ORGANIC SYNTHESES

ORGANIC SYNTHESES

Collective Volume 11

A Revised Edition of Annual Volumes 80–84

The procedures in this text are intended for use only by persons with prior training in the field of organic chemistry. In the checking and editing of these procedures, every effort has been made to identify potentially hazardous steps and to eliminate as much as possible the handling of potentially dangerous materials; safety precautions have been inserted where appropriate. If performed with the materials and equipment specified, in careful accordance with the instructions and methods in this text, the Editors believe the procedures to be very useful tools. However, these procedures must be conducted at one's own risk. Organic Syntheses, Inc., its Editors, who act as checkers, and its Board of Directors do not warrant or guarantee the safety of individuals using these procedures and hereby disclaim any liability for any injuries or damages claimed to have resulted from or related in any way to the procedures herein.

Published by John Wiley & Sons, Inc., Hoboken, New Jersey.

Published simultaneously in Canada.

"John Wiley & Sons, Inc. is pleased to publish this volume of Organic Syntheses on behalf of Organic Syntheses, Inc. Although Organic Syntheses, Inc. has assured us that each preparation contained in this volume has been checked in an independent laboratory and that any hazards that were uncovered are clearly set forth in the write-up of each preparation, John Wiley & Sons, Inc. does not warrant the preparations against any safety hazards and assumes no liability with respect to the use of the preparations."

Library of Congress Catalog Card Number: 21-17747
ISBN 978-0-470-39548-6

Printed in the United States of America
10 9 8 7 6 5 4 3 2 1

ORGANIC SYNTHESES

VOLUME	VOLUME EDITOR	PAGES
I*	†ROGER ADAMS	84
II*	†JAMES BRYANT CONANT	100
III*	†HANS THACHER CLARKE	105
IV*	†OLIVER KAMM	89
V*	†CARL SHIPP MARVEL	110
VI*	†HENRY GILMAN	120
VII*	†FRANK C. WHITMORE	105
VIII*	†ROGER ADAMS	139
IX*	†JAMES BRYANT CONANT	108
Collective Vol. I	A revised edition of Annual Volumes I–IX †HENRY GILMAN, *Editor-in-Chief* 2nd Edition revised by †A. H. BLATT	580
X*	†HANS THACHER CLARKE	119
XI*	†CARL SHIPP MARVEL	106
XII*	†FRANK C. WHITMORE	96
XIII*	†WALLACE H. CAROTHERS	119
XIV*	†WILLIAM W. HARTMAN	100
XV*	†CARL R. NOLLER	104
XVI*	†JOHN R. JOHNSON	104
XVII*	†L. F. FIESER	112
XVIII*	†REYNOLD C. FUSON	103
XIX*	†JOHN R. JOHNSON	105
Collective Vol. II	A revised edition of Annual Volumes X–XIX †A. H. BLATT, *Editor-in-Chief*	654
20*	†CHARLES F. H. ALLEN	113
21*	†NATHAN L. DRAKE	120
22*	†LEE IRVIN SMITH	114
23*	†LEE IRVIN SMITH	124
24*	†NATHAN L. DRAKE	119
25*	†WERNER E. BACHMANN	120
26*	†HOMER ADKINS	124
27*	†R. L. SHRINER	121
28*	†H. R. SNYDER	121
29*	†CLIFF S. HAMILTON	119
Collective Vol. III	A revised edition of Annual Volumes 20–29 †E. C. HORNING, *Editor-in-Chief*	890
30*	†ARTHUR C. COPE	115
31*	†R. S. SCHREIBER	122

* *Out of print.*
† *Deceased.*

VOLUME	VOLUME EDITOR	PAGES
32*	†Richard Arnold	119
33*	†Charles Price	115
34*	†William S. Johnson	121
35*	†T. L. Cairns	122
36*	†N. J. Leonard	120
37*	†James Cason	109
38*	†John C. Sheehan	120
39*	†Max Tishler	114
Collective Vol. IV	A revised edition of Annual Volumes 30–39 †Norman Rabjohn, *Editor-in-Chief*	1036
40*	†Melvin S. Newman	114
41*	John D. Roberts	118
42*	†Virgil Boekelheide	118
43*	†B. C. McKusick	124
44*	†William E. Parham	131
45*	†William G. Dauben	118
46*	E. J. Corey	146
47*	†William D. Emmons	140
48*	†Peter Yates	164
49*	Kenneth B. Wiberg	124
Collective Vol. V	A revised edition of Annual Volumes 40–49 †Henry E. Baumgarten, *Editor-in-Chief*	1234
Cumulative Indices to Collective Volumes, I, II, III, IV, V	†Ralph L. and †Rachel H. Shriner, *Editors*	
50*	Ronald Breslow	136
51*	†Richard E. Benson	209
52*	Herbert O. House	192
53*	Arnold Brossi	193
54*	Robert E. Ireland	155
55*	†Satoru Masamune	150
56*	†George H. Büchi	144
57*	Carl R. Johnson	135
58*	†William A. Sheppard	216
59*	Robert M. Coates	267
Collective Vol. VI	A revised edition of Annual Volumes 50–59 Wayland E. Noland, *Editor-in-Chief*	1208
60*	†Orville L. Chapman	140
61*	†Robert V. Stevens	165

* *Out of print.*

† *Deceased.*

VOLUME	VOLUME EDITOR	PAGES
62*	Martin F. Semmelhack	269
63*	Gabriel Saucy	291
64*	Andrew S. Kende	308
Collective Vol. VII	A revised edition of Annual Volumes 60–64 Jeremiah P. Freeman, *Editor-in-Chief*	602
65*	Edwin Vedejs	278
66*	Clayton H. Heathcock	265
67*	Bruce E. Smart	289
68*	James D. White	318
69*	Leo A. Paquette	328
Reaction Guide to Collective Volumes I–VII and Annual Volumes 65–68	Dennis C. Liotta and Mark Volmer, *Editors*	854
Collective Vol. VIII	A revised edition of Annual Volumes 65–69	696
Cumulative Indices to Collective Volumes, I, II, III, IV, V, VI, VII, VIII, VIII	Jeremiah P. Freeman, *Editor-in-Chief*	
70*	†Albert I. Meyers	305
71*	Larry E. Overman	285
72*	†David L. Coffen	333
73*	Robert K. Boeckman, Jr.	352
74*	Ichiro Shinkai	341
Collective Vol. IX	A revised edition of Annual Volumes 70–74 Jeremiah P. Freeman, *Editor-in-Chief*	840
75	Amos B. Smith, III	257
76	Stephen Martin	340
77	David J. Hart	312
78	William R. Roush	326
79	Louis S. Hegedus	328
Collective Vol. X	A revised edition of Annual Volumes, 75–79 Jeremiah P. Freeman, *Editor-in-Chief*	809

* *Out of print.*
† *Deceased.*

VOLUME	VOLUME EDITOR	PAGES
80	STEVEN WOLFF	259
81	RICK L. DANHEISER	296
82	EDWARD J. J. GRABOWSKI	195
83	DENNIS P. CURRAN	221
84	MARVIN J. MILLER	410
Collective Vol. XI	A revised edition of Annual Volumes 80–84 CHARLES K. ZERCHER, *Editor-in-Chief*	1138
85	SCOTT E. DENMARK	321

Collective Volumes, Collective Indices to Collective Volumes I–IX, Annual Volumes, 75–79 and Reaction Guide are available from John Wiley & Sons, Inc.

**Out of print.*
†Deceased.

HANDLING HAZARDOUS CHEMICALS

A Brief Introduction

General Reference: *Prudent Practices in the Laboratory*; National Academy Press; Washington, DC, 1995.

Physical Hazards

Fire. Avoid open flames by use of electric heaters. Limit the quantity of flammable liquids stored in the laboratory. Motors should be of the nonsparking induction type.

Explosion. Use shielding when working with explosive classes such as acetylides, azides, ozonides, and peroxides. Peroxidizable substances such as ethers and alkenes, when stored for a long time, should be tested for peroxides before use. Only sparkless "flammable storage" refrigerators should be used in laboratories.

Electric Shock. Use 3-prong grounded electrical equipment if possible.

Chemical Hazards

Because all chemicals are toxic under some conditions, and relatively few have been thoroughly tested, it is good strategy to minimize exposure to all chemicals. In practice this means having a good, properly installed hood; checking its performance periodically; using it properly; carrying out all operations in the hood; protecting the eyes; and, since many chemicals can penetrate the skin, avoiding skin contact by use of gloves and other protective clothing at all times.

a. Acute Effects. These effects occur soon after exposure. The effects include burn, inflammation, allergic responses, damage to the eyes, lungs, or nervous system (e.g., dizziness), and unconsciousness or death (as from overexposure to HCN). The effect and its cause are usually obvious and so are the methods to prevent it. They generally arise from inhalation or skin contact, so should not be a problem if one follows the admonition “work in a hood and keep chemicals off your hands”. Ingestion is a rare route, being generally the result of eating in the laboratory or not washing hands before eating.

b. Chronic Effects. These effects occur after a long period of exposure or after a long latency period and may show up in any of numerous organs. Of the chronic effects of chemicals, cancer has received the most attention lately. Several dozen chemicals have been demonstrated to be carcinogenic in man and hundreds to be carcinogenic to animals. Although there is no simple correlation between carcinogenicity in animals and in man, there is little doubt that a significant proportion of the chemicals used in laboratories have some potential for carcinogenicity in man. For this and other reasons, chemists should employ good practices at all times.

The key to safe handling of chemicals is a good, properly installed hood, and the referenced book devotes many pages to hoods and ventilation. It recommends that in a laboratory where people spend much of their time working with chemicals there should be a hood for each two people, and each should have at least 2.5 linear feet (0.75 meter) of working space at it. Hoods are more than just devices to keep undesirable vapors from the laboratory atmosphere. When closed they provide a protective barrier between chemists and chemical operations, and they are a good containment device for spills. Portable shields can be a useful supplement to hoods, or can be an alternative for hazards of limited severity, e.g., for small-scale operations with oxidizing or explosive chemicals.

Specialized equipment can minimize exposure to the hazards of laboratory operations. Impact resistant safety glasses are basic equipment and should be worn at all times. They may be supplemented by face shields or goggles for particular operations, such as pouring corrosive liquids. Because skin contact with chemicals can lead to skin irritation or sensitization or, through absorption, to effects on internal organs, protective gloves should be worn at all times.

Laboratories should have fire extinguishers and safety showers. Respirators should be available for emergencies. Emergency equipment should be kept in a central location and must be inspected periodically.

MSDS (Materials Safety Data Sheets) sheets are available from the suppliers of commercially available reagents, solvents, and other chemical materials; anyone performing an experiment should check these data sheets before initiating an experiment to learn of any specific hazards associated with the chemicals being used in that experiment.

DISPOSAL OF CHEMICAL WASTE

General Reference: *Prudent Practices in the Laboratory* National Academy Press, Washington, D.C. 1996

Effluents from synthetic organic chemistry fall into the following categories:

1. **Gases**
   - 1a. Gaseous materials either used or generated in an organic reaction.
   - 1b. Solvent vapors generated in reactions swept with an inert gas and during solvent stripping operations.
   - 1c. Vapors from volatile reagents, intermediates and products.
2. **Liquids**
   - 2a. Waste solvents and solvent solutions of organic solids (see item 3b).
   - 2b. Aqueous layers from reaction work-up containing volatile organic solvents.
   - 2c. Aqueous waste containing non-volatile organic materials.
   - 2d. Aqueous waste containing inorganic materials.
3. **Solids**
   - 3a. Metal salts and other inorganic materials.
   - 3b. Organic residues (tars) and other unwanted organic materials.
   - 3c. Used silica gel, charcoal, filter aids, spent catalysts and the like.

The operation of industrial scale synthetic organic chemistry in an environmentally acceptable manner* requires that all these effluent categories be dealt with properly. In small scale operations in a research or

*An environmentally acceptable manner may be defined as being both in compliance with all relevant state and federal environmental regulations *and* in accord with the common sense and good judgment of an environmentally aware professional.

academic setting, provision should be made for dealing with the more environmentally offensive categories.

1a. Gaseous materials that are toxic or noxious, e.g., halogens, hydrogen halides, hydrogen sulfide, ammonia, hydrogen cyanide, phosphine, nitrogen oxides, metal carbonyls, and the like.
1c. Vapors from noxious volatile organic compounds, e.g., mercaptans, sulfides, volatile amines, acrolein, acrylates, and the like.
2a. All waste solvents and solvent solutions of organic waste.
2c. Aqueous waste containing dissolved organic material known to be toxic.
2d. Aqueous waste containing dissolved inorganic material known to be toxic, particularly compounds of metals such as arsenic, beryllium, chromium, lead, manganese, mercury, nickel, and selenium.
3. All types of solid chemical waste.

Statutory procedures for waste and effluent management take precedence over any other methods. However, for operations in which compliance with statutory regulations is exempt or inapplicable because of scale or other circumstances, the following suggestions may be helpful.

Gases

Noxious gases and vapors from volatile compounds are best dealt with at the point of generation by "scrubbing" the effluent gas. The gas being swept from a reaction set-up is led through tubing to a large trap to prevent suck-back and into a sintered glass gas dispersion tube immersed in the scrubbing fluid. A bleach container can be conveniently used as a vessel for the scrubbing fluid. The nature of the effluent determines which of four common fluids should be used: dilute sulfuric acid, dilute alkali or sodium carbonate solution, laundry bleach when an oxidizing scrubber is needed, and sodium thiosulfate solution or diluted alkaline sodium borohydride when a reducing scrubber is needed. Ice should be added if an exotherm is anticipated.

Larger scale operations may require the use of a pH meter or starch/iodide test paper to ensure that the scrubbing capacity is not being exceeded.

When the operation is complete, the contents of the scrubber can be poured down the laboratory sink with a large excess (10–100 volumes) of water. If the solution is a large volume of dilute acid or base, it should be neutralized before being poured down the sink.

Liquids

Every laboratory should be equipped with a waste solvent container in which *all* waste organic solvents and solutions are collected. The contents of these containers should be periodically transferred to properly labeled waste solvent drums and arrangements made for contracted disposal in a regulated and licensed incineration facility.**

Aqueous waste containing dissolved toxic organic material should be decomposed *in situ*, when feasible, by adding acid, base, oxidant, or reductant. Otherwise, the material should be concentrated to a minimum volume and added to the contents of a waste solvent drum.

Aqueous waste containing dissolved toxic inorganic material should be evaporated to dryness and the residue handled as a solid chemical waste.

Solids

Soluble organic solid waste can usually be transferred into a waste solvent drum, provided near-term incineration of the contents is assured.

Inorganic solid wastes, particularly those containing toxic metals and toxic metal compounds, used Raney nickel, manganese dioxide, etc. should be placed in glass bottles or lined fiber drums, sealed, properly labeled, and arrangements made for disposal in a secure landfill.** Used mercury is particularly pernicious and small amounts should first be amalgamated with zinc or combined with excess sulfur to solidify the material.

Other types of solid laboratory waste including used silica gel and charcoal should also be packed, labeled, and sent for disposal in a secure landfill.

Special Note

Since local ordinances may vary widely from one locale to another, one should always check with appropriate authorities. Also, professional disposal services differ in their requirements for segregating and packaging waste.

**If arrangements for incineration of waste solvent and disposal of solid chemical waste by licensed contract disposal services are not in place, a list of providers of such services should be available from a state or local office of environmental protection.

INFORMATION FOR AUTHORS OF PROCEDURES

Organic Syntheses welcomes and encourages submissions of experimental procedures that lead to compounds of wide interest or that illustrate important new developments in methodology. Proposals for *Organic Syntheses* procedures will be considered by the Editorial Board upon receipt of an outline proposal as described below. A full procedure will then be invited for those proposals determined to be of sufficient interest. These full procedures will be evaluated by the Editorial Board, and if approved, assigned to a member of the Board for checking. In order for a procedure to be accepted for publication, each reaction must be successfully repeated in the laboratory of a member of the Editorial Board at least twice, with similar yields (generally ±5%) and selectivity to that reported by the submitters.

Organic Syntheses Proposals

A cover sheet should be included providing full contact information for the principal author and including a scheme outlining the proposed reactions (an Organic Syntheses Proposal Cover Sheet can be downloaded at orgsyn.org). Attach an outline proposal describing the utility of the methodology and/or the usefulness of the product. Identify and reference the best current alternatives. For each step, indicate the proposed scale, yield, method of isolation and purification, and how the purity of the product is determined. Describe any unusual apparatus or techniques required, and any special hazards associated with the procedure. Identify the source of starting materials. Enclose copies of relevant publications (attach pdf files if an electronic submission is used).

Submit proposals by mail or as e-mail attachments to:

Professor Charles K. Zercher
Associate Editor, Organic Syntheses
Department of Chemistry
University of New Hampshire
23 Academic Way, Parsons Hall
Durham, NH 03824

For electronic submissions: *org.syn@unh.edu*

Submission of Procedures

Authors invited by the Editorial Board to submit full procedures should prepare their manuscripts in accord with the Instructions to Authors which are described below or may be downloaded at orgsyn.org. Submitters are also encouraged to consult earlier volumes of *Organic Syntheses* for models with regard to style, format, and the level of experimental detail expected in *Organic Syntheses* procedures. Manuscripts should be submitted to the Associate Editor. Electronic submissions are encouraged; procedures will be accepted as e-mail attachments in the form of Microsoft Word files with all schemes and graphics also sent separately as ChemDraw files.

Procedures that do not conform to the Instructions to Authors with regard to experimental style and detail will be returned to authors for correction. Authors will be notified when their manuscript is approved for checking by the Editorial Board, and it is the goal of the Board to complete the checking of procedures within a period of no more than six months.

Additions, corrections, and improvements to the preparations previously published are welcomed; these should be directed to the Associate Editor. However, checking of such improvements will only be undertaken when new methodology is involved.

NOMENCLATURE

Both common and systematic names of compounds are used throughout this volume, depending on which the Volume Editor felt was more appropriate. The Chemical Abstracts indexing name for each title compound, if it differs from the title name, is given as a subtitle. Systematic

Chemical Abstracts nomenclature, used in the Collective Indexes for the title compound and a selection of other compounds mentioned in the procedure, is provided in an appendix at the end of each preparation. Chemical Abstracts Registry numbers, which are useful in computer searching and identification, are also provided in these appendices. Whenever two names are concurrently in use and one name is the correct Chemical Abstracts name, that name is preferred.

ACKNOWLEDGMENT

Organic Syntheses wishes to acknowledge the contributions of Merck & Co. and Pfizer, Inc. to the success of this enterprise through their support, in the form of time and expenses, of members of the Board of Editors.

PREFACE

In keeping with the established tradition, Collective Volume XI of Organic Syntheses is a compilation of procedures that were originally published in annual volumes 80–84 (2003–2007). The five-year cycle of Collective Volumes is reflective of the rapid pace of development in synthetic organic chemistry. Our preparation of this Collective Volume, which emphasizes the organic chemistry transformations of most recent interest and impact to the organic chemistry community, is one of our efforts to support this rapid development. Many of the procedures in this volume have been revised from the version in the annual volume to include updated references and modified discussions. Some changes to the experimental text of procedures have been made for the purpose of clarity, but no modifications of the experimental procedures have been incorporated, since this would require a rechecking of the original procedure.

A survey of the procedures found in this Collective Volume reveals the vibrancy of modern organic chemistry. In this volume are included new oxidation and reduction processes, creative methods for the formation of heterocyclic systems, and transformative organometallic reactions. Traditional carbon-carbon-bond formation, which is central to organic synthesis, is well represented, and asymmetric synthetic methods abound. The generation of organometallic catalysts, the formation of biological tools, and the assembly of structures with unusual architecture attest to the richness and diversity of the procedures found in this volume.

The contents of this volume are organized by reaction class or compound type, with the precise classification made by the predilection of the editor, who has attempted to ascertain the primary purpose or utility of the procedure. For example, methods that describe the use of reagents for the purpose of oxidation are grouped together, although procedures that include an oxidative transformation might be found elsewhere in the volume. Methods that describe reductions, heterocycle formation, alkene formation, cycloadditions, and halogenation reactions, among others, are also organized as groups of procedures. A more useful tool for searching the volume is likely to be the indices found at the end of the Collective Volume. In particular, the reaction index and compound index should be of utility in finding transformations and functional groups of interest to the reader. An author index, a molecular formula index, and a general (subject) index are also available for searching this volume. If one has a literature citation to the procedure in an annual volume, the concordance index at the end of the volume allows the reader to find the latest version.

Organic Syntheses, Inc. is committed to providing reliable methods for the preparation of organic compounds. The editors of *Organic Syntheses* welcome corrections, suggestions, and procedures being submitted for consideration by the Board of Editors. Prospective submitters should consult the section entitled "Information for Authors of Procedures" found at the front of any recent annual volume or at the Organic Syntheses website (www.orgsyn.org). Correspondence should be addressed to the Associate Editor of Organic Syntheses, Dr. Charles K. Zercher, Department of Chemistry, University of New Hampshire, Durham, NH 03824.

The Editor would like to thank all of the submitters, who have responded enthusiastically to requests for procedures and additional information. The names of the Checking Editors and their coworkers are listed with each procedure. Without their dedicated efforts and expertise, *Organic Syntheses* would cease to exist. The editor of this volume is grateful to the assistance and encouragement offered by a number of individuals. Professor Jeremiah Freeman, who served as the Editor of the most recent four Collective Volumes, established a standard for the day-to-day operations of the organization. His willingness to answer a multitude of questions during the assembly of this Collective Volume has been greatly appreciated. Professor Rick Danheiser, whose leadership for the past five years as the Editor-in-Chief of *Organic Syntheses* reinvigorated the organization, has provided encouragement. A large portion of this project has fallen to the able hands of my assistant, Peg Torch. Her tireless efforts, agreeable spirit, and commitment to excellence have made this task far less onerous than once feared.

CHARLES K. ZERCHER

Durham, NH
June 2009

CONTENTS

CATALYTIC REDUCTION OF AMIDES TO AMINES WITH HYDROSILANES USING A TRIRUTHENIUM CARBONYL CLUSTER AS THE CATALYST 1

SYNTHESIS OF 4-TRIISOPROPYLSILYL-3-BUTYN-2-OL BY ASYMMETRIC TRANSFER HYDROGENATION 9

PREPARATION OF OPTICALLY ACTIVE (*R,R*)-HYDROBENZOIN FROM BENZOIN OR BENZIL 17

SYNTHESIS OF (+)-(1*R*,2*S*,9*S*)-11-METHYL-7,11-DIAZATRICYCLO [7.3.1.0^{2,7}]TRIDECANE, A (+)-SPARTEINE SURROGATE 25

SYNTHESIS AND RU(II)-BINAP REDUCTION OF A KETOESTER DERIVED FROM HYDROXYPROLINE 38

2-(2′, 2′-DIMETHYLPROPOXY)-2,3-DIHYDRO-1*H*-INDENE 46

PREPARATION OF α-ACETOXY ETHERS BY THE REDUCTIVE ACETYLATION OF ESTERS: *endo*-1-BORNYLOXYETHYL ACETATE 52

INDIUM/AMMONIUM CHLORIDE-MEDIATED SELECTIVE REDUCTION OF AROMATIC NITRO COMPOUNDS: ETHYL 4-AMINOBENZOATE 57

PREPARATION OF 1,4:5,8-DIMETHANO-1,2,3,4,5,6,7,8-OCTAHYDRO-9,10-DIMETHOXYANTHRACENIUM HEXACHLOROANTIMONATE ($4^{+\cdot}SbCl_6^-$): A HIGHLY ROBUST RADICAL-CATION SALT 62

A PRACTICAL PROCEDURE FOR CARBONYL α-OXIDATION: SYNTHESIS OF (2-BENZOYLOXY)-1,4-CYCLOHEXANEDIONE *MONO*-ETHYLENE KETAL . . . 70

OXIDATION OF NEROL TO NERAL WITH IODOBENZENE AND TEMPO 77

AU(I)-CATALYZED HYDRATION OF ALKYNES: 2,8-NONANEDIONE 82

D-RIBONOLACTONE AND 2,3-ISOPROPYLIDENE (D-RIBONOLACTONE) 88

PREPARATION OF 4-ACETYLAMINO-2, 2, 6, 6-TETRAMETHYL-PIPERIDINE-1-OXOAMMONIUM TETRAFLUOROBORATE, AND THE OXIDATION OF GERANIOL TO GERANIAL 93
(*R*)-(+)-3,4-DIMETHYLCYCLOHEX-2-EN-1-ONE 100
OXIDATION OF PRIMARY ALCOHOLS TO CARBOXYLIC ACIDS WITH SODIUM CHLORITE CATALYZED BY TEMPO AND BLEACH: 4-METHOXYPHENYLACETIC ACID 107
METHYLTRIOXORHENIUM CATALYZED OXIDATION OF SECONDARY AMINES TO NITRONES: *N*-BENZYLIDENE-BENZYLAMINE *N*-OXIDE 114
PHENYLSULFENYLATION OF NONACTIVATED δ-CARBON ATOM BY PHOTOLYSIS OF ALKYL BENZENESULFENATES: PREPARATION OF 2-PHENYLTHIO-5-HEPTANOL 121
(5*S*)-(d-MENTHYLOXY)-2(5*H*)-FURANONE 128
MILD AND SELECTIVE OXIDATION OF SULFUR COMPOUNDS IN TRIFLUOROETHANOL: DIPHENYL DISULFIDE AND METHYL PHENYL SULFOXIDE 135
TRANSFORMATION OF PRIMARY AMINES TO *N*-MONOALKYLHYDROXYLAMINES: *N*-HYDROXY-(*S*)-1-PHENYLETHYLAMINE OXALATE 138
SYNTHESIS OF AN *ANTI*-α-AMINO EPOXIDE BY ONE-CARBON HOMOLOGATION OF AN α-AMINO ESTER: (2*S*,3*S*)-1,2-EPOXY-3-(BOC-AMINO)-4-PHENYLBUTANE 149
PREPARATION OF (*S*)-METHYL GLYCIDATE VIA HYDROLYTIC KINETIC RESOLUTION 157
SYNTHESIS OF 1,2:5,6-DIANHYDRO-3, 4-O-ISOPROPYLIDENE-L-MANNITOL 164
SYNTHESIS AND USE OF GLYCOSYL PHOSPHATES AS GLYCOSYL DONORS 169
SYNTHESIS OF 1,2:4,5-DI-*O*-ISOPROPYLIDENE-D-erythro-2,3-HEXODIULO-2,6-PYRANOSE. A HIGHLY ENANTIOSELECTIVE KETONE CATALYST FOR EPOXIDATION 177
ASYMMETRIC EPOXIDATION OF *trans*-β-METHYLSTYRENE AND 1-PHENYLCYCLOHEXENE USING A D-FRUCTOSE-DERIVED KETONE: (*R,R*)-*trans*-β-METHYLSTYRENE OXIDE AND (*R,R*)-1-PHENYLCYCLOHEXENE OXIDE 183

N,N-DIBENZYL-*N*-[1-CYCLOHEXYL-3-(TRIMETHYLSILYL)-2-PROPYNYL]-AMINE FROM CYCLOHEXANECARBALDEHYDE, TRIMETHYLSILYLACETYLENE AND DIBENZYLAMINE 189

SYNTHESIS OF ACYLSILANES FROM MORPHOLINE AMIDES. SYNTHESIS OF 1-(DIMETHYL(PHENYL)SILYL)PROPAN-1-ON 197

(*S*)-(+)-NEOMENTHYLDIPHENYLPHOSPHINE IN NICKEL-CATALYZED ASYMMETRIC REDUCTIVE COUPLING OF ALKYNES AND ALDEHYDES: ENANTIOSELECTIVE SYNTHESIS OF ALLYLIC ALCOHOLS AND α-HYDROXY KETONES 205

A XANTHATE-TRANSFER APPROACH TO α-TRIFLUOROMETHYLAMINES [2-(*N*-ACETYLAMINO)-4, 4-DIMETHOXY-1,1,1-TRIFLUOROBUTANE] 212

STEREOSELECTIVE ISOPRENOID CHAIN EXTENSION WITH ACETOACETATE DIANION: (*E, E, E*)-GERANYLGERANIOL FROM (*E, E*)-FARNESOL 221

SYNTHESIS OF TERMINAL 1,3-DIYNES VIA SONOGASHIRA COUPLING OF VINYLIDENE CHLORIDE FOLLOWED BY ELIMINATION. PREPARATION OF 1,3-DECADIYNE 234

SYNTHESIS OF YNAMIDES BY *N*-ALKYNYLATION OF AMINE DERIVATIVES. PREPARATION OF *N*-ALLYL-*N*-(METHOXYCARBONYL)-1, 3-DECADIYNYLAMINE 244

1,3-DICHLOROACETONE AS A CYCLOPROPANONE EQUIVALENT: 5-OXASPIRO [3.4]OCTAN-1-ONE 256

COPPER(I) IODIDE DIMETHYL SULFIDE CATALYZED ADDITION OF A VINYLZIRCONIUM REAGENT. PREPARATION OF 4-PHENYL-5(E)-DECEN-2-ONE 261

ORTHO-FORMYLATION OF PHENOLS; PREPARATION OF 3-BROMOSALICYLALDEHYDE 267

HIGHLY EFFICIENT ADDITION OF ACTIVATED METHYLENE COMPOUNDS TO ALKENES CATALYZED BY GOLD AND SILVER 272

SYNTHESIS OF MULTI-FUNCTIONALIZED KETONES THROUGH THE FUKUYAMA COUPLING REACTION CATALYZED BY PEARLMAN'S CATALYST: PREPARATION OF ETHYL 6-OXOTRIDECANOATE 281

1-(4-ACETYLPHENYL)-2-PHENYLETHANE FROM POTASSIUM 2-PHENETHYLTRIFLUOROBORATE AND 4-BROMOACETOPHENONE . . . 289
ONE-FLASK SYNTHESIS OF METHYL ARYLVINYLDIAZOACETATES AND THEIR APPLICATION IN ENANTIOSELECTIVE C–H FUNCTIONALIZATION: SYNTHESIS OF (*E*)-METHYL 2-DIAZO-4-PHENYLBUT-3-ENOATE AND (*S,E*)-METHYL 2-((*R*)-4-METHYL-1, 2-DIHYDRONAPHTHALEN-2-YL)-4-PHENYLBUT-3-ENOATE . . . 296
RADICAL ALLYLATION OF *B*-ALKYLCATECHOLBORANES . . . 308
SUZUKI-MIYAURA CROSS-COUPLING: PREPARATION OF 2′-VINYLACETANILIDE . . . 315
SYNTHESIS OF 6,9,12,15,18-PENTAMETHYL-1, 6,9,12,15,18-HEXAHYDRO(C_{60}-I_h)[5,6]FULLERENE . . . 319
NICKEL-CATALYZED HOMOALLYLATION OF ALDEHYDES WITH 1,3-DIENES . . . 327
TRIFLUOROMETHYLATION AT THE α-POSITION OF α, β-UNSATURATED KETONES: 4-PHENYL-3-(TRIFLUOROMETHYL)BUTAN-2-ONE . . . 336
A PRACTICAL AND SAFE PREPARATION OF 3,5-BIS (TRIFLUOROMETHYL) ACETOPHENONE . . . 343
dl-SELECTIVE PINACOL-TYPE COUPLING USING ZINC, CHLOROSILANE, AND CATALYTIC AMOUNTS OF Cp_2VCl_2; *dl*-1,2-DICYCLOHEXYLETHANEDIOL . . . 348
4-NONYLBENZOIC ACID . . . 353
PALLADIUM CATALYZED CROSS-COUPLING OF (*Z*)-1-HEPTENYLDIMETHYLSILANOL WITH 4-IODOANISOLE: (*Z*)-1-HEPTENYL-4-METHOXYBENZENE . . . 359
PLATINUM CATALYZED HYDROSILYLATION AND PALLADIUM CATALYZED CROSS-COUPLING: ONE-POT HYDROARYLATION OF 1-HEPTYNE TO (*E*)-1-(1-HEPTENYL)-4-METHOXYBENZENE . . . 367
HECK REACTIONS OF ARYL CHLORIDES CATALYZED BY PALLADIUM/TRI-*TERT*-BUTYLPHOSPHINE: (*E*)-2-METHYL-3-PHENYLACRYLIC ACID BUTYL ESTER AND (*E*)-4-(2-PHENYLETHENYL) BENZONITRILE . . . 372

SYNTHESIS OF N-(*TERT*-BUTOXYCARBONYL)-β-IODOALANINE METHYL ESTER: A USEFUL BUILDING BLOCK IN THE SYNTHESIS OF NONNATURAL α-AMINO ACIDS VIA PALLADIUM CATALYZED CROSS COUPLING REACTIONS 384
SYNTHESIS OF 3-PYRIDYLBORONIC ACID AND ITS PINACOL ESTER. APPLICATION OF 3-PYRIDYLBORONIC ACID IN SUZUKI COUPLING TO PREPARE 3-PYRIDIN-3-YLQUINOLINE 393
EFFICIENT ASYMMETRIC SYNTHESIS OF *N-TERT*-BUTOXYCARBONYL α-AMINO ACIDS USING 4-*TERT*-BUTOXYCARBONYL-5,6-DIPHENYLMORPHOLIN-2-ONE: (*R*)-(*N-TERT*-BUTOXYCARBONYL) ALLYLGLYCINE 399
PREPARATION OF *O*-ALLYL-*N*-(9-ANTHRACENYLMETHYL) CINCHONIDINIUM BROMIDE AS A PHASE TRANSFER CATALYST FOR THE ENANTIOSELECTIVE ALKYLATION OF GLYCINE BENZOPHENONE IMINE *TERT*-BUTYL ESTER: (4*S*)-2-(BENZHYDRYLIDENAMINO) PENTANEDIOIC ACID, 1-*TERT*-BUTYL ESTER-5-METHYL ESTER 404
RUTHENIUM-CATALYZED ALKYLATION OF AROMATIC KETONES WITH OLEFINS: 8-[2-(TRIETHOXYSILYL)ETHYL]-1-TETRALONE 410
DIETHYLAMINOTRIMETHYLSILANE-CATALYZED 1,4-ADDITION OF ALDEHYDES TO VINYL KETONES: (3*R*)-3,7-DIMETHYL-2-(3-OXOBUTYL)-6-OCTENAL 416
FRAGMENTATION-RECOMBINATION NAZAROV CYCLIZATION: 3,4-DIMETHYLCYCLOPENT-2-EN-1-ONE 420
FACILE SYNTHESES OF AMINOCYCLOPROPANES: *N,N*-DIBENZYL-*N*-(2-ETHENYLCYCLOPROPYL) AMINE 425
GENERATION AND CYCLIZATION OF 5-HEXENYLLITHIUM: 2-CYCLOPENTYLACETOPHENONE 433
THE CATALYTIC INTRAMOLECULAR PAUSON-KHAND REACTION: 2,3,3α,4-TETRAHYDRO-2-[(4-METHYLBENZENE) SULFONYL] CYCLOPENTA[C] PYRROL-5(1H)-ONE 442

INTRA- AND INTERMOLECULAR KULINKOVICH CYCLOPROPANATION REACTIONS OF CARBOXYLIC ESTERS WITH OLEFINS: BICYCLO[3.1.0]HEXAN-1-OL AND *TRANS*-2-BENZYL-1-METHYLCYCLOPROPAN-1-OL 450
SYNTHESIS AND [3 + 2] CYCLOADDITION OF A 1,1-DIALKOXY-2-METHYLENECYCLOPROPANE: 6,6-DIMETHYL-1-METHYLENE-4,8-DIOXASPIRO [2.5] OCTANE AND *CIS*-5-(5,5-DIMETHYL-1,3-DIOXAN-2-YLIDENE)HEXAHYDRO-1(2H)-PENTALEN-2-ONE 457
PREPARATION AND USE OF A NEW DIFLUOROCARBENE REAGENT. TRIMETHYLSILYL 2-FLUOROSULFONYL-2, 2-DIFLUOROACETATE: *n*-BUTYL 2,2-DIFLUOROCYCLOPROPANECARBOXYLATE 469
SYNTHESIS OF 2a-BENZYLOXY-8-OXABICYCLO[3.2.1]OCT-6-EN-3-ONE BY [4 + 3] CYCLOADDITION 473
TRIFLUOROMETHANESULFONIMIDE-CATALYZED (2 + 2)-CYCLOADDITION OF SILYL ENOL ETHERS WITH α, β-UNSATURATED ESTERS: 1-(*TERT*-BUTYLDIMETHYLSILOXY)-8-(METHOXYCARBONYL)-6-METHYLBICYCLO [4.2.0]OCTANE 481
PREPARATION OF BENZOCYCLOBUTENONE DERIVATIVES BASED ON AN EFFICIENT GENERATION OF BENZYNES 488
AN EFFICIENT, HIGHLY DIASTEREO- AND ENANTIOSELECTIVE HETERO-DIELS-ALDER CATALYST. PREPARATION OF (2*S*,6*R*)-6-(*TERT*-BUTYLDIMETHYL-SILYLOXYMETHYL)-2-METHOXY-2, 5-DIHYDROPYRAN 498
CATALYTIC ASYMMETRIC ACYL HALIDE-ALDEHYDE CYCLOCONDENSATION REACTION 506
SYNTHESIS OF 9-SPIROEPOXY-*ENDO*-TRICYCLO[5.2.2.0^{2,6}]UNDECA-4, 10-DIEN-8-ONE 514
PREPARATION OF 9,10-DIMETHOXYPHENANTHRENE AND 3,6-DIACETYL-9,10-DIMETHOXYPHENANTHREN 519
HELICENEBISQUINONES: SYNTHESIS OF A [7] HELICENEBISQUINONE 524
PREPARATION OF A TRISAMIDOMOLYBDENUM(VI) PROPYLIDYNE COMPLEX –A HIGHLY ACTIVE CATALYST PRECURSOR FOR ALKYNE METATHESIS . . . 529
PREPARATION OF A CARBAZOLE-BASED MACROCYCLE VIA PRECIPITATION-DRIVEN ALKYNE METATHESIS 540

2-SUBSTITUTED-1,3-CYCLOHEXADIENES BY INTERMOLECULAR, METHYLENE-FREE TANDEM ENYNE METATHESIS 552
SYNTHESIS OF 1,3-DIENES FROM ALKYNES AND ETHYLENE:ACETIC ACID 2-METHYLENE-3-PHENETHYLBUT-3-ENYL ESTER 561
RING-CLOSING METATHESIS SYNTHESIS OF *N*-BOC-3-PYRROLINE 570
GENERATION OF YNOLATE AND *Z*-SELECTIVE OLEFINATION OF ACYLSILANES: (*Z*)-2-METHYL-3-TRIMETHYLSILYL-2-BUTENOIC ACID 577
SYNTHESIS OF 2-CHLOROACROLEIN DIETHYL ACETAL 586
CONVERSION OF ARYLALKYLKETONES INTO DICHLOROALKENES: 1-CHLORO-4-(2,2-DICHLORO-1-METHYLVINYL)BENZENE 594
GENERATION OF AN ACETYLENE-TITANIUM ALKOXIDE COMPLEX: PREPARATION OF (*Z*)-1,2-DIDEUTERIO-1-(TRIMETHYLSILYL)-1-HEXENE 600
ALKYNE *VIA* SOLID-LIQUID PHASE-TRANSFER CATALYZED DEHYDROHALOGENATION: ACETYLENE DICARBOXALDEHYDE TETRAMETHYL ACETAL AND ACETYLENE DICARBOXALDEHYDE DIMETHYL ACETAL 607
GENERATION OF NONRACEMIC 2-(*t*-BUTYLDIMETHYLSILYLOXY)-3-BUTYNYLLITHIUM FROM (*S*)-ETHYL LACTATE: (*S*)-4-(*t*-BUTYLDIMETHYLSILYLOXY)-2-PENTYN-1-OL . . 615
PREPARATION OF 1-METHOXY-2-(4-METHOXYPHENOXY) BENZENE 626
IRIDIUM-CATALYZED SYNTHESIS OF VINYL ETHERS FROM ALCOHOLS AND VINYL ACETATE 631
THE USE OF POLYSTYRYLSULFONYL CHLORIDE RESIN AS A SOLID SUPPORTED CONDENSATION REAGENT FOR THE FORMATION OF ESTERS: SYNTHESIS OF *N*-[(9-FLUORENYLMETHOXY)CARBONYL]-L-ASPARTIC ACID; α-*TERT*-BUTYL ESTER, β-(2-ETHYL[(1E)-(4-NITROPHENYL)AZO]PHENYL] AMINO]ETHYL ESTER 635
1-HYDROXY-3-PHENYL-2-PROPANONE 642
DIRECT CHLORINATION OF ALCOHOLS: SYNTHESIS OF ETHYL 3-CHLORO-3-PHENYLPROPANOATE 646
PREPARATION OF *HEXAKIS*(4-BROMOPHENYL) BENZENE (HBB) 652

SYNTHESIS OF 5-BROMOISOQUINOLINE AND 5-BROMO-8-NITROISOQUINOLINE 656
COPPER(I)-CATALYZED PREPARATION OF (*E*)-3-IODOPROP-2-ENOIC ACID 662
SELECTIVE Pd-CATALYZED AMINATIONS ON CHLORO-IODOPYRIDINES: SYNTHESIS OF ETHYL 4-[(6-CHLOROPYRIDIN-3-YL)AMINO]BENZOATE 665
A NEW PRACTICAL ONE-POT CONVERSION OF PHENOLS TO ANILINES: 6-AMINO-3,4-DIHYDRO-1(2*H*)-NAPHTHALENONE 670
PRACTICAL SYNTHESIS OF A CHIRAL YNAMIDE: (*R*)-4-PHENYL-3-(2-TRIISOPROPYLSILYL-ETHYNYL) OXAZOLIDIN-2-ONE 677
COPPER-CATALYZED ELECTROPHILIC AMINATION OF DIORGANOZINC REAGENTS: 4-PHENYLMORPHOLINE 684
ALLYLBORONATION OF IMINES: 1-PHENYLHEX-5-EN-3-AMINE 691
IRIDIUM-CATALYZED *N*-HETEROCYCLIZATION OF PRIMARY AMINES WITH DIOLS: *N*-BENZYLPIPERIDINE 697
(2*S*)-(−)-3-exo-(MORPHOLINO)ISOBORNEOL[(−)-MIB] 702
PREPARATION OF (η^5-(*S*)-2-(4-METHYLETHYL)-OXAZOLINYLCYCLOPENTADIENYL)-(η^4-TETRAPHENYLCYCLO-BUTADIENE)COBALT 708
PREPARATION OF THE COP CATALYSTS: $[(S)\text{-COP-OAc}]_2$, $[(S)\text{-COP-Cl}]_2$, AND (*S*)-COP-hfacac 714
ASYMMETRIC REARRANGEMENT OF ALLYLIC TRICHLOROACETIMIDATES: PREPARATION OF (*S*)-2,2,2-TRICHLORO-*N*-(1-PROPYLALLYL) ACETAMIDE 720
PRACTICAL SYNTHESIS OF NOVEL CHIRAL ALLENAMIDES: (*R*)-4-PHENYL-3-(1,2-PROPADIENYL) OXAZOLIDIN-2-ONE 726
BORIC ACID CATALYZED AMIDE FORMATION FROM CARBOXYLIC ACIDS AND AMINES: *N*-BENZYL-4-PHENYLBUTYRAMIDE 733
PREPARATION OF *N*-*p*-TOLYLSULFONYL-(*E*)-1-PHENYLETHYLIDENEIMINE 741
PREPARATION OF ENANTIOMERICALLY ENRICHED (1*S*)-1-PHENYLPROPAN-1-AMINE HYDROCHLORIDE BY A CATALYTIC ADDITION OF DIORGANOZINC REAGENTS TO IMINES 749
(*S*)-(+)-2,4,6-TRIMETHYLBENZENESULFINAMIDE 761

(R_S)-(+)-2-METHYL-2-PROPANESULFINAMIDE 770
2,2-DIETHOXY-1-ISOCYANOETHANE 778
1-(*TERT*-BUTYLIMINO-METHYL)-1,3-DIMETHYL-UREA HYDROCHLORIDE 782
THIONATION: GLCNAC-THIAZOLINE TRIACETATE {(3A*R*, 5*R*, 6*S*, 7*R*, 7A*R*)-5-ACETOXYMETHYL-6, 7-DIACETOXY-2-METHYL-5,6,7,7A-TETRAHYDRO-3A*H*-PYRANO[3,2-*D*]THIAZOLE} 787
ELECTROPHILIC CYCLIZATION WITH *N*-IODOSUCCINIMIDE: PREPARATION OF 5-(4-BROMOPHENYL)-3-IODO-2-(4-METHYL-PHENYL)FURAN 794
PREPARATION OF SUBSTITUTED 5-AZAINDOLES: METHYL 4-CHLORO-1*H*-PYRROLO[3,2-*C*] PYRIDINE-2-CARBOXYLATE 802
SYNTHESIS OF AN *N*-SUBSTITUTED LACTAM USING AN INTRAMOLECULAR SCHMIDT REACTION: FORMATION OF 2,3,11,11A-TETRAHYDRO-1*H*-BENZO[D]PYRROLO[1,2-A]AZEPIN-5(6*H*)-ONE 811
SYNTHESIS OF (−)-(*S,S*)- BIS(4-ISOPROPYLOXAZOLINE) 821
INDIUM-CATALYZED CYCLOISOMERIZATION: PREPARATION OF 4-METHYLPYRROLO[1,2-*A*] QUINOLINE 826
5-*ENDO-TRIG* CYCLIZATION OF 1,1-DIFLUORO-1-ALKENES: SYNTHESIS OF 3-BUTYL-2-FLUORO-1-TOSYLINDOLE . . 834
SYNTHESIS OF DITHIENO[3,2-*B*:2′, 3′-*D*]THIOPHENE 843
1,4-DIOXENE 850
PREPARATION OF 1-BUTYL-3-METHYLIMIDAZOLIUM TETRAFLUOROBORATE 859
PREPARATION OF 2,4-DISUBSTITUTED IMIDAZOLES: 4-(4-METHOXYPHENYL)-2-PHENYL-1H-IMIDAZOLE . . . 862
PREPARATION OF *N*-ARYL-5*R*-HYDROXYMETHYL-2-OXAZOLIDINONES FROM *N*-ARYL CARBAMATES: *N*-PHENYL-(5*R*)-HYDROXYMETHYL-2-OXAZOLIDINONE 867
PREPARATION OF PIVALOYL HYDRAZIDE IN WATER 874
ASYMMETRIC SYNTHESIS OF *N-TERT*-BUTOXYCARBONYL α-AMINO ACIDS. SYNTHESIS OF (5*S*,6*R*)-4-*TERT*-BUTOXYCARBONYL-5,6-DIPHENYLMORPHOLIN-2-ONE 879
SYNTHESIS OF (*R,R*)-4,6-DIBENZOFURANDIYL-2,2′-BIS (4-PHENYLOXAZOLINE) (DBFOX/PH) – A NOVEL TRIDENTATE LIGAND 888
(4*S*)-4-(1-METHYLETHYL)-5,5-DIPHENYL-2-OXAZOLIDINONE . . 896

SYNTHESIS OF INDOLES BY PALLADIUM-CATALYZED REDUCTIVE *N*-HETEROANNULATION OF 2-NITROSTYRENES: METHYL INDOLE-4-CARBOXYLATE 903
PREPARATION OF SUBSTITUTED PYRIDINES VIA REGIOCONTROLLED [4 + 2] CYCLOADDITIONS OF OXIMINOSULFONATES: METHYL 5-METHYLPYRIDINE-2-CARBOXYLATE 911
GENERATION AND [2 + 2] CYCLOADDITIONS OF THIO-SUBSTITUTED KETENES: *TRANS*-1-(4-METHOXYPHENYL)-4-PHENYL-3-(PHENYLTHIO)AZETIDIN-2-ONE 920
PREPARATION OF 2-CHLORO-1,3 BIS(DIMETHYLAMINO) TRIMETHINIUM HEXAFLUOROPHOSPHATE 929
SYNTHESIS OF ALPHA-HALO ETHERS FROM SYMMETRIC ACETALS AND *IN SITU* METHOXYMETHYLATION OF AN ALCOHOL 934
MONO CARBAMATE PROTECTION OF ALIPHATIC DIAMINES USING ALKYL PHENYL CARBONATES 942
PROTECTION OF ALCOHOLS USING 2-BENZYLOXY-1-METHYLPYRIDINIUM TRIFLUOROMETHANESULFONATE: METHYL (*R*)-(−)-3-BENZYLOXY-2-METHYL PROPANOATE 947
A CONVENIENT PREPARATION OF AN ORTHOGONALLY PROTECTED C^{α}, C^{α}-DISUBSTITUTED AMINO ACID ANALOG OF LYSINE: 1-*TERT*-BUTYLOXYCARBONYL-4-((9-FLUORENYLMETHYLOXYCARBONYL) AMINO)-PIPERIDINE-4-CARBOXYLIC ACID 956
SYNTHESIS OF ELECTRON-DEFICIENT SECONDARY PHOSPHINE OXIDES AND SECONDARY PHOSPHINES: BIS[3,5-BIS(TRIFLUOROMETHYL) PHENYL] PHOSPHINE OXIDE AND BIS[3,5-BIS (TRIFLUOROMETHYL)PHENYL] PHOSPHINE 965
PREPARATION OF ENANTIOMERICALLY PURE (*R,R*)-BOZPHOS 982
PREPARATION OF (TRIPHENYLPHOSPHORANYLIDENE)-KETENE FROM (METHOXYCARBONYLMETHYLENE)-TRIPHENYLPHOSPHORANE 986
PREPARATION OF [1-(METHOXYMETHYLCARBAMOYL) ETHYL]PHOSPHONIC ACID BIS-(2,2,2-TRIFLUOROETHYL) ESTER: A USEFUL INTERMEDIATE IN THE SYNTHESIS OF *Z*-UNSATURATED *N*-METHOXY-*N*-METHYLAMIDES 992
TETRABENZYL PYROPHOSPHATE 1001

IRIDIUM-CATALYZED C-H BORYLATION OF ARENES AND HETEROARENES: 1-CHLORO-3-IODO-5-(4,4,5,5-TETRAMETHYL-1,3,2-DIOXABOROLAN-2-YL) BENZENE AND 2-(4,4,5,5,-TETRAMETHYL-1,3,2-DIOXABOROLAN-2-YL) INDOLE 1007
SYNTHESIS OF ORTHO SUBSTITUTED ARYLBORONIC ESTERS BY IN SITU TRAPPING OF UNSTABLE LITHIO INTERMEDIATES: 2-(5,5-DIMETHYL-1,3,2-DIOXABORINAN-2-YL)BENZOIC ACID ETHYL ESTER 1015
PREPARATION OF (*S,S*)-1,2-BIS(*TERT*-BUTYLMETHYLPHOSPHINO)ETHANE ((*S,S*)-*t*-Bu-BISP*) AS A RHODIUM COMPLEX 1020
CHIRAL LITHIUM AMIDE BASE DESYMMETRIZATION OF A RING FUSED IMIDE: FORMATION OF (3a*S*,7a*S*)-2-[2-(3,4-DIMETHOXYPHENYL)-ETHYL]-1,3-DIOXO-OCTAHYDRO-ISOINDOLE-3a-CARBOXYLIC ACID METHYL ESTER 1028
CATALYTIC ASYMMETRIC ACYLATION OF ALCOHOLS USING A CHIRAL 1,2-DIAMINE DERIVED FROM (*S*)-PROLINE: (1*S*,2*S*)-*TRANS*-1-BENZOYLOXY-2-BROMOCYCLOHEXANE 1037
(*R,R*)-2,2′-BISPYRROLIDINE AND (*S,S*)-2,2′-BISPYRROLIDINE: USEFUL LIGANDS FOR ASYMMETRIC SYNTHESIS . . . 1046
LIPASE-CATALYZED RESOLUTION OF 4-TRIMETHYLSILYL-3-BUTYN-2-OL AND CONVERSION OF THE (*R*)-ENANTIOMER TO (*R*)-3-BUTYN-2-YL MESYLATE AND (*P*)-1-TRIBUTYLSTANNYL-1,2-BUTADIENE 1056
ASYMMETRIC ALCOHOLYSIS OF MESO-ANHYDRIDES MEDIATED BY ALKALOIDS 1068

TYPE OF REACTION INDEX 1074
TYPE OF COMPOUND INDEX 1088
FORMULA INDEX 1104
AUTHOR INDEX 1111
GENERAL INDEX 1115
CONCORDANCE INDEX 1137

CATALYTIC REDUCTION OF AMIDES TO AMINES WITH HYDROSILANES USING A TRIRUTHENIUM CARBONYL CLUSTER AS THE CATALYST

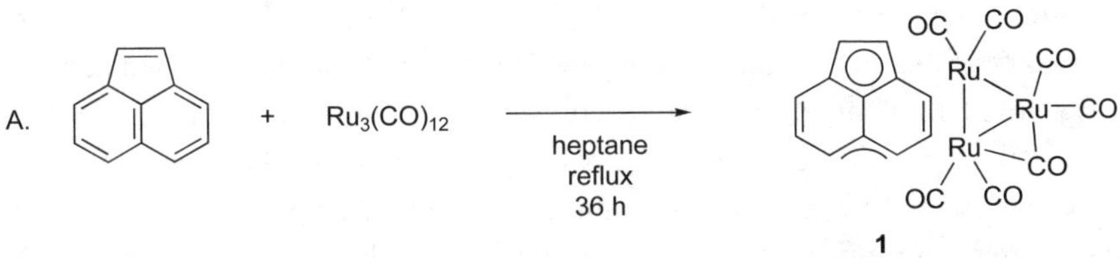

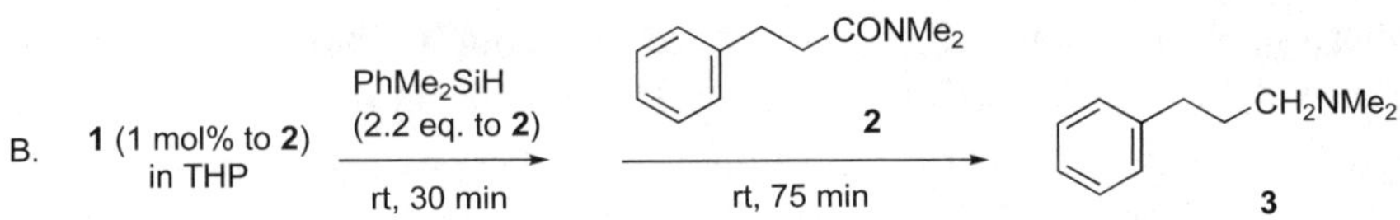

Submitted by Yukihiro Motoyama, Chikara Itonaga, Toshiki Ishida, Mikihiro Takasaki, and Hideo Nagashima.[1]
Checked by Joshua G. Pierce and Peter Wipf.

1. Procedure

Caution! Since reaction intermediates are potentially unstable to air and moisture, preparation of the ruthenium cluster should be carried out under an inert gas atmosphere.

A. (μ_3, η^2: η^3: η^5-Acenaphthylene)$Ru_3(CO)_7$ (**1**). A two-necked, 200-mL round-bottomed flask is equipped with a magnetic stir bar, a reflux condenser, and a stopcock. The top of the condenser is fitted with a three-way stopcock, of which one way is connected to an argon flow line and the other is connected to a vacuum line. The apparatus is flame-dried while under vacuum, and allowed to cool to room temperature under an argon purge. The stopcock is removed, and the flask is charged with $Ru_3(CO)_{12}$ (Note 1) (639 mg, 1 mmol), acenaphthylene (Note 2) (182 mg, 1.2 mmol), and heptane (Note 3) (75 mL) under an argon stream. The stopcock is again placed on the flask, and the mixture is heated under reflux. At the initial stage, the solution contains a yellow supernatant and an undissolved orange solid of $Ru_3(CO)_{12}$. The color of the solution gradually turns to red and all of the solid materials are dissolved after 30 min (Note 4). After 36 h, a dark red solution and some red precipitates have formed. The mixture is cooled to room temperature under an argon atmosphere. From this stage, subsequent work can be performed in the air. The precipitates are collected on a glass filter, washed with *n*-hexane (30 mL), and dissolved in dichloromethane (20 mL) (Note 5). Celite® (*ca.* 3 g) (Note 6) is added to this solution and stirred for 1 min, then the solvent is removed under reduced pressure. The crude ruthenium complex adsorbed to Celite® is applied to a pre-packed silica gel column (20 mm x 100 mm) (Note 7), then sea sand (10 mm) (Note 8) is loaded onto the top of the Celite® for preventing the powder from diffusing into the eluent. The silica gel is first washed with 80 mL of *n*-hexane to remove a small amount of unreacted acenaphthylene (yellow band), then the product (red-brown band) is eluted with 2.5 L of 2:1 ether:*n*-hexane (Note 9, 10) to afford 620 mg (95%) of (μ_3, η^2: η^3: η^5-acenaphthylene)$Ru_3(CO)_7$ (**1**) (Note 11), which can be used for reduction of amides without further purification. Dark red crystals of **1** are obtained by recrystallization of the product obtained as above from a mixture of dichloromethane and *n*-hexane at -30 °C (Note 12).

*Caution! The cluster **1** is first treated with excess hydrosilane for activation. Treatment of the activated catalyst with amides in the presence of excess hydrosilane is exothermic. The reaction apparatus should be equipped with an efficient reflux condenser, and the amide should be added carefully to avoid overheating of the reaction mixture.*

B. N,N-Dimethyl-3-phenylpropylamine (**3**). A 200-mL, two-necked round-bottomed flask is equipped with an efficient reflux condenser and a septum. The top of the reflux condenser is fitted with a three-way stopcock, of which one way is connected to an argon flow line and the other is connected to a vacuum line. The apparatus is flame-dried *in vacuo*, and allowed to cool to room temperature under an argon purge. The septum is removed, and the flask is charged with **1** (365 mg, 0.557 mmol, 1 mol% based on *N,N*-dimethyl-3-phenylpropionamide (**2**) (Note 13)) under an argon flow. The septum is again placed on the flask, and tetrahydropyran (THP) (Note 14, 15) (4.5 mL) is added to dissolve the catalyst. Phenyldimethylsilane (Note 16) (19 mL, 123 mmol) is charged *via* syringe through the septum, and the mixture is stirred for 30 min at room temperature. At this point, the initial dark-orange color of the catalyst solution changed to light orange. *N,N*-Dimethyl-3-phenylpropionamide (**2**) (10 mL, 56 mmol) (Note 17) is added drop wise over a period of 75 min *via* syringe (Note 18). *Caution! The reaction is exothermic* (Note 19). The solution is stirred for an additional 75 min, resulting in consumption of the amide as determined by TLC analysis (Note 20). The solvent is removed by rotary evaporator at room temperature (70 Torr) and the residue is poured into a 500 mL Erlenmeyer flask containing conc. HCl (84 mL, 12 M) at 0 °C. The reaction flask is rinsed with ether (3 x 84 mL), and the liquid is added to the Erlenmeyer flask. The mixture is moved to a separatory funnel and shaken well. The resulting aqueous solution is washed three times with 84 mL of ether (totally 252 mL). All ether layers are discarded and the aqueous phase is moved to a 500 mL Erlenmeyer flask, cooled in an ice bath, and basified with KOH pellets (84 g) with stirring to form white solids. Ether (164 mL) is added to the mixture, stirred for 20 min, and the organic layer is separated. This ether extraction is repeated four times (84 mL each) (Note 21). The combined organic phases are dried over anhydrous Na_2SO_4 and concentrated under reduced pressure at room temperature (50 Torr). Distillation of the crude product at 57 °C (2.1 Torr) gives *N,N*-dimethyl-3-phenylpropylamine (**3**) as a colorless oil (Note 22, 23) (8.32 g, 91%).

2. Notes

1. $Ru_3(CO)_{12}$ (99%) is commercially available from Aldrich Chemical Company, Inc., although the submitters have prepared it according to a reported procedure.[2]

2. Acenaphthylene was purchased from Aldrich Chemical Company, Inc. (99+%) or Tokyo Kasei Kogyo Company, Ltd. (99+%) and used as received.

3. Heptane (99+%) was purchased from Kanto Chemical Company, Inc. and distilled from calcium hydride before use. Checkers purchased heptane from Aldrich Chemical Company.

4. The progress of the reaction was followed by TLC analysis (Merck Art No. 5715 : Silica gel 60 F_{254}, 0.25 mm thickness) with 2:1 ether:*n*-hexane as eluent and visualization with a UV lamp at 254 nm. The acenaphthylene and $Ru_3(CO)_{12}$ starting materials have an R_f= 0.89 and 0.95, respectively, and the complex **1** has an R_f = 0.42.

5. *n*-Hexane (96%) and dichloromethane (99%) were purchased from Kishida Chemical Co., Ltd. (submitters) or VWR (checkers), and used as received.

6. Celite 545 was purchased from Kishida Chemical Company, Ltd. (submitters) or Fisher Scientific (checkers), and used as received.

7. Silica gel 60 (70-230 mesh) (Merck Art No. 7734) was purchased from Merck (submitters) or Sorbent Technologies (checkers), and used as received.

8. Sea sand (425-850 μm; 20-35 mesh) was purchased from Wako Pure Chemical Industries, Ltd. (submitters) or Aldrich (checkers), and used as received.

9. Ether (99.5%) was purchased from Kishida Chemical Co., Ltd. (submitters) or VWR (checkers), and used as received.

10. Although the submitters prefer ether to dichloromethane for environmental reasons, a 1:1 mixture of dichloromethane:*n*-hexane (600 mL) can alternatively be used as the eluent for collection of a second red-brown band. Use of dichloromethane instead of ether can reduce the total amounts of eluent. TLC analysis (Merck Art No. 5715 : Silica gel 60 F_{254}, 0.25 mm thickness) was performed with 50% dichloromethane-hexane as eluent and visualization with UV lamp at 254 nm. The acenaphthylene and $Ru_3(CO)_{12}$ starting materials have $R_f = 0.86$ and 0.91, and the complex **1** has $R_f = 0.22$.

11. The product exhibits the following physical and spectral properties: mp 231.2-232.1 °C (dec). ^{1}H NMR (500 MHz, $CDCl_3$) δ: 6.48 (d, 1 H, $J = 6.3$ Hz), 6.25 (dd, 1 H, $J = 9.2$, 6.1 Hz), 6.01 (d, 1 H, $J = 2.5$ Hz), 5.81 (d, 1 H, $J = 9.2$ Hz), 5.58 (t, 1 H, $J = 6.3$ Hz), 5.48 (d, 1 H, $J = 6.0$ Hz), 4.90 (d, 1 H, $J = 2.5$ Hz), 2.52 (d, 1 H, $J = 6.1$ Hz). ^{13}C NMR (76 MHz, THF-d_8, -60 °C) δ: 265.9 (μ-CO), 207.4 (CO), 204.7 (CO), 204.0 (CO), 195.7 (CO), 192.2 (CO), 188.7 (CO), 130.4, 116.1, 99.1, 88.3, 84.8, 82.2, 80.1, 78.5, 69.0, 67.0, 66.1, 37.7. IR (KBr) cm^{-1} ν_{CO} 2039 (s), 2028 (s), 2012 (s), 1996 (s), 1986 (s), 1946 (s), 1768 (m: μ-CO). The purity (>98%) was determined by ^{1}H NMR and mp.

12. The complex (624 mg) is dissolved in 30 mL of dichloromethane at 40 °C, then cooled at room temperature. To this solution, 70 mL of *n*-hexane is carefully added and the two-phase solution is stored at –37 °C overnight. Orange crystals are collected by suction filtration on a Buchner funnel, washed three times with 20 mL of *n*-hexane (total of 60 mL), and then transferred to a 20 mL, round-bottomed flask and dried for 4 h at 0.01 Torr to provide 497 mg (80%) of **1** as orange microcrystals, which can be stored in the air for years without decomposition.

13. At lower catalyst concentration, formation of an unidentified byproduct decreases the yield of the product. For example, use of 0.1 mol% of the catalyst to **2** affords the desired amine in about 60% yield.

14. Tetrahydropyran (THP) (98+%) was purchased from Kanto Chemical Company, Inc. (submitters) or Aldrich (checkers), and distilled from benzophenone ketyl before use. Anhydrous toluene, diethyl ether, and 1,4-dioxane (a cancer suspected reagent) can alternatively be used as the

solvent.

15. The ruthenium cluster activated by hydrosilane is an efficient catalyst for conversion of hydrosilanes to silanols and siloxanes. The reduction should be performed in anhydrous THP under an inert gas atmosphere to avoid contact with moisture.

16. Phenyldimethylsilane (98+%) was purchased from Aldrich Chemical Company, Inc. and distilled before use.

17. *N,N*-Dimethyl-3-phenylpropionamide is commercially available in milligram quantities (TimTec Overseas Stock, Ambinter Screening Library, MicroChemistry Screening Collection), but it can prepared from dihydrocinnamoyl chloride (95+%) (purchased from Tokyo Kasei Kogyo Company, Ltd. (submitters) or Aldrich (checkers)) and dimethylamine (99+%) (purchased from Aldrich Chemical Company, Inc.) by application of a standard method and distilled before use (bp 108 °C / 2 Torr). The purity (>99%) was determined by ^{1}H NMR, and the submitters also used capillary GLC with a TC-WAX column (0.25 mm x 30 m) (oven temperature: 230 °C; head pressure: 60 kPa; retention time: 11.8 min).

18. Although a syringe was used for the addition of amide **2**, a syringe pump or an addition funnel can also be used.

19. Although the submitters controlled the exothermic reaction by the addition rate of the amide, a water bath can also be used to cool the reaction mixture.

20. The progress of the reaction was followed by TLC analysis (Merck Art No. 5715: Silica gel 60 F_{254}, 0.25 mm thickness) with ethyl acetate as eluent and visualization with UV lamp at 254 nm. The amide starting material has an $R_f = 0.48$ and the amine product has an $R_f = 0.19$.

21. The submitters reported the reduction gave a single product (determined by ^{1}H NMR) in an earlier paper,[3] but isolated yields were somewhat lower (75%). This can be attributed to a loss of the product during the chromatographic removal of siloxane byproducts. In the improved procedure reported herein, separation of the silane products from the desired amine is achieved by simple extraction, which contributes to an improvement in the yield of the product.

22. The product **3**[4] exhibits the following physical and spectral properties: ^{1}H NMR (500 MHz, $CDCl_3$) δ: 7.32-7.27 (m, 2 H), 7.22-7.18 (m, 3 H), 2.65 (t, 2 H, J = 7.7 Hz), 2.31 (t, 2 H, J = 7.3 Hz), 2.24 (s, 6 H), 1.45-1.78 (m, 2 H). ^{13}C NMR (76 MHz, $CDCl_3$) δ: 142.2, 128.3, 128.2, 125.6, 59.2, 45.4, 33.5, 29.4. IR (neat) cm^{-1} 3062, 3026, 2942, 2764, 1603, 1496, 1454, 1265, 1030. EIMS *m/z* 163 (M^+, 40), 91 (60), 84 (65), 74 (60),

59 (100), 58 (90); HRMS (EI) *m/z* calcd for $C_{11}H_{17}N$ 163.1376, found 163.1361. The purity (>99%) was determined by ^{1}H NMR, and the submitters also used capillary GLC with a TC-WAX column (0.25 mm x 30 m) (oven temperature: 120 °C; head pressure: 60 kPa; retention time: 16.1 min).

23. The submitters measured the amount of residual ruthenium species in the product by ICP-MS to be less than *ca.* 3 ppm.

Safety and Waste Disposal Information

All hazardous materials should be handled and disposed of in accordance with "Prudent Practices in the Laboratory"; National Academy Press; Washington, DC, 1995.

3. Discussion

The reduction of amides has been generally accomplished by aluminum hydrides such as $LiAlH_4$ in THF or Et_2O, DIBAL-H, and Red-Al in benzene at room temperature.[5,6] Alanes, AlH_3 in Et_2O and $AlH_3 \cdot Et_3N$, or boranes, $BH_3 \cdot THF$ and $BH_3 \cdot SMe_2$ are also useful.[6] However, these reagents are air- and moisture sensitive, and careful handling is required. It is also well known that separation of the product from aluminum or boron wastes is often troublesome. Alkali borohydrides are rather stable and their handling is easy; however, reduction of amides with $NaBH_4$ requires application of high temperatures or addition of activators. For example, reduction of amides is achieved by $NaBH_4$ in the presence of Lewis acids such as $TiCl_4$ and $ZrCl_4$.[5] Activation of amides is an alternative method for successful borohydride reduction; a two step procedure consisting of iminium salt formation from amides by treatment with Et_3OBF_4 or $POCl_3$ followed by reduction with $NaBH_4$ is reported.[7]

Hydrosilanes are stable reducing agents and their handling is easy. Although a number of catalytic hydrosilylations of ketones and aldehydes have been reported, in which addition of the H-Si bond to the C=O moiety is catalyzed by transition metal salts or complexes, there have been few reports for the reduction of amides. The $RhCl(PPh_3)_3$-catalyzed reduction of amides has recently been reported by Ito and coworkers; the reaction requires the use of dihydrosilanes or trihydrosilanes.[8] Removal of the silicon waste from the reaction mixture is not easy. Fuchikami reported the reduction of amides with trialkylsilanes, which are rather convenient to manipulate compared to dihydrosilanes or trihydrosilanes, in the presence of several ruthenium catalysts; however, the reaction proceeds under drastic conditions.[9] Thus, the procedure using trialkylsilanes, which are stable and easy to work with, deserves further attention, especially when the reduction of amides can be achieved under mild conditions.

The ruthenium complex **1**[10] is a useful catalyst for hydrosilylation of ketones and aldehydes,[10] reduction of carboxylic acids, esters, and amides with hydrosilanes,[3] and reduction of acetals and cyclic ethers with hydrosilanes.[11] It is also a good catalyst for silane-induced ring-opening polymerization of cyclic ethers,[11] cyclic siloxanes,[12] and vinyl ethers.[13] Treatment of **1** with molecular hydrogen followed by treatment with carbon monoxide provides 4,5-dihydroacenaphthylene, which has never been synthesized by other methods.[14]

The amide reduction is accomplished under mild conditions by the use of **1** as the catalyst. Phenyldimethylsilane as the reducing reagent has the advantage of stability towards air and moisture, and it is easy to handle. It is known that the siloxane byproducts are sometimes difficult to separate in many silane-mediated reductions of carbonyl compounds. In the present procedure, the siloxane byproducts are easily removed by extraction. Thus, the present method is an easy way to synthesize amines from amides.

1. Institute for Materials Chemistry and Engineering, Graduate School of Engineering Sciences, Kyushu University, Kasuga, Fukuoka 816-8580, Japan.

2. Bruce, M. I.; Jensen, C. M.; Jones, N. L. *Inorg. Synth.* **1990**, *28* (Reagents Transition Met. Complex Organomet. Synth.), 216-220, and references cited therein.
3. (a) Matsubara, K.; Iura, T.; Maki, T.; Nagashima, H. *J. Org. Chem.* **2002**, *67*, 4985. (b) Motoyama, Y.; Mitsui, K.; Ishida, T.; Nagashima, H. *J. Am. Chem. Soc.* **2005**, *127*, 13150. (c) Hanada, S.; Ishida, T.; Motoyama, Y.; Nagashima. H. *J. Org. Chem.* **2007**, *72*, 7551.
4. Kuehne, M. E.; Shannon, P. J. *J. Org. Chem.* **1977**, *42*, 2082.
5. Cope, A. C.; Ciganek, E. *Org. Synth. Coll. Vol.* **1963**, *4*, 339.
6. Seyden-Penne, J. *Reductions by the Alumino- and Borohydrides in Organic Synthesis*; Wiley-VCH (1997).
7. Borch, R. F. *Tetrahedron Lett.* **1968**, 61.
8. Kuwano, R.; Takahashi, M.; Ito, Y. *Tetrahedron Lett.* **1998**, *39*, 1017.
9. Igarashi, M.; Fuchikami, T. *Tetrahedron Lett.* **2001**, *42*, 1945.
10. Nagashima, H.; Fukahori, T.; Aoki, K.; Itoh, K. *J. Am. Chem. Soc.* **1993**, *115*, 10430.
11. Nagashima, H.; Suzuki, A.; Iura, T.; Ryu, K.; Matsubara, K. *Organometallics* **2000**, *19*, 3579.

12. Matsubara, K.; Terasawa, J.; Nagashima, H. *J. Organomet. Chem.* **2002**, *660*, 145.
13. Nagashima, H.; Itonaga, C.; Yasuhara, J.; Motoyama, Y.; Matsubara, K. *Organometallics* **2004**, *23*, 5779.
14. Nagashima, H.; Suzuki, A.; Nobata, K.; Itoh, K. *J. Am. Chem. Soc.* **1996**, *118*, 687.

SYNTHESIS OF 4-TRIISOPROPYLSILYL-3-BUTYN-2-OL BY ASYMMETRIC TRANSFER HYDROGENATION

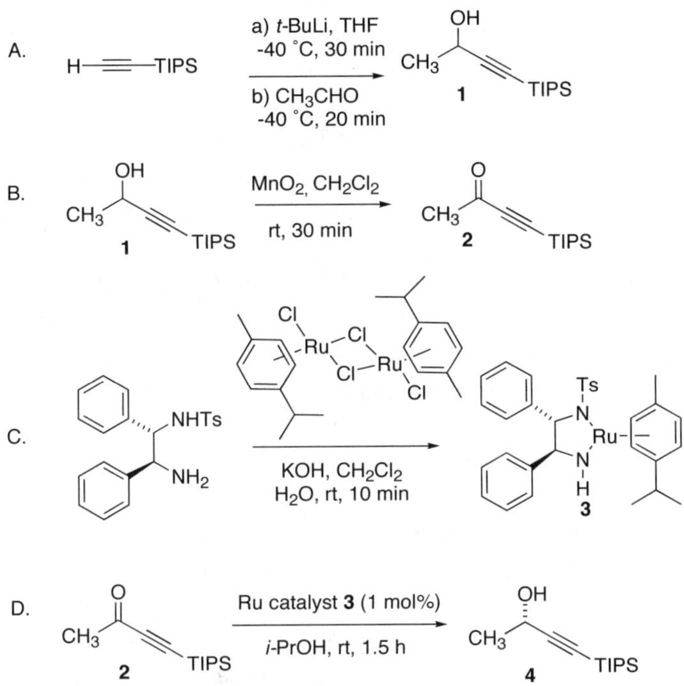

Submitted by James A. Marshall, Patrick Eidam, and Hilary Schenck Eidam.[1a]
Checked by Kai Zhang and Dennis P. Curran.[1b]

1. Procedure

A. *(rac)-4-Triisopropylsilyl-3-butyn-2-ol (1).* A single-necked, flame-dried, 250-mL round-bottomed flask equipped with a magnetic stir bar and a rubber septum fitted with an argon inlet needle is flushed with argon and charged with triisopropylsilyl acetylene (95%, 6.1 mL, 4.75 g, 26.0 mmol) (Note 1) and dry THF (54 mL) (Note 2) by syringe through the septum. The reaction mixture is cooled to –40 °C (bath temperature) by means of a dry ice-acetonitrile bath and *tert*-BuLi (1.7 M in pentane, 18 mL, 32.3 mmol) is added dropwise by means of a syringe (Note 3). The resulting bright yellow mixture is stirred at –40 °C for 30 min and then ACS reagent grade acetaldehyde (>99.5%, 2.3 mL, 39.2 mmol) is added in one portion by means of a syringe. The reaction mixture is stirred for 20 min at –40 °C and then poured over a rapidly stirring solution of saturated aqueous NH_4Cl (75 mL) at room temperature. After 15 min, the phases are separated and the aqueous layer is extracted with diethyl ether (2 x 40 mL). The combined organic extracts are washed with brine, dried over $MgSO_4$, filtered, and the solvent is removed with a rotary evaporator (40 °C, water aspirator pressure). The residue is purified by bulb-to-bulb distillation (95 °C, 0.2 mmHg; Note 4) to yield the racemic alcohol (5.8 g, 98%) as a clear oil (Notes 5, 6, 7). This material is used in the next step without further purification.

B. 4-Triisopropylsilyl-3-butyn-2-one (2). A one-necked, flame-dried, 250-mL round-bottomed flask equipped with a magnetic stir bar and a rubber septum fitted with an argon inlet needle is flushed with argon. The septum is temporarily removed and the flask is charged sequentially with CH_2Cl_2 (54 mL), 4-triisopropylsilyl-3-butyn-2-ol (**1**) (5.8 g, 25.6 mmol) and technical grade MnO_2 (85%, 34.4 g, 336.9 mmol) (Notes 8, 9), which is added in one portion. The reaction mixture is stirred at room temperature for 30 min and then filtered through a short pad of Celite. The Celite is washed with CH_2Cl_2 (2 x 30 mL) and the solvent is removed on a rotary evaporator (40 °C, water aspirator pressure). The resulting yellow oil is purified by bulb-to-bulb distillation (80 °C, 0.2 mmHg; Note 4) to yield ketone **2** (5.0–5.1 g, 87–89%) as a clear oil (Notes 10, 11). This material is used in the next step without further purification.

C. RuCl[(S,S)-NTsCH(C_6H_5)CH(C_6H_5)NH_2(η^6-cymene) (3). (Note 12). A one-necked, flame-dried, 50-mL round-bottomed flask equipped with a magnetic stir bar and a rubber septum fitted with an argon inlet needle is flushed with argon. The septum is temporarily removed and the flask is charged with CH_2Cl_2 (6 mL), (1*S*,2*S*)-(+)-*N*-*p*-toluenesulfonyl-1,2-diphenylethylene-diamine (126 mg, 0.34 mmol), dichloro(*p*-cymene)ruthenium(II) dimer (107 mg, 0.17 mmol), and powdered KOH (141 mg, 2.5 mmol) (Note 1). The resulting orange mixture is stirred for 5 min at room temperature and then distilled H_2O (6 mL) is added in one portion by means of a syringe. The resulting biphasic mixture is stirred for 10 min during which time the organic phase turns dark purple. The mixture is transferred to a separatory funnel, diluted with 10 mL of H_2O and the layers are separated. The aqueous phase is extracted with CH_2Cl_2 (2 x 10 mL). The combined organic extracts are dried over CaH_2 (Note 1), filtered and solvent is removed with a rotary evaporator (40 °C, water aspirator pressure) to furnish the catalyst **3** (180–182 mg, 86–87%) as a dark purple solid that is used in the subsequent reduction without additional purification.

D. *(S)-4-Triisopropylsilyl-3-butyn-2-ol (4)*. A one-necked, flame-dried, 500-mL round-bottomed flask equipped with a magnetic stir bar and a rubber septum fitted with an argon inlet needle is flushed with argon. Isopropyl alcohol (250 mL) and 4-triisopropylsilyl-3-butyn-2-one (**2**) (5.1 g, 22.7 mmol) are added to the flask through the septum by syringe. The ruthenium catalyst **3** (180 mg, 0.3 mmol) is taken up in a minimal amount of CH_2Cl_2 (~5 mL) and added to the reaction mixture in one portion by means of a syringe. The mixture is stirred for 1.5 h and then the solvent is removed with a rotary evaporator (40 °C, water aspirator pressure). The brown residue is purified by bulb-to-bulb distillation (95 °C, 0.2 mmHg; Note 4) to yield the alcohol **4** (5.0 g, 85% for three steps) as a clear oil (Notes 13, 14, 15).

2. Notes

1. All chemicals were purchased from Aldrich Chemical Co. and were used without further purification. The submitters purchased THF and CH_2Cl_2 as optidry Fischer Paks from Fischer Scientific. The checkers purchased dry CH_2Cl_2 from Aldrich and used THF from a sodium/benzophenone still.

2. The submitters reported that the use of diethyl ether in place of THF resulted in diminished yields and the formation of impurities not separable by distillation.

3. The submitters reported that the use of *n*-BuLi (20.2 mL of 1.6 M *n*-BuLi in hexanes) afforded the adduct **1** in quantitative yield. The spectra of the sample prepared in this way were identical to those of the sample from the *tert*-BuLi experiment.

4. Bulb-to-bulb distillations were performed with an Aldrich Kugelrohr apparatus.

5. Physical characteristics for 4-triisopropylsilyl-3-butyn-2-ol (**1**): R_f = 0.41 (15% EtOAc-hexane) on Merck silica gel 60 F 254 precoated 250 μm plates; IR (film): 3325, 2943, 2866, 2173, 1464 cm^{-1}; ^{1}H NMR (300 MHz, $CDCl_3$) δ: 1.08 (m, 21 H), 1.48 (d, J = 6.6 Hz, 3 H), 1.78 (s, 1 H), 4.55 (q, J = 6.6 Hz, 1 H); ^{13}C NMR (75 MHz, $CDCl_3$) δ: 11.1, 18.5, 24.5, 58.7, 84.3, 109.8.

6. The submitters reported that the product was contaminated by up to 3% triisopropylacetylene in some runs. This was analyzed by integration of the alkynyl H signal at 2.35 ppm in the ^{1}H NMR spectrum.

7. An analytical sample was prepared by chromatography on silica gel. A 1.5 cm x 20 cm glass column packed with 9 g of EMD silica gel 60 was used to purify 20 mg of alcohol by elution with 10% diethyl ether in hexanes. Anal. Calcd for $C_{13}H_{26}OSi$: C, 68.96; H, 11.57. Found: C, 68.96; H, 11.66.

8. MnO_2 was purchased from Aldrich Chemicals and dried in an oven at 110 °C overnight prior to use.

9. The submitters reported that exposure to less than 13 equiv of MnO_2 resulted in incomplete conversion and formation of byproducts upon prolonged reaction times.

10. Physical characteristics for 4-triisopropylsilyl-3-butyn-2-one **(2)**: R_f = 0.76 (15% EtOAc-hexane) on Merck silica gel 60 F 254 precoated 250 μm plates; IR (film): 2946, 2868, 2147, 1684, 1464, 1196 cm^{-1}; ^1H NMR (300 MHz, $CDCl_3$) δ: 1.10 (m, 21 H), 2.37 (s, 3 H); ^{13}C NMR (75 MHz, $CDCl_3$) δ: 11.3, 18.4, 32.8, 95.2, 104.6, 184.4.

11. An analytical sample was prepared by chromatography on silica gel. (A 1.5 cm x 20 cm glass column packed with 9 g of of EMD silica gel 60 was used to purify 20 mg of ketone by elution with 5% diethyl ether in hexanes. Anal. Calcd for $C_{13}H_{24}OSi$: C, 69.58; H, 10.78. Found: C, 69.56; H. 10.89.

12. The procedure described for the preparation of RuCl[(*S*,*S*)-NTsCH(C_6H_5)CH(C_6H_5)NH_2(η^6-cymene) was taken from reference 9.

13. Physical characteristics for *(S)*-4-triisopropylsilyl-3-butyn-2-ol (**4**): Same as for racemic 4-triisopropylsilyl-3-butyn-2-ol; $[\alpha]_D^{20}$ –21.3 (c = 1.58, $CHCl_3$). The er of this material was found to be >95:5 by ^{19}F NMR analysis [δ–73.68 ppm (major) and –73.02 ppm (minor), in $CDCl_3$ versus benzotrifluoride as an internal standard at –63.72 ppm] of the *(R)*-Mosher ester. To prepare the Mosher ester, 4-triisopropylsilyl-3-butyn-2-ol (4 mg), (*S*)-(+)-methoxy-α-trifluoromethylphenylacetyl chloride (5 μL) and dry pyridine (0.15 mL) were added to a one-necked, flame-dried, 5-mL round-bottomed flask. The flask was equipped with a magnetic stir bar and a rubber septum and was flushed with argon. The reaction mixture was stirred at room temperature for 2 h, then the volatiles were evaporated to yield the crude (*R*)-Mosher ester, which was directly analyzed by ^{19}F NMR spectroscopy without purification.

14. GC analysis on a β-Dex column temperature programmed from 130 to 180 °C gave a single peak at 22.35 min (submitters). GC analysis on an Agilent MP-1 column (1909/Z-413E, 30 m x 0.32 mm), temperature ramp 10 °C per min from 50 °C to 315 °C, gave a single peak at 10.61 min

(checkers). Anal. Calcd for $C_{13}H_{26}OSi$: C, 68.96; H, 11.57. Found: C, 69.03; H, 11.74.

15. The submitters measured $[\alpha]_D^{20}$ –19.5 (c = 1.55, $CHCl_3$) for a run on the same scale and concentration. When they performed the reduction at a ten-fold higher concentration of ketone, the derived alcohol **4** exhibited $[\alpha]_D^{20}$ –16.5 (c = 1.81, $CHCl_3$).

Safety and Waste Disposal Information

All hazardous materials should be handled and disposed of in accordance with "Prudent Practices in the Laboratory"; National Academy Press; Washington, DC, 1995.

3. Discussion

The palladium-catalyzed conversion of enantioenriched propargylic mesylates such as **2** to enantioenriched allenylzinc (**3**) or indium (**4**) reagents and in situ additions to aldehydes provides a convenient route to homopropargylic alcohol intermediates related to polyketide natural products (Scheme 1).[2,3,4]

Scheme 1. Synthesis of Enantioenriched Homopropargylic Alcohols from Propargylic Mesylates

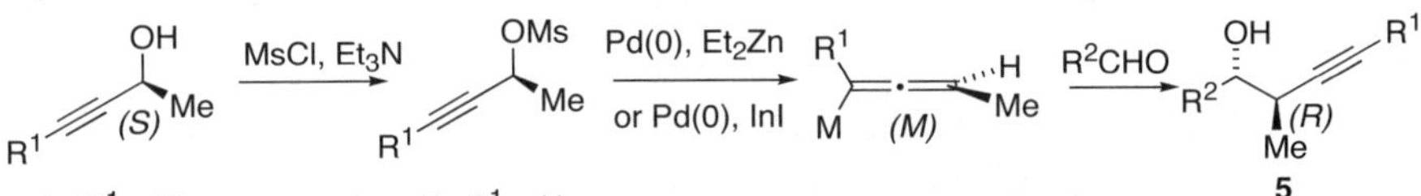

1a R¹ = H
1b R¹ = TMS
1c R¹ = TIPS

2a R¹ = H
2b R¹ = TMS
2c R¹ = TIPS

3a/b/c M = Zn(OMs)
4a/b/c M = InI(OMs)

Although the alcohol precursor of mesylate (*S*)-**2a** and its enantiomer are commercially available,[5] their high cost discourages widespread use of the methodology. Accordingly, we have explored simple and economical preparations through use of chiral pool starting materials[6] or lipase resolution of racemic propargylic alcohols for a more cost effective synthesis.[7,8] The TMS reagents **2b**, prepared through lipase resolution of racemic **1b**, show significantly higher diastereoselectivity than the parent alkyne **2a**. While this alternative represents an improvement, the intrinsic inefficiency of optical resolution coupled with the volatility of the resolved alcohols represent significant drawbacks to the methodology. The present procedure employs Noyori asymmetric hydrogenation of alkynyl ketones to prepare the alkynyl TIPS derivative **1c** of 3-butyn-2-ol.[9] This route offers several major advantages compared with the previous approaches to the parent **1a** and the TMS alcohol **1b**:

(1) Unlike alcohol **1a** and its alkynyl TMS analog **1b**, the TIPS compound **1c** is nonvolatile and completely insoluble in water, thereby allowing facile isolation by simple extraction;
(2) all intermediates can be vacuum-dried and purified by short path distillation;
(3) both enantiomers of the chiral catalyst are readily available;
(4) the catalyst precursor can be stored without loss of activity; and,
(5) addition reactions of the derived TIPS allenylzinc reagent to aliphatic aldehydes proceeds readily and with comparable diastereoselectivity to additions of the parent reagent or the TMS derivative.

The outcome of addition reactions of the three allenylzinc reagents generated in situ from mesylates **2a**, **2b**, and **2c** to cyclohexanecarboxaldehyde is summarized in Table 1.

Table 1. Comparison of Addition Reactions of Allenylzinc Reagents Derived in situ from the Mesylate of *(S)*-3-Butyn-2-ol (**2a**) and the Silylated Analogs **2b** and **2c** to Cyclohexanecarboxaldehyde.

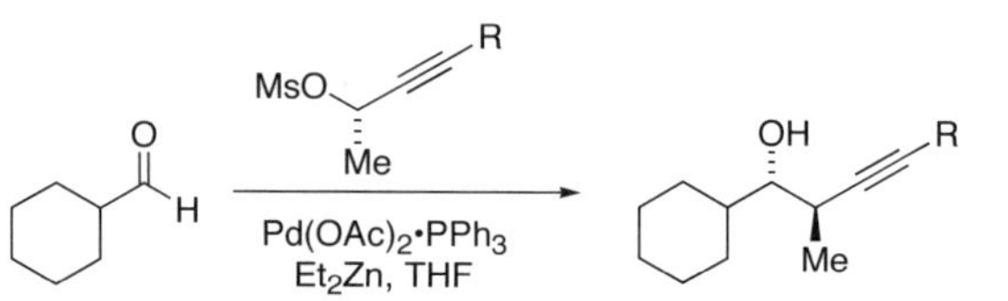

R	yield	dr	er
H (**2a**)	86	19:1	99:1
TMS (**2b**)	84	24:1	97:3
TIPS (**2c**)	89	19:1	98:2

An even larger difference in diastereoselectivity for the three reagents was observed with an α-methylated enal (Table 2). Removal of the TIPS group from the adducts is easily effected in minutes with tetrabutylammonium fluoride in THF.

Table 2. Comparison of Addition Reactions of Allenylzinc Reagents Derived in situ from the Mesylate of *(S)*-3-Butyn-2-ol (**2a**) and the Silylated Analogs **2b** and **2c** to a Conjugated Aldehyde.

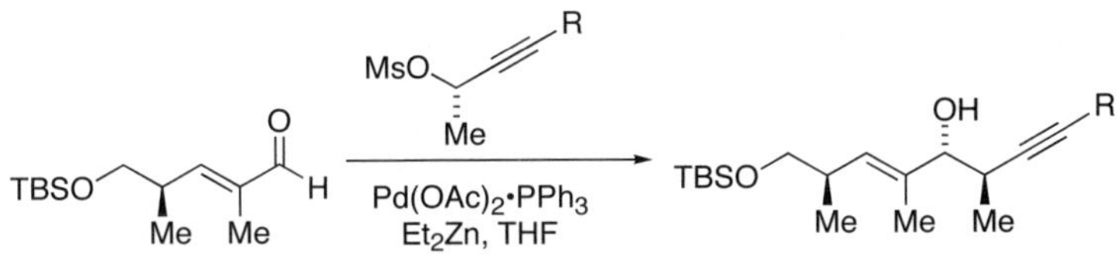

R	yield	dr
H (**2a**)	71	85:15
TMS (**2b**)	56	88:12
TIPS (**2c**)	80	92:8

1. (a) Department of Chemistry, University of Virginia, PO Box 400319, Charlottesville VA 22904, e-mail jam5x@virginia.edu. (b) Department of Chemistry, University of Pittsburgh, Pittsburgh, PA 15260, curran@pitt.edu.

2. Marshall, J. A. *Chem. Rev.* **1996**, *96*, 31. (b) Marshall, J. A. *Chem. Rev.* **2000**, *100*, 3163.

3. (a) Marshall, J. A.; Adams, N. D. *J. Org. Chem.* **1999**, *64*, 5201. (b) Marshall, J. A.; Adams, N. D. *J. Org. Chem.* **1998**, *63,* 3812. (c) Marshall, J. A.; Grant, C. M. *J. Org. Chem.* **1999**, *64*, 8214.

4. (a) Zincophorin. Marshall, J. A.; Palovich, M. R. *J. Org. Chem.* **1998**, *63*, 3701. (b) Discodermolide. Marshall, J. A.; Johns. B. A. *J. Org. Chem.* **1998**, *63*, 7885. (c) Callystatin A. Marshall, J. A.; Fitzgerald, R. N. *J. Org. Chem.* **1999**, *64*, 4477. (d) Aplyronine A. Marshall, J. A.; Johns, B. A. *J. Org. Chem.* **2000**, *65*, 1501. (e) Balifomycin. Marshall, J. A.; Adams, N. D. *Organic Lett.* **2000**, *2*, 2897. (f) Tautomycin. Marshall, J. A.; Yanik, M. *J. Org. Chem.* **2001**, *66,* 1373. (g) (–)-Callystatin A. Marshall, J. A.; Bourbeau, M. P. *J. Org. Chem.* **2002**, *67,* 2751. Marshall, J. A.; Bourbeau, M. P. *Org. Lett.* **2002** *4,* 3931. (h) Membrenone C. Marshall, J. A.; Ellis, K. C. *Org. Lett.* **2003**, *5,* 1729. (i) Leptofuranin D. Marshall, J. A.; Schaaf, G. M. *J. Org. Chem.* **2003**, *68,* 7428. (j) Cytostatin. Marshall, J. A.; Ellis, K. *Tetrahedron Lett.* **2004**, *45,* 1351.

5. Aldrich Chemical Co., Inc. 1001 West St. Paul Avenue, Milwaukee, WI 53233.

6. Marshall, J. A.; Yanik, M. M.; Adams, N. D.; Ellis, K. C.; Chobanian, H. R. *Org Synth.* **2004**, *81,* 157.

7. Marshall, J. A.; Chobanian, H. R.; Yanik, M. M. *Org. Lett.* **2001**, *3,* 3369.

8. Burgess, K.; Jennings, L. D. *J. Am. Chem. Soc.* **1991**, *113*, 6129. For an application of the lipase resolution methodology to (*E*)-4-diphenylsilyl-3-buten-2-ol and 4-diphenylsilyl-3-butyn-2-ol, see , respectively: Beresis, R. T.; Solomon, J. S.; Yang, M. G.; Jain, N. F.; Panek, J. S. *Org.Synth.* **1997**, *75*, 78.

9. Matsumura, K.; Hashiguchi, S.; Ikariya, T.; Noyori, R. *J. Am. Chem. Soc.* **1997**, *119,* 8738.

PREPARATION OF OPTICALLY ACTIVE (*R,R*)-HYDROBENZOIN FROM BENZOIN OR BENZIL

1,2-Ethanediol, 1,2-diphenyl-, (1*R*,2*R*)-

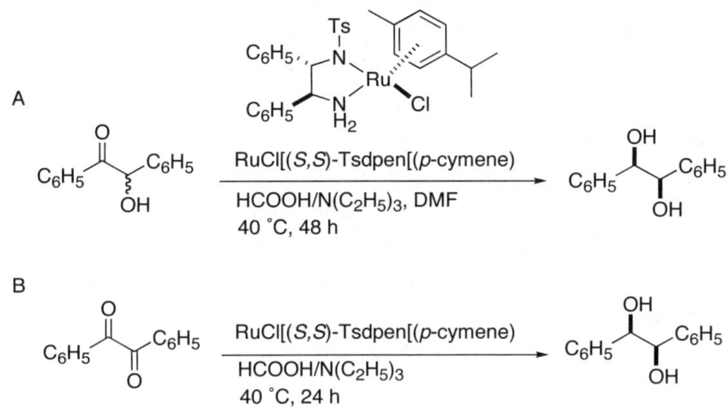

Submitted by Takao Ikariya,[1] Shohei Hashiguchi,[2] Kunihiko Murata,[3] and Ryoji Noyori.[4]
Checked by Peter Wipf and David Amantini.

1. Procedure

A. *A 100 g scale synthesis from rac-Benzoin:* A 1 L four-necked, round-bottomed flask equipped with a mechanical stirrer, a reflux condenser bearing an inert gas inlet tube, a thermometer and a dropping funnel is charged with 290 mL (2.08 mol) of triethylamine (Note 1). The triethylamine is cooled to 4 °C in an ice bath and formic acid (97.0 mL, 2.57 mol) is added slowly (Note 2). To the mixture of formic acid and triethylamine at ambient temperature is added *rac*-benzoin (Note 3) (170 g, 0.801 mol), RuCl[1*S*,2*S*)-*N*-*p*-toluenesulfonyl-1,2-diphenylethanediamine]-(η^6-*p*-cymene) (Notes 4 and 5) (0.204 g, 0.321 mmol), and dry DMF (Notes 1 and 6) (80 mL) . After the reaction mixture is stirred at 40 °C for 48 h, 300 mL of water is added at 0 °C with stirring (Note 7). The pale pink precipitate is filtered through a Büchner funnel, washed with water (2 x 500 mL), and dried in *vacuo* to give a white solid in 97% yield (Note 8). The crude product is dissolved in hot methanol (700 mL) at 60 °C. A small amount of insoluble material is removed through filtration and the filtrate is cooled initially to room temperature and then to 0 to 5 °C to provide white crystals. The crystalline product is isolated by filtration, washed with cooled (ice bath) 2-propanol (400 mL), and dried to provide 129.7 g of optically pure (*R*,*R*)-hydrobenzoin as white crystals (*dl* > 99%, 99.9% ee, Note 9). Concentration of the mother liquors and another recrystallization from methanol (100 mL) provides a second crop of the product, 19.1 g (*dl* > 99%, 99.9% ee, Note 9). The overall yield is 148.8 g (87%).

B. *A 10 g scale synthesis from benzil:* A 100 mL four-necked, round-bottomed flask equipped with a mechanical stirrer, a reflux condenser bearing an inert gas inlet tube, a thermometer, and a dropping funnel is charged with a mixture of formic acid (8.70 mL, 230 mmol) and triethylamine (19.0 mL, 136 mmol) in a similar manner to Procedure A. To this formic acid-triethylamine mixture at ambient temperature is added benzil (Note 3) (11.0 g, 52.3 mmol), and RuCl[(1*S*,2*S*)-*N*-p-toluenesulfonyl-1,2-diphenylethanediamine](η^6-p-cymene) (33.3 mg, 0.0524 mmol) (Note 6). After the reaction mixture is stirred at 40 °C for 24 h, 50 mL of water is added at 0 °C with stirring. The pale pink precipitate is filtered through a Büchner funnel, washed with water (50 mL), and dried in *vacuo* to give a white solid in 95% yield. The crude product is dissolved in hot methanol (50 mL) at 60 °C. The filtrate is cooled to room temperature and then to -40 °C to give white crystals. The crystalline product is isolated by filtration, washed with cooled (ice-bath) 2-propanol (10 mL), and dried to provide 9.3 g of optically pure (*R*,*R*)-hydrobenzoin as white crystals (82%, 100% ee).

2. Notes

1. Triethylamine and formic acid were purchased from Kanto Chemical Company and used without further purification. The Checkers used chemicals from Fisher and Fluka. An azeotropic mixture of triethylamine and formic acid is commercially available but could not be used for this reaction (see discussion section). Dry DMF was purchased from J. T. Baker.

2. The reaction of triethylamine with formic acid is exothermic and may proceed violently unless performed by controlled addition.

3. Benzoin and benzil were purchased from Kanto Chemical Company and used without further purification. The Checkers used chemicals from TCI and Acros, respectively. Substituted benzoins were prepared by benzoin condensation of the corresponding ring-substituted benzaldehydes.[5]

4. The Checkers used the following procedure for the preparation of (1*S*,2*S*)-*N*-*p*-toluenesulfonyl-1,2-diphenylethanediamine (TsDPEN): a dry CH_2Cl_2 solution (10 mL) of *p*-toluenesulfonyl chloride (0.893 g, 4.69 mmol) was added dropwise over 5 h (syringe pump addition) to a mixture of (*S*,*S*)-DPEN (0.995 g, 4.69 mmol) and triethylamine (0.69 mL, 4.5 mmol) in dry CH_2Cl_2 (30 mL) at 0 °C. After the reaction mixture was stirred at 0 °C for 6 h, the solution was washed with water (10 mL x 2) and saturated NaCl solution (10 mL) and then dried with Na_2SO_4. The solvent was removed under reduced pressure to give 1.659 g of crude white solid product. Recrystallization from ethyl acetate (8 mL) gave 1.146 g (67% yield) of the desired product as white crystals.

5. Commercially available chiral Ru complexes from Kanto Chemical Company were used by the Submitters; however, the complexes were prepared by the Checkers following a modified literature procedure.[6] (*R*)-RuCl[(1*S*,2*S*)-p-TsNCH(C_6H_5)CH(C_6H_5)NH_2](η^6-p-cymene): A mixture of [$RuCl_2$(η^6-p-cymene)]$_2$[7] (0.651 g, 1.06 mmol), (S,S)-TsDPEN[8] (0.780 g, 2.13 mmol), and triethylamine (0.60 mL, 4.3 mmol) in 2-propanol (21 mL) was stirred at 80 °C for 1 h. The orange solution was concentrated and the resulting solid was collected by filtration, washed with a small amount of water and dried under reduced pressure to give the chiral Ru complex. After recrystallization from methanol (20 mL), 0.552 g of pure Ru (II) catalyst were collected as bright orange crystals. After two additional recrystallizations of the concentrated mother liquor from methanol (8.0 and 5.0 mL, respectively), an additional 0.327 g of pure catalyst were collected. The overall yield was 0.879 g (65%). mp > 100 °C (dec.); IR (KBr) [cm^{-1}]: 3468, 3277, 3220 (H-N), 3062, 3029 (H–$C_{aromat.}$), 2961, 2872 (H–$C_{aliphat.}$); MS (EI): *m/z* (%) = 603 (63), 601 (100), 600 (66), 599 (58); 1H NMR (300 MHz, $CDCl_3$, 25 °C, TMS): δ 1.37 (m, 6H, CH(CH_3)$_2$), 2.21 (s, 3H, CH_3 in p-cymene), 2.35 (s, 3H, CH_3 in *p*-Ts), 3.12 (m, 1H, CH(CH_3)$_2$), 3.52 (m, 2H, N*H*H and *H*CNH_2), 3.73 (d, 1H, *J* = 10.5 Hz, *H*CN-*p*-Ts), 5.72 (m, 4H, $CH_{aromat.}$ in p-cymene), 6.04 (m, 1H, NH*H*), 6.34–7.05 (m, 14H, *p*-$CH_3C_6H_4$-SO_2NCH(C_6H_5)CH(C_6H_5)NH_2).

6. DMF was used to maintain the homogeneity of the reaction mixture, but it is not crucial for the catalysis to be efficient and practical. On small scale the addition of DMF is not needed. In fact, the reaction of benzil with a substrate to catalyst ratio (S/C) of 1,000 (4.7 M) in a mixture of HCOOH and N(C_2H_5)$_3$ containing the (S,S)-Ru catalyst (benzil:HCOOH:N(C_2H_5)$_3$ = 1:4.4:2.6) proceeded heterogeneously at the early stages of the reaction because of the low solubility of benzil. After

about ten minutes, the reaction mixture changed to a completely homogenous solution, giving almost the same results as with DMF as solvent.

7. On large scale, the reaction flask should be connected to an Ar gas inlet to allow CO_2 to escape.

8. The diastereoselectivity of the product, *dl*:*meso* = 95.0:5.0, was determined by integration in the 1H NMR (300 MHz, $CDCl_3$); (*R*,*R*)-hydrobenzoin: δ 3.07 (s, 2H, OH), 4.86 (s, 2H, CH–OH), 7.25–7.40 (m, 10H, aromatic ring protons), *meso*-hydrobenzoin: δ 2.33 (s, 2H, OH), 4.98 (s, 2H, CHOH), 7.30–7.45 (m, 10H, aromatic ring protons).

9. (*R*,*R*)-Hydrobenzoin: $[\alpha]_D^{25}$ + 91.6 (c 1.05, ethanol), (lit.[9a] $[\alpha]_D^{23}$, +95 (c 0.87 ethanol), 99% ee (*R*,*R*)). HPLC separation conditions, (column: CHIRALCEL OJ (4.6 mm i.d. x 250 mm), eluent: hexane/2-propanol = 90/10, flow rate: 1.0 mL/min, temp: 25 °C, detection UV 254 nm); retention time, (*S*,*S*)-hydrobenzoin, 14.2 min, (*R*,*R*)-hydrobenzoin, 16.5 min.

Safety and Waste Disposal Information

All hazardous materials should be handled and disposed of in accordance with "Prudent Practices in the Laboratory"; National Academy Press; Washington, DC, 1995.

3. Discussion

Optically active hydrobenzoins are useful building blocks for the stereoselective synthesis of various biologically active compounds, as well as chiral ligands and auxiliaries. The preparation of these chiral hydrobenzoins by Sharpless asymmetric dihydroxylation of trans-stilbene is one of the most convenient and well established methods.[9] Asymmetric reduction of readily available benzils or benzoins would be a promising and widely applicable approach; however, no practical reduction systems have been reported except for oxazaborolidine-catalyzed reductions of benzils with borane/methylsulfide.[10] Direct asymmetric hydrogenation of benzils or benzoins catalyzed by well-established Ru-BINAP complexes could potentially lead to optically active 1,2-diols. However, *meso*-isomers are obtained as the major products, because the substrate control of the hydroxy ketone intermediate favors *meso*-diol formation.[11] The procedure described herein provides a highly efficient method accessing the desired chiral hydrobenzoins in high enantiomeric excess using commercially available chiral (Ru(II) catalysts, RuCl(TsDPEN) (η^6-arene),[6,12] and easily handled reagents such as benzils or benzoins as substrates and a formic acid and triethylamine mixture as the hydrogen source.[13]

A mixture of formic acid and triethylamine is the best hydrogen donor for this reduction. In the absence of triethylamine, no conversion of benzoins or benzils was observed. The addition of triethylamine to the reaction mixture causes a significant increase in the conversion of the substrates. In the reaction of benzoin, a formic acid:triethylamine molar ratio of 3.2:2.6 to 3.2:4.4 gives the best catalyst performance in terms of both reactivity and stereoselectivity. The reduction of benzil requires a molar ratio of 4.4:2.6 to 4.4:4.4. The reaction with an azeotropic mixture of formic acid and triethylamine (5:2) gave no conversion under otherwise identical conditions as described in the Procedure.

The success of this asymmetric reduction of benzil or benzoin leading to the optically active hydrobenzoin with the formic acid and triethylamine mixture relies strongly on the nature of benzoin with a configurationally labile stereogenic center and the enantiomer discrimination ability of the chiral Ru complexes. Due to a sufficiently rapid stereomutation of benzoins under the basic reaction conditions, the dynamic kinetic resolution of benzoins allows the diastereo- and enantioselective synthesis of optically active hydrobenzoins.[13] Reduction of (*R*)-benzoin with the (*S,S*)-Ru catalyst

Table 1. Asymmetic Reduction of Benzils with (*S,S*)-Ru Catalyst

R	S/C	temp, °C	time, h	yield, %	*dl:meso*	ee, %
H	1000	40	24	100	98.4:1.6	>99
p-CH_3	1000	40	48	67	96.7:3.3	>99
p-OCH_3[a]	200	35	48	75	94.4:5.6	>99
p-F	1000	40	24	100	94.2:5.8	>99

Conditions: (*S,S*)-Ru cat 0.005 mmol, ketone/HCOOH/$N(C_2H_5)_3$ = 1/4.4/2.6.
[a] ketone/HCOOH/$N(C_2H_5)_3$ = 4.4/4.4 in 1.2 M DMF.

in DMF under the same conditions gave (*R,R*)-hydrobenzoin quantitatively and in 100% ee, indicating that the (*S,S*)-Ru catalyst favors the reaction of (*R*)-benzoin.[13a] The rate of the reduction of (*R*)-benzoin with the (*S,S*)-Ru catalyst proceeds 55 times faster than the *S*-isomer. The slow-reacting *S*-isomer undergoes a rapid racemization.

Various benzil derivatives bearing substituents on aromatic rings can be reduced stereoselectively to the chiral hydrobenzoins in high ee's and in good yields (Table 1). The benzils with electron-donating substituents such as methyl or methoxy groups are reduced with excellent enantioselectivity but more slowly, while the reduction of *p*-fluorobenzil proceeded rapidly, as expected, giving a product with a high ee.[13]

The described, chiral Ru catalyst-promoted asymmetric transfer hydrogenation with a formic acid and triethylamine mixture is also applicable to the enantioselective reduction of acetophenone,[12] ring-

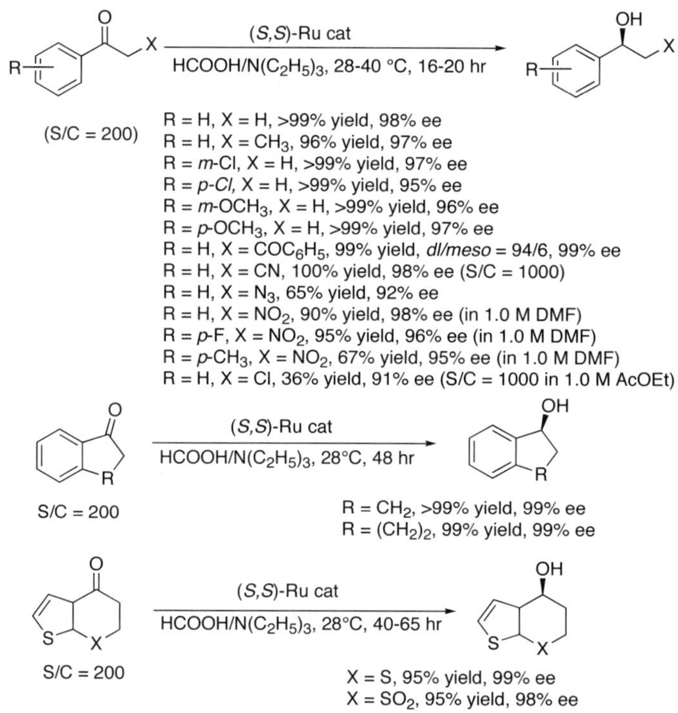

(*S,S*)-Ru cat: RuCl[(*S,S*)-Tsdpen](η^6-mesitylene) or RuCl[(*S,S*)-Tsdpen](*p*-cymene)

Figure 1. Examples of Chiral Ru Catalyzed Asymmetric Reduction of Ketones with $HCOOH/N(C_2H_5)_3$

substituted acetophenone derivatives,[12] α-substituted acetophenones,[14,15] acetylpyridine derivatives,[16] and functionalized ketones[17,18] leading to the corresponding optically active alcohols in excellent ee. These asymmetric reductions with the chiral Ru catalyst are characterized by a rapid, carbonyl group-selective transformation because of the coordinatively saturated nature of the diamine-based Ru hydride complexes.[6,17] The neighboring groups at the α-position of the carbonyl group do not interact with the metal center, leading to excellent reactivity and enantioselectivity. Some representative examples are listed in Figure 1.

1. Graduate School of Science and Engineering, Frontier Collaborative Research Center, Tokyo Institute of Technology, Meguro-ku, Tokyo 152-8552 Japan.
2. Takeda Chemical Industries, ltd., Yodogawa-ku, Osaka, Japan.
3. Kanto Chemical Corp. Inc., Central Research Laboratory, Soka, Saitama, Japan.
4. Department of Chemistry and Research Center for Materials Science, Nagoya University, Chikusa-ku, Nagoya, Japan.
5. Ide, W. S.; Buck, J. S. In "Organic Reactions", Adams, M.; Bachmann, W. E.; Blatt, A. H.; Fieser, L. F.; Johnson, J. R.; Snyder, H. R. ed. John Wiley & Sons, New York, **1948**, *4*, pp. 269-304.
6. Haack, K.-J.; Hashiguchi, S.; Fujii, A.; Ikariya, A.; Noyori, R. *Angew. Chem., Int. Ed. Engl.* **1997**, *36*, 285-288.
7. Bennett, M. W.; Huang, T.-N.,; Matheson, T. W.; Smith, A. K. *Inorg. Synth.* **1982**, *21*, 74-78.
8. Oda, T.; Irie, R.; Katuski, T.; Okawa, H. *Synlett* **1992**, 641-643.
9. (a) Wang, Z.-M.; Sharpless, K. B. *J. Org. Chem.* **1994**, *59*, 8302-8303. (b) Kolb, H. C.; VanNieuwenhze, M. S.; Sharpless, K. B. *Chem. Rev.* **1994**, *94*, 2483-2547.
10. Quallich, G. J.; Keavey, K. N.; Woodall, T. M.; *Tetrahedron Lett.* **1995**, *36*, 4729-4732.
11. Kitamura, M.; Ohkuma, T.; Inoue, S.; Sayo, N.; Kumobayashi, H.; Akutagawa, S.; Ohta, T.; Takaya, H.; Noyori, R. *J. Am. Chem. Soc.* **1988**, *110*, 629-631.
12. Hashiguchi, S.; Fujii, A.; Takehara, J.; Ikariya, T.; Noyori, R. *J. Am. Chem. Soc.* **1995**, *117*, 7562-7563.
13. (a) Murata, K.; Okano, K.; Miyagi, M.; Iwane, H.; Noyori, R.; Ikariya, T. *Org. Lett.* **1999**, *1*, 1119-1121. (b) Koike, T.; Murata, K.; Ikariya, T. *Org. Lett.* **2000**, *2*, 3833-3836.
14. (a) Hamada, T.; Torii, T.; Izawa, K.; Noyori, R.; Ikariya, T. *Org. Lett.* **2002**, *4*, 4373-4376. (b) Hamada, T.; Torii, T.; Izawa, K.; Ikariya, T. *Tetrahedron* **2004**, *60*, 7411-7417. (c) Hamada, T.; Torii, T.; Onishi, T.; Izawa, K.; Ikariya, T. *J. Org. Chem.* **2004**, *69*, 7391-7394.
15. Watanabe, M.; Murata, K.; Ikariya, T. *J. Org. Chem.* **2002**, *67*, 1712-1715.
16. Okano, K.; Murata, K.; Ikariya, T. *Tetrahedron Lett.* **2000**, *41*, 9277-9280.

17. (a) Noyori, R.; Hashiguchi, S. *Acc. Chem. Res.* **1997**, *30*, 97-102; (b) Ikariya, T.; Murata, K. and Noyori, R. Org. Biomol. Chem. **2006**, *4*, 393-406; (c) Ikariya, T. Blacker, A. J., *Acc. Chem. Res.* in press.

18. (a) Palmer, M. J.; Wills, M. *Tetrahedron: Asymmetry* **1999**, *10*, 2045-2061. (b) Cross, D. J.; Kenny, J. A.; Houson, I.; Cambell, L.; Walsgrove, T.; Wills, M. *Tetrahedron: Asymmetry* **2001**, *12*, 1801-1806. (c) Mohar, B.; Valleix, A.; Desmurs, J.-R.; Felemez, M.; Wagner, A.; Mioskowski, C. *Chem. Commun.* **2001**, 2572-2573. (d) Eustache, F.; Dalko, P. I.; Cossy, J. *Org. Lett.* **2002**, *4*, 1263-1265.

SYNTHESIS OF (+)-(1*R*,2*S*,9*S*)-11-METHYL-7,11-DIAZATRICYCLO[7.3.1.0^{2,7}]TRIDECANE, A (+)-SPARTEINE SURROGATE

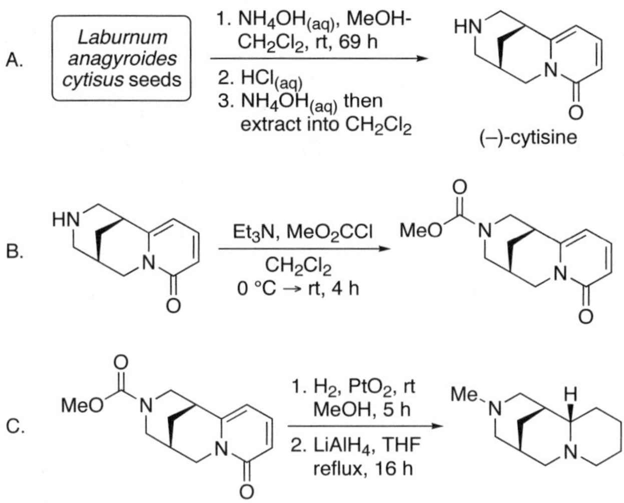

Submitted by Amanda J. Dixon, Matthew J. McGrath, and Peter O'Brien.[1]
Checked by Sigrid Holle and Alois Fürstner.

1. Procedure

A. (–)-Cytisine. A 2-L, three-necked round-bottomed flask equipped with an overhead mechanical stirrer with large Teflon-coated blades (Note 1) and two glass stoppers is charged with finely ground *Laburnum anagyroides* seeds (598 g) (Notes 2 and 3), dichloromethane (837 mL), methanol (239 mL) and aqueous 25% w/v ammonium hydroxide (90 mL) (Note 4). The resulting mixture is stirred vigorously (Note 1) at room temperature. After a short induction period, a slight exotherm is observed lasting for the first ~24 h, accompanied with the formation of a thick gel (Note 1), which slowly disappears on prolonged stirring. After stirring the suspension for a total of 69 h, the mixture is filtered in three separate batches and the filter cake is washed with CH_2Cl_2 until the filtrate is colorless (Note 5). The filtrate is transferred to a 5-L separatory funnel and shaken with 3.3 M HCl (500 mL). After 2 h (Note 6), the two layers are separated and the aqueous layer is transferred to a 2-L conical flask equipped with a magnetic stirring bar. The stirred aqueous solution is basified to pH 9-10 by the careful, portionwise addition of aqueous 25% w/v ammonium hydroxide (~150 mL) over 1 h (Note 7), followed by stirring for an additional 2 h (Note 8). The resulting solution is extracted with CH_2Cl_2 (10 × 100 mL) (Note 9). The CH_2Cl_2 extracts are combined, dried ($MgSO_4$), filtered and evaporated under reduced pressure to give crude (–)-cytisine (8.31 g) (Note 10) as a yellow-brown solid. The crude (–)-cytisine is purified by recrystallization (Note 11) from toluene (25-30 mL) to afford 5.25 g (0.9 % mass yield) of pure (–)-cytisine (Note 12) as a yellow-brown solid, having greater than 99.5% ee (Note 13).

B. (–)-Methyl (1R,9R)-6-Oxo-7,11-diazatricyclo[7.3.1.0^{2,7}]trideca-2,4-diene-11-carboxylate. A flame-dried 250-mL, two-necked round-bottomed flask equipped with a Teflon-coated magnetic stir bar, a glass stopper and a vacuum take-off adapter attached to the argon line is charged with (–)-cytisine (5.25 g, 27.5 mmol), CH_2Cl_2 (80 mL) (Note 4) and triethylamine (4.21 mL, 30.2 mmol) (Note 14). The resulting magnetically stirred solution is immersed in an ice bath and methyl chloroformate (2.34 mL, 30.2 mmol) (Note 14) is added dropwise via syringe over 10 min at 0 °C. The resulting mixture is stirred for 1 h at 0 °C and then for 3 h at room temperature before the solvent is evaporated under reduced pressure. Ethyl acetate (EtOAc) (40 mL) is added to the residue and the solids are removed by filtration through Celite (Note 15). The filter cake is washed with EtOAc (3 × 15 mL) and the filtrate is evaporated under reduced pressure. The residue is purified by column chromatography over a short plug of silica (Note 16) with CH_2Cl_2:MeOH (9:1) as eluent. The fractions containing the product (R_f = 0.51; 9:1, CH_2Cl_2:MeOH) are combined and evaporated under reduced pressure followed by removal of the last traces of solvent by high vacuum drying (10^{-3} mbar) for 2 h to afford 6.45 g (94%) of pure cytisine methyl carbamate (Note 17) as a thick colorless oil.

C. (+)-(1R,2S,9S)-11-Methyl-7,11-diazatricyclo[7.3.1.0^{2,7}]tridecane. A flame-dried, 250-mL, three-necked round-bottomed flask equipped with a magnetic stir bar, a glass stopper, and a vacuum take-off adapter that can be attached either to the argon line or to the H_2 supply is charged with cytisine methyl carbamate (6.50 g, 26.2 mmol), MeOH (100 mL) and platinum(IV) oxide (600 mg, 2.7 mmol) (Note 18). The resulting magnetically-stirred suspension is carefully evacuated and backfilled with argon (three times) before evacuating and backfilling with hydrogen (*via* hydrogen balloons attached to the two-tap adaptor). The cloudy black mixture is stirred vigorously under a hydrogen atmosphere for 5-12 h (Note 19). The solids are removed by filtration through Celite (Note 20) and the filter cake is washed with CH_2Cl_2:MeOH (9:1) (50 mL). The filtrate is evaporated under reduced pressure followed by removal of the last traces of solvent by high vacuum drying (10^{-3} mbar) to afford 5.93 g (90 %) of crude hydrogenation product (Note 21) as an off-white solid.

A flame-dried, 500-mL, two-necked round-bottomed flask equipped with a magnetic stir bar, a reflux condenser attached to the argon line, and a glass stopper is charged with lithium aluminium hydride (5.12 g, 134.6 mmol) and tetrahydrofuran (THF) (150 mL) (Note 22). The resulting magnetically-stirred suspension is immersed in an ice bath and a solution of the crude hydrogenation product (5.93 g, 23.5 mmol) in THF (150 mL) is added dropwise over 10 min *via* a cannula at 0 °C. The mixture is allowed to warm to room temperature and then refluxed under argon for 16 h (Note 23). After cooling to 0 °C, diethyl ether (150 mL) is added, followed by the careful, portionwise addition of solid hydrated sodium sulfate ($Na_2SO_4 \cdot 10H_2O$) (17 g) (Note 22), which causes a vigorous evolution of hydrogen gas. The resulting viscous mixture is stirred for 30 min until gas evolution has ceased and the solids are removed by filtration through Celite (Note 24). The filter cake is washed with CH_2Cl_2:MeOH (9:1) (150 mL) and the washings are combined, dried (Na_2SO_4), filtered and evaporated under reduced pressure to give crude diamine (3.95 g). The crude diamine is purified by Kugelrohr distillation (oven temperature 150–160 °C at 0.07 mbar) to afford 2.65 g (52% over two steps) of pure diamine (Note 25) as a colorless oil, having greater than 95% ee (Note 26).

2. Notes

1. Effective stirring is essential for the success of the isolation process. Because the reaction mixture temporarily forms a thick gel for several hours that slowly disappears, the use of a strong mechanical stirrer is required. The checkers used a mechanical stirrer with approximately 9 x 2 cm Teflon-coated blades. An effective extraction was ensured at 300 turns/min of this stirrer.

2. *Laburnum Anagyroides* seeds were purchased from Vilmorin, Division Semences d'Arbres, Route du Manoir, 49250 La Ménitré, France.

3. The seeds are ground to a fine (< 2 mm), non-uniform powder by using a standard coffee grinder. Typically, 30-50 g batches of the seeds are ground for ~30 sec. The use of coarse particles results in an inefficient extraction and/or extended extraction times.

4. Dichloromethane, methanol, toluene and 12 M hydrochloric acid (diluted with water to 3.3 M) were purchased from Aldrich and used as received. Ammonium hydroxide (25% w/v) purchased from Reininghaus was used as received.

5. The following procedure is used for the filtration process: (i) approximately one third of the stirred reaction mixture slurry is rapidly poured into a 14 cm-diameter Büchner funnel (fitted with a 12.5 cm-diameter Whatman No. 1 filter paper) attached to a 2-L Büchner flask. The filter cake is sucked dry and then washed with CH_2Cl_2 (~100-150 mL) until the filtrate runs through colorless. The filter cake is discarded. (ii) CH_2Cl_2 (150 mL) is added to the extraction mixture in the round-bottomed flask and approximately half of the remaining stirred mixture slurry is rapidly poured into the Büchner funnel/flask set-up. The filter cake is sucked dry and then washed with CH_2Cl_2 (150 mL) until the filtrate runs through colorless. The filter cake is discarded. (iii) The final portion of the reaction mixture slurry is filtered by repeating step (ii). The total volume of the filtrate at this stage is typically 1.8 L.

6. The mixture is left for 2 h to complete acidification and is periodically shaken with frequent venting to ensure good mixing. The pH is 1–2 as tested by pH paper (E. Merck, Darmstadt).

7. The basification is exothermic and a slow addition of the aqueous 25% w/v ammonium hydroxide is necessary.

8. The solution is left for at least 2 h so that a stable pH of 9-10 (pH paper, E. Merck, Darmstadt) is obtained. The solution can be left overnight at this stage without any detrimental effect on the isolated yield of (–)-cytisine.

9. If the mixture is not extracted with 10 × 100 mL portions of CH_2Cl_2, lower yields of (–)-cytisine are obtained.

10. The yield of crude (–)-cytisine can vary widely (8-15 g) depending on the particular crop of the seeds that is used, but ≥ 5 g of pure (–)-cytisine is typically obtained after recrystallization from 25-30 mL of toluene.

11. For recrystallization, the crude product is dissolved in 25-30 mL of boiling toluene in a 50-mL conical flask and then allowed to cool slowly by standing overnight at room temperature. The flask is cooled at 0 °C for 1 h before collecting the crystals by filtration. The submitters report that on some occasions it is necessary to collect a second crop of crystals to ensure that >5 g of pure (–)-cytisine is obtained.

12. The properties are as follows: ^1H NMR ($CDCl_3$, 300 MHz) δ: 1.94 (br s, 2 H), 2.24 (br s, 1 H), 2.32 (br s, 1 H), 2.89 (br s, 1 H), 2.97–3.13 (m, 4

H), 3.90 (dd, 1 H, *J* = 15.5, 6.5 Hz), 4.10 (d, 1 H, *J* = 15.5 Hz), 5.98 (dd, 1 H, *J* = 6.8, 1.4 Hz), 6.42 (dd, 1 H, *J* = 9.1, 1.4 Hz), 7.27 (dd, 1 H, *J* = 9.1, 6.9 Hz); ^{13}C NMR ($CDCl_3$, 75.5 MHz) δ: 26.1, 27.6, 35.3, 49.6, 52.6, 53.6, 105.1, 116.8, 138.8, 150.7, 163.6; IR (film): 1649, 1546 cm^{-1}; $[\alpha]_D^{20}$ –59.3 ($CHCl_3$, *c* 0.84); mp 153–154 °C; R_f = 0.16 (9:1, CH_2Cl_2:MeOH).

13. The submitters determined the enantiomeric excess by conversion into *N*-benzyl cytisine and analysis by chiral HPLC: A 10-mL, one-necked round-bottomed flask equipped with a magnetic stirrer bar and reflux condenser (fitted with a rubber septum, attached to a nitrogen line) is charged with (–)-cytisine (100 mg, 0.52 mmol), benzyl bromide (0.14 mL, 1.04 mmol), potassium carbonate (380 mg, 2.6 mmol) and acetonitrile (3 mL). The resulting magnetically-stirred suspension is heated at reflux under nitrogen for 5 h. After cooling to room temperature, the solvent is evaporated under reduced pressure. Dichloromethane (10 mL) is added to the residue and the solids are removed by filtration through Celite. The filtrate is evaporated under reduced pressure and the residue is purified by column chromatography over silica with CH_2Cl_2 and then CH_2Cl_2:MeOH (97:3) as eluent. The fractions containing the product (R_f = 0.6; 9:1, CH_2Cl_2:MeOH) are combined and evaporated under reduced pressure followed by removal of the last traces of solvent by high vacuum drying to afford 137 mg (94%) of pure *N*-benzyl cytisine as a white solid. The properties are as follows: ^{1}H NMR ($CDCl_3$, 400 MHz) δ: 1.77–1.94 (m, 2 H), 2.32 (dd, 1 H, *J* = 11.0, 2.0), 2.37 (br d, 1 H, *J* = 11.0), 2.40–2.48 (m, 1 H), 2.83–2.96 (m, 3 H), 3.39 (d, 1 H, *J* = 14.0), 3.46 (d, 1 H, *J* = 14.0), 3.89 (dd, 1H, *J* = 15.0, 7.0), 4.12 (d, 1 H, *J* = 15.0), 5.84 (dd, 1 H, *J* = 7.0, 1.0), 6.50 (dd, 1 H, *J* = 9.0, 1.0), 6.98–7.01 (m, 2 H), 7.17–7.23 (m, 3 H), 7.29 (dd, 1 H, *J* = 9.0, 7.0); ^{13}C NMR ($CDCl_3$, 100.6 MHz) δ: 25.9, 28.1, 35.4, 49.9, 59.9, 60.0, 61.9, 104.6, 116.5, 126.8, 128.1, 138.0, 138.5, 151.4, 163.6 (one aromatic resonance not resolved); IR (CH_2Cl_2) cm^{-1} 1650, 1560, 1545; $[\alpha]_D^{20}$ –302 ($CHCl_3$, *c* 0.5); mp 139-141 °C; R_f = 0.6 (9:1 CH_2Cl_2-MeOH). HPLC analysis using a Chiralcel-OD column with 20% 2-propanol in heptane containing 0.1% diethylamine as eluent, and a flow rate of 0.5 mL/min indicates a single peak with retention time of 18 min. HPLC analysis of the corresponding racemic *N*-benzyl cytisine affords two peaks at 18 min and 23 min corresponding to (–)- and (+)-cytisine, respectively.

14. Triethylamine was purchased from Aldrich and distilled over potassium hydroxide before use. Methyl chloroformate was purchased from Aldrich Chemical Company, Inc. and used as received. Ethyl acetate (EtOAc) was purchased from Fisher Scientific and used as received.

15. Filtration is through a 1 cm-depth of Celite in a 5 cm-diameter Büchner funnel (fitted with a 42.5 mm-diameter Whatman No. 1 filter paper) attached to a 250-mL Büchner flask.

16. Flash silica gel 60 (220-440 mesh) purchased from E. Merck, Darmstadt, is placed in a 4 cm-diameter column (8 cm-depth of silica) and eluted with ~300-400 mL of solvent.

17. The submitters reported a yield of 99%. The properties are as follows: ^{1}H NMR ($CDCl_3$, 400 MHz) δ: 1.85–2.00 (m, 2 H), 2.42 (br s, 1 H), 3.02 (br s, 3 H), 3.41–3.60 (m, 3 H), 3.83 (dd, 1 H, J = 15.6, 6.6 Hz), 4.09 (d, 1 H, J = 15.6 Hz), 4.01–4.32 (m, 2 H), 6.02 (d, 1 H, J = 5.6 Hz), 6.40 (dd, 1 H, J = 9.1, 1.5 Hz), 7.25 (dd, 1 H, J = 9.1, 6.5 Hz); ^{13}C NMR ($CDCl_3$, 100.6 MHz) rotamers observed, δ for major rotamer: 25.7, 27.1, 34.3, 48.9, 50.1, 51.0, 52.7, 105.6, 117.3, 138.9, 148.8, 156.1, 163.4; IR (film): 1700, 1656, 1546 cm^{-1}; $[\alpha]_D^{20}$ –207.9 ($CHCl_3$, c 0.47); R_f = 0.51 (9:1, CH_2Cl_2:MeOH).

18. Platinum(IV) oxide (surface area ≥60 m^2/g, 81-83% Pt) was purchased from Heraeus and used as received.

19. The time required for the hydrogenation reaction to reach completion varies from 5-12 h (even when the same batch of platinum(IV) oxide was employed). The disappearance of starting methyl carbamate can be identified qualitatively by TLC: R_f = 0.5 (9:1, CH_2Cl_2:MeOH) for methyl carbamate.

20. Filtration is through a 2-cm depth of Celite in a 5-cm diameter Büchner funnel (fitted with a 42.5 mm-diameter Whatman No. 1 filter paper) attached to a 250-mL Büchner flask.

21. The crude hydrogenation product is of sufficient purity for direct use in the next step. It can be purified by column chromatography over silica with CH_2Cl_2:MeOH:aqueous 25% w/v ammonium hydroxide (97:2:1) as eluent. The fractions containing the product (R_f = 0.3; 97:2:1 CH_2Cl_2:MeOH:aqueous 25% w/v ammonium hydroxide) are combined and evaporated under reduced pressure followed by removal of the last traces of solvent by high vacuum drying (10^{-3} mbar) to afford pure hydrogenation product as a white solid. The physical properties are as follows: ^{1}H NMR ($CDCl_3$, 400 MHz) approx. 4:1 mixture of rotamers, δ: 1.52–1.67 (m, 2 H), 1.73–1.98 (m, 5 H), 2.06–2.21 (m, 1 H), 2.27–2.49 (m, 2 H), 2.77 (br d, 1 H, J = 13.5 Hz), 2.84 (dd, 0.85 H, J = 13.9, 2.0 Hz), 2.92 (br d, 0.15 H, J = 13.9 Hz), 2.98 (br d, 0.15H, J = 13.0 Hz), 3.04 (br dt, 0.85H, J = 13.5, 2.1 Hz), 3.40–3.47 (m, 1 H), 3.57 (s, 2.5 H), 3.64 (s, 0.5 H), 4.17 (d, 0.85 H, J = 13.5 Hz), 4.26 (br d, 0.15 H, J = 13.0 Hz), 4.34 (br d, 0.15 H, J = 13.5 Hz), 4.58

(d, 0.85 H, J = 13.9 Hz), 4.71 (br, 0.15 H), 4.74 (d, 0.85 H, J = 13.5 Hz); ^{13}C NMR ($CDCl_3$, 100.6 MHz) δ: 20.1, 27.7, 27.8. 32.7, 33.1, 33.3, 44.3, 45.8, 49.0, 52.6, 59.5, 156.1, 169.6; IR (film): 1695, 1635 cm^{-1}; $[\alpha]_D^{20}$ –169.9 ($CHCl_3$, c 1.05); mp 118–120 °C; R_f = 0.2 (97:2:1, CH_2Cl_2:MeOH:aq. 25% w/v ammonium hydroxide).

22. Lithium aluminum hydride and hydrated sodium sulfate were purchased from Aldrich Chemical Company, Inc. and used as received. Tetrahydrofuran was purchased from Fisher Scientific and distilled from sodium/benzophenone ketyl under nitrogen. Diethyl ether was purchased from Fisher Scientific and used as received.

23. The reaction mixture should not be heated at reflux for more than 16 h as a lower yield of diamine is obtained.

24. Filtration is through a 2-cm depth of Celite in a 5-cm diameter Büchner funnel (fitted with a 42.5-mm diameter Whatman No. 1 filter paper) attached to a 1-L Büchner flask.

25. The submitters report a yield of 62% over two steps. A similar yield (61%) was obtained by the checkers when the reaction was performed on half-scale. The physical properties of the product are as follows: ^{1}H NMR ($CDCl_3$, 400 MHz) δ: 1.18–1.31 (m, 2 H), 1.41–1.80 (m, 9 H), 1.85 (br d, 1 H, J = 11.0 Hz), 1.92 (dd, 1 H, J = 11.5, 3.1 Hz), 2.10 (s, 3 H), 2.08–2.13 (m, 1 H), 2.19 (ddd, 1 H, J = 11.2, 3.5, 1.6 Hz), 2.79–2.87 (m, 2 H), 2.92–2.98 (m, 2 H); ^{13}C NMR ($CDCl_3$, 100.6 MHz) δ: 25.1, 25.7, 30.6, 30.8, 33.9, 35.2, 47.4, 56.3, 57.6, 60.4, 60.5, 66.4; IR (film): 2930 cm^{-1}; $[\alpha]_D^{20}$ + 29.7 (c = 1.10, EtOH).

26. The enantiomeric excess was determined by high resolution ^{1}H NMR spectroscopy (400 MHz, $CDCl_3$) in the presence of 3.0 equivalents of (R)- or (S)-2,2,2-trifluoro-1-(9-anthryl)ethanol: A 0.12 M solution of the diamine in $CDCl_3$ is prepared by dissolving the diamine (46 mg, 0.24 mmol) in $CDCl_3$ (2.0 mL) and a 0.06 M solution of (R)-2,2,2-trifluoro-1-(9-anthryl)ethanol in $CDCl_3$ is prepared by dissolving (R)-2,2,2-trifluoro-1-(9-anthryl)ethanol (33 mg, 0.12 mmol) in $CDCl_3$ (2.0 mL). The sample for ^{1}H NMR spectroscopy analysis is then prepared by using 0.06 mL of the 0.12 M solution of the diamine in $CDCl_3$ (0.006 mmol), 0.36 mL of the 0.06 M solution of (R)-2,2,2-trifluoro-1-(9-anthryl)ethanol in $CDCl_3$ (0.018 mmol, 3.0 equiv.) and 0.18 mL of $CDCl_3$ (total volume of NMR sample ~ 0.6 mL). Key signals: ^{1}H NMR ($CDCl_3$, 400 MHz) δ: 2.11 (s, 3 H), 2.53 (br s, 1 H), 2.66 (br s, 1 H), 2.87 (br d, 1 H, J = 11.0 Hz), 2.95 (br d, 1 H, J = 11.0 Hz). In a similar fashion, a sample for ^{1}H NMR spectroscopy analysis is prepared using 0.05 mL of the 0.12 M solution of the diamine in $CDCl_3$ (0.006

mmol), 0.30 mL of a 0.06 M solution of (*S*)-2,2,2-trifluoro-1-(9-anthryl)ethanol in $CDCl_3$ (0.018 mmol, 3.0 equiv) and 0.15 mL of $CDCl_3$ (total volume of NMR sample ~ 0.5 mL). Key signals: ^{1}H NMR ($CDCl_3$, 400 MHz) δ: 1.97 (s, 3 H), 2.65–2.85 (m, 2 H), 2.93 (br d, 1 H, J = 11.0). The absence of any signals due to the other diastereomeric complex in each of these ^{1}H NMR spectra indicates that the diamine is present in greater than 95% ee. (*R*)- and (*S*)-2,2,2-Trifluoro-1-(9-anthryl)ethanol were purchased from Aldrich Chemical Company, Inc. and used as received.

Safety and Waste Disposal Information

All hazardous materials should be handled and disposed of in accordance with "Prudent Practices in the Laboratory"; National Academy Press; Washington, DC, 1995.

3. Discussion

(–)-Sparteine is a widely used ligand in asymmetric synthesis[2] but suffers from the drawback that it is only commercially available in one enantiomeric form. Diamine (+)-**1** was designed as a (+)-sparteine surrogate because it possesses most of the three-dimensional architecture of (+)-sparteine. The procedure described here is a simple, three-step synthesis of diamine (+)-**1** from *Laburnum anagyroides cytisus* seeds.[3,4] The route has been successfully used by other research groups.[5-7] The extraction process is a modified version of a protocol reported by Rouden, Lasne and coworkers[8] and is a simple and high yielding alkaloid isolation. Subsequent *N*-protection (as a methyl carbamate), pyridone hydrogenation (completely diastereoselective in the sense depicted, as established by X-ray crystallography)[4] and lithium aluminum hydride reduction furnishes diamine (+)-**1** in good overall yield. Distillation (Kugelrohr) of the diamine immediately before use is recommended (as is usual when using (–)-sparteine with organolithium reagents).[9] Two multi-step, asymmetric syntheses of (–)-**1**[10] and (+)-**1**[11] have also been described. Other analogues of which **2-4**[5,12] are representative, have been prepared by using the appropriate acid chloride in the first step of the synthesis and analogue **5** was prepared by using a modified route.[5,13] In terms of applications, we[12,13] and others[5,6] have found that diamine (+)-**1** is the most effective and versatile (+)-sparteine surrogate of those diamines prepared from (–)-cytisine.

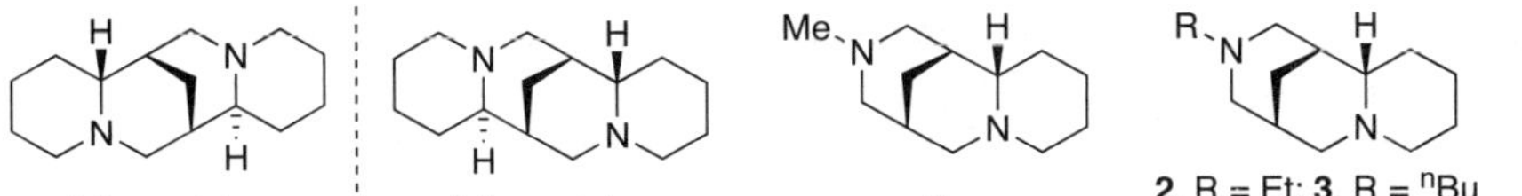

(–)-sparteine (+)-sparteine **1** **2**, R = Et; **3**, R = ^{n}Bu **4**, R = $CH_2{}^tBu$; **5**, R = ^{i}Pr

A diverse range of examples that utilizes diamine (+)-**1** in asymmetric synthesis with organolithium reagents is presented in Table 1. All of the products show opposite enantioselectivity to those obtained with (–)-sparteine and a similarly high degree of enantioselection, thus demonstrating that (+)-**1** is an excellent (+)-sparteine surrogate. Lithiations and subsequent rearrangement (Entry 3) or electrophilic trapping (Entries 1-2 and 4-8), including three examples reported by other groups (Entries 6-8),[5,6] are particularly successful. Recently, (+)-**1** was used to control the *regioselectivity* of deprotonation of an enantiomerically enriched functionalized *N*-Boc pyrrolidine used in a route to (–)-kainic acid.[7] In a similar fashion, inter- and intramolecular carbolithiations work well (Entries 9-10).

The use of diamine (+)-**1** is not limited to organolithium-mediated processes; reactions including magnesium (Entry 1), copper (Entry 2) and palladium (Entry 3) are also successful (Table 2). The results in Tables 1 and 2 cover a wide range of mechanistic pathways and processes (namely asymmetric deprotonation and substitution, carbometallation, anhydride ring opening, dynamic thermodynamic resolution and kinetic resolution).

Examples where diamine (+)-**1** actually outperforms (–)-sparteine include phosphine borane lithiation (Table 1, entry 6), benzylic organolithium functionalization (Table 1, entries 7-8) and copper(II)-mediated dynamic thermodynamic resolution of racemic BINOL (Table 2, Entry 2). In recent work, we have introduced a new ligand exchange catalytic approach for asymmetric deprotonation using sub-stoichiometric amounts of (–)-sparteine and (+)-**1**. In all cases, (+)-**1** gives higher enantioselectivity than (–)-sparteine.[14]

In contrast, we are aware of three processes where diamine (+)-**1** is significantly inferior to (–)-sparteine. First, dynamic kinetic resolution of *N*,*N*-diisopropyl-*o*-ethylbenzamide gave an enantiomeric ratio of 68:32 with (+)-**1** (Table 1, entry 11) but an enantiomeric ratio of 94:6 with (–)-sparteine. Second, attempted kinetic resolution of racemic indanol using palladium(II) and molecular oxygen in the presence of (+)-**1** gave a selectivity factor (s) of 6.8 whereas s = 8.3 was obtained using (–)-sparteine (Table 2, entry 3).

Table 1. Asymmetric Synthesis Using (+)-Sparteine Surrogate **1** and Organolithium Reagents

Entry	Substrate	Product	Yield (%)	er	Ref
1	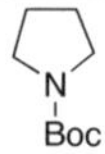	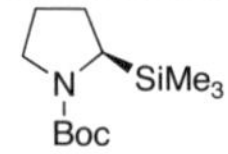	84[a]	95:5	3, 4, 16
2	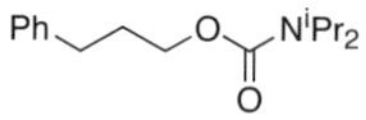	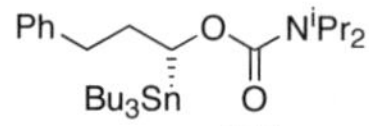	84[b]	96:4	3, 17
3	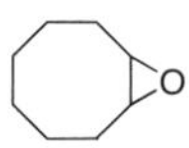	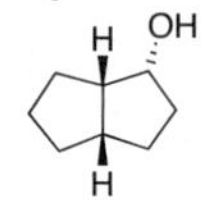	70[c]	81:19	3, 12, 18
4	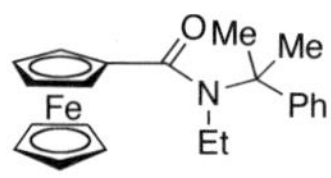	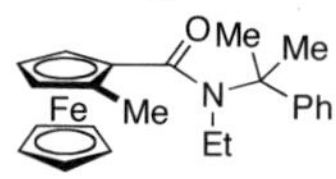	78[d]	96:4	19, 20
5	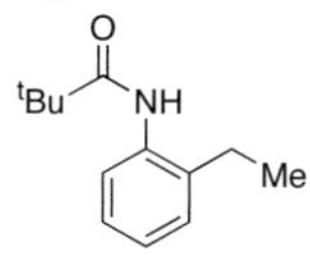	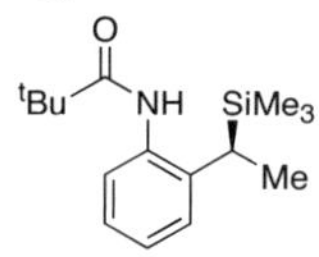	58[e]	93:7	12, 21
6	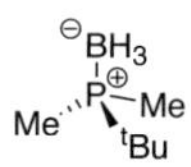	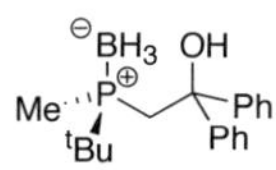	78[f]	96:4	5, 22
7	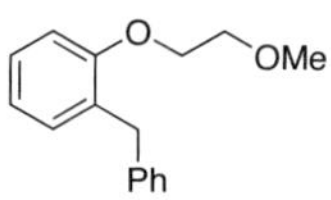	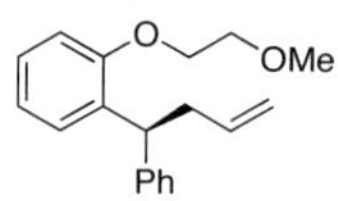	90[g]	96:4	6
8	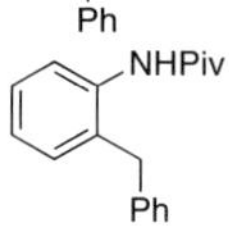	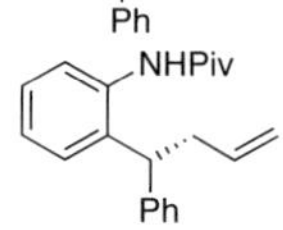	89[h]	90:10	6
9	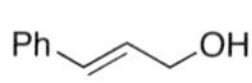	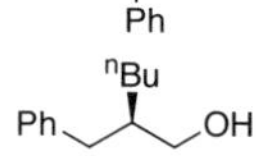	71[i]	87:13	12, 23
10	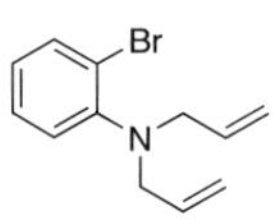	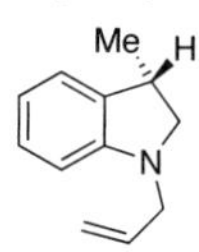	84[j]	85:15	24, 25
11	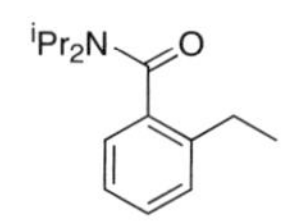	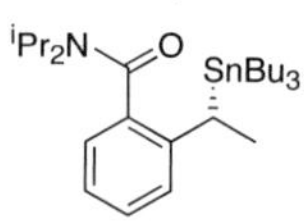	41[k]	68:32	21, 26

[a] (i) ^{s}BuLi, (+)-**1**, –78 °C, Et_2O, 5 h; (ii) Me_3SiCl. [b] (i) ^{s}BuLi, (+)-**1**, –78 °C, Et_2O, 5 h; (ii) Bu_3SnCl. [c] ^{s}BuLi, (+)-**1**, –78 °C, Et_2O, 5 h. [d] (i) ^{n}BuLi, (+)-**1**, –78 °C, 6:1 Et_2O-toluene, 2 h; (ii) MeI. [e] ^{s}BuLi, Et_2O, –25 °C, 2 h; (ii) (+)-**1**, –25 °C, 45 min; (iii) –78 °C then Me_3SiCl. [f] (i) ^{s}BuLi, (+)-**1**, –78 °C, 3 h; (ii) Ph_2CO; (iii) $HCl_{(aq)}$. [g] (i) ^{s}BuLi, (+)-**1**, –78 °C, Et_2O; (ii) –20 °C, 1 h; (iii) –78 °C then 0.5 eq allyl tosylate; (iv) –20 °C, 1 h; (v) –78 °C then 0.5 eq allyl tosylate, warm to rt. [h] ^{s}BuLi, Et_2O, 0 °C, 3 h; (ii) (+)-**1**, –78 °C, 2 h; (iii) allyl bromide, warm to rt . [i] ^{n}BuLi, (+)-**1**, 0 °C, cumene, 1 h. [j] (i) ^{t}BuLi, –78 °C, Et_2O-pentane, 10 min; (ii) (+)-**1**, –40 °C, 1.5 h; (iii) MeOH. [k] (i) ^{s}BuLi, (+)-**1**, –78 °C, pentane, 90 min; (ii) Bu_3SnCl, –78 °C, 1 h.

Finally, the attempted dynamic thermodynamic resolution of lithiated *tert*-butylphenylphosphine borane using (+)-**1** generated a racemic adduct (95% ee with (–)-sparteine).[12] Crucial to the success of this protocol is the formation of a "voluminous precipitate" during the equilibration, and this was not observed with (+)-**1**.[15]

To summarize, the procedure presented here is a simple synthesis of diamine (+)-**1**. Evaluation of diamine (+)-**1** by us and others has demonstrated that it is a very good mimic of (+)-sparteine and can be employed in reactions where (–)-sparteine is a successful ligand. It thus allows access to a range of products of opposite absolute configuration to those obtained by using (–)-sparteine.

Table 2. Asymmetric Synthesis Using (+)-Sparteine Surrogate **1** and Magnesium, Copper and Palladium.

Entry	Substrate	Product	Yield (%)	er	Ref
1	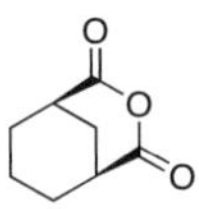	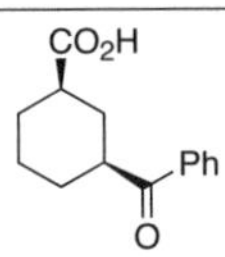	78[a]	89:11	12, 27
2	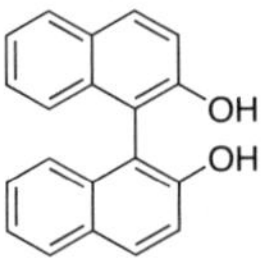 *racemic*	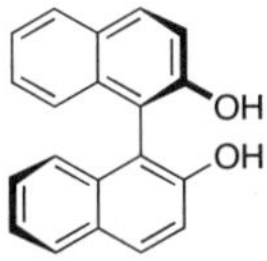	86[b]	99:1	12, 28
3	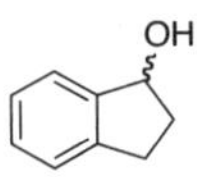 *racemic*	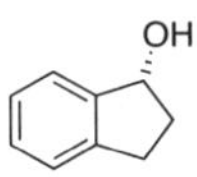	–[c,d]	–	12, 29

[a] PhMgCl, (+)-**1**, –78 °C, toluene, 20 h. [b] (i) CuCl, air, MeOH, (+)-**1**, rt, sonicate, 30 min; (ii) Ar, rt, sonicate, 1 h; (iii) *rac*-BINOL, CH_2Cl_2-MeOH, rt, 8 h; (iv) –25 °C, 16 h; (v) –25 °C, conc. HCl.
[c] (+)-**1**, Pd(nbd)Cl_2, O_2, 3Å MS, 60 °C, toluene, 54 h. [d] Selectivity factor for kinetic resolution, s =6.8.

1. Department of Chemistry, University of York, Heslington, York YO10 5DD UK.
2. For reviews, see: (a) Hoppe, D.; Hense, T. *Angew. Chem., Int. Ed. Engl.* **1997**, *36*, 2282. (b) Beak, P.; Basu, A.; Gallagher, D. J.; Park, Y. S.; Thayumanavan, S. *Acc. Chem. Res.* **1996**, 29, 552. (c) Clayden, J. "Organolithiums: Selectivity for Synthesis", Pergamon, **2002**. (d) Schütz, T. *Synlett*, **2003**, 901. (e) Iyengar, R.; Gracias, V. *Chemtracts* **2004**, *17*, 92. (f) Hoppe, D. Christoph, G. "The Chemistry of Organolithium Compounds. In *The Chemistry of Functional Groups*; Rappoport, Z., Marek, I., Ed.; Wiley: Chichester **2004**; 1077.

3. Dearden, M. J.; Firkin, C. R.; Hermet, J.-P. R.; O'Brien, P. *J. Am. Chem. Soc.* **2002**, *124*, 11870.

4. Hermet, J.-P. R.; Porter, D. W.; Dearden, M. J.; Harrison, J. R.; Koplin, T.; O'Brien, P.; Parmene, J.; Tyurin, V.; Whitwood, A. C.; Gilday, J.; Smith, N. M. *Org. Biomol. Chem.* **2003**, *1*, 3977.

5. (a) Johansson, M. J.; Schwartz, L. O.; Amedjkouh, M.; Kann, N. C. *Eur. J. Org. Chem.* **2004**, 1894. (b) Johansson, M. J.: Schwartz, L.; Amedjkouh, M.; Kann, N. *Tetrahedron: Asymmetry* **2004**, *15*, 3531.

6. (a) Wilkinson, J. A.; Rossington, S. B.; Ducki, S.; Leonard, J.; Hussain, N. *Tetrahedron: Asymmetry* **2004**, *15*, 3011. (b) Wilkinson, J. A.; Rossington, S. B.; Ducki, S.; Leonard, J.; Hussain, N. *Tetrahedron* **2006**, *62*, 1833.

7. Morita, Y.; Tokuyama, H. Fukuyama, T. *Org. Lett.* **2005**, *7*, 4337.

8. Marrière, E.; Rouden, J.; Tadino, V.; Lasne, M.-C. *Org. Lett.* **2000**, *2*, 1121.

9. Nikolic, N. A.; Beak, P. *Org. Synth.* **1997**, *74*, 23.

10. Danieli, B.; Lesma, G.; Passarella, D.; Piacenti, P.; Sacchetti, A.; Silvani, A.; Virdis, A. *Tetrahedron Lett.* **2002**, *43*, 7155.

11. Danieli, B.; Lesma, G.; Passarella, D.; Sacchetti, A.; Silvani, A. *Tetrahedron Lett.* **2005**, *46*, 7121.

12. Dearden, M. J.; McGrath, M. J.; O'Brien, P. *J. Org. Chem.* **2004**, *69*, 5789.

13. O'Brien, P.; Wiberg, K. B.; Bailey, W. F.; Hermet, J.-P. R.; McGrath, M. J. *J. Am. Chem. Soc.* **2004**, *126*, 15480.

14. McGrath, M. J.; O'Brien, P. *J. Am. Chem. Soc.* **2005**, *127*, 16378.

15. Wolfe, B; Livinghouse, T. *J. Am. Chem. Soc.* **1998**, *120*, 5116.

16. Beak, P.; Kerrick, S. T.; Wu, S.; Chu, J. *J. Am. Chem. Soc.* **1994**, *116*, 3231.

17. Tomooka, K.; Shimizu, H.; Inoue, T.; Shibata, H.; Nakai, T. *Chem. Lett.* **1999**, 759. See also: Hoppe, D.; Hintze, F.; Tebben, P. *Angew. Chem. Int. Ed. Engl.* **1990**, *29*, 1422.

18. Hodgson, D. M.; Lee, G. P.; Marriott, R. E.; Thompson, A. J.; Wisedale, R.; Witherington, J. *J. Chem. Soc., Perkin Trans. 1* **1998**, 2151.

19. Genet, C.; Canipa, S. J.; O'Brien, P.; Taylor, S. *J. Am. Chem. Soc.*, **2006**, *128*, 9336.

20. (a) Metallinos, C.; Szillat, H.; Taylor, N. J.; Snieckus, V. *Adv. Catal. Synth.* **2003**, *345*, 370. (b) Tsukazaki, M.; Tinkl, M.; Roglans, A.; Chapell, B. J.; Taylor, N. J.; Snieckus, V. *J. Am. Chem. Soc.* **1996**, *118*, 685.

21. Thayumanavan, S.; Basu, A.; Beak, P. *J. Am. Chem. Soc.* **1997**, *119*, 8209.

22. Imamoto, T.; Watanabe, J.; Wada, Y.; Masuda, H.; Yamada, H.; Tsurata, H.; Matsukawa, S.; Yamaguchi, K. *J. Am. Chem. Soc.* **1998**, *120*, 1635.

23. Norsikian, S.; Marek, I.; Klein, S.; Poisson, J. F.; Normant, J. F. *Chem. Eur. J.* **1999**, *5*, 2055.

24. Bailey, W. F.; Luderer, M. R.; O'Brien, P. unpublished results.

25. Mealy, M. J.; Luderer, M. R.; Bailey, W. F.; Sommer, M. B. *J. Org. Chem.* **2004**, *69*, 6042.

26. Dearden, M. J.; O'Brien, P. unpublished results.

27. Shintani, R.; Fu, G. C. *Angew. Chem. Int. Ed.* **2002**, *41*, 1057.

28. Zhang, Y.; Yeung, S.-M.; Wu, H.; Heller, D. P.; Wu, C.; Wulff, W. D. *Org. Lett.* **2003**, *5*, 1813.

29. Ferraira, E. M.; Stoltz, B. M. *J. Am. Chem. Soc.* **2001**, *123*, 7725.

SYNTHESIS AND RU(II)-BINAP REDUCTION OF A KETOESTER DERIVED FROM HYDROXYPROLINE: 2(*S*)-(β-*tert*-BUTOXYCARBONYL-α-(*S*) and α-(*R*)-HYDROXYETHYL)-4(*R*)-HYDROXYPYRROLIDINE-1-CARBOXYLIC ACID, *tert*-BUTYL ESTER

(2-Pyrrolidinepropanoic acid, 1-[(1,1-dimethylethoxy)carbonyl]-β,4-dihydroxy-, 1,1-dimethylethyl ester, [2*S*-[2α(*S),4β]]-)**

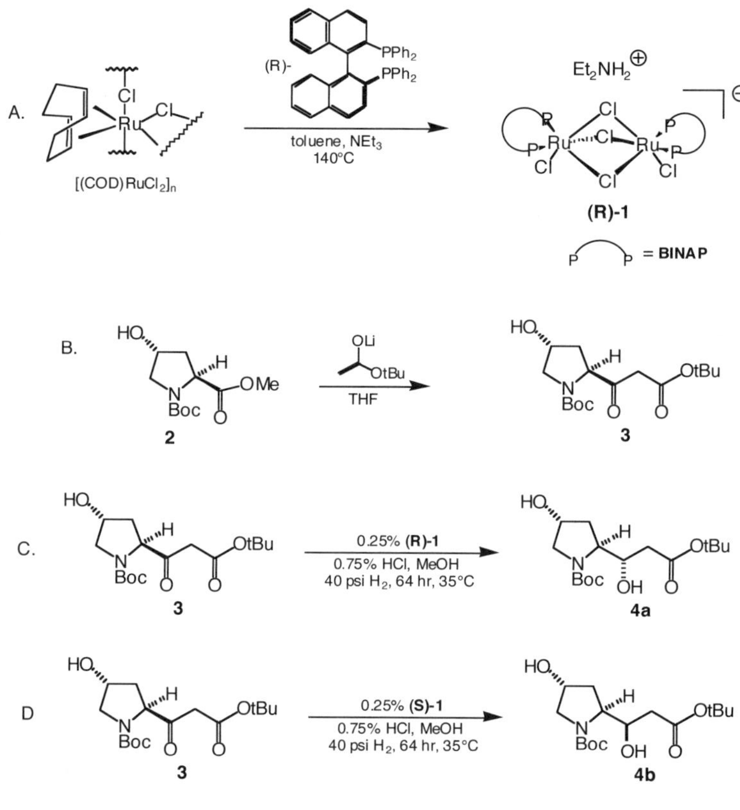

Submitted by Steven A. King, Joseph Armstrong, and Jennifer Keller.[1]
Checked by Amos B. Smith, III and Meinrad Brenner.

1. Procedure

A. *Preparation of catalyst.* Two 100-mL, round-bottomed flasks are connected to a double-ended filter which consists of two chambers, each equipped with a side arm, separated by a glass frit (Note 1). Vacuum grease is used to ensure an air-tight seal and rubber bands are used to secure the flasks to the filter assembly. A magnetic stirbar, (cyclooctadienyl)ruthenium dichloride polymer (0.426 g, 1.5 mmol), and (*R*)-BINAP (1.00 g, 1.60 mmol) (Note 2) are placed in one flask, and the entire apparatus is evacuated and filled with nitrogen. Toluene (34 mL) and triethylamine (3.4 mL) (Note 2), which have been deoxygenated by nitrogen sparging for several min, are added via the lower side arm. The vessel is sealed and the mixture heated to 140 °C producing a deep brick red colored solution (Note 3). After 4 h, the apparatus is allowed to cool to room temperature and the reaction mixture is vigorously stirred while the catalyst precipitates. The apparatus is vented to nitrogen and inverted to filter the product using vacuum on the lower side arm and nitrogen on the upper side arm (Note 4). If the frit cakes with product, slowing down filtration, momentarily reversing the flow of nitrogen will lift the cake away from the frit. The precipitate is washed with 34 mL of deoxygenated toluene and the flask containing the filtrate is replaced with an empty one. The entire apparatus is placed under vacuum and the product is dried overnight to give 0.850 g (67%) of a dark red solid (Notes 5, 6).

B. *(2S,4R)-2-tert-Butoxycarbonylacetyl-4-hydroxypyrrolidine-1-carboxylic acid, tert-butyl ester.* A 12-L, four-necked, round-bottomed flask equipped with a mechanical stirrer, addition funnel, thermocouple, and nitrogen inlet is charged with 3.3 L (3.82 mol) of 1.15M lithium hexamethyldisilylamide (LHS) solution in THF (Note 7). The solution is cooled to –40 °C, and 514 mL (3.82 mol) of *tert*-butyl acetate is added via the addition funnel over 15 min. A solution of *N*-Boc hydroxyproline methyl ester **2** (187 g, 0.763 mol) in 940 mL of THF (Note 8) is cooled to 0°C and then added slowly via the addition funnel over 20 min while the temperature of the reaction mixture is maintained below –30 °C. The resulting solution is stirred until the reaction is complete (30 min)(Note 9) and then poured over 20 min into an ice cold mixture of 3.8 L of 1M citric acid and 1.7 L of heptanes (Note 10). After stirring for an additional 30 min, the layers are separated (Note 11). The organic layer is concentrated by distillation under reduced pressure to a volume of 1.1 L (Note 12). Heptanes (3.6 L) are added and the mixture is concentrated by distillation under reduced pressure at 30 °C to a volume of 1.1 L (Note 13). The resulting solution is diluted with 800 mL of heptanes and cooled at 0 °C overnight. The resulting precipitate is filtered, washed with two 800-mL portions of heptanes, and dried overnight in a vacuum oven to provide 227 g (90%) of β-ketoester **3** as white crystals (Note 14).

C. *2(S)-(β-tert-Butoxycarbonyl-α-(S)-hydroxyethyl)-4-(R)-hydroxy-pyrrolidine-1-carboxylic acid, tert-butyl ester.* A 50-mL Parr shaker bottle capped with a rubber septum was charged with β-ketoester **3** (10.0 g, 30 mmol) and 15 mL of methanol and the solution was deoxygenated by sparging with nitrogen for 15 min. (*R*)-BINAP catalyst ***(R)*-1** prepared as described above (0.064 g, 0.035 mmol) was added followed by 8.7N HCl in methanol (26 μL, 0.23 mmol). The vessel was transferred to a standard Parr shaker apparatus and flushed by evacuating and refilling with nitrogen and then hydrogen several times (Note 15). The apparatus was heated at 35 °C with shaking under 40 psi of hydrogen (Note 16). After 20 min, the reaction mixture became a homogeneous clear yellow solution. After 11 hr, the reaction mixture is transferred to a 250-mL, round-bottomed flask and concentrated to dryness. The residue is suspended in 50 mL of toluene and concentrated to dryness to ensure complete removal of methanol. The product is slurried in 25 mL of toluene/hexane/isopropyl acetate solution (90:5:5) and stirred overnight at room temperature. The precipitate is filtered, washed with 10 mL of toluene, and dried overnight in a vacuum oven (35 °C, <0.5 mmHg) to provide 8.5 g (85%) of (*S*)-hydroxy ester **4a** (Notes 17, 18).

C. *2(S)-(β-tert-Butoxycarbonyl-α-(R)-hydroxyethyl)-4-(R)-hydroxy-pyrrolidine- 1-carboxylic acid, tert-butyl ester.* The identical procedure was followed, in this case using the (*S*)-BINAP catalyst **(*S*)-1**. Hydrogenation is conducted for 64 h, and the reaction mixture is then transferred to a 250-mL, round-bottomed flask and concentrated to dryness. The residue is dissolved in 17 mL of methanol and cooled to 15°C. After the slow addition of 7 mL of DI water, the solution is aged for 15 min gradually forming a thin slurry. More DI water (75 mL) is added over 1 hr and the mixture is allowed to stand for an additional 1 hr at 15 °C. The resulting crystals (Note 19) are filtered at 15 °C, washed with 10 mL of 1:4-MeOH:water, and then dried overnight in a vacuum oven (35 °C, 686 mm) to yield 7.0 g (70%) 0f (*R*)-hydroxy ester **4b** (Note 20).

2. Notes

1. A double-ended filler was purchased from Kontes (catalog #215500-6044) and is shown below.

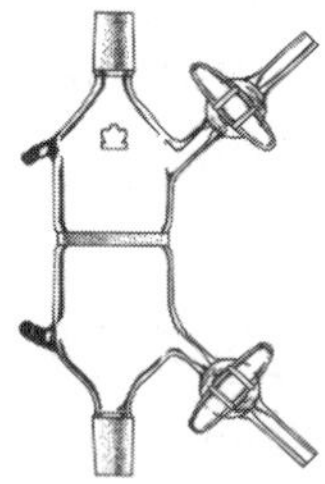

2. The checkers purchased $[(COD)RuCl_2]_n$ from Fluka (purum quality), BINAP from Aldrich (97 %), and toluene (HPLC grade) and triethylamine (reagent grade) from Fisher Scientific; the latter was distilled from CaH_2 under Ar prior to use. The submitters dried toluene and triethylamine over 4 Å molecular sieves. Karl-Fischer titration indicated <200 µg/mL water.

3. A dark green color of the solution indicates decomposition of the reaction mixture by oxygen.

4. ^{31}P NMR showed that the filtrate contained none of the desired product.

5. The checkers obtained 0.849-0.875 g of catalyst, but ^{1}H-NMR analysis revealed that the product was of low purity. Reactions run on a 0.75 mmol scale afforded the catalyst in 47-51% yield and in higher purity. Since the reaction is run in a sealed apparatus, the pressure built up in the apparatus (and therefore the boiling point and reaction temperature) may depend on the reaction scale and how far the flask is immersed into the oil bath.

6. ^{1}H-NMR (CD_2Cl_2, 500 MHz) δ: 1.47 (t, J = 7.4, 6 H), 3.20-3.30 (m, 4 H), 6.52-6.60 (m, 12 H), 6.65 (d, J = 8.6, 2 H), 6.69 (t, J = 7.4, 2 H), 6.74-6.81 (m, 4 H), 6.86 (t, J = 7.4, 2 H); 6.96 (t, J = 7.7, 2 H), 7.18-7.34 (m, 14 H), 7.38 (t, J = 7.2, 2 H), 7.47-7.61 (m, 12 H), 7.63-7.68 (m, 6 H), 7.84 (t, J = 8.3, 2 H), 8.08 (t, J = 8.4, 4 H), 8.57 (s br., 2 H); ^{31}P NMR (CD_2Cl_2, 101 MHz) δ: 52.3 (d, J = 38), 56.5 (d, J = 38).

7. The submitters obtained LHS in THF from Callery Chemical Co. and determined its concentration to be 1.15M by titration according to the method of Ireland (Ireland, R.; Meissner, R. *J. Org. Chem.* **1991**, *56*, 4566-

4568). The checkers purchased LHS as a 1M solution in THF from Aldrich Chemical Co.

8. The submitters obtained *N*-Boc-4-hydroxyproline methyl ester from Synthetech, Inc. [Albany, OR, (503) 967-6575] or Bachem California [Torrance, CA, (310) 530-1571] and dried THF over 4Å molecular sieves for two days prior to use (Karl-Fischer titration gave 145 μg/mL water). The checkers purchased N-Boc hydroxyproline methyl ester (97%) and *tert*-butyl acetate (99+%) from Aldrich Chemical Co. and obtained THF (HPLC grade) from Fisher Scientific.

9. The progress of the reaction was followed by HPLC (Zorbax RX-C8 column, 1.5 mL/min, 50:50 CH_3CN:0.01M H_3PO_4 in water, room temperature, detection at 200 nm; retention times: methyl ester **2** 2.343 min, β-ketoester **3** 3.987 min. After 30 min, less than 0.5 area% methyl ester was found to be remaining.

10. There is an exotherm to 10 °C during the quench.

11. GC assay of the organic layer showed no $HN(TMS)_2$ remaining after 15 min of stirring (GC conditions: Restek RTX-1 column (30 m x 0.53 mm, 1 m film thickness), 2.53 mL/min, initial temperature 50 °C, final temperature 300 °C, rate 20 deg/min, injection temperature 200 °C, detector temperature 350 °C, injection volume 1 μL, inject sample neat; retention times: *tert*-butyl alcohol 1.4 min, THF 1.7 min, heptane 2.1 min, $HN(TMS)_2$ 2.6 min, ethylbenzene (present in commercial LHS) 3.1 min, *tert*-butyl acetate 4.0 min). Volume percents were determined based on standard solution counts.

12. The checkers found the product to contain small amounts of citrates. Drying ($MgSO_4$) or washing (H_2O) the organic phase prior to concentration prevents this contamination.

13. The distillation is carried out at ca. 60 mm and at a temperature below 40°C due to the instability of the β-ketoester in the crude solution. It is important to remove all THF prior to crystallization.

14. In smaller scale runs, the checkers obtained the product in 88-93% yield. The product exhibited the following properties: mp 98-99 °C, $[\alpha]_D^{23°}$: –73.2 (c = 0.95, MeOH); ^{13}C NMR (CD_3OD, 500 MHz, rotamers) δ: 28.4, 28.6, 28.7, 38.3, 39.1, 56.0, 56.3, 65.3, 65.6, 70.0, 70.8, 81.7, 82.2, 83.0, 155.9, 156.7, 168.1, 168.3, 204.7, 204.9. IR (film) cm^{-1}: 3439 (br. s), 2979 (s), 2934 (m), 1707 (s), 1395 (s), 1368 (s), 1326 (m), 1258 (m), 1162 (s), 1077 (m), 983 (m), 963 (m), 853 (m), 772 (m). Anal. Calcd for $C_{16}H_{27}NO_6$: C, 58.34; H, 8.26; N, 4.24. Found: C, 58.71; H, 8.52; N, 4.16.

15. Thorough removal of oxygen is imperative to prevent destruction of the catalyst.

16. The progress of the reaction is followed by HPLC (Zorbax RX-C8 column, 1.5 mL/min, gradient elution of 20:80 to 70:30 CH_3CN:0.01 M H_3PO_4 in water over 10 min, held for 20 min, room temperature, detection at 200 nm, retention times: β-keto *t*-butyl ester **2** 8.31 min, hydroxy *tert*-butyl ester **3** 7.10 min, hydroxy methyl ester 4.45 min, β-keto methyl ester 5.47 min).

17. The checkers obtained the product in 73-90% yield. The product exhibited the following properties: mp 99-100 °C; $[\alpha]_D^{23°}$: –50.5 (c = 0.93, MeOH); ^1H NMR ($CDCl_3$, 500 MHz) δ: 1.46 (s, 9 H), 1.47 (s, 9 H), 1.90-2.00 (m, 1 H), 2.00-2.10 (m, 1 H), 2.34 (dd, *J* = 15.7, 9.0, 1 H), 2.41 (dd, *J* = 15.7, 3.3, 1 H), 3.39 (dd, *J* = 12.1, 4.1, 1 H), 3.55-3.70 (m, 1 H), 4.11-4.17 (m, 1 H), 4.40-4.50 (m, 1 H). ^{13}C NMR (CD_3OD, 500 MHz, rotamers) δ: 28.4, 28.7, 36.7, 37.1, 39.3, 40.8, 56.6, 56.9, 61.0, 61.4, 70.1, 70.3, 70.7, 71.9, 81.3, 81.5, 81.8, 156.7, 157.2, 172.9. IR (film) cm^{-1}: 3418 (br. s), 2978 (s), 2933 (s), 1703 (s), 1403 (s), 1367 (s), 1257 (s), 1159 (s), 999 (m), 969 (m), 848 (m), 774 (m), 737 (m). Anal. Calcd for $C_{16}H_{29}NO_6$: C, 57.99; H, 8.82; N, 4.23. Found: C, 58.18; H, 9.09; N, 3.98.

18. The submitters determined the crystalline hydroxy esters to be >99.9% diastereomerically pure by supercritical fluid chromatography (EMdiol silica column and a Chiralcel (+) OD-(H) column (Chiral Technologies, Inc.) in tandem (100 bar CO_2; 1.0 mL/min; 8% MeOH, modifier 35°C; retention times: β-ketoester (keto form) **3** 15.83 min, β-ketoester (enol form) **3** 18.01 min, (*R*)-hydroxy ester **4b** 18.78 min, (*S*)-hydroxy ester **4a** 19.70 min).

19. The checkers found that the product was still slightly brown, indicating that the purification procedure is not effective in removing the catalyst.

20. The product exhibits the following properties: mp 135-137 °C; $[\alpha]_D^{23°}$: –70.2 (c = 0.34, MeOH); ^1H NMR ($CDCl_3$, 500 MHz) δ: 1.46 (s, 9 H), 1.47 (s, 9 H), 1.89-1.94 (m, 1 H), 2.00-2.20 (m br., 1 H), 2.25 (dd, *J* = 15.8, 9.5, 1 H), 2.32 (dd, *J* = 15.8, 3.5, 1 H), 3.40 (dd, *J* = 11.9, 4.0, 1 H), 3.45-3.80 (m br., 1 H), 3.80-4.15 (m br., 1 H), 4.30-4.45 (m, 1 H), 4.45-4.55 (m, 1 H); ^{13}C NMR (CD_3OD, 500 MHz, rotamers) δ: 28.4, 28.8, 34.4, 34.8, 41.1, 41.7, 56.1, 56.7, 61.8, 69.1, 69.2, 70.4, 70.7, 81.0, 81.3, 81.8, 156.7, 157.2, 172.3, 172.6. IR (film) cm^{-1}: 3419 (br. s), 2980 (s), 2919 (s), 1733 (s), 1673 (s), 1477 (m), 1418 (s), 1366 (s), 1244 (m), 1161 (s), 1080 (m), 1049 (m), 1002 (m), 952 (m), 876 (m), 852 (m), 775 (m), 749 (m); Anal. Calcd for

$C_{16}H_{29}NO_6$: C, 57.99; H, 8.82; N, 4.23. Found: C, 58.33; H, 9.01; N, 3.81.

3. Discussion

The Ru(II)-BINAP complex,[2] $[Et_2NH_2]^+[Ru_2Cl_5(BINAP)_2]^-$,[3] is prepared as a toluene solvate in nearly pure form by this procedure. Typical crystallized product shows no other signals in the phosphorus NMR and gives a good combustion analysis. The material is quite stable and can be routinely handled in air. Storage under nitrogen will extend its shelf life, however.

The Claisen condensation of *t*-butyl acetate with a methyl ester is a general route for the preparation of complex β-ketoesters.[4] The reaction requires an excess of the enolate of *t*-butyl acetate to rapidly deprotonate the product and prevent tertiary alcohol formation. Some workers have also used excess LDA or *t*-butoxide for this purpose.

The reduction of β-ketoesters with Ru(II)-BINAP is the most efficient man-made catalytic asymmetric reaction.[5] Enantioselectivity is nearly always greater than 97% and 1,000 turnovers are typically accomplished in 2-8 hr. In the case of the optically active ketoester shown here, diastereoselectivity is >99:1 in the matched (C.) reaction and 94:6 in the mismatched (D.) case. Not surprisingly, the matched reaction is also considerably faster. Ketoester reductions are best run as a concentrated solution in methanol. (Methylene chloride/methanol mixtures have also been used; the presence of an alcoholic cosolvent is mandatory.) Reaction mixtures are extremely susceptible to poisoning by the simultaneous presence of hydrogen and oxygen. Thus, thorough degassing is necessary prior to reaction. The addition of a small amount of strong acid (e.g., HCl, H_2SO_4, CH_3SO_3H) is necessary to activate the catalyst; in its absence nearly no reduction occurs. The particular case of *t*-butyl esters demonstrates that the conditions are still quite mild, since acid-catalyzed dealkylation and transesterification are possible side reactions.

1. Merck Research Laboratories, Rahway, NJ 07065.

2. King, S. A.; Thompson, A. S.; King, A. O.; Verhoeven, T. R. *J. Org. Chem.* **1992**, *57*, 6689.

3. This structure is a revision of the originally assigned structure which was $[Ru_2Cl_4(BINAP)_2]\cdot Et_3N$. King, S. A.; DiMichele, L. In "Catalysis of Organic Reactions"; Scaraos, M. G.; Prunier, M. L., Ed.; **1995**, 157.
4. See, for example: Ohta, S.; Shimabayashi, A.; Hayakawa, S.; Sumino, S.; Okamoto, M. *Synthesis* **1985**, 45; Yamaguchi, M.; Nakamura, S.; Okuma, T.; Minami, T. *Tetrahedron Lett.* **1990**, *31*, 3913; Brower, P. L.; Butler, D. E.; Deering, C. F.; Le, T. V.; Millar, A.; Nanninga, T. N.; Roth, B. D. *Tetrahedron Lett.* **1992**, *33*, 2279.
5. Noyori, R.; Ohkuma, T.; Kitamura, M.; Takaya, H.; Sayo, N.; Kumobayashi, H.; Akutagawa, S. *J. Am. Chem. Soc.* **1987**, *109*, 5856; Noyori, R.; Takaya, H. *Acc. Chem. Res.* **1990**, 23, 345.

2-(2′,2′-DIMETHYLPROPOXY)-2,3-DIHYDRO-1*H*-INDENE

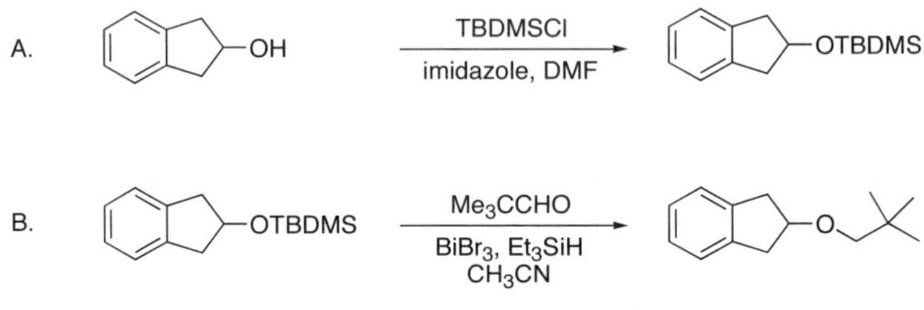

Submitted by Joginder S. Bajwa,[1] Kapa Prasad and Oljan Repič.
Checked by Sigrid Holle and Alois Fürstner.

1. Procedure

A. *2,3-Dihydro-2-tert-butyldimethysiloxy-1H-indene.* A flame-dried 500-mL, four-necked round-bottomed flask equipped with a Teflon-coated magnetic stir bar, a pressure-equalizing dropping funnel, a nitrogen inlet, an internal thermometer and a glass stopper is charged successively under nitrogen atmosphere with 2-indanol (13.42 g, 100 mmol), imidazole (8.17 g, 120 mmol) and 60 mL of dry *N,N*-dimethylformamide (DMF) (Note 1). The mixture is stirred until a clear solution has formed (Note 2). The flask is immersed in an ice bath and a solution of *tert*-butyldimethylsilyl chloride (16.32 g, 108 mmol) in 40 mL of dry DMF (Note 3) is added over a period of 10 min at such a rate that the reaction temperature remains below 10 °C. The ice-water bath is removed, the mixture is allowed to warm to ambient temperature and stirring is continued for 45 min (Note 4). Deionized water (150 mL) is added slowly via the dropping funnel, causing a slight exotherm (Note 5), followed by the addition of 100 mL of *n*-heptane. The resulting mixture is filtered through a pad of Celite (Note 6), the organic layer is separated, and the aqueous phase is extracted with 100 mL of *n*-heptane. The combined organic phases are washed with 100 mL of water, dried over anhydrous magnesium sulfate, filtered and concentrated by rotary evaporation (40 °C/10 mmHg). The crude product is purified by flash chromatography on 100 g of silica gel (Note 1) packed in a 4.5 x 38 cm column. Elution with *n*-heptane (650 mL) followed by 2% ethyl acetate in *n*-heptane (about 1 L) provides 22.5 g (91 %) of 2,3-dihydro-2-*tert*-butyldimethysiloxy-1*H*-indene as a colorless oil (Notes 7, 8).

B. 2-(2,2-Dimethylpropoxy)-2,3-dihydro-1H-indene. A 250-mL, flame-dried three-necked round-bottomed flask equipped with a Teflon-coated magnetic stir bar, a distillation head attached to a Liebig condenser, a nitrogen inlet and an internal thermometer is charged under nitrogen with 2,3-dihydro-2-*tert*-butyldimethysiloxy-1*H*-indene (10.6 g, 42.66 mmol) and 75 mL of dry acetonitrile (Note 1). The flask is immersed in a preheated oil bath and the resulting solution is refluxed such that approximately 25 mL of the solvent is distilled off at atmospheric pressure (Note 9). The solution is allowed to reach ambient temperature and the distillation head is replaced by a glass stopper. Bismuth tribromide (1.32 g, 2.94 mmol) (Notes 1, 10) is added in one portion and stirring is continued for 10 min. The glass stopper is then replaced by a rubber septum. Triethylsilane (8.2 mL, 51.34 mmol) (Note 1) is added via syringe over 5 min causing the immediate formation of a black suspension. The mixture is cooled in an ice-water bath to 0–5 °C before trimethylacetaldehyde (5.1 mL, 46.96 mmol) (Note 1) is added via syringe at such a rate (Note 11) as to maintain the internal temperature below 15 °C. The ice-water bath is removed, the mixture is allowed to warm to ambient temperature over 30 min and stirring is continued for an additional 30 min. (Note 12). The mixture is filtered through a pad of Celite (Note 13) to remove the black precipitate. The filtrate is concentrated by rotary evaporation (40 °C/10 mmHg) to give a mixture of clear and orange oils. To this residue are added 100 mL of *n*-heptane and 40 g of silica gel (Notes 1, 14). The mixture is concentrated by rotary evaporation (40 °C/10 mmHg) to afford a free-flowing powder that is placed at the top of a column (4.5 x 38 cm) of 100 g of silica gel. The column is eluted with *n*-heptane (1 L) followed by 2% ethyl acetate in *n*-heptane (about 1.2 L) to give 7.51 g (86 %) of 2,3-dihydro-2-(2′,2′-dimethylethoxy)-1*H*-indene as a colorless oil (Notes 15, 16).

2. Notes

1. Triethylsilane, 2-indanol and trimethylacetaldehyde were purchased from Acros Organics and used as received. Imidazole, *tert*-butyldimethylsilyl chloride and bismuth tribromide were purchased from Aldrich Chemical Company, Inc. and used as received. Dimethylformamide (water content < 0.02%), acetonitrile (water content, 0.003%), ethyl acetate, *n*-heptane and silica gel (200-425 mesh) were purchased from Fisher Scientific and were used as such. The checkers used DMF dried by distillation over CaH_2.

2. The dissolution of 2-indanol and imidazole in dimethylformamide is endothermic. The temperature of the solution changes from 22 °C to 12 °C over 5 min.

3. *tert*-Butyldimethylsilyl chloride was dissolved in DMF by heating the mixture in warm water bath (40 °C).

4. The progress of the reaction is monitored by TLC on silica gel (elution with 10% ethyl acetate in *n*-heptane; visualized with 254-nm UV lamp and phosphomolybdic acid). The product TBS ether has an R_f = 0.81 and the alcohol starting material an R_f = 0.19. TLC analysis indicates that the reaction is complete after stirring for 45 min at room temperature.

5. The temperature rises from 21 °C to 35 °C upon addition of water.

6. An orange-brown rag layer is formed at the interface of the phases and is removed by filtration through a pad of Celite (3.0 g in 4.5 cm diameter Büchner funnel) to facilitate separation of the layers.

7. The submitters reported a yield of 23.7 g (95%). The spectral data for 2,3-dihydro-2-*tert*-butyldimethylsiloxy-1*H*-indene are as follows: ^{1}H NMR (400 MHz, $CDCl_3$) δ: 0.08 (s, 6 H), 0.89 (s, 9 H), 2.87 (dd, J = 5.9, 15.6 Hz, 2 H), 3.11 (dd, J = 6.8, 15.6 Hz, 2 H), 4.63–4.69 (m, 1 H), 7.11–7.18 (m, 4 H); ^{13}C NMR (100 MHz, $CDCl_3$) δ: –4.7, 18.2, 25.9, 42.6, 73.9, 124.5, 126.4, 141.2. IR (neat): 2952, 2929, 2856, 1471, 1461, 1252, 1101, 1064, 987 cm^{-1}.

8. The product is > 96% pure as determined by GC analysis on a 0.53 mm i.d. x 30 m capillary column (DB-1) at 50-260 °C, heating increment 25 °C/min. Retention time is 7.4 min.

9. Partial distillation azeotropically removes traces of water.

10. Bismuth tribromide is hygroscopic and should be handled in a glove box. Use of wet $BiBr_3$ in this reaction significantly lowers the yield of the product.

11. The reaction is highly exothermic. However, the exotherm is easily controlled by external cooling and slow (10 min) addition of trimethylacetaldehyde.

12. The reaction is monitored by TLC on silica gel (2% ethyl acetate in *n*-heptane; product R_f = 0.56, TBS ether R_f = 0.39) and is complete after 30 min at room temperature.

13. Celite (3.0 g) is used on top of a filter paper in a 4.5 cm diameter Büchner funnel.

14. The evaporation of *n*-heptane removes any residual acetonitrile that causes poor separation of the product from the silicon by-products.

15. The spectral data for 2,3-dihydro-2-(2′,2′-dimethylethoxy)-1*H*-indene are as follows: ^{1}H NMR (400 MHz, $CDCl_3$) δ: 0.90 (s, 9 H), 2.95 (dd, J = 5.2, 16.0 Hz, 2 H), 3.13 (s, 2 H), 3.16 (dd, J = 6.8, 16.0 Hz, 2 H), 4.26–4.33 (m, 1 H), 7.13–7.16 (m, 2 H), 7.17-7.21 (m, 2 H); ^{13}C NMR (100 MHz, $CDCl_3$) δ: 26.8, 32.0, 39.4, 79.6, 80.7, 124.6, 126.3, 141.2; IR (neat): 2952, 2866, 1480, 1362, 1110, 1091 cm^{-1}; Calcd for $C_{14}H_{20}O$: 204.1514; found: 204.1516 ([M^+]).

16. The product is >99% pure as determined by GC analysis performed under the same conditions as described in Note 8. The retention time of the product is 6.7 min.

Safety and Waste Disposal Information

All hazardous materials should be handled and disposed of in accordance with "Prudent Practices in the Laboratory"; National Academy Press; Washington, DC, 1995.

3. Discussion

Ethers are usually prepared by the Williamson method whereby an alkoxy anion is reacted with an alkyl halide or sulfonate.[2] The presence of base-labile functionalities in the reactants and competing eliminations with the use of secondary and tertiary halides can limit the use of this method. Alternatively, ethers can be synthesized by triethylsilane-mediated reductive alkylation of carbonyl compounds catalyzed by various Lewis acids such as $TrClO_4$,[3] BF_3OEt,[4] TMSOTf,[5] and metal triflates.[6] As compared to these catalysts, bismuth tribromide was found to work more efficiently.[7]

The present method for the preparation of ethers is a variation of the procedure originally described by Komatsu,[7] who used only TMS ethers as the substrates. Since TMS ethers are hydrolytically rather labile, the reaction was successfully extended[8] to more stable silyl ethers such as triethylsilyl (TES), triisopropylsilyl (TIPS) and *tert*-butyldimethylsilyl (TBDMS) ethers (Table 1). The reaction is very sensitive to moisture. Even a trace amount of water in the reaction mixture leads to a substantial amount of desilylated by-product alcohol.[9] Therefore, it is necessary that the solvent and the reagents are rigorously anhydrous. The reaction is highly exothermic, however, the exotherm is easily controlled by slow addition of aldehyde and by external

Table 1. Transformation of Silyl Ethers into Dialkyl Ethers with $BiBr_3$ and Et_3SiH

Silyl Ether	Aldehyde or Ketone	Product	Yield (%)
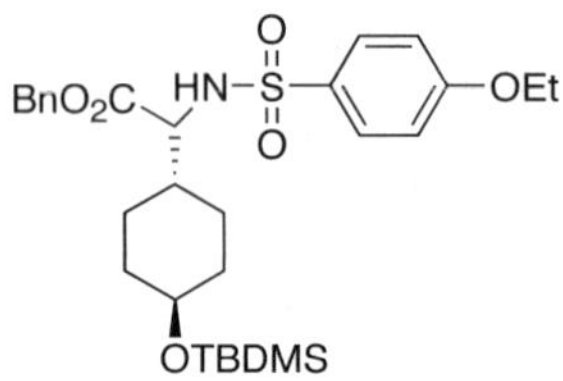	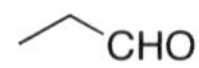	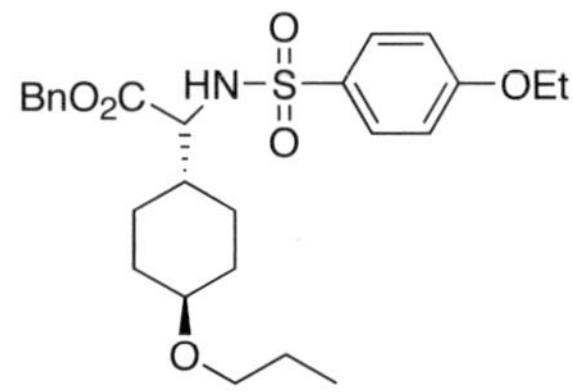	88[a]
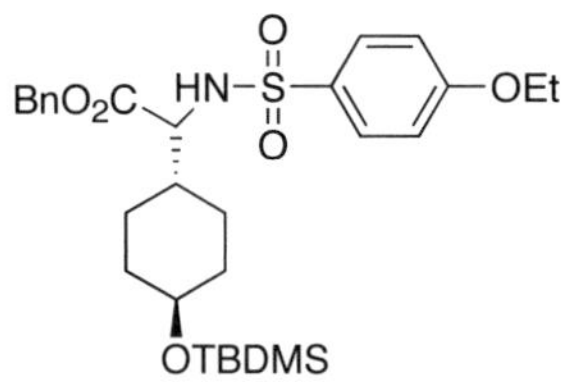		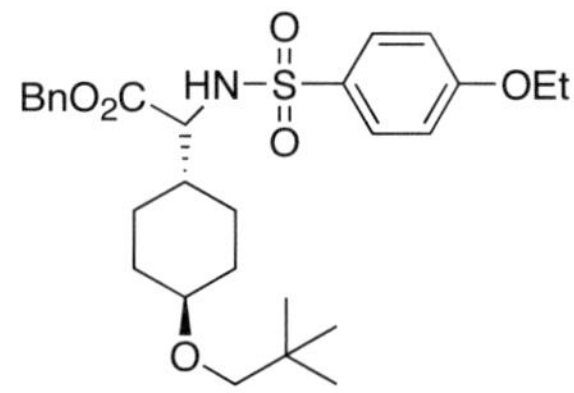	92[a]
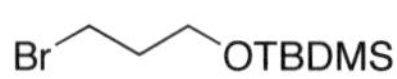	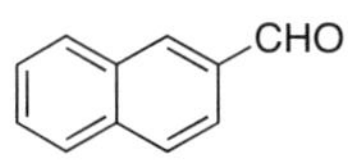	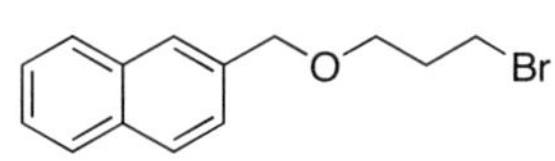	81[a]
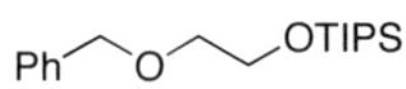		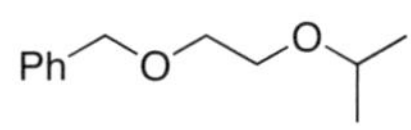	85[a]
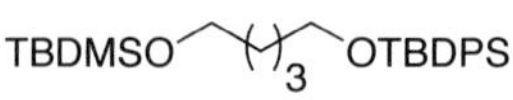		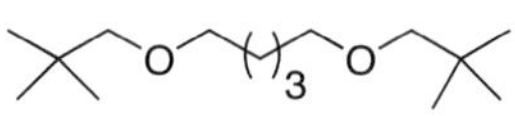	45[a]
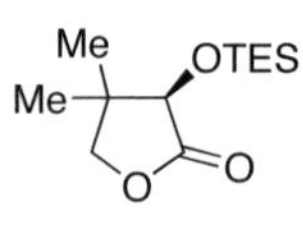	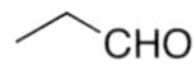	No Reaction	0[a]
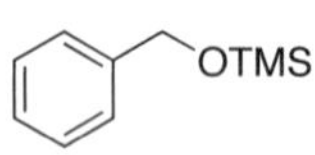	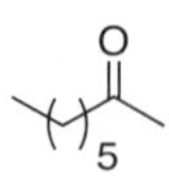	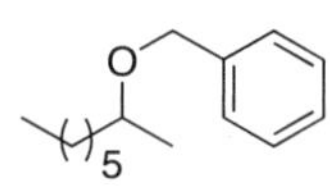	96[b]
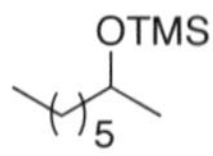	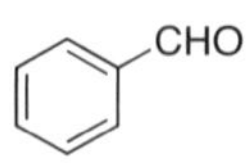	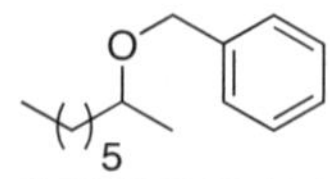	88[b]

[a] Reference 8; [b] Reference 7.

cooling. We believe that $BiBr_3$ by itself does not act as a Lewis acid. Instead, it reacts with Et_3SiH to generate elemental bismuth, hydrogen bromide and triethylsilylbromide.[8]

Overall, the procedure described above constitutes a general and effective method for the synthesis of dialkyl ethers. Its utility and scope are evident from the examples compiled in Table 1, some of which are not easily accessible by the classical Williamson's method.

1. Process R&D, Chemical and Analytical Development, Novartis Institute for Biomedical Research, One Health Plaza, East Hanover, NJ 07936.
2. Smith, M. B.; March, J. March's *Advanced Organic Chemistry: Reactions, Mechanisms, and Structure*, 5th ed.; Wiley: New York, **2001**; pp 477-478 and references cited therein.
3. Kato, J.; Iwasawa, N.; Mukaiyama, T. *Chem. Lett.* **1985**, 743-746.
4. Doyle, M. P.; West, C. T.; Donnelly, S. J.; McOsker, C. C. *J. Organomet. Chem.* **1976**, *117*, 129-140.
5. Hatakeyama, S.; Mori, H.; Kitano, K.; Yamada, H.; Nishizawa, M. *Tetrahedron Lett.* **1994**, *35*, 4367-4370.
6. Yang, W.-C.; Lu, X.-A.; Kulkarni, S. S.; Hung, S.-C. *Tetrahedron Lett.* **2003**, *44*, 7837-7840.
7. Komatsu, N.; Ishida, J.; Suzuki, H. *Tetrahedron Lett.* **1997**, *38*, 7219-7222.
8. Bajwa, J. S.; Jiang, X.; Slade, J.; Prasad, K.; Repič, O. Blacklock, T. J. *Tetrahedron Lett.* **2002**, *43*, 6709-6713
9. Bajwa, J. S.; Vivelo, J.; Slade, J.; Repič, O.; Blacklock, T. *Tetrahedron Lett.* **2000**, *41*, 6021-6024.

PREPARATION OF α-ACETOXY ETHERS BY THE REDUCTIVE ACETYLATION OF ESTERS: *endo*-1-BORNYLOXYETHYL ACETATE

[(Ethanol, 1-[[(1*R*,2*S*,4*R*)-1,7,7-trimethylbicyclo[2.2.1]hept-2-yl]oxy]-, acetate)]

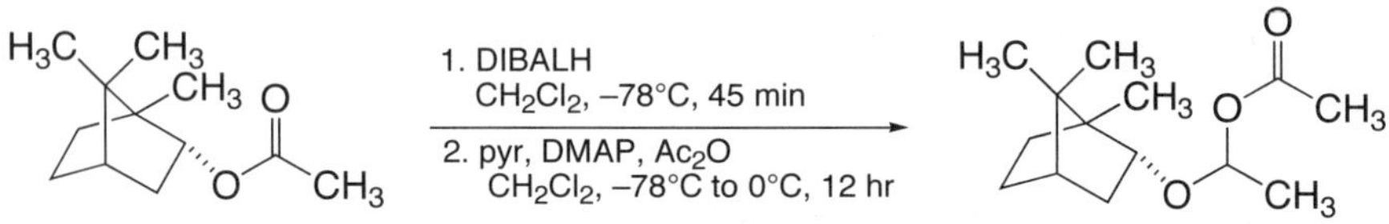

Submitted by David J. Kopecky and Scott D. Rychnovsky.[1]
Checked by Guillaume L. N. Péron, Cédric R. N. Fischmeister, and Andrew B. Holmes.

1. Procedure

A flame-dried, 2-L, three-necked, round-bottomed flask, equipped with a large Teflon magnetic stirring bar and three rubber septa, is charged with 5.5 g (28.1 mmol) of (−)-bornyl acetate (Note 1) in 162 mL of dichloromethane (Note 2) under an argon atmosphere. A thermocouple (Note 3) is immersed in the solution through one of the rubber septa. The flask is placed in a dry ice/acetone bath and cooled to an internal solution temperature of −78 °C. Stirring is begun and diisobutylaluminum hydride, 54.4 mL of a 1.0M solution in hexanes (54.4 mmol, 2 eq), is added as a slow stream via a glass syringe over a period of 13 min; during this addition, the internal solution temperature does not exceed −72 °C (Notes 1,4). The clear solution is stirred for an additional 45 min at −78 °C, then 6.6 mL of pyridine (6.5 g, 81.6 mmol, 3.0 eq) is added dropwise via a glass syringe, maintaining the internal temperature below −76 °C. Immediately following, a solution of 6.5 g of 4-dimethylaminopyridine (DMAP) (54.5 mmol, 2.0 eq) in 80 mL of dichloromethane (Note 2) is added slowly via an 18-gauge stainless steel cannula (Note 5). The internal temperature is not allowed to rise above −72 °C during this addition. Acetic anhydride (15.4 mL, 16.6 g, 163.2 mmol, 6.0 eq) is then added dropwise via a glass syringe over 5.5 min at a rate to maintain the internal temperature below −72 °C.

The resultant light yellow solution is stirred at −78 °C for 12 hr (Note 6). During this time, the color of the solution intensifies considerably. The bright yellow solution is warmed to 0 °C in an ice water bath and stirred for

35 min, during which time the color of the solution becomes bright yellow-orange. The reaction is then quenched by the sequential addition of 270 mL of saturated aqueous ammonium chloride and 200 mL of saturated aqueous sodium potassium tartrate. The emulsion is removed from the ice water bath and stirred vigorously for 50 min, by which time adequate phase separation is observed (Note 7). The biphasic mixture is transferred to a 2-L separatory funnel and is extracted with four 175-mL portions of dichloromethane. The combined extracts are washed sequentially with two 500-mL portions of an ice-cooled 1M aqueous solution of sodium bisulfate, three 500-mL portions of saturated aqueous sodium bicarbonate and 500 mL of saturated aqueous sodium chloride. The solution is dried over anhydrous sodium sulfate, filtered, and is concentrated under reduced pressure to afford 8.5 g of a yellow oil.

The residue is purified using flash chromatography with 145 g of triethylamine-deactivated Merck 9385 silica gel 60 (230-400) (Note 8) eluting with 5% ethyl ether in hexanes (Note 9). Fractions of 50-mL are collected, and product-containing fractions are identified by TLC analysis (Note 10). The product (6.4 g, 95%) is obtained as a clear oil (Notes 11,12). ^{1}H NMR and ^{13}C NMR spectra indicate that the product is a 2.8:1 mixture of diastereomers. Short path vacuum distillation of a portion of the product afforded material of comparable purity, bp 73.5−75 °C (2 mm).

2. Notes

1. (−)-Bornyl acetate, purchased from Aldrich Chemical Company, Inc., was of 97% purity and was used as received. Diisobutylaluminum hydride (DIBALH) (1.0M solution in hexanes) and 4-dimethylaminopyridine were purchased from Aldrich Chemical Company, Inc. and used as received. Pyridine was purchased from Aldrich Chemical Company, Inc. and was distilled from calcium hydride at atmospheric pressure prior to use. Acetic anhydride was purchased from Fisher Scientific, Inc. and was distilled at atmospheric pressure prior to use.

2. The checkers obtained dry dichloromethane by distillation from calcium hydride. The submitters dried dichloromethane through an alumina filtration system.[2]

3. A Digi-Sense digital thermometer was used. The internal temperature needs to be monitored, and the additions should be slow enough to avoid significant exotherms. The checkers found that temperature control is the major factor influencing the outcome of the reaction. The initial

procedure was submitted as double scale [i.e. 11 g (56.2 mmol) (−)-bornyl acetate]. However, the checkers found temperature control on full scale to be difficult, resulting in incomplete reaction and/or over-reduction of the bornyl acetate. The checkers found that the procedure is reliably reproducible on half scale (5.5 g (−)-bornyl acetate).

4. The checkers found that reproducible results were obtained when a freshly opened bottle of diisobutylaluminium hydride was used.

5. The cannula transfer was accomplished by briefly evacuating the reaction flask using a vacuum line connected to a vacuum pump, and then pressurizing the flask containing the DMAP solution with just over 1 atmosphere of argon.

6. A bath temperature of −78 °C was maintained during the 12 hr period by packing the bath with dry ice and covering the bath and reaction flask with aluminum foil.

7. Although the two phases separated, the aqueous phase remained cloudy.

8. The submitters used ICN 230-400 mesh silica gel. The silica gel was deactivated by mixing it with 325 mL of 2% (v/v) triethylamine in hexanes. The resultant slurry was poured onto the column, packed, and then flushed with 200 mL of 5% ethyl ether in hexanes to remove residual triethylamine before product loading.

9. The head pressure of the flash column during chromatography was 2.5 psi.

10. The α-acetoxy ether elutes with an R_f of 0.32 on Merck silica gel F_{254} in 5% (v/v) ethyl ether in hexanes (product stains yellow-green in an acidic solution of *p*-anisaldehyde).

11. The submitters obtained the equivalent of 6.4 g (95%). Physical and spectroscopic properties are as follows: IR (mixture of isomers, neat) cm^{-1}: 2952, 2879, 1737, 1452, 1372, 1247, 1175, 1141, 1072, 1033, 1009, 929, 833; ^{1}H NMR (500 MHz, $CDCl_3$, mixture of isomers) δ: 0.81-0.85 (m, 9 H), 0.99 (dd, 1 H, *J* = 13.1, 3.4, minor isomer), 1.10 (dd, 1 H, *J* = 13.3, 3.4, major isomer), 1.16-1.28 (m, 2 H), 1.38 (d, 3 H, *J* = 5.4, minor), 1.40 (d, 3 H, *J* = 5.3, major), 1.59 (t, 1 H, *J* = 4.5, major). 1.64 (t, 1 H, *J* = 4.5, minor), 1.63-1.74 (m, 1 H), 1.90-2.01 (m, 1 H), 2.04 (s, 3 H, major), 2.05 (s, 3 H, minor), 2.07-2.19 (m, 1 H), 3.81 (ddd, 1 H, *J* = 9.9, 3.2, 2.0), 5.93 (q, 1 H, *J* = 5.2, minor), 5.94 (q, 1 H, *J* = 5.2, major); ^{13}C NMR (125 MHz, $CDCl_3$, mixture of isomers) δ: 13.4 (minor isomer), 13.6 (major isomer), 18.8 (major), 18.8 (minor), 19.7 (major), 19.7 (minor), 20.9 (major), 21.2 (minor), 21.4 (minor), 21.6 (major), 26.4 (minor), 26.6 (major), 28.1

(major), 28.2 (minor), 36.1 (minor), 37.0 (major), 44.9 (minor), 45.0 (major), 47.2 (major), 47.6 (minor), 48.8 (minor), 49.3 (major), 82.6 (minor), 85.9 (major), 95.3 (minor), 97.7 (major), 170.6 (major), 170.9 (minor). Anal. Calcd for $C_{14}H_{24}O_3$: C, 70.0; H, 10.1. Found: C, 70.0; H, 10.0. The submitters obtained C, 70.0; H, 10.0.

12. The product contained a 3% impurity of (−)-bornyl acetate as indicated by the presence of the following diagnostic peaks in the ^{1}H NMR spectrum: 0.87 (s, 3 H), 0.90 (s, 3 H), 0.96 (buried dd, 1 H), 2.06 (s, 3 H). This impurity originates from incomplete reduction and/or over-reduction.

Waste Disposal Information

All toxic materials were disposed of in accordance with "Prudent Practices in the Laboratory"; National Academy Press; Washington, DC, 1995.

3. Discussion

The reductive acetylation of esters and lactones can be accomplished by treatment of the ester with a two-fold excess of DIBALH at −78 °C followed by trapping of the resulting aluminum hemiacetal intermediate with acetic anhydride in the presence of pyridine and DMAP at low temperature.3 This transformation has been successfully applied to a wide range of esters and lactones, including sterically-hindered acyclic esters, β-alkoxy esters, benzyl esters and macrolactones. A compilation of representative examples is presented in Table I.[3a] Upon exposure to mild acid, the α-acetoxy ether products undergo facile conversion to the corresponding oxacarbenium ions, which can be trapped by a wide variety of carbon- or heteroatom-based nucleophiles.[3b] For example, (*endo*)-1-bornyloxyethyl acetate can be efficiently allylated by treatment with boron trifluoride etherate and allyltrimethylsilane (82% yield; d.r. = 3.7:1), or ethylated with trimethylsilyl triflate and diethylzinc (87% yield; d.r. = 1.9:1).[4] α-Acetoxy ethers have also been utilized as Prins cyclization precursors in our studies toward the diastereoselective synthesis of polyfunctionalized tetrahydropyran ring systems.[5]

1. Department of Chemistry, University of California, Irvine, CA 92697-2025.
2. Pangborn, A. B.; Giardello, M. A.; Grubbs, R. H.; Rosen, R. K.; Timmers, F. J. *Organometallics* **1996**, *15*, 1518.
3. (a) Kopecky, D. J.; Rychnovsky, S. D. *J. Org. Chem.* **2000**, *65*, 191; (b) Dahanukar, V. H.; Rychnovsky, S. D. *J. Org. Chem.* **1996**, *61*, 8317.
4. Both reactions were performed in dichloromethane at −78 °C.
5. (a) Rychnovsky, S. D.; Hu, Y. ; Ellsworth, B. *Tetrahedron Lett.* **1998**, *39*, 7271 (b) Rychnovsky, S. D.; Thomas, C. R. *Org. Lett.* **2000**, *2*, 1217 (c) Jaber, J. J.; Mitsui, K.; Rychnovsky, S. D. *J. Org. Chem.* **2001**, *66*, 4679.

TABLE 1[3a]
DIBALH REDUCTIVE ACETYLATION OF VARIOUS ESTERS

Entry	Substrate	Product	Yield(%)
1	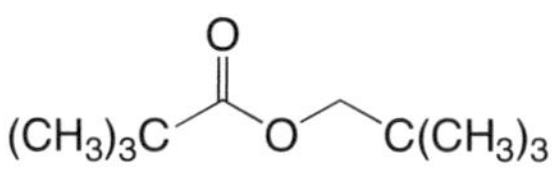	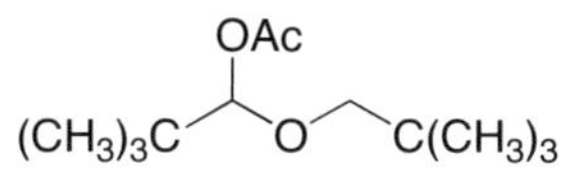	87
2	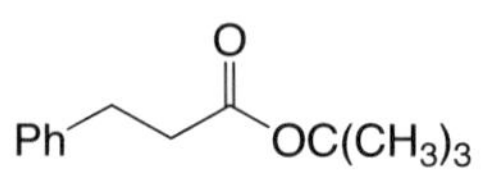	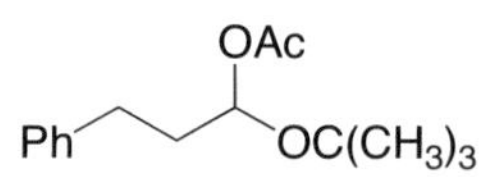	90
3	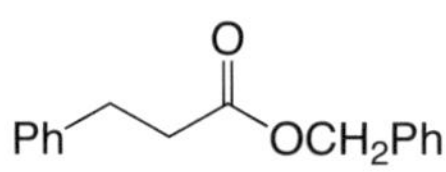	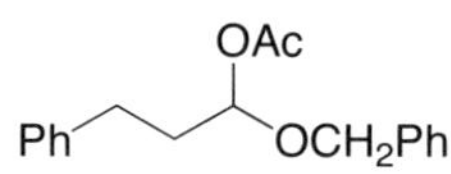	78
4	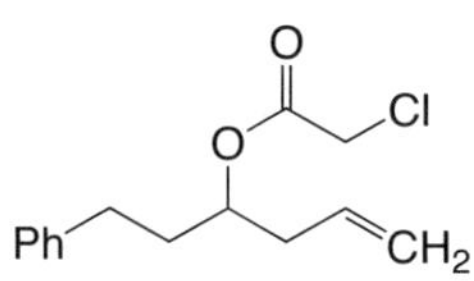	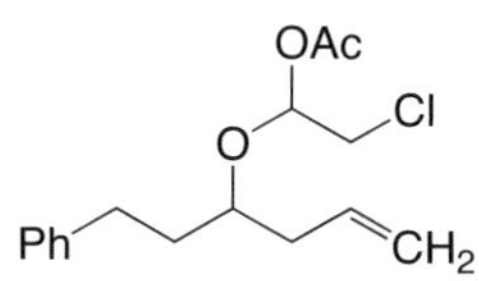	92 (d.r. = 1.1:1)
5	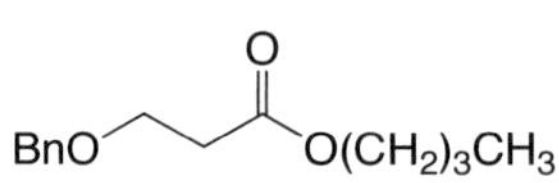	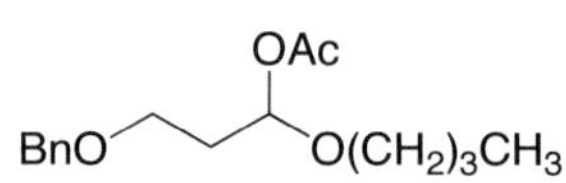	81
6	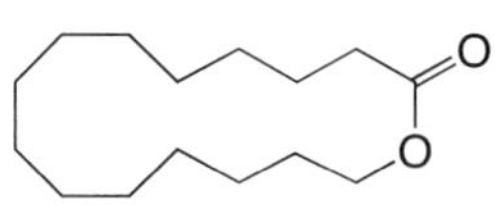	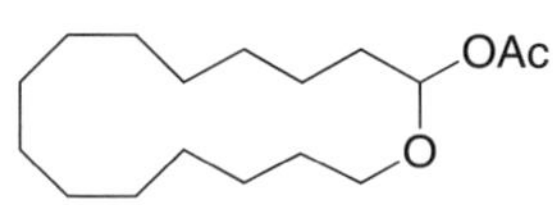	84
7	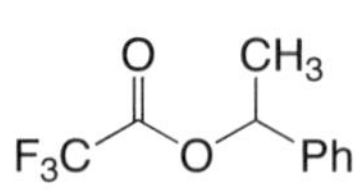	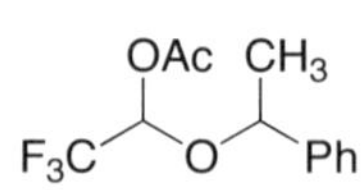	93 (d.r. = 2.2:1)

INDIUM/AMMONIUM CHLORIDE-MEDIATED SELECTIVE REDUCTION OF AROMATIC NITRO COMPOUNDS: ETHYL 4-AMINOBENZOATE

(Benzoic acid, 4-amino-, ethyl ester)

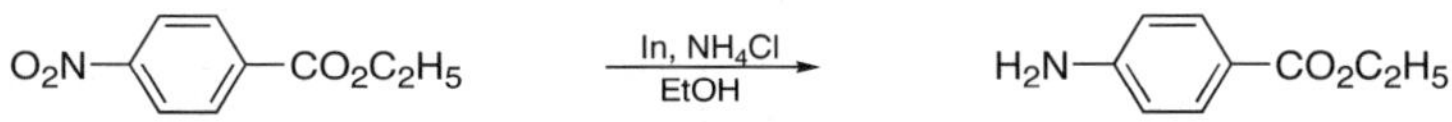

Submitted by Bimal K. Banik[1], Indrani Banik, and Frederick F. Becker.
Checked by Weiqiang Huang and Marvin J. Miller.

1. Procedure

Ethyl 4-aminobenzoate. A 1000-mL, round-bottomed flask equipped with a magnetic stirbar is charged with a suspension of 10 g (51 mmol) of ethyl 4-nitrobenzoate in 250 mL of ethanol, and a solution of 27.4 g (510 mmol) of ammonium chloride in 125 mL of water is then added (Note 1). Indium powder (23.5 g, 205 mmol) (Note 2) is added, and the resulting mixture is heated at reflux for 2.5 hr. The reaction mixture is allowed to cool to room temperature, diluted with 350-400 mL of water, and filtered under vacuum. The filtrate is extracted with 6-8 portions of 50-60 mL of dichloromethane, and the combined organic phases are washed with 100 mL of brine and dried over anhydrous sodium sulfate. The resulting solution is concentrated under reduced pressure and the crude product is dissolved in 100 mL of dichloromethane. The solution is concentrated by warming, and 50 mL of hexane is then added. The resulting solution is allowed to stand in a refrigerator overnight and then filtered under vacuum to give 7.63 g (90%) of ethyl 4-aminobenzoate (Note 3).

2. Notes

1. Ethyl 4-nitrobenzoate, ammonium chloride, ethanol were purchased from Aldrich Chemical Company and used as received.

2. Indium powder (99.99%) was obtained from Aldrich Chemical Company.

3. The spectral properties of ethyl 4-aminobenzoate are as follows: IR (film) cm^{-1}: 3424, 3345, 3224, 1685, 1636, 1598, 1515, 1367, 1312, 1281, 1173, 773; ^{1}H NMR (300 MHz, $CDCl_3$) δ: 1.36 (3 H, t, J = 6), 4.04 (2 H,

brs), 4.31 (2 H, q, J = 7), 6.63 (2 H, d, J = 9), 7.85 (2 H, d, J = 9); ^{13}C NMR (75 MHz, $CDCl_3$) δ: 14.6 (CH_3), 60.4 (CH_2), 114.0 (Ar-CH), 120.5 (Ar-q), 131.7 (Ar-CH), 151.0 (Ar-q), 166.9 (Ar-q = quaternary carbon atom); *m/e*: 166 (M+H); exact mass: m/e (M^+): 165.0764 (measured), 165.0790 (theoretical).

Waste Disposal Information

All toxic materials were disposed of in accordance with the policy of UTMDACC to handle and dispose of hazardous waste, which is in accordance with the regulations of the Environmental Protection Agency, Occupational Safety and Health Administration, Federal Department of Transportation, Texas Department of Health, and the Texas Water Commission.

3. Discussion

The synthesis of aromatic amines is an active and important area of research.[2] Many methods are available in the literature for the synthesis of these compounds. Though some of these are widely used, still they have limitations based on safety or handling considerations. For example, catalytic hydrogenation[3] of nitro or azido compounds in the presence of metals such as palladium on carbon or Raney nickel require stringent precautions because of their flammable nature in the presence of air. In addition, these methods require compressed hydrogen gas and a vacuum pump to create high pressure within the reaction flask. To overcome these difficulties, several new methods have been reported in the recent literature[4] involving such reducing agents as decaborane,[5] electrochemically generated Raney nickel,[6] dimethyl hydrazine/ferric chloride,[7] hydrazine hydrate/ferric oxide-magnesium oxide,[8] diethyl chlorophosphite,[9] and sodium borohydride-sodium-methoxide in methanol[10] In general, the main drawbacks of these methods are long reaction time and non-chemoselectivity. The submitters[11] have also described new methods for the reduction of aromatic nitro compounds and imines to aromatic amines by novel samarium-induced iodine catalyzed and ammonium chloride mediated reduction.[12] While our samarium-induced reduction of aromatic nitro compounds works well in the polycarbocyclic series, similar reaction with several heteroaromatic nitro compounds results in a mixture of products under identical conditions. Moreover, samarium-induced reduction of the nitro compounds requires anhydrous reaction conditions.

The submitters have been actively involved in the use of polyaromatic amines for the development of anticancer agents.[13] Therefore, the submitters began a research program aimed at developing methods to synthesize several aromatic amines rapidly and in high selectivity, by using ecologically friendly reagents. The present study describes a method for the selective reduction of aromatic nitro compounds to the corresponding amines by indium in the presence of ammonium chloride in aqueous ethanol.[14]

The chemistry of indium metal is the subject of current investigation, especially since the reactions induced by it can be performed in aqueous solution.[15] The selective reductions of ethyl 4-nitrobenzoate (entry 1), 2-nitrobenzyl alcohol (entry 2), 1-bromo-4-nitrobenzene (entry 3), 4-nitrocinnamyl alcohol (entry 4), 4-nitrobenzonitrile (entry 5), 4-nitrobenzamide (entry 6), 4-nitroanisole (entry 7), and 2-nitrofluorenone (entry 8) with indium metal in the presence of ammonium chloride using aqueous ethanol were performed and the corresponding amines were produced in good yield. These results indicate a useful selectivity in the reduction procedure. For example, ester, nitrile, bromo, amide, benzylic ketone, benzylic alcohol, aromatic ether, and unsaturated bonds remained unaffected during this transformation. Many of the previous methods produce a mixture of compounds. Other metals like zinc, tin, and iron usually require acid-catalysts for the activation process, with resultant problems of waste disposal.

Because of the non-flammable nature of the process and ready availability of indium and ammonium chloride, the submitters believe this method is practical for the preparation of several aromatic amines. Further, this method is performed in aqueous ethanol, is extremely safe from the environmental point of view, and should prove useful in organic chemistry.

TABLE 1
SELECTIVE REDUCTION OF AROMATIC NITRO COMPOUNDS BY INDIUM METAL IN THE PRESENCE OF $NH_4Cl/H_2O/EtOH$

Entry	Nitro compound	Product	Reduction time (h)	%Yield	mp °C
1	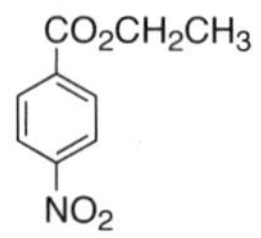	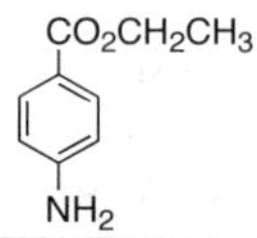	2.5	94	88 (lit. 88-90)
2	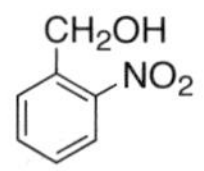	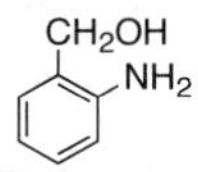	2h, 5 min	68.5	84 (lit. 83-85)
3	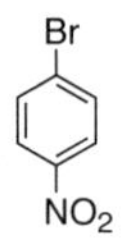	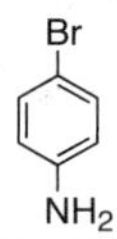	1.5	80	59 (lit. 60-64)
4	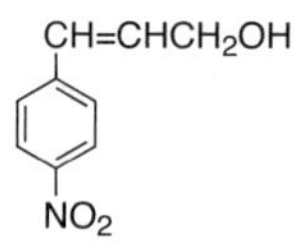	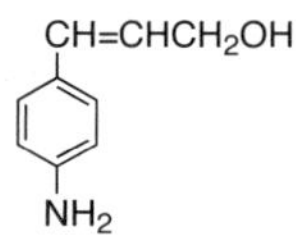	1	oil	oil
5	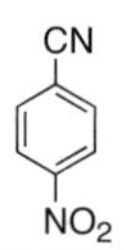	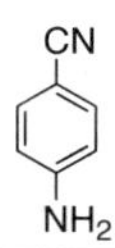	2	75	85 (lit 83-85)
6	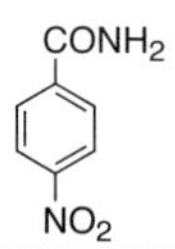	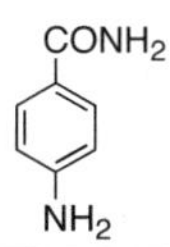	1.75	71	182 (lit. 181-183)
7	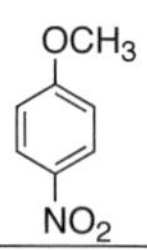	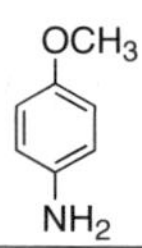	5	90	58-60 (lit. 57-60)
8	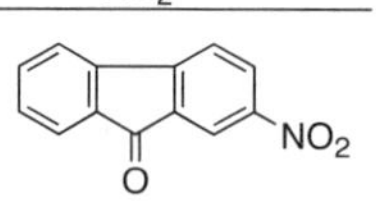	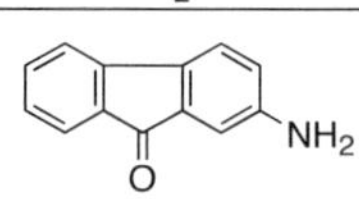	8	80	157-160 (lit. 160)

1. The University of Texas, M. D. Anderson Cancer Center, Department of Molecular Pathology, Box 89, 1515 Holcombe Blvd., Houston, TX 77030.

2. For a recent reference, see: Main, B. G.; Tucker, H. In *Medicinal Chemistry*, 2nd Ed.; Genellin, C. R.; Roberts, S. M.; Academic Press: New York, **1993**, p187.

3. (a) Rylander, P. N. *Hydrogenation Methods;* Academic Press: New York, **1985**; Chapter 8 (b) Johnstone, R. A. W.; Wilby, A. H.; Entistle, I. D. *Chem. Rev.* **1985**, *85*, 129.

4. (a) Banik, B. K.; Barakat, K. J.; Wagle, D. R.; Manhas, M. S.; Bose, A. K. *J. Org. Chem.* **1999**, *64*, 5746. (b) Weiner, H.; Blum, J.; Sasson, Y. *J. Org. Chem.* **1991**, *56,* 4481.

5. Bae, J. W.; Cho, Y. J.; Lee, S. H.; Yoon, C. M. *Tetrahedron Lett.* **2000,** *41*, 175.

6. Yasuhara, A.; Kasano, A.; Sakamoto, T. J. *Org. Chem.* **1999,** *64*, 2301.

7. Boothroyd, S. R.; Kerr, M. A. *Tetrahedron Lett.* **1995,** *36*, 2411.

8. Kumbhar, P. S.; Sanchez-Valente, J.; Figueras, F. *Tetrahedron Lett.* **1998,** *39*, 2573.

9. Fischer, B.; Sheihet, L. J. *Org. Chem.* **1998,** *63*, 393.

10. Suwinski, J.; Wagner, P.; Holt, E. M. *Tetrahedron* **1996,** *52*, 9541.

11. (a) Banik, B. K.; Mukhopadhyay, C.; Venkatraman, M. S.; Becker, F. F. *Tetrahedron Lett.* **1998**, *39*, 7243. (b) Banik, B. K., Zegrocka, O., Banik, I.; Hackfeld, L.; Becker, F. F. *Tetrahedron Lett.* **1999**, *40*, 6731.

12. Basu, M. K.; Becker, F. F.; Banik, B. K. *Tetrahedron Lett.* **2000,** *41*, 5603.

13. (a) Becker, F. F.; Banik, B. K. *Bioorg. Med. Chem. Lett.* **1998,** *8*, 2877. (b) Becker, F. F., Mukhopadhyay, C., Banik, I., Hackfeld, L., Banik, B. K. *Bioorg. & Med. Chem.* **2000**, *8*, 2693. (c) Banik, B. K. Becker, F. F. *Bioorg. & Med. Chem.* **2001**, *9*, 593. (d) Banik, B. K. Becker, F. F. *Current Med. Chem.* **2001**, *8*, 1513.

14. Banik, B. K.; Suhendra, M.; Banik, I.; Becker, F. F. *Synth. Commun.* **2000**, *30*, 3745. Also see: (a) Moody, C. J.; Pitts, M. R. *Synlett* **1998**, 1028. (b) Moody, C. J.; Pitts, M. R. *Synlett* **1998**, 1029.

15. Li has carried out a significant work on indium-mediated reactions, for example, see: (a) Li, C. J.; Chan, T.-H. *Organic Reactions in Aqueous Media*; J. Wiley & Sons: New York, **1997**. (b) Li, C. J. *Tetrahedron* **1996**, *52*, 5643. (c) Li, C. J. *Chem. Rev.* **1993**, *93*, 2023.

PREPARATION OF 1,4:5,8-DIMETHANO-1,2,3,4,5,6,7,8-OCTAHYDRO-9,10-DIMETHOXYANTHRACENIUM HEXACHLOROANTIMONATE ($4^{+\bullet}SbCl_6^-$): A HIGHLY ROBUST RADICAL-CATION SALT

[1,4:5,8-Dimethanoanthracene, 1,2,3,4,5,6,7,8-octahydro-9,10-dimethoxy-, radical ion(1+), hexachloroantimonate(1-)]

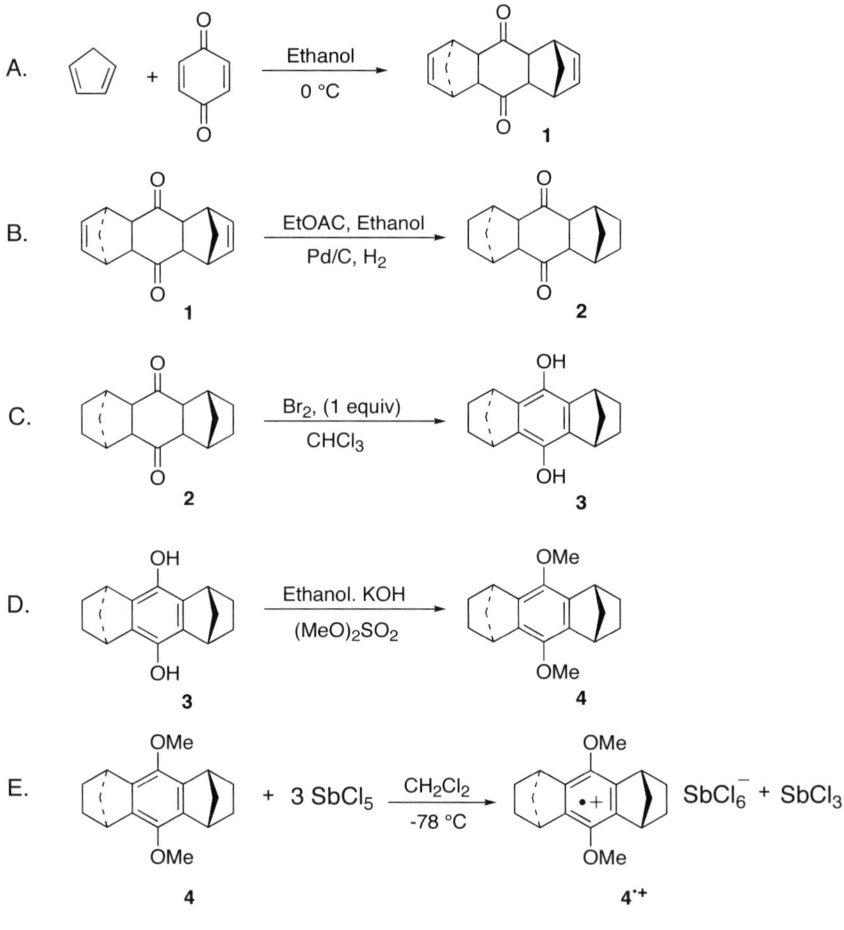

Submitted by Rajendra Rathore,[1] Carrie L. Burns, and Mihaela I. Deselnicu.
Checked by Scott E. Denmark and Tommy Bui.

1. Procedure

*A. 1,4:5,8-Dimethano-1,4,4a,5,8,8a,9a,10a-octahydroanthracene-9,10-dione (**1**).* To a 500-mL conical flask is added *p*-benzoquinone (10.8 g, 100 mmol) [Note 1] and absolute ethanol (100 mL). The mixture is stirred at room temperature for 10 min and then is cooled in an ice-salt bath (ca. –5 °C, external). [Note 2] Freshly cracked cyclopentadiene (13.2 g, 200 mmol) [Note 3] is added slowly via syringe over 3-5 min and the resulting mixture is stirred in the ice-salt bath for 20 min and then for an additional 30 min at room temperature. A large amount of the adduct **1** precipitates during the course of the reaction. [Note 4] The resulting slurry is cooled in an ice bath (ca. 0 °C, external) and the precipitate is filtered using a Büchner funnel and is washed with ice-cold ethanol to yield 20.9-21.6 g (87-90%) of cycloadduct **1** as a nearly colorless, crystalline solid [Note 5].

*B. 1,4:5,8-Dimethanododecahydroanthracene-9,10-dione (**2**).* A solution of diketone **1** (19.2 g, 80 mmol) in a mixture of ethanol (50 mL) and ethyl acetate (250 mL) is combined with 10% Pd-C (200 mg) in a 500-mL pressure bottle [Note 6]. The bottle is pressured to 60 psi of hydrogen and shaken on a Parr apparatus for three hours (when the uptake of hydrogen ceases) [Note 7]. The mixture is then diluted with dichloromethane (~200 mL) to dissolve the precipitate and the solution is filtered *in vacuo* through a short pad of Celite (100 g) in a sintered glass funnel. The Celite pad is washed with dichloromethane (3 x 150 mL) and the colorless solution is concentrated on a rotary evaporator to afford diketone **2** (18.8 g, 96%) as a white solid. [Note 8]

*C. 1,4:5,8-Dimethano-1,2,3,4,5,6,7,8-octahydroanthracene-9,10-diol (**3**).* In a 500-mL, three-necked, round-bottomed flask equipped with a 125-mL pressure-equalizing addition funnel and nitrogen inlet and outlet adapters, is placed diketone **2** (18.3 g, 75 mmol) and anhydrous chloroform (75 mL). [Note 9] A solution of Br_2 (12.0 g, 75 mmol) in chloroform (75 mL) was placed in the dropping funnel and was added dropwise over 0.5 h under a slow stream of nitrogen at room temperature. [Note 10] Stirring the reaction mixture while purging the flask with nitrogen is continued for another hour to remove the gaseous HBr. The resulting suspension is cooled in an ice-salt bath (approx. -5 °C) and is filtered *in vacuo* using a Büchner funnel. The solid is washed with ice-cold chloroform (2 x 25 mL) and is suction dried to afford 17.4 g (96 %) of hydroquinone **3** as a colorless powder, which is used in the next step without further purification.

*D. 1,4:5,8-Dimethano-1,2,3,4,5,6,7,8-octahydro-9,10-dimethoxy-anthracene (**4**).* A solution of hydroquinone **3** (16.9 g, 70 mmol) in absolute ethanol (175 mL) is placed in a three-necked, 500-mL, round-bottomed flask equipped with a pressure-equalizing addition funnel and nitrogen inlet and outlet adapters. To this mixture is added a solution of KOH (8.7 g, 155 mmol) in water (8.7 mL) followed by a solution of sodium hydrosulfite (2 g) in water (10 mL) under an inert atmosphere. Dimethyl sulfate (19.6 g, 155 mmol) is placed in the addition funnel and is then added dropwise over 30 min. After the addition is complete, the addition funnel is replaced with a condenser and the mixture is heated to reflux in a 120°C oil bath for 3 h and then is cooled to room temperature. A second portion of KOH (2.0 g, 35 mmol, 0.5 equiv) is added to the flask followed by the dropwise addition of a second portion of dimethyl sulfate (4.4 g, 35 mmol, 0.5 equiv) through a 25-mL addition funnel. The addition funnel is replaced with a condenser and the mixture is heated for one hour and then is allowed to cool to room temperature. The reaction mixture is diluted with water (250 mL) and then is extracted with dichloromethane (3 x 100 mL). The combined dichloromethane extracts are successively washed with 10% aqueous NaOH solution (50 mL), water (100 mL), brine (50 mL), and then are dried over anhydrous magnesium sulfate. Removal of solvent *in vacuo* affords a crude material that is purified by filtration through a short column of silica gel (200 g) as follows. A solution of the crude product in a 25:1 mixture of hexanes and ethyl acetate (90 mL) is eluted through the silica gel column followed by an additional 900 mL of hexane/ethyl acetate, 25:1. The product-containing fractions are concentrated on a rotary evaporator to afford dimethyl ether **4** (15.3 g, 81%) as a cream colored solid [Notes 11 and 12].

E. Preparation of $\boldsymbol{4^{+\bullet}SbCl_6^-}$. A solution of hydroquinone ether **4** (2.7 g, 10 mmol) in anhydrous dichloromethane (20 mL) [Note 13] is placed in a 100-mL, three-necked, round-bottomed flask equipped with a 50-mL pressure-equalizing addition funnel and nitrogen inlet and outlet adapters. The addition funnel is charged with a 1.0 M solution of $SbCl_5$ in dichloromethane (15 mL,15 mmol) [Note 14] and the flask is cooled in a dry ice-acetone bath (approximately –75 °C, external). The $SbCl_5$ solution is slowly added (3-5 min) under a flow of nitrogen. The reaction mixture immediately turns red and a large amount of dark-red material precipitates.

The resulting mixture is warmed to 0 °C over 5-10 min, and anhydrous diethyl ether (50 mL) is added to precipitate the dissolved $\mathbf{4}^{+\bullet}SbCl_6^-$ salt. The dark-red microcrystalline precipitate is suction filtered using a medium-porosity sintered-glass funnel under a blanket of dry nitrogen and is washed with dry diethyl ether (2 x 20 mL). [Note 15] The salt is dried *in vacuo* at room temperature to afford $\mathbf{4}^{+\bullet}SbCl_6^-$ (5.6 g, 93%) as a red, crystalline solid. [Notes 16 and 17]

2. Notes

1. *p*-Benzoquinone, 98% was purchased from Aldrich Chemical Co. and was used without further purification.

2. *p*-Benzoquinone, generally, precipitates upon cooling and it dissolves eventually as the cycloaddition reaction proceeds.

3. Cyclopentadiene is prepared by heating commercial dicyclopentadiene (available from Aldrich Chemical Company, Inc.) at 160 °C in a distillation apparatus. Cyclopentadiene distills smoothly at 39–45 °C. For a detailed procedure, see: Moffett, R. B. *Org. Synth.*, *Coll. Vol. IV*, **1963**, 238.

4. This adduct can be directly hydrogenated (using a Parr hydrogenation apparatus) after dissolving the precipitate by adding 200-mL ethyl acetate and 10% Pd-C (200 mg, Aldrich Chemical Co.); however, a longer hydrogenation time is required. Furthermore, a less pure product which required recrystallization was afforded. A simple filtration of the adduct **1** followed by its hydrogenation (as described in the procedure above) yields a colorless solid of high purity.

5. The spectral data for 1,4:5,8-dimethano-1,4,4a,5,8,8a,9a,10a-octahydroanthra-cene-9,10-dione (**1**) are as follows: mp 157-158 °C; ^{1}H NMR ($CDCl_3$) δ 1.29 (br d, 2 H, J = 8.6), 1.45 (dt, 2 H, J = 8.6, 1.72), 2.87 (br, 4 H), 3.35 (br, 4 H), 6.18 (t, 4 H, J = 1.7); ^{13}C NMR ($CDCl_3$) δ 48.2, 49.5, 53.2, 136.4, 212.7.

6. A standard hydrogenation bottle purchased from ACE glass Inc. was used.

7. The hydrogenation can also be carried out at atmospheric pressure as well.

8. The spectral data for 1,4:5,8-dimethanododecahydroanthracene-9,10-dione (**2**) are as follows: mp 226-229 °C (decomp.); ^{1}H NMR ($CDCl_3$) δ

1.25-1.30 (m, 4 H), 1.38-1.45 (m, 4 H), 1.38-1.58, (m, 4 H), 2.80 (br, 4 H), 2.86 (br, 4 H); ^{13}C NMR ($CDCl_3$) δ 25.0, 39.2, 43.7, 53.6, 214.5.

9. Ethanol-free, anhydrous chloroform, which can be readily prepared according to the procedure from Perrin,[9] is recommended for this step. Anhydrous dichloromethane can also be used as the solvent for the bromination.

10. The nitrogen outlet was connected to gas-wash bottle containing a 20% (w/w) solution of sodium hydroxide.

11. The spectral data for 1,4:5,8-dimethano-1,2,3,4,5,6,7,8-octahydro-9,10-dimethoxyanthracene (**4**) are as follows: mp 116-117 °C (decomp.); ^{1}H NMR ($CDCl_3$) δ 1.15-1.19 (m, 4 H), 1.42-1.44 (br d, 2 H, J = 8.55), 1.66-1.69 (m, 2 H), 1.87 (br d, 4 H, J = 7.33), 3.54 (m, 4 H), 3.83 (s, 6 H); ^{13}C NMR ($CDCl_3$) δ 27.0, 40.6, 48.9, 61.3, 137.5, 143.5.

12. The submitters described recrystallization of **4** by dissolving 10 g in boiling hexanes (150 mL) and allowing the product to precipitate by standing at room temperature overnight.

13. Anhydrous dichloromethane was obtained according to the procedure of Perrin.[9]

14. A 1 M solution of antimony pentachloride in dichloromethane can be purchased from Aldrich Chemical Co. This solution can also be prepared by dissolving neat $SbCl_5$ in anhydrous dichloromethane.

15. A large inverted funnel, connected to a nitrogen outlet, positioned above the sintered glass funnel is generally sufficient for maintaining an inert atmosphere during filtration of the cation radical salt (see sketch of the apparatus below).

16. The spectral data of [$4^{+\cdot}$ $SbCl_6^-$]: UV-vis (dichloromethane), λ_{max} =

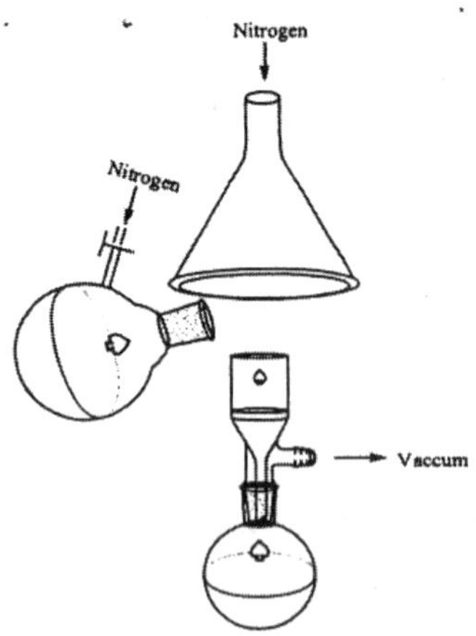

518 and 486 (shoulder) nm, extinction coefficient, ε_{518} = 7300 M^{-1} cm^{-1}. Anal. calcd. for $C_{18}H_{22}O_2$ $SbCl_6$: C, 35.74, H, 3.67, Cl, 35.17. Found: C, 35.64, H, 3.54, Cl, 34.16. The pure cation-radical salt (**4**$^{+\bullet}$$SbCl_6^-$) is >99% can be stored in a glass bottle at ambient temperatures indefinitely. The purity of the **4**$^{+\bullet}$$SbCl_6^-$ salt can be confirmed by iodimetric titration as follows. Thus, a solution of **4**$^{+\bullet}$ $SbCl_6^-$ (60.45 mg, 0.01 M) in anhydrous dichloromethane was added to a dichloromethane solution containing excess tetra-*n*-butylammonium iodide (1 mmol, 0.1 M) at 22 °C, under an argon atmosphere, to afford a dark brown solution. The mixture was stirred for 5 min and was titrated (with rapid stirring) by a slow addition of a standard aqueous sodium thiosulfate solution (0.005 M) in the presence of a starch solution as an internal indicator. Based on the amount of thiosulfate solution consumed (58.6 mL), purity of the cation radical was determined to be ca. 98.5%. Moreover, the purity can also be estimated by utilizing the spectrophotometric data as follows. Thus, a known quantity of **4**$^{+\bullet}$$SbCl_6^-$ was dissolved in anhydrous dichloromethane and the absorbance was monitored at 518 nm using a 1-cm quartz cuvette. For example, a solution of 3.0 mg of **4**$^{+\bullet}$$SbCl_6^-$ in 30 mL anhydrous dichloromethane showed an absorbance of 1.20 at 518 nm (actual concentration = 1.64×10^{-4} M, 99.6% pure). The actual concentration of the **4**$^{+\bullet}$$SbCl_6^-$ was calculated by using the equation: Absorbance at 518 nm/7300 = actual concentration.

17. The radical cation salt **4**$^{+\bullet}$$SbCl_6^-$ was also characterized by ESR spectroscopy. A solution of the salt (2 mg, 3.3 μmol) in 30 mL of dichloromethane was prepared in a dry box in a flame-dried, 50-mL round-bottomed flask. The solution was filtered through filter paper to remove particulates. About 0.25-0.35 mL of this solution was placed in a dry ESR tube under argon. The tube was sealed with a septum and the spectrum was recorded. G = 2.0046; A_1 = 3.3 Gauss (6H), A_2 = 2.1 Gauss (2H), A_3 = 1.2 Gauss (2H).

Safety and Waste Disposal Information

All hazardous materials should be handled and disposed of in accordance with "Prudent Practices in the Laboratory"; National Academy Press; Washington, DC, 1995.

3. Discussion

The hydroquinone ether **4** is an electron-rich molecule (E^{o}_{ox} = 1.11 V vs. SCE) that is readily oxidized to a remarkably stable cation-radical salt, which can be stored as a crystalline solid for an indefinite period. Stable radical-cation salts are of fundamental importance to organic materials science because they constitute the smallest units that carry both a delocalized positive charge and an unpaired electron. Thus, radical-cation salts provide the basis for experimentation in conductivity and ferromagnetism, etc. Radical cations are important intermediates in a rich menu of organic transformations.[2] They are also useful as electron-transfer catalysts[3] in a variety of organic and organometallic transformations, as well as tools for elucidation of electron transfer mechanisms.

Removal of an electron from a neutral molecule leads to an "umpolung" of the normal reactivity character. The cation radical possesses greatly enhanced electrophilic reactivity (due to the cationic charge), as well as homolytic reactivity (due to the radical character). Radical cations can promote a variety of unimolecular and bimolecular reactions as illustrated by following examples.[2]

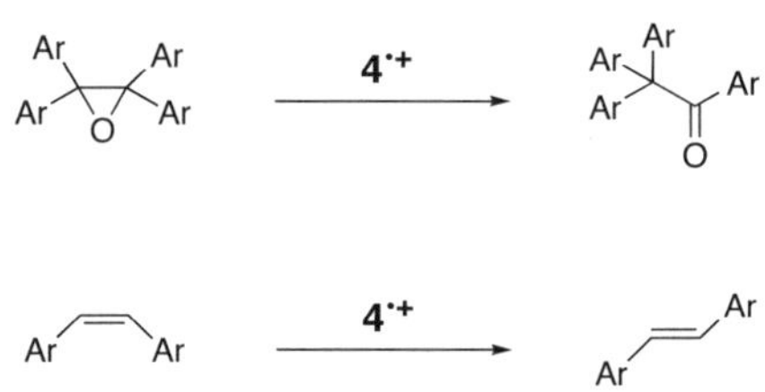

(Ar = Anisyl or 2,5-dimethoxy-4-methylphenyl)

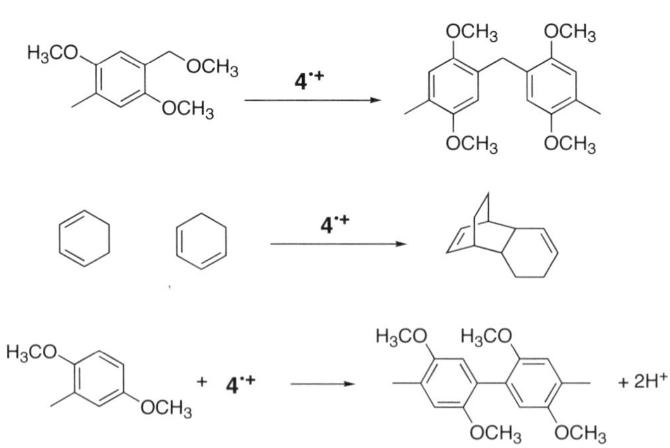

In addition, the ready availability of $\mathbf{4}^{+\bullet}$ allows its utilization for the formation of cation radicals from a variety of organic donors for spectroscopic monitoring, as well as for the elucidation of electron transfer mechanisms.[3f,3g]

The method described here for the preparation of $\mathbf{4}^{+\bullet}$ is essentially a detailed description of our recently published procedure.[4] To date, the only radical cation salts readily accessible to organic chemists are the heteroatom-centered *tris*(4-bromophenylaminium)[5] and thianthrenium[6] salts. It is envisioned that the ready availability of **4** and its cation radical $\mathbf{4}^{+\bullet}$ will foster its use in a variety of organic and organometallic syntheses,[3,7] as well as in the exploration of the role of single electron transfer (SET) in various chemical processes.[2,8]

1. Department of Chemistry, Marquette University, P.O. Box 1881, Milwaukee, WI 53201-1881.

2. (a) Rathore, R.; Kochi J. K. *Adv. Phys. Org. Chem.* **2000**, *35*, 196. (b) Rathore, R.; Kochi, J. K. *Acta. Chem. Scand.* **1998**, *52*, 14.
3. (a) Chanon, M. *Bull. Soc. Chim. Fr.* **1985**, 209. (b) Bauld, N. L.; Bellville, D. J.; Harirchian, B.; Lorenz, K. T.; Pabon, P. A., Jr.; Raynolds, D. W.; Wirth, D. D.; Chiou, H. S.; Marsh, B. K. *Acc. Chem. Res.* **1987**, *20*, 371. (c) Rathore, R.; Burns, C. L.; Deselnicu, M. I. *Org. Lett.* **2001**, *3*, 2887. (d) Rathore, R.; Kochi, J. K. *J. Org. Chem.* **1995**, *60*, 7479. (e) Bard, A. J.; Ledwith, A.; Shine, H. J. *Adv. Phys. Org. Chem.* **1976**, *13*, 155 and references cited therein. (f) Rathore, R.; Burns, C. L. *J. Org. Chem.* **2003**, *68*, 4071. (g) Rathore, R.; Burns, C. L.; Deselnicu, M. I. *Org. Lett.* **2001**, *3*, 2887.
4. Rathore, R.; Kochi, J .K. *J. Org. Chem.* **1994**, *60*, 4399.
5. Bell, F. A.; Ledwith, A.; Sherrington, D. C. *J. Chem. Soc. C.* **1969**, 2719.
6. Bandish, B. K.; Shine, H. J. *J. Org. Chem.* **1977**, *42*, 561.
7. Connelly, N. G.; Geiger, W. E. *Chem. Rev.* **1996**, *96*, 877 and references therein.
8. Eberson, L. *Electron Transfer Reactions in Organic Chemistry;* Springer: New York, 1987 and references cited therein.
9. Perrin, D. D.; Armarego, W. L. F.; Perrin, D.R. *Purification of Laboratory Chemicals, 2nd Ed.*; Pergamon: New York, 1980.

A PRACTICAL PROCEDURE FOR CARBONYL α-OXIDATION: SYNTHESIS OF (2-BENZOYLOXY)-1,4-CYCLOHEXANEDIONE *MONO*-ETHYLENE KETAL

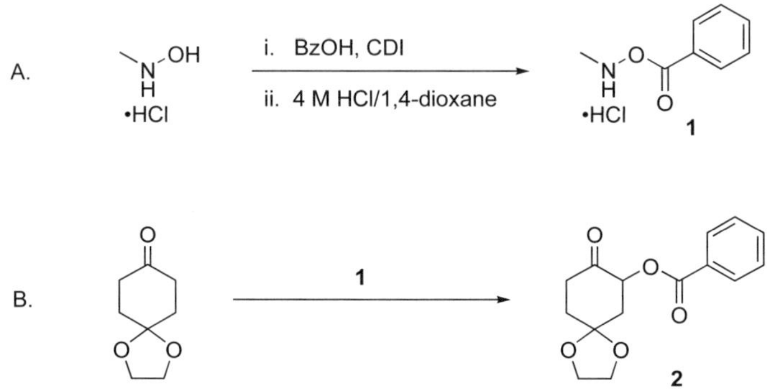

Submitted by Teyrnon C. Jones and Nicholas C. O. Tomkinson.[1]
Checked by Kalyani Patil and Peter Wipf.[2]

1. Procedure

A. *N-Methyl-O-benzoylhydroxylamine hydrochloride* **1** (Note 1). Benzoic acid (18.32 g, 150.0 mmol, 1 eq.) (Note 2) is placed in an open single-necked 500-mL round-bottomed flask equipped with a large magnetic stirrer and dissolved in 200 mL of methylene chloride (Note 3). This clear solution is cooled to 0 °C in an ice-water bath before the controlled addition of carbonyldiimidazole (24.32 g, 150.0 mmol, 1 equiv.) over a period of five min (Note 4). The ice bath is removed and the resultant yellow solution is stirred at ambient temperature until all effervescence has ceased (15 minutes). After this period, *N*-methylhydroxylamine hydrochloride (15.87 g, 190.0 mmol, 1.27 equiv.) is added in one portion to the reaction mixture and stirring is continued for a further 40 min at 25 °C. The reaction broth is diluted with 100 mL of methylene chloride, transferred to a 1-L separatory funnel and washed sequentially with cold (0 °C) 1 M hydrochloric acid (300 mL) (Note 5) and saturated sodium hydrogen carbonate solution (300 mL). The organic layer is separated, dried using sodium sulfate (40 g), filtered through a Büchner funnel and concentrated to dryness in a 500-mL round-bottomed flask on a rotary evaporator (25 °C, 20 mmHg). Diethyl ether (200 mL) is added to the resulting liquid followed by 4 M HCl in dioxane (75 mL, 0.30 mol, 2 equiv.) (Note 6). After 30 minutes, the solid precipitate is collected in a Büchner funnel (Note 7), washed with 100 mL of ice-cold diethyl ether and dried under high vacuum (0.5 mmHg) for 8 h to give *N*-methyl-*O*-benzoylhydroxylamine hydrochloride **1** as a white microcrystalline solid (17.45 g, 62%) (Note 8) that is used in Step B without further purification (Note 9).

B. *(2-Benzoyloxy)-1,4-cyclohexanedione (mono)ethylene ketal* **2**. An open 25-mL single-necked round-bottomed flask (Note 10) equipped with a magnetic stirring bar is charged at ambient temperature with 1,4-cyclohexanedione mono-ethylene ketal (4.68 g, 30.0 mmol, 1 equiv.) (Note 11) and dimethyl sulfoxide (7.5 mL). *N*-Methyl-*O*-benzoylhydroxylamine hydrochloride **1** (5.63 g, 30.0 mmol, 1 equiv., as prepared above) is added to the flask over a period of 15 min (Note 12) and the reaction mixture is stirred for 1 h (Note 13). The solution is subsequently poured into 300 mL of ethyl acetate and transferred to a 1-L separatory funnel where it is washed with 250 mL of water. The layers are separated and retained: the organic layer is washed further with two 250 mL portions of water, which are discarded; and the original aqueous layer is washed with two 100 mL portions of ethyl acetate which are added to the original organic layer. The combined organic fractions are dried using magnesium sulfate (25 g), filtered through a Büchner funnel and concentrated on a rotary evaporator (40 °C, 20 mmHg) to give a brown solid as a crude product (Note 14). The crude solid is transferred to a 25-mL Erlenmeyer flask and dissolved in 5 mL of hot (80 °C) isopropyl alcohol, then allowed to slowly cool to room temperature. The resultant solid is collected in a Büchner funnel and sequentially washed with ice-cold isopropyl alcohol (50 mL) and ice-cold 35-60 petroleum ether (50 mL) (Note 11) before being dried under high vacuum (0.5 mmHg) for 8 h to yield (2-benzoyloxy)-1,4-cyclohexanedione mono-ethylene ketal **2** (Note 15) as pale yellow needles (6.38 g, 77%) (Note 16).

2. Notes

1. Procedure A is a modification of the method of Geffken: Geffken, D. *Chem. Ber.* **1986**, *119*, 744.

2. Benzoic acid (99+%), carbonyldiimidazole, *N*-methylhydroxylamine hydrochloride (98%), and 4 M hydrochloric acid in 1,4-dioxane were obtained from Aldrich Chemical Company, Inc. Methylene chloride and diethyl ether were obtained from Fisher Scientific Ltd. Submitters purchased benzoic acid (98%) from Avocado Research

Chemicals Ltd. and carbonyldiimidazole (97%) from Lancaster Synthesis Ltd. All chemicals were used as received.

3. There was no need to oven-dry the vessels used in the reaction, and all steps were carried out in flasks open to the atmosphere.

4. The reaction produced carbon dioxide gas, resulting in vigorous effervescence.

5. The hydrochloric acid was prepared by dilution of 12 M hydrochloric acid purchased from EMD Chemicals Inc.

6. The submitters generated hydrochloride acid gas from a slow addition of sulfuric acid to ammonium chloride and bubbled the gas directly into the diethyl ether solution.

7. Filtration should be carried out in a fume hood as the mother liquor contains dissolved hydrogen chloride.

8. The checkers performed the reaction also at 50%, 25%, and 10% scale. The yields ranged from 58-62%.

9. Physical properties and spectral data for **1** are as follows: mp 135–137 °C; IR (nujol) 2923, 1770, 1599, 1475, 1260 cm^{-1}; ^1H NMR (300 MHz, DMSO-d^6) δ: 2.90 (s, 3 H), 7.57 (t, J = 7.8 Hz, 2 H), 7.72 (t, J = 7.5 Hz, 1 H), 7.95 (d, J = 7.2 Hz, 2 H), 11.38 (brs, 2 H); ^{13}C NMR (75 MHz, DMSO-d^6) δ: 36.7, 126.4, 129.6, 129.8, 135.1, 163.7; MS (EI) m/z (rel intensity) 151 ($[M\text{-}HCl]^+$, 11), 136 (25), 123 (24), 122 (87), 107 (8), 106 (70), 105 (56); HRMS (EI) calculated for $C_8H_9NO_2$ 151.0633, found 151.0631; Anal. calcd. for $C_8H_{10}NO_2Cl$ C, 51.2, H, 5.4, N, 7.5, Cl, 18.9; found C, 51.3, H, 5.5, N, 7.5, Cl, 18.9.

10. There was no need to oven-dry the reaction flask, and the reaction was carried out open to the atmosphere.

11. 1,4-Cyclohexanedione mono-ethylene ketal (97%) was obtained from Aldrich Chemical Company, Inc. Dimethyl sulfoxide (99.9%) was obtained from Fisher Scientific Ltd. Isopropyl alcohol and 35-60 petroleum ether was obtained from J. T. Baker Chemical Co. Submitters used dimethyl sulfoxide (99%) purchased from Aldrich Chemical Company, Inc., and, isopropyl alcohol and 40-60 petroleum ether purchased from Fisher Scientific Ltd. All chemicals were used as received.

12. It was noted that this reaction was exothermic. Measuring of the internal reaction temperature with a thermometer showed that during the addition of **1** the temperature rose to a maximum of 70 °C.

13. The progress of the reaction was monitored by TLC analysis on EMD Silicagel 60 F_{254} 0.25 mm silica plates, using 50% ethyl acetate/35-60 petroleum ether as eluent and visualising with potassium permanganate

solution. The ketone starting material had an R_f = 0.37 and the oxidized product had an R_f = 0.51.

14. Occasionally, the crude product was obtained as a brown oil. In these cases the oil may solidify on standing or can be induced to crystallize by the addition of a small amount of 35-60 petroleum ether, which was subsequently removed by rotary evaporation.

15. The checkers performed the reaction also at 50%, 25%, and 10% scale. The yields ranged from 78-81%.

16. Physical properties and spectral data for **2** are as follows: mp 114-116 °C; IR (dichloromethane) 2964, 1720, 1451, 1274, 1111, 1043 cm^{-1}; ^{1}H NMR (500 MHz, $CDCl_3$) δ: 2.01–2.12 (m, 2 H), 2.27 (t, J = 12.8 Hz, 1 H), 2.45–2.51 (m, 2 H), 2.80 (dt, J = 14.2, 6.6 Hz, 1 H), 4.04–4.13 (m, 4 H), 5.68 (q, J = 6.6 Hz, 1 H), 7.44 (t, J = 7.7 Hz, 2 H), 7.57 (t, J = 7.5 Hz, 1 H), 8.08 (d, J = 7.8 Hz, 2 H); ^{13}C NMR (75 MHz, $CDCl_3$) δ: 34.5, 35.8, 40.3, 64.9, 65.0, 73.6, 107.3, 128.4, 129.5, 129.8, 133.2, 165.3, 203.3; MS (EI) *m/z* (rel intensity) 276 (M^+, 16), 219 (25), 171 (22), 155 (23), 126 (16), 122 (20), 115 (12), 105 (100); HRMS (EI) calculated for $C_{15}H_{16}O_5$ 276.0997, found 276.1000; Anal. calcd. for $C_{15}H_{16}O_5$ C, 65.2, H, 5.8; found C, 65.1, H, 5.8.

3. Discussion

α-Oxyacylated carbonyl compounds are important functional groups present in many natural products, pharmaceuticals and synthetic intermediates of broad utility. Shi and coworkers have reported the catalytic asymmetric introduction of the oxybenzoyl group α- to ketone carbonyl groups by the low temperature epoxidation of preformed enol esters followed by rearrangement under acidic conditions, which furnishes products in good yields and excellent enantioselectivity.[3] The preparation of α-acetoxy ketones catalyzed by iodobenzene via a hypervalent iodine intermediate has also been described.[4] Despite the effectiveness of this method, it has not been extended to aldehyde substrates and is only effective for the introduction of acetoxy groups. Discrimination between primary and non-sterically encumbered secondary carbon centers is also poor using this method.

The procedure described above provides a very practical alternative for the α-acyloxylation of carbonyl compounds. The reagents are bench stable and react under mild conditions with both aldehydes and ketones in good yields. Not only do these reagents avoid air-sensitive intermediates and therefore obviate the need for purified solvents and specialized equipment; but the reagents display a remarkable functional group tolerance and useful chemo- and regioselectivity.

N-Methyl-*O*-benzoylhydroxylamine hydrochloride **1** is a general reagent for the oxygenation of both aldehydes and ketones.[5] The related *N*-*tert*-butyl-*O*-benzoylhydroxylamine hydrochloride **3** allows the chemoselective α-oxybenzoylation of aldehyde substrates in good to excellent yield, since it is completely inert to ketones.[6] Both classes of reagents can be used for the introduction of the oxyacyl group of choice by simple modification of the carboxylic acid used in the reagent synthesis. Representative examples for the α-functionalization of aldehydes and ketones by *N*-alkyl-*O*-acylhydroxylamine hydrochlorides are given in Tables 1 and 2.

TABLE 1
α-Oxybenzoylation of Aldehydes and Ketones using *N*-Methyl-*O*-Benzoylhydroxylamine Hydrochloride (**1**)

Entry	Substrate	Product	Yield (%)	Entry	Substrate	Product	Yield (%)
1	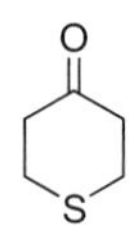	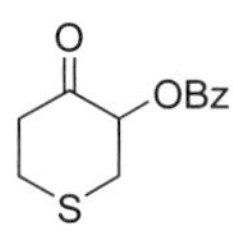	75[a]	6	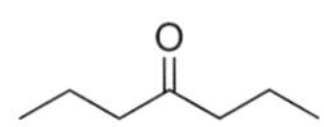	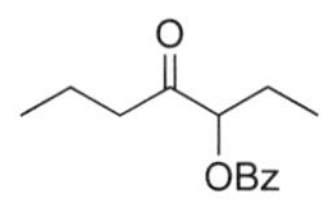	90[b]
2	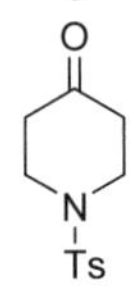	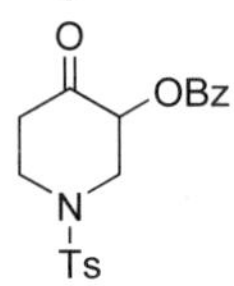	74[a]	7	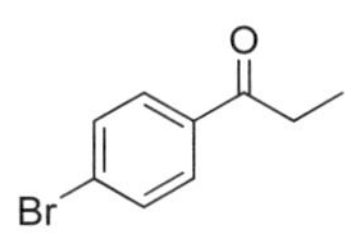	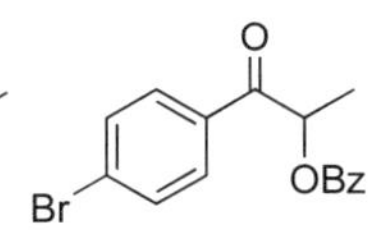	82[c]
3	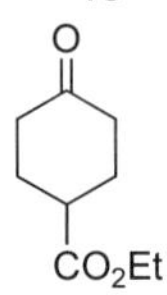	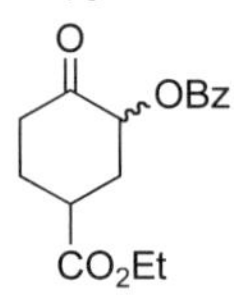	75[a]	8	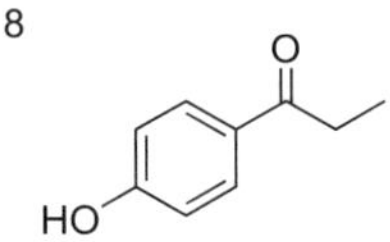	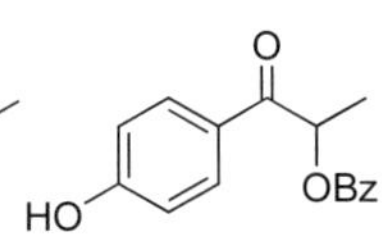	85[c]
4	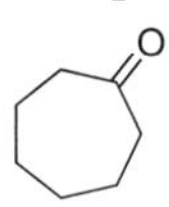	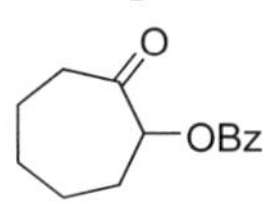	69[a]	9	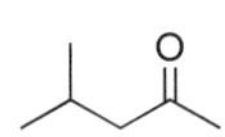	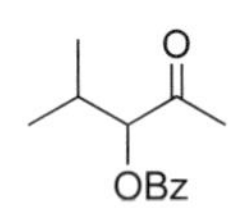	73[b]
5	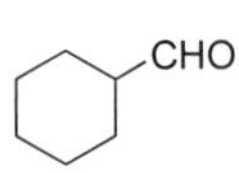	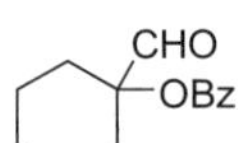	74[a]	10	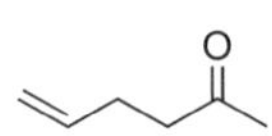	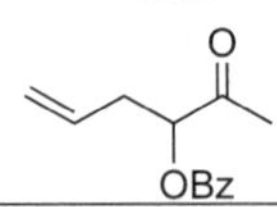	81[b]

[a] DMSO, rt, 4-24 h, 1 equiv. **1**; [b] DMSO, 50 °C, 24 h, 1 equiv. **1**; [c] DMSO, 50 °C, 48 h, 2 equiv. **1**

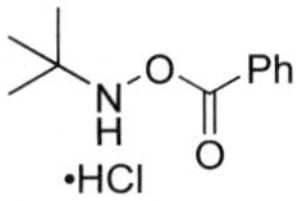
3

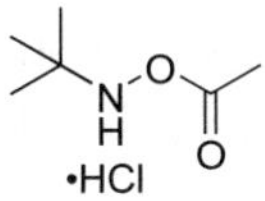
4

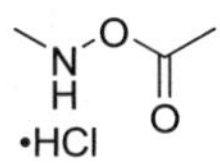
5

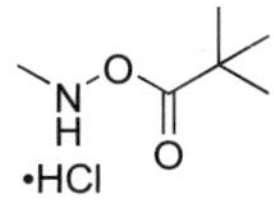
6

TABLE 2
α-Acyloxylaton of Carbonyl Compounds using *N*-Alkyl-*O*-Acyl Hydroxylamine Hydrochlorides (**3-6**)

Entry	Reagent	Substrate	Product	Yield (%)
1	**3**	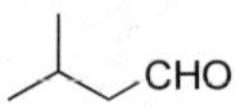	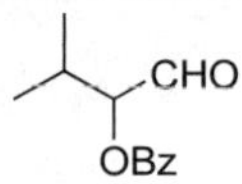	79[a]
2	**3**	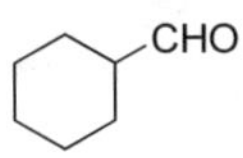	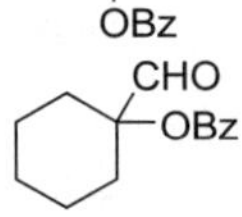	82[a]
3	**3**	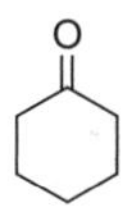	None	0[a]
4	**4**	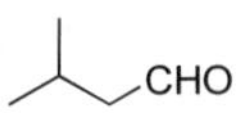	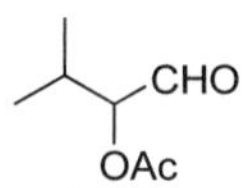	72[a]
5	**5**	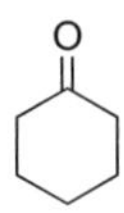	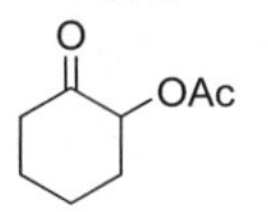	67[b]
6	**6**	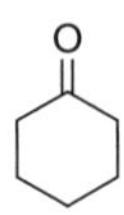	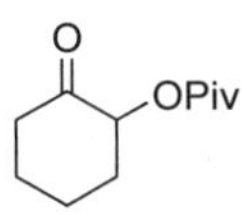	69

[a] THF/H_2O (9:1), 50 °C, 24 h; [b] DMSO, rt, 24 h

1. School of Chemistry, Cardiff University, Main Building, Park Place, Cardiff, CF10 3AT, UK; tomkinsonnc@cardiff.ac.uk.
2. Department of Chemistry, University of Pittsburgh, Pittsburgh, PA 15260, USA; pwipf@pitt.edu.
3. (a) Zhu, Y.; Shu, L.; Tu, Y.; Shi, Y. *J. Org. Chem.* **2001**, *66*, 1818; (b) Feng, X.; Shu, L.; Shi, Y. *J. Org. Chem.* **2002**, *67*, 2831.
4. Ochiai, M.; Takeuchi, Y.; Katayama, T.; Sueda, T.; Miyamoto, K. *J. Am. Chem. Soc.* **2005**, *127*, 12244.

5. Beshara, C. S.; Hall, A.; Jenkins, R. L.; Jones, K. L.; Jones, T. C.; Killeen, N. M.; Taylor, P. H.; Thomas, S. P.; Tomkinson, N. C. O. *Org. Lett.* **2005**, *7*, 5729.

6. Beshara, C. S.; Hall, A.; Jenkins, R. L.; Jones, T. C.; Parry, R. T.; Thomas, S. P.; Tomkinson, N. C. O. *Chem. Commun.* **2005**, 1478.

OXIDATION OF NEROL TO NERAL WITH IODOBENZENE AND TEMPO
[(*Z*)-3,7-Dimethyl-2,6-octadienal]

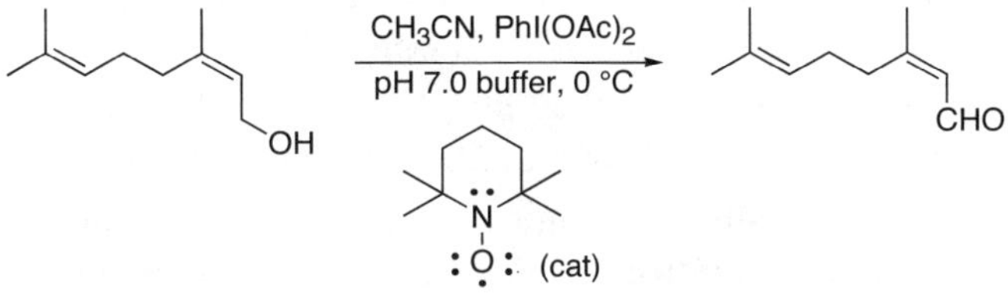

Submitted by Giovanni Piancatelli and Francesca Leonelli.[1a]
Checked by Nga Do and John Ragan.[1b]

1. Procedure

(Z)-3,7-Dimethyl-2,6-octadienal. A 250-mL round-bottomed flask equipped with a Teflon-coated magnetic stirring bar is charged with the following order of the reagents: acetonitrile (28 mL) (Note 1), (*Z*)-3,7-dimethyl-2,6-octadien-1-ol (nerol) (5.70 mL, 32.5 mmol) (Note 2), aqueous pH 7.0 buffer solution (8 mL) (Note 3), 2,2,6,6-tetramethylpiperidin-1-oxyl (TEMPO) (490 mg, 3.24 mmol) (Note 4), and iodobenzene diacetate (IBD) (11.49 g, 35.71 mmol) (Note 5). The reaction mixture is stirred at 0 °C (Note 6) until nerol is no longer detectable by TLC analysis (Note 7). The reaction mixture is diluted with diethyl ether (100 mL) and transferred to a 500-mL separatory funnel. The orange mixture is washed with saturated aqueous sodium thiosulfate (2 x 50 mL) (Note 8). The aqueous phase is separated and extracted with diethyl ether (3 x 35 mL). The combined organic layers are washed with saturated aqueous sodium hydrogen carbonate (40 mL) and then with saturated aqueous sodium chloride (40 mL) (Note 9). The organic layer is dried over anhydrous sodium sulfate, filtered, and concentrated with a rotary evaporator (35 °C, 70 mmHg). The residue is purified by column chromatography on silica gel (Note 10), using a 1:9 mixture of diethyl ether and hexanes as eluent (Note 11) to afford 4.30–4.39 g (87–89%) of (*Z*)-3,7-dimethyl-2,6-octadien-1-al (neral) as a colorless oil (Note 12). The material is homogenous by TLC, IR, ^{1}H and ^{13}C-NMR (Note 13).

2. Notes

1. "RPE-For analysis"-grade CH_3CN, as supplied by Carlo Erba Reagents (Italy), was used.

2. Nerol (97%) was purchased from Aldrich and used directly without purification. The compound revealed traces of geraniol (< 3%) from GCMS analysis (Shimadzu GCMS-QP5000; EQUITY^{TM}-5 FUSED SILICA Capillary Column 30 m x 0.32 mm x 0.25 μm film thickness; 80 °C (1 min), 80-240 °C (16 min), 240 °C (1 min)); t_{nerol} = 8.43, $t_{r geraniol}$ = 8.75. The checkers used an Agilent HP-5 fused silica capillary column, 30 m x 0.32 mm x 0.25 μm; 250 °C detector, 100 °C isothermal program at 2 mL/min. This system gave the following retention times: t_{nerol} = 7.3 min, t_{neral} = 7.8 min, $t_{geraniol}$ = 8.4 min, $t_{geranial}$ = 9.1 min. The checkers' lot of nerol (also purchased from Aldrich) showed <0.5% geraniol contamination.

3. The pH 7.0 buffer solution was purchased from Fluka.

4. TEMPO (98%) was purchased from Aldrich and used as received.

5. IBD (98%) was purchased from Aldrich and used as received.

6. An ice-bath is placed under the reaction flask after the dissolution of IBD (≈ 3 min) because the reaction is slightly exothermic. On a smaller scale (1-5 mmol) this precaution is not necessary. The checkers placed the reaction flask in an ice bath prior to addition of the TEMPO and IBD; as the IBD began dissolving, an internal temperature probe showed a temperature rise of 6–9 °C over the course of the reaction. This was not an issue on the scale run, but should be carefully monitored if the reaction were to be significantly increased in scale.

7. The oxidation reaction is usually very fast and is complete after 20 min on the scale described above. The reaction was checked by TLC to confirm completion. Thin-layer chromatography analysis was carried out on E.Merck silica gel F254 plates by elution with Et_2O/hexanes (3/7), then sprayed with 2 N H_2SO_4 solution and heated with a hot plate for 1 min. The alcohol starting material has an R_f = 0.29 (light brown) and the aldehyde product has an R_f = 0.66 (deep purple). The reaction can also be monitored by GC/MS, as indicated by the checkers in Note 3.

8. Washing with saturated aqueous sodium thiosulfate removes TEMPO from the organic phase that becomes light yellow. If the organic phase should be still orange after the two washings, the solution is shaken in the separatory funnel with 50 mL of saturated aqueous sodium thiosulfate and 15 mL of 0.5 N HCl until it becomes light yellow (the checkers found this extra thiosulfate-HCl wash to be necessary). TEMPO must be removed

carefully at this stage, because it cannot be removed in the chromatographic purification.

9. The aqueous phase must be neutral. Acidic impurities can catalyze (*E*)/(Z) isomerization of the aldehyde in the purification stage.

10. Silica gel 60 (0.063-0.2 mm/ 70-230 mesh ASTM), purchased from Macherey-Nagel, was used. The column chromatography was performed using a 1/20 ratio between the crude product and silica gel.

11. The solvent required for the purification of the product varies from 1.4 to 1.6 L. The chromatographic solvent is removed with a rotary evaporator (35 °C, 70 mmHg).

12. The checkers used diethyl ether:hexane (2:8) for column elution. Attempted purification by distillation (bp 118–120 °C, 20 mmHg) led to significant olefin isomerization (the pot residue was a 50:50 mixture of neral and geranial).

13. The following analytical data have been obtained for (*Z*)-3,7-dimethyl-2,6-octadien-1-al: IR ($CHCl_3$) cm^{-1}: 1675 (C=O), 1635 (C=C); ^1H NMR (400 MHz, $CDCl_3$) δ: 1.58 (s, 3 H). 1.67 (s, 3 H), 1.97 (d, J = 1.2 Hz, 3 H), 2.15–2.25 (m, 2 H), 2.57 (t, J = 7.5 Hz, 2 H), 5.06–5.11 (m, 1 H), 5.86 (d, J = 8.3 Hz, 1 H), 9.88 (d, J = 8.3 Hz, 1 H); ^{13}C NMR (100 MHz, $CDCl_3$) δ: 17.9, 25.3, 25.9, 27.2, 32.7, 122.4, 128.8, 133.9, 164.1, 191.0; GCMS purity >97%, t_r=8.56. The compound revealed traces of geranial (< 3%, t_r=8.98) from GCMS analysis (See Note 3).

Safety and Waste Disposal Information

All hazardous materials should be handled and disposed of in accordance with "Prudent Practices in the Laboratory"; National Academy Press; Washington, DC, 1995.

3. Discussion

The catalytic procedure described here illustrates a fast and inexpensive conversion of a (*Z*)-configured primary allylic alcohol into the corresponding (*Z*)-configured α,β-unsaturated aldehyde in high yields. The result demonstrates the chemoselectivity of the process, in that the easily isomerizable *Z*-configuration of the nerol is maintained.

Various methods for the oxidation of nerol to neral are available. Although individually having some synthetic advantages, most methods suffer from one or more experimental drawbacks, such as severe or delicate reaction conditions, complicated reaction procedures, and the need to use toxic or unstable reagents.[2]

Oxammonium salts **1** are the effective oxidant species derived from TEMPO, and have been used extensively either in stoichiometric or in catalytic amounts for the oxidation of primary and secondary alcohols to the corresponding carbonyl compounds.[3] Compound **1** has been generated *in situ* from nitroxyl radicals, such as TEMPO, in combination with a number of secondary oxidants.[4]

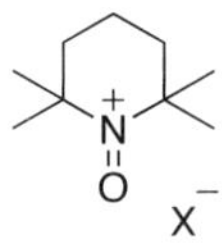

1

Hypervalent iodine reagents have been used recently for a variety of organic transformations.[4] *Inter alia*, IBD in combination with catalytic amounts of TEMPO is used as a stoichiometric oxidant in the conversion of primary and secondary alcohols to carbonyl compounds.[5] This oxidation protocol works efficiently at room temperature in dichloromethane (and also in most common organic solvents and neat in some cases) and can be

Table 1. Oxidation of Primary Alcohols to Aldehydes[5]

Entry	Substrate	Product	Time (h)	yield (%)
1	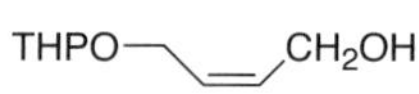	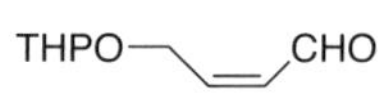	0.2	95
2	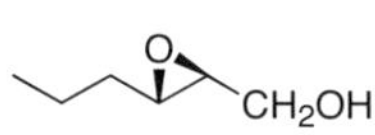		1	70
3	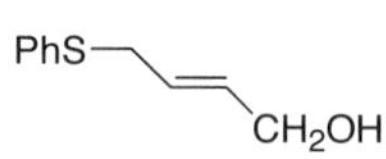	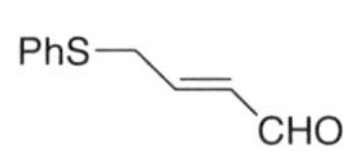	1	70
4	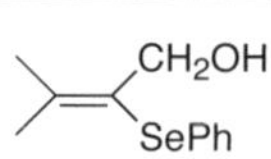	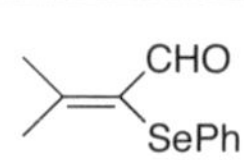	15	55
5	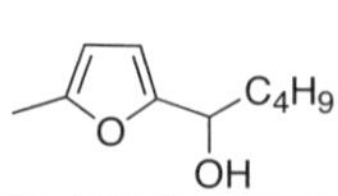	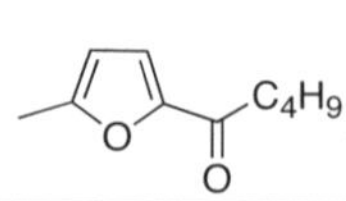	3	95

performed in an open flask without special precautions (*e.g.* inert atmosphere or dry solvent). This process exhibits a very high degree of selectivity for the oxidation of primary alcohols to aldehydes, without noticeable over-oxidation to carboxyl compounds, and a high chemoselectivity in the presence of either secondary alcohols or of other oxidizable moieties.[5]

Many sensitive protective groups are not affected by this method. Some examples of carbonyl compounds synthesized with this method are reported in Table 1.[5]

1. (a) Dipartimento di Chimica, Università "La Sapienza", Piazzale Aldo Moro 5, 00185 Roma, Italy. (b) Chemical Research & Development, Pfizer Global Research & Development, Eastern Point Road, Groton, CT, USA.
2. For example, see: (a) Takahashi, M.; Oshima, K.; Matsubara, S. *Tetrahedron Lett.* **2003**, *44*, 9201-9203. (b) Matsuo, J-i.; Iida, D.; Yamanaka, H; Mukaiyama, T. *Tetrahedron* **2003**, *59*, 6739-6750. (c) Bhar, S.; Chaudhuri, S. K. *Tetrahedron* **2003**, *59*, 3493-3498. (d) Matano, Y.; Nomura, H. *Angew. Chem., Int. Ed.* **2002**, *41*, 3028-3031. (e) Muldoon, J.; Brown, S. N. *Org. Lett.* **2002**, *4*, 1043-1045. (f) De Luca, L.; Giacomelli, G.; Porcheddu, A. *Org. Lett.* **2001**, *3*, 3041-3043. (g) Kakiuchi, N.; Maeda, Y.; Nishimura, T.; Uemura, S. *J. Org. Chem.* **2001**, *66*, 6620-6625.
3. For recent reviews of TEMPO-catalyzed alcohol oxidation, see: (a) de Nooy, A. E. J.; Besemer, A. C.; van Bekkum, H. *Synthesis* **1996**, 1153-1174. (b) Adam, W.; Saha-Moller, C. R.; Ganeshpure, P. A. *Chem. Rev.* **2001**, *101*, 3499-3548. (c) Sheldon, R. A.; Arends, I. W. C. E.; ten Brink, G.-J.; Dijksman, A. *Acc. Chem. Res.* **2002**, *35*, 774-781.
4. For recent reviews, see: (a) Stang, P. J.; Zhdankin, V. V. *Chem. Rev.* **1996**, *96*, 1123-1178. (b) Kita, Y.; Takada, T.; Tohma, H. *Pure Appl. Chem.* **1996**, *68*, 627-630. (c) Varvoglis, A. *Hypervalent Iodine in Organic Synthesis, Academic Press, San Diego,* **1997**. (d) Ochiai, M. in *Chemistry in Hypervalent Compounds* (Ed.: K. Akiba), Wiley-VCH, New York, **1999**, chap. 12. (e) Zhdankin, V. V.; Stang, P. J. *Chem. Rev.* **2002**, *102*, 2523-2584. (f) Moriarty, R. *J. Org. Chem.* **2005**, *70*, 2893-2903.
5. De Mico, A.; Margarita, R.; Parlanti, L.; Vescovi, A.; Piancatelli, G. *J. Org. Chem.* **1997**, *62*, 6974 -6977.

AU(I)-CATALYZED HYDRATION OF ALKYNES: 2,8-NONANEDIONE

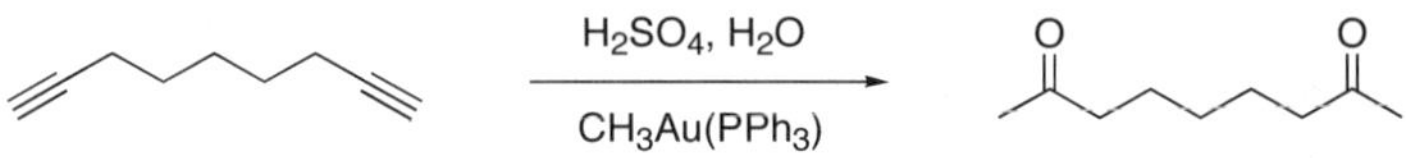

Submitted by Eiichiro Mizushima,[1] Dong-Mei Cui,[1] Dilip Chandra Deb Nath,[1] Teruyuki Hayashi,*[1,2a] and Masato Tanaka.[1,2b]
Checked by Rick L. Danheiser and Tin Yiu Lam.

1. Procedure[3]

2,8-Nonanedione. A 250-mL, three-necked, round-bottomed flask fitted with a rubber septum, a glass stopper, a reflux condenser equipped with an argon inlet, and a magnetic stirbar is evacuated and filled with argon, and then charged with 1,8-nonadiyne (6.3 mL, 5.0 g, 42 mmol, 1 equiv) (Note 1), methyl(triphenylphosphine)gold[4] (CH_3AuPPh_3) (0.042 g, 0.088 mmol, 0.2 mol%) (Note 2), sulfuric acid (prepared by diluting 2.08 g (21.2 mmol, 0.5 equiv) of concentrated H_2SO_4 with 21 mL of water) (Note 3), and methanol (120 mL) (Note 4). The mixture is heated and stirred at reflux (oil bath temperature 70–75 °C) for 2 h, at which time analysis by gas chromatography (Note 5) shows that the conversion of 1,8-nonadiyne is complete. The reaction mixture is allowed to cool to room temperature and then transferred to a 250-mL, round-bottomed flask and most of the methanol is removed by rotary evaporation (30–40 °C, 20 mmHg). The resulting viscous, colorless, and cloudy oil (about 20 mL) is diluted with water (70 mL) and diethyl ether (50 mL). The mixture is stirred for 5 min and then the aqueous layer is separated and extracted with ether (3 x 50 mL). The combined ether layers are washed with saturated $NaHCO_3$ solution (50 mL) and saturated NaCl solution (50 mL), dried over anhydrous magnesium sulfate, filtered, and concentrated by rotary evaporation (20 °C, 20 mmHg) to give 5.80 g of a colorless solid. This material is dissolved in boiling pentane (about 30 mL) in a 50-mL Erlenmeyer flask heated on a hot plate and the solution is cooled to 0 °C. The resulting solid is collected by vacuum filtration on a Büchner funnel, washed with ice-cold pentane (10 mL), and then dried at 20 mmHg to afford 5.57 g (86%) of 2,8-nonanedione as colorless crystals (Note 6).

2. Notes

1. 1,8-Nonadiyne (98% purity) was purchased by the submitters from Acros Organics and used as received. The checkers obtained the diyne (98% purity) from Aldrich Chemical Co.

2. Methyl(triphenylphosphine)gold(I) was prepared as follows according to the procedure of ref. 4. A 100-mL, two-necked, round-bottomed flask equipped with a rubber septum, a reflux condenser fitted with an argon inlet, and a magnetic stirbar was evacuated and filled with argon. Chloro(triphenylphosphine)gold(I) (Strem Chemicals, 98+% purity) (0.5 g, 1.0 mmol, 1 equiv) and 6 mL of diethyl ether (purified by pressure filtration through activated alumina) were added and the resulting colorless suspension was cooled to 0 °C. Methylmagnesium iodide solution (3M in diethyl ether, Aldrich Chemical Co.) (1.7 mL, 5.1 mmol, 5.1 equiv) was added dropwise via syringe over 10 min. The resulting mixture was heated at reflux (oil bath temperature 50–55 °C) for 1 h and then allowed to cool to room temperature. Ice-cold 0.5% H_2SO_4 solution (24 mL, degassed by argon purging for 30 min before use) was added, resulting in the formation of a purple precipitate. Diethyl ether (3 mL, purified by pressure filtration through activated alumina) was added and the resulting mixture of two liquid phases and the precipitate was vacuum filtered through a Büchner funnel (washing with three 10-mL portions of toluene that had been purified by pressure filtration through activated alumina). The combined filtrates were transferred to a separatory funnel and the layers were separated. The organic phase was washed with 50 mL of water, dried over anhydrous Na_2SO_4, filtered, and concentrated to a volume of about 10 mL. This solution was transferred to a 50-mL Erlenmeyer flask, diluted with 30 mL of hexane (degassed by purging with argon for 30 min prior to use), and then slowly cooled to –18 °C. The resulting crystals were collected on a Büchner funnel by vacuum filtration (washing with ca. 15 mL of ice-cold hexane) and air-dried to give 0.248 g (52%) of methyl(triphenylphosphine)gold(I) as colorless crystals. The product exhibits the following physical properties: mp: 164–167 °C (lit.[4] 167–168 °C); ^{1}H NMR (300 MHz, $CDCl_3$) δ: 0.55 (d, J_{PH} = 7.8 Hz, 3 H), 7.43–7.58 (m, 15 H); ^{31}P NMR (121 MHz, $CDCl_3$) δ: 47.6; Anal. Calcd for $C_{19}H_{18}AuP$: C, 48.12; H, 3.83; Found: C, 48.11; H, 3.51.

3. Concentrated sulfuric acid was obtained by the checkers from Mallinckrodt Chemicals and was diluted with deionized water. The aqueous acid was degassed by purging with argon for 30 min prior to use.

4. Methanol was purchased by the submitters from Wako Pure Chemical Industries, Ltd. and used as received. The checkers obtained anhydrous methanol from Mallinckrodt Chemicals. The methanol was degassed by purging with argon for 30 min prior to use.

5. The submitters employed a Shimadzu GC-17A gas chromatograph equipped with a capillary column of Ultra Alloy UA1-30M-0.25F with a stationary phase of poly(dimethylsiloxane) using the following conditions: initial temperature at 50 °C, then increased at a rate of 10 °C/min, and held at 250 °C for 30 min. The checkers used a HP 6890 Series GC system equipped with a capillary column HP-5 with 5% phenyl dimethylpolysiloxane as the stationary phase under the following conditions: initial temperature 50 °C, then increased at a rate of 10 °C/min, and held at 240 °C for 30 min. Retention time of 1,8-nonadiyne: 4.37 min.

6. The product has the following characteristics: mp 47–48 °C (lit.[5] 47–48 °C); ^{1}H NMR (500 MHz, $CDCl_3$) δ: 1.22–1.28 (m, 2 H), 1.54 (apparent quintet, J = 7.5 Hz, 4 H), 2.10 (s, 6 H), 2.40 (t, J = 7.5 Hz, 4 H); ^{13}C NMR (125 MHz, $CDCl_3$) δ: 23.6, 28.7, 30.1, 43.5, 209.1; IR (KBr): 2936, 2866, 1716, 1702, 1464, 1408, 1378, 1360, 1230, 1163, 958 cm^{-1}; Anal. Calcd for $C_9H_{16}O_2$: C, 69.19; H, 10.32. Found: C, 68.92; H, 10.56.

Safety and Waste Disposal Information

All hazardous materials should be handled and disposed of in accordance with "Prudent Practices in the Laboratory"; National Academy Press; Washington, DC, 1995.

3. Discussion

2,8-Nonanedione is a useful synthetic intermediate for the synthesis of rootworm sex pheromones, among other applications. The methods for the synthesis of 2,8-nonanedione have been recently summarized by Keinan and coworkers[6] as follows: (1) mercury-catalyzed hydration of 1,8-nonadiyne,[7] (2) palladium-catalyzed Wacker oxidation of 1,8-nonadiene,[8] (3) reaction of 1,5-pentylene biscadmium halide with acetyl chloride,[9] (4) oxidative cleavage of 2-methylcycloheptanone,[5] and (5) reaction of bromohexanone ketal and acetoacetate.[6] Other synthetic routes include (6) alkylation of acetone dimethylhydrazone followed by hydrolysis,[10] (7) condensation of an alkoxyalkyl phenyl sulfone and a dihalopentane,[11] (8) imidazolylation of pimelonamide followed by alkylation and elimination,[12] and (9) alkylation of a phosphonium salt of pimeloyl chloride followed by hydrolysis.[13]

Table 1 Preparation of Various Methyl Ketones[a]

Alkyne	Product	$[(PPh_3)AuCH_3]$/mmol	Time	Yield[b]
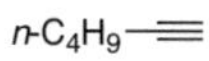	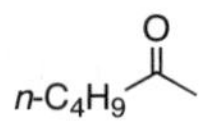	0.002	2 h[c]	99%
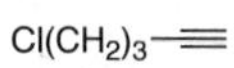	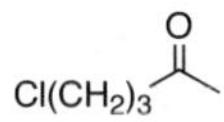	0.01	4 h[d]	72%
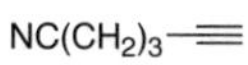	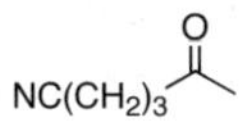	0.002	1 h	83%
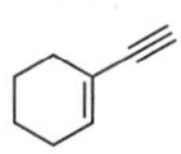	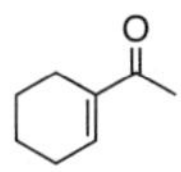	0.01	1 h	90%
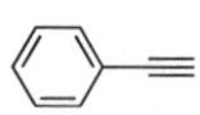	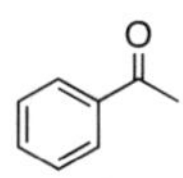	0.002[e]	1 h	98%
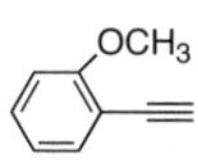	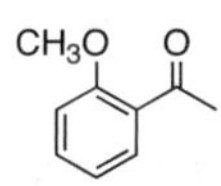	0.002	1 h	95%
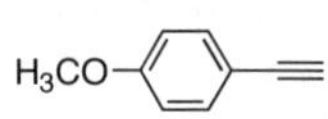	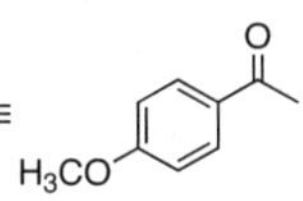	0.002	1 h	93%
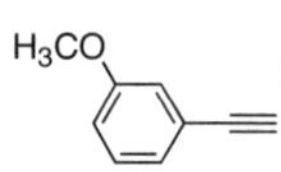	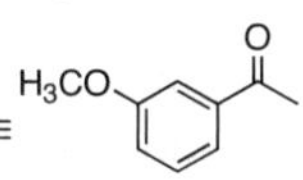	0.01	1 h	77%
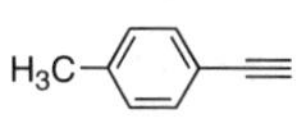	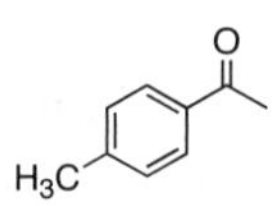	0.002[e]	1 h	96%
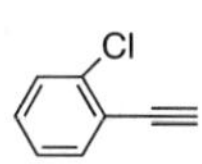	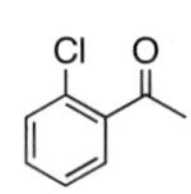	0.002	1 h	66%
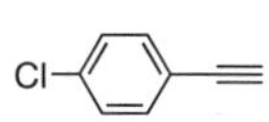	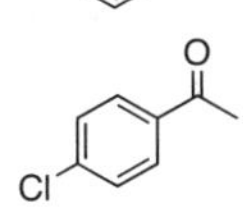	0.01	1 h	54%
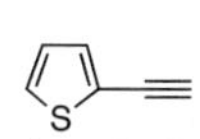	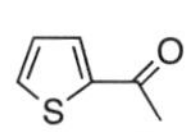	0.01	1h	92%

a) Catalyst $[(Ph_3P)AuCH_3]$, alkyne 1 mmol, water 0.5 mL, H_2SO_4 0.5 mmol, methanol 1-3 mL, 70 °C. b) Yields by GC c) 60 °C. d) Under CO atmosphere. e) CF_3SO_3H instead of H_2SO_4.

However, reactions (3) - (9) are multi-step procedures that give massive salt waste. Wacker oxidation (2) is principally the best reaction, while the oxidation of 1,8-nonadiene[8] itself is not selective.

On the other hand, acid-catalyzed hydration of alkynes to give carbonyl compounds is one of the most straightforward methods to form the carbon-oxygen bond and is long known.[14] However, only electron-rich alkynes react satisfactorily. The reaction of simple alkynes is sluggish and requires a co-catalyst such as mercury(II) salts[6,15] as in reaction (1). To avoid the use of toxic mercury(II) salts, the use of various transition metal complexes such as Ru(II),[16] Ru(III),[17] Rh(III),[18] Pt(II),[19] and Au(III)[20] have been examined. However, the efficiencies of these catalyst systems have remained insufficient and the highest turnover frequency is in the 100/h range. Recently, Teles and coworkers[21] have claimed in a patent application that hydration of propargyl alcohol is catalyzed by an Au(I)-acid catalyst system. However, the efficiency is still low.

We have discovered that simple alkynes can be hydrated by the CH_3AuPPh_3-acid catalyst system to give the corresponding methyl ketones in high yields.[3] The reaction is significantly affected by the reaction media and methanol is the best solvent. Turnover frequency increases in methanol to 15600/h.[22]

The reaction is applied not only to simple alkynes and diynes, but also to various alkynes as shown in Table 1. Aliphatic terminal alkynes give corresponding methyl ketones in good yields, even when electron withdrawing substituents such as chloro and cyano groups are attached. The use of CO atmosphere is required in some cases. Though C=C and C≡N groups can be potentially hydrated or hydrolyzed, both groups remain intact in the presence of this CH_3AuPPh_3-acid catalyst system. Aromatic alkynes give the corresponding acetophenones in high yields, while the use of CF_3SO_3H instead of H_2SO_4 is required in some cases. The presence of electronegative *o*- and *p*-chloro- and *m*-methoxy-substituents results in lower yields. The present catalyst CH_3AuPPh_3 is also applicable to hydroamination of alkynes.[23] A timely review on recent explosive progress in gold catalysis has appeared.[24]

1. National Institute of Advanced Industrial Science and Technology.
2. Present affiliation: (a) Department of Chemistry, The University of Tokyo; (b) Tokyo Institute of Technology.

3. A less detailed procedure has been published: Mizushima, E.; Sato, K.; Hayashi, T.; Tanaka, M. *Angew. Chem. Int. Ed.*, **2002**, *41*, 4563.

4. Tamaki, A.; Kochi, J. K. *J. Organomet. Chem.*, **1973**, *61*, 441.

5. Nishinaga, A.; Rindo, K.; Matsuura, T. *Synthesis*, **1986**, 1038.

6. Keinan, E.; Sinha, S. C.; Sinha-Bagchi, A. *J. Org. Chem.*, **1992**, *57*, 3631.

7. Vo-Quang, Y. *Ann. Chim. (Paris)*, **1962**, *7*, 785.

8. Zahalka, H. A.; Januszkiewicz, K.; Alper, H. *J. Mol. Catal.*, **1986**, *35*, 249.

9. Soussan, G; Freon, P. *Bull. Soc. Chim. Fr.*, **1972**, 4228.

10. Yamashita, M.; Matsumiya, K.; Morimoto, H., Suemitsu, R. *Bull. Chem. Soc. Jpn.*, **1989**, *62*, 1668.

11. Tanaka, T.; Matsui, S., Kaji, A. *Bull. Chem. Soc. Jpn.*, **1980**, *53*, 3619.

12. Hayakawa, S.; Baba, J.; Okamoto, M.; Ohta, S. *Chem. Pharm. Bull.*, **1988**, *36*, 22.

13. Maeda, H.; Hino, N.; Yamauchi, Y.; Ohmori, H. *Chem. Pharm. Bull.*, **2000**, *48*, 1196.

14. Smith, M. B.; March, J. "Advanced Organic Chemistry," 5th ed., Wiley: New York, **2001**, p.995.

15. Jäger, V.; Viehe, H. G. in "Houben-Weyl, Methoden der Organischen Chemie," Vol. 5/2a, Ed.: Müller, E., Thieme Verlag: Stuttgart, **1977**, p.726.

16. Tokunaga, M; Wakatsuki, Y. *Angew. Chem. Int. Ed.*, **1998**, *37*, 2867.

17. Halpern, J.; James, B. R.; Kemp, A. L. W. *J. Am. Chem. Soc.*, **1961**, *83*, 4097.

18. James, B. R.; Rempel, G. L. *J. Am. Chem. Soc.*, **1969**, *91*, 863.

19. Francisco, L. W.; Moreno, D. A.; Atwood, J. D., *Organometallics*, **2001**, *20*, 4237.

20. Fukuda, Y.; Utimoto, K. *J. Org. Chem.*, **1991**, *56,* 3729.

21. Schulz, M.; Teles, J. H. U. S. Patent 6,037,482 for BASF, Mar. 14, 2000.

22. Reaction conditions: 1-octyne 20 mmol, CH_3AuPPh_3 0.001 mmol, CF_3SO_3H 0.5 mmol, H_2O 1 mL, methanol 10 mL, 70°C, 1 h, under CO atmosphere.

23. Mizushima, E.; Hayashi, T.; Tanaka, M. *Org. Lett.,* **2003**, *5*, 3349.

24. Stephen, A.; Hashmi, K.; Hutchings, G. *J. Angew. Chem. Int. Ed.*, **2006**, *45*, 7896.

D-RIBONOLACTONE AND 2,3-ISOPROPYLIDENE(D-RIBONOLACTONE)

(D-Ribonic acid, 2,3-*O*-(1-methylethylidene)-, γ-lactone)

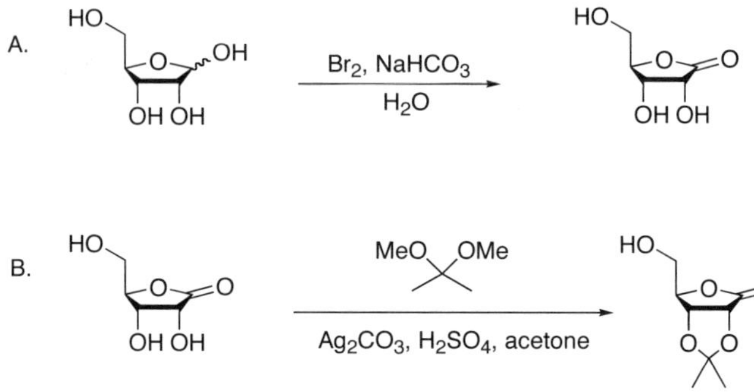

Submitted by John D. Williams,[1] Vivekanand P. Kamath,[2] Philip E. Morris,[2] and Leroy B. Townsend.[1]
Checked by Anthony Cuzzupe and Peter Wipf.

1. Procedure

Caution! Bromine is volatile and corrosive, and causes severe burns upon contact with skin. Proper protective equipment and an efficient fume hood are required.

A. ***D**-Ribonolactone.* A 1-L three-necked, round-bottomed flask fitted with a mechanical stirrer, a 100-mL pressure-equalizing addition funnel and an internal thermometer is charged with **D**-ribose (100 g, 0.67 mol), sodium bicarbonate (112 g, 1.3 mol, 2 equiv) (Note 1) and water (600 mL). The mixture is stirred at room temperature for 15 min, during which time most of the solids dissolve (Note 2). The flask is then immersed in an ice-water bath. The addition funnel is filled with bromine (112 g, 0.70 mol, 1.04 equiv) (Note 1) and the bromine is added to the vigorously stirred aqueous solution at a rate of about 2 drops/sec such that the reaction temperature does not exceed 5 °C. When the addition is complete (about 1 hr), the funnel is replaced with a stopper and the resulting orange solution is stirred for an additional 50 min. Sodium bisulfite (6.5 g, 62.5 mmol) is added in order to completely discharge the orange color (Note 3). The clear aqueous solution is transferred to a 2-L flask and evaporated on a rotary evaporator (bath temperature 60 °C – 70 °C, water aspirator pressure) until a wet slurry remains. Absolute EtOH (400 mL) and toluene (100 mL) are added to give a cloudy suspension and the solvent is removed by rotary evaporation (bath temperature 50 °C, water aspirator pressure) to provide a damp solid. Absolute EtOH (400 mL) is added and the mixture is heated on a steam bath for 30 min. The hot ethanolic suspension is filtered and the solids are rinsed with hot absolute EtOH (100 mL). The filtrate is cooled to room temperature, and then refrigerated for 16 h. The crystalline product is filtered, rinsed first with cold absolute EtOH (100 mL) then with Et_2O (100 mL), and dried under vacuum (room temperature, 0.25 mmHg) to yield 125 g of crude product (Note 4). The filtrate is concentrated (200 mL) and refrigerated to obtain additional product, which is filtered, washed with cold EtOH (25 mL) and Et_2O (25 mL) and dried under vacuum (room temperature, 0.25 mmHg) to provide 35 g of additional crude product.

*2,3-Isopropylidene(**D**-ribonolactone).* In a 2-L round-bottom flask, the crude ribonolactone (160 g) from above is suspended in dry acetone (700 mL), 2,2-dimethoxypropane (100 mL) and conc. H_2SO_4 (1 mL, 20 mmol) (Note 1). The solution is stirred vigorously at room temperature for 50 min, then silver carbonate (20 g, 73 mmol) is added (Notes 1 and 5). The resulting suspension is stirred at room temperature for 50 min, then the suspension is filtered through a 2 cm Celite pad. The solids are rinsed with acetone (100 mL), and the filtrate is evaporated to dryness. The crude acetonide is dissolved in EtOAc (250 mL) with heating on a steam bath. The resulting suspension is filtered through a 2 cm Celite pad, the solids are rinsed with hot EtOAc (50 mL), and the filtrate is allowed to cool to room temperature. Crystals of 2,3-isopropylidene(**D**-ribonolactone) form spontaneously as the solution cools. The resulting crystalline product is filtered and dried under vacuum (room temperature, 0.25 mmHg) to yield 68.7 g of product. The mother liquor is concentrated to 50 mL to yield 22.5 g of additional product after filtration and drying as above. The solids are combined to afford 91.2 g (73% overall yield from ribose) of white crystalline solid (Note 6, 7).

2. Notes

1. Ribose, bromine and silver carbonate were purchased from Acros Organics. Sodium bicarbonate was obtained from Mallinckrodt Baker, Inc. 2,2-Dimethoxypropane was obtained from Avocado Research Chemicals,

Ltd. Deionized water was used as the solvent. All reagents were used as purchased.

2. Some of the sodium bicarbonate remains undissolved, but this does not affect the reaction yield, as the excess will be consumed during the course of the reaction.

3. If any orange color remains at this stage, the product will develop a pale brown color that cannot be removed by activated carbon.

4. The major contaminant is ~40-45% sodium bromide. The submitters recrystallized the crude product twice from hot *n*-BuOH (800 mL) to obtain 80 g of pure material. However, the crude material is of sufficient quality for the subsequent reaction. Pure ribonolactone has the following physical properties: mp 85-87 °C; $[a]_D^{25}$+11.9 (*c* 0.99, H_2O); IR (KBr) cm^{-1}: 3513, 3373, 3159, 1761, 1627, 1397, 1197, 1143; R_f 0.56 (30% v/v MeOH/$CHCl_3$, *p*-anisaldehyde in ethanol stain); ^1H NMR (300 MHz, DMSO-d_6) d 5.73 (d, 1H, D_2O exch, J = 7.7 Hz), 5.35 (d, 1H, D_2O exch, J = 3.8 Hz), 5.15 (t, 1H, D_2O exch, J = 5.4 Hz), 4.40 (dd, 1H, J = 7.7, 5.4 Hz), 4.21 (t, 1H, J = 3.5 Hz), 4.11 (app t, 1H, J = 4.9 Hz), 3.56 (dd, 2H, J = 5.4, 3.6 Hz); ^{13}C NMR (75 MHz, DMSO-d_6) d 176.5, 85.4, 69.3, 68.6, 60.5.

5. As an alternative to the expensive silver carbonate, the submitters also used 30 g of the weakly basic resin Amberlyst A-21. In this instance, additional acetone (200 mL total) is used to rinse the filtered solids, and care must be taken upon evaporation of the solvent to remove the residual water.

6. The recrystallized **D**-ribonolactone acetonide has the following physical properties: mp 134-137 °C; $[a]_D^{24}$-66.7 (*c* 1.03, $CHCl_3$); IR (KBr) cm^{-1}: 3469, 2991, 2952, 2932, 1767, 1467, 1389, 1379, 1224; R_f 0.51 (1:3, hexane:EtOAc; *p*-anisaldehyde in ethanol stain); ^1H NMR (300 MHz, DMSO-d_6) d 5.30 (t, 1H, D_2O exch, J = 5.0 Hz), 4.76 (s, 2H), 4.60 (t, 1H, J = 2.1 Hz), 3.67-3.55 (m, 2H), 1.34 (s, 3H), 1.30 (s, 3H); ^{13}C NMR (75 MHz, DMSO-d_6) d 174.3, 111.6, 82.3, 78.1, 75.0, 60.4, 26.6, 25.1; Anal. Calcd for $C_8H_{12}O_5$: C, 51.06; H, 6.43. Found: C, 51.26; H, 6.46.

7. Extremely pure ribonolactone acetonide melts at >140 °C, but material with a melting point >130 °C is suitable for most reactions. Very pure material can be obtained by 1-2 additional recrystallizations from hot EtOAc.

Safety and Waste Disposal Information

All hazardous materials should be handled and disposed of in accordance with "Prudent Practices in the Laboratory"; National Academy Press; Washington, DC, 1995.

3. Discussion

The present procedure represents a modification of a previously published procedure for the bromine oxidation of ribose to ribonolactone.[3] The improved procedure allows for the addition of a liquid reagent (bromine) instead of the solid reagent (sodium carbonate) used in the previous preparation, and thus requires neither a powder addition apparatus nor constant attention. Additionally, the present procedure does not require the careful monitoring of pH, the absence of which can lead to the complete failure of the previous preparation.

Industrial preparations of ribonolactone involve the epimerization of arabinonic acid salts under very caustic and harsh conditions, followed by fractional crystallization of the resulting ribonic acid salts from the arabinonic/ribonic acid salt mixture, and the cyclization of the resulting salts.[4-6] An alternative rhenium-catalyzed oxidation procedure requires the preparation of an expensive catalyst and the removal of substantial amounts of benzalacetone (used as a terminal oxidant) and its reduction products.[7]

D-Ribonolactone is no longer commercially available in large quantities, and is very expensive. The acetonide is also very expensive. Although the above preparation of **D**-ribonolactone and **D**-ribonolactone acetonide is not high-yielding, the starting materials are inexpensive and the preparation is quite convenient. **D**-Ribonolactone and its derivatives have been used for the syntheses of many nucleoside analogs[8,9] and natural products.[10,11]

1. Department of Medicinal Chemistry, University of Michigan, 428 Church St., Ann Arbor, MI 48105.
2. BioCryst Pharmaceuticals, 2190 Parkway Lake Dr., Birmingham, AL 35244.
3. Pudlo, J. S.; Townsend, L. B. In *Nucleic Acid Chemistry: Improved and New Synthetic Procedures, Methods, and Techniques, Part 4.*; Townsend, L. B., Tipson, R. S., Eds.; John Wiley & Sons: New York, **1991**, p. 51-53.
4. Flexser, L. A.; Hoffmann-La Roche, Inc.: US Patent 2,438,883, **1948**.
5. Sternbach, L. H.; Hoffmann-La Roche, Inc.: US Patent 2,438,881, **1948**.

6. Schmidt, W.; Paust, J.; BASF Aktiengesellschaft: US Patent 4,294,766, **1981**.
7. Isaac, I.; Stasik, I.; Beaupere, D.; Uzan, R. *Tetrahedron Lett.* **1995**, *36*, 383-386.
8. Pankiewicz, K. W.; Sochacka, E.; Kabat, M. M.; Ciszewski, L. A.; Watanabe, K. A. *J. Org. Chem.* **1988**, *53*, 3473-3479.
9. Cheng, J. C.-Y.; Hacksell, U.; Daves, G. D. J. *J. Org. Chem.* **1985**, *50*, 2778-2780.
10. Fürstner, A.; Radkowski, K.; Wirtz, C.; Goddard, R.; Lehmann, C. W.; Mynott, R. *J. Am. Chem. Soc.* **2002**, *124*, 7061-7069.
11. Jiang, S.; Mekki, B.; Singh, G.; Wightman, R. H. *Tetrahedron Lett.* **1994**, *35*, 5505-5508.

PREPARATION OF 4-ACETYLAMINO-2, 2, 6, 6-TETRAMETHYLPIPERIDINE-1-OXOAMMONIUM TETRAFLUOROBORATE, AND THE OXIDATION OF GERANIOL TO GERANIAL

(2,6-Octadienal, 3,7-dimethyl-, (2E)-)

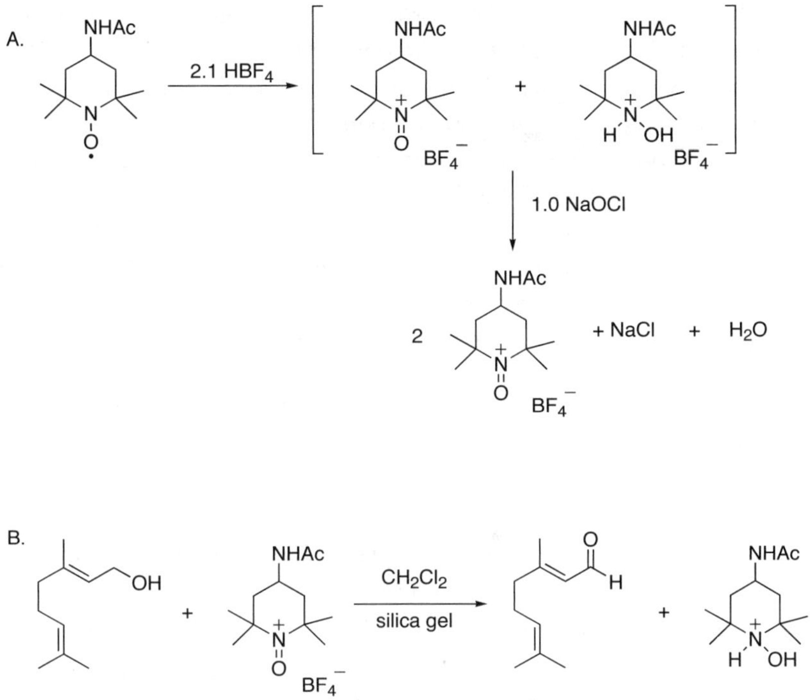

Submitted by James M. Bobbitt and Nabyl Merbouh.[1]
Checked by Peter Wipf and David Amantini.

1. Procedures (Note 1)

A. 4-Acetylamino-2,2,6,6-tetramethylpiperidine-1-oxoammonium tetrafluoroborate. In a 250-mL one-necked, round-bottomed flask equipped with a 3-cm magnetic stirring bar, 4-acetylamino-2,2,6,6-tetramethyl-1-piperidinyloxy (4-acetamido-TEMPO) (25.0 g, 0.117 mol) (Note 2) and water (50.0 mL) are added. An aqueous solution of HBF_4 (48% aqueous solution, 17.7 mL, 0.135 mol) (Note 3) is charged into a 25-mL dropping funnel and added dropwise over 30 min to the vigorous stirring orange mixture. A dark brown solution is formed initially, followed by formation of a yellow precipitate. The yellow slurry is stirred for an additional 30 min. Commercial sodium hypochlorite (Chlorox® bleach, 6.00 % NaOCl) (65.5 mL, 0.058 mol) is transferred into a 100-mL dropping funnel and added dropwise over a 1 hr period to the heterogeneous mixture (Note 4). The yellow slurry is then cooled to 0 °C and stirred at this temperature for 2 hrs. The mixture is filtered and the yellow solid washed with cooled water (4 °C, 2 x 20 mL) (Note 5) and dichloromethane (3 x 20 mL) to remove sodium chloride and unreacted 4-acetamido-TEMPO, respectively. The bright yellow salt is compressed with a spatula to remove additional solvent, transferred into a 100-mL round-bottomed flask and dried under high vacuum at room temperature overnight (Note 6). The product is obtained as a bright yellow solid (27.4 g, 78%, 98% purity) (Note 7).

B. The oxidations of a number of alcohols with 4-acetylamino-2,2,6,6-tetramethylpiperidine-1-oxoammonium perchlorate have been described on a 10 mmol scale.[2] The oxoammonium perchlorate salt detonated when a 9 g sample was dried at 40 °C under high vacuum, after being apparently stable for several years.[3] Thus, attention has been shifted to development of the oxoammonium tetrafluoroborate salt. In all cases investigated, the oxidative properties of the two salts are identical.

Oxidation of geraniol to geranial. Into a 500-mL one-necked, round-bottomed flask are added geraniol (7.70 g, 50.0 mmol) (Note 8) and dichloromethane (400 mL). 4-Acetylamino-2,2,6,6-tetramethylpiperidine-1-oxoammonium tetrafluoroborate (16.5 g, 55.0 mmole, 1.10 eq) and 2 g of silica gel are added to the solution (Note 9). The flask is equipped with a mechanical stirrer and the reaction mixture is vigorously stirred at room temperature for 6 h. The reaction begins immediately, and the yellow slurry turns gradually to the off-white color of the reduced oxidant. A 1-cm thick pad of silica gel is placed in a fritted glass funnel (*ca.* 7 cm in diameter), wetted with dichloromethane and covered with a piece of filter paper. The slurry is carefully poured onto the silica gel pad and filtered (Note 10). The pad is washed with four successive 50 mL portions of dichloromethane (Note 11). The solvent is evaporated under vacuum (room temperature) to give 7.08 g (93%) of geranial (98% purity by GC analysis) as a colorless oil (Note 12). The hydroxylamine-silica gel precipitate can be processed to obtain 4-acetamido-TEMPO for recycling (Note 13).

2. Notes

1. No professional safety check has been carried out to determine the stability of 4-acetylamino-2,2,6,6-tetramethylpiperidine-1-oxoammonium tetrafluoroborate salt. All the procedures herein described have been carried out behind a safety shield.

2. 4-Acetamido-TEMPO was purchased from TCI and used without further purification; however, submitters report that it can readily be prepared from 4-amino-2,2,6,6-tetramethylpiperidine (Fluka). In their original procedure for the preparation of 4-acetamido-TEMPO,[2] potassium carbonate was used to basify 4-acetylamino-2,2,6,6-tetramethylpiperidinium acetate; however, due to its insolubility the potassium tetrafluoroborate can accumulate in the salt. Therefore, sodium carbonate should be used and all potassium salts should be avoided. Sodium tetrafluoroborate is quite water-soluble.

3. Tetrafluoroboric acid (48% aqueous solution) was purchased from ACROS.

4. Slow addition is necessary since the tetrafluoroborate oxoammonium salt reacts with excess bleach to produce undesired byproducts.

5. The oxoammonium tetrafluoroborate salt is fairly soluble in pure water (6 g/100 mL at 0 °C, 8 g/100 mL at 20 °C and >100 g/100 mL at 100 °C), so care must be taken during the washes.

6. The oxoammonium tetrafluoroborate salt reacts, albeit slowly, with hot water. If the wet salt is dried in an oven, some decomposition will take place.

7. The melting point of the oxoammonium tetrafluoroborate salt is a vigorous, instantaneous decomposition in a capillary tube: mp 184-185 °C (dec.). The submitters report that the observed melting point can vary from 180 °C to 194 °C, even though the purity is about the same. Since the melting points are ambiguous, the purity of the salt was measured by oxidation of a known quantity of 1-decanol with a known, limited amount of salt. The conversion of 1-decanol to 1-decanal was quantified using ^{1}H NMR (300 MHz, $CDCl_3$). The C-2 proton NMR signal (triplet) at 2.43 ppm (decanal) and the C-1 proton signal (triplet) at 3.65 ppm (1-decanol) are integrated and the percentage of conversion is compared with the theoretical value. In a 25-mL round-bottomed flask 1-decanol (0.079 g, 0.50 mmol), dichloromethane (7.0 mL), silica gel (0.10 g) and 4-acetylamino-2,2,6,6-tetramethylpiperidine-1-oxoammonium tetrafluoroborate (0.075 g, 0.25

mmol, 0.50 eq) were added and the resulting slurry left under stirring at room temperature for 12 h. The mixture was filtered through a 1-cm silica gel pad and subsequently washed with dichloromethane (25 mL). The solvent was removed under vacuum at room temperature to give a mixture of 1-decanol and decanal as a colorless oil.

8. Geraniol (99% purity) was purchased from ACROS and used without further purification.

9. The amount of silica gel and the reaction time depend on the alcohol being oxidized. Geraniol, an allylic alcohol, requires less silica gel and takes less time. Primary aliphatic alcohols will require twice as much silica gel and longer reaction times.[2]

10. If the slurry is filtered without the pad of clean silica gel, a small amount of nitroxide will contaminate the product; however, some product loss occurs due to its incomplete removal from the silica gel.

11. The submitters report that this solution can be used for various reactions, without actual isolation of the carbonyl compound. Reactions reported include Wittig, Grignard, and Baylis-Hillman reactions, as well as cyanohydrin and acetal formation.

12. The crude aldehyde can be distilled at 20 mm and 115-120 °C (lit.[4] 117 °C at 20 mm) to give 6.53 g (89 %) of geranial with spectral and chromatographic properties identical with the undistilled product: colorless oil, IR (KBr) [cm^{-1}]: 2967, 2918, 2836, 1676, 1633, 1444, 1381, 1194, 1121; EI-MS: *m/z* (%) = 153 (7), 152 (24), 137 (19), 123 (19), 109 (29), 94 (39), 84 (51), 69 (100); ^{1}H NMR (300 MHz, $CDCl_3$, TMS) δ 1.62, (s, 3H), 1.69 (s, 3H), 2.18 (s, 3H), 2.21-2.28 (m, 4H), 5.05-5.10 (m, 1H), 5.87-5.91 (m, 1H), 10.00 (d, J = 8.0 Hz, 1H); ^{13}C NMR (75 MHz, $CDCl_3$) δ 17.4, 17.6, 25.5, 25.6, 40.5, 122.5, 127.3, 132.7, 163.5, 191.0. GC-Analysis was performed with a EC-1 column of 30 m /0.32 mm ID and a T-ramp of 70 °C to 240 °C in 15 °C/min, providing a retention time for geraniol of 6.94 min.

13. The submitters report that the spent oxidant consisting of a mixture of hydroxylamine tetrafluoroborate and silica gel can be collected from several oxidations and processed as follows to give 4-acetamido-TEMPO for recycling: The mixture is washed with several portions of warm water to extract hydroxylamine tetrafluoroborate salt from the silica gel. The aqueous solution is basified with $NaHCO_3$ to pH 7 and treated with an excess of H_2O_2 (30% aqueous solution). The 4-acetamido-TEMPO can be extracted with dichloromethane. Recrystallization from boiling water or boiling ethyl acetate provides purified 4-acetamido-TEMPO, which tends to be quite soluble at the boiling point of either solvent and precipitates quickly

upon cooling. The resulting 4-acctamido-TEMPO should melt above 144 °C to be judged suitable for further use.

Safety and Waste Disposal Information

All hazardous materials should be handled and disposed of in accordance with "Prudent Practices in the Laboratory"; National Academy Press; Washington, DC, 1995. If all of the materials are recycled as described, the only waste products are NaCl and $NaBF_4$.

3. Discussion

Oxoammonium salts are stable, non heavy-metal, specific oxidizing agents for the preparation of aldehydes or ketones from alcohols.[5] Under normal conditions,[2] the reactions are nearly quantitative; the reactions are colorimetric; and product isolation is simple. Anhydrous conditions may be used, but are not necessary.

Oxoammonium salt oxidations are actually the stoichiometric version of nitroxide-catalyzed oxidations using a secondary oxidant such as bleach, as described in a previous *Organic Syntheses* procedure.[6] Disadvantages of the catalyzed oxidations are that a two phase system is generally used and that the secondary oxidant can cause undesired reactions. Catalytic reactions are, however, more suitable for large-scale reactions. On the other hand, stoichiometric oxidations undergo fewer side reactions, but are more appropriate on a 1-50 mmol scale.

Other than convenience, two major advantages of oxoammonium salt reactions are that phenolic alcohols can be oxidized without protecting the phenol[2,7] and allylic alcohols can be oxidized without isomerization of double bonds.[2, 8, 9]

Many oxoammonium salts are known with various substitution patterns and different anions.[5] The properties vary tremendously, being especially important with respect to their hygroscopic nature and their rates of reaction. 4-Acetylamino-2,2,6,6-tetramethylpiperidine-1-oxoammonium perchlorate and tetrafluoroborate salts were described in our preliminary publication,[2] and the perchlorate was used exclusively for the oxidations. However, the perchlorate proved to be unstable and its use was discontinued.[3] All further work has been done with the tetrafluoroborate oxoammonium salt, which is a bright yellow crystalline material, completely non-hygroscopic and stable indefinitely. Its solubility and oxidation properties appear to be identical with those of the perchlorate.

The tetrafluoroborate salt possesses sufficient solubility in dichloromethane to react easily, and its reduced form, 4-acetylamino-2,2,6,6-N-hydroxypiperidinium tetrafluoroborate is colorless and essentially insoluble in this solvent. Thus, it can be removed by filtration, leaving a quantitative yield of aldehyde or ketone in solution.

Reaction rates vary, allylic and benzylic alcohols being fast (1-3 hr), acetylenic and secondary alcohols next (4-6 hr) and primary aliphatic alcohols being slowest (2-3 days without silica gel).[2] Silica gel is an effective catalyst and may also assist in the purification of the products. When silica gel is used, primary aliphatic alcohols are oxidized in 10-15 hr and allylic and benzylic alcohols require a lower catalyst loading. Completion of the oxidation can be easily monitored colorimetrically, since the yellow tetrafluoroborate oxoammonium salt starting material is converted to its white reduced form.

The most serious side reaction is the rapid oxidation of amines, a reaction that is not well understood.[2,5] Two slow competing reactions, the oxidation of benzyloxy groups[10] and the addition to activated double bonds,[2,11] can result in side product formation, particularly when slowly oxidized substrates (primary aliphatic alcohols) are used. Several other side reactions have been documented for oxoammonium salts,[5] but most are slow and of lesser importance, especially when the anion is tetrafluoroborate.

For reasons that are not understood, alcohols with a b-oxygen (in any functional group) or b-nitrogen (as amide) react so slowly with the oxidant as to be useless. Interestingly, this limitation is apparently not true for nitroxide-catalyzed reactions.[6]

Oxidations in dichloromethane are slightly acidic, due to the reduced oxidant, which has a pKa of about 5.6.[2] For this reason, simple acetal groups and such acid labile protecting groups as the *tert*-butyldimethylsiloxy group are slowly cleaved, although the *tert*-butyldiphenylsiloxy group seems to be stable.

Stoichiometric oxidations can be carried out in the presence of a base such as pyridine, although only one paper has appeared on the subject.[12] The reaction takes place in a different manner with two equivalents of oxidant and pyridine being required. The products are the desired oxidation product, pyridine tetrafluoroborate, and nitroxide. Under these conditions, acetal groups are stable, and b-oxygenated materials are oxidized.

1. Department of Chemistry, University of Connecticut, Storrs, CT 06269-3060.
2. Bobbitt, J. M. *J. Org. Chem.* 1998, *63*, 9367.
3. Bobbitt, J. M. *Chem. & Eng. News*, July 19, 1999, *77*, 6.
4. Cardillo, G.; Orena, M.; Sandri, S. *Synthesis*, 1976, 394.
5. For reviews, see (a) Bobbitt, J. M.; Flores, M. C. L. Heterocycles 1988, *27*, 509. (b) de Nooy, A. E. J.; Besemer, A. C.; van Bekkum, H. Synthesis 1996, 1153. (c) Merbouh, N.; Bobbitt, J. M.; and Brückner, C. *Org. Prep. Proced. Int.* 2004, *36,* 1.
6. Anelli, P. L.; Montanari, F.; Quici, S. *Org. Synth.* 1990, *69*, 212.
7. Abad, A.; Agullo, C.; Cunat, A. C.; Perni, R. H. *Tetrahedron: Asymmetry*, 2000, *11*, 1607.
8. Koch, T.; Hoskovec, M.; Boland, W. *Tetrahedron* 2002, *58*, 3271.
9. Hoye, T. R.; Hu, M. *J. Am. Chem. Soc.* 2003, *125*, 9576.
10. Miyazawa, T.; Endo T. *Tetrahedron Lett.* 1986, *27*, 3395.
11. Takata, T.; Tsujino, Y.; Nakanishi, S.; Nakamura, K.; Yoshida, E.; Endo, T. *Chem. Lett.* 1999, 937.
12. Merbouh, N.; Bobbitt, J. M.; Brückner C. *Tetrahedron Lett.* 2001, *42*, 8793.

(*R*)-(+)-3,4-DIMETHYLCYCLOHEX-2-EN-1-ONE
((*R*)-(+)-3,4-Dimethyl-2-cyclohexen-1-one)

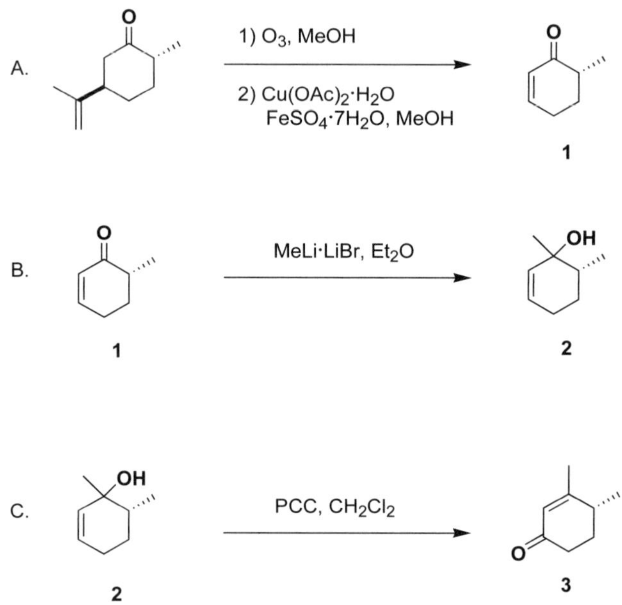

Submitted by James D. White, Uwe M. Grether, and Chang-Sun Lee.[1]
Checked by Rick L. Danheiser and Charnsak Thongsornkleeb.

1. Procedure

A. (R)-(+)-6-Methylcyclohex-2-en-1-one (**1**). A 500-mL, three-necked, round-bottomed flask is equipped with a mechanical stirrer, thermometer, and a disposable pipette attached via tygon tubing to an ozone generator (Note 1). The flask is charged with (2*R*,5*R*)-(+)-*trans*-dihydrocarvone (10.4 g, 68 mmol) (Note 2) and 200 mL of methanol (Note 3). The solution is cooled to –30 °C in a dry ice-acetone bath and ozone is passed through the solution until TLC analysis shows that no dihydrocarvone remains (Note 4). The reaction mixture is flushed with argon for 7 min and then allowed to warm to –20 °C. Copper(II) acetate monohydrate (27.2 g,136 mmol) (Note 5) is added in one portion. The resulting suspension is stirred for 15 min, and ferrous sulfate heptahydrate (22.7 g, 82 mmol) (Note 5) is added in 1 g portions over a 20 min period. The dark green suspension is stirred at –20 °C for 7 hr and then at ambient temperature for 3 hr. Water (200 mL) is added and the resulting mixture is extracted with six 125-mL portions of diethyl ether. The combined organic layers are washed with 100 mL of saturated aq sodium hydrogen carbonate (Note 6) and 50 mL of brine, and dried over anhydrous magnesium sulfate. The solvent is removed by rotary evaporation at ambient temperature (50 mmHg) and the residue is purified by column chromatography on 350 g of silica (elution with 6:1 to 4:1 pentane/diethyl ether) (Notes 7, 8) to give 4.26 g (57%) of (*R*)-(+)-6-methylcyclohex-2-en-1-one as a colorless oil (Notes 9, 10).

B. (6R)-(+)-1,6-Dimethylcyclohex-2-en-1-ol (**2**). A flame-dried, 250-mL, three-necked, round-bottomed flask is equipped with a magnetic stir bar, rubber septum, glass stopper, and an argon inlet. The flask is charged with a solution of (*R*)-(+)-6-methylcyclohex-2-en-1-one (4.16 g, 38 mmol) in 70 mL of diethyl ether (Note 3) and cooled at -78 °C. A solution of methyllithium-lithium bromide complex in diethyl ether (27.4 mL, 1.5M, 41 mmol) (Note 5) is then added via syringe during 20 min. The cooling bath is removed and the resulting solution is stirred for 3 hr at ambient temperature. After 3 hr, the reaction mixture is cooled at 0 °C and 40 mL of water is very slowly added to the yellow solution (Note 11). The aqueous phase is separated and extracted with two 40-mL portions of diethyl ether. The combined organic layers are washed with 40 mL of water, dried over anhydrous magnesium sulfate, and concentrated by rotary evaporation at room temperature (50 mmHg) to afford 4.79 g (100%) of crude (6*R*)-(+)-1,6-dimethylcyclohex-2-en-1-ol (**2**) as a yellow oil which is used in the next step without further purification (Note 12).

C. (R)-(+)-3,4-Dimethylcyclohex-2-en-1-one (**3**). A flame-dried, 250-mL, three-necked, round-bottomed flask is equipped with a magnetic stir bar, rubber septum, glass stopper, and an argon inlet. The flask is charged with pyridinium chlorochromate (16.4 g, 76 mmol) (Note 5) and 75 mL of dichloromethane (Note 3). A solution of crude (6*R*)-(+)-1,6-dimethylcyclohex-2-en-1-ol (4.79 g, 38 mmol)prepared as described above in 25 mL of dichloromethane (Note 3) is transferred into the reaction mixture via cannula over 5 min, and the resulting mixture is stirred at ambient temperature for 3 hr. The reaction mixture is then diluted with 120 mL of diethyl ether (Note 7), the solution is decanted, and the remaining black resinous polymer is thoroughly washed with three 50-mL portions of diethyl ether. The combined dark brown/black ether solution is washed successively with two 100-mL portions of 5% aqueous sodium hydroxide solution, 100 mL of 5% aqueous hydrochloric acid, and two 50-mL portions of saturated aqueous $NaHCO_3$ solution, dried over anhydrous magnesium sulfate, filtered, and concentrated by rotary evaporation at room temperature (50 mmHg) to give 4.42 g of the crude product as a yellow oil. Purification by column chromatography (Note 13) yields 3.71-3.85 g (79-82% overall from **1**) of (*R*)-(+)-3,4-dimethylcyclohex-2-en-1-one (**3**) as a colorless oil (Note 14).[4]

2. Notes

1. The checkers used a Welsbach model T-816 ozone generator and introduced the ozone via a disposable pipette in an open neck of the flask fitted with a reducing adapter. The submitters vented the reaction flask through a trap filled with 40% aqueous sodium bisulfite solution.

2. (+)-Dihydrocarvone was purchased from Aldrich Chemical Company, Inc. as a ca. 80:20 mixture of (2*R*,5*R*)-(+)-*trans*-dihydrocarvone and (2*S*,5*R*)-(+)-*cis*-dihydrocarvone. Prior to use, this material (23 ml, 21 g, 136 mmol) was purified by column chromatography on 450 g of silica (elution with 8:1 to 6:1 hexane/diethyl ether) to afford 13.9 g (91 mmol) of pure (2*R*,5*R*)-(+)-*trans*-dihydrocarvone.

3. The submitters purchased absolute methanol from J. T. Baker, Inc. and used it without further purification. Anhydrous methanol was purchased by the checkers from Mallinckrodt Chemical Company, Inc., and used as received. The submitters freshly distilled dichloromethane from calcium hydride under argon before use and distilled diethyl ether from sodium and benzophenone under argon. Dichloromethane and diethyl ether were purified by the checkers by pressure filtration through activated alumina .

4. TLC analysis is carried out on silica gel (elution with 50% ethyl acetate-hexane, visualization with $KMnO_4$). Dihydrocarvone exhibits R_f = 0.67. The reaction mixture is a pale blue color after ozonolysis is complete.

5. Copper(II) acetate monohydrate (≥ 99.0%) and ferrous sulfate heptahydrate (≥ 99.5%) were purchased from Fluka Chemical Company, Inc., and were used without further purification. Methyllithium, as a complex with lithium bromide, and pyridinium chlorochromate (98%) were purchased from Aldrich Chemical Company, Inc., and were used without further purification.

6. Caution: a large amount of carbon dioxide is liberated!

7. The crude product consists of a mixture of α,β and β,γ enone isomers in a ratio of 74:26. The isomers have R_f values of 0.60 and 0.70, respectively (silica gel TLC, elution with 50% ethyl acetate-hexane and visualization with $KMnO_4$). For purification, the product is charged on a column (24 x 50 cm) of 350 g of silica gel (Sorbent Technologies, 32-63 μm) and eluted with 500 mL of 6:1 pentane-ethyl ether. At that point, fraction collection (30-mL fractions) is begun, and elution is continued with 1500 mL of 6:1 pentane-ethyl ether, 1500 mL of 5:1 pentane-ether, and then 1500 mL of 4:1 pentane-ether. The desired product is obtained in fractions 63-116 and the corresponding β,γ-unsaturated isomer is obtained in fractions 38-55 (17% yield). Pentane, purchased from J.T. Baker, and diethyl ether, purchased from Mallinckrodt Chemical Company, Inc. (certified ACS), were used without further purification.

8. The fractions containing the product were concentrated by careful fractional distillation at room temperature and atmospheric pressure through a 10-inch Vigreux column to a volume of ca. 40-50 mL. This step is necessary to avoid loss of the relatively volatile product. Further concentration was conducted by rotary evaporation at room temperature at 50 mmHg.

9. In other runs the product was obtained in 48-63% yield.

10. (*R*)-(+)-6-Methylcyclohex-2-en-1-one has the following physical properties: ^{1}H NMR (500 MHz, $CDCl_3$): δ 1.15 (d, J = 7 Hz, 3 H), 1.70-1.79 (m, 1 H), 2.05-2.11 (m, 1 H), 2.37-2.45 (m, 3 H), 5.99 (td, J = 10 Hz, 3 Hz, 1 H), 6.92-6.96 (m, 1 H); ^{13}C NMR (125 MHz, $CDCl_3$): δ 15.2, 25.7, 31.0, 41.8, 129.6, 149.9, 202.6; IR (neat) cm^{-1}: 3033, 2964, 2932, 1682, 1389, 1215; $[\alpha]_D^{20}$ + 83 (*c* 1.31, MeOH); + 86 (*c* 1.46, $CHCl_3$); lit[2] + 91 (*c* 1.1 $CHCl_3$); lit[3] + 70 (*c* 3.0, $CHCl_3$)); Anal. Calcd for $C_7H_{10}O$: C, 76.33; H, 9.15. Found: C, 76.13; H, 9.13. Enantiomeric excess of the product was determined by the Submitters to be >99:1 (OD column, 92:8 Hexane/*i*-PrOH, 0.6 mL/min).

11. Water is carefully added dropwise from a disposable pipette. Addition of the first ca. 5 mL is accompanied by violent bubbling.

12. (6*R*)-(+)-1,6-Dimethylcyclohex-2-en-1-ol (**2**) is a pungent liquid. ^{1}H NMR analysis of the crude product indicated that it consists of a 67:33 mixture of diastereomers (the configuration of the isomers was not assigned.)

13. The product is charged on a column (19 x 43 cm) of 150 g of silica gel (Sorbent Technologies, 32-63 μm) and eluted with 300 mL of 6:1 pentane-ether. At that point, fraction collection (30-mL fractions) is begun,

and elution is continued with 1200 mL of 6:1 pentane-ether, 500 mL of 4:1 pentane-ether, 500 mL of 2:1 pentane-ether, and then 700 mL of 1:1 pentane-ether. The desired product is obtained in fractions 38-72 and has R_f = 0.47 (silica gel TLC, elution with 50% ethyl acetate-hexane, visualization with *p*-anisaldehyde).

14. The product has the following physical properties: ^{1}H NMR (500 MHz, $CDCl_3$): δ 1.20 (d, *J* = 7 Hz, 3 H), 1.73-1.80 (m, 1 H), 1.96 (s, 3 H), 2.09-2.15 (m, 1 H), 2.32 (ddd, *J* = 5, 8, 17 Hz, 1 H), 2.38-2.50 (m, 2 H), 5.83 (app s, 1 H); ^{13}C NMR (125 MHz, $CDCl_3$): δ 17.9, 22.9, 30.5, 34.5, 34.7, 126.5, 166.8, 199.8; IR (neat) cm^{-1}: 3028, 2966, 2877, 1671, 1626, 1378, 1255; $[\alpha]_D^{20}$ + 108 (*c* 1.28, $CHCl_3$; lit[4] + 111 (*c* 1.06)). The purity of the product was determined to be 95% by GC analysis. The enantiomeric excess of the product was determined by the submitters to be 100% (OD column, 95:5 hexane/*i*-PrOH, 0.6 mL/min; OJ Column, 97:3 Hexane/*i*-PrOH, 0.6 mL/min).

Safety and Waste Disposal Information

All hazardous materials should be handled and disposed of in accordance with "Prudent Practices in the Laboratory"; National Academy Press; Washington, DC, 1995.

3. Discussion

The method described for the preparation of (*R*)-(+)-6-methylcyclohex-2-en-1-one (**1**) was first reported by Schreiber[5] and was improved by Solladié and Hutt.[3] An alternative approach to (*R*)-(+)-6-methylcyclohex-2-en-1-one (**1**) commences from cyclohex-2-en-1-one using Enders' methodology.[2,6] This route delivers **1** via a four-step sequence and in an overall yield of 46%, including two steps for attachment and removal of the chiral auxiliary, (*S*)-1-amino-2-methoxymethylpyrrolidine (SAMP).[6] Since (+)-dihydrocarvone is inexpensive and the *trans*- and *cis*- isomers are conveniently separated by either column chromatography on silica[7] or on 100 g scale by fractional crystallization of the corresponding oximes,[8] the route to **1** from (+)-dihydrocarvone is superior for the large scale preparation of (*R*)-(+)-6-methylcyclohex-2-en-1-one (**1**).

The transformation of (*R*)-(+)-6-methylcyclohex-2-en-1-one (**1**) to (*R*)-(+)-3,4-dimethylcyclohex-2-en-1-one (**3**) via (6*R*)-(+)-1,6-dimethylcyclohex-2-en-1-ol (**2**) was first reported by Tokoroyama *et al*[2] and is based upon a procedure by Dauben and Michno.[9] The optical purity of **3** derived from the Tokoroyama synthesis ($[\alpha]_D^{20}$ + 111 (c 1.06, $CHCl_3$))[2] is comparable to that of the present route starting from (+)-dihydrocarvone ($[\alpha]_D^{20}$ + 104 (c 1.04, $CHCl_3$)). Alternatively, **3** can be prepared via a five-step sequence starting from comparatively expensive (*R*)-(+)-pulegone.[4,10-12]

(*R*)-(+)-3,4-Dimethylcyclohex-2-en-1-one (**3**) is a valuable starting material for the asymmetric synthesis of a variety of natural products and their analogues, many of which possess important properties. Examples include irones, e.g. (2*R*,6*S*)-(+)-*cis*-γ-irone (**4**),[7] which are constituents of essential oils that are highly prized ingredients of certain perfumes; the epiaflavinine derivative 3-demethylaflavinine (**5**);[12] the sponge metabolite (+)-agelasimine A (**6**);[8] the clerodane diterpenoid (-)-methyl kolavenate (**7**);[2] and the antiviral antibiotic (-)-ascochlorin (**8**).[4]

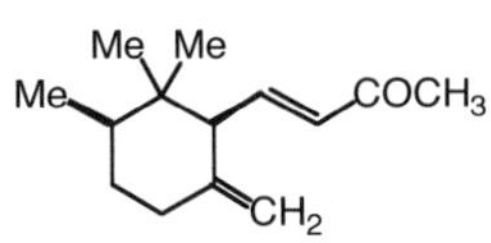

(2*R*,6*S*)-(+)-cis-γ-Irone (**4**)

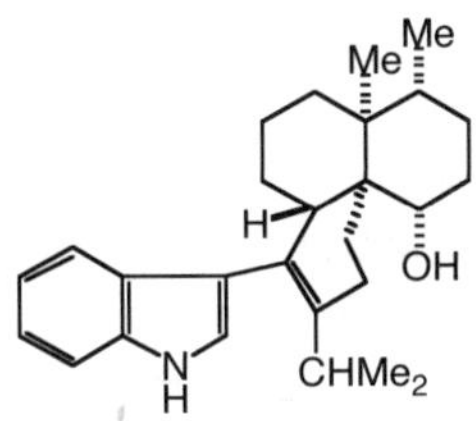

3-Demethylaflavinine (**5**)

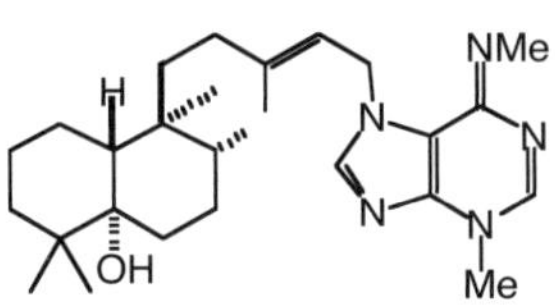

(+)-Agelasimine A (**6**)

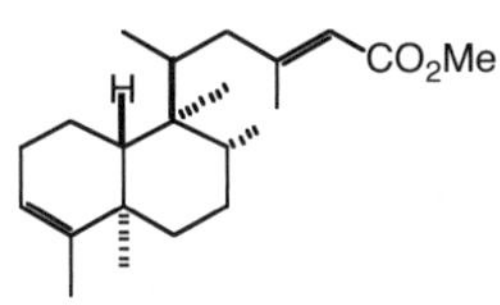

(-)-Methyl Kolavenate (**7**)

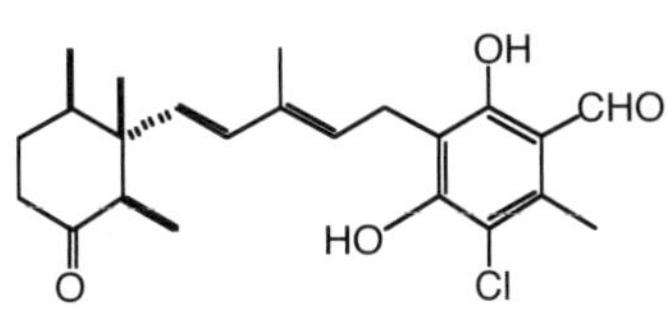

(-)-Ascochlorin (**8**)

1. Department of Chemistry, Oregon State University, Corvallis, Oregon 97331-4003.
2. Iio, H.; Monden, M.; Okada, K.; Tokoroyama, T. *J. Chem. Soc., Chem. Commun.* **1987**, 358-359.

3. Solladié, G.; Hutt, J. *J. Org. Chem.* **1987**, *52*, 3560-3566.
4. Mori, K.; Takechi, S. *Tetrahedron* **1985**, *41*, 3049-3062.
5. Schreiber, S. L. *J. Am. Chem. Soc.* **1980**, *102,* 6163-6165.
6. Enders, D.; Eichenauer, H. *Chem. Ber.* **1979**, *112,* 2933-2960.
7. Laval, G.; Audran, G.; Galano, J. -M.; Monti, H. *J. Org. Chem.* **2000**, *65*, 3551-3554.
8. Ohba, M.; lizuka, K.; Ishibashi, H.; Fujii, T. *Tetrahedron* **1997**, *53*, 16977-16986.
9. Dauben, W. G.; Michno, D. M. *J. Org. Chem.* **1977**, *42,* 682-685.
10. Silvestri, M. G. *J. Org. Chem.* **1983**, *48,* 2419-2420.
11. Danishefsky, S.; Harrison, P.; Silvestri, M.; Segmuller, B. *J. Org. Chem.* **1984**, *49,* 1319-1321.
12. Danishefsky, S.; Chackalamannil, S.; Harrison, P.; Silvestri, M.; Cole, P. *J. Am. Chem. Soc.* **1985**, *107,* 2474-2484.

OXIDATION OF PRIMARY ALCOHOLS TO CARBOXYLIC ACIDS WITH SODIUM CHLORITE CATALYZED BY TEMPO AND BLEACH: 4-METHOXYPHENYLACETIC ACID

(Benzeneacetic acid, 4-methoxy-)

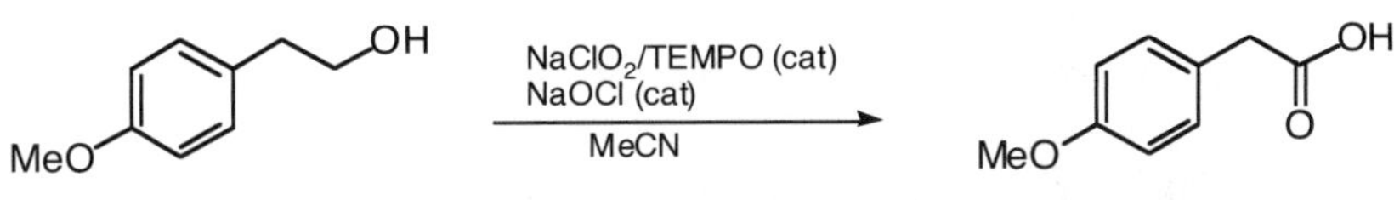

1f **2f**

Submitted by Matthew M. Zhao,[1] Jing Li,[1] Eiichi Mano,[2] Zhiguo J. Song,[1] and David M. Tschaen.[1]
Checked by Arun Ghosh, Jamie Sieser, Weiling Cai, and Sarah E. Kelly.

1. Procedure

A. 1-L, three-necked, round-bottomed flask equipped with a mechanical stirrer and two 100-mL addition funnels is charged with 4-methoxyphenethyl alcohol (**1f**) (6.09 g, 40 mmol), TEMPO (2,2,6,6-tetramethyl-1-piperidinyloxy free radical) (0.436 g, 2.8 mmol), 200 mL of acetonitrile, and 150 mL of 0.67M sodium phosphate buffer (pH 6.7) (Notes 1,2). A solution of sodium chlorite is prepared by dissolving 80% $NaClO_2$ (9.14 g, 80.0 mmol) in 40 mL of water and a solution of dilute sodium hypochlorite (NaOCl) is prepared by diluting household bleach (5.25% NaOCl, 1.06 mL, ca. 2.0 mol%) with 19 mL of water. The reaction mixture is heated to 35 °C with stirring and approximately 20% of the $NaClO_2$ solution is added via one addition funnel followed by 20% of the dilute bleach solution via the other funnel. The remaining portions of both reagents are then added simultaneously over 2 h (Note 3). The resulting mixture is stirred at 35 °C until the reaction is complete (Note 4) (usually 6-10 h). After cooling the reaction mixture to 25 °C, 300 mL of water is added and the pH is adjusted to 8.0 by addition of ca. 48 mL of 2.0N NaOH. The reaction mixture is then poured into ice-cold sodium sulfite solution (12.2 g in 200 mL of water) maintained below 20 °C with an ice-water bath (Note 5). After stirring for 0.5 h at rt, 200 mL of methyl t-butyl ether (MTBE) is added and the resulting mixture is stirred for 15 min. The organic layer is separated and discarded. More MTBE (100 mL) is added and the rapidly stirred mixture is acidified with 2.0N HCl to pH 3-4. The organic layer is separated and the aqueous layer is extracted with two 100-mL portions of MTBE. The combined organic phases are washed with two 100-mL portions of water, 100 mL of brine, and then concentrated to give 6.16-6.22 g (93-94%) of 4-methoxyphenylacetic acid (**2f**) as a colorless solid (Note 6).

2. Notes

1. All reactants and reagents were obtained from Sigma-Aldrich and used without purification.
2. The sodium phosphate buffer consisted of a 1:1 mixture of 0.67M NaH_2PO_4 and 0.67M Na_2HPO_4 (pH = 6.5 at 22 °C). For substrates not prone to chlorination, lower pH (3-4) can be employed to speed up the reaction.
3. *Caution*: It is not advisable to mix sodium chlorite solution and bleach prior to the addition since the mixture appears to be unstable. On a small scale, it is acceptable to mix the substrate, sodium chlorite, TEMPO, acetonitrile, and buffer first and then add bleach in one portion. The submitters report that the appearance of a greenish off-gas appeared in some instances. The use of a nitrogen or carbon dioxide sweep to carry the off-gas to an aqueous Na_2SO_3 trap can be used in these cases.
4. The progress of the reaction is monitored by HPLC: YMC ODS-AM column, 4.6x250mm, flow rate 1.00mL/min, 30 °C, linear gradient 20-80% MeCN/15 min, 0.1% H_3PO_4; retention time **1f**: 9.09 min, **2f**: 9.25 min.
5. The pH of the aqueous layer should be 8.5 – 9.0.
6. The product exhibits the following properties: mp 86-88.5 °C (lit.[9] 86 °C); ^{1}H NMR (400 MHz, $CDCl_3$) δ: 3.59 (s, 2 H), 3.79 (s, 3 H), 6.88 (d, J = 8.5, 2 H), 7.19 (d, J = 8.5, 2 H); ^{13}C NMR (100 MHz, $CDCl_3$) δ: 178.4, 159.1, 130.7, 125.5, 114.3, 55.5, 40.4; HPLC purity >99 area %.

Waste Disposal Information

All toxic material were disposed of in accordance with "Prudent Practices in the Laboratory"; National Academic Press; Washington, DC, 1996.

3. Discussion

The procedure described here provides an efficient and environmentally benign method for oxidizing primary alcohols to carboxylic acids using stoichiometric $NaClO_2$, catalytic TEMPO, and NaOCl. Compared with the previously reported TEMPO/NaOCl/CH_2Cl_2 protocol,[5c] this new procedure gives significantly improved yields and purity of the desired product by reducing the chlorination of the aromatic groups. No racemization or epimerization is observed for substrates with labile chiral centers. Additionally, no chlorinated solvent is required. Similar to TEMPO/NaOCl, this procedure is not applicable to alkenic alcohols and substrates with exceedingly electron-rich aromatic groups.

The development of this procedure stems from our recent work involving the oxidation of primary alcohol **1** to the carboxylic acid **2**.[3]

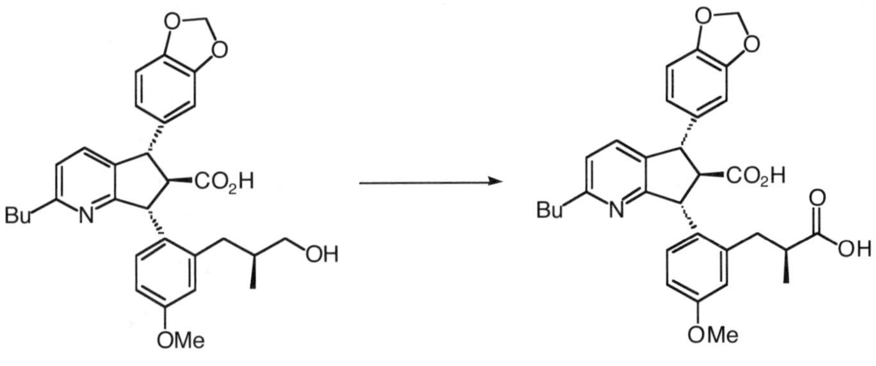

1 **2**

The submitters found that the $RuCl_3/H_5IO_6$ protocol[4] gave low yields (<50%) due to the destruction of the electron-rich aromatic rings. TEMPO catalyzed oxidation[5] with NaOCl also gave unsatisfactory results because of the chlorination of the aromatic groups. Other oxidants, such as H_2O_2, $MeCO_3H$, t-BuO_2H, etc., were examined without much success. In the submitters' recent report[7] on a CrO_3 (1 mol%) catalyzed oxidation, although it is much improved over the Jones oxidation in terms of environmental impact, the heavy metal issue is not completely eliminated. Investigation of the TEMPO/NaOCl protocol using sodium chlorite ($NaClO_2$)[8] as the oxidant resulted in the development of this new procedure.[3b] The reaction appeared to be very slow (~2%/h) initially. Subsequently, it was found that NaOCl dramatically accelerated the reaction. The conversion reached >50% in 1 h and went to completion in 2-4 h.

The remarkable acceleration can be explained by the proposed catalytic cycle shown in Scheme 1. TEMPO radical is first oxidized by NaOCl to the *N*-oxoammonium ion[5] **A**, which rapidly oxidizes the primary alcohol (**1**) to the aldehyde (**3**) and gives a molecule of the hydroxylamine **D**.[5] The aldehyde (**3**) is then oxidized by $NaClO_2$ to the carboxylic acid (**2**)[6] and regenerates a molecule of NaOCl. The hydroxylamine **D** can either be directly oxidized to **A** or undergo a syn proportionation with a molecule of **A** to give two molecules of TEMPO radical which can be oxidized back to **A**[5a,b] Although the exact mechanism of TEMPO-catalyzed oxidation of alcohols is still unclear, previous work[5] has shown that *N*-oxoammonium ion **A** and hydroxylamine **D** are involved. It is also known that $NaClO_2$ can readily oxidize aldehydes to the carboxylic acids in the absence of TEMPO.[6] The long induction period of the reaction without a catalytic amount of bleach is likely due to the relatively slow oxidation of TEMPO radical or the hydroxylamine **D** by $NaClO_2$. Once the reaction is initiated, it becomes self-sustaining as NaOCl is continuously regenerated. The chlorination problem is greatly reduced because the concentration of NaOCl remained low throughout the reaction. Risks for epimerization of the neighboring chiral center are also reduced since the labile aldehyde intermediate is rapidly oxidized to the carboxylic acid by sodium chlorite.

Scheme 1. Catalytic Cycle for the TEMPO/NaOCl Catalyzed Oxidation

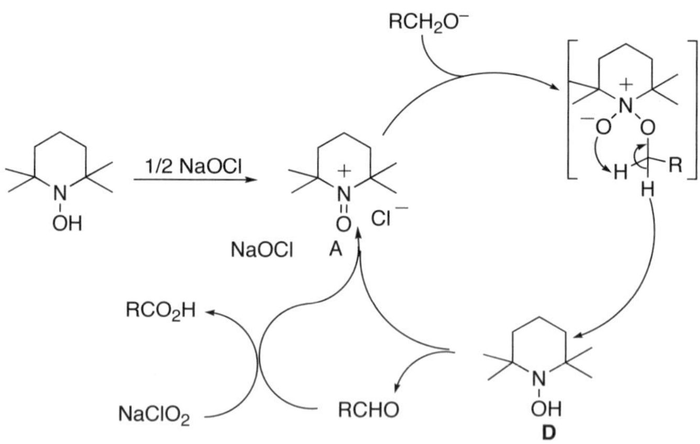

This method is mild and efficient and has been demonstrated on a variety of primary alcohols (Table 1). For comparison purposes, substrates with electron-rich aromatic rings were also subjected to the TEMPO/NaOCl oxidation[5c] (entries 3, 5-7, 10). In all of these cases, much improved yields were obtained with this new procedure. The most striking example is **1g**. The yield of the desired product **2g** was only 42% using stoichiometric NaOCl in contrast to the quantitative yield obtained using our procedure (entry 7). One of the major side products in the oxidation with stoichiometric NaOCl was isolated and identified as the chlorinated compound **4** (Figure 1) based on NMR studies (NOE). Similarly, 3-phenylpropargyl alcohol (**1h**) was oxidized to the acid **2h** in 90% yield using our procedure vs. < 5% with NaOCl. It appeared that carbon-carbon triple bonds can be tolerated, but substrates with ordinary carbon-carbon double bonds such as cinnamyl alcohol failed to react. This was likely due to quenching of the catalytic NaOCl, which shut down the catalytic cycle. Substrates with very electron-rich aromatic groups such as **1n-1o** (Figure 1) also failed for similar reasons. Surprisingly, oxidation of 4-methoxybenzyl alcohol (**1p**) was very sluggish. Substrate **1i,** which contained a cyclopropyl group, posed no problem (entry 9). Chiral alcohols, including protected amino alcohol **1l**, were oxidized to the corresponding carboxylic acids without racemization (or epimerization) of the labile chiral centers (entries 10-12).

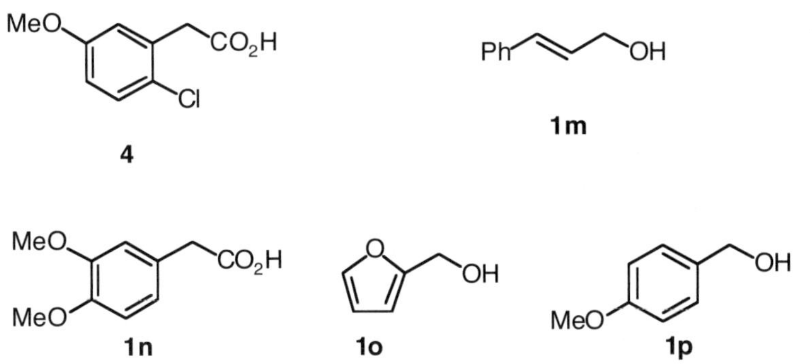

Figure 1

Table 1. **TEMPO** Catalyzed Oxidation of Primary Alcohols

Entry	Substrate	Product	Yield % ($NaClO_2$)	Yield % (NaOCl)
1	 **1a**	$Ph\text{-}CO_2H$ **2a**	98	–
2	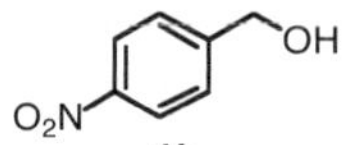 **1b**	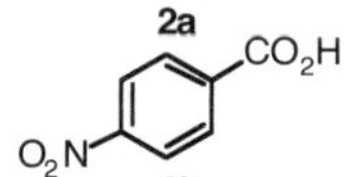 **2b**	100	–
3	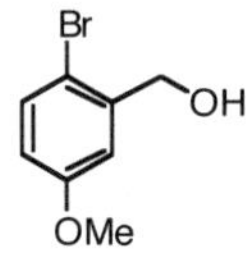 **1c**	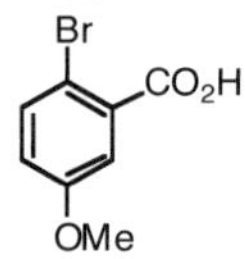 **2c**	96	80
4	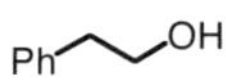 **1d**	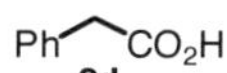 **2d**	100	–
5	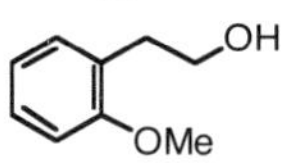 **1e**	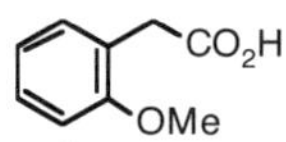 **2e**	99	65
6	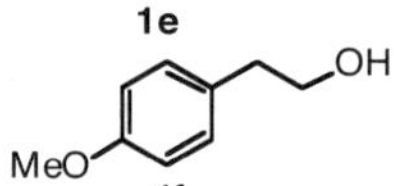 **1f**	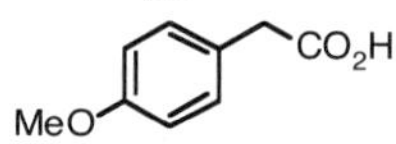 **2f**	100	86
7	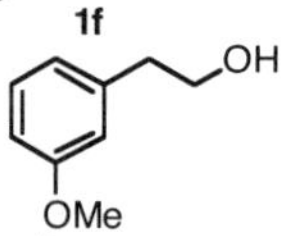 **1g**	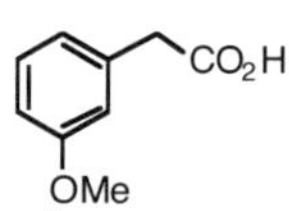 **2g**	96	42
8	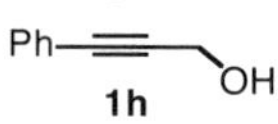 **1h**	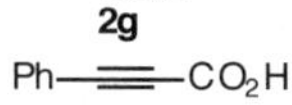 **2h**	90	< 5
9	 **1i**	 **2i**	95	–
10	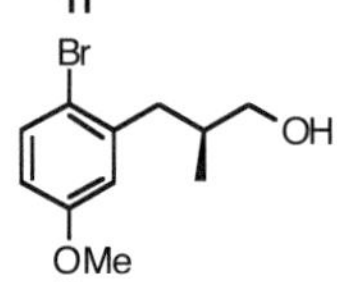 **1j**	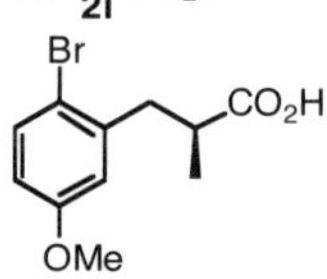 **2j**	92	60
11	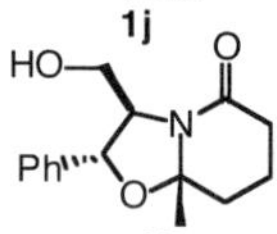 **1k**	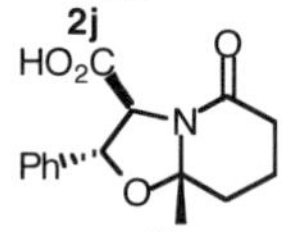 **2k**	95	–
12	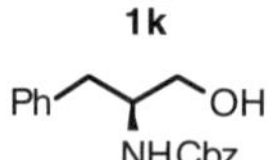 **1l**	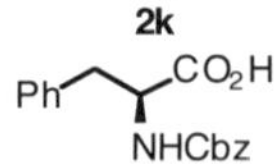 **2l**	85	–

[a] All entries except **3**, **9** and **10** were checked. The reactions took 6-10 h for completion in all cases. Additional charges of both $NaOCl_2$ and NaOCl, however, showed faster rate. Entry **11** was particularly sluggish towards oxidation. Yields and purities of products were generally comparable to those reported in the Table. For substrates not prone to chlorination, lower pH (3-4) can be employed to speed up the reaction.

1. Process Research, Merck Research Laboratories, Merck & Co. Inc., P.O. Box 2000, Rahway, NJ 07065

2. Pharm. R & D, Banyu Pharmaceutical Co. Ltd. Menuma, Saitama, Japan

3. (a) Song, Z. J.; Zhao, M.; Desmond, R.; Devine, P.; Tschaen, D. M.; Tillyer, R.; Frey, L.; Heid, R.; Xu, F.; Foster, B.; Li, J.; Reamer, R.; Volante, R.; Grabowski, E. J. J.; Dolling, U. H.; Reider, P. J.; Okada, S.; Kato, Y.; Mano, E. *J. Org. Chem.* **1999**, *64*, 9658-9667. (b) Zhao, M.; Li, J.; Mano, E.; Song, Z.; Tschaen, D. M.; Grabowski, E. J. J.; Reider, P. J. *J. Org. Chem.* **1999**, *64*, 2564-2566.

4. Carlsen, P. H. J.; Katsuki, T.; Martin V. S.; Sharpless, K. B. *J. Org. Chem.* **1981**, *46*, 3936–3938.

5. (a) Nooy, A. E. J. de; Besemer, A. C.; Bekkum, H. V. *Synthesis* **1996**, 1153–1174. (b) Bobbitt, J. M.; Flores, M. C. L. *Heterocycles* **1988**, *27*, 509-533. (c) Anelli, P. L.; Biffi, C.; Montanari, F.; Quici, S. *J. Org. Chem.* **1987**, *52*, 2559–2562. (d) Miyazawa, T.; Endo, T.; Shiihashi, S.; Okawara, M. *J. Org. Chem.* **1985**, *50*, 1332–1334. (e) Rychnovsky, S. D.; Vaidyanathan, R. *J. Org. Chem.* **1999**, *64*, 310–312.

6. (a) Lindgren, B. O.; Nilsson, T. *Acta Chem. Scand.* **1973**, *27*, 888–890. (b) Dalcanale, E.; Montanari, F. *J. Org. Chem.* **1986**, *51*, 567–569.

7. Zhao, M.; Li, J.; Song, Z.; Desmond, R.; Tschaen, D. M.; Grabowski, E. J. J.; Reider, P. J. *Tetrahedron Lett.* **1998,** *39,* 5323-5326.

8. The use of $NaBrO_2$ has been reported; however, $NaClO_2$ was reported to be unsuccessful. Additionally, $NaBrO_2$ is not readily available commercially. Inokuchi, T; Matsumoto, S.; Nishiyama, T.; Torii, S. *J. Org. Chem.* **1990**, *55*, 462-466.

9. CRC Handbook of Data on Organic Compounds, **1985**, HODOC No:04247.

METHYLTRIOXORHENIUM CATALYZED OXIDATION OF SECONDARY AMINES TO NITRONES: *N*-BENZYLIDENE-BENZYLAMINE *N*-OXIDE

[Benzenemethanamine, (*N*-phenylmethylene)-, *N*-oxide]

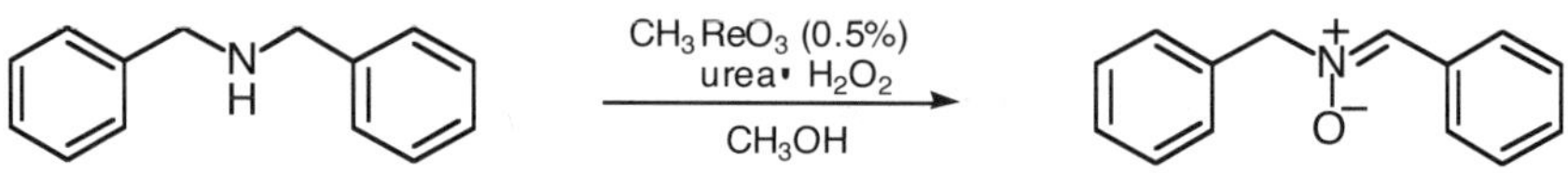

Submitted by Andrea Goti,[1] Francesca Cardona, and Gianluca Soldaini.
Checked by Xiaowu Jiang and Marvin J. Miller.

1. Procedure

A. 250-mL, two-necked, round-bottomed flask equipped with a magnetic stirbar, thermometer, and a reflux condenser fitted with a rubber septum and balloon of argon is charged with a solution of methyltrioxorhenium (MTO) (0.013 g, 0.05 mmol, 0.1% mol equiv) in 100 mL of methanol (Note 1). Urea hydrogen peroxide (UHP) (14.3 g, 152 mmol) is added (Notes 1, 2, 3, 4), the flask is cooled in an ice bath, and dibenzylamine (9.7 mL, 50.7 mmol) is then added dropwise via syringe over 10 min (Notes 1, 5). After completion of the addition, the ice bath is removed and the mixture is stirred at room temperature (Note 6). A white precipitate forms after approximately 5 min (Note 7) and the yellow color disappears within 20 min (Note 8). Another four portions of MTO (0.1% mol equiv, 0.013 g each) are added at 30-min intervals (2.5 hr total reaction time). After each addition, the reaction mixture develops a yellow color, which then disappears; only after the last addition does the mixture remain pale yellow (Note 9). The reaction flask is cooled in an ice bath and solid sodium thiosulfate pentahydrate (12.6 g, 50.7 mmol) is added in portions over 20 min in order to destroy excess hydrogen peroxide (Note 10). The cooled solution is stirred for 1 h further, at which point a KI paper assay indicates that the excess oxidant has been completely consumed. The solution is decanted into a 500-mL flask to remove small amounts of undissolved thiosulfate. The solid is washed with 50 mL of MeOH and the methanol extract is added to the reaction solution which is then concentrated under reduced pressure by rotary evaporation. Dichloromethane (250 mL) is added to the residue and the urea is removed by filtration through cotton and celite. Concentration of the filtrate affords 10.3 g (97%) of the nitrone as a yellow solid (Note 11).

If desired, the nitrone can be recrystallized from diisopropyl ether (200 mL) (Notes 12, 13), affording 9.0 g (84%) of a white solid (Note 14). Concentration of the mother liquor and recrystallization of the residue from 40 mL of diisopropyl ether provides 0.600 g of additional nitrone. Total yield: 9.6 g (89%) of the nitrone as a white solid (Note 15).

2. Notes

1. Dibenzylamine, MTO, and UHP were purchased from Aldrich Chemical Company and used as received.

2. *Caution:* Potential hazards are associated with the use of the urea hydrogen peroxide complex (UHP). Due to the relatively high hydrogen peroxide content (ca. 36%) of this reagent, care should be exercised in its use. Additionally, concentrated solutions of H_2O_2 are potentially explosive, especially when contaminants such as metals are present. However, it has been reported that UHP is a safe source of hydrogen peroxide,[2] which makes it a good alternative to anhydrous or highly concentrated aqueous solutions of H_2O_2. UHP has been judged stable by negative impact and pressure-time tests for explosiveness on small samples[2] and the submitters have never experienced problems in its use.

3. Although the oxidation requires 2 mol equiv of oxidant with respect to amine, use of 3 mol equiv of UHP is advised in order to speed up the reaction and to reduce the amount of catalyst required. For example, use of a smaller excess of UHP (2.5 mol equiv) required an increase to 0.7% mol equiv of MTO and a 4 h reaction time in order for the oxidation to proceed to completion.

4. Immediately upon addition of the urea-hydrogen peroxide adduct to the solution containing methyltrioxorhenium, a yellow color develops due to formation of the catalytically active rhenium peroxo complexes.[3]

5. The internal temperature of the solution rises from ca. 4 °C to 12-13 °C after all of the amine has been added.

6. The internal temperature varies from 24 °C to 32 °C during the whole reaction time, with the highest values being reached after each addition of the catalyst.

7. The white precipitate is presumably *N,N*-dibenzylhydroxylamine, an intermediate oxidation product, which disappears after the third catalyst addition.

8. The disappearance of the yellow color indicates a low concentration of the active catalytic species due to loss in efficiency of the cycle.

9. Monitoring of the reaction by TLC on silica gel showed complete consumption of the starting amine (elution with 20:1 CH_2Cl_2/CH_3OH, visualization by UV and by iodine; dibenzylamine: R_f = 0.4, nitrone: R_f = 0.7).

10. The submitters found that on small scale (1 mmol) this treatment is unnecessary. On a larger scale, it is recommended to destroy excess hydrogen peroxide for safety, although oxidations on larger scale without this treatment have been carried out without experiencing any problem. It was also observed that product of higher purity is obtained when hydrogen peroxide is destroyed.

11. The crude nitrone has the following physical data: mp 75-77 °C; Anal. Calcd for $C_{14}H_{13}NO$: C, 79.59; H, 6.20; N, 6.63. Found: C, 79.39; H, 6.18; N, 6.78.

12. *Caution:* Use of diisopropyl ether from a freshly opened bottle is advised, due to possible formation of peroxides in aged ethereal solvents. Other solvents were less effective for this recrystallization.

13. The crude solid is readily dissolved in a smaller amount of solvent, but employing the indicated volumes is recommended in order to achieve better yields of recrystallization. With this volume of solvent, heating at 74 °C is sufficient to have complete solubilization of the solid.

14. The nitrone exhibits the following spectral and physical data: mp (uncorr.) 79-81 °C (lit.[4] 81-83 °C); ^{1}H NMR (300 MHz, $CDCl_3$) δ: 5.09 (s, 2 H), 7.38-7.49 (m, 9 H), 8.17-8.20 (m, 2 H); ^{13}C NMR (75 MHz, $CDCl_3$) δ: 71.29, 128.41, 128.57, 128.93, 129.17, 130.35, 130.51, 133.35, 134.11; IR (KBr) cm^{-1}: 3061, 3030, 2995, 1581, 1497, 1459, 1153; MS (EI, 70 eV) m/z (%): 211 ($M^{+\cdot}$, 27), 195 (7), 181 (16), 91 (100), 83 (78), 66 (70), 65 (52); exact mass (FAB): Calcd $(M+H)^+$ 212.1075, found 212.1064. Anal. Calcd for $C_{14}H_{13}NO$: C, 79.59; H, 6.20; N, 6.63. Found: C, 79.57; H, 5.92; N, 6.51.

15. The submitters report a yield of 86% and the checkers obtained the product in 85-90% yield in several runs.

Waste Disposal Information

All toxic materials were disposed of in accordance with "Prudent Practices in the Laboratory"; National Academy Press; Washington, DC, 1995.

3. Discussion

Nitrones are very useful tools in the construction of structurally complex molecules and particularly nitrogen-containing biologically active compounds, usually allowing a high degree of diastereocontrol. In this context, both the [3+2] cycloaddition of nitrones to alkenes[5,6] and the alkylation of nitrones by organometallic reagents[7] have been extensively developed and have become extremely reliable synthetic procedures. In addition, nitrones are useful spin trap reagents, widely employed in biological systems.[8]

The most widely employed methods for the synthesis of nitrones are the condensation of carbonyl compounds with *N*-hydroxylamines[5] and the oxidation of *N,N*-disubstituted hydroxylamines.[5,9] Practical and reliable methods for the oxidation of more easily available secondary amines have become available only recently.[10-13] These include reactions with stoichiometric oxidants not readily available, such as dimethyldioxirane[10] or *N*-phenylsulfonyl-*C*-phenyloxaziridine,[11] and oxidations with hydrogen peroxide catalyzed by Na_2WO_4[4,12] or SeO_2.[13] All these methods suffer from limitations in scope and substrate tolerance. For example, oxidations with dimethyldioxirane seem to be limited to arylmethanamines and the above mentioned catalytic oxidations have been reported (and we have experienced as well) to give somewhat erratic and unreproducible results in some cases. Particularly, functionalized pyrrolidines and piperidines behaved as capricious substrates in those protocols. For example, a chiral optically active pyrrolidine underwent racemization to variable degrees upon oxidation to the corresponding nitrone with H_2O_2 catalyzed by Na_2WO_4[14] and oxidation with dimethyldioxirane of a chiral 2-morpholinone gave scarcely reproducible results.[15] In these cases, the described oxidation method using methyltrioxorhenium/urea·H_2O_2 complex was demonstrated to be a better alternative, furnishing consistently good and reproducible yields (see Table, entries 9 and 11, respectively), without loss of enantiomeric purity, also on multi-gram scale. The examples reported in the Table are illustrative of the scope of the oxidation method described here. Most of the reactions reported in the Table have been carried out on a 0.5-1 mmol scale and are unoptimized, with a 2% mol equiv of MTO being employed. The procedure described above has been scaled-up and optimized, also in terms of minimal catalyst use. The amount of catalyst to be used in order to achieve complete conversion is not greatly influenced by the reaction scale.

Indeed, essentially the same results as those reported above have been obtained on a 10 mmol scale with 0.4% mol equiv of MTO and in a further scale-up of the process on 0.1 mol of dibenzylamine with a minimal increase to 0.6% mol equiv of MTO. On a 10-12 mmol scale, 0.7% mol equiv of MTO were necessary for oxidizing completely dibutylamine and 1% mol equiv was sufficient to oxidize the more sterically congested *N*-(*tert*-butyl)-*N*-benzylhydroxylamine.

Due to its broad scope, as well as to its favorable features (commercial availability of the catalyst, use of a "green" oxidant, economy, extremely simple procedure and work-up), this method has been rapidly accepted, as demonstrated by its use by several different research groups, despite its recent disclosure.[16,17] This procedure employing CH_3ReO_3 and UHP appears to be the method of choice when optically active nitrones are prepared by oxidation of the corresponding amines.[14,15,18]

Table

Entry	Amine	Nitrone	Yield (%)
1[a]	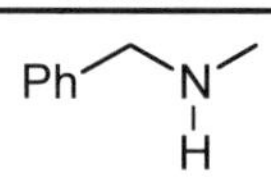	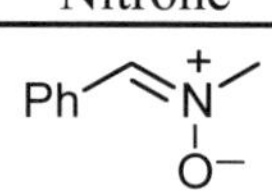	70
2[a]	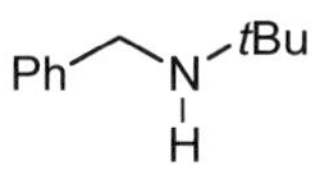	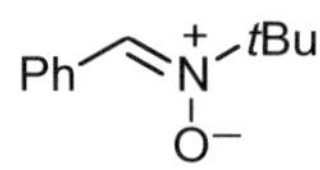	86
3[a]	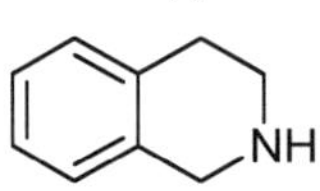	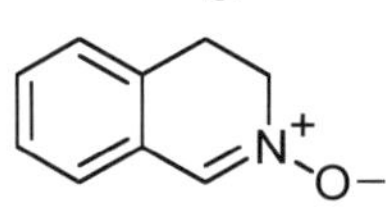	82
4[a]	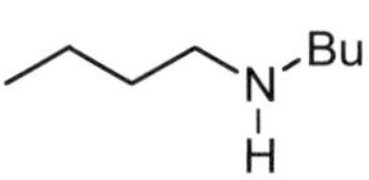	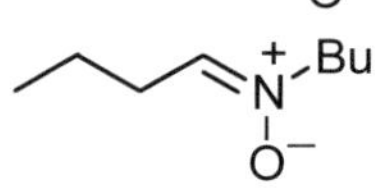	60
5[a]	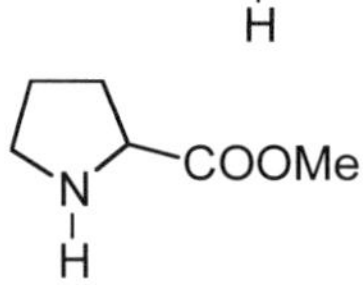	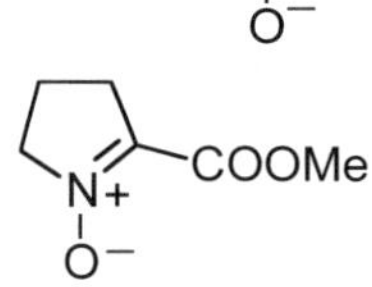	65
6[a]	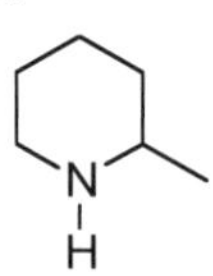	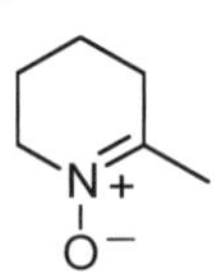	60[b]

Entry	Amine	Nitrone	Yield
7[a,c]	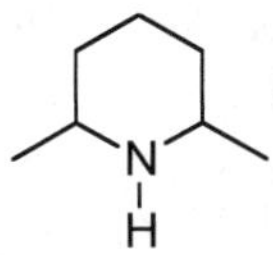	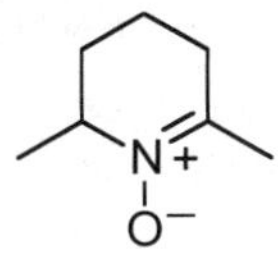	91
8[d]	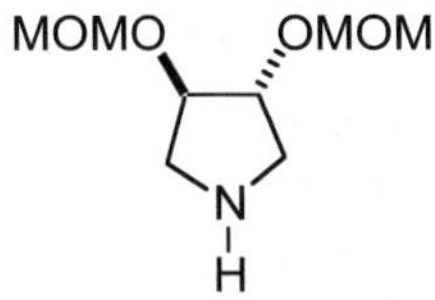	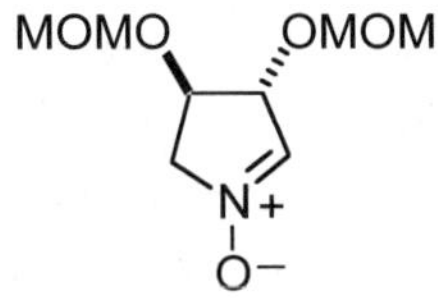	80
9[e]	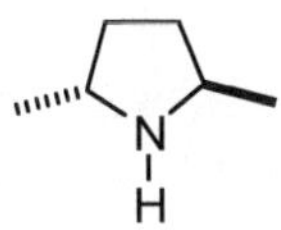	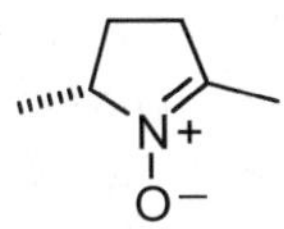	51[f]
10[g]	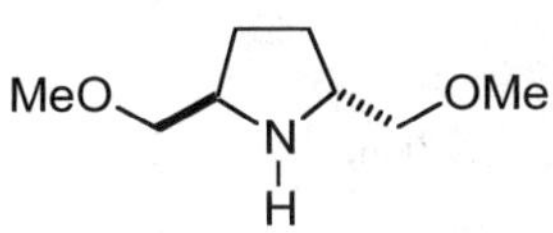	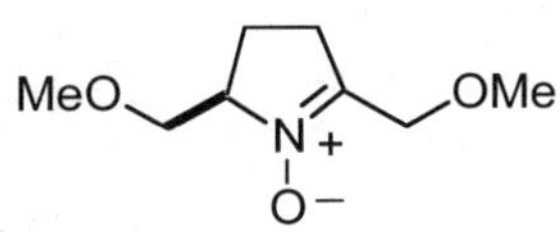	94[h]
11[i]	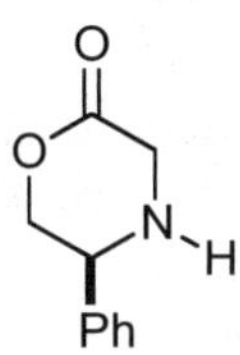	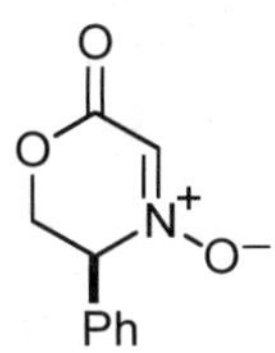	70-80[l]
12[m]	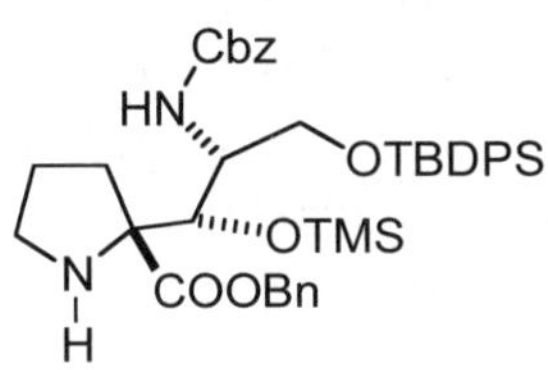	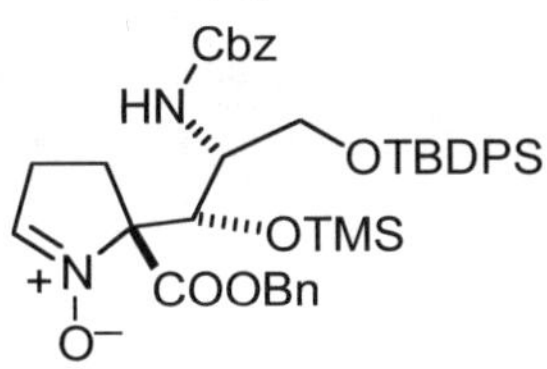	85

[a]Reference 16.
[b]About 10% of the regioisomeric aldonitrone was observed in the ^{1}H NMR spectrum of the crude mixture.
[c]Reference 18a.
[d]Reference 17a.
[e]Reference 14.
[f]Calculated over two steps starting from the *N*-benzylamine derivative.
[g]Reference 18b.
[h]Yield of crude product, used without further purification.
[i]Reference 15.
[l]On multigram scale.
[m]Reference 18c.

1. Dipartimento di Chimica Organica “Ugo Schiff”, Polo Scientifico, Università degli Studi di Firenze, via della Lastruccia 13, I-50019 Sesto Fiorentino, Firenze, Italy.
2. Cooper, M. S.; Heaney, H.; Newbold, A. J.; Sanderson, W. R. *Synlett* **1990**, 533.

3. (a) Herrmann, W. A.; Fischer, R. W.; Scherer, W.; Rauch, M. U. *Angew. Chem. Int. Ed. Engl.* **1993**, *32*, 1157; (b) Espenson, J. H. *Chem. Commun.* **1999**, 479.
4. Murahashi, S.-I.; Mitsui, H.; Shiota, T.; Tsuda, T.; Watanabe, S. *J. Org. Chem.* **1990**, *55,* 1736.
5. (a) Tufariello, J. J. In *1,3-Dipolar Cycloaddition Chemistry*; Padwa, A.; Ed.; John Wiley & Sons: New York, 1984; (b) Confalone, P. N.; Huie, E. M. *Org. React.* **1988**, *36*, 1; (c) Torssell, K. B. G. *Nitrile Oxides, Nitrones, and Nitronates in Organic Synthesis*; Feuer, H., Ed.; VCH Publishers: New York, 1988; (d) Frederickson, M. *Tetrahedron* **1997**, *53*, 403.
6. For a recent use of nitrones in a different [3+2] cycloaddition, see: Hanessian, S.; Bayrakdarian, M. *Tetrahedron Lett.* **2002**, *43*, 967.
7. (a) Bloch, R. *Chem. Rev.* **1998**, *98*, 1407; (b) Enders, D.; Reinhold, U. *Tetrahedron: Asymmetry* **1997**, *8*, 1895; (c) Lombardo, M.; Trombini, C. *Synthesis* **2000**, 759; (d) Merino, P.; Franco, S.; Merchan, F. L.; Tejero, T. *Synlett* **2000**, 442.
8. (a) Janzen, E. G. In *Free Radicals in Biology*; Pryor, W. A., Ed.; Academic Press: New York, 1980; vol. 4, pp. 115-154; (b) Rosen, G. M.; Finkelstein, E. *Adv. Free Rad. Biol. Med.* **1985**, *1*, 345-375; (c) Janzen, E. G.; Haire, D. L. In *Advances in Free Radical Chemistry*; Tanner, D. D., Ed.; Jai Press: Greenwich, CT, 1990.
9. (a) Cicchi, S.; Corsi, M.; Goti, A. *J. Org. Chem.* **1999**, *64,* 7243; (b) Cicchi, S.; Marradi, M.; Goti, A.; Brandi, A. *Tetrahedron Lett.* **2001**, *42*, 6503.
10. Murray, R. W.; Singh, M. *J. Org. Chem.* **1990**, *55,* 2954.
11. Zajac, W. W., Jr.; Walters, T. R.; Darcy, M. G. *J. Org. Chem.* **1988**, *53,* 5856.
12. Murahashi, S.-I.; Shiota, T.; Imada, Y. *Org. Synth., Coll. Vol. IX* **1998**. 632.
13. Murahashi, S.-I.; Shiota, T. *Tetrahedron Lett.* **1987**, *28*, 2383.
14. Einhorn, J.; Einhorn, C.; Ratajczak, F.; Gautier-Luneau, I.; Pierre, J.-L. *J. Org. Chem.* **1997**, *62*, 9385.
15. Long, A.; Baldwin, S. W. *Tetrahedron Lett.* **2001**, *42*, 5343.
16. Goti, A.; Nannelli, L. *Tetrahedron Lett.* **1996**, *37*, 6025.
17. Related methods employing H_2O_2 instead of UHP have been published: (a) Murray, R. W.; Iyanar, K.; Chen, J.; Wearing, J. T. *J. Org. Chem.* **1996**, *61*, 8099; (b) Yamazaki, S. *Bull. Chem. Soc. Jpn.* **1997**, *70*, 877.
18. (a) Einhorn, J.; Einhorn, C.; Ratajczak, F.; Durif, A.; Averbuch, M.-T.; Pierre, J.-L. *Tetrahedron Lett.* **1998**, *39*, 2565; (b) Shibata, T.; Uemae, K.; Yamamoto, Y. *Tetrahedron: Asymmetry* **2000**, *11*, 2339; (c) Okue, M.; Kobayashi, H.; Shin-ya, K.; Furihata, K.; Hayakawa, Y.; Seto, H.; Watanabe, H. Kitahara, T. *Tetrahedron Lett.* **2002**, *43*, 857; (d) Revuelta, J.; Cicchi, S.; Goti, A.; Brandi, A. *Synthesis* **2007**, 485.
19. It has been recently discovered that imines also undergo oxidation to nitrones under similar conditions: Soldaini, G.; Cardona, F.; Goti, A. *Org. Lett.* **2007**, *9*, 473.

PHENYLSULFENYLATION OF NONACTIVATED δ-CARBON ATOM BY PHOTOLYSIS OF ALKYL BENZENESULFENATES: PREPARATION OF 2-PHENYLTHIO-5-HEPTANOL
(Heptane, 2-phenylthio-5-hydroxy)

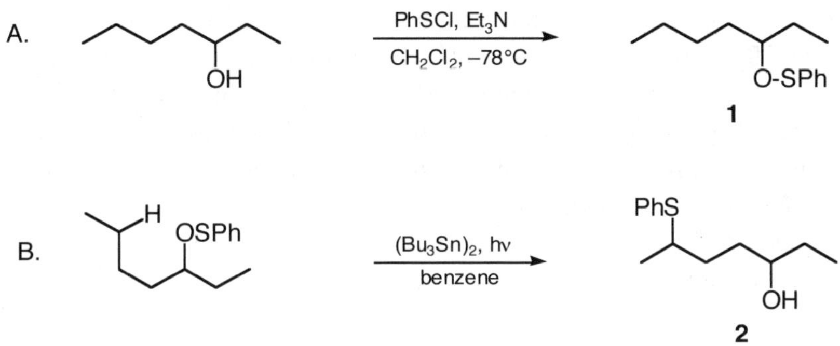

Submitted by Goran Petrovic, Radomir N. Saicic and Zivorad Cekovic.[1]
Checked by Fangzheng Li and Marvin J. Miller.

1. Procedure

A. 3-Heptyl benzenesulfenate. A 1-L, three-necked, round-bottomed flask equipped with a 50-mL pressure-equalizing addition funnel fitted with an argon inlet, an internal thermometer, a drying tube, and a magnetic stir bar is charged with 10.0 g (86.1 mmol) of 3-heptanol (Note 1), 420 mL of anhydrous methylene chloride (Note 2), and 30 mL (22 g, 215 mmol) of freshly distilled triethylamine (Note 3). The flask is flushed with argon and the solution is cooled in a acetone/dry ice bath to an internal temperature of −72 °C. Benzenesulfenyl chloride (10.93 mL, 13.7 g, 94.73 mmol) (Note 4) is added over 20 min to the efficiently stirred solution. The reaction mixture is stirred at −72 °C for 45 min and then protected from light (Note 5) and allowed to warm to room temperature. The resulting mixture is diluted with 200 mL of methylene chloride and then washed successively with 200 mL of deionized water, 200 mL of 1.5M hydrochloric acid, 200 mL of deionized water, 200 mL of saturated aqueous sodium hydrogen carbonate solution, and 200 mL of deionized water. The organic solution is dried over anhydrous sodium sulfate, filtered, and then concentrated under reduced pressure from a flask wrapped with aluminum foil. The residual oil is purified by vacuum distillation (bp 131-132 °C, 3 mmHg) to give 17.2 g (89% yield) of the title compound **1** as a green-yellow oil (Notes 6, 7, 8).

B. 2-Phenylthio-5-heptanol. A photochemical reactor consisting of a tubular pyrex flask, a magnetic stirbar, a water-cooled high pressure mercury lamp, and an argon inlet tube (Note 9) is charged with 12.3 g (54.8 mmol) of 3-heptyl benzenesulfenate, 1.6 g (2.8 mmol) of hexabutylditin (Note 10), and 220 mL of benzene (Note 11). The solution is purged with argon for 10 min to remove oxygen and then irradiated for 1 h with water cooling (Note 12). The solvent is then removed under reduced pressure to afford a pale yellow oil which is purified by dry flash chromatography on silica gel (Note 13). The fractions containing the desired product are combined and concentrated by rotary evaporation under reduced pressure to afford 6.5 g (53%) of 2-phenylthio-5-heptanol as a colorless oil (Note 14).

2. Notes

1. 3-Heptanol was purchased from Fluka Chemika and distilled (bp 155 °C) before use. The checkers purchased 3-heptanol from Aldrich Chemical Company and used it without purification.
2. Methylene chloride was distilled from phosphorus pentoxide and stored over activated molecular sieves (4Å).
3. Triethylamine was purchased from the Merck & Company, Inc. and distilled from calcium hydride. The checkers used triethylamine purchased from Aldrich Chemical Company.
4. Benzenesulfenyl chloride was prepared from the reaction of thiophenol with sulfuryl chloride in the presence of triethylamine in petroleum ether (bp 35-50 °C) according to the procedure of Barrett, A. G. M.; Dhanak, D.; Graboski, G. G.; Taylor, S. J. *Org. Synth. Coll. Vol. VIII*, **1993**, 550. This reagent was stored under argon in a refrigerator and protected from light, under which conditions it can be stored for one month without significant change (G. Zelčans, *Encyclopedia of Reagents for Organic Synthesis,* Ed. L. A. Paquette, Wiley, New York, 1995, Vol. 1, p. 272).
5. Alkyl benzenesulfenates are sensitive to light and their preparation should be carried out in a darkened hood.
6. The crude product is sufficiently pure for use in the next step. The submitters preferred to perform the distillation at a lower pressure (89 °C, 0.3 mmHg) in order to minimize decomposition and obtained the

product in 91% yield. The checkers obtained the product in 84-89% yield in several runs.

7. The preparation of alkyl benzenesulfenates is accompanied by the formation of products of higher oxidation states of sulfur. These products have higher boiling points and are separated by careful distillation of the product under reduced pressure.

8. The product exhibits the following spectroscopic properties: IR (KBr) cm^{-1}: 3061, 2960, 2934, 2874, 2000-1600, 1583, 1477, 1464, 1439, 1113, 1069, 1024, 943, 916, 893, 811, 737, 690; ^1H NMR (300 MHz, $CDCl_3$) δ: 0.88 (t, J = 7.5 Hz, 6H), 1.24-1.31 (m, 4H), 1.56-1.70 (m, 4H), 3.52 (q, J = 5.7 Hz, 1H), 7.16-7.52 (m, 5H); ^{13}C NMR (75 MHz, $CDCl_3$) δ: 9.5, 14.1, 22.9, 26.9, 27.5, 33.5, 89.3, 125.3, 128.9, 141.8.

9. The reaction apparatus used by the submitters for irradiation on a 12-g scale was 18 cm high and 6 cm in diameter and had a 55/50 joint at the top for an immersion well. The apparatus is fitted with an argon inlet tube and water-cooled condenser which is connected to a mineral oil bubbler. The light source was a 125W high pressure Hanovia mercury lamp. The checkers used a tubular pyrex flask 15 cm high and 7.5 cm in diameter. The light source was a 450W high pressure ACE mercury lamp (7825-34 lamp) with filter to provide 250 nm light.

10. Hexabutylditin was purchased from Aldrich Chemical Company, Inc. and used without purification.

11. Anhydrous benzene was purchased from Aldrich Chemical Company and used without purification.

12. The course of the reaction was followed by TLC (silica gel 60 A, 2.5 x 7.5 cm plates, elution with 95:5 ethyl acetate/hexane, visualization with 50% sulfuric acid followed by heating) by monitoring the disappearance of benzenesulfenate starting material **1**.

13. A chromatography column of 4.5-cm diameter was charged with 250 g of silica gel (ICN Silica 10/18 60 A). The viscous oily crude product is dissolved in 10 mL of methylene chloride, 10 g of silica gel is added, and the solvent is evaporated. The resulting dry powder is applied on the top of the column which is then successively eluted with petroleum ether, 95:5 petroleum ether/acetone, and 90:10 petroleum ether/acetone.

14. The submitters obtained the product in 61% yield; the checkers isolated the purified product in 53-56% yield in several runs. The product exhibits the following spectroscopic properties: IR (KBr) cm^{-1}: 3400, 3074, 2961, 2930, 2874, 2000-1600, 1584, 1480, 1458, 1439, 1376, 1303, 1263,

1178, 1092, 1068, 1025, 1000, 971, 923, 746, 692; ^{1}H NMR (300 MHz, $CDCl_3$) δ: 0.92 (t, J = 7.5 Hz, 3H), 1.27 (d, J = 6.9 Hz, 3H), 1.40-1.80 (m, 6H), 3.22 (m, 1H), 3.48 (m, 1H), 7.18-7.29 (m, 3H), 7.37-7.40 (m, 2H); ^{13}C NMR (75 MHz, $CDCl_3$) δ: 9.9, 21.5, 21.5, 30.4, 30.4, 32.9, 33.1, 34.3, 34.5, 43.7, 43.8, 73.2, 73.3, 126.9, 128.9, 132.3, 135.7.

Waste Disposal Information

All toxic materials were disposed of in accordance with "Prudent Practices in the Laboratory"; National Academy Press; Washington, DC, 1995.

3. Discussion

The introduction of various functional groups onto a remote nonactivated carbon atom is of great synthetic importance.[2] The regioselective functionalization of a nonactivated δ-carbon atom involves a free radical 1,5-hydrogen transfer from δ-carbon atom to oxygen or nitrogen centered radicals.[3] Transposition of a radical center from oxygen or nitrogen to the remote carbon atom offers possibilities for introduction of different functional groups onto the nonactivated carbon atom or for formation of a new carbon-carbon bond. Diversity of this type of reaction offers possibilities for introduction of different oxygen,[2] nitrogen,[4] halogen,[5] and sulfur[6] functional groups as well as an olefinic bond[7] onto the δ-carbon atom. The present procedure is based on the original results[6] of the author and co-workers.

Formation of δ-phenylthio alcohols from alkyl benzenesulfenates **3** formally represents a simple interchange of the positions of δ-hydrogen and phenylthio group. The sequence of radical reactions is initiated by tributyltin radical and is based on its thiophilic addition to the sulfur of an alkyl benzenesulfenate.[6] The formed alkoxyl radical **4** upon intramolecular 1,5-hydrogen migration gives the δ-carbon radical **5**. In the absence of any other reactive species the carbon radical undergoes homolytic substitution at the sulfur atom in the alkyl benzenesulfenate **3** to give the δ-phenylthio alcohol **6** and to generate the alkoxyl radical **4** as a transfer radical which continues the chain (see Scheme 1).

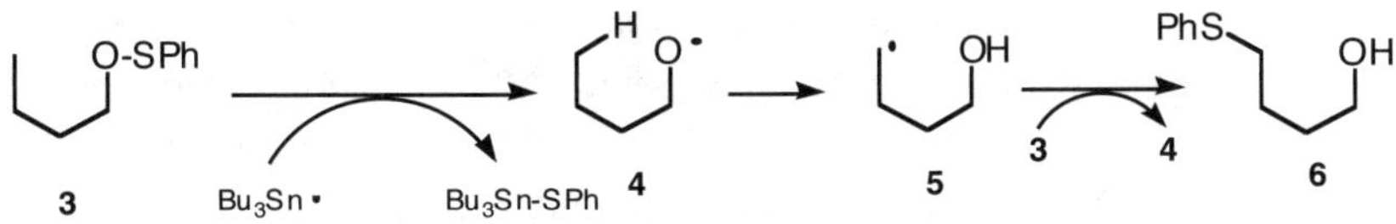

Scheme 1

δ-Phenylsulfenylation is also conceivable in a tin-free variant; however, when alkyl benzenesulfenates were irradiated in the absence of hexabutylditin, the reaction proceeded at a lower rate and gave 10-15% lower yields of δ-phenylthio alcohols.[6]

This method for introduction of the thioether functional group tolerates the presence of a broad range of functional groups, such as alkene, ester, carbonyl, and cyano groups.

Introduction of the phenylthio group onto the δ-carbon atom of alcohols can have valuable synthetic applications. δ-Phenylthio alcohols can be oxidized to the corresponding δ-sulfoxides and sulfones (with their versatile reactivities) or they can be deprotonated by strong base converting the δ-carbon atom to a nucleophilic species. Conversion of δ-phenylthio alcohols to the corresponding δ-carbonyl compounds can be achieved via halogenation followed by subsequent hydrolysis. In this way an inversion of the reactivity of the δ-carbon atom may be accomplished and it can react as an electron acceptor.

The procedure of phenylsulfenylation of δ-carbon atom was applied to a variety of other substrates as summarized in Table 1.

1. Faculty of Chemistry, University of Belgrade, Studentski trg 16, PO Box 158, 11000 Belgrade, Serbia&Montenegro

2. (a) Heusler, K; Kalvoda, J. *Angew. Chem., Int. Ed. Engl.* **1964**, *3,* 525; (b) Mihailovic, M. Lj.; Cekovic, Z. *Synthesis,* **1970**, 209; (c) Barton, D. H. R., *Pure and Appl. Chem.* **1968**, *16,* 1; (d) Feray, N.; Kuznetsov, P.; Renaud, P., in *Radicals in Organic Synthesis,* Eds. Renaud, P. and Sibi, M. P., Vol. 2. Ch. 3.6. p. 246, Wiley-VCH, Weinheim, 2001; (e) Cekovic, Z.,*Tetrahedron* **2003**, *59,* 8073

3. Majetich, G.; Wheless, K. *Tetrahedron* **1995**, *51,* 7095.

4. (a) Barton, D. H. R.; Parakh, S. I. *Half a Century of Free Radical Chemistry,* Cambridge University Press, Cambridge, 1993; (b) Barton, D. H. R.; Beaton, J. M.; Geller, L. E.; Pechet, M. M. *J. Am. Chem. Soc.* **1960**, *82*, 2640; (c) Barton, D. H. R.; Beaton, J. M.; Geller, L. E.; Pechet, M. M. *J. Am. Chem. Soc.* **1961**, *83*, 4076.

5. Walling, C.; Padwa, A. *J. Am. Chem. Soc.* **1961**, *83*, 2207;

6. Petrovic, G.; Saicic, R. N.; Cekovic, Z. *Tetrahedron Lett.* **1997**, *38*, 7107; (b) Petrovic, G.; Saicic, R. N.; Cekovic, Z. *Tetrahedron* **2003**, *59,* 187; (c) Petrovic, G.; Saicic, R. N.; Cekovic, Z. *Synlett,* **1999**, 635; (d) Pasto, D. J.; Cottard, F. *Tetrahedron Lett.* **1994**, *35*, 4303.

7. (a) Cekovic, Z.; Green, M. M. *J. Am. Chem. Soc.* **1974**, *96*, 3000; (b) Cekovic, Z.; Dimitrijevic, Lj.; Djokic, G.; Srnic, T. *Tetrahedron,* **1979**, *35*, 2021

TABLE 1.
δ-PHENYLTHIO ALCOHOLS PREPARED BY PHOTOLYSIS OF ALKYL BENZENESULFENATES IN THE PRESENCE OF HEXABUTYLDITIN

Alkyl benzenesulfenates	δ-Phenylthio alcohols	Yields (%)
		17
	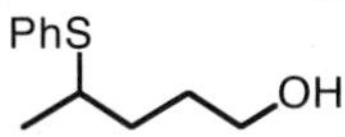	80
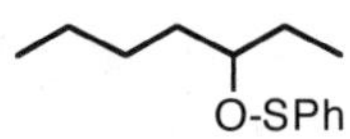	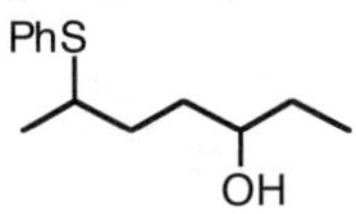	61
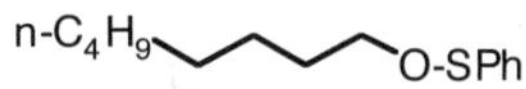	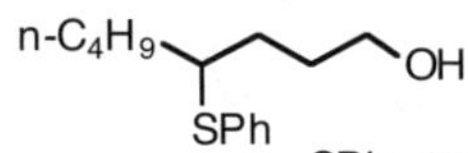	47
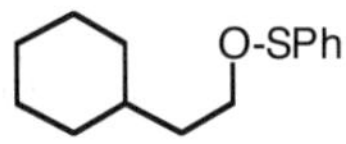	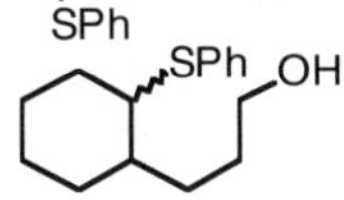	51
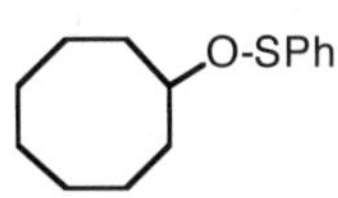	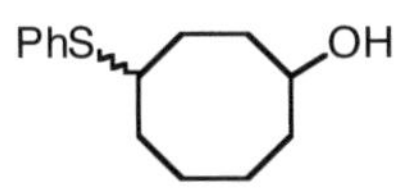	35
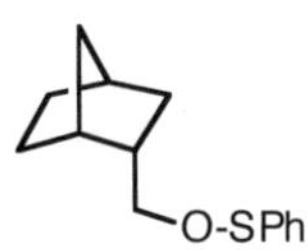	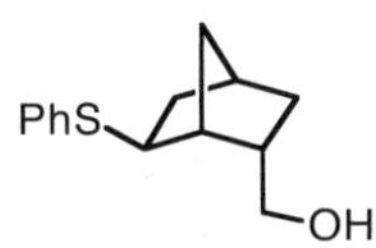	64
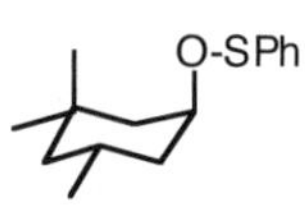	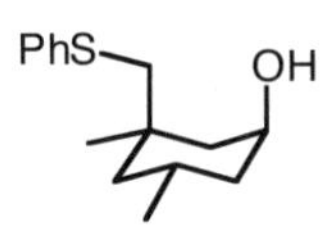	62
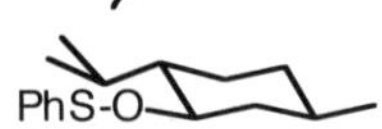	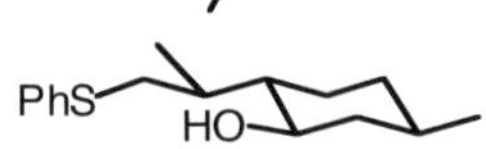	43
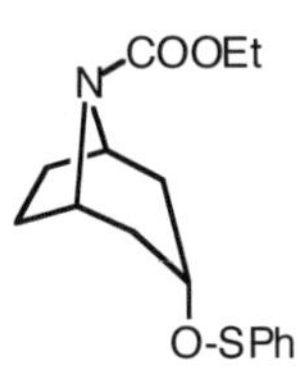	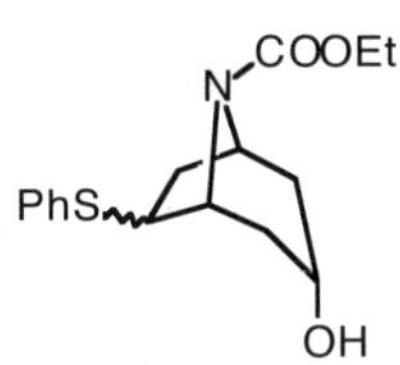	45
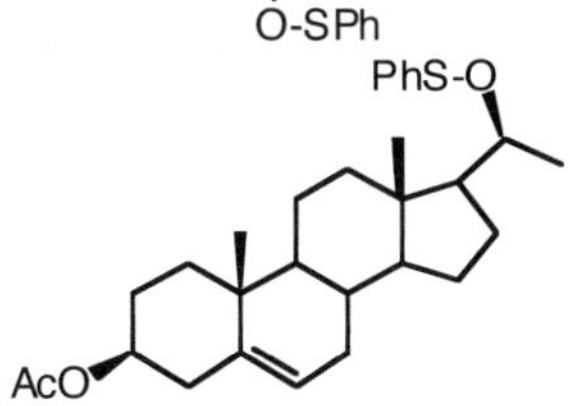	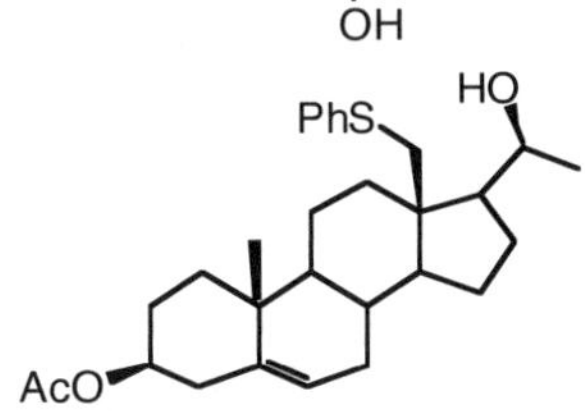	40

(5*S*)-(d-MENTHYLOXY)-2(5*H*)-FURANONE [(2(5*H*)-Furanone, 5[[(1*S*,2*R*,5*S*)-5-methyl-2-(1-methylethyl)cyclohexyl]oxy]-, (5*S*)-)]

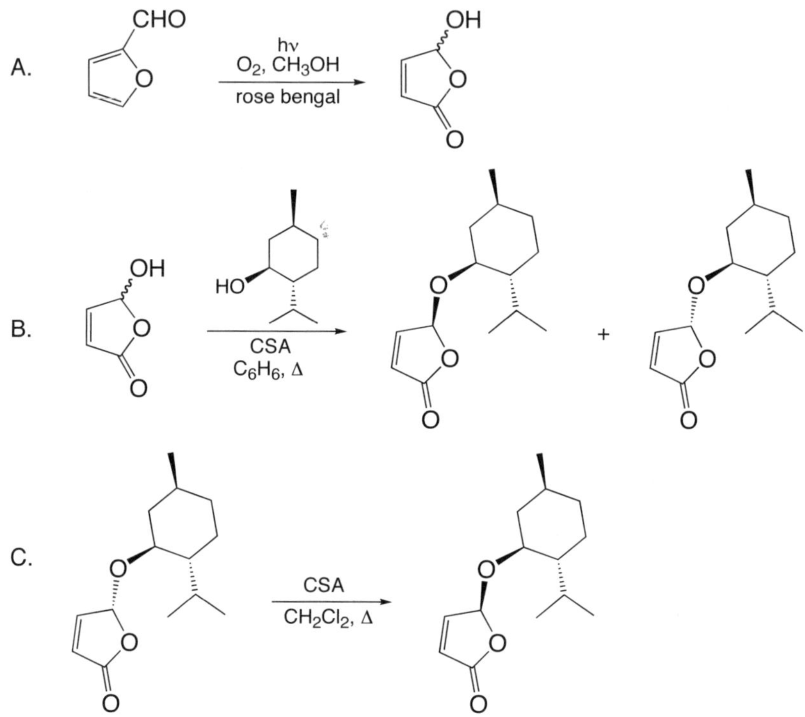

Submitted by Oscar M. Moradei and Leo A. Paquette.[1]
Checked by Christian Peschko and Rick L. Danheiser.

1. Procedure

A. 5-Hydroxy-2(5H)-furanone. A solution consisting of freshly distilled furfural (85.84 g, 0.893 mol) (Note 1) and rose bengal (1.82 g, 1.8 mmol, 0.2% mol/eq) (Note 2) in dry methanol (450 mL) is placed in a 600-mL Pyrex photochemical reactor. The outer vessel, which is fitted at its base with a fritted disc, is also equipped with a thermometer, magnetic stirring bar, and reflux condenser. With oxygen bubbling from the bottom of the reactor (Note 3) and appropriate (water) cooling in the reactor well, the stirred solution is irradiated with a tungsten halogen lamp (Note 4). The reaction temperature is maintained at or below 33 °C at all times. After 18-22 h, when no residual furfural can be detected by thin layer chromatography (TLC) (Note 5), the solution is carefully concentrated at 35-38 °C (bath temperature) (Note 6). Following standing overnight, the orange solid that forms is taken up in 70 mL of cold (−78 °C) chloroform and the resulting crystals are collected by vacuum filtration, washed once with 70 mL of chloroform, and dried under high vacuum to afford 72.70-75.65 g (81-85%) of 5-hydroxy-2(5*H*)-furanone as sticky, yellowish crystals, mp 52 °C (Note 7).

B. (5S)-(d-Menthyloxy)-2(5H)-furanone. A 500-mL, round-bottomed flask equipped with a magnetic stirring bar, 10-mL Dean-Stark trap, and reflux condenser is charged with d-menthol (51.9 g, 0.330 mol), 5-hydroxy-2(5*H*)-furanone (37.3 g, 0.370 mol), D-(+)-camphorsulfonic acid (3.96 g, 0.170 mol), and 190 mL of dry benzene (Note 8). The stirred suspension is heated to reflux under argon with an oil bath preheated to 100 °C (Note 9). After 1-2 h, a total of 5.1 mL of water is collected and no residual menthol is apparent by TLC analysis (Note 10). The reaction mixture is cooled in an ice bath and treated carefully with 100 mL of saturated sodium bicarbonate solution. After completion of the addition, stirring is maintained for 90 min as the mixture is allowed to warm to room temperature. The product is extracted with dichloromethane (300 mL) and the organic phase is washed with three 70-mL portions of brine, dried over anhydrous magnesium sulfate, filtered and concentrated by rotary evaporation to provide 66.7 g (85%) of a 1:1 diastereomeric mixture of the menthyloxybutenolides (Note 11). This mixture is dissolved in 400 mL of hot petroleum ether (bp 35-60 °C) and subsequently cooled to room temperature and eventually to −10 °C (Note 12). Two additional recrystallizations from petroleum ether (200 mL on each occasion) with slow cooling affords 10.7-17.33 g (14-22%) of pure (5*S*)-(d-menthyloxy)-2(5*H*)-furanone, mp 79-80 °C (Note 13).

C. Acid-catalyzed epimerization. The combined mother liquors from Step B are concentrated to dryness, dissolved in 50 mL of benzene, and concentrated again under reduced pressure. The resulting solid is dissolved in 150 mL of dichloromethane, treated with D-(+)-camphorsulfonic acid (2.0 g, 8.6 mmol) and the mixture is heated to reflux under argon with magnetic stirring for 19 h. The resulting solution is cooled to 0 °C and saturated sodium bicarbonate solution (40 mL) is carefully added. The mixture is stirred for 2 h at 20 °C. The organic phase is separated and washed with two 50-mL portions of water, dried over anhydrous magnesium sulfate, filtered, and concentrated by rotary evaporation. The residue is crystallized three times from petroleum ether as described in Step B to give an additional 13.04-17.60 g (17-22%) of furanone. The total yield of pure (5*S*)-(d-menthyloxy)-2(5*H*)-furanone (based on the menthol used in Step B) is 24.75-34.93 g (32-44%) (Note 14).

2. Notes

1. Furfural was purchased from Alfa Aesar and was distilled at reduced pressure (47 mm) from hydroquinone before use.

2. Rose bengal was obtained by the submitters from Acros (certified 80%) and by the checkers from the Aldrich Chemical Company, Inc. (certified 89%) and used as received. Anhydrous methanol was obtained from Malinckrodt and used as received.

3. A gentle flow of oxygen is sufficient. A vigorous stream of the gas results in rapid loss of solvent, followed by an undesirable increase in temperature and decomposition of the product.

4. The submitters used a conventional slide projector bulb (Radiac DYS/SYV/BHC 120V, 600W) and the checkers employed a General Electric DYS 600W/120V bulb. The bulb was welded to lead wires contained within porcelain tubing that set properly within the inner well of the photoreactor. The bulb was attached to a Variac at full power (the checkers used a Variac set at 480W). Importantly, the bulb was additionally cooled with a strong flow of air that was introduced two-thirds of the way down in the well by means of Tygon tubing appropriately fastened to the porcelain.

5. TLC was carried out by the checkers on silica gel plates using 50:1 chloroform:methanol as eluent and $KMnO_4$ for visualization. Under these conditions, the R_fs of furfural and the furanone are 0.6 and 0.0, respectively.

6. It is convenient to divide the reaction mixture into three portions (ca. 170 mL each) and to transfer them into three 1-L, round-bottomed flasks to effect concentration separately. In each case, some chloroform was introduced and concentration was carried out at 35-38 °C. This process was repeated a second time and the residual oils were combined in a 300-mL beaker, this transfer being effected with the aid of a minimum amount of chloroform.

7. The submitters, starting with 266.3 g (2.77 mol) of furfural, obtained the product as colorless crystals, mp 57-59 °C and found that, upon stirring the mother liquors at −78 °C and seeding, an additional 70.3 g (total

yield of 77%) of the butenolide could be obtained. The checkers noted that their product could be decolorized by charcoal treatment (twice) of a chloroform solution heated at reflux to furnish an analytically pure sample of the furanone with mp 54 °C. The product displays the following spectral data: ^{1}H NMR (300 MHz, $CDCl_3$) δ: 5.24 (br s, 1H), 6.23 (dd, J= 1.2, 5.4, 1H), 6.27 (m, 1H), 7.33 (dd, J = 1.2, 5.4, 1H); ^{13}C NMR (75 MHz, $CDCl_3$) δ: 99.4, 124.7, 152.5, 172.0; IR (NaCl) cm^{-1}: 3363, 3114, 2929, 2739, 1792, 1756, 1611, 1583, 1442, 1341, 1283, 1183, 1131, 1083, 996; Anal. Calcd for $C_4H_4O_3$: C, 48.01; H, 4.03. Found: C, 48.11; H, 4.15.

8. The submitters purchased d-menthol from TCI and the checkers obtained this compound (99%, 98% ee) from Aldrich Chemical Company, Inc. Menthol was used as received. D-(+)-camphorsulfonic acid (99%) was purchased from Avocado and used as received. Benzene was purchased from EM Science and used as received.

9. The submitters observed that the two liquid phases that are present when refluxing begins form a homogeneous solution within a few minutes.

10. The theoretical amount of water is 5.9 mL. TLC analysis was conducted by the checkers on silica gel plates using 50:1 chloroform:methanol as eluent and iodine and $KMnO_4$ for visualization. Under these conditions, the R_fs of the product ($KMnO_4$) and menthol (I_2) are 0.59 and 0.46, respectively.

11. This mixture solidifies spontaneously on standing. ^{1}H NMR analysis (in $CDCl_3$ solution) clearly shows the α- (δ 3.53) and β-diastereomers (δ 3.66) to be present in equal amounts.

12. In order to maximize the purity of the product, rapid cooling should be avoided. The submitters allowed the solution to cool slowly to room temperature in an open beaker in the hood, followed by storage in a refrigerator at +5 °C and ultimately at −10 °C.

13. The product exhibits the following spectral data: ^{1}H NMR (300 MHz, $CDCl_3$) δ: 0.79 (d, J= 7.0, 3H), 0.84 (d, J = 7.0 Hz, 3H), 0.94 (d, J = 6.5, 3H), 0.96-1.50 (m, 3H), 1.61-1.72 (m, 2H), 2.08-2.17 (m, 2H), 3.66 (dt, J = 4.5, 10.8, 1H), 6.09 (t, J = 1.2, 1H), 6.20 (dd, J = 1.2, 5.9, 1H), 7.17 (dd, J = 1.2, 5.9, 1H); ^{13}C NMR (75 MHz, $CDCl_3$) δ: 15.9, 21.1, 22.4, 23.3, 25.5, 31.7, 34.4, 40.5, 47.9, 79.3, 100.7, 125.0, 151.1, 171.0; IR (NaCl) cm^{-1}: 2930, 1795, 1752, 1451, 1387, 1347, 1312, 1173, 1132, 1016, 923, 825; Anal. Calcd for $C_{14}H_{22}O_3$: C, 70.56; H, 9.30. Found: C, 70.52: H, 9.16. The submitters found $[\alpha]^{20}_{D}$ +136.9 (EtOH, c 8.2) and the checkers observed $[\alpha]^{20}_{D}$ +135.12 ($CHCl_3$, c 8.2).

14. The submitters obtained 14.2 g of product in Step C. Based on the amount of furanone theoretically present in the mother liquors subjected to Step C, this represents a 35% yield of the desired diastereomer in the epimerization step.

Waste Disposal Information

All toxic materials were disposed of in accordance with "Prudent Practices in the Laboratory"; National Academy Press; Washington, DC, 1995.

3. Discussion

The synthetic utility of (5*S*)- and (5*R*)-menthyloxy-2(5*H*)-furanones has been extensively explored and is well documented. The ready availability of these enantiomerically pure butenolides via the singlet photooxygenation of furfural[2] and their wide range of reactivity contribute to their popularity as chiral building blocks. The present procedure for the preparation of the title compound and the useful epimerization step are modifications of methods originally reported by Feringa and co-workers.[3]

The double bond of these enantiomerically pure butenolides participates readily in asymmetric Diels-Alder reactions,[4,5,6] 1,4-conjugate additions,[7,8,9,10,11,12,13,14,15] [2+2] photochemical reactions,[16,17,18] 1,3-dipolar cycloadditions,[3,19,20] diastereoselective dihydroxylation,[21] and tandem double Michael addition/intramolecular nucleophilic substitution processes.[22,23] As a consequence, they have played a key strategic role in the total syntheses of such structurally varied enantiopure targets as dibenzylbutyrolactone[24,25,26,27] and aryltetralin lignans,[28] podophyllotoxin[29,30,31] and isomers thereof,[32,33] β-lactams,[34] grandisol,[17] terebic acid,[35] isostegane derivatives,[36] and chiral nitronic esters[37] among others.

The current procedure provides material of high enantiomeric purity (>98% ee) starting with inexpensive and easily handled reagents. The two steps proceed in reasonable overall yield and require no chromatographic separation.

1. Evans Chemical Laboratories, The Ohio State University, Columbus, OH 43210.

2. (a) Doerr, I. L.; Willette, R. E. *J. Org. Chem.* **1973**, *38*, 3878; (b) Yuste, F.; Sánchez-Obregón, R. *J. Org. Chem.* **1982**, *47*, 3665.
3. Feringa, B. L.; de Lange, B.; de Jong, J. C. *J. Org. Chem.* **1989**, *54*, 2471.

4. Feringa, B. L.; de Jong, J. C. *J. Org. Chem.* **1988**, *53*, 1125.
5. de Jong, J. C.; Jansen, J. F. G. A.; Feringa, B. L. *Tetrahedron Lett.* **1990**, *31*, 3047.
6. de Jong, J. C.; van Bolhuis, F.; Feringa, B. L. *Tetrahedron: Asymmetry* **1991**, *2*, 1247.
7. Feringa, B. L.; de Lange, B. *Tetrahedron Lett.* **1988**, *29*, 1303.
8. Feringa, B. L.; de Lange, B. *Tetrahedron* **1988**, *44*, 7213.
9. Jansen, J. F. G. A.; Feringa, B. L. *Tetrahedron Lett.* **1989**, *30*, 5481.
10. de Lange, B.; van Bolhuis, F.; Feringa, B. L. *Tetrahedron* **1989**, *45*, 6799.
11. Jansen, J. F. G. A.; Feringa, B. L. *Tetrahedron: Asymmetry* **1990**, *1*, 719.
12. Jansen, J. F. G. A.; Jansen, C.; Feringa, B. L. *Tetrahedron: Asymmetry* **1991**, *2*, 109.
13. Jansen, J. F. G. A.; Feringa, B. L. *Synth. Commun.* **1992**, *22*, 1367.
14. Kang, F.-A.; Yin, H.-Y.; Yin, C.-L. Chin. *Chem. Lett.* **1997**, *8*, 365.
15. Kang, F.-A.; Yu, Z.-Q.; Yin, H.-Y.; Yin, C.-L. *Tetrahedron: Asymmetry* **1997**, *8*, 3591.
16. Hoffmann, N.; Scharf, H.-D.; Runsink J. *Tetrahedron Lett.* **1989**, *30*, 2637.
17. Hoffmann, N.; Scharf, H.-D. *Liebigs Ann. Chem.* **1991**, *12*, 1273.
18. Bertrand, S.; Hoffmann, N.; Pete, J.-P. *Tetrahedron* **1998**, *54*, 4873.
19. de Lange, B.; Feringa, B. L. *Tetrahedron Lett.* **1988**, *29*, 5317.
20. Rispens, M. T.; Keller, E.; de Lange, B.; Zijlstra, R. W. J.; Feringa, B. L. *Tetrahedron: Asymmetry* **1994**, *5*, 607.
21. Sundermann, B.; Scharf, H.-D. *Tetrahedron: Asymmetry* **1996**, *7*, 1995.
22. Huang, H.; Chen, Q. *Tetrahedron: Asymmetry* **1998**, *9*, 4103.
23. Huang, H.; Chen, Q. *Tetrahedron: Asymmetry* **1999**, *10*, 1295.
24. Pelter, A.; Ward, R. S.; Jones, D. M.; Maddocks, P. *Tetrahedron: Asymmetry* **1990**, *1*, 857.
25. Pelter, A.; Ward, R. S.; Jones, D. M.; Maddocks, P. *Tetrahedron: Asymmetry* **1992**, *3*, 239.
26. Pelter, A.; Ward, R. S.; Jones, D. M.; Maddocks, P. *J. Chem. Soc., Perkin Trans. 1* **1993**, 2631.
27. van Oeveren, A.; Jansen, J. F. G. A.; Feringa, B. L. *J. Org. Chem.* **1994**, *59*, 5999.
28. Pelter, A.; Ward, R. S.; Jones, D. M.; Maddocks, P. *J. Chem. Soc., Perkin Trans. 1* **1993**, 2621.
29. van Speybroeck, R.; Guo, H.; Van der Eycken, J.; Vandewalle, M. *Tetrahedron* **1991**, *47*, 4675.
30. Bush, E. J.; Jones, D. W. *J. Chem. Soc., Chem. Commun.* **1993**, 1200.
31. Bush, E. J.; Jones, D. W. *J. Chem. Soc., Perkin Trans. 1* **1996**, 151.
32. Pelter, A.; Ward, R. S.; Storer, N. P. *Tetrahedron* **1994**, *50*, 10829.
33. Pelter, A.; Ward, R. S.; Li, Q.; Pis, J. *Tetrahedron: Asymmetry* **1994**, *5*, 909.

34. Lubben, M.; Feringa, B. L. *Tetrahedron: Asymmetry* **1991**, *2*, 775.

35. Hoffmann, N. *Tetrahedron: Asymmetry* **1994**, *5*, 879.

36. Pelter, A.; Ward, R. S.; Abd-el-Ghani, A. *J. Chem. Soc., Perkin Trans. 1* **1996**, 1353.

37. Kang, F.-A.; Yin, C.-L.; She, S.-W. *J. Org. Chem.* **1996**, *61*, 5523.

MILD AND SELECTIVE OXIDATION OF SULFUR COMPOUNDS IN TRIFLUOROETHANOL: DIPHENYL DISULFIDE AND METHYL PHENYL SULFOXIDE

[(Disulfide, diphenyl- and Benzene, (methylsulfinyl)-)]

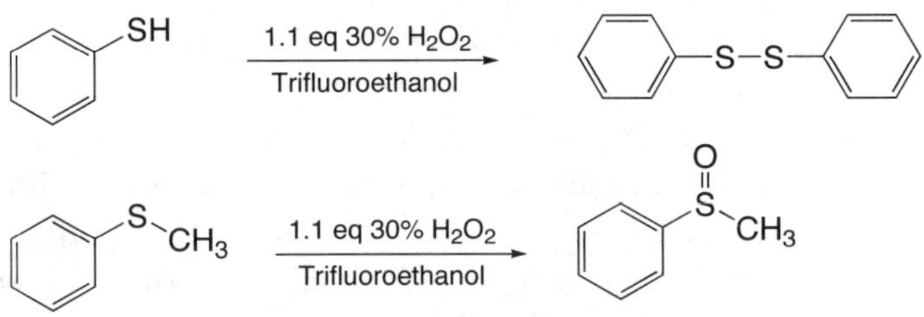

Submitted by Kabayadi S. Ravikumar, Venkitasamy Kesavan, Benoit Crousse, Danièle Bonnet-Delpon, and Jean-Pierre Bégué.[1]
Checked by Jonathan Tripp and Dennis P. Curran.

1. Procedure

A. Diphenyl disulfide. In a 100-mL, round-bottomed flask equipped with a magnetic stirrer are placed 11.0 g (10.25 ml, 0.1 mol) of benzenethiol and 50 mL of trifluoroethanol (Note 1). The mixture is stirred and cooled in an ice bath (Note 2) and 12.5 mL (3.73 g, 0.11 mol) of 30% aqueous hydrogen peroxide (Note 3) is added dropwise over a period of 15 min through an addition funnel. After completion of the addition, the ice bath is removed and the reaction mixture is allowed to stir at room temperature for 24 h. Diphenyl disulfide is sparingly soluble in trifluoroethanol and precipitates out of solution. The solids are collected on a Buchner funnel and dried under vacuum to afford 10.6 g of diphenyl disulfide (97%) (Note 4). Sodium sulfite 2.52 g (0.02 mol) is added to the mother liquor to decompose the excess peroxide and the mixture is heated in a water bath at 50 °C for 30 min. A starch iodide test is negative. The liquid is transferred to a 100-mL, round-bottomed flask, fitted with a distillation unit having a Vigreux column (5 cm). The flask is heated in an oil bath and the solvent is distilled to recover the trifluoroethanol (Note 5).

B. Methyl phenyl sulfoxide. To a 100-mL, round-bottomed flask equipped with a magnetic stirrer are added 12.4 g (0.1 mol) of thioanisole and 50 mL of trifluoroethanol (Note 1). The mixture is stirred and cooled in an ice bath (Note 6) and 12.5 mL (3.73 g, 0.11 mol) of 30% aqueous hydrogen peroxide is added dropwise over 30 min using an addition funnel

(Note 7). After completion of the addition, the ice bath is removed and the reaction mixture is stirred at room temperature for 8 h (Note 8). Sodium sulfite (2.52 g, 0.02 mol) is added to decompose the excess hydrogen peroxide and the mixture is heated in a water bath at 50 °C for 30 min (a starch iodide test is negative). The flask is fitted with a distillation unit having a Vigreux column (5 cm) and heated with an oil bath to remove the solvent by distillation (Note 5). Ether (100 mL) is added to the residue and the mixture is dried over 28 g of magnesium sulfate. Solids are removed by filtration and washed with ether (2 × 50 mL). The combined filtrate and washings are concentrated under vacuum to afford 14.9 g of a slightly yellow oil that crystallizes upon cooling in an ice bath. The crude sulfoxide is transferred to a 25-mL distillation flask with the aid of a small amount of dichloromethane. After removal of solvent, simple short path distillation (Note 9) of the crude product affords 12.8 g (91%) of pure methyl phenyl sulfoxide (Notes 10, 11).

2. Notes

1. Both benzenethiol and phenyl methyl sulfide are sparingly soluble in trifluoroethanol.

2. During the addition of hydrogen peroxide, the temperature of the reaction mixture increases. The submitters report that the addition can be done carefully without an ice bath and that no overoxidation is observed.

3. 30% Aqueous hydrogen peroxide is purchased from Fluka Chemical Company and handled with caution.

4. The product's physical and spectroscopic properties are: mp 52-53 °C; 1H NMR δ: 7.5-7.6 (m, 10 H); MS m/z 218 (M+), 109.

5. Initially trifluoroethanol is obtained (37-38 mL, bp 77-79 °C). Subsequently, a fraction (11-13 mL), containing a mixture of trifluoroethanol and water, is collected at 85-87 °C. The 19F NMR spectrum of this fraction was identical to that of trifluoroethanol. The recovered trifluoroethanol can be reused for the oxidation of sulfides.

6. When the addition of hydrogen peroxide is carried out at room temperature, the submitters reported that the reaction is exothermic and that 3-4% of the sulfone is formed.

7. When 2 equivalents of hydrogen peroxide are used, the submitters report that 0-2% of the sulfone is obtained. When phenyl methyl sulfoxide is oxidized under similar conditions, no sulfone is observed, even after 24 h.

8. The submitters report that oxidation of phenyl methyl sulfide to

phenyl methyl sulfoxide also can be achieved selectively by using hexafluoroisopropyl alcohol (HFIP) as the solvent; the reaction is complete after 10 min.

9. The technique of adding a small amount of activated carbon to the distillation pot afforded a colorless distillate.

10. A fore-fraction is not obtained.

11. The product's physical and spectroscopic properties are: bp 78-79 °C (0.1 mm); 1H NMR (CDCl3) δ: 2.7 (s, 3 H); 7.5-7.7 (m, 5 H); MS m/z 140 (M+), 125, 97, 77.

3. Discussion

Methods of preparation. Diphenyl disulfide can be prepared by the action of hydrogen peroxide in acidic or basic media,[2] iodine-hydrogen iodide,[3] Sm/BiCl3 in water,[4b] FeCl3/NaI,[4c] and by using bromine-aqueous potassium hydrogen carbonate.[4] Methyl phenyl sulfoxide has also been prepared from thioanisole by the action of hydrogen peroxide in methanol,[5] mCPBA,[5b] peroxytrifluoroacetic acid[6] and sodium periodate.[7]

Merits of preparation. Selective oxidative transformation of thiols to disulfides is of interest for both biological[8] and synthetic reasons.[2] Most of the existing methods involve the use of metal catalysts or halogenated reagents, which lead to the formation of side-products.

Organic sulfoxides are useful intermediates for the construction of many chemically and biologically significant molecules.[9] Most of the current methods for the oxidation of sulfides to sulfoxides[10] suffer from over-oxidation to sulfones. Even when controlling the reaction temperature, it is difficult to avoid over-oxidation in most of the reported procedures.
The present procedure is based on the use of fluoroalkyl alcohols as solvents in oxidation reactions. The method is efficient and versatile, and produces disulfides and sulfoxides under mild conditions. These reactions have also been developed using hexafluoroisopropyl alcohol (HFIP) as solvent with a large variety of substrates (alkyl sulfides, alkyl thiols, vinyl sulfides, fluorinated vinyl sulfides, thioglucosides) by Bégué et al.[11,12] Replacement of HFIP with trifluoroethanol, a more common and less expensive solvent, also allows the use of mild conditions and affords high yields of disulfides and sulfoxides without contamination. 30% Aqueous hydrogen peroxide is inexpensive and, since water is the sole byproduct, this method is environmentally friendly.

Cooling is advisable during the addition of hydrogen peroxide, but the remainder of the reaction can be carried out at ambient temperature. This method may be useful for large-scale reactions since trifluoroethanol can be recovered.

1. Laboratoire BIOCIS associé au CNRS, Centre d'Etudes Pharmaceutiques, Rue J.B. Clément, 92296 Chatenay-Malabry, France.

2. Capozzi, G.; Modena, G. In "Chemistry of the Thiol Group"; Patai, S., Ed.; John Wiley & Sons, New York, 1974; Part 2, p. 785.

3. Aida, T.; Akasaka, T.; Furukawa, N.; Oae, S. *Bull. Chem. Soc., Jpn.* **1976**, *49*, 1441.

4. (a) Drabowicz, J.; Mikolajczyk, M. *Synthesis* **1980**, 32; (b) Wu, X.; Rieke, R. D.; Zhu, L. *Synth. Commun.* **1996**, *26*, 191; (c) Iranpoor, N.; Zeynizadeh, B. *Synthesis* **1999**, 1, 49.

5. (a) Drabowicz, J.; Lyzwa, P.; Popielarczyk M.; Mikolajczyk, M. *Synthesis* **1990**, 937; (b) Drabowicz, J.; Kielbasinski, P.; Mikolajczyk, M. In "Syntheses of Sulphones, Sulphoxides and Cyclic Sulphides"; Patai, S.; Rappoport, Z., Eds.; John Wiley and Sons: West Sussex, 1994; p. 109.

6. Venier, C. G.; Squires, T. G.; Chen, Y.-Y.; Hussmann, G. P.; Shei, J. C.; Smith, B. F. *J. Org. Chem.* **1982**, *47*, 3773.

7. Johnson, C. R.; Keiser, J. E. *Org. Synth., Coll. Vol. V* **1973**, 791.

8, Jocelyn, P. C. In "Biochemistry of the SH Group"; Academic Press: London, New York, 1972.

9. (a) Block, E. In "Reactions of Organosulfur Compounds"; Academic Press: New York, 1978; (b) Durst, T. In "Comprehensive Organic Chemistry"; Barton, D.; Ollis, W. D., Eds.; Pergamon: Oxford, 1979; Vol. 3.

10. Drabowicz, J.; Kielbasinski, P.; Mikolajczyk, M. In the "The Chemistry of Sulfones and Sulfoxides; Patai, S.; Rappoport, Z.; Stirling, C. J. M., Eds.; Wiley: New York, 1988; p. 233.

11. (a) Ravikumar, K. S.; Bégué, J.-P.; Bonnet-Delpon, D. *Tetrahedron Lett.* **1998**, *39*, 3141; (b) Ravikumar, K. S.; Zhang, Y. M.; Bégué, J.-P.; Bonnet-Delpon, D. *Eur. J. Org. Chem.* **1998**, 2937; (c) Ravikumar, K. S.; Barbier, F.; Bégué, J.-P.; Bonnet-Delpon, D. *J. Fluorine Chem.* **1999**, *95*, 123.

12. Kesavan, V.; Bégué, J.-P.; Bonnet-Delpon, D. Synthesis 2000, 223-225.

TRANSFORMATION OF PRIMARY AMINES TO *N*-MONOALKYLHYDROXYLAMINES: *N*-HYDROXY-(*S*)-1-PHENYLETHYLAMINE OXALATE

[(Benzenemethanamine, *N*-hydroxy-α-methyl-, (*S*)-, ethanedioate)]

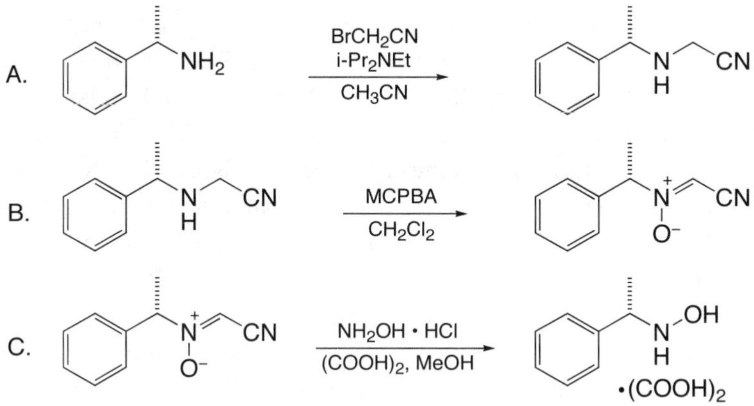

Submitted by Hidetoshi Tokuyama, Takeshi Kuboyama, and Tohru Fukuyama.[1]
Checked by Scott E. Denmark and Jeromy J. Cottell.

1. Procedure

Caution! Hydroxylamines may cause explosions under certain conditions. Careful handling and proper safety equipment are needed especially when they are heated or have not been converted to salts with acids.

A. (S)-[(1-Phenylethyl)amino]acetonitrile. A 1-L, round-bottomed flask equipped with a Teflon-coated magnetic stirrer bar, rubber septum and argon gas inlet is charged with (*S*)-1-phenylethylamine (10.17 g, 83.92 mmol) (Note 1), acetonitrile (MeCN) (170 mL) (Note 2), and diisopropylethylamine (i-Pr_2NEt) (29.2 mL, 168 mmol) (Note 3). After stirring the solution for 5 min, bromoacetonitrile (6.43 mL, 92.3 mmol) (Note 3) is added via syringe over 10 min. The reaction mixture is stirred at ambient temperature until completion of the reaction (Note 4). The mixture is then concentrated on a rotary evaporator to give a white solid, to which is added saturated aqueous sodium bicarbonate ($NaHCO_3$) solution (100 mL); the suspension is extracted with dichloromethane (CH_2Cl_2) (100 mL). The organic phase is washed with brine (100 mL), and the combined aqueous phases are back-extracted with CH_2Cl_2 (3 × 100 mL). The combined organic extracts are dried over anhydrous sodium sulfate, filtered, and concentrated under reduced pressure to give a slightly yellow oil (21.6 g). To the crude oil is added CH_2Cl_2 (10 mL) and the resulting solution is passed through a short column of silica gel (5 cm i. d. × 10 cm, 100 g of SiO_2) (Note 5). A forerun was obtained by elution with 400 mL of 9:1 hexanes:ethyl acetate, followed by collection of the desired fractions with 100 mL of 9:1 hexanes:ethyl acetate, then 700 mL of 7:3 hexanes:ethyl acetate (Note 6). Concentration of the appropriate fractions on a rotary evaporator and then under high vacuum gives (*S*)-[(1-phenylethyl)amino]acetonitrile as a colorless oil (13.62 g, 101%) (Note 7).

B. [(1S)-1-Phenylethyl]imino]acetonitrile N-oxide. A 1-L, three-necked flask equipped with a mechanical stirrer, argon gas inlet, and thermometer is charged with (*S*)-[(1-phenylethyl)amino]acetonitrile (10.34 g, 64.54 mmol) and CH_2Cl_2 (200 mL) (Note 8), and the resulting solution is cooled in an ice bath. *m*-Chloroperbenzoic acid (MCPBA) (ca. 77%, 34.3 g, 153 mmol) (Note 9) is added in portions over 30 min (Notes 10-12). After completion of the addition, the ice bath is removed and the mixture is stirred at ambient temperature. When the reaction is complete (Note 13), the mixture is cooled in an ice bath again, and aqueous sodium thiosulfate ($Na_2S_2O_3$) (16.0 g, 129 mmol, in 60 mL of H_2O) and saturated $NaHCO_3$ (200 mL) are added. The resulting slurry is stirred vigorously for 15 min until the white solid completely dissolves. The two-phase solution is separated using a separatory funnel. The aqueous phase is extracted with CH_2Cl_2 (100 mL). The organic layer is washed with brine (100 mL), and the combined aqueous phases are back-extracted with CH_2Cl_2 (3 × 100 mL). The combined organic extracts are dried over anhydrous sodium sulfate, filtered, and concentrated on a rotary evaporator to afford crude nitrone (11.5 g) as slightly yellow crystals (Note 14).

C. N-Hydroxy-(S)-1-phenylethylamine oxalate. A 1-L, round-bottomed flask equipped with a reflux condenser, Teflon-coated magnetic stirrer bar, and argon gas inlet is charged with crude nitrone (11.5 g) and methanol (MeOH) (130 mL) (Note 15). After addition of hydroxylamine hydrochloride (20.8 g, 323 mmol) (Note 16) in one portion at ambient temperature, the mixture is warmed to 60 °C (55 °C internal) and is stirred at that temperature for 2 h (Note 17). The reaction mixture is cooled to room temperature and diluted with CH_2Cl_2 (260 mL). After stirring for 5 min, the resulting precipitate is collected by filtration and the filter cake is washed with CH_2Cl_2 (40 mL). The filtrate is neutralized with saturated $NaHCO_3$ (150 mL) and partitioned. The aqueous phase is extracted with CH_2Cl_2 (100 mL). The organic phase is washed with brine (100 mL), and the combined aqueous phases are back-extracted with CH_2Cl_2 (3 × 100 mL). The combined organic extracts are dried over anhydrous sodium sulfate, filtered and concentrated to a minimum volume at a temperature below 25 °C (Note 18) under reduced pressure. To the residue is added a MeOH (25 mL) solution of oxalic acid (11.6 g, 129 mmol) (Note 19) and ether (Et_2O) (100 mL), precipitating white crystals, which are collected by filtration and washed with Et_2O (3 × 20 mL). After drying under reduced pressure, (*S*)-*N*-1- phenylethylhydroxylamine oxalate (12.6 g, 86% for two steps) is obtained as white crystals. The oxalate salt was recrystallized from hot ethanol (110 mL) and washed with Et_2O (50 mL) to provide analytically pure material (10.45, 71% for two steps) (Notes 20, 21). Concentration and recrystallization of the mother liquor provides additional (*S*)-*N*-1-phenylethylhydroxylamine oxalate (0.85 g, 6%).

2. Notes

1. (*S*)-1-Phenylethylamine was purchased from Aldrich Chemical Co., Inc. and used as received. The enantiomeric purity of (*S*)-1-phenylethylamine was determined as the dinitrobenzoyl derivative to be 95% ee using a chiral stationary phase HPLC column and a chiral stationary phase SFC column (see (Note 21)). The submitters used reverse phase CSP HPLC analysis for this purpose (Daicel CROWNPAK CR, aq. $HClO_4$ (pH 1.5), 0.8 mL/min, 25 °C, 210 nm).

2. Reagent grade acetonitrile was purchased from Fisher Scientific Co. and used as received.

3. Diisopropylethylamine and bromoacetonitrile were purchased from Aldrich Chemical Co., Inc. and used as received.

4. There is a slight exotherm (≈10 °C) over the first 30 min. The reaction typically takes 10-12 h to complete and was monitored by TLC analysis on Merck silica gel 60 F-254 plates eluting with isopropylamine:hexane:EtOAc (5:45:50), $R_f = 0.68$ (visualized with 254-nm UV lamp and ninhydrin spray).

5. Silica gel was purchased from Merck (40-100 mesh, spherical, neutral).

6. The submitters found that the crude product can be purified by distillation (110-120 °C, 1.5 mm). Starting from 24.3 g of amine, 21.8 g (68%) of the desired product was isolated. The reduced yield was due to dialkylation of the product with unreacted bromoacetonitrile during distillation.

7. While the product was sufficiently pure for the subsequent step, ^1H NMR analysis indicated the presence of 4 wt.% of bromoacetonitrile, which was absent after the Step B work-up. To secure analytically pure material, the checkers distilled chromatographically purified product to obtain 9.93 g (74%) of (*S*)-[(1-phenylethyl)-amino]acetonitrile. The pure product exhibits the following physical properties: bp 91-93 °C (1 mm); $[\alpha]_D^{27}$ −217 (c 1.23, $CHCl_3$), $[\alpha]_D^{23}$ −199.3 (c 1.08, EtOH), lit.[2] $[\alpha]_D^{25}$ −248.8 (c 4.7, benzene); IR (film) cm^{-1}: 3335, 2968, 2235, 1957, 1887, 1818, 1494, 1452, 1207, 1132, 870, 763; ^1H NMR (500 MHz, $CDCl_3$) δ: 1.42 (d, *J* = 6.4, 3 H), 1.67 (s (br), 1H), 3.28 (d, *J* = 17.6, 1 H), 3.58 (d, *J* = 17.6, 1 H), 4.04 (q, *J* = 6.5, 1 H), 7.26-7.34 (m, 5H); ^{13}C NMR (166 MHz, $CDCl_3$) δ: 23.9, 35.0, 56.7, 117.8, 126.9, 127.7, 128.7, 142.8. Anal. Calcd for $C_{10}H_{12}N_2$: C, 74.97; H, 7.55; N, 17.48. Found: C, 74.94; H, 7.70; N, 17.32.

8. Reagent grade dichloromethane was purchased from Fisher Scientific Co. and used as received.

9. MCPBA (ca. 65%) was purchased from Aldrich Chemical Co., Inc. and used as received.

10. The reaction temperature was maintained below 7 °C.

11. Alternatively, conversion to the nitrone could be carried out with a catalytic amount of sodium tungstate and hydrogen peroxide[3] as described below: A 200-mL, round-bottomed flask equipped with a 30-mL, dropping funnel, Teflon-coated magnetic stirbar and argon gas inlet is charged with (*S*)-[(1-phenylethyl)amino]acetonitrile (6.10 g, 38.1 mmol), MeOH (64 mL), and sodium tungstate dihydrate ($Na_2WO_4{\cdot}2H_2O$) (504 mg, 1.53 mmol) in one portion. With cooling with an ice-bath, 30% aqueous hydrogen peroxide (14.7 mL, 153 mmol) is added to the solution over 20 min. The reaction mixture is then allowed to warm to ambient temperature. After TLC analysis shows completion of the reaction (usually 10 to 20 h), aqueous $Na_2S_2O_3$ (15 mL) is added slowly with cooling (ice-bath). The resulting suspension is extracted with CH_2Cl_2 (100 mL). The organic extracts are washed with brine (100 mL), and the combined aqueous phases are back-extracted with CH_2Cl_2 (3 × 100 mL). The combined extracts are dried over anhydrous sodium sulfate, filtered, and concentrated to dryness on a rotary evaporator to give crude nitrone (6.1 g) as a slightly yellow solid. The nitrone is directly

converted to hydroxylamine oxalate without further purification (5.40 g, 62% yield for 2 steps). The enantiomeric purity of the hydroxylamine was shown to be >99% ee upon reduction with Zn/AcOH and derivatization (see (Note 21)).

12. Conversion to the nitrone could also be carried out with magnesium monoperoxyphthalate (MMPP)[4] as follows: A 1-L, three-necked, flask equipped with a 30-mL addition funnel, Teflon-coated magnetic stirbar, thermometer and argon gas inlet, is charged with an aqueous solution (153 mL) of magnesium monoperoxyphthalate hexahydrate (26.0 g, 80%, 42 mmol), which is then cooled in an ice-bath. A solution of (*S*)-[(1-phenylethyl)amino]acetonitrile (6.11 g, 38.1 mmol) in MeOH (30 mL) is subsequently added dropwise over a period of 20 min (the addition funnel is washed with 8 mL of MeOH). The resulting mixture is then stirred at ambient temperature for 30 min. After the reaction is complete, the mixture is cooled in an ice-bath and is diluted with CH_2Cl_2 (200 mL). With vigorous stirring, 4.8 g (19 mmol) of $Na_2S_2O_3{\cdot}5H_2O$ and 10 g (96 mmol) of sodium carbonate are added portionwise and the resulting two-phase mixture is partitioned. The aqueous phase is extracted with CH_2Cl_2 (100 mL). The organic phase is washed with brine (100 mL) and the combined aqueous layers are back-extracted with CH_2Cl_2 (3 × 100 mL). The combined organic extracts are dried over anhydrous sodium sulfate, filtered, and concentrated to dryness on a rotary evaporator to give crude nitrone (6.4 g) as a slightly yellow solid. The nitrone is directly converted to hydroxylamine oxalate without further purification (7.51 g, 87% yield for 2 steps). The enantiomeric purity of the hydroxylamine was shown to be >99% ee upon reduction with Zn/AcOH and derivatization (see Note 21).

13. The reaction typically takes 30 min to complete. TLC analysis on Merck silica gel 60 F-254 plates eluting with MeOH:CH_2Cl_2 (2.5:97.5) shows formation of the (Z)-nitrone (R_f = 0.68) (visualized by a 254-nm UV lamp and ethanolic phosphomolybdic acid), with only a trace amount of the (E)-nitrone (R_f = 0.73).

14. Judging by ^{1}H NMR analysis, the crude product is highly pure (Z)-nitrone (92/8, Z/E), which can be used for the next step. Recrystallization from EtOAc-hexane affords pure (Z)-nitrone, which exhibits the following physical properties: mp 89.5-91.0; $[\alpha]_D^{28}$ +83 (c 0.494, $CHCl_3$), $[\alpha]_D^{23}$ +135.8 (c 0.995, EtOH); IR (KBr) cm^{-1}: 3098, 2222, 1541, 1452, 1442, 1377, 1295, 1181, 1074, 1007, 748, 701; ^{1}H NMR (500 MHz, $CDCl_3$) δ: 1.83 (d, *J* = 7.0, 3 H), 5.18 (quint, *J* = 6.9, 1 H), 6.67 (s, 1 H), 7.43 (br s, 5 H); ^{13}C NMR (166 MHz, $CDCl_3$) δ: 19.0, 79.5, 105.8, 112.2,

128.9, 129.2, 129.8, 136.2. Anal. Calcd for $C_{10}H_{10}N_2O$: C, 68.95; H, 5.79; N, 16.08. Found: C, 68.83; H, 5.68; N, 15.90.

15. Reagent grade MeOH was purchased from Aldrich Chemical Co., Inc. and used as received.

16. Hydroxylamine hydrochloride was purchased from Aldrich Chemical Co., Inc. and used as received.

17. *Caution! The reaction should be carried out in a well-ventilated hood because of the potential of generating HCN gas.*

18. Care must be exercised not to allow the bath temperature to rise above 25 °C and not to concentrate the solution to complete dryness, both of which would increase the risk of explosion of the hydroxylamine. The checkers found that some of the hydroxylamine would precipitate, resulting in a white slurry that would dissolve upon addition of the next reagent.

19. Oxalic acid was purchased from Aldrich Chemical Co., Inc. and used as received. The checkers found that sonication was helpful in dissolving the oxalic acid in MeOH.

20. The pure product exhibits the following physical properties: mp 177-180 °C (dec.); $[\alpha]_D^{28}$ −2.2 (c 1.06, MeOH); IR (KBr) cm^{-1}: 3220, 2987, 2578, 1761, 1610, 1483, 1458, 1210, 986, 961, 775, 714, 702; ^1H NMR (500 MHz, CD_3OD) δ: 1.68 (d, J = 6.8, 3 H), 4.52 (q, J = 6.9, 1 H), 7.39-7.50 (m, 5 H); ^{13}C NMR (166 MHz, CD_3OD) δ: 16.1, 62.9, 129.9, 130.1, 130.6, 136.0, 166.4. Anal. Calcd for $C_8H_{11}NO\cdot(COOH)_2$: C, 52.86; H, 5.77; N, 6.16. Found: C, 52.72; H, 5.68; N, 6.18.

21. The enantiomeric excess of the product was determined by reduction with Zn in AcOH, followed by derivatization with 3,5-dinitrobenzoyl chloride. (*S*)-*N*-1-Phenylethylhydroxylamine oxalate (0.50 g, 2.20 mmol) was placed in a 50-mL, single-necked, round-bottomed flask equipped with an argon inlet, rubber septum, and Teflon-coated stirbar, followed by acetic acid (AcOH) (glacial, 6 mL), HCl (1M, 12 mL), and Zn dust (5 g). The slurry was heated to 80 °C for 6 h, during which time the Zn began to conglomerate. The reaction mixture was cooled to room temperature and filtered through glass wool. The Zn metal was washed with H_2O (10 mL) and CH_2Cl_2 (20 mL) and the filtrate was concentrated under reduced pressure. The residue was partitioned between CH_2Cl_2 (40 mL) and 1M NaOH (40 mL). The organic phase was washed with H_2O (20 mL) and brine (20 mL). The combined aqueous phases were back-extracted with CH_2Cl_2 (2 × 20 mL). The organic extracts were combined, dried over magnesium sulfate, and concentrated to give crude amine. Bulb-to-bulb distillation of the amine [150 °C (air bath temp), 100 mm] provided (*S*)-1-

phenylethylamine (0.21 g, 78%). In a 25-mL, two-necked, round-bottomed flask, with an argon inlet, septum, and Teflon-coated stirbar, was placed (*S*)-1-phenylethylamine (0.18 g, 1.5 mmol) and THF (6 mL). The solution was cooled in an ice bath to 0 °C; triethylamine (0.3 mL, 2.2 mmol) and subsequently 3,5-dinitrobenzoyl chloride (0.38 g, 1.7 mmol) were added. The ice bath was removed and the solution was stirred for 2.5 h at ambient temperature, during which time the solution turned yellow and a white precipitate formed. The slurry was then partitioned between H_2O (15 mL) and Et_2O (15 mL). The organic phase was separated, then washed with brine (10 mL). The combined aqueous phases were back-extracted with Et_2O (10 mL). The combined organic extracts were dried over magnesium sulfate, filtered, and concentrated under reduced pressure. The crude amide was purified by column chromatography (30 mm × 150 mm SiO_2, hexane/EtOAc, 4/1→2/1) to provide (*S*)-*N*-(3,5-dinitrobenzoyl)-1-phenylethylamine (0.47 g, 99%). The enantiomeric purity of the derivative was shown to be >99% ee by chiral stationary phase HPLC and SFC. (HPLC: Pirkle S-N1N-Naphthylleucine column, hexane/*i*-PrOH, 3/21 mL/min, t_R 16.54 min (R isomer: t_R = 14.22 min), SFC: Chiralcel OD column, 20% MeOH, 2 mL/min, t_R = 9.136 (R isomer: 8.539 min). The submitters used reverse phase CSP HPLC for this analysis (see Note 1).

Waste Disposal Information

All toxic materials were disposed of in accordance with "Prudent Practices in the Laboratory"; National Academy Press; Washington, DC, 1995.

3. Discussion

Although *N*-monoalkylhydroxylamines are important precursors for *N*-hydroxyl-α-amino acid and hydroxamic acids,[5] there are few practical and general methods for the preparation of this class of compounds.[6] Direct oxidation of primary amines with hydrogen peroxide or MCPBA is generally ineffective, often leading to over-oxidation products. A widely used method for the conversion of primary amines to hydroxylamines employs oxidation of the corresponding Schiff bases, although the intermediacy of relatively acid-labile Schiff bases is one of its disadvantages.[7]

The present procedure provides a novel protocol for transformation of primary amines to *N*-monoalkylhydroxylamines in a three-step sequence involving selective mono-cyanomethylation, regioselective nitrone formation using MCPBA, and hydroxylaminolysis.[8] The procedure in Step A is representative of selective mono-cyanomethylation. Depending upon the steric bulk of the alkyl substituents, three conditions can be employed to achieve selective mono-cyanomethylation (Table 1). Since cyanomethylated amines are relatively stable, they can be used as reliable precursors for hydroxylamines. The procedure in Step B illustrates the regioselective formation of nitrone. The significance of the cyano group in directing the formation of nitrones was also demonstrated by the successful application of this method to a wide variety of substrates including α-aminoester derivatives[7b,7d,9] (Table 1). For the conversion of nitrones to hydroxylamines (Step C), we have modified the literature procedure.[7] The use of excess hydroxylamine hydrochloride at higher temperature reduced the reaction time and improved the yields. One of the advantages of the present procedure is that the oxime by-product generated after hydroxylaminolysis, namely NC-CH=NOH, can be easily removed by aqueous extraction.

Table 1

Entry	Amine	Cyanomethylation[a] Conditions	time (h)	Yield (%)	Overall Yield (%) of Hydroxylamine from Cyanomethylated Amine
1	$PhCH_2NH_2$	A	24	95	82
2	$PhCH_2CH_2NH_2$	A	24	96	75
3	$PhCH_2CH_2CH_2NH_2$	A	24	93	74
4	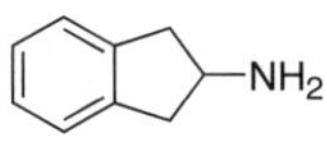	B[b,c]	22	97	79
5	Ph_2CHNH_2	B	15	98	55
6	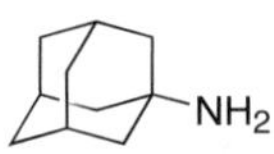	C	1	89	76
7	$BnO_2CCH_2NH_2$	B[d]	18	91	62
8	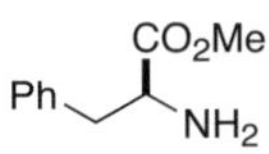	B[b,e]	26	92	61

[a]Conditions A. $ClCH_2CN$ (1.5 eq), K_2CO_3 (2.0 eq), CH_3CN, 60°C; Conditions B. $BrCH_2CN$ (1.5) eq), *i*-Pr_2NEt (2.0 eq), CH_3CN, rt; Conditions C. ICH_2CN (2 eq), K_2CO_3 (2.5 eq), DMF, rt.
[b]HCl salt of amine was used.
[c]$BrCH_2CN$ (1.3 eq), *i*-Pr_2NEt (3.0 eq).
[d]$BrCH_2CN$ (1.2 eq), *i*-Pr_2NEt (2.0 eq).
[e]$BrCH_2CN$ (2.0 eq), *i*-Pr_2NEt (3.0 eq).

1. Graduate School of Pharmaceutical Sciences, The University of Tokyo, 7-3-1 Hongo, Bunkyo-ku, Tokyo 113-0033, Japan.
2. Okawara, T.; Harada, K. *J. Org. Chem.* **1972**, *37*, 3286.
3. (a) Murahashi, S.-I.; Mitsui, H.; Shiota, T.; Tsuda, T.; Watanabe, S. *J. Org. Chem.*, **1990**, *55*, 1736-1744. (b) Murahashi, S.-I.; Shiota, T.; Imada, Y. *Org, Synth. Coll. Vol. IX*, **1998**, 632-636.
4. For a review on magnesium monoperoxyphthalate, see: Heaney, H. *Aldrichimica Acta*, **1993**, *26*, 35.

5. (a) Ottenheijm, H. C. J.; Herscheid, J. D. M. *Chem. Rev.* **1986**, *86*, 697. (b) Akiyama, A. *Yuki Gosei Kagaku* **1982**, *40*, 1189.

6. For synthesis of hydroxylamines, see: Bowman, W. R.; Marmon, R. J. In *Comprehensive Organic Functional Group Transformations*; Katritzky, A. R., FRS; Meth-Cohn, O.; Rees, C. W., FRS, Eds., Pergamon, Elsevier, Oxford, **1995**; Vol. 2, Chapter 2.06 and references cited therein.

7. (a) Emmons, W. D. *J. Am. Chem. Soc.* **1957**, *79*, 5739. (b) Polonski, T.; Chimiak, A. *Tetrahedron Lett.* **1974**, 2453. (c) Wovkulich, P. M.; Uskokovic, M. R. *Tetrahedron* **1985**, *41*, 3455. (d) Grundke, G.; Keese, W.; Rimpler, M. *Synthesis* **1987**, 1115.

8. Tokuyama, H.; Kuboyama, T.; Amano, A.; Yamashita, T.; Fukuyama, T. *Synthesis*, **2000**, 1299.

9. Wittman, M. D.; Halcomb, R. L.; Danishefsky, S. J. *J. Org. Chem.* **1990**, *55*, 1981.

SYNTHESIS OF AN *ANTI*-α-AMINO EPOXIDE BY ONE-CARBON HOMOLOGATION OF AN α-AMINO ESTER: (2*S*,3*S*)-1,2-EPOXY-3-(BOC-AMINO)-4-PHENYLBUTANE

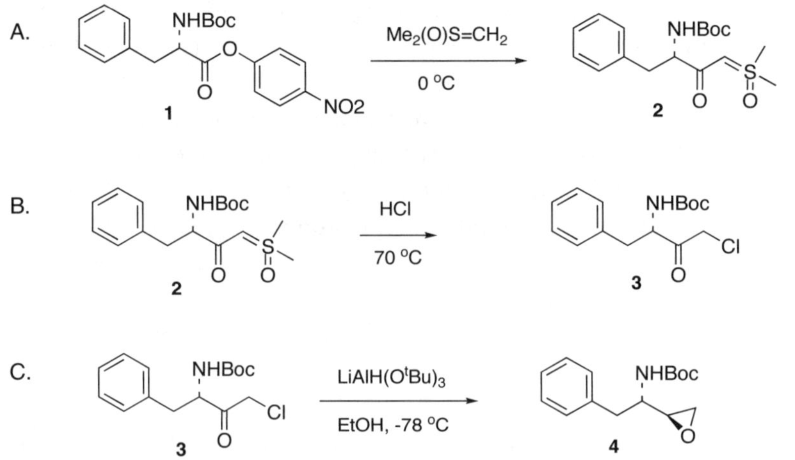

Submitted by Dengjin Wang[1a] and William A. Nugent.[1b]
Checked by Jason J. Kowal and David J. Mathre.

1. Procedure

A. Dimethylsulfoxonium (S)-2-oxo-3-(Boc-amino)-4-phenylbutylide (**2**). A 500-mL three-necked, round-bottomed flask is equipped with a large stir bar, a nitrogen inlet, a reflux condenser capped with a nitrogen outlet, and a rubber septum through which is inserted a temperature probe. The flask is flushed with nitrogen for 5 min and is then charged with trimethylsulfoxonium iodide (33.0 g, 150 mmol) (Note 1), which is added through a neck that is temporarily opened, and 150 mL of anhydrous tetrahydrofuran (Note 2), which is added by syringe through a septum. A 1.0 M solution of potassium *tert*-butoxide in tetrahydrofuran (150 mL, 150 mmol) (Note 3) is added via syringe through a septum, after which the mixture is heated at reflux for 2 h (Note 4). The resulting solution of dimethylsulfoxonium methylide (containing precipitated potassium iodide) is cooled to 0 °C. The reflux condenser is replaced with a 125-mL pressure-equalizing addition funnel containing a solution of Boc-L-phenylalanine 4-nitrophenyl ester (**1**) (19.3 g, 50.0 mmol) (Note 5) in 100 mL of anhydrous tetrahydrofuran and capped with a nitrogen outlet. The solution is added dropwise over 20 min so that the internal temperature does not exceed 5 °C. The mixture is stirred for 1 h at 0 °C. The flask is briefly opened by temporary removal of the addition funnel to allow addition of 50 mL of water and stirring is continued for 15 min at 0 °C. After warming to room temperature, the mixture is filtered through a short pad (3 mm) of Celite on a 100-mL fritted-glass Büchner funnel and the solvent is removed by rotary evaporation (25 °C, 20 mmHg). The concentrated mixture is rinsed into a 1-L separatory funnel with 200 mL of water and extracted with one 200-mL portion and two 100-mL portions of ethyl acetate (Note 6). The combined extracts are washed with one 50-mL portion of water and two 50-mL portions of brine. (In some cases solid product separates during the first extraction, but it redissolves when the organic phases are combined.) The solution is dried over sodium sulfate, filtered and concentrated by rotary evaporation (25 °C, 20 mmHg) to afford **2** as a light yellow solid (15.3–15.4 g, 90–91%) (Note 7), which is carried on to the next step without further purification.

B. (S)-1-Chloro-3-(Boc-amino)-4-phenyl-2-butanone (**3**). A 500-mL three-necked, round-bottomed flask is equipped with a stir bar, a nitrogen inlet, a reflux condenser capped with a nitrogen outlet, and a rubber septum through which is inserted a temperature probe. The flask is flushed with nitrogen for 5 min and is then charged with dimethylsulfoxonium (*S*)-2-oxo-3-(Boc-amino)-4-phenylbutylide (**2**) (13.6 g, 40 mmol), which is added through temporary removal of the reflux condenser. Anhydrous tetrahydrofuran (250 mL) is added by syringe through a septum. A 4.0 M solution of hydrogen chloride in 1,4-dioxane, (11.6 mL, 46.0 mmol) (Note 8) is added via syringe through the septum over 5 min at room temperature during which time a solid sulfoxonium salt separates. The mixture is then heated with stirring at reflux (70 °C) for 4 h. The reaction mixture stirs as a very thick slurry for the first 30 min of this period. After cooling to room temperature, the resulting homogeneous solution is transferred to a 1-L separatory funnel, to which is added 100 mL of ethyl acetate and 200 mL of hexanes (Note 9). The mixture is washed with two 200-mL portions of water, one 100-mL portion of saturated sodium bicarbonate solution, four 200-mL portions of water and one 100-mL portion of brine. (Note 10) The solution is dried over sodium sulfate and filtered through a 100-mL fritted-glass Büchner funnel. The filtrate is concentrated by rotary evaporation (25 °C, 20 mmHg) to obtain 13–14 g of a light yellow solid. This material is dissolved in 40 mL of hot hexanes (70 °C). Upon cooling, a solid separates which is collected by filtration on a 100-mL fritted-glass Büchner funnel and dried overnight at 25 °C and 0.5 mmHg to afford **3** (9.61–9.71 g, 81% yield) (Note 11) as an off-white solid.

C. (2S,3S)-1,2-Epoxy-3-(Boc-amino)-4-phenylbutane (**4**). A 500-mL three-necked, round-bottomed flask is equipped with a large stir bar (Note 12), a nitrogen inlet, a 125-mL pressure-equalizing addition funnel capped with a nitrogen outlet, and a rubber septum through which is inserted a temperature probe. The flask is flushed with nitrogen for 5 min and is then charged with (*S*)-1-chloro-3-(Boc-amino)-4-phenyl-2-butanone (**3**) (7.5 g, 25 mmol), which is added through temporary removal of the addition funnel. Anhydrous ethanol (200 mL, Note 13) is added by syringe through a septum. The addition funnel is opened slightly to allow delivery via syringe of a 1.0 M solution of lithium tri-*tert*-butoxyaluminum hydride in tetrahydrofuran (52.5 mL, 52.5 mmol) (Note 14). The flask is cooled to –78 °C (internal temperature) in a dry ice-acetone bath. The contents of the addition funnel are then added dropwise over 20 minutes so that the internal temperature does not exceed –70 °C. The mixture is stirred at this temperature for 2 h. The cooling bath is removed and the internal temperature is allowed to rise to –20 °C. The reaction is quenched through the addition of 50 mL of Rochelle's salt solution in one portion (Note 15). The bath is removed and stirring is continued for 15 min. The reaction mixture is concentrated by rotary evaporation (25 °C, 20 mmHg) to provide two liquid phases of ~80 mL combined volume. The concentrated mixture is transferred to a 1-L separatory funnel. An additional 200 mL of Rochelle's salt solution is added and the mixture is extracted with one 500-mL portion and one 150-mL portion of ethyl acetate. The combined extracts are washed with two 200-mL portions of water and one 200-mL portion of brine. The solution is dried over sodium sulfate, filtered and concentrated by rotary evaporation (25 °C, 20 mmHg) to obtain the crude product as a light yellow solid. This material is dissolved with heating in a mixture of 30 mL of ethyl acetate and 10 mL of hexanes (70 °C). Upon cooling, a solid separates which is collected by filtration on a 100-mL fritted-glass Büchner funnel and dried overnight at 25 °C and 0.5 mmHg to afford **4** (5.44–5.56 g, 82–84% yield) (Note 16) as a white solid.

2. Notes

1. Trimethylsulfoxonium iodide was purchased from the Aldrich Chemical Company, Inc. and used as received.

2. Anhydrous tetrahydrofuran (<50 ppm water) was purchased from EM Science and used as received.

3. Potassium *tert*-butoxide (1.0 M in tetrahydrofuran) was purchased from Aldrich Chemical Company, Inc. and used as received.

4. This heating step was essential when the starting ester was potentially epimerizable. If this step was omitted in the case of sulfur ylide **2**, good chemical yields were obtained but the product was nearly racemic.

5. Boc-L-Phenylalanine 4-nitrophenyl ester was purchased from Bachem Bioscience, Inc. and used as received.

6. Ethyl acetate was purchased from EMD Chemicals, Inc. and was used as received.

7. An analytical sample of compound **2** that was dried for 24 h at room temperature at 0.5 mmHg had the following properties: mp (DSC) 163 °C. $[\alpha]_D^{25}$ –31.8 (c = 1.00, MeOH). ^{1}H NMR ($CDCl_3$, 400 MHz) δ: 1.41 (s, 9 H), 2.98 (m, 2 H), 3.25 (s, 3 H), 3.35 (s, 3 H), 4.28 (m, 2 H), 5.22 (d, *J* = 8.8, 1 H), 7.15–7.31 (m, 5 H). ^{13}C NMR ($CDCl_3$, 100 MHz) δ: 28.3, 39.7, 41.7, 42.0, 57.8, 69.4, 79.2, 126.4, 128.1, 129.5, 137.6, 155.1, 186.5. Anal. Calcd for $C_{17}H_{25}NO_4S$: C, 60.15; H, 7.42; N, 4.13; S, 9.44 Found: C, 60.21; H, 7.61; N, 4.06. TLC: dichloromethane/methanol (95:5) R_f = 0.35.

8. Hydrogen chloride (4.0 M in dioxane) was purchased from Aldrich Chemical Company, Inc. and was used as received.

9. Hexanes was purchased from EMD Chemicals, Inc. and was used as received.

10. The multiple washes were required to completely remove an impurity (presumably co-product DMSO) that interfered with the crystallization of **3**.

11. The enantiomeric excess was determined to be >99% by supercritical fluid chromatography (Chiralcel OD-H, 250 x 4.6 mm, 5 μm particle size, 3% methanol in CO_2 mobile phase, 40 °C, 2 mL/min, 150 bar). Retention times for the (*R*)- and (*S*)-enantiomers were 6.2 and 6.8 min, respectively. Properties of **3** are as follows: mp (DSC) 103 °C. $[\alpha]_D^{25}$ = –43.3 (c = 1.00, MeOH). ^{1}H NMR (400 MHz, $CDCl_3$) δ: 1.40 (s, 9 H), 2.94–3.12 (m, 2 H), 4.00 (d, *J* = 16.2, 1 H), 4.18 (d, *J* = 16.2, 1 H), 4.68 (m, 1 H), 5.10 (b, 1 H), 7.14–7.47 (m, 5 H). ^{13}C NMR ($CDCl_3$, 100 MHz) δ: 28.3, 37.8, 47.5, 58.5, 80.6, 126.2, 129.0, 129.2, 135.7, 155.3, 201.5. Anal.

Calcd for $C_{15}H_{20}ClNO_3$: C, 60.50; H, 6.77; N, 4.70. Found: C, 60.15; H, 6.81; N, 4.72. TLC: hexanes/ethyl acetate (2:1) R_f = 0.7.

12. The checkers used overhead mechanical stirring.

13. Anhydrous ethanol (<0.005% water) was purchased from Aldrich Chemical Company, Inc. and used as received.

14. Lithium tri-*tert*-butoxyaluminum hydride (1.0M in tetrahydrofuran) was purchased from Aldrich Chemical Company, Inc. and used as received.

15. Rochelle's salt solution was prepared by dissolving potassium sodium tartrate (30 g) and potassium carbonate (3.0 g) in 270 mL of water.

16. The enantiomeric excess of the epoxide **4** was >99% by chiral HPLC. Conditions: Chiralcel OD NP, 250 x 4.6 mm, 10 μm particle size, solvent (isocratic) 98% heptane, 1% methanol, 1% ethanol, flow rate 1.0 mL/min. Retention times were: *(R,R)*-epoxide, 10.2 min; *(S,S)*-epoxide (**4**), 11.0 min. Compound **4** had the following properties: mp (DSC) 125 °C. $[\alpha]_D^{25}$ -8.3 (c = 1.00, MeOH). ^1H NMR (400 MHz, $CDCl_3$) δ: 1.38 (s, 9 H), 2.75 (br m, 1 H), 2.80 (m, 1 H), 2.87 (m, 1 H), 2.95 (m, 2 H), 3.69 (m, 1 H), 4.46 (br s, 1 H), 7.2–7.4 (m, 5 H). ^{13}C NMR (100 MHz, $CDCl_3$) δ: 28.2, 37.5, 46.7, 52.5, 53.2, 79.5, 126.6, 128.4, 129.4, 137.7, 155.2. Anal. Calcd for $C_{15}H_{21}NO_3$: C, 68.42; H, 8.04; N, 5.32. Found: C, 68.44; H, 8.19; N, 5.30.

Safety and Waste Disposal Information

All hazardous materials should be handled and disposed of in accordance with "Prudent Practices in the Laboratory"; National Academy Press; Washington, DC, 1995.

3. Discussion

N-Protected α-amino epoxides are broadly useful synthetic intermediates and in particular are ideal precursors for the hydroxyethylamine class of protease inhibitors. These compounds are frequently synthesized by one-carbon homologation of α-amino acids using diazomethane.[2] There are significant hazards associated with the use of diazomethane since this gaseous reagent is highly toxic, shock-sensitive, and explosive.[3] The procedure reported here, which is based on an earlier synthesis of α-fluorinated chloroketones,[4] was developed to provide an alternative to the use of diazomethane.[5]

Efficient reaction of Corey's reagent, dimethylsulfoxonium methylide,[6] with esters requires an electron-withdrawing group adjacent to the carbonyl, provided in the case of **1** by the α-BOC-amino substituent. Several examples of chloroketone synthesis via this method are shown in the Table. Esters of simple aliphatic acids react only sluggishly. In the presence of a suitable activating group, methyl esters as well as *p*-nitrophenyl esters will react. However, for substrates that are subject to epimerization such as **1**, the more reactive aryl ester is needed to preclude partial racemization.

Our earlier procedure[5] for reduction of chloroketone **3** utilized sodium borohydride as the reductant. With that reducing agent, the *anti/syn* selectivity was only 4:1, necessitating the isolation and crystallization of the intermediate β-chlorohydrin prior to epoxide ring-closure. The current procedure takes advantage of Hoffman's report[7] that high *anti*-selectivity could be achieved with related substrates using lithium tri-*tert*-butoxyaluminum hydride in ethanol. This protocol improved selectivity for reduction of **3** to 92:8 (*anti/syn*). The intermediate β-chlorohydrins underwent epoxide ring-closure during work-up with the basic Rochelle salt solution. The desired *anti* isomer **4** was isolated from the mixed diastereomers via a simple crystallization. The Rochelle's salt solution also served to sequester the aluminum-containing co-products, simplifying the extractive workup.

As an alternative to this "telescoped" procedure, it is also possible to isolate the β-chlorohydrin intermediate. This is accomplished by replacing the Rochelle salt solution with water during the extractive work-up.[5] The β-chlorohydrin is recovered as a mixture of diastereomers as shown below. The pure *anti*-diastereomer is a crystalline solid (mp 122 °C) and can be isolated by crystallization from hot ethyl acetate.[5]

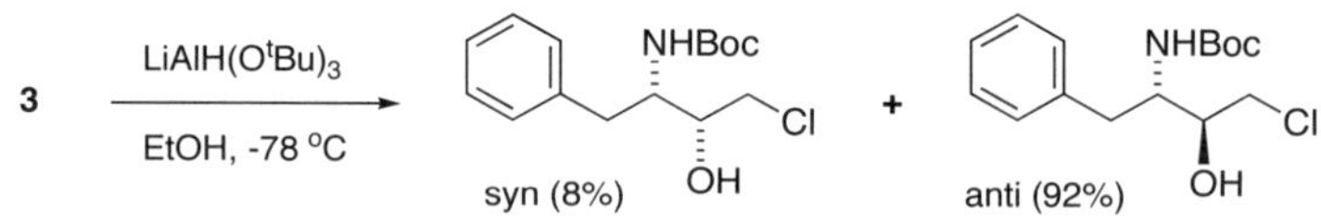

The best method for this reduction step varies depending on the nature of the α-chloroketone. When the chloroketone is prepared from an α-amino acid with a sterically large R group (e.g., valine), simple borohydride reduction provides high diastereoselectivity.[8] When the R group is small (e.g., alanine), hydrogenation using a "matched" ruthenium-BINAP catalyst is advantageous.[5] For asymmetric reduction of achiral chloroketones, the Noyori catalytic transfer hydrogenation procedure is recommended.[9] It should also be noted that, depending on which enantiomer of the Noyori catalyst is employed, this procedure can also be used to convert **3** into *either* the *syn* or the *anti* β-chlorohydrin, in each case with 90:10 diastereoselectivity.[10] In cases where the final epoxide is not crystalline, it will generally be desirable to isolate and crystallize the intermediate β-chlorohydrin.

1. (a) Bristol Myers Squibb Co., Process Research and Development Department, P. O. Box 4000, Princeton, NJ 08543-4000. (b) Vertex Pharmaceutical Co., 130 Waverly Street, Cambridge, MA 02139.

2. (a) Penke, B.; Czombos, J.; Balaspiri, L.; Petres, J.; Kovaks, K. *Helv. Chim. Acta* **1970**, *53*, 1057. (b) Albeck, A.; Persky, R. *Tetrahedron* **1994**, *50*, 6333.

3. (a) *Dangerous Prop. Ind. Mater. Rep.* **1992**, *12*, 530. (b) de Boer, Th. J.; Backer, H. J. *Organic Syntheses*; Wiley: New York, 1963; Collect. Vol. IV, 250.

4. Elkik, E.; Imbeaux-Oudette, M. *Bull. Soc. Chim. Fr.* **1987**, 861.

5. (a) Wang, D.; Schwinden, M. D.; Radesca, L.; Patel, B.; Kronenthal, D.; Huang, M.-H.; Nugent, W. A. *J. Org. Chem.* **2004**, *69*, 1629. (b) Schwinden, M. D.; Kronenthal, D., U. S. Pat. 6,399,793 (**2002**).

6. Corey, E. J.; Chaykovsky, M. *J. Am. Chem. Soc.* **1965**, *87*, 1353.

7. Hoffman, R. V.; Maslouh, N.; Cervantes-Lee, F. *J. Org. Chem.* **2002**, *67*, 1045.

8. Rotella, D. P. *Tetrahedron Lett.* **1995**, *36*, 5453.

9. Hamada, T.; Torii, T.; Izawa, K.; Ikariya, T. *Tetrahedron*, **2004**, *60*, 7099.

10. Hamada, T.; Torii, T.; Onishi, T.; Izawa, K.; Ikariya, T. *J. Org. Chem.* **2004**, *69*, 2186.

11. Wang, D.; Nugent, W. A., unpublished results.

Table. Synthesis of Chloroketones by One-Carbon Homologation of Esters

Ester	α-Chloroketone	Equiv/time/temp[a]	Yield (%)[b]	Ref.
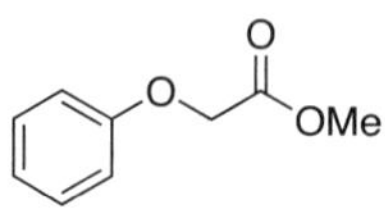	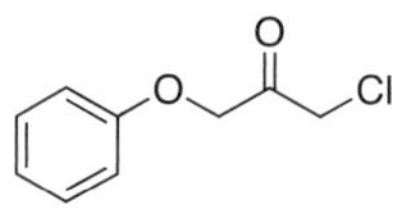	3/16/25	63[c]	5
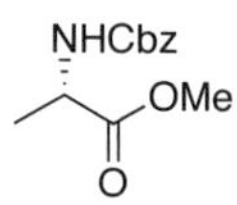	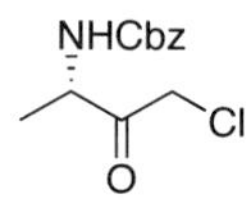	3/4/0	68[c]	5
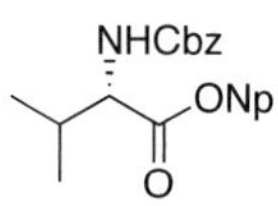	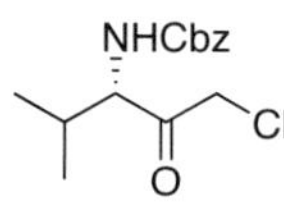	5/4/0	74[d]	5
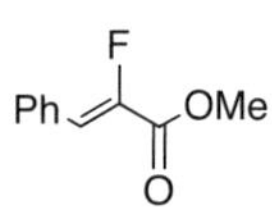	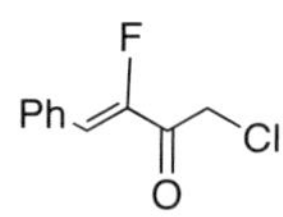	1/2/15	54	4
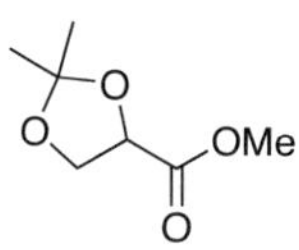	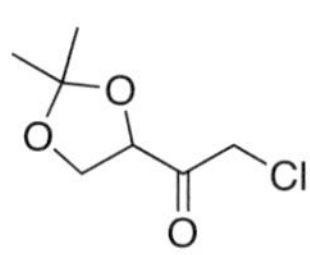	3/16/25	71[e]	11
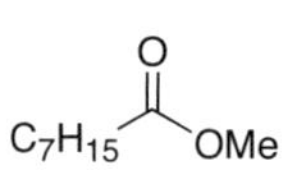	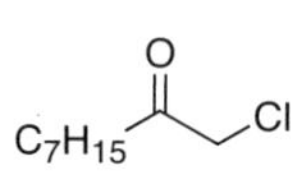	5/72/0	25[f]	5

a) Reaction conditions: equiv of dimethylsulfoxonium methylide, time (h), temperature (°C).
b) Combined yield for ylide formation and HCl cleavage steps. c) Reagent generated from trimethylsulfoxonium chloride; other cases use the iodide. d) Yield after flash chromatography.
e) Cleavage step carried out with pyridinium hydrochloride (70 °C, 3 h) rather than HCl.
f) Yield for sulfur ylide step after flash chromatography; not converted to chloroketone.

PREPARATION OF (*S*)-METHYL GLYCIDATE VIA HYDROLYTIC KINETIC RESOLUTION

(Oxiranecarboxylic acid, methyl ester, (2*S*)-)

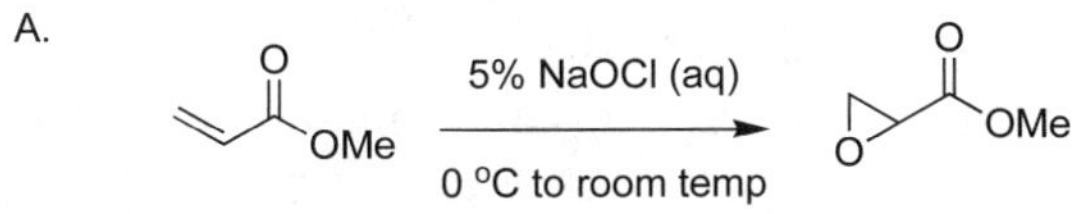

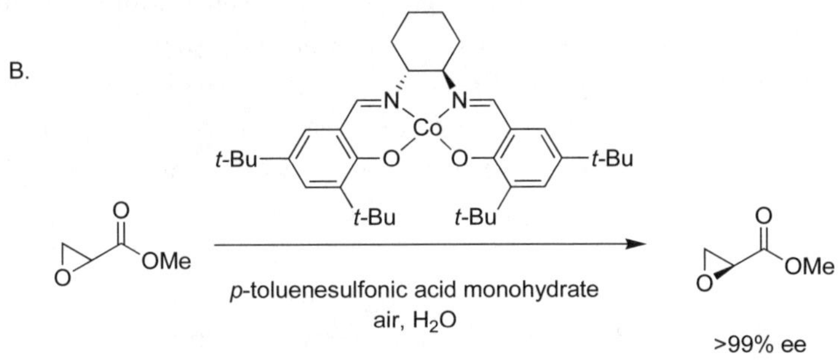

Submitted by Christian P. Stevenson, Lars P. C. Nielsen and Eric N. Jacobsen.[1]

Checked by Jason D. McKinley, Timothy D. White, Michel A. Couturier, and John Ragan.

1. Procedure

A. (±)-Methyl glycidate (Note 1). A 3-L, round-bottomed flask equipped with a magnetic stir bar and an internal thermometer is charged with aqueous sodium hypochlorite solution (6.0 wt%, 940 mL, 0.755 mol, Note 2) and cooled to 0 °C in an ice bath. Methyl acrylate (58.5 g, 0.680 mol, 1 equiv) is added in one portion, and the flask is capped loosely under air (Note 3). The biphasic mixture is stirred vigorously at 0 °C (Note 4). After 30 min, the ice bath is removed, and the solution is stirred for an additional 1.5 h (Note 5). During this time, the internal temperature gradually rises to 30–37 °C, and the biphasic, yellow mixture becomes turbid and colorless. The reaction mixture is then cooled to 20–25 °C in an ice bath, transferred to a separatory funnel, and extracted with dichloromethane (4 x 150 mL). The organic extracts are dried over sodium sulfate, filtered, and concentrated to a volume of approximately 50 mL by rotary evaporation (Note 6). The solution is transferred to a 100-mL recovery flask and purified by fractional vacuum distillation (bp 84–87 °C/70 mmHg, Notes 7-9) to afford racemic methyl glycidate (30.3–31.6 g, 44–46%) as a colorless liquid (Note 10).

B. (S)-Methyl glycidate. A 200-mL, round-bottomed flask equipped with a magnetic stir bar is charged with (*R*,*R*)-(–)-*N*,*N'*-bis(3,5-di-*tert*-butylsalicylidene)-1,2-cyclohexanediaminocobalt(II) (0.916 g, 1.52 mmol, 0.50 mol%), *p*-toluenesulfonic acid monohydrate (0.304 g, 1.60 mmol, 0.53 mol%, Note 11), and dichloromethane (20 mL). The solution rapidly turns from bright red to dark green/brown and is stirred open to air for 30 min (Note 12). After removing the stir bar, the solution is concentrated by rotary evaporation, then the residue is dried under vacuum (<1 mmHg) for 30 min (Note 13). The stir bar is returned to the flask, racemic methyl glycidate (31.0 g, 0.304 mol, 1 equiv) is added, and the flask is loosely capped with a greased glass stopper (Note 14). After dissolving most of the catalyst residue by swirling the solution for 1-2 min, distilled water (3.85 mL, 0.214 mol, 0.70 eq) is added, the reaction vessel is immersed in a room temperature water bath and stirred vigorously for 24 hours. The stopper and water bath are removed, and the reaction flask is fitted with a reflux condenser. The mixture is heated at 85-90 °C (oil bath temperature) for two h. Upon cooling to room temperature and stirring for 30 min, the precipitated red (salen)Co(II) complex is recovered by vacuum filtration, and washed with distilled water (3 x 15 mL, Notes 15 and 16). The filtrate is extracted with dichloromethane (3 x 60 mL). The combined organic extracts are dried over sodium sulfate, filtered, and concentrated to a volume of less than 20 mL by rotary evaporation (Note 6). This brown solution is transferred to a 50-mL recovery flask, and short path vacuum distillation (bp 54–57 °C/33 mmHg, Notes 7-9) affords (*S*)-methyl glycidate (11.1-11.4 g, 36-37%) as a colorless liquid in >99% enantiomeric excess (Notes 10 and 17).

2. Notes

1. This procedure is a modification of that described by Nemes and coworkers,[2] first developed in these laboratories by K. B. Hansen.[3]

2. The checkers used 6 wt% NaOCl (Aldrich). The submitters used Clorox® brand household bleach solution (5.25 wt% NaOCl), which required 1.07 L to achieve 0.755 mol.

3. For both procedures, reagents were purchased from Aldrich Chemical Company, Inc. (the submitters purchased (salen)Co(II) pre-catalyst from Strem Chemicals, Inc.). All reagents were used as received without further purification.

4. Rapid stirring must be maintained to ensure sufficient mixing of the biphasic reaction.

5. Longer reaction times lead to diminished yields due to saponification of methyl glycidate under the basic reaction conditions (pH = 11−12).

6. Care should be taken to minimize losses of the volatile methyl glycidate. The submitters concentrated the solution at 20 °C/40 mmHg and lost <2% of the material.

7. Upon reaching 115 mmHg, the pressure of the distillation apparatus is lowered slowly. Vigorous stirring is maintained to keep the solution from foaming over into the receiving flask. Upon reaching the desired distillation pressure, the apparatus is maintained at room temperature with vigorous stirring for one hour to ensure removal of all residual solvent.

8. The receiving flask is maintained at 0 °C once the product starts to distill. At the end of the distillation, when the distillation head temperature starts to fall, the head is warmed gently with a heat gun to maximize product recovery.

9. The submitters reported a boiling point of 59−61 °C/20 mmHg. The checkers found that distillation at slightly higher pressures provided a more distinct boiling point separation between residual solvent/starting material and the product epoxide.

10. The product was found to be >98% pure by gas chromatography on a 30 m HP5 capillary column from Agilent Technologies, Inc. (65 °C isothermal, t_R = 1.31 min). The product has been characterized as follows: ^{1}H NMR (400 MHz, $CDCl_3$) δ: 2.92 (dd, *J* = 6.0, 4.0, 1 H), 2.95 (dd, *J* = 6.0, 2.5, 1 H), 3.43 (dd, *J* = 4.0, 2.5, 1 H), 3.77 (s, 3 H); ^{13}C NMR (100 MHz, $CDCl_3$) δ: 46.5, 47.1, 52.3, 169.9; IR (neat) cm^{-1}: 1740, 1441, 1394, 1298, 1209, 1027, 883; HRMS (*m/z*) (CI NH_3) calc. for $C_4H_{10}NO_3$ $(M+NH_4)^+$ 120.0661, found 120.0661; *d* = 1.16 g/mL. Distilled racemic methyl glycidate stored for two years at 0 °C under air showed no decomposition by ^{1}H NMR and GC analysis. However, this racemate developed an undetectable contaminant that significantly retarded the rate of the resolution (14 h, 25% ee vs 93% ee with freshly prepared epoxide). Even redistilled, the old epoxide was never as reactive as freshly prepared

material. Enantioenriched product isolated from procedure B did not develop this contaminant upon storage.

11. Similar results can also be obtained by replacing *p*-toluenesulfonic acid with electron-deficient benzoic acids, although these catalysts require somewhat higher loadings (for example, 0.7 mol% (salen)Co(II) with 3,5-bis(trifluoromethyl)benzoic acid, 0.8 mol% (salen)Co(II) with 3,5-dinitrobenzoic acid, or 0.9 mol% (salen)Co(II) with 4-nitrobenzoic acid). Such electronic tuning of the catalyst may be beneficial for certain epoxides (see Discussion).

12. Oxidation to the catalytically active Co(III) complex is dependent on maintaining sufficient oxygenation of the solution. Use of a smaller flask or slow stirring can lead to incomplete oxidation and substantially slower kinetic resolution reactions.

13. If desired, this catalyst may be further purified by suspension in pentane, filtration, and drying under vacuum. This provides the (salen)Co(III)OTs•H_2O catalyst as a free-flowing, green powder. This catalyst remains fully active for at least four months when stored under air on the benchtop. Reactions with this catalyst are initiated by simply adding the (salen)Co(III) species to a stirring mixture of epoxide and water at room temperature.[4d]

14. A glass stopper is preferable to a septum, which can absorb solvent and epoxide.

15. Heating the mixture to reflux induces catalyst reduction and allows removal of most of the Co-containing species by filtration. The aqueous extraction serves to remove most of the diol product. This workup procedure was developed to eliminate catalyst- and diol-induced foaming during distillation of the epoxide. This is not a significant problem in the hydrolytic kinetic resolution of many other epoxides, in which case direct distillation of the epoxide from the reaction mixture remains the most convenient isolation method.[4]

16. After drying under vacuum for 1 h, the recovered (salen)Co(II) complex (0.758-0.789 g, 83-86% recovery) can be recycled without any further purification. The reuse of this catalyst has been demonstrated by resolving an additional 31.0 g of methyl glycidate by using recovered (salen)Co(II) (plus a small amount of fresh complex to reach the desired 0.5 mol% catalyst loading).

17. The enantiomeric purity was determined by chiral gas chromatography on a 20 m Chiraldex γ-TA capillary column from Advanced Separation Technologies, Inc. Using an 80°C isothermal method, the (*R*)-

and (*S*)- enantiomers had retention times of 4.1 and 7.3 min, respectively. The optical rotations of (*S*)-methyl glycidate were found to be $[\alpha]_D^{27}$ –17.3 (neat) and $[\alpha]_D^{26}$ –10.3 (methanol, c 5.34).

Waste Disposal Information

All hazardous materials should be handled and disposed of in accordance with "Prudent Practices in the Laboratory"; National Academy Press; Washington, DC, 1995.

3. Discussion

The hydrolytic kinetic resolution (HKR) of terminal epoxides catalyzed by chiral (salen)Co complexes affords highly enantioenriched unreacted epoxides and 1,2-diol products.[4] The HKR displays extraordinary scope, as a wide assortment of sterically and electronically varied epoxides can be resolved to ≥99% ee (Figure 1). The general availability of racemic

Figure 1: Epoxides Resolved to >99% ee via the HKR[4c]

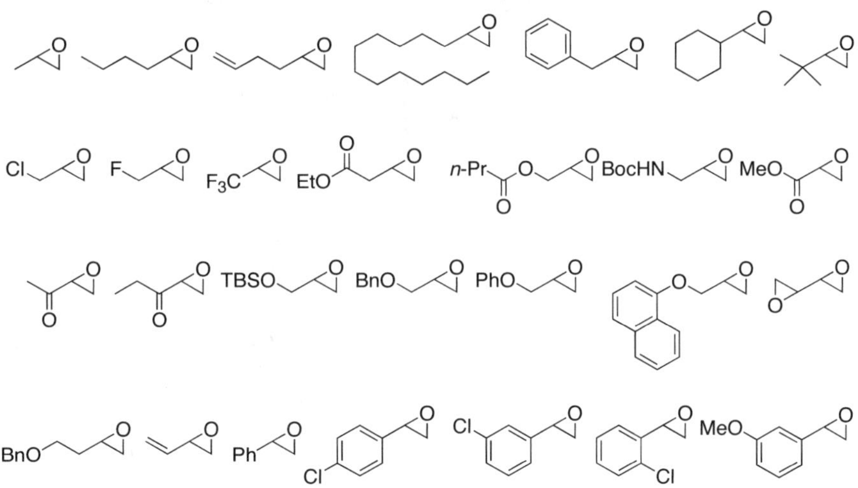

epoxides, the use of water as the resolving agent, and the low loadings of a recyclable, commercially available catalyst make this methodology particularly practical. Accordingly, the utility of the HKR has already been widely demonstrated in organic synthesis.[5] Recent mechanistic studies[4d] have revealed that the more electrophilic (salen)Co(III)OTs catalyst is significantly more reactive than the (salen)Co(III)OAc catalyst originally reported.[4a-c] The increased Lewis acidity of the (salen)Co(III)OTs catalyst has, however, been shown to be detrimental to the HKR's of styrene oxide and propargyl epoxides, presumably due to participation of less selective S_N1 pathways.

While the broad utility of epoxides as synthetic intermediates has been well documented,[6] methyl glycidate is a particularly versatile chiral building block. A range of reactions has been developed that selectively transform either the epoxide or ester functionality (Scheme 1). The epoxide can be converted to an α-hydroxyester by carbonylation[7] or by organocuprate,[8] phenol,[9] or indole[10] addition. The resulting hydroxyl group can then be displaced under Mitsunobu conditions[11] or as the sulfonate.[12] The epoxide can also be opened with 1-morpholino-2-trimethylsilyl acetylene yielding a γ-lactone.[13] Alternatively, methyl glycidate can be converted to a variety of epoxyketones by addition of Grignard or organolithium reagents.[14] The resulting epoxyketones undergo diastereoselective reductive amination,[15] or can be converted to tri-substituted epoxyalcohols by using a Grignard addition/Payne rearrangement protocol.[16]

Scheme 1: Synthetic Uses of Methyl Glycidate

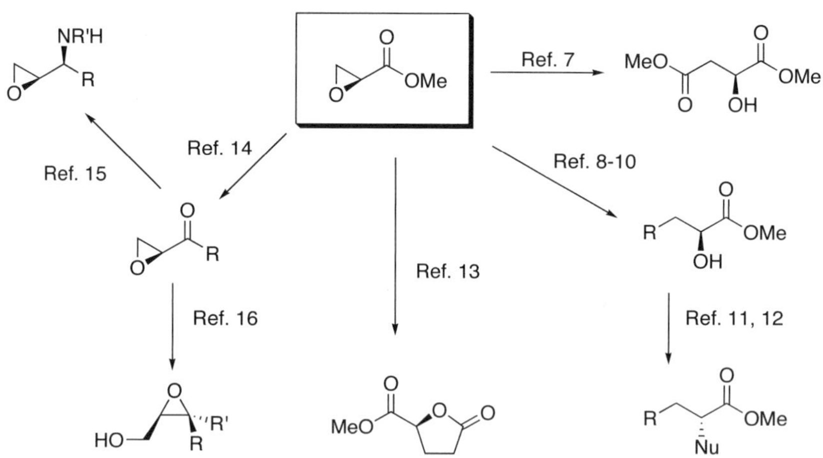

1. Department of Chemistry and Chemical Biology, Harvard University, Cambridge, MA, 02138.

2. Nemes, A.; Czibula, L.; Visky, G.; Farkas, M.; Kreidl, J. *Heterocycles* **1991**, *32*, 2329.
3. Hansen, K. B., Ph. D. Thesis, Harvard University; Cambridge, Massachusetts, 1998.
4. (a) Tokunaga, M.; Larrow, J. F.; Kakiuchi, F.; Jacobsen, E. N. *Science* **1997**, *277*, 936; (b) Furrow, M. E.; Schaus, S. E.; Jacobsen, E. N. *J. Org. Chem.* **1998**, *63*, 6776; (c) Schaus, S. E.; Brandes, B. D.; Larrow, J. F.; Tokunaga, M.; Hansen, K. B.; Gould, A. E.; Furrow, M. E.; Jacobsen, E. N. *J. Am. Chem. Soc.* **2002**, *124*, 1307; (d) Nielsen, L. P. C.; Stevenson, C. P.; Blackmond, D. G.; Jacobsen, E. N. *J. Am. Chem. Soc.* **2004**, *126*, 1360.
5. Selected examples: (a) Schaus, S. E.; Brånalt, J. E.; Jacobsen, E. N. *J. Org. Chem.* **1998**, *63*, 4876; (b) Liu, P.; Panek, J. S. *J. Am. Chem. Soc.* **2000**, *122*, 1235; (c) Chavez, D. E.; Jacobsen, E. N. *Angew. Chem., Int. Ed. Engl.* **2001**, *40*, 3667; (d) Holland, J. M.; Lewis, M.; Nelson, A. *Angew. Chem., Int. Ed. Engl.* **2001**, *40,* 4082; (e) Bode, J. W.; Carreira, E. M. *J. Org. Chem.* **2001**, *66*, 6410; (f) Fürstner, A.; Thiel, O. R.; Ackermann, L. *Org. Lett.* **2001**, *3*, 449.
6. For reviews, see: (a) Winstein, S.; Henderson, R. B. In *Heterocyclic Compounds*, Vol. 1; Elderfield, R.C., Ed.; Wiley: New York, 1950; Chapter 1; (b) Parker, R. E.; Isaacs, N. S. *Chem. Rev.* **1959**, *59*, 737; (c) Bartók, M.; Láng, K. L. Small Ring Heterocycles. In *The Chemistry of Heterocyclic Compounds*, Vol. 42, Part 3; Hassner, A., Ed.; Wiley: New York, 1985; Chapter 1; (d) Rao, A. S.; Paknikar, S. K.; Kirtane, J. G. *Tetrahedron* **1983**, *39*, 2323; (e) Smith, J. G. *Synthesis* **1984**, 629.
7. Hinterding, K.; Jacobsen, E. N. *J. Org. Chem.* **1999**, *64*, 2164.
8. Mori, K.; Takikawa, H. *Tetrahedron* **1991**, *47*, 2163.
9. Ready, J. M.; Jacobsen, E. N. *J. Am. Chem. Soc.* **1999**, *121*, 6086.
10. Entzeroth, M.; Kunczik, T.; Jaenicke, L. *Liebigs Ann. Chem.* **1983**, 226.
11. For a review, see: Mitsunobu, O. *Synthesis* **1981**, 1.
12. Hoffman, R. V.; Kim, H.-O. *Tetrahedron Lett.* **1993**, *34*, 2051; and references therein.
13. Movassaghi, M.; Jacobsen, E. N. *J. Am. Chem. Soc.* **2002**, *124*, 2456.
14. Pegorier, L.; Petit, Y.; Mambu, A.; Larchevêque, M. *Synthesis* **1994**, 1403
15. (a) Pegorier, L.; Petit, Y.; Larchevêque, M. *J. Chem. Soc., Chem. Commun.* **1994**, 633; (b) Haddad, M.; Botuha, C.; Larchevêque, M. *Synlett* **1999**, 1118.
16. Luchi, J. T.; Jacobsen, E. N., Unpublished results.

SYNTHESIS OF 1,2:5,6-DIANHYDRO-3,4-O-ISOPROPYLIDENE-L-MANNITOL

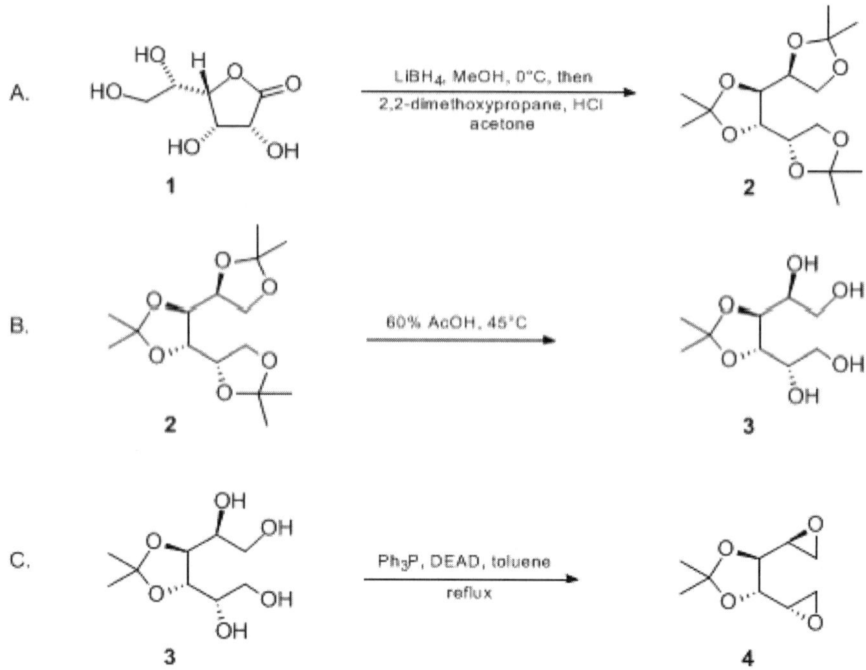

Submitted by David A. Nugiel,[1] Kim Jacobs, A. Christine Tabaka, and Chris A. Teleha.[2]
Checked by Peter A. Orahovats, Jason S. Newcom, and William R. Roush.

1. Procedure

A. 1,2:3,4:5,6-Tri-O-isopropylidene-L-mannitol (**2**). A 2-L, two-necked, round-bottomed flask equipped with a powder funnel and a magnetic stirbar is charged with L-mannonic acid γ-lactone (20 g, 0.11 mol) (Note 1) and 300 mL of methanol. The resulting suspension is cooled in an ice bath and lithium borohydride (4.12 g, 0.193 mol) is added in several portions via the powder funnel (Note 2) over 30 min. *Caution! Extensive foaming due to hydrogen gas evolution.* After the addition is complete, the ice bath is removed and the mixture is stirred for an additional 30 min. The reaction mixture is then recooled in an ice-bath and quenched by the addition of 4N HCl in dioxane (Note 3) until the mixture stops bubbling (1-2 mL). The mixture is then transferred to a 2-L, single-necked, round-bottomed flask and the solvent is concentrated at reduced pressure until a glassy solid is obtained (Note 4). The round-bottomed flask is fitted with a three-way stopcock to which a nitrogen-filled balloon is attached. The solid is then suspended in 100 mL of acetone and 2,2-dimethoxypropane (78 mL, 0.66 mol) (Note 5) is added in one portion. Next, 4N HCl in dioxane (82.5 mL, 0.33 mol) is added slowly with stirring. The reaction mixture is stirred at ambient temperature for 16 h, during which time it becomes bright red in color. If TLC analysis indicates the reaction is not complete (Note 6), then an additional portion of 2,2-dimethoxypropane (13 mL, 0.11 mol) is added and stirring is continued for an additional 6 h. The solvent volume is then reduced by 80% at reduced pressure and the reaction mixture is slowly poured into 400 mL of saturated sodium bicarbonate solution. The product precipitates from solution, and after 10-12 h the solid is collected on a Büchner funnel and air-dried (Note 7). The waxy solid is dissolved in 350 mL of absolute ethanol and cooled to −78 °C to give 23.1-24.3 g (69%) of the product as a white solid. A second crop of crystals is collected (5.7-5.9 g) and combined to give a final yield of 29.0-30.0 g (86-87%) (Notes 8, 9).

B. 3,4-O-Isopropylidene-L-mannitol (**3**). A 2-L, single-necked, round-bottomed flask equipped with a magnetic stirbar is charged with 1,2:3,4:5,6-tri-*O*-isopropylidene-L-mannitol (2) (20 g, 0.066 mol) and 300 mL of 60% acetic acid. The flask is attached to a rotary evaporator (at atmospheric pressure) and the reaction mixture is stirred at 45 °C for ca. 1.5 h. At that point, monitoring by TLC indicates that the reaction has progressed approximately halfway to completion. The pressure in the rotary evaporator is then reduced to 1 mmHg (by attaching it to a vacuum pump in a hood), and the bath temperature is reduced to 40 °C. Under these conditions, full removal of the solvent is achieved in ca. 60 min to yield a highly viscous slurry. This residue is taken up in 200 mL of dichloromethane, stirred for 10 min, and any precipitate is removed by filtration through Celite (Note 11). The filtrate is concentrated at reduced pressure, and the residue is crystallized from 120 mL of diethyl ether and dried at 1 mmHg to give 9.8 g (67%) of the product as a white solid. The filtrate is concentrated to a volume of 50 mL and a second crop of 1.1 g (7%) is collected by filtration and combined to give a final yield of 10.9 g (74%) (Notes 12, 13).

C. 1,2:5,6-Dianhydro-3,4-O-isopropylidene-L-mannitol (**4**). A 1-L, two-necked, round-bottomed flask equipped with a nitrogen inlet, addition funnel, and a magnetic stirbar is charged with 3,4-*O*-isopropylidene-L-mannitol (14.5 g, 0.065 mol), 160 mL of dry toluene (Note 14), and triphenylphosphine (42.9 g, 0.163 mol). The stirred suspension is cooled under a nitrogen atmosphere in an ice bath while diethyl azodicarboxylate (25.9 mL, 0.163 mol) (Note 15) is added dropwise over 20 min. During the addition the reaction mixture becomes homogeneous. After the addition is complete, the ice bath is replaced with an oil bath, the addition funnel is replaced with a reflux condenser, and the reaction mixture is heated at reflux with continued stirring for 1-2 h (Note 16). The resulting pink reaction mixture is allowed to cool to room temperature, applied directly to a dry silica gel column, and eluted with 30-50% ether/hexane to give 9.4 g (78%) of the desired product as a volatile oil (Note 17).

2. Notes

1. L-Mannonic acid γ-lactone was purchased from Sigma Chemical Company, St. Louis, MO.

2. If the lithium borohydride sticks to the spatula or powder funnel, additional methanol can be used to wash the material into the reaction mixture without affecting the yield. On this reaction scale, an additional 25-50 mL of methanol could be used. Using 2.0M $LiBH_4$ in THF instead of solid $LiBH_4$ complicates the following step resulting in lower yields.

3. 4N HCl in dioxane was purchased from Aldrich Chemical Company, Inc.

4. Caution should be used evaporating the solvent, as the residue tends to foam upon concentration. Concentration is carried out by rotary evaporation and then for 30 min at high vacuum (1 mmHg) while warming with a heat gun to remove all of the methanol.

5. Purchased from Aldrich Chemical Company, Inc. and used as received.

6. The reaction was monitored using TLC with 60% ether/hexane as the eluent. The desired product has $R_f = 0.95$ and the undesired diacetonide has $R_f = 0.25$. Visualization was done using *p*-anisaldehyde.

7. The checkers found that best results were obtained when the trisacetonide was allowed to precipitate from the sodium bicarbonate solution over 10-12 h.

8. The checkers found that it was necessary to cool the crystallization solution to −78 °C to induce crystallization.

9. The physical and spectral properties are as follows: mp 72-74 °C; $[\alpha]_D = -13.4$ ($CHCl_3$, c 1.2); ^1H NMR (400 MHz, $CDCl_3$) δ: 1.4 (s, 6 H),

1.45 (s, 6 H), 1.5 (s, 6 H), 4.05 (m, 4 H), 4.15 (m, 2 H), 4.25 (m, 2 H); CIMS (NH_3) *m/z*: 303 ($M+H^+$, 100%).

10. The reaction was monitored using TLC with 5% $MeOH/CH_2C_2$ as the eluent. The desired monoacetonide 3 has a R_f = 0.25. It was desirable to run the reaction to the point where most of the higher R_f components were consumed. This will produce some unwanted L-mannitol that can be filtered off as described above.

11. The L-mannitol recovered in this manner could be recycled if desired.

12. The checkers found that further hydrolysis of mannitol diacetonide can occur during the removal of solvent, and that removal of all acetic acid from the product was problematic. Residual acetic acid in the product complicates the next step. The procedure described reduces these problems.

13. The physical and spectral properties are as follows: mp 83-86 °C; $[\alpha]_D = -26.4$ (H_2O, c 3); ^1H NMR (400 MHz, DMSO-d_6) δ: 1.25 (s, 6 H), 3.35 (m, 2 H), 3.5 (m, 4 H), 3.85 (m, 2 H), 4.45 (t, $J = 5.9$ Hz, 2 H), 5.1 (d, $J = 4.4$ Hz, 2 H); ^{13}C NMR (100 MHz, DMSO-d_6) δ: 29.1, 64.8, 74.7, 80.9, 110.1; CIMS (NH_3) *m/z*: 240 ($M+NH_4^+$, 100%); Anal. calcd. for $C_9H_{14}O_4$: C, 48.64; H, 8.16. Found: C, 48.54; H, 8.16.

14. The toluene used was from a freshly opened bottle obtained from EM Science.

15. Neat DEAD was unavailable from commercial sources at the time the procedure was being checked. Therefore, the checkers used a commercially available 40% solution of DEAD in toluene with results comparable to that described in the original procedure.

16. The reaction was initially monitored using 10% $MeOH/CH_2Cl_2$ to follow the disappearance of tetraol **3**. The appearance of the diepoxide **4** is monitored using 50% ether/hexane.

17. The physical and spectral properties are as follows: $[\alpha]_D = +0.38$ ($CHCl_3$, c 0.8); ^1H NMR (400 MHz, $CDCl_3$) δ: 1.40 (s, 6 H), 3.77 (dd, 2 H, $J = 2.9, 1.4$), 2.65 (dd, 2 H, $J = 5.2, 3.0$), 2.80 (t, 2 H, $J = 9.5$), 3.10 (m, 2 H); ^{13}C NMR (100 MHz, $CDCl_3$) δ: 16.6, 27.9, 46.3, 52.6, 79.5, 111.5; CIMS (NH_3 m/z: 204 ($M+NH_4^+$, 100%); Anal calcd. for $C_9H_{14}O_4$: C, 58.05; H, 7.58. Found: C, 57.78; H, 7.52.

3. Discussion

Preparing the title compound from L-mannitol using methodology developed for the D-isomer[2] would be prohibitively expensive. Our approach uses a much less expensive, commercially available starting material. Current synthetic approaches to the D-isomer involve selective conversion of the two primary hydroxyl groups of 3,4-*O*-isopropylidene-D-mannitol into good leaving groups followed by base treatment to facilitate epoxide formation. We found prolonged exposure of the epoxide to base reduces the reaction's overall yield. To overcome this liability we used a one-pot conversion of 3,4-*O*-isopropylidene-L-mannitol to the title compound using Mitsunobu-based technology. This approach was found to be more reproducible and consistently gave yields in the 60-80% range.

The title compound is a key C_6 building block. Several labs have prepared novel α-amino acids, biological probes and other interesting compounds using the D-diepoxide as a key intermediate.[3] An efficient route to the L-enantiomer provides a pathway to compounds with the opposite configuration, one not readily available from commercial sources, and a valuable probe of stereochemistry in biological systems and reaction mechanism.

1. Astrazeneca, Inc., B312, CNS Discovery, 1800 Concord Pike, Wilmington, DE, USA 19850-5437

2. Johnson and Johnson Pharmaceutical Research and Development, LLC, Welsh and McKean Roads, P.O. Box 776, Spring House, PA 19477-0776.

3. (a) Wiggins, L.F. *J. Chem. Soc.* **1946**, 384. (b) Le Merrer, Y.; Dureault, A.; Greck, C.; Micas-Languin, D.; Gravier, C.; Depezay, J. *Heterocycles*, **1987**, *25*, 541. (c) Ghosh, A.K.; McKee, S.P.; Thompson, W.J. *Tetrahedron Lett.* **1991**, *32(41)*, 5729.

SYNTHESIS AND USE OF GLYCOSYL PHOSPHATES AS GLYCOSYL DONORS

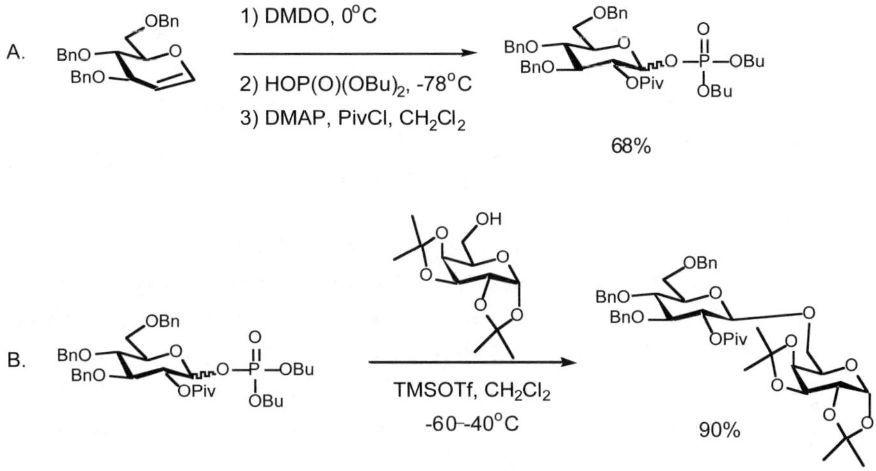

Submitted by Kerry R. Love and Peter H. Seeberger.[1]
Checked by Youseung Shin and Dennis P. Curran.

1. Procedure

A. Dibutyl 3,4,6-tri-O-benzyl-2-O-pivaloyl-D-glucopyranosyl phosphate. A dry 500-mL, round-bottomed flask is charged with 5.00 g (12.0 mmol) of 3,4,6-tri-*O*-benzyl-D-glucal (Note 1) and 7 mL of toluene (Note 2). The toluene is removed azeotropically on a rotary evaporator and this process is repeated two more times. The residue is dried under vacuum for 30 min. Under a flow of nitrogen, the flask is charged with a magnetic stirbar and 50 mL of dichloromethane (Note 2) and the resulting solution is cooled to 0 °C. A solution of dimethyldioxirane (200 mL of a 0.08M solution, 16.0 mmol) (Note 3) is added to the flask via cannula and the reaction mixture is stirred at 0 °C for 10 min (Note 4). Volatile materials are then removed by rotary evaporation at 0 °C (Note 5) and the resulting white residue is dried at 2 mm for 5 min. The flask is equipped with a rubber septum fitted with a nitrogen inlet needle and the residue is dissolved in 40 mL of CH_2Cl_2 and the solution is cooled to –78 °C. A solution of dibutyl phosphate (2.85 mL, 14.4 mmol) (Note 6) in 10 mL of dichloromethane is added via cannula. After 10 min at –78 °C, the reaction mixture is warmed to 0 °C and 4-(dimethylamino)pyridine (DMAP) (5.86 g, 48.0 mmol) and pivaloyl chloride (2.96 mL, 24.0 mmol) are added (Note 7). The solution is stirred for 30 min at 0 °C, then allowed to warm to room temperature and stirred for 16 h (Note 8). A solution of 25% ethyl acetate in hexanes is added and the resulting suspension is filtered through a pad of silica gel. The filtrate is concentrated to a clear oil and purified by column chromatography through a short plug of 80 g of silica gel (elution with 35% ethyl acetate-hexanes) (Note 9) to afford 5.86 g (67%) of dibutyl 3,4,6-tri-*O*-benzyl-2-*O*-pivaloyl-D-glucopyranosyl phosphate as a colorless oil (Notes 10, 11).

B. 3,4,6-Tri-O-benzyl-2-O-pivaloyl-β-D-glucopyranosyl-(1→6)-1,2:3,4-di-O-isopropylidene-α-D-galactopyranoside. A dry 500-mL, round-bottomed flask is charged with dibutyl 3,4,6-tri-*O*-benzyl-2-*O*-pivaloyl-D-glucopyranosyl phosphate (5.89 g, 8.11 mmol), 1,2:3,4-di-*O*-isopropylidene-α-D-galactopyranose (1.92 g, 7.37 mmol) (Note 12), and 7 mL of toluene (Note 2). The toluene is removed azeotropically on a rotary evaporator and this process is repeated two more times. The residue is dried under vacuum for 1 hr. The flask is equipped with a rubber septum fitted with a nitrogen inlet needle and a magnetic stirbar, and 100 mL of CH_2Cl_2 is added. The resulting solution is cooled to –60 °C. Trimethylsilyl triflate (TMSOTf) (1.47 mL, 8.11 mmol) (Note 13) is added and the solution is allowed to warm slowly to –40 °C (Note 14). After 30 min, the reaction is quenched by addition of 15 mL of triethylamine (Note 15) and allowed to warm to room temperature. The reaction mixture is concentrated by rotary evaporation and the residue is purified by column chromatography on 140 g of silica gel (elution with 30% ethyl acetate-hexanes) to afford 5.09 g (85%) of 3,4,6-tri-*O*-benzyl-2-*O*-pivaloyl-D-glucopyranosyl-(1→6)-1,2:3,4-di-*O*-isopropylidene-α-D-galactopyranoside as a colorless oil (Notes 16,17).

2. Notes

1. Tri-*O*-benzyl-D-glucal (97%) was purchased from Aldrich Chemical Company, Inc. and was used as supplied without further purification.

2. Toluene and dichloromethane (CH_2Cl_2) were purified by a JT Baker Cycle-Tainer Solvent Delivery System.

3. Dimethyldioxirane was prepared according to the literature procedure[2] as an 0.08M solution and was dried over 4Å beaded molecular sieves for 24 h prior to use.

4. Analytical thin-layer chromatography was performed on E. Merck silica gel 60 F_{254} plates (0.25 mm) and compounds were visualized by dipping the plates in a cerium sulfate-ammonium molybdate solution followed by heating.

5. Expedient removal of the volatiles after the epoxidation with dimethyldioxirane is crucial to achieve reproducible yields because the epoxide is extremely water sensitive.

6. Dibutyl phosphate (97%) was purchased from Fluka and was used as supplied without further purification.

7. DMAP (99%) and pivaloyl chloride (99%) were purchased from Aldrich Chemical Company, Inc. and were used as supplied without further purification.

8. The submitters recommend that the reaction be worked up immediately upon reaching room temperature, but the checkers found that it was not detrimental to allow the reaction to stir overnight at room temperature prior to workup.

9. Prolonged chromatography when purifying glycosyl phosphates will lead to hydrolysis and decomposition of the product.

10. Early fractions were mixtures of α/β anomers while later fractions contained pure β-anomer. The fractions were combined to give the product as a 9:1 mixture of β/α-anomers. Analytical data for this compound are as follows: (β-phosphate) R_f 0.29 (ethyl acetate:hexane, 1:3); $[\alpha]^{24}_{D}$ −1.9 (c = 1.50, CH_2Cl_2); IR (thin film) cm^{-1}: 2946, 1740, 1454, 1282, 1016; ^1H NMR (500 MHz, $CDCl_3$) δ: 0.96-0.88 (m, 6H), 1.20 (s, 9H), 1.40-1.34 (m, 4H), 1.64-1.59 (m, 4H), 3.64-3.61 (m, 1H), 3.78-3.70 (m, 3H), 3.82 (t, J = 9.5 Hz, 1H), 4.08-4.00 (m, 4H), 4.51 (d, J = 11.0 Hz, 1H), 4.69-4.54 (m, 2H), 4.70 (d, J = 11.0 Hz, 1H), 4.80-4.75 (m, 2H), 5.17 (app t, J = 8.5 Hz, 1H), 5.24 (app t, J = 7.3 Hz, 1H), 7.16-7.14 (m, 2H), 7.33-7.25 (m, 13H); ^{13}C NMR (125 MHz, $CDCl_3$) δ: 14.0, 19.1, 26.9, 32.7, 39.2, 68.1, 68.2, 68.4, 73.3, 73.9, 75.9, 76.2, 83.1, 97.0, 127.6, 128.0, 128.1, 128.2, 128.3, 128.7, 138.1, 138.2, 177.2, (d, J_{C-P} = 5.0 Hz); ^{31}P NMR (200 MHz, $CDCl_3$) δ: -2.2; FAB MS m/z $(M)^+$ calcd 726.3532, obsd 726.3537. (α-Phosphate) R_f 0.36 (ethyl acetate:hexane, 1:3); $[\alpha]^{24}_D$ +50.5 (c = 0.63, CH_2Cl_2); IR (thin film) cm^{-1}: 2960, 2872, 1736, 1454, 1282; ^1H NMR (500 MHz, $CDCl_3$) δ: 0.97-0.91 (m, 6H), 1.24 (s, 9H), 1.44-1.36 (m, 4H), 1.86-1.61 (m, 4H), 3.68 (d, J = 11.0 Hz, 1H), 3.86-3.79 (m, 2H), 4.10-4.02 (m, 5H), 4.56-4.50 (m, 3H), 4.63 (d, J = 11.5 Hz, 1H), 4.83-4.80 (m, 3H), 4.99-4.97 (m, 1H), 5.85 (dd, J = 1.8, 6.4 Hz, 1H), 7.18-7.15 (m, 2H), 7.35-7.27 (m, 13H); ^{13}C NMR

(125 MHz, $CDCl_3$) δ: 13.8, 18.8, 27.3, 32.4, 32.5, 39.0, 67.8, 67.9, 68.0, 68.2, 72.6, 72.7, 73.7, 75.4, 75.6, 79.5, 94.7, 127.6, 127.7, 127.9, 128.0, 128.1, 128.2, 128.3, 128.5, 128.6, 138.0, 138.1, 138.3, 177.7, (d, $J_{C\text{-}P}$ = 5.5 Hz); ^{31}P NMR (200 MHz, $CDCl_3$) δ: -2.5; FAB MS *m/z* $(M)^+$ calcd 726.3532, obsd 726.3537.

11. Practical considerations limit the scalability of this reaction due to the highly reactive and water sensitive intermediates formed. Furthermore, the time required for removal of large amounts of solvent *in vacuo* allows for the opening of the intermediate epoxide leading to diol formation.

12. 1,2:3,4-Di-*O*-isopropylidene-α-D-galactopyranoside (97%) was purchased from Aldrich Chemical Company, Inc. and was used as supplied without further purification.

13. TMSOTf (99%) was purchased from Acros Organics and was used as supplied without further purification.

14. The submitters report that the less reactive α-phosphate can be completely consumed by allowing the reaction mixture to warm to –20 °C prior to workup.

15. Triethylamine (98%) was purchased from J. T. Baker and was used as supplied without further purification.

16. The product was obtained in 90-91% yield in other runs. Analytical data for this compound are as follows: $[\alpha]^{24}{}_D$–45.2 (*c* = 2.34, CH_2Cl_2); IR (thin film) cm^{-1}: 3029, 2978, 2933, 2904, 1741, 1134, 1028; ^{1}H NMR (500 MHz, $CDCl_3$) δ: 1.21 (s, 9H), 1.31 (s, 3H), 1.32 (s,3H), 1.43 (s, 3H), 1.51 (s, 3H), 3.53-3.50 (m, 1H), 3.63-3.59 (m,1H), 3.76-3.69 (m, 4H), 3.97-3.94 (m, 1H), 4.10-4.07 (m, 1H), 4.29-4.25 (m, 2H), 4.46 (d, *J* = 8.0 Hz, 1H), 4.58-4.53 (m, 3H), 4.64 (d, *J* = 8.0 Hz, 1H), 4.79-4.69 (m, 3H), 5.10 (app t, *J* = 8.5 Hz), 5.49 (d, *J* = 5.0 Hz, 1H), 1H), 7.19-7.15 (m, 2H), 7.39-7.24 (m, 13H); ^{13}C NMR (125 MHz, $CDCl_3$) δ: 24.7, 25.4, 26.3, 26.4, 27.5, 39.1, 67.4, 69.0, 70.9, 71.5, 73.3, 73.9, 75.2, 75.7, 78.1, 83.6, 96.6, 101.8, 108.8, 109.5, 127.7, 127.9, 127.8, 128.0, 128.1, 128.2, 128.6, 128.7, 138.4, 138.5, 177.1; FAB MS *m/z* $(M)^+$ calcd 776.3772, obsd 776.3770.

17. Continued elution gave unreacted α-phosphate (0.125 g, 2%).

Waste Disposal Information

All toxic materials were disposed of in accordance with "Prudent Practices in the Laboratory"; National Academy Press; Washington, D.C., 1995.

3. Discussion

The role of complex carbohydrates in biological processes is now widely appreciated. Oligosaccharides and glycoconjugates are essential for many cellular events, such as recognition, adhesion and signaling between cells.[3] Carbohydrates have also been implicated in a variety of disease states including cancers[4] and in a host of bacterial and viral infections.[5] The widespread biological implications of carbohydrates have rendered them targets of intense study. Microheterogeneity of naturally derived oligosaccharides, however, limits the access to pure materials in appreciable quantity. Chemical synthesis, therefore, remains the best way procure material for biological investigations.

Advances in carbohydrate chemistry, particularly in the development of powerful glycosylating agents, have provided access to molecules of biological interest. In particular, glycosyl trichloroacetimidates,[6] glycosyl fluorides,[7] thioglycosides[8] and n-pentenyl glycosides[9] have each been used in the construction of complex oligosaccharides. Limitations of these glycosyl donors, including lengthy syntheses, difficult activation conditions, and long reaction times, have lead us to develop glycosyl phosphates for the facile construction of a variety of glycosidic linkages.

Glycosyl phosphates can be synthesized from a variety of intermediates; anomeric lactols[10,11] and 1,2-anhydrosugars[12] are most typically used. Additionally, other glycosyl donors, such as glycosyl trichloroacetimidates[13] and *n*-pentenyl glycosides,[14] may be converted into glycosyl phosphates. The aforementioned procedure generates the 1,2-anhydrosugar from a suitably protected glycal with dimethyldioxirane. Opening of the anhydrosugar at low temperature using a phosphate diester, followed by acylation of the C^2-hydroxyl group leads to fully protected glycosyl phosphates (Figure 1).[12]

This method is attractive for three reasons: First, glycal precursors are desirable starting materials because they require the differentiation of only three hydroxyl groups, each having unique reactivity. Generation of phosphates from anomeric lactols or other glycosyl donors requires many synthetic steps for the differential protection of the five hydroxyl groups present. Second, the procedure can be carried out in one pot providing quick access to the desired glycosyl phosphates in high yields. Finally, choice of solvent for the opening of the anhydrosugar can lead to either alpha- or beta-enriched phosphates.[12]

Initial work by Ikegami and coworkers showed that glycosyl phosphates are highly reactive glycosylating agents.[15] Activation of the glycosyl phosphate by addition of a stoichiometric amount of a Lewis acid in the presence of an acceptor quickly forms the desired disaccharide in good yield (Table 1). Glycosyl phosphates have been successfully employed in the synthesis of both *O*- and *C*-glycosides,[16] including biologically relevant compounds such as the Leishmania antigenic tetrasaccharide[17] and the H-type II blood group pentasaccharide.[18] Use of phosphates on solid-support in an automated fashion further demonstrates their utility as glycosyl donors.[19]

Figure 1
Glycosyl Phosphates Prepared from Glycals

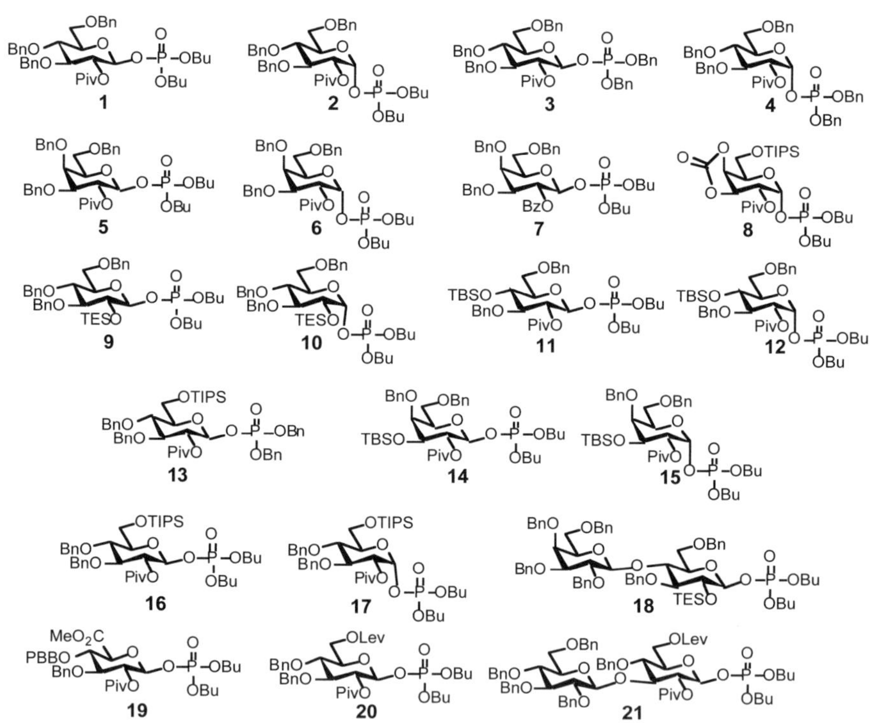

Table 1
Glycosylation with Glycal Derived Glycosyl Phosphates

Donor	Acceptor	Product	Yield (%)
1	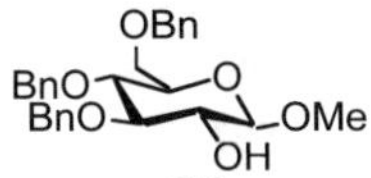 **24**	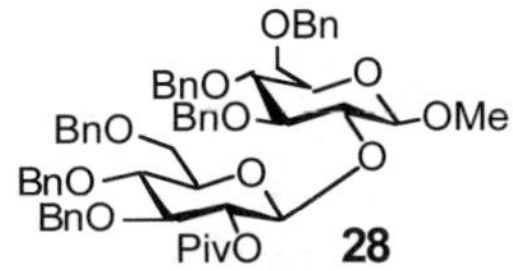 **28**	83
2	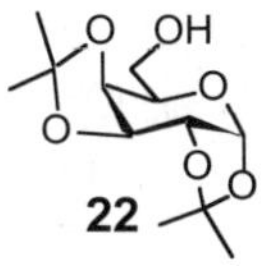 **22**	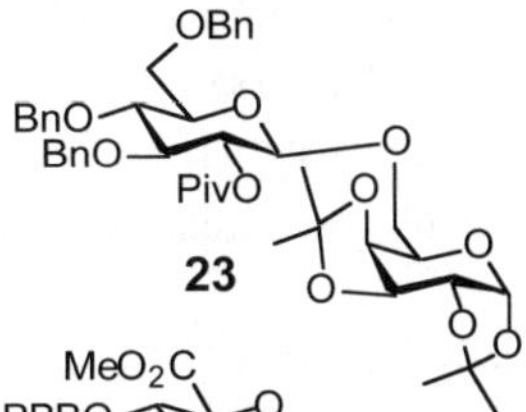 **23**	87
19	**22**	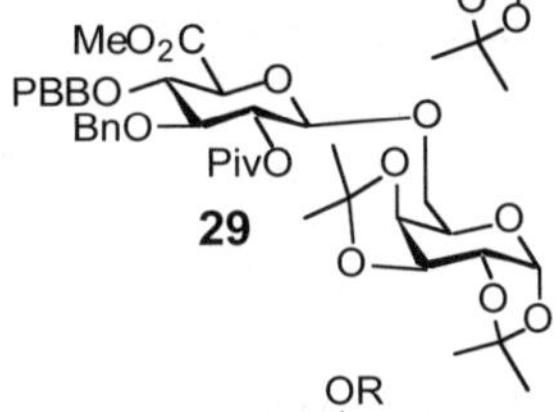 **29**	72
19	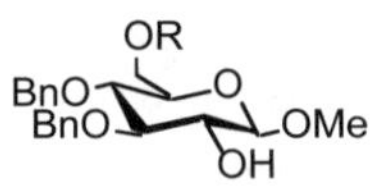 **25**	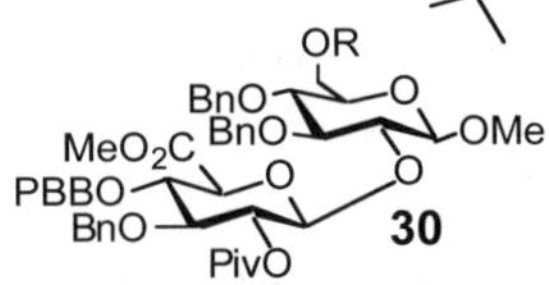 **30**	84
3	EtSH **26**	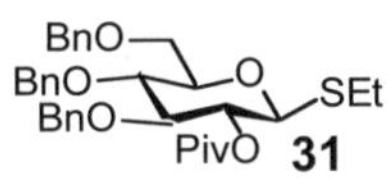 **31**	90
7	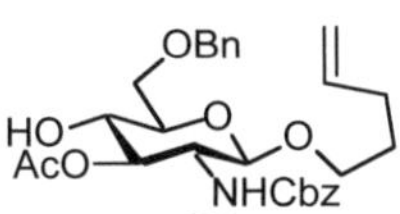 **27**	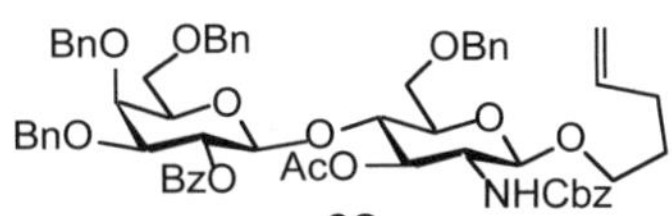 **32**	96
9	**22**	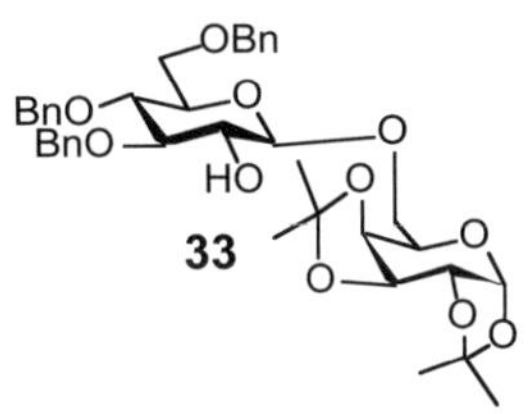 **33**	71

1. Department of Chemistry, Massachusetts Institute of Technology, Cambridge, MA 02139. Present address: Laboratory of Organic Chemistry, ETH Hönggerberg/HCL, CH-8093, Zürich, Switzerland, e-mail: seeberger@org.chem.ehtz.ch.

2. Murray, R. W.; Singh, M. *Org. Synth.* **1996**, *73*, 91.

3. Varki, A. *Glycobiology* **1993**, *3*, 97.
4. Hakomori, S. *Adv. Cancer Res.* **1989**, *52*, 257.
5. Rudd, P. M.; Eliott, T.; Cresswell, P.; Wilson, I. A.; Dwek, R. A. *Science* **2001**, *291*, 2370.
6. Schmidt, R. R. *Adv. Carbohydr. Chem. Biochem.* **1994**, *50*, 21.
7. Mukaiyama, T.; Murai, Y.; Shida, S. *Chem. Lett.* **1981**, 431.
8. Garegg, P. J. *Adv. Carbohydr. Chem. Biochem.* **1997**, *52*, 179.
9. Mootoo, D. R.; Date, V.; Fraser-Reid, B. J. *J. Am. Chem. Soc.* **1988**, *110*, 2662.
10. Via the chlorophosphate: (a) Sabesan, S.; Neira, S. *Carbohydr. Res.* **1992**, *223*, 169. (b) Inage, M.; Chaki, H.; Kusumoto, S.; Shiba, T. *Chem Lett.* **1982**, 1281.
11. Via dehydrative glycosylation: Garcia, B. A.; Gin, D. Y. *Org. Lett.* **2000**, *2*, 2135.
12. Plante, O. J.; Palmacci, E. R.; Andrade, R. B.; Seeberger, P. H. *J. Am. Chem. Soc.* **2001**, *123*, 9545.
13. Hoch, M.; Heinz, E.; Schmidt, R. R. *Carbohydr. Res.* **1989**, *191*, 21.
14. Pale, P.; Whitesides, G. M. *J. Org. Chem.* **1991**, *56*, 4547.
15. Hashimoto, S.; Honda, T.; Ikegami, S. *J. Chem. Soc. Comm.* **1989**, 685.
16. Palmacci, E. R.; Seeberger, P. H. *Org. Lett.* **2001**, *3*, 1547.
17. Hewitt, M. C.; Seeberger, P. H. *J. Org. Chem.* **2001**, *66*, 4233.
18. Love, K. R.; Andrade, R. B.; Seeberger, P. H. *J. Org. Chem.* **2001**, *66*, 8165.
19. Plante, O. J.; Palmacci, E. R.; Seeberger, P. H. *Science*, **2001**, *291*, 1523.

SYNTHESIS OF 1,2:4,5-DI-*O*-ISOPROPYLIDENE-D-erythro-2,3-HEXODIULO-2,6-PYRANOSE. A HIGHLY ENANTIOSELECTIVE KETONE CATALYST FOR EPOXIDATION

[[β-D-erythro-2,3-Hexodiulo-2,6-pyranose, 1,2:4,5-bis-*O*-(1-methylethylidene)-]]

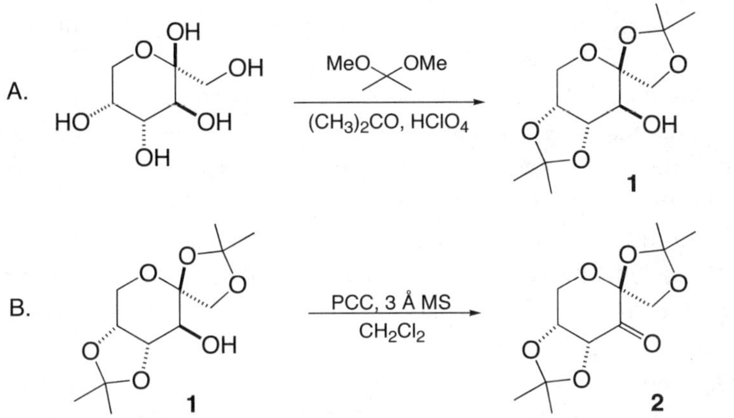

Submitted by Yong Tu, Michael Frohn, Zhi-Xian Wang, and Yian Shi.[1]
Checked by Jason M. Diffendal and Rick L. Danheiser.

1. Procedure

A. 1,2:4,5-Di-O-isopropylidene-β-D-fructopyranose (**1**).[2] D-Fructose (18.0 g, 100 mmol) (Note 1) and 2,2-dimethoxypropane (7.4 mL, 60 mmol) (Note 1) are added to 350 mL of acetone (Note 1) in a 1-L, round-bottomed flask equipped with a Teflon-coated magnetic stir bar. The flask is cooled in an ice bath for 15 min, then 4.3 mL of 70% perchloric acid (Note 2) is added in one portion. The resulting suspension is stirred for 6 h at 0 °C (Note 3). Concentrated ammonium hydroxide (4.8 mL) is then added to neutralize the acid and the solvent is removed by rotary evaporation at 25 °C to give a white solid. This solid is dissolved in 200 mL of dichloromethane (CH_2Cl_2) and washed with two 50-mL portions of saturated sodium chloride solution, dried over sodium sulfate (Na_2SO_4), filtered, and concentrated by rotary evaporation (25 °C) until the total volume is about 40 mL (Note 4). Boiling hexane (100 mL) is then added (Note 5) and the flask is allowed to cool to room temperature, during which time the bulk of the product crystallizes out of solution. Additional product crystallizes upon cooling to −25 °C for 4 h.

Isolation of the solid by vacuum filtration and careful washing with three 25-mL portions of cold (−25 °C) hexane gives 13.4-13.6 g (51-52%) of the title alcohol as fine white needles (Note 6).

B. *1,2:4,5-Di-O-isopropylidene-D-erythro-2,3-hexodiulo-2,6-pyranose* (**2**).[2] A 500-mL, round-bottomed flask equipped with a 4.5-cm, egg-shaped Teflon-coated magnetic stir bar is charged with 130 mL of CH_2Cl_2 (Note 1), the alcohol prepared in Step A (10.4 g, 40.0 mmol), and 15 g of freshly powdered 3 Å molecular sieves (Note 7). Pyridinium chlorochromate (21.5 g, 100 mmol) (Note 1) is added portionwise over 10 min and the resulting mixture is stirred at room temperature for 15 h (Note 8). Ether (200 mL) is added slowly with vigorous stirring and the solution is filtered under vacuum through a pad of 35 g of Celite (Note 9). The solids remaining in the reaction flask are transferred to the Celite pad by scraping with a spatula and washing with three 50-mL portions of ether. The resulting cloudy brown filtrate is concentrated by rotary evaporation at room temperature to give a brown solid. To this solid is added 25 mL of 1:1 ether:hexane and the solids are scraped with a spatula. The mixture is then poured onto 60 g of Whatman 60 Å (230-400 mesh) silica gel packed in a 4-cm diameter chromatography column and the liquid is adsorbed onto the silica gel by gravity (Note 10). The material remaining in the flask is further washed with 1:1 ether:hexane and transferred onto the silica gel; this process is repeated until all the material has been loaded onto the silica gel. The ketone is eluted using 500 mL of 1:1 ether:hexane and the eluent is concentrated by rotary evaporation to afford the crude ketone as a white solid. This material is dissolved in 40-45 mL of boiling hexane. Upon cooling the solution to room temperature, the ketone begins to crystallize. The flask is then cooled to −25 °C for 2 h. The resulting solids are collected by filtration, washed with three 25-mL portions of cold (−25 °C) hexane, and dried to afford 8.84-9.08 g (86-88%) of the ketone as a white solid (Note 11).

2. Notes

1. D-Fructose (98%), 2,2-dimethoxypropane (98%), and pyridinium chlorochromate (98%) were obtained from Aldrich Chemical Company, Inc. and used as received. ACS grade acetone and dichloromethane were purchased from Fisher Scientific and used as received.

2. Perchloric acid (70%) was obtained from J. T. Baker Company. Reaction of 70% perchloric acid with organic materials can lead to fires and

explosions, and anhydrous $HClO_4$ is potentially explosive. Although no incidents occurred in the experience of the submitters, care should be taken in handling this compound.

3. The suspension turns into a clear, colorless solution after 1-2 h. The title compound is the kinetic product of the reaction, and can readily isomerize to 2,3:4,5-di-O-isopropylidene-β-D-fructopyranose (the thermodynamic product; see ref. 2c). Control of the reaction time is important to minimize the formation of the thermodynamic product.

4. The solvent volume can vary slightly without much effect on the yield of the recrystallization step.

5. The white, crystalline product begins to precipitate in the first 5 min after the addition of hexane.

6. The product exhibits the following physical and spectral properties: mp 118.5-119.5 °C; IR (KBr) cm^{-1}: 3547; ^{1}H NMR (500 MHz, $CDCl_3$) δ: 1.37 (s, 3 H), 1.44 (s, 3 H), 1.52 (s, 3 H), 1.54 (s, 3 H), 1.99 (d, 1 H, J = 8.1), 3.67 (dd, 1 H, J = 8.1, 6.8), 3.98 (d, 1 H, J = 9.0), 4.01 (dd, 1 H, J = 13.2, 0.9), 4.12 (dd, 1 H, J = 13.2, 2.7), 4.13 (dd, 1 H, J = 6.8, 5.7), 4.19 (d, 1 H, J = 9.0), 4.22 (ddd, 1 H, J = 5.7, 2.7, 0.9); ^{13}C NMR (125 MHz, $CDCl_3$) δ: 26.1, 26.5, 26.6, 28.1, 61.0, 70.6, 73.52, 73.53, 77.5, 104.7, 109.6, 112.0; HRMS (ESI) Calcd for $C_{12}H_{20}O_6$ $[M+Na]^+$: 283.1152; Found: 283.1149; $[\alpha]_D^{25}$ −144.2 ($CHCl_3$, c 1.0).

7. Mallinckrodt Grade 564 CCGT3A molecular sieves are used without further activation.

8. The mixture turns from orange-brown to a dark brown color during the course of the reaction, indicating the reduction of Cr(VI) to Cr(III).

9. Slow addition of ether is necessary for a high yield. The addition of ether leads to the precipitation of only a small amount of the reduced chromium. This filtration mainly removes the molecular sieves and chromium species adsorbed during stirring.

10. The yield of the ketone will be reduced if all the brown solids are not loaded onto the column (these solids contain some of the ketone).

11. The product exhibits the following physical and spectral properties: mp 101-103 °C; IR (KBr) cm^{-1}: 1749; ^{1}H NMR (500 MHz, $CDCl_3$) δ: 1.40 (s, 6 H), 1.46 (s, 3 H), 1.55 (s, 3 H), 4.00 (d, 1 H, J = 9.5), 4.12 (d, 1 H, J = 13.4), 4.39 (dd, 1 H, J = 13.4, 2.2), 4.55 (ddd, 1 H, J = 5.7, 2.2, 1.0), 4.61 (d, 1 H, J = 9.5), 4.73 (d, 1 H, J = 5.7); ^{13}C NMR (125 MHz, $CDCl_3$) δ: 26.20, 26.24, 26.7, 27.3, 60.3, 70.2, 76.1, 78.1, 104.3, 110.8,

114.0, 197.1; HRMS (ESI) Calcd for $C_{12}H_{18}O_6[M+Na]^+$: 281.0996; Found: 281.0985; $[\alpha]_D^{25}$ −125.4 ($CHCl_3$, c 1.0).

Waste Disposal Information

All toxic materials were disposed of in accordance with "Prudent Practices in the Laboratory"; National Academy Press; Washington, DC, 1995.

3. Discussion

Dioxiranes are remarkably versatile oxidizing agents which show encouraging potential for asymmetric synthesis, particularly asymmetric epoxidation. Dioxiranes can be generated in situ from Oxone ($KHSO_5$) and ketones (Scheme 1).[3] In principle, only a catalytic amount of ketone is required, so with a chiral ketone there exists the opportunity for catalytic asymmetric epoxidation.[4,5] Since the first asymmetric epoxidation of olefins with a chiral dioxirane reported by Curci in 1984,[4a] this area has received intensive interest and significant progress has been made.[4,5]

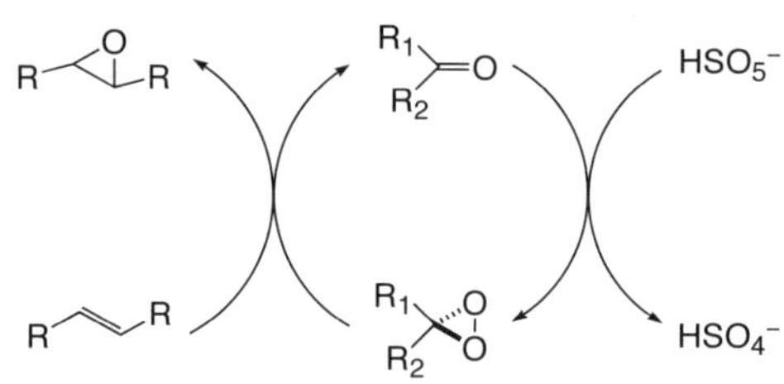

The fructose-derived ketone (**2**) described herein is a member of a class of ketones designed to contain the following general features: (1) The stereogenic centers are close to the reacting center, resulting in efficient transfer of stereochemistry between substrate and catalyst. (2) The presence of fused ring(s) or a quaternary center α to the carbonyl group minimizes epimerization of the stereogenic centers. (3) Approach of an olefin to the reacting dioxirane can be controlled by sterically blocking one face or using a C2- or pseudo-C2-symmetric element. (4) Electron-withdrawing (by induction) substitutents are introduced to activate the carbonyl.

Ketone **2** gives very high enantioselectivities for a variety of trans-disubstituted and trisubstituted olefins.[5] The ketone catalyst can be readily synthesized from very incxpensive D-fructose by ketalization with acetone[2e,h] and subsequent oxidation of the remaining alcohol to the ketone. Other acids such as H_2SO_4 can also be used for ketalization.[2c,f,i] Although the present procedure uses PCC for the oxidation, many other oxidants such as PDC-Ac_2O,[2f] DMSO-Ac_2O,[2a,b,d] DMSO-DCC,[2e] DMSO-$(COCl_2)$,[2g] $RuCl_3$-$NaIO_4$,[2h] Ru-TBHP,[2j] etc. are also effective. The enantiomer of catalyst **2** (**ent-2**) can be prepared in the same fashion from L-fructose, which in turn can be prepared from readily available L-sorbose.[6,5c] As expected, the enantiomeric catalyst shows the same enantioselectivity in epoxidation reactions.

1. Department of Chemistry, Colorado State University, Fort Collins, CO 80523.
2. For leading references see: (a) McDonald, E. *J. Carbohydr. Res.* **1967**, *5*, 106; (b) James, K.; Tatchell, A. R.; Ray, P. K. *J. Chem. Soc. C* **1967**, 2681; (c) Brady, R. F., Jr. *Carbohydr. Res.* **1970**, *15*, 35; (d) Tipson, R. S.; Brady, R. F. Jr.; West, B. F. *Carbohydr. Res.* **1971**, *16*, 383; (e) Prisbe, E. J.; Smejkal, J.; Verheyden, J. P. H.; Moffatt, J. G. *J. Org. Chem.*, **1976**, *41*, 1836; (f) Fayet, C.; Gelas, J. *Carbohydr. Res.* **1986**, *155*, 99; (g) Maryanoff, B. E.; Reitz, A. B.; Nortey, S. O. *Tetrahedron* **1988**, *44*, 3093; (h) Mio, S.; Kumagawa, Y.; Sugai, S. *Tetrahedron* **1991**, *47*, 2133; (i) Kang, J.; Lim, G. J.; Yoon, S. K.; Kim, M. Y. *J. Org. Chem.* **1995**, *60*, 564; (j) Fung, W.-H.; Yu, W.-Y.; Che, C.-M. *J. Org. Chem.* **1998**, *63*, 2873.
3. For general leading references on dioxiranes see: (a) Adam, W.; Curci, R.; Edwards, J. O. *Acc. Chem. Res.* **1989**, *22*, 205; (b) Murray, R. W. *Chem. Rev.* **1989**, *89*, 1187; (c) Curci, R.; Dinoi, A.; Rubino, M. F. *Pure & Appl. Chem.* **1995**, *67*, 811; (d) Clennan, E. L. *Trends Org. Chem.* **1995**, *5*, 231; (e) Adam, W.; Smerz, A. K. *Bull. Soc. Chim. Belg.* **1996**, *105*, 581; (f) Denmark, S. E.; Wu, Z. *Synlett* **1999**, 847.
4. For leading references on asymmetric epoxidation mediated by chiral ketones see: (a) Curci, R.; Fiorentino, M.; Serio, M. R. *J. Chem. Soc., Chem. Commun.* **1984**, *155*; (b) Curci, R.; D'Accolti, L.; Fiorentino, M.; Rosa, A. *Tetrahedron Lett.* **1995**, *36*, 5831; (c) Denmark, S. E.; Forbes, D. C.; Hays, D. S.; DePue, J. S.; Wilde, R. G. *J. Org. Chem.* **1995**, *60*, 1391; (d) Brown, D. S.; Marples, B. A.; Smith, P.; Walton, L. *Tetrahedron* **1995**, *51*, 3587; (e) Yang, D.; Yip, Y.-C.; Tang, M.-W.; Wong, M.-K.; Zheng, J.-H.; Cheung, K.-K. *J. Am. Chem. Soc.* **1996**, *118*, 491; (f) Yang, D.; Wang, X.-C.; Wong, M.-K.; Yip, Y.-C.; Tang, M.-W. *J. Am. Chem. Soc.* **1996**, *118*, 11311; (g) Song, C. E.; Kim, Y. H.; Lee, K. C.; Lee, S.; Jin, B. W. *Tetrahedron: Asymmetry* **1997**, *8*, 2921; (h) Adam, W.; Zhao, C.-G. *Tetrahedron: Asymmetry* **1997**, *8*, 3995; (i) Denmark, S. E.; Wu, Z.; Crudden, C.

M.; Matsuhashi, H. *J. Org. Chem.* **1997**, *62*, 8288; (j) Wang, Z.-X.; Shi, Y. *J. Org. Chem.* **1997**, *62*, 8622; (k) Armstrong, A.; Hayter, B. R. *Chem. Commun.* **1998**, *621*; (l) Yang, D.; Wong, M.-K.; Yip, Y.-C.; Wang, X.-C; Tang, M.-W.; Zheng, J.-H.; Cheung, K.-K. *J. Am. Chem. Soc.* **1998**, *120*, 5943; (m) Yang, D.; Yip, Y.-C.; Chen, J.; Cheung, K.-K. *J. Am. Chem. Soc.* **1998**, *120*, 7659; (n) Adam, W.; Saha-Möller, C. R.; Zhao, C.-G. *Tetrahedron: Asymmetry* **1999**, *10*, 2749; (o) Wang, Z.-X.; Miller, S. M.; Anderson, O. P.; Shi, Y. *J. Org. Chem.* **1999**, *64*, 6443.

5. For examples of asymmetric epoxidation mediated by fructose-derived ketones see: (a) Tu, Y.; Wang, Z.-X.; Shi, Y. *J. Am. Chem. Soc.* **1996**, *118*, 9806; (b) Wang, Z.-X.; Tu, Y.; Frohn, M.; Shi, Y. *J. Org. Chem.* **1997**, *62*, 2328; (c) Wang, Z.-X.; Tu, Y.; Frohn, M.; Zhang, J.-R.; Shi, Y. *J. Am. Chem. Soc.* **1997**, *119*, 11224; (d) Frohn, M.; Dalkiewicz, M.; Tu, Y.; Wang, Z.-X.; Shi, Y. *J. Org. Chem.* **1998**, *63*, 2948; (e) Wang, Z.-X.; Shi, Y. *J. Org. Chem.* **1998**, *63*, 3099; (f) Cao, G.-A.; Wang, Z.-X.; Tu, Y.; Shi, Y. *Tetrahedron Lett.* **1998**, *39*, 4425; (g) Zhu, Y.; Tu, Y.; Yu, H.; Shi, Y. *Tetrahedron Lett.* **1998**, *39*, 7819; (h) Tu, Y.; Wang, Z.-X.; Frohn, M.; He, M.; Yu, H.; Tang, Y.; Shi, Y. *J. Org. Chem.* **1998**, *63*, 8475; (i) Wang, Z.-X.; Cao, G.-A.; Shi, Y. *J. Org. Chem.* **1999**, *64*, 7646; (j) Warren, J. D.; Shi, Y. *J. Org. Chem.* **1999**, *64*, 7675; (k) Shu, L.; Shi, Y. *Tetrahedron Lett.* **1999**, *40*, 8721.

6. Chen, C.-C.; Whistler, R. L. *Carbohydr. Res.* **1988**, *175*, 265.

ASYMMETRIC EPOXIDATION OF *trans*-β-METHYLSTYRENE AND 1-PHENYLCYCLOHEXENE USING A D-FRUCTOSE-DERIVED KETONE: (*R,R*)-*trans*-β-METHYLSTYRENE OXIDE AND (*R,R*)-1-PHENYLCYCLOHEXENE OXIDE [(Oxirane, 2-methyl-3-phenyl-, (2*R*,3*R*)- and 7-Oxabicyclo[4.1.0]heptane,1-phenyl-)]

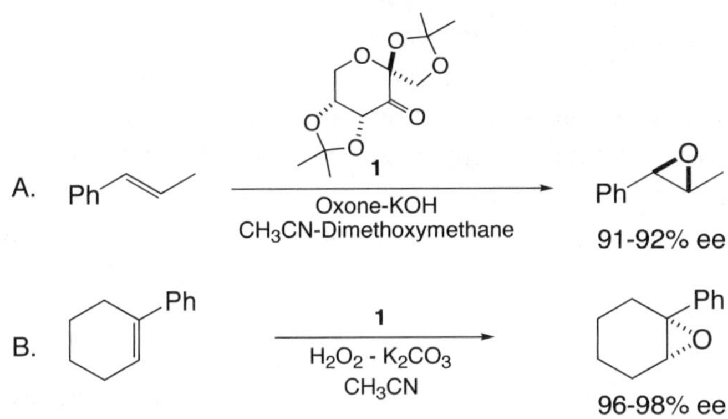

Submitted by Zhi-Xian Wang, Lianhe Shu, Michael Frohn, Yong Tu, and Yian Shi.[1]
Checked by Jason M. Diffendal and Rick L. Danheiser.

1. Procedure

A. (R,R)-trans-β-Methylstyrene oxide.[2] A 2-L, three-necked, round-bottomed flask (Note 1) equipped with a 5-cm, egg-shaped, Teflon-coated stir bar and two addition funnels is cooled in an ice-bath. The flask is charged with *trans*-β-methylstyrene (5.91 g, 50.0 mmol) (Note 2), 500 mL of a 2:1 mixture of dimethoxymethane (Note 2) and acetonitrile (CH_3CN) (Note 2), 300 mL of potassium carbonate-acetic acid buffer solution (Note 3), tetrabutylammonium hydrogen sulfate (0.375 g, 1.1 mmol), and the chiral ketone **1** (4.52 g, 17.5 mmol, 35 mol%) (Note 4). One addition funnel is charged with a solution of Oxone (46.1 g, 75.0 mmol) in 170 mL of aqueous 4×10^{-4}M disodium ethylenediaminetetraacetate (Na_2EDTA) solution (Note 5) and the other addition funnel is charged with 170 mL of 1.47M aqueous potassium hydroxide (KOH) solution. The two solutions in the addition funnels are added dropwise at the same rate over 2.5 h to the cooled reaction mixture which is stirred vigorously at 0 °C (Notes 6, 7). The resulting suspension is stirred at 0 °C for an additional hour and then 250 mL of pentane is added. The aqueous phase is separated and extracted with two 250-mL portions of pentane, and the combined organic phases are dried over Na_2SO_4, filtered, and concentrated by rotary evaporation at 0 °C (Note 8). The resulting oil is loaded onto 50 g of Whatman 60 Å (230-400 mesh) silica gel (Note 9) packed in a 5-cm diameter column. The silica gel is first washed with 200 mL of hexane to remove trace amounts of unreacted olefin, then the product is eluted with 200 mL of 10:1 hexane:ether to afford 6.02-6.31 g (90-94%) of *trans*-β-methylstyrene oxide (Notes 10-12).

B. (R,R)-1-Phenylcyclohexene oxide.[3] A 250-mL, round-bottomed flask (Note 1) equipped with a 4.5-cm, egg-shaped Teflon-coated magnetic stir bar is charged with 1-phenylcyclohexene (7.91 g, 50.0 mmol) (Note 2) and the chiral ketone **1** (1.29 g, 5.00 mmol, 10 mol%) (Note 4). The flask is cooled in an ice-bath, and 75 mL of CH_3CN and 75 mL of a solution which is 2.0M potassium carbonate and 4×10^{-4}M EDTA are added. The reaction mixture is cooled to 0 °C, and 20 mL (200 mmol) of 30% hydrogen peroxide (H_2O_2) is added. The resulting mixture is vigorously stirred at 0 °C for 6 h (Note 13), then diluted with 50 mL of hexane. The aqueous phase is separated and extracted with three 200-mL portions of hexane, and the combined organic phases are washed with two 50-mL portions of 1M aqueous sodium thiosulfate solution and 100 mL of brine, dried over Na_2SO_4, filtered, and concentrated by rotary evaporation at 0 °C. The resulting oil is applied to 180 g of Whatman 60 Å (230-400 mesh) silica gel (Note 9) packed in a 5-cm diameter column, then the product is eluted with 600 mL of hexane and finally 1 L of 20:1 hexane:Et_2O to afford 6.88-8.01 g (79-92%) of (*R*,*R*)-1-phenylcyclohexene oxide as a colorless oil (Notes 14, 15).

2. Notes

1. All glassware used for the epoxidation reaction is carefully washed to remove trace metals, which may catalyze the decomposition of Oxone and H_2O_2. The checkers used Alconox, followed by water, and then acetone.

2. β-Methylstyrene (99%) and dimethoxymethane (99%) were obtained from Aldrich Chemical Company, Inc. and used as received. ACS grade acetonitrile and 30% H_2O_2 were purchased from Fisher Scientific and used as received. 1-Phenylcyclohexene (99%) was obtained from Alfa Aesar.

3. The buffer solution is prepared by adding 4.5 mL of glacial acetic acid (Fisher Scientific) to 1 L of a 0.1M solution of K_2CO_3 (Fisher Scientific).

4. The chiral ketone **1** was prepared as described in *Org. Synth.* **2003**, *80*, 1.

5. Oxone was obtained from the Aldrich Chemical Company, Inc. The activity of commercial Oxone in oxidation reactions occasionally varies with different batches. Na_2EDTA was purchased from Fisher Scientific.

6. The concentration of Oxone in the reaction mixture and the reaction pH are very important factors in determining the epoxidation efficiency. Both the Oxone and KOH solutions must be added to the reaction mixture in a steady, uniform manner over 2.5 h.

7. As the reaction progresses, the organic and aqueous phases separate. Salts precipitate during the first 10-20 min of addition. Vigorous stirring is required to sufficiently mix the two phases; however, excessive splashing of the reaction mixture must be avoided in order to maximize conversion.

8. The epoxide product is volatile. Care should be taken to minimize the loss of the epoxide during concentration.

9. The silica gel is buffered by packing the column with hexane containing 1% triethylamine.

10. The product exhibits the following physical and spectral properties: ^{1}H NMR (500 MHz, $CDCl_3$) δ: 1.45 (d, 3 H, *J* = 5.1), 3.03 (qd, 1 H, *J* = 5.1, 2.1), 3.57 (d, 1 H, *J* = 2.1), 7.20-7.40 (m, 5 H); ^{13}C NMR (125 MHz, $CDCl_3$) δ: 18.1, 59.2, 59.7, 125.7, 128.2, 128.6, 137.9; Anal. Calcd C, 80.56; H, 7.51. Found: C, 80.39; H, 7.37; $[\alpha]^{25}_{D}$ +45.3 ($CHCl_3$, c 1.84).

11. The submitters obtained the product in 91-92% ee as determined by chiral GC with a Chiraldex G-TA column (25 μm × 30 m) (oven temperature: 80 °C; head pressure: 20 psi; retention time: minor isomer at 11.8 min, major isomer at 15.8 min). The checkers obtained epoxide with 89% ee.

12. The submitters report that the ee can be increased to 94% if the epoxidation is carried out at −8 to −10 °C.

13. The reaction was monitored by TLC and was complete after 6 hr. The reaction rate is affected by the rate of stirring.

14. The product exhibits the following physical and spectral properties: ^{1}H NMR (500 MHz, $CDCl_3$) δ: 1.25-1.65 (m, 4 H), 1.95-2.03 (m,

2 H), 2.12 (dt, 1 H, J = 15.0, 5.1), 2.29 (ddd, 1 H, J = 15.0, 8.4, 5.1), 3.09 (br s, 1 H), 7.20-7.40 (m, 5 H); ^{13}C NMR (125 MHz, $CDCl_3$) δ: 19.8, 20.1, 24.7, 28.9, 60.2, 61.9, 125.3, 127.1, 128.2, 142.5. Anal. Calcd: C, 82.72; H, 8.10. Found: C, 82.76; H, 8.13; $[\alpha]_D^{25}$ +113.0 (benzene, c 0.56).

15. The submitters obtained the product in 96-98% ee as determined by chiral GC with a Chiraldex G-TA column (25 μm × 30 m) (oven temperature: 80 °C; head pressure: 25 psi; retention time: minor isomer at 68.0 min, major isomer at 71.8 min). The checkers obtained the product in 94-95% ee.

Waste Disposal Information

All toxic materials were disposed of in accordance with "Prudent Practices in the Laboratory"; National Academy Press; Washington, DC, 1995.

3. Discussion

The epoxidation procedure described herein utilizes the fructose-derived ketone (1) as catalyst and Oxone[4a-j] or H_2O_2[4k,l] as oxidant. The procedure provides a valuable method for the preparation of enantiomerically-enriched epoxides from *trans*- and trisubstituted olefins. High enantioselectivies have been obtained for a variety of unfunctionalized trans-disubstituted and trisubstituted olefins,[4a-c] vinylsilanes,[4j] hydroxyalkenes,[4e] conjugated dienes,[4d] conjugated enynes,[4f,i] and enol derivatives.[4g] Representative examples are shown in Table I.

Previously, the generation of dioxiranes almost exclusively used potassium peroxomonosulfate ($KHSO_5$) as oxidant.[5,6] Recently, we found that hydrogen peroxide (H_2O_2) could be used as primary oxidant in combination with a nitrile for the epoxidation catalyzed by **1**.[4k,l] In this epoxidation, the peroxyimidic acid generated from the addition of H_2O_2 to CH_3CN is likely to be the active oxidant for the formation of the dioxirane.[7] The Oxone procedure described herein can be applied to a wide variety of olefins without the need for extensive optimization. The more recent H_2O_2 procedure uses much less solvent and salts and is operationally simpler. However, this procedure is somewhat sensitive to the reactivity and

Table 1. Asymmetric Epoxidation of Representative Olefins by Ketone **1** (0.3 equiv) with Oxone

Entry	Substrate	Product	Yield (%)	ee (%)	Ref.
1.	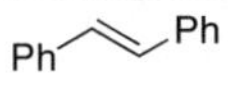	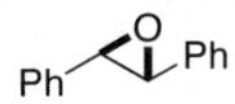	85	98	4c
2.	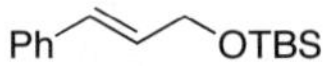	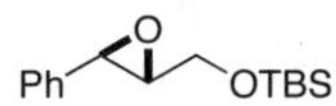	87	94	4c
3.	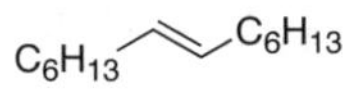	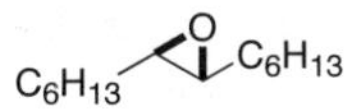	89	95	4c
4.	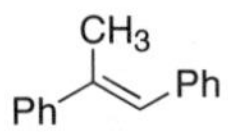	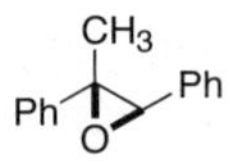	89	95	4c
5.	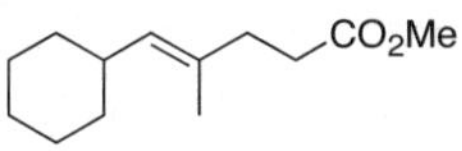	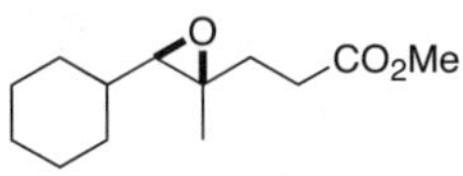	89	94	4c
6.	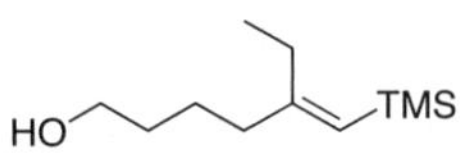	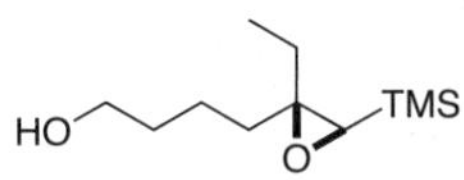	71	93	4j
7.	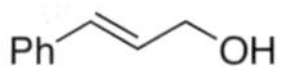	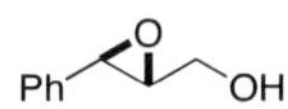	85	94	4e
8.	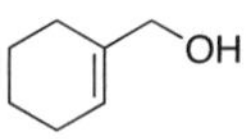	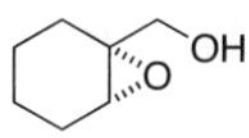	93	94	4e
9.	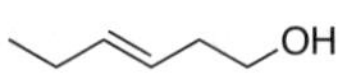	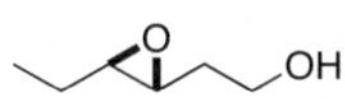	82	90	4e
10.	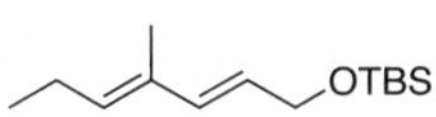	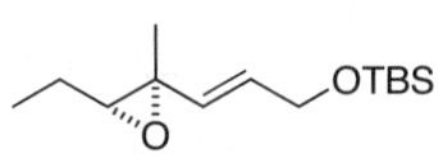	81	96	4d
11.	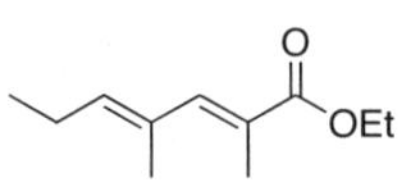	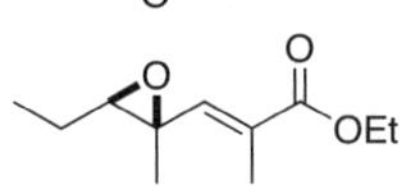	89	94	4d
12.	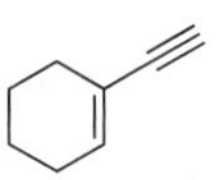	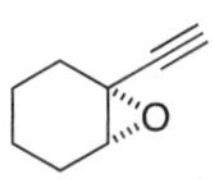	78	93	4i
13.	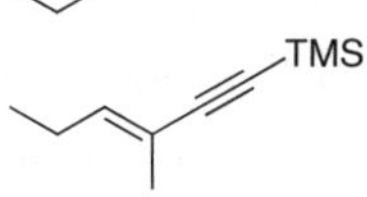	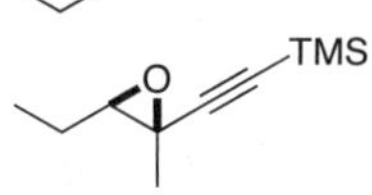	84	95	4i
14.	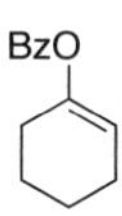	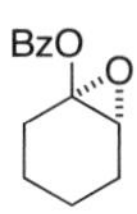	82	93	4g
15.	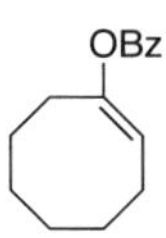	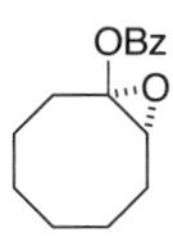	82	95	4g

solubility of olefins; some optimization (varying catalyst loading, reaction time, reaction temperature, amount of H_2O_2 and solvent) may be required for different substrates.[41]

1. Department of Chemistry, Colorado State University, Fort Collins, CO 80523.
2. Witkop, B.; Foltz, C. M. *J. Am. Chem. Soc.* **1957**, *79*, 197.
3. Berti, G.; Macchia, B.; Macchia, F.; Monti, L. *J. Org. Chem.* **1968**, *33*, 4045.
4. For examples of asymmetric epoxidation mediated by fructose-derived ketones see: (a) Tu, Y.; Wang, Z.-X.; Shi, Y. *J. Am. Chem. Soc.* **1996**, *118*, 9806; (b) Wang, Z.-X.; Tu, Y.; Frohn, M.; Shi, Y. *J. Org. Chem.* **1997**, *62*, 2328; (c) Wang, Z.-X.; Tu, Y.; Frohn, M.; Zhang, J.-R.; Shi, Y. *J. Am. Chem. Soc.* **1997**, *119*, 11224; (d) Frohn, M.; Dalkiewicz, M.; Tu, Y.; Wang, Z.-X.; Shi, Y. *J. Org. Chem.* **1998**, *63*, 2948; (e) Wang, Z.-X.; Shi, Y. *J. Org. Chem.* **1998**, *63*, 3099; (f) Cao, G.-A.; Wang, Z.-X.; Tu, Y.; Shi, Y. *Tetrahedron Lett.* **1998**, *39*, 4425; (g) Zhu, Y.; Tu, Y.; Yu, H.; Shi, Y. *Tetrahedron Lett.* **1998**, *39*, 7819; (h) Tu, Y.; Wang, Z.-X.; Frohn, M.; He, M.; Yu, H.; Tang, Y.; Shi, Y. *J. Org. Chem.* **1998**, *63*, 8475; (i) Wang, Z.-X.; Cao, G.-A.; Shi, Y. *J. Org. Chem.* **1999**, *64*, 7646; (j) Warren, J. D.; Shi, Y. *J. Org. Chem.* **1999**, *64*, 7675; (k) Shu, L.; Shi, Y. *Tetrahedron Lett.* **1999**, *40*, 8721; (l) Shu, L.; Shi, Y. *Tetrahedron* **2001**, *57*, 5213.
5. Oxone ($2KHSO_5{\cdot}KHSO_4{\cdot}K_2SO_4$) is currently the common source of potassium peroxomonosulfate ($KHSO_5$).
6. As close analogues of potassium peroxomonosulfate, arenesulfonic peracids generated from (arenesulfonyl)imidazole/H_2O_2/NaOH have also been shown to produce dioxiranes from acetone and trifluoroacetone as illustrated by ^{18}O-labeling experiments see: Schulz, M.; Liebsch, S.; Kluge, R.; Adam, W. *J. Org. Chem.* **1997**, *62*, 188.
7. For leading references on epoxidation using H_2O_2 and RCN see: (a) Payne, G. B.; Deming, P. H.; Williams, P. H. *J. Org. Chem.* **1961**, *26*, 659; (b) Payne, G. B. *Tetrahedron* **1962**, *18*, 763; (c) McIsaac, J. E. Jr.; Ball, R. E.; Behrman, E. J. *J. Org. Chem.* **1971**, *36*, 3048; (d) Bach, R. D.; Knight, J. W. *Org. Synth. Coll. Vol. VII* **1990**, 126; (e) Arias, L. A.; Adkins, S.; Nagel, C. J.; Bach, R. D. *J. Org. Chem.* **1983**, *48*, 888.

N,N-DIBENZYL-N-[1-CYCLOHEXYL-3-(TRIMETHYLSILYL)-2-PROPYNYL]-AMINE FROM CYCLOHEXANECARBALDEHYDE, TRIMETHYLSILYLACETYLENE AND DIBENZYLAMINE

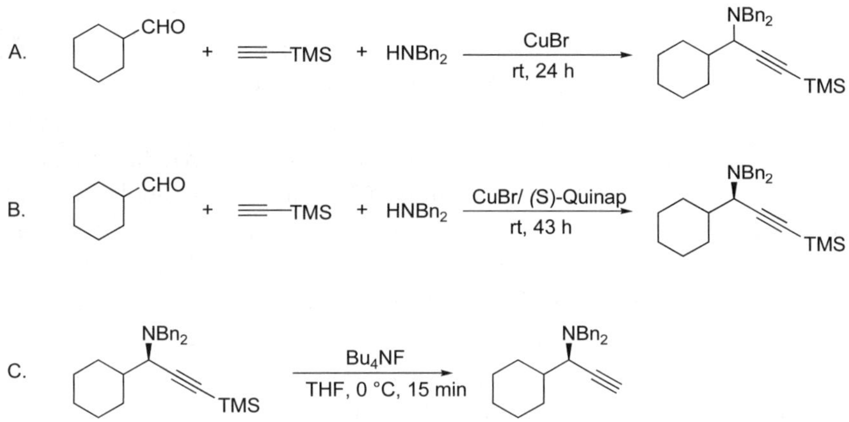

Submitted by Nina Gommermann and Paul Knochel.[1]
Checked by Jason C. Rech and Jonathan A. Ellman.

1. Procedure

A. (±)-N,N-Dibenzyl-N-[1-cyclohexyl-3-(trimethylsilyl)-2-propynyl]-amine. A 100-mL Schlenk flask equipped with a magnetic stirring bar and a rubber septum is flame-dried under vacuum (5 x 10^{-2} mbar), cooled to rt, flushed with argon and charged with copper(I) bromide (143 mg, 1.0 mmol) (Note 1). The flask is evacuated (5 x 10^{-2} mbar) for 15 min, flushed with argon and charged with dry toluene (40 mL) (Note 2) giving a white suspension. Molecular sieves 4 Å (10 g) (Note 3) is added under a flow of argon. Trimethylsilylacetylene (1.96 g, 2.85 mL, 20.0 mmol) (Note 4) followed by cyclohexanecarbaldehyde (2.24 g, 2.42 mL, 20.0 mmol) (Note 5) are added via syringe (Note 6) in one portion. Dibenzylamine (3.95 g, 3.87 mL, 20.0 mmol) (Note 7) is added during 5 min via syringe. During the addition, the solution becomes clear and turns green at the end of addition. The reaction is slightly exothermic during the addition of dibenzylamine, and subsequently turns milky again. The reaction mixture is stirred for 24 h. After this time, GC-analysis (Note 8) shows more than 96% conversion to the desired product. The solution is filtered (Note 9), and the molecular sieves 4 Å are washed with diethyl ether (150 mL). The solution is concentrated by rotary evaporation (55 °C, 760 mmHg, then 38 mmHg). The resulting light green oil is purified by column chromatography (Note 10) affording *N,N*-dibenzyl-*N*-[1-cyclohexyl-3-(trimethylsilyl)-2-propynyl]-amine (6.86 g, 17.6 mmol) as a clear, colorless oil which solidifies upon standing at rt (88%) (Note 11).

B. (R)-(+)-N,N-Dibenzyl-N-[1-cyclohexyl-3-(trimethylsilyl)-2-propynyl]-amine. A 25-mL Schlenk flask equipped with a magnetic stirring bar and a rubber septum is flame-dried under vacuum (5 x 10^{-2} mbar), cooled to rt, flushed with argon and charged with copper(I) bromide (21.5 mg, 0.150 mmol) (Note 1) and (*S*)-Quinap (72.5 mg, 0.165 mmol) (Note 12). The flask is evacuated (5 x 10^{-2} mbar) for 15 min, flushed with argon and charged with dry toluene (9 mL) (Note 2) giving a yellow solution with white particles. The mixture is stirred for 30 min at rt, during this time, the solution turns milky. Molecular sieves 4 Å (1.5 g) (Note 3) are added under a flow of argon. Trimethylsilylacetylene (295 mg, 0.43 mL, 3.0 mmol) (Note 4) followed by cyclohexanecarbaldehyde (337 mg, 0.36 mL, 3.0 mmol) (Note 5) are added via syringe (Note 6) in one portion. Dibenzylamine (592 mg, 0.58 mL, 3.0 mmol) (Note 7) is added during one minute via syringe. The suspension is stirred for 43 h at rt, after which time GC-analysis (Note 8) shows more than 96% conversion to the desired product. The solution is filtered (Note 9) and the molecular sieves 4 Å are washed with diethyl ether (50 mL). The solution is concentrated by rotary evaporation (55 °C, 760 mmHg, then 38 mmHg). The resulting yellow oil is dissolved in pentane:diethyl ether (99:1) and filtered from the precipitated catalyst (Note 13). The clear filtrate is subjected to column chromatographic purification (Note 14) affording (*R*)-(+)-*N,N*-dibenzyl-*N*-[1-cyclohexyl-3-(trimethylsilyl)-2-propynyl]-amine (1.07 g, 2.7 mmol) as a clear, colorless oil which solidifies upon standing at rt (91%, 95 % ee) (Note 15).

C. (R)-(+)-N,N-Dibenzyl-N-(1-cyclohexyl-2-propynyl)-amine. (*R*)-(+)-*N,N*-Dibenzyl-*N*-[1-cyclohexyl-3-(trimethylsilyl)-2-propynyl]-amine (390 mg, 1.00 mmol), placed in a 25-mL round-bottomed flask equipped with a magnetic stirring bar and a rubber septum, is dissolved in dry THF (5 mL) (Note 16). The solution is cooled to 0 °C for 10 min and Bu_4NF (0.30 mmol, 0.30 mL, 1 M solution in THF) (Note 17, 18) is added dropwise. During the addition, the solution turns slightly yellow. After 15 min, GC analysis (Note 8) shows full conversion. The reaction mixture is quenched with water (30 mL), and the aqueous phase is extracted with diethyl ether (3 x 40 mL). The combined organic fractions are dried over Na_2SO_4 (Note 19) and concentrated by rotary evaporation (55 °C, 760 mmHg, then 38 mmHg). The resulting light yellow oil is purified by column chromatography (Note 20) affording *N,N*-dibenzyl-N-(1-cyclohexyl-2-propynyl)-amine (307 mg, 0.97 mmol) as a clear, colorless oil. (97%) (Note 21).

2. Notes

1. CuBr was prepared as described by Taylor:[2] $CuBr_2$ (5.00 g, 22.4 mmol) was placed in a 100-mL round bottomed flask equipped with a magnetic stirring bar and dissolved in water (5 mL). A freshly prepared solution of Na_2SO_3 (3.80 g, 30.1 mmol) in water (25 mL) was added slowly via an addition funnel during 10 min. The initially dark brown mixture turned white and rapidly deposited a white precipitate of CuBr. After the addition was complete, the mixture was stirred for an additional 15 min. Afterwards, the suspension was poured into water (400 mL) which contained Na_2SO_3 (0.50 g) and concentrated hydrochloric acid (1 mL). The mixture was stirred vigorously for 10 min, after which the CuBr was allowed to settle. The supernatant liquid was decanted and the precipitate was transferred to a sintered-glass filter (pore size P4; 75 mL) with diluted sulfuric acid (100 mL, 0.1 M) to prevent oxidation. For the transfer, as well as for the following washing steps, it was ensured that liquid covered the precipitate at all times. With application of gentle suction (375 mmHg), the precipitate was washed with glacial acetic acid (4 x 15 mL), absolute ethanol (3 x 15 mL) and anhydrous diethyl ether (6 x 10 mL). The white crystalline CuBr was immediately transferred to a 50-mL Schlenk flask and dried under vacuum (4×10^{-2} mmHg). CuBr (3.08 g, 21.5 mmol, 96%) was obtained as a white solid and stored under argon, but the solid can be handled routinely in air.

2. Toluene was dried by alumina column (checkers) or dried by distillation under argon from sodium (submitters).

3. Molecular sieves 4 Å, 8 to 12 mesh, were obtained from Acros Organics and used as obtained.

4. Trimethylsilylacetylene was purchased from Lancaster (checkers) or obtained as a generous gift from Wacker Chemie GmbH, Burghausen (submitters), and used as obtained.

5. Cyclohexanecarbaldehyde (98%) was obtained from Aldrich and freshly distilled under reduced pressure (24 mmHg) before use. Reduced

yields (65–80%), but not reduced enantioselectivities, were observed when aged starting materials were employed in this procedure.

6. Although the liquid starting materials were transferred via syringe, weighing of the compounds in the syringes instead of measuring the volumes was found to be advantageous. Especially for the smaller scale of the reaction in presence of the chiral ligand (Procedure B), yields were found to be more reliable when the amounts of starting materials were determined by weighing. Therefore, the use of the given mass-values instead of volumes is strongly recommended.

7. Dibenzylamine (98%) was obtained from Acros Organics and freshly distilled under reduced pressure (0.04 mmHg) before use.

8. GC-analysis was carried out using a Hewlett&Packard 5890 Series 2 machine equipped with a HP Ultra-2.5 %-phenylmethylpolysiloxane column (12m x 0.2 mm x 0.33 μm). Oven program for GC-Analysis: Starting temperature, 90 °C for 1 min; heating to 250 °C by a rate of 50 °C per min; 8 min at 250 °C. Retention time of *N*,*N*-Dibenzyl-*N*-[1-cyclohexyl-3-(trimethylsilyl)-2-propynyl]-amine: 7.34 min. The reaction was monitored by the disappearance of the enamine (retention time: 5.83 min).

9. A glass filter with pore size P3 and a volume of 75 mL was used. Filtration was carried out under reduced pressure (300 mmHg).

10. The oil is taken up in 20 mL of the solvent mixture (pentane/diethyl ether, 99:1) and applied to a 5-cm diameter column packed with 200 g Merck silica gel 60 mesh (0.063–0.200 mm), R_f= 0.35 (TLC, aluminum sheets, silica gel 60 F254, obtained from Merck). Approximately 1.5 L of the solvent mixture is used.

11. The submitters reported yields ranging from 6.97–7.01g (88–89%). Spectral data are as follows: checkers mp 68–70 °C, submitters mp 81–82 °C; ^{1}H-NMR (300 MHz, $CDCl_3$) δ: 0.28 (s, 9 H). 0.67–0.89 (m, 2 H), 1.01–1.28 (m, 3 H), 1.52–1.71 (m, 4 H), 1.96 (m, 1H), 2.27 (m, 1 H), 3.01 (d, J = 10.5 Hz, 1 H), 3.34 (d, J = 13.8 Hz, 2 H), 3.78 (d, J = 13.8 Hz, 2 H), 7.19–7.23 (m, 2 H), 7.26–7.34 (m, 4 H), 7.37–7.45 (m, 4 H). ^{13}C-NMR (75 MHz, $CDCl_3$) δ: 0.5, 25.9, 26.1, 26.6, 30.2, 31.2, 39.4, 54.8, 58.5, 90.1, 103.5, 126.8, 128.2, 128.8, 139.8. MS (70 eV, EI): 307 (27), 306 (M^+–c-Hex, 100), 91 (34). HRMS (EI): Calcd. for $C_{26}H_{36}NSi$ [M+H]: 390.2617, found: 390.2628. IR (film): 2924, 2850, 2160, 1494, 1450, 1249, 1004, 841, 746, 698 cm^{-1}. Anal Calcd. for $C_{26}H_{35}NSi$: C: 80.14, H: 9.05, N: 3.59, Found: C: 79.98, H: 8.82, N: 3.79.

12. (S)-Quinap was obtained from Strem Chemicals, Inc. and used as obtained.

13. The oil was mixed with approximately 5 mL of the solvent mixture (pentane/diethyl ether, 99:1) and a light yellow precipitate was formed. The solid was collected by filtration and washed with the solvent mixture (10 mL).

14. Column chromatography was carried out on a 2.5-cm diameter column packed with 50 g Merck silica gel 60 mesh (0.063–0.200 mm) using pentane/diethyl ether, 99:1 as eluent, (R_f = 0.35, TLC aluminum sheets, Silica gel 60 F254, obtained from Merck). Approximately 0.5 L of the solvent mixture is used.

15. The submitters report yields in the range of 1.02–1.04g (88–89%, 95% ee). The enantiomeric excess was determined by HPLC analysis using a Chiracel OD-H column and n-heptane as the eluent after deprotection to form the desilylated derivative (see procedure C), retention times: 36.6 min (*S*), 41.9 min (*R*) (submitters). The enantiomeric excess can also be determined using a Chiracel OD column with hexane as the eluent, retention times: 7.45 min (*S*), 7.89 min (*R*) (checkers). Analytical data: $[\alpha]_D^{20}$ = +168 (c = 1.00, $CHCl_3$) (checkers). The purity of the product was established by ^{1}H-NMR analysis to be >98 %, showing the same data as reported in Note 10.

16. THF was dried by alumina column (checkers) or by distillation from sodium/benzophenone under nitrogen (submitters).

17. Bu_4NF was obtained as a 1M solution in THF from Alfa Aesar and used as obtained.

18. The use of only 0.4 equiv. of Bu_4NF for the cleavage of a TMS-group from acetylenes has been reported by Nishikawa.[3]

19. Na_2SO_4 was obtained from Acros Organics and used as obtained.

20. Column chromatography was carried out on a 2.5-cm diameter column packed with 20 g Merck silica gel 60 mesh (0.063–0.200 mm) using pentane/diethyl ether, 99:1 as eluent, (R_f= 0.30, TLC aluminum sheets, silica gel 60 F254, obtained from Merck).

21. The submitters reported a yield of 307 mg (99%). Spectral data as follows: $[\alpha]_D^{20}$ = +135 (c = 1.10, $CHCl_3$). ^{1}H-NMR (300 MHz, $CDCl_3$) δ: 0.67–0.89 (m, 2 H), 1.01–1.28 (m, 3 H), 1.52–1.71 (m 4 H), 2.00 (m, 1 H), 2.29 (m, 1 H), 2.34 (d, *J* = 2.4 Hz, 1 H), 3.03 (dd, *J* = 10.5, 2.1 Hz, 1 H), 3.37 (d, *J* = 13.5 Hz, 2 H), 3.81 (d, *J* = 13.5 Hz, 2 H), 7.18-7.25 (m, 2 H), 7.26–7.35 (m, 4 H), 7.36–7.45 (m, 4 H). ^{13}C-NMR (75 MHz, $CDCl_3$) δ: 25.9, 26.1, 26.5, 30.2, 31.2, 39.5, 54.8, 57.6, 73.4, 81.0, 126.8, 128.2, 128.8, 139.7. MS (70 eV, EI): 235 (21), 234 (M+–c-Hex, 100), 91 (88). HRMS

(EI): Calcd. for $C_{23}H_{28}N$ [M+H]: 318.2222, found: 318.2222. IR (film): 3300, 2923, 2850, 1490, 1450, 745, 698 cm^{-1}. Anal Calcd. for $C_{23}H_{27}N$: C: 87.02, H: 8.57, N: 4.41, Found: C: 86.78, H: 8.63, N: 4.55. The checkers observed that the colorless oil solidifies upon standing: mp 75–76 °C.

Safety and Waste Disposal Information

All hazardous materials should be handled and disposed of in accordance with "Prudent Practices in the Laboratory"; National Academy Press; Washington, DC, 1995.

3. Discussion

The catalytic, enantioselective formation of new carbon-carbon bonds is an important class of reactions. Especially attractive are multi-component reactions that allow the formation of several bonds including new carbon-carbon bonds in a one-pot procedure. The present procedure describes a convenient method for the preparation of enantiomerically enriched propargylamines in good selectivities and excellent yields.[4] The reaction is catalyzed by the chiral complex formed from Cu(I)Br and commercially available (*R*)- or (*S*)-Quinap[5] (5 mol%). The reaction tolerates a broad range of different aldehydes, including aromatic and heteroaromatic aldehydes, as well as branched and non-branched aliphatic aldehydes. Many types of terminal alkynes are tolerated, but the use of trimethylsilylacetylene provides the best enantiomeric excesses. Furthermore, desilylation to the terminal alkyne potentiates many different transformations (Scheme 1). As the third component, only aliphatic amines can be used; amides or anilines do not undergo the reaction. A major advantage of this reaction is the convenience of the procedure. The reaction is carried out at room temperature with equimolar amounts of starting materials and only 5 mol% of catalyst is necessary, which demonstrates a very atom economical reaction.[6] The propargylamines obtained by this procedure are valuable building blocks for organic synthesis. A related synthesis involving imines that uses CuOTf/pybox as the catalytic system is also known, but the reaction is limited to aromatic aldehydes and anilines.[7]

Scheme 1. Transformations of terminal propargylamines.

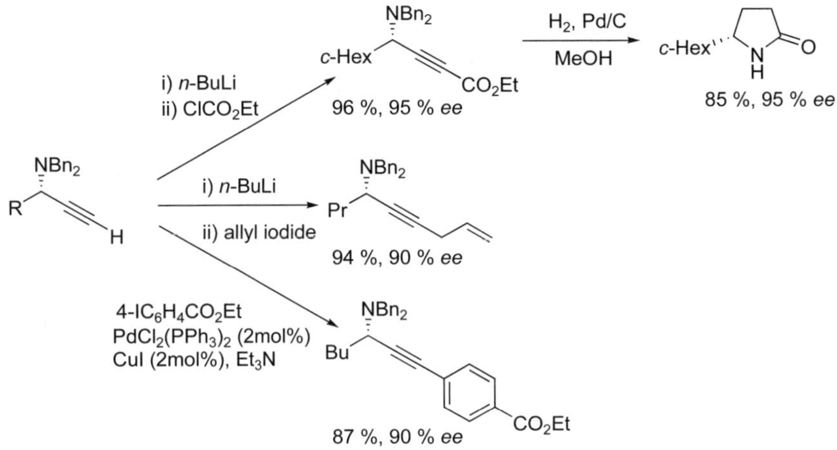

1. Department of Chemistry, Ludwig-Maximilians-University Munich, Butenandtstr. 5-13, D-81377 München, Germany. Paul.Knochel@cup.uni-muenchen.de
2. Taylor, R. J. *Organocopper Reagents*, *Oxford Univ. Press*, **1994**.
3. Nishikawa, T.; Ino, A.; Isobe, M. *Tetrahedron* **1994**, *50*, 1449.
4. (a) Gommermann, N.; Koradin, C.; Polborn, K.; Knochel, P. *Angew. Chem. Int. Ed.* **2003**, *42*, 5763; (b) Dube, H.; Gommermann, N.; Knochel. P. *Synthesis* **2004**, 2015; (c) Gommermann, N.; Knochel, P. *Chem. Commun.* **2004**, 2324.
5. (a) Valk, J. M.; Whitlock, G. A.; Layzell, T. P.; Brown, J. M. *Tetrahedron: Asymmetry* **1995**, *6*, 2593; (b) Fernandez, E.; Maeda, K.; Hooper, M. W.; Brown, J. M. *Chem. Eur. J.* **2000**, *6*, 1840.
6. (a) Trost, B. M. *Angew. Chem. Int. Ed.* **1995**, *34*, 259; (b) Trost, B. M. *Science* **1991**, *254*, 1471.
7. Wei, C.; Li, C.-J. *J. Am. Chem. Soc.* **2002**, *124*, 5638.

Table 1: Enantiomerically enriched propargylamines obtained by the Cu(I)-Quinap catalyzed three-component coupling of aldehydes, secondary amines and terminal alkynes.

Entry	RCHO	Product of type 3	Yield (%)[a]	%ee[b]
		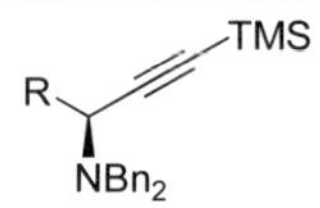		
1	*n*-Pr	R= n-Pr	88	90
2	*n*-Bu	R= n-Bu	75	90
3	*i*-Bu	R= i-Bu	84	94
4	*i*-Pr	R= i-Pr	84	96
5	*s*-Pent	R= s-Pent	97	98
6	*c*-Pr	R= c-Pr	97	92
7	*c*-Pent	R= c-Pent	97	96
8	3-Benzothiophene	R=3-Benzothiophene	92	82
9	$(C_6H_5)CH=CH$	R= $(C_6H_5)CH=CH$	93	82
10	$(C_6H_5)_2C=CH$	R= $(C_6H_5)_2C=CH$	80	84
		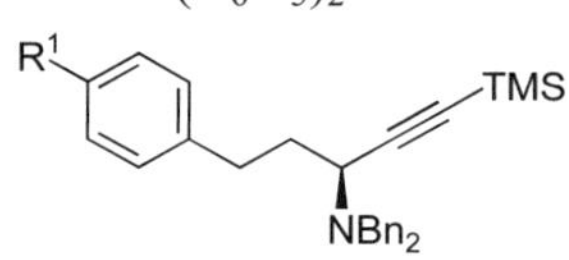		
11	$C_6H_5(CH_2)_2$	R^1= H	76	88
12	4-Br-$C_6H_4(CH_2)_2$	R^1= Br	72	88
13	4-CO_2Et-$C_6H_4(CH_2)_2$	R^1= CO_2Et	83	87
14	Ph	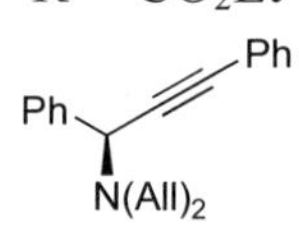	91	70
15	*i*-Bu	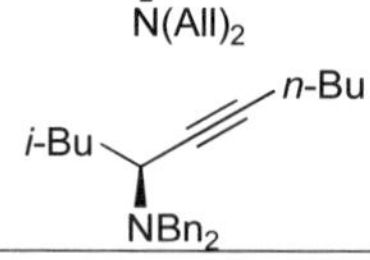	85	82

[*a*] Isolated yield of analytically pure product; [*b*] Enantiomeric excess determined by HPLC using Chiracel OD-H column (*n*-heptane : *i*-PrOH).

SYNTHESIS OF ACYLSILANES FROM MORPHOLINE AMIDES. SYNTHESIS OF 1-(DIMETHYL(PHENYL)SILYL)PROPAN-1-ONE [Dimethyl(1-oxopropyl)phenylsilane]

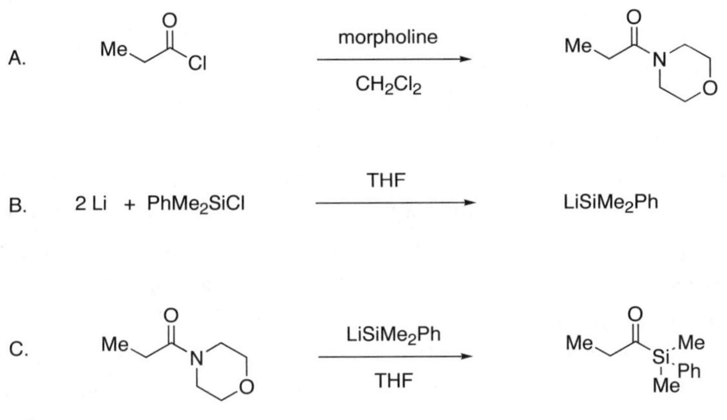

Submitted by Robert B. Lettan II, Benjamin C. Milgram, and Karl A. Scheidt.[1]
Checked by Scott E. Denmark and Ramil Y. Baiazitov.

1. Procedure

A. 1-Morpholinopropan-1-one. A flame-dried, 1-L, three-necked, round-bottomed flask, equipped with a magnetic stirring bar, 50-mL addition funnel (capped with a rubber septum), a nitrogen inlet and Teflon-coated thermocouple (Note 1) is charged with 32.9 mL (32.9 g, 378 mmol, 1.1 equiv) of morpholine (Note 2), which is added by syringe to the addition funnel with the stopcock open. Methylene chloride (175 mL) (Note 3) is added by syringe through the same septum and is used to rinse the addition funnel. The stopcock is closed and solution is placed under a positive pressure of nitrogen from a manifold. While stirring the solution, the flask and its contents are cooled in a NaCl/ice bath (internal temperature –8 °C). To the stirred solution is added propanoyl chloride (15.0 mL, 15.9 g, 172 mmol) (Note 4) through the addition funnel over 20 min (Note 5). Following the addition, the mixture is allowed to warm to ambient temperature. The mixture is stirred for 23 h at ambient temperature (Note 6). Then the solution is diluted with 70 mL of methylene chloride and transferred to a 1-L separatory funnel. The mixture is subsequently washed with 1N HCl (2 x 50 mL), and a mixture of saturated aqueous sodium bicarbonate (25 mL) and brine (50 mL) in a 1-L separatory funnel. The organic phase is dried over 10 g of anhydrous magnesium sulfate, filtered through glass wool, and concentrated by rotary evaporation (10 mmHg) to provide 27.4 g of a colorless liquid. The product is distilled with a short path distillation head (50 mm Vigreaux column) (oil bath temperature 110–120 °C) to afford 21.3 g (86%) of 1-morpholinopropan-1-one as a clear liquid (bp 83–86 °C, 1.1 mmHg) (Notes 7, 8).

B. Dimethylphenysilyllithium. This reagent is prepared according to the procedure of Fleming and coworkers.[2] A flame-dried, 300-mL, two-necked, Schlenk flask equipped with a magnetic stirring bar, nitrogen inlet and a rubber septum is charged with 33 cm (45 mg/cm, 1.49 g, 214 mmol, 1.5 equiv) of lithium wire (Note 9) and is filled with nitrogen. Tetrahydrofuran (60 mL, Note 10) is added by syringe through the septum. To the suspension of lithium wire is added chlorodimethylphenylsilane (12.0 mL, 71.5 mmol) (Note 11) by syringe at ambient temperature with stirring (Note 12). Following the addition of chlorodimethylphenylsilane, the reaction flask is tightly sealed and then transferred to a –15 to – 20 °C freezer and is allowed to stand for 59 h. The resulting, purple solution of dimethylphenylsilyllithium in THF is titrated to 0.97 M against phenolphthalein with 0.1 N HCl (Note 13).

C. 1-(Dimethyl(phenyl)silyl)propan-1-one. A flame-dried, 250-mL, 3-necked, round-bottomed flask, equipped with a magnetic stirring bar, a nitrogen inlet, a Teflon-coated thermocouple and a rubber septum, is charged with of 1-morpholinopropan-1-one (6.0 g, 41.9 mmol) by syringe under a nitrogen atmosphere from a manifold. To the flask is added 45 mL of THF (Note 10) by syringe. The solution is then cooled to –74 °C (internal) with a dry ice/acetone bath and dimethylphenylsilyllithium in THF (0.97 M solution, 52 mL, 50.4 mmol, 1.2 equiv) is added dropwise by syringe over 8 min such that the internal temperature never rises above –60 °C. The purple reaction mixture is stirred for 70 min at –75 °C (Note 14), quenched with 100 mL of saturated aqueous ammonium chloride solution, and then is allowed to warm to ambient temperature. The mixture is diluted with 40 mL of water in a 1-L separatory funnel and is extracted with hexanes (150 mL) and ethyl acetate (2 x 50 mL). The combined organic extracts are washed with saturated aqueous ammonium chloride (50 mL) and brine (100 mL), then are dried over 11 g of anhydrous magnesium sulfate, filtered through glass wool, and concentrated by rotary evaporation under vacuum (10 mm Hg). The residue is further dried by azeotropic evaporation with hexanes (2 x 50 mL, 10 mmHg) to afford 10.0 g of a light-yellow oil. The product is purified by chromatography on silica gel (50 mm column diameter, 150 g silica gel, packed with precooled hexanes (–15 °C)) using a gradient of precooled solvents (hexanes/CH_2Cl_2: 1/0 (350 mL), 10/1 (500 mL), 5/1 (550 mL), 1/1 (450 mL), 0/1 (250 mL), 60 mL fractions) to provide the acylsilane (5.30 g, 66%) as a light yellow oil and 0.63 g (8%) of a less pure fraction. The larger portion was distilled using a 65-mm Vigreaux column (oil bath, 130 °C) to afford 4.73 g (59%) of 1-(dimethyl(phenyl)silyl)propan-1-one as a light-yellow oil (bp 69–72 °C at 0.94 mmHg) (Notes 15, 16, 17).

2. Notes

1. A PFA-coated thermocouple probe, Type K (Omega Engineering, Inc.) was inserted through a teflon thermometer adapter sealed with an O-ring.

2. Morpholine (99+%) was purchased from Sigma-Aldrich Chemical Co., and was used as received.

3. Anhydrous methylene chloride was obtained by passage through a bed of activated alumina.[3]

4. Propanoyl chloride (98%) was purchased from Sigma-Aldrich Chemical Co., and was used as received.

5. The reaction temperature was monitored internally with a thermocouple during addition of propanoyl chloride, such that the temperature never rose above +10 °C. The reaction was very vigorous and a thick suspension rapidly formed which was difficult to stir magnetically. The use of a mechanical stirrer is highly recommended for larger scale runs.

6. The submitters monitored the progress of the reaction by ^1H NMR spectroscopy. A small aliquot of the reaction mixture was removed by syringe and quenched with 1 mL of saturated aqueous NH_4Cl. The mixture was extracted with 3 mL of ethyl acetate. The organic extraction was dried over Na_2SO_4, filtered through glass wool, and concentrated by vacuum. The remaining material was dissolved in $CDCl_3$, and monitored by ^1H NMR. Reaction completion was monitored by observance of product by ^1H NMR analysis.

7. The submitters used the 1-(dimethyl(phenyl)silyl)propan-1-one without purification for the next step. The amide is somewhat volatile and considerable amounts can be lost by extended exposure to high vacuum.

8. The physical properties are as follows: bp 83–86 °C, (1.1 mmHg); ^{1}H NMR (400 MHz, $CDCl_3$) δ: 1.13 (t, J = 7.4 Hz, 3 H), 2.32 (q, J = 7.5 Hz, 2 H), 3.42–3.46 (t, J = 4.9 Hz, 2 H), 3.58–3.62 (m, 2 H), 3.63–3.66 (m, 4 H). ^{13}C NMR (101 MHz, $CDCl_3$) δ: 9.3, 26.2, 41.8, 45.7, 66.5, 66.9, 172.4. IR (film): 3493, 2977, 2920, 2854, 1644, 1463, 1435, 1380, 1362, 1300, 1274, 1233, 1116, 1069, 1035, 1008, 990, 942, 905 cm $^{-1}$; MS (EI, 70 eV) *m/z* (rel. intensity) 144 (11), 143 [M^+, 98], 128 (131), 114 (21). Anal. Calcd for $C_7H_{13}NO_2$: C, 58.72; H, 9.15; N, 9.78. Found: C, 58.39; H, 9.32; N, 9.66. The submitters found their product to be > 99% pure by GC (Agilent 6890 Series, column: HP-5; 5% phenyl methyl siloxane, 30 m x 320 μm x 0.25 μm); T-program: 70 °C (1.0 min)→285 °C (25 °C/min); 6.03 min retention time) without distillation.

9. The lithium wire (99.9% in mineral oil) was purchased from Sigma-Aldrich Chemical Co., and was pre-washed with pentanes (2 x 30 mL) to remove the oil and was charged in 1–3 mm segments. If larger (5–10 mm) segments were used the reaction took considerably longer time.

10. Anhydrous tetrahydrofuran was obtained by passage through a bed of activated alumina.[3]

11. A moderate heat evolution was noted in the upper regions of the flask where the lithium floated.

12. The reagent was unstable at room temperature and was kept in an ice bath (0 °C) when not stored in the freezer.

13. Chlorodimethylphenylsilane (98%) was purchased from Gelest and was purified by distillation under vacuum through a short path distillation apparatus, topped with a thermometer, and attached to a receiving flask. (bp 75–77 °C (13–15 mmHg).

14. The submitters monitored the progress of the reaction by TLC analysis on silica gel with 20% EtOAc-hexanes as eluent and visualization with potassium permanganate stain. The amide starting material had R_f = 0.14 and the acylsilane product had R_f = 0.62. Silyl impurities could be visualized at R_f = 0.36 and R_f = 0.72.

15. The checkers found that chromatographic purification was necessary to remove the silicon-containing impurities (mostly disilane and disiloxane). However, chromatography with room temperature solvents led to a significant exotherm (and decomposition) when loading the acylsilane on silica gel (Silicycle Inc. Ultra Pure Silica Gel Silia*Flash*® P60. Grade: 230–400 mesh (40–63 microns), 60 Å, catalog number R12030B, pH=6.9, water content 6.0%). The acylsilane had an R_f = 0.22 (hexane/CH_2Cl_2, 1/1, UV, I_2, CAN) and the major impurities had R_f = 0.43, 0.63, and 0.68.

16. Analytical and spectral data are as follows: ^{1}H NMR (500 MHz, $CDCl_3$) δ: 0.51 (s, 6 H), 0.92 (t, J = 7.3 Hz, 3 H), 2.61 (q, J = 7.2 Hz, 2 H), 7.36–7.44 (m, 3 H), 7.54–7.58 (m, 2 H). ^{13}C NMR (125 MHz, $CDCl_3$) δ: –4.8, 6.1, 41.9, 128.1, 129.8, 133.9, 134.6, 246.2. ^{29}Si NMR (60 MHz, $CDCl_3$) δ: –17.01. IR (film) 2972, 2937, 2898, 1644, 1429, 1403, 1373, 1250, 1111, 1032, 834 cm^{-1} MS (EI, 70 eV) *m/z* (rel. intensity) 191 (4), 163 (20), 137 (12), 136 (19), 135 (100), 105 (11). Anal. Calcd for $C_{11}H_{16}OSi$: C, 68.69; H, 8.39. Found: C, 68.52; H, 8.25.

17. A second run on this scale provided 5.46 g (68%) of the acylsilane. The submitters obtained an 85% yield of material that was still contaminated with disilane and disiloxane.

Safety and Waste Disposal Information

All hazardous materials should be handled and disposed of in accordance with "Prudent Practices in the Laboratory"; National Academy Press; Washington, DC, 1995.

3. Discussion

In the presence of a strong nucleophile, acylsilanes have the ability undergo a Brook rearrangement (1,2-silyl shift from carbon to oxygen), which has a firmly established synthetic utility.[4] This tactic has been used in various umpolung approaches to carbonyl anions[5] by generating nucleophilicity on the normally electrophilic carbonyl carbon. To exploit this reactivity, acylsilanes have been employed in the synthesis of enolsilanes[6] and alcohols[7] in addition to providing direct access to homoenolates upon addition of vinyl or alkynyl organometallic reagents.[8] More recently, acylsilanes that undergo the Brook rearrangement have been utilized in tandem annulation reactions,[9] and new catalytic acyl anion addition reactions have been developed taking advantage of the unique benefits of the resulting acyl anions.[10] More specifically, acylsilanes in our laboratory have been employed as acyl anion equivalents in the presence of catalytic quantities of thiazolium salts (20-30 mol%) to carry out sila-Stetter reactions,[11] tandem one-pot sila-Stetter/Paal-Knoor transformations,[12] and the direct syntheses of α-amino ketones.[13] However, although acylsilanes are useful entities, standard procedures for accessing these compounds have numerous limitations such as low overall efficiency and a reliance on stoichiometric quantities of a transition metal.

There are many synthetic approaches to construct the carbonyl carbon-silicon bond.[14] Unconventional approaches include the conversion of α,α-dibromobenzylsilanes to acylsilanes on silica gel[15] and the hydroboration of silylalkynes.[16] The trapping of deprotonated dithianes with chlorosilanes has been employed for many years, but this strategy always requires the unmasking of the α-silyl dithiane, which is not always compatible with the substrate.[17] Additionally, benzotriazol-1-yl phenoxyalkanes are useful precursors to a variety of acylsilane structures.[18] Not surprisingly, many of these processes are suboptimal because of poor availability of starting materials or their lack of atom economy.[19] α,β-Unsaturated acylsilanes can be prepared either by a Horner-Wadsworth-Emmons reaction of α-(phosphonoacyl)-silanes[20] or the silyl-Wittig rearrangement of allylic alcohols followed by oxidation.[21] Interestingly, the palladium-catalyzed conversion of aryl acid chlorides to aryl acylsilanes has been reported, but this reaction is limited to the trimethylsilyl group and undergoes decarbonylation with electron deficient aromatic systems.[22]

The addition of anionic silyl nucleophiles to acid chlorides is typically the most direct method for the synthesis of acylsilanes, but this method requires at least two equivalents of the silyllithium reagent and suffers largely from the need for a stoichiometric amount of copper(I) cyanide required for the reaction to proceed in high yield.[23] Unfortunately on a preparative scale, this process becomes prohibitive. As an alternate approach, Fleming and coworkers first noted that the addition of silyl anions to amides afforded acylsilanes, but this process was not adequately explored as a method to construct acylsilanes.[24]

Because the direct addition of organometallic reagents to morpholine amides without over-addition is possible,[25] and more economical than the corresponding Weinreb amides, a direct and efficient synthesis of acylsilanes from amides could be developed.[26] The use of the morpholine amide minimizes over-addition of the silyl nucleophile and also provides the highest yields of the amides surveyed.

The extension of this method to the use of aromatic morpholine amides does not yield the desired acylsilanes. Unlike with the alkyl morpholine amides, additions of silyl anions to aromatic amides usually results in a Brook rearrangement, presumably due to stabilization of the resulting carbanion by the aromatic moiety. Surprisingly, this anion subsequently undergoes an α-elimination of the silyloxy substituent and the ensuing carbene either encounters an additional equivalent of the silyllithium species or dimerizes.[27]

The overall process to prepare alkyl substituted acylsilanes from morpholine amides demonstrated in this procedure has been optimized for a preparative scale, avoids the use of transition metals, and tolerates varied alkyl amide structures, including various protecting groups.

1. Department of Chemistry, Northwestern University, 2145 Sheridan Road, Evanston, Illinois, 60208.
2. Fleming I.; Newton, T. W.; Roessler, F. *J. Chem. Soc., Perkin Trans I* **1981**, 2527-2532.
3. Pangborn, A. B.; Giardello, M. A.; Grubbs, R. H.; Rosen, R. K.; Timmers, F. J. *Organometallics* **1996**, *15*, 1518-1520.
4. (a) Brook, A. G.; LeGrow, G. E.; MacRae, D. M.; *Can. J. Chem.* **1967**, *45*, 239-253. (b) Brook, A. G. *Acc. Chem. Res.* **1974**, *7*, 77-84. (c) Ricci, A.; Degl'Innocenti, A. *Synthesis* **1989**, 647-661. (d) Cirillo, P. F.; Panek, J. S. *Org. Prep. Proced. Int.* **1992**, *24*, 553-582. (e) Moser, W. H. *Tetrahedron* **2001**, *57*, 2065-2084.
5. (a) Grobel, B. T.; Seebach, D. *Synthesis* **1977**, 357-402. (b) Seebach, D. *Angew. Chem., Int. Ed. Engl.* **1979**, *18*, 239-258. (c) Johnson, J. S. *Angew. Chem., Int. Ed.* **2004**, *43*, 1326-1328.
6. (a) Reich, H. J.; Holtan, R. C.; Bolm, C. *J. Am. Chem. Soc.* **1990**, *112*, 5609-5617. (b) Lefebvre, O.; Brigaud, T.; Portella, C. *J. Org. Chem.* **2001**, *66*, 4348-4351.
7. (a) Buynak, J. D.; Geng, B.; Uang, S.; Strickland, J. B. *Tetrahedron Lett.* **1994**, *35*, 985-988. (b) Takeda, K.; Ohnishi, Y.; Koizumi, T. *Org. Lett.* **1999**, *1*, 237-240. (c) Patrocínio, A. F.; Corrêa, I. R.; Moran, P. J. S. *J. Chem. Soc., Perk. Trans. 1* **1999**, 3133-3137. (d) Bolm, C.; Kasyan, A.; Heider, P.; Saladin, S.; Drauz, K.; Gunther, K.; Wagner, C. *Org. Lett.* **2002**, *4*, 2265-2267. (e) Cossrow, J.; Rychnovsky, S. D. *Org. Lett.* **2002**, *4*, 147-150.
8. (a) Enda, J.; Kuwajima, I. *J. Am. Chem. Soc.* **1985**, *107*, 5495-5501. (b) Mi, Y.; Schreiber, J. V.; Corey, E. J. *J. Am. Chem. Soc.* **2002**, *124*, 11290-11291 and references cited therein. (c) Reich, H. J.; Olson, R. E.; Clark, M. C. *J. Am. Chem. Soc.* **1980**, *102*, 1423-1424.
9. (a) Takeda, K.; Haraguchi, H.; Okamoto, Y. *Org. Lett.* **2003**, *5*, 3705-3707. (b) Sasaki, M.; Kawanishi, E.; Nakai, Y.; Matsumoto, T.; Yamaguchi, K.; Takeda, K. *J. Org. Chem.* **2003**, *68*, 9330-9339.
10. (a) Degl'Innocenti, A.; Pike, S.; Walton, D. R. M.; Seconi, G.; Ricci, A.; Fiorenza, M. *J. Chem. Soc., Chem. Commun.* **1980**, 1201-1202. (b) Degl'Innocenti, A.; Ricci, A.; Mordini, A.; Reginato, G.; Colotta, V. *Gazz. Chim. Ital.* **1987**, *117*, 645-648. (c) Xin, L. H.; Johnson, J. S. *Angew. Chem., Int. Ed.* **2003**, *42*, 2534-2536.

11. Mattson, A. E.; Bharadwaj, A. R.; Scheidt, K. A. *J. Am. Chem. Soc.* **2004**, *126*, 2314-2315.
12. Bharadwaj, A. R.; Scheidt, K. A. *Org. Lett.* **2004**, *6*, 2465-2468.
13. Mattson, A. E.; Scheidt, K. A. *Org. Lett.* **2004**, *6*, 4363-4366.
14. (a) Colvin, E. W. *Silicon Reagents in Organic Synthesis*; Academic Press: London, 1988. (b) Page, P. C. B.; Klair, S. S.; Rosenthal, S. *Chem. Soc. Rev.* **1990**, *19*, 147-195.
15. Degl'Innocenti, A.; Walton, D. R. M.; Seconi, G.; Pirazzini, G.; Ricci, A.*Tetrahedron Lett.* **1980**, *21*, 3927-3928.
16. (a) Miller, J. A.; Zweifel, G. *J. Am. Chem. Soc.* **1981**, *103*, 6217-6219. (b) Miller, J. A.; Zweifel, G. *Synthesis* **1981**, 288-289.
17. (a) Brook, A. G.; Duff, J. M.; Jones, P. F.; Davis, N. R. *J. Am. Chem. Soc.* **1967**, *89*, 431-434. (b) Corey, E. J.; Seebach, D.; Freedman, R. *J. Am. Chem. Soc.* **1967**, *89*, 434-436.
18. (a) Katritzky, A. R.; Yang, Z. J.; Hong, Q. M. *J. Org. Chem.* **1994**, *59*, 5097-5099. (b) Katritzky, A. R.; Wang, Z. Q.; Lang, H. Y. *Organometallics* **1996**, *15*, 486-490. (c) Katritzky, A. R.; Lang, H. Y.; Wang, Z. Q.; Lie, Z. *J. Org. Chem.* **1996**, *61*, 7551-7557.
19. Trost, B. M. *Angew. Chem., Int. Ed.* **1995**, *34*, 259-281.
20. Nowick, J. S.; Danheiser, R. L. *J. Org. Chem.* **1989**, *54*, 2798-2802.
21. Danheiser, R. L.; Fink, D. M.; Okano, K.; Tsai, Y. M.; Szczepanski, S. W. *J. Org. Chem.* **1985**, *50*, 5393-5396.
22. (a) Picard, J. P.; Calas, R.; Dunogues, J.; Duffaut, N.; Gerval, J.; Lapouyade, P. *J. Org. Chem.* **1979**, *44*, 420-424. (b) Rich, J. D. *J. Am. Chem. Soc.* **1989**, *111*, 5886-5893.
23. (a) Capperucci, A.; Degl'Innocenti, A.; Faggi, C.; Ricci, A.; Dembech, P.; Seconi, G. *J. Org. Chem.* **1988**, *53*, 3612-3614. (b) Bonini, B. F.; Comes-Franchini, M.; Mazzanti, G.; Passamonti, U.; Ricci, A.; Zani, P. *Synthesis* **1995**, 92-96.
24. Fleming, I.; Ghosh, U. *J. Chem. Soc., Perkin Trans. 1* **1994**, 257-262.
25. (a) Martín, R.; Romea, P.; Tey, C.; Urpí, F.; Vilarrasa, J. *Synlett* **1997**, 1414-1416. (b) Douat, C.; Heitz, A.; Martinez, J.; Fehrentz, J.-A. *Tetrahedron Lett.* **2000**, *41*, 37-40.
26. Clark, C. T.; Milgram, B. C.; Scheidt, K. A. *Org. Lett.* **2004**, *6*, 3977-3980.
27. For the first reported observation of this behavior, see: Fleming, I.; Ghosh, U. *J. Chem. Soc., Perkin Trans. 1* **1994**, 257-262. (b) Buswell, M.; Fleming, I. *Chem. Commun.* **2003**, 202-203 and references cited therein.

(*S*)-(+)-NEOMENTHYLDIPHENYLPHOSPHINE IN NICKEL-CATALYZED ASYMMETRIC REDUCTIVE COUPLING OF ALKYNES AND ALDEHYDES: ENANTIOSELECTIVE SYNTHESIS OF ALLYLIC ALCOHOLS AND α-HYDROXY KETONES

[(*R*)-Butan-2-one, 1-cyclohexyl, 1-hydroxy-]

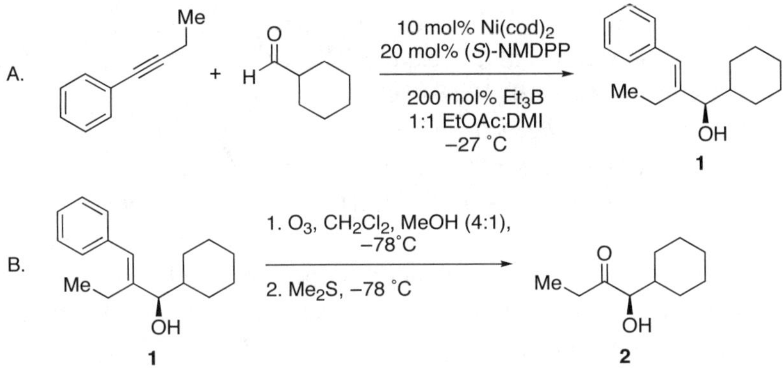

Submitted by Aaron R. Van Dyke, Karen M. Miller, and Timothy F. Jamison.[1]
Checked by Matthew J. Fleming and Mark Lautens.

1. Procedure

A. (R)-(E)-2-Benzylidiene-1-cyclohexyl-butan-1-ol. In a glovebox, a flame-dried 500-mL round-bottomed flask is charged with Ni(cod)$_2$ (1.38 g, 5.00 mmol) (Note 1) and (*S*)-(+)-neomenthyldiphenylphosphine (3.24 g, 10.0 mmol) (Note 2). The flask is sealed with a septum, removed from the glovebox, and transferred to a fume-hood. An argon inlet (Note 3) is then attached to the flask and degassed EtOAc (50 mL) (Note 4), freshly distilled, degassed DMI (50 mL) (Note 5), and triethylborane (14.5 mL, 100 mmol) (Note 6) are added sequentially via syringe. *Caution! Triethylborane is extremely pyrophoric.* The solution is allowed to stir 15 min at room temperature and then placed in a –27 °C bath (Note 7) for 30 min. 1-Phenyl-1-butyne (7.1 mL, 50 mmol) (Note 8) is added in one portion via syringe followed by addition of cyclohexanecarboxaldehyde (9.1 mL, 75 mmol) (Note 9) via syringe pump over 9 h to the solution in a –27 °C bath. The reaction is stirred at –27 °C for 36 h, then quenched with saturated NH_4Cl

(100 mL) (Note 10), 1M HCl (40 mL) at –27 °C. The solution is allowed to warm to room temperature and then stirred 15 min at room temperature. Air is vigorously bubbled through the reaction using a pasteur pipet for 30 min, which results in a light yellow emulsion (Note 11). The mixture is extracted with EtOAc (2 x 200 mL). The combined organic layers are washed twice with saturated NH_4Cl (300 mL), once with brine (300 mL) (Note 12), dried over $MgSO_4$ (10 g) (Note 13), filtered and concentrated on a rotary evaporator (20 °C, 11 mmHg, then 3 mmHg) to remove trace EtOAc. The resulting yellow oil is purified by flash chromatography on silica gel with a hexanes to 9:1 hexanes: ethyl acetate gradient (Note 14) to yield 10.87 g (89%) of **1** as a colorless oil (Note 15).

B. (R)-1-Cyclohexyl-1-hydroxy-butan-2-one. In a 1-L round-bottomed, one-necked flask with a magnetic stirbar, **1** (10.87 g, 44.5 mmol) is dissolved in methanol (60 mL) (Note 16) and dichloromethane (240 mL) (Note 17). The vessel is cooled to –78 ° C and ozone is bubbled through the solution using a Pasteur pipet until a persistent blue color appears (approximately two hours) (Note 18). Argon is then bubbled through the solution for 30 min and dimethylsulfide (131 mL, 1780 mmol) is added (Note 19). The reaction is warmed slowly to ambient temperature and stirred for 13 h. The solvent and excess dimethylsulfide are removed by rotary evaporation (20 °C, 11 mmHg). The crude oil is purified by flash chromatography on silica gel, eluting with 50:1 hexanes: ethyl acetate (Note 20) to yield 5.30 g (70%) of **2** as a colorless oil (Note 21).

2. Notes

1. $Ni(cod)_2$ (98%) was purchased from Strem and used as received. The crystals should appear yellow in color, approximately 1.5 mm in diameter. $Ni(cod)_2$ appearing as a yellow powder or black should not be used. $Ni(cod)_2$ from other vendors should not be used. Silver colored deposits on the inside of $Ni(cod)_2$ vials were indicative of formation of Nickel mirror and other undesired species. Such samples should not be used.

2. (*S*)-(+)-Neomenthyldiphenylphosphine was purchased from Strem and used as received. NMDPP is moderately air sensitive and will oxidize unless opened, manipulated, and stored in a glovebox.

3. All "argon inlet" references are defined as inserting a 16-gauge needle attached to a positive flow of argon through the top of a rubber septum that is sealed around the flask neck with electrical tape.

4. Ethyl acetate (99.9%) was purchased from Burdick and Jackson, distilled over $MgSO_4$ (77 °C) and degassed by bubbling argon through it for 20 min. The checkers used ethyl acetate (99.5%) purchased from EMP Chemicals, Inc.

5. Dimethylimidizolidinone (99.5%) was purchased from Fluka, freshly distilled from CaH_2 (98 °C, 11.7 mmHg) prior to use. DMI is **extremely hygroscopic**, and trace water led to alkylative coupling (transfer of an ethyl group from Et_3B) instead of reductive coupling. Alkylative coupling was inseparable from the product by chromatography but identified by its carbinol proton ^1H NMR (400 MHz, $CDCl_3$) δ: 4.35 (d, J = 9, 1 H). Failure to use freshly distilled DMI gave approximately 10% alkylative coupling, whereas freshly distilled DMI gave 2% or less.

6. Triethylborane (98%) was purchased from Aldrich and used as received. The headspace of the Et_3B canister must be thoroughly purged with argon before opening. Triethylborane (Et_3B) in the syringe should be kept under argon at all times. Any Et_3B remaining in the syringe should be diluted with CH_2Cl_2 and quenched by stirring the solution open to air overnight in a hood. If a spark is observed at the needle tip, fresh Et_3B must be used. The checkers used triethylborane (98%) purchased from Strem.

7. Isopropyl alcohol bath temperature –27 $\pm$ 2 °C was maintained by Neslab Cryocool CC-60IIA. Temperature control was critical for maximum ee and yield. Due to the air sensitive nature of the reaction, internal temperature readings were not taken. All readings indicate the isopropanol bath temperature. The checkers used a Neslab Cryobath CB-80.

8. 1-Phenyl-1-propyne (99%) was purchased from Aldrich Chemical Co. and distilled from $MgSO_4$ (74 °C, 4 mmHg).

9. Cyclohexanecarboxaldehyde (98%) was purchased from Aldrich Chemical Co. and distilled from $MgSO_4$ (65 °C, 28 mmHg).

10. Ammonium chloride (reagent grade) was purchased from J.T. Baker and added to distilled tap water until saturated.

11. Air was vigorously bubbled through the quenched reaction to oxidize all nickel species to Ni(II) salts and to facilitate cleavage of the RO-BEt_2 bond, liberating the alcohol. An alternative to the NH_4Cl/HCl quench involved slow addition of a basic hydrogen peroxide solution via syringe (50 mL 30% H_2O_2 (Aldrich) in 200 mL 0.75M NaOH (Aldrich)). An 18-gauge needle was inserted into the septum with this workup to avoid pressure buildup.

12. Sodium chloride (reagent grade) was purchased from Aldrich Chemical Co. and added to distilled tap water until saturated.

13. Magnesium sulfate (reagent grade) was purchased from Aldrich Chemical Co and used as received.

14. A column (10 cm by 22 cm) was prepared from a slurry of 600 g of silica gel (40 – 63μm, obtained from Silicycle) in hexanes. The oil was loaded onto this column and eluted with 2 L of hexanes, 2 L 95:5 hexanes : EtOAc, then 9:1 hexanes : EtOAc (approximately 3 L). Two silica gel columns were required for optimal purity.

15. The reaction was monitored by TLC. The product was visualized with UV followed by phosphomolybdic acid (10 g phosphomolybdic acid in 100 mL EtOH) stain, $R_f = 0.20$ (5:1, hexanes : EtOAc); ^{1}H NMR (400 MHz, $CDCl_3$) δ: 0.99–1.30 (m, 8 H) 1.55–1.82 (m, 6 H), 2.00 (d, J = 12.5 Hz, 1 H), 2.21 (dq, J = 19, 7.5 Hz, 1 H), 2.36 (dq, J = 19, 7.5 Hz, 1 H), 3.93 (d, J = 7 Hz, 1 H), 6.45 (s, 1 H), 7.22–7.36 (m, 5 H); ^{13}C NMR (100 MHz, $CDCl_3$) δ: 14.0, 21.4, 26.1, 26.3, 26.5, 28.3, 30.1, 41.6, 81.4, 126.3, 126.4, 128.1, 128.6, 137.6, 145.5; IR (thin film NaCl): 3395, 3055, 3023, 2927, 2851, 1599, 1493, 1448, 1308, 1261, 1173. Anal. Calcd for $C_{17}H_{24}O$: C, 83.55; H, 9.90. Found: C, 83.30; H, 9.81. Enantiomeric excess (87%) was established by chiral HPLC (Chiralcel OD, hexanes: 2-propanol, 98:2, 1 mL/min): t_R[(R)-**1**] = 14.5 min, t_R[(S)-**1**] = 16.5 min. Repeating the reaction on a half-scale gave 90% yield and 88% ee.

16. Methanol (99.9%) was purchased from VWR and used as received. The checkers used methanol (99.8%) purchased from Caledon, Inc.

17. Dichloromethane (HPLC grade) was purchased from Burdick and Jackson and used as received. The checkers used dichloromethane (99.5%) purchased from Caledon, Inc.

18. Clear Water Tech Ozone generator, model CD1500.

19. Dimethylsulfide (99+%) was purchased from Aldrich Chemical Co. and used as received. With the exception of rotary evaporation, all manipulations involving dimethylsulfide and dimethylsulfide-containing solutions were performed in a well-ventilated fume hood. The distillate collected during rotary evaporation was treated with 1:1 solution of commercial-grade (Clorox®) and water (2 x 200 mL) prior to disposal. All reaction glassware and the collection bulb on the rotary evaporator were washed with a bleach solution (1:1, bleach:water) after use.

20. A column (6.5 cm by 12 cm) was prepared from 225 g of silica gel (40 – 63μm, obtained from Silicycle).

21. Product was visualized with PMA stain, $R_f = 0.25$ (8:2, hexanes: ethyl acetate); ^{1}H NMR (400 MHz, $CDCl_3$) δ: 1.13–1.35 (m, 8 H), 1.48 (dq, J = 12.5, 4 Hz, 1 H), 1.64–1.83 (m, 5 H), 2.41–2.57 (overlapping dq, J = 19,

7.5 Hz, 2 H), 3.41 (d, J = 10 Hz, 1 H), 4.06 (br s, 1H); ^{13}C NMR (100 MHz, $CDCl_3$) δ: 7.6, 25.1, 25.8, 26.0, 26.5, 30.1, 31.4, 41.4, 80.5, 213.0; IR (thin film NaCl): 3474, 2976, 2931, 2853, 1709, 1450, 1406, 1349, 1260, 1105; Anal. Calcd for $C_{10}H_{18}O_2$: C, 70.55; H, 10.66. Found: C, 70.35; H, 10.55. Enantiomeric excess 88%, established by chiral GC (Alltech B-PH, column = 95 °C, injector = 200 °C, flow (H_2) = 2 mL/min): t_R[(R)-**2**] = 30.7 min, t_R[(S)-**2**] = 31.5 min. The checkers could not get separation of peaks on Alltech B-PH column (10 years old) using the submitter's conditions. Changing the column temp to 80 °C gave two peaks, but without baseline separation. Enantiomeric excess could be estimated to be between 85–90%: t_R[(R)-**2**] = 70.6 min, t_R[(S)-**2**] = 74.2 min. Repeating the reaction on a half-scale gave 66% yield and 85–90% ee.

Safety and Waste Disposal Information

All hazardous materials should be handled and disposed of in accordance with "Prudent Practices in the Laboratory"; National Academy Press; Washington, DC, 1995.

3. Discussion

Allylic alcohols are synthetically useful intermediates and are present in a variety of natural products. An established method for accessing these compounds involves the addition of alkenyl halides (Nozaki-Hiyama-Kishi reaction[2]) or alkenyl metals (prepared via hydroboration[3] or hydrozirconation[4] of alkynes) to aldehydes. All of these involve the use of a stoichiometric amount of transition metal. Intramolecular[5] and intermolecular[6] nickel-catalyzed reductive coupling of alkynes and aldehydes have been reported in the literature, and that described here is the only one that affords products with high levels of enantioselectivity.[7]

The nickel-catalyzed asymmetric reductive coupling of alkynes and aldehydes reported herein allows rapid access to enantiomerically enriched (*E*)-trisubstituted allylic alcohols.[8] The catalyst is derived from $Ni(cod)_2$ and(*S*)-(+)-neomenthyldiphenylphosphine (NMDPP), a commercially available phosphine, and the terminal reductant is Et_3B. The enantioselective, catalytic reductive coupling exhibits excellent regioselectivity (>95:5) and exclusive *cis* addition across the alkyne. Ozonolysis yields an α-hydroxyketone with complete preservation of enantiomeric purity. In addition to being an example of a useful family of building blocks, the TBS ether of **2** was developed by Masamune for use in asymmetric aldol reactions.[9]

Table 1. Asymmetric catalytic reductive coupling of alkynes and aldehydes.[a]

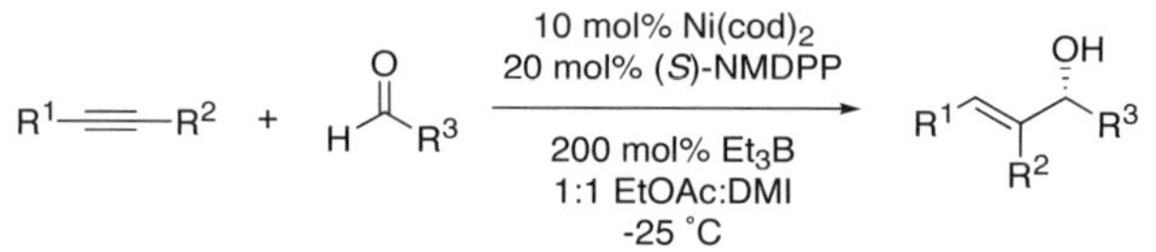

entry	R^1	R^2	R^3	yield (%),[b] regioselectivity[c]	ee (%)[d]
1	Ph	Me	*i*–Pr	95 (>95:5)	90[e]
2	"	"	Cy	97 (>95:5)	90
3	"	"	Ph	79 (91:9)	73
4	"	"	*n*–Pr	82 (>95:5)	65
5	(*p*-MeO)Ph	"	*i*–Pr	80 (>95:5)	88
6	(*p*-Cl)Ph	"	"	75 (>95:5)	83
7	(*p*-CF_3)Ph	"	"	78 (>95:5)	81
8	(*o*-Me)Ph	"	"	86 (>95:5)	84
9	1-naphthyl	"	"	93 (>95:5)	90
10	Ph	Et	"	81 (>95:5)	93
11[f]	Ph	Et	Cy	78 (>95:5)	89
12	"	*n*–Pr	*i*–Pr	74 (>95:5)	92
13	"	*i*–Pr	"	58[g] (>95:5)	92
14	"	cyclopropyl	"	67 (>95:5)	92
15	"	CH_2OTBS	"	59 (>95:5)	85
16	"	CH_2NHBoc	"	60 (>95:5)	96
17	"	$SiMe_3$	*n*–Pr	43[g] (>95:5)	92
18	*n*–Pr	*n*–Pr	*i*–Pr	35[g] (–)	42
19	Ph	H	"	15 (>95:5)	75

[a] Experimental procedure: A solution of Ni(cod)$_2$ (0.05 mmol), (*S*)-(+)-NMDPP (0.10 mmol), and Et_3B (1.0 mmol) in EtOAc/DMI (1:1, total volume 0.50 mL) was cooled to –25 °C. The alkyne (0.50 mmol) was added, and then the aldehyde (1.0 mmol) was added dropwise via syringe over 8 h. After 36 h, saturated aqueous NH_4Cl (2 mL) and 1 M HCl (0.5 mL) were added and the mixture was extracted with EtOAc (3 x 10 mL). Crude material was purified by silica gel chromatography. [b] Yield of isolated product. [c] Determined by ^{1}H NMR. [d] Determined by chiral HPLC or GC analysis. [e] Absolute configuration determined to be (*R*) by Mosher's ester analysis. [f] Performed on 5.0 mmol scale. [g] Some alkylative coupling (transfer of Et from Et_3B instead of H) was observed.

1. Department of Chemistry, Massachusetts Institute of Technology, Cambridge, MA, 02139. e-mail: tfj@mit.edu.

2. (a) Jin, H.; Uenishi, J.; Christ, W.J.; Kishi, Y. *J. Am. Chem. Soc.* **1986**, *108*, 5644; (b) Takai, K.; Tagashira, M.; Kuroda, T.; Oshima, K.; Utimoto, K.; Nozaki, H. *J. Am. Chem. Soc.* **1986**, *108*, 6048.

3. Oppolzer, W.; Radinov, R. N. *J. Am. Chem. Soc.* **1993**, 115, 1593.

4. Wipf, P.; Xu, W. *Tetrahedron Lett.* **1994**, *35*, 5197.

5. (a) Oblinger, E.; Montgomery, J. *J. Am. Chem. Soc.* **1997**, *119*, 9065; (b) Crowe, W.E.; Rachita, M. J. *J. Am. Chem. Soc.* **1995**, *117*, 6787.

6. (a) Huang, W. S.; Chan, J.; Jamison, T. F. *Org Lett.* **2000**, *2*, 4221; (b) Colby, E. A.; Jamison, T. F. *J. Org. Chem.* **2003**, *68*, 156; (c) Mahandru, G. M.; Liu, G.; Montgomery, J. *J. Am. Chem. Soc.* **2004**, *126*, 3698.

7. Colby, E. A.; Jamison, T. F. *J. Org. Chem.* **2003**, *68*, 156.

8. Miller, K. M.; Huang, W. S.; Jamison, T. F. *J. Am. Chem. Soc.* **2003**, *125*, 3442.

9. (a) Masamune, S.; Choy, W.; Kerdesky, F. A. J.; Imperiali, B. *J. Am. Chem. Soc.* **1981**, *103*, 1566; (b) Masamune, S.; Hirama, M.; Mori, S.; Ali, S. A.; Garvey, D. S. *J. Am. Chem. Soc.* **1981**, *103*, 1568.

A XANTHATE-TRANSFER APPROACH TO α-TRIFLUOROMETHYLAMINES [2-(*N*-ACETYLAMINO)-4,4-DIMETHOXY-1,1,1-TRIFLUOROBUTANE]

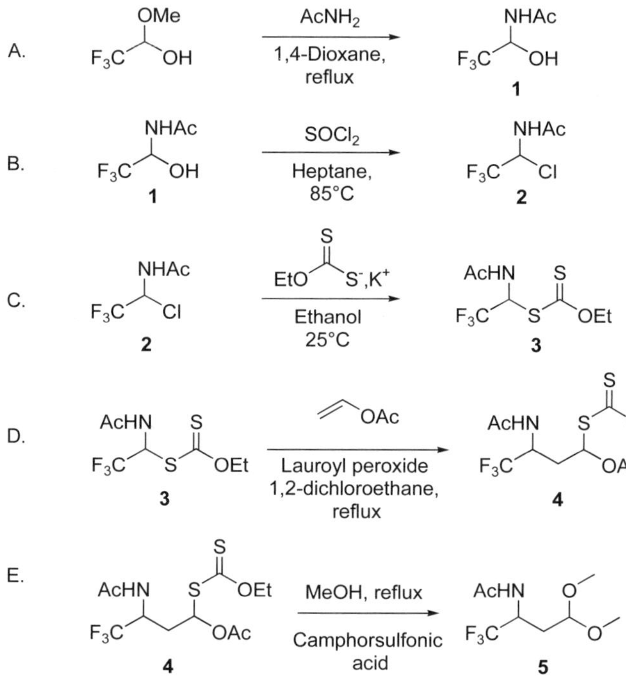

Submitted by Fabien Gagosz and Samir Z. Zard.[1]
Checked by Daniel Laurich and Alois Fürstner.

1. Procedure

A. N-(2,2,2-Trifluoro-1-hydroxyethyl)-acetamide (**1**). A 250-mL, two-necked, round-bottomed flask, equipped with a magnetic stir bar and fitted with a glass stopper and a condenser, is charged with 2,2,2-trifluoro-1-methoxyethanol (14.75 g, 102 mmol) (Note 1), acetamide (6.0 g, 102 mmol) (Note 1) and 1,4-dioxane (100 mL) (Note 2). The resulting solution is heated to reflux and stirred in an oil bath under argon for 2 h. The reaction mixture is then cooled to ambient temperature and transferred to a single-necked flask. The solvent is evaporated under reduced pressure and the residue dried under vacuum (25 °C, 1 mmHg) for 1 h to give 16.8 g of a colorless solid residue. This product is dissolved in *tert*-butyl methyl ether (20 mL) in a one-necked, round-bottomed flask. Silica gel (10 g) is added to the solution and the solvent is evaporated under reduced pressure (Note 3). The resulting adsorbate is added on top of a silica gel column (approximately 180 g), which is eluted with hexanes/ethyl acetate (3:2). The fractions containing the product are combined and evaporated to give *N*-(2,2,2-trifluoro-1-hydroxyethyl)acetamide (**1**) as a colorless solid (8.98 g, 56%) (Notes 4, 5).

B. N-1-(Chloro-2,2,2-trifluoro-ethyl)acetamide (**2**). A 250-mL, two-necked, round-bottomed flask is equipped with a magnetic stir bar and fitted with a glass stopper and a condenser. The top of the condenser is fitted with a T-joint allowing argon to sweep the effluent gases (HCl and sulfur dioxide) from the reaction into the exit duct of a well ventilated fume hood. The flask is charged with compound **1** (8.98 g, 57 mmol), thionyl chloride (7.1 g, 60 mmol) (Note 6) and heptane (65 mL) (Note 2). The resulting suspension is stirred and heated to 85 °C in an oil bath under argon until a clear solution has formed. Stirring is continued for 15 min before the mixture is allowed to cool to ambient temperature, whereupon the product starts to crystallize. After standing for 30 min, the crystals are collected by filtration, rinsed with heptane (2 x 25 mL) and dried under vacuum (25 °C, 1 mmHg) for 1 h to provide product **2** (9.3 g, 93%) as colorless crystals, which were used without further purification in the next step (Note 7).

C. S-(1-Acetylamino-2,2,2-trifluoroethyl) O-ethyl dithiocarbonate (**3**). A 500-mL, one-necked, round-bottomed flask, equipped with a magnetic stir bar, is charged with *N*-(1-chloro-2,2,2-trifluoro-ethyl)acetamide **2** (9.2 g, 52 mmol) and EtOH (100 mL) (Note 2). Potassium *O*-ethyl xanthate (9.2 g, 57 mmol) (Note 8) is added in portions over a period of 5 min. The resulting mixture is stirred at ambient temperature for 15 min (Note 9) before the reaction is quenched with water (100 mL). The solution is extracted with ether/hexanes solution (7:3, 3 x 150 mL), the combined organic layers are dried over anhydrous magnesium sulfate, filtered and evaporated under reduced pressure to provide a solid residue which is dried under vacuum (25 °C, 1 mmHg) for 1 h to give xanthate **3** (12.5 g, 91%) as a white solid (Note 10).

D. 3-Acetylamino-1-ethoxythiocarbonylsulfanyl - 4, 4, 4-trifluorobutyl acetate (**4**). A flame-dried, 100-mL, two-necked, round-bottomed flask is fitted with magnetic stir bar, a glass stopper and a condenser connected to the argon line. The flask is charged with xanthate **3** (12.4 g, 47.5 mmol), vinyl acetate (5.1 mL, 54.6 mmol) (Note 11) and 1,2-dichloroethane (50 mL) (Note 11) and the resulting solution is heated under reflux in an oil bath for

15 min under argon. Four equal portions of lauroyl peroxide (473 mg each, 1.9 g overall, 10 mol%) (Notes 11–13) are added in intervals of 1.5 h to the refluxing solution. After 7 h under reflux, the solution is allowed to reach ambient temperature before the solvent is evaporated under reduced pressure to give adduct **4** as a pale yellow oil (17.8 g), which was used in the next step without further purification (Note 14).

E. N-(3,3-Dimethoxy-1-trifluoromethyl-propyl)-acetamide (**5**). A flame-dried, 250-mL, two-necked, round-bottomed flask is fitted with a large magnetic stir bar, a glass stopper and a reflux condenser connected to the argon line. The flask is charged with the crude radical adduct **4** (17.8 g), MeOH (100 mL) and (±)-10-camphorsulfonic acid (12.5 mol%, 1.37 g, 5.9 mmol) (Note 15). The mixture is refluxed in an oil bath for 24 h before it is cooled to ambient temperature. The solvent is evaporated under reduced pressure. The residue is dissolved in ethyl acetate (80 mL) and the organic phase is washed successively with saturated aqueous sodium bicarbonate (20 mL) and saturated aqueous sodium chloride (20 mL). The organic solvent is dried over $MgSO_4$, filtered, and evaporated under reduced pressure. The resulting brown syrup (11.8 g) is purified by flash chromatography on silica gel (Notes 16, 17) using hexanes/ethyl acetate (2:3) as the eluent to give *N*-(3,3-dimethoxy-1-trifluoromethyl-propyl)-acetamide (**5**) as colorless crystals (4.43 g, 41%) (Note 18).

2. Notes

1. 2,2,2-Trifluoro-1-methoxyethanol was obtained from Avocado (trifluoroacetaldehyde methyl hemiacetal), tech. 90% and acetamide was obtained from Acros, 99%. The checkers used 2,2,2-trifluoro-1-methoxyethanol purchased from ABCR and acetamide purchased from Riedel de Haën.

2. 1,4-Dioxane was obtained from SDS Carlo Erba and used as received. All the solvents (petroleum ether, ethyl acetate, heptane, acetone, ethanol, methanol, *tert*-butyl methyl ether) were obtained from SDS Carlo Erba and used as received. The checkers used reagent-grade solvents purchased from Acros.

3. (E. Merck, Darmstadt, 230-240 mesh) was used. The progress of the reaction was monitored by TLC on silica gel using hexanes/ethyl acetate (3:2) as eluent. The product has an R_f = 0.15 (stained with potassium permanganate solution [300 mL of water, 3 g of $KMnO_4$, 20 g of K_2CO_3, 0.25 mL of acetic acid]).

4. The submitters reported purification of the crude product by recrystallization: Dichloromethane (60 mL) is added to the crude product and the resulting mixture is vigorously stirred for 10 min. The resulting white precipitate is filtered, washed twice with 30 mL of dichloromethane and dried under vacuum (25 °C, 1 mmHg) for 1 h to give a crop of pure product. The corresponding filtrate is concentrated under reduced pressure, diluted in 80 mL of ethyl acetate and washed twice with 20 mL of a half-saturated aqueous sodium chloride solution. The organic solvent is then dried over anhydrous magnesium sulfate, filtered and evaporated to give a semi-solid residue. Dichloromethane (30 mL) is added and the mixture stirred vigorously for 10 min. The resulting white precipitate is filtered, washed twice with 10 mL of dichloromethane and dried under vacuum (25 °C, 1 mmHg) for 1 h to provide a second crop of pure product. The two crops are combined to give amide **1** (53% yield). The checkers obtained 41% of product using this purification method.

5. The product exhibited the following properties: mp 117–119 °C; ^{1}H NMR (MeOD, 400 MHz) δ: 2.01 (s, 3 H), 4.83 (s, 2 H), 5.71 (m, 1 H).

6. Thionyl chloride (99.5%) was obtained from Acros and used as received; the checkers used thionyl chloride (99%) purchased from Fluka.

7. The product exhibited the following properties: mp 85–87 °C (heptane). ^{1}H NMR ($CDCl_3$, 400 MHz) δ: 2.15 (s, 3 H), 6.33 (qd, J = 5.3, 11.0 Hz, 1 H), 7.31 (br. m, 1 H); ^{13}C NMR ($CDCl_3$, 100 MHz) δ: 23.0, 60.7 (q, J = 38 Hz), 121.8 (q, J = 277 Hz), 170.4; IR (film): 3279, 3037, 1684, 1530, 1375, 1349, 1244, 1192, 1150, 1133, 857, 788, 690 cm^{-1}; HRMS (EI): Calcd for $C_4H_5NOF_3Cl$: 175.0011; Found: 175.0010.

8. Potassium *O*-ethyl xanthate (99%) was obtained from Aldrich (listed under ethylxanthic acid potassium salt) and recrystallized from hot ethanol before use. The checkers used the commercial product as received.

9. The progress of the reaction was monitored by TLC analysis on silica gel using hexanes/ethyl acetate (4:1) as eluent; the product had an R_f = 0.23 (stained with *p*-anisaldehyde solution [(950 mL of ethanol, 95%, 35 mL of concentrated sulfuric acid, 26 mL of *p*-anisaldehyde, 10.5 mL of acetic acid)].

10. The product exhibited the following properties: mp 84–86 °C; ^{1}H NMR ($CDCl_3$, 400 MHz) δ: 1.45 (t, J = 7.0 Hz, 3 H), 2.11 (s, 3 H), 4.70 (q, J = 7.0 Hz, 2 H), 6.61 (qd, J = 7.6, 9.7 Hz, 1 H), 7.43 (d, J = 9.7 Hz, 1 H); ^{13}C NMR ($CDCl_3$, 100 MHz) δ: 13.6, 22.8, 57.9 (q, J = 38 Hz), 71.4, 123.5 (q, J = 279 Hz), 170.9, 207.1; IR (film): 3278, 2990, 2933, 1667, 1514, 1370, 1334, 1289, 1246, 1210, 1186, 1127, 1115, 1102, 1049, 857, 819, 684 cm^{-1}.

The product obtained following the procedure was pure enough for use in the next step. However, it could be recrystallized by dissolving 1 g of the compound in a hot mixture of 1 mL of ethyl acetate and 10 mL of heptane and allowing the solution to cool to room temperature whereupon the product crystallized (yield of crystallization > 85%). Anal. Calcd for $C_7H_{10}F_3NO_2S_2$: C, 32.18; H, 3.86. Found: C, 32.57; H, 3.91.

11. Vinyl acetate (99+%) and lauroyl peroxide (DLP, 97%) were obtained from Aldrich, and used as received. 1,2-Dichloroethane (99+%) was purchased from SDS Carlo Erba and used as received. The checkers used 1,2-dichloroethane (99%) purchased from KMF.

12. When performed on a smaller scale (4.28 g of xanthate **3**) complete conversion was reached with only 3 portions of lauroyl peroxide with a reaction time of only 4.5 h.

13. The progress of the reaction was followed by TLC analysis on silica gel using hexanes/ethyl acetate (4:1) as eluent and visualization was performed with *p*-anisaldehyde solution. The product had an $R_f = 0.075$.

14. The product exhibited the following properties: ^1H NMR ($CDCl_3$, 400 MHz, mixture of diastereomers) δ: 1.30 (br. t, 3 H), 1.41 (t, $J = 7.0$ Hz, 3 H), 2.02 (s, 6 H), 2.07 (s, 6 H), 2.10–2.27 (m, 2 H), 2.44–2.51 (m, 2 H), 4.61–4.67 (m, 4 H), 4.79–4.85 (m, 2 H), 6.54 (dd, $J = 3.0, 9.7$ Hz, 2 H), 6.65 (m, 2 H); IR (CCl_4): 3429, 2982, 1767, 1706, 1503, 1369, 1235, 1188, 1137, 1049 cm^{-1}. MS (ESI): 370 ($[M+Na]^+$).

15. (±)-10-Camphorsulfonic acid was purchased from Acros, 98%, and used as received.

16. Approximately 150 g of silica (E. Merck, Darmstadt, 230-240 mesh) was used with hexanes/ethyl acetate (2:3) as the eluent, $R_f = 0.16$ (stained with potassium permanganate solution).

17. The submitters reported that the crude product could be purified by crystallization by cooling a solution in Et_2O (20 mL) to −78 °C (52% yield). The checkers, however, experienced problems caused by gelation of the mixture and therefore utilized chromatographic purification.

18. The product exhibited the following properties: mp 61–63 °C (ether). ^1H NMR ($CDCl_3$, 400 MHz) δ: 1.83 (ddd, $J = 3.9, 9.8, 14.2$ Hz, 1 H), 2.03 (s, 3 H), 2.04 (ddd, $J = 3.3, 7.6, 14.2$ Hz, 1 H), 3.34 (s, 3 H), 3.36 (s, 3 H), 4.46 (dd, $J = 7.6, 3.9$ Hz, 1 H), 4.75 (m, 1 H), 6.89 (d, $J = 9.6$ Hz, 1 H); ^{13}C NMR ($CDCl_3$, 100 MHz) δ: 22.9, 31.7, 47.6 (q, $J = 32$ Hz), 53.1, 53.9, 101.6, 125.2 (q, $J = 280$ Hz), 170.6; IR (CCl_4): 3285, 3072, 2939, 2834, 1665, 1551, 1437, 1375, 1299, 1265, 1181, 1135, 1064 cm^{-1}. MS (EI) *m/z* (rel. intensity): 198 (45) 139 (58), 124 (56), 75 (100); HRMS (ESI): 252

($[M+Na]^+$). Anal. Calcd for $C_8H_{14}F_3NO_3$: C, 41.92; H, 6.16. Found: C, 41.84; H, 6.09.

Safety and Waste Disposal Information

All hazardous materials should be handled and disposed of in accordance with "Prudent Practices in the Laboratory"; National Academy Press; Washington, DC, 1995.

3. Discussion

The introduction of fluorine atoms in a given molecule often dramatically alters its chemical properties and its pharmacological profile in the case of a biologically active substance.[2] As a consequence, much ongoing effort has been devoted to the development of practical synthetic routes to the various classes of fluorinated compounds.[3] Radical reactions have proved quite useful in this respect. Xanthates, with their unique ability to mediate difficult radical processes such as intermolecular additions to non-activated olefins,[4] provide a collection of highly efficient reagents for the assembly of fluorinated derivatives. Some of these reagents are displayed in Figure 1. The transformation described above is representative of the use of xanthate **3**.[5] Vinyl acetate can be replaced by a number of other olefinic traps, as shown by the examples in the Table (entries 1-5).[5] Xanthates **6** and **7** can be prepared by a modification of the route devised for the synthesis of compound **3**.[6] They also add efficiently to various olefins as indicated by the examples in entries 6-10 in the Table.[6] A simple trifluoromethyl group can be introduced by the use of xanthate **8** (entry 11),[7] whereas reagent **9** gives directly a trifluoromethyl ketone (entry 12).[8]

Figure 1: Reagents for the assembly of fluorinated derivatives

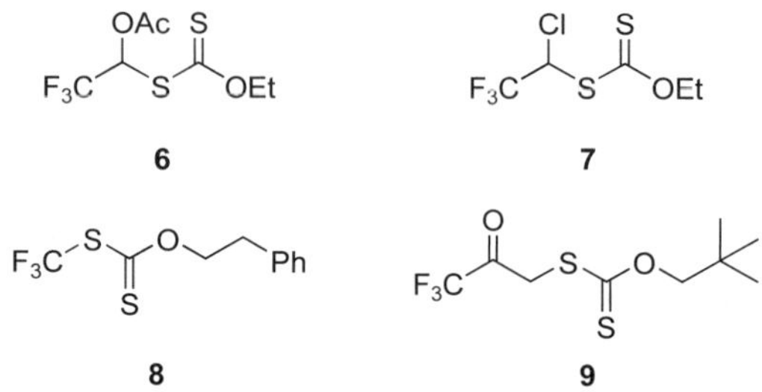

This approach to fluorinated derivatives combines efficiency with flexibility. The presence of the xanthate in the product can be exploited in many ways, since it provides an entry into the extremely rich chemistry of sulfur. It also allows the implementation of a second radical transformation, as shown by the three examples given in Scheme 1. The first example involves radical addition to allyl trimethylsilane leading to the densely functionalized structure **11**.[5] The second transformation illustrates a process for the tin-free reductive removal of the xanthate group (**10f** to **12**).[6] The last example (**13**) highlights the possibility of performing ring closures onto aromatic rings, providing a simple and direct entry into indolines.[6]

Table 1: Radical Additions of Xanthates (Xa = S-C(=S)OEt).

Entry	Xanthate	Olefin	Reaction time (h)	Adduct	Yield %	Isomeric ratio
1	3	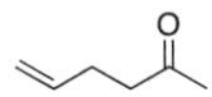	1.5	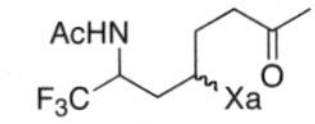 **10a**	88	1/1
2	3	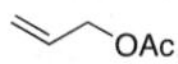	1.5	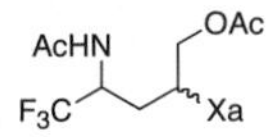 **10b**	84	1/1
3	3	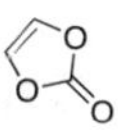	6	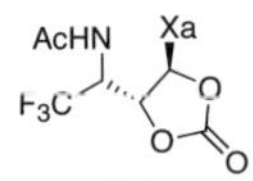 **10c**	72	1/1
4	3	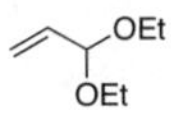	1.5	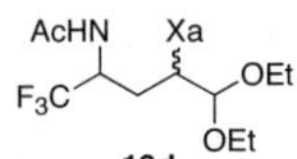 **10d**	95	6/43
5	3	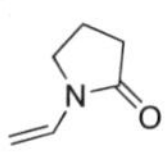	3	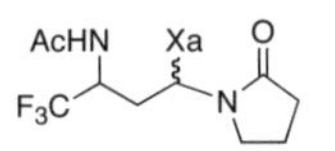 **10e**	62	6/4
6	6	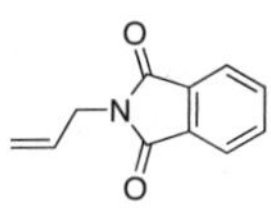	4.5	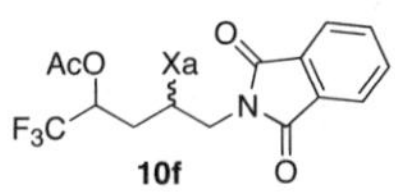 **10f**	78	1/1
7	6	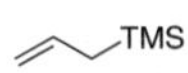	1.5	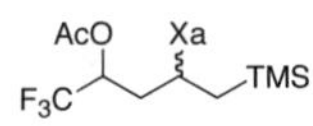 **10g**	80	2/3
8	6	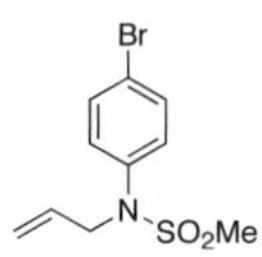	8	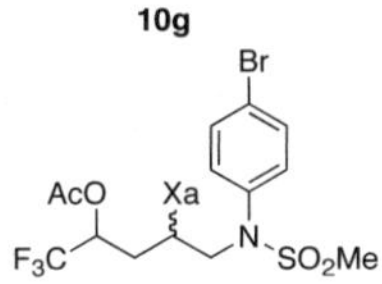 **10h**	82	1/1
9	7	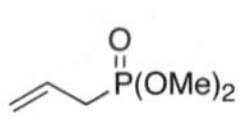	3	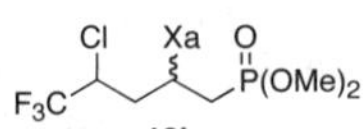 **10i**	27*	3/2
10	7	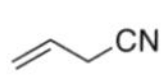	3	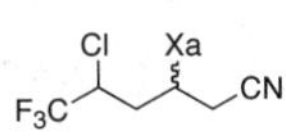 **10j**	52	2/1
11	8	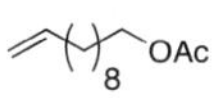	4.5	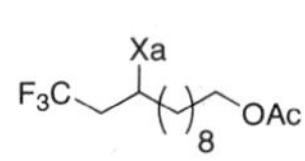 **10k**	84	
12	9	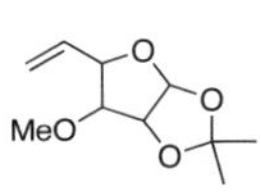	3	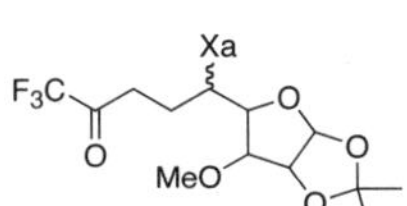 **10l**	80	2/3

* yield based on the starting trifluoroacetaldehyde hemiacetal.

Scheme 1: Radical transformations of adducts (DLP = dilauroyl peroxide)

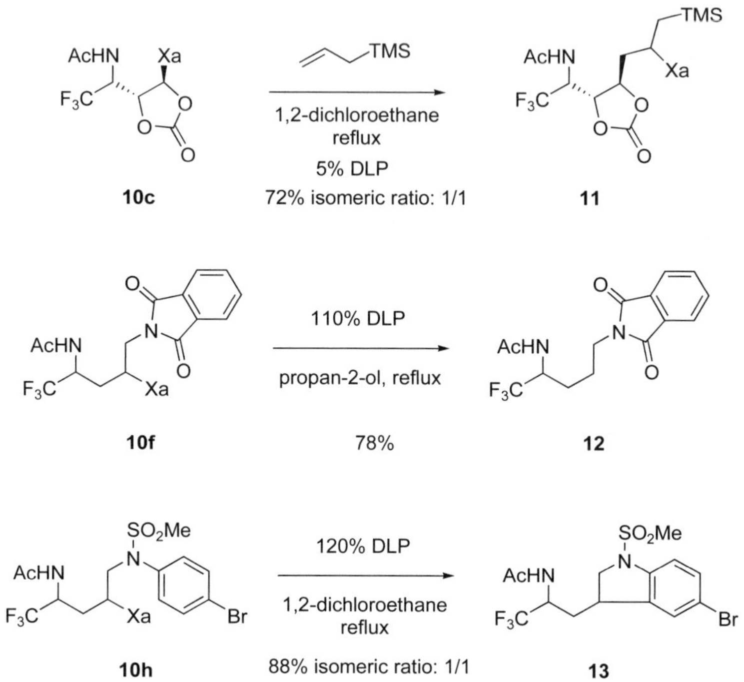

1. Laboratoire de Synthèse Organique, Ecole Polytechnique, 91128 Palaiseau, France.
2. (a) *Organofluorine in Medicinal Chemistry and Biochemical Applications*; Filler, R., Kobayashi, Y., Yagupolskii, L. M., Eds.; Elsevier: Amsterdam, The Netherlands, 1993; Stud. Org. Chem., Vol. 48. (b) *Fluorine-Containing Amino Acids: Synthesis and Properties*; Kukhar, V. P., Soloshonok, V. A., Eds.; Wiley: Chichester, UK, 1995. (c) *Biomedical Frontiers of Fluorine Chemistry*; Ojima, I., McCarthy, J. R., Welch, J. T., Eds.; American Chemical Society: Washington, DC, 1996; ACS Symp. Ser., Vol. 639. (d) Ismail, F. M. D. *J. Fluorine Chem.* **2002**, *118*, 27.
3. (a) *Fluoroorganic Chemistry: Synthetic Challenges and Biomedicinal Rewards*; Resnati, G., Soloshonok, V. A., Eds.; Pergamon: Oxford, UK, 1996. (b) *Organofluorine Chemistry: Techniques and Synthons*; Chambers, R. D., Ed.; Springer: Berlin, Germany, 1997; *Top. Curr. Chem.*, Vol. 193.
4. (a) Zard, S. Z. *Angew. Chem., Int. Ed. Engl.* **1997**, *36*, 672. (b) Quiclet-Sire, B.; Zard, S. Z. *Phosphorus, Sulfur Silicon Relat. Elem.* **1999**, *153-154*, 137. (c) Quiclet-Sire, B.; Zard, S. Z. *Top. Curr. Chem.* **2006**, 264, 201. (d) Quiclet-Sire, B.; Zard, S. Z. *Chem. Eur. J.* **2002**, *12*, 6002.
5. Gagosz, F.; Zard, S. Z. *Org. Lett.* **2003**, *5*, 2655.
6. Tournier, L.; Zard, S.Z. *Tetrahedron Lett.* **2004**, *46*, 455.
7. Bertrand, F.; Pevere, V.; Quiclet-Sire, B.; Zard, S. Z. *Org. Lett.* **2001**, *3*, 1069.
8. Denieul, M.-P.; Quiclet-Sire, B.; Zard, S. Z. *J. Chem. Commun.* **1996**, 2511.

STEREOSELECTIVE ISOPRENOID CHAIN EXTENSION WITH ACETOACETATE DIANION: (*E*, *E*, *E*)-GERANYLGERANIOL FROM (*E*, *E*)-FARNESOL

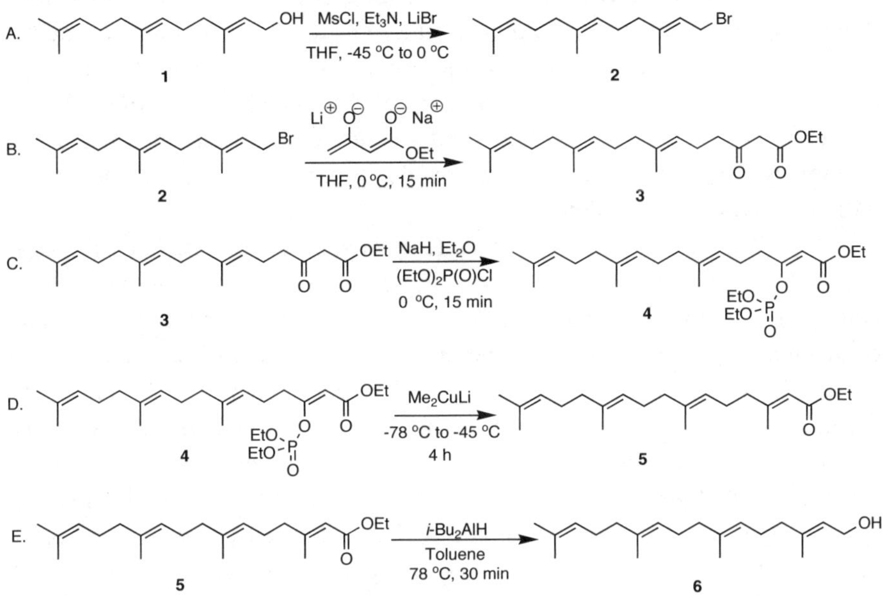

Submitted by Yinghua Jin, Frank G. Roberts, and Robert M. Coates.[1]
Checked by Hillary M. Peltier and Jonathan A. Ellman.

1. Procedure

A. (E, E)-Farnesyl bromide. A flame-dried, 500-mL, three-necked, round-bottomed flask, equipped with a magnetic stirring bar, two rubber septa, and a nitrogen inlet connected to a mineral oil bubbler, is charged with 10.0 g (11.3 mL, 45.0 mmol) of (*E*, *E*)-farnesol (Note 1) and 150 mL of THF (Note 2). The solution is stirred and cooled in a dry ice/acetonitrile bath (temperature –45 °C to –47 °C), while methanesulfonyl chloride (6.70 g, 4.53 mL, 58.5 mmol) (Note 3) is added through one septum via syringe over 5 min. Triethylamine (9.10 g, 12.6 mL, 89.9 mmol) (Note 4) is then added through one septum by syringe over 5 min. A white solid precipitates during the addition, and the resulting suspension is stirred at –45 °C to –47 °C for 45 min. A room temperature solution of lithium bromide (15.6 g, 180 mmol) (Note 5) in 50 mL of THF is prepared in a separate flask and added by cannula transfer over 5 min. The dry ice/acetonitrile bath is removed, and the suspension is allowed to warm to 0 °C with an ice-bath. The yellowish mixture is stirred at 0°C for 1 h. The reaction mixture is poured into 300 mL of ice water. The aqueous layer is separated and extracted with three 150-mL portions of cold diethyl ether (Note 6). The combined organic extracts are sequentially washed with ice-cold saturated $NaHCO_3$ (200 mL) and brine (200 mL), dried over anhydrous $MgSO_4$ (Note 7), filtered, and concentrated by rotary evaporation and vacuum drying to give 11.6 g of (*E*, *E*)-farnesyl bromide as a light yellow oil which is used in the next step without purification (Note 8).

B. (6E, 10E)-Ethyl 7,11,15-Trimethyl-3-oxohexa-deca-6,10,14-trienoate. A flame-dried, 500-mL, three-necked, round-bottomed flask, equipped with a magnetic stirring bar, two rubber septa, and a nitrogen inlet connected to a mineral oil bubbler, is charged with a 60% dispersion of sodium hydride in mineral oil (5.32 g, 133 mmol) (Note 9). The mineral oil is removed by washing with three 10-mL portions of hexanes, allowing the sodium hydride to settle and withdrawing the supernatant solvent with a syringe. Tetrahydrofuran (100 mL) (Note 2) is added, and the hydride suspension is stirred and cooled in a 0 °C ice-bath as ethyl acetoacetate (15.8 g, 15.5 mL, 121 mmol) (Note 10) is added dropwise via syringe over 15 min. The light yellow solution of the sodium enolate is stirred at 0 °C for an additional 15 min (Note 11). A solution of *n*-BuLi in hexane (59.2 mL, 2.15 M, 127 mmol) (Note 12) is then added by syringe over 15 min. The resulting orange solution of acetoacetate dianion is stirred at 0 °C for 15 min. A room temperature solution of farnesyl bromide (11.6 g, 40.6 mmol) in 20 mL of THF is added by cannula transfer over 10 min. The resulting orange suspension is stirred at 0 °C for 15 min, aqueous HCl (1N, 300 mL) is added, and the aqueous layer is separated and extracted with diethyl ether (3 x 150 mL). The combined organic extracts are washed sequentially with saturated aqueous $NaHCO_3$ (2 x 200 mL) and brine (2 x 200 mL), dried over anhydrous $MgSO_4$, filtered, and concentrated by rotary evaporation and vacuum drying to give 21.6 g of crude product. The excess ethyl acetoacetate is removed by Kugelrohr distillation (2 h, 60 °C, 0.4 mmHg) to afford 16.0 g of β-keto ester **3** as an orange oil. The residue is loaded onto a column of 220 g of silica gel (Note 13) and eluted with 300 mL of hexanes. At that point, collection of 50-mL fractions in glass tubes (25 x 150 mm) is begun, while elution is continued with 500 mL of 2% EtOAc-hexanes, 500 mL of 5% EtOAc-hexanes, 500 mL of 10% EtOAc-hexanes and then 500 mL of 15 % EtOAc-hexanes. The fractions are analyzed by TLC on silica gel with 30% EtOAc-hexanes as eluent and visualized with *p*-anisaldehyde staining reagent (Note 14). The product is obtained in fractions 20-33, which are combined and concentrated under reduced pressure using a rotary evaporator to yield β-keto ester **3** (11.8 g, overall 78% from farnesol) as a light yellow oil (Note 15).

C. (2Z,6E,10E)-Ethyl 7,11,15-Trimethyl-3-diethylphosphoryloxy-6,10, 14-trienoate. A flame-dried, 500-mL, three-necked, round-bottomed flask, equipped with a magnetic stirring bar, two rubber septa, and a nitrogen inlet connected to a mineral oil bubbler, is charged with a 60% dispersion of sodium hydride in mineral oil (1.62 g, 40.6 mmol) (Note 9). The mineral oil is removed by washing with hexanes (3 x 10 mL) allowing the sodium hydride to settle and withdrawing the supernatant solvent with a syringe. Diethyl ether (150 mL) (Note 16) is added, and the mixture is stirred and cooled at 0 °C with an ice-bath. β-Keto ester **3** (11.7 g, 35.3 mmol) in 50 mL of diethyl ether is added slowly via cannula over 20 min, and the reaction mixture is stirred at rt for 20 min or until gas evolution ceases. The clear orange solution is then cooled to 0 °C, and neat diethyl chlorophosphate (9.13 g, 7.67 mL, 52.9 mmol) (Note 17) is added by syringe over 5 min. The mixture is stirred at 0 °C for 15 min, after which time the reaction is quenched by adding saturated aqueous NH_4Cl (300 mL). The aqueous layer is separated and extracted with diethyl ether (3 x 150 mL). The combined organic extracts are washed sequentially with saturated aqueous $NaHCO_3$ (2 x 200 mL) and brine (2 x 200 mL), dried over anhydrous $MgSO_4$, filtered, and concentrated by rotary evaporator and vacuum drying to give the crude enol phosphate **4** (16.9 g) as a yellow oil (Note 18).

D. (2E,6E,10E)-Ethyl 3, 7, 11, 15-Tetramethylhexadeca-2, 6, 10, 14-tetraenoate. A flame-dried, 500-mL three-necked, round-bottomed flask, equipped with a magnetic stirring bar, one rubber septum, a thermometer, and a nitrogen inlet connected to a mineral oil bubbler, is charged with copper(I) iodide (12.1 g, 63.5 mmol) (Note 19) and 40 mL of diethyl ether (Note 16). The suspension is stirred and cooled at 0 °C in an ice bath as methyllithium in diethyl ether (98.8 mL, 1.29 M, 127 mmol) (Note 20) is added via syringe over 5 min. A yellow solid precipitates immediately during the addition, and the precipitate disappears after stirring at 0 °C for 30 min. The colorless or light tan solution is cooled to –78 °C using a dry ice/acetone bath and stirred magnetically, as a room temperature solution of enol phosphate **4** (16.9 g) in 50 mL of diethyl ether is added slowly over 20 min via cannula while maintaining the internal temperature below –60 °C

(Note 21). The orange-red solution is stirred at –78 °C for 2 h. The reaction mixture is then allowed to warm to –45 °C using a dry ice/acetonitrile bath and stirred for 2 h, resulting in a dark purple mixture. Methyl iodide (4.50 mL, 10.3 g, 72.3 mmol) is added while stirring at –45 °C for 10 min (Note 22). The mixture is then poured into 300 mL of ice-cold 4:1 (v:v) saturated NH_4Cl and concentrated NH_4OH mixture, and stirred for approximately 1 h until all the copper salts disappear. The aqueous layer is separated and extracted with ether (3 x 150 mL). The combined organic extracts are washed sequentially with 10 % NH_4OH (2 x 200 mL), water (2 x 200 mL), and brine (2 x 200 mL), dried over anhydrous $MgSO_4$, filtered, and concentrated by rotary evaporation and vacuum drying to give crude ester **5** (9.58 g) as a 19:1–20:1 mixture of trans and cis isomers according to GC analysis (Note 23).

E. (E, E, E)-Geranylgeraniol. A flame-dried, 500-mL, three-necked, round-bottomed flask, equipped with a magnetic stirring bar, two rubber septa, and a nitrogen inlet connected to a mineral oil bubbler, is charged with ester **5** (9.58 g, ~ 28.8 mmol) and toluene (100 mL) (Note 24). The solution is cooled in a dry ice/acetone bath (temperature –78 °C), and diisobutylaluminum hydride in hexanes (86.4 mL, 86.4 mmol) (Note 25) is added slowly via syringe through one of the septa over 20 min. After 30 min at –78 °C, the reaction is quenched by careful addition of methanol (10 mL) via a glass pipette over approximately 10 min (Note 26). The mixture is allowed to warm to room temperature and poured into a mixture of saturated NH_4Cl (250 mL) and 1N HCl (250 mL) with stirring, and the stirring is continued for 30 min. The aqueous layer is separated and extracted with ether (3 x 200 mL). The combined organic extracts are sequentially washed with water (2 x 200 mL) and brine (2 x 200 mL), dried over anhydrous $MgSO_4$, filtered, and concentrated by rotary evaporator and vacuum drying. The concentrated residue is loaded onto a column (5.0 x 21 cm) of 200 g of silica gel (Note 13) and eluted with 300 mL of hexanes. At that point, collection of 50-mL fractions in glass tubes (25 x 150 mm) is begun, while elution is continued with 500 mL of 5% EtOAc-hexanes, 500 mL of 10% EtOAc-hexanes, 500 mL of 15% EtOAc-hexanes, and then 500 mL of 20 % EtOAc-hexanes. The fractions are analyzed by TLC on silica gel with 30% EtOAc-hexanes as eluent and visualized with *p*-anisaldehyde staining reagent (Note 14). The product is obtained in fractions 24-42, which are combined and concentrated under reduced pressure using a rotary evaporator to yield geranylgeraniol (7.20 g, overall 70% from β-keto ester **3**) (Note 27).

2. Notes

1. *Trans*, *trans*-farnesol (~95%) was purchased from the Fluka Chemical Corp. and was used without further purification.

2. Tetrahydrofuran was purchased from Fisher Scientific Co. and passed through an alumina column (checkers) or purchased from Burdick and Jackson and freshly distilled from sodium/benzophenone under nitrogen (submitters).

3. Methanesulfonyl chloride (>99.5 % purity) was purchased from Aldrich Chemical Co. and used as received. When this reagent was taken from a previously opened container, it was distilled from P_2O_5 before use.

4. Triethylamine (reagent grade) was purchased from Fisher Scientific Co., distilled from calcium hydride, and stored over potassium hydroxide pellets.

5. Lithium bromide was purchased from Fisher Scientific Co. and flame-dried under vacuum before use.

6. Unless otherwise indicated, the solvents used were of reagent grade quality and were used without further purification.

7. Anhydrous $MgSO_4$ was purchased from Fisher Scientific Co.

8. Bromide **2** is prone to decomposition and should be kept cold by adding ice during the extractions and washings. It should be stored in a freezer (approximately –20 °C) no longer than one day. The physical data for (*E*, *E*)-farnesyl bromide (provided by submitters) follow: TLC R_f = 0.72 (7:3 hexanes/ethyl acetate); ^{1}H NMR (500 MHz, $CDCl_3$) δ: 1.61 (br s, 6 H, 2CH_3), 1.69 (s, 3 H, CH_3), 1.74 (s, 3 H, CH_3), 1.97–2.13 (m, 8 H, 4CH_2), 4.03 (d, J = 8.3 Hz, 2 H, CH_2Br), 5.08–5.11 (m, 2 H, vinyl *H*), 5.54 (br t, J = 8.3 Hz, 1 H, vinyl *H*); ^{13}C NMR (100 MHz, $CDCl_3$) δ: 16.2, 16.3, 17.9, 25.9, 26.3, 26.9, 30.0, 39.7, 39.9, 120.7, 123.6, 124.5, 131.6, 135.8, 143.8.

9. Sodium hydride was purchased from Acros Organics and used as received.

10 Ethyl acetoacetate was purchased from Aldrich Chemical Co. and purified by washing with saturated aqueous $NaHCO_3$ and water, drying over $MgSO_4$, and distilling under reduced pressure (bp 30–32 °C, 1 mmHg). A slower, incomplete reaction (89 % conversion after 5 h at 0 °C, 68% yield based on conversion) occurred when only 1.2 equiv of ethyl acetoacetate was used, instead of 3.0 equiv, under otherwise identical conditions.

11. ***Caution:*** Vigorous evolution of hydrogen occurs. The mixture should be stirred until gas evolution ceases.

12. *n*-Butyllithium in hexanes was purchased from Aldrich Chemical Co. and titrated using 2,6-di-*tert*-butyl-4-methylphenol and 1,10-phenanthroline (checkers).

13. Silica gel 60 230-400 mesh ASTM was purchased from Merck and Co. The 6.2-cm diameter glass column was packed with silica gel (6.2 x 14.5 cm) and hexanes.

14. Anisaldehyde staining reagent was prepared by adding, in small portions with stirring, 6.5 mL of concentrated H_2SO_4, 2.5 mL of glacial acetic acid, and 5 mL of *p*-anisaldehyde in the order listed to 186 mL of cold 95% EtOH.

15. The absolute yield of β-keto ester **3** was determined in one of the submitters' runs prior to chromatography (17.3 g of crude **3**) by ^{1}H NMR analysis using mesitylene as a standard. A solution of crude product (47 mg) and mesitylene (11 mg) in chloroform-*d* was prepared. Integrations of the ^{1}H NMR spectrum showed a 1.00:0.79 molar ratio of the product and the standard, indicating the yield was 13.7 g (91 %). The submitters obtained 11.7–12.2 g of β-keto ester **3** (overall 78–81% from farnesol) after chromatography as mixtures of keto and enol tautomeric forms. The physical data (supplied by the checkers) for pure β-keto ester **3** (6.7:1 mixture of keto and enol tautomeric forms) are as follows: TLC: R_f = 0.60 (7:3, hexanes/ethyl acetate); IR: 1721, 1745, 2920 cm^{-1}. ^{1}H NMR (500 MHz, $CDCl_3$) δ: 1.27 (t, *J* = 7.2 Hz, 3 H, CH_3), 1.59 (br s, 3 H, CH_3), 1.60 (s, 3 H, CH_3), 1.61 (s, 3 H, CH_3), 1.68 (br s, 3 H, CH_3), 2.00–2.13 (m, 8 H, 4 CH_2), 2.28 (q, *J* = 7.5 Hz, 2 H, CH_2), 2.56 (t, *J* = 7.5, 2 H, CH_2), 3.42 (s, 2 H, CH_2), 4.19 (q, *J* = 7.2 Hz, 2 H, CH_2), 4.97 (s, 0.164 H, -C(OH)=C*H*-), 5.06–5.09 (m, 3 H, vinyl *H*), 12.1 (s, 0.146 H, -C(O*H*)=CH-).; ^{13}C NMR (125 MHz, $CDCl_3$) δ: 14.4, 16.33, 16.35, 18.0, 22.5, 26.0, 26.9, 27.1, 39.9, 40.0, 43.4, 49.7, 61.7, 122.4, 124.4, 124.7, 131.6, 135.4, 137.1, 167.6, 202.9. Anal. Calcd. for $C_{21}H_{34}O_3$: C, 75.41; H, 10.25. Found: C, 75.63; H, 10.47.

16. Diethyl ether was purchased from Fisher Scientific Co. and passed through an alumina column (checkers) or freshly distilled from sodium/benzophenone under nitrogen (submitters).

17. Diethyl chlorophosphate (97%) was purchased from Aldrich Chemical Co. and was used as received.

18. The submitters recommend using the unpurified enol phosphate since an attempt to purify the intermediate by large-scale flash column chromatography resulted in reduced yield, probably owing to the decomposition of the product. The crude enol phosphate **4** (18.7 g) was analyzed by the submitters by ^{1}H NMR spectroscopy using mesitylene as a

standard. A solution of 20 mg of enol phosphate **4** and 10 mg of mesitylene in chloroform-*d* was prepared. Integrations of the ^{1}H NMR spectrum showed a 1.00:2.08 molar ratio of the product and the standard, indicating the purity of the crude product was 90 %. A 70-mg sample of enol phosphate was purified by the submitters with flash column chromatography using 30 % ethyl acetate in hexanes as eluent to yield 27 mg of pure enol phosphate **4**: TLC: R_f = 0.26 (7:3 hexanes/ethyl acetate); ^{1}H NMR (400 MHz, $CDCl_3$) δ: 1.27 (t, *J* = 7.1 Hz, 3 H, CH_3), 1.36 (td, *J* = 7.1, 1.2 Hz, 6 H, $2CH_3$), 1.59 (s, 3 H, CH_3), 1.60 (s, 3 H, CH_3), 1.61 (s, 3 H, CH_3), 1.68 (s, 3 H, CH_3), 1.95–2.07 (m, 8 H, $4CH_2$), 2.27 (q, *J* = 7.3 Hz, 2 H, CH_2), 2.45 (t, *J* = 7.6 Hz, 2 H, CH_2), 4.15 (q, *J* = 7.2 Hz, 2 H, CH_2), 4.26 (dq, *J* = 7.8, 7.1 Hz, 4 H, $2CH_2$), 5.07–5.11 (m, 3 H, vinyl *H*), 5.35 (s, 1 H, vinyl *H*); ^{13}C NMR ($CDCl_3$, 125 MHz) δ: 14.4, 16.16, 16.23, 16.28, 16.29, 17.9, 25.1, 25.9, 26.8, 26.9, 35.4, 39.9 (d, *J* = 7.6 Hz), 60.1, 64.9 (d, *J* = 6.1 Hz), 105.5 (d, *J* = 7.4 Hz), 122.0, 124.2, 124.5, 131.5, 135.3, 137.0, 161.6 (d, *J* = 7.3 Hz), 164.0; ^{31}P NMR ($CDCl_3$, 202 MHz) δ –7.6.

19. Copper(I) iodide (98%) purchased from Aldrich Chemical Co. was used as received.

20. Methyllithium in diethyl ether was purchased from Aldrich Chemical Co. and titrated using 2,6-di-*tert*-butyl-4-methylphenol and 1,10-phenanthroline (checkers).

21. Enol phosphate **4** should be added slowly. The coupling reaction was exothermic; fast addition of enol phosphate **4** caused a temperature rise that led to a lower *trans*/*cis* ratio.

22. Direct hydrolysis of the crude product with a mixture of ice-cold 5:1 (*v*/*v*) saturated NH_4Cl and concentrated NH_4OH afforded all *trans* ester **5** contaminated by approximately 5% of the simple reduction byproduct, (2*E*,6*E*,10*E*) ethyl 7,11,15-trimethylhexadeca-2,6,10,14-tetraenoate, which was inseparable by chromatography. Prior addition of methyl iodide decreased the amount of the reduction byproduct to ≤ 1%.

23. The unpurified ester **5** (11.1 g) was analyzed by the submitters in one run by ^{1}H NMR spectroscopy using mesitylene as internal standard. Integrations of the spectrum of the ester (26 mg) and mesitylene (10 mg) in chloroform-*d* showed a 1.00:1.17 molar ratio of the product and standard, indicating a purity of 90 %. In a different run a small sample (approximately 30 mg) of the crude product was purified by the submitters with flash column chromatography using 10 % ethyl acetate in hexanes as eluent to afford a characterization sample of pure ester **5**: TLC R_f = 0.72 (7:3 hexane/ethyl acetate); ^{1}H NMR (400 MHz, $CDCl_3$) δ: 1.28 (t, *J* = 7.2 Hz, 3

H, CH_3), 1.60–1.61 (m, 9 H, 3CH_3), 1.69 (d, J = 1.1 Hz, 3 H, CH_3), 1.96–2.17 (m, 12 H, 6CH_2), 2.17 (d, J = 1.1 Hz, 3 H, CH_3), 4.15 (q, J = 7.2 Hz, 2 H, CH_2), 5.08–5.13 (m, 3 H, vinyl H), 5.67 (m, 1 H, vinyl H); ^{13}C NMR (100 MHz, $CDCl_3$) δ: 14.6, 16.22, 16.25, 17.9, 19.0, 25.9, 26.2, 26.8, 27.0, 39.88, 39.92, 41.2, 59.7, 115.8, 123.1, 124.3, 124.6, 131.5, 135.3, 136.4, 160.1, 167.1.

24. Toluene was purchased from Fisher Scientific Co. and passed through a column of alumina (checkers) or dried over 4Å molecular sieves (submitters).

25. Diisobutylaluminum hydride (1.0 M in hexanes) was purchased from Aldrich Chemical Co.

26. ***Caution:*** Methanol should be added very slowly to avoid vigorous gas evolution.

27. The submitters obtained 6.90–7.94 g of geranylgeraniol (68–77% yield from β-keto ester **3**). Physical data for geranylgeraniol (**6**) (supplied by the checkers): TLC: R_f = 0.43 (7:3 hexanes/ethyl acetate); IR: 1445, 1668, 2921, 3319 cm^{-1}. ^{1}H NMR (500 MHz, $CDCl_3$) δ: 1.59 (br s, 9 H, 3CH_3), 1.67 (br s, 6 H, 2CH_3), 1.98–2.13 (m, 12 H, 6CH_2), 4.13 (d, 2 H, J = 7.0), 5.07–5.10 (m, 3 H, vinyl H), 5.40 (t, J = 7.0 Hz, 1 H, vinyl H); ^{13}C NMR ($CDCl_3$, 125 MHz) δ: 16.2, 16.3, 16.6, 18.0, 26.0, 26.6, 26.9, 27.0, 39.9, 39.99, 40.0, 59.6, 123.7, 124.1, 124.5, 124.7, 131.5, 135.2, 135.6, 139.9. Anal. Calcd. for $C_{20}H_{34}O$: C, 82.69; H, 11.80. Found: C, 82.58; H, 12.16. The purity of geranylgeraniol (**6**) was determined to be 96 % according to GC analysis. GC analysis was performed on a Shimadzu GC-14A with 30-m Restek Rtx-5 column, and the sample was analyzed using the following parameters, initial temperature 40 °C, hold for 5 min, 20 °C/min until 300 °C, hold for 5 min.

Safety and Waste Disposal Information

All hazardous materials should be handled and disposed of in accordance with "Prudent Practices in the Laboratory"; National Academy Press; Washington, DC, 1995.

3. Discussion

(*E, E, E*)-Geranylgeranyl diphosphate (GGPP) is considered to be the universal substrate for diterpene cyclases, prephytoene synthase, and many prenyl diphosphate synthases,[2] as well as cosubstrate for protein geranylgeranyl transferases I and II.[3] The diphosphate ester is readily prepared from (*E, E, E*)-geranylgeraniol (**6**, GGOH) by conversion to the corresponding allylic chloride, bromide, or diethyl phosphate followed by S_N2 displacement with tetrabutylammonium pyrophosphate.[4,5] Although both GGOH and GGPP are commercially available, the prices are quite high, and there is need for reliable methods to access different labeled forms and analogues as pure all trans isomers for research on diterpene synthases and transferases. The present five-step procedure (**1** → **6**) effects isoprene chain extension of (*E, E*)-farnesol to give 6.9–7.9 g of the isoprenylogue **6** in 53–61% overall yield and with high E selectivity using Weiler's acetoacetate dianion alkylation-enol phosphate coupling methods.[6,7] 6-FluoroGGOH was prepared in the same way from 2-fluorofarnesol.[8] A similar route to **6** utilizes a Suzuki reaction of the enol triflate related to enol phosphate **4** to obtain α,β-enoate **5**.[9]

The most common syntheses of (*E, E, E*)-geranylgeraniol (**6**) in the literature proceed by C_2 chain extensions of (*E, E*)-farnesylacetone, reactions that invariably afford mixtures of 2E and 2Z isomers of esters or alcohols. Farnesylacetone has been synthesized from farnesol derivatives by traditional α-alkylation of acetoacetate[10a,b] and by alkylation of lithiated acetone *N,N*-dimethylhydrazone,[10c] and by Wittig olefination of the keto aldehyde generated from ozonolysis of isoprene cyclic trimer.[11] Conversion to isomeric mixtures of geranylgeranoates or geranylgeraniol has been accomplished by an acetylide addition-reduction-allylic rearrangement sequence,[10a] by Horner-Wadsworth-Emmons olefinations,[12] by ethoxyacetylene addition and hydrolysis,[10b,13] and by vinylmetal addition followed by allylic rearrangement.[10c,11] Reductions of ester **5** to the GGOH have been carried out with $LiAlH_4$, $i Bu_2AlH$, and $LiAlH_4$-$AlCl_3$.

Geranylgeraniol has also been obtained by Peterson and Wittig olefinations to complete the isoprene units at the opposite ends of the chain.[14] A convergent approach involves allylic couplings of two C_{10} geraniol derivatives,[15] geranylbarium alkylation,[16a] and an allyl silane-acetal S_E' method.[16b] Palladium-mediated couplings of an α-sulfonylfarnesyl-1-carboxylate with isoprene 3,4-epoxide and of methyl farnesoate with α-sulfonyl-γ,γ-dimethylallylcarboxylate have been reported.[17] Synthesis of the 2E, 6Z, 10E isomer has been achieved by sequential Pd-catalyzed couplings of zincates with vinyl iodides followed by zirconocene-mediated addition of Me_3Al to a trienyne intermediate to form the hydroxyl-bearing isoprene unit.[18] An iterative C_5 isoprenoid chain extension approach to (*E, E, E*,)geranylgeraniol by means of Biellmann couplings with sulfonyl-stabilized anions has been reported recently.[19a] A moderately selective method (E/Z 91:9) for effecting palladium-catalyzed allylic rearrangement of geranyllinalyl acetate followed by cleavage of the resulting amine to geranylgernaiol has been disclosed.[19b] Patents describe the separation of geranylgeranyl sterate isomers by low temperature crystallizations,[19c] and isolation of high purity (*E*, *E*, *E*)-geranylgeraniol after hydrolysis of naturally occurring esters extracted from annatto seeds.[19d]

Although numerous methods for isoprenoid chain extensions have been investigated,[20] the present approach[6a,b,7] has the important advantages of high E-selectivity at both the 2,3 and 6,7 double bonds, satisfactory overall yield, and the potential to form 2Z,6E, 2E,6Z, and 2Z,6Z isomers.[21] The 5-step sequence begins with conversion of (*E*,*E*)-farnesol (**1**) to the corresponding bromide (**2**) with methanesulfonyl chloride, triethylamine, and lithium bromide in THF at –45 to 0 °C.[22] The Li/Na dianion is generated by lithiation of sodio acetoacetate with *n*-butyllithium in THF at 0 °C, and subsequent γ-alkylation with the unpurified bromide affords β-keto ester **3**. Phosphorylation of the sodium enolate with diethyl chlorophosphate in ether gives rise to all trans enol phosphate **4** which undergoes vinylic coupling with lithiodimethyl cuprate in ether solution at –78 to –45 °C to give (*E*,*E*,*E*)-ethyl geranylgeranoate (**5**) in good yield and ~20:1 E-selectivity. Reduction of the ester with isobutylaluminum hydride in toluene provides (*E,E,E*)-geranylgeraniol (**6**) in 53-61% overall yield after purification by chromatography on silica gel.

Acetoacetate dianion is a very useful carbon nucleophile for C-C bond formation at the γ-position and for accessing a wide variety of β-keto esters.[23] It is sufficiently stable to undergo alkylations with unactivated primary halides and epoxides,[24] and aziridines bearing *N*-sulfonyl or *N*-phosphoryl groups.[25] Reactions with carbonyl electrophiles afford δ-hydroxy-β-keto esters and δ, β-diketo esters (Scheme 1).[26]

Scheme 1

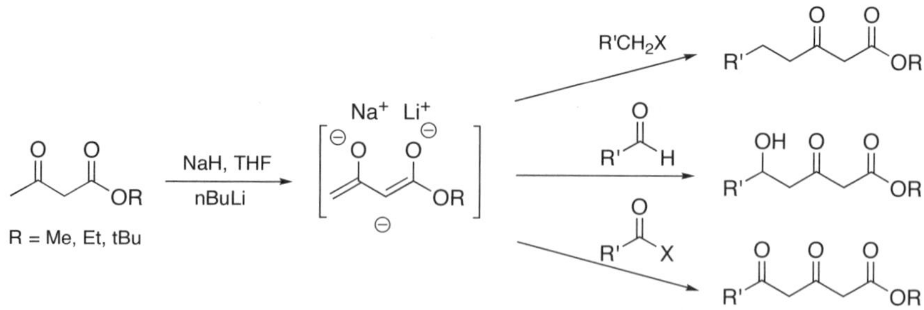

Coupling of enol phosphates with organocopper reagents offers a versatile, stereoselective synthesis of tri- and tetra-substituted α,β-unsaturated esters from β-keto esters with either double bond configuration. *O*-Phosphorylation of chelated sodium enolates of β-keto esters with diethyl or diphenyl phosphoryl chlorides in diethyl ether give rise to the Z (trans) enol phosphate isomers[6a] whereas the E (cis) isomers can be obtained through reactions with tertiary amine bases in polar solvents, eg Et_3N or *i*Pr_2NEt in HMPA (Scheme 2).[21,27] Vinylic couplings with methyl, primary alkyl, and *t*-butyl cuprates or Grignard reagents in the presence of Cu(I) salts usually take place efficiently with retention of the double bond geometry. Palladium-catalyzed Suzuki reactions of Z and E enol triflates with methyl-, *t*-butyl-, cyclopropyl-, and phenylboronic acids have been employed in similar syntheses of E and Z isoprenoid esters and their C-3 analogues.[9]

Scheme 2

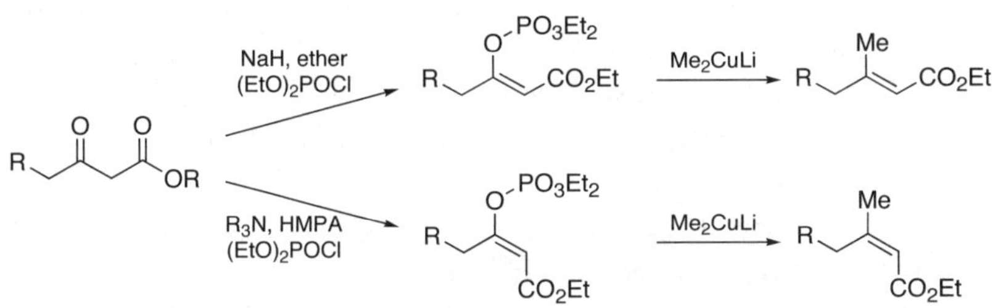

(*E,E,E*)-Geranylgeraniol (**6**) has other research applications beyond its conversion to the diphosphate. The forms labeled with deuterium and tritium at C-1 [1,1-$^2H_2$]- and [1-^{3}H]**6** can be oxidized to the corresponding labeled aldehydes which are reduced to (*R*)- and (*S*)-[1-$^2H_1$]- and [1-^{3}H]**6** for studies on stereochemistry of diterpene cyclase and transferase reactions.[5,28] The terminal double bonds of the benzyl ether and acetate derivatives of **6** are converted to the 14,15 bromohydrins by hypobromination reactions with *N*-bromosuccinimide in aqueous THF or DME.[5,10b,29] The bromohydrins are readily cyclized to the epoxides and then converted to 14,15-oxidogeranylgeraniol.[10b] Acid-catalyzed hydrolysis to the 14,15-diol and oxidative cleavage with periodate affords the trisnoraldehyde that is a key intermediate in synthesis of **6** bearing deuterium in the terminal methyl positions by E- or Z-selective olefinations and deuteride reductions.[28a] Reductive amination of the trisnoraldehyde acetate and hydrolysis gives 14-

aza-14,15-dihydro **6**,[5] the diphosphate of which is a potent inhibitor abietadiene synthase.[30]

1. Department of Chemistry, University of Illinois, 600 South Mathews Avenue, Urbana, IL 61801

2. (a) MacMillan, J.; Beale, M. H. *Comprehensive Natural Products Chemistry*; Elsevier: New York, 1999; Vol. 2, Chap. 8, pp 217-243. (b) Davis, E. M.; Croteau, R. *Top. Curr. Chem.* **2000**, *209*, 53-95.

3. (a) Zhang, F. L.; Casey, P. J. *Annu. Rev. Biochem.* **1996**, *65*, 241-269. (b) Harris, C. M.; Poulter, C. D. *Nat. Prod. Rep.* **2000**, *17*, 137-144.

4. (a) Davisson, V. J.; Woodside, A. B.; Neal, T. R.; Stremler, K. E.; Muehlbacher, M.; Poulter, C. D. *J. Org. Chem.* **1986**, *51*, 4768-4779. (b) Woodside, A. B.; Huang, Z.; Poulter, C. D. *Org. Synth. Coll. Vol. 8*, 616-620.

5. Ravn, M. M.; Jin. Q.; Coates, R. M. *Eur. J. Org. Chem.* **2000**, 1401-1410.

6. (a) Sum, F. W.; Weiler, L. *Can. J. Chem.* **1979**, *57*, 1431-1441. (b) Sum, F. W.; Weiler, L. *Tetrahedron*, **1981**, *37*, 303-317. (c) Alderdice, M.; Sum, F. W.; Weiler, L. *Org. Synth. Coll. Vol. 7*, 351-356.

7. Eis, K.; Schmalz, H.-G. *Synthesis* **1997**, 202-206.

8. Jin, Y.-H.; Williams, D. C.; Croteau, R.; Coates, R. M. *J. Am. Chem. Soc.* **2005**, *127*, 7834-7842.

9. Mu, Y.-Q.; Eubanks, L. M.; Poulter, C. D.; Gibbs, R. A. *Bioorg. Med. Chem.* **2002**, *10*, 1207-1219, and references cited therein.

10. (a) Ruzicka, L.; Firmenich, G. *Helv. Chim. Acta* **1939**, *22*, 392-396. (b) van Tamelen, E. E.; Nadeau, R. G. *Bioorg. Chem.* **1982**, *11*, 197-218. (c) Boland, W.; Gäbler, A. *Helv. Chim. Acta* **1989**, *72*, 247-253.

11. Odinokov, V. N.; Akhmetova, V. R.; Savchenko, R. G.; Mallyabaeva, M. I.; Lobanova, N. V. Russ. *J. Org. Chem.* **1998**, *34*, 1099-1101.

12. Klinge, S; Demuth, M. *Synlett* **1993**, 783-784.

13. Caliezi, A.; Schinz, H. *Helv. Chim. Acta* **1952**, *35*, 1637-1649; 1649-1655.

14. (a) Odinokov, V. N.; Kukovinets, O. S.; Sakharova, N. I.; Tolstikov, G. A. Russ. *J. Org. Chem.* **1984**, *20*, 1702-1707. (b) Odinokov, V. N.; Kukovinets, O. S.; Sakharova, N. I.; Tolstikov, G. A. Russ. *J. Org. Chem.* **1993**, *29*, 24-30.

15. (a) Altman, L. J.; Ash, L.; Marson, S. *Synthesis* **1974**, 129-131. (b) Coates, R. M.; Ley, D. A.; Cavender, P. L. *J. Org. Chem.* **1978**, *43*, 4915-4922. (c) Sato, K.; Inoue, S.; Onishi, A.; Uchida, N.; Minowa, N. *J. Chem. Soc., Perkin Trans. 1* **1981**, 761-769.

16. (a) Corey, E. J.; Shieh, W.-C. *Tetrahedron Lett.* **1992**, *33*, 6435-6438. (b) Radetich, B.; Corey E. J. *J. Am. Chem. Soc.* **2002**, *124*, 2430-2431.

17. (a) Bouzbouz, S.; Kirschleger, B. *Synlett* **1994**, 763-764. (b) Trost, B. M.; Weber, L. *J. Org. Chem.* **1975**, *40*, 3617-3619.

18. Negishi, E.-i.; Liou, S.-Y.; Xu, C.; Huo, S. *Org. Lett.* **2002**, *4*, 261-264.

19. (a) Kuk, J.; Kim, B. S.; Jung, H.; Choi, S.; Park, J.-Y.; Koo, S. *J. Org. Chem.* **2008**, ASAP Article; February 2, 2008. (b) Yuasa, Y.; Yuasa, Y. *Synth. Commun.* **2006**, *36*, 1671-1677. (c) Mori, T.; Sato, J.; Fukumoto, T.; Nakao, K.; Tamai, Y. "Process for producing geranylgeraniol." (Kuraray Co., Ltd., Japan). Eur. Pat. Appl. 711749, May 15, 1996; (d) Nara, T.; Iwama, T.; Nakadate, (Tama Biochemical Co., Ltd., Japan). "Manufacture of high-purity geranylgeraniol from annatto seeds," *Jpn. Kokai Tokkyo Koho* **2007**, Japanese Patent 2007238508; September 20, 2007; CAN 147:340018M.

20. Cainelli, G.; Cardillo, G. *Acc. Chem. Res.* **1981**, *14*, 89-94.

21. Alderdice, M.; Spino, C.; Weiler, L. *Can. J. Chem.* **1993**, *71*, 1955-1963.

22. Corey, E. J.; Luo, G.; Lin, L. S. *J. Am. Chem. Soc.* **1997**, *119*, 9927-9928.

23. (a) Thompson, C. M. *Dianion Chemistry in Organic Synthesis*; CRC Press: Boca Raton, Florida, 1994. (b) Thompson, C. M.; Green, D. L. C. *Tetrahedron* **1991**, *47*, 4223-4285.

24. (a) Bryson, T. A. *J. Org. Chem.* **1973**, *38*, 3428-3429. (b) Langer, P.; Freifeld, I. *Chem. Eur. J.* **2001**, *7*, 565-572.

25. (a) Lygo, B. *Synlett* **1993**, 764-766. (b) Osowska-Pacewicka, K.; Zwierzak, A. *Synth. Commun.* **1998**, *28*, 1127-1137.

26. (a) Huckin, S. N.; Weiler, L. *Can. J. Chem.* **1974**, *52*, 2157-2164. (b) Huckin, S. N.; Weiler, L. *Can. J. Chem.* **1974**, *52*, 1343-1351.

27. White, J. D.; Kawasaki, M. *J. Org. Chem.* **1992**, *57*, 5292-5300.

28. (a) Jin, Q.; Williams, D. C.; Hezari, M.; Croteau, R.; Coates, R. M., *J. Org. Chem.* **2005**, *70*, 4667-4675. (b) Ravn, M. M.; Coates, R. M.; Flory, J. E.; Peters, R. J.; Croteau, R. *Org. Lett.* **2000**, *2*, 573-576.

29. (a) Takanashi, S.-i.; Mori, K. *Ann. Chem./Recueil* **1997**, 825-838. (b) Grigor'eva, N. Y.; Avrutov, I. M.; Semenovskii, A. V.; Odinokov, V. N.; Akhunova, V. R.; Tolstikov, G. A. *Akad. Nauk. Sci. Bull. (Engl.)* **1979**, *28*, 353-358.

30. Peters, R. J.; Ravn, M. M.; Coates, R. M.; Croteau, R. B. *J. Am. Chem. Soc.* **2001**, *123*, 8974-8978.

SYNTHESIS OF TERMINAL 1,3-DIYNES VIA SONOGASHIRA COUPLING OF VINYLIDENE CHLORIDE FOLLOWED BY ELIMINATION. PREPARATION OF 1,3-DECADIYNE

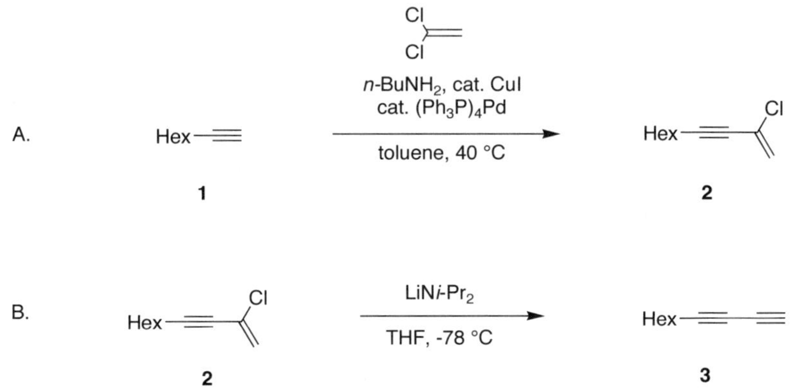

Submitted by Amanda L. Kohnen and Rick L. Danheiser.[1]
Checked by Scott E. Denmark and Xiaorong Liu.

1. Procedure

A. 2-Chloro-1-decen-3-yne (**2**). A 350-mL, round-bottomed pressure flask (Note 1) equipped with a magnetic stir bar, rubber septum, and argon inlet needle (Note 2) is charged with $(Ph_3P)_4Pd$ (3.46 g, 2.99 mmol, 0.03 equiv) (Note 3) and 150 mL of toluene (Note 4), and then vinylidene chloride (11.98 mL, 14.54 g, 150 mmol, 1.5 equiv) (Note 5) is added via syringe over 2 min. The resulting yellow mixture is stirred at room temperature for 20 min and then a mixture of 1-octyne (14.75 mL, 11.02 g, 100.0 mmol) (Note 6) and *n*-butylamine (14.82 mL, 10.97 g, 150 mmol, 1.5 equiv) (Note 7) is transferred into the reaction mixture via cannula over 5 min from a 50-mL, round-bottomed flask fitted with a septum and an argon inlet needle. The flask is rinsed with two 3-mL portions of toluene, which are transferred into the pressure flask by cannula. The septum is temporarily removed from the pressure flask and copper(I) iodide (0.571 g, 2.99 mmol, 0.03 equiv) (Note 8) is added in one portion. Material adhering to the sides of the pressure flask is rinsed into the reaction mixture with 5 mL of toluene. The pressure flask is then sealed under argon with a threaded Teflon cap and submerged in a preheated oil bath at 40 °C behind a safety shield. (Note 9) The color of the clear yellow solution darkens to red over 4 h. After 8 h (Note 10), the cloudy red mixture is allowed to cool to room temperature and then filtered through a medium frit sintered glass funnel to remove the precipitated ammonium salt. The solid is washed with 100 mL of hexanes, and the filtrate is concentrated by rotary evaporation (40 °C, 20 mmHg). The resulting red oil (28.0 g) is diluted with 20 mL of CH_2Cl_2 and concentrated onto 40 g of silica gel (40 °C, 16 mmHg, Note 11). The clumpy powder is transferred to a column (25 x 8 cm) of 600 g of silica gel (Note 11) and is eluted with hexanes. After elution with 1000 mL of hexanes, 50-mL fractions are collected. Fractions 6 to 20 are combined and concentrated by rotary evaporation (40 °C, 20 mmHg) to afford 12.80–12.97 g (75–76%) of the desired enyne as a yellow liquid (Note 12).

B. 1,3-Decadiyne (**3**). A 1-L, three-necked, round-bottomed flask equipped with a magnetic stir bar, rubber septum, a 125-mL pressure-equalizing addition funnel fitted with a rubber septum, and a Claisen adapter fitted with an argon inlet adapter and a rubber septum penetrated by the Teflon probe of a thermometer (Notes 2, 13) is charged via syringe with diisopropylamine (23.4 mL, 16.9 g, 167 mmol, 2.6 equiv) (Note 14) followed by 200 mL of THF (Note 15). The solution is cooled at 0 °C in an ice-water bath and *n*-butyllithium solution (2.39 M in hexanes, 67 mL, 160 mmol, 2.5 equiv) (Note 16) is added dropwise via the addition funnel over 15 min. The resulting yellow solution is stirred at 0 °C for 30 min and then cooled to –78 °C (internal temperature) in a dry ice-acetone bath with the surface of the reaction mixture approximately 3 cm below the level of the cold acetone. A 250-mL, three-necked, round-bottomed flask equipped with a magnetic stir bar, rubber septum, glass stopper, and argon inlet adapter is charged via syringe with a solution of enynyl chloride **2** (11.0 g, 64.4 mmol, 1.0 equiv) in 80 mL of THF and cooled to –78 °C (external temperature). The solution of enynyl chloride is transferred via cannula into the solution of LDA at such a rate so as to maintain the internal temperature between –70 and –78 °C (Notes 17, 18). The 250-mL flask is washed with two 5-mL portions of THF and the washings are transferred by cannula to the flask containing the LDA. The reaction mixture, which immediately darkens to a reddish color, is allowed to stir at –78 °C for 5–10 min and then poured into a 1-L separatory funnel containing 100 mL of half-saturated NH_4Cl solution. The resulting mixture is diluted with 150 mL of pentanes and extracted with 100 mL of 10% HCl solution. The organic phase is washed with 10% HCl

(2 x 100 mL) and the combined aqueous phases are back-extracted with pentanes (2 x 100 mL). The combined organic layers are washed with 100 mL of saturated NaCl solution, dried over $MgSO_4$, filtered, and concentrated by rotary evaporation (23 °C, 16 mmHg) to afford 9.40 g of a red liquid. This material is dissolved in 20 mL of CH_2Cl_2 and deposited onto 10 g of silica gel (Note 11) by rotary evaporation (23 °C, 16 mmHg). The resulting clumpy powder is transferred onto a column (12 x 8 cm) of 300 g of silica gel (Note 11) and is eluted with pentanes. Fractions 15 to 33 (50 mL each) are combined and concentrated under reduced pressure (23 °C, 16 mmHg) to provide 5.95–6.26 g (69–73%) of diyne **3** as an orange liquid (Note 19).

2. Notes

1. The 350-mL, threaded, heavy-walled, round-bottomed pressure vessel (142 mm length x 82.5 mm OD) was purchased from Chemglass, Inc. The vessel was equipped with a Teflon cap (thread size 15 mm). Carrying out the reaction in a sealed vessel minimized the loss of vinylidene chloride (bp 30–32 °C).

2. The apparatus was flame-dried under reduced pressure (0.08 mmHg) and then maintained under an atmosphere of argon during the course of the reaction.

3. Tetrakis(triphenylphosphine)palladium(0) (99.9%) was purchased from Strem Chemicals, Inc. and used as received. The catalyst was stored at –18 °C under argon and was weighed in the glove box. The submitters observed no difference in the yield of **2** when the catalyst was stored and transferred in a dry box.

4. Toluene was purified by pressure filtration under argon through activated alumina.

5. *Caution: Vinylidene chloride is light sensitive and highly susceptible to polymerization and peroxide formation.* Vinylidene chloride (99%) was purchased from Aldrich Chemical Company, Inc. and stored in the dark at 0 °C under argon. To remove any polymers and autoxidation products, the reagent was purified immediately before use by filtration through alumina according to the following procedure. A jacketed glass column (46 cm x 2.5 cm) cooled with cold running water and equipped with a needle at the bottom was charged with 30 g of Activity I basic alumina (EM Science, 70-230 mesh). A flame-dried, 50-mL, two-necked pear-shaped flask equipped with two rubber septa was flushed with argon and attached to the needle at the bottom of the alumina column. Vinylidene

chloride (approximately 25 mL) was added to the column and an exit needle was attached to the other septum of the flask. Nitrogen pressure was then applied to the top of the column to drive the vinylidene chloride through the column into the flask.

6. 1-Octyne (98%) was purchased from Alfa Aesar and used as received.

7. *n*-Butylamine (99.5%) was purchased from Aldrich Chemical Company, Inc. and purified by distillation at atmospheric pressure under argon from CaH_2.

8. Copper(I) iodide (98%) was purchased from Alfa Aesar and purified according to the following procedure. A 500-mL, round-bottomed flask was fitted with a Soxhlet extractor equipped with a reflux condenser fitted with a septum and argon inlet needle. The apparatus was flame-dried under vacuum (0.08 mmHg) and purged with argon. A Soxhlet thimble filled with 25 g of CuI was inserted into the extractor and 200 mL of THF was added via syringe. The extractor was wrapped with aluminum foil to shield the CuI from light and the THF in the flask was heated in an oil bath at reflux for 24 h. The apparatus was allowed to cool to room temperature and the CuI remaining in the thimble was transferred to a flame-dried, 100-mL, round-bottomed flask and dried at 0.08 mmHg for 24 h. The resulting CuI was ground up with a glass rod to afford a white, free-flowing powder.

9. The checkers submerged the flask such that the surface of the solution was 1 cm below the oil bath level. However, the checkers found that the oil level had no effect on the yield so long as good stirring was maintained.

10. The reaction was terminated when TLC analysis indicated that all of the octyne was consumed. TLC analysis was conducted using silica gel plates (elution with hexanes, visualization by UV and vanillin or $KMnO_4$); R_f values: octyne = 0.42, enyne product (**2**) = 0.55, bis-coupling product **4** = 0.30.

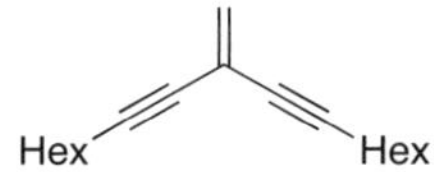

4

11. Silica gel was purchased from Silicycle (230-400 mesh).

12. Enyne **2** is light sensitive and decomposes slowly at room temperature. This compound is best stored at 0 °C in the dark under argon.

The product exhibits the following physical properties: R_f = 0.55 (hexanes); IR (neat) cm^{-1}: 2932, 2215, 1602, 1195; ^1H NMR (400 MHz, $CDCl_3$) δ: 0.89 (t, *J* = 7.1 Hz, 3 H), 1.28 – 1.33 (m, 4 H), 1.41 (app quint, *J* = 7.1 Hz, 2 H), 1.55 (app quint, *J* = 7.3 Hz, 2 H), 2.34 (t, *J* = 7.1 Hz, 2 H), 5.51 (s, 1 H), 5.55 (s, 1 H); ^{13}C NMR (101 MHz, $CDCl_3$) δ: 14.3, 19.4, 22.7, 28.3, 28.7, 31.5, 77.7, 93.0, 120.6, 121.0. Anal. Calcd. for $C_{10}H_{15}Cl$: C, 70.37; H, 8.86. Found: C, 70.55; H, 8.84.

13. An Omegaette HH3087 mini-thermometer type K with a Teflon-coated thermocouple was used to monitor internal temperature.

14. Diisopropylamine (99.5%) was purchased from Aldrich Chemical Company, Inc. and distilled at atmospheric pressure under argon from CaH_2.

15. Tetrahydrofuran was purified by pressure filtration under argon through activated alumina.

16. *n*-Butyllithium was purchased from Alfa Aesar and titrated by the method of Watson and Eastham.[2]

17. The position of the exit needle of the cannula was adjusted so that it is at the level of the dry ice-acetone mixture in the external cooling bath. In this fashion, the solution of enynyl chloride ran down the cold wall of the flask approximately 3 cm above the surface of the solution of LDA. If the temperature of the reaction mixture rose above –70 °C, lower yields of diyne were obtained due to isomerization (see Discussion Section).

18. Approximately 20 min was required for the addition.

19. 1,3-Decadiyne was stored as a solution in pentanes at –20 ºC under argon. The diyne exhibits the following physical properties: R_f = 0.41 (hexanes); IR (neat) cm^{-1}: 3310, 2957, 2298, 2226, 1466; ^1H NMR (400 MHz, $CDCl_3$) δ: 0.88 (t, *J* = 7.0 Hz, 3 H), 1.24–1.33 (m, 4 H), 1.38 (app quint, *J* = 7.0 Hz, 2 H), 1.53 (app quint, *J* = 7.2 Hz, 2 H), 1.96 (s, 1 H), 2.25 (t, *J* = 7.0 Hz, 2 H); ^{13}C NMR (101 MHz, $CDCl_3$) δ: 14.2, 19.2, 22.7, 28.2, 28.7, 31.5, 64.6, 64.8, 68.7, 78.8. Anal. Calcd. for $C_{10}H_{14}$: C, 89.49; H, 10.51. Found: C, 89.11; H, 10.84.

Safety and Waste Disposal Information

All hazardous materials should be handled and disposed of in accordance with "Prudent Practices in the Laboratory"; National Academy Press; Washington, DC, 1995.

3. Discussion

1,3-Diynes serve as valuable building blocks for the synthesis of natural products,[3] supramolecular constructs, and a variety of oligomeric alkynes and advanced materials with interesting electronic and optical properties. The most popular approaches to the synthesis of conjugated diynes involve alkynyl coupling reactions.[4] For example, the homocoupling of acetylenes can conveniently be achieved with copper catalysis using the Glaser reaction and its various modifications While unsymmetrical 1,3-diynes are available via the Cadiot-Chodkiewicz reaction,[6] the latter method involves the copper-catalyzed coupling of an alkyne with a 1-bromoalkyne; a variant of this process employs the coupling of alkynes with 1-haloalkynes in the presence of a combination of copper and palladium catalysts.[7,8] Several strategies have been developed for the synthesis of *terminal* 1,3-diynes based on the Cadiot-Chodkiewicz reaction. One approach employs the coupling of (bromoethynyl)trialkylsilanes followed by desilylation of the resultant 1-trialkylsilyl-1,3-diynes,[9] while an alternative strategy utilizes coupling reactions of propargylic alcohol derivatives followed by base-promoted fragmentation to unmask the terminal triple bond.[10,11] The application of the former approach to the preparation of the title compound, 1,3-decadiyne, has been reported by Zweifel[b] and is outlined in Scheme 1.[12]

While approaches such as this provide access to a variety of internal and terminal conjugated diynes, in many cases these coupling reactions are complicated by the formation of alkyne homodimers as significant byproducts. The cost of bis(trimethylsilyl)acetylene and the need for at least three synthetic operations further limits the attractiveness of this approach as a practical method for the synthesis of terminal diynes. These limitations have motivated the development of several alternative strategies for the preparation of terminal 1,3-diynes which are based on the Sonogashira coupling[13] of an alkyne with a dihaloethylene followed by b-elimination.

Scheme 1

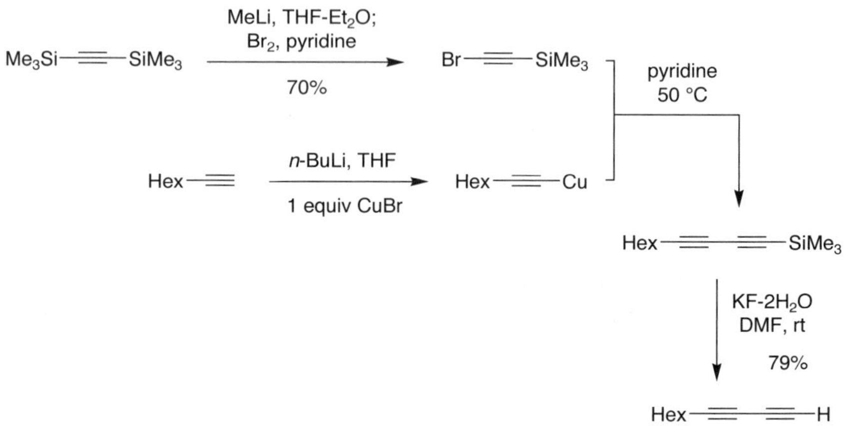

The selective mono-Sonogashira coupling of 1,2-dichloroethylene with alkynes was first demonstrated by Linstrumelle in 1981,[14] and subsequently several laboratories reported the synthesis of 1,3-diynes by Sonogashira coupling of various 1,2-dihaloethylenes followed by base-promoted dehydrohalogenation.[15] The requisite 1,2-dihaloalkenes are quite expensive, however, and Negishi[16] consequently introduced a more economical approach in 2003 based on the Sonogashira coupling of inexpensive 1,1-dichloroethylene (vinylidene chloride).[17]

The method of Negishi provides the basis for the optimized procedure described here for the synthesis of 1,3-decadiyne. In the case of the first step, the Sonogashira coupling of vinylidene chloride and 1-octyne, the catalyst loading has been reduced relative to the original Negishi protocol (from 5 to 3 mol% each for $(Ph_3P)_4Pd$ and CuI), and the amount of vinylidene chloride has been reduced from 5 to 1.5 equivalents. Our initial attempts to reduce the amount of vinylidene chloride below 5 equivalents were not successful. As the amount of vinylidene chloride was reduced, the octyne homodimer (7,9-hexadecadiyne) and enediyne **4** (the product of bis-coupling) appeared as major byproducts and the desired chloroenyne was obtained in greatly diminished yield. We hypothesized that the volatile vinylidene chloride (bp 30–32 °C) was escaping during the course of the reaction, and indeed, the formation of these byproducts could be minimized by simply carrying out the reaction in a sealed flask. Under the optimal conditions described here using 1.5 equiv of vinylidene chloride, the desired enyne was obtained in 75–76% yield and the crude coupling product was found to contain only 5-10% of enediyne **4** and less than 5% of the homodimer.

As reported by Negishi, exposure of enynyl chloride **2** to LDA at –78 °C effected β-elimination leading to the desired diyne **3**. We have noted that temperature control was critical to the success of this transformation, particularly when the reaction was conducted on scales greater than 500 mg. Specifically, the allenyne **8** was formed as a byproduct in yields of up to 40% if the reaction temperature was allowed to rise above –70 °C during the addition of the chloroenyne to the solution of LDA. Scheme 2 outlines the likely mechanism for the formation of this byproduct. 1,4-Elimination[18] of **2** leads to the tetraene **5**, which is further metalated by LDA to afford the organolithium intermediates **6** and/or **7**. Protonation then furnishes the allenyne **8**. The possibility that **8** arises from isomerization of 1,3-decadiyne was ruled out by control experiments that demonstrated that the diyne was not converted to **8** upon exposure to LDA under the conditions of the reaction.

Scheme 2

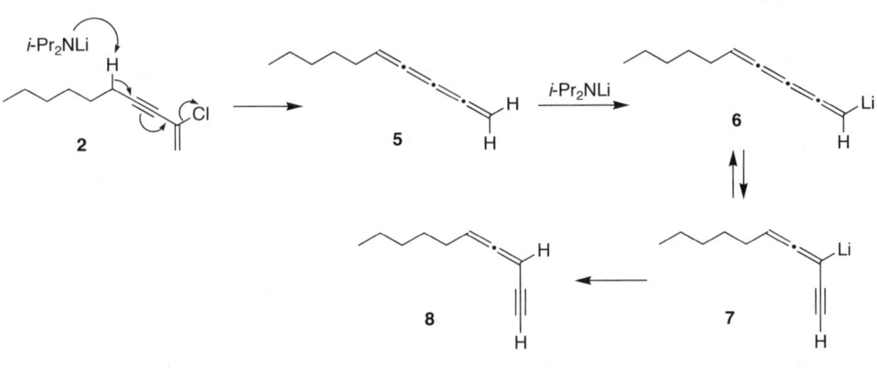

The formation of the allenyne byproduct could be minimized by carefully controlling the temperature of the reaction mixture, especially during the addition of the chloroenyne to LDA, which resulted in an exothermic reaction. Temperature control was best achieved by precooling the solution of chloroenyne in THF to –78 °C and employing a solution of less than 1.0 M concentration. The solution was added slowly at such a rate that the internal temperature of the reaction mixture did not exceed –70 °C. Under these conditions the diyne was obtained in 66–72% yield contaminated with <2% of the allenyne.

In summary, an improved protocol for the preparation of 1,3-decadiyne (**3**) by the method of Negishi has been developed. This approach employs the Sonogashira coupling of 1-octyne with vinylidene chloride followed by base-promoted dehydrochlorination with LDA. The optimized conditions described here provide 1,3-decadiyne in good yield on multigram-scale and are considerably more practical and efficient than previous methods reported for the preparation of this diyne.[19]

1. Department of Chemistry, Massachusetts Institute of Technology, Cambridge, MA 02139. Financial support by the National Institutes of Health (GM 28273) is gratefully acknowledged. A.L.K was supported in part by a Massachusetts Institute of Technology Leventhal Presidential Fellowship and by the American Council of Independent Laboratories.
2. Watson, S. C.; Eastham, J. F. *J. Organomet. Chem.* **1967**, *9*, 165.
3. The structure of a number of marine natural products incorporate conjugated diynes, for a review, see: Faulkner, D. J. *Nat. Prod. Rep.* **2002**, *19*, 1.
4. Siemsen, P.; Livingston, R. C.; Diedrich, F. *Angew. Chem. Int. Ed.* **2000**, *39*, 2632.
5. Niedballa, U. In *Methoden der Organischen Chemie (Houben-Weyl)*; Muller, E., Ed.; Thieme, Stuttgart, **1977**; Vol. 5, pp 925-928.
6. Cadiot, P.; Chodkiewicz, W. In *Chemistry of Acetylenes*; Viehe, H. G., Ed.; Marcel Dekker: New York, 1969; pp 597-647.
7. Negishi, E.; Anastasia, L. *Chem. Rev.* **2003**, *103*, 1979.
8. For recent examples, see (a) Alami, M.; Ferri, F. *Tetrahedron Lett.* **1996**, *37*, 2763. (b) Kim, S.; Kim, S.; Lee, T.; Ko, H.; Kim, D. *Org. Lett.* **2004**, *6*, 3601.
9. (a) Eastmond, R.; Walton, D. R. M. *Tetrahedron* **1972**, *28*, 4591. (b) Miller, J. A.; Zweifel, G. *Synthesis* **1983**, 128.
10. (a) Shostakovskii, N.; Bogdanova, A. V. In *The Chemistry of Diacetylenes*; John Wiley & Sons: New York, 1974. (b) For some recent examples, see Dabdoub, M. J.; Dabdoub, V. B.; Lenardão, E. J. *Tetrahedron Lett.* **2001**, *42*, 1807.
11. Vereshchagin, L. I.; Buzilova, S. R.; Bol'shdvorskaya, R. L.; Kirillova, L. P. *J. Org. Chem. USSR (Engl. Transl.)* **1976**, *12*, 1174.
12. 1,3-Decadiyne has also been generated by fragmentation (NaOH, toluene, reflux) of the tertiary, propargylic alcohol prepared by Cadiot-Chodkiewicz coupling of octyne with 4-bromo-2-methyl-3-butyn-2-ol. See Dabdoub, M. J.; Baroni, A. C. M.; Lenardão, E. J.; Gianeti, T. R.; Hurtado, G. R. *Tetrahedron* **2001**, 57, 4271.
13. For recent reviews, see: (a) Sonogashira, K. In *Metal-Catalyzed Cross-Coupling Reactions*; Diederich, F., Stang, P. J., Eds.; Wiley-VCH: Weinheim, **1998**; 203-229. (b) Sonogashira, K. Sonogashira Alkyne Synthesis. In *Handbook of*

Organopalladium Chemistry for Organic Synthesis; Negishi, E., de Meijere, A., Eds.; Wiley-Interscience: New York, 2002; Vol. I, 493. (b) Negishi, E.; Xu, C. Palladium-Catalyzed Alkynylation with Alkynylmetals and Alkynyl Electrophiles. In *Handbook of Organopalladium Chemistry for Organic Synthesis*; Negishi, E., de Meijere, A., Eds.; Wiley-Interscience: New York, 2002; Vol. I, 531.

14. Linstrumelle, G.; Ratovelomanana, V. *Tetrahedron Lett.* **1981**, *22*, 315.

15. (a) Negishi, E.; Okukado, N.; Lovich, S. F.; Luo, F.-T. *J. Org. Chem.* **1984**, *49*, 2629. (b) Kende, A. S.; Smith, C. A. *J. Org. Chem.* **1988**, *53*, 2655. (c) Alami, M.; Crousse, B.; Linstrumelle, G. *Tetrahedron Lett.* **1995**, *36*, 3687.

16. Qian, M.; Negishi, E. *Org. Proc. Res. & Dev.* **2003**, *7*, 412.

17. For the first report of mono-Sonogashira coupling reactions of vinylidene chloride, see Ratovelomanana, V.; Hammoud, A.; Linstrumelle, G. *Tetrahedron Lett.* **1987**, *28*, 1649.

18. For examples of related 1,4-elimination reactions, see Leclerc, E.; Tius, M. A. *Org. Lett.* **2003**, *5*, 1171 and references cited therein.

19. (a) Brandsma, L.; Verkruijsse, H. D.; Fossatelli, M. *Recl. Trav. Chim. Pays-Bas* **1991**, *110*, 131. (b) Stracker, E. C.; Zweifel, G. *Tetrahedron Lett.* **1990**, *31*, 6815.

SYNTHESIS OF YNAMIDES BY *N*-ALKYNYLATION OF AMINE DERIVATIVES. PREPARATION OF *N*-ALLYL-*N*-(METHOXYCARBONYL)-1,3-DECADIYNYLAMINE

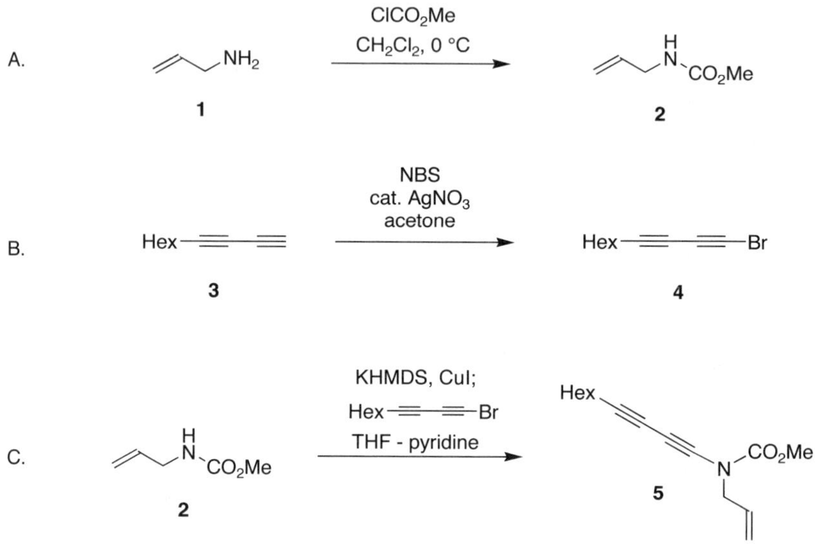

Submitted by Amanda L. Kohnen, Joshua R. Dunetz, and Rick L. Danheiser.[1]
Checked by Scott E. Denmark and Xiaorong Liu.

1. Procedure

A. N-(Methoxycarbonyl)-2-propenylamine (**2**) (Note 1). A 250-mL, three-necked, round-bottomed flask equipped with a magnetic stir bar, an argon gas inlet tube, a rubber septum, which is pierced with a Teflon-coated thermocouple, and a 200-mL pressure-equalizing addition funnel fitted with a rubber septum (Note 2) is charged via syringe through the septum with allylamine (**1**) (14.0 mL, 10.68 g, 187 mmol, 2.1 equiv) (Note 3) and 100 mL of dichloromethane (Note 4). The solution is cooled to 0 °C (internal temperature) in an ice-water bath and a solution of methyl chloroformate (7.0 mL, 8.6 g, 91 mmol) (Note 5) in 30 mL of dichloromethane is added dropwise via the addition funnel over 10 min. The resulting white mixture is allowed to warm to room temperature over 30 min and then transferred to a 500-mL separatory funnel and extracted with 50 mL of aqueous 1M HCl solution, 50 mL of saturated aqueous $NaHCO_3$ solution, and 50 mL of saturated aqueous NaCl solution. The organic phase is dried over $MgSO_4$, filtered, and concentrated by rotary evaporation (23 °C, 16 mmHg) to afford 10.12–10.26 g (97–98%) of carbamate **2** as a colorless oil (Note 6).

B. 1-Bromo-1,3-decadiyne (**4**). Into a 200-mL, round-bottomed flask equipped with a magnetic stir bar (Note 2) is added 1,3-decadiyne (**3**) (5.66 g, 42.2 mmol) and 100 mL of acetone (Note 7). The flask is equipped with a rubber septum and argon inlet needle. *N*-Bromosuccinimide (8.26 g, 46.4 mmol, 1.1 equiv) (Note 8) and $AgNO_3$ (0.717 g, 4.22 mmol, 0.10 equiv) (Note 9) are each added in one portion through temporary removal of the septum. The rubber septum is replaced with an argon inlet adapter, the flask is wrapped with aluminum foil (Note 10), and the reaction mixture is stirred at room temperature for 22 h (Note 11). The resulting cloudy orange mixture is transferred to a 500-mL separatory funnel with the aid of 100 mL of pentanes and washed with two 50-mL portions of saturated aqueous $Na_2S_2O_3$ solution (Note 12). The combined aqueous layers are extracted with two 50-mL portions of pentanes and the combined organic phases are washed with 100 mL of saturated NaCl solution, dried over $MgSO_4$, filtered, and concentrated by rotary evaporation (23 °C, 16 mmHg) to provide 8.18–8.25 g (91–92%) of bromodiyne **4** as a red oil (Notes 13, 14).

C. N-(Methoxycarbonyl)-N-(2-propenyl)-1,3-decadiynylamide (**5**). Into a 500-mL, three-necked, round-bottomed flask is weighed carbamate **2** (3.98 g, 34.6 mmol). The flask is equipped with a large magnetic stir bar, an argon inlet adapter, rubber septum, and a 125-mL pressure-equalizing addition funnel fitted with a rubber septum (Note 1). Tetrahydrofuran (130 mL) (Note 15) and pyridine (70 mL, 68.0 g, 867 mmol, 25 equiv) (Note 16) are added via syringe through the septum. The solution is cooled to 0 °C (internal temperature) in an ice-water bath and 38 mL of potassium hexamethyldisilazide solution (0.91 M in THF, 34.6 mmol, 1.0 equiv) (Note 17) is added via syringe over 10 min. The peach-colored slurry is allowed to warm to room temperature over 15 min and then copper(I) iodide (6.58 g, 34.6 mmol, 1.0 equiv) (Notes 18, 19) is added in one portion by temporarily removing the septum. The resulting yellow-green mixture is stirred at room temperature for 2 h. A solution of bromodiyne **4** (8.47 g, 39.8 mmol, 1.15 equiv) in 40 mL of THF is then added via the addition funnel over 45 min (5 mL THF rinse) (Note 20). The flask and addition funnel are wrapped with aluminum foil to protect the bromodiyne **4** from light and the resulting mixture is stirred at room temperature for 20 h (Note 21). The dark red-brown reaction mixture is transferred to a 1-L separatory funnel, diluted with

150 mL of diethyl ether, and washed with four 100-mL portions of a 3:1 mixture of saturated NaCl solution and concentrated NH_4OH solution (Note 22). The blue aqueous layers are extracted with two 100-mL portions of Et_2O, and the combined organic layers are washed with two 200-mL portions of 1M HCl solution, 200 mL of saturated NaCl solution, dried over $MgSO_4$, filtered, and concentrated by rotary evaporation (40 °C, 16 mmHg) to afford 18.5 g of dark red oil. This material is transferred to a column (12 x 8 cm) of 300 g of silica gel and eluted with 500 mL of hexanes. Elution is then continued with 5% EtOAc-hexanes and 50-mL fractions are collected. Fractions 14 to 51 are combined and concentrated by rotary evaporation (23 °C, 16 mmHg) to provide 6.88 g (80%) of diynamide **5** as a red oil (Notes 23, 24).

2. Notes

1. *t*-Butyl and ethyl *N*-allylcarbamate are commercially available. The method described here for the synthesis of the methyl derivative **2** illustrates a general procedure for the preparation of carbamates which usually delivers the desired compounds in near-quantitative yield and in a state of high purity, suitable for use in the *N*-alkynylation reaction without further purification.

2. The apparatus was flame-dried under reduced pressure (0.08 mmHg) and then maintained under an atmosphere of argon during the course of the reaction.

3. Allylamine (98%) was purchased from Alfa Aesar and used without purification.

4. Dichloromethane was purified by pressure filtration under argon through activated alumina.

5. Methyl chloroformate (99%) was obtained from Aldrich Chemical Company, Inc. and used as received.

6. The product exhibits the following spectroscopic properties: R_f = 0.17 (hexanes/EtOAc, 4/1); IR (neat) cm^{-1}: 3337, 3084, 2952, 1702, 1534; ^{1}H NMR (400 MHz, $CDCl_3$) δ: 3.68 (s, 3 H), 3.81 (app s, 2 H), 4.73 (br s, 1 H), 5.13 (dd, J = 10.2, 1.2 Hz, 1 H), 5.19 (dd, J = 17.1, 1.2 Hz, 1 H), 5.84 (ddt, J = 17.1, 10.5, 5.6 Hz, 1H); ^{13}C NMR (101 MHz, $CDCl_3$) δ: 43.6, 52.4, 116.2, 134.7, 157.1. MS (EI) *m/z* 115 (M^+, 91), 100 (100). Anal. Calcd. for $C_5H_9NO_2$: C, 52.16; H, 7.88; N, 12.17. Found: C, 51.77; H, 7.89; N, 11.89.

7. 1,3-Decadiyne was prepared via the *Organic Syntheses* procedure directly preceding this procedure. Acetone (99.5%) was purchased from Aldrich Chemical Company and was used as received.

8. *N*-Bromosuccinimide (NBS) (99%) was purchased from Aldrich Chemical Company, Inc. and purified by recrystallization from water according to the following procedure.[2] A 500-mL Erlenmeyer flask equipped with a magnetic stir bar was charged with NBS (30 g, 169 mmol) and 300 mL of deionized water and heated on a hot plate until the yellow mixture became homogeneous (approximately 90 °C). The resulting dark orange solution was cooled in an ice bath over 3 h and the white crystals that formed were collected by vacuum filtration on a Büchner funnel, washed with 100 mL of water, and then dried over P_2O_5 in a vacuum dessicator at 22 mmHg for 12 h to afford 21.2 g of NBS as white crystals.

9. Silver(I) nitrate (99.9%) was obtained from Alfa Aesar and used as received.

10. The reaction mixture was shielded from light due to the photosensitivity of NBS and bromodiyne **4**.

11. The progress of the reaction was monitored by thin layer chromatography. TLC analysis was conducted using silica gel plates (elution with hexanes, visualization by UV and vanillin); R_f values: diyne **3** = 0.41, bromodiyne **4** = 0.49.

12. Washing with saturated sodium thiosulfate solution removed traces of bromine.

13. The product exhibits the following spectroscopic properties: R_f = 0.49 (hexanes); IR (neat) cm^{-1}: 2956, 2930, 2858, 2141, 1466; ^1H NMR (400 MHz, $CDCl_3$) δ: 0.89 (t, J = 7.1 Hz, 3 H), 1.24–1.32 (m, 4 H), 1.38 (app quint, J = 7.1 Hz, 2 H), 1.52 (app quint, J = 7.2 Hz, 2 H), 2.24 (t, J = 7.0 Hz, 2 H); ^{13}C NMR (125 MHz, $CDCl_3$) δ: 14.2, 19.2, 22.7, 28.2, 28.7, 31.4, 37.0, 65.5, 65.9, 77.7. Anal. Calcd. for $C_{10}H_{13}Br$: C, 56.36; H, 6.15. Found: C, 56.35; H, 6.29.

14. In contrast to many other alkynyl bromides we have worked with, the bromodiyne **4** does not exhibit lachrymatory properties. As with all alkynyl halides, however, we nonetheless recommend avoiding any skin contact with this compound and carrying out all operations in a fume hood.

15. Tetrahydrofuran was purified by pressure filtration through activated alumina under argon.

16. Pyridine was purchased from Aldrich Chemical Company and purified by distillation under argon from CaH_2. The submitters found that lower yields were obtained in these coupling reactions if pyridine (Alfa

Aesar, 99%, 0.1% water content) from a freshly opened bottle was used instead (see Discussion Section).

17. Potassium hexamethyldisilazide solution was purchased from Alfa Aesar and used as received. KHMDS in toluene (0.5 M, Aldrich) can be employed with similar results.

18. Copper(I) iodide (98%) was purchased from Alfa Aesar and purified according to the following procedure. A 500-mL, round-bottomed flask was fitted with a Soxhlet extractor equipped with a reflux condenser fitted with a rubber septum and argon inlet needle. The reaction apparatus was flame-dried under vacuum (0.08 mmHg) and purged with argon. A Soxhlet thimble filled with 25 g of CuI was inserted into the extractor and 200 mL of THF was added via syringe. The extractor was wrapped with aluminum foil to shield the CuI from light and the THF was heated in an oil bath at reflux for 24 h. The apparatus was allowed to cool to room temperature and the CuI was transferred to a flame-dried, 100-mL, one-necked, round-bottomed flask and dried at 0.08 mmHg for 24 h. The resulting CuI was ground up with glass rod to afford a white, free-flowing powder, which was stored under argon shielded from light.

19. The yield of ynamide product in a related coupling reaction was unchanged when CuI was used as received without purification.

20. Increasing the rate of addition resulted in reduced yields (see Discussion Section).

21. The progress of the reaction was monitored by thin layer chromatography. TLC analysis was conducted using silica gel plates (elution with 20% EtOAc-hexanes, visualization by UV and vanillin or $KMnO_4$); R_f values: bromodiyne **4** = 0.69, carbamate **2** = 0.17, diynamide **5** = 0.48.

22. The aqueous phase in the fourth wash should be colorless. Extraction of copper salts from the organic phase was found to be important in order to obtain the ynamide product in good yield.

23. The yield of ynamide varied from 73–80% in several runs carried out between one-half-scale and the scale described here.

24. The product exhibits the following spectroscopic properties: R_f = 0.48 (hexanes/EtOAc, 4/1); IR (neat) cm^{-1}: 2956, 2261, 2170, 174, 1442; ^1H NMR (400 MHz, $CDCl_3$) δ: 0.88 (t, J = 7.0 Hz, 3 H), 1.23–1.32 (m, 4 H), 1.38 (app quint, J = 7.2 Hz, 2 H), 1.52 (app quint, J = 7.1 Hz, 2 H), 2.29 (t, J = 7.1 Hz, 2 H), 3.82 (s, 3 H), 4.05 (d, J = 6.1 Hz, 2 H), 5.25 (dd, J = 1.2, 10.3 Hz, 1 H), 5.29 (dd, J = 1.3, 17.3 Hz, 1 H), 5.85 (ddt, J = 6.0, 10.3, 18.8 Hz, 1 H); ^{13}C NMR (101 MHz, $CDCl_3$) δ: 14.3, 19.8, 22.7, 28.5, 28.7, 31.5,

52.7, 54.5, 58.0, 64.3, 67.5, 84.3, 119.3, 131.3, 155.9; MS (EI) *m/z*: 247 (M^+, 100). Anal. Calcd. for $C_{15}H_{21}NO_2$: C, 72.84; H, 8.56; N, 5.66. Found: C, 72.59; H, 8.40; N, 5.31.

Safety and Waste Disposal Information

All hazardous materials should be handled and disposed of in accordance with "Prudent Practices in the Laboratory"; National Academy Press; Washington, DC, 1995.

3. Discussion

In recent years, ynamides have emerged as exceptionally useful building blocks for organic synthesis.[3] Considerably more robust than simple ynamines, ynamides are more easily stored and handled, and more resistant to hydrolysis and polymerization. Ynamides tolerate a variety of reaction conditions incompatible with simple ynamines and have thus proved to be versatile substrates for a variety of synthetic transformations. Recent representative examples include Pauson-Khand reactions,[4] thermal and metal-promoted [4 + 2] cycloadditions,[5] [2 + 2 + 2] cyclotrimerizations,[6] Heck and other carbometalation reactions,[7] ring-closing metatheses,[8] radical-π cyclizations,[9] Pt-catalyzed cycloisomerizations,[10] ketene [2 + 2] cycloadditions,[11] and Pictet-Spengler-type cyclizations.[12]

Several strategies are available for the synthesis of ynamides.[3] While some success has been realized using the base-catalyzed isomerization of propargyl amides[13] and the dehydrohalogenation of halo enamides,[14] the most attractive approach to the synthesis of ynamides employs the direct *N*-alkynylation of amine derivatives. The first effective means for achieving this transformation was based on the addition of metalated amides to alkynyl(phenyl)iodonium salts.[15] Unfortunately, this method is subject to several serious limitations. While *sulfonamide* derivatives participate smoothly in these reactions, lactams, acyclic carbamates, and other carboxamide derivatives do not undergo alkynylation in good yield. Most significantly, this strategy can only be applied to the synthesis of ynamides in which the substituent on the alkynyl carbon is a hydrogen atom, or trialkylsilyl or aryl group.

Recent advances in copper-promoted coupling reactions of amine derivatives have provided the basis for a more general and efficient method that is applicable to the synthesis of a wide variety of ynamides. The preparation of diynamide **5** described here illustrates the method introduced by our laboratory in 2003 for the *N*-alkynylation of several classes of amine derivatives.[16] In our initial studies, we attempted to effect the coupling of carbamates with alkynyl halides using the conditions developed by Buchwald[17] for the amidation of aryl halides. Unfortunately, attempted application of Buchwald's catalyst system to the coupling of acyclic carbamates with 1-bromo-2-phenylacetylene gave only trace amounts of the desired ynamide, with the predominant product being the 1,3-diyne generated from "homocoupling" of the alkynyl halide. After considerable experimentation, success in effecting the desired alkynylation was finally achieved when we turned our attention to protocols in which complete conversion of the amide substrate to its copper derivative was carried out *prior* to addition of the alkynyl halide.[18] Under these conditions, the rate of alkynylation is significantly faster, copper-promoted dimerization of the alkynyl halide is greatly diminished, and the desired ynamides emerge as the major product of the reaction.

Both alkynyl bromides and alkynyl iodides participate smoothly in the desired reaction.[19] The procedure described here for the preparation of 1-bromo-1,3-decadiyne is based on the method of Hofmeister and coworkers[20] and provides the alkynyl bromide in near-quantitative yield and in a state of high purity. If extended storage is required, we recommend storing this compound as a solution in pentane at –20 °C.

The alkynylation procedure described here is a slightly modified version of our original protocol. As reported previously, a solution of the amide derivative is first deprotonated by reaction with 1 equiv of commercially available KHMDS (solution in THF) and then treated with 1 equiv of CuI. Somewhat diminished yields result if CuCN is employed in place of CuI. Although a full equivalent of CuI is required for the reaction, this salt is inexpensive and can be used as received without purification. After 2 h, the alkynyl bromide is added dropwise and the reaction is allowed to proceed at room temperature overnight. Slow addition of the bromide is important in minimizing the homocoupling side reaction. Only a slight excess of **4** (1.15 equiv) is employed in the optimal procedure described here, although in the case of other alkynyl bromides we occasionally found that the yield of coupling product is improved if 1.5 or even 2 equiv of the alkyne is employed. Originally we carried out the alkynylation reactions in a mixture of pyridine, benzene (from the alkynyl bromide stock solution), and THF (60:30:10 ratio by volume). In the optimized procedure described here, however, the amount of pyridine has been reduced to 25 equivalents, and the final solvent system consists of a mixture of THF and pyridine in a ratio of approximately 3:1.

The examples shown in Table 1 illustrate the scope of this method for the synthesis of ynamides. As reported previously, a broad range of alkynyl bromides participate in the reaction, including compounds that are especially prone to homocoupling such as halo derivatives of conjugated enynes and aryl acetylenes. Acyclic carbamates, sulfonamides, oxazolidinones, and cyclic ureas all undergo *N*-alkynylation in good yield. Significantly, the alkynylation reaction proceeds smoothly at room temperature, allowing its application to the synthesis of several classes of thermally sensitive ynamides, which are not available by other methods that require reaction at elevated temperature (vide infra).

Concurrent with our studies, Hsung and co-workers developed a complementary method for the synthesis of ynamides that also involves the coupling of amine derivatives with halo acetylenes. In 2003, Hsung reported that the copper catalyst system originally developed by Buchwald for the *arylation* and *vinylation* of amides, lactams, and carbamates[17] is also effective for the alkynylation of oxazolidinones and lactams.[21] The Hsung procedure requires reaction at elevated temperatures (110–150 °C) and employs catalytic CuCN in conjunction with a 1,2-diamine ligand in the presence of K_3PO_4. Unfortunately, under these conditions acyclic carbamates, sulfonamides, and cyclic ureas undergo alkynylation in poor yield. In 2004, Sato and co-workers reported a modified Buchwald protocol that is applicable to the alkynylation of acyclic and cyclic sulfonamide derivatives,[22] and the same year Hsung described a "second-generation catalyst system"[23] employing catalytic copper sulfate and 1,10-phenanthroline as ligand that is applicable to a broad range of amine derivatives including sulfonamides, imidazolidinones, and acyclic carbamates. This procedure does require reaction at elevated temperature (60–95 °C), however, and therefore is not applicable to the preparation of thermally sensitive ynamides. In the case of ynamide **5**, we found that the desired diynamide is produced in only 34–38% yield using the Hsung "second-generation" protocol as compared to 74–82% employing our room temperature procedure.

Table 1. Synthesis of Ynamides by Copper-Mediated N-Alkynylation

entry	amine derivative	alkyne	ynamide	yield (%)[a]
1[b]	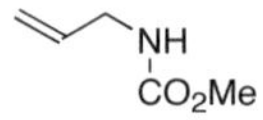		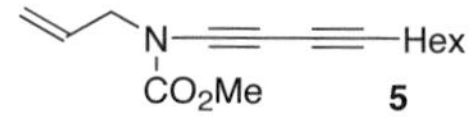 **5**	82[c]
2[b]	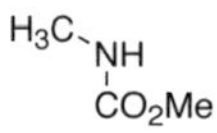		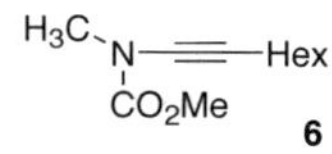 **6**	61[d]
3[b]	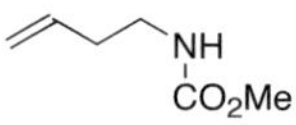		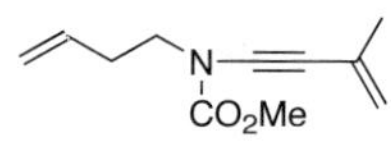 **7**	68[d]
4[b]	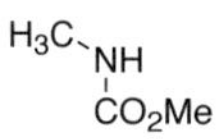	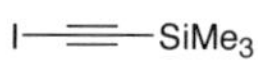	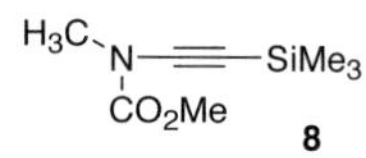 **8**	64[d]
5[f]	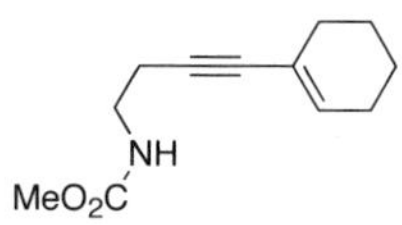	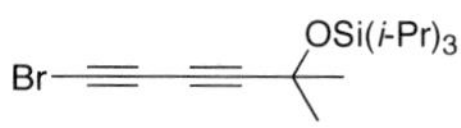	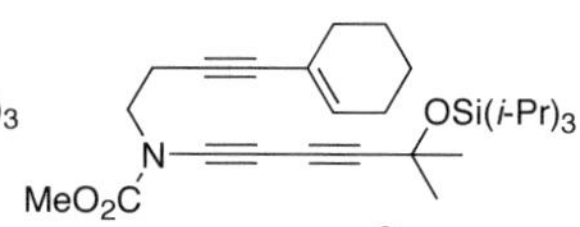 **9**	51[e]
6[f]	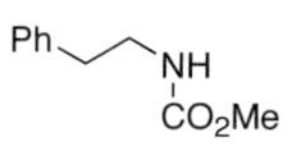	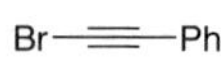	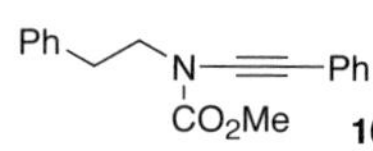 **10**	76[g]
7[f]	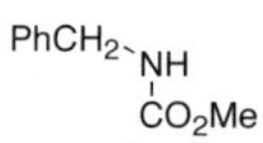	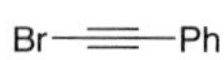	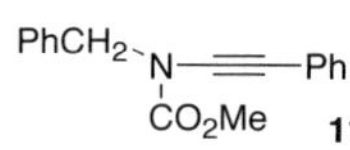 **11**	63[d]
8[f]	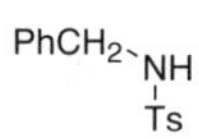	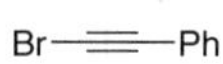	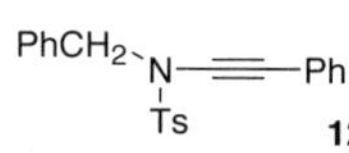 **12**	78[g]
9[f]	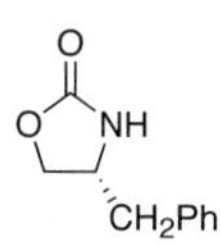	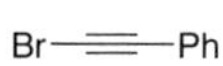	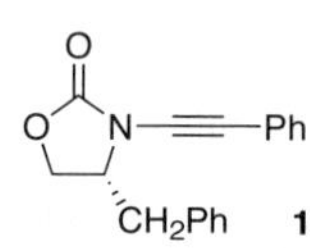 **13**	74[g]
10[f]	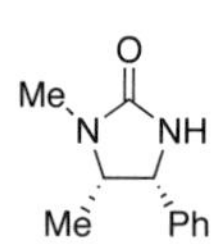	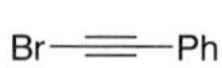	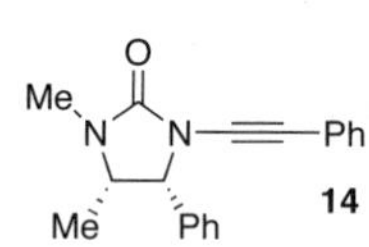 **14**	60[g]

[a]Isolated yields of products purified by column chromatrography. [b]Reactions conducted using improved protocol with 1.15-1.5 equiv of alkynyl halide and 25 equiv of pyridine. [c]This work. [d]See ref. 11. [e]See ref. 5b. [f]Reaction conducted using original protocol with 1.5-2 equiv of alkynyl bromide. [g]See ref. 16.

For the synthesis of most ynamides, we have found both our method and the Hsung "second-generation" procedure to be reliable and reproducible on both small and large (i.e., multigram) scales. Recently, Tam and co-workers reported an alternative *N*-alkynylation protocol[24] that involves a melding of the procedures previously developed in our laboratory and that of Hsung. Tam's conditions involve the use of catalytic (8–40 mol%) CuI and 1,10-phenanthroline (12–46 mol%), with slow addition of 1.0–1.7 equiv of KHMDS over several hours to a hot (90 °C) solution of the copper salt, ligand, amine derivative, and alkynyl bromide in toluene. Tam was motivated to introduce this modified procedure when he encountered difficulties in reproducing results previously reported by both our group and that of Hsung. For example, Tam reports obtaining none of the desired ynamide from the reaction of 1-bromo-2-phenylacetylene with $BnNHCO_2Me$ using either our method or the method of Hsung. Hsung had previously reported obtaining this ynamide in 73% yield,[23] and we had reported the synthesis of the corresponding Boc derivative in 61% yield.[16]

We investigated the coupling of $BnNHCO_2Me$ with 1-bromo-2-phenylacetylene in some detail in an attempt to identify the experimental variables that might be responsible for Tam's inability to reproduce our results and the results of Hsung and co-workers.[11] We obtained the expected ynamide **11** in 63% yield using our method (Table 1, entry 7), and in 79% yield by employing the "second-generation" procedure of Hsung. In our experience, both methods have proved to be highly reproducible. We have noted, however, that results employing our method are affected by the quality of the pyridine used in the reaction. Thus, the yield reported here for ynamide **11** was obtained using pyridine freshly distilled from CaH_2 or KOH as recommended in our original report.[16] When pyridine (Alfa Aesar, 99%, 0.1% water content) from a freshly opened bottle was used (without distillation) instead, the yield of ynamide product declined to 52%. Most significantly, when distilled pyridine that had then been exposed to the atmosphere for several days was used for the *N*-alkynylation, none of the desired ynamide was formed.

1. Department of Chemistry, Massachusetts Institute of Technology, Cambridge, MA 02139. Financial support by the National Institutes of Health (GM 28273) and Merck Research Laboratories is gratefully acknowledged. A.L.K was supported in part by a Massachusetts Institute of Technology Leventhal Presidential Fellowship and by the American Council of Independent Laboratories.

2. Dauben, H. J.; McCoy, L. L. *J. Am. Chem. Soc.* **1959**, *81*, 4863.
3. For reviews, see (a) Himbert, G. In *Methoden der Organischen Chemie (Houben-Weyl)*; Kropf, H.; Schaumann, E., Eds.; Georg Thieme Verlag: Stuttgart, 1993; Vol.

E15, Part 3, pp 3267–3443. (b) Zificsak, C. A.; Mulder, J. A.; Hsung, R. P.; Rameshkumar, C.; Wei, L.-L. *Tetrahedron* **2001**, *57*, 7575.

4. (a) Witulski, B.; Gössmann, M. *Synlett* **2000**, 1793. (b) Rainier, J. D.; Imbriglio, J. E. *J. Org. Chem.* **2000**, *65*, 7272 and references cited.

5. (a) Witulski, B.; Lumtscher, J.; Bergsträsser, U. *Synlett* **2003**, 708. (b) Dunetz, J. R.; Danheiser, R. L. *J. Am. Chem. Soc.* **2005**, *127*, 5776. (c) Martínez-Esperón, M. F.; Rodríguez, D.; Castedo, L.; Saá, C. *Org. Lett.* **2005**, *7*, 2213.

6. Witulski, B.; Alayrac, C. *Angew. Chem. Int. Ed.* **2002**, *41*, 3281 and references cited.

7. (a) Witulski, B.; Alayrac, C.; Tevzadze-Saeftel, L. *Angew. Chem. Int. Ed.* **2003**, *42*, 4257. (b) Couty, S.; Liégault, B.; Meyer, C.; Cossy, J. *Org. Lett.* **2004**, *6*, 2511. (c) Chechik-Lankin, H.; Livshin, S.; Marek, I. *Synlett* **2005**, 2098 and references cited.

8. (a) Saito, N.; Sato, Y.; Mori, M. *Org. Lett.* **2002**, *4*, 803. (b) Huang, J.; Xiong, H.; Hsung, R. P.; Rameshkumar, C.; Mulder, J. A.; Grebe, T. P. *Org. Lett.* **2002**, *4*, 2417.

9. Marion, F.; Courillon, C.; Malacria, M. *Org. Lett.* **2003**, *5*, 5095.

10. Marion, F.; Coulomb, J.; Courillon, C.; Fensterbank, L.; Malacria, M. *Org. Lett.* **2004**, *6*, 1509.

11. Kohnen, A. L.; Mak, X. Y.; Lam. T. Y.; Dunetz, J. R.; Danheiser, R. L. *Tetrahedron* **2006**, *62*, 3815.

12. Zhang, Y.; Hsung, R. P.; Zhang, X.; Huang, J.; Slafer, B. W.; Davis, A. *Org. Lett.* **2005**, *7*, 1047.

13. (a) Wei, L.-L.; Mulder, J. A.; Xiong, H.; Zificsak, C. A.; Douglas, C. J.; Hsung, R. P. *Tetrahedron* **2001**, *57*, 459. (b) Gravestock, D.; Dovey, M. C. *Synthesis* **2003**, 523 and references cited.

14. (a) Brückner, D. *Synlett* **2000**, 1402. (b) Couty, S.; Barbazanges, M.; Meyer, C.; Cossy, J. *Synlett* **2005**, 905 and references cited.

15. (a) Murch, P.; Williamson, B. L.; Stang, P. J. *Synthesis* **1994**, 1255. (b) Feldman, K. S.; Bruendl, M. M.; Schildknegt, K.; Bohnstedt, A. C. *J. Org. Chem.* **1996**, *61*, 5440. (c) Witulski, B.; Stengel, T. *Angew. Chem., Int. Ed. Engl.* **1998**, *37*, 489. (d) Rainier, J. D.; Imbriglio, J. E. *Org. Lett.* **1999**, *1*, 2037. (e) Witulski, B.; Gössmann, M. *J. Chem. Soc., Chem. Commun.* **1999**, 1879. (f) Witulski, B.; Stengel, T. *Angew. Chem., Int. Ed. Engl.* **1999**, *38*, 2426.

16. Dunetz, J. R.; Danheiser, R. L. *Org. Lett.* **2003**, *5*, 4011.

17. (a) Klapars, A.; Huang, X.; Buchwald, S. L. *J. Am. Chem. Soc.* **2002**, *124*, 7421. (b) Jiang, L.; Job, G. E.; Klapars, A.; Buchwald, S. L. *Org. Lett.* **2003**, *5*, 3667.

18. Copper promoted vinylation of amides has previously been achieved by reaction of alkenyl halides with the potassium salts of amides. See Ogawa, T.; Kiji, T.; Hayami, K.; Suzuki, H. *Chem. Lett.* **1991**, 1443.

19. For reviews of the chemistry of halo acetylenes, see (a) Hopf, H.; Witulski, B. In *Modern Acetylene Chemistry*; Stang, P. J., Diederich, F., Eds.; VCH: Weinheim, 1995; pp 33-66. (b) Brandsma, L. *Synthesis of Acetylenes, Allenes, and Cumulenes, Methods and Techniques*; Elsevier, Ltd.: Oxford, 2004; pp 191-202.

20. Hofmeister, H.; Annen, K.; Laurent, H.; Wiechert, R. *Angew. Chem., Int. Ed. Engl.* **1984**, *23*, 720.

21. Frederick, M. O.; Mulder, J. A.; Tracey, M. R.; Hsung, R. P.; Huang, J.; Kurtz, K. C. M.; Shen, L.; Douglas, C. J. *J. Am. Chem. Soc.* **2003**, *125*, 2368.

22. Hirano, S.; Tanaka, R.; Urabe,H.; Sato, F. *Org. Lett.* **2004**, *5*, 727.

23. Zhang, Y.; Hsung, R. P.; Tracey, M. R.; Kurtz, K. C. M.; Vera, E. L. *Org. Lett.* **2004**, *6*, 1151.

24. Riddell, N.; Villenueve, K.; Tam, W. *Org. Lett.* **2005**, *7*, 3681.

1,3-DICHLOROACETONE AS A CYCLOPROPANONE EQUIVALENT: 5-OXASPIRO[3.4]OCTAN-1-ONE

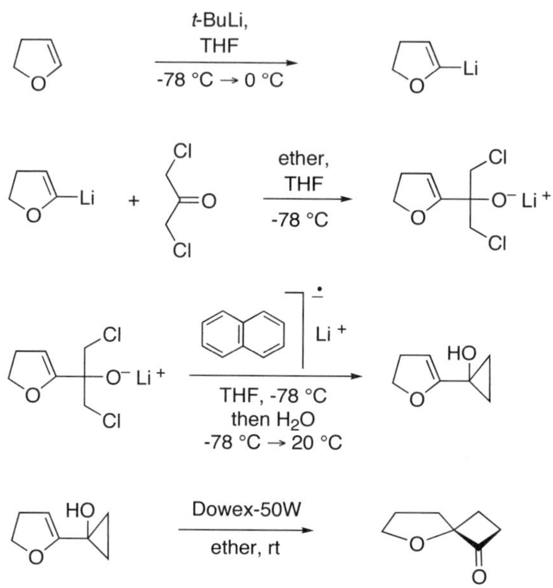

Submitted by David G. Hilmey and Leo A. Paquette.[1a]
Checked by Richard Todd Bibart, Danny G. Lafrance and John A. Ragan.[1b]

1. Procedure

5-Oxaspiro[3.4]octan-1-one. Anhydrous tetrahydrofuran (100 mL) (Note 1) and 2,3-dihydrofuran (7.8 mL, 102 mmol, 1.3 equiv) (Note 2) are added to a single-necked 500-mL round-bottomed flask equipped with a rubber septum and previously purged with nitrogen. The magnetically stirred reaction mixture is cooled to –78 °C and 15 min later treated via cannula with *t*-butyllithium in pentane (1.7 M, 60 mL, 1.3 equiv) (Note 3) over 10 min. The resulting yellow solution is stirred in the cold for 30 min, at which point the acetone/dry ice bath is replaced by an ice-water bath and the reaction mixture is allowed to warm to 0 °C and kept at this temperature for 30 min (Note 4). Concurrently, dry ether (200 mL) and 1,3-dichloroacetone (10.0 g, 78.8 mmol, 1.0 equiv) (Note 5) are placed in a rubber septum-fitted 1-L flask and cooled to –78 °C under nitrogen with stirring (Note 6). The lithiated dihydrofuran solution is next introduced via cannula over 20 min (Note 7). After 2 h at –78 °C, a solution of lithium naphthalenide in tetrahydrofuran (197 mL of 1 M, 2.5 equiv) (Note 8) is introduced via cannula during 20 min to give a dark green solution that is stirred at –78°C for 4–5 h and at room temperature overnight (Note 9). The reaction mixture is then cooled to 0 °C and quenched with saturated aqueous sodium bicarbonate solution (250 mL). The separated aqueous phase is extracted with ether (3 x 100 mL) and the combined organic layers are dried over $MgSO_4$ and filtered. The solution that contains the crude, unstable cyclopropanol is charged with 10 g of Dowex-50W (Note 10) and a stir bar. The solution is stirred overnight. Solvent evaporation is performed with care on a rotary evaporator at 20 °C and 50 mmHg until solvent evaporation ceased. The residue is chromatographed on ~450 grams of silica gel with 5-25% ether in hexanes as eluent (Note 11). The fractions containing the product are carefully concentrated through rotary evaporation as previously described. 5-Oxaspiro[3.4]octan-1-one is obtained (6.9 g, 70%) as a faint yellow oil (Note 12).

2. Notes

1. Anhydrous tetrahydrofuran (fresh Sure-Seal bottle) and ether (fresh 500 mL bottle) were obtained from Aldrich and J. T. Baker, respectively, and used as provided.

2. 2,3-Dihydrofuran (99%+ purity) was obtained from Aldrich and used as provided.

3. *t*-Butyllithium was obtained from Aldrich and used as provided. An atmosphere of nitrogen was maintained in all vessels throughout the reaction.

4. A color change from yellow to almost colorless was witnessed during the warming process.

5. 1,3-Dichloroacetone of 95% purity was obtained from Aldrich and recrystallized from 1:3 ether/hexane, mp 43 °C.

6. 1,3-Dichloroacetone is soluble in ether at room temperature, but crystallizes from this medium at –78 °C. Accordingly, the addition of the lithiated 2,3-dihydrofuran was to a white suspension in ether.

7. The white solid dissolved during the addition leaving a yellow solution. If the solution went turbid again before the addition of lithiated 2,3-dihydrofuran was completed, the product yield was significantly

decreased. The use of a fresh bottle of *t*-butyllithium proved to be quite helpful in eliminating appearance of turbidity and ensuring good conversion to product.

8. Lithium naphthalenide was prepared as follows: To an oven-dried, 500-mL, one-necked, round-bottomed, septum-stoppered flask was added dry tetrahydrofuran (200 mL), naphthalene (25.3 g, 197 mmol), and a teflon-covered stir bar. The flask was purged with argon. To the resulting solution was added lithium wire (1.37 g, 197 mmol containing 0.6-0.8% sodium) in small chunks (3–5 mm long) that had previously been scraped under oil and rinsed in hexanes to remove the oxide coating. The mixture, which developed a deep-green color during 30 min, was allowed to stir overnight and was considered to constitute a 1 M solution.

9. The progress of the reaction was monitored by TLC on glass-backed plates. 1,3-Dichloroacetone was not visible by UV or *p*-anisaldehyde staining. However, the product, which had an R_f value of 0.38 in 3:1 hexanes/ether, stained faintly under these conditions.

10. The ion exchange resin of 100 or 400 mesh was washed in turn with 150 mL of 5% HCl, 100 mL of water, 300 mL of methanol, and 500 mL of ether, then dried overnight under high vacuum (0.1 mmHg) to give dried Dowex resin.

11. When solvent evaporation slowed, the solid that resulted was taken up in toluene (25 mL) and the residual liquid was loaded onto a column of silica gel. With 5% ether in petroleum ether as the eluent, naphthalene eluted in fractions 4-6 (each of 200 mL volume). The desired cyclobutanone eluted in fractions 9-19 following an increase in solvent polarity to 25% ether in hexanes. The purity of the product was quite acceptable at this stage. Attempts at distillation led to decomposition.

12. The analytical properties of the volatile title compound were as follows: R_f = 0.38 in 25% ether in hexanes; IR (neat, cm^{-1}) 2954 (s), 2876(s), 1788 (s), 1058 (s), 1021(s); ^1H NMR (400 MHz, $CDCl_3$) δ: 1.87–1.95 (m, 3 H), 2.01–2.17 (m, 3 H), 2.60–2.77 (m, 2 H), 3.87 (t, J = 6.4 Hz, 2 H); ^{13}C NMR (100 MHz, $CDCl_3$) δ: 25.8, 26.8, 33.9, 39.8, 70.2, 97.0, 211.9; MS (EI) *m/z* 127.06 $(M + H)^+$ calcd 144.05; Anal. Calcd. for $C_7H_{10}O_2$: C, 66.65; H, 7.99. Found: C, 66.24; H, 8.35.

Safety and Waste Disposal Information

All hazardous materials should be handled and disposed of in accordance with "Prudent Practices in the Laboratory"; National Academy Press; Washington, DC, 1995.

3. Discussion

As a consequence of its highly reactive nature, cyclopropanone cannot conveniently be generated in preparative quantities in a laboratory setting.[2,3] Rather, the handling of this energetic compound has most often been made possible by virtue of its ability to undergo conversion to addition products from which the three-membered cyclic ketone may be released under relatively mild conditions. These precursors include the ethyl hemiketal,[4] 1-acetoxycyclopropanol,[5] and select carbinolamines.[6]

With the advent of 1,3-dichloroacetone as an article of commerce, an alternative cyclopropanone equivalent became available, although it has

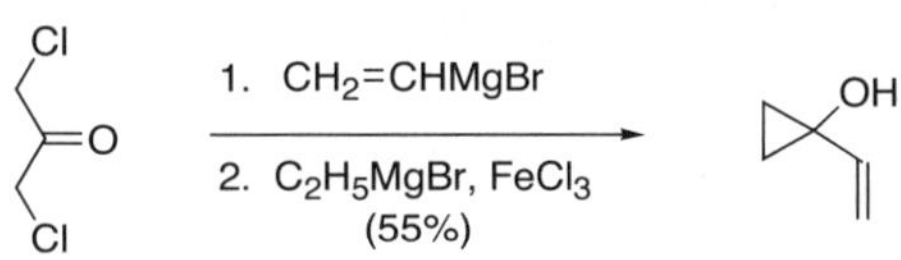

been little recognized as such. Several decades ago, the capacity of the halogenated acyclic ketone to undergo 1,2-addition reactions with Grignard reagents such as vinylmagnesium bromide was recognized.[7] The subsequent introduction of C_2H_5MgBr and a catalytic quantity of $FeCl_3$ delivered 1-vinylcyclopropanol in 55% yield. Barluenga's subsequent contribution was to recognize that the reductive cyclization of the initial Grignard adducts could be carried out as well with anhydrous $MgBr_2$ and lithium powder, or more conveniently with lithium naphthalenide in THF.[8] The second and third steps of the present procedure constitute a modification of the latter protocol. The lithiation of 2,3-dihydrofuran is adapted from the literature,[9] while the final step has its foundation in the many oxonium ion-initiated pinacolic ring expansion reactions explored in this laboratory.[10]

The use of 1,3-dichloroacetone as a cyclpropanone equivalent leads exclusively to 1-substituted cyclopropanols. Consequently, competing processes open to the parent ketone such as decarbonylation,[11] ring opening,[12] polymerization,[5,13] and twofold addition to the carbonyl group[3] do not operate. Advantageously, the present methodology is considered to be scalable, the only limitation being the need to remove naphthalene chromatographically. Finally, the chemistry here described is open to Grignard and organolithium reagents alike, and carries no restrictions as do cyclopropanone hemiacetals.[4]

1. (a) Evans Chemical Laboratories, The Ohio State University, Columbus, Ohio 43210. (b) Pfizer Global Research & Development, Groton, Connecticut 06340.
2. Review: Turro, N. J. *Acc. Chem. Res.* **1969**, *2*, 25.
3. Review: Wasserman, H. H.; Clark, G. M.; Turley, P. C. *Top. Curr. Chem.* **1974**, *47*, 73.
4. (a) Review: Salaün, J. *Chem. Rev.* **1983**, *83*, 619. (b) Salaün, J.; Marguerite, J. *Org. Synth.* **1985**, *63*, 147.
5. Turro, N. J.; Hammond, W. B. *Tetrahedron* **1968**, *24*, 6029.
6. (a) Wasserman, H. H.; Adickes, H. W.; Espejo de Ochoa, O. *J. Am. Chem. Soc.* **1971**, *93*, 5586. (b) Wasserman, H. H.; Glazer, E. A. *J. Org. Chem.* **1975**, *40*, 1505.
7. (a) De Puy, C. H.; Dappen, G. M.; Eilers, K. L.; Klein, R. A. *J. Org. Chem.* **1964**, *29*, 2813. (b) Salaün, J.; Conia, J. M. *Tetrahedron Lett.* **1972**, 2849.
8. Barluenga, J.; Florez, J.; Yus, M. *J. Chem. Soc., Chem. Commun.* **1982**, 1153.
9. (a) Boeckman, R. K., Jr.; Bruza, K. J. *Tetrahedron* **1981**, *37*, 3997. (b) Boeckman, R. K., Jr.; Bruza, K. *Tetrahedron Lett.* **1977**, 4187.
10. (a) Paquette, L. A. *Recent Res. Devel. in Chemical Sciences* **1997**, *1*, 1. (b) Paquette, L. A.; Negri, J. T.; Rogers, R. D. *J. Org. Chem.* **1992**, *57*, 3947. (c) Paquette, L. A.; Andrews, J. F. P.; Vanucci, C.; Lawhorn, D. E.; Negri, J. T.; Rogers, R. D. *J. Org. Chem.* **1992**, *57*, 3956. (d) Paquette, L. A.; Lanter, J. C.; Johnston, J. N. *J. Org. Chem.* **1997**, *62*, 1702. (e) Paquette, L. A.; Hilmey, D. G.; Selvaraj, P. R. *Heterocycles*, **2005**, *57*.
11. (a) Sclove, D. B.; Pazos, J. F.; Camp, R. L.; Greene, F. D. *J. Am. Chem. Soc.* **1970**, *92*, 7488. (b) Breslow, R.; Oda, M. *J. Am. Chem. Soc.* **1972**, *94*, 4787.
12. (a) Turro, N. J.; Hammond, W. B. *J. Am. Chem. Soc.* **1966**, *88*, 3672. (b) Barber, L. L.; Chapman, O. L.; Lassila, J. D. *J. Am. Chem. Soc.* **1969**, *91*, 3664.
13. (a) Schaafsma, S. E.; Steinberg, H.; de Boer, T. J. *Rec. Trav. Chim.* **1967**, *86*, 651. (b) Schaafsma, S. E.; Steinberg, H.; de Boer, T. J. *Rec. Trav. Chim.* **1966**, *85*, 1170.

COPPER(I) IODIDE DIMETHYL SULFIDE CATALYZED ADDITION OF A VINYLZIRCONIUM REAGENT. PREPARATION OF 4-PHENYL-5(E)-DECEN-2-ONE

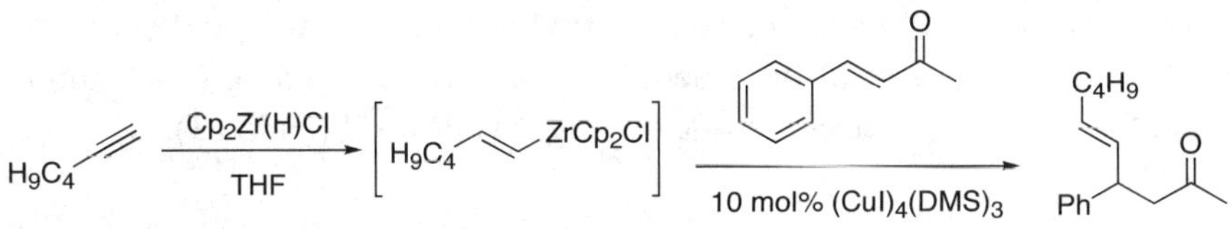

Submitted by Amer El-Batta and Mikael Bergdahl.[1]
Checked by Monica Trincado, Yazmin Rosa-Bauza, and Jonathan A. Ellman.

1. Procedure

An oven-dried 250-mL, two-necked, round-bottomed flask equipped with a magnetic stirring bar, rubber septum and a water condenser provided with an argon atmosphere inlet, is charged with Schwartz's reagent (9.36 g, 36.3 mmol) (Note 1) by temporarily removing the septum. The flask is immersed in an ice/water cooling bath equipped with an external thermometer (Note 2), and placed on a magnetic stir plate. Under an atmosphere of argon, 25.0 mL of tetrahydrofuran (Note 3) and 1-hexyne (3.71 mL, 33.0 mmol) (Note 4) are added to the flask by syringe through the septum. After stirring for 5 min, the ice/water cooling bath is replaced by an oil bath equipped with an external thermometer. Stirring is maintained and the reaction flask is heated to 40 °C for 6 h. The resulting yellow solution is cooled to room temperature (Note 5), *trans*-4-phenyl-3-buten-2-one (4.82 g, 33.0 mmol) is added (Note 6) and 5 min later copper(I) iodide-dimethyl sulfide complex (782 mg, 3.30 mmol) (Note 7) is added by temporarily removing the septum. The reaction temperature is increased to 40 °C, and the stirring is continued under an argon atmosphere for 12 h. The heating is interrupted and the oil bath is replaced by an ice/water bath. After removing the argon inlet and the water condenser, the reaction mixture is quenched with ether saturated with water (50 mL) at the ice/water bath temperature. The resulting heterogeneous mixture is stirred for 10 min and filtered in *vacuo* through a Celite® pad (Note 8) in a Büchner funnel. The Celite is carefully washed with 3 x 100 mL portions of ether, and the resulting ethereal phase is transferred to a 500-mL separatory funnel. The organic layer is washed with 2 x 20 mL of saturated sodium bicarbonate solution, dried over sodium sulfate (Note 9), filtered through a second Celite pad in a Büchner funnel, and concentrated under reduced pressure. The resulting crude product is purified on a silica gel 60Å chromatography column (38 cm x 4 cm) (Note 10) using a hexanes-ethyl acetate 1:0-10:1 gradient (Note 11) (Note 12). Removal of the eluent in *vacuo* afforded 5.55 g (24.1 mmol, 73%) of 4-phenyl-5(*E*)-decen-2-one as a light yellow oil (Note 13).

2. Notes

1. The submitters prepared bis(cyclopentadienyl)zirconium chloride hydride by lithium aluminum hydride reduction of bis(cyclopentadienyl)zirconium dichloride employing the procedure of Buchwald and his coworkers.[2] The checkers purchased Schwartz's reagent from Strem Chemicals, Inc. (Cat. No. 40-1040), and it was used as received.

2. The reaction has not been conducted on a larger scale than 36.3 mmol. As the reaction was mildly exothermic,[3] use of an ice/water bath is recommended on a larger scale.

3. Tetrahydrofuran was freshly distilled from sodium/ benzophenone under an argon atmosphere (submitters) or dried through an alumina column (checkers).

4. 1-Hexyne was purchased from Aldrich Chemical Company, Inc. (Cat. No. 24,442-2) and was used as received. The alkyne was slowly added (~1 mL/min) using a gas tight syringe.

5. The flask was placed in a water bath and the mixture was stirred for 10 min.

6. Benzalacetone was purchased from Aldrich Chemical Company, Inc. (Cat. No. 24,109-1) and was used as received.

7. Copper(I) iodide (submitters used 98%, checkers used 99.999%), commercially available from Aldrich Chemical Company, Inc., was purified via its dimethyl sulfide complex following the House protocol.[4] Copper(I) iodide (5.0 g, 26 mmol) in 15 mL of anhydrous ether (ACS grade) was mixed with 20 mL (380 mmol) of dimethyl sulfide (Aldrich Chemical Company, Inc., Cat. No. 47,157-7). The resulting yellow solution was filtered through a cotton plug to remove heterogeneous material. The filtrate was diluted with 50 mL of cold hexanes. The crystalline material obtained was filtered with suction and washed with cold hexanes (3 x 50 mL) to afford 5.1 g of the copper complex as white prisms. The CuI/DMS complex was subsequently dried under vacuum (ca. 0.1 mbar) at room temperature

where it rapidly lost DMS and reached a constant weight within two hours to afford 4.5 of white crystals. Elemental analysis of this material indicated a stoichiometry of CuI/0.75 DMS. Anal. Calcd. for $C_6H_{18}Cu_4I_4S_3$: C, 7.60; H, 1.91; S, 10.15. Found: C, 7.53; H, 1.72; S, 10.13.

8. Celite 545® was purchased from Fisher Scientific Co. (Cat. No. C212-500).

9. Sodium sulfate was purchased from EMD Chemicals, Inc. (Cat. No. SX0763-3).

10. Silica gel was purchased from Whatman (Cat. No. 4745-250), 230-400 Mesh.

11. Ethyl acetate and hexanes were purchased from Fisher Scientific, ACS reagent grade. The submitters applied flash-chromatography conditions with 10% ethyl acetate in hexanes.

12. The product had a TLC R_f value of 0.47 [ethyl acetate: hexanes, 1:10].

13. The submitters reported a yield of 6.23 g (82%). Full characterization of the product was as follows: ^{1}H NMR (400 MHz, $CDCl_3$) δ: 0.86 (t, J = 6.8 Hz, 3 H), 1.26–1.42 (m, 4 H), 2.03–2.09 (m, 2 H), 2.09 (s, 3 H), 2.85 (dd, J = 7.6, 7.2 Hz, 2 H), 3.91 (q, J = 7.2 Hz, 1 H), 5.53 (dtd, J = 15.2, 7.2, 0.8 Hz, 1 H), 5.62 (ddt, J = 15.2, 7.2, 1.2 Hz, 1 H), 7.20–7.35 (m, 5 H); ^{13}C NMR (100 MHz, $CDCl_3$) δ: 13.9 (CH_3), 22.3 (CH_2), 30.6 (CH_3), 31.6 (CH_2), 32.3 (CH_2), 44.0 (CH), 49.8 (CH_2), 126.4 (CH), 127.5 (2xCH), 128.6 (2xCH), 130.9 (CH), 132.3 (CH), 143.8 (C), 207.0 (CO); IR (thin film) cm^{-1}: 1716, 1602, 1494, 1453, 968, 699; HRMS (EI) *m/z* Calcd for $C_{16}H_{22}O$: 230.1671. Found: 230.1671. The purity of the product was determined to be >98% by ^{1}H NMR and elemental analysis. Anal. Calcd. for $C_{16}H_{22}O$: C, 83.43; H, 9.63. Found: C, 83.46; H, 9.72.

Safety and Waste Disposal Information

All hazardous materials should be handled and disposed of in accordance with "Prudent Practices in the Laboratory"; National Academy Press; Washington, DC, 1995.

3. Discussion

The CuI•0.75DMS complex is an excellent catalyst for the direct conjugate addition of alkenyl groups from vinylzirconocene reagents to α,β-unsaturated aldehydes and ketones.[5] The presence of the catalyst with an alkenylzirconocene, at 40 °C in THF, circumvents the need for making discrete alkenylcopper reagents. The catalyst is superior in terms of product yields and alkene flexibility in comparison to other copper(I) sources as well as the nickel(II)[6] catalyzed conjugate addition. This simple one-pot procedure shows that one equivalent of the vinylzirconocene relative to the enone is needed to afford a high yield of the 1,4-addition of the alkenyl group. Some representative examples of the present reaction are shown in Table 1.

Copper(I) catalyzed conjugate additions of vinylzirconocenes, using the rather unappreciated CuI·DMS complex, circumvent problems associated with the formation of dienes. Our preliminary studies show that the dimerization of the alkenyl groups is minimal as long as freshly prepared $(CuI)_4(Me_2S)_3$ complex[7,8] is employed. A vinylzirconocene reagent, in the presence of $(CuI)_4(Me_2S)_3$, is an extremely mild and considerably less basic system than the lithium alkenyl compounds or the corresponding discrete vinylcuprate reagents commonly used in 1,4-addition reactions.

Table 1. Conjugate additions of vinyl groups.

Enone	Alkyne[a]	Cond.[b]	Product	Yield[c]
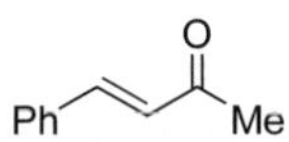		12 h	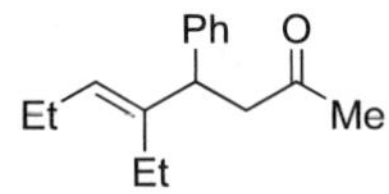	86[d]
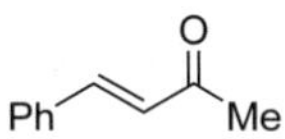	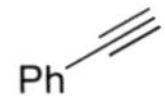	12 h	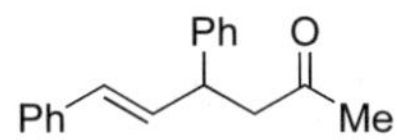	79[d]
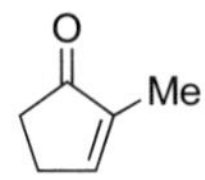	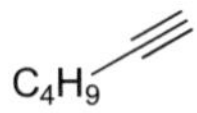	5 h	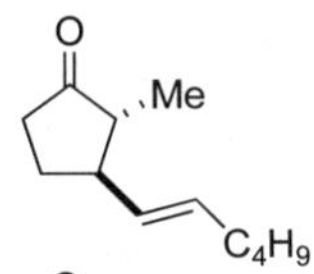	54[d] 87[e]
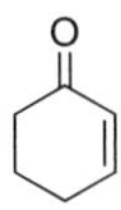	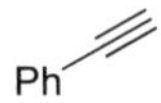	5 h	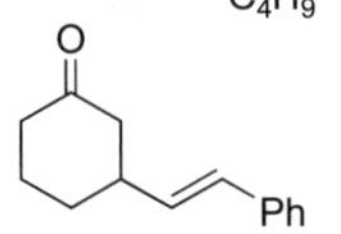	78[d]
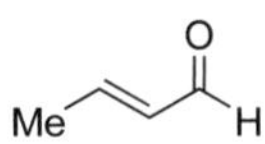	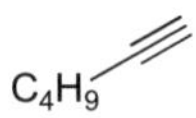	2 h	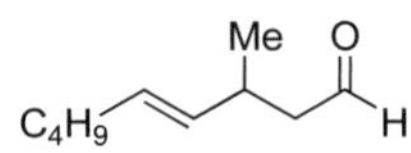	90[d]
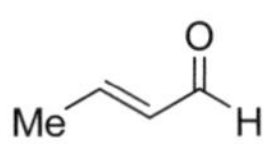	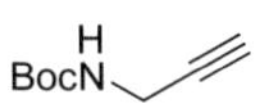	2 h	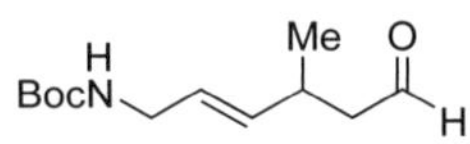	88[d]
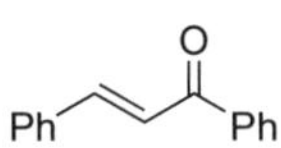	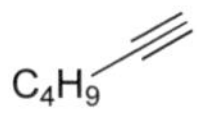	10 h	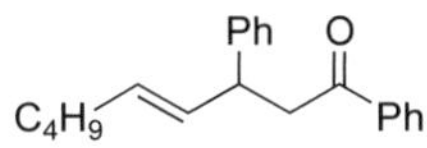	84[d]
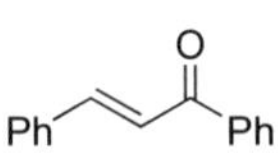	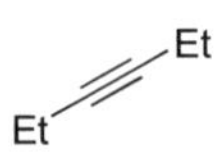	10 h	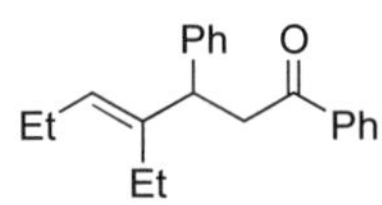	71[d]

[a] 1.0 Equiv alkyne *vs*. enone. [b] 40 °C. [c] Based on isolated and purified material (%). [d] 10 mol% CuI•0.75DMS *vs*. alkyne [e] 100 mol% CuI•0.75DMS, *trans-cis* ratio, 20:1.

1. San Diego State University, Department of Chemistry and Biochemistry, San Diego, CA 92182-1030.
2. Buchwald, S. L.; LaMaire, S. J.; Nielsen, R. B.; Watson, B. T.; King, S. M. *Org. Synth., Coll. Vol. IX* **1998**, 162–165. See also: Buchwald, S. L.; LaMaire, S. J.; Nielsen, R. B.; Watson, B. T.; King, S. M. *Tetrahedron Lett.* **1987**, *28*, 3895–3898.

3. Sun, C. R.; Okabe, M.; Coffen, D. L.; Schwartz, J. *Org. Synth., Coll. Vol. IX* **1998**, 640–643.
4. House, H. O.; Chu, C.-Y.; Wilkins, J. M.; Umen, M. J. *J. Org. Chem.* **1975**, *40*, 1460–1469.
5. El-Batta, A.; Hage, T. R.; Plotkin, S.; Bergdahl, M. *Org. Lett.* **2004**, *6*, 107–110.
6. Loots, M. J.; Schwartz, J. *J. Am. Chem. Soc.* **1977**, *99*, 8045–8046.
7. Eriksson, M.; Hjelmencrantz, A.; Nilsson, M.; Olsson, T. *Tetrahedron* **1995**, *51*, 12631–12644.
8. Eriksson, M.; Iliefski, T.; Nilsson, M.; Olsson, T. *J. Org. Chem.* **1997**, *62*, 182–187.

ORTHO-FORMYLATION OF PHENOLS; PREPARATION OF 3-BROMOSALICYLALDEHYDE

(3-Bromo-2-hydroxybenzaldehyde)

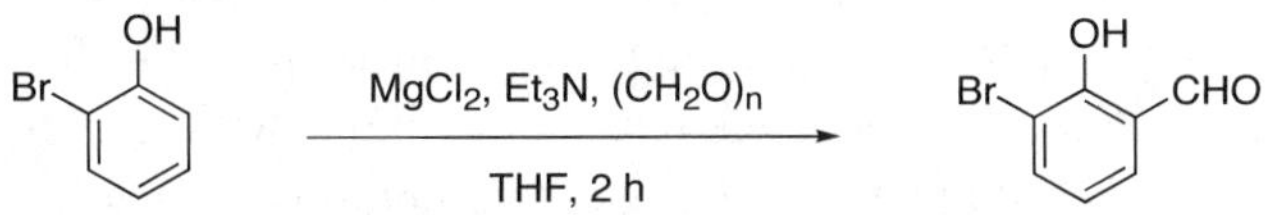

Submitted by Trond Vidar Hansen[1] and Lars Skattebøl.[2]
Checked by David Guthrie and Dennis P. Curran.

1. Procedure

A dry 500-mL, three-necked round-bottomed flask equipped with a stirring bar, reflux condenser and rubber septa is purged with argon gas. Anhydrous magnesium dichloride (Notes 1 and 2) (9.52 g, 100 mmol) and solid paraformaldehyde (Notes 3 and 4) (4.50 g, 150 mmol) are added, while a slight positive pressure of argon is maintained. Dry tetrahydrofuran (Note 5) (250 mL) is added by syringe. Triethylamine (Notes 6 and 7) (10.12 g, 100 mmol) is added dropwise by syringe and the mixture is stirred for 10 min. 2-Bromophenol (Note 8) (8.65 g, 50 mmol) is added dropwise by syringe, resulting in an opaque, light pink mixture. This mixture is immersed in an oil bath at about 75 °C (bath temperature) and soon turns a bright orange-yellow color. Heating at gentle reflux is maintained for 4 hr (Note 9).

The reaction mixture is cooled to room temperature and 100 mL of ether is added. The resulting organic phase is transferred to a 1-L separatory funnel and washed successively with 1N HCl (3 x 100 mL) (*Caution! gas evolves*) and water (3 x 100 mL) (Note 10), dried over anhydrous magnesium sulfate ($MgSO_4$), and filtered. The solvent is removed by rotatory evaporation leaving a pale yellow oil that solidifies on further vacuum drying at 1-2 mmHg. The resulting yellow solid (8.09-8.11 g, 80-81%) consisting mainly (≥95 %) of 3-bromosalicylaldehyde (Note 11) is sufficiently pure for further synthetic use. Recrystallization from hexane (50 mL) gives 6.80-6.94 g (68-69%, Note 12) of pure aldehyde as pale yellow needles (Note 13).

2. Notes

1. Magnesium dichloride, anhydrous beads, –10 mesh, 99.9% was purchased from the Aldrich Chemical Company, Inc. The submitters dried the beads over phosphorus pentoxide under reduced pressure for 24 hr prior to use. The checkers purchased the beads in ampules with less than 100 ppm water and used these directly. The use of anhydrous beads is crucial; the checkers observed little or no reaction when using anhydrous magnesium chloride powder dried over phosphorous pentoxide.

2. The submitters report that the reaction proceeds at a slower rate with less than two equivalents of magnesium dichloride and gives a lower yield of aldehyde.

3. Paraformaldehyde powder was purchased from Aldrich Chemical Company, Inc. and dried over phosphorus pentoxide under reduced pressure for 24 hr prior to use.

4. The reaction requires two equivalents of paraformaldehyde; however, the submitters report that an excess of this reagent results in a faster reaction and higher yield of aldehyde.

5. Tetrahydrofuran was purchased from Aldrich Chemical Company, Inc. The submitters emphasize the use of dry solvent in this reaction and the THF was distilled from sodium benzophenone ketyl prior to use. The checkers passed the THF through a column of dry, activated aluminum oxide.

6. Triethylamine was purchased from Aldrich Chemical Company, Inc., distilled from calcium hydride and stored over 4Å molecular sieves prior to use.

7. Equimolar amounts of triethylamine and magnesium dichloride are required.

8. 2-Bromophenol was purchased from Aldrich Chemical Company, Inc. and used as received.

9. Progress of the reaction was followed by thin layer chromatography using Merck silica gel 60 F254 aluminum-backed plates, eluting with hexane/EtOAc (10:1) and visualizing with a 254 nm UV lamp: 2-bromophenol $Rf \approx 0.58$; 3-bromsalicylaldehyde $Rf \approx 0.45$.

10. Vigorous shaking of the separatory funnel should be avoided since emulsions can result. The mixture is swirled and tilted back and forth gently for 3-4 min per washing. This results in rapid separation of the layers on standing.

11. The main impurity is the starting bromophenol. The product can be analyzed by GLC on a Varian GC 3300 instrument equipped with a 25 m SP2100 capillary column or by ^{1}H NMR spectroscopy.

12. The submitters obtained 9.05 g (90 %) of the recrystallized product.

13. 3-Bromosalicylaldehyde exhibited the following physical and spectral properties: mp (uncorr.) 52-53 °C (lit.[3] 52 °C); ^{1}H NMR (300 MHz, $CDCl_3$) δ: 6.94 (t, J = 7.8 Hz), 7.55 (dd, J = 1.3 Hz, 7.7 Hz), 7.75 (dd, J = 1.3 Hz, 7.7 Hz), 9.85 (s, 1H), 11.60 (s, 1H); ^{13}C NMR (75 MHz, $CDCl_3$) δ: 111.0, 120.7, 121.2, 132.9, 139.9, 157.9, 196.0; FTIR spectrum ($CHCl_3$) cm^{-1}: 3152, 2852, 1662, 906; LRMS (EI), *m/z* 202/200 (M^+), 159, 149, 117, 115, 71, 57, 55; HRMS, Calcd. for $C_7H_5BrO_2$: 199.9473; Found: 199.9475.

Safety and Waste Disposal Information

All hazardous materials should be handled and disposed of in accordance with "Prudent Practices in the Laboratory"; National Academy Press; Washington, DC, 1995.

3. Discussion

Surprisingly few substituted salicylaldehydes are commercially available even though they are important intermediates in organic synthesis for the preparation of a variety of oxygen-containing heterocyclic compounds and as a source for salen ligands.[5] Salicylaldehydes are accessible from the corresponding phenols by several classical formylation reactions;[4] however, the yields are often moderate and the lack of regioselectivity is problematic.

The present procedure is a simple, efficient and regioselective method for the preparation of substituted salicylaldehydes that is based on the work reported by Hofsløkken and Skattebøl.[6] It is applicable for large-scale preparations; on a 0.25 molar scale, 3-bromosalicylaldehyde was obtained in 90% yield. The method gives exclusively *ortho*-formylation of phenols and naphthols; no bis-formylation has been observed. Tetrahydrofuran may be replaced by acetonitrile with a negligible effect on yields. Reactions usually require 2-4 hours for completion. Electron releasing substituents enhance the reaction rate while the opposite effect is observed for electron withdrawing substituents. Prolonged reaction times increase the amounts of byproducts, particularly formation of the corresponding 2-methoxymethylphenol derivatives.[6]

TABLE
ORTHO FORMYLATION OF PHENOLS AND NAPHTHOLS WITH $MgCl_2$/PARAFORMALDEHYDE

Phenol	Salicylaldehyde	Yield[Ref]
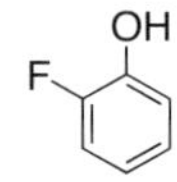	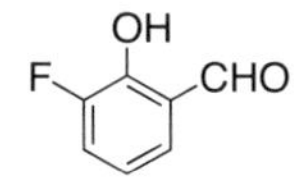	80%[7]
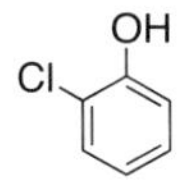	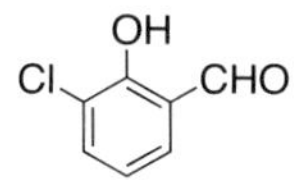	87%[6]
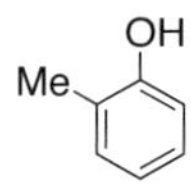	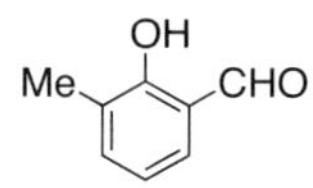	99%[6]
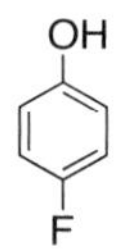	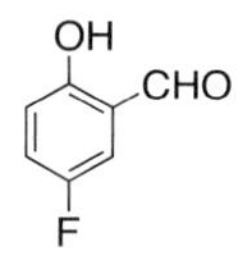	85%[7]
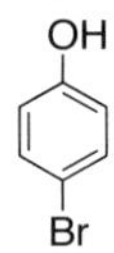	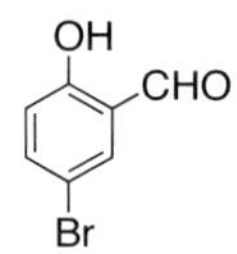	91%[7]
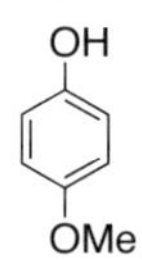	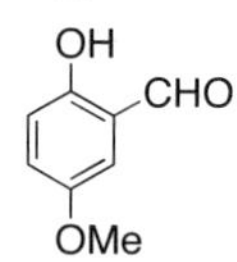	87%[6]
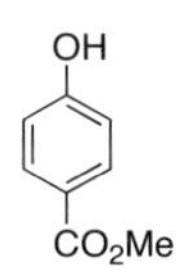	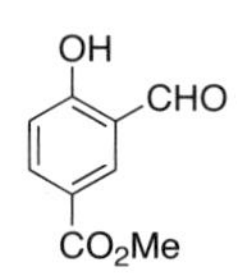	88%[6]
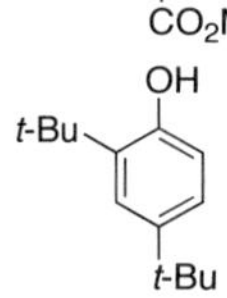	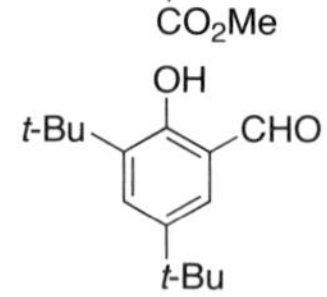	84%[8]
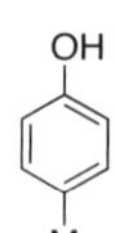	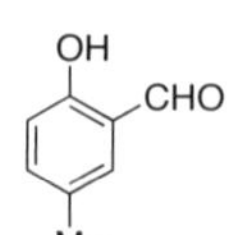	90%[6]

CHO

Some representative examples of salicylaldehydes prepared by this method are compiled in the Table, and a recent reference presents further examples.[9]

1. Department of Medicinal Chemistry, School of Pharmacy, University of Oslo, POB 1185, Blindern, N-0316 Oslo, Norway.
2. Department of Chemistry, University of Oslo, POB 1033, Blindern, N-0315 Oslo, Norway.
3. Müller, J. *Ber.* **1909**, *42*, 1166.
4. For a review, see: Laird, T. in *Comprehensive Organic Chemistry*, Stoddart, J. F., Ed., Pergamon, Oxford 1979, Vol. 1, p 1105.
5. Larrow, J. F.; Jacobsen, E. N.; Gao, Y.; Hong, Y., Nie; X. Zepp, C. M. *J. Org. Chem.* **1994**, *59*, 1939.
6. Hofsløkken, N. U.; Skattebøl, L. *Acta Chem. Scand.* **1999**, *53*, 258.
7. Hansen, T. V.; Skattebøl, L., *unpublished work.*
8. Hansen, T. V.; Skattebøl, L. *Tetrahedron Lett.* **2005**, *46*, 3829.
9. Doshi, J. M. ; Tian, D.; Xing, C. *J. Med. Chem.* **2006**, *49*, 7731.

HIGHLY EFFICIENT ADDITION OF ACTIVATED METHYLENE COMPOUNDS TO ALKENES CATALYZED BY GOLD AND SILVER

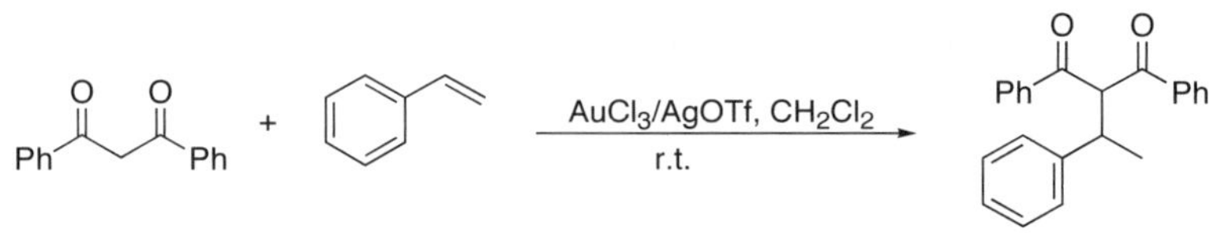

1

Submitted by Chao-Jun Li[1] and Xiaoquan Yao.
Checked by Brian P. Jones and John A. Ragan.

1. Procedure

1,3-Diphenyl-2-(1-phenylethyl)propane-1,3-dione. An oven-dried, 300-mL, three-necked, round-bottomed flask equipped with a nitrogen inlet adapter, a rubber septum, a stopcock and a Teflon-coated magnetic stir bar is cooled to room temperature under a flow of nitrogen. While temporarily removing the septum, the flask is charged with gold(III) chloride ($AuCl_3$) (0.303 g, 1 mmol, 5 mol% Au) and silver trifluoromethanesulfonate (AgOTf) (0.77 g, 3 mmol, 15 mol%) (Note 1) under N_2. Dichloromethane (150 mL) (Note 2) is added, and the catalyst mixture is stirred at room temperature for 1.5–2 h, resulting in a brown suspension (Note 3). Dibenzoylmethane (4.48 g, 20 mmol) (Note 4) is added to the reaction flask over 2 min while temporarily removing the septum and the color of the reaction mixture turns green. Styrene (3.45 mL, 30 mmol) (Notes 5, 6) is diluted to 25 mL with CH_2Cl_2 and added dropwise to the reaction mixture by syringe pump through the septum over 6–7 h at room temperature (Note 7). The mixture is then stirred for another 30 min (Note 8). A deep-green solution is obtained (Note 9). The solution is filtered through approximately 40 mL of silica gel in a 60-mL glass Büchner filter funnel to remove the catalyst (Note 10). The silica gel is washed with ether (6 x 50 mL), and the filtrate is concentrated by rotary evaporation at room temperature. The residue, a clear, orange-brown oil, is then purified by flash column chromatography (Note 11) to provide the crude product as a pale yellow solid (6.44 g, 98 %) (Note 12). Further purification is achieved by recrystallization. The crude product is dissolved in 30 mL of hot EtOAc, and diluted with 90 mL of hexanes to the point of cloudiness. The mixture is allowed to cool slowly to room temperature overnight, and the resulting solids are collected by filtration, rinsing with 10 mL of hexanes. The solids are air-dried, then placed in a vacuum oven for 4 h to provide analytically pure product as a white solid (4.16 g, 12.7 mmol, 63%) (Notes 13, 14).

2. Notes

1. Gold(III) chloride (99.99%+) and silver trifluoromethanesulfonate (99%+) were purchased from the Aldrich Chemical Company and used as received.

2. Dichloromethane was purchased from J. T. Baker (HPLC grade) and used as received. The submitters distilled their dichloromethane from CaH_2.

3. Silver trifluoromethanesulfonate was stirred with gold chloride in dichloromethane, which results in a brown (in some cases orange) precipitate (the clear solution was almost colorless). The clear solution had no catalytic activity.

4. Dibenzoylmethane (98%) was purchased from the Aldrich Chemical Company and used as received.

5. Styrene (99%, inhibited with 10–15 ppm 4-*tert*-butylcatechol) was purchased from the Aldrich Chemical Company and used as received.

6. The main side reaction was the dimerization of styrene, resulting in the formation of 1,3-diphenyl-but-1-ene as the by-product. In order to increase the yield of the major product, excess styrene had to be used. When enlarging the scale of the reaction, the dimerization of styrene became more serious. Therefore, in this case, 1.5 to 2 equiv. of styrene was required.

7. A competition exists between the addition of dibenzoylmethane to styrene and the dimerization of styrene, therefore introduction of styrene in one portion or too quickly resulted in a decreased yield.

8. The submitters stirred the reaction for an additional 4 h. The checkers found that a post-addition stir time ranging from 30 min to 16 h had no impact on product yield or quality.

9. The product was detected by TLC (hexanes/EtOAc = 10/1). R_f (dimer of styrene): 0.91; R_f (diketone): 0.49; R_f (product): 0.29.

10. Length: 3.5 cm diameter × 4 cm. If the catalyst was not removed before evaporation, the product tended to decompose to the diketone and the dimer of styrene. In a control experiment, when the catalyst was not removed before evaporation (bath temperature of the rotary evaporator: 50 °C), a lower yield was obtained (4.14 g, 63% yield before recrystallization).

11. The oil was diluted with 50 mL of CH_2Cl_2, and 13 g of SiO_2 (Baker, 40 μm, flash chromatography grade) was added. The resulting slurry was concentrated on the rotary evaporator to a tan powder, which was charged to the top of a column prepared from flash silica gel slurried in 2% EtOAc-hexanes (column dimensions 3.5 cm diameter x 18 cm height). Elution with 2% EtOAc-hexanes (1.5 L, removed unreacted dibenzoylmethane) followed by 10% EtOAc-hexanes (6 L) afforded the product in fractions 19-74 (75 mL fractions were collected). The product-containing fractions were combined and concentrated to provide the crude product.

12. The product at this stage was probably suitable for most subsequent applications (ca. 90% purity), but two impurities were observed by ^{1}H NMR and LC/MS and the product failed elemental analysis. LC/MS showed the impurities to be present at ca. 5% each. The impurities appeared to be the two regioisomers (each a mixture of two diastereomers) generated from the addition of dibenzoylmethane to the styrene dimer ($(PhCO)_2CHCH(Ph)CH_2CH(Ph)Me$ and $(PhCO)_2CHCH(CH_2Ph)CH$-(Ph)Me): $C_{31}H_{28}O_2$ requires *m/z* 432, M+Na = 455 observed.

13. The submitters recrystallized a sample from 10:1 hexanes-CH_2Cl_2 for elemental analysis and melting point.

14. IR (thin film): ν_{max} 3062, 3028, 2968, 1693, 1662, 1595, 1579, 1494, 1447, 1321, 1264, 1215, 1194, 1180, 979, 907, 756, 732, 686 cm^{-1}; ^{1}H NMR ($CDCl_3$, 400 MHz) δ: 1.34 (d, *J* = 7.0 Hz, 3 H), 4.07 (dq, *J* = 10.0, 7.0 Hz, 1 H), 5.60 (d, *J* = 10.0 Hz, 1 H), 7.05–7.57 (m, 11 H), 7.73 (d, *J* = 7.5 Hz, 2 H), 8.03 (d, *J* = 7.5 Hz, 2 H); ^{13}C NMR ($CDCl_3$, 100 MHz) δ: 20.1, 41.1, 64.4, 126.5, 127.6, 128.3, 128.3, 128.6, 128.7, 132.9, 133.4, 136.7, 137.0, 143.6, 194.5, 194.8. mp 123–124 °C (lit. 126–127 °C).[2] Anal. Calcd for $C_{23}H_{20}O_2$: C, 84.12; H, 6.14; Found: C, 84.10; H, 6.06.

Safety and Waste Disposal Information

All hazardous materials should be handled and disposed of in accordance with "Prudent Practices in the Laboratory"; National Academy Press; Washington, DC, 1995.

3. Discussion

As one of the most common methodologies for the formation of carbon-carbon bonds, the alkylation of 1,3-dicarbonyl compounds usually requires the use of a stoichiometric amount of base and an organic halide. An alternative reaction via transition metal-catalyzed addition of 1,3-dicarbonyl compounds to alkenes would provide a more atom-economical approach and has attracted much interest in recent years.

Recently, Widenhoefer and co-workers reported an elegant intramolecular hydroalkylation of alkenes by carbonyl compounds catalyzed by palladium.[3] Compared with intramolecular addition, intermolecular hydroalkylations of alkenes involving such activated methylene C-H bonds were rarely reported. Widenhoefer reported a platinum- or palladium-catalyzed intermolecular addition of ethylene with β-diketones.[4] Hartwig also reported a palladium-catalyzed addition of mono- and dicarbonyl compounds to conjugated dienes.[5]

Using the addition of 2,4-pentanedione to styrene as a prototype, we found:[6] (1) AgOTf or $AuCl_3$ provided only trace amounts of the desired product, (2) the combination of $AuCl_3$ and AgOTf provided the desired product in good yields, (3) changing the counter ion of the silver salt from OTf lowered the yield, but no difference was observed when the counter ion of gold was changed, and (4) CH_2Cl_2 was the solvent of choice relative to DCE and nitromethane. It should be noted that since the catalytic system is very strongly Lewis acidic, THF was not stable in the solution of the gold/silver catalyst. THF decomposed almost immediately, and a semi-solid was obtained in the tube.

A side reaction observed in this system was the polymerization of styrene, in which the dimerization promoted by metal triflate species was found to be the main reaction and gave 1,3-diphenyl-but-1-ene as the main by-product.[7] Since styrene readily dimerized in the reaction, the use of catalytic $AuCl_3$/AgOTf in CH_2Cl_2 with an excess of styrene, which was diluted and slowly added to the reaction, provided the best yield. It should be mentioned that dimerization became more serious with an increase in the reaction scale, especially for the less reactive 1,3-diketone substrates, such as 2,4-pentandione. In these cases, increasing the excess of styrene and reaction temperature was beneficial. Alternatively, use of a more diluted solution and slower addition helped to inhibit the dimerization of styrene. The results of various 1,3-diketones and styrenes are summarized in Table 1.

Since gold(III) chloride is expensive, decreasing the amount of catalyst was also attempted. For the less active diketones, such as 2,4-pentandione, reduction of the amount of gold(III) resulted in a decrease in yield. However, for the procedure described above, there was only ca. 3% decrease in yield.

This reaction was reversible at higher temperatures when catalyzed by gold or silver catalyst.[8] Dibenzoylmethane derivatives were most prone to the reverse reaction, which gave the diketone and the styrene dimer. In small scale reactions (1 mmol or less), the reversibility was not a serious problem, since it was not necessary to remove CH_2Cl_2 before separation. However, for larger scale reactions, the catalyst had to be removed by a short silica gel column before concentration, especially for the dibenzoylmethane derivatives (although it was not necessary for the 2,4-pentanedione derivatives).

Under the same reaction conditions, various cyclic dienes and cyclic enol ethers were reacted with dibenzoylmethane and 1-benzoylacetone (Table 2).[9] Addition to a triene (entry 2) was also selective in the generation of a diene product. The reaction had to be carried out at low temperature in some cases in order to prevent polymerization or dimerization of the diene (entries 3 and 4). The use of cyclic alkenes bearing a heteroatom (entries 5-8) resulted in selective product formation. However, addition of dibenzoylmethane to an ether-substituted DHP (entries 6 and 7) led to a significant amount of bis-alkylation products. The use of simple cyclic and acyclic alkenes led to very low conversions, while use of acyclic conjugated dienes resulted in the formation of a complicated unidentified mixture. The use of sterically hindered dienes or less activated methylenes (such as 2,4-pentadione and cyclic diketones) also led to low conversions under the present conditions.

Table 1 Addition of β-Diketone to Alkenes Catalyzed by Gold Catalyst[a]

Entry	β–Diketone	Alkenes[b]	Product	Yield (%)[c]
1	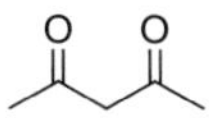	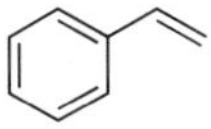	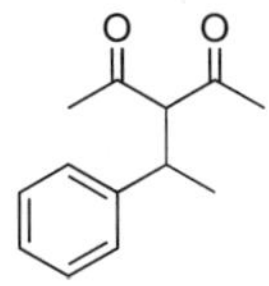	89[d]
2	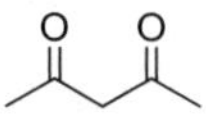	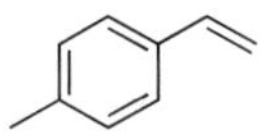	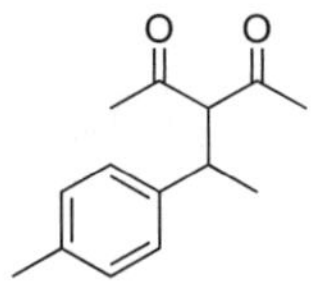	94
3	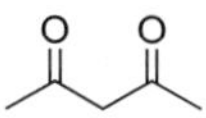	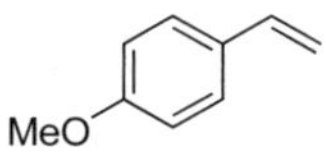	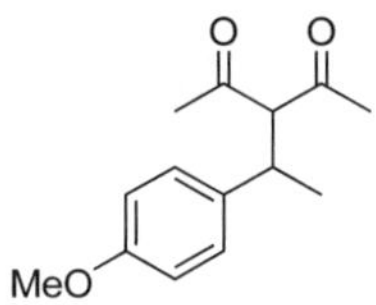	62
4	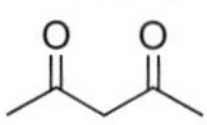	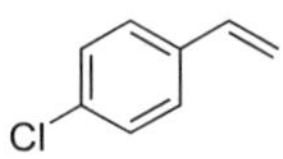	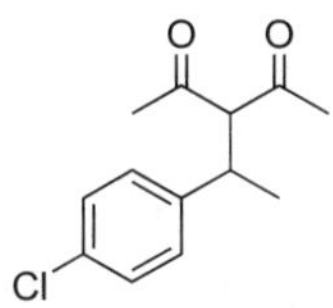	97
5[e]	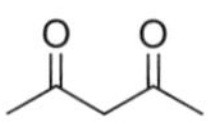	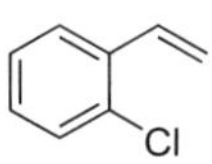	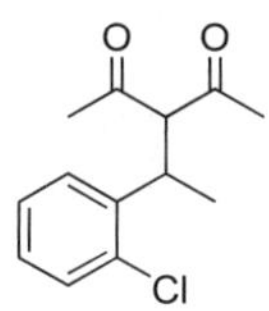	50
6	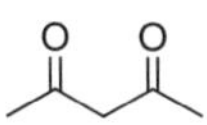	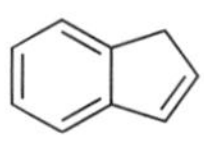	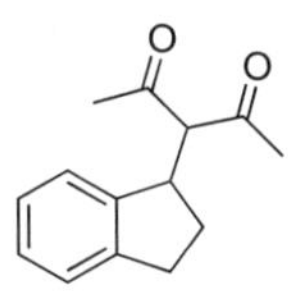	44
7	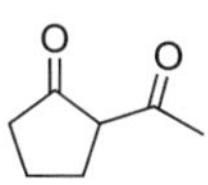	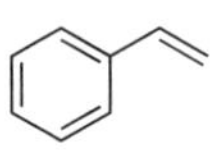	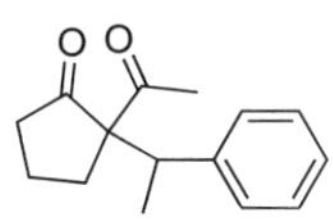	70 (1:1)[f]
8	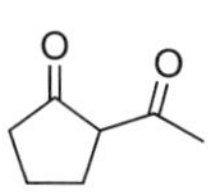	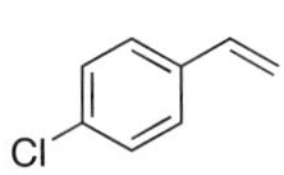	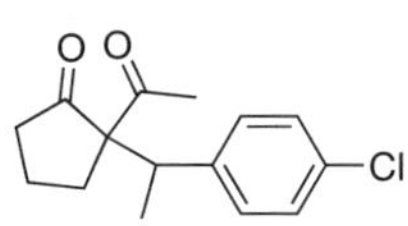	55 (1:1)[f]

Table 1 (continued)

Entry	β–Diketone	Alkenes[b]	Product	Yield (%)[c]
9	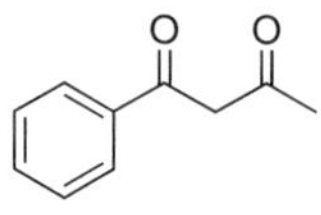	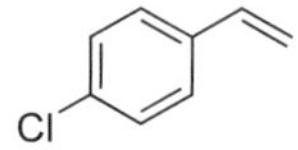	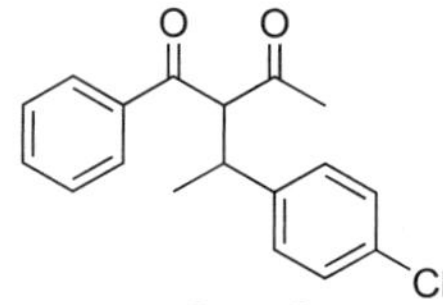	89 (1:1)[f]
10	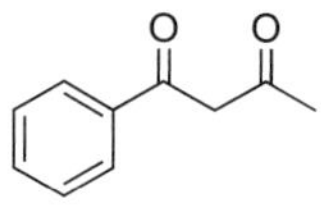	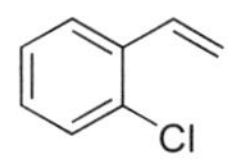	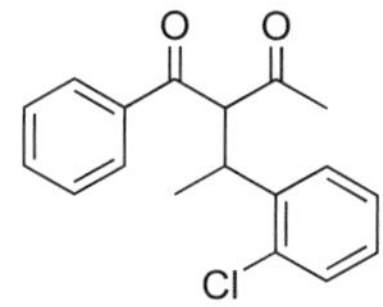	67 (1:1)[f]
11	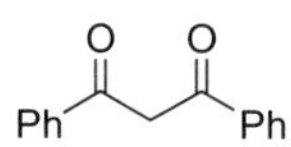	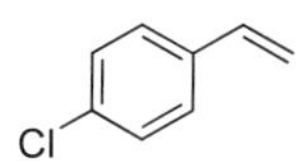	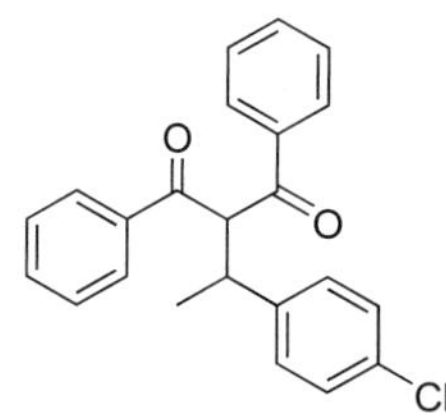	98
12	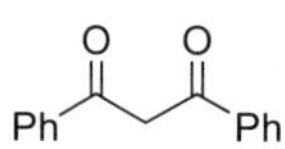	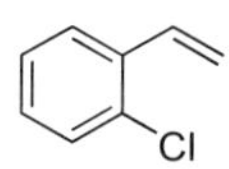	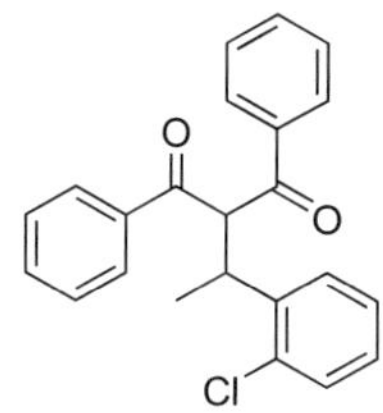	91
13	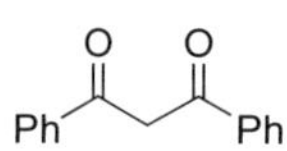	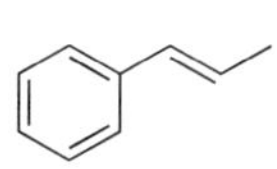	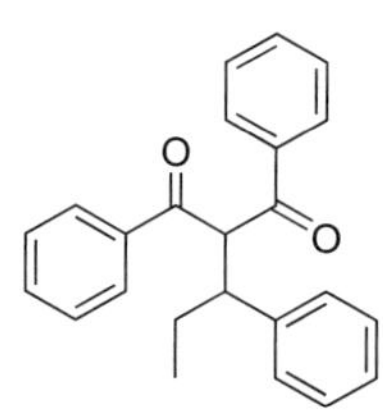	39[g]
14	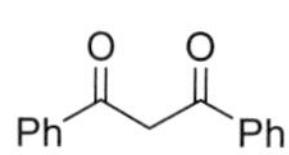		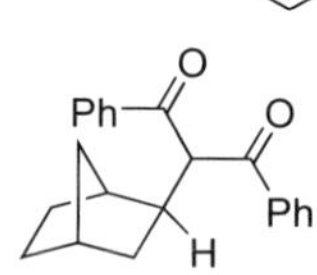	81

[a] Reaction conditions: 1 mmol scale in dichloromethane. [b] The alkenes (1.5 equiv., diluted to 3 mL) were added by syringe pump over 5 h. [c] Isolated yield. [d] This reaction was also carried out on a 30 mmol scale (2 equiv. of styrene was used), and 70–74% yields were achieved. [e] The reaction was carried out in refluxing nitromethane. [f] The ratio of two diastereomers was determined by ^{1}H-NMR. [g] Approximately 56% of the diketone was recovered.

Table 2 Addition of β-diketone to cyclic alkenes catalyzed by gold catalyst[a]

Entry	β–Diketone	Alkenes[b]	Product	Yield (%)[c]
1	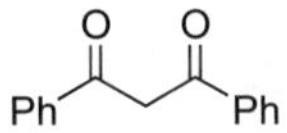	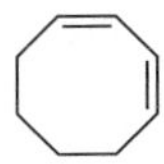	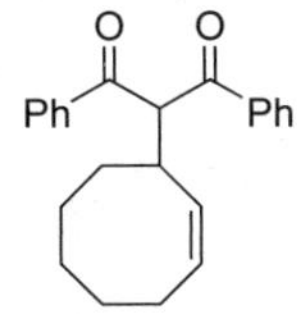	65
2[d]	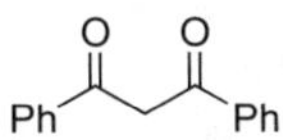		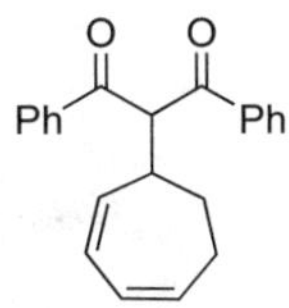	42
3[e]	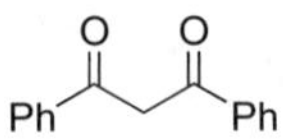		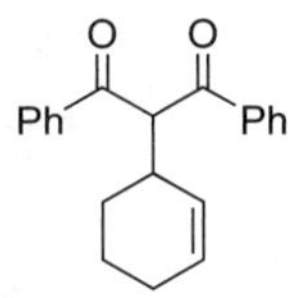	51
4[e]	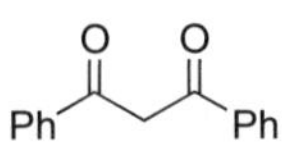		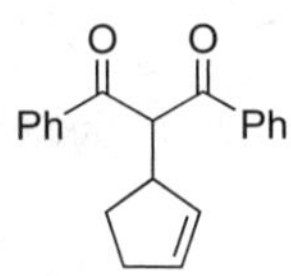	50
5[e]	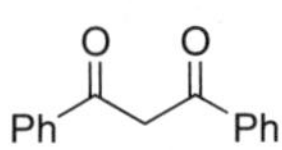	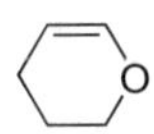	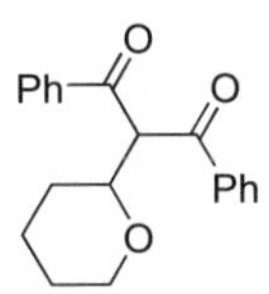	68
6	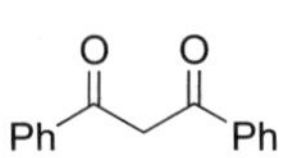	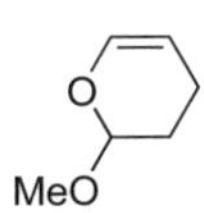	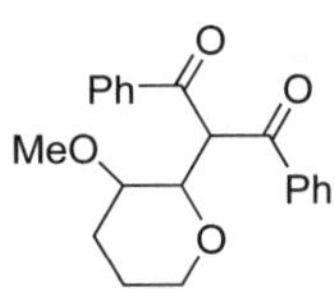 + 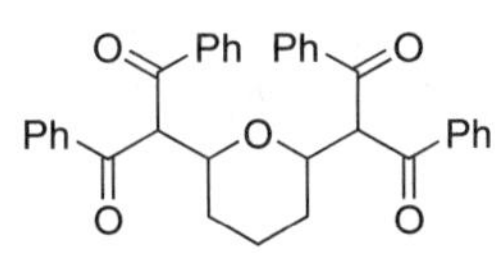	47 + 15
7	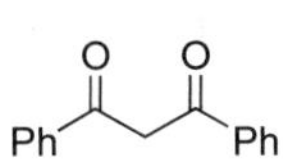	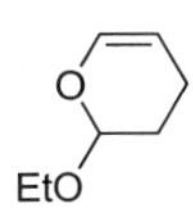	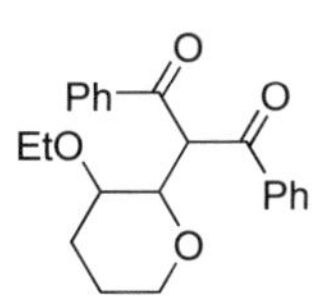 + 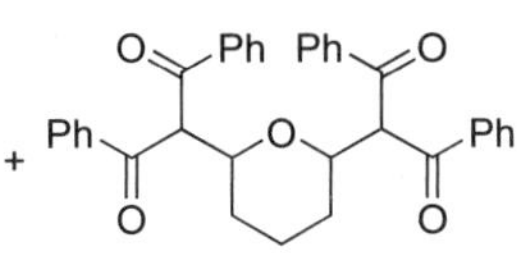	44 + 15

Table 2 (continued)

Entry	β–Diketone	Alkenes[b]	Product	Yield (%)[c]
8	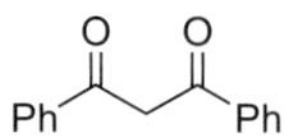	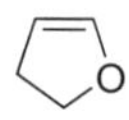	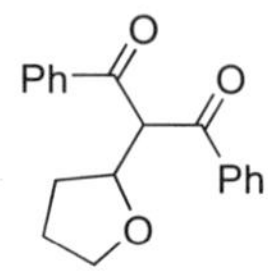	58
9	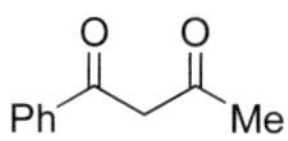	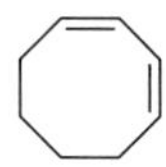	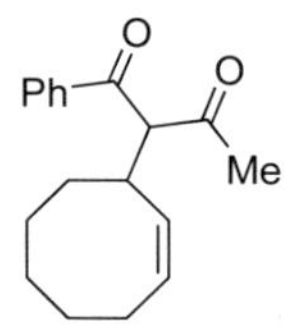	38 (1.4:1)[f]
10	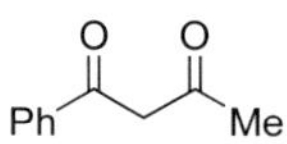	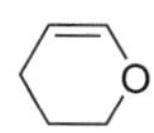	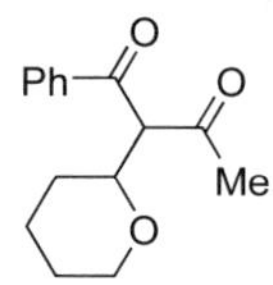	35 (1.2:1)[f]

[a] Reactions were performed in dry dichloromethane. [b] Alkenes were added by syringe pump. [c] Isolated yield. [d] The position of the double bonds was confirmed by COSY-NMR. [e] Reactions were run at 0 °C. [f] The ratio of the two diastereoisomers was determined by ^{1}H NMR.

1. Department of Chemistry, McGill University, 801 Sherbrooke St. West, Montreal, Quebec H3A 2K6, Canada
2. Gonzalez, A.; Marquet, J.; Moreno-Manas, M. *Tetrahedron Lett.* **1988**, *29*, 1469.
3. (a) Pei, T.; Widenhoefer, R. A. *J. Am. Chem. Soc.* **2001**, *123*, 11290; (b) Pei, T.; Widenhoefer, R. A. *Chem. Commun.* **2002**, 650; (c) Qian, H.; Widenhoefer, R. A. *J. Am. Chem. Soc.* **2003**, *125*, 2056. (d) Pei, T.; Wang, X.; Widenhoefer, R. A. *J. Am. Chem. Soc.* **2003**, *125*, 648. (e) Wang, X.; Pei, T.; Han, X.; Widenhoefer, R. A. *Org. Lett.* **2003**, *5*, 2699. See also (f) Yang, D.; Li, J. H.; Gao, Q.; Yan, Y. L. *Org. Lett.* **2003**, *5*, 2869.
4. Wang, X.; Widenhoefer, R. A. *Chem. Commun.* **2004**, 660.
5. Leitner, A.; Larsen, J.; Steffens, C.; Hartwig, J. F. *J. Org. Chem.* **2004**, *69*, 7552.
6. Yao, X.; Li, C.-J. *J. Am. Chem. Soc.* **2004**, *126*, 6884.
7. Tsuchimoto, T., Kamiyama, S.; Negoro, R.; Shirakawa, E.; Kawakami, Y. *Chem. Commun.* **2003**, 852.
8. Yao X.; Li, C.-J. *J. Org. Chem.* **2005**, *70*, 5752.
9. Nguyen, R.-V.; Yao, X.; Bohle, D. S.; Li, C.-J. *Org. Lett.* **2005**, *7*, 673.

SYNTHESIS OF MULTI-FUNCTIONALIZED KETONES THROUGH THE FUKUYAMA COUPLING REACTION CATALYZED BY PEARLMAN'S CATALYST: PREPARATION OF ETHYL 6-OXOTRIDECANOATE

(Tridecanoic acid, 6-oxo-, ethyl ester)

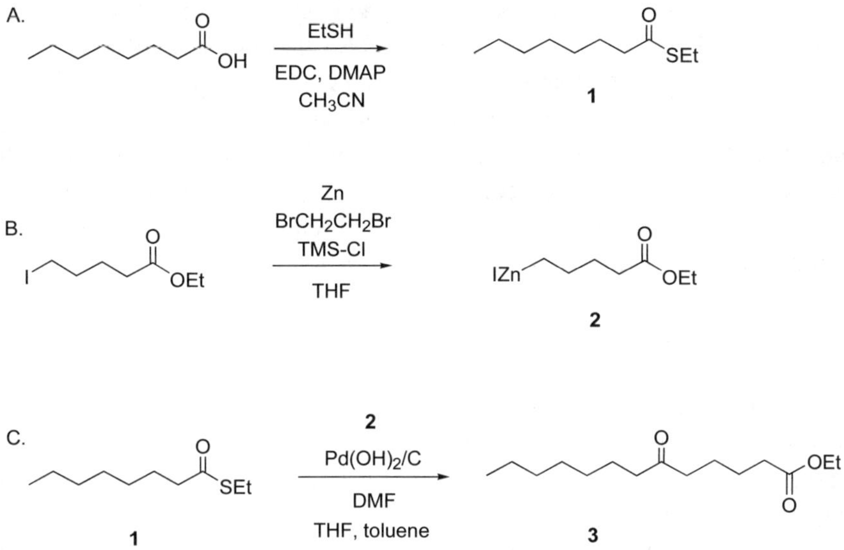

Submitted by Yoshikazu Mori and Masahiko Seki.[1]
Checked by Naohiro Aoyama and Masakatsu Shibasaki

1. Procedure

Caution! All reactions should be conducted in a well-ventilated hood because of the stench of EtSH.

A. S-Ethyl Octanethioate (**1**). A 200-mL, three–necked, round-bottomed flask equipped with a rubber septum, a thermometer, a three-way stopcock connected to an argon balloon (Note 1), and a Teflon-coated magnetic stirring bar is charged with octanoic acid (Note 2) (7.50 g, 52.0 mmol) by a pipette. After being purged with argon in the flask, acetonitrile (CH_3CN) (Note 2) (80 mL) is added by using a measuring cylinder with temporary removal of the septum, while keeping a positive flow of argon. The flask is immersed in an ice-water bath. *N,N*-Dimethylaminopyridine (DMAP) (Note 2) (0.60 g, 4.9 mmol) and

1-ethyl-3-(3-dimethylaminopropyl)carbodiimide hydrochloride (EDC) (Note 3) (10.0 g, 52.2 mmol) are added successively to the solution while the temperature is kept below 10 °C. The mixture is stirred below 10 °C for 10 min and ethanethiol (EtSH) (Note 2) (3.9 mL, 53 mmol) is added via syringe. The ice-water bath is removed and the mixture is stirred at 25 °C for 22 h (Note 4). The mixture is evaporated (Note 5). The residue is partitioned after adding ethyl acetate (50 mL) and 3.5 % hydrochloric acid (HCl) (50 mL). The organic layer is successively washed with water (50 mL), saturated aqueous sodium hydrogen carbonate (50 mL) and water (50 mL). The organic layer is dried over anhydrous magnesium sulfate ($MgSO_4$) (5.0 g) and filtered. The solid $MgSO_4$ is washed with ethyl acetate (40 mL). The filtrate and the washings are combined and evaporated. The crude material obtained as the residue is purified by column chromatography with a 5-cm diameter column containing 200 g of silica gel (Note 6). The column is eluted with a mixture of hexanes and ethyl acetate (20:1) to give 9.44–9.49 g (50.1–50.4 mmol, 96–97 %) of **1** as a colorless oil (Note 7).

B. 5-Ethoxy-5-oxopentylzinc iodide (**2**). A 100-mL, three–necked, round-bottomed flask is equipped with a reflux condenser (fitted with a three-way stopcock connected to an argon balloon), a thermometer, a rubber septum, and a Teflon-coated magnetic stirring bar. After the apparatus is purged with argon gas, the flask is charged with zinc powder (7.09 g, 108 mmol) (Note 8) with temporary removal of the septum, while keeping a positive flow of argon and anhydrous tetrahydrofuran (THF) (16 mL) (Note 9) via syringe under an argon atmosphere. 1,2-Dibromoethane (240 μL, 2.79 mmol) (Note 10) is added to the stirred suspension via syringe and the mixture is heated to 65 °C and kept at 65 °C for 3 min (Note 11). After cooling the mixture to 25 °C, chlorotrimethylsilane (TMS-Cl; 240 μL, 1.88 mmol) (Note 12) is added via syringe and the slurry is stirred for 15 min at 25 °C (Note 13). The mixture is warmed up to 55–60 °C using an oil bath. Ethyl 5-iodopentanoate (13.8 g, 53.9 mmol) (Note 14) is added to the mixture at 55–60 °C over 90 min via syringe with vigorous stirring (Note 15). The mixture is stirred at 55–60 °C for 30 min and the oil bath is removed. The reaction mixture is allowed to cool to 30–35 °C to give **2** (Note 16). The suspension is used for the subsequent step without further purification.

C. Ethyl 6-Oxotridecanoate (**3**). To the mixture obtained above are added at 30–35 °C, under an argon atmosphere, *S*-ethyl octanethioate (5.04 g, 26.8 mmol) and anhydrous toluene (16 mL) (Note 17) via syringe, Pearlman's catalyst (5% palladium(II) hydroxide on activated carbon) (345 mg, 0.162 mmol) (Note 18) with temporary removal of the septum, while keeping a positive flow of argon and anhydrous *N,N*-dimethylformamide (DMF) (2 mL, 25.8 mmol) (Note 17) via syringe. An initial exotherm (from 30–35 °C to approximately 45 °C) is observed. The mixture is stirred at 25 °C for 14 h (Note 19). The reaction mixture is cooled to 5 °C, and 3 M HCl (30 mL) is added to the mixture keeping the temperature below 25 °C (Note 20). The reaction mixture is stirred at 25 °C for 30 min and is filtered. The filtrate is partitioned after addition of ethyl acetate (120 mL). The organic layer is washed with water (60 mL x 2) and dried over anhydrous $MgSO_4$ (6 g). The mixture is filtered and the $MgSO_4$ is washed with ethyl acetate (45 mL). The filtrate and the washings are combined and evaporated to afford 10.2 g of yellow oil (Note 21). The crude material is purified by chromatography on a 5-cm diameter column containing 200 g of silica gel (Note 6). The column is eluted with a mixture of hexanse and ethyl acetate (15:1) to give 5.35 g (20.9 mmol, 78 %) of **3** as a pale yellow oil (Note 22).

2. Notes

1. The submitters performed the reaction under a nitrogen atmosphere.

2. Octanoic acid, acetonitrile, DMAP, and EtSH were purchased from Aldrich Chemical Company, Inc., and used as received.

3. EDC was purchased from Mitsubishi Tanabe Pharma Corp., Fine Chemicals Department and used as received.

4. The conversion of the reaction was monitored by TLC (hexanes/ethyl acetate = 8:1, phosphomolybdic acid was used as stain, silica gel was used as stationary phase, octanoic acid: R_f = 0.2, product (*S*-ethyl octanethioate): R_f = 0.85.

5. Checkers conducted the evaporation at 50 °C under approximately 30 mmHg. In order to control the odor, the vapor output from evaporator was passed through a trap cooled at –78 °C by dry ice-acetone bath.

6. Silica gel 60, particle size 230–400 mesh ASTM, Merck Ltd. was used.

7. The physical properties of the product were as follows: ^{1}H NMR (500 MHz, $CDCl_3$) δ: 0.88 (t, 3 H, J = 7.1 Hz). 1.34–1.21 (m, 11 H), 1.69–1.62 (m, 2 H), 2.55–2.51 (m, 2 H), 2.87 (q, 2 H, J = 7.4 Hz); ^{13}C NMR (125 MHz, $CDCl_3$) δ: 14.0, 14.8, 22.6, 23.2, 25.7, 28.9, 28.9, 31.6, 44.1, 199.8; IR (neat): 2928, 2857, 1692 cm^{-1}; MS (ESI) *m/z* 211 $[M + Na]^+$.

HRMS (EI) *m/z* $[M–SEt]^+$: Calcd for $C_8H_{15}O$, 127.1117; Found, 127.1119. Anal. Calcd for $C_{10}H_{20}OS$: C, 63.77; H, 10.70. Found: C, 63.52; H, 10.44.

8. Zinc powder was purchased from Merck (GR for analysis, particle size <45 μm, Merck 108789), and used as received.

9. The submitters used anhydrous THF distilled from lithium aluminum hydride and stored over molecular sieves 4Å. The checkers used anhydrous THF freshly distilled from sodium/benzophenone ketyl under argon prior to use.

10. 1,2-Dibromoethane was purchased from Aldrich Chemical Company, Inc. (submitters) or Tokyo Kasei Kogyo Co., Ltd. (checkers) and used as received.

11. The checkers kept the internal temperature at 65 °C for 3 min after the internal temperature reached 65 °C (oil bath temperature was approximately 75 °C).

12. TMS-Cl was purchased from Aldrich Chemical Company, Inc., and used as received.

13. Checkers observed evolution of gas after addition of TMS-Cl.

14. Ethyl 5-iodopentanoate was purchased from Mitsubishi Tanabe Pharma Corp., Fine Chemicals Department and was freshly distilled before use (bp 91–95 °C/7 mmHg).

15. The addition period depended on the scale of the reaction. An extended addition time of up to 2 h did not affect the reaction. Consistent addition of the iodide with control of the internal temperature at the indicated range (55–60 °C) was critical to ensure the completion of the reaction. The checkers kept the inner temperature at 60 °C.

16. The conversion of the reaction was assessed by the following procedures: (A, submitters' procedure) an aliquot of the reaction mixture was quenched with 1 M HCl and extracted with ethyl acetate and subjected to the HPLC analysis (Inertsil ODS-2, 40 °C, 50 mM KH_2PO_4 (pH 3.0)/MeOH = 30:70, 0.8 mL/min, 210 nm, ethyl pentanoate: 5.4 min, ethyl 5-iodopentanoate: 6.4 min, the intensity of ethyl 5-iodopentanoate is 1.9 times higher than that of ethyl pentanoate). (B, checkers' procedure) an aliquot of the reaction mixture was quenched with 1 M HCl and extracted with $CDCl_3$ (chloroform-d) and subjected to the ^1H NMR (500 MHz, $CDCl_3$) analysis (ethyl pentanoate: δ: 0.92 ($\underline{CH_3}(CH_2)_3CO_2Et$, t, *J* = 7.3 Hz), ethyl 5-iodopentanoate: δ: 3.19 ($I\underline{CH_2}(CH_2)_3CO_2Et$, t, *J* = 7.1 Hz)). Integration of these signals showed less than 1% of ethyl 5-iodopentanoate and more than 99% of ethyl pentanoate was detected.

17. Toluene and DMF were purchased from Aldrich Chemical Company, Inc., and dried over molecular sieves 4Å before use.

18. Pearlman's catalyst (Degussa nomenclature, E101NE/D) was purchased from Degussa Japan Co., Ltd. Pearlman's catalyst was added as a solid under argon flow.

19. The checkers immersed the reaction flask in the oil bath to keep the outer temperature at 25 °C. The conversion of the reaction was monitored by TLC (hexanes/ethyl acetate = 8 : 1, phosphomolybdic acid was used as an indicator, SM (*S*-ethyl octanethioate): R_f = 0.85, product (ethyl 6-oxotridecanoate): R_f = 0.4).

20. The reaction mixture produced an unpleasant odor due to the formation of EtSH.

21. The checkers conducted the evaporation at 50 °C under 36 mmHg. The submitters reported that the crude material weighed 8.1 g.

22. The physical properties of the product were as follows: ^{1}H NMR (500 MHz, $CDCl_3$) δ: 0.88 (t, 3H, *J* = 6.8 Hz), 1.20–1.34 (m, 11 H), 1.52–1.65 (m, 6 H), 2.31 (t, 2 H, *J* = 7.1 Hz), 2.36–2.44 (m, 4 H), 4.12 (q, 2 H, *J* = 7.4 Hz); ^{13}C NMR (125 MHz, $CDCl_3$) δ: 14.0, 14.2, 22.6, 23.2, 23.9, 24.5, 29.1, 29.2, 31.7, 34.1, 42.3, 42.9, 60.3, 173.4, 211.0. IR (neat): 2929, 2857, 1736, 1179 cm^{-1}. HRMS (FAB) *m/z* $[M+H]^+$: Calcd for $C_{15}H_{29}O_3$, 257.2111; Found, 257.2110. Anal. Calcd for $C_{15}H_{28}O_3$: C, 70.27; H, 11.01. Found: C, 69.99; H, 10.86.

Waste Disposal Information

All toxic materials were disposed of in accordance with "Prudent Practices in the laboratory"; National Academy Press; Washington, DC, 1995.

3. Discussion

Palladium-catalyzed coupling reactions have found wide application in the preparation of valuable compounds such as drugs and natural products. One of the most attractive features of palladium-catalyzed coupling reactions is its wide applicability to the substrates carrying various functional groups. Tedious protection-deprotection sequences, as well as functional group transformations are often avoided. Fukuyama and co-workers have recently developed a highly efficient synthetic method for the generation of ketones.[2] Treatment of thiol esters with zinc reagents in the presence of dichlorobis(triphenylphosphine) palladium(II) $[PdCl_2(PPh_3)_2]$ provides highly functionalized ketones in excellent yields. The reaction is featured by unusually high chemoselectivity, mild reaction conditions and the use of less-toxic reagents. However, requirement of the expensive palladium catalyst has hindered the commercial application of the technology.

The use of heterogeneous palladium catalysts to perform the coupling chemistry described above enables facile recovery of palladium, as well as eliminates metal contamination of the product. The heterogeneous catalytic process thus provides a reaction that is much more cost conscious, compared with the homogeneous counterparts. The present procedure employs heterogeneous Pearlman's catalyst (palladium(II) hydroxide on activated carbon, $Pd(OH)_2/C$).[3] Use of this catalyst is advantageous in terms of low catalyst loading and safe handling of the catalyst, which is less flammable.

The method was tested in the reaction of various functionalized thiol esters with the zinc reagent (Table 1). The reaction between the zinc reagent and primary or secondary alkyl thiol esters with or without an ester or a keto group afforded the corresponding ketones in good yields (Table 1, Entries 2-5). The reaction using an aromatic thiol ester carrying a chloro or a sulfide group provided a good yield, as well (Table 1, Entries 7, 8).

In the workup as originally performed, the reaction mixture was filtered directly through Celite to recover the palladium catalyst. However, viscous zinc salts in the reaction mixture often required a prolonged period for filtration. As an improvement, prior to filtration, aqueous hydrochloric acid was added to the reaction mixture. The aqueous hydrochloric acid dissolved the zinc salts and permitted separation of the zinc-free Pd catalyst as a filter cake. No palladium was detected in the filtrate and recovery of palladium from the filter cake was more than 95%.

The present procedure was applied to the introduction of the side chain of (+)-biotin[4] (Scheme 1).[5] The zinc reagent was allowed to react with thiolactone {(*4S,5R*)-1,3-dibenzyl-3,3a,6,6a-tetrahydro-1*H*-thieno[3,4-*d*]-imidazol-2,4-dione} to afford, after dehydration with hydrochloric acid, a compound carrying the required side chain in high yield. The product is readily converted to (+)-biotin by hydrogenation and subsequent removal of protective groups.[6] Through the present process, (+)-biotin is obtained in three steps from thiolactone, which is three steps shorter than the conventional method that employs a Grignard reaction.[7] The reduced number of steps, high yield, operational simplicity, ready availability of the reagents and mild reaction conditions of the process provides a much more practical route to (+)-biotin.

Table 1. Fukuyama coupling reaction in the presence of $Pd(OH)_2/C$.

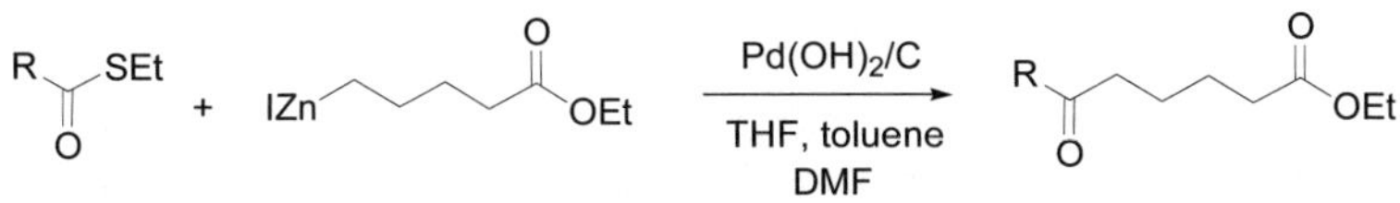

Entry	Thiol Ester	Catalyst Loading (mol % Pd)	Reaction Time (h)	Yield (%)[a]
1	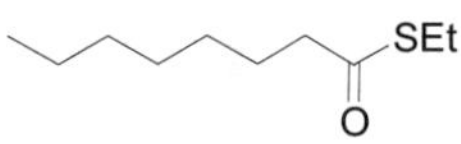	0.6	14	78
2	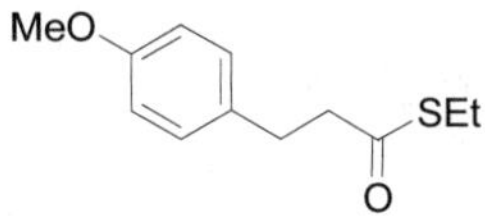	0.15	50	84
3	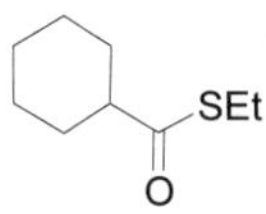	0.6	72	79
4	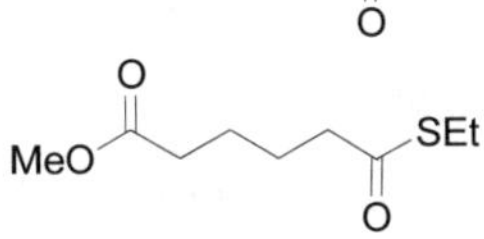	0.9	17	78
5	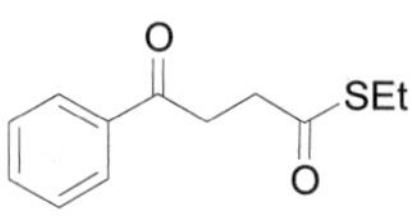	0.9	23	70
6	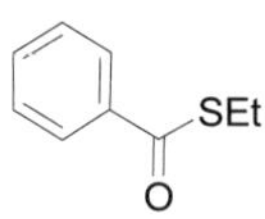	0.9	71	72
7	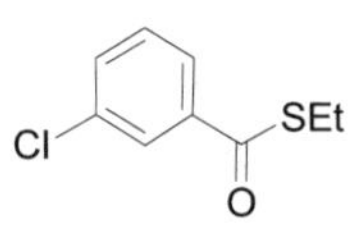	0.6	25	73
8	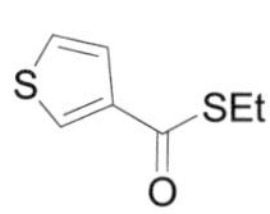	0.9	68	61

[a]Isolated Yield.

Scheme 1

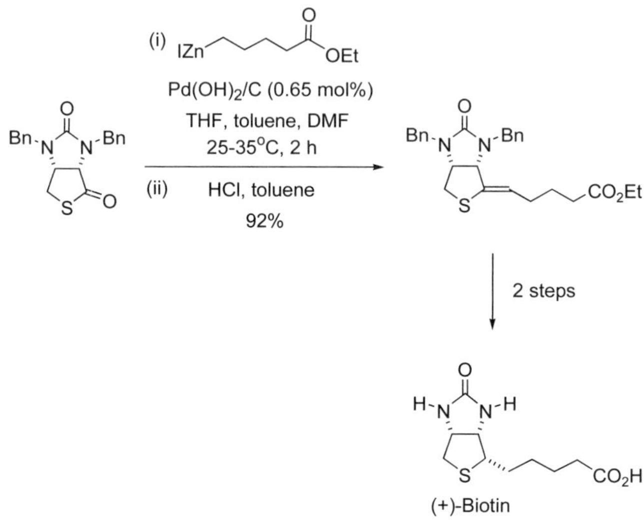

1. Mitsubishi Tanabe Pharma Corp., Fine Chemicals Department, 3-16-89, Kashima, Yodogawa-Ku, Osaka 532-8505, Japan.
2. (a) Tokuyama; H.; Yokoshima; S.; Yamashita; T.; Fukuyama, T. *Tetrahedron Lett.* **1998**, *39*, 3189; (b) Miyazaki, T.; Han-ya, Y.; Tokuyama, H.; Fukuyama, T. *Synlett* **2004**, 477; (c) Tokuyama, H.; Miyazaki, T.; Yokoshima, S.; Fukuyama, T. *Synlett* **2003**, 1512; (d) Fukuyama, T.; Tokuyama, H. *Aldrichimica Acta* **2004**, *37*, 87.
3. Mori, Y.; Seki, M. *J. Org. Chem.* **2003**, *68*, 1571.
4. For reviews, see: (a) De Clercq, P. J. *Chem. Rev.* **1997**, *97*, 1775. (b) Seki, M. *Med. Res. Rev.* **2006**, *26*, 434. (c) Seki, M. *Synthesis* **2006**, 2975. (d) Seki, M. "Studies in Natural Products Chemistry". Ed. by Atta-ur-Rahman, Elsevier, Amsterdam, in press.
5. Mori, Y.; Seki, M. *Heterocycles* **2002**, *58*, 125.
6. Seki, M.; Hatsuda, M.; Mori, Y.; Yoshida, S.; Yamada, S.; Shimizu, T. *Chem. Eur. J.* **2004**, *10*, 6102.
7. (a) Goldberg, M. W.; Sternbach, L. H. US Pat 2489232, **1949**; *Chem. Abst.* **1951**, *45*, 184. (b) Goldberg, M. W.; Sternbach, L. H. US Pat 2489235, **1949**; *Chem. Abst.* **1951**, *45*, 186s. (c) Goldberg, M. W.; Sternbach, L. H. US Pat 2489238, **1949**; *Chem. Abst.* **1951**, *45*, 186g. (d) Gerecke, M.; Zimmermann, J.-P.; Ashwanden, W. *Hel. Chim. Acta* **1970**, *53*, 991.

1-(4-ACETYLPHENYL)-2-PHENYLETHANE FROM POTASSIUM 2-PHENETHYLTRIFLUOROBORATE AND 4-BROMOACETOPHENONE

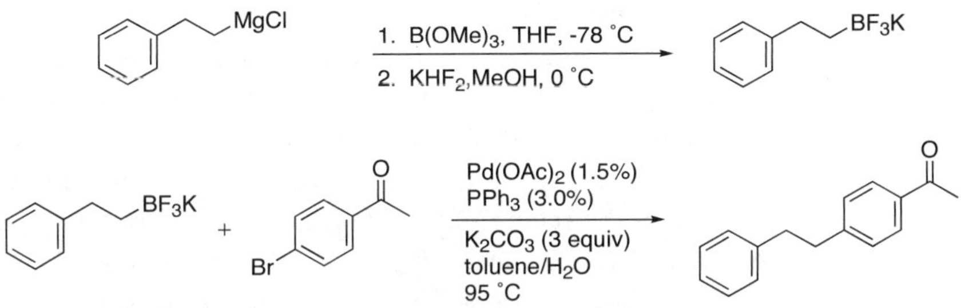

Submitted by Gary A. Molander and Daniel E. Petrillo.[1]
Checked by Scott E. Denmark and Christophe Eggertswyler.

1. Procedure

A. Preparation of Potassium 2-Phenethyltrifluoroborate. A single-necked, 500-mL round-bottomed flask, equipped with a magnetic stirring bar, a 125-mL pressure-equalizing addition funnel, and connected by a three-way stopcock fitted with a rubber septum to a vacuum-argon manifold, is flame-dried under vacuum (0.3 mm Hg), then is flushed with dry argon. THF (90 mL) is added via syringe to the addition funnel, followed by trimethyl borate (10.0 mL, 89.2 mmol, 1.5 equiv) (Note 1). After the addition of the trimethyl borate to the round-bottomed flask, additional THF (30 mL) (Note 1) is added via syringe to rinse the addition funnel. The solution is then cooled in a dry ice/isopropyl alcohol bath. Phenethylmagnesium chloride (60.0 mL of a 1.0 M solution in THF, 60.0 mmol) (Note 2) is then added dropwise over 1 h via the addition funnel. The mixture is stirred in the dry ice/isopropyl alcohol bath for 1.5 h, then is allowed to warm to room temperature over 1 h. The resulting white slurry is then cooled in an ice/water bath and methanol (60 mL) (Note 3) is charged into the addition funnel by syringe. The methanol is added over 15 min to dilute the slurry (Note 4). The argon inlet is removed and a solution of potassium hydrogen fluoride (80.0 mL of 4.5 M aqueous solution, 360 mmol, 6.0 equiv) (Note 5) is then charged into the addition funnel and is added dropwise over 15 min. The resulting solution is stirred for 1 h in the ice/water bath, and then is concentrated on a rotary evaporator (60 °C water bath, 9 mm Hg). The resulting white solids are placed on a high-vacuum line (0.3 mm Hg) for 36 h until the solids appear dry (Note 6). Acetone (120 mL) (Note 7) is added and the solids are broken up into a fine powder with a spatula. The suspension is then placed on a rotary evaporator and heated (40 °C) with rotation under atmospheric pressure for 5–10 min and filtered over a 14 g pad of Celite in a glass-fritted filter funnel. Acetone (120 mL) is added to the solid remaining in the flask, the suspension is placed on a rotary evaporator and heated (40 °C) with rotation under atmospheric pressure for 5–10 min and filtered over the same Celite pad. This process is repeated one additional time (Note 8). The combined filtrates are concentrated on a rotary evaporator (60 °C water bath) to a white solid, which is dissolved in a minimal amount of dry acetone (~25 mL). Diethyl ether (30 mL) is then added to precipitate the product. The white slurry is allowed to cool first to 23 °C, and then is cooled in an ice/water bath for 15 min. The slurry is placed in a refrigerator (4 °C) for 1 h, and is then filtered to give 9.84 g (77%) of a white solid (Note 9).

B. Preparation of 1-(4-Acetylphenyl)-2-phenylethane. A three-necked, 250-mL, round-bottomed flask, equipped with a reflux condenser, a rubber septum, a glass stopper, a magnetic stirring bar, and connected to a vacuum-argon manifold, is charged sequentially with potassium 2-phenethyltrifluoroborate (7.77 g, 36.6 mmol, 1.05 equiv), 4-bromoacetophenone (6.97 g, 35.0 mmol), potassium carbonate (14.5 g, 105 mmol, 3.0 equiv), triphenylphosphine (0.275 g, 1.05 mmol, 0.03 equiv) and palladium acetate (0.118 g, 0.53 mmol, 0.015 equiv) (Note 10). The flask is evacuated and filled with dry argon three times. Toluene (105 mL) (Note 11) and H_2O (21 mL) are added, and the resulting two-phase mixture is stirred in a 95 °C oil bath until consumption of the 4-bromoacetophenone is indicated by GC analysis (Note 12). The mixture is cooled to room temperature and then is transferred to a 250-mL separatory funnel. The flask is rinsed with toluene (20 mL) and H_2O (20 mL). The layers are shaken and separated. The organic layer is washed with 10% aq. citric acid solution (20 mL) (Note 13) and saturated aq. NaCl solution (20 mL). The organic layer is dried over anhydrous Na_2SO_4 (0.780 g) and filtered through a medium porosity glass-fritted funnel, rinsing the drying agent with toluene (20 mL). The resulting solution is concentrated (50 °C water bath, 9 mmHg) to a light yellow solid (Note 14), which is crystallized from methanol to give 5.79 g (74%, two crops) (Notes 15, 16) of a light tan solid. Recrystallization of 0.3

g of this solid from hot methanol (2 mL) provided white, crystalline material (0.25 g) of analytical purity (Note 17).

2. Notes

1. Trimethyl borate (98%) was obtained from Aldrich Chemical Company and was distilled from sodium. THF (Fischer, HPLC Grade) was dried by percolation through a column packed with neutral alumina and a column packed with Q5 (a supported copper catalyst for removing oxygen) under a positive pressure of argon.

2. Phenethylmagnesium chloride (1.0 M in THF) was obtained from Aldrich Chemical Company and was used as received.

3. Methanol (ACS Grade) was obtained from Aldrich Chemical Company and was distilled continuously from $Mg(OMe)_2$.

4. At this stage the solution may be transferred to a Nalgene vessel to avoid etching of the glass flask by the KHF_2.

5. Potassium hydrogen fluoride (p.a. >99%) was obtained from Fluka and was used as received.

6. Care was taken to ensure that the solids are free of water or else inorganic fluoride contaminated the final product.

7. Acetone (Spectroscopic Grade) was obtained from Aldrich Chemical Company and was used as received. Diethyl ether (ACS Grade) was obtained from Fischer Scientific and was used as received.

8. The desired product was soluble in acetone whereas the inorganic salts were not. This extractive process ensured that any product trapped within the inorganic solids would be dissolved.

9. Analytical data: mp: 332–335 °C (sealed tube). IR cm^{-1}: 3026, 2921, 2847; ^{1}H NMR (600 MHz, acetone-d^6) δ: 0.51 (br, 2 H), 2.56 (dt, J = 8.8, 4.7 Hz, 2 H), 7.00–7.05 (m, 1 H), 7.14–7.17 (m, 4 H); ^{13}C NMR (150 MHz, acetone-d^6) δ: 32.1, 124.5, 127.9, 128.0, 148.5; ^{11}B NMR (192.4 MHz, acetone-d^6) δ 5.48; ^{19}F NMR (470 MHz, acetone-d^6) δ –142.32; MS (ES): *m/z* 173.1 (100) Anal. Calcd for $C_9H_9BF_3K$: C, 45.31; H, 4.28; F, 26.88. Found: C, 45.03; H, 4.29; F, 25.38.

10. 4-Bromoacetophenone (98%) was obtained from Aldrich Chemical Company and was used as received. Potassium carbonate (ACS Grade) was purchased from Fisher Scientific and was used as received. Triphenylphosphine (ReagentPlus Grade, 99%) was obtained from Aldrich Chemical Company and was used as received. Palladium acetate was obtained from Frontier Scientific and was used as received.

11. Toluene (ACS Grade, 99.5+%) was obtained from Aldrich Chemical Company and was used as received.

12. The reaction was assayed as follows: an aliquot (20 μL) was removed and filtered through a small pad of silica gel in a Pasteur pipette, rinsing with EtOAc (1.5 mL). The sample was then injected on an HP 5890 GC with an HP 1 column (100 °C for 1 min., ramp to 250 °C at 25 °C/min.). Typical retention times were: 4-bromoacetophenone: 4.1 min, 1-(4-acetylphenyl)-2-phenylethane: 7.4 min. The reaction generally required 16–20 h.

13. Citric acid monohydrate (Baker Analyzed Reagent) was obtained from J. T. Baker Chemical and was used as received.

14. The submitters found that the crude product was a dark yellow solid that needed to be decolorized with activated carbon in MeOH prior to recrystallization.

15. The crude product was slurried in MeOH (25 mL) in a 100-mL flask and heated to 50 °C in an oil bath. After dissolution, the solution was allowed to cool to room temperature. It was then placed in an ice/water bath for 1 h, after which crystals were removed by filtration. The filtrate was concentrated (50 °C water bath) to ~ 10 mL and cooled to 0 °C, after which a second crop of the crystals was collected via filtration. This material had a melting point of 66–68 °C.

16. The submitters noted that the product may also be purified by silica gel column chromatography (10% EtOAc/hexanes as eluent, $R_f = 0.34$) to give an 89% yield of product of unspecified purity.

17. Analytical data: mp: 68–72 ºC (lit. mp: 68–70 ºC).[2] IR cm^{-1}: 1676, 1601; ^1H NMR (600 MHz, $CDCl_3$) δ: 2.58 (s, 3 H), 2.92–3.01, (m, 4 H), 7.15-7.30 (m, 7 H), 7.88 (d, J = 8.3 Hz, 2 H); ^{13}C NMR (150 MHz, $CDCl_3$) δ: 26.5, 37.3, 37.8, 126.1, 128.3, 128.4, 128.5, 128.7, 135.1, 141.0, 147.4, 197.7. MS (EI, 70 eV): *m/z* 224 (M^+, 44), 209 (12), 181 (8), 133 (12), 105 (8), 91 (100), 65 (7); Anal. Calcd for $C_{16}H_{16}O$: C, 85.68; H, 7.19. Found: C, 85.59; H, 7.31.

Safety and Waste Disposal Information

All hazardous materials should be handled and disposed of in accordance with "Prudent Practices in the Laboratory"; National Academy Press; Washington, DC, 1995.

3. Discussion

The metal-catalyzed cross-coupling reaction between organometallic reagents with electrophiles has emerged as one of the most powerful methods for carbon-carbon bond formation.[3] The Suzuki-Miyaura[4] reaction using organoboron reagents is among the most important, due its tolerance of a broad range of functional groups as well as its formation of non-toxic byproducts. The use of alkylboron reagents has received special attention because of their utility in natural product synthesis.[5]

The coupling of alkylboron reagents has generally been performed with trialkylboranes (R_3B), alkylborinate esters (R_2BOR'), alkylboronic acids [$RB(OH)_2$] or boronate esters [$RB(OR')_2$]. Trialkylboranes such as 9-BBN[6] derivatives are among the most common, although they present problems such as functional group intolerance, air sensitivity, and a lack of atom economy. The use of alkylboronate esters often results in low yields in coupling reactions unless highly toxic thallium bases are used.[7] Additionally, these species are often hydrolytically unstable. The coupling of alkylboronic acids[8] has shown more promise, although their tendency to form cyclic anhydrides (boroxines) often makes determining their precise stoichiometry difficult.

Recently, potassium organotrifluoroborates (RBF_3K) have been shown to be highly effective coupling partners in the Suzuki-Miyaura reaction.[9] They are easily prepared on large-scale (>200 g) by the addition of KHF_2 (an inexpensive fluoride source) to various organoboron reagents such as boronic acids, boronate esters, and organodihaloboranes. Their advantages over traditional organoboron reagents include their greater nucleophilicity, air stability, and atom economy. Initial reports described their use in the palladium-catalyzed coupling with arenediazonium[10] or diaryliodonium[11] salts. Previous reports from our laboratory have shown the successful cross-coupling of potassium alkyl-,[12] alkenyl-,[13] aryl-,[14] and alkynyltrifluoroborates[15] with aryl- and alkenyl halides as well as triflates using catalytic $PdCl_2(dppf){\bullet}CH_2Cl_2$ and Cs_2CO_3 or Et_3N as a base. In the case of aryltrifluoroborates, it was also shown that ligandless conditions using $Pd(OAc)_2$ and K_2CO_3 can be used.[16] Batey and Quach have reported similar cross-coupling reactions using tetraalkylammonium trifluoroborates.[17] Recently, functionalized organotrifluoroborates such as epoxidized[18] and dihydroxylated[19] potassium alkyltrifluoroborates have been coupled to electrophiles under palladium catalysis. The procedure described above represents a typical coupling reaction in which a commercially available, easily-handled potassium alkyltrifluoroborate is coupled to a common aryl electrophile.

1. Department of Chemistry, University of Pennsylvania, Philadelphia, Pennsylvania, 19104-6323.
2. Lutz, R.E.; Allison, R.K.; Ashburn, G.; Bailey, P.S.; Clark, M.T.; Codington, J.F.; Deinet, A.J.; Freek, J.A.; Jordan, R.H.; Leake, N.H.; Martin, T.A.; Nicodemus, K.C.; Rowlett, R.J.; Shearer, N.H.; Smith, J.D.; Wilson, J.W. *J. Org. Chem.* **1947**, *12*, 617-703.
3. *Metal-Catalyzed Cross-Coupling Reactions,* 2nd ed; de Meijere, A.; Diederich, F., Eds.; Wiley-VCH: Weinheim, 2004.
4. (a) Suzuki, A.; Brown, H.C. *Organic Syntheses via Boranes: Volume 3 Suzuki Coupling*: Aldrich Chemical Co.: Milwaukee, 2003. (b) Miyaura, N. *Top. Curr. Chem.* **2002**, *219*, 11. (c) Miyaura, N.; Suzuki, A. *Chem. Rev.* **1995**, *95*, 2457.
5. Chemler, S.R.; Trauner, D.; Danishefsky, S.J. *Angew. Chem., Int. Ed.* **2001**, *40*, 4545-4568.
6. (a) Miyaura, N.; Ishiyama, T.; Ishikawa, M.; Suzuki, A. *Tetrahedron Lett.* **1986**, *27*, 6369-6372. (b) Miyaura, N. Ishiyama, T.; Sasaki, H.; Ishikawa, M.; Satoh, M.; Suzuki, A. *J. Am. Chem. Soc.* **1989**, *111*, 314-321. (c) Old, D.W.; Wolfe, J.P.; Buchwald, S.L.; High, A. *J. Am. Chem. Soc.* **1998**, *120*, 9722-9723. (d) Wolfe, J.P.; Buchwald, S.L. *J. Am. Chem. Soc.* **1999**, *121*, 9550-9561.
7. Sato, M.; Miyaura, N.; Suzuki, A. *Chem. Lett.* **1989**, 1405-1408.
8. (a) Zou, G.; Reddy, K.; Falck, J.R. *Tetrahedron Lett.* **2001**, *42*, 7213-7215. (b) Molander, G.A.; Yun, C.-P. *Tetrahedron* **2002**, *58*, 1465-1470.
9. (a) Darses, S.; Genêt, J.-P. *Eur. J. Org. Chem.* **2003**, 4313-4327. (b) Molander, G.A.; Figueroa, R. *Aldrichimica Acta* **2005**, *38*, 49-56. (c) Molander, G. A. Ellis, N. *Acc. Chem. Res.* **2007**, *40*, 275. (d) Darses, S.; Genêt, J.-P. *Chem. Rev.* **2008**, *108*, 288-325.
10. Darses, S.; Michaud, G.; Genêt, J.-P. *Eur. J. Org. Chem.* **1999**, 1875-1883.
11. Xia, M.; Chen, Z.-C. *Synth. Commun.* **1999**, *29*, 2457-2465.
12. (a) Molander, G.A.; Ito, T. *Org. Lett.* **2001**, *3*, 393-396. (b) Molander, G.A.; Yun, C.-P.; Ribagorda, M.; Biolatto, B. *J. Org. Chem.* **2003**, *68*, 5534-5539. (c) Molander, G.A.; Ham, J.; Seapy, D.G. *Tetrahedron* **2007**, *63*, 768-775. (d) Molander, G.A.; Vargas, F. *Org. Lett.* **2007**, *9*, 203-206. (e) Molander, G.A.; Sandrock, D.L. *Org. Lett.* **2007**, *9*, 1597-1600.

13. (a) Molander, G.A.; Rodriguez Rivero, M. *Org. Lett.* **2002**, 4, 107-109. (b) Molander, G.A.; Bernardi, C.R. *J. Org. Chem.* **2002**, 67, 8424-8429. (c) Molander, G.A.; Felix, L.A. *J. Org. Chem.* **2005**, *70*, 3950-3956.

14. (a) Molander, G.A.; Biolatto, B. *Org Lett.* **2002**, *4*, 1867-1870. (b) Molander, G.A.; Biolatto, B. *J. Org. Chem.* **2003**, *68*, 4302-4314. (c) Molander, G.A.; Petrillo, D.E.; Landzberg, N.R.; Rohanna, J.C.; Biolatto, B. *Synlett* **2005**, *11*, 1763-1766. (d) Molander, G.A.; Fumagalli, T. *J. Org. Chem.* **2006**, *71*, 5743-5747.

15. Molander, G.A.; Katona, B.W.; Machrouhi, F. *J. Org. Chem.* **2002**, *67*, 8416-8423.

16. Molander, G.A.; Biolatto, B. *Org Lett.* **2002**, *4*, 1867-1870.

17. Batey, R.A.; Quach, T.D. *Tetrahedron Lett.* **2001**, *42*, 9099-9103.

18. Molander, G.A.; Ribagorda, M. *J. Am. Chem. Soc.* **2003**, *125*, 11148-11149.

19. Molander, G.A.; Figueroa, R. *Org. Lett.* **2006**, *8*, 75-78.

ONE-FLASK SYNTHESIS OF METHYL ARYLVINYLDIAZOACETATES AND THEIR APPLICATION IN ENANTIOSELECTIVE C–H FUNCTIONALIZATION: SYNTHESIS OF (*E*)-METHYL 2-DIAZO-4-PHENYLBUT-3-ENOATE AND (*S,E*)-METHYL 2-((*R*)-4-METHYL-1,2-DIHYDRONAPHTHALEN-2-YL)-4-PHENYLBUT-3-ENOATE

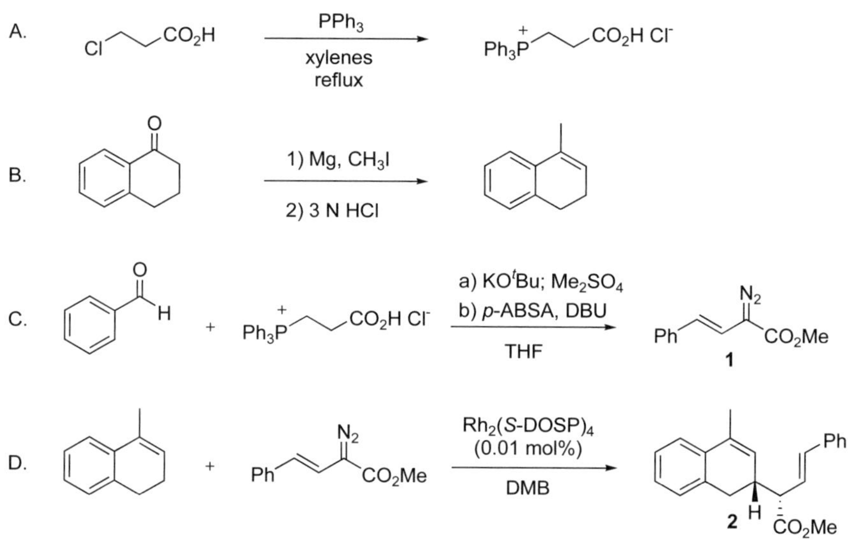

Submitted by James R. Manning and Huw M. L. Davies.[1]
Checked by Scott E. Denmark and William R. Collins.

1. Procedure

A. 2-Carboxyethyltriphenylphosphonium chloride. 3-Chloropropionic acid (16.3 g, 150 mmol) (Note 1) and triphenylphosphine (43.3 g, 165 mmol, 1.1 equiv) (Note 2) are added to a 250-mL, round-bottomed flask equipped with a magnetic stir bar. Xylenes (75 mL) (Note 3) are added and a water-cooled condenser is attached to the flask. Vigorous stirring is commenced and the solution is heated to reflux in an oil bath for 3 h during which time two layers form. The flask is then removed from the oil bath and the flask is fitted with a thermometer. When the solution has cooled to 60–70 °C, the thermometer is removed and acetonitrile (50 mL) (Note 4) is slowly added with vigorous stirring and scraping using a metal spatula to prevent the product from sticking to the sides of the flask (Note 5). The product, which is eventually converted to a white solid, is collected by vacuum filtration in a Büchner funnel. The product is washed with diethyl ether (300 mL) (Note 6) and then dried under vacuum (0.05 mmHg) for 2 h to afford 46.6–46.7g (84%) of pure 2-carboxyethyltriphenylphosphonium chloride as a white solid (Note 7).

B. 1-Methyl-3,4-dihydronaphthalene. A 500-mL, 3-necked round-bottomed flask equipped with a stir bar is fitted with an argon inlet adaptor, a 100-mL pressure-equalized addition funnel and a reflux condenser. The tops of the addition funnel and reflux condenser are sealed with septa, the apparatus is flame dried under vacuum (0.05 mmHg) and then is backfilled with argon. Magnesium turnings (3.65 g, 150 mmol, 1.50 equiv) (Note 8) are then added and the flask is evacuated and filled with argon twice. Dry diethyl ether (100 mL) (Note 9) is added through the condenser via cannula. Iodomethane (9.98 mL, 22.7 g, 160 mmol, 1.60 equiv) (Note 10) is then added by syringe to the addition funnel. Stirring is commenced and the iodomethane is added dropwise to the magnesium turnings at a rate sufficient to maintain a gentle reflux (13–17 min is required to complete the addition). The solution is then stirred for another 20 min after the addition is complete. A solution of α-tetralone (14.6 g, 100 mmol) (Note 11) in dry diethyl ether (10 mL) is transferred via syringe from a flame-dried 100-mL round bottom flask to the addition funnel and then added dropwise to the methyl magnesium iodide solution at a rate sufficient to maintain a gentle reflux (12–16 min is required for the addition). The solution is stirred for another 15 min after completion of the addition. The rubber septa are then removed from the tops of both the condenser and the addition funnel. An aqueous solution of 3 N HCl (60 mL) is placed in the addition funnel and then is added dropwise to the solution over 30 min. The mixture is then vigorously stirred for 1 h at room temperature, after which time the mixture is poured into a 250-mL separatory funnel and the layers are separated. The aqueous layer is extracted with diethyl ether (3 x 30 mL). The combined ether extracts are washed with brine (50 mL), dried over anhydrous $MgSO_4$, vacuum filtered through a plug of silica gel (24 g) in a Büchner funnel and then concentrated by rotary evaporation at room temperature. The residue is purified by flash column chromatography on 130 g of silica gel (Note 12) eluting with pentane (450 mL). The fractions containing product are collected and concentrated by rotary evaporation (11 mmHg, room temperature) to afford 13.1–13.2 g (91–92%) of pure 1-methyl-3,4-dihydronaphthalene as a clear oil (Notes 13, 14).

C. (E)-Methyl 2-diazo-4-phenylbut-3-enoate. 2-Carboxyethyltriphenylphosphonium chloride (26.7 g, 72.0 mmol, 1.20 equiv) is added to a flame-dried, 2-necked 500-mL, round-bottomed flask equipped with a magnetic stir bar and fitted with a rubber septum and an argon inlet adaptor. The flask is evacuated (0.05 mmHg) and filled with argon two times after which a positive argon pressure is maintained on the flask. Benzaldehyde (6.07 mL, 6.37 g, 60.0 mmol) (Note 15) and THF (130 mL) (Note 16) are then added by syringe and the flask is externally cooled to 0 °C in an ice-water bath and vigorous stirring is begun. A solution of potassium *tert*-butoxide (16.8 g, 150 mmol, 2.5 equiv) in THF (80 mL) under an argon atmosphere and at 0 °C is then added *via* cannula under positive argon pressure over 30 min (Notes 16, 17). After the addition is complete, the solution is stirred at 0 °C for 30 min, then the ice-water bath is removed and stirring is continued for another 20 min. Dimethyl sulfate (11.4 mL, 15.1 g, 120 mmol, 2.0 equiv) (Note 18) is then added rapidly by syringe and stirring is continued at ambient temperature for 2.5 h. The septum is then briefly removed from the flask and *para*-acetamidobenzenesulfonyl azide (18.7 g, 78.0 mmol, 1.30 equiv) (Note 19) is added in one portion. The septum is replaced and the flask, under a positive argon pressure, is externally cooled to 0 °C in an ice-water bath. DBU (11.7 mL, 11.9 g, 78.0 mmol) (Note 20) is then added rapidly by syringe and the solution quickly becomes red. Stirring is continued at 0 °C for 4 h and then the solution is allowed to warm to room temperature and then is transferred to a 1-L, single-necked, round-bottomed flask and the solution is concentrated by rotary evaporation (27 mmHg, ambient temperature) (Note 21). The residue is treated with saturated aqueous ammonium chloride solution (150 mL) and diethyl ether (400 mL) and the mixture is poured into a 1-L separatory funnel and vigorously shaken. The aqueous layer is drained and the organic layer is washed with saturated ammonium chloride solution (150 mL) and then is dried over anhydrous $MgSO_4$ (5 g) (Note 22). The solvent is removed by rotary evaporation (11 mmHg, room temperature) to give a dark-red oil. The oil is then adsorbed onto 50 g of silica gel, which is then placed at the top of a packed column of silica gel (350 g) (Note 23). The column is eluted with pentanes/diethyl ether, 12:1. The red fractions containing the product are collected and concentrated by rotary evaporation at room temperature to afford 9.3–9.7 g (77–80%) of **1** as a red solid (Note 24).

D. (S,E)-Methyl 2-((R)-4-methyl-1,2-dihydronaphthalen-2-yl)-4-phenylbut-3-enoate. 1-Methyl-3,4-dihydronaphthalene (5.34 g, 37.0 mmol) and Rh_2(*S*-DOSP)$_4$ (7.0 mg, 0.0037 mmol, 0.0001 equiv) (Note 25) are added to a flame-dried, 250-mL 2-necked, round-bottomed flask equipped with a magnetic stir bar and fitted with an argon inlet adaptor and a rubber septum. The flask is evacuated (0.05 mmHg) and purged with argon twice and then maintained under a positive argon pressure. 2,2-Dimethylbutane (40 mL) (Note 26) is added by syringe and vigorous stirring is begun. A solution of (*E*)-methyl 2-diazo-4-phenylbut-3-enoate **1** (8.23 g, 40.7 mmol, 1.10 equiv) (Note 27) in 2,2-dimethylbutane (60 mL) is added by syringe pump over 4 h (Note 28). The solution is stirred for an additional 10 h and the solvent is then removed by rotary evaporation (11 mmHg, room temperature) to give a light-yellow solid. Absolute ethanol (20 mL) is added to the crude material and a reflux condenser is attached to the flask. The solution is heated to reflux to dissolve the material and then allowed to cool to room temperature, whereupon the product crystallizes as an off-white solid. The solution is then cooled to 0 °C for 2 h and the product is collected by vacuum filtration on a Büchner funnel to give 8.0–8.3 g (68–70%) of pure **2** as an off-white solid (Note 29).

2. Notes

1. 3-Chloropropionic acid was purchased from Acros Organics and used as received.
2. Triphenylphosphine was purchased from Acros Organics and used as received.
3. Xylenes were purchased from Mallinckrodt Inc. and used as received.
4. Acetonitrile was purchased from J. T. Baker and used as received.
5. Scraping the precipitated material was crucial to induce solidification of the product, which initially separated out of the xylene solution as a taffy-like material.
6. Washing should be continued with diethyl ether until the wash solvent no longer appears white after filtration.
7. The product displayed the following properties: mp 201–204 °C (sealed tube); IR (neat): cm^{-1} 2900, 1742, 1421, 1586, 1489, 1439, 1421, 1388, 1352, 1220, 1165, 1116, 1037, 996, 950, 910, 859, 816, 752, 698, 613;

^{1}H NMR (500 MHz, $CDCl_3$) δ: 3.09–3.15 (m, 2 H), 3.69–3.75 (m, 2 H), 7.69–7.84 (m, 15 H); ^{13}C NMR (125 MHz, $CDCl_3$) δ: 19.1, 28.1, 117.2, 130.4, 133.4, 171.3, 135.2; LRMS EI (relative intensity) *m/z*: 369 ($[M]^+$, 6), 317 (5), 262 (100), 183 (25), 152 (4); Anal. Calcd for $C_{21}H_{20}ClO_2P$: C, 68.02; H, 5.44. Found: C, 68.03; H, 5.48. The spectroscopic data were consistent with previously reported literature values.[2a,b] The procedure described herein is a modified version of a literature procedure.[2b]

8. Magnesium turnings were purchased from Fisher Scientific and used as received.

9. Diethyl ether was purchased from EMD Chemicals Inc., dried over 4Å molecular sieves overnight and further dried by pressure filtration through activated alumina.

10. Iodomethane (99%) was purchased from Aldrich Chemical Company, Inc. and used as received. It is classified as a carcinogen and should be handled with care

11. α-Tetralone was purchased from Aldrich Chemical Company, Inc. (98%) and used as received.

12. Silica gel was purchased from Silicycle (40-63 μm particle size, 60 Å pore diameter) and loaded as a slurry onto a 30-mm diameter column. Fractions were collected (19 mL) and elution monitored by TLC (silica gel, R_f 0.51, pentane).

13. The submitters occasionally found that the product was contaminated with the isomeric 1-methylene-1,2,3,4-tetrahydronaphthalene in a 25:1 ratio on some of the trials. This impurity was inseparable by chromatography but was converted quantitatively to the desired product by refluxing the mixture in toluene with 1 mol % toluenesulfonic acid monohydrate for 1 h.

14. The product displayed the following properties: R_f 0.51 (pentanes); IR (neat): cm^{-1} 3097, 3060, 2933, 2883, 2830, 2856, 1487, 1450, 1438, 1427, 1378, 1239, 1069, 1039, 1020, 936, 886, 868, 812, 791, 756, 731, 704, 633; ^{1}H NMR (500 MHz, $CDCl_3$) δ: 2.09 (s, 3 H), 2.27–2.31 (m, 2 H), 2.79 (t, J=8.5 Hz, 2 H), 5.88 (m, 1 H), 7.17–7.22 (m, 2 H), 7.22–7.29 (m, 2 H); ^{13}C NMR (125 MHz, $CDCl_3$) δ: 23.4, 28.6, 122.9, 125.6, 126.5, 126.9, 127.6, 132.4, 136.1, 136.5; LRMS EI (relative intensity) *m/z*: 144 ($[M]^+$, 100), 129 (47), 115 (10), 102 (4), 89 (5), 77 (3), 63 (5); Anal. Calcd for $C_{11}H_{12}$: C, 91.61; H, 8.39. Found: C, 91.57; H, 8.22. The spectroscopic data were consistent with previously reported literature values.[3] The procedure described herein is a modified version of a literature procedure.[3]

15. Benzaldehyde (redistilled, 99.5+%) was purchased from Aldrich Chemical Company, Inc. and used as received.

16. THF was purchased from J. T. Baker Chemical Co., dried over 4 Å molecular sieves overnight and further dried by pressure filtration through activated alumina.

17. Potassium *tert*-butoxide (98+%) was purchased from Acros Organics and used as received. The solution in THF was cooled externally to 0 °C in an ice-water bath before and during addition.

18. Dimethyl sulfate was purchased from Acros Organics and used as received. It is classified as a carcinogen and should be handled with care.

19. *p*-Acetamidobenzenesulfonyl azide (*p*-ABSA) is commercially available (Aldrich Chemical Company, Inc.), but was prepared by the submitters using the published procedure, See *Org. Synth., Coll. Vol. IX* **1998**, 422. Although the compound has not exhibited shock sensitivity,[4] *all azide compounds should be handled with care.*

20. DBU was purchased from Acros Organics and used as received.

21. The reaction mixture was concentrated on a rotary evaporator in a fume hood because of the excess dimethyl sulfate that was collected in the receiver.

22. Anhydrous $MgSO_4$ was purchased from Fisher Scientific Company.

23. Silica gel was purchased from Silicycle (40–63 μm particle size, 60 Å pore diameter) and loaded as a slurry onto a 60-mm diameter column.

24. The product displayed the following properties: R_f 0.35 (9:1 pentane/ether); IR (neat): cm^{-1} 2960, 2168, 2088($C=N_2$), 1706, 1626, 1492, 1449, 1438, 1405, 1364, 1318, 1286, 1243, 1194, 1113, 1010, 950, 752, 734, 692; ^1H NMR (500 MHz, $CDCl_3$) δ: 3.86 (s, 3 H), 6.20 (d, *J* =16.0 Hz, 1 H), 6.49 (d, *J*= 16.0 Hz, 1 H), 7.20 (t, *J* = 7.5 Hz, 1 H), 7.30–7.37 (m, 4 H); ^{13}C NMR (125 MHz, $CDCl_3$) δ: 52.05, 111.0, 122.8, 125.6, 126.8, 128.4, 136.5, 165.0; LRMS EI (relative intensity) *m/z*: 202.1 ($[M]^+$, 38), 170.0 (10), 159 (4), 142 (12), 129 (16)); Anal. Calcd for $C_{11}H_{10}N_2O_2$ C, 65.34; H, 4.98; N, 13.85, Found: C, 65.40; H, 4.94; N, 13.70 The procedure described herein is a modified version of a literature procedure.[5] The product should be stored at 0 °C and immediately purified before use if stored for more than ~2 days as it will slowly undergo an electrocyclization reaction to form a pyrazole. The pyrazole is observed as a white crystalline solid. *All diazo compounds should be handled carefully,* although in this case, the only thermal decomposition pathway exhibited by this compound in the absence of

catalyst is electrocyclization to the pyrazole. The compound should not be stored in solution as this accelerates the decomposition reaction.

25. Rh_2(*S*-DOSP)$_4$ is commercially available from Strem Chemicals, Inc. and Aldrich Chemical Company, Inc. and is air stable. Alternatively, it can be readily made by the published procedure.[6]

26. 2,2-Dimethylbutane (DMB) was purchased from Lancaster Synthesis and distilled from sodium under argon prior to use.

27. (*E*)-Methyl 2-diazo-4-phenyl-3-butenoate **1** was freshly purified when used. The checkers found that 500 μL of diethyl ether was needed to wet the solid before it was diluted with DMB. If the diethyl ether was not used, the starting material was not completely soluble in the DMB.

28. The checkers found that carrying out the syringe pump addition with a 50-mL syringe led to clogging from formation of the insoluble pyrazole. The best results were obtained by carrying out the addition in six portions with a 10-mL syringe with a 16-gauge needle (40 min per addition for the total of a 4-h addition period).

29. The product was obtained in 87–92% yield when purified by silica gel column chromatography (See Table 2), but contained a slight greenish tinge. The product displayed the following properties: mp 77–79 °C (sealed tube), R_f 0.35 (10:1 pentane/ether); IR (KBr): cm^{-1} 2946, 1956, 1886, 1735, 1638, 1487, 1450, 1427, 1380, 1331, 1298, 1264, 1215, 1151, 1098, 1024, 993, 972, 944, 902, 842, 807, 761, 723, 659; ^{1}H NMR (500 MHz, $CDCl_3$) δ: 2.07 (s, 3 H), 2.72 (m, 1 H), 2.84–2.88 (m, 2 H), 3.11 (t, *J*=9.5 Hz, 1 H), 3.69 (s, 3 H), 5.73 (d, *J*=3.5 Hz, 1 H), 6.14 (dd, *J*=16.0, 9.5 Hz, 1 H), 6.42 (d, *J*=16.0 Hz, 1 H), 7.08 (d, *J*= 7.0 Hz, 1 H), 7.15 (td, *J*= 7.0, 2.0 Hz, 1 H), 7.23–7.24 (m, 4 H), 7.30–7.36 (m, 4 H); ^{13}C NMR (125 MHz, $CDCl_3$) δ: 19.4 (CH_3), 31.7, 36.1, 51.9, 53.2, 123.0, 126.0, 126.2, 126.4, 126.6, 127.2, 127.7, 127.8, 128.5, 134.4, 135.3, 136.6, 173.6; LRMS ESI (relative intensity) *m/z*: 319.2 ($[M+H]^+$,7), 341.1 ($[M+Na]^+$, 0.4), 287 (3), 259.1 (3), 176 (100), 143 (94); HPLC analysis: >99.8:0.2 (unable to accurately integrate minor enantiomer) (Chiralcel OD-H, 2% *i*-PrOH in hexane, 0.8 mL/min, λ = 254 nm, t_R = 7.25 min , major; t_R = 7.99 min, minor); Anal. Calcd for $C_{22}H_{22}O_2$: C, 82.99; H, 6.96. Found: C, 82.62; H, 6.94. $[\alpha]_D$ = –21.4 (c = 1.0, ethanol). The procedure described herein is a modified version of a literature procedure.[7]

Safety and Waste Disposal Information

All hazardous materials should be handled and disposed of in accordance with "Prudent Practices in the Laboratory"; National Academy Press; Washington, DC, 1995.

3. Discussion

Dirhodium tetracarboxylate based catalysts efficiently decompose diazo compounds to generate reactive metal carbenoids that are capable of undergoing extraordinarily regio- and stereoselective reactions.[8] This is especially true when such catalysts are used in conjunction with diazo compounds that contain both electron-donating groups (EDG) and electron-withdrawing groups (EWG). Our work over the last decade in this area has shown that the prolinate based catalyst Rh_2(*S*-DOSP)$_4$ and its second generation relative, the bridged catalyst Rh_2(*S*-biTISP)$_2$ can routinely produce highly diastereo- and enantioselective intermolecular cyclopropanation reactions with the latter catalyst capable of achieving very high turnover numbers (92,000).[9]

While highly stereoselective cyclopropanation reactions have been known for some time, asymmetric intermolecular C–H activation reactions are a more recent development. Extensive work in this area in recent years has led to carbenoid surrogates for many classic organic transformations including the aldol addition,[10] the Mannich reaction,[11] and the Claisen rearrangement.[12] Rh_2(*S*-DOSP)$_4$ has been an exceptional catalyst for these transformations and has been found to perform best with diazo compounds substituted with both a methyl ester (EWG) and an aryl, heteroaryl, or vinyl group (EDG). C–H activation products are regularly formed as single diastereomers and with >90% ee. One of the more spectacular transformations discovered in recent years using this catalyst has been the combined C–H activation/Cope rearrangement reaction.[7,13] It occurs when a vinyl-substituted carbenoid undergoes a C–H activation of an allylic C–H bond that is interrupted by a Cope rearrangement. The product of this transformation will then normally undergo a retro-Cope rearrangement at room temperature or upon heating in toluene to give the thermodynamically more stable "formal" C–H activation product. This transformation is extremely efficient with dihydronaphthalenes as substrates and routinely produces products in high yields and with >98% ee. The process is amenable to a variety of vinyldiazoacetates and substituted dihydronaphthalenes. The procedure described here can be used to synthesize many different vinyldiazoacetates starting from readily available aromatic aldehydes. Several examples of this are shown in Table 1,[5] performed on a 5-mmol scale.

Table 1.

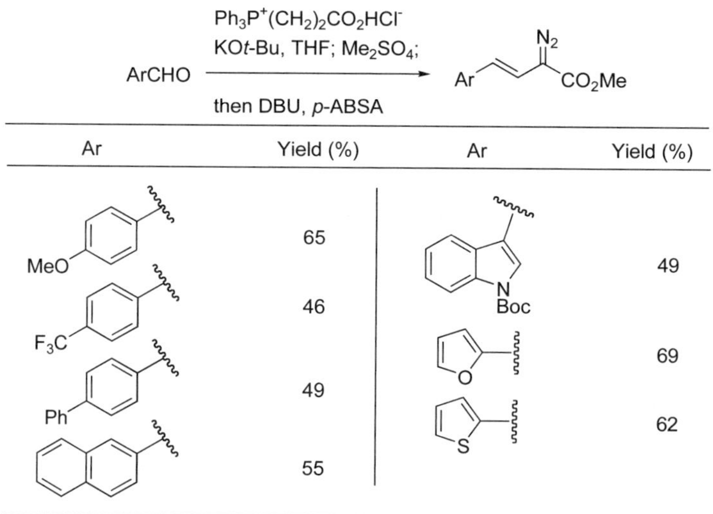

The combined C–H activation/Cope rearrangement reaction has proven to be a very favorable transformation and catalyst loadings as low as 0.01 mol% can be used with no significant drop in yield or stereoselectivity (Table 2).

Table 2.

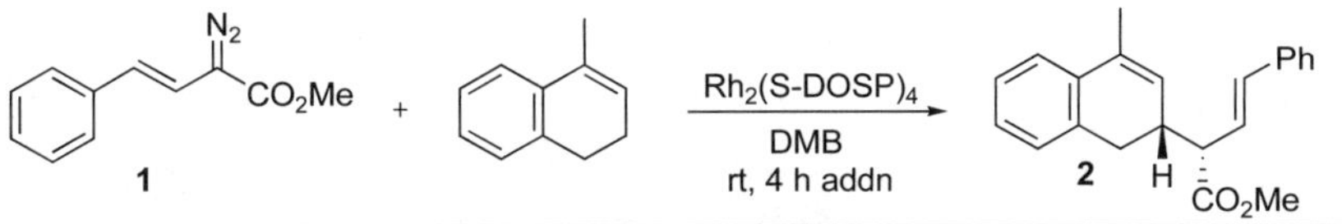

diazo (mmole)	substrate (mmole)	catalyst (mole %)	catalyst T.O.	yield[a] (%)	de (%)	ee (%)
1.20	1.00	0.10	920	92	>94	99.3
12.0	10.0	0.050	1740	87	>94	99.4
12.0	10.0	0.040	2300	92	>94	99.5
12.0	10.0	0.030	2967	89	>94	99.7
12.0	10.0	0.020	4550	91	>94	99.2
12.0	10.0	0.010	9100	91[b]	>94	99.5
12.0	10.0	0.0075	-	73[c]	>94	97.3

[a] isolated yield after flash column chromatography

[b] solution was green at the end of addition-TOF at least 2275 h^{-1}

[c] reaction incomplete-yield from crude ^{1}H NMR ratio of product:sm (2.7:1)

The procedure described here is readily performed on a large scale, as the amount of catalyst required is small compared to the product generated and the crude reaction mixture is easily recrystallized from ethanol to give the pure product, requiring no other purification steps. Although the sequence is illustrated with a methyl-substituted dihydronaphthalene, unsubstituted substrates and other substituents, including siloxy and acetoxy, can be used.[7] This methodology has been successfully used in a variety of applications, including total synthesis (Scheme 1).[7,13]

Scheme 1.

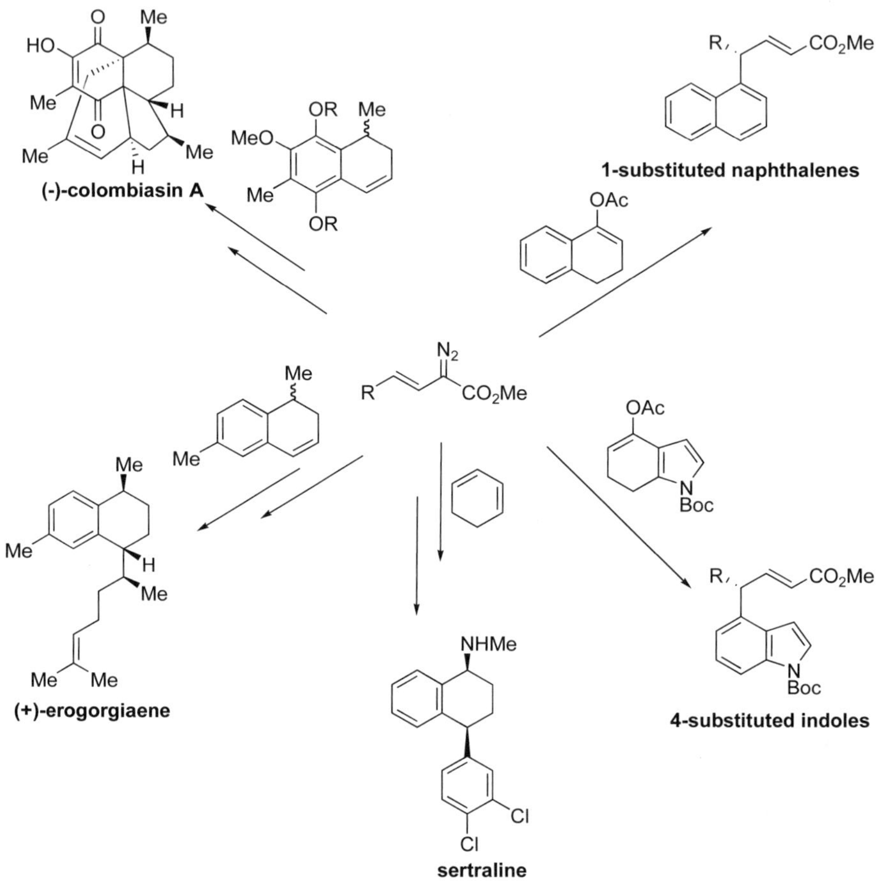

1. Department of Chemistry, University at Buffalo, The State University of New York, Buffalo, NY 14260, E-mail: hdavies@buffalo.edu.
2. (a) Denney, D. B.; Smith, L. C. *J. Org. Chem.* **1962**, *27*, 3404-3408. (b) Narayanan, K. S.; Berlin, K. D. *J. Org. Chem.* **1980**, *45*, 2240-2243.
3. Adamczyk, M.; Watt, D. S.; Netzel, D. A. *J. Org. Chem.* **1984**, *49*, 4226-4237.
4. Baum, J. S.; Shook, D.A.; Davies, H. M. L.; Smith, H. D. *Synth. Commun.* **1987**, *17*, 1709-1716.
5. Davies, H. M. L.; Yang, J.; Manning, J. R. *Tetrahedron: Asymmetry* **2006**, *17*, 665-673.

6. Davies, H. M. L.; Bruzinski, P. R.; Lake, D. H.; Kong, N.; Fall, M. J. *J. Am. Chem. Soc.* **1996**, *118*, 6897-6907.

7. Davies, H. M. L.; Jin, Q. *J. Am. Chem. Soc.* **2004**, *126*, 10862-10863.

8. (a) Davies, H. M. L.; Beckwith, R. E. *Chem. Rev.* **2003**, *103*, 2861-2903. (b) Doyle, M.; McKervey, M.; Ye, T. *Modern Catalytic Methods for Organic Synthesis with Diazo Compounds: From Cyclopropanes to Ylides*; Wiley: New York, 1998.

9. Davies, H. M. L; Venkataramani, C. *Org. Lett.* **2003**, *5*, 1403-1406.

10. Davies, H. M. L; Beckwith, R. E. J.; Antoulinakis, E. G.; Jin, Q. *J. Org. Chem.* **2003**, *68*, 6126-6132.

11. Davies, H. M. L.; Venkataramani, C.; Hansen, T.; Hopper, D. W. *J. Am. Chem. Soc.* **2003**, *125*, 6462-6468.

12. Davies, H. M. L; Yang, J.; Nikolai, J. *J. Organomet. Chem.* **2005**, *690*, 6111-6124.

13. (a) Davies, H. M. L.; Stafford, D. G.; Hansen, T. *Org. Lett.* **1999**, *1*, 233-236. (b) Davies, H. M. L.; Jin, Q. *Proc. Nat. Acad. Sci. U.S.A.* **2004**, *101*, 5472-5475. (c) Davies, H. M. L.; Walji, A. M. *Angew. Chem., Int. Ed.* **2005**, *44*, 1733-1735. (d) Davies, H. M. L.; Dai, X.; Long, M. S. *J. Am. Chem. Soc.* **2006**, *128*, 2485-2490. (e) Davies, H. M. L; Manning, J. R. *J. Am. Chem. Soc.* **2006**, *128*, 1060-1061.

RADICAL ALLYLATION OF *B*-ALKYLCATECHOLBORANES [Ethyl 2-{[(1*S*,2*R*,3*R*,5*S*)-2,6,6-trimethylbicyclo[3.1.1]hept-3-yl]methyl}acrylate]

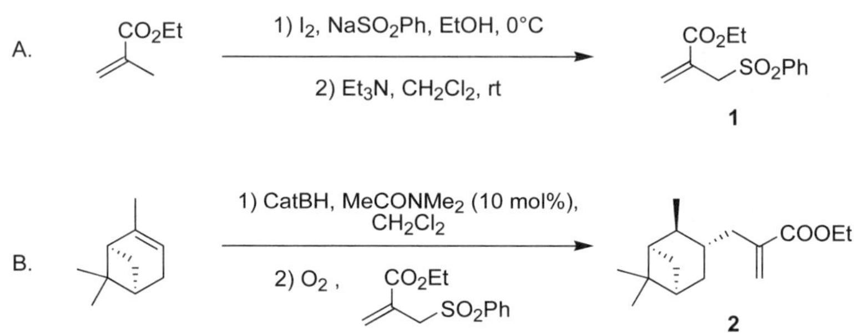

Submitted by Vincent Darmency, Eoin Martin Scanlan, Arnaud Pierre Schaffner and Philippe Renaud.[1]
Checked by Bin Sui and Dennis P. Curran.

1. Procedure

A. 2-(Ethoxycarbonyl)prop-2-en-1-yl phenyl sulfone (**1**).[2] Ethyl methacrylate (7.50 mL, 60 mmol) (Note 1) and absolute ethanol (120 mL) are added to a 250-mL, three-necked, round-bottomed flask (Note 2) equipped with a magnetic stir bar, a glass stopper, a thermometer, and an argon inlet adaptor. The mixture is cooled to 0 °C in an ice bath, the stopper is removed, iodine (18.1 g, 71 mmol) and benzenesulfinic acid sodium salt (22.7 g, 126 mmol) are added with the aid of a plastic funnel, and the stopper is replaced. The temperature of the mixture increases to 6–7 °C upon addition of the sodium salt. The reaction mixture is recooled to 0 °C and stirred for 5 h, then allowed to warm to room temperature. Dichloromethane (100 mL) is added and the dark brown reaction mixture is transferred to a 500-mL separatory funnel. The organic layer is washed with water (2 x 100 mL) (Note 3), and the water is back-extracted with dichloromethane (2 x 25 mL). The combined organic layer is washed with saturated sodium hydrogen carbonate solution (100 mL) and sodium dithionite (5% solution, 2 x 100 mL). The resulting yellow solution is dried over $MgSO_4$, filtered into a 250-mL, one-neck, round-bottomed flask and concentrated to dryness by rotary evaporation (40 °C, 500 mbar) to yield 22.58–22.68 g (98-99%) crude adduct as a dark oil (Note 4).

After addition of a magnetic stir bar, the flask is capped with a septum, placed under an argon atmosphere, and the crude adduct is diluted with dichloromethane (35 mL). Triethylamine (16.7 mL, 120 mmol) is added dropwise by syringe over 10 min (Note 5). The resulting orange-brown mixture is stirred for 9 h at room temperature (Note 6). The reaction mixture is concentrated by rotary evaporation (35 °C, bath temperature) to about half of its original volume (Note 7), and the concentrate is charged onto a column (8 x 18 cm) of 350 g of silica gel (Note 8). The column is eluted with 500 mL of *tert*-butyl methyl ether/cyclohexane (1:4), at which time the eluent is changed to *tert*-butyl methyl ether/cyclohexane (3:7). After elution with about 1 L of the solvent mixture, the product begins to emerge, as identified by TLC analysis (Note 6). The eluting solvent is changed to *tert*-butyl methyl ether/cyclohexane (1:1), and the product fractions are collected. Combination of the fractions, concentration by rotary evaporation (35 °C, bath temperature), and vacuum drying afforded 13.45–13.85 g of **1** as a colorless oil (88–90% yield) (Note 9).

B. Ethyl 2-{[(1S,2R,3R,5S)-2,6,6-trimethylbicyclo[3.1.1]hept-3-yl]methyl}acrylate (**2**).[3,4] (+)-α-Pinene (7.14 mL, 45 mmol), *N,N*-dimethylacetamide (0.42 mL, 4.5 mmol) (Note 10) and dichloromethane (30 mL) are added to a 100-mL, three-necked, round-bottomed flask equipped with a large, oblong magnetic stir bar, a reflux condenser with an argon inlet, a glass stopper, and a straight glass stopcock fitted with rubber septum. The stirred mixture is cooled to 0 °C in an ice bath, and catecholborane (12.0 mL, 113 mmol) (Note 11) is added dropwise via syringe by placing the syringe needle through the septum on the open stopcock. The stopcock is closed, the ice bath is removed and colorless reaction mixture is heated to reflux for 5 h (Note 12). The mixture is cooled to 0 °C, and methanol (3.0 mL, 74 mmol) is carefully added [*CAUTION: vigorous evolution of hydrogen gas*] to quench the excess of catecholborane, and the resulting mixture is warmed to room temperature and stirred for 15 min.

2-(Ethoxycarbonyl)prop-2-en-1-yl phenyl sulfone **1** from Step A (13.7 g, 54 mmol) is added via syringe, and the mixture is heated to reflux. Air (40 mL, 0.33 mmol O_2) (Note 13) is introduced over 1 h by a syringe pump with the needle of the syringe placed through the stopcock and immersed just under the surface of the liquid. After 15–20 min, the clear yellow mixture turns black (Notes 12 and 14). After 1 h, the mixture is cooled and the solvent is removed by rotary evaporation (35 °C, bath temperature). The resulting black oil is dissolved in a minimum of dichloromethane (~20 mL). The solution is charged onto a column (8 x 9 cm) of 200 g silica gel, and the column is eluted with *tert*-butyl methyl ether/cyclohexane 1:19. Fractions containing the product (Note 15) are concentrated by rotary evaporation (35 °C, bath temperature) and dried under high vacuum to provide 8.70–8.75 g compound **2** (77–78%) as a colorless oil (Note 16).

2. Notes

1. The submitters purchased ethyl methacrylate (99%), iodine (99%) and benzenesulfinic acid sodium salt (97%) from Fluka Chemie GmbH. The checkers purchased ethyl methacrylate (99%) and benzenesulfinic acid sodium salt (98%) from Aldrich, and iodine (Certified A.C.S.) from Fischer Scientific. All reagents were used as received.

2. The apparatus is dried in an oven at 130 °C for 24 h, cooled in a dessicator, and maintained under an argon atmosphere during the course of the reaction.

3. Both layers are dark. Use of a flashlight helps to identify the phase boundary. Small additional amounts of water and dichloromethane can be added if the phases do not separate after the first extraction.

4. The intermediate adduct is $PhSO_2CH_2C(I)(CH_3)CO_2Et$: $R_f = 0.58$, 1/1 *tert*-butyl methyl ether/cyclohexane. ^1H NMR (300 MHz, $CDCl_3$): δ: 1.34 (t, $J = 7.2$ Hz, 3 H), 2.45 (s, 3 H), 3.95 (d, $J = 13.8$ Hz, 1H), 4.22−4.32 (m, 2 H), 4.51 (d, $J = 13.8$ Hz, 1 H), 7.60 (m, 2 H), 7.68 (m, 1 H), 7.92 (m, 2 H). The sample becomes a dark solid upon standing in the freezer.

5. The reaction mixture warms during the addition, so the rate of addition must be slow enough to keep the temperature well below the reflux point.

6. Reaction progress can be monitored by TLC analysis on silica eluting with 1/1 *tert*-butyl methyl ether/cyclohexane (UV lamp visualization): starting adduct $R_f = 0.58$; product **1** $R_f = 0.35$.

7. Salts may begin to precipitate towards the end of the concentration. These are loaded onto the column along with the residual liquid.

8. The submitters used SDS silica gel (40-63 mm). The checkers used standard grade silica gel (40-63 mm) from Sorbent Technologies, Inc.

9. ^1H NMR (300 MHz, $CDCl_3$)[2] δ: 1.15 (t, $J = 7.1$ Hz, 3 H), 4.00 (q, $J = 7.1$ Hz, 2 H), 4.15 (d, $J = 0.8$ Hz, 2 H), 5.89 (narrow q, $J = 0.8$, 0.6 Hz, 1 H), 6.49 (d, $J = 1$ Hz, 0.6 H), 7.52 (m, 2 H), 7.62 (m, 1 H), 7.84 (m, 2 H);

^{13}C NMR (75 MHz, $CDCl_3$) δ: 13.9, 57.4, 61.4, 128.6, 128.9, 129.0, 133.2, 133.8, 138.2, 164.6. The submitters report that the product can be further purified by distillation in a Kugelrohr oven; bp 90–95 °C (6 mmHg).

10. The submitters purchased α-pinene (97%) and *N,N*-dimethylacetamide (99%) from Fluka Chemie GmbH. The checkers purchased (*R*)-(+)-α-pinene (98%, 91% ee) from Aldrich and *N,N*-dimethylacetamide (99%) from J. T. Baker. These were used as received.

11. The quality of the catecholborane is crucial for success. The submitters purchased catecholborane from BASF corporation, Mount Olive, NJ, USA and distilled it (bp 50 °C, 50 mmHg) prior to use. The checkers used catechol borane (98%) directly from a fresh bottle purchased from Aldrich.

12. The progress of the hydroboration reaction and the radical allylation reaction can be monitored by GC analysis on an Agilent MP-1 methyl siloxane capillary column (19091Z-413E, 30 m x 0.32 mm); temperature ramp, 10 °C per min from 50 °C to 315 °C. T_R; pinene, 3.59 min; hydroboration adduct, 15.21 min; **1**, 15.29 min; **2**, 12.68 min. The submitters used a CE instrument, MEGA Series HRGC fitted with an Optima delta-3 0.25 μm fused silica capillary column from Macherey-Nagel, 30 m x 0.25 mm; temperature ramp, 6 °C per min from 60 °C to 280 °C; T_R pinene, 7.38 min; **1**, 31.53 min; **2** = 24.86 min.

13. The submitters report that initiation with di-*tert*-butyl hyponitrite in refluxing dichloromethane afforded a similar reaction yield.[3]

14. GC analysis at this point (Note 12) indicates that the reaction is complete. The checkers observed the black color after 15–20 min, while the submitters observed the color change after about 1 h.

15. Thin layer chromatography (TLC) analysis is used to identify product fractions. TLCs are performed by using Merck silica gel 60 F_{254} analytical plates; detection with UV or by dipping in a solution of $KMnO_4$ (3 g), K_2CO_3 (20 g), NaOH 5% (5 mL) in H_2O (300 mL) and subsequent heating. Product R_f 0.62 in *tert*-butyl methyl ether/cyclohexane (1:9).

16. ^{1}H NMR (300 MHz, $CDCl_3$): δ: 0.74 (d, *J* = 9.5 Hz, 1 H), 0.97 (s, 3 H), 1.00 (d, *J* = 7.1 Hz, 3H), 1.16 (s, 3 H), 1.28 (t, *J* = 7.1 Hz, 3 H), 1.40 (ddd, *J* = 13.3, 5.9, 2.6 Hz, 1 H), 1.66 (quintd, *J* = 7.1, 1.9 Hz, 1 H), 1.75 (td, *J* = 5.7, 1.9 Hz, 1 H), 1.82–1.97 (m, 2 H), 2.04 (tdd, *J* = 11.4, 3.5, 2.1 Hz, 1 H), 2.13 (ddd, *J* = 13.5, 9.3, 0.7 Hz, 1 H), 2.27 (dtd, *J* = 9.5, 6.2, 2.1 Hz, 1 H), 2.51 (ddd, *J* = 13.5, 5.1, 1.0 Hz, 1 H), 4.19 (m, 2 H), 5.52 (dd, *J* = 1.7, 1.0 Hz, 1 H), 6.11 (d, *J* = 1.7 Hz, 1 H); ^{13}C NMR (75 MHz, $CDCl_3$): δ: 14.1,

21.4, 22.9, 28.0, 34.0, 34.2, 34.6, 38.8, 41.9, 43.4, 43.5, 48.2, 60.4, 125.3, 139.8, 167.3; MS (EI): *m/z* (%): 250 [M^+], 235, 207, 194, 176, 137, 121, 107, 93, 81, 67, 55; $[\alpha]_D^{20}$ –31.5 (c, 1.0, $CHCl_3$). If desired, the product can be further purified by vacuum distillation (bp 113–114 °C, 0.45 mmHg) to provide a clear oil; Anal. calcd for $C_{16}H_{26}O_2$: C, 76.75; H, 10.47. Found: C, 76.77; H, 10.41.

Safety and Waste Disposal Information

All hazardous materials should be handled and disposed of in accordance with "Prudent Practices in the Laboratory"; National Academy Press; Washington, DC, 1995.

3. Discussion

Here we describe an efficient one-pot radical hydroallylation of an alkene under tin-free conditions.[3,4] *B*-Alkylcatecholboranes are easily prepared by hydroboration of alkenes with catecholborane according to the efficient and cost-effective conditions developed by Fu.[5] These are used as radical precursors in a fragmentation process where the resulting benzenesulfonyl radical can sustain the radical chain. This methodology has proven to be an effective alternative to radical reactions mediated by toxic tin reagents. Interestingly, this radical approach complements nicely Knochel's procedure for the allylation of organoboranes, which requires transmetalation to organozinc derivatives.[6]

Oxygen-centered radicals react efficiently with *B*-alkylcatecholboranes.[7] The use of oxygen from air as the radical initiator is efficient for a large-scale reaction. Care is always required when oxygenating warm organic solvents. In this reaction, slow syringe pump addition of air is a suitable way to continuously provide the small amounts of oxygen needed for initiation. This transformation can be carried out under mild conditions and is tolerant of a wide range of organic functionalities.

Typical results of this reductive allylation (hydroallylation) of alkenes are shown in Table 1. The allylated products were obtained in satisfactory to excellent yields by using only 1.2 equivalents of the allyl sulfones with primary, secondary and tertiary alkyl radicals (Entries 1-3). The unsubstituted allyl sulfone also reacts under these conditions and provides the volatile allylated product in moderate isolated yield (Entry 4). Finally, allylic sulfones bearing a sulfonyl group (Y = $PhSO_2$) and a bromine atom

(Y = Br) reacted equally well (Entries 5-6). The stereochemical outcome of all these reactions is rationalized in a straightforward manner, both the hydroboration and the radical reaction occur from the less hindered face of the alkene and of the radical, respectively.

The allylation chemistry described here has been extended to a tandem radical process involving conjugate addition of a *B*-alkylcatecholborane to an activated alkene followed by allylation. The whole process can be considered as a unique and selective coupling of three different alkenes.[8]

Table 1

Entry	Alkene	Trap	Product	Yield	dr
1	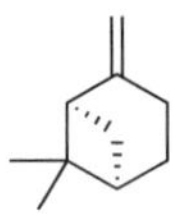	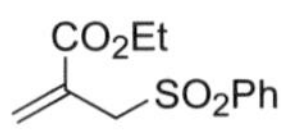	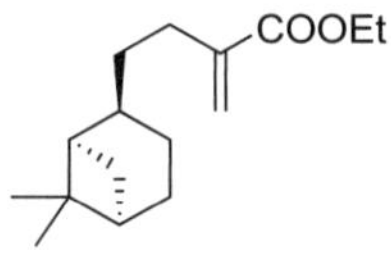	62 %[a]	90/10
2	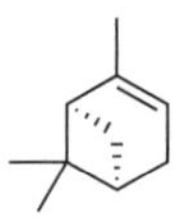	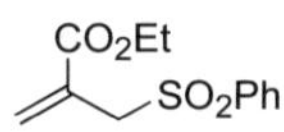	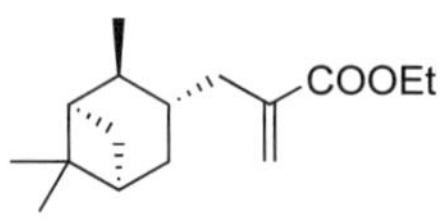	80 %[b]	>98/2
3		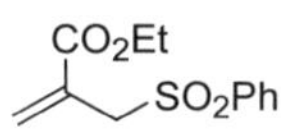	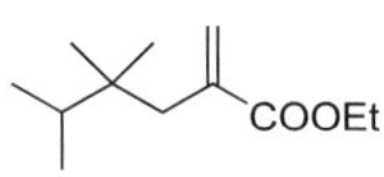	65 %[a]	-
4	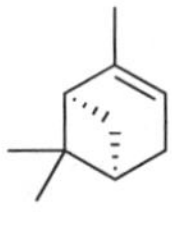	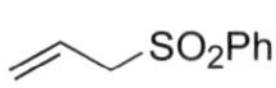	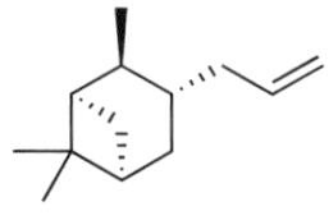	52 %[a]	95/5
5		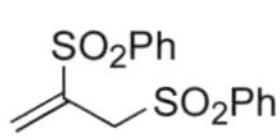	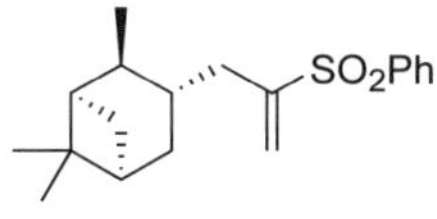	89 %[a]	96/4
6	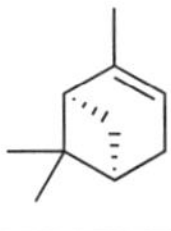	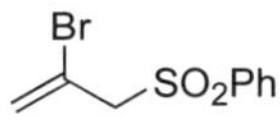	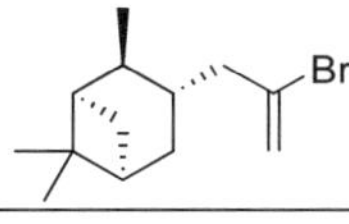	58 %[a]	96/4

a) 2-3 mmol scale b) 45 mmol scale

1. University of Bern, Department of Chemistry and Biochemistry, Freiestrasse 3, 3000 Bern 9 (Switzerland).
2. Ghera, E.; Yechezkel, T.; Hassner, A. *J. Org. Chem.* **1990**, *55*, 5977.
3. Schaffner, A.-P.; Renaud, P. *Angew. Chem. Int. Ed.* **2003**, *42*, 2658.
4. Schaffner, A.-P.; Becattini, B.; Ollivier, C.; Weber, V.; Renaud, P. *Synthesis* **2003**, 17, 2740.
5. Garett, C. E.; Fu, G. C. *J. Org. Chem.* **1996**, *61*, 3224.
6. Boudier, A.; Darcel, C.; Flachsmann, F.; Micouin, L.; Oestreich, M.; Knochel, P. *Chem. Eur. J.* **2000**, *6*, 2748.
7. Schaffner, A.-P.; Renaud, P. *Eur. J. Org. Chem.* **2004**, 2291.
8. Schaffner, A. P.; Sarkunum, K.; Renaud, P. *Helv. Chim. Acta* **2006**, *89*, 2450.

SUZUKI-MIYAURA CROSS-COUPLING: PREPARATION OF 2'-VINYLACETANILIDE
[*N*-(2-Ethenylphenyl)acetamide]

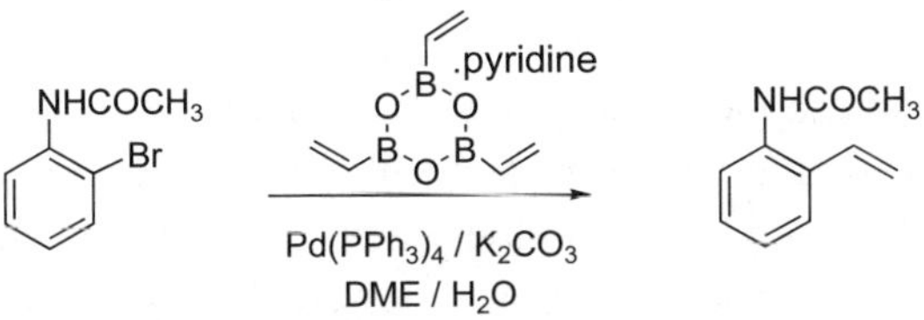

Submitted by Bertrand Cottineau, Albane Kessler and Donal F. O'Shea.[1]
Checked by Anuradha Gupta and Marvin J. Miller.

1. Procedure

A 500-mL, two-necked, round-bottomed flask covered in tin-foil and is equipped with a stirring bar, a reflux condenser fitted with a nitrogen inlet, and a stopper. The flask is charged with *N*-(2-bromophenyl)acetamide (10.0 g, 46.6 mmol, Note 1), ethylene glycol dimethyl ether (180 mL) (Note 2), and tetrakis(triphenylphosphine)palladium(0) (1.08 g, 0.9 mmol) (Note 3). The apparatus is maintained under an atmosphere of nitrogen during the course of the reaction. The mixture is stirred at room temperature for 20 min then potassium carbonate (6.44 g, 46.6 mmol) (Note 4) dissolved in distilled water (55 mL) is added via funnel, followed by 2,4,6-trivinylcyclotriboroxane-pyridine complex (5.6 g, 23.3 mmol) (Note 5). The reaction mixture is stirred and heated at reflux in an oil bath for 20 h, then cooled to ambient temperature. Distilled water (75 mL) is added via a funnel, and the resulting mixture is filtered on a Büchner funnel. The filtrate is transferred to a separatory funnel and extracted with diethyl ether (3 x 100 mL) (Note 6). The combined organic phases are dried over sodium sulfate, filtered on filter paper and concentrated to dryness by rotory evaporation (30 °C, 25 mmHg). The resulting yellow solid is purified by column chromatography (Note 7) affording *N*-(2-vinylphenyl)acetamide (5.9 g, 37.3 mmol) as a pale yellow solid. The solid is dissolved in a hot mixture of cyclohexane:dicholoromethane (4:1) (55 mL) and the warm mixture is filtered through a Büchner funnel. The filtrate is allowed to cool to room temperature for 20 min. The flask is immersed for 30 min in an ice bath in order to complete precipitation. The resulting crystals are collected by suction filtration on a Büchner funnel, washed with cyclohexane (10 mL), and dried under reduced pressure (15 h at 0.1 mmHg) to provide *N*-(2-vinylphenyl)acetamide (5.30 g, 71%) as a white solid (Note 8).

2. Notes

1. *N*-(2-Bromophenyl)acetamide (96%) was purchased from Aldrich Chemical Company, Inc. and was used without purification.

2. Ethylene glycol dimethyl ether (DME) was purchased from Aldrich Chemical Company, Inc. and passed through a column of Merck aluminium oxide 90 (30 g) immediately prior to use.

3. Tetrakis(triphenylphosphine)palladium(0) was used as received from Aldrich Chemical Company, Inc.

4. Potassium carbonate (99%) was used as received from Aldrich Chemical Company, Inc.

5. 2,4,6-Trivinylcyclotriboroxane-pyridine complex (purity not indicated) was used as received from Frontier Scientific, Inc.

6. Diethyl ether was used as received from Aldrich Chemical Company, Inc. (HPLC grade).

7. The yellow solid, dissolved in 20 mL of dichloromethane, was applied to 4.5 cm diameter column packed with 600 g of flash silica gel (Merck silica 60 mesh 0.040–0.063 mm). The product was eluted with diethyl ether/cyclohexane (9:1) (3.5 L). R_f of 2'-vinylacetanilide: 0.26 using the same eluting solvent and TLC aluminium sheets with silica gel $60F_{254}$ (received from Aldrich Chemical Company, Inc.). Dichloromethane, diethyl ether and cyclohexane were used as received from Aldrich Chemical Company, Inc. (HPLC grade).

8. Analytical data: ^{1}H NMR (500 MHz, DMSO-d_6) δ: 2.05 (s, 3 H), 5.30 (dd, *J* = 11.0, 1.1 Hz, 1 H), 5.77 (dd, *J* = 17.5, 1.1 Hz, 1 H), 6.89 (dd, *J* = 17.5, 11.0 Hz, 1 H), 7.16–7.27 (m, 2 H), 7.37 (d, *J* = 7.9 Hz, 1 H), 7.61 (d, *J* = 7.8 Hz, 1H), 9.52 (bs, 1H); ^{13}C NMR (125 MHz, DMSO-d_6) δ: 23.1, 115.3, 125.2, 125.4, 126.2, 127.9, 131.5, 132.4, 135.1, 168.5; mp: 94–95 °C (lit.[3a] 89–90 °C); IR (KBr) cm^{-1}: 3228.8, 1648. MS (ES^-) *m/z* 160.1; Anal. required for $C_{10}H_{11}NO$: C, 74.51; H, 6.88; N, 8.69; Found: C, 74.44; H, 6.99; N, 8.61. HPLC analysis [column: Atlantis™ C_{18} 5 μm 4.6 x 250 mm, elution: 80% CH_3CN 20% H_2O (pH = 4, adjusted with 100 μL of HCO_2H in 1.1 L of water), 1 mL/min, 1300 psi, retention time 3.6 min] gives 100% purity at 200 and 254 nm.

Waste Disposal Information

All hazardous materials should be handled and disposed of in accordance with "Prudent Practices in the Laboratory"; National Academy Press; Washington, DC, 1995.

3. Discussion

We have previously reported a synthetic methodology for the generation of substituted styrene derivatives using 2,4,6-trivinylcyclotriboroxane-pyridine complex as the vinyl source in a Suzuki-Miyaura cross-coupling protocol.[2] Functionalized styrenes are of considerable synthetic importance as key synthetic intermediates and for the generation of new polymeric materials. Specifically *N*-(2-vinylphenyl)-acetamide is an intermediate used for the synthesis of the indole ring system.[3]

Numerous palladium-catalyzed methods have been described in which aryl halides are utilized as starting substrates in reactions with a variety of reagents as sources of the alkene functionality. To date the most commonly utilized approaches have been the Heck[4] and Stille[5] methodologies, which use ethene and tributyl(vinyl)tin, respectively, as the vinyl provider. Vinylmagnesium bromide,[6] trimethylvinylsilane,[7] vinyltrimethylsiloxane,[8] vinylpolysiloxanes,[9] potassium vinyltrifluoroborate,[10] vinylboronic acid dibutyl ester,[11] trivinylindium,[12] vinylbromide,[13] and divinylaluminium-2-dimethylaminoethanolate[14] have also been demonstrated as coupling reagents.

Aryl halides with *ortho*-subsituents are known to be challenging substrates for cross-coupling reactions of boronic acids, therefore we chose *N*-(2-bromophenyl)acetamide as a test substrate.[15] The target product has been previously synthesised by the coupling of ethene with *N*-(2-bromophenyl)acetamide using a high pressure Heck protocol.[3a]

In our reaction the cross-coupling proceeds efficiently by using a standard coupling procedure without the need of specialized catalysts. The reaction is successful with 0.5 mole equivalent of the trivinylboronic anhydride indicating that the anhydride provides more than one of the vinyl groups for reaction. In summary, 2,4,6-trivinylcyclotriboroxane-pyridine complex serves as a versatile coupling partner with aryl halides for the generation of substituted styrenes.

1. Centre for Synthesis and Chemical Biology, Conway Institute, Department of Chemistry, University College Dublin, Belfield, Dublin 4, Ireland.
2. Kerins F.; O'Shea D. F. *J. Org. Chem.* **2002**, *67*, 4968.
3. (a) Kasahara, A.; Izumi, T.; Murakami, S.; Miyamoto, K.; Toshimi, H. *J. Heterocyc. Chem.* **1989**, *26*, 1405. (b) Izumi, T.; Sugano, M.; Konno, T. *J Heterocyc. Chem.* **1992**, *29*, 899.
4. Plevyak, J. E.; Heck, R. F., *J. Org. Chem.* **1978**, *43*, 2454.
5. McKean, D. R.; Parrinello, G.; Renaldo, A. F.; Stille, J. K. *J. Org. Chem.* **1987**, *52*, 422.
6. Bumagin, N. A.; Luzikova, E. V. *J. Organomet. Chem.* **1997**, *532*, 271.
7. (a) Hatanaka, Y.; Hiyama, T. *J. Org. Chem.* **1988**, *53*, 918. (b) Jeffery, T. *Tetrahedron Lett.*, **1999**, *40*, 1673.
8. Mowery, M. E.; DeShong, P. *J. Org. Chem.* **1999**, *64*, 1684.
9. (a) Denmark, S. E.; Wang, Z. *Synthesis* **2000**, 999. (b) Denmark, S. E.; Wang, Z. *J. Organomet. Chem.*, **2001**, *624*, 372. (c) Denmark, S. E.; Sweis, R.F. *Acc. Chem. Res.* **2002**, *35*, 835.
10. (a) Darses, S.; Michaud, G.; Genêt, J.-P. *Tetrahedron Lett.* **1998**, *39*, 5045. (b) Molander, G. A.; Rivero, M. R. *Org. Lett.* **2002**, *4*, 107. (c) Molander, G. A.; Bernardi, C. R. *J. Org. Chem.* **2002**, *67*, 8424.
11. (a) Hunt, A. R.; Stewart, S. K.; Whiting, A. *Tetrahedron Lett.* **1993**, *34*, 3599. (b) Lo Galbo, F.; Occhiato, E. G.; Guarna, A.; Faggi, C. *J. Org. Chem.* **2003**, *68*, 6360.
12. Pérez, I.; Pérez Sestelo, J.; Sarandeses, L. A. *J. Am. Chem. Soc.* **2001**, *124*, 4155.
13. Lando, V. R.; Monteiro, A. L. *Org. Lett.* **2003**, *16*, 2891.
14. Schumann, H.; Kaufmann, J.; Schmalz, H.-G.; Böttcher, A.; Gotov, B. *Synlett*, **2003**, 1783.
15. Suzuki, A. *Pure Appl. Chem.* **1994**, *66*, 213.

SYNTHESIS OF 6,9,12,15,18-PENTAMETHYL-1,6,9,12,15,18-HEXAHYDRO(C_{60}-I_h)[5,6]FULLERENE

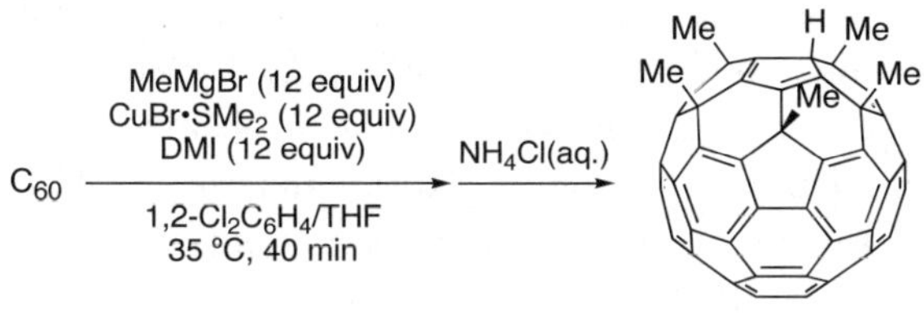

Submitted by Yutaka Matsuo, Ayako Muramatsu, Kazukuni Tahara, Madoka Koide, and Eiichi Nakamura.[1]
Checked by Peter Wipf and David L. Waller.[2]

1. Procedure

A 200-mL two-necked, round-bottomed flask (Note 1) connected to a vacuum/nitrogen manifold through a three-way stopcock is equipped with a Teflon-coated magnetic stirring bar, a vacuum/nitrogen inlet, and a glass stopper. A microcrystalline sample of [60]fullerene (2.00 g, 2.78 mmol) (Note 2) is placed in the 200-mL flask and the apparatus is flushed with nitrogen. The glass stopper is replaced with a rubber septum, and 1,2-dichlorobenzene (90 mL) (Note 3) is introduced into the flask via a syringe under nitrogen. The rubber septum is again replaced with a glass stopper. The reaction mixture is cooled with an ice/water bath and stirred under reduced pressure (1 mmHg) for 30 min to remove dissolved oxygen. Then the flask is flushed with nitrogen and warmed to room temperature (approximately 23 °C).

A 500-mL three-necked, round-bottomed flask (Note 1), connected to a vacuum/nitrogen manifold through a three-way stopcock, is equipped with a 3-cm Teflon-coated magnetic stir bar, a vacuum/nitrogen inlet, a 100-mL pressure-equalizing dropping funnel with a glass stopper, and a glass stopper. The flask is charged with copper(I) bromide-dimethyl sulfide complex (6.84 g, 33.3 mmol) (Note 4) and flushed with nitrogen (Note 5). The glass stopper on the top of the pressure-equalizing dropping funnel is replaced with a rubber septum, and the 1,2-dichlorobenzene solution of [60]fullerene is transferred to the dropping funnel via a Teflon cannula. The glass stopper on the 500-mL flask is replaced with a rubber septum and tetrahydrofuran (47 mL) (Note 6) is added to the copper complex via syringe.

The resulting white suspension is stirred and the flask is placed into an oil bath (approximately 23 °C). A tetrahydrofuran solution of methylmagnesium bromide (3 M, 11.1 mL, 33.3 mmol) (Note 7) is added to this white suspension by syringe, followed by 1,3-dimethyl-2-imidazolidinone (DMI, 3.62 mL, 33.3 mmol) (Note 8). The organocopper reagent is quickly warmed in an oil bath to 35 °C within 5 min. The 1,2-dichlorobenzene solution of [60]fullerene in the dropping funnel is added at once to the organocopper reagent (Note 9). The color of the suspension changes to dark brown. The dropping funnel is replaced with a glass stopper.

After stirring for 40 min at 35 °C (Note 10), 3.0 mL of degassed saturated aqueous ammonium chloride solution is added quickly via syringe. The color of the mixture turns from dark brown to reddish brown. The septum is replaced with a glass stopper and the reaction mixture is stirred under reduced pressure (1 mmHg) at 23 °C to remove tetrahydrofuran and dimethyl sulfide until the volume is halved.

The concentrated mixture is diluted with degassed toluene (200 mL) and filtered through a pad of silica gel (Note 11) with toluene as eluent (Notes 12 and 13). The vermilion filtrate is collected into a 1-L, round-bottomed flask and concentrated on a rotary evaporator under reduced pressure (10 mmHg) at 40 °C. To remove 1,2-dichlorobenzene the bath temperature is raised to 80 °C until solid just begins to appear on the inside wall of the flask (Note 14). Nitrogen is added while the vacuum is broken, and the flask is removed from the rotary evaporator. Degassed methanol is added gently against the inside wall of the flask (Note 15). As methanol mixes with the product, the solution becomes cloudy and a precipitate forms. The addition is continued until precipitation is complete. The total amount of methanol required is approximately 400 mL. The precipitate is collected by filtration on a Buchner funnel (Note 16) and washed with methanol (5 x 10 mL) and hexane (3 x 10 mL). The product is dried under reduced pressure (1 mmHg) for 12 h at room temperature to obtain red microcrystals of the title compound (2.08 g, 2.61 mmol) in 94% overall yield and 91% purity (determined by HPLC analysis) (Note 17). The product can be used as is for further transformations, or it can be purified by preparative HPLC (Notes 18 and 19).

2. Notes

1. All glassware was dried in an oven (110 °C), assembled, and allowed to cool under nitrogen atmosphere. All solvents used for workup

need to be degassed by N_2-bubbling since the final product as well as the intermediate are susceptible to air oxidation.

2. [60]Fullerene was provided by Frontier Carbon Corporation.

3. The checkers purchased 1,2-dichlorobenzene from Aldrich Inc. (99%) and used it without further purification. The submitters purchased 1,2-dichlorobenzene from Kanto Kagaku (>99.0%) and used it without further purification. The water content was checked with a Karl-Fischer Moisture Titrator (MK-210, Kyoto Electronics Company) to be less than 30 ppm.

4. Although the copper bromide dimethyl sulfide complex may be purchased from commercial suppliers and may work well, it is advisable to prepare it freshly because a successful reaction crucially depends on the purity of the copper complex. Copper bromide dimethyl sulfide complex, $CuBr{\bullet}SMe_2$, was synthesized from commercially available CuBr (Aldrich Inc. (98%, checkers); Wako Pure Chemical Industries Ltd., (submitters)) according to the following literature procedures.[3,4] CuBr (25 g) is washed with methanol (4 x 50 mL) to remove colored impurities, and dried under reduced pressure for 1 h. The resulting green-tinged white powder of CuBr is dissolved in dimethyl sulfide (60 mL), and insoluble impurities are removed by filtration. The filtrate is diluted with hexane (200 mL) and the precipitate is collected by filtration under suction. After washing the colorless crystals five times with hexane, the light-sensitive white crystals are dried under a stream of nitrogen for 3 h. Drying under reduced pressure causes loss of the dimethyl sulfide ligand, which adversely affects the reaction. CAUTION: All procedures need to be performed in a well-ventilated hood because of the foul odor of dimethylsulfide.

5. The dimethyl sulfide complex $CuBr{\bullet}SMe_2$ should not be kept under reduced pressure, as it will lose the sulfide ligand.[5]

6. Anhydrous tetrahydrofuran was purchased from EMD Chemicals Inc. (checkers) or Wako Pure Chemical Industries Ltd. (submitters) and used as received. The water content was checked with a Karl-Fischer Moisture Titrator (MK-210, Kyoto Electronics Company) to be less than 30 ppm.

7. A tetrahydrofuran solution of MeMgBr was purchased from Aldrich Inc. (3 M, checkers) or Tokyo Kasei Kogyo Co., Ltd. (1 M, submitters) and used as received. Excess Grignard reagent is necessary to ensure completion of the reaction, since the copper reagent decomposes in competition with the desired reaction. The amount of excess Grignard reagent necessary is therefore reagent-dependent, and, for instance, the

corresponding phenylation reaction under similar conditions may need about 8 equiv of PhMgBr and copper salt.

8. 1,3-Dimethyl-2-imidazolidinone (>98.0%) was purchased from Aldrich Inc. (checkers) or Tokyo Kasei Kogyo Co., Ltd. (submitters) and used without purification.

9. All operations from the addition of the solution of MeMgBr to the addition of the 1,2-dichlorobenzene solution of [60]fullerene are performed in about 15 min.

10. The flask is connected to a nitrogen-inlet/outlet during the reaction so that positive pressure due to ethane formation can be released.

11. Filtration is performed by silica gel flash chromatography (45 x 200 mm size) with the use of 90 g of silica gel (Sorbent Technologies Silica gel, particle size 230 – 450 mesh (checkers); Kanto Kagaku silica gel 60, spherical neutral, particle size 40 – 100 mm (submitters)).

12. The flask is washed with toluene three to ten times (the total amount is approximately 100 mL). The silica gel is repeatedly washed with toluene until the color of the filtrate turns to pale yellow.

13. The product gradually reacts with oxygen in air to afford $C_{60}Me_5O_nH$ (n = 2-3) in solution, therefore filtration needs to be performed as quickly as possible.

14. Care must be taken to stop evaporation when solid material just appears on the wall.

15. To minimize air oxidization of the product, methanol must be added immediately after the evaporation is complete.

16. If a part of the fine powder of the precipitate passes through the filter paper, the mother liquor can be filtered using the same filter paper.

17. HPLC analysis (flow rate: 1 mL/min) was performed on a Buckyprep column (Nacalai Tesque. Co. Ltd., 4.6 x 250 mm) with toluene/isopropanol (7/3) as eluent or an ODS column (4.6 x 250 mm) with toluene/isopropyl alcohol (3/7) as eluent.

18. The product shows the following spectroscopic properties: IR (neat) cm^{-1}: 2973, 2912, 2858, 1572, 1546, 1518, 1459, 1370, 1324, 1286, 1265, 1234, 1169, 1145, 1128, 792, 747, 684; ^1H NMR (500 MHz, $CS_2/CD_3COCD_3(5\%)$) δ: 2.34 (s, 6 H), 2.35 (s, 6 H), 2.46 (s, 3 H), 4.50 (s, 1 H, C_{60}-*H*); ^{13}C NMR (100 MHz, $CS_2/CD_3COCD_3(5\%)$) δ: 27.23 (2C), 27.61 (2C), 33.14 (1C), 51.37 (1C), 51.45 (2C), 53.51 (2C), 59.75 (1C), 143.61 (2C), 144.12 (2C), 144.48 (2C), 144.53 (2C), 144.77 (2C), 144.89 (4C), 145.20 (2C), 145.42 (2C), 145.87 (2C), 145.96 (2C), 146.98 (1C), 147.14 (2C), 147.24 (2C), 148.03 (2C), 148.12 (2C), 148.29 (3C), 148.54 (2C),

148.60 (2C), 148.91 (4C), 149.10 (2C), 150.12 (2C), 154.27 (2C), 154.37 (2C), 154.62 (2C), 157.76 (2C).

19. The purity of the product in solution stored in the presence of air decreased to 80% in 24 h. Although the product is rather stable as a solid, its storage under air for several months may lead to a decrease of the purity.

Safety and Waste Disposal Information

All hazardous materials should be handled and disposed of in accordance with "Prudent Practices in the Laboratory"; National Academy Press; Washington, DC, 1995.

3. Discussion

Among numerous procedures to install organic residues into fullerenes, the procedure described here is unique for its selectivity and efficiency, as well as for the versatility of the product. The reaction can install five organic groups of considerable diversity into [60]fullerene in a single step in essentially quantitative yield.[6] The same reaction puts three organic groups onto [70]fullerenes.[7] This process represents one of the two reactions that are known for fullerenes to take place in nearly quantitative yield.[8] The examples are shown in Table 1. For the addition of arylmagnesium bromide/CuBr•SMe_2 reagents, the 1,3-dimethyl-2-imidazolidinone additive is not necessary. However, the additive is mandatory for the MeMgBr addition.

The penta-addition product has a cyclopentadienyl moiety surrounded by the five organic groups, and it can be derivatized to a number of interesting compounds. As in standard cyclopentadiene chemistry, the cyclopentadienyl moiety can be converted to metal complexes.[9] One striking application is the synthesis of a hybrid of ferrocene and the fullerene (bucky ferrocene).[8b,c] The penta-organo groups can serve to protect from oxidation low-valent metal atoms attached to the center cyclopentadienide group through steric and electronic protective effects.[8i] Crystal structures of some representative metal complexes are shown below.

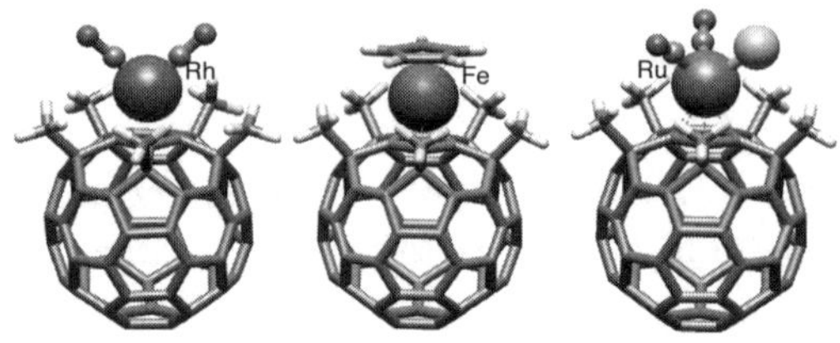

$Rh(C_{60}Me_5)(CO)_2$ $Fe(C_{60}Me_5)Cp$ $Ru(C_{60}Me_5)Cl(CO)_2$

Table 1. Addition of Organocopper Reagent to Fullerenes

Fullerene	R	Product	RMgBr (equiv.)	CuBr·SMe$_2$ (equiv.)	Yield (%)	reference
C_{60}	Me[a]	$C_{60}R_5H$	12	12	93	this work
C_{60}	Me[a]	$C_{60}R_5H$	30	30	92	6c
C_{60}	Ph	$C_{60}R_5H$	15	15	99.5	6b
C_{60}	4-$CF_3C_6H_4$	$C_{60}R_5H$	12	12	99	6b
C_{60}	4-$MeOC_6H_4$	$C_{60}R_5H$	16	16	99	6b
C_{60}	4-ClC_6H_4	$C_{60}R_5H$	15	15	99.6	6b
C_{60}	4-PhC_6H_4	$C_{60}R_5H$	15	15	99.8	6b
C_{60}	(*E*)-1-propenyl[b]	$C_{60}R_5H$	25	50	85	6d
C_{60}	(*Z*)-1-propenyl[b]	$C_{60}R_5H$	25	50	77	6d
C_{60}	(*E*)-2-phenylethenyl[b]	$C_{60}R_5H$	25	50	83	6d
C_{70}	Me[a]	$C_{70}R_3H$	30	30	90	7b
C_{70}	Ph	$C_{70}R_3H$	30	30	95	7b
C_{70}	4-PhC_6H_4	$C_{70}R_3H$	30	30	99	7b
C_{70}	4-$CF_3C_6H_4$	$C_{70}R_3H$	16	16	94	7b
C_{70}	4-ClC_6H_4	$C_{70}R_3H$	16	16	94	7b
C_{70}	1-naphthyl	$C_{70}R_3H$	16	16	95	7b

[a]1,3-Dimethyl-2-imidazolidinone was used as an additive. [b]Substituted ethenyllithium reagents were used instead of Grignard reagents.

The fullerene compounds serve as useful building blocks in supramolecular chemistry.[10] Another application of the metal complexes is the formation of bilayer vesicles upon dissolution of a potassium salt in an aqueous solution.[9a] Upon installation of five suitable side chains, the shuttlecock-shaped penta-adduct exhibits a columnar liquid crystalline phase.[9d,e,f]

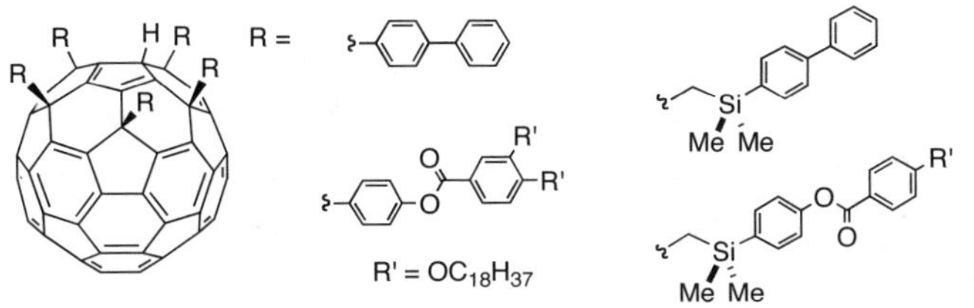

1. Department of Chemistry, The University of Tokyo, 7-3-1 Hongo, Bunkyo-ku, Tokyo 113-0033, Japan.
2. Department of Chemistry, University of Pittsburgh, Pittsburgh, PA 15260, USA.
3. Wuts, P. G. M. *Synth. Commun.* **1981**, *11*, 139.
4. "Organocopper Reagents", Taylor, R. J. K., Ed., Oxford, **1994**, pp 39.
5. Lipshutz, B. H.; Kozlowski, J. A.; Wilhelm, R. S. *J. Org. Chem.* **1983**, *48*, 546.
6. (a) Sawamura, M.; Iikura, H.; Nakamura, E. *J. Am. Chem. Soc.* **1996**, *118*, 12850. (b) Sawamura, M.; Iikura, H.; Ohama, T.; Hackler, U. E.; Nakamura, E. *J. Organomet. Chem.* **2000**, *599*, 32. (c) Sawamura, M.; Toganoh, M.; Kuninobu, Y.; Kato, S.; Nakamura, E. *Chem. Lett.* **2000**, 270. (d) Sawamura, M; Nagahama, N.; Toganoh, M.; Nakamura, E. *J. Organomet. Chem.* **2002**, *652*, 31. (e) Nakamura, E.; Sawamura, M. *Pure Appl. Chem.* **2001**, *73*, 355.
7. (a) Sawamura, M.; Iikura, H.; Hirai, A.; Nakamura, E. *J. Am. Chem. Soc.* **1998**, *120*, 8285. (b) Sawamura, M.; Toganoh, M.; Iikura, H.; Matsuo, Y.; Hirai, A.; Nakamura, E. *J. Mater. Chem.* **2002**, *12*, 2109.
8. Isobe, H.; Tomita, N.; Nakamura, E. *Org. Lett.* **2000**, *2*, 3663.
9. (a) Sawamura, M.; Kuninobu, Y.; Nakamura, E. *J. Am. Chem. Soc.* **2000**, *122*, 12407. (b) Sawamura, M.; Kuninobu, Y.; Toganoh, M.; Matsuo, Y.; Yamanaka, M.; Nakamura, E. *J. Am. Chem. Soc.* **2002**, *124*, 9354. (c) Nakamura, E. *Pure Appl. Chem.* **2003**, *75*, 427. (d) Matsuo, Y.; Nakamura, E. *Organometallics* **2003**, *22*, 2554. (e) Toganoh, M.; Matsuo, Y.; Nakamura, E. *Angew. Chem. Int. Ed.* **2003**, *42*, 3530. (f) Toganoh, M.; Matsuo, Y.; Nakamura, E. *J. Organomet. Chem.* **2003**, *683*, 295. (g) Toganoh, M.; Matsuo, Y.; Nakamura, E. *J. Am. Chem. Soc.* **2003**, *125,* 13974. (h) Matsuo, Y.; Kuninobu, Y.; Ito, S.; Nakamura, E. *Chem. Lett.* **2004**, *33*, 68. (i) Kuninobu, Y.; Matsuo, Y.; Toganoh, M.; Sawamura, M.; Nakamura, E. *Organometallics* **2004**, *23*, 3259. (j) Matsuo, Y.; Iwashita, A.; Nakamura, E. *Organometallics* **2005**, *24*, 89.
10. (a) Sawamura, M.; Nagahama, N.; Toganoh, M.; Hackler, U. E.; Isobe, H.; Nakamura, E.; Zhou, S.-Q.; Chu, B. *Chem. Lett.* **2000**, 1098. (b) Zhou, S.; Burger, C.; Chu, B.; Sawamura, M.; Nagahama, N.; Toganoh, M.; Hackler, U. E.; Isobe, H.;

Nakamura, E. *Science* **2001**, *291*, 1944. (c) Burger, C.; Hao, J.; Ying, Q.; Isobe, H.; Sawamura, M.; Nakamura, E.; Chu, B. *J. Colloid Interface Sci.* **2004**, *275*, 632. (d) Sawamura, M.; Kawai, K.; Matsuo, Y.; Kanie, K.; Kato, T.; Nakamura, E. *Nature* **2002**, *419*, 702. (e) Nakamura, E.; Isobe, H. *Acc. Chem. Res.* **2003**, *36*, 807. (f) Matsuo, Y.; Muramatsu, A.; Hamasaki, R.; Mizoshita, N.; Kato T.; Nakamura, E. *J. Am. Chem. Soc.* **2004**, *126*, 432.

NICKEL-CATALYZED HOMOALLYLATION OF ALDEHYDES WITH 1,3-DIENES (*anti*-3-Methyl-1-phenyl-4-penten-1-ol and *anti*-5-Methyl-1-phenyl-6-hepten-3-ol)

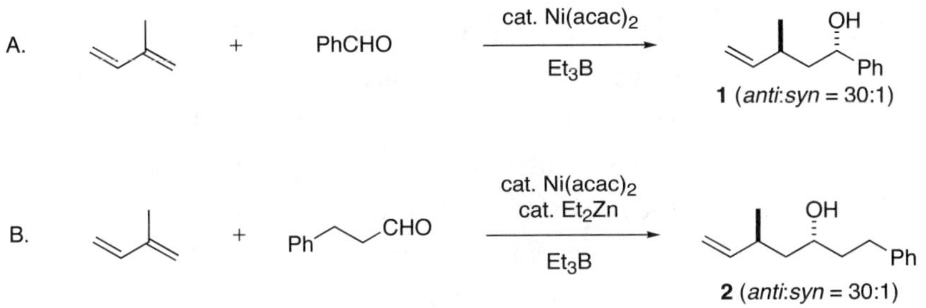

1 (*anti*:*syn* = 30:1)

2 (*anti*:*syn* = 30:1)

Submitted by Yoshinao Tamaru[1a] and Masanari Kimura.[1b]
Checked by Masakatsu Shibasaki and Noriyuki Yamagiwa.

1. Procedure

A. 3-Methyl-1-pheny-4-penten-1-ol (1).[2] An oven-dried, 300-mL, two-necked round-bottomed flask, equipped with a rubber septum, an air condenser (fitted with a three-way stopcock connected to a nitrogen balloon), and a Teflon-coated magnetic stir bar is charged with $Ni(acac)_2$ (180 mg, 0.7 mmol) (Note 1). The apparatus is purged with nitrogen and the flask is charged successively via syringe with freshly distilled THF (100 mL), isoprene (5 mL, 50 mmol), and benzaldehyde (2.65 g, 25 mmol) (Note 2). Triethylborane (50 mL, 1 M hexane solution; 50 mmol) (Note 3) is added via syringe over 5 min while the solution is stirred with a magnetic stirrer. Stirring is continued for 30 h at ambient temperature. After the reaction is completed (Note 4), most of the solvents and the remaining reagents (isoprene and Et_3B) are removed with a rotary evaporator. The residual mixture is diluted with diethyl ether (60 mL). The organic phase is washed with 4 M KOH (2 x 50 mL) and brine (2 x 30 mL), and then dried over magnesium sulfate, filtered, and concentrated by rotary evaporation (Note 5). The residue is distilled by means of a Kugelrohr apparatus (90–100 °C/0.03

mmHg) to give 3-methyl-1-phenyl-4-penten-3-ol (**1**) (3.89–4.05 g, 88–92%) consisting of *anti*- and *syn*-isomers in a ratio of 30:1 (Note 6).

B. 5-Methyl-1-phenyl-6-hepten-3-ol (2).[3] An oven-dried, 300-mL, two-necked round-bottomed flask equipped with a rubber septum, an air condenser (fitted with a three-way stopcock connected to a nitrogen balloon), and a Teflon-coated magnetic stir bar is charged with $Ni(acac)_2$ (192 mg, 0.75 mmol). The apparatus is purged with nitrogen and the flask is charged with freshly distilled THF (100 mL). Diethylzinc (0.75 mL, 1 M hexane solution; 0.75 mmol) is added into this solution at ambient temperature (Note 7). The color of the reaction mixture changes from yellow-green to brown. Isoprene (7.5 mL, 75 mmol) and dihydrocinnamaldehyde (3.4 g, 25 mmol) are added by syringe. Triethylborane (60 mL, 1 M hexane solution; 60 mmol) is then added via syringe over 5 min while stirring the solution with a magnetic stirrer (Note 3). Stirring is continued for 8 h at ambient temperature. After the reaction is complete (Note 8), most of the solvents and the remaining reagents (isoprene and Et_3B) are removed with a rotary evaporator. The residual solution is diluted with diethyl ether (60 mL). The organic phase is washed with 4 M KOH (2 x 40 mL) and brine (2 x 30 mL), and then dried over magnesium sulfate, filtered, and concentrated by rotary evaporation. The residue is distilled by means of Kugelrohr apparatus under reduced pressure (100–120 °C/0.03 mmHg) to give 5-methyl-1-phenyl-6-hepten-3-ol (**2**) (3.50–3.59 g, 68–70%) (Note 9) consisting of the *anti*- and *syn*-isomers in a ratio of 30:1 (Note 10).

2. Notes

1. $Ni(acac)_2$ is used as received from Aldrich Chemical Company, Inc.

2. THF (tetrahydrofuran) is distilled from sodium/benzophenone ketyl under N_2 prior to use. Benzaldehyde was used as received from Wako Pure Chemical Industry, Ltd.

3. Et_3B (1 M hexane solution) is used as received from Aldrich Chemical Company, Inc. Hydrocinnamaldehyde was used as received from Wako Pure Chemical Industry, Ltd.

4. The reaction is monitored by TLC (Merck, Silica gel 60F254). R_f (**1**) = 0.67; R_f (benzaldehyde) = 0.72: hexane/ethyl acetate = 2/1, v/v.:

visualized by a 254-nm UV lamp as well as by iodine.

5. It is crucial to wash the organic phase with a strong base, such as 4 M KOH prior to distillation. The concentrated residue obtained by washing with sat. $NaHCO_3$ did not withstand distillation (100 °C/0.05 mmHg) and decomposed to give an intractable mixture of products.

6. ***anti*-3-Methyl-1-phenyl-4-penten-1-ol (*anti*-1)**: IR (neat) 3357 (s), 2960 (s), 2926 (s), 1494 (s), 1454 (s), 997 (s), 911 (s), 756 (s), 700 (s) cm^{-1}; ^1H NMR (400 MHz, $CDCl_3$) δ: 1.01 (d, J = 6.7 Hz, 3 H), 1.63 (ddd, J = 13.7, 6.1, 6.1 Hz, 1 H), 1.83 (ddd, J = 13.7, 7.9, 7.9 Hz, 1 H), 1.89 (brs, 1 H), 2.20 (ddt, J = 6.1, 7.9, 6.7 Hz, 1 H), 4.71 (dd, J = 6.1, 7.9 Hz, 1 H), 4.96 (brd, J = 10.5 Hz, 1 H), 5.00 (brd, J = 17.7 Hz, 1 H), 5.77 (ddd, J = 7.9, 10.5, 17.7 Hz, 1 H), 7.25 - 7.34 (m, 5 H); ^{13}C NMR (100 MHz, $CDCl_3$) δ: 20.4, 35.2, 45.9, 73.0, 113.2, 126.0, 127.6, 128.5, 144.5, 144.6. Anal. calcd. for $C_{12}H_{16}O$: C, 81.77; H, 9.15. Found: C, 81.21; H, 9.21. ***syn*-3-Methyl-1-phenyl-4-penten-1-ol (*syn*-1)** was prepared from the *anti*-isomer by Mitsunobu reactions for reference (see discussion): IR (neat) 3356 (s), 1640 (w), 1043 (s), 912 (s), 700 (s) cm^{-1}m; ^1H NMR (400 MHz, $CDCl_3$) δ: 1.05 (d, J = 6.8 Hz, 3 H), 1.61 (ddd, J = 3.6, 9.3, 13.9 Hz, 1 H), 1.80 (ddd, J = 4.7, 9.3, 13.9 Hz, 1 H), 1.83 (d, J = 3.6 Hz, 1 H), 2.42 (dqm, J = 8.0, 6.8 Hz, 1 H), 4.72 (dt, J = 9.3, 3.6 Hz, 1 H), 5.01 (br d, J = 10.0 Hz, 1 H), 5.06 (br d, J = 17.1 Hz, 1 H), 5.73 (ddd, J = 8.0, 10.0, 17.1 Hz, 1 H), 7.23 - 7.38 (m, 5 H); ^{13}C NMR (100 MHz, $CDCl_3$) δ: 21.0, 34.9, 46.1, 72.4, 113.7, 125.7, 127.4, 128.4, 143.8, 145.0. Anal. calcd. for $C_{12}H_{16}O$: C, 81.77; H, 9.15. Found: C, 81.21; H, 9.21. The diastereomeric ratio (*anti*-**1**/*syn*-**1**) was determined from the ratio of resonance integrations at 5.77 ppm (*anti*) and 5.73 ppm (*syn*) in ^1H-NMR spectra.

7. Et_2Zn (1 M hexane solution) is used as received from Aldrich Chemical Company, Inc.

8. The reaction is monitored with TLC (Merck, Silica gel 60F254). R_f(**2**)= 0.58: R_f(hydrocinnamaldehyde) = 0.50: hexane/ethyl acetate = 2/1, v/v.: visualized by a 254-nm UV lamp as well as by iodine.

9. The submitters report an isolated yield of 80% yield.

10. ***anti*-5-Methyl-1-phenyl-6-hepten-3-ol (*anti*-2)**: IR (neat) 3373 (s), 2927 (s), 1639 (s), 1603 (s), 1495 (s), 1454 (s), 913 (s), 699 (s) cm^{-1};^1H NMR (500 MHz, $CDCl_3$) δ: 0.99 (d, J = 6.9 Hz, 3 H), 1.42 (ddd, J = 4.0, 6.9,

13.8 Hz, 1 H), 1.52 (ddd, J = 5.8, 9.8, 13.8 Hz, 1 H), 1.65 (brs, 1 H), 1.68-1.81 (m, 2 H), 2.28-2.34 (m, 1 H), 2.65 (ddd, J = 6.3, 9.8, 13.8 Hz, 1 H), 2.78 (ddd, J = 5.8, 9.8, 13.8 Hz, 1 H), 3.68-3.73 (m, 1 H), 4.92 (dd, J = 1.7, 10.3 Hz, 1 H), 5.00 (brd, J = 1.7, 17.2 Hz, 1 H), 5.76 (ddd, J = 8.0, 10.3, 17.2 Hz, 1 H), 7.16-7.29 (m, 5 H); ^{13}C NMR (125 MHz, $CDCl_3$) δ: 20.3, 31.9, 35.6, 39.3, 44.5, 70.0, 112.9, 125.8, 128.4, 128.4, 142.2, 145.0. Anal. calcd. for $C_{14}H_{20}O$: C, 82.30; H, 9.87. Found: C, 82.03; H, 9.58. ***syn*-5-Methyl-1-phenyl-6-hepten-3-ol** (*syn*-**2**) was prepared from the anti-isomer by Mitsunobu reactions for reference (see discussion): IR (neat) 3354 (s), 1641 (m), 1030 (m), 997 (s), 914 (s), 698 (s) cm^{-1}; ^{1}H NMR (400 MHz, $CDCl_3$) δ: 1.02 (d, J = 6.4 Hz, 3 H), 1.43 (dm, J = 13.9 Hz, 1 H), 1.50 (dm, J = 13.9 Hz, 1 H), 1.70-1.78 (m, 2 H), 2.40 (dqm, J = 8.0, 6.4 Hz, 1 H), 2.66 (dm, J = 13.9 Hz, 1 H), 2.78 (dm, J = 13.9 Hz, 1 H), 3.68 (br s, 1 H), 4.96 (br d, J = 10.1 Hz, 1 H), 5.03 (d, J = 17.2 Hz, 1 H), 5.67 (ddd, J = 8.0, 10.1, 17.2 Hz, 1 H), 7.15-7.29 (m, 5 H); ^{13}C NMR (120 MHz, $CDCl_3$) δ: 21.1, 32.1, 34.8, 39.6, 44.3, 69.3, 113.3, 125.6, 128.2, 142.0, 143.9; Anal. calcd. for $C_{14}H_{20}O$: C, 82.30; H, 9.87. Found: C, 82.03; H, 9.58. The diastereomeric ratio (*anti*-**2**/*syn*-**2**) was determined from the ratio of resonance integrations at 5.76 ppm (*anti*) and 5.67 ppm (*syn*) in ^{1}H-NMR spectra.

Safety and Waste Disposal Information

All hazardous materials should be handled and disposed of in accordance with "Prudent Practices in the Laboratory"; National Academy Press; Washington, DC, 1995.

3. Discussion

The structures of the minor components of **1** and **2** were confirmed to be the diastereomeric isomers, *syn*-**1** and *syn*-**2**, respectively, by comparison of the spectral data with those of authentic samples prepared by the Mitsunobu inversion ($PhCO_2H$, diethyl azodicarboxylate, Ph_3P) followed by hydrolysis (NaOH, EtOH-H_2O): *anti*-**1**:*syn*-**1** = 1:30 in 79% overall yield from *anti*-**1**:*syn*-**1** = 30:1; *anti*-**3**:*syn*-**3** = 1:30 in 72% overall yield from *anti*-**3**:*syn*-**3** = 30:1. The structure determination of **1** and **2**[4] relied on the ^{13}C NMR chemical shifts of their cyclic derivatives **3** and **4**, respectively (Scheme 1).

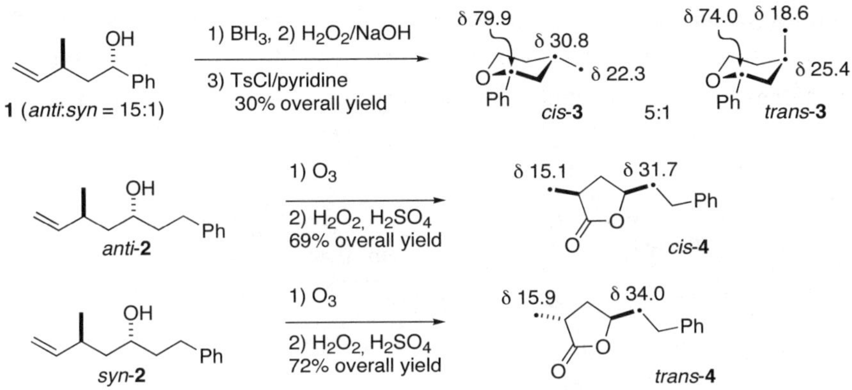

Scheme 1

Transformation of **1** and **2** into Heterocycles **3** and **4** and their ^{13}C Chemical Shifts (δ in ppm, 100 MHz in $CDCl_3$).

Bishomoallyl alcohols and ethers are ubiquitous structural motifs of natural products, especially of polyether antibiotics and acetogenin-derived natural products.[5] The present nickel-catalyzed homoallylation of aldehydes with isoprene (and other 1,3-dienes)[6] provides a powerful method for constructing such classes of compounds.[7] The reaction shows high 1,3-diastereoselectivity furnishing 1,3-*anti*-4-penten-1-ol products with as high as 95 – 100% stereoselectivity.

Three variations on the homoallylation procedure are available and appropriate choice of the method to be applied requires consideration of the reactivity of the aldehyde. For reactive aldehydes (aromatic and unsaturated aldehydes), procedure A is the method of choice.[2] A second (unchecked) procedure is particularly effective for the homoallylation of relatively unreactive and sterically hindered aliphatic aldehydes, as well as ketones.[3] This second procedure is sometimes plagued with a side reaction resulting from ethylation of the aldehyde. This problem is most apparent when reactive primary and secondary aliphatic aldehydes are used. For such cases, procedure B is recommended, because ethylation can be avoided almost completely. Thus, these three methods complement one another.

A variation on these procedures has been successfully applied to the homoallylation of aldimines formed *in situ* from aromatic aldehydes and anisidine. Good yields of 1,3-*syn*-bishomoallyl amines are furnished with excellent stereoselectivity. Aldimines are generally by far less reactive than aldehydes. Interestingly, the sense of stereoselection is opposite to that observed for aldehydes (Eq 1).[8]

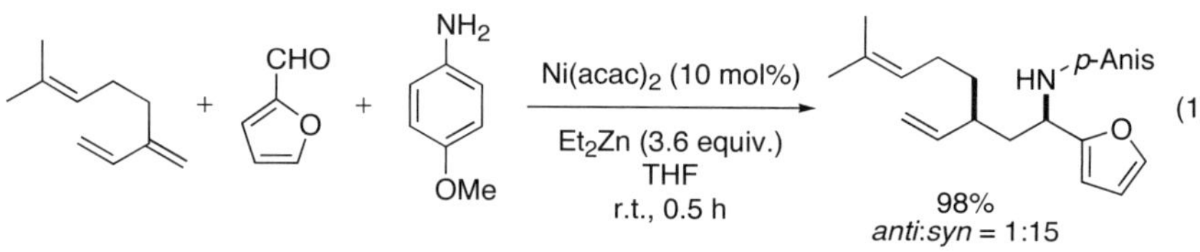

Sorbic acid methyl ester, an electron-deficient diene, engages by (Eq 2) reacting regioselectively at the C2 position with high 1,2-*anti* stereoselectivity. All results to date suggest that unsymmetrically substituted 1,3-dienes react with aldehydes regioselectively at the terminus bearing the highest electron density.

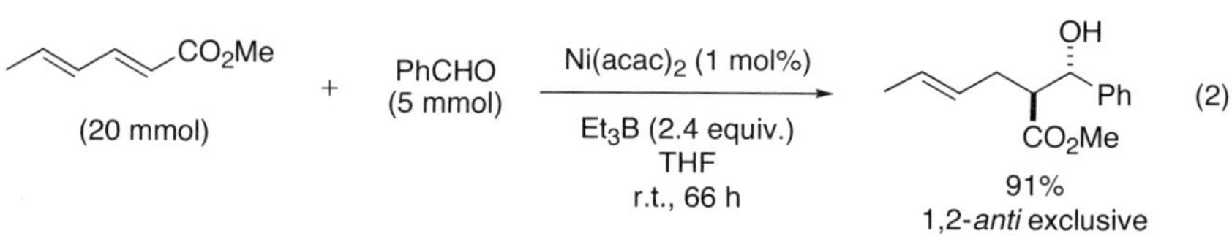

Triethyl borane (Et_3B) is compatible with water; hence, the method described in procedure A can be applied to the homoallylation of aldehydes that are stable only in water.[9] For example, in the presence of a catalytic amount of $Ni(acac)_2$ and a stoichiometric amount of Et_3B, isoprene reacts regioselectively at the C1 position with a commercial 50% aqueous solution of glutaraldehyde to provide a 1,3-*anti*-bishomoallyl alcohol. Cyclic hemiacetals (n = 1 or 2) undergo the homoallylation with isoprene to form bishomoallylic diols with excellent 1,3-asymmetric induction (Eq 4).

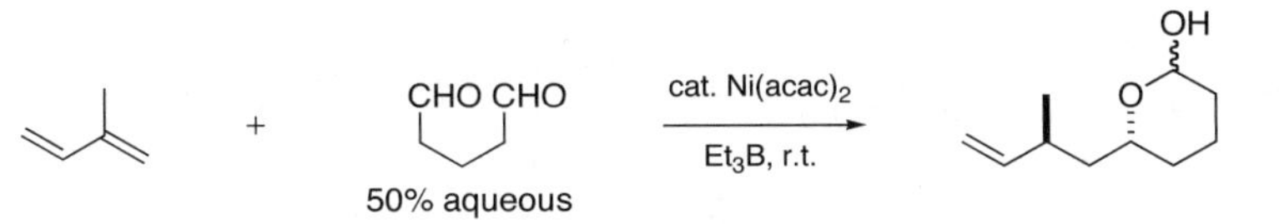

(3)

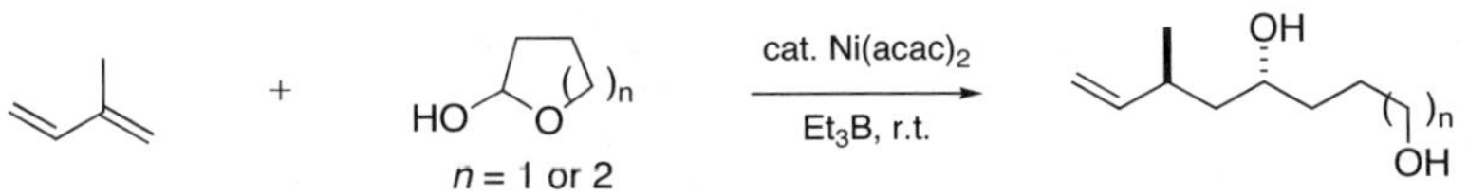

(4)

The protocol for the intermolecular homoallylation of aldehydes with 1,3-dienes can be applied successfully to an intramolecular version of ω-dienyl aldehydes forming five- and six-membered cycloalkanols (Eq 5).[10]

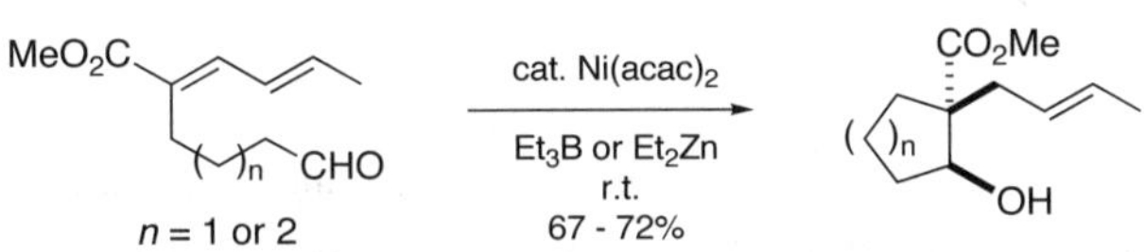

(5)

The combination of nickel and Et_3B has been proved to be applicable to the reductive coupling of alkynes and aldehydes to provide allylic alcohols with high regio- and stereoselectivities (Eq 6).[11] The procedure is extended to the asymmetric synthesis of allyl alcohols and is applied to the total synthesis of natural products.[12] Under similar conditions, enantioenriched epoxides undergo the reductive coupling with alkynes and provide homoallyl alcohols with complete preservation of the original chirality (Eq 7).[13] In contrast to the reactions with aldehydes and epoxides, the reaction with aldimines delivers the ethyl group of Et_3B, instead of hydrogen, at the distal carbon of alkynes to give rise to stereochemically defined allylamines (Eq 8).[14]

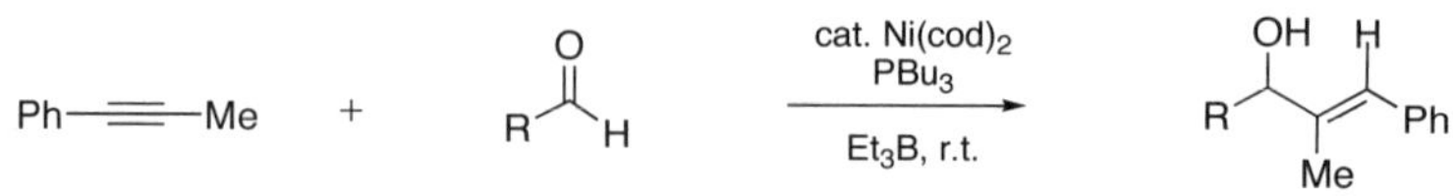

(6)

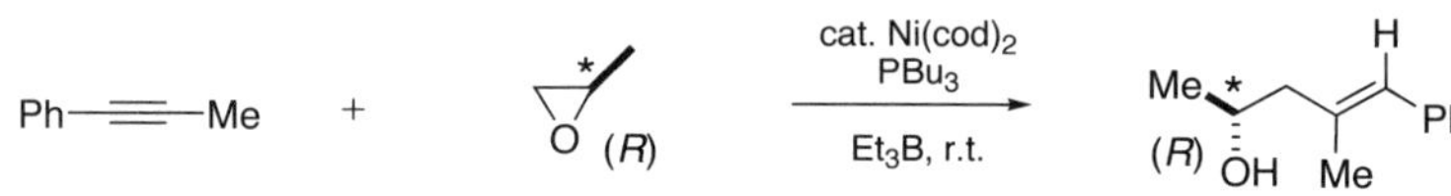

(7)

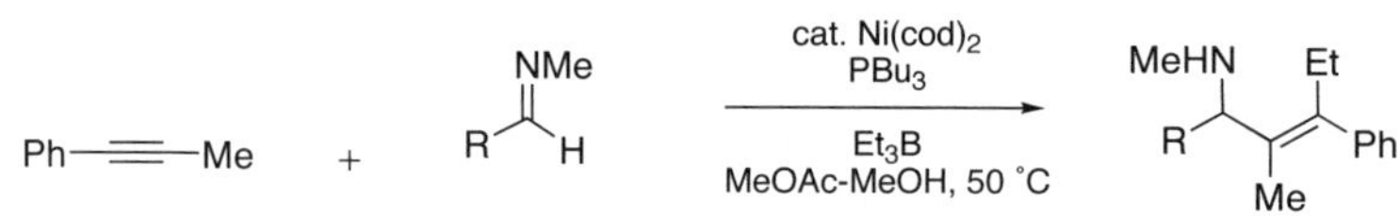

(8)

1. (a) Department of Applied Chemistry, Faculty of Engineering, Nagasaki University, 1-14 Bunkyo-machi, Nagasaki 852-8521, Japan. tamaru@net.nagasaki-u.ac.jp; (b) Graduate School of Science and Technology, Nagasaki University, 1-14 Bunkyo-machi, Nagasaki 852-8521, Japan. masanari@net.nagasaki-u.ac.jp

2. (a) Kimura, M.; Ezoe, A.; Shibata, K.; Tamaru, Y. *J. Am. Chem. Soc.* **1998**, *120*, 4033. (b) Kimura, M.; Ezoe, A.; Mori, M.; Iwata, K.; Tamaru, Y. *J. Am. Chem. Soc.* **2006**, *128*, 8559.

3. Kimura, M.; Fujimatsu, H.; Akihiro, E.; Shibata, K.; Shimizu, M.; Matsumoto, S.; Tamaru Y. *Angew. Chem. Int. Ed.* **1999**, *38*, 397.

4. (a) E. Nakamura, H. Oshio, I. Kuwajima, *J. Am. Chem. Soc.* **1986**, *108*, 3745; (b) G. Fronza, C. Fuganti, P. Grasselli, M. Terreni, *Tetrahedron* **1992**, *48*, 7363.

5. (a) Gottwald, T.; Greb, M.; Hartung, J. *Synlett*, **2004**, 65; (b) Hartung, J.; Greb, M. *J. Organometal. Chem.* **2002**, *661*, 67; (c) Brandl, M.; Wu, X.; Liu, Y.; Pease, J.; Holper, M.; Hooijmaaijer, E.; Lu, Y.; Wu, P. *J. Pharm. Sci.* **2003**, *92*, 1981.

6. We have noticed that 1,3-cyclohexadiene is unusual and selectively undergoes allylation (not homoallylation) under the conditions used in reference 3. Accordingly, the report that claims 1,3-cyclohexadiene selectively undergoes homoallylation of aldehydes should be re-examined: Loh, T. P.; Song, H. Y.; Zhou, Y. *Org. Lett.* **2002**, *4*, 2715. Our results differ from the observation of Loh and co-workers.

7. (a) Sato, Y.; Takimoto, M.; Mori, M. *Tetrahedron Lett.* **1996**, *37*, 887; (b) Sato, Y.;

Takimoto, M.; Mori, M. *Synlett*, **1997**, 734; (c) Sato, Y.; Takahashi, T.; Hoshida, M.; Mori, M. *Tetrahedron Lett.* **1998**, *39*, 5579; (d) Sato, Y.; Takimoto, M.; Mori, M. *J. Am. Chem. Soc.* **2000**, *122*, 1624; (e) Sato, Y.; Sawaki, R.; Saito, N.; Mori, M. *J. Org. Chem.* **2002**, *67*, 656.

8. Kimura, M.; Miyachi, A.; Kojima, K.; Tanaka, S.; Tamaru, Y. *J. Am. Chem. Soc.* **2004**, *126*, 14360.

9. Kimura, M.; Ezoe, A.; Tanaka, S.; Tamaru, Y. *Angew. Chem. Int. Ed.* **2001**, *40*, 3600.

10. Shibata, K.; Kimura, M.; Shimizu, M.; Tamaru, Y. *Org. Lett.* **2001**, *3*, 2181.

11. (a) Huang, W.-S.; Chan, J.; Jamison, T. F. *Org. Lett.* **2000**, *2*, 4221; (b) Miller, K. M.; Luanphaisarnnont, T.; Molinaro, C.; Jamison, T. F. *J. Am. Chem. Soc.* **2004**, *126*, 4130.

12. (a) Miller, K. M.; Huang, W.-S.; Jamison, T. F. *J. Am. Chem. Soc.* **2003**, *125*, 3442; (b) Colby, E. A.; Jamison, T. F. *J. Org. Chem.* **2003**, *68*, 156; (c) Chan, J.; Jamison, T. F. *J. Am. Chem. Soc.* **2003**, *125*, 11514.

13. (a) Molinaro, C.; Jamison, T. F. *J. Am. Chem. Soc.* **2003**, *125*, 8076; (b) Colby, E. A.; O'Brien, K. C.; Jamison, T. F. *J. Am. Chem. Soc.* **2004**, *126*, 998.

14. Patel, S. J.; Jamison, T. F. *Angew. Chem. Int. Ed.* **2003**, *42*, 1364.

TRIFLUOROMETHYLATION AT THE α-POSITION OF α,β-UNSATURATED KETONES: 4-PHENYL-3-(TRIFLUOROMETHYL)BUTAN-2-ONE

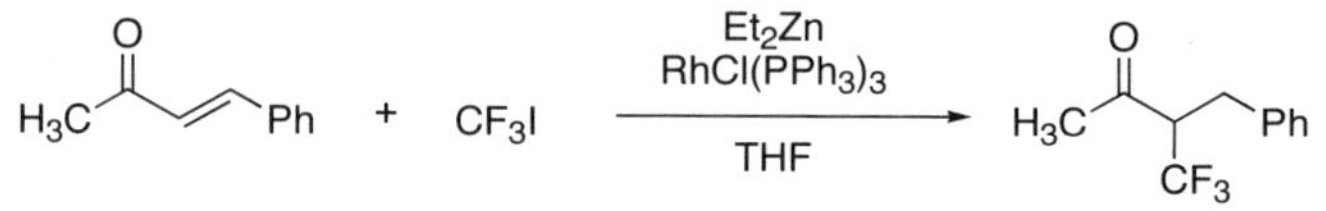

Kazuyuki Sato,[1] Masaaki Omote, Akira Ando, Itsumaro Kumadaki.*
Checked by Scott E. Denmark and Won-jin Chung.

1. Procedure

4-Phenyl-3-trifluoromethyl-2-butanone. A flame-dried, 500-mL, four-necked, round-bottomed flask is equipped with a rubber septum, a dry ice/isopropyl alcohol cooled cold-finger condenser topped with a nitrogen stopcock inlet, a pressure-equalizing dropping funnel topped with a rubber septum, a Teflon-coated thermocouple (Note 1), and a Teflon-coated magnetic stirring bar. The reaction vessel is charged with 4-phenylbut-3-en-2-one (7.310 g, 50 mmol), $RhCl(PPh_3)_3$ (925 mg, 1.0 mmol) (Note 2), and 190 mL of tetrahydrofuran (THF) is added under nitrogen (Note 2). The brown solution is cooled in an isopropyl alcohol bath to –30 °C (Note 3). In a graduated tube equipped with a three-way stopcock attached to a nitrogen inlet from a Schlenk line, 7 mL (ca. 86 mmol) of trifluoroiodomethane (CF_3I) (Note 4) is condensed at –78 °C by using a dry ice/isopropyl alcohol bath. After the CF_3I condenses, the three-way stopcock is closed and the top of the stopcock is fitted with a rubber septum. The condensed CF_3I is added to the reaction mixture quickly through a cannula by warming the graduated tube with a water bath. After almost all of CF_3I is transferred to the reaction vessel, 10 mL of THF is added to the graduated test-tube through the septum, and the THF is transferred to the reaction vessel through the cannula by briefly evacuating the reaction vessel through a hose to the Schlenk line. The reaction mixture is immersed in an ice-bath, and then diethylzinc in hexane solution (75 mL 1.0 M) (Note 5) is gradually added to the solution through the presser-equalizing dropping funnel over 2 h. During the addition, the temperature of the solution is kept below 5 °C by cooling with an ice bath. The reaction mixture is allowed to warm to room temperature over 30 min after removing the ice-bath, whereupon it is stirred at room temperature for 1 h (Note 6). The progress of the reaction is followed by thin layer chromatographic (TLC) analysis (Note 7), which confirms that the reaction is complete (Note 8). The resulting mixture is carefully poured into a mixture of 10% HCl (150 mL) and ice (about 50 g) in a 1-L Erlenmeyer flask equipped with a Teflon-coated magnetic stirring bar (Note 9). The ice melts in a few minutes, then the mixture is extracted with Et_2O (4 x 150 mL). The combined organic layers are washed with brine (100 mL) and then are dried over $MgSO_4$ (about 20 g). The $MgSO_4$ is removed by filtration through a sintered glass-filter, and the solvent is removed under reduced pressure on a rotary evaporator (23 °C, 30 mm Hg). The residue is purified by silica gel (Note 10) column chromatography (190 g, diameter = 50 mm, eluted with hexane/ethyl acetate, 30/1; 1.2 L) to remove small amounts of impurities and most of the black color. The product-containing portions are concentrated under reduced pressure, and are purified by reduced-pressure distillation using short path distillation apparatus (66–68 °C / 3 mmHg) to afford 7.610 g (72 %) of 4-phenyl-3-(trifluoromethyl)butan-2-one as a clear, colorless oil (Notes 10 and 11).

2. Notes

1. A PFA coated thermocouple probe, Type K (Omega Engineering, Inc.) was inserted through the septum.

2. 4-Phenylbut-3-en-2-one (>99%) and $RhCl(PPh_3)_3$ (>99.99%) were purchased from Aldrich Chemical Co. and were used without further purification. The reaction apparatus is shown in Figure 1. The submitters used argon.

3. Anhydrous THF was obtained by filtration through an alumina drying column on a GlassContour system (Irvine, CA).

4. CF_3I was purchased from Aldrich Chemical Co. (99%). Its boiling point is –22.1 °C, and density is about 2.4 g/mL. CF_3I was transferred from the container to the graduated test-tube by using the apparatus shown in Figure 2. The tube was evacuated and cooled below –78 °C in a dry ice/isopropyl alcohol bath. Then the vacuum line was closed, stopcock **A** was opened and the correct amount of CF_3I was collected in the graduated test-tube.

5. Diethylzinc (1.0 M in hexane solution) was purchased from Aldrich Chemical Co. It was transferred to the pressure-equalizing addition funnel via cannula by briefly evacuating the reaction flask through the Schlenk line.

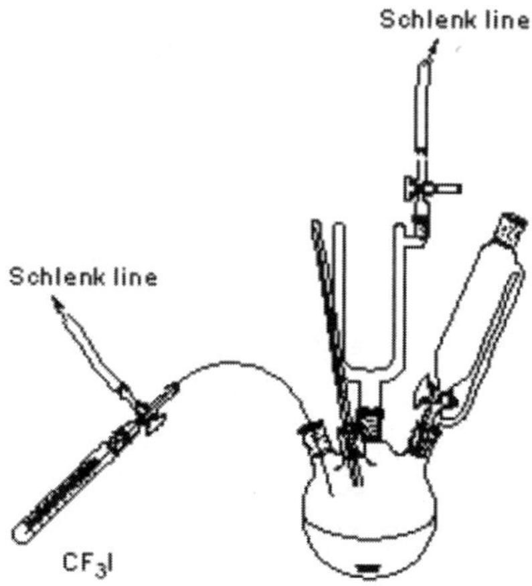

Figure 1. Reaction Apparatus for the Synthesis of α-CF_3 Ketones

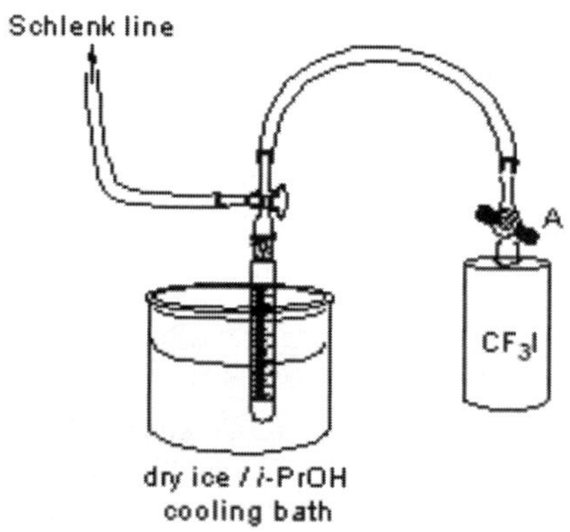

Figure 2. Apparatus for Collecting CF_3I

6. A mild evolution of gas was observed upon addition of diethylzinc. The composition of the gas was not determined, but it is most likely trifluoromethyl iodide and/or ethylene. As the reaction mixture was warmed to room temperature, slight gas evolution was observed, and the color of the solution changed to dark brown.

7. TLC analysis was performed on Merck silica gel plates with QF-254 indicator and hexane/ethyl acetate, 9/1 as eluent. Visualization was accomplished with UV light and $KMnO_4$ staining solution.

8. An aliquot of the reaction mixture was removed and quenched with 10% aq. HCl, then Et_2O was added. The organic phase was analyzed by TLC (see Note 7). 4-Phenylbut-3-en-2-one (R_f = 0.23) almost disappeared after 0.5 h, but 4-phenylbutan-2-one, (R_f= 0.29) which can be generated by quenching the corresponding rhodium enolate was still present. After 1.0 h, 4-phenylbutan-2-one disappeared and only 4-phenyl-3-(trifluoromethyl) butan-2-one (R_f = 0.50) was detected. The submitters used gas chromatography to monitor the reaction progress.

9. Vigorous evolution of gas was observed upon quenching.

10. The submitters found that 4-phenyl-3-(trifluoromethyl)butan-2-one is stable at room temperature under an argon atmosphere over three months.

11. The final product had the following characteristics; ^{1}H NMR ($CDCl_3$, 500 MHz) δ: 2.08 (s, 3 H), 3.06 (dd, 1 H, J= 13.8, 4.3 Hz), 3.18 (dd, 1 H, J= 13.8, 10.8 Hz), 3.56 (dqd, 1 H, J= 10.8, 8.6, 4.1 Hz), 7.15–7.32 (m, 5 H); ^{13}C NMR ($CDCl_3$, 125 MHz) δ: 31.6 (q, $J_{C\text{-}F}$ = 1.8 Hz), 31.8 (q, $J_{C\text{-}F}$ = 2.8 Hz), 57.4 (q, $J_{C\text{-}F}$ = 24.9 Hz), 124.4 (q, $J_{C\text{-}F}$ = 279.5 Hz), 127.0, 128.7, 128.8, 136.4, 201.3 (q, $J_{C\text{-}F}$ = 1.5 Hz); ^{19}F NMR ($CDCl_3$, 396 MHz) δ: –67.72 (d, 3 F, J = 7.5 Hz); IR (KBr) cm^{-1}: 3033, 2942, 1731, 1606, 1499, 1457, 1422, 1362, 1301, 1263, 1161, 1115, 1081, 1505, 874, 747, 704; MS *m/z*: 216 (M^+); HRMS Calc. $C_{11}H_{11}OF_3$: 216.0762 (M^+), Found: 216.0763; Elemental Analysis Calc. for $C_{11}H_{11}OF_3$: C, 61.11; H, 5.13; F, 26.36. Found: C, 60.87; H, 5.04; F, 26.68.

Safety and Waste Disposal Information

All hazardous materials should be handled and disposed of in accordance with "Prudent Practices in the Laboratory"; National Academy Press; Washington, DC, 1995.

3. Discussion

Trifluoromethylated (CF_3) compounds constitute one of the most important classes of fluorine-containing compounds, and there are many reports on trifluoromethylation reactions. However, the reactions for introducing a CF_3 group at the α-position of ketones are limited to trifluoromethylation of metal enolates of carbonyl compounds or imides,[2] electrophilic trifluoromethylation with calcogenium reagents,[3] and photochemical[4a] and ultrasonic[4b] reactions of enamines with CF_3I.

Recently, our group reported that α,β-unsaturated ketones react with ethyl bromodifluoroacetate and Et_2Zn in the presence of a rhodium catalyst to give unexpected products, in which either the CF_2COOEt group was introduced on the α-carbon of α,β-unsaturated ketones or the expected 1,2-addition product proceeded in good yields depending on the solvent.[5] On the basis of this result, we treated α,β-unsaturated ketones with CF_3I in the presence of Et_2Zn and $RhCl(PPh_3)_3$ to prepare α-CF_3 ketones.[6] This method was found to be superior to previous routes for synthesis of α-CF_3 ketones because it did not need special reagents, could be carried out under mild conditions, and provided the product in good yield. Moreover, trifluoromethylation at the α-position of α,β-unsaturated ketones had not been reported.

A mechanistic proposal is shown below (Figure 3).[6] A β-hydrogen atom in Et_2Zn appears to play an important role in this reaction because dimethylzinc does not work at all.

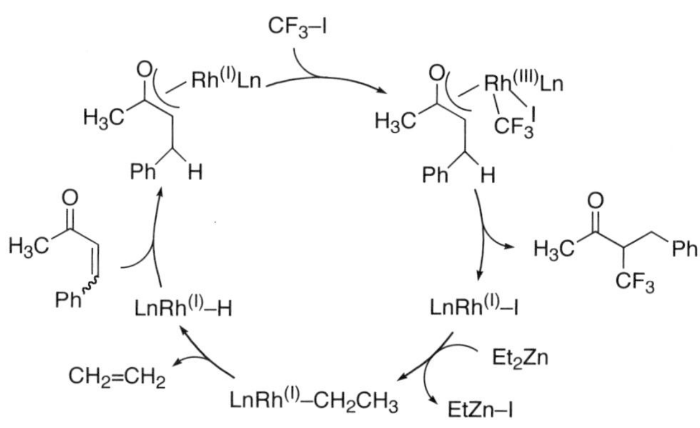

Figure 3. Proposed Mechanism

Following the procedure described above, various α-CF_3 ketones were obtained in good yields as shown in Table 1. The yields reported in the Table are based on reactions performed on a 2-mmol scale.

Table 1. Synthesis of Various α-CF_3 Ketones

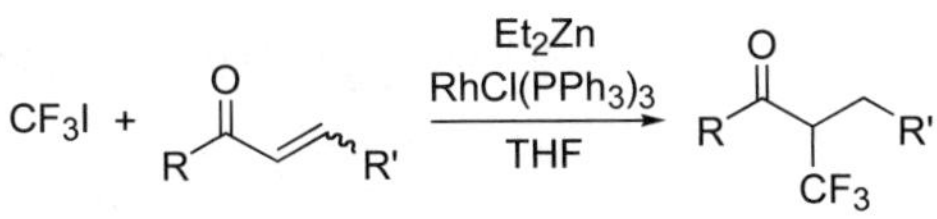

entry	enone	time (h)	yield (%)[a]
1	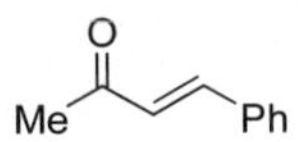	0.5	77
2	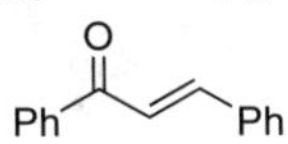	1	31
3	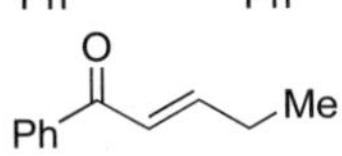	1	35
4	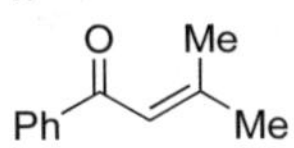	3	0
5	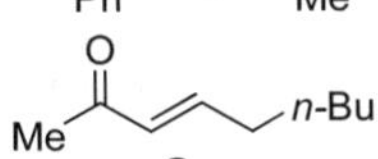	1	67
6	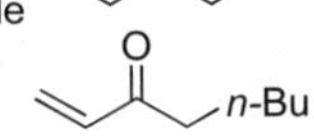	1	59
7	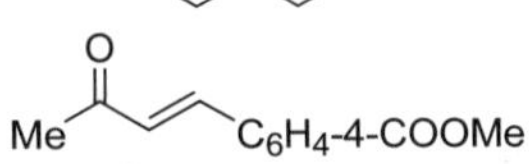	1	65
8	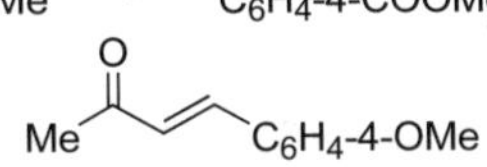	1	65
9	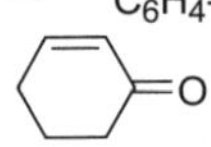	0.5	55
10	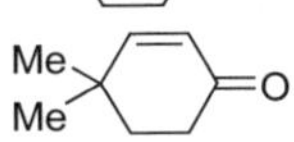	1	53
11	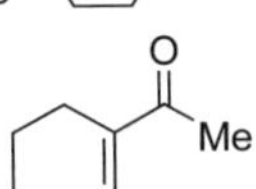	0.5	0[b]

(a) Isolated yield. (b) 1,4-adduct was obtained in 54%.

1. Faculty of Pharmaceutical Sciences, Setsunan University, 45-1, Nagaotoge-cho, Hirakata, Osaka 573-0101, Japan

2. (a) Itoh, Y.; Mikami, K. *Org. Lett.* **2005**, *7*, 649-651. (b) Iseki, K.; Nagai, T.; Kobayashi, Y. *Tetrahedron: Asymmetry* **1994**, *5*, 961-974.

3. (a) Ma, J. -A.; Cahard, D. *J. Org. Chem.* **2003**, *68*, 8726-8729. (b) Umemoto, T.; Adachi, K. *J. Org. Chem.* **1994**, *59*, 5692-5699.

4. (a) Crusiani, G.; Margaretha, P. *J. Fluorine Chem.* **1987**, *37*, 95-101. (b) Kitazume, T.; Ishikawa, N. *J. Am. Chem. Soc.* **1985**, *107*, 5186-5191.

5. Sato, K.; Tarui, A; Kita, T.; Ishida, Y.; Tamura, H.; Omote, M.; Ando, A.; Kumadaki, I. *Tetrahedron Lett.* **2004**, *45*, 5735-5737.

6. Sato, K.; Omote, M.; Ando, A.; Kumadaki, I. *Org. Lett.* **2004**, *6*, 4359-4361.

A PRACTICAL AND SAFE PREPARATION OF 3,5-BIS(TRIFLUOROMETHYL)ACETOPHENONE (1-[3,5-Bis(trifluoromethyl)phenyl]-ethanone)

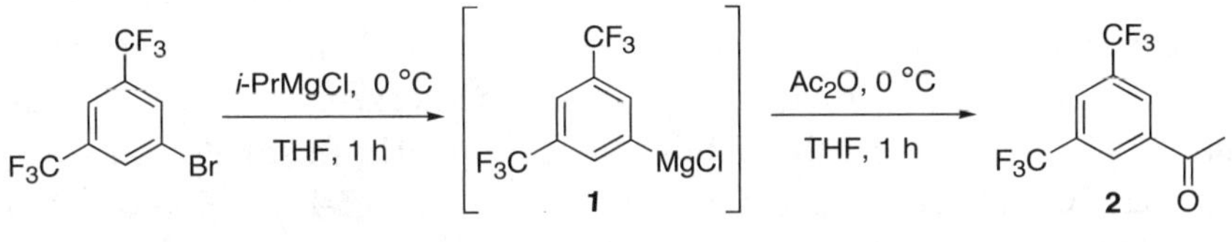

Submitted by Johnnie L. Leazer Jr. and Raymond Cvetovich.[1]
Checked by Kevin M. Maloney and Rick L. Danheiser.

1. Procedure[2]

A 100-mL, three-necked, round-bottom flask equipped with a teflon-coated thermocouple probe, inert gas inlet, a 100-mL pressure-equalizing addition funnel fitted with a rubber septum, and a magnetic stirbar is charged with 3,5-bis(trifluoromethyl)bromobenzene (20.0 g, 11.8 mL, 68.3 mmol) and 35 mL of anhydrous THF (Notes 1, 2, 3). The solution is cooled to -5 °C in a ice-salt water bath and a solution of *i*-PrMgCl in THF (37.6 mL, 2M, 75.2 mmol) is added dropwise over 1 hr at a rate such that the internal reaction temperature does not exceed 0 °C (Notes 4, 5). Upon completion of the addition, the reaction mixture is stirred for 1 hr at 0 to -10 °C.

A 250-mL, three-necked, round-bottomed flask equipped with a teflon-coated magnetic stir bar, a reflux condenser fitted with an inert gas inlet, a glass stopper, and a septum pierced by a teflon-coated thermocouple probe (Note 1) is charged with acetic anhydride (24.4 mL, 258 mmol) (Note 6) and then cooled to -5 °C in a ice-salt water bath. The solution of 3,5-bis(trifluoromethyl)phenylmagnesium chloride (at -5 °C, prepared as described above) is transferred into the flask containing acetic anhydride via an 18-gauge, 2-ft, double-ended needle at a rate such that the internal reaction temperature does not exceed 0 °C (ca. 2 hr) (Note 7). The resulting pale yellow solution is stirred for 30 min at 0 °C. The flask is removed from the ice-water bath and 35 mL of deionized water is added dropwise over 1 hr (Note 8). The resulting biphasic mixture is heated at 60 °C in an oil bath for 15 min (Note 9). The reaction mixture is then allowed to cool to room temperature. The organic layer is separated (Note 10) and diluted with 35 mL of MTBE (Note 11). The resulting solution is transferred to a 250-mL Erlenmeyer flask and vigorously stirred at 15 °C while 2 mL of aqueous 2.5N NaOH is added dropwise so that the pH of the aqueous phase reaches 7.1. The organic phase is separated and washed with two 20-mL portions of saturated aqueous $NaHCO_3$ solution and 20 mL of brine, dried over Na_2SO_4, and concentrated by rotary evaporation (20 mmHg) at room temperature. The residue is transferred to a 50-mL, one-necked, pear-shaped flask. Purification by bulb-to-bulb distillation at 25 mmHg (distillate collected at 122-132 °C) affords 15.1-15.2 g (86-87%) of 3,5-bis(trifluoromethyl)-acetophenone as a clear, colorless oil (Note 12).

2. Notes

1. The apparatus was flame-dried and maintained under an atmosphere of argon (checkers) or nitrogen (submitters) throughout the course of the reaction.

2. 3,5-Bis(trifluoromethyl)bromobenzene was purchased from Aldrich Chemical Company, Inc., and used without further purification. Use of 3,5-bis(trifluoromethyl)bromobenzene prepared as described by Leazer et al.[2] gave identical results.

3. The submitters purchased anhydrous THF (HPLC grade) from EM Scientific and used it without further purification. Karl Fischer titration indicated there was 34 μg of water per mL of solvent. The checkers dried THF by pressure filtration under argon through activated alumina.

4. All temperatures refer to internal reaction temperatures.

5. A solution of *i*-PrMgCl (2M in THF) was purchased from Aldrich Chemical Company, Inc., and used as received.

6. Acetic anhydride (certified) was purchased from Fisher Scientific and used without further purification.

7. Addition of acetic anhydride to a freshly prepared solution of **1** gave bis-adduct **3** as the exclusive product. On the other hand, addition of Grignard reagent **1** to an excess of acetic anhydride gave acetophenone **2** in 90% isolated yield. Thus, excess acetic anhydride is required to obtain desired acetophenone.

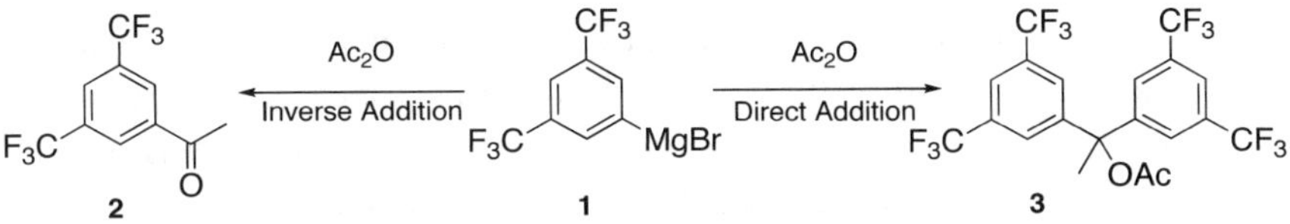

8. The reaction mixture warms to room temperature during the addition of water.

9. The 60 °C age is necessary in order to destroy residual Ac_2O.

10. The separation of layers is facile. No emulsion problems are encountered. The organic layer is clear and yellow. The aqueous layer is clear and colorless. The reaction mixture turns bright cloudy yellow after addition of 5 mL of water. A gummy residue is observed coating the flask wall after addition of 15 mL of water; however, this residue completely dissolves upon addition of 25 mL of water. Vigorous stirring is used during the addition of water.

11. HPLC grade methyl *tert*-butyl ether (MTBE) was purchased from Fisher Scientific Co. and used without further purification.

12. Product **2** has the following physical properties: ^{1}H NMR ($CDCl_3$, 400 MHz) δ 2.72 (s, 3H), 8.09 (s, 1H), 8.40 (s, 2H); ^{13}C NMR ($CDCl_3$ 100 MHz) δ 26.7 (s), 123.2 (q, J = 272.9 Hz), 126.3 (sept, J = 3.6 Hz), 128.3, (q, J = 3.4 Hz), 132.5 (q, J = 34.0 Hz), 138.6 (s), 195.1 (s); ^{19}F NMR ($CDCl_3$, 376 MHz) δ -63.67; IR (neat) 1705, 1610, 1005, 720 cm^{-1}; Anal. Calcd for $C_{10}H_6F_6O$: C, 46.89; H, 2.36. Found: C, 46.49; H, 2.37.

Safety and Waste Disposal Information

All hazardous materials should be handled and disposed of in accordance with "Prudent Practices in the Laboratory"; National Academy Press; Washington, DC, 1995.

3. Discussion

The trifluoromethylphenyl moiety is a functionality frequently encountered in pharmaceutical drugs,[3] catalysts,[4] and synthetic intermediates.[5] Trifluoromethylphenyl ketones can be classically prepared *via* addition of dimethyl copper lithium to an acid chloride in ethereal solvent;[6] however, the cryogenics make this method unsuitable for large scale synthesis. A more convenient method involves direct acetylation of Grignard reagent **1**. Preparation of the requisite 3,5-bis(trifluoromethyl)-

phenyl Grignard reagent is not straightforward, as there are references to detonations of trifluoromethylphenyl Grignard reagents.[7] One such detonation resulted in loss of life.[8] The chemical community is thus in need of a safe and reliable preparation of these valuable reagents.

Our data indicates that detonations associated with formation of trifluoromethylphenyl Grignards using Mg^0 may be attributed to loss of contact with solvent, runaway exothermic side reactions, and potentially, the presence of a highly activated form of Mg. Use of Mg to generate trifluoromethylphenyl Grignard reagents should be avoided if possible. If these reagents are to be generated from Mg, the reaction should be done in THF, and well below reflux, since loss of solvent contact can cause a runaway exothermic reaction.[2] The Knochel procedure should be used whenever possible as it bypasses the factors leading to runaway reactions.[2,9] Knochel's procedure for the low- temperature halogen-magnesium exchange has exhibited no propensity toward runaway reaction. The exchange is rapid (<1h) at cold temperatures (0 °C) and suitable for multi-kilogram scale-up.

1. Department of Process Research, Merck Research Laboratories, PO Box 2000, Rahway, NJ 07065, USA.

2. A less detailed version of this procedure has been published: Leazer Jr., J. L.; Cvetovich, R.; Tsay, F.-R.; Dolling, U.; Vickery, T.; Bachert, D. *J. Org. Chem.* **2003**, *68*, 3695.

3. (a) Brands, K. M. J.; Payack, J. F.; Rosen, J. D.; Nelson, T. D.; Candelario, M. A.; Huffman, M. A.; Zhao, M. M.; Li, J.; Craig, B.; Song, Z. J.; Tschaen, D. M.; Hansen, K.; Devine, P. N.; Pye, P. J.; Rossen, K.; Dormer, P. G.; Reamer, R.A.; Welch, C. J.; Mathre, D. J.; Tsou, N. N.; McNamara, J. M.; Reider, P. J. *J. Am. Chem. Soc.* **2003**, *125*, 2129. (b) Houlihan, W.J.; Gogerty, J.H.; Ryan, E.A.; and Schmitt, G. *J. Med. Chem.* **1985**, *28*, 28. (c) John T. Welch and Seetha Eswarakrishnan, Fluorine in Bioorganic Chemistry. John Wiley & Sons, p. 246, NY, 1991. (d) Desai, R.C.; Cicala, P.; Meurer, L.C.; Finke, P.E. *Tetrahedron Lett.* **2002**, *43*, 4569. (e) Kuethe, J.T.; Wong, A.; Wu, J.; Davies, I.W.; Dormer, P.G.; Welch, J.W.; Hillier, M.C.; Hughes, D.L.; Reider, P.J. *J. Org. Chem.* **2002**, *67,* 5993. (f) Riachi, N.J.; Arora, P.K.; Sayre, L.M.; and Harik, S.I. *J. Neurochem.* **1988**, 1319.

4. Kaul, F.A.R; Puchta, G.T.; Schneider, H.; Grosche, M.; Mihalios, D.; Herrmann, W.A.; *J. Organomet. Chem.* **2000**, *621*, 184.

5. Pinho, P.; Guijarro, D.; Andersson, P.G.; *Tetrahedron* **1998**, *54*, 7897. (b) Broeke, J.; Deelman, B-J.; Koten, G. *Tetrahedron Lett.* **2001**, *42*, 8085.

6. Posner, G.H.; Whitten, C.E. *Tetrahedron Lett.* **1970**, *11,* 4647.

7. (a) Broeke, J.; Deelman, B-J.; Koten, G. *Tetrahedron. Lett.* **2001**, *42,* 8085. (b) Beck, C.; Park, Y-J.; Crabtree, R. *Chem. Commun.* **1998**, 693. (c) Pinho, P.; Guijarro, D.; Andersson, P. *Tetrahedron* **1998**, *54*, 7897. (d) Kaul, F.; Puchta, G.; Schneider, H.; Grosche, M.; Mihalios, D.; Herrmann, W. *J. Organomet. Chem.* **2001**, *621*, 184. (e) Li, N-S.; Yu, S.; Kabalka, G. *J. Organomet. Chem.* **1997,** *531,* 101. (f) Doctorvich, F.; Deshpande, A.; Ashby, E. *Tetrahedron* **1994,** *50,* 5945.

8. (a) Appleby, I.C. *Chemistry and Industry*. **1971**, 120. (b) Post Tribune; Gary, Indiana; November 6, 1981.

9. Abarbri, M.; Dehmel, F.; Knochel, P. *Tetrahedron Lett.* **1999**, *40*, 7449.

dl-SELECTIVE PINACOL-TYPE COUPLING USING ZINC, CHLOROSILANE, AND CATALYTIC AMOUNTS OF Cp_2VCl_2; *dl*-1,2-DICYCLOHEXYLETHANEDIOL

(1,2-Ethanediol,1,2-dicyclohexyl-)

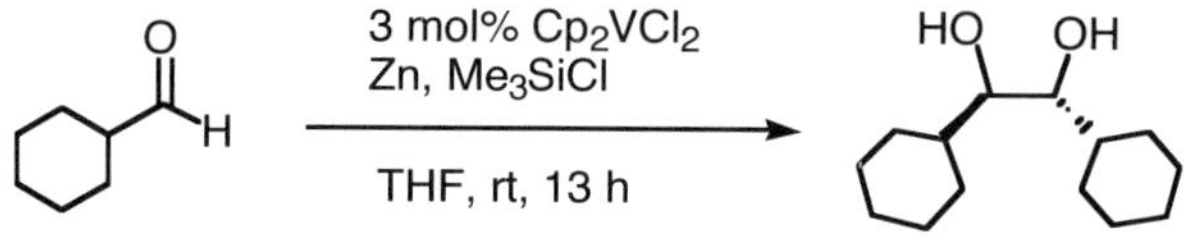

Submitted by Toshikazu Hirao, Akiya Ogawa, Motoki Asahara, Yasuaki Muguruma, and Hidehiro Sakurai.[1]
Checked by Helga Krause and Alois Fürstner.

1. Procedure

A 500-mL, two-necked, round-bottomed flask equipped with a rubber septum fitted with an argon inlet needle, a 200-mL, pressure-equalizing addition funnel capped with a rubber septum, and a magnetic stirbar is charged with zinc powder (13 g, 200 mmol) (Note 1) and dichlorodicyclopentadienylvanadium (0.756 g, 3.0 mmol) (Note 2). The flask is flame-dried and purged with argon, allowed to cool to room temperature, and then a solution of chlorotrimethylsilane (21.7 g, 200 mmol) (Note 3) in 200 mL of tetrahydrofuran (THF) (Note 4) is added dropwise over ca. 15 min via the addition funnel. The reaction mixture is stirred at room temperature for 1 hr during which time the color of the solution changes from red purple to light blue. Cyclohexanecarboxaldehyde (11.2 g, 100 mmol) (Note 5) is added via syringe over 5 min and the reaction mixture is stirred at room temperature for 13 hr. Diethyl ether (100 mL) and 100 mL of 2M aqueous HCl solution are added to the resulting mixture. The organic phase is separated and washed with 50 mL of saturated aqueous sodium bicarbonate solution, two 50-mL portions of water, and 50 mL of brine, dried over sodium sulfate, and concentrated by rotary evaporation to give 9.86-10.79 g (87-95%) of 1,2-dicyclohexylethanediol as a mixture of diastereomers (*dl*/meso = 87/13, Note 6). Three to four consecutive recrystallizations from ethanol (10 mL/g) affords 2.3-3.4 g (20-30%) of the *dl*-isomer in pure form (Notes 7,8).

2. Notes

1. Zinc powder was purchased from Wako Pure Chemical Industries and used as received.

2. Dichlorodicyclopentadienylvanadium was purchased from Aldrich Chemical Company, Inc. and used as received.

3. Chlorotrimethylsilane was purchased from Wako Pure Chemical Industries and freshly distilled from CaH_2 before use.

4. THF was obtained from Kanto Chemical Co., Inc. as dehydrated stabilizer-free grade.

5. Cyclohexanecarboxaldehyde was purchased from Tokyo Chemical Industry Co., Ltd. and used as received.

6. The diastereomeric ratio was determined by the NMR integral ratio of the methine protons adjacent to the hydroxyl group (3.33 ppm for the *dl*-isomer; 3.44 ppm for the *meso*-isomer).

7. The checkers obtained 6.7 g (60%) after the first recrystallization; this sample showed a *dl*/meso ratio of 95:5. The submitters report obtaining 5.5-7.0 g (49-62%) of the pure *dl* isomer after three recrystallizations.

8. The physical properties of the *dl*-isomer are as follows: mp 136-138°C; IR (KBr) 3320, 2915, 2850, 1450, 1410, 1270, 1105, 1020, 890, 735 cm^{-1}; ^{1}HNMR (400 MHz, $CDCl_3$) δ 1.0-1.3 (m, 10H), 1.5-1.9 (m, 14H), 3.33 (br, 2H); ^{13}CNMR (100 MHz, $CDCl_3$) δ 26.1, 26.3, 26.5, 28.3, 29.7, 40.5, 75.2; MS (EI) *m/z* (rel. intensity): 226 (0.7, [M^+]), 143 (17), 113 (50), 112 (40), 96 (13), 95 (100), 81 (10), 67 (10), 55 (16), 41 (11); Anal. Found: C, 74.18; H, 11.52%. Calcd for $C_{14}H_{26}O_2$: C, 74.29; H, 11.58%.

Waste Disposal Information

All toxic materials were disposed of in accordance with "Prudent Practices in the Laboratory"; National Academy Press; Washington, DC, 1995.

3. Discussion

Metal-induced reductive dimerization of carbonyl compounds is a useful synthetic method for the formation of vicinally functionalized carbon-carbon bonds. For stoichiometric reductive dimerizations, low-valent metals such as aluminum amalgam, titanium, vanadium, zinc, and samarium have been employed. Alternatively, ternary systems consisting of *catalytic* amounts of a metal salt or metal complex, a chlorosilane, and a stoichiometric co-reductant provide a catalytic method for the formation of pinacols based on reversible redox couples.[2] The homocoupling of aldehydes is effected by vanadium or titanium catalysts in the presence of Me_3SiCl and Zn or Al to give the 1,2-diol derivatives; high selectivity for the *dl*-isomer is observed in the case of secondary aliphatic or aromatic aldehydes.

A variety of such ternary catalytic systems has been developed for diastereoselective carbon-carbon bond formations (Table). A Cp-substituted vanadium catalyst is superior to the unsubstituted one,[3] whereas a reduced species generated from $VOCl_3$ and a co-reductant is an excellent catalyst for the reductive coupling of aromatic aldehydes.[4] A trinuclear complex derived from Cp_2TiCl_2 and $MgBr_2$ is similarly effective for *dl*-selective pinacol coupling.[5] The observed *dl*-selectivity may be explained by minimization of steric effects through *anti*-orientation of the bulky substituents in the intermediate.

Chlorosilanes appear to contribute to the catalytic reactions in various ways. Importantly, they are necessary to liberate the catalyst from the primary product formed and they are also thought to facilitate the electron transfer to the carbonyl moiety, generating the stabilized silyloxyalkyl radicals that subsequently undergo dimerization. Moreover, the observed diastereoselectivity partly depends on the substituents of the chlorosilanes. Similar ligand and additive effects are observed in diastereoselective titanium-catalyzed coupling reactions of aromatic aldehydes.[6,7] Use of water as a solvent does not require the chlorosilanes as an additive.[8]

The cat. $Cp_2VCl_2/R_3SiCl/Zn$ system outlined above can also be used for the reductive coupling of aldimines with *meso*-diastereoselectivity.[8] The observed selectivity depends on the substituents on the nitrogen as well as the silicon atoms. 1,5- and 1,6-dialdehydes undergo intramolecular pinacol coupling to give cyclic *vic*-diols with excellent selectivity.[3,6] The reductive coupling has also been applied to the catalytic diastereoselective cyclization of arylidene malononitriles and ketonitriles to give the corresponding cyclopentene and cyclopentenol derivatives, respectively.[9]

Table. Cat. $Cp_2VCl_2/Me_3SiCl/Zn$-Induced Pinacol Coupling of Aliphatic Aldehydes[3]

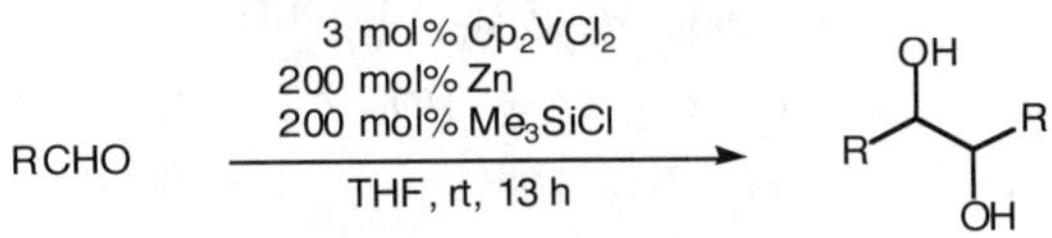

aldehyde	yield / % [*dl* / *meso*]
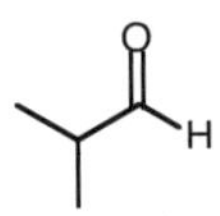	89 [91/9]
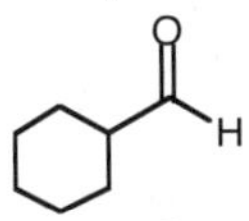	86 [87/13]
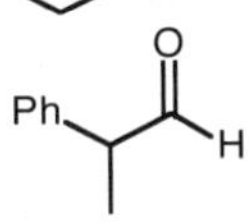	66 [94/6]
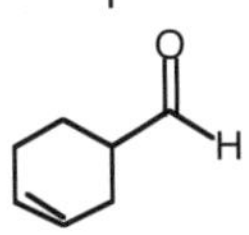	97 [85/15]
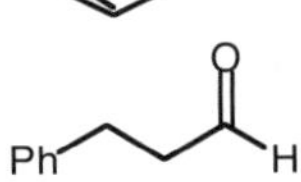	97 [64/36]

1. Department of Applied Chemistry, Graduate School of Engineering, Osaka University, Yamada-oka, Suita, Osaka 565-0871, Japan.
2. Hirao, T. *Chem. Rev.* **1997**, *97*, 2707; Hirao, T. *Synlett* **1999**, 175; A. Fürstner, *Chem. Eur. J.* **1998**, *4*, 567; Hirao, T. in *Catalytic Reductive C-C Bond Formation*, Krische, M. J., Ed., Springer: Heidelberg, **2007**, 53.
3. Hirao, T.; Hasegawa, T.; Muguruma, Y.; Ikeda, I. *Abstracts for the 6th International Conference on New Aspects of Organic Chemistry*, 1994; p 175;

Hirao, T.; Asahara, M.; Muguruma, Y.; Ogawa, A. *J. Org. Chem.* **1998**, *63*, 2812; Hirao, T.; Hasegawa, T.; Muguruma, Y.; Ikeda, I. *J. Org. Chem.* **1996**, *61*, 366; Hirao, T.; Hatano, B.; Asahara, M.; Muguruma, Y.; Ogawa, A. *Tetrahedron Lett.* **1998**, *39*, 5247.

4. Hirao, T.; Hatano, B.; Imamoto, Y.; Ogawa, A. *J. Org. Chem.* **1999**, *64*, 7665.

5. Gansäuer, A. *Chem. Commun.* **1997**, 457.

6. Lipski, T. A.; Hilfiker, M. A.; Nelson, S. G. *J. Org. Chem.* **1997**, *62*, 4566.

7. Gansäuer, A.; Bauer, D. *J. Org. Chem.* **1998**, *63*, 2070.

8. Xu, X.; Hirao, T., *J. Org. Chem.* **2005**, *70*, 8594.

9. Hatano, B.; Ogawa, A.; Hirao, T. *J. Org. Chem.* **1998**, *63*, 9421.

10. Zhou, L.; Hirao, T. *Tetrahedron Lett.* **2000**, *41*, 8517; Zhou, L.; Hirao, T. *Tetrahedron* **2001**, *57*, 6927.

4-NONYLBENZOIC ACID
(Benzoic acid, 4-nonyl-)

A.

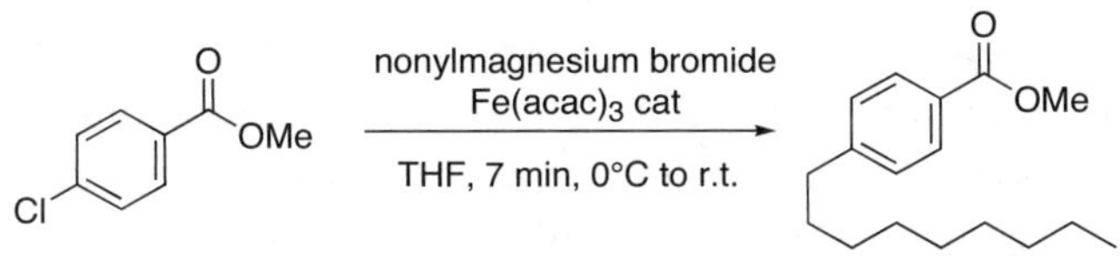

B.

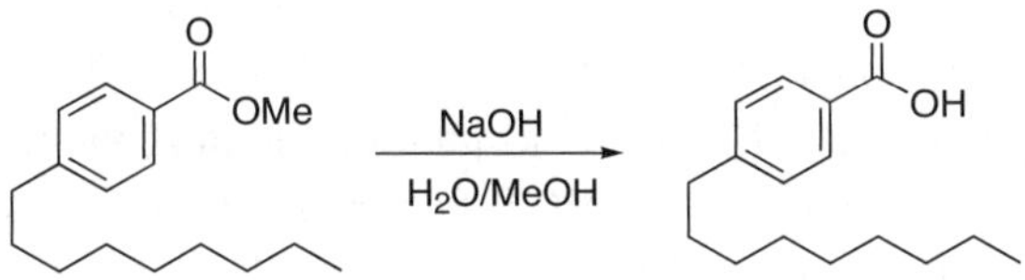

Submitted by Alois Fürstner, Andreas Leitner, and Günter Seidel.
Checked by Renee Kontnik and Steven Wolff.

1. Procedure

A. 4-Nonylbenzoic acid methyl ester. A 250-ml, three-necked, round-bottomed flask equipped with a Teflon-coated magnetic stirbar, glass stopper, a reflux condenser fitted with an argon stopcock inlet, and a pressure-equalizing dropping funnel is charged with magnesium turnings (2.95 g, 121.0 mmol). The flask is evacuated and flame-dried while the magnesium turnings are gently stirred. After reaching ambient temperature, the apparatus is flushed with argon, the magnesium turnings are suspended in 20 mL of tetrahydrofuan (Note 1), and 1,2-dibromoethane (0.3 ml, 3.6 mmol) (Note 2) is introduced. A solution of 1-bromononane (20.52 g, 97.0 mmol) (Note 2) in 100 mL of THF (Note 1) is then added via the addition funnel to the suspension over a period of ca. 45 min at such a rate as to maintain gentle reflux. After the addition is complete, the mixture is refluxed for another 20 min and then the resulting solution of nonylmagnesium bromide is allowed to cool to ambient temperature. In the meantime, an oven-dried, 2-L, two-necked, round-bottomed flask equipped with a magnetic stirbar, rubber septum, and an argon inlet is flushed with argon and charged with 4-chlorobenzoic acid methyl ester (13.0 g, 76.2 mmol) (Note 2), ferric acetylacetonate [Fe(acac)$_3$] (1.35 g, 3.82 mmol) (Note 2), 450 ml of THF (Note 1), and 25 mL of N-methylpyrrolidinone (NMP) (Note 1). The flask is immersed in an ice bath and the solution of nonylmagnesium bromide prepared above is immediately added within one min via a polyethylene cannula. This causes an immediate color change from red to black-violet. The ice bath is removed and the resulting dark mixture is stirred for 7-10 min at ambient temperature, diluted with 200 mL of diethyl ether, and then carefully quenched by addition of 300 mL of 1M HCl with stirring. The mixture is transferred into a separatory funnel (the flask is rinsed with 200 mL of Et_2O), and the aqueous phase is separated and extracted with 200 mL of Et_2O. The combined organic phases are washed with 300 mL of saturated aq $NaHCO_3$, dried over Na_2SO_4, and concentrated by rotary evaporation at reduced pressure (Note 3). The resulting crude orange-red residue is purified by short path distillation under high vacuum (1×10^{-4} torr) to give 15.81-16.85 g (79-84%) of 4-nonylbenzoic acid methyl ester as a colorless syrup, bp 103-105 °C (Notes 4, 5).

B. 4-Nonylbenzoic acid. A 500-mL, round-bottomed flask equipped with a Teflon-coated magnetic stirbar and a reflux condenser is charged with 4-nonylbenzoic acid methyl ester (10.07 g, 38.37 mmol), 100 mL of methanol (Note 1), and 96 mL of 1M aqueous NaOH. The resulting mixture is heated at reflux for 18 hr and then allowed to cool to room temperature. The reaction mixture is carefully acidified by addition of 200 mL of 1M aqueous HCl, and the resulting solution is transferred to a separatory funnel and extracted with four 250-mL portions of ethyl acetate. The combined organic layers are dried over Na_2SO_4, filtered, and concentrated by rotary evaporation at reduced pressure. The residue (ca. 9.5 g) is recrystallized from 70 mL of hexanes to give 8.32-8.35 g (87-88%) of 4-nonylbenzoic acid as a white solid (Notes 6, 7).

2. Notes

1. The submitters used THF that was freshly distilled over Na/K alloy; NMP was distilled over CaH_2. The checkers used commercially available anhydrous THF and NMP obtained from Aldrich Chemical Co. There appears to be a slight increase in yield when THF distilled from sodium/benzophenone ketyl is used instead of the commercial THF. All other solvents used were of reagent grade quality and were used without further purification.

2. 1,2-Dibromoethane, 1-bromononane (98%), 4-chlorobenzoic acid methyl ester (99%), and $Fe(acac)_3$ (99.9%) were purchased from Aldrich Chemical Co. and used as received.

3. The checkers found that a dark red oil separated upon concentration of the ethereal solution of the crude ester. In this instance, the ester was diluted with 200 mL of ether, washed with 200 mL of brine, and then again concentrated.

4. The submitters obtained the product in 87-90% yield.

5. The product is ≥ 95% pure by GC (Agilent 6890 Series, column: HP-5MS; 5% phenyl methyl siloxane, 30 m x 250 μm 0.25 μm); the remainder is octadecane formed by oxidative coupling of the Grignard reagent; T-program: 70 °C (3.5 min)→280 °C (20 °C/min); 12.62 min retention time; ^{1}H NMR (300 MHz, $CDCl_3$): δ 0.86 (t, J = 6.7 Hz, 3H), 1.2-1.36 (m, 12H), 1.60 (m, 2H), 2.63 (t, J = 7.7 Hz, 2H), 3.88 (s, 3H), 7.22 (d, J = 8.4 Hz, 2H), 7.93 (d, J = 8.3 Hz, 2H); ^{13}C NMR (75 MHz, $CDCl_3$): δ 14.1, 22.7, 29.2, 29.3, 29.4, 29.7, 31.1, 31.9, 36.0, 51.9, 127.6, 128.4, 129.6, 148.5, 167.2,; IR (film): 3032, 2953, 2926, 2855, 1725, 1611, 1465, 1435, 1278, 1178, 1109, 1021, 854 cm $^{-1}$; MS (El): *m/z* (rel. Intensity) 262 (63, [M^+]), 231 (24), 163 (13), 150 (100), 121 (8), 105, (9), 91 (41), 57 (8), 43 (13). Anal. Calcd for $C_{17}H_{26}O_2$: C, 77.82; H, 9.99. Found C, 78.30; H, 10.45.

6. The submitters obtained the product in 94% yield.

7. The product has the following physical properties: mp 92.5-94.3 °C; ^{1}H NMR (400 MHz, $CDCl_3$): δ 0.86 (t, J = 6.9 Hz, 3H), 1.2-1.38 (m, 12H), 1.62 (m, 2H), 2.66 (t, J = 7.7 Hz, 2H), 7.26 (d, J = 8.2 Hz, 2H), 8.01 (d, J = 8.2 Hz, 2H); ^{13}C NMR (100 MHz, $CDCl_3$): δ 14.1, 22.7, 29.2, 29.3, 29.4, 29.5, 31.1, 31.8, 36.1, 126.7, 128.6, 130.4, 149.6, 171.9; IR (film): 3072, 2924, 2852, 2669, 2554, 1683, 1609, 1575, 1469, 1424, 1321, 1290, 945, 859, 758 cm $^{-1}$; MS (El): *m/z* (rel. intensity): 248 (56, [M^+]), 177 (7), 149 (9), 136 (100), 107 (9), 92 (38), 57 (14); Anal. Calcd for $C_{16}H_{24}O_2$-H_2O: C, 72.14; H, 9.84. Found C, 72.53; H, 9.76.

Waste Disposal Information

All toxic materials were disposed in accordance with "Prudent Practices in the Laboratory"; National Academy Press; Washington, DC 1995.

3. Discussion

Modern cross coupling chemistry is heavily dominated by the use of palladium and nickel complexes as the catalysts, which show an impressively wide scope and an excellent compatibility with many functional groups.[2] This favorable application profile usually overcompensates the disadvantages resulting from the high price of the palladium precursors, the concerns about the toxicity of nickel salts, the need for ancillary ligands to render the complexes sufficiently active and stable, and the extended reaction times that are necessary in certain cases.

Our group has recently developed an alternative method for alkyl-(hetero)aryl- as well as aryl-heteroaryl cross coupling reactions catalyzed by iron salts. [3,4] This methodology was inspired by early reports of Kochi et al. [5,6] on iron-catalyzed cross coupling of vinyl halides and is distinguished by several notable advantages.

(1) The expensive noble metal catalysts are replaced by cheap, air stable, commercially available and toxicologically benign iron salts without any loss in efficiency. The reactions are usually carried out under "ligand free" conditions using inexpensive Grignard reagents as the preferred coupling partners.

(2) Aryl chlorides, which are the most attractive type of substrates due to their low cost and ready availability, perform inherently better than the corresponding aryl bromides or iodides. Moreover, aryl triflates and even electron-deficient aryl tosylates can be used as the starting materials.

(3) Most iron-catalyzed reactions proceed at unprecedentedly high rates and are finished within a few minutes even when carried out at or below ambient temperature.

(4) Due to the efficiency with which the iron catalysts activate the C-Cl bond, several functional groups are tolerated that normally would react with a Grignard reagent.

(5) When applied to polyfunctional substrates, iron catalyzed reactions allow either for selective, exhaustive, or consecutive cross-coupling processes to be carried out in "one pot."

Most of these chemical features are evident from the preparation of 4-nonylbenzoic acid described above which requires less than 10 min reaction time at 0 °C→ r.t. even when performed on a > 15 g scale. No competing attack of the Grignard reagent onto the methyl ester of the substrate can be detected under the reaction conditions. It should be noticed that 4-nonylbenzoic acid, like other alkyl benzoic acid derivatives that are

TABLE. IRON CATALYZED CROSS COUPLING REACTIONS OF GRIGNARD REAGENTS WITH (HETERO)ARYL CHLORIDES OR -TRIFLATES

Substrate	Product	Yield
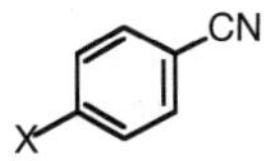	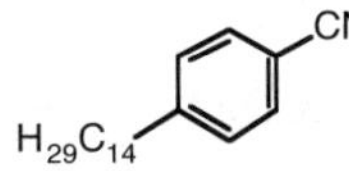	91% (X = Cl) 80% (X = OTf)
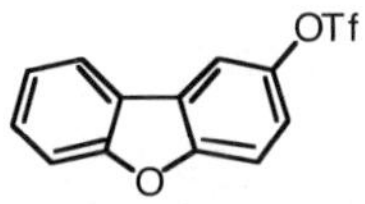	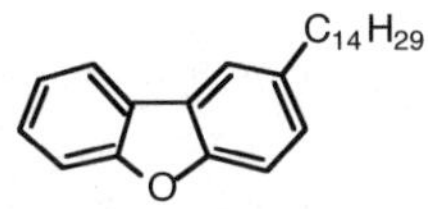	81%
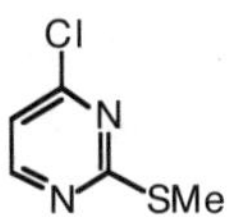	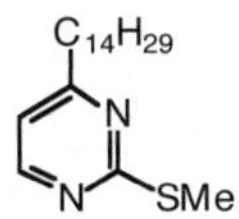	89%
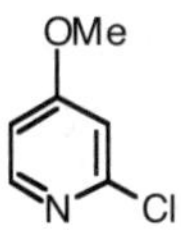	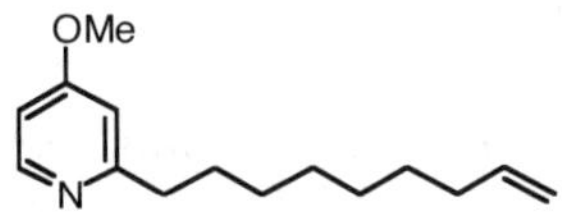	81%
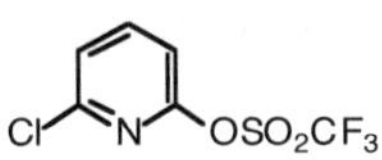	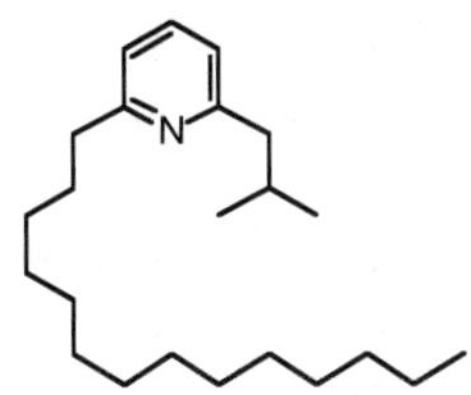	71%[a]
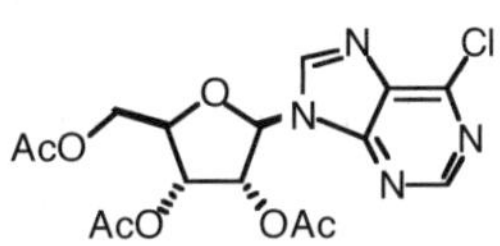	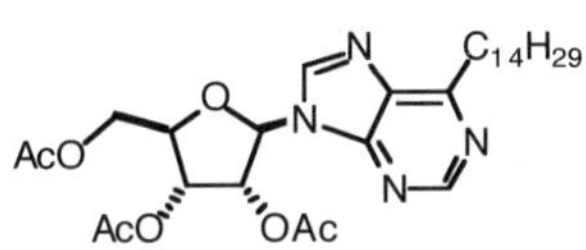	72%
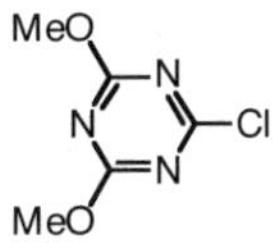	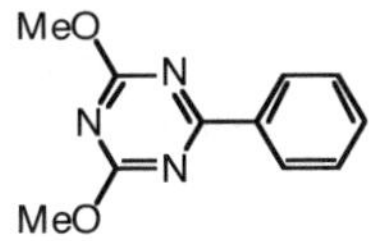	63%
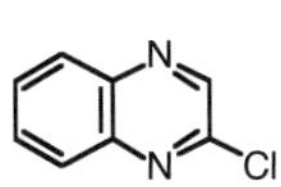	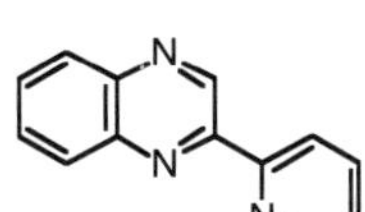	82%

a) Consecutive addition of isobutylmagnesium bromide and tetradecylmagnesium bromide within an overall reactioin time of 8 min.

equally available by this iron-catalyzed cross coupling method, is of some practical relevance as liquid crystalline material or a component thereof.[7,8] The additional examples compiled in the Table illustrate the scope and performance of this new protocol in more detail.[3,4,9,10]

1. Max-Planck-Institut für Kohlenforschung, Kaiser-Wilhelm-Platz 1, D-45470 Mülheim/Ruhr, Germany.

2. *Metal Catalyzed Cross-coupling Reactions;* Diederich, F.; Stang, P. J., Eds.; Wiley-VCH: Weinheim, **1998**.

3. Fürstner, A.; Leitner, A. *Angew. Chem., Int. Ed.* **2002**, *41*, 609.

4. (a) Fürstner, A.; Leitner, A.; Méndez, M.; Krause, H. *J. Am. Chem. Soc.* **2002**, *124*, 13856. (b) Scheiper, B.; Bonnekessel, M.; Krause, H.; Fürstner, A. *J. Org. Chem.* **2004**, *69*, 3943. (c) Short review: Fürstner, A.; Martin, R. *Chem. Lett.* **2005**, *34*, 624.

5. (a) Tamara, M. Kochi, J. K. *J. Am. Chem. Soc.* **1971**, *93*, 1487. (b) Tamura, M.; Kochi J. *Synthesis* **1971**, 303. (c) Neumann, S. M.; Kochi, J. K. *J. Org. Chem.* **1976**, *41*, 502.

6. See also: Cahiez, G; Avedissian, H. *Synthesis* **1998**, 1199.

7. Sage, I. In *Ullmann's Encyclopedia of Industrial Chemistry*, VCH, Weinheim, **1990**, Vol. A15, pp. 359.

8. See also: (a) Kato, T.; Fukumasa, M.; Frechet, J. M. J. *Chem. Mater.* **1995**, *7*, 368. (b) Kato, T.; Frechet, J. M. J.; Wilson, P. G.; Saito, T.; Uryu, T.; Fujishima, A.; Jin, C.; Kaneuchi, F. *Chem. Mater.* **1993**, *5*, 1094. (c) Ojha, D. P.; Kumar, D.; Pisipati, V. G. K. M. *Z. Naturforsch.* **2002**, *57a*, 189.

9. Scheiper, B.; Bonnekessel, M.; Krause, H.; Fürstner, A.; *J. Org. Chem.* **2004**, *69*, 3943.

10. Short review: Fürstner, A.; Martin, R. *Chem. Lett.* **2005**, *34*, 624.

PALLADIUM CATALYZED CROSS-COUPLING OF (*Z*)-1-HEPTENYLDIMETHYLSILANOL WITH 4-IODOANISOLE: (*Z*)-1-HEPTENYL-4-METHOXYBENZENE [Benzene,1-(1Z)-1-heptenyl-4-methoxy-]

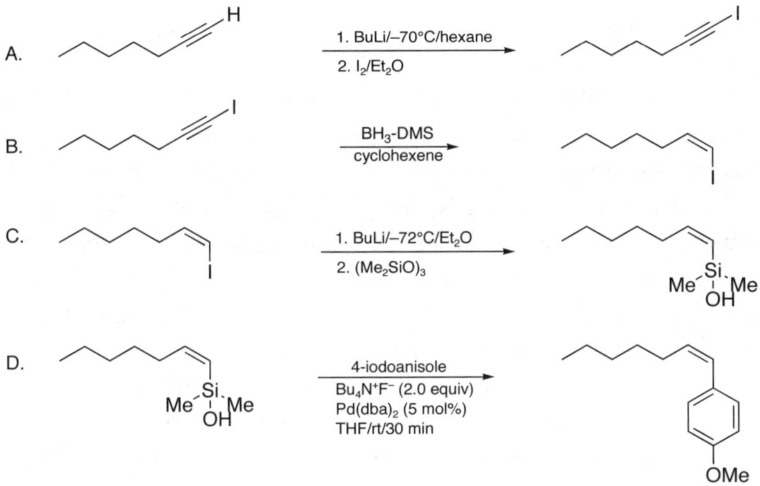

Submitted by Scott E. Denmark and Zhigang Wang.[1]
Checked by Matthew Campbell and Dennis P. Curran.

1. Procedure

A. 1-Iodo-1-heptyne.[2] A solution of 9.62 g (100 mmol) of 1-heptyne (Note 1) in 50 mL of hexane (Note 2) is placed in a flame-dried, three-necked, 500-mL, round-bottomed flask equipped with a large stirbar, a pressure equalizing addition funnel, and an N_2 inlet. The solution is cooled to –50 to –55 °C (Note 3) and a solution of *n*-BuLi in hexane (61.0 mL, 1.65M, 100 mmol) (Note 4) is then added dropwise over 30 min. The resulting thick suspension is stirred at –70 °C for 1.5 h. A solution of 25.4 g (100 mmol) of Iodine (Note 5) in 130 mL of diethyl ether (Note 6) is added over 45 min. The cooling bath is removed and the reaction solution is allowed to warm to room temperature. Water (100 mL) is added and the resulting mixture is stirred at room temperature for 10 min. The aqueous layer is separated and extracted with 100 mL of hexane, and the combined organic phases are washed with two 200-mL portions of water, 50 mL of 20% aqueous sodium thiosulfate ($Na_2S_2O_3$) solution, and then dried over anhydrous $MgSO_4$ (Note 7). The solvents are removed by rotary evaporation (Note 8) and the residue is dissolved in 30 mL of pentane (Note 9) and the solution is then passed through a short column of silica gel (Note 10) followed by further elution with 250 mL of pentane. The solvent is removed by rotary evaporation and the residue is purified by Kugelrohr distillation to afford 17.9-18.5 g (81-83%) of 1-iodo-1-heptyne as a colorless liquid, bp 60-65 °C at 1.2 mm (lit.[2] 55-60 °C at 5 mm) (Notes 11, 12).

B. (Z)-1-Iodo-1-heptene.[2] A solution of 8.52 g of (112 mmol) of borane-dimethylsulfide complex (Note 13) in 100 mL of ether is added to a flame-dried, three-necked, 300-mL, round-bottomed flask equipped with stirbar, temperature probe, and N_2 inlet. The solution is cooled to 5 °C with an ice-bath. Cyclohexene (18.4 g, 224 mmol) (Note 14) is then added by syringe over 10 min while keeping the temperature below 15 °C. The mixture is stirred at 5 °C for 15 min. A white solid precipitates either towards the end of the addition or during the subsequent stirring period. The reaction mixture is allowed to warm to room temperature and is stirred for 1 h. The non-homogeneous solution is cooled to 2-3 °C and 22.7 g (102 mmol) of 1-iodo-1-heptyne is added by syringe over 10 min. The reaction mixture is stirred at 2-3 °C for 30 min, and then the cooling bath is removed and the mixture is stirred at room temperature for 1 h. The solution is cooled to 2-3 °C and 51 mL of acetic acid (AcOH) (Note 15) is slowly added. The ice-bath is removed after the addition of AcOH is completed and the mixture is stirred at room temperature for 2 h. Ether (100 mL) is added and the solution is washed with four 75-mL portions of water, dried over anhydrous $MgSO_4$, filtered, and concentrated by rotary evaporation to give a pale yellow liquid which is dissolved in 50 mL of pentane. The solution is passed through a short column of silica gel (Note 10) followed by further elution with 250 mL of pentane. The solvent is removed by rotary evaporation to give the crude product which is purified by Kugelrohr distillation to afford 15.7 g (69%) of (*Z*)-1-iodo-1-heptene as a colorless liquid, bp 75-80 °C at 1.5 mm (lit.[2] 60-63 °C at 5.5 mm) (Notes 11, 16).

C. (Z)-1-(1-Heptenyl)-1,1-dimethylsilanol. A solution of 13.0 g (58 mmol) of (*Z*)-1-iodo-1-heptene in 50 mL of ether is placed in a flame-dried, three-necked, 300-mL, round-bottomed flask equipped with stirbar, septum, temperature probe, and N_2 inlet. The solution is cooled to –72 °C and a solution of *n*-BuLi in hexane (Note 4) (35.6 mL, 1.63M, 58 mmol) is then added by syringe over 15 min. The mixture is stirred at –72 °C for 30 min and then a solution of 4.30 g (19.3 mmol) of hexamethylcyclotrisiloxane (Note 17) in 50 mL of ether is added over 15 min. The cooling bath is removed, and the reaction solution is allowed to warm to room temperature and stirred for 14 h. The solution is then cooled to 0 °C and 50 mL of water is slowly added to quench the reaction. The aqueous layer is separated and extracted with three 50-mL portions of diethyl ether. The combined organic phases are washed with 30 mL of water and two 30-mL portions of brine, dried over anhydrous $MgSO_4$, filtered, and concentrated by rotary evaporator and vacuum drying to give a crude product which is distilled to afford 7.34 g (73%) of (*Z*)-1-(1-heptenyl)-1,1-dimethylsilanol as a colorless liquid, bp 54-55 °C at 0.15 mm (lit.[4] 120 °C at 0.9 mm) (Notes 18, 19).

D. (Z)-1-(1-Heptenyl)-4-methoxybenzene. A solution of 4.14 g (24 mmol) of (*Z*)-1-(1-heptenyl)-1,1-dimethylsilanol in 2.0 mL of THF is added to a flame-dried, 250 mL, three-necked, round-bottomed flask equipped with a stirbar, addition funnel, temperature probe, and argon inlet. A solution of tetrabutylammonium fluoride in THF (40 mL, 1.0M, 40 mmol) (Note 20) is then added over 5 min. The mixture is stirred at room temperature for 10 min and then 0.575 g (1.0 mmol) of $Pd(dba)_2$ (Note 21) is added. A solution of 4-iodoanisole in THF (20 mL, 1.0M, 20 mmol) (Note 22) is added slowly through the addition funnel (Note 23) such that the temperature of the solution is kept between 30-33 °C (addition time 30 min). The mixture is stirred for 30 min after complete addition of 4-iodoanisole and then 70 mL of diethyl ether is added and the mixture is stirred for an additional 10 min. The mixture is passed through a short column of silica gel (Note 10) followed by further elution with 200 mL of diethyl ether. The combined eluate is concentrated by rotary evaporation and vacuum drying to give a residue which is purified by Kugelrohr distillation to afford a colorless or light yellow liquid. This liquid is further purified by chromatography (Note 24) followed by Kugelrohr distillation to afford 3.29 g (80%) of (*Z*)-1-(1-heptenyl)-4-methoxybenzene as a colorless oil, bp 90-95 °C at 0.1 mm (lit.[4] 80-85 °C at 1.3 mm) (Notes 11, 25).

2. Notes

1. 1-Heptyne was purchased from GFS Chemicals Inc. and was used without further purification.

2. Hexane was purchased from Fisher Scientific Company and was freshly distilled from sodium.

3. Unless otherwise noted, all temperatures refer to internal temperatures measured with Teflon-coated thermocouples.

4. Butyllithium was purchased from FMC Corporation Lithium Division and its concentration was determined by double titration by the Gilman method.[3]

5. I_2 was purchased from Mallinckrodt Inc. and was used as received.

6. Ether was freshly distilled from sodium/benzophenone.

7. Anhydrous $MgSO_4$ was purchased from Fisher Scientific Company.

8. The vacuum for the rotary evaporator is about 15 mm.

9. Pentane was purchased from Fisher Scientific Company and was distilled from $CaCl_2$.

10. Silica gel was purchased from VWR Scientific (230-400 mesh). A 45-mm diameter column was employed and 50 g silica gel was loaded as a pentane slurry.

11. Boiling points (bp) correspond to uncorrected air-bath temperatures in the Buchi GKR-50 Kugelrohr.

12. The analytical data are as follows: ^1H NMR (500 MHz, $CDCl_3$) δ: 0.90 (t, $J = 7.1$, 3 H), 1.34 (m, 4 H), 1.51 (m, 2 H), 2.35 (t, $J = 7.1$, 2 H); ^{13}C NMR (125.6 MHz, $CDCl_3$) δ: –7.2, 13.9, 20.7, 22.1, 28.2, 30.9, 94.6; IR (NaCl) cm^{-1}: 2956, 2932, 2859, 1462; Anal. Calcd. for $C_7H_{11}I$: C, 37.86; H, 4.99; I, 57.15; Found: C, 37.90; H, 5.08; I, 57.50.

13. Borane-methyl sulfide complex (neat) was purchased from Aldrich Chemical Company, Inc. and was used as received.

14. Cyclohexene was purchased from Aldrich Chemical Company, Inc. and was used as received.

15. Acetic acid (100%) was purchased from J. T. Baker Company and was used as received.

16. The checkers obtained the product in 65-74% yield. The analytical data are as follows: ^1H NMR (500 MHz, $CDCl_3$) δ: 0.90 (t, J = 7.1, 3 H), 1.32 (m, 4 H), 1.34 (m, 2 H), 2.13 (m, 2 H), 6.17 (m, 2 H); ^{13}C NMR (125.6 MHz, $CDCl_3$) δ: 14.0, 22.5, 27.6, 31.3, 34.6, 82.1, 141.5; IR (NaCl) cm^{-1}: 2956, 2927, 2857, 1608, 1463, 1276; GC analysis: t_R (*Z*) 6.34 min (99.3%), t_R (*E*) 6.57 min (0.7%) (HP-5, 150 °C, 15 psi); Anal. Calcd. for $C_7H_{13}I$: C, 37.52; H, 5.85; I, 56.63; Found: C, 37.72; H, 5.96; I, 56.81.

17. Hexamethylcyclotrisiloxane was purchased from Gelest Inc. and was used as received.

18. The (*Z*)-1,1-dimethyl-1-heptenylsilanol easily forms 1,1,1',1'-tetramethyl-1,1'-diheptenyldisiloxane. The crude product was distilled (60-62 °C at 0.5 mm) to afford 8.02 g (80%) of (*Z*)-1,1-dimethyl-1-

heptenylsilanol containing 2–5% of 1,1,1',1'-tetramethyl-1,1'-diheptenyldisiloxane. To obtain analytically pure product, the silanol was distilled twice again (bp 54-55 °C, 0.15 mm).

19. The checkers obtained the product in 67-89% yield. The analytical data are as follows: ^{1}H NMR (500 MHz, $CDCl_3$) δ: 0.24 (s, 6 H), 0.89 (t, J = 7.1, 3 H), 1.30 (m, 4 H), 1.39 (m, 2 H), 1.69 (br, 1 H), 2.19 (m, 2 H), 5.47 (dt, J = 14.1, 1.3, 1 H), 6.36 (dt, J = 14.1, 7.5, 1 H); ^{13}C NMR (125.6 MHz, $CDCl_3$) δ: 1.7, 14.0, 22.6, 29.3, 31.5, 33.6, 127.8, 151.6; IR (NaCl) cm^{-1}: 3274 (br), 2959, 2928, 2858, 1607, 1465, 1253, 1067, 864, 785; GC analysis: t_R 7.15 min (HP-1, 100 °C, 10 psi); Anal. Calcd. for $C_9H_{20}OSi$: C, 62.72; H, 11.70; Found: C, 62.56; H, 12.04.

20. A solution of tetrabutylammonium fluoride (TBAF) in THF (1.0M) is prepared from colorless, crystalline tetrabutylammonium fluoride trihydrate purchased from Fluka Chemical Corp.

21. $Pd(dba)_2$ was prepared by the literature procedure without recrystallization from $CHCl_3$;[5] See Note 8, *Org. Synth.*, **2004**, *81*, 57.

22. 4-Iodoanisole was purchased from Aldrich Chemical Company, Inc., and was purified by chromatography on silica gel (hexane/EtOAc, 50/1) prior to use.

23. The coupling reaction is very exothermic. The reaction temperature can be controlled by the rate of addition of 4-iodoanisole.

24. A 55 mm diameter column is employed and 200 g silica gel is used (pentane/EtOAc, 50/1).

25. The checkers obtained the product in 90-97% yield. The analytical data are as follows: ^{1}H NMR (400 MHz, $CDCl_3$) δ: 0.89 (t, J = 7.1, 3 H), 1.31 (m, 4 H), 1.45 (qn, J = 7.3, 2 H), 2.31 (qd, J = 7.3, 2 H), 3.81 (s, 3 H), 5.57 (dt, J = 11.7, 7.1, 1 H), 6.33 (d, J = 11.7, 1 H), 6.87 (d, J = 8.8, 2 H), 7.22 (d, J = 8.8, 2 H); ^{13}C NMR (100.6 MHz, $CDCl_3$) δ: 13.9, 22.6, 28.6, 29.7, 31.6, 55.0, 113.5, 128.0, 129.9, 130.5, 131.5, 158.1; IR (NaCl) cm^{-1}: 3005, 2956, 2927, 2856, 1608, 1511, 1463, 1301, 1249, 1175, 1038, 837; GC analysis: t_R (*Z*) 6.05 min (99.1%), t_R (*E*) 6.53 min (0.9%) (HP-5, 250 °C, 15psi); Anal. Calcd. for $C_{14}H_{20}O$: C, 82.30; H, 9.87; Found: C, 82.25; H, 9.96.

Waste Disposal Information

All toxic materials were disposed of in accordance with "Prudent Practices in the Laboratory", National Academy Press; Washington, DC, 1995.

3. Discussion

Palladium-catalyzed, cross-coupling reactions of organosilicon reagents with aryl heteroaryl and alkenyl halides have emerged in recent years as a synthetically viable alternative to the traditional organotin (Stille) and organoboron (Suzuki) cross-coupling reactions. The most extensively investigated of these processes involves the use of mono-, di-, and trifluorosilanes as the coupling partners activated by a fluoride source.[6] Recently, aryl trialkoxysiliconates have also shown promise for carbon-carbon bond formation under similar conditions.[7]

Our recent demonstration that a silicon-oxygen bond is the key structural feature that lends high reactivity of organosilicon compounds toward cross-coupling has led to the development of a wide range of silicon-based substrates for this process.[4,8] The most thoroughly investigated of these derivatives are the dialkylsilanols. These compounds are easily prepared by a number of different methods, the most common of which illustrated here is the reaction of organolithium compounds with hexamethylcyclotrisiloxane (D_3).[9,10] The addition is general for alkenyl-, aryl-, heteroaryl- and alkynyllithium agents and all three silicon groups in D_3 are available.

The palladium-catalyzed cross-coupling of alkenylsilanols has been extensively studied with respect to the structure of both the silicon component and the acceptor halide. The preferred catalyst for coupling of aryl iodides is $Pd(dba)_2$ and for aryl bromides it is $[allylPdCl]_2$. The most effective promoter is tetrabutylammonium fluoride used as a 1.0M solution in THF. In general the coupling reactions occur under mild conditions (room temperature, in 10 min to 12 h) and some are even exothermic.

Varying substitution patterns have been studied and while the rate of the coupling is attenuated by steric crowding of the dimethylsilanol, the yields and stereospecificities are still high. In all cases studied to date, the coupling reactions proceed with retention of configuration. The examples compiled in Figure 1 represent successful couplings of a number of classes of alkenylsilanols in combination with both aryl and alkenyl halides.[8]

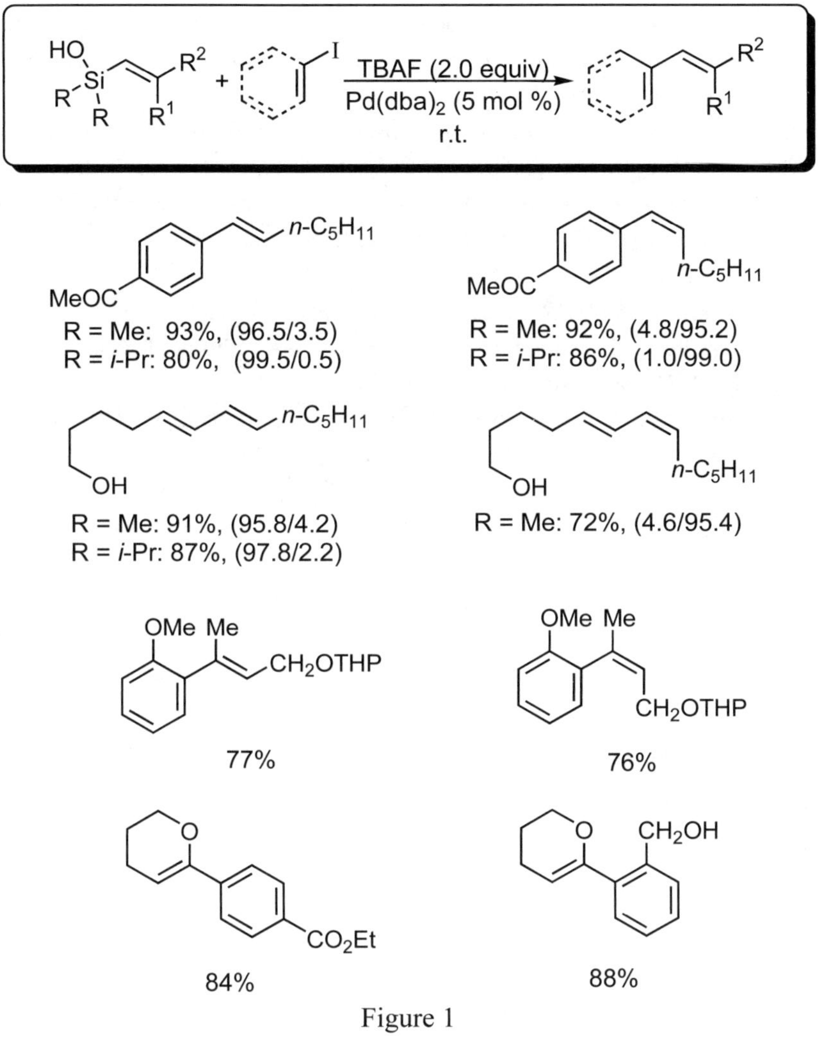

Figure 1

1. Department of Chemistry, Roger Adams Lab., University of Illinois at Urbana-Champaign. Urbana, IL, 61801, E-mail: denmark@scs.uiuc.edu.
2. Ravid, U.; Silverstein, R. M.; Smith, L. R. *Tetrahedron* **1978**, *34*, 1449-1452.
3. Gilman, H.; Cartledge, F. K.; Sin, S.-Y. *J. Organomet. Chem.* **1963**, *1*, 8-14.
4. Denmark, S. E.; Wehrli, D. *Org. Lett.* **2000**, *2*, 565-568.
5. Ukai, T.; Kawazura, H.; Ishii Y. *J. Organomet. Chem.* **1974**, *65*, 253-266.

6. Hiyama, T. Organosilicon Compounds in Cross-Coupling Reactions. In *Metal Catalyzed Cross-Coupling Reactions*; Diederich, F., Stang, P. J., Eds.; Wiley-VCH: Weinhein, 1998; Chapter 10.

7. Tamao, K.; Kobayashi, K.; Ito, Y. *Tetrahedron Lett.* **1989**, *30*, 6051-6054. (b) Mowery, M. E.; DeShong, P. *J. Org. Chem.* **1999**, *64*, 1684-1688. (c) Mowery, M. E.; DeShong, P. *J. Org. Chem.* **1999**, *64*, 3266-3270. (d) Mowery, M. E.; DeShong, P. *Org. Lett.* **1999**, *1*, 2137-2140. (e) Lee, H. M.; Nolan, S. P. *Org. Lett.* **2000**, *2*, 2053-2055.

8. Denmark, S. E.; Sweis, R. F. *Acc. Chem. Res.* **2002**, *35,* 835-846.

9. Hirabayashi, K.; Takahisa, E.; Nishihara, Y.; Mori, A.; Hiyama, T. *Bull. Chem. Soc. Jpn*. **1998**, *71*, 2409-2417.

10. Sieburth, S. McN.; Fensterbank, L. *J. Org. Chem.* **1993**, *58*, 6314-6318. (b) Sieburth, S. McN.; Mu, W. *J. Org. Chem.* **1993**, *58*, 7584-7586.

PLATINUM CATALYZED HYDROSILYLATION AND PALLADIUM CATALYZED CROSS-COUPLING: ONE-POT HYDROARYLATION OF 1-HEPTYNE TO (E)-1-(1-HEPTENYL)-4-METHOXYBENZENE

[Benzene, 1-(1E)-1-heptenyl-4-methoxy-]

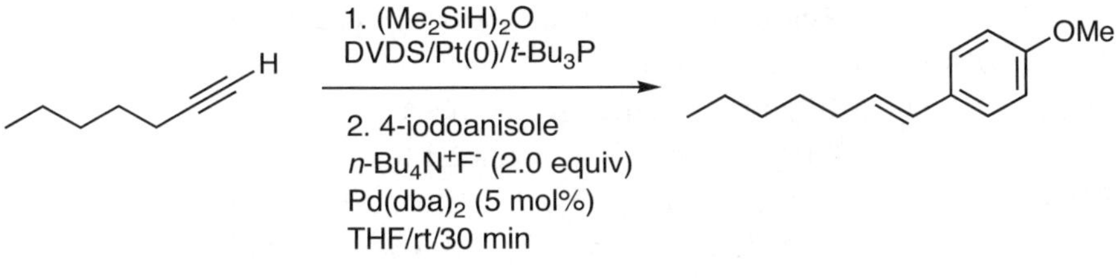

Submitted by Scott E. Denmark and Zhigang Wang.[1]
Checked by Adam I. Keller and Dennis P. Curran.

1. Procedure

(E)-1-(1-Heptenyl)-4-methoxybenzene. A solution of 2.79 mL (2.12 g, 15.6 mmol) of 1,1,3,3-tetramethyldisiloxane in 3.0 mL of tetrahydrofuran (THF) is added to a flame-dried, three-necked, 250-mL, round-bottomed flask equipped with a temperature probe, addition funnel, argon inlet, and stirbar (Note 1). t-Bu_3P-Pt(0) complex (200 μL) (Notes 2 and 3) is added by syringe. 1-Heptyne (3.43 mL, 2.51 g, 26.0 mmol) (Note 4) is then slowly added by syringe with external cooling in a water bath so that the temperature of the reaction solution is not allowed to exceed 30 °C (Note 5). The mixture is stirred at room temperature for 60 min after complete addition of 1-heptyne. A solution of tetrabutylammonium fluoride in THF (40 mL, 1.0M, 40 mmol) is then added over 10 min (Notes 6 and 7). After complete addition of TBAF, the solution is stirred for 10 min and then 0.575 g (1.0 mmol) of $Pd(dba)_2$ is added in one portion (Note 8). A solution of 4-iodoanisole (Note 9) in THF (20 mL, 1.0M, 20 mmol) is then slowly added via the addition funnel such that the temperature of the reaction solution is kept between 30-33 °C (addition time 45 min). The mixture is stirred at room temperature for 30 min after the complete addition of 4-iodoanisole. Diethyl ether (50 mL) is added and the mixture is stirred for an additional 5 min. The mixture is filtered through a short column of silica gel (Note 10), eluting with 200 mL of diethyl ether. The filtrate is concentrated by rotary evaporation and vacuum drying to give a residue that is purified by Kugelrohr distillation, chromatography, and

Kugelrohr distillation to afford 3.38-3.42 g (83-84%) of (*E*)-1-(1-heptenyl)-4-methoxybenzene as a colorless oil, bp 90-95 °C/0.1 mm (lit.[4] 85-90 °C/ 1.2 mm) (Notes 11, 12, and 13).

2. Notes

1. 1,1,3,3-Tetramethydisiloxane was purchased from Lancaster Synthesis Inc. and used as received. THF was obtained from Mallinckrodt Inc. and freshly distilled from sodium/benzophenone ketyl.

2. *t*-Bu_3P-Pt(0) complex was prepared by the literature procedure:[2] *t*-Bu_3P (32 mg, 0.158 mmol) (Strem Chemicals) was dissolved in platinum(0)-1,3-divinyl-1,1,3,3-tetramethyldisiloxane complex (1.5 mL of xylene solution obtained from Aldrich Chemical Company, Inc., catalog number 47-951-9). The mixture was stirred at 65 °C (oil bath) for 5 min and then was slowly cooled to room temperature. This solution could be stored under N_2 in the freezer indefinitely.

3. *t*-Bu_3P is pyrophoric and should be handled and stored under an inert atmosphere of nitrogen or argon.

4. 1-Heptyne was purchased from GFS Chemicals Inc. and was used without further purification.

5. The hydrosilylation of 1-heptyne with 1,1,3,3-tetramethyldisiloxane is exothermic. If the temperature of the reaction exceeds 30 °C, the amount of the undesired α-isomer will increase. However, the solution should not be cooled to below 10 °C as the reaction becomes very sluggish.

6. A solution of tetrabutylammonium fluoride (TBAF) in THF (1.0M) was prepared from colorless, crystalline tetrabutylammonium fluoride trihydrate purchased from Fluka Chemical Corp.

7. Caution: TBAF reacts exothermally with the disiloxane mixture and the first few drops must be added very slowly and carefully.

8. Pd(dba)$_2$ was prepared by the literature procedure:[3] palladium chloride (1.05 g, 5.92 mmol) was added to hot (ca. 50 °C) methanol (150 mL) containing dibenzylideneacetone (DBA) (4.60 g, 19.6 mmol) and sodium acetate (3.90 g, 47.5 mmol). The mixture was stirred for 4 h at 40 °C to give a reddish-purple precipitate and allowed to cool to complete the precipitation. The precipitate was isolated by filtration, washed successively with water and acetone, and dried in vacuo.

9. 4-Iodoanisole was purchased from Aldrich Chemical Company, Inc., and was purified by chromatography on silica gel (elution with hexane/EtOAc, 50/1) prior to use.

10. Silica gel was purchased from VWR Scientific (230-400 mesh). A 45-mm diameter column was employed and 60 g of silica gel was loaded as a pentane slurry.

11. The boiling point (bp) corresponds to uncorrected air-bath temperature of a Büchi GKR-50 Kugelrohr.

12. The crude product was purified by Kugelrohr distillation (98-103°/0.2 mm) followed by column chromatography (55-mm diameter column, 200 g of silica gel, elution with pentane/EtOAc, 50/1), and another Kugelrohr distillation.

13. The analytical data are as follows: ^{1}H NMR (500 MHz, $CDCl_3$) δ: 0.91 (t, J = 7.1, 3 H), 1.33 (m, 4 H), 1.46 (m, 2 H), 2.18 (qd, J = 7.1, 1.5, 2 H), 3.80 (s, 3 H), 6.09 (dt, J = 15.9, 6.9, 1 H), 6.33 (d, J = 15.9, 1 H), 6.84 (d, J = 8.8, 2 H), 7.28 (d, J = 8.8, 2 H); ^{13}C NMR (100.6 MHz, $CDCl_3$) δ: 14.1, 22.6, 29.2, 31.4, 33.0, 55.2, 113.8, 126.9, 128.9, 129.1, 130.8, 158.5; IR (NaCl) cm^{-1}: 2956, 2927, 2855, 1608, 1511, 1464, 1291, 1248, 1175, 1037, 964, 833; Anal. calcd for $C_{14}H_{20}O$: C, 82.30; H, 9.87; Found: C, 82.10; H, 10.07; GC analysis: $t_R(Z)$ = 6.07 min (1.6%), $t_R(E)$ = 6.56 min (95.3%), tR(α) = 5.91 min (3.1%) (HP-5, 250 °C, 15 psi).

Waste Disposal Information

All toxic materials were disposed of in accordance with "Prudent Practices in the Laboratory", National Academy Press: Washington, DC, 1995.

3. Discussion

One-pot coupling reactions of organoboron- and organotin reagents with aryl or alkenyl halides and their applications in organic synthesis have been well documented.[5,6] There were only a few examples of hydrosilylation/cross-coupling applied to specific synthetic target molecules.[7,8] The procedures presented here provide a general route to (*E*)-olefins by palladium-catalyzed cross-coupling of alkenyl disiloxanes with aryl or alkenyl iodides.[9]

Tetramethyldisiloxane is an inexpensive, commercially available reagent. It reacts with alkynes to generate the alkenyldisiloxanes in situ, which can then react with a variety of iodides bearing different substituent patterns. The processes display very good functional group compatibility and afford products in high yield and with excellent regio- and stereoselectivity (Table 1). A notable characteristic of this protocol is the extremely mild reaction conditions employed. The hydrosilylation was complete at room temperature after 30 min. Most coupling reactions proceeded in 10 min to a few hours at room temperature.

The generality and functional group compatibility were also excellent with regard to the alkynes (Table 2). Free hydroxyl groups and even a remote double bond were compatible and give very good yields and purities.

The couplings of reactive iodides (with electron-withdrawing groups) were exothermic. However, the reaction temperature can be modulated by the slow addition of iodides last. These protocols are suitable for larger scale applications as well.

Table 1. One-Pot Hydrosilyation/Cross-Coupling of 1-Heptyne with Aryl Iodides

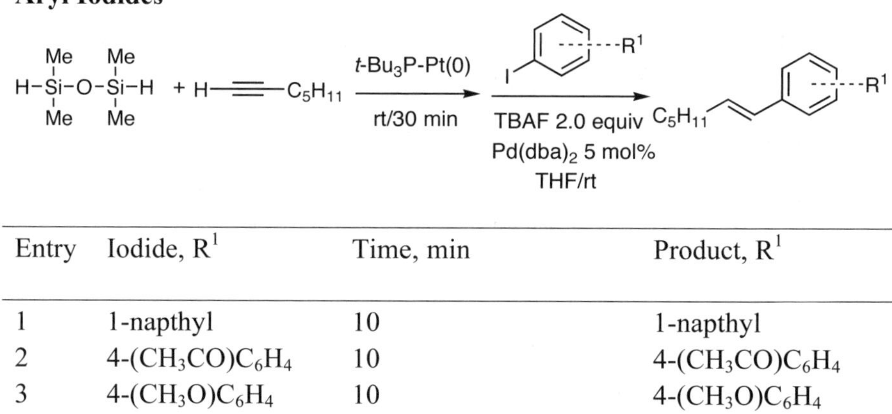

Entry	Iodide, R[1]	Time, min	Product, R[1]	Yie d,
1	1-napthyl	10	1-napthyl	82
2	4-$(CH_3CO)C_6H_4$	10	4-$(CH_3CO)C_6H_4$	94
3	4-$(CH_3O)C_6H_4$	10	4-$(CH_3O)C_6H_4$	89
4	3-$(NO_2)C_6H_4$	10	3-$(NO_2)C_6H_4$	85
5	3-$(NO_2)C_6H_4$	60 (1 mol% Pd)	3-$(NO_2)C_6H_4$	89
6	3-$(CH_3)C_6H_4$	10	3-$(CH_3)C_6H_4$	78
7	2-$(CH_3OCO)C_6H_4$	20 h (10 mol% Ph_3As)	2-$(CH_3OCO)C_6H_4$	88
8	2-$(CH_3O)C_6H_4$	10	2-$(CH_3O)C_6H_4$	82

Table 2. One-Pot Hydrosilyation/Cross-Coupling of 1-Alkynes with Aryl Iodides

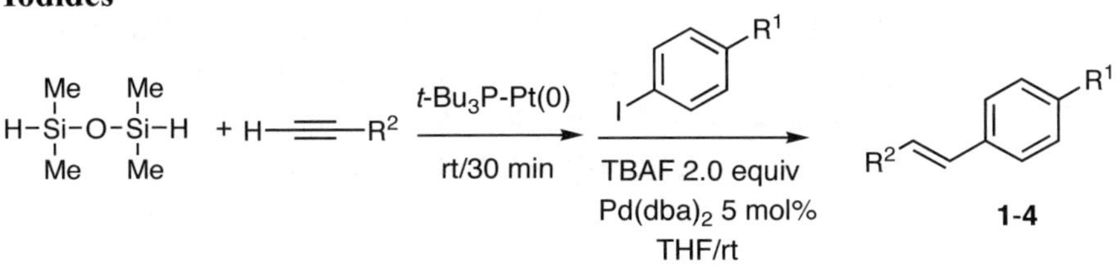

Entry	Alkyne, R^2	Iodide, R^1	Time, min	Product	Yield, %
1	C_6H_5	4-$(CH_3CO)C_6H_4$	10	**1a**	89
2	C_6H_5	4-$(CH_3O)C_6H_4$	10	**1b**	74
3	$HO(CH_2)_3$	4-$(CH_3CO)C_6H_4$	30	**2a**	82
4	$HO(CH_2)_3$	4-$(CH_3O)C_6H_4$	60	**2b**	89
5	$C_6H_5C(OH)(CH_3)$	4-$(CH_3CO)C_6H_4$	24 h	**3a**	72
6	$C_6H_5C(OH)(CH_3)$	4-$(CH_3O)C_6H_4$	24 h	**3b**	79
7	$CH_2{=}CHCH_2O(CH_2)_3$	4-$(CH_3CO)C_6H_4$	10	**4a**	78
8	$CH_2{=}CHCH_2O(CH_2)_3$	4-$(CH_3O)C_6H_4$	10	**4b**	76

1. Department of Chemistry, Roger Adams Laboratory, University of Illinois at Urbana-Champaign, Urbana, IL 61801, E-mail: denmark@MITscs.uiuc.edu.
2. Chandra, G.; Lo, P. Y.; Hitchcock, P. B.; Lappert, M. F. *Organometallics* **1987**, *6*, 191-192.
3. Ukai, T.; Kawazura, H.; Ishii Y. *J. Organomet. Chem.* **1974**, *65*, 253-266.
4. Denmark, S. E.; Wehrli, D. *Org. Lett.* **2000**, *2*, 565-568.
5. (a) Miyaura, N.; Yano, T.; Suzuki, A. *Tetrahedron Lett.* **1980**, *21*, 2865-2868. (b) Miyaura, N.; Ishiyama, T.; Sasaki, H.; Ishikawa, M.; Satoh, M.; Suzuki, A. *J. Am. Chem. Soc.* **1989**, *111*, 314-321. (c) Ishiyama, T.; Nishijima, K.-i.; Miyaura, N.; Suzuki, A. *J. Am. Chem. Soc.* **1993**, *115*, 7219-7225. (d) Baudoin, O.; Guenard, D.; Gueritte, F. *J. Org. Chem.* **2000**, *65*, 9268-9271.
6. Maleczka, R. E., Jr.; Gallagher, W. P.; Terstiege, I. *J. Am. Chem. Soc.* **2000**, *122*, 384-385.
7. Tamao, K.; Kobayashi, K.; Ito, Y. *Tetrahedron Lett.* **1989**, *30*, 6051-6054.
8. Takahashi, K.; Minami, T.; Ohara, Y.; Hiyama, T. *Tetrahedron Lett.* **1993**, *34*, 8263-8266.
9. Denmark, S. E.; Wang, Z. *Org. Lett.* **2001**, *3*, 1073-1076.

HECK REACTIONS OF ARYL CHLORIDES CATALYZED BY PALLADIUM/TRI-*tert*-BUTYLPHOSPHINE: (E)-2-METHYL-3-PHENYLACRYLIC ACID BUTYL ESTER AND (E)-4-(2-PHENYLETHENYL)BENZONITRILE

[2-Propenoic acid, 2-methyl-3-phenyl-, butyl ester and Benzonitrile, 4-[(1E)-2-phenylethenyl]]

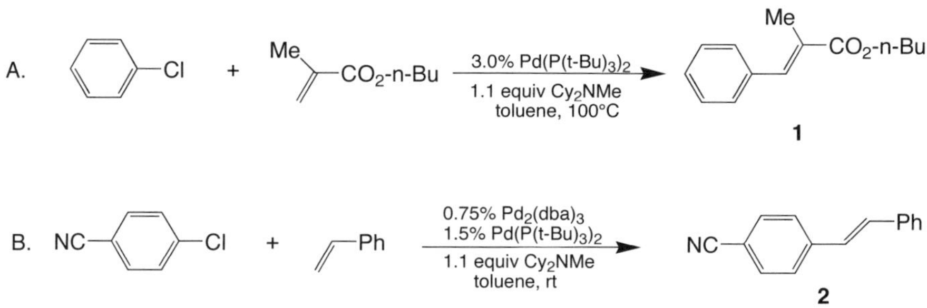

Submitted by Adam F. Littke and Gregory C. Fu.[1]
Checked by Michael H. Ober and Scott E. Denmark.

1. Procedure

A. (E)-2-Methyl-3-phenylacrylic acid butyl ester (**1**). An oven-dried, 250-mL, three-necked, round-bottomed flask equipped with a reflux condenser (fitted with an argon inlet adapter), rubber septum, glass stopper, and a teflon-coated magnetic stir bar is cooled to room temperature under a flow of argon. The flask is charged with bis(tri-*tert*-butylphosphine)palladium ($Pd(P(t\text{-}Bu)_3)_2$) (0.482 g, 0.943 mmol, 3.0 mol% Pd) (Notes 1, 2) and again purged with argon. Toluene (32 mL) (Notes 9, 10) is added, and the mixture is stirred at room temperature, resulting in a homogeneous brown-orange solution. Chlorobenzene (3.20 mL, 31.5 mmol) (Note 11), *N*-methyldicyclohexylamine (Cy_2NMe) (7.50 mL, 35.0 mmol) (Note 12), and butyl methacrylate (5.50 mL, 34.6 mmol) (Note 12) are then added successively via syringe. The resulting mixture is allowed to stir at room temperature for 5 min, resulting in a homogeneous light-orange solution. The rubber septum is then replaced with a glass stopper, and the flask is heated in a 100 °C oil bath under a positive pressure of argon for 22 hr (Note 13). Upon heating, the solution becomes bright canary-yellow in color, and within 10-15 min the formation of a white precipitate (the amine hydrochloride salt) is observed. Upon completion of the reaction, shiny deposits of palladium metal form on the sides of the flask, and a large quantity of white precipitate is present. The reaction mixture is allowed to cool to room temperature and then diluted with 100 mL of diethyl ether. The resulting solution is washed with 100 mL of H_2O, and the aqueous layer is extracted with three 50-mL portions of diethyl ether. The combined organic phases are washed with 100 mL of brine and then concentrated by rotary evaporation. Any residual solvent is removed at 0.5 mm. The crude product, a dark-brown oil, is then purified by flash column chromatography (Note 14) to afford 6.67 - 6.72 g (95%) of **1** as a pale red-orange liquid. This liquid appears to be pure by ^1H and ^{13}C NMR spectroscopy; however, if desired, the discoloration can be removed by filtering the product through a small column of silica gel (3 cm diameter x 10 cm height), which furnishes 6.49-6.62 g (95-96%) of **1** as a clear, colorless liquid (Notes 15 and 16).

B. (E)-4-(2-Phenylethenyl)benzonitrile (**2**). An oven-dried, 250-mL, three-necked, round-bottomed flask equipped with an argon inlet adapter, rubber septum, glass stopper, and a teflon-coated magnetic stir bar is cooled to room temperature under a flow of argon. The flask is charged successively with bis(tri-*tert*-butylphosphine)palladium [($Pd(P(t\text{-}Bu)_3)_2$] (0.238 g, 0.466 mmol, 1.5 mol% Pd) (Notes 1, 2), tris(dibenzylideneacetone)dipalladium(0) ($Pd_2(dba)_3$) (0.213 g, 0.233 mmol, 1.5 mol% Pd) (Note 3), and 4-chlorobenzonitrile (4.25 g, 30.9 mmol) (Note 17). The flask is purged with argon, and 62 mL of toluene is added (Notes 9, 10, 18). The resulting mixture is stirred at room temperature, resulting in a dark, red-purple solution. *N*-Methyldicyclohexylamine (7.5 mL, 35.0 mmol) (Note 12) and styrene (3.8 mL, 33.2 mmol) (Note 19) are then added via syringe. The reaction mixture is allowed to stir at room temperature under a positive pressure of argon for 72 hr (Note 13). Within the first 1-2 hr, the color changes from deep red-purple to dark brown, and a precipitate (the amine hydrochloride salt) begins to form. Upon completion of the reaction, 100 mL of ethyl acetate is added, and the resulting solution is washed with 100 mL of water. The aqueous layer is separated and extracted with three 50-mL portions of diethyl ether, and the combined organic phases are concentrated by rotary evaporation. Any residual solvent is removed at 0.5 mm. The crude product, a yellow solid, is purified via flash column chromatography (Note 20) to afford 5.30-5.34 g (83-84%) of **2** as white crystalline spheres (Notes 21, 22, 23).

2. Notes

1. $Pd(P(t\text{-}Bu)_3)_2$ was prepared according to the following procedure. In a nitrogen-filled Vacuum Atmospheres glovebox, a 100-mL, one-neck, round-bottomed flask equipped with a teflon-coated magnetic stir bar is charged with tris(dibenzylideneacetone)dipalladium(0) ($Pd_2(dba)_3$) (2.98 g, 3.25 mmol) (Note 3). A solution of tri-*tert*-butylphosphine ($P(t\text{-}Bu)_3$) (2.88 g, 14.2 mmol) (Note 4) in 43 mL of *N*, *N*-dimethylformamide (DMF) (Note 5) is then added to the reaction flask via a glass pipette, and the resulting dark green-brown solution is stirred at room temperature for 23 hr. The reaction mixture is then filtered through a 30-mL medium-porosity glass frit to collect the unpurified product, $Pd(P(t\text{-}Bu)_3)_2$, as a gray solid. The reaction flask is rinsed with three 6-mL portions of DMF and 5 mL of methanol (Note 6), which are passed through the glass frit (the final rinses should be colorless). The reaction product is then dissolved in 100 mL of toluene (Note 7), and the resulting solution is filtered through a 3-cm diameter 0.45 μm Gelman acrodisc (to remove some insoluble black material) into a 250-mL Schlenk tube, affording a homogeneous orange-yellow solution. The Schlenk tube is removed from the glovebox, and the toluene solution is concentrated under high vacuum (0.5mm) to a volume of approximately 25 mL, at which point a white crystalline solid begins to precipitate. The Schlenk tube is taken into the glovebox, and the toluene solution and crystalline solid are transferred to a 250-mL Erlenmeyer flask via a glass pipette. MeOH (100 mL) is then added slowly via pipette over 10 min, resulting in the precipitation of additional white crystalline solid. The solution is allowed to stand for one hour, and then the mother liquor is separated from the solid via pipette. The solid is washed with two 10-mL portions of MeOH, transferred to a tared 20-mL glass vial, and dried under high vacuum, affording 2.41 g (72%) of $Pd(P(t\text{-}Bu)_3)_2$ as a white, crystalline solid (Note 8). The $Pd(P(t\text{-}Bu)_3)_2$ can be stored indefinitely under nitrogen in a Vacuum Atmospheres glovebox. Although $Pd[P(t\text{-}Bu)_3]_2$ has been reported to be "stable in air in the solid state,"[2] if a glovebox is not available, it is recommended that $Pd[P(t\text{-}Bu)_3]_2$ be stored in a tightly capped vial in a desiccator, preferably under argon or nitrogen. **Alternative workup and crystallization**: After the reaction mixture has stirred for 23 hours, the resulting precipitate is collected by filtration, washed thoroughly with DMF, and then dissolved in hexane. The hexane solution is filtered and concentrated, yielding $Pd(P(t\text{-}Bu)_3)_2$ as white crystals.

2. Alternatively, $Pd[P(t\text{-}Bu)_3]_2$ may be purchased from Strem Chemicals.

3. $Pd_2(dba)_3$ was purchased from the Aldrich Chemical Company and used as received.

4. $P(t\text{-}Bu)_3$ (99%) was purchased from Strem Chemicals and used as received.

5. DMF (anhydrous, DriSolv) was purchased from EM Science and degassed under high vacuum for 10-15 min prior to use.

6. Methanol (certified A.C.S., purchased from Fisher Scientific) was distilled from $Mg(OMe)_2$ and was degassed by three freeze-pump-thaw cycles prior to use.

7. Toluene (J. T. Baker; CYCLE-TRAINER solvent delivery kegs) was vigorously purged with argon for 2 h and then passed through two packed columns of neutral alumina and copper(II) oxide under argon pressure.[3]

8. $^{31}P\{^1H\}$ NMR (C_6D_6, 202 MHz): δ 85.3; ^1H NMR (C_6D_6, 500 MHz): δ 1.51 (t, J = 5.4 Hz). The NMR sample must be prepared under inert atmosphere to avoid aerobic oxidation of the catalyst as evidenced by free $P(t\text{-}Bu)_3$ at δ1.27 ppm in the ^1H NMR spectrum. Even with these precautions, ca. 1% of $P(t\text{-}Bu)_3$ is observed.

9. Toluene (anhydrous, 99.8%, Sure/Seal™ bottle) was purchased from the Aldrich Chemical Company and used as received.

10. 1,4-Dioxane is an equally suitable solvent for these Heck couplings (and is the solvent used in the published procedures); however, due to the lower cost and the lower toxicity of toluene, it was chosen as the solvent for these reactions.

11. Chlorobenzene (anhydrous, 99.8%, Sure/Seal™ bottle) was purchased from the Aldrich Chemical Company and used as received.

12. *N*-Methyldicyclohexylamine (97%) and butyl methacrylate (99%) were purchased from the Aldrich Chemical Company and gently sparged with argon for 5-10 min prior to use.

13. The progress of the reaction was monitored by GC.

14. Flash column chromatography was performed using silica gel (6 cm diameter x 35 cm height), eluting with 19/1 hexane/diethyl ether.

15. Compound **1** has the following properties: bp 111 °C (1 mm) ^1H NMR ($CDCl_3$, 500 MHz) δ: 0.98 (t, J = 7.5, 3H), 1.45 (sext, J = 1.0, 2H), 1.71 (qn, J = 1.0, 2H), 2.12 (d, J = 1.5, 3H), 4.22 (t, J = 6.6, 2H), 7.33 (m, 1H), 7.39 (m, 4H), 7.68 (apparent d, J = 1.5, 1H); ^{13}C NMR ($CDCl_3$ 126 126MHz): δ13.7, 14.0, 19.3, 30.7, 64.8, 128.2, 128.4, 128.7, 129.6, 136.0,

138.8, 168.8; IR (neat, cm^{-1}): 3059, 3026, 2960, 2933, 2873, 1709, 1635, 1492, 1448, 1388, 1356, 1254, 1201, 1115, 1074, 1003, 931, 766, 704; Anal. Calcd for $C_{14}H_{18}O_2$: C, 77.03; H, 8.31. Found: C, 77.02; H, 8.30.

16. The checkers found that the product could be further purified by distillation (110 °C/1 mm).

17. 4-Chlorobenzonitrile (99%) was purchased from Aldrich Chemical Company and used as received.

18. To allow more efficient stirring, it was beneficial to run this reaction at half of the concentration (2 mL solvent per mmol of aryl chloride) of the original published procedure (1 mL solvent per mmol of aryl chloride).

19. Styrene (99+%) was purchased from Aldrich Chemical Company and gently sparged with argon for 5-10 min prior to use.

20. Flash column chromatography was performed using silica gel (10 cm diameter x 27 cm height), eluting with 4/1 toluene/hexane. A small amount of aryl chloride that remained unreacted after 72 hr was recovered mixed with a small quantity (<5%) of the desired product.

21. Compound **2** has the following properties: mp (corr.) 115-117 °C (lit.[4] mp 115 °C); ^{1}H NMR ($CDCl_3$, 500 MHz): δ7.09 (d, J = 16.5, 1H), 7.22 (d, J = 16.5, 1H), 7.33 (d, J = 7.1, 1H), 7.40 (t, J = 7.1, 2H), 7.54 (d, J = 7.5, 2H), 7.59 (d, J = 8.3, 2H), 7.64 (d, J = 8.4, 2H); ^{13}C NMR ($CDCl_3$, 126 MHz): δ110.8, 119.3, 126.9, 127.0, 127.1, 128.9, 129.1, 132.6, 132.7, 136.5, 142.1; IR (neat, cm^{-1}): 3029, 2964, 2225, 1648, 1604, 1531, 1450, 1278, 1226, 1174, 1095, 966, 821, 769; Anal. Calcd for $C_{15}H_{11}N$: C, 87.78; H, 5.40; N, 6.82. Found: C, 87.43; H, 5.26; N, 6.86.

22. The checkers found that the product could be further purified by sublimation (127 °C/0.5 mm).

23. The submitters obtained the product in 89% yield.

Waste Disposal Information

All toxic materials were disposed of in accordance with "Prudent Practices in the Laboratory"; National Academy Press; Washington, DC, 1995.

3. Discussion

Since its discovery in the early 1970's, the palladium-catalyzed arylation of olefins (Heck reaction; eq 1 and Figure 1)[5,6] has been applied to a diverse array of fields, ranging from natural products synthesis[7,8] to materials science[9] to bioorganic chemistry.[10] This powerful carbon-carbon bond-forming process has been practiced on an industrial scale for the production of compounds such as naproxen[11] and octyl methoxycinnamate.[12] The Heck reaction is typically performed in the presence of a palladium/tertiary phosphine catalyst and a stoichiometric amount of an inorganic or organic base. High functional-group tolerance and the ready availability and low cost of simple olefins contribute to the exceptional utility of the Heck arylation.

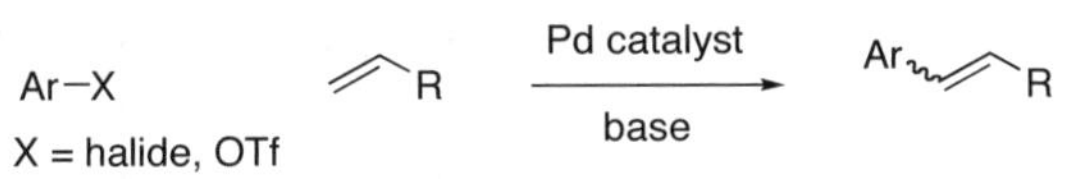

(1)

Figure 1
Outline of the Catalytic Cycle for the Heck Coupling Reaction

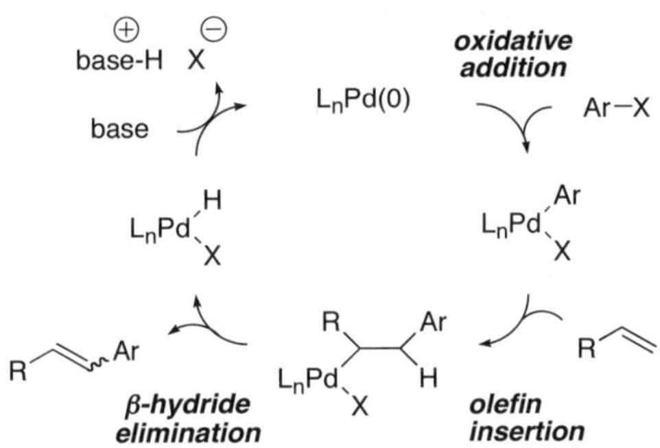

Until recently, one important unsolved problem for the Heck reaction was the poor reactivity of aryl chlorides, which are arguably the most attractive class of aryl halides, due to their lower price and greater availability as compared with the corresponding bromides and iodides.[13,14] For the few catalyst systems that have displayed activity for Heck couplings of aryl chlorides (e.g., those of Milstein (bulky, electron-rich chelating bisphosphines),[15] Herrmann (palladacycles, *N*-heterocyclic carbenes),[16] Reetz (tetraphenylphosphonium salts),[17] and Beller (phosphites),[18,19,20] the scope of the reactions has been quite narrow and the reaction temperatures have been rather high ($\geq$120 °C). This need for elevated temperatures can be problematic for a variety of reasons, including decomposition of thermally unstable substrates and decreased regio- and stereoselectivities.

Pd/P(*t*-Bu)$_3$, in the presence of Cy$_2$NMe, is an unusually mild and versatile catalyst for Heck reactions of aryl chlorides (Tables 1 and 2) (as well as for room-temperature reactions of aryl bromides).[21,22,23] Example A, the coupling of chlorobenzene with butyl methacrylate, illustrates the application of this method to the stereoselective synthesis of a trisubstituted olefin; α-methylcinnamic acid derivatives are an important family of compounds that possess biological activity (e.g., hypolipidemic[24] and antibiotic[25]) and serve as intermediates in the synthesis of pharmaceuticals (e.g., Sulindac, a non-steroidal anti-inflammatory drug[26]). Example B, the coupling of 4-chlorobenzonitrile with styrene, demonstrates that Pd/P(*t*-Bu)$_3$ can catalyze the Heck reaction of activated aryl chlorides *at room temperature*.

From a practical point of view, it is worth noting that Heck reactions catalyzed by Pd/P(*t*-Bu)$_3$ do not typically require rigorously purified reagents or solvents. In addition, the palladium and phosphine sources, Pd[P(*t*-Bu)$_3$]$_2$ and Pd$_2$(dba)$_3$, are commercially available and can be handled in air.

Thus, in terms of scope, mildness, and convenience, Pd/P(*t*-Bu)$_3$ provides an attractive method for achieving Heck couplings of aryl chlorides.

Table 1.
Heck Couplings of Aryl Chlorides at Elevated Temperature

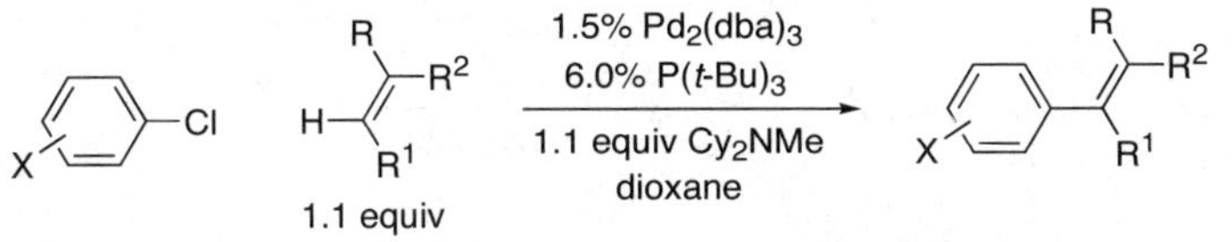

Entry	Aryl Chloride	Olefin	Temperature	Product	Yield[a,b]
1[c]	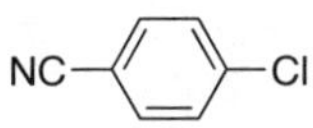	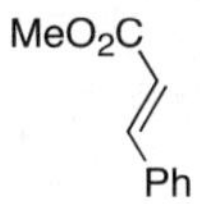	70 °C	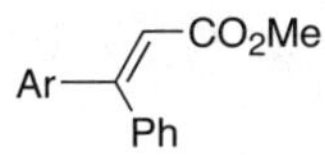	67%[d]
2	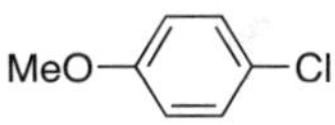		120 °C		72%
3	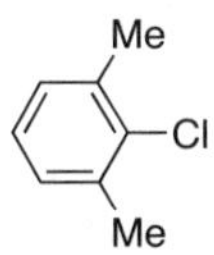	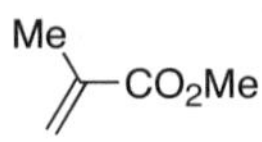	120 °C	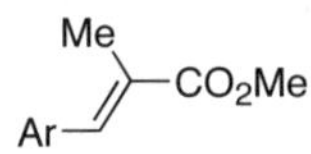	80%
4	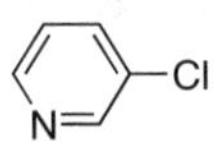	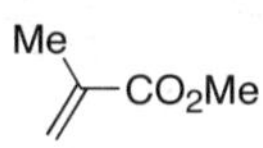	100 °C	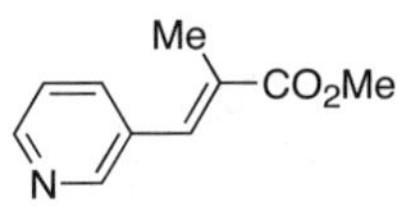	76%
5	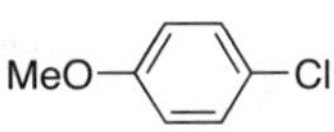	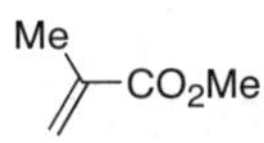	120 °C	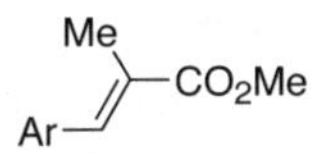	72%

[a]Isolated yield, average of two runs. [b]E:Z ratio is >20:1, as determined by ^{1}H NMR. [c]3.6% P(t-Bu)$_3$ was used. [d]Product includes 2.5% P(t-Bu)$_3$/OP(t-Bu)$_3$.

Table 2.
Heck Couplings of Activated Aryl Chlorides at Room Temperature

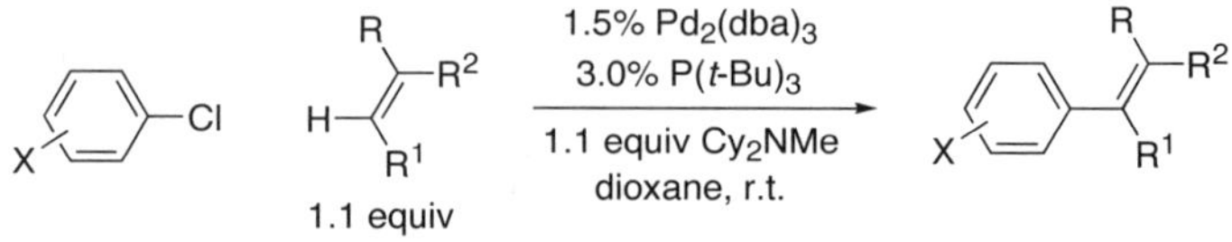

Entry	Aryl Chloride	Olefin	Product	Yield[a,b]
1	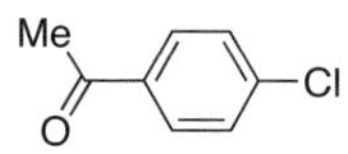	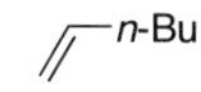	 6:1 E:Z	70%
2	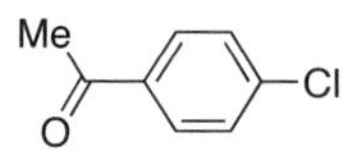	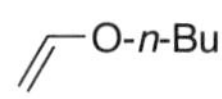	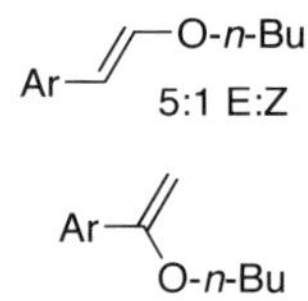 5:1 E:Z; 10 : 1	87%[c]
3	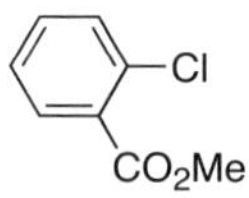	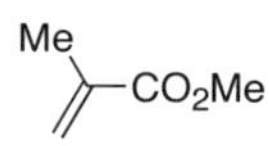	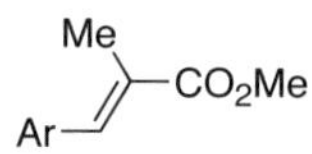	90%
4	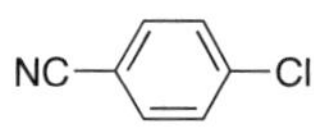	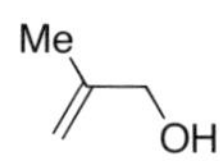	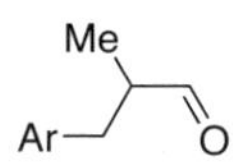	79%
5[d]	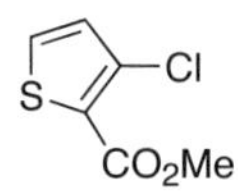	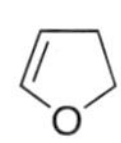	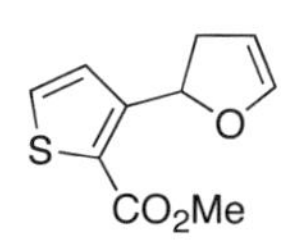	87%
6[e]	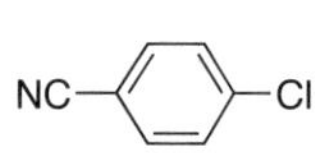	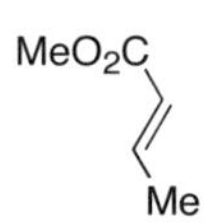	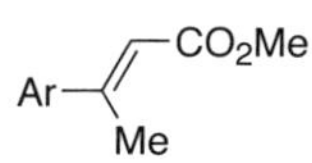	57%[f]

[a]Isolated yield, average of two runs. [b]Unless otherwise indicated, the E:Z ratio is >20:1, as determined by ^{1}H NMR. [c]Product includes 5% 4'-chloroacetophenone. [d]3 equiv. of olefin was used. [e]2 equiv. of olefin was used. [f]Product includes 2% P(t-Bu)$_3$/OP(t-Bu)$_3$.

1. Department of Chemistry, Massachusetts Institute of Technology, Cambridge, MA 02139.

2. (a) Otsuka, S.; Yoshida, T.; Matsumoto, M.; Nakatsu, K. *J. Am. Chem. Soc.* **1976**, *98*, 5850-5858. (b) Yoshida, T.; Otsuka, S. *J. Am. Chem. Soc.* **1977**, *99*, 2134-2140. (c) Yoshida, T.; Otsuka, S. *Inorg. Synth.* **1990**, *28*, 113-119.
3. (a) Pangborn, A. B.; Giardello, M. A.; Grubbs, R. H.; Rosen, R. K.; Timmers, F. J. *Organometallics* **1996**, *15*, 1518-1520. (b) Alaimo, P. J.; Peters, D. W.; Arnold, J.; Bergman, R. J. *J. Chem. Ed.* **2001**, *78*, 64-64.
4. Gusten, H.; Salzwedel, M. *Tetrahedron* **1967**, *23*, 173-185.
5. (a) Mizoroki, T.; Mori, K.; Ozaki, A. *Bull. Chem. Soc. Jpn.* **1971**, *44*, 581. (b) Heck, R. F.; Nolley, J. P., Jr. *J. Org. Chem.* **1972**, *37*, 2320-2322.
6. For reviews of the Heck reaction, see: (a) Bräse, S.; de Meijere, A. In *Metal-Catalyzed Cross-Coupling Reactions*; Diederich, F., Stang, P. J., Eds.; Wiley: New York, 1998; Chapter 3. (b) Beletskaya, I. P.; Cheprakov, A. V. *Chem. Rev.* **2000**, *100*, 3009-3066. (c) Heck, R. F. In *Comprehensive Organic Synthesis*; Trost, B. M., Ed.; Pergamon: New York, 1991; Vol. 4, Chapter 4.3. (d) Heck R. F. *Org. React.* **1982**, *27*, 345-390. (e) Crisp, G. T. *Chem. Soc. Rev.* **1998**, *27*, 427-436. (f) de Meijere, A.; Meyer, F. E. *Angew. Chem., Int. Ed. Engl.* **1994**, *33*, 2379-2411. (g) Jeffery, T. In *Advances in Metal-Organic Chemistry*; Liebeskind, L. S., Ed.; JAI: London, 1996; Vol. 5, pp. 153-260. (h) Cabri, W.; Candiani, I. *Acc. Chem. Res.* **1995**, *28*, 2-7.
7. For example, see: (a) Taxol: Danishefsky, S. J.; Masters, J. J.; Young, W. B.; Link, J. T.; Snyder, L. B.; Magee, T. V.; Jung, D. K.; Isaacs, R. C. A.; Bornmann, W. G.; Alaimo, C. A.; Coburn, C. A.; Di Grandi, M. J. *J. Am. Chem. Soc.* **1996**, *118*, 2843-2859. (b) Scopadulcic acid: Overman, L. E.; Ricca, D. J.; Tran, V. D. *J. Am. Chem. Soc.* **1993**, *115*, 2042-2044.
8. For overviews of applications of the Heck reaction in natural products synthesis, see: (a) Link, J. T.; Overman, L. E. In *Metal-Catalyzed Cross-Coupling Reactions*; Diederich, F., Stang, P. J., Eds.; Wiley-VCH: New York, 1998; Chapter 6. (b) Bräse, S.; de Meijere, A. In *Metal-Catalyzed Cross-Coupling Reactions*; Diederich, F., Stang, P. J., Eds.; Wiley: New York, 1998; Chapter 3.6. (c) Nicolaou, K. C.; Sorensen, E. J. *Classics in Total Synthesis*; VCH: New York, 1996; Chapter 31. These authors refer to the Heck reaction as "one of the true "power tools" of contemporary organic synthesis" (p. 566).
9. For example, see: (a) *Step-Growth Polymers for High-Performance Materials*; Hedrick, J. L., Labadie, J. W., Eds.; ACS Symp. Ser. 624; American Chemical Society: Washington, DC, 1996; Chapters 1, 2, and 4. (b) DeVries, R. A.; Vosejpka, P. C.; Ash, M. L. *Catalysis of Organic Reactions*; Herkes, F. E., Ed.;

Marcel Dekker: New York, 1998; Chapter 37. (c) Tietze, L. F.; Kettschau, G.; Heuschert, U.; Nordmann, G. *Chem. Eur. J.* **2001**, *7*, 368-373.

10. For some recent examples, see: (a) Haberli, A.; Leumann, C. J. *Org. Lett.* **2001**, *3*, 489-492. (b) Burke, T. R., Jr.; Liu, D.-G.; Gao, Y. *J. Org. Chem.* **2000**, *65*, 6288-6292.
11. Stinson, S. C. *Chem. Eng. News* January 18, 1999, p. 81.
12. Octyl methoxycinnamate (OMC) is the most common UV-B sunscreen that is on the market: Eisenstadt, A. In *Catalysis of Organic Reactions*; Herkes, F. E., Ed.; Marcel Dekker: New York, 1998; Chapter 33.
13. For discussions of coupling reactions of aryl chlorides, see: (a) Grushin, V. V.; Alper, H. In *Activation of Unreactive Bonds and Organic Synthesis*; Murai, S., Ed.; Springer-Verlag: Berlin, 1999; pp. 193-226. (b) Grushin, V. V.; Alper, H. *Chem. Rev.* **1994**, *94*, 1047-1062.
14. The low reactivity of aryl chlorides is usually attributed to the strength of the C-Cl bond (bond dissociation energies for aryl-X: Cl = 96 kcal/mol; Br = 81 kcal/mol; I = 65 kcal/mol).
15. (a) Ben-David, Y.; Portnoy, M.; Gozin, M.; Milstein, D. *Organometallics* **1992**, *11*, 1995-1996. (b) Portnoy, M.; Ben-David, Y.; Milstein, D. *Organometallics* **1993**, *12*, 4734-4735. (c) Portnoy, M.; Ben-David, Y.; Rousso, I.; Milstein, D. *Organometallics* **1994**, *13*, 3465-3479.
16. (a) Herrmann, W. A.; Brossmer, C.; Ofele, K.; Reisinger, C.-P.; Priermeier, T.; Beller, M.; Fischer, H. *Angew. Chem., Int. Ed. Engl.* **1995**, *34*, 1844-1848. (b) Herrmann, W. A.; Brossmer, C.; Reisinger, C.-P.; Reirmeier, T. H.; Ofele, K.; Beller, M. *Chem. Eur. J.* **1997**, *3*, 1357-1364. (c) Herrmann, W. A.; Elison, M.; Fischer, J.; Köcher, C.; Artus, G. R. J. *Angew. Chem., Int. Ed. Engl.* **1995**, *34*, 2371-2374. (d) See also: Herrmann, W. A.; Brossmer, C.; Ofele, K.; Beller, M.; Fischer, H. *J. Mol. Catal. A* **1995**, *103*, 133-146.
17. Reetz, M. T.; Lohmer, G.; Schwickardi, R. *Angew. Chem. Int. Ed.* **1998**, *37*, 481-483.
18. Beller, M.; Zapf, A. *Synlett* **1998**, 792-793.
19. See also: Kaufmann, D. E.; Nouroozian, M.; Henze, H. *Synlett* **1996**, 1091-1092.
20. For very early work on Heck reactions of aryl chlorides, see: (a) Davison, J. B.; Simon, N. M.; Sojka, S. A. *J. Mol. Cat.* **1984**, *22*, 349-352. (b) Spencer, A. *J. Organomet. Chem.* **1984**, *270*, 115-120.
21. Littke, A. F.; Fu, G. C. *J. Am. Chem. Soc.* **2001**, *123*, 6989-7000. See also: Littke, A. F.; Fu, G. C. *J. Org. Chem.* **1999**, *64*, 10-11.

22. See also: Shaughnessy, K. H.; Kim, P.; Hartwig, J. F. *J. Am. Chem. Soc.* **1999**, *121*, 2123-2132.

23. Generally, for Pd/P(t-Bu)$_3$-catalyzed Heck couplings that proceed at elevated temperature, a 1:2 Pd:phosphine ratio is preferred. For reactions that occur at room temperature, a 1:1 Pd:phosphine ratio is usually desirable (2:1 mixture of Pd(P(t-Bu)$_3$)$_2$:Pd$_2$(dba)$_3$).

24. For example, see: Watanabe, T.; Hayashi, K.; Yoshimatsu, S.; Sakai, K.; Takeyama, S.; Takashima, K. *J. Med. Chem.* **1980**, *23*, 50-59.

25. For example, see: Buchanan, J. G.; Hill, D. G.; Wightman, R. H.; Boddy, I. K.; Hewitt, B. D. *Tetrahedron* **1995**, *51*, 6033-6050.

26. Eisenstadt, A. In *Catalysis of Organic Reactions*; Herkes, F. E., Ed.; Marcel Dekker: New York, 1998; Chapter 33.

SYNTHESIS OF N-(*tert*-BUTOXYCARBONYL)-β-IODOALANINE METHYL ESTER: A USEFUL BUILDING BLOCK IN THE SYNTHESIS OF NONNATURAL α-AMINO ACIDS VIA PALLADIUM CATALYZED CROSS COUPLING REACTIONS
(L-Alanine, N-[(1,1-dimethylethoxy)carbonyl]-3-iodo-, methyl ester)

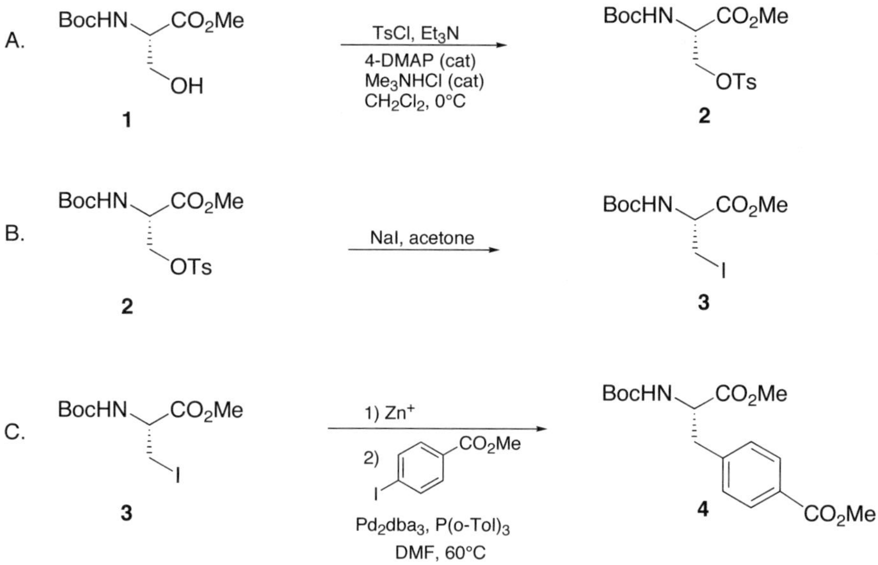

Submitted by Richard F. W. Jackson and Manuel Perez-Gonzalez.[1]
Checked by Rick L. Danheiser and Aimee L. Crombie.

1. Procedure

A. N-(tert-Butoxycarbonyl)-O-(p-toluenesulfonyl)-L-serine methyl ester (**2**). A one-necked, 500-mL, round-bottomed flask equipped with a rubber septum and magnetic stirbar was charged with 26.1 g (119 mmol) of N-(*tert*-butoxycarbonyl)-L-serine methyl ester (**1**) (Note 1) and 200 mL of CH_2Cl_2 (Note 2). The solution is cooled in an ice bath at 0 °C while 0.700 g (6.0 mmol) of 4-dimethylaminopyridine (4-DMAP), 1.1 g (12 mmol) of Me3NHCl (Note 3), and 22.7 g (119 mmol) of freshly recrystallized *p*-toluenesulfonyl chloride (TsCl) (Note 4) are added. The septum is replaced with a dropping funnel charged with 17 mL (119 mmol) of triethylamine (Et_3N) in 50 ml of CH_2Cl_2 which is added dropwise to the reaction mixture at 0 °C over 40 min (Note 5). The resulting slurry is stirred at 0 °C for 2 h and then poured into a mixture of 100 mL of ice, 100 mL of water, and 50 mL of 2M HCl solution. The aqueous layer is extracted with 100 mL of CH_2Cl_2, and the combined organic layers are washed with two 60-mL portions of brine, dried over magnesium sulfate, and concentrated by rotary evaporation to yield 59.4 g of a light yellow solid. This product may contain ca. 15% of starting material and TsCl that can be efficiently removed by crystallization according to the following procedure. The solid is dissolved in 140 mL of hot diethyl ether, filtered, and the filtrate is allowed to cool to room temperature and then to 0 °C. Once crystallization begins (Note 6), a total of 250 mL of petroleum ether is added in five portions over 2 h and then crystallization is allowed to proceed at –20 °C overnight. The crystals are collected by suction filtration on a Büchner funnel and air-dried to give 28.3-30.8 g (64-69%) of **2** as a white solid (Note 7).

B. N-(tert-Butoxycarbonyl)-β-iodoalanine methyl ester (**3**). A one-necked, 250-mL, round-bottomed flask equipped with a rubber septum and magnetic stirbar is charged with 27.8 g (74.0 mmol) of N-(*tert*-butoxycarbonyl)-O-(*p*-toluenesulfonyl)-L-serine methyl ester (**2**) and 160 mL of acetone (Note 8). The solution is stirred at room temperature and 13.4 g (89.0 mmol) of NaI (Note 9) is added in one portion. The reaction mixture is stirred in the dark for 3 days, after which an additional 3.3 g (22 mmol) of NaI is added and stirring is continued for an additional day (Note 10). The reaction mixture is then suction filtered through a sintered glass funnel and the filtrate is collected in a one-necked, 500-mL, round-bottomed flask. The solid is washed with acetone until it is colorless. The solid is discarded and the filtrate is concentrated by rotary evaporation under reduced pressure (Note 11). The residual yellow oil is partitioned between 150 mL of diethyl ether and 60 mL of 1M sodium thiosulfate ($Na_2S_2O_3$) solution. The organic layer is separated and washed with 40 mL of 1M $Na_2S_2O_3$ 50 mL of brine, dried over magnesium sulfate, filtered, and concentrated by rotary evaporation (Note 11) to afford 23.2 g of a colorless oil that solidifies on standing at 0 °C. The solid is dissolved in 30 mL of hot (40 °C) petroleum ether (bp 35-45 °C), cooled to room temperature and then to 0 °C. Once a precipitate appears (Note 6), the mixture is cooled at –20 °C for 1 h and the white solid is collected on a Büchner funnel and washed with cold petroleum ether to yield 19.4-20.0 g (80-82%) of N-(*tert*-butoxycarbonyl)-β-iodoalanine methyl ester **3** as white to pale yellow crystals (Note 12).

C. N-(tert-Butoxycarbonyl)-β-[4-(methoxycarbonyl)phenyl]alanine methyl ester (**4**). A three-necked, 100-mL, round-bottomed flask equipped with an argon inlet adapter, reflux condenser, rubber septum, and magnetic stirbar (Note 13) is charged with 2.7 g (42 mmol) of zinc dust and 0.039 g (0.15 mmol) of I_2. The flask is evacuated to 0.03 mm and heated with a heat gun for 10 min, flushed three times with argon, and then allowed to cool to room temperature. Dimethylformamide (1 mL) (Note 14) is added *via* syringe, and then a solution of 10.8 g (33 mmol) of iodide **3** in 14 mL of DMF is added dropwise *via* syringe to the well-stirred suspension of zinc dust. The reaction mixture is stirred for 30 min at 0 °C to produce a solution of the zinc reagent and a suspension of excess zinc dust. The ice bath is removed and 7.9 g (30 mmol) of methyl 4-iodobenzoate (Note 15), 0.137 g (0.15 mmol) of tris(dibenzylideneacetone)dipalladium (0), and 0.183 g (0.6 mmol) of tri-*o*-tolylphosphine are added. The reaction mixture is then stirred at 60 °C for 5 h. The resulting mixture is poured into a conical flask containing 300 mL of water. An additional 100 mL of 10% citric acid solution is added in order to break the black emulsion. The aqueous layer is extracted with two 150-mL portions of diethyl ether, and the combined organic layers are washed with two 100-mL portions of water and 100 mL of brine. At this point, the organic extracts are decanted from a black solid [consisting of palladium and zinc(0), together with a bright white solid; (Note 16)], dried over magnesium sulfate, filtered, and concentrated by rotary evaporation to yield 9.1 g of a pale yellow solid. This material is purified by column chromatography (Note 17). The product is charged on a column of 200 g of silica gel (Sorbent 60A, 32-63 μm) and eluted with 1 L of 20% EtOAc-hexane. At that point, fraction collection (25-mL fractions) is begun, and elution is continued with 500 mL of 25% EtOAc-hexane and then 200 mL of 30% EtOAc-hexane. The desired product is collected in fractions 13-59 which are concentrated by rotary evaporation to yield 3.5-3.9 g (35-39%) of **4** as a white solid (Note18).

2. Notes

1. The checkers purchased *N*-(*tert*-butoxycarbonyl)-L-serine methyl ester from Aldrich Chemical Company, Inc. The submitters prepared this compound, a thick amber oil, according to the procedure described in Dondoni, A.; Perrone, D. *Org. Synth.* Coll. Vol. X, **2004**, 320.

2. The checkers purified dichloromethane by pressure filtration through activated alumina.

3. The checkers purchased 4-DMAP and $Me_3N{\cdot}HCl$ from Aldrich Chemical Company, Inc. Et_3N was purchased from Alfa Aesar Chemical Company and was distilled from CaH_2. The submitters obtained 4-DMAP and Et_3N from Lancaster and Me3N·HCl from Aldrich Chemical Company, Inc.

4. TsCl was obtained from Aldrich Chemical Company, Inc. and purified by recrystallization according to the following procedure. p-Toluenesulfonyl chloride (85 g) is dissolved in 150 mL of hot $CHCl_3$ and 200 mL of petroleum ether (room temperature) is added in one portion to the clear, colorless solution. The resulting cloudy solution is clarified by addition of ca. 5 g of charcoal, stirred for 1 min, and filtered on a Büchner funnel. The filtrate is concentrated to ca. 1/5th of its original volume by rotary evaporation, and the solid which appears is collected by filtration and dried under reduced pressure (25 °C, 0.03 mm) to afford 68 g of TsCl as bright white crystals.

5. Triethylamine addition must be carried out slowly to avoid base-promoted elimination of *p*-toluenesulfonate in the final product.

6. It may be necessary to scratch the flask in order to start crystallization.

7. The product exhibits the following properties: mp 74-76 °C, lit.[7a] 74-75 °C; $[\alpha]_D^{20}$ + 3.0 (methanol, c 2.0), lit.[7b] +4.6 (methanol, c 2); Rf 0.24 (petroleum ether/EtOAc 3:1); IR (CH_2Cl_2) cm^{-1}: 3384, 2979, 1753, 1714, 1511, 1367, 1191, 1177; ^1H NMR (500 MHz, $CDCl_3$) δ: 1.43 (s, 9 H), 2.46 (s, 3 H), 3.71 (s, 3 H), 4.29 (dd, J = 10.1, 3.1 Hz, 1 H), 4.40 (dd, J = 10.1, 3.1 Hz, 1 H), 4.50-4.53 (m, 1 H), 5.30 (d, J = 7.9 Hz, NH), 7.37 (app d, J = 7.9 Hz, 2 H), 7.77 (app d, J = 8.2 Hz, 2 H); ^{13}C NMR (125 MHz, $CDCl_3$) δ: 21.9, 28.4 (3 C), 53.1 (2 C), 69.7, 80.7, 128.2 (2 C), 130.1 (2 C), 132.5, 145.4, 155.1, 169.2; MS (EI) *m/z* 314 (M^+-CO_2Me, 20%), 300 [(M^+-O^tBu), 8%], 258 [(M^+-$BocNH_2$), 13%], 215 (11%), 155 ($TolSO_2$, 43%), 57 (100%); HRMS (EI) *m/z*: $[M - CO_2Me]^+$ calcd for $C_{14}H_{20}NO_5S$, 314.1057; found, 314.1059.

8. A.C.S. Reagent grade acetone (99.5%) was purchased from Aldrich Chemical Company, Inc.

9. The checkers purchased sodium iodide from Mallinckrodt Chemical Co. and dried it under vacuum (0.03 mm) at 100 °C for 2 days.

10. The reaction can be monitored by TLC: elution with 3:1 petroleum ether/ethyl acetate, Rf = 0.24 (O-tosylserine), Rf = 0.60 (iodoalanine).

11. Due to the unstable nature of the iodide, the solution should be cooled in an ice-water bath at 0 °C during concentration.

12. Iodoalanine derivative **3** is also available from Aldrich Chemical Company, Inc. Compound **3** exhibits the following spectroscopic and physical properties: mp 45-47 °C, lit.[6b] mp = 51 °C; $[\alpha]_D^{20}$ −3.7 (MeOH), c 3.0, lit.[6b] $[\alpha]_D^{20}$ = −4.0 (c 3 methanol); Rf 0.60 (petroleum ether/AcOEt 3:1); IR (CH_2Cl_2) cm^{-1}: 1163, 1259, 1422, 1714, 1748, 2986, 3054, 3424; ^1H NMR (500 MHz, CDCl3) δ: 1.47 (s, 9 H), 3.55-3.67 (m, 2 H), 3.81 (s, 3 H), 4.53-4.54 (m, 1 H), 5.36 (d, $J = 6.4$ Hz, 1 H); C NMR (125 MHz, $CDCl_3$ δ: 8.6, 29.0 (3 C), 53.7, 54.4, 81.2, 155.5, 170.8,; MS (EI) *m/z* 270 (M^+-CO_2Me 25%), 214 [(M^+-Boc), 8%], 57 (100%); HRMS (ESI) *m/z*: $[M + Na]^+$ calcd for $C_9H_{16}INO_4$, 352.0016; found, 352.0021; Anal. Calcd for $C_9H_{16}INO_4$: C, 32.8; H 4.9; N, 4.3. Found: C, 33.2; H 4.8; N, 4.1.

13. The submitters used a two-necked, round-bottomed flask equipped with a magnetic stirbar, three-way stopcock connected to vacuum and a N_2 source, and a solid addition tube. The addition tube was charged with methyl 4-iodobenzoate, tris(dibenzylideneacetone)dipalladium(0), and tri-*o*-tolylphosphine. After the zinc reagent was formed and the ice bath was removed, the solids contained in the addition tube were added in one portion by inverting the tube.

14. DMF was distilled from CaH_2 and stored over activated 4Å molecular sieves.

15. Methyl 4-iodobenzoate was purchased from Avocado Chemical Co. Zinc dust (< 10 microns, 95% purity) was purchased from Aldrich Chemical Ccompany, Inc. The checkers purchased Pd_2dba_3 and P(*o*-Tol)$_3$ from Strem Chemicals, Inc. and I_2 from Mallinckrodt. The submitters purchased Pd_2dba_3 and P(*o*-Tol)$_3$ from Aldrich Chemical Company, Inc.

16. This solid was identified as dimethyl biphenyl-4,4'-dicarboxylate: ^1H NMR (200MHz, $CDCl_3$) δ: 3.96 (s, 6H), 7.67-7.72 (BB', 4H), 8.10-8.16 (AA', 4H); mp 219-220 °C, (lit. mp 214-217 °C: Catalogue handbook of Fine Chemicals. Aldrich 1999-2000).

17. The submitters purified the product by the following procedure. The residual pale yellow solid is dissolved in 50 ml of diethyl ether and the remaining solid is filtered off (Note 16). The filtrate is concentrated to a volume of ca. 25 mL, and the solution is allowed to crystallize at 0 °C. Once crystallization begins, 50 mL of petroleum ether is added in two portions over 10 h, and then crystallization is allowed to proceed overnight at 0 °C. The white solid is collected by filtration and washed with a mixture of 3:1

petroleum ether-diethyl ether to afford 3.8 g of **4**. Chromatographic purification of the mother liquor (5.5 x 18 cm of DSH silica gel 40-63 mm, elution with 1 L of petroleum ether/ethyl acetate 4:1 followed by 1.5 L of 3:1 petroleum ether-ethyl acetate) gives 2.5 g of **4** as a pale yellow solid. All the material is combined and recrystallized from diethyl ether/petrol as above to yield 5.2 g (47%) of **4** in two crops.

18. The enantiomeric purity of compound **4** was higher than 99.5% as determined by HPLC (2-cm Supelco Si precolumn, 250 x 4.6 mm Chiralpak AS, 10% isopropanol in heptane, 1mL/min, t_r = 14.5 min). The D enantiomer (t_r = 11.5 min) was not detected. Alternatively, the Boc protecting group of compound **4** (0.100 g) was removed by treatment with a solution prepared by addition of 100 μL of AcCl to 1.5 mL of methanol. Mosher's amides were formed using *i*-Pr_2NEt (1.1 equiv), EDCI (1.1 equiv), BtOH (1.1 equiv) and (R) or (S)-α-methoxy-α-(trifluoromethyl)phenylacetic acid (1.5 equiv), and ^1H NMR (500 MHz) analysis of both amides showed ee > 98%. Compound **4** exhibits the following spectroscopic and analytical properties: mp 82-83 °C; $[\alpha]_D^{20}$ = –5.0 (acetone, c 1.0); IR (CH_2Cl_2) cm^{-1}: 3367, 2979, 2954, 1720, 1613, 1512, 1282, 1167; 1H NMR (500 MHz, $CDCl_3$) δ: 1.42 (s, 9 H), 3.10 (dd, *J* = 13.7, 6.4 Hz, 1 H), 3.20 (dd, *J* = 13.7, 5.8 Hz, 1 H), 3.91 (s, 3 H), 3.72 (s, 3 H), 4.61-4.63 (m, 1 H), 5.00 (d, *J* = 7.9 Hz, NH), 7.21 (app d, *J* = 7.9 Hz, 2 H), 7.98 (app d, *J* = 8.5 Hz, 2 H); ^{13}C NMR (125 MHz, $CDCl_3$) δ: 29.0 (3 C), 39.1, 52.8, 53.1, 54.9, 80.8, 129.6, 130.1 (2 C), 130.5 (2 C), 142.2, 155.7, 167.6, 172.7; MS (EI) m/z 337 (M^+, 0.3%), 306 [(M^+–OMe), 7%], 264 [(M^+–O^tBu), 4%], 220 [(M^+–$BocNH_2$), 83%], 150 (79%), 57 (100%); HRMS (ESI) *m/z*: $[M + Na]^+$ calcd for $C_{17}H_{23}NO_6$ 360.1418; found, 360.1407; Anal. Calcd for $C_{17}H_{23}NO_6$: C, 60.51; H, 6.88; N, 4.15. Found: C, 60.3; H, 6.6; N, 4.0.

Waste Disposal Information

All toxic materials were disposed of in accordance with "Prudent Practices in the Laboratory"; National Academy Press; Washington, DC, 1995.

3. Discussion

The development of synthetic equivalents for the alanine β-anion is an important target for the synthesis of α-amino acids. A useful solution was provided by the discovery that organozinc reagents could be prepared from the iodoalanine derivative (**3**) by insertion of activated zinc into the carbon-iodine bond.[2] The resulting organometallic reagent could then be coupled with carbon electrophiles in the presence of a suitable palladium catalyst. Furthermore, treatment of the organozinc reagent with copper cyanide gave the organocuprate, which could then couple directly with reactive electrophiles.[3]

The iodoalanine derivative is available in both enantiomeric forms and this method offers an extremely simple route to large numbers of non-natural amino acids.[4] Therefore a reliable and practical method for the synthesis of fully protected iodoalanine **3** in a multigram scale is highly desirable.

Although the tosylation reaction of fully protected serine has already been described,[5] we have found that extended reaction times are required when the reaction is carried out on 100-200 mmol scale, taking from 3 to 5 days to go to completion. As a result of longer reaction times, the amount of by-products is increased making the purification step more difficult.

Recently Tanabe and co-workers have found that several alcohols were smoothly and efficiently tosylated using tosyl chloride/ triethylamine and a catalytic amount of trimethylamine hydrochloride as reagents.[6] Compared with the traditional method using pyridine as solvent, this procedure has the merit of much higher reaction rates, and it avoids the side reaction in which the desired tosylate is converted into the corresponding chloride.

Entry	Electrophile	Catalyst	T(°C), t	Product	Yield[a]
1	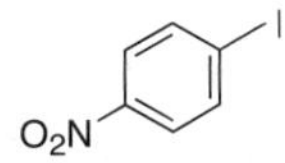	$Pd_2(dba)_3$ 1% $P(oTol)_3$ 4%	rt, 2 h	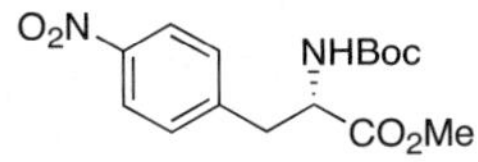	72%[b]
2	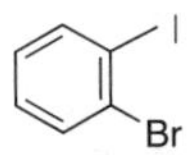	$Pd_2(dba)_3$ 1.2% $P(oTol)_3$ 5%	40 °C, 5.5h	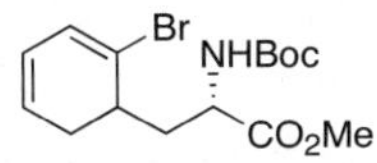	47%[b]
3	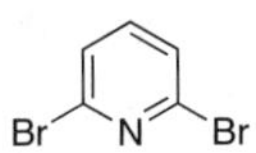	$Cl_2Pd(PPh_3)_2$ 5%	50 °C, 5h	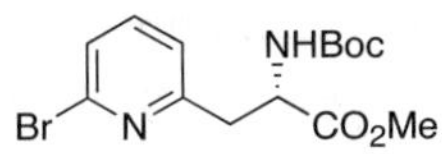	52%[c]
4	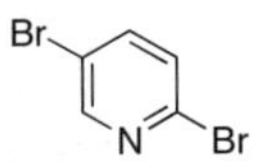	$Cl_2Pd(PPh_3)_2$ 2%	50 °C, 5.5h	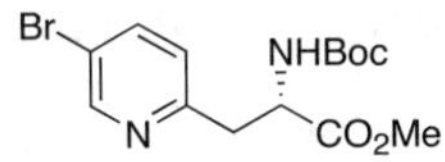	64%[d]
5	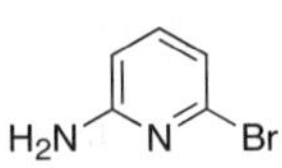	$Cl_2Pd(PPh_3)_2$ 5%	40 °C, 6h	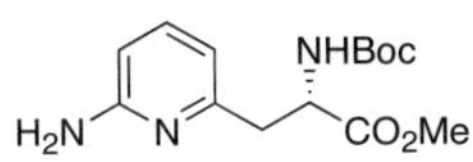	34%[e]

a) Isolated yield. b) 5 mmol scale. c) 10 mmol scale. d) 7 mmol scale. e) 2 mmol scale.

Stirring a mixture of tosylate and NaI in acetone at room temperature for four days produced the iodoalanine derivative **3** via a *pseudo*-Finkenstein reaction. Although this reaction has been reported to proceed in shorter reaction times when carried out at reflux,[7] in our hands, an increase in the temperature of the reaction produced significant amounts of by-products and a poor mass balance. For instance, when the experimental procedure stated in step B was carried out at reflux of acetone only 4 g of crude material was recovered. Three water-soluble products account for the rest of the mass balance, and one of them has been identified as 4-methoxycarbonyloxazolidin-2-one.[8] Better yields can be obtained when the reaction was carried out at reflux on a smaller scale (2 g of compound **2**) and under much more dilute conditions as reported.[7] It is our contention that due to the instability of the iodoalanine derivative, high temperatures during the reaction or the work-up promote decomposition of the product *via* intramolecular nucleophilic attack by the carbamate moiety. Nevertheless, compound **3** could be kept in the refrigerator for one year without evidence of decomposition.

The procedure in Section C is representative of the synthesis of non-natural α-amino acids featuring the palladium cross coupling reaction of a β-alanine organozinc derivative with aromatic electrophiles. This methodology has been successfully extended with modifications to both the electrophile and the catalyst as shown in the Table.

Applications of the zinc reagent derived from iodide **3** continue be reported, and a recent comprehensive review, which updates reference 4, has been published.[9] A range of other amino acid-derived organozinc reagents have been prepared (**5** and **6** from aspartic acid, and **7** and **8** from glutamic acid) and been shown to undergo Negishi coupling with aromatic iodides under similar conditions to those reported in the procedure. In general, reactions of zinc reagents **5-8** proceed at lower temperatures, and with significantly higher yields, than the zinc reagent derived from iodide **3**. This is largely due to the greater reactivity of zinc reagents **5-8**.

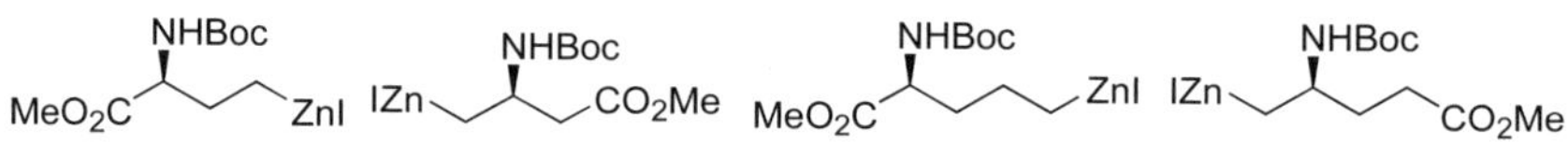

5 **6** **7** **8**

The method for zinc activation now routinely used for step C involves addition of the iodine to a rapidly stirred suspension of zinc dust in dimethylformamide under an atmosphere of nitrogen, followed by stirring at room temperature until the iodine decolorizes (typically < 5min).[10] This procedure is simpler than the reported method and allows very similar results to be obtained.

1. Department of Chemistry, The University of Sheffield, Sheffield, S3 7HF, UK
2. Jackson, R. F. W.; Wishart, N.; Wood, A.; James, K.; Wythes, M. J. *J. Org. Chem.* **1992**, *57*, 3397.
3. Dunn, M. J.; Jackson, R. F. W.; Pietruszka, J.; Turner, D. *J. Org. Chem.* **1995**, *60*, 2210.
4. For a review see: Gair, S.; Jackson, R. F. W. *Current Org. Chem.* **1998**, *2*, 527.
5. Hermkens, P. H. H.; Maarseveen, J. H. v.; Ottenheijm, H. C. J.; Kruse, C. G.; Scheeren, H. W. *J. Org. Chem.* **1990**, *55*, 3998.
6. Yoshida, Y.; Sakakura, Y.; Aso, N.; Okada, S.; Tanabe, Y. *Tetrahedron* **1999**, *55*, 2183.
7. a) Stocking, E. M.; Schwarz, J. N.; Senn H.; Salzmann, M.; Silks, L. A. *J. Chem. Soc., Perkin Trans. 1* **1997**, 2443. b) Bajgrowicz, J. A.; A. El Hallaoui; Jacquier, R.; Pigiere, Ch.; Viallefont, P. *Tetrahedron* **1985**, *41*, 1833.
8. Sibi, M. P.; Rutherford, D.; Sharma, R. *J. Chem. Soc., Perkin Trans.1* **1994**, 1675.
9. Rilatt, I. Caggiano, L.; Jackson, R.F.W. *Synlett* **2005**, 2701.
10. Huo, S. *Org. Lett.* **2003**, *5*, 423.

SYNTHESIS OF 3-PYRIDYLBORONIC ACID AND ITS PINACOL ESTER. APPLICATION OF 3-PYRIDYLBORONIC ACID IN SUZUKI COUPLING TO PREPARE 3-PYRIDIN-3-YLQUINOLINE [Quinoline, 3-(3-pyridinyl)-]

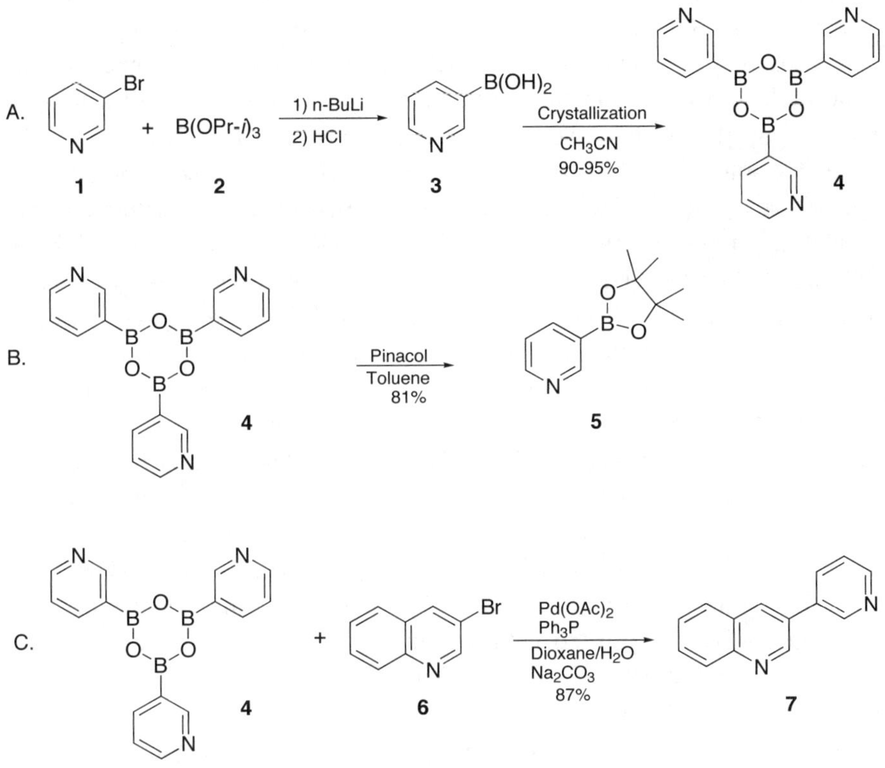

Submitted by Wenjie Li, Dorian P. Nelson, Mark S. Jensen, R. Scott Hoerrner, Dongwei Cai and Robert D. Larsen.[1]
Checked by Scott E. Denmark, Geoff T. Halvorsen, and Jeffrey M. Kallemeyn.

1. Procedure

A. 3-Pyridylboronic acid [tris(3-pyridyl)boroxin]. A 1-L, 3-necked flask equipped with a temperature probe, overhead stirrer, and a nitrogen inlet adaptor capped with a rubber septum is charged with 320 mL of toluene, 80 mL of THF, triisopropyl borate (55.4 mL, 240 mmol), and 3-bromopyridine (19.3 mL, 200 mmol) (Notes 1, 2). The mixture is cooled to

–40 °C using a dry ice/acetone bath and 96 mL of *n*-butyllithium solution (2.5M in hexanes, 240 mmol) (Note 3) is added dropwise with a syringe pump over 1 h. The reaction mixture is stirred for an additional 30 min maintaining the temperature at –40 °C. The acetone/dry ice bath is then removed, and the reaction mixture is allowed to warm to –20 °C whereupon a solution of 200 mL of 2N HCl solution is added. When the mixture reaches room temperature, it is transferred to a 1-L separatory funnel and the aqueous layer (pH ≈ 1) is drained into a 500-mL Erlenmeyer flask equipped with a magnetic stir bar. The pH of the aqueous layer is adjusted to 7.6-7.7 using 5N aqueous NaOH (ca. 30 mL) (Note 4). A white solid precipitates out as the pH approaches 7. The aqueous mixture is then saturated with solid NaCl, transferred to a 1-L separatory funnel, and extracted with three 250-mL portions of THF. The combined organic phases are concentrated on a rotary evaporator to leave a solid residue which is suspended in 80 mL of acetonitrile for crystallization. The mixture is heated to 70 °C in an oil bath, stirred for 30 min, and then allowed to cool slowly to room temperature and then to 0 °C in an ice bath. After being stirred at 0 °C for 30 min, the mixture is filtered through a fritted-glass funnel. The solid is washed with 15 mL of cold (5 °C) acetonitrile, and then dried under vacuum to afford 18.9 g of tris(3-pyridyl)boroxin·1.0 H_2O as a white solid (Note 5).

B. 3-(4,4,5,5-Tetramethyl-[1,3,2]dioxaborolan-2-yl)pyridine. A 250-mL, one-necked, round-bottomed flask equipped with a magnetic stirbar and a Dean-Stark trap fitted with a condenser capped with a nitrogen inlet adaptor is charged with tris(3-pyridyl)boroxin·0.85 H_2O (3.0 g, 9.1 mmol), pinacol (4.07 g, 34.4 mmol) (Note 6), and 120 mL of toluene. The solution is heated at reflux for 2.5 h in a 120 °C oil bath. The reaction is complete when the mixture changes from cloudy-white to clear. The solution is then concentrated under reduced pressure on a rotary evaporator to afford a solid residue. This solid is suspended in 15 mL of cyclohexane (Note 7) and the slurry is heated to 85 °C, stirred at this temperature for 30 min, and then allowed to cool slowly to room temperature. The slurry is filtered, rinsed twice using the mother liquors, washed with 3 mL of cyclohexane, and dried under vacuum to afford 4.59 g (82%) of 3-pyridylboronic acid pinacol ester as a white solid (Note 8).

C. 3-Pyridin-3-yl-quinoline. A 100-mL, round-bottomed Schlenk flask equipped with a magnetic stir bar is charged with tris(3-pyridyl)boroxin·0.85 H_2O (3.80 g, 11.5 mmol), 3-bromoquinoline (6.24 g, 30.0 mmol) (Note 9), 30 mL of a 2M aqueous solution of Na_2CO_3, and 30 mL of 1,4-dioxane (Note 10). Palladium(II) acetate (0.336 g, 1.5 mmol) and triphenylphosphine (1.57 g, 6.0 mmol) are added (Note 10), the mixture is degassed using five vacuum/nitrogen back-fill cycles, and then is heated to 95 °C for 2.5 h with vigorous stirring (Note 11). The mixture is allowed to cool to room temperature, transferred to a 500-mL separatory funnel, and diluted with 100 mL of water and 150 mL of ethyl acetate. The aqueous layer is separated and back-extracted with 50 mL of ethyl acetate, and the combined organic layers are extracted with three 50-mL portions of 1M HCl solution. The combined acidic aqueous layers are treated with 40 mL of 5M aqueous NaOH resulting in a pH of ca. 9. The cloudy aqueous layer is then extracted with three 50-mL portions of ethyl acetate. The combined organic layers are dried over $MgSO_4$, filtered into a 500-mL, round-bottomed flask, and concentrated on a rotary evaporator under reduced pressure to afford a brown solid. Isopropyl acetate (10 mL) is added and the slurry is heated to reflux at which point 40 mL of heptane is added. The mixture is allowed to cool to room temperature over 3 h, and then is stirred at room temperature for 2 h. The solid is isolated by filtration, washed with 10 mL of 4:1 heptane/isopropyl acetate, and dried under vacuum at room temperature to give 5.36-5.40 g (87%) of 3-pyridin-3-ylquinoline as a white solid (Note 12).

2. Notes

1. Toluene and THF were purchased from Fisher Scientific and dried over 4Å molecular sieves overnight prior to use. The water content of the solvents was < 50 μg/mL by Karl Fischer titration.
2. Triisopropyl borate was purchased from Aldrich Chemical Company, Inc. 3-Bromopyridine was purchased from Lancaster Synthesis. Both compounds were used without further purification.
3. *n*-Butyllithium (2.5M in hexanes) was purchased from Aldrich Chemical Company, Inc. and was titrated prior to use.
4. Measurement of pH was performed using a Metrohm model 691 pH meter equipped with a Metrohm combined LL micro pH glass electrode calibrated prior to use with pH = 2 and 9 buffers. The checkers found that adjustment to a lower pH led to product with higher amounts of inorganic impurities. The checkers also found that the use of pH paper results in different pH values as compared to the pH meter.
5. The checkers obtained trispyridylboroxin in various hydration levels (0.85 – 1.0 H_2O). A satisfactory melting point for this solid could not be obtained. ^1H NMR (400 MHz, CD_3OD): δ 7.66 (br s, 1H), 8.38 (d, J =

6.6, 1H), 8.51 (dd, J = 1.2, 4.4, 1H), 8.61 (br s, 1H). ^{1}H NMR spectra were complicated in other solvents such as $CDCl_3$ and DMSO-d_6. Anal Calcd. for $C_{15}H_{12}B_3O_3N_3$ • $1.0H_2O$: C, 54.15; H, 4.24; N, 12.63. Found: C, 53.95; H, 3.91; N, 12.35. Yield based on this formula is 85%. The submitters report obtaining the product in 91% yield.

6. Pinacol was purchased from Aldrich Chemical Company, Inc. and was used without further purification.

7. Cyclohexane was purchased from Aldrich Chemical Company, Inc. and was used without further purification.

8. The checkers obtained the product in 74-82% yield in different runs. The product exhibits the following physical properties: mp 102–105 °C; IR (KBr pellet) cm^{-1} 2994, 2968, 2932, 1609, 1572, 1476, 1410, 1361, 1209, 1154, 1063, 1017, 953, 926, 859, 833, 800, 759, 705; ^{1}H NMR (500 MHz, $CDCl_3$): δ 1.33 (s, 12H), 7.25 (ddd, J = 1.1, 4.9, 7.5, 1H), 8.03 (dt, J = 1.8, 7.5, 1H), 8.64 (dd, J = 1.9, 4.9, 1H), 8.93 (d, J = 1.1, 1H); ^{13}C NMR (126 MHz, $CDCl_3$) δ 24.8, 84.1, 123.0, 142.2, 152.0, 155.5; MS (EI, 70 eV): 205 (M+, 46), 204 (15), 191 (12), 190 (100), 189 (25), 162 (10), 148 (44), 147 (14), 120 (18), 119 (11), 106 (100), 105 (35), 85 (15), 59 (17), 58 (19); HRMS (EI) *m/z* 205.1280, calcd for $C_{11}H_{16}NO_2B$ 205.1274). Anal. Calcd for : $C_{11}H_{16}BO_2N$: C, 64.43; H, 7.86; N, 6.83. Found: C, 64.23; H, 7.99; N, 6.88.

9. 3-Bromoquinoline was purchased from Acros Organics and was used without further purification.

10. 1,4-Dioxane, palladium (II) acetate, and triphenylphosphine were purchased from Aldrich Chemical Company, Inc. and were used without further purification.

11. A heterogeneous mixture that is difficult to stir may be formed during the reaction. The checkers found that rapid stirring from the onset of the reaction prevented loss of stirring. Inefficient stirring was found to lower the yield of the reaction.

12. The following characterization data was obtained: mp: 122–125 °C; IR (KBr pellet): cm^{-1} 3044, 1568, 1495, 1410, 1338, 1298, 1187, 1126, 1059, 1022, 952, 932, 816, 784, 758, 709 909, 808, 786, 752, 709, 61; ^{1}H NMR (500 MHz, $CDCl_3$): δ 7.45 (ddd, J = 0.9, 4.9, 7.9, 1H), 7.60 (dt, J = 0.9, 7.9, 1H), 7.75 (ddd, J = 1.2, 6.8, 8.3, 1H), 7.90 (dd, J = 0.9, 8.1, 1H), 8.00 (dt, J = 2.2, 7.8, 1H), 8.15 (d, J = 8.4, 1H), 8.32 (d, J = 2.3, 1H), 8.68 (dd, J = 1.5, 4.9, 1H), 8.97 (d, J = 2.1, 1H), 9.15 (d, J = 2.1, 1H); ^{13}C NMR (126 MHz, $CDCl_3$) δ 123.8, 127.3, 127.8, 128.0, 129.3, 129.9, 130.6, 133.57, 133.63, 134.6, 147.7, 148.5, 149.27, 149.29; MS (EI, 70 eV): 207

(16), 206 (M+, 100), 205 (40); HRMS (EI) *m/z* 206.0847, calcd for $C_{14}H_{10}N_2$ 206.0944). Anal. Calcd for $C_{14}H_{10}N_2$: C, 81.53; H, 4.89; N, 13.58. Found: C, 81.43; H, 4.86; N, 13.40.

Waste Disposal Information

All toxic materials were disposed of in accordance with "Prudent Practices in the Laboratory"; National Academy Press; Washington, DC, 1995.

3. Discussion

3-Pyridylboronic acid is a useful compound for the introdution of a 3-pyridyl moiety into a molecule by the Suzuki reaction.[2-3] It is commercially available only in small quantities at high cost. Typical preparation of this class of compounds involves reaction between boronic esters and an organometallic reagent (Li or Mg)[4-7] usually prepared by magnesium insertion[4] or lithium-halogen exchange[5-7] of 3-bromopyridine. The literature protocols, however, require conditions that are not suitable for large scale, including using dibromoethane as the solvent, running the reaction at very low temperature (–78 °C), and affording poor to modest yields. Recently, we reported a revised procedure for the preparation of 3-pyridylboronic acid via lithium-halogen exchange and *in situ* quench with triisopropyl borate.[8] In this protocol, *n*-butyllithium is added to a solution of 3-bromopyridine and triisopropyl borate in THF/toluene based on the reasoning that lithium-halogen exchange is much faster than the reaction between *n*-butyllithium and triisopropyl borate. The 3-lithiopyridine intermediate thus generated, reacts rapidly with the borate in the reaction mixture, thereby minimizing the chance for 3-lithiopyridine to undergo side reactions.[9] The new procedure allows the reaction to be run at much higher temperatures, giving the best yields (90–95%) at –40 °C and a respectable 80% yield even at 0 °C. The reaction can be carried out at kilogram scale. The product is isolated by crystallization from acetonitrile in the form of boroxin **4**. The characterization of **4** is difficult, however, due to the presence of varying amounts of water. Therefore, boroxin **4** is converted to its pinacol ester **5**, which is fully characterized by spectroscopic data and elemental analysis. The utility of **4** in palladium-catalyzed cross-coupling reaction is demonstrated in a Suzuki reaction with 3-bromoquinoline.[10]

1. Process Research Department, Merck Research Laboratories, P. O. Box 2000, Rahway, NJ 07065.
2. For reviews of Suzuki reaction, see: (a) Miyaura, N.; Suzuki A. *Chem. Rev.* **1995**, *95*, 2457-2483. (b) Suzuki, A. *J. Organomet. Chem.* **1999**, *576*, 147-168.
3. For examples of 3-pyridylboronic acid in cross-coupling reactions, see: (a) Enguehard, C.; Hervet, M.; Allouchi, H.; Debouzy, J.-C.; Leger, J.-M.; Gueiffier, A. *Synthesis*, **2001**, *4*, 595-600. (b) Li, J. J.; Yue, W. S. *Tetrahedron Lett.* **1999**, *40*, 4507-4510. (c) Bower, J. F.; Guillaneux, D.; Nguyen, T.; Wong, P. L.; Snieckus, V. *J. Org. Chem.* **1998**, *63*, 1514-1518.
4. Fischer, F. C.; Havinga, E. *Recl. Trav. Chim. Pays-Bas* **1965**, *84*, 439-440.
5. Terashima, M.; Kakimi, H.; Ishikura, M.; Kamata, K. *Chem. Pharm. Bull.* **1983**, *31*, 4573-4577.
6. Ishikura, M.; Kamada, M.; Oda, I.; Terashima, M. *Heterocycles* **1985**, *23* (1), 117-120.
7. Coudret, C. *Synth. Comm.* **1996**, *26*, 3543-3547.
8. (a) Li, W.; Nelson, D. P.; Jensen, M. S.; Hoerrner, R. S.; Cai, D.; Larsen, R. D.; Reider, P. J. *J. Org. Chem.* **2002**, *67*, 5394-5397. (b) Cai, D.; Larsen, R. D.; Reider, P. J. *Tetrahedron Lett.* **2002**, *43*, 4285-4287.
9. Gilman, H.; Spatz, S. M. *J. Org. Chem.* **1951**, *16*, 1485-1494.
10. (a) Ishikura, M.; Oda, I.; Terashima, M. *Heterocycles*, **1985**, *23 (9)*, 2375-2386. (b) Ishikura, M.; Kamada, M.; Terashima, M. *Synthesis*, **1984**, *11*, 936-938.

EFFICIENT ASYMMETRIC SYNTHESIS OF *N-tert*-BUTOXYCARBONYL α-AMINOACIDS USING 4-*tert*-BUTOXYCARBONYL- 5,6-DIPHENYLMORPHOLIN-2-ONE: (*R*)-(*N*-*tert*-BUTOXYCARBONYL)ALLYLGLYCINE

[(4-Pentenoic acid, 2-[[(1,1-dimethylethoxy)carbonyl]amino]-, (2*R*)-)]

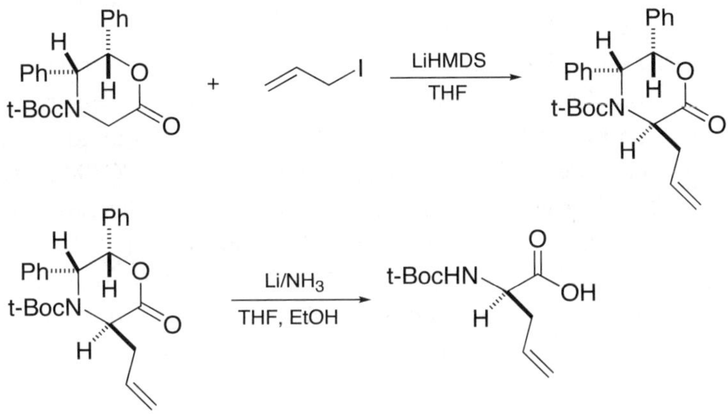

Submitted by Robert M.Williams,[1] Peter J. Sinclair, and Duane E. DeMong. Checked by Wenlin Lee and Marvin J. Miller.

1. Procedure

A. (3R,5R,6S)-4-tert-Butoxycarbonyl-5,6-diphenyl-3-(1'-prop-2'-enyl) morpholin-2-one. A 250-mL, single neck, round-bottomed flask equipped with a magnetic stirring bar is charged with 1.0 g (2.83 mmol) of (5*R*,6*S*)-4-*tert*-butoxycarbonyl-5,6-diphenylmorpholin-2-one and placed under vacuum for 3 h. The evacuated flask is purged with argon, a rubber septum with an argon inlet is attached to the neck, and tetrahydrofuran (THF) (50 mL) is added via syringe (Note 1). After dissolution of the solid by stirring, 2 mL (10 mmol) of allyl iodide (Note 2) is added via syringe. With stirring, the flask and its contents are cooled to −78 °C with a dry ice/acetone bath, then 2.8 mL (2.83 mmol) of lithium bis(trimethylsilyl)amide (1M in THF) is added dropwise (Note 3). The mixture is stirred for 1 h at −78 °C (Note 4), then the solution is quenched by the addition of 100 mL of water. The mixture is extracted with 250 mL of ethyl acetate in a separatory funnel. After separation of the aqueous phase, the organic phase is washed with brine, dried over anhydrous magnesium sulfate ($MgSO_4$), and concentrated by rotary evaporation under vacuum to afford an orange oil. The crude product is dissolved in a minimum amount of 4:1 hexanes:ethyl acetate, and purified by flash chromatography using 70 g of 35-75 mesh silica gel (eluting with 4:1 hexane:ethyl acetate followed by 2:1 hexanes:ethyl acetate). Removal of solvent from the appropriate fractions by rotary evaporation at room temperature affords 997 mg (90%) of (3*R*,5*R*,6*S*)-4-*tert*-butoxycarbonyl-5,6-diphenyl-3-(1'-prop-2'-enyl)-morpholin-2-one as a white solid (Notes 5, 6).

B. (R)-(N-tert-Butoxycarbonyl)allylglycine. A flame-dried, 100-mL, two-necked, round-bottomed flask equipped with a magnetic stirring bar is charged with 35 mL of liquid ammonia and 124 mg (17.8 mmol) of lithium metal (Note 7). The resulting solution is stirred at −33 °C. Meanwhile, a flame-dried 25-mL, round-bottomed flask fitted with a rubber stopper and an argon inlet is charged with 539 mg (1.37 mmol) of (3*R*,5*R*,6*S*)-4-*tert*-butoxycarbonyl-5,6-diphenyl-3-(1'-prop-2'enyl)morpholin-2-one 0.150 mL of absolute ethanol (Note 8), and 5 mL of anhydrous THF (Note 1). The resulting mixture is then transferred via cannula to the lithium/liquid ammonia solution. After 20 min, the reaction is carefully quenched with solid ammonium chloride and allowed to warm to room temperature (Note 9). After evaporation of the remaining ammonia, the residue is diluted with 35 mL of water and extracted twice with 35-mL portions of diethyl ether (Note 10). The aqueous phase is carefully acidified with 1N HCl to a pH of 2 while stirring with 35 mL of ethyl acetate (Note 11). Following separation of the organic phase, the aqueous phase is extracted three times with 20-mL portions of ethyl acetate. The organic extracts are combined, dried over $MgSO_4$, filtered, and concentrated under vacuum to yield 178 mg (60%) of crude (*R*)-(*N-tert*-butoxycarbonyl)allylglycine (% ee ≥ 96%) as a viscous oil (Notes 12-15).

2. Notes

1. Anhydrous tetrahydrofuran is obtained by distilling from sodium benzophenone ketyl.

2. Allyl iodide (98%) can be purchased from Aldrich Chemical Co., Inc.

3. Lithium bis(trimethylsilyl)amide (1M solution in tetrahydrofuran) can be purchased from Aldrich Chemical Co., Inc.

4. Phosphomolybdic acid (PMA) dissolved in ethanol is a suitable reagent for visualizing the product on TLC. After dipping the TLC plate in the PMA solution, the chromatogram is developed by heating on a hot plate.

5. Trace amounts of impurities can be removed by recrystallizing the product from either diethyl ether/hexanes or dichloromethane/hexanes. The yield obtained by the checker after the first recrystallization was 65%, although an additional crop of the product could be obtained by recrystallization of the mother liquor.

6. Analytical and spectral data are as follows: mp. 177-178 °C, $[\alpha]^{25}_{D}$ +45.7 (c 1.34, CH_2Cl_2); ^{1}H NMR (300 MHz) (DMSO-d_6 vs. DMSO) (120 °C) δ: 1.22 (9H, s broad); 2.87 (2H, m); 4.90 (1H, dd, *J* = 7.3, 7.0); 5.17-5.29 (3H, m); 5.89-6.04 (1H, m); 6.20 (1H, d, *J* = 3.3); 6.59-6.62 (2H, m); 7.1-7.3 (8H, m). IR (NaCl, CH_2Cl_2) cm^{-1}: 3050, 2970, 2920, 1755, 1690, 1375, 1350, 1260, 1155, 1110. Analysis (recrystallized from diethyl ether/hexanes or dichloromethane/hexanes); calcd for $C_{28}H_{27}NO_4$: C, 73.26; H, 6.92; N, 3.56. Found: C, 72.37; H, 6.85; N, 3.68. Note that the NMR spectrum must be recorded at high temperature due to slow conformational exchange of the urethane group on the NMR time scale.

7. Ammonia is distilled from sodium metal immediately prior to use.

8. Absolute ethanol is used as received from Pharmco Products Incorporated.

9. When quenching the reaction and warming to room temperature, caution must be exercised in order to prevent the reaction mixture from overflowing the flask if ammonium chloride is added too quickly or the reaction is warmed too rapidly.

10. Reagent grade diethyl ether is used as received from Fisher Scientific.

11. An Orion Research model 301 analog pH meter was used.

12. The crude product is typically analytically pure; however, the product can be further purified by flash chromatography using 15 g of 35-75 mesh silica gel (elution with 60 mL of 5% methanol in dichloromethane).

13. Analytical and spectral data are as follows: $[\alpha]^{25}_{D}$ = +3.3 (c, 1.5, CH_2Cl_2); ^{1}H NMR (300 MHz) (DMSO-d_6) (60 °C) δ TMS: 1.39 (9H, s); 2.29-2.46 (2H, m); 3.92-4.0 (1H, m); 5.02-5.13 (2H, m); 5.77 (1H, dddd, *J* = 17.2, 10.3, 7.0, 7.0); 6.73 (1H, br s); 12.13 (1H, br s). IR (neat) cm^{-1}: 3430, 3050, 2980, 1715, 1500, 1370, 1265, 1155; mass spectrum (NH_3, Cl) *m/e* 232.9 (M^+ +18, 1.8), 215.9 (M^+ +1, 2.1), 214.9 (M^+, 0.3), 116.0 (62.9).

14. Optical rotations were measured on a Rudolf Research Autopol III automatic polarimeter operating at 589 nm, corresponding to the sodium D line.

15. The enantiomeric excess of the product is determined by Mosher amide formation and NMR analysis as follows: (*R*)-(*N-tert*-butoxycarbonyl)allylglycine is dissolved in 3 mL of 1M ethanolic HCl and heated to reflux for 2 h. The reaction is cooled and the solvent removed under vacuum. The crude amino ester hydrochloride is dissolved in 0.3 mL of pyridine and 0.3 mL of CCl_4. To this mixture is added 25 mg (0.10 mmol) of (+)-α-methoxy-α-(trifluoromethyl)phenylacetyl chloride and the reaction is stirred for 6 h at room temperature. The reaction is quenched with 1 mL water, taken up in diethyl ether, and washed consecutively with 1M HCl, saturated aqueous $NaHCO_3$, and water. The ether layer is dried over anhydrous $MgSO_4$, filtered and evaporated. The Mosher amide formation is repeated with (±)-α-methoxy-α-(trifluoromethyl)phenylacetyl chloride, and, now with this reference standard, the two products are compared by ^{1}H NMR.

Waste Disposal Information

All toxic materials were disposed of in accordance with "Prudent Practices in the Laboratory"; National Academy Press; Washington, DC, 1995.

3. Discussion

The morpholinone glycine synthon, which can be obtained in both enantiomeric forms, can be employed as a versatile scaffold for the preparation of a wide range of *N-t*-Boc protected *R*- and *S*-α-amino acids.[2,3,4] Allylglycine, the example selected here to demonstrate this methodology, is useful for the construction of several amino acid derivatives, such as 2,7-diaminosuberic acid[5,6] and mechanism-based inhibitors of pyridoxal-dependent enzymes.[7] Allylglycine has also served as a starting material for the synthesis of several natural products.[8,9,10,11] The current procedure is the only direct method of synthesizing this *N-t*-Boc-protected amino acid. Other reports have appeared for preparing this amino acid[12,13] including an *Organic Syntheses* procedure by Myers.[14] The use of the dissolving metal reduction to remove the chiral auxiliary allows the direct isolation of the *N-t*-Boc-protected amino acid, thus eliminating the need to protect the free amino acid in a subsequent step. The by-product of the reduction of the diphenylamino alcohol moiety is bibenzyl, which is insoluble in water and conveniently removed by extraction as described above. The current procedure enables these alkylations to be carried out on a preparative scale in good yield and excellent ee.

1. Department of Chemistry, Colorado State University, Fort Collins, Colorado 80523. This material is based upon work supported by the National Science Foundation under Grant No. 0202827.
2. Williams, R. M., *Aldrichimica Acta* **1992**, *25*, 11.
3. Williams, R. M., *Advances in Asymmetric Synthesis*, A. Hassner, Ed., JAI Press **1995**, Vol. 1, p 45.
4. Williams, R. M, *Peptidomimetics Protocols*, Kazmierski, W., Ed., *Methods in Molecular Medicine* **1999**, Vol. 23, Chapter 19, p. 339, Humana Press.
5, Williams, R. M.; Liu, J. *J. Org. Chem.* **1998**, *63*, 2130.
6. Gao, Y.; Lane-Bell, P.; Vederas, J. C. *J. Org. Chem.* **1998**, *63*, 2133.
7. Johnston, M.; Raines, R.; Walsh, C.; Firestone, R. A. *J. Am. Chem. Soc.* **1980**, *102*, 4241.
8. Hutton, C. A.; White, J. M. *Tetrahedron Lett.* **1997**, *38*, 1643.
9. Kurokawa, N.; Ohfune, Y. *Tetrahedron* **1993**, *49*, 6195.
10. Baldwin, J. E.; Bradley, M.; Turner, N. J.; Adlington, R.M.; Pitt, A. R.; Sheridan, H. *Tetrahedron* **1991**, *47*, 8203.
11. Schneider, H.; Sigmund, G.; Schricker, B.; Thirring, K.; Berner, H. *J. Org. Chem.* **1993**, *58*, 683.
12. Hamon, D. P. G.; Massy-Westropp, R. A.; Razzino, P. *Tetrahedron* **1995**, *51*, 4183.
13. R. M. Williams, *Synthesis of Optically Active α-Amino Acids* (Organic Chemistry Series, J. E. Baldwin, Series Editor) **1989**, p. 410, Pergamon Press, New York.
14. Myers, A. G.; Gleason, J.L. *Org. Synth.* **1999**, *76*, 57.

PREPARATION OF *O*-ALLYL-*N*-(9-ANTHRACENYLMETHYL)CINCHONIDINIUM BROMIDE AS A PHASE TRANSFER CATALYST FOR THE ENANTIOSELECTIVE ALKYLATION OF GLYCINE BENZOPHENONE IMINE *tert*-BUTYL ESTER: (4*S*)-2-(BENZHYDRYLIDENAMINO)PENTANEDIOIC ACID, 1-*tert*-BUTYL ESTER-5-METHYL ESTER

[[Cinchonanium, 1-(9-anthracenylmethyl)-9-(2-propenyloxy)-, bromide, (8α,9*R*)-and L-Glutamic acid, *N*-(diphenylmethylene)-, 1-(1,1-dimethylethyl) 5-methyl ester]]

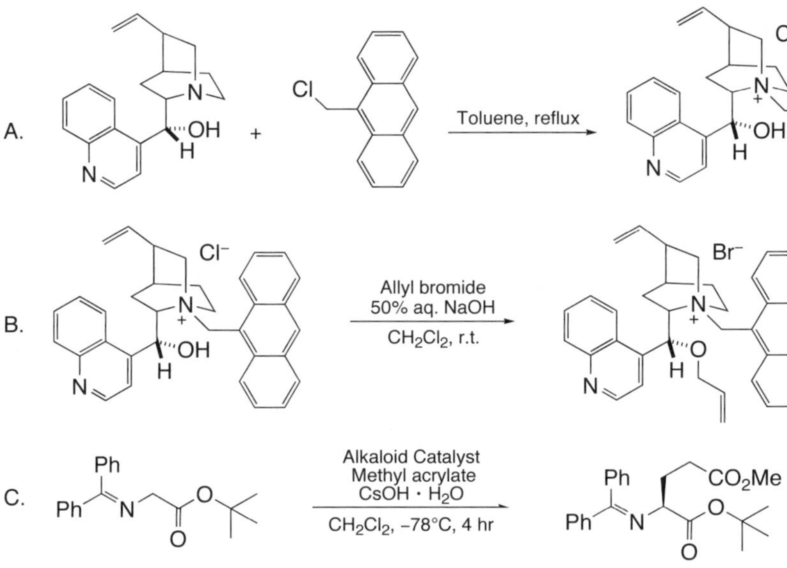

Submitted by E. J. Corey[1] and Mark C. Noe.[2]
Checked by S. Springfield, J. Amato, and E. J. J. Grabowski.

1. Procedure

A. N-(9-Anthracenylmethyl)cinchonidinium chloride (Note 1). A 2-L, three-necked flask, equipped with an overhead stirrer, reflux condenser and nitrogen inlet, is charged with 78.2 g (266 mmol) of cinchonidine (Note 2), 63.0 g (278 mmol) of 9-chloromethylanthracene (Note 3), and 800 mL of toluene (Note 4). The mixture is stirred and heated to reflux employing a bath temperature of ~130 °C for 3 h. Within 30 min all of the material dissolves, then the product begins to crystallize. At the end of the reflux period the mixture is cooled to ~30 °C and 800 mL of diethyl ether is added over 10 min. The slurry is stirred for 20 min, filtered through a coarse sintered glass, fritted funnel, washed with 250 mL of diethyl ether and dried at 50 °C under vacuum overnight to afford 144.4 g (235 mmol, 88%) of *N*-(9-anthracenylmethyl)cinchonidinium chloride as a crystalline, light yellow solid (Notes 1, 5), which is sufficiently pure for use in the next step.

B. O-Allyl-N-(9-Anthracenylmethyl)cinchonidinium bromide. A 1-L, three-necked flask, equipped with an overhead stirrer and nitrogen inlet, is charged with 49.5 g (80.7 mmol) of *N*-(9-anthracenylmethyl)cinchonidinium chloride-toluene solvate, 400 mL of methylene chloride (CH_2Cl_2), 25 mL of allyl bromide (35.0 g, 289 mmol) (Note 6) and 50 mL of 50% aqueous potassium hydroxide (Note 7). The mixture is stirred vigorously for 4 h, then diluted with 400 mL of water and stirred for 5 min. After separation of the phases, the organic phase is washed with a solution of 25 g of sodium bromide in 250 mL of water and dried over 50 g of sodium sulfate (Na_2SO_4). The solution is filtered using 50 mL of CH_2Cl_2 as a rinse, and 275 mL of ethyl ether is added over 15 min producing a heavy crystalline mass. After stirring for 1 h, the solids are collected by filtration, washed with 250 mL of 1:4 methylene chloride:ethyl ether and dried at 50 °C under vacuum for 1 h to afford 38.1 g (62.9 mmol, 78%) of *O*-allyl-*N*-(9-anthracenylmethyl)cinchonidinium bromide as a crystalline yellow solid (Note 8).

C. Enantioselective Reaction of N-(Diphenylmethylene)glycine tert-Butyl Ester with Methyl Acrylate – (4S)-2-(Benzhydrylidenamino) pentanedioic Acid, 1-tert-Butyl Ester-5-Methyl Ester. A flame-dried, 200-mL, three-necked flask, equipped with a large magnetic stirrer, 50 mL addition funnel, septum and nitrogen inlet, is charged with 5.0 g (16.9 mmol) of *N*-(diphenylmethylene)glycine *tert*-butyl ester (Note 9), 0.95 g (1.7 mmol) of *O*-allyl-*N*-(9-anthracenylmethyl)cinchonidinium bromide and 40 mL of anhydrous CH_2Cl_2. The mixture is cooled in a dry ice-acetone bath to −78 °C and 29 g (170 mmol) of cesium hydroxide monohydrate ($CsOH{\cdot}H_2O$) is added to the well-stirred mixture (Note 10). After stirring for 5 min, a solution of 5 mL of methyl acrylate (Note 11) in 10 mL of CH_2Cl_2 is added dropwise over 10 min through the addition funnel. The mixture is stirred vigorously for 3 h, at which point TLC analysis (Note 12) indicates complete reaction. The mixture is diluted with 150 mL of ethyl ether and 50 mL of water and the cooling bath is removed. The mixture is stirred for 5 min, transferred to a separatory funnel, and 650 mL of ethyl ether is added. The organic phase is washed twice with 200 mL of water and once with 50 mL of brine, dried over magnesium sulfate, filtered, and concentrated on a rotary evaporator. The residue is loaded onto a 6-cm column wet-packed with 80 g of silica gel and hexane. Elution with 1 L of 20:1 hexane:ethyl acetate followed by 1 L of 5:1 hexane:ethyl acetate affords 5.8 g of product as a colorless syrup (90% yield, 95.2% ee for the *S*-enantiomer, (Note 13). Recovery of the catalyst as the chloride salt is accomplished by extraction of the aqueous washes with two 50-mL portions of CH_2Cl_2, drying over Na_2SO_4, filtration, and concentration. Trituration of the residue with ethyl ether affords 0.89 g (94% recovery) of the catalyst as a light yellow solid.

2. Notes

1. The cinchona alkaloids and their quaternary salts are photosensitive and should be stored in brown bottles.

2. (−)-Cinchonidine (96%) was purchased from the Aldrich Chemical Co., Inc. and used as received. It typically contains ~5% of the dihydro analog which behaves like cinchonidine throughout this procedure.

3. 9-Chloromethylanthracene (98+%) was purchased from the Aldrich Chemical Co., Inc. and used as received.

4. The toluene was stored over 4 Å molecular sieves prior to use.

5. The product is isolated as a toluene solvate (mp 154-156 °C) and is sufficiently pure for use in the following step. A reference sample of the solvent-free salt can be prepared by dissolving the toluene solvate in CH_2Cl_2 (1 g/5 mL) followed by filtration, washing and drying of the resulting crystals of *N*-(9-anthracenylmethyl)cinchonidinium chloride: $[\alpha]_D^{23}$ −363 (c 0.81, $CHCl_3$); mp 168-171 °C (dec.); FTIR (film) cm^{-1}: 3500-2500, 1625, 1589, 1571, 1509, 1478, 1462, 1451, 1422, 1265, 1062; ^1H NMR (400 MHz, $CDCl_3$): δ 0.99 (m, 1H), 1.09 (m, 1H), 1.68 (bs, 1H), 1.80 (m, 2H), 2.12 (bs, 1H), 2.40 (app. t, 1H, *J* = 11.1), 2.56 (dd, 1H, *J* = 10.7, 12.8), 4.07 (bd, 1H, *J* = 12.9), 4.71 (m, 2H), 4.89 (dd, 1H, *J* = 1.3, 10.5), 5.25 (dd, 1H, *J* = 1.0, 17.3), 5.42 (m, 1H), 6.67 (d, 1H, *J* = 13.6), 6.83 (d, 1H, *J* = 13.5), 7.24 (m, 4H), 7.20 (m, 2H), 7.38 (m, 1H), 7.56 (d, 1H, *J* = 8.2), 7.60 (m, 1 H), 7.63 (d, 1H, *J* = 8.2), 7.96 (s, 1H), 8.02 (d, 1H, *J* = 4.4), 8.21 (d, 1H, *J* = 5.2), 8.71 (d, 1H, *J* = 8.2), 8.85 (m, 2H), 9.06 (d, 1H, *J* = 9.0); ^{13}C NMR (100 MHz, $CDCl_3$): δ 23.4, 25.7, 25.9, 38.5, 50.4, 54.8, 61.3, 67.0, 67.3, 117.7, 118.3, 120.1, 124.1, 124.2, 124.7, 124.8, 125.6, 126.3, 126.9, 127.4, 127.6,

128.3, 128.5, 128.6, 129.2, 130.2, 130.4, 131.1, 132.7, 133.2, 136.4, 145.7, 147.1, 149.4; FABMS: 485 [M-Cl]$^-$; HRMS calcd for [$C_{34}H_{33}N_2OCl$-Cl]$^-$: 485.2593, found: 485.2575. Anal. Calcd for $C_{34}H_{33}ClN_2O$: C, 78.37; H, 6.38; Cl, 6.80; N, 5.38. Found: C, 78.04; H, 6.42; Cl, 6.81; N, 5.50.

6. Allyl bromide (99%) was purchased from the Aldrich Chemical Co., Inc. and used as received. Excess reagent is used, as hydrolysis of the allyl bromide is competitive with the *O*-alkylation of *N*-(9-anthracenylmethyl)cinchonidinium chloride.

7. The 50% (w/w) aqueous potassium hydroxide (KOH) solution is prepared immediately before use in a neoprene bottle. *Caution: Cooling with a water bath is necessary since dissolution of KOH is very exothermic.*

8. *O*-Allyl-*N*-(9-anthracenylmethyl)cinchonidinium bromide has the following properties: $[\alpha]_D^{23}$ −320 (c 0.45, $CHCl_3$); mp 194-197 °C; FTIR (film) cm^{-1}: 3504, 3082, 2950, 2907, 2884, 1646, 1641, 1625, 1588, 1509, 1450, 1067, 996; ^1H NMR (400 MHz, CD_3OD); δ 1.60 (m, 2H), 1.96 (d, 1H, *J* = 2.9), 2.17 (m, 1H), 2.48 (m, 2H), 2.61 (m, 2H), 7.79-7.77 (m, 2H), 7.95-7.92 (m, 3H), 8.25-81.9 (m, 3H), 8.44 (d, 1H, *J* = 9.0), 8.57 (m, 1H), 8.76 (d, 1H, *J* = 9.0), 8.89 (s, 1H), 9.02 (d, 1H, *J* = 4.6); ^{13}C NMR (100 MHz, CD_3OD): δ 23.4, 26.2, 27.3, 39.5, 49.9, 53.6, 57.4, 63.4, 69.9, 71.4, 117.8, 119.0, 121.8, 125.7 (2C), 126.5, 126.6, 127.1, 129.2, 129.5, 130.5, 131.1, 131.3, 131.5, 133.0, 133.1, 133.8, 133.9, 134.6, 134.7, 134.8, 138.6, 143.0, 149.3, 151.1; FABMS: 525 [M-Br]$^-$; HRMS calcd for [$C_{37}H_{37}N_2OBr$-Br]$^-$: 525.2906, found: 525.2930. Anal. Calcd for $C_{37}H_{37}BrN_2O$: C, 73.38; H, 6.16; Br, 13.19; N, 4.63. Found: C, 73.40; H, 6.12; Br, 13.19; N 4.47.

9. *N*-(Diphenylmethylene)glycine *tert*-butyl ester was purchased from the Aldrich Chemical Co., Inc. and used as received.

10. Cesium hydroxide monohydrate was purchased from the Aldrich Chemical Co., Inc. and used as received.

11. Methyl acrylate was purchased from the Aldrich Chemical Co., Inc. and used as received.

12. The TLC analysis was performed using silica gel plates, 10% ethyl acetate in hexane and UV detection.

13. Characterization data for the Michael adduct: $[\alpha]_D^{23}$ −100 (c 1.35, CH_2Cl_2); mp 194-196 °C; FTIR (film) cm^{-1}: 3061, 3056, 2977, 2950, 2933, 1735, 1624, 1446, 1368, 1316, 1277, 1254, 1195, 1150; ^1H NMR (400 MHz, $CDCl_3$): δ 1.43 (s, 9H), 2.24-2.18 (m, 2H), 2.39-2.35 (m, 2H), 3.59 (s, 3H), 3.95 (dd, 1H, *J* = 7.2, 5.6), 7.18-7.16 (m, 2H), 7.45-7.25 (m, 6H), 7.63 (d,

2H, J = 7.2); ^{13}C NMR (125 MHz, $CDCl_3$): δ 28.3, 28.8, 30.7, 51.8, 65.0, 81.5, 128.1, 128.3, 128.7, 128.9, 129.1, 130.6, 136.6, 139.6, 170.5, 170.9, 173.8; CIMS: 382 $[M+H]^+$, 280, 134; HRMS calcd for $[C_{23}H_{27}NO_4+H]^+$: 382.2018, found: 382.2017. The enantioselectivity was determined by chiral HPLC analysis (Regis Whelk-O1 column, 20% 2-propanol-hexane, 0.5 mL/min, λ = 254 nm, retention times: R (minor): 16.1 min, S (major): 19.1 min).

Waste Disposal Information

All toxic materials were disposed of in accordance with "Prudent Practices in the Laboratory"; National Academic Press; Washington, DC 1998.

3. Discussion

O-Allyl-*N*-(9-anthracenylmethyl)cinchonidinium bromide can be used as a versatile phase transfer catalyst for highly enantioselective alkylation and Michael addition reactions of the benzophenone Schiff base of glycine tert-butyl ester.[3,4,5] The reaction is performed with 10 mol% of the cinchona alkaloid catalyst, 1.5 to 5 equivalents of the electrophile and 10 equivalents of $CsOH{\cdot}H_2O$ in methylene chloride at −60 to −78 °C.[3] The reaction may be performed using 50% aqueous KOH (w/v) at 0 °C instead of $CsOH{\cdot}H_2O$ with a slight reduction in enantioselectivity. The catalyst itself is prepared from inexpensive, readily available starting materials in two steps without chromatography. It is stable at room temperature for many months and can be recovered from the asymmetric alkylation reaction and reused. The alkylation reaction can be performed with a variety of electrophiles, including aliphatic, allylic and benzylic halides as well as cyclic and acyclic α,β-unsaturated esters and ketones, affording convenient access to orthogonally protected, enantiomerically pure amino acid derivatives useful as chiral building blocks in asymmetric synthesis. The use of the 9-anthracenylmethyl group on the cationic nitrogen rigidifies the catalyst and favors reaction through a highly structured contact ion pair.[3,4,5] Similar reactions using previously prepared catalysts are less enantioselective.[6,7,8]

1. Harvard University, Department of Chemistry and Chemical Biology, 12 Oxford Street, Cambridge, MA 02138.
2. Pfizer Inc., Central Research, Eastern Point Road, Groton, CT 06340.
3. Corey, E. J.; Xu, F.; Noe, M. C. *J. Am. Chem. Soc.* **1997**, *119*, 12414.
4. Corey, E. J.; Bo, Y.; Busch-Petersen, J. *J. Am. Chem. Soc.* **1998**,*120*, 13000.
5. Corey, E. J.; Noe, M. C.; Xu, F. *Tetrahedron Lett.* **1998**, *39*, 5347.
6. O'Donnell, M. J.; Bennett, W. D.; Wu, S. *J. Am. Chem. Soc.* **1989**, *111*, 2353.
7. O'Donnell, M. J.; Wu, S.; Huffman, J. C. *Tetrahedron* **1994**, *50*, 4507.
8. Lipkowitz, K. B.; Cavanaugh, M. W.; Baker, B.; O'Donnell, M. J. *J. Org. Chem.* **1991**, *56*, 5181.

RUTHENIUM-CATALYZED ALKYLATION OF AROMATIC KETONES WITH OLEFINS: 8-[2-(TRIETHOXYSILYL)ETHYL]-1-TETRALONE

[(1(2*H*)-Naphthalenone, 3,4-dihydro-8-[2-(triethoxysilyl)ethyl]-)]

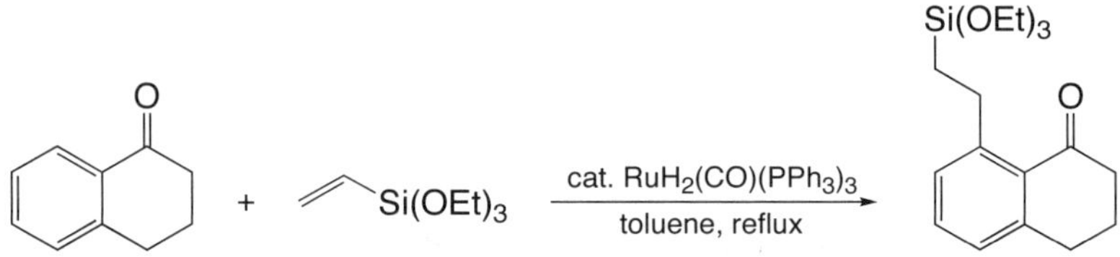

Submitted by Fumitoshi Kakiuchi and Shinji Murai.[1]
Checked by Dennis P. Curran and Andre Lapierre.

1. Procedure

8-[2-(Triethoxysilyl)ethyl]-1-tetralone (Note 1). An apparatus, consisting of a 100-mL, two-necked, round-bottomed flask, reflux condenser connected to a vacuum/N_2 line, inlet tube sealed with a rubber septum, and magnetic stirring bar, is evacuated, then flushed with nitrogen. This cycle is repeated four times. The apparatus is flame-dried under a flow of nitrogen, then cooled to room temperature under a nitrogen atmosphere. Carbonyldihydridotris(triphenylphosphine)ruthenium(II), $RuH_2(CO)(PPh_3)_3$, (0.918 g, 1.00 mmol, 0.010 eq) (Notes 2, 3) is placed in the flask under a slow flow of nitrogen. Addition of 20 mL of toluene (Note 4) to the flask affords a suspension of white solids (Note 5). To this suspension are added triethoxyvinylsilane (20.93 g, 110 mmol, 1.10 eq) (Notes 6, 7), then 1-tetralone (14.62 g, 100 mmol, 1.00 eq) (Notes 6, 8) at room temperature through the rubber septum using a syringe. The resulting mixture containing the white solids is heated to reflux in an oil bath (oil bath temperature: 135 °C) (Notes 9, 10). The reaction mixture becomes colorless within 1 min, then changes to a dark wine red color within 5 min (Note 11). After heating for 30 min, the reaction mixture is cooled to room temperature. About half of the reaction mixture is transferred to a 50-mL, round-bottomed flask (Note 12) for concentration and distillation. Volatile materials (toluene and triethoxyvinylsilane) are removed by rotary evaporation under reduced pressure (2 mm) at 40 °C. After almost all volatile materials are removed, the other half of the reaction mixture is transferred to the flask. The reaction vessel is rinsed with two 5-mL portions of toluene, then the rinses are transferred to the distillation flask. The combined reaction mixture and rinses are concentrated. Solvent and vinylsilane are removed by rotary evaporation under reduced pressure (2 mmHg at 40 °C) (Note 13). Distillation of the residue under reduced pressure gives 31-32.5 g (92-96%) of 8-[2-(triethoxysilyl)ethyl]-1-tetralone as a pale yellow liquid, bp 133-135 °C/0.2 mmHg (Notes 14-18).

2. Notes

1. This procedure is a modification of one published by our group.[2,3]

2. The ruthenium complex, carbonyldihydridotris (triphenylphosphine)ruthenium(II), can be synthesized according to the literature method.[3,4] This complex is commercially available from Aldrich Chemical Company, Inc. and Strem Chemicals, Inc.

3. The commercially available ruthenium complex (Strem Chemicals, Inc.) and the complex prepared in the submitters' laboratory showed comparable catalytic activity. The checkers purchased the complex from Strem Chemicals, Inc.

4. Toluene was dried over CaH_2, then distilled under a N_2 atmosphere.

5. The ruthenium complex is only slightly soluble in toluene at room temperature.

6. Triethoxyvinylsilane, 1-tetralone, and toluene were purchased from commercial suppliers (Aldrich Chemical Company, Inc., Strem Chemicals, Inc., Chisso Co., or Wako Pure Chemical Industries, Ltd.).

7. The silane was dried over CaH_2, then distilled under reduced pressure (bp = 70 °C/6 mm).

8. 1-Tetralone was dried over $CaSO_4$, then distilled under reduced pressure (bp 85 °C/2 mm).

9. Maintaining the oil bath temperature above 130 °C (preferentially around 135 °C) is necessary to effect the catalytic reaction with good reproducibility.

10. The actual temperature of the reaction mixture was around 125 °C throughout the reaction.

11. The color of the reaction mixture changes to dark green, wine red, and finally dark wine red within 5 min.

12. The use of a small flask is recommended to minimize loss of the product.

13. The checkers transferred the entire reaction mixture to a 100-mL, round-bottomed flask with the aid of 2 × 5-mL washes with toluene. A distillation apparatus with a three-way adapter and three collection vessels was attached, and the solvent was removed by gentle heating at 3 mmHg. The residue was then distilled as described.

14. The second fraction is collected. The first fraction (0.20-0.28 g; 65-100 °C/0.2 mmHg) containing unknown impurities and a small amount of the desired product is discarded.

15. The residue is prone to bump during the late stages of the distillation. Wine red splashes on the wall of the distillation head results in contamination of the product with impurities, including triphenylphosphine, triphenylphosphine oxide, and the ruthenium complex. If the product is contaminated with these impurities, it can be redistilled.

16. Heating the distillation head with a heat gun is recommended to avoid reducing the yield of the product.

17. The spectral properties are as follows: ^1H NMR (300 MHz, $CDCl_3$) δ: 0.94-1.00 (m, 2 H, $SiCH_2$), 1.26 (t, 9 H, J = 6.97, CH_3), 2.05 (quintet, 2 H, J = 6.00, $CH_2\underline{CH_2}CH_2$), 2.61 (t, 2 H, J = 6.33, $ArCH_2$), 2.90 (t, 2 H, J = 5.66, $C(O)CH_2$), 3.08-3.14 (m, 2 H, $Ar\underline{CH_2}CH_2Si$), 3.87 (q, 6 H, J = 5.58, OCH_2), 7.05 (d, 1 H, J = 7.17, ArH), 7.11 (d, 1 H, J = 7.47, ArH), 7.26-7.31 (dd, 1 H, J = 7.56, ArH); ^{13}C NMR (75 Hz, $CDCl_3$) δ: 12.18 ($SiCH_2$), 18.21(CH_3), 22.81 (CH_2), 28.57, 30.82 [($SiCCH_2$) and ($ArCH_2$)], 40.96 ($C(O)CH_2$), 58.18 (OCH_2), 126.73, 129.15, 130.31, 132.33, 145.68, 147.95 (Ar), 199.60 (C=O); MS (m/z): 336 (M^+), 290 (M^+ - EtOH), 189, 173, 163, 135, 115, 79, 63; IR (neat) cm^{-1}: 1680 (C=O). Anal. Calcd for $C_{18}H_{28}O_4Si$: C, 64.25; H, 8.39. Found: C, 64.24; H, 8.29.

18. The checkers obtained bp 160-168 °C/2 mmHg.

Waste Disposal Information

All toxic materials were disposed of in accordance with "Prudent Practices in the Laboratory"; National Academy Press; Washington, DC, 1995.

3. Discussion

The procedure described here is typical for the catalytic alkylation of aromatic ketones at the ortho position by alkenes. Aromatic ketones are readily available by Friedel-Crafts acylation and many other methods,[5] and many of these ketones are suitable substrates for the present catalytic alkylation with alkenes affording the corresponding ortho-alkylated ketones.[3,6] The present method provides a direct way to alkylate aromatics with olefins. Moreover, the C-C bond formation takes place with exclusive ortho selectivity, while mixtures of o-, m-, p-isomers are usually obtained in the conventional Friedel-Crafts alkylation of aromatic compounds.

In the present ruthenium-catalyzed reaction, a C-C bond is formed directly from a C-H bond without prior conversion of the C-H bond to another functional group such as halogen. The preparation of 8-[2-(triethoxysilyl)ethyl]-1-tetralone described above can be modified for a variety of substituted aromatic ketones. Some additional, representative examples of the $RuH_2(CO)(PPh_3)_3$-catalyzed addition of C-H bonds in aromatic ketones to olefins are shown in the Table.[3,6] Naphthalene derivatives and heteroaromatic ketones can be employed in the present coupling reaction. Functional group compatibility of this reaction is broad and both electron-donating and electron-withdrawing substituents are tolerated. In many cases, simple bulb-to-bulb distillation of the reaction mixture gives an analytically pure product. This simple one-pot procedure provides a new opportunity for site-selective alkylations of aromatic and heteroaromatic compounds.

TABLE. RUTHENIUM-CATALYZED REACTION OF AROMATIC KETONES WITH OLEFINS

Entry	Ketone	Product	Time	Isolated Yield	GLC Yield
1 [a,b]	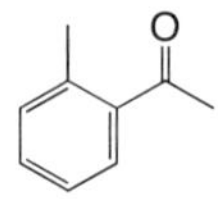	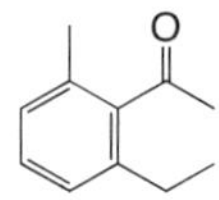	24 h	88%	100%
2 [a,c]	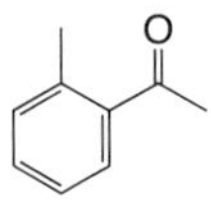	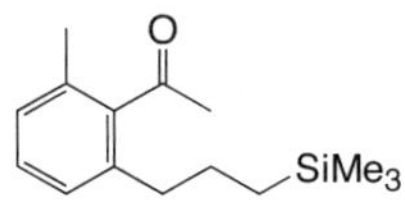	4 h	97%	100%
3 [a,d]	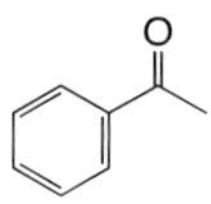	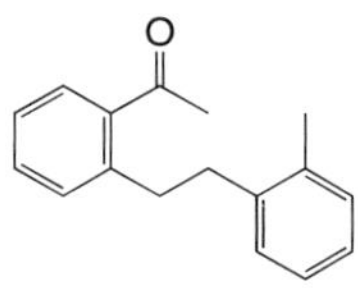	4 h	89%	—
4 [a,e]	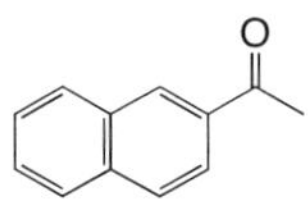	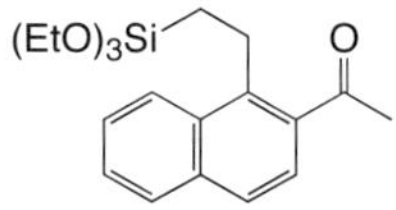	6 h	88%	100%
5 [a,e]	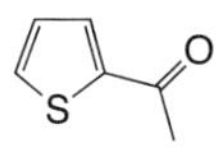	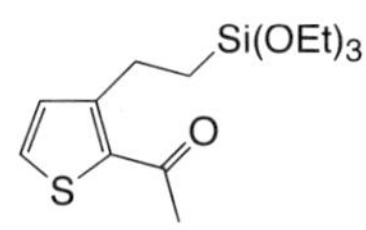	1 h	93%	100%
6 [e,f]	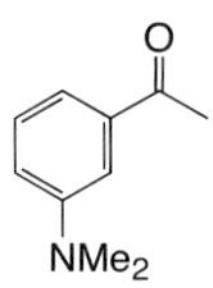	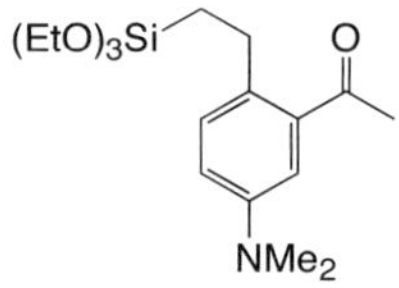	8 h	72%	85%
7 [e,f]	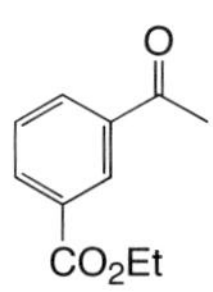	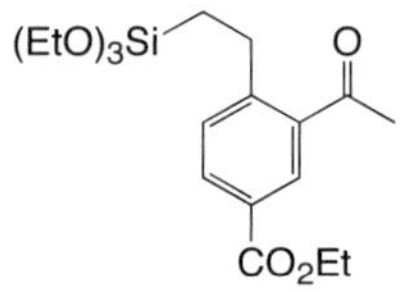	4 h	91%	91%
8 [e,f]	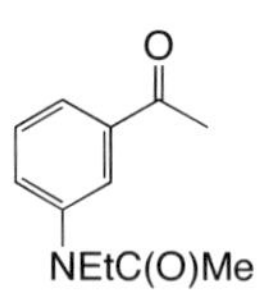	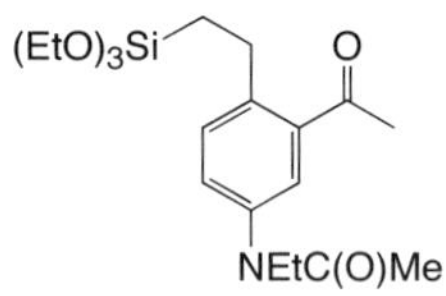	8 h	92%	96%
9 [e,f]	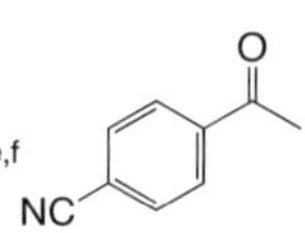	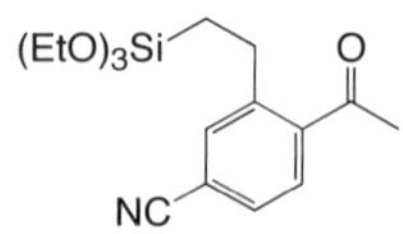	5 h	73%	88%

[a]See ref 3. [b]Ethylene was used. [c]Allyltrimethylsilane was used. [d]2-Methylstyrene was used. [e]Triethoxyvinylsilane was used. [f]See ref 6.

1. Department of Applied Chemistry, Faculty of Engineering, Osaka University, Suita, Osaka 565-0871, Japan.

2. Murai, S.; Kakiuchi, F.; Sekine, S.; Tanaka, Y.; Kamatani, A.; Sonoda, M.; Chatani, N. *Nature* **1993**, *366*, 529.

3. Kakiuchi, F.; Sekine, S.; Tanaka, Y.; Kamatani, A.; Sonoda, M.; Chatani, N.; Murai, S. *Bull. Chem. Soc. Jpn.* **1995**, *68*, 62.

4. Ahmad, N.; Levison, J. J.; Robinson, S. D.; Uttley, M. F. *Inorg. Synth.* **1974**, *15*, 45.

5. Walter, D. S. In "Comprehensive Organic Functional Group Transformations"; Katritzky, A. R.; Meth-Cohn, O.; Rees, C. W., Eds.; Pergamon: Cambridge, U.K., 1995; Vol. 3, Chapter 3.06.

6. (a) Sonoda, M.; Kakiuchi, F.; Chatani, N.; Murai, S. *Bull. Chem. Soc. Jpn.* **1997**, *70*, 3117. (b) Kakiuchi, F.; Murai, S. *Acc. Chem. Res.* **2002**, *35*, 826.

DIETHYLAMINOTRIMETHYLSILANE-CATALYZED 1,4-ADDITION OF ALDEHYDES TO VINYL KETONES: (3*R*)-3,7-DIMETHYL-2-(3-OXOBUTYL)-6-OCTENAL [(6-Octenal, 3,7-dimethyl-2-(3-oxobutyl)-, (3*R*)-)]

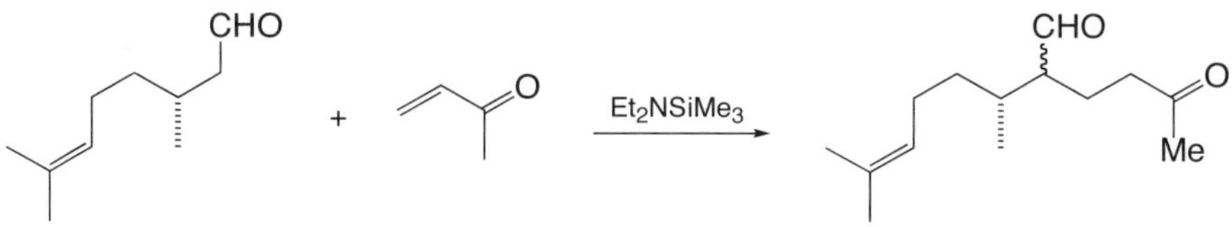

Submitted by Hisahiro Hagiwara, Hiroki Ono, and Takashi Hoshi.[1]
Checked by Kwame Nti-Addae and Steven Wolff.

1. Procedure

(3R)-3,7-Dimethyl-2-(3-oxobutyl)-6-octenal. To a two-necked, 300-mL flask, equipped with a magnetic stirrer, addition funnel, and condenser with a nitrogen balloon on the top, and immersed in an ice-cooled bath is added 18.1 mL of citronellal (100 mmol) (Note 1), 12.5 mL of 3-buten-2-one (150 mmol) (Note 2), 1.9 mL of diethylaminotrimethylsilane (10 mmol) (Notes 3, 4) and 400 mL of acetonitrile (Note 5). Following the addition, the resulting homogeneous solution is heated at reflux in an oil bath and the progress of the reaction is monitored by TLC analysis of the product using Merck silica gel 60 F_{254} plates: $R_f = 0.39$ (5:1 hexane:ethyl acetate). After 46 h, when the citronellal has been consumed, the resulting solution is concentrated to dryness under reduced pressure and the residue is distilled using a Kugelrohr apparatus twice (oven temperature 120~140 °C at 1.9 mm) to give (3*R*)-3,7-dimethyl-2-(3-oxobutyl)-6-octenal (21.6 g, 96%) (Notes 6-8) as a pale yellow liquid.

2. Notes

1. *R*-(+)-Citronellal was obtained from Tokyo Chemical Industry Co. Ltd. and used as received.
2. 3-Buten-2-one was obtained from Merck & Co., Inc. and used as received.
3. Diethylaminotrimethylsilane was obtained from Shin-Etsu Chemical Co., Ltd. and used as received.

4. Diethylaminotrimethylsilane can be prepared according to the procedure by Middleton, W. J.; Bingham, E. M. *Org. Synth., Coll. Vol. VI,* **1988**, 440.

5. Acetonitrile (Kanto Chemical Co., Ltd. special grade) was used as received.

6. The purity of the product was determined to be 100% by medium pressure LC (silica gel packed column, eluent: 5:1 n-hexane:ethyl acetate).

7. The physical properties of (3*R*)-3,7-dimethyl-2-(3-oxobutyl)-6-octenal, a 1:1 mixture of diastereoisomers, are as follows: IR (thin film) cm^{-1}: 2930, 2709, 1722, 1450, 1376, 1240 and 1164; ^{1}H NMR ($CDCl_3$, 200 MHz) δ: 0.89 (d, 1.5 H, J = 6.9), 0.99 (d, 1.5 H, J = 6.9), 1.60 (s, 3 H), 1.69 (s, 3 H), 1.08-2.10 (m, 7 H), 2.13 (s, 3 H), 2.26-2.65 (m, 3 H), 5.08 (m, 1 H), 9.60 (d, 0.5 H, J = 2.4) and 9.64 (0.5 H, J = 2.9); ^{13}C (50 MHz) δ: 15.9, 16.7, 17.6, 18.0, 19.7, 25.1, 25.6, 29.9, 32.2, 33.3, 33.8, 34.4, 41.36, 41.42, 55.9, 56.2, 123.7, 131.9, 204.9, 205.2, 207.9 and 208.0; m/z 224 (M^+, 0.1%), 148 (32), 109 (26), 95 (38), 82 (32), 71 (25), 69 (52), 58 (28), 55 (37), 43 (100) and 41 (68); Calcd for $C_{14}H_{24}O_2$; M^+, 224.1776. Found: M^+, 224.1798.

8. When 0.5 eq of Et_2NTMS was used, the reaction is complete in 19 h in 100% yield as shown in the Table, entry 8.

Waste Disposal Information

All toxic materials were disposed of in accordance with "Prudent Practices in the Laboratory"; National Academy Press; Washington, DC, 1995.

3. Discussion

This is a mild, simple and practical procedure for 1,4-addition of an aldehyde to methyl vinyl ketone,[2] without converting the aldehyde into an enamine or a silyl enol ether. The products, substituted 5-ketoaldehydes, are important compounds, especially for the preparation of substituted 2-cyclohexen-1-one derivatives, which have been versatile starting materials for syntheses of natural products such as terpenoids.[3] These 5-ketoaldehydes have been prepared previously by the 1,4-addition of modified aldehydes, i.e., morpholinoenamines of aldehydes,[4,5] trimethylsilyl enol ethers of aldehydes in the presence of a Lewis acid,[6] or diethylallylamine in the presence of a catalytic amount of a Ru complex,[7] to methyl vinyl ketones.

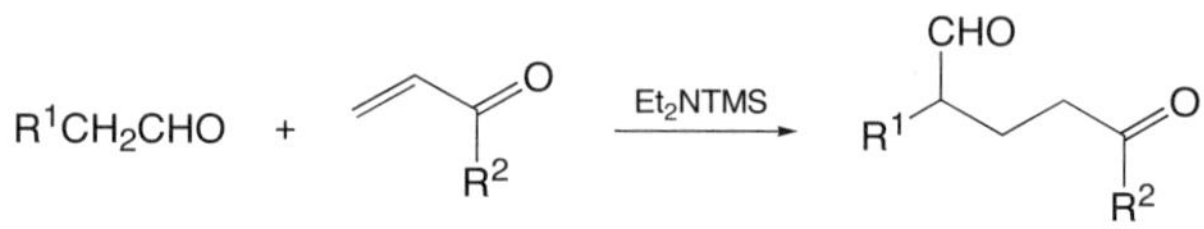

Table 1,4-Addition of aldehydes with vinyl ketones

Entry	Aldehyde (R^1)	Vinyl ketone (R^2)	Time (hr)	Product	Isolated yield (%
1	C_8H_{17}	Me	20	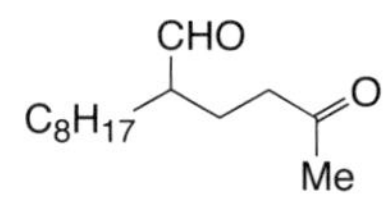	78
2	i-Pr	Me	9	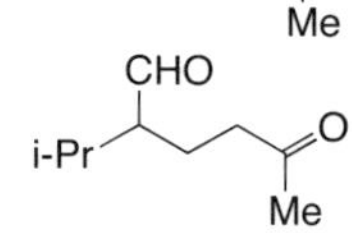	64
3	$PhCH_2$	Me	6	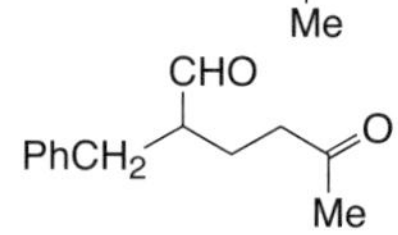	59
4	$THPOC_3H_6$	Me	6	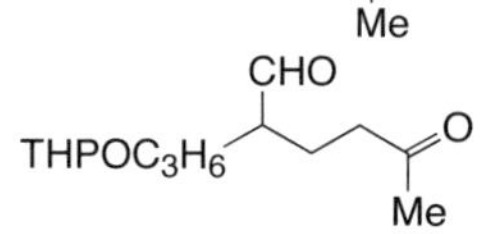	89
5	$AcOC_8H_{16}$	Me	9	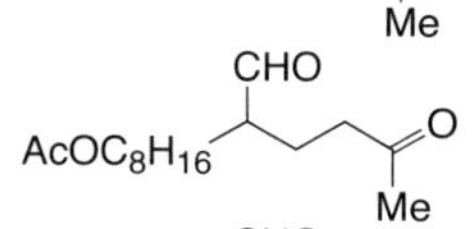	76
6	C_8H_{17}	Et	10.5	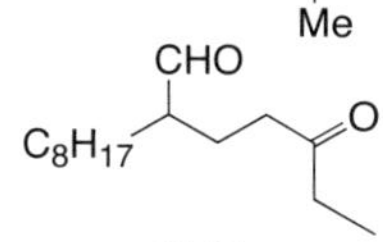	76
7	$PhCH_2$	Et	4	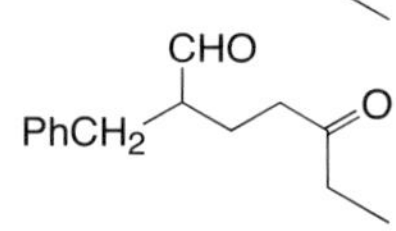	68
8[a]	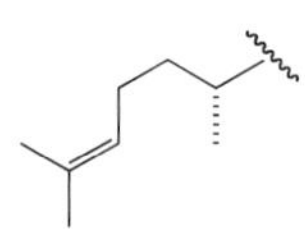	Me	19	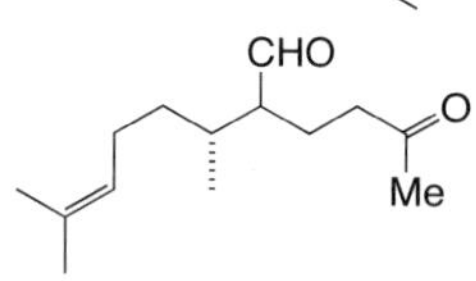	100

[a]Et_2NTMS (0.5 equiv.) was used.

The reaction is carried out simply by refluxing an acetonitrile solution of the aldehyde, vinyl ketone and 10 mol% of diethylaminotrimethylsilane until disappearance of the aldehyde (monitored by TLC analysis). Bulb-to-bulb distillation provides the 5-ketoaldehydes without aqueous work up. When the amount of diethylaminotrimethylsilane is increased, the reaction proceeds faster (Table, entry 8), although the resulting reaction mixture is not colorless. Some representative examples of the present reaction are shown in the Table. The mildness of the reaction is well exemplified by obtaining satisfactory yields in the reactions with acid- or base-sensitive aldehydes (having a THP or acetyl protecting group) (Table, entries 4 and 5).

The reaction also can be carried out without solvent, although the yields using isobutyraldehyde and citronellal were relatively low.

1. Graduate School of Science and Technology, Niigata University, 8050, 2-Nocho, Ikarashi, Nishi-ku, Niigata 950-2181, Japan.
2. Hagiwara, H.; Kato, M. *Tetrahedron Lett.* **1996**, *37*, 5139; Hagiwara, H.; Komatsubara, N.; Ono, H.; Okabe, T.; Hoshi, T.; Suzuki, T.; Ando, M.; Kato, M. *J. Chem. Soc., Perkin Trans. 1* **2001**, 316.
3. Hagiwara, H.; Okabe, T.; Ono, H.; Kamat, V. P.; Hoshi, T.; Suzuki, T.; Ando, M. *J. Chem. Soc., Perkin Trans. 1* **2002**, 895.
4. Stork, G.; Brizzolara, A.; Landesman, H.; Szmuszkovicz, J.; Terrell, R. *J. Am. Chem. Soc.* **1963**, *85*, 207.
5. Brown, M. *J. Org. Chem.* **1968**, *33*, 162.
6. Duhamel, P.; Hennequin, L.; Poirier, J. M.; Tavel, G.; Vottero, C. *Tetrahedron* **1986**, *42*, 4777.
7. Nakajima, T.; Maruyama, Y.; Shimizu, I. In "Abstracts of Papers of 65th Annual Meeting of Chemical Society of Japan," Tokyo, **1993**; Vol. II, p. 335.

FRAGMENTATION-RECOMBINATION NAZAROV CYCLIZATION: 3,4-DIMETHYLCYCLOPENT-2-EN-1-ONE

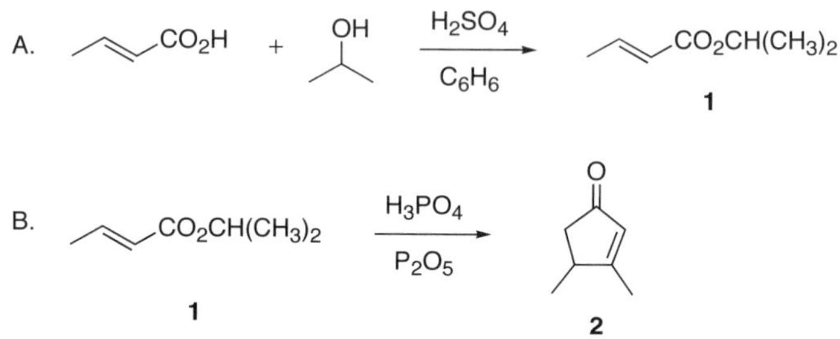

Submitted by Keith D. Schwartz and James D. White.[1]
Checked by Shin-ya Tosaki and Masakatsu Shibasaki.

1. Procedure

A. Isopropyl (E)-but-2-enoate (**1**). A 500-mL round-bottomed flask is fitted with a Soxhlet extractor containing magnesium sulfate (10 g) in a cellulose extraction thimble. Atop the extractor is a reflux condenser fitted with a rubber septum and an argon line. The flask is charged with crotonic acid (17.00 g, 197.5 mmol), isopropanol (106 mL, 1.39 mol), concentrated sulfuric acid (2.5 mL) and benzene (30 mL) (Note 1). The mixture is heated to reflux over the $MgSO_4$ plug for 18 h and is then cooled to room temperature. Benzene (60 mL) is added, followed by the dropwise addition of a 10% aqueous solution of sodium bicarbonate (100 mL) while stirring. The mixture is transferred to a separatory funnel, the layers are separated, and the aqueous portion is extracted with benzene (3 x 25 mL). The organic layers are combined, washed with brine (50 mL), dried over anhydrous sodium sulfate, and filtered. The solvent is evaporated under vacuum and the resulting yellow oil is purified by distillation (46–48 °C/25 mmHg) to give 17.98–18.00 g (71%) of isopropyl (*E*)-but-2-enoate (**1**) (Note 2) as a colorless oil.

B. 3,4-Dimethylcyclopent-2-enone (**2**). A three-necked, 1-L round-bottomed flask is fitted with a mechanical stirrer, a glass stopper, a rubber septum, and an argon line. The flask is charged with phosphoric acid (32.0 g) and phosphorus pentoxide (48.0 g) (Note 1). The mixture is stirred at 100 °C for 1 h, at which time it becomes homogenous. To this mixture is added the neat ester **1** (10.0 g, 78.0 mmol), and the mixture is stirred for 3 min at 100 °C. The mixture is cooled to 0 °C with an ice bath. Diethyl ether

(Et_2O, 100 mL) is added followed by the slow addition of saturated aqueous $NaHCO_3$ (150 mL) with vigorous stirring. Solid $NaHCO_3$ is then cautiously added to the mixture in small portions until foaming subsides. The contents of the flask are transferred to a separatory funnel with an additional quantity (50 mL) of Et_2O, the layers are separated, and the aqueous layer is extracted with Et_2O (3 x 60 mL). The organic layers are combined, washed sequentially with water (50 mL) and brine (50 mL), dried over anhydrous sodium sulfate, and filtered. The solvent is evaporated under vacuum and the resulting orange colored oil is purified by fractional distillation (68–69 °C/20 mmHg) to give 5.61–5.63 g (65–66 %) of 3,4-dimethylcyclopent-2-enone (**2**) (Note 3) as a colorless oil.

2. Notes

1. Crotonic acid was purchased from TCI and was used as received. Isopropanol was purchased from KANTO CHEMICAL and was used as received. Phosphorus pentoxide powder, phosphoric acid (85 wt. % solution in water), concentrated sulfuric acid, and benzene were purchased from Wako and were used as received. Diethyl ether was purchased from Showa-ether and was used as received.

2. Physical data for isopropyl (*E*)-but-2-enoate: IR (neat, cm^{-1}) 2981, 2940, 1718, 1661; ^{1}H NMR (500 MHz, $CDCl_3$) δ: 1.25 (d, J = 6.4 Hz, 6 H), 1.86 (dd, J = 7.0, 1.8 Hz, 3 H), 5.05 (septet, J = 6.4 Hz, 1 H), 5.81 (dq, J = 15.6, 1.8 Hz, 1 H), 6.95 (dq, J = 15.6, 7.0 Hz 1 H); ^{13}C NMR (125 MHz, $CDCl_3$) δ: 17.9, 21.9, 67.3, 123.3, 144.1, 166.1. Purity was established by gas chromatographic analysis (Restek Rtx-20 column, flow rate 1 mL/min, temperature 45–280 °C ramped at 60 °C/min, retention time 6.49 min).

3. Physical data for 3,4-dimethylcyclopent-2-enone: IR (neat, cm^{-1}) 2965, 2927, 1712, 1684, 1619; ^{1}H NMR (500 MHz, $CDCl_3$) δ: 1.19 (d, J = 7.3 Hz, 3 H), 2.00 (dd, J = 18.6, 2.1 Hz, 1 H), 2.08 (s, 3 H), 2.64 (dd, J = 18.6, 6.5 Hz, 1 H), 2.81 (ddt, J = 7.3, 6.5, 2.1 Hz, 1 H), 5.88 (s, 1 H); ^{13}C NMR (125 MHz, $CDCl_3$) δ: 17.0, 18.8, 38.8, 44.2, 130.3, 182.6, 208.9. Purity was established by gas chromatographic analysis (Restek Rtx-20 column, flow rate 1 mL/min, temperature 120–250 °C ramped at 50 °C/min, retention time 7.87 min).

Safety and Waste Disposal Information

All hazardous materials should be handled and disposed of in accordance with "Prudent Practices in the Laboratory"; National Academy Press; Washington, DC, 1995.

3. Discussion

The transformation of isopropyl (*E*)-but-2-enoate to 3,4-dimethylcyclopent-2-en-1-one represents an example of a general rearrangement of esters of α,β-unsaturated acids to cyclopentenones catalyzed by polyphosphoric acid at elevated temperature. The reaction was first reported by Conia and Leriverend,[2] and presumably takes place via a fragmentation-recombination-Nazarov cyclization pathway, as shown in Scheme 1. A variety of substituted cyclopent-2-en-1-ones can be prepared by this method; esters of benzoic acid give 1-indanones under the same conditions (see Table 1).

Scheme 1

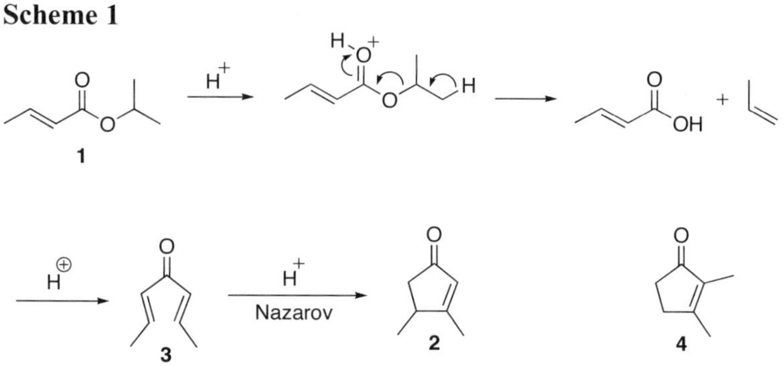

3,4-Dimethylcyclopent-2-en-1-one (**2**) is reported to be formed in 60% yield when hepta-2,5-dien-4-one (**3**) is treated with phosphoric acid and 98% formic acid.[3] However, a later report suggests that the major product from **3** under "Nazarov conditions" (phosphoric acid and formic acid at 90 °C) is 2,3-dimethylcyclopent-2-en-1-one (**4**),[4] a result in accord with an observation made by Nozaki on Nazarov cyclization of nona-3,6-dien-5-one, which gave 2,3-diethylcyclopent-2-en-1-one.[5] Exposure of **3** to fluorosulfonic acid at 0 °C gives **2** in 44% yield.[4]

3,4-Dimethylcyclopent-2-en-1-one reacts with an alkyllithium reagent and *p*-toluenesulfonic acid to give 1,2-dimethylcyclopenta-1,3-diene which, as its lithio or potassio derivative, has been employed in a variety of coupling reactions to produce substituted 1,2-dimethylcyclopentadienes.[6]

Bromination of 3,4-dimethylcyclopent-2-en-1-one with *N*-bromo-succinimide gives 4-bromo-3,4-dimethylcyclopent-2-en-1-one (Scheme 2).[7] This compound, upon treatment with triethylamine, generates unstable 3,4-dimethylcyclopentadienone, which undergoes spontaneous self-Diels-Alder cycloaddition to give *endo* adduct **5**.[8]

Table 1

Conversion of α,β-Unsaturated Esters to Cyclopent-2-en-1-ones under Nazarov Conditions[a]

Ester	Product	Yield (%)
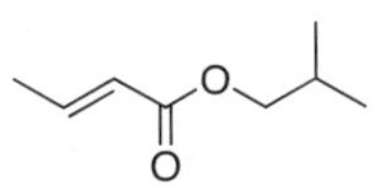	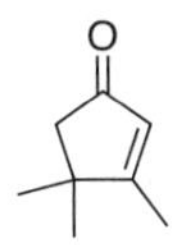	60
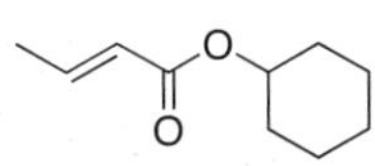	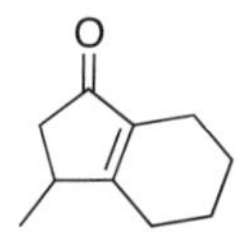	60
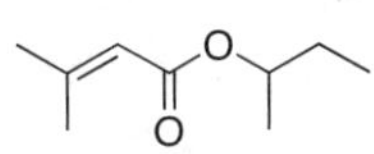	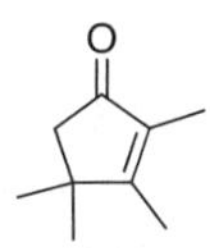	60
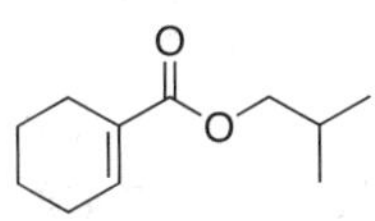	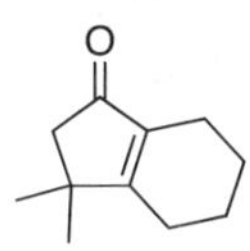	59
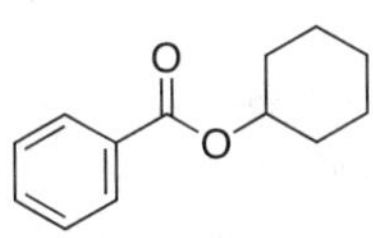	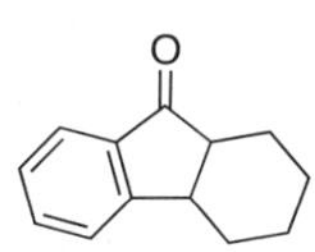	40

a) H_3PO_4, P_2O_5

3,4-Dimethylcyclopent-2-en-1-one reacts with sodium azide in the presence of trifluoroacetic acid to give 5,6-dihydro-4,5-dimethyl-2-pyridone, a formal example of a Schmidt reaction leading to ring expansion.[9]

Scheme 2

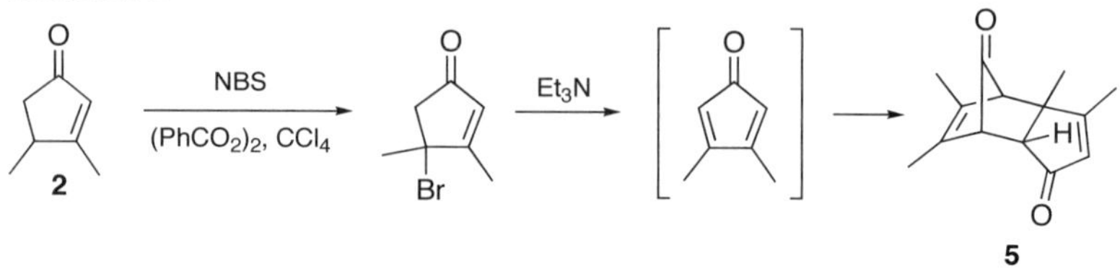

Finally, conjugate addition of lithium divinylcuprate to 3,4-dimethylcyclopent-2-en-1-one in the presence of trimethylsilylchloride is stereoselective and yields 3-alkyl-3,4-dimethylcyclopentanone **6** after hydrolysis of the silyl enol ether (Scheme 3).[10]

Scheme 3

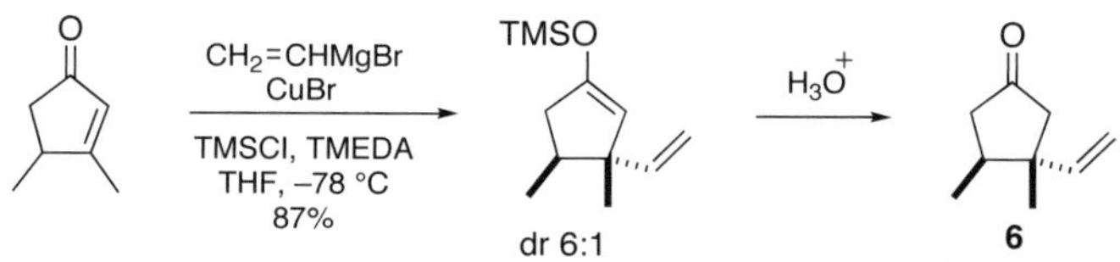

1. Department of Chemistry, Oregon State University, Corvallis, OR 97331, USA
2. Conia, J.-M.; Leriverend, M.-L. *Bull. Soc. Chim. Fr.* **1970**, 2981.
3. Jones, N.; Taylor, H. T. *J. Chem. Soc.* **1961**, 1345.
4. Howell, J. A. S.; O'Leary, P.; Yates, P. C. *Tetrahedron* **1995**, *51*, 7231.
5. Hirano, S.; Tagaki, S.; Hiyama, T.; Nozaki, H. *Bull. Chem. Soc. Jpn.* **1980**, *53*, 169.
6. (a) Bunel, E. E.; Campos, P.; Ruz, J.; Valle, L.; Chadwick, I.; Santa Ana, M.; Gonzalez, G.; Manriquez, J. M. *Organometallics* **1988**, *7*, 474. (b) Bunel, E. E.; Valle, L.; Jones, N. L.; Carroll, P. J.; Gonzalez, M.; Munoz, N.; Manriquez, J. M. *Organometallics* **1988**, *7*, 789. (c) Huttenloch, M. E.; Dielbold, J.; Rief, U.; Brintzinger, H. H.; Gilbert, A. M.; Katz, T. J. *Organometallics* **1992**, *11*, 3600. (d) Juengling, S.; Muelhaupt, R.; Plenio, H. *J. Organometall. Chem.* **1993**, *460*, 191. (e) Rouzaud, J.; Joudat, M.; Castel, A.; Delpech, F.; Riviere, P.; Gomitzka, H.; Manriquez, J. M.; Chavez, I. *J. Organometall. Chem.* **2002**, *651*, 44.
7. Dowd, P.; Schappert, R.; Garner, P. *Tetrahedron Lett.* **1982**, *23*, 3.
8. Dowd, P.; Schappert, R.; Garner, P.; Go, C. L. *J. Org. Chem.* **1985**, *50*, 44.
9. (a) Yasufumi, K.; Kobayashi, K.; Shinya, K.; Tadashi, T.; Masanori, M.; Toshihiko, N.; Katsuya, H.; Atsuko, F.; Keisuke, H.; Masao, H.; Masahuru, K.; Hisao, N.; Masaaki. T. *Bioorg. Med. Chem. Lett.* **2002**, *12*, 2291.
10. Benbow, J.; Katoch, R.; Martinez, B.; Shetzline, S. *Tetrahedron Lett.* **1997**, *38*, 4017.

FACILE SYNTHESES OF AMINOCYCLOPROPANES: *N,N*-DIBENZYL-*N*-(2-ETHENYLCYCLOPROPYL)AMINE [(Benzenemethanamine, *N*-(2-ethenylcyclopropyl)-*N*-(phenylmethyl)-]

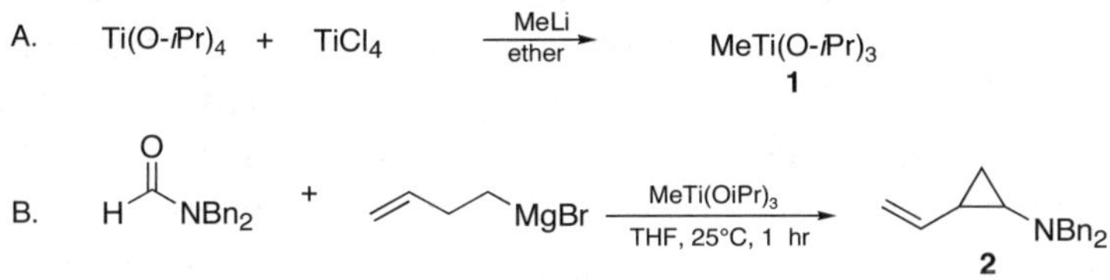

Submitted by Armin de Meijere, Harald Winsel, and Björn Stecker.[1]
Checked by Fanglong Yang, Sivaraman Dandapani, and Dennis P. Curran.

1. Procedure

A. Methyl tris(isopropoxy)titanium.[2] A 500-mL, three-necked, round-bottomed flask equipped with two rubber septa and a dropping funnel fitted with an argon inlet is charged with a mixture of 44.4 mL (150 mmol) of titanium tetraisopropoxide (Note 1) in 50 mL of anhydrous ether and cooled to 0° C in an ice bath. Titanium tetrachloride (5.48 mL, 50 mmol) (Note 1) is added dropwise via syringe pump over 30 min. The resulting mixture is allowed to warm to room temperature and stirred for 2 h. The reaction mixture is then cooled to 0° C and 123 mL (197 mmol) of a 1.6M solution of methyllithium in ether (Note 1) is added via the dropping funnel over 40 min. The mixture is allowed to warm to room temperature and stirred for 1 h. The dropping funnel is replaced with a short path distillation head, and the ether is removed by distillation at ambient pressure. Distillation of the crude product at reduced pressure (bp 56-57° C, 60 mm) gives 38.5 g (80%) of $MeTi(O\textit{i}Pr)_3$ as a yellow oil (Note 2).

B. N,N-Dibenzyl-N-(2-ethenylcyclopropyl)amine. An oven-dried, 250-mL, three-necked, round-bottomed flask containing a magnetic stirbar and equipped with two rubber septa and an addition funnel fitted with a nitrogen inlet is charged with a solution of 9.0 g (40 mmol) of *N,N*-dibenzylformamide (Notes 3 and 4) in 10 mL of anhydrous tetrahydrofuran (Note 5). $MeTi(O\textit{i}Pr)_3$ (11.5 g, 48 mmol, 1.2 equiv) is added in one portion by syringe, and then 3-butenylmagnesium bromide (1.48M in THF, 40.5 mL, 60 mmol, 1.5 equiv) (Notes 6, 7) is added dropwise via the addition funnel over 30 min. The resulting black suspension is stirred at ambient temperature for an additional hour (Note 8), and the reaction is then quenched by slow addition of 50 mL of diethyl ether followed by 5 mL of water. The mixture is vigorously stirred for 1 h (Note 9), filtered, and the colorless precipitate is then washed with two 20-mL portions of ether. The combined organic phases are washed with 30 mL of brine and dried over anhydrous $MgSO_4$, filtered, and concentrated at reduced pressure (Note 10) to afford 9.4 g of **2** (1:7 ratio of *cis*- and *trans*-isomers) as a deep yellow oil (Notes 11, 12). This crude product is purified by flash chromatography on 350 g of silica gel (200-400 mesh, elution with 100:1 pentane/ether) (Notes 13, 14) to yield 0.73 g (7%) of the *cis*-diastereomer, 0.51 g (5%) of a mixed fraction, and 6.2 g (59%) of the *trans*-diastereomer of 2, each as a pale yellow oil (Notes 15, 16).

2. Notes

1. [MeTi(O*i*Pr)$_3$] must be used in pure form rather than as a solution in hexane. Titanium tetraisopropoxide (95%) was obtained from ABCR, Karlsruhe by the submitters and from Aldrich Chemical Company by the checkers and was distilled under nitrogen before use. Titanium tetrachloride (>99%) was purchased from VWR by the submitters and from Aldrich Chemical Company by the checkers and was used without further purification. The submitters obtained a solution of methyllithium in ether (1.6M, 5 wt%) from Fluka. The amount of titanium tetrachloride was incorrectly given as 6.48 mL and the concentration of methylithium solution was incorrectly given as 1.0 M in the original published version of this *Organic Syntheses* procedure due to typographical errors.

2. The product is air and moisture sensitive, and the checkers used it immediately after preparation. The submitters report that MeTi(O*i*Pr)$_3$ can be stored for several months in a freezer under an inert atmosphere.

3. Other *N,N*-dialkylformamides including commercially available *N*-methylformanilide, *N*-formylmorpholine, *N*-formylpiperidine, *N,N*-diethylformamide and *N,N*-dimethylformamide (DMF) can be employed with essentially the same procedure. With *N,N*-dimethylformamide, care has to be exercised because of the volatility of the product, which can be purified by distillation or chromatography of its hydrochloride salt.

4. *N,N*-Dibenzylformamide was prepared as follows. A 250-mL, round-bottomed flask equipped with reflux condenser fitted with an argon inlet was charged with 100 mL of DMF, 40.4 mL (210 mmol) of dibenzylamine, and 12.8 mL (300 mmol) of formic acid. The solution was heated at reflux overnight, allowed to cool to room temperature, and then

quenched by carefully adding 40 mL of saturated Na_2CO_3 solution. The aqueous phase was separated and extracted with four 50-mL portions of ether, and the combined organic layers were washed with two 20-mL portions of brine, dried over $MgSO_4$, and concentrated by rotary evaporation at reduced pressure. The crude product was crystallized from 100 mL of 70:30 ether/pentane to yield 40.2 g (85%) of *N,N*-dibenzylformamide, mp 50-51° C; it is hygroscopic and must be dried before use under reduced pressure (0.007 mm) overnight.

5. The submitters used THF that was freshly distilled from sodium, while the checkers used THF that was dried by pressure filtration through activated alumina.

6. 3-Butenylmagnesium bromide is prepared as follows. A 250-mL, three-necked, round-bottomed flask containing a magnetic stirbar and equipped with two rubber septa and an addition funnel fitted with an argon inlet was charged with 4.8 g (200 mmol) of magnesium turnings (Aldrich) suspended in 70 mL of anhydrous THF. A solution of 18.3 mL (180 mmol) of 4-bromo-1-butene (Aldrich) in 60 mL of anhydrous THF was added by dropping funnel at a rate adjusted to maintain a gentle reflux (total addition time, ca. 3 h). The reaction mixture was then stirred for an additional hour. To determine the concentration of product, a 1-mL aliquot of the mixture was quenched with 0.1N HCl and then back-titrated with 0.1N NaOH (phenolphthalein indicator) which indicated a concentration of 1.5M.

The concentration of the Grignard reagent should not be higher than 1.5M. The use of more concentrated solutions leads to lower yields. Because of the rather high concentration of the reagents, it is important to stir the reaction mixture vigorously during and after the addition of the Grignard reagent. If the reaction is carried out on a larger scale than that described here, then the use of a mechanical stirrer is recommended.

7. The submitters used a syringe pump to add the Grignard solution. However, the checkers experienced problems with precipitation of the Grignard reagent and recommend the use of an addition funnel. If the reagent begins to precipitate in the funnel during the addition period, then gentle warming with a heat gun accomplishes redissolution.

8. The reductive cyclopropanation reaction proceeds very rapidly. For example, the submitters report that hydrolysis of the reaction mixture after only 1 min gives the product in 75% yield. However, stirring for an additional 1 h is recommended, and this should be sufficient for analogous transformations with other Grignard reagents and formamides as well.

Higher *N,N*-dialkylcarboxamides may require longer reaction times (up to 5 h).

9. The hydrolysis should be carried out with access of air so that the black titanium(II) derivatives are oxidized rapidly and completely to colorless titanium(IV) compounds.

10. The solvents are removed within 1 h at a bath temperature of 50° C at 15 mm.

11. The *cis/trans* diastereomeric ratio is determined by integration of resonances in the ^{1}H NMR spectrum of the crude mixture and is usually found in the range of 1/6.0 to 1/7.5.

12. TLC analysis of the crude product (elution with 50:1 pentane:ether, visualization with iodine) showed three non-baseline spots: R_f 0.65 (*cis* isomer), R_f 0.52 (unknown impurity), and R_f 0.32 (*trans* isomer). The unknown impurity is intensely sensitive to iodine and largely coelutes with the *cis*-isomer in the subsequent column chromatography. However, the ^{1}H NMR spectrum of this isomer shows excellent purity despite the presence of this spot on TLC. In 100:1 pentane:ether, R_f values of the *cis* and *trans* isomers are about 0.50 and 0.15, respectively.

13. It is important to wet-pack the column with the eluting solvent. Dry packing results in strong retention of the sample and poor separation, possibly because the small amount of ether in the eluting solvent is adsorbed by the dry silica gel.

14. If separation of the diastereomers is not required, then filtration of the crude product through a pad of 50 g of silica eluting with dichloromethane followed by concentration affords a mixture of diastereomers separated from baseline materials.

15. The submitters report obtaining 1.29 (12%) of the *cis*-diastereoisomer and 8.39 g (80%) of the *trans*-diastereoisomer. The submitters recommend that the purified diastereomers be stored in the freezer (-18° C). Both pure isomers slowly equilibrate to a 21:79 mixture of cis/trans isomers upon standing at ambient temperature for several weeks.

16. The products exhibit the following spectroscopic properties: *trans* isomer: ^{1}H NMR (300 MHz, $CDCl_3$) δ 0.60 (ddd, *J* = 6.7, 5.5, 4.9 Hz, 1 H), 0.76 (ddd, *J* = 9.0, 4.6, 4.6 Hz, 1 H), 1.35-1.26 (m, 1 H), 1.79 (ddd, *J* = 7.1, 4.2, 3.0 Hz, 1 H), 3.60 (d, *J* = 13.5 Hz, 2 H), 3.70 (d, *J* = 13.5 Hz, 2 H), 4.83-4.76 (m, 2 H), 5.33 (ddd, *J* = 17.0, 10.3, 8.5 Hz, 1 H), 7.34-7.20 (m, 10 H); ^{13}C NMR (75 MHz, $CDCl_3$) δ 15.9, 25.5, 45.1, 58.2, 112.2, 126.8, 128.0, 129.4, 138.4, 139.6; IR cm^{-1}: 3062, 3027, 2921, 2809, 1635, 1493, 1453, 1364. *cis* isomer: ^{1}H NMR (300 MHz, $CDCl_3$) δ 0.45 (ddd, *J* = 4.8, 4.8, 4.8

Hz, 1 H), 0.87 (ddd, J = 8.6, 6.8, 4.9 Hz, 1 H), 1.57-1.51 (m, 1 H), 2.06 (ddd, J = 6.8, 6.8, 4.6 Hz, 1H), 7.34-7.22 (m, 10 H), 3.51 (d, J = 13.7 Hz, 2 H), 3.74 (d, J = 13.7 Hz, 2 H), 4.97 (dd, J = 10.3, 2.0 Hz, 1 H), 5.14 (dd, J = 17.3, 2.0 Hz, 1 H), 5.88 (ddd, J = 17.3, 10.3, 9.5 Hz, 1 H); ^{13}C NMR (75 MHz, $CDCl_3$) δ 15.6, 23.3, 42.4, 57.2, 113.60, 126.8, 128.0, 129.5, 138.3, 138.4; IR cm^{-1}: 3062, 3026, 2922, 2807, 1634, 1493, 1453, 1364.

3. Discussion

The transformation of *N,N*-dialkylcarboxamides with low valent titanium compounds formed in situ from Grignard reagents and titanium alkoxides of the type $XTi(OiPr)_3$ (X = O*i*Pr, Cl, Me) to correspondingly substituted *N,N*-dialkylcyclopropylamines has been thoroughly examined in recent years since it was first discovered.[3] Related to the conversion of esters to cyclopropanols,[5] this new and reasonably general access to this important class of compounds[4] is now easily carried out and generally furnishes high yields.[6] For example, unsubstituted *N,N*-dialkylcyclopropylamines are obtained from *N,N*-dialkylformamides and ethylmagnesium bromide following this protocol in up to 98% yield.

Scheme 1. Generality of the titanium mediated reductive cyclopropanation of *N,N*-diakylcarboxamides. More than 60 compounds have been prepared by this route.

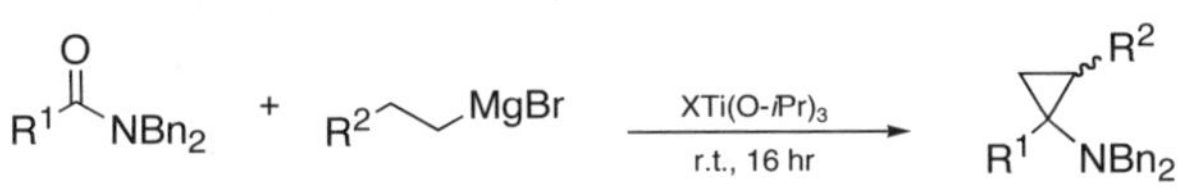

entry	R^1	R^2	yield (%)	*cis/trans*
1	H	H	98	—
2	H	*n*Bu	90	1 : 2.0
3	H		85	1 : 2.2
4	H	Ph	92	1 : 2.0
5	H	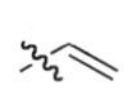	92	1 : 7.0
6	Me	H	77	—
7	Et	H	70	—

R^1 = R^2 = H, Alkyl, Alkenyl, Aryl, functionally substituted Alkyl, Alkenyl, Aryl
X = Me, Cl, O*i*Pr

The products can be purified by distillation or crystallization as hydrochloride salts, but they can often be used for subsequent transformations without purification.

The best titanium mediator appears to be methyltitanium triisopropoxide, yet good yields are also obtained with titanium tetraisopropoxide and chlorotitanium triisopropoxide. The methyl group on titanium serves as a dummy alkyl ligand which is eliminated as methane after hydride transfer from the other alkyl group introduced from the Grignard reagent so that essentially only one equivalent of the latter would be required. However, yields do increase from 80-85% to over 90% in most cases by using an excess of the Grignard reagent (up to 2.5 equiv.). In the presence of titanium tetraisopropoxide and with the use of more than two equivalents of the Grignard reagent, the desired product is usually obtained from an *N,N*-dialkylformamide in over 80% yield along with 10-15% of the corresponding dialkylamine.

The reaction is best performed in THF. In less polar solvents such as diethyl ether or benzene, the chemical yields are significantly lower.

In the presence of 1 equiv. of chlorotrimethylsilane, the transformation can be performed with substoichiometric amounts of titanium tetraisopropoxide (25 mol%) to yield 73% of the diastereomeric mixture of the corresponding *N,N*-dialkylaminocyclopropane.[7]

Ethenylcyclopropylamines such as the described one are essentially functionally substituted cyclopropylamines, since the ethenyl group can easily be converted to functionally substituted ethyl groups, e. g., by hydroboration/oxidation or hydroboration/amination, and eventually the benzyl groups can be removed by catalytic hydrogenation to yield the corresponding primary cyclopropylamines. Ethenylcyclopropylamines can also easily be transformed to cyclopenten-4-ylamines by thermal rearrangement.[3c] Functionality can also directly be introduced with the carboxamide,[8a] the Grignard reagent,[8b] or the use of functionalized organozinc reagents.[8c]

A wide range of alkenyl-, aryl-, alkyl-, and dialkyl-substituted cyclopropylamines can be made by reaction of an *N,N*-dialkylformamide (or higher *N,N*-dialkylcarboxamide) with cyclohexylmagnesium halide in the presence of methyltitanium triisopropoxide (or titanium tetraisopropoxide) and a 1,3-diene, a styrene, a terminal or an internal alkene following this protocol with some minor modifications.[9] In these cases the reacting low valent titanium intermediate is generated by ligand exchange,[10] and the overall transformation corresponds to an aminocyclopropanation of the added alkene. For example, a solution of *N*-Boc-protected pyrroline (4.36 g, 25.8 mmol) and $Ti(OiPr)_4$ (9.06 mL, 30.9 mmol) in THF (100 mL) is treated with MeMgCl (10.3 mL of a 3M solution in THF, 30.9 mmol, added within 10 min) at 0° C and the mixture is warmed to ambient temperature. *N,N*-Dibenzylformamide (6.95 g, 30.9 mmol) is added in one portion, then dropwise with stirring a solution of cyclohexylmagnesium bromide (28.1 mL of a 2.2M solution in diethyl ether, 61.9 mmol) within 2 h. The mixture is heated under reflux for an additional 1 h, then cooled to ambient temperature. Addition of 100 mL of water and 100 mL of pentane, filtration and further workup with filtration through a pad of 10 g of basic aluminum oxide and subsequent crystallization from pentane yields 8.75 g (90%) of crystalline *N*-Boc-protected *exo*-6-*N,N*-dibenzylamino-3-azabicyclo [3.1.0] hexane.[9b]

1. Institut für Organische und Biomoleckulare Chemie der Georg-August- Universität Göttingen, Tammannstrasse 2, D-37077 Göttingen, Germany; Fax: + 49 (0)551 399475; E-mail: armin.deMeijere @chemie.uni-goettingen.de

2. Reetz, M. T.; J. Westermann, J.; Steinbach, R.; Wenderoth, B.; Peter, R.; Ostarek, R.; Maus, S. *Chem. Ber.* **1985**, *118*, 1421-1440.

3. (a) Chaplinski, V.; de Meijere, A. *Angew. Chem. Int. Ed. Engl.* **1996**, *35*, 413-414. (b) Chaplinski, V.; Winsel, H.; Kordes, M.; de Meijere, A. *Synlett* **1997**, 111-114. (c) Willliams, C. M.; de Meijere, A. *J. Chem. Soc., Perkin Trans.* **1998**, 3699-3702. (d) Kordes, M.; Winsel, H.; de Meijere, A. *Eur. J. Org. Chem.* **2000**, 3235-3245.

4. Salaün, J. *Topics in Current Chemistry* **2000**, *207*, 1-67.

5. Kulinkovich, O. G.; Sviridov, S. V.; Vasilevski, D. A.; Pritytskaya, T. S. *Zh. Org. Khim.* **1989**, *25*, 2244-2245; *J. Org. Chem. USSR (Engl. Transl.)* **1989**, *25*, 2027-2028.

6. Reviews: (a) Kulinkovich, O. G.; de Meijere, A. *Chem. Rev.* **2000**, *100*, 2789-2834. (b) de Meijere, A.; Kozhushkov, S. I.; Savchenko, A. I. in *Titanium and Zirconium in Organic Synthesis*, (Ed.: I. Marek), Wiley-VGH, Weinheim, **2002**, 390-434.

7. de Meijere, A.; Stecker, B., to be published. Stecker, B., Dissertation, Universität Göttingen, 2002.

8. (a) Winsel, H.; Gazizova, V.; Kulinkovich, O.; Pavlov, V.; de Meijere, A. *Synlett* **1999**, 1999-2003. (b) Winsel, H. Dissertation Universität Göttingen, 2000. (c) Wiedemann, S.; Marek, I.; de Meijere, A. *Synlett.* **2002**, 879-882.

9. (a) Williams, C. M.; Chaplinski, V.; Schreiner, P. R.; de Meijere, A. *Tetrahedron Lett.* **1998**, *39*, 7695-7698. (b) de Meijere, A.; Williams, C. M.; Chaplinski, V.;

Kourdioukov, A.; Sviridov, S. V.; Savtchenko, A.; Kordes, M.; Stratmann, C. *Chem. Eur. J.* **2002**, *8*, 3789-3801.

10. (a) Lee, J. C.; Cha, J. K. *J. Org. Chem.* **1997**, *62*, 1584-1585. (b) Lee, J. C.; U, J. S.; Blackstock, S. C.; Cha, J. K. *J. Am. Chem. Soc.* **1997**, *119*, 10241-10242. (c) Lee, J. C.; Sung, M. J.; Cha, J. K. *Tetrahedron Lett.* **2001**, *42*, 2059-2061.

GENERATION AND CYCLIZATION OF 5-HEXENYLLITHIUM: 2-CYCLOPENTYLACETOPHENONE
(Ethanone, 2-cyclopentyl-1-phenyl-)

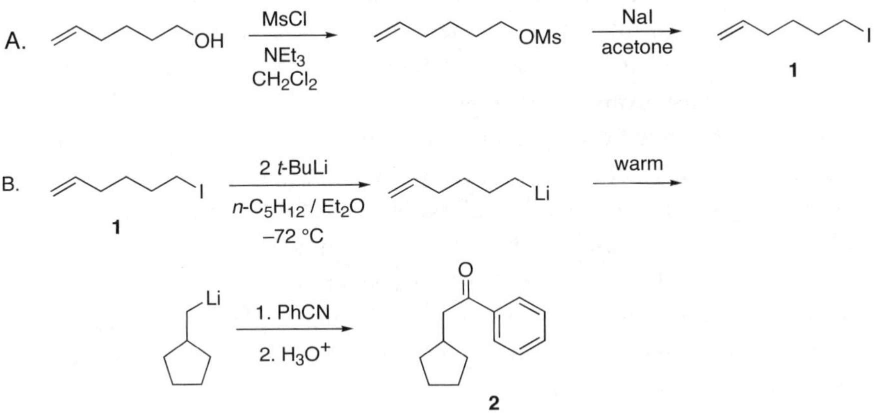

Submitted by William F. Bailey, Matthew R. Luderer, Michael J. Mealy, and Eric R. Punzalan.[1]
Checked by Scott E. Denmark and Stephen L. MacNeil.

1. Procedure

A. 6-Iodo-1-hexene (**1**). A flame-dried, 1-L, one-necked, round-bottomed flask equipped with a magnetic stirbar and a pressure-equalizing addition funnel fitted with a nitrogen inlet adapter is charged with 10.0 g (0.100 mol) of 5-hexen-1-ol, 15.2 g (0.151 mol) of triethylamine, and 500 mL of dichloromethane (Note 1). The flask is cooled in an ice-salt bath (0 to –5 °C) for 30 min and then 9.3 mL (0.12 mol) of methanesulfonyl chloride (Note 2) is added dropwise via the addition funnel. The reaction mixture is stirred at –5 to –10 °C for an additional 1 h before being transferred to a cold 1-L separatory funnel and washed successively with 150 mL of cold water, 150 mL of cold 10% (ca. 3.3N) aqueous hydrochloric acid, 150 mL of cold, saturated, aqueous sodium bicarbonate solution, and 150 mL of cold brine (Note 3). The organic layer is dried over $MgSO_4$ and then divided into three portions (Note 4). One portion is filtered into a 300-mL, round-bottomed flask and concentrated by rotary evaporation at 8-10 mm in a 20-25 °C water bath, and this is repeated with the remaining two portions to give the desired mesylate as a clear, pale-yellow oil (Note 5).

Dry acetone (200 mL) and then 18.5 g (0.12 mol) of anhydrous sodium iodide (Note 6) are added to the above 300-mL flask, which is then equipped with a magnetic stirbar and a Friedrichs condenser fitted with a nitrogen inlet adapter. The pale-yellow solution is stirred in the dark under a positive pressure of nitrogen at gentle reflux for 4 h (Note 7). The resulting mixture is allowed to cool to room temperature and the acetone is then removed by rotary evaporation at 8–10 mm in a 20-25 °C water bath. The residue is partitioned between 50 mL of pentane and 50 mL of 10% aqueous sodium thiosulfate solution by swirling the flask until all of the precipitate dissolves. The aqueous phase is discarded and the organic layer is washed with 50 mL of brine, dried over $MgSO_4$, filtered, and concentrated by rotary evaporation at 8–10 mm in a 20-25 °C water bath. The residue is then passed through a 60-mL, medium porosity, sintered-glass funnel containing ca. 20 g of alumina (Note 8) using ca. 100 mL of pentane as eluent. The pentane is removed by rotary evaporation at 8–10 mm in a 20-25 °C water bath to give 17.44-17.97 g (83–86% overall from the alcohol) of **1** as a clear, colorless oil (Notes 9 and 10).

B. 2-Cyclopentylacetophenone (**2**). A 500-mL, two-necked, round-bottomed flask (Note 11) equipped with an egg-shaped magnetic stir bar (4 cm x 1.5 cm), argon inlet adapter, and a rubber septum is flame-dried and allowed to cool to room temperature under a positive pressure of argon. A thermometer is inserted through the septum (Note 12) and the flask is charged with 114 mL of dry, alkene-free pentane and 76 mL of anhydrous diethyl ether (Notes 13, 14). The solution is cooled to approximately –72 °C using a 2-propanol-dry ice bath and 52 mL of a 1.90M solution of *tert*-butyllithium (*t*-BuLi) in heptane (98.8 mmol) is added via teflon cannula at a rate of approximately 1.7 mL/min so as to maintain an internal temperature below ca. –68 °C (Notes 15, 16). A two-necked, 100-mL, round-bottomed flask equipped with a rubber septum and argon inlet adapter is flame-dried under an atmosphere of argon, allowed to cool to room temperature, and then charged with 10.0 g (47.6 mmol) of oxygen-free 6-iodo-1-hexene (**1**) (Note 17) and 50 mL of dry, alkene-free pentane (Note 14). The iodide solution is cooled to –72 °C using a 2-propanol-dry ice bath and then transferred via a teflon cannula under a positive pressure of argon to the stirred, –72 °C solution of *t*-BuLi at a rate of approximately 1.4 mL/min so as to maintain an internal temperature below –65 °C (Note 18). The last of the iodide solution is transferred through the cannula by addition of 5 mL of dry pentane to the flask. Upon completion of the addition, the mixture is stirred at –72 °C for 15 min and the cooling bath is then removed. After 3.5

h (Note 19), the opaque, pale-yellow solution is cooled to ca. 0 °C in an ice-water bath, and 4.85 g (47.0 mmol) of benzonitrile (Note 20) is added dropwise via syringe (Note 21). The cooling bath is removed, and the bright-orange solution is allowed to stir for 1 h. The reaction mixture is then re-cooled in an ice-water bath and 35.0 mL of 10% (ca. 3.3N) aqueous hydrochloric acid is added rapidly via syringe (Note 22). Once the exotherm abates, the cooling bath is removed, and stirring is continued for 2 h. The contents of the flask are completely transferred to a 500-mL separatory funnel by repeated washings with pentane and ether, shaken vigorously, and the aqueous layer is discarded. The organic phase is washed with 50 mL of water and 50 mL of brine, dried over $MgSO_4$. filtered, and concentrated by rotary evaporation at 8–10 mm. Kugelrohr distillation of the residue (bath temperature ca. 180 °C, 5 mm) affords 6.64-6.78 g (75-77 %) of pure **2** as a clear, colorless oil (Notes 10, 23, 24).

2. Notes

1. 5-Hexen-1-ol and triethylamine were purchased from Acros Organics and used without further purification. Alternatively, 5-hexen-1-ol may be prepared from 2-(chloromethyl)tetrahydropyran according to the literature procedure for the preparation of 4-penten-1-ol (Brooks, L. A.; Snyder, H. R. *Org. Synth. Coll. Vol. III* **1955**, 698). Dichloromethane (certified ACS) was purchased from Fisher Scientific and was used as received.

2. Methanesulfonyl chloride was purchased from Aldrich Chemical Co. and used as received. This reagent, when taken from a previously opened container, is distilled from phosphorus pentoxide prior to use.

3. The separatory funnel, water, 10% hydrochloric acid, saturated sodium bicarbonate solution, and brine are cooled to ca. 0 °C in a freezer prior to use.

4. The dried organic layer is filtered and concentrated in ca. 150-mL portions.

5. The mesylate is used in the next step without further purification and is essentially pure; a small quantity of residual dichloromethane has no effect on the yield of the Finkelstein reaction. The mesylate has the following spectroscopic properties: ^{1}H NMR ($CDCl_3$): δ 1.46 - 1.52 (m, 2H), 1.71 - 1.77 (m, 2H), 2.06 - 2.10 (q, J = 7.1 Hz, 2H), 2.98 (s, 3H), 4.21 (t, J = 6.6 Hz, 2H), 4.95 - 5.02 (m, 2H), 5.72 - 5.80 (m, 1H); ^{13}C NMR ($CDCl_3$): δ 24.5, 28.4, 32.9, 37.2, 69.9, 115.1, 137.8.

6. Acetone (certified ACS) was purchased from Fisher Scientific and was dried over calcium sulfate and filtered prior to use. Anhydrous sodium iodide was purchased from Acros Organics and used as received.

7. Aluminum foil is wrapped around the flask and lower part of the condenser during the reaction.

8. Alumina, adsorption, ca. 150 mesh, was purchased from Aldrich Chemical Co. and used as received.

9. The known iodide **1**[2] is isolated in high purity (> 99 % by GC analysis) and may be used without further purification. The material displays the following properties: n_D^{20} = 1.5121 (lit.[3] n_D^{20} = 1.5106); ^{1}H NMR ($CDCl_3$: δ 1.46 - 1.55 (m, 2H), 1.80 - 1.87 (m, 2H), 2.05 - 2.11 (m, 2H), 3.19 (t, J = 7.0, 2H), 4.95 - 5.04 (m, 2H), 5.74–5.84 (m, 1H); ^{1}H NMR ($CDCl_3$: δ 6.9, 29.6, 32.6, 32.9, 115.0, 138.1; IR (neat) 3076, 2931, 2856, 1641, 1441, 1215, 1176, 991, 912 cm^{-1}. The iodide may be stored indefinitely in a freezer if first saturated with dry argon.

10. The purity of **1** and **2** is assessed by analytical gas-liquid chromatography (GC) on a Hewlett-Packard 5890 gas chromatograph equipped with a flame-ionization detector and fitted with a 50 m x 0.2 mm HP-5 fused silica glass capillary column using linear temperature programming from an initial temperature of 150 °C for 5 min to a final temperature of 200 °C for 10 min at a rate of 5 °C/min.

11. A one-necked Ace Glass 6935 flask with style-C side well (ordering code 6935-72) was used by the submitters, but the checkers found the use of a two-neck flask to be equally satisfactory.

12. An Omega model HH22 type J-K digital thermometer, connected to a type K thermocouple probe inserted directly into the flask, was used to measure the temperature.

13. Pentane (HPLC grade) was purchased from Fisher Scientific and used as received. Anhydrous diethyl ether was purchased from Mallinckrodt Inc. and distilled under nitrogen from a dark blue solution of sodium and benzophenone.

14. The appropriate volume of solvent is most conveniently transferred under argon from flame-dried graduated cylinders fitted with rubber septa using stainless steel double-tipped cannulas. Although the proportions of pentane and diethyl ether used as solvent are not crucial to the success of the lithium-iodine exchange, it is necessary to run the reaction in a solvent system that contains enough diethyl ether to render the *t*-BuLi

dimeric.[4] The submitters have found that a solvent mixture composed of 3:2 (by volume) of pentane and ether gives reproducible results.

15. The concentration of solutions of *t*-BuLi in heptane, purchased from FMC, Lithium Division, were determined immediately prior to use by titration with 2-butanol in xylene using 1,10-phenanthroline as the indicator following the procedure described by Watson and Eastham.[5] A typical procedure is as follows: a flame-dried, 25-mL, round-bottomed flask, equipped with a magnetic stirbar, is charged under a blanket of argon with 3 mL of a 0.10 % (w/v) solution of 1,10-phenanthroline in dry benzene and 0.40-0.50 mL of *t*-BuLi in heptane; the violet solution is titrated with a ca. 1.0M solution of 2-butanol in xylenes until the pale yellow end point is reached. It should be noted that the nominal concentration recorded on commercial samples of an organolithium is often in error, particularly once the original seal is breached, or if the material has been stored for an extended period, and it is essential that the actual concentration be determined immediately prior to use. At least two molar equivalents of *t*-BuLi must be used in the exchange reaction: a full equivalent of *t*-BuLi is consumed in reaction with the *tert*-butyl iodide, generated as a by-product of the exchange, to give lithium iodide, isobutylene, and isobutane.[4] A slight surplus of *t*-BuLi in excess of 2 molar equiv is not deleterious, since residual *t*-BuLi is rapidly consumed by proton abstraction from diethyl ether when the reaction mixture is warmed to room temperature,[4] but a large excess should be avoided.

16. A commercial solution of *t*-BuLi in pentane may be substituted for *t*-BuLi in heptane. However, the lower volatility and higher flash-point of the heptane solvent makes *t*-BuLi in heptane *much* easier and safer to handle than this alternative.

17. The 6-iodo-1-hexene is rendered free of oxygen by bubbling argon through the neat iodide for at least 5 min prior to use.

18. The exchange reaction is exothermic and too rapid an addition of the iodide will cause a temperature rise that may lead to the formation of unwanted side-products. A white precipitate of LiI is observed when approximately two-thirds of the iodide solution has been added.

19. The solution warms slowly to room temperature; after 1.5 h the internal temperature is approximately 15-20 °C.

20. Benzonitrile, purchased from Aldrich Chemical Co., was distilled from calcium hydride and saturated with dry argon for at least 5 min prior to use.

21. An approximately 5 °C exotherm was noted during the addition.

22. The addition of the hydrochloric acid solution results in an exotherm of approximately 20 °C, accompanied by the appearance of a white precipitate and a viscous, orange oil.

23. The known ketone **2**,[6] which is isolated in high purity (>99% by GC analysis), exhibits the following physical and spectroscopic properties: n_D^{20} = 1.5362; ^{1}H NMR ($CDCl_3$): δ 1.13 - 1.22 (m, 2H), 1.50 - 1.68 (m, 4H), 1.84 - 1.91 (m, 2H), 2.38 (7-line pattern, J = 7.7 Hz, 1H), 2.98 (d, J = 7.1 Hz, 2H), 7.43 - 7.46 (m, 2H), 7.52 - 7.56 (m, 1H), 7.94 -7.97 (m, 2H); ^{13}C NMR ($CDCl_3$): δ 24.9, 32.7, 36.0, 44.7, 128.0, 128.5, 132.8, 137.2 200.31; IR (neat) 3062, 2951, 2868, 1685, 1448, 1209, 752, 690 cm^{-1}; Anal. calcd for $C_{13}H_{16}O$: C, 82.94; H, 8.57, Found: C, 82.71; H, 8.50

24. Purification by short-path distillation (138-139 °C/4.5 mm) resulted in a slightly diminished yield (6.25 g, 71%).

Waste Disposal Information

All toxic materials were disposed of in accordance with "Prudent Practices for Disposal of Chemicals from Laboratories"; National Academy Press; Washington, DC, 1995.

3. Discussion

The procedure described above may be used for the generation and 5-*exo* cyclization of a variety of substituted 5-hexenyllithiums and, with appropriate modification, provides a general route to a variety of other olefinic[7] and acetylenic[8] organolithiums that cyclize upon warming.[9]

Primary alkyllithiums may be prepared rapidly and in virtually quantitative yield by treatment of a primary alkyl iodide with slightly more than two molar equiv of *t*-BuLi in a hydrocarbon–ether solvent system at low temperature.[4] The exchange protocol outlined above for the preparation of 5-hexenyllithium (**1**) is a general one but its success often depends crucially on the appropriate choice of both halide and solvent. Primary alkyl iodides, rather than bromides, must be used to ensure that the reaction proceeds cleanly, most likely via a 10-I-2 ate-complex:[10] addition of *t*-BuLi to a primary alkyl bromide often initiates radical-mediated processes. The lithium-iodine exchange appears to involve a dimeric *t*-BuLi solvate and, for this reason, the best medium for the preparation of primary alkyllithiums by lithium-iodine exchange is a predominantly hydrocarbon solvent system that contains a quantity of a simple alkyl ether, such as Et_2O, MTBE, dibutyl ether, or the like, in which the *t*-BuLi is predominantly dimeric. Solvent systems containing THF, TMEDA, or other Lewis bases that render the *t*-BuLi monomeric should be avoided since elimination and coupling reactions often compete with exchange under these conditions.

Cyclization of an organolithium tethered to a suitably positioned carbon-carbon π-bond is a thermodynamically favorable process that proceeds in a totally regioselective *exo*-fashion with a high degree of stereocontrol via a transition state in which the lithium atom is intramolecularly coordinated with the remote π-bond.[9] The stereochemical outcome of the cyclization of a substituted 5-hexenyllithium follows from the preference of the substituent to occupy a pseudoequatorial position in the chair-like transition state depicted below.[7]

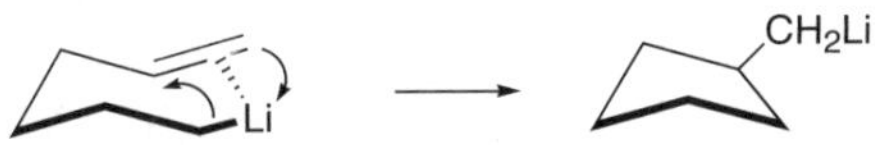

Despite the favorable thermodynamics associated with the cyclization of unsaturated organolithiums, the isomerization is often sluggish when the ring closure involves generation of a quaternary center or formation of a strained framework. In such cases it has been found that addition of lithiophilic Lewis bases such as THF or TMEDA facilitate the reaction.[7,9] The preparation of cuparene, a sterically congested sesquiterpene possessing two adjacent quaternary centers, illustrates the methodology.[11]

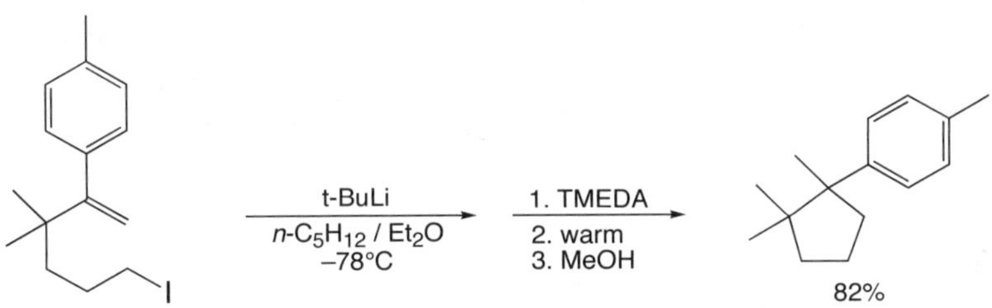

The predictable stereochemistry of the ring closure of substituted 5-hexenyllithiums, coupled with the ease with which the product organolithium may be functionalized, permits rational design of synthetic routes to polycyclic systems by sequential anionic cyclizations of polyolefinic alkyllithiums.[12] The preparation of stereoisomerically pure *endo*-2-substituted bicyclo[2.2.1]heptanes is illustrative of this approach to polycyclic systems.[12]

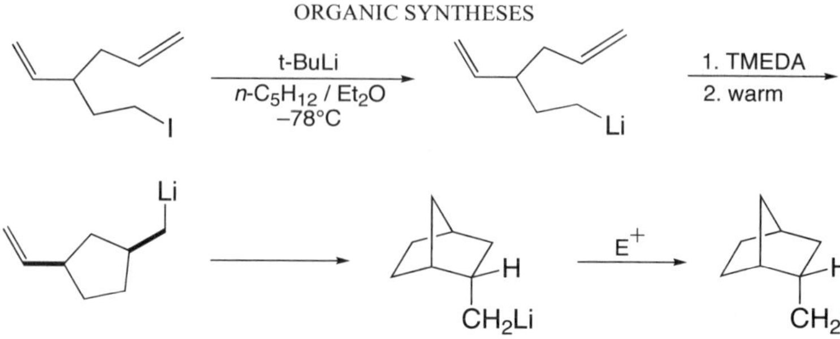

Heterocyclic systems, such as the substituted indoline illustrated below,[13] may also be constructed via cyclization of unsaturated organolithiums and a recent review of the preparation of nitrogen- and oxygen-containing heterocycles via this approach is available.[14]

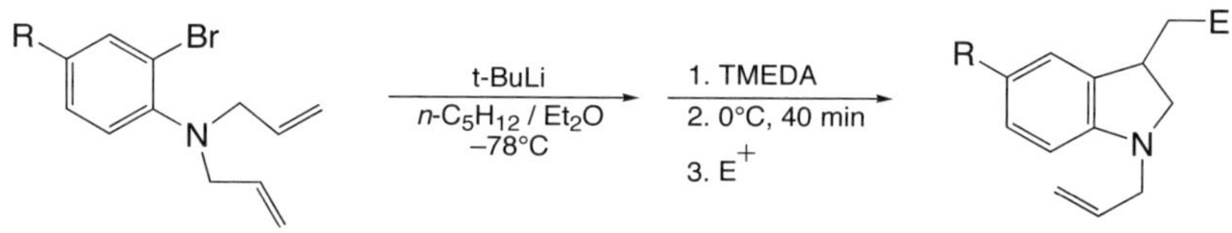

It should also be noted that the 5-exo-trig cyclization of achiral olefinic organolithiums has been found to proceed enantioselectively when conducted in the presence of a chiral ligand that serves to render the lithium atom stereogenic. Thus, for example, (*R*)-1-allyl-3-methylindoline has been prepared in 86 % ee by cyclization of an achiral aryllithium in the presence of an equivalent of (–)-sparteine.[15]

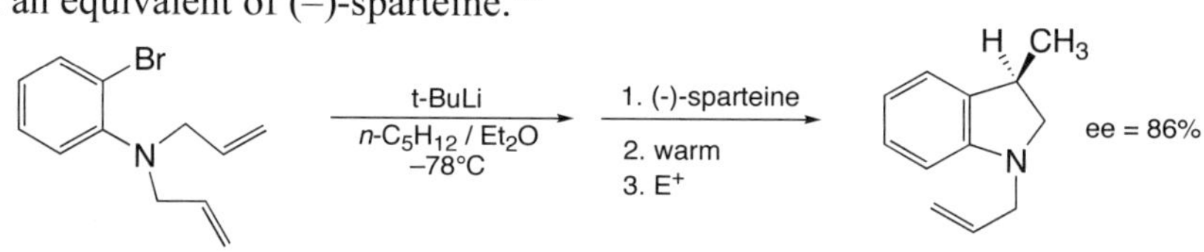

1. Department of Chemistry, University of Connecticut, Storrs, CT, 06269.
2. Bailey, W. F.; Gagnier, R. P.; Patricia, J. J. *J. Org. Chem.* **1984**, *49*, 2098.
3. Ashby, E. C.; DePriest, R. N. Goel, A. B.; Wenderoth, B.; Pham, T. N. *J. Org. Chem.* **1984**, *49*, 3545.
4. Bailey, W. F.; Punzalan, E. R. *J. Org. Chem.* **1990**, *55*, 5404.
5. Watson, S. C.; Eastham, J. F. *J. Organomet. Chem.* **1967**, *9*, 165.
6. Moureu, H.; Chovin, P; Bloch, G.; Rivoal, G. *Bull. Soc. Chim. Fr.* **1949**, *457*, 475.

7. Bailey, W. F.; Khanolkar, A. D.; Gavaskar, K.; Ovaska, T. V.; Rossi, K.; Thiel, Y.; Wiberg, K. B. *J. Am. Chem. Soc.* **1991**, *113*, 5720.

8. Bailey, W. F.; Ovaska, T. V. *J. Am. Chem. Soc.* **1993**, *115*, 3080.

9. Bailey, W. F.; Ovaska, T. V. In *Advances in Detailed Reaction Mechanisms*; Coxon, J. M., Ed.; JAI Press: Greenwich, CT, 1994; Vol. 3, *Mechanisms of Importance in Synthesis*; p. 251-273.

10. Wiberg, K. B.; Sklenak, S.; Bailey, W. F. *J. Org. Chem.* **2000**, *65*, 2014.

11. Bailey, W. F.; Khanolkar, A. D. *Tetrahedron* **1991**, *47*, 7727.

12. Bailey, W. F.; Khanolkar, A. D.; Gavaskar, K. V. *J. Am. Chem. Soc.* **1992**, *114,* 8053.

13. Bailey, W. F.; Jiang, X.-L. *J. Org. Chem.* **1996**, *61*, 2596.

14. Mealy, M. J.; Bailey, W. F. *J. Organomet. Chem.* **2002**, *646*, 59.

15. Bailey, W. F.; Mealy, M. J. *J. Am. Chem. Soc.* **2000**, *122*, 6787.

THE CATALYTIC INTRAMOLECULAR PAUSON-KHAND REACTION: 2,3,3α,4-TETRAHYDRO-2-[(4-METHYLBENZENE) SULFONYL]CYCLOPENTA[C]PYRROL-5(1*H*)-ONE [(Cyclopenta[b]pyrrol-5(1*H*)-one, 2,3,3a,4-tetrahydro-1-[(4-methylphenyl)sulfonyl]-)]

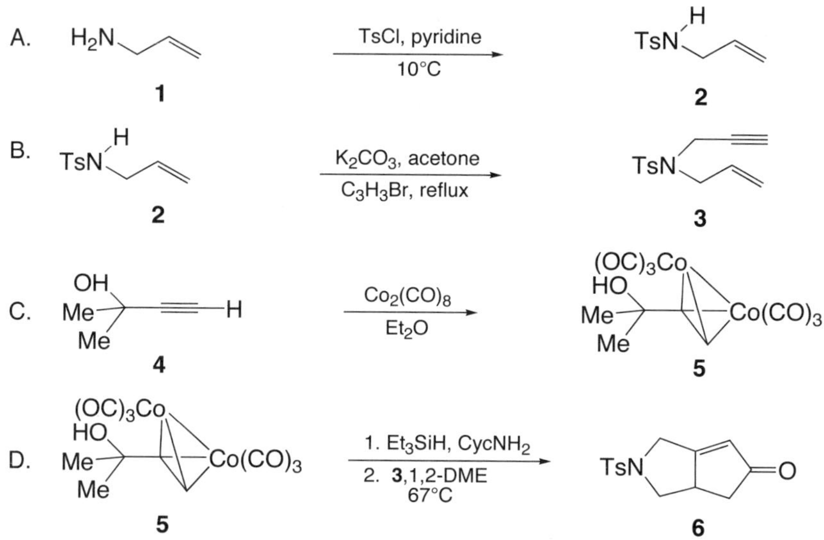

Submitted by Mittun C. Patel,[1] Tom Livinghouse,[2] and Brian L. Pagenkopf.[1]
Checked by Kyung-Hee Kim and Marvin J. Miller.

1. Procedure

A. N-(2-Propenyl)-4-methylbenzenesulfonamide (**2**). A 1-L, two-necked, round-bottomed flask equipped with a magnetic stirring bar, internal thermometer and powder funnel is charged with *p*-toluenesulfonyl chloride (97.2 g, 0.51 mol) (Note 1). The powder funnel is replaced with a rubber septum connected to a positive pressure of argon and an oil bubbler. The apparatus is flushed with argon and charged with tetrahydrofuran (THF) (400 mL, (Note 2)) and pyridine (42.9 mL, 0.53 mol). The flask is placed in an ice bath and, after the reaction mixture has cooled to ca. 10 °C, allylamine (37.5 mL, 0.50 mol) (Note 3) is added portion-wise by syringe over ca. 40 min (exothermic) maintaining the internal temperature below 15 °C. The ice bath is removed and the resulting solution is allowed to warm to room temperature. After 4 h the rubber septum is removed and the mixture is treated with 75 mL of a 2M aqueous solution of sodium hydroxide. After another 4 h, the reaction mixture is transferred to a separatory funnel, the organic phase is separated, and the aqueous phase is extracted with two 100-mL portions of ethyl acetate (EtOAc). The combined organic phases are washed with brine (50 mL) and dried with magnesium sulfate ($MgSO_4$) in the presence of activated carbon (4 g). The solution is filtered through a plug of silica gel (diameter: 5 cm; height: 3 cm) and the cake is washed with EtOAc (300 mL). The combined filtrate and washes are concentrated under reduced pressure. The crude product is recrystallized from 275 mL of 30% EtOAc/hexanes to afford 65 g (62%) of allyl tosylamide as a first crop. A second crop of 27 g (26%, 88% in total) is obtained from the mother liquor (Note 4).

B. N-(2-Propenyl)-N-(2-propynyl)-4-methylbenzenesulfonamide (**3**). A 500-mL, single-necked, round-bottomed flask equipped with a Teflon-coated stirring bar (Note 1) is charged with allyl tosylamide (31.7 g, 150 mmol), anhydrous potassium carbonate (K_2CO_3) (24.8 g, 1.2 equiv, 180 mmol) (Note 5), 1-bromo-2-propyne (20.0 mL, 1.2 equiv, 180 mmol), and acetone (300 mL). The flask is equipped with a water-cooled condenser fitted with a rubber septum. The apparatus is flushed with argon introduced through the septum and a positive pressure of argon is maintained with an argon-filled balloon (Note 6). The reaction mixture is heated to reflux with stirring for 24 h. After complete consumption of starting material, monitored by thin layer chromatography (TLC, (Note 7)), the reaction mixture is allowed to cool and is concentrated under reduced pressure on a rotary evaporator. The residue is diluted with EtOAc (250 mL) and water (125 mL) and the organic phase is separated. The aqueous phase is extracted with 200 mL of EtOAc and the combined organic phases are washed with brine (50 mL), dried ($MgSO_4$), filtered and concentrated under reduced pressure with a rotary evaporator. The residue is recrystallized from 2.5 mL of ether and 250 mL of 20% EtOAc/hexanes to afford ca. 30 g (80%) of **3** as nearly colorless crystals. A second crop totaling 5.9 g (16%, 96% in total) is also obtained (Note 8).

C. Hexacarbonyl[μ-[(3,4-η:3,4-η)-2-methyl-3-butyn-2-ol]]dicobalt (5). A single-necked, 50-mL, round-bottomed flask (Note 9), equipped with a Teflon-coated stirring bar, is charged with dicobalt octacarbonyl ($Co_2(CO)_8$) (1.7 g, 5.0 mmol, 1.0 equiv) (Note 10). After attaching a rubber septum, the flask is flushed with carbon monoxide (CO), then charged with 25 mL of degassed diethyl ether (Note 11). The flask is placed in an ice bath and, after 10 min, a bubbler is connected through the septum, followed immediately by the addition of 2-methyl-3-butyn-2-ol (0.51 mL, 5.3 mmol, 1.05 eq). After removing the ice bath and stirring for 3 h at ambient temperature, the resultant solution is filtered through a pad of Celite (diameter: 3 cm; height: 1 cm) (Note 12), which is washed with ether until the washes are clear (ca. 75 mL).

The alkyne cobalt complex is purified by silica gel chromatography. To the red solution is added 2 g of silica gel and the mixture is concentrated on a rotary evaporator (Note 13). The silica gel is then placed under vacuum (1 mm) until effervescence ceases (about 10 min). The dried silica gel is placed at the top of a column of silica gel (diameter: 3 cm; height: 9 cm), eluting first with hexane to remove non-polar by-products and subsequently with 15% ether/hexane. The desired product elutes as a red band. Concentration of the appropriate fractions on a rotary evaporator (Note 13) and under high vacuum (0.7 mm for 10 min) affords 1.4 g (79%) of 5 as a fine red powder (Note 14).

D. 2,3,3a,4-Tetrahydro-2-[(4-methylbenzene)sulfonyl]cyclopenta[c] pyrrol-5(1H)-one (**6**). A 250-mL, two-necked, round-bottomed Schlenk flask, equipped with a thermometer and Teflon-coated stirring bar, is charged with 6.0 g (24.1 mmol) of enyne 3 prepared in Step B (Note 1). A 12-cm, air-cooled condenser is attached to the flask. The top of the condenser is fitted with a rubber septum, which is connected to a CO supply and the apparatus is flushed with CO through the Schlenk valve (Note 15) for several min. After the CO purge is completed, the flask is charged with 120 mL of 1,2-dimethoxyethane (1,2-DME) (Note 16). Meanwhile, another 50-mL, round-bottomed Schlenk flask is charged with the $Co_2(CO)_6$-alkyne complex **5** (445 mg, 1.20 mmol, 5 mol%) (Note 9) and flushed with CO with a CO-filled balloon as described above. To the flask containing the $Co_2(CO)_6$-alkyne complex **5** are added sequentially 22.5 mL of 1,2-DME, cyclohexylamine (0.413 mL, 3.61 mmol, 15 mol%), 3 eq of amine/cobalt dimer) and triethylsilane (Et_3SiH) (2.40 mL of a 0.5M solution in xylenes, 1.20 mmol, 5 mol%). The assembly is heated in an oil bath at 67 °C for 15 min (Note 17), then transferred via cannula under an atmosphere of CO to the Schlenk flask containing enyne **3** (Note 18). The resulting mixture is heated at 67 °C for 24 h, when the reaction is usually complete as determined by TLC analysis (Note 19). After complete carbocyclization, the reaction mixture is transferred (in air) to a single-necked flask and concentrated on a rotary evaporator. The dark residue is dissolved in 150 mL of dichloromethane (CH_2Cl_2), washed with 10% aqueous sulfuric acid (2 × 15 mL), water (15 mL), and brine (45 mL) and treated with ca. 4 g of activated carbon and anhydrous sodium sulfate (Na_2SO_4). After 4 h the solution is filtered through a pad of Celite (diameter: 5 cm; height: 2 cm) (Note 12) and concentrated on a rotary evaporator to afford a brown solid. Purification by flash chromatography with silica gel (Note 20) affords 5.8-6.2 g (86-93% yield) of **6** as a slightly yellow solid. Colorless microcrystalline material may be obtained by recrystallization from a cold (−20 °C) solution of 1:1 2-propanol:CH_2Cl_2 (4.6-5.1 g, 69-77%) (Notes 21, 22).

2. Notes

1. All glassware is dried in a 200 °C oven overnight prior to use, assembled while hot, and allowed to cool to room temperature under dry argon.

2. Mallinckrodt ChromAR® HLPC grade THF was used as received.

3. Allylamine was purchased from Acros Organics and distilled from calcium sulfate before use.

4. The checkers obtained 80 g (76%). Properties of **2:**[10] mp 58.0-59.0 °C (EtOAc/hexanes); ^1H NMR ($CDCl_3$) δ: 2.34 (s, 3H), 3.49 (app t, J = 5.95, 2H), 4.96-5.12 (m, 2H), 5.23 (s, 1H), 5.57-5.68 (m, 1H), 7.21 (d, J = 8.2, 2H), 7.69 (d, J = 8.2, 2H); ^{13}C NMR ($CDCl_3$) δ: 21.2, 45.5, 117.2, 126.9, 129.5, 132.9, 136.9, 143.2.

5. Unless otherwise noted, all solvents and reagents were purchased from VWR Scientific or Aldrich Chemical Co., Inc. and used as received. $CDCl_3$ was filtered through basic alumina (Aldrich Chemical Co., Inc.) prior to use.

6. The checkers used a double needle in the septum, one for argon supply and the other attached to an argon-filled balloon.

7. The R_f of the starting material and product is 0.30 and 0.52, respectively (1:1 EtOAc:hexanes).

8. The checkers obtained 33.2 g (89%). Properties of **3**: mp 60.8-61.5 °C (EtOAc/hexanes); ^1H NMR ($CDCl_3$) δ: 1.96 (t, J = 2.4, 1 H), 2.35 (s, 3H), 3.76 (d, J = 6.5, 2H), 4.02 (d, J = 2.4, 2H), 5.25-5.14 (m, 2H), 5.72-5.59 (m, 1 H), 7.23 (d, J = 8.1, 2H), 7.66 (dd, J = 8.3, 1.5, 2H); ^{13}C NMR ($CDCl_3$) δ: 21.3 (CH_3), 35.7 (CH_2), 48.9 (CH_2), 73.6 (CH_2), 119.7 (CH_2), 127.6 (CH), 129.3 (CH), 131.8 (CH), 136.0 (C), 143.4 (C); IR (thin film) cm^{-1}: 3257, 3049, 2987, 2885, 2117, 1646, 1601, 1425, 1330, 1165.

9. The glassware must be cooled to room temperature before use.

10. *Caution. Dry $Co_2(CO)_8$ can be pyrophoric.* $Co_2(CO)_8$ is toxic and should be handled in a glove box or well-ventilated hood. $Co_2(CO)_8$ was purchased from Strem Chemical Co., Inc. High purity $Co_2(CO)_8$ is critical for successful catalytic reactions. A freshly opened sample of $Co_2(CO)_8$ stored in a dry box is often sufficiently pure to be used directly in this reaction provided three eq of cyclohexylamine per $Co_2(CO)_8$ are added to the reaction mixture. However, a more reliable approach is to generate the active cobalt catalyst in situ as described herein.

11. Ether was distilled from a purple solution of sodium/benzophenone ketyl and degassed by bubbling with argon.

12. Celite 545 was purchased from Fluka.

13. The bath temperature of the rotary evaporator must be kept below 35 °C.

14. The checkers obtained 1.35 g (75%) Properties of **5:**[10] ^{1}H NMR ($CDCl_3$) δ: 1.55 (s, 6H), 1.70 (s, 1 H), 6.01 (s, 1 H); ^{13}C NMR ($CDCl_3$) δ: 33.4, 71.7, 72.9, 76.8, 107.3, 200.0 (m).

15. Carbon monoxide is highly toxic and should be handled in a well-ventilated hood.

16. All solvents must be rigorously degassed for the Pauson-Khand step. 1,2-DME obtained from VWR Scientific was freshly distilled from a deep blue solution of sodium/benzophenone ketyl and degassed with argon prior to use.

17. The internal temperature of the mixture is 64 °C. Temperatures above 70 °C lead to catalyst decomposition.

18. To facilitate transfer of the catalyst, the CO source connected to the large reaction vessel is removed and replaced with a line leading to a mineral oil bubbler. When the transfer is complete, the bubbler is disconnected and a CO-filled double-walled balloon is connected to the top of the condenser and used to maintain a CO atmosphere throughout the reaction. Strict exclusion of air during these transfers is critical.

19. The course of the reaction may be monitored by TLC: R_f of **3** and **6** is 0.82 and 0.36, respectively (1:1 EtOAc:hexanes). The color of the reaction mixture typically turns from burgundy to dark brown at the end of the reaction.

20. The solid residue is dissolved in a minimum of dichloromethane and purified by flash chromatography on silica gel (8 cm diameter by 15 cm long column), eluting with 1.5 % EtOH in CH_2Cl_2.

21. The crystals are collected by filtration on a fritted glass filter funnel, rinsed with ether and dried under reduced pressure.

22. The checkers obtained 4.3 g (65%). Properties of **6**: mp 147.0-148.0 °C (2-PrOH/CH_2Cl_2); ^{1}H NMR ($CDCl_3$) δ: 1.96-2.03 (dd, *J* = 17.9, 3.7, 1H), 2.37 (s, 3H), 2.48-2.60 (m, 2H), 3.04-3.10 (m, 1H), 3.93-4.00 (dd, *J* = 14.3, 5.1, 2H), 4.27 (d, *J* = 16.5, 1H), 5.92 (s, 1H), 7.29 (d, *J* = 8.1, 2H), 7.66 (d, *J* = 8.2, 2H); ^{13}C NMR ($CDCl_3$) δ: 21.0 (CH_3), 39.3 (CH_2), 43.4 (CH), 47.1 (CH_2), 51.9 (CH_2), 125.6 (CH), 126.9(CH), 129.5 (CH), 133.0 (C), 143.6 (C), 178.2 (C), 206.8 (C); IR (KBr) cm^{-1}: 3067, 1711, 1648, 1343, 1162, 1090.

Waste Disposal Information

All toxic materials were disposed of in accordance with "Prudent Practices in the Laboratory"; National Academy Press; Washington, DC, 1995.

3. Discussion

The Pauson-Khand reaction is one of the most convenient methods for the synthesis of cyclopentenones, and both inter- and intramolecular variations exist.[3,4] In the classical procedure, the enyne is mixed with one equivalent of $Co_2(CO)_8$ and the mixture is heated (either at reflux or in a sealed tube) for several hours to days. The synthetic ease of the Pauson-Khand reaction was enhanced when it was discovered that silica gel, tertiary amine *N*-oxides and DMSO all markedly accelerated the reaction.[4] The Pauson-Khand reaction is compatible with enynes containing esters, ethers, amines, sulfides, 1,2-disubstituted olefins and substituted or terminal alkynes. A major drawback to large scale, classical Pauson-Khand reactions is the requirement for stoichiometric quantities of $Co_2(CO)_8$. We recently reported the first catalytic Pauson-Khand reaction conveniently operative at 1 atmosphere of CO.[5,6] The key factors of successful catalytic cyclizations are the use of highly purified $Co_2(CO)_8$, strictly anaerobic conditions, and temperature control. To circumvent the requirement for very pure $Co_2(CO)_8$ we developed a chemically robust, air-stable $Co_2(CO)_6$-alkyne complex as a source of pure $Co_2(CO)_8$ when released by silylative decomplexation.[7] This procedure employs cyclohexylamine, which has been shown to accelerate stoichiometric Pauson-Khand reactions[8] and minimizes the sensitivity of the catalyst to oxygen. The scope of the catalytic Pauson-Khand has been further enhanced by the discovery that 2,2,2-trifluoroethanol as a co-solvent minimizes some non-carbonylative isomerization processes.[9] The utility of the method is demonstrated by the diverse range of substrates shown in the Table.

TABLE

THERMALLY PROMOTED CATALYTIC PAUSON-KHAND CYCLIZATIONS

Entry	Enyne	Product	Yield (diastereomer ratio)
1	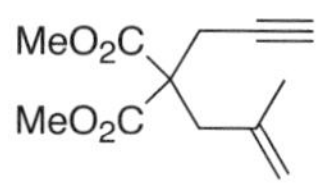	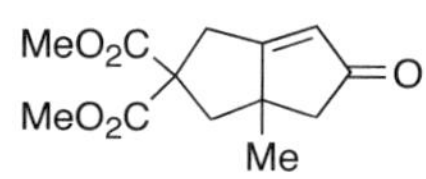	95[a]
2	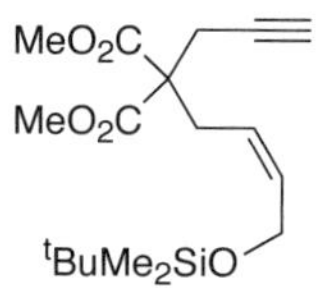	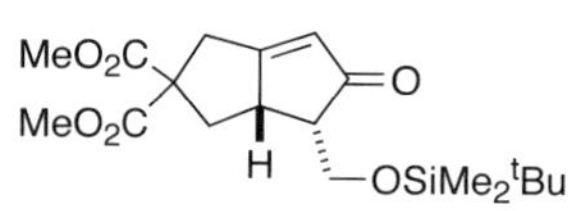	90[b] (>30:1)
3	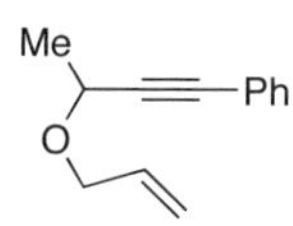	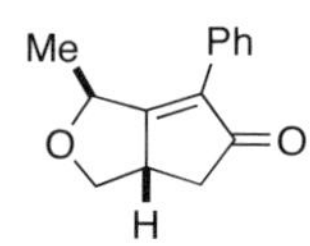	96[a,c] (30:1)
4	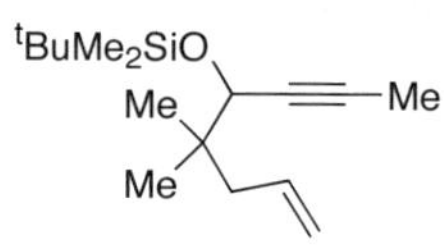	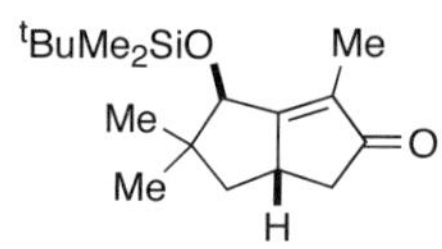	74[a,d] (2.5:1)
5	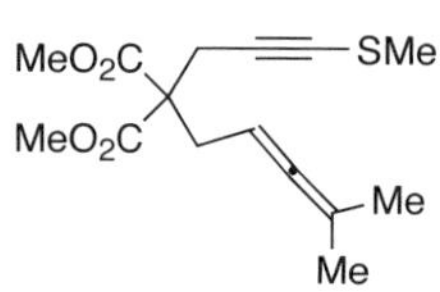	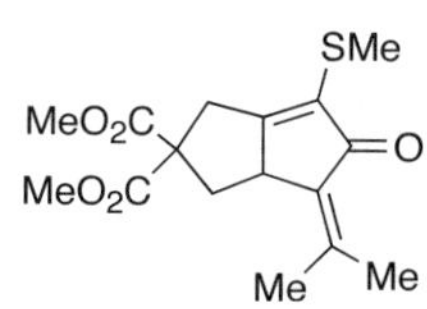	84[b]

[a]5 mol% $Co_2(CO)_8$. [b]10 mol% $Co_2(CO)_8$. [c]Reaction solvent was 2,2,2-trifluoroethanol; see ref.9. [d]Reaction solvent was 2,2,2-trifluoroethanol:1,2-DME (2:1); see ref.9.

1. The University of Texas at Austin, Department of Chemistry and Biochemistry, Austin, TX 78712.
2. Montana State University, Department of Chemistry and Biochemistry, Bozeman, MT 59717.

3. Piper, J. R.; Rose, L. M.; Johnston, T. P.; Grennan, M. M. *J. Med. Chem.* **1975**, *18*, 803.

4. The Pauson-Khand reaction has been extensively reviewed: (a) Schore, N.E. *Chem. Rev.* **1988**, *88*, 1081; (b) Schore, N.E., *Org. React.* (NY) **1991**, *40*, 1; (c) Schore, N.E. in *Comprehensive Organic Synthesis*; Trost, B.M.; Fleming, I., Eds.; Pergamon: Oxford, **1991**, Vol 5, p. 1037; (d) Schore, E.E. in *Comprehensive Organometallic Chemistry II*; Hegedus, L.S.; Ed. Elsevier: Oxford, **1995**; Vol. 12, p. 703; (e) Brummond, K. M.; Kent, J. L. *Tetrahedron* **2000**, *56*, 3263.

5. Pagenkopf, B. L.; Livinghouse, T. *J. Am. Chem. Soc.* **1996**, *118*, 2285.

6. Belanger, D. B.; O'Mahony, D. J. R.; Livinghouse, T. *Tetrahedron Lett.* **1998**, *39*, 7641.

7. Belanger, D. B.; Livinghouse, T. *Tetrahedron Lett.* **1998**, *39*, 7641.

8. Sugihara, T.; Yamada, M.; Ban, H.; Yamaguchi, M.; Kaneko, C. *Angew. Chem. Int. Ed. Engl.* **1997**, *36*, 2801.

9. Pagenkopf, B. L.; Belanger, D. B.; O'Mahony, D. J. R.; Livinghouse, T. *Synthesis* **2000**,1009.

10. Piper, J. R.; Rose, L. M.; Johnston, T. P.; Grennan, M. M. *J. Med. Chem.* **1975**, *18*, 803.

11. Giordano, R.; Sappa, E.; Predieri, G. *Inorg. Chim. Acta* **1995**, *228*, 139.

INTRA- AND INTERMOLECULAR KULINKOVICH CYCLOPROPANATION REACTIONS OF CARBOXYLIC ESTERS WITH OLEFINS: BICYCLO[3.1.0]HEXAN-1-OL AND *trans*-2-BENZYL-1-METHYLCYCLOPROPAN-1-OL

[(Bicyclo[3.1.0]hexan-1-ol and 1-Cyclopropanol, 1-methyl-2-phenylmethyl-)]

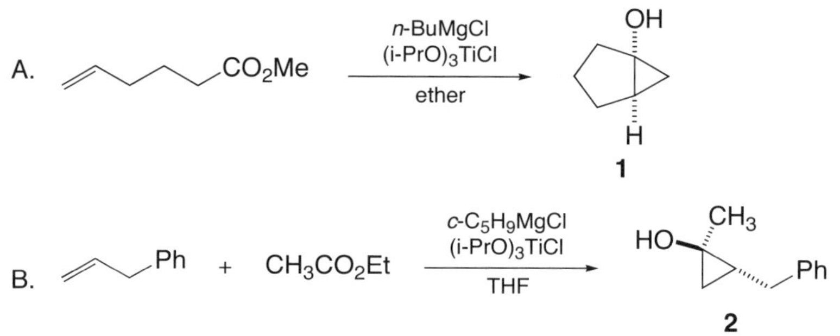

Submitted by Se-Ho Kim, Moo Je Sung, and Jin Kun Cha.[1]
Checked by Fanglong Yang and Dennis P. Curran.

1. Procedure

A. Bicyclo[3.1.0]hexan-1-ol (**1**). To a 500-mL, round-bottomed flask, equipped with a magnetic stirring bar and rubber septum (Note 1), is added at room temperature a mixture of 2.0 g (15.6 mmol) of methyl 5-hexenoate, 11.2 mL (11.2 mmol) of a 1M solution of chlorotitanium triisopropoxide in hexane, and 54 mL of anhydrous ether (Note 2) under a nitrogen atmosphere. A 1M solution of *n*-butylmagnesium chloride in ether (52 mL, 52 mmol; (Note 3)) is added over a period of 6.5 h via a syringe pump (Note 4) at room temperature. After the addition is complete, the resulting black reaction mixture is stirred for an additional 20 min. The mixture is cooled to 0 °C with an ice bath, diluted with 50 mL of ether and then quenched by slow addition of water (14 mL). The resulting mixture is stirred for an additional 3 h at room temperature. The organic phase is separated and the aqueous phase is extracted with ether (3 × 100 mL). The combined organic extracts are washed with brine (2 × 50 mL), dried over anhydrous magnesium sulfate, filtered, and concentrated under reduced pressure using a rotary evaporator (Note 5). Purification of the crude product (Note 6) by column chromatography on 40 g of silica gel (Note 7) using a gradient of

5% to 10% ether/pentane as the eluent provides 1.09 g (71%) of bicyclo[3.1.0]hexan-1-ol (**1**)[2] as a colorless oil (Note 8).

B. trans-2-Benzyl-1-methylcyclopropan-1-ol (**2**). To a 500-mL, round-bottomed flask, equipped with a magnetic stirring bar and rubber septum, is added a mixture of 2.5 g (21 mmol) of allylbenzene, 2 mL (20 mmol) of ethyl acetate, and 20 mL of a 1M solution of chlorotitanium triisopropoxide in hexane (Note 2), and 160 mL of anhydrous tetrahydrofuran (THF). After the mixture has been cooled to 0 °C with an ice bath under a nitrogen atmosphere, a 1M solution of cyclopentylmagnesium chloride in ether (80 mL, 80 mmol; (Note 9)) is added over a period of 2.5 h via a syringe pump (Note 4). After the addition is complete, the resulting black reaction mixture is stirred for 30 min at 0 °C, then is quenched by the cautious addition of water (15 mL). The resulting mixture is stirred for an additional 1 h at room temperature and filtered through a pad of Celite, which is rinsed thoroughly with ether (4 × 50 mL). The combined filtrate and rinsings are poured into a separatory funnel containing 50 mL of water and shaken thoroughly. The organic phase is separated, washed with brine (50 mL), dried over anhydrous magnesium sulfate, filtered, and concentrated under reduced pressure using a rotary evaporator. Purification of the crude product (obtained as a pale yellow oil) by column chromatography on 80 g of silica gel (Note 7) using 1:20 ethyl acetate:hexane as the eluent provides 2.6 g (80%) of *trans*-2-benzyl-1-methylcyclopropan-1-ol (**2**) as a colorless oil (Note 10).

2. Notes

1. All glassware, needles, syringes were oven-dried (>100 °C) and quickly assembled prior to use.

2. (a) Methyl 5-hexenoate was purchased from TCI America and used as received. A 1M solution of chlorotitanium triisopropoxide in hexane was purchased from Aldrich Chemical Company, Inc. and used as obtained. (b) Ether and THF were dried and purified by distillation from sodium/benzophenone. Unless stated otherwise, all solvents and reagents in this procedure were obtained commercially and used as received.

3. *n*-Butylmagnesium chloride (a 2M solution in ether) was available from Aldrich Chemical Company, Inc. Alternatively, it was freshly prepared from 1-chlorobutane and magnesium.

4. (a) The tip of a syringe needle was placed right above the surface of the reaction mixture, rather than near the neck of the reaction flask. This placement was found to minimize the amount of the Grignard reagent to be

added. An excess of the Grignard reagent was used to ensure complete conversion. (b) To prevent the glass syringe plunger from sticking in the barrel, vacuum grease was applied around the edge of the exposed portion of the plunger.

5. The solvent was removed under water (aspirator) vacuum, while the bath temperature was kept below 20 °C. Otherwise, considerable loss occurred due to the volatility of the cyclopropanol product.

6. Immediate purification of the crude product is recommended. Otherwise, it should be stored in a cold (−30 °C) freezer to avoid decomposition.

7. Flash chromatography was performed on E. Merck silica gel (70-230 mesh). For Procedure A, the checkers used 60 g silica gel, eluting first with 150 mL 10% ether/pentane, then with 200 mL 20% ether/pentane. For Procedure B, the checkers used 100 g silica gel and eluted first with 100 mL 9% ether/hexane, then with 200 mL 17% ether/hexane.

8. Yields in several runs were obtained in the range of 68-75% by the submitters and 67-71% by the checkers. Spectral data for **1** [R_f = 0.27 (1:4 ethyl acetate:hexane)] are as follows: bp 53-55 °C (15 mm, Kugelrohr) [lit.[2b] bp 63-65 °C (20 mm)]; IR (neat) cm^{-1}: 3300; ^1H NMR (360 MHz, acetone-d6) δ 0.47 (t, J = 4.6, 1 H), 0.67 (dd, J = 4.6, 8.9, 1 H), 1.02-1.15 (m, 2 H), 1.43 (dd, J = 7.9, 12.1, 1 H), 1.55 (m, 1 H), 1.75-1.89 (m, 3 H), 4.25 (s, 1 H, -OH); ^{13}C NMR (90 MHz, acetone-d_6) δ 14.9, 21.8, 24.3, 27.4, 34.3, 64.2; Mass spectrum: *m/z* 98 (M^+, 29), 97 (47), 83 (56), 70 (100), 55 (54).

9. Cyclopentylmagnesium chloride (a 2M solution in ether) was available from Aldrich Chemical Company, Inc. Alternatively, it was freshly prepared from chlorocyclopentane and magnesium.

10. Yields in several runs were obtained in the range of 78-86% by the submitters and 78-82% by the checkers. Spectral data for **2** [R_f = 0.47 (1:3 ethyl acetate:hexane)] are as follows: IR (neat) cm^{-1}: 3332, 1228; ^1H NMR (360 MHz, $CDCl_3$) δ 0.29 (dd, J = 5.4, 6.0, 1 H), 0.98 (dd, J = 5.4, 10.0, 1 H), 1.30-1.39 (m, 1 H), 1.49 (s, 3 H), 1.88 (s, 1 H, -OH), 2.55 (dd, J = 7.5, 15.3, 1 H), 2.67 (dd, J = 7.3, 15.3, 1 H), 7.19-7.33 (m, 5 H); ^{13}C NMR (90 MHz, $CDCl_3$) δ 20.4, 20.8, 26.1, 35.6, 55.8, 125.9, 128.1, 128.4, 141.6; Mass spectrum (EI): *m/z* 162 (M^+, 3), 104 (76), 92 (58), 71 (100).

Waste Disposal Information

All toxic materials were disposed of in accordance with "Prudent Practices in the Laboratory"; National Academy Press: Washington, DC, 1995.

3. Discussion

Kulinkovich and co-workers recently developed a new method for preparing cyclopropanols (the subject of recent reviews[3]) by treating carboxylic esters with a suitable Grignard reagent (i.e., containing β-hydrogen atoms) in the presence of titanium tetraisopropoxide.[4] The present procedures represent a convenient variant of the Kulinkovich cyclopropanation reaction by using a sacrificial Grignard reagent (n-butylmagnesium chloride or cyclopentylmagnesium chloride) to generate the putative titanacyclopropane intermediate for subsequent ligand exchange with a monosubstituted olefin. A mixture of the two rapidly interconverting titanacyclopropane intermediates is thus formed, followed by the intramolecular cyclopropanation reaction, to afford the corresponding bicyclic cyclopropanol products in good to excellent yield.[5,6] Because formation of five- and six-membered rings proceeds much faster than the intermolecular reaction, the precise position of equilibrium is inconsequential for successful cyclization, as demonstrated by the facile preparation of bicyclo[3.1.0]hexan-1-ol (**1**). Consequently, various Grignard reagents have been shown to be effective for the intramolecular cyclopropanation reactions with insignificant differences in yields. Whereas several titanium alkoxides and aryloxides can also be employed, chlorotitanium triisopropoxide and methyltitanium triisopropoxide have often been found to be the titanium reagent of choice.[7] Ether, THF, toluene, or even dichloromethane are generally appropriate reaction solvents.

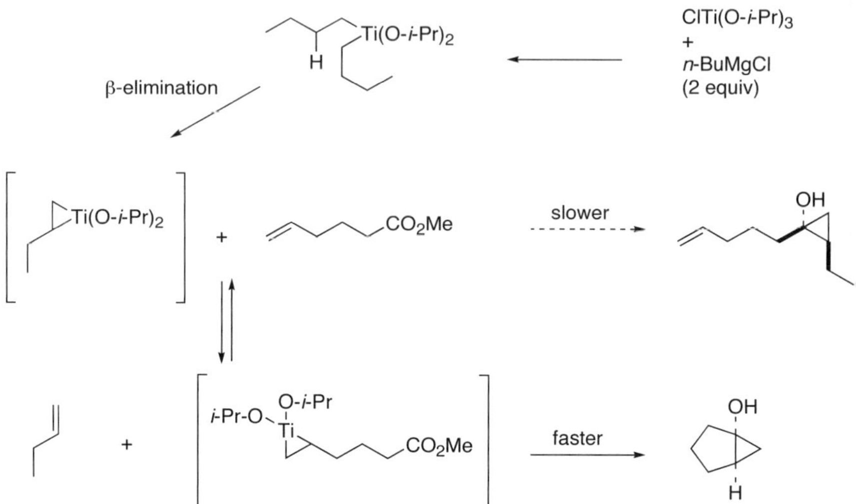

1

Central to the successful extension of the olefin exchange strategy to intermolecular coupling reactions of an ester and a monosubstituted olefin, as illustrated by the preparation of trans-2-benzyl-1-methylcyclopropan-1-ol (**2**), is the use of cyclohexyl or cyclopentyl Grignard reagents.[8] These secondary Grignard reagents are believed to allow the position of equilibrium to be shifted to the desired titanacyclopropane intermediate (less substituted, more stable). Notable characteristics of the ligand exchange-mediated variant of the Kulinkovich cyclopropanation include the simplicity and ease of operation using inexpensive reagents and also excellent compatibility with several common functional groups (for example, alkoxy, bromo, and alkenyl substituents).

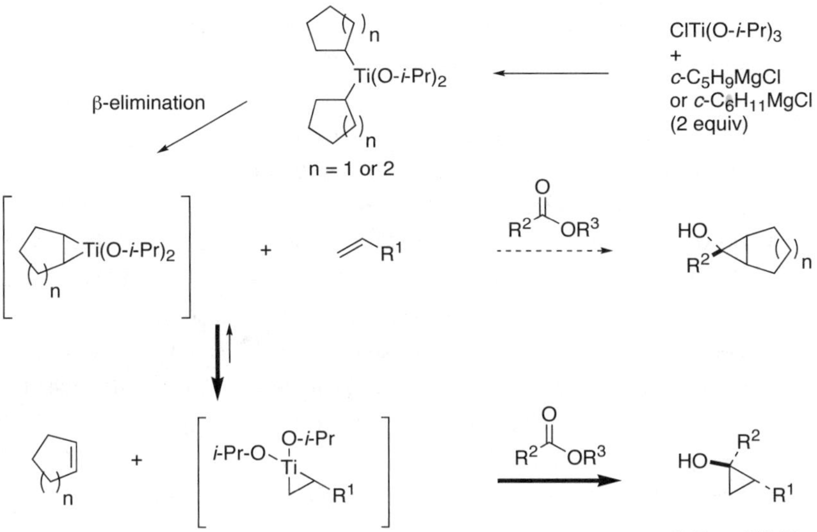

2: $R_1 = CH_2Ph$
$R_2 = CH_3$

In addition, other carboxylic acid derivatives, such as tertiary carboxamides and cyclic carbonates, readily undergo cyclopropanation reactions under similar conditions to provide cyclopropylamines[9,10] and cyclopropanone hemiacetals,[11] respectively. The cognate titanium-mediated coupling of imides has been shown to afford synthetically useful N-acylhemiaminals.[12]

1. Current address: Department of Chemistry, Wayne State University, Detroit, MI 48202.
2. Bicyclo[n.1.0]alkan-1-ols were previously prepared by the Simmons-Smith cyclopropanation reaction of 1-(trimethylsilyloxy)cycloalkenes and subsequent hydrolysis: (a) Conia, J. M.; Girard, C. *Tetrahedron Lett.* **1973**, 2767. (b) Murai, S.; Aya, T.; Sonoda, N. *J. Org. Chem.* **1973**, 38, 4354.
3. (a) Kulinkovich, O. G.; Sviridov, S. V.; Vasilevskii, D. A.; Pritytskaya, T. S. *Zh. Org. Khim.* **1989**, *25*, 2244. (b) Kulinkovich, O. G.; Sviridov, S. V.; Vasilevskii, D. A.; Savchenko, A. I.; Pritytskaya, T. S. *Zh. Org. Khim.* **1991**, *27*, 294. (c) Kulinkovich, O. G.; Sorokin, V. L.; Kel'in, A. V. *Zh. Org. Khim.* **1993**, *29*, 66. (d) Kulinkovich, O. G.; Savchenko, A. I.; Sviridov, S. V.; Vasilevskii, D. A. *Mendeleev Commun.* **1993**, 230.

4. (a) Kulinkovich, O. G.; de Meijere, A. *Chem. Rev.* **2000**, *100*, 2789. (b) Sato, F.; Urabe, H.; Okamoto, S. *Chem. Rev.* **2000**, *100*, 2835. (c) For a recent theoretical study on mechanism, see: Wu, Y.-D.; Yu, Z.-X. *J. Am. Chem Soc.* **2001**, *123*, 5777.

5. Lee, J.; Kang, C. H.; Kim, H.; Cha, J. K. *J. Am. Chem. Soc.* **1996**, *118*, 291.

6. (a) Kasatkin, A.; Sato, F. *Tetrahedron Lett.* **1995**, *36*, 6079. (b) Kasatkin, A.; Kobayashi, K.; Okamoto, S.; Sato, F. *Tetrahedron Lett.* **1996**, *37*, 1849.

7. (a) Corey, E. J.; Rao, S. A.; Noe, M. C. *J. Am. Chem. Soc.* **1994**, *116*, 9345. (b) Chaplinski, V.; Winsel, H.; Kordes, M.; de Meijere, A. *Synlett* **1997**, 111. (c) Lee, J. C.; Sung, M. J.; Cha, J. K. *Tetrahedron Lett.* 2001, **42**, 2059.

8. Lee, J.; Kim, H.; Cha, J. K. *J. Am. Chem. Soc.* **1996**, *118*, 4198.

9. Chaplinski, V.; de Meijere, A. *Angew. Chem., Int. Ed. Engl.* **1996**, *35*, 413.

10. Lee, J.; Cha, J. K. *J. Org. Chem.* **1997**, *62*, 1584.

11. Lee, J.; Kim, Y. G.; Bae, J. G.; Cha, J. K. *J. Org. Chem.* **1996**, *61*, 4878.

12. Lee, J.; Ha, J. D.; Cha, J. K. *J. Am. Chem. Soc.* **1997**, *119*, 8127.

SYNTHESIS AND [3+2] CYCLOADDITION OF A 1,1-DIALKOXY-2-METHYLENECYCLOPROPANE: 6,6-DIMETHYL-1-METHYLENE-4,8-DIOXASPIRO[2.5]OCTANE and *cis*-5-(5,5-DIMETHYL-1,3-DIOXAN-2-YLIDENE)HEXAHYDRO-1(2H)-PENTALEN-2-ONE [(4,8-Dioxaspiro[2.5]octane, 6,6-dimethyl-1-methylene-)]

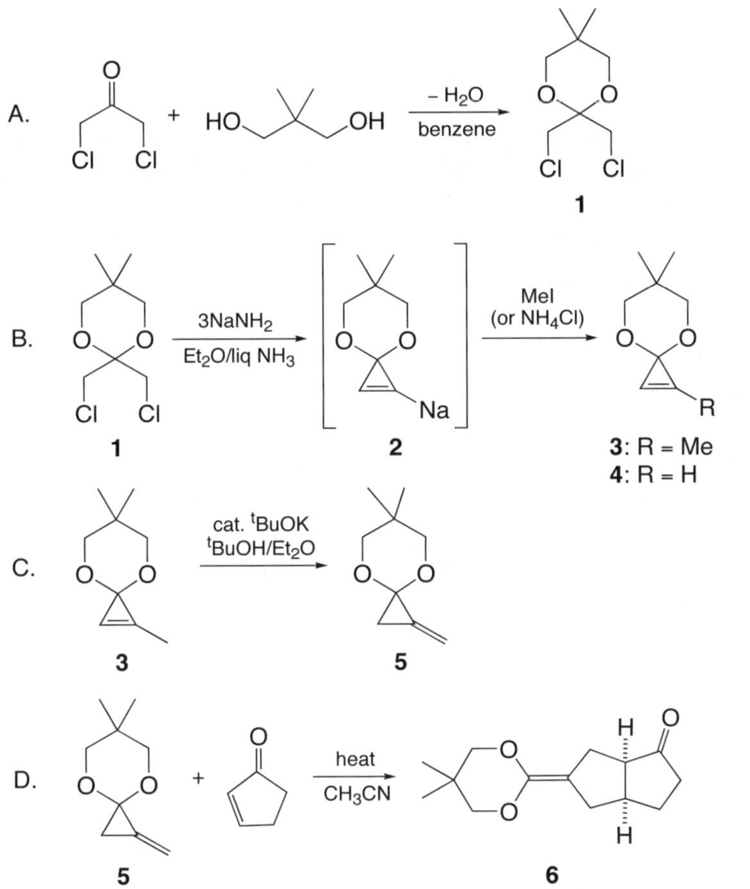

Submitted by Masaharu Nakamura,[1] Xiao Qun Wang,[1] Masahiko Isaka,[1] Shigeru Yamago,[2] and Eiichi Nakamura.[1]
Checked by Jimmy Wu and Edward J. J. Grabowski.

1. Procedure

Caution! All operations should be performed in a well-ventilated hood, and care should be taken to avoid skin contact with 1,3-dichloro-2-propanone and cyclopropenones.

A. 2,2-Bis-(chloromethyl)-5,5-dimethyl-1,3-dioxane (**1**). A mixture of 1,3-dichloroacetone (152 g, 1.20 mol, an irritant), neopentyl glycol (138 g, 1.32 mol), p-toluenesulfonic acid (4.6 g, 0.024 mol) and benzene (100 mL) (Note 1) is heated to reflux for 19 h in a 500-mL, round-bottomed flask equipped with a Dean-Stark trap and a condenser with azeotropic removal of water. After separation of water ceases, the reaction mixture is partitioned between hexane (500 mL) and saturated sodium bicarbonate ($NaHCO_3$) (200 mL). The organic phase is washed with water (100 mL) and saturated sodium chloride (NaCl) (100 mL), dried over magnesium sulfate ($MgSO_4$), and concentrated on a rotary evaporator. Distillation of the residue yields, after about 5 g of forerun, 249 g (97%) of the acetal **1** as a colorless oil (bp 99-100 °C, 3.5 mm) (Note 2).

B. 1,6,6-Trimethyl-4,8-dioxaspiro[2.5]oct-1-ene (**3**) (Note 3). A solution of sodium amide in liquid ammonia is prepared according to a procedure previously described (Notes 4, 5). An oven-dried, 2-L, three-necked, round-bottomed flask is equipped with a mechanical stirrer, nitrogen gas inlet, and dry ice/acetone condenser protected with a drying tube containing potassium hydroxide pellets. The flask is flushed with nitrogen introduced through the gas inlet, then is placed in a dry ice/acetone bath. The nitrogen source is replaced with a hose connected to a cylinder of ammonia (NH_3). Gaseous NH_3 is introduced to the flask condensing ca. 400 mL of NH_3 and gentle stirring is started. The NH_3 inlet is replaced with a glass stopper and the dry ice/acetone bath is replaced with a −35 °C bath (electronically controlled or a dry ice/trichloroethylene bath). Crystals of $Fe(NO_3)_3 \cdot 9H_2O$ (0.3 g) are added through the neck having a stopper. To the resulting orange solution is added a small (about 5 mm3) cube of sodium. The solution is stirred (Note 6) until the blue color disappears and fine black particles appear. Pieces of sodium (21.44 g, 0.930 mol) are then added over 35 min. After 20 min, the solution turns into a dark gray suspension with a white precipitate. The cooling bath is replaced with a dry ice/acetone bath, and the gas inlet is replaced with a pressure-equalizing addition funnel containing a solution of 2,2-bis-(chloromethyl)-5,5-dimethyl-1,3-dioxane **1** (63.93 g, 0.300 mol) in 150 mL of dry ether (Et_2O). This solution is added dropwise to the slurry of sodium amide in liquid NH_3 over 1 h. The addition funnel is rinsed with 20 mL of dry Et_2O. The cooling bath is removed, and the mixture is stirred for 3 h (Note 7), then is cooled again with a dry ice/acetone bath. After 10 min, a solution of freshly distilled iodomethane (44.71 g, 0.315 mol) in 80 mL of dry Et_2O is added via the addition funnel over 1 h (Notes 8, 9); the funnel is rinsed with 20 mL of dry Et_2O. After stirring for 15 min, the cooling bath is removed and the mixture is stirred for 1 h. The mixture is cooled with a dry ice/acetone bath again and solid ammonium chloride (NH_4Cl) (20.24 g, 0.378 mol) is added in several portions over 5 min. The dry ice condenser is removed and the ammonia is allowed to evaporate. The cooling bath is replaced with a water bath (ca. 30 °C), and a 1:1 mixture of dry Et_2O and dry pentane (400 mL) is added through the addition funnel over 10 min with vigorous stirring. The water bath temperature is maintained between 25-30 °C. After evaporation of most of the ammonia (1.5-2 h), the ethereal solution is filtered by suction through a pad of Hyflo Super Cel to remove the inorganic salts. The filter cake is washed three times with 80 mL of Et_2O. The combined filtrate and washes are concentrated under reduced pressure (30-40 mm, 25 °C) (Note 10) and the residue is distilled to yield **3** as a colorless oil (32.3 g, 70%; bp 58-61 °C, 6-7 mm) (Notes 11-14). While this material is adequate for use in Step C, a pure sample can be obtained by flash chromatography of the crude product followed by distillation. For 36.1 g of crude product (obtained on a 0.25 mol scale), chromatography is performed with 290 g of silica gel (Merck, Kieselgel 60) and Et_2O:pentane (5:95) as eluent, collecting 160-mL fractions. Fractions 3-16 containing the product are combined. Removal of the solvent under reduced pressure followed by vacuum distillation afforded 29.3 g of **3** in 76% yield (99% pure by GC).

C. 6,6-Dimethyl-1-methylene-4,8-dioxaspiro[2.5]octane (**5**). An oven-dried, 100-mL, round-bottomed flask, equipped with a magnetic stirring bar and three-way stopcock, which is connected to a vacuum/nitrogen line, is flushed with nitrogen. The flask is placed in a plastic bag filled with nitrogen, and powdered potassium *tert*-butoxide (1.32 g, 11.8 mmol) (Note 15) is introduced quickly. The flask is connected to a nitrogen source, *tert*-butyl alcohol (7.11 g, 96 mmol) and 40 mL of Et_2O (Note 16) are introduced by syringe, and the solution is stirred for 10 min at room temperature.

An oven-dried 300-mL, round-bottomed flask, equipped with a magnetic stirring bar and three-way stopcock, connected to a vacuum/nitrogen line, is charged with 1,6,6-trimethyl-4,8-dioxaspiro[2.5]oct-1-ene **3** (46.3 g, 0.30 mol) under nitrogen. Ether (100 mL) is introduced via syringe, and the solution is stirred with external cooling at −78 °C (dry ice/acetone bath). The solution of potassium *tert*-butoxide is added to this solution through a 1.5 mm i.d. cannula over 5 min. After stirring for 5 min at this temperature, the cooling bath is removed and the solution is allowed to warm to room temperature (Note 17). After stirring for 4 h (Note 18), 15 mL of 1N HCl is added in one portion. The mixture is transferred to a separatory funnel with the aid of 30 mL of Et_2O, and the aqueous phase is separated. The organic phase is washed successively with 5 mL each of saturated $NaHCO_3$ and saturated NaCl, and is dried over Na_2SO_4. After concentration on a rotary evaporator, a pale yellow oil is obtained. Distillation under reduced pressure (Note 19) (57-59 °C/7.3 mm) affords 36.49 g (79% yield) of 5 (Notes 19, 20) after ca. 1.5 g of a forerun. The material is sufficiently pure for the subsequent step.

D. cis-5-(5,5-Dimethyl-1,3-dioxan-2-ylidene)hexahydro-1(2H)-pentalen-2-one (**6**). An oven-dried, 50-mL, round-bottomed flask, equipped with a magnetic stirring bar and three-way stopcock, is flushed with nitrogen. A mixture of 6,6-dimethyl-1-methylene-4,8-dioxaspiro[2.5]octane, **5**, (8.48 g, 55 mmol) and 2-cyclopenten-1-one (4.52 g, 55 mmol) in 15 mL of acetonitrile is introduced via syringe and the solution is heated at 60 °C for 12 h. The three-way stopcock is replaced with a distillation head. Solvent is removed by distillation (ca. 30-120 °C, ca. 20-1.4 mm) and the residue is distilled under reduced pressure (142-143 °C, 1.4 mm) to afford 6 (10.0 g, 42.3 mmol, 77%), which crystallizes upon standing at room temperature (Notes 21, 22).

2. Notes

1. The following chemicals were purchased from Kanto Chemical Co. Inc. and used as received: neopentyl glycol, *p*-toluenesulfonic acid, sodium, and *n*-BuLi. N,N,N',N'-Tetramethylethylenediamine was purchased from Aldrich Chemical Company, Inc. and distilled before use. 1,3-Dichloro-2-propanone was obtained from Wacker Chemicals East Asia and used as received. Reagent grade benzene, pentane, Et_2O, THF, *tert*-butyl alcohol, acetonitrile, and toluene were purchased from Wako Chemicals Industries Ltd. Benzene, pentane, toluene, and *tert*-butyl alcohol were distilled from CaH_2; Et_2O and THF from sodium benzophenone ketyl immediately before use; acetonitrile successively from P_2O_5 and anhydrous K_2CO_3. Potassium *tert*-butoxide and 2-cyclopenten-1-one were purchased from Tokyo Kasei Kogyo Co. Ltd.; the ketone was distilled before use.

2. The product was pure by 270 MHz ^{1}H NMR analysis. Spectral properties: IR (neat) cm^{-1}: 2950, 2860, 1105, 1025, 770; ^{1}H NMR (270 MHz, $CDCl_3$) δ: 1.00 (s, 6 H), 3.57 (s, 4 H), 3.80 (s, 4 H).

3. Important notes: (1) Since the reactions are sensitive to temperature effects, care must be taken to allow the reaction mixture to reach thermal equilibrium with the cooling bath at each stage of the operation. (2) Because of the rather unstable nature of the product in its impure form, the entire procedure except the final distillation must be carried out within a day. This will take at least 12 h. See also (Note 10).

4. One of the procedures for converting sodium to sodium amide described in *Organic Syntheses* is used.[3]

5. The use of commercial $NaNH_2$ is not recommended since the yield and the purity of the cyclization product are heavily dependent on the purity of $NaNH_2$.

6. Gentle stirring (e.g., 120 rpm) is recommended to avoid loss of the amide base which may adhere to the upper part of the flask. This was found to be a good practice to achieve the most accurate base/substrate molar ratio.

7. The reaction may be complete in about 2 h. An aliquot may be removed and analyzed by GC for the intermediate 2-chlorocyclopropanone acetal. In case of a deficiency of sodium amide, a small amount of the chloride may remain but can be removed by distillation. The analysis may be performed on a 0.25 mm i.d. × 25 m capillary GC column (HR-1, Shinwa Chemical Industries, Ltd.) at 80 °C. Typical retention times for the 2-chlorocyclopropanone acetal and **3** are 9.9 and 4.6 min, respectively. The amount of the former may be estimated by comparison of its peak area with that of 3 assuming equal detector response factors of the two compounds.

8. Rapid addition of iodomethane causes the formation of a significant amount of 2,3-dialkylated compound and 2-methylenecyclopropanone acetal (by isomerization of the olefin).

9. If iodomethane is replaced with solid NH_4Cl (3-4 eq to **1**, added very carefully in several portions), the unsubstituted cyclopropenone acetal **4** can be obtained in the same manner. The submitters carried out this reaction on a 0.3-1 mol scale in 70-85% distilled yield.

10. Care must be taken not to lose material upon concentration. In case of a lack of time, the crude product may be kept under nitrogen in a freezer (−20 °C) and distilled later.

11. In order to obtain a good yield, it is important to carry out the distillation as fast as possible (<25 min), but with the bath temperature below 100 °C.

12. This product may contain 6~8% of the unsubstituted cyclopropenone acetal **4** and 3-5% of 2,3-dimethylcyclopropenone acetal and ca. 1% of methylene-cyclopropanone acetal **5** as determined by capillary GC analysis on a 0.25 mm i.d. × 25 m capillary column (HR-1) at 80 °C. Typical retention times for **3**, **4**, **5**, and 2,3-dimethyl cyclopropenone acetal are 6.4, 4.3, 5.8 and 9.6 min, respectively. The checkers used material of this purity for Step C.

13. The product has the following physical properties: IR (neat) cm^{-1}: 2950, 1840 (w), 1735, 1280, 1070; ^1H NMR (270 MHz, $CDCl_3$) δ: 0.99 (s, 3 H, $C\underline{H}_3$), 1.07 (s, 3 H, $C\underline{H}_3$), 2.19 (3 H, cyclopropenyl CH_3), 3.63 (s, 4 H, $OC\underline{H}_2C$), 7.34 (s, 1 H, C=C$\underline{H}$); ^{13}C NMR (100 MHz, $CDCl_3$) δ: 10.27, 22.14, 22.45, 30.47, 77.10, (2 C), 83.24, 116.04, 125.42.

14. The cyclization in Step B is an improvement of Butler's procedure for the synthesis of **4**,[4,5] which employs less convenient reagents, KNH_2 and an acetal of 1-bromo-3-chloro-2-propenone. Beside the acetals derived from neopentyl glycol, those derived from ethanol, 1,3-propanediol and 2,4-pentanediol have been synthesized by the present method.[8a] The second part of Step B involves the formation and the electrophilic trapping of cyclopropenyl anion **2**, which is the key element of the present preparations. Step B provides a simple route to substituted cyclopropenones, but the reaction is limited to alkylation with alkyl halides. The use of lithiated and zincated cyclopropenone acetal,[8] on the other hand, is more general and permits the reaction with a variety of electrophiles; alkyl, aryl and vinyl halides, Me_3SiCl, Bu_3SnCl, aldehydes, ketones, and epoxides. Repetition of the lithiation/alkylation sequence provides disubstituted cyclopropenone acetals.

15. High purity of potassium *tert*-butoxide is crucial for a clean reaction. The *tert*-butanol must be anhydrous.

16. A 1:1 mixture of dry dimethylsulfoxide/Et_2O was successfully used for other higher homologues of alkylidenecyclopropanone acetals.

17. The reaction can be monitored either by TLC or GC. TLC analysis may be performed with a plate (Merck No. 5765), eluting with a 1:9 mixture of ethyl acetate and hexane. The R_f values of **3** and **5** are 0.22 and 0.44, respectively. GC analysis performed under the same conditions as described in (Note 12) separates these two compounds, which have retention times of 6.4 and 5.8 min, respectively.

18. In the event that the reaction does not start or has not been completed, another 0.6-0.9 g portion of potassium *tert*-butoxide dissolved in *t*-BuOH/ether may be added.

19. It is important to keep the bath temperature as low as possible. Distillation above 100 °C would result in greatly reduced yield due to thermal decomposition of **5**.

20. The product has the following physical properties: IR (neat) cm^{-1}: 3150 (w), 1270, 1255, 1245, 1150, 1140, 1070, 1030, 1015, 970, 905; ^{1}H NMR (200 MHz, $CDCl_3$) δ: 1.04 (s, 3 H, CH_3), 1.06 (s, 3 H, CH_3), 1.62 (dd, 2 H, J = 3.1, 2.4, cyclopropyl C$\underline{H}_2$), 3.63 (s, 4 H, OC$\underline{H}_2$C), 5.45 (t, 1 H, J = 2.4, C = C$\underline{H}$H), 5.81 (t, 1 H, J = 3.1, C = CH$\underline{H}$); Anal. Calcd for $C_9H_{14}O_2$: C, 70.10; H, 9.15. Found: C, 70.02; H, 9.25.

21. The cycloadduct has the following spectral properties: ^{1}H NMR (500 MHz, CD_3CN) δ: 0.96 (s, 3 H), 0.99 (s, 3 H), 1.64-1.70 (m, 1 H), 2.00 (br dd, 1 H, J = 15.6, 4.6), 2.02-2.10 (m, 1 H), 2.21 (distorted t, 2 H, J = 7.6), 2.33 (br dd, 1 H, J = 16.5, 3.7), 2.39-2.49 (m, 2 H), 2.52 (ddd, 1 H, J = 9.6, 4.7, 2.1) 2.74-2.81 (m, 1 H), 3.59 (d, 1 H, J = 7.3), 3.60 (d, 1 H, J = 7.3), 3.61 (s, 2 H); ^{13}C NMR (125 MHz, CD_3CN) δ: 22.13 (q), 22.18 (q), 26.57 (t), 29.80 (t), 31.26 (s), 33.19 (t), 37.46 (t), 42.21 (d), 53.25 (d), 77.42 (t), 77.44 (t), 97.92 (s), 148.69 (s), 222.44 (s).

The ketene acetal functionality of the cycloadduct is moisture sensitive. It can be readily hydrolyzed to the cyclopentanone ester **7** upon treatment with aqueous acetic acid. Spectral properties of **7** are as follows: the major, more polar product-IR (neat) cm^{-1}: 3440, 1730; ^{1}H NMR (200 MHz, $CDCl_3$) 0.91 (s, 6 H), 1.60 (dt, 1 H, J = 13.3, 9.7), 1.74-2.49 (m, 8 H), 2.61 (dt, 1 H, J = 9.5, 5.9), 2.71-2.99 (m, 2 H), 3.29 (br s, 2 H), 3.92 (s, 2 H); Anal. Calcd for $C_{14}H_{22}O_4$: C, 66.11; H, 8.72. Found: C, 66.00; H, 8.72; the minor, less polar product-IR (neat) cm^{-1}: 3440, 1730, 1720; ^{1}H NMR (200 MHz, $CDCl_3$) δ: 0.92 (s, 6 H), 1.49-1.86 (m, 2 H), 2.03-2.37 (m, 7 H), 2.57-2.84 (m, 2 H), 2.85-3.06 (m, 1 H), 3.29 (br d, 2 H, J = 2.3), 3.94 (s, 2 H); Anal. Calcd for $C_{14}H_{22}O_4$: C, 66.11; H, 8.72. Found: C, 66.00; H, 8.86.

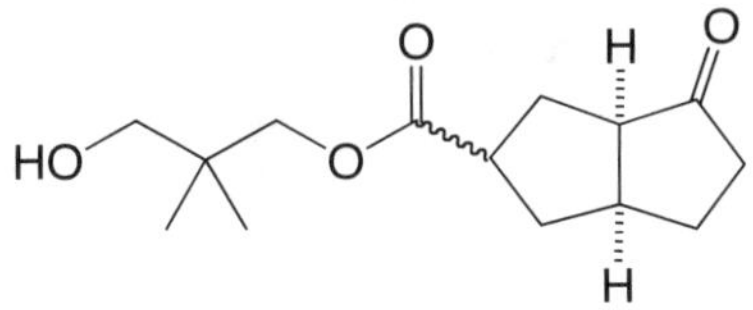

7

22. This product may contain ca. 5% of impurities as judged by capillary GC analysis on a 0.25 mm i.d. × 25 m capillary column (HR-1) at 170 °C. Typical retention times for the title compound and the side product are 8.8 and 4.5 min respectively.

3. Discussion

The procedures described herein illustrate the preparation of a substituted cyclopropenone acetal and an alkylidene cyclopropanone acetal.[6] The latter compound has been used to generate a dipolar trimethylenemethane (TMM) species that undergoes [3+2] cycloaddition with electron-deficient 2π-electron C=C and C=X compounds.[7]

The substituted cyclopropenone acetal synthesized in Step B can be readily hydrolyzed to the corresponding cyclopropenone.[8] This synthetic sequence provides the best and most versatile current synthetic route to substituted cyclopropenones.[9] The deficiencies of conventional procedures are precisely the synthesis of cyclopropenones with aliphatic substituents and functional groups, for which the present method has proven to be particularly useful.[10]

Cyclopropenones show considerable biological activity,[11,12] and have recently been employed as a key structural unit for a novel inhibitor of a cysteine protease.[13] The utility of cyclopropenone acetals has recently been recognized for vinylcarbene formation,[14,15] asymmetric synthesis,[16] and other processes.[17] Cycloaddition reactions of cyclopropenone acetals and congeners have also proven to be useful for chiral functionalization of buckminsterfullerenes.[18]

Step C describes a synthesis of the 1,1-dialkoxy-2-methylenecyclopropane (**5**) from 1,1-dialkoxy-2-methylenecyclopropene (**3**).[19] As described in Step D, upon thermolysis under mild conditions, **5** undergoes a [3+2] cycloaddition reaction by the reversible generation of a reactive dipolar TMM **8** (eq 1).[20] Various 1,1-dialkoxy-2-alkylidenecyclopropanes can also be prepared by the same procedure from the corresponding l-1-dialkoxy-2-alkylcyclopropenes, and have been shown to undergo regio- and stereoselective [3+2] cycloaddition reactions.[21] The use of a nitrogen atmosphere and dry solvent is not essential, and the reactions can be carried out under air in the presence of a small amount of water (the use of a slight excess of the methylenecyclopropane removes any water without affecting the yield of the cycloaddition product). Solvent is also not essential; the reaction rate and course may be affected by the choice of solvent (or absence of solvent).

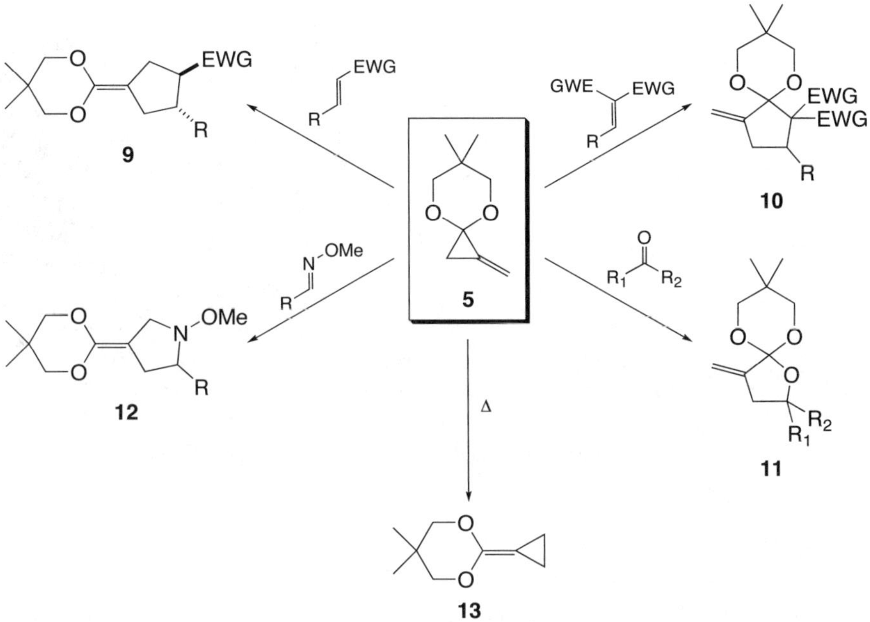

Olefins substituted with a single ester, nitrile, or ketone group function as good TMM acceptors to afford ketene acetals **9** (Table 1, entries 1-4). The cycloaddition reaction takes place stereoselectively with retention of the olefin geometry and "endo" orientation of the directing groups.[21] In contrast, the reaction with olefins whose reduction potential is larger than −1.8 eV (vs. SCE) gives mainly the acetal of an α-methylenecyclopentanone 10 via a single electron transfer process.[22] This cycloaddition can be used for the functionalization of buckminsterfullerene (C_{60} and C_{70})[23,24] whose reduction potential is −0.42 eV (see below). A radioactive version of **3** and **5** have been prepared by using $^{14}CH_3I$ in Step B and the latter has been employed for pharmacokinetic studies of water soluble C_{60}.[25] Cycloaddition to electron-deficient acetylenes takes place smoothly to give cyclopentene carboxylic acid esters in one step (entries 7 and 8).[26] Cycloaddition to carbonyl compounds gives tetrahydrofuran derivatives (entries 9-11),[27] and to an O-alkyloxime affords a pyrrolidine (entry 12).[28] Prolonged heating gives a highly reactive ketene acetal **13**, which serves as a useful synthon of cyclopropanecarboxylate enolate.[29]

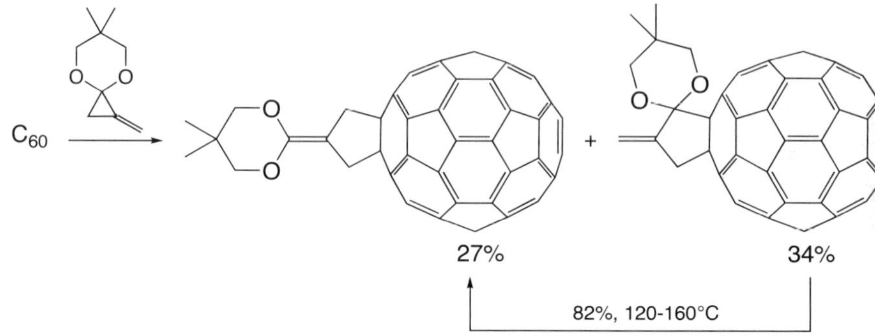

Table I. Cycloaddition of **2** with 2π Electron Acceptors.[a]

entry	substrate	product	% yield[b]	entry	substrate	product	% yield[b]
1	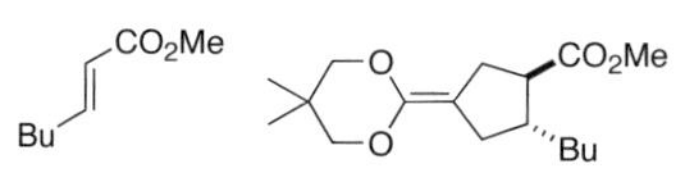		89	2	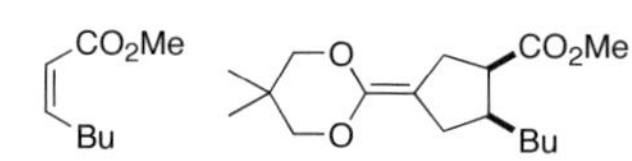		86[c]
3	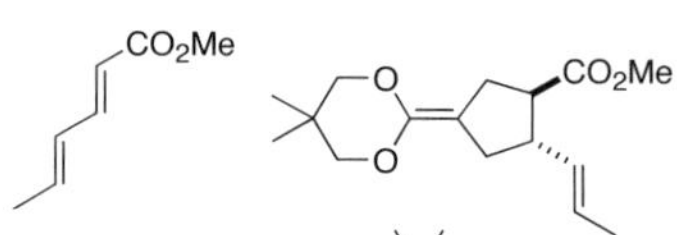		90	4	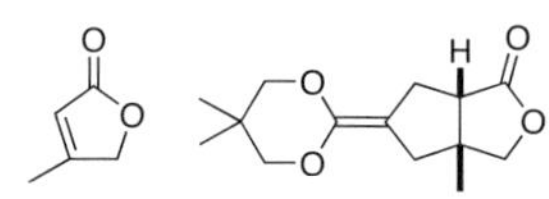		84
5	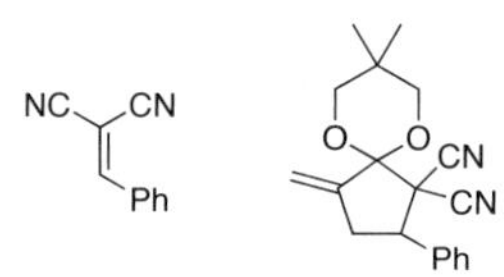		77	6	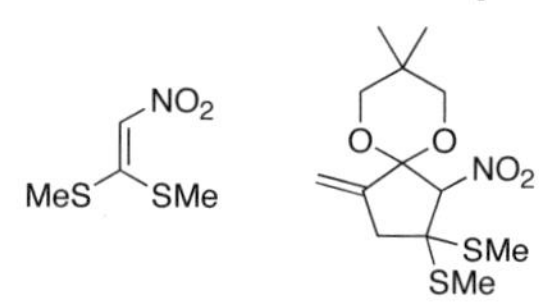		42
7	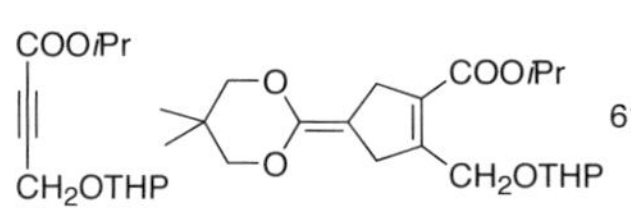		61	8	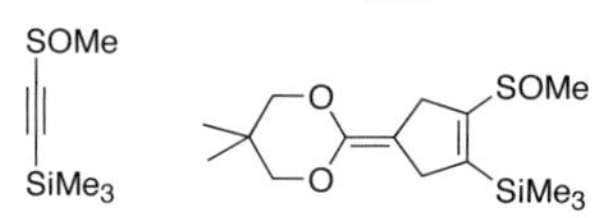		88
9	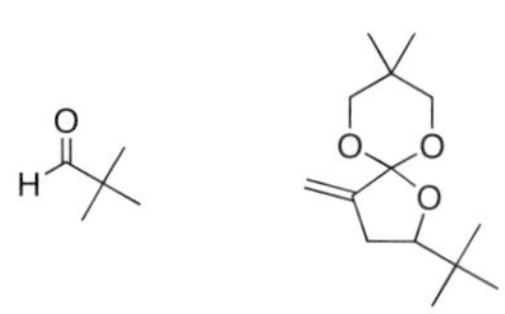		86	10	$(CH_2O)n$[d]	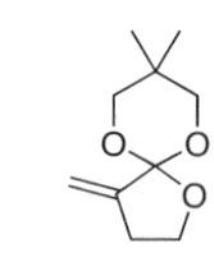	62
11	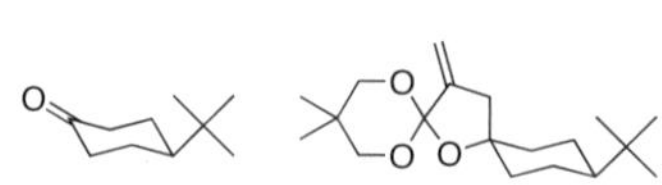		81[e]	12	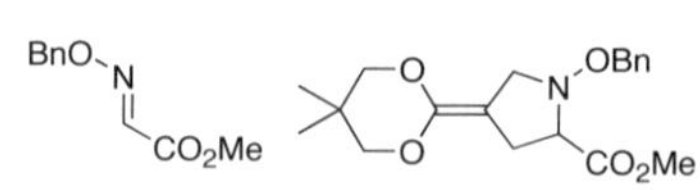		89

[a]The reactions were carried out at 40-130 °C. [b]Isolated yield. The initially formed ketene acetal products were isolated as cyclopentane carboxylic acid esters after acid hydrolysis. [c]The starting olefin was 100%*Z* and the product was 98.3%*Z*. [d]$ZnCl_2$ (20 mol%) was added to the reaction mixture. [e]The product consisted of a 82:18 mixture of two diastereomers. The major isomer is shown.

1. Institute for Chemical Research, Kyoto University, Uji, Kyoto 611-0011, Japan.
2. Department of Chemistry, The University of Tokyo, Bunkyo-ku, Hongo, Tokyo 113-0033, Japan.
3. Hancock, E. M.; Cope, A. C. *Org. Synth., Coll. Vol. III* **1965**, 219.
4. Butler, G. B.; Herring, K. H.; Lewis, P. L.; Sharpe, III, V. V.; Veazey, R. L. *J. Org. Chem.* **1977**, *42*, 679; Baucom, K. B.; Butler, G. B. *J. Org. Chem.* **1972**, *37*, 1730.
5. Boger, D. L.; Brotherton, C. E.; Georg, G. I. *Org. Synth.* **1987**, *65*, 32; Breslow, R.; Pecoraro, J.; Sugimoto, T. *Org. Synth., Coll. Vol. VI* **1988**, 361.
6. Review: Nakamura, M.; Isobe, H.; Nakamura, E. *Chem. Rev.* **2003**, *103*, 1295.
7. Review: Nakamura, E. In *Organic Synthesis in Japan. Past, Present and Future*; Noyori, R., Ed., Tokyo Kagaku Dojin, 1992.
8. (a) Isaka, M.; Ejiri, S.; Nakamura, E. *Tetrahedron* **1992**, *48*, 2045. (b) Isaka, M.; Ando, R.; Morinaka, Y. Nakamura, E. *Tetrahedron Lett.* **1991**, *32*, 1339. (c) Isaka, M.; Matsuzawa, S.; Yamago, S.; Ejiri, S.; Miyachi, Y.; Nakamura, E. *J. Org. Chem.* **1989**, *54*, 4727.
9. Review: Potts, K. T.; Baum, J. S. *Chem. Rev.* **1974**, *74*, 189; Eicher, T.; Weber J. L. *Top. Curr. Chem.* **1975**, *57*, 1.
10. Review: Nakamura, E. *J. Synth. Org. Chem., Jpn.* **1994**, *52*, 935-945.
11. Tokuyama, H.; Isaka, M.; Nakamura, E.; Ando, R.; Morinaka, Y. J. *Antibiotics* **1992**, *45*, 1148.
12. Okuda, T.; Shimma, N.; Furumai, T. *J. Antibiotics* **1984**, *37*, 723; Okuda, T.; Furumai, T. *J. Antibiotics* **1985**, *38*, 631.
13. (a) Ando, R.; Morinaka, Y.; Tokuyama, H.; Isaka, M.; Nakamura, E. *J. Am. Chem. Soc.* **1993**, *115*, 1174. (b) Tokuyama, H.; Isaka, M.; Nakamura, E. *Synth. Commun.* **1995**, *25*, 2005.
14. Boger, D. L.; Brotherton C. E. In *Advances in Cycloaddition*; JAI Press: Greenwich, **1990**; Vol. II, p. 147. Boger, D. L.; Brotherton, C. E. *J. Am. Chem. Soc.* **1986**, *108*, 6695.
15. Tokuyama, H.; Isaka, M.; Nakamura, E. *J. Am. Chem. Soc.* **1992**, *114*, 5523; Tokuyama, H.; Yamada, T.; Nakamura, E. *Synlett* **1993**, 589.
16. Isaka, M.; Nakamura, E. *J. Am. Chem. Soc.* **1990**, *112*, 7428; Nakamura, E.; Isaka, M.; Matsuzawa, S. *J. Am. Chem. Soc.* **1988**, *110*, 1297; Nakamura, M.; Arai, M.; Nakamura, E. *J. Am. Chem. Soc.* **1995**, *117*, 1179.
17. Albert, R. M.; Butler, G. B. *J. Org. Chem.* **1977**, *42*, 674.
18. Nakamura, E.; Isobe, H.; Tokuyama, H.; Sawamura, M. *J. Chem. Soc., Chem. Commun.* **1996**, 1747. Isobe, H.; Tokuyama, H.; Sawamura, M.; Nakamura, E. *J. Org. Chem.* **1997**, *62*, 5034.
19. Yamago, S.; Nakamura, E. *J. Am. Chem. Soc.* **1989**, *111*, 7285.

20. Nakamura, E.; Yamago, S.; Ejiri, S.; Dorigo, A. E.; Morokuma, K. J. *Am. Chem. Soc.* **1991**, *113*, 3183.

21. Ejiri, S.; Yamago, S.; Nakamura, E. *J. Am. Chem. Soc.* **1992**, *114*, 8707.

22. Yamago, S.; Ejiri, S.; Nakamura, M.; Nakamura, E. *J. Am. Chem. Soc.* **1993**, *115*, 5344.

23. Prato, M.; Suzuki, T.; Foroudian, H.; Li, Q.; Khemani, K.; Wudl, F.; Leonetti, J.; Little, R. D.; White, T.; Rickborn, B.; Yamago, S.; Nakamura, E. *J. Am. Chem. Soc.* **1993**, *115*, 1594.

24. Yamago, S.; Nakamura, E. *Chem. Lett.* **1996**, 395.

25. Yamago, S.; Tokuyama, H.; Nakamura, E.; Kikuchi, K.; Kananishi, S.; Sueki, K.; Nakahara, H.; Enomoto, S.; Ambe, F. *Chemistry & Biology*, **1995**, *2*, 385.

26. Yamago, S.; Ejiri, S.; Nakamura, E. *Angew. Chem., Int. Ed. Engl.* **1995**, *34*, 2154.

27. Yamago, S.; Nakamura, E. *J. Org. Chem.* **1990**, *55*, 5553.

28. Yamago, S.; Nakamura, M.; Wang, X. Q.; Yanagawa, M.; Tokumitsu, S.; Nakamura, E. *J. Org. Chem.* **1998**, *63*, 1694.

29. Yamago, S.; Takeichi, A.; Nakamura, E. *J. Am. Chem. Soc.* **1994**, *116*, 1123; Yamago, S.; Takeichi, A.; Nakamura, E. *Synthesis*, **1996**, 1380.

PREPARATION AND USE OF A NEW DIFLUOROCARBENE REAGENT. TRIMETHYLSILYL 2-FLUOROSULFONYL-2,2-DIFLUOROACETATE: *n*-BUTYL 2,2-DIFLUOROCYCLOPROPANECARBOXYLATE

[(Cyclopropanecarboxylic acid, 2,2-difluoro-, butyl ester and Acetic acid, difluoro(fluorosulfonyl)-, trimethylsilyl ester)]

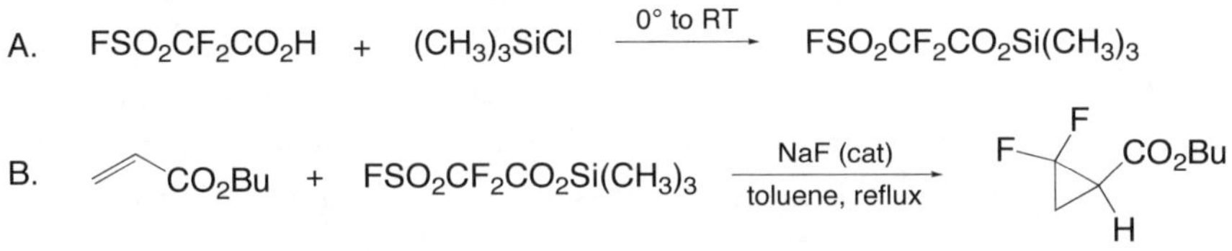

Submitted by W. R. Dolbier, Jr.1, F. Tian, J.-X. Duan, and Q.-Y. Chen.
Checked by Victoria D. Bock, David P. Richardson, and Steven Wolff.

1. Procedure

Caution: Trimethylsilyl 2-fluorosulfonyl-2,2-difluoroacetate should be handled with care. Skin contact, which will cause painful burns and blistering, should be avoided. If such contact occurs, the affected area should be washed thoroughly with water and with sodium bicarbonate solution.

A. Trimethylsilyl 2-fluorosulfonyl-2,2-difluoroacetate. All glassware is oven-dried for about 1 h prior to use. A 1-L, three-necked, round-bottomed flask is equipped with a magnetic stirrer, addition funnel with a nitrogen (N_2) inlet, and water-cooled condenser with gas outlet. The gas outlet is connected by Tygon tubing to an empty 500-mL back-up trap, and then to an inverted 75-mm glass funnel outlet positioned just above a 1-L beaker containing 75 g of sodium bicarbonate ($NaHCO_3$) in 500 mL water (to neutralize HCl and chlorotrimethylsilane from the reaction vessel) (Note 1). The flask is charged with 150 g of 2-fluorosulfonyl-2,2-difluoroacetic acid (0.84 mol) (Note 2). Then, with a slow N_2 flow and cooling with an ice bath, 363 g of chlorotrimethylsilane (3.2 mol) (Notes 3, 4) is added dropwise with stirring over a 2 h period. Upon completion of the addition, the mixture is allowed to warm to room temperature and is stirred with N_2 bubbling for 24 h. The product mixture is distilled to give 175 grams (83%) of trimethylsilyl 2-fluorosulfonyl-2,2-difluoroacetate (bp 52 °C at 15 mm) (Notes 5, 6).

B. n-Butyl 2,2-difluorocyclopropanecarboxylate. A 500-mL, round-bottomed, three-necked flask is fitted with a magnetic stirrer, condenser, addition funnel, gas dispersion tube extending to the bottom of the flask, and gas outlet with a paraffin oil bubbler. The flask is charged with 200 mL of toluene, 0.4 g of sodium fluoride (0.06 eq), and 20 g of *n*-butyl acrylate (0.156 mol) (Notes 7, 8). The solution is heated to reflux and slow N_2 bubbling is initiated with stirring for 1 h. Trimethylsilyl 2-fluorosulfonyl-2,2-difluoroacetate (62.5 g, 0.31 mol, 1.6 eq) (Note 5) is added dropwise (Note 9). The mixture is heated for 8 h, then cooled and filtered under vacuum filtration through a Celite pad (Note 10). Toluene is removed by simple distillation at atmospheric pressure, and the residue distilled at reduced pressure to obtain 15.4 g of *n*-butyl 2,2-difluorocyclopropanecarboxylate (55%) (bp 99-101 °C at 58 mm) as a colorless liquid (Note 11).

2. Notes

1. In order to avoid suction of the $NaHCO_3$ solution into the trap, the mouth of the funnel should not be submerged in the $NaHCO_3$ solution.

2. 2-Fluorosulfonyl-2,2-difluoroacetic acid can be obtained from Aldrich Chemical Co., Inc., Oakwood Products, Inc., or from FluoroTech, LLC (PO Box 13135, Gainesville, FL 32604).

3. A 3.5-4 fold excess of chlorotrimethylsilane is required for full conversion of the 2-fluorosulfonyl-2,2-difluoroacetic acid. Residual acid can be difficult to separate from the ester.

4. Chlorotrimethylsilane (bp 55 °C) was obtained from Aldrich Chemical Co., Inc. and used as received.

5. Trimethylsilyl 2-fluorosulfonyl-2,2-difluoroacetate is moisture-sensitive and, if possible, should be prepared immediately before use in the next step.

6. Trimethylsilyl 2-fluorosulfonyl-2,2-difluoroacetate is characterized as follows: ^{1}H NMR ($CDCl_3$) δ: 0.40 ppm (s); ^{13}C NMR, δ: −1.05 ppm (s), 112.22 (dt, J = 31.5, 299.0), 155.13 (t, J = 27.0); ^{19}F NMR, δ: −103.74 ppm (2F, s), 40.58 (1F, s).[2]

7. Toluene (anhydrous), sodium fluoride (Reagent grade), and *n*-butyl acrylate (99+%) were obtained from Aldrich Chemical Co., Inc. and used as received.

8. Other acrylate esters should function equally well in this synthesis. The *n*-butyl ester was specifically chosen because the boiling point of the

product is sufficiently different from that of toluene, permitting convenient separation by distillation.

9. Addition of the trimethylsilyl 2-fluorosulfonyl-2,2-difluoroacetate should be slow at the beginning of the reaction, but can be gradually increased. A vigorous evolution of gas (CO_2 and SO_2) accompanies the early stages of addition, but will be barely distinguishable from the nitrogen flow at the end.

10. Filtration through the Celite pad is recommended in order to remove small amounts of acrylate polymer.

11. *n*-Butyl 2,2-difluorocyclopropanecarboxylate is characterized as follows: ^{1}H NMR ($CDCl_3$) δ: 0.91 (3H, t, J = 7.32), 1.39 (2H, m), 1.62 (2H, m), 1.72 (1H, m), 2.04 (1H, m), 2.41 (1H, m), 4.15 (2H, q, J = 6.75); ^{19}F NMR, δ: −126.50 (1F, m), −114.70 (1F, m); ^{13}C NMR, δ: 13.49, 16.30 (t, J = 10.6), 18.92, 25.51 (t, J = 10.6), 30.43, 65.31, 110.6 (m), 166.57 ppm. The product obtained by the checkers was contaminated with small amounts of *n*-butyl acrylate, which could not be removed by distillation.

Waste Disposal Information

All waste materials were disposed of in accordance with "Prudent Practices in the Laboratory"; National Academy Press; Washington, DC, 1995.

3. Discussion

The present example provides a typical procedure for difluorocyclopropanation utilizing a new and highly versatile source of difluorocarbene.[3,4] This novel reagent, trimethylsilyl 2-fluorosulfonyl-2,2-difluoroacetate, provides a convenient procedure for the synthesis of difluorocyclopropane compounds via addition of difluorocarbene to alkenes of highly variable nucleophilicity.[2] A relatively small excess of reagent is required for high yields, and even electrophilic alkenes, such as n-butyl acrylate, provide acceptable yields of the difluorocyclopropane adduct. The reactivity and preparative utility of our reagent is probably comparable to Seyferth's reagent, phenyl(trifluoromethyl)mercury,[5] but this latter substance is no longer commercially available and is both tedious and hazardous to prepare.

1. Department of Chemistry, University of Florida, PO Box 117200, Gainesville, FL 32611-7200 (email: wrd@chem.ufl.edu).
2. Tian, F.; Kruger, V.; Bautista, O.; Duan, J.-X.; Li, A.-R.; Dolbier, W. R., Jr.; Chen, Q.-Y. *Org. Lett.* **2000**, *2*, 563.
3. Brahms, D. L. S.; Dailey, W. P. *Chem. Rev.* **1996**, *96*, 1585.
4. Burton, D. J.; Hahnfeld, J. L. *Fluorine Chem. Rev.* **1977**, *8*, 119.
5. Seyferth, D.; Hopper, S. P. *J. Org. Chem.* **1972**, *37*, 4070.

SYNTHESIS OF 2α-BENZYLOXY-8-OXABICYCLO[3.2.1]OCT-6-EN-3-ONE BY [4+3] CYCLOADDITION

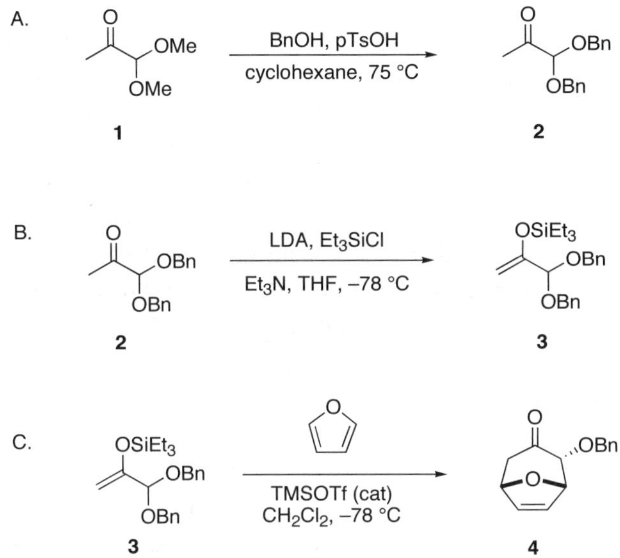

Submitted by María Vidal-Pascual, Carolina Martínez-Lamenca and H. M. R. Hoffmann.[1]
Checked by Timothy E. Long and Marvin J. Miller.

1. Procedure

A. 1,1-Bis(benzyloxy)propan-2-one (**2**). A one-necked, 100-mL, round-bottomed flask equipped with a magnetic stirring bar is charged with pyruvic aldehyde dimethyl acetal (12.1 mL, 100 mmol) in cyclohexane (50 mL), benzyl alcohol (22.8 mL, 220 mmol) and *p*-toluenesulfonic acid monohydrate (0.95 g, 5 mmol) (Note 1). The resulting mixture is heated at reflux for 2 h using a Dean–Stark separator for the removal of methanol. When the reaction is complete (approximately 2 h), approximately 8.1 mL (200 mmol) of MeOH is obtained. The reaction mixture is cooled to room temperature and washed with saturated potassium carbonate solution (25 mL) and water (20 mL). The aqueous layer is extracted twice with cyclohexane (2 x 50 mL). The combined organic phase is dried (Na_2SO_4), filtered, evaporated and the crude black oil is purified by column chromatography using a 10-cm diameter column packed with 900 g silica gel (Note 2) and eluting with 2.1 L of MTBE/cyclohexane (1:20) to afford keto acetal **2** as a yellowish oil (22.3 g, 83%) (Note 3).

B. [1,1-(Bis-benzyloxymethyl)vinyloxy]triethylsilane (**3**) (Note 4). An LDA solution is prepared in a 250-mL, two-necked, round-bottomed flask equipped with a nitrogen inlet and nitrogen balloon, magnetic stirring bar and rubber septum by adding BuLi (1.6 M solution in hexane, 22.5 mL, 36 mmol) (Note 5) into a solution containing diisopropylamine (5.1 mL, 36 mmol) (Note 6) in THF (36 mL) (Note 7) at −78 °C (Note 8). The resulting mixture is stirred for an additional 15 min at room temperature. A separate two-necked, 250-mL round-bottomed flask is equipped with two septa and a nitrogen balloon. Dibenzyl acetal **2** (8.1 g, 30 mmol) and chloro-triethylsilane (7.5 mL, 45 mmol) (Note 9) are dissolved in THF (30 mL) under a nitrogen atmosphere. This mixture is cooled to −78 °C and the LDA solution is added by cannula over the course of 10 min. Triethylamine (18.8 mL, 135 mmol) (Note 10) was immediately added by syringe over the course of 10 min. The resulting reaction mixture is stirred for 16 h at −78 °C. Water (25 mL) is added, the cooling bath is removed, and the mixture is stirred until it reaches room temperature. The aqueous phase is extracted twice with cyclohexane. After being dried over sodium sulfate (Na_2SO_4) the organic solution is concentrated under vacuum and purified by column chromatography (6-cm diameter column, 360 g silica gel, MTBE/cyclohexane (1:100, 9 L) with 0.1% Et_3N) (Note 11) to give **3** as a yellow oil (8.9 g, 77%) (Notes 12, 13).

C. 2α-Benzyloxy-8-oxabicyclo[3.2.1]oct-6-en-3-one (**4**). A 250-mL, one-necked, round-bottomed flask capped with a rubber septum and equipped with a nitrogen balloon and magnetic stirring bar is charged under nitrogen atmosphere with silyl enol ether **3** (9.2 g, 24 mmol) (Note 14) and dichloromethane (25 mL) (Note 15). The mixture is cooled to −78 °C, and furan (1.8 mL, 24 mmol) (Note 16) is added by syringe. After 15 min at −78 °C, TMSOTf (0.46 mL, 2.4 mmol) is added by syringe (Note 17). The mixture is stirred for 30 min at −78 °C and then a saturated solution of sodium bicarbonate (25 mL) is added. The cooling bath is removed and the flask is shaken thoroughly until the mixture reaches room temperature. The aqueous layer is extracted with dichloromethane (3 × 25 mL) and the combined organic phase is dried (Na_2SO_4). After removal of the solvent under reduced pressure, the crude product is purified by column chromatography (5-cm diameter column, 250 g silica gel,

MTBE/cyclohexane (1:3)) giving cycloadduct **4** as a colorless oil (3.72 g, 70%) (Note 18). On storage in the freezer at –25 °C, the product crystallizes from *tert*-butyl methyl ether to give a white solid (2.87 g, 54%) mp 64–65 °C (Notes 19-20).

2. Notes

1. Pyruvic aldehyde dimethyl acetal was purchased from Acros Organics. Benzyl alcohol was purchased from Lancaster. *p*-Toluenesulfonic acid was purchased from Aldrich.
2. Silica gel (230-400 mesh) was obtained from Macherey Nagel.
3. Spectral data for 1,1-bis(benzyloxy)propan-2-one **2**: ^{1}H NMR (500 MHz, $CDCl_3$) δ: 2.33 (s, 3 H), 4.68 (d, J = 12 Hz, 2 H), 4.78 (d, J = 12 Hz, 2 H), 4.84 (s, 1 H), 7.41–7.45 (m, 10 H); ^{13}C NMR (75 MHz, $CDCl_3$) δ: 25.0, 69.2, 101.0, 127.4, 128.0, 128.5, 137.0, 203.7. EI-MS (*m/z*): 271 (5), 259 (15), 228 (10), 182 (18), 181 (100), 165 (14); HRMS (FAB) (*m/z*): calcd for $C_{17}H_{19}O_2$ (M^++1) 271.1334, observed 271.1308.
4. The preparation of [(1,1-bis(benzyloxy)methyl)vinyloxy]-triethylsilane (**3**) is appropriate for scales of 30 mmol of dibenzyl acetal or less. See Note 20 for information regarding larger scale reactions.
5. A commercial solution of 1.6 M butyllithium in hexane from Acros Organics was used.
6. Diisopropylamine was purchased from Acros and was distilled from KOH pellets and stored over solid KOH, purchased from Fisher Scientific.
7. Tetrahydrofuran supplied by Mallinckrodt is dried by distillation from sodium and benzophenone under an argon atmosphere.
8. The reaction mixture is cooled to –78 °C using dry ice-acetone bath or a cryostat.
9. Chlorotriethylsilane from both Acros Organics and Gelest, Inc. was used.
10. Triethylamine was dried and distilled from and stored over KOH pellets purchased from Fisher Scientific.
11. The checkers found that addition of triethylamine to the eluent is required to prevent decomposition of the silyl enol ether during chromatography.
12. Spectral data for [(1,1-bis(benzyloxy)methyl)-vinyloxy] triethylsilane **3**: ^{1}H NMR (300 MHz, $CDCl_3$) δ: 0.69 (q, J = 7.8 Hz, 6 H), 0.95 (t, J = 8.1 Hz, 9 H), 4.37 (d, J = 1.2 Hz, 1 H), 4.57 (d, J = 12 Hz, 2 H);

4.63 (d, $J = 11.7$ Hz, 2 H), 4.70 (s, 1 H), 4.89 (s, 1 H), 7.31–7.26 (m, 10 H); ^{13}C NMR (125 MHz, $CDCl_3$) δ: 4.9, 6.7, 67.6, 92.3, 99.2, 127.5, 127.8, 128.4, 138.2, 153.8. EI-MS (*m/z*): no M^+, 279 (9), 249 (4), 248 (3), 193 (4), 187 (6), 181 (7), 159 (14), 157 (13), 115 (17), 91 (100).

13. The submitters report that this procedure can also be used for the preparation of 3,3-bis-benzyloxy-2-trimethylsilyloxypropene by using chlorotrimethylsilane (from Acros Organics) instead of chlorotriethylsilane. In that case, the resultant trimethylsilyl enol ether cannot be purified by column chromatography due to its acid sensitivity, so the crude product is used directly in the [4+3] cycloaddition reaction (Note 20).

14. An alternative to silyl enol ether **3** as oxyallylic cation precursor is 3,3-bis-benzyloxy-2-trimethylsilyloxypropene (Notes 13, 20).

15. Commercial dichloromethane was freshly distilled from CaH_2.

16. Commercial furan (99%, stabilized with BHT) from Acros Organics was used without further purification. 2,5-Dimethylfuran (99%) from Acros Organics was used by the submitters in the unchecked [4+3] cycloaddition (85% yield of cycloadduct) following the same procedure as for furan.

17. Commercial trimethylsilyl trifluoromethanesulfonate (99%) from Acros Organics was used. Ten mole % (0.1 equiv) of TMSOTf was sufficient to catalyze the [4+3] cycloaddition.

18. The checkers observed that the oil partially solidifies to a white solid upon standing. The submitters report that residual benzyl alcohol can be removed by heating the oil at 100 °C for two hours using a Kugelrohr apparatus. The checkers did not identify the presence of benzyl alcohol by NMR analysis and did not perform the heating.

19. Spectral data for 2α-benzyloxy-8-oxabicyclo[3.2.1]oct-6-en-3-one **4**: ^{1}H NMR (500 MHz, $CDCl_3$) δ: 2.37 (d, $J = 15.5$ Hz, 1 H), 2.75 (dd, $J = 15, 5$ Hz, 1 H), 4.13 (d, $J = 5$ Hz, 1 H), 4.64 (d, $J = 12$ Hz, 1 H), 4.91 (dd, $J = 5, 1.5$ Hz. 1 H), 4.99 (m, 2 H), 6.30 (dd, $J = 6.5, 1.5$ Hz, 1 H), 6.34 (dd, $J = 6, 2$ Hz, 1 H), 7.39–7.31 (m, 5 H); ^{13}C NMR (125 MHz, $CDCl_3$) δ: 46.1, 73.7, 78.6, 80.0, 84.3, 128.1, 128.7, 132.0, 134.8, 137.8, 205.2. IR ($CHCl_3$) cm^{-1}: 1724 (very strong), 1112 (very strong), 731 (strong), 697 (strong; EI-MS (rt): 230 (6, M^+), 201 (4), 158 (38), 139 (31, M^+-Bn), 121 (10), 108 (25), 91 (100), 81 (30), 77 (14), 69 (23). Anal. Calcd for $C_{14}H_{14}O_3$: C, 72.89; H, 6.05. Found: C, 73.03; H, 6.13.

20. Alternatively, the [4+3] cycloadduct has been prepared by the submitters using inexpensive chlorotrimethylsilane (Note 13) as follows:

A two-necked, 250-mL, round-bottomed flask equipped with a magnetic stirring bar is charged with dibenzyl acetal **2** (27 g, 100 mmol) in anhydrous DMF (33 mL) (Dimethylformamide (DMF) is dried over powdered BaO for 2 weeks, followed by decanting and then distilling under reduced pressure.) and chlorotrimethylsilane (28.6 mL, 225 mmol) (Note 13) and heated at 75 °C. Triethylamine (38.9 mL, 280 mmol) (Note 10) is added slowly by syringe pump (60 mL/h). The reaction mixture is refluxed for 16 h at 75 °C and the reaction mixture becomes thicker and dark brown. The reaction mixture is cooled to 0 °C, washed with 30 mL of a cold saturated solution of ammonium chloride (30 mL) and water (approximately 15 mL). The aqueous phase is extracted with cyclohexane (4 x 50 mL). The combined organic phase is dried (Na_2SO_4) and the solvent removed with a rotatory evaporator. The resulting brown viscous oil is used directly in the [4+3] cycloaddition as described in procedure C.

When the 3,3-*bis*-benzyloxy-2-trimethylsilyloxypropene is used as the oxyallylic cation precursor, the yield of the [4+3] cycloadduct with furan is 41 % (two steps). The yield of the [4+3] cycloadduct with 2,5-dimethylfuran is 53 % (two steps). The procedure has been used in a reaction using 130-mmol of starting dibenzyl acetal.

Safety and Waste Disposal Information

All hazardous materials should be handled and disposed of in accordance with "Prudent Practices in the Laboratory"; National Academy Press; Washington, DC, 1995.

3. Discussion

In view of the wide synthetic interest of [4+3] cycloadducts of oxyallyl cations and dienes, a variety of methods have been developed for the generation and capture of appropriate oxyallyl cations. The classical routes to generate these reactive intermediates start from halogenated precursors:[2] α,α′-Dihalo ketones or α-monohalo ketones. For example, 2,4-dimethyl-8-oxabicyclo[3.2.1]oct-6-en-3-one (5α,α) has been prepared from both a dihalo-[3a] and a monohaloketone[3b] and has served as a *meso*-configured workhorse for exploring desymmetrization of the seven-carbon backbone with its three pro-stereogenic sp^2 centers. Allylic halides and silver salts have been used also.[2a, 4] Further, the use of trialkylsilyloxyallyl cations from 1,1-dimethoxyacetone in [4+3] cycloadditions with furans has been reported.[5]

In the procedure described here, a triethylsilyloxyallylic cation with a π-donating benzyloxy substituent at the allylic terminus is generated via treatment of triethylsilyl enol ether **3** with TMSOTf in catalytic amount and, in presence of the 4π-component, the [4+3] cycloaddition proceeds smoothly giving 2α-benzyloxy-8-oxabicyclo[3.2.1]oct-6-en-3-ones. If furan is replaced by 2,5-dimethylfuran, then cycloaddition yields are higher (leading to **6**). Increasing the equivalents of diene does not alter the yield appreciably. The availability of oxabicycle **6** opens a short route to the dioxatricyclic core of the marine metabolite dictyoxetane (**6** → **7**)[6a] and to functionalized scopolines.[6b] An asymmetric variant of the [4+3] cycloaddition involving chiral allyl cations has also been accomplished affording **8** with high *ee*.[7]

Oxabicycle **8** is a useful building block and is endowed with all the chiral information for the *de novo* construction of C-glycosides. For example, axial hydroxylation (**8** → **9**) proceeds smoothly by treatment of the derived TES enol ether with *m*-chloroperbenzoic acid provided that *wet* THF is used as solvent.[8] The use of dilute solutions of dimethyldioxirane as oxidant which is toxic and dangerous, is obviated.[9] Epimerization of **9** affords the diequatorial epimer **10**. Starting from **8** and parent 8-oxabicyclo[3.2.1]oct-6-en-3-one, a complete series of C-glycosides with fully resolved seven-carbon backbone polyol stereochemistry and with complete anomeric control has been prepared (Scheme 1).[8]

Scheme 1

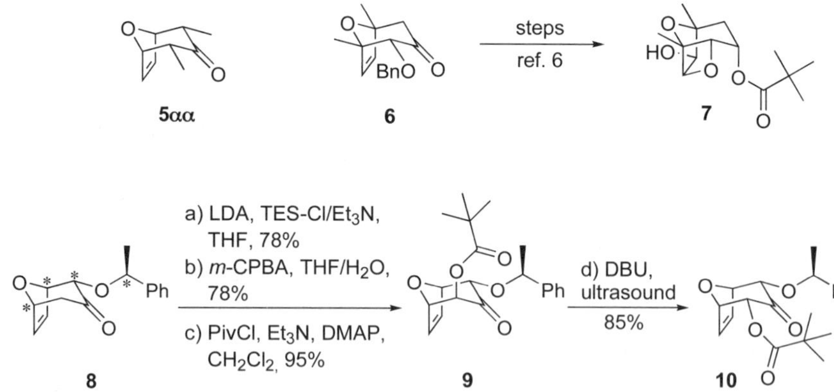

8-Oxabicyclo[3.2.1]oct-6-en-3-ones have been converted into [3.3.1]lactone acetals, which are versatile synthetic intermediates and allow further stereocontrolled C–C, C–S, C–$P^+Ph_3X^-$ bond formation. C–C Bonds and extended carbon chains are formed with complete anomeric control in the presence of a Lewis acid (Me_3SiOTf) and silylated nucleophile in acetonitrile (Scheme 2).[10] 8-Oxabicyclo[3.2.1]oct-6-en-3-ones have been used as key precursors in natural product syntheses, as accomplished by the total synthesis of hinokitiol[11] and callistatin A.[12]

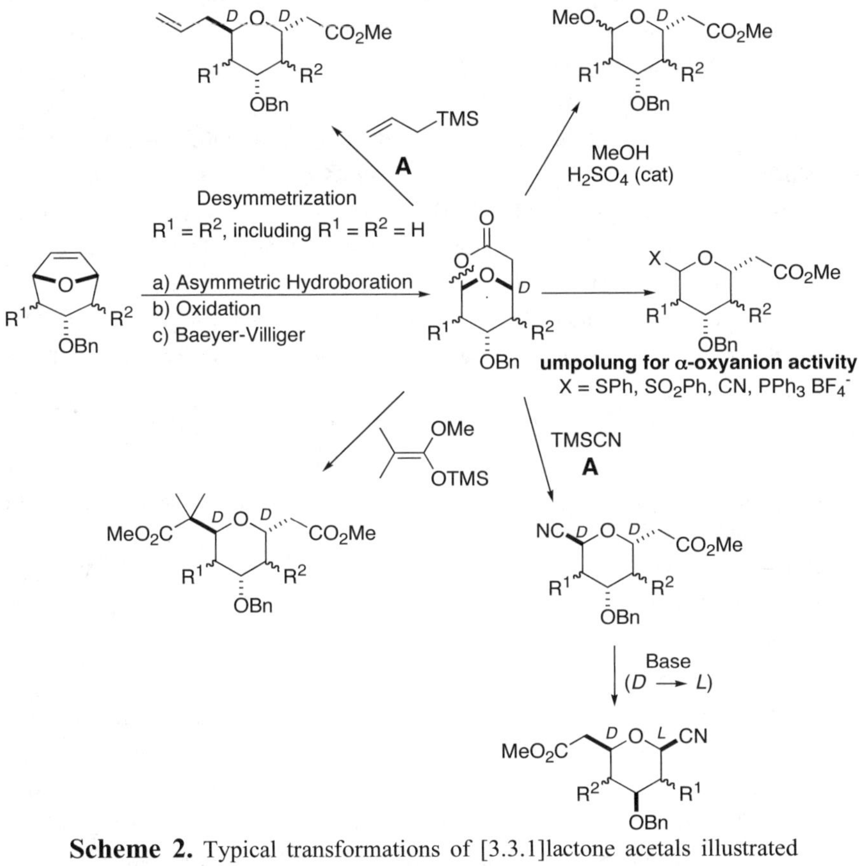

Scheme 2. Typical transformations of [3.3.1]lactone acetals illustrated for C-glycoside synthesis with generalized *D-D* type configuration. **A**: All reactions were run with TMSOTf (ca. 1 equiv) in MeCN, –40 °C to 0 °C; then addition of MeOH, r.t. followed by *in situ* esterification. Yields are uniformly about 95%.

Intramolecular [4+3] cycloadditions have also been studied.[13] Recently, metathesis of substituted oxabicyclics combined with spiroannulation has been reported.[14] Ru-Catalyzed asymmetric ring-opening/cross-metathesis provides another efficient enantioselective route to functionalized tetrahydropyrans.[15] Vinylic nonaflates from 8-oxabicyclo [3.2.1] oct-6-en-3-ones have been used to elaborate the bicyclic scaffold.[16]

1. Institut für Organische Chemie, Universität Hannover, Schneiderberg 1B, D-30167 Hannover, Germany.

2. a) Hoffmann, H. M. R. *Angew. Chem. Int. Ed. Engl.* **1984**, *23*, 1-19; b) Sarhan, A. A. O. *Curr. Org. Chem.* **2001**, *5*, 827-844; c) Rigby, J. H.; Pigge, F. C. *Org. React.* **1997**, *51*, 351-478.

3. a) Ashcroft, M. R.; Hoffmann, H. M. R. *Org. Synth. Coll. Vol. VI*, 512; b) Lautens, M.; Bouchain, G. *Org. Synth.* **2002**, *79*, 251.

4. Oh, J.; Ziani-Cherif, C.; Choi, J. -R.; Cha, J. K. *Org. Synth.* **2002**, *78*, 212.

5. Albizati, K. F.; Murray, D. H. *Tetrahedron Lett.* **1990**, *31*, 4109; for the use of Lewis acids and 3-alkoxy-2-oxo-3-alkenenitriles in cycloadditions to cyclopentadiene see Hoffmann, H. M. R.; Giesel, K.; Lies, R.; Ismail, Z.M. *Synthesis* **1986**, 548; see also Niess, B.; Hoffmann, H.M.R. *Angew.Chem.Int.Ed.* **2005**, *44*, 26.

6. a) Proemmel, S.; Wartchow, R.; Hoffmann, H. M. R. *Tetrahedron* **2002**, *31*, 6199; b) Vidal-Pascual, M.; Proemmel, S.; Beil, W.; Wartchow, R.; Hoffmann, H. M. R. *Org. Lett.* **2004**, *6*, 4155.

7. a) Stark, C. B. W.; Eggert, U.; Hoffmann, H. M. R. *Angew. Chem. Int. Ed. Engl.* **1998**, *37*, 1266; b) Stark, C. B. W.; Pierau, S.; Wartchow, R.; Hoffmann, H. M. R. *Chem. Eur. J.* **2000**, *6*, 684; c) Beck, H.; Stark, C. B. W.; Hoffmann, H. M. R. *Org. Lett.* **2000**, *2*, 883.

8. Hoffmann, H. M. R.; Dunkel, R.; Mentzel, M.; Reuter, H.; Stark, C. B. W. *Chem. Eur. J.* **2001**, *7*, 4771; Hartung, I. V.; Hoffmann, H. M. R. *Angew. Chem. Int. Ed.* **2004**, *43*, 1934.

9. Murray, R. W.; Singh, M. *Org. Synth. Coll. Vol.* **9**, 288.

10. Gaertzen, O.; Misske, A. M.; Wolbers, P.; Hoffmann, H. M. R. *Tetrahedron Lett.* **1999**, *40*, 6359; Gaertzen, O.; Misske, A. M.; Wolbers, P.; Hoffmann, H. M. R. *Synlett.* **1999**, 1041; Dunkel, R.; Hoffmann, H. M. R. *Tetrahedron* **1999**, *55*, 8385.

11. Lee, J. C.; Cho, S. Y.; Cha, J. K. *Tetrahedron Lett.* **1999**, *40*, 7675.

12. Lautens, M.; Stammers, T. A. *Synthesis* **2002**, 1993 and references therein.

13. Harmata, M. *Acc. Chem. Res.* **2001**, *34*, 595; Harmata, M.; Rashatasakhon, P. *Tetrahedron* **2003,** *59***,** 2379**.**

14. Usher, L. C.; Estrella-Jiménez, M.; Ghiviriga, I.; Wright, D. L. *Angew. Chem. Int. Ed.* **2002**, *41*, 4560.

15. Gillingham, D. G.; Kataoka, O.; Garber, S. B.; Hoveyda, A. H. *J. Am. Chem. Soc.* **2004**, *126*, 12288.

16. Högermeier, J.; Reissig, H.-U.; Brüdgam, I.; Hartl, H. *Adv. Synth. Catal.* **2004**, *346*, 1868.

TRIFLUOROMETHANESULFONIMIDE-CATALYZED (2 + 2)-CYCLOADDITION OF SILYL ENOL ETHERS WITH α,β-UNSATURATED ESTERS: 1-(*tert*-BUTYLDIMETHYLSILOXY)-8-(METHOXYCARBONYL)-6-METHYLBICYCLO[4.2.0]OCTANE

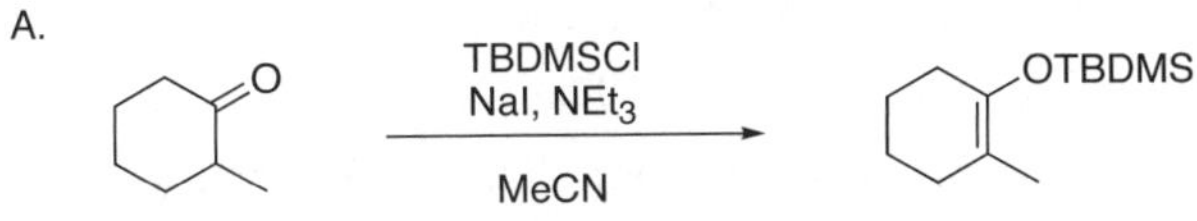

1

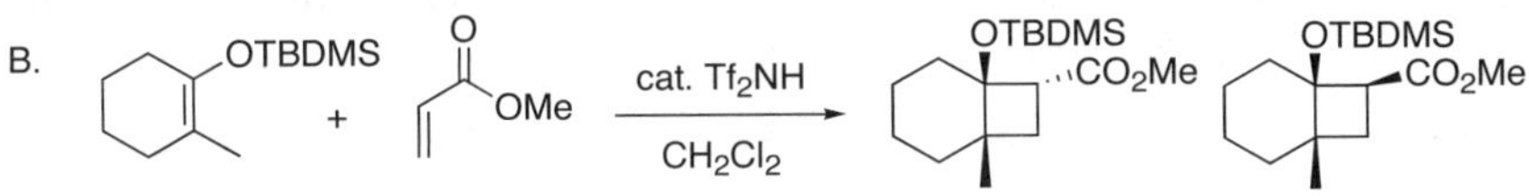

trans-**2** *cis*-**2**

Submitted by Kiyosei Takasu, Takayuki Ishii, Kazato Inanaga and Masataka Ihara.[1]
Checked by Katarzyna Kowalczuk and John A. Ragan.

1. Procedure

A. 1-tert-Butyldimethylsilyloxy-2-methyl-1-cyclohexene. A nitrogen-purged, 300-mL, three-necked, round-bottomed flask is equipped with a magnetic stirring bar, a rubber septum, a glass stopper, a temperature probe, and a nitrogen inlet adapter. The flask is charged with 2-methylcyclohexanone (10.0 mL, 82.6 mmol) (Note 1), triethylamine (13.9 mL, 100 mmol) (Note 2), and *t*-butyldimethylsilyl chloride (TBDMSCl) (15.1 g, 100 mmol) (Note 1). To the flask is added a solution of sodium iodide (15.0 g, 100 mmol) (Note 1) in acetonitrile (100 mL) (Note 3) via syringe over 30 min at ambient temperature. The reaction solution is stirred at ambient temperature for 18 h. The resulting mixture is quenched by addition of saturated sodium bicarbonate solution (100 mL). The mixture is extracted with hexane twice (2 x 200 mL). The combined organic phases are washed with brine (40 mL) and dried over $MgSO_4$, filtered and concentrated at reduced pressure (15–25 mmHg, 25–35 °C) to afford crude product **1**

(20.2 g) as a pale yellow oil. This crude product is purified by filtration through a silica gel pad (200 g of silica in a 10-cm diameter fritted glass funnel, height of silica was 20 cm), rinsing with 1 L of hexanes (Note 4) to provide 17.75–17.80 g (95%) of **1** as a colorless oil (Notes 5, 6).

B. 1-(tert-Butyldimethylsilyloxy)-8-(methoxycarbonyl)-6-methylbicyclo[4.2.0]octane. A flame-dried, 300-mL, three-necked, round-bottomed flask is equipped with a magnetic stirring bar, a rubber septum, an internal temperature probe and an argon inlet adapter. The flask is charged with a solution of **1** (4.98 g, 22.0 mmol) in dichloromethane (100 mL) (CH_2Cl_2) (Note 7) at ambient temperature, and the mixture is cooled in a dry ice-acetone bath to –78 °C while methyl acrylate (1.80 mL, 20.0 mmol) (Note 8) is added. To the mixture is added dropwise a solution of trifluoromethanesulfonimide (80 mM solution; 2.50 mL, 0.20 mmol) in toluene (Notes 9, 10) via syringe over 30 min at –78 °C (Note 11). The mixture is stirred at –78 °C for 30 min (Note 12), and is then quenched with saturated sodium bicarbonate solution (100 mL). The mixture is extracted with methyl *tert*-butyl ether (MTBE) (2 x 100 mL) (Note 13). The combined organic phases are dried over $MgSO_4$, filtered and concentrated at reduced pressure (15–25 mmHg, 25–35 °C) to afford crude product **2** (7.80 g) as a colorless oil. This crude product is purified by chromatography with 400 g of silica gel (6.5 cm i.d. column, eluent with 50:1 hexane/Et_2O) (Note 14) to provide 5.13–5.27 g (82–84%) of the *trans*-diastereomer of **2** as a colorless oil (Notes 15, 16).

2. Notes

1. 2-Methylcyclohexanone (>98%), *t*-butyldimethylsilyl chloride (97%), and sodium iodide (>99.5%) were purchased from Aldrich Chemical Company and used as received.

2. Triethylamine (anhydrous) was purchased from J. T. Baker and used as received.

3. Acetonitrile (anhydrous, Sure-Seal) was purchased from Aldrich Chemical Company and used as received.

4. The checkers used 40-μm silica gel for the filtration (J. T. Baker). The submitters purified the product by column chromatography (9.5 cm i.d. column, elution with hexane) on 400 g of Silica gel 60 N (spherical, neutral,

63–210 mesh), purchased from Kanto Chemical Co. Inc. The R_f value for **1** is 0.75 (hexane).

5. The product from step A exhibits the following data: IR (neat): 2928, 2857, 1687, 1348, 1253, 1168, 831 cm^{-1}; ^{1}H NMR (400 MHz, $CDCl_3$) δ: 0.10 (s, 6 H) 0.94 (s, 9 H), 1.51–1.55 (m, 2 H), 1.56 (s, 3 H), 1.60–1.66 (m, 2 H), 1.92–1.95 (m, 2 H), 1.98–2.02 (m, 2 H); ^{13}C NMR (100 MHz, $CDCl_3$) δ: –3.6, 16.6, 18.4, 23.2, 24.1, 26.1, 30.5, 30.6, 111.7, 143.2; MS (EI) *m/z*: 226 (50%), 169 (80%), 75 (100%); Anal. Calcd for $C_{13}H_{26}OSi$: C, 68.96; H 11.57. Found: C, 68.64; H, 11.64.

6. A trace amount (<3%) of the regioisomeric silyl enol ether (1-*tert*-butyldimethylsiloxy-6-methyl-1-cyclohexene) was observed by ^{1}H NMR spectroscopy. Separation of the isomer is not required for the next reaction.

7. Dichloromethane (anhydrous, Sure-Seal) was purchased from Aldrich Chemical Company and used as received.

8. Methyl acrylate (97%) was purchased from Aldrich Chemical Company and used as received. The submitters purchased this material from Tokyo Kasei Kogyo Co., Ltd., and distilled under reduced pressure before use.

9. Trifluoromethanesulfonimide (95%) was purchased from Aldrich Chemical Company, Inc. A 80 mM solution of trifluoromethanesulfonimide in toluene was prepared as follows. A flame-dried, 50-mL, round-bottomed flask equipped with gas inlet is charged with trifluoromethanesufonimide (675 mg, 2.40 mmol) under an atmosphere of argon, and toluene (anhydrous, Aldrich Sure-Seal) (30 mL) is quickly added under an atmosphere of argon. The solution can be stored for more than 1 month in the dark at ambient temperature.

10. Trifluoromethanesulfonimide (neat) is moisture-sensitive and, if possible, should be dissolved in toluene under an argon atmosphere after a fresh bottle is opened.

11. The solution temperature inside the reaction vessel was monitored by the internal thermometer and kept within –76 to –78 °C.

12. Since the (2 + 2)-cycloaddition reaction is reversible, the effects of temperature, reaction time, concentration, and solvent are important. The *trans*-isomer of **2** corresponds to the kinetic product. Interestingly, on

smaller scale reactions (~100 mg of silyl enol ether), the submitters obtained *trans*- and *cis*-isomers of **2** in 98% and 1% yields, respectively, under the same conditions.

13. The submitters performed these extractions with diethyl ether. The checkers found that MTBE worked equally well for this extraction.

14. Chromatography was performed on 40-μm silica gel purchased from J. T. Baker. The R_f values for *trans*- and *cis*-isomers of **2** are 0.26 and 0.22, respectively (50:1; hexane/Et_2O), as reported by the submitters. The checkers were unable to cleanly discern the minor isomer on TLC.

15. The *trans*-isomer exhibits the following spectroscopic properties: IR (neat) ν: 2929, 2857, 1734, 1462, 1220, 1095, 834 cm^{-1}; ^1H NMR (400 MHz, $CDCl_3$) δ: 0.05 (s, 3 H), 0.10 (s, 3 H), 0.84 (s, 9 H), 1.03 (s, 3 H), 1.14–1.60 (m, 8 H), 1.62–1.64 (m, 1 H), 1.78 (t, *J* = 10, 1 H), 3.10 (t, *J* = 10, 1 H), 3.62 (s, 3 H); ^{13}C NMR (100 MHz, $CDCl_3$) δ: –3.2, –2.9, 18.6, 20.3, 21.8, 24.9, 26.0, 26.9, 32.3, 33.4, 41.3, 48.5, 51.1, 77.6, 173.6; LRMS (EI) *m/z*: 255 (M^+ – 57, 100%); Anal. Calcd for $C_{17}H_{32}O_3Si$: C, 65.33; H 10.32. Found: C, 65.22; H, 10.78.

16. The checkers were unable to isolate a clean sample of the minor isomer, although its presence was confirmed in the crude ^1H NMR spectrum (approximately a 10:1 ratio of major/minor isomers, which is in agreement with the submitter's report). The submitters reported the isolation of the *cis*-isomer of **2** (8%), which exhibited the following characterization data: IR (neat): 1746 cm^{-1}; ^1H NMR (400 MHz, $CDCl_3$) δ: 0.05 (s, 3 H), 0.09 (s, 3 H), 0.85 (s, 9 H), 0.97 (s, 3 H), 1.08–1.24 (m, 1 H), 1.31–1.61 (m, 5 H), 1.75 (m, 2 H), 2.03 (d, *J* = 12.2 Hz, 1 H), 2.22 (t, *J* = 10.0 Hz, 1 H), 3.16 (t, *J* = 7.8 Hz, 1 H), 3.62 (s, 3 H); ^{13}C NMR (100 MHz, $CDCl_3$) δ: –1.64, –1.58, 18.6, 21.0, 21.6, 24.0, 26.1, 33.4, 33.7, 38.8, 41.0, 43.1, 51.1, 83.7, 172.6; LRMS (EI) *m/z*: 255 (M^+ – 57); Anal. Calcd for $C_{17}H_{32}O_3Si$: C, 65.33; H 10.32. Found: C, 65.36; H, 9.98.

Safety and Waste Disposal Information

All hazardous materials should be handled and disposed of in accordance with "Prudent Practices in the Laboratory"; National Academy Press; Washington, DC, 1995.

3. Discussion

The procedure described here is typical for the catalytic (2 + 2)-cycloaddition reaction of silyl enol ethers with α,β-unsaturated esters to produce multi-substituted silyloxycyclobutanes.[2] Silyl enol ethers are readily available by enol silylation of the corresponding ketones. Previously, we have reported that the (2 + 2)-cycloaddition of silyl enol ethers is catalyzed by a hard Lewis acid such as $EtAlCl_2$.[3,4,5] Compared with that method, the procedure using trifluoromethanesulfonimide (Tf_2NH) provides high chemical yield and stereoselectivity under practical and environmentally benign conditions, with broader substrate-applicability (Table).

We have found the Tf_2NH-catalyzed (2 + 2)-cycloadditions of silyl enol ethers with α,β-unsaturated esters are eventually reversible. Although the kinetic product in the reaction possesses the *trans*-configuration of silyloxy and ester functionalities, longer reaction times or higher reaction temperatures allow the retro (2 + 2)-cycloaddition to occur, leading to the thermodynamically more stable *cis*-isomer. The isomerization can be monitored by careful TLC analysis.

Yamamoto and coworkers reported that Tf_2NH-catalyzed aldol reactions between TMS enol ethers and aldehydes are promoted by highly reactive, *in situ* generated $TMSNTf_2$.[6] We also observed that a catalytic amount of pre-assembled $TBSNTf_2$ promoted the (2 + 2)-cycloaddition reactions. Thus, Tf_2NH appears to act similarly as a pre-catalyst to produce the real catalyst $TBSNTf_2$ through reaction with the *t*-butyldimethylsilyl enol ethers. Importantly, decomposition of $TBSNTf_2$ to form Tf_2NH during the course of these processes is reversed by reaction of the latter with the TBS enol ether substrates. As a result, high turnover numbers are achieved in this catalytic system. Thus, a reaction using a stoichiometric amount of Tf_2NH results in low yields of the desired product and decomposition of the silyl enol ether.

Table 1. Catalytic (2 + 2)-Cycloaddition Reactions[a]

entry	silyl enol ether	α,β-unsaturated esters	products: *trans*	products: *cis*		% yield[g] (ratio)
1[b] 2[b] 3[b]	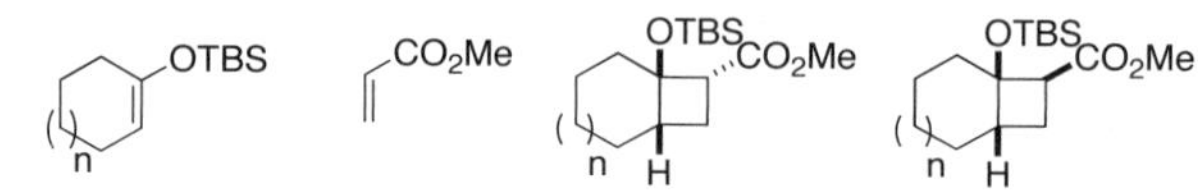				(n = 0) (n = 1) (n = 2)	77 (>99 : 1) 70 (80 : 20) 91 (93 : 7)
4[b]	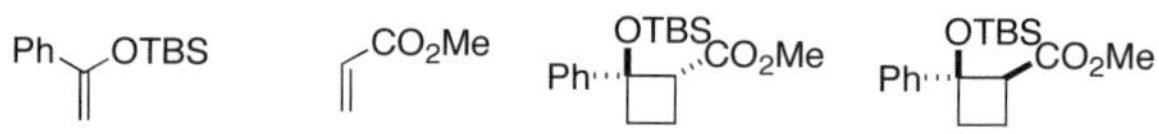					75 (72 : 28)
5[b]	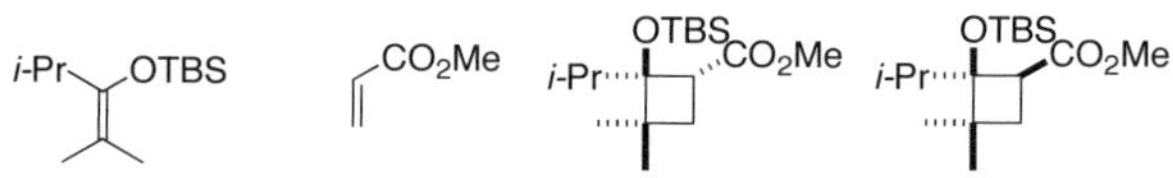					93 (81 : 19)
6[c]	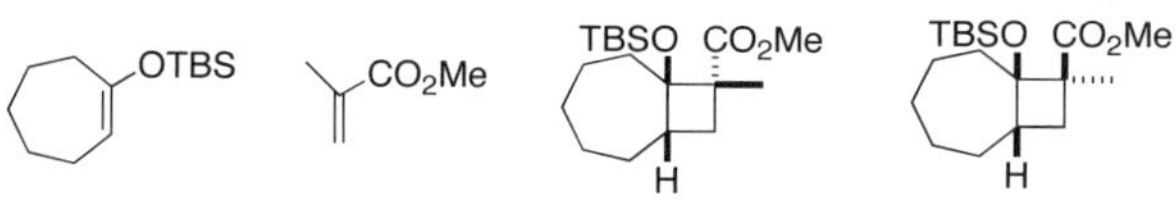					75 (67 : 33)
7[c]	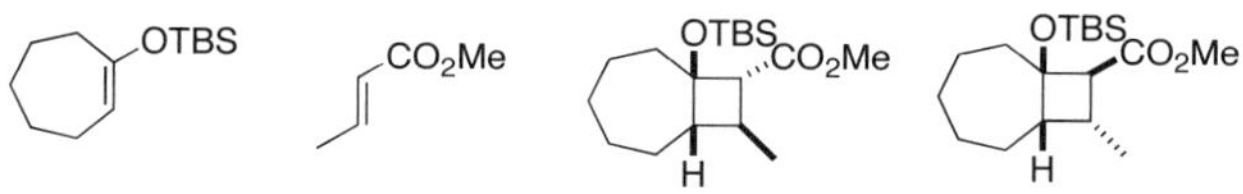					71 (87 : 13)
8[d] 9[d] 10[d]					(n = 0, R = H) (n = 1, R = Me) (n = 2, R = H)	66 (-) 53 (-) 80 (-)
11[e]	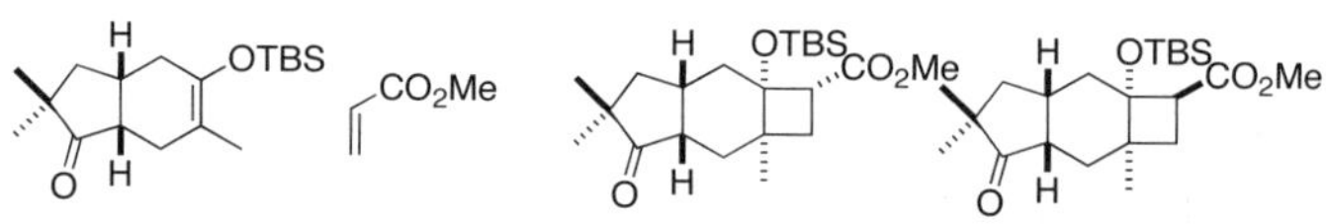					78 (80 : 20)[h]
12[f]	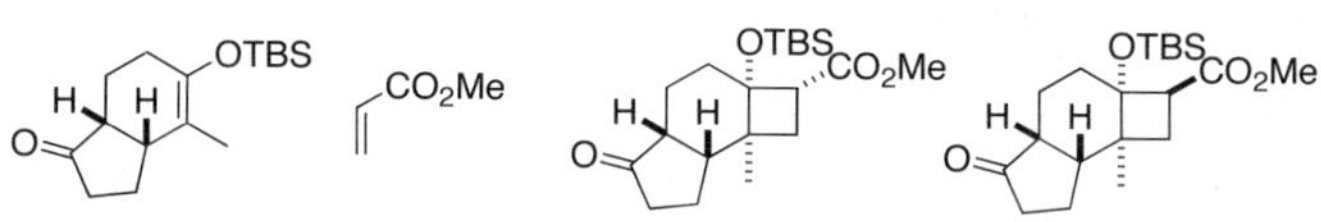					43 (80 : 20)[h]

[a] Reactions are performed using the α,β-unsaturated ester (1 equiv), silyl enol ethers (1.1 equiv) and catalytic amounts of Tf_2NH in CH_2Cl_2 (0.1-0.3 M). [b] Tf_2NH (1.0 mol%), -78 °C, 2 h. [c] Tf_2NH (1.0 mol%), -20 °C to -40 °C, 3 h. [d] Tf_2NH (2.0 mol%), rt, 0.5 h. [e] α,β-unsaturated esters (1.5 equiv), silyl enol ethers (1 equiv), Tf_2NH (1 mol%), -40 °C, 0.5 h. [f] α,β-unsaturated ester (2.0 equiv), silyl enol ethers (1 equiv), Tf_2NH (1.0 mol%), -40 °C, 3 h. [g] Chemical yields were calculated based on α,β-unsaturated esters except for entries 11 and 12. [h] Chemical yields were calculated based on silyl enol ethers.

1. Institute of Medicinal Chemistry, Hoshi University, 2-4-41 Ebara, Shinagawa-ku, Tokyo 142-8501, Japan.
2. Inanaga, K.; Takasu, K.; Ihara, M. *J. Am. Chem. Soc.* **2005**, *127*, 3668.
3. Takasu, K.; Ueno, M.; Inanaga, K.; Ihara, M. *J. Org. Chem.* **2004**, *69*, 517.
4. Takasu, K.; Nagao, S.; Ueno, M.; Ihara, M. *Tetrahedron* **2004**, *60*, 2071.
5. Inanaga, K.; Takasu, K.; Ihara, M. *J. Am. Chem. Soc.* **2004**, *126*, 1352.
6. Ishihara, K.; Hiraiwa, Y.; Yamamoto, H. *Synlett* **2001**, 1851-1854.

PREPARATION OF BENZOCYCLOBUTENONE DERIVATIVES BASED ON AN EFFICIENT GENERATION OF BENZYNES (Bicyclo[4.2.0]octa-1,3,5-trien-7-one, 5-(phenylmethoxy)-)

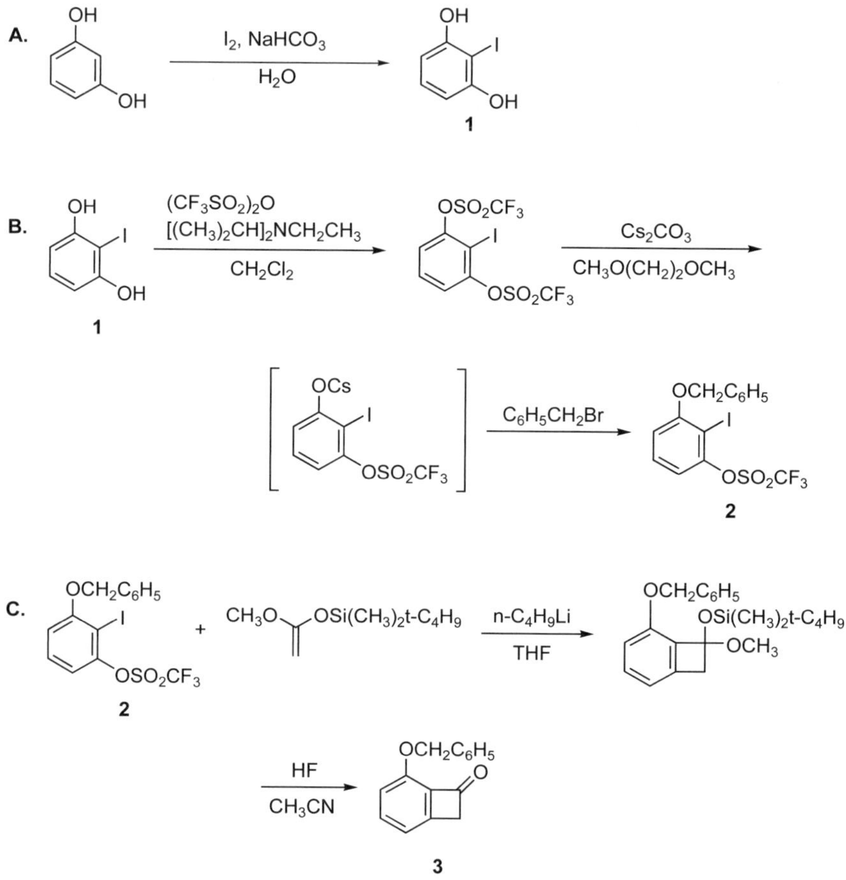

Submitted by Shin-ichiro Tsujiyama and Keisuke Suzuki.[1a]
Checked by David B. Guthrie, Heather M. Gibney and Dennis P. Curran.[1b]

1. Procedure

Caution! Iodine, trifluoromethanesulfonic anhydride, benzyl bromide and hydrofluoric acid are toxic and/or corrosive reagents and should be handled in a well-ventilated hood. Calcium gluconate gel should be present for first aid in case of accidental exposure of skin to HF.

A. 2-Iodoresorcinol (**1**) (Note 1). A 250-mL, one-necked, round-bottomed flask equipped with a large, elliptical stir bar and open to the atmosphere is charged with distilled water (46 mL), resorcinol (7.27 g, 66.0 mmol) and iodine (17.92 g, 70.6 mmol), and then placed in an ice-water bath (Note 2). To this open flask is slowly added sodium bicarbonate (6.16 g, 73.3 mmol) in portions by spatula over 5 min at 0 °C with vigorous stirring (Notes 2, 3). During the addition, vigorous gas evolution (carbon dioxide) is observed. The ice bath is removed and the mixture is allowed to warm to room temperature over 20 min, then stirred for a further 10 min, during which time it becomes a brown slurry (Note 4). The products are extracted with ethyl acetate (3 x 50 mL). The combined organic extract is successively washed with 10% aqueous sodium thiosulfate solution and brine, dried over anhydrous sodium sulfate, filtered, and concentrated under reduced pressure (Note 5). The solid residue is triturated in chloroform (20 mL) for 10 min at –10 °C (ice/ethanol bath), filtered, and washed with cold (–10 °C) chloroform (20 mL) to give 9.12 g (58%) of 2-iodoresorcinol (**1**) as a cream-colored solid. The filtrate is concentrated under reduced pressure, and the residue is triturated in chloroform (7 mL) for 30 min at –10 °C, filtered, and washed with 5 mL of chloroform (0 °C) to give an additional 1.18 g (8%, 66% combined yield) of 2-iodoresorcinol (**1**) as a cream-colored solid. (Notes 6, 7, 8, 9).

B. 3-Benzyloxy-2-iodophenyl trifluoromethanesulfonate (**2**). An oven-dried, 500-mL, one-necked, round-bottomed flask equipped with a magnetic stirring bar, an argon gas inlet and a rubber septum is charged with 2-iodoresorcinol (**1**) (11.49 g, 48.7 mmol) and dichloromethane (115 mL) (Note 10), and the mixture is cooled to –78 °C with a dry ice–acetone bath. *N,N*-Diisopropylethylamine (20.4 mL, 117 mmol) is added by syringe. Then trifluoromethanesulfonic anhydride (18.0 mL, 107 mmol) is slowly added over 15 min by syringe pump while maintaining the reaction mixture at –78 °C (Note 11). After stirring for 10 min at –78 °C, the reaction mixture is warmed to 0 °C by replacing the dry ice bath with an ice water bath, and stirred for 1 h (Note 12). The reaction is quenched by the slow addition of water (100 mL, added over 10 min), and the products are extracted with dichloromethane (2 x 50 mL). The combined organic extract is successively washed with saturated aqueous sodium bicarbonate solution (2 x 50 mL) and brine (2 x 50 mL), dried over anhydrous sodium sulfate, filtered, and concentrated under reduced pressure. The residue is dissolved in diethyl ether (50 mL) and hexanes (50 mL), and then silica gel (12.0 g)

(Note 13) is added. After stirring for 30 min at room temperature, the mixture is filtered, and the silica cake is washed with a mixture of diethyl ether (75 mL) and hexanes (75 mL). The filtrate is concentrated under reduced pressure and vacuum dried to give 24.31–24.33 g of crude 2-iodoresorcinol bis(trifluoromethanesulfonate) as a dark brown oil. This is used directly for the next step without further purification.

A 500-mL, one-necked round-bottomed flask equipped with a magnetic stirring bar and a reflux condenser attached to an inert atmosphere inlet is placed under argon and charged with the crude 2-iodoresorcinol bis(trifluoromethanesulfonate) (24.3 g), 1,2-dimethoxyethane (100 mL) and cesium carbonate (19.06 g, 58.5 mmol) (Note 14). The resulting mixture is stirred for 4 h in an 80 °C oil bath (Note 15). After cooling to room temperature, benzyl bromide (6.96 mL, 58.5 mmol) (Note 16) is added via syringe over 2 min, then the reaction mixture is stirred for 5 h at room temperature (Note 17). Diethylamine (2.5 mL, 24.4 mmol) (Note 18) is added via syringe through the top of the condenser, and the mixture is stirred for 1 h at room temperature. The mixture is cooled in an ice bath and 1 M hydrochloric acid (200 mL) is added, dropwise at first, then more rapidly as the exotherm subsides. The mixture is warmed to room temperature and the products are extracted with ethyl acetate (3 x 100 mL). The combined organic extract is successively washed with brine (2 x 100 mL), saturated aqueous sodium bicarbonate (2 x 100 mL), and brine (2 x 100 mL). The extracts are dried over anhydrous sodium sulfate, filtered, and concentrated under reduced pressure to give 26.3–26.5 g of a brown oil. The residue is purified by column chromatography on silica gel (Note 19) to give 16.5–17.4 g (74–78%) of 3-benzyloxy-2-iodophenyl trifluoromethanesulfonate (**2**) as yellow crystals (Note 20), which are used directly for the next step.

C. *5-Benzyloxybicyclo[4.2.0]octa-1,3,5-trien-7-one* (**3**). An oven-dried, single-necked 500-mL round-bottomed flask equipped with a magnetic stirring bar, an argon inlet and a rubber septum is charged with 3-benzyloxy-2-iodophenyl trifluoromethanesulfonate (**2**) (12.6 g, 27.5 mmol). The flask is flushed with argon, tetrahydrofuran (76 mL) (Note 21) and 1-(*tert*-butyldimethylsilyloxy)-1-methoxyethene (8.99 mL, 41.2 mmol) (Note 21) are added by syringe, and the mixture is cooled to –78 °C (bath temperature) with a dry ice–acetone bath. A solution of butyllithium in hexanes (20.63 mL of a 1.6 M solution, 33.0 mmol) (Note 22) is added to the reaction mixture over 20 min via syringe pump using a needle inserted through the septum. The reaction mixture is further stirred for 10 min at

−78 °C (Note 23), then water (2 mL) is added dropwise. After warming to room temperature, water (100 mL) is added, and the products are extracted with ethyl acetate (3 x 100 mL). The combined organic extract is successively washed with brine (3 x 100 mL), dried over anhydrous sodium sulfate, filtered, and concentrated under reduced pressure to give 10.5–10.8 g of crude 5-benzyloxy-7-*tert*-butyldimethylsilyloxy-7-methoxy-bicyclo-[4.2.0]octa-1,3,5-triene as a brown oil. This is used directly for the next step without further purification.

A 250-mL, three-necked, round-bottomed flask (Note 24) equipped with a magnetic stirring bar, an addition funnel, a stopper and an argon inlet is charged with 5-benzyloxy- 7-*tert*-butyldimethylsilyloxy-7-methoxybicyclo[4.2.0]octa-1,3,5-triene (10.5–10.8 g), acetonitrile (82 mL), and the mixture is cooled to 0 °C in an ice-water bath. Aqueous hydrofluoric acid (48%, 12.5 mL, 345 mmol) (Note 25) is slowly added during 2 min via addition funnel. The mixture is warmed to room temperature over 20 min, then stirred an additional 20 min (Note 26). The reaction mixture is carefully poured into saturated aqueous sodium bicarbonate solution (300 mL) (Note 27), and the products are extracted with ethyl acetate (3 x 100 mL). The combined organic extract is successively washed with saturated aqueous sodium bicarbonate solution (3 x 100 mL) and brine (3 x 100 mL), dried over anhydrous sodium sulfate, filtered, and concentrated under reduced pressure to give 6.24–7.08 g of a yellow solid. The solid is taken up in hot hexanes (40 mL warmed to 60 °C) and the resulting mixture is quickly filtered to remove small amounts of insoluble brown solid. Cooling to 0 °C in an ice bath followed by filtration of the precipitate and washing with cold hexanes gives 3.72–4.34 g (60–70%) of 5-benzyloxybicyclo[4.2.0]octa-1,3,5-trien-7-one (**3**) as a white solid (Notes 28, 29).

2. Notes

1. The iodination process was based on the procedure by Thomsen, I.; Torssell, K. B. G. *Acta Chem. Scand.* **1991**, *45*, 539.

2. The submitters used distilled water, iodine and sodium bicarbonate purchased from Kanto Chemical Co., Inc. Resorcinol was purchased from Tokyo Kasei Kogyo Co., Inc. The checkers used Aldrich HPLC grade distilled water. Sodium bicarbonate was purchased from EMD, resorcinol from Lancaster (99%), and iodine from Aldrich or Fisher (ACS reagent grade, >99.8%). All reagents were used without further purification.

3. The checkers used an elliptical stir bar shaped like an American football, 32 mm long and 17 mm in diameter at the widest point. The submitters used a mechanical stirrer. Rapid stirring was essential for the success of the reaction, and a good vortex was required to prevent a foam from forming on the surface of the mixture during the addition of bicarbonate. When the stirring was not efficient, less 2-iodoresorcinol was formed along with increased amounts of 4-iodoresorcinol, 2,4-diiodoresorcinol and unreacted resorcinol. Sodium bicarbonate was added every few seconds in portions of about 55–65 mg by spatula.

4. Reaction progress was monitored by TLC analysis on Merck silica gel 60 F254 plates, visualized by a 254-nm UV lamp and stained with an aqueous solution of phosphomolybdic acid. TLC analysis showed the formation of 2-iodoresorcinol (**1**) (3/2 hexanes/ethyl acetate, R_f = 0.48). The side products (4-iodoresorcinol and 2,4-diidoresorcinol) had similar R_f values to the target product, while the R_f of resorcinol was 0.28.

5. The residue usually solidified spontaneously at this point. If not, chloroform (20 mL) was added and the mixture was again evaporated.

6. The submitters obtained a first crop from trituration of 9.49 g (61%) and a second crop of 2.18 g (14%) for a total yield of 2-iodoresorcinol (**1**) of 75%.

7. A reaction conducted at twice the indicated scale gave 21.86 g (70%) 2-iodoresorcinol (**1**) after two triturations (20.11 g, then 1.75 g).

8. The product was stored in a dark bottle at 5 °C because slow decomposition occurred on exposure to ambient light at room temperature.

9. Mp 99–101 °C; ^{1}H NMR (300 MHz, acetone-d_6) δ: 6.46 (d, J = 8.1 Hz, 2 H), 7.00 (t, J = 8.1 Hz, 1 H), 8.83 (s, 2 H); ^{13}C NMR (75 MHz, acetone-d_6) δ: 75.4, 107.1, 130.4, 158.8; IR (thin film, $CHCl_3$): 3456, 3363 (broad), 1593, 1582, 1491, 1458, 1304, 1278, 1250, 1185, 1159, 1022, 994, 784, 704 cm^{-1}; EIMS (*m/z*): M^+ 237 (6), 236 (100), 218 (23), 127 (16), 81 (13), 63 (8), 53 (17); HRMS (*m/z*): $[M]^+$ calcd for $C_6H_5IO_2$, 235.9334; found, 235.9333. Anal. calcd. for $C_6H_5IO_2$: C, 30.54; H, 2.14. Found: C, 30.27; H, 2.14.

10. Dichloromethane, purchased from Asahi Glass Co., Inc. or Aldrich Chemical Co, was distilled from phosphorus pentoxide and calcium hydride (submitters) or dried by passing through an activated alumina column (checkers).

11. *N,N*-Diisopropylethylamine was purchased by the submitters from Tokyo Kasei Kogyo Co., Inc., and trifluoromethanesulfonic anhydride was donated from Central Glass Co., Inc. These reagents were used without

further purification. The submitters purchased these chemicals from Aldrich Chemical Co. and redistilled the amine before use.

12. TLC analysis showed the formation of 2-iodoresorcinol bis(trifluoromethanesulfonate) (4/1 hexanes/ethyl acetate, $R_f = 0.57$).

13. Silica gel (60 Å, for column chromatography) was purchased from Kanto Chemical Co., Inc. (submitters) or Sorbent Technologies (60 Å, 230 x 400 mesh) (checkers).

14. 1,2-Dimethoxyethane was purchased from Tokyo Kasei Kogyo Co., Inc. (submitters) or J. T. Baker (checkers). Cesium carbonate (Reagent Plus, 99%) was purchased from Aldrich Chemical Co. These reagents were used without further purification.

15. TLC analysis showed the formation of 3-hydroxy-2-iodophenyl trifluoromethanesulfonate (4/1 hexanes/ethyl acetate, $R_f = 0.24$).

16. Benzyl bromide purchased from Tokyo Kasei Kogyo Co., Inc. (submitters) or Aldrich Chemical Co. (Reagent grade, 98%) (checkers) was used without further purification.

17. TLC analysis showed the formation of 3-benzyloxy-2-iodophenyl trifluoromethanesulfonate (**2**) (4/1 hexanes/ethyl acetate, $R_f = 0.45$).

18. Diethylamine was used to quench the excess benzyl bromide. Diethylamine was purchased from Kanto Chemical Co., Inc. (submitters) or Aldrich Chemical Co. (Reagent Plus, >99.5%) (checkers) and was used without further purification.

19. The product was taken up in 3/1 hexanes/ethyl acetate and stirred for 1 h. The organic solution was decanted from a small amount of dark brown solid directly onto the column. The chromatography was performed on 300 g of silica gel eluting with 10/1 hexanes/ethyl acetate.

20. Mp 86–87 °C; ^{1}H NMR (300 MHz, $CDCl_3$) δ: 5.21 (s, 2 H), 6.86 (d, J = 8.1 Hz, 1 H), 6.98 (d, J = 8.4 Hz, 1 H), 7.32–7.40 (m, 2 H), 7.40–7.48 (m, 2 H), 7.48–7.53 (m, 2 H); ^{13}C NMR (75 MHz, $CDCl_3$) δ: 71.4, 83.3, 111.6, 114.3, 118.7 (q, $J_{C\text{-}F}$ = 318.8 Hz), 126.9, 128.1, 128.6, 130.3, 135.6, 151.2, 159.4; IR (thin film, $CHCl_3$): 1589, 1450, 1425, 1273, 1224, 1209, 1139, 1056, 1024, 947, 850, 826, 732 cm^{-1}; EIMS (*m/z*): M^+ 459 (6), 458 (36), 382 (24), 368 (11), 367 (7), 249 (11), 197 (52), 107 (64), 92 (98), 79 (44), 69 (78), 65 (100); HRMS (*m/z*): $[M]^+$ calcd for $C_{14}H_{10}F_3IO_4S$, 457.9297; found, 457.9316. Anal. calcd. for $C_{14}H_{10}F_3IO_4S$: C, 36.70; H, 2.20. Found: C, 36.77; H, 2.19.

21. 1-(*tert*-Butyldimethylsilyloxy)-1-methoxyethene was obtained from Aldrich Chemical Co., Inc., and used without further purification. The submitters observed that distillation of 1-(*tert*-butyldimethylsilyloxy)-

1-methoxyethene under argon did not result in increased product yield or purity. The submitters used tetrahydrofuran (dehydrated, stabilizer free) purchased from Tokyo Kasei Kogyo Co., Inc. without further purification. The checkers used THF from Aldrich Chemical Co. (99%, inhibitor free), which was dried by passing through activated alumina.

22. Butyllithium in hexanes was purchased from Kanto Chemical Co., Inc. (submitters) or Aldrich Chemical Co. (checkers).

23. TLC analysis showed consumption of starting material and formation of 5-benzyloxy-7-*tert*-butyldimethylsilyloxy-7-methoxybicyclo-[4.2.0]octa-1,3,5-triene (4/1 hexanes/ethyl acetate, $R_f = 0.74$).

24. The submitters and checkers used a glass vessel for this reaction, but the submitters reported that a polypropylene flask could also be used to avoid corrosion of the flask by hydrofluoric acid.

25. The submitters used acetonitrile and a 46% aqueous solution of hydrofluoric acid, purchased from Kanto Chemical Co., Inc. without further purification. The checkers used acetonitrile from J. T. Baker and 48% HF from Mallinckrodt.

26. TLC analysis showed the formation of 5-benzyloxybicyclo-[4.2.0]octa-1,3,5-trien-7-one (**3**) (4/1 hexanes/ethyl acetate: $R_f = 0.55$).

27. Vigorous evolution of carbon dioxide gas was observed.

28. Mp 80–81 °C; ^{1}H NMR (300 MHz, $CDCl_3$) δ: 3.95 (s, 2 H); 5.47 (s, 2 H), 6.90 (d, $J = 8.4$ Hz, 1 H), 7.06 (d, $J = 6.9$ Hz, 1 H), 7.30-7.43 (m, 3 H), 7.43-7.51 (m, 3 H); ^{13}C NMR (75 MHz, $CDCl_3$) δ: 51.1, 73.8, 115.1, 116.3, 127.7, 128.0, 128.3, 132.4, 136.3, 137.6, 150.4, 152.1, 184.7; IR (thin film, $CHCl_3$): 3033, 2929, 1761, 1604, 1573, 71475, 1453, 1386, 1274, 1160, 1129, 1052, 787, 755 cm^{-1}; EIMS (*m/z*): M^+ 224 (20), 105 (6), 91 (100), 65 (13), 51 (14). HRMS (*m/z*): $[M]^+$ calcd for $C_{15}H_{12}O_2$: 224.0837; found: 224.0845. Anal. calcd. for $C_{15}H_{12}O_2$: C, 80.34; H, 5.39. Found: C, 80.36; H, 5.10.

29. The submitters obtained a second crop as follows. The filtrate was concentrated under reduced pressure to give 1.62 g of a yellowish solid, which was recrystallized from hexanes (6 mL, at 60 °C followed by cooling down to 0 °C using an ice bath) to give the second crop of **3** (0.60 g, 10%). In the checkers' hands, the second crop product was about 90–95% pure according to ^{1}H NMR spectroscopy.

Safety and Waste Disposal Information

All hazardous materials should be handled and disposed of in accordance with "Prudent Practices in the Laboratory"; National Academy Press; Washington, DC, 1995.

3. Discussion

Benzocyclobutenes have long attracted much attention due to their molecular structure, chemical reactivity and synthetic potential.[2] They have served as important building blocks for syntheses of various natural products by exploiting their unique reactivity. Due to their inherent strain, benzocyclobutenes are capable of isomerizing to quinodimethanes that can in turn be used to derive various polycyclic skeletons via Diels-Alder reactions or electrocyclizations. Described here is an effective procedure for the synthesis of highly oxygenated benzocyclobutenones, which have great synthetic potential, since a carbonyl group in the four-membered ring can serve as a platform to introduce various functionalities. The present protocol is particularly attractive because it provides benzocyclobutene analogues in assorted oxidation states by the use of various ketene silyl acetals (Scheme 1).[3] These products can be important synthetic intermediates for various polyaromatic natural products, such as gilvocarcin,[4] pradimicin,[5] aquayamycin,[6] and TAN-1085.[7]

Scheme 1

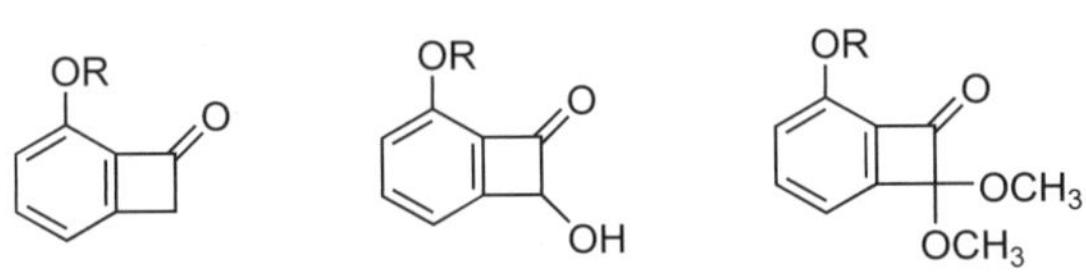

Arynes are important reactive intermediates, and many studies of their generation and reactions have been undertaken.[8] Several protocols for generation of arynes have been known basically relying on the 1,2-elimination on aromatic rings. These include treatment of aryl halides with a strong base such as sodium amide,[8] decomposition of benzenediazonium-2-carboxylate,[9] halogen–metal exchange on *ortho*-dihalosubstituted benzenes,[10] and treatment of *ortho*-trimethylsilylphenyl triflates with fluoride.[11] However, all methods require high reaction temperatures and/or long reaction times, and often accompanied by side reactions.

In the present procedure, an *ortho*-iodoaryl triflate is employed as an efficient benzyne precursor.[12] This is easily prepared by mono-cleavage of a *bis*-triflate by cesium carbonate[13] and in situ-benzylation of the resulting cesium phenoxide. Benzynes are immediately generated by treatment of the *ortho*-iodoaryl triflate with butyllithium at low temperature (typically −78 °C). Compared with former methods, the present protocol enables faster generation of benzynes at much lower temperature because of the rapid halogen–lithium exchange and the ability of the triflate to serve as an excellent leaving group. The benzyne intermediate thus generated combines with arynophiles with high-lying HOMOs such as ketene silyl acetals,[3] furans,[12] and nitrones[14] in reactions that afford the corresponding cycloadducts in high yields (Scheme 2).

Scheme 2

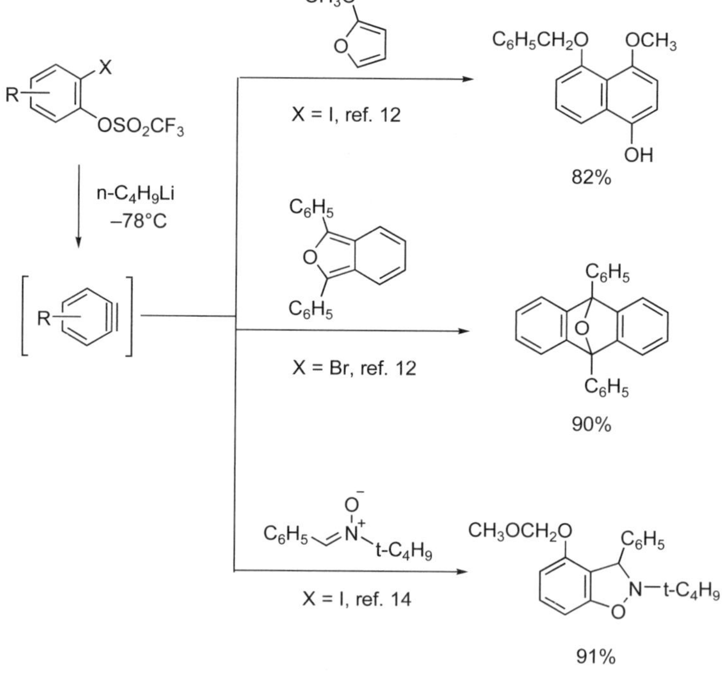

1. (a) Department of Chemistry, Tokyo Institute of Technology and CREST, Japan Science and Technology Agency (JST), O-okayama, Meguro-ku, Tokyo 152-8551, Japan. Email: ksuzuki@chem.titech.ac.jp. (b) Department of Chemistry,

University of Pittsburgh, Pittsburgh, PA 15260, USA. Email: curran@pitt.edu.

2. For reviews on benzocyclobutenes, see: (a) Wada, Y.; Tago, T.; Nishimura, J. *J. Synth. Org. Chem. Jpn.* **1992**, *50*, 616; (b) Farona, M. F. *Prog. Polym. Sci.* **1996**, *21*, 505; (c) Michellys, P.-Y.; Pellissier, H.; Santelli, M. *Org. Prep. Proced. Int.* **1996**, *28*, 545; (d) Segura, J. L.; Martín, N. *Chem. Rev.* **1999**, *99*, 3199; (e) Mehta, G.; Kotha, S. *Tetrahedron* **2001**, *57*, 625; (f) Sadana, A. K.; Saini, R. K.; Billups, W. E. *Chem. Rev.* **2003**, *103*, 1539.

3. (a) Hosoya, T.; Hasegawa, T.; Kuriyama, Y.; Matsumoto, T.; Suzuki, K. *Synlett* **1995**, 177; (b) Hamura, T.; Hosoya, T.; Yamaguchi, H.; Kuriyama, Y.; Tanabe, M.; Miyamoto, M.; Yasui, Y.; Matsumoto, T.; Suzuki, K. *Helv. Chim. Acta* **2002**, *85*, 3589.

4. (a) Matsumoto, T.; Hosoya, T.; Suzuki, K. *J. Am. Chem. Soc.* **1992**, *114*, 3568; (b) Hosoya, T.; Takashiro, E.; Matsumoto, T.; Suzuki, K. *J. Am. Chem. Soc.* **1994**, *116*, 1004.

5. (a) Ohmori, K.; Kitamura, M.; Suzuki, K. *Angew. Chem. Int. Ed.* **1999**, *38*, 1226; (b) Kitamura, M.; Ohmori, K.; Kawase, T.; Suzuki, K. *Angew. Chem. Int. Ed.* **1999**, *38*, 1229.

6. (a) Matsumoto, T.; Yamaguchi, H.; Hamura, T.; Tanabe, M.; Kuriyama, Y.; Suzuki, K. *Tetrahedron Lett.* **2000**, *41*, 8383; (b) Yamaguchi, H.; Konegawa, T.; Tanabe, M.; Nakamura, T.; Matsumoto, T.; Suzuki, K. *Tetrahedron Lett.* **2000**, *41*, 8389; (c) Matsumoto, T.; Yamaguchi, H.; Tanabe, M.; Yasui, Y.; Suzuki, K. *Tetrahedron Lett.* **2000**, *41*, 8393.

7. Ohmori, K.; Mori, K.; Ishikawa, Y.; Tsuruta, H.; Kuwahara, S.; Harada, N.; Suzuki, K. *Angew. Chem. Int. Ed.* **2004**, *43*, 3167.

8. For reviews on arynes, see: (a) Kessar, S. V. In *Comprehensive Organic Synthesis*; Trost, B. M. Ed., Pergamon: Oxford, U. K., 1991, Vol. 4, pp 483–515; (b) Pellissier, H.; Santelli, M. *Tetrahedron* **2003**, *59*, 701.

9. Friedman, L.; Logullo, F. M. *J. Am. Chem. Soc.* **1963**, *85*, 1549.

10. Wittig, G. *Org. Synth.* **1959**, *39*, 75 (Coll. Vol. 4, **1963**, 964)

11. Himeshima, Y.; Sonoda, T.; Kobayashi, H. *Chem. Lett.* **1983**, 1211.

12. Matsumoto, T.; Hosoya, T.; Katsuki, M.; Suzuki, K. *Tetrahedron Lett.* **1991**, *32*, 6735.

13. The mono-cleavage process is analogous to mono-saponification of bis-benzoate: Zaugg, H. E. *J. Org. Chem.* **1976**, *41*, 3419.

14. Matsumoto, T.; Sohma, T.; Hatazaki, S.; Suzuki, K. *Synlett* **1993**, 843.

AN EFFICIENT, HIGHLY DIASTEREO- AND ENANTIOSELECTIVE HETERO-DIELS-ALDER CATALYST. PREPARATION OF (2*S*,6*R*)-6-(*tert*-BUTYLDIMETHYL-SILYLOXYMETHYL)-2-METHOXY-2,5-DIHYDROPYRAN

(Silane, [[(2*R*,6*S*)-3,6-dihydro-6-methoxy-2H-pyran-2-yl]methoxy]-(1,1-dimethylethyl)dimethyl)-

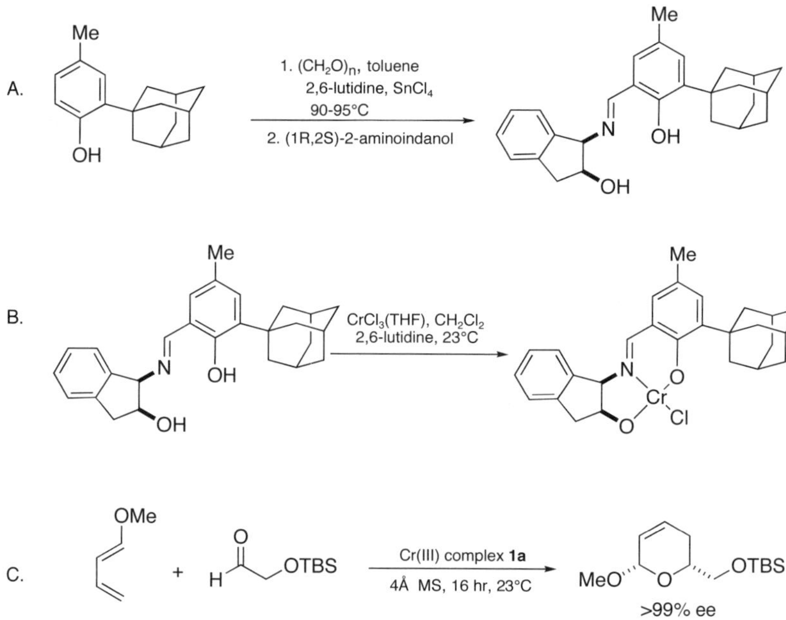

Submitted by David E. Chavez and Eric N. Jacobsen.[1]
Checked by E. J. J. Grabowski and Michele Kubryk.

1. Procedure

A. (1R,2S)-1-[3-Adamantyl)-2-hydroxy-5-methylbenzylidenamino]indan-2-ol. An oven-dried, 300-mL, three-necked, round-bottomed flask is equipped with a magnetic stirbar, fitted with a reflux condenser and thermometer, and purged with a nitrogen atmosphere by means of an inlet fitted to the condenser. The flask is charged with 2-adamantyl-4-methylphenol (12.1 g, 50.0 mmol, 1 eq) (Note 1), freshly distilled toluene (110 mL) (Note 2), and 2,6-lutidine (4.28 g, 4.67 mL, 40.00 mmol, 0.8 eq); the open neck of the flask is capped with a septum. Neat stannic chloride ($SnCl_4$) (2.60 g, 1.17 mL, 10.00 mmol, 0.2 eq) is added by syringe over 10 min (Note 3). The solution turns pale yellow in color, and a pale yellow precipitate is also observable. The mixture is allowed to stir at room temperature for 20 min, then the septum is removed and solid paraformaldehyde (6.00 g, 200 mmol) is added in one portion against a gentle nitrogen counterflow (Note 4). The mixture is stirred an additional 10 min, the nitrogen inlet is replaced with a nitrogen balloon, the reaction flask is placed in a 90-95 °C bath, and heating is maintained at this temperature for 6 h. The reaction mixture is then allowed to cool to room temperature and filtered through a pad of premixed Celite® and silica gel (1:1, 12 g). The filter pad is washed with ethyl acetate (200 mL), and the combined organic filtrates are washed with water (350 mL), 1N HCl (350 mL), and brine (350 mL), and then dried over anhydrous Na_2SO_4. Concentration is effected by rotary evaporation, followed by removal of trace solvent on a high vacuum pump (0.5 mm) (13.4 g crude, 99.5%) (Note 5). Absolute ethanol (200 mL) is added and the mixture is heated gently until complete dissolution occurs (Note 6). (1*R*,2*S*)-1-Amino-2-indanol (7.83 g, 52.50 mmol, 1.05 equiv. Note 2) is added in one portion. The reaction mixture is then heated at 80 °C for 45 min, cooled to room temperature, and allowed to stand for 3-5 hours. The yellow solid product is isolated filtration, washed with cold ethanol (50 mL), and dried in the air (15.1 g, 75.2% over 2 steps) (Note 7).

*B. Chromium(III) Cl complex (**1a**).* To a 200-mL round-bottomed flask is added chromium(III) chloride-tetrahydrofuran complex (1:3) (2.80 g, 7.48 mmol, 1 equiv.) and (1*R*,2*S*)-1-[3-adamantyl)-2-hydroxy-5-methylbenzyliden-amino]indan-2-ol (3.00 g, 7.48 mmol, 1 equiv.). The reaction mixture is placed under a nitrogen atmosphere, and dichloromethane (CH_2Cl_2, 60 mL) is added followed by dropwise addition of 2,6-lutidine (1.74 mL, 14.96 mmol, 2 equiv). The solution is stirred for 3 h, diluted with CH_2Cl_2 (300 mL), and washed with water (3 x 180 mL), then brine (180 mL) (Note 8). The organic phase is dried over anhydrous Na_2SO_4, filtered, and concentrated by rotary evaporation. The resulting solid is triturated with acetone (10 mL), filtered, washed with an additional portion of acetone (10 mL), and air-dried to give the chromium complex (**1a**) as a brown solid (2.3 g). Water (2 mL) is added to the filtrate (Note 9) and the solution allowed to stand uncovered at 23 °C overnight. The resulting precipitate is filtered and washed with cold acetone to give an additional 600-800 mg of the chromium complex (**1a**) (combined yield 2.9-3.1 g, 80 – 85%) (Notes 10, 11).

C. (2S,6R)-6-(tert-Butyldimethylsilyloxymethyl)-2-methoxy-2,5-dihydro-pyran. 1-Methoxybutadiene (2.40 g, 2.89 mL, 28.7 mmol, 1.11

equiv.) is added dropwise to a stirring mixture of (*tert*-butyldimethylsilyloxy)acetaldehyde (90%, 5.00 g, 5.46 mL, 25.8 mmol, 1 equiv.), (1*R*, 2*S*) chromium(III) chloride complex (**1a**) (200 mg, 0.19 mmol, 1.5 mol% (Note 12) and 4Å molecular sieves (Note 13) under N_2 at 0 °C. The reaction mixture is allowed to stir at 0 °C for 1 h and then warmed to room temperature and allowed to stir for an additional 16 h. Distillation of this mixture (Kügelrohr, 110 °C, 0.5 mm) affords the cycloadduct (6.0 g, 90%) as a colorless oil (Note 14) in >99% ee (Note 15).

2. Notes

1. The purity of the 2-adamantyl-4-methylphenol is important; in particular, the material should be free of 2,6-diadamantyl-4-methylphenol.
2. All reagents were obtained from commercial suppliers (Acros, Aldrich Chemical Company, Inc., or Strem Chemicals, Inc.). Toluene was distilled from sodium, and dichloromethane was distilled from calcium hydride. All other reagents were used as received without further purification.
3. The use of a syringe containing a teflon plunger prevents clogging during the addition of $SnCl_4$.
4. Caution must be taken to prevent the fluffy solid paraformaldehyde from dispersing outside of the flask during this addition process.
5. This procedure for the synthesis of 2-adamantyl-5-methylsalicylaldehyde is a modification of the method reported by Casiraghi.[2] The aldehyde can be recrystallized from hexanes, but purification is not essential for successful formation of the Schiff base. The purified aldehyde has the following spectral and physical properties: mp 151.5-152 °C; IR (KBr) 3200-2500, 1649, 1607, 1524, 1447, 1416, 1356, 1312, 1244, 1221, 1163, 1105, 1084, 1040, 963, 864 cm^{-1}; ^{1}H NMR (500 MHz, $CDCl_3$) δ 1.78 (s, 6H), 2.08 (s, 3H), 2.12 (s, 6H), 2.31 (s, 3H), 7.14 (d, J = 1.5 Hz, 1H), 7.26 (d, J = 1.5 Hz, 1H), 9.8 (s, 1H), 11.65 (s, 1H); ^{13}C NMR (100 MHz, $CDCl_3$) δ 20.5, 28.9, 36.9, 40.1, 120.3, 128.2, 131.2, 135.4, 138.1, 159.3, 197.1; Calcd for $C_{18}H_{22}O_2$: C, 79.96; H, 8.20. Found: C, 79.70; H, 8.16.
6. The aldehyde is observed to dissolve completely between 60-70 °C.
7. The product exhibits the following physical and spectroscopic properties: mp 219-221 °C; $[\alpha]^{26}_{D}$ +70.0 (c .100, THF); IR (KBr disk) 3584,

2905, 2849, 1624, 1597cm^{-1}; ^{1}H NMR (500 MHz, DMSO- d_{6}.) δ 1.69 (m, 6H), 1.99 (m, 3H), 2.05 (m, 6H), 2.23 (s, 3H), 2.95 (dd, J = 6.0, 15.5 Hz, 1H), 3.11 (dd, J = 6.1, 15.5 Hz, 1H), 4.54 ('q', J = 5.7 Hz, 1H), 4.73, (d, J = 5.5 hz, 1H), 5.23, (d, J = 4.9 hz, 1H), 7.01 (s, 1H), 7.09 (s,1H), 7.18-7.31 (m, 4H), 8.61 (s, 1H), 10.94 (s, 1H); ^{13}C-NMR (ppm): 20.2, 28.3, 36.2, 36.5, 39.0, 39.7, 73.9 (2 carbons), 118.2, 124.7, 125.0, 125.7, 126.6, 127.4, 127.9, 129.6, 130.0, 136.4, 141.0, 142.0, 158.5, 166.5; HRMS (*m/z*) (Cl NH_3) calcd for $C_{27}H_{35}NO_2(M)^+$ 401.2355, found 401.2341.

8. The water washes should be carried out with gentle shaking in order to avoid formation of intractable emulsions.

9. If partial concentration occurs during filtration, the filtrate should be diluted with acetone prior to addition of water such that the total volume is 20 mL. Upon addition of water, a small amount of precipitate may form. This should be redissolved by gently warming the solution or by addition of a minimal amount of acetone.

10. X-Ray quality crystals are obtained by recrystallization from acetone/water. The solid state structure of complex **1** is that of a dimer bearing a bridging water molecule and one terminal water molecule on each metal center.[3] This dimeric complex exhibits the following spectral properties: IR (KBr): 3414, 2903, 2847, 1618, 1537, 1433, 1340, 1305, 1228, 1168, 1078 cm^{-1}. LRMS (FAB): calcd for dimer $C_{54}H_{68}Cl_2N_2O_7Cr_2$, $(M\text{-}2Cl\text{-}2H_2O)^+$, 920, found 919. A dehydrated sample suitable for elemental analysis was prepared as follows: Chlorotrimethylsilane (39.0 μL, 0.31 mmol) was added to a solution of Cr(III)Cl complex (50.0 mg, 0.048 mmol) in dry *tert*-butyl methyl ether (2 mL). The mixture was stirred for 2 h under nitrogen to give a green precipitate. The mixture was concentrated *in vacuo*, suspended in dry *tert*-butyl methyl ether (2 mL), filtered and the residue washed with dry *tert*-butyl methyl ether. The residue was then dried under high vacuum (0.5 mm). Anal. Calcd for [$C_{27}H_{29}ClCrNO_2$+2HCl]: C, 57.92; H, 5.58; Cr, 9.29; N 2.50. Found: C, 57.49; H, 5.73; Cr, 9.00; N, 2.48.

11. For certain applications (see, for example, the first entry in Table 1), superior results in HDA reactions are obtained with catalyst **1b**, wherein the chloride counterion of **1a** is replaced with SbF_6. Preparation of catalyst **1b** is achieved as follows: A flame-dried, 50-mL, foil wrapped round-bottomed flask equipped with a stirbar was charged with complex **1a** (100 mg, 0.97 mmol, 1 equiv) and silver hexafluoroantimonate (66.8 mg, 0.19 mmol, 2 equiv). The flask was placed under a nitrogen atmosphere, *tert*-butyl methyl ether (30 mL) was added, and the mixture allowed to stir

for 3 h. The reaction mixture was then filtered through Celite® and the isolated solids are washed with *tert*-butyl methyl ether (20 mL). The filtrates were combined and concentrated by rotary evaporation to afford the desired SbF_6 complex **1b** as a brown solid (165 mg). IR (KBr) 3378, 2973, 2905, 1615, 1538, 1229, 1069 cm^{-1}. LRMS (m/z) (FAB) mass calcd for $C_{27}H_{35}CrNO_{2,}$ $(M)^+$ 451; found 451; calcd for $2[C_{27}H_{35}CrNO_{2,}]$, $(2M)^+$, 902; found 902; calcd for 2 $C_{27}H_{35}CrNO_{2,}$ + $H_2O]$, $(2M + H_2O)^+$, 920; found 921.

12. The catalyst loading was calculated based on the number of equivalents of chromium relative to the limiting aldehyde substrate.

13. The molecular sieves (1.6 mm pellets) are powdered with a mortar and pestle and activated in a vacuum oven (130 °C) overnight before use. Alternatively, commercially available finely powdered 4Å molecular sieves (<5 micron) may be used.

14. The product has the following spectral and physical properties: $[\alpha]_D^{26}$ +55.3 (c 1.14, $CDCl_3$); R_f = 0.70 (1:1 ether/hexanes); IR (thin film) 2955, 2934, 2888, 2858, 1471, 1400, 1339, 1255, 1204, 1129, 1112, 1080, 1057 cm^{-1}; ^1H NMR (500 MHz, $CDCl_3$) δ 0.07 (s, 6H), 0.89 (s, 9H), 2.08 (m, 2H), 3.47 (s, 3H), 3.65 (dd, *J* = 6.5, 10.4 Hz, 1H), 3.76 (dd, *J* = 5.6, 10.4 Hz, 1H), 3.85 ('q', *J* = 6.3 Hz, 1H), 5.02 (m, 1H), 5.65 ('dq', *J* = 3.7, 10.2 Hz, 1H), 5.97 ('dq', *J* = 5.3, 10.2 Hz, 1H); ^{13}C NMR (125 MHz, $CDCl_3$) δ 5.2, 5.3, 18.4, 25.9, 26.8, 55.2, 65.5, 72.6, 97.7, 127.0 128.5; HRMS (*m/z*) (Cl) calc. for $C_{13}H_{30}NO_3Si$ $(M+NH_4)^+$ 276.1995, found 276.2003.

15. Enantiomeric excess was determined by GC analysis following conversion to (*R*)-6-(*tert*-butyldimethylsilyoxymethyl)-5,6-dihydropyran-2-one, according to the following procedure: Pyridinium dichromate (1.04 g, 2.75 mmol) was added to a solution of the acetal (256 mg, 1.38 mmmol) and acetic acid (3 mL) in CH_2Cl_2 (20 mL) at 23 °C. The mixture was stirred overnight, diluted with 1:1 ether/hexanes (20 mL), and filtered through a pad of $MgSO_4$. The residue remaining in the reaction flask was washed thoroughly with 1:1 ether/hexanes (4 x 20 mL) and the extracts were filtered. The combined filtrates were filtered once more through a fresh pad of $MgSO_4$ and concentrated *in vacuo*. Kügelrohr distillation (210-220 °C, 10 mm) afforded the product lactone (267 mg, 57.0%). GC analysis using a commercial chiral column (Cyclodex β. 135 °C, isothermal) revealed the product to be in >99% ee (t_R(major) = 50.23 min). $[\alpha]_D^{26}$ +79 (c 1.00, $CDCl_3$). R_f = 0.17 (10% ether/hexanes). IR (thin film) 2955, 2930, 2859, 1732, 1471, 1407, 251, 1136, 1093, 1043 cm^{-1}. ^1H NMR (500 MHz, $CDCl_3$) δ 0.06 (s, 6H), 0.87 (s, 9H), 2.40 ('dt', *J* = 4.6, 18.6 Hz, 1H), 2.51 (ddd, *J* =

2.6, 11.1, 18.6 Hz, 1H), 3.78 (dd, $J = 5.4, 10.9$ Hz, 1H), 3.80 (dd, $J = 4.64, 10.9$ Hz, 1H), 4.45 (dddd, $J = 4.4, 4.6, 5.4, 11.1$ Hz, 1H), 5.99 (d, $J = 9.7$ Hz, 1H), 6.89 (ddd, $J = 2.6, 5.8, 9.7$ Hz, 1H). ^{13}C NMR (125 MHz, $CDCl_3$) δ -5.4, 18.3, 25.8, 64.2, 77.8, 121.1, 145.0, 163.9. HRMS (*m/z*) (Cl) Calcd for $C_{12}H_{26}NO_3Si$ $(M+NH_4)^+$ 260.1682. Found 260.1679.

Waste Disposal Information

All hazardous materials should be handled and disposed of in accordance with "Prudent Practices in the Laboratory"; National Academy Press; Washington, DC, 1995.

3. Discussion

This procedure describes a practical synthesis of the chiral tridentate Schiff base complex **1a**, and the use of this complex to catalyze an efficient, highly diastereo- and enantioselective hetero-Diels-Alder (HDA) reaction. The unique characteristic of this catalyst, and the derived SbF_6 complex **1b** (see Note 11), lies in their demonstrated ability to promote asymmetric hetero-Diels-Alder reactions between aldehydes and dienes bearing a single oxygen substituent.[4] Reactions proceed generally with excellent diastereo- and enantioselectivity, and provide access to enantiomerically enriched dihydropyran derivatives from simple achiral starting materials (Table 1). This HDA methodology has already been showcased in several natural product syntheses.[5] More recently, the same catalyst system has been applied to highly enantioselective inverse demand hetero-Diels-Alder reactions between conjugated aldehydes and ethyl vinyl ether.[3]

The method for the synthesis of complex **1a** described herein represents a significant improvement over the procedure first reported in 1999.[4] The use of air- and moisture-sensitive $CrCl_2$ is now avoided, and the necessity of conducting the metal-insertion step in a glove box is thereby precluded. Instead, the use of the $(Cr(III)Cl_3 \cdot [C_4H_8O]_3)$ complex allows the reaction to be conducted in a fume hood. Additionally, the procedure for the formylation of 2-adamantyl-4-methylphenol has been adapted such that purification of the resulting aldehyde by recrystallization is no longer necessary. Finally, and perhaps most important, catalysts prepared by the new procedure displays measurably higher enantioselectivity in a variety of HDA reactions.[3]

Table 1

Aldehyde	Diene	Product	Cat	ee (%)	Yield (%)	Ref
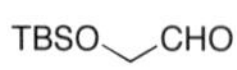	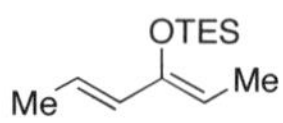	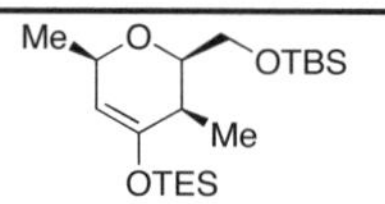	**1a** **1b**	99 >99	90 97	4 4
PhCHO	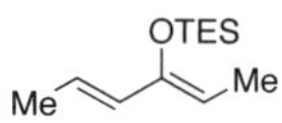	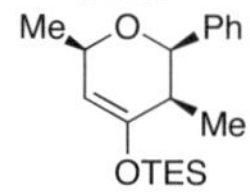	**1b**	90	72	4
n-C_5H_{11}CHO	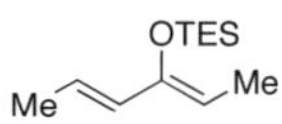	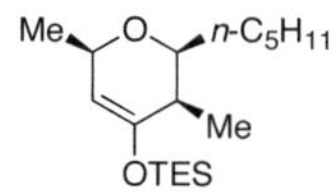	**1b**	98	85	4
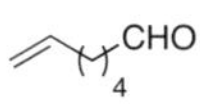	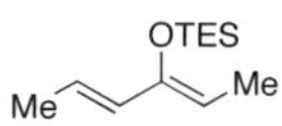	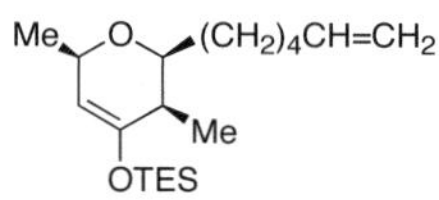	**1b**	98	85	4
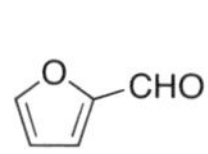	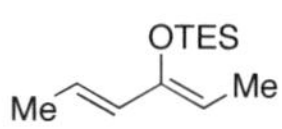	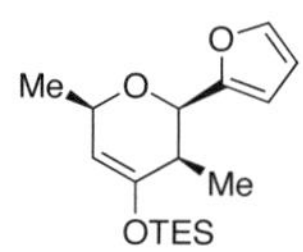	**1b**	95	77	4
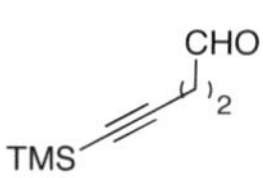	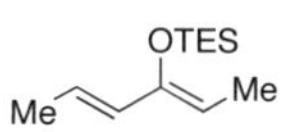	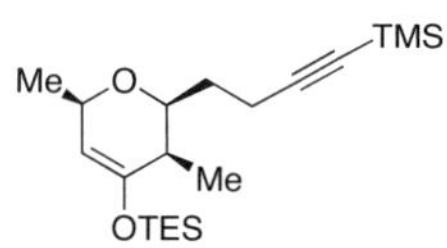	**1b**	95	92	5a
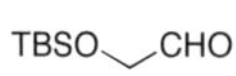	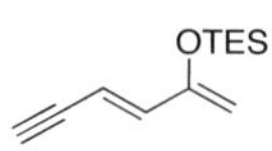	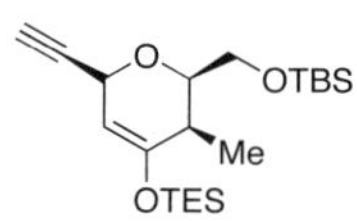	**1b**	98	61	5a
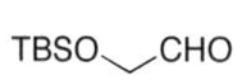	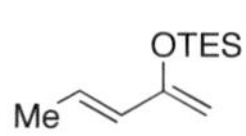	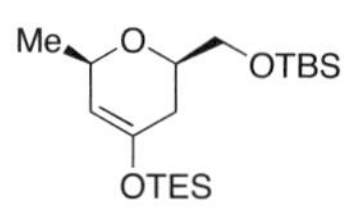	**1b**	98	78	4
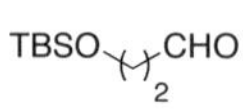	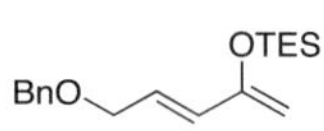	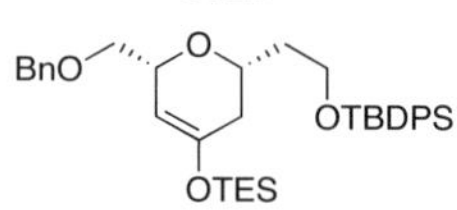	**1a**	97	64	5c
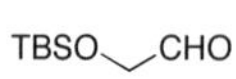	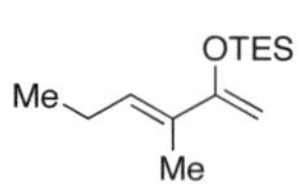	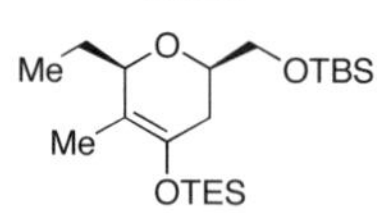	**1a**	>99	87	5c
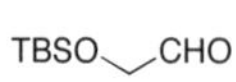	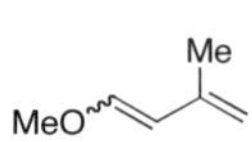	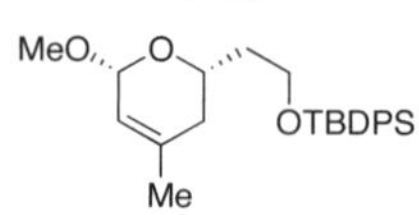	**1a**	>95	71	5e
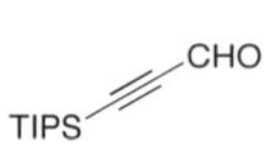	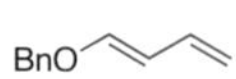	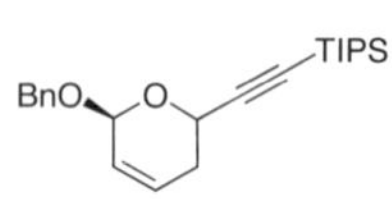	**1a**	94	80	5b, 3

The hetero-Diels-Alder reaction illustrated in this procedure utilizes commercially available 1-methoxy-1,3-butadiene and (*t*-butyldimethylsilyloxy)acetaldehyde. The reaction is carried out with 1.5 mol% catalyst under solvent-free conditions. The dihydropyran is isolated in 90% yield, >97:3 dr, and >99% ee by direct distillation of the reaction mixture. The product can be oxidized to the corresponding lactone readily and in one step,providing efficient access to a substructure that occurs in several interesting natural products (i.e., fostriecin[5b], callystatin A[6a], ratjadone[6b]).

1. Department of Chemistry and Chemical Biology, Harvard University, Cambridge, MA 02138.

2. Casiraghi, G.; Casnati, G.; Puglia, G.; Sartori, G.; Terenghi, G. *J. Chem. Soc., Perkin Trans.* **1980**, 1862.

3. Gademann, K.; Chavez, D. E.; Jacobsen, E. N. *Angew. Chem. Int. Ed.* **2002**, *41*, in press.

4. Dossetter, A. G.; Jamison, T. F.; Jacobsen, E. N. *Angew. Chem. Int. Ed.* **1999**, *38*, 2398.

5. (a) FR901464: Thompson, C. F.; Jamison, T. F.; Jacobsen, E. N. *J. Am. Chem. Soc.* **2000**, *122*, 10482; *J. Am. Chem. Soc.* **2001**, *121*, 9974 (b) Fostriecin: Chavez, D. E.; Jacobsen, E. N. *Angew. Chem., Int. Ed. Engl.* **2001**, *40*, 3667; c) Ambruticin: Liu, P.; Jacobsen, E. N. *J. Am. Chem. Soc.* **2001**, *123*, 10772; (d) Apicularin A: Bhattacharjee, A.; De Brabander, J. K. *Tetrahedron Lett.* **2000**, 41, 8069. (e) Laulilamide: Paterson, I.; De Savi, C.; Tutdge, M. *Org. Lett.* **2001**, *3*, 3149.

6. (a) Callystatin A: Crimmins, M. T.; King, B. W. *J. Am. Chem. Soc.* **1998**, *120*, 9084; (b) Ratjadone: Christman, M.; Bhatt, U.; Quitschalle, E.: Claus, E.; Kalesse, M. *Angew. Chem. Int. Ed.* **2000**, *39*, 4364.

CATALYTIC ASYMMETRIC ACYL HALIDE-ALDEHYDE CYCLOCONDENSATION REACTION
[(4*S*)-4-(2-Phenylethyl)-2-oxetanone]

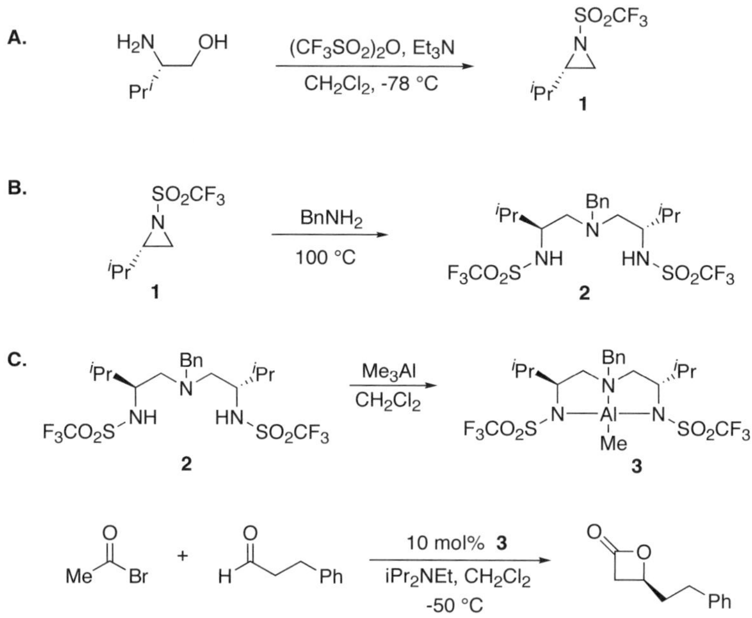

Submitted by Scott G. Nelson and Paul M. Mills.[1]
Checked by Takashi Ohshima and Masakatsu Shibasaki.

1. Procedure

A. (S)-N-Trifluoromethylsulfonyl-2-isopropylaziridine. An oven-dried, 500 mL, round-bottomed flask is equipped with Teflon-coated magnetic stirring bar and sealed with a rubber septum containing a needle adapter to a N_2 source. The flask is charged with (*S*)-valinol (8.30 g, 80.5 mmol) (Note 1), triethylamine (26.3 mL, 189 mmol) (Note 2) and dichloromethane (135 mL) (Note 3). The flask is placed in a –78 °C dry ice-acetone bath (Note 4) and, to the well-stirred solution, is added of trifluoromethanesulfonic anhydride (31.8 mL, 189 mmol) (Note 5) via syringe over 20 min (Note 6). The resulting reaction mixture is held at –78 °C for 5 h. The reaction mixture is transferred to a 500 mL separatory funnel containing 200 mL of 0.1 N HCl and the mixture is thoroughly shaken and the layers are separated. The organic portion is washed successively with one portion of 0.1 M HCl (200 mL), two portions of saturated aqueous $NaHCO_3$ (200 mL each) and one portion of brine (200 mL). The organic portion is dried over anhydrous magnesium sulfate (Note 7), filtered and concentrated under reduced pressure on a rotary evaporator (Note 8) to afford 16.8-16.9 g (96-97%) of (*S*)-*N*-trifluoromethylsulfonyl-2-isopropylaziridine (**1**) as a pale yellow oil (Note 9). The (*S*)-*N*-trifluoromethylsulfonyl-2-isopropylaziridine is used in the next transformation without further purification.

B. (2S,6S)-4-Benzyl-1,7-bis(trifluoromethylsulfonyl)-2,6-diisopropyl-1,4,7-triazaheptane. A flame-dried, 250 mL round-bottomed flask containing a Teflon-coated magnetic stirring bar is charged with (*S*)-*N*-trifluoromethylsulfonyl-2-isopropylaziridine (**1**) (16.9 g, 77.8 mmol). Benzylamine (4.03 mL, 37.0 mmol) (Note 10) is added resulting in a mild exotherm (Note 11). Once the exotherm subsided (approx. 10 min), the reaction flask is placed in a 100 °C oil bath and held at this temperature for 12 h. The crude product mixture is separated by flash chromatography on silica gel (approx. 300 g) (Note 12) using 10% ethyl acetate in hexane as the eluent (Note 13) to give 20.4 - 21.4 g of pale yellow sticky solid, which included 18.1-18.5 g of **2** (90-92% yield based on ^{1}H NMR analysis), ethyl acetate, and trace amount of yellow material (Note 14). This sticky solid is further purified by trituration with hexane (20 mL, vigorous stirring) and after filtration 15.6-16.0 g (78-80%) of the triamide **2** is obtained as a white powder (Note 15).

C. Acyl halide-aldehyde cyclocondensation: (4S)-4-(2- Phenethyl)-oxetan-2-one. An oven-dried, 500 mL, three neck round-bottomed flask is equipped with a Teflon-coated magnetic stirring bar and a 125 mL addition funnel and is sealed with rubber septa containing a needle adapter to a N_2 source and a Teflon-coated thermocouple probe attached to a digital thermometer. The flask is charged with triamine **2** (4.00 g, 7.40 mmol) and dichloromethane (150 mL), whereupon a 2 M hexanes solution of trimethylaluminum (3.70 mL) (Note 16) is added at ambient temperature (23 °C) (CAUTION: Methane gas evolution). The resulting solution is stirred at 23 °C for 2 h before being cooled to –50 °C (Note 17). Once the reaction mixture has reached –50 °C, diisopropylethylamine (22.0 mL, 126 mmol) (Note 18) and acetyl bromide (10.4 mL, 141 mmol) (Note 19) are added consecutively via syringe at a rate that maintains the internal reaction temperature ≤ –42 °C. The addition funnel is charged with hydrocinnamaldehyde (9.75 mL, 74.0 mmol) and CH_2Cl_2 (10 mL) and this solution is added dropwise to the reaction mixture at a rate that maintains the internal temperature at ≤ –46 °C (Note 20). Once addition is complete, the reaction mixture is stirred at –50 °C for 16 h. The reaction mixture is diluted with 350 mL CH_2Cl_2 then transferred to a 2 L separatory funnel containing 800 mL of 0.1 M HCl and the mixture is thoroughly shaken and the layers are separated. The organic portion is washed consecutively with two portions of 0.1 M HCl (800 mL each), three portions of saturated aqueous $NaHCO_3$ (800 mL each) and two portions of brine (800 mL each). After each washing, the separated aqueous payer is extracted with one portion of diethyl ether (400 mL) using the same portion of ether for each extraction; the ethereal extract is reserved until washing of the CH_2Cl_2 layer is complete. The CH_2Cl_2 solution and the ether extract are combined, dried over anhydrous sodium sulfate (Note 21), filtered and concentrated under reduced pressure on a rotary evaporator. The crude product mixture is distilled under reduced pressure (100 °C at 5 Pa using turbo-molecular pump) (Note 22) yielding 10.4 g (80%) of (4*S*)-4-(2-phenethyl)oxetan-2-one (Notes 23 and 24).

2. Notes

1. The submitters prepared (*S*)-valinol according to an *Organic Syntheses* procedure employing 200 g (1.71 mol) of (*S*)-valine, 100 g (2.64 mol) of lithium aluminum hydride and 6 L of THF; the THF is used directly from a freshly opened bottle and is not distilled. The crude product mixture was purified by vacuum distillation (80 °C at 7 mm Hg) to afford 118 g (67%) of (*S*)-valinol. See: Dickman, D. A.; Meyers, A. I. *Org. Synth., Coll. Vol. VII* **1990**, 530. The checker used commercially available (*S*)-valinol, which was purchased from Aldrich Chemical Company, after distillation (70 °C at 1.3 kPa).

2. Triethylamine was purchased from Fisher Scientific Company and was freshly distilled over calcium hydride.

3. Methylene chloride was purchased from EM Science and was freshly distilled over calcium hydride.

4. Bath temperature was achieved using a dry ice-acetone slurry.

5. Trifluoromethanesulfonic anhydride was purchased from Aldrich Chemical Company and was used as received. The use of only 2 mol equivalent of trifluoromethanesulfonic anhydride (27.0 mL, 161 mmol) gave almost identical result.

6. Internal temperature was gradually increased to ca. –60 °C.

7. Anhydrous magnesium sulfate was purchased from EM Science.

8. The residue was concentrated using diaphragm pump (2 kPa). Concentration using oil-pump (0.5 kPa) caused loss of the product due to the low boiling point of the product.

9. Spectral data for the crude aziridine: ^1H NMR (500 MHz, $CDCl_3$) δ 2.90 (td, *J* = 7.0, 5.0 Hz, 1 H), 2.86 (d, *J* = 7.0 Hz, 1 H), 2.45 (d, *J* = 5.0 Hz, 1 H), 1.65 (octet, *J* = 7.0 Hz, 1 H), 1.05 (d, *J* = 7.0 Hz, 3 H), 1.03 (d, *J* = 7.0 Hz, 3 H).

10. Benzylamine was purchased from Aldrich Chemical Company and freshly distilled before use.

11. Internal temperature reached to 65 °C even when benzylamine was added very slowly (0.5 mL/min).

12. Silica gel was purchased from Merck (Silica gel 60, 230-400 mesh ASTM).

13. The submitters used 80% hexanes-20% ethyl acetate mixture as the eluent for flash chromatography on silica gel purchased from Bodman Industries (70-239 mesh). Using this eluent system, however, the checker obtained unsatisfactory separation.

14. The submitters used the pale yellow solid for the next reaction without further purification (90% yield). To maintain the purity of the triamide **2**, the checkers used the trituration technique.

15. The analytical data are as follows: mp 118-119 °C (lit. 112 °C); TLC R_f = 0.50 (hexane:EtOAc = 3:1) $[\alpha]_D$ –50° (*c* 1.6, MeOH); IR (NaCl) cm^{-1}: 3308, 2968, 2880, 2840, 1435, 1371, 1229, 1195, 1148, 1025; IR (KBr) cm^{-1}: 3262, 2969, 1449, 1379, 1231, 1195, 1151, 617; ^1H NMR (500 MHz, $CDCl_3$) δ 7.39-7.27 (m, 5 H), 5.40 (br s, 2 H), 3.94 (d, *J* = 13.5 Hz, 1 H), 3.61-3.56 (m, 2 H), 3.39 (d, *J* = 13.5 Hz, 1 H), 2.66 (dd, *J* = 13.5, 8.5 Hz,

2 H), 2.50 (dd, J = 13.5, 5.5 Hz, 2 H), 2.00-1.93 (m, 2 H), 0.87 (d, J = 7.0 Hz, 6 H), 0.84 (d, J = 6.9 Hz, 6 H); ^{13}C NMR (125 MHz, $CDCl_3$) δ 137.0, 129.6, 128.6, 127.7, 119.3 (q, J_{CF} = 319 Hz), 58.60, 58.56, 55.4, 29.7, 18.1, 17.2; MS (EI, 70 eV): *m/z* 542 $(M+1)^+$, 540, 337, 91; HRMS *m/z* calcd for $C_{19}H_{29}F_6N_3O_4S_2$: 541.1504; found: 541.1516; Elemental Anal. calcd for $C_{19}H_{29}F_6N_3O_4S_2$: C, 42.14; H, 5.40; N, 7.76; found: C, 42.14; H, 5.40; N, 7.76.

16. Trimethylaluminum (2.0 M solution in hexanes) was purchased from Aldrich Chemical Company and was use as received.

17. Bath temperature was controlled using a circulating chiller equipped with a submersible cooling probe.

18. *N,N*-Diisopropylethylamine was purchased from Aldrich Chemical Company and was freshly distilled over calcium hydride.

19. Acetyl bromide was purchased from Aldrich Chemical Company and was freshly distilled over P_2O_5.

20. Hydrocinnamaldehyde was purchased from Aldrich Chemical Company and was freshly distilled over calcium hydride.

21. Anhydrous sodium sulfate was purchased from L. T. Baker.

22. The checkers experienced some decomposition of the β-lactone to 3-butenylbenzene during distillation, potentially caused by trace acidic impurities (c.f., *i*-Pr_2NEt•HBr) remaining in the crude product mixture). Alternatively, the crude product mixture can be purified by silica-gel column chromatography (10% EtOAc in hexane) to afford the product in 90% yield.

23. The analytical data are as follows: TLC R_f = 0.33 (hexane:EtOAc = 3:1) $[\alpha]_D$ –43.1° (*c* 1.12, CH_2Cl_2) –47.4° (*c* 1.6, $CHCl_3$, 92% ee); IR (NaCl): 3085, 2987, 1828, 1545, 1455, 1135, 700 cm^{-1}; IR (neat): 3027, 2930, 1825, 1496, 1455, , 1412, 1133, 829, 750, 700 cm^{-1}; ^{1}H NMR (500 MHz, $CDCl_3$) δ 7.34-7.18 (m, 5 H), 4.51-4.48 (m, 1 H), 3.48 (dd, J = 16.5, 5.5 Hz, 1 H), 3.03 (dd, J = 16.5, 4.5 Hz, 1 H), 2.84-2.80 (m, 1 H), 2.76-2.70 (m, 1 H), 2.22-2.18 (m, 1 H), 2.12-2.07 (m, 1 H); ^{13}C NMR (125 MHz, $CDCl_3$) δ 168.0, 140.0, 128.5, 128.2, 126.2, 70.3, 42.7, 36.2, 31.1; MS (EI, 70 eV): *m/z* 176 (M^+), 158, 131, 117, 104, 91, 84; HRMS *m/z* calcd for $C_{11}H_{12}O_2$: 176.0838; found: 176.0839; Anal. calcd for $C_{11}H_{12}O_2$: C, 74.98; H, 6.86; found: C, 74.97; H, 7.07.

24. Separation of the enantiomers by chiral HPLC (Daicel Chiracel™ OD-H column, flow rate 1.0 mL/min, 10% ^{i}PrOH, 90% hexane, T_r 14.3 (*S*) and 16.5 (*R*) min) provided the enantiomer ratio: 4(*S*):4(*R*) = 96.1:3.9 (92% ee).

Safety and Waste Disposal Information

All hazardous materials should be handled and disposed of in accordance with "Prudent Practices in the Laboratory"; National Academy Press; Washington, DC, 1995.

3. Discussion

Recent success in developing *de novo* asymmetric syntheses of enantioenriched β-lactones has created renewed interest in these heterocycles as versatile platforms for asymmetric organic synthesis.[3,4] β-Lactones are direct progenitors of numerous useful building blocks including enantioenriched β-amino acids,[5] allenes[6] and β,β-disubstituted carboxylic acids.[7] β-Lactones are also functional equivalents of ester aldol addition products.[8] In this latter context, we developed acyl halide-aldehyde cyclocondensation (AAC) reactions that deliver enantioenriched β-lactone acetate aldol surrogates from commercially available starting materials (eq 1).

We have recently described Al(III)-catalyzed cyclocondensations of acyl halides and aldehyde electrophiles as an operationally simple route to highly enantioenriched 4-substituted 2-oxetanones. The easily prepared Al(III)-triamine catalyst **3** is uniquely effective in mediating highly enantioselective [2+2] cycloadditions of *in situ* generated ketene and aldehydes. The enantioenriched triamine ligand **2** required for preparing the AAC catalyst is obtained in two high yielding steps from (*S*)-valinol. The AAC catalyst **3** is prepared from the triamine ligand *in situ* by reacting **2** with $AlMe_3$; the acetyl bromide, aldehyde and diisopropylethylamine required for the AAC reaction are then simply added to the resulting catalyst solution. The enantioenriched β-lactones emerging from the AAC reactions are typically sufficiently pure to be used in subsequent transformations without purification.

The catalyst complex **3** renders a variety of structurally diverse aldehydes as effective electrophiles for the catalyzed asymmetric AAC reactions. The procedure described herein highlights the reactivity of enolizable aliphatic aldehydes under the AAC reactions. The examples compiled in Table 1 are indicative of other aldehyde substrates that participate in efficient AAC reactions. Aromatic aldehydes bearing alkyl or electron-withdrawing substituents, functionalized aldehydes bearing common oxygen protecting groups and conjugated ynals are all very reactive electrophiles in the asymmetric AAC reactions. Aldehydes that are not useful AAC substrates include conjugated enals and α-branched aldehydes (c.f., cyclohexanecarboxaldehyde); these types of aldehydes afford little to no β-lactone product under the AAC reaction conditions.

Table 1

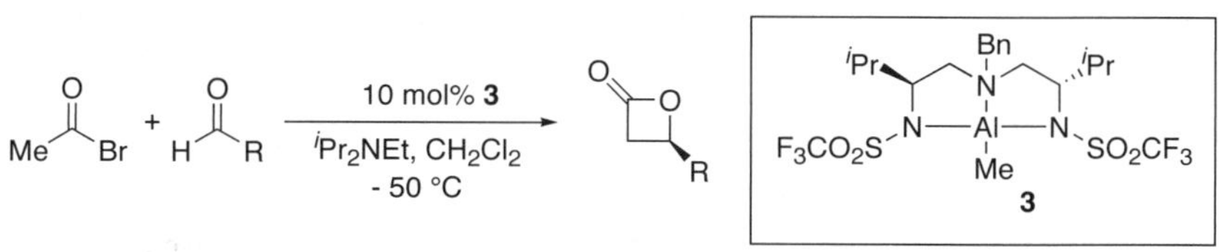

Aldehyde (R)	% yield	% ee
$-(CH_2)_8CH{=}CH_2$	91	91
$-CH_2CH(CH_3)_2$	80	93
$-CH_2OCH_2Ph$	91	92
$-CH_2OSiPh_2tBu$	74	89
$-CH_2CH_2OCH_2Ph$	90	91
$-C{\equiv}CCH_2OCH_2Ph$	86	93
$-C_6H_4NO_2$	93	95

1. Department of Chemistry, University of Pittsburgh, Pittsburgh, Pennsylvania, 15260.

2. Cernerud, M.; Skrinning, A.; Bérgère, I.; Moberg, C. *Tetrahedron: Asymm.* **1997**, *8*, 3437-3441.

3. Acyl halide-aldehyde cyclocondensations: (a) Nelson, S. G. Peelen, T. J.; Wan, Z. *J. Am. Chem. Soc.* **1999**, *121*, 9742. (b) Nelson, S. G.; Wan, Z. *Org. Lett.* **2000**, *2*, 1883. (c) Nelson, S. G.; Zhu, C.; Shen, X. *J. Am. Chem. Soc.* **2004**, *126*, 14-15.

4. Leading references to catalytic asymmetric β-lactone preparation: (a) Wynberg, H.; Staring, E. G. J. *J. Am. Chem. Soc.* **1982**, *104*, 166. (b) Romo, D.; Harrison, P. H. M.; Jenkins, S. I.; Riddoch, R. W.; Park, K.; Yang, H. W.; Zhao, C.; Wright, G. D. *Bioorg. Med. Chem.* **1998**, *6*, 1255. (c) Tennyson, R.; Romo, D. *J. Org. Chem.* **2000**, *65*, 7248. (d) Evans, D. A.; Janey, J. M. *Org. Lett.* **2001**, *3*, 2125. (e) Cortez, G. S.; Tennyson, R. L.; Romo, D. *J. Am. Chem. Soc.* **2001**, *123*, 7945. (f) Calter, M. A.; Liao, W. *J. Am. Chem. Soc.* **2002**, *124*, 13127 and references therein.

5. Nelson, S. G.; Spencer, K. L. Angew. *Chem. Int. Ed.* **2000**, 39, 1323.

6. Wan, Z.; Nelson, S. G. *J. Am. Chem. Soc.* **2000**, *122*, 10470.

7. Nelson, S. G.; Wan, Z.; Stan, M. A. *J. Org. Chem.* **2002**, *67*, 4680.

8. Nelson, S. G.; Wan, Z.; Peelen, T. J.; Spencer, K. L. *Tetrahedron Lett.* **1999**, *40*, 6535.

SYNTHESIS OF 9-SPIROEPOXY-*endo*-TRICYCLO[5.2.2.0^{2,6}]UNDECA-4, 10-DIEN-8-ONE
(Spiro[4,7-ethano-1H-indene-8,2'-oxiran]-9-one, 3a,4,7a-tetrahydro-)

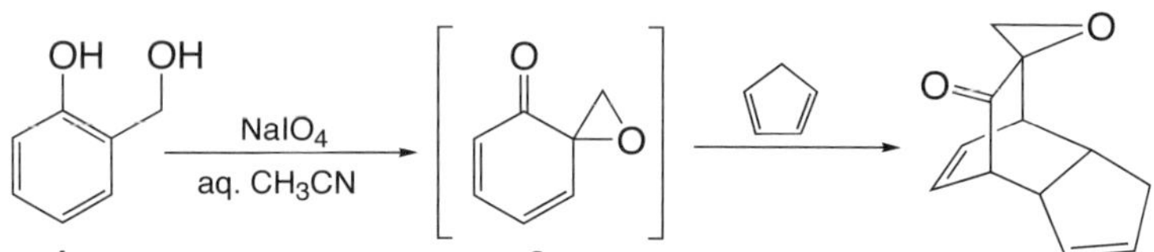

1 **2** **3**

Submitted by Vishwakarma Singh*, Mini Porinchu, Punitha Vedantham and Pramod K. Sahu.[1]
Checked by Aaron Murray and Marvin J. Miller.

1. Procedure

9-Spiroepoxy-endo-tricyclo[5.2.2.0^{2,6}]undeca-4,10-dien-8-one. A 2-L, single-necked, round-bottomed flask equipped with a magnetic stirbar is charged with salicyl alcohol **1** (20.0 g, 161 mmol) (Note 1) and 325 mL of acetonitrile (Note 2). The resulting solution is cooled at 0 °C in an ice bath while freshly generated cyclopentadiene (20 mL, 15.6 g, 236 mmol) (Note 3) is added in one portion. A pressure-equalizing addition funnel is then attached to the flask and charged with a solution of sodium metaperiodate (60 g, 280 mmol) (Note 4) in 220 mL of water. This solution is added dropwise to the ice-cooled, vigorously stirred reaction mixture over a period of 2 h (Note 5). The color of the reaction mixture initially turns yellow-orange. After the reaction mixture is stirred at 0 °C for an additional 1 h, additional cyclopentadiene (10 mL) is added and stirring is continued for 30 min at 0 °C. The ice bath is then removed and after 30 min an additional 10-mL portion of cyclopentadiene (Note 6) is added and the reaction mixture is stirred overnight at room temperature (Note 7).

Sodium chloride (50 g) is added to the reaction mixture and the suspension is stirred for 1 h. The reaction mixture is then filtered under vacuum through a 30 x 35 mm pad of Celite in a sintered glass funnel washing with two 50-mL portions of diethyl ether. Two phases form in the filtrate. The organic phase is separated and concentrated by rotary evaporation. The residue is combined with the aqueous phase and extracted with four 100-mL portions of ether, and the combined organic phases are washed with 50 mL of brine, dried over anhydrous sodium sulfate, filtered, and concentrated to afford 32.5 g of an orange-brown oil which is adsorbed onto 25 g of silica gel and charged onto a 430 mm x 45 mm column of 350 g of 60-120 mesh silica gel. The column is eluted with ca. 700 mL of petroleum ether (bp 60-80 °C) to remove hydrocarbon impurities, and elution is then continued with 500 mL of 98:2 petroleum ether and then with 95:5 petroleum ether:ethyl acetate to give a small amount of salicylaldehyde (Note 8). Further elution with 3.0 L of 95:5 petroleum ether:ethyl acetate gives pure **3**. The column is further eluted with 90:10 petroleum ether-ethyl acetate to give 3.5 g of starting material (**1**). Recrystallization of **3** from a mixture of 25 mL of ether and 15 mL of petroleum ether gives 18.57-19.5 g (61-64%) of pure **3** as a white solid (Note 9).

2. Notes

1. Salicyl alcohol was purchased from Lancaster, UK or from Fluka and used as received.

2. The submitters purchased acetonitrile (AR grade) from Sisco Research Laboratories, India and used it as received. The checkers obtained acetonitrile from Aldrich Chemical Company.

3. Cyclopentadiene was obtained by pyrolysis of dicyclopentadiene (pract., 90%) which was purchased from Fluka. Pyrolysis was carried out following the previously described general procedure.[2] Dicyclopentadiene (80 mL) was placed in a round-bottomed flask equipped with a magnetic stirbar and a Vigreux column fitted with a distillation head through which cold water was circulated. The contents of the flask were slowly heated with stirring at 160 °C in an oil bath, and ca. 60 mL of cyclopentadiene (bp 38-42 °C) was collected in a receiver cooled in an ice-salt bath. The cyclopentadiene was used immediately and the residue left in the distillation flask was discarded.

4. Sodium metaperiodate was purchased by the submitters from S.D. Fine Chemicals, India and by the checkers from Aldrich Chemical Company. Occasionally a small amount of some material remains insoluble and is removed by filtration; the checkers observed that it begins to precipitate during the addition. An excess of the reagent was used so that oxidation is as complete as possible.

5. A powerful magnetic stirrer is required since a large amount of inorganic material precipitates after the addition of sodium metaperiodate. The submitters used a 35-mm stirbar to achieve vigorous stirring.

6. An excess of cyclopentadiene was used because of its volatility and tendency to dimerize. Cyclopentadiene was added intermittently in portions so that a sufficient amount is always present in the reaction mixture to efficiently intercept the spiroepoxycyclohexadienone (**2**) generated *in situ*.

7. Analysis by thin layer chromatography (TLC) indicates product formation and a small amount of starting material. TLC was performed on Merck Silica gel 60 F_{254} coated on aluminum sheets. Product **3** has R_f = 0.44 (elution with 70:30 petroleum ether-ethyl acetate; visualization with iodine vapor).

8. Some salicylaldehyde is also formed during the oxidation.

9. The physical properties of **3** were as follows: mp 40-41°C; IR (neat) cm^{-1}: 3058, 2931, 1734; ^1H NMR (300 MHz, $CDCl_3$): δ 2.01 (m, 1 H), 2.57 (m, 1 H), 2.63 (m, 1 H), 2.81 (m, 1 H), 3.03 (m, 1 H), 3.08 (m, 1 H), 3.34 (m, 2 H), 5.44 (m, 1 H), 5.69 (m, 1 H), 6.10 (t, superimposed dd, $J_1 = J_2$ =7.2, 1 H), 6.38 (t, superimposed dd, $J_1 = J_2$=7.2, 1 H); ^{13}C NMR (75 MHz, $CDCl_3$) δ: 36.3, 38.5, 44.1, 50.4, 52.4, 52.8, 58.1, 126.4, 129.6, 132.2, 133.5, 205.3. These spectral features match data reported earlier.[4a] The orientation of the oxirane ring was suggested on the basis of general tendency of cyclohexa-2,4-dienones during their cycloaddition and comparison with other similar adducts.[3,4]

Waste Disposal Information

All toxic materials were disposed of in accordance with "Prudent Practices in the Laboratory"; National Academy Press; Washington, DC, 1995.

3. Discussion

This preparation is a modification of a general method developed in our laboratory for syntheses of a variety of tricycloundecadienones of type (**3**) which involves *in-situ* generation of spiroepoxycyclohexa-2,4-dienones and their interception with dienes.[3] This method is fairly versatile and a number of tricycloundecadienones have been prepared by oxidation of hydroxymethyl phenols and interception of the resulting cyclohexadienones with cyclopentadiene, spiro[4,2]cycloheptadiene and dimethylfulvene.[4] This synthesis also constitutes an example of generation of molecular complexity in a single step from simple starting materials.[5] Recently, interception of cyclohexa-2,4-dienones with electron-rich olefins has also been reported.[6]

Oxidation of hydroxymethyl phenols with sodium metaperiodate has been extensively studied and is known to involve spiroepoxycyclohexa-2,4-dienones which are highly unstable and undergo dimerization instantaneously.[7] Recently, substituted spiroepoxycyclohexadienones were isolated and used as precursors in synthesis.[8,9] Oxidation of 5-bromosalicyl alcohol was also reported to give a stable 4-bromo-6-spiroepoxycyclohexa-2,4-dienone.[10]

Though the cycloaddition described here gives the adduct **3** in a regio- and stereoselective fashion, there exists a mechanistic dichotomy regarding the mode of the pericyclic reaction between cyclohexa-2,4-dienones with dienes.[11,12] The spiroepoxycyclohexa-2,4-dienone **2** may participate as a 4-π component (diene) and cyclopentadiene as a 2-π partner (dienophile) to give the adduct **3** directly. Alternatively, the cyclohexa-2,4-dienone may react as a 2-π partner (dienophile) and cyclopentadiene as a 4-π component (diene) to give the adduct of type **I** which may undergo a 3,3-shift[12] to give the product **3**. However, we were unable to isolate adducts of type **I** under our experimental conditions.

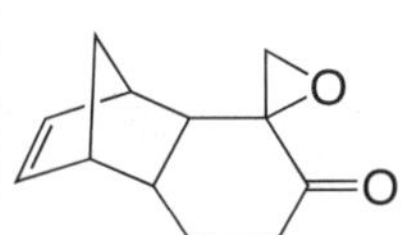

I

The epoxyketone **3** is a versatile precursor for a variety of tricycloundecane systems having a β,γ-unsaturated carbonyl chromophore, which are not so readily accessible. The contiguous epoxy ketone functionality and the double bond present in the five-membered ring provide opportunities for further manipulation. Adduct **3** may be transformed into a variety of molecular frameworks such as linearly fused *cis:anti:cis* tricyclopentanoids, protoilludanes, and marasmanes in a stereoselective fashion after suitable chemical and photochemical manipulation.[13]

1. Department of Chemistry, Indian Institute of Technology, Bombay, Mumbai, 400 076, India.

2. Moffett, R. B. *Org. Synth., Coll. Vol. IV,* **1963**, 238-241. (b) Partridge, J. J.; Chadha, N. K.; Uskokovic, M. R. *Org. Synth., Coll. Vol. VII,* **1990**, 339-345.
3. (a) Singh, V.; Porinchu, M. *J. C. S. Chem. Commun.* **1993**, 134-136. (b) Singh, V.; Thomas, B. *J. C. S. Chem. Commun.* **1992**,1211-1213.

4. (a) Singh, V.; Porinchu, M. *Tetrahedron.* **1996**, *52*, 7087-7126. (b) Singh, V.; Thomas, B. *J. Org. Chem.* **1997**, *62*, 5310-5320.

5. (a) Chanon, M.; Barone, R.; Baralotto, C.; Julliard, M.; Hendrickson, J. B. *Synthesis* **1998**, 1559-1583. (b) Corey, E. J.; Cheng, X.-M., *The Logic of Chemical Synthesis;* John Wiley & sons: New York, 1989.

6. Bonnarme, V.; Bachmann, C.; Cousson, A.; Mondon, M.; Gesson, J. -P. *Tetrahedron,* **1999**, *55*, 433-448.

7. (a)Adler, E.; Brasen, S.; Miyake, H. *Acta. Chem. Scand.* **1971**, *25*, 2055-2069. (b) Adler, E.; Holmberg, K. *Acta. Chem. Scand.* **1974**, *28B*,465-472.

8. Haseltine, J. N.; Cabal, M. P.; Mantlo, N. B.; Iwasawa, N.; Yamashita, D. S.; Coleman, R. S.; Danishefsky, S. J.; Schulte, G. K. *J. Am. Chem. Soc.* **1991**, *113*, 3850-3866.

9. Corey, E. J.; Dittami, J. P. *J. Am. Chem. Soc.* **1985**, *107,* 256-257.

10. Mondon, M.; Boeker, N.; Gesson, J. -P. *J. Chem. Research(S)* **1999**, 484-485.

11. Gesson, J. -P.; Hervaud, L.; Mondon, M. *Tetrahedron Lett.* **1993**, *34*, 2941-2944.

12. Singh, V.; Sharma, U.; Prasanna, V.; Porinchu, M. *Tetrahedron,* **1995**, *51*, 6015-6032.

13. Singh, V. *Acc. Chem. Res.* **1999**, *32*, 324-333.

PREPARATION OF 9,10-DIMETHOXYPHENANTHRENE AND 3,6-DIACETYL-9,10-DIMETHOXYPHENANTHRENE

[(Phenanthrene, 9,10-dimethoxy- and Ethanone, 1,1'-(9,10-dimethoxy-3,6-phenanthrenediyl)bis-)]

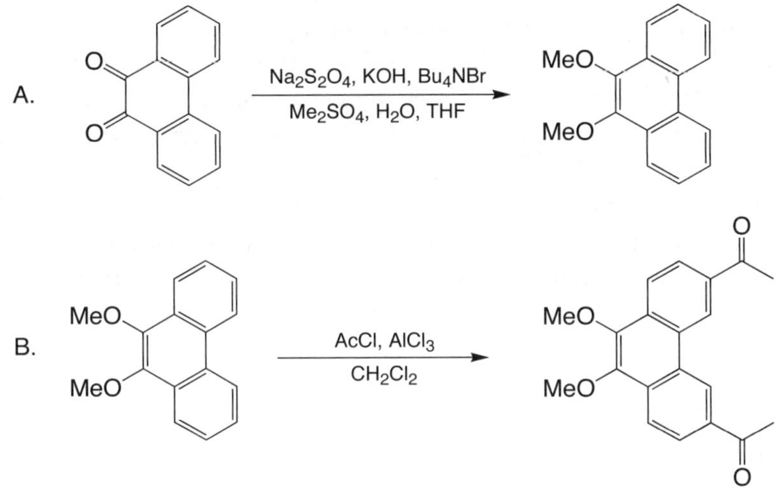

Submitted by Kamil Paruch, Libor Vyklicky, and Thomas J. Katz.[1]
Checked by Mitsuru Kitamura and Koichi Narasaka.

1. Procedure

Caution: Dimethyl sulfate is highly toxic and a potential carcinogen. It should be handled with appropriate safeguards in a well-ventilated fume hood.

A. 9,10-Dimethoxyphenanthrene. A mixture of 26 g (0.125 mol) of 9,10-phenanthrenequinone (Note 1), 13 g (0.04 mol) of tetrabutylammonium bromide (Bu_4NBr), 65 g (0.37 mol) of sodium dithionite ($Na_2S_2O_4$), 250 mL of tetrahydrofuran (THF), and 250 mL of water is shaken for 5 min in a 2-L separatory funnel. Dimethyl sulfate (62 mL, 0.65 mol) is added, followed by an aqueous solution of sodium hydroxide (64 g, 1.6 mol, in 125 mL of water) and 200 g of ice. The mixture is shaken for 5 min and, after another 200 g of ice has been added, shaken for another 10 min. Ethyl acetate (EtOAc, 100 mL) is added and the mixture is shaken. The aqueous phase is separated and extracted with EtOAc (2 × 100 mL). The combined organic extracts are washed with water (3 × 100 mL), 15% aqueous ammonia (2 × 100 mL), water (3 × 100 mL), and finally brine (100 mL). The solution is dried over sodium sulfate (Na_2SO_4) and filtered. The solvents are removed under vacuum (initially using a water aspirator, then at 0.5 torr). The residue, a thick brown oil. is dissolved in 80 mL of a ca. 2:1 mixture of dichloromethane (CH_2Cl_2) and hexane and poured onto a plug of neutral alumina, from which it is eluted with a 170-mL portion of the solvent mixture. After the solvent has been removed, the residue is dried under vacuum (ca. 0.5 torr) to give 23.7 g (80%) of 9,10-dimethoxyphenanthrene as a yellow oil (Note 2).

B. 3,6-Diacetyl-9,10-dimethoxyphenanthrene. A 1-L, three-necked, round-bottomed flask, fitted with an addition funnel, mechanical stirrer, and hydrogen chloride trap, is charged with 23.7 g of 9,10-dimethoxyphenanthrene and 120 mL of CH_2Cl_2 (Note 3). The solution is cooled in an ice-bath, and acetyl chloride (120 mL, (Note 4)) is added slowly. The cooling bath is removed, and over the course of 5 min, 44 g (0.33 mol) of aluminum chloride (Note 5) is added in portions to the stirred solution. The mixture is stirred for 15 min at ambient temperature and then carefully poured onto 1 L of crushed ice. The organic phase is separated, and the aqueous phase is extracted three times with 120-mL portions of CH_2Cl_2. The combined organic phases are washed with 120 mL of water, then with 120 mL of saturated aqueous sodium bicarbonate ($NaHCO_3$). The solution is dried over Na_2SO_4 and filtered. The solvent is removed, and the residual solid is suspended in 120 mL of methanol (MeOH), filtered, washed with an additional 120 mL of MeOH, and then dried (80 °C/ca. 0.5 torr) to afford 24.8 g (77% yield based on dimethoxyphenanthrene, 62% yield based on phenanthrenequinone) of 3,6-diacetyl-9,10-dimethoxyphenanthrene as a slightly pale yellow solid (Note 6).

2. Notes

1. 9,10-Phenanthrenequinone (95%), purchased from Acros, was used as received. The checkers purchased it from Tokyo Chemical Industry.

2. In two runs, the checkers obtained 27.0 and 25.7 g (91% and 90% yields). IR (neat film between NaCl plates) cm^{-1}: 2936, 1602, 1327, 1070. ^1H NMR ($CDCl_3$, 500 MHz in accord with the literature[2]) δ: 4.10 (s, 6H), 7.59-7.65 (m, 4H), 8.24 (dd, J = 8.0, J = 1.5, 2H), 8.64 (dd, J = 7.0, J = 1.5, 2H); ^{13}C NMR ($CDCl_3$, 75 MHz) δ: 60.9, 122.1, 122.6, 125.8, 126.8, 128.6,

129.1, 143.9. HRMS (FAB) *m/z* calcd for $C_{16}H_{14}O_2$: $[M]^+$ 238.0994; found 238.1003.

3. Dichloromethane was distilled from CaH_2.

4. Acetyl chloride (98%), purchased from Aldrich Chemical Company, Inc., was used as received. The checkers purchased it from Tokyo Chemical Industry.

5. Aluminum chloride (98%), purchased from Aldrich Chemical Company, Inc., was used as received. The checkers purchased it from Wako Pure Chemical Industry.

6. In two runs, the checkers obtained 26.6 and 28.8 g (82% and 83% yields). Samples can be purified further by crystallization from MeOH, but no significant amounts of impurities are detected when the material is analyzed before crystallization by means of ^{1}H NMR and ^{13}C NMR spectroscopy and HPLC (reverse phase, 5 µm LiChrospher(r) 100 from Hewlett-Packard, acetonitrile-H_2O gradients varying from 80:20 to 100:0). The pure product is a white solid, mp 160-161 °C. IR (CCl_4) cm^{-1}: 2939, 1687, 1611, 1318, 1059. ^{1}H NMR ($CDCl_3$, 400 MHz) δ: 2.82 (s, 6 H), 4.13 (s, 6 H), 8.20 (dd, J = 8.5, 1.3, 2 H), 8.32 (d, J = 8.5, 2 H), 9.31 (d, J = 1.3, 2 H); ^{13}C NMR ($CDCl_3$, 75 MHz) δ: 27.0, 61.1, 122.8, 123.6, 126.3, 128.5, 132.6, 134.7, 145.5, 198.0. UV-vis (CH_3CN, $c = 1.81 \times 10^{-4}$ M): λ_{max} (log ε) 266 (4.05), 325 nm (4.06). Anal. Calcd for $C_{20}H_{18}O_4$: C, 74.52; H, 5.63. Found: C, 74.52; H; 5.66.

Waste Disposal Information

All toxic materials were disposed of in accordance with "Prudent Practices in the Laboratory"; National Academy Press; Washington, DC, 1995.

3. Discussion

Almost no 9,10-dialkoxyphenanthrenes are known, and the quantities of those few made by other methods are either small or unstated.[2,3] The best alternative procedure gave 9,10-dimethoxyphenanthrene when 9,10-phenanthrenequinone was combined first with sodium in diglyme and then with dimethyl sulfate, but the yield after two chromatographic purifications was 51%.[2b] The only other reported alkylations of phenanthrene-9,10-diol are by 1-chloro-2-diethylaminoethane in 10% yield,[3f] and by 1,2-

di(bromomethyl)benzene in 22% yield.[3g] Unlike the procedure recorded here, none of these employ two-phase alkylations catalyzed by phase transfer agents.[4]

3,6-Diacetyl-9,10-dimethoxyphenanthrene has not been prepared by any other method. However, two related compounds, 9,10-di-TBDMSO- and 9,10-diphenyl-methylenedioxyphenanthrene, have been prepared by brominating phenanthrene-9,10-dione, converting the diones to the ethers, coupling the dibromides with tributyl(1-ethoxyethenyl)stannane by the Stille method, and hydrolyzing the enol ethers.[5] The procedure described here circumvents three disadvantages of the tin reagents used in the Stille procedure: expense, toxicity, and the need for chromatography to remove side products, mainly Bu_3SnBr. 3,6-Diacetyl-9,10-dialkoxyphenanthrene, like the 9,10-di-TBDMSO- and 9,10-diphenylmethylenedioxy-derivatives, combines with 1,4-benzoquinone to give a [7]helicenebisquinone.[6]

There are no reported alternatives to the current procedure for the acylation of a phenanthrene using the Friedel-Crafts reaction.[7,8] Indeed, alternative methods to cleanly prepare 3,6-disubstituted derivatives of phenanthrene by means of electrophilic substitution are not known,[9,10,11] nor is there a precedent for the electrophilic substitution of any 9,10-dialkoxyphenanthrene.

1. Department of Chemistry, Columbia University, New York, NY 10027.

2. (a) Eichner, M.; Merz, A. *Tetrahedron Lett.* **1981**, *22*, 1315; (b) Rio, G.; Berthelot, J. *Bull. Soc. Chim. France* **1972**, 822.

3. (a) Dannenberg, H.; Keller, H.-H. *Chem. Ber.* **1967**, *100*, 23; (b) Adams, C.; Kamkar, N. M.; Utley, J. H. P. *J. Chem. Soc., Perkin Trans. 2* **1979**, 1767; (c) Sucharda-Sobczyk, A.; Syper, L. *Rocz. Chem.* **1975**, *49*, 749; (d) Santamaria, J.; Ouchabane, R. *Tetrahedron* **1986**, *42*, 5559; (e) Goldschmidt, S.; Schmidt, W. *Chem. Ber.* **1922**, *55*, 3197; (f) Fourneau, E.; Matti, J. *Bull. Soc. Chim. France* **1942**, 633; (g) Kurebayashi, H.; Mine, T.; Harada, K.; Usui, S.; Okajima, T.; Fukazawa, Y. *Tetrahedron* **1998**, 54, 13495.

4. Dehmlow, E. V.; Dehmlow, S. S. "Phase Transfer Catalysis"; VCH: New York, 1993.

5. Fox, J. M.; Goldberg, N. R.; Katz, T. J. *J. Org. Chem.* **1998**, *63*, 7456.

6. Paruch, K.; Vyklicky, L.; Katz, T. J. *Org. Synth.* **2003**, *80*, 233.

7. Monoacylations: (a) Mosettig, E.; van de Kamp, J. *J. Am. Chem. Soc.* **1930**, *52*, 3704; (b) Gore, P. H. *J. Org. Chem.* **1957**, *22*, 135; (c) Blin, P.; Bunel, C.; Maréchal, E. *J. Chem. Res. (S)* **1978**, 206; (d) Fernández, F.; Gómez, G.; López, C.; Santos, A. *J. Prakt. Chem.* **1989**, *331*, 15.

8. The only diacylations are by phthalic anhydride with aluminum chloride in "tetrachloroethane" and by benzoyl chloride with $AlCl_3$ in the absence of additional solvent. The diketones were isolated in 15% and unspecified yields, respectively, and their structures were not determined. Clar, E.; Kelly, W. *J. Am. Chem. Soc.* **1954**, *76*, 3502.

9. Upon sulfonation, phenanthrene-3-sulfonic acid gives mainly (but not exclusively) the 3,6-disulfonic acid (Fieser, L. F. *J. Am. Chem. Soc.* **1929**, *51*, 2471), but the starting phenanthrene-3-sulfonic acid can be obtained from phenanthrene in only ca. 25% yield, and then isolated only from a mixture including a comparable amount of the 2-sulfonic acid (Fieser, L. F. *Org. Synth., Coll. Vol. II* **1943**, 482).

10. Cerfontain, H.; Koeberg-Telder, A.; Laali, K.; Lambrechts, H. J. A. *J. Org. Chem.* **1982**, *47*, 4069.

11. Schmidt, J.; Schairer, O. *Chem. Ber.* **1923**, *56*, 1331.

HELICENEBISQUINONES: SYNTHESIS OF A [7]HELICENEBISQUINONE [(Dinaphtho[2,1-c:1',2'-g]phenanthrene-1,4,15,18-tetrone, 9,10-dimethoxy-6,13-bis[[tris(1-methylethyl)silyl]oxy]-)]

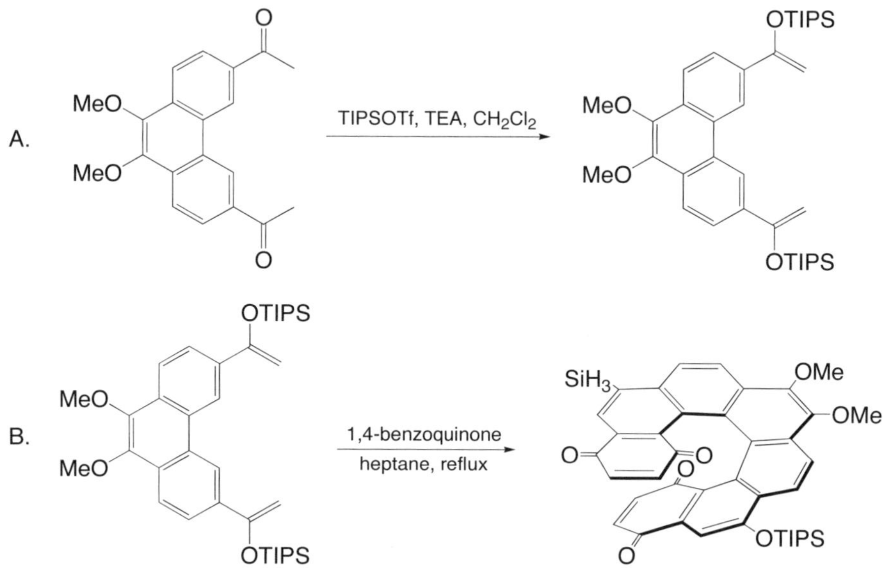

Submitted by Kamil Paruch, Libor Vyklicky, and Thomas J. Katz.[1]
Checked by Mitsuru Kitamura and Koichi Narasaka.

1. Procedure

A. 3,6-Bis[1-(triisopropylsiloxy)ethenyl]-9,10-dimethoxyphenanthrene. A solution of 8 g (0.025 mol) of 3,6-diacetyl-9,10-dimethoxyphenanthrene (Note 1) and 28 mL of triethylamine (Et_3N) (Note 2) in 80 mL of dry dichloromethane (CH_2Cl_2) (Note 2) is prepared under nitrogen in a 250-mL, round-bottomed flask equipped with a magnetic stir bar. The flask is cooled in an ice bath and 14 mL (0.052 mol) of triisopropylsilyl triflate (Note 3) is slowly added by syringe. The mixture is stirred for 15 min while cooling in the ice bath, then for 1 h at ambient temperature. Hexane (200 mL) is added, and the organic phase is washed twice with 100-mL portions of 1M aqueous potassium hydroxide (KOH), dried quickly over potassium carbonate (K_2CO_3), and filtered. The solvent is removed, and the oily residue is dissolved in ca. 40 mL of hexane. The solution is poured onto a 4 × 4-cm plug of neutral alumina and washed with 200 mL of hexane. The solvent is removed and the residue is dried under vacuum (0.5 torr/80 °C). 3,6-Bis[1-(triisopropylsiloxy)ethenyl]-9,10-dimethoxyphenanthrene (14.4 g, 91%) is obtained as a yellow oil (Notes 4, 5) and used in the subsequent step without further purification.

B. 6,13-Bis(triisopropylsiloxy)-9,10-dimethoxy[7]helicenebisquinone. A solution of the material obtained in Step A in 100 mL of heptane (Notes 6, 7) is added under nitrogen to a 500-mL, round-bottomed flask fitted with a reflux condenser. Thirty-four grams (0.31 mol) of 1,4-benzoquinone is added (Notes 8, 9), and the mixture is heated with an oil bath at 120 °C for 3.5 days with stirring (Note 10). The reaction mixture is then cooled to 25 °C and 25 mL of CH_2Cl_2 is added. The solids are broken into small pieces with the aid of a spatula, and the mixture is shaken for 15 min with 50 g of sand (Note 11). The supernatant liquid is decanted, the residue is extracted three times with 70-mL portions of 1:1 hexane:CH_2Cl_2, and the combined extracts are filtered through a pad of Celite. The solution is concentrated, and 1,4-benzoquinone is removed by sublimation by heating the residue to 100 °C and applying a vacuum of ca. 0.5 torr. Methanol (MeOH, 100 mL) and ca. 20 g of sand are added, and the mixture is shaken until the solids are finely suspended (ca. 30 min). Water (20 mL) is added and the mixture is shaken for 10 min. The solids are collected by filtration and washed with ca. 250 mL of 5:1 MeOH:H_2O. The powder obtained after drying under vacuum at 100 °C is dissolved in a small amount of toluene and purified by flash chromatography[2] on a 12 × 8-cm column of silica gel. A purple impurity is eluted with toluene, and the product is eluted with 1-3% tetrahydrofuran (THF) in toluene. After removal of the solvents and drying under vacuum at 100 °C, 4.5 g of the helicenebisquinone (a 25% yield based on the enol ether, 22% based on 3,6-diacetyl-9,10-dimethoxyphenanthrene) is obtained as a dark red solid (Note 12).

2. Notes

1. Prepared according to the accompanying procedure (Paruch, K.; Vyklicky, L.; Katz, T. J. *Org. Synth.* **2003**, *80*, 227).
2. Dichloromethane and triethylamine were distilled from CaH_2.
3. Triisopropyl triflate (97%) was purchased from GSF Chemicals and used as received. The checkers purchased it from Tokyo Chemical Industry.
4. The product solidifies occasionally.

5. In two runs, the checkers obtained 15.8 and 15.2 g (99% and 98% yields). The product exhibits the following properties: ^{1}H NMR ($CDCl_3$, 400 MHz) δ: 1.18 (d, 36 H, *J* = 7.3), 1.35 (m, 6 H), 4.11 (s, 6 H), 4.57 (d, 2 H, *J* = 1.8), 5.06 (d, 2 H, *J* = 1.8), 7.90 (dd, 2 H, *J* = 8.6, 1.6), 8.17 (d, 2 H, *J* = 8.6), 8.96 (d, 2 H, *J* = 1.5); ^{13}C NMR ($CDCl_3$, 75 MHz) δ: 12.8, 18.2, 61.0, 90.8, 119.5, 122.0, 124.3, 128.6, 129.1, 135.4, 144.1, 156.4; IR (CCl4) cm^{-1}: 2946, 2868, 1607, 1464, 1322, 1292, 1113, 1015. The checkers reported the following elemental analysis. Calcd. for $C_{30}H_{58}O_4Si_2$: C, 71.87; H, 9.21. Found: C, 71.65; H, 9.20.

6. Heptane was distilled from sodium/benzophenone.

7. The yield is lower when toluene is used as solvent.

8. 1,4-Benzoquinone (98%, Aldrich) is purified by slurrying it in CH_2Cl_2 with two times its weight of basic alumina, filtering the mixture through Celite, concentrating the filtrate, and drying the residue under vacuum.

9. When seven molar equivalents of 1,4-benzoquinone are used, the yield drops to 12%.

10. Sometimes sublimation of 1,4-benzoquinone plugs the condenser, but turning off the cooling water ameliorates the problem.

11. The 20-30 mesh sand is obtained from Fisher Scientific.

12. In two runs the checkers obtained 8.4 and 8.5 g (40% and 42% yields). They found that the product can be purified further by recrystallization from MeOH/EtOAc; mp 272-274 °C. IR (CCl_4) cm^{-1}: 2948, 2870, 1665, 1610, 1573, 1471, 1385, 1295, 1096. ^{1}H NMR ($CDCl_3$, 400 MHz) δ: 1.20 (d, 18 H, *J* = 7.5), 1.24 (d, 18 H, *J* = 7.5), 1.51 (m, 6 H), 4.21 (s, 6 H), 5.92 (d, 2 H, *J* = 10.1), 6.48 (d, 2 H, *J* = 10.1), 7.38 (s, 2 H), 8.40 (d, 2 H, *J* = 8.9), 8.46 (d, 2 H, *J* = 8.9); ^{13}C NMR ($CDCl_3$, 75 MHz) δ: 13.0, 18.1, 61.2, 107.7, 121.1, 123.3, 125.0, 126.1, 128.0, 129.6, 129.9, 133.0, 134.1, 140.4, 145.3, 157.4, 183.6, 184.6. UV-vis (CH_3CN, c = 5.50 × 10^{-5} M): λ_{max} nm (log ε) 241 (4.54), 285 (4.48), 342 (4.18), 417 (3.76). Anal. Calcd for $C_{50}H_{58}O_8Si_2$: C, 71.22; H, 6.93. Found: C, 70.99; H, 6.87. The checkers obtained C, 71.20; H, 6.93.

Waste Disposal Information

All toxic materials were disposed of in accordance with "Prudent Practices in the Laboratory"; National Academy Press; Washington, DC, 1995.

3. Discussion

The procedure illustrates one that has been used to prepare significant quantities of a variety of helicenebisquinones.[3] The majority of helicenes have been synthesized by photocyclization of derivatives of stilbene,[4] but in most cases only small amounts can be prepared practically. Moreover, the functional groups that have been incorporated in these helicenes have not been generally useful. Using our procedure, helicenebisquinones having functional groups suitable for subsequent transformations can be prepared easily in significant amounts. Among the structures synthesized from these helicenes are helical polymers,[5] helical discotic columnar aggregates[3e,6] and liquid crystals,[7] helical ligands,[8] nonlinear optical materials,[9] asymmetric catalysts,[10] and substances that probe remote chirality.[11]

Similar procedures have been used to prepare [7]helicenes that have different side chains,[3a,b] [5]- and [6]carbohelicenes,[3c,d,e,g] and [7] heterohelicenes.[3f,h]

1. Department of Chemistry, Columbia University, New York, NY 10027.
2. Still, W. C.; Kahn, M.; Mitra, A. *J. Org. Chem.* **1978**, *43*, 2923.
3. (a) Fox, J. M.; Goldberg, N. R.; Katz, T. J. *J. Org. Chem.* **1998**, *63*, 7456; (b) Paruch, K.; Katz, T. J.; Incarvito, C.; Lam, K.-C.; Rhatigan, B.; Rheingold, A. L. *J. Org. Chem.* **2000**, *65*, 7602; (c) Katz, T. J.; Liu, L.; Willmore, N. D.; Fox, J. M.; Rheingold, A. L.; Shi, S.; Nuckolls, C.; Rickman, B. H. *J. Am. Chem. Soc.* **1997**, *119*, 10054; (d) Dreher, S. D.; Katz, T. J.; Lam, K.-C.; Rheingold, A. L. *J. Org. Chem.*, **2000**, *65*, 815; (e) Nuckolls, C.; Katz, T. J.; Katz, G.; Collings, P. J.; Castellanos, L. *J. Am. Chem. Soc.* **1999**, *121*, 79; (f) Dreher, S. D.; Weix, D. J.; Katz, T. J. *J. Org. Chem.* **1999**, *64*, 3671; (g) Paruch, K.; Vyklicky, L.; Katz, T. J.; Incarvito, C. D.; Rheingold, A. L. *J. Org. Chem.* **2000**, *65*, 8774; (h) Phillips, K. E. S.; Katz, T. J.; Jockush, S.; Lovinger, A. J.; Turro, N. J. *J. Am. Chem. Soc.* **2001**, *123*, 11899.
4. (a) Laarhoven, W. H.; Prinsen, W. J. *Top. Curr. Chem.* **1984**, *125*, 63. (b) Mallory, F. B.; Mallory, C. W. *Organic Reactions*; Wiley: New York, **1984**; Vol. 30, p. 1.
5. Dai, Y.; Katz, T. J. *J. Org. Chem.* **1997**, *62*, 1274.
6. (a) Lovinger, A. J.; Nuckolls, C.; Katz, T. J. *J. Am. Chem. Soc.* **1998,** *120*, 264; (b) Nuckolls, C.; Katz, T. J.; Verbiest, T.; Van Elshocht, S.; Kuball, H.-G.; Kiesewalter, S.; Lovinger, A. L.; Persoons, A. *J. Am. Chem. Soc.* **1998**, *120*, 8656; (c) Fox, J. M.; Katz, T. J.; Van Elshocht, S.; Verbiest, T.; Kauranen, M.;

Persoons, A.; Thongpanchang, T.; Krauss, T.; Brus, L. *J. Am. Chem. Soc.* **1999**, *121*, 3453.

7. (a) Nuckolls, C.; Katz, T. J. *J. Am. Chem. Soc.* **1998**, *120*, 9541; (b) Nukolls, C,; Shao, R.; Jang, W.-G.; Clark, N. A.; Walba, D.; Katz, T. *J. Chem. Mater.* **2000**, *14*, 773.

8. Fox, J. M.; Katz, T. J. *J. Org. Chem.* **1999**, *64*, 302.

9. (a) Verbiest, T.; Van Elshocht, S.; Kauranen, M.; Hellemans, L.; Snauwaert, J.; Nuckolls, C.; Katz, T. J.; Persoons, A. *Science* (Washington, D. C.) **1998**, *282*, 913; (b) Busson, B.; Kauranen, M.; Nuckolls, C.; Katz, T. J. Persoons, A. *Phys. Rev. Lett.,* **2000**, *84*, 79; (c) Verbiest, T.; Sioncke, S.; Persoons, A.; Vyklicky, L.; Katz, T. *J. Angew. Chem. Int. Ed.* **2002**, *14*, 3882.

10. Dreher, S. D.; Katz, T. J.; Lam, K.-C.; Rheingold, A. L. *J. Org. Chem.,* **2000**, *65*, 815.

11. Weix, D. J.; Dreher, S. D.; Katz, T. J. *J. Am. Chem. Soc.* **2000**, *122*, 10027.

PREPARATION OF A TRISAMIDOMOLYBDENUM(VI) PROPYLIDYNE COMPLEX – A HIGHLY ACTIVE CATALYST PRECURSOR FOR ALKYNE METATHESIS

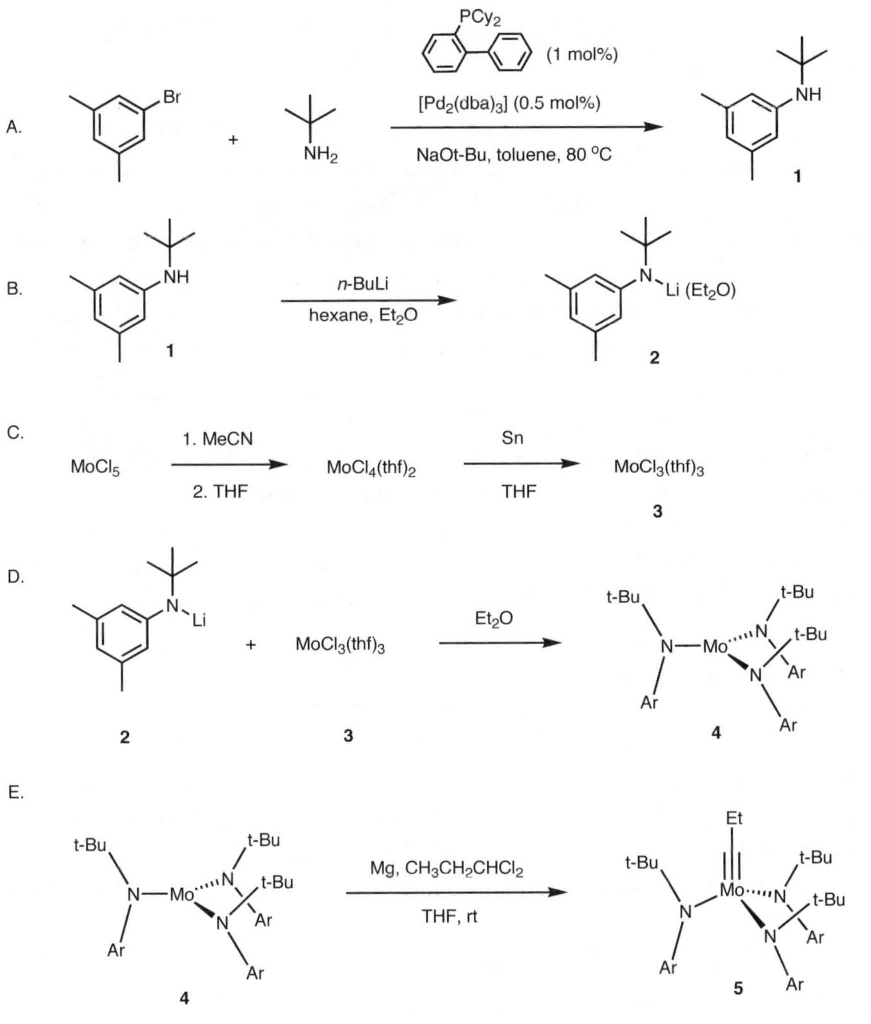

Submitted by Wei Zhang, Yunyi Lu and Jeffrey S. Moore.[1]
Checked and modified by Günter Seidel and Alois Fürstner.

1. Procedure

*CAUTION: Procedures involving molybdenum trisamide **4** were conducted under an atmosphere of argon. It has been reported that **4** is able to activate molecular nitrogen at or below room temperature.[2] Therefore, N_2 must not be used as a protecting atmosphere for any experiment involving this reagent (Note 1).*

*A. N-(3,5-Dimethylphenyl)-tert-butylamine (**1**).* A 500-mL, three-necked, round-bottomed flask is equipped with two glass stoppers, a magnetic stir bar and a reflux condenser fitted with a gas inlet connected to an argon-vacuum manifold. The flask is flame-dried under vacuum, then filled with argon. After evacuating and backfilling with argon three times, the flask, in which a positive flow of argon is maintained throughout the entire procedure, is sequentially charged with sodium *tert*-butoxide (19.22 g, 0.20 mol) (Note 2), 2-(dicyclohexylphosphino)biphenyl (0.50 g, 1.4 mmol), [$Pd_2(dba)_3$] (0.65 g, 0.71 mmol) (Note 3), toluene (200 mL) (Note 4), *tert*-butylamine (18.0 mL, 0.17 mol) and 5-bromo-1,3-dimethylbenzene (19.4 mL, 0.14 mol) (Note 2). The resulting viscous red reaction mixture is heated at 80 °C in an oil bath for 12 h while maintaining positive argon pressure. The solution color changes from dark red to orange red with white solids precipitating out. The mixture is transferred to a 500-mL, one-necked, round-bottomed flask and concentrated to a residue by rotary evaporation (water bath temperature 50–55 °C) before cooling down to room temperature (Note 5). *tert*-Butyl methyl ether (250 mL) (Note 2) and brine (80 mL) are added to the residue. After separation, the aqueous layer is extracted with *tert*-butyl methyl ether (2 x 250 mL), the combined organic layers are dried ($MgSO_4$), filtered and concentrated by rotary evaporation, and the crude product is purified by distillation under vacuum (bp 50–53 °C, 10^{-3} mbar, oil bath temperature 70 °C) affording aniline **1** as colorless liquid (21.8 g, 86%) (Note 6).

*B. Lithium N-(3,5-dimethylphenyl)-tert-butylamide etherate (**2**).* A 500-mL, three-necked, round-bottomed flask is equipped with a magnetic stirring bar, two glass stoppers and a pressure-equalizing dropping funnel fitted with a gas inlet connected to an argon-vacuum manifold. The flask is flame-dried under vacuum, then filled with argon. After evacuating and backfilling with argon three times (Note 1), the flask is charged with substituted aniline **1** (11.3 g, 63.7 mmol) and *n*-hexane (220 mL) (Note 4) under argon purge. The solution is cooled to –78 °C in a dry ice-acetone bath and a solution of *n*-BuLi (1.6 M in hexanes, 48 mL, 76.8 mmol) (Note 2) is added via the dropping funnel over 10 min. After the addition is complete, the mixture is stirred for 12 h during which time it is gradually warmed to room temperature; a white precipitate forms upon warming. The dropping funnel is replaced by a distillation head and condenser and all volatiles are removed under vacuum (approximately 7.5 mmHg) at room temperature without exposing the contents to the open atmosphere (Notes 1, 7). Then ether (25 mL) and *n*-hexane (50 mL) (Note 4) are added to the remaining viscous liquid causing further precipitation of a white solid. The resulting suspension is kept at –30 °C under argon for 12 h (Note 1, 7). The white solids of lithium anilide etherate (**2**•Et_2O) formed are collected by filtration under argon, washed with cold *n*-hexane (2 x 10 mL) and dried under vacuum (10^{-3} mbar) to give lithium *N*-(3,5-dimethylphenyl)-*tert*-butylamide etherate as colorless crystals (9.41 g, 57%). The filtrate is concentrated under vacuum to approximately half of its volume and slowly cooled to –78 °C to give a second crop of the product which is isolated by filtration under argon as described above (3.48 g, 21%) (Note 8).

*C. [$MoCl_3(THF)_3$](**3**).* A 500-mL, two-necked, round-bottomed flask equipped with a magnetic stir bar, a glass stopper and a gas inlet connected to the argon/vacuum line is flame-dried and back-filled with argon. While maintaining a positive argon pressure, the flask is charged with acetonitrile (250 mL) (Note 4). Molybdenum pentacarbonyl ($MoCl_5$) (51.17 g, 160 mmol) is then added in portions over a period of 20 min, causing a gentle warming of the mixture. The resulting suspension is stirred for 20 h at ambient temperature. The solid material is filtered off and is carefully rinsed with acetonitrile (50 mL) before it is dried under vacuum (10^{-3} mbar) to give $MoCl_4(MeCN)_2$ as a red-brown solid (47.5 g, 93%). This material is re-dispersed in THF (200 mL) and the suspension is stirred for 3.5 h under argon. The resulting complex is filtered off under argon, the solid is carefully washed with pentane (100 mL) and dried under vacuum (10^{-3} mbar) to give $MoCl_4(THF)_2$ as orange-brownish crystals (53.30 g, 94%). A 500-mL, two-necked, round-bottomed flask equipped with a magnetic stir bar, a glass stopper and a gas inlet connected to the argon/vacuum line is flame-dried and back-filled with argon. This flask is charged with the $MoCl_4(THF)_2$ complex (53.30 g, 140 mmol), THF (300 mL) and tin shot (107.3 g, 0.904 mol), and the resulting suspension is stirred for 30 min, resulting in a slight exotherm (a temperature of approximately 30 °C is reached). Stirring is then discontinued to allow the remaining tin to settle. Upon gentle manual swirling, leaving the metallic tin at the bottom of the flask, the supernatant suspension (liquid and all solids except metallic tin) is decanted, the solid material is collected by filtration under argon, the product is carefully rinsed with pentane (100 mL) and dried under vacuum (10^{-3} mbar), affording $MoCl_3(THF)_3$ as an orange solid (51.2 g, 88%) (Note 9).

D. *Tris[N-(tert-butyl)(3,5-dimethylphenyl)-amido]molybdenum(III) (4).* A 500-mL two-necked, round-bottomed flask equipped with a magnetic stir bar, a glass stopper and a gas inlet connected to the argon/vacuum manifold is flame-dried under vacuum and back-filled with argon (Note 1). The flask is then charged with $[MoCl_3(THF)_3]$ **3** (7.96 g, 19.0 mmol) by removing a stopper and adding the solid, followed by Et_2O (300 mL) (Note 4), which is transferred by syringe. The resulting suspension is cooled to –30 °C with an acetone/dry ice bath before anilide **2** (9.79 g, 38.0 mmol) is added through removal of the stopper. The resulting mixture is stirred for 3 h during which time it is allowed to reach room temperature and the solution turns dark red. The suspension is filtered under argon into a flame-dried 500-mL, two-necked, round-bottomed flask to remove LiCl; the solid is rinsed with ether (15 mL). With one neck equipped with a rubber septum and the other one connected via a cold-trap to the vacuum line, the filtrate is concentrated under reduced pressure (7.5 mmHg) to approximately 1/6 of its original volume (50 mL) (Note 10). The remaining solution is cooled to – 30 °C for 16 h, causing the separation of dark red crystals. The supernatant liquid is decanted off while keeping a positive argon pressure, and the remaining dark-red, crystalline molybdenum trisamide **4** is dried under vacuum (10^{-3} mbar) (4.28 g, 55%) (Note 11).

E. *Tris[N-(tert-butyl)(3,5-dimethylphenyl)-amido]molybdenum(VI) propylidyne (5).* A 250-mL, two-necked, round-bottomed flask equipped with a rubber septum, a magnetic stirring bar, and a gas inlet connected to the argon/vacuum manifold is flame-dried and back-filled with argon (Note 1, 10). Molybdenum trisamide **4** (5.31 g, 8.48 mmol), magnesium turnings (2.54 g, 106 mmol) (Note 2) and THF (160 mL) (Note 4) are added to the flask by removal of the septum. To the stirred solution, 1,1-dichloropropane (1.7 mL, 17 mmol) (Note 2, Note 12) is added dropwise via syringe within 5 min, resulting in a color change from red to dark amber. The resulting mixture is stirred for 2.5 h at room temperature under argon. The rubber septum is replaced by a distillation head connected to the vacuum line, the solvent is removed under reduced pressure (approximately 7.5 mmHg) (Note 10), and the residue is redissolved in pentane (100 mL) (Note 4). After vigorous stirring/manual shaking, the solid precipitate ($MgCl_2$) and excess magnesium turnings are removed by filtration under argon. The filtrate is concentrated to approximately 1/10 of its volume and slowly cooled to –78 °C, causing the molybdenum propylidyne complex to precipitate (Note 13). The supernatant is removed via cannula and the remaining beige crystals are dried under vacuum (10^{-3} mbar) (3.08 g, 54%) (Notes 1, 14, 15).

2. Notes

1. All compounds described in this procedure, except *N*-(3,5-dimethylphenyl)-*tert*-butylamine (**1**), are highly sensitive to air and moisture. Therefore, proper protection from atmospheric influences must be maintained throughout the preparation and handling of these products. The checkers ensured this by applying Schlenk techniques, whereas the submitters largely relied on the use of an argon-filled glove box.

2. Sodium *tert*-butoxide (97%), *tert*-butylamine (98%), 5-bromo-1,3-dimethylbenzene (97%), *tert*-butyl methyl ether (≥ 99%), *n*-butyllithium (1.6 M in hexanes), magnesium turnings (~3/16 in., 99.98%), 1,1-dichloropropane (97%, dried using activated, powdered 3Å molecular sieves overnight before use), *p*-nitrophenol (≥ 99%), 1-phenyl-1-butyne (99%, passed through a short plug of basic alumina and transferred under vacuum before use) were purchased from Aldrich Chemical Company, Inc. and used as received unless further purification indicated.

3. 2-(Dicyclohexylphosphino)biphenyl (98%), [$Pd_2(dba)_3$], $MoCl_5$ (99.6%) and tin shot (99.8%) were purchased from Strem Chemicals, Inc. and used as received.

4. The checkers used solvents purchased from Acros which were purified and deoxygenated by distillation under argon over the drying agents indicated: toluene, hexane, pentane (Na/K alloy), Et_2O, THF, *tert*-butyl methyl ether (Mg-anthracene), MeCN (CaH_2). The submitters used toluene, hexane, Et_2O, THF and pentane deoxygenated by bubbling argon through them for a period of 20 min and dried using 3 Å molecular sieves overnight before use.

5. When cooling down to room temperature, the reaction mixture turned into a viscous solution, making the direct solvent removal under vacuum impractical. In case this happens, *tert*-butyl methyl ether (250 mL) and H_2O (100 mL) should be added, and the aqueous layer removed. The organic layer should be concentrated under vacuum and the residue subjected to further workup as described in the text.

6. The procedure for synthesizing aniline **1** follows that reported by Fürstner, et al.[3a] The spectral characteristics of compound **1** are as follows:

^{1}H NMR ($CDCl_3$, 400 MHz) δ: 1.31 (s, 9 H) 2.22 (s, 6 H), 3.30 (s, 1 H), 6.38 (s, 2 H), 6.41 (s, 1 H). ^{13}C NMR ($CDCl_3$, 100 MHz) δ: 21.5, 30.1, 51.3, 115.4, 120.2, 138.3, 146.8. According to ^{1}H NMR and GC/MS, the product contains approximately 3% of 5-bromo-1,3-dimethylbenzene as an impurity that cannot be separated at this stage. The spectroscopic data of **1** are in agreement with those reported in the literature.[3a]

7. With identical results, the reaction can be set up in the glove box.

8. The procedure for synthesizing anilide **2** follows that reported by Fürstner.[3a] The spectral characteristics of compound **2** are as follows: ^{1}H NMR (300 MHz, d_8-toluene) δ: 0.91 (t, J = 6.8 Hz, 6 H), 1.57 (s, 9 H), 2.29 (s, 6 H), 3.13 (q, J = 6.8 Hz, 4 H), 6.11 (s, 1 H), 6.52 (s, 2 H). These data are in agreement with those reported in the literature.[3a]

9. $MoCl_3(THF)_3$ prepared according to this procedure shows the following analytical properties: Elemental analysis calcd. for $C_{12}H_{24}Cl_3MoO_3$ (418.62): C 34.43, H 5.78; found: C 34.54, H 5.59. The submitters reported a "one-pot" preparation of **3** following a procedure reported by Poli.[4] Under these conditions, however, the checkers got erratic results with the appearance of the samples obtained in different runs ranging in color from orange to brown to lilac; in none of the cases was the elemental analysis correct. Therefore, the stepwise procedure outlined in the text is highly recommended.

10. The submitters performed this procedure in a glove box. The vacuum source was obtained by connecting an oil pump and a liquid nitrogen cold trap to a gas inlet with a hose in the glove box. The flask can be refilled with argon by closing the vacuum line and punching the septum with a needle. During concentration under vacuum, the flask was gently heated with a heat gun occasionally to speed the evaporation.

11. The procedure for synthesizing molybdenum triamide **4** follows that reported by Fürstner.[3a] The spectral characteristics of compound **4** are as follows: ^{1}H NMR (300 MHz, d_8-toluene) δ: –50.4 (br. s, 3 H), –20 (br. s, 6 H), –9.3 (br. s, 18 H), 61.7 (br, 27 H). Trisamide **4** can be used as such as an efficient alkyne metathesis precatalyst upon *in situ* activation with CH_2Cl_2 in toluene.[3a, 9, 10]

12. Trisamide **4** was treated with 2 equiv of 1,1-dichloropropane instead of higher equiv in order to minimize a side product that presumably arises from a Grignard reaction.

13. The title compound's spectral characteristics are as follows: ^{1}H NMR (300 MHz, C_6D_6) δ: 1.45 (br. s, 30 H), 2.12 (s, 18 H), 3.13–3.95 (m, 2 H), 6.05 (br. s, 6 H), 6.66 (s, 3 H); ^{13}C NMR (150 MHz, THF-d_8) δ: 14.3

(br), 21.7, 33.9 (br), 45.6 (br), 61.2 (br), 127.7 (br), 130.9 (br), 137.3, 151.6 (br), 303.2; MS (EI, vaporization temperature = 130 °C): *m/z* (%): 665 (76) [M^+], 610 (100), 552 (80), 490 (41), 432 (32), 391 (44), 375 (70), 335 (48), 306 (19); HRMS (EI) ($C_{39}H_{59}MoN_3$): calcd for ^{98}Mo: 667.3764, found 667.3761. The deuterated solvents used were dried and deoxygenated by distillation over potassium metal chips before use. The product contained small amounts (< 5 wt% based on ^{1}H NMR integration) of *N-tert*-butyl aniline **1**,[7] which could not be removed even upon repeated recrystallization from pentane. Therefore, the product was used for alkyne metathesis without any further purification.

14. The submitters report testing of the metathesis activity of catalyst precursor **5** as follows: In the glove box under an atmosphere of argon (Note 1), toluene-d_8 (0.7 mL) was added to a mixture of molybdenum(VI) propylidyne **5** (3.7 mg, 0.0056 mmol) and *p*-nitrophenol (2.3 mg, 0.017 mmol) (Note 2) in a sealable NMR tube (Note 16). The color of the resulting catalyst solution changed from yellow to dark red within 4 min of mixing (Note 17). Then 1-phenyl-1-butyne (**6**) (16.0 μL, 0.112 mmol) (Note 2) was added to the catalyst solution. After mixing well, the metathesis reaction was monitored in this sealed NMR tube at 20 °C. The ^{1}H NMR (d1 = 5 s) (Note 18) spectrum was recorded until the ratio of 3-hexyne to **6** was constant. The reaction reached equilibrium within 10 min with 46% conversion obtained under closed system conditions (Note 19).

15. The submitters reported isolation of the molybdenum propylidyne complex **5** (5.05 g, 89%) as a light yellow powder by simple evaporation of the pentane under vacuum in the glove box. The product can be further purified by recrystallization from pentanes as originally reported.[5] The submitters did not observe noticeable difference in the catalytic activity of the catalyst generated from **5** after recrystallization or from **5** without recrystallization.

16. J. Young valve NMR tube (Wilmad-Labglass, product No. 528-JY-8) was used.

17. A red precipitate was observed during mixing, which was presumably due to the low solubility of the generated species in toluene. In the case of chloroform, ethyl acetate or THF as solvent, no precipitation was observed.

18. A long relaxation time (d1 = 5 s) was employed to ensure the accurate ^{1}H NMR integration on the signals of both starting compound **6** and products.

19. Ferrocene was used as internal standard and the mass balance was higher than 95%. Conversion was calculated from the integrated intensities of product and starting material. The equilibrium position was established by conducting the metathesis reaction in both forward (starting from **6**) and backward (starting from 3-hexyne and diphenylacetylene) directions. In both cases, the same molar ratio (1.0:1.7) of dimer to monomer was obtained.

Safety and Waste Disposal Information

All hazardous materials should be handled and disposed of in accordance with "Prudent Practices in the Laboratory"; National Academy Press; Washington, DC, 1995.

1. Discussion

Alkyne metathesis has attracted great interest in the past two decades due to its powerful application in both organic synthesis and polymer chemistry.[9] Ring closing alkyne metathesis (RCAM) has been successfully applied to natural products synthesis,[10] and acyclic diyne metathesis (ADIMET) has been employed in preparation of high molecular weight polymers.[11] More recently, one-step gram-scale synthesis of shape-persistent arylene ethynylene macrocycles from simple monomer precursors have also been accomplished.[8]

The application of alkyne metathesis depends on the availability and performance of the alkyne metathesis catalysts, especially with regard to convenience of catalyst synthesis, substrate compatibility, and temperature required for catalytic activity. The most readily available of these catalysts is a poorly characterized species.[12] Many of the known catalysts require high temperature ($\geq$ 130 °C) for metathesis activity, and exhibit limited functional group tolerance. The first well-defined alkyne metathesis catalyst was the tungsten alkylidyne complex $(t\text{-BuO})_3W{\equiv}C(t\text{-Bu})$ developed by Schrock in the early 1980's.[13] Recently, Cummins has reported a new route to synthesize analogous trialkoxymolybdenum(VI) alkylidyne complexes $(R^3O)_3Mo{\equiv}CR^1$, which are active alkyne metathesis catalysts even at room temperature.[14] Although trialkoxy alkylidyne **8** can be conveniently obtained from alcoholysis of the metathesis inactive triamide **7** (eq 1), the starting complex **7** [15] requires a time-consuming, multistep synthetic procedure and presents a practical limitation to this potentially useful catalyst.

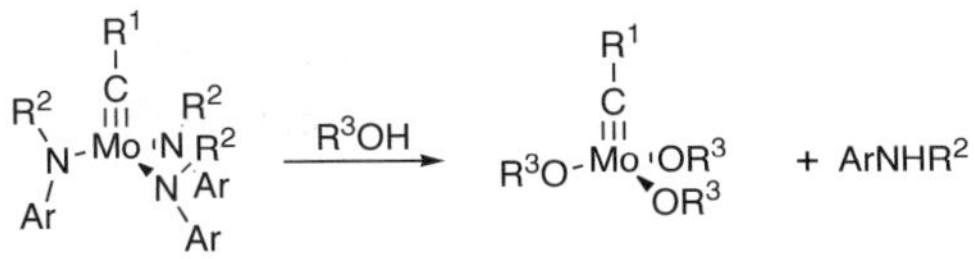 (eq 1)

7 8

A related alkyne metathesis catalyst was developed by Fürstner who combined molybdenum trisamide **4** with methylene chloride to produce a mixture of ClMo[N(*t*-Bu)Ar]$_3$ (**9**) and HC≡Mo[N(*t*-Bu)Ar]$_3$ (**10**) (eq 2). This mixture exhibited alkyne metathesis activity at 80 °C with good functional group compatibility.[3a,b] The catalytically active species is derived from monochloride **9**, whereas methylidyne **10** exhibits poor metathesis activity.[16]

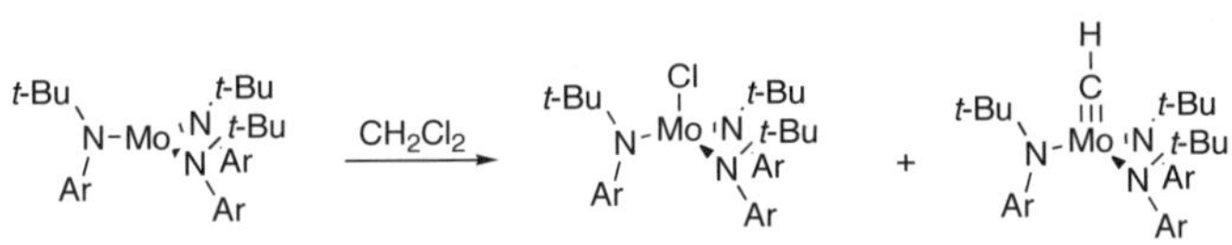 (eq 2)

4 9 10

Given that Fürstner's method directly produces alkylidyne **10**, which is the precursor to Cummins' active trialkoxyalkylidyne **8**, an expedient synthesis of complex **7** is envisioned to involve redirecting the reaction in eq 2 to favor production of HC≡Mo[N(*t*-Bu)Ar]$_3$. Magnesium selectively reacts with monochloride **9** returning it to starting complex **4**, while complex **10** remains unaffected. In the presence of excess methylene chloride, the net result is a reductive recycle approach to selectively generate the desired complex **10** in one pot (eq 3).

Reductive recycle appoach to Mo(VI) alkylidyne complexes.

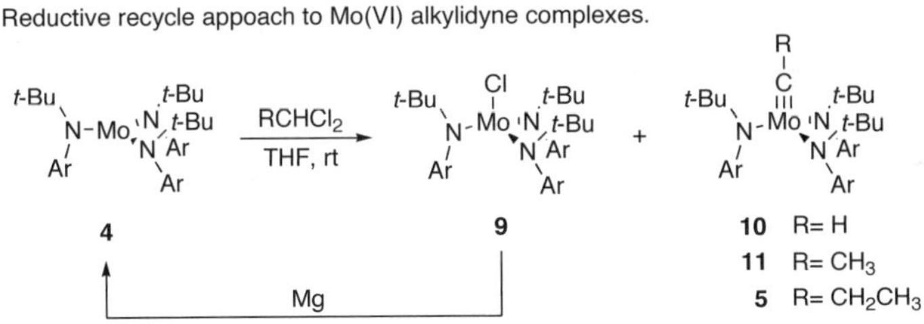 (eq 3)

Using this method, trisamidomolybdenum(VI) ethylidyne **11** and –propylidyne **5** were prepared in high yield by the treatment of trisamide **4** with 1,1-dichloroethane or 1,1-dichloropropane, respectively, in the presence of magnesium. Treatment of trisamidomolybdenum(VI) propylidyne (**5**) –

the catalyst precursor – with various phenols/alcohols produced alkyne metathesis catalysts active at room temperature. The Mo(VI) catalyst with *p*-nitrophenol as the ligand showed the highest catalytic activity.[17] In contrast to the reported incompatibilities of the tungsten alkylidyne complex ($(t\text{-BuO})_3W{\equiv}C(t\text{-Bu})$) with amines or polyether chains,[3a,10] the molybdenum(VI) propylidyne catalyst prepared through the reductive recycle strategy displayed good compatibility with these and other functional groups, as well as a wide variety of solvents.[17] This catalyst has been successfully used in synthesis of small molecules,[17] defect-free poly(thienyleneethynylene)s (PTEs) (**12**) and poly(phenyleneethynylene)s PPEs (**13**) with high molecular weight.[18] A variety of arylene ethynylene macrocycles (**14**, **15**) have also been successfully prepared in one step starting from monomers catalyzed by this Mo(VI) propylidyne catalyst.[8,19]

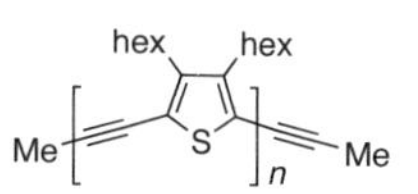

12

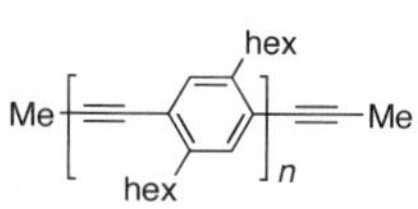

13

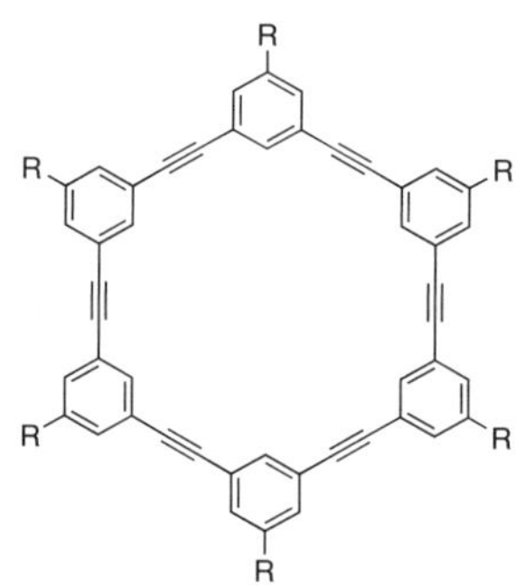

14

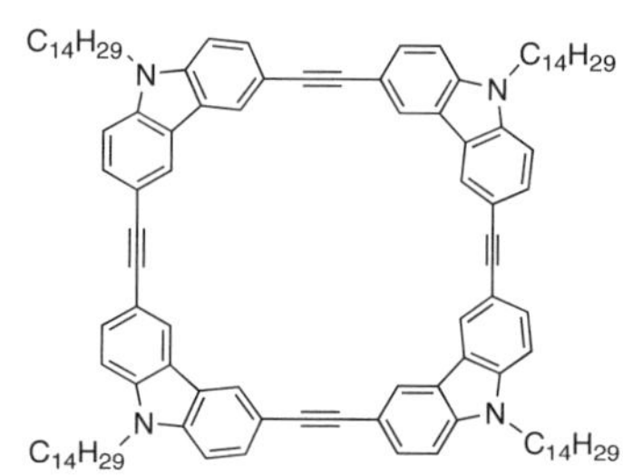

15

R = *t*-Bu, OTg, CH_2OTg, CO_2Tg, CO_2t-Bu
Tg = $-(CH_2CH_2O)_3CH_3$

1. Roger Adams Laboratory, Departments of Chemistry and Materials Science & Engineering, University of Illinois at Urbana-Champaign, Urbana, IL 61801, USA. E-mail: jsmoore@uiuc.edu.

2. (a) Cummins, C. C. *Chem. Commun.* **1998**, 1777-1786. (b) Laplaza, C. E.; Cummins, C. C. *Science* **1995**, *268*, 861-863.

3. (a) Fürstner, A.; Mathes, C.; Lehmann, C. W.; *Chem. Eur. J.* **2001**, *7*, 5299-5317. (b) Fürstner, A.; Mathes, C.; Lehmann, C. W. *J. Am. Chem. Soc.* **1999**, *121*, 9453-9454. Cummins pioneered the synthesis and study of $Mo[NAr(t\text{-Bu})]_3$ (**4**); see: (c) Laplaza, C. E.; Odom, A. L.; Davis, W. M.; Cummins, C. C.; Protasiewicz, J. D. *J. Am. Chem. Soc.* **1995**, *117*, 4999-5000. (d) Johnson, A. R.; Cummins, C. C. *Inorg. Synth.* **1998**, *32*, 123-132. (e) Tsai, Y.-C.; Stephens, F. H.; Meyer, K.; Mendiratta, A.; Gheorghiu, M. D.; Cummins, C. C. *Organometallics* **2003**, *22*, 2902-2913.

4. Stoffelbach, F.; Saurenz, D.; Poli, R. *Eur. J. Inorg. Chem.* **2001**, 2699-2703.

5. Zhang, W.; Kraft. S.; Moore, J. S. *Chem. Commun.* **2003**, 832-833.

6. To avoid the overlap of toluene–d_8 aromatic peaks with those of compound **8**, THF–d_8 was used in acquiring ^{13}C NMR data.

7. A similar problem was also previously encountered in separation of trialkoxy Mo(VI) complex from aniline **1**, see Tsai, Y.-C.; Diaconescu, P. L.; Cummins, C. C. *Organometallics* **2000**, *19*, 5260-5262.

8. Zhang, W.; Moore, J. S. *J. Am. Chem. Soc.* **2004**, *126*, 12796.

9. For reviews, see (a) Fürstner, A.; Davies, P. W. *Chem. Commun.* **2005**, 2307-2320. (b) Schrock, R. R.; Hoveyda, A. H. *Angew. Chem., Int. Ed.* **2003**, *42*, 4592-4633. (c) Schrock, R. R. *Chem. Rev.* **2002**, *102*, 145-179. (d) Fürstner, A. in *Handbook of Metathesis*, ed. Grubbs, R. H., Wiley-VCH, Weinheim, **2003**, vol. 2, ch. 2.12. (e) Bunz, U. H. F. *Acc. Chem. Res.* **2001**, *34*, 998-1010.

10. (a) Fürstner, A.; Guth, O.; Rumbo, A.; Seidel, G. *J. Am. Chem. Soc.* **1999**, *121*, 11108-11113. (b) Fürstner, A.; Seidel, G. *Angew., Chem. Int. Ed.* **1998**, *37*, 1734-1736. (c) Fürstner, A.; Grela, K.; Mathes, C.; Lehmann, C. W. *J. Am. Chem. Soc.* **2000**, *122*, 11799-11805. (d) Fürstner, A.; Mathes, C.; Grela, K. *Chem. Commun.* **2001**, 1057-1059. (e) Fürstner, A.; Stelzer, F.; Rumbo, A.; Krause, H. *Chem. Eur. J.* **2002**, *8*, 1856-1871. (f) Fürstner, A.; Castanet, A.-S.; Radkowski, K.; Lehmann, C. W. *J. Org. Chem.* **2003**, *68*, 1521-1528. (g) Fürstner, A.; Turet, L. *Angew. Chem., Int. Ed.* **2005**, *44*, 3462-3466.

11. (a) Kloppenburg, L.; Song, D.; Bunz, U. H. F. *J. Am. Chem. Soc.* **1998**, 120, 7973-7974. (b) Kloppenburg, L.; Jones, D.; Bunz, U. H. F. *Macromolecules* **1999**, *32*, 4194-4203.

12. Mortreux, A.; Blanchard, M. *J. Chem. Soc., Chem. Commun.* **1974**, 786-787.

13. Schrock, R. R.; Clark, D. N.; Sancho, J.; Wengrovius, J. H.; Rocklage, S. M.; Pedersen, S. F. *Organometallics*, **1982**, *1*, 1645-1651.

14. Blackwell, J. M.; Figueroa, J. S.; Stephens, F. H.; Cummins, C. C. *Organometallics* **2003**, *22*, 3351-3353.

15. (a) Freudenberger, J. H.; Schrock, R. R.; Churchill, M. R.; Rheingold, A. L.; Ziller, J. W. *Organometallics* **1984**, *3*, 1563-1573. (b) Listemann, M. L.; Schrock, R. R. *Organometallics* **1985**, *4*, 74-83. (c) Agapie, T.; Diaconescu, P. L.; Cummins, C. C. *J. Am. Chem. Soc.* **2002**, *124*, 2412-2413 and ref 7, 13.

16. Schrock reported that terminal alkynes are unsuitable in alkyne metathesis, which may be explained by the propensity of terminal alkylidynes to suffer ligand loss along the reaction pathway, see: Schrock, R. R. *Polyhedron* **1995**, *14*, 3177-3195.

17. Zhang, W.; Kraft, S.; Moore, J. S. *J. Am. Chem. Soc.* **2004**, *126*, 329-335.

18. Zhang, W.; Moore, J. S. *Macromolecules* **2004**, *37*, 3973-3975.

19. Zhang, W.; Moore, J. S. *J. Am. Chem. Soc.* **2005**, *127*, 11863-11870.

PREPARATION OF A CARBAZOLE-BASED MACROCYCLE VIA PRECIPITATION-DRIVEN ALKYNE METATHESIS

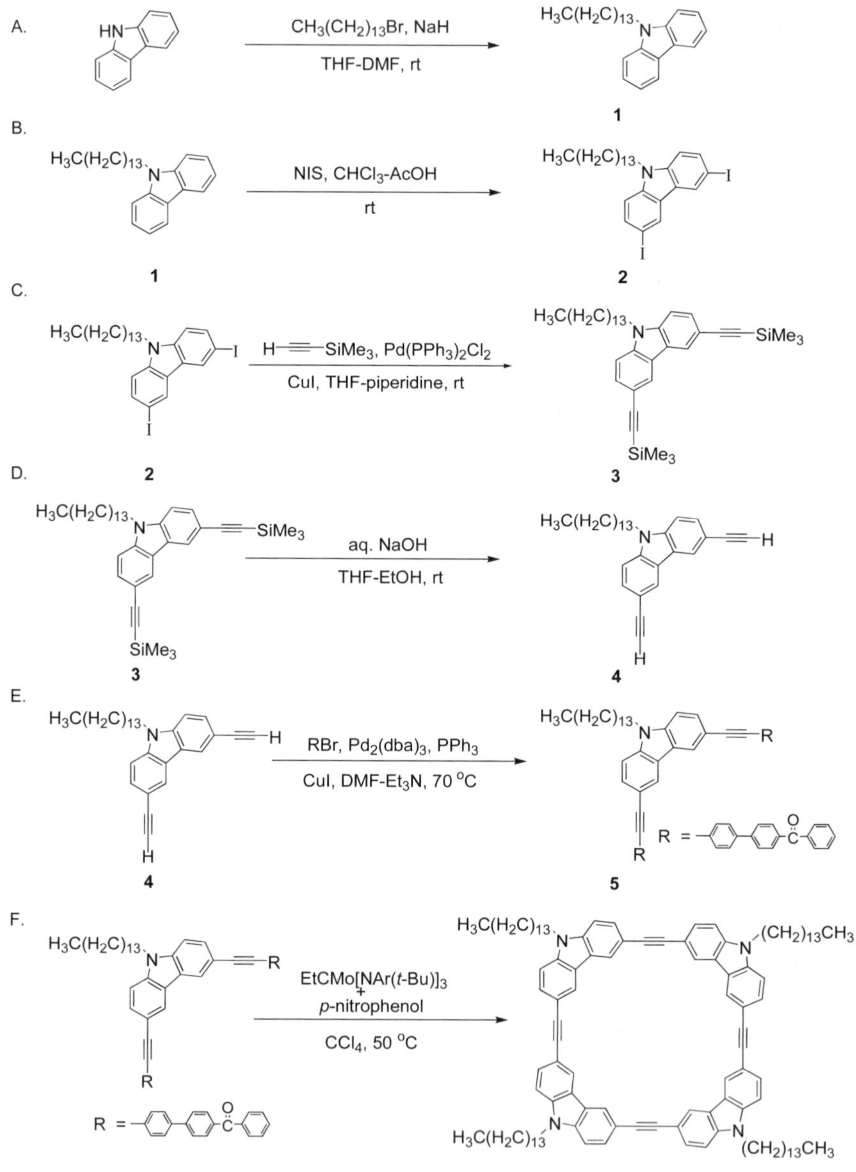

Submitted by Wei Zhang, Hyeon Mo Cho and Jeffrey S. Moore.[1]
Checked by Scott E. Denmark and Christopher R. Butler.

1. Procedure

A. 9-Tetradecylcarbazole (1). A single-necked 500-mL, round-bottomed, side-armed flask (Note 1) equipped with a glass stopper and a magnetic stirring bar is connected to a nitrogen-vacuum manifold. After evacuating and backfilling with nitrogen three times, the flask (with stopper removed), in which a positive flow of nitrogen is maintained throughout the entire procedure, is sequentially charged with carbazole (12.5 g, 0.072 mol) (Note 2), 1-bromotetradecane (24.3 mL, 0.079 mol) (Note 3), THF (60 mL) and DMF (20 mL) (Note 4). To the stirred solution, NaH (60% in oil, 4.32 g, 0.11 mol) (Note 3) is added to the solution in small portions within 20 min, avoiding vigorous bubbling and heating. The flask is capped with a glass stopper and the resulting suspension is stirred for 2 h at room temperature under nitrogen. Methanol (30 mL) is slowly added to the reaction mixture within 5 min to quench the remaining NaH. The solution is transferred to a 1-L, round-bottomed flask and then is concentrated by rotary evaporation (15–20 mmHg, water bath temperature 50–55 °C) to afford a solid residue, which is extracted with CH_2Cl_2 (3 x 100 mL). The combined organic layers are sequentially washed with 2 N aq HCl (2 x 100 mL), H_2O (2 x 100 mL), then are dried (Na_2SO_4) and concentrated by rotary evaporation. The crude product is purified by column chromatography [silica gel, 80 x 250 mm, hexanes] to afford carbazole **1** as a white solid (25.8 g, 98%) (Note 5).

B. 3,6-Diiodo-9-tetradecylcarbazole (2). A 500-mL, one-necked, round-bottomed flask equipped with a magnetic stirring bar is sequentially charged with carbazole **1** (14.5 g, 0.040 mol), *N*-iodosuccinimide (19.1 g, 0.081 mol) (Note 6), $CHCl_3$ (280 mL) and acetic acid (100 mL). The flask is capped with a nitrogen inlet adapter and wrapped in aluminum foil. The reaction mixture is stirred under nitrogen for 16 h at room temperature. Chloroform and acetic acid are removed by rotary evaporation. A pink precipitate forms upon pouring water (450 mL) into the residue and then transferring the slurry into a 1-L Erlenmeyer flask containing more water (200 mL). The solids are collected by filtration, air dried and dissolved in $CHCl_3$ (300 mL). The organic solution is sequentially washed with saturated $Na_2S_2O_3$ (3 x 70 mL), brine (2 x 100 mL), dried (Na_2SO_4), filtered and concentrated by rotary evaporation (15–20 mmHg, water bath temperature 30–35 °C). Iodinated carbazole **2** is obtained as a white solid (23.8 g, 97%) and used without further purification (Note 7).

C. 3,6-Bis[(trimethylsilyl)ethynyl]-9-tetradecylcarbazole (3). A 500-mL single-necked round-bottomed, side-arm flask equipped with a glass stopper and a magnetic stirring bar is connected to a nitrogen-vacuum manifold. After evacuating and backfilling with nitrogen three times, the flask, in which a positive flow of nitrogen is maintained throughout the procedure, is sequentially charged with iodinated carbazole **2** (23.7 g, 0.038 mol), CuI (0.147 g, 0.77 mmol) (Note 3), $Pd(PPh_3)_2Cl_2$ (1.35 g, 1.9 mmol) (Note 3), THF (90 mL), and piperidine (55 mL) (Note 3). The flask is cooled in an ambient temperature water bath before 1-trimethylsilylacetylene (45.7 mL, 0.38 mol) (Note 8) is added in two portions over 5 min (Note 9). The reaction mixture is then stirred for 23 h at room temperature. The solution changes from green to yellow with the formation of a white precipitate, and then the suspension changes to green and then black. The reaction mixture is then filtered through a coarse fritted funnel into a 1-L, one-necked, round-bottomed flask and is concentrated by rotary evaporation (15–20 mmHg, water bath temperature 40–45 °C). The crude product is purified by column chromatography [silica gel, 80 x 250 mm, hexanes 1 L, then hexanes/EtOAc 50/1 (2 L), then hexanes/EtOAc, 20/1 (1 L)] to afford diyne-substituted carbazole **3** as a yellow oil (19.9 g, 93%) (Note 10).

*D. 3,6-Diethynyl-9-tetradecylcarbazole (**4**).* A 1-L, one-necked, round-bottomed flask equipped with a magnetic stirring bar is sequentially charged with diyne substituted carbazole **3** (19.5 g, 0.035 mol), THF (100 mL) and ethanol (400 mL). As the resulting solution is stirred, a solution of NaOH (4.21 g, 0.11 mol) in H_2O (30 mL) is added slowly over 5 min. The solution is stirred under nitrogen at room temperature for an additional 2 h during which time the solution changes from orange to red. The solvent is removed by rotary evaporation (water bath temperature 45–50 °C) to afford a residue, which is partitioned between CH_2Cl_2 (450 mL) and H_2O (100 mL) in a 1-L separatory funnel. The separated organic layer is washed with H_2O (2 x 100 mL) and brine (2 x 100 mL), and then is dried (Na_2SO_4), filtered and concentrated by rotary evaporation (20–25 mmHg, water bath temperature 30–35 °C). Diyne-substituted carbazole **4** is obtained as a yellow-red oil that becomes a tan solid upon standing (14.0 g, 97%) and is used without further purification (Notes 11, 12).

*E. 3,6-Bis(benzoylbiphenyl)ethynyl-9-tetradecylcarbazole (**5**).* A 500-mL, single-necked round-bottomed, side-arm flask equipped with a Teflon thermometer adapter and a magnetic stirring bar is connected to a nitrogen-vacuum manifold. The flask is flame-dried under vacuum, then filled with nitrogen. After evacuating and backfilling with nitrogen three times, the flask, in which a positive flow of nitrogen is maintained throughout the entire procedure, is sequentially charged with CuI (0.058 g, 0.31 mmol)

(Note 3), $Pd_2(dba)_3$ (1.68 g, 1.8 mmol) (Note 13), PPh_3 (3.21 g, 12 mmol) (Note 13) and 4-benzoyl-4'-bromobiphenyl (21.7 g, 0.064 mol) (Note 14). The flask containing the solid mixture is evacuated and backfilled with nitrogen three times. Under a nitrogen purge, triethylamine (70 mL) (Note 3, Note 4) and a solution of diyne-substituted carbazole **4** (12.6 g, 0.031 mol) in DMF (150 mL) (Note 4) are sequentially added. Oxygen is removed from the mixture by three freeze-pump-thaw cycles. The reaction mixture is then stirred for 16 h at 70 °C (internal temperature). The solution changes from yellow to a dark purple. When the reaction mixture cools to room temperature, yellow precipitates are observed. The mixture is transferred to a 1-L, single-necked, round-bottomed flask with the aid of dichloromethane washes (2 x 250 mL). Dichloromethane and Et_3N are removed by rotary evaporation (15–20 mmHg, water bath temperature 40–45 °C) and DMF is removed under high vacuum by using a short-path distillation apparatus (0.50–1.0 mmHg, air bath temperature 45–55 °C) (Note 15). Dichloromethane (400 mL) is added and the organic layer is washed with saturated aq. NH_4Cl solution (2 x 200 mL), brine (2 x 200 mL) and then was dried (Na_2SO_4) and filtered. Dichloromethane is removed by rotary evaporation and the residual solid is treated with methanol (200 mL), resulting in the formation of a yellow solid. The solid bisarylethynyl-substituted carbazole **5** is collected by filtration, then is washed with methanol (3 x 50 mL), dried in vacuo (26.9 g, 95%) and used without further purification (Notes 16, 17).

*F. Carbazole-based tetrakismacrocycle (**6**).* A 400-mL Schlenk tube equipped with a magnetic stirring bar and a glass stopper is introduced to a glove box. The flask is sequentially charged with 4-nitrophenol (0.680 g, 4.89 mmol) (Note 3), trisamidomolybdenum(VI) propylidyne (1.09 g, 1.63 mmol) (Note 18) and CCl_4 (150 mL) (Note 4). The resulting solution is stirred for 4 min and the solution changes from light yellow to orange (Note 19). A solution of monomer **5** (15.1 g, 16.3 mmol) in CCl_4 (200 mL) (Notes 20, 21) is added to the stirred catalyst solution. The flask is sealed with grease and electrical tape and is removed from the glove box. The resulting mixture is stirred at 50 °C (Note 22) for 24 h during which time a light-yellow precipitate forms and the color of the reaction mixture changes from yellow to brown. The mixture is transferred to a 1-L Erlenmeyer flask, along with CH_2Cl_2 (200 mL). The solids are removed by suction filtration through a medium-porosity fritted-glass funnel and are washed with CH_2Cl_2 (2 x 50 mL). The filtrate is concentrated by rotary evaporation (20–25 mmHg, water bath temperature 40–45 °C). Methanol (200 mL) is added to the residue and the resulting suspension is filtered, washed with methanol (3 x 75 mL) and dried in vacuo. The crude product is further purified by silica gel column chromatography (Note 3) [silica gel, 44 x 572 mm, $CHCl_3/CH_2Cl_2$, 3/1 to 1/1] to afford macrocyclic product **6** as a pale-yellow solid (3.80 g, 61%) (Notes 23, 24).

2. Notes

1. All glassware was oven-dried at 140 °C overnight.

2. Carbazole (96%) was purchased from Acros and used as received.

3. 1-Bromotetradecane (97%), sodium hydride (60% dispersion in mineral oil), CuI (99.999%), $Pd(PPh_3)_2Cl_2$ (98%), piperidine (redistilled, 99.5+%), triethylamine (99.5%), *p*-nitrophenol (≥ 99%) and grade 62 silica gel with pore size of 150 Å were purchased from the Aldrich Chemical Company, Inc. and used as received unless further purification is indicated.

4. THF and DMF were dried by standing over 4 Å molecular sieves over night before use. Carbon tetrachloride was distilled over P_2O_5 and stored over 3 Å molecular sieves.

5. A small portion of compound **1** was further purified by recrystallization from ethanol to afford analytically pure material. The analytical data from compound **1** were as follows: mp 42–44 °C; TLC: R_f = 0.22 (hexanes); ^{1}H NMR ($CDCl_3$, 500 MHz) δ: 0.87 (t, *J* = 7.4 Hz, 3 H), 1.23–1.40 (m, 22 H), 1.87 (tt, *J* = 7.3 and 7.3 Hz, 2 H), 4.30 (t, *J* = 7.3 Hz, 2 H), 7.22 (dt, *J* = 7.8 and 0.9 Hz, 2 H), 7.40–7.48 (m, 4 H), 8.10 (d, *J* = 7.8 Hz, 2 H); ^{13}C NMR ($CDCl_3$, 125 MHz) δ: 14.1, 22.7, 27.3, 28.9, 29.3, 29.4, 29.47, 29.54, 29.57, 29.62, 29.64, 31.9, 43.0, 108.6, 118.6, 120.3, 122.8, 125.5, 140.4. IR (KBr): 3048, 2947, 2919, 2869, 2848, 2341, 2360, 1628, 1600, 1485, 1465, 1453, 1371, 1351, 1328, 1235, 1153, 1122 cm^{-1}; LRMS (EI): *m/z* 363.3, 180.1. Anal. Calcd for $C_{26}H_{37}NO$: C, 85.89; H, 10.26; N, 3.85. Found: C, 86.12; H, 10.66; N, 4.03. The spectroscopic data were in agreement with those previously reported.[2]

6. *N*-Iodosuccinimide (95%) was purchased from the Aldrich Chemical Company, Inc. and used as received.

7. The analytical data from compound **2** were as follows: mp 79–80 °C; ^{1}H NMR ($CDCl_3$, 500 MHz) δ: 0.88 (t, *J* = 7.1 Hz, 3 H), 1.22–1.30 (m, 22 H), 1.81 (tt, *J* = 7.1 and 7.1 Hz, 2 H), 4.21 (t, *J* = 7.4 Hz, 2 H), 7.16 (d, *J* = 8.6 Hz, 2 H), 7.70 (dd, *J* = 8.6 and 1.7 Hz, 2 H), 8.32 (d, *J* = 1.5 Hz, 2 H); ^{13}C NMR ($CDCl_3$, 125 MHz) δ: 14.1, 22.7, 27.2, 28.8, 29.31, 29.35, 29.35,

29.42, 29.51, 29.56, 29.61, 29.63, 29.65, 31.9, 43.2, 81.6, 110.9, 124.0, 129.3, 134.5, 139.5; IR (KBr): 3062, 2921, 2847 1585, 1466, 1428, 1374, 1348, 1312, 1290, 1236, 1216, 1152, 1047, 1014 cm^{-1}; LRMS (EI): *m/z* 615.1, 431.9, 208.1. These spectroscopic data were in agreement with those previously reported.[2]

8. 1-Trimethylsilylacetylene (98+%) was purchased from GFS Chemicals and used as received.

9. Upon addition of 1-trimethylsilylacetylene, a significant exotherm (15 °C) was observed, and stirring became difficult as the piperidine hydroiodide was formed.

10. The analytical data from compound **3** were as follows: TLC R_f = 0.32 (*n*-hexane/EtOAc, 20/1); ^{1}H NMR ($CDCl_3$, 500 MHz) δ: 0.28 (s, 18 H), 0.87 (t, J = 7.1 Hz, 3 H), 1.22–1.31 (m, 22 H), 1.83 (p, J = 7.1 Hz, 2 H), 4.25 (t, J = 7.2 Hz, 2 H), 7.29 (d, J = 8.1 Hz, 2 H), 7.56 (dd, J = 8.5 and 1.6 Hz, 2 H), 8.19 (d, J = 1.0 Hz, 2 H); ^{13}C NMR ($CDCl_3$, 125 MHz) δ: 0.15, 14.1, 22.7, 27.2, 28.9, 29.32, 29.34, 29.45, 29.51, 29.57, 29.61, 29.63, 29.65, 31.9, 43.3, 92.0, 106.3, 108.8, 113.7, 122.3, 124.7, 130.0, 140.5; IR (film): 2926, 2854, 2153, 1869, 1629, 1598, 1483, 1406, 1382, 1350, 1318, 1287, 1248, 1202, 1149, 1133, 1059 cm^{-1}; MS (EI): *m/z* 555.4, 372.1, 73.1. These spectroscopic data were in agreement with those previously reported.[2]

11. A small portion of compound **4** was further purified by column chromatography (*n*-hexanes/EtOAc, 100/1 to 20/1) to afford analytically pure material. The analytical data from compound **4** were as follows: TLC R_f = 0.29 (*n*-hexanes/EtOAc, 20/1); ^{1}H NMR ($CDCl_3$, 500 MHz) δ: 0.88 (t, J = 7.3 Hz, 3 H), 1.22–1.35 (m, 22 H), 1.84 (tt, J = 7.1 and 7.1 Hz, 2 H), 3.00 (s, 2 H), 4.25 (t, J = 7.3 Hz, 2 H), 7.32 (d, J = 8.6 Hz, 2 H), 7.60 (dd, J = 8.3, 1.5 Hz, 2 H), 8.21 (d, J = 1.1 Hz, 2 H); ^{13}C NMR ($CDCl_3$, 125 MHz) δ: 14.1, 22.7, 27.2, 28.9, 29.30, 29.33, 29.43, 29.50, 29.56, 29.60, 29.62, 29.64, 31.9, 43.3, 75.4, 84.7, 108.9, 112.7, 122.2, 124.7, 130.1, 140.6; IR (film): 3279 (s), 2924 (s), 2849 (s), 2103 (s), 1630 (m), 1598 (s), 1483 (s), 1381 (m), 1349 (s), 1321 (w), 1290 (s), 1234 (m), 1153 (s), 1134 (m), 882 (s), 812 (s), 728 (w), 681 (m), 653 (s), 614 (s) cm^{-1}; LRMS (EI): *m/z* 411.3, 228.1, 75.0. Anal. Calcd for $C_{30}H_{37}N$: C, 87.54; H, 9.06; N, 3.40. Found: C, 87.38; H, 8.78; N, 3.63. These spectroscopic data were in agreement with those previously reported.[2]

12. The appearance of more than one ^{1}H NMR singlet around δ 3.1 ppm (terminal acetylene) indicated the presence of mono-desilylation side product. In this case, **4** was further purified by column chromatography [*n*-hexanes/EtOAc, 100/1 to 20/1, TLC R_f = 0.29 (*n*-hexanes/EtOAc, 20/1].

13. $Pd_2(dba)_3$ and PPh_3 (99%) were purchased from Strem Chemicals, Inc. and used as received.

14. 4-Benzoyl-4'-bromobiphenyl (99%) was purchased from Lancaster and used as received.

15. The submitters reported the use of a Kugelrohr short-path distillation apparatus (Cat. No. Z401137-1SET), which was purchased from Aldrich Chemical Company, Inc.

16. A small portion (500 mg) of compound **5** was further purified by recrystallization from hot isopropyl acetate (ca. 20 mL) to provide analytically pure material. The analytical data from compound **5** were as follows: mp 162–167 °C; TLC R_f = 0.26 (CH_2Cl_2/*n*-hexanes, 1/1); ^1H NMR ($CDCl_3$, 500 MHz) δ: 0.88 (t, *J* = 7.1 Hz, 3 H, CH_3), 1.19–1.41 (m, 22 H, CH_2), 1.89 (tt, *J* = 7.1, 7.1 Hz, 2 H), 4.31 (t, *J* = 7.3 Hz, 2 H), 7.40 (d, *J* = 8.4 Hz, 2 H), 7.52 (m, 4 H), 7.61 (tt, *J* = 7.5, 1.4 Hz, 2 H), 7.66–7.71 (m, 10H), 7.74 (d, *J* = 8.6 Hz, 4 H), 7.85 (m, 4 H), 7.92 (d, *J* = 8.4 Hz, 4 H), 8.32 (d, *J* = 1.4 Hz, 2 H); ^{13}C NMR ($CDCl_3$, 125 MHz) δ: 14.1, 22.7, 27.2, 28.9, 29.3, 29.46, 29.51, 29.57, 29.61, 29.62, 29.64, 31.9, 43.4, 87.6, 92.0, 109.1, 113.7, 122.5, 123.8, 124.2, 126.8, 127.2, 128.3, 129.8, 130.0, 130.8, 132.1, 132.4, 136.4, 137.7, 139.2, 140.6, 144.4, 196.3; IR (KBr): 2924, 2851, 2201, 1648, 1600, 1522, 1481, 1384, 1317, 1277, 1150 cm^{-1}. LRMS (MALDI): *m/z* 923.94, 825.5, 809.5, 755.9, 740.8, 567.4, 551.4, 545.4. Anal. Calcd for $C_{68}H_{61}NO_2$: C, 88.38; H, 6.65; N, 1.52. Found: C, 88.06; H, 6.41; N, 1.70. These spectroscopic data were in agreement with those previously reported.[2]

17. The submitters found that the appearance of more than one ^1H NMR triplet around δ 4.3 ppm (α-methylene protons on the tetradecyl side chain) indicated the presence of unreacted starting material and/or mono-alkynylated product. In this case, **5** could be further purified by column chromatography [*n*-hexanes/CH_2Cl_2, 1/1 v/v, TLC R_f = 0.26 (CH_2Cl_2/*n*-hexane, 1/1)].

18. Gram-scale preparation of trisamidomolybdenum(VI) propylidyne is reported in the preceding *Organic Syntheses* report.

19. A red precipitate was observed during mixing, and was presumably due to the low solubility of the species generated in CCl_4. In the case of chloroform, ethyl acetate or THF as solvent, no precipitate was observed.

20. Only a portion of monomer **5** dissolved in CCl_4 at the start of the reaction. However, the poor solubility of **5** did not affect the macrocycle synthesis. As the metathesis reaction proceeded, monomer **5** dissolved and was transformed into soluble macrocycles.

21. In addition to CCl_4, 1,2,4-trichlorobenzene and toluene are alternative solvents for the macrocycle synthesis. The amount of aromatic impurities observed in the reaction (based on ^1H NMR analysis) conducted in CCl_4 was less than that observed in toluene.

22. In a previous report on the small-scale synthesis of macrocycle **6** via precipitation-driven alkyne metathesis,[2] the reaction was conducted at 30 °C. However, for gram-scale preparation, the low solubility of oligomeric intermediates dramatically increased the solution viscosity and made stirring difficult. Raising the reaction temperature from 30 °C to 50 °C greatly improved the intermediate solubility and allowed efficient stirring.

23. The crude product **6** (after methanol wash, before column separation) was obtained in 98% yield (6.13 g) and contained a small amount of aromatic impurities (mainly 4-benzoyl-4'-bromobiphenyl). Macrocycles of high purity were obtained after column chromatography. The analytical data from compound **6** were as follows: mp 222–224 °C; TLC; $R_f = 0.89$ ($CHCl_3/CH_2Cl_2$, 3/1); ^1H NMR ($CDCl_3$, 500 MHz) δ: 0.88 (t, $J = 7.1$ Hz, 12 H), 1.20–1.45 (m, 88 H), 1.90 (tt, $J = 7.1, 7.1$, 8 H), 4.31 (t, $J = 7.1$ Hz, 8 H), 7.39 (d, $J = 8.5$ Hz, 8 H), 7.71 (dd, $J = 8.3, 1.3$ Hz, 8 H), 8.42 (d, $J = 0.7$ Hz, 8 H); ^{13}C NMR ($CDCl_3$, 125 MHz) δ: 14.1, 22.7, 27.3, 29.0, 29.35, 29.50, 29.55, 29.60, 29.64, 29.67, 31.9, 43.3, 89.0, 108.9, 114.4, 122.6, 123.9, 129.3, 140.1; IR (KBr): 2922, 2848, 1628, 1599, 1570, 1490, 1382, 1352, 1307, 1283, 1214, 1148, 1131 cm^{-1}; MS (MALDI): *m/z* 1542.7, 1370.1. Anal. Calcd for $C_{112}H_{140}N_4$: C, 87.22; H, 9.15; N, 3.63; Found: C, 86.84; H, 9.19; N, 3.26. These spectroscopic data were in agreement with those previously reported.[2]

24. The checkers found that accurate analytical data could be obtained from **6** only after thorough drying at 0.3 mm Hg in an Abderhalden apparatus heated with refluxing xylenes.

Safety and Waste Disposal Information

All hazardous materials should be handled and disposed of in accordance with "Prudent Practices in the Laboratory"; National Academy Press; Washington, DC, 1995.

3. Discussion

Shape-persistent arylene ethynylene macrocycles have attracted attention in the fields of supramolecular chemistry and materials science over the past decade due to their novel properties and potential applications.[3] However, the study and application of shape-persistent macrocycles are impeded by inefficient macrocycle syntheses. The usual preparation involves a large number of synthetic steps, requires dilute conditions (< 1 mM), and affords low overall yields.[4] Considerable efforts have been devoted to the development of methodologies allowing selectively functionalized structures to be obtained in high yields.

Conventional synthetic routes for arylene ethynylene macrocycles are dominated by Sonogashira coupling[5] between aryl iodides and terminal acetylenes and Glaser-type couplings[6] between terminal acetylenes. These cross-coupling approaches have advantages such as high tolerance of functional groups, with generally high yields in the individual steps. However, to obtain high yields of macrocycles, precursor oligomers must be pre-synthesized in a stepwise fashion and subsequently subjected to cross-coupling reactions under pseudo-high dilution condition (< 1 mM) to form macrocycles via intramolecular cyclization. Such tedious and time-consuming synthesis of the precursors impedes the efficient large-scale preparation of macrocycles via this route. Random cyclization of three or more monomer units has also been applied to preparing target macrocycles, but usually only low yields (1–18% on cyclization step) were obtained.[7]

Inspired by the successful applications of thermodynamically controlled reversible reactions in preparing unique molecular architectures[8] such as macrocyclic compounds, molecular capsules and interlocked structures, we envisioned that a convenient synthesis of arylene ethynylene macrocycles could be accomplished by using dynamic covalent chemistry.[9] In the past two decades, there has been rapid progress in alkyne metathesis catalyzed by either defined carbyne complexes[10] or catalysts generated in situ.[11] The ready availability of highly active Mo(VI) alkylidyne catalysts synthesized by a reductive recycle strategy[12] and successful examples of using dynamic covalent chemistry in organic syntheses prompted us to develop a synthetic approach to macrocycle preparation directly from a monomer.

We reported a successful approach involving precipitation-driven alkyne metathesis for convenient, multi-gram synthesis of arylene ethynylene macrocycles near room temperature.[2] Driven by the precipitation of a diarylacetylene byproduct, the desired macrocycles were obtained in one step from monomers in high yields. Macrocycle formation is a thermodynamically favored process under equilibrium control.[13]

We have utilized precipitation-driven alkyne metathesis to prepare other carbazole-based tetrameric macrocycles, such as **10c**. The four side chains are easily removed via thermolysis.[14] The resulting macrocycles may exhibit interesting electronic properties such as mixed-valence (MV) states.[15]

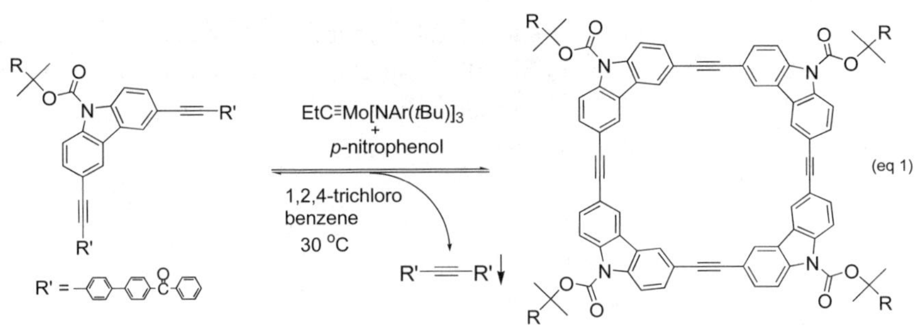

9a: R = Methyl
9b: R = Heptyl
9c: R = dodecyl

10a: R = Methyl < 5%
10b: R = Heptyl < 5%
10c: R = dodecyl 81%

Three dialkyne monomers **9a-c**, which are functionalized with *t*-butyl carboxylate, 1,1-dimethyloctyl carboxylate and 1,1-dimethyltridecyl carboxylate, respectively, were prepared and subjected to standard metathesis conditions (eq 1). In the case of **9a-b**, at the end of the reaction nearly quantitative conversion of starting monomers into insoluble materials was observed. Based on elemental analysis, the insoluble materials are believed to be a mixture of intermediate oligomers and diarylacetylene byproduct. In great contrast, metathesis of monomer **9c** proceeded well in 1,2,4-trichlorobenzene at 30 °C, and tetrameric macrocycle **10c** was obtained in 81% isolated yield.[16]

The successful generation of tetrameric macrocycle **6** and **10c**, as well as those failed cases when using monomer **9a-b**, illustrate the importance of balancing the solubility through the course of macrocycle formation.[13] In the case of metathesis of **9a-b**, the generated oligomeric species, either open-chain or cyclic versions, presumably precipitate out of solution due to their insufficient solubility. In great contrast, carbazole oligomers with 1,1-dimethyltridecyl carboxylate side chains (from metathesis of **9c**) have better solubility and intermediate species remain in solution, thus allowing for high yield formation of tetracycles via alkyne metathesis—a dynamic covalent approach.

This finding has important practical significance in that it demonstrates that the solubility of longer arylene ethynylene oligomers, especially those overshooting the target macrocycle length, must be taken into consideration to accomplish high-yield macrocycle syntheses. All the intermediate compounds should be sufficiently soluble along the reaction pathway to allow the thermodynamically stable product to be obtained through reversible alkyne metathesis.

It is envisioned that the successful preparation of macrocycles using the precipitation-driven, reversible alkyne-metathesis may open the way to other two-dimensional or three-dimensional arylene ethynylene structures, as well as alkyne-bridged oligomers and polymers.

1. Roger Adams Laboratory, Departments of Chemistry and Materials Science & Engineering, University of Illinois at Urbana-Champaign, Urbana, IL 61801, USA. E-mail: jsmoore@uiuc.edu.

2. Zhang, W.; Moore, J. S. *J. Am. Chem. Soc.* **2004**, *126*, 12796.

3. (a) Zhang, W.; Moore, J.S. *Angew. Chem. Int. Ed.* **2006**, *45*, 4416–4439. (b) Höger, S. *Chem. Eur. J.* **2004**, *10*, 1320–1329. (c) Zhao, D.; Moore, J. S. *Chem. Commun.* **2003**, 807–818. (d) Yamaguchi, Y.; Yoshida, Z. *Chem. Eur. J.* **2003**, *9*, 5430–5440. (e) Bunz, U. H. F.; Rubin, Y.; Tobe, Y. *Chem. Soc. Rev.* **1999**, *28*, 107–119. (f) Haley, M. M.; Pak, J. J.; Brand, S. C. *Top. Curr. Chem.* **1999**, *201*, 81–130. (g) Moore, J. S. *Acc. Chem. Res.* **1997**, *30*, 402–413. (h) Young, J. K.; Moore, J. S. in Modern Acetylene Chemistry (Eds.: P. J. Stang, F. Diederich), VCH, Weinheim, **1995**, chap. 12

4. (a) Grave, C.; Schlüter, A. D. *Eur. J. Org. Chem.* **2002**, 3075–3098. (b) Höger, S. *J. Polym. Sci. Part A: Polym. Chem.* **1999**, *37*, 2685–2698. (c) Prautzsch, V.; Ibach, S.; Vögtle, F. J. *Incl. Phenom. Macrocycl. Chem.* **1999**, *33*, 427–457.

5. (a) Sonogashira, K.; Tohda, Y.; Hagihara, N. *Tetrahedron Lett.* **1975**, 4467–4470. (b) Sonogashira, K.; Yatake, T.; Tohda, Y.; Takahashi, S.; Hagihara, N. *J. Chem. Soc. Chem. Commun.* **1977**, 291–292.

6. (a) Glaser, C. *Ber. Dtsch. Chem. Ges.* **1869**, *2*, 422–424. (b) Eglinton, G. *J. Chem. Soc.* **1959**, 889–892. (c) Behr, O. M.; Eglinton, G.; Galbraith, A. R.; Raphael, R. A. *J. Chem. Soc.* **1960**, 3614–3625.

7. (a) Höger, S.; Meckenstock, A. D.; Pellen, H. *J. Org. Chem.* **1997**, *62*, 4556–4557. (b) Tobe, Y.; Fuji, T.; Matsumoto, H.; Tsumuraya, K.; Noguchi, D.; Nakagawa, N.; Sonoda, M.; Naemura, K.; Achiba, Y.; Wakabayashi, T. *J. Am. Chem. Soc.* **2000**, *122*, 1762–1775. (c) Maruyama, S.; Hokari, H.; Wada, T.; Sasabe, H. *Synthesis*

2001, 1794–1799. (d) Tobe, Y.; Utsumi, N.; Nagano, A.; Sonoda, M.; Naemura, K. *Tetrahedron* **2001**, *57*, 8075–8083. (e) Ohkita, M.; Ando, K.; Tsuji, T. *Chem. Commun.* **2001**, 2570–2571. (f) Krömer, J.; Rios-Carreras, I.; Fuhrmann, G.; Musch, C.; Wunderlin, M.; Debaerdemaeker, T.; Mena-Osteritz, E.; Bäuerle, P. *Angew. Chem. Int. Ed.* **2000**, *39*, 3481–3486. (g) Fuhrmann, G.; Krömer, J.; Bäuerle, P. *Synth. Met.* **2001**, *119*, 125–126. (h) Li, J.; Ambroise, A.; Yang, S. I.; Diers, J. R.; Seth, J.; Wack, C. R.; Bocian, D. F.; Holten, D.; Lindsey, J. S. *J. Am. Chem. Soc.* **1999**, *121*, 8927–8940.

8. (a) Rowan, S. J.; Sanders, J. K. M. *J. Org. Chem.* **1998**, *63*, 1536–1546. (b) Tam-Chang, S.-W.; Stehouwer, J. S.; Hao, J. *J. Org. Chem.* **1999**, *64*, 334–335. (c) Hamilton, D. G.; Montalti, M.; Prodi, L.; Fontani, M.; Zanello, P.; Sanders, J. K. M. *Chem. Eur. J.* **2000**, *6*, 608–617.

9. Rowan, S. J.; Cantrill, S. J.; Cousins, G. R. L.; Sanders, J. K. M.; Stoddart, J. F. *Angew. Chem. Int. Ed.* **2002**, *41*, 898–952.

10. (a) Weiss, K.; Michel, A.; Auth, E. M.; Bunz, U. H. F.; Mangel, T.; Müllen, K. *Angew. Chem. Int. Ed.* **1997**, *36*, 506–509. (b) Fürstner, A.; Guth, O.; Rumbo, A.; Seidel, G. *J. Am. Chem. Soc.* **1999**, *121*, 11108–11113.

11. (a) Kloppenburg, L.; Jones, D.; Bunz, U. H. F. *Macromolecules* **1999**, *32*, 4194–4203. (b) Ge, P.-H.; Fu, W.; Herrmann, W. A.; Herdtweck, E.; Campana, C.; Adams, R. D.; Bunz, U. H. F. *Angew. Chem. Int. Ed.* **2000**, *39*, 3607–3610.

12. (a) Zhang, W.; Kraft, S.; Moore, J. S. *Chem. Commun.* **2003**, 832–833. (b) Zhang, W.; Kraft, S.; Moore, J. S. *J. Am. Chem. Soc.* **2004**, *126*, 329–335.

13. Zhang, W.; Moore, J. S. *J. Am. Chem. Soc.* **2005**, *127*, 11863–11870.

14. (a) Sadighi, J. P.; Singer, R. A.; Buchwald, S. L. *J. Am. Chem. Soc.* **1998**, *120*, 4960–4976; (b) Rawal, V. H.; Jones, R. J.; Cava, M. P. *J. Org. Chem.* **1987**, *52*, 19–28.

15. Coropceanu, V.; Gruhn, N. E.; Barlow, S.; Lambert, C.; Durivage, J. C.; Bill, T. G.; Nöll, G.; Marder, S. R.; Brédas, J.-L. *J. Am. Chem. Soc.* **2004**, *126*, 2727–2731.

16. The characterization data for macrocycle **10c** were as follows: ^{1}H NMR ($CDCl_3$, 500 MHz) δ: 0.87 (t, J = 6.6 Hz, 12 H, $OC(CH_3)_2CH_2(CH_2)_{10}CH_3$], 1.23–1.54 (m, 80 H, $OC(CH_3)_2CH_2(CH_2)_{10}CH_3$], 1.80 [s, 24 H, $OC(CH_3)_2CH_2(CH_2)_{10}CH_3$], 2.11 [m, 8 H, $OC(CH_3)_2CH_2(CH_2)_{10}CH_3$], 7.44 (d, J = 8.6 Hz, 8 H), 8.00 (s, 8 H), 8.10 (d, J = 8.6 Hz, 8H); ^{13}C NMR ($CDCl_3$, 125 MHz) δ: 14.1, 22.7, 24.2, 26.4, 29.4, 29.5, 29.6, 29.7 (br, signal overlap), 30.0, 31.9, 40.8, 87.0 [$OC(CH_3)_3$], 89.1 [Ar*CC*Ar], 116.0, 118.2, 122.4, 125.1, 130.7, 138.0, 150.5 (NCOO); MS (FD): m/z (%): 1773.3 (21), 887.0 (100), 525.3 (66); Anal. Calcd for $C_{120}H_{148}N_4O_8$ (1773.1): C, 81.22; H, 8.41; N, 3.16; Found: C, 80.93; H, 8.05; N, 3.24; GPC 2030 (M_n), 1.02 (M_w/M_n).

2-SUBSTITUTED-1,3-CYCLOHEXADIENES BY INTERMOLECULAR, METHYLENE-FREE TANDEM ENYNE METATHESIS

[2-(Benzoyloxymethyl)-1,3-cyclohexadiene]

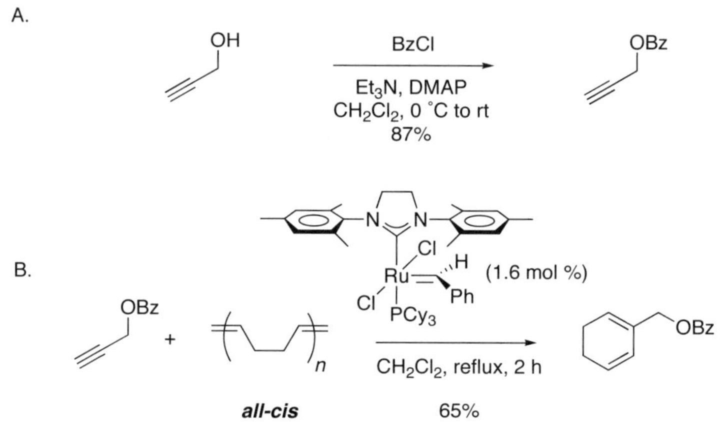

Submitted by Amol A. Kulkarni and Steven T. Diver.[1]
Checked by Masakatsu Shibasaki and Keisuke Maki.

1. Procedure

A. Propargyl benzoate. Into an oven-dried, argon-cooled 1-L round-bottomed flask equipped with magnetic stirbar and rubber septum is added CH_2Cl_2 (250 mL) (Note 1), which is maintained under an atmosphere of argon (Note 2). Propargyl alcohol (13.6 mL, 240 mmol, 1.2 equiv) and triethylamine (Note 3) (42 mL, 300 mmol, 1.5 equiv) are added by gas-tight syringe, followed by the addition of DMAP (Note 4) (2.4 g, 20 mmol, 0.1 equiv). The reaction mixture is subsequently cooled in an ice bath for 30 min. Benzoyl chloride (Note 5) (23.2 mL, 200 mmol, 1.0 equiv) is then added slowly by syringe over 15 min. The reaction mixture is stirred in the ice bath for an additional 30 min and then warmed to room temperature and stirred for 4 h. The reaction is quenched by addition of ice cold water (300 mL). The contents are transferred into a 1-L separatory funnel. The organic layer is washed with 1 M HCl (2 x 300 mL), followed by distilled water (300 mL) and brine (300 mL). The organic layer is separated, dried (Na_2SO_4) and concentrated under vacuum (rotary evaporator, 8–9 mmHg, 24 °C) (Note 6). The crude yellow oil is purified by bulb-to-bulb distillation (80

°C pot temperature) under reduced pressure (1–2 mmHg) to yield the desired ester (28.4–28.9 g, 89–90% yield) as a colorless oil (Note 7).

B. 2-(Benzoyloxymethyl)-1,3-cyclohexadiene. Into an oven-dried, argon-cooled 100-mL Schlenk tube (Note 8) equipped with a magnetic stirbar, rubber septum and a cold finger condenser (Note 9) under an atmosphere of argon (Note 2) is added all-*cis* polybutadiene (8.66 g, 160 mmol, 8 equiv) (Note 10) followed by 40 mL CH_2Cl_2 (Note 1). To the Schlenk tube is added Grubbs second generation catalyst (Note 11) (271 mg, 0.32 mmol, 1.6 mol %) and the flask is then immersed in an oil bath maintained at 45 °C. Propargyl benzoate (3.2 g, 20 mmole) is dissolved in about 1 mL of CH_2Cl_2 and this solution is added to the suspension of polybutadiene in CH_2Cl_2 over a period of 2 h by means of syringe pump addition with a gas-tight syringe connected to a short section of teflon tubing (Note 12). The reaction flask is maintained at 45 °C during the addition. After the addition is complete, the syringe is rinsed with about 1 mL of CH_2Cl_2 and the contents are added to the reaction mixture over about 10 min. The reaction mixture is stirred at 45 °C for another 30 min. At this stage, ethyl vinyl ether (Note 13) (612 μL, 0.64 mmol, 2 equiv based on ruthenium catalyst) is added and heating is continued for an additional 30 min. The reaction mixture is concentrated under vacuum (rotary evaporator, 8–9 mmHg, 23–25 °C) to give a brown oil which is then partially purified by passing through a plug of silica gel (50 g; 2 x 10 cm) (Note 14) eluting with 0.8 % ethyl acetate in hexanes. The first 250 mL of the eluent is discarded (Note 15). Then 700 mL of 4% ethyl acetate in hexanes is used to elute the desired cyclohexadiene along with other oligomeric materials. The eluate is concentrated under vacuum (rotary evaporator, 10 mmHg, 25–40 °C). The viscous yellow oil thus obtained is further purified by means of bulb-to-bulb distillation (checker; 120–130 °C pot temperature/submitter; 124-128 °C) at reduced pressure (1–2 mmHg) to afford the desired cyclohexadiene derivative as a colorless oil (2.64–2.69 g, 62–63 % yield) (Notes 16,17,18). The ^1H and ^{13}C NMR spectra of the isolated compound match previously reported characterization data.[2]

2. Notes

1. The submitters used argon-sparged dichloromethane (CH_2Cl_2, Mallinckrodt, 99%) drawn from a solvent purifier (two 0.05 m x 1 m stainless steel columns packed with activated alumina) maintained under argon immediately prior to use. The checkers used anhydrous

dichloromethane, which was used as received from Kanto Chemical Co. No degassing of the solvent was performed.

2. Argon was supplied statically through a gas manifold with a mineral oil bubbler to generate slight back-pressure.

3. The submitters used triethylamine (Acros, 99%) distilled from CaH_2. The checkers used triethylamine (Aldrich, 99.5%) as received.

4. DMAP (99%) was used as received from Aldrich Chemical Co.

5. The submitters used benzoyl chloride (Baker, 99.8%) as received. The checkers used benzoyl chloride (Aldrich, 99%) as received.

6. The submitters used a diaphragm-type vacuum pump equipped with a primary dry ice-acetone cooled condensor (V-shaped cold finger-type) and a secondary in-line vacuum trap (dry ice-acetone cooling) to collect solvent. The checkers used a diaphragm-type vacuum pump equipped with –20 °C cooling circulator.

7. Further purification was not necessary; material obtained by bulb-to-bulb distillation was 99.5% pure (GC), and was used in the subsequent step. GC retention time of 11.90 min (capillary GC, HP-5 column, 0.25 μm x 30 m; temperature program: 40 °C for 4 min, then increased 25 °C/min to 280 °C. To obtain an accurate boiling point, the ester was distilled at 80–81 °C (1.1 mmHg) through a short path column. Analytical data is reported for the distillate obtained from the bulb-to-bulb distillation: Analytical TLC: R_f 0.36 (10 % ethyl acetate in hexanes); ^{1}H NMR (500 MHz, $CDCl_3$) δ: 2.53 (t, J = 2.4 Hz, 1 H), 4.92 (d, J = 2.4 Hz, 2 H), 7.46 (t, J = 7.9 Hz, 2 H), 7.55 (tt, J = 7.4, 1.2 Hz, 1 H), 8.06-8.08 (m, 2 H); ^{13}C NMR (125 MHz, $CDCl_3$) δ: 52.4, 75.0, 77.7, 128.4, 129.3, 129.7, 133.3, 165.7; FT-IR (thin film, cm^{-1}) 3296, 1725, 1268, 1108; HRMS (FAB; M^++H, *m/z*) calcd for $C_{10}H_9O_2$ 161.0602, found 161.0604.

8. The Schlenk tube dimensions were 3.7 cm x 18 cm with a sidearm of 9 mm OD. The septum was fitted over the 9 mm tube with argon supplied through an 18-gauge needle. The teflon cannula is threaded through the same septum. Use of a round-bottomed flask in place of the Schlenk tube resulted in incomplete conversion.

9. The submitters reported the use of a cold-finger condenser in the Schlenk tube. The checkers used a standard reflux condenser and obtained identical results.

10. All-*cis* polybutadiene (M_w *ca.* 2–3 x 10^6, 98% *cis*) was purchased from Aldrich (18137-4) and used without further purification. The polymer was kept in a polyethylene bag and no special precautions were taken to exclude air. The use of fewer equivalents of polybutadiene led to incomplete

conversions. For example, 4.0 equiv of polybutadiene delivered under the same reaction conditions described in this procedure produced only 64% yield of cyclohexadiene (NMR yield against mesitylene internal standard).

11. Ruthenium [1,3-bis-(2,4,6-trimethylphenyl)-2-imidazolidinylidine]-dichloro(phenylmethylene)(tricyclohexylphosphine) (Grubbs second generation catalyst) was obtained from Aldrich Chemical Co. (Milwaukee, WI) or from Materia, Inc. (Pasadena, CA).

12. A Teflon cannula was used (ID = 0.8 mm and 0.4 mm thickness of about 20 cm length). One end of the cannula was attached to the needle by slipping over the 22 gauge stainless steel needle until snug; the other end was cut with a razor blade at a 45° angle and pierced through the rubber septum. The tip of the cannula was touched to the cooled glass surface just below the level of solvent reflux.

13. Ethyl vinyl ether (99%) was used as received from Aldrich Chemical Co.

14. The filtration was carried out using Merck silica gel 60 (230–400 mesh). The viscous crude mixture required high pressure to elute the solvent. The use of a smaller column (<2 cm) was not suitable.

15. This fraction contains oligomeric byproducts including isomeric 1,5,9-cyclododecatrienes. This fraction should be checked for cyclohexadiene product by TLC analysis (R_f 0.42; 10% ethyl acetate in hexanes) before discarding.

16. The submitters reported obtaining a pale yellow oil. The cyclohexadiene is not stable as the neat liquid, even when stored at –20 °C. Partial decomposition by polymerization was variable, and depended significantly on sample storage conditions. When BHT (5 % by weight) was added to the distillate, polymerization was retarded somewhat. Extended storage of the neat cyclohexadiene is not recommended. Pure cyclohexadiene could be obtained from degraded samples by redistillation with loss of material.

17. Further purification was not necessary. In order to obtain an accurate boiling point, the dienylester was distilled at 124–125 °C (1.1 mmHg) through a short path column.

18. The analytical data are provided for the distillate from the bulb-to-bulb distillation: Analytical TLC: R_f 0.42 (10 % ethyl acetate in hexanes). ^{1}H-NMR (500 MHz, $CDCl_3$) δ: 2.15–2.26 (m, 4 H), 4.80 (m, 2 H), 5.87–5.93 (m, 2 H), 5.99 (ddd, J = 9.5, 3.4, 1.5 Hz, 1 H), 7.44 (t, J = 7.6 Hz, 2 H), 7.55 (tt, J = 7.4, 1.2 Hz, 1 H), 8.05–8.08 (m, 2 H); ^{13}C-NMR (125 MHz, CDCl) δ: 22.0, 22.1, 67.0, 124.3, 124.9, 127.6, 128.3, 129.6, 130.3, 131.1,

132.9, 166.5; FT-IR (thin film, cm^{-1}) 2935, 1719, 1450, 1271, 1111, 712; HRMS (FAB; M^+, *m/z*) calcd for $C_{14}H_{15}O_2$ 215.1072, found 215.1069. GC was used to assess purity of the cyclohexadiene, which was found to be >98.5%. GC retention time of 15.07 min (capilliary GC, HP-5 column, 0.25 μm x 30 m; temperature program 40 °C for 5 minutes, then increased 25 °C/min to 280 °C).

Safety and Waste Disposal Information

All hazardous materials should be handled and disposed of in accordance with "Prudent Practices in the Laboratory"; National Academy Press; Washington, DC, 1995.

3. Discussion

In recent years, alkene metathesis has emerged as a powerful method for the synthesis of the carbon-carbon double bond. Synthetic applications increased dramatically with the advent of the functional group tolerant ruthenium carbenes developed by Grubbs.[3-11] Enyne metathesis is a closely related catalytic reaction which involves the coupling of an alkene with an alkyne to give a conjugated 1,3-diene.[12] This is a particularly attractive reaction because the conjugated diene product participates in a host of transformations and can thus be used as a building block for the synthesis of more complex organic molecules.

Recent metathesis applications involving tandem transformations are exciting due to their potential to build rings from simple acyclic precursors. These applications are very attractive because they permit the rapid assembly of complex structures. Tandem metathesis transformations had previously been limited to intramolecular reactions.[13,14] Though attractive, these tandem reactions require the synthesis of polyenyne reactants. Ring synthesis from unsaturated reactants by an *inter*molecular reaction is advantageous because the reactants are comparatively simple. The difficulty in intermolecular ring building reactions is the initial, non-stereoselective metathesis between the alkene and alkyne component. Control of alkene geometry is a current, general problem in intermolecular metathesis research. The tandem diene-alkyne metathesis described in this procedure is challenging because it is triggered by an intermolecular bond formation which lacks stereocontrol.

The challenge posed by nonstereoselective cross metathesis was addressed by development of "methylene-free" conditions. In general, the low stereoselectivity in cross enyne metathesis of alkenes and alkynes limits its utility in synthesis.[15] Low selectivity was also evident in our initial work on the tandem dienyne metathesis.[16] In the case of 1,3-cyclohexadiene synthesis, only the *Z*-alkene would lead to the ring-closed product. This problem led us to consider interconversion of vinyl carbene intermediates to produce more of the *Z*-intermediate, which would then give ring-closing metathesis. This process would produce more of the desired cyclohexadiene product. Our analysis led us to consider conditions that would favor interconversion of vinyl carbenes by omitting CH_2 sources. These conditions were expected to decelerate the rate of catalyst turnover, giving the isomeric vinyl carbene intermediates time to interconvert. Since there are no CH_2 sources in the reaction and no $L_nRu{=}CH_2$ intermediates, we termed these conditions to be "methylene-free."[2] The term is used to differentiate these conditions from typical cross metatheses, which had previously required the use of 1-alkene reactants. (However, the catalytic enyne metathesis mechanism is complex,[17-19] and the hypothesis that methylene-free conditions permit the equilibration of vinyl carbene intermediates requires further experimentation.)

The procedure described here illustrates that the methylene-free conditions provide an effective ring synthesis of 1,3-cyclohexadienes. In the best cases, the cyclohexadienes are produced in greater than 80% yield (as assessed by NMR). The isolation procedure requires quenching of carbene reactivity and separation of oligomeric materials derived from polybutadiene. In this procedure, carbene reactivity is quenched upon addition of ethyl vinyl ether (EVE). However, our group has recently developed an improved protocol for the quenching and removal of ruthenium carbene, which can also be used in the methylene-free synthesis of 1,3-cyclohexadienes (see eq 8, Scheme 3 in ref. 20).[20] In terms of alkene reactant, the reported procedure employs polybutadiene (PB) as the alkene reactant, yet similar results are obtained using 1,5-cyclooctadiene (1,5-COD) as the alkene. In the 20-mmole scale procedure above using 4 equiv of 1,5-COD (8 alkene equivalents), 61% yield of the corresponding 1,3-cyclohexadiene was obtained (in comparison to 65% with 8 equiv PB). However, use of fewer equivalents of polybutadiene (4 equiv) led to incomplete conversion of alkyne reactant (64% conversion of propargyl benzoate as assessed by ^{1}H NMR comparison to the mesitylene internal standard).

Table 1 shows results using 4 equiv of 1,5-COD in place of polybutadiene on a 1-mmole scale. On the smaller scale, catalyst loading was not optimized lower than 5 mole %, though it is anticipated that catalyst loading can be decreased during scale-up.[21] We have found that either 8 equiv PB or 4 equiv 1,5-COD can be used with equal efficiency.

Table 1. Ring Synthesis by Methylene-Free Enyne Metathesis[a,b]

entry	alkyne	X	Y	yield(NMR)	yield
	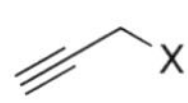			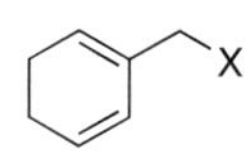	
1		OBz	-	83	71
2		OTBDPS	-	62	51
3		OCH_2Ph	-	69	59
	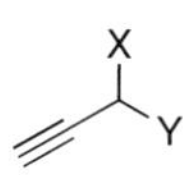			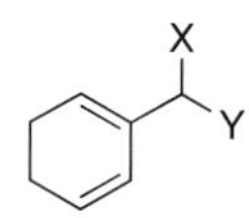	
4		CH_3	OBz	88	74
5		CH_2Ph	OAc	80	66
	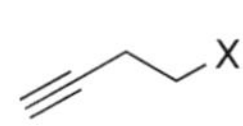			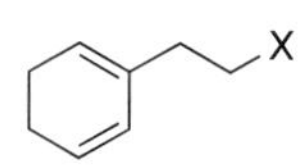	
6[c]		OC(O)β-Nap	-	50	41
7		N(Ts)Boc	-	77	62
8		CO_2Et	-	72	68
	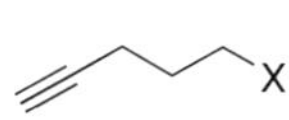			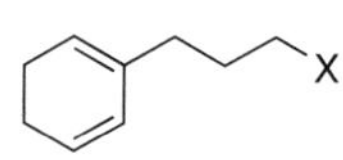	
9[c,d]		ONap	-	49	42
10		Cl	-	62	48

(a) All reactions were conducted on a 1 mmole scale and used the Grubbs second generation catalyst. (b) A solution of alkyne (1 equiv) and 1,5-COD (2 equiv) was added to a solution of ruthenium catalyst (5 mol %) and 1,5-COD (2 equiv) in refluxing CH_2Cl_2 over a period of 4 h. (c) Incomplete conversion. (d) 10 mol % ruthenium catalyst was used.

A variety of terminal alkynes participated efficiently in this reaction. Esters at the propargylic and homopropargylic position were well tolerated (entries 1, 4, 5, 6). Protected propargyl ethers gave reasonable isolated yields (entries 2, 3). Nitrogen-containing homopropargylic functionality was tolerated if suitably protected (entry 7), and a remote ester posed no difficulty (entry 8). The more remote naphthoate ester required higher catalyst loading and still proceeded to only partial conversion (entry 9). This is presumably due to a coordination problem in the intermediate vinyl carbenes, which may both slow the reaction and result in catalyst decomposition. Coordinating ethers in the homopropargylic position were similarly not well tolerated under the conditions reported in this procedure; however, conditions using higher concentrations of 1,5-COD have been successful with these substrates.[2b]. Since internal alkenes coordinate more weakly to metals than 1-alkenes, the putative slow step, vinyl carbene reaction with alkene, is slower than a typical enyne metathesis involving 1-alkenes. As a result, the reaction proves more sensitive to potentially coordinating functional groups than typical ring-closing alkene or enyne metathesis reactions. In these unusual cases, higher catalyst loading and higher alkene concentrations are often necessary. With higher concentration of alkene (polybutadiene or 1,5-COD), internal alkynes have been successfully used in the 1,3-cyclohexadiene ring synthesis.[2b] Methylene-free conditions have also been used for the ring expansion of cyclopentene to give 1,3-cycloheptadienes.[22]

In conclusion, a practical synthesis of 1,3-cyclohexadienes is possible from simple, readily available unsaturated reactants. The optimal 20-mmole scale procedure employs the second generation Grubbs carbene as a functional group tolerant catalyst. A moderately low catalyst loading was achieved despite the demands of a slow intermolecular reaction. The key to this method is the use of methylene-free conditions, achieved with either polybutadiene or 1,5-cyclooctadiene as the alkene.

1. Department of Chemistry, 572 Natural Sciences Complex, University at Buffalo, The State University of New York, Amherst, NY 14260-3000, (716) 645-6800 x2201, email: diver@buffalo.edu.

2. (a) Kulkarni, A. A.; Diver, S. T., *J. Am. Chem. Soc.* **2004**, *126*, 8110-8111. (b) Peppers, B. P.; Kulkarni, A. A.; Diver, S. T. *Org. Lett.* **2006**, *8*, 2539-2542.
3. Sanford, M. S.; Love, J. A.; Grubbs, R. H., *J. Am. Chem. Soc.* **2001**, *123*, 6543-6554.
4. Sanford, M. S.; Ulman, M.; Grubbs, R. H., *J. Am. Chem. Soc.* **2001**, *123*, 749-750.
5. Scholl, M.; Ding, S.; Lee, C. W.; Grubbs, R. H., *Org. Lett.* **1999**, *1*, 953-956.
6. Schwab, P.; France, M. B.; Ziller, J. W.; Grubbs, R. H., *Angew. Chem. Int. Ed.* **1995**, *34*, 2039-2041.
7. Nguyen, S. T.; Grubbs, R. H.; Ziller, J. W., *J. Am. Chem. Soc.* **1993**, *115*, 9858-9859.
8. Trnka, T. M.; Grubbs, R. H., *Acc. Chem. Res.* **2001**, *34*, 18-29.

9. Grubbs, R. H., *Tetrahedron* **2004**, *60*, 7117-7140.

10. Love, J. A.; Sanford, M. S.; Day, M. W.; Grubbs, R. H., *J. Am. Chem. Soc.* **2003**, *125*, 10103-10109.

11. Chatterjee, A. K.; Choi, T.-L.; Sanders, D. P.; Grubbs, R. H., *J. Am. Chem. Soc.* **2003**, *125*, 11360-11370.

12. Diver, S. T.; Giessert, A. J., *Chem. Rev.* **2004**, *104*, 1317-1382.

13. Kim, S.-H.; Bowden, N.; Grubbs, R. H., *J. Am. Chem. Soc.* **1994**, *116*, 10801-10802.

14. Zuercher, W. J.; Scholl, M.; Grubbs, R. H., *J. Org. Chem.* **1998**, *63*, 4291-4298.

15. Stragies, R.; Schuster, M.; Blechert, S., *Angew. Chem. Int. Ed.* **1997**, *36*, 2518-2520.

16. Smulik, J. A.; Diver, S. T., *Tetrahedron Lett* **2001**, *42*, 171-174.

17. Galan, B. R.; Giessert, A. J.; Keister, J. B.; Diver, S. T., *J. Am. Chem. Soc.* **2005**, *127*, 5762-5763.

18. Giessert, A. J.; Diver, S. T., *Org. Lett.* **2005**, *7*, 351-354.

19. Giessert, A. J.; Diver, S. T., *J. Org. Chem.* **2005**, *70*, 1046-1049.

20. Galan, B. R.; Kalbarczyk, K. P.; Szczepankiewicz, S.; Keister, J. B.; Diver, S. T. *Org. Lett.* **2007**, *9*, 1203-1206.

21. Diver, S. T.; Giessert, A. J., *Synthesis* **2004**, 466-471.

22. Kulkarni, A. A.; Diver, S. T., *Org. Lett.* **2003**, *5*, 3463-3466.

SYNTHESIS OF 1,3-DIENES FROM ALKYNES AND ETHYLENE: ACETIC ACID 2-METHYLENE-3-PHENETHYLBUT-3-ENYL ESTER

[Benzenepentanol, β,γ-bis(methylene)-, acetate]

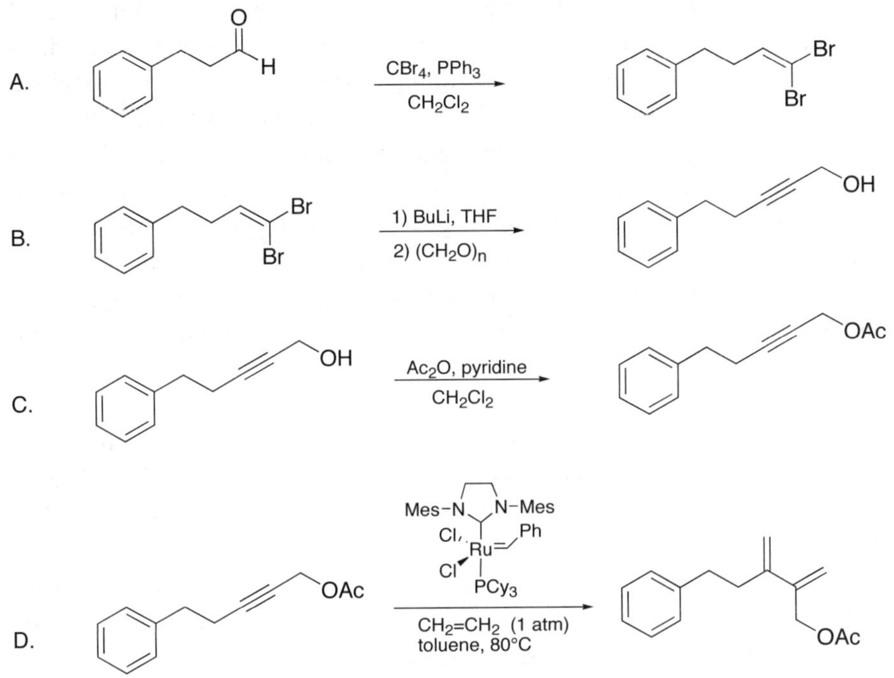

Submitted by Miwako Mori, Keisuke Tonogaki, and Atsushi Kinoshita.[1]
Checked by Scott E. Denmark and Tetsuya Kobayashi.

1. Procedure

A. (4,4-Dibromobut-3-enyl)benzene. A flame-dried, 2-L, three-necked flask, equipped with a magnetic stirbar, a rubber septum, an addition funnel fitted with a rubber septum, and an argon inlet is charged with 82.46 g (248.64 mmol) of carbon tetrabromide (Note 1) and then 400 mL of anhydrous dichloromethane (CH_2Cl_2, Note 2) is added by syringe. The solution is cooled in an ice-water bath and a solution of 130.43 g (497.28 mmol) of triphenylphosphine (Note 3) in 400 mL of anhydrous CH_2Cl_2 is then added dropwise via the addition funnel over 30 min. The reaction mixture is stirred at 0 °C for 10 min, and then a solution of 16.4 mL (16.7 g, 124.3 mmol) of 3-phenylpropionaldehyde (Note 4) in 266 mL of anhydrous

CH_2Cl_2 is added over 10 min via cannula. The solution is stirred at 0 °C for 1 h, and then 500 mL of H_2O is added. The resulting mixture is transferred to a 4-L separatory funnel and the aqueous layer is separated and extracted with three 200-mL portions of CH_2Cl_2. The combined organic layers are washed with 600 mL of brine, dried over Na_2SO_4, filtered, and concentrated under reduced pressure using a rotary evaporator. To the residue is added 500 mL of diethyl ether, and the resulting suspension is filtered to remove triphenylphosphine oxide. The solid that is collected is washed with three 300-mL portions of diethyl ether, and the combined ethereal filtrate is concentrated under reduced pressure using a rotary evaporator. The residue is purified by flash column chromatography on 500 g of silica gel (elution with hexane/ether, 10:1). The fractions containing the product are collected and concentrated under reduced pressure using a rotary evaporator to yield 34.75 g (96%) of (4,4-dibromobut-3-enyl)benzene as a colorless oil (Note 5).

B. 5-Phenylpent-2-yn-1-ol. A flame-dried, two-necked, 1-L, round-bottomed flask equipped with an argon inlet, rubber septum, and a magnetic stirbar is charged with (4,4-dibromobut-3-enyl)benzene (30.00 g, 103.45 mmol) and then anhydrous tetrahydrofuran (345 mL) (Note 6) is added to the flask by syringe. The reaction mixture is cooled in a –78 °C dry ice-acetone bath while a solution of *n*-butyllithium in hexane (1.6M, 148 mL, 237 mmol) (Note 7) is added via syringe over 30 min. The solution is stirred at –78 °C for 30 min, and then paraformaldehyde (9.32 g, 310.35 mmol) (Note 8) is added in one portion. The solution is stirred at –78 °C for 30 min, and then allowed to warm to room temperature and stirred for 1 h. Saturated aqueous NH_4Cl (400 mL) is added and the aqueous layer is separated and extracted with three 200-mL portions of ethyl acetate. The combined organic layers are washed with 400 mL of brine, dried over Na_2SO_4, filtered, and concentrated under reduced pressure using a rotary evaporator. The residue is purified by flash column chromatography on 500 g of silica gel (elution with hexane/ethyl acetate, from 7:1 to 1:1). The fractions containing the product are collected and concentrated under reduced pressure using a rotary evaporator to yield 14.32 g (86%) of 5-phenylpent-2-yn-1-ol (Note 9) as a colorless oil.

C. Acetic acid 5-phenylpent-2-ynyl ester. A flame-dried, 500-mL, three-necked, round-bottomed flask equipped with an argon inlet, two rubber septa, and a magnetic stirbar is charged with 5-phenylpent-2-yn-1-ol (13.51 g, 84.33 mmol) and then 280 mL of anhydrous CH_2Cl_2 is added by syringe. The solution is cooled in an ice-water bath and pyridine (68 mL, 843 mmol)

(Note 10) is added in one portion by syringe, followed by the addition of acetic anhydride (32 mL, 337 mmol) (Note 11) over 10 min. The reaction mixture is stirred at room temperature overnight. The mixture is cooled to 0 °C, 10.3 mL of MeOH is added, and the resulting solution is stirred at 0 °C for 10 min. 4N Aqueous HCl (300 mL) is added, and the aqueous layer is separated and extracted with three 200-mL portions of ethyl acetate. The combined organic layers are washed with 400 mL of saturated aqueous $NaHCO_3$ solution, 400 mL of brine, dried over Na_2SO_4, filtered, and concentrated under reduced pressure using a rotary evaporator. The residue is purified by flash column chromatography on 500 g of silica gel (elution with hexane/ethyl acetate, from 7:1 to 3:1). The fractions containing the product are collected and concentrated under reduced pressure using a rotary evaporator to yield 16.77 g (98%) of acetic acid 5-phenylpent-2-ynyl ester (Note 12) as a colorless oil.

D. Acetic acid 2-methylene-3-phenethylbut-3-enyl ester.[3] A flame-dried, 250-mL, two-necked flask equipped with a magnetic stirbar, a rubber septum, and a reflux condenser capped with a three-way stopcock (used as the ethylene inlet adapter) is charged with acetic acid 5-phenylpent-2-ynyl ester (5.16 g, 25.52 mmol) and then 51 mL of anhydrous toluene (Note 13) is added to the flask by syringe. A solution of the second-generation ruthenium carbene complex[4e] (0.217 g, 0.26 mmol, 1 mol %, Note 14) in 5 mL of toluene is then added, and the resulting solution is degassed by repeated freeze-pump-thaw cycles. A 1-gallon (3.8 L) gasbag (Note 15) filled with purified ethylene gas (Notes 16, 17) is connected to the ethylene inlet adapter and the solution is stirred at 80 °C for 2 h under an atmosphere of ethylene gas. After allowing the solution to cool, 6 mL of ethyl vinyl ether is added (Note 18). The solution is concentrated under reduced pressure using a rotary evaporator, and the residue is purified by flash column chromatography on 250 g of silica gel (elution with hexane/ethyl acetate, 7:1). The fractions containing the product are collected and concentrated under reduced pressure using a rotary evaporator to yield 5.68 g (97%) of acetic acid 2-methylene-3-phenethylbut-3-enyl ester (Note 19) as a light yellow oil. Colorless material can be obtained by subjecting this product to Kugelrohr distillation to afford 5.55 g (94%) of the product as a colorless oil (Note 20).

2. Notes

1. Carbon tetrabromide (99%) was purchased from the Aldrich Chemical Company Inc. and was used without further purification.

2. Dichloromethane (unstabilized HPLC grade) was purchased from Fisher Scientific and was dried by percolation through two columns packed with neutral alumina under a positive pressure of argon.

3. Triphenylphosphine (99%) was purchased from the Aldrich Chemical Company Inc. and was used without further purification.

4. 3-Phenylpropionaldehyde (>90%) was purchased from Fluka and was purified by distillation.

5. The product displayed the following properties: R_f 0.73 (10:1 hexane-ether); IR ($CHCl_3$): cm^{-1} 3087, 3066, 3030, 3015, 2928, 2861, 1604, 1496, 1454, 1083, 1030, 907; ^1H NMR (500 MHz, $CDCl_3$): δ 2.42 (q, *J* = 7.7 Hz, 2 H), 2.74 (t, *J* = 7.7 Hz, 2 H), 6.42 (t, *J* = 7.7 Hz, 2 H), 7.19 (br d, *J* = 7.6 Hz, 2 H), 7.22 (br t, *J* = 7.6 Hz, 1 H), 7.31 (br t, *J* = 7.6 Hz, 2 H); ^{13}C NMR (125.7 MHz, $CDCl_3$): δ 33.8, 34.6, 89.5, 126.3, 128.4 (2C), 128.5 (2C), 137.6, 140.6. LRMS EI (relative intensity) *m/z*: 292 (M^++2, 1.0), 290 (M^+, 1.9), 288 (M^+-2, 1.2), 227 (0.9), 225 (1.0), 221 (49), 209 (52), 129 (23), 128 (20), 92 (43), 91 (100), 77 (13). Anal. Calcd for $C_{10}H_{10}$ Br_2: C; 41.42, H; 3.48. Found: C; 41.33, H; 3.40. The spectroscopic data were identical with literature values for material previously prepared by this procedure.[2]

6. Dry HPLC grade tetrahydrofuran was purchased from Fisher and was dried by percolation through two columns packed with neutral alumina under a positive pressure of argon.

7. A hexane solution of *n*-butyllithium was purchased from the Aldrich Chemical Company Inc. and was titrated prior to use.

8. Paraformaldehyde was purchased from Fisher and was used without further purification.

9. 5-Phenylpent-2-yn-1-ol exhibits the following properties: R_f 0.25 (3:1 hexane-ethyl acetate); IR ($CHCl_3$): cm^{-1} 3610 (OH), 3088, 3066, 3029, 3013, 2932, 2874, 2360, 2223, 1604, 1496, 1455, 1430, 1385, 1342, 1233, 1133, 1078, 1006, 968; ^1H NMR (500 MHz, $CDCl_3$): δ 1.60 (t, *J* = 6.0 Hz, 1 H), 2.52 (tt, *J* = 7.5, 2.1 Hz, 2 H), 2.84 (t, *J* = 7.5 Hz, 2 H), 4.23 (dt, *J* = 6.0, 2.1 Hz, 2 H), 7.18-7.26 (m, 3 H), 7.31 (br t, *J* = 7.3 Hz, 2 H); ^{13}C NMR (125.7 MHz, $CDCl_3$): δ 20.9, 34.9, 51.3, 79.0, 85.7, 126.3, 128.38 (2C), 128.40 (2C), 140.5; LRMS EI (relative intensity) *m/z*: 160 (M^+, 1.2), 159 (5), 142 (32), 129 (15), 92 (10), 91 (100), 65 (14). Anal. Calcd for $C_{11}H_{12}O$: C, 82.46; H, 7.55. Found: C, 82.27; H, 7.68.

10. Pyridine (99.9%) was purchased from Fisher and was purified by distillation over CaH_2.

11. Acetic anhydride (99.7%) was purchased from Fisher and was purified by distillation.

12. Acetic acid 5-phenylpent-2-ynyl ester has the following properties: R_f 0.5 (3:1 hexane-ethyl acetate); IR ($CHCl_3$): cm^{-1} 3066, 3028, 2948, 2865, 2237 (alkyne), 1737 (CO), 1604, 1496, 1454, 1438, 1380, 1360, 1224, 1210, 1147, 1078, 1026, 967; ^1H NMR (500 MHz, $CDCl_3$): δ 2.09 (s, 3 H), 2.52 (tt, J = 7.7, 2.1 Hz, 2 H), 2.84 (t, J = 7.7 Hz, 2 H), 4.65 (t, J = 2.1 Hz, 2 H), 7.18-7.24 (m, 3 H), 7.30 (br t, J = 7.1 Hz, 2 H); ^{13}C NMR (125.7 MHz, $CDCl_3$): δ 20.8, 20.9, 34.8, 52.7, 74.7, 86.8, 126.4, 128.4 (4C), 140.4, 170.4; LRMS EI (relative intensity) *m/z*: 202 (M^+, 0.14), 201 (0.85), 160 (14), 142 (33), 141 (18), 92 (10), 91 (100), 65 (13). Anal. Calcd for $C_{13}H_{14}O_2$: C, 77.20; H, 6.98. Found: C, 77.37; H, 7.20.

13. Toluene (99.9%) is purchased from Fisher and dried by distillation under argon from sodium benzophenone ketyl.

14. Second-generation ruthenium-carbene complex, (tricyclohexyl-phosphine-[1,3-bis(2,4,6-trimethylphenyl)-4,5-dihydroimidazol- 2- ylidene] [benzylidene]ruthenium (IV) dichloride),[4e] was purchased from Strem Chemicals.

15. The ethylene gasbag was purchased from the Aldrich Chemical Company, Inc., catalog number Z 18,674-0.

16. Ethylene was purchased from Matheson (>99.5%) and was purified by bubbling it sequentially through an aqueous solution of CuCl (2 g of CuCl in 200 mL of saturated aqueous NH_4Cl), concentrated H_2SO_4, and then a KOH tube as shown in Figure 1. The reaction apparatus for the synthesis of the 1,3-diene is shown in Figure 2.

17. The checkers found that this purification of ethylene was not necessary. Without the purification, the checkers obtained the product quantitatively after purification by chromatography and in 97% yield after distillation. Ethyl vinyl ether is added to quench the reaction.[5]

Figure 1. Purification of Ethylene Gas

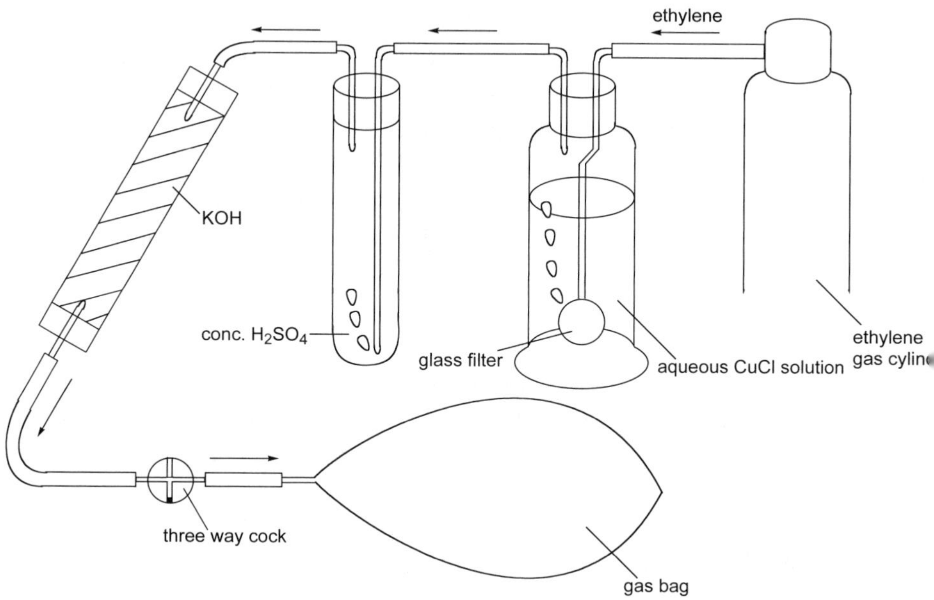

Figure 2. Reaction Apparatus for the Synthesis of 1,3-Diene

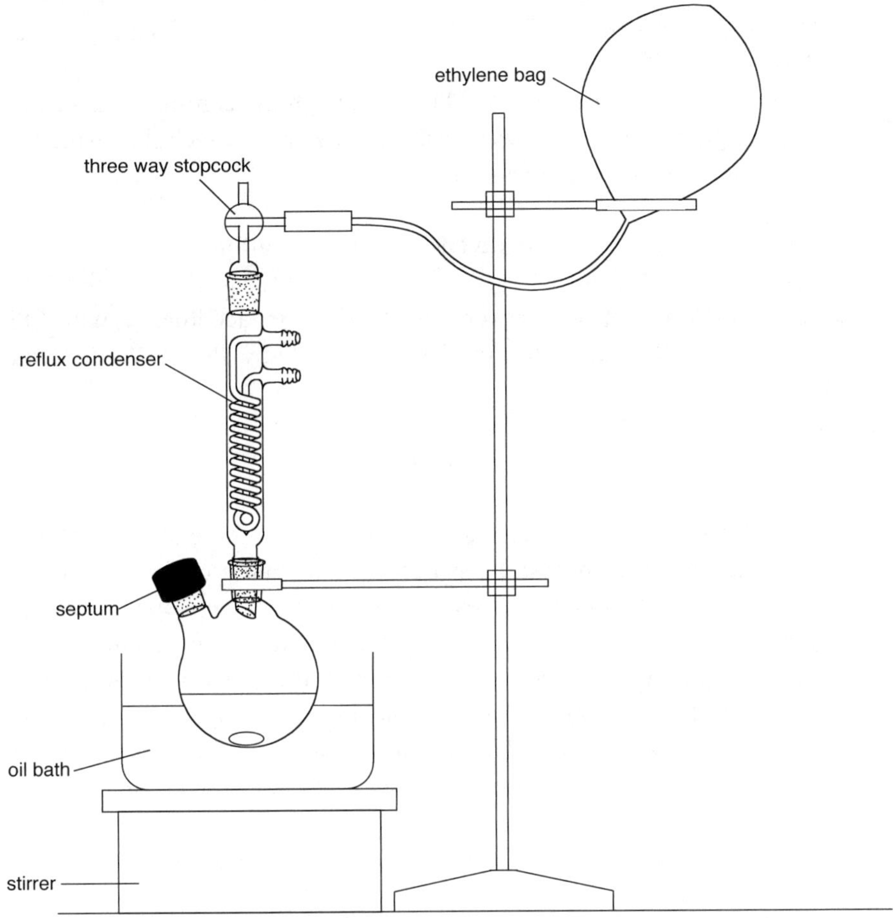

18. Ethyl vinyl ether was purchased from the Aldrich Chemical Company Inc. and was used without purification.

19. 2-Methylene-3-phenethylbut-3-enyl acetate has the following properties: R_f 0.55 (3:1 hexane-ethyl acetate); IR ($CHCl_3$): cm^{-1} 3690, 3087, 3065, 3020, 2945, 2865, 2360, 2342, 1736, 1635, 1601, 1496, 1455, 1370, 1222, 1046, 1030, 905; ^{1}H NMR (500 MHz, $CDCl_3$): δ 2.11 (s, 3 H), 2.58 (br t, J = 8.1 Hz, 2 H), 2.80 (br t, J = 8.1 Hz, 2 H), 4.80 (s, 2 H), 5.03 (br s, 1 H), 5.10 (s, 1 H), 5.31 (br s, 1 H), 5.37 (s, 1 H), 7.16-7.23 (m, 3 H), 7.29 (br t, J = 7.3 Hz, 2 H); ^{13}C NMR (125.7 MHz, $CDCl_3$): δ 21.0, 34.7, 35.9, 65.2, 113.1, 114.8, 125.9, 128.34 (2C), 128.38 (2C), 141.4, 141.9, 144.2, 170.8;

LRMS EI (relative intensity) *m/z*: 230 (M^+, 0.04), 170 (30), 155 (28), 142 (37), 141 (17), 130 (21), 129 (20), 128 (14), 104 (10), 92 (18), 91 (100), 65 (15). Anal. Calcd for $C_{13}H_{18}O_2$: C, 78.23; H, 7.88. Found: C, 78.26; H, 7.91.

20. The checkers found that although the chromatographed product was a light yellow oil, it was analytically pure. Kugelrohr distillation was required to obtain a colorless oil.

Waste Disposal Information

All toxic materials were disposed of in accordance with "Prudent Practices in the Laboratory"; National Academy Press; Washington, DC, 1995.

3. Discussion

This procedure describes the synthesis of 1,3-dienes from alkynes and ethylene (at atmospheric pressure) by ruthenium-catalyzed, cross-enyne metathesis.[3,6] Although this reaction was initially developed using the first-generation ruthenium-carbene complex,[7] the second-generation ruthenium-carbene complex[4] is more effective and the reaction proceeded in the presence of even 1 mol % of the ruthenium catalyst. A number of different functional groups tolerate this reaction. Various 1,3-dienes are obtained in high yields following the procedure described in the text (Table).[3]

1. Graduate School of Pharmaceutical Sciences, Hokkaido University, Sapporo 060-0812, Japan.

2. (a) Renaud, J.-L; Aubert, C.; Malacria, M. *Tetrahedron* **1999**, *55*, 5113. (b) Harada, T.; Katsuhira, T.; Hara, D.; Kotani, Y.; Maejima, K.; Kaji, R.; Oku, A. *J. Org. Chem.* **1993**, *58*, 4897.
3. Tonogaki, K.; Mori, M. *Tetrahedron Lett.* **2002**, *43*, 2235.
4. (a) Weskamp, T.; Schattenmann, W. C.; Spiegler, M.; Herrmann, W. A. *Angew. Chem. Int. Ed.* **1998**, *37*, 2490. (b) Huang, J.; Stevens, E. D.; Nolan, S. P.; Peterson, J. L. *J. Am. Chem. Soc.* **1999**, *121*, 2674. (c) Scholl, M.; Trnka, T. M.; Morgan, J. P.; Grubbs, R. H. *Tetrahedron Lett.* **1999**, *40*, 2247. (d) Jafarpour, L.; Nolan, S. P. *Organometallics* **2000**, *19*, 2055. (e) Scholl, M.; Ding, S.; Lee, C. W.; Grubbs, R. H. *Org. Lett.* **1999**, *1*, 953.
5. Louie, J.; Grubbs, R. H. *Organometallics* **2002**, *21*, 2153.
6. (a) Kinoshita, A.; Sakakibara, N.; Mori, M. *J. Am. Chem. Soc.* **1997**, *119*, 12388.

(b) Kinoshita, A.; Sakakibara, N.; Mori, M. *Tetrahedron* **1999**, *55*, 8155. (c) Mori, M.; Tonogaki, K.; Nishiguchi, N. *J. Org. Chem.* **2002**, *67*, 224.

7. Schwab, P.; France, M. B.; Ziller, J. W.; Grubbs, R. H. *Angew. Chem., Int. Ed. Engl.* **1995**, *34,* 2039.

TABLE
SYNTHESIS OF VARIOUS 1,3-DIENES[a]

run	alkyne	diene	time (hr)	yield of diene (%)[b]
1	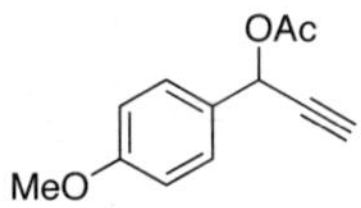	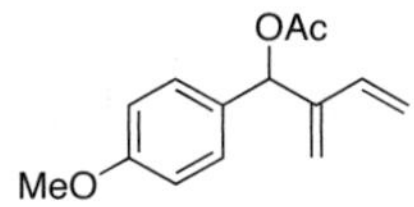	0.5	100
2	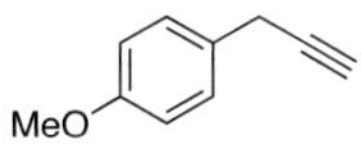	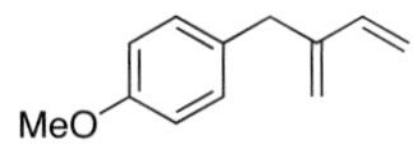	0.5	83
3	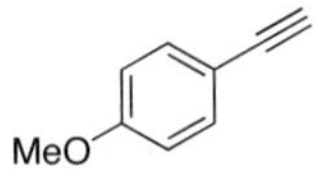	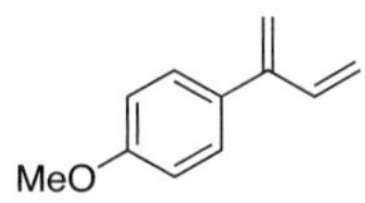	0.5	88
4	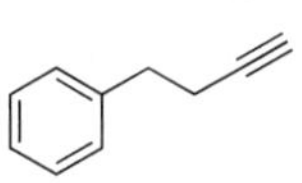	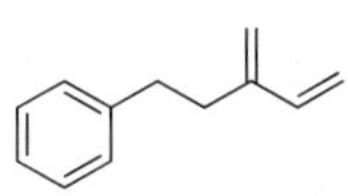	0.5	71
5	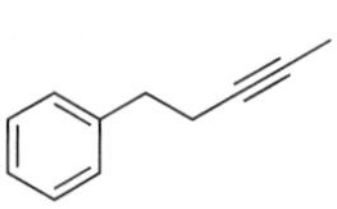	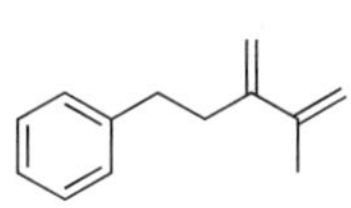	0.5	85
6[c]	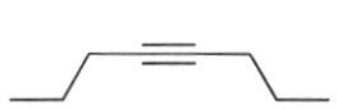	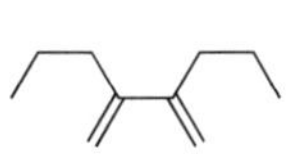	24	80
7	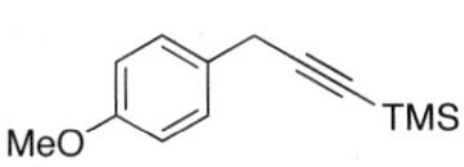	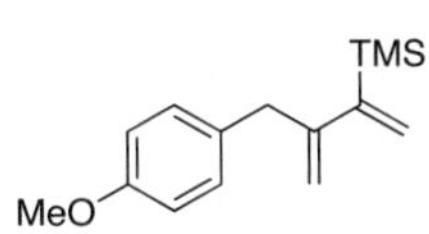	16	87 (10)
8[d]	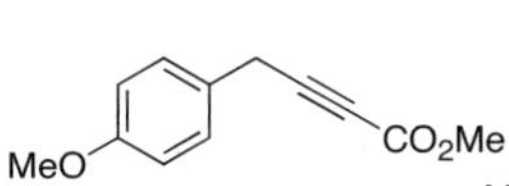	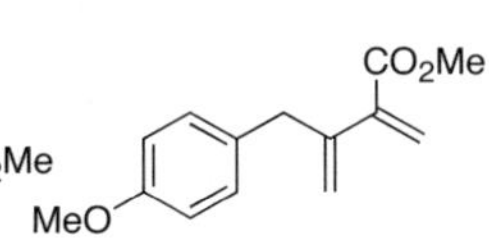	16	43 (34)

a) All reactions were carried out in toluene at 80°C using 5 mol % of ruthenium carbene complex under an atmosphere of ethylene gas by the representative procedure described in the text.
b) The number in parentheses shows the yield of alkyne recovered.
c) This reaction was carried out in CH_2Cl_2 under reflux condition.
d) This reaction was carried out using 10 mol % of ruthenium carbene complex.

RING-CLOSING METATHESIS SYNTHESIS OF *N*-Boc-3-PYRROLINE

[(1*H*-Pyrrole-1-carboxylic acid, 2,5-dihydro-, 1,1-dimethylethyl ester)]

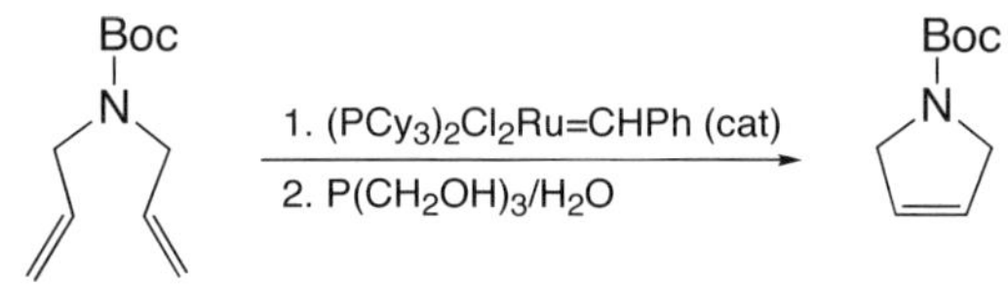

Submitted by Marcelle L. Ferguson,[1] Daniel J. O'Leary,[1] and Robert H. Grubbs.[2]
Checked by Louise M. Stamp and Andrew B. Holmes.

1. Procedure

N-Boc-3-pyrroline.[3] An oven-dried, 3-L, three-necked, round-bottomed flask equipped with an overhead mechanical stirrer, rubber septum with nitrogen inlet, and condenser with a bubbler-sealed outlet, is charged with 732 mg (0.89 mmol, 0.5 mol%) of bis(tricyclohexyl-phosphine)benzylidine ruthenium dichloride (Note 1) and 445 mL of dry dichloromethane (Note 2). The mixture is stirred under a nitrogen atmosphere to give a burgundy-colored solution. *N*-Boc-diallylamine (38.3 mL, 178 mmol) (Note 3) is transferred to the flask by syringe. After addition of the *N*-Boc-diallylamine, ethylene gas is evolved vigorously and the color of the solution changes from burgundy to dark brown. The solution is heated to reflux with stirring for 2.5 h (Note 4), then cooled to room temperature.

An aqueous methanolic solution of tris(hydroxymethyl)phosphine is prepared (Note 5). This solution is transferred to the flask containing the *N*-Boc-3-pyrroline, using 5 mL of methanol as a rinse. Triethylamine (0.247 mL, 1.78 mmol) is added to the mixture, which is vigorously stirred at room temperature overnight under a nitrogen atmosphere (Note 9).

Deionized water (400 mL) is added to the flask and the mixture is stirred vigorously for 30 min. Using a 2-L separatory funnel, the dichloromethane phase is separated from the aqueous phase, which is discarded. The organic phase is then washed with 350 mL of deionized water and with 200 mL of 50% v/v water/saturated brine solution. The dichloromethane phase is dried with 10 g of $MgSO_4$, then filtered through a sintered glass funnel. The filtrate is concentrated under reduced pressure by rotary evaporation (water aspirator) using a bath temperature of ca. 40 °C.

The crude product is divided in half and each portion is separately distilled in a Kugelrohr apparatus (50 mL distillation and receiver flasks) (Notes 10, 11) to give a combined yield of 27.2-28.3 g (90-94%) of *N*-Boc-3-pyrroline as a white, crystalline, low-melting solid (Note 12).

2. Notes

1. Bis(tricyclohexylphosphine)benzylidine ruthenium dichloride was obtained from Strem Chemical Company and used as received.

2. Dichloromethane was distilled over calcium hydride before use. The submitters found that identical reaction yields were obtained using commercial amylene-free HPLC-grade dichloromethane.

3. *N*-Boc-diallylamine was obtained from Aldrich Chemical Company, Inc. and used as received.

4. The progress of the reaction can be conveniently monitored by gas chromatography. GC analyses were performed on a Hewlett-Packard 6890 series gas chromatograph equipped with a 6890 series mass selective detector and an HP-5MS 30-m × 0.25-mm × 0.25-μm column under the following conditions: injector temp 180 °C; detector temp 150 °C; oven temp 70 °C, 2 min; ramp 5.75 °C/min; final temp 150 °C; helium gas flow 20.0 mL/min; t_R = 9.73-9.75 min (the submitters observed t_R 9.71 min). Complete conversion was usually observed after 2 h of reflux.

5. Preparation of tris(hydroxymethyl)phosphine. A 250-mL, three-necked, round-bottomed flask, equipped with a condenser having a bubbler-sealed outlet, rubber septum with nitrogen inlet, and magnetic stir bar, is charged with 43 mL (74.1-79.1 mmol) of a 1.72-1.84M aqueous solution of tetrakis(hydroxymethyl)phosphonium sulfate (Note 6). The flask is immersed in an oil bath. Methanol (25 mL) is added to the flask (Note 7), and the contents are heated to a gentle reflux under a nitrogen atmosphere. Sodium hydroxide pellets (3.1 g, 76.5 mmol) are added to the flask over the course of 30 min, accompanied by the gradual addition of 40 mL of methanol (Note 8). The mixture is stirred for an additional 10 min, then cooled to room temperature with stirring.

6. Tetrakis(hydroxymethyl)phosphonium sulfate was obtained as a 70-75% w/v (ca. 1.72-1.84M) aqueous solution from Fluka and was not purified before use. The submitters used 30.5 mL (76.5 mmol) of a 2.56M aqueous solution (11% active phosphorus by weight) obtained from Cytec, Inc.

7. Methanol was added to the tetrakis(hydroxymethyl)phosphonium sulfate by syringe through the rubber septum and this resulted in a cloudy, white solution.

8. A white precipitate forms on the bottom of the flask as NaOH is added over the 30 min period, causing the solution to become viscous. Additional methanol is added to enable stirring to be maintained.

9. The brown color of the *N*-Boc-3-pyrroline solution begins to fade immediately upon addition of tris(hydroxymethyl)phosphine. Approximately 45 min later, a viscous, brown substance is observed to adhere to the walls of the flask. This initially insoluble material dissolves after stirring overnight. A pale yellow solution results.

10. The submitters used a Kugelrohr distillation apparatus fitted with 24/40 ground glass joints and the *N*-Boc-3-pyrroline was distilled from a 500-mL, round-bottomed flask directly into a 250-mL receiving flask.

11. The Kugelrohr oven was heated from room temperature to 80 °C over the course of 45 min. The submitters noted that the pressure was 280 mtorr as the residual solvent was removed. As the oven temperature approached 50 °C, an acetone and dry ice bath was placed under the collection bulb. The submitters noted that the system pressure then dropped to 70 mtorr and the product distilled as a clear and colorless liquid that rapidly crystallized. The checkers observed similar results. The distillation was terminated when the distilling flask was nearly empty.

12. The submitters obtained 28.71 g (96%) of *N*-Boc-3-pyrroline. The analytical properties are as follows: mp 36-38 °C; ^{1}H NMR (400 MHz, $CDCl_3$) δ: 1.46 (s, 9 H), 4.10 (s, 4 H), 5.75 (s, 2 H); ^{13}C NMR (100 MHz, $CDCl_3$) δ: 28.5, 53.1, 79.2, 125.8, 154.3; IR ($CHCl_3$) cm^{-1} 2937, 1680, 1623, 1522, 1477, 1405, 1208, 1124, 995, 950, 871, 742, 624; MS (EI) *m/z* (rel intensity) 169 (20, M^+), 114 (30), 96 (45), 68 (100); HRMS (ES^+) *m/z* 192.1004 ($[M + Na]^+$, calcd. for $C_9H_{15}NO_2Na$ 192.1000). Anal. Calcd. for $C_9H_{15}NO_2$: C, 63.9; H, 8.9; N, 8.3. Found C, 63.3; H, 8.8; N, 8.1.

Waste Disposal Information

All toxic materials were disposed of in accordance with "Prudent Practices in the Laboratory"; National Academy Press; Washington, DC, 1995.

3. Discussion

The preparation of 3-pyrroline was recently reported in *Organic Syntheses*.[4] The protected form, *N*-Boc-3-pyrroline, has been used effectively in Heck arylations for the preparation of various 4-aryl endocyclic enecarbamates.[5] It has also been used in the preparation of regioisomeric 3-hydroxyisoxazolinyl prolines which are medicinally active.6 The catalytic asymmetric hydroformylation of *N*-Boc-pyrroline has been investigated.7 N-Boc-pyrroline has also been utilized in the synthesis of various antibacterial compounds.[8,9,10]

A procedure to remove colored ruthenium impurities using lead tetraacetate oxidation followed by filtration through a silica gel plug has been described.[11]

Table 1. Examples of RCM using Grubb's Catalyst, **1**

Entry	Substrate	Conditions (yield)	Product	Reference
1	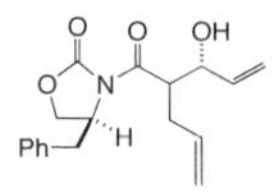 **3**	1 mol% **1**, 30 min, 25 °C, CH_2Cl_2 (97%)	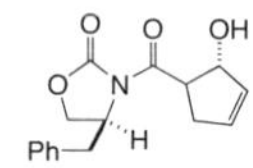 **4**	12
2	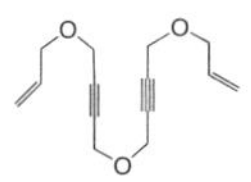 **5**	8 mol% **1**, 4 h, 45 °C, C_6H_6, (60%)	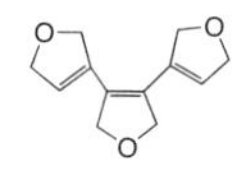 **6**	13
3	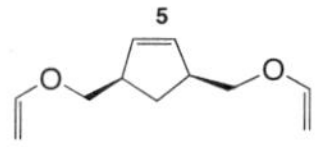 **7**	5 mol% **1**, 2 h, 60 °C, C_6,H_6, (90%)	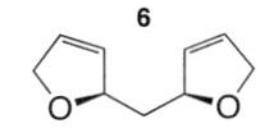 **8**	14
4	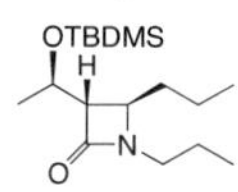 **9**	5 mol% **1**, 6 h, 25 °C, CH_2Cl_2, (78%)	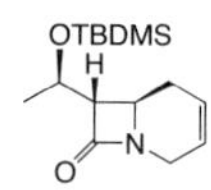 **10**	15
5	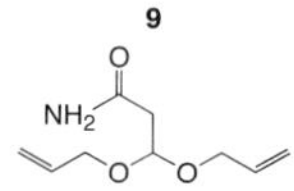 **11**	5 mol% **1**, 21 h, 25 °C, CH_2Cl_2, (83%)	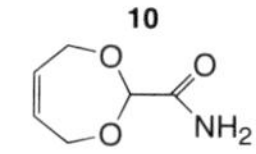 **12**	16
6	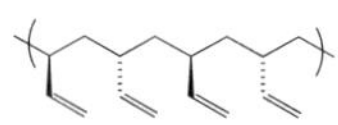 **13**	2 mol% **1**, 21 h, 35 °C, CH_2Cl_2, (>97%)	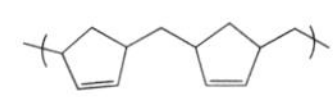 **14**	17
7	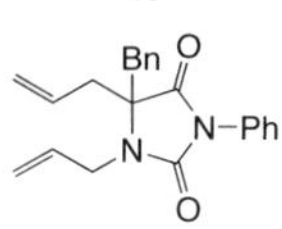 **15**	10 mol% **1**, overnight, reflux, CH_2Cl_2, (>95%)	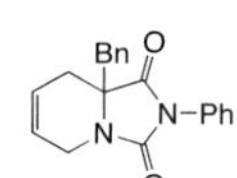 **16**	18
8	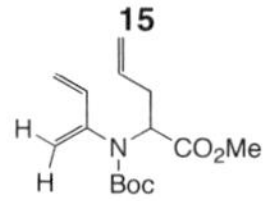 **17**	5 mol% **1**, 30 min, reflux, CH_2Cl_2, (93%)	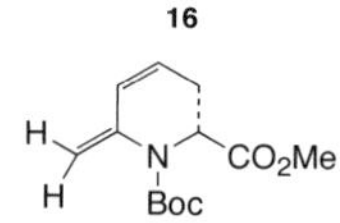 **18**	19
9	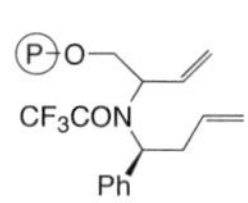 **19**	13 mol% **1**, 12 h, reflux, CH_2Cl_2, (70%) (tritylpolystyrol resin)	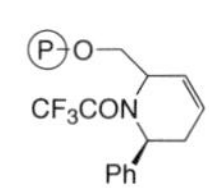 **20**	20
10	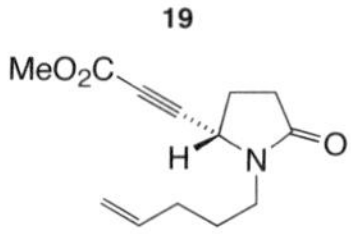 **21**	4 mol% **1**, 5 h, 25 °C, CH_2Cl_2, (87%)	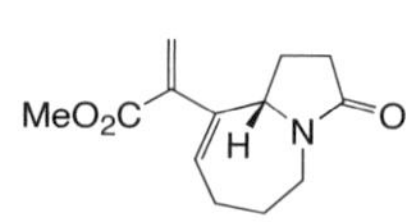 **22**	21

An efficient large-scale preparation of *N*-Boc-3-pyrroline using 0.1 mol% bis(tricyclohexylphosphine)ethylidene ruthenium dichloride was published[22] during the time that this procedure was in press. For this particular transformation, even lower catalyst loadings could likely be realized by use of the more active and more stable *N*-heterocyclic carbene-based ruthenium catalysts (e.g., the second-generation Grubbs catalyst).[23, 24]

Newer methods for the removal of ruthenium-derived by-products have also been disclosed, including the use of a polar isocyanide,[25] supercritical CO_2,[26] functionalized silicates,[27] polymer-bound triphenylphosphine oxide/DMSO,[28] polymer bound chelating phosphines,[29] and silica gel/activated carbon.[30]

1. Department of Chemistry, Pomona College, Claremont, CA 91711.
2. Division of Chemistry and Chemical Engineering, California Institute of Technology, Pasadena, CA 91125.
3. For the original reports from which this procedure was adapted, see: (a) Fu, G. C.; Nguyen, S. T.; Grubbs, R. H. *J. Am. Chem. Soc.* **1993**, *115*, 9856; (b) Maynard, H. D.; Grubbs, R. H. *Tetrahedron Lett.* **1999**, *40*, 4137.
4. Meyers, A. I.; Warmus, J. S.; Dilley, G. *J. Org. Synth.*, Coll. Vol. IX, **1998**, *666*.
5. Carpes, M. J. S.; Correia, C. R. D. *Synlett* **2000**, *7*, 1037.
6. Conti, P.; Dallanoce, C.; De Amici, M.; De Micheli, C.; Fruttero, R. *Tetrahedron* **1999**, *55*, 5623.
7. Horiuchi, T.; Ohta, T.; Shirakawa, E.; Nozaki, K.; Takaya, H. *J. Org. Chem.* **1997**, *62*, 4285.
8. Hong, C. Y.; Kim, Y. K.; Lee, Y. H.; Kwak, J. H. *Bioorg. Med. Chem. Lett.* **1998**, *8*, 221.
9. Lee, J. W.; Son, H. J.; Lee, K. S.; Yu, Y. H.; Yoon, G. J. *Yakhak Hoechi* **1994**, *38*, 677. *Chem. Abstr.* **1994**, *122*, 235095.
10. Okada, T.; Sato, H.; Tsuji, T.; Tsushima, T.; Nakai, H.; Yoshida, T.; Matsuura, S. *Chem. Pharm. Bull.* **1993**, *41*, 132.
11. Paquette, L. A.; Schloss, J. D.; Efremov, I.; Fabris, F.; Gallou, F.; Mendez-Andino, J.; Yang, *J. Org. Lett.* **2000**, *2*, 1259.
12. Crimmins, M. T.; King, B. W. *J. Org. Chem.* **1996**, *61*, 4192.
13. Zuercher, W. J.; Scholl, M.; Grubbs, R. H. *J. Org. Chem.* **1998**, *63*, 4291.
14. Zuercher, W. J.; Hashimoto, M.; Grubbs, R. H. *J. Am. Chem. Soc.* **1996**, *118*, 6634.
15. Barrett, A. G. M.; Baugh, S. P. D.; Gibson, V. C.; Giles, M. R.; Marshall, E. L.; Procopiou, P. A. *Chem. Commun.* **1997**, *155*.

16. Barrett, A. G. M.; Baugh, S. P. D.; Gibson, V. C.; Giles, M. R.; Marshall, E. L.; Procopiou, P. A. *Chem. Commun.* **1996**, *2231.*

17. Coates, G. W.; Grubbs, R. H. *J. Am. Chem. Soc.* **1996**, *118*, 229.

18. Dyatkin, A. B. *Tetrahedron Lett.* **1997**, *38*, 2065.

19. Rutjes, F. P. J. T.; Schoemaker, H. E. *Tetrahedron Lett.* **1997**, *38*, 677.

20. Schuster, M.; Pernerstorfer, J.; Blechert, S. *Angew. Chem. Int. Ed. Eng.* **1996**, *35*, 1979.

21. Kinoshita, A.; Mori, M. *J. Org. Chem.* **1996**, *61*, 8356.

22. Sturmer, R.; Schafer, B.; Wolfart, V.; Stahr, H.; Kazmaier, U.; Helmchen, G. *Synthesis* **2001**, *46.*

23. Ritter, T.; Hejl, A.; Wenzel, A. G.; Funk, T. W.; Grubbs, R. H. *Organometallics* **2006**, *25*, 5740.

24. Schrodi, Y.; Pederson, R. L. *Aldrichimica Acta* **2007**, *40*, 45.

25. Galan, B. R.; Kalbarczyk, K. P.; Szczepankiewicz, S.; Keister, J. B.; Diver, S. T. *Org. Lett.* **2007**, *9*, 1203.

26. Gallou, F.; Saim, S.; Koenig, K. J.; Bochniak, D.; Horhota, S. T.; Yee, N. K.; Senanayake, C. H. *Org. Proc. Res. Dev.* **2006**, *10*, 937.

27. McEleney, K.; Allen, D. P.; Holliday, A. E.; Crudden, C. M. *Org. Lett.* **2006**, *8*, 2663.

28. Haack, K. L.; Ahn, Y. M.; Georg, G. I. *Mol. Diversity*, **2005**, *9*, 301.

29. Westhus, M.; Gonthier, E.; Brohm, D.; Breinbauer, R. *Tetrahedron Lett.* **2004**, *45*, 3141.

30. Cho, J. H.; Moon, K. B. *Org. Lett.* **2003**, *5*, 531.

GENERATION OF YNOLATE AND *Z*-SELECTIVE OLEFINATION OF ACYLSILANES: (*Z*)-2-METHYL-3-TRIMETHYLSILYL-2-BUTENOIC ACID

((*Z*)-2-Butenoic acid, 2-methyl-3-trimethylsilyl-)

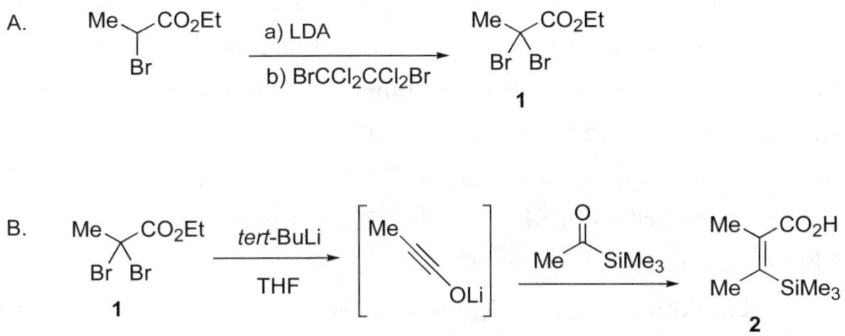

Submitted by Mitsuru Shindo,[1] Kenji Matsumoto,[2] and Kozo Shishido.[2]
Checked by Scott E. Denmark and Brian M. Eklov.

1. Procedure

A. Ethyl 2,2-dibromopropionate (**1**).[3] A 1-L, three-necked, round-bottomed flask is equipped with a magnetic stirring bar and fitted with a thermocouple temperature probe (Note 1) inserted through an adapter, a rubber septum, and a 100-mL, pressure-equalizing addition funnel fitted with a vacuum adapter which is capped with a rubber septum and the sidearm of which is connected to a vacuum/argon gas line. The flask is flame-dried under reduced pressure and then is maintained under an atmosphere of argon during the course of the reaction. The flask is charged with diisopropylamine (23.0 mL, 164 mmol, 1.1 equiv) (Note 2) and anhydrous THF (350 mL) (Note 3), and then is cooled to –71 °C (internal) in a dry ice/isopropyl alcohol bath (Note 4) while a 1.62 M solution of *n*-butyllithium in hexane (102 mL, 165 mmol, 1.1 equiv) (Note 5) is added through the addition funnel over 10 min. After an additional 16 min, a solution of ethyl 2-bromopropanoate (20.0 mL, 27.8 g, 154 mmol) (Note 6) in anhydrous THF (20 mL) is added dropwise through the addition funnel over 18 min (the addition funnel is rinsed with 10 mL of anhydrous THF).

After an additional 34 min, a solution of 1,2-dibromo-1,1,2,2-tetrachloroethane (53.7 g, 165 mmol, 1.1 equiv) (Note 7) in THF (60 mL) is added rapidly in one portion through the addition funnel with vigorous stirring (Note 8). After being stirred in the cold bath for 40 min (Note 9), the mixture is poured into 150 mL of saturated aqueous sodium bicarbonate solution in a 1-L separatory funnel (Note 10) and the mixture is extracted with hexanes (2 x 150 mL). The combined organic layers are washed with 150 mL of water and 150 mL of brine, then are dried over $MgSO_4$ (6 g), filtered under vacuum, and concentrated by rotary evaporation (40 °C, 50–100 mmHg) to afford a dark brown oil. The oil is transferred to a 100-mL, round-bottomed flask equipped with a magnetic stirbar and fitted with a short-path distillation unit. Distillation of the product under reduced pressure using an oil bath affords 36.1 g (90% yield) of ethyl 2,2-dibromopropionate (**1**) as a colorless liquid (71–80 °C / 17 mmHg) (Notes 11, 12) after an approximate 5 g forerun.

B. (Z)-2-Methyl-3-trimethylsilyl-2-butenoic acid (**2**).[4] A 500-mL, three-necked, round-bottomed flask is equipped with a magnetic stirring bar and fitted with a thermocouple temperature probe adapter, a rubber septum, and a pressure-equalizing, 200-mL addition funnel fitted with a vacuum adapter capped with a rubber septum. The side arm of the vacuum adapter is connected to a vacuum/argon gas system. The flask is flame-dried under reduced pressure and then maintained under an atmosphere of argon during the course of the reaction. The thermocouple is inserted through the adapter and the flask is charged with ethyl 2,2-dibromopropionate (**1**, 17.6 g, 68.0 mmol, 1.2 equiv) and anhydrous THF (180 mL) (Note 3). The solution is cooled with stirring to –72 °C (internal) in a dry ice/isopropyl alcohol bath, then a 1.72 M solution of *tert*-butyllithium in pentane (158 mL, 272 mmol, 4.9 equiv) (Note 13) is added dropwise over 67 min through the addition funnel such that the internal temperature is kept below –64 °C (Note 14). The addition funnel is rinsed with 5 mL of THF. The yellow solution is stirred at –72 °C for an additional 110 min (3 h total reaction time), and then the cooling bath is replaced by an ice bath. After being stirred at 2 °C (internal) for 30 min, the ice bath is removed, and the mixture is allowed to warm to room temperature. A solution of acetyltrimethylsilane (6.50 g, 56 mmol, Note 15) in anhydrous THF (15 mL) is added through the addition funnel over 15 min (Note 16). The addition funnel is rinsed with 5 mL of anhydrous THF, and the mixture is stirred for 45 min at room temperature. The reaction mixture is poured onto 1 M aqueous NaOH (120 mL) in a 1-L separatory funnel. The organic phase is separated and the aqueous phase (pH 8) is washed with ethyl acetate (2 x 120 mL). Thc combined organic extracts are washed with 1 M aqueous sodium hydroxide solution (120 mL) and set aside. The combined aqueous extracts (pH 9) are acidified with 6 M HCl (100 mL) to pH 1. The resulting cloudy suspension is extracted with ethyl acetate (3 x 120 mL). These organic extracts are combined, washed with brine (120 mL), dried over $MgSO_4$ (4 g), filtered through a coarse glass frit and then are concentrated on a rotary evaporator (23 °C, 10 mmHg) to yield 7.64 g of a viscous yellow-orange oil that solidifies upon standing in the freezer (–20 °C) (Note 17). The crude product is transferred to a sublimation chamber (chamber: inside diameter, 57 mm; height, 90 mm; condenser: outside diameter, 27 mm; height 70 mm). The bottom of the chamber is cooled in an ice bath and the apparatus is attached to a vacuum line. The bath is removed once the majority of the volatile materials are evaporated as judged by stabilization of the pressure. When the pressure inside the chamber drops below 0.1 mmHg, the condenser is charged with a dry ice/isopropyl alcohol mixture and the chamber is placed in an oil bath, pre-heated to 40 °C. The sublimate was removed from the cold finger three times to afford 6.12 – 6.60 g (64–69%) of **2** as a white crystalline solid (mp 48.5– 50.5 °C, Note 19).

2. Notes

1. A PFA-coated thermocouple probe, Type K (Omega Engineering, Inc.) was inserted through the septum.

2. Diisopropylamine was purchased from Sigma-Aldrich Company (>98.0%) and was distilled from calcium hydride.

3. Anhydrous THF (>99.5%, stabilizer free) was purchased from Fischer Scientific and was used as received (Karl Fischer analysis gave < 120 μg/mL).

4. For a small scale (smaller than 500-mL flask), the submitters used a low temperature bath with magnetic stirrer (Tokyo Rikakikai Co., Ltd., PSL-

1800).

5. *n*-Butyllithium solution was purchased from Sigma-Aldrich and was titrated by both No-D NMR spectroscopy[5a] and Gilman double titration.[5b]

6. Ethyl 2-bromopropanoate (>98%) was purchased from Tokyo Kasei Kogyo Co. Ltd. (TCI) and was distilled.

7. 1,2-Dibromo-1,1,2,2-tetrachloroethane (98%) was purchased from Alfa-Aesar and was used as received. 1,2-Dibromo-1,1,2,2-tetrafluoroethane (>99.0%, Tokyo Kasei Kogyo Co. Ltd.) is more convenient as a bromination reagent, but is not available in the USA.

8. Addition required approximately 30 sec during which time the temperature rose from –71 °C to –35 °C. If this reagent was added slowly, a significant amount of side products were generated, probably due to self-condensation.

9. The progress of the reaction was monitored by gas chromatography. GC-MS analysis was performed on a JMS-Automass SUN200 instrument equipped with a TC-17 30-m x 0.32-mm x 0.25-μm column under the following conditions: injector temp 250 °C; oven temp 60 °C, 3 min; ramp 8 °C/min; final temp 250 °C; helium gas flow 0.8 mL/min; t_R = 4.56 min (dibromo ester), t_R = 2.13 min (bromo ester). The Checkers did not monitor the reaction, but found that the product could be analyzed by GC-MS (Model HP 5890A GC equipped with an HP 5970 MS detector; column HP-1).

10. A large portion of the aqueous solution froze on contact with the reaction mixture. The frozen material was allowed to thaw before performing the extraction.

11. A single distillation afforded material that was contaminated by the starting bromo ester (which could be detected by NMR spectroscopy) and the brominating agent (1,2-dibromo-1,1,2,2-tetrachloroethane). The latter impurity was difficult to identify by TLC (not UV active and did not readily stain), GC (did not combust nor did it ionize easily), IR (C-Cl stretches were weak and were obscured by the dibromo ester), or NMR (no protons to observe and the ^{13}C resonance was suppressed by the long relaxation time and the lack of a proton nOe). Combustion analysis proved to be the only reliable method to determine the purity of the distillate. The combustion analysis results clearly demonstrated that the primary distillation did not

satisfactorily remove the haloalkane. Both of these impurities were removed by a second distillation (65–67 °C, 9 mmHg) to afford 29.3–29.4 g (74–75%) of analytically pure material. The dibromo ester was a stable compound that can be distilled, purified by silica gel column chromatography, and stored for months in a refrigerator.

12. Physical and spectroscopic properties of **1** are as follows: ^{1}H-NMR (500 MHz, $CDCl_3$) δ: 1.36 (t, 3 H, J = 7.0), 2.65 (s, 3 H), 4.34 (q, 2 H, J = 7.0). ^{13}C-NMR (125 MHz, $CDCl_3$) δ: 13.7 (q), 37.3, 52.2, 63.7, 166.8. IR (film) cm^{-1}: 1737. EI-MS *m/z*: 260 (M^{+}+2), 258 (M^{+}). Anal. Calcd for $C_5H_8Br_2O_2$: C, 23.10; H, 3.10; Br, 61.48. Found: C, 22.79; H, 2.91; Br, 61.53.

13. A solution of *tert*-butyllithium in pentane was purchased from Sigma Aldrich Company and was titrated by both No-D NMR spectroscopy[5a] and Gilman double titration.[5b] This lithium-halogen exchange procedure for the preparation of ynolates[6] is suitable for a laboratory scale because of its convenience and simplicity. Reductive lithiation with lithium metal catalyzed by naphthalene was better for larger scale.[7]

14. The exothermicity of the reaction subsided substantially after one half of the alkyllithium reagent was added. Thus, the rate of addition was increased after this point without a concomitant increase in internal temperature.

15. Acetyltrimethylsilane (97%) was purchased from Aldrich Chemical Company and was distilled prior to use (bp 110–116 °C).

16. The reaction of the acylsilane with the lithium ynolate was exothermic and reached a maximum temperature of 39 °C. The addition rate was adjusted to minimize the loss of pentanes from the solution.

17. TLC analysis was performed on Merck silica gel 60 F-254 plates eluting with hexanes/EtOAc, 4:1, R_f = 0.43 (visualized with 254 nm UV lamp).

18. The product was obtained in analytically pure form by distillation on a half scale. The submitters reported that distillation of the product on the full scale led to polymerization.

19. Physical and spectroscopic properties of **2** are as follows: ^{1}H-NMR (500 MHz, $CDCl_3$) δ: 0.16 (s, 9 H), 1.90 (q, 3 H, J = 1.0), 1.97 (q, 3 H, J = 1.0), 11.6–12.6 (br s, 1 H). ^{13}C-NMR (125 MHz, $CDCl_3$) δ: 0.2, 15.4, 20.4,

136.0, 155.6, 175.1. ^{29}Si-NMR (119 MHz, $CDCl_3$) δ: -5.15. IR (film) cm^{-1}: 2966, 1679. MS (EI) *m/z* 157 (M^+ –Me). Anal. Calcd. for. $C_8H_{16}O_2Si$ C, 55.77; H 9.36. Found: C, 55.78, H. 9.63.

Safety and Waste Disposal Information

All hazardous materials should be handled and disposed of in accordance with "Prudent Practices in the Laboratory"; National Academy Press; Washington, DC, 1995.

3. Discussion

Ynolates[8] are ketene anion equivalents and are expected to act as multifunctional carbanions for highly reactive ketenes. Although about 10 reports on the syntheses of ynolates have been published, studies on the versatility of ynolates have remained at a rather basic level, probably due to a lack of convenient and general methods for their synthesis. In *Organic Syntheses*, only one report on ynolates has been published.[9]

The procedure described here demonstrates a practical and convenient method for the generation of ynolates, and a (*Z*)-selective olefination of acylsilanes leading to (*Z*)-2-methyl-3-trimethylsilyl-2-butenoic acid.

The method for the generation of ynolates involves thermal cleavage of ester dianions derived from α,α-dibromo esters. We have synthesized aliphatic dibromo esters (R = Me, Bu, *i*-Pr, cyclohexyl, *t*-Bu, phenylethyl, trimethylsilyl, etc.) by the method described herein. Aromatic dibromo esters are synthesized via radical bromination. Two procedures for the generation of ester dianions from the dibromo esters have been developed. Herein, lithium-halogen exchange of dibromo esters via *tert*-butyllithium giving the ester dianions is described. This procedure is very convenient, simple and suitable for laboratory scale. In an alternative procedure, the naphthalene-catalyzed reductive lithiation[6] of the dibromo esters with lithium metal is employed. As outlined in the following scheme ~~of~~ for the generation of ynolates using reductive lithiation catalyzed by naphthalene,[10] the dibromo ester reacts with lithium naphthalenide at –78 °C to give an α-bromo ester enolate, which is successively lithiated to form the ester dianion.

The dianion is cleaved at 0 °C to generate an ynolate by release of lithium ethoxide.

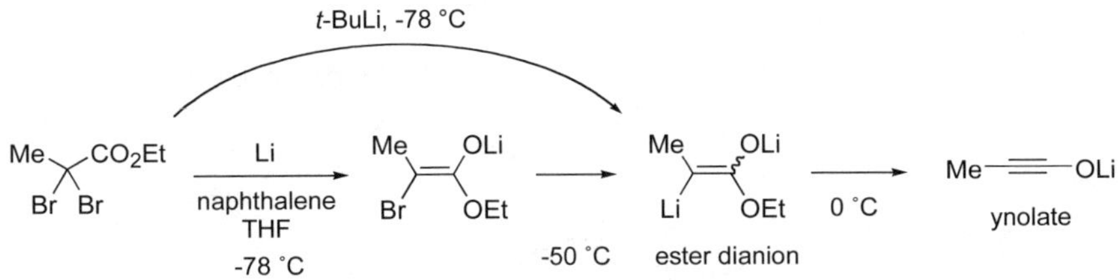

Vinylsilanes are powerful tools in synthetic organic chemistry. Although olefination of acylsilanes is expected to be a useful method for the preparation of vinylsilanes, few reports on this process have been reported due to the unsuitability of conventional olefination reagents. We have developed a method in which olefination of ketones via ynolates provides α,β-unsaturated carboxylates with good to moderate *E/Z* selectivities.[11] Using acylsilanes as a ketone in this process, however, extremely high *Z*-selectivity is achieved in the synthesis of various substituted vinylsilanes. The *E/Z* selectivity is controlled in the electrocyclic ring-opening of the intermediate β-lactone enolates, prepared by cycloaddition of carbonyl groups with ynolates. In this step, the transition state would be stabilized by the orbital interaction between a breaking C-O s orbital and Si vacant orbitals.[12]

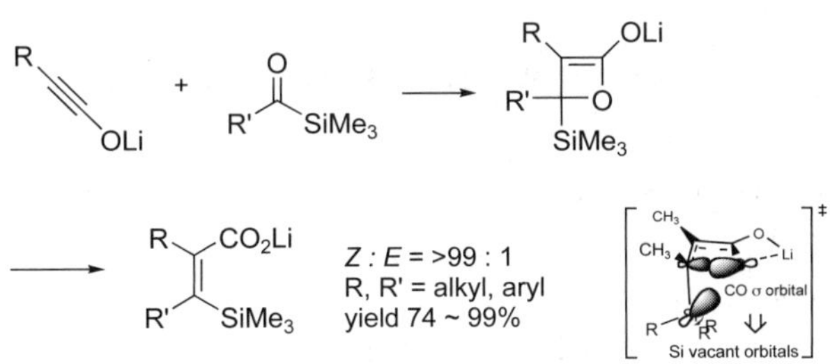

The vinylsilanes produced can be converted into a variety of functionalized multisubstituted alkenes.

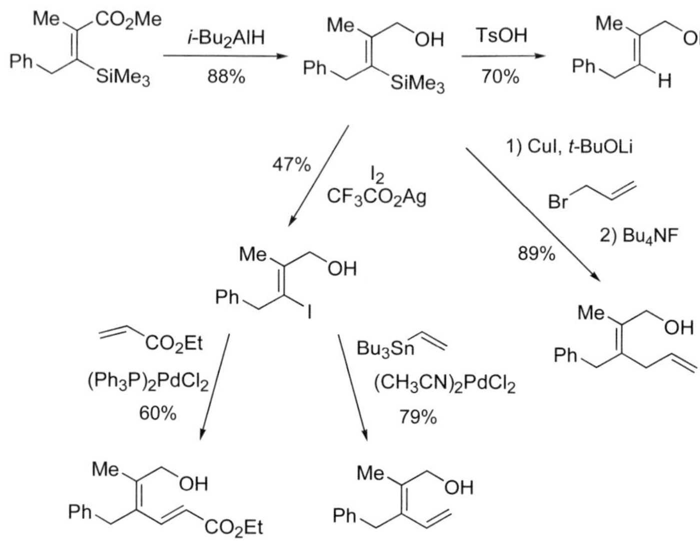

1. Institute for Materials Chemistry and Engineering, Kyushu University, 6-1 Kasugakoen, Kasuga, Fukuoka 816-8580, Japan. E-mail: shindo@cm.kyushu-u.ac.jp
2. Institute of Health Biosciences, The University of Tokushima Graduate School, Sho-machi 1, Tokushima 770-8505, Japan.
3. Shindo, M.; Sato, Y.; Koretsune, R.; Yoshikawa, T.; Matsumoto, K.; Itoh, K.; Shishido, K. *Chem. Pharm. Bull.* **2003**, *51*, 477-478.
4. Shindo, M.; Matsumoto, K.; Mori, S.; Shishido, K. *J. Am. Chem. Soc.* **2002**, *124*, 6840-6841
5. (a) Hoye, T. R.; Aspaas, A. W.; Eklov, B. M.; Voloshin, M. *Org. Lett.* **2004**, *6*, 2567-2570. (b) Gilman, H.; Cartledge, F. K. *J. Organomet. Chem.* **1964**, *2*, 447.
6. Shindo, M.; Sato, Y.; Shishido, K. *Tetrahedron* **1998**, *54*, 2411-2422.
7. Shindo, M.; Koretsune, R.; Yokota, W.; Itoh, K.; Shishido, K. *Tetrahedron Lett.*, **2001**, *42*, 8357-8360.
8. (a) Shindo, M. *Synthesis* **2003**, 2275-2288; (b) Shindo, M. *Tetrahedron* **2007**, *63*, 10-36.
9. Reddy, R.; Kowalski, C. *Org. Syn. Coll. Vol. 9* **1998**, 426-432.
10. Yus, M.; Herrera, R. P.; Guijarro, A. *Chem. Eur. J.* **2002**, *8*, 2574-2584.
11. (a) Shindo, M.; Sato, Y.; Shishido, K. *J. Org. Chem.*, **2000**, *65*, 5443-5445. (b) Shindo, M.; Sato, Y.; Yoshikawa, T.; Koretsune, R.; Shishido, K. *J. Org. Chem.*

2004, *69*, 3912-3916.

12. For related effects on torquoselectivity, see: (a) Murakami, M.; Miyamoto, Y.; Ito, Y. *Angew. Chem., Int. Ed.* **2001**, *40*, 189-190. (b) Lee, P. S.; Zhang, X.; Houk, K. N. *J. Am. Chem. Soc.* **2003**,*125*, 5072-5079. (c) Mori, S.; Shindo, M. *Org. Lett.* **2004**, *6*, 3945-3948.

SYNTHESIS OF 2-CHLOROACROLEIN DIETHYL ACETAL (2-Chloroprop-2-enal, diethyl acetal)

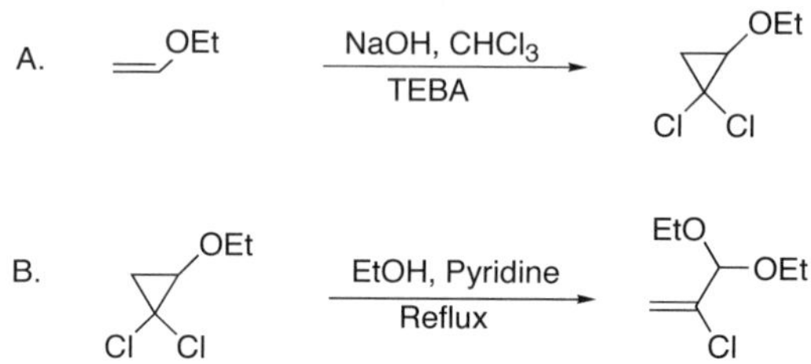

Submitted by Ole H. Kvernenes and Leiv K. Sydnes.[1]
Checked by David M. Arnold and Peter Wipf.[2]

1. Procedure

A. 1,1-Dichloro-2-ethoxycyclopropane. A 250-mL, three-necked, round-bottomed flask equipped with a septum, a 3.8 cm Teflon-coated magnetic stir bar and a condenser connected to a drying tube containing Drierite® is charged with ethyl vinyl ether (9.55 mL, 7.20 g, 0.10 mol) (Note 1), chloroform (31.9 mL, 47.8 g, 0.40 mol) (Note 2), and benzyltriethylammonium chloride (TEBA) (0.100 g, 0.44 mmol) (Note 3) and placed in an ice bath (Note 4). After the reaction flask is cooled for 10 min, the mixture is stirred vigorously (Note 5) and 24 g of a 50% aqueous solution of sodium hydroxide (NaOH) (12.0 g, 0.30 mol in 12 mL of H_2O) (Note 6) is added dropwise via a syringe in 10 min (Note 7). The reaction mixture is stirred vigorously at bath temperature (approximately 0 °C) for 2 h and at room temperature for 22 h. The reaction mixture is then cooled in an ice bath and quenched by injecting 6 M hydrochloric acid (30 mL) dropwise over a 10 min period (Note 8). The hydrolysate is transferred to a 500-mL separatory funnel, and the flask is rinsed with water (2 x 20 mL) and dichloromethane (30 mL), which is also added to the funnel. After addition of more water (60 mL) to the funnel, the mixture is shaken and the organic layer is separated from the aqueous layer. The aqueous layer is extracted with dichloromethane (3 x 60 mL) (Note 9). The combined organic extracts are dried with approximately 2 g of anhydrous magnesium sulfate ($MgSO_4$) (Note 10), vacuum filtered, and concentrated under reduced pressure on a rotary evaporator (25 °C at 70 mmHg) to afford a pale yellow residue. This residue is transferred to a 50-mL round bottom flask using approximately 2 mL of dichloromethane (Note 11). The dichloromethane is removed at 30–35 °C (oil bath temperature) at 50 mmHg, and the product is distilled at 65–70 °C (oil bath temperature) at 30 mmHg to give 12.94 g (84%) (Note 12) of 1,1-dichloro-2-ethoxycyclopropane (Note 13) as a colorless liquid, bp 55 °C/30 mmHg.

B. 2-Chloro-3,3-diethoxyprop-1-ene (Note 14). A 250-mL, single-necked, round-bottomed flask equipped with a 3.8 cm Teflon-coated magnetic stir bar and a condenser connected to a Drierite® drying tube is charged with absolute ethanol (50 mL) (Note 15), pyridine (7.33 mL, 7.20 g, 0.09 mol) (Note 16) and 1,1-dichloro-2-ethoxycyclopropane (10.86 g, 0.070 mol) (Note 17). The resulting mixture is stirred at a moderate speed at reflux for 48 h in an oil bath at 95 °C (bath temperature), and is then allowed to cool to room temperature before being concentrated under reduced pressure on a rotary evaporator (40 °C at 70 mmHg vacuum) until the volume is approximately 20 mL (Note 18). The residue is transferred to a 250-mL separatory funnel, and the flask is rinsed with water (2 x 50 mL) and dichloromethane (60 mL), which are also added to the separatory funnel. The mixture is shaken, and the organic layer is separated from the aqueous layer. The aqueous phase is extracted with dichloromethane (2 x 50 mL), and the combined organic phases are washed with a 0.7 M aqueous solution (2 x 100 mL) of copper sulfate ($CuSO_4$) (Note 19), and dried with approximately 2 g of anhydrous magnesium sulfate (Note 10). The organic phase is then vacuum filtered into a 500-mL round-bottomed flask through a plug of aluminum oxide (Note 20). Dichloromethane (15 mL) is added to the flask containing the $MgSO_4$ and after mixing the contents for approximately 1 min, the dichloromethane solution is filtered through the plug as well. The combined filtrates are then concentrated under reduced pressure on a rotary evaporator (25 °C at 70 mmHg) and transferred with approximately 2 mL of dichloromethane to a 50-mL round-bottomed flask for vacuum distillation (Notes 11, 21). The dichloromethane is removed at 21 °C and 30 mmHg. Following the removal of the dichloromethane, the apparatus is placed again under vacuum (30 mmHg) and the 25-mL receiving flask is submerged in an acetone/dry ice bath. The product is distilled with an oil bath temperature of 80–85 °C. Upon completion of the distillation, nitrogen gas is bled into the apparatus from the vacuum adaptor as the vacuum is released. The receiving flask is warmed to room temperature under a nitrogen atmosphere before being removed. The distillation yields 6.75 g (59 %) of a clear liquid, bp 70–71 °C/30 mmHg (Note 22). The product, 2-chloro-3,3-diethoxyprop-1-ene (Note 23), is found by GC analysis to be 96.0–97.2% pure (Notes 24, 25, 26).

2. Notes

1. Ethyl vinyl ether of 99% purity was purchased from Aldrich Chemical Co. and used as obtained.

2. Chloroform, with a purity of better than 99%, was purchased from J. T. Baker (Mallinckrodt Baker) and used as obtained.

3. Checkers used TEBA, a phase-transfer catalyst, which was 98% pure, purchased from Acros Organics, and used as obtained. Submitters used 99% pure TEBA purchased from Aldrich Chemical Co.

4. The bath was a Pyrex dish (125 mm x 65 mm), to which were added about 300 mL of ice and 300 mL of cold water.

5. It is essential that the stirring is vigorous so that the organic and aqueous phases are thoroughly mixed. The submitters used a mechanical stirrer with a stirring speed of approximately 500 revolutions per min.

6. Sodium hydroxide, with a purity of 98.4%, was purchased from J. T. Baker and used as obtained. A 50% stock solution of sodium hydroxide was made by dissolving 20.0 g of sodium hydroxide in 20.0 g of water.

7. The submitters report that the sodium hydroxide solution can be added much faster without any consequences at the scale used here. At larger scales, however, it is important to add the solution slowly to avoid excessive heating. When the reaction was carried out on a 2-mole scale, the addition of 240 g of 50% aqueous sodium hydroxide required 30 min.

8. The hydrochloric acid was prepared by dilution of concentrated (36-38%, 1.19 g/mL, 12 M) hydrochloric acid of Baker-analyzed quality purchased from J. T. Baker (Mallinckrodt Baker). Rapid addition of the hydrochloric acid solution at room temperature generated a lot of heat and discoloration (yellowing) of the reaction mixture was noted.

9. Dichloromethane with a purity of 99.5% was purchased from EMD Chemicals Inc. and used as obtained.

10. Checkers used anhydrous magnesium sulfate that was purchased from EMD Chemicals Inc. and used as obtained. Submitters used dehydrated but not completely anhydrous magnesium sulfate from several suppliers.

11. Checkers carried out the distillation with a 50-mL round-bottomed flask equipped with a Teflon-coated magnetic stirring bar and a short path distillation head fitted with a 25-mL receiving flask. Submitters used standard equipment consisting of a 100-mL pear-shaped flask with a magnetic stirrer, a connecting adapter with thermometer, a Liebig condenser, a vacuum adapter, and a 25-mL receiving flask, all with 14/23 joint(s).

12. The reaction was also run at a 50% scale and yielded 5.64 g of product. The yield range is 73–84%. The submitters report that the reaction has been reproduced up to a 2-mole scale following the procedure described here. At such a large scale, the yield is considerably better, usually in the 82-96% range, provided proper care is taken to avoid excessive heating of the reaction mixture (see Notes 7 and 8). When the reaction is carried out on a larger scale, the amount of product remaining in the distillation flask, the distillation head and the condenser is relatively smaller.

13. The product exhibits the following spectroscopic properties: ^1H NMR ($CDCl_3$, 300 MHz) δ: 1.29 (t, 3 H, J = 7.0 Hz), 1.53 (dd, 1 H, J = 8.5, 5.1 Hz), 1.66 (t, 1 H, J = 8.5 Hz), 3.54 (dd, 1 H, J = 8.1, 5.1 Hz), 3.68–3.89 (m, 2 H); ^{13}C NMR ($CDCl_3$, 75 MHz) δ: 14.9, 27.9, 58.6, 63.1, 67.3; IR (neat) 3095 (w), 2980 (s), 2931 (m), 2875 (m), 1481 (w), 1444 (w), 1425 (m), 1398 (m), 1374 (m), 1345 (m), 1300 (w), 1235 (m), 1179 (m), 1115 (m), 1065 (s), 1019 (m), 977 (w), 905 (m), 890 (w), 800 (m), 766 (s) cm^{-1}; EI-MS *m/z* 158 ($[M+4]^+$, 0.1), 156 ($[M+2]^+$, 0.5), 154 (M^+, 0.8), 119 (54), 109 (17), 97 (34), 91 (100), 84 (9), 73 (8), 61 (62), 55 (8). The purity of the product was >99% based on GC analysis (assuming identical response factors) using a HP 6890 Series GC system (100% dimethylpolysiloxane coated column (30 m x 320 µm x 0.25 µm); column flow rate 1 mL/min; 70 °C - start; temperature increased at 15 °C/min until 240 °C; temperature kept at 240 °C for 5 min; temperature increased at 25 °C/min until 300 °C; retention time: 3.975 min.

14. The procedure for this step is a modification of that described by Skattebøl.[3]

15. Checkers used absolute ethanol containing at least 99.95% ethanol without further purification as purchased from Pharmco Products Inc. Submitters used at least 99.98% pure absolute ethanol without further purification as purchased from Arcus, Norway.

16. Anhydrous pyridine, with a purity of 99.8%, was purchased from Aldrich Chemical Co. and was used without further purification. Submitters report that at least 1 equiv of pyridine has to be used; usually 1.3 equiv are used. If less than 1 equiv is used, ethanol addition to the product takes place, furnishing 2-chloro-1,1,3-triethoxypropane.

17. The order of addition does not affect the reaction yield.

18. When the volume is approximately 20 mL, the liquid turns turbid upon cooling. A slight yellowing of the solution was noted.

19. Checkers used copper sulfate pentahydrate, with a purity of >99%, purchased from Fisher Scientific to prepare the aqueous copper

sulfate solution by dissolving 35.0 g of $CuSO_4{\bullet}5H_2O$ in 200 mL of water. Submitters used copper sulfate pentahydrate, with a purity of 98 %, purchased from Acros Organics.

20. Checkers used activated basic aluminum oxide (150 mesh, 58 Å) purchased from Aldrich Chemical Co. Submitters used the Fluka Chromatography product 6300, neutral 120–125 mesh aluminum oxide. The filtration was carried out using a column with a diameter of 2.0 cm and a height of 10 cm containing a vacuum adaptor connected to a water aspirator; upon addition of 6.3 g of aluminum oxide the alumina plug was 2.0 cm high.

21. Distillation of the product has to be carried out with care to avoid loss of material due to the high volatility of this alkene.

22. The reaction was also reproduced on a 50% scale and yielded 2.86 g of product for a yield range of 50-59%.

23. The product exhibits the following spectroscopic properties: ^1H NMR ($CDCl_3$, 300 MHz) δ: 1.23 (t, 6 H, J = 7.0 Hz), 3.47–3.68 (m, 4 H), 4.84 (s, 1 H), 5.45 (dd, 1 H, J = 1.2, 0.5 Hz), 5.66 (t, 1 H, J = 1.2 Hz); ^{13}C NMR ($CDCl_3$, 75 MHz) δ: 15.1, 61.8, 100.7, 115.1, 138.2; IR (neat) 2978 (s), 2886 (s), 1635 (m), 1481 (w), 1445 (m), 1372 (m), 1336 (m), 1296 (w), 1206 (m), 1062 (s), 964 (m), 904 (m), 765 (w), 747 (w) 703 (w), 677 (w) cm^{-1}; EI-MS *m/z* 166 ($[M+2]^+$, 4), 164 (M^+, 13), 129 (10), 119 (84), 103 (60), 91 (100), 75 (14), 63 (9), 55 (8); HRMS *m/z* calculated for $C_7H_{13}O_2Cl$ 164.0604, found 164.0612. Analysis calcd. for $C_7H_{13}O_2Cl$: C, 51.07; H, 7.96; found C, 50.90; H, 7.88.

24. The purity of the product was 96.0–97.2% according to GC analyses. The GC analyses (assuming identical response factors) were performed with a HP 6890 Series GC system (100% dimethylpolysiloxane column (30 m x 320 μm x 0.25 μm); column flow rate 1 mL/min; 70 °C - start; temperature increased at 15 °C/min until 240 °C; temperature kept at 240 °C for 5 min; temperature increased at 25 °C/min until 300 °C; retention time: 4.410 min.

25. ^1H NMR and GC analyses determined that the product was contaminated with 2.8–4.0% of 1,1-dichloro-2-ethoxycyclopropane. Subsequent distillations did not improve the purity of the product above 97.2%.

26. An aliquot of the product mixture was resubjected to the reaction conditions, heated at reflux for an additional 24 h and worked up according to the outlined procedure. A GC analysis performed under identical conditions as above showed this product to be >99% pure.

Safety and Waste Disposal Information

All hazardous materials should be handled and disposed of in accordance with "Prudent Practices in the Laboratory"; National Academy Press; Washington, DC, 1995.

3. Discussion

2-Haloacrolein diethyl acetals are densely functionalized compounds with a considerable potential in organic synthesis. This potential has been partly realized with 2-bromoacrolein diethyl acetal, which has proven to be useful for attachment of the 1-(diethoxymethyl)ethenyl group to a variety of substrates in multistep syntheses.[4-15] The chloro analogue, however, has barely been used,[16] probably due to its tedious synthesis.

Both 2-chloroacrolein diethyl acetal and 2-bromoacrolein diethyl acetal can be prepared from acrolein,[9,17,18] but the syntheses are hampered by the ease of polymerization of this unsaturated aldehyde.[19] Thus, the starting material should be thoroughly purified to facilitate product formation,[20] and an elaborate work-up procedure must be applied to obtain a pure product.[9] Furthermore, it is a disadvantage that hazardous chlorine gas and bromine liquid are used as reagents.

An alternative synthesis of 2-chloroacrolein diethyl acetal, involving ring opening of 1,1-dichloro-2-ethoxycyclopropane, has been reported by Skattebøl.[3] The cyclopropane is prepared from ethyl vinyl ether following the Doering-Hoffman procedure ($CHCl_3$, *t*-BuOK, pentane, low temperature), which is rather laborious and requires absolutely dry conditions.[21] An alternative synthesis of the cyclopropane, using CCl_4, CH_2Br_2, 50% aqueous NaOH, and tetrabutylammonium hydrogensulfate, has also been published,[22] but the method is rather unattractive due to the use of toxic CCl_4.[23] A third procedure involves vinyl ether, chloroform and magnesium,[24] but this synthesis is conceivably difficult to run on a large scale.

Inspired by Skattebøl's synthesis, we have worked out a new and simpler procedure for the preparation of the title compound. 1,1-Dichloro-2-ethoxycyclopropane is synthesized from ethyl vinyl ether by using Makosza's method with chloroform instead of CCl_4.[25] This method has been used to prepare other 2-alkoxy-1,1-dichlorocyclopropanes,[25-27] albeit in lower yields than obtained in this case. The purification of the cyclopropane is straightforward, and it is rewarding that the yield is improved when the reaction is preformed on larger scales.[28] The ring opening of the cyclopropane has also been modified compared to Skattebøl's method. Pyridine, added to remove the hydrogen chloride formed during the ring opening, is removed effectively by washing with a solution of copper(II) sulfate, which forms water-soluble complexes with pyridine.[29] Subsequent filtration through an aluminum-oxide pad was also introduced to remove any pyridine-copper complex remaining in the combined organic phases.

The yield of 2-chloroacrolein diethyl acetal in the second step is lower than what has been reported in the literature.[3] This is probably because we have performed the distillation in such a way that contamination of the corresponding acetylenic acetal, 1,1-diethoxy-2-propyne, was avoided.

1. Department of Chemistry, University of Bergen, Allégt. 41, NO-5007 Bergen, Norway.
2. Department of Chemistry, University of Pittsburgh, Pittsburgh, PA 15260, USA.
3. Skattebøl, L. *J. Org. Chem.* **1966**, *31*, 1554.
4. Depezay, J.-C.; Merrer, Y. L. *Tetrahedron Lett.* **1974**, 2751.
5. Marino, J. P.; Farina, J. S. *Tetrahedron Lett.* **1975**, 3901.
6. Grieco, P. A.; Wang, C.-L. J.; Majetich, G. *J. Org. Chem.* **1976**, *41*, 726.
7. Marino, J. P.; Farina, J. S. *J. Org. Chem.* **1976**, *41*, 3213.
8. Boeckman, Jr., R. K.; Ramaiah, M. *J. Org. Chem.* **1977**, *42*, 1581.
9. Smith, III, A. B.; Levenberg, P. A.; Jerris; P. J.; Scarborough, Jr., R. M.; Wovkulich; P. M. *J. Org. Chem.* **1981**, *103*, 1501.
10. Marino, J. P.; Linderman, R. J. *J. Org. Chem.* **1981**, *46*, 3696
11. Roush, W. R.; Peseckis, S. M. *J. Am. Chem. Soc.* **1981**, *103*, 6696.
12. Taylor, M. D.; Smith, III, A. B. *Tetrahedron Lett.* **1983**, *24*, 1867.
13. Caine, D.; Venkataramu, S. D.; Kois, A. *J. Org. Chem.* **1992**, *57*, 2960.
14. Paquette, L. A.; Montgomery, F. J.; Wang, T.-Z. *J. Org. Chem.* **1995**, *60*, 7857.
15. Paterson, I.; Fessner, K.; Finlay, M. R. V.; Jacobs, M. F. *Tetrahedron Lett.* **1996**, *37*, 8803.
16. An example is the synthesis of 3,3,4,4-tetraethoxybut-1-yne from 2-chloroacrolein diethyl acetal; see Sydnes, L. K. *Eur. J. Org. Chem.* **2000**, 3511; see also Sydnes, L. K.; Kvernenes, O. H.; Valdersnes, S. *Pure Appl. Chem.* **2005**, *77*, 119.
17. Mesnard, D.; Hanaï, N.; Miginiac, L. *J. Organomet. Chem.* **1996**, *519*, 125.
18. Shostakovskii, M. F.; Annenkova, V. Z.; Ivanova, L. T.; Ugryumova, G. S. *Izv. Sib. Otd. Akad. Nauk SSSR, Ser. Khim. Nauk* **1967**, 104; *Chem. Abstr.* **1968**, *69*, 51483s.

19. (a) Mark, H. F.; Bikales, N. M.; Overberger, C. G.; Menges, G. "Encyclopedia of Polymer Science and Engineering", 2nd edition; John Wiley & Sons: New York,1985, pp. 160-169; (b) Redtenbacher, J. *Ann. Chem. Pharm.* **1843**, *47*, 113.

20. Andreeva, I. V.; Koton, M. M.; Akopova, A. N.; Kukarkina, N. V. *Zh. Org. Khim.* **1975**, *11*, 954; *Chem. Abstr.* **1975**, *83*, 42773y.

21. For a detailed description of the Doering-Hoffman procedure, see Skattebøl, L.; Solomon, S. *Org. Synth. Collect. Vol. V* **1973**, 306.

22. (a) Jonczyk, A.; Balcerzak, P. "Prepartion of substituted *gem*-dichlorocyclopropanes", Pol. Patent PL156,825, 30 Apr 1992; *Chem. Abstr.* **1993**, *119*, 225616f; (b) Jonczyk, A.; Balcerzak, P. *Tetrahedron Lett.* **1989**, *30*, 4697.

23. Elkins, H. B. "The Chemistry of Industrial Toxicology"; John Wiley & Sons, New York, 1950, p. 132.

24. Huang, N.; Xu, L.-H. *Chinese Science Bulletin* **1991**, *36*, 831.

25. Makosza, M.; Wawrzyniewicz, M. *Tetrahedron Lett.* **1969**, 4659.

26. Kajigaeshi, S.; Kuroda, N.; Matsumoto, G.; Wada, E.; Nagashima, A., A. *Tetrahedron Lett.* **1971**, 4887.

27. Tao, F.; Zhu, Q.; Fang, Y.; Al, H.; Xu, L.; Zhang, X. *Youiji Huaxue* **1993,** *13*, 603; *Chem. Abstr.* **1994**, *121*, 34842h.

28. See note 11.

29. (a) Bjerrum, J. *Acta Chem. Scand.* **1964**, *18*, 843; (b) Libus, W.; Uruska, I. *Inorg. Chem.* **1966**, *5*, 256; (c) Biagetti, R. V.; Bottjer, W. G.; Haendler, H. M. *Inorg. Chem.* **1966**, *5*, 379.

CONVERSION OF ARYLALKYLKETONES INTO DICHLOROALKENES: 1-CHLORO-4-(2,2-DICHLORO-1-METHYLVINYL)BENZENE (1-Chloro-4-(2,2-dichloro-1-methylethenyl)-benzene)

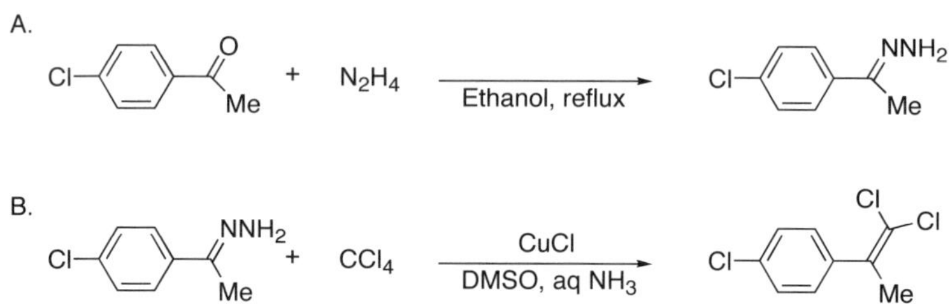

Submitted by Valentine G. Nenajdenko,[1] Vasily N. Korotchenko,[1] Alexey V. Shastin,[2] and Elisabeth S. Balenkova.[1]
Checked by Kristin Brinner and Jonathan A. Ellman.

1. Procedure

A. 1-(4-Chlorophenyl)ethanone hydrazone. A single-necked, 250 mL round-bottomed flask equipped with a reflux condenser is charged with absolute ethanol (100 mL) (Note 1), 100% hydrazine hydrate (9.7 mL, 0.2 mol) (Note 2) and 1-(4-chlorophenyl)ethanone (13.0 mL, 0.1 mol) (Note 3). The reaction solution is refluxed for 3 h. The ethanol and excess of hydrazine are removed by evaporation at ca. 50 °C/20 mm (Note 4). The slight-yellow crude 1-(4-chlorophenyl)ethanone hydrazone is redissolved in 10 mL of hot ethanol (Note 1). The mixture is cooled to –18 °C, kept at that temperature overnight and filtered. The solid is washed with hexane (2 x 20 mL) (Note 5) and dried under vacuum (2-3 mm) for 2 h to afford 13.6-14.8 g (81-88%) of colorless crystals (Note 6). The product is stored at –18 °C until used in step B (Note 7).

B. 1-Chloro-4-(2,2-dichloro-1-methylvinyl)benzene. A single-necked, 250 mL round-bottomed flask equipped with a magnetic stirring bar and a pressure-equalizing dropping funnel is charged with dimethyl sulfoxide (DMSO) (50 mL) (Note 8), 1-(4-chlorophenyl)ethanone hydrazone (8.4 g, 0.05 mol) (Note 6) and 28% aqueous ammonium hydroxide (16.8 mL, 0.25 mol) (Note 9). Freshly purified copper (I) chloride (500 mg, 0.005 mol) (Note 10) was added in one portion. After 5 min, carbon tetrachloride (24 mL, 0.25 mol) (Note 11) is added dropwise over 30 min (Note 12), while the temperature is maintained between 15-20 °C (water bath). After the addition is complete, the reaction mixture is stirred at room temperature overnight and quenched with 500 mL of 5% aqueous hydrochloric acid. The reaction products are extracted with hexane (4 x 50 mL) (Note 5). The combined extracts are dried over anhydrous sodium sulfate. After filtration and removal of the solvent by rotary evaporation (Note 13), the yellow crude product is purified by flash chromatography (5.0 x 5.0 cm) through silica gel (30 g, 63-200 mesh) using hexane (4-5 50 mL portions) (Notes 13 and 14) to afford 8.64-8.85 g (78-80% yield) of 1-chloro-4-(2,2-dichloro-1-methylvinyl)benzene as a colorless viscous oil that is > 98% pure by ^{1}H NMR analysis (Note 15).

2. Notes

1. Reagent grade ethanol was used without further purification by the checkers.

2. 100% Hydrazine hydrate was purchased from E. Merck.

3. 1-(4-Chlorophenyl)ethanone (98% purity) was purchased from E. Merck. The checkers purchased 97% purity material from Aldrich.

4. The checkers report that ethanol was removed by rotary evaporation at 30 mm with the bath temperature at 30 °C. Hydrazine was then removed at 10 mm with the bath temperature at 25 °C. The submitters report that the ethanol–hydrazine mixture collected in a trapping flask can be used for the preparation of next portion of hydrazone.

5. Reagent grade hexane was used without further purification.

6. mp 48.4-50.1 °C (the same melting point was observed after repeated recrystallizations; however, the submitters report a mp of 55-56 °C). ^{1}H NMR (400 MHz, DMSO-d_6) δ: 2.00 (s, 3H), 6.46 (s, 2H), 7.32 (d, $^3J_{HH}$ = 8.8, 2H), 7.61 (d, $^3J_{HH}$ = 8.8, 2H). ^{13}C NMR (100 MHz, DMSO-d_6) δ: 11.1, 126.7, 128.4, 131.9, 139.2, 141.1. IR (nujol) cm^{-1}: 3365, 3310, 3225, 2970, 2945, 2865, 1640, 1600, 1490, 1430, 1380, 1285, 1135, 1100, 1070, 1015, 970, 840.[3]

7. The submitters do not recommend the use of hydrazone stored more than 7 days as this results in decreased yields.

8. Reagent grade DMSO (99.6% purity) purchased from Sigma Aldrich was used without further purification.

9. Ammonium hydroxide (28% NH_3 in water) was purchased from the Sigma-Aldrich Company.

10. Commercially available copper (I) chloride (98% purity) purchased from Sigma-Aldrich (10 g) was dissolved in 100 mL of 10% HCl, precipitated by dilution with 1000 mL of water and filtered. The solid was washed with water (100 mL), ethanol (100 mL), acetone (100 mL) and ethyl ether (100 mL), and then dried under vacuum (2-3 mm) in a desiccator.

11. Reagent grade CCl_4 (99.5% purity) purchased from Sigma-Aldrich was used without further purification.

12. Intense gas evolution is observed. A characteristic color change from slight-green to blue occurs.

13. Care should be taken to keep the bath temperature at 25 °C during rotary evaporation of hexanes because the product is moderately volatile.

14. The main impurity (TLC (hexane:CH_2Cl_2; 2:1): R_f 0.45) was identified as the azine of 1-(4-chlorophenyl)ethanone.

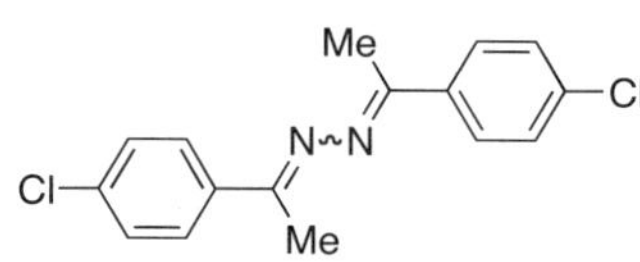

15. ^{1}H NMR (400 MHz, $CDCl_3$) δ: 2.17 (s, 3H), 7.19 (d, $^3J_{HH}$ = 8.8, 2H), 7.32 (d, $^3J_{HH}$ = 8.8, 2H). ^{13}C NMR (100 MHz, $CDCl_3$) δ: 22.9, 117.6, 128.6, 129.2, 133.6, 134.5, 138.4. IR (neat) cm^{-1}: 2930, 2860, 1915, 1650, 1615, 1595, 1490, 1430, 1400, 1380, 1265, 1100, 1070, 1040, 1020, 920, 840. TLC (R_f (hexane) 0.55). Found: C, 49.02; H, 3.14. Calcd for $C_9H_7Cl_3$: C, 48.80; H, 3.18 %.

Safety and Waste Disposal Information

All hazardous materials should be handled and disposed of in accordance with "Prudent Practices in the Laboratory"; National Academy Press; Washington, DC, 1995.

3. Discussion

Development of new catalytic methods for the construction of double carbon-carbon bonds is a goal of modern synthetic chemistry. Olefination of carbonyl compounds, that is, conversion of the carbonyl group into C=C fragment, is a widely used approach for the formation of alkenes. Wittig-type reactions, Peterson and Julia olefination and reductive couplings of carbonyl compounds are the most common olefination methods.[4,5] However, the necessity to use equimolar quantities or a large excess of phosphorus, silicon, sulfur reagents or metals is a significant disadvantage of these techniques.

Table 1

Preparation of Dichloroalkenes

Substrate	Product	Yield
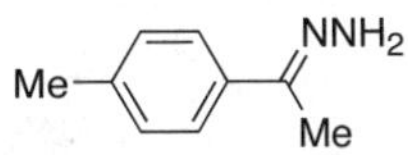	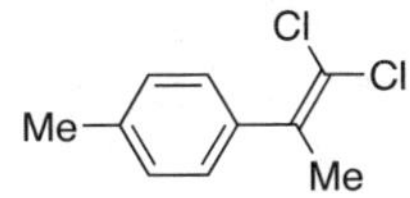	65
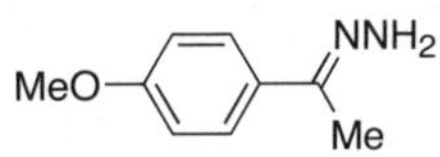	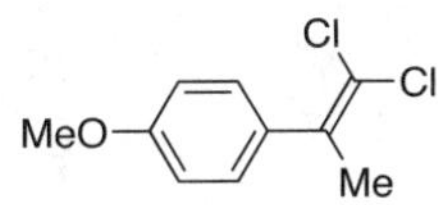	70
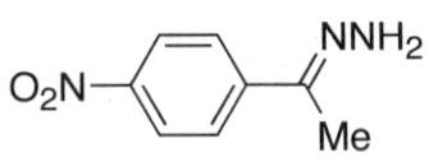	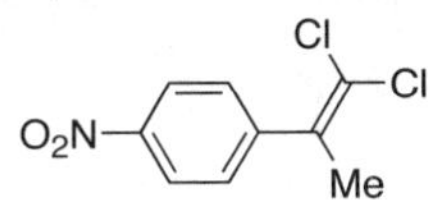	58
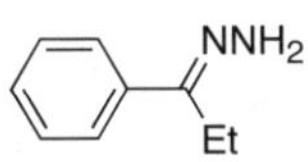	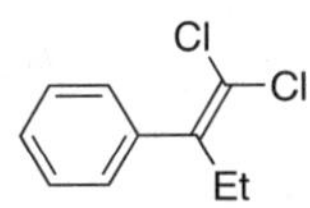	57

The present procedure describes a convenient preparation of dichloroalkenes from aryl alkyl ketones based on a novel catalytic olefination reaction (COR) of carbonyl compounds.[6] Some representative examples of dichloromethylenation of ketones are compiled in Table 1.

Recently we found that *N*-unsubstituted hydrazones of aldehydes and ketones could be transformed into alkenes when treated with polyhalogenated alkanes in the presence of copper (I) chloride. This method has already found applications in the synthesis of dihaloalkenes,[6] monohaloalkenes,[7] fluorinated olefins,[8] and a range of alkenes containing functional groups[9] from the corresponding carbonyl precursors (Figure 1). Mild reaction conditions and operational simplicity are important features of this catalytic olefination procedure.

Figure 1

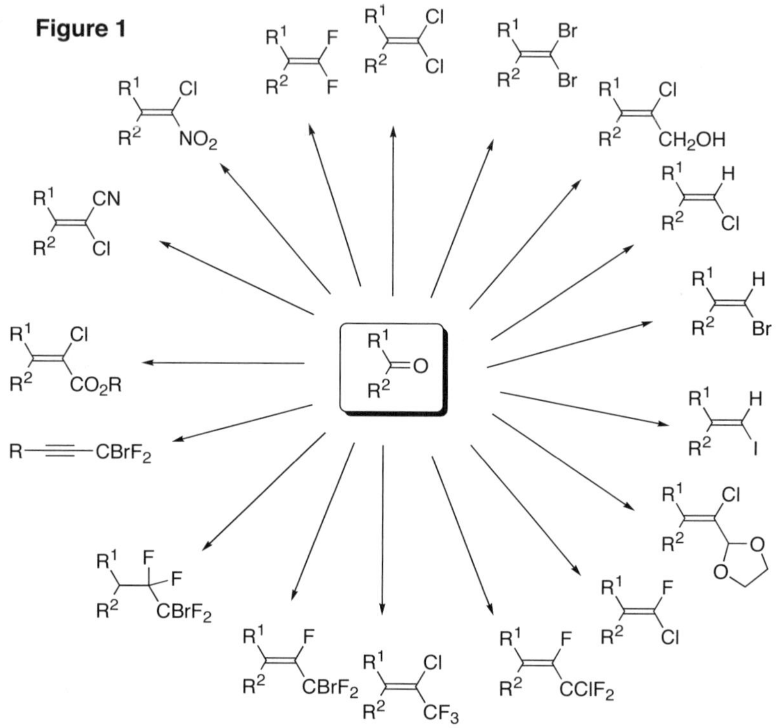

R^1 = Alkyl, Aryl, HetAryl
R^2 = H, Alkyl

1. Moscow State University, Department of Chemistry, Leninskie Gory, Moscow 119992, Russia. E-mail: nen@acylium.chem.msu.ru
2. Institute of Problems of Chemical Physics, Chernogolovka, Moscow region, 142432, Russia
3. Newkome, G. R.; Fishel, D. L. *J. Org. Chem.* **1966**, *31*, 677.
4. Lawrence, N. L. In *Preparation of Alkenes*; Williams, J. M. J., Ed.; Oxford University Press: Oxford, **1996.**
5. Kelly, S. E. In *Comprehensive Organic Synthesis*, eds. Trost, B. M.; Fleming, I. Ed.; Pergamon Press, Oxford, **1992**, pp. 730 – 810.
6. (a) Shastin, A. V.; Korotchenko, V. N.; Nenajdenko, V. G.; Balenkova, E. S. *Tetrahedron* **2000**, *56*, 6557; (b) Shastin, A. V.; Korotchenko, V. N.; Nenajdenko, V. G.; Balenkova, E. S. *Synthesis* **2001**, 2081; (c) Korotchenko, V. N.; Shastin, A. V.; Nenajdenko, V. G.; Balenkova, E. S. *J. Chem. Soc. Perkin Trans. 1* **2002**, 883;

(d) Korotchenko, V. N.; Shastin, A. V.; Nenajdenko, V. G.; Balenkova, E. S. *Org. Biomol. Chem.* **2003**, *1*, 1906.

7. (a) Shastin, A. V.; Korotchenko, V. N.; Nenajdenko, V. G.; Balenkova, E. S. *Russ. Chem. Bull.* **2001**, *50*, 1401; (b) Shastin, A. V.; Korotchenko, V. N.; Varseev G. N.; Nenajdenko, V. G.; Balenkova, E. S. *Russ. J. Org. Chem.* **2003**, *39*, 403.
8. (a) Korotchenko, V. N.; Shastin, A. V.; Nenajdenko, V. G.; Balenkova, E. S. *Tetrahedron* **2001**, *57*, 7519; (b) Nenajdenko, V. G.; Shastin, A. V.; Korotchenko, V. N.; Varseev G. N.; Balenkova, E. S. *Eur. J. Org. Chem.* **2003**, 302; (c) Nenajdenko, V. G.; Varseev G. N.; Korotchenko, V. N.; Shastin, A. V.; Balenkova, E. S. *J. Fluorine Chem.* **2003**, *124*, 115.
9. (a) Nenajdenko, V. G.; Lenkova O. N.; Shastin, A. V.; Balenkova, E. S. *Synthesis* **2004**, 573; (b) Nenajdenko, V. G.; Shastin, A. V.; Golubinskii I. V.; Lenkova O. N.; Balenkova, E. S. *Russ. Chem. Bull.* **2004**, *1*, 218.

GENERATION OF AN ACETYLENE-TITANIUM ALKOXIDE COMPLEX: PREPARATION OF (*Z*)-1,2-DIDEUTERIO-1-(TRIMETHYLSILYL)-1-HEXENE

[(Silane, (1,2-dideuterio-1-hexenyl)trimethyl-, (*Z*)-)]

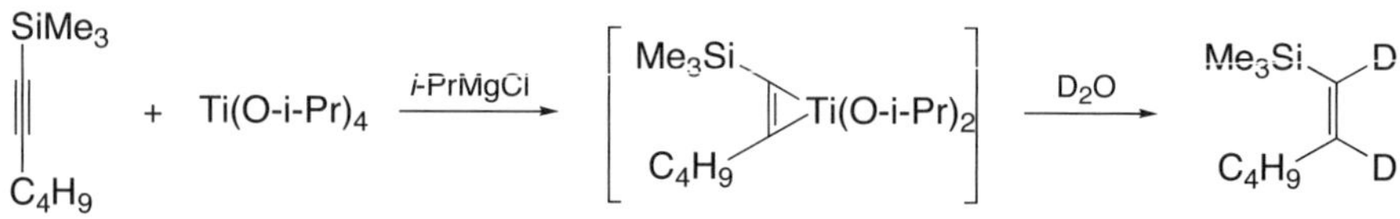

Submitted by Hirokazu Urabe,[1] Daisuke Suzuki,[1] and Fumie Sato.[1]
Checked by Katherine Hervert and Louis Hegedus.

1. Procedure

An oven-dried, 1-L, two-necked, round-bottomed flask is equipped with a stirring bar and 100-mL, pressure-equalizing addition funnel, the top of which is fitted with a rubber septum. The other neck of the flask is fitted with a three-way stopcock, one outlet of which is capped with a rubber septum and the other is connected alternately to a vacuum/argon gas system. The flask is evacuated, then flushed with argon; this operation is repeated twice. After 400 mL of anhydrous diethyl ether (Note 1), 1-(trimethylsilyl)-1-hexyne (Note 2) (5.00 g, 32.40 mmol) and freshly distilled titanium tetraisopropoxide (Note 3) (11.95 mL, 40.50 mmol) are added in this order through the septum of the three-way stopcock via syringe. The resulting solution is stirred and cooled in a dry ice-hexane bath to ca. −78 °C. The addition funnel is charged via syringe through the rubber septum with a solution of isopropylmagnesium chloride in diethyl ether (Note 4) (50.62 mL of a 2.0M solution in ether, 101.24 mmol), which then is added dropwise to the cold mixture over 30 min. The mixture is warmed to −50 °C over 30 min (and turns black) and is aged for 3 h at −50 °C to insure completion of reaction (Note 5). Deuterium oxide (12 mL) (Note 6) is rapidly added via the addition funnel to the cooled reaction mixture with stirring. The cooling bath is then removed and the reaction mixture is allowed to warm to room temperature. At this point, the addition funnel and three-way stopcock are removed. After anhydrous sodium sulfate (100 g) is added to the heterogeneous mixture, the suspension is stirred at room temperature for 10 min and filtered under vacuum through a short pad of Celite, which is washed with small portions of ether. The combined filtrate and washings are concentrated with a rotary evaporator (Note 7) to approximately 30 mL, which is transferred to a 50-mL distillation flask with the aid of small portions of ether. The crude product is purified by distillation under partial vacuum to furnish 3.56-4.47 g (69-87%) of the olefin, bp 128-129 °C (350 mm), as a colorless liquid (Notes 8, 9).

2. Notes

1. Anhydrous diethyl ether was purchased from Kanto Chemical Co., Ltd. (Japan).

2. 1-(Trimethylsilyl)-1-hexyne was purchased from Aldrich Chemical Co., Inc. and used as received. Alternatively, this can be prepared according to a method described in Organic Syntheses.[2]

3. Titanium isopropoxide was purchased from Tokyo Kasei Kogyo Co., Ltd. (Japan), distilled before use, and stored under an argon atmosphere.

4. A 2.0M solution of isopropylmagnesium chloride in diethyl ether was purchased from Aldrich Chemical Co., Inc. and used as received.

5. Reduction of the amount of the ethyl ether (initially 400 mL) increases the formation of the 2:1 acetylene-titanium alkoxide complex, i.e., a titanacyclopentadiene. However, under these conditions, hydrolysis of the reaction mixture reveals that the formation of 2,3-dibutyl-1,4-bis(trimethylsilyl)-1,3-butadiene arising from the titanacyclopentadiene is less than 3%. In any event, after distillation, this diene does not contaminate the desired product.

6. Deuterium oxide (100.0 atom% D) was purchased from Aldrich Chemical Co., Inc.

7. The product is moderately volatile so that extensive concentration under reduced pressure decreases the yield. When most of the solvent is removed under atmospheric pressure with a standard distillation apparatus (rather than with a rotary evaporator), the product is obtained in 87% yield.

8. The spectra are as follows: IR (neat) cm^{-1}: 2957, 2927, 2862, 1585 (C=C), 1459, 1248 (Si-Me), 839, 756. ^1H NMR (300 MHz, $CDCl_3$): δ 0.14 (s, 9 H, Me_3Si), 0.91 (t, J = 7.2, 3 H, Me), 1.35-1.55 (m, 4 H, $(CH_2)_2$), 2.22 (t, J = 7.2, 2 H, allylic CH_2). ^{13}C NMR (75 MHz, $CDCl_3$): δ 0.08 ($SiMe_3$), 13.91, 22.34, 31.90, 33.04, 128.23 (t, $J\ {}^{13}C\text{-}{}^2H$ = 20.5), 148.96 (t, $J\ {}^{13}C\text{-}{}^2H$ = 22.8). Anal. Calcd for $C_9H_{18}D_2Si$: C, 68.26. Found: C, 68.11. Deuterium incorporation α and β to the silyl group is determined by ^1H NMR analysis to be >99.5%.

9. Structural identification, except for the deuterium incorporation, is secured through (Z)-1-(trimethylsilyl)-1-hexene, a known compound obtained by hydrolysis of the acetylene-titanium complex with H_2O instead of D_2O using the same procedure. Spectra for the unlabeled product follow: IR (neat) cm^{-1}: 2958, 2925, 2859, 1607 (C=C), 1458, 1249 (Si-Me), 837, 763; ^1H NMR (300 MHz, $CDCl_3$): δ 0.11 (s, 9 H, Me_3Si), 0.90 (t, J = 7.2, 3 H, Me), 1.25-1.45 (m, 4 H, $(CH_2)_2$), 2.12 (symmetrical m, 2 H, allylic CH_2), 5.47 (d/t, J = 13.8, 1.2, 1 H, Me_3SiCH=C), 6.30 (d/t, J = 13.8, 7.2, 1 H, CH=$CHSiMe_3$). ^{13}C NMR (75 MHz, $CDCl_3$): δ 0.09 ($SiMe_3$), 13.90, 22.31, 31.99, 33.18, 128.80, 149.41. These data are in good agreement with reported values.[3] The ratio of Z and E isomers is determined to be >99.5:0.5 by ^1H NMR analysis of this sample. No contamination by the saturated silane is also confirmed by NMR analysis.

Waste Disposal Information

All toxic materials were disposed of in accordance with "Prudent Practices in the Laboratory"; National Academy Press; Washington, DC 1995.

3. Discussion

Isomerically pure 1-alkenylsilanes are useful intermediates in organic synthesis.[4] Thus, their deuterium-labeled counterparts are versatile precursors for the preparation of labeled organic molecules. Catalytic reduction of acetylenes using deuterium gas (D_2) is a convenient method to prepare D-labeled olefins.[5] However, some drawbacks have been observed, including scrambling between hydrogen atoms in the substrate and D_2 (which decreases the degree of deuterium uptake as well as the positional selectivity[6]), the variable E/Z ratio of the resulting olefinic bond,[7] and over-reduction of the acetylenic bond.[5] In addition, handling (and indeed availability) of the required amount of D_2 may be often problematic in some laboratories. The method described herein[8] is a convenient alternative to reduction with D_2 because (1) high deuterium incorporation of the vinylic positions and (2) formation of the Z-olefin is insured, and finally (3) inexpensive reagents (including D_2O as the source of deuterium) are utilized. The labeled olefins shown in the Table have been prepared by this method.[8,9]

Table. Preparation of (Z)-Dideuterio Alkenes

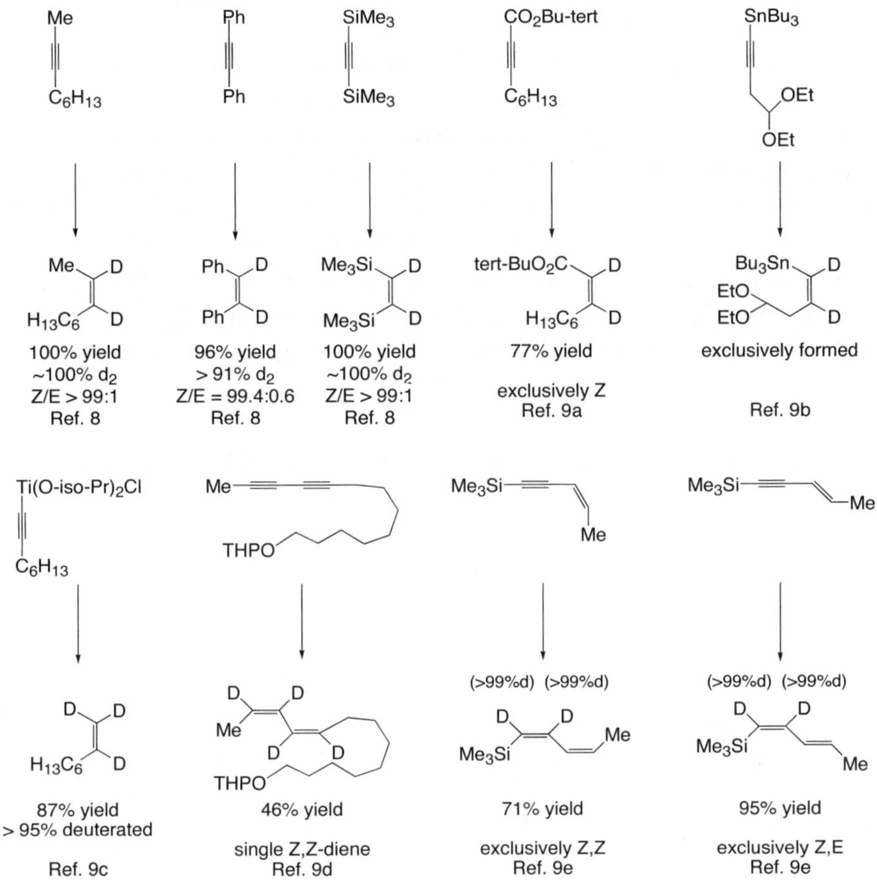

More importantly, the procedure represents a novel method for the generation of a new type of acetylene-Group 4 transition metal complex (**1** in the Scheme[10]). Acetylene-metal complexes often have been utilized as useful intermediates in organic synthesis, even though their structures may not be known.[11] The Scheme summarizes the four types of transformations available to the titanium complex **1**[10]: i) **1** can serve as a cis-1,2-bis-carbanionic species, which are otherwise difficult to generate, and affords adducts upon reaction with a variety of electrophiles (route a); ii) allenyl- or propargyltitanium reagents are generated from acetylenes having an appropriate leaving group at the propargylic position (route b); iii) the carbon-titanium bond of the complex reacts intramolecularly with an ester carbonyl group to give unsaturated carbonyl compounds (route c); iv)

complex **1** participates in coupling reactions with other olefins or acetylenes to give the corresponding titanacycles, enabling a variety of subsequent synthetic applications (route d).

The synthetic advantages of the titanium alkoxide complex **1** over other Group 4 metal complexes such as titanocene or zirconocene-based ones are: i) the titanium alkoxide is very inexpensive compared with metallocene reagents; ii) the reaction work up is very simple, since titanium is completely hydrolyzed to give water-soluble inorganic salts, readily separable from the desired organic product(s); iii) various functionalized acetylene complexes have been generated for the first time; and finally, iv) several reactions, some of which cannot be accomplished by conventional Group 4 metal complexes, are now feasible through these new acetylene complexes.[10]

Scheme. Generation of and Synthetically Useful Reactions of Acetylene-Titanium Complexes 1

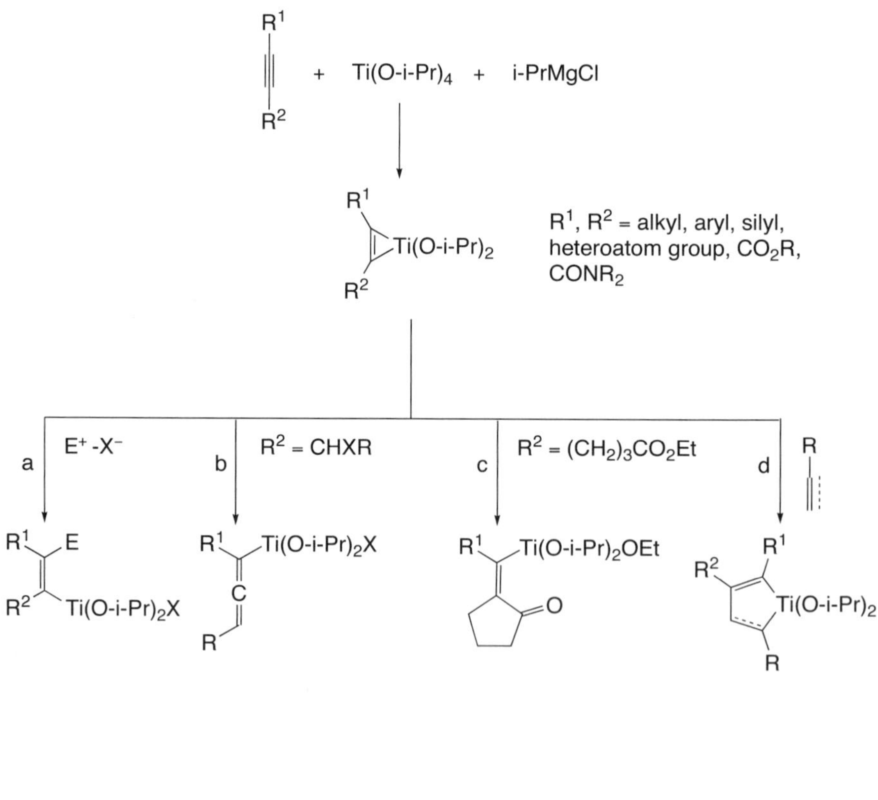

1. Department of Biomolecular Engineering, Graduate School of Bioscience and Biotechnology, Tokyo Institute of Technology, 4259-B-59 Nagatsuta-cho, Midori-ku, Yokohama, Kanagawa 226-8501 Japan. Correspondence should be addressed to H. U. (hurabe@bio.titech.ac.jp).
2. Stang, P. J.; Kitamura, T. *Org. Synth., Coll. Vol. IX*, **1998**, 477.
3. (a) Soderquist, J. A. ; Santiago, B. *Tetrahedron Lett.* **1990**, *31*, 5113. (b) Doyle, M. M.; Jackson, W R.; Perlmutter, P. *Aust. J. Chem.* **1989**, *42*, 1907. See also: (c) Sato, F.; Ishikawa, H.; Sato, M. *Tetrahedron Lett.* **1981**, *22*, 85. (d) Miller R. B.; McGarvey, G. *J. Org. Chem.* **1978**, *43*, 4424. (e) Miller, R. B.; Reichenbach, T. *Tetrahedron Lett.* **1974**, *15*, 543.
4. (a) Weber, W. P. In *Silicon Reagents for Organic Synthesis*; Springer-Verlag: Berlin, **1983**; p. 79. (b) Colvin, E. W. In *Silicon in Organic Synthesis*; Butterworth, Sevenoaks, **1981**; p. 44. (c) Fleming, I.; Dunogués, J.; Smithers, R. In *Organic Reactions*; Kende, A. S., Ed.; Wiley: New York, **1989**; Vol. *37*, p. 57. (d) Fleming, I.; Barbero, A.; Walter, D. *Chem. Rev.* **1997**, *97*, 2063.
5. Freifelder, M. *Catalytic Hydrogenation in Organic Synthesis. Procedures and Commentary*; Wiley: New York, 1978; p. 10. Rylander, P. N. In *Selections from the Aldrichimica Acta*; Aldrich: Milwaukee, 1984; p. 195; *Aldrichimica Acta*, **1979**, *12*, 53.
6. March J. *Advanced Organic Chemistry*; 4th ed., Wiley: New York **1992**; p. 776.
7. Ref. 4b, p. 48. Ref. 3d. Overman, L. E.; Brown, M. J.; McCann, S. F. *Org. Synth. Coll. Vol. VIII*, **1993**, 609.
8. Harada, K.; Urabe, H.; Sato, F. *Tetrahedron Lett.* **1995**, *36*, 3203.
9. (a) Hamada, T.; Suzuki, D.; Urabe, H.; Sato F. *J. Am. Chem. Soc.* **1999**, *121*, 7342. (b) Launay, V.; Beaudet, I.; Quintard, J.-P. *Synlett* **1997**, 821. (c) Averbuj, C.; Kaftanov, J.; Marek, I. *Synlett* **1999**, 1939. (d) Hungerford, N. L.; Kitching, W. *J. Chem. Soc., Perkin Trans. 1* **1998**, 1839. (e) Hamada, T.; Mizojiri, R.; Urabe, H.; Sato F. *J. Am. Chem. Soc.* **2000**, *122*, 7138. See also: (f) Sawada, D.; Kanai, M.; Shibasaki, M. *J. Am. Chem. Soc.* **2000**, *122*, 10521.
10. (a) Sato, F.; Urabe, H.; Okamoto, S. *Pure Appl. Chem.* **1999**, 71, 1511. (b) Sato, F.; Urabe, H.; Okamoto, S. *Chem. Rev.* **2000**, *100*, 2835. (c) Sato, F.; Urabe, H.; Okamoto, S. *Synlett* **2000**, 753. (d) Sato, F.; Okamoto, S. *Adv. Synth. Catal.* **2001**, *343*, 759-784. (e) Sato, F.; Urabe, H. In *Titanium and Zirconium in Organic Synthesis*; Marek, I., Ed.; Wiley-VCH: Weinheim, 2002, 319-354. (f) See also: Kulinkovich, O. G.; de Meijere, *A. Chem. Rev.* **2000**, *100*, 2789. (g) Eisch, J. J. *J. Organomet. Chem.* **2001**, *617-618*, 148.
11. (a) Buchwald, S. L.; Nielsen, R. B. *Chem. Rev.* **1988**, *88*, 1047. (b) Negishi, E.; Takahashi, T. *Acc. Chem. Res.* **1994**, *27*, 124. (c) Ohff, A.; Pulst, S.; Lefeber, C.;

Peulecke, N.; Arndt, P.; Burlakov, V. V.; Rosenthal, U. *Synlett* **1996**, *111*. (d) Negishi, E.; Takahashi, T. *Bull. Chem. Soc. Jpn.* **1998**, *71*, 755. (e) Kataoka, Y.; Miyai, J.; Oshima, K.; Takai, K.; Utimoto, K. *J. Org. Chem.* **1992**, *57*, 1973. (f) Hartung, J. B., Jr.; Pedersen, S. F. *J. Am. Chem. Soc.* **1989**, *111*, 5468. (g) Strickler, J. R.; Bruck, M. A.; Wexler, P. A.; Wigley, D. E. *Organometallics* **1990**, *9*, 266.

ALKYNE *via* SOLID-LIQUID PHASE-TRANSFER CATALYZED DEHYDROHALOGENATION: ACETYLENE DICARBOXALDEHYDE TETRAMETHYL ACETAL AND ACETYLENE DICARBOXALDEHYDE DIMETHYL ACETAL

[1,1,4,4-Tetramethoxy-2-butyne and 4,4-Dimethoxy-2-Butynal]

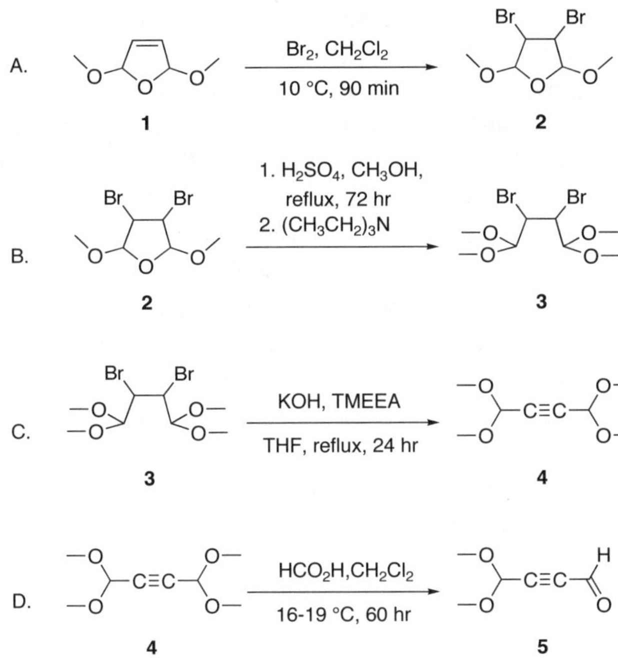

Submitted by Rufine Akué-Gédu and Benoît Rigo.[1]
Checked by Shigeki Matsunaga and Masakatsu Shibasaki.

1. Procedure

A. 3,4-Dibromo-2,5-dimethoxytetrahydrofuran (**2**). A 1-L three-necked, round-bottomed flask, equipped with a magnetic stirring bar, thermometer, pressure-equalizing addition funnel and water condenser connected to nitrogen, is loaded with dichloromethane (150 mL) (Note 2) and 2,5-dimethoxy-2,5-dihydrofuran (**1**) (50 g, 0.384 mol) (Note 3). The flask is cooled to 10 °C by using an ice bath. Bromine (61.4 g, 0.384 mol) is added over 120 min, while keeping the reaction temperature below 15 °C.

After addition is complete, the reaction mixture is stirred at 10 °C until the bromine color disappears. The solution is concentrated on a rotary evaporator while keeping the bath temperature below 20 °C (Note 4). 3,4-Dibromo-2,5-dimethoxytetrahydrofuran (**2**) is obtained as a light colored semi-solid mass (110-111 g, 99%, mixtures of isomers) (Note 5, 6).

B. 2,3-Dibromo-1,1,4,4-tetramethoxybutane (**3**). A 3-L three-necked, round-bottomed flask, equipped with a heating mantle, a pressure-equalizing addition funnel and a water condenser connected to nitrogen, is loaded with crude 3,4-dibromo-2,5-dimethoxytetrahydrofuran (**1**) prepared in Step A (110-111 g), methanol (2 L) and boiling chips. The solution is heated to reflux, at which time 96% sulfuric acid (37.7 g, 0.384 mol) is added over 30 min (Notes 7, 8). The solution is refluxed for 72 h, then cooled at room temperature. Triethylamine (49 mL, 0.352 mol) is added (Notes 9, 10), then the solution is concentrated on a rotary evaporator. Heptane (250 mL) is added and the residue is stirred and refluxed for 20 min. The lightly colored heptane phase is separated using a separatory funnel. The heptane extraction step is repeated three times. The combined heptane solutions are concentrated on a rotary evaporator to give 2,3-dibromo-1,1,4,4-tetramethoxybutane (**3**) as a yellow oil (99-100 g including inpurities, < 79%) (Note 11).

C. 1,1,4,4-Tetramethoxybut-2-yne (**4**). A 1-L one-necked, round-bottomed flask is equipped with a powerful magnetic stirrer, a large rugby ball-shaped stirring bar, and a water condenser equipped with a nitrogen inlet. The flask is charged with tetrahydrofuran (300 mL) (Note 3), tris[2-(2-methoxyethoxy)-ethyl]amine (TMEEA, 10.1 g, 31.2 mmol) (Note 3) and crude 2,3-dibromo-1,1,4,4-tetramethoxybutane (**3**) prepared in Step B (99 g). Potassium hydroxide pellets (85%) (70 g, 1.06 mol) (Note 12) are then added to the homogenous solution, and the stirred mixture is refluxed for 24 h using an oil-bath. Solvent is evaporated on a rotary evaporator, water (250 mL) is added, and the solution is partitioned three times with ether (250 mL) (Note 13). The combined ether phases are washed three times with water (100 mL). After the solution is dried over $MgSO_4$, the solution is concentrated on a rotary evaporator to give 1,1,4,4-tetramethoxybut-2-yne (**4**) as a yellow oil (50-51 g, 97-99%) (Note 14, 15, 16).

D. 4,4-Dimethoxybut-2-ynal (**5**). A 1-L round-bottomed flask, charged with dichloromethane (150 mL) (Note 2) and crude 1,1,4,4-tetramethoxybut-2-yne (**4**) (50 g, 287 mmol) prepared in Step C is cooled to 0 °C in an ice bath. Another flask charged with formic acid (280 g, 6 mol), dichloromethane (150 mL), and H_2O (1.5 mL) is also cooled to 0 °C. The cooled solution of formic acid is then rapidly added to the diacetal solution. The neck is sealed with a rubber septum equipped with a stainless steel syringe needle (Note 17). The flask is rapidly covered by a large opaque towel, and the reaction is stirred in the dark at 16-19 °C using a temperature regulated water bath for 60 h (Note 18). The dark brown solution is partitioned three times with ice cooled water (200 mL x 1, then 150 mL x 2). The aqueous phases are combined and extracted three times with dichloromethane (100 mL). The dichloromethane extracts are combined, washed with ice-cooled water (100 mL), and then dried with a mixture of Na_2SO_4/$NaHCO_3$ (10/1). The solution is concentrated on a rotary evaporator while keeping the bath temperature below 25 °C (Note 4), to give 4,4-dimethoxybut-2-ynal (**5**) (25.5 g, 69% from **4**, 52% from **1**) as a brown oil (Note 19, 20, 21, 22, 23).

2. Notes

1. This procedure is an adaptation of a procedure published earlier by our group.[2]

2. Dichloromethane was distilled from P_2O_5.

3. 2,5-Dimethoxy-2,5-dihydrofuran (99%) was purchased from Alfa Aesar; triethylamine (99.5%) was purchased from Aldrich; tetrahydrofuran (99+%) was purchased from Acros; TMEEA, (tris[2-(2-methoxyethoxy)ethyl]amine) (99%) was purchased from Alfa Aesar; formic acid (95-97%) was purchased from Aldrich. These compounds were used as received.

4. Heating the oil bath of the rotary evaporator leads to a more colored, less pure product.

5. While the product was sufficiently pure for the subsequent step, ^{1}H NMR indicated the presence of undetermined impurities and showed that a mixture of the three isomers of 3,4-dibromo-2,5-dimethoxytetrahydrofuran was obtained in a ratio depending mainly on the temperature and length of the evaporation of solvent (The ratio of products as determined by ^{1}H NMR 43/17/40 (submitters), 46/12/42 (checkers); Isomer 1 (46% of the mass, mp 88 °C) ^{1}H NMR ($CDCl_3$) δ: 3.48 (s, 3H), 3.50 (s, 3H), 4.16 (dd, J = 9.6, 3.9 Hz, 1H), 4.26 (dd, J =9.6, 3.9 Hz, 1H), 4.94 (d, J = 3.9 Hz, 1H), 5.21 (d, J = 3.9 Hz, 1H); Isomer 2 (12% of the mass, mp 56 °C) ^{1}H NMR ($CDCl_3$) δ: 3.47 (s, 6H), 4.42 (dd, J = 1.9, 0.7 Hz, 2H), 5.23 (dd, J = 1.9, 0.7 Hz, 2H);

Isomer 3 ^{1}H NMR ($CDCl_3$) δ: (42% of the mass, mp 72 °C) 3.50 (s, 6H), 4.18 (dd, J = 1.9, 1.3 Hz, 2H), 5.29 (dd, J = 1.9, 1.3 Hz, 2H).

6. The submitter describes the following unchecked procedure to separate three isomers; the mixture of three isomers of **2** can be partially separated by cooling the crude semisolid mass at 0 °C overnight. Filtration of the solid and washing with cold (0 °C) heptane, gave 78 g of isomer with mp 88-89 °C. Heptane was added to the residue, and the solution was cooled at 0 °C. Filtration and washing of the solid with cold (0 °C) heptane, gave isomer with mp 72-74 °C. Chromatography of the residue on alumina can give the third isomer (mp 56-58 °C).[3]

7. If a lesser amount of sulfuric acid is used, the reaction takes a longer time and leads to less pure product.

8. Slow addition of sulfuric acid to the hot reaction mixture allows for better control of the exothermicity of the addition, and provides a less colored final product.

9. Neutralization of sulfuric acid before evaporation of methanol avoids decomposition and colorization of diacetal **3**.

10. Triethylamine was selected for the neutralization because it decomposes the methyl sulfate formed from reaction of sulfuric acid with methanol.

11. Product **3** is difficult to purify, but remaining impurities are removed during the treatment of the next step. ^{1}H NMR ($CDCl_3$) δ: 3.44 (two singlet peaks overlapped with impurities, 12 H), 4.34 (d, J = 7.9 Hz, 2H), 4.58 (d, J = 7.9 Hz, 2H).

12. Potassium hydroxide pellets are better than finely ground powder, because the powder readily solidifies into a mass causing the reaction to slow down.

13. It was necessary to use diethyl ether instead of CH_2Cl_2 to avoid extraction of the phase transfer agent.

14. It was possible to perform the dehydrobromination step with potassium *tert*-butoxide in DMSO (2 h, 20 °C in a smaller scale), but treatment of the reaction mixture was less efficient and the purity of the product was lower.

15. While the product was sufficiently pure for the subsequent step, ^{1}H NMR indicated the presence of a small amount of undetermined impurities. If very pure product is needed for another use, compound **4** can be purified by distillation through a 10-cm Vigreux column: bp 55-59 °C (0.5 mmHg) [bp = 70-73 °C (1 mmHg),[4] bp = 101-103 °C (13 mmHg)[5]] to give a colorless liquid. ^{1}H NMR ($CDCl_3$) δ ppm: 3.40 (s, 12H), 5.21 (s, 2H).

16. Alternatively, it was not necessary to perform heptane extraction of compound **3**: the solution obtained after neutralization **3** with triethylamine (Step B) was evaporated on a rotary evaporator. Tetrahydrofuran (250 mL) (Note 2), TMEEA (0.03 mol, 10.1 g) (Note 2) and pellets of 85% potassium hydroxide (110 g, 1065 mol) (Note 11) were added. The mixture was refluxed for 24 h, and then part of THF was evaporated. Water was added and the mixture was extracted with ether, leading to 56 g of **4**. The weight of diacetal **4,** including impurities, is higher with no heptane extraction (56 g for Steps B and C); however, the purity of **4** is higher using the heptane extraction procedure.

17. The syringe needle is used for carbon dioxide evolution from the reaction mixture containing formic acid and acetal **4**.

18. It is important to perform the deprotection step below 20 °C in the dark, because exposure of the reaction mixture to a higher temperature or to light can lead to a lower yield through formation of a highly colored by-product. The reaction time and yield of Step D strongly depended on the quality of formic acid used. 95-97% formic acid from Aldrich gave best results. Drier formic acids: 99% from Acros Organics (Cat. No. 14793), 98% (Cat. No. 8841) from Lancaster, 97% from Avocado (Cat. No. 13285), gave decomposition of the monoacetal and formation of black tars. Even using 95-97% formic acid from Aldrich, difference in lot number may cause different yield and reaction time. Although addition of a small amount of H_2O (1.5 mL) was not a part of the submitter's original procedure, the isolated yield of **5** by the checkers is slightly improved by addition of a small amount of H_2O to the reaction mixture. Loss of product during work-up can also be problematic. It is necessary to wash the organic extracts with ice-cooled water quickly to avoid decomposition of product during extraction and washing. The temperature of the water bath also affected the reaction time significantly. The reaction was judged to be complete after 60 h at 16-19 °C (by the checkers), after 32 h at 17-20 °C (by the checkers), and after 20 h at 20-21 °C (by the submitters).

19. Because the formic acid quality, the reaction temperature, the quality of crude **4**, and the amount of water greatly influence the reaction time and isolated yield, it is essential to follow the reaction by NMR analysis of the crude reaction mixture. A 0.5 mL sample of the reaction mixture is evaporated under reduced pressure (2 mmHg) below 20 °C and dissolved in $CDCl_3$. A resonance at 5.20 ppm ($C\mathit{H}(OMe)_2$ of dialdehyde) is compared

with resonances at 5.31 ppm {C*H*(OMe)$_2$} and 9.27 ppm {(C*H*O) of the monoaldehyde}.

20. ^{1}H NMR spectrum of 4,4-dimethoxybut-2-ynal (**5**) indicated the presence of tiny amount of undetermined impurities. ^{1}H NMR ($CDCl_3$) δ: 3.41 (s, 6H), 5.31 (s, 1H), 9.27 (s, 1H); ^{13}C NMR ($CDCl_3$) δ: 53.0, 83.0, 88.3, 92.7, 176.0. Purity of this crude compound is sufficient for participation in Michael reactions with nucleophiles. Aldehyde **5** remains unaltered when kept for years at -40 °C.

21. While the brown color of the crude compound can be removed by filtration on silica, most impurities remain in the product. [A solution of the product in CH_2Cl_2 was applied to the top of a short column of flash-grade silica gel (200 g, 10-cm diam x 4 cm) loaded with CH_2Cl_2. The brown color remains on the top of the column upon washing with CH_2Cl_2. Last traces of solvent were removed at 0.5 mmHg to afford **5** as a slighly yellow liquid.]

22. If one needs a purer product, compound **5** can be purified by distillation: 24 g of aldehyde **5** and a rugby ball-shaped stirring bar were introduced in a 50-mL one-neck flask immersed in an oil-bath, and fitted with a 10-cm Vigreux column, a water condenser cooled at –10 to –20 °C by a cooling regulator, and a receiving flask cooled at –40 °C by cooling bath. Aldehyde **5** was distilled under vacuum (0.2 mmHg), while the temperature of the oil-bath was slowly raised from 25 °C to 60 °C by using a temperature regulator. Aldehyde **5** (17.3 g (72 %), bp 34-37 °C (0.2 mmHg) [bp = 75-78 °C (14 mm)].)[4] was obtained as a slightly yellow liquid.

23. While working in a 20 to 50 g scale of diacetal **4**, compound **5** was obtained in 85-92 % yields (by the submitters) and in 72% (by the checkers in 25 g scale).

Safety and Waste Disposal Information

All hazardous materials should be handled and disposed of in accordance with "Prudent Practices in the Laboratory"; National Academy Press; Washington, DC, 1995.

3. Discussion

This procedure describes the synthesis of acetylene dicarboxaldehyde mono and bis *dimethyl* acetals. It was reported[4,6] that acetylene dicarboxaldehyde *dimethyl* acetals are less reactive than their *diethyl* analogs. The new procedure takes advantages of this slightly lower reactivity and higher stability of the *dimethyl* acetals (Reaction time for the formic acid hydrolysis described in Step D is 20-60 h when using *dimethyl* acetals whereas it is 10 h when using *diethyl* acetals). Acetylene dicarboxaldehyde[4-15] and its precursors mono and bis *diethyl* acetals are useful starting materials, which have found many applications in dipolar cycloaddition,[7-11] Diels-Alder,[4,6,7,12,13] Michael,[4,6,7,13] or Wittig[15] reactions.

A major drawback for a larger use of these compounds is their high price[16] or difficulties in their preparation, which preclude a large-scale synthesis. Two main procedures for the synthesis of these compounds are described in the literature. In a first reaction sequence, addition of bromine to acrolein followed by a dehydrobromination step leads to acetylene carboxaldehyde diethyl acetal. Reaction of this compound with triethyl orthoformate using $ZnCl_2$ afforded the ethyl analog of acetal **4**.[2] In order to obtain 50 g of diacetal using this method,[2,6,17] it would be necessary to use 175 g of triethyl orthoformate, 345 g of KOH and 575 g of tetrabutylammonium hydrogen sulfate. In another approach, acetylene was reacted with ethyl magnesium bromide to give a bis magnesium salt. Reaction of this salt with triethyl orthoformate yields the ethyl analog of acetal **4**. In order to obtain 50 g of the diacetal using this second method,[4,6] it would be necessary to use 200 g of triethyl orthoformate, 19 g of magnesium and 260 L of gaseous acetylene. During this last procedure, it is necessary to avoid precipitation of the intermediate bis-magnesium salt of acetylene, which due to difficulties in stirring can lead to poor yields.[4]

The use of 99% formic acid to perform hydrolysis of one functional group of acetylene dicarboxaldehyde tetramethyl acetal is an adaptation of a method from Gorgues.[4,5,18]

The procedure described above reports an easy, high yielding, large-scale preparation of dimethyl acetals **4** from inexpensive 2,5-dimethoxy-2,5-dihydrofuran (**1**), a compound that contains all the carbon atoms of acetylene dicarboxaldehyde. The use of the phase transfer agent TMEEA is very important for this reaction,[19] because TMEEA enables the use of KOH and THF instead of DMSO and potassium *tert*-butoxide in the dehydrobromination step.

1. Groupe de Recherche sur l'Inhibition de la Prolifération Cellulaire (EA 2692), Ecole des Hautes Etudes Industrielles, 13 rue de Toul, 59046 Lille, France.

2. Akué-Gédu, R.; Rigo B. *Tetrahedron Lett.,* **2004**, *45*, 1829.
3. Gagnaire, D.; Vottero, P. *Bull. Soc. Chim. Fr.* **1963**, 2779.

4. Gorgues, A.; Stephan, D.; Belyasmine, A.; Khanous, A.; Le Coq, A. *Tetrahedron* **1990,** *46*, 2817.

5. Wohl, A.; Bernreuter, E. *Ann. Chem.* **1930**, *481*, 1.

6. Gorgues, A.; Simon, A.; Le Coq, A.; Hercouet, A.; Corre, F. *Tetrahedron* **1986,** *42*, 351.

7. Gorgues, A. *Janssen Chim. Acta*, **1986,** *4*, 21.

8. Gorgues, A.; Le Coq, A. *Tetrahedron Lett.* **1979,** *20*, 4829.

9. Henkel, K.; Weygand, F. *Chem. Ber.* **1943**, *76B*, 812.

10. Ramanaiah, K. C. V.; Stevens, E. D.; Trudell, M. L.; Pagoria, P.F. *J. Heterocyclic Chem.* **2000**, *37*, 1597.

11. Frère, P.; Belyasmine, A.; Gouriou, Y.; Jubault, M.; Gorgues, A.; Duguay, G.; Wood, S.; Reynolds, C.D.; Bryce, M.R. *Bull. Soc. Chim. Fr.* **1995**, *132*, 975.

12. Gorgues, A.; Le Coq, A. *Chem. Commun.* **1979**, 767.

13. Stephan, D.; Gorgues, A.; Belyasmine, A.; Le Coq, A. *Chem. Commun.* **1988,** 263.

14. Blanco, L.; Bloch, R.; Bugnet, E.; Deloisy, S. *Tetrahedron Lett.* **2000**, *41*, 7875.

15. Barrett, A. G. M.; Hamprecht, D.; Ohkubo, M. *J. Org. Chem.* **1997**, *62*, 9376.

16. 3,3-Diethoxy-1-propyne: 329 €/25 g (Lancaster); acetylene-dicarboxaldehyde tetraethyldiacetals: 75 €/1 g (Acros); acetylene- dicarboxaldehyde monodiethylacetal: 78 €/1 g (Acros); 2,5-dimethoxy-2,5-dihydrofuran (**1**): 191 €/1000 g (Alfa Aesar).

17. (a) Le Coq, A.; Gorgues, A. *Org. Synth. Coll. Vol. VI*, **1988**, 954; (b) Montaña, A. M.; Fernandez, D.; Pagès, R.; Filippou, A.C.; Kociok-Köhn, G. *Tetrahedron* **2000**, *56*, 425.

18. Gorgues, A. *Bull. Soc. Chim. Fr.* **1974**, 529.

19. (a) Soula, G. *J. Org. Chem.* **1985**, *50*, 3717; (b) Stafford, A.; McMurry, J. E. *Tetrahedron Lett.,* **1988**, *29*, 2531. This phase transfer agent strongly accelerates dehydrobromination of alkyl halides in the presence of KOH. (c) Tamao, K.; Nakagawa, Y.; Ito, Y. *Org. Synth. Coll. Vol. IX* **1998**, 539.

GENERATION OF NONRACEMIC 2-(*t*-BUTYLDIMETHYLSILYLOXY)-3-BUTYNYLLITHIUM FROM (*S*)-ETHYL LACTATE: (*S*)-4-(*t*-BUTYLDIMETHYLSILYLOXY)-2-PENTYN-1-OL

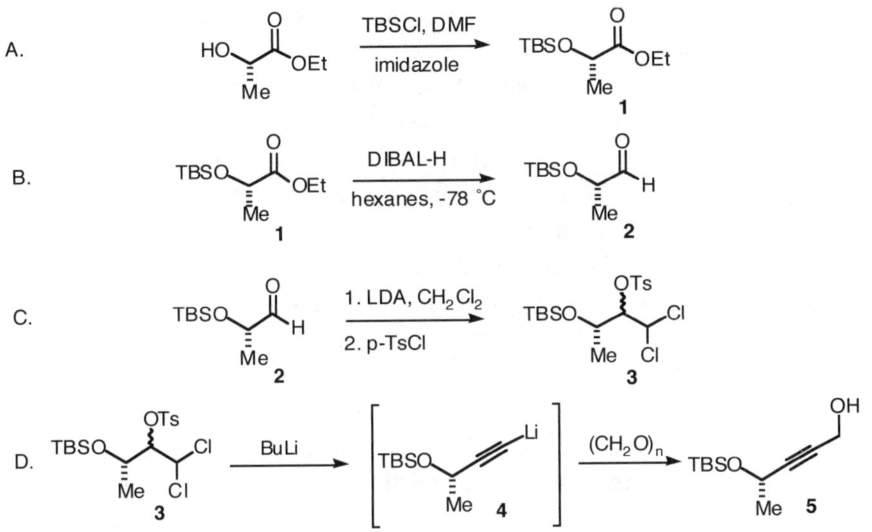

Submitted by James A. Marshall, Mathew M. Yanik, Nicholas D. Adams, Keith C. Ellis, and Harry R. Chobanian.[1]
Checked by Venugopal Gudipati, Tiffany Turner, and Dennis P. Curran.

1. Procedure

A. (S)-Ethyl 2-(t-Butyldimethylsilyloxy)propanoate (**1**). A 2-L, two-necked, round-bottomed flask equipped with a mechanical stirrer and inert gas inlet (Note 1) is charged with (*S*)-ethyl lactate (118 g, 1.0 mol), 500 mL of dimethylformamide (DMF), and imidazole (102 g, 1.5 mol) (Note 2). The solution is cooled in an ice bath and *tert*-butyldimethylsilyl chloride (TBDMSCl) (150 g, 1.0 mol) is added in three 50-g portions, at intervals of 30 min between each addition. After the addition of the third portion, a white precipitate forms. The ice bath is allowed to melt gradually overnight. After 18 h, the reaction mixture is diluted with 300 mL of water and 500 mL of hexanes. The aqueous phase is separated and extracted with 300 mL of hexanes, and the combined hexane extracts are washed with three 50-mL portions of saturated brine, dried over $MgSO_4$, filtered, and concentrated by rotary evaporation to afford 240 g (103%) of the TBDMS ether as a colorless liquid. The product is distilled under vacuum (bp 70-78 °C, 0.5 mmHg; bath temperature 95-105 °C) (Note 3) to afford 222 g (96%) of ester **1** as a colorless liquid (Notes 4, 5).

B. (S)-2-(t-Butyldimethylsilyloxy)propanal (**2**). A 2-L, single-necked, round-bottomed flask equipped with a large magnetic stirbar, rubber septum, and inert gas inlet (Note 1) is charged with (*S*)-ethyl 2-(*t*-butyldimethylsilyloxy)propanoate (**1**) (69.9 g, 300 mmol) and 600 mL of hexanes, and cooled in a dry ice-acetone bath at –78 °C. A 500-mL, round-bottomed flask equipped with a rubber septum and inert gas inlet is charged with 310 mL of DIBAL-H (1.0M in hexanes, 310 mmol) (Note 6) and cooled to –78 °C. The DIBAL-H solution is transferred by cannula into the well-stirred solution of ester over 20 to 25 min. After completion of the addition, the reaction mixture is stirred for 1 h at –78 °C and then quenched by addition of 30 mL of MeOH and stirred for 15 min at –78 °C. The cold solution is transferred to a 2-L flask equipped with a mechanical stirrer containing 600 mL of saturated Rochelle salt and the resulting mixture is vigorously stirred for 3.5 h (Note 7). The aqueous phase is separated and extracted with 200 mL of hexanes, and the combined organic extracts are washed with 100 mL of saturated brine, dried over $MgSO_4$, filtered, and concentrated by rotary evaporation (Note 8). After removal of the solvent, the residue is distilled (bp 45-52 °C, 0.5 mmHg, bath temperature 65-80 °C) to afford 50.3-50.9 g (91-92%) of aldehyde **2** as a colorless oil (Notes 9, 10).

C. 1,1-Dichloro-(3S)-(t-butyldimethylsilyloxy)-2-butyl p-toluenesulfonate (**3**). A 2-L, single-necked, round-bottomed flask equipped with a large magnetic stirbar and a 500-mL pressure equalizing addition funnel (Note 1) is charged with aldehyde **2** (44.3 g, 235 mmol), dichloromethane (45.0 mL, 705 mmol), and 500 mL of tetrahydrofuran (Note 11) and the mixture is cooled in a dry ice-acetone bath at –78 °C. A 500-mL, three-necked, round-bottomed flask equipped with a magnetic stirbar, a graduated 150-mL pressure-equalizing addition funnel, and two rubber septa is charged with 250 mL of THF and diisopropylamine (49.3 mL, 376 mmol) and then cooled with an ice bath. The addition funnel on the 500-mL flask is charged with 141 mL of BuLi solution (2.5M in hexanes, 352 mmol) which is added dropwise to the solution of amine over 30 min. The resulting solution of LDA is transferred by cannula to the addition funnel on the 2-L flask, and this solution is added dropwise over 1 h to the reaction mixture, resulting in a color change to light yellow. After 30 min, the mixture is warmed to 0 °C for 30 min with a change in color to dark brown. The addition funnel is replaced with a rubber septum, and *p*-toluenesulfonyl chloride (44.8 g, 235 mmol) is added in one portion through a funnel. After 10 min, the ice bath is removed and the mixture is stirred for 1.5 h. The reaction is quenched by addition of 10 mL of water, stirred for 30 min, and then transferred to a separatory funnel and extracted with 250 mL of 10% HCl and 150 mL of 1N NaOH solution. Each of these aqueous extracts is separately back-extracted with 500 mL of Et_2O, and the combined organic phases are washed with 100 mL of brine, dried over $MgSO_4$, and filtered into a 1-L, round-bottomed flask for use in the next step. The solvent is removed under reduced pressure whereupon 80-85 g of the dichloro tosylate **3** is obtained as a dark brown oil which is used in the next step without purification (Note 12).

D. (S)-4-(t-Butyldimethylsilyloxy)-2-pentynol (**5**). THF (500 mL) is added to tosylate **3** (76 g, *ca* 177 mmol) prepared as described above, and the 1-L flask is equipped with a large magnetic stirbar and a pressure-equalizing addition funnel fitted with a nitrogen inlet. The solution is cooled in a dry ice-acetone bath at –78 °C for 20 min and the addition funnel is charged with 219 mL of BuLi solution (2.5M in hexanes, 548 mmol) which is then added dropwise over ca. 1 h to the reaction mixture. When the addition is complete, the solution is stirred for 30 min at –78 °C and warmed over 1 h to 0 °C. The reaction mixture is then recooled to –78 °C. To the resulting black solution of lithium acetylide **4** is then added paraformaldehyde (10.6 g, 354 mmol) (Note 13) in one portion. After 15 min, the mixture is allowed to warm to room temperature over 4 h during which time the suspension of paraformaldehyde gradually dissolves. The reaction is quenched by addition of 200 mL of aqueous ammonium chloride solution, and the aqueous layer is separated and extracted with three 200-mL portions of Et_2O. The combined organic extracts are washed with 100 mL of 1N NaOH and 50 mL of brine, dried over $MgSO_4$, filtered, and concentrated under reduced pressure to give the desired product contaminated with *tert*-butyldimethylsilanol. The product is purified by fractional distillation (Note 14) to give 21-24 g (44-53% overall from aldehyde **2**) of the alcohol **5** as a colorless oil (Notes 15, 16).

2. Notes

1. All glassware and needles were dried in an oven at 120 °C overnight and assembled under a nitrogen purge or flame-dried immediately

prior to use. All reactions were performed under nitrogen (submitters) or argon (checkers).

2. (*S*)-Ethyl lactate (98%) and imidazole (99+%) were purchased from Aldrich Chemical Co. TBDMSCl was purchased from FMC Corporation (submitters) or Aldrich (checkers). A newly opened bottle of dimethylformamide (ACS Reagent grade, 0.02% water) was used as received.

3. The distillation was conducted in a 500-mL, round bottomed flask equipped with a magnetic stirring bar and a variable take-off distillation head.

4. The ester **1** displayed the following properties: $[\alpha]_D$ –25.9 (c 1.56, $CHCl_3$); IR (thin film): cm^{-1} 1753; ^1H NMR (300 MHz, $CDCl_3$) δ: 0.05 (s, 3H), 0.08 (s, 3H), 0.88 (s, 9H), 1.23 (t, J = 7.2 Hz, 3H), 1.36 (d, J = 6.6 Hz, 3H), 4.11 (m, 1H), 4.25 (q, J = 7.2 Hz, 2H); ^{13}C NMR (75 MHz,$CDCl_3$) δ: –5.3, –4.9, 14.2, 18.3, 21.3, 25.7, 60.7, 68.4, 174.1.

5. The submitters report that by the same procedures without modifications the (*R*)-enantiomer can be prepared from (*R*)-isobutyl lactate, available from Sigma Chemical Co. The physical properties for (*R*)-isobutyl 2-(*t*-butyldimethylsilyloxy)-propanoate are: bp 85-88 °C, 0.1 mmHg; $[\alpha]_D$ +28.7 (c 1.61, $CHCl_3$); IR (thin film) 1763 cm^{-1}; ^1H NMR (300 MHz, $CDCl_3$) δ: 0.06 (s, 3H), 0.09 (s, 3H), 0.89 (s, 9H), 0.91 (d, J = 6.6 Hz, 6H), 1.38 (d, J = 6.6 Hz, 3H), 1.94 (septet, 1H), 3.88 (dq, J = 10.5, 6.9 Hz, 2H), 4.28 (q, J = 6.6 Hz, 1H); ^{13}C NMR (125 MHz, $CDCl_3$) δ: –5.4, –5.0, 18.3, 19.0, 21.4, 25.7, 27.7, 68.4, 70.8, 174.1

6. DIBAL-H (1.0M in hexanes) was purchased from Aldrich Chemical Co. Hexanes from a freshly opened bottle (ACS Reagent Grade) was used as solvent.

7. Potassium sodium tartrate (Rochelle salt) was purchased from Fluka (purum p.a. grade). The quantity of salt solution specified was found to be optimal (2 mL/mmol of DIBAL-H). Use of less salt resulted in incomplete complexation. The submitters reported similar yields with stirring overnight.

8. Concentration is carried out without heating. Heating the rotary evaporator bath above 35 °C results in lower yields due to the volatility of the aldehyde.

9. The aldehyde contains small amounts of the starting ester and the overreduced alcohol along with other minor impurities. It can be stored for short periods of time (1-2 days) in a freezer at –20 °C without significant deterioration. However long term storage is not recommended.

10. The enantiomeric excess of aldehyde **2** was estimated to be >96% by derivatization as the Schiff bases with (*S* and *R*)-α-methylbenzylamine as described below. Spectral characteristics for aldehyde **2**: $[\alpha]_D$ –12.1 (c 1.96, $CHCl_3$); IR (thin film): cm^{-1} 1741; ^1H NMR (300 MHz, $CDCl_3$) δ: 0.07 (s, 3H), 0.08 (s, 3H), 0.9 (s, 9H), 1.26 (d, *J* = 6.9 Hz, 3H), 4.07 (dq, *J* = 6.9, 1.3 Hz, 1H), 9.59 (d, *J* = 1.3 Hz, 1H); ^{13}C NMR (125 MHz, $CDCl_3$) δ: –4.8, 18.1, 18.5, 25.6, 73.8, 204.2.

(S) and (R)-α-Methylbenzylimines of (S)-4-(t-Butyldimethylsilyloxy) propanal. To a solution of (*S*)-aldehyde **2** (0.150 g, 0.74 mmol), 4Å molecular sieves (0.100 g), in 1 mL of CH_2Cl_2 was added (*S*)-α-methylbenzylamine (100 μL, 0.77 mmol, Aldrich, 98% ee by GLC analysis). After 2 h, the mixture was filtered through Celite, rinsed with CH_2Cl_2, and concentrated affording the (*S*,*S*)-imine as a colorless oil (0.215 g, 95%). The imine from (*R*)-α-methylbenzylamine (96% ee by GLC analysis) was prepared in identical fashion. The ratio of diastereoisomers determined through integration of the ^1H NMR spectra was 98:2 for the (*S*,*S*) derivative and 96.5:3.5 for the (*S*,*R*) derivative. **(*S*,*S*) Imine**: $[\alpha]_D$ –55.2 (c 1.20, $CHCl_3$); IR (thin film) 2968, 2951, 2864, 1675 cm^{-1}; ^1H NMR ($CDCl_3$, 300 MHz) δ: 0.11 (s, 6H), 0.93 (s, 9H), 1.26, (d, *J* = 6.3 Hz, 3H), 1.50 (d, *J* = 6.6 Hz, 3H), 4.29 (q, *J* = 6.6 Hz, 1H), 4.38 (q, *J* = 6.3 Hz, 1H), 7.24-7.34 (m, 5H), 7.62 (d, *J* = 5.4 Hz, 1H); ^{13}C NMR (125 MHz, $CDCl_3$) δ: –4.7, –4.6, 18.2, 21.8, 24.3, 25.8, 68.9, 70.7, 126.5, 126.8, 128.4, 144.6, 166.4. **(*S*,*R*) Imine**: $[\alpha]_D$ +46.2 (c 1.20 $CHCl_3$); ^1H NMR δ: –0.03 (s, 3H), 0.03 (s, 3H), 0.85 (s, 9H), 1.29 (d, *J* = 6.6 Hz, 3H), 1.48 (d, *J* = 6.6 Hz, 3H), 4.27 (q, *J* = 6.6 Hz, 1H), 4.32 (qd, *J* = 6.6 Hz, 5.4 Hz, 1H), 7.23-7.40 (m, 5H), 7.59 (d, *J* = 5.4 Hz, 1H); ^{13}C NMR (125 MHz, $CDCl_3$) δ: –4.8, –4.7, 18.1, 21.7, 24.2, 25.8, 68.9, 70.7, 126.5, 126.8, 128.3, 144.4, 166.2.

11. THF (99.9% anhydrous, inhibitor free) and CH_2Cl_2 (99.8% anhydrous) were obtained from Aldrich Chemical Co. and used as received. The submitters distilled diisopropylamine and stored it over KOH. The checkers used 99.5% diisopropylamine as received from Aldrich. *p*-Toluenesulfonyl chloride (99%) was obtained from Acros Chemical Co. (submitters) or Avocado (checkers).

12. This material consisted of a 45:55 mixture of diastereomers based on ^1H NMR analysis. R_f = 0.53 (10% EtOAc/hexanes, phosphomolybdic acid stain). Spectral characteristics for tosylate **3**: ^1H NMR (300 MHz, $CDCl_3$) δ: 0.08 (s, 6H), 0.09 (s, 6H), 0.88 (s, 18H), 1.24 (d, *J* = 6.3 Hz, 3H), 1.29 (d, *J* = 6.0 Hz, 3H), 2.44 (s, 6H), 4.04 (dq,

J = 6.0, 7.5 Hz, 1H), 4.34 (dq, J = 3.3, 6.3 Hz, 1H), 4.70 (dd, J = 3.3, 6.0 Hz, 1H), 4.74 (dd, J = 2.1, 7.2 Hz, 1H), 5.81 (d, J = 6.3 Hz, 1H), 6.00 (d, J = 2.1 Hz, 1H),7.31 (d, J = 8.4 Hz, 4H), 7.82 (dd, J = 5.4, 8.4 Hz, 4H).

13. Paraformaldehyde (95%) was obtained from Aldrich Chemical Co. (submitters) or Baker (checkers) and was dried azeotropically with benzene by concentrating a benzene solution (100 mL per gram of paraformaldehyde) at 45 °C by rotary evaporation, repeating this process, and then drying the residue under high vacuum overnight. Dry paraformaldehyde was stored in a sealed container under argon. Butyllithium was obtained from Acros Chemical Co (submitters) or Aldrich (checkers).

14. The submitters reported that an immediate preliminary distillation is advisable to minimize contact time with the dark polymeric byproducts which results in decomposition and lowered yields. This is achieved by means of a Kugelrohr distillation apparatus preheated to 80 °C. Care should be exercised to prevent bumping in the early stage of this distillation. However, the checkers had problems with bumping and preferred direct fractional distillation according to the following procedure. After concentration, the flask containing the crude product was equipped with a 30 x 1.5 cm distillation column and evacuated at 0.1 mm. After an initial period of distillation at room temperature to remove residual solvent, a first fraction of bp up to 45 °C was collected; this is believed to be TBSOH. The main product fraction (21-24 g) was collected at bp 79-82 °C.

15. The ee of alcohol **5** was determined to be >97% by derivatization with a chiral silylating reagent (Note 16). Physical characteristics for alcohol **5**: R_f = 0.42 (25% EtOAc/hexanes, phosphomolybdic acid); $[\alpha]_D$ –53.0 (c 1.42, $CHCl_3$); IR (thin film): cm^{-1} 3370; ^1H NMR (300 MHz, $CDCl_3$): δ 0.09 (s, 3H), 0.11 (s, 3H), 0.89 (s, 9H), 1.39 (d, J = 6.5 Hz, 3H), 2.44 (t, J = 6.2 Hz, 1H), 4.27 (dd, J = 6.2, 1.7 Hz, 2H), 4.54 (tq, J= 6.5, 1.7 Hz, 1H); ^{13}C NMR (75 MHz, $CDCl_3$) δ: –5.0, –4.7, 18.2, 25.2, 25.7, 50.8, 59.0, 81.4, 88.0.

16. The enantiomeric purity of alcohol 5 was determined by conversion to the silyl ether 8 via the following sequence:

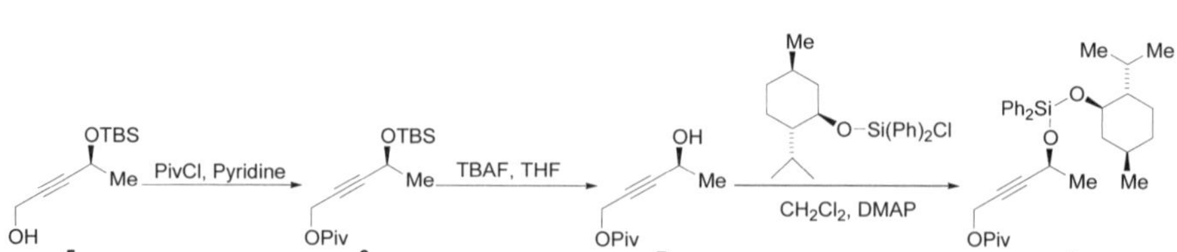

To a solution of (*S*)-pentynol **5** (0.107 g, 0.50 mmol) in pyridine (2.4 mL) is added trimethylacetyl chloride (PivCl, 0.1 mL, 0.8 mmol) (Note 17). After 3 h, ice (*ca* 3 g) is added and the mixture is stirred vigorously. After 1 h, 10% aqueous HCl (5 mL) is added and the resulting mixture is extracted with three 10-mL portions of ethyl acetate. The combined organic extracts are washed with 20 mL of brine, dried over $MgSO_4$, filtered, and concentrated under reduced pressure. The residue is purified by chromatography on silica gel (elution with 1% EtOAc/hexanes) to afford 0.129 g (87%) of ester **6** (Note 18).

To a solution of (*S*)-ester **6** (0.566 g, 1.90 mmol) in 20 mL of THF at 0 °C is added tetrabutylammonium fluoride (3.0 mL, 3.0 mmol) (Note 19). After 15 min, the reaction mixture is warmed to room temperature. After 3 h, 20 mL of aqueous saturated NH_4Cl is added and the mixture is extracted with three 10-mL portions of Et_2O. The combined organic extracts are washed with 20 mL of brine, dried over $MgSO_4$, filtered, and concentrated under reduced pressure. The residue is purified by chromatography on silica gel (elution with 20% EtOAc/hexanes) to afford 0.271 g (78%) of alcohol **7** (Note 20).

To a solution of (*S*)-alcohol **7** (0.052 g, 0.28 mmol) in 2.8 mL of CH_2Cl_2 is added (–)-chloromenthyloxydiphenylsilane (0.105 g, 0.28 mmol) (Note 21) followed by DMAP (0.035 g, 0.29 mmol). After 2 h, the mixture is concentrated and the residue is purified by chromatography on silica gel (elution with 1% EtOAc/hexanes) to afford 0.112 g (76%) of silylmenthol derivative **8**. The identical procedures are employed with the (*R*)-pentynol (Note 22).

17. Pyridine was freshly distilled from calcium hydride and stored under nitrogen over potassium hydroxide. Trimethylacetyl chloride (99%) was obtained from Aldrich Chemical Company and used as received.

18. Physical characteristics of (*S*)-ester **6**: R_f = 0.83 (25% EtOAc/hexanes, cerium molybdate); $[\alpha]_D$ –36.0 (*c* 0.79, $CHCl_3$); IR (thin film): cm^{-1} 2968, 1745; ^1H NMR (500 MHz, $CDCl_3$) δ: 0.11 (s, 3H), 0.13 (s, 3H), 0.90 (s, 9H), 1.21 (s, 9H), 1.40 (d, *J* = 6.5 Hz, 3H), 4.55 (qt, *J* = 6.0, 2.0 Hz, 1H), 4.68 (d, *J* = 2.0 Hz, 2H); ^{13}C NMR (125 MHz, $CDCl_3$) δ: –5.0, –4.6, 18.2, 25.1, 25.8, 27.1, 38.7, 52.3, 58.9, 77.5, 88.9, 177.7.

19. Tetrabutylammonium fluoride (1.0M in THF) was obtained from Aldrich Chemical Company and used as received.

20. Physical characteristics of (*S*)-alcohol **7**: R_f = 0.30 (25% EtOAc/hexanes, cerium molybdate); IR (thin film): cm^{-1} 3440, 2986, 1745; $[\alpha]_D$ –18.1 (*c* 1.2, $CHCl_3$); ^1H NMR (500 MHz, $CDCl_3$) δ: 1.22 (s, 9H), 1.46 (d, *J* = 7.0 Hz, 3H), 4.57 (dq, *J* = 1.5 Hz, 6.5 Hz, 1H), 4.69 (s, 2H); ^{13}C NMR (125 MHz, $CDCl_3$) δ: 24.0, 27.0, 38.7, 52.2, 58.3, 78.3, 88.2, 177.9.

21. The (–)-chloromenthyloxydiphenylsilane was prepared from (–) -menthol (Aldrich, >99% ee) according to the published procedure: Weibel, D. B.; Walker, T. R.; Schroeder, F. C.; Meinwald, J. *Org. Lett.* **2000**, *2*, 2381.

22. Diagnostic ^1H NMR peaks (singlets) are located at δ 4.62 (*S*) and δ 4.57 (*R*) ppm, respectively, in the ^1H NMR spectra of the crude products prior to chromatography. Physical characteristics of the (*S*)-silylmenthyl derivative **8**: R_f = 0.67 (25% EtOAc/hexanes, cerium molybdate); $[\alpha]_D$ –73.5 (*c* 0.63, $CHCl_3$); IR (thin film): cm^{-1} 2960, 1745; ^1H NMR (500 MHz, $CDCl_3$) δ: 0.56 (d, *J* = 6.9 Hz, 3H), 0.87 (d, *J* = 6.3 Hz, 3H), 0.90 (d, *J* = 7.2 Hz, 3H), 0.80-0.94 (m, 2H), 1.22 (s, 9H), 1.12-1.33 (m, 3H), 1.42 (d, *J* = 6.6 Hz, 3H), 1.56-1.62 (m, 2H), 2.09 (m, 1H), 2.33 (m, 1H), 3.63 (dt, *J* = 10.2, 4.2 Hz, 1H), 4.62 (d, *J* = 1.5 Hz, 2H), 4.72 (dq, *J* = 6.6, 1.5 Hz, 1H), 7.33-7.46 (m, 6H), 7.65-7.70 (m, 4H); ^{13}C NMR (125 MHz $CDCl_3$) δ: 15.6, 21.3, 22.3, 22.6, 24.9, 25.3, 27.1, 31.5, 34.4, 38.7, 45.2, 50.0, 52.3, 59.2, 73.5, 77.8, 88.4, 127.6, 130.1, 133.1, 133.5, 135.1, 177.8. (*R*)-silylmenthyl derivative: $[\alpha]_D$ +1.1 (*c* 0.94, $CHCl_3$); IR (thin film): cm^{-1} 2957, 1739; ^1H NMR (300 MHz, $CDCl_3$) δ: 0.59 (d, *J* = 7.2 Hz, 3H), 0.85 (d, *J* = 6.3 Hz, 3H), 0.91 (d, *J* = 7.2 Hz, 3H), 0.79-0.92 (m, 2H), 1.12 (apparent q, *J* = 12.0 Hz, 1H) 1.21 (s, 9H), 1.24-1.34 (m, 2H), 1.46 (d, *J* = 6.3 Hz, 3H), 1.56-1.62 (m, 2H), 2.02 (m, 1H), 2.38 (m, 1H), 3.64 (dt, *J* = 10.2, 4.2 Hz, 1H), 4.57 (d, *J* = 1.5 Hz, 2H), 4.73 (qt, *J* = 6.6, 1.5 Hz, 1H), 7.33-7.45 (m, 6H), 7.64-7.72 (m, 4H); ^{13}C NMR (125 MHz $CDCl_3$) δ: 15.6, 21.3, 22.2, 22.6, 24.9, 25.3, 27.1, 31.5, 34.4, 38.7, 45.2, 50.0, 52.3, 59.2, 73.6, 77.8, 88.3, 127.6, 130.1, 133.3, 135.1, 177.7.

Waste Disposal Information

All toxic materials were disposed of in accordance with "Prudent Practices in the Laboratory"; National Academy Press; Washington, DC, 1995.

3. Discussion

The present route to (*S*)-4-(*t*-butyldimethylsilyloxy)-2-pentyn-1-ol is based on a procedure for the preparation of terminal alkynes reported by a group from the Chemical Process Department at the DuPont Pharmaceutical Company.[2] This alcohol serves as a convenient starting material for the preparation of 1-acyloxy 4-mesylates **10** (eq 1).

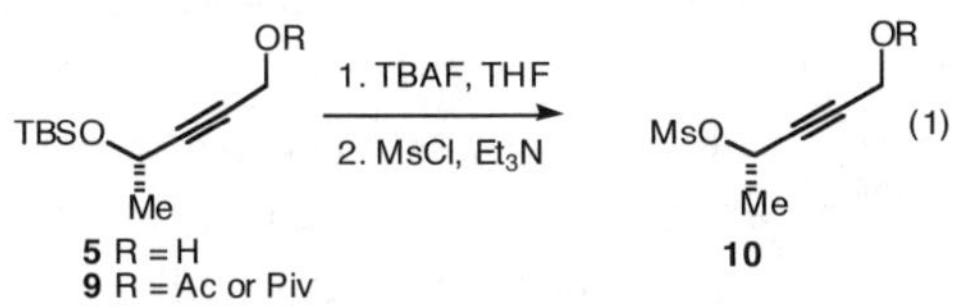

These mesylates, in turn, can be converted to enantioenriched allenyltin, zinc, and indium reagents which add to aldehydes with excellent diastereo-and enantioselectivity to afford either syn- or anti-homopropargylic alcohols or allenylcarbinols (eq 2, 3, and 4).[3,4] Adducts of this type serve as useful intermediates for the synthesis of polyketide and hydrofuran natural products.[5]

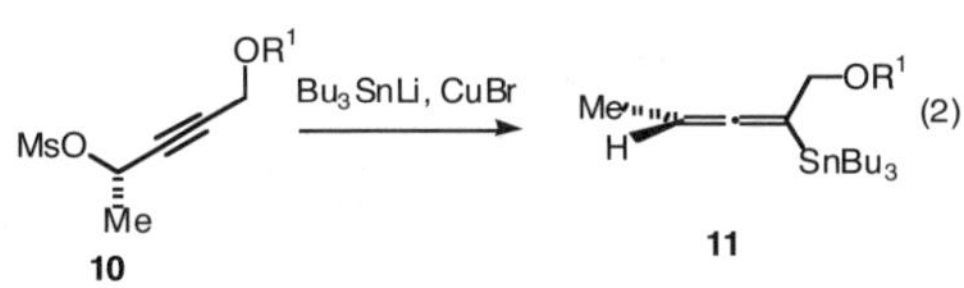

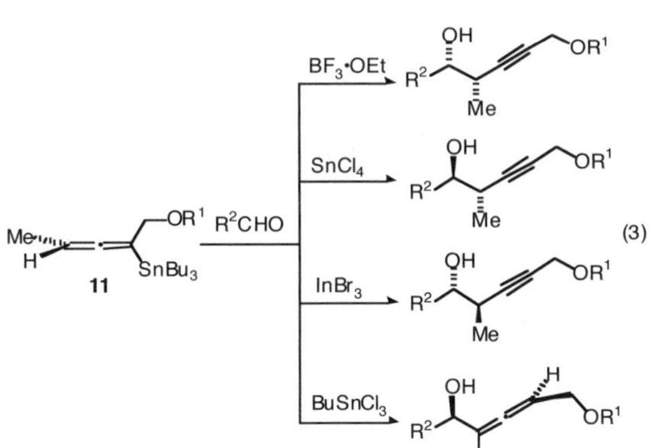

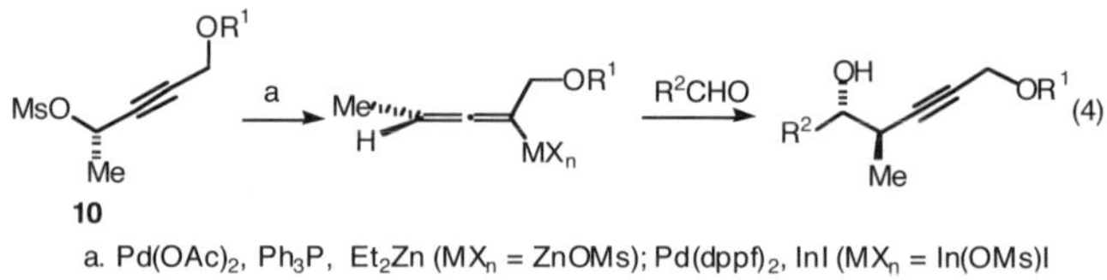

Previous syntheses of terminal alkynes from aldehydes employed Wittig methodology with phosphonium ylides and phosphonates.[6,7] The DuPont procedure circumvents the use of phosphorus compounds by using lithiated dichloromethane as the source of the terminal carbon. The intermediate lithioalkyne **4** can be quenched with water to provide the terminal alkyne or with various electrophiles, as in the present case, to yield propargylic alcohols, alkynylsilanes, or internal alkynes. Enantioenriched terminal alkynylcarbinols can also be prepared from allylic alcohols by Sharpless epoxidation and subsequent basic elimination of the derived chloro- or bromomethyl epoxide (eq 5). A related method entails Sharpless asymmetric dihydroxylation of an allylic chloride and base treatment of the acetonide derivative.[8] In these approaches the product and starting material contain the same number of carbons.

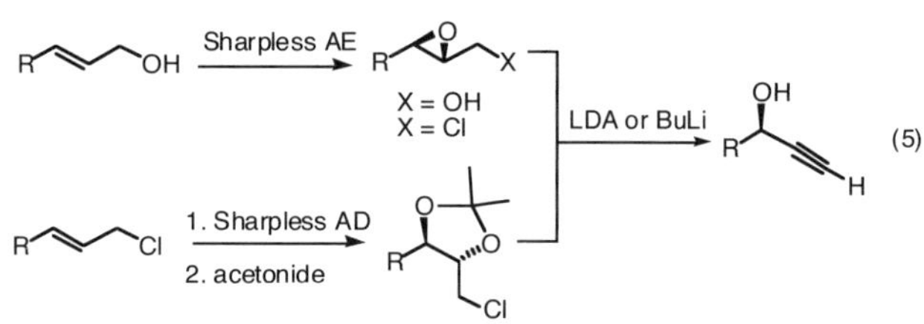

1. Department of Chemistry, McCormick Road, P.O. Box 400319, University of Virginia, Charlottesville, VA 2290, E-mail: jam5x@virginia.edu.
2. Wang, Z.; Yin, J.; Campagna, S.; Pesti, J. A.; Fotunak, J. M. *J. Org. Chem.* **1999**, *64*, 6918.
3. a) Marshall, J. A. *Chem. Rev.* **1996**, *96*, 31. (b) Marshall, J. A. in "Lewis Acids in Organic Synthesis," H. Yamamoto Ed., Vol. 1, Wiley-VCH, Weinheim, (**2000**). Chapter 10, pp. 453-520.
4. Marshall, J. A. *Chem. Rev.* **2000**, *100*, 3163.
5. a) Marshall, J. A.; Lu, S.-H.; Johns, B. A. *J. Org. Chem.* **1998**, *63*, 817. b) Marshall, J. A.; Palovich, M. R. *J. Org. Chem.* **1998**, *63*, 3701. c) Marshall, J. A.; Johns, B. A. *J. Org. Chem.* **1998**, *63*, 7885. d) Marshall, J. A.; Fitzgerald, R. A. *J.*

Org. Chem. **1999**, *64*, 4477. e) Marshall, J. A.; Johns, B. A. *J. Org. Chem.* **2000**, *65*, 1501. f) Marshall, J. A.; Yanik, M. M. *J. Org. Chem.* **2001**, *66*, 1373.

6. a) Horner, L.; Hoffmann, H.; Wippel, H. G.; Klaher, G. *Chem. Ber.* **1959**, 2499. b) Wadsworth, W. S. Jr.; Emmons, W. D. *J. Am. Chem. Soc.* **1961**, *83*, 1733. c) Corey, E. J.; Achiwa,K.; Katzenellenbogen, J. A. *J. Am. Chem. Soc.* **1969**, *91*, 4318. d) Corey, E. J.; Fuchs, P. L. *Tetrahedron Lett.* **1972**, 3769. e) Gilbert, J. C.; Weerasooriya, U. *J. Org. Chem.* **1982**, *47*, 1837. f) Muller, S.; Liepold, B.; Roth, G. J.; Bestmann, H. J. *Syn Lett* **1996**, 521.

7. a) For a previous synthesis of the pivalic ester related to mesylate **10** from *(R)*-methyl lactate and $Ph_3P{=}CBr_2$, see Marshall, J. A.; Xie, S. *J. Org. Chem.* **1995**, *60*, 7230. b) An analogous synthesis of *(S)*-3-butyn-2-ol derivatives has also been reported; Ku, Y.-Y.; Patel, R. R.; Elisseou, E. M.; Sawick, D. P. *Tetrahedron Lett.* **1995**, 36*, 2738*.

8. a) Yadav, J. S.; Chander, M. C.; Rao, C. S. *Tetrahedron Lett.* **1989**, *30*, 5455. b) Yadav, J. S.; Chander, M. C.; Joshi, R. V. *Tetrahedron Lett.* **1988**, *29*, 2737.

PREPARATION OF 1-METHOXY-2-(4-METHOXYPHENOXY)BENZENE
(Benzene, 1-methoxy-2-(4-methoxyphenoxy)-)

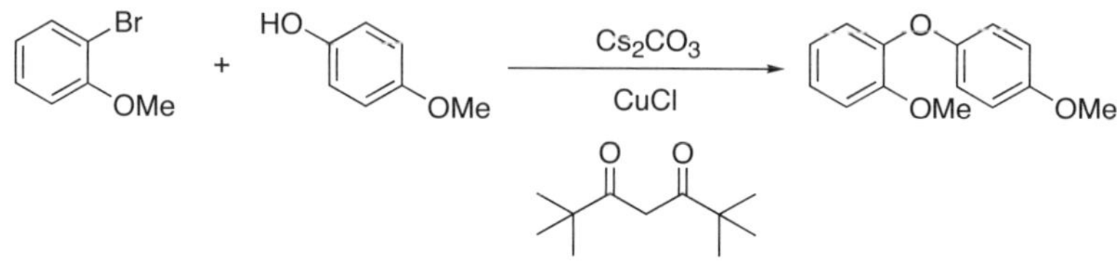

Submitted by Elizabeth Buck and Zhiguo J. Song.[1]
Checked by Scott E. Denmark and John D. Baird.

1. Procedure

To a 500-mL, three-necked, round-bottomed flask equipped with reflux condenser, nitrogen inlet, thermocouple and mechanical stirrer is added 14.0 g (113 mmol, 2.0 equiv) of 4-methoxyphenol (Note 1) and 95 mL of 1-methyl-2-pyrrolidinone (NMP) (Note 2). To this solution is then added 36.8 g (113 mmol, 2.0 equiv) of cesium carbonate (Note 3). The resulting slurry is degassed by evacuation and filling with nitrogen three times. Then 10.5 g (56.1 mmol) of 2-bromoanisole (Note 4) and 1.0 g (5.4 mmol, 0.096 equiv)) of 2,2,6,6-tetramethylheptane-3,5-dione (Note 5) are added followed by 2.8 g (28 mmol, 0.5 equiv) of copper (I) chloride (Note 6). The reaction mixture is degassed by evacuation and filling with nitrogen three times and then is heated to 120 °C under nitrogen. The reaction is monitored by HPLC analysis (Note 8) and is stopped after 25 h when >98 % conversion (by peak area) is obtained. The reaction mixture is cooled to room temperature and is diluted with 125 mL of methyl *t*-butyl ether (MTBE). The slurry is filtered and filter cake is washed with 125 mL of MTBE. The combined filtrates are washed subsequently with 175 mL of aqueous 2N HCl solution, 175 mL of 0.6N aqueous HCl solution, 150 mL of 2M aqueous NaOH solution, and 150 mL of 10% aqueous NaCl solution. The resulting organic layer is dried over $MgSO_4$ and is concentrated by rotary evaporation under vacuum at $\leq$ 30 °C to remove all solvents. The crude, brown solid is triturated by magnetic stirring in 55 mL of hexanes (Note 9), then is aged for 4 h and filtered to give 9.9-10.1 g (78% average yield) of an off-white solid (mp 77 – 78 °C) of 99% purity as determined by HPLC analysis (Note 10).

2. Notes

1. 4-Methoxyphenol was purchased from Aldrich Chemical Co. and was used as received. Use of less than two equivalents resulted in a longer reaction time.

2. Anhydrous 1-methyl-2-pyrrolidinone (NMP) was purchased from Aldrich Chemical Co. and was used as received. NMP that was dried over 4Å molecular sieves with water content $\leq$100 μg/mL (Karl Fisher titration) gave slightly faster reactions.

3. Cesium carbonate was purchased from Alfa Aesar and was used as received. It is hygroscopic and should be weighed with minimum exposure to the air. Use of less than two equivalents resulted in slower reactions.

4. 2-Bromoanisole was purchased from Aldrich Chemical Co. and was used as received.

5. 2,2,6,6-Tetramethylheptane-3,5-dione was purchased from Aldrich Chemical Co. and was used as received.

6. Copper (I) chloride was purchased from Acros and was used as received.

7. MTBE (methyl t-butyl ether) was purchased from EM Science and was used as received.

8. The HPLC instrument used was a HP1100. HPLC conditions: Perkin Elmer 3X3 CR C18; MeCN/0.1% H_3PO_4; 2 mL/min; gradient: time 0 min 25/75, 1 min 25/75, 9 min 95/5, 10 min 95/5; UV detector 210 nm. Retention times: 4-methoxyphenol 0.6 min, 2-bromoanisole 3.4 min, product 4.3 min.

9. Hexanes was purchased from Fisher Scientific Co. and used directly as received.

10. Purity is reported as HPLC area percent (A%). Assay yields are reported using isolated material as standard. Analytical Data: ^{1}H NMR ($CDCl_3$, 400 MHz) δ: 7.09-7.05 (m, 1 H), 7.00-6.98 (m, 1 H), 6.97-6.93 (m, 2 H), 6.90-6.84 (m, 4 H), 3.88 (s, 3 H), 3.79 (s, 3H); ^{13}C NMR ($CDCl_3$, 100 MHz) δ:155.3, 150.8, 150.7, 146.6, 123.7, 120.9, 119.3, 119.0, 114.6, 112.5, 55.9, 55.6. Anal. Calcd. for $C_{14}H_{14}O_3$: C, 73.03; H, 6.13. Found: C, 73.27; H, 6.05. MS (EI) *m/z* 230 (M^+, 100).

Safety and Waste Disposal Information

All hazardous materials should be handled and disposed of in accordance with "Prudent Practices in the Laboratory"; National Academy Press; Washington, DC, 1995.

3. Discussion

The Ullmann ether formation has traditionally been carried out under rather harsh conditions, usually at high temperatures in pyridine as the solvent.[1-3] The yields are typically low to moderate and reactions between electron-rich aryl halides and electron-deficient phenols typically do not work well. Other milder conditions, employing more expensive reagents or catalysts such as palladium or copper (II) triflate, are also available.[5,6] The major advantages of the process reported here are low cost of the copper additive and the simplicity of operation. Relative to other less expensive ligands for copper such as pyridine-based ligands or phosphine-based ligands,[5a,b] 2,2,6,6-tetramethylheptane-3,5-dione (TMHD) accelerates the Ullmann coupling reaction more effectively, allowing some of the electronically unfavorable substrates to undergo ether formation readily, as listed in Table 1. Nevertheless, these reactions exhibit the general trend that electron-donating groups on the phenol and electron-withdrawing groups on the halide make the reaction more favorable. Some tolerance of electron-withdrawing groups on the phenols was observed. The major side reaction in these Ullmann ether formations is the reduction of the aryl halides to the arene (Aryl-Br to Aryl-H). The reaction of an aryl iodide is faster and higher yielding than the corresponding bromide, as indicated by a comparison of entries 12 and 1.

Table 1. 2,2,6,6-Tetramethylheptane-3,5-dione Catalyzed Ullmann Ether Formation

Ar-X	ArOH	ether product	reaction time (h)	assay yield (%)	isolated yield (%)
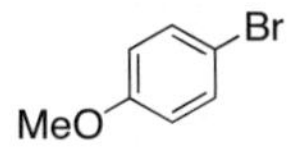	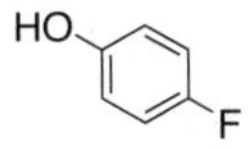	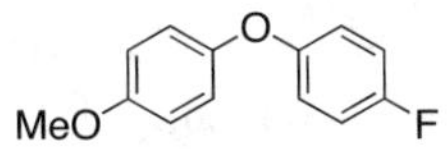	10	70	55
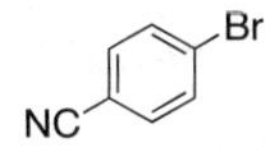	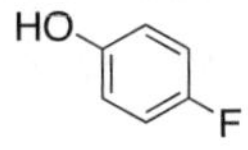	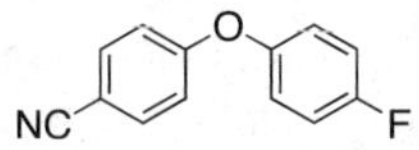	1.5	90	82
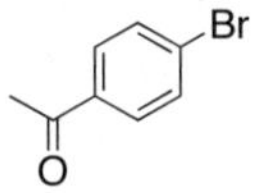	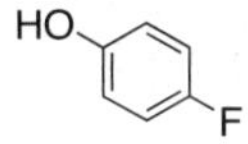	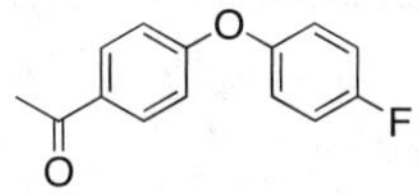	4	83	60
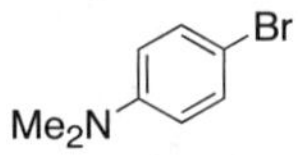	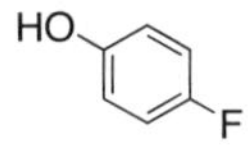	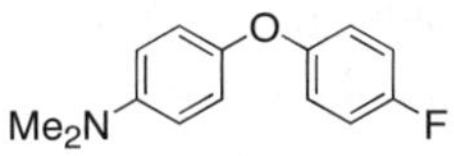	15	79	59
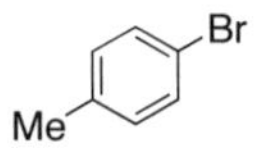	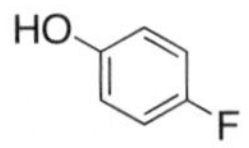	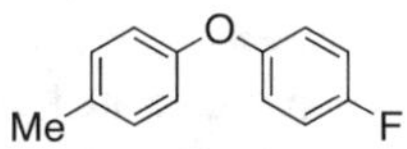	7.5	86	63
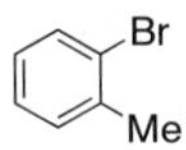	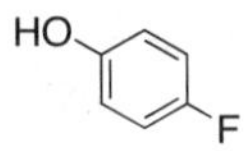	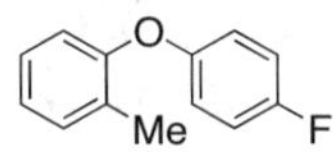	15	97	85
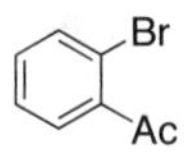	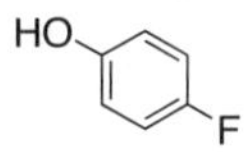	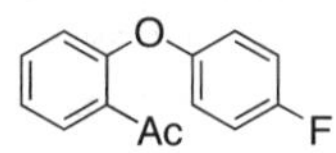	24	67	61
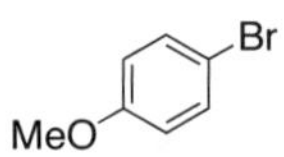	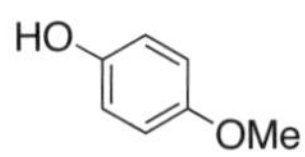	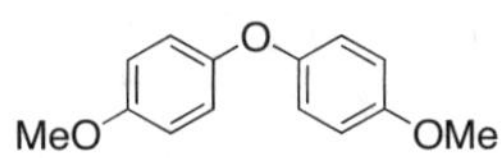	10	85	77
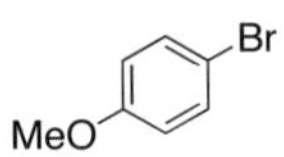	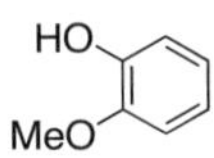	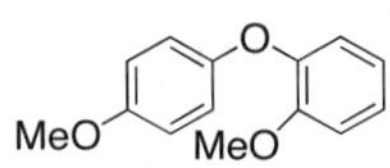	10	80	66
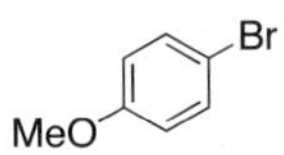	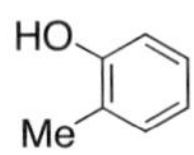	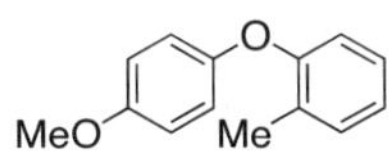	10	92	79
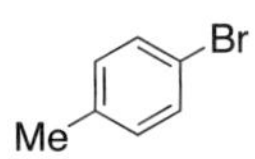	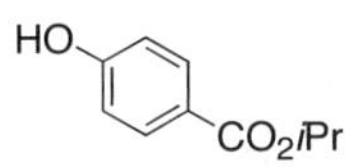	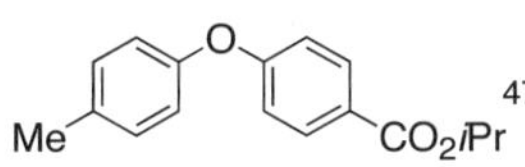	47	56	51
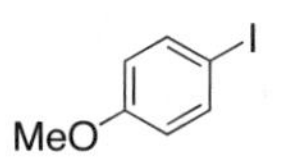	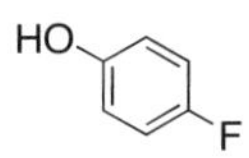	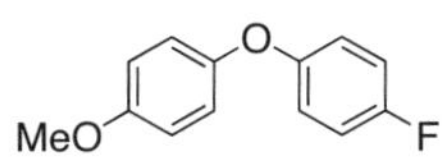	6	77	68

1. Department of Process Research, Merck Research Laboratories, P.O. Box 2000, Rahway, NJ 07065.
2. This procedure was first reported by: Buck, E.; Song, Z. J.; Tschaen, D.; Dormer, P. G.; Volantc, R. P.; Reider, P. J. Org. Lett. 2002, 4, 1623.
3. For reviews: (a) Sawyer, J. S. Tetrahedron 2000, 56, 5045. (b) Theil, F. Angew Chem., Int. Ed. 1999, 38, 2345. (c) Lindley, J. Tetrahedron 1984, 40, 1433.
4. Some earlier reports: (a) Williams, A. L.; Kinney, R. E.; Bridger, R. F. *J. Org. Chem.* **1967**, *32,* 2501. (b) Bacon, R. G. R.; Rennison, S. C. *J. Chem. Soc.(C)* **1969**, 312. (c) Bacon, R. G. R.; Stewart, O. J. *J. Chem. Soc.* **1965**, 4953.
5. For palladium mediated ether synthesis: (a) Torraca, K. E.; Huang, X.; Parrish, C. A.; Buchwald, S. L. *J. Am. Chem. Soc.* **2001**, *123*, 10770. (b) Aranyos, A.; Old, D. W.; Kiyomori, A.; Wolfe, J. P.; Sadighi, J. P.; Buchwald, S. L. *J. Am. Chem. Soc.* **1999**, *121*, 4369. (c) Mann, G.; Hartwig, J. F. *Tetrahedron Lett.* **1997***, 38*, 8005. (d) Mann, G.; Incarvito, C.; Rheingold, A. L.; Hartwig, J. F. *J. Am. Chem. Soc.* **1999**, *121*, 3224. (e) Shelby, Q.; Kataoka, N.; Mann, G.; Hartwig, J. *J. Am. Chem. Soc.* **2000**, *122*, 10718.
6. For some recent reports of copper mediated ether formations: (a) Gujadhur, R. K.; Bates, C. G.; Venkataraman, D. *Org. Lett.* **2001**, *3,* 4315. (b) Fagan, P. J.; Hauptman, E.; Shapiro, R.; Casalnuovo, A. *J. Am. Chem. Soc.* **2000**, *122*, 5043. (c) Marcoux, J.-F.; Doye, S.; Buchwald, S. L. *J. Am. Chem. Soc.* **1997**, *119*, 10539. (d) Kalinin, A. V.; Bower, J. F.; Riebel, P.; Snieckus, V. *J. Org. Chem.* **1999**, *64*, 2986. (e) Evans, D. A.; Katz, J. L.; West, T. R. *Tetrahedron Lett.* **1998**, *39,* 2937. (f) Palomo, C.; Oiarbide, M.; Lopez, R.; Gomez-Bengoa, E. *Chem. Commun.* **1998***,* 2091. (g) Smith, K.; Jones, D. *J. Chem. Soc. Perkin Trans. I* **1992**, 407. (g) Wolter, M.; Nordmann, G.; Job, G. E.; Buchwald, S. L. *Org. Lett.* **2002**, *4,* 973.

IRIDIUM-CATALYZED SYNTHESIS OF VINYL ETHERS FROM ALCOHOLS AND VINYL ACETATE
(1-Methoxy-4-vinyloxybenzene)

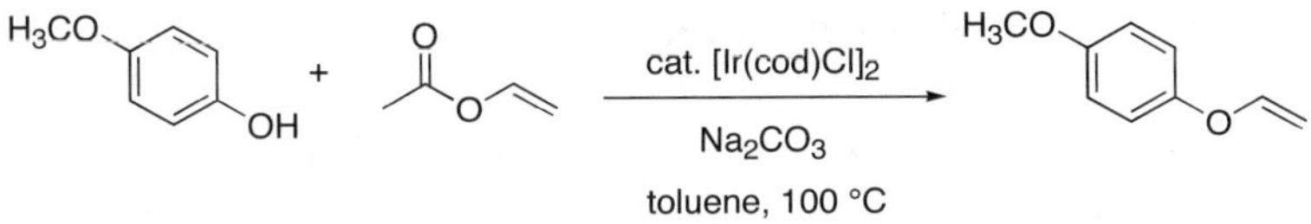

Submitted by Tomotaka Hirabayashi, Satoshi Sakaguchi and Yasutaka Ishii.[1,2]
Checked by Paul W. Davies and Alois Fürstner.

1. Procedure

A 100-mL, two-necked round-bottomed flask is fitted with a magnetic stirbar, a reflux condenser connected to an argon/vacuum line, and a rubber septum. The equipment is flame dried under vacuum and then flushed with argon. Di-μ-chloro-bis(1,5-cyclooctadiene)diiridium(I), $[Ir(cod)Cl]_2$, (0.34 g, 0.01 eq., 0.5 mmol) (Note 1) and sodium carbonate (3.18 g, 30 mmol, 0.6 eq.) (Note 2) are rapidly weighed in air and added to the flask which is resealed, evacuated and backfilled with argon. Toluene (50 mL) (Note 3), *p*-methoxyphenol (6.21 g, 50 mmol) (Note 4) and vinyl acetate (8.61 g, 100 mmol) (Note 5) are successively introduced and the flask is placed into a preheated oil-bath at 100 °C with magnetic stirring. The reaction mixture changes from a yellow color after the addition of the vinyl acetate to a wine red color within 30 minutes of heating (Note 6) and continues to darken over the course of the reaction. After heating for 2 h, the reaction mixture is allowed to cool to ambient temperature before being transferred to a separatory funnel. Portions of EtOAc used to rinse the flask (3 × 50 mL) are added to the separatory funnel. The combined organic fractions are washed with water (3 × 100 mL) and brine (50 mL) before being dried over $MgSO_4$. The solution is filtered, the cake is washed with EtOAc (2 × 20 mL), and the combined filtrates are evaporated (Note 7). The resulting crude product is applied to a pre-packed silica gel column (50 × 3.5 cm) (Note 8) and eluted with hexanes/EtOAc (4:1) giving 1-methoxy-4-vinyloxybenzene as a pale yellow liquid after evaporation of the solvent (6.85 g, 91%) (Notes 7, 9, 10).

2. Notes

1. $[Ir(cod)Cl]_2$ can be prepared according to the literature method.[3] This complex is commercially available from Aldrich Chemical Company, Inc. The checkers used $[Ir(cod)Cl]_2$ purchased from Strem without further purification.

2. Na_2CO_3 purchased from Wako Pure Chemical Industries Ltd. was used as received. The checkers, however, found that the use of moist Na_2CO_3 led to lower yields of the desired product and increased amounts of the acetate side-product. Well reproducible results were obtained by using Na_2CO_3 dried under high vacuum (10^{-4} Torr) at 80 °C overnight before use.

3. Toluene was dried by distillation over Na under Ar atmosphere.

4. *p*-Methoxyphenol was purchased from Wako Pure Chemical Industried Ltd. or Acros and used as received.

5. Vinyl acetate purchased from commercial suppliers (Wako Pure Chemical Industries Ltd., or Acros) was dried over MS 4Å and distilled at normal pressure (bp 72 °C) under Ar.

6. This characteristic color change is much slower if insufficiently dried Na_2CO_3 is used, thus leading to longer overall reaction times.

7. To avoid any loss of product during the evaporation of the solvent, the vacuum was set to ≥ 20 mbar and the temperature of the water bath should not exceed 40 °C.

8. Silica gel (230-400 mesh) from Kanto Kagaku Reagent Division was used. The checkers used Merck silica gel (230-400 mesh).

9. A small amount (< 3%) of 1-acetoxy-4-methoxybenzene was observed as a side product. The checkers obtained somewhat larger amounts of this compound (0.52 g, 6%).

10. Analytical data: ^1H NMR (400 MHz, $CDCl_3$) δ 6.97-6.84 (m, 4H), 6.59 (dd, J = 6.2, 13.8 Hz, 1H), 4.64 (dd, J = 1.7, 13.8 Hz, 1H), 4.34 (dd, J = 1.7, 6.2 Hz, 1H), 3.79 (s, 3H); ^{13}C NMR (101 MHz, $CDCl_3$) δ 155.8, 150.7, 149.6, 118.8, 114.8, 93.7, 55.8; IR (film) 2953, 2835, 1639, 1500, 1209, 1147, 1034, 954, 830 cm^{-1}. Anal. Calcd for $C_9H_{10}O_2$: C, 71.98; H 6.71; Found: C, 71.85; H, 6.67.

Safety and Waste Disposal Information

All hazardous materials should be handled and disposed of in accordance with "Prudent Practices in the Laboratory"; National Academy Press; Washington, DC, 1995.

3. Discussion

Vinyl ethers are conventionally prepared by mercury-catalyzed transfer vinylation,[4] base- or transition metal-catalyzed isomerization of allyl

Table 1. Iridium-Catalyzed Synthesis of Various Vinyl Ethers

Entry	Alcohol	Product	Time	GLC Yield	Isolated Yield
1	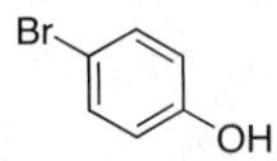	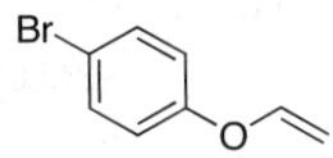	3 h	91%	88%
2	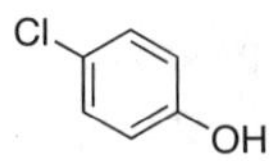	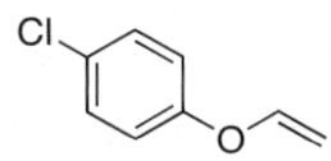	3 h	92%	49%
3	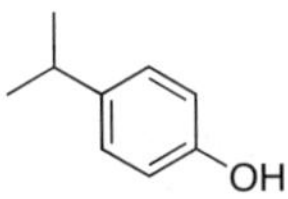	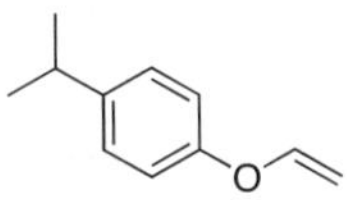	3 h	98%	92%
4	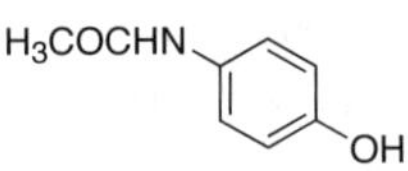	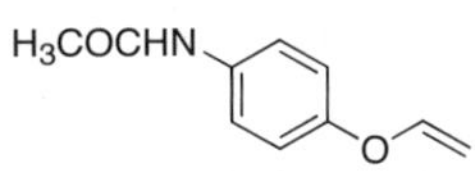	15 h	84%	79%
5[a]	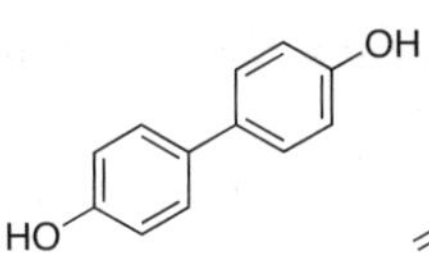	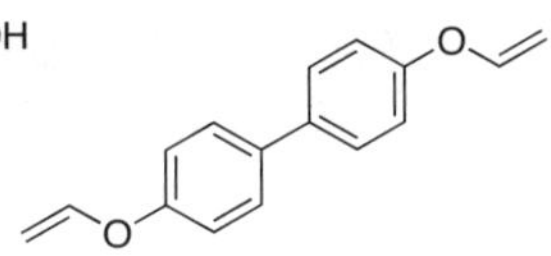	6 h	88%	80%
6[a]	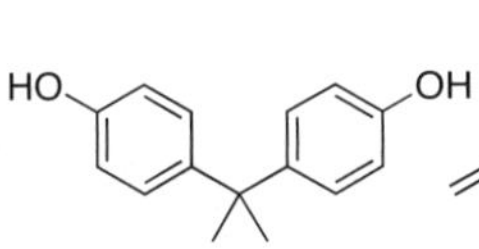	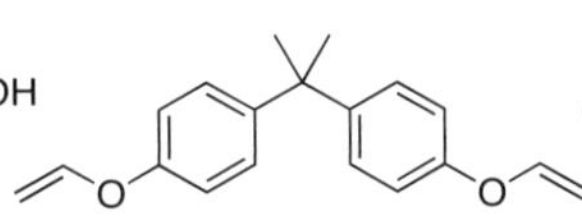	6 h	90%	86%
7	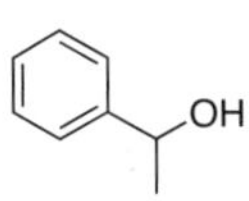	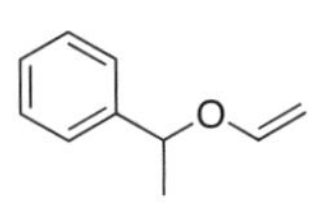	2 h	95%	-
8[b]	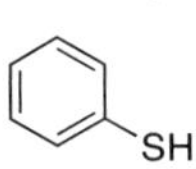	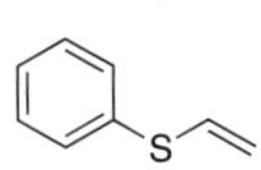	24 h	92%	-
9[a]	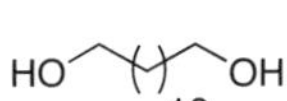	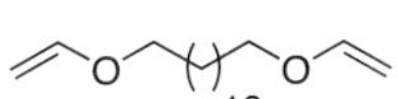	6 h	95%	83%

[a] 4 eq. of vinyl acetate and 2 eq. of Na_2CO_3 were used. [b] See text.

ethers,[5] or elimination reactions.[6] Quite recently, the synthesis of allyl and alkyl vinyl ethers by the transfer vinylation of alcohols with butyl vinyl ether in the presence of a palladium catalyst has been reported.[7]

The present method uses commercially available $[Ir(cod)Cl]_2$ as the precatalyst for the synthesis of vinyl ethers which are very difficult to prepare otherwise.[2] The reaction is thought to proceed through an addition-elimination sequence of alcohol and acetic acid in the presence of the iridium complex as a catalyst and Na_2CO_3 as the base. Representative examples are compiled in the Table. The method can also be applied to the synthesis of vinyl ethers derived from secondary alcohols (entry 7) and holds promise for the preparation of alkyl vinyl ethers (entry 9). Although sulfur compounds frequently inhibit transition metal-catalyzed reactions, thiophenol reacts with vinyl acetate to form phenylvinyl thioether in excellent yield (entry 8); this compound, however, is unstable in air and is therefore difficult to isolate in analytically pure form (~15 %).

1. Department of Applied Chemistry, Faculty of Engineering, Kansai University, Suita, Osaka 564-8680, Japan.
2. Okimoto, Y.; Sakaguchi, S.; Ishii, Y. *J. Am. Chem. Soc.* **2002**, *124*, 1590.
3. Herde, J. L.; Lambert, J. C.; Senoff, C. V. *Inorg. Synth.* **1974**, *15*, 18.
4. Watanabe, W. H.; Conlon, L. E. *J. Am. Chem. Soc.* **1956**, *79*, 2828.
5. Larock, R. C. *Comprehensive Organic Transformations,* 2nd ed.; Wiley-VCH: New York, **1999**, pp. 222 and 225-226.
6. Nerdel, F.; Buddrus, J.; Brodowski, W.; Hentschel, P.; Klamann, D.; Weyerstahl, P. *Justus Liebigs Ann. Chem.* **1967**, *710*, 36.
7. Bosch, M.; Schlaf, M. *J. Org. Chem.* **2003**, *68*, 5225.

THE USE OF POLYSTYRYLSULFONYL CHLORIDE RESIN AS A SOLID SUPPORTED CONDENSATION REAGENT FOR THE FORMATION OF ESTERS: SYNTHESIS OF *N*-[(9-FLUORENYLMETHOXY)CARBONYL]-L-ASPARTIC ACID; α-*tert*-BUTYL ESTER, β-(2-ETHYL[(1E)-(4-NITROPHENYL)AZO]PHENYL]AMINO]ETHYL ESTER

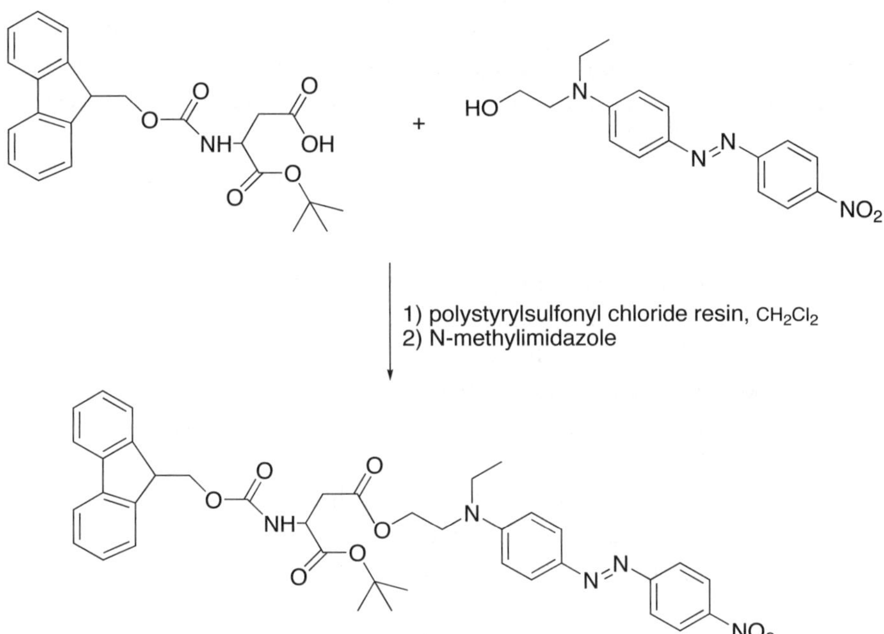

1

Submitted by Norbert Zander[1] and Ronald Frank.[2]
Checked by Richard V. Coelha, Jr., Klaas Schildknegt, and Sarah E. Kelly.

1. Procedure

To a 50-mL polypropylene vial (Note 1) are added 0.839 g (2.67 mmol) of 2-[ethyl[4-[(1*E*)-(4-nitrophenyl)azo]phenyl]amino]ethanol (Disperse Red 1, Note 2), 0.985 g of (2.39 mmol) *N*-[(9*H*-fluoren-9-ylmethoxy)carbonyl]-L-aspartic acid,1-(1,1-dimethylethyl) ester (Fmoc-L-Asp-OtBu, Note 3), 3.26 g (4.73 mmol) of polystyrylsulfonyl chloride resin (Note 4), and 30 mL anhydrous methylene chloride (Note 5). The vial is capped and the mixture is shaken for five min (Note 6). *N*-Methylimidazole

(0.764 mL, 9.58 mmol) is then added to the deep red mixture (Note 7) and the resulting mixture is shaken for 2 h (Note 8).

N-Methylimidazole is then removed from the reaction mixture with Amberlyst 15 ion exchange resin (Note 9) using the following procedure. To a 2-cm diameter column equipped with a glass frit (Note 1) is added 7 g of Amberlyst 15, which is rinsed with 25 mL of methylene chloride. The reaction mixture is added and the resin mixture is rinsed with 50 mL of methylene chloride. The filtrate is collected in a 250-mL flask (Note 10) and the solvent is removed on a rotary evaporator to afford 1.63-1.66 g (96-98%) of the desired ester **1** as a deep red foam. HPLC analysis showed a purity of 98% (Notes 11, 12, 13, 14).

2. Notes

1. Polystyrene resin strongly sticks to glass, and consequently either polypropylene reaction vessels or silanized glassware should be used. Glassware was silanized by the following procedure. The glassware and a small beaker containing neat chlorotrimethylsilane were placed in a desiccator. Vacuum was applied until the chlorotrimethylsilane started to boil and then the desiccator was closed. The next day, the glassware was removed, rinsed with methylene chloride, and dried at 80 °C.

2. Disperse Red 1, dye content ca. 95%, is available from Aldrich, Sigma-Aldrich Chemie GmbH. The checkers obtained Disperse Red 1 from Sigma-Aldrich Corporation.

3. The submitters obtained Fmoc-L-Asp-OtBu from Calbiochem-Novabiochem AG. The checkers obtained Fmoc-L-Asp-OtBu from EMD Biosciences, Inc.

4. Polystyrylsulfonyl chloride resin is available from Argonaut Technologies. The batch used by the submitters had a substitution of 1.44 mmol/g (the checker's batch had a substitution of 1.45 mmol/g). The use of polystyrylsulfonyl chloride resin from Novabiochem in this type of reaction resulted in drastically longer reaction times and less pure products.

5. Methylene chloride was the solvent most compatible with the reagents and resin. Tetrahydrofuran or a 1: 1 mixture of dimethylformamide (DMF) and methylene chloride could also be employed, but longer reaction times were necessary. 1,4-Dioxane, toluene, *N*-methylpyrrolidinone, and DMF alone were not suitable as solvents.

6. Mixtures containing polystyrene resins should either be shaken or stirred with a mechanical stirrer. Stirring with a magnetic stir bar results in

destruction of resin beads and the resulting debris can clog frits during filtrations.

7. In place of *N*-methylimidazole (MeIm), only dimethylaminopyridine (DMAP) could be substituted. The solid-supported amines piperidinomethyl- or morpholinomethyl polystyrene resins, pyridine, and tertiary amines like triethylamine and *N*-methylmorpholine were not effective.

8. HPLC analysis showed complete conversion after 90 min. The use of less base resulted in longer reaction times.

9. Amberlyst 15 is available from Fluka, Sigma-Aldrich Chemie GmbH. The checkers obtained Amberlyst 15 from Sigma-Aldrich Corporation. The use of only 6 g of Amberlyst 15 resulted in incomplete removal of MeIm.

10. Further rinsing with methylene chloride yielded trace quantities of additional product, which was less pure by HPLC analysis. The main impurity was **2** (see Discussion) produced by cleavage of the *t*-butyl ester by the Amberlyst 15 resin.

11. Analytical reversed-phase HPLC was performed using a 50 mm x 2 mm i.d. 3 µm C18(2) column (LUNA, Phenomenex, Germany); solvent A was 0.1% TFA in HPLC-grade water; solvent B was 0.1% TFA in HPLC-grade acetonitrile; UV-detection was at 220 nm; flow rate was 0.8 mL/min; gradient elution: 0 min 95% A, 0-11 min 5% A, 11-12.5 min 5% A, 12.5-13 min 95% A, 13.5-17 min 95% A.

12. The submitters performed the reaction using a 0.12 mmol excess of Fmoc-L-Asp-O*t*-Bu, under which conditions aminomethylated polystyrene resin was required to remove the excess carboxylic acid (Note 13). The checkers modified the reaction to use 0.28 mmol excess Disperse Red 1. The initial Amberlyst-15 filtration removes this material.

13. To remove carboxylic acid, the crude product is redissolved in 20 mL of methylene chloride and is shaken for 30 min with 1 g of aminomethylated polystyrene resin with a substitution of 1.02 mmol/g, available from Novabiochem (Note 3). After filtration and washing of the resin with 50 mL of methylene chloride, the filtrates were collected together in a 250-mL flask and the solvent was removed on a rotary evaporator.

14. Spectral properties of the product are as follows: IR (KBr): cm^{-1} 2975, 1733, 1600, 1549, 1514, 1433, 1391, 1336, 1253, 1133, 853, 741; ^1H NMR (400 MHz, DMSO-d_6) δ: 1.11 (t, 3H, J = 6.8 Hz), 1.30 (s, 9H), 2.61 (dd, 1H, J = 16.4, 8.1 Hz), 2.71 (dd, 1H, J = 16.2, 6.6 Hz), 3.50 (q, 2H, J = 6.9 Hz), 3.68 (m, 2H), 4.18 (t, 1H, J = 6.6 Hz), 4.23 – 4.32 (m, 5H), 6.88 (d,

2H, J = 9.5 Hz), 7.28 (t, 2H, J = 7.5 Hz), 7.37 (t, 2H, J = 7.3 Hz), 7.66 (d, 2H, J = 7.5 Hz), 7.74 (d, 1H, J = 8.3 Hz), 7.81 (d, 2H, J = 9.1 Hz), 7.85 (d, 2H, J = 7.5 Hz), 7.88 (d, 2H, J = 9.1 Hz), 8.32 (d, 2H, J = 9.1 Hz; ^{13}C NMR (100 MHz, DMSO-d_6) δ: 12.6, 28.1, 36.1, 36.7, 45.7, 47.2, 48.8, 51.6, 62.6, 66.3, 81.8, 112.4, 120.2, 120.8, 123.1, 123.8, 125.6, 125.8, 126.8, 127.7, 128.3, 136.3, 141.4, 143.4, 144.4, 147.5, 152.3, 156.5, 156.7, 170.5, 170.7; HRMS (ESI): Calcd for $C_{39}H_{41}N_5O_8$+Na^+: 730.2853, found 730.2854.

Waste Disposal Information

All toxic materials were disposed in accordance with "Prudent Practices in the Laboratory"; National Academy Press; Washington, DC 1995.

3. Discussion

Although solid-supported reagents and scavengers have been used in organic synthesis for decades, it was the development of combinatorial and parallel high throughput synthesis techniques that brought this class of reagents to wider attention. While there are numerous applications of solid-supported reagents and scavengers only a few examples for the formation of esters have been described.[3] Carboxylates, generated with solid-supported organic bases, were alkylated with alkyl halides.[4] Solid-supported organic bases were also used as scavenger resins in the esterification of benzyl alcohol with benzoyl chlorides, giving clean benzyl esters in high yields.[5] This approach requires the acid chloride to be available. A recent report describes the alkylation of carboxylic acids with carbenium ions, generated from polymer-supported triazines.[6] This approach, however, requires the preformation of the polymer-supported triazine for each alkyl group to be transferred and a relatively high excess of the alkylating resin. No examples for the formation of aryl esters are given.

We report here a convenient and general procedure for the formation of esters directly from the easily available carboxylic acids and alcohols or phenols.[7] The reaction uses polystyrylsulfonyl chloride resin as an efficient dehydration reagent and *N*-methylimidazole as basic catalyst. The method employs only commercially available supported reagents and scavengers and allows compounds to be obtained in excellent yields and high purity by simple filtration. It should therefore be especially suitable for an automated parallel synthesis. The scope of our procedure was examined by treating six carboxylic acids with Fmoc-glycinol and seven alcohols with Fmoc-glycine. The results are summarized in Table 1. The purity of all the different esters derived from Fmoc-glycinol (entry 1-6) was very good (>93%), while the reaction time necessary for a complete conversion varied considerably. All products of the esterification of Fmoc-glycine (entry 7-13) with different alcohols and donor or acceptor substituted phenols were essentially 100% pure, except for *tert*-butyl alcohol, which reacted only after the addition of 0.25 equivalents of DMAP. The conversion was slow, but the resulting ester was of good purity.

Table 1

Entry	**Carboxylic acid**	**Alcohol**	**Purity [%]**[c, d]	**Reaction Time**
1	isobutyric acid	Fmoc-Gly-ol[a]	100	60 min
2	pivalic acid	"	98 (2)	22 h
3	6-bromohexanoic acid	"	100	30 min
4	benzoic acid	"	94	60 min
5	4-acetamidobenzoic acid	"	93 (7)	22 h
6	3-nitrobenzoic acid	"	100	15 min
7	Fmoc-Gly-OH[b]	ethanol	100	30 min
8	"	isobutyl alcohol	100	30 min
9	"	tert-butyl alcohol	87	22 h[e]
10	"	benzyl alcohol	100	30 min
11	"	phenol	100	30 min
12	"	4-Ethoxycarbonylphenol	98	30 min
13	"	4-methoxyphenol	100	30 min

a) 0.0706 mmol Fmoc-Gly-ol, 1.3 equiv. carboxylic acid, 1.3 equiv. sulfonyl chloride resin, 4 eq. MeIm 1 mL abs. DCM; b) 0.0673 mmol Fmoc-Gly-OH, 1.3 eq. alcohol, 1.3 eq sulfonyl chloride resin, 4 equiv. MeIm 1 mL abs. DCM; c) HPLC of the crude product, percent starting material in parentheses; d) all compounds were characterized by HPLC-ESI-MS or ^{1}H-NMR, e) 0.25 eq. DMAP added

Scheme 1

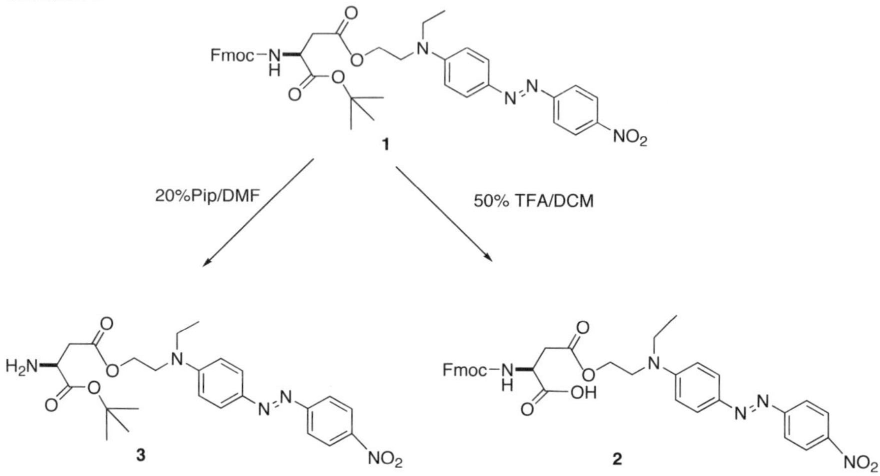

The example described in the experiment illustrates the compatibility of this reaction and the work up procedure with the base-sensitive Fmoc- as well as the acid-sensitive *tert*-Bu-ester protecting group. No by-product **3** due to Fmoc cleavage by *N*-methylimidazole was found. The small amount of by-product **2** resulting from the cleavage of the *tert*-butyl ester by the acidic ion exchange resin Amberlyst 15 during the removal of *N*-methylimidazole is easily separated during this process (Note 13). The product **1** has potential in the synthesis of color-labelled peptides[8] after cleavage of either of the protecting groups as illustrated in Scheme 1. The amine **3** should be useful for peptide synthesis in solution while the carboxylic acid **2** could be applied for solid phase synthesis. The Fmoc group was easily removed in 10 min with 20% piperidine in DMF. LC-MS analysis of **3** showed no by-products resulting from the cleavage of the two ester groups. The *tert*-butyl ester was cleaved in 1 h with 50% TFA in methylene chloride without side reactions.

1. AIMS Scientific Products GmbH, Inhoffenstrasse 7, Building Y, D-38124 Braunschweig, Germany.

2. Helmholtz Centre for Infection Research, Department of Chemical Biology, Inhoffenstrasse 7, D-38124 Braunschweig. Germany.

3. Ley, S. V.; Baxendale, I. R.; Bream, R. N.; Jackson, P. S.; Leach, A. G.; Longbottom, D. A.; Nesi, N.; Scott, J. S.; Storer, R. I.; Taylor, S. J.; *J. Chem. Soc., Perkin Trans. 1*, **2000**, 3815-4195.

4. Cainelli, G.; Manescalchi, F.; *Synthesis* **1975**, 723-724.
5. Gayo, L. M.; Suto, M. J.; *Tetrahedron Lett.* **1997**, *38*, 513-516.
6. Rademann, J.; Smerdka, J.; Jung, G.; Grosche, P.; Schmid, D.; *Angew. Chem.* **2001**, *113*, 390-393.
7. Zander, N.; Frank, R.; *Tetrahedron Lett.* **2001**, *42*, 7783-7785.
8. For N-α-dye labelled amino acid derivatives see: Sameiro, M.; Gonçalves, T.; Maia, H. L. S.; *Tetrahedron Lett.* **2001**, *42*, 7775-7777.

1-HYDROXY-3-PHENYL-2-PROPANONE
[2-Propanone, 1-hydroxy-3-phenyl-]

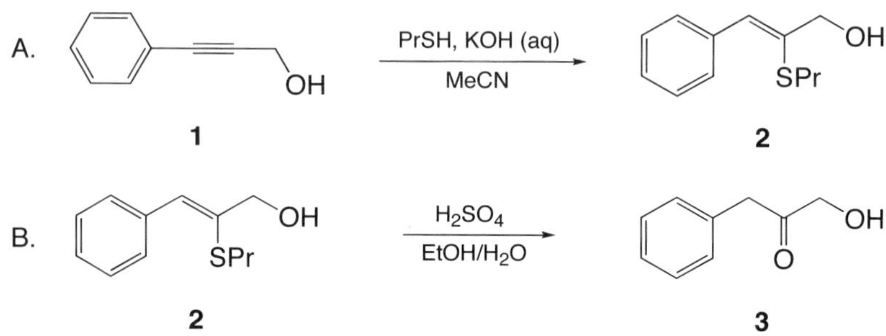

Submitted by Marjorie See Waters, Kelley Snelgrove, and Peter Maligres.[1]
Checked by Günter Seidel and Alois Fürstner.

1. Procedure

Caution: The procedures should be carried out in a well-ventilated hood because of the extreme stench of the mercaptan products. All glassware used in the procedures should be rinsed with bleach solution prior to removal from the fume hood.

A. 3-Phenyl-2-(propylthio)-2-propen-1-ol (**2**). A 250-mL, one-necked, round-bottomed flask, equipped with a magnetic stirring bar, thermocouple probe and nitrogen inlet, is charged with acetonitrile (150 mL), 3-phenyl-2-propyn-1-ol (19.8 g, 0.15 mol), 30 wt% KOH (2.84 g, 15.2 mmol) and 1-propanethiol (14.0 mL, 0.16 mol) (Note 1). The reaction mixture is heated to 60 °C with a heating mantle and stirred for 6 h (Note 2). After cooling to room temperature, the bulk of the solvent is removed under reduced pressure. The resulting yellow oil is taken up in 100 mL of *tert*-butyl methyl ether (TBME) and filtered through 50 g of silica gel (Note 3). The silica gel plug is washed with 300 mL of TBME. The filtrates are combined and concentrated under reduced pressure to give 30.83 g (99%) of 3-phenyl-2-(propylthio)-2-propen-1-ol as a pale yellow oil which can be used without further purification (Note 4).

B. 1-Hydroxy-3-phenyl-2-propanone (**3**). A 1-L, three-necked, round-bottomed flask, equipped with a magnetic stirring bar, thermocouple probe, rubber septum and nitrogen inlet and outlet that leads to a scrubber containing bleach, is charged with 3-phenyl-2-(propylthio)-2-propen-1-ol (30.1 g, 0.14 mol), 150 mL of ethanol and 150 mL of 1M sulfuric acid. The suspension, which becomes a homogeneous yellow solution, is heated at 65 °C with a heating mantle for 4 h with vigorous stirring and a nitrogen sweep (Notes 5 and 6).

After cooling to room temperature, the reaction mixture is transferred to a 1-L separatory funnel with the aid of 50 mL of water and washed with 250 mL of hexanes, which is discarded (Note 7). After adding 150 mL of brine to the aqueous phase, the product is extracted with ethyl acetate (3 × 250 mL). The combined organic extracts are dried over magnesium sulfate, filtered, and concentrated under reduced pressure (Note 8).

The residue is dissolved in 40 mL of ethyl acetate and filtered through a pad of Celite into a 500-mL round-bottomed flask. With stirring, 240 mL of *n*-heptane is added over 20 min to the ethyl acetate solution via a pressure-equalizing addition funnel. During the addition, the product begins to crystallize as a white solid. The volume of the mixture is reduced by distillation at reduced pressure (20 °C, 10-20 mm) to about 150 mL. The mixture is then cooled to −5 °C and aged for 1 h. The solids are collected via vacuum filtration, washed with heptane (100 mL), and dried under vacuum at room temperature to afford 1-hydroxy-3-phenyl-2-propanone (18.9 g, 87% yield) as white crystals (Note 9).

2. Notes

1. 3-Phenyl-2-propyn-1-ol was purchased from Lancaster Synthesis Ltd. and used as received. 1-Propanethiol was purchased from Aldrich Chemical Company, Inc., and used as received.

2. The reaction progress can be monitored by GC analysis (HP-1701, 15 m × 0.53 mm capillary column, 1.00 μm film thickness, 100 °C increase at 15°/min to 250 °C); the retention time was 4.6 min for the alkynol (**1**) and 7.9 min for the vinyl sulfide (**2**).

3. Silica gel is added to a 300 mL fritted glass funnel and pre-washed with TBME under vacuum, discarding the filtrate. The reaction mixture is poured onto the pad and filtered under vacuum, followed by the TBME washes. An orange layer of by-products remains at the top of the silica gel.

4. The crude product typically contains 2-3% of a by-product resulting from addition to the 3-position of the alkynol, 2-3% of the E-addition product, and residual propanethiol. Spectral data for **2** are as follows: IR cm^{-1}: 3358, 2962, 2930, 2871, 1598, 1491, 1445, 1378, 1291, 1237, 1098, 1078, 1009, 753, 694. ^{1}H NMR (400 MHz, $CDCl_3$) δ: 0.96 (t, 3H, J = 7.3), 1.60 (m, 2H), 2.76 (t, 2H, J = 7.3), 4.37 (s, 2H), 6.83 (s, 1H),

7.2-7.4 (m, 3H), 7.63 (d, 2H, J =7.6); ^{13}C NMR (100 MHz, $CDCl_3$) δ: 13.4, 23.4, 33.4, 67.0, 127.4, 128.1, 129.5, 129.9, 135.6, 136.3. The checkers obtained **2** in 95% yield. The checkers noted that compound **2** may eventually isomerize to 3-phenyl-2-(propylthio)propanal when kept in $CDCl_3$ which was not rigorously dried prior to use.

5. A moderate nitrogen purge is necessary to remove the propanethiol, which is formed in the reaction. Care should be taken to treat any distillate collected in the trap with bleach.

6. The reaction rate will depend on the stirring and nitrogen purge rates. The reaction should be monitored for completion (>98% conversion) by GC using the method described in (Note 2); the retention time for hydroxyketone **3** is 4.9 min.

7. The hexanes wash removes thiol-containing by-products and should be treated with bleach prior to disposal.

8. The concentration is done at 20-25 °C at the lowest possible pressure. The hydroxyketone (**3**) is reactive and all product-containing solutions should be maintained below 25 °C during the work up.

9. The physical properties of 3 are as follows: mp 47-48 °C; IR (KBr) cm^{-1}: 3378, 1721, 1496, 1454, 1402, 1291, 1076, 1043, 755, 701; ^{1}H NMR (400 MHz, $CDCl_3$) δ: 2.94 (s, 1H), 3.74 (s, 2H), 4.30 (s, 2H), 7.2-7.4 (m, 5H); ^{13}C NMR (100 MHz, $CDCl_3$) δ: 45.9, 67.7, 127.6, 129.0, 129.4, 132.8, 207.3. Anal. Calcd for $C_9H_{10}O_2$: C, 71.98; H, 6.71. Found: C, 71.79; H, 6.70. The checkers obtained material having mp 45-46 °C in 83% yield.

Waste Disposal Information

All toxic materials were disposed of in accordance with "Prudent Practices in the Laboratory"; National Academy Press; Washington, DC, 1995.

3. Discussion

Hydroxyketones are versatile intermediates in the synthesis of pharmaceutical intermediates and heterocyclic molecules.[2] α-Aryl hydroxyketones have been prepared by reaction of aryl aldehydes with 1,4-dioxane followed by reduction with lithium aluminum hydride (LAH)[3] and by the selective LAH reduction of α-silyloxy α,β-unsaturated esters.[4] Wissner[5] has shown that treatment of acid chlorides with tris(trimethylsilyloxy)ethylene affords alkyl and aryl hydroxymethyl ketones. 1-Hydroxy-3-phenyl-2-propanone (**3**) has been generated by the osmium-catalyzed oxidation of phenylpropene[6] and by the palladium-catalyzed rearrangement of phenyl epoxy alcohol[7] both in 62% yield.

The present method is superior in that it requires neither oxidation nor reduction and does not use expensive and toxic heavy metal catalysts. The two step process offers mild conditions and high yields. Recovery of the thiol is possible by using fractional distillation during the hydrolysis, making the process efficient on large scale. The procedure is general for a variety of α-aryl hydroxyketones.[8]

1. Department of Process Research, Merck Research Laboratories, Merck & Co., Inc., Rahway, NJ 07065

2. (a) Horikawa, M.; Nakajima, T.; Ohfune, Y. *Synlett* **1997**, 253-254; (b) Liu, K.; Wood, H. B.; Jones, A. B. *Tetrahedron Lett.* **1999**, *40*, 5119-5122; (c) Jerris, P. J.; Wovkulich, P. M.; Smith, A. B., III *Tetrahedron Lett.* **1979**, *20*, 4517-4520.

3. Fetizon, M.; Hanna, I.; Rens, J. *Tetrahedron Lett.* **1985**, *26*, 3453-3456.

4. Dalla, V.; Catteau, J. P *Tetrahedron* **1999**, *55*, 6497-6510.

5. Wissner, A. *Tetrahedron Lett.* **1978**, *19*, 2749-2752.

6. Murahashi, S.; Naota, T.; Hanaoka, H. *Chem. Lett.* **1993**, *10*, 1767-1770.

7. Vankar, Y. D.; Chaudhuri, N. C.; Singh, S. P. *Synth. Comm.* **1986**, *16*, 1621-1626.

8. Waters, M. S.; Cowen, J. A.; McWilliams, J. C.; Maligres, P. E.; Askin, D. *Tetrahedron Lett.* **2000**, *41*, 141-144.

DIRECT CHLORINATION OF ALCOHOLS: SYNTHESIS OF ETHYL 3-CHLORO-3-PHENYLPROPANOATE
[Benzenepropanoic acid, β-chloro-, ethyl ester]

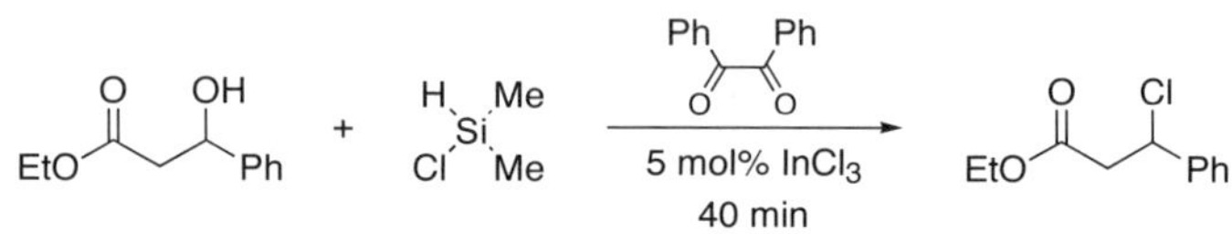

Submitted by Makoto Yasuda, Satoshi Yamasaki, Yoshiyuki Onishi, and Akio Baba.[1]
Checked by John R. Heemstra, Jr. and Scott E. Denmark.

1. Procedure

Caution! This reaction generates hydrogen (H_2) gas and therefore should be conducted in an efficient hood.

A 500-mL, three-necked, round-bottomed flask is equipped with a magnetic stir bar, a glass stopper, a Teflon-coated thermocouple, and a Leibig reflux condenser fitted with an inert gas inlet tube that is attached to a nitrogen manifold. The flask is flame-dried under vacuum, (0.2 mmHg) flushed with nitrogen, cooled to room temperature, and maintained under a slightly positive pressure of nitrogen. The flask is charged with $InCl_3$ (0.775 g, 3.5 mmol) (Note 1), which is then dried by heating the bottom of the flask with a hot air gun under vacuum (0.2 mmHg) for 2 min (Note 2). The flask is then refilled with nitrogen and kept under a slightly positive pressure. The glass stopper is removed and to the flask are added benzil (14.7 g, 70 mmol) (Note 3), dichloromethane (140 mL) (by syringe) (Note 4), and ethyl 3-hydroxy-3-phenylpropanoate (by syringe) (13.6 g, 70 mmol) (Note 5). Chlorodimethylsilane ($HSiMe_2Cl$) (7.28 g, 77 mmol) is added to the resulting mixture by syringe through an open neck while stirring (Note 6). The stopper is replaced and gentle evolution of gas (H_2) starts in a few minutes (Note 7). The extent of the gas evolution gradually increases and the reaction temperature slowly increases (Note 8).

After 35 min, gas evolution ceases (Note 9), indicating the end of the reaction. The solution is stirred for 5 min further (40 min total after addition of chlorodimethylsilane), whereupon water (100 mL) is added to the flask and the mixture is transferred into a 1-L separatory funnel containing diethyl ether (200 mL). After the mixture is shaken, the organic layer is separated and the aqueous layer is extracted with diethyl ether (1 x 100 mL, 1 x 60 mL). The combined organic extracts (yellow solution) are dried over anhydrous magnesium sulfate ($MgSO_4$), filtered and concentrated at reduced pressure by rotary evaporation (30 °C/14 mmHg). The residue is diluted with hexane (20 mL) and cooled in an ice-bath to give a precipitate, which is filtered (Note 10). The filtrate is concentrated at reduced pressure by rotary evaporation (30 °C/14 mmHg) to afford a yellow oil.

The oil is loaded onto a column (4 cm diam x 14 cm height) of 100 g of silica gel (Note 11) with hexane. Using compressed air, the column is eluted with hexane (250 mL) (Note 12) and then ethyl acetate (300 mL) (Note 13). The ethyl acetate fraction (yellow solution) is concentrated by rotary evaporation (30 °C/14 mmHg) to afford 19.08 g of yellow liquid (Note 14). The liquid is distilled at reduced pressure to give 14.0 g (94%) of ethyl 3-chloro-3-phenylpropanoate (90 °C/0.2 mmHg) as a yellow oil (Notes 15-18).

2. Notes

1. $InCl_3$ (98%) was obtained from Aldrich Chemical Company, Inc., and used as received.
2. $InCl_3$ is hygroscopic and this procedure gives dry $InCl_3$.
3. Benzil (98%) was obtained from Aldrich Chemical Company, Inc., and used as received.
4. Dry dichloromethane was obtained from Wako Pure Chemical Industries, Ltd. (Japan), and used as received.
5. The reaction temperature does not change during this process. A yellow solution with a precipitate is observed.
6. Chlorodimethylsilane (98%) was obtained from Aldrich Chemical Company, Inc., and used as received. The bottle of the reagent should be cooled at –30 °C in freezer before the syringe transfer operation.
7. The evolution of H_2 can be checked by bubbler and is seen in the flask as bubbles.
8. Internal temperatures: 19 °C (0-6 min), 21 °C (10 min), 26 °C (20

min), 30 °C (25 min), 31 °C (30 min), 30-29 °C (35-40 min).

9. The gas (H_2) evolves constantly for about 30 min and ceases at 35 min. The period of the time is shorter for smaller scale reactions.

10. The precipitate is separated by vacuum filtration using 60 mm Büchner funnel. The solid is carefully washed with cold hexane (50 mL). The separated pale yellow solid is pure benzil (10.1 g).

11. Silica gel was purchased from Silicycle (SiliaFlash P60 (40-63 μ, 60 Å pore size)). For TLC analysis, EMD Chemicals, Inc. silica gel 60 F_{254} TLC plates were used, with hexane/Et_2O, 7:3 as eluent. The starting alcohol (ethyl 3-hydroxy-3-phenylpropanoate), the chlorinated product (ethyl 3-chloro-3-phenylpropanoate), and benzil have R_f values of ca. 0.28, 0.66, and 0.57, respectively.

12. The column is eluted with hexane until a few drops of yellow solution come out. This process effectively separates silyl by-product.

13. A yellow-colored solution comes out.

14. The ^{1}H NMR spectrum shows that the liquid contains the desired chloride and benzil (79:21), along with a small amount of silyl species.

15. A forerun (10 drops) is collected and discarded to avoid contamination by silyl species.

16. The purity of the product was found to be 94%. The major contaminants detected by ^{1}H NMR (500 MHz) spectroscopic analysis are benzil (2%) and ethyl 3-phenyl-2-propenoate (4%).

17. The residue after distillation contains mostly benzil. The pure benzil (2.47 g) is recovered from the residue by vacuum filtration and is washed with cold hexane. A total of 12.57 g (86%) of benzil is recovered (Note 11).

18. The checkers obtained an analytically pure sample of the product as follows: a 300-mg portion of the product was purified by silica gel column chromatography (2 cm diam x 11 cm height, 20 g) (Note 12) with hexane/CH_2Cl_2, 2:1 (200-300 mL). In this solvent system the desired product has an R_f = 0.33 and the elimination product an R_f = 0.23. Concentration of the product-containing fractions (30 °C/14 mmHg) provided 275 mg of ethyl 3-chloro-3-phenylpropanoate, which was purified by bulb-to-bulb distillation to afford 133 mg of analytically pure material. Physical properties of the purified product are as follows: ^{1}H NMR (500 MHz, $CDCl_3$) δ: 1.24 (t, J = 7.1 Hz, 3 H), 3.03 (A<u>B</u>X, J_{AB} = 15.9 and J_{BX} =

5.8 Hz, 1 H), 3.17 (ABX, J_{AB} = 15.9 and J_{AX} = 9.1 Hz, 1 H), 4.16 (m, 2 H), 5.35 (dd, J = 5.9 and 9.0 Hz, 1 H), 7.30–7.43 (m, 5 H); ^{13}C NMR (125 MHz, $CDCl_3$) δ: 14.0, 44.8, 58.1, 60.9, 126.9, 128.67, 128.74, 140.2, 169.5; IR (neat) cm^{-1}: 3454 (m), 3066 (m), 3034 (m), 2983 (s), 2938 (s), 2906 (m), 2360 (w), 1955 (w), 1883 (w), 1740 (s), 1604 (m), 1587 (m), 1495 (m), 1455 (m), 1367 (s), 1333 (s), 1272 (s), 1196 (s), 1159 (s), 1097 (m), 1021 (s), 948 (s); MS (EI, 70 eV) m/z (%): 212 (M^+, 43), 183 (28), 167 (17), 149 (19), 138 (84), 125 (89), 105 (100), 91 (14), 77 (47), 63 (11); HRMS (EI, M^+) *m/z* calcd for $C_{11}H_{13}ClO_2$: 212.0604, found: 212.0596. Anal. calcd for $C_{11}H_{13}ClO_2$: C, 62.12; H, 6.16; Cl, 16.67. Found: C, 62.17; H, 6.16; Cl, 16.56.

Safety and Waste Disposal Information

All hazardous materials should be handled and disposed of in accordance with "Prudent Practices in the Laboratory"; National Academy Press; Washington, DC, 1995.

3. Discussion

The procedure described herein provides a useful method for chlorination of alcohols under mild and neutral conditions.[3] Even an acid sensitive hydroxy ester (ethyl 3-hydroxy-3-phenylpropanoate) is cleanly converted into the desired chloride. For this substrate, conventional chlorination methods[4] in which strong protic acids are generated *in situ* using PCl_3 or PCl_5 give low yield of the chloride contaminated with ethyl 3-phenyl-2-propenoate that forms through β-elimination.[3]

Indium trichloride ($InCl_3$) has moderate Lewis acidity and oxophilicity as compared with typical Lewis acids such as aluminum halides or boron halides.[5-7] This character enables $InCl_3$ to catalyze the reactions of carbonyls or alcohols.[8,9]

In the alcohol/$HSiMe_2Cl$/$InCl_3$/benzil system, role of benzil is significant. When benzil is not included, hydro-dehydroxylation (reduction) occurs through the formation of the hydrodimethylsilyl ether with release of HCl.[10] The addition of benzil changes the reaction pathway to form the chlorodimethylsilyl ether with release of H_2. Although benzil is used as an equimolar additive, about 80% of the benzil can easily be recovered. The amount of benzil can be reduced for certain substrates, but the use of a stoichiometric amount leads to faster reaction rates, higher yields, and broader substrate scope.

The generality of this chlorination method is summarized in Table 1. Various secondary and tertiary alcohols are converted into the corresponding

Table 1. Chlorination of Various Alcohols

entry	alcohol	time/ h	product	yield/ %
1	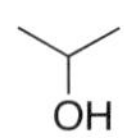	6	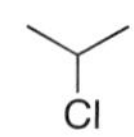	100
2	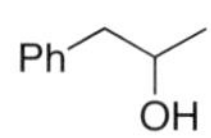	15	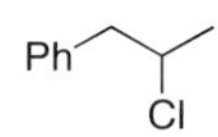	68
3	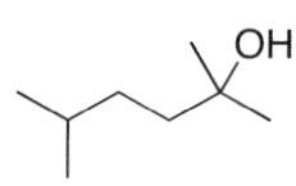	9	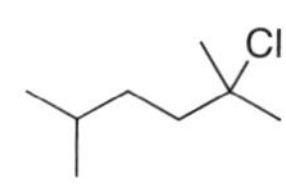	100
4		23		71
5	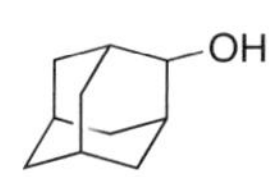	91	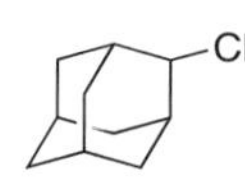	98
6	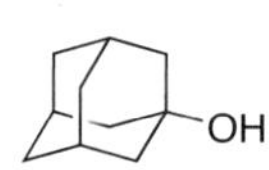	82		93
7	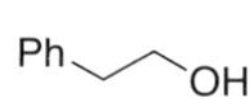	24	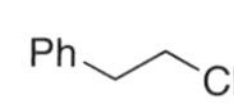	0
8	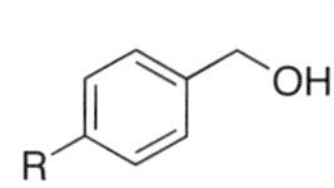 R = H	1	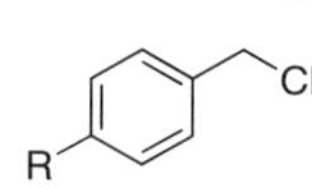	80
9	R = Cl	2.5		92
10	R = Me	0.7		83
11	R = NO_2	24		97
12	R = COOEt	1.5		77
13	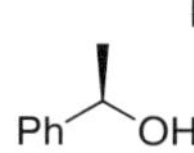	2.5	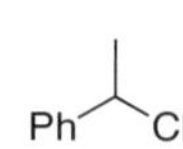	91 (<2% ee)

chlorides in high yields (entries 1-6). A primary alcohol (2-phenylethanol) does not give the desired product (entry 7). However, effective transformation proceeds in the reaction with benzylic alcohols which bear electron-withdrawing or donating substituents (entries 8-10). Nitro and ester groups tolerate these reaction conditions to furnish the corresponding chlorides (entries 11 and 12). The enantiomerically pure alcohol (1-phenylethanol) gives racemic 1-phenylethyl chloride (entry 13). These results suggests that the reaction proceeds via a carbocationic intermediate.

Selective chlorination at the tertiary site of a diol that contains both primary and tertiary alcohols illustrates the unique selectivity of this chlorination system (Scheme 1). On the contrary, conventional chlorination systems such as PPh_3/CCl_4 or PCl_5 afford the primary chloride.

Scheme 1

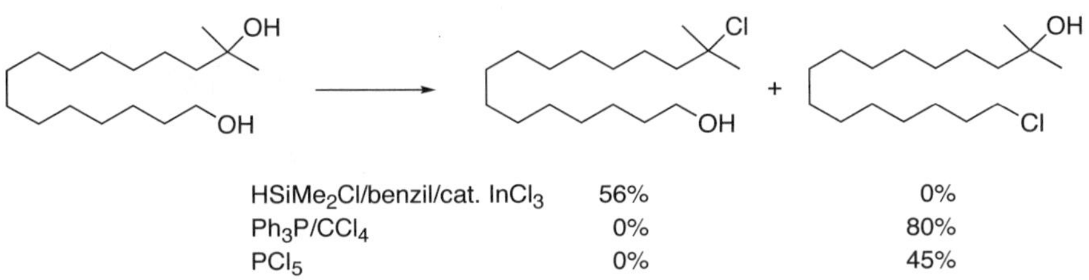

1. Department of Applied Chemistry, Graduate School of Engineering, Osaka University, Suita, Osaka 565-0871, Japan.
2. Hauser, C. R.; Breslow, D. S. *Org. Synth., Coll. Vol. III*, 408-410.
3. Yasuda, M.; Yamasaki, S.; Onishi, Y.; Baba, A. *J. Am. Chem. Soc.* **2004**, *126*, 7186-7187.
4. (a) *Comprehensive Organic Transformations*; 2nd ed.; Larock, R. C., Ed.; Wiley-VCH: New York, 1999; pp. 689-693. (b) *Comprehensive Organic Syntheses*; Trost, B. M., Ed.; Pergamon Press: Oxford, U.K., 1991; Vol. 6, pp. 204-206.
5. Olah, G. A.; Kobayashi, S.; Tashiro, M. *J. Am. Chem. Soc.* **1972**, *94*, 7448-7461.
6. Babu, G.; Perumal, P. T. *Aldrichimica Acta*, **2000**, *33*, 16-22.
7. Ranu, B. C. *Eur. J. Org. Chem.* **2000**, 2347-2356.
8. Yasuda, M.; Saito, T.; Ueba, M.; Baba, A. *Angew. Chem. Int. Ed.,* **2004**, *43*, 1414-1416.
9. Onishi, Y.; Ogawa, D.; Yasuda, M.; Baba, A. *J. Am. Chem. Soc.*, **2002**, *124*, 13690-13691.
10. Yasuda, M.; Onishi, Y.; Ueba, M.; Miyai, T.; Baba, A. *J. Org. Chem.* **2001**, *66*, 7741-7744.

PREPARATION OF *HEXAKIS*(4-BROMOPHENYL)BENZENE (HBB) [1,1':2',1''-Terphenyl, 4,4''-dibromo-3',4',5',6'-tetrakis(4-bromophenyl)-]

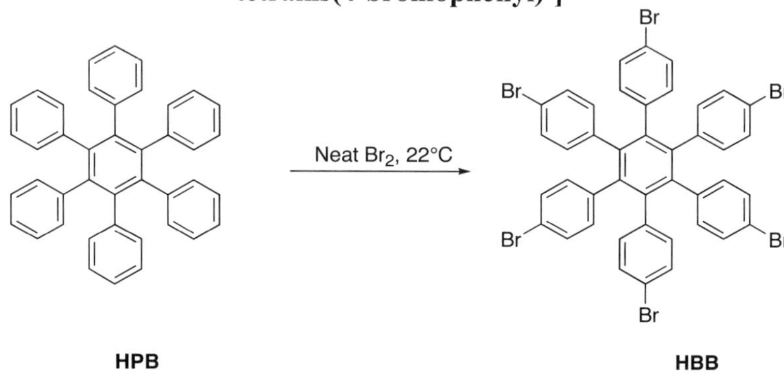

Submitted by Rajendra Rathore[1] and Carrie L. Burns.
Checked by Scott E. Denmark and Shinji Fujimori.

1. Procedure

A 500-mL, three-necked flask, equipped with a mechanical stirrer (fitted with an 11 cm Teflon paddle), a septum and an outlet adapter connected *via* rubber tubing to a pipette which is immersed in a 10% aqueous sodium hydroxide solution (250 mL) in a 500 mL Erlenmeyer flask is charged with 70 mL of bromine (1.37 mol) [Note 1]. The septum is replaced with a 250-mL powder addition funnel charged with 26.7 g (50 mmol) of hexaphenylbenzene (HPB) [Note 2]. The flask is placed in a water bath at ambient temperature to control the heat evolved from the reaction. To the slowly stirred bromine, hexaphenylbenzene is added slowly over 1 hour [Note 3]. The reaction starts immediately as judged by an evolution of gaseous hydrobromic acid [Note 4]. After the addition of hexaphenylbenzene is complete, the dark-orange slurry is stirred for an additional 20 min. [Note 5]

The bromine slurry of resulting product is carefully poured into 500 mL of pre-chilled (approx. –78 °C) ethanol in a 1-L Erlenmeyer flask with stirring by a magnetic stir bar [Note 6]. To the three-necked flask is added cold (–78 °C) ethanol (100 mL x 2) and the remaining precipitate is transferred to the Erlenmeyer flask. The suspension of the product in ethanol is allowed to warm to room temperature over 2 h with stirring, and the suspension is filtered using a Büchner funnel. The pale yellow precipitate is washed with ethanol (50 mL), aqueous sodium bisulfite (5%, 100 mL), and ethanol (2 X 50 mL) successively. After being dried overnight *in vacuo* (0.5 mmHg) at room temperature to a constant weight, 47.8 g (96% yield) of *hexakis*(4-bromophenyl)benzene (HBB) is obtained with greater than >95% purity as judged by ^{1}H NMR spectroscopy.

The precipitated product is sufficiently pure for most purposes; however, it can be further purified by re-precipitation from tetrahydrofuran. Thus, 10 g of HBB was dissolved in refluxing tetrahydrofuran [Note 7] (350 mL) in a beaker. Upon slow evaporation at room temperature, the solution yields a colorless (microcrystalline) precipitate (9.4 g) of *hexakis*(4-bromophenyl)benzene (HBB). [Note 8] [Note 9].

2. Notes

1. Bromine was obtained from Aldrich Chemical Co. was used as received.
2. Hexaphenylbenzene (HPB) was obtained following the *Organic Syntheses* procedure (Fieser, L. F. *Org. Synth.*, **1973**, *46*, 44; *CV 5*, 604). Commercially available HPB (Aldrich Chemical Co.) can also be used.
3. The reaction is carried out without added solvent and thus for a thorough mixing of reagents an excess of bromine is required for a complete conversion of HPB to HBB. It is critical to maintain the reaction mixture as a slurry for complete conversion.
4. Gaseous HBr was trapped in an aqueous solution of sodium hydroxide.
5. At this point the HBr evolution completely ceases.
6. The slurry should be poured into cold ethanol (-78 °C) to prevent an exothermic reaction between excess bromine and ethanol.
7. Tetrahydrofuran (Optima grade) was obtained from Fischer Inc. and was used as received.
8. The spectral data for analytically pure HBB: mp 358-359 °C; ^{1}H NMR ($CDCl_3$) δ: 6.61 (d, J = 8.5 Hz, 12 H), 7.06 (d, J = 8.6 Hz, 12 H); ^{13}C NMR ($CDCl_3$) δ: 120.3, 130.5, 132.6, 138.4, 139.6. Anal. Calcd for $C_{42}H_{24}Br_6$: C, 50.04; H, 2.40; Br, 47.56. Found: C, 49.92; H, 2.27.
9. Crystallization can be performed in refrigerator (-15 °C) over 3 days to provide larger size crystals.

Waste Disposal Information

All hazardous materials should be handled and disposed of in accordance with "Prudent Practices in the Laboratory"; National Academy Press; Washington, DC, 1995.

3. Discussion

The hexaphenylbenzene core is being extensively investigated as a platform for the preparation of nanometer-size macromolecules and supramolecular assemblies owing to their importance as materials that can be used as molecular devices such as sensors, switches, ferromagnets, and other electronic and optoelectronic devices.[2-5] As we recently demonstrated, the elaboration of the hexaphenylbenzene core can be readily achieved using HBB for the preparation of a hexacation-radical salt[6] for use as a (multi)electron-transfer catalyst in a variety of organic and organometallic transformations.[7] The hexaphenylbenzene core is also being utilized for the preparation of well-defined graphite-like structures by Müllen and others and the progress in the area has been reviewed in two recent *Chemical Review* articles.[8,9]

There is one reported procedure for the preparation of HBB, which utilizes the Diels-Alder approach using *tetrakis*(4-bromophenyl) cyclopentadienone and *bis*(4-bromophenyl)acetylene (also known as 4,4'-dibromotolan) as the starting materials.[10] Both of these starting materials are prepared *via* multi-step syntheses. We have also discovered that trimerization of 4,4'-dibromotolan using *bis*(acetonitrile)palladium dichloride affords HBB in fair yield.[11]

A similar bromination procedure (as described above for the preparation of HBB) can be employed for the preparation of *tetrakis*(4-bromophenyl)methane using tetraphenylmethane and bromine.[6]

The method described here for the preparation of HBB is essentially a detailed description of our recently published procedure[6] using hexaphenylbenzene and neat bromine. It is believed that the ready availability of HBB from hexaphenylbenzene will facilitate the synthesis of a variety of materials, which were otherwise not readily accessible.[12]

1. Department of Chemistry, Marquette University, P.O. Box 1881, Milwaukee, WI 53201-1881.

2. Lambert, C.; Noll, G. *Angew. Chem. Int. Ed. Engl.* **1998**, *37*, 2107.

3. Praefcke, K.; Khone, B.; Singer, D. *Angew. Chem. Intl. Ed. Engl.* **1990,** *29*, 177.

4. Laschewsky, A. *Angew. Chem. Intl. Ed. Engl.* **1989**, *28*, 1574.

5. Kobayashi, K.; Shiraska, T.; Horn, E.; Furukawa, N. *Tetrahedron Lett.* **2000**, *41*, 89 and references therein.

6. Rathore, R.; Burns, C.L.; Deselnicu, M. I. *Org. Lett.* **2001**, *3*, 2887.

7. Connelly, N.G.; Geiger, W.E. *Chem. Rev.* **1996,** *96*, 877 and references therein.

8. Watson, M.D.; Fechtenkotter, A.; Mullen, K. *Chem. Rev.* **2001**, *101*, 1267 (review) and references therein.

9. Berresheim, A.J.; Muller, M.; Mullen, K. *Chem. Rev.* **1999**, *99*, 1747 (review) and references therein.

10. Broser, W.; Siegle, P.; Curreck, H. *Chem. Ber.* **1968**, *101*, 69-83.

11. Rathore, R. (unpublished results).

12. Rathore, R., Burns, C. L., Guzei, I. A. *J. Org. Chem.* **2004**, *69*. 1524-1530.

SYNTHESIS OF 5-BROMOISOQUINOLINE AND 5-BROMO-8-NITROISOQUINOLINE

(Isoquinoline, 5-bromo- and Isoquinoline, 5-bromo-8-nitro-)

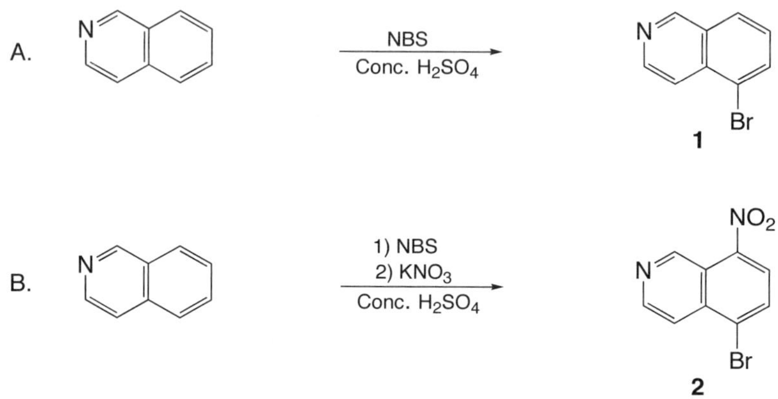

Submitted by William Dalby Brown[1a] and Alex Haahr Gouliaev.[1b]
Checked by Steven Wolff and Walter Burger.

1. Procedure

A. 5-Bromoisoquinoline. A 1-L, three-necked, round-bottomed flask equipped with an internal thermometer, mechanical stirrer, and an addition funnel fitted with a nitrogen inlet is charged with concentrated sulfuric acid (96%, 340 mL, Note 1) and cooled to 0 °C. Isoquinoline (40 mL, 44.0 g, 330 mmol) is slowly added to the well-stirred acid at a rate such that the internal temperature is maintained below 30 °C. The solution is cooled to –25 °C in a dry ice-acetone bath and *N*-bromosuccinimide (64.6 g, 363 mmol; Note 2) is added to the vigorously stirred solution in portions such that the internal temperature is maintained between –22 and –26 °C (Note 3). The suspension is efficiently stirred for 2 hr at –22 ± 1 °C and then for 3 hr at –18 ± 1 °C. The resulting homogeneous reaction mixture is poured onto 1.0 kg of crushed ice in a 5-L flask placed in an ice water bath. The reaction flask is quickly washed with ice-cold water, which is added to the 5-L flask. The resulting mixture is stirred while the pH is adjusted to 9.0 using 25% aq NH_3 with the internal temperature maintained below 25 °C. The resulting alkaline suspension is added to 800 mL of diethyl ether and the biphasic system is vigorously mixed (Note 4). The two clear phases are separated and the aqueous phase is extracted with two 200-mL portions of diethyl ether. The combined organic phases are washed with 200 mL of 1M NaOH (aq) and 200 mL of H_2O, dried over anhydrous $MgSO_4$, filtered, and concentrated to afford 47 g of a light brown solid. Fractional distillation under reduced pressure (bp 145-149 °C at 14 mm, Note 5) furnishes 34-36 g (47-49%) of 5-bromoisoquinoline as a white solid (Notes 6 and 7).

B. 5-Bromo-8-nitroisoquinoline. A 1-L, three-necked, round-bottomed flask fitted with an internal thermometer, mechanical stirrer, and an addition funnel fitted with a nitrogen inlet is charged with concentrated sulfuric acid (96%, 340 mL, Note 1) and cooled to 0 °C. Isoquinoline (40 mL, 44.0 g, 330 mmol) is slowly added from the addition funnel to the well-stirred acid such that the internal temperature is maintained below 30 °C. The solution is cooled to –25 °C in a dry ice-acetone bath and N-bromosuccinimide (76.4 g, 429 mmol; Note 2) is added to the vigorously stirred solution in portions such that the internal temperature is maintained between –22 and –26 °C (Note 3). The suspension is efficiently stirred for 2 hr at –22 °C ± 1 °C and then for 3 hr at –18 °C ± 1 °C. Potassium nitrate (35.0 g, 346 mmol) is then added at a rate such as to maintain the internal temperature below –10 °C and the mixture is then stirred at 10 °C for 1 hr. The cooling bath is removed and the solution is stirred overnight. The resulting homogeneous reaction mixture is poured onto 1.0 kg of crushed ice in a 5-L flask and the reaction flask is quickly washed with ice-cold water, which is added to the 5-L flask. The resulting mixture is stirred while the pH is adjusted to 8.0 using 25% aq NH_3 with the internal temperature maintained below 30 °C. The resulting suspension is stirred in an ice water bath for 2 hr and the precipitated solids are isolated by filtration using a glass filter funnel. The solids are thoroughly washed three times with 1-L portions of ice-cold water and then air-dried to constant weight to afford 65 g of a slightly yellow solid. This material is suspended in 1000 mL of heptane and 250 mL of toluene (Note 8) in a 2-L, round-bottomed flask and heated at reflux for 1.5 hr with stirring. The hot solution is then filtered through Celite using vacuum suction (Note 9). The volume of the filtrate is reduced by distillation to 1000 mL and the resulting orange solution is allowed to slowly cool with stirring overnight. The solids are isolated by filtration, washed with 350 mL of ice-cold heptane, and air-dried to constant weight to afford 40-44 g (47-51%) of 5-bromo-8-nitroisoquinoline (Notes 10 and 11).

2. Notes

1. Sulfuric acid 96% (technical quality) and diethyl ether (technical quality) were purchased from Bie & Berntsen A/S, Sandbaekvej 7, DK-2610 Roedovre, Denmark and used without further purification. Isoquinoline (97%) and potassium nitrate (99%) were purchased from Aldrich Chemical Company, Inc. and used without further purification.

2. Isoquinoline must be completely dissolved prior to the addition of *N*-bromosuccinimide. *N*-Bromosuccinimide (99%) was purchased from Aldrich Chemical Company, Inc. and recrystallized[2] and air-dried prior to use. Recrystallization is essential in order to obtain high yield and pure product. The use of more NBS than stated (i.e., more than 1.1 equiv for the synthesis of 5-bromoisoquinoline and 1.3 equiv for the synthesis of 5-bromo-8-nitroisoquinoline) to obtain complete transformation of isoquinoline should be avoided as this leads to formation of 5,8-dibromoisoquinoline, which cannot easily be separated from 5-bromoisoquinoline and which will also lead to a lower yield of 5-bromo-8-nitroisoquinoline.

3. The temperature is controlled throughout the reaction by intermittently adding additional pieces of dry ice to the dry ice-acetone bath. Strict temperature control throughout the bromination reaction is important to obtain high regioselectivity and purity of the product as the side products cannot be removed with ease.

4. Vigorous shaking for at least 5 min is required.

5. A 40-cm column insulated with cotton and aluminum foil is used for the fractional distillation. The condenser is heated to a constant temperature of 80 °C by attaching an 80 °C water bath and a circulatory pump. As distillation progresses, the temperature rises smoothly, settling at 126-128 °C for some time, and then rising to 144 °C (at 14 mm), at which temperature and pressure the product starts to solidify in the condenser. Increased heating (heating the water bath to 85-90 °C) allows for a steady flow of product through the condenser. All fractions boiling below 144 °C/14 mm are discarded. As an alternative to the preheated condenser, a heat gun may be used to melt the solidifying product.

6. The isolated product typically contains 0-3% isoquinoline. Physical data for 5-bromoisoquinoline: TLC: Rf = 0.30 (9:1 dichloromethane/diethyl ether); IR ($CHCl_3$) cm^{-1}: 3053, 1582, 1509, 1484, 1352, 1263, 1222, 1112, 1000; ^1H NMR (500 MHz, DMSO-d_6) δ: 7.62 (t, 1 H, *J* = 7.8), 7.91 (d, 1 H, *J* = 6.0), 8.15 (d, 1 H, *J* = 7.5), 8.19 (d, 1 H, *J* = 8.2),

8.66 (d, 1 H, $J = 5.9$), 9.37 (s, 1 H); ^{13}C NMR (DMSO-d_6) δ: 118.5, 120.3, 127.9, 128.4, 129.3, 133.9, 134.3, 144.6, 152.9.

7. Alternatively, the product can be isolated in >99% purity by column chromatography on 63-200 μm silica gel eluting with 9:1 → 5:1 dichloromethane/ethyl acetate and then 9:1 → 4:1 dichloromethane/diethyl ether. The product obtained in this manner has mp 81-82 °C (lit. 82-84 °C,[3] 79.5-80.5 °C,[4] 82-83 °C[5]). Anal. Calcd for C_9H_6BrN: C, 51.96; H, 2.91; Br, 38.40; N, 6.73. Found: C, 52.06; H, 2.74; Br, 38.21; N, 6.65.

8. Heptane was purchased from Fisher Chemicals and toluene (technical quality) was purchased from SvedaKemi A/S, Rosenoerns Allé 9, DK-1970 Frederiksberg, Denmark.

9. The glass filter funnel containing Celite was preheated using a heat gun to avoid precipitation of product in the funnel. After filtration, an additional preheated mixture of 120 mL of heptane and 30 mL of toluene was passed through the Celite using suction. A precipitate forms in the filtrate, which is redissolved by heating to reflux.

10. The isolated product typically contains 0-1% 8-bromo-5-nitroisoquinoline, 0-1% 5,8-dibromoisoquinoline, and 0-1% 5-nitroisoquinoline. Physical data for 5-bromo-8-nitroisoquinoline: mp 137-139 °C (lit. 139-141 °C,[3] 138-140 °C,[4] 139-141 °C[5]); TLC $R_f = 0.57$ (9:1 dichloromethane/ethyl acetate); IR ($CHCl_3$) cm^{-1}: 3053, 1619, 1580, 1485, 1374, 1265, 1201; ^{1}H NMR (500 MHz, DMSO-d_6) δ: 8.12 (dd, 1 H, $J_A = 0.8$, $J_B = 6.0$), 8.33† (d, 1 H, $J = 8.3$), 8.35 (AB, 1 H, $J = 8.2$), 8.84 (d, 1 H, $J = 5.9$), 9.78 (s, 1 H); ^{13}C NMR δ: 118.9, 120.0, 125.8, 127.8, 133.4, 134.5, 145.3, 145.7, 148.1.

11. Alternatively, the product can be isolated in >99% purity by column chromatography on 63-200 μm silica gel, elution with 9:1 → 6:1 dichloromethane/diethyl ether followed by recrystallization to yield 5-bromo-8-nitroisoquinoline with mp 139 -141 °C. Anal. Calcd for C_9H_6BrN: C, 42.72; H, 1.99; Br, 31.58; N, 11.07. Found: C, 43.03; H, 1.82; Br, 31.19; N, 10.94.

Waste Disposal Information

All toxic materials were disposed of in accordance with "Prudent Practices in the Laboratory"; National Academy Press; Washington, DC, 1995.

3. Discussion

Previously published methods for electrophilic bromination of isoquinoline[3-5] lead to mixtures of isomers only separable with difficulty, use expensive additives or large excesses of reactants, or involve multistep procedures.

The present procedure describes conditions, which allow for the formation of 5-bromoisoquinoline in good yield and high purity using easily available and inexpensive starting materials. In order to obtain the desired product, it is important to ensure careful temperature control to suppress the formation of 8-bromoisoquinoline, which is difficult to remove. By choosing sulfuric acid as solvent for the bromination, a convenient one-pot procedure to prepare 5-bromo-8-nitroisoquinoline, without prior isolation of 5-bromoisoquinoline, has been developed. Finally, the method can easily be scaled up from grams to kilograms of the title compounds.

Many pharmacologically active compounds have been synthesized using 5-bromoisoquinoline or 5-bromo-8-nitroisoquinoline as building blocks.[6-11] The haloaromatics participate in transition-metal couplings [8,10,12] and Grignard reactions. The readily reduced nitro group of 5-bromo-8-nitroisoquinoline provides access to an aromatic amine, one of the most versatile functional groups. In addition to *N*-alkylation, *N*-acylation and diazotiation, the amine may be utilized to direct electrophiles into the *ortho*-position.

The heterocyclic ring may be reduced under very mild conditions after *N*-alkylation, giving access to bicyclic amines[7-10,13] or enamines[5] Use of 5-bromoisoquinoline in a metalation reaction yielded 6-aminoisoquinoline, a compound otherwise accessed with difficulty.[14]

1. (a) NeuroSearch A/S, DK-2750 Ballerup, Denmark; (b) Nuevolution A/S, DK-2100 Copenhagen.

2. Perrin, D. D.; Armarego, W. L. F. *Purification of Laboratory Chemicals,* 3rd Edition, Pergamon Press: Oxford, England, **1988**, 105.
3. Osborn, A. R.; Schofield, K.; Short, L. N. *J. Chem. Soc.* **1956,** 4191.
4. Gordon, M.; Pearson, D. E. *J. Org. Chem.* **1964,** *29*, 329.
5. Rey, M.; Vergnani, T.; Dreiding, A. S. *Helv. Chim. Acta* **1985**, *68*, 1828.
6. Ortwine, D. F.; Malone, T. C.; Bigge, C. F.; Drummond, J. T.; Humblet, C.; Johnson, G.; Pinter, G. W. *J. Med. Chem.* **1992**, *35*, 1345-1370.
7. Bigge, C. F.; Humblet, C.; Johnson, G.; Malone, T.; Ortwine, D. F.; Pinter, G. W. *Trends Med. Chem.'90,* Proc. Int. Symp. Med. Chem., 11th **1992**, 153-159.

8. Moldt, P.; Wätjen, F. (NeuroSearch A/S, DK) **1996** WO pat. Appl. WO96/08495.

9. Bigge, C. F.; Malone, T.; Schelkun, R. M.; Yi, C. S. (Warner-Lambert Company, US) **1996** WO pat. Appl. WO96/28445.

10. Wätjen, F.; Drejer, J. (NeuroSearch A/S, DK) **1994** WO pat. Appl. WO94/26747.

11. Srivastava, S. K.; Chauhan, P. M. S.; Agarwal, S. K.; Bhaduri, A. P.; Singh, S. N.; Fatma, N.; Chatterjee, R. K.; Bose, C.; Srivastava, V. M. L. *Bio. Med. Chem. Lett.* **1996**, *6*, 2623.

12. Pridgen, L. N. *J. Heterocycl. Chem.* **1980**, *17*, 1289.

13. Mathison, I. W.; Morgan, P. H. *J. Org. Chem.* **1974**, *39*, 3210-3214.

14. Poradowska, H.; Huczkowska, E.; Czuba, W. *Synthesis* **1975**, *11*, 733.

COPPER(I)-CATALYZED PREPARATION OF (*E*)-3-IODOPROP-2-ENOIC ACID
[(2-Propenoic acid, 3-iodo-, (2*E*)-)]

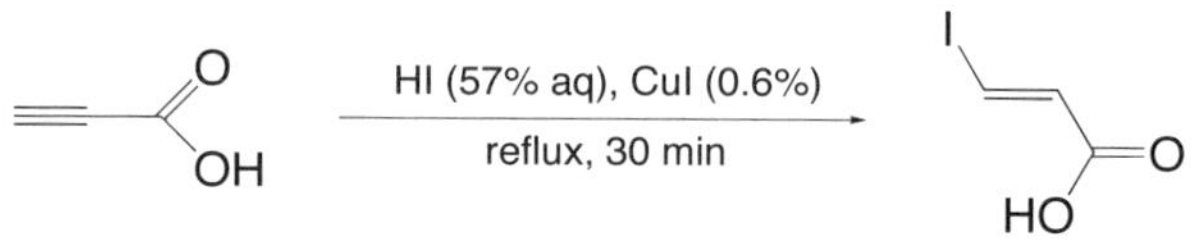

Submitted by Darren J. Dixon, Steven V. Ley, and Deborah A. Longbottom.[1]
Checked by Michelle DeRitter and Steven Wolff.

1. Procedure

A 250-mL, three-necked, round-bottomed flask, equipped with a magnetic stirring bar, thermometer and water condenser, is charged with copper(I) iodide (200 mg, 1.1 mmol) (Note 1) and hydriodic acid (57% aq, 40 mL) (Notes 2, 3). Propiolic acid (10 ml, 162 mmol) (Note 4) is added over 1 min via syringe (Note 5), during which time the reaction temperature rises to ~100 °C. The reaction mixture is immediately immersed in an oil bath, preheated with a thermostatically controlled stirrer-hotplate to 130 °C (Note 6) and a gentle reflux (110 °C) is reached after ~4 min. The mixture is heated to reflux for a further 30 min, then the oil bath is removed and replaced with a room temperature water bath. The vigorously stirred reaction mixture is allowed to cool to 28 °C over 15 min (Note 7), during which time a large quantity of white needles crystallizes from the reaction mixture (Note 8). The solution is stirred for a further 15 min at room temperature, then filtered through a sintered glass funnel. The crystals are washed with 3 × 70 mL of distilled water and dried with suction for ~1 hr, then over phosphorus pentoxide in a vacuum desiccator to constant weight. Analytically and isomerically pure (*E*)-3-iodoprop-2-enoic acid [25.3 g (79%)] is obtained as white needles (Note 9).

2. Notes

1. Copper(I) iodide (99%) was purchased from Aldrich Chemical Company, Inc. (Cat. No. 20,554-0) and used as received.

2. Hydriodic acid was purchased from Aldrich Chemical Company, Inc. (Cat. No. 21,002-1) and used as received.

3. Propiolic acid was purchased from Lancaster Synthesis Ltd. (Cat. No. 0669) and used as received.

4. Commercially available HI must be colorless to straw-yellow colored (not brown or darker) for a successful reaction.

5. On larger scales (up to 50 grams of propiolic acid) the addition should still be complete within 1 min to allow the internal reaction temperature to rise to ~100 °C.

6. The temperature of the reaction mixture must be raised to the reflux temperature (110 °C) immediately after the addition of propiolic acid.

7. On larger scales the water bath may be replaced with an ice bath to reduce the temperature of the reaction mixture to 25-30 °C over 15 min.

8. The exact ratio of reagents for this procedure is important for successful formation of pure product in this reaction.

9. Full characterization of the product is as follows: mp 146-149 °C; ν_{max} (CH_2Cl_2) cm^{-1}: 3500-2500 (OH, CH), 1701 (C=O); ^{1}H NMR (400 MHz, CD_3OD) δ: 6.90 (1H, d, *J* =14.9, ICH=C<u>H</u>), 8.07 (1H, d, *J* =14.9, ICH); ^{13}C NMR (100 MHz, CD_3OD) δ: 103.0 (I<u>C</u>H), 135.7 (ICH=<u>C</u>H), 169.2 (C=O); *m/z* (EI) 198 (100, $[M]^+$), 181 (11), 153 (12), 127 (53), 71 (15). Anal. Calcd for $C_3H_3IO_2$: C, 18.20; H, 1.52%. Found: C, 18.13; H, 1.49%.

Waste Disposal Information

Aqueous washings from the filtration and isolation of desired product were extracted with ether and disposed of as halogenated waste.

3. Discussion

Pure (*E*)-3-iodoprop-2-enoic acid can be obtained in one step by heating propiolic acid and HI at 130 °C in a sealed tube for extended periods. It can also be formed by thermal isomerization of the pure (*Z*)-isomer under similar conditions.[2] In a recent improvement to this protocol, (*Z*)-3-iodoacrylic acid was isomerized by extended periods of heating in hydriodic acid, thus obviating the use of a sealed tube.[3,4,5]

However, in order to obtain isomerically pure (*E*)-3-iodoprop-2-enoic acid, a two-step protocol is still required and reactions must be heated for 24 hr in both steps; work-up and purification of the reaction mixture is also necessary in each case. We were hopeful that we could improve upon these procedures by the use of copper(I) salts to catalyze the reaction.

Copper(I) halides have been shown previously to accelerate the formation of (*Z*)-3-iodoprop-2-enoic acid.[6,7,8,9,10] The formation of the (E)-isomer during these low temperature investigations indicated that, at higher temperature, copper catalysis could accelerate the formation of the thermodynamically favored (E)-isomer. Therefore, we decided to investigate the effect of temperature on the copper-catalyzed addition of HI to propiolic acid, with a view to creating a copper-catalyzed, one-step synthesis of the (E)-isomer.

Following optimization studies, we have now found that our method is a dramatic improvement of existing methods and relies on the copper(I)-catalyzed (0.6%) addition-isomerization of hydriodic acid to propiolic acid. The reaction time is short (30 min) and the analytically pure crystalline product is filtered directly from the reaction mixture, providing the desired material in excellent yield (79%). No further purification is necessary, all of which make this procedure the most attractive yet for the synthesis of (*E*)-3-iodoprop-2-enoic acid.

1. University of Cambridge, Department of Chemistry, Lensfield Road, Cambridge, CB2 1EW, United Kingdom.

2. Abarbri, M.; Parrain, J.-L.; Cintrat J.-C.; Duchene, A. *Synthesis* **1996**, 82.
3. Zoller, T.; Uguen, D. *Tetrahedron Lett.*, **1998**, *39*, 6719.
4. Takeuchi, R.; Tanabe, K.; Tanaka, S. *J. Org. Chem.* **2000**, *65*, 1558.
5. Paterson, I.; Lombart, H.-G.; Allerton, C. *Org. Lett.* **1999**, *1*, 19.
6. Topek, K.; Vsetecka, V.; Procházka, M. *Collect. Czech. Chem. Commun.* **1978**, *43*, 2395.
7. Moss, R. A.; Wilk, B.; Krogh-Jespersen, K.; Westbrook, J. D. *J. Am. Chem. Soc.* **1989**, *111*, 6729.
8. Grandjean, D.; Pale, P.; Chuche, J. *Tetrahedron Lett.* **1992**, *33*, 5355.
9. Grandjean, D.; Pale, P.; Chuche, J. *Tetrahedron* **1993**, *49*, 5225.
10. Keen, S. P.; Weinreb, S. M. *J. Org. Chem.* **1998**, *63*, 6739.

SELECTIVE Pd-CATALYZED AMINATIONS ON CHLORO-IODOPYRIDINES: SYNTHESIS OF ETHYL 4-[(6-CHLOROPYRIDIN-3-YL)AMINO]BENZOATE

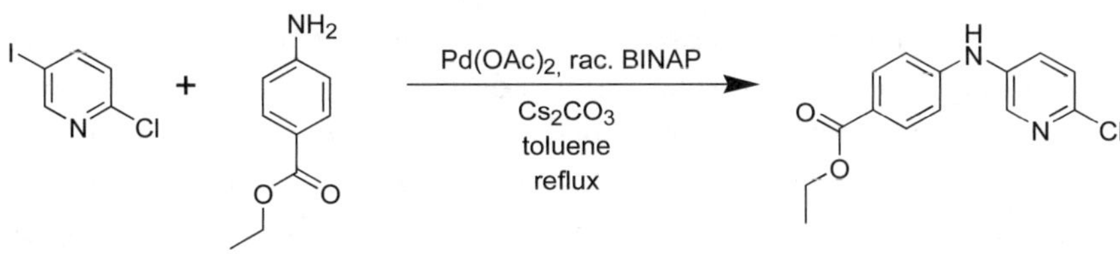

Submitted by Geert Rombouts,[1] Bert U.W. Maes,[1] Tim H.M. Jonckers,[2] and Kristof T.J. Loones.[1]
Checked by Praew Thansandote[3] and Mark Lautens.[3]

1. Procedure

Ethyl 4-[(6-chloropyridin-3-yl)amino]benzoate. To a 250-mL, one-necked, round-bottomed flask, fitted with a septum pierced with nitrogen inlet and outlet needles, is added $Pd(OAc)_2$ (0.135 g, 0.6 mmol) (Note 1), rac. BINAP (0.374 g, 0.6 mmol) (Note 2) and 100 mL of toluene (Note 3). The mixture is stirred under a nitrogen flow for 15 min. In a 1000 mL round-bottomed, three-necked flask is added 2-chloro-5-iodopyridine (7.183 g, 30 mmol) (Note 4), ethyl 4-aminobenzoate (5.947 g, 36 mmol) (Note 5) and Cs_2CO_3 (48.873 g, 150 mmol) (Note 6). Then, a reflux condenser equipped with a bubbler filled with oil is placed on the three-necked flask and one of the side necks is stoppered. Subsequently, a nitrogen inlet is connected to the remaining neck and the apparatus is flushed with nitrogen for 5 min. Next, under a low flow of nitrogen, the stoppered side neck is opened and the catalyst solution is added through a funnel using 200 mL of toluene (Note 3) to rinse the flask. Under magnetic stirring the resulting mixture is flushed with nitrogen for 10 min. Then, the nitrogen inlet is removed, the neck stoppered and the 1000 mL, three-necked flask placed in a preheated oil bath at 130 °C for 20 h under a nitrogen atmosphere (Note 7). **Vigorous** magnetic stirring is required to prevent precipitation of the base (Note 8). When the reaction mixture is cooled down to room temperature, it is filtered over a sintered glass filter funnel containing Celite® (Note 9) using 1000 mL of EtOAc (Note 10). The filtrate is concentrated at reduced pressure by rotary evaporation at 45 °C (75 mmHg) to afford a brown oil. The residue is dissolved in a minimal amount of 99:1 CH_2Cl_2:MeOH (Notes 11 and 12) and loaded on a pre-packed silica gel column (40 x 6 cm, ± 500 g) (Notes 13 and 14). The 99:1 CH_2Cl_2:MeOH eluent is used to isolate ethyl 4-[(6-chloropyridin-3-yl)amino]benzoate, in addition to fractions mixed with ethyl-4-aminobenzoate (Note 15). The impure fractions are collected and concentrated at reduced pressure by rotary evaporation at 45 °C (75 mmHg). The resulting dark beige crystals are then dissolved in a minimum amount of dichloromethane (Note 11) and loaded onto a second pre-packed silica gel column (40 x 6 cm, ± 500 g) (Notes 13 and 14). Pure dichloromethane (Note 11) is used as an eluent until the product begins to elute and subsequently, 99:1 CH_2Cl_2:MeOH is used to afford ethyl 4-[(6-chloropyridin-3-yl)amino]benzoate (7.068 g combined from two purifications, 25.5 mmol, 85%) (Notes 16-20) as a light beige solid.

2. Notes

1. Palladium(II)acetate (47.5 % Pd) was used as received from Acros.

2. (rac)-2,2'-Bis(diphenylphosphino)-1,1'-binaphtyl (98%) was used as received from Acros.

3. Toluene (for analysis, ACS) was used as received from Acros.

4. 2-Chloro-5-iodopyridine (98+%) was used as received from Alfa Aesar. The submitters used this compound as received from Acros.

5. Ethyl 4-aminobenzoate (98%) was used as received from Acros.

6. Cesium carbonate (99.5%) was used as received from Acros for the first trial, and cesium carbonate (99.9%) from Aldrich was used as received for the second trial. No difference in the reactions was observed when using the different sources. Since a large excess (5 equiv) of both bases (Acros and Aldrich) were used and the reaction was only checked after 20 hours at reflux, no differences in conversion between both bases were expected.

7. The checkers used an argon atmosphere with no change in the results of the procedure.

8. In a 500-mL round-bottomed, three-necked flask, a magnetic stirring rate of 700 rpm was used; for the 1000-mL round-bottomed, three-necked flask, the stirring rate was 1000 rpm.

9. A sintered glass filter funnel with a volume of 350 mL, a diameter of 9 cm and a pore size of 0.010–0.016 mm was used. The funnel was filled with Celite® to a height of 6 cm.

10. Ethyl acetate (min. 99%) was used as received from VWR Scientific.

11. Dichloromethane (99.9%) was used as received from VWR Scientific. The submitters used dichloromethane (technical) from Caldic, which was distilled before use.

12. MeOH (99+%) was used as received from Acros.

13. Silica gel (0.040–0.063 mm, pore diameter ca 6 nm) from Silicycle was used. The submitters used silica gel (0.035-0.070 mm, pore diameter approximately 6 nm) from Acros.

14. The pre-packed silica gel column was made by using a slurry of silica gel and the eluent.

15. At this stage, 62% of pure ethyl 4-[(6-chloropyridin-3-yl)amino]benzoate was isolated. Chromatography was performed at ambient pressure to ensure good separation.

16. Before calculating the yield, the reaction product was dried under vacuum (0.5 mmHg) for 20 h.

17. The R_f value of ethyl 4-[(6-chloropyridin-3-yl)amino]benzoate in 94:6 CH_2Cl_2:MeOH was 0.67. The submitters reported an R_f value of 0.16 using TLC plates Alugram® Sil G/UV_{254} from Marcherey-Nagel). Hence, the submitters used this solvent system as a starting eluent for column chromatography, and finished with 92:8 CH_2Cl_2:MeOH once the product began to elute. See Notes 11 and 12 for solvent quality and Note 13 for silica type.

The R_f of ethyl 4-[(6-chloropyridin-3-yl)amino]benzoate in 99:1 CH_2Cl_2:MeOH was 0.31. Thin layer chromatography (TLC) by the checkers was conducted using plates cut from 20 x 20 cm TLC aluminum sheets (Silica gel 60 F_{254} from Merck). See Notes 11 and 12 for solvent quality and Note 13 for silica type.

18. Ethyl 4-[(6-chloropyridin-3-yl)amino]benzoate: ^{1}H NMR ($CDCl_3$) δ: 1.38 (t, J = 7.4 Hz, 3 H, CH_3), 4.35 (q, J = 7.6 Hz, 2 H, CH_2), 6.12 (br s, 1 H, NH), 6.99 (d, J = 8.8 Hz, 2 H, H-2',6'), 7.27 (d, J = 8.0 Hz, 1 H, H-3), 7.50 (dd, J = 8.4 Hz, 2.4 Hz, 1 H, H-4), 7.96 (d, J = 8.4 Hz, 2 H, H-3',5'), 8.24 (d, J = 3.2 Hz, 1 H, H-6); ^{13}C NMR ($CDCl_3$) δ: 14.4; 60.7, 115.3, 123.1, 124.5, 129.3, 131.6, 137.0, 141.2, 144.2, 146.3, 166.2; HRMS (ESI) for $C_{14}H_{14}ClN_2O_2$ [$M+H^+$] found: 277.0738, calcd: 277.0744; Calcd for $C_{14}H_{13}ClN_2O_2$: C, 60.85; H, 4.75; Cl: 12.67; N, 10.14; O, 11.59. Found: C, 60.72; H, 4.81; N, 10.0.

19. Elemental analyses were recorded for the checkers on a 2400 Series II CHNS Analyzer. Elemental analysis was recorded for the

submitters in the "Hungarian Academy of Sciences" on a Vario EL III (CNHS) (project: GVOP-3.2.1-2004-04-0311/3.0).

20. The procedure as reported enabled the recovery of starting material (ethyl 4-aminobenzoate). Alternatively, the submitters suggested the following procedure to remove residual aniline prior to purification:

Derivatization of the excess aniline to the corresponding acetamide was required in some cases to simplify the purification of the desired reaction product from the excess of aniline. The need for derivatization depended on the chloro-iodopyridine substrate and the specific aniline used. For the synthesis of ethyl 4-[(2-chloropyridin-3-yl)amino]benzoate, derivatization of the excess aniline was used to separate the reaction product from ethyl 4-aminobenzoate. Acetamide formation through the reaction of 4-aminobenzoate with acetyl chloride in refluxing CH_2Cl_2 with triethylamine as base was used to effect derivative formation.

Safety and Waste Disposal Information

All hazardous materials should be handled and disposed of in accordance with "Prudent Practices in the Laboratory"; National Academy Press; Washington, DC, 1995.

3. Discussion

The procedure described above is a typical example of the Pd-catalyzed C-5 arylamination of 2-chloro-5-iodopyridine.[4] The observed selectivity is due to chemoselective oxidative addition on the substrate. The weak non-nucleophilic base used in this protocol is essential to obtain broad functional group compatibility.[5] For instance, the traditionally-used MO*t*Bu, which is often employed as the base in Buchwald-Hartwig reactions, is incompatible with the presence of an ester group. If weak bases are used in Buchwald-Hartwig reactions, rate-limiting deprotonation of the Pd(II)-amine intermediate formed in the catalytic cycle is most likely and consequently often observed. Interestingly, the use of a large excess of weak base, although mostly insoluble in the standard solvents used (tetrahydrofuran, dioxane and toluene),[5] speeds up the reactions in those cases.[6] In 2004, our laboratory gained experimental evidence that this remarkable "*base effect*" is due to an interphase deprotonation mechanism of the Pd(II)-amine intermediate formed.[6] Vigorous magnetic stirring of the reaction mixture is, in these cases, essential in order to keep the specific surface area of the base maximal. Interestingly, hitherto there are no other straightforward methods known to allow access to *N*-aryl-2-chloropyridin-5-amines. The selective Buchwald-Hartwig protocol on 2-chloro-5-iodopyridine can also be extended to other chloro-iodopyridines such as 2-chloro-3-iodopyridine, as was recently shown by us and scientists from Merck.[4,6,7]

1. Department of Chemistry, University of Antwerp, Groenenborgerlaan 171, B-2020 Antwerp, Belgium (email: bert.maes@ua.ac.be)

2. Tibotec BVBA, Generaal de Wittelaan L11 B3, B-2800 Mechelen, Belgium

3. Department of Chemistry, University of Toronto, 80 St. George Street, Toronto, Ontario, Canada, M5S 3H6 (email: mlautens@chem.utoronto.ca)

4. Maes, B.U.W.; Loones, K.T.J.; Jonckers, T.H.M.; Lemière, G.L.F.; Dommisse, R.A.; Haemers, A. *Synlett* **2002**, 1995.

5. (a) Muci, A.R.; Buchwald, S.L. *Top. Curr. Chem.* 2002, 219, 131. (b) Schlummer, B.; Scholz, U. Adv. Synth. Catal. **2004**, *346*, 1599.

6. Meyers, C.; Maes, B.U.W.; Loones, K.T.J.; Bal, G.; Lemière, G.L.F.; Dommisse, R.A. *J. Org. Chem.* **2004**, *69*, 6010.

7. Kuethe, J.T.; Wong, A.; Davies, I.W. *J. Org. Chem.* **2004**, *69*, 7752.

A NEW PRACTICAL ONE-POT CONVERSION OF PHENOLS TO ANILINES: 6-AMINO-3,4-DIHYDRO-1(2*H*)-NAPHTHALENONE
[1(2*H*)-Naphthalenone, 6-amino-3,4-dihydro-]

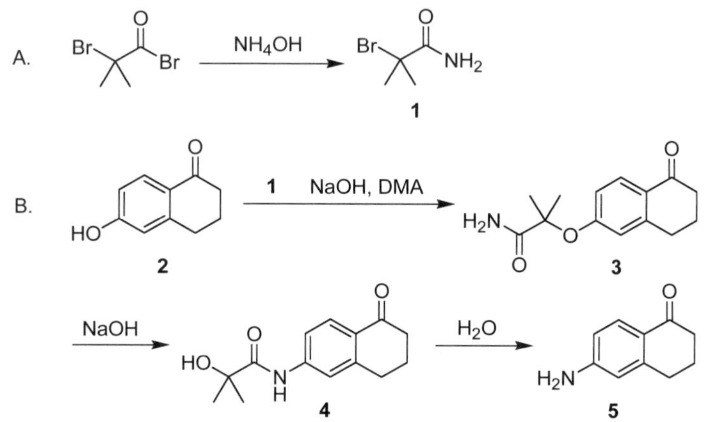

Submitted by Masahiro Mizuno[1] and Mitsuhisa Yamano.[1]
Checked by Jun Takeuchi and Jonathan A. Ellman.

1. Procedure

A. 2-Bromo-2-methylpropanamide (***1***): A 250-mL, three-necked, round-bottomed flask, fitted with a thermometer, pressure-equalizing dropping funnel and equipped with a mechanical stirrer is charged with 28.0 -30.0% aqueous ammonium hydroxide (35.7 mL, approximately 600 mmol) (Note 1) and water (44.3 mL). The dropping funnel is charged with 2-bromo-2-methylpropanoyl bromide (50.0 g, 217.5 mmol) (Note 1), and the mixture is cooled below 5 °C (internal temperature) using a brine/ice bath with stirring. The 2-bromo-2-methylpropanoyl bromide is added dropwise slowly with stirring so that the internal temperature does not rise above 15 °C with cooling using a brine/ice bath (Caution: Exothermic reaction!). Stirring is continued for 1 h after the addition is complete, with the temperature of the reaction mixture maintained between 0 and 5 °C. The resulting white precipitate is collected by filtration on a glass filter funnel and washed three times with 40 mL of water to give **1** (33.0-33.5 g, 91.4-93.8 %) as a white crystalline powder (Note 2).

B. 6-Amino-3,4-dihydro-1(2H)-naphthalenone (***5***): A 500-mL, three-necked, round-bottomed flask, fitted with a reflux condenser, thermometer, addition funnel and equipped with a large magnetic stirring bar is charged with 3,4-dihydro-6-hydroxy-1(2*H*)-naphthalenone (**2**) (10.0 g, 61.7 mmol) (Note 1) and *N*,*N*-dimethylacetamide (90 mL) (Note 1). Sodium hydroxide (7.40 g, 185 mmol) (Note 1) is added into the flask by temporarily removing the addition funnel and the resulting mixture is stirred at 20–30 °C (Note 3) for 1 h. 2-Bromo-2-methylpropanamide (**1**) (30.7 g, 185 mmol) is added to the reaction mixture (Note 4) through the addition funnel, and the resulting mixture is stirred at 25–35 °C (Note 3) for 5 h. After the reaction period (Note 5), sodium hydroxide (22.2 g, 555 mmol) is added, and the resulting mixture is stirred for 1 h with heating to 50–60 °C (internal temperature) using an oil bath. After the reaction period (Note 6), water (90 mL) is added by use of an addition funnel, and the mixture is heated at reflux using an oil bath for 1 h (internal temperature is 85–95 °C). Water (180 mL) is added to the reaction solution using an addition funnel, and the resulting precipitated mixture is allowed to cool slowly to 20–30 °C. The precipitated crystalline powder is collected by filtration on a glass filter funnel (Note 7) and washed three times with 90 mL of water to give **5** (6.28 g, 63.2%) as a brown crystalline powder (Notes 8 and 9).

2. Notes

1. 2-Bromo-2-methylpropanoyl bromide and 3,4-dihydro-6-hydroxy-1(2*H*)-naphthalenone (**2**) were purchased from Aldrich Chemical Company, Inc. The submitters purchased 25–27.9% aqueous ammonium hydroxide, *N*,*N*-dimethylacetamide and sodium hydroxide from Wako Pure Chemical Industries, Ltd. The checkers purchased *N*,*N*-dimethylacetamide and sodium hydroxide from Aldrich Chemical Company, Inc. The checkers purchased 28.0 – 30.0% aqueous ammonium hydroxide from EMD Chemicals, Inc. All of the reagents were used without further purification.

2. 2-Bromo-2-methylpropanamide (**1**) showed the following physical data: mp 146–148 °C; IR (film): 3382, 3187, 1652, 1625, 1111 cm^{-1}; ^{1}H NMR (400 MHz, $CDCl_3$) δ: 1.97 (s, 6 H), 6.20 (br, 1 H), 6.60 (br, 1 H); ^{13}C NMR (100 MHz, $CDCl_3$) δ: 32.3, 61.1, 174.9; MS (EI): *m/z* 167 (3), 165 (3), 123 (23), 121 (23), 86 (100); Anal. Calcd. for C_4H_8BrNO: C, 28.94; H, 4.86; N, 8.44; Br, 48.13. Found: C, 28.71; H, 4.96; N, 8.33. The submitters reported that HPLC analysis was performed on YMC-Pack ODS-A302 column (150 mm×4.6 mm i.d.) with 50 mM aqueous KH_2PO_4 solution-MeCN (65:35) at 25 °C. Detected at 254 nm(UV) t_R: 2.7 (area 100.0%)

3. After the addition of the reagents, the internal temperature gradually rose to approximately 30 °C and then over time returned to ambient temperature.

4. This reaction was slightly endothermic.

5. Although the alkylation reaction was not complete after 5 h, the procedure could be continued with no problem. The submitters reported that the intermediate formed at this stage, 2-methyl-2-[(5-oxo-5,6,7,8-tetrahydro-2-naphthalenyl)oxy]propanamide (**3**), can be isolated without further purification by the addition of water to the reaction mixture. Intermediate **3** isolated by the submitters in this fashion showed the following physical data: HPLC (YMC-Pack ODS-A302 column (150 mm x 4.6 mm i.d.) with 50 mM aqueous KH_2PO_4 solution-MeCN (80:20) at 25 °C. Detected at 254 nm (UV)) t_R: 4.5(**1**), 11.1(**2**), 20.4(**3**, area 99.4%); mp 117.5-119.5 °C; IR (KBr): 3357.5, 3164.6, 1687.4, 1664.3, 1604.5 cm^{-1}; ^1H NMR (300 MHz, $CDCl_3$) δ: 1.61 (s, 6 H), 2.12 (quint, *J* = 6.1 Hz, 2 H), 2.61 (t, *J* = 6.2 Hz, 2 H), 2.91 (t, *J* = 6.1 Hz, 2 H), 5.65 (br, 1 H), 6.41 (br, 1 H), 6.75 (d, *J* = 2.3 Hz, 1 H), 6.84 (dd, *J* = 8.6 Hz, 2.4Hz, 1 H), 7.98 (d, *J* = 8.7 Hz, 1 H); ^{13}C NMR (75 MHz, $CDCl_3$) δ: 23.7, 25.5(2C), 30.4, 39.3, 81.8, 118.7, 119.0, 128.1, 129.6, 146.9, 159.2, 177.4, 197.6; MS (EI): *m/z* 247 $[M]^+$; Anal. Calcd. for $C_{14}H_{17}NO_3$: C, 68.00; H, 6.93; N, 5.66. Found: C, 67.71; H, 6.76; N, 5.71.

6. The submitters reported that this intermediate, 2-hydroxy-2-methyl-*N*-(5-oxo-5,6,7,8-tetrahydro-2-naphthalenyl)propanamide (**4**), can be isolated without further purification by the addition of water to the reaction mixture. Intermediate **4** isolated by the submitters in this fashion showed the following physical data: HPLC (YMC-Pack ODS-A302 column (150 mm x 4.6 mm i.d.) with 50 mM aqueous KH_2PO_4 solution-MeCN (80:20) at 25 °C. Detected at 254 nm (UV)) t_R: 4.5(**1**), 11.1(**2**), 17.7(**4**, area 99.6%), 20.4(**3**); mp 162.5-163 °C; IR (KBr): 3305.4, 1672.0, 1600.6, 1540.9 cm^{-1}; ^1H NMR (300 MHz, $CDCl_3$) δ: 1.58 (s, 6 H), 2.14 (quint, *J* = 6.2 Hz, 2 H), 2.65 (t, *J* = 6.2 Hz, 2 H), 2.73(s, 1 H), 2.96 (t, *J* = 6.2 Hz, 2 H), 7.31 (dd, *J* = 8.5, 2.1 Hz, 1 H), 7.78 (d, *J* = 1.6 Hz, 1 H), 8.01 (d, *J* = 8.5 Hz, 1 H), 8.90 (br, 1 H); ^{13}C NMR (75 MHz, $CDCl_3$) δ: 23.3, 27.8(2C), 30.0, 39.0, 74.2, 117.6, 118.4, 128.6, 128.7, 142.1, 146.4, 175.1, 197.8; MS (FAB): *m/z* 248 $[M+H]^+$; Anal. Calcd. for $C_{14}H_{17}NO_3$: C, 68.00; H, 6.93; N, 5.66. Found: C, 68.19; H, 7.06; N, 5.64.

7. After the reaction mixture was stirred for 14 h, the checkers collected the precipitated crystalline powder.

8. The checkers obtained a 60.4% yield when the reaction was run

on half-scale. 6-Amino-3,4-dihydro-1(2*H*)-naphthalenone (**5**) showed the following physical data: mp 128–130 °C; IR(film): 3423, 3349, 3239, 1567, 1553, 1356, 1325, 1292 cm^{-1}; ^{1}H NMR (400 MHz, $CDCl_3$) δ: 2.05 (quint, J = 6.4 Hz, 2 H), 2.56 (t, J = 6.4 Hz, 2 H), 2.81 (t, J = 6.4 Hz, 2 H), 4.28 (br, 2 H), 6.41 (d, J = 2.0 Hz, 1 H), 6.53 (dd, J = 8.4, 2.0 Hz, 1 H), 7.87 (d, J = 8.4 Hz, 1 H); ^{13}C NMR (100 MHz, $CDCl_3$) δ: 23.2, 29.9, 38.7, 112.3, 113.0, 123.6, 129.6, 147.0, 151.4, 197.0; MS (EI): *m/z* 161 (76, $[M]^{+}$), 133 (100), 105 (28); Anal. Calcd. for $C_{10}H_{11}NO$: C, 74.51; H, 6.88; N, 8.69. Found: C, 74.24; H, 6.93; N, 8.67.

9. The submitters reported the isolation of an off-white crystalline powder. The submitters reported that HPLC analysis was performed on a YMC-Pack ODS-A302 column (150 mm x 4.6 mm i.d.) with 50 mM aqueous KH_2PO_4 solution-MeCN (80:20) at 25 °C. Detected at 254 nm (UV) t_R: 4.5(**1**), 8.6(**5**, area 100.0%), 11.1(**2**), 17.7(**4**), 20.4(**3**).

Waste Disposal Information

All toxic materials were disposed of in accordance with "Prudent Practices in the Laboratory"; National Academy Press; Washington, DC, 1995.

3. Discussion

Since more phenols are generally more available than the corresponding anilines, several methods have been developed for the direct conversion of phenols to anilines.

These methods include the Bucherer reaction[2] and activation of phenols with 4-chloro-2-phenylquinazoline [3] or diethyl chlorophosphate;[4] however, all of these methods suffer from drawbacks. The Bucherer reaction is restricted to naphthalenes and related heterocycles, whereas benzene derivatives are much less reactive. Rearrangement of aryloxyquinazoline, prepared from 4-chloro-2-phenylquinazoline, requires extremely high temperatures (approximately 300 °C) and strongly basic conditions. Preparation of aryl diethyl phosphate esters requires the use of toxic diethyl chlorophosphate and potassium metal in liquid ammonia.

Another method, which has been used for the conversion of phenols to anilines, is an alkylation-Smiles rearrangement-hydrolysis[5] sequence. However, this method also has some disadvantages. The three-step process requires the use of sodium hydride and *N,N*-dimethylformamide, which has the risk of causing an exothermic, uncontrollable reaction.[6]

Recently, palladium-catalyzed amination of aryl sulfonates, e.g., aryl triflates and aryl nonaflates, has been investigated.[7] However, this method requires that aryl sulfonates be synthesized from the corresponding phenols. The high cost of palladium and the potential for metal contamination of the final product are issues associated with this method.

6-Amino-1-tetralone (**5**), which is an important synthetic intermediate, has been synthesized by two general synthetic methods. The first of these involves a Friedel-Crafts ring-closure step near the end of the synthetic sequence. The starting material, β-benzoylpropionic acid, was nitrated to give the *meta*-nitro compound, which was then reduced to provide the amine. Acetylation of the amino group and hydrogenolysis of the ketone carbonyl generated the acetamido acid. Friedel-Crafts cyclization provided 6-acetamido-1-tetralone, which was hydrolyzed to aniline **5**.[8] An alternative synthetic method involved the introduction of the ketone carbonyl into the saturated side chain as a key step. The tetralin starting material was acetylated to give 6-acetyltetralin, which then was converted to the oxime. Beckmann rearrangement of the oxime gave 6-acetamidotetralin. Selective oxidation gave 6-acetamido-1-tetralone, which was hydrolyzed to aniline **5**.[8b, 9]

These two general methods required many steps to produce aniline **5**. Our novel three-step, one-pot Smiles rearrangement process[10] is a practical and useful alternative route to this compound.

ONE-POT CONVERSION OF PHENOLS (THIOPHENOLS) TO ANILINES

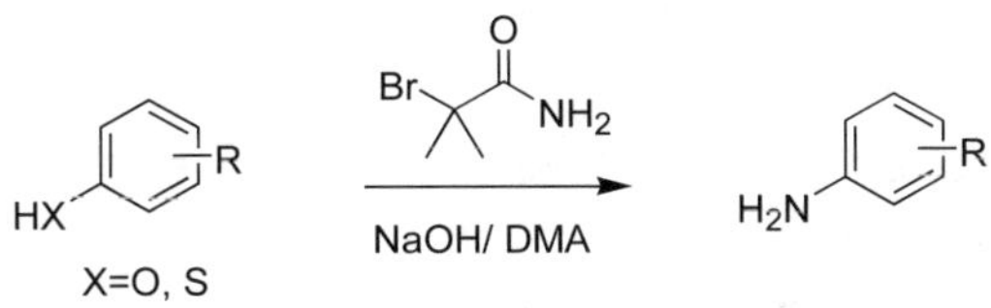

Entry	Substrate	Product	Conv[a]	Yield[b]
1	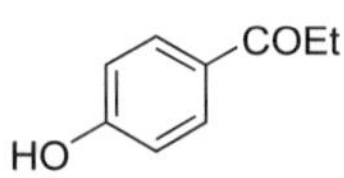	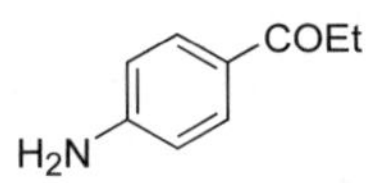	92	86
2	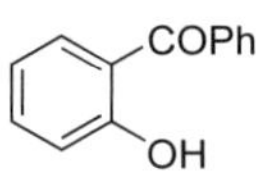	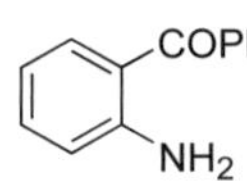	72	77
3	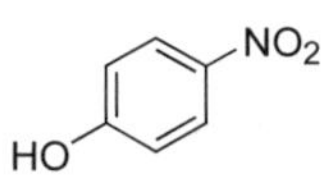	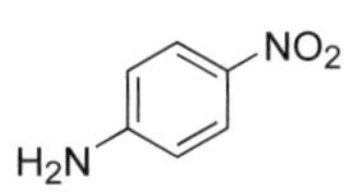	83	38
4	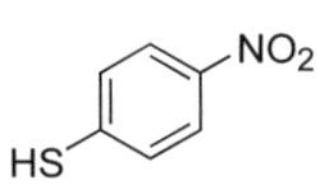	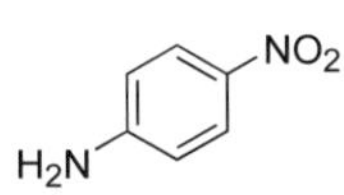	99	49

[a] Conversion based on quantitative HPLC analysis (not optimized)
[b] Isolated yield (products separated as crystals from reaction mixture by filtration)

1. Chemical Development Laboratories, Pharmaceutical Production Division, Takeda Pharmaceutical Company Limited., 17-85, Jusohonmachi 2-chome, Yodogawa-ku, Osaka, 532-8686, Japan., Mizuno_Masahiro@takeda.co.jp.
2. (a) Bucherer, H. T. *J. Prakt. Chem.* **1904**, *69*, 49. (b) Wagner, R. B.; Zook, H. D. *Synthetic Organic Chemistry, John Wiley and Sons: New York, NY,* **1953**; pp. 670. (c) Schroter, R. *Houben-Weyl-Muller.* **1957**, *11*, 143. (d) Rieche, A.; Seeboth, H. *Ann.* **1960**, *638*, 66.
3. (a) Morrow, D. F.; Hofer, R. M. *J. Med. Chem.* **1966**, *9*, 249. (b) Conrow, R. B.; Bernstein, S. *Steroids.* **1968**, *11*, 151. (c) Scherrer, R. A.; Beatty, H. R. *J. Org. Chem.* **1972**, *37*, 1681. (d) Matsumoto, K.; Stark, P.; Meister, R. G. *J. Med. Chem.* **1977**, *20*, 17. (e) Sadek, S. A.; Shaw, S. M.; Kessler, W. V.; Wolf, G. C. *J. Org. Chem.* **1981**, *46*, 3259.

4. Rossi, R. A.; Bunnett, J. F. *J. Org. Chem.* **1972**, *37*, 3570.

5. (a) Bayles, R.; Johnson, M. C.; Maisey, R. F.; Turner, R. W. *Synthesis* **1977**, *31*, 33. (b) Fitzgerald, L. R.; Blakeley, R. L.; Zerner, B. *Chem. Lett.* **1984**, 29. (c) Coutts, I. G. C.; Southcott, M. R. *J. Chem. Res. (Synopses)* **1988**, 241. (d) Coutts, I. G. C.; Southcott, M. R. *J. Chem. Soc., Perkin Trans., 1* **1990**, 767. (e) Weidner, J. J.; Weintraub, P. M.; Schnettler, R. A.; Peet, N. P. *Tetrahedron* **1997**, *53*, 6303. (f) Beier, P.; Mindl, J.; Sterba, V.; Hanusek, J. *Org. Biomol. Chem.* **2004**, *2*, 562.

6. (a) Buckley, J.; Webb, R. L.; Laird, T.; Ward, R. J. *Chem. Eng. News*, **1982**, *60*(28), 5. (b) De Wall, G. *Chem. Eng. News*, **1982**, *60*(37), 5, 43.

7. (a) Wolfe, J. P.; Buchwald, S. L. *J. Org. Chem.* **1997**, *62*, 1264. (b) Louie, J.; Driver, M. S.; Hamann, B. C.; Hartwig, J. *J. Org. Chem.* **1997**, *62*, 1268. (c) Ahman, J.; Buchwald, S. L. *Tetrahedron Lett.* **1997**, *38*, 6363. (d) Wolfe, J. P.; Ahman, J.; Sadighi, J. P.; Singer, R. A.; Buchwald, S. L. *Tetrahedron Lett.* **1997**, *38*, 6367. (e) Wolfe, J. P.; Tomori, H.; Sadighi, J. P.; Yin, J.; Buchwald, S. L. *J. Org. Chem.* **2000**, *65*, 1158. (f) Ullrich, T.; Giraud, F. *Tetrahedron Lett.* **2003**, *44*, 4207. (g) Anderson, K. W.; Mendez-Perez, M.; Priego, J.; Buchwald, S. L. *J. Org. Chem.* **2003**, *68*, 9563. (h) Rahman, O.; Kihlberg, T.; Langstroem, B. *Org. Biomol. Chem.* **2004**, *2*, 1612.

8. (a) Somerville, L. F.; Allen, C. F. H. *Org. Syntheses, Coll. Vol. II, John Wiley and Sons: New York*, **1943**; p. 81. (b) Allinger, N. L.; Jones, E. S. *J. Org. Chem.* **1962**, *27*, 70.

9. (a) Newman, M. S.; Zahm, H. V. *J. Am. Chem. Soc.* **1943**, *65*, 1097. (b) Schofield, K.; Swain, T; Theobald, R. S. *J. Chem. Soc.* **1949**, 2399. (c) Biggs, D. F.; Casy, A. F.; Chu, Ih; Coutts, R. T. *J. Med. Chem.* **1976**, *19*, 472.

10. Mizuno, M.; Yamano, M. *Org. Lett.* **2005**, *7*, 3629.

PRACTICAL SYNTHESIS OF A CHIRAL YNAMIDE: (*R*)-4-PHENYL-3-(2-TRIISOPROPYLSILYL-ETHYNYL)OXAZOLIDIN-2-ONE

[2-Oxazolidinone, 4-Phenyl-3-(2-triisopropylsilyl-ethynyl)-, (4*R*)-]

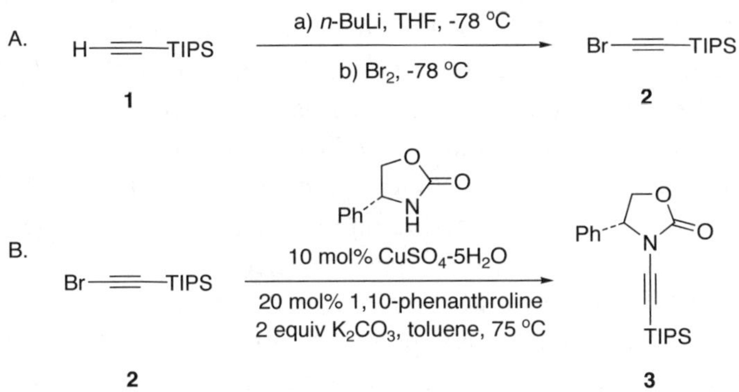

Submitted by I. K. Sagamanova, K. C. M. Kurtz, and R. P. Hsung.[1]
Checked by Karen M. Marcantonio and David J. Mathre.

1. Procedures

A. 1-Bromo-2-triisopropylsilyl-ethyne (**2**). To a flame-dried single-necked 1-L round-bottomed flask equipped with a magnetic stir bar is added a solution of triisopropylsilylacetylene (22.0 g, 120.6 mmol) (Note 1) in anhydrous THF (500 mL) (Note 2). The solution is cooled to – 78 °C, and *n*-BuLi (50.7 mL, 126.7 mmol, 1.05 equiv) (Note 3) is added by syringe through the septum. The reaction is stirred for 30 min at – 78 °C, and Br_2 (6.80 mL, 132.7 mmol, 1.10 equiv) (Note 4) is added slowly through the septum using a syringe. The reddish brown color from Br_2 disappears as it is consumed upon addition. The solution remains reddish brown when the addition is complete. The mixture is stirred for 15 min at –78 °C and then quenched by addition of saturated aqueous $Na_2S_2O_3$ (150 mL) after removal of the septum. The reaction mixture is transferred to a separatory funnel and the layers separated. The aqueous layer is further extracted with methyl *tert*-butyl ether (MTBE) (3 × 50 mL), and the combined organic extracts are washed with saturated aqueous NaCl (50 mL), dried over Na_2SO_4, filtered and concentrated on a rotary evaporator (20 – 30 mmHg, 50 °C) to yield the crude alkynyl bromide **2** (30.22–30.24 g, 96%) as a pale yellow oil (Note 5). Alkynyl bromide **2** is used without further purification (Note 6).

B. (R)-4-Phenyl-3-(2-triisopropylsilyl-ethynyl)oxazolidin-2-one (**3**). To a solution of 1-bromo-2-triisopropylsilylacetylene (**2**) (28.19 g, 107.9 mmol) (Note 7) in freshly distilled anhydrous toluene (100 mL) (Note 2) in a 250-mL, single-necked, round-bottomed flask fitted with a magnet stir bar are added *R*-phenyloxazolidinone (17.60 g, 107.9 mmol, 1.00 equiv) (Note 8), K_2CO_3 (29.81 g, 215.7 mmol, 2.00 equiv) (Note 9), $CuSO_4$•$5H_2O$ (2.69 g, 10.8 mmol, 0.10 equiv) (Note 10), and 1,10-phenanthroline (3.89 g, 21.6 mmol, 0.20 equiv) (Note 11). The flask is fitted with a reflux condenser topped with a septum and N_2 inlet, and heated in an oil bath at 75 °C (bath temperature) for 48 h. The reaction is monitored using TLC analysis (Note 12). Upon completion, the reaction mixture is cooled to room temperature and filtered through a 200-mL coarse-fritted vacuum filtration funnel containing a 5-cm layer of silica gel covered with a 1-cm layer of Celite. The mixture is washed through with 50% EtOAc/hexanes (400 mL), and the filtrate is concentrated on a rotary evaporator (20 – 30 mmHg, 70 °C). The crude residue is purified using silica gel column flash chromatography (Note 13) to give ynamide **3** (29.5–32.7 g, 81–88%) (Notes 14 and 15) as a yellow oil.

2. Notes

1. Triisopropylacetylene (97%) was purchased from GFS Chemicals.
2. Anhydrous solvents were obtained from an MBraun solvent purification system. Unstabilized THF was purchased from JT Baker, and toluene was purchased from Aldrich.
3. *N*-Butyllithium (2.5 *M* solution in hexanes) was purchased from Aldrich Chemical Co.
4. Bromine (99+%) was purchased from Fisher Scientific.
5. Characterization of **2**: R_f = 0.72 [25% EtOAc in hexanes]; yellow oil; ^{1}H NMR (300 MHz, $CDCl_3$) δ 1.06–1.10 (m, 21 H); ^{13}C NMR (75 MHz, $CDCl_3$) δ 11.2, 18.4, 61.6, 83.4.
6. This purification step was not necessary, because, in most cases, high purity was obtained judging from ^{1}H NMR. However, product **2** was stable to silica gel column flash chromatography and could be eluted with hexanes. Other alkynyl bromides may require silica gel column flash chromatography for purification.
7. The amidation reaction was performed successfully on a variety of scales [see Discussion section].

8. *R*-Phenyloxazolidinone (98%) was purchased from Aldrich and used as received.

9. Potassium carbonate (98%; ~ 325 mesh powder) was purchased from Aldrich.

10. $CuSO_4 \cdot 5H_2O$ (99.3%) was purchased from J. T. Baker. The solid was ground into a powder with a mortar and pestle before use.

11. 1,10-Phenanthroline (99+%) was purchased from Aldrich. The solid was ground into a powder with a mortar and pestle before use.

12. TLC analysis (25% EtOAc/hexanes): R_f(oxazolidinone) = 0.03, R_f(**2**) = 0.72, R_f(**3**) = 0.39.

13. The crude product was dissolved in 200 mL of CH_2Cl_2 in a round-bottomed-flask, and 75 g silica gel was added to the flask. The CH_2Cl_2 was removed on a rotary evaporator (20–30 mmHg, 35 °C) equipped with a bump trap. The dried contents of the flask were loaded onto the column (size $l \times d$ = 30 cm × 5 cm) containing a slurry of silica gel and hexanes layered with sand. Gradient elution [EtOAc in hexanes]: 800 mL 0%, 800 mL 2%, 1000 mL 5%, 1500 mL 7%, 2000 mL 10%.

14. GC analysis of ynamide **3** shows its purity to be ≥98.0%.

15. Characterization of **3**: R_f = 0.39 [25% EtOAc in hexanes]; clear oil; $[\alpha]_D^{23}$ – 133.7 (c 1.21, CH_2Cl_2); Chiralcel OD-H (250 x 4.6 mm), 5% IPA/heptane, 1.0 mL/min, 215 nm, t_R (*R*) 10.77 min, t_R (*S*) 13.66 min: >99.5 % ee R. ^{1}H NMR (500 MHz, $CDCl_3$) δ: 0.90–0.93 (m, 21 H), 4.28 (dd, 1 H, *J* = 7.6, 8.8 Hz), 4.73 (t, 1 H, *J* = 8.8 Hz), 5.06 (dd, 1 H, *J* = 7.6, 8.8 Hz), 7.34–7.45 (m, 5 H); ^{13}C NMR (75 MHz, $CDCl_3$) δ: 11.0, 18.3, 62.2, 70.4, 71.9, 91.8, 127.1, 129.1, 129.4, 135.7, 155.2; IR (thin film) cm^{-1} 2943 (m), 2185 (w), 1782 (s), 1394 (m), 883 (m); mass spectrum (APCI): *m/z* (% relative intensity) 344 (13) $(M + H)^+$, 334 (33), 318 (100); *m/z* calcd for $C_{20}H_{29}NO_2Si$ 344.2040, found 344.2047

Safety and Waste Disposal Information

All hazardous materials should be handled and disposed of in accordance with "Prudent Practices in the Laboratory"; National Academy Press; Washington, DC, 1995.

3. Discussion

It is noteworthy that this cooper-catalyzed amidation of alkynyl bromide can be carried out with scales ranging from 48 to 107 mmol. The respective ynamide products were isolated with yields ranging from 86–89%.

Table 1

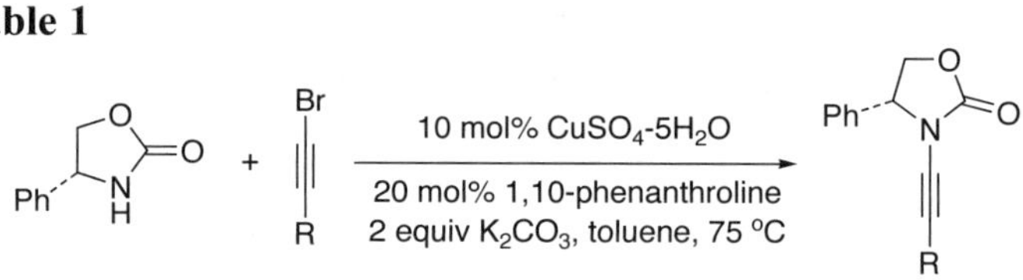

halide	R =	scale [mmols]	product	isolated yield [%, g]
2a	TIPS	48	**3a**	89 [14.63 g]
	TIPS	107		86 [32.07 g]
2b	Ph	18	**3b**	60 [4.66 g]
	Ph	35		71 [6.72 g]

The level of purity of the ynamides prepared using this procedure can be unambiguously established using GC/HPLC analysis for ynamide **3**. The level of optical purity of the product is dependent upon the optical purity of the chiral auxiliaries acquired from commercial sources. The $[\alpha]_D^{20}$ value for the commercial (*R*)-2-phenyloxazolidinone used in this preparation was –52.3 [c 2.0, $CHCl_3$]. The $[\alpha]_D^{20}$ values reported by Aldrich for (*R*)-2-phenyloxazolidinone and (*S*)-2-phenyloxazolidinone are –48.0 [c 2.0, $CHCl_3$], which indicates that the *ee* or optical integrity of the Evans' auxiliary used for this work was very high. Severe erosion of the auxiliary's *ee* is unlikely under the amidation conditions described herein. The anticipated high *ee* of chiral ynamide (*R*)-**3** was confirmed by chiral HPLC.

Ynamides[2-9] have become a highly attractive building block for developing synthetic methodologies.[10-11] As illustrated in Figure 1, there have been at least 30 reports in the last few years describing dozens of different strategies employing ynamides. Particularly attractive are those in which the nitrogen atom of ynamides becomes an integral part of various products that possess potential in alkaloid synthesis. These efforts demonstrate that there can be a distinct advantage in utilizing ynamides over simple alkynes. Therefore, the key issue, in part being addressed in this *Organic Syntheses* procedure, is their preparation. We, as well as others, have exerted much effort toward this goal.[3-9]

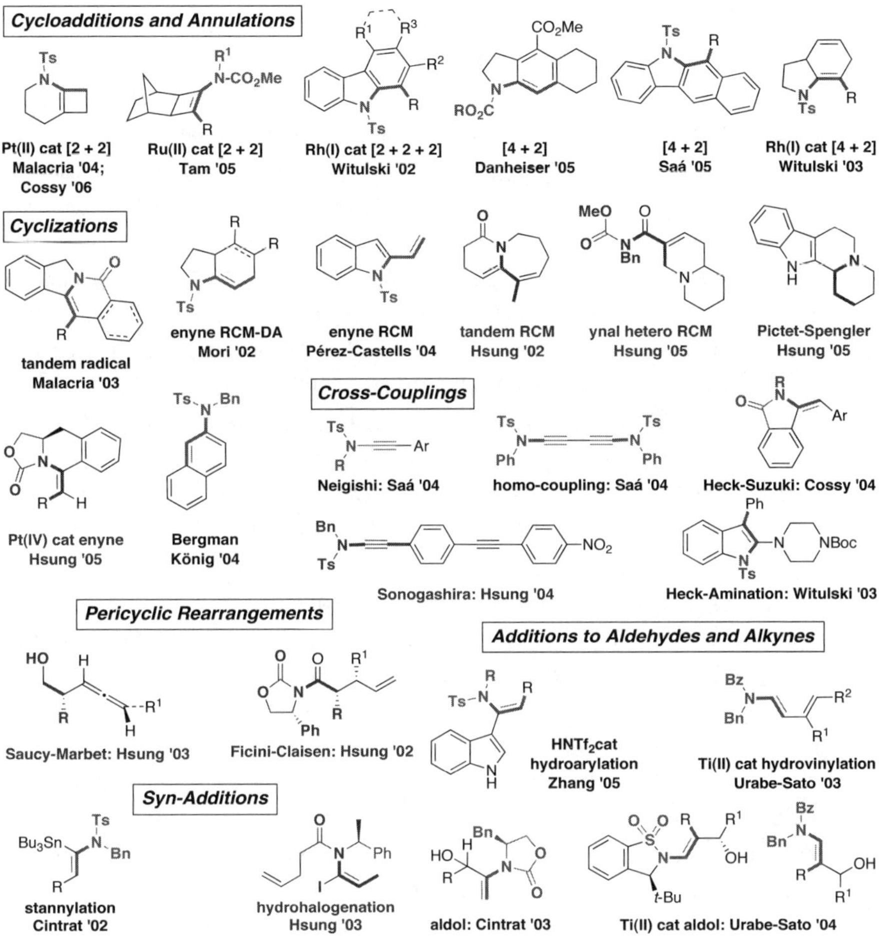

1. School of Pharmacy, University of Wisconsin, 777 Highland Ave., Madison, Wisconsin 53705-2222 USA.

2. For reviews on ynamides, see: (a) Zificsak, C. A.; Mulder, J. A.; Hsung, R. P.; Rameshkumar, C.; Wei, L.-L. *Tetrahedron* **2001**, *57*, 7575. (b) Zhang, Y.; Hsung, R. P. *ChemTracts* **2004**, *17*, 442. (c) Katritzky, A. R.; Jiang, R.; Singh, S. K. *Heterocycles* **2004**, *63*, 1455.

3. For reviews on the synthesis of ynamides, see: (a) Tracey, M. R.; Hsung, R. P.; Antoline, J.; Kurtz, K. C. M.; Shen, L.; Slafer, B. W.; Zhang, Y., in *Science of Synthesis, Houben-Weyl Methods of Molecular Transformations*, Steve M. Weinreb, Ed. Georg Thieme Verlag KG: Chapter 21.4, **2005**. (b) Mulder, J. A.; Kurtz, K. C. M.; Hsung, R. P. *Synlett* **2003**, 1379.

4. For the first preparations of ynamides, see: Janousek, Z.; Collard, J.; Viehe, H. G. *Angew. Chem., Int. Ed.* **1972**, *11*, 917.

5. For some examples of alkynyl iodonium triflate salts, see: (a) Feldman, K. S.; Bruendl, M. M.; Schildknegt, K.; Bohnstedt, A. C. *J. Org. Chem.* **1996**, *61*, 5440. (b) Witulski, B.; Stengel, T. *Angew. Chem. Int. Ed.* **1998**, *37*, 489. (c) Witulski, B.; Stengel, T.; Fernàndez-Hernàndez, J. M. *Chem. Commun.* **2000**, 1965. (d) Witulski, B.; Buschmann, N.; Bergsträßer, U. *Tetrahedron* **2000**, *56*, 8473. (e) Rainier, J. D.; Imbriglio, J. E. *J. Org. Chem.* **2000**, *65*, 7272. (f) Brückner, D. *Synlett* **2000**,1402. (g) Fromont, C.; Masson, S. *Tetrahedron* **1999**, *55*, 5405.

6. For the synthesis of ynamides using copper catalyzed amidations, see: (a) Frederick, M. O.; Mulder, J. A.; Tracey, M. R.; Hsung, R. P.; Huang, J.; Kurtz, K. C. M.; Shen, L.; Douglas, C. J. *J. Am. Chem. Soc.* **2003**, *125*, 2368. (b) Zhang, Y.; Hsung R. P.; Tracey, M. R.; Kurtz, K. C. M.; Vera, E. L. *Org. Lett.* **2004**, *6*, 1151.

7. Also see: (a) Dunetz, J. R.; Danheiser, R. L. *Org. Lett.* **2003**, *5*, 4011. (b) Hirano, S.; Tanaka, R.; Urabe, H.; Sato, F. *Org. Lett.* **2004**, *6*, 727. (c) Riddell, N.; Villeneuve, K.; Tam, W. *Org. Lett.* **2005**, *7*, 3681.

8. For the synthesis of ynamides using an based promoted elimination *Z*-beta-bromo enamides, see: Wei, L.-L.; Mulder, J. A.; Xiong, H.; Zificsak, C. A.; Douglas, C. J.; Hsung, R. P. *Tetrahedron* **2001**, *57*, 459.

9. Also see: Couty, S.; Barbazanges, M.; Meyer, C.; Cossy, J. *Synlett* **2005**, 906.

10. For an impressive array of reports on the chemistry of ynamides just in the last three years, see: (a) For a special issue dedicated to the chemistry of ynamides, see: *Tetrahedron-Symposium-In-Print:* "Chemistry of Electron-Deficient Ynamines and Ynamides." *Tetrahedron* **2006**, *62*, Issue No.16. (b) Tanaka, K.; Takeishi, K.; Noguchi, K. *J. Am. Chem. Soc.* **2006**, *128*, 4586. (c) Couty, S.; Meyer, C.; Cossy, J. *Angew. Chem. Int. Ed.* **2006**, *45*, 6726. (d) Dunetz, J. R.; Danheiser, R. L. *J. Am. Chem. Soc.* **2005**, *127*, 5776. (e) Zhang, Y. *Tetrahedron Lett.* **2005**, *46*, 6483. (f) Riddell, N.; Villeneuve, K.; Tam, W. *Org. Lett.* **2005**, *7*, 3681. (g) Martinez-Esperon, M. F.; Rodriguez, D.; Castedo, L.; Saá, C. *Org. Lett.* **2005**, *7*, 2213. (h) Bendikov, M.; Duong, H. M.; Bolanos, E.; Wudl, F. *Org. Lett.* **2005**, *7*, 783. (i) Marion, F.; Coulomb, J.; Courillon, C.; Fensterbank, L.; Malacria, M. *Org. Lett.* **2004**, *6*, 1509. (j) Rosillo, M.; Domínguez, G.; Casarrubios,L.; Amador, U.; Pérez-Castells, J. *J. Org. Chem.* **2004**, *69*, 2084. (k) Couty, S.; Liégault, B.; Meyer, C.; Cossy, J. *Org. Lett.* **2004**, *6*, 2511. (l) Rodríguez, D.; Castedo, L.; Saá, C. *Synlett* **2004**, 783. (m) Rodríguez, D.; Castedo, L.; Saá, C. *Synlett* **2004**, 377. (n) Hirano, S.; Tanaka, R.; Urabe, H.; Sato, F. *Org. Lett.* **2004**, *6*, 727. (o) Klein, M.; König, B. *Tetrahedron* **2004**, *60*, 1087. (p) Marion, F.; Courillon, C.; Malacria, M. *Org. Lett.* **2003**, *5*, 5095. (q) Witulski, B.; Alayrac, C.; Tevzaadze-Saeftel, L. *Angew. Chem., Int. Ed.* **2003**, *42*, 4257. (r) Tanaka, R.; Hirano, S.; Urabe, H.; Sato, F. *Org. Lett.*

2003, *5*, 67. (s) Witulski, B.; Lumtscher, J.; Bergsträber, U. *Synlett* **2003**, 708. (t) Naud, S.; Cintrat, J.-C. *Synthesis* **2003**, 1391. (u) Witulski, B.; Alayrac, C. *Angew. Chem. Int. Ed.* **2002**, *41*, 3281. (v) Saito, N.; Sato, Y.; Mori, M. *Org. Lett.* **2002**, *4*, 803. (w) Timbart, J.-C.; Cintrat. J.-C. *Chem. Eur. J.* **2002**, *8*, 1637. (x) For many other contributions before 2001, see references 2 and 3.

11. For our own recent applications of ynamides, see: (a) Zhang, X.; Li, H.; You, L.; Tang, Y.; Hsung, R. P. *Chem. Commun.* **2007**, *ASAP*. (b) Oppenheimer, J.; Johnson, W. L.; Tracey, M. R.; Hsung, R. P.; Yao, P.-Y.; Liu, R.; Zhao, K. *Org. Lett.* **2007**, *9*, *ASAP*. (c) You, L.; Al-Rashid, Z. F.; Figueroa, R.; Ghosh, S. K.; Li, G.; Lu, T.; Hsung, R. P. *Synlett* **2007**, *in press*. (d) Tracey, M. R.; Oppenheimer, J.; Hsung, R. P. *J. Org. Chem.* **2006**, *71*, 8629. (e) Zhang, X.; Li, H.; You, L.; Tang, Y.; Hsung, R. P. *Adv. Syn. Cat.* **2006**, *348*, 2437. (f) Zhang, X.; Hsung, R. P.; You, L. *Organic Biomol. Chem.* **2006**, *6*, 2679. (g) Zhang, X.; Zhang, Y.; Huang, J.; Hsung, R. P.; Kurtz, K. C. M.; Oppenheimer, J.; Petersen, M. E.; Sagamanova, I. K.; Tracey, M. R. *J. Org. Chem.* **2006**, *71*, 4170. (h) Kurtz, K. C. M.; Frederick, M. O.; Lambeth, R. H.; Mulder, J. A.; Tracey, M. R.; Hsung, R. P. *Tetrahedron* **2006**, *62*, 3928. (i) Kurtz, K. C. M.; Hsung, R. P.; Zhang, Y. *Organic Lett.* **2006**, *8*, 231. (j) Zhang, Y.; Hsung, R. P.; Zhang, X.; Huang, J.; Slafer, B. W.; Davis, A.; *Org. Lett.* **2005**, *7*, 1047. (k) Tracey, M. R.; Zhang, Y.; Frederick, M. O.; Mulder, J. A.; Hsung, R. P. *Org. Lett.* **2004**, *6*, 2209. (l) Shen, L.; Hsung, R. P. *Tetrahedron Lett.* **2003**, *44*, 9353. (m) Frederick, M. O.; Hsung, R. P.; Lambeth, R. H.; Mulder, J. A.; Tracey, M. R. *Org. Lett.* **2003**, *5*, 2663. (n) Mulder, J. A.; Kurtz, K. C. M.; Hsung, R. P.; Coverdale, H. A.; Frederick, M. O.; Shen, L.; Zificsak, C. A. *Org. Lett.* **2003**, *5*, 1547. (o) Huang, J.; Xiong, H.; Hsung, R. P.; Rameshkumar. C.; Mulder, J. A.; Grebe, T. P. *Org. Lett.* **2002**, *4*, 2417. (p) Mulder, J. A.; Hsung, R. P.; Frederick, M. O.; Tracey, M. R.; Zificsak, C. A. *Org. Lett.* **2002**, *4*, 1383.

COPPER-CATALYZED ELECTROPHILIC AMINATION OF DIORGANOZINC REAGENTS: 4-PHENYLMORPHOLINE

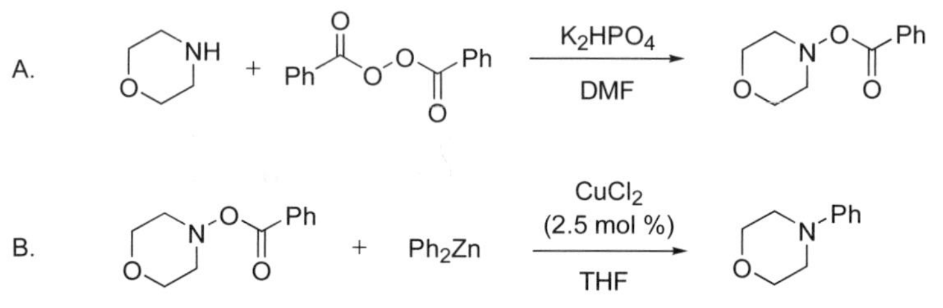

Submitted by Ashley M. Berman and Jeffrey S. Johnson.[1]
Checked by George Nora and Marvin J. Miller.

1. Procedure

A. 4-Benzoyloxymorpholine. A 500-mL, one-necked, round-bottomed flask equipped with a Teflon-coated magnetic stir bar is charged with benzoyl peroxide (12.11 g, 50 mmol), dipotassium hydrogen phosphate (13.06 g, 75 mmol), and *N,N*-dimethylformamide (125 mL). The suspension is stirred and morpholine (5.20 mL, 59.4 mmol) (Note 1) is added via syringe in one portion (Note 2). The flask is capped with a septum and the suspension is stirred at ambient temperature for 1 h, during which time a gradual discoloration of the reaction mixture from clear to yellow occurs. Deionized water (200 mL) is added and the contents are stirred vigorously for several min until all solid has dissolved. The reaction mixture is transferred to a 1-L separatory funnel and extracted with ethyl acetate (1 x 150 mL). The organic phase is collected and washed with saturated aqueous $NaHCO_3$ solution (2 x 100 mL). All of the aqueous fractions are combined and extracted with ethyl acetate (3 x 100 mL). All of the organic fractions are combined and washed with three 100-mL portions of deionized water, 100 mL of brine, dried over $MgSO_4$, filtered, and concentrated by rotary evaporation. The resulting pale-yellow solid is purified by flash column chromatography (Note 3), to yield 7.28–8.03 g (70.3–77.5%) of 4-benzoyloxymorpholine as a white, crystalline solid (Note 4).

B. 4-Phenylmorpholine. An oven-dried 250-mL, one-necked, round-bottomed flask equipped with a Teflon-coated magnetic stir bar and a septum is maintained under an inert atmosphere of argon and charged with zinc chloride (2.04 g, 15 mmol), and anhydrous tetrahydrofuran (27 mL) (Note 5). The solution is stirred and cooled in an ice-bath. Phenylmagnesium bromide (30 mL, 30 mmol) is added via syringe in one portion (Note 6). The resulting orange-colored solution is stirred for an additional 30 min in the ice-bath prior to use.

An oven-dried 1-L, two-necked, round-bottomed flask equipped with a thermometer, a septum, and a Teflon-coated magnetic stir bar is maintained under an inert atmosphere of argon and charged with 4-benzoyloxy-morpholine (5.18 g, 25 mmol), copper(II) chloride (0.084 g, 0.625 mmol), and anhydrous tetrahydrofuran (250 mL) (Note 7). The solution is stirred and cooled in an ice-bath. The previously generated diphenylzinc solution is added via cannula over the course of 5–10 min at a rate such that the internal temperature does not exceed 5 °C. The resulting light brown solution is stirred for 1.5 h in the ice-bath. Diethyl ether (250 mL) is added and the reaction mixture is transferred to a 1-L separatory funnel. The reaction mixture is washed with saturated aqueous $NaHCO_3$ solution (3 x 200 mL) (Note 8), brine (1 x 100 mL), and concentrated by rotary evaporation. The resulting yellow oil is dissolved in diethyl ether (100 mL) and the solution is extracted with 10 % aqueous HCl solution (3 x 50 mL). The aqueous extracts are basified with 10 % aqueous NaOH solution (200 mL) and extracted with dichloromethane (3 x 100 mL). The organic fraction is washed with 150-mL of brine, dried over Na_2SO_4, filtered, and concentrated by rotary evaporation to yield 2.73 g (67%) of 4-phenylmorpholine as an off-white, crystalline solid (Note 9).

2. Notes

1. Benzoyl peroxide (97%) was purchased from Aldrich Chemical Company. Dipotassium hydrogen phosphate was purchased from Fisher Scientific. *N,N*-Dimethylformamide (99%) and morpholine (99%) (d = 0.996 g/mL @ 25 °C) were purchased from Acros. All reagents were used as received.

2. A slight exotherm develops approximately 30 seconds after the completion of addition of morpholine.

3. Flash column chromatography was performed on a silica gel column (20 cm length x 18 cm width, 180 g of silica gel) eluting with 50% EtOAc:hexanes. Collected fractions were analyzed by thin layer chromatography (TLC), eluting with 50% EtOAc:hexanes (R_f = 0.35 for 4-benzoyloxymorpholine). Visualization was accomplished with UV light and

on spraying with aqueous ceric ammonium molybdate solution followed by heating.

4. Analytical data for 4-benzoyloxymorpholine: mp 82–84 °C, lit mp 81-82 °C;[2] IR (Nujol, cm^{-1}) 2924, 2852, 1730, 1599, 1456, 1377, 1315, 1269, 1248, 1178, 1165, 1101, 1084, 1066, 1049, 1007, 922, 858, 712; ^{1}H NMR (400 MHz, $CDCl_3$) δ: 3.03 (br t, J = 9.4 Hz, 2 H), 3.43 (br d, J = 9.3 Hz, 2 H), 3.85 (br t, J = 11.2 Hz, 2 H), 3.96 (br d, J = 10.7 Hz, 2 H), 7.45–7.41 (m, 2 H), 7.58–7.53 (m, 1 H), 8.00–7.98 (m, 2 H); ^{13}C NMR (100 MHz, $CDCl_3$) δ: 57.0, 65.8, 128.4, 129.2, 129.4, 133.1, 164.6. Anal. Calcd for $C_{11}H_{13}NO_3$: C, 63.76; H, 6.32; N, 6.76. Found: C, 63.96; H, 6.40; N, 6.67. The checkers noted that storage of this compound in the freezer (-10 °C) or at room temperature resulted in discoloration. The checkers recommend that it be used immediately upon formation.

5. Zinc chloride (98%) was purchased from Riedel-de-Haën and because it is very hydroscopic, it was dried under vacuum (0.1 mmHg) at 150 °C for 12 h prior to use. Tetrahydrofuran was dried by passage through a column of neutral alumina under nitrogen prior to use.

6. Phenylmagnesium bromide was purchased from Aldrich Chemical Company as a 1.0 M solution in tetrahydrofuran, and was used as received. The checkers noted that if the color does not change upon addition of the Grignard reagent, then the reaction does not proceed.

7. Copper(II) chloride (97%) was purchased from Aldrich Chemical Company and was used as received. Upon addition of the copper (II) chloride the solution turned a light green.

8. Copious formation of a white precipitate accompanied the initial wash with saturated aqueous $NaHCO_3$ solution.

9. Analytical data for 4-phenylmorpholine: mp 50–52 °C, (lit mp 53-54 °C);[3] IR (Nujol, cm^{-1}) 2922, 2854, 1601, 1498, 1458, 1377, 1231, 1126, 928, 771, 758, 690; ^{1}H NMR (400 MHz, $CDCl_3$) δ: 3.16–3.13 (m, 4 H), 3.86–3.84 (m, 4 H), 6.92–6.85 (m, 3 H), 7.30–7.24 (m, 2 H); ^{13}C NMR (100 MHz, $CDCl_3$) δ: 49.4, 66.9, 115.7, 120.0, 129.2, 151.3. Anal. Calcd for $C_{10}H_{13}NO$: C, 73.59; H, 8.03; N, 8.58. Found: C, 73.71; H, 8.03; N, 8.52. The checkers report that the procedure, when checked at half-scale, provided a 71% yield.

Waste Disposal Information

All hazardous materials should be handled and disposed of in accordance with "Prudent Practices in the Laboratory"; National Academy Press; Washington, DC, 1995.

3. Discussion

Amines are of fundamental interest in many areas of chemistry. Consequently, the development of methodology for the facile preparation of these compounds continues to be a topical area of research.[4] Among the modern methods for the preparation of amines, the Buchwald-Hartwig coupling is a convenient and general route to aryl amines, and represents the benchmark in catalyzed nucleophilic amination.[5,6] Electrophilic amination of nonstabilized carbanions is an alternative approach, and is noteworthy for its use of the *umpolung* strategy for C-N bond construction.[7,8] Generally, electrophilic amination has been limited to only the most reactive carbon donors (RLi and RMgX reagents), and yields of amine products are modest. Reports of transition-metal catalyzed methods present a possible solution to these problems, but have as yet been limited to the preparation of primary aniline derivatives.[9,10]

The copper-catalyzed electrophilic amination of diorganozinc reagents presented here allows for the expeditious preparation of tertiary amines under mild reaction conditions (eq 1).[11]

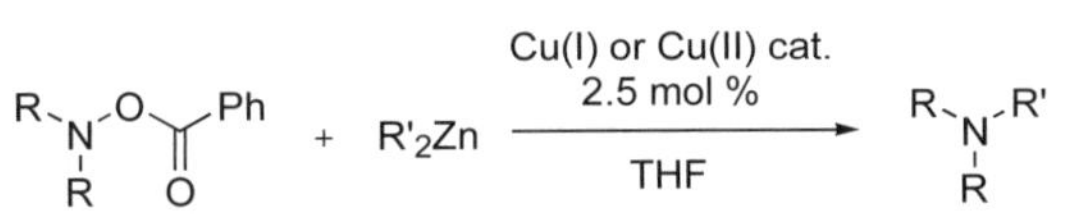

(1)

The *O*-benzoyl hydroxylamines employed are easily prepared from the corresponding secondary amines upon treatment with benzoyl peroxide, and show good stability: they are crystalline solids which can be purified via column chromatography and stored indefinitely in a freezer without decomposition or loss of reactivity. These reagents consequently represent a convenient source of sp^3-hydbridized nitrogen electrophiles for the direct delivery of R_2N^+ synthons. The secondary amine oxidation that is used here is a modification of a protocol originally reported by Ganem.[2] Elevated temperatures were found to be unnecessary and the use of DMF as reaction solvent provided higher yields than diethyl ether.

The amination reaction is quite general in the nucleophilic component, with both sp^2- (aryl, heteroaryl) and sp^3-hybridized (alkyl, benzyl) carbanions undergoing coupling in uniformly high yields (Table 1). On a scale of <1 mmol, it is convenient to conduct the reaction at ambient temperature; however, the 25-mmol preparation conducted in this procedure should be performed in an ice bath to minimize a deleterious exotherm.

The R_2Zn reagents employed are prepared via transmetallation from the corresponding RMgX or RLi with 0.5 equivalents of $ZnCl_2$, and are used *in situ* without isolation or purification.

Table 1. Scope of the Copper-Catalyzed Amination of Diorganozinc Reagents[a]

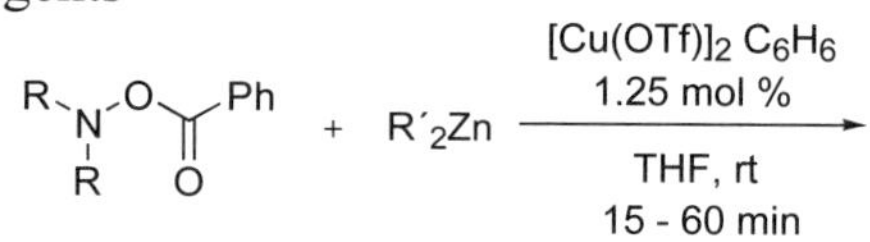

entry	R_2N-OC(O)Ph	R´	% yield[b]
1		*p*-$MeOC_6H_4$	93
2		*o*-MeC_6H_4	94
3	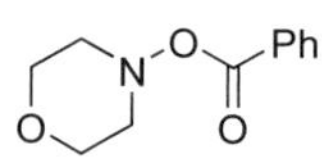	2-pyridyl	71
4		Bn	80
5		Ph	91
6	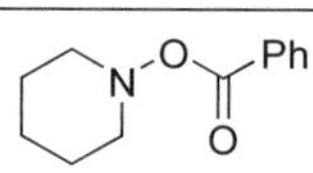	Ph	94
7		Bn	91
8	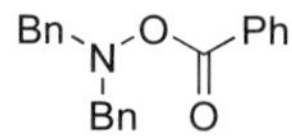	Et	91
9		*t*-Bu	98

[a] Unless otherwise noted, 1.1 equiv of R_2Zn were employed. Reactions were run on a 0.50 mmol scale of the starting R_2N-OC(O)Ph. [b] Isolated yield of product (average of at least two experiments). Yield is based on the starting R_2N-OC(O)Ph.

Functionalized tertiary aryl amines can also be prepared when functionalized Ar_2Zn reagents are employed.[12] The latter are generated *in situ* from the corresponding aryl iodides via a simple Mg/I exchange/transmetallation sequence.[13] The reaction shows good functional group tolerance, with nitrile, ester, halide, triflate, and nitro functionalities all accommodated (Table 2).

In conclusion, the copper-catalyzed amination of diorganozinc reagents represents a convenient and general route to tertiary amines. The reaction is noteworthy for the mild reaction conditions employed and ease of product purification (acid/base extractive workup). The catalyst is cheap and requires no exogenous supporting ligand. Such procedures may find utility as a complement to Buchwald-Hartwig coupling reactions.

Table 2. Scope of the Copper-Catalyzed Amination of Functionalized Diarylzinc Reagents[a]

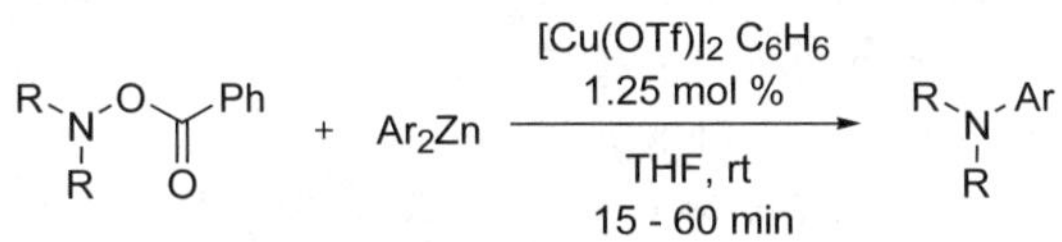

entry	R_2N-OC(O)Ph	Ar	% yield[b]
1		*p*-NCC_6H_4	76
2		*p*-$EtO_2CC_6H_4$	77
3	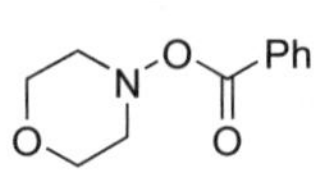	*p*-ClC_6H_4	93
4		*p*-$TfOC_6H_4$	95
5		*o*-$O_2NC_6H_4$	83
6		*p*-NCC_6H_4	95
7	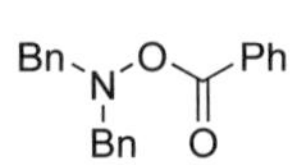	*p*-$EtO_2CC_6H_4$	99
8		*o*-$O_2NC_6H_4$	97

[a] 1.1equiv of Ar_2Zn were employed. Reactions were run on a 0.25 mmol scale of the starting R_2N-OC(O)Ph. [b] Isolated yield of product (average of at least two experiments). Yield is based on the starting R_2N-OC(O)Ph.

1. Department of Chemistry, University of North Carolina, Chapel Hill, NC 27599-3290, E-mail: jsj@unc.edu.

2. Biloski, A. J.; Ganem, B. *Synthesis* **1983**, 537-538.
3. Kawaguchi, M.; Ohashi, J.; Kawakami, Y.; Yamamoto, Y.; Oda, J. *Synthesis* **1985**, 701-3.
4. Ley, S. V.; Thomas, A. W. *Angew. Chem. Int. Ed.* **2003**, *42*, 5400-5449.
5. Muci, A. R.; Buchwald, S. L. *Top. Curr. Chem.* **2002**, *219*, 131-209.
6. Hartwig, J. F. In *Handbook of Organopalladium Chemistry for Organic Synthesis*; Negishi, E.-i., Ed.; Wiley: New York, 2002; Vol. 1, 1051-1096.
7. Dembech, P.; Seconi, G.; Ricci, A. *Chem. Eur. J.* **2000**, *6*, 1281-1286.
8. Erdik, E.; Ay, M. *Chem. Rev.* **1989**, *89*, 1947-1980.
9. Erdik, E.; Daskapan, T. *J. Chem. Soc. Perkin Trans. 1* **1999**, 3139-3142.
10. Narasaka, K. *Pure App. Chem.* **2002**, *74*, 143-149.
11. Berman, A. M.; Johnson, J. S. *J. Am. Chem. Soc.* **2004**, *126*, 5680-5681.
12. Berman, A. M.; Johnson, J. S. *J. Org. Chem.* **2005**, *70*, 364-366.
13. Knochel, P.; Dohle, W.; Gommermann, N.; Kneisel, F. F.; Kopp, F.; Korn, T.; Sapountzis, I.; Vu, V. A. *Angew. Chem. Int. Ed.* **2003**, *42*, 4302-4320.

ALLYLBORONATION OF IMINES: 1-PHENYLHEX-5-EN-3-AMINE

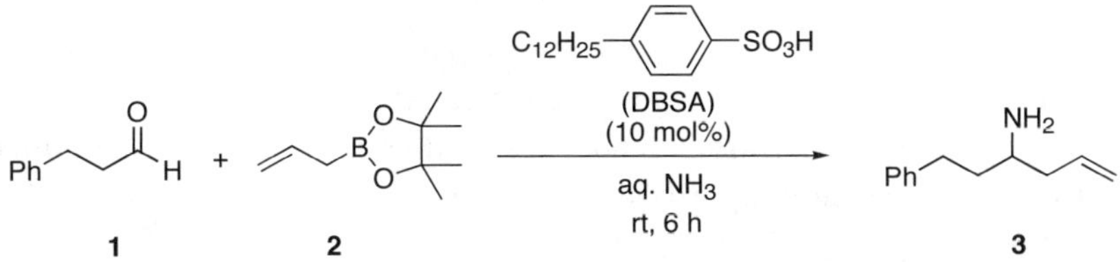

Submitted by Masaharu Sugiura, Keiichi Hirano, and Shū Kobayashi.[1]
Checked by Nai-Wen Tseng and Mark Lautens.

1. Procedure

Caution! The reaction should be conducted in a well-ventilated hood. An oven-dried, 250-mL, two-necked, round-bottomed flask is charged with dodecylbenzenesulfonic acid (1.307 g, 4.0 mmol) (Note 1), flushed with argon, and equipped with a magnetic stirring bar, a rubber septum, and an argon inlet. The flask is charged with 80 mL of 28% aqueous ammonia (Note 2) by syringe. After the gas evolution ceases, the mixture is stirred at 23–25 °C to give a clear solution, then cooled to 0 °C in an ice-water bath. 2-Allyl-4,4,5,5-tetramethyl-1,3,2-dioxaborolane (**2**) (8.120 g, 9.0 mL, 48.3 mmol) (Note 3) is added dropwise over 5 min via a syringe (Note 4) and the mixture is stirred at 23–25 °C for 30 min. 3-Phenylpropionaldehyde (**1**) (5.328 g, 5.3 mL, 39.7 mmol) (Note 5) is added dropwise via a syringe over 5 min at 23–25 °C. The slurry mixture is vigorously stirred at 23–25 °C for 6 h (Note 6) and transferred to a 1-L separatory funnel with water (100 mL) and saturated aqueous NaCl solution (200 mL). The aqueous layer is extracted three times with diethyl ether (200, 150, and 150 mL), and the combined organic layers are washed with saturated aqueous NaCl solution (200 mL), dried over anhydrous sodium carbonate (Na_2CO_3), filtered, and concentrated under reduced pressure to afford crude amine **3** as a yellow oil (Note 8). The crude material is chromatographed on a 5.5 x 45 cm column containing 300 g of silica gel (Note 9). A mixture of hexane/isopropyl amine (20/1) is used as the eluent. Fractions (27 mL each) are collected in 30-mL test tubes (Note 10). Fractions containing the product are evaporated under reduced pressure (30 °C, 200 mmHg) to afford the amine as yellow oil. In order to completely remove the pinacol, the amine is diluted with diethyl ether (200 mL) and extracted three times with 1 M hydrochloric acid (3 x 50 mL). The combined aqueous layers are washed twice with diethyl ether (2 x 100 mL), carefully basified with 6 M sodium hydroxide (60 mL) (Note 7), and extracted three times with dichloromethane (200, 150, and 150 mL). The combined organic layers are dried over anhydrous sodium carbonate (Na_2CO_3), filtered, concentrated under reduced pressure (30 °C, 200 mmHg) and dried under vacuum (approximately 0.2 mmHg, 15 min) to afford 6.680 g of **3** (96%) as yellow oil (Note 11).

2. Notes

1. Dodecylbenzenesulfonic acid (soft type) (>90.0%) was purchased from Tokyo Kasei Kogyo Co., Ltd. or TCI America, and used without purification. This viscous liquid material was measured directly in the reaction flask using a 5-mL pipette.

2. Aqueous ammonia solution (25.0-27.9%) was purchased from Wako Pure Chemical Industries, Ltd. and used as received. The checkers used aqueous ammonia solution (28.0-30.0%) purchased from Aldrich Chemical Company, Inc.

3. 2-Allyl-4,4,5,5-tetramethyl-1,3,2-dioxaborolane (**2**) (95%) was purchased from Aldrich Chemical Company, Inc. This material included white precipitates. Therefore, it was filtered with diethyl ether, evaporated (30 °C, 200 mmHg), and distilled (72-73 °C, 27 mmHg, the boiling point is uncorrected) prior to use.

4. On addition of **2**, colorless precipitates were formed to give a slurry mixture.

5. 3-Phenylpropionaldehyde (**1**) (>90% GC) was purchased from Tokyo Kasei Kogyo Co., Ltd. and used after distillation. The checkers purchased this reagent from Aldrich Chemical Company, Inc.

6. As the reaction progressed, the precipitates gradually dissolved to give a cloudy solution. The reaction could be monitored by TLC analysis on Merck silica gel 60 F_{254} plates and visualization with UV and phosphomolybdic acid [sodium phosphomolybdate hydrate (9.7 g) in 85% phosphoric acid (6.0 mL), concentrated sulfuric acid (20 mL), and water (400 mL)]. The checkers used 5% phosphomolybidic acid in ethanol as the TLC indicator. Disappearance of **1** was observed within 30 min. Compound **1** has R_f = 0.68 with hexane/ethyl acetate (3/1) as eluent (UV absorption, gray spot). Compound **3** has R_f = 0.13 with hexane/ethyl acetate (1/1) as eluent and R_f = 0.63 with hexane/isopropyl amine (10/1) as eluent (UV

absorption, white spot). Pinacol has $R_f = 0.30$ with hexane/ethyl acetate (1/1) as eluent and $R_f = 0.33$ with hexane/isopropyl amine (10/1) as eluent (UV inactive, blue spot).

7. After addition of sodium hydroxide, the pH was checked by a pH-test paper to be approximately 9.

8. The crude material was a mixture of **3** and pinacol (ca. 10/1 molar ratio). Distillation under reduced pressure (88 °C, 3 mmHg) gave almost pure **3**, but could not separate pinacol completely.

9. Silica gel 60 70-230 mesh ASTM (Merck Ltd.) was used. Checkers used Silica gel 60 40-63 mm (EMD Chemicals, Inc.).

10. The column fractions were checked by TLC analysis on Merck silica gel 60 F_{254} plates with hexane/isopropyl amine (20/1) as eluent and visualization with UV and phosphomolybdic acid. R_f values are given in Note 6.

11. The physical properties of **3** are as follows: ^1H NMR (300 MHz, $CDCl_3$) δ: 1.26 (brs, 2 H), 1.61 (dddd, *J* = 13.6, 10.1, 7.9, 5.7 Hz, 1 H), 1.76 (dddd, *J* = 13.5, 10.3, 6.2, 5.4 Hz, 1 H), 2.03 (dddt, *J* = 13.8, 7.9, 7.9, 0.9 Hz, 1 H), 2.27 (dddt, *J* = 13.7, 6.3, 4.8, 1.4 Hz, 1 H), 2.64 (ddd, *J* = 13.6, 10.1, 6.2 Hz, 1 H), 2.75 (ddd, *J* = 13.7, 10.2, 5.7 Hz, 1 H), 2.82 (tt, *J* = 7.9, 4.6 Hz, 1 H), 5.09 (dm, *J* = 10.0 Hz, 1 H), 5.10 (dm, *J* = 17.2 Hz, 1H), 5.79 (dddd, *J* = 17.2, 10.0, 7.9, 6.4 Hz, 1 H), 7.15–7.21 (m, 3 H), 7.25–7.30 (m, 2 H); ^{13}C NMR (75 MHz, $CDCl_3$) δ: 32.6, 39.4, 42.7, 50.1, 117.4, 125.7, 128.31, 128.33, 135.6, 142.2; HR-ESIMS calcd for $C_{12}H_{18}N$ $(M+H^+)$ 176.1439, found 176.1435; Anal. Calcd for $C_{12}H_{17}N$: C, 82.23; H, 9.78; N, 7.99. Found: C, 81.92; H, 9.77; N, 7.89.

Safety and Waste Disposal Information

All hazardous materials should be handled and disposed of in accordance with "Prudent Practices in the Laboratory"; National Academy Press; Washington, DC, 1995.

3. Discussion

Addition of allylmetals to C=N double bonds is a useful synthetic method for formation of homoallylic amine derivatives.[2] However, in order to obtain synthetically versatile homoallylic primary amines, removal of substituents on the nitrogen is often necessary. To address this issue, hydrolytically labile *N*-silyl, *N*-boryl, or *N*-metalloimines have been utilized frequently as precursors to the primary amines; however, pre-formation of those imines are required. In contrast, the method in this procedure is based on in situ formation of ammonia-derived imines and subsequent chemoselective allylation.[4,5] Thus, homoallylic primary amines are directly obtained from the three components without incorporation and cleavage of *N*-substituents. Additionally, when (*E*)- and (*Z*)-crotylboronates are used, *anti*- and *syn*-crotylated products can be obtained with high diastereoselectivities, respectively (Scheme 1).

Scheme 1.

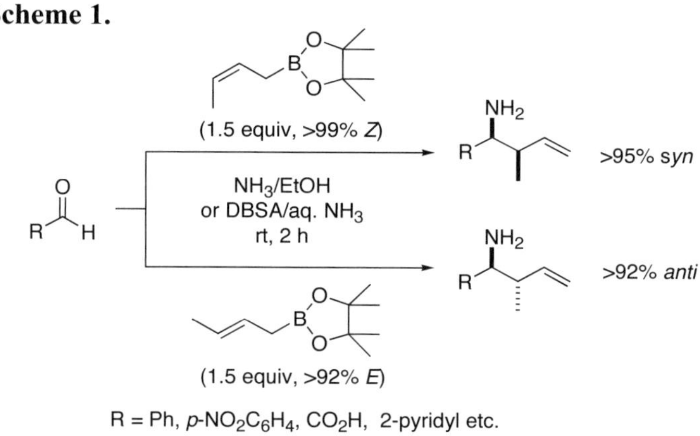

Use of ammonia as the nitrogen source is key in this process. In contrast to ammonia, primary amines scarcely undergo the reaction, and hence highly selective formation of homoallylic primary amines (suppression of over-reaction of the products) has been attained. Although ammonia in ethanol was utilized previously,[4] use of aqueous ammonia in the presence of DBSA has made the reaction more practical.[5] A variety of additives were examined for the reaction of 3-phenylpropionaldehyde (**1**) with allylboronate (**2**) in aqueous ammonia. Among them, DBSA was found to be the best, whereas lauric acid, sodium dodecyl sulfate, and sodium dodecylbenzenesulfonate were effective as well. In the absence of any additive, the yield of **3** and the chemoselectivity of the reaction (formation of **3** vs the corresponding homoallylic alcohol) were low. The hydrophobic alkyl chain of DBSA plays an important role, since *p*-toluenesulfonic acid showed much lower activity. In reactions performed with DBSA, a variety of aliphatic, aromatic, heteroaromatic, and a,b-unsaturated aldehydes afford the corresponding homoallylic primary amines in good to excellent yields (Table 1).[5]

However, addition of DBSA is not always necessary; hydrophilic carbonyl compounds such as a-oxo carboxylic acids and hydroxy aldehydes or ketones, including carbohydrates, were found to undergo the reactions smoothly with high selectivity as shown in Schemes 2 and 3.[6]

Table 1. Amine Formation

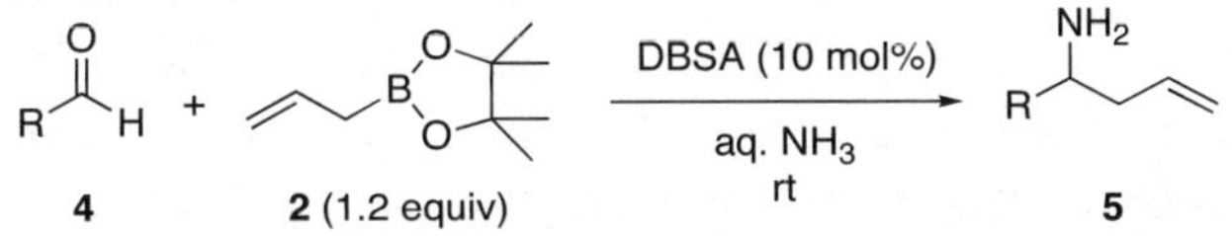

entry	R (**4**)	time/h	%yield of **5**[a]
1	Ph	6	61
2	*p*-$NO_2C_6H_4$	6	60
3	*p*-$MeOC_6H_4$	6	60
4	*o*-HOC_6H_4	6	75
5	2-Pyridyl	6	88
6	3-Pyridyl	2	85
7	4-Pyridyl	2	83
8	2-Thienyl	6	60
9	3-Thienyl	2	95
10	2-Furyl	2	53
11	3-Furyl	2	93
12	(*E*)-PhCH=CH	2	78
13	*c*-C_6H_{11}	6	68
14	*n*-C_5H_{11}	2	49

[a] Isolated yields.

Scheme 2.

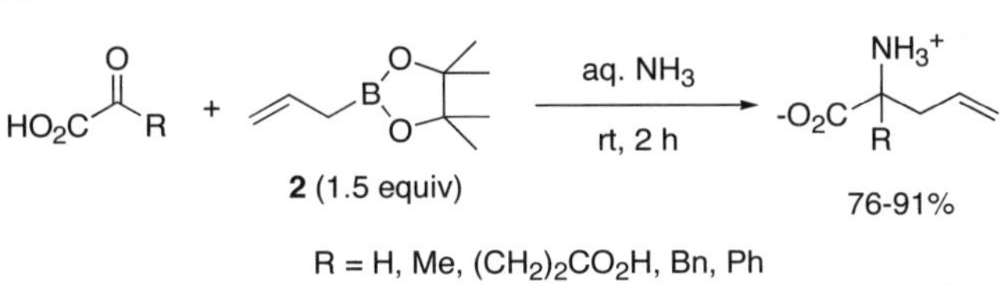

Scheme 3.

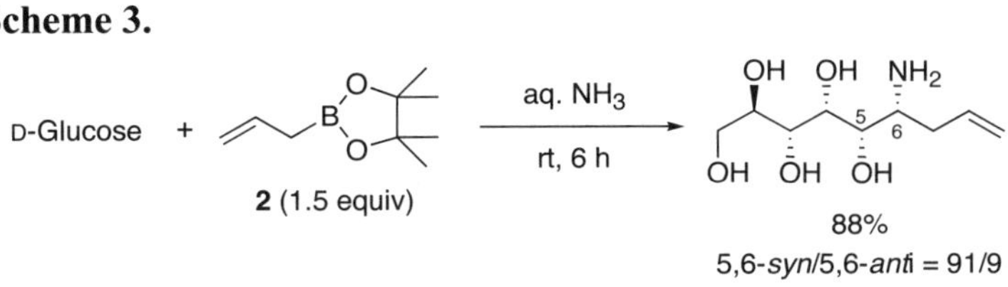

1. Graduate School of Pharmaceutical Sciences, The University of Tokyo, Hongo, Bunkyo-ku, Tokyo 113-0033, Japan
2. For reviews: (a) Kleinman, E. F.; Volkmann, R. A. In *Comprehensive Organic Synthesis*; Trost, B. M.; Fleming, I., Eds.; Pergamon Press, Oxford, **1991**, Vol. 2, p 975. (b) Yamamoto, Y.; Asao, N. *Chem. Rev.* **1993**, *93*, 2207. (c) Enders, D.; Reinhold, U. *Tetrahedron: Asymmetry* **1997**, *8*, 1895. (d) Bloch, R. *Chem. Rev.* **1998**, *98*, 1407. (e) Kobayashi, S.; Ishitani, H. *Chem. Rev.* **1999**, *99*, 1069. (f) Puentes, C. O.; Kouznetsov, V. *J. Heterocyclic Chem.* **2002**, *39*, 595. (g) Kobayashi, S.; Sugiura, M.; Ogawa, C. *Adv. Synth. Cat.* **2004**, *346*, 1023.
3. (a) Watanabe, K.; Ito, K.; Itsuno, S. *Tetrahedron: Asymmetry* **1995**, *6*, 1531. (b) Itsuno, S.; Watanabe, K.; Ito, K.; El-Shehawy, A. A.; Sarhan, A. A. *Angew. Chem. Int. Ed. Engl.* **1997**, *36*, 109. (c) Watanabe, K.; Kuroda, S.; Yokoi, A.; Ito, K.; Itsuno, S. *J. Organomet. Chem.* **1999**, *581*, 103. (d) Itsuno, S.; Watanabe, K.; Matsumoto, T.; Kuroda, S.; Yokoi, A.; El-Shehawy, A. *J. Chem. Soc., Perkin Trans. 1,* **1999**, 2011. (e) Itsuno, S.; Yokoi, A.; Kuroda, S. *Synlett* **1999**, 1987. (f) Chen, G.-M.; Ramachandran, P. V.; Brown, H. C. *Angew. Chem. Int. Ed.* **1999**, *38*, 825. (g) Chen, G.-M.; Brown, H. C. *J. Am. Chem. Soc.* **2000**, *122*, 4217.
4. Sugiura, M.; Hirano, K.; Kobayashi, S. *J. Am. Chem. Soc.* **2004**, *126*, 7182.
5. Kobayashi, S.; Hirano, K.; Sugiura, M. *Chem. Commun.* **2005**, 104.
6. Unpublished results.

IRIDIUM-CATALYZED *N*-HETEROCYCLIZATION OF PRIMARY AMINES WITH DIOLS: *N*-BENZYLPIPERIDINE

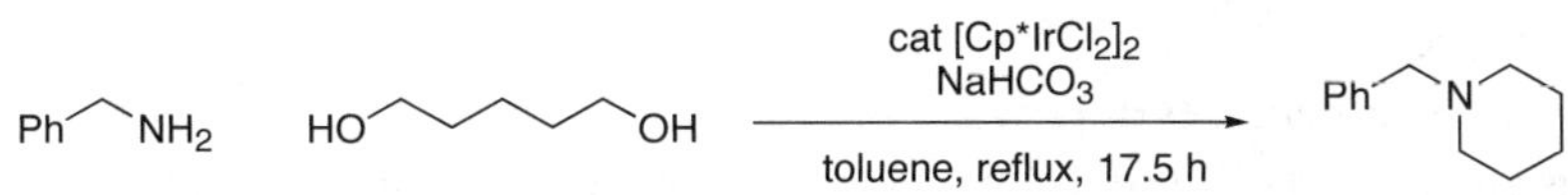

Submitted by Ken-ichi Fujita,[1a] Youichiro Enoki, and Ryohei Yamaguchi.[1b]
Checked by Gustavo Moura-Letts and Dennis P. Curran.[1c]

1. Procedure

N-Benzylpiperidine (Note 1). A 100-mL, two-necked, round-bottomed flask fitted with a magnetic stirring bar, a rubber septum, and a reflux condenser with a bubbler-sealed outlet is charged with di-μ-chloro-dichlorobis(η^5-pentamethylcyclopentadienyl)diiridium $[Cp^*IrCl_2]_2$ (199 mg, 0.25 mmol) (Note 2) and sodium bicarbonate (41 mg, 0.48 mmol) (Note 3) under an argon atmosphere. Addition of 10 mL of toluene (Note 4) by syringe to the flask affords an orange suspension. Benzylamine (10.70 g, 10.91 mL, 99.86 mmol) (Note 5) is added by syringe through the rubber septum over 10 sec. During the addition, the color of the suspension changes to yellow. Then 1,5-pentanediol (10.39 g, 10.45 mL, 99.76 mmol) (Note 5) is added by syringe through the septum over 30 sec. Under an argon flow, the rubber septum is replaced with a glass stopper. The black suspension (Note 6) is heated at reflux in an oil bath (oil bath temperature: 120 °C) for 17.5 h, then the reaction mixture is cooled to room temperature (Note 7). The reflux condenser is removed, and a short-path vacuum distillation apparatus is mounted to the flask. Distillation at 21.0 mmHg yields a fraction boiling between 123–125 °C; the clear, very pale yellow liquid is *N*-benzylpiperidine (14.20–14.30 g, 81–82% based on 1,5-pentanediol) (Notes 8, 9).

2. Notes

1. This is a modification of a published procedure.[2]
2. The submitters prepared the iridium complex di-μ-chloro-dichlorobis(η^5-pentamethylcyclopentadienyl)diiridium according to the literature method.[3] They also report that the commercially available

complex from Aldrich Chemical Company, Inc. or Strem Chemicals, Inc. can be substituted. The checkers purchased the catalyst from Strem.

3. Sodium bicarbonate (EP) was purchased from Wako Pure Chemical Industries, Ltd. (submitters) or Fisher Scientific (checkers) and used as received.

4. Toluene (GR) was purchased from Wako Pure Chemical Industries, Ltd. (submitters) or Aldrich (checkers) and distilled from sodium benzophenone ketyl under an argon atmosphere before use.

5. Benzylamine (GR, >99%) and 1,5-pentanediol (EP, >97%) were purchased from Tokyo Kasei Kogyo Co., Ltd. (submitters) or Aldrich (checkers) and used as received.

6. The submitters reported that the mixture remained yellow, but the checkers observed that it turned black after about 5 min.

7. At this time, a small quantity of water was observed in the bottom of the flask.

8. The submitters reported that the distillate was colorless. The checkers observed that the clear, very pale yellow liquid could be redistilled to provide a clear, colorless liquid, if desired. Alternatively, in a separate run where special care was taken to collect only clear distillate, 13.30 g (76%) product was obtained.

9. The spectral data and elemental analysis are as follows: ^{1}H NMR (300 MHz, $CDCl_3$) δ: 1.42 (m, 2 H), 1.56 (m, 4 H), 2.36 (m, 4 H), 3.46 (s, 2 H), 7.31–7.23 (m, 5 H); ^{13}C NMR (75 MHz, $CDCl_3$) δ: 24.3, 25.9, 54.4, 63.8, 126.7, 127.9, 129.0, 138.6; IR (thin film) 3062, 2934, 2852, 2684, 1493, 1369, 1154, 1113 cm^{-1}; EIMS *m/z*: 91(100), 175(95), 98(92), 84(91), 65(75), 176(58). Purity (>99.5%) was assessed by Gas Chromatography (GC) with a retention time (R_t) of 11.16 min: (Agilent 19091Z-413E, HP-1 Methyl siloxane, 0.32mm x 30m x 0.25μm; temperature 50 °C, ramp 10°/min to 350 °C).

Safety and Waste Disposal Information

All hazardous materials should be handled and disposed of in accordance with "Prudent Practices in the Laboratory"; National Academy Press; Washington, DC, 1995.

3. Discussion

N-Heterocyclic compounds are important intermediates in medicinal chemistry, material chemistry, and synthetic organic chemistry. In particular, pyrrolidine, piperidine and morpholine derivatives are present in large classes of biologically active natural products. For the past several decades, much effort has been devoted to development of efficient methods for the synthesis of these *N*-heterocycles. Recently, a variety of transition metal-catalyzed reactions for the synthesis of *N*-heterocyclic compounds have been disclosed and reviewed.[4,5] From an environmental point of view, *N*-heterocyclization of primary amines with diols is an attractive method because an *N*-heterocyclic product can be obtained from easily available starting materials in one pot without generation of wasteful or harmful byproducts (H_2O is the only byproduct). Although some ruthenium-catalyzed systems for *N*-heterocyclization of primary amines with diols have been reported, most of them require high reaction temperature (>150 °C) and applicable substrates are rather restricted.[6-9]

The method outlined here represents a convenient and environmentally benign *N*-heterocyclization of primary amines with diols catalyzed by a Cp*(C_5Me_5) iridium complex. The reaction can be conducted under relatively mild conditions (reflux in toluene as a solvent), and it does not generate any wasteful byproducts. Additional examples of the $[Cp^*IrCl_2]_2$-catalyzed *N*-heterocyclization of primary amines with diols are shown in the Table.[2] Using this protocol, a variety of pyrrolidine, piperidine, and morpholine derivatives can be synthesized in good to excellent yields. A seven-membered cyclic amine (azepane) can be also synthesized in a satisfactory yield (entry 2).

Table. Cp*Ir Complex-Catalyzed *N*-Heterocyclization of Primary Amines with a Variety of Diols[a]

entry	amine	diol	cat (%Ir)	yield[b] (%)	product
1	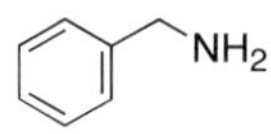	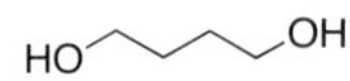	1.0	72	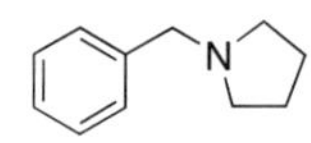
2[c]	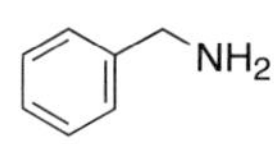	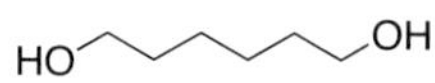	2.0	73	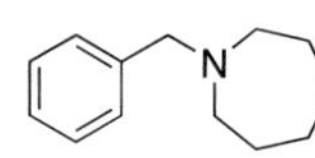
3[d]	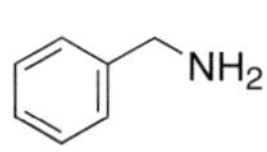	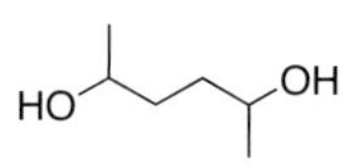	1.0	94[e]	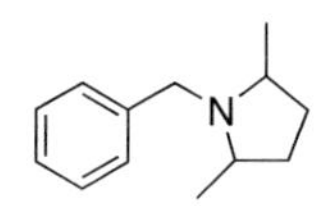
4	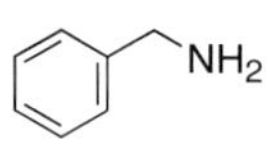	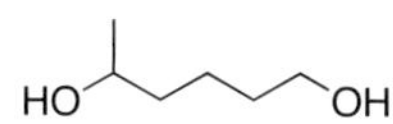	1.0	79	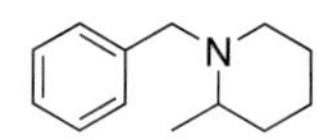
5	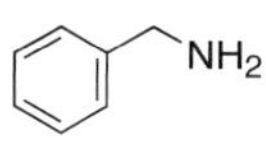	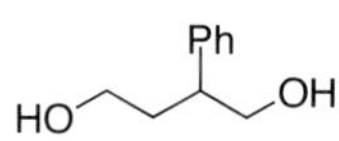	2.0	90	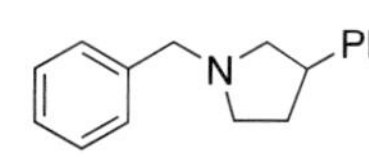
6	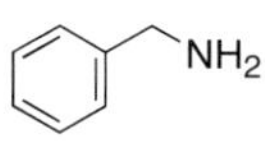	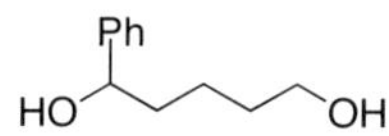	4.0	78[f]	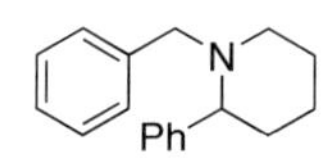
7[g,h]	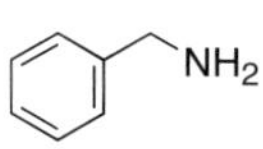	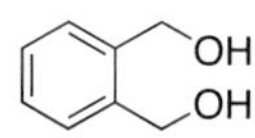	2.0	63	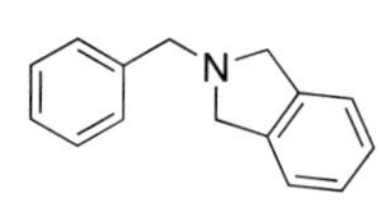
8	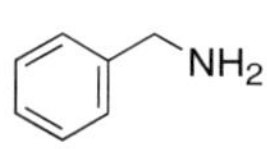	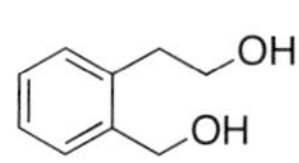	2.0	76	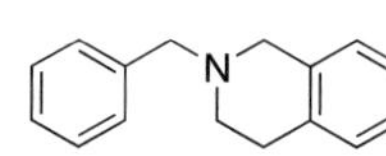
9	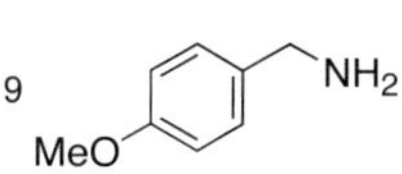	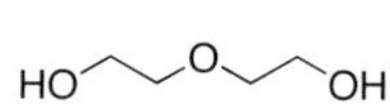	2.0	76	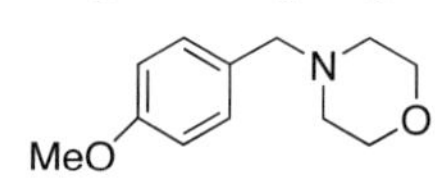
10[i,j]	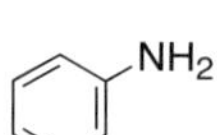	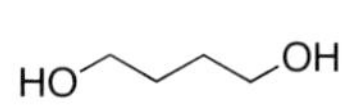	5.0	70	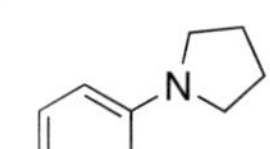
11[i]	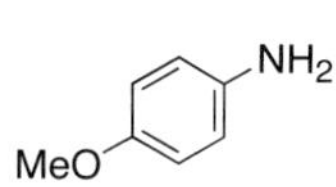	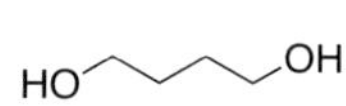	5.0	90	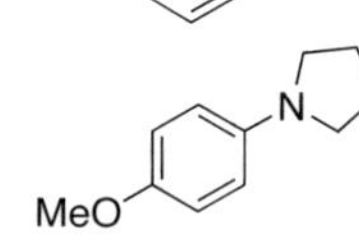
12	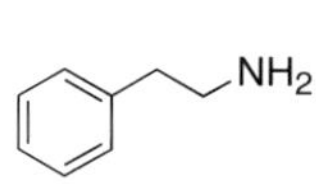	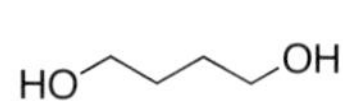	4.0	73	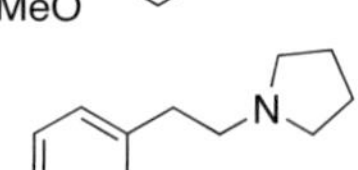
13	n-$C_8H_{17}NH_2$	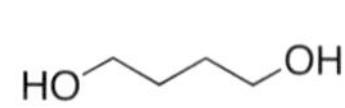	4.0	81[f]	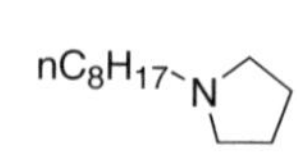

[a]The reaction was carried out at 110 °C for 17 h with amine (3.0 mmol), diol (2.0 mmol), $[Cp^*IrCl_2]_2$ (1.0 –5.0 % Ir), and $NaHCO_3$ (same number of equiv as Ir catalyst) in toluene (1 mL). [b]Isolated yield. [c] Toluene (3 mL) was used. [d]Na_2CO_3 was used as base. [e]*Cis/trans* = 73/27 (determined by ^1H NMR analysis. [f]GC yield. [g]Amine (2.0 mmol) was used. [h]Base was not added. [i]Reaction temperature was 130 °C. [j]Reaction time was 40 h.

1. a) Graduate School of Human and Environmental Studies, Kyoto University, Kyoto 606-8501, Japan; email: fujitak@kagaku.mbox.media.kyoto-u.ac.jp
b) email: yama@kagaku.mbox.media.kyoto-u.ac.jp c) Department of Chemistry, University of Pittsburgh, Pittsburgh, PA 15260, email: curran@pitt.edu

2. Fujita, K.; Fujii, T.; Yamaguchi, R. *Org. Lett.* **2004**, *6*, 3525.

3. White, C.; Yates, A.; Maitlis, P. M. *Inorg. Synth.* **1992**, *29*, 228.

4. Felpin, F.-X.; Lebreton, J. *Eur. J. Org. Chem.* **2003**, 3693.

5. Nakamura, I.; Yamamoto, Y. *Chem. Rev.* **2004**, *104*, 2127.

6. Murahashi, S.-I.; Kondo, K.; Hakata, T. *Tetrahedron Lett.* **1982**, *23*, 229.

7. Tsuji, Y.; Huh, K.-T.; Ohsugi, Y.; Watanabe, Y. *J. Org. Chem.* **1985**, *50*, 1365.

8. Felföldi, K.; Klyavlin, M. S.; Bartók, M. *J. Organomet. Chem.* **1989**, *362*, 193.

9. Abbenhuis, R. A. T. M.; Boersma, J.; van Koten, G. *J. Org. Chem.* **1998**, *63*, 4282.

(2*S*)-(–)-3-exo-(MORPHOLINO)ISOBORNEOL [(–)-MIB] ([1*R*-(exo,exo)]-1,7,7-Trimethyl-3-morpholin-4-yl-bicyclo[2.2.1] heptan-2-ol)

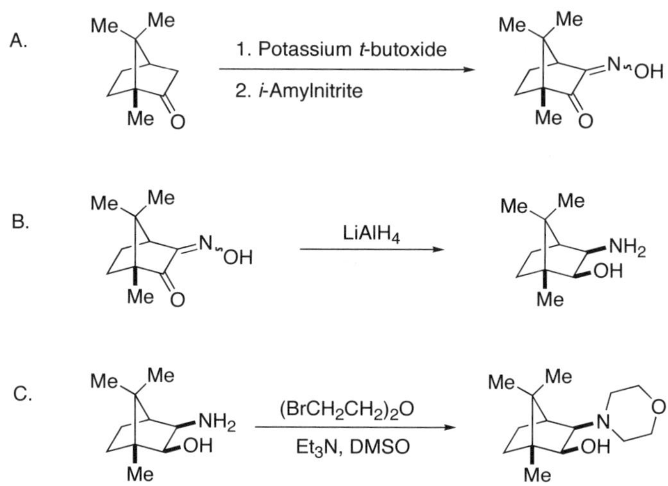

Submitted by Young K. Chen,[1] Sang-Jin Jeon,[1] Patrick J. Walsh,[1] and William A. Nugent.[2]
Checked by Christopher Kendall and Peter Wipf.

1. Procedure

A. Camphorquinone oxime. A flame-dried 1-L three-necked, round-bottomed flask equipped with a mechanical stirrer, a thermometer and septum is charged with potassium *t*-butoxide (46.0 g, 410 mmol) (Note 1). The contents of the flask are thoroughly purged with a stream of N_2 exhausted through an oil bubbler. After 15 min, 500 mL of Et_2O (Note 2) is added *via* cannula and the flask is submerged in a cold bath cooled to -30 °C (Note 3). A second 250-mL round-bottom flask is charged with (*R*)-Camphor (50.0 g, 328 mmol) (Note 4) and 100 mL of Et_2O. The resulting clear solution is added *via* cannula into the first flask over 10 min while keeping the internal temperature below -30 °C. The second flask is thoroughly rinsed with 20 mL of Et_2O, which is also transferred into the first flask. The cooling bath is removed and the reaction mixture is allowed to warm to room temperature. After stirring at room temperature for 30 min, the reaction mixture is cooled to -30 °C. Isoamyl nitrite (55.0 mL, 409 mmol) (Note 5) is added by syringe over 20 min while keeping the internal temperature below -30 °C. An orange to red color appears during the addition of isoamyl nitrite. The reaction mixture is allowed to warm to room temperature and is stirred for 16 h at ambient temperature under a N_2 atmosphere. The solution is extracted with water (3 x 150 mL), and the combined aqueous layers (approximately 450 mL) are cooled in an ice bath with magnetic stirring. The pH is adjusted to 4 by dropwise addition of approximately 30 mL of conc. HCl. After the pH is adjusted, an off-white solid precipitates from solution. The biphasic mixture is extracted with CH_2Cl_2 (3 x 150 mL). The combined organic layers are washed successively with 50 mL of saturated $NaHCO_3$, 50 mL of water, 50 mL of brine, and then dried over anhydrous $MgSO_4$. The solvent is removed *in vacuo* to afford 52.8 g (89%) of the title compound. The solid is a 10/90 mixture of *syn*- and *anti*-camphorquinone oximes (Note 6). This material is suitable for the next reaction without further purification.

B. (2S)-(-)-3-exo-Aminoisoborneol. A 250-mL two-necked, round-bottomed flask is charged with camphorquinone oximes (12.1 g, 66.8 mmol). The flask is fitted with a condenser and a septum, and is thoroughly purged with a steady stream of N_2. After 15 min, the N_2 flow is reduced to a slow bleed and 30 mL of anhydrous THF (Note 2) is added via syringe. The homogeneous solution is cooled in an ice bath to 0 °C. A 1.0 M solution of $LiAlH_4$ in THF (100 mL, 100 mmol) (Note 7) is slowly transferred *via* cannula to the mixture over 30 min (Note 8). After vigorous H_2 gas evolution ceased, the reaction mixture is allowed to warm to ambient temperature and then heated at reflux for 30 min. The solution is cooled to room temperature, diluted with 65 mL of Et_2O, cooled to 0 °C and quenched by the successive dropwise addition of 3.8 mL of water, 3.8 mL of 10% NaOH solution, and 11.4 mL of water (Note 9). The colorless precipitate was vacuum filtered through Celite, and the filter cake was washed with THF (3 x 20 mL) (Note 10). The combined filtrate was concentrated to give 10.4 g (92%) of a waxy solid. This material was used in the next step without further purification.

C. (2S)-(-)-exo-(Morpholino)isoborneol. To a 150-mL round-bottomed flask charged with (2*S*)-(-)-3-*exo*-aminoisoborneol (6.53 g, 38.6 mmol) is added 40 mL of reagent grade DMSO followed by Et_3N (16.3 mL, 117 mmol) (Note 11). Bis(2-bromoethyl) ether (6.50 mL, 46.5 mmol) (Note 11) in 30 mL of DMSO is added dropwise over 10 min and the reaction mixture is stirred at ambient temperature under N_2 for 72 h. The solution is poured into 400 mL of water, and the aqueous mixture is extracted with

Et_2O (3 x 150 mL). The combined organic layers are washed successively with 100 mL of water, 50 mL of brine and then dried over anhydrous $MgSO_4$. The solvent is removed *in vacuo* and the residue is purified by flash column chromatography (Notes 12, 13 and 14) to give 5.23 g (57%) of (-)-MIB as a colorless solid.

2. Notes

1. Potassium *t*-butoxide was purchased from Acros and used under a stream of N_2.

2. Et_2O and THF were purchased from Fisher and dried by passage through an activated alumina column under N_2.

3. The reaction was successfully performed at temperatures ranging from –30 °C to –50 °C.

4. (1*R*)-(+)-Camphor 98%, was purchased from Aldrich Chemical. The optical purity of (1*R*)-(+)-Camphor varies with the natural source. For analysis of the optical purity of commercial (+)- and (–)-camphor see: Armstrong, D. W.; Lee, J. T.; Chang, L. W. *Tetrahedron: Asymmetry* **1998**, *9*, 2043.

5. Isoamyl nitrite was purchased form Aldrich Chemical Company, Inc. Because of the appalling odor of the reagent, it was used without prior distillation.

6. *Anti*-(1*S*)-(–)-camphorquinone 3-oxime ^1H NMR (500 MHz, $CDCl_3$) δ: 9.70-8.80 (br, 1 H, N-OH), 3.25 (d, $J = 4.3$, 1 H), 2.08-2.00 (m, 1 H), 1.83-1.74 (m, 1 H), 1.62-1.52 (m, 2 H), 1.00 (s, 3 H), 1.03 (s, 3 H), 0.89 (s, 3 H) ppm; ^{13}C {^1H} NMR (125 MHz, $CDCl_3$) δ: 204.5, 160.1, 58.8, 47.1, 45.3, 31.1, 24.2, 21.1, 18.0, 9.3, 18.0 ppm. *Syn*-(1*S*)-(–)-camphorquinone 3-oxime ^1H NMR (500 MHz, $CDCl_3$) δ: 9.70-8.80 (br, 1 H, N-OH), 2.70 (d, $J = 4.2$, 1 H), 2.14-2.08 (m, 1 H), 1.83-1.74 (m, 1 H), 1.66-1.55 (m, 2 H), 1.03 (s, 3 H), 1.01 (s, 3 H), 0.93 (s, 3 H) ppm; ^{13}C {^1H} NMR (125 MHz, $CDCl_3$) δ: 205.3, 156.6, 60.0, 50.0, 47.4, 30.4, 25.4, 21.0, 18.4, 8.8 ppm.[3,4a]

7. 1.0 M $LiAlH_4$ in THF was purchased from Aldrich. The yields and diastereoselectivity were much higher when homogeneous solutions of $LiAlH_4$ were used instead of the powder form.

8. H_2 gas evolution was vigorous at the beginning of the addition.

9. Addition of the initial 3.8 mL of water is accompanied by vigorous generation of H_2 and is very exothermic.

10. The best product recovery was accomplished when the filter cake was washed with THF. For spectral properties see reference 4b.

11. Di(bromoethyl) ether and Et_3N were purchased from Aldrich. Reagent grade DMSO was purchased from Fisher.

12. ~200 g of silica, 10% to 15% EtOAc in hexanes. TLC in 20% EtOAc in hexanes; R_f 0.25; Stained brown in I_2 Chamber.

13. Recrystallization is also possible from hexanes 4 mL/g at –30 °C.[5]

14. Characterization data for (2*S*)-(–)-exo-(morpholino)isoborneol: $[\alpha]^{20}_D$ = -6.9 (*c* = 1.0, MeOH); mp = 65-66 °C (hexane); ^{1}H NMR (400 MHz, C_6D_6) δ: 3.91 (s, 1 H), 3.43 (d, 1 H, *J* = 7.1 Hz), 3.40 (bs, 4 H), 2.30 (bs, 2 H), 2.10 (bs, 2 H), 2.07 (d, 1 H, *J* = 7.1 Hz), 1.66 (d, 1 H, *J* = 4.7 Hz), 1.51 (tt, 1 H, *J* = 12.0, 4.6 Hz), 1.35 (td, 1 H, *J* = 12.2, 3.6), 1.13 (s, 3 H), 1.02 (s, 3 H), 0.92-0.82 (m, 1 H), 0.77-0.70 (m, 1 H), 0.68 (s, 3 H); ^{13}C {^{1}H} NMR (76 MHz, C_6D_6, d_1 = 5 sec) δ 79.6, 74.0, 67.4, 50.1, 47.2, 46.0, 33.2, 28.6, 22.8, 21.7, 12.4; IR (KBr) 3460, 3367, 1478, 1448, 1396, 1360, 1284, 1261, 1202, and 1200 cm^{-1}; EIMS *m/z* 239 (M^+, 13), 154 (100); HRMS (EI) *m/z* calcd for $C_{14}H_{25}NO_3$: 239.1885, found 239.1889. Enantiomeric excess was determined as followed: To a screw cap vial (1 dram) charged with (–)-MIB (24 mg, 0.1 mmol) was added dichloromethane (1 mL), followed by Et_3N (17 μL, 0.12 mmol), and DMAP (~2 mg). *p*-Bromobenzoyl chloride (22 mg, 0.1 mmol) was added to the clear solution and stirred for 10 min. The reaction mixture was concentrated under reduced pressure and the residue purified by column chromatography (5% ethyl acetate in hexanes). HPLC analysis of the resultant *p*-bromobenzyl ester established the enantiomeric excess as 96.0% ee (Chiralcel OD column, flow 1 mL/min, 254 nm, 2% isopropanol in hexanes; minor isomer 4.6 min, and major isomer 5.3 min).

Safety and Waste Disposal Information

All hazardous materials should be handled and disposed of in accordance with "Prudent Practices in the Laboratory"; National Academy Press; Washington, DC, 1995.

3. Discussion

(–)-MIB has been shown to be an excellent chiral ligand in the asymmetric alkylation of aldehydes. In comparison, MIB was equal or better than DAIB in Et_2Zn addition to aromatic aldehydes with ee's > 95. Highly enantioselective Et_2Zn additions to alkyl substituted aldehydes are also possible with (–)-MIB.[5] Remarkably, the generation of (*S*)-1-phenylpropan-1-ol with 90% ee can be achieved when MIB of only 10% ee was used in the Et_2Zn addition with benzaldehyde. This result shows a large positive non-linear effect that parallels DAIB.[6] Recently, MIB has been shown to be equally effective with DAIB in the alkenylzinc addition to aldehydes pioneered by Oppolzer.[7] A number of terminal alkynes with various substituents were used in the study. Using MIB in this reaction constitutes a powerful and practical method to access both enantiomers of allylic alcohols in high optical purity.[8] Furthermore, it has recently been shown that MIB can be used in a one-pot tandem asymmetric addition/diastereoselective epoxidation sequence to generate epoxy alcohols with up to three stereocenters and the asymmetric addition/diastereoselective cyclopropanation with up to four stereogenic centers with high enantio- and diastereoselectivity.[9]

The distinct advantage of MIB is its ease of preparation. Gram quantities of both enantiomers of MIB can be made in only three steps and one purification, while the most efficient synthesis of (–)-DAIB was achieved in six steps and involved a low yielding and laborious purification step to remove the undesired diastereomer.[4] Furthermore, MIB is a crystalline solid and can be stored for months in the presence of air without noticeable decomposition.[5]

1. Department of Chemistry, University of Pennsylvania, Philadelphia, PA, 19104-6323, USA.

2. Bristol-Myers Squibb Company, Process Research and Development Department, P. O. Box 4000, Princeton, NJ, 08543-4000.

3. Roy, S.; Chakraborti, A. K. *Tetrahedron Lett.* **1998**, 6355-6356.

4. (a) White, J. D.; Wardrop, D. J.; Sundermann, K. F. *Org. Synth.* **2002**, 79, 125-129. (b) White, J. D.; Wardrop, D. J.; Sundermann, K. F. *Org. Synth.* **2002**, *79*, 130-138.

5. Nugent, W. A. *J. Chem. Soc., Chem. Comm.* **1999**, 1369-1370.

6. Chen, Y. K.; Costa, A. M.; Walsh, P. J. *J. Am. Chem. Soc.* **2001**, *123*, 5378-5379.

7. Oppolzer, W.; Radinov, R. N. *Helv. Chim. Acta* **1992**, *75*, 170-173.

8. (a) Chen, Y. K.; Lurain, A. E.; Walsh, P. J. *J. Am. Chem. Soc.* **2002**, *124*, 12225-12231. (b) Lurain, A. E.; Walsh, P. J. *J. Am. Chem. Soc.* **2003**, *125*, 10677-10683.

9. (a) Lurain, A. E.; Maestri, A.; Kelly, A. R.; Carroll, P. J.; Walsh, P. J. *J. Am. Chem. Soc.* **2004**, *126*, 13608-13609. (b) Lurain, A. L.; Carroll, P. J.; Walsh, P. J. *J. Org. Chem.* **2005**, *70*, 1262-1268. (c) Rowley Kelly, A.; Lurain, A.E.; Walsh, P.J. *J. Am. Chem. Soc.* **2005**, *127*, 14668-14674. (d) Kim, H.Y.; Lurain, A.E.; García-García, P.; Carroll, P.J.; Walsh, P.J. *J. Am. Chem. Soc.* **2005**, *127*, 13138-13139.

PREPARATION OF (η^5-(*S*)-2-(4-METHYLETHYL)-OXAZOLINYLCYCLOPENTADIENYL)-(η^4-TETRAPHENYLCYCLO-BUTADIENE)COBALT (Cobalt, [1,1',1'',1'''-(η^4-1,3-cyclobutadiene-1,2,3,4-tetrayl)tetrakis[benzene]][(1,2,3,4,5-η)-1-[(4*S*)-4,5-dihydro-4-(1-methylethyl)-2-oxazolyl]-2,4-cyclopentadien-1-yl])

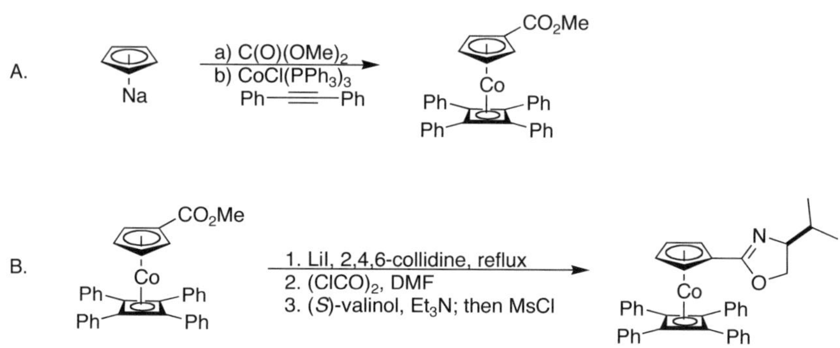

Submitted by Carolyn E. Anderson, Larry E. Overman,*[1] Christopher J. Richards, Mary P. Watson and Nicole White.
Checked by Joshua Bolger and Marvin J. Miller.

1. Procedure

A. Preparation of (η^5-carbomethoxycyclopentadienyl)-(η^4-tetraphenyl-cyclobutadiene)cobalt. A flame-dried, 500-mL, three-necked, round-bottomed flask is fitted with a stirring bar, reflux condenser, rubber septa, and an argon inlet. The flask is flushed with dry argon and charged with sodium cyclopentadienide (Note 1) (20 mL, 2.0 M in THF, 40 mmol) and 20 mL of anhydrous tetrahydrofuran (Note 2) via syringe. Dimethyl carbonate (Note 1) (10.2 mL, 120 mmol) is added via syringe and the resulting solution is warmed to reflux for 4 h before cooling to room temperature (Note 3).

Once cooled, 160 mL of toluene (Note 2) is added via syringe, followed by chlorotris(triphenylphosphine)cobalt(I) (Note 4) (30.8 g, 34.8 mmol) and diphenylacetylene (Note 5) (14.2 g, 80 mmol) by temporarily removing a rubber septum. The resulting mixture is heated at reflux for 5 h before cooling to room temperature, transferring to a single-neck flask, and concentrating under reduced pressure. The residue is suspended in hexanes (150 mL) and filtered through a Büchner funnel. The solids are washed with hexanes (~3.5 L) until the filtrate is colorless. The resultant dark mustard-colored filter cake is dissolved from the filter into a new collection flask with methylene chloride. Washing is continued until the filtrate is colorless. A black insoluble solid is left after washing and is discarded. The organic solution is concentrated *in vacuo* to provide 12.7 g (23.6 mmol, 67%) of (η^5-carbomethoxycyclopentadienyl)-(η^4-tetraphenylcyclobutadiene)cobalt as a mustard-colored solid (Notes 6, 7).

B. Preparation of (η^5-(S)-2-(4-methylethyl)oxazolinylcyclopentadienyl)-(η^4-tetraphenylcyclobutadiene)cobalt. A flame-dried, 250-mL, round-bottomed flask is fitted with an air condenser, stirring bar and an argon inlet. The flask is flushed with dry argon and then charged with (η^5-carbomethoxycyclopentadienyl)-(η^4-tetraphenylcyclobutadiene)cobalt (10 g, 18.6 mmol), lithium iodide (Note 8) (4.95 g, 37.2 mmol), and 100 mL of 2,4,6-collidine (Note 1) by temporarily removing the condenser. The resulting solution is warmed to reflux. After 16 h, the solution is cooled to room temperature (Note 9) and diluted with 50 mL of methylene chloride. This solution is washed with 150 mL of 2 N aqueous hydrochloric acid, and the aqueous layer is extracted with 150 mL of methylene chloride. The combined organic phases are then washed with 2 N aqueous hydrochloric acid (4 x 150 mL), dried over anhydrous sodium sulfate, filtered, and concentrated under reduced pressure to provide (η^5-carboxycyclopentadienyl)(η^4-tetraphenylcyclobutadiene)cobalt as an orange solid (Note 10), which is used directly in the next transformation.

A flame-dried, 250-mL, round-bottomed flask is fitted with a stirring bar and an argon inlet. The flask is flushed with argon and then charged with crude (η^5-carboxycyclopentadienyl)(η^4-tetraphenylcyclobutadiene)-cobalt and 124 mL of methylene chloride (Note 2). Oxalyl chloride (Note 1) (3.25 mL, 37.2 mmol) and dimethylformamide (Note 11) (3 drops) are added sequentially. Upon addition of the latter, gas evolution is observed. The resulting solution is maintained at room temperature. After 30 min, the solution is concentrated using a rotary evaporator. Volatile byproducts and excess starting materials are removed by dissolving the residue in 100 mL of methylene chloride and concentrating this solution using a rotary evaporator. The evaporation procedure is repeating three times to provide the corresponding acid chloride as a red-brown residue, which is used directly in the next transformation.

A flame-dried, 500-mL, round-bottomed flask is fitted with a stirring bar and a septum containing argon inlet and outlet needles. The flask is flushed with argon and then charged with (*S*)-valinol·HCl salt (Note 12) (3.6 g, 26 mmol) by temporarily removing the septum, triethylamine (Note 11) (15.5 mL, 112 mmol) and 86 mL of methylene chloride (Note 2) via syringe. A solution of crude acid chloride and 100 mL of methylene chloride (Note 2) is added to the flask via cannula. The resulting solution is maintained at room temperature, and, after 2 h, the solution is cooled to 0 °C using an ice/water bath. Methanesulfonyl chloride (Note 1) (3.6 mL, 47 mmol) is then added in one portion via syringe, and the resulting solution is allowed to warm to room temperature. After 16 h, the solution is transferred to a separatory funnel and washed with 150 mL of saturated aqueous sodium bicarbonate and 150 mL of brine. The organic layer is then dried over anhydrous sodium sulfate, filtered, and concentrated using a rotary evaporator. The residue is dissolved in a minimum volume of methylene chloride (~50 mL) and purified through a column of silica gel using 9:1 hexanes-ethyl acetate as eluent (Note 13). Evaporation of solvent gives 8.2 g (13.9 mmol, 75% yield) of (η^5-(*S*)-2-(4-methylethyl)oxazolinylcyclopentadienyl)(η^4-tetraphenylcyclobutadiene)cobalt as a yellow solid (Note 14).

2. Notes

1. Sodium cyclopentadienide was purchased from Aldrich Chemical Company or Strem Chemical. This reagent should be light red/orange in color. Commercial reagent that is black in color and/or contains some solid residue should not be used. Dimethyl carbonate, oxalyl chloride, methanesulfonyl chloride, and 2,4,6-collidine were purchased from Acros Organics or Aldrich Chemical Company. These chemicals were used as received.

2. Tetrahydrofuran, toluene and methylene chloride were purified by passage through activated alumina using a GlassContour solvent purification system.[2] The checkers used anhydrous Acroseal solvents.

3. Sodium carbomethoxycyclopentadienide can be isolated,[3] but its isolation and purification were unnecessary for this procedure.

4. Chlorotris(triphenylphosphine)cobalt(I) was prepared according to a detailed literature procedure.[4] This reagent could be purchased from Aldrich Chemical Company. However, it is quite expensive; the use of the commercial material typically resulted in lower yields of product.

5. Diphenylacetylene was purchased from Acros Organics and was recrystallized from ethanol prior to use.

6. Somewhat higher yield (13.8 g, 25.6 mmol, 74%) was observed when the solution of sodium carbomethoxycyclopentadienide and THF was transferred by cannula to a 500-mL round-bottomed flask charged with dichlorotris(triphenylphosphine)cobalt(I), diphenylacetylene and toluene. However, this procedure was less convenient.

7. The product, (η^5-carbomethoxycyclopentadienyl)(η^4-tetraphenyl-cyclobutadiene)cobalt, showed the following characterization data; ^{1}H NMR (500 MHz, $CDCl_3$) δ: 3.23 (s, 3 H), 4.79 (t, J = 2.0 Hz, 2 H), 5.21 (t, J = 2.0 Hz, 2 H), 7.22–7.30 (m, 12 H), 7.43–7.45 (m, 8 H); ^{13}C NMR (125 MHz, $CDCl_3$) δ: 51.4, 76.5, 84.7, 86.6, 86.8, 126.9, 128.2, 129.0, 135.3, 166.6; Infrared spectrum (thin film) cm^{-1}: 3058, 2953, 1713, 1596, 1498, 1467, 1281, 1140; HRMS (FAB+) m/z 538.1320 [538.1343 calcd for $C_{35}H_{27}O_2Co$ (M^+)]; Anal. Calcd. for $C_{35}H_{27}O_2Co$: C, 78.06; H, 5.05. Found: C, 78.30; H, 5.43.

8. Lithium iodide was purchased from Acros Organics. In some instances, the use of old bottles of this reagent resulted in incomplete conversion.

9. The disappearance of starting material could be monitored by TLC analysis. Using 20% ethyl acetate: hexanes as eluent, (η^5-carbomethoxycyclopentadienyl)(η^4-tetraphenylcyclobutadiene)cobalt had an R_f of 0.8.

10. The crude product, (η^5-carboxycyclopentadienyl)(η^4-tetraphenyl-cyclobutadiene)cobalt, showed the following ^{1}H NMR spectrum; (500 MHz, $CDCl_3$) δ: 4.84 (2d, J = 2.0 Hz, 2 H), 5.23 (2d, J = 2.0 Hz, 2 H), 7.20–7.24 (m, 12 H), 7.42–7.44 (m, 8 H).

11. Dimethylformamide and triethylamine were purified by passage through GlassContour solvent purification columns.[2] The checkers used anhydrous dimethylformamide purchased from Acros Organics. Triethylamine was purchased from Aldrich Chemical Company and used as received.

12. (S)-Valinol was prepared by reduction of (S)-valine.[5] The checkers used (S)-valinol purchased from Aldrich Chemical Company, Inc. The salt was formed by treatment of the freebase with 2 N HCl in diethylether.

13. The submitters reported dissolving the crude residue in 9:1 hexanes:EtOAc prior to loading on the column, but the checkers isolated 5.6 g (9.5 mmol, 51% yield) of pure product by filtration of the resulting slurry. The residue remaining after concentration of the filtrate was dissolved in a

minimum of methylene chloride and purified as described to give an additional 2.6 g (4.4 mmol, 24% yield) of the pure product. Subsequent reactions were purified exclusively by column chromatography.

14. The product, (η^5-(*S*)-2-(4-methylethyl)oxazolinylcyclopentadienyl)-(η^4-tetraphenylcyclobutadiene)cobalt, shows the following characterization data; ^{1}H NMR (500 MHz, $CDCl_3$) δ: 0.76 (d, *J* = 7.0 Hz, 3 H), 0.98 (d, *J* = 6.5 Hz, 3 H), 1.37–1.44 (m, 1 H), 3.41–3.55 (m, 3 H), 4.71 (2d, *J* = 2.5 Hz, 2 H), 4.8 (2d, *J* = 2.5 Hz, 2 H), 5.09 (s, 1 H), 5.25 (brs, 1 H) 7.20–7.29 (m, 12 H), 7.45–7.47 (m, 8 H); ^{13}C NMR (125 MHz, $CDCl_3$) δ: 18.4, 19.6, 33.0, 69.5, 72.7, 76.0, 82.0, 84.4, 85.0, 86.4, 126.4, 128.0, 129.0, 135.4, 160.5; Infrared spectrum (thin film) cm^{-1}: 3058, 2958, 1652, 1597, 1499, 1373, 1113, 1024; HRMS (FAB+) *m/z* 592.2055 [592.2051 calcd for $C_{39}H_{35}CoNO$ (M+H)]. Anal. Calcd. for $C_{39}H_{34}ONCo$: C, 79.17; H, 5.79; N, 2.37. Found: C, 79.12; H, 5.83; N, 2.33.

Safety and Waste Disposal Information

All hazardous materials should be handled and disposed of in accordance with "Prudent Practices in the Laboratory"; National Academy Press; Washington, DC, 1995.

3. Discussion

This synthesis of (η^5-(*S*)-2-(4-methylethyl)oxazolinylcyclopentadienyl)-(η^4-tetraphenylcyclobutadiene)cobalt is a modification of a procedure originally reported by Richards and Stevens.[6] The procedure has been streamlined and modified to enable large-scale preparation of this oxazoline. Nucleophilic conditions, rather than hydrolytic, are used to cleave the ester to the acid in Step B, thereby eliminating the use of DMSO, which is difficult to remove. Removal of DMSO is an essential feature of the original procedure as oxalyl chloride is employed in the following step.[6] In the present procedure, the oxazoline is formed directly from the crude valinol amide by reaction with methanesulfonyl chloride and triethylamine in methylene chloride. These conditions eliminate the need for toxic carbon tetrachloride; previously the isolated valinol amide was cyclized using triphenylphosphine, carbon tetrachloride and triethylamine.[6] In this optimized procedure, conversion of the ester to the oxazoline is accomplished in one pot.

(η^5-(*S*)-2-(4-methylethyl)oxazolinylcyclopentadienyl)-(η^4-tetraphenylcyclobutadiene)cobalt is an intermediate in the synthesis of the planar, chiral palladium(II) catalysts, $[COP\text{-}Cl]_2$,[7] $[COP\text{-}OAc]_2$[7] and COP-hfacac,[7] which have been used in the catalytic asymmetric synthesis of chiral allylic amides[8,9] and allylic esters.[10] This procedure illustrates a general synthesis of enantiopure (η^5-(*S*)-2-oxazolinylcyclopentadienyl)-(η^4-tetraphenylcyclobutadiene)cobalt complexes. Analogous oxazoline complexes prepared from (*S*)-serine have been used as ligands for catalytic asymmetric addition of organozinc reagents to aldehydes and palladium-catalyzed allylic alkylations.[11] As enantioenriched monodentate oxazolines have found use as chiral ligands,[12] and η^5-(cyclopentadienyl)-(η^4-tetraphenylcyclobutadiene)cobalt complexes have been used to prepare a variety of novel materials,[13] the oxazoline product of this procedure could find use in its own right.

1. Department of Chemistry, University of California, Irvine, California, 92697-2025; E-mail: leoverma@uci.edu

2. (a) Pangborn, A. B.; Giardello, M. A.; Grubbs, R. H.; Rosen, R. K.; Timmers, F. J. *Organometallics* **1996**, *15*, 1518–1520. (b) http://www.glasscontour.com/

3. Hart, W. P.; Shihua, D.; Rausch, M. D. *J. Organomet. Chem.* **1985**, *282*, 111–121.

4. Watatsuki, Y.; Yamazaki, H. *Inorg. Syn.* **1989**, *26*, 189–200.

5. McKennon, M. J.; Meyers, A. I.; Drauz, K.; Schwarm, M. *J. Org. Chem.* **1993**, *58*, 3568–3571.

6. Stevens, A. M.; Richards, C. J. *Organometallics* **1999**, *18*, 1346–1348.

7. Anderson, C. E; Kirsch, S. F.; Overman, L. E.; Richards, C. J.; Watson, M. P. *Org. Synth.* **2007**, *84*, 148-155.

8. Kirsch, S. F.; Overman, L. E. *J. Org. Chem.* **2005**, *70,* 2859–2861.

9. (a) Kirsch, S. F.; Overman, L. E.; Watson, M. P. *J. Org. Chem.* **2004**, *69*, 8101–8104. (b) Anderson, C. E.; Overman, L. E. *J. Am. Chem. Soc.* **2003**, *125*, 12412.

10. Kirsch, S. F.; Overman, L. E. *J. Am. Chem.* Soc. **2005**, *127*, 2866–2867.

11. (a) Jones, G.; Richards, C. J. *Tetrahedron Lett.* **2001**, *42*, 5553–5555. (b) Jones, G.; Richards, C. J. *Tetrahedron: Assymetry* **2004**, *15*, 653–664.

12. See, inter alia.: Dakovic, S.; Liscic-Tumir, L.; Kirin, S. I.; Vinkovic, V.; Raza, Z.; Suste, A.; Sunjic, V. *J. Mol. Catal. A: Chem.* **1997**, *118*, 27–31.

13. For illustrative examples, see: (a) Lee, J.-C.; Nishio, A.; Tomita, I.; Endo, T. *Macromolecules* **1997**, *30*, 5205–5212. (b) Harrison, R. M.; Brotin, T.; Noll, B. C.; Michl, J. *Organometallics* **1997**, *16*, 3401–3412.

PREPARATION OF THE COP CATALYSTS: [(*S*)-COP-OAc]$_2$, [(*S*)-COP-Cl]$_2$, AND (*S*)-COP-hfacac

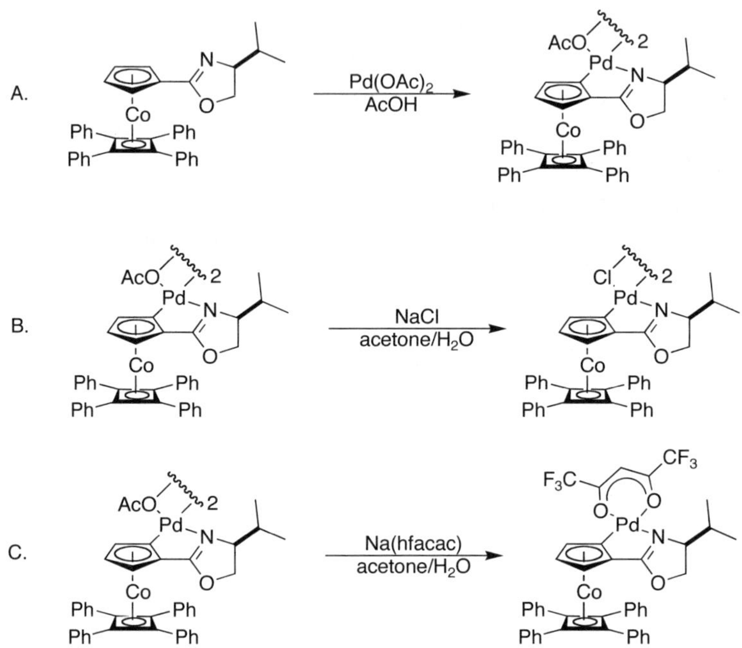

Submitted by Carolyn E. Anderson, Stefan F. Kirsch, Larry E. Overman,*[1] Christopher J. Richards and Mary P. Watson.
Checked by Brian S. Bodnar and Marvin J. Miller.

1. Procedure

A. Preparation of di-μ-acetatobis[(η^5-(S)-(pR)-2-(2'-(4'-methylethyl)oxazolinyl)cyclopentadienyl, 1-C, 3'-N)(η^4-tetraphenylcyclobutadiene)cobalt]dipalladium ([COP-OAc]$_2$). A flame-dried, single-necked, 250-mL round-bottomed flask is fitted with a stirring bar and an argon inlet. The flask is evacuated, refilled with argon and then, while temporarily removing the argon inlet, charged with (η^5-(*S*)-2-(4-methylethyl)oxazolinylcyclopentadienyl)-(η^4-tetraphenylcyclobutadiene)-cobalt (9.6 g, 16.2 mmol)[2] and 96 mL of glacial acetic acid. Palladium(II) acetate (Note 1) (3.6 g, 16.2 mmol) is then added. The red solution is then heated at 95 °C (oil bath temperature) and the formation of an orange precipitate is observed. After 30 min, the solution is cooled to room temperature and filtered to provide an orange solid. This solid is washed with 50 mL of glacial acetic acid and dried under vacuum to provide 8.9 g (5.9 mmol, 73% yield) of di-μ-acetatobis[(η^5-(*S*)-(*pR*)-2-(2'-(4'-methylethyl)oxazolinyl)cyclopentadienyl, 1-*C*, 3'-*N*)(η^4-tetraphenylcyclobutadiene)cobalt]dipalladium as a mustard-colored solid (Note 2).

B. Preparation of di-μ-chlorobis[(η^5-(S)-(pR)-2-(2'-(4'-methylethyl)oxazolinyl)cyclopentadienyl, 1-C, 3'-N)(η^4-tetraphenylcyclobutadiene)cobalt]dipalladium ([COP-Cl]$_2$). A single-necked, 250-mL round-bottomed flask is fitted with a stirring bar. The flask is charged with di-μ-acetatobis[(η^5-(*S*)-(*pR*)-2-(2'-(4'-methylethyl)oxazolinyl) cyclopentadienyl, 1-*C*, 3'-*N*) (η^4-tetraphenylcyclobutadiene)cobalt] dipalladium (8.9 g, 5.9 mmol) and 59 mL of acetone. Aqueous sodium chloride (2 M, 30 mL) is then added, and the resulting heterogeneous mixture is stirred at room temperature (Note 3). After 4 h, the yellow solid is collected by filtration of the reaction mixture. This solid is washed with 125 mL of water followed by 20 mL of acetone and then dried under vacuum to provide 8.2 g (5.6 mmol, 95% yield) of di-μ-chlorobis[(η^5-(*S*)-(*pR*)-2-(2'-(4'-methylethyl)oxazolinyl)cyclopentadienyl, 1-*C*, 3'-*N*)(η^4-tetraphenylcyclobutadiene)cobalt]dipalladium, [COP-Cl]$_2$, as a mustard-colored solid (Notes 4, 5, 6).

C. Preparation of hexafluoroacetylacetonate[(η^5-(S)-(pR)-2-(2'-(4'-methylethyl)oxazolinyl)cyclopentadienyl, 1-C, 3'-N)(η^4-tetraphenylcyclobutadiene)cobalt]palladium (COP-hfacac). A single-necked, 25-mL round-bottomed flask is equipped with a stir bar and charged with [COP-OAc]$_2$ (1.0 g, 0.66 mmol), sodium hexafluoroacetylacetonate (Note 7) (1.6 g, 6.8 mmol), 6.6 mL of acetone and 3.3 mL of water. The mixture is stirred vigorously at room temperature (Note 3). After 9 h, the mixture is filtered, and the filter cake is washed with 10–20 mL of water and then dried over P_2O_5 in a vacuum dessicator for 12 h to provide 1.1 g (1.2 mmol, 91%) of hexafluoroacetylacetonate[(η^5-(*S*)-(*pR*)-2-(2'-(4'-methylethyl)oxazolinyl)-cyclopentadienyl, 1-*C*, 3'-*N*)(η^4-tetraphenylcyclobutadiene)cobalt]-palladium, COP-hfacac, as an orange solid (Notes 8, 9).

2. Notes

1. Palladium acetate should be an orange/red solid. It can be obtained from either Strem Chemicals or Aldrich Chemical Co. In one instance, the submitters report that a commercial sample gave poor results; good results were obtained when this sample of palladium acetate was purified by crystallization from benzene.

2. Characterization data: ^{1}H NMR (500 MHz, $CDCl_3$) δ: –0.01 (d, *J* = 6.6 Hz, 3 H, CH_3), 0.46 (d, *J* = 7.1 Hz, 3 H, CH_3), 1.72–1.81 (m, 1 H, *CH*), 1.96 (s, 3 H, CH_3), 2.98 (td, *J* = 9.0, 3.1 Hz, 1 H, *CH*), 3.36 (t, *J* = 9.0 Hz, 1 H, CH_2), 4.08 (dd, *J* = 8.6, 3.9 Hz, 1 H, CH_2), 4.23 (t, *J* = 2.4 Hz, 1 H, *CH*), 4.62 (d, *J* = 1.4 Hz, 1 H, *CH*), 4.68 (d, *J* = 2.0 Hz, 1 H, *CH*), 7.20–7.29 (m, 12 H, Ar*H*), 7.64 (m, 8 H, Ar*H*); ^{13}C NMR (125 MHz, $CDCl_3$) δ: 13.3, 18.8, 24.2, 29.1, 65.0, 71.2, 76.2, 79.4, 85.0, 85.4, 86.8, 98.1, 126.2, 128.1, 129.5, 136.1, 170.8, 180.9; Infrared spectrum (thin film) cm^{-1}: 3061, 2961, 1583, 1502, 1417, 1366, 1181, 1069. Anal. Calcd. for $C_{82}H_{72}O_6N_2Co_2Pd_2$: C, 65.04; H, 4.93; N, 1.85. Found: C, 64.81; H, 4.74; N, 1.98.

3. An aliquot was removed and filtered. The solid was analyzed by ^{1}H NMR spectroscopy to determine when the reaction was complete, which typically required 4 h.

4. The checkers found that the sample could be purified by filtration through a short plug of silica gel using methylene chloride as the eluent to afford the product as an orange solid (99% yield before purification, 97% yield after purification). The product, di-μ-chlorobis[(η^5-(*S*)-(*pR*)-2-(2'-(4'-methylethyl)oxazolinyl)cyclopentadienyl, 1-*C*, 3'-*N*)(η^4-tetraphenylcyclobutadiene)cobalt]dipalladium, exists as a 1.0:0.7 mixture of dimers in $CDCl_3$. ^{1}H NMR (500 MHz, $CDCl_3$) δ: 0.70–0.74 (m, 6 H, CH_3), 0.76 (d, *J* = 6.8 Hz, 3 H, CH_3), 0.80 (d, *J* = 7.0 Hz, 3 H, CH_3), 2.20–2.33 (m, 2 H, CH), 3.03–3.15 (m, 2 H, CH), 3.34 (t, *J* = 9.0 Hz, 1 H, CH_2), 3.43 (t, *J* = 8.9 Hz, 1 H, CH_2), 4.16–4.22 (m, 2 H, CH_2), 4.28 (t, *J* = 2.5 Hz, 1 H, CH), 4.40 (t, *J* = 2.5 Hz, 1 H, CH), 4.70 (d, *J* = 2.0 Hz, 1 H, CH), 4.73 (d, *J* = 2.0 Hz, 1 H, CH), 4.98 (d, *J* = 1.5 Hz, 1 H, CH), 4.99 (d, *J* = 1.5 Hz, 1 H, CH), 7.16–7.22 (m, 12 H, ArH), 7.23–7.30 (m, 12 H, ArH), 7.58–7.62 (m, 8 H, ArH), 7.66–7.71 (m, 8 H, ArH); ^{13}C NMR (125 MHz, $CDCl_3$) δ: 14.3, 14.5, 19.0 (2C), 29.0, 29.2, 65.6, 66.0, 71.4, 71.5, 76.5, 76.8, 81.0, 84.2, 84.5, 84.8, 84.9, 85.7, 87.1, 98.6, 98.7, 126.4, 126.5, 128.2, 128.3, 129.4, 129.5, 135.5, 135.6, 171.1, 171.2; Infrared spectrum (thin film) cm^{-1}: 3061, 2961, 1602, 1502, 1370, 1185.

5. The enantiomeric purity of [(*S*)-COP-Cl]$_2$ is best determined after conversion to the acetylacetonate derivative. A mixture of [COP-Cl]$_2$ (15 mg, 0.010 mmol), sodium acetylacetonate (70 mg, 0.573 mmol), acetone (0.4 mL) and water (0.2 mL) is stirred vigorously at room temperature for 20 h. Methylene chloride (0.5 mL) and water (0.5 mL) are added, and the layers were separated. The aqueous layer is extracted with methylene chloride (0.5 mL x 3). The combined organic layers is dried ($MgSO_4$), filtered and concentrated to provide an orange residue. This residue is filtered through a short plug of silica gel (diameter = 0.25 cm, height = 4 cm) using 4% *i*-PrOH/hexanes as eluent. The enantiopurity of this sample was determined to be >99% by HPLC analysis (AD column, 95:5 hexane:*i*-PrOH, 230 nm); retention times were: 5.8 min for (*S*)-COP-acac and 3.8 min for (*R*)-COP-acac. The checkers found that a sample of (*S*)-COP-acac gave one signal at 5.1 min that was determined to be 97% pure by HPLC analysis (Diacel Chiralpak AD-H column (0.46 cm x 25 cm), 1 mg/mL sample concentration, flow rate of 1.0 mL/min, 95:5 hexanes: *i*-PrOH).

6. The enantiomer of [COP-OAc]$_2$ and [COP-Cl]$_2$ can be prepared in an analogous fashion from (η^5-(*R*)-2-(4-methylethyl)oxazolinylcyclopentadienyl)-(η^4-tetraphenylcyclobutadiene)cobalt prepared from (*R*)-valinol.[2]

7. Sodium hexafluoroacetylacetonate was purchased from Aldrich Chemical Company, Inc., and was used as received.

8. The checkers found that in one case it was necessary to stir the reaction for 11 h in order to achieve full conversion to the product. COP-hfacac prepared by this method was of sufficient purity to catalyze allylic imidate rearrangements in polar solvents such as tetrahydrofuran or methylene chloride. COP-hfacac can be purified further by filtration through a short plug of silica gel using methylene chloride as eluent. The checkers found that purification of a sample of (*S*)-COP-hfacac yielded an orange solid (89% yield after purification).

9. The product, hexafluoroacetylacetonate[(η^5-(*S*)-(*pR*)-2-(2'-(4'-methylethyl)oxazolinyl)cyclopentadienyl, 1-*C*, 3'-*N*)(η^4-tetraphenylcyclobutadiene)cobalt]palladium, displayed the following characterization data: ^{1}H NMR (500 MHz, $CDCl_3$) δ: 0.79 (d, J = 6.9 Hz, 3 H, CH_3) 0.83 (d, J = 7.0 Hz, 3 H, CH_3), 2.05–2.08 (m, 1 H, C*H*), 3.45 (td, J = 9.4, 5.1 Hz, 1 H, C*H*), 3.72 (t, J = 9.4 Hz, 1 H, CH_2), 4.33 (dd, J = 8.6, 5.3 Hz, 1 H, CH_2), 4.53 (t, J = 2.3 Hz, 1 H, C*H*), 4.90 (d, J = 2.3 Hz, 1 H, C*H*), 4.68 (d, J = 2.3 Hz, 1 H, C*H*), 5.95 (s, 1 H, C*H*), 7.19–7.29 (m, 12 H, Ar*H*), 7.54–7.56 (m, 8 H, Ar*H*); ^{13}C NMR (125 MHz, $CDCl_3$) δ: 14.9, 18.4, 29.3, 65.3, 72.3, 76.7,

78.9, 84.1, 84.8, 87.7, 90.1, 97.6, 116.5, 118.8, 126.5, 127.9, 128.9, 135.5, 173.3, 173.6, 174.1; Infrared spectrum (thin film) cm^{-1}: 3061, 2964, 1629, 1598, 1509, 1475, 1258, 1208, 1150.

Safety and Waste Disposal Information

All hazardous materials should be handled and disposed of in accordance with "Prudent Practices in the Laboratory"; National Academy Press; Washington, DC, 1995.

3. Discussion

This synthesis of (*S*)-COP-OAc dimer, is a modification of the procedure originally reported by Richards and Stevens.[3] The procedure has been streamlined and modified to enable large-scale preparation of this and related COP catalysts. Using the procedure detailed here and in the accompanying procedure,[2] 8.9 g $[(S)\text{-COP-OAc}]_2$ or 8.2 g of $[(S)\text{-COP-Cl}]_2$ can be made in 40% overall yield from sodium cyclopentadienide with only one chromatographic purification being required.

Palladation of (η^5-(*S*)-2-(4-methylethyl)oxazolinylcyclopentadienyl)-(η^4-tetraphenylcyclobutadiene)cobalt occurs in a diastereoselective fashion because of the large steric floor provided by the tetraphenylcyclobutadiene moiety. This floor forces the oxazoline into a single diastereomeric conformation, allowing *ortho*-palladation to proceed on only one of the diastereotopic positions of the cyclopentadienyl ring.[3]

$[\text{COP-OAc}]_2$ is a planar, chiral palladium(II) complex, which catalyzes the reaction of carboxylic acids with prochiral primary (*Z*)-allylic trichloroacetimidates to provide enantioenriched 3-acyloxy-1-alkenes.[4] These reactions proceed with predictable high stereoinduction and are accomplished at room temperature using low catalyst loadings. This acetate-bridged dimer serves also as a convenient precursor of enantiopure $[\text{COP-Cl}]_2$ and the enantiopure monomeric hexafluoroacetonylacetonate complex, COP-hfacac.

$[\text{COP-Cl}]_2$ is a planar chiral palladium(II) complex, which catalyzes the [3,3]-sigmatropic rearrangement of (*E*)-allylic trichloacetimidates to the corresponding trichloacetamides in high yields and enantioselectivities for a wide range of substituted trihaloacetimidates.[5] The use of $[\text{COP-Cl}]_2$ for the catalytic asymmetric rearrangement of allylic trihaloacetimidates is described in an *Organic Syntheses* procedure.[6] The corresponding hexafluoroacetylacetonate complex, COP-hfacac, also catalyzes asymmetric allylic trichloroacetimidate rearrangements.[7] It is more soluble than [COP-Cl]$_2$ and can be used in a wider variety of solvents. This enhanced solubility allows catalytic asymmetric allylic trichloroacetimidate rearrangements to be conducted at high substrate concentrations using as little as 1 mol % of COP-hfacac.

1. Department of Chemistry, University of California, Irvine, California, 92697-2025; E-mail: leoverma@uci.edu
2. Anderson, C. E.; Kirsch, S. F.; Overman, L. E.; Richards, C. J.; and Watson, M. P. *Org. Synth.* **2007**, *84*, 139-147.
3. Stevens, A. M.; Richards, C. J. *Organometallics* **1999**, *18*, 1346–1348.
4. Kirsch, S. F.; Overman, L. E. *J. Am. Chem.* Soc. **2005**, *127*, 2866–2867.
5. Anderson, C. E.; Overman, L. E. *J. Am. Chem. Soc.* **2003**, *125*, 12412–12413.
6. Anderson, C. E.; Overman, L. E.; Watson, M. P. *Org. Synth.* **2005**, *82*, 134–139.
7. Kirsch, S. F.; Overman, L. E.; Watson, M. P. *J. Org. Chem.* **2004**, 69, 8101–8104.

ASYMMETRIC REARRANGEMENT OF ALLYLIC TRICHLOROACETIMIDATES: PREPARATION OF (*S*)-2,2,2-TRICHLORO-*N*-(1-PROPYLALLYL)ACETAMIDE (Acetamide, 2,2,2-trichloro-*N*-[(1*S*)-1-ethenylbutyl]-)

A.

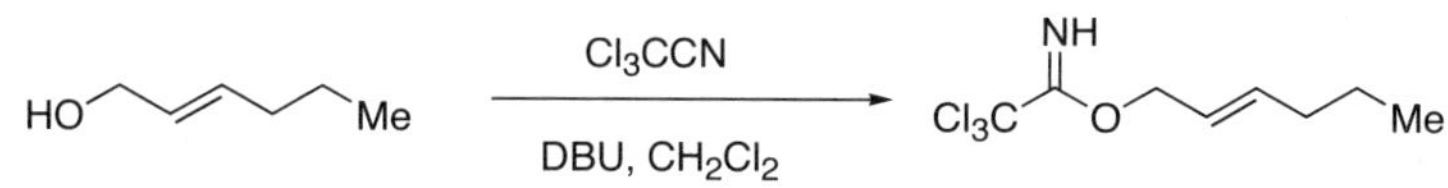

B.

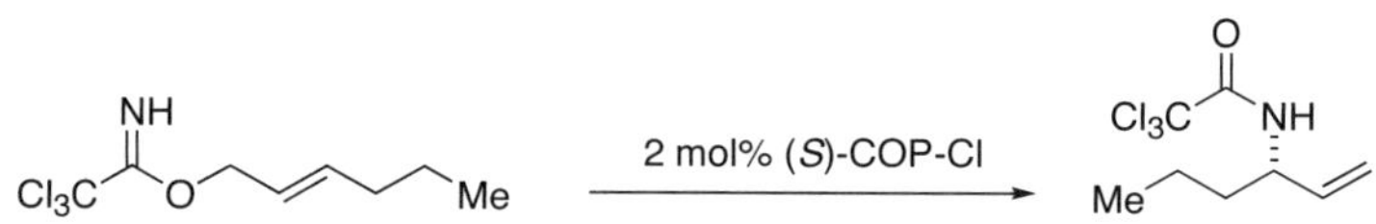

Submitted by Carolyn E. Anderson, Larry E. Overman* and Mary P. Watson.
Checked by Matthew L. Maddess and Mark Lautens.

1. Procedure

Caution! Part A should be carried out in a well-ventilated hood to avoid exposure to trichloroacetonitrile vapors.

A. Preparation of (E)-2,2,2-trichloroacetimidic acid hex-2-enyl ester. A 500 mL round-bottomed flask equipped with a stirring bar is flame dried under a stream of nitrogen and allowed to cool to room temperature. The flask is then charged with *trans*-2-hexen-1-ol (Note 1) (3.3 mL, 27.8 mmol), 1,8-diazabicyclo[5.4.0]undec-7-ene (DBU) (Note 2) (0.84 mL, 5.6 mmol) and 170 mL of methylene chloride (Note 3). The solution is cooled to 4 °C using an ice/water bath before trichloroacetonitrile (Note 2) (4.2 mL, 42 mmol) is added over five minutes by syringe. With time, the reaction solution is observed to change from clear to orange in color, and within 1 hour the starting materials are consumed (Note 4). The stir bar is removed with a magnetic rod and the solution is then concentrated under reduced pressure using a rotary evaporator to give a brown oil (Note 5). This oil is purified by flash chromatography through a plug of silica gel (Silicycle® 230-400 mesh, 6 cm tall, 5 cm diameter) using 2% ethyl acetate in hexanes (600 mL, 60 mL fractions) to yield 6.48 g (95%) of nearly pure (*E*)-2,2,2-trichloroacetimidic acid hex-2-enyl ester as a colorless oil (Notes 6, 7).

B. (S)-COP-Cl catalyzed rearrangement of (E)-2,2,2-trichloroacetimidic acid hex-2-enyl ester to (S)-2,2,2-trichloro-N-(1-propylallyl)acetamide. A 150-mL, round-bottomed flask is fitted with a stirring bar and then charged with (*E*)-2,2,2-trichloroacetimidic acid hex-2-enyl ester (6.81 g, 27.8 mmol), Di-μ-chlorobis[η^5-(*S*)-(p*R*)-2-(2'-(4'-isopropyl)oxazolinylcycloentadienyl, 1-*C*, 3'-*N*))-(η^4-tetraphenylcyclobutadiene)cobalt]dipalladium [(*S*)-COP-Cl] (Note 8) (816 mg, 0.56 mmol) and 9.3 mL of methylene chloride (Note 3). The flask is sealed with a polyethylene cap, the cap is secured to the flask with Parafilm, and the flask is placed in an oil bath preheated to 38 °C +/- 2 °C. After 24 h, the solution is cooled to room temperature, the stir bar is removed using a magnetic rod, and the solution is concentrated using a rotary evaporator to yield a brown oil. This oil is purified by flash chromatography through a column of silica gel (Silicycle® 230-400 mesh, 20 cm tall, 5 cm diameter) using 0.5% to 2% ethyl acetate:hexanes as eluent (3 L 0.5% ethyl acetate:hexanes, 1 L 1% ethyl acetate:hexanes, 1 L 2% ethyl acetate: hexanes). Evaporation of solvent provides 6.61 g (97% yield) of (*S*)-2,2,2-trichloro-*N*-(1-propylallyl)acetamide, 94% ee, as a pale yellow oil (Notes 9, 10, 11, 12).

2. Notes

1. The checkers used *trans*-2-hexen-1-ol purchased from Aldrich Chemical Company, Inc. Although of sufficient purity for this series of transformations, the commercial reagent is contaminated with 3% of 1-hexanol.[1] The submitters used (*E*)-2-Hexen-1-ol (>99% *E*) prepared from butanal by Horner-Wadsworth-Emmons reaction with trimethyl phosphonoacetate to form hex-2-enoic acid methyl ester, followed by reduction of this product with diisobutylaluminum hydride (–78 °C to room temperature in THF).

2. 1,8-Diazabicyclo[5.4.0]undec-7-ene and trichloroacetonitrile were purchased from Aldrich Chemical Company, Inc. These chemicals were used as received.

3. Methylene chloride was purified by passage through a solvent purification system. The submitters used a GlassContour alumina solvent purification columns and the checkers used a MBRAUN® solvent purification system.[2]

4. The reaction progress can be analyzed by thin layer chromatography (Silicycle®, plastic backed, 250 μm thickness). Using 10% ethyl acetate:hexanes as eluent, the product (*E*)-2,2,2-trichloroacetimidic

acid hex-2-enyl ester has an R_f of 0.45, whereas the starting alcohol has an R_f of 0.12. Both the imidate and starting alcohol can be visualized by potassium permanganate stain.

5. Upon concentrating the solution, black semi-solids sometimes form. These unwanted byproducts are insoluble in 2% ethyl acetate:hexanes.

6. The product, (*E*)-2,2,2-trichloroacetimidic acid hex-2-enyl ester, exhibits a spectrum that matches that which is reported in the literature.[3] ^{1}H NMR (300 MHz, $CDCl_3$) δ: 8.27 (broad s, 1H, N*H*), 5.86 (dt, *J* = 15.6, 6.3 Hz, 1H, C*H*), 5.68 (dt, *J* = 15.6 Hz, 6.9 Hz, 1H, C*H*), 4.74 (d, *J* = 6.3 Hz, 2H, CH_2), 2.06 (tq, *J* = 6.9, 6.9 Hz, 2H, CH_2), 1.43 (tq, *J* = 6.9, 7.2 Hz, 2H, CH_2), 0.91 (t, *J* = 7.2 Hz, 3H, CH_3). ^{13}C NMR (75 MHz, $CDCl_3$) δ 162.6, 136.9, 123.3, 91.7, 70.0, 34.5, 22.1, 13.7.

7. The product is contaminated with 2,2,2-trichloroacetimidic acid hexyl ester (3%) arising from the reaction of hexyl alcohol with trichloroacetonitrile.

8. (*S*)-COP-Cl was purchased from Aldrich Chemical Company, Inc. and was used as received. The specific rotation of this catalyst was measured to be: $[\alpha]_D^{23.8}$ = +1169 (c 0.25, $CHCl_3$). A detailed procedure for the synthesis of (*S*)-COP-Cl has been published: Anderson, C.E.; Kirsch, S.F.; Overman, L.E., Richards, C.J.; Watson, M.P. *Org. Synth.* **2007**, *84*, 148-155.

9. The enantiomeric ratio was determined by HPLC analysis by comparison to a racemic sample:[3a] Hewlett-Packard HP Series 1100, Chiracel OD with guard column, 99:1 hexanes:isopropyl alcohol, 0.4 mL/min, 30 °C, R_t (major) = 15.3 min, R_t (minor) = 16.4 min.

10. ^{1}H NMR (400 MHz, $CDCl_3$) δ: 6.60 (broad s, 1H, N*H*), 5.81 (ddd, *J* = 16.4, 10.4, 5.6 Hz, 1H, C*H*), 5.17–5.27 (m, 2H, CH_2), 4.39–4.47 (m, 1H, C*H*), 1.55–1.71 (m, 2H, CH_2), 1.35–1.46 (m, 2H, CH_2), 0.96 (t, *J* = 7.2 Hz, 3H, CH_3); ^{13}C NMR (100 MHz, $CDCl_3$) d 161.2, 136.7, 115.9, 92.8, 53.4, 36.5, 18.9, 13.7; IR (neat) cm^{-1}: 3424, 3322, 2961, 2935, 1714, 1520, 1249, 926, 821; HRMS (EI) *m/z* 244.0063 [244.0063 calcd for $C_8H_{13}Cl_3NO$ (M+H)]; R_f = 0.36 (10% ethyl acetate:hexanes).

11. The submitters reported, that when commercially available starting material is used, the product is contaminated with observable quantities of 2,2,2-trichloroacetimidic acid hexyl ester. The checkers, however, report that 2,2,2-trichloroacetimidic acid hexyl ester is not observed to the detection limits of ^{1}H or ^{13}C NMR.

12. On half-scale to that described, with prolonged drying under high vacuum the product was observed to solidify to a white crystalline solid (mp = 28 - 29 °C). On the scale described above, the product remained an oil but could be rapidly induced to crystallize by addition of a seed crystal.

Safety and Waste Disposal Information

All hazardous materials should be handled and disposed of in accordance with "Prudent Practices in the Laboratory"; National Academy Press; Washington, DC, 1995.

3. Discussion

This procedure illustrates a general method for the preparation of enantioenriched chiral allylic trichloroacetamides from readily available prochiral (*E*)-allylic alcohols by catalytic asymmetric rearrangement of trichloroacetimidate intermediates.[4] The rearrangement tolerates a variety of alkyl and Lewis basic substituents at C3 of the starting allylic alcohol; however, substitution at C2 is not permitted (Table 1).[4,5] Although 5 mol % COP-Cl is convenient to use for small scale reactions, the catalyst loading

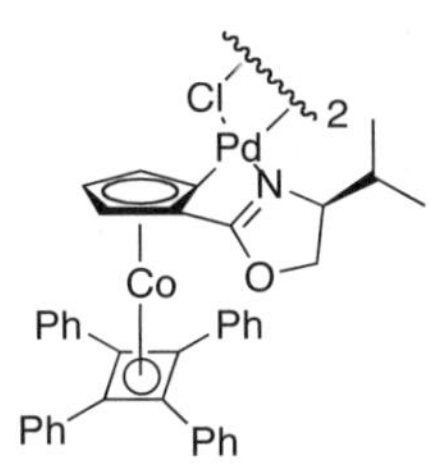

(*S*)-COP-Cl

can be decreased to as low as 1 mol % if the concentration is simultaneously increased (Entry 3). The reaction conditions exemplified in this example (2 mol % COP-Cl, CH_2Cl_2 (3 M), 38 °C, 24 h) were chosen to insure complete conversion of the allylic imidate intermediate within 24 hours. (*Z*)-Allylic trichloroacetimidates do not undergo COP-Cl catalyzed rearrangement at a practical rate. However, allylic trichloroacetamides of opposite absolute configuration can be prepared using (*R*)-COP-Cl, synthesized from (*R*)-valinol.

The COP-Cl catalyzed transformation of prochiral allylic alcohols to chiral allylic trichloroacetamides is technically simple as allylic trichloroacetimidate intermediates require minimal purification.

Additionally, no special precautions are required to protect the rearrangement reaction from light, air or traces of moisture. Moreover, the trichloroacetyl group of the product amide can be readily cleaved or this functional group can be converted to other functional arrays.[6]

Table 1. Enantioselective Synthesis of Allylic Trichloroacetimidates from (*E*)-Allylic Trichloroacetimidates.[a]

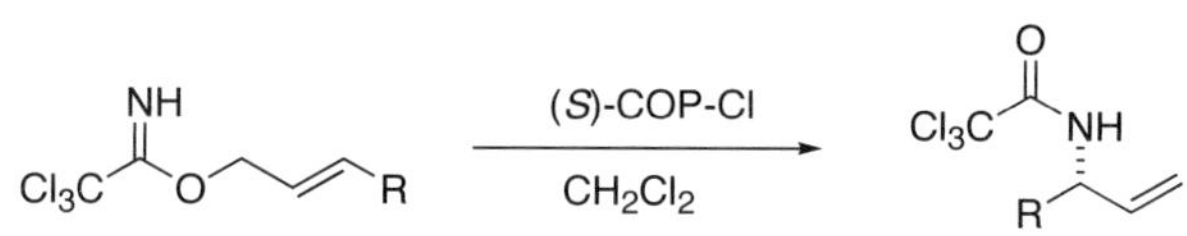

entry	R	amide yield (%)[b]	amide % ee[c]/conf
1	*n*-Pr	99	95/*S*
2	*i*-Bu	95	96/*S*
3[d]	*i*-Bu	92	98/*S*
4	CH_2CH_2Ph	93	93/*S*
5	$(CH_2)_3OAc$	97	92/*S*
6	$(CH_2)_2COMe$	98	95[e]/*S*
7	$CH_2OTBDMS$	98	96/*S*
8	$(CH_2)_3NBn(Boc)$	96	95/*S*

(a) Conditions: 5 mol % (*S*)-COP-Cl, CH_2Cl_2 (0.6 M), 38 °C, 18 h. (b) Duplicate experiments (±3%). (c) Determined by HPLC analysis of duplicate experiments (±2%). (d) 1 mol % (*S*)-COP-Cl, CH_2Cl_2 (1.2 M). (e) Determined by chiral GC analysis of duplicate experiments (±2%).

1. Hill, J. G.; Sharpless, K. B.; Exon, C. M.; Regenye, R. *Org. Syn., Coll. Vol. 7*, 461.
2. (a) Pangborn, A. B.; Giardello, M. A.; Grubbs, R. H.; Rosen, R. K.; Timmers, F. J. *Organometallics* **1996**, *15*, 1518–1520. (b) http://www.glasscontour.com/ or http://www.mbraunusa.com/.
3. (a) Overman, L. E. *J. Am. Chem. Soc.* **1976**, *98*, 2901–2910. (b) Bongini, A.; Cardillo, G.; Orena, M.; Sandri, S.; Tomasini, C. *J. Org. Chem.* **1986**, *51*, 4905–4910.
4. Anderson, C. E.; Overman, L. E. *J. Am. Chem. Soc.* **2003**, *125*, 12412–12413.
5. (a) Overman, L. E. *Acc. Chem. Res.* **1980**, *13*, 218–224. (b) Overman, L. E. *Angew. Chem., Int. Ed. Engl.* **1984**, *23*, 579–586.

6. For examples, see: (a) Cardillo, G.; Orena, M.; Sandri, S. *J. Chem. Soc., Chem. Commun.* **1983**, 1489-1490. (b) Nagashima, H.; Wakamatsu, H.; Itoh, K. *J. Chem. Soc., Chem. Commun.* **1984**, 652-653. (c) Atanassova, I. A.; Petrov, J. S.; Mollov, N. M. *Synthesis* **1987**, 734–736. (d) Yamamoto, N.; Isobe, M. *Chem. Lett.* **1994**, 2299–2302. (e) Urabe, D.; Sugino, K.; Nishikawa, T.; Isobe, M. *Tetrahedron Lett.* **2004**, *45*, 9405-9407.

PRACTICAL SYNTHESIS OF NOVEL CHIRAL ALLENAMIDES: (*R*)-4-PHENYL-3-(1,2-PROPADIENYL)OXAZOLIDIN-2-ONE (2-Oxazolidinone, 4-phenyl-3-(1,2-propadienyl)–, (4*R*)–)

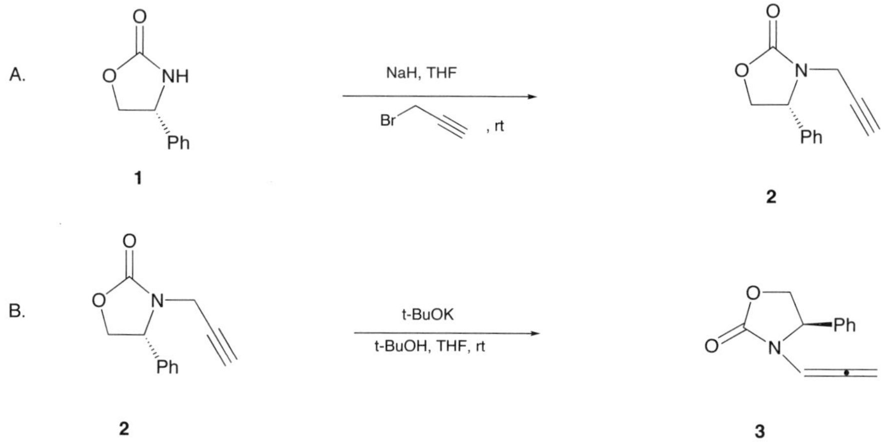

Submitted by H. Xiong, M. R. Tracey, T. Grebe, J. A. Mulder, and R. P. Hsung.[1]
Checked by Peter Wipf and Jennifer Smotryski.

1. Procedure

A. (R)-4-Phenyl-3-(2-propynyl)oxazolidin-2-one (2). A flame-dried, 500-mL, round-bottomed flask equipped with a magnetic stirbar and a rubber septum fitted with a nitrogen inlet is charged with 9.90 g (61 mmol) of (*R*)-4-phenyl-2-oxazolidinone (Note 1) and 200 mL of anhydrous tetrahydrofuran (THF) under a nitrogen atmosphere (Note 2). Sodium hydride (NaH) (2.90 g, 60% w/w in mineral oil, 1.20 equiv, 73 mmol) is added in small portions (Note 3), and the resulting white slush is stirred for 1 h at room temperature before carefully adding 8.00 mL of a solution of propargyl bromide in toluene (80% w/w in toluene, 72 mmol, 1.18 equiv) (Note 4) dropwise over ca. 10 min via syringe. Precipitation of sodium bromide is observed and does not affect the progress of the reaction. The reaction mixture is stirred at room temperature for 24 h and then concentrated by rotary evaporation under reduced pressure. The residue is dissolved in 100 mL of anhydrous ether and filtered through a small bed of Celite washing with 1:1 ethyl acetate-hexane. The filtrate is concentrated by rotary evaporation, and the resulting residue is purified using silica gel column chromatography (gradient eluent with 0-33% ethyl acetate-hexane) to give 5.71-6.09 g (47-50%) of the desired propargyl amide **2** as a lightly yellow-colored oil (Note 5).

B. (R)-4-Phenyl-3-(1,2–propadienyl)-2-oxazolidinone (3). A flame-dried, 500-mL, round-bottomed flask equipped with a magnetic stirbar and a rubber septum fitted with a nitrogen inlet is charged with a solution of 6.67 g (33 mmol) of propargyl amide **2** in 300 mL of anhydrous THF (Note 2). Potassium *tert*-butoxide (1.26 g, 11 mmol, 0.33 equiv) (Notes 6, 7) is added in portions to the reaction mixture over ca. 10 min. The reaction mixture is stirred at room temperature for 24 h and the progress of the reaction is monitored by TLC (elution with 50% ethyl acetate-hexane) and ^{1}H NMR analysis. Upon completion of the reaction, the solvent is removed by rotary evaporation under reduced pressure. The resulting crude residue is dissolved in 50 mL of ethyl acetate and vacuum filtered through a small bed of silica gel. The solids are washed with two 40-mL portions of 25-50% ethyl acetate-hexane, and the filtrate is then concentrated by rotary evaporation. The residue is purified using silica gel column chromatography (gradient elution with 0% to 25% ethyl acetate-hexanes) to give 1.41-1.59 g (38-41%) of the desired allenamide **3** as a yellow brownish-red oil (Note 8).

2. Notes

1. (*R*)-4-Phenyl-2-oxazolidinone was purchased from Sigma-Aldrich or Urquima, S.A. Arnau de Vilanove, Barcelona, Spain and used as received.

2. Anhydrous THF was freshly distilled from sodium/ benzophenone under nitrogen.

3. *Caution: evolution of large amount of H_2 gas.* NaH was obtained from Sigma-Aldrich and used as received. NaH free of oil can also be used with no difference in yield and was prepared via four cycles of washing and decanting (via pipette) with anhydrous pentane.

4. Propargyl bromide was obtained as an 80% w/w solution in toluene from Sigma-Aldrich.

5. The submitters report obtaining the product in 66% yield prior to purification at 0.050 mole-scale and suggest that the purification step is not necessary because in most cases simple filtration through a small bed of Celite or silica gel provided the desired propargyl amide with high purity as

determined by GC analysis. Specifically, GC analysis of propargyl amide **2** shows its purify to be 98.2% prior to column chromatography, and 98.5% after chromatography. After column chromatography, the submitters obtained 6.40 g (52%) of the desired amide. The amide displays the following physical properties: R_f = 0.49 (50% EtOAc in hexane); $[\alpha]^{20}_{D}$ 148.8 (c 10.2, EtOH); ^{1}H NMR (300 MHz, $CDCl_3$) δ: 2.25 (t, 1H, *J* = 2.4 Hz), 3.35 (dd, 1H, *J* = 2.4, 17.7), 4.12 (dd, 1H, *J* = 7.8, 8.7), 4.36 (dd, 1H, *J* = 2.4, 17.7), 4.65 (t, 1H, *J* = 8.7), 4.95 (dd, 1H, *J* = 7.8, 8.7), 7.32-7.44 (m, 5H); ^{13}C NMR (75 MHz, $CDCl_3$) δ: 32.0, 59.0, 70.0, 73.4, 76.6, 127.3, 129.3, 129.4, 136.6, 157.8; IR (neat) cm^{-1}: 3285, 2913, 2100, 1733, 1494, 1177, 860; mass spectrum (EI): *m/e* (%relative intensity) 201 (20) M^+, 156 (51), 143 (24), 124 (30), 116 (66), 104 (100), 91 (22), 77 (21); *m/e* calcd for $C_{12}H_{12}NO_2$: 202.086; found 202.0871.

6. Potassium *tert*-butoxide was obtained from Sigma-Aldrich and used as received. Alternatively, equivalent results were obtained using *t*-BuOK that was prepared from potassium metal and anhydrous *tert*-butyl alcohol (*t*-BuOH) followed by removal of excess *t*-BuOH. In this case, the molecular weight of *t*-BuOK was calculated based on a 1:1 ratio of *t*-BuOK to *t*-BuOH (i.e., 186.34 for $C_8H_{19}O_2K$).

7. The submitters report that equivalent results were obtained using between 0.20-0.35 equiv of *t*-BuOK.

8. The submitters report obtaining the allenamide in 63% yield on this scale with its purity determined by GC analysis to be between 94%-96% in different runs. The lower yield obtained by the checkers may be explained by partial decomposition of product during the chromatographic purification on standard (40-60 nm) silica gel. The physical properties of 3 are as follows: R_f = 0.59 (50% ethyl acetate-hexane); $[\alpha]^{20}_{D}$ –156.4 (c 0.225, $CHCl_3$); ^{1}H NMR (500 MHz, $CDCl_3$)) δ: 4.14 (dd, 1H, *J* = 5.9, 9.8), 4.69 (t, 1H, *J* = 9.0), 4.87 (dd, 1H, *J* = 6.4, 9.5), 4.88 (dd, 1H, *J* = 6.4, 9.5), 5.16 (dd, 1H, *J* = 5.9, 9.8), 6.79 (t, 1H, *J* = 6.4), 7.23 – 7.37 (m, 5H); ^{13}C NMR (75 MHz, $CDCl_3$) δ: 59.0, 70.6, 87.7, 95.6, 126.5, 128.7. 129.0, 138.4, 155.5, 201.9; IR (neat) cm^{-1} 3063, 3035, 2979, 1963, 1767, 1494, 1462, 1216, 966, 911, 881; mass spectrum (EI): *m/e* (%relative intensity) 201 (17) M^+, 156 (100), 129 (17), 115 (20), 104 (54), 91 (17), 77 (17); *m/e* calcd for $C_{12}H_{11}NO_2$ 201.0790, found 201.0784.

Waste Disposal Information

All toxic materials were disposed of in accordance with "Prudent Praactices in the Laboratory"; National Academy Press; Washington, DC, 1995.

3. Discussion

A. Preparation. The level of purity during preparations of allenamides was unambiguously established using GC analysis for both the propargyl amide intermediates and the allenamides. The level of optical purity was established based on optical purity of the chiral auxiliaries from commercial sources. The $[\alpha]_D^{20}$ values for the commercial (*R*)-2-phenyloxazolidinone and (*S*)-2-phenyloxazolidinone that were used for this preparation are –52.33 (c 2.0, $CHCl_3$) and +52.95 (2.0, $CHCl_3$), respectively. Based on the known $[\alpha]_D^{20}$ values for (*R*)-2-phenyloxazolidinone and (*S*)-2-phenyloxazolidinone from Aldrich (-48.0 (c 2.0, $CHCl_3$) and + 48.0 (c 2.0, $CHCl_3$), respectively), the *ee* or optical integrity of these two enantiomeric auxiliaries should be very high. Since it is not likely to severely erode the *ee* of the auxiliary under the rather mild reaction conditions described in these two procedures, the level of *ee* for propargyl amide (*R*)-2 and allenamide (*R*)-3 should be comparable to that of the starting (*R*)-2-phenyloxazolidinone.

Furthermore, using (*S*)-2-phenyloxazolidinone led to the propargyl amide (*S*)-**2a** that has an opposite optical rotation of + 157.97 (c 10.0, EtOH). After based-induced isomerization, the allenamide (*S*)-**3a** also possesses an opposite optical rotation of + 157.13 (c 10.0, CH_2Cl_2).

B. Applications. Stereoselective inverse-demand hetero (4 + 2) cycloadditions. A Chiral Template for *C*-Aryl Glycoside Synthesis. Chiral allenamides[2,3,4] had been used in highly stereoselective inverse-demand hetero (4 + 2) cycloaddition reactions with heterodienes.[5] These reactions lead to stereoselective synthesis of highly functionalized pyranyl heterocycles. Further elaboration of these cycloadducts provides a unique entry to C-aryl-glycosides and pyranyl structures that are common in other natural products (**Scheme 1**).

Scheme 1

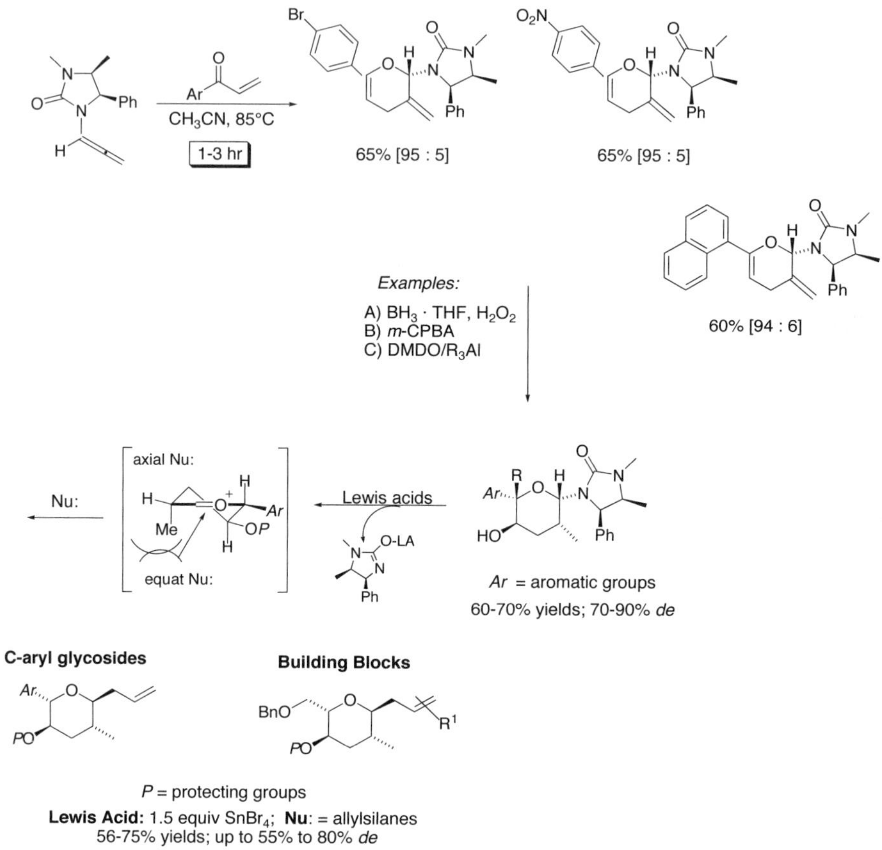

2. *Highly stereoselective [4 + 3] oxyallyl cycloadditions.* An *endo*-Selective Sequential Epoxidation-Oxyallyl Cycloaddition and the First Nitrogen-Stabilized Oxyallyl Cations.

Epoxidations of chiral allenamides lead to chiral nitrogen-stabilized oxyallyl catioins that undergo highly stereoselective (4 + 3) cycloaddition reactions with electron-rich dienes.[6] These are the first examples of epoxidations of allenes, and the first examples of chiral nitrogen-stabilized oxyallyl cations. Further elaboration of the cycloadducts leads to interesting chiral amino alcohols that can be useful as ligands in asymmetric catalysis (**Scheme 2**).

Scheme 2

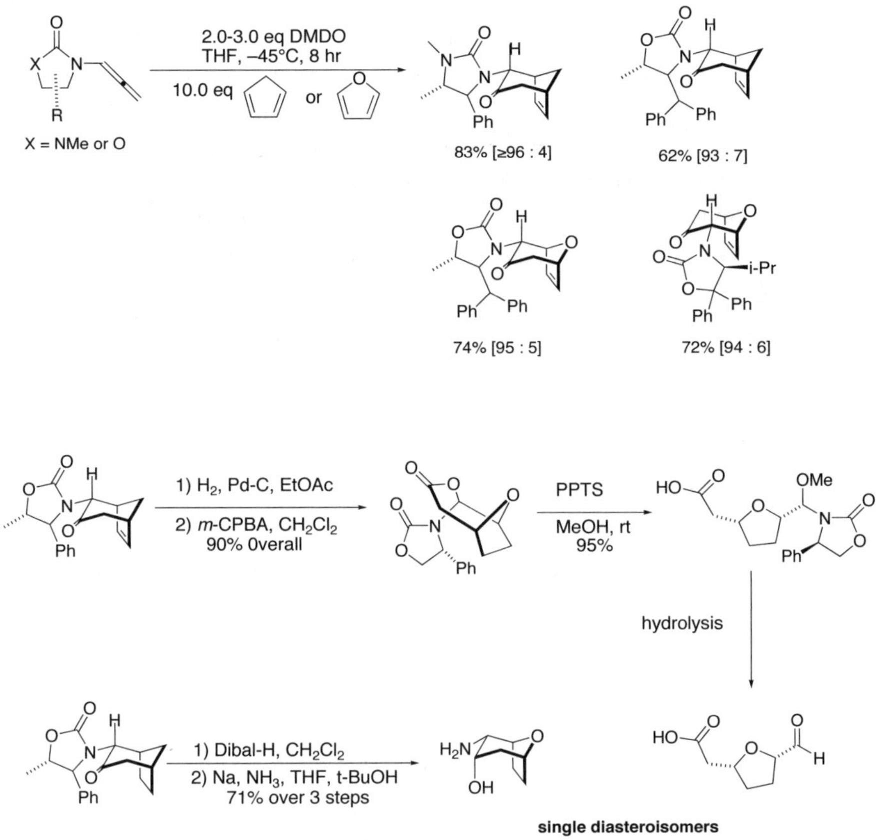

1. School of Pharmacy, University of Wisconsin, 777 Highland Ave., Madison, Wisconsin 53705-2222 USA.

2. (a) Saalfrank, R. W.; Lurz, C. J., in *Methoden Der Organischen Chemie* (*Houben-Weyl*), Kropf, H. and Schaumann, E., Eds. Georg thieme Verlag: Stuttgart, **1993**, 3093. (b) Schuster, H. E.; Coppola, G. M. *Allenes in Organic Synthesis*, John Wiley and Sons: New York, 1984.

3. First preparations of allenamides or electron deficient allenamines: (a) Dickinson, W. B.; Lang, P. C. *Tetrahedron Lett.* **1967**, *8*, 3035. (b) Overman, L. E.; Marlowe, C. K.; Clizbe, L. A. *Tetrahedron Lett.* **1979**, *20*, 599.

4. (a) Gardiner, M.; Grigg, R.; Sridharan, V.; Vicker, N. *Tetrahedron Lett.* **1998**, *39*, 435, and references cited therein. (b) Noguchi, M.; Okada, H.; Watanabe, M.; Okuda, K.; Nakamura, O. *Tetrahedron* **1996**, *52*, 6581. (c) Farina, V.; Kant, J.

Tetrahedron Lett. **1992**, *33*, 3559 and 3563. (d) Kimura, M.; Horino, Y.; Wakamiya, Y.; Okajima, T.; Tamaru, Y. *J. Am. Chem. Soc.* **1997**, *119*, 10869.

5. (a) Wei, L.-L.; Xiong, H.; Douglas, C. J.; Hsung, R. P. *Tetrahedron Lett.* **1999**, *40*, 6903. (b) Wei, L.-L.; Hsung, R. P.; Xiong, H.; Mulder, J. A.; Nkansah, N. T. *Organic Lett.* **1999**, *1*, 2145. (c) Xiong, H.; Hsung, R. P.; Wei, L.-L.; Berry, C. R.; Mulder, J. A.; Stockwell, B. *Organic Lett.* **2000**, *2*, 2869.

6. (a) Xiong, H.; Hsung, R. P.; Berry, C. R.; Rameshkumar, C. *J. Am. Chem. Soc.* **2001**, *123*, 7174. (b) Rameshkumar, C.; Xiong, H.; Tracey, M. R.; Berry, C. R.; Yao, L. J.; Hsung, R. P. *J. Org. Chem.* **2002**, *67*, 1339.

BORIC ACID CATALYZED AMIDE FORMATION FROM CARBOXYLIC ACIDS AND AMINES: *N*-BENZYL-4-PHENYLBUTYRAMIDE (Benzenebutanamide, *N*-(phenylmethyl)-)

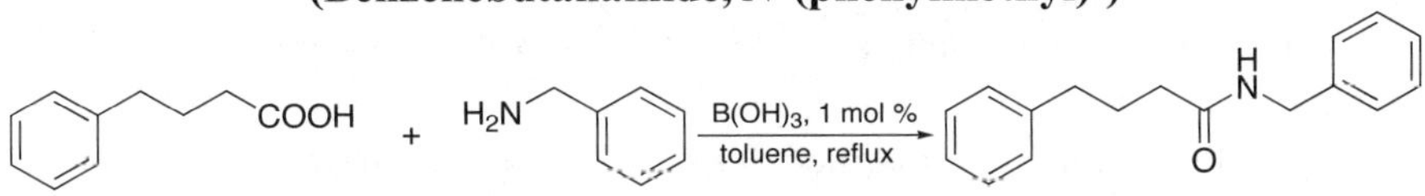

Submitted by Pingwah Tang.[1]
Checked by Helga Krause and Alois Fürstner.

1. Procedure

N-Benzyl-4-phenylbutyramide. A flame-dried, 200-mL, three-necked, round-bottomed flask is equipped with two glass stoppers, a vacuum-jacketed Dean-Stark trap topped with a reflux condenser fitted with a nitrogen inlet, and a Teflon-coated magnetic stirring bar. The reaction vessel is charged with 4-phenylbutyric acid (4.92 g, 30 mmol) (Note 1), boric acid (0.020 g, 0.30 mmol) (Note 2), and 88 mL of toluene (Note 3). To the stirred colorless reaction mixture is added benzylamine (3.32 g, 31 mmol) (Note 4) in one portion. The reaction mixture is heated at reflux for 16 h and ca. 0.6 mL of water is collected in the Dean-Stark trap (Notes 5, 6, 7, 8). The mixture is allowed to cool to ambient temperature and then is poured with stirring into 500 mL of hexanes leading to the immediate precipitation of a white solid. Stirring is continued for an additional 30 min and then the precipitate is filtered off with suction through a sintered glass filter funnel. The collected solid is successively washed with two 60-mL portions of hexanes and two 60-mL portions of distilled water (Note 9) and then is dried in vacuo at room temperature for 12 h to afford 6.90 g (91%) of *N*-benzyl-4-phenylbutyramide as a white solid (Notes 10, 11).

2. Notes

1. 4-Phenylbutyric acid (99%) was purchased from Aldrich Chemical Company and used as received.

2. Boric acid (99.999%) was purchased from Aldrich Chemical Company and used as received.

3. Toluene (99.8% anhydrous, water <0.001%, evaporation residue <0.005%) was purchased from Aldrich Chemical Company and used as

received. The capacity of the receiver in the Dean-Stark trap was 28 mL. The initial amount of toluene placed in the flask was 88 mL. Upon heating to reflux, 28 mL of toluene was distilled from the flask and collected in the receiver. The remaining volume of toluene in the reaction flask was ca. 60 mL, corresponding to an approximately 0.50M concentration of the reactants.

4. Benzylamine (redistilled 99.5+ %) was purchased from Aldrich Chemical Company and used without further purification.

5. The theoretical amount of water is 0.54 mL.

6. TLC analysis indicated that the reaction was complete. TLC was performed on pre-coated plates with silica gel 60 F_{254} purchased from EM Sciences (Merck) using hexanes/ethyl acetate (1/1) as the eluent.

7. HPLC analysis indicated that the reaction was complete. HPLC was conducted utilizing Perkin-Elmer's Series 200 autosampler and pump with either a Perkin-Elmer Series 200 or Applied Biosystems 785A UV/Vis detector (both monitoring at λ = 220 nm). Phenomenex's Kromasil 5 micron C18 column (50 mm x 4.6 mm) fitted with Phenomenex's C18 (ODS, Octadecyl) security guard cartridges (4 mm x 3 mm) were utilized with a 3 mL/min flow-rate. The solvent system consisted of mixtures of acetonitrile and water, containing 0.1% (v/v) trifluoroacetic acid, with the gradient starting at 5% acetonitrile and increasing to 95% over 10.5 min.

8. GC analysis indicated that the reaction was complete. Gas chromatographic analyses were performed on a Agilent 6890N GC system equipped with a 30-m 5% polyphenyl methyl siloxane capillary column (0.32 mm i.d.). Helium was used as the carrier gas, and the flow rate was kept constant at 7.7 mL/min. The retention time was measured under the following conditions: injector temperature: 250 °C; initial temperature of the column: 150 °C; increment rate 8 °C/min to 200 °C. After being kept at 200 °C for 8.0 min, the temperature was raised to 290 °C at an increment rate of 8 °C/min.

9. The water wash removes residual boric acid. Elemental analysis of the final product indicated that boron content was less than 0.005%.

10. The filtrate (wash liquor) contained some white solid which was collected, triturated with ca. 5 mL of heptane, and dried to provide additional 0.35-0.50 g of pure product as a white solid (0.4-0.7%). Analytical data (TLC, GC, HPLC and elemental analysis) indicated that this crop has a purity of ≥ 95%.

11. The amide was obtained in 88-91% yield in different runs. The product has the following physico-chemical properties: GC: R_t: 10.31 min

(Note 8); HPLC: R_t: 5.63 min (Note 7); TLC R_f: 0.40 (Note 6); mp 78-80 °C; IR (Kbr) cm^{-1} 3293, 3085, 3029, 2944, 1632, 1550, 1453, 1236, 1155, 1080, 1027, 724, 695; ^1H NMR (300 MHz, DMSO-d_6) δ: 1.76-1.87 (m, 2 H), 2.15 (t, 2 H, $J = 7.2$), 2.55 (t, 2 H, $J = 7.5$), 4.26 (d, 2 H, $J = 6.0$), 7.16-7.33 (m, 10 H), 8.30 (br, 1 H); ^{13}C NMR (75.6 MHz, DMSO-d_6) δ: 27.1, 34.6, 34.7, 41.9, 125.7, 126.6, 127.1, 128.20, 128.23, 128.27, 139.7, 141.7, 171.7. Anal. calcd for $C_{17}H_{19}NO$: C, 80.60; H, 7.56; N, 5.53. Found: C, 80.29; H, 7.53; N, 5.47; B < 0.005.

Waste Disposal Information

All toxic materials were disposed in accordance with "Prudent Practices in the Laboratories"; National Academy Press; Washington, DC, 1965

3. Discussion

Many procedures for the formation of carboxylic acid amides are known in the literature. The most widely practiced method employs carboxylic acid chlorides as the electrophiles which react with the amine in the presence of an acid scavenger. Despite its wide scope, this protocol suffers from several drawbacks. Most notable are the limited stability of many acid· chlorides and the need for hazardous reagents for their preparation (thionyl chloride, oxalyl chloride, phosgene, etc.) which release corrosive and volatile by-products. Moreover, almost any other functional group in either reaction partner needs to be protected to ensure chemoselective amide formation.[2] The procedure outlined above presents a convenient and catalytic alternative to this standard protocol.

Although there are several reports in the literature on boron-mediated amide formations, the boron reagents had to be used in stoichiometric amounts.[3-9] Recently, Yamamoto et al. presented the first truly catalytic method allowing for a direct amide formation from free carboxylic acids and amines as the reaction partners.[10-12] Best results were obtained by using phenylboronic acids bearing electron withdrawing substituents in the *meta*- and/or *para*-positions such as 3,4,5-trifluorophenylboronic acid or 3,5-bis(trifluoromethyl)boronic acid as the catalysts.

During the course of our discovery program directed to small molecules for drug delivery, it was discovered that cheap, readily available, non-toxic, and environmentally benign boric acid, $B(OH)_3$, also constitutes a highly effective catalyst for direct amide formation. Benzylamines and cyclic aliphatic amines such as piperidines react smoothly. In most cases, the use of 5 mol% of $B(OH)_3$ is sufficient for obtaining excellent yields. Likewise, aniline derivatives afford the corresponding amides without incident even if they are hardly nucleophilic due to the presence of electron withdrawing substituents on the arene ring; in such cases, however, the amount of $B(OH)_3$ has to be increased to ca. 25% to ensure complete conversion. The scope of the method is illustrated by the examples compiled in Tables 1 and 2. Particularly noteworthy are the operational simplicity of this new method which might therefore qualify for large-scale preparations, as well as the excellent chemoselectivity profile that can make protection/deprotection sequences obsolete. This is illustrated by the examples shown below.

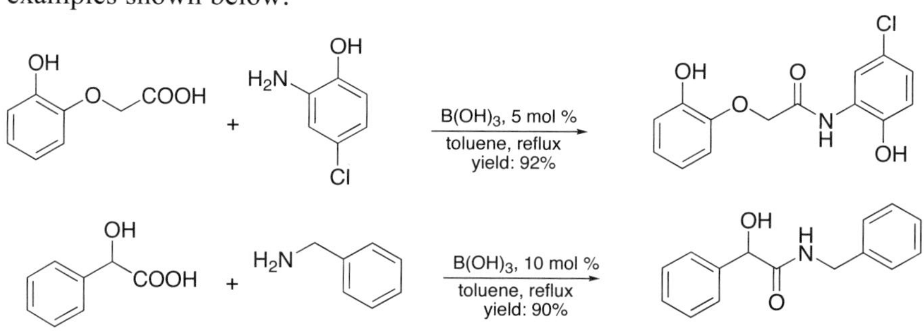

It is proposed that the boric acid reacts with the carboxylic acid to form a mixed anhydride as the actual acylating agent.[9,13] Upon reaction with an amine, this intermediate forms the desired carboxamide and regenerates the catalytically active boric acid.

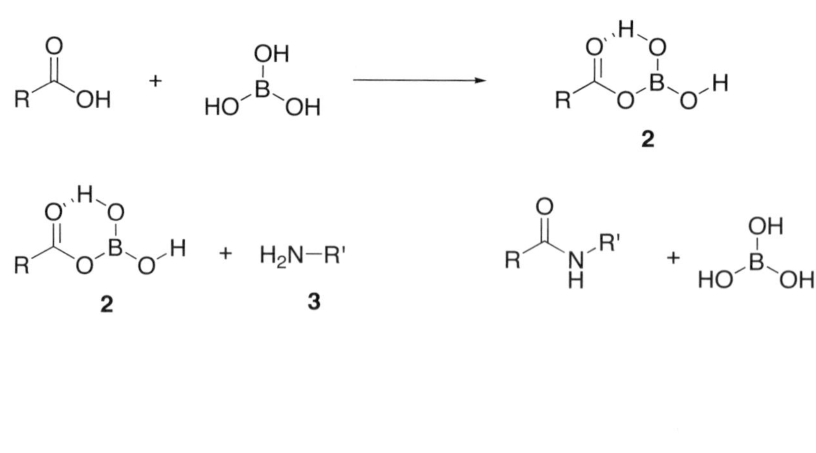

Table 1

Carboxylic acid + Amine $\xrightarrow[\text{reflux}]{B(OH)_3}$ Carboxamide

Entry	Carboxylic acid	Amine	Carboxamide	Solvent	Time (hr)	Isolated Yield
1	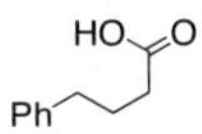		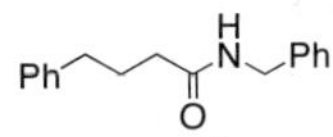	Heptane	15	99
2	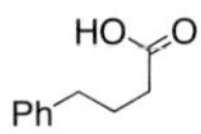	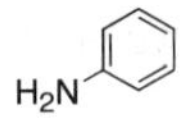	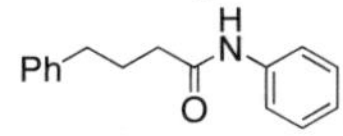	Toluene	15	92
3	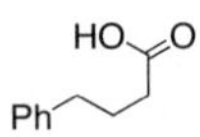	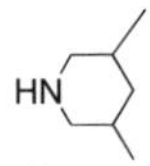	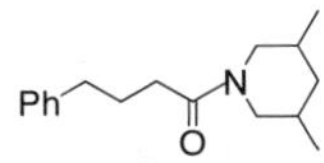	Toluene	15	95
4	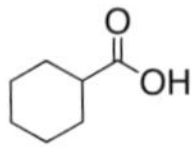	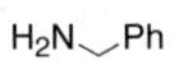	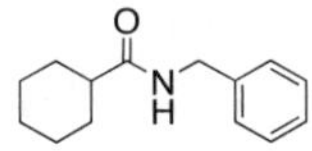	Toluene	15	95
5	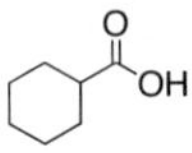	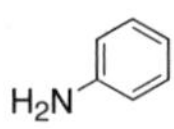	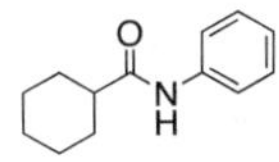	Toluene	15	83
6	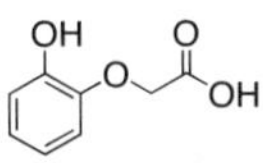	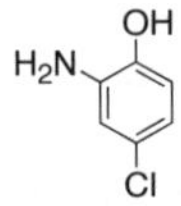	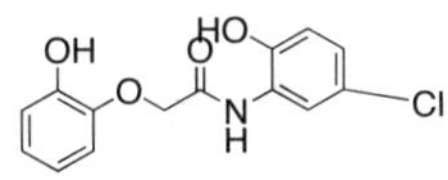	Toluene	16	92
7	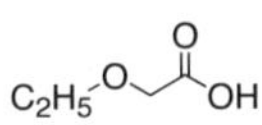	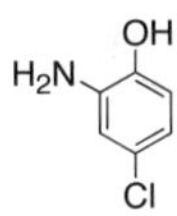	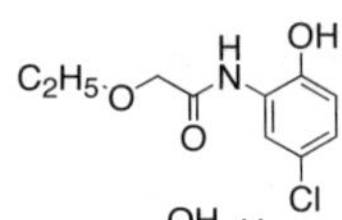	Toluene	4	90
8	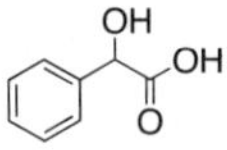	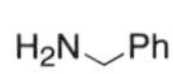	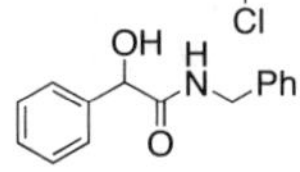	Xylene	15	90
9	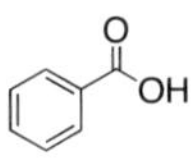	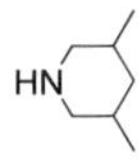	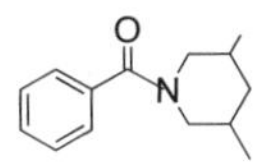	o-Xylene	20	85
10	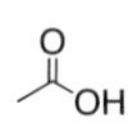	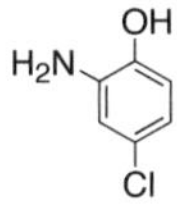	**No carboxamide produced**	Toluene	16	0
11	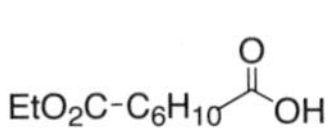	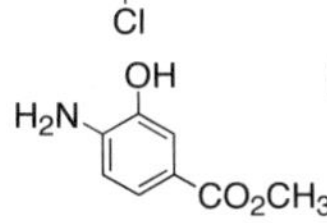	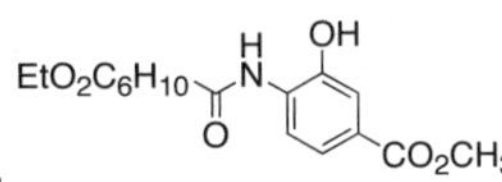	Toluene	16	80
12	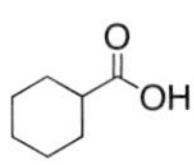	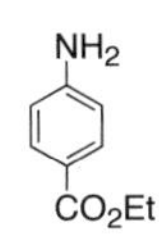	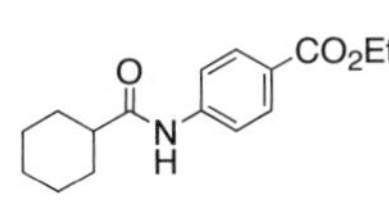	Toluene	16	60

Table II

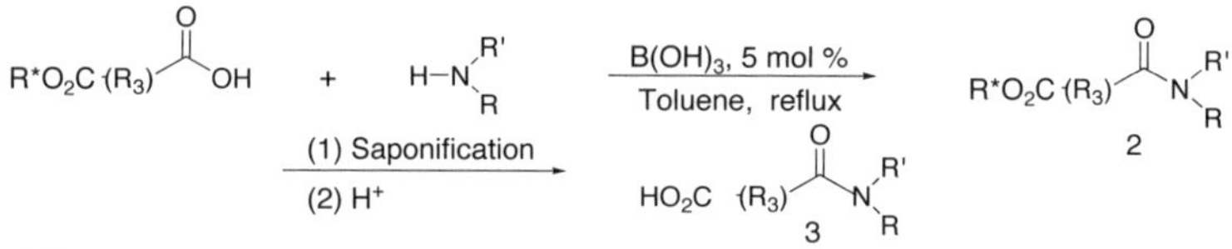

Entry	Carboxylic acid	Amine	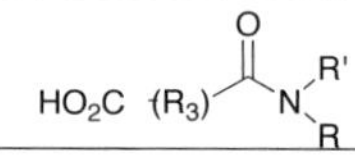	Reaction Time (hr)	Yield % of 3
1	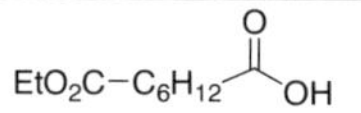	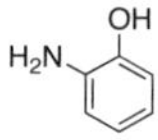	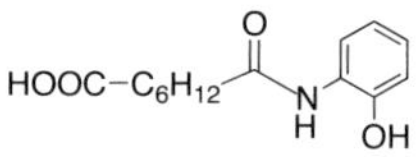	16	75
2	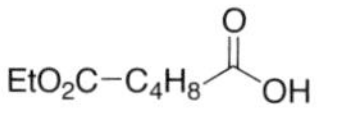	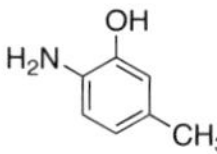	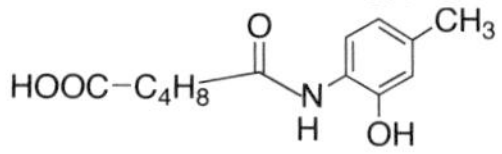	4	77
3	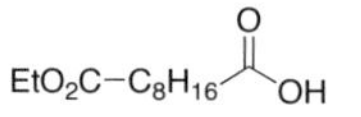	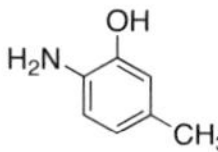	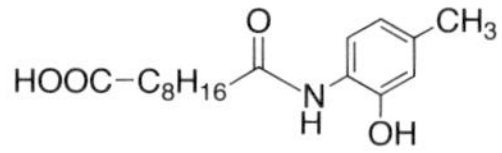	4	78
4	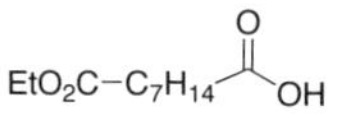	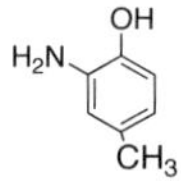	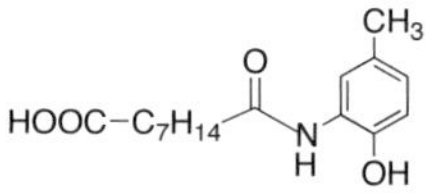	4	77
5	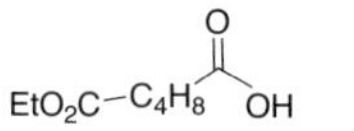	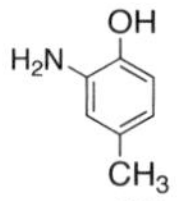	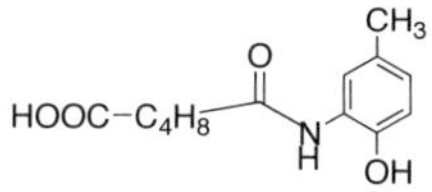	4	78
6	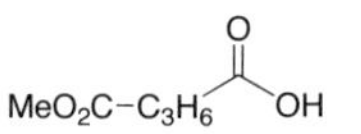	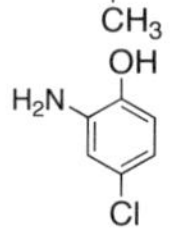	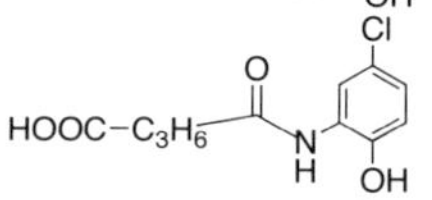	16	72
7	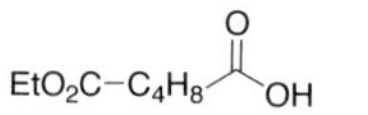	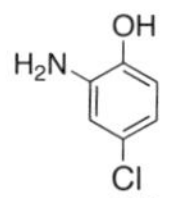	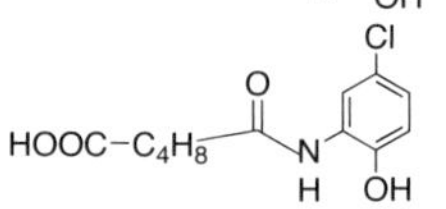	16	70
8	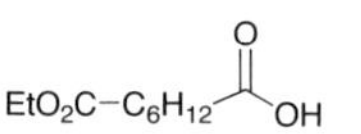	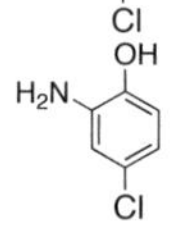	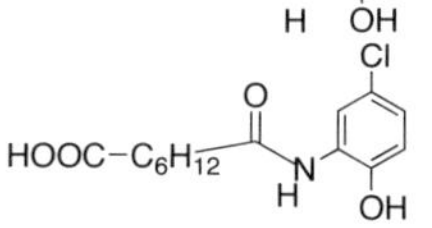	16	57
9	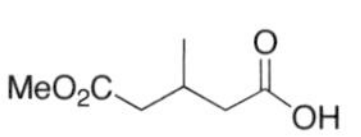	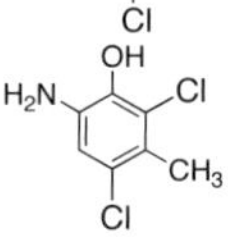	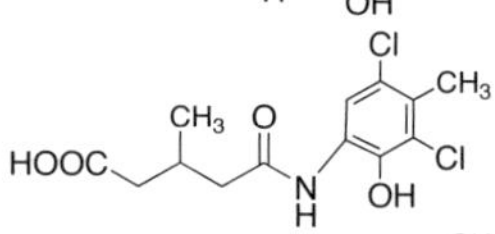	16	68
10	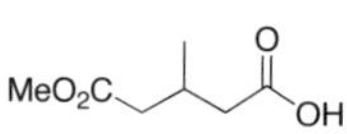	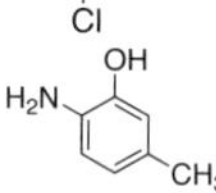	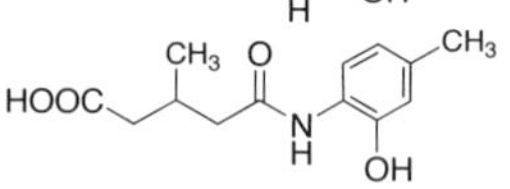	16	70

Several important generalizations emerge from the study of the boric acid catalyzed amidation of carboxylic acids and amines.

(1) In most cases, 5 mol% of the catalyst is sufficient to catalyze the amidation. In all cases, the reactions proceed cleanly in high yields to the expected carboxamides.

(2) No or little side reactions either with the unprotected hydroxy group present in the phenylamines, or with the unprotected hydroxyl group present in the carboxylic acids are observed. The catalytic amidation of unprotected α-hydroxycarboxylic acids also proceeds well under similar conditions.

(3) The present method is successfully applicable to aromatic amines, which are less nucleophilic and less reactive. It works well with phenylamines, even bearing deactivating groups such as carboxylic acid ester moieties; in these cases we use 25 mol% of boric acid to drive the reaction to completion.

(4) The catalyst employed in the reactions, boric acid, is inexpensive and commercially available. Boric acid is a "green" catalyst. By virtue of the simplicity of the process, the operation is easy to conduct. Therefore, it is amenable for large-scale preparations.

(5) The catalytic amidation is an atom-economical process because it maximizes the incorporation of all materials used in the process into the final product. Therefore, it allows organic molecule architects for the quick building of molecular complexity.

1. Department of Chemistry, Emisphere Technologies, Inc., Tarrytown, NY 10591-6715

2. Kelly, S. E.; LaCour, T. G. *Synth. Commun.* **1992,** *22*, 859.

3. Tani, J.; Oine, T.; Inoue, I. *Synthesis* **1975**, 714.

4. Trapani, G.; Reho, A.; Latrofa, A. *Synthesis* **1983**, 1013.

5. Pelter, A.; Levitt, T. E.; Nelson, P. *Tetrahedron* **1970**, *26*, 1539.

6. Collum, D. B.; Chen, S.-C.; Ganem, B. *J. Org. Chem.* **1978**, *43*, 4393.

7. Pelter, A.; Levitt, T. E. *Tetrahedron* **1970**, *26*, 1545.

8. Carlson, R.; Lundstedt, T.; Nordahl, A.; Prochazka, M. *Acta Chem. Scand., Ser. B* **1986**, *40*, 522.

9. For mixed anhydride: Srinivas, P.; Gentry, E. J.; Mitscher, L. A. Poster session 237, 223rd ACS National Meeting, Division of Organic Chemistry, Orlando, FL, U.S.A., April 7-11, 2002.

10. Ishihara, K.; Ohara, S.; Yamamoto, H. *J. Org. Chem.* **1996,** *61*, 4196.

11. Ishihara, K.; Kondo, S.; Yamamoto, H. *Synlett* **2001**, *9*, 1371.

12. Ishihara, K.; Ohara, S.; Yamamoto, H. *Org. Synth., Coll. Vol. X* **2004**, 80.

13. Ishihara, K. In *Lewis Acids in Organic Synthesis*; Yamamoto, H., Ed.; Wiley-VCH: Weinheim, Germany, **2000**, Vol 1, p. 89.

PREPARATION OF *N-p*-TOLYLSULFONYL-(*E*)-1-PHENYLETHYLIDENEIMINE

A.

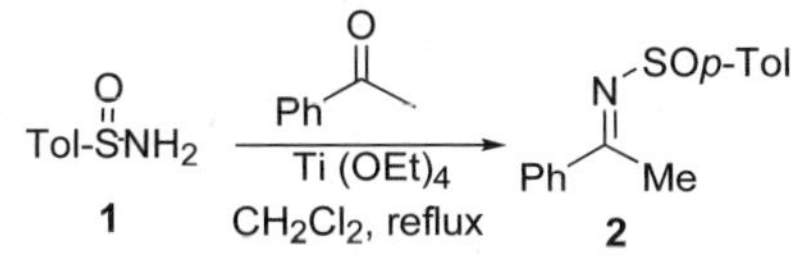

B.

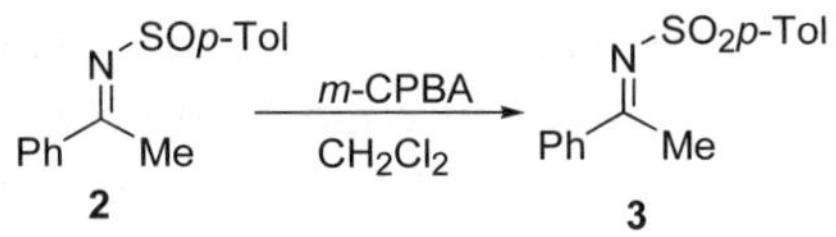

Submitted by José Luis García Ruano, José Alemán, Alejandro Parra, and M. Belén Cid.[1]
Checked by Kounosuke Oisaki and Masakatsu Shibasaki.

1. Procedure

A. N-p-Tolylsulfinyl-(E)-1-phenylethylideneimine (**2**). An oven-dried 300-mL two-necked round-bottomed flask, equipped with a reflux condenser, a magnetic stirring bar, and a rubber septum, is charged with *N*-*p*-toluenesulfinamide **1** (Note 1) (5.2 g, 33.5 mmol, 1.0 equiv), acetophenone (8.0 mL, 68.5 mmol, 2.0 equiv) (Note 2), and $Ti(OEt)_4$ (35 mL, 168 mmol, 5.0 equiv) (Note 3) by temporarily removing the septum. A vacuum adaptor is attached to the condenser and the reaction mixture is stirred under vacuum (1 mmHg) for 15 min. A balloon of argon is attached to the adapter at the top of the condensor. The solid is dissolved in CH_2Cl_2 (75 mL), which is added via syringe through the septum (Note 4). The solution is heated under reflux for 36 h (external bath temperature 80 °C and internal temperature 40 °C), and the reaction is monitored by TLC (Note 5). After cooling the solution to 0 °C using an ice bath, MeOH (50 mL), CH_2Cl_2 (50 mL) and a saturated solution of $NaHCO_3$ (7 mL) are sequentially added with vigorous stirring until precipitation of titanium salts is observed. The solid is vacuum filtered (25 mmHg) through a Büchner funnel (Note 6) and washed with CH_2Cl_2 (3 x 75 mL). The organic phase is dried over anhydrous Na_2SO_4, filtered into a single-necked, 500-mL round-bottomed flask and concentrated by rotary evaporation (25 °C, 10 mmHg) to afford a yellow oil, which is charged onto a silica gel column (8 cm x 18 cm) (Note 7). While collecting fractions of 50 mL, the column is eluted with AcOEt/hexanes (10:90) until acetophenone has completely emerged (volumes of eluents used = 0.6 L). The elution is then continued with AcOEt/hexanes (50:50). The product-containing fractions (total volume of eluent used for this product = 1.7 L) are concentrated and successively recrystallized from CH_2Cl_2/hexanes to afford *N*-sulfinylketimine **2** (5.78 g, 67%) as a white solid (Notes 8 and 9).

B. N-p-Tolylsulfonyl (E)-1-phenylethylideneimine (**3**). Into a round-bottomed 300-mL flask equipped with a magnetic stirring bar is added *N-p*-tolylsulfinylketimine **2** (5.2 g, 20 mmol, 1.0 equiv) and CH_2Cl_2 (75 mL) (Note 10). The solution is cooled to 0 °C (Note 11) and commercially available *m*-CPBA (4.5 g, 20 mmol, 1.0 equiv) (Note 12) is added slowly (portions of 1.0 g per min) to the solution, which is open to the atmosphere. The mixture is allowed to warm to room temperature and the reaction is monitored by TLC (Note 13). After five min, the mixture is diluted with CH_2Cl_2 (50 mL), transferred to a separatory funnel, washed with a saturated solution of $NaHCO_3$ (5 x 25 mL) (Note 14), and the organic phase is dried over anhydrous Na_2SO_4. The solution is filtered into a single-necked, 300-mL round-bottomed flask and concentrated by rotary evaporation (25 °C, 10 mmHg) to afford crude **3** as a white solid, which is diluted with 7 mL of CH_2Cl_2 and charged on a silica gel column (150 g, 5.5 cm x 18 cm) (Note 7). The column is eluted initially with AcOEt/hexanes (10:90) until the *N*-sulfinylketimine **2** emerges, and then the elution is then continued with AcOEt/hexanes (10:50). The product-containing fractions are concentrated to afford *N*-sulfonylketimine **3** (4.23 g, 77%) as a white solid (Notes 14, 15 and 16).

2. Notes

1. *N*-Sulfinamide **1** could be purchased from Aldrich Chemical Company, Inc. in optically pure form. However, (±)-**1** can be obtained at much lower cost following a known procedure,[11] which involves two steps from *p*-tolyldisulfide.

2. Acetophenone, 99% was purchased from Aldrich Chemical company, Inc. and was used without further purification.

3. Titanium (IV) ethoxide, technical grade was purchased from Aldrich Chemical Company, Inc. and was used without further purification.

4. No difference in the yield was observed when using either dried CH_2Cl_2 (from P_2O_5) or the directly purchased CH_2Cl_2 from SDS (purex analytical grade).

5. The reaction was monitored by TLC on silica gel using hexanes/EtOAc (2:1) as eluent and visualization with phosphomolibdic acid. The starting material **1** had R_f= 0.1 (green), acetophenone had R_f= 0.77 (green), and the final product **2** had R_f= 0.38 (brown).

6. A double filter paper of medium porosity was used.

7. Silica gel (230–400 mesh) from Merck was used.

8. The physical properties of *N*-sulfinylketimine **2** were as follows: white solid, mp 95.0–96.0 ºC (lit.[2] mp 99–100 ºC). ^{1}H NMR (500 MHz) δ: 2.40 (s, 3 H) 2.78 (s, 3 H), 7.32 (d, *J* = 8.2 Hz, 2 H), 7.39 (t, *J* = 8.0 Hz, 2 H), 7.47 (t, *J* = 6.7 Hz, 1 H), 7.72 (d, *J* = 8.2 Hz, 2 H), 7.88 (d, *J* = 7.6 Hz, 2 H), ^{13}C NMR (125.7 MHz) δ: 20.2, 21.4, 125.2, 127.5, 128.4, 129.8, 131.8, 138.2, 141.8, 143.3, 173.9; IR (KBr, cm^{-1}): 3430, 1637, 1565, 1096; LRMS (ESI-MS) *m/z* 258 $(M+H)^+$, 280 $(M+Na)^+$; HRMS (FAB-MS) calcd for $C_{15}H_{16}NOS$ $(M+H)^+$ 258.0947, found 258.1008.

9. *N*-Sulfinylketimine **2** and *N*-sulfonylketimine **3** are hydrolyzed upon extended contact with silica gel; therefore the flash chromatography should be carried out rapidly.

10. In this case, dry CH_2Cl_2 (distilled from P_2O_5) was used, since traces of water can hydrolyze the starting material **2** or *N*-sulfonylketimine **3**.

11. This reaction was exothermic, therefore the addition should be carried out at 0 °C. Without an ice bath (0 °C) an increase in temperature from 21 to 40 °C was detected.

12. 3-Chloro-perbenzoic acid (>77 %) was purchased from Aldrich Chemical Company, Inc. (Lot number S22939 was used, which contained 18.03 % of water and 71.70 % of 3-chloroperbenzoic acid). The reagent was directly used without removal of the water, which resulted in a 78% yield. The yield of the reaction could be increased to 84% when water was removed from this reagent. The procedure for removing water was as follows: Prior to its use, water was removed by dissolving the commercial product in CH_2Cl_2, drying the organic solution over anhydrous Na_2SO_4, filtering and concentrating the reagent to dryness under vacuum. Titration following the procedure reported in the literature (McDonald, R. N.; Steppel, R. N.; Dorsey, J. E. *Org. Synth.* **1970**, *50*, 15) indicated a purity of 90% (using $Na_2S_2O_3$ 0.1004 N). A purity of 91% was obtained when the purity

determination was carried out by HPLC: performed on ZORBAX RX-C8 (4.6 mm x 25 cm) with 50:50 water:methanol, with a flow rate of 0.5 mL/min at 210 nm (3-chlorobenzoic acid t_R= 5.5 min, 3-Chloroperbenzoic acid t_R= 21.5 min). When a calibration standard was used (3-chlorobenzoic acid at different concentrations) under these conditions, the purity was determined to be 90%.

13. The reaction was monitored by TLC on silica gel using hexanes/EtOAc (2:1) as eluent and visualization with a phosphomolibdic acid. The starting material **2** has an R_f= 0.38 while R_f= 0.70 for the final product **3**.

14. *N*-Sulfonylketimine **3** is hydrolyzed upon extended contact with water; therefore, the washes should be carried out rapidly.

15. The physical properties of *N*-sulfonylketimine **3** were as follows: white solid, mp 88.5–90.5 ºC (lit.[10k] mp 88–90 ºC). ^{1}H NMR (500 MHz) δ: 2.45 (s, 3 H) 2.99 (s, 3 H), 7.34 (d, *J* = 8.2 Hz, 2 H), 7.41 (t, *J* = 8.0 Hz, 2H), 7.53 (t, *J* = 7.2 Hz, 1H), 7.90 (d, *J* = 8.2 Hz, 2 H), 7.93 (d, *J* = 8.0 Hz, 2 H); ^{13}C NMR (125.7 MHz, $CDCl_3$) δ: 21.1, 21.5, 127.0, 128.2, 128.6, 129.4, 133.1, 137.5, 138.7, 143.5, 179.8; IR (KBr, cm^{-1}): 3435, 1590, 1570, 1286, 1180, 1083; LRMS (ESI-MS) *m/z* 274 $(M+H)^+$, 296 $(M+Na)^+$; HRMS (FAB-MS) calcd for $C_{15}H_{16}NO_2S$ $(M+H)^+$ 274.0896, found 274.0775. The purity of the sample was determined by HPLC (>98%), performed on Chiral AS (Daicel) with 60:40 hexane:2-propanol and a flow rate of 0.6 mL/min.

16. The *N*-sulfonylketimine **3** (0.007 mol, 2 g) could be crystallized from Et_2O (80 mL). The solution was heated until the solid dissolved, and filtered through filter paper of medium porosity. The solution was kept for 12 h at –20 °C. White needles were collected using a Büchner funnel under vacuum (25 mmHg), which resulted in the isolation of 1.90 g of pure **3** (95% yield).

Safety and Waste Disposal Information

All hazardous materials should be handled and disposed of in accordance with "Prudent Practices in the Laboratory"; National Academy Press; Washington, DC, 1995.

3. Discussion

As the sulfonyl moiety has proven to be a powerful activating group of iminic C=N bonds, *N*-sulfonyl imines have been widely used in organic synthesis.[3] They are excellent substrates in aza Diels-Alder reactions,[4] nucleophilic additions,[5] reductions,[6] as well as in radical[7] or ene[8] reactions. They have been also used in the synthesis of aziridines.[9]

Despite the plethora of methods reported for obtaining *N*-sulfonyl imines,[10] most of the methods are only efficient in preparing those compounds derived from aromatic or non-enolizable aldehydes. Therefore, it was necessary to find a new general protocol for synthesizing any *N*-sulfonyl aldimine or ketimine, even those containing enolizable α-hydrogens. We have recently reported a new procedure that fulfills all of these requirements.[11] The procedure involves the condensation of carbonyl compounds with *p*-tolyl (Table 1) and *tert*-butyl (Table 2) sulfinamides followed by oxidation of the resulting *N*-sulfinylimines with *m*-CPBA. The method is applicable to aldehydes (aliphatics and aromatics) and ketones (dialkyl, diaryl, and alkyl aryl), even those containing enolizable protons. The configurations of α-stereogenic centers are not affected and the method is highly selective in the presence of C=N and C=C (deactivated or not) double bonds. The new procedure has been applied to the preparation of compounds collected in Table 1. The main advantages of this procedure derive from the near absence of structural limitations for the starting carbonyls used for preparing sulfinylimines, as well as the higher stability of these sulfinylimines (they can be stored without significant decomposition for long periods of time) with respect to their corresponding sulfonylimines.

Table 1. Preparation of *N-p*-Tolylsulfonylimines.

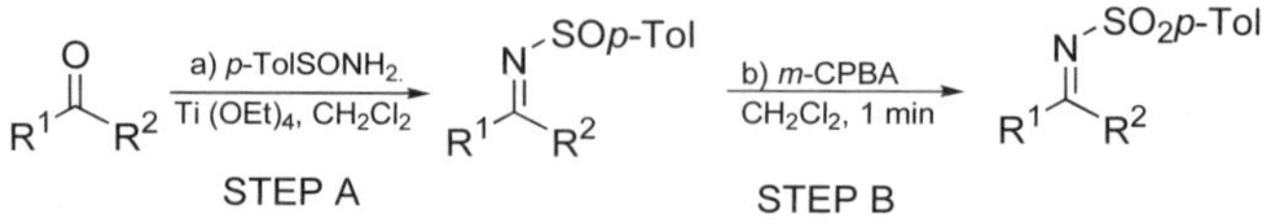

Entry	R^1	R^2	Reaction Yields (%)		
			Step A	Step B	Overall
1	Ph	H	93	100	93
2	3-MeOC_6H_4	H	92	100	92
3	2-Napht	H	94	95	89
4	2-BrC_6H_4	H	91	100	91
5	4-CNC_6H_4	H	81	85[a]	69
6	PhCH=CH_2	H	80	95	76
7	*n*-Pr	H	78	96[b]	75
8	*i*-Pr	H	65	100[b]	65
9	*t*-Bu	H	79	96	76
10	Ph	Ph	79	88	69
11	4-MeOC_6H_4	Me	73	84	61
12	4-CNC_6H_4	Me	77	90	69
13	*t*-Bu	Me	48	90	43
14	*i*-Pr	Me	64	90	58

[a] In this case the reaction time was 15 min. [b] To avoid oxaziridine formation, this reaction was carried out at 45 ºC.

Table 2. Preparation of *N-t*-Butylsulfonylimines.

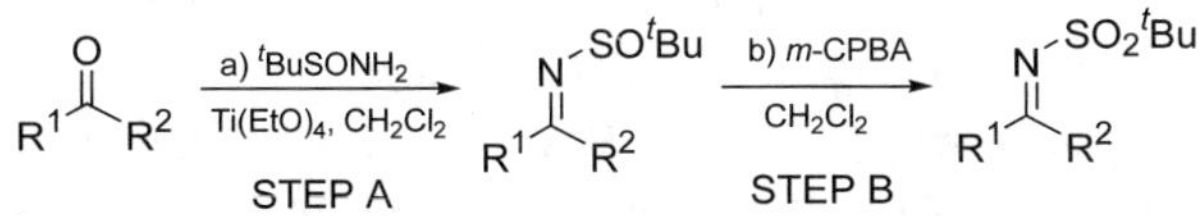

Entry	R^1	R^2	Reaction yields (%)		
			Step A	Step B	Overall
1	Ph	H	96	100	96
2	2-furyl	H	63	100	63
3	*n*-Pr	H	100	84	84
4	Ph	Ph	75	100	75
5	Ph	Me	63	100	63
6	4-MeOC$_6$H$_4$	Me	62	100	62
7	4-CNC$_6$H$_4$	Me	73	100	73
8	-(CH$_2$)$_5$-		65	90	59
9	Et	Me	73[a]	78	57
10	2-pyridyl	Me	59[b]	80	47

[a] A 4:1 mixture of the E/Z isomers was formed. [b] In this case the reaction time was 15 m.

1. Departamento de Química Orgánica, Universidad Autónoma de Madrid, Cantoblanco, 28049-Madrid, Spain.
2. Davis, F. A.; Zhang, Y.; Andemichael, Y.; Fang, T.; Fanelli, D. L.; Zhang, H. *J. Org. Chem.* **1999**, *64*, 1403.
3. (a) Weinreb, S.M. *Top. Curr. Chem.* **1997**, *190*, 131. (b) Gohain, M. *Synlett*, **2003**, *13*, 2097. (c) Bloch, R. *Chem. Rev.* **1998**, *98*, 1407.
4. (a) Sisko, J.: Weinreb, S. M. *Tetrahedron Lett.* **1989**, *30*, 3037. (b) Boger, D. L ; Corbett, W. L; Curran, T.; Kasper, A. M. *J. Am. Chem. Soc.* **1991**, *113*, 1713. (c) García-Mancheño, O; Gómez-Arrayás, R.; Carretero, J.C. *J. Am. Chem. Soc.* **2004**, *126*, 456 and references cited therein.
5. (a) Melnick, M. J. ; Freyes, A. J. ; Weinreb, S. M. *Tetrahedron Lett.*, **1988**, *29*, 3891. (b) Yamada, K; Fujihara, H.; Yamamoto, Y.; Miwa, Y.; Taba, T.; Tomioka, K. *Org. Lett.* **2002**, *4*, 3509. (c) Aggarwal, V.; Alonso, E.; Ferrar, M.; Spey, S. E. *J. Org. Chem.* **2002**, *67*, 2335. (d) Yanmaka, M.; Nishida, A.; Nagawaka, M. *Org. Lett.* **2002**, *2*, *159*. 3891. (e) Soeta, T.; Nagai, K.; Fujihara, H.; Kuriyama, M.; Tomioka,

K. *J. Org. Chem.* **2003**, *68*, 9723. (f) Wipf, P.; Kendall, C.; Stephenson, C. R. J. *J. Am. Chem. Soc*, **2003**, *125*, 761. (g) Günter, M.; Gais, H.-J. *J. Org. Chem*, **2003**, *68*, 8037. (h) Yim, H.-K.; Wong, H. N. C. *J. Org. Chem.* **2004**, *69*, 2892.

6. (a) Hojo, M.; Murakami, C.; Fujii, A.; Hosomi, A. *Tetrahedron Lett.*, **1999**, *40*, 911. (b) Nisikori, H.; Yoshihara, R.; Hosomi, A. *Synlett*, **2003**, *4*, 561. (c) Chen, Y.-C.; Wu, T.-F.; Deng, J.-G.; Liu, H.; Cui, X.; Zhu, J.; Jiang, Y.-Z.; Choi, M. C. K.; Chan, A. S. C. *J. Org, Chem.* **2002**, *67*, 5301.

7. Yamada, K.-I.; Fujihara, H.; Yamamoto, Y.; Miwa, Y.; Taga, T.; Tomioka, K. *Org. Lett.* **2002,** *4*, 3509.

8. Yamanaka, M.; Nishida, A.; Nakagwa, M. *Org. Lett.* **2002**, *2*, 159.

9. (a) Hori, R. Aoayama, T.; Shoiri, T. *Tetrahedron Lett.* **2000**, *41*, 9455. (b) Arini, L. G.; Sinclair, A.; Szeto, P.; Stockan, R. A. *Tetrahedron Lett.* **2004**, *45*, 1589.

10. (a) Davis, F. A.; Lamendola Jr., J.; Nadir, U.; Kluger, E. W.; Sedergran, T. C.; Panunto, T. W.; Billmers, R.; Jenkins Jr., R.; Turchi, I. J.; Watson, W. H.; Chen, J. S.; Kimura, M. *J. Am. Chem. Soc.* **1980**, *102*, 2000. (b) Trost, B. M.; Christopher, M. *J. Org. Chem.* **1991**, *56*, 6468. (c) Jennings, W.B.; Lovely, C. J. *Tetrahedron*, **1991**, *47*, 5561. (d) Boger, D. L.; Corbett, W. L. *J. Org. Chem.* **1992**, *57*, 4777. (e) Love, B. E.; Raje, P. S.; Willimas, II, T. C. *Synlett*, **1994**, 493. (f) Georg, G. I.; Harriman, G. C. B.; Peterson, S.A. *J. Org. Chem.* **1995**, *60*, 7366. (g) Chemla, F.; Hebbe, V.; Normant, J.-F. *Synthesis*, **2000**, *1*, 75. (h) Artman, G. D.; Bartolozzi, A.; Franck, R. W.; Weinreb, S. M. *Synlett* **2001**, 232. (i) Ram, R. N.; Khan, A.A. *Synth. Commun.* **2001**, *31*, 841. (j) Lee, K. Y.; Lee, C. C.; Kim, J. N. *Tetrahedron Lett.* **2003**, *44*, 1231. (k) Wolfe, J.; Ney, J. E. *Org. Lett.* **2003**, *5*, 4607. (l) Jin, T.; Feng, G.; Yang, M.; Li, T. *Synth. Commun.* **2004**, *34*, 1277.

11. García Ruano, J. L.; Alemán, J.; Cid, M. B.; Parra, A. *Org. Lett.* **2005**, 7, 179.

PREPARATION OF ENANTIOMERICALLY ENRICHED (1*S*)-1-PHENYLPROPAN-1-AMINE HYDROCHLORIDE BY A CATALYTIC ADDITION OF DIORGANOZINC REAGENTS TO IMINES

[Benzenemethanamine, α-ethyl-, hydrochloride, (α*S*)-]

A.

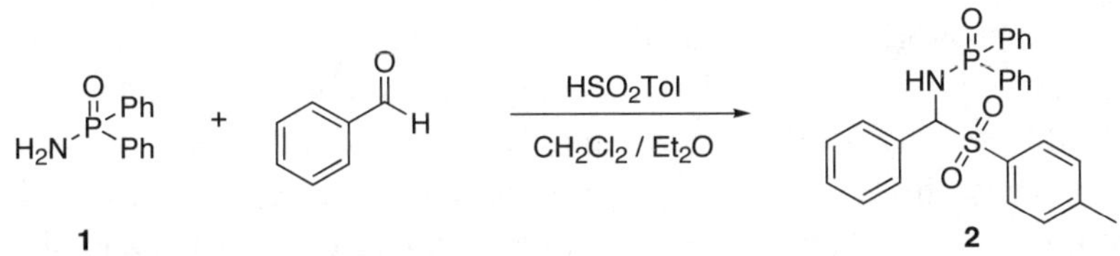

B.

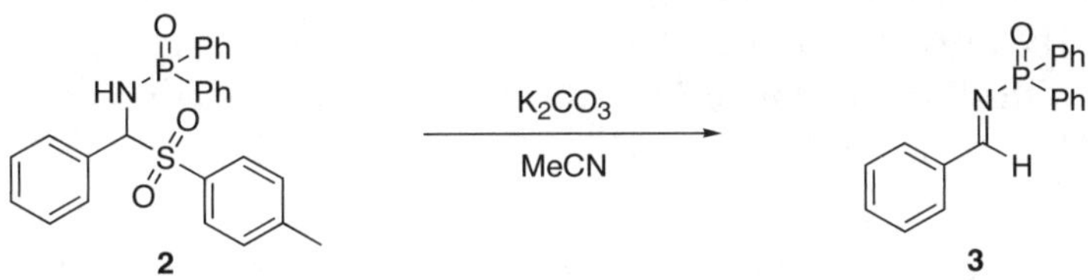

C.

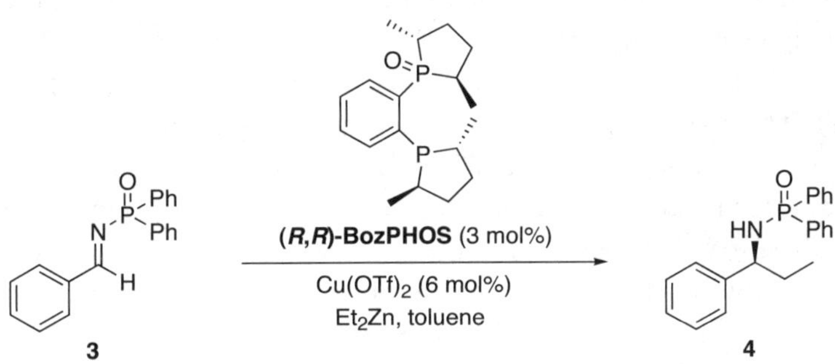

D.

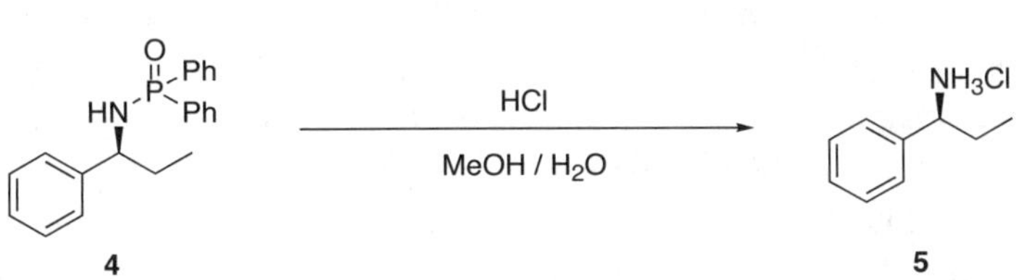

Submitted by Jean-Nicolas Desrosiers, Alexandre Côté, Alessandro A. Boezio and André B. Charette.[1]
Checked by Scott E. Denmark and Justin I. Montgomery.

1. Procedure

A. N-[(4-Methylphenyl)sulfonyl(phenyl)methyl]-P,P-diphenylphosphinic amide (**2**). A 1-L, three-necked, round-bottomed flask equipped with a mechanical stirrer (fitted with a 7-cm Teflon paddle) and two rubber septa is charged with **1** (9.55 g, 43.97 mmol) (Notes 1, 2). Dichloromethane (75 mL) (Note 3) and diethyl ether (365 mL) (Note 4) are added successively to the flask. The resulting suspension is stirred for 5 min and benzaldehyde (6.7 mL, 65.96 mmol) (Note 5) is added via a syringe. *p*-Toluenesulfinic acid (10.3 g, 65.96 mmol) (Note 6) is then added in one portion at room temperature. The reaction mixture is capped and allowed to stir for 48 h (200 rpm), during which time a white precipitate is slowly formed. The mixture is filtered through a sintered glass funnel and the white solid is washed with diethyl ether (100 mL) and dried under vacuum (25 °C, 0.1 mmHg) to afford 18.12–18.45 g (89–91% yield) (Note 7) of the title compound.

B. N-[(Phenyl)methylene]-P,P-diphenylphosphinic amide (**3**). A flame-dried 1-L, one-necked, round-bottomed flask equipped with an egg-shaped magnetic stirring bar and a rubber septum is charged with **2** (18.00 g, 39.00 mmol) and anhydrous potassium carbonate (26.95 g, 195.0 mmol) (Note 8). The flask is purged with argon and anhydrous acetonitrile (365 mL) (Note 9) is added via cannula under argon at room temperature. The resulting suspension is vigorously stirred under argon at room temperature for 12 h. The reaction mixture is filtered through a sintered glass funnel and the residue is washed with acetonitrile (50 mL). The filtrate is concentrated at 30 °C by rotary evaporation (30–50 mmHg) to afford 10.88 g (91% yield) (Note 10) of the title compound as an off-white solid.

C. N-[(1S)-1-Phenylpropyl]-P,P-diphenylphosphinic amide (**4**). A flame-dried, 500-mL, one-necked, round-bottomed flask equipped with an egg-shaped magnetic stirring bar and a rubber septum is charged with $Cu(OTf)_2$ (745 mg, 2.06 mmol) (Note 11) and ***(R,R)*-BozPHOS** (332 mg, 1.03 mmol) (Note 12) in a glove-box under argon. Anhydrous toluene (172 mL) (Note 13) is added to the flask at room temperature via cannula. The resulting dark green heterogeneous solution is stirred for 1 h at room temperature and neat diethylzinc (7.05 mL, 68.78 mmol) (Note 14) is added at room temperature under argon via a 10-mL gas-tight syringe (*CAUTION: NEAT DIETHYLZINC IS HIGHLY PYROPHORIC*). The resulting dark brown suspension is stirred for 20 min at room temperature and then is diluted by the addition of anhydrous toluene (172 mL) via cannula under argon. The mixture is cooled to 0 °C (internal temperature measured with a thermocouple probe (Note 15)) with a cryostat (Note 16) and the system is opened to add the solid **3** (10.50 g, 34.39 mmol) quickly in one portion. The flask is sealed with a septum and purged with argon over a period of 5 min. The reaction mixture is allowed to stir for 21 h at 0 °C under argon. Aqueous saturated ammonium chloride (100 mL) is added dropwise via cannula at 0 °C to quench the reaction. The mixture is poured into a 2-L separatory funnel containing aqueous saturated ammonium chloride (400 mL). The biphasic mixture is extracted with dichloromethane (3 x 500 mL). The combined extracts are dried over Na_2SO_4 (150 g), filtered and concentrated at 40 °C by rotary evaporation (100–150 mmHg) and then at 20 mmHg to afford 11.82 g of an off-white solid containing the title compound and the catalyst. This mixture is used without purification for the next step. A 200.0 mg sample of the crude product is purified by silica gel chromatography (Note 17) to afford 185.7 mg (92.9% recovery, 97% ee) (Note 18) of the title compound as a white solid.

D. (1S)-1-Phenylpropan-1-amine hydrochloride (**5**). A 1-L, one-necked, round-bottomed flask equipped with an egg-shaped magnetic stirring bar is charged with **4** (11.62 g, 92.9% purity, 32.19 mmol). A mixture of methanol (270 mL) (Note 19) and concentrated aqueous HCl (60.9 mL) (Note 21) is added to the flask. The flask is capped and the resulting clear, bright yellow solution is allowed to stir at room temperature for 16.5 h (Note 20). The reaction mixture is then concentrated at 40 °C by rotary evaporation (10 mmHg) and the residue is dissolved in 1.21 M aqueous HCl (300 mL). The precipitate (Note 22) is removed by filtration on a sintered-glass funnel. The acidic filtrate is extracted with diethyl ether (3 x 300 mL) (Note 23), then it is basified (pH ≥ 12, Note 24) by adding 2 M NaOH (~300 mL). The resulting milky solution is extracted with dichloromethane (4 x 300 mL). The combined extracts are dried over Na_2SO_4 (160 g) and filtered. A solution of HCl in diethyl ether (1 M, 67.6 mL, 67.6 mmol) (Note 25) is added to the organic layer and the mixture is concentrated at 30 °C by rotary evaporation (200–250 mmHg), then at 50–100 mmHg. The resulting off-white to yellow solid (5.60 g) is triturated with ethyl acetate (50 mL) and filtered to afford 4.89 g (88% yield, 97% ee) (Notes 26-28) of the pure title compound as a white solid.

2. Notes

1. *P,P*-Diphenylphosphinic amide (**1**) was purchased from Alfa Aesar as diphenylphosphinamide (98+%).

2. *P,P*-Diphenylphosphinic amide (**1**) can be prepared by the following procedure (not attempted by the checkers): A flame-dried 2-L, one-necked, round-bottomed flask, equipped with an egg-shaped magnetic stirring bar, a rubber septum and an internal thermocouple probe is purged with argon. The flask is then charged with 1 L of anhydrous dichloromethane under argon. The septum is removed and acetone oxime (16.56 g, 227 mmol) is added quickly in one portion and the flask is sealed and purged again with argon. Freshly distilled triethylamine (31.6 mL, 227 mmol) is added via a syringe and the resulting colorless solution is cooled to –78 °C (internal temperature) with an acetone/dry ice bath (approximately 30 min are needed to reach this temperature). Freshly distilled, colorless chlorodiphenylphosphine (42 mL, 227 mmol) is added dropwise (2 mL/min) via a syringe at such a rate that the internal temperature does not exceed –70 °C. After the addition is complete, the resulting milky solution is stirred 15 min at –78 °C under argon. The cooling bath is removed, and the solution is allowed to warm to room temperature over a period of 1.5 h and stirred 1 h at that temperature. The milky solution becomes clear. The solvent is removed under reduced pressure and the resulting off-white to yellow solid is dried under vacuum for 12 h. The residue is dissolved in acetone (200 mL) and dried under reduced pressure. This last step is repeated once again. (This step makes the removal of the solid from the flask easier. Furthermore, the solid can be ground to a finer powder, which facilitates the next step. This step can be avoided but problems may occur, such as the formation of a gummy yellow solid during the addition to the aqueous ammonium hydroxide solution. This problem can be related to the presence of dichloromethane in the solid. Heating the mixture with a heat gun to dissolve everything can solve this problem, but lower yields and side-products are obtained.) The solid is removed from the flask and ground to a powder using a mortar and pestle. The powder is added in one portion into an Erlenmeyer flask containing a stirring solution (magnetic stirring bar) of concentrated aqueous ammonium hydroxide (660 mL) and distilled water (330 mL). The mixture is usually heterogeneous and the precipitate is a white solid. Sometimes the solid added dissolves rapidly to form a clear yellow solution and, after few seconds, a white solid starts to precipitate. The heterogeneous mixture is stirred for 45 min and **1** is obtained as a white solid by filtration

through a sintered glass funnel. The aqueous filtrate is extracted with dichloromethane (3 x 600 mL). The combined extracts are dried over Na_2SO_4, filtered and evaporated under reduced pressure. The white solid obtained is combined with the white precipitate obtained above after filtration. The combined solids are dissolved in benzene (300 mL) and evaporated. This step is repeated three times to remove traces of water to afford 46.31–47.87 g (94–97% yield) of **1**. The purity can be increased by crystallization in ethyl acetate.

3. Dichloromethane (ACS grade) was purchased from Fisher Scientific and was used as received.

4. Diethyl ether (ACS grade) was purchased from Fisher Scientific and was used as received.

5. Benzaldehyde was purchased from Aldrich Chemical Company, Inc. and was freshly distilled (70–72 °C, 20 mmHg) prior to use.

6. *p*-Toluenesulfinic acid was prepared according to the procedure described in *Organic Syntheses*, see ref 2.

7. The physical properties are as follows: mp 153–155 °C (sealed tube); IR (neat) cm^{-1}: 725, 748, 756, 888, 1030, 1086, 1109, 1124, 1148, 1191, 1289, 1301, 1312, 1436, 1459, 1597, 2954, 3056, 3205; ^{31}P NMR (202 MHz, DMSO) δ: 24.98. Due to the very low solubility of the compound in DMSO, clean NMR spectral data are difficult to obtain because minor impurities are much more soluble than the compound of interest.

8. Granular potassium carbonate was purchased from Aldrich Chemical Company, Inc. and was ground to a powder using a mortar and pestle then dried under vacuum (100 °C, 1 mmHg).

9. Anhydrous acetonitrile was obtained by filtration through a drying column on a GlassContour system (Irvine, CA).

10. The purity is greater than 97% according to the ^{31}P and ^{1}H NMR spectra and the only observed impurities are benzaldehyde and **1**. These impurities can be removed on a short pad of silica gel (EtOAc 100%). The physical properties of the purified material are as follows: mp 144–146 °C; R_f 0.47 (EtOAc); MS (ESI+) *m/z* 307.1 (20%), 306.1 (M^++H, 100%), 233.1 (37%), 219.1 (22%), 201.2 (14%); Anal. calcd for $C_{19}H_{16}NOP$: C, 74.74; H, 5.28; N, 4.59. found: C, 74.45; H, 5.27; N, 4.71; IR (KBr) cm^{-1}: 704, 729, 752, 832, 848, 926, 998, 1074, 1110, 1127, 1199, 1311, 1368, 1443, 1577, 1597, 1626, 1663, 1698, 2881, 3024, 3056; ^{1}H NMR (400 MHz, $CDCl_3$) δ: 7.39–7.56 (m, 9 H), 7.90–8.02 (m, 6 H), 9.32 (d, J = 32.2 Hz, 1 H); ^{13}C NMR (100 MHz, $CHCl_3$) δ: 128.3 (d, $J_{C\text{-}P}$ = 12.5 Hz), 128.8, 130.0, 131.4 (d, $J_{C\text{-}P}$ = 9.2 Hz), 131.7 (d, $J_{C\text{-}P}$ = 3.6 Hz), 132.8 (d, $J_{C\text{-}P}$ = 127.5 Hz), 133.5,

135.6 (d, $J_{C\text{-}P}$ = 24.2 Hz), 173.5 (dd, $J_{C\text{-}P}$ = 7.7, 3.0 Hz); ^{31}P NMR (162 MHz, $CDCl_3$) δ: 24.78.

11. $Cu(OTf)_2$ was purchased from Strem Chemicals, Inc. It was stored under argon atmosphere in a glove-box and was used without prior purification.

12. (*R,R*)-BozPHOS was prepared according to the preceeding procedure and was stored under argon in a glove box.

13. Anhydrous toluene was obtained by filtration through a drying column on a GlassContour system (Irvine, CA).

14. Diethylzinc is a moisture sensitive and pyrophoric liquid and must be manipulated in an inert atmosphere. Neat diethylzinc was purchased from Aldrich Chemical Co. Inc. and was used without prior purification.

15. A PFA coated thermocouple probe, Type K (Omega Engineering, Inc.) was inserted through the septum after the addition of diethylzinc to monitor the internal temperature of the reaction solution.

16. Approximately 35 min are needed to reach that temperature. A Neslab, model CC-65II, cryostat was used to maintain the reaction mixture at 0 °C.

17. The mixture was dissolved in a minimum amount of dichloromethane and then was charged onto a column (diameter = 2 cm, height = 15 cm) of 200 g of UltraPure silica gel (40–63 μm) purchased from Silicycle. The column was eluted with EtOAc and 8-mL fractions were collected. Fractions 15-33 were combined and concentrated by rotary evaporation (30 °C, 50 mmHg). The desired product can be visualized on TLC with a UV lamp or by spraying with a phosphomolybdic acid solution. The R_f value of the title compound in EtOAc is 0.37.

18. The physical properties are as follows: mp 127–129 °C; MS (ESI+) *m/z*: 337.2 (M^+ + H, 21%), 336.1 (M^+, 100%), 233.2 (17%), 219.2 (18%), 218.1 (28%), 201.2 (9%). Anal. calcd. for $C_{21}H_{22}NOP$: C, 75.21; H, 6.61; N, 4.18; found: C, 74.90; H, 6.55; N, 4.35; IR (KBr) cm^{-1}: 722, 752, 904, 933, 1057, 1090, 1109, 1122, 1182, 1198, 1438, 1460, 2873, 2926, 2963, 3055, 3135; ^{1}H NMR (500 MHz, $CDCl_3$) δ: 0.78 (t, *J* = 7.5 Hz, 3 H), 1.78–1.88 (m, 1 H), 1.96–2.06 (m, 1 H, exchanges with D_2O), 3.27 (br, 1 H), 4.09 (pent, *J* = 8.4 Hz, 1 H), 7.14–7.16 (m, 2 H), 7.21–7.34 (m, 5 H), 7.39–7.50 (m, 4 H), 7.75 (ddt, *J* = 12.0, 6.9, 1.5 Hz, 2 H), 7.86 (ddt, *J* = 12.0, 7.1, 1.3 Hz, 2 H); ^{13}C NMR (126 MHz, $CHCl_3$) δ: 10.5, 32.5 (d, $J_{C\text{-}P}$ = 3.9 Hz), 57.1, 126.5, 127.1, 128.3 (d, $J_{C\text{-}P}$ = 12.9 Hz), 128.4, 128.4 (d, $J_{C\text{-}P}$ = 12.9 Hz), 131.7 (d, $J_{C\text{-}P}$ = 2.7 Hz), 131.8 (d, $J_{C\text{-}P}$ = 9.2 Hz), 131.8 (d, $J_{C\text{-}P}$ = 2.8 Hz), 131.9 (d, $J_{C\text{-}P}$ = 131.0 Hz), 132.6 (d, $J_{C\text{-}P}$ = 10.2 Hz), 133.2 (d, $J_{C\text{-}P}$ =

128.0 Hz), 143.5 (d, J_{C-P} = 5.5 Hz); ^{31}P NMR (202 MHz, $CDCl_3$) δ: 22.74; $[\alpha]_D^{20}$ –41.6 (c = 2.14, MeOH). The enantiomeric excess of the product is determined by HPLC analysis at 254 nm [Chiralpak, AD, 85:15 hexanes: *i*-PrOH, 1mL/min: (*R*) t_r (minor) = 10.2 min, (*S*) t_r (major) = 13.4 min)].

19. Methanol (ACS grade) was purchased from Fisher Scientific and was used as received.

20. The reaction was followed by ^{31}P NMR spectroscopic analysis. The rate of disappearance of the signal at 23.2 ppm is proportional to the rate of formation of the desired product. The new signals at 34.3 ppm and at 36.6 ppm correspond, respectively, to diphenylphosphinic acid and methyl diphenylphosphinate.

21. Aqueous concentrated HCl (ACS grade) was purchased from Fisher Scientific and was used as received.

22. The precipitate is diphenylphosphinic acid.

23. This step is used to extract the methyl diphenylphosphinate.

24. An ColorpHast indicator strip purchased from EM Science, Inc. was used to measure the pH.

25. 1 M Hydrogen chloride solution in diethyl ether was purchased from Aldrich Chemical Company, Inc. and was used as received.

26. The physical properties are as follows: mp 231–234 °C; MS (EI) *m/z*: 135.1, (4%), 106.1 (100%), 79.1 (18%), 74.1 (75%), 59.0, (100%). Anal. calcd. for $C_9H_{14}ClN$: C, 62.97; H, 8.22; N, 8.16; found: C, 62.71; H, 8.33; N, 8.18; IR (KBr) cm^{-1}: 754, 764, 1386, 1394, 1458, 1509, 1600, 2536, 2618, 2685, 2914, 2963, 3035; ^{1}H NMR (400 MHz, MeOD) δ: 0.88 (t, *J* = 7.3 Hz, 3 H), 1.90–2.10 (m, 2 H), 4.15 (dd, *J* = 9.5, 5.8 Hz, 1 H), 7.39–7.49 (m, 5 H); ^{13}C NMR (100 MHz, MeOD) δ: 10.5, 28.7, 58.3, 128.4, 130.3, 130.3, 138.1; $[\alpha]_D^{20}$ +16.2 (c = 1.03, MeOH).

27. The enantiomeric excess of the product was determined on the trifluoroacetyl derivative of the amine by GC analysis with an FID detector (Astec Chiraldex GT-A (30 m x 0.32 mm) 10 psi, 90 °C isothermal): (*R*) t_r (minor) = 19.59 min, (*S*) t_r (major) = 20.53 min). To prepare the trifluoroacetamide derivative, 1 mL of 2 M aqueous NaOH solution was added to 10 mg of the amine hydrochloride salt in a 5 mL conical vial. The aqueous layer was extracted with 1 mL of dichloromethane. The organic layer was dried over Na_2SO_4 (0.5 g), and was decanted into a 5-mL round-bottomed flask. An egg-shaped magnetic stir bar and 250 μL of trifluoroacetic anhydride were added and the mixture was stirred at room temperature under argon for 5 min. The solvent and excess reagents were removed by rotary evaporation (80 mmHg, 25 °C, 5 min; then 5 mmHg, 25

°C, 10 min). The solid residue was dissolved in 1 mL of dichloromethane and the liquid was filtered through a syringe filter for GC analysis. For the enantioenriched compounds, the sample was concentrated to ca. 0.1 mL to observe the minor isomer.

28. The submitters were able to determine the enantiomeric excess of the free amine directly by GC analysis with an FID detector [Beta Dex™ 120; 30 °C to 85 °C in 11 min and isothermal thereafter: (*R*) t_r (minor) = 43.0 min, (*S*) t_r (major) = 43.6 min)].

Safety and Waste Disposal Information

All hazardous materials should be handled and disposed of in accordance with "Prudent Practices in the Laboratory"; National Academy Press; Washington, DC, 1995.

3. Discussion

This procedure provides an attractive route to α-substituted chiral amines by using a catalytic asymmetric addition of a diorganozinc reagent. Numerous activating groups have been reported for the catalytic asymmetric additions of diorganozinc reagents to imines. Among them, *N*-tosylimines,[4] *N*-arylimines[5] and *N*-acylimines[6] have been used as precursors. Herein, we use the *N*-phosphinoyl group because of its facile cleavage under mildly acidic conditions. The other activating groups require harsh oxidizing, reducing or basic conditions for cleavage. Another advantage of using the *N*-phosphinoyl activating group is that a crystalline product is generally obtained. This product can often be recrystallized from low polarity solvents to increase enantiopurity.[7]

Earlier syntheses of *N*-diphenylphosphinoylimines involved either a condensation between an aldehyde and *N*-diphenylphosphinoyl amide promoted by $TiCl_4$ and Et_3N[8] or a radical rearrangement between an oxime and chlorodiphenylphosphine with the presence of Et_3N.[9] These methods are not useful for the synthesis of *N*-phosphinoylimines on large scale because of the low yields that result from tedious workup procedures or extensive purification. However, the present synthesis of *N*-phosphinoylimines via a stable sulfinyl adduct is very effective for production of multi-gram quantities of material.[10] The procedure, which consists of deprotonating the sulfinyl adduct **2** with K_2CO_3 to induce imine formation, is general, straightforward, inexpensive, and high yielding.

Several *N*-phosphinoylimines derived from aromatic or aliphatic aldehydes can be generated by this sequence. For instance, acetaldehyde, heptanal, isovaleraldehyde, hydrocinnamaldehyde, isobutyraldehyde, cyclopentanecarboxaldehyde and cyclohexanecarboxaldehyde afford the desired sulfinic acid adducts of the *N*-phosphinoylimine on a small scale (Table 1).[11] These adducts are generated in diethyl ether and precipitate from the reaction mixture as white solids. In the present case, a mixture of dichloromethane and diethyl ether gives a better yield and a higher purity of the sulfinyl adduct **2**. Because of the higher solubility of its adduct, 1-naphthaldehyde gives a poor yield (30%). To avoid problems of handling unstable imines of aliphatic aldehydes, it is also possible to carry out the asymmetric addition directly on the sulfinic acid adduct.[11,12] The same reaction conditions are used except that an excess of diethylzinc (2.5 equiv) is needed to generate the imine *in situ*.

Table 1. Sulfinyl adducts synthesis[a,b]

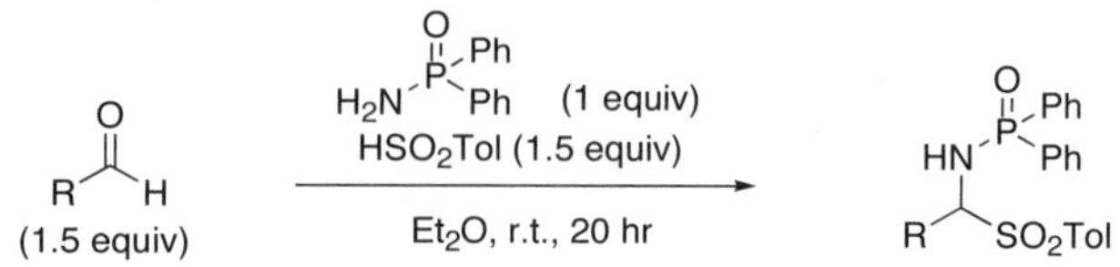

Entry	R	Yield (%)
1		91
2	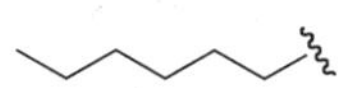	87
3	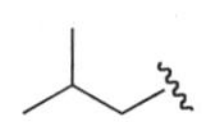	84
4[c]	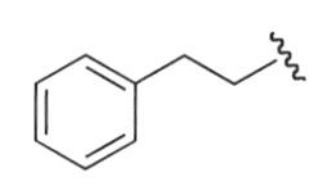	97
5		88
6	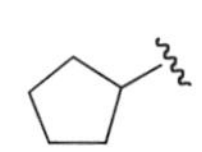	95
7	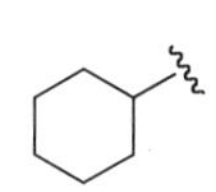	72

[a] See reference 11 [b] 5 mmol scale [c] 10 mmol scale.

The copper-catalyzed addition of diorganozinc reagents to alkyl- or aryl-substituted *N*-phosphinoylimines proceeds in high yields, with excellent enantioselectivities and without using a large excess of organometallic reagents. Furthermore, both enantiomers of the BozPHOS ligand synthesized from the commercially available enantiomers of Me-DuPHOS. The reaction is proceeds well with different dialkyl- (Me_2Zn, *n*-Bu_2Zn and *i*-Pr_2Zn) or functionalized zinc reagents (Table 2).[7] This system can also tolerate a wide variety of *N*-phosphinoylimines obtained from either electron-poor or electron-rich aldehydes.[7, 13]

Table 2. Enantioselective Addition of Dialkylzinc Reagents to Imines[a,b]

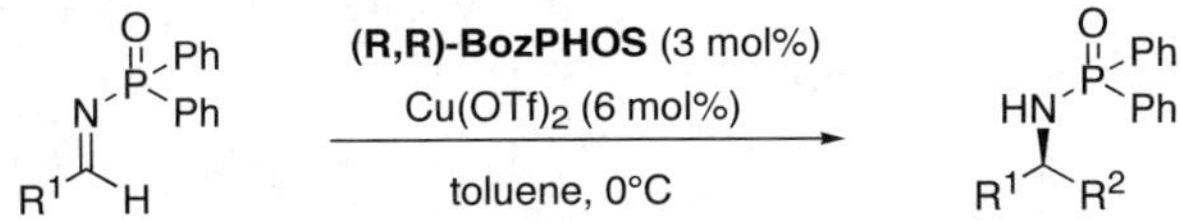

Entry	R^1	$R^2{}_2Zn$	Yield (%)	ee (%)
1	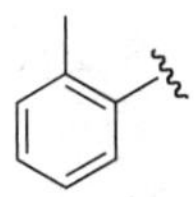	Et_2Zn	97	99
2	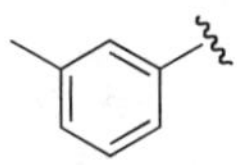	Et_2Zn	96	97
3	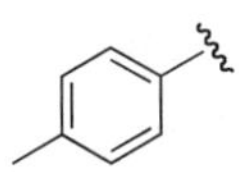	Et_2Zn	94	98
4	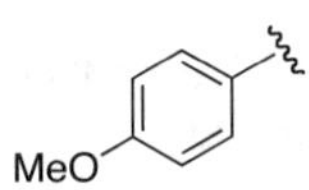	Et_2Zn	91	98
5	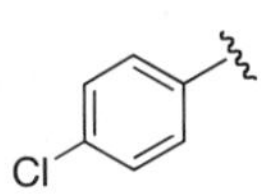	Et_2Zn	97	97
6	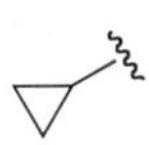	Et_2Zn	95	94
7[c]	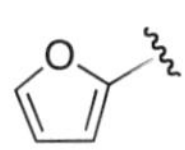	Et_2Zn	97	96
8[d]	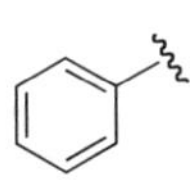	Me_2Zn	87	97
9	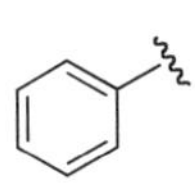	$n\text{Bu}_2Zn$	92	96
10[e]	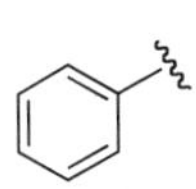	$i\text{Pr}_2Zn$	84	95

[a] See reference 7 [b] 2 mmol scale [c] Specific conditions: 2.8 mol% $Cu(OTf)_2$ / 3.0 mol% **(R,R)-BozPHOS** at -15 °C [d] Specific conditions: 5.0 mol% $Cu(OTf)_2$• Toluene / 5.0 mol% **(R,R)-BozPHOS** at r.t. [e] 3 equiv of $i\text{Pr}_2Zn$.

1. Département de Chimie, Université de Montréal, P.O. Box 6128, Station Downtown, Montréal (Québec) Canada, H3C 3J7.
2. Sisko, J.; Mellinger, M.; Sheldrake, P. W.; Baine, N. H. *Org. Synth.* **2000**, *77*, 198-205.
3. Côté, A.; Desrosiers, J.-N.; Boezio, A. A.; Charette, A. B. *Org. Synth.* preceding preparation in this volume.
4. Fujihara, H.; Nagai, K.; Tomioka, K. *J. Am. Chem. Soc.* **2000**, *122*, 12055–12056.
5. Porter, J. R.; Traverse, J. F.; Hoveyda, A. H.; Snapper, M. L. *J. Am. Chem. Soc.* **2001**, *123,* 10409–10410.
6. Hermanns, N.; Dahmen, S.; Bolm, C.; Bräse, S. *Angew. Chem. Int. Ed.* **2002**, *41*, 3692–3694.
7. Boezio, A. A.; Pytkowicz, J.; Côté, A.; Charette, A. B. *J. Am. Chem. Soc.* **2003**, *125*, 14260–14261.
8. Jennings, W. B.; Lovely, C. J. *Tetrahedron Lett.* **1988**, , 3725-3728.
9. Krzyzanowska, B.; Stec, W.J. *Synthesis* **1978**, 521–524.
10. (a) Mecozzi, T.; Petrini, M. *J. Org. Chem.* **1999**, *64,* 8970–8972. (b) Chemla, F.; Hebbe, V.; Normant, J.-F. *Synthesis* **2000**, 75–77. (c) Kanazawa, A. M.; Denis, J.-N.; Greene, A. E. *J. Org. Chem.* **1994**, *59*, 1238–1240.
11. Côté, A.; Boezio, A. A. ; Charette, A. *Proc. Natl. Acad. Sci. USA* **2004**, *101*, 5405–5410.
12. Dahmen, S.; Bräse, S. *J. Am. Chem. Soc.* **2002**, *124,* 5940–5941.
13. Boezio, A. A.; Charette, A. B. *J. Am. Chem. Soc.* **2003***, 125,* 1692–1693.

(*S*)-(+)-2,4,6-TRIMETHYLBENZENESULFINAMIDE

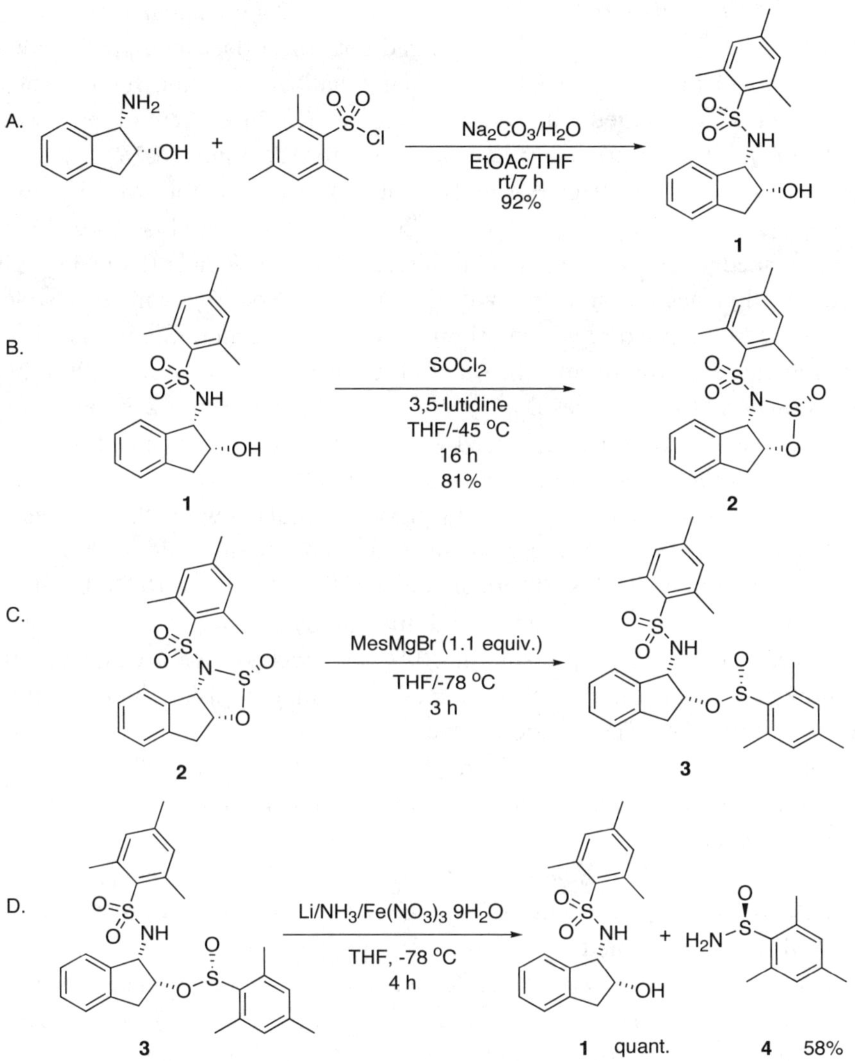

Submitted by Tokala Ramachandar, Yongzhong Wu, Junyi Zhang, and Franklin A. Davis.[1]
Checked by Shinji Harada, Riichiro Tsuji, and Masakatsu Shibasaki.

1. Procedure

*A. (1S,2R)-(+)-N-(2-Hydroxyindan-1-yl)-2,4,6-trimethylbenzenesulfonamide (**1**).*[2] A 1-L, 2-necked, round-bottomed flask equipped with a magnetic stirring bar, a 100-mL pressure-equalizing addition funnel, and a glass stopper is charged with sodium carbonate (21.34 g, 0.2 mol) and water (100 mL). The reaction mixture is stirred for 20 min. (1*S*,2*R*)-(–)-*cis*-1-Amino-2-indanol (15.0 g, 0.1 mol) (Note 1) is added, followed by ethyl acetate (250 mL) (Note 2). After 30 min, the addition funnel is charged with 2,4,6-trimethylbenzenesulfonyl chloride (21.3 g, 0.097 mol) (Note 1) in 50 mL of ethyl acetate and tetrahydrofuran (1:1) (Note 3), which is added dropwise to the mixture within 20 min. After the reaction mixture is stirred at ambient temperature for 7 h, water (100 mL) and ethyl acetate (200 mL) are slowly added. The resulting mixture is transferred to a 2 L separatory funnel. The reaction flask is rinsed with ethyl acetate (50 mL) and the ethyl acetate solution is added to the separatory funnel. The mixture is shaken, the phases are separated and the aqueous phase is extracted with ethyl acetate (2 × 200 mL). The combined organic phase is washed with water (100 mL), 1 N hydrochloric acid (2 × 100 mL), water (100 mL), brine (50 mL), dried over anhydrous sodium sulfate, and filtered by gravity. The solvent is removed under reduced pressure at 30 °C or lower to give a white/slightly yellow solid, which is dissolved in a 1 L Erlenmeyer flask containing ethyl acetate (150 mL). The solution is warmed to 40 °C in a water bath, and *n*-hexane (300 mL) (Note 4) is added. The solution is kept at –20 °C (Note 5) for 16 h and the resulting 30.9 g (92%) of white crystalline solid of **1** (Note 6) is collected.

*B. 3-(2,4,6-Trimethylbenzenesulfonyl)-3,3a,8,8a-tetrahydro-2H-1-oxa-2λ⁴-thia-3-aza-cyclopenta[a]inden-2-ol (**2**).*[2] An oven-dried, 1-L three-necked, round-bottomed flask (Note 7) equipped with a magnetic stirring bar, temperature probe, an argon inlet, and a rubber septum is charged with **1** (30.0 g, 0.090 mol). Tetrahydrofuran (180 mL) (Note 3) is added via syringe at room temperature and the solution is cooled to –45 °C (Note 8). To the colorless solution is added thionyl chloride (9.9 mL, 0.135 mol) (Note 9) *via* syringe over a 5 min period. A tetrahydrofuran (150 mL) solution containing 3,5-lutidine (25.8 mL, 0.226 mol) (Note 9) is added to the reaction mixture over 60 minutes *via* cannula at –45 °C. A white precipitate appears almost immediately. After stirring for 16 h at –45 °C, the reaction mixture is quenched at this temperature by dropwise addition of a saturated aqueous sodium bicarbonate solution (120 mL) using a 250-mL dropping funnel. The mixture is diluted with ethyl acetate (150 mL) and warmed to room temperature with stirring. The mixture is shaken, the phases are separated and the aqueous phase is extracted with ethyl acetate (2 × 200 mL). The reaction flask is rinsed with ethyl acetate (50 mL) and the combined organic phases are transferred to a 2-L separatory funnel. The organic phase is washed with brine (100 mL), dried over anhydrous sodium sulfate, and filtered into a 500-mL round-bottomed flask. The solvent is removed under reduced pressure at 30 °C or lower. The resulting light yellow solid is stirred with *n*-heptane (250 mL) (Note 10) for 2 h affording a white precipitate. The slurry is filtered and the white cake is washed with *n*-heptane (75 mL) to give 33.0 g (yield 96.5%) of white powder. The white powder is dissolved in ethyl acetate (450 mL) in a 1-L Erlenmeyer flask at 40 °C and *n*-hexane (200 mL) is added. The solution is cooled to –20 °C (Note 6) for 16 h and filtered to give 27.5 g (81%) of a white crystalline solid **2** (Note 11).

C. (R$_S$,1S,2R)-(–)-2,4,6-Trimethylbenzenesulfinic acid 1-(2,4,6-trimethylbenzenesulfonylamino)-2-indan-2-yl ester (3).[2] An oven-dried, 1-L three-necked, round-bottomed flask (Note 7) equipped with a magnetic stirring bar, a temperature probe and an argon inlet is charged with white crystalline **2** (20.0 g, 0.053 mol). Tetrahydrofuran (400 mL) is added via syringe at room temperature and the solution is cooled to –78 °C. After 5 min, 2-mesitylmagnesium bromide (58.4 mL, 1.0 M solution in tetrahydrofuran, 0.058 mol) (Note 12) is added to the solution over a 5 min period. The slightly yellow reaction mixture is stirred at –78 °C for 3 h and quenched at this temperature by addition of a saturated sodium bicarbonate aqueous solution (100 mL) via a 250-mL dropping funnel. The mixture is diluted with ethyl acetate (150 mL), warmed to room temperature with stirring, and transferred to a 2-L separatory funnel. The mixture is shaken, the phases are separated, and the aqueous phase is extracted with ethyl acetate (2 × 200 mL). The combined organic phases are washed with brine (100 mL), dried over anhydrous sodium sulfate, and filtered. The solvent is removed under reduced pressure at 30 °C or lower to give 26.8 g (quantitative) of a white foam, which is impure **3** (Note 13). The resulting white foam is used without further purification.

D. (S)-(+)-2,4,6-Trimethylphenylsulfinamide (4).[2] An oven-dried, 1-L 3-necked round-bottomed flask (Note 7) is equipped with a magnetic stirring bar. One neck of the reaction flask is fitted with an acetone-dry ice-cooled cold finger condenser bearing an argon inlet/outlet vented through a mineral oil bubbler. Another neck is fitted with a gas inlet valve, and the third neck is fitted with a dropping funnel. The flask is cooled to –78 °C using an external acetone-dry ice bath. Ammonia (400 mL) is condensed into the flask through the gas inlet valve. The gas inlet is replaced with a stopper and crystalline iron(III) nitrate (0.1 g, 0.25 mmol) (Note 14) is added. The solution develops a brown (Note 15) color immediately. Then lithium wire (10.0 g, 1.443 mol) (Note 16) is added portion wise over 30 min and the resulting gray (Note 17) solution is stirred at –78 °C for 90 min. A solution containing the white foam **3** (20.0 g, 0.039 mol) in THF (100 mL) is added over a 30 min period via dropping funnel. The reaction mixture is stirred at this temperature for 4 h, and quenched at this temperature by addition of solid NH_4Cl (25.0 g) portion-wise over 20 min. The mixture is warmed to room temperature and the ammonia is allowed to evaporate to give a dark solid residue, to which is added Et_2O (200 mL). The mixture is filtered under vacuum into a 1-L round-bottomed flask. The solid residue is thoroughly washed with excess Et_2O (2 × 250 mL) and the combined organic solutions are concentrated. Flash chromatography (SiO_2, EtOAc:hexanes, 1:1) (Note 18) affords 13.3 g (100%) of chiral auxiliary **1** (product in *Part A*) and 4.95 g of crude **4**. This material is dissolved in EtOAc (30 mL) and *n*-hexane (30 mL) is added. The solution is kept at –20 °C (Note 6) for 24 h and then filtered to give 4.25 g (58%) of white crystalline **4** (Note 19 and 20).

2. Notes

1. (1*S*,2*R*)-(–)-*cis*-1-Amino-2-indanol and 2,4,6-trimethylbenzenesulfonyl chloride, were purchased from Aldrich Chemical Company, Inc.

2. The checkers purchased ethyl acetate from Wako Pure Chemical Industries, Ltd. and used it as received. The submitters purchased ethyl acetate from Fisher Scientific Company.

3. Reagent grade tetrahydrofuran was purified by passing through columns packed with activated alumina and supported copper catalyst (Glass Contour, Irvine, CA).

4. The checkers purchased *n*-hexane from Wako Pure Chemical Industries, Ltd. and used it as received. The submitters purchased *n*-hexane from Fisher Scientific Company.

5. A refrigerator having a freezer compartment at –20 °C was used.

6. Compound (+)-**1** exhibits the following physical and spectroscopic properties: mp 149–150 °C; $[\alpha]_D$ +17.9 (*c* 1.09, $CHCl_3$); IR (KBr): 3476, 3331, 2949, 1601, 1331, 1154 cm^{-1}; ^1H NMR (500 MHz, $CDCl_3$) δ: 2.33 (s, 3 H), 2.72 (s, 6 H), 2.91 (d, *J* = 17.0 Hz, 1 H), 3.08 (dd, *J* = 5.0, 16.5 Hz, 1 H), 4.40–4.43 (m, 1 H), 4.62 (dd, *J* = 5.0, 10.0 Hz, 1 H), 5.34 (d, *J* = 9.5 Hz, 1 H), 7.00 (s, 2 H), 7.12–7.24 (m, 4 H); ^{13}C NMR (125 MHz, $CDCl_3$) δ: 20.9, 23.0, 39.4, 61.1, 72.9, 124.6, 125.3, 127.2, 128.5, 132.0, 133.7, 139.3, 139.4, 139.6, 142.6; Anal. Calcd for $C_{18}H_{21}NO_3S$: C, 65.23; H, 6.39; N, 4.23. Found: C, 65.16; H, 6.38; N, 4.11.

7. All glassware was dried at 120 °C for at least 4 h and cooled to room temperature in a desiccator prior to use.

8. An acetonitrile/dry ice bath was used.

9. Thionyl chloride and 3,5-lutidine were purchased from Aldrich Chemical Company, Inc and used as received. The checkers found that 1.5 equiv of $SOCl_2$ is necessary to drive the reaction to completion.

10. The checkers purchased *n*-heptane from Wako Pure Chemical Industries, Ltd. and used it as received. The submitters purchased it from Fisher Scientific Company.

11. The known compound (–)-**2**[3] exhibits the following physical and spectroscopic properties: mp 170–171 °C; $[\alpha]_D$ –2.6 (*c* 1.0, $CHCl_3$), IR (KBr): 2937, 1601, 1335, 1158 cm^{-1}; ^1H NMR (500 MHz, $CDCl_3$) δ: 2.38 (s, 3 H), 2.73 (s, 6 H), 3.42 (dd, *J* = 7.0, 18.5 Hz, 1 H), 3.57 (d, *J* = 17.5 Hz, 1 H), 5.54 (d, *J* = 6.5 Hz, 1 H), 5.85 (dt, *J* = 1.5, 7.0 Hz, 1 H), 6.59 (d, *J* = 8.0 Hz, 1 H), 7.07–7.27 (m, 3 H); ^{13}C NMR (125 MHz, $CDCl_3$) δ: 21.1, 23.1, 39.3, 66.2, 95.6, 124.7, 125.4, 127.8, 129.4, 131.9, 132.5, 138.41, 138.45, 140.9, 144.7.

12. 2-Mesitylmagnesium bromide (1.0 M solution in tetrahydrofuran) was purchased from Aldrich Chemical Company, Inc.

13. The submitter's report an isolated yield of 23.7 g (90%). A portion of compound **3** was purified for the purpose of characterization by flash chromatography (SiO_2, EtOAc:hexanes, 3:7) as follows: The product (approximately 0.1 g) is charged on a column (40 x 2.5 cm) of 50 g of silica gel and eluted with 200 mL of hexanes. At this point, fraction collection (25-mL fractions) begins, and elution is continued with 400 mL of 15% EtOAc-hexane until compound **3** is obtained in fractions 5-10. These fractions are concentrated by rotary evaporation (25 °C, 15 mmHg) to give 0.112 g (85%) of a white crystalline solid. The R_f of compound **3** is 0.57. The known compound (–)-**3**[3] exhibits the following physical and

spectroscopic properties: mp 145–146 °C; $[\alpha]_D$ –58.8 (*c* 1.0, $CHCl_3$), IR (KBr): 2970, 1602, 1332, 1154 cm^{-1}; ^1H NMR (500 MHz, $CDCl_3$) δ: 2.25 (s, 3 H), 2.32 (s, 3 H), 2.43 (s, 6 H), 2.70 (s, 6 H), 3.08 (dd, *J* = 4.5, 17.0 Hz, 1 H), 3.16 (d, J = 16.5 Hz, 1 H), 4.85 (dd, *J* = 4.5, 9.0 Hz, 1 H), 4.96–4.98 (m, 1 H), 5.53 (d, *J* = 9.5 Hz, 1 H), 6.79 (s, 2 H), 6.98 (s, 2 H), 7.13–7.25 (m, 4 H); ^{13}C NMR (125 MHz, $CDCl_3$) δ: 18.8, 20.9, 21.1, 23.1, 37.2, 60.2, 81.0, 124.4, 124.7, 127.4, 128.2, 130.7, 132.0, 134.2, 137.4, 137.6, 137.8, 139.3, 140.0, 142.3.

14. Iron(III) nitrate was purchased from Acros Organics, Inc. and was used as received.

15. The submitters reported that a gray color is formed.

16. Lithium wire (3.2 mm diameter in mineral oil, 99.9%, Catalog No: 220914) was purchased from Aldrich Chemical Company, Inc., and was cut into small pieces.

17. The submitter reported that a blue color is formed.

18. The product is charged on a column (60 x 7.5 cm) of 400 g of silica gel and eluted with 500 mL of hexane. At that point, fraction collection (75-mL fractions) begins, and elution is continued with 1 L of 20% EtOAc-hexane until auxiliary **1** is obtained in fractions 10-15. At this time the column is eluted with 1.5 L of 80% of EtOAc-hexane until sulfinamide **4** was obtained in fractions 15-25. The products fractions are concentrated by rotary evaporation (25 °C, 15 mmHg). The R_fs of the two compounds are 0.51 (for chiral auxiliary **1**) and 0.24 (for **4**). The resulting chiral auxiliary **1** was obtained in 9.2 g (69%) by submitters. Compound **1** was used without further purification in *Part B* and the purity of the recovered auxiliary was evaluated by comparison of its physical properties with an authentic sample: mp 149–150 °C; $[\alpha]_D$ +17.9 (*c* 1.09, $CHCl_3$).

19. The known compound (+)-**4**[2] exhibits the following physical and spectroscopic properties: mp 120–122 °C; $[\alpha]_D$ + 303 (*c* 0.46, $CHCl_3$), IR (KBr): 3287, 3103, 1601, 1450, 1062, 1027, 848 cm^{-1}; ^1H NMR (500 MHz, $CDCl_3$) δ: 2.27 (s, 3 H), 2.59 (s, 6 H), 4.46 (brs, 2 H), 6.85 (s, 2 H); ^{13}C NMR (125 MHz, $CDCl_3$) δ: 19.1, 20.9, 130.8, 136.1, 138.7, 140.7. The enantiomeric excess (97%) was determined by using a Chiralcel OD, 4-6 x 250 mm, 10 μm; 9:1 (hexane/*i*-PrOH), 1.0 mL/min, 250 nm; (*S*)-**4**, r_t = 17.5 min; the minor isomer, (*R*)-**4**, r_t = 23.5 min.

20. The submitters report the use of LiHMDS to effect the conversion of **3** to **4** on a 22.4 gram scale resulted in 11.2 g (50%) of starting material **3**, 6.8 g (46%) of the chiral auxiliary **1** and 3.8 g (46%) sulfinamide **4**. On a 1.0 g scale the LiHMDS procedure gave **4** in 70% yield.[3]

Waste Disposal Information

All hazardous materials should be handled and disposed of in accordance with "Prudent Practices in the Laboratory"; National Academy Press; Washington, DC, 1995.

3. Discussion

The most direct and reliable method for the asymmetric synthesis of diverse amine derivatives is the addition of an organometallic reagent across the C-N bondof an imine having a chiral *N*-sulfinyl auxiliary.[4] Sulfinimines (*N*-sulfinylimines) provide a general solution to the problem of addition of organometallic reagents across the C-N bond of chiral imines because the sulfinyl group activates the imine for addition, is highly stereodirecting, and easily removed in the sulfinamide product giving the enantiomerically pure amine product of known stereochemistry. Generally sulfinimines are prepared by condensation of an enantiopure (*S*)- or (*R*)-sulfinamide (RS(O)-NH_2) with an aldehyde or ketone in the presence of a weak Lewis acid dehydrating agent such as $Ti(OEt)_4$.[4] For this purpose the commercially available and easily prepared *p*-toluenesulfinamide and *N*-*tert*-butane-sulfinamide are generally employed. However, a useful addition to the sulfinimine protocols would be the ability to "tune" the reactivity of the sulfinimine by varying the steric and electronic properties of the *N*-sulfinyl auxiliary.[5] The method described here for the asymmetric synthesis of (*S*)-(+)-2,4,6-trimethylphenylsulfinamide **4** is a representative general procedure for the asymmetric synthesis of structurally diverse sulfinamides developed by Senanayake and co-workers.[2] In this procedure *N*-sulfonyl-1,2-3-oxathiazolidines-2-oxide **2** is treated with Grignard reagents to afford the sulfenate ester **3**, which on reaction with Li/NH_3 or LiHMDS affords the corresponding enantiopure sulfinamide **4**.[2,3] The reaction of Grignard reagents with **3** has been used to prepared enantiomerically pure sulfoxides.[6]

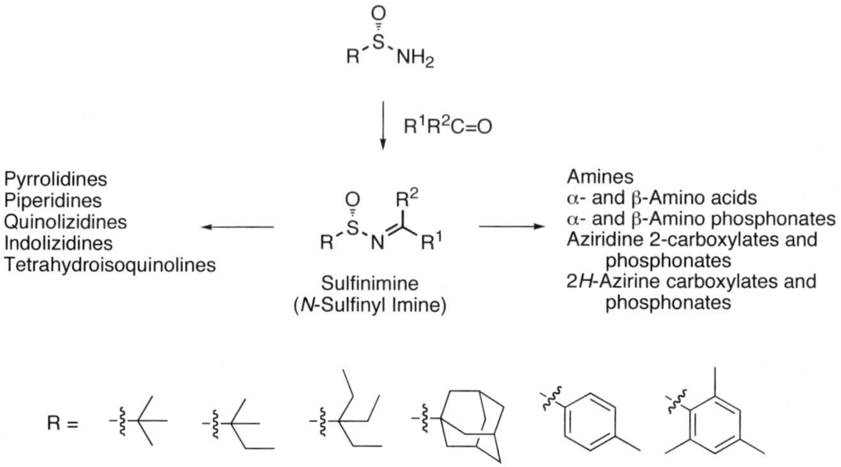

In the aza-Darzens reaction of (*S*)-(+)-*N*-(benzylidene)-2,4,6-trimethyphenylsulfinamide **5** (prepared using (*S*)-(+)-2,4,6-trimethylphenylsulfinamide (**4**) with lithium diethyl iodomethylphosphonate) a single diastereomeric aziridine 2-phosphonate (+)-**6** was obtained.[3] Other *N*-sulfinyl auxiliaries gave mixtures of products or could not be removed with MeMgBr to give the corresponding NH-aziridine 2-phosphonate (–)-**7**. NH-Aziridine 2-phosphonates are valuable chiral building block for the synthesis of α-amino phosphonates[3,7] and the 2*H*-azirine 3-phosphonate which are chiral imino dienophiles.[8] An example of the application of 2,4,6-trimethylsulfinamide has been reported.[9]

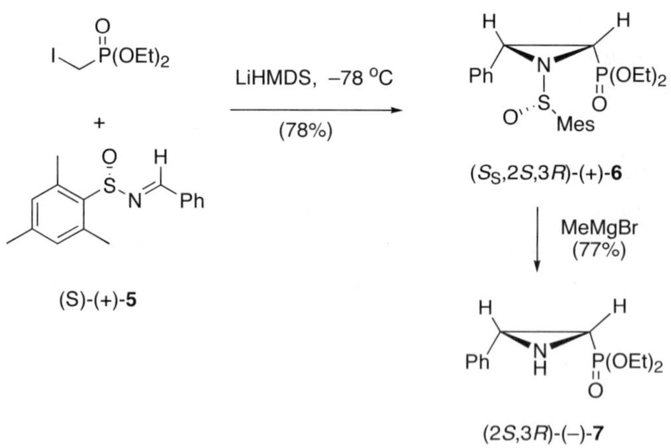

1. Department of Chemistry, Temple University, Philadelphia, Pennsylvania 19122.

2. Han, Z.; Krishnamurthy, D.; Grover, P.; Fang, Q. P.; Senanayake, C. H. *J. Am. Chem. Soc.* **2002**, *124*, 7880.

3. Davis, F. A.; Ramachandar, T.; Wu, Y. *J. Org. Chem.* **2003**, *68*, 6894.

4. For recent reviews see: (a) Senanayake, C. H.; Krishnamurthy, D.; Lu, Z.-H.; Han, Z.; Gallou, I. *Aldrichimica Acta* **2005**, *38*, 93. (b) Zhou, P.; Chen, B.-C.; Davis, F. A. *Tetrahedron* **2004**, *60*, 8003. (c) Ellman, J. A.; Owens, T. D.; Tang, T. P. *Acc. Chem. Res.* **2002**, *35*, 984.

5. For examples see: (a) Davis, F. A.; Szewczyk, J. M.; Reddy, R. E.; *J. Org. Chem.* **1996**, *61*, 2222. (b) Garcia-Ruano, J. L.; Fernandez, I.; del Prado-Catalina, M.; Hermoso, J. A.; Sanz-Aparicio, J.; Martinez-Ripoll, M. *J. Org. Chem.* **1998**, *63*, 7157. (c) Kawecki, R. *Tetrahedron* **2001**, *57*, 8385. (d) Davis, F. A.; Mohanty, P. K. *J. Org. Chem.* **2002**, *67*, 1290. (e) Han, Z.; Krishnamurthy, D.; Pflum, D.; Grover, P.; Wald, S. A.; Senanayake, C. H. *Org. Lett.* **2002**, *4*, 4025. (f) Kawecki, R. *Tetrahedron: Asymmetry* **2003**, *14*, 2827. (g) Han, Z.; Krishnamurthy, D.; Grover, P.; Fang, Q. K.; Pflum, D. A.; Senanayake, C. H. *Tetrahedron Lett.* **2003**, *44*, 4195. (h) Fernandez, I.; Valdivia, V.; Gori, B.; Alcudia, F.; Alvarez, E.; Khiar, N. *Org. Lett.* **2005**, *7*, 1307.

6. Han, Z.; Krishnamurthy, D.; Grover, P.; Wilkinson, H. A.; Fang, Q. P.; Su, X.; Lu, Z.-H.; Magiera, D.; Senanayake, C. H. *Angew. Chem. Int. Ed.* **2003**, *42*, 2032.

7. Davis, F. A.; Wu, Y.; Yan, H.; McCoull, W.; Prasad, K. R. *J. Org. Chem.* **2003**, *68*, 2410.

8. Davis, F. A.; Wu, Y.; Yan, H.; Prasad, K. R.; McCoull, W. *Org. Lett.* **2002**, *4*, 655.

9. Davis, F. A.; Song, M. *Org Lett.* **2007**, *9*, 2413.

(R_S)-(+)-2-METHYL-2-PROPANESULFINAMIDE
[*tert*-Butanesulfinamide]

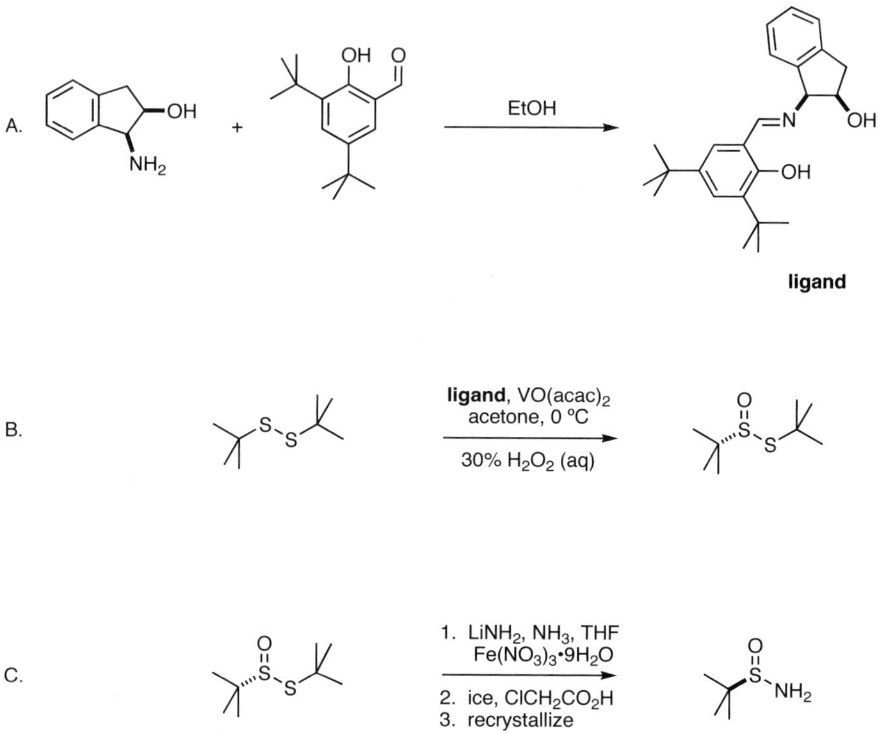

Submitted by Daniel J. Weix and Jonathan A. Ellman.[1]
Checked by Xiao Wang and Dennis P. Curran.

1. Procedures

A. (1S,2R)-1-[(2-Hydroxy-3,5-di-tert-butylbenzylidene)amino]indan-2-ol (Note 1). A 100-mL round-bottomed flask equipped with a magnetic stir bar is charged with (1*S*,2*R*)-(−)-*cis*-1-amino-2-indanol (0.77 g, 5.17 mmol) (Note 2), 3,5-di-*tert*-butyl salicylaldehyde (1.21 g, 5.16 mmol), and 33 mL of absolute ethanol. The yellow reaction mixture is stirred under argon for 2 h, and the solvent is removed under reduced pressure to give a yellow oil. The oil is dissolved in 33 mL of dichloromethane and the solvent is again removed under reduced pressure. The dissolution and evaporation are repeated. After careful vacuum drying (the substance foams as the

vacuum is initially applied), the resulting solid (1.85 g, 98%) is crushed to a yellow powder (Note 3) that is used without purification in Step B.
CAUTION: 30% hydrogen peroxide can cause burns.

B. *(R_S)-(+)-tert-Butyl tert-butanethiosulfinate.* A three-neck 1-L round-bottomed flask fitted with an overhead stirrer and two rubber septa is charged with the ligand from *Step A* (1.85 g, 5.06 mmol) and vanadyl *bis*-acetylacetonate (1.33 g, 5.00 mmol) (Note 4). Acetone (250 mL) is added and the dark green reaction mixture is stirred vigorously for 30 min while open to the air. Di-*tert*-butyl disulfide (178 g, 1.00 mol) (Note 5) is slowly poured in from a graduated cylinder, and the resulting mixture is cooled to 0 °C (Note 6). The reaction mixture is stirred vigorously and 30% aq hydrogen peroxide (110 mL, 1.10 mol) (Notes 7 and 8) is added over 20 h by using two syringe pumps (Note 9). Upon the addition of the first few drops of hydrogen peroxide, the color of the mixture changes from dark green to purple, and the purple color deepens during the addition. Reaction progress can be monitored by ^{1}H NMR spectroscopy (Note 10). After 20 h, the reaction is quenched at 0 °C by adding 50 mL of saturated aqueous sodium thiosulfate by syringe pump over 30 min (Note 11). The purple color fades through red to light blue-green. The resulting mixture is transferred to a 2-L separatory funnel. The reaction flask is rinsed with hexanes (250 mL) and the hexanes wash is added to the separatory funnel. The mixture is shaken, the layers are separated and the aqueous layer is extracted with hexanes (2 x 250 mL). The combined organic layers are then washed with brine (50 mL), dried over sodium sulfate and filtered. The solvent is removed under reduced pressure at 30 °C or lower. The resulting yellow or brown oil (216-231 g) (Notes 10 and 12) is used without purification for *part C* (Notes 13 and 14).
CAUTION: Ammonia is toxic and corrosive and should be used only in a well-ventilated fume hood.
CAUTION: This procedure generates hydrogen gas and should be performed in a well-ventilated fume hood.
CAUTION: tert-Butanethiol, although efficiently scavenged, is produced in stoichiometric quantities. It has a foul odor and is the odorant used for natural gas.

C. *(R_S)-(+)-tert-Butanesulfinamide.* A 5-L three-necked round-bottomed flask equipped with a mechanical stirrer (glass rod and glass paddle), a cold finger fitted with a gas inlet, and a Claisen adapter fitted with a glass stopper and a gas outlet connected to a mineral oil bubbler is placed under argon and cooled in a bucket of dry ice/acetone. The cold finger is filled with dry ice/acetone, and the gas inlet is changed from argon to an ammonia cylinder (Note 15). The ammonia gas pressure is adjusted so that most of the ammonia is condensed into the vessel as it passes through the cold finger (Note 16). After about 2 L of ammonia is condensed (about 2.5 h), the mineral oil bubbler is removed from the Claisen adapter gas outlet and replaced with an argon line. The ammonia line is removed from the cold finger gas inlet and replaced with a mineral oil bubbler.

The reaction mixture at –75°C (internal temperature) is flushed gently with an argon stream, and $Fe(NO_3)_3{\bullet}9H_2O$ (1.0 g, 2.48 mmol) (Note 17) is added. The mixture turns brown as dissolution occurs on stirring. The Claisen adaptor is replaced with an adaptor containing a low temperature thermometer. The cooling bath is lowered and pieces of lithium ribbon (17.4 g, 2.50 mol) (Note 18) are added with tweezers in ~1 g portions (Note 19). Each piece is cut in half just prior to addition to ensure fast dissolution and reaction. As each piece of lithium is added, the mixture briefly turns a dark blue color. This fades to give a grey suspension of lithium amide with concomitant evolution of hydrogen gas (Note 19). Once the reaction returns to the grey color, another piece of lithium is added. The dry ice/acetone bath is periodically raised to contact the bottom of the flask to slow the refluxing caused by the formation of $LiNH_2$ (Note 19). After the addition of the lithium is complete (about 1.5 h), the reaction mixture is allowed to reflux for 15 minutes (Note 20), and then submerged in the dry ice/acetone bath.

After cooling to –78 °C (about 30 minutes) and replacement of the thermometer adaptor by an addition funnel, a solution of crude (R_S)-(+)-*tert*-butyl *tert*-butanethiosulfinate (entire crude product from Step B, 1.00 mol) in 320 mL of freshly distilled, dry THF is added to the vigorously stirring reaction mixture over 45 minutes via the addition funnel (Note 21). The addition funnel is rinsed with a small amount (5-10 mL) of THF and this is added to the reaction mixture. The addition funnel is replaced by the thermometer adaptor. After one hour at –78 to –72 °C, the cooling bath is removed, stirring is slowed to the lowest possible rate and the reaction mixture is allowed to slowly warm to room temperature overnight (15 h) under a stream of argon (Note 22).

The resulting thick gray mixture, still under an argon stream, is cooled in an ice-water bath. Ice (540 g, 30 mmol) is added in one portion, and the dark green mixture is stirred until it is mostly homogeneous (Note 23). Chloroacetic acid (104 g, 1.10 mol) (Note 24) is added to the ice-cold mixture in five portions over 20 minutes. The resulting mixture is stirred overnight (12 h) at ambient temperature under argon. The claret solution is poured into a 2 L separatory funnel and extracted with methylene chloride (6 x 1 L). The combined organic layers are dried over Na_2SO_4, and filtered through a pad of celite. The solvent is removed under reduced pressure. Hexanes (200 mL) are added and the solvent is removed under reduced pressure. More hexanes (50 mL) are added and the evaporation is repeated. The resulting yellow-orange solid is triturated (Note 25) with hexanes (250 mL) in a 500 mL Erlenmeyer flask. After the solids are crushed, a magnetic stir bar is added and the slurry is stirred for 30 minutes. The solids are collected on a suction filter, washed with hexanes (15 mL), and suction-dried (Note 26). The triturated material is then recrystallized by dissolving it in boiling hexanes (150 mL), followed by slow cooling to room temperature with fast stirring. The crystals are collected by suction filtration and washed with cold hexanes (300 mL, precooled to –30 °C) and suction-dried to provide 84.9-86.9 g (70-72% from di-*tert*-butyl disulfide) of the desired product (Note 27).

2. Notes

1. This ligand has been reported several times.[2] This procedure is adapted from the procedure of Ruck and Jacobsen.[2d] (1*S*,2*R*)-(–)-*cis*-1-Amino-2-indanol leads to (R_S)-*tert*-butanesulfinamide while (1*R*,2*S*)-(+)-*cis*-1-amino-2-indanol leads to (S_S)-*tert*-butanesulfinamide.

2. (1*S*,2*R*)-(–)-*cis*-1-Amino-2-indanol was purchased from Strem (submitters) or Aldrich (checkers). 3,5-Di-*tert*-butyl-2-hydroxy-benz-aldehyde was purchased from Aldrich.

3. The yellow powder exhibited the following features: mp 64 - 68 °C (65 - 68 °C lit.[2c]); ^{1}H NMR ($CDCl_3$, 300 MHz) δ 1.33 (s, 9 H), 1.43 (s, 9 H), 2.17 (br s, 1 H), 3.13 (dd, J = 4.9, 15.9 Hz, 1 H), 3.26 (dd, J = 5.8, 15.9 Hz, 1 H), 4.69 (m, 1 H), 4.81 (d, J = 5.3 Hz, 1 H), 7.19 (d, J = 2.2 Hz, 1 H), 7.20-7.32 (m, 4 H), 7.44 (d, J = 2.3 Hz, 1 H), 8.63 (s, 1 H), 13.12 (br s, 1 H); ^{13}C NMR ($CDCl_3$, 75 MHz) δ 29.4, 31.5, 34.1, 35.0, 39.7, 75.2, 75.7, 117.8, 124.9, 125.4, 126.5, 127.0, 127.6, 128.5, 136.9, 140.4, 140.8, 140.9, 157.9, 168.3; IR (neat) 739, 755, 1626, 2957 cm^{-1}; HRMS-FAB (*m/z*) calcd for $C_{24}H_{31}NO_2$, 365.2355, found, 365.2353. Anal. Calcd for $C_{24}H_{31}NO_2$: C, 78.86; H, 8.55; N, 3.83. Found: C, 79.15; H, 8.69; N, 3.69.

4. Vanadium (IV) bis(acetylacetonato)oxide was purchased from Strem (submitters) or Aldrich (checkers). Industrial Grade acetone was used as received.

5. Di-*tert*-butyl disulfide was purchased from Acros and distilled (91-96 ºC/10 mmHg). The distillation proceeds smoothly if the pot is not allowed to boil vigorously, otherwise excessive foaming and bumping can result. Avoid Boileezers® or stir bars in the pot.

6. A recirculating chiller (Lauda RM-6B or FTS System) with an external cooling coil and a 2-propanol bath was used. Slightly higher temperatures result in a small decrease in enantioselectivity, but no decrease in conversion.

7. Hydrogen peroxide (30% aq, stabilized with sodium stannate) was purchased from Fisher Scientific.

8. This mixture of hydrogen peroxide, acetone, and water is considered safe. See the "hydrogen peroxide product information manual" available from Eka chemicals (www.eka.com).

9. The submitters used a model NE-1600 programmable six-syringe pump from New Era Pump Systems, Inc., while checkers used a two separate syringe pumps model M361 from Thermo Orion.

10. ^{1}H NMR ($CDCl_3$, 300 MHz) δ 1.39 (s, 9 H), 1.54 (s, 9 H). When a reaction is incomplete, a starting material resonance, is visible at 1.31 ppm; ^{13}C NMR ($CDCl_3$, 75 MHz) δ 24.2, 32.3, 48.6, 59.3. Analysis by chiral phase HPLC (Daicel Chiralpak AS-H column, 25 cm x 0.46 cm; 97:3 hexanes:2-propanol; 1 mL/min; 254 nm; (*S*) r_t = 6.6 min; (*R*) r_t = 7.8 min) shows that the product has an 86-87% ee.

11. It is best to quench any remaining hydrogen peroxide before allowing the reaction to warm to room temperature because a mild exotherm can occur due to hydrogen peroxide decomposition upon warming. When saturated aq sodium thiosulfate was added in one portion with cooling, an exotherm occurred which raised the internal temperature of the reaction to 32 ºC; therefore, slow addition with adequate cooling is preferred.

12. By ^{1}H NMR spectroscopic analysis, the product contained less than 1% acetone (peak at 2.17 ppm in $CDCl_3$). If acetone is still present, it is removed by adding more hexanes and removing the azeotrope under reduced pressure. The checkers sometimes observed small amounts of an insoluble yellow solid in the late stages of solvent removal. This is elemental sulfur (mp 111-113 ºC, lit. 112.8 ºC), and was removed by decantation. The decomposition of $S_2O_3^{-2}$ to S and SO_3^{-2} is a well-known reaction.

13. *tert*-Butyl *tert*-butanethiosulfinate is a useful starting material for the synthesis of *tert*-butyl sulfoxides. It can be purified by distillation or crystallization if needed.[2a]

14. The thiosulfinate should be stored at –20 °C. At this temperature it is a solid and no decomposition has been observed over several months. Gradual decomposition and erosion of enantiopurity occur on storage at room temperature.

15. Ammonia gas (99.99% pure) was purchased from Matheson or Praxair.

16. The dry ice evaporates rapidly from the cold finger if the ammonia is passed in too quickly.

17. $Fe(NO_3)_3{\bullet}9H_2O$ was purchased from Mallinckrodt (submitters) or Aldrich (checkers).

18. The submitters used lithium bars (12.7 mm x 165 mm, 99.9 % pure) purchased from Alfa Aesar. The checkers used lithium ribbon (dissolves more rapidly) purchased from Aldrich (45 mm x 0.75 mm, 99.9%). The lithium was washed with hexanes to remove storage oil prior to rapid weighing in a tared beaker of hexanes. The thermometer adaptor was partially removed to drop in the lithium.

19. Lithium amide does not form rapidly at –78 °C, but the blue color begins to form as the mixture is warmed to about –45 °C. During the addition, the cooling bath is raised and lowered periodically to keep the internal temperature between –50 °C and –45 °C.

20. All the lithium must be dissolved and converted to lithium amide to prevent an exotherm from occurring during the subsequent evaporation of the ammonia.

21. The internal temperature is kept at –73 °C or below. The thiosulfinate is added directly into the reaction mixture to prevent it from solidifying on the sidewalls of the vessel. Adjusting the mixing speed to wash any solidified material back into the reaction mixture may be necessary.

22. The submitters report that stirring is not needed during this stage.

23. The initially thick precipitate largely dissolves, but a small amount of powder remains suspended. If the mixture is not stirred during the ammonia evaporation or if the stirring stops, it can be difficult to restart the stirrer. Churning the glass rod by hand breaks up the mass so that stirring can be reinitiated.

24. Chloroacetic acid was purchased from Acros or Aldrich.

25. Trituration improves the enantiopurity and removes most of the colored impurities. The recovery in the trituration will decrease if any CH_2Cl_2 remains in the crude material because *tert*-butanesulfinamide is very soluble in this solvent.

26. This material has a 95% ee and represents a 75% yield from di-*tert*-butyl disulfide.

27. The product is a white to off-white crystalline solid; mp 102 - 105 °C (102 - 103 °C lit.[2a]); ^{1}H NMR ($CDCl_3$, 300 MHz) δ 1.22 (s, 9 H), 3.73 (s, 2 H). ^{13}C NMR ($CDCl_3$, 75 MHz) δ 22.1, 55.2; IR (neat) 1030, 1363, 1460, 1588, 3119, 3216 cm^{-1}; Anal. Calcd for $C_4H_{11}NOS$: C, 39.64; H, 9.15; N, 11.56. Found: C, 39.62; H, 9.12; N, 11.29. Chiral phase HPLC (Daicel Chiralpak AS-H column, 25 cm x 0.46 cm; 90:10 hexanes:ethanol; 1.2 mL/min; 220 nm; (*R*) r_t = 7.9 min; (*S*) r_t = 10.6 min) shows that the product has ≥ 99% ee; The submitters report that the analysis can also be conducted on a Daicel Chiralpak OD column; 93/7, hexanes:i-PrOH, 1 mL/min, 220 nm; (S) r_t = 23.7 min, (R) r_t = 27.7 min;

Waste Disposal Information

All toxic materials were disposed of in accordance with "Prudent Practice in the Laboratory", National Academy Press; Washington, D. C., 1995.

3. Discussion

Chiral amines are key components of many pharmaceutical agents, materials, and catalysts. Since its introduction in 1997,[3] enantiopure *tert*-butanesulfinamide has proven to be an extremely versatile chiral ammonia equivalent for the asymmetric synthesis of amines.[4]

The original synthesis included an oxidation that did not scale well and did not provide procedures for removing the offensive *tert*-butane thiol odor. Recent work has improved the scalability of the oxidation, allowed easy access to both (R_S)- and (S_S)-*tert*-butanesulfinamide, and eliminated the tedious distillation of the *tert*-butyl *tert*-butanethiosulfinate intermediate.[2g] This procedure used sodium hydroxide solutions to trap the thiol; however, the thiol removal was incomplete. *tert*-Butane thiol is the primary odorant added to natural gas because its odor can be detected at very low levels, and because it is heavier than air and resists oxidation well.

We initially explored trapping the thiol by the use of gas washing bottles filled with bleach solution. While this was effective at controlling odor, it presents an unnecessary danger when combined with ammonia (possible formation of chlorine gas, nitrogen trichloride, or hydrazine; explosion hazard). After a survey of various methods, we found that evaporation of the ammonia before quenching the reaction with ice eliminated thiol odor during the ammonia removal process. Furthermore, if chloroacetic acid is then added to the crude aqueous solution and the mixture stirred, the residual lithium *tert*-butanethiolate reacts with the lithium chloroacetate to form the highly water soluble lithium (*tert*-butylthio)acetate. At this stage, the reaction mixture has no discernible thiol odor. Simple extraction with methylene chloride then separates the sulfinamide from the lithium (*tert*-butylthio)acetate. Trituration followed by crystallization yields highly enantioenriched *tert*-butanesulfinamide. This procedure allows for the preparation of either enantiomer of *tert*-butanesulfinamide in good yield on mole scale with no thiol odor and only trituration and crystallization for purification.

1. Department of Chemistry, University of California, Berkeley, CA 94720; jellman@uclink.berkeley.edu.

2. (a) Cogan, D. A.; Liu, G.; Kim, K.; Backes, B. J.; Ellman J. A. *J. Am. Chem. Soc.* **1998**, *120*, 8011-8019. (b) Zhen, L.; Fernandez, M.; Jacobsen, E. N. *Org. Lett.* **1999**, *1*, 1611-1613. (c) Flores-Lopez, L. Z.; Parra-Hake, M.; Somanathan, R.; Walsh, P. J. *Organometallics* **2000**, *19*, 2153-2160. (d) Ruck, R. T.; Jacobsen, E. N. *J. Am. Chem. Soc.* **2002**, *124*, 2882-2883. (e) Gama, A.; Flores-Lopez, L. Z.; Aguirre, G.; Parra-Hake, M.; Somanathan, R.; Walsh, P. J. *Tetrahedron Asymm.* **2002**, *13*, 149-154. (f) Beatrice, P.; Anson, M. S.; Campbell, I. B.; MacDonald, S. J. F.; Priem, G.; Jackson, R. F. W. *Synlett* **2002**, 1055-1060. (g) Weix, D. J.; Ellman, J. A. *Org. Lett.* **2003**, *5*, 1317-1320.

3. Liu, G.; Cogan, D.; Ellman, J. A. *J. Am. Chem. Soc.* **1997**, *119*, 9913-9914.

4. (a) Ellman, J. A.; Owens, T. D.; Tang, T. P. *Acc. Chem. Res.* **2002**, *35*, 984-995. (b) Ellman, J. A. *Pure Appl. Chem.* **2003**, *75*, 39-46. (c) Mukade, T.; Dragoli, D. R.; Ellman, J. A. *J. Comb. Chem.* **2003**, 590-596. (d) Kochi, T.; Tang, T. P.; Ellman, J. A. *J. Am. Chem. Soc.* **2003**, 11276-11282. (e) Evans, J. W.; Ellman, J. A. *J. Org. Chem.* **2003**, 9948-9957. (f) Kells, K. W.; Chong, J. M. *Org. Lett.* **2003**, *5*, 4215-4218. (g) Cooper, I. R.; Grigg, R.; Hardie, M. J.; MacLachlan, W. S.; Sridharan, V.; Thomas, W. A. *Tetrahedron Lett.* **2003**, *44*, 2283-2285 (h) Rech, J. R.; Floreancig, P. E. *Org. Lett.* **2003**, *5*, 1495-1498. (i) Jung, P. M.; Beaudegnies, R.; De Mesmaeker, A.; Wendeborn, S. *Tetrahedron Lett.* **2003**, *44*, 293-297.

2,2-DIETHOXY-1-ISOCYANOETHANE
(2,2-Diethoxyethyl isocyanide, Isocyanoacetaldehyde diethyl acetal)

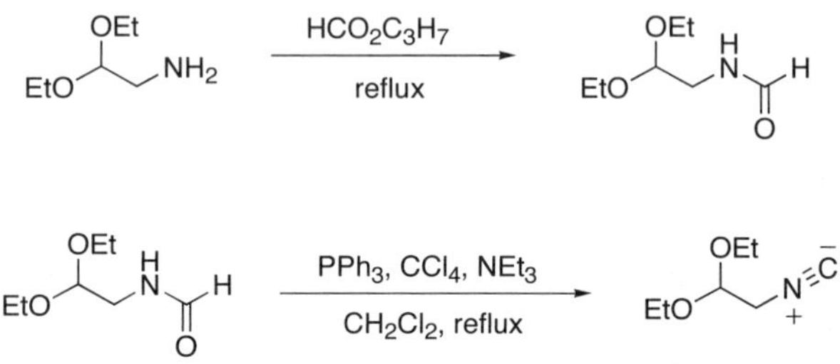

Submitted by Francesco Amato and Stefano Marcaccini.[1]
Checked by Raghuram S. Tangirala and Dennis P. Curran.

1. Procedure

CAUTION: All the operations must be conducted in an efficient hood because the isocyanide has an obnoxious odor.

A. N-(2,2-Diethoxy)ethyl formamide. A 100-mL round-bottomed flask equipped with a magnetic stir bar and fitted with a reflux condenser is charged with aminoacetaldehyde diethyl acetal (28.34 g, 213 mmol, Note 1) and propyl formate (22.48 g, 255 mmol, Note 2). The resulting clear solution is heated at reflux in an oil bath for 3 h (Note 3). After cooling, the reaction mixture is transferred to a 250-mL round-bottomed flask and freed from the 1-propanol and the unreacted propyl formate by rotary evaporation. The residue is transferred to a distillation apparatus equipped with a 10 cm Vigreux column and a two-necked receiver, and distilled under reduced pressure. After a short forerun, the fraction boiling at 110-111 °C (0.5 mmHg) is collected to give 29.3–29.5 g (86% yield) of *N*-(2,2-diethoxy)ethyl formamide (Notes 4, 5).

B. 2,2-Diethoxy-1-isocyanoethane. A 500-mL one-necked flask equipped with a reflux condenser bearing a $CaCl_2$ trap at the upper end is charged with *N*-(2,2-diethoxyethyl formamide (24.20 g, 150 mmol), tetrachloromethane (24.61 g, 160 mmol, Note 6), triphenyl phosphine (44.59 g,170 mmol, Note 7), triethylamine (17.20 g, 150 mmol, Note 8), and 150 mL dichloromethane (Note 9). The clear mixture is heated at reflux in an oil bath, and a precipitate (triphenyl phosphine oxide) begins to appear after 20-30 min. After 3.5 h at reflux, the suspension is cooled to 5 °C and filtered through a Büchner funnel under vacuum from a water aspirator. The collected solid is washed with 50 mL diethyl ether. The filtrate and the washings are combined and evaporated to dryness, and the residue is stirred with a mixture of 100 mL ethyl ether and 100 mL pentanes (Note 10). The resulting suspension is allowed to stand overnight in the freezer (Note 11), and then filtered through a fritted funnel under vacuum with chilling of the collected filtrate in an ice-sodium chloride bath. The solid residue is washed with 60 mL of pentanes. The filtrate is concentrated on a rotary evaporator in a fume hood and the residue is transferred to a flask equipped with a short path distillation head and a two-necked receiver. Distillation under reduced pressure gives 13.5–13.8 g (63-64%) of 2,2-diethoxy-1-isocyanoethane (Note 12), bp 60-61 °C at 1 mm Hg, as a colorless, vile-smelling liquid (Notes 13, 14).

2. Notes

1. Aminoacetaldehyde diethyl acetal was purchased from Aldrich and used as supplied.

2. Propyl formate (40 mL, purchased from Aldrich) is stirred with 20 mL aq. 5% $NaHCO_3$ for 2 min. The layers are separated, and the propyl formate layer is washed with 3 x 20 mL of distilled water then dried over magnesium sulfate. After filtration, the filtrate is distilled at atmospheric pressure and the fraction boiling at 80-81 °C is collected for use in Step A.

3. The submitters used an electric shell for heating.

4. The distillation tends to bump, but the Vigreux column prevents overflow into the receiver.

5. IR spectrum (neat) 3304, 1666 cm^{-1}; ^{1}H NMR (300 MHz, $CDCl_3$), the formamide is an 6/1 ratio of amide rotamers in this solvent, major rotamer resonances: δ 8.21 (s, 1 H, NCHO), 5.81 (broad s, 1 H, NH), 4.52 (t, J = 5.1 Hz, 1 H, O-CH-O), 3.50-3.77 (m, 4 H, $C\underline{H_2}CH_3$), 3.45 (t, J = 5.4 Hz, 2 H, CH_2N), 1.29 (t, J = 7.1 Hz, 3 H, $CH_2C\underline{H_3}$), minor rotamer resonances: δ8.05 (d, J = 13.5 Hz, 1 H, NCHO), 4.45 (t, J = 6 Hz, 1 H, O-CH-O), 3.31 (t, J = 6 Hz, 2 H, CH_2N); ^{13}C NMR (75 MHz, $CDCl_3$) major rotamer resonances: δ 161.3, 100.2, 62.4, 40.1, 14.9, minor rotamer resonances: δ 165.0, 101.3, 63.0, 44.3; LRMS (EI) *m/z* 117 (M – CH_3NO, 10%), 103 (100%), 91 (17%), 84 (55%), 75 (77%); HRMS (EI) *m/z* calcd for $C_7H_{16}NO_2$ (M + H) 162.1130, found, 162.1126.

6. Tetrachloromethane was purchased from Baker or Fisher and dried over molecular sieves before use.

7. Triphenyl phosphine (99%) was purchased from Aldrich or Acros and used as supplied.

8. Reagent grade triethylamine (Fluka) was dried over calcium hydride pellets and distilled. The fraction boiling at 89 °C was employed.

9. Dichloromethane was stored overnight on 4 Å molecular sieves (submitters) or distilled from calcium hydride (checkers) prior to use.

10. The lumps that formed were carefully broken with a spherical-ended glass rod.

11. The freezer temperature is about –15 °C. If this operation is omitted, then additional solid that precipitates during the distillation makes this process difficult.

12. IR (neat) 2156 cm^{-1}; ^1H NMR (300 MHz, $CDCl_3$) δ 4.64 (t, J = 5.4 Hz, 1 H, O-CH-O), 3.48-3.71 (m, 2 H, $C\underline{H}_2CH_3$), 3.43 (d, J = 5.4 Hz, 2 H, CH_2NC), 1.17 (t, J = 7.0 Hz, 3 H, $CH_2C\underline{H}_3$); ^{13}C NMR (75 MHz, $CDCl_3$) δ 157.8, 99.1, 62.8, 44.3, 14.8.

13. The submitters obtained 71-75% yields. The submitters report that 2,2-diethoxy-1-isocyanoethane can be stored for at least two years at –30 °C under nitrogen without appreciable decomposition. The checkers stored a sample at –20 °C for three months, and the resulting liquid was still clear and exhibited a ^1H NMR spectrum identical to that recorded on the starting sample.

14. The checkers had the impression that this isonitrile smells fouler than phenyl isonitrile and related aryl isonitriles. Glassware can be freed from the isonitrile odor by rinsing with a 1:10 mixture of 37% hydrochloric acid/ethanol.

Waste Disposal Information

All hazardous materials should be handled and disposed of in accordance with "Prudent Practices in the Laboratory"; National Academy Press; Washington, DC, 1995.

3. Discusssion

This synthesis of 2,2-diethoxy-1-isocyanoethane is based on the dehydration of *N*-substituted formamides, which is the most important route to isocyanides.[2] The combination of triphenylphosphine, carbon tetrachloride and triethylamine allows a smooth dehydration and a facile workup. The formylation of aminoacetaldehyde diethyl acetal employs propyl formate, because of the instability of acetals towards acidic reagents such as formic acid and formic-acetic anhydride that are usually employed in *N*-formylations.

Hartke[3] reported a synthesis of 2,2-diethoxy-1-isocyanoethane in which amino acetaldehyde acetal was transformed initially into the corresponding thioformamide. The thioformamide was then converted into the isocyanide by treatment with diphenylacetyl chloride/diisopropyl carbodiimide/triethylamine. The present method appears to be more convenient, because the experimental procedures are simpler, the yields are higher (62-68% overall) and the reagents are easily available and cheap.

This isocyanide has been employed as C–C–N–C unit in the synthesis of imidazoles, imidazo-imidazoles and aminoisoxazoles.[4] 2,2-Dimethoxy-1-isocyanoethane can be prepared starting from aminoacetaldehyde dimethyl acetal following this procedure. The yields are similar and the product shows the same reactivity of its ethyl analogues. A procedure for the synthesis of 2,2-dimethoxy-1-isocyanoethane based on the dehydration of the corresponding formamide with $POCl_3/NEt_3$ has recently been reported.[5] The dehydration with $PPh_3/CCl_4/NEt_3$ appears to be a superior method, giving better yields without employing a tedious aqueous work-up. 2,2-Dimethoxy-1-isocyanoethane has been used for the synthesis of 4,5-dihydrothiazoles and thiazoles.[5]

1. Dipartimento di Chimica Organica "Ugo Schiff", Università di Firenze, via della Lastruccia 13, I-50019 Sesto Fiorentino (FI), Italy.
2. Ugi, I.; Dömling, A. *Angew, Chem. Int. Ed.* **2000**, *39*, 3168.
3. Hartke, K. *Chem. Ber.* **1966**, *99*, 3163.
4. (a) Bossio, R.; Marcaccini, S.; Pepino, R.; Polo, C.; Torroba, T. *Heterocycles* **1990**, *31*, 1287. (b) Bossio, R.; Marcaccini, S.; Pepino, R. *Liebigs Ann. Chem.* **1993**, 1289. (c) Bossio, R.; Marcaccini, S.; Pepino, R.; Torroba, T. *J. Org. Chem.* **1996**, *61*, 2202. (d) Buron, R.; El Kaïm, L.; Uslu, A. *Tetrahedron Lett.* **1997**, *46*, 8027.
5. Kazmaier, U.; Ackermann, S. *Org. Biomol. Chem.* **2005**, *3*, 3184.

1-(*TERT*-BUTYLIMINO-METHYL)-1,3-DIMETHYL-UREA HYDROCHLORIDE

[Urea, N-[[(1,1-dimethylethyl)imino]methyl]-N,N'-dimethyl-, monohydrochloride]

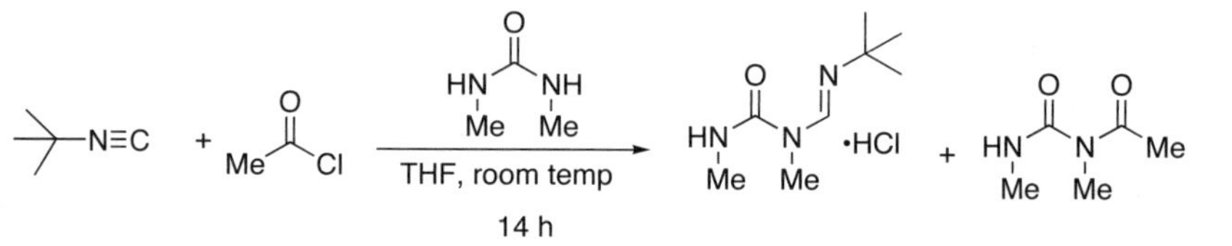

Submitted by David D. Díaz[1], Amy S. Ripka,[1] and M.G. Finn.[1]
Checked by Daniel Laurich, Günter Seidel, and Alois Fürstner.

1. Procedure

Caution! Use safety glasses and nitrile gloves under a well-ventilated hood since isocyanides have pungent odors and some are known to be toxic. A 1-L, three-necked, round-bottomed flask is equipped with a mechanical stirrer (Note 1), a vacuum take-off adapter attached to a nitrogen source, and a pressure equalizing dropping funnel closed with a glass stopper. The flask is flame-dried, cooled under nitrogen, and charged with tetrahydrofuran (THF) (250 mL) (Note 2) and *tert*-butyl isocyanide (12.4 mL, 110 mmol) (Note 3). To this homogeneous solution is added acetyl chloride (8.6 mL, 121 mmol) (Note 4) and the mixture is vigorously stirred for 15 min. A solution of 1,3-dimethylurea (24.2 g, 275 mmol) (Note 3) in 150 mL of THF is then introduced slowly and continuously through the dropping funnel. After 15-20 min, a white precipitate appears (Note 5). The reaction mixture is stirred for 14 h at room temperature, after which time the heterogeneous mixture is filtered through a Buchner funnel connected to vacuum. The white precipitate is washed with cold THF (Note 6) to remove excess urea, the acetylurea byproduct and colored impurities (Note 7). The product is dried in a vacuum oven at 50 °C for 20 h to give 19-20 g (85-90%) of analytically pure product as a white powder (Notes 8-10). The hygroscopic product should be stored under nitrogen.

2. Notes

1. The submitters used a magnetic stir bar (50 mm in length x 7 mm in diameter) instead of the mechanical stirrer but noted that stirring may become difficult as the reaction proceeds. This problem is avoided when a mechanical stirrer is used.

2. The submitters used commercial tetrahydrofuran, grade OPTIMA (Fisher). The checkers used THF dried by distillation over Mg-anthracene under argon.

3. *tert*-Butyl isocyanide (≥ 98%, Fluka) and 1,3-dimethylurea, 98% (Acros) were used as received.

4. Acetyl chloride, p.a., was used as supplied by Acros, and dispensed by syringe. For the scale proposed it is not necessary to distill this reagent, although this is recommended for smaller scale reactions.

5. The submitters noted that stirring becomes difficult shortly after the precipitation starts if a magnetic stir bar is used.

6. THF at 0 °C is necessary for the washing process as follows: suction is stopped and the white solid is suspended completely in cold THF, the suction is turned on to remove the liquid, and the operation is repeated several times until 1 L of cold THF is used. After washing, air is pulled through the solid for 20 min to dry the product, and clumps are broken manually with a spatula. The solid remaining on the walls of the funnel is dissolved in CH_2Cl_2 and the solvent evaporated under vacuum. The two fractions of solid are combined.

7. The byproduct *N*-acylurea may be isolated in equimolar amounts to formamidine urea by column chromatography on silica gel (7:3 EtOAc:hexanes, R_f ca. 0.5) of the residue from the combined filtrates.

8. The checkers dried the product *in vacuo* (10^{-3} Torr) at ambient temperature for 3 h.

9. Spectral and analytical data: mp 172-173 °C; ^{1}H NMR (400 MHz, $CDCl_3$): δ 11.09 (d, J = 14.8 Hz, 1 H), [8.72 (d, J = 15.2 Hz), 8.7 (m); 2 H], 3.85 (s, 3 H), 2.94 (d, J = 4.4 Hz, 3 H), 1.59 (s, 9 H); ^{13}C NMR (100 MHz, $CDCl_3$): δ 152.8, 150.7, 58.7, 34.7, 29.1, 28.0; IR (film): 3209, 2984, 1722, 1671, 1535 cm^{-1}. Anal. calcd for $C_8H_{18}ClN_3O$: C, 46.26; H, 8.74; Cl, 17.09; N, 20.23. Found: C, 46.14; H, 8.63; Cl, 17.10; N, 20.29.

10. As a precaution, the formamidine urea should be handled with care in a well-ventilated hood.

Safety and Waste Disposal Information

All hazardous materials should be handled and disposed of in accordance with "Prudent Practices in the Laboratory"; National Academy Press; Washington, DC, 1995.

3. Discussion

Formamidines are of interest in synthetic chemistry[2] and have been used extensively as pesticides. The present procedure is an example of a novel process in which the addition of a substituted urea to a mixture of isocyanide and acid chloride gives formamidine urea salts **1** in pure form.[3] The yields of **1** are maximized by the use of 2.5-3.0 equiv of urea, and THF provides the best results in general. The process was found to be tolerant of substantial variations in the nature of the components, but restricted almost exclusively to acid chlorides. A proposed mechanism for the formation of **1** is shown in Scheme 1.[3] The first key intermediate is the electrophilic adduct of isonitrile and acid chloride, **2**. Substitution of chloride by urea at the chloroiminium carbon gives **3**. The unique aspect of the mechanism is then proposed to be capture of **3** at the active carbonyl carbon by another equivalent of urea to form the soluble acylurea **4** and intermediate **5**, which may be formulated as the betaine shown or a stabilized carbenoid. Structure **5** should undergo rapid proton transfer to give the formamidine **6**, and an equivalent of HCl is extracted by precipitation of the hydrochloride salt **1**. The efficiency and convenience of the reaction is governed by both the generation of the reactive adducts and the precipitation of the final product. The overall transformation is made possible by the ability of the acid chloride to change the nature of the isonitrile carbon from nucleophilic to electrophilic. The entry to the formamidine urea skeleton exemplified here is quite a bit more convenient than previous methods.

Scheme 1

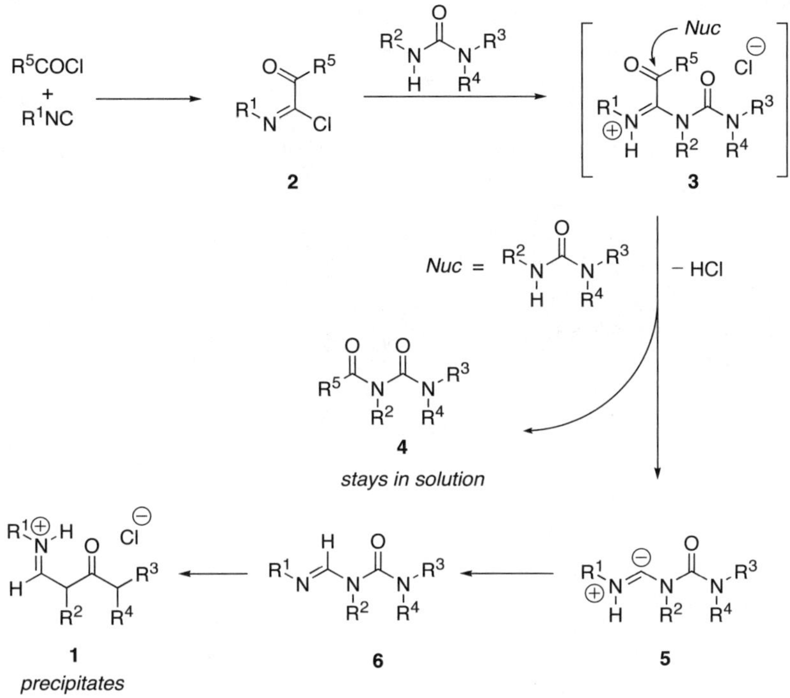
R⁵COCl
+
R¹NC
2
3
Nuc
Nuc =
− HCl
4
stays in solution
1
precipitates
6
5

Scheme 2

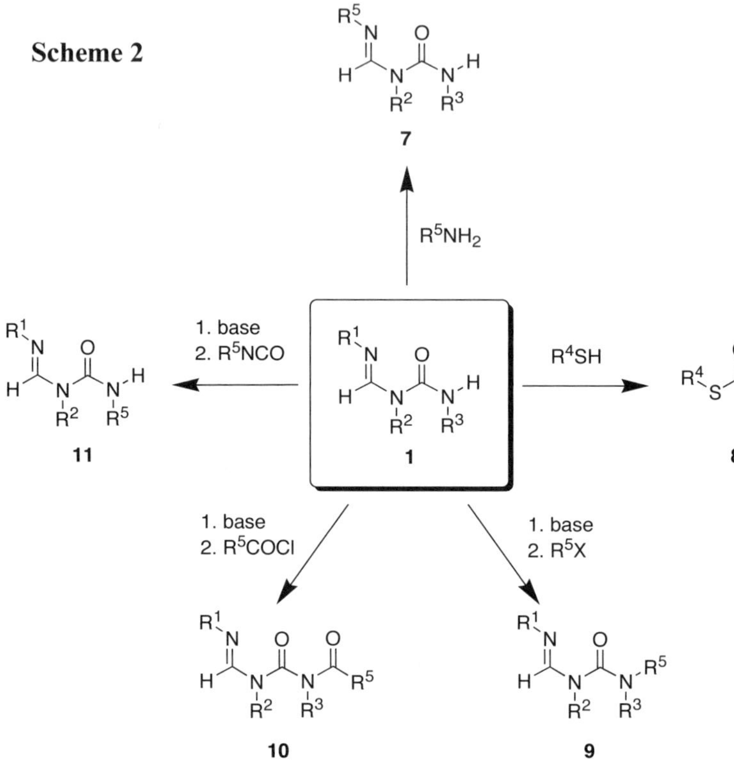

The *tert*-butylamino fragment of **1** may be readily exchanged with a wide variety of primary nitrogen nucleophiles,[4] allowing the easy preparation of a variety of formamidine ureas (**7**, Scheme 2). Formamidine ureas also undergo substitution at the carbonyl carbon to give thiolcarbamates (**8**) and the formamidine urea nucleus can be elaborated by base-mediated alkylation and acylation (**9**, **10**, **11**).[5]

1. Department of Chemistry and The Skaggs Institute for Chemical Biology, The Scripps Research Institute, 10550 N. Torrey Pines Rd., La Jolla, CA, 92037, USA. We thank Dr. Chunmei Li for her valuable assistance.

2. (a) Meyers, A. I.; Hutchings, R. H., *Tetrahedron* **1993**, *49*, 1807-1820. (b) Meyers, A. I.; Elworthy, T. R. *J. Org. Chem.* **1992**, *57*, 4732-4740, and references cited therein.
3. Ripka, A. S.; Díaz, D. D.; Sharpless, K. B.; Finn, M. G. *Org. Lett.* **2003**, *5*, 1531-1533.
4. Díaz, D. D.; Finn, M. G. *Chem. Eur. J.* **2004**, *10*, 303-309.
5. Díaz, D. D.; Finn, M. G. *Org. Lett.* **2004**, *6*, 43-46.

THIONATION: GlcNAc-THIAZOLINE TRIACETATE {(3a*R*,5*R*,6*S*,7*R*,7a*R*)-5-ACETOXYMETHYL-6,7-DIACETOXY-2-METHYL-5,6,7,7a-TETRAHYDRO-3a*H*-PYRANO[3,2-*d*]THIAZOLE}

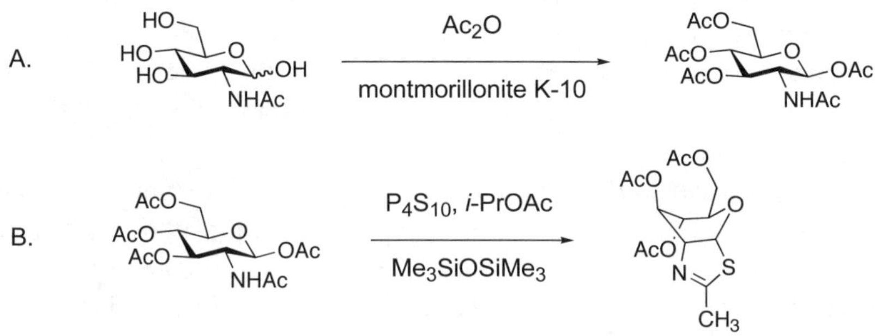

Submitted by Spencer Knapp, Richard A. Huhn, and Benjamin Amorelli.[1]
Checked by Erikah E. Englund and Peter Wipf.[2]

1. Procedure

Caution! The following operations produce lachrymatory and corrosive vapors and should be carried out in a well-ventilated fume hood.

A. 2-Acetamido-2-deoxy-1,3,4,6-tetra-O-acetyl-β-D-glucopyranose. A dry, 2000-mL, three-necked, round-bottomed flask is equipped with a Teflon-coated magnetic stir bar (41 × 19 mm) and an ice-salt bath, and at the respective necks with a thermometer, a septum, and a rubber stopper. A nitrogen atmosphere is incorporated by a needle inserted through the septum. The stopper is temporarily removed, replaced with a funnel and acetic anhydride (900 mL, 9.52 mol, Note 1) is added from a bottle that had been stored in a refrigerator at 5 °C. The stopper is replaced and the acetic anhydride is cooled to 0 °C (internal temperature) with stirring, the stopper is again replaced by a funnel, and then 80.0 g (362 mmol) of *N*-acetyl-D-glucosamine and 270 g of montmorillonite K-10 (Note 1) are sequentially added over a 10-min period such that the internal temperature does not rise above ambient. The stopper is replaced, the ice bath is removed, and the suspension is stirred for 24 h.

The reaction mixture is filtered through a medium porosity sintered glass funnel (600 mL capacity) pre-coated with a pad of 20 g of Celite moistened with methyl acetate (Note 1). The flask and filtered solids are washed with methyl acetate, about 1 L total, and then the combined filtrate is concentrated on a rotary evaporator (40 °C, 5 mmHg). The resulting orange residue is dislodged with a spatula, and then dissolved in 200 mL of hot methanol. The methanolic solution is placed in a 5 °C explosion proof refrigerator and a precipitate forms over 24 h. The solid is collected by filtration and washed with 100 mL of diethyl ether, then dried (23 °C, 1 mm) to afford 46.3 g (33%) of product as a white solid (mp 179–180 °C) (Notes 2 and 3).

B. GlcNAc-thiazoline triacetate. A 1000-mL, two-necked, round-bottomed flask under a nitrogen atmosphere immersed in an oil bath is fitted with Teflon-coated magnetic stir bar (41 × 19 mm), a rubber septum and a reflux condenser capped with a septum, through which nitrogen is affixed by use of a needle. The septum on the flask is temporarily removed, a funnel inserted and the vessel is charged with 300 mL of isopropyl acetate and 18.3 g (40 mmol) of phosphorus pentasulfide (P_4S_{10}). The resulting suspension is stirred rapidly while 76.3 mL (360 mmol) of hexamethyldisiloxane (Note 4) is added via syringe over a 5-min period. The mixture is heated at reflux for 10 min while vigorous stirring is maintained. The solution is allowed to cool to just below reflux, the septum on the flask is replaced with a funnel and 2-acetamido-2-deoxy-1,3,4,6-tetra-*O*-acetyl-β-D-glucopyranose (40.0 g, 103 mmol) is added in four portions with continuous stirring over a 5-min period. The septum is replaced and the reaction mixture is then brought back to reflux, and held at that temperature for 2–3 h, during which period it becomes homogeneous and orange. TLC analysis (3:1, dichloromethane / ethyl acetate) indicates complete conversion to product (R_f = 0.40 on silica).

The reaction mixture is allowed to cool to room temperature and then stirred for an additional 12 h (Note 5). The septum is removed and acetone (150 mL) is added. The flask is cooled to 0 °C in an ice-salt bath (Note 6), and then 52 mL of 5.3 M aqueous K_2CO_3 is added. The quenched reaction mixture is stirred at 0 °C for 30 min, then water and isopropyl acetate are used to transfer the reaction mixture to a 2-L separatory funnel (total 300 mL of water and 200 mL of isopropyl acetate). The organic layer is separated and washed with 0.53 M aqueous K_2CO_3 (3 × 200 mL or until the aqueous layer is colorless), and then dried over 30 g of anhydrous magnesium sulfate. Filtration followed by concentration on the rotary evaporator (23 °C, 5 mmHg) gives an orange oil, which is mixed with 50 mL of isopropyl acetate and 55 g of silica, and then concentrated to give an orange powder. A plug of 170 g of silica moistened with isopropyl acetate is layered onto a coarse porosity sintered glass funnel (600 mL capacity). The orange powder is placed on the silica plug and tamped with a piece of filter paper, which is left in place. A mixture of heptanes / isopropyl acetate (70:30, 1.25 L, then 40:60, 125 mL) is used to wash through high R_f impurities, which are discarded as waste. Elution with 1.5 L of 40:60 heptanes / isopropyl acetate is followed by pure isopropyl acetate (~1 L) until the product is washed off the silica according to TLC analysis. Concentration of this latter solution on the rotary evaporator (23 °C, 5 mmHg) followed by a high vacuum (27 °C, 0.05 mmHg), until a constant weight is obtained, provides 31.1 g (88%) of GlcNAc-thiazoline triacetate as an orange oil (Note 7).

Notes

1. Acetic anhydride, *N*-acetyl-D-glucosamine, and montmorillonite K-10 were purchased from Aldrich Chemical Company, Inc., and are used as received. All solvents throughout this procedure were reagent grade and were used as received. The submitters report that a nitrogen atmosphere is not required for either Step A or Step B. Similar results are obtained in the absence of a nitrogen atmosphere.

2. Lit. mp 187 °C (ethanol),[3] 185–186 °C (water),[4] 186–186.5 °C (methanol/ether),[5] 188–189 °C (ethanol, repeated).[6]

3. The product was >99% pure according to ^{1}H NMR analysis. The α-anomer, when present, gives a signal in the ^{1}H NMR spectrum at δ 6.17 (d, *J* = 3.2 Hz, 1 H). The spectral properties of the β-pentaacetate are as follows: $[\alpha]_D^{20}$ 7.2 (*c* 1.0, $CHCl_3$); IR (KBr) cm^{-1} 3257, 3074, 1747, 1656, 1555, 1444, 1370, 1220, 1069; ^{1}H NMR ($CDCl_3$, 300 MHz) δ: 1.93 (s, 3 H), 2.04 (s, 3 H), 2.05 (s, 3 H), 2.09 (s, 3 H), 2.12 (s, 3 H), 3.81 (ddd, *J* = 2.4, 4.5, 9.0 Hz, 1 H), 4.12 (dd, *J* = 2.1, 12.5 Hz, 1 H), 4.27 (dd, *J* = 4.6, 12.6 Hz, 1 H), 4.26–4.35 (m, 1 H), 5.09–5.17 (m, 2 H), 5.56 (d, *J* = 9.5 Hz, 1 H), 5.69 (d, *J* = 8.8 Hz, 1 H); ^{13}C NMR ($CDCl_3$, 75 MHz) δ: 20.5, 20.6, 20.7, 20.8, 23.1, 53.1, 61.7, 67.8, 72.6, 72.9, 92.6, 169.2, 169.5, 170.1, 170.6, 171.2; MS *m/z* (relative intensity) 413 (13), 412 (100, $[M+Na]^+$), 353 (5), 352 (40), 330 (3); HRMS Calcd for $C_{16}H_{23}NO_{10}Na$ (M+Na) 412.1220, Found 412.1216; Anal. Calcd for $C_{16}H_{23}NO_{10}$: C, 49.36; H, 5.95. Found: C, 48.98; H, 5.79.

4. Phosphorus pentasulfide and hexamethyldisiloxane were purchased from Aldrich Chemical Company, Inc., and used as received. The submitters utilized a drying tube as opposed to the nitrogen atmosphere and obtained similar results.

5. Cleaner reaction mixtures were obtained when the reaction mixture

was allowed to rest for an additional period after reflux.

6. At quench temperatures higher than 0 °C, some hydrolysis of product during the carbonate treatment was observed by TLC.

7. The product was >95% pure according to ^{1}H NMR analysis. The spectral and analytical properties are as follows: $[\alpha]_D^{20}$ –28.3 (*c* 0.4, $CHCl_3$); IR (neat) cm^{-1} 3308, 2958, 1747, 1629, 1433, 1371, 1232, 1040; ^{1}H NMR ($CDCl_3$, 300 MHz) δ: 2.08 (s, 6 H), 2.13 (s, 3 H), 2.31 (d, *J* = 2.3 Hz, 3 H), 3.53 (dt, *J* = 9.5, 4.2 Hz, 1 H), 4.11 (app d, *J* = 4.1 Hz, 2 H), 4.44–4.48 (m, 1 H), 4.95 (dt, *J* = 9.5, 1.6 Hz, 1 H), 5.56 (dd, *J* = 1.7, 3.3 Hz, 1 H), 6.23 (d, *J* = 7.1 Hz, 1 H); ^{13}C NMR ($CDCl_3$, 75 MHz) δ: 20.5, 20.6, 20.7, 20.8, 63.2, 68.3, 69.2, 70.5, 76.5, 88.6, 168.2, 169.2, 169.5, 170.5; MS *m/z* (relative intensity): 346 (9), 345 (23, $[M]^+$), 312 (15), 286 (29), 285 (24), 272 (25), 225 (25), 183 (26), 166 (44), 165 (100), 152 (53), 141 (60); HRMS Calcd for $C_{14}H_{19}NO_7S$ 345.0882; Found 345.0878; Anal. Calcd for $C_{14}H_{19}NO_7S$: C, 48.69; H, 5.55; N, 4.06; S, 9.29. Found: C, 48.44; H, 5.68; N, 4.00; S, 9.26.

Safety and Waste Disposal Information

All hazardous materials should be handled and disposed of in accordance with "Prudent Practices in the Laboratory"; National Academy Press; Washington, DC, 1995.

3. Discussion

The preparation of 2-acetamido-2-deoxy-tetra-*O*-acetyl-β-D-glucopyranose is based on the report by Bhaskar and Loganathan[7] that describes the use of montmorillonite K-10 clay as an acidic catalyst for per-*O*-acetylation of sugars. The conditions have been modified to facilitate scale-up and isolation of pure β-anomer. This product has most commonly been made from D-glucosamine by a four-step route involving the intermediacy of an *N*-(*p*-methoxybenzylidene) derivative,[3] and by acetylation with anhydrous zinc(II) chloride as promoter.[4] In our hands neither procedure proved as convenient on the 0.1 mole scale as the present one. The β-peracetate may also obtained commercially from Aldrich Chemical Company ($87 for 5 g in 2005).

GlcNAc-thiazoline triacetate may be synthesized from 2-acetamido-2-deoxy-tetra-*O*-acetyl-β-D-glucopyranose in quantitative yield on a 2 g scale by using freshly prepared Lawesson's reagent, 2,4-bis(4-methoxyphenyl)-

1,3-dithia-2,4-diphosphetane 2,4-disulfide, followed by column chromatography.[8] The corresponding GalNAc-thiazoline triacetate has been synthesized analogously.[9] The thionation/cyclization reaction described herein is based on the P_4S_{10}/hexamethyldisiloxane procedure of Curphey,[10] which minimizes the formation of P/S byproducts that must be removed chromatographically. The amount of P_4S_{10} relative to starting material is increased to 0.4 equivalents in this procedure to accommodate the thionation of the byproduct, acetic acid, and to ensure that the reaction runs to completion.

GlcNAc-thiazoline triacetate has been hydrolyzed to the corresponding triol (Scheme 1), which is an effective inhibitor[11] and mechanistic/structural probe [12,13,14,15] of a family of 20 *N*-acetylhexosaminidases and related enzymes.[16,17,18] Alternatively, hydrolysis of the thiazoline ring gives the α-mercaptan stereoselectively,[8] and this has been used in turn to prepare α-GlcNAc thioconjugates,[8,19] including glycopeptide analogues,[20,21] mycothiol analogues,[22,23] and α-GlcNAc-UDP analogues.[24] Oxidation of the thiazoline in the presence of ethanol leads to the α-GlcNAc 1-C-sulfinate and 1-C-sulfonate *O*-ethyl esters.[25]

The thiazoline has been modified on the methyl group by buffer- or acylation-induced imine-to-enamine conversion and then electrophile or radical addition. Several of the methyl-functionalized GlcNAc-thiazolines show highly selective inhibition of human *O*-GlcNAcase.[26] The α-mercaptan has been converted to an α-GlcNAc thiolsulfonate, which shows distinctive inhibition of the short isoform, and irreversible inhibition of the long isoform, of human *O*-GlcNAcase.[27]

Scheme 1

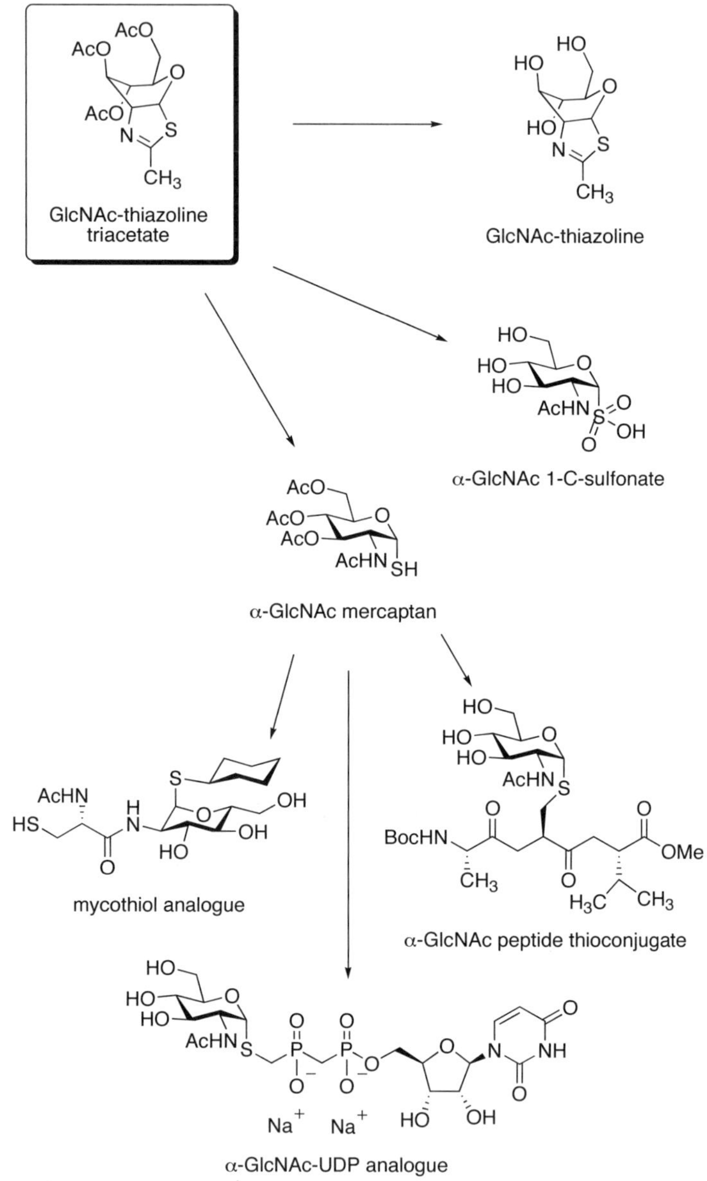

1. Department of Chemistry & Chemical Biology, Rutgers The State University of New Jersey, Piscataway, NJ 08554-8087, USA.

2. Department of Chemistry, University of Pittsburgh, Pittsburgh, PA 15260, USA.

3. Bergmann, M.; Zervas, L. *Chem. Ber.* **1931**, *64*, 975–980.
4. Findlay, J.; Levvy, G. A.; Marsh, C. A. *Biochem. J.* **1958**, *69*, 467–476.
5. Horton, D. *J. Org. Chem.* **1964**, *29*, 1776–1782.
6. Hudson, C. S.; Dale, J. K. *J. Am. Chem. Soc.* **1916**, *38*, 1431–1436.
7. Bhaskar, P. M.; Loganathan, D. *Tetrahedron Lett.* **1998**, *39*, 2215–2218.
8. Knapp, S.; Myers, D. S. *J. Org. Chem.* **2001**, *66*, 3636–3638.
9. Knapp, S.; Myers, D. S. *J. Org. Chem.* **2002**, *67*, 2995–2999.
10. Curphey, T. J. *J. Org. Chem.* **2002**, *67*, 6461–6473.
11. Knapp, S.; Vocadlo, D.; Gao, Z.; Kirk, B.; Lou, J.; Withers, S. G. *J. Am. Chem. Soc.* **1996**, *118*, 6804–6805.
12. Mark, B. L.; Vocadlo, D. J.; Knapp, S.; Triggs-Raine, B. L.; Withers, S. G.; James, M. N. G. *J. Biol. Chem.* **2001**, *276*, 10330–10337.
13. Tropak, M. B.; Reid, S. P.; Guiral, M.; Withers, S. G.; Mahuran, D. *J. Biol. Chem.* **2004**, *279*, 13478–13487.
14. Mark, B. L.; Mahuran, D. J.; Cherney, M. M.; Zhao, D.; Knapp, S.; James, M. N. G. *J. Mol. Biol.* **2003**, *327*, 1093-1109.
15. Mark, B. L.; James, M. N. G. *Can. J. Chem.* **2002**, *80*, 1064–1074.
16. Macauley, M. S.; Whitworth, G. E.; Debowski, A.; Chin, D.; Vocadlo, D. J. *J. Biol. Chem.* **2005**, *280*, 25313–25322.
17. Reid, C. W.; Blackburn, N. T.; Legaree, B. A.; Auzanneau, F.-I.; Clark, A. J. *FEBS Lett.* **2004**, *574*, 73–79.
18. Ritter, T. K.; Wong, C.-H. *Tetrahedron Lett.* **2001**, *42*, 615–618.
19. Naus, P.; Leseticky, L.; Smrcek, S.; Tislerova, I.; Sticha, M. *Synlett* **2003**, 2117–2122.
20. Zhu, X. M.; Schmidt, R. R. *Tetrahedron Lett.* **2003**, *44*, 6063–6067.
21. Zhu, X. M.; Schmidt, R. R. *Chem. Eur. J.* **2004**, *10*, 875–887.
22. Knapp, S.; Gonzalez, S.; Myers, D. S.; Eckman, L. L.; Bewley, C. A. *Org. Lett.* **2002**, *4*, 4337–4339.
23. Knapp, S.; Amorelli, B.; Darout, E.; Ventocilla, C. C.; Goldman, L. M.; Huhn, R. A.; Minnihan, E. C. J. Carbohydr. Chem. **2005**, *24*, 103–130.
24. Zhu, X. M.; Stolz, F.; Schmid, R. R. *J. Org. Chem.* **2004**, *69*, 7367–7370.
25. Knapp, S.; Darout, E. *Tetrahedron Lett.* **2002**, *43*, 6075–6078.
26. Knapp, S.; Abdo, M.; Ajayi, K.; Huhn, R. A.; Emge, T. J.; Kim, E. J.; Hanover, J. A. *Org. Lett.* **2007**, *9*, 2321-2324.
27. Kim, E. J.; Amorelli, B.; Abdo, M.; Thomas, C. J.; Love, D. C.; Knapp, S.; Hanover, J. A. *J. Am. Chem. Soc.* **2007**, *129*, 14854-14855.

ELECTROPHILIC CYCLIZATION WITH *N*-IODOSUCCINIMIDE: PREPARATION OF 5-(4-BROMOPHENYL)-3-IODO-2-(4-METHYLPHENYL)FURAN

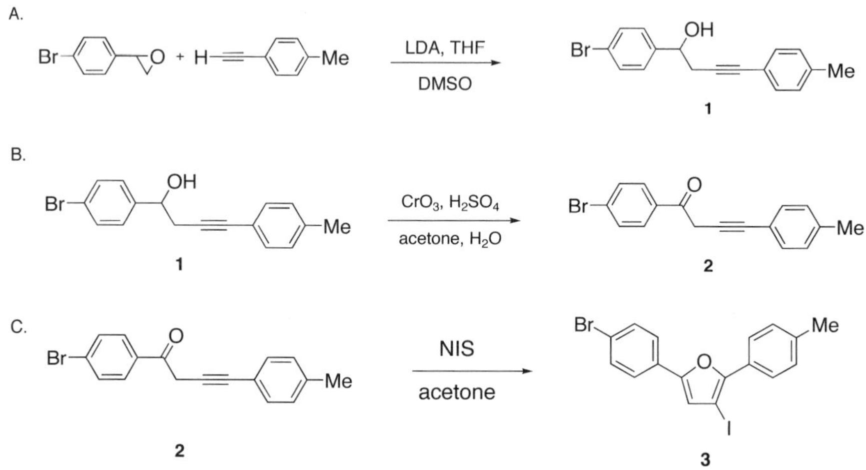

Submitted by Adam Sniady, Marco S. Morreale, and Roman Dembinski.[1]
Checked by Cody Timmons and Peter Wipf.[2]

1. Procedure

A. 1-(4-Bromophenyl)-4-(4-methylphenyl)but-3-yn-1-ol (**1**). A 500-mL, three-necked, round-bottomed flask (Note 1), equipped with a magnetic stirring bar and fitted with a nitrogen inlet adapter and two rubber septa, is charged with THF (40 mL) (Note 2) and 4-ethynyltoluene (4.08 mL, 32.2 mmol) (Note 3) via syringes. The reaction vessel is placed in a dry ice-acetone bath. After 10 min, lithium diisopropylamide solution (LDA) (16.1 mL, 32.2 mmol) (Note 4) is added within 5 min via syringe to the stirred reaction mixture. The mixture is allowed to stir for 10 min after which time the bath is removed and the reaction vessel is allowed to warm to room temperature. Dimethyl sulfoxide (DMSO) (75 mL) (Note 5) is added over 5 min via syringe to the reaction vessel, followed by a solution of 4-bromostyrene oxide (4.94 g, 24.8 mmol) (Note 6) in 5 mL of THF over 1 min via syringe. The reaction mixture acquires a dark red wine color and is allowed to stir at room temperature for 2 h (Note 7). Then, the reaction flask is placed in an ice-bath and quenched with saturated aqueous ammonium chloride (100 mL). The mixture is transferred to a 500-mL separatory funnel using saturated aqueous ammonium chloride (50 mL) as a rinse. The aqueous phase is extracted with ether (3 × 150 mL). The combined ether extracts are transferred to a 1-L separatory funnel and washed with saturated aqueous sodium chloride (2 × 200 mL). The organic phase is dried with anhydrous sodium sulfate (10 g) and filtered through a fritted-glass funnel of medium porosity. The solvent is removed via rotary evaporation (23 °C, 25 mmHg) followed by high vacuum (23 °C, 0.6 mmHg, 3 h). The crude product is purified by recrystallization and chromatography of the mother liquor on SiO_2 (Note 8) to yield **1** as a white crystalline solid (6.01 g, 19.1 mmol, 77%) (Note 9).

B. 1-(4-Bromophenyl)-4-(4-methylphenyl)but-3-yn-1-one (**2**). A 500-mL, single-necked, round-bottomed flask equipped with a rubber septum and a magnetic stirring bar is charged with **1** (6.01 g, 19.1 mmol) and acetone (200 mL) (Note 10). Jones reagent (8.0 mL, 3.0 M solution, 24 mmol) (Note 11) is added by syringe over 5 min at 0 °C (ice-water bath) to the stirred mixture (Notes 12, 13). The bath is removed and the reaction mixture is allowed to stir for an additional 15 min (Note 7), then immediately ether (130 mL) and isopropanol (4 mL) are added. The blue-green solid is filtered off through a medium porosity fritted-glass funnel and the filtrate is transferred to a 1-L separatory funnel filled with 100 mL of ice water. Three extractions with ether (3 × 100 mL) are carried out. The combined organic phases are washed twice with saturated sodium bicarbonate (2 × 150 mL), dried over sodium sulfate (10 g), filtered, and concentrated at reduced pressure on a rotary evaporator (23 °C, 25 mmHg). The resulting light yellow solid is transferred to a 50-mL, single neck round-bottomed flask and dried for 6 h under vacuum (23 °C, 0.6 mmHg) to give **2** (5.56 g, 17.8 mmol, 93%), which is used in the next step without further purification (Note 14).

C. 5-(4-Bromophenyl)-3-iodo-2-(4-methylphenyl)furan (**3**). A 500-mL, three-necked, round-bottomed flask equipped with a magnetic stirring bar and a rubber septum is charged with **2** (5.10 g, 16.3 mmol). An additional rubber septum with a nitrogen inlet is attached to the remaining neck, and acetone (150 mL) (Note 10) is added via syringe. In a separate flask *N*-iodosuccinimide (NIS) (3.85 g, 17.1 mmol) (Note 15) is dissolved in acetone (60 mL). The solution of NIS is then injected via syringe over 5 min into the reaction mixture that is stirred at room temperature under a nitrogen atmosphere for 2 h. TLC analysis (Note 7) indicates a complete conversion. The solvent is removed by rotary evaporation (23 °C, 25 mmHg) and the residue is extracted with ether (150 mL) (Note 16). The solid (succinimide) is filtered off through a medium porosity fritted-glass funnel, and additionally washed with ether (60 mL) (Note 16). The etheral extracts are combined and transferred to a 500-mL separatory funnel containing 50 mL of a 2.0 M aqueous sodium bisulfite solution. The phases are separated and the aqueous phase is further extracted with ether (3 × 50 mL). The combined organic extracts are washed with brine (1 × 100 mL) and dried over Na_2SO_4 (10 g). The mixture is filtered through a medium porosity fritted glass funnel and the solvent is removed by rotary evaporation (23 °C, 25 mmHg). The resulting material is charged on a column (3 × 40 cm) of silica gel (35 g) (Note 17) and eluted with hexanes/ethyl acetate (8:1) (approximately 600 mL) (Note 18). The product is concentrated by rotary evaporation (23 °C, 25 mmHg) and then dried at 0.6 mmHg (23 °C, 6 h) to yield **3** as a colorless solid (5.80 g, 13.2 mmol, 81%) (Note 19).

2. Notes

1. The apparatus was oven-dried at 135 °C for 2 h and maintained under an atmosphere of dry nitrogen during the course of the reaction. The flasks had neck sizes of 24/40.

2. THF was dried over sodium/benzophenone and freshly distilled under an atmosphere of dry nitrogen.

3. 4-Ethynyltoluene (98%) was purchased from GFS Chemicals and used as received. The checkers used chemicals from Alfa Aesar without further purification.

4. Lithium diisopropylamide (2.0 M solution in THF/*n*-heptane) was purchased from Acros Organics and used as received.

5. Methylsulfoxide, anhydrous, DriSolv was purchased from EMD Chemicals, Inc. and used as received. The checkers used 99.7% anhydrous DMSO from Acros Organics.

6. 4-Bromostyrene oxide (96%) was purchased from Aldrich Chemical Company, Inc. and used as received.

7. The progress of the reaction was monitored by TLC analysis on silica gel eluted with hexanes/ethyl acetate (8:2) and visualized with phosphomolybdic acid. The R_f values were: 0.53 (blue) for 4-bromostyrene oxide, 0.32 (dark blue) for alcohol **1**, 0.59 (blue) for ketone **2**, and 0.74 (orange-brown) for furan **3**.

8. Ethyl acetate (15 mL) was added to the flask containing crude **1** (10.23 g). The flask was heated gently with a heat-gun under hand-stirring until the all of the crude material was dissolved. Then, the solution was

layered with hexanes (30 mL) and placed in the refrigerator (–6 °C) for 12 h to induce crystallization. The flask was placed in an ice-bath and the mother liquor was removed via syringe. The crystalline residue was rinsed with cold hexanes (2 × 10 mL) (placed in an ice bath for at least 45 min prior to use) and the rinses were removed via syringe; then the crystalline material was dried (23 °C, 0.6 mmHg, 3 h) to afford alcohol **1** (4.45 g, 14.1 mmol, 57%). The rinses were combined with the mother liquor and the solvent was removed via rotary evaporation (23 °C, 25 mmHg) followed by high vacuum (23 °C, 0.6 mmHg, 3 h) to yield 4.12 g of crude material. This material was dissolved in ethyl acetate (10 mL) with hand-stirring and the help of a heat-gun. The solution was layered with hexanes (20 mL) and placed in the refrigerator for 48 h. Then the crystallization flask was placed into an ice bath and the mother liquor was removed via syringe. The crystalline material was rinsed with cold hexanes (10 mL) and the rinse was removed via syringe; then the crystalline material was dried (23 °C, 0.6 mmHg, 3 h) to afford an additional 6% of alcohol **1** (0.46 g, 1.5 mmol). The rinses and mother liquor were combined and concentrated by rotary evaporation. The resulting material was charged on a column (3 × 40 cm) of silica gel (35 g) and eluted with hexanes/ethyl acetate (6:1) (approximately 400 mL) to provide an additional 14% of alcohol **1** (1.10 g, 3.49 mmol). The total amount of alcohol **1** obtained in this reaction was 6.01 g (19.1 mmol, 77%).

9. mp 133–135 °C (ethyl acetate/hexanes). Calcd for $C_{17}H_{15}BrO$: C, 64.78; H, 4.80. Found: C, 64.59; H, 4.85. IR (cm^{-1}, NaCl disk) 3411 (m), 1509 (m), 1046 (m), 817 (s). UV-Vis (ε, $M^{-1}cm^{-1}$; ether; 3.5×10^{-5} M) 244 (29000), 254 (24000). MS (rel intensity) 185 (88), 130 (100), 115 (74), 77 (96), 63 (34). ^{1}H NMR (300 MHz, $CDCl_3$) δ: 2.36 (s, 3 H), 2.61 (d, 1 H, J = 3.0 Hz), 2.80–2.84 (m, 2 H), 4.90 (td, 1 H, J = 6.3, 1.5 Hz), 7.12 (d, 2 H, J = 7.8 Hz), 7.28–7.33 (m, 4 H), 7.47 (d, 2 H, J = 8.4 Hz); ^{13}C NMR δ: 21.4, 30.6, 71.9, 83.6, 84.5, 119.9, 121.6, 127.5, 129.0, 131.4, 131.5, 138.2, 141.6.

10. High Purity Solvent grade purchased from Burdick & Jackson was used as received. The checkers used OmniSolv High Purity Solvent grade acetone from EMD Chemicals. For procedure B reagent grade acetone can be used.

11. Prepared according to ref. 11.

12. Precipitation of blue-green chromium(III) salts was observed.

13. Reaction can be carried out in an open flask.

14. The light yellow solid had a purity of approximately 95% as determined by NMR. A portion of the material (0.46 g) was retained for recrystallization as described in Note 8 to afford colorless crystals for

analytical data: mp 121–123 °C (ethyl acetate /hexanes). Calcd for $C_{17}H_{13}BrO$: C, 65.19; H, 4.18. Found: C, 65.33; H, 4.14. IR (cm^{-1}, NaCl disk) 1677 (s), 809 (s). UV-Vis (ε, $M^{-1}cm^{-1}$; ether; 4.5 × 10^{-5} M) 254 (35000). MS (rel intensity) 312 (M^+, 53), 183 (100), 155 (60), 129 (74). ^1H NMR ($CDCl_3$) δ: 2.33 (s, 3 H), 4.02 (s, 2 H), 7.09 (d, 2 H, J = 7.8 Hz), 7.31 (d, 2 H, J = 8.1 Hz), 7.61 (d, 2 H, J = 8.4 Hz), 7.91 (d, 2 H, J = 8.4 Hz); ^{13}C NMR δ: 21.3, 31.5, 80.9, 85.7, 119.7, 128.7, 128.9, 130.1, 131.4, 131.9, 134.1, 138.3, 192.1.

15. NIS was obtained from Simafex and used as received. The checkers used NIS from Alfa Aesar.

16. Ether (unstabilized) was purchased from EMD Chemicals, Inc. and used as received. The checkers used diethyl ether that was freshly distilled from sodium/benzophenone. Extraction with ether stabilized with ethanol led to "leaching" of the succinimide.

17. Sorbent Technologies 60 Å, 230-400 mesh.

18. Flash column chromatography was performed in order to remove residues of succinimide. The solid was dissolved in a minimum amount of hexanes/ethyl acetate (8:1) and loaded onto the column. Pressured air was applied and one fraction was collected.

19. mp 98–100 °C (ethyl acetate /hexanes). Calcd for $C_{17}H_{12}BrIO$: C, 46.50; H, 2.75. Found: C, 46.72; H, 2.74. IR (cm^{-1}, NaCl disk) 1492 (s), 1475 (s), 1400 (w), 946 (s), 816 (s). UV-Vis (ε, $M^{-1}cm^{-1}$; ether, 2.5 × 10^{-5} M) 237 (23000), 253 (22000), 330 (35000). MS (rel intensity) 438 (M^+, 100), 283 (M – C_6H_4Br, 70). ^1H NMR ($CDCl_3$) δ: 2.42 (s, 3 H), 6.83 (s, 1 H), 7.28 (d, 2 H, J = 8.1 Hz), 7.48–7.59 (m, 4 H), 7.94 (d, 2 H, J = 8.1 Hz); ^{13}C NMR δ: 21.4, 62.1, 116.1, 121.7, 125.2, 126.1, 127.2, 128.5, 129.1, 131.9, 138.4, 151.7, 152.6.

Safety and Waste Disposal Information

All hazardous materials should be handled and disposed of in accordance with "Prudent Practices in the Laboratory"; National Academy Press; Washington, DC, 1995.

3. Discussion

The furan ring is a key component of many pharmaceutical agents, fragrances, natural products, and organic materials. Halofurans allow for further functionalization of the ring, usually via transition metal-catalyzed coupling reactions; they serve as starting materials for combinatorial chemistry.[3] The title molecule includes differently halogenated furan and phenyl rings that can be further functionalized.

The method described herein applies electrophilic cyclization for the synthesis of 3- (*beta*) iodofurans, with different, but electronically similar substituents in the 2- and 5- (*alpha*) positions. The method is of general use; other 3-halo (iodo- and bromo-) furans can also be synthesized in a regioselective manner (Table 1),[4] using other electrophilic reagents (ICl, NBS, and Br_2). This approach provides a facile, non-acidic, non-basic, non-transition metal based protocol, in which an easy to handle reagent[5] facilitates cyclization and serves simultaneously as a halogen donor. The reaction conditions are mild (room temperature in acetone). The yield is high, the reaction time is relatively short (0.5–2 h), and the isolation procedure is simple.

The best current alternatives for the synthesis of 2,5-unsymmetrically substituted 3-iodofurans include: a) Hydriodic acid cyclization of 4-hydroxybut-2-yn-1-ones.[3,6] This procedure features a strong acid. b) Iodocyclization (3 equiv of iodine) of alk-3-yn-1,2-diols that requires the use of a base.[7] c) Sonogashira coupling-addition-cyclocondensation one-pot procedure that starts from acyl chlorides and terahydropyranyl ethers of propargyl alcohols.[8]

Alkynol **1** was prepared by a known method.[9] It was observed by ^{1}H NMR that the ring opening of the 2-(4-bromophenyl)oxirane proceeded in a completely regioselective manner. $BF_3{\cdot}Et_2O$ was not used as an additive, as referenced; although such a procedure may be carried out in THF.[10] THF and DMF were found to be inefficient solvents for this protocol. Alkynone **2** was prepared by oxidation of the alkynol **1** with CrO_3 (Jones reagent).[11] Among other oxidation methods screened, Dess-Martin periodinane[12] was comparable. Jones oxidation for relevant compounds has been reported with low,[13] reasonable,[14,15] and high yields.[16] No experimental procedure, however, was provided for the high yielding example. A catalytic oxidation protocol has also been described.[17] Alk-3-yn-1-ones are reportedly prone to isomerization to the corresponding allenes;[17,18] this does not pertain to the procedure for **2**, presumably due to stabilization of the triple bond by conjugation to the phenyl ring. Alkynone **2** was found to be stable as a solid (stored in a freezer); cycloisomerization was observed when **2** was left for a prolonged time in solution. Cycloisomerization of homologous alk-3-yn-1-ones to the non-halogenated furans has been reported to proceed efficiently with the use of metal catalysts such as Hg,[16] Au,[19] Pd,[15,20] or Zn[21] derivatives.

Table 1. Preparation of other halofurans via electrophilic cyclization.[4]

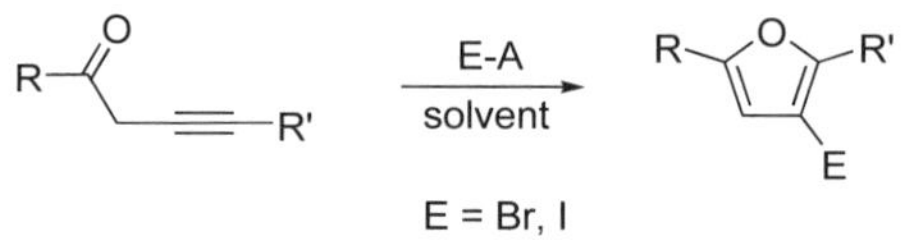

Entry	*R*	*R'*	*E-A, solvent*	*Yield (%)*
1	C_6H_5	*p*-$CH_3C_6H_4$	NBS, acetone	89
2			NIS, acetone	84
3			ICl, CH_2Cl_2	82
4	*p*-BrC_6H_4	*p*-$CH_3C_6H_4$	NBS, acetone	89
5			NIS, acetone	86
6	*p*-ClC_6H_4	*p*-$CH_3C_6H_4$	NBS, acetone	83
7			NIS, acetone	94
8	*p*-BrC_6H_4	*p*-*t*-BuC_6H_4	NBS, acetone	87
9			NIS, acetone	81
10	*p*-ClC_6H_4	*p*-*t*-BuC_6H_4	NBS, acetone	86
11			NIS, acetone	88

1. Department of Chemistry, Oakland University, Rochester, MI 48309-4477, USA; dembinsk@oakland.edu

2. Department of Chemistry, University of Pittsburgh, Pittsburgh, PA 15260, USA; pwipf@pitt.edu

3. Sauers, R. R.; Van Arnum, S. D. *J. Comb. Chem.* **2004**, *6*, 350.

4. For reaction leading to 3-chlorofurans see: Sniady, A.; Morreale, M. S.; Wheeler, K. A.; Dembinski, R. *Eur. J. Org. Chem.* **2008**, 3449.

5. Sniady, A. *Synlett* **2006**, 960.

6. Obrecht, D. *Helv. Chim. Acta* **1989**, *72*, 447.

7. a) El-Taeb, G. M. M.; Evans, A. B.; Jones, S.; Knight, D. W. *Tetrahedron Lett.* **2001**, *42*, 5945. b) Bew, S. P.; El-Taeb, G. M. M.; Jones, S.; Knight, D. W.; Tan,

W.-F. *Eur. J. Org. Chem.* **2007**, 5759.

8. a) Karpov, A. S.; Merkul, E.; Oeser, T.; Müller, T. J. *J. Chem. Commun.* **2005**, 2581. b) Karpov, A. S.; Merkul, E.; Oeser, T.; Müller, T. J. J. *Eur. J. Org. Chem.* **2006**, 2991.

9. Brandsma, L. *Preparative Acetylenic Chemistry*, 2nd Ed.; Studies in Organic Chemistry 34; Elsevier: Amsterdam, 1988; p 67.; a relevant procedure that uses ethylenediamine complex of lithium acetylide is also available: McDonald, F. E.; White, B. H. *Org. Synth.* Coll. Vol. 10, 621 (2004); Vol. *79*, 27 (2002).

10. Yamaguchi, M.; Hirao, I. *Tetrahedron Lett.* **1983**, *24*, 391.

11. Eisenbraun, E. J. *Org. Synth.* Coll. Vol. 5, 310 (1973); Vol. *45*, 28 (1965).

12. a) Boeckman, Jr., R. K.; Shao, P. ; Mullins, J. J. *Org. Synth.* Coll. Vol. 10, 696 (2004); Vol. *77*, 141 (2000). b) Ireland, R. E.; Liu, L. *J. Org. Chem.* **1993**, *58*, 2899.

13. Bol'shedvorskaya, R. L.; Pavlova, G. A.; Gavrilov, L. D.; Alekseeva, N. V.; Vereshchagin, L. I. *Zh. Org. Khim.* **1972**, *8*, 1879; *J. Org. Chem. USSR* **1972**, *8*, 1927.

14. Padwa, A.; Crumrine, D.; Hartman, R.; Layton, R. *J. Am. Chem.* Soc. **1967**, *89*, 4435.

15. Fukuda, Y.; Shiragami, H.; Utimoto, K.; Nozaki, H. *J. Org. Chem.* **1991**, *56*, 5816.

16. Ménard, D.; Vidal, A.; Barthomeuf, C.; Lebreton, J.; Gosselin, P. *Synlett* **2006**, 57.

17. Schmieder-van de Vondervoort, L.; Bouttemy, S.; Padrón, J. M.; Le Bras, J.; Muzart, J.; Alsters, P. L. *Synlett* **2002**, 243.

18. Krause, N.; Hashmi, A. S. K. *Modern Allene Chemistry,* Wiley: Weinheim, 2004; p 3-50.

19. Hashmi, A. S. K.; Schwarz, L.; Choi, J.-H.; Frost, T. M. *Angew. Chem., Int. Ed.* **2000**, *39*, 2285.

20. a) Utimoto, K. *Pure Appl. Chem.* **1983**, *55*, 1845. b) Sheng, H.; Lin, S.; Huang, Y. *Synthesis* **1987**, 1022.

21. a) Sniady, A.; Durham, A.; Morreale, M. S.; Wheeler, K. A.; Dembinski, R. *Org. Lett.* **2007**, 9, 1175. b) Sniady, A.; Durham, A.; Morreale, M. S.; Marcinek, A.; Szafert, S.; Lis, T.; Brzezinska, K. R.; Iwasaki, T.; Ohshima, T.; Mashima, K.; Dembinski, R. *J. Org. Chem.* **2008**, *73*, 5881.

PREPARATION OF SUBSTITUTED 5-AZAINDOLES: METHYL 4-CHLORO-1*H*-PYRROLO[3,2-*C*]PYRIDINE-2-CARBOXYLATE

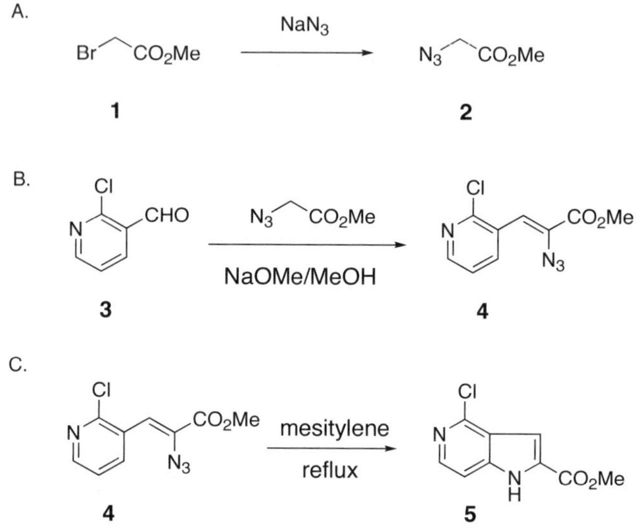

Submitted by Patrick Roy[1], Michel Boisvert, and Yves Leblanc.
Checked by Akiyoshi Kuramochi and Masakatsu Shibasaki.

1. Procedure

A. Methyl azidoacetate (**2**). *This reaction should be performed behind a safety shield.* To a solution of methyl bromoacetate (**1**) (51.5 mL, 0.528 mol) in MeOH (86 mL) in a 500-mL, three-necked, round-bottomed flask equipped with a magnetic stirrer, reflux condenser, a drying tube, a septum through which is passed a metal temperature probe and a 20 °C water bath is added a slurry of sodium azide (42.3 g, 0.649 mol, 1.23 equiv) in H_2O (40 mL) via a funnel in one portion. The residual sodium azide is rinsed into the round-bottomed flask with water (2 x 6 mL). The suspension is stirred for 20 min in the water bath and a mild exotherm occurs (Note 2). The solution is then brought to gentle reflux with a heating mantle for 2 h. After cooling to room temperature, the solvent is removed carefully on the rotary evaporator (40 mmHg, 25 °C water bath), as the heterogeneous mixture has a tendency to bump. The residue is partitioned between Et_2O (100 mL) and H_2O (100 mL), transferred to a separatory funnel and the aqueous phase is further extracted with of Et_2O (3 x 30 mL). The combined organic layer is dried with $MgSO_4$, filtered, and evaporated to give methyl azidoacetate (53.9–54.5 g, 89–90%) as a slightly yellow oil.

B. Methyl (2Z)-2-azido-3-(2-chloropyridin-3-yl)acrylate (**4**). *This reaction should be performed behind a safety shield.* A 2-L three-necked, round-bottomed flask (24/40 ground glass joints) is fitted with a drying tube, a 125-mL addition funnel, a magnetic stir bar (Note 3), and a septum through which is passed a metal temperature probe. To the flask is added 2-chloro-3-pyridine carboxaldehyde (20.0 g, 141 mmol) (Note 1), MeOH (200 mL), and methyl azidoacetate (32 mL, 349 mmol, 2.5 equiv) (Note 4). A slightly positive pressure of nitrogen is placed on the system via a needle through the septum. The flask is placed in an acetone bath, and dry ice is added slowly to the bath until the internal temperature reaches –10 to –12 °C (Note 5). At that point, a 25% solution of NaOMe in MeOH (80 mL, 4.37 M, 349 mmol, 2.5 equiv) is slowly added using the addition funnel over a period of 25 to 30 min. Dry ice is occasionally added to the cold bath during this time to maintain the internal temperature at approximately –9 to –12 °C. The reaction color changes from clear pale brown to cloudy orange/yellow as the addition proceeds, and mild bubbling is noted. When the addition is completed, the mixture is stirred an additional 15 min at –10 °C. The flask is then transferred to an ice bath, and stirring is continued for 2.5 h, during which time the internal temperature rises to 1.5 to 2 °C (Note 6). A tan-colored precipitate gradually forms (product), and slow bubbling is observed. The reaction is then placed in an ice bath in a 4 °C cold room, and is stirred overnight (16 h) with the drying tube in place. The flask is covered in aluminum foil to exclude light. The next morning, a 4-L Erlenmeyer flask is placed in an ice bath and the flask is then charged with 500 g of ice and 1.5 L of saturated NH_4Cl solution. A stir bar is transferred to the Erlenmeyer flask, and the tan-colored suspension is poured onto the stirring mixture over approximately 20 sec. The reaction flask is rinsed with 3 x 50 mL of ice water to ensure that any remaining solid is transferred. The final pH of the suspension is approximately 7.5. After stirring for an additional 5 min, the product is collected on a 12-cm Büchner funnel using a standard vacuum filtration setup, washing with ice-cold water (3 x 100 mL) (Note 7). The tan-colored solid is air-dried for 10 min on the Büchner funnel under suction from an aspirator, and is then transferred to a 1-L single-necked round-bottomed flask and placed under high vacuum for 30 min. Dichloromethane (250 mL) is then added to dissolve the product, and $MgSO_4$ (approximately 3–4 g) is added to dry the solution. The pale brown solution is filtered through a 2-cm thick pad of silica gel 60 on a 4-cm sintered glass funnel, which removes some dark brown baseline material. The pale yellow solution is concentrated to dryness on the rotary-evaporator (approximately 15 mmHg) at 25 °C to give the product as a pale yellow solid (17.7–18.8 g, 53–56%) (Note 8).

C. Methyl 4-chloro-1H-pyrrolo[3,2-c]pyridine-2-carboxylate (**5**). *This reaction should be performed behind a safety shield.* The product (**4**) from Step B (17.7 g, 74.2 mmol) is suspended in 740 mL of mesitylene (0.1 M) in a 2–L, single-necked round-bottomed flask equipped with a magnetic stir bar and a reflux condenser with a drying tube. The reaction mixture is stirred while being heated in an oil bath at 185 °C (Note 9), and after approximately 3 min, the starting material dissolves. At 5 min, bubbling occurs for several min as N_2 is extruded from **4**. After 30 min, the mixture is gently refluxing, which is maintained for 1 h. The clear deep yellow solution is then allowed to stand at room temperature for 1 h, during which time a thick pale yellow precipitate forms. After cooling to 0 °C for 1 h, the product is collected on a 12–cm Büchner funnel, washing with 4 x 50 mL of cold hexanes. The solid is then suspended with vigorous stirring overnight in 200 mL of EtOAc:hexanes (1:20, vol/vol). The product is then collected on a 12–cm Büchner funnel and washed with 2 x 50 mL of hexanes to give a pale yellow solid (10.7 g, 69%) (Note 10).

2. Notes

1. Methyl bromoacetate (97% purity) was obtained from Lancaster Synthesis. Sodium azide (>99.5% purity), 2-chloro-3-pyridine carboxaldehyde (97% purity), 25% NaOMe/MeOH and mesitylene (98% purity) were obtained from the Aldrich Chemical Co. MeOH was Optima D.I.G. grade (4 L, 0.07% H_2O) from Fisher.

2. A mild exotherm to approximately 45 °C was observed, although this exothermic behaviour never showed any sign of loss of control. Differential scanning calorimetry indicated an endotherm at 58 °C. This reaction was carried out at this scale or greater without incident. The characterization of **2** is as follows: ^{1}H NMR (500 MHz, acetone-d_6) δ: 3.77 (s, 3 H), 4.04 (s, 2 H). ^{13}C NMR (126 MHz, acetone-d_6) δ: 50.47. 52.64, 169.87; IR (neat): cm^{-1} 2108, 1751.

3. Mechanical stirring was not necessary, although for other substrates such as 2-methoxy-3-pyridine carboxaldehyde[2] the reaction became a thick

paste overnight at 0 °C. This thickening, however, did not appear to affect the yield.

4. In the original papers using methyl azidoacetate for the condensation step (Monatshefte Chemie **1969**, 100(5) 1599), the authors used 4 equivalents of the reagent, which was reduced to 2.5 equivalents in the above procedure. The excess was felt to be necessary since the reagent slowly decomposed with time, as indicated by the gentle bubbling during the condensation step.

5. The addition was also carried out successfully at –30 °C for this substrate. However, for 2,4-dichloro-3-pyridine carboxaldehyde[2] addition at –30 °C followed by warming to 0 °C resulted in an exotherm to >12 °C, which led to the mixture frothing out of the flask. No product was recovered. As a *precaution* when carrying out the reaction, the flaps on the septa are not folded down in order to minimize problems arising from any build-up of pressure in the event of an exotherm.

6. If a non-insulated container such as a plastic bowl is used for the ice bath, care must be taken to maintain enough ice to prevent the internal temperature from rising too high. Internal temperatures of 4 °C have been noted with ice floating in water in the bath.

7. The filtration can require 15–20 min due to the fine nature of the crystals. A larger Büchner funnel reduces this filtration time. The aqueous/MeOH filtrate soon turned a dark red after filtration.

8. The product was best stored in the refrigerator or freezer, covered in aluminum foil, although we noted no change in the NMR spectra after several days at room temperature. Depending on the substitution pattern on the pyridine ring, the intermediate **4** gradually darkened over several weeks/months in the refrigerator. The material was repurified by dissolution in CH_2Cl_2 and filtration through a small plug of silica gel to remove baseline material. Differential scanning calorimetry of this material indicated that an exotherm occurred at approximately 120 ºC. The characterization for **4** is as follows: mp = 107–109 ºC (dec, gas bubbles); ^{1}H NMR (500 MHz, $CDCl_3$) δ: 3.93 (s, 3 H). 7.16 (s, 1 H), 7.26 (dd, J = 7.9, 4.8 Hz, 1 H), 8.29 (dd, J = 4.8, 1.7 Hz, 1 H), 8.50 (dd, J = 7.9, 1.7 Hz, 1 H); ^{13}C NMR (125 MHz, $CDCl_3$) δ: 53.54, 118.65, 122.50, 128.29, 129.22, 139.48, 149.35, 151.33, 151.33. IR (KBr): cm^{-1} 2129, 1708. MS (ESI+): *m/z* 260.7 (M+1) (100%)

9. The reaction was also carried out through use of a heating mantle, but some dark brown solid material formed on the walls of the flask. The yield was not greatly affected.

10. The characterization for **5** prior to chromatography was as follows: mp = 210–212 °C (dec); ^{1}H NMR (500 MHz, DMSO-d_6) δ: 3.91 (s, 3 H), 7.19 (s, 1 H), 7.43 (d, J = 5.8 Hz, 1 H), 8.09 (d, J = 5.8 Hz, 1 H), 12.77 (br s, 1 H); ^{13}C NMR (126 MHz, DMSO-d_6) δ: 52.29, 105.97, 107.87, 122.08, 129.15, 141.51, 141.85, 144.23, 160.81; IR (KBr): cm^{-1} 1718. MS (ESI+) *m/z* 210.8 (M+H) (58%), M+Na 232.7 (100%). Anal. Calcd. for $C_9H_7ClN_2O_2$: C, 51.32; H, 3.35; N, 13.30. Found: C, 51.31; H, 3.41; N, 13.11. Although further purification was not necessary, the product was purified on small scale (100 mg) by flash chromatography on silica gel 60 (EMD), eluting with 30 % EtOAc/hexanes. The crude solution in mesitylene was applied directly to the column before the product precipitated.

Safety and Waste Disposal Information

All hazardous materials should be handled and disposed of in accordance with "Prudent Practices in the Laboratory"; National Academy Press; Washington, DC, 1995.

3. Discussion

This procedure describes the preparation of methyl 4-chloro-1H-pyrrolo[3,2-c]pyridine-2-carboxylate, a 4-Cl-5-azaindole, via thermolysis of an intermediate azidopyridine acrylate (Hemetsberger-Knittel reaction).[3] The product is isosteric with the analogous indole template, and represents an interesting scaffold from the point of view of medicinal chemistry. Furthermore, with four sites available for selective functionalization, one could envisage a wide range of chemical diversity in the construction of libraries based on this core structure. Table 1 shows the results obtained for the preparation of a variety of 5, 6, and 7-azaindoles. Procedure A (100 mg of pyridine acrylate at 0.1 M) involves bulb-to-bulb distillation of the mesitylene followed by chromatography. Procedure B is used for amounts greater than 500 mg of the pyridine acrylate and is performed as described in the current preparation. Procedure C is for the synthesis of 6-azaindoles (entries 13 and 14) and involves adding the pyridine acrylate to a refluxing solution of decalin (190 °C) at a concentration of 2 g/ 500 mL and refluxing 10 min before cooling and filtering.

Of the unsubstituted pyridine acrylates (entries 1-3), only the 7-azaindole was obtained, albeit in low yield (25%). For the two 6-azaindole precursors listed in entries 13 and 14, decomposition/poor yields were observed upon thermolysis at 160 °C, whereas use of refluxing decalin gave acceptable yields (44%, 51%). There appears to be a minimum threshold below which the desired cyclization does not readily take place for these substrates. The best results were obtained with 5-azaindoles (entries 4-12). The thermolysis proceeds smoothly at 160 °C with halogen, ether, thioether, methylsulfone, and alkyl substituents on the pyridine ring to give good yields (66–93%) with the exception of the methoxychloro-substituted entry 12 (32%).

Table 1. Thermolysis of 2-Azido-3-pyridine Acrylates

Entry	Azidopyridine acrylate	Procedure	Product	Yield(%)
1	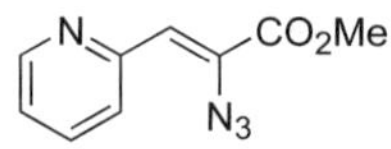	B	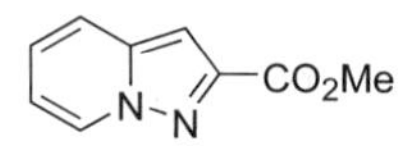	67
2	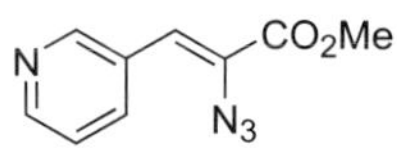	B	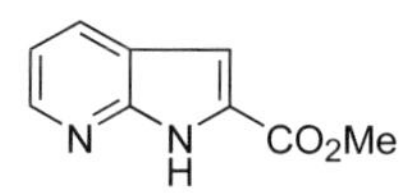	25
3	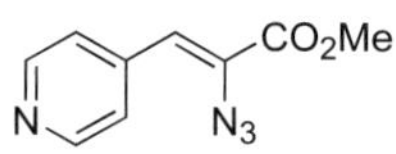	B	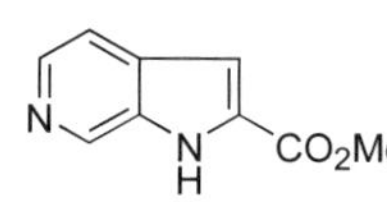	0
4	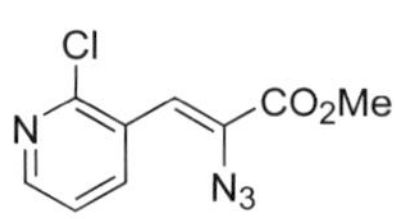	B	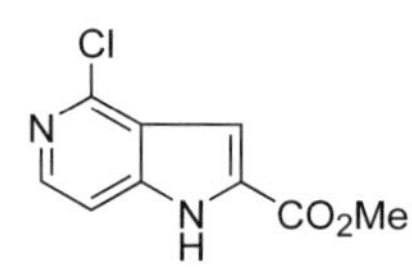	72
5	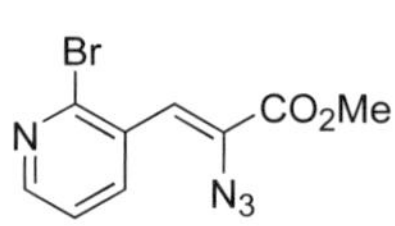	A	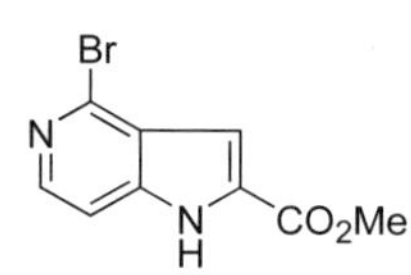	78
6	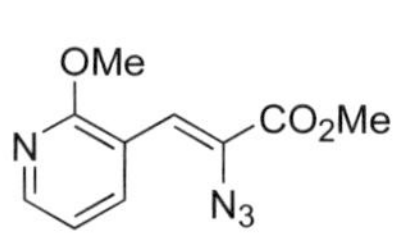	A	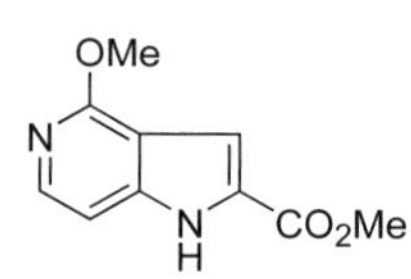	93
7	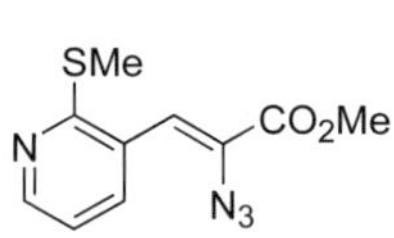	A	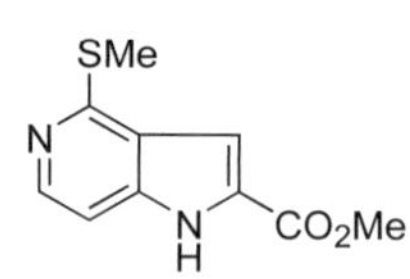	84
8	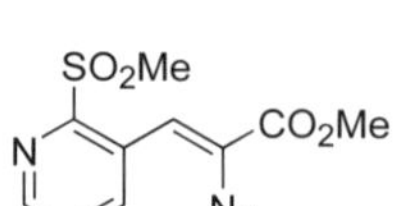	A	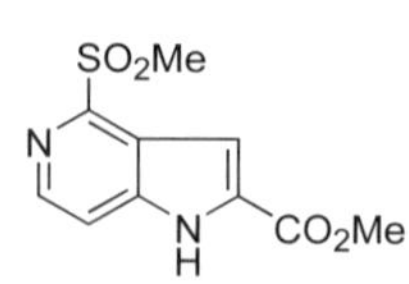	69

Table 1. (continued) Thermolysis of 2-Azido-3-pyridine Acrylates

Entry	Azidopyridine acrylate	Procedure	Product	Yield(%)
9	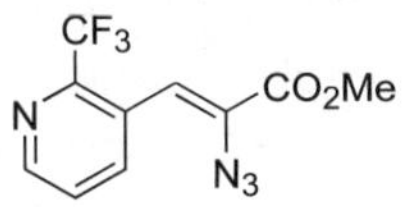	B	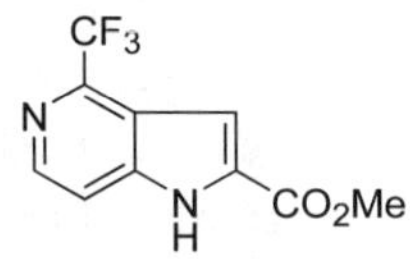	66
10	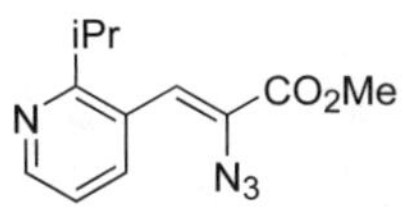	B	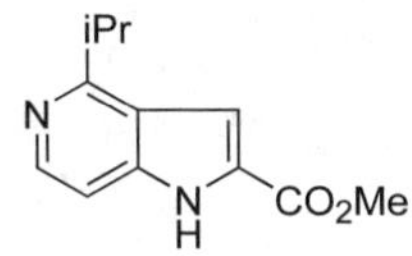	75
11	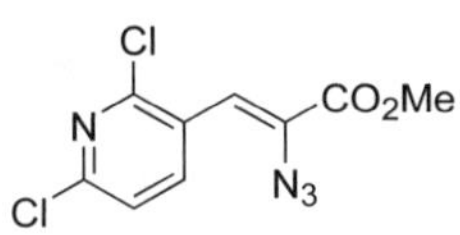	A	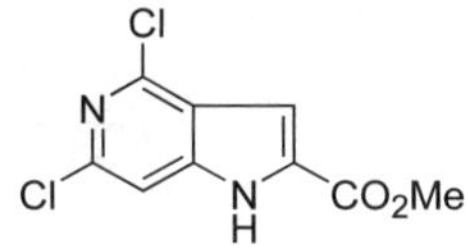	77
12	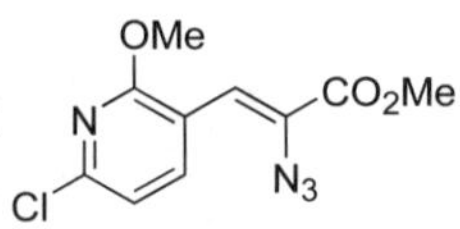	A	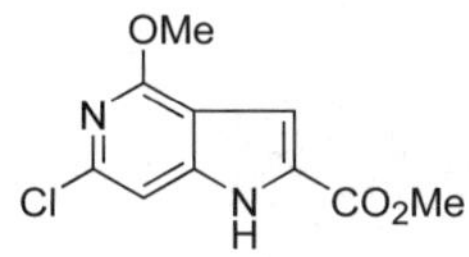	32
13	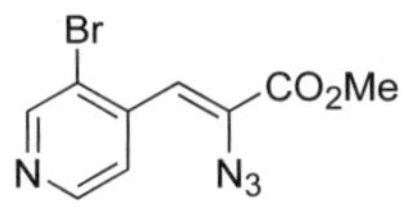	C	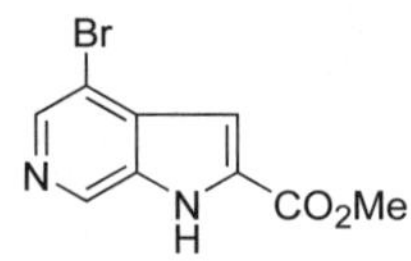	44
14	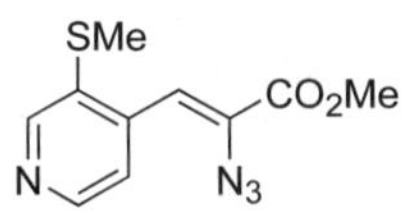	C	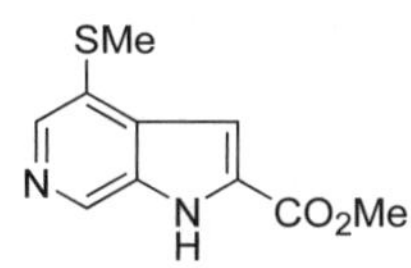	51
15	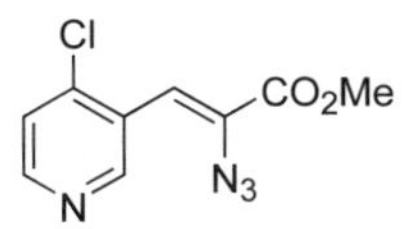	A	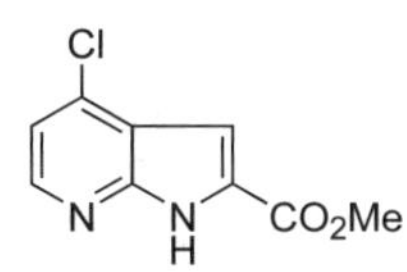	40
16	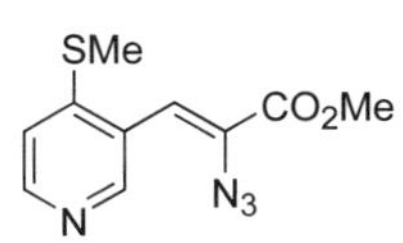	B	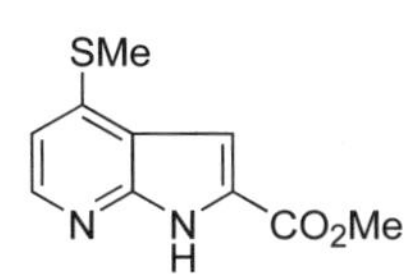	56

1. Merck Frosst Centre for Therapeutic Research, C.P. 1005, Pointe Claire-Dorval , Québec H9R 4P8, Canada.

2. Roy, P.J.; Dufresne, C.; Lachance, N.; Leclerc, J-P.; Boisvert, M.; Wang, Z.; Leblanc, Y. *Synthesis* **2005**, *16*, 2751-2757.

3. a) Hemtsberger, H.; Knittel, D.; Wiedmann, H. *Monatshefte Chemie* **1970**, *101*, 161-165. For references on the preparation of azaindoles via this reaction, see b) Fresneda, P. M.; Molina, P.; Delgado, S.; Bleda, J. A. *Tetrahedron Lett.* **2000**, *41,* 4777-4780 and c) Lomberget, T.; Radix, S.; Barret, R. *Synlett* **2005**, *13*, 2080-2082.

SYNTHESIS OF AN *N*-SUBSTITUTED LACTAM USING AN INTRAMOLECULAR SCHMIDT REACTION: FORMATION OF 2,3,11,11a-TETRAHYDRO-1*H*-BENZO[d]PYRROLO[1,2-a]AZEPIN-5(6*H*)-ONE

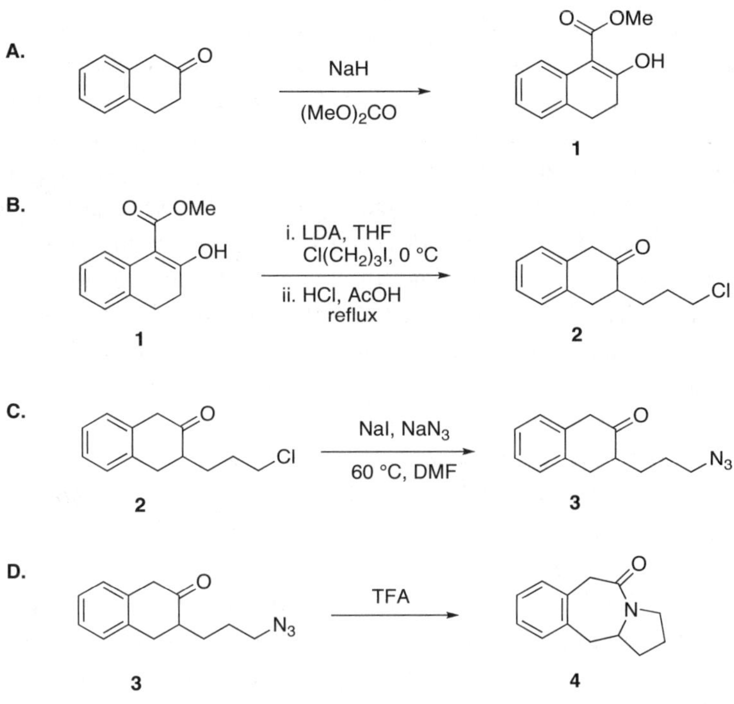

Submitted by Scott Grecian and Jeff Aubé[1].
Checked by Dalian Zhao and David J. Mathre.

Caution! *Sodium azide is toxic as well as heat and shock sensitive. Metallic and sharp surfaces should be avoided.*

1. Procedure

A. Methyl 2-oxo-1,2,3,4-tetrahydronaphthalene-1-carboxylate (**1**). An oven-dried 250-mL three-necked, round-bottomed flask equipped with a glass stopper, thermometer, rubber septum, large magnetic stirring bar, and two venting needles is charged with 8.0 g (200 mmol) of 60% sodium hydride in mineral oil by temporarily removing the glass stopper. The suspension is then washed with three 30 mL portions of anhydrous hexanes using a syringe through the septum. Anhydrous dimethyl carbonate is added (Note 1) (120 mL) and the mixture is cooled to 5 °C with an ice bath. β-Tetralone (Note 2) (13.3 mL, 100 mmol) is added via syringe over 20 min. The resulting green/brown mixture is then stirred for 1 h, at which time the ice bath is removed and the reaction is warmed to 23 °C. After 12 h, the glass stopper is removed and the reaction flask is fitted with a reflux condenser and heated to reflux for 1 h in an oil bath (Note 3). The brown mixture is cooled to room temperature, the condenser is removed, 20 mL of water is added dropwise and the mixture is poured into a 1-L separatory funnel followed by the addition of 200 g of ice and the slow addition of 250 mL of 1.0 M HCl. The mixture is extracted with diethyl ether (3 x 100 mL). The ethereal extracts are combined and washed once with 150 mL of brine and dried with magnesium sulfate, filtered and concentrated to give 20–25 g of a red oil. The oil is transferred to a 100-mL round-bottomed flask and topped with a short path distillation head fitted with a thermometer and four receiving flasks submerged in an ice bath. Vacuum distillation is performed at 0.9 mm Hg while the oil bath temperature is increased gradually to 185 °C. The first several drops are collected between 136 and 141 °C and discarded. Methyl 2-oxo-1,2,3,4-tetrahydronaphthalene-1-carboxylate (**1**) is collected between 138 and 141 °C as a pale yellow oil (16.9 – 17.8 g, 83–87%) (Notes 4 and 5).

B. 3-(3'-Chloropropyl)-3,4-dihydronaphthalen-2(1H)-one (**2**). To an oven-dried 250-mL, three-necked, round-bottomed flask equipped with a rubber septum, a thermometer, a glass stopper, and a magnetic stir bar is added 120 mL of freshly distilled THF and diisopropylamine (15.1 mL, 108 mmol) (Note 6) by syringe through the septum. After cooling the flask to 0 °C in an ice bath, 45 mL of *n*-butyllithium (105 mmol, 2.32 *M* /hexanes) are added dropwise over 10 min while maintaining the temperature at or below 10 °C (Note 7). After an additional 45 min, methyl 2-hydroxy-3,4-dihydronaphthalene-1-carboxylate (**1**) (10.0 g, 49 mmol) dissolved in 25 mL of THF is added via cannula through the syringe over 15 min while maintaining the reaction temperature below 10 °C (Note 8). After 15 min, the ice bath is removed and the reaction is stirred for 1 h at which time it is cooled to 0 °C in an ice bath. After an additional 30 min, 1-chloro-3-iodopropane (6.22 mL, 59 mmol) (Note 9) is added in several portions via syringe over 25 min while maintaining the reaction temperature below 10 °C. The ice bath is removed after 30 min, and the reaction is stirred for an additional h, at which time the reaction is quenched by the slow addition of 50 g of ice. The mixture is transferred to a separatory funnel and 150 mL of

1.0 M HCl is added slowly. The organic layer is separated and the mixture is extracted with diethyl ether (2 x 100 mL). The combined organic layers are washed once with 175 mL of brine, dried with magnesium sulfate, filtered and concentrated to give a red oil (15.2–15.8 g) (Note 10). The oil is transferred to a 100-mL, single-necked, round-bottomed flask equipped with a stir bar and a reflux condenser, and the oil is dissolved in 35 mL of glacial acetic acid. Concentrated HCl (5 mL) is added. The mixture is stirred at room temperature for 30 min, then placed in an oil bath and heated to 110 °C for 3 h, then cooled to room temperature and poured into a separatory funnel followed by the addition of 100 mL of water and 75 mL of diethyl ether. The aqueous layer is extracted with diethyl ether (2 x 75 mL). The combined extracts are washed with a saturated solution of sodium bicarbonate (3 x 100 mL) (Note 11). The organic layer is then washed once with 150 mL of brine, dried over magnesium sulfate, filtered and concentrated to a red oil (11.7–12.0 g). The oil is chromatographed over 100 g of silica gel (25 × 5 cm) starting with 500 mL of 5% ethyl acetate/hexanes followed by 500 mL of 10% ethyl acetate/hexanes. The desired product is collected in fractions 28 to 40 (25 mL fractions), which upon concentration, gives 8.1–8.2 g (73–74% from **1**) of 3-(3′-chloropropyl)-3,4-dihydronaphthalen-2(1*H*)-one (**2**) as a light yellow oil (Note 12).

C. 3-(3′-Azidopropyl)-3,4-dihydronaphthalen-2(1H)-one (**3**). In a 100-mL single-necked round-bottomed flask equipped with a stir bar is dissolved 3-(3-chloropropyl)-3,4-dihydronaphthalen-2(1H)-one (**2**) (5.0 g, 22 mmol) in 60 mL of anhydrous *N,N*-dimethylformamide. A blasting shield is placed around the reaction flask and sodium iodide (3.96 g, 26.4 mmol) (Note 13) and sodium azide (2.86 g, 44.0 mmol) (Note 14) are added. The mixture is stirred while immersed in an oil bath heated to 60 °C for 12 h, after which time the reaction mixture is poured into 150 mL of water and extracted with diethyl ether (2 x 100 mL) (Notes 15 and 16). The organic layer is washed with brine (3 x 150 mL), dried with magnesium sulfate, filtered and concentrated at or below 30 °C to a red oil (4.9–5.2 g). The oil is loaded on a column of 60 g of silica gel (18 × 5 cm) and eluted with 750 mL of 10% ethyl acetate/hexanes. The desired product is collected in fractions 18-28 (25 mL fractions) which, after concentration, gives 4.10–4.24 g (81–84%) of 3-(3′-azidopropyl)-3,4-dihydronaphthalen-2(1*H*)-one (**3**) as a red oil (Note 17).

D. 2,3,11,11a-Tetrahydro-1H-benzo[d]pyrrolo[1,2-a]azepin-5(6H)-one (**4**). An oven-dried, three-necked 100-mL round-bottomed flask is equipped with a thermometer, rubber septum, an elbow adapter connected to a tygon tube with an in-line bubbler, and a stir bar. 3-(3′-Azidopropyl)-3,4-

dihydronaphthalen-2(1H)-one (4.0 g, 17.44 mmol) of (**3**) is added to the flask by temporarily removing the septum and cooled to 0 °C in an ice-bath. Trifluoroacetic acid (Note 18) (20 mL) is added dropwise through the septum by syringe over 15 min (Note 19). After 30 min, the ice bath is removed and the reaction is allowed to warms to room temperature. After an additional 30 min, 50 mL of water is added to the reaction, which is then poured into a separatory funnel. The aqueous layer is extracted with chloroform (3 x 40 mL). The organic extracts are combined and washed with 100 mL of a saturated solution of sodium bicarbonate, dried with magnesium sulfate, filtered and concentrated to an orange solid (3.7–4.0 g). The solid is dissolved in 5 mL of hot chloroform, loaded onto a column of 40 g of silica gel (10 × 5 cm) and eluted with 350 mL of 20% ethyl acetate/hexanes followed by 500 mL of ethyl acetate. The desired product is collected in fractions 19-34 (25 mL fractions) which, after concentration, gives 2.87–3.09 g (82–88%) of 2,3,11,11a-tetrahydro-1H-benzo[d]-pyrrolo[1,2-a]azepin-5(6*H*)-one (**4**) as an off-white solid (Note 20).

2. Notes

1. Sodium hydride and dimethyl carbonate were purchased from Aldrich Chemical Company, Inc.

2. 3,4-Dihydronaphthalen-2(1*H*)-one (β-tetralone) was purchased from Aldrich Chemical Company, Inc.

3. The reaction progress was monitored by silica gel thin layer chromatography.

4. This compound was prepared according to the procedure described in Oommen, P. K. *Aust. J. Chem.* **1976**, *29*, 1393-1396.

5. The product exhibited the following properties: R_f = 0.48 (1:9, ethyl acetate/hexanes); IR (neat): 1640, 1595, 1225 cm^{-1}; ^1H NMR (400 MHz, $CDCl_3$) δ: 2.59 (t, *J* = 7.3 Hz, 2 H), 2.86 (t, *J* = 7.3 Hz, 2 H), 3.96 (s, 3 H), 7.18–7.12 (m, 2 H), 7.25 (t, *J* = 10.2 Hz, 1 H), 7.75 (d, *J* = 6.7 Hz, 1 H), 13.39 (s, 1 H); ^{13}C NMR (100 MHz, $CDCl_3$) δ: 27.8, 29.6, 51.8, 99.9, 125.0, 125.9, 126.5, 127.3, 131.4, 133.2, 172.5, 178.5; MS (FAB) *m/z* (relative intensity): 204 (M^+, 99), 172 (100); HRMS calcd for $C_{12}H_{13}O_3$ (MH^+): 205.0865, found: 205.0853.

6. Anhydrous diisopropylamine was purchased from the Aldrich Chemical Company, Inc., and used without further purification.

7. *n*-Butyllithium was purchased from the Aldrich Chemical Company, Inc., and titrated prior to use. For the titration procedure used, see: Kofron, W. G.; Baclawski, L. M. *J. Org. Chem.* **1976**, *41*, 1879-1880.

8. A voluminous light yellow precipitate began to form near the end of the addition.

9. 1-Chloro-3-iodopropane was purchased from Aldrich Chemical Company, Inc.

10. An aliquot (285 mg) was loaded on 10 g of silica gel and eluted with 5% ethyl acetate/hexanes. Concentration of the fractions containing the product gave 178 mg of methyl 3-(3-chloropropyl)-2-oxo-1,2,3,4-tetrahydronaphthalene-1-carboxylate as a light yellow oil. This product exhibited the following physical properties: R_f = 0.40 (1:9, ethyl acetate/hexanes); IR (neat): 3440, 1735, 1640, 1595 cm^{-1}; ^1H NMR (400 MHz, $CDCl_3$) δ: 1.51–1.57 (m, 1 H), 1.67–1.71 (m, 1 H), 1.84–1.93 (m, 2 H), 2.55–2.59 (m, 1 H), 2.66 (dd, J = 4.9, 15.1 Hz, 1 H), 3.06 (dd, J = 5.6, 15.1 Hz, 1 H), 3.53 (sextet, J = 6.9 Hz, 2 H), 3.95 (s, 3 H), 7.09–7.16 (m, 2 H), 7.22 (dt, J = 1.5, 7.1 Hz, 1 H), 7.73 (d, J = 7.8 Hz, 1 H), 13.49 (s, 1 H); ^{13}C NMR (100 MHz, $CDCl_3$) δ: 27.0, 30.3, 33.2, 38.5, 44.7, 51.9, 99.3, 125.2, 125.8, 126.5, 128.0, 130.6, 131.5, 172.7, 180.4; MS (FAB) *m/z* (relative intensity): 280 (M^+, 100), 248 (77); HRMS calcd for $C_{15}H_{17}ClO_3$ (M^+): 280.0866, found: 280.0873.

11. Significant gas evolution was observed upon addition of a saturated solution of sodium bicarbonate.

12. The product exhibited the following properties: R_f = 0.41 (1:4, ethyl acetate/hexanes); IR (neat): 2955, 1710, 750 cm^{-1}; ^1H NMR (500 MHz, $CDCl_3$) δ: 1.58–1.62 (m, 1 H), 1.87–1.97 (m, 3 H), 2.48 (sextet, J = 6.4 Hz, 1 H), 2.89 (dd, J =10.3 Hz, 15.4, 1 H), 3.13 (dd, J = 5.5 Hz, 15.4, 1 H), 3.58 (sextet, J = 4.4 Hz, 2 H), 3.62 (d, J = 8.1 Hz, 2 H), 7.14 (m, 1 H), 7.24 (m, 3 H); ^{13}C NMR (125 MHz, $CDCl_3$) δ: 27.1, 30.1, 34.4, 44.4, 44.8, 46.7, 126.8, 126.9, 127.8, 128.0, 133.1, 135.6, 210.9; MS (FAB) *m/z* (relative intensity): 223 (MH^+, 100); HRMS calcd for $C_{13}H_{16}ClO$ (MH^+): 223.0890, found: 223.0890.

13. Sodium iodide and sodium azide were purchased from the Aldrich Chemical Company Inc.

14. **Caution**: *Sodium azide is explosive and toxic. Metallic and sharp surfaces should be avoided.*

15. Although the submitters have not experienced any explosions of alkyl azides, the substances should be treated with extreme caution.

16. The reaction was monitored by silica gel thin layer chromatography.

17. The product exhibited the following properties: R_f = 0.39 (1:4, ethyl acetate/hexanes); IR (neat): 2920, 2080, 1700 cm^{-1}; ^1H NMR (300 MHz, $CDCl_3$) δ: 1.49 (ddd, J = 6.8, 9.2, 13.3 Hz, 1 H), 1.69 (ddd, J = 6.8, 6.8, 14.0 Hz, 1 H), 1.79–1.93 (m, 1 H), 2.46 (sextet, J = 6.4 Hz, 1 H), 2.85 (dd, J = 10.3, 15.4 Hz, 1 H), 3.11 (dd, J = 5.6, 15.5 Hz, 1 H), 7.15–7.23 (m, 3 H), 3.29 (dt, J = 1.4, 6.8 Hz, 2 H), 3.60 (d, J = 6.4 Hz, 2 H) 7.05–7.15 (m, 1 H); ^{13}C NMR (100 MHz, $CDCl_3$) δ: 26.5, 27.0, 34.5, 44.5, 47.0, 51.4, 126.9, 127.0, 127.9, 128.1, 133.2, 135.7, 211.0; Anal calcd for $C_{13}H_{15}N_3O$: C, 68.10; H, 6.59; N, 18.33; found C, 68.38; H, 6.71; N, 17.98.

18. Trifluoroacetic acid was purchased from the Aldrich Chemical Company, Inc.

19. The reaction turned purple, significant gas evolution was observed and heat was evolved during the addition of trifluoroacetic acid.

20. The product exhibited the following properties: mp 120–121 °C; R_f = 0.17 (2:3, ethyl acetate/hexanes); IR (neat): 2955, 1625 cm^{-1}; ^1H NMR (300 MHz, $CDCl_3$) δ: 1.69–2.25 (m, 3 H), 2.18–2.31 (m, 1 H), 2.95–3.14 (m, 2 H), 3.47–3.66 (m, 3 H), 4.08–4.22 (m, 2 H), 7.02–7.22 (m, 4 H); ^{13}C NMR (100 MHz, $CDCl_3$) δ: 23.0, 33.7, 39.4, 44.1, 46.7, 57.1, 127.0, 127.5, 130.3, 131.0, 132.5, 136.5, 170.0; MS (EI) *m/z* (relative intensity): 201 (M^+, 100); HRMS calcd for $C_{13}H_{15}NO$ (M^+): 201.1154, found: 201.1144. Anal. Calcd for $C_{13}H_{15}NO$: C, 77.58; H, 7.51; N, 6.96. Found: C, 77.55; H, 7.38; N, 7.12.

Waste Disposal Information

All toxic materials were disposed of in accordance with "Prudent Practices in the Laboratory"; National Academy Press; Washington, DC, 1995.

3. Discussion

Although the Schmidt reaction of hydrazoic acid with ketones has been known for many years, the corresponding reaction of ketones with alkyl azides, affording *N*-substituted lactams, is of much more recent vintage.[1-4] In the 1940s, several researchers reported that the reaction of methyl azide with simple ketones using standard Schmidt reaction conditions (concentrated H_2SO_4) did not lead to any lactam products.[5,6] In the following decade, Boyer found that amides could be formed in modest yields (≤30%) when aromatic aldehydes were allowed to react with *n*-hexyl azide under H_2SO_4 promotion, but reported that these conditions failed when applied to ketones.[7] In the early 1990s, it was learned that Schmidt reactions of alkyl azides and ketones could be actuated by carrying out intramolecular reactions in δ- or ε-azido ketones.[8]

The intramolecular Schmidt reaction requires that the ketone and azide group be separated by four or five atoms (usually carbons), with the former being significantly more favorable.[8,9] Thus, Schmidt reactions of δ-azido ketones can be promoted by strong protic acids, as exemplified in the present procedure, while the homologous substrates require stronger Lewis acid promotion and are generally much more sluggish (Table 1). This is presumably because the former reactions proceed through a six-membered azidohydrin intermediate en route to products, whereas the seven-membered ring version of this intermediate is less favored (Scheme 1). In a specialized example, it was shown that advantage could be taken of the dependence of reactivity on tether length to effect a regioselective reaction in a complex example of the reaction on a diketone reactant.[10] Subsequently, it was found that the more stringent conditions used to carry out Schmidt reactions of ε-azido ketones could also be applied to intermolecular reactions of simple alkyl azides and ketones, although this reaction is of severely limited scope and frequently plagued by complicating (if sometimes interesting) side reactions.[11,12] Conversely, the intramolecular Schmidt reaction is very forgiving with respect to substitution pattern and to the nature of the reacting ketone, with successful reactions being observed to date in ketones containing from 3 to 12 members. Aldehydes also react, but can give rise to either *N*-substituted or -unsubstituted lactams depending on whether an alkyl group or a hydrogen atom migrates.

Some other features of the intramolecular Schmidt reaction are that it is highly regioselective and, where relevant, stereoselective. Thus, the product formed in a typical example of this reaction is almost always the fused lactam formed by migration of the α-carbon between the carbonyl and azide. However, a few examples of bridged adducts have been observed in specialized substrates.[10,13,14] As is common for related processes like the classical Schmidt reaction and the Baeyer–Villiger reaction, the intramolecular Schmidt reaction is stereoselective insofar as the α-carbon migrates with retention of configuration.[9] An important practical consideration, however, is that apparent loss of stereoselectivity can be observed due to the fact that the α-hydrogen is acidic and therefore this center is subject to enolization and epimerization. This is particularly problematic for sluggish substrates that require heating or extended reaction times.[15]

Scheme 1

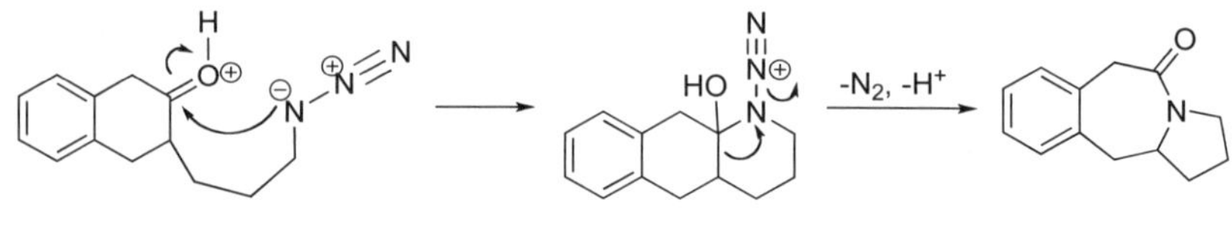

Table 1. Intramolecular Schmidt Reactions of Azido Ketones

Entry	Ketone	Product	Yield (%)
1	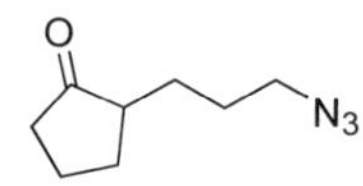	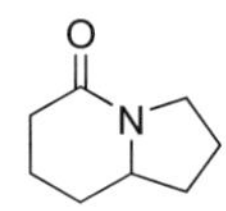	83
2	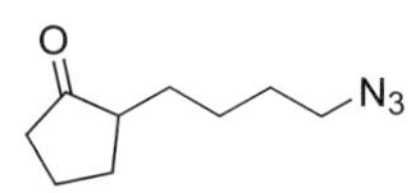	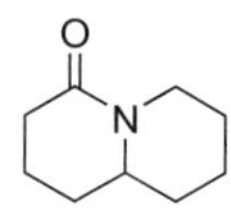	56
3	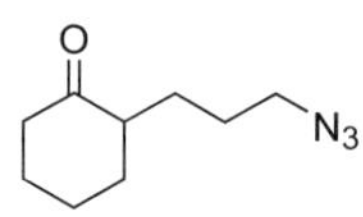	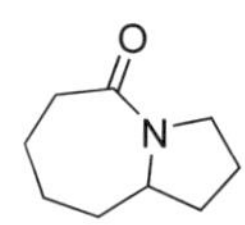	85
4	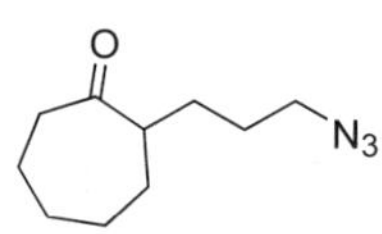	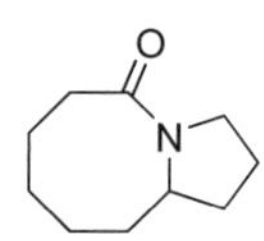	80
5	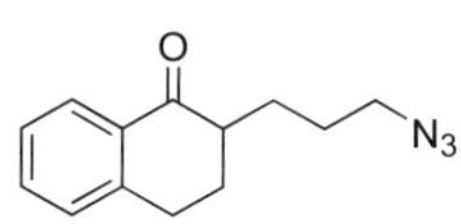	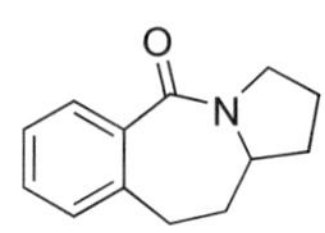	45
6	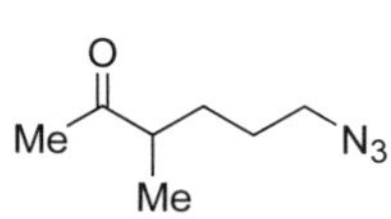	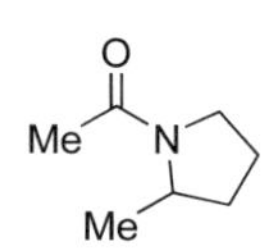	75
7	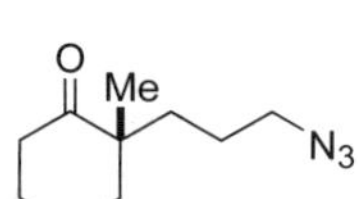 91% ee	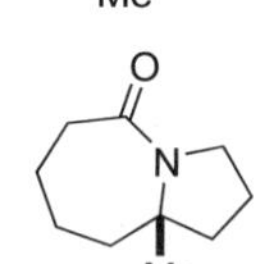 89% ee	87
8	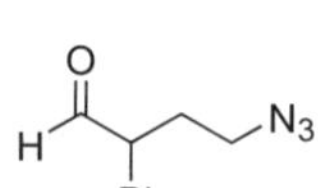	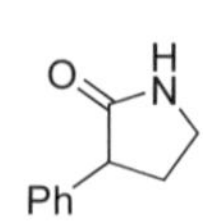	29

Over the same time period that saw the development of the intramolecular Schmidt reaction, a number of related processes were also reported. These included the reactions of alkyl azides with carbocations generated from alkenes[16-19] or other functional groups,[20,21] principally developed by Pearson but explored by other researchers as well. Subsequent to the discovery of the intramolecular Schmidt reaction, this laboratory has also explored the related reactions of 2- or 3-hydroxyalkyl azides with ketones, some of which can be used to effect an asymmetric nitrogen insertion reaction.[22-25] These, along with other important developments in modern alkyl azide chemistry, have been summarized in an exhaustive review article.[26]

The main utility of the present reaction is in the preparation of multicyclic nitrogen containing rings in complex alkaloid synthesis. Some targets that have been explored using this chemistry include homoerythrina alkaloids,[27] indolizidines isolated from South American frogs,[28,29] gephyrotoxin,[30] sparteine,[31] stenine,[13] and aspidospermidine.[10]

1. Department of Medicinal Chemistry, 1251 Wescoe Hall Drive, Room 4070 Malott Hall, University of Kansas, Lawrence, KS 66045-7582; jaube@ku.edu. Support for this work from the National Institutes of Health (GM-49093) is gratefully acknowledged.

2. (a) Schmidt, K. F. *Ber.* **1924**, *57B*, 704-706. (b) Wolff, H. *Org. React.* **1946**, *3*, 307-336.
3. Smith, P. A. S. In *Molecular Rearrangements*; de Mayo, P., Ed.; John Wiley & Sons: New York, 1963; Vol. 1, pp 457-591.
4. Abramovich, R. A.; Kyba, E. P. In *The Chemistry of the Azido Group*; Patai, S., Ed.; John Wiley & Sons: London, 1971, pp 221-329.
5. Briggs, L. H.; De Ath, G. C.; Ellis, S. R. *J. Chem. Soc.* **1942**, 61-63.
6. Smith, P. A. S. *J. Am. Chem. Soc.* **1948**, *70*, 320-323.
7. Boyer, J. H.; Hamer, J. *J. Am. Chem. Soc.* **1955**, *77*, 951-954.
8. Aubé, J.; Milligan, G. L. *J. Am. Chem. Soc.* **1991**, *113*, 8965-8966.
9. Milligan, G. L.; Mossman, C. J.; Aubé, J. *J. Am. Chem. Soc.* **1995**, *117*, 10449-10459.
10. Iyengar, R.; Schildknegt, K.; Aubé, J. *Org. Lett.* **2000**, *2*, 1625-1627.
11. Aubé, J.; Milligan, G. L.; Mossman, C. J. *J. Org. Chem.* **1992**, *57*, 1635-1637.
12. Desai, P.; Schildknegt, K.; Agrios, K. A.; Mossman, C.; Milligan, G. L.; Aubé, J. *J. Am. Chem. Soc.* **2000**, *122*, 7226-7232.
13. Golden, J. E.; Aubé, J. *Angew. Chem., Int. Ed.* **2002**, *41*, 4316-4318.

14. Lei, Y.; Wrobleski, A. D.; Golden, J. E.; Powell, D. R.; Aubé, J. *J. Am. Chem. Soc.* **2005**, *127*, 4552-4553.

15. Gracias, V.; Zeng, Y.; Desai, P.; Aubé, J. *Org. Lett.* **2003**, *5*, 4999-5001.

16. Pearson, W. H.; Schkeryantz, J. M. *Tetrahedron Lett.* **1992**, *33*, 5291-5294.

17. Pearson, W. H.; Walavalkar, R.; Schkeryantz, J. M.; Fang, W.-k.; Blickensdorf, J. D. *J. Am. Chem. Soc.* **1993**, *115*, 10183-10194.

18. Pearson, W. H.; Fang, W.-k. *J. Org. Chem.* **1995**, *60*, 4960-4961.

19. Pearson, W. H.; Gallagher, B. M. *Tetrahedron* **1996**, *52*, 12039-12048.

20. Reddy, P. G.; Varghese, B.; Baskaran, S. *Org. Lett.* **2003**, *5*, 583-585.

21. Reddy, P. G.; Baskaran, S. *J. Org. Chem.* **2004**, *69*, 3093-3101.

22. Gracias, V.; Milligan, G. L.; Aubé, J. *J. Am. Chem. Soc.* **1995**, *117*, 8047-8048.

23. Gracias, V.; Frank, K. E.; Milligan, G. L.; Aubé, J. *Tetrahedron* **1997**, *53*, 16241-16252.

24. Forsee, J. E.; Aubé, J. *J. Org. Chem.* **1999**, *64*, 4381-4385.

25. Sahasrabudhe, K.; Gracias, V.; Furness, K.; Smith, B. T.; Katz, C. E.; Reddy, D. S.; Aubé, J. *J. Am. Chem. Soc.* **2003**, *125*, 7914-7922.

26. Bräse, S.; Gil, C.; Knepper, K.; Zimmermann, V. *Angew. Chem., Int. Ed.* **2005**, *44*, 5188-5240.

27. Le Dréau, M.-A.; Desmaële, D.; Dumas, F.; d'Angelo, J. *J. Org. Chem.* **1993**, *58*, 2933-2935.

28. Aubé, J.; Rafferty, P. S.; Milligan, G. L. *Heterocycles* **1993**, *35*, 1141-1147.

29. Wrobleski, A.; Sahasrabudhe, K.; Aubé, J. *J. Am. Chem. Soc.* **2002**, *124*, 9974-9975.

30. Pearson, W. H.; Fang, W.-k. *J. Org. Chem.* **2000**, *65*, 7158-7174.

31. Smith, B. T.; Wendt, J. A.; Aubé, J. *Org. Lett.* **2002**, *4*, 2577-2579.

SYNTHESIS OF (–)-(*S,S*)- BIS(4-ISOPROPYLOXAZOLINE)
(4*S*, 4*S'*)-2,2'-(Propane-2,2-diyl)bis(4-isopropyl-4,5-dihydrooxazole)

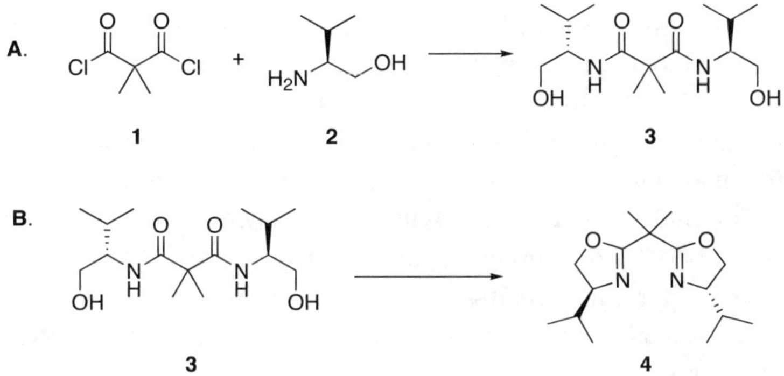

Submitted by David A. Evans,[1] Keith A. Woerpel,[1,2] Bernd Nosse,[3] Andreas Schall,[3] Yogesh Shinde,[3] Eva Jezek,[3] Mohammad Mahbubul Haque,[3] R. B. Chhor,[3] and Oliver Reiser.[3]
Checked by Peter Wipf and Nilukshi Jayasuriya.[4]

1. Procedure

A. (–)-(S,S)-N,N'-Bis(1-hydroxymethyl-2-methylpropyl)-2,2-dimethylmalonamide (**3**): An oven-dried 250 mL, 3-necked round-bottom flask equipped with a stirring bar and two 50 mL pressure-equalizing addition funnels connected to a mineral oil bubbler is purged with nitrogen and charged with (L)-valinol (**2**, 5.13 g, 0.050 mol, Note 1). The flask is immersed in an ice bath at 0 °C and triethylamine (17.4 mL, 0.124 mol, Note 2) is added dropwise via the first addition funnel. 2,2-Dimethylpropanedioyl dichloride (3.3 mL, 0.25 mol, Note 3) in dry dichloromethane (25 mL, Note 4) is then added dropwise over 25 minutes via the second addition funnel. The internal temperature increases from 0 °C to 10 °C during the addition. Subsequently, the ice bath is removed and the reaction mixture is allowed to warm to room temperature. Stirring is continued for 45 min, resulting in a colorless precipitate that is dissolved by addition of dry dichloromethane (120 mL). After addition of aqueous HCl (1 N, 30 mL), the aqueous layer is separated and extracted with dichloromethane (3 x 15 mL). The combined organic layers are washed with saturated $NaHCO_3$ solution (30 mL) and brine (30 mL), dried over $MgSO_4$, filtered and concentrated *in vacuo* to afford crude **3** as a pale yellow solid. Recrystallization of the crude product from ethyl acetate (40 mL) yields **3** (4.30 g, 14.2 mmol, 57%) as white crystals. The mother liquor is concentrated and the residue recrystallized from ethyl acetate (10 mL) to yield a second crop of **3** (1.60 g, 5.27 mmol, 21%); the process is repeated to yield a third crop of **3** (0.440 g, 1.45 mmol, 6%, total yield: 6.40 g, 21.1 mmol, 84%, Note 5).

*B. (–)-(S,S)-Bis(4-isopropyloxazoline) (**4**):* An oven-dried 500 mL, 2-necked round-bottom flask equipped with a stirring bar and a 50 mL, pressure-equalizing addition funnel connected to a mineral oil bubbler is purged with nitrogen and charged with (–)-(*S*,*S*)-*N*,*N'*-bis-(1-hydroxymethyl-2-methylpropyl)-2,2-dimethylmalonamide (**3**, 5.5 g, 18.4 mmol), 4-dimethylaminopyridine (0.204 g, 1.67 mmol, Note 6) and dry dichloromethane (130 mL, Note 4). The flask is immersed in a water bath at room temperature and triethylamine (10.25 mL, 73.4 mmol, Note 2) is added slowly via syringe. Subsequently, tosyl chloride (7.10 g, 37 mmol, 2.0 equiv., Note 7), dissolved in dry dichloromethane (15 mL), is added dropwise over 30 minutes via the addition funnel. After completion of the addition, the funnel is rinsed with dry dichloromethane (2.5 mL) and the reaction mixture is stirred for an additional 27 h at room temperature (Note 8). The reaction mixture is treated with saturated NH_4Cl solution (70 mL) followed by water (40 mL). The aqueous layer is separated and extracted with dichloromethane (3 x 55 mL), and the combined organic layers are dried over $MgSO_4$. The organic solution is filtered and concentrated under vacuum. The oily residue is treated with hot pentane (40 mL, Note 9), stirred for 5 min and the supernatant liquid is decanted. This procedure is repeated three times and the collected pentane layers are combined and concentrated under vacuum to yield **4** (4.05 g, 15.2 mmol, 83%, Note 10) as a colorless oil.

2. Notes

1. (L)-Valinol was prepared from (L)-valine in 81% yield in an analogous way to the reduction of (*S*)-*tert*-leucine to (*S*)-*tert*-leucinol according to Evans, D. A.; Peterson, G. S.; Johnson, J. S.; Barnes, D. M.; Campos, K. R.; Woerpel, K. A. *J. Org. Chem.* **1998**, *63*, 4541, or purchased from Aldrich Chemical Company, Inc.

2. Triethylamine was obtained from Alfa Aesar and was distilled prior to use from calcium hydride under nitrogen.

3. 2,2-Dimethylpropanedioyl dichloride was purchased from Aldrich Chemical Company, Inc. or synthesized from 2,2-dimethylmalonic acid in

90% yield according to Evans, D. A.; Peterson, G. S.; Johnson, J. S.; Barnes, D. M.; Campos, K. R.; Woerpel, K. A. *J. Org. Chem.* **1998**, *63*, 4541.

4. Dichloromethane was purified by filtration through activated alumina.

5. Physical properties and spectral data for **3** are as follows: R_f 0.25 (EtOAc:MeOH, 95:5); $[\alpha]_D^{24}$ −6.0 (c 0.50, CH_2Cl_2); mp 98−99 °C; ^1H NMR (300 MHz, $CDCl_3$) δ: 0.92 (d, J = 6.8 Hz, 6 H), 0.96 (d, J = 6.8 Hz, 6 H), 1.50 (s, 6 H), 1.82 (oct, J = 6.8 Hz, 2 H), 2.66 (bs, 2 H), 3.52 (m, 2 H), 3.69−3.86 (m, 4 H), 6.41 (d, J = 8.6 Hz, 2 H); ^{13}C NMR (75 MHz, $CDCl_3$) δ: 18.8, 19.6, 23.7, 29.1, 50.2, 57.1, 63.5, 174.5; IR (KBr) cm^{-1} 3349, 3380, 2962, 2886, 1658, 1530, 1049, 1033; MS (ES) *m/z (%)* 405.2 (20), 325.2 ($[M+Na]^+$, 100), 285.2 (10); HRMS (ES) calcd. for $C_{15}H_{30}N_2O_4Na$ 325.2103, found 325.2084; Anal. calcd. for $C_{15}H_{30}N_2O_4$: C, 59.57; H, 10.00; N, 9.26. found: C, 59.01; H, 10.12; N, 9.11.

6. 4-Dimethylaminopyridine was obtained from the Aldrich Chemical Company, Inc and used as received.

7. Tosyl chloride was obtained from the Aldrich Chemical Company, Inc and used as received.

8. After the reaction mixture was stirred for 27 h, the submitters noticed crystalline solid precipitate. The solid precipitate was dissolved by adding dichloromethane (50 mL) prior to workup. The checkers, however, did not observe any precipitate.

9. In some experiments, a cloudy precipitate was formed, which was then dissolved in dichloromethane (2.5 mL per mmol of **3**) prior to workup.

10. An analytically pure sample for characterization purposes was obtained by Kugelrohr distillation (95-100 °C, 0.5 mmHg) of the crude material. Physical properties and spectral data for **4** are as follows R_f 0.25 (CH_2Cl_2/MeOH, 19:1); $[\alpha]_D^{24}$ −107.5 (c 1.0, CH_2Cl_2); ^1H NMR (300 MHz, $CDCl_3$) δ: 0.85 (d, J = 6.8 Hz, 6 H), 0.91 (d, J = 6.8 Hz, 6 H), 1.51 (s, 6 H), 1.88−1.73 (m, 2 H), 4.06−3.93 (m, 4 H), 4.26−4.15 (m, 2 H); ^{13}C NMR (75 MHz, $CDCl_3$) δ: 17.3, 18.5, 24.4, 32.2, 38.5, 69.9, 71.5, 168.7; IR (film) cm^{-1} 3308, 2960, 2874, 1746, 1525, 1353, 1303, 1037, 1017, 895, 815, 714; MS (EI) *m/z (%)* 266 (M^+, 15), 265 (30), 223 (100), 195 (30), 155 (65), 137 (97), 110 (50); HRMS (EI) calcd for $C_{15}H_{26}N_2O_2$ 266.1994, found 266.1987; Anal. calcd for $C_{15}H_{26}N_2O_2$: C, 67.63; H, 9.84; N, 10.52. found C, 66.65; H, 9.81; N, 10.08.

Safety and Waste Disposal Information

All hazardous materials should be handled and disposed of in accordance with "Prudent Practices in the Laboratory"; National Academy Press; Washington, DC, 1995.

3. Discussion

Since the discovery of oxazolines as ligands for catalysis,[5] bis(oxazolines)[6] have proved to be privileged structures because they promote a great number of metal-catalyzed transformations with unprecedented selectivity. Most commonly, bis(oxazolines) can be obtained from amino alcohols, either by a two-step condensation/cyclization sequence with acid chlorides as described here, or by condensation with dinitriles in the presence of a metal salt such as zinc(II) chloride[7], trifluoromethanesulfonic acid[8] or stepwise formation of the corresponding imidates[9] followed by cyclization. While the commercially available *tert*-butyl-substituted bis(oxazoline) **5**[10,11] often gives rise to the highest selectivities, a number of applications[12] have been developed in which the bis(4-isopropyloxazoline) **4** or the phenyl-substituted bis(oxazoline) **6** will give similar or even better results. Moreover, for the synthesis of **5** the unnatural and therefore expensive amino acid *tert*-leucine is required. Consequently, bis(4-isopropyloxazoline) **4**, available as either enantiomer from inexpensive (*S*)- or (*R*)-valine, is an attractive alternative for large scale applications if equally good enantioselectivity can be achieved.

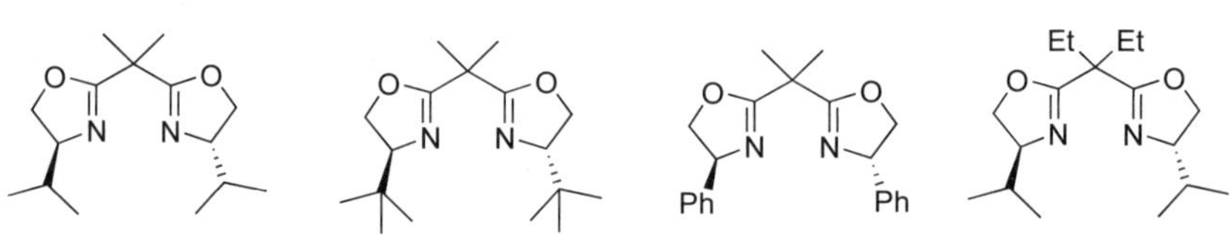

4 **5** **6** **7**

The procedure described here for bis(4-isopropyloxazoline) **4** closely resembles the previously reported protocols for the bis(4-ethyloxazoline) **7**[13] and the bis(4-*tert*-butyloxazoline) ligand **5**.[10] The main differences can be found in the workup conditions, which accommodate the quite different solubility and crystallization properties of intermediates and final ligands.

1. Department of Chemistry and Chemical Biology, Harvard University, 12 Oxford Street, MA 02139, USA
2. New address: Department of Chemistry, University of California, Irvine, CA 92697, USA
3. Institut für Organische Chemie, Universität Regensburg, Universitätsstr. 31, 93053 Regensburg, Germany.
4. Department of Chemistry, University of Pittsburgh, Pittsburgh, PA 15260, USA.
5. (a) Brunner, H.; Obermann, U.; Wimmer, P. J. *Organomet. Chem.* **1986**, *316*, C1; (b) Brunner, H.; Obermann, U. *Chem. Ber.* **1989**, *122*, 499; (c) Brunner, H.; Obermann, U.; Wimmer, P. *Organometallics* **1989**, *8*, 821; (d) Nishiyama, H.; Sakaguchi, H.; Nakamura, T.; Horihata, M.; Kondo, M.; Itoh, K. *Organometallics* **1989**, *8,* 846; (e) Balavoine, G.; Clinet, J.; Lellouche, 1. *Tetrahedron Lett.* **1989**, *30,* 5141.
6. *Reviews:* (a) Johnson, J. S.; Evans, D. A. *Acc. Chem. Res.* **2000,** *33*, 325; (b) Jørgensen, K. A.; Johannsen, M.; Yao, S.; Audrain, H.; Thornauge, J. *Acc. Chem. Res.* **1999**, *32*, 605; (c) Ghosh, A. K.; Mathivanan, P.; Cappiello, J. *Tetrahedron: Asymmetry* **1998,** *9*, 1; (d) Pfaltz, A. *Acta Chem. Scand.* **1996,** *50*, 189; (e) Reiser, O. *Nachr. Chem. Tech. Lab.* **1996,** *44*, 744.
7. (a) Bolm, C.; Weickhardt, K.; Zehnder, M.; Ranff, T. *Chem. Ber.* **1991**, *124*, 1173; (b) Witte, H.; Seeliger, W. *Liebigs Ann. Chem.* **1974**, 996.
8. Davies, I. W.; Senanayake, C. H.; Larsen, R. D.; Verhoeven, T. R.; Reider, P. J. *Tetrahedron Lett.* **1996**, *37*, 813.
9. Hall, J.; Lehn, J. M.; DeCian, A.; Fischer, J. *Helv. Chim. Acta* **1991**, *74*, 1.
10. (a) Evans, D. A.; Peterson, G. S.; Johnson, J. S.; Barnes, D. M.; Campos, K. R.; Woerpel, K. A. *J. Org. Chem.* **1998**, *63*, 4541; (b) Evans, D. A.; Woerpel, K. A.; Hinman, M. M.; Faul, M. M. *J. Am. Chem. Soc.* **1991**, *113*, 726.
11. Fluka GmbH:Catalog No. 59427; Aldrich Inc.: Catalog No. 40,614-7.
12. (a) Matsumura, Y.; Maki, T.; Murakami, S.; Onomura, O. *J. Am. Chem. Soc.* **2003**, *125*, 2052; (b) Jezek, E.; Schall, A.; Kreitmeier, P.; Reiser, O. *Synlett* **2005**, 915; (c) Chhor, R. B.; Nosse, B.; Soergel, S.; Böhm, C.; Seitz, M.; Reiser, O. *Chem. Eur. J.* **2003,** 9, 260; (d) Nosse, B.; Chhor, R. B.; Jeong, W. B.; Böhm, C.; Reiser, O. *Org. Lett.* **2003,** 5, 941.
13. Denmark, S. E.; Nakajima, N.; Nicaise, O. J.-C.; Faucher, A.-M.; Edwards, J. P. *J. Org. Chem.* **1995**, *60,* 4884.

INDIUM-CATALYZED CYCLOISOMERIZATION: PREPARATION OF 4-METHYLPYRROLO[1,2-*a*]QUINOLINE

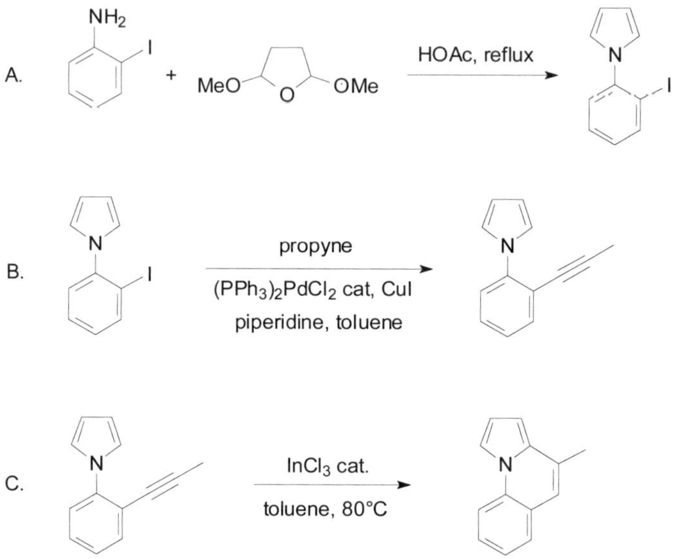

Submitted by Alois Fürstner, Victor Mamane, Günter Seidel, and Daniel Laurich.[1]
Checked by Scott E. Denmark and Jack Hung-Chang Liu.

1. Procedure

A. 1-(2-Iodophenyl)pyrrole. A 50-mL, three-necked, round-bottomed, flame-dried flask equipped with a Teflon-coated stirbar, a glass stopper, an addition funnel and a reflux condenser is charged with 2-iodoaniline (12.5 g, 57.1 mmol) (Notes 1, 2) and glacial acetic acid (12.5 mL, Note 1). The resulting solution is heated to reflux before 2,5-dimethoxytetrahydrofuran (7.8 mL, 59.0 mmol, Note 1) is added over a period of 10 min via the dropping funnel and reflux is continued for 5 min once the addition is complete. The addition funnel and reflux condenser are replaced by an inlet for a thermocouple and a distillation head with condenser, respectively, and the acetic acid is slowly removed by distillation under reduced pressure (15 mmHg, bath temperature about 50 °C) over a period of approximately 4 h. During this time, the internal temperature slowly reaches 50 °C. The remaining brown residue is transferred to a 25-mL round-bottomed flask and is purified by short-path distillation in vacuum (0.04 mmHg, Note 3). The fraction distilling at 90−102 °C is collected to give 1-(2-iodophenyl)pyrrole as a brown liquid (9.78–10.12 g, 64–66%) (Notes 4, 5).

B. 1-(2-(1-Propynyl)phenyl)pyrrole. A flame-dried 250-mL, two-necked, round-bottomed flask equipped with a magnetic stirbar, a gas dispersion tube (5 mm OD, Note 6), and a bubbler-sealed gas outlet is charged sequentially with 1-(2-iodophenyl)pyrrole (8.98 g, 33.4 mmol), piperidine (10 mL, Note 1), $(PPh_3)_2PdCl_2$ (1.20 g, 1.67 mmol, Note 1), CuI (318 mg, 1.67 mmol, Note 1) and toluene (90 mL, Note 1) to give a homogeneous reddish solution. Propyne is bubbled through this solution for 90 min at 0.1 L/min via the dispersion tube until the reaction mixture turns black (Notes 7, 8). For workup, the flask is vented, the reaction mixture is filtered through a short pad of silica (approximately 30 g) which is carefully rinsed with *tert*-butyl methyl ether (350 mL). The combined filtrates are transferred into a 1-L separatory funnel and are successively washed with water (2 x 200-mL portions) and brine (350 mL) before being dried over Na_2SO_4 and evaporated. The remaining orange liquid is transferred to a 25-mL round-bottomed flask and is purified by short-path distillation in vacuum (Note 3) to give 1-(2-(1-propynyl)phenyl)pyrrole as a pale yellow liquid (bp 70−75 °C, 0.005 mm Hg) (4.56–4.72 g, 75–78%, Notes 5, 9).

C. 4-Methylpyrrolo[1,2-a]quinoline. A 250-mL, three-necked, round-bottomed flask equipped with a reflux condenser fitted with an argon inlet, a glass stopper, a stopcock gas inlet and a Teflon-coated magnetic stirbar is evacuated and flame-dried before it is flushed with argon and allowed to cool to ambient temperature. The flask is charged with 1-(2-(1-propynyl)phenyl)pyrrole (4.53 g, 25 mmol), toluene (100 mL, Note 1) and $InCl_3$ (277 mg, 1.25 mmol, Note 10) and the resulting mixture is stirred at 80 °C bath temperature for 4 h until GC analysis shows complete conversion of the substrate (Note 11). For workup, the dark mixture is filtered through a pad of silica gel (5 cm x 3 cm, approximately 30 g), which is then carefully rinsed with toluene (150 mL). The crude product was obtained as a yellow solid after removal of the solvent under reduced pressure. The product is dissolved in toluene (20 mL) and this solution is adsorbed on silica gel (approximately 10 g) and the resulting slurry is evaporated to dryness. The adsorbate is placed in a Soxhlet thimble (8.0 cm x 3.5 cm) and is extracted with pentane (80 mL, Note 1) for 18 h using a Soxhlet apparatus of ca. 50 mL inner volume (Note 12). Evaporation of the resulting pentane extract provides 4-methylpyrrolo[1,2-a]quinoline as a pale yellow solid, which is pure enough for most applications (mp 63−65°C, ≥ 95% by GC, Note 11)

(4.12 g, 91%). Analytically pure material is obtained by subsequent sublimation (60 °C bath temperature, 5 x 10^{-3} mmHg for 3.5 h, Note 13) as a white solid (3.05 g, 67%, Notes 14-16), mp 67–68 °C.

2. Notes

1. 2-Iodoaniline (>98%) was purchased from Aldrich Chemical Co. or from Lancaster. 2-Iodoaniline was recrystallized by dissolving the commercial material in benzene with stirring in a round-bottomed flask, in a 50 °C oil bath. For each gram of 2-iodoaniline, 1 mL of benzene was used. To the dark brown solution of 2-iodoaniline was then slowly added petroleum ether with stirring. Approximately four times the volume of benzene was added. This mixture was allowed to cool to room temperature and was then cooled in –20 °C for approximately 4 hours. The dark brown crystals were filtered off through a coarse fritted funnel, and were further purified by sublimation under high vacuum for approximately 2 to 2.5 h (oil bath temperature: 50–55 °C, cold finger of the sublimator cooled by a dry ice-acetone bath). The melting point of the purified material was 56–57 °C and was significantly less colored. Copper(I) iodide (CuI, 99.999%), $(PPh_3)_2PdCl_2$ (99.99%) (Aldrich Chemical Co.), glacial acetic acid (99%, Fluka), 2,5-dimethoxytetrahydrofuran (99%, *cis/trans* mixture, Acros Organics), and piperidine (99%, Aldrich Chemical Co.) were purchased from the suppliers indicated. 2,5-Dimethoxytetrahydrofuran was purified by short path distillation. Other chemicals were used as received. Toluene and pentane were dried by distillation over Na/benzophenone prior to use.

2. For Step A, the checkers were not able to reproduce the yield when the reaction was carried out on 114.0 mmol scale; considerably more unreacted 2-iodoaniline remained than when the reaction was carried out on 57.0 mmol scale. This was most likely due to the less efficient heat transfer into a 100-mL flask. Consequently, the distillate was a mixture of starting material and product.

3. The short path distillation was performed with a one-piece distillation apparatus consisting of a vacuum-jacketed distillation head and short (approximately 3 cm) condenser arm. This equipment enables adiabatic distillation and is commercially available from Ace Glass (catalog no. 9317-42).

4. ^{1}H NMR (500 MHz, $CDCl_3$) δ: 6.35 (t, J = 2.1 Hz, 2 H), 6.82 (t, J = 2.1 Hz, 2 H), 7.11 (ddd, J = 9.0, 7.9, 1.7 Hz, 1 H), 7.31 (dd, J = 7.8, 1.6 Hz, 1 H), 7.42 (ddd, J = 8.8, 7.7, 1.4 Hz, 1 H), 7.95 (dd, J = 8.0, 1.4 Hz, 1 H);

^{13}C NMR (126 MHz, $CDCl_3$) δ: 96.0, 109.3, 122.3, 128.2, 129.1, 129.6, 140.1, 144.2; MS (EI) *m/z* (relative intensity): 269 (100%, [M^+]), 142 (35%), 115 (62%); HRMS (EI) *m/z*, calcd for $C_{10}H_8NI$: 268.9702; found: 268.9701; IR (film): 3101, 1582, 1494, 1438, 1072, 1012, 924, 760, 724 cm^{-1}. Anal. calcd for $C_{10}H_8IN$ (269.08): C, 44.64; H, 3.00; N, 5.21. Found: C, 44.65; H, 2.77; N, 5.02.

5. The product is stable for extended periods of time when kept in a refrigerator.

6. The gas dispersion tube was purchased from ACE Glass (catalog no. 9435-78).

7. Propyne purchased from Matheson Inc. was used without further purification.

8. A flow-meter (Matheson Tri-Gas: PM-1000, part no. MN11E101N201) showed that a total of approximately 9 L of propyne was passed through the solution. If a less vigorous stream of propyne is chosen, then the reaction time will increase accordingly. In all cases, however, the formation of a black precipitate indicated complete conversion.

9. ^1H NMR (500 MHz, $CDCl_3$) δ: 2.03 (s, 3 H), 6.36 (dt, J = 2.5, 1.3 Hz, 2 H), 7.16 (dt, J = 2.2, 1.1 Hz, 2 H), 7.25 (dt, J = 7.5, 1.4 Hz, 1 H), 7.31 (dd, J = 8.0, 1.0 Hz, 1 H), 7.36 (dt, J = 8.1, 1.5 Hz, 1 H), 7.55 (d, J = 7.8 Hz, 1 H); ^{13}C NMR (126 MHz, $CDCl_3$) δ: 4.7, 76.8, 91.0, 109.2, 118.8, 121.8, 125.0, 126.3, 128.7, 134.2, 142.0; MS (EI) *m/z* (relative intensity): 181 (100%, [M^+]), 180 (95%), 166 (3%), 154 (18%), 152 (11%), 140 (2%), 127 (4%), 115 (3%), 89 (5%), 77 (7%); HRMS (EI) *m/z*, calcd for $C_{13}H_{11}N$: 181.0891, found: 181.0886; IR (film): 3102, 2914, 2849, 2226, 1599, 1501, 1478, 1332, 1102, 1070, 761, 725 cm^{-1}. Anal. calcd for $C_{13}H_{11}N$ (181.24): C, 86.15; H, 6.12; N, 7.73; Found: C, 86.02; H, 6.12; N, 8.04.

10. $InCl_3$ (99.999%, Strem Chemicals) was used as received. Due to its hygroscopic character, this compound was kept in a dry box.

11. Occasionally, the reaction time was found to be somewhat longer. The progress of the reaction and the purity of the product can be monitored by GC (Hewlett-Packard HP-5, 5% phenylmethylsiloxane column: 50 m, 0.25 mm), temperature program: 70 °C → 270 °C, 20 °C/min, then 270 °C for 6.0 min; retention time: 11.81 min (starting material), 14.24 min (product).

12. In a second run, the checkers found that a higher yield (4.20 g, 93%) could be obtained by exhaustive Soxhlet extraction for 48 h. Sublimation of this material (oil bath: 50–60 °C, 7 h), afforded 3.97 g (88%) of the product as a white solid.

13. The sublimation was performed by immersing a large side-armed test tube (42 mm x 175 mm) containing the crude product in an oil bath and subliming the product under reduced pressure onto the cold wall of a smaller test tube (25 mm x 150 mm) cooled with a dry-ice/acetone slush. The smaller test tube, which served as a cold finger, was secured by a large rubber stopper. During the sublimation, the bath temperature should not exceed 60 °C to avoid undue decomposition. The purity of the material can be checked by GC (Note 11).

14. The submitters found that analytically pure material (2.49 g, 55%) could also be obtained by recrystallization from hexanes (15 mL).

15. In solid form, the product can be stored in a refrigerator for extended periods of time. However, the product is sensitive in solution to traces of acid present in $CDCl_3$ that has not been rigorously purified.

16. ^{1}H NMR (500 MHz, $CDCl_3$) δ: 2.45 (d, J = 1.3 Hz, 3 H), 6.54 (dd, J = 3.8, 1.4 Hz, 1 H), 6.81 (dd, J = 3.8, 3.7 Hz, 1 H), 6.83 (s, 1H), 7.30 (dt, J = 7.5, 1.1 Hz, 1 H), 7.46 (dt, J = 7.8, 1.5 Hz, 1 H), 7.60 (dd, J = 7.8, 1.4, 1 H), 7.87 (d, J = 1.3 Hz, 1 H), 7.87 (d, J = 7.7 Hz, 1 H); ^{13}C NMR (126 MHz, $CDCl_3$) δ: 18.5, 101.3, 112.5, 112.6, 114.2, 117.7, 123.7, 124.6, 127.0, 128.0, 128.1, 129.4, 132.7; MS (EI) *m/z* (relative intensity): 181 (100%, [M+]), 180 (42%), 152 (8%), 91 (7%), 77 (7%); HRMS (EI) *m/z*, calcd for $C_{13}H_{11}N$: 181.0891; found: 181.0890; IR (film): 3142, 1608, 1541, 1486, 1458, 1419, 1366, 1192, 1088, 867, 840, 774, 753, 739, 703 cm^{-1}. Anal. calcd for $C_{13}H_{11}N$ (181.24): C, 86.15; H, 6.12; N, 7.73. Found: C, 86.03; H, 6.15; N, 7.85.

Safety and Waste Disposal Information

All hazardous materials should be handled and disposed of in accordance with "Prudent Practices in the Laboratory"; National Academy Press; Washington, DC, 1995.

3. Discussion

The ability of "soft" metal salts such as $PtCl_2$, $AuCl_3$ or $InCl_3$ to render alkynes susceptible to attack by (tethered) nucleophiles such as alkenes, allyl ethers, or aromatic systems has been recognized only recently.[2-4] The ensuing skeletal rearrangements are inherently attractive for increasing molecular complexity. In this context, it has been shown that readily available biaryl derivatives containing an alkyne unit at one of their *ortho*-positions are converted into substituted phenanthrenes, or heterocyclic congeners thereof, on exposure to catalytic amounts of these salts in an inert solvent.[5-7] The most widely applied catalyst for this purpose is $PtCl_2$, although $GaCl_3$ and $InCl_3$ tend to give higher yields with heteroaromatic substrates as well as with biaryl derivatives bearing halo-alkynes. Such metal-catalyzed transformations accommodate substantial structural variations as can be seen from the selected examples compiled in the Table. In addition to a host of phenanthrene derivatives, substituted helicenes, benzoindoles, benzocarbazoles, napthothiophenes, or pyrrolo[1,2-*a*]quinolines can be obtained in good to excellent yields.[5-7] Since the latter class of heterocycles is endowed with promising biological activities but is difficult to make by more conventional methodology,[8] the procedure detailed above provides a good illustration for the advantages associated with this novel catalytic approach.

TABLE. Selected Examples of Phenanthrenes and Polycyclic Heteroarenes Formed by Metal-Catalyzed Cycloisomerization Reactions[5-7]

Substrate	Catalyst	Product	Yield
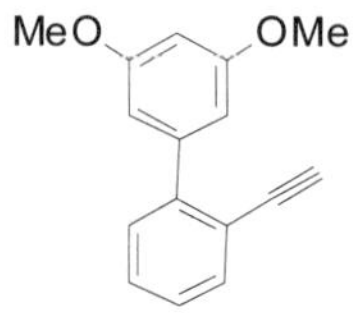	$PtCl_2$	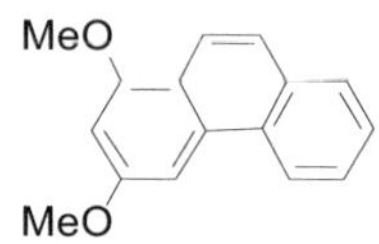	73%
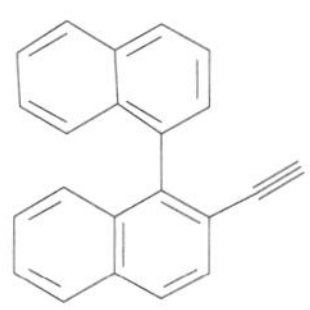	$PtCl_2$	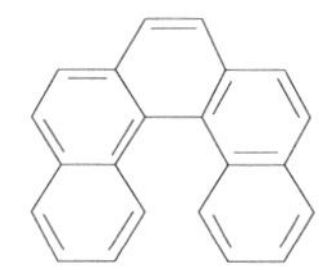	56%
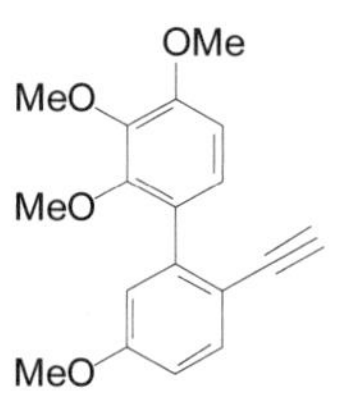	$PtCl_2$	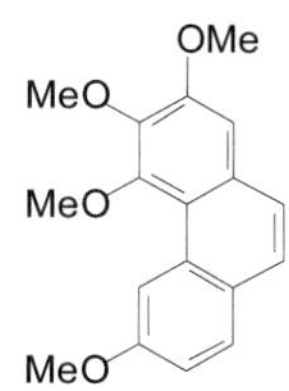	86%
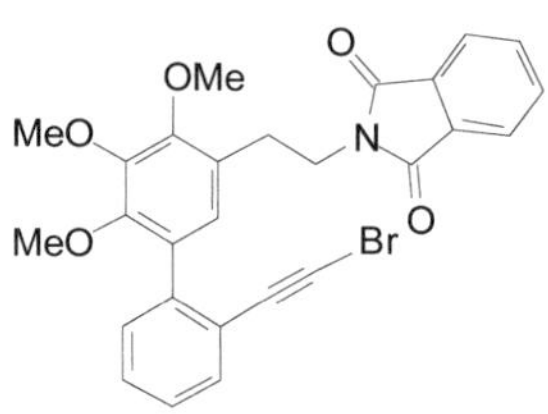	$PtCl_2$	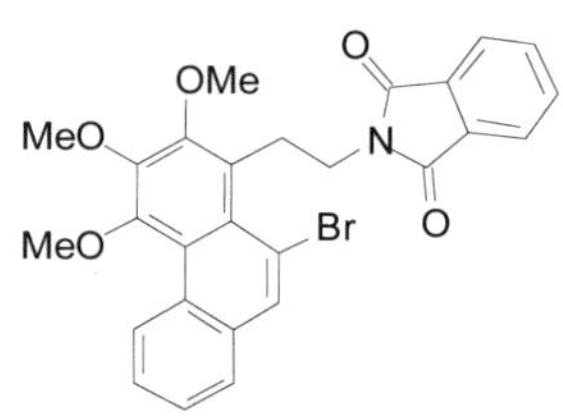	87%
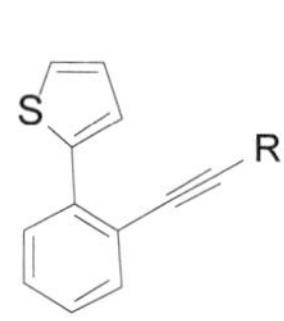	$PtCl_2$ $InCl_3$	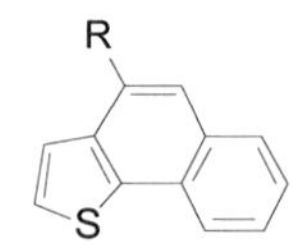	54% (R = H) 83% (R = Ph)
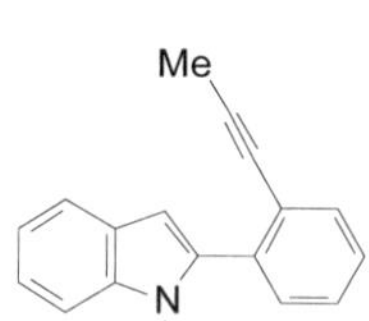	$PtCl_2$	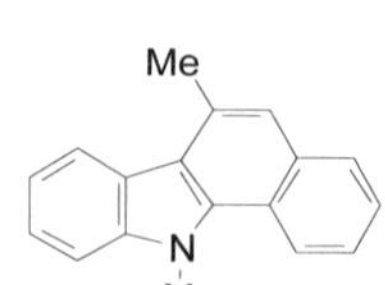	93%

1. Max-Planck-Institut für Kohlenforschung, Kaiser-Wilhelm-Platz 1, D-45470 Mülheim, Germany.
2. Reviews: (a) Méndez, M.; Mamane, V.; Fürstner, A. *Chemtracts – Org. Chem.* **2003**, *16*, 397. (b) Aubert, C.; Buisine, O.; Malacria, M. *Chem. Rev.* **2002**, *102*, 813. (c) Méndez, M.; Echavarren, A. M. *Eur. J. Org. Chem.* **2002**, 15. (d) Lloyd-Jones, G. C. *Org. Biomol. Chem.* **2003**, *1*, 215. (e) Fürstner, A.; Davies, P. W. Angew. *Chem. Int. Ed.* **2007**, *46*, 3410.
3. Pioneering studies: (a) Chatani, N.; Furukawa, N.; Sakurai, H.; Murai, S. *Organometallics* **1996**, *15*, 901. (b) Blum, J.; Beer-Kraft, H.; Badrieh, Y. *J. Org. Chem.* **1995**, *60*, 5567. (c) Trost, B. M.; Chang, V. K. *Synthesis* **1993**, 824.
4. (a) Fürstner, A.; Szillat, H.; Gabor, B.; Mynott, R. *J. Am. Chem. Soc.* **1998**, *120*, 8305. (b) Fürstner, A.; Szillat, H.; Stelzer, F. *J. Am. Chem. Soc.* **2000**, *122*, 6785. (c) Fürstner, A.; Stelzer, F.; Szillat, H. *J. Am. Chem. Soc.* **2001**, *123*, 11863. (d) Mamane, V.; Gress, T.; Krause, H.; Fürstner, A. *J. Am. Chem. Soc.* **2004**, *126*, 8654. (e) Fürstner, A.; Hannen, P. *Chem. Commun.* **2004**, 2546.
5. Fürstner, A.; Mamane, V. *J. Org. Chem.* **2002**, *67*, 6264.
6. (a) Fürstner, A.; Mamane, V. *Chem. Commun.* **2003**, 2112. (b) Fürstner, A.; Kennedy, J. W. *J. Chem. Eur. J.* **2006**, *12*, 7398.
7. Mamane, V.; Hannen, P.; Fürstner, A. *Chem. Eur. J.* **2004**, *10*, 4556.
8. Illustrative examples: (a) Anderson, W. K.; DeRuiter, J.; Heider, A. R. *J. Org. Chem.* **1985**, *50*, 722. (b) Bode, M. L.; Kaye, P. T. *J. Chem. Soc., Perkin Trans. 1* **1993**, 1809. (c) Garcia, E. E. *Org. Prep. Proc. Int.* **1974**, *6*, 11.

5-*endo-trig* CYCLIZATION OF 1,1-DIFLUORO-1-ALKENES: SYNTHESIS OF 3-BUTYL-2-FLUORO-1-TOSYLINDOLE (1*H*-Indole, 3-butyl-2-fluoro-1-[(4-methylphenyl)sulfonyl]-)

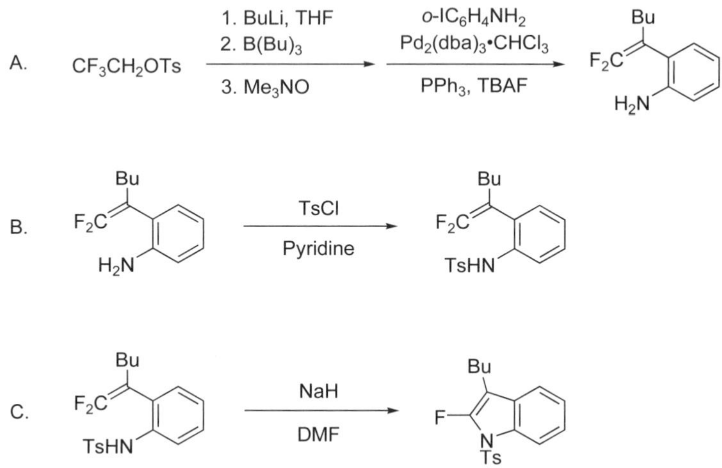

Submitted by Junji Ichikawa, Ryo Nadano, Takashi Mori, and Yukinori Wada.[1]
Checked by Sigrid Holle and Alois Fürstner.

1. Procedure

A. o-(1,1-Difluorohex-1-en-2-yl)aniline. A 1-L, three-necked, round-bottomed flask is equipped with a Teflon-coated magnetic stirbar, two pressure-equalizing dropping funnels, and a reflux condenser fitted with an argon stopcock inlet. The system is flame-dried and flushed with argon. The flask is charged with anhydrous tetrahydrofuran (200 mL) (Note 1) and 2,2,2-trifluoroethyl *p*-toluenesulfonate (15.3 g, 60 mmol) (Note 2), and the solution is cooled to –78 °C in a dry ice–acetone bath. Butyllithium (76.8 mL, 1.64 M in hexane, 126 mmol) (Note 3) is added via one of the dropping funnels over 10 min (Note 4) and the resulting mixture is stirred for an additional 20 min. Then tributylborane (66.0 mL, 1.0 M in THF, 66 mmol) (Note 5) is added via the other dropping funnel at –78 °C over 3 min. After being stirred for 1 h, the reaction mixture is warmed to room temperature and stirred for an additional 3 h to generate 2,2-difluorovinylborane. This solution of 2,2-difluorovinylborane is cooled in an ice bath prior to addition of trimethylamine oxide (10.9 g, 145 mmol) (Note 6). The reaction mixture is stirred at 0 °C for 2 h, and then allowed to warm to room temperature. Triphenylphosphine (PPh_3, 2.52 g, 9.60 mmol) (Note 7) and tris(dibenzylideneacetone)dipalladium-chloroform (1/1) ($Pd_2(dba)_3 \cdot CHCl_3$, 2.48 g, 2.40 mmol) (Note 8) are added in this order, and the mixture is stirred for 15 min. To the mixture are added *o*-iodoaniline (9.20 g, 42.0 mmol) (Note 9) and tetrabutylammonium fluoride (168 mL, 1.0 M in tetrahydrofuran, 168 mmol) (Note 10). After being stirred at room temperature for 15 min, the resulting mixture is heated in a 60 °C oil bath for 13 h, at which time TLC analysis indicates the reaction to be complete (Notes 11, 12). Then the reaction mixture is cooled to ambient temperature, and phosphate buffer (200 mL) (Note 13) is added. Organic materials are extracted with ethyl acetate (AcOEt, 3 x 200 mL), the combined extracts are washed successively with water (3 x 200 mL) and brine (200 mL), and then dried over anhydrous magnesium sulfate (30 g). After filtration through a filter paper and removal of the solvent under reduced pressure (Note 14), the palladium catalyst and ammonium salt are removed by short column chromatography on silica gel (Note 15). The eluent is concentrated under reduced pressure (Note 14) to give the crude product as a brown oil. This oil is distilled twice under reduced pressure (74–77 °C/1.2–1.4 mmHg) to give *o*-(1,1-difluorohex-1-en-2-yl)aniline (5.63 g, 63%) (Note 16) as a pale yellow liquid that was pure enough (90–95%) for use in the next step.

B. o′-(1,1-Difluorohex-1-en-2-yl)-p-toluenesulfonanilide. A 100-mL, two-necked, round-bottomed flask containing *o*-(1,1-difluorohex-1-en-2-yl)aniline (4.94 g, 23.4 mmol) is equipped with a glass stopper, a Teflon-coated magnetic stirbar, and a stopcock inlet connected to the argon line. Pyridine (50 mL) (Note 17) is introduced and the solution is cooled to 0 °C. *p*-Toluenesulfonyl chloride (TsCl, 6.69 g, 35.1 mmol) (Note 18) is added, and the mixture is stirred at room temperature for 15 h. Water (5 mL) is added and the reaction mixture is stirred for 0.5 h. Water (50 mL) and ether (100 mL) are added to the mixture before 2 M aqueous HCl solution (400 mL) is introduced. Organic materials are extracted with ether (4 x 100 mL). The combined organic extracts are washed successively with water (100 mL) and brine (100 mL), and then dried over anhydrous magnesium sulfate (30 g). After filtration through filter paper and removal of the solvent under reduced pressure (Note 14), a brown liquid (8.77 g) (Note 19) is obtained. This is used in the next step without further purification.

C. 3-Butyl-2-fluoro-1-tosylindole. A 300-mL, two-necked, round-bottomed flask is equipped with a glass stopper, a Teflon-coated magnetic stirbar, and a reflux condenser connected to the argon line. The system is flame-dried and flushed with argon. The flask is charged with anhydrous *N,N*-dimethylformamide (50 mL) (Note 1) and crude *o′*-(1,1-difluorohex-1-en-2-yl)-*p*-toluenesulfonanilide (8.77 g) (Note 19). Sodium hydride (0.65 g, 27 mmol) (Note 20) is added in portions to the solution at 0 °C. After the reaction mixture is stirred at 0 °C for 30 min, the flask is placed in an 80 °C oil bath. The mixture is stirred at this temperature for 7 h, at which time TLC analysis indicates the reaction to be complete (Notes 11, 21). The reaction mixture is cooled to ambient temperature, and phosphate buffer (100 mL) (Note 13) is added. Organic materials are extracted with ethyl acetate (3 x 150 mL), and the combined extracts are washed with water (4 x 100 mL) and brine (100 mL), and then dried over anhydrous magnesium sulfate (30 g). After filtration through a filter paper and removal of the solvent under reduced pressure (Note 14), the resulting brown residue is purified by column chromatography on silica gel (Note 22). The eluent containing the product is concentrated under reduced pressure (Note 14) to provide pure 3-butyl-2-fluoro-1-tosylindole as a yellow liquid (6.43 g, 80% over two steps) (Note 23).

2. Notes

1. Anhydrous tetrahydrofuran (THF) and *N,N*-dimethylformamide (DMF) (organic synthesis grade) were purchased from Kanto Chemical Co., Inc. and used as supplied. The checkers used anhydrous THF dried by distillation over Mg-anthracene, and DMF (Fluka) that was dried by storing over molecular sieves (3 Å) and subsequent distillation under reduced pressure.

2. Trifluoroethyl *p*-toluenesulfonate (Guaranteed Reagent grade) was purchased from Tokyo Kasei Kogyo Co., Ltd. and used as supplied. The checkers purchased this compound from Lancaster.

3. Butyllithium (ca. 1.6 M solution in hexane, organic synthesis grade) was purchased from Kanto Chemical Co., Inc. and used as supplied. The checkers purchased BuLi (1.64 M in hexanes) from Fischer Science.

4. The submitters introduced the butyllithium via syringe, adding the solution along the wall of the reaction flask that was cooled by a dry ice–acetone bath.

5. Tributylborane (1.0 M solution in tetrahydrofuran) was purchased from Aldrich Chemical Company, Inc. and used as supplied.

6. Anhydrous trimethylamine oxide was prepared by sublimation of

trimethylamine oxide dihydrate under reduced pressure (150 °C, 2.0 mmHg). Trimethylamine oxide dihydrate (Guaranteed Reagent grade) was purchased from Tokyo Kasei Kogyo Co., Ltd. The checkers used anhydrous trimethylamine oxide purchased from Aldrich.

7. Triphenylphosphine (Extra Pure grade) was purchased from Tokyo Kasei Kogyo Co., Ltd. and recrystallized from methanol prior to use. The checkers used triphenylphosphine (Aldrich) recrystallized from acetone prior to use.

8. The palladium catalyst was prepared from palladium(II) diacetate and dibenzylideneacetone according to the literature method.[2]

9. *o*-Iodoaniline (Extra Pure grade) was purchased from Tokyo Kasei Kogyo Co., Ltd. and used as supplied. The checkers used the commercial sample purchased from Acros as received.

10. Tetrabutylammonium fluoride (1.0 M solution in tetrahydrofuran) was purchased from Tokyo Kasei Kogyo Co., Ltd. and used as supplied. The checkers purchased the solution of this reagent from Aldrich.

11. Silica gel TLC plates ($60F_{254}$) were purchased from Merck Ltd. Japan, and visualized with 3% aqueous $KMnO_4$.

12. The R_f value of iodoaniline was 0.30 (ethyl acetate:hexane (1:5)); the product possessed an R_f = 0.42 in this solvent system. The submitters reported that the reaction was complete after only 6 h.

13. Phosphate buffer (pH 7, 1 L) was prepared by dissolving KH_2PO_4 9.1 g and $Na_2HPO_4 \cdot 12H_2O$ (47.7 g) in distilled water.

14. Rotary evaporation was conducted at 100 mmHg in a 40 °C water bath.

15. Short column chromatography was performed by using a 3.5-cm x 40-cm column packed with 250 mL of silica gel (Fuji Silysia Chemical Ltd., PSQ100B, >100 µm). The product was eluted with 500 mL of ethyl acetate:hexane (1:5). The checkers used silica gel purchased from E. Merck, Darmstadt (230–400 mesh).

16. *o*-(1,1-Difluorohex-1-en-2-yl)aniline has the following physical properties: R_f = 0.42 (ethyl acetate:hexane (1:5); ^1H NMR (300 MHz, $CDCl_3$) δ: 0.84–0.89 (m, 3 H), 1.30–1.35 (m, 4 H), 2.24–2.32 (m, 2 H), 3.59 (br. s, 2 H), 6.70–6.77 (m, 2 H), 6.99 (dd, 1 H, J = 1.6, 7.6 Hz), 7.11 (ddd, 1 H, J = 1.6, 7.5, 7.8 Hz); ^{13}C NMR (75 MHz, $CDCl_3$) δ: 13.8, 22.3, 27.7, 29.8 (dd, J = 2, 3 Hz), 89.0 (dd, J = 17, 22 Hz), 115.6, 118.4, 119.1 (dd, J = 1, 4 Hz), 128.9, 130.4 (dd, J = 2, 3 Hz), 144.2 (dd, J = 1, 3 Hz), 152.8 (dd, J = 287, 288 Hz); ^{19}F NMR (282 MHz, $CDCl_3/CFCl_3$) δ: −89.2 (d, 1 F, J = 43 Hz), −93.2 (d, 1 F, J = 43 Hz); IR (KAP) 3475, 3385, 2958, 2930, 2862,

1737, 1617, 1496, 1232 cm^{-1}; MS (70 eV) *m/z* (rel intensity) 211 (M^+; 80), 168 (98), 148 (100); Anal. Calcd for $C_{12}H_{15}NF_2$: C, 68.23; H, 7.16; N, 6.63. Found: C, 66.95; H, 7.86; N, 5.78.

17. Pyridine (JIS special grade) was purchased from Kokusan Chemical Works, Ltd. and used as supplied. The checkers used pyridine (E. Merck, Darmstadt) distilled over KOH prior to use.

18. *p*-Toluenesulfonyl chloride (Extra Pure grade) was purchased from Kanto Chemical Co., Inc. and recrystallized from toluene prior to use. The checkers used tosyl chloride as received from Aldrich.

19. An analytically pure sample of the toluenesulfonanilide can be isolated by column chromatography on silica gel (Fuji Silysia Chemical Ltd., PSQ100B, >100 μm). The product was eluted with ethyl acetate:hexane (1:5) and the eluent was concentrated with a rotary evaporator (Note 14) to give *o′*-(1,1-difluorohex-1-en-2-yl)-*p*-toluenesulfonanilide as a white powder. The product exhibits the following physical properties: R_f = 0.31 (ethyl acetate:hexane (1:5)) (Note 11); ^{1}H NMR (400 MHz, $CDCl_3$) δ: 0.79 (t, 3 H, *J* = 7.1 Hz), 1.05–1.23 (m, 4 H), 1.98 (br. s, 2 H), 2.35 (s, 3 H), 6.52 (s, 1 H), 6.98–7.06 (m, 2 H), 7.19–7.27 (m, 3 H), 7.60 (d, 1 H, *J* = 7.9 Hz), 7.68 (d, 2 H, *J* = 8.4 Hz); ^{13}C NMR (100 MHz, $CDCl_3$) δ: 13.6, 21.5, 22.2, 28.0, 29.4, 88.0 (dd, *J* = 16, 23 Hz), 119.7, 124.1 (d, *J* = 4 Hz), 124.3, 127.2, 129.1, 129.7, 130.6, 135.0, 136.4, 144.1, 153.0 (dd, *J* = 288, 291 Hz); ^{19}F NMR (282 MHz, $CDCl_3/CFCl_3$) δ: –86.9 (d, 1 F, *J* = 39 Hz), –90.3 (d, 1 F, *J* = 41 Hz); IR (KAP) 3274, 2958, 1742, 1494, 1401, 1339, 1247, 1167, 1092, 920, 666 cm^{-1}; MS (70 eV) *m/z* (rel intensity) 365 (M^+; 0.5), 210 (100), 148 (62); HRMS calcd for $C_{19}H_{21}NO_2F_2SNa$ 388.11533 (M^+ + Na); found 388.11525.

20. The checkers used pure NaH by removing the mineral oil from the commercial sample (Aldrich) with pentane (three washings). The submitters used sodium hydride (NaH) in mineral oil (1.06 g, 60% dispersion in mineral oil, 27 mmol, organic synthesis grade) purchased from Kanto Chemical Co., Inc. without further purification.

21. The product possessed an R_f = 0.39 (ethyl acetate:hexane (1:5)).

22. Column chromatography was performed by using a 3-cm x 50-cm column packed with 300 mL of silica gel (Fuji Silysia Chemical Ltd., PSQ100B, >100 μm). The product was eluted with ethyl acetate:hexane (1:10). After collection of 200 mL of eluent, 50 mL-fractions were collected. The $KMnO_4$-active product was eluted in the fractions 6–11 (Note 11). The checkers used silica purchased from E. Merck, Darmstadt (230–400 mesh).

23. 3-Butyl-2-fluoro-1-tosylindole exhibits the following physical properties: ^{1}H NMR (400 MHz, $CDCl_3$) δ: 0.84 (t, 3 H, *J* = 7.3 Hz), 1.16−1.25 (m, 2 H), 1.48−1.56 (m, 2 H), 2.33 (s, 3 H), 2.51 (dt, 2 H, *J* = 0.8, 7.4 Hz), 7.19 (d, 2 H, *J* = 8.0 Hz), 7.23 (ddd, 1 H, *J* = 1.1, 7.6, 7.6 Hz), 7.25−7.29 (m, 1 H), 7.33 (d, 1 H, *J* = 7.1 Hz), 7.71 (d, 2 H, *J* = 8.3 Hz), 8.07 (d, 1 H, *J* = 8.1 Hz); ^{13}C NMR (100 MHz, $CDCl_3$) δ: 13.7, 21.3 (d, *J* = 2 Hz), 21.6, 22.1, 30.5 (d, *J* = 2Hz), 99.7 (d, *J* = 11 Hz), 114.4, 118.9 (d, *J* = 7 Hz), 123.9, 124.0 (d, *J* = 4 Hz), 126.8, 128.1 (d, *J* = 6 Hz), 129.8, 130.6, 134.7, 145.2, 147.4 (d, *J* = 277 Hz); ^{19}F NMR (282 MHz, $CDCl_3/CFCl_3$) δ: −132.8; IR (KAP) 2957, 2931, 2861, 1659, 1453, 1393, 1190, 1179, 746, 689, 663 cm^{-1}; MS (70 eV) *m/z* (rel intensity) 345 (M^+; 91), 190 (100), 148 (71); HRMS calcd for $C_{19}H_{20}NO_2FSNa$ 368.10910 (M^+ + Na); found 368.10906; Anal. Calcd for $C_{19}H_{20}NO_2FS$: C, 66.06; H, 5.84; N, 4.05. Found: C, 65.97; H, 5.90; N, 4.10.

Waste Disposal Information

All hazardous materials should be handled and disposed of in accordance with “Prudent Practices in the Laboratory”; National Academy Press; Washington, DC, 1995.

3. Discussion

Synthetic methods for the preparation of 2-fluoroindoles have been limited to difluorination of indole derivatives followed by elimination[3a,b] and electrophilic fluorination of stannylindoles.[3c] The procedures described herein illustrate an efficient construction of 2-fluoroindoles via intramolecular substitution of the vinylic fluorine in β,β-difluorostyrenes bearing an *ortho*-amido substituent.[4] This reaction can be classified as a 5-*endo-trig* ring closure, a disfavored process according to Baldwin’s rules.[5-8] While having only rarely been observed in synthetic chemistry,[6] the nucleophilic 5-*endo-trig* cyclization was successfully achieved by taking advantage of the unique properties of fluorine: (i) the highly polarized difluorovinylidene double bond (^{13}C NMR: ca. 150 ppm and 90 ppm for CF_2=C) aids initial ring formation by electrostatic attraction between the CF_2 carbon and the internal nucleophile, and (ii) the successive elimination of the fluoride ion suppresses the reverse ring opening, thus functioning as a “lock”.[4]

Difluorostyrene derivatives **4** including *o*-(2,2-difluorovinyl)anilines are easily prepared by using a one-pot sequence as outlined below.[9a,c] The sequence comprises the following two processes starting from 2,2,2-trifluoroethyl *p*-toluenesulfonate (**1**): (i) a boron-mediated alkylation via 1,2-migration, leading to 2,2-difluorovinylboranes **2** and (ii) subsequent palladium-catalyzed coupling reaction with aryl iodides via 2,2-difluorovinylboronates **3**, which affords difluorostyrenes **4** in good yield. The substituent R at the vinylic position of **4** is derived from a trialkylborane (BR_3), which is readily generated by hydroboration of the corresponding alkene. The selective oxidation of boron–alkyl bonds with trimethylamine oxide prevents the coupling reaction of B-alkyl groups in **2**.

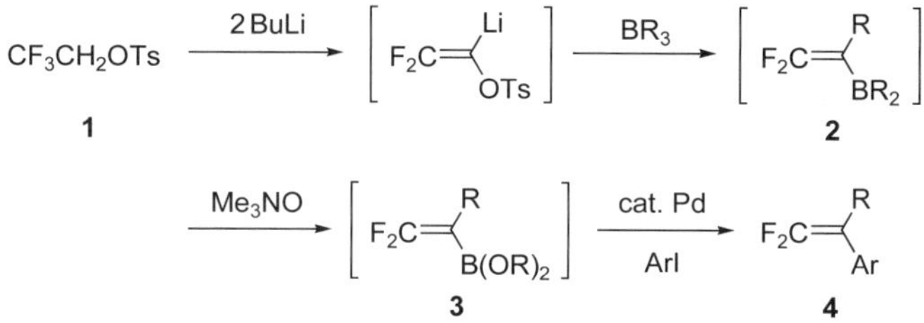

Another vinyl-selective coupling reaction of **2** with aryl iodides is accomplished via the transmetalation to 2,2-difluorovinylcopper species by adding cuprous iodide.[4a,9b,c] β,β-Difluorostyrenes **4** (R = H) without a vinylic substituent can also be prepared via 2,2-difluorovinylzirconocene from **1**.[10]

After tosylation of *o*-(2,2-difluorovinyl)anilines, treatment of the obtained sulfonanilides with 1.1 equiv of NaH in DMF promotes the normally "disfavored" 5-*endo-trig* cyclization to afford the corresponding 2-fluoroindoles in high yield. High-dilution conditions are not required in this ring closure.

This methodology for constructing five-membered rings is widely applicable to the cyclization of (i) β,β-difluorostyrene derivatives and (ii) 1,1-difluoro-1-butene derivatives bearing a nucleophilic nitrogen, oxygen, or sulfur atom at the *ortho* or the homoallylic position, respectively. Under similar conditions, these substrates undergo cyclization to afford ring-fluorinated heterocycles such as pyrrolines, furans, and thiophenes in high yield as shown in the Table. Five-membered carbocycles bearing a fluorine on the ring can also be obtained by this method.[4a,11] Moreover, six-membered ring-fluorinated heterocyclic and carbocyclic compounds can be provided in line with the "intramolecular substitution" concept.[11b,12]

Table. Nucleophilic 5-*endo-trig*-Cyclization of 1,1-Difluoro-1-alkenes

Substrate	Product	Conditions[a]	Yield
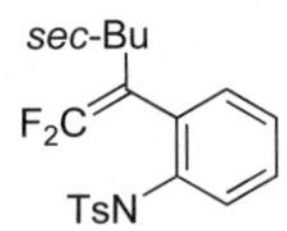	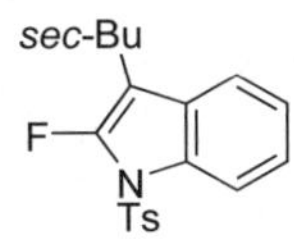	80 °C, 5 h	81%
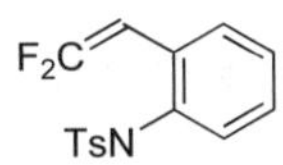	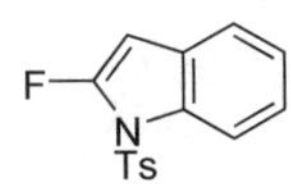	70 °C, 23 h	73%
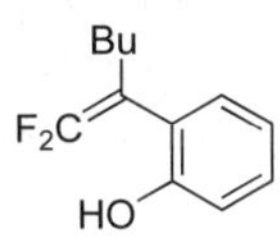	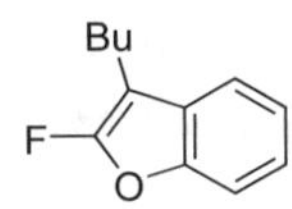	60 °C, 2 h	80%
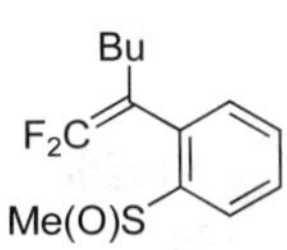	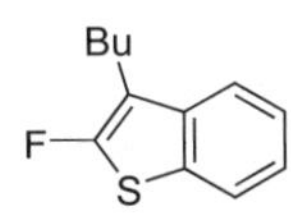	reflux, 3 h[b]	82%
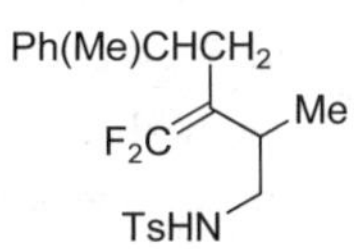	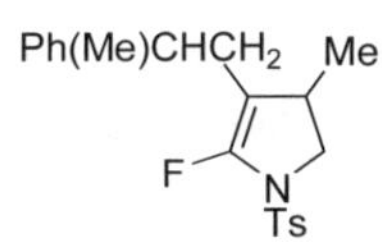	90 °C, 4 d	80%
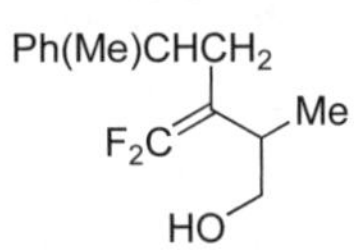	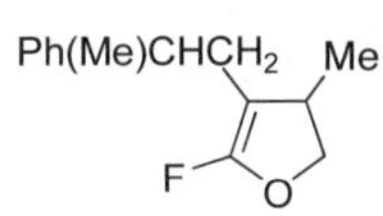	90 °C, 7 h	67%
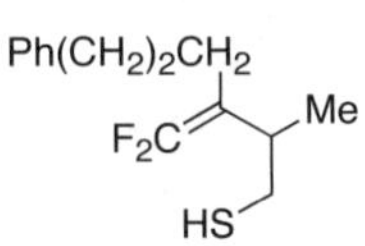	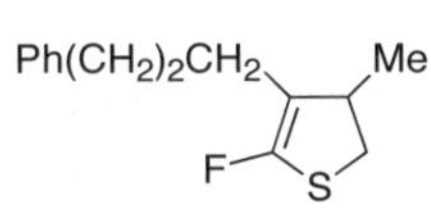	90 °C, 4 h	76%

a. NaH (1.1–1.2 equiv), DMF. b. (i) $(CF_3CO)_2O$ (3 equiv), NEt_3(3 equiv), CH_2Cl_2, 0 °C, 0.5 h. (ii) K_2CO_3 (6 equiv), MeOH, 0 °C–rt, 1 h then reflux.

1. Department of Chemistry, Graduate School of Science, The University of Tokyo, Hongo, Bunkyo-ku, Tokyo 113-0033, Japan.
2. Coulson, D. R. *Inorg. Synth.* **1972**, *13*, 121.
3. For the synthesis of 2- and 3-fluoroindoles, see: (a) Torres, J. C.; Garden, S. J.; Pinto, A. C.; da Silva, F. S. Q.; Boechat, N. *Tetrahedron* **1999**, *55*, 1881. (b) Barton, D. H.

R.; Hesse, R. H.; Jackman, G. P.; Pechet, M. M. *J. Chem. Soc., Perkin Trans. I* **1977**, 2604. (c) Hodson, H. F.; Madge, D. J.; Slawin, A. N. Z.; Widdowson, D. A.; Williams, D. J. *Tetrahedron* **1994**, *50*, 1899.

4. For the original reports on the synthesis of 2-fluoroindoles, see: (a) Ichikawa, J.; Wada, Y.; Fujiwara, M.; Sakoda, K. *Synthesis* **2002**, 1917. (b) Ichikawa, J.; Wada, Y.; Okauchi, T.; Minami, T. *Chem. Commun.* **1997**, 1537. See also: (c) Ichikawa, J.; Fujiwara, M.; Wada, Y.; Okauchi, T.; Minami, T. *Chem. Commun.* **2000**, 1887.
5. (a) Baldwin, J. E. *J. Chem. Soc., Chem. Commun.* **1976**, 734. (b) Baldwin, J. E.; Cutting, J.; Dupont, W.; Kruse, L.; Silberman, L.; Thomas, R. C. *J. Chem. Soc., Chem. Commun.* **1976**, 736. (c) Baldwin, J. E.; Thomas, R. C.; Kruse, L. I.; Silberman, L. *J. Org. Chem.* **1977**, *42*, 3846.
6. For reports on nucleophile-driven 5-*endo-trig* cyclizations, see: (a) Ref. 4a and references therein. (b) Ichikawa, J.; Mori, T.; Iwai, Y. *Chem. Lett.* **2004**, *33*, 1354. (c) Clayden, J.; Watson, D. W.; Helliwell, M.; Chambers, M. *Chem. Commun.* **2003**, 2582.
7. For a recent review on electrophile-driven 5-*endo-trig* cyclizations, see: Knight, D. W. *Prog. Heterocycl. Chem.* **2002**, *14*, 19.
8. For recent reviews on radical-initiated 5-*endo-trig* cyclizations, see: (a) Chatgilialoglu, C.; Ferreri, C.; Guerra, M.; Timokhin, V.; Froudakis, G.; Gimisis, T. *J. Am. Chem. Soc.* **2002**, *124*, 10765. (b) Ishibashi, H.; Sato, T.; Ikeda, M. *Synthesis* **2002**, 695. (c) Parsons, A. F. *C. R. Acad. Sci., Ser. IIc: Chem.* **2001**, *4*, 391.
9. For the original reports on the synthesis of 1,1-difluoro-1-alkenes, see: (a) Ichikawa, J.; Moriya, T.; Sonoda, T.; Kobayashi, H. *Chem. Lett.* **1991**, 961. (b) Ichikawa, J.; Minami, T.; Sonoda, T.; Kobayashi, H. *Tetrahedron Lett.* **1992**, *33*, 3779. (c) Ichikawa, J. *J. Fluorine Chem.* **2000**, *105*, 257 and references cited therein.
10. Ichikawa, J.; Fujiwara, M.; Nawata, H.; Okauchi, T.; Minami, T. *Tetrahedron Lett.* **1996**, *37*, 8799.
11. (a) Ichikawa, J.; Sakoda, K.; Wada, Y. *Chem. Lett.* **2002**, 282. (b) Ichikawa, J.; Miyazaki, H.; Sakoda, K.; Wada, Y. *J. Fluorine Chem.* **2004**, *125*, 585.
12. Ichikawa, J. In *Fluorine-Containing Synthons*, Soloshonok, V. A., Ed.; *ACS Symp. Ser.*, Vol. 911, American Chemical Society and Oxford University Press, Washington, D.C., **2005**; Chap. 14.

SYNTHESIS OF DITHIENO[3,2-*b*:2′,3′-*d*]THIOPHENE

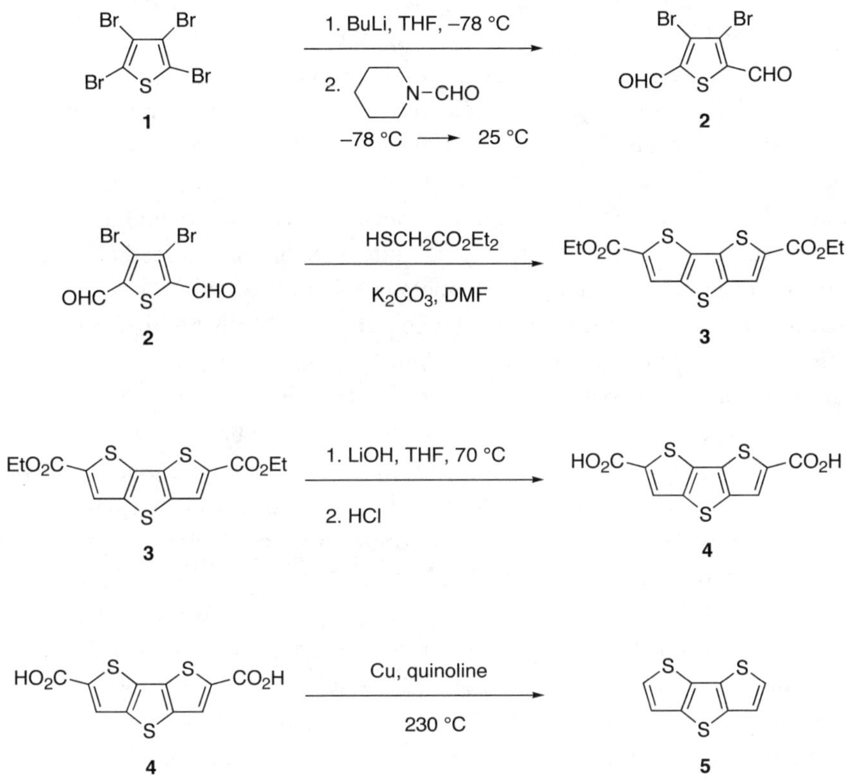

Submitted by Joseph Frey, Steffen Proemmel, Michael A. Armitage, and Andrew B. Holmes.[1]
Checked by Ryo Takita and Masakatsu Shibasaki.

1. Procedure

A. 3,4-Dibromo-2,5-diformylthiophene (**2**). An oven-dried, 1-L, three-necked, round-bottomed flask is equipped with a magnetic stir bar, a low-temperature thermometer, a rubber septum and a three-way stopcock to which an argon balloon is attached. Under a gentle flow of argon, the flask is charged with tetrabromothiophene **1** (16.0 g, 40.0 mmol) (Note 1) and freshly-distilled tetrahydrofuran (200 mL) (Note 2). The solution is cooled with a dry ice/2-propanol bath to an internal temperature of less than –65 °C and a solution of butyllithium (50 mL, 1.6 M in hexanes, 80 mmol, 2 equiv) is added via syringe. The addition proceeds at a rate that keeps the internal temperature below –60 °C, usually taking 30 min. When the addition is complete, the brown solution is stirred at about –70 °C for 30 min. Dry *N*-formylpiperidine (10 mL, 90 mmol, 2.25 equiv) (Note 3) is added quickly (in about 10 s) by syringe to the reaction mixture, which is then allowed to warm slowly to ambient temperature overnight. The reaction mixture is cooled to 0 °C in an ice-water bath, and hydrochloric acid (100 mL, 6 M) is added slowly to the reaction mixture, causing a yellow precipitate to form. The mixture is stirred at 0 °C for 45 min then filtered immediately under vacuum through a sintered-glass funnel. The solid is washed with water (150 mL) and dried in the funnel under vacuum for 30 min before being dried further in a vacuum dessicator (25 mmHg, overnight, silica gel dessicant with moisture indicator) to give the crude dialdehyde **2** (8.8 g, 74%) (Notes 4, 5). The material is carried forward without further purification.

B. Dithieno[3,2-b:2',3'-d]thiophene-2,6-dicarboxylic acid diethyl ester (**3**). In an oven-dried 500-mL, two-necked, round-bottomed flask equipped with a magnetic stir bar, nitrogen inlet and rubber septum, 3,4-dibromo-2,5-diformylthiophene **2** (8.0 g, 26.9 mmol) is suspended in anhydrous *N,N*-dimethylformamide (250 mL) (Note 6). Anhydrous potassium carbonate (9.65 g, 69.8 mmol, 2.6 equiv) (Note 7) and ethyl 2-mercaptoacetate (6 mL, 55 mmol, 2.05 equiv) are added to the slurry, causing a slight exotherm, and the reaction mixture is stirred under nitrogen at ambient temperature for three days. The dark solution is poured into a beaker containing 500 mL of water stirred with a magnetic stir bar, and a yellow precipitate forms. The aqueous suspension is extracted with dichloromethane (3 x 250 mL). The red organic extracts are combined, washed with brine (4 x 500 mL) and dried over anhydrous magnesium sulfate (Note 8). After filtration through fluted filter paper, the solvents are evaporated on a rotary evaporator and a yellow solid begins to form. When the volume of solvent remaining is about 100 mL, the solid is filtered under vacuum on a Büchner funnel, washed with water (100 mL) and dried in a vacuum dessicator (7–8 mmHg, overnight, silica gel dessicant with moisture indicator) to give the crude diester **3** (6.8 g, 74%) (Note 9). The material is carried forward without further purification.

C. Dithieno[3,2-b:2',3'-d]thiophene-2,6-dicarboxylic acid (**4**). In a 500-mL, one-necked, round-bottomed flask equipped with a magnetic stir bar and reflux condenser under a static argon atmosphere (Note 10), dithieno[3,2-*b*:2',3'-*d*]thiophene-2,6-dicarboxylic acid diethyl ester **3** (5.00 g,

14.7 mmol) is suspended in tetrahydrofuran (75 mL). An aqueous solution of 1 M lithium hydroxide (75 mL, 75 mmol, 5.1 equiv.) is added (Note 11), causing the color of the suspension to change from yellow to terracotta. The mixture is heated to 70 °C (oil bath temperature) and stirring is maintained at this temperature for 4 h to give an orange solution. The solvent is evaporated on a rotary evaporator until the volume remaining is about 75 mL. A small amount of precipitate is observed. The solution is acidified with 1 M hydrochloric acid (100 mL) to give a white precipitate with some slight foam. The solid is filtered slowly under vacuum on a Büchner funnel (Note 12) and washed sequentially with water (200 mL), methanol (100 mL) and diethyl ether (100 mL). Finally, the solid is dried overnight in a high vacuum oven (40 °C, approx. 0.02 mmHg) to afford the crude diacid **4** (3.8 g, 91%) (Notes 13, 14). The material is carried forward without further purification.

D. Dithieno[3,2-b:2′,3′-d]thiophene (**5**). In a 50-mL, two-necked, round-bottomed flask equipped with a magnetic stir bar, a reflux condenser and an internal thermometer, dithieno[3,2-*b*:2′,3′-*d*]thiophene-2,6-dicarboxylic acid **4** (3.80 g, 13.4 mmol) and copper powder (0.81 g, 12.7 mmol, 0.95 equiv) (Note 15) are suspended in quinoline (25 mL). The mixture is heated to reflux at 230 °C for 1 h using a heating mantle. Petroleum ether (40–60) is added to the cooled reaction mixture and the organic layer is decanted into a 500-mL separatory funnel (Note 16). This process is repeated several times until the volume of the organic layer is about 150 mL. The organic phase is washed with 2 M hydrochloric acid (3 x 70 mL) and brine (3 x 100 mL). The organic layer is then dried over anhydrous magnesium sulfate, filtered through a fluted filter paper and the solvent is removed on a rotary evaporator to give a pale yellow oil. The residue is dissolved in a small amount of dichloromethane and filtered through a short (about 8 cm length x 5.5 cm diameter) column of silica (Note 17), using petroleum ether (40-60) as eluent. The eluent is concentrated on a rotary evaporator to give a white crystalline solid. The solid is recrystallized by dissolving in 2 mL of dichloromethane at 30 °C, adding methanol dropwise at a rate of one drop every 5 seconds to the solution until it becomes cloudy (about 5 mL), adding another drop of dichloromethane to remove the cloudiness and allowing the solution to cool to ambient temperature for 30 min. The resulting crystals are collected under vacuum on a Büchner funnel, washed with ice-cold methanol (10 mL), dried in the funnel under vacuum and transferred to an appropriate vial to be further dried in a vacuum dessicator (25 mmHg, overnight, silica gel dessicant with moisture indicator) to give the purified dithieno[3,2-*b*:2′,3′-*d*]thiophene **5** (1.9 g, 72%) (Note 18).

2. Notes

1. Tetrabromothiophene (99+%) and ethyl 2-mercaptoacetate (99%) were purchased from Acros Organics and were used as received. Butyllithium (1.6 M solution in hexanes) was purchased from Aldrich.

2. Tetrahydrofuran was distilled from sodium benzophenone ketyl.

3. *N*-Formylpiperidine (98%) was purchased from Lancaster Synthesis and dried over activated 4Å molecular sieves before use. The molecular sieves were activated in a round-bottomed flask that was heated in a microwave oven (50 W) at full power for 5 min, then evacuated under vacuum (7–8 mmHg) on a vacuum line for 5 min while being allowed to cool to room temperature. This procedure was repeated three times. The flask was sealed with a three-way cock until the activated sieves were used. *N*-Formylpiperidine was dried by standing in a sealed bottle over activated MS 4Å at room temperature overnight. The submitters found that some mono-aldehyde [^{1}H NMR (400 MHz, $CDCl_3$) δ: 7.75 (s, 1 H), 9.94 (s, 1 H)] was formed if the *N*-formylpiperidine was insufficiently dry.

4. All analytical thin layer chromatography analyses (TLC) were carried out on Merck pre-coated 0.25 mm thick plates of silica gel 60 F_{254}. The plates were visualized with 254 nm UV light followed by a phosphomolybdic acid solution (10% w/v in MeOH) stain.

5. The submitters report yields that range from 69–80%. The crude product has the following properties; R_f = 0.30 (5:1, hexane:ethyl acetate); mp 225 °C (sub.) (lit.,[2] 227 °C); IR (solid): 1663 cm^{-1}; ^{1}H NMR (500 MHz, $CDCl_3$) δ: 10.07 (s, 2 H); ^{13}C NMR (125 MHz, $CDCl_3$) δ: 123.7, 142.3, 183.3; LRMS (ESI) *m/z* 425 [M + 2(*i*-PrOH) + Li], 723 [2M + 2(*i*-PrOH) + Li]; Anal. Calcd for $C_6H_2O_2SBr_2$: C, 24.19; H, 0.68. Found: C, 24.25; H, 0.90.

6. Anhydrous *N,N*-dimethylformamide (99.8%) and lithium hydroxide monohydrate were purchased from Aldrich Chemical Company and were used as received.

7. Anhydrous potassium carbonate (>99%) was purchased from Wako Pure Chemical Industries and was used as received. In the submitter's case, anhydrous potassium carbonate (>99%) was purchased from Fisher Scientific UK and was used as received.

8. Anhydrous magnesium sulfate (99%) was purchased from Wako Pure Chemical Industries and was used as received. In the submitter's case, anhydrous magnesium sulfate (99%) was purchased from BDH and was used as received.

9. The submitters report yields that range from 69–71%. The crude product has the following properties; R_f = 0.40 (5:1, hexane:ethyl acetate); mp 189–190 °C (lit.,[3] 196 °C); IR (solid): 2983, 2934, 1715, 1691, 1228 cm^{-1}; ^{1}H NMR (500 MHz, $CDCl_3$) δ: 1.41 (t, *J* = 7.5 Hz, 6 H), 4.40 (q, *J* = 7.5 Hz, 4 H), 8.00 (s, 2 H); ^{13}C NMR (125 MHz, $CDCl_3$) δ: 14.5, 61.8, 126.8, 135.4, 136.0, 143.4, 162.2; LRMS (ESI (+)) *m/z* 363 [M + Na], 703 [2M + Na]; Anal. Calcd for $C_{14}H_{12}O_4S_3$: C, 49.39; H, 3.55. Found: C, 49.23; H, 3.44.

10. A static inert atmosphere was provided by attaching a three-way cock with an argon balloon at the top of the condenser. The flask was flushed with argon after the addition of reagents.

11. The aqueous solution of lithium hydroxide was prepared by dissolving 3.15 g (75 mmol) of lithium hydroxide monohydrate (Aldrich) in 75 mL of distilled water.

12. Filter paper purchased from Toyo Roshi Kaisha (No.2) was used. The checkers observed some clogging or a partial leak of a small amount of solid. The submitters used Whatman filter paper No. 54 (hardened, ashless).

13. The submitters report yields that range from 95–96%. The crude product has the following properties; mp >300 °C; IR (solid) cm^{-1}: 3083-2557 br, 1692, 1661; ^{1}H NMR (500 MHz, DMSO-d_6) δ: 8.18 (s, 2 H); ^{13}C NMR (126 MHz, DMSO-d_6) δ: 127.7, 134.3, 137.0, 143.8, 163.3. LRMS (ESI (-)) *m/z* 283 [M-H], 567 [2M-H]; Anal. Calcd for $C_{10}H_4O_4S_3$: C, 42.24; H, 1.42. Found: C, 42.32; H, 1.61.

14. Because the product is only soluble in DMSO, TLC analysis gives R_f = 0 in MeOH. TLC may therefore be used to verify the loss of starting material but will not confirm the formation of the product.

15. Copper powder (99+%, irregular 300 mesh) and quinoline (98+%) were purchased from Lancaster Synthesis and were used as received.

16. The organic phases are decanted to separate the product from the insoluble copper salts.

17. Column chromatography was performed using silica gel (Merck silica gel 60 (230–400 mesh)).

18. The submitters report yields that range from 59–71%. The purified product has the following properties; R_f = 0.33 (hexane); mp 66–67 °C (lit.,[4] 66–67 °C); IR (solid) cm^{-1}: 3096, 3074, 1464, 1357; ^{1}H NMR

(500 MHz, $CDCl_3$) δ: 7.29 (d, J = 5.5 Hz, 2 H), 7.36 (d, J = 5.5 Hz, 2 H); ^{13}C NMR (126 MHz, $CDCl_3$) δ: 120.9, 125.9, 131.0, 141.7; LRMS (ESI (+)) *m/z* 305 [M+Ag], 501 [2M+Ag]; Anal. Calcd for $C_8H_4S_3$: C, 48.95; H, 2.05. Found: C, 48.69; H, 2.20.

Safety and Waste Disposal Information

All hazardous materials should be handled and disposed of in accordance with "Prudent Practices in the Laboratory"; National Academy Press; Washington, DC, 1995.

3. Discussion

In recent years much interest has been directed to dithieno[3,2-*b*:2′,3′-*d*]thiophene (DTT) **5** as a unit in semiconducting organic materials.[5-8] In particular, the dimer α,α′-bis(dithieno[3,2-*b*:2′,3′-*d*]thiophene)[9] and its derivatives[10] have been used as high mobility semiconductors in field-effect transistors, with field-effect mobilities of up to 0.05 $cm^2V^{-1}s^{-1}$ being recorded.[9]

The most recent syntheses of DTT were reported in 2002, by Hellberg and coworkers[11] in Sweden, and by Holmes and Frey[3] in Cambridge, UK. Both of these routes showed improvements over the previous method, which was published by De Jong in 1971.[4] Although the route reported by Hellberg appears to be more convenient in having fewer synthetic steps, the present method has several advantages. Each of the intermediates is a solid that requires no purification before taking forward, thus providing a quick and efficient synthetic route. The final product, DTT **5**, is quickly purified by flash column chromatography followed by recrystallization. Finally, the synthesis may be carried out on scales of over 30 g, although the lithiation step is limited on a laboratory scale due to the difficulty in cooling large-scale glassware.

DTT can be brominated in the 2- and 6- positions with NBS. This dibromide can be used in a variety of cross coupling reactions to form interesting conjugated oligomers and polymers. In addition, a novel silylation procedure can introduce TMS groups to the 3- and 5- positions.[3]

1. Melville Laboratory for Polymer Synthesis, Department of Chemistry, University of Cambridge, Lensfield Road, Cambridge, CB2 1EW, U.K. and School of Chemistry,

University of Melbourne, Vic. 3010, Australia.; Bio21 Institute, University of Melbourne, Vic. 3010, Australia.

2. Steinkopf, K. *Justus Liebigs Ann. Chem.* **1937**, *532*, 250-279.

3. Frey, J.; Bond, A. D.; Holmes, A. B. *Chem. Commun.* **2002**, 2424-2425.

4. De Jong, F.; Janssen, M. J. *J. Org. Chem.* **1971**, *36*, 1645-1648.

5. Kim, O. K.; Lee, K. S.; Woo, H. Y.; Kim, K. S.; He, G. S.; Swiatkiewicz, J.; Prassad, P. N. *Chem. Mater.* **2000**, *12*, 284.

6. Ventelon, L.; Moreaux, L.; Mertz, J.; Blanchard-Desce, M. *Chem. Commun.* **1999**, 2055-2056.

7. Kim, O. K.; Fort, A.; Barzoukas, M.; Blanchard-Desce, M.; Lehn, J. M. *J. Mater. Chem.* **1999**, *9*, 2227-2232.

8. Tsivgoulis, G. M.; Lehn, J. M. *Angew. Chem., Int. Ed. Engl.* **1995**, *34*, 1119-1122.

9. Li, X. C.; Sirringhaus, H.; Garnier, F.; Holmes, A. B.; Moratti, S. C.; Feeder, N.; Clegg, W.; Teat, S. J.; Friend, R. H. *J. Am. Chem. Soc.* **1998**, *120*, 2206-2207.

10. Morrison, J. J.; Murray, M. M.; Li, X. C.; Holmes, A. B.; Moratti, S. C.; Friend, R. H.; Sirringhaus, H. *Synth. Met.* **1999**, *102*, 987-988.

11. Allared, F.; Hellberg, J.; Remonen, T. *Tetrahedron Lett.* **2002**, *43*, 1553-1554.

1,4-DIOXENE
(2,3-Dihydro-1,4-dioxin)

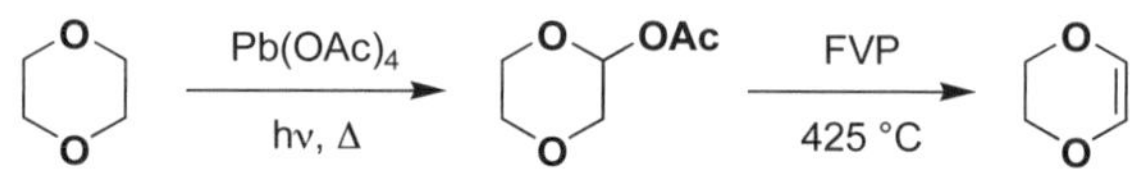

Submitted by Matthew M. Kreilein, James C. Eppich and Leo A. Paquette.[1]
Checked by Christopher P. Davie and Rick L. Danheiser.

1. Procedure

A. 2-Acetoxy-1,4-dioxane. An oven-dried, 500-mL, three-necked, round-bottomed flask is equipped with a magnetic stirbar, a reflux condenser fitted with an argon inlet, a thermometer, and a 20-cm length of black rubber tubing (1/8 in thick, 1 in diameter) attached to a dry 125-mL Erlenmeyer flask wrapped in aluminum foil and containing 91.9 g (207 mmol) of lead tetraacetate (Note 1). The flask is charged with 200 mL (2.35 mol) of dioxane (Note 2) and heated with stirring at 80 °C with a "Power Light" 500 W lamp (Note 3). The lead tetraacetate is added portionwise over approximately 30 min while the temperature is increased to the reflux point. A cloudy white or tan solution develops. The reaction mixture is heated at reflux for 1 h (Note 4) and then allowed to cool to rt. Saturated $NaHCO_3$ solution (200 mL) is carefully introduced, and the resulting mixture is filtered through a pad of Celite. The filtrate is extracted with three 100-mL portions of CH_2Cl_2, and the combined organic fractions are washed with 100 mL of saturated $NaHCO_3$ solution, dried over Na_2SO_4, filtered, and concentrated by rotary evaporation (20 mmHg) at room temperature. The resulting oil (27.3 g) is fractionally distilled at reduced pressure (Note 5) to remove dioxane and deliver 22.0 g (73%) of 2-acetoxy-1,4-dioxane as a colorless liquid, bp 54-55 °C (0.4 mmHg) (Note 6).

B. 1,4-Dioxene. A pyrolysis apparatus is assembled as shown in the photographs below. A 50-mL, two-necked, round-bottomed flask containing 18.0-18.1 g (0.123-0.124 mol) of 2-acetoxy-1,4-dioxane is attached to a quartz tube (Note 7) packed with coarse quartz chips (ca. 1/4") and heated at 425 °C in a horizontal furnace (Notes 8, 9). The tube is attached to a U-tube that is charged with ca. 11 g of NaOH pellets. A Dewar (cold finger) condenser with a 105° angled side joint is attached to the other end of the U-tube and filled with dry ice/acetone (–78 °C). A vacuum trap, cooled in liquid nitrogen, is positioned between the cold finger and the vacuum source. A gentle flow of N_2 (constricted through a needle valve) is initiated at the side neck of the flask and a gentle vacuum (ca. 265 mmHg, controlled by a digital vacuum regulator) is applied via the angled side joint of the Dewar condenser (Note 10).

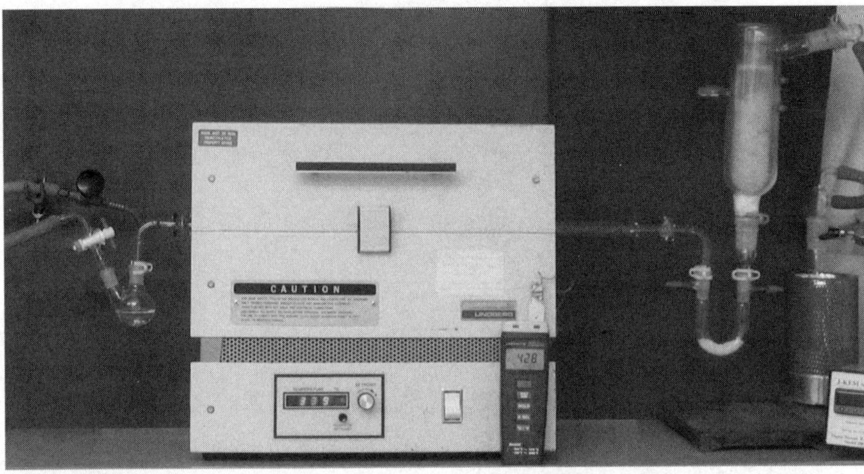

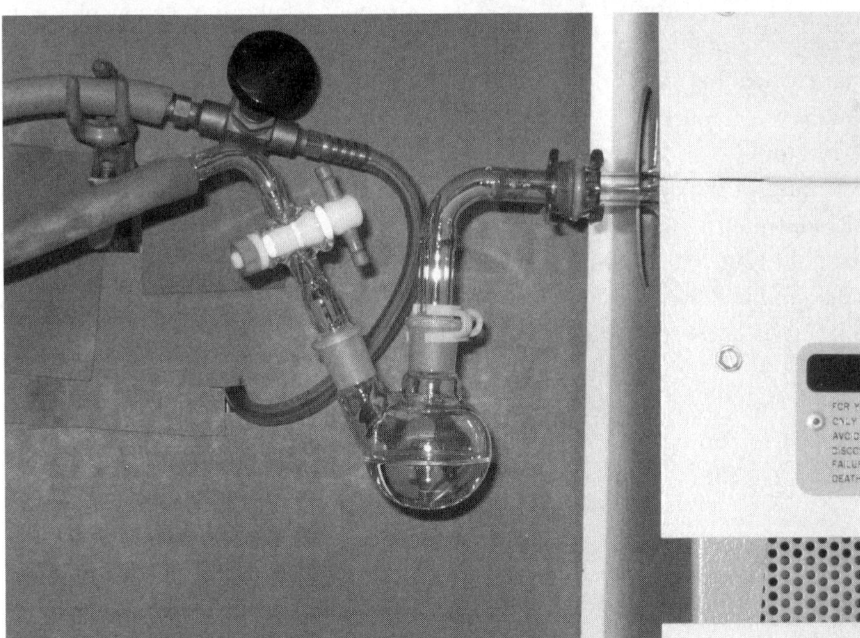

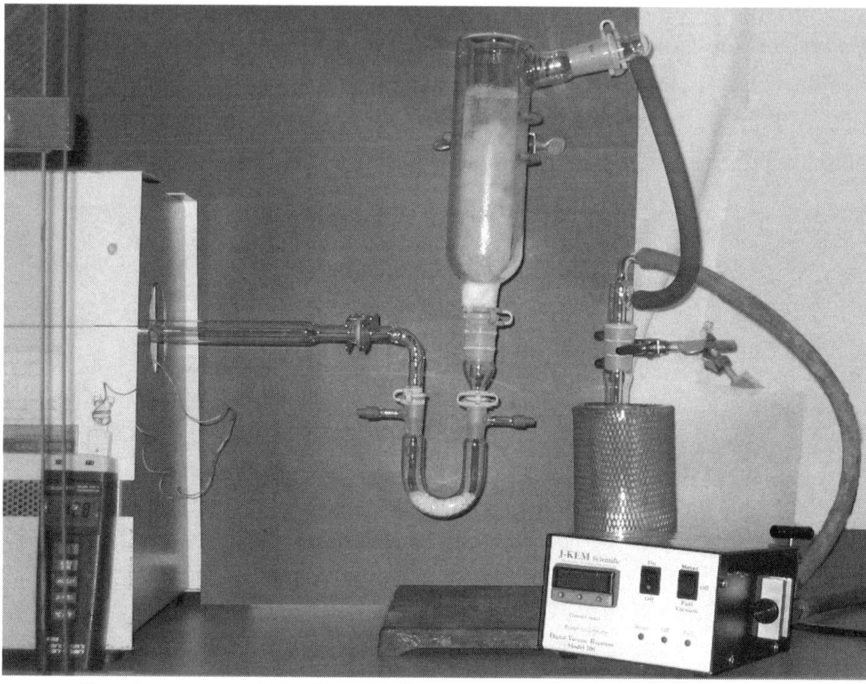

The 2-acetoxy-1,4-dioxane is introduced into the vapor phase by appropriate application of a modest amount of heat from a commercial heat gun (Note 11). After the introduction of the acetate is completed (25 min), the Dewar condenser is mounted above a 25-mL, round-bottomed flask and a mixture of 1,4-dioxene and acetic acid is collected therein following warming of the cold finger bath. The liquids contained in the U-tube (Note 12) and liquid nitrogen trap are combined with the material collected from the cold finger to afford 9.4-9.8 g of a colorless to pale yellow liquid (Note 13). Sodium hydroxide pellets (4.0 g) are added, and the resulting gel-like mixture is allowed to stand for 1 h and then distilled at atmospheric pressure (Note 14). The distillate of water and 1,4-dioxene is placed in a –18 °C freezer overnight and 1,4-dioxene is obtained by decantation from the ice crystals to furnish 4.7-6.5 g (44-61%) of 1,4-dioxene as a colorless liquid (Notes 15, 16).

2. Notes

1. Lead tetraacetate was obtained from Acros Organics (95% purity, stabilized with ca. 4% acetic acid) and used as received. If lightly colored, the reagent was used directly. If dark-brown in color, the lead tetraacetate was recrystallized from acetic acid according to the procedure described in Perrin, D. D.; Armarego, W. L. F.; Perrin, D. R. *Purification of Laboratory Chemicals,* 2nd ed.; Pergamon Press: Oxford, 1980; pp 497-498.

2. Dioxane (anhydrous, 99.8%) was purchased from Alfa Aesar and distilled from calcium hydride under argon before use.

3. The lamp was an inexpensive model purchased from a discount distributor (Sam's Club). The manufacturer was The Designer's Edge, 11730 N.E. 12th St., Bellevue, WA 98006.

4. In the event that gas evolution (as evidenced by briefly turning off the Ar source and watching the bubbler) persists after the reaction mixture is heated at reflux for 1 h, refluxing is continued until gas evolution ceases.

5. A short path distillation head attached to a receiver cow was employed with the receiving flask cooled in a dry ice/acetone bath.

6. In different runs, the yield ranged from 68-73%. The product exhibits the following spectroscopic properties: IR (film): 2976, 2860, 1748, 1454, 1374, 1225 cm^{-1}; ^{1}H NMR (500 MHz, $CDCl_3$) δ 2.16 (s, 3H), 3.64 (app dt, J = 11.8, 2.7 Hz, 1H), 3.71-3.82 (m, 4H), 4.09-4.17 (m, 1H), 5.85 (t, J = 2.1 Hz, 1H); ^{13}C NMR (125 MHz, $CDCl_3$) δ 21.3, 61.9, 66.3, 67.9, 89.4, 170.0; HRMS (ESI) Calcd for $C_6H_{10}O_4$ $[M + Na]^+$: 169.0471; Found: 169.0478.

7. The dimensions of the quartz tube used by the checkers were as follows: overall length, 590 mm; length of tube between insulation inside oven, 308 mm; outside diameter, 19 mm. The tube used by the submitters had the following dimensions: overall length, 490 mm; length of tube inside oven between insulation, 280 mm; inside diameter, 13 mm.

8. Any type of furnace may be employed, although "hot spots" should be avoided if possible. Should the bore of the oven aperture be significantly larger than the diameter of the quartz tube, the ends of the tube may be wrapped with glass wool tape to fit. The checkers used an oven manufactured by Lindberg (a unit of General Signal), Watertown, WI 53094. The submitters used an oven manufactured by Hevi Duty Electric Company, Milwaukee, WI. The checkers monitored the oven temperature using an Omegaette® Model HH308 digital thermometer equipped with a

thermocouple probe that was inserted into the center of the oven. The submitters monitored the oven temperature using a Keithley Model 871 digital thermometer equipped with a ceramic temperature probe that was inserted into the center of the oven.

9. The quartz tube was preheated at 425 °C for ≥1h before the 2-acetoxy-1,4-dioxane was introduced.

10. The pyrolysis setup used by the submitters did not include a liquid nitrogen trap. Their U-tube contained ca. 5 g of NaOH, and they controlled the flow of nitrogen with a fine capillary.

11. The checkers used a heat gun with a 260-399 °C range (14 amp) manufactured by the Master Appliance Corporation, Racine, WI 53403. The submitters used a heat gun made by the same manufacturer with a 149-260°C range (12 amp).

12. In one run, the checkers found that gently heating the U-tube with a heat gun (after the apparatus had been disassembled) allowed more liquid to be decanted from the U-tube.

13. ^{1}H NMR analysis indicated that this material is a ca. 75:25 mixture of 1,4-dioxene and AcOH.

14. A short path distillation head attached to a 25-mL, round-bottomed receiver flask were used. The mixture of water and 1,4-dioxene distilled at 58-94 °C (760 mmHg).

15. The submitters report that they obtained 10-20% pure dioxene from the Dewar condenser cold finger and additional product by distillation of the liquid decanted from the U-tube; total yield: 65%.

16. The product exhibits the following spectroscopic properties: IR (film): 3099, 2982, 2933, 2879, 2023, 1654, 1458, 1395, 1268, 1128, 1067, 954, 898, 739 cm^{-1}; ^{1}H NMR (500 MHz, $CDCl_3$) δ 4.07 (s, 4H), 5.96 (s, 2H); ^{13}C NMR (125 MHz, $CDCl_3$) δ 64.9, 127.2; Anal. Calcd for $C_4H_6O_2$: C, 55.81; H, 7.02; N; Found: C, 55.69; H, 7.01.

Safety and Waste Disposal Information

All hazardous materials should be handled and disposed of in accordance with "Prudent Practices in the Laboratory"; National Academy Press; Washington, DC, 1995.

3. Discussion

The 1,4-dioxene molecule (**1**) has attracted attention for a number of years since its initial preparation by Summerbell and Bauer in 1935.[2] The symmetrical nature of its double bond and the cis orientation of the two oxygen atoms are features not commonly resident in other structural contexts. One consequence is the lowering of the first ionization potential of

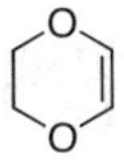

1

1 to 8.43 eV relative to the corresponding value in dihydropyran (8.84 eV).[3] As a result, this heterocyclic building block undergoes successful [2+2] photocycloaddition to conjugated enones,[4] 1,2-diketones,[5] and benzene.[6] The facility with which **1** enters into the Paterno-Büchi reaction has also been documented.[7] Comparable interest has surrounded the involvement of 1,4-dioxene in thermal inverse electron demand Diels-Alder processes,[8,9] trapping with molybdenum and chromium carbene complexes,[10] [2+2] ketene cycloadditions,[11] and cyclopropanations with diazo compounds.[12]

The enol ether constitution of **1** has caused it to be regarded as a protecting group for alcohols.[13] Beyond this, conversion to 2-dioxenyllithium (**2**) can be efficiently accomplished by exposure to *tert*-butyllithium in THF at low temperature.[14] This organometallic intermediate has been broadly exploited by Fétizon and Hanna in synthesis,[15] and the derived stannane **3** can be smoothly acylated with acyl chlorides under conditions of palladium catalysis.[16] Added scope is provided by the higher order cuprate **4** whose reactivity is well suited to electrophilic capture.[17]

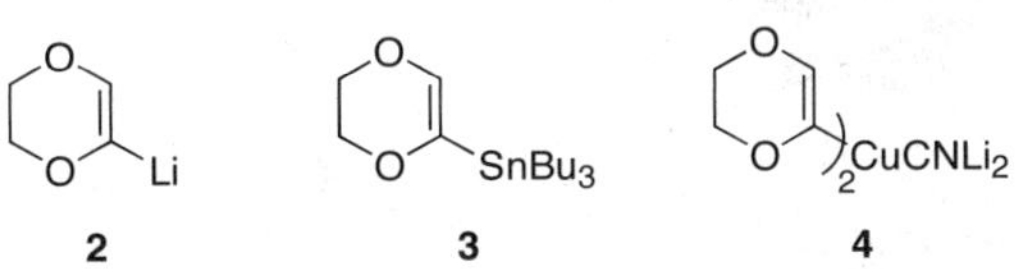

2 **3** **4**

Despite the considerable promise of 1,4-dioxene in organic synthesis, only three preparative routes to **1** have been reported. More astonishing yet was the unsuitability of all three routes for the laboratory-scale acquisition of reasonable amounts of *pure* reagent (Scheme 1). The original pathway proceeds via the photochlorination of *p*-dioxane to the 2,3-dichloro derivative **5** (70%) followed by reductive dehalogenation with magnesium and iodine (49%).[2] Entry has also been made from diethylene glycol (**6**), heating of which with a copper chromite catalyst and $KHSO_4$ in the liquid phase proceeds with oxidation and cyclodehydration (67%).[18] The complication here is the co-production of 2-*p*-dioxanone, a side reaction highly dependent on the proportion of $KHSO_4$ present. The third strategy involves the photoaddition of phenanthrenequinone to *p*-dioxane and thermal activation of the resulting **7** at 230-250 °C to liberate **1**.[19]

The two-step process described here results in the clean formation of 1,4-dioxene free of contaminants and is therefore expected to find serviceable application in the synthesis of this useful heterocyclic intermediate. In this more practical and convenient route, advantage is taken of the rarely exploited capability[20,21] of lead tetraacetate to engage in the acetoxylation of C-H bonds positioned at benzylic[22] and allylic sites,[23] as well as adjacent to ethereal oxygen centers.[24] In the specific case of *p*-dioxane, the eight available C-H bonds are equivalent by virtue of symmetry, thus simplifying matters considerably. The result of irradiating a refluxing solution of lead tetraacetate in dioxane with an inexpensive commercial 500 W light source for ca. 1 h and subsequent fractional distillation is to provide the 2-acetoxy derivative in 68-73% yield. The boiling point of this colorless liquid is sufficiently higher than that of the starting material to permit its isolation in a pure state. The flash vacuum pyrolysis step results in the smooth thermal extrusion of acetic acid to deliver **1**. When this experiment is conducted in that manner where no acetate remains unreacted, the yield of volatile 1,4-dioxene is 44-61%.

Scheme 1

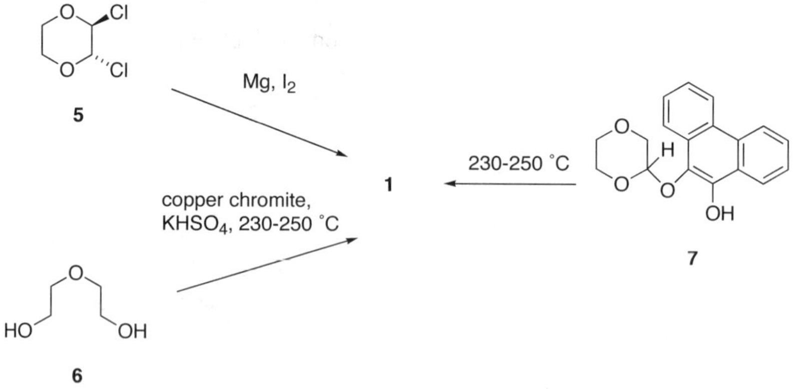

1. Department of Chemistry, The Ohio State University, Columbus, Ohio 43210-1106.

2. Summerbell, R. K.; Bauer, L. N. *J. Am. Chem. Soc.* **1935**, 57, 2364. See also Summerbell, R. K.; Umhoefer, R. R. *J. Am. Chem. Soc.* **1939**, *61*, 3016.

3. Bloch, M.; Brogli, F.; Heilbronner, E.; Jones, T. B.; Prinzbach. H.; Schweikert, O. *Helv. Chim. Acta* **1978**, *61*, 1388.

4. (a) Bernassau, J. M.; Bouillot, A.; Fétizon, M.; Hanna, I.; Maia, E. R.; Prangé, T. *J. Org. Chem.* **1987**, *52*, 1993. (b) Blechert, S.; Jansen, R.; Velder, J. *Tetrahedron* **1994**, *50*, 9649.

5. Horspool, W. M.; Khandelwal, G. D. *Chem. Commun.* **1967**, 1203.

6. (a) Atkins, R. J.; Fray, G. I.; Drew, M. G. B.; Gilbert, A.; Taylor, G. N. *Tetrahedron Lett.* **1978**, *19*, 2945. (b) Mattay, J.; Leismann, H.; Scharf, H. D. *Chem. Ber.* **1979**, *112*, 577. (c) Atkins, R. J.; Fray, G. I.; Gilbert, A.; bin Samsudin, M. W.; Steward, A. J. K.; Taylor, G. N. *J. Chem. Soc., Perkin Trans. 1* **1979**, 3196. (d) Gilbert, A.; Taylor, G. N.; bin Samsudin, M. W. *J. Chem. Soc., Perkin Trans. 1* **1980**, 869. (e) Gilbert, A. *Pure Appl. Chem.* **1980**, *52*, 2669.

7. (a) Lazear, N. R.; Schauble, J. H. *J. Org. Chem.* **1974**, *39*, 2069. (b) Araki, Y.; Nagasawa, J.; Ishido, Y. *J. Chem. Soc., Perkin Trans. 1* **1981**, 12. (c) Freilich, S. C.; Peters, K. S. *J. Am. Chem. Soc.* **1981**, *103*, 6255. (d) Adam, W.; Kliem, U.; Lucchini, V. *Tetrahedron Lett.* **1986**, *27*, 2953. (e) Buschmann, H.; Hoffmann, N.; Scharf, H. D. *Tetrahedron: Asymmetry* **1991**, *2*, 1429.

8. Thalhammer, F.; Wallfahrer, U.; Sauer, J. *Tetrahedron Lett.* **1990**, *31*, 6851.

9. 1,4-Dioxene is also known to undergo thermal [2+2] cycloadditions to *N*-phenyltriazolinedione [Koerner von Gustorf, E.; White, D. V; Kim, B.; Hess, D.; Leitich, J. *J. Org. Chem.* **1970**, *35*, 1155] and singlet oxygen [Schaap, A. P. *Tetrahedron Lett.* **1971**, *12*, 1757], but not with tetracyanoethylene [Huisgen, R.; Steiner, G. *Tetrahedron Lett.* **1973**, *14*, 3763]. The latter reagent gives rise to a violet charge-transfer complex in $CHCl_3$.

10. Harvey, D. F.; Grenzer, E. M.; Gantzel, P. K. *J. Am. Chem. Soc.* **1994**, *116*, 6719.

11. (a) Huisgen, R.; Feiler, L. A.; Otto, P. *Chem. Ber.* **1969**, *102*, 3444. (b) Fétizon, M.; Hanna, I. *Synthesis* **1990**, 583.

12. (a) Huisgen, R.; Feiler, L. A.; Otto, P. *Chem. Ber.* **1969**, *102*, 3405. (b) Shatzmiller, S.; Neidlein, R. *Justus Liebigs Ann. Chem.* **1977**, 910. (c) Jendralla, H.; Pflaumbaum, W. *Chem. Ber.* **1982**, *115*, 229. (d) Wenkert, E.; Greenberg, R. S.; Raju, M. S. *J. Org. Chem.* **1985**, *50*, 4681.

13. Fétizon, M.; Hanna, I. *Synthesis* **1985**, 806.

14. Saylor, R. W.; Sebastian, J. F. *Synth. Commun.* **1982**, *12*, 579.

15. (a) Fétizon, M.; Hanna, I.; Rens, J. *Tetrahedron Lett.* **1985**, *26*, 3453. (b) Fétizon, M.; Goulaouic, P.; Hanna, I. *Tetrahedron Lett.* **1985**, *26*, 4925. (c) Fétizon, M.;

Goulaouic, P.; Hanna, I.; Prangé, T. *J. Org. Chem.* **1988**, *53*, 5672. (d) Fétizon, M.; Goulaouic, P.; Hanna, I. *Heterocycles* **1989**, *28*, 521. (e) Fétizon, M.; Goulaouic, P.; Hanna, I. *J. Chem. Soc., Perkin Trans. 1* **1990**, 1107. (f) Hanna, I. *Tetrahedron Lett.* **1995**, *36*, 889.

16. Blanchot, V.; Fétizon, M.; Hanna, I. *Synthesis* **1990**, 755.

17. (a) Blanchot-Courtois, V.; Hanna, I. *Tetrahedron Lett.* **1992**, *33*, 8087. (b) Boger, D. L.; Zhu, Y. *J. Org. Chem.* **1994**, *59*, 3453.

18. Moss, R. D.; Paige, J. N. *J. Chem. Eng. Data* **1967**, *12*, 452.

19. Rubin, M. B. *Synthesis* **1977**, 266.

20. Butler, R. N. In *Synthetic Reagents;* Pizey, J. S., Ed.; Ellis Horwood Limited: Chichester, 1977; Vol. 3, pp 277-419.

21. Mihailovic, M. Lj.; Cerkovic, Z. In *Encyclopedia of Reagents for Organic Synthesis*; Paquette, L. A., Ed.; Wiley: Chichester, 1995; Vol. 5, pp 2949-2954.

22. (a) Heiba, E. I.; Dessau, R. M.; Koehl, W. J., Jr. *J. Am Chem. Soc.* **1968**, *90*, 1082. (b) Cavill, G. W. K.; Solomon, D. H. *J. Chem. Soc.* **1954**, 3943.

23. Crilley, M. M. L.; Larsen, D. S.; Stoodley, R. J.; Tomé, F. *Tetrahedron Lett.* **1993**, *34*, 3305.

24. Hill, R. K.; Morton, G. H.; Peterson, J. R.; Walsh, J. A.; Paquette, L. A. *J. Org. Chem.* **1985**, *50*, 5528.

PREPARATION OF 1-BUTYL-3-METHYLIMIDAZOLIUM TETRAFLUOROBORATE
[1H-Imidazolium, 1-butyl-3-methyl, tetrafluoroborate (1–)]

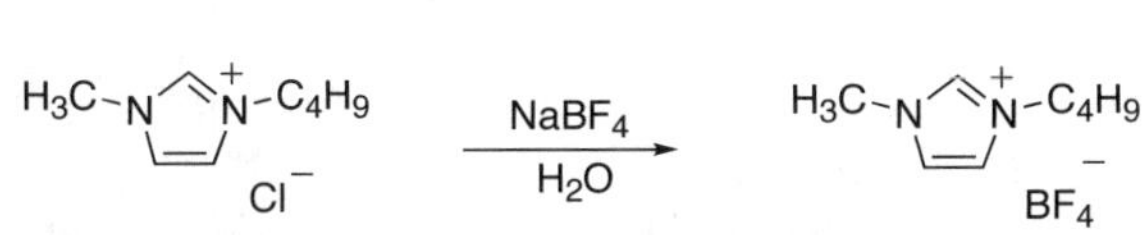

Submitted by Xavier Creary and Elizabeth D. Willis.[1]
Checked by Gustavo Moura-Letts and Dennis P. Curran.

1. Procedure

1-Butyl-3-methylimidazolium chloride (30.00 g, 172 mmol) (Note 1) is placed in a 125 mL Erlenmeyer flask containing a stir bar and a thermometer. This salt is dissolved in 35 mL of distilled water and $NaBF_4$ (20.00 g, 182 mmol) (Note 2) is added in portions with stirring over 10-15 min. The $NaBF_4$ dissolves as the mixture emulsifies and cools to 14 °C (Note 3). After the mixture warms back to ambient temperature, 30 mL of CH_2Cl_2 is added and the contents are transferred to a 125 mL separatory funnel. The bottom CH_2Cl_2 phase is separated (Note 4). The aqueous phase is extracted with an additional 20 mL of CH_2Cl_2. The combined CH_2Cl_2 phases are shaken in a separatory funnel with a solution of $NaBF_4$ (10.0 g, 91 mmol) in 20 mL of water. The CH_2Cl_2 phase is separated and dried over a mixture of 1.0 g of Na_2SO_4 and 3.0 g of $MgSO_4$. The mixture is filtered through a Büchner funnel and the salts are washed with an additional 15 mL of CH_2Cl_2. A short path distillation head is attached to the 250 mL round-bottom-flask containing the filtrate and most of the CH_2Cl_2 is removed by distillation at 30 mm pressure. Care is taken not to heat the product above 50°C. The solvent is condensed in a receiver flask cooled in an ice-water slurry (Note 5). The last traces of CH_2Cl_2 are removed using a rotary evaporator at 15 mm and 45 °C followed by vacuum drying at ambient temperature until the weight remains constant. The pure 1-butyl-3-methylimidazolium tetrafluoroborate (34.2–34.3 g; 89% yield) is a colorless to pale yellow, viscous liquid (Notes 6-8).

2. Notes

1. The submitters prepared 1-butyl-3-methylimidazolium chloride as previously described.[2] The checkers purchased this from Aldrich Chemical Co.

2. Sodium tetrafluoroborate (98%) was purchased from Acros Organics and used as received.

3. If the emulsion is allowed to stand for 30-60 min, then two phases results. However, there is no need to do this since addition of CH_2Cl_2 induces rapid phase separation.

4. The product at this point is not completely chloride free. In a separate analysis by the submitters, removal of the CH_2Cl_2 from this solution and analysis of 1.53 g of product by titration with 0.100 M $AgNO_3$ (Mohr titration; K_2CrO_4 indicator)[3] required 0.6 mL of the silver nitrate solution to precipitate all of the chloride ion. This corresponds to a product containing 0.7% $BMIM^+$ Cl^- and 99.3% $BMIM^+$ BF_4^-.

5. The distilled CH_2Cl_2 has droplets of water from the azeotrope with CH_2Cl_2. This distillation process helps to dry the product and allows recovery of most of the CH_2Cl_2 used in the procedure.

6. Dissolution of 1.50 g of this liquid in 3 mL of water followed by addition of 0.100 M $AgNO_3$ gave no precipitation or cloudiness.

7. The submitters report that the water content of the product was 0.17% as determined from the ^1H NMR spectrum of the neat liquid by integration of the water signal at d 2.9 and the CH_2 signal at d 1.93. The product is hygroscopic and water content increases with exposure to air. The checkers could not detect water in their spectrum in CD_2Cl_2.

8. The product exhibits the following spectroscopic properties: IR (thin film) 3646, 3162, 3122, 2964, 2877, 1575, 1467, 1431, 1171, 1048, 850 cm^{-1}; ^1H NMR (500 MHz, CD_2Cl_2) δ 8.66 (s, 1 H), 7.37 (s, 2 H), 4.14 (t, 2 H, J = 7.3 Hz), 3.89 (s, 3 H), 1.81 (quintet, 2 H, J = 7.4 Hz), 1.32 (sextet, 2 H, J = 7.3 Hz), 0.91 (t, 3 H, J = 7.3 Hz); ^{13}C NMR (125 MHz, CD_2Cl_2) δ 136.2, 123.8, 122.5, 49.7, 36.1, 31.9, 19.3, 13.1. Anal. Calcd for $C_8H_{15}BF_4N_2$: C, 42.51; H, 6.69; N, 12.37; Found: C 41.94; H, 6.85; N, 12.37; trace analysis for Cl, 636 ppm.

Safety and Waste Disposal Information

All hazardous materials should be handled and disposed of in accordance with "Prudent Practices in the Laboratory"; National Academy Press; Washington, DC, 1995.

3. Discussion

This procedure is based on the general method recently reported in *Organic Syntheses*.[2] The preparation of 1-butyl-3-methylimidazolium hexafluorophosphate ($BMIM^+ PF_6^-$) proceeds as described.[2] However, in our hands, the reported method for the preparation of 1-butyl-3-methylimidazolium tetrafluoroborate ($BMIM^+ BF_4^-$; the most commonly used of the ionic liquids) was not reproducible. The product was contaminated with varying (and significant) amounts of $BMIM^+ Cl^-$[4]. The use of the relatively insoluble KBF_4 (0.44 g/100 mL) in the original preparation makes dissolution and subsequent reaction problematic. Thus, analysis of 0.716 g of a typical ionic liquid product produced from KBF_4 by titration with 0.100 M $AgNO_3$ required 11.1 mL of the silver nitrate solution to precipitate all of the chloride ion. This corresponds to a product containing 27% $BMIM^+ Cl^-$ and 73% $BMIM^+ BF_4^-$.

The procedure described herein uses the much more soluble $NaBF_4$ (97.3 g/100 mL). After the first cycle, the chloride content is only 0.7%, and the chloride can no longer be detected after the second cycle. In addition, the tedious process of removing water by distillation under reduced pressure is eliminated. This procedure is also useful for the preparation of the ionic liquid *n*-butyl pyridinium tetrafluoroborate, which is subject to the same chloride contamination problems when prepared from KBF_4. Finally, the amount of residual water in the product can be readily determined from the ^1H NMR spectrum of the neat product.

1. Department of Chemistry and Biochemistry, University of Notre Dame, Notre Dame, IN 46556.
2. Dupont, J.; Consorti, C. S.; Suarez, P.A.Z.; de Sousa, R. F. *Org. Synth.* **2002** *79*, 236.
3. Christian, G. D. "Analytical Chemistry, Fifth Edition", John Wiley and Sons, Inc., New York, 1994, p. 278.
4. Contamination of $BMIM^+ BF_4^-$ and other ionic liquids with chloride in various preparations has been noted in the literature. See Seddon, K. R.; Stark, A.; Torres, M.-J. *Pure Appl. Chem.* **2000**, *72*, 2275.

PREPARATION OF 2,4-DISUBSTITUTED IMIDAZOLES: 4-(4-METHOXYPHENYL)-2-PHENYL-1H-IMIDAZOLE [1H-Imidazole, 4-(4-methoxyphenyl)-2-phenyl-]

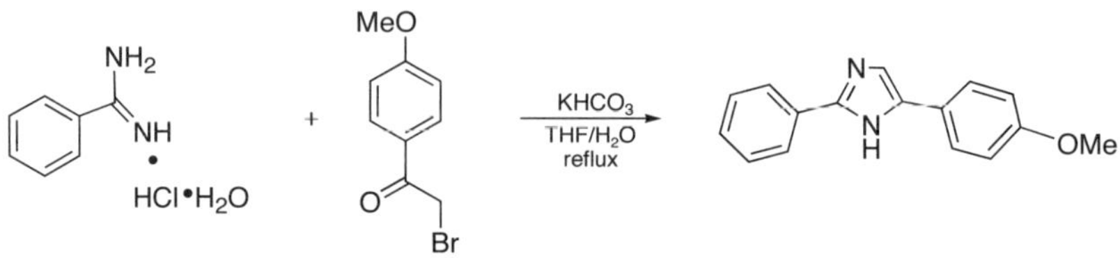

Submitted by Bryan Li,[1] Charles K-F Chiu, Richard F. Hank, Jerry Murry, Joshua Roth, and Harry Tobiassen.
Checked by Renee Kontnik and Steven Wolff.

1. Procedure

4-(4-Methoxyphenyl)-2-phenyl-1H-imidazole. A 2-L, three-necked, round-bottomed flask equipped with an addition funnel, reflux condenser, and mechanical stirrer is charged with 500 mL of tetrahydrofuran (THF) and 125 mL of water. Benzamidine hydrochloride monohydrate (50 g, 0.29 mol) (Note 1) is added, followed by the slow, portionwise addition of potassium bicarbonate (54.4 g, 0.57 mol) (Note 2). The reaction mixture is vigorously heated to reflux. A solution of 4-methoxyphenacyl bromide (65.3 g, 0.29 mol) in 325 mL of THF is then added dropwise via the addition funnel over a period of 30 min while the reaction is maintained at reflux. After completion of the addition, the mixture is heated at reflux for 18-20 hr (Note 3), then cooled in an ice bath (Note 4), and THF is removed under reduced pressure using a rotary evaporator. An additional 100 mL of water is added, and the resulting suspension is stirred at 50-60 °C for 30 min. The mixture is cooled in an ice bath and the solids are collected by filtration. The filter cake is rinsed with two 100-mL portions of water and air-dried in the filter funnel for 2 h. The crude product is transferred to a 500-mL flask and 150 mL of diisopropyl ether and 150 mL of hexanes are added. The mixture is stirred for 2 h at room temperature, and the solids are again collected by filtration. The filter cake is dried in a vacuum oven for 48 h (68 °C/ca. 100 mm) to give 68.6 g (96%) of the desired imidazole as an off-white solid (Notes 5, 6).

2. Notes

1. All reagents were purchased from Aldrich Chemical Company and used without further purification.

2. Caution: liberation of carbon dioxide.

3. TLC analysis (elution with 1:1 hexanes:ethyl acetate) indicated that the reaction was not complete after 5 h; heating was continued overnight.

4. Inorganic salts are observed upon cooling and are removed by decantation.

5. The dried product contained 0.10-0.54% H_2O w/w; and has the following physical properties: mp 144.5-145.6 °C [lit. mp 178-179 °C[2]]; IR (KBr): cm^{-1} 3003, 2836, 1617, 1567, 1499, 1461, 1403, 1298, 1248, 1176; ^{1}H NMR (300 MHz, DMSO-d^6): δ 3.76 (s, 3H), 6.94 (d, J = 8.4 Hz, 2H), 7.46-7.30 (m, 3H), 7.58 (s, 1H), 7.74 (d, J = 8.7 Hz, 2H), δ 7.97 (d, J = 7.8 Hz, 2H); ^{13}C NMR (75 MHz, DMSO-d^6): δ 55.0, 114.0, 115.8, 124.9, 125.7, 126.4, 127.9, 128.6, 130.7, 139.2, 145.8, 158.1. MS *m/z* 250 ($[M]^+$); 235 ($[M-CH_3]$); 77 ($[C_6H_6]$); Anal. Calcd for $C_{16}H_{14}N_2O$: C, 76.78; H, 5.64; N, 11.19. Found C, 76.49; H, 5.62; N, 11.03.

6. HPLC analysis indicated 97.1 area % purity. HPLC conditions: Zorbax XDB-C8 column (3.0 x 100mm) eluting with 5-100% MeCN + 0.1%TFA (0.5 mL/min); 220 nm wavelength.

Waste Disposal Information

All toxic materials were disposed of in accordance with "Prudent Practice in the Laboratory"; National Academy Press; Washington, DC, 1995.

3. Discussion

The imidazole nucleus is often found in biologically active molecules,[3] and a large variety of methods have been employed for their synthesis.[4] We recently needed to develop a more viable process for the preparation of kilogram quantities of 2,4-disubstituted imidazoles. The condensation of amidines, which are readily accessible from nitriles,[5] with α-halo ketones has become a widely used method for the synthesis of 2,4-disubstituted imidazoles. A literature survey indicated that chloroform was the most commonly used solvent for this reaction.[6] In addition to the use of a toxic solvent, yields of the reaction varied from poor to moderate, and column chromatography was often required for product isolation. Use of other solvents such as alcohols,[7] DMF,[8] and acetonitrile[9] have also been utilized in this reaction, but yields are also frequently been reported as poor.

Our initial attempts to optimize this reaction focused on utilizing anhydrous reaction conditions due to stability concerns of α-bromo ketones under basic aqueous conditions. Condensations using a variety of bases (potassium *t*-butoxide, potassium carbonate, cesium carbonate, etc.) in THF, DMF, CH_3CN or CH_2Cl_2 gave low yields. Reactions in alcohols (ethanol, 2-propanol and *t*-butyl alcohol) were equally unsatisfactory. We then investigated mixed organic/aqueous reaction media, since we reasoned that amidines are stronger nucleophiles than water, and therefore the condensation rate of α-bromo ketones with amidines should be faster than the decomposition rate of the bromo ketones in water. A series of reactions using THF, DMF and alcohols as the organic solvent were conducted, and from these experiments we made a number of observations.

(1) Aqueous THF is a suitable solvent system to solubilize the very polar amidines and non-polar α-bromo ketones, and it is superior to aqueous DMF or alcohol.
(2) Higher reaction temperatures in aqueous THF accelerate the condensation.
(3) Bicarbonate is the base of choice, since it only scavenges the acid produced during the condensation reaction.
(4) Since α-bromo ketones decompose under the reaction conditions, their concentration in the reaction should be minimized.

We found that the optimal reaction protocol was to add a solution of α-bromo ketone in THF to the amidine in aqueous THF in the presence of potassium bicarbonate under vigorous reflux. Using this procedure, 2,4-disubstituted imidazoles were isolated in excellent yields with >95% purity without column chromatography. Aromatic and aliphatic α-halo ketones participate in this reaction with a variety of aromatic amidines, as indicated in Table 1. Particularly noteworthy is that reactions involving pyridylamidines or chloroacetone are substantially more robust using this process (entries 3 and 4). We have successfully used this protocol on a multi-kilogram scale.

In conclusion, a scaleable process for the preparation of 2,4-subsituted imidazole from amidines and α-halo ketones is described. This method avoids the use of chloroform as solvent and affords the desired products in consistently good to excellent yields.

Table 1. Amidine and α-Halo ketone Condensations[a]

Amidine	α-Halo ketone	Product	Isolated Yields	Lit. Yields
Benzamidine	2-Bromo-acetophenone	2,5-Diphenyl-1*H*-imidazole	86%	62% [b 10]
Nicotinamidine	Chloroacetone	3-(5-Methyl-1*H*-imidazol-2-yl)-pyridine	83%	15% [11]
Thiophene-2-carboxamidine	2-Bromo-4'-methoxy-acetophenone	5-(4-Methoxyphenyl)-2-thiophen-2-yl-1*H*-imidazole	87%	80%[c 12]
Benzamidine	Chloroacetone	5-Methyl-2-phenyl-1*H*-imidazole	87%	59%[d 13]

[a] The 2,4-disubstituted imidazoles in Table 1 were previously characterized in the literature. The spectroscopic data of all products are consistent to that originally reported.

[b] Irreproducible results were reported.

[c] 3 eq. of 2-thiophenylamidine was used; yield was based on the α-bromo ketone.

[d] From the condensation of amidine and an α-halo oxime.

1. Chemical Research and Development, Pfizer Global Research and Development, Groton Laboratories, Groton, CT 06340, USA. Jerry Murry is currently located at Process Research Department, Merck Research Laboratories, Merck and Co. Inc., Rahway, NJ, 07065.
2. Lombardino, J. G.; Wiseman, E. H. *J. Med. Chem.* **1974**, *17*, 1182.
3. (a) Kudzma, l. V.; Turnvull, S. P. Jr. *Synthesis*, **1991**, 1021; (b) Compagnone, R. S.; Rapoport, H. *J. Org. Chem.* **1986**, *51*, 1713 and references cited therein; (c) Shapiro, S.; Enz, A. *Drugs of the Future*, **1992**, *17*, 489.
4. For examples, see (a) Grimmett, M. R. in *Comprehensive Heterocycle Chemistry*, Katritzky, A. R.; Rees, C. Ed. Pergamon Press, vol. 5, 1984 and references therein; (b) Grimmet, M. R. in *Comperhensive Heterocyclic Chemistry II*, Katritzky, A. R.; Rees, C.; Scriven, E. F.V. Ed. Pergamon Press, vol. 3, 1996 and references cited therein; (c) Varma, R. S.; Kumar, D. *Tetrahedron Lett.* **1999**, *40*, 7665; ((e) Lengeler, D.; Weisz, K. *Nuclesides Nucleotides*, **1999**, *18*, 1657; (f) Batanero, B.; Escudero, J.; Barba, F. *Org. Lett.* **1999**, *1*, 1521; (g) Bergemann, M.; Neidlein, R. *Helv. Chim. Acta* **1999**, *82*, 909.
5. (a) Boeré, R. T.; Oakley, R. T.; Reed, R. W. *J. Organomet. Chem.* **1987**, *331*, 161; (b) Thurkauf, A.; Hutchison, A.; Peterson, J.; Cornfield, L; Meade, R. *J. Med. Chem.* **1995**, *38*, 2251.

6. (a) Kempter, G.; Spindler, J.; Fiebig, H. J.; Sarodnick, G. *J. Prakt. Chem.* **1971**, *313*, 977; (b) Baldwin, J. J.; Christy, M. E.; Denny, G. H.; Habecker, C. N.; Freedman, M. B. *J. Med. Chem.* **1986**, *29*, 1065; (c) Nagao, Y.; Takahashi, K.; Torisu, K.; Kondo, K.; Hamanaka, N. *Heterocycles*, **1996**, *42*, 517; (d) Baldwin, J. J.; Engelhardt, E. L.; Hirschmann, R.; Lundell, G. F.; Ponticello, G. S. *J. Med. Chem.* **1979**, *22*, 687.

7. Baldwin, J. E.; Fryer, A. M.; Pritchard, G. J. *J. Org. Chem.* **2001**, *66*, 2588.

8. Kikuchi, K.; Hibi, S.; Yoshimura, H.; Tokuhara, N.; Tai, K.; Hida, T.; Yamauchi, T.; Nagai, M.; *J. Med. Chem.* **2000**, *43*, 409.

9. Moody, C. J.; Roffey, J. R. A. *Chem. Abstr.* **2000**, 134:71748.

10. Burtles, R. J.; Pyman, F. L. *J. Chem. Soc.* **1923**, *123*, 362.

11. Baldwin, J. J.; Lumma, P. K.; Novello, F. C.; Ponticello, G. S.; Sprague, J. M.; Duggan, D. E., *J. Med. Chem.* **1977**, *20*, 1189.

12. Baldwin, J. J.; Engelhardt, E. L.; Hirschmann, R.; Lundell, G. S.; Ponticello, G. S. *J. Med. Chem.* **1979**, *22*, 687.

13. Nakanish, S.; Nantaku, J.; Otsuji, Y. *Chem. Lett.* **1983**, *3*, 341.

PREPARATION OF *N*-ARYL-5*R*-HYDROXYMETHYL-2-OXAZOLIDINONES FROM *N*-ARYL CARBAMATES: *N*-PHENYL-(5*R*)-HYDROXYMETHYL-2-OXAZOLIDINONE
[2-Oxazolidinone, 5-(hydroxymethyl)-3-phenyl-, (5R)-]

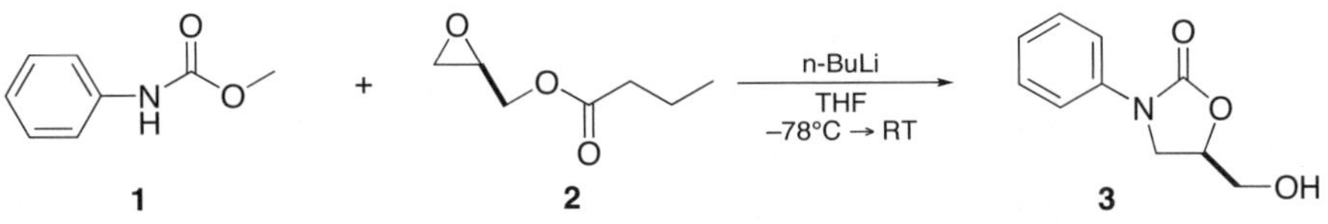

Submitted by Peter R. Manninen[1] and Steven J. Brickner.[2]
Checked by Ossama Darwish and Marvin J. Miller.

1. Procedure

N-Phenyl-(5R)-hydroxymethyl-2-oxazolidinone. A 2-L, three-necked, round-bottomed flask equipped with an addition funnel, rubber septum, nitrogen inlet, and a large magnetic stirbar is charged with 24.81 g (164.1 mmol) of *N*-phenylcarbamic acid methyl ester (**1**) (Notes 1, 2, 3). Freshly distilled tetrahydrofuran (THF, 750 mL) (Note 4) is added *via* syringe in 50-mL portions. The resulting solution is cooled to –78 °C in a dry ice/acetone bath while 103 mL of butyllithium solution (1.6M in hexanes, 164.8 mmol) (Note 5) is added dropwise via the addition funnel over 60 min. The addition funnel is rinsed with 10 mL of distilled THF and the rinse is added to the reaction mixture. The reaction mixture is stirred for 38 min while the flask is cooled in the dry ice/acetone bath and then 23.4 mL (164.8 mmol) of *R*-(–)-glycidyl butyrate **2** (Note 6) is added dropwise via syringe over 6 min. After 15 min, the dry ice/acetone bath is removed and the reaction mixture is allowed to warm to room temperature and stirred for 22 h. To the resulting thick slurry is then added 750 mL of saturated aqueous ammonium chloride solution and 20 mL of water (Note 7). The aqueous layer is separated and extracted with three 350-mL portions of ethyl acetate, and the combined organic layers are dried over magnesium sulfate, filtered, and concentrated under reduced pressure. The residual solid is dried in a vacuum oven at 80 °C for 72 h to provide 29.87 g (95%) of **3** as an off-white crystalline solid (Notes 8, 9).

2. Notes

1. All glassware was either dried in an oven or flame-dried and cooled under nitrogen.

2. A large football-shaped stir bar is recommended because the mixture becomes a thick slurry.

3. The *N*-phenylcarbamic acid methyl ester was purchased from TCI America.

4. Tetrahydrofuran was distilled from sodium/benzophenone under nitrogen.

5. Butyllithium was purchased from Aldrich Chemical Company, Inc.

6. *R*-(–)-Glycidyl butyrate was purchased from Lonza.

7. A little water is added to dissolve the precipitate that results in the aqueous layer.

8. The submitters report obtaining the product in 99% yield. The enantiomeric excess of the Mosher ester of **3** was measured to be 98% using a Chiralcel OD column (40% 2-propanol/hexane). This optical purity measurement substantiated the optical purity assessment made by ^{1}H NMR studies of **3** and racemic **3** prepared using a different method.[3] Addition of the chiral shift reagent tris[3-(heptafluoropropylhydroxymethylene)-(+)-camphorato]europium (III) resulted in clear resolution of the respective aromatic proton signals for the two enantiomers, which was demonstrated with the racemate. Under similar conditions, NMR analysis of **3** showed that within the detectable limits of the experiment (ca. <3%), there was none of the disfavored enantiomer.

9. The product has the following physical properties: mp 139-141 °C; ^{1}H NMR (300 MHz, $CDCl_3$) δ: 2.05 (s, broad, 1H), 3.77 (dd, J = 12.6 Hz, J' = 4.1 Hz, 1H), 4.02 (m, 3H), 4.76 (m, 1H), 7.15 (t, J = 7.15 Hz, 1H), 7.38 (t, J = 7.4 Hz, 2H), 7.54 (d, J = 9.0 Hz, 2H); ^{13}C NMR (125 MHz, $CDCl_3$) δ: 154.6, 138.1, 129.1, 124.2, 118.3, 72.7, 62.9, 46.3; IR (mineral oil mull): cm^{-1} 3391 (m), 1716 (s), 1424 (m), 1380 (m), 1307 (m), 1232 (m), 1146 (m); mass spectrum (EI): *m/z* (rel. abundance) 193 (100, M^+), 106 (42.9), 77 (38.0), (FAB): 194[100,$(MTH)^+$], 136(62.5), 106(50.1). Anal. Calcd for $C_{10}H_{11}NO_3$: C, 62.17; H, 5.74; N, 7.25. Found: C, 62.09; H, 5.80; N, 7.06. $[\alpha]_D$= –61 (c = 0.969, $CHCl_3$). TLC: ethyl acetate:hexane (1:1); R_f = 0.13.

Waste Disposal Information

All toxic materials were disposed of in accordance with "Prudent Practices in the Laboratory", National Academy Press; Washington, DC, 1995.

3. Discussion

N-Aryl-5-*R*-hydroxymethyl-2-oxazolidinones, represented by **3**, are important intermediates for the synthesis of oxazolidinone antibacterial agents. The procedure previously reported[3] and illustrated in Scheme 1 applied the method of Herweh–Kauffmann[4] in which an aryl isocyanate is reacted with kinetically resolved (*R*)-glycidyl butyrate in the presence of solubilized lithium bromide catalyst. While this method works well to provide the butyrate ester oxazolidinone **5** in high yield, an important limitation when utilized in a more general sense is the need to prepare non-commercially available isocyanate. This typically involves the reaction of anilines with phosgene, a reaction that generally terminates at 50% conversion, due to the formation of an equivalent of the aniline hydrochloride salt. In addition to the hazardous nature of phosgene, other limitations are the elevated temperatures required for the oxazolidinone ring formation, and the need for an additional step to cleave the butyrate ester to provide the requisite 5-(hydroxymethyl)oxazolidinone.

Scheme 1

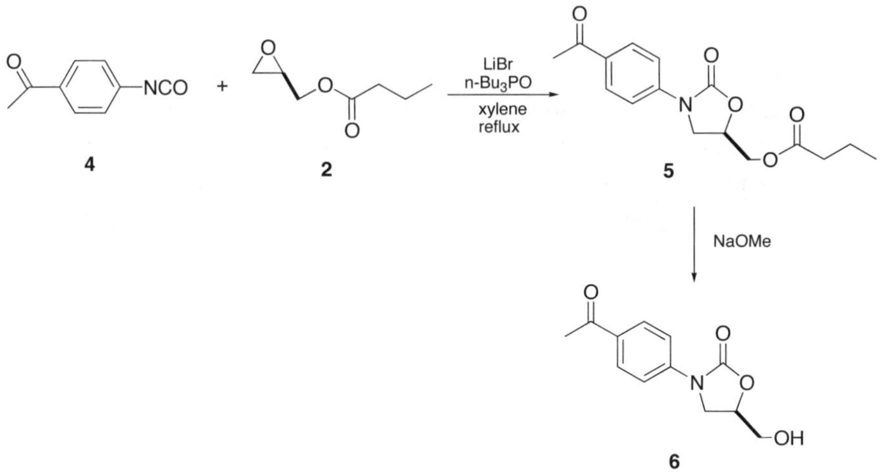

We have developed a novel and mild general approach to 5-(*R*)-hydroxymethyl-2-oxazolidinones that involves the alkylation of commercially available (*R*)-glycidyl butyrate with *N*-lithio-*N*-aryl carbamates generated by the deprotonation of aryl carbamates with *n*-butyllithium at –78 °C. The *N*-aryl-5-hydroxymethyl oxazolidinone is directly obtained from this reaction, by virtue of the *in situ* transesterification of the oxazolidinone butyrate ester with the lithium alkoxide generated in the cyclization. This transesterification equilibrium between the two esters and alkoxides is driven toward the desired pathway as a consequence of the precipitation of the lithium salt **11** of the 5-(hydroxymethyl) oxazolidinone (Scheme 2). In the case of compound **3** derived from *O*-methyl *N*-phenyl carbamate, a very clean product (with purity assessed by an acceptable combustion analysis) is obtained simply by evaporative removal of the solvent following extractive aqueous workup, which conveniently removes the by-product methyl butyrate. When the *O*-benzyl *N*-phenyl carbamate is employed, **3** is obtained in high purity by taking the crude product derived from aqueous workup and triturating with ethyl acetate–hexane (1:1), which removes the benzyl butyrate. In other examples, it is usually necessary to chromatograph the product to attain analytical purity. Table 1 lists examples of *N*-aryl-(5*R*)-hydroxymethyl-2-oxazolidinones prepared in high yield and purity from the respective benzyl carbamate.

We have found in our studies that the use of the lithium counter ion in the base is essential for successful reaction with regiochemical control and allows cyclization to proceed under mild thermal conditions. In contrast, use of sodium (NaH, $NaN(SiMe_3)_2$) or potassium (KH, $KN(SiMe_3)_2$) bases require elevated temperatures, and results in poor yields of the desired product, and a mixture of several by-products, including the regioisomeric 4-hydroxymethy-2-oxazolidinone,[6] resulting from alternate processes. Thus, the lithium ion plays a very important role in the mechanism of this reaction.

Scheme 2

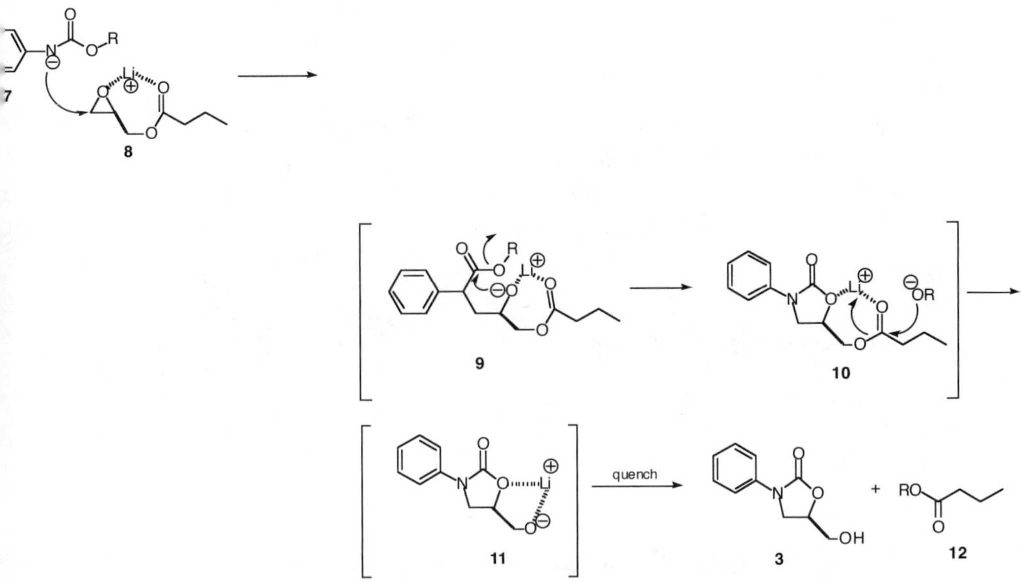

The preparation reported here represents a very general method[7] for the asymmetric synthesis of *N*-aryl-(5*R*)-hydroxymethyl-2-oxazolidinones with several distinct advantages over the previous method: (1) the oxazolidinone ring can be formed from readily available *N*-aryl carbamates, and avoids the hazards and limitations associated with the preparation and isolation of aryl isocyanates. This method significantly broadens the scope of aryl or heteroaryl substitution found in the oxazolidinone. The requisite carbamates are easily prepared from substituted anilines or heteroaryl amines and alkoxy carbonyl chlorides.[5] (2) The desired 5-(hydroxymethyl)oxazolidinone is directly obtained without need for a subsequent saponification step. (3) The reaction conditions are very mild, high-yielding, and provide the 5-(hydroxymethyl)oxazolidinone product in high enantiomeric excess, often with simple work up conditions.

Table 1.

Carbamate	Product	Yield	Ref.
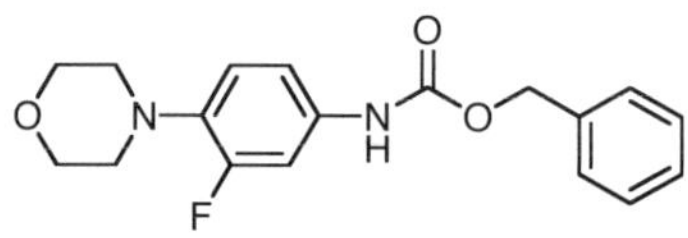	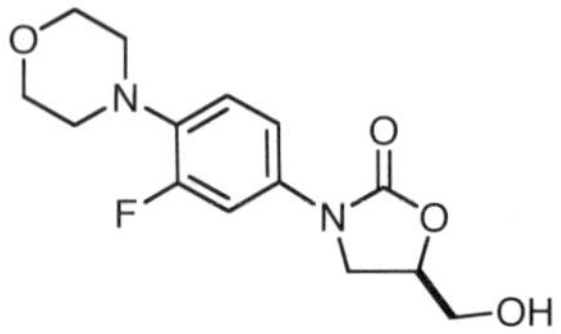	83%	5
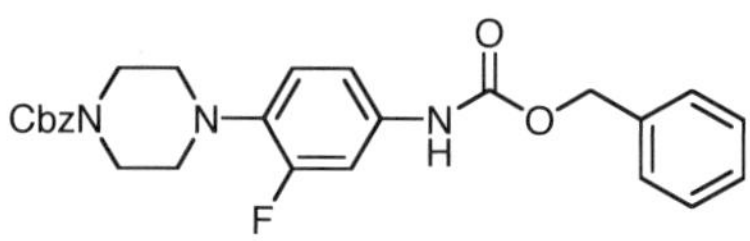	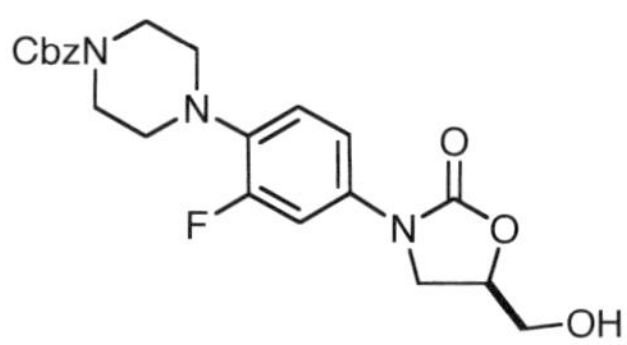	85%	5

1. This work was carried out in Medicinal Chemistry Research, Pharmacia, Kalamazoo MI 49001. Author's present address: Eli Lilly Corporation, Lilly Corporate Center, Drop Code 1523, Indianapolis, IN 46285.
2. Co-author's present address: Pfizer Global Research & Development, Groton, CT 06340.
3. Gregory, W. A.; Brittelli, D. R.; Wang, C. L. J.; Wounola, M. A.; McRipley, R. J.; Eustice D. C.; Eberly, V. S.; Bartholomew, P. T.; Slee, A. M.; Forbes, M. *J. Med. Chem.* **1989**, *32*, 1673.
4. Herweh, J. E.; Kauffman, W. J.; *Tetrahedron Lett.* **1971**, *12*, 809.
5. Brickner, S. J.; Hutchinson, D. K.; Barbachyn, M. R.; Manninen P. R.; Ulanowicz, D. A.; Garmon, S. A.; Grega, K. C.; Hendges, S. K.; Toops, D. S.; Ford. C. W.; Zurenko, G. E. *J. Med. Chem.* **1996**, *39*, 673.
6. Manninen, P. R.; Little, H. A.; Brickner, S. J.; Division of Organic Chemistry, American Chemical Society, 212th ACS National Meeting, Orlando, FL, August, 1996.
7. Citations that have used this method: Barbachyn, M. R.; Hutchinson, D. K.; Brickner, S. J.; Cynamon, M. H.; Kilburn, J. O.; Klemens, S. P.; Glickman, E. S.; Grega, K. C.; Hendges, S. K.; Toops, D. S.; Ford. C. W.; Zurenko, G. E. *J. Med. Chem.* **1996**, *39*, 680.; Genin, M. J.; Allwine, D. A.; Anderson, D. J.; Barbachyn, M. R.; Emmert, D. E.; Garmon, S. A.; Graber, D. R.; Grega, K. C.; Hester, J. B.; Hutchinson, D. K.; Morris, J.; Reischer, R. J.; Ford. C. W.; Zurenko, G. E.; Hamel, J. C.; Schaadt, R. D.; Stapert, D.; Yagi, B. H. *J. Med. Chem.* **2000**, *43*, 953.; Tucker, J. A.; Allwine, D. A.;

Grega, K. C.; Barbachyn, M. R.; Klock, J. L.; Adamski, J. L.; Brickner, S. J.; Hutchinson, D. K.; Ford. C. W.; Zurenko, G. E.; Conradi, R. A.; Burton, P. S.; Jensen, R. M. *J. Med. Chem.* **1998**, *41*, 3727.

PREPARATION OF PIVALOYL HYDRAZIDE IN WATER
(Propanoic acid, 2,2-dimethyl-, hydrazide)

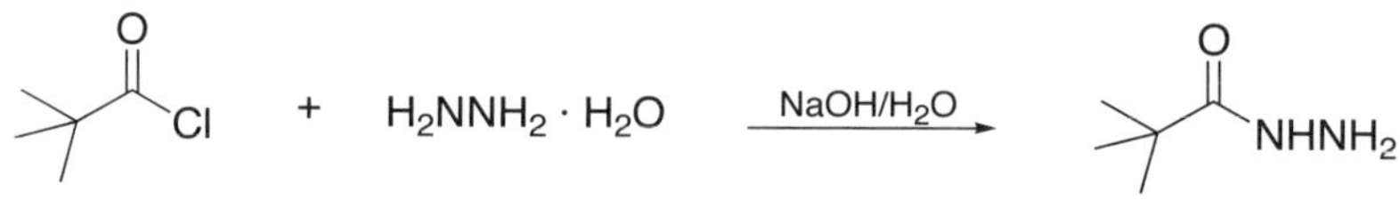

Submitted by Bryan Li,[1] Raymond J. Bemish, David R. Bill, Steven Brenek, Richard A. Buzon, Charles K-F Chiu and Lisa Newell.
Checked by Claire Coleman and Peter A. Wipf.

1. Procedure

Pivaloyl hydrazide. A 1-L, three-necked, round-bottomed flask equipped with a Teflon-coated thermocouple and mechanical stirrer is charged with 400 mL of water and sodium hydroxide (12.87 g, 322 mmol), and the resulting mixture is stirred until all of the solids dissolve (Note 1). Hydrazine (35% aqueous solution, 36.83 g, 400 mmol) is then added in one portion. The mixture is cooled in an ice-water/acetone bath to an internal temperature of –5 to 0 °C, and trimethylacetyl chloride (38.6 mL, 320 mmol) is added dropwise (Note 2) over a period of 40-60 min while maintaining the reaction temperature between –5 and 0 °C (Note 3). The reaction mixture is then transferred to a 1-L, pear-shaped flask and concentrated to a volume of ca. 100 mL by rotary evaporation (at ca. 100 mm) (Notes 4, 5) and the resulting suspension is filtered (Note 6). The filtrate is further concentrated to a volume of ca. 40 mL (Note 7) and 100 mL of toluene is then added. The resulting solution is transferred to a three-necked, round-bottomed flask equipped with a Dean-Stark water separator, thermometer, and rubber septum. Distillation is continued at atmospheric pressure until a constant boiling point is reached (Note 8) and then the pot volume is further reduced to ca. 40 mL. The resulting heterogenous mixture is filtered (Note 9) and the filtrate is concentrated by rotary evaporation under reduced pressure to provide the desired product as a colorless oil, which on standing solidifies to a white semi-solid. This material is recrystallized from 100 mL of isopropyl ether to afford 18.6-20.4 g (50-55%) of pivaloyl hydrazide (Notes 10, 11).

2. Notes

1. All reagents were purchased from Aldrich Chemical Company (except for trimethylacetyl chloride, which the checkers obtained from Acros) and were used without further purification. The checkers used a low temperature alcohol thermometer in place of a Teflon-coated themocouple. The third neck of the flask was left open to the atmosphere.

2. A syringe pump was used for the addition of acid chloride in order to achieve a steady flow rate. The tip of the syringe needle (gauge 20) was submerged in the reaction mixture. Dropwise addition of trimethylacetyl chloride at 0-5 °C resulted in the immediate formation of a precipitate.

3. The reaction was complete at the end of the pivaloyl chloride addition. On 5-L or larger scale, the reaction was conducted at temperatures of 10-15 °C without loss of selectivity.

4. A small amount of hydrazine hydrate was present in the reaction mixture at this point, but a safety evaluation indicated the final reaction mixture had a very low thermal potential (DH=15.3 J/g). This poses a minimum thermal hazard for vacuum distillation.

5. The submitters concentrated the reaction mixture by vacuum distillation (100 mm, bath temperature 70 °C, vapor temperature 51 °C). The weight after concentration was ca. 120 g. The checkers used rotary evaporation with a bath temperature of 65 to 70 °C without any problems, and employed an explosion shield as a safety precaution.

6. The bis-acylation byproduct ($Me_3CCONHNHCOCMe_3$) was removed by filtration; 20 mL of water was used for washing the filter cake.

7. The submitters removed solvent by vacuum distillation (100 mm, bath temperature 70 °C, vapor temperature 51 °C).

8. Azeotropic removal of water was complete when the vapor temperature reached 111 °C.

9. Sodium chloride was removed by filtration.

10. The submitters obtained the product in 72% yield without recrystallization and determined the product to be >97% pure by HPLC (by area; conditions: 250 mm Kromasil C4 column using acetonitrile (A)/water (B) and 0.1% TFA in water (C), 0:90:10 A:B:C ramp to 90:0:10 A:B:C over 15 min and hold for 5 min. Flow rate 1 mL/min and detection wavelength 210 nm. The checkers found the purity prior to recrystallization to be typically 85-90% with the impurity determined to be the bis-acylation product by LC-MS. The checkers employed C8 microsorb-MW 100 (250

mm) and Alltech C18 (100 mm) columns using acetonitrile (A)/water (B) and 0.1% TFA in water (C), 0:90:10 A:B:C ramp to 90:0:10 A:B:C over 15 min and hold for 5 min. Flow rate 1 mL/min and detection wavelength 210 nm.

11. The product exhibits the following physical properties: mp 67.1-69.2 °C; IR (NaCl) cm^{-1} 3471, 3327, 2968, 1660; ^1H NMR ($CDCl_3$, 400 MHz) δ: 1.19 (s, 9H), 4.40-4.85 (br, 3 H); ^{13}C NMR ($CDCl_3$, 75 MHz) δ: 27.4, 38.1, 179.3; MS (*m/z*): 117 ([M + 1] $^+$). Anal. Calcd for $C_5H_{12}N_2O$: C, 51.70; H, 10.41; N, 24.12. Found C, 51.62; H, 10.78; N, 24.22.

Waste Disposal Information

All toxic materials were disposed of in accordance with "Prudent Practice in the Laboratory"; National Academy Press; Washington, DC, 1995.

3. Discussion

Hydrazides ($RCONHNH_2$) are highly useful starting materials and intermediates in the synthesis of heterocyclic molecules.[2] They can be synthesized by hydrazinolysis of amides, esters and thioesters.[3] The reaction of hydrazine with acyl chlorides or anhydrides is also well known,[4] but it is complicated by the formation of 1,2-diacylhydrazines, and often requires the use of anhydrous hydrazine which presents a high thermal hazard. Diacylation products predominate when hydrazine reacts with low molecular weight aliphatic acyl chlorides, which makes the reaction impractical for preparatory purposes.[5]

Recently we needed to prepare large amounts of pivaloyl hydrazide (**1**). A literature survey indicated several approaches: (1) heating pivalic acid with hydrazine hydrate with a Lewis acid catalyst such as activated alumina[6] or titanium oxide;[7] (2) heating hydrazine hydrate at high temperature (140 °C) with ethyl pivalate;[8] (3) condensing phthaloyl hydrazine with pivaloyl chloride, followed by deprotection of the phthaloyl group;[9] and (4) reaction of ethyl thiopivalate with hydrazine hydrate. Reaction safety evaluations revealed that hydrazine monohydrate has an onset temperature of *ca.* 125 °C in a Differential Scanning Calorimetry (DSC) experiment, and possesses a very high thermal potential (ΔH = 2500 J/g),[10,11] which prompted us to develop a method for the synthesis of **1** that did not require heating. After some experimentation we determined that the reaction of pivaloyl chloride with hydrazine proceeds most efficiently in water to give a 4:1 ratio[12] of **1** to $Me_3CCONHNHCOCMe_3$ (**2**). The use of organic solvents (MeOH, THF, 2-propanol) with water[13] invariably led to formation of biphasic mixtures and predominant formation of **2**.[14] Reaction workup is also simplified using water as solvent. Upon partial concentration the bis-acylhydrazide by-product **2** precipitated out of the reaction mixture and is conveniently removed by filtration. Removal of the remainder of the water by displacement with toluene leads to precipitation of NaCl, which is also easily removed by filtration. The filtrate is then further concentrated to provide **1** in >97% purity, typically in 55–75% yield. This procedure has been employed to prepare 10 Kg batches of **1** with no difficulty.

The protocol is effective in preparation of hydrazides of 5 carbons or less. Cyclopropanecarboxylic acid hydrazide[15] and isobutyric acid hydrazide[16] were prepared from their corresponding acid chlorides in 64% and 71% yields, respectively. However, when this method was applied to cyclohexanecarboxylic acid chloride, the bis-acylhydrazide was the predominant product, and the mono-acylhydrazide[17] was isolated in 25% yield.[18]

1. Chemical Research and Development, Pfizer Global Research and Development, Groton Laboratories, Groton, CT 06340, USA
2. (a) Monge, A.; Aldana, I.; Lezamiz, I.; Fernandez-Alvarez, E. *Synthesis*, **1984**, *2*, 160; (b) Rao, C. S.; Ramachandra, R.; Kshirsagar, S.; Vashi, D. M.; Murty, V. S. N. *Indian J. Chem. Sect. B,* **1983**, *22*, 230; (c) Essassi, E. M.; Fifani, J. *Bull. Soc. Chim. Belg.* **1987**, *96*, 63.
3. Edwards, L. H. US 4500539, US Appl. 83-514073.
4. Paul H.; Stoye, D. *Chap. 10 The Chemistry of Hydrazide, In The Chemistry of Amides*; Zabicky J.; Ed.; John Wiley & Sons, **1970**, 515.
5. Smith, P. A. S. *Org. React.* **1946**, *3*, 366.
6. Flowers, W. T.; Robinson, J. F.; Taylor, D. R.; Tipping, A. E. *J. Chem. Soc. Perkin Trans.* 1, **1981**, 349.
7. Verbrugge, P. A.; De Waal, J. *Eur. Pat. Appl.* **1995**, EP 653419.
8. (a) Struve, G. *J. Prakt. Chem.* **1894**, *50*, 295; (b) Quast, H.; Kees, F. *Chem. Ber.* **1981**, *114*, 802; (c) Wieland, H.; Dennstedt, J. *Justus Liebigs Ann. Chem.* **1927**, *16*, 452.

9. Brosse, N.; Pinto, M-F.; Jamart-Gregoire, B. *J. Chem. Soc., Perkin Trans. 1* **1998**, *22*, 3685.
10. Under nitrogen atmosphere, the thermal potential is slightly lower (DH = 2150 J/g), and the onset temperature is 150 °C.
11. For perspective, 1 g of hydrazine monohydrate is equivalent to ~0.5 g of 2,4,6-trinitrotoluene (TNT) in terms of thermal potential; T. Grewer, *Thermal Hazards of Chemical Reactions*; Elsevier:Amsterdam, **1994**, Vol. 4.
12. By integration of the ^1H NMR spectrum of the crude reaction product.
13. Water is still introduced from the use of hydrazine hydrate when no additional water is added.
14. Pivaloyl chloride and **1** are both preferentially soluble in the organic phase, which gives rise to **2** as the major product.
15. Szmuszkovicz, J. *Ger. Offen.* (**1972**), DE 2228044; CAN 78:72230.
16. Galons, H.; Cave, C.; Miocque, M.; Rinjard, P.; Tran, G.; Binet, P. *Eur. J. Med. Chem.* **1990**, *25*, 785.
17. McElvain, S. M.; Starn, R. E., Jr. *J. Am. Chem. Soc.* **1955**, *77*, 4571.
18. Cyclohexanecarboxylic acid hydrazide has low solubility in water. A biphasic mixture developed during the reaction, which led to the predominant formation of the bis-acylated material.

ASYMMETRIC SYNTHESIS OF *N-tert*-BUTOXYCARBONYL α-AMINO ACIDS. SYNTHESIS OF (5*S*,6*R*)-4-*tert*-BUTOXYCARBONYL-5,6-DIPHENYLMORPHOLIN-2-ONE [(4-Morpholinecarboxylic acid, 6-oxo-2,3-diphenyl-, 1,1-dimethylethyl ester, (2*S*,3*R*)-)]

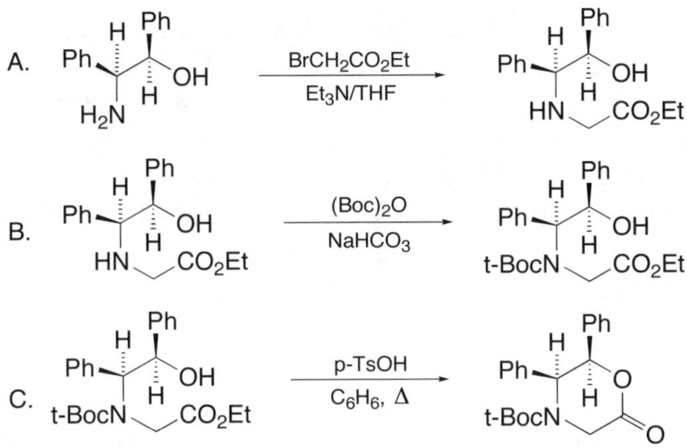

Submitted by Robert M. Williams, Peter J. Sinclair, Duane E. DeMong, Daimo Chen, and Dongguan Zhai.[1]
Checked by Wenlin Lee and Marvin J. Miller.

1. Procedure

A. Ethyl (1'S,2'R)-N-(1',2'-diphenyl-2'-hydroxyethyl)glycinate. A 1-L, three-necked, round-bottomed flask, equipped with a magnetic stirring bar, pressure-equalizing addition funnel, and two stoppers, is charged with 25 g (0.117 mol) of (1*S*,2*R*)-1,2-diphenyl-2-hydroxyethylamine (Note 1), 29 g (0.176 mol) of ethyl bromoacetate and 455 mL of anhydrous tetrahydrofuran (THF) (Note 2). To the stirred mixture, 24 g (0.234 mol) of triethylamine (Note 3) is added dropwise via the addition funnel. After stirring vigorously for 20 h at room temperature, the solids are removed by filtration and washed three times with THF (Note 4). The clear yellowish filtrate and washes are concentrated on a rotary evaporator (to remove solvent, excess triethylamine, and ethyl bromoacetate) to afford an off-white solid, which is collected with a Büchner funnel, then washed with water (Note 5) to dissolve residual salts. The residue is air-dried in the filter funnel with water aspirator vacuum. The crude product is dissolved in 117 mL of boiling absolute ethanol (EtOH) and stored at room temperature overnight. White crystals are collected by filtration, washed with ice-cold (0 °C) absolute EtOH carefully, and dried at 50 °C (Note 6). The above described procedure is done in duplicate, providing ethyl (1'*S*,2'*R*)-*N*-(1',2'-diphenyl-2'-hydroxyethyl)glycinate in yields of 29.01 g and 28.96 g, (average yield: 83%) (Notes 7, 8).

B. Ethyl (1'S,2'R)-N-tert-butyloxycarbonyl-N-(1',2'-diphenyl-2'-hydroxyethyl-glycinate.[3] A 1-L, three-necked, round-bottomed flask, equipped with a magnetic stirring bar, reflux condenser, and two stoppers, is charged with 167 mL of a solution of saturated aqueous sodium bicarbonate ($NaHCO_3$), 34.2 g (0.585 mol) of sodium chloride (NaCl), 350 mL of chloroform ($CHCl_3$) (Note 9), 25 g (0.084 mol) of ethyl (1'*S*,2'*R*)-*N*-(1',2'-diphenyl-2'-hydroxyethyl)glycinate, and 36.5 g (0.167 mol) of di-*tert*-butyl dicarbonate (Note 10). The resulting mixture is then heated to reflux for 20 h and vigorously stirred.

The reaction mixture is placed in a separatory funnel; the aqueous phase is separated and extracted twice with chloroform ($CHCl_3$). The organic phases are combined, washed twice with water, and dried with anhydrous sodium sulfate overnight. After filtration to remove sodium sulfate, concentration under vacuum affords a residue, which is heated (5 mm, 130 °C oil bath) to remove and recover excess di-*tert*-butyl dicarbonate (Note 11). A total of 33.2 g (83 mmol) of crude ethyl (1'*S*,2'*R*)-*N*-*tert*-butyloxycarbonyl-*N*-(1',2'-diphenyl-2'-hydroxyethyl)glycinate, is obtained as an oil and used directly in the subsequent lactonization reaction.

C. (5S,6R)-4-tert-Butoxycarbonyl-5,6-diphenylmorpholin-2-one. A 2-L, round-bottomed flask, equipped with a magnetic stirring bar, is charged with 33.2 g (0.083 mol) of crude ethyl (1'*S*,2'*R*)-*N*-*tert*-butyloxycarbonyl-*N*-(1',2'-diphenyl-2'-hydroxyethyl) glycinate, 1.90 g (0.010 mol) of *p*-toluenesulfonic acid monohydrate (Note 12), and 670 mL of benzene (Notes 13, 14). The reaction flask is fitted with a Soxhlet extractor, which is packed with 65 g of calcium chloride ($CaCl_2$, anhydrous, 30 mesh) and a condenser with a drying tube. The reaction mixture is heated to reflux with stirring for 10 h (Note 15). After the first 4 h, the $CaCl_2$ in the Soxhlet extractor is replaced; the total amount of $CaCl_2$ used in the reaction is 130 g. Ultimately, the reaction mixture becomes a suspension. Removal of benzene under reduced pressure (Note 16) yields white solids that are dissolved in 310 mL of dichloromethane (CH_2Cl_2). The solution is washed with water and 5% aqueous $NaHCO_3$ in a separatory funnel until the CH_2Cl_2 solution becomes clear. After removal of CH_2Cl_2 under vacuum, the resulting white crystals are dissolved by boiling in 600 mL of absolute EtOH; the resulting solution is allowed to stand at room temperature overnight. The white crystals are collected by filtration, washed with ice-cold EtOH, and dried at 60 °C. Concentration of the mother liquor by rotary evaporation affords a residue that can also be recrystallized in absolute EtOH to provide additional product. The above described procedure is performed in duplicate, providing (5*S*,6*R*)-4-*tert*-butyloxycarbonyl-5,6-diphenylmorpholin-2-one in yields of 21.7 g and 20.3 g (average yield for the two steps: 71%) (Notes 17-19).

2. Notes

1. (1*S*,2*R*)- and (1*R*,2*S*)-1,2-Diphenyl-2-hydroxyethylamine (>98% ee) were prepared according to the procedure of Tishler, et al.[2] The checkers obtained the amino alcohols from Aldrich Chemical Co., Inc.

2. Tetrahydrofuran was dried by distillation from sodium-benzophenone ketyl.

3. Triethylamine was dried by distillation from CaH_2.

4. Reagent grade or technical grade (i.e. undistilled) THF can be used for this wash.

5. Extensive washing in a Büchner funnel with water is recommended.

6. The product was dried in a crystallizing dish in a vacuum oven.

7. The physical and spectroscopic properties for ethyl (1'*S*,2'*R*)-*N*-(1',2'-hydroxyethyl)glycinate are as follows: mp. 127-128 °C, $[\alpha]^{25}_{D}$ +24.4 (c 5.5, CH_2Cl_2); IR (NaCl, $CDCl_3$) cm^{-1}: 3840-3430, 3330, 3080, 3045, 2995, 2940, 1750, 1460, 1385, 1210, 1035, 915, 740; ^1H NMR (270 MHz, $CDCl_3$) δ: 1.20 (3H, t, J = 7.1), 2.2 (2H, br s), 3.15 (1H, 1/2 AB q, J = 17.5), 3.29 (1H, 1/2 AB q, J = 17.5), 3.95 (1H, d, J = 6.0), 4.11 (2H, q, J = 7.1), 4.80 (1H, d, J = 6.0), 7.17-7.32 (10H, m); Anal. Calcd. for $C_{18}H_{21}NO_3$: C, 72.22; H, 7.07; N, 4.68. Found: C, 72.31; H, 7.16; N, 4.56.

8. From the antipode (1*R*,2*S*)-1,2-diphenyl-2-hydroxyethylamine is obtained the corresponding enantiomer ethyl (1'*R*,2'*S*)-*N*-(1',2'-diphenyl-2'-hydroxyethyl)-glycinate. mp 127-128 °C, $[\alpha]^{25}_{D}$ −24.4 (c 5.5, CH_2Cl_2). This series, also performed in duplicate, provides yields of 29.2 g and 29.5 g (average yield: 84%).

9. Technical grade chloroform was used.

10. Di-*tert*-butyl dicarbonate was purchased from Aldrich Chemical Co., Inc.

11. The recovered di-*tert*-butyl dicarbonate is very pure and can be reused.

12. *p*-Toluenesulfonic acid monohydrate was purchased from Aldrich Chemical Co., Inc.

13. ACS grade benzene was purchased from Fisher Scientific.

14. The checkers found that 770 mL of cyclohexane could also be used as the solvent for this step, avoiding the use of large amounts of toxic benzene. The crude Step B product, a white solid, very slowly dissolves in refluxing cyclohexane; upon conversion to the morpholinone, which is insoluble in cyclohexane, the product precipitates out of solution. The reaction mixture remains a suspension throughout the reaction, but the texture of the solids shows a subtle change over time, indicating the progress of the reaction. The color of the solution also turns slightly yellow. The reaction time is extended to 17 h and the $CaCl_2$ pellets in the Soxhlet extractor are changed after the first 5-6 h. Work up of the reaction is identical to that described in the procedure.

15. The Soxhlet extractor does not require the standard Soxhlet filter thimbles; cotton above and below the $CaCl_2$ pellets is sufficient.

16. *Caution! Care should be exercised in performing the reaction and concentrating benzene in a fume hood.*

17. The physical and spectroscopic properties for (5*S*,6*R*)-4-*tert*-butyloxycarbonyl-5,6-diphenylmorpholin-2-one are as follows: mp 207 °C; $[\alpha]^{25}_{D}$ +86.3 (c 5.6, CH_2Cl_2); IR (NaCl, CH_2Cl_2) cm^{-1}: 3050, 2975, 1755, 1690, 1380, 1255, 1150, 1100, 1045; ^1H NMR (200 MHz) (DMSO-d_6 vs. DMSO) (120 °C) δ: 1.25 (9H, s), 4.52 (2H, d, J = 1.1), 5.16 (1H, d, J = 3.0), 6.17 (1H, d, J = 3.0), 6.63-6.68 (2H, m), 7.0-7.3 (8H, m); Anal. calcd. for $C_{21}H_{23}NO_4$: C, 71.37; H, 6.56; N, 3.96. Found: C, 71.08; H, 6.44; N, 3.98. The checkers obtained 22.7 g, (77%) after the first recrystallization; mp 205-207 °C; $[\alpha]^{25}_{D}$ −86.0 (c 1.0, CH_2Cl_2).

18. From the antipode, ethyl (1'*R*,2'*S*)-*N*-(1',2'-diphenyl-2'-hydroxyethyl)-glycinate, is obtained the corresponding enantiomer, (5*R*,6*S*)-4-*tert*-butyloxycarbonyl-5,6-diphenylmorpholin-2-one, mp 207 °C, $[\alpha]^{25}_{D}$ +85.8 (c 5.7, CH_2Cl_2). This series, also performed in duplicate, provides yields of 24.0 g and 23.6 g, respectively (average yield for the two steps: 80%).

19. Optical rotations were measured on a Rudolf Research Autopol III automatic polarimeter operating at 589 nm, corresponding to the sodium D line.

Waste Disposal Information

All toxic materials were disposed of in accordance with "Prudent Practices in the Laboratory"; National Academy Press; Washington, DC, 1995.

3. Discussion

(+)- and (−)-4-*tert*-Butoxycarbonyl-5,6-diphenylmorpholin-2-one can be used for the asymmetric synthesis of α-amino acids.[3] The methylene carbon of the morpholinone can be efficiently brominated with NBS in CCl_4 to give a single diastereomeric bromide (anti). This bromo lactone serves as a useful electrophilic glycine synthon that undergoes stereoselective coupling with a variety of organometallic reagents. In general, coupling proceeds with net retention of stereochemistry furnishing the crystalline anti-homologation products. Dissolving metal reduction (lithium or sodium in liquid NH_3/EtOH/THF) directly furnishes the corresponding *N-tert*-butoxycarbonyl-protected amino acid derivative. Alternatively, the *N-t*-Boc group can be removed (TFA or HCO_2H) and catalytic hydrogenation (PdCl2/EtOH/THF, 25 °C, 40 psi) furnishes the free zwitterionic amino acid. In some cases, coupling has been observed to proceed with net inversion of stereochemistry to furnish syn-homologation products as oils. In cases where modest stereoselectivity is observed, the syn- and anti-isomers can be readily separated by chromatography. If the coupling is anti-selective (which is typical), a simple recrystallization will consistently give the final amino acid in >98% ee.

The lactones are quite stable to storage and handling (shelf life at room temperature >16 years). Treatment of these substances with either lithium, sodium or potassium hexamethyldisilazane in THF at −78 °C generates the corresponding enolate that can be stereoselectively alkylated with an appropriate electrophile to give anti-homologated products. Similar reductive processing of these materials provides amino acid derivatives in high enantiomeric excess.

The present method offers a versatile and practical protocol for preparing either D- or L- amino acids and offers the unique advantage of allowing direct access to the *N-t*-Boc derivatives. Conversion of the diphenyl amino alcohol moiety into the desired product and the water-insoluble bibenzyl, which is removed by simple extraction into ether or pentane, by either reductive protocol (1 and 2, shown below) is noteworthy. The water-soluble zwitterions, or in the case of the *N-t*-Boc amino acids, the *N-t*-Boc carboxylate salts, remain in the aqueous phase and are conveniently isolated in high chemical purity without, in most cases, the need for additional time-consuming and expensive chromatographic purification steps. With the oxidative methods (3 and 4), the diphenyl amino alcohol moiety is converted into two equivalents of benzaldehyde which can generally be separated from the amino acid by extraction. The ease and convenience of processing the derivatized morpolinones to the desired amino acids is testament to the flexibility and reliability of this methodology.

Method	**Conditions**	**Type of 'R' Group**	**Amino Acid**
1) Birch Reduction	(Li° or Na°, $NH_{3(l)}$)	Unsaturated; Saturated, Aliphatic	*N-t*-Boc
2) Catalytic hydrogenation (after removal of the Boc group)	(H_2 / Pd-C, EtOH)	Saturated, Aliphatic	Zwitterion
3) Oxidation	1) H^+; 2) $NaIO_4$	Aromatic; Saturated, Aliphatic	Zwitterion
4) Oxidation	1) H^+; 2) $Pb(OAc)_4$	Aromatic β-Arylcyclopropyl Saturated, Aliphatic	Methyl ester

As illustrated in the accompanying procedure, (*N-t*-Boc)allylglycine, this morpholinone has proven to be one of the most versatile and useful chiral glycine equivalents from which a large variety of amino acids can be prepared. Several other, recent asymmetric electrophilic glycine enolate and cation equivalents have been devised.[4]

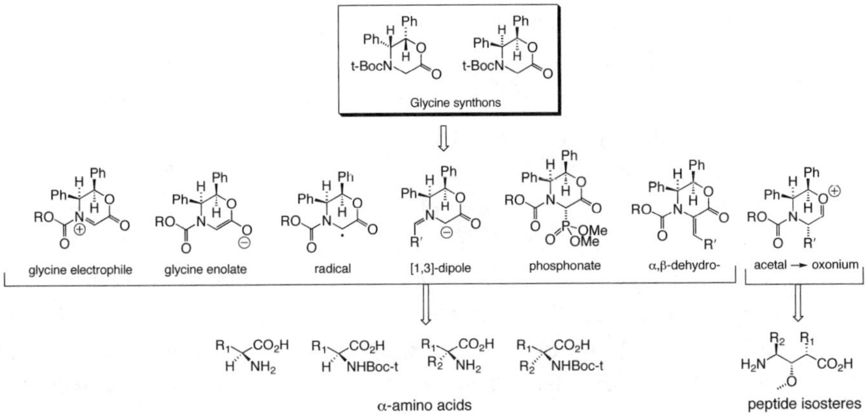

As shown in the scheme above, no other chiral glycine equivalent (from the laboratories of Seebach, Oppolzer, Evans, Myers, Schöllkopf, etc.) can provide the range of reactivity of this substance.

The chemistry of these compounds was reviewed in 1992[3j] and again in 1995[3q] and there are numerous publications and patent citations in the literature from other laboratories that have employed these synthons; a few examples are cited.[5,6,7,8,9,10,11,12,13,14,15,16,17] Finally, the *N-t*-Boc-protected morpholinones have been used extensively in our laboratory to prepare a variety of amino acids and peptide isosteres as shown below.[3] These examples serve to illustrate the great versatility that these glycine derivatives offer for preparing structurally diverse amino acids in high optical purity.

1. Department of Chemistry, Colorado State University, Fort Collins, Colorado 80523. This material is based upon work supported by the National Science Foundation under Grant No. 0202827.
2. (a) Sinclair, P. J.; Zhai, D.; Reibenspies, J.; Williams, R. M. *J. Am. Chem. Soc.* **1986**, *108*, 1103; (b) Williams, R. M.; Zhai, D.; Sinclair, P. J. *J. Org. Chem.* **1986**, *51*, 5021; (c) Williams, R. M.; Sinclair, P. J.; Zhai, W. *J. Am. Chem. Soc.* **1988**, *110*, 482; (d) Williams, R. M.; Im, M.-N. *Tetrahedron Lett.* **1988**, *29*, 6075 (e) Williams, R. M.; Sinclair, P. J.; Zhai, D.; Chen, D. *J. Am. Chem. Soc.* **1988**, *110*, 1547; (e) Williams, R. M.; Zhai, W. *Tetrahedron* **1988**, *44*, 5425; (f) Williams, R. M.; Hendrix, J. A. *J. Org. Chem.* **1990**, *55*, 3723; (g) Williams, R. M.; Im, M.-N.; Cao, J. *J. Am. Chem. Soc.* **1991**, *113*, 6976; (h) Williams, R. M.; Fegley, G. J. *J. Am. Chem. Soc.* **1991**, *113*, 8796; (i) Williams, R. M.; Im, M.-N. *J. Am. Chem. Soc.* **1991**, *113*, 9276; (j) Williams, R. M. *Aldrichimica Acta* **1992**, *25*, 11; (k) Williams,

R. M.; Yuan, C. *J. Org. Chem.* **1992**, *57*, 6519; (l) Williams, R. M.; Fegley, G. J. *Tetrahedron Lett.* **1992**, *33*, 6755; (m) Williams, R. M.; Zhai, W.; Aldous, D. J.; Aldous, S. C. *J. Org. Chem.* **1992**, *57*, 6527; (n) Williams, R. M.; Fegley, G. J. *J. Org. Chem.* **1993**, *58*, 6933; (o) Williams, R. M.; Yuan, C. *J. Org. Chem.* **1994**, *59*, 6190; (p) Williams, R. M.; Colson, P. J.; Zhai, W. *Tetrahedron Lett.* **1994**, *35*, 9371; (q) Williams, R. M., *Advances in Asymmetric Synthesis*, JAI Press, A. Hassner, Ed. **1995**, Vol. 1, p. 45; (r) Williams, R. M.; Fegley, G. J.; Gallegos, R.; Schaeffer, F.; Pruess, D. L. *Tetrahedron* **1996**, *52*, 1149; (s) Bender, D. M.; Williams, R. M. *J. Org. Chem.* **1997**, *62*, 6690; (t) Williams, R. M.; Liu, J. *J. Org. Chem.* **1998**, *63*, 2130; (u) Williams, R. M.; Aoyagi, Y. *Tetrahedron* **1998**, *54*, 10419; (v) Scott, J. D.; Tippie, T. N.; Williams, R. M. *Tetrahedron Lett.* **1998**, *39*, 3659; (w) Aoyagi, Y.; Williams, R.M. *Synlett* **1998**, 1099; (x) Aoyagi, Y.; Williams, R. M. *Tetrahedron* **1998**, *54*, 13045; (y) Williams, R. M, *Peptidomimetics Protocols*, Kazmierski, W., Ed., *Methods in Molecular Medicine* **1999**, Vol. 23, Chapter 19, p. 339, Humana Press; (z) Sebahar, P. R.; Williams, R. M. *J. Am. Chem. Soc.* **2000**, *122*, 5666.

3. Weijlard, J.; Pfister, K.; Swanezy, E. F.; Robinson, C. A.; Tishler, M. *J. Am. Chem. Soc.* **1951**, *73*, 1216.

4. For reviews, see: (a) Synthesis of Optically Active α-Amino Acids, Robert M. Williams, (*Organic Chemistry Series*, J. E. Baldwin, Series Editor); **1989**, 410, Pergamon Press, New York; (b) Duthaler, R. O. *Tetrahedron* **1994**, *50*, 1539; (c) O'Donnell, M. J., Ed., *Tetrahedron Symposia in Print Tetrahedron* **1988**, *44*, 5253; (d) Cativiela, C.; Diaz-de-Villegas, M. D. *Tetrahedron: Asymm.* **1998**, *9*, 3517; (e) Cativiela, C.; Diaz-de-Villegas, M. D. *Tetrahedron: Asymm.* **2000**, *11*, 645.

5. Ramer, S. E.; Cheng, H.; Vederas, J. C. *Pure Appl. Chem.* **1989**, *61*, 489.

6. Reno, D. S.; Lotz, B. T.; Miller, M. J. *Tetrahedron Lett.* **1990**, *31*, 827.

7. Dudycz, L. W. *Nucleosides and Nucleotides* **1991**, *10*, 329.

8. Dong, Z. *Tetrahedron Lett.* **1992**, *33*, 7725.

9. Baldwin, J. E.; Lee, V.; Schofield, C. J. *Heterocycles* **1992**, *34*, 903.

10. Baldwin, J. E.; Lee, V.; Schofield, C. J. *Synlett* **1992**, 249.

11. Schow, S. R.; DeJoy, S. Q.; Wick, M. M.; Kerwar, S. S. *J. Org. Chem.* **1994**, *59*, 6850.

12. Maggini, M.; Scorrano, G.; Bianco, A.; Toniolo, C.; Sijbesma, R. P.; Wudl, F.; Prato, M. J. *Chem. Soc. Chem. Comm.* **1994**, 305.

13. Schuerman, M. A.; Keverline, K. I.; Hiskey, R. G. *Tetrahedron Lett.* **1995**, *36*, 825.

14. Solas, D.; Hale, R. L.; Patel, D. V. *J. Org. Chem.* **1996**, *61*, 1537.

15. Hutton, C. A.; White, J. M. *Tetrahedron Lett.* **1997**, *38*, 1643.

16. Lee, S-H.; Nam, S-W. Bull. *Korean Chem. Soc.* **1998**, *19*, 613.

17. van den Nieuwendijk, A. M. C. H.; Benningshof, J. C. J.; Wegmann, V.; Bank, R. A.; te Koppele, J. M.; Brussee, J.; van der Gen, A. *Bioorg. Med. Chem. Lett.* **1999**, *9*, 1673.

SYNTHESIS OF (*R,R*)-4,6-DIBENZOFURANDIYL-2,2'-BIS (4-PHENYLOXAZOLINE) (DBFOX/PH) – A NOVEL TRIDENTATE LIGAND

[[Oxazole, 2,2'-(4,6-dibenzofurandiyl)bis(4,5-dihydro-4-phenyl-, (4*R*,4'*R*)-]]

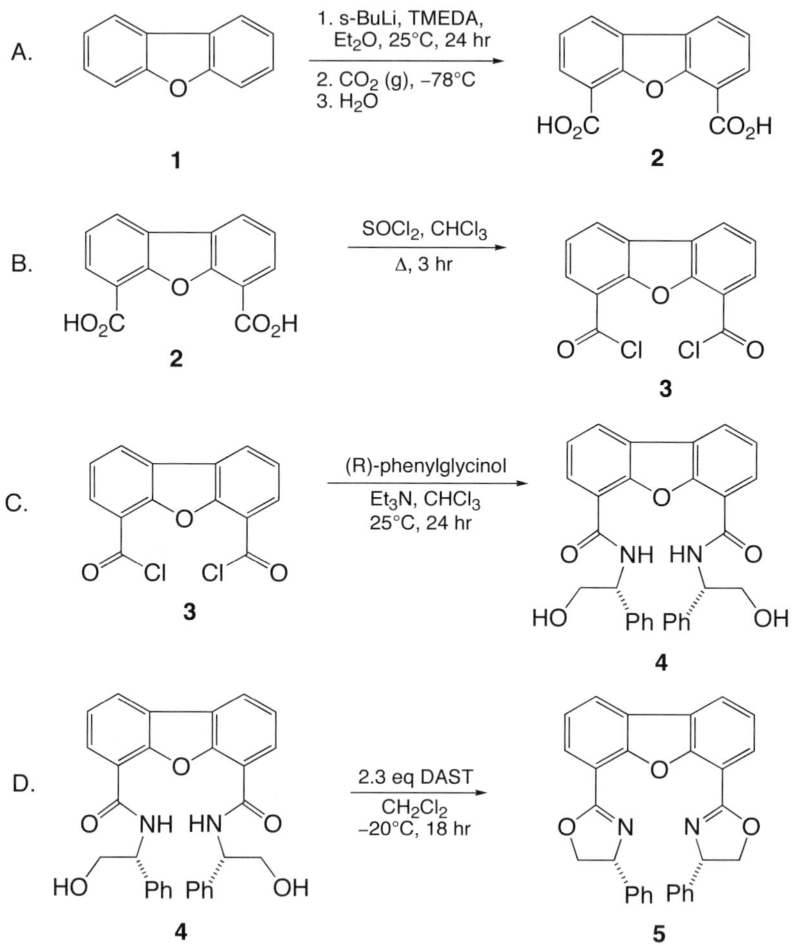

Submitted by Ulrich Iserloh, Yoji Oderaotoshi, Shuji Kanemasa, and Dennis P. Curran.[1]
Checked by Richard S. Gordon and Andrew B. Holmes.

1. Procedure

*A. Dibenzofuran-4,6-dicarboxylic acid (**2**).*[2] A 1-L, three-necked, round-bottomed flask (Note 1) is fitted with a large magnetic stirbar. Under argon, the flask is charged with 350 mL of dry, freshly distilled diethyl ether, 10.0 g (59.4 mmol) of dibenzofuran (Notes 2, 3) and 27 mL (179 mmol) of tetramethylethylenediamine (TMEDA). The mixture is cooled to −78 °C (bath temperature) with a dry ice/acetone bath, and stirred (Note 4) while 137 mL (178 mmol) of *sec*-butyllithium is added over 1 h via cannula and syringe (Note 5). After warming the reaction mixture to 25 °C, the suspension is stirred for 24 h. The mixture is cooled again to −78 °C under vigorous stirring and gaseous carbon dioxide (CO_2) (Note 6) is introduced over 1 h; the reaction mixture is then warmed to 25 °C over 4 h under a constant CO_2 stream (Note 7). After decanting the supernatant liquid, a brown solid is isolated by filtration using a Buchner funnel (Note 8). The residue is washed with 300 mL of diethyl ether and then suspended in 200 mL water, acidified with aqueous 2N HCl to pH 3, and stirred for 1 h. After filtration, the beige solids are washed first with 400 mL of water and then with 200 mL of diethyl ether. The solids are dried under vacuum over phosphorus pentoxide (P_2O_5) for three days to yield 14.5 g (96%) of the title compound **2** (Note 9).

B. Dibenzofuran-4,6-dicarbonyl chloride (**3**): A 250-mL, three-necked flask (Note 1), equipped with a magnetic stirbar, condenser, thermometer, and addition funnel, is charged with 5.0 g (19.6 mmol) of diacid **2** and 68 mL of chloroform ($CHCl_3$) (Note 10). At 25 °C, 44 mL (600 mmol) of thionyl chloride (Note 2) and 1 drop of dimethylformamide (DMF) are added, followed by heating the mixture at 68 °C for 3 h. After the initial suspension turns into a yellow solution (Note 11), the heating source is removed and the diacid chloride **3** precipitates as a white solid. After cooling the reaction mixture to 25 °C, the solid is collected via filtration using a Buchner funnel, washed with $CHCl_3$ (Note 12), and dried in a vacuum desiccator for 15 h to give 5.12 g (90%) of **3** as a white powder (Note 13).

C. (R,R)-Dibenzofuran-4,6-dicarboxylic acid bis(2-hydroxy-1-phenylethyl) amide (**4**): A 250-mL, three-necked flask (Note 1), equipped with a magnetic stirbar, is charged with 5.00 g (17.07 mmol) of diacid chloride **3** and 135 mL of freshly distilled $CHCl_3$. After cooling the suspension to 0 °C under nitrogen, a solution of 5.18 g (37.74 mmol) of (*R*)-phenylglycinol (Note 2) and 5.26 mL (37.74 mmol) of triethylamine (Et_3N) in 25 mL of $CHCl_3$ (Note 14) is added over 2 h via an addition funnel (Note

15). After the reaction mixture is stirred for 15 h at 25 °C, $CHCl_3$ (100 mL) and solid ammonium chloride (5 g) is added and the suspension is stirred for a further 0.5 h (Note 16). The suspension is filtered and the residue is washed with $CHCl_3$ (3 × 50 mL). The residue is put aside for suspension in tetrahydrofuran (THF). The combined filtrate and washings are evaporated to dryness under reduced pressure. The resulting residue is purified by two recrystallizations from ethyl acetate/hexane (ca. 9:1) to give the amide **4** (combined yield 5.2 g, 61.0%). The mother liquor from the recrystallizations is chromatographed using silica gel (eluting with ethyl acetate/hexanes initially in a 1:1 ratio and increasing to 4:1, then 1:0) to give more amide **4** (0.6 g, 7.0%). The original solid residue, obtained from the filtration, is suspended in THF (200 mL); the slurry is stirred using a magnetic stirbar for 30 min. The product is filtered. The THF extraction of this residue is repeated (Note 17) and the filtrates are combined. Removal of the solvent under reduced pressure afforded additional amide **4**, which was recrystallized as before (1.7 g, 20.0%). The overall yield of amide **4** is 5.9-7.5 g (11.9-14.9 mmol, 70-88%) (Note 18).

D. Caution! Diethylaminosulfur trifluoride is a powerful corrosive and reacts violently with water. Appropriate precautions should be taken whenever it is handled, since one of its decomposition products is hydrofluoric acid.

(R,R)-4,6-Dibenzofurandiyl-2,2'-bis(4-phenyloxazoline), DBFOX/Ph (**5**): A 100-mL, one-necked flask (Note 1) equipped with a magnetic stirbar is charged with 2.40 g (4.85 mmol) of **4** and 50 mL of dichloromethane (CH_2Cl_2). After the suspension is cooled (Note 19) to −20 °C, 1.49 mL (11.28 mmol) of diethylaminosulfur trifluoride (DAST, (Note 2)) is slowly added via syringe. After 18 h at −20 °C, 4 mL of aqueous 4N ammonium hydroxide (NH_4OH) is added and the resulting solution is stirred for 15 min at −20 °C. Subsequently, the cooling bath is removed and 10 mL of water is added. The orange aqueous phase is extracted with CH_2Cl_2 (3 × 15 mL) and the combined organic phases are dried over magnesium sulfate ($MgSO_4$) and concentrated under vacuum. The resulting material is chromatographed using silica gel (20% EtOAc/hexanes, (Note 20)) to yield 2.10 g (4.59 mmol, 94%) of **5** as an off-white foam. This foam can be recrystallized from EtOAc/hexanes to yield white crystals (Notes 21-23).

2. Notes

1. The checkers dried all glassware at 120 °C for at least 2 h, and cooled under argon where appropriate. The submitters dried all glassware at 140 °C for at least 4 h and cooled to 25 °C under a flow of argon. Solvents were distilled from the indicated drying agent and were used immediately: dichloromethane (CaH_2), diethyl ether and tetrahydrofuran (sodium/benzophenone), chloroform (potassium carbonate), and triethylamine (KOH).

2. Dibenzofuran (99%), sec-butyllithium (1.3 M in cyclohexane), *N,N,N',N'*-tetramethylethylenediamine (99%), *N,N*-dimethylformamide (anhydrous, 99.8%), (*R*)-(−)-2-phenylglycinol (98%) and (diethylamino)sulfur trifluoride (DAST) were purchased from Aldrich Chemical Company, Inc. Thionyl chloride (99.5%) was purchased from Acros Organics and carbon dioxide (commercial grade) was purchased from Matheson Gas Products Inc. Except for dibenzofuran, all reagents were used as received.

3. Commercial dibenzofuran was not purified. The checkers used recrystallized and commercial dibenzofuran as received in separate experiments with no significant difference in yield.

4. The submitters used a large stirbar followed by a mechanical stirrer (upon introduction of CO_2) to ensure efficient stirring throughout. After initial experiments, the checkers carried out the whole procedure using a powerful magnetic stirrer (and a large magnetic stirbar) with no adverse effect on yield. The checkers observed no improvement in yield when a half-scale reaction was carried out using a mechanical stirrer.

5. To transfer 137 mL of sec-butyllithium, the contents of an entire 100-mL Sure/Seal™ bottle were transferred via cannula, while the remaining 37 mL were transferred via syringe from another bottle.

6. Carbon dioxide was dried by passing the gas through a gas-washing bottle (equipped with a fritted cylindrical gas-dispersion tube) containing concentrated sulfuric acid. For safety reasons, empty gas-washing bottles were placed before and after the H_2SO_4-containing bottle. The gas was introduced into the flask through a wide pipette (0.8 cm i. d.) which was submerged below the level of the reaction mixture. The checkers observed very little difference in yield irrespective of whether the inlet tube was above or below the surface of the reaction mixture.

7. The beige-colored suspension turned white upon introduction of carbon dioxide. The submitters adjusted the rate of addition such that the

internal temperature remained below −65 °C. At −25 °C, the white suspension turned red-brown. The checkers did not monitor the internal temperature of the reaction, but ensured that the CO_2 addition was slow and constant.

8. The yellow supernatant/filtrate comprises TMEDA in Et_2O, while the remaining solid is the lithiocarboxylate of diacid **2**. No precaution was taken during filtration to exclude air or moisture.

9. The submitters obtained 15.1 g (99%). Data for **2**: ^{1}H NMR (250 MHz, DMSO-d_6) δ 7.53 (dd, 2 H, J = 7.5 and 7.5), 8.04 (d, 2 H, J = 7.5), 8.44 (d, 2 H, J = 7.5), 13.30 (bs, 2 H); ^{13}C NMR (100 MHz, DMSO-d_6) δ 116.9, 123.7, 124.9, 126.2, 130.4, 154.6, 165.7; IR (solid phase) cm^{-1}: 2858, 1686, 1431, 1305, 1288, 1154; MS (EI) *m/z* 256 (M^+, 15), 181 (15), 69 (100); HRMS (ES, M + Na^+) *m/z* Calcd for $C_{14}H_8O_5Na$ 279.0264. Found 279.0269, Anal. Calcd for $C_{14}H_8O_5$: C, 65.6; H, 3.1. Found: C, 65.1; H, 3.2. The submitters reported mp 324-325 °C. The checkers could not obtain a mp for this material, but observed decomposition at temperatures > 320 °C (lit.[4] > 300 °C).

10. The checkers carried out the reaction with undistilled chloroform, with no difference in yields.

11. Although the solution is yellow, the reaction mixture remained cloudy throughout.

12. It is important not to use too much chloroform because the product is sparingly soluble.

13. The submitters obtained 5.0 g (87%). Data for **3**: mp 242-244 °C, ^{1}H NMR (400 MHz, $CDCl_3$) δ 7.59 (dd, 2 H, J = 5 and 5), 8.31 (dd, 2 H, J = 5 and 1), 8.33 (dd, 2 H, J = 5 and 1); ^{13}C NMR (100 MHz, DMSO-d_6) δ 119.3, 123.9, 124.8, 127.7, 132.6, 163.4, 168.2; IR (solid phase) cm^{-1}: 1757, 1622, 1587, 1474, 1425, 1236, 1184, 1142; MS (EI) *m/z* 292 (M^+, 40), 257 (40), 169 (20), 119 (20); HRMS (EI, M^+) *m/z* Calcd for $C_{14}H_6Cl_2O_3$ 291.9694. Found: 291.9699; Anal. Calcd for $C_{14}H_6Cl_2O_3$: C, 57.5; H, 2.1. Found: C, 56.7; H, 2.1.

14. The checkers had to carefully warm the contents in the addition funnel (using a heat gun) to completely dissolve the (*R*)-(−)-2-phenylglycinol. The contents of the funnel were allowed to cool before addition to the reaction mixture.

15. The submitters observed that slow addition is necessary to avoid formation of a thick gel-like slurry and noted that adding additional $CHCl_3$ is beneficial if the slurry[3] is too viscous to stir. The submitters observed (over the course of 15 h) the formation of a thick slurry which in turn became a

free flowing white suspension. The checkers observed the immediate formation of a clear yellow solution, which became cloudy, and sometimes a thick slurry, during the course of the reaction.

16. The product is partially soluble in $CHCl_3$ and additional $CHCl_3$ is added to recover as much product as possible. However, significant product still remains undissolved.

17. The submitters omitted the extraction of the first residue (which consists of triethylamine hydrochloride, excess ammonium chloride (NH_4Cl) and amide **4**) with THF, but the checkers found this essential in order to obtain the higher yield (88%) reported. To recover all amide **4**, it is necessary to repeat the procedure, because the product requires extended periods to dissolve completely. As an alternative work up procedure, the checkers removed the $CHCl_3$ under reduced pressure and stirred the reaction mixture in THF (250 mL containing ca. 5 g of NH_4Cl) for 1 h. After filtration of the suspension, recovery of **4** was poor (3.6 g from two recrystallizations).

18. The submitters obtained 7.4 g (87%) of **4**. Data for **4**: mp 185-186 °C; ^{1}H NMR (250 MHz, DMSO-d_6) δ 3.66-3.81 (m, 4 H, C<u>H</u>$_2$OH), 4.98 (t, 2 H, *J* = 5, CH_2O<u>H</u>), 5.20 (dd, 2 H, *J* = 7 and 14, C<u>H</u>Ph), 7.16-7.27 (m, 6 H), 7.42-7.60 (m, 6 H), 8.00 (dd, 2 H, *J* = 1 and 8), 8.36 (dd, 2 H, *J* = 1 and 8), 8.81 (d, 2 H, *J* = 8, PhCHN<u>H</u>); ^{13}C NMR (63 MHz, DMSO-d_6) δ 56.5, 65.3, 120.1, 123.9, 124.4, 124.7, 127.4, 127.6, 128.2, 128.7, 141.3, 153.1, 163.8; IR (solid phase) cm^{-1}: 3365, 3278, 1636, 1571, 1530, 1339, 1290, 1190, 1029; MS (ES) *m/z* 517 (M+Na$^+$, 40), 495 (M+H, 100), 375 (50), 255 (25); HRMS (ES, M + Na$^+$) *m/z* Calcd for $C_{30}H_{26}N_2O_5Na$: 517.1739. Found: 517.1732; Anal. Calcd for $C_{30}H_{26}N_2O_5$: C, 72.9; H, 5.3; N, 6.0. Found: C, 72.5; H, 5.4; N, 5.7.

19. The checkers used an electrical refrigeration bath to maintain this reaction temperature.

20. The silica gel is stirred for 30 min in 20% EtOAc/hexanes containing 5% (v/v) Et_3N prior to packing the column. Before applying the crude material, the column was equilibrated with 20% EtOAc/hexanes to remove excess Et_3N.

21. Data for recrystallized **5**: R_f (**5**) = 0.19 (40% EtOAc/hexanes), mp 126-127 °C (submitters observed mp 125.0-126 °C, lit.[4] 134.0-135 °C), $[\alpha]_D^{23}$ +46.92 (c 0.97, $CHCl_3$) [submitters observed $[\alpha]_D^{23}$ +50.75 (c 1.07, $CHCl_3$)]; [lit.[4] +47.75 (c 1.07 $CHCl_3$)], ^{1}H NMR (250 MHz, $CDCl_3$) δ 4.38 (br t, 2 H, *J* = 8), 4.96 (dd, 2 H, *J* = 10 and 10), 5.54 (dd, 2 H, *J* = 9.5 and 9.5), 7.23-7.45 (m, 12 H), 8.10-8.19 (m, 4 H); ^{13}C NMR (63 MHz, $CDCl_3$) δ 69.9,

74.9, 113.1, 123.2, 124.0, 124.9, 126.9, 127.6, 128.8, 142.4, 154.4, 162.4; IR (solid phase) cm^{-1}: 1642, 1424, 1412, 1283, 1184, 1120, 1072; MS (ES) *m/z* 459 (M+H, 95), 253 (10); HRMS (ES, M + H) *m/z* Calcd for $C_{30}H_{22}N_2O_3Na$: 459.1708. Found: 459.1721. Anal. Calcd for $C_{30}H_{22}N_2O_3$: C, 78.6; H, 4.8; N, 6.1. Found: C, 78.5; H, 4.8; N, 6.1.

22. The checkers also prepared (*S,S*)-**5** over two steps according to the same procedure from the reaction of **3** with (*S*)-(−)-2-phenylglycinol purchased from Aldrich Chemical Co., Inc. In addition to the spectroscopic properties of (*S,S*)-**5** which were identical to those reported above for the (*R,R*)-enantiomer, the checkers observed mp 126.0-127 °C, $[\alpha]_D^{23}$ −43.20 (c 1.00, $CHCl_3$).

23. The enantiomeric purities of (*R,R*)-**5** and (*S,S*)-**5** were measured separately by courtesy of Mr. Eric Hortense (GlaxoSmithKline, Stevenage). Chiral HPLC analysis [25 cm Chiracel OD-H, column no. ODHOCE-IF029, mobile phase ethanol/heptane 1:4 (v/v), UV detector at 215 nm, flow rate 1.0 mL/min at 25 °C] of (*R,R*)-**5** yielded a retention time of 7.4 min, while that of enantiomer (*S,S*)-**5** showed a retention time of 8.7 min. The enantiomeric ratio of (*R,R*)-**5** was in excess of 99.5:0.5, while that of (*S,S*)-**5** was also >99.5:0.5. The submitters determined the enantiomeric purity of (*R,R*)-5 as 99% ee by chiral HPLC on a Chiracel OD column (7% 2-propanol/hexanes, 1 mL/min, t(R) = 45 min, t(S) = 65 min).[4]

Waste Disposal Information

All toxic materials were disposed of in accordance with "Prudent Practices in the Laboratory"; National Academic Press; Washington, DC, 1995.

3. Discussion

(*R,R*)-4,6-Dibenzofurandiyl-2,2'-bis(4-phenyloxazoline), DBFOX/Ph (**5**), is a novel tridentate bisoxazoline ligand developed by Kanemasa and coworkers[5] that has been successfully used as a chiral Lewis acid in enantioselective Diels-Alder-reactions, nitrone cycloadditions and conjugate additions of radicals[6] and thiols[7] to 3-(2-alkenoyl)-2-oxazolidinones. Representative examples for cycloadditions using the $Ni(ClO_4)_2·6H_2O$ derived complex are shown below.

The procedure described here is a variant of the original preparation[5] that offers the following advantages:

(1) The overall yield of the four-step ligand preparation is increased from 28% to 71%. The improved dicarboxylation of 4,6-dilithiodibenzofuran significantly contributes to this yield increase. Originally, the dilithiated dibenzofuran was quenched with dry ice (solid CO_2), which presumably led to competitive protonation and ketone formation. In this procedure, dry CO_2 is introduced as a gas and gives a quantitative yield of dicarboxylated material **2**.

(2) The original protocol was not amenable for scale-up. The current procedure performs reliably on a 2-5 gram scale.

(3) All intermediates are purified by recrystallization. This provides for clean bis(2-hydroxy-1-phenylethyl) amide **4**, which is essential for the excellent yields obtained in the final cyclodehydration.

(4) The DAST-mediated cyclodehydration gives higher yields of DBFOX than the original base-mediated cyclization.

1. Department of Chemistry, University of Pittsburgh, Pittsburgh, PA 15260, e-mail: curran@pitt.edu.

2. We acknowledge Professor X. Zhang, The Pennsylvania State University, for helpful suggestions and we also thank Dr. Ana Gabarda for helping with Step A. The dilithiation of dibenzofuran has been previously reported: (a) Haenel, M. W.; Jakubik, D.; Rothenberger, E.; Schroth, G. *Chem. Ber.* **1991**, *124*, 1705; (b) Jean, F.; Melnyk, O.; Tartar, A. *Tetrahedron Lett.* **1995**, *36*, 765.

3. Kanemasa, S.; Oderaotoshi, Y.; Sakaguchi, S.; Yamamoto, H.; Tanaka, J.; Wada, E.; Curran, D. P. *J. Am. Chem. Soc.* **1998**, *120*, 3074.

4. Gel formation was also observed by other researchers: Takacs, J. M.; Quincy, D. A.; Shay, W.; Jones, B. E.; Ross, C. R. *Tetrahedron: Asymmetry* **1997**, *8*, 3079.

5. Kanemasa, S.; Oderaotoshi, Y.; Tanaka, J.; Wada, E. *J. Am. Chem. Soc.* **1998**, *120*, 12355.

6. Iserloh, U.; Curran, D. P.; Kanemasa, S. *Tetrahedron: Asymmetry* **1999**, *10*, 2417.

7. Kanemasa, S.; Oderaotoshi, Y.; Wada, E. *J. Am. Chem. Soc.* **1999**, *121*, 8675.

(4*S*)-4-(1-METHYLETHYL)-5,5-DIPHENYL-2-OXAZOLIDINONE [(2-Oxazolidinone, 4-(1-methylethyl)-5,5-diphenyl-, (4*S*)-)]

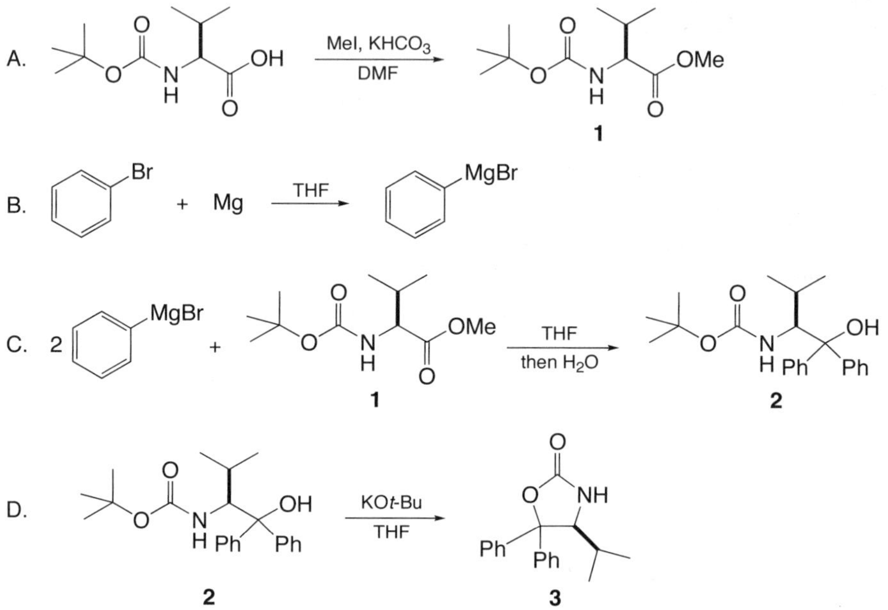

Submitted by Meinrad Brenner,[1] Luigi La Vecchia,[2] Thomas Leutert,[2] and Dieter Seebach.[1]
Checked by Mitsuru Kitamura and Koichi Narasaka.

1. Procedure

A. N-(tert-Butoxycarbonyl)-L-valine methyl ester (**1**) (Note 1). A 1-L, three-necked flask, fitted with a pressure-equalizing, 50-mL addition funnel, thermometer, and magnetic stirring bar is charged under an argon atmosphere (Note 2) with (*S*)-*N*-(*tert*-butoxycarbonyl)valine (65.2 g, 0.300 mol) (Note 3), dimethylformamide (DMF) (440 mL) (Note 4) and potassium hydrogen carbonate (60.1 g, 0.600 mol) (Note 5). Methyl iodide (29.9 mL, 0.480 mol) (Note 6) is added dropwise to this suspension over a period of 30 min. After the addition is complete, stirring is continued for 3 h. The reaction mixture is poured into a 2-L separatory funnel, diluted with water (1.2 L) and extracted with a 1:1 mixture of ethyl acetate and hexane (3 × 250 mL). The organic phases are combined and successively washed with water (2 × 250 mL), aqueous 5% sodium sulfite (2 × 250 mL), and brine (250 mL).

After drying over anhydrous magnesium sulfate, the solvent is removed and the oily residue is dried under vacuum (10 mbar) at room temperature. Crude **1** (69.4 g) is used in Step C without further purification.

B. Phenylmagnesium bromide. Under an argon atmosphere (Note 2) a 2-L, three-necked flask, fitted with a reflux condenser, pressure-equalizing, 500-mL addition funnel, thermometer, and magnetic stirring bar, is charged with magnesium turnings (25.5 g, 1.05 mol) and tetrahydrofuran (THF) (130 mL) (Note 7). Bromobenzene (15.7 g, 0.100 mol) (Note 8) is then added, and the reaction is initiated by warming with a heat gun. A solution of bromobenzene (149 g, 0.95 mol) in THF (500 mL) is added at such a rate that gentle refluxing is maintained. After the addition is complete (Note 9), the reaction mixture is heated to reflux for 1 h with an oil bath.

C. (S)-[1-(Hydroxydiphenylmethyl)-2-methylpropyl]carbamic acid, tert-butyl ester (**2**). The Grignard solution obtained in Step B is cooled with an ice bath (Note 10) and a solution of *N*-(*tert*-butoxycarbonyl)-L-valine methyl ester (**1**) (69.4 g, 0.30 mol) in THF (300 mL) is added at a rate to maintain the internal temperature below 6 °C (Note 11). After the addition is complete, the reaction mixture is allowed to warm to room temperature and stirring is continued for another 15 h (Note 12).

The reaction mixture is poured into an ice-cold aqueous saturated ammonium chloride solution (700 mL). The aqueous phase is separated and extracted with ethyl acetate (2 × 500 mL). The organic phases are combined, washed with brine (500 mL), and dried over anhydrous magnesium sulfate. After removal of the solvent under reduced pressure, crude **2** (115 g) is obtained as a white solid.

The crude product is dissolved in boiling ethyl acetate (600 mL) (Note 13) in a 1-L flask fitted with a reflux condenser and a magnetic stirring bar; then hexane (230 mL) is added. A few minutes after cessation of heating, a white solid starts to precipitate. The suspension is allowed to cool to room temperature, then is cooled for an additional hour with an ice bath. The solids are collected, washed with ice-cold hexane (100 mL), and dried under high vacuum (0.1 mbar) at room temperature for 12 h to give purified **2** (77.1 g, 72%) (Note 14) as a white solid. The mother liquor is concentrated under reduced pressure (to 220 g) and hexane (150 mL) is added. A few minutes after the addition of hexane, colorless crystals start to form. After standing for several hours at room temperature, the crystals are collected, washed with ice-cold hexane (100 mL) and dried as described above to yield a second crop of **2** (12.5 g, 12%). A third crop (11.3 g, 11%) is obtained after concentration (to 70 g) and treating of the mother liquor as described above (Note 15).

D. (4S)-4-(1-Methylethyl)-5,5-diphenyl-2-oxazolidinone (**3**). Under an argon atmosphere (Note 2), a 2-L, three-necked flask, fitted with a thermometer and magnetic stirring bar (Note 16) is charged with (*S*)-[1-(hydroxydiphenylmethyl)-2-methylpropyl]carbamic acid, *tert*-butyl ester (**2**) (97.8 g, 0.275 mol) (Note 17) and THF (1.8 L). The resulting solution is cooled to an internal temperature below 5 °C with an ice bath. Potassium *tert*-butoxide (37.0 g, 0.330 mol) (Note 18) is then added in one portion (Note 19). A few minutes after the addition of potassium *tert*-butoxide a white solid begins to precipitate. Stirring below 5 °C is continued for 2 h. The resulting suspension is poured into a 10% aqueous solution of ammonium chloride (2 L) and stirred for 10 min. The white solids are collected and washed with water (4 × 400 mL) (Note 20). The solids are transferred to a 1-L flask fitted with a reflux condenser and a magnetic stirring bar. Methanol (650 mL) is added and the resulting suspension is heated to reflux for 1 h with stirring. The mixture is allowed to cool to room temperature and stirring is continued for 1 h in an ice bath. The white solids are collected and washed with ice-cold methanol (100 mL). After drying for 24 h under high vacuum (0.1 mbar) at room temperature, oxazolidinone **3** (72.5 g, 94%) (Note 21) is obtained as a white solid.

2. Notes

1. *N*-(*tert*-Butoxycarbonyl)-L-valine methyl ester is also commercially available (Aldrich Chemical Co., Inc). The procedure described here is taken from the literature[3] and is readily reproducible.

2. The flask was flushed with argon and the argon atmosphere was maintained with an argon balloon.

3. (*S*)-*N*-(*tert*-Butoxycarbonyl)valine (puriss grade) was obtained from Fluka Chemie AG. (the submitters) or from Tokyo Chemical Industry (the checkers).

4. DMF (p. a. grade) was used as received from Merck AG. The checkers obtained DMF (99%) from Kokusan Chemical Co., Ltd.

5. Powdered potassium hydrogen carbonate (purum grade) was purchased from Fluka Chemie AG (the submitters) or Kokusan Chemical Co., Ltd. (the checkers).

6. Methyl iodide (purum grade) was obtained from Fluka Chemie AG (the submitters) or Tokyo Chemical Industry (the checkers).

7. THF (p. a. grade) was used as received from Merck AG. The checkers obtained THF from Kanto Chemical Co. (<0.005% water).

8. Bromobenzene (puriss grade) was obtained from Fluka Chemie AG (the submitters) or Tokyo Chemical Industry (the checkers).

9. The bromobenzene solution is added over ca. 1.5 h.

10. The clear dark Grignard solution becomes a suspension after cooling with an ice bath.

11. The protected valine ester is added over ca. 1.5 h.

12. TLC analysis indicates almost complete reaction after stirring for 3 h at room temperature. Stirring is continued overnight to ensure complete reaction.

13. The resulting light yellow solution is slightly turbid.

14. The product exhibits the following properties: mp 188.5-189 °C. $[\alpha]_D^{r.t.}$ = −65 (c 1.0, CH_2Cl_2) [The checkers obtained $[\alpha]_D^{29°C}$ = −60 (c 1.0, CH_2Cl_2)]. IR ($CHCl_3$) cm^{-1}: 3446, 2965, 1703, 1498, 1368. ^1H NMR (400 MHz, $CDCl_3$) δ: 0.88 (d, J = 6.8, Me), 0.90 (d, J = 6.9, Me), 1.32 (s, t-Bu), 1.43 (br. s, t-Bu), 1.80 (septd, J = 6.8, 2.2, CH), 2.63 (s, OH), 4.42-4.50 (m, CH), 4.58-4.62 (m, CH), 5.01 (d, J = 10.2, NH), 7.13-7.32 (m, 6 H), 7.44-7.55 (m, 4 H). ^{13}C NMR (100 MHz, $CDCl_3$) δ: 17.4, 22.7, 28.3, 28.8, 59.1, 79.0, 82.4, 125.3, 125.7, 126.7, 126.8, 128.2, 128.3, 145.6, 146.3, 156.3. Exact mass calc. for $C_{22}H_{29}NO_3Na^+$ (M + Na^+): 378.2040. Found (MALDI) *m/z* 378.2042. For comparison with literature data, see Ref. 4 and 5.

15. The third crop contains (4*S*)-4-(1-methylethyl)-5,5-diphenyloxazolidin-2-one (**3**) as a byproduct (ca. 7 mol% as determined from the ^1H NMR spectrum).

16. Since the reaction mixture becomes a viscous suspension, a large magnetic stirring bar or mechanical stirrer has to be used.

17. The three crops obtained in Step B are used.

18. Potassium *tert*-butoxide (purum grade) was purchased from Fluka Chemie AG.

19. The temperature rises to 8 °C after the addition of potassium *tert*-butoxide.

20. The third and fourth washes are negative when tested for chloride with silver nitrate.

21. The enantiomer ratio was determined by HPLC (column: Chiralcel OD-H, 4.6 × 15 mm, 5 µm, Daicel Chemical Industries; eluent: hexane/2-PrOH 10:1; flow: 1 mL/min; detection: UV at 254 nm; (S)-**3**: t_R = 4.3 min; (R)-**3**: t_R = 12.8 min) to be >99:1. The product exhibits the

following properties: mp 253.5-254 °C, $[\alpha]_D^{r.t.} = -244$ (c 0.61, CH_2Cl_2). IR ($CHCl_3$) cm^{-1}: 3460, 2967, 1759. ^1H NMR (400 MHz, $CDCl_3$) δ: 0.69 (d, J = 6.6, Me), 0.90 (d, J = 7.0, Me), 1.81-1.90 (m, CH), 4.36 (d, J = 3.7, CH), 6.61 (br. s, NH), 7.23-7.40 (m, 8 H), 7.52-7.55 (m, 2 H). ^{13}C NMR (100 MHz, $CDCl_3$) δ: 15.6, 20.8, 29.6, 65.9, 89.4, 125.7, 126.3, 127.7, 128.1, 128.2, 128.5, 139.2, 143.9, 158.8. Exact mass calc. for $C_{18}H_{19}NO_2Na^+$ (M + Na^+): 304.1308. Found (MALDI) *m/z* 304.1308. For comparison with literature data see Ref. 5, 6, 7, 8.

Waste Disposal Information

All toxic materials were disposed of in accordance with "Prudent Practices in the Laboratory"; National Academy Press; Washington, DC, 1995.

3. Discussion

The present procedure for the preparation of oxazolidinone **3** is a variation of the procedures described by Luche[4] and Davies[5]. Yields have been substantially enhanced by improving the purification procedures. The preparation of **3** starting from valine methyl or ethyl ester hydrochloride has been described by several authors.[6,7,8,9] These procedures suffer from moderate yields for the Grignard addition step and some of them use hazardous reagents like phosgene.

Oxazolidinones, such as **4** and **5**, have been introduced by Evans as very useful chiral auxiliaries.[10] These compounds are prepared from readily available chiral amino alcohols and are successfully employed in a variety of different stereoselective reactions.[11]

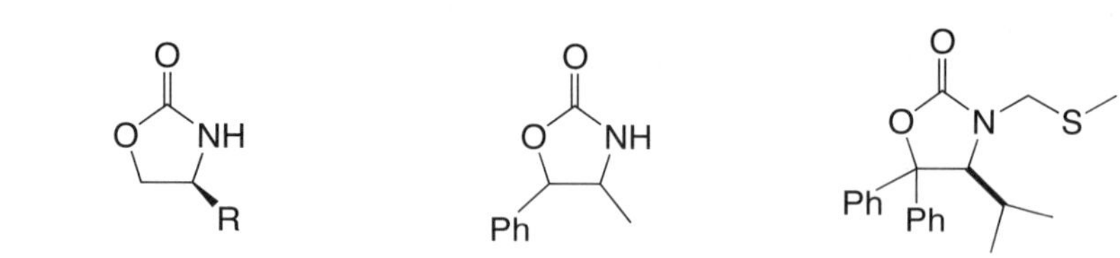

R = $CHMe_2$, CH_2Ph, Ph **4**   (4R,5S) or (4S,5R) **5**   **6**

(4*S*)-4-(1-Methylethyl)-5,5-diphenyl-2-oxazolidinone (**3**), whose preparation is described here, has several advantages over Evans' original auxiliaries:[8] i) Derivatives of **3** are more likely to crystallize. In many cases the separation and purification of diastereoisomers can be achieved by simple recrystallization rather than by expensive and time-consuming chromatography. ii) Acylation of **3** can be carried out at 0 °C (instead of −78 °C for **4** and **5**) by deprotonation with BuLi, followed by treatment with an activated carboxylic acid derivative. iii) Lithium enolates of *N*-acyl derivatives of **3** can be obtained directly by treatment with BuLi at −78 °C, in comparison to **4** and **5** when the more expensive lithium diisopropylamide or lithium hexamethyldisilazane is required. iv) The *N*-acyl derivatives of **3** can be cleaved by using sodium hydroxide without any detectable nucleophilic attack on the oxazolidinone carbonyl group. Expensive and hazardous (in large scale) $LiOH/H_2O_2$ has to be used for the cleavage of *N*-acyl derivatives of **4** and **5**. v) Due to its low solubility in most solvents, oxazolidinone **3** is easily recovered in high purity by simple filtration after the cleavage step.

Enolates of *N*-acyl derivatives of oxazolidinone **3** have been used for asymmetric alkylations,[5,7,8,9] aldol reactions,[8] Mannich reactions,[8] Michael additions to nitroolefins,[8,12] azide transfer reactions,[9] and samarium-Reformatsky reactions.[13] Furthermore, α,β-unsaturated *N*-acyl derivatives of **3** have been employed in Diels-Alder reactions and in Michael addition reactions of cuprates.[8] The *N*-methyl-thiomethyl derivative **6** of oxazolidinone **3** can be lithiated to provide a reagent which is synthetically equivalent to a chiral formyl anion.[14]

1. Laboratorium für Organische Chemie, ETH Hönggerberg, Wolfgang-Pauli-Str. 10, HCI, CH-8093 Zürich, Switzerland
2. Novartis Pharma Ltd., Basel, Switzerland.
3. Hamada, Y.; Shibata, M.; Sugiura, T.; Kato, S.; Shioiri, T. *J. Org. Chem.* **1987**, *52*, 1252.
4. Delair, P.; Einhorn, C.; Einhorn, J.; Luche, J. L. *J. Org. Chem.* **1994**, *59*, 4680.
5. Bull, S. D.; Davies, S. G.; Jones, S.; Sanganee, H. J. *J. Chem. Soc., Perkin Trans. 1*, **1999**, *387*.
6. Gawley, R. E.; Zhang, P. *J. Org. Chem.* **1996**, *61*, 8103.
7. Isobe, T.; Fukuda, K. Jap. Pat. JP 09,143,173 (*Chem. Abstr.* **1997**, *127*, 50635x)
8. (a) Hintermann, T.; Seebach, D. *Helv. Chim. Acta* **1998**, *81*, 2093. (b) Seebach, D.; Schaeffer, L.; Gessier, F.; Bindschädler, P.; Jäger, C.; Josien, D.; Kopp, S.; Lelais, G.; Mahajan, Y.; Micuch, P.; Sebesta, R.; Schweizer, B. W. *Helv. Chim. Acta* **2003**,

86, 1852; (c) Lelais, G.; Micuch, P.; Josien-Lefebvre, D.; Rossi, F.; Seebach, D. *Helv. Chim. Acta* **2004**, *87*, 3131.

9. Gibson, C. L.; Gillon, K.; Cook, S. *Tetrahedron Lett.* **1998**, *39*, 6733.

10. For a review, see: Evans, D. A. *Aldrichimica Acta*, **1982**, *15*, 23.

11. For reviews, see: (a) Ager, D. J.; Prakash, I.; Schaad, D. R. *Chem. Rev.* **1996**, *96*, 835; (b) Ager, D. J.; Prakash, I.; Schaad, D. R. *Aldrichimica Acta*, **1997**, *30*, 3.

12. Brenner, M.; Seebach, D. *Helv. Chim. Acta* **1999**, *82*, 2365.

13. Fukuzawa, S.; Matsuzawa, H.; Yoshimitsu, S. *J. Org. Chem.* **2000**, *65*, 1702.

14. (a) Gaul, C.; Seebach, D. *Org. Lett.* **2000**, *2*, 1501; (b) Gaul, C.; Schärer, K.; Seebach, D. *J. Org. Chem.* **2001**, *66*, 3059; (c) Gaul, C.; Schweizer, B. W.; Seiler, P.; Seebach, D. *Helv. Chim. Acta* **2002**, *85*, 1546.

SYNTHESIS OF INDOLES BY PALLADIUM-CATALYZED REDUCTIVE *N*-HETEROANNULATION OF 2-NITROSTYRENES: METHYL INDOLE-4-CARBOXYLATE

[(1*H*-Indole-4-carboxylic acid, methyl ester)]

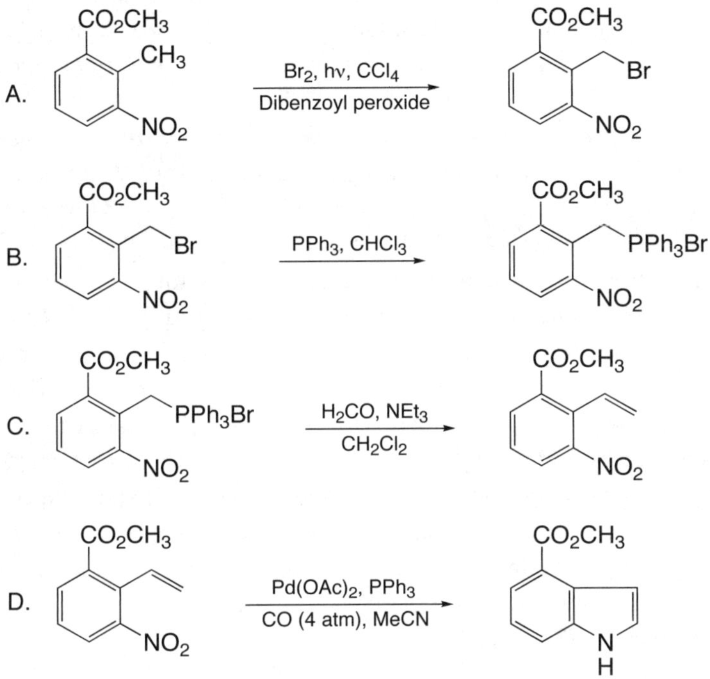

Submitted by Björn C. Söderberg[1], James A. Shriver, and Jeffery M. Wallace.
Checked by J. David Warren and Louis S. Hegedus.

1. Procedure

A. Methyl 2-bromomethyl-3-nitrobenzoate. To a 250-mL, two-necked, round-bottomed flask, equipped with a condenser and addition funnel, is added methyl 2-methyl-3-nitrobenzoate (19.1 g, 97.9 mmol, (Note 1)), dibenzoyl peroxide (1.21 g, 5.00 mmol, (Note 2)), and 100 mL of carbon tetrachloride (Note 3). The mixture is heated to reflux, and a clear, pale yellow solution is formed. A solution of bromine (16.1 g, 100.6 mmol) (Note 2) in 20 mL of carbon tetrachloride is added over 10 min to the boiling solution under irradiation using a 100-W flood lamp. The reaction mixture is heated and irradiated for 24 h. The resulting orange solution is allowed to cool to ambient temperature; 50 mL of dichloromethane is added, and the solution is washed with three 50-mL portions of saturated aqueous sodium bicarbonate. The organic phase is dried ($MgSO_4$), filtered, and the solvents are removed at water aspirator pressure on a rotary evaporator affording 25.90 g (96.5%) of methyl 2-bromomethyl-3-nitrobenzoate as pale yellow crystals. The material is used in the next step without further purification (Note 4).

B. (2-Carbomethoxy-6-nitrobenzyl)triphenylphosphonium bromide. To a 500-mL, round-bottomed flask is added methyl 2-bromomethyl-3-nitrobenzoate (25.90 g, 94.5 mmol) and 150 mL of chloroform (Note 5). Triphenylphosphine (28.44 g, 108.4 mmol) (Note 2) is added in one portion and the resulting yellow solution is heated to reflux for 1.5 h. After cooling to ambient temperature, the orange solution is poured into 400 mL of anhydrous diethyl ether (Note 6) with vigorous stirring to precipitate the Wittig salt. The slurry is cooled in a freezer (−20 °C, 1 h). The white solids are collected by filtration, washed with 4 × 100 mL of anhydrous diethyl ether, and dried under high-vacuum to give crude (2-carbomethoxy-6-nitrobenzyl)triphenylphosphonium bromide. A ^{1}H NMR spectrum indicated that the product contained some triphenylphosphine, thus the salt was washed again with diethyl ether (2 × 100 mL). After drying, a considerably purer product (50.91 g) was obtained. The salt is used in the next step without further purification (Note 7).

C. Methyl 2-ethenyl-3-nitrobenzoate. Paraformaldehyde (30 g) (Note 8) is placed in a 250-mL, two-necked, round-bottomed flask. An argon inlet is connected to one of the necks. The other neck is connected to a 1-L, two-necked, round-bottomed flask via Tygon tubing (2 cm id) and two male 24/40 joints (Note 9). In the 1-L flask, 2-carbomethoxy-6-nitrobenzyl)triphenylphosphonium bromide (50.91 g) is dissolved in 500 mL of dichloromethane. A glass tube is attached to the Tygon tubing such that the end extends several cm into the solution. Triethylamine (39.13 g, 53.8 mL, 386.7 mmol) is added to the 1-L flask, resulting in the immediate formation of a deep blue/purple solution; a condenser is attached to the second neck of the flask. Argon is then passed over the paraformaldehyde so that a steady stream of gas is constantly bubbling through the purple ylide solution. The paraformaldehyde is heated to 160 °C (oil bath) in order to generate formaldehyde, which flows through the solution (Note 10). Upon completion of the addition of formaldehyde, the purple color of the ylide slowly changes from deep purple to brown over a period of 1-2 h, indicating completion of the reaction. The resulting solution is poured into 500 mL of hexanes, forming a precipitate, which is removed by filtration. The precipitate is washed with 100 mL of hexanes and the combined filtrate and wash are concentrated at water aspirator pressure on a rotary evaporator. The crude yellow-brown solid is dissolved in 100 mL of dichloromethane; 30 g of silica gel is added and the mixture is concentrated again. The resulting powder applied to a pre-packed silica gel column (40 × 4 cm) (Note 11), and eluted with 7:3 hexanes:ethyl acetate to give methyl 2-ethenyl-3-nitrobenzoate (15.87 g, 76.6 mmol, 81%, Notes 12-14) as pale yellow crystals.

D. Caution! Due to the toxicity of carbon monoxide and the risk of an explosion in handling pressurized glassware, this transformation should be carried out in a well vented fume hood with a blast shield and the hood sash down. Users should exercise appropriate caution at all times.

Methyl indole-4-carboxylate. To a threaded glass, 200-mL reaction vessel, equipped with a Teflon stirring bar, is added methyl 2-ethenyl-3-nitrobenzoate (10.35 g, 50.0 mmol), triphenylphosphine (3.23 g, 12.3 mmol) and 100 mL of acetonitrile (Note 15). The mixture is stirred for 10 min to dissolve the reagents. Palladium acetate (0.673 g, 3.00 mmol) (Note 16) is added; a yellow precipitate is immediately formed. The tube is attached to a pressure head (Note 17) and the solution is saturated with carbon monoxide (four cycles to 59 psi of CO) (Note 18). The reaction mixture is heated to 90 °C (oil bath temperature) under CO (59 psi) for 50 h. A red-brown solution forms after heating for a few minutes. In order to remove carbon dioxide that is formed and to monitor the progress of the reaction, the vessel is removed from the oil bath for 15 min, carefully vented, and repressurized with CO every 10-12 h (Note 19). After 50 h, the reaction mixture is cooled and concentrated using a rotary evaporator at water aspirator pressure, giving a dark brown/black oil. The crude product is purified by chromatography on a pre-packed 40 × 4 cm silica column (Note 11) using 7:3 hexanes:CH_2Cl_2 (0.80 L), then 1:1 hexanes:CH_2Cl_2 (2.80 L) as eluents affording methyl indole-4-carboxylate (7.95 g, 44.75 mmol, 91%) as a pale yellow solid (Note 20).

2. Notes

1. Methyl 2-methyl-3-nitrobenzoate was used as received from Aldrich Chemical Company, Inc., or prepared from 2-methyl-3-nitrobenzoic acid in >97% yield according to the literature procedure.[2]

2. Dibenzoyl peroxide, bromine, and triphenylphosphine were used as received from Aldrich Chemical Company, Inc.

3. Carbon tetrachloride (anhydrous) was used as received from Aldrich Chemical Company, Inc.

4. The checkers did not observe complete conversion to the bromide. Analytical data: mp 63-66 °C (uncorrected). ^{1}H NMR (300 MHz, $CDCl_3$) δ: 3.97 (s, 3 H), 5.13 (s, 2 H), 7.52 (t, J = 8.1, 1 H), 7.93 (dd, J = 8.1, 1.2, 1 H), 8.08 (dd, J = 7.7, 1.2, 1 H); ^{13}C NMR (75 MHz, $CDCl_3$) δ: 23.2, 53.4, 128.0, 129.4, 132.5, 132.8, 134.9, 150.7, 166.0; IR (neat) cm^{-1}: 1726, 1532, 1270. Anal. Calcd for $C_9H_8BrNO_4$: C, 39.44; H, 2.94. Found: C, 39.33; H, 2.94.

5. Chloroform (anhydrous) was used as received from Aldrich Chemical Company, Inc.

6. Diethyl ether (anhydrous) was used as received from Fisher Scientific.

7. Analytical data: mp 217-229 °C (sealed capillary, uncorrected); ^{1}H NMR (300 MHz, $CDCl_3$) δ: 3.67 (s, 3 H), 5.69 (d, J_{PH} = 14.6, 2 H), 7.55-7.82 (overlapping multiplets, 16 H), 7.95 (d, J = 8.1, 1 H), 8.09 (d, J = 7.9, 1 H); ^{13}C NMR (75 MHz, $CDCl_3$) δ: 26.3 (d, J_{CP} = 51.7), 53.5, 117.7 (d, J_{CP} = 86.4), 124.8 (d, J_{CP} = 8.4), 129.1 (d, J_{CP} = 2.8), 130.0 (d, J_{CP} = 12.5), 130.8, 133.8 (d, J_{CP} = 9.8), 135.2 (d, J_{CP} = 2.9), 135.7, 150.9, 165.5; IR (neat) cm^{-1}: 1717, 1532, 1437, 1275, 1108.

8. Paraformaldehyde was used as received from Fisher Scientific.

9. Formaldehyde is generated according to the procedure of Smith, A. B., III; Branca, S. J.; Guaciaro, M. A.; Wovkulich, P. M. *Org. Synth., Coll. Vol. VII* **1990**, *271*; see Figure 1, p. 272.

10. The glass inlet tube has a tendency to clog. This problem can be remedied by lifting the tube out of the solution, discontinuing heating, and removing the obstruction. It is important to closely follow the pyrolysis. If insufficient heat is supplied, the ylide solution may be sucked into the flask containing paraformaldehyde. If too much heat is supplied, the joints may separate, usually resulting in a small flame at the joint.

11. Silica gel (200-400 mesh) from Natland International Corp. was used.

12. The yield is based on methyl-2-bromomethyl-3-nitrobenzoate.

13. A small amount (1.5%) of hydrolysis product, methyl 2-methyl-3-nitrobenzoate, was observed at times.

14. Analytical data: mp 39-40 °C (uncorrected); ^{1}H NMR (300 MHz, $CDCl_3$) δ: 3.89 (s, 3 H), 5.23 (dd, J = 17.6, 0.8, 1 H), 5.43 (dd, J = 11.3, 0.8, 1 H), 7.18 (dd, J = 17.6, 11.5, 1 H), 7.47 (t, J = 7.9, 1 H), 7.88 (dd, J = 8.3,

1, 1 H), 7.98 (dd, J = 7.7, 1.2, 1 H); ^{13}C NMR (75 MHz, $CDCl_3$) δ: 52.9, 119.7, 126.5, 128.1, 132.1, 132.2, 133.2, 134.2, 150.4, 166.8; IR (neat) cm^{-1}: 1730, 1531, 1292, 1264, 1124, 707. Anal. Calcd for $C_{10}H_9NO_4$: C, 57.97; H, 4.38. Found: C, 57.72; H, 4.47.

15. Acetonitrile was distilled from calcium hydride prior to use.

16. Palladium acetate was used as received from Pressure Chemical Co.

17. The pressure head is assembled as follows using, if possible, stainless steel parts: A tee, equipped with a pressure relief valve (preset to open at 120 psi) and a pressure gauge (0-400 psi), is connected to a cross via a nipple. A ball valve (for pressure release and sample removal) is attached to the top of the cross, a valve (for introduction of CO) on one side of the cross, and a nipple connected to a Swagelok Teflon adapter attached to the bottom of the cross. An autoclave or a metal reaction vessel can be substituted for this reaction assembly.

18. UHP-grade carbon monoxide is used as purchased from Matheson Gas Products, Inc.

19. The reaction is monitored by thin layer chromatography on 60 Å silica gel (R_f = 0.15; 4:6 hexanes:CH_2Cl_2).

20. After ca. 500 mL of solvent, 20-mL fractions are collected and analyzed by thin layer chromatography. The plates are visualized at 254 nm, and the product appears as a bright fluorescent blue spot. Test tubes containing the product are combined into two major fractions the first (1.173 g) contaminated by a small amount of an unknown impurity (higher R_f) and the second (6.663 g) as pure methyl indole-4-carboxylate. The ^{1}H NMR spectra of the two fractions are identical; however; a slight melting point depression is observed for the first fraction. Uncorrected melting points: first fraction 64-66 °C; second fraction 68-69 °C. Note: A range of melting points has been reported for this compound: Lit. mp 63 °C;[3] 64-65 °C;[4] 67-69 °C;[2] 69-71 °C.[5] ^{1}H NMR (300 MHz, $CDCl_3$) δ: 3.99 (s, 3 H), overlapping 7.19 (br d, J = 3.0, 1 H), 7.23 (t, J = 7.7, 1 H), 7.34 (t, J = 3.0, 1 H), 7.59 (d, J = 8.1, 1 H), 7.93 (d, J = 7.5, 1 H), 8.48 (br s, 1 H); ^{13}C NMR (75 MHz, $CDCl_3$) δ: 52.2, 103.9, 116.5, 121.3, 121.6, 123.6, 126.8, 127.6, 136.9, 168.4; IR (neat) cm^{-1}: 3355, 1699, 1276.

Waste Disposal Information

All toxic materials were disposed of in accordance with "Prudent Practices in the Laboratory"; National Academy Press; Washington, DC, 1995.

3. Discussion

The procedure described here illustrates an efficient and relatively mild synthesis of methyl indole-4-carboxylate in 72% overall yield starting from commercially available materials. The synthetic sequence compares favorably to the previously described Batcho-Leimgruber indole syntheses of this substance.[2,3] In general, the palladium/phosphine-catalyzed *N*-heteroannulation of 2-nitrostyrenes offers a very flexible entry to functionalized indoles.[6] A few examples of this reaction, performed on a 1-2 mmol scale, are shown in the Table below. A number of palladium reagents, both palladium(0) and palladium(II), can be used as precatalyst for the reaction. For example, palladium diacetate, bis(acetonitrile)palladium dichloride, bis(dibenzylidenacetone)palladium and palladium on carbon (10% Pd) have all been shown to produce indoles in the presence of carbon monoxide and a catalytic amount of phosphine.[6a] Perhaps the most important feature of the reaction is its compatibility with other functional groups. For example, functionalities such as bromides, triflates, alcohols, ethers, esters, ketones, nitriles, and additional nitro groups all remain unaffected in this reaction. In contrast to the Batcho-Leimgruber route, indoles substituted in the 2- and/or 3-position can readily be prepared from 2-nitrostyrenes. More complex, fused indoles, can be obtained by annulation of bicyclic nitrostyrene derivatives (entries 7-8).

Table
Indoles By Reductive *N*-Heteroannulation of 2-Nitrostyrenes

Entry	Styrene	Indole	Yield (%)
1	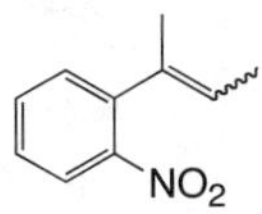	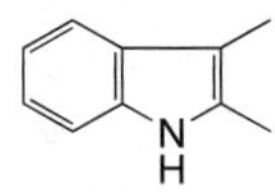	97
2	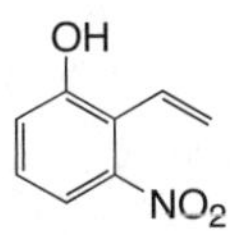	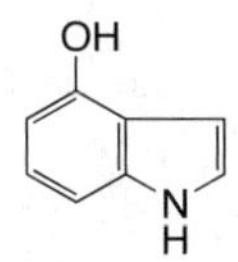	96
3	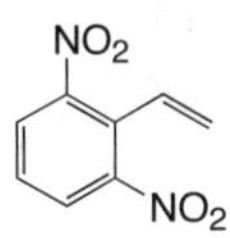	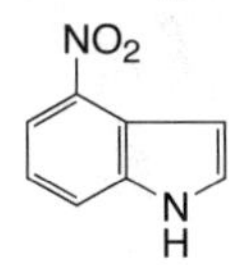	89
4	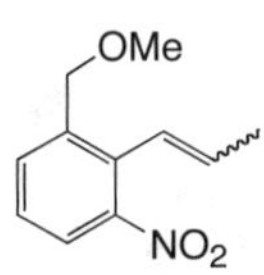	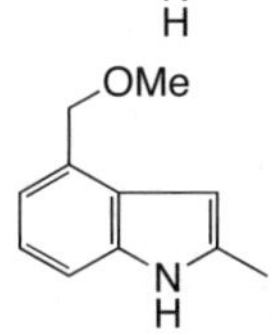	76
5	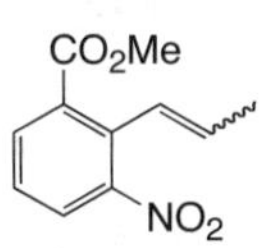	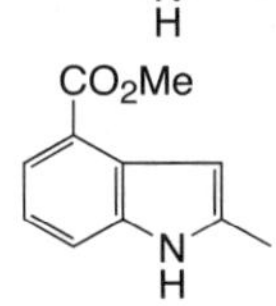	90
6	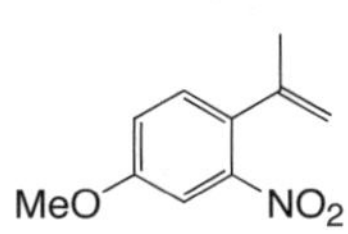	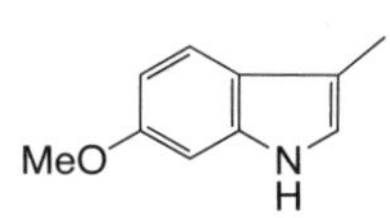	81
7	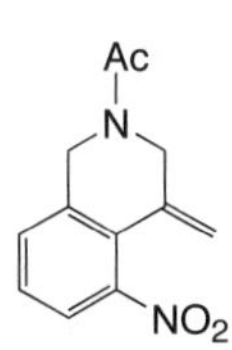	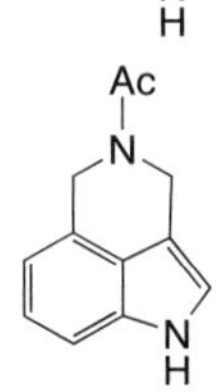	41
8	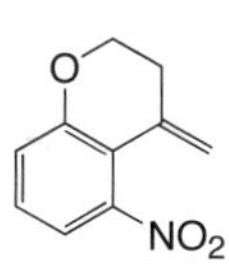	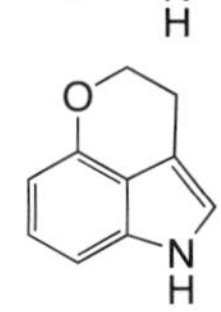	63

1. Department of Chemistry, West Virginia University, Morgantown, WV 26506.

2. Hoffmann-La Roche and Co., British Patent 1 276 966; **1972**; Batcho, A. D. and Leimgruber, W. German Patent 2 057 840; **1971**.

3. (a) Ponticello, G. S.; Baldwin, J. J. *J. Org. Chem.* **1979**, *44*, 4003; (b) Kozikowski, A. P.; Ishida, H.; Chen, Y.-Y. *J. Org. Chem.* **1980**, *45*, 3350.

4. Watanabe, T.; Hamaguchi, F.; Ohki, S. *Chem. Pharm. Bull.* **1972**, *20*, 2123.

5. Aldrich Handbook of Fine Chemicals and Laboratory Equipment, **2000-2001**, 1126.

6. (a) Söderberg, B. C.; Rector, S. R.; O'Neil, S. N. *Tetrahedron Lett.* **1999**, *40*, 3657; (b) Söderberg, B. C.; Shriver, J. A. *J. Org. Chem.* **1997**, *62*, 5838 and references therein. (c) Söderberg, B. C.; Chisnell, A. C.; O'Neil, S. N.; Shriver, J. A. *J. Org. Chem.* **1999**, *64*, 9731.

PREPARATION OF SUBSTITUTED PYRIDINES VIA REGIOCONTROLLED [4 + 2] CYCLOADDITIONS OF OXIMINOSULFONATES: METHYL 5-METHYLPYRIDINE-2-CARBOXYLATE

[(2-Pyridinecarboxylic acid, 5-methyl-, methyl ester)]

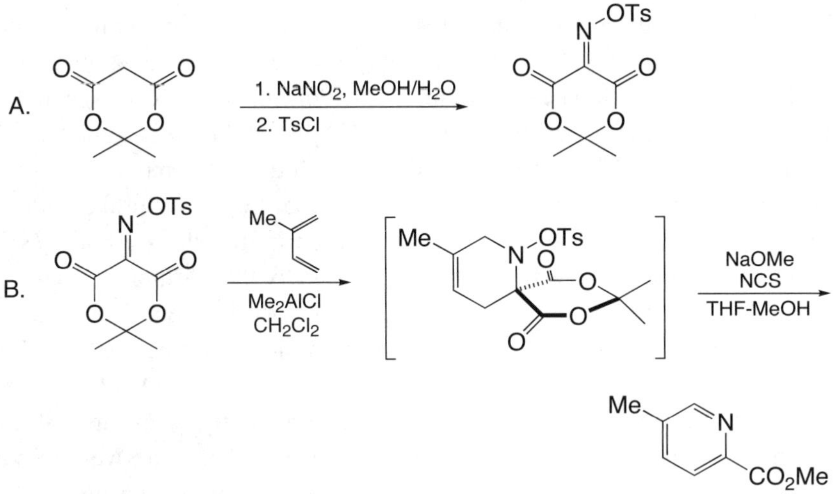

Submitted by Rick L. Danheiser, Adam R. Renslo, David T. Amos, and Graham T. Wright.[1]
Checked by Helga Krause and Alois Fürstner.

1. Procedure

A. 5-(Tosyloxyimino)-2,2-dimethyl-1,3-dioxane-4,6-dione. A 100-mL, three-necked, round-bottomed flask equipped with an argon inlet adapter, magnetic stirring bar, rubber septum, and solid addition funnel fitted with a rubber septum (Note 1) is charged with 4.00 g (27.8 mmol) of Meldrum's acid (Note 2) and 23 mL of methanol (Note 3). To this suspension is added in one portion a solution of 1.91 g (27.8 mmol) of sodium nitrite (Note 4) in 15 mL of water. The reaction mixture is stirred for 2 h at room temperature to give a deep red solution which is then treated with 4 mL of pH 7 phosphate buffer (Note 5) in one portion and then cooled to 0 °C. *p*-Toluenesulfonyl chloride (4.81 g, 25.2 mmol) (Note 6) is added over 3 min via the solid addition funnel and the cooling bath is removed. The resulting peach-colored mixture is vigorously stirred for 30 min, then the solids are collected by filtration on a Büchner funnel with the aid of 40 mL of cold methanol. The resulting solids are dried at 0.2 mm in a desiccator over phosphorus pentoxide (P_2O_5) for 4 h to provide 4.85 g (56-59% based on TsCl) of the oximinosulfonate as a white solid, mp 155-156 °C (Note 7).

B. Methyl-5-methylpyridine-2-carboxylate. A 250-mL, three-necked, round-bottomed flask, equipped with an argon inlet adapter, magnetic stirring bar, rubber septum, and pressure-equalizing addition funnel fitted with a rubber septum (Note 8), is charged with a solution of 4.50 g (13.7 mmol) of oximinosulfonate in 100 mL of dichloromethane (Note 9) and 4.10 mL (2.81 g, 41.1 mmol) of isoprene (Note 10). The solution is cooled at −78 °C while 27.5 mL (27.5 mmol) of a 1.0M solution of dimethylaluminum chloride (Me_2AlCl) in hexane (Note 11) is added dropwise via the addition funnel over 9 min. The resulting orange solution is stirred for 3 h at −78 °C to give a yellow solution which is then quenched by the addition of 60 mL of saturated sodium potassium tartrate solution in one portion. The resulting mixture is allowed to warm to 0 °C over 30 min, then is transferred to a 500-mL separatory funnel. Dichloromethane (80 mL) and 150 mL of water are added, and the aqueous phase is separated and extracted with three 80-mL portions of dichloromethane. The combined organic phases are washed with 80 mL of saturated sodium chloride solution, dried over magnesium sulfate ($MgSO_4$), filtered, and concentrated at reduced pressure on a rotary evaporator to provide 5.48-5.54 g of cycloadduct as an orange foam which is used in the next step without further purification (Notes 12, 13).

A 250-mL, round-bottomed flask equipped with an argon inlet adapter and magnetic stirring bar (Note 8) is charged with a solution of the crude cycloadduct prepared above in 80 mL of tetrahydrofuran (Note 14) and 80 mL of methanol (Note 3). The solution is cooled at 0 °C while 26.3 mL (41.1 mmol) of a 1.56M solution of sodium methoxide (NaOMe) in methanol (Note 15) is added via syringe over 4 min followed by the addition of 1.83 g (13.7 mmol) of *N*-chlorosuccinimide (Note 16) in one portion. The cooling bath is removed, and the resulting yellow solution is stirred in the dark for 16 h (Note 17). The reaction mixture is concentrated to a volume of ca. 20 mL by rotary evaporation, then diluted with 150 mL of ethyl acetate and 150 mL of pH 7 phosphate buffer (Note 5). The aqueous phase is separated and extracted with three 100-mL portions of ethyl acetate, and the combined organic phases are extracted with three 100-mL portions of 1.0N HCl. The combined aqueous extracts are neutralized by the slow addition of solid sodium bicarbonate (Note 18), then extracted with three 100-mL portions of ethyl acetate. The combined organic phases are washed with 100 mL of saturated sodium chloride solution, dried over Na_2SO_4, filtered, and concentrated at reduced pressure on a rotary evaporator to afford 1.61-1.76 g (78-85%) of methyl 5-methylpyridine-2-carboxylate as a yellow solid, mp 47-49 °C (Notes 19-22).

2. Notes

1. The apparatus is purged with argon and maintained under an atmosphere of argon during the course of the reaction.

2. 2,2-Dimethyl-1,3-dioxane-4,6-dione (Meldrum's acid) was purchased from Aldrich Chemical Co., Inc. and recrystallized before use according to the following procedure. Meldrum's acid (10 g) was dissolved in a minimum amount of warm (40-45 °C) acetone (ca. 20 mL). Room temperature water (15 mL) was then added in one portion, resulting in the immediate precipitation of white needles which were collected by filtration on a Büchner funnel and dried (ca. 0.05 mm) in a desiccator over P_2O_5 overnight.

3. Anhydrous grade methanol was purchased from Mallinkrodt, Inc. and used as received.

4. Sodium nitrite was purchased from Mallinkrodt, Inc. and used as received.

5. Phosphate buffer (pH 7.00) was purchased from VWR Scientific Products, Inc; the checkers used phosphate buffer purchased from Riedel-de Haen.

6. *p*-Toluenesulfonyl chloride (TsCl) was purchased from Fluka Corporation and recrystallized before use according to the following procedure. TsCl (30 g) was dissolved in a minimum amount (ca. 50 mL) of warm (55-60 °C) chloroform to give a yellow solution. To this solution was added 70 ml of warm petroleum ether and 3 g of activated charcoal. After stirring for 10 min, the mixture was filtered and concentrated to half volume. Crystals of TsCl precipitated and after 30 min were collected by filtration on a Büchner funnel. Drying at 0.05 mm for 1 h afforded 22 g of TsCl as white crystals, mp 67-68 °C.

7. The oximinosulfonate has the following spectroscopic properties: IR ($CHCl_3$) cm^{-1}: 3020, 1790, 1765, 1596, 1400, 1290; ^1H NMR (300 MHz, $CDCl_3$) δ: 1.78 (s, 6 H), 2.46 (s, 3 H), 7.40 (d, 2 H, $J = 8.4$), 7.93 (d, 2 H, $J = 8.4$); ^{13}C NMR (75 MHz, $CDCl_3$) δ: 21.8, 28.1, 106.8, 129.5, 130.1, 130.2, 138.8, 147.1, 149.6, 154.7. Anal. Calcd for $C_{13}H_{13}NO_7S$: C, 47.70; H, 4.00; N, 4.28; Found: C, 47.76; H, 4.02; N, 4.22.

8. The apparatus is flame-dried under reduced pressure and then maintained under an atmosphere of argon during the course of the reaction.

9. Dichloromethane was purchased from J.T. Baker, Inc. and purified by pressure filtration through activated alumina.

10. Isoprene was purchased from Aldrich Chemical Co., Inc. and distilled under argon at atmospheric pressure immediately before use.

11. Dimethylaluminum chloride in hexanes (1.0M) was purchased from Aldrich Chemical Company, Inc.

12. A pure sample of the intermediate cycloadduct can be obtained by column chromatography on silica gel (elution with 1% methanol/dichloromethane). The cycloadduct exhibits the following spectroscopic properties: IR ($CHCl_3$) cm^{-1}: 3020, 1780, 1750, 1385, 1300; ^1H NMR (300 MHz, $CDCl_3$) δ: 1.67 (s, 3 H), 1.69 (s, 3 H), 1.88 (s, 3 H), 2.48 (s, 3 H), 2.71 (br dd, 2 H, J = 1.2, 3.3), 3.93 (s, 2 H), 5.33 (br., 1H), 7.36 (d, 2H, J = 7.9), 7.81 (d, 2 H, J = 8.4); ^{13}C NMR (75 MHz, $CDCl_3$) δ: 20.3, 21.7, 28.5, 29.3, 32.8, 57.4, 66.3, 106.2, 113.9, 129.2, 129.55, 129.62, 131.2, 145.9, 164.0.

13. *CAUTION:* The checkers experienced a rather vigorous decomposition of this product upon evaporating the solvent in the rotary evaporator and drying of the crude material at 39 °C (bath temperature) and 15 mm. No hazards, however, were encountered when this operation was carried out at ambient temperature (22 °C) at 38 mm. The resulting crude material was directly used in the next step.

14. Tetrahydrofuran was purchased from J.T. Baker, Inc. and purified by pressure filtration through activated alumina.

15. A 1.56M solution of sodium methoxide was prepared by careful addition of 2.69 g of sodium metal (cut into ca. 15 pieces) to 75 mL of methanol at 0 °C under an atmosphere of argon.

16. *N*-Chlorosuccinimide (NCS) was purchased from Aldrich Chemical Company, Inc. and recrystallized before use according to the following procedure. NCS (25 g) was dissolved in 125 mL of glacial acetic acid at 60-65 °C in a 250-mL, Erlenmeyer flask and the resulting solution was allowed to cool to room temperature. The Erlenmeyer flask was then placed in an ice-water bath (15-20 °C), and white flakes of NCS immediately precipitated and were collected by filtration on a Buchner funnel (washing with 20-mL of glacial acetic acid and then with two 20-mL portions of hexanes) and then dried at 0.5 mm for 6 h and stored in the dark until use.

17. For this purpose, the flask was wrapped in aluminum foil.

18. Sodium bicarbonate was added until gas evolution was no longer observed.

19. The purity of this material was confirmed by spectroscopic and elemental analysis: IR (KBr) cm^{-1}: 3091, 3048, 3004, 2956, 1730, 1591, 1448, 1320, 1251, 1126, 1032, 781, 704; ^1H NMR (300 MHz, $CDCl_3$) δ: 2.43 (s, 3 H), 4.00 (s, 3H), 7.64 (dd, 1 H, J = 7.8, 2.0), 8.05 (d, 1 H, J = 8.3), 8.57 (d, 1 H, J = 2.0); ^{13}C NMR (75 MHz, $CDCl_3$) δ: 18.5, 52.6, 124.6, 137.1, 137.3, 145.2, 150.2, 165.6. Anal. Calcd for $C_8H_9NO_2$: C, 63.56; H, 6.00; N, 9.27; Found: C, 63.21; H, 6.02; N, 9.43; TLC: R_f = 0.11 (elution with 25% ethyl acetate-hexanes; phosphomolybdic acid in ethanol stain).

20. If desired, further purification can be effected by chromatography (Note 21) or, alternatively, by recrystallization according to the following procedure: the crude product is dissolved in 20 mL of warm (50 °C) hexanes and the solution is decanted from an insoluble orange solid and transferred to a 50-mL, round-bottomed flask. Concentration provides 1.34 g (65%) of methyl 5-methylpyridine-2-carboxylate as a colorless solid, mp 53-54 °C.

21. Purification by column chromatography was carried out according to the following procedure: silica gel (10 g) is added to the dried solution of crude product obtained from the work up, and the solution is concentrated by rotary evaporation to afford a free-flowing powder which is placed at the top of a column (4.5 cm diameter) of 40 g of silica gel (230-400 mesh) and eluted with 25% ethyl acetate/hexane (containing 1% triethylamine) to afford 1.47 g (71%) of methyl 5-methylpyridine-2-carboxylate as a pale yellow solid, mp 54-55 °C.

22. The checkers obtained the crude material as a yellow-brown syrup (1.86 g, 89%). Column chromatography as described in (Note 21) afforded 1.29 g (63%) of the title compound as a pale yellow solid, mp 52-53 °C.

Waste Disposal Information

All toxic materials were disposed of in accordance with "Prudent Practices in the Laboratory"; National Academy Press; Washington, DC, 1995.

3. Discussion

The structures of a number of natural products incorporate the pyridine ring as do important commercial compounds including herbicides, insecticides, fungicides, and a variety of medicinal agents. Among the numerous strategies that have been developed for pyridine synthesis,[2] methods based on the Diels-Alder reaction are particularly attractive due to their intrinsic convergent character and regiochemical features. Although a variety of azadienophiles[3] have been used for the synthesis of tetrahydropyridines, few examples of their application in the synthesis of pyridines have previously been reported. The current procedure illustrates a versatile method for the synthesis of substituted pyridines via [4+2] cycloaddition of an oximinosulfonate derived from Meldrum's acid with conjugated dienes. Diels-Alder reactions of this new azadienophile are subject to Lewis acid promotion, permitting highly regioselective cycloadditions with a wide range of diene substrates under relatively mild reaction conditions.

As described here, we have developed a simple procedure that delivers multigram quantities of the key oximinosulfonate in good yield and excellent purity in one operation without the need for additional purification.[4] The oximinosulfonate is an easily handled solid that can be transferred in air and is stable for months when stored under argon in a refrigerator. Reaction of the oximinosulfonate with a conjugated diene and 2 equiv of Me_2AlCl leads to an efficient Diels-Alder reaction, providing cycloadducts which can then be transformed to substituted pyridines in a single synthetic operation. Significantly, the regiochemical course of the cycloaddition is opposite to that observed with conventional imino dienophiles, thus producing heterocycles with substitution patterns that cannot be accessed using prior imino-dienophile Diels-Alder methodology. A series of experiments in which the Me_2AlCl stoichiometry was varied from 0.1-2.5 equivalents established the requirement that a full 2.0 equiv of the Lewis acid be employed. As shown below, it is likely that the second equiv of Me_2AlCl serves to promote the ionization of chloride from an initial 1:1 Lewis acid complex and thereby generate the more reactive ionic 2:1 complex.[5,6]

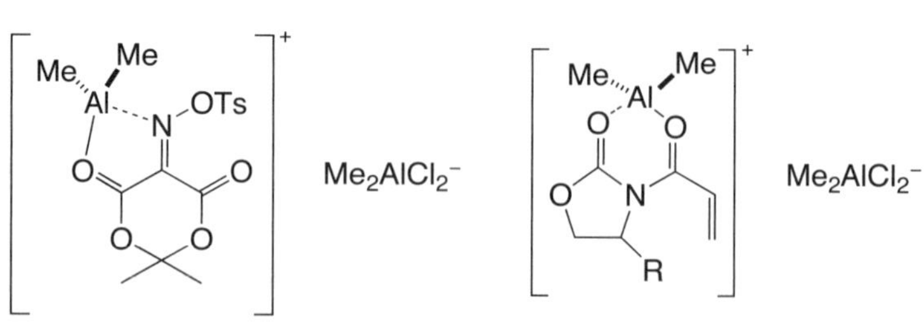

The cycloadducts obtained in the oximinosulfonate Diels-Alder reaction are best converted directly to pyridines without purification. Exposure of the spiro-fused cycloadducts to a combination of NCS and sodium methoxide brings about cleavage of the dioxanedione ring with concomitant elimination of acetone and carbon dioxide. Elimination of tosylate from the resulting ester enolate then generates a dihydropyridine, and subsequent chlorination by NCS and elimination of HCl finally provides the desired aromatic pyridine product.

The Lewis acid-promoted reaction of our oximinosulfonate with dienes and the conversion of the resulting cycloadducts to pyridines comprises a new annulation method for the synthesis of substituted pyridines from conjugated dienes. As illustrated in the Table, very good overall yields are obtained in reactions of 2-substituted dienes, providing 5-substituted pyridine-2-carboxylates. Reactions with 1-substituted dienes yield 3-substituted pyridines, and disubstituted dienes react smoothly to afford trisubstituted pyridines in good yield. Polycyclic systems are obtained when dienes such as 1-vinyl-1-cyclohexene are employed in the annulation.

TABLE
[4+2] Pyridine Annulations[a]

diene[b]	pyridine	% yield[c]
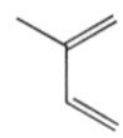	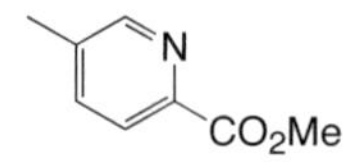	78-85
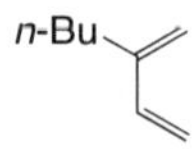	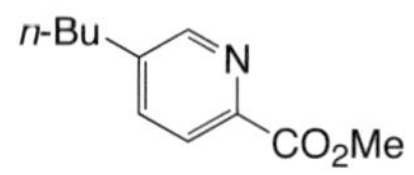	72
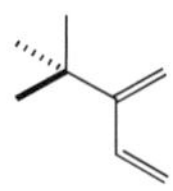	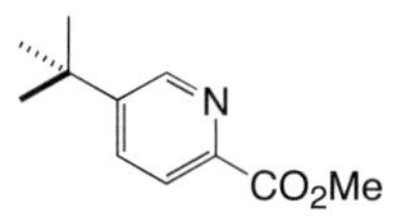	73
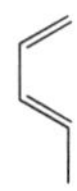	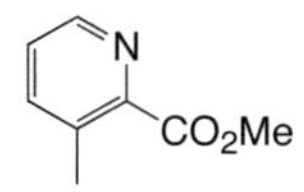	56-57
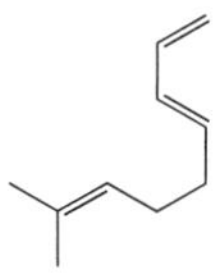	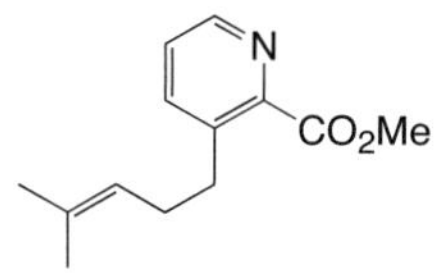	60
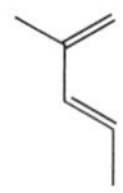	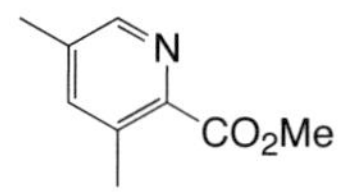	70
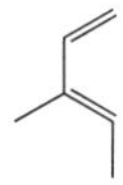	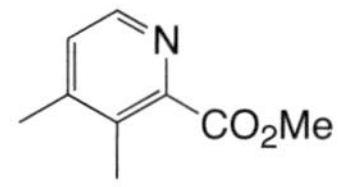	40
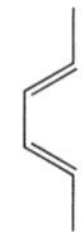	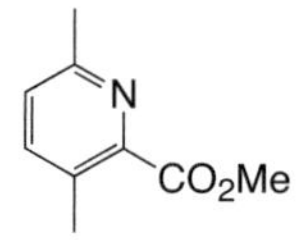	40[d]
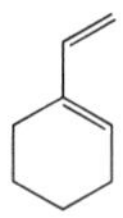	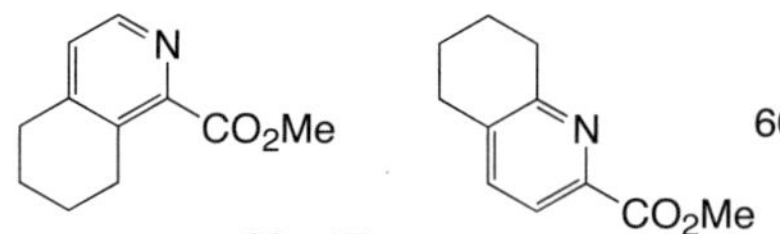 83 : 17	60

[a]Conditions: (a) oximinosulfonate, 2.0 equiv Me_2AlCl, CH_2Cl_2, −78 °C, 2-4 h; (b) 3.0 equiv MeONa, 1.0 equiv NCS, MeOH-THF (1:1), rt, 14-16 h. [b]1.5-3.0 equiv of diene was employed. [c]Isolated over all yield for two steps. [d]5.0 equiv of NaOMe and 4.0 equiv of NCS were used.

1. Department of Chemistry, Massachusetts Institute of Technology, Cambridge, MA 02139. We thank the National Institutes of Health (GM 28273) for generous financial support. D.T.A. was supported in part by NIH training grant CA 09112.
2. For general reviews of the synthesis and chemistry of pyridines, see: (a) Scriven, E. F. V. in *Comprehensive Heterocyclic Chemistry*; Katritzky, A. R., Rees, C. W., Eds.; Pergamon: New York, **1984**; Vol. 2, Part 2A, 165-314. (b) Yates, F. S. in *Comprehensive Heterocyclic Chemistry*; Katritzky, A. R., Rees, C. W., Eds.; Pergamon: New York, **1984**; Vol. 2, Part 2A, 511-524. (c) Tomasik, P.; Ratajewicz, Z. in *Pyridine-Metal Complexes*; Newkome, G. R., Strekowski, L., Eds.; *The Chemistry of Heterocyclic Compounds*; Wiley: New York, **1985**; Vol. 14, Part 6A. (d) Jones, G. in *Comprehensive Heterocyclic Chemistry*; Katritzky, A. R., Rees, C. W., Eds.; Pergamon: New York, **1984**; Vol. 2, Part 2A, 395-510. (e) *Pyridine and its Derivatives*; Newkome, G. R., Ed; *The Chemistry of Heterocyclic Compounds*; Wiley: New York, **1984**; Vol. 14, Part 5.
3. For reviews of the application of azadienophiles in the Diels-Alder reaction, see:(a) Weinreb, S. M. in *Comprehensive Organic Synthesis*; Trost, B. M.; Fleming, I., Eds.; Pergamon Press: Oxford, **1991**, Vol. 5, 401-449. (b) Weinreb, S. M.; Levin, J. I. *Heterocycles* **1979**, 12, 949. (c) Weinreb, S. M.; Staib, R. R. *Tetrahedron* **1982**, 38, 3087. (d) Boger, D. L.; Weinreb, S. M. *Hetero Diels-Alder Methodology in Organic Synthesis*; Academic: San Diego, **1987**; Chapter 2. (e) Tietze, L. F.; Kettschau, G. Top. *Curr. Chem.* **1997**, *189*, 1.
4. Renslo, A. R.; Danheiser, R. L. *J. Org. Chem.* **1998**, *63*, 7840.
5. Lehmkuhl, H.; Kobs, H.-D. *Liebigs Ann. Chem.* **1968**, *719*, 11.
6. Evans, D. A.; Chapman, K. T.; Bisaha, J. *J. Am. Chem. Soc.* **1988**, *110*, 1238.

GENERATION AND [2+2] CYCLOADDITIONS OF THIO-SUBSTITUTED KETENES: *trans*-1-(4-METHOXYPHENYL)-4-PHENYL-3-(PHENYLTHIO)AZETIDIN-2-ONE

[(2-Azetidinone, 1-(4-methoxyphenyl)-4-phenyl-3-(phenylthio)-, *trans*-)]

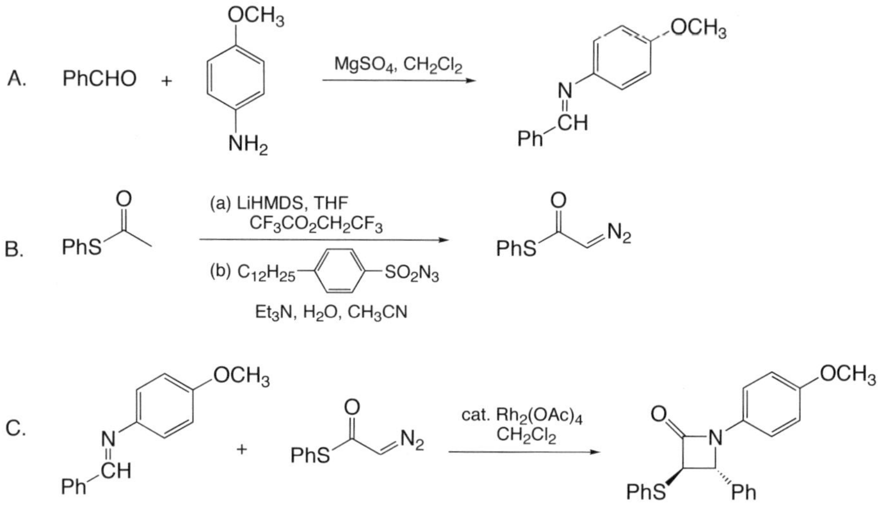

Submitted by Rick L. Danheiser, Iwao Okamoto, Michael D. Lawlor, and Thomas W. Lee.[1]

Checked by Scott E. Denmark and Michael H. Ober.

1. Procedure

Caution! Diazo compounds are presumed to be toxic and potentially explosive and therefore should be handled with caution in a fume hood. Although in carrying out this reaction numerous times we have never observed an explosion, we recommend that these reactions be conducted behind a safety shield.

A. N-Benzylidene-p-anisidine. A 100-mL, three-necked, round-bottomed flask is equipped with an argon inlet adapter, rubber septum, glass stopper, and a magnetic stirring bar (Note 1). The flask is charged with 45 mL of dichloromethane (CH_2Cl_2) (Note 2) and 3.00 mL (0.030 mol) of benzaldehyde (Note 3), and then is cooled in an ice-water bath while a solution of 3.50 g (0.028 mol) of *p*-anisidine (Note 4) in 5 mL of CH_2Cl_2 is added dropwise via syringe over 15 min. After 30 min, 7.5 g of anhydrous magnesium sulfate is added in one portion. The ice-water bath is removed, and the reaction mixture is stirred at room temperature for 2 h. The resulting mixture is then filtered through a sintered glass funnel with the aid of two 5-mL portions of CH_2Cl_2, and the filtrate is concentrated at reduced pressure by rotary evaporation at room temperature to afford a pale brown powder. This material is dissolved in 150 mL of ethanol heated in an 80 °C water bath while 270 mL of hot water is added with stirring. The resulting solution is allowed to cool to room temperature and then is cooled in an ice-water bath for 2 h. Filtration provides 5.31 g (88%) of *N*-benzylidene-*p*-anisidine as brown metallic plates (Note 5).

B. S-Phenyl diazothioacetate. A 250-mL, three-necked, round-bottomed flask is equipped with a magnetic stirring bar, argon inlet adapter, rubber septum, and a 50-mL pressure-equalizing dropping funnel fitted with a rubber septum (Note 1). The flask is charged with 50 mL of dry tetrahydrofuran (THF) (Note 6) and 12.0 mL (0.057 mol) of 1,1,1,3,3,3-hexamethyldisilazane (Note 7), and then is cooled in an ice-water bath while 20.7 mL (0.052 mol) of a 2.58M solution of *n*-butyllithium in hexane (Note 8) is added dropwise over 5 min by syringe. After 10 min, the resulting solution is cooled at −78 °C in a dry ice-acetone bath, and a solution of 7.00 mL (0.052 mol) of *S*-phenyl thioacetate (Note 9) in 40 mL of dry tetrahydrofuran is added dropwise via the addition funnel over 30 min (the funnel is rinsed with two 2-mL portions of dry THF). The reaction mixture is allowed to stir for 30 min at −78 °C, and then 8.50 mL (0.063 mol) of 2,2,2-trifluoroethyl trifluoroacetate (Note 10) is added rapidly in one portion via syringe. After 10 min, the reaction mixture is poured into 200 mL of 5% aqueous hydrochloric acid and is extracted with three portions (250, 100, and 100 mL) of ether. The combined organic phases are washed with 100 mL of saturated sodium chloride solution, dried over anhydrous sodium sulfate, filtered, and concentrated under reduced pressure with a rotary evaporator, then under high vacuum to afford a colorless solid. This material is immediately dissolved in 100 mL of acetonitrile (Note 11) and is transferred into a 500-mL, three-necked, round-bottomed flask equipped with a magnetic stirring bar, argon inlet adapter, rubber septum, and 125-mL pressure-equalizing dropping funnel fitted with a rubber septum. Water (0.95 mL, 0.053 mol) and 10.0 mL (0.072 mol) of triethylamine (Note 12) are added via syringe, and a solution of 26.9 g (0.076 mol) of dodecylbenzenesulfonyl azide[2] (Note 13) in 45 mL of acetonitrile is then added through the dropping funnel over 15 min. After stirring at room temperature for 16 h, the solution is concentrated at reduced pressure with a rotary evaporator and then a vacuum pump to give a brown oil. Silica gel (20

g) is added, and the resulting slurry is loaded onto a column (60 mm) of 200 g of silica gel (230-400 mesh) and is eluted with 5% ethyl acetate-hexane to afford 7.16 g (77%) of *S*-phenyl diazothioacetate as a pale yellow oil (Note 14).

C. trans-1-(4-Methoxyphenyl)-4-phenyl-3-(phenylthio)azetidin-2-one. A 500-mL, three-necked, round-bottomed flask is equipped with a magnetic stirring bar, reflux condenser fitted with an argon inlet adapter, glass stopper, and 50-mL, pressure-equalizing dropping funnel fitted with a glass stopper (Note 1). The flask is charged with 3.50 g (0.016 mol) of *N*-benzylidene-*p*-anisidine, 200 mL of CH_2Cl_2 (Note 2), and 0.045 g (0.10 mmol) of rhodium(II) acetate dimer (Note 15), and the resulting green solution is heated at reflux while a solution of 4.43 g (0.025 mol) of *S*-phenyl diazothioacetate in 40 mL of CH_2Cl_2 is added via the dropping funnel over 1 h (the dropping funnel is rinsed with 2 mL of CH_2Cl_2). The reaction mixture is further heated at reflux for 5 min, then allowed to cool to room temperature. After transfer to a 500-mL, round-bottomed flask, the mixture is concentrated by rotary evaporation at reduced pressure to provide a brown oil. This material is filtered through a column of 50 g of silica gel (elution with 750 mL of CH_2Cl_2) to remove polar impurities and is concentrated under reduced pressure to afford a brown solid, which is washed on a sintered glass funnel with 10 mL of ethyl acetate and then 20 mL of hexane. The resulting pale yellow powder is dissolved in 40 mL of ethyl acetate at 80 °C, and 400 mL of hexane (pre-heated in a water bath at 80 °C) is then added in one portion. The resulting solution is allowed to cool to room temperature and then is cooled further at −20 °C for 2 h to afford 5.50 g (91%) of *trans*-1-(4-methoxyphenyl)-4-phenyl-3-(phenylthio) azetidin-2-one as off-white crystals (Notes 16 and 17).

2. Notes

1. The apparatus is flame-dried under reduced pressure and then maintained under an atmosphere of argon during the course of the reaction.

2. Dichloromethane was distilled from calcium hydride immediately before use.

3. Benzaldehyde was purchased from Aldrich Chemical Company, Inc., and was distilled before use.

4. *p*-Anisidine was purchased from Aldrich Chemical Company, Inc. and used as received.

5. The imine has the following physical properties: mp 70-71 °C (lit.[3] 70-71 °C); ^{1}H NMR (500 MHz, $CDCl_3$) δ: 3.83 (s, 3 H), 6.93 (d, 2 H, J = 9.0), 7.24 (d, 2 H, J = 8.8), 7.46 (m, 3 H), 8.48 (s, 1 H), 7.89 (m, 2 H); ^{13}C NMR (166 MHz, $CDCl_3$) δ: 55.7, 114.6, 122.4, 128.8, 128.9, 131.3, 136.6, 145.2, 158.5, 158.7; IR (CCl_4) cm^{-1}: 3065, 3030, 3002, 2959, 2935, 2909, 2880, 2835, 1626, 1529, 1505, 1464, 1366, 1289, 1245, 1191, 1106, 1040, 920, 882, 828, 761, 691, 542; Anal. Calcd for $C_{14}H_{13}NO$: C, 79.59; H, 6.20; N, 6.63. Found: C, 79.50; H, 6.05; N, 6.74.

6. Tetrahydrofuran was distilled from sodium benzophenone ketyl immediately before use.

7. 1,1,1,3,3,3-Hexamethyldisilazane was purchased from Aldrich Chemical Company, Inc., and was distilled from calcium hydride prior to use.

8. *n*-Butyllithium was purchased from Fisher Scientific Company, and was titrated prior to use according to the method of Watson and Gilman.[4]

9. *S*-Phenyl thioacetate was purchased from Aldrich Chemical Company, Inc. and used as received.

10. 2,2,2-Trifluoroethyl trifluoroacetate was purchased from Aldrich Chemical Company, Inc., and was distilled from calcium hydride prior to use.

11. Acetonitrile was distilled from calcium hydride prior to use.

12. Triethylamine was purchased from EM Science, and was distilled from calcium hydride prior to use.

13. Dodecylbenzenesulfonyl azide was purchased by the checkers from TCI Americas Inc. and was used as received.

14. *S*-Phenyl diazothioacetate decomposes slowly on storage at −20 °C with the generation of nitrogen gas. The diazo compound is best used immediately in the next step or purified by chromatography immediately before use. Physical properties are as follows: ^{1}H NMR (500 MHz, $CDCl_3$) δ: 5.25 (s, 1 H), 7.40-7.45 (m, 3 H), 7.49-7.51 (m, 2 H); ^{13}C NMR (166 MHz, $CDCl_3$) δ: 54.5, 127.6, 129.6, 130.1, 135.6, 184.3; IR (thin film) cm^{-1}: 3099, 3077, 2466, 2354, 2269, 2108, 1641, 1477, 1441, 1334, 1139, 1022, 854, 748, 690, 638.

15. Rhodium(II) acetate dimer was purchased from Alfa Aesar Company and used as received.

16. Physical properties of the β-lactam are as follows: mp 144-146 °C (lit. 142 °C[5], 147-148 °C[6]); ^{1}H NMR (500 MHz, $CDCl_3$) δ: 3.72 (s, 3 H), 4.25 (d, 1 H, J = 2.4), 4.77 (d, 1 H, J = 2.4), 7.52-7.51 (m, 2 H), 7.36-7.34

(m, 3 H), 7.31-7.26 (m, 5 H), 7.16 (d, 2 H, J = 9.2), 6.75 (d, 2 H, J = 9.2); ^{13}C NMR (166 MHz, $CDCl_3$) δ: 162.9, 156.5, 136.6, 132.5, 130.9, 129.6, 129.5, 129.4, 129.2, 128.2, 126.3, 118.8, 114.5, 63.3, 61.7, 55.6; IR (CCl_4) cm^{-1}: 3065, 3033, 3003, 2953, 2934, 2909, 2835, 1760, 1512, 1455, 1381, 1248, 1179, 1137, 1067, 1040, 828, 736, 696, 590; Anal. Calcd for $C_{22}H_{19}NO_2S$: C, 73.10; H, 5.30; N, 3.88; S, 8.87. Found: C, 73.03; H, 5.29; N, 4.02; S, 9.20.

17. Cleavage of the *p*-methoxyphenyl group can be achieved in 65% yield by the following procedure. A three-necked, 250-mL, round-bottom flask, equipped with a magnetic stirbar, rubber septum, argon inlet adapter, and glass stopper, is charged with 0.512 g (1.42 mmol) of *trans*-1-(4-methoxyphenyl)-3-(phenylthio)azetidin-2-one and 55 mL of acetonitrile. After cooling the flask in an ice-water bath, a solution of 2.380 g (4.34 mmol) of ceric ammonium nitrate in 20 mL of water is added by cannula over 5 min followed by stirring for 1 h. The reaction mixture is poured into 500 mL of water and is extracted with three 200-mL portions of ethyl acetate. The organic phases are combined and washed with 400 mL of 5% aqueous $NaHCO_3$ solution. The washes are back extracted with 100 mL of ethyl acetate. The combined organic phases are washed additionally with six 200-mL portions of saturated aqueous Na_2SO_3 solution, then two 150-mL portions of saturated aqueous NaCl solution. After drying with Na_2SO_4, the solution is concentrated to give a brown oil. Silica gel (1.5 g) is added, and the resulting slurry is applied to a column of 10 g of silica gel and eluted with 10% ethyl acetate-hexane. The eluent is concentrated to provide a yellow solid which is dissolved in 125 mL of boiling hexane, which is then reduced to a total volume of 50 mL. The resulting solution is allowed to cool to room temperature, then is cooled further to −20 °C to afford 0.234 g (65%) of 4-phenyl-3-(phenylthio)-2-azetidinone as a pale yellow solid, mp 111-112 °C (lit.[7] 110.5-111.5 °C).

Waste Disposal Information

All toxic materials were disposed of in accordance with "Prudent Practices in the Laboratory"; National Academy Press; Washington, DC, 1995.

3. Discussion

[2+2] Cycloadditions of ketenes with alkenes and alkynes constitute the most popular method for the synthesis of cyclobutanones and cyclobutenones. Unfortunately, however, this process is truly general only for highly nucleophilic ketenophiles such as conjugated dienes and enol ethers. In general, unactivated alkenes and alkynes fail to react in good yield with either alkyl- or aryl-substituted ketenes, or with ketene itself. To circumvent this limitation, dichloroketene is usually employed as a ketene equivalent, since this electrophilic ketene reacts well with many types of unactivated multiple bonds, and the resultant cycloadducts undergo facile dechlorination under mild conditions.[8]

The procedure described here illustrates a practical and convenient method for the generation of thio-substituted ketenes which participate in surprisingly facile cycloadditions with both activated and unactivated alkenes, alkynes, and imines to form four-membered carbocycles and heterocycles. As outlined in the following scheme, exposure of α-diazo thiol esters to the action of catalytic rhodium(II) acetate leads to a remarkably facile "thia-Wolff rearrangement," producing thio-substituted ketenes which combine with a variety of ketenophiles to provide access to α-thiocyclobutanones, cyclobutenones, and β-lactams.

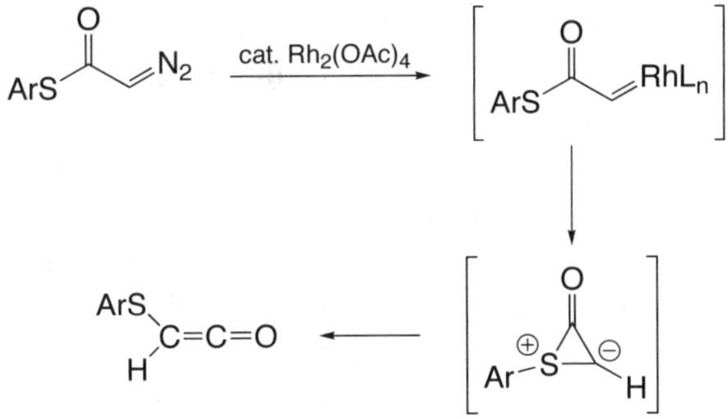

As illustrated below, desulfurization of α-thiocyclobutanone cycloadducts is easily achieved in high yield upon exposure to either tributyltin hydride or activated zinc dust, and the overall sequence thus represents a useful new alternative to the existing dichloroketene-based methodology for the synthesis of a variety of four-membered carbocycles and heterocycles.[9]

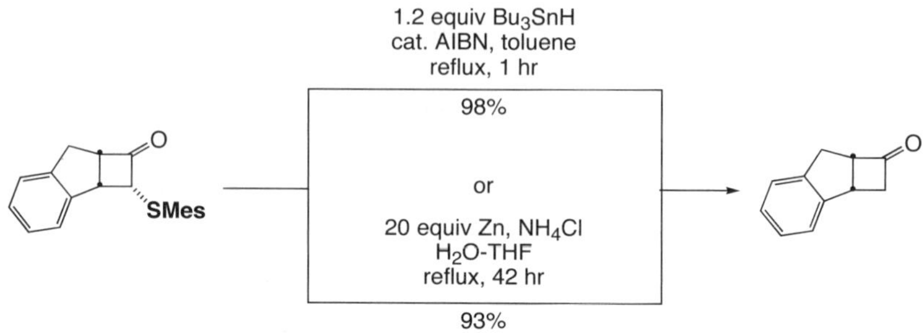

The requisite α-diazo thiol esters are conveniently prepared by using the "detrifluoroacetylative" diazo transfer strategy previously developed in our laboratory.[10] Cycloadditions are best carried out by using as little as 0.006 equiv of rhodium(II) acetate to promote the thia-Wolff rearrangement. Reactions involving the more nucleophilic ketenophiles proceed smoothly in refluxing dichloromethane (40 °C), while cycloadditions with less reactive partners are best accomplished in 1,2-dichloroethane (83 °C). As is standard for ketene cycloadditions, the optimal protocol involves slowly adding a solution of the diazo thiol ester to a solution of the ketenophile and catalyst in order to minimize competitive ketene dimerization.

Examples of the application of this chemistry to the preparation of cyclobutanones, cyclobutenones, and β-lactams are presented in the Table. The mesityl thiol ester has proven to be particularly effective in reactions with less ketenophilic alkenes, although with the more reactive ketenophiles nearly identical results are obtained using either the mesityl α-diazo thiol ester or the more readily available thiophenyl ester. In the case of readily available ketenophiles, the reaction is best conducted using excess alkene, alkyne, or imine, but in other cases the cycloaddition can be carried out with excess diazo thiol ester. The efficiency of the reaction with unactivated alkenes is especially notable, and compares favorably with results obtained previously employing dichloroketene. For example, addition of dichloroketene to methylenecyclohexane is reported to proceed in 55% yield,[11] while up to 81% of the desired [2+2] cycloadduct is produced in the reaction of (mesitylthio)ketene with this olefin under our conditions.

TABLE

Preparation of Cyclobutanones and β-Lactams from α-Diazo Thiol Esters

ketenophile	diazo thiol ester	reaction conditions[a]	product	% yield (ratio)[b]
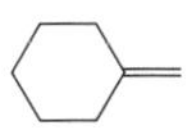	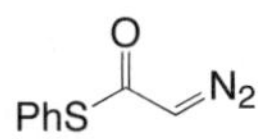	A	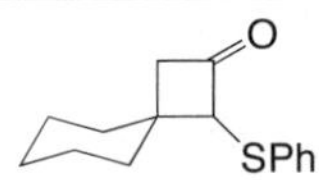	73-78
	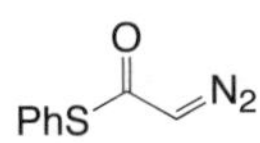	B	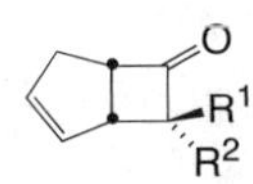 R^1 = H, R^2 = SPh; R^1 = SPh, R^2 = H	90-96 (5-11:1)
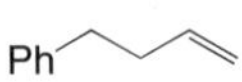	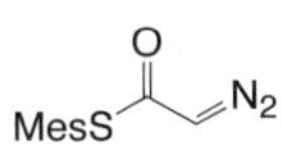	A	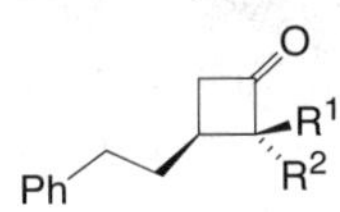 R^1 = H, R^2 = SMes; R^1 = SMes, R^2 = H	62 (83:17)
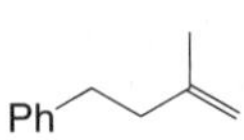	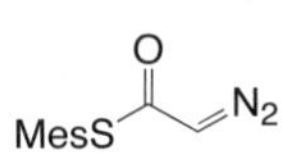	A	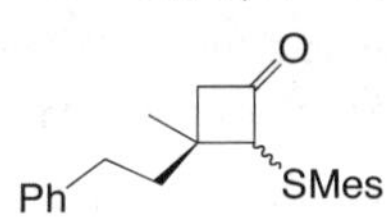	59 (57:43)
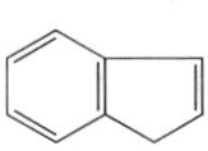	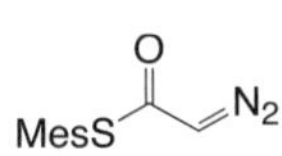	A	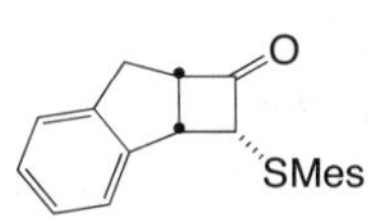	58
PhCH=NPh	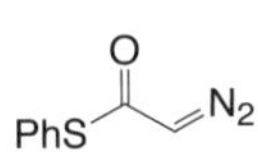	B	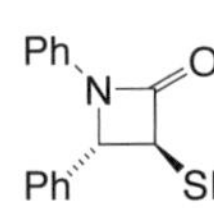	85
	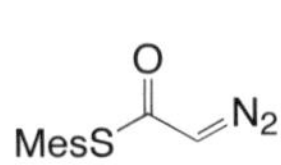	A	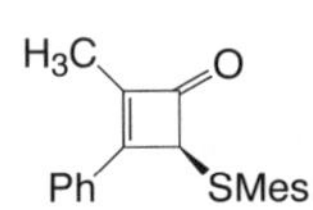	50

[a]The diazo thiol ester is treated with 0.02 equiv of $Rh_2(OAc)_4$ and 3.5-10 equiv of ketenophile. Procedure A: reaction in 1,2-dichloroethane at 83 °C for 3 h. Procedure B: reaction in dichloromethane at 40 °C for 3 h. [b]Isolated yield and ratio of diastereoisomers.

1. Department of Chemistry, Massachusetts Institute of Technology, Cambridge, MA 02139. We thank the National Institutes of Health (GM 28273) for generous financial support. M.D.L. was supported in part by NIH training grant CA 09112 and I. O. was supported as a J.S.P.S. Research Fellow. T.W.L. was a James Flack

Norris Summer Undergraduate Research Scholar of the Northeastern Section of the American Chemical Society and a Pfizer Undergraduate Summer Research Fellow.

2. Hazen, G. G.; Bollinger, F. W.; Roberts, F. E.; Russ, W. K.; Seman, J. J.; Staskiewicz, S. *Org. Synth. Coll. Vol. IX* **1998**, 400.

3. Roe, A.; Montgomery, J. A. *J. Am. Chem. Soc.* **1953**, *75*, 910.

4. Gilman, H.; Cartledge, F. K. *J. Organomet. Chem.* **1964**, *2*, 447.

5. van der Veen, J. M.; Bari, S. S.; Krishnan, L.; Manhas, M. S.; Bose, A. K. *J. Org. Chem.* **1989**, *54*, 5758.

6. Cossío, F. P.; Ganboa, I.; García, J. M.; Lecea, B.; Palomo, C. *Tetrahedron Lett.* **1987**, *28*, 1945.

7. Hart, D. J.; Kanai, K.; Thomas, D. G.; Yang, T.-K. *J. Org. Chem.* **1983**, *48*, 289.

8. For recent reviews, see: (a) Hyatt, J. A.; Raynolds, P. W. *Org. Reactions* **1994**, *45*, 159. (b) Tidwell, T. T. Ketenes; Wiley: New York, 1995. (c) Schaumann, E.; Scheiblich, S. in *Methoden der Organischen Chemie* (Houben-Weyl); Kropf, H., Schaumann, E., Eds.; Thieme: Stuttgart, Germany, **1993**; Vol. E15, Parts 2 and 3. (d) Lee-Ruff, E. in *Methods of Organic Chemistry* (Houben-Weyl); de Meijere, A., Ed.; Thieme: Stuttgart, Germany, **1997**; Vol. E17e, pp 190-219.

9. Lawlor, M. D.; Lee, T. W.; Danheiser, R. L. *J. Org. Chem.* **2000**, *65*, 4375.

10. (a) Danheiser, R. L.; Miller, R. F.; Brisbois, R. G.; Park, S. Z. *J. Org. Chem.* **1990**, *55*, 1959. (b) Danheiser, R. L.; Miller, R. F.; Brisbois, R. G. *Org. Synth. Coll. Vol. IX* **1998**, *197*.

11. Dunkelblum, E. *Tetrahedron* **1976**, *32*, 975.

PREPARATION OF 2-CHLORO-1,3 BIS(DIMETHYLAMINO) TRIMETHINIUM HEXAFLUOROPHOSPHATE

[[Methanaminium, *N*-[2-chloro-3-(dimethylamino)-2-propenylidene]-*N*-methyl-, hexafluorophosphate(1-)]]

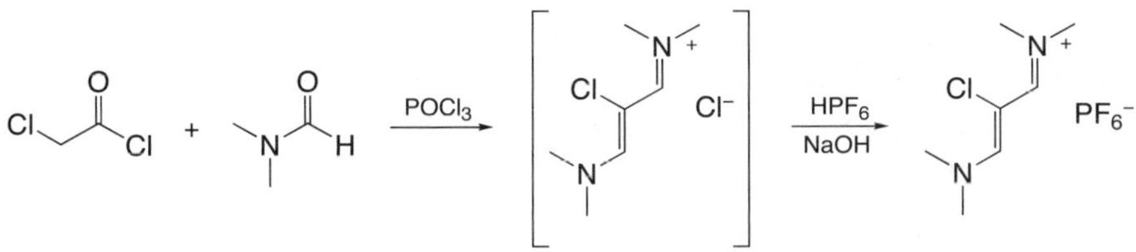

Submitted by Ian W. Davies, Jean-Francois Marcoux, and Jeremy Taylor.[1]
Checked by Yu Fan and Scott E. Denmark.

1. Procedure

A 250-mL, three-necked, round-bottomed flask, equipped with a stir bar, nitrogen bubbler (Note 1), thermocouple, and pressure-equalizing addition funnel, is charged with dimethylformamide (DMF) (60 mL) (Note 2). Chloroacetyl chloride (14.13 g, 0.125 mol) (Notes 3-5) is added over 5 min at 50 °C. The mixture is then heated to 65-70 °C to give a clear, yellow solution. Phosphorus oxychloride (19.20 g, 0.125 mol) (Notes 6, 7) is added at ≈5 mL/h maintaining the temperature of the mixture at 70 °C. After completion of the addition, the mixture is heated for 3 h to give a deep red solution. The reaction mixture is cooled to 20-30 °C and transferred to a pressure-equalizing addition funnel. The reaction flask is rinsed with DMF (10 mL), which is added to the addition funnel. A solution of sodium hexafluorophosphate is prepared by the addition of 5N aqueous sodium hydroxide (NaOH) solution (37 mL) to a solution of hexafluorophosphoric acid *(Caution: Hexafluorophosphoric acid is toxic and corrosive)* (33.15 g of 60 wt%, 0.135 mol) in water (150 mL) in a 1-L, three-necked, round-bottomed flask fitted with a mechanical stirrer (Note 8). The vinamidinium chloride solution (≈80 mL) and 5N aqueous NaOH solution (70 mL) are added concurrently over 1 h to the solution of sodium hexafluorophosphate in water (150 mL) with stirring at a temperature of <10 °C to give a yellow slurry (Note 9). The mixture is aged for 1 h, then the solids are collected under vacuum. The crude solids are washed with water (75 mL) (Note 10), then transferred to a 500-mL, three-necked, round-bottomed flask equipped with a mechanical stirrer and thermocouple. Water (225 mL) and 2-propanol

(60 mL) are added to the flask. The mixture is heated to 70 °C to dissolve the solids (Note 11). The mixture is then cooled to 0-5 °C over 1 h. The light-yellow solids are collected by filtration, washed with cold water/2-propanol (50 mL, 20/1), and dried under reduced pressure at <40 °C (Note 12) to give 2-chloro-1,3-bis(dimethylamino)trimethinium hexafluorophosphate as a light-yellow solid (29.1 g, 77%) (Notes 13, 14). *Caution: The salt is a skin and eye irritant and was positive in the Ames mutagenesis assay. It should be handled in a hood with adequate personal protective equipment.* The product is >99.9% pure by HPLC analysis (Note 15).

2. Notes

1. The reaction is run under dry nitrogen.
2. A fresh bottle of anhydrous DMF from Aldrich Chemical Co., Inc. was used. The water content of the DMF was 0.024 g/L by Karl Fischer titration.
3. The checkers purchased chloroacetyl chloride from Aldrich Chemical Co., Inc.
4. Chloroacetic acid may be used in place of chloroacetyl chloride by adjusting the charge of phosphorus oxychloride to 0.25 mol. The submitters purchased chloroacetic acid from Säurefabrik Schweizerhall (CH). The material is also available from Aldrich Chemical Co., Inc.
5. The addition is mildly exothermic.
6. A syringe pump was used for this addition by the checkers.
7. When chloroacetic acid is used in place of chloroacetyl chloride, the charge of phosphorus oxychloride is increased to 0.25 mol.
8. The checkers purchased hexafluorophosphoric acid from Aldrich Chemical Co., Inc.
9. The addition is mildly exothermic. Sodium hexafluorophosphate is commercially available although significantly more expensive. The sodium salt may be used in place of the hexafluorophosphoric acid by omitting the initial sodium hydroxide charge.
10. The crude solids may be dried and were ≈98 area% pure by HPLC analysis according to the submitters. The crude material may be used in a number of applications without recrystallization.
11. The submitters have observed that aging at elevated temperature for long periods (>8 h) will lead to a 3-5% decrease in yield. Over-heating the solids on the walls of the flask may discolor the product.

12. According to the submitters there is a minor exotherm initiating at 60 °C by differential scanning (Reactive System Screening Tool) (dT/dt °C/min 2.0, dP/dt °C/min <1). The exotherm is non-hazardous but will lead to deterioration of product quality presumably due to hydrolysis.

13. The submitters have run this procedure at various scales up to 300 kg with yields ranging from 73-79%.

14. The product is fully characterized: mp 125-126 °C; ^{1}H NMR (400 MHz, DMF-d_7) δ: 3.44 (s, 6 H), 3.60 (s, 6 H), 7.97 (s, 2 H); ^{13}C NMR (125 MHz, DMF-d_7) δ: 39.9, 49.7, 92.6, 161.1; ^{19}F NMR (470 MHz DMF-d7) δ: −72.3 (d, J = 708); ^{31}P NMR (202 MHz, DMF-d_7) 143.2 (septet, J = 709); Anal. Calcd. for $C_7H_{14}ClN_2PF_6$: C, 27.42; H, 4.60; N, 9.14; Cl, 11.56; F, 37.18; P, 10.10. Found: C, 27.09; H, 4.86; N, 8.85; Cl, 11.47; F, 37.14; P, 10.09%. IR (KBr) cm^{-1}: 1610(s), 1492(w), 1424(s), 1404(s), 1291(s), 1200(s), 1117(s), 1045(m), 963(w), 840(s), 740(w); MS (FAB 70 eV): 161(100), 162(3), 163(32), 164(3). X-ray crystallographic analysis reveals a "W"-conformation.

15. A reverse phase ion-pairing HPLC method was developed by the submitters for analysis. Chromatographic conditions: A 10-μL sample (0.1 mg/mL in acetonitrile) is injected onto a suitable liquid chromatograph equipped with a Waters Symmetry Shield RP18 column, 250 × 4.6 mm, 5 μm particle size at 40 °C with a mobile phase of 0.404 g/L heptanesulfonic acid, sodium salt + 0.1% phosphoric acid (Component A, pH 2.2) and acetonitrile (Component B) at a flow rate of 1.0 mL/min, programmed with a linear gradient from 95:5 A:B (v/v) to 30:70 A:B (v/v) over 20 min. Detection is achieved by UV at 300 nm. The retention time is approximately 10 min.

Waste Disposal Information

All toxic materials were disposed of in accordance with "Prudent Practices in the Laboratory"; National Academy Press, Washington, DC, 1995.

3. Discussion

Vinamidinium salts are important intermediates in the synthesis of heterocycles.[2] The 2-chloro-1,3-bis(dimethylamino)trimethinium hexafluorophosphate salt has been used in the preparation of the highly selective Cox-2 inhibitor etoricoxib (Scheme 1).[3] This method describes a straightforward preparation of the hexafluorophosphate salt which is a crystalline, thermally and shock-stable, non-hygroscopic solid. The submitters have extensively studied the preparation of vinamidinium salts and demonstrated that the method is applicable to substituted acetic acids that contain an electron-withdrawing group (Table 1).[4] The annulation reaction is also general and useful for the preparation of pyridines, pyridones and pyridine *N*-oxides.[5]

Scheme 1

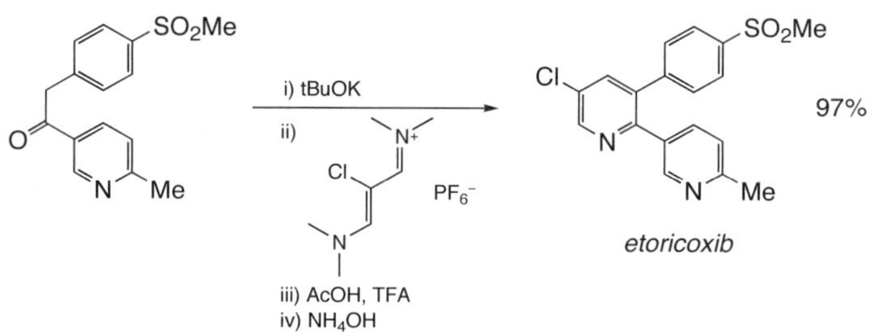

Table 1

Entry	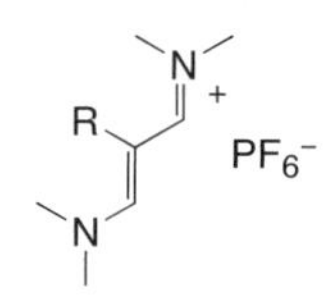	Yield
1	R = Cl	78 %
2	R = Br	78 %
3	R = I	60 %
4	R = CF_3	68 %
5	R = Ph	90 %
6	R = 4-NO_2-C_6H_4-	80 %
7	R = 4-MeO-C_6H_4-	75 %
8	R = 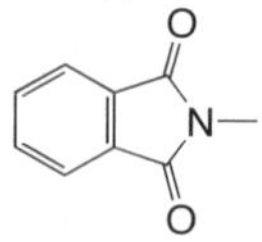	68 %

1. Department of Process Research, Merck & Co., Inc., PO Box 2000, Rahway, NJ 07065-0900.

2. (a) For examples see: Gupton, J. T.; Petrich, S. A.; Hicks, F. A.; Wilkinson, D. R.; Vargas, M.; Hosein, K. N.; Sikorski, J. A. *Heterocycles* **1998**, *47*, 689. (b) Kase, K.; Katayama, M.; Ishihara, T.; Yamanaka, H.; Gupton, J. T. *J. Fluor. Chem.* **1998**, *90*, 29.

3. Davies, I. W.; Marcoux, J. -F.; Corley, E. G.; Journet, M.; Cai, D.-W.; Palucki, M.; Wu, J.; Larsen, R. D.; Rossen, K.; Pye, P. J.; DiMichele, L.; Dormer, P.; Reider, P. J. *J. Org. Chem.* **2000**, *65*, 8415.

4. (a) Davies, I. W.; Marcoux, J.-F.; Wu, J.; Palucki, M.; Corley, E. G.; Robbins, M. A.; Tsou, N.; Ball, R. G.; Dormer, P.; Larsen, R. D.; Reider, P. *J. Org. Chem.* **2000**, *65*, 4571. (b) The method is general, but the reaction times/temperatures may require some slight modification to give optimal yields of the desired vinamidinium. In the case of the trifluoromethylvinamidinium 60 °C is optimal: Davies, I. W.; Tellers, D. M.; Shultz, C. S.; Fleitz, F. J.; Cai, D. -W.; Sun, Y. *Org. Lett.* **2002**, *4*, 2969.

5. (a) Marcoux, J.-F.; Corley, E. G.; Rossen, K.; Pye, P.; Wu, J.; Robbins, M. A.; Davies, I. W.; Larsen, R. D.; Reider, P. J. *Org. Lett.* **2000**, *2*, 2339. (b) Marcoux, J.-F.; Marcotte, F.-A.; Wu, J.; Dormer, P. G.; Davies, I. W.; Hughes, D.; Reider, P. J. *J. Org. Chem.* **2001**, *66*, 4194. (c) Davies, I. W.; Marcoux, J.-F.; Reider, P. J. *Org. Lett.* **2001**, *3*, 209.

SYNTHESIS OF ALPHA-HALO ETHERS FROM SYMMETRIC ACETALS AND *in situ* METHOXYMETHYLATION OF AN ALCOHOL

[1-Methoxymethoxy-1-phenylethane]

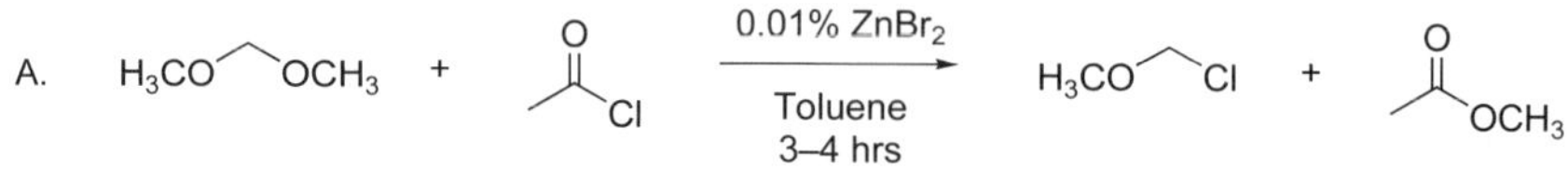

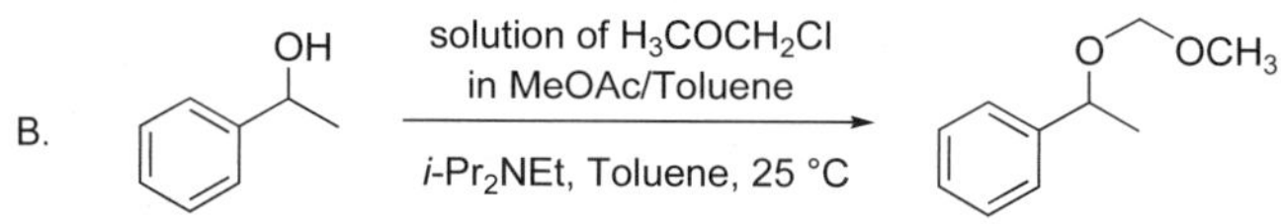

Submitted by Martin Berliner[1] and Katherine Belecki.
Checked by William D. Paquette and Peter Wipf.[2]

1. Procedure

CAUTION: Chloromethyl methyl ether is an OSHA registered carcinogen and is listed as an extremely hazardous substance by the EPA and the European Community. Other α-halo ethers are likely to be carcinogenic. Exercise due caution to minimize handling of these compounds.

A. Chloromethyl methyl ether as a solution in toluene. A three-necked 500-mL flask is fitted with a magnetic stirbar, thermometer, reflux condenser with nitrogen inlet, and addition funnel (Note 1). The vessel is charged with dimethoxymethane (22.1 mL, 250 mmol, 1 equiv) (Note 2), toluene (66 mL, 3 volumes) (Notes 3, 4) and $ZnBr_2$ (5.6 mg, 0.01 mol%) (Notes 5, 6). The zinc bromide dissolves shortly after addition. Acetyl chloride (17.8 mL, 250 mmol, 1 equiv) (Note 7) is placed in the addition funnel and then introduced into the reaction mixture at a constant rate over 5–15 minutes (Note 8). The addition funnel is rinsed with a minimum volume of toluene (5 mL) directly into the reaction mixture. The reaction self-heats slowly to 40–45 °C and then cools to ambient temperature over 2–3 h, at which time the exchange reaction is typically complete (Note 9). This clear, colorless solution of chloromethyl methyl ether in toluene is employed directly in the subsequent step without further manipulation (Note 10).

B. 1-Methoxymethyl-1-phenylethane. The reaction flask containing a toluene solution of chloromethyl methyl ether (250 mmol, 1.5 equiv) prepared above is immersed in a cold water bath and the contents cooled to 5–10 °C. α-Phenethyl alcohol (20 mL, 166 mmol, 1 equiv) is charged to the addition funnel and added in one portion (Note 11). Diisopropylethylamine (36.0 mL, 208 mmol, 1.25 equiv) is charged to the addition funnel and added to the reaction mixture dropwise over 30 min (Notes 12, 13). The internal temperature of the mixture is maintained below 25 °C during and immediately after the amine addition, and once heat evolution has ceased, the reaction mixture is allowed to come to ambient temperature and then stirred for 12 h. At this time the starting material is consumed (Note 14). The orange solution is diluted with ethyl acetate (100 mL) and a saturated aqueous NH_4Cl solution (100 mL), and the biphasic mixture stirred vigorously for a minimum of five minutes to ensure all residual chloromethyl methyl ether is decomposed (Note 15). The layers are separated and the resulting clear, colorless organic layer is washed once with water (100 mL), once with brine (100 mL), dried with $MgSO_4$ and concentrated by rotary evaporation to a clear, colorless oil (Note 16). The material is transferred to a 50-mL round-bottomed flask equipped with a magnetic stirbar and a short-path distillation head. The product is purified by distillation, collecting a single fraction (bp 60–63 °C, 0.5 mmHg) of 1-methoxymethoxy-1-phenylethane as a clear, colorless oil (26.2 g, 157 mmol, 95% yield) (Notes 17, 18, 19).

2. Notes

1. The success of the reaction does not depend on having previously dried the glassware.
2. Dimethoxymethane was obtained from Acros Organics at 99.5+% purity and used as received. The submitters used dimethoxymethane, obtained from Aldrich Chemical Co. at 99% purity, as received.
3. ACS grade toluene was employed without additional purification. A similar grade of dichloromethane, dichloroethane and alkyl acetate esters (i.e. ethyl acetate, isopropyl acetate) can be substituted for toluene with no change in reaction behavior or yield.
4. Although the reaction can be conducted neat, the authors strongly recommend using a co-solvent to mitigate the strongly exothermic nature of the process. The volume of solvent employed in this reaction was chosen to provide a good balance of reaction rate and thermal control. If less solvent is employed, active cooling should be utilized to keep the reaction temperature below 45 °C.

5. Zinc bromide (anhydrous powder, 98+%) was obtained from Acros Organics and used as received.

6. For small- and medium-scale reactions with dimethoxymethane, such small amounts of catalyst are required that the authors found it easiest to dispense the correct amount of catalyst from standard solutions of $ZnBr_2$ in dimethoxymethane (20–50 mg/mL).

7. The submitters obtained acetyl chloride from Aldrich Chemical Co. and Acros Organics at 97–99% purity and used it as received.

8. The reaction temperature increases slightly upon addition of acetyl chloride; the reaction is not dose controlled at this temperature. For a dose-controlled addition, the submitters recommend pre-heating the solvent/acetal/catalyst solution to 40 °C before dropwise addition of the halogenating agent, since the catalyzed exchange reaction is very rapid once the temperature exceeds 40 °C.

9. A small amount of the reaction mixture (20 mL) is dissolved in 0.6 mL of $CDCl_3$ for the analysis. The ratio of AcCl (δ 2.6 ppm):MeOAc (δ 2.0 ppm) is utilized to evaluate reaction progress as a small excess of dimethoxymethane is usually present once the exchange is complete (typical ratio MeOAc:AcCl ≥ 99:1). ^{1}H NMR ($CDCl_3$) δ: 5.44 (s, 2 H, MOMCl), 3.64 (s, 3 H, MeOAc), 3.49 (s, 3 H, MOMCl), 2.03 (s, 3 H, MeOAc). *In situ* IR can also be employed to monitor reaction progress without requiring sampling of the reaction contents.

10. The solution of MOMCl in toluene prepared using this stoichiometry have a density of 0.91 g/mL, are approx 2.1 M (18% w/w), and are stable for months if adequately sealed.

11. No heat evolution is observed during the addition.

12. Addition of the amine base results in a strong and extended exotherm as the reaction proceeds. In the absence of external cooling the reaction temperature can exceed 60 °C.

13. The order of addition of the amine and alcohol can be interchanged without affecting the purity or yield of the product. The following procedure is suggested: The solution of MOMCl is cooled to 10 °C and the base is added, slowly at first, keeping the internal temperature below 25 °C. Alcohol is added in one portion and the solution allowed to warm to ambient temperature and stirred until complete. Workup is as described in the main procedure.

14. Analysis is by HPLC. Starting material t_R 4.46 min, product t_R 7.25 min. Agilent Zorbax® SB-C8 column 4.6 x 150 mm, 2 mL/min flow, using acetonitrile and 0.1% $HClO_4/H_2O$ as the eluent. The following gradient

profile is used: 2 min isocratic (10:90, organic:aqueous), ramp to 90:10 (organic:aqueous) over 6 min, then 2 min isocratic at 90:10 (organic:aqueous). Analysis can also be done by TLC (5:1, hexanes:EtOAc; α-phenethyl alcohol, R_f= 0.13; 1-methoxy-1-phenylethane, R_f= 0.5).

15. The quench is mildly exothermic as residual MOMCl decomposes. Typically all remaining α-chloro ether has been quenched five minutes after addition of the aqueous solution. Water and saturated carbonate solutions are also effective for quenching the reaction.

16. Typical mass of crude product is 27 g (98% yield) with purity >95%.

17. The submitters observed yields from 77–80%. Some decomposition of the product occurs during extended batch distillation, resulting in decreased yields. This is not observed when the purification is conducted on small scale in a Kugelrohr apparatus, or when a chromatographic purification is employed.

18. The checkers obtained a yield of 90% (12.5 g) on half-scale.

19. Physical data for the product: R_f 0.5 (5:1, hexanes:EtOAc). ^{1}H NMR (300 MHz, $CDCl_3$) δ: 1.49 (d, J = 6.6 Hz, 3 H), 3.39 (s, 3 H), 4.55 (d, J = 6.9 Hz, 1 H), 4.59 (d, J = 6.6 Hz, 1 H), 4.76 (q, J = 6.6 Hz, 1 H), 7.27–7.37 (m, 5 H). ^{13}C NMR (75 MHz, $CDCl_3$) δ: 23.6, 55.3, 73.7, 94.0, 126.3, 127.5, 128.4, 143.2. IR (neat, cm^{-1}): 2976, 2931, 2887, 1451, 1155, 1099, 1036, 919, 759, 700. Anal. Calcd for $C_{10}H_{14}O_2$: C, 72.26; H, 8.49. Found: C, 72.22; H, 8.64.

Safety and Waste Disposal Information

All hazardous materials should be handled and disposed of in accordance with "Prudent Practices in the Laboratory"; National Academy Press; Washington, DC, 1995.

3. Discussion

Chloromethyl methyl ether (MOMCl) and other α-halo ethers are versatile reagents for protecting a variety of functional groups[3] and as precursors of carbenes, electrophiles and α-stannyl ethers that have wide use in organic synthesis.[4] This procedure describes an improved method for the generation of α-halo ethers from symmetric aliphatic acetals, including MOMCl and higher substituted analogs.[5] The original preparation of MOMCl in this series[6] and current commercial preparations utilize the reaction between methanol, formaldehyde and hydrogen chloride and produce substantial amounts of the very highly carcinogenic bis(chloromethyl) ether which cannot be readily separated by distillation. More recent procedures[7] have employed the exchange reaction between dimethoxymethane and a halide donor catalyzed by protic acid; these minimize formation of bis(chloromethyl)ether and are preferred to the original preparation. These methods employ different halide sources and different acid catalysts, but all require extended (2–24 hours or more) reflux to effect full exchange, and due to the high vapor pressure of the reactants and products, efficient condensers are required to minimize material loss.

The present method utilizes catalytic zinc(II) salts (0.001 to 0.01 mol%) to accelerate the exchange reaction between halide donors and acetals. Reactions typically are complete in one to four hours irrespective of the scale of the preparation (mmol through mole scale) without the need for external heating. A range of zinc(II) salts can be utilized, with $Zn(OTf)_2$, $ZnBr_2$ and $Zn(OAc)_2$ being the easiest to manipulate. Acid halides and thionyl chloride (but not thionyl bromide) are effective as the halide source (Table 1). A variety of symmetric aliphatic acetals participate in the reaction to give synthetically useful α-halo ethers (Table 2). The solutions of α-halo ethers are employed without isolation in base-mediated reactions where the presence of the ester by-product does not interfere, including alkylations of acids, alcohols, phenols and stabilized enolates under the usual conditions (Table 3). Since α-halo ethers are hazardous (some are carcinogenic; others are possible carcinogens), this procedure illustrates a convenient on-demand synthesis and use of these reagents without isolation or unnecessary handling. The improved safety of this telescoped procedure addresses concerns of both academic and industrial chemists who are under increasing pressure to reduce laboratory hazards.

Table 1. Halogen Sources

$$(MeO)_2CH_2 + X\text{–}R \xrightarrow[25\ °C]{0.01\%\ ZnBr_2} MeOCH_2X + MeOR$$

Entry	Halide Source	Equiv required	Time
1	$(COCl)_2$	0.5	< 5 min
2	RCOCl	1.0	30 min
3	$SOCl_2$	0.5	30 min
4	CH_3COBr	1.0	45 min

Table 2. Acetals Employed in the Exchange Reaction

Entry	Acetal	α-Halo Ether	Halide Source	Catalyst	Time[a]
1	$(EtO)_2CH_2$	$EtOCH_2Cl$	AcCl	$ZnBr_2$	4 hr
2	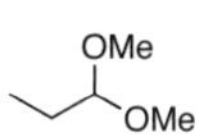	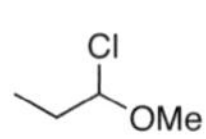	$(COCl)_2$	$Zn(OTf)_2$	1 hr
3	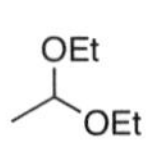	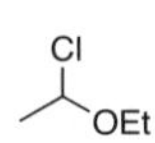	AcCl	$ZnBr_2$	1 hr
4	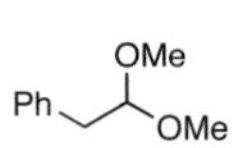	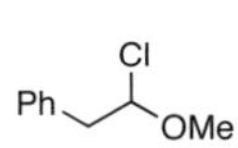	AcCl	$Zn(OTf)_2$	30 min
5	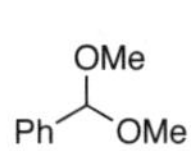	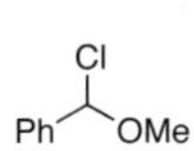	AcCl	$Zn(OTf)_2$	30 min

(*a*) Time to completion of exchange reaction by NMR spectroscopy.

Table 3: Alkylation Reactions of In Situ Generated α-Halo Ethers[a]

Entry	Starting Material	Product	Yield[b] (%)
1	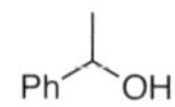	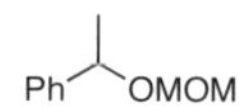	92
2	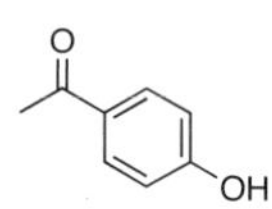	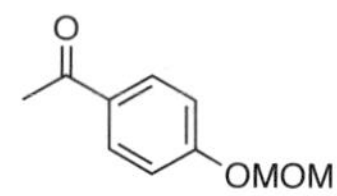	86
3[c]	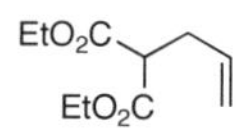	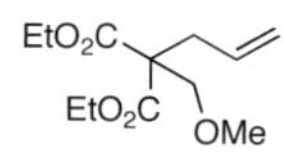	80
4	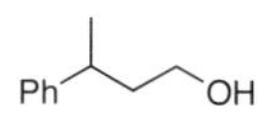	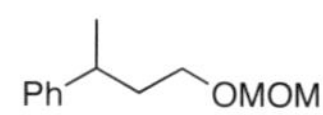	92
5	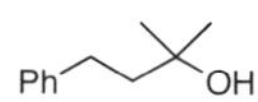		92
6	$PhCO_2H$	$PhCO_2MOM$	92
7	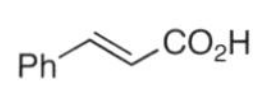	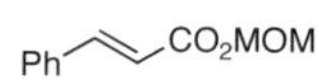	96
8	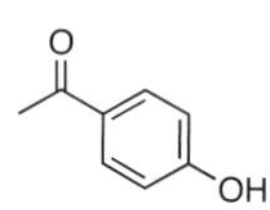	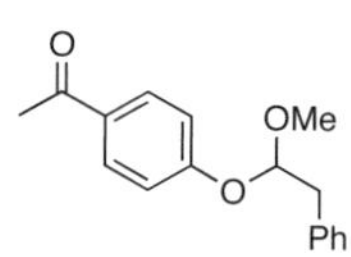	93
9	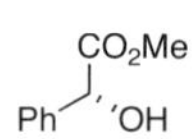	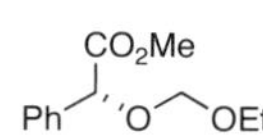	85

(*a*) Reactions are conducted on 10-20 mmol scale with 1.5–2.0 equivalents of α-halo ether and 1.25 equivalents of *i*-Pr_2NEt in toluene for 12–16 hours followed by a standard aqueous workup as described in this procedure. (*b*) Yield of purified product. (*c*) Malonate anion was prepared using NaH in THF followed by addition to a MOMCl solution prepared in toluene.

1. Chemical Research and Development, Pfizer Global Research and Development, Groton, CT 06340, *martin.a.berliner@pfizer.com.*
2. Department of Chemistry, University of Pittsburgh, Pittsburgh, PA 15260, USA.
3. Greene, T. W.; Wuts, P. G. M. *Protective Groups in Organic Synthesis*, 3rd ed.; John Wiley & Sons: New York, 1999.
4. For a review, see: Benneche, T. *Synthesis* **1995**, 1 and references therein.
5. Berliner, M. A.; Belecki, K. *J. Org. Chem.* **2005**, *70*, 9618.
6. Marvel, C. S.; Porter, P. K. *Organic Syntheses*; John Wiley & Sons: New York, 1941. Coll. Vol. I, p. 377.

7. With HCl: (a) Weinstock, L.M.; Karady, S.; Sletzinger, M. US 3,972,947 (*Chem. Abstr* **1976**, *85*, 142633) (b) Amato, J. S.; Karady, S.; Sletzinger, M.; Weinstock, L. M. *Synthesis* **1979**, 970. (c) Linderman, R. J.; Jaber, M.; Griedel, B. D. *J. Org. Chem.* **1994**, *59*, 6499. With H_2SO_4: (d) Chong, J. M.; Shen, L. *Synth. Commun.* **1998**, *28*, 2801. (e) Williams, A. G. WO 02/059070 A1 (*Chem. Abstr.* **2002**, *137*, 124930). (f) Reggelin, M.; Doerr, S. *Synlett* **2004**, 1117. By other methods: (g) Stadlwieser, J. *Synthesis* **1985**, 490. (h) Jones, M. *Synthesis* **1984**, *9*, 727.

MONO CARBAMATE PROTECTION OF ALIPHATIC DIAMINES USING ALKYL PHENYL CARBONATES
[(2-Aminoethyl)carbamic acid *tert*-butyl ester]

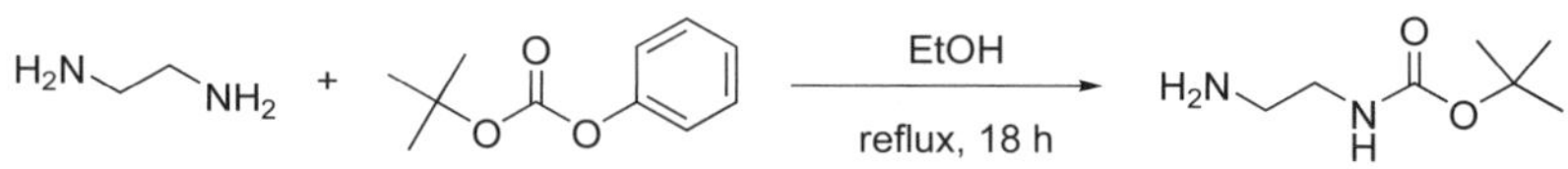

Submitted by Michael Pittelkow, Rasmus Lewinsky and Jørn B. Christensen.[1]
Checked by Cara Cesario and Marvin J. Miller.

1. Procedure

(2-Aminoethyl)carbamic acid tert-butyl ester. tert-Butyl phenyl carbonate (Note 1) (64.62 g, 61.6 mL, 0.33 mol) is added to a solution of 1,2-ethanediamine (Note 2) (20.0 g, 22.3 mL, 0.33 mol) in absolute EtOH (Note 3)(200 mL) in a 500-mL, single-necked, round-bottomed flask equipped with a stirring bar and a reflux condenser. The reaction mixture is heated gently to reflux overnight (18 h) (Note 4) ensuring that the temperature of the oil bath is at a maximum of 80 °C resulting in a yellow solution. The reaction mixture is cooled to room temperature and the solution concentrated to approximately 150 mL using a rotary evaporator (20 °C, 20 mmHg), which leaves a yellow solution. Water (300 mL) is added and the pH is adjusted to approximately 3 (Note 5) by careful addition of aqueous HCl (2M, approximately 110 mL) followed by extraction with CH_2Cl_2 (3 × 400 mL) (Note 6). The aqueous phase is adjusted to pH 12 by addition of aqueous NaOH (2M, 150 mL) and extracted with CH_2Cl_2 (5 × 500 mL). The combined organic extracts are dried (Na_2SO_4), filtered and concentrated using a rotary evaporator (20 °C, 20 mmHg) to afford 27.0 g (51%) of the title compound as a yellow oil (Notes 7, 8, 9, and 10).

2. Notes

1. *tert*-Butyl phenyl carbonate (98%) was purchased from Aldrich Chemical Company, Inc. and used as received. *tert*-Butyl phenyl carbonate is a low melting solid that should be stored in a refrigerator. It is convenient to allow it to melt before handling.

2. 1,2-Ethanediamine was purchased from Aldrich Chemical Company, Inc. (99%). Distillation before use is recommended if the compound has a deep yellow color. Otherwise it can be used as received.

3. Absolute Ethanol (99.9%) was purchased from De Danske Spritfabrikker and was used without further efforts to remove water. The checkers used 200 proof ethanol from Aaper Alcohol and Chemical Company; the solvent was used as received.

4. The Boc protection group is known to be thermally unstable and temperatures above 85–90 °C for prolonged periods should be avoided.[2]

5. The acid labile protective group should not be exposed to acidic aqueous media for prolonged periods of time, otherwise unwanted cleavage of the group may occur.

6. Thorough extraction was needed at this stage to remove unreacted alkyl phenyl carbonate and side products including the bis-carbonate-protected diamine and phenol. Also, it was important to conduct this extraction immediately after addition of 2M HCl, as the Boc group is known to be acid sensitive.

7. The product had the following physical data: ^{1}H NMR (300 MHz, $CDCl_3$) δ: 1.23 (br s, 2 H, NH_2), 1.37 (s, 9 H, $(CH_3)_3$), 2.78 (t, 2 H, $J = 6.0$ Hz), 3.09 (q, 2 H, $J = 6.0$ Hz), 5.18 (br s, 1 H, NHCO); ^{13}C NMR (125 MHz, $CDCl_3$) δ: 28.2, 41.7, 42.2, 78.9, 156.1. MS (FAB): *m/z* 161 (MH^+). IR (film) 3350, 2977, 1693, 1524, 1391, 1366, 1252, 1172. Anal. Calcd for $C_7H_{16}N_2O_2$: C, 52.48; H, 10.07; N, 17.48. Found: C, 52.54; H, 9.83; N, 17.28.

8. The level of phenol in the product was below the level that could be detected by the human nose. No other products were detected by NMR and TLC (visualized either by UV light at 256 nm or 316 nm, by oxidation with I_2 or by ninhydrin test).

9. If long-term storage of the product is required, it should be stored under an inert atmosphere as it is prone to react with CO_2 from the air to produce a solid compound.

10. The checkers also performed the procedure on half-scale and obtained 14.4 g (54%) of the title compound as a yellow oil. The submitters report yields ranging from 59-65% when the reaction is performed on the scale of the procedure.

Safety and Waste Disposal Information

All hazardous materials should be handled and disposed of in accordance with "Prudent Practices in the Laboratory"; National Academy Press; Washington, DC, 1995.

3. Discussion

A challenge often encountered in the synthesis, or in the synthetic use of polyamines, is selective protection of the amino groups. This synthetic challenge has been dealt with in a number of ways and elegant multi-step synthetic procedures have been developed to strategically incorporate protection groups in polyamines.[2-4]

The most widely used procedure for mono carbamate protection of aliphatic diamines is that of Krapcho and co-workers that uses an excess of diamine as compared to di-*tert*-butyl dicarbonate to achieve high yield with respect to the di-*tert*-butyl dicarbonate.[5,6] This procedure, however, is not attractive if the diamine is a valuable intermediate; in general, the use of a large excess of a reagent should be avoided if possible. Other reagents such as 2-(*tert*-butoxycarbonyloxyimino)-2-phenylacetonitrile (BOC-ON)[7] and *O*-alkyl-*O'*-(*N*-succinimidyl) carbonates[8] have also been successful for similar transformations, but issues with strict temperature control and pH control complicates these procedures. The use of alkyl chloroformates has also been described, but these generally give statistical mixtures of products.

This procedure, which uses alkyl phenyl carbonates as the electrophiles, is a simple, efficient and selective method for the preparation of mono-carbamate-protected diamines. It is applicable for large laboratory scale preparations and the purification of the product proceeds without the need for column chromatography, distillation or recrystallization. The procedure has been shown to work almost equally well for the introduction of Boc, Cbz and Alloc protecting groups.[9]

The selectivity in the introduction of carbamate protecting groups has further been tested by reaction with unsymmetrical aliphatic diamines.[9] Selectivity towards reaction with primary amines located on a primary carbon in the presence of a primary amine located on either a secondary or a tertiary carbon was studied by reaction of the alkyl phenyl carbonates with the two diamines, 2-methyl-1,2-propanediamine and 1,2-propanediamine. Both of these diamines were successfully mono-carbamate-protected on the primary amino groups located on the primary carbon atom in high yields with the three carbonates tested (Table 1).

Some representative examples of mono-carbamate-protected diamines prepared by this method are compiled in Table 1, which illustrates the possibility to incorporate the Boc, Cbz and Alloc protecting groups by the same procedure.

Table 1 Selective Carbamate Protection of Aliphatic Diamines.[9]

Amine	Protected Amine	Yield[a]	
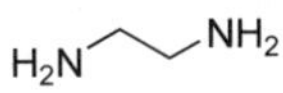	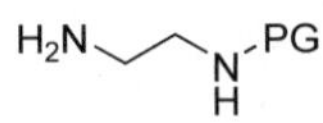	PG = Boc PG = Cbz PG = Alloc	65% 64% 67%
	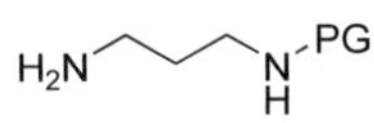	PG = Boc PG = Cbz PG = Alloc	68% 72% 86%
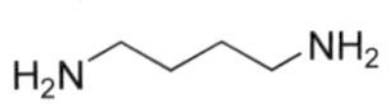	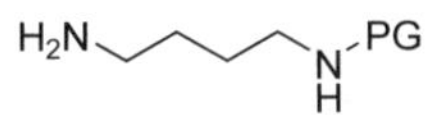	PG = Boc PG = Cbz PG = Alloc	63% 63% 77%
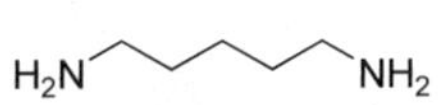	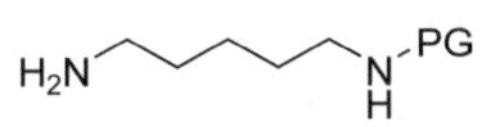	PG = Boc PG = Cbz PG = Alloc	50% 56% 51%
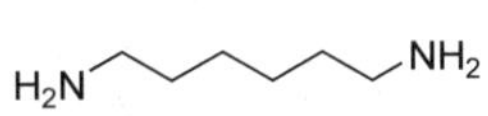	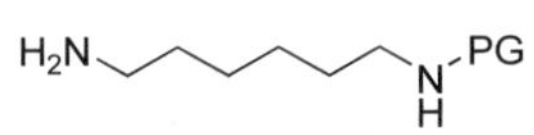	PG = Boc PG = Cbz PG = Alloc	49% 48% 46%
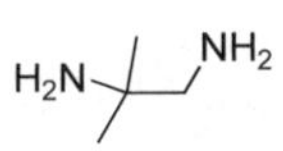	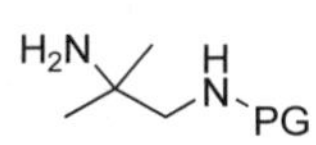	PG = Boc PG = Cbz PG = Alloc	91% 97% 96%
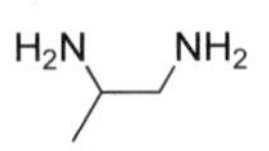	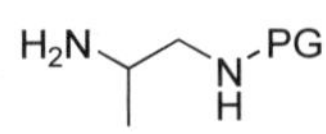	PG = Boc PG = Cbz PG = Alloc	70%[b] 69%[b] 75%[b]

a) PG =Protection Group: Boc = $(CH_3)_3COCO$, Cbz = $PhCH_2OCO$, Alloc = CH_2CHCH_2OCO
b) Yields are based on the polyamine; two equivalents of alkyl phenyl carbonate were used.

1. Department of Chemistry, University of Copenhagen, Universitetsparken 5, DK-2100, Copenhagen Ø, Denmark. E-mail: jbc@kiku.dk
2. Greene, T. W.; Wuts, P. G. M. *Protective Groups in Organic Synthesis*, 3rd ed.; Wiley: New York, **1999**.
3. Kuksa, V.; Buchan, R.; Lin, P. K. T. *Synthesis* **2000**, 1189.
4. Bergeron, R. J.; Stolowich, N. J.; Porter, C. W. *Synthesis* **1982**, 689.
5. Krapcho, A. P.; Maresch, M. J.; Lunn, J. *Synth. Commun.* **1993**, *23*, 2443.
6. Krapcho, A. P.; Kuell, C. S. *Synth. Commun.* **1990**, *20*, 2559.

7. Phanstiel, O.; Price, H. L.; Wang, L.; Juusola, J.; Kline, M.; Shah, S. M. *J. Org. Chem.* **2000**, *65*, 5590.

8. Adamczyk, M.; Fishpaugh, J. R.; Heuser, K. J. *Org. Prep. Proced. Int.* **1998**, *30*, 339.

9. Pittelkow, M.; Lewinsky, R.; Christensen, J. B. *Synthesis* **2002**, 2195.

PROTECTION OF ALCOHOLS USING 2-BENZYLOXY-1-METHYLPYRIDINIUM TRIFLUOROMETHANESULFONATE: METHYL (*R*)-(–)-3-BENZYLOXY-2-METHYL PROPANOATE

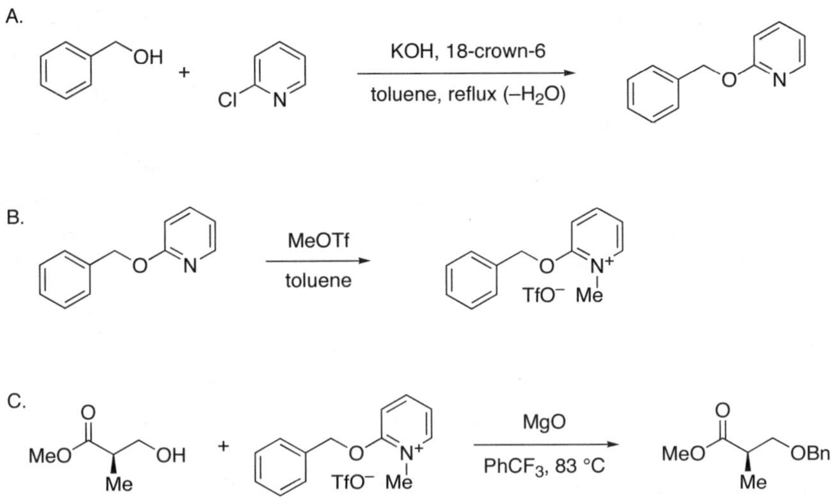

Submitted by Kevin W. C. Poon, Philip A. Albiniak, and Gregory B. Dudley.[1]
Checked by Martin A. Berliner and John A. Ragan.[2]

1. Procedure

Caution: *Exercise care in the handling of 18-Crown-6 (harmful, target organ: nerves) and methyl triflate (corrosive, causes burns, target organs: eyes, skin, mucous membranes).*

A. 2-Benzyloxypyridine. A 500-mL, three-necked, round-bottomed flask equipped with a magnetic stir bar, a thermocouple probe, a glass stopper, and a Dean-Stark trap with reflux condenser and a nitrogen bubbler (Note 1) is charged with benzyl alcohol (11.7 g, 0.108 mol, 1.0 equiv), 2-chloropyridine (13.5 g, 0.119 mol, 1.1 equiv) (Note 2), potassium hydroxide (20.0 g, 0.356 mol, 3.3 equiv) (Note 3), and toluene (210 mL) (Note 4) by temporary removal of the glass stopper. The suspension is stirred and 18-crown-6 (1.43 g, 5.40 mmol, 0.05 equiv) (Note 2) is added in one portion. The reaction mixture is heated to reflux (Note 5) with azeotropic removal of water (Note 6) for 1 h. The reaction mixture is then allowed to cool to 25 °C. The fittings of the flask are removed. Water (100 mL) is added to the cooled reaction mixture and the resulting biphasic solution is transferred to a 500-

mL separatory funnel. The reaction flask is rinsed with 50 mL of toluene and the rinse is combined with the reaction mixture. The lower, strongly basic aqueous layer (pH 14) is removed. The organic phase is washed a second time with water (100 mL) and the aqueous layer (pH ~7) is removed. The cloudy, colorless organic phase is transferred to a 500-mL flask and concentrated under reduced pressure on a rotary evaporator (25–45 mmHg, bath temperature 45–50 °C). Residual water remaining in the organic phase is azeotropically removed during the concentration, to provide the crude product as a clear and colorless to pale yellow oil. This material is transferred to a 100-mL single-necked round-bottomed flask with magnetic stir bar and purified by vacuum distillation using a short-path distillation apparatus. A single fraction is collected (bp 93–95 °C, 1.0 mmHg) providing 19.0–19.2 g (95–96%) of 2-benzyloxypyridine as a colorless liquid (Notes 7,8).

B. 2-Benzyloxy-1-methylpyridinium trifluoromethanesulfonate. A three-necked, 250-mL, round-bottomed flask equipped with a nitrogen bubbler inlet (Note 1), thermocouple probe, rubber septum and overhead mechanical stirrer (Note 9) is charged with 2-benzyloxypyridine (10.6 g, 57.3 mmol, 1.0 equiv) and toluene (60 mL) (Note 4) by temporary removal of the septum. The resulting solution is cooled to 0 °C in an ice bath. Methyl trifluoromethanesulfonate (6.80 mL, 60.1 mmol, 1.05 equiv) (Note 10) is added dropwise via syringe over 15 min, during which time white solids crystallize and the internal temperature increases to approximately 15 °C. The ice bath is removed and the viscous slurry is allowed to warm to ambient temperature. After 1 h (Note 11), the heterogeneous reaction mixture is diluted with 50 mL of hexanes (Note 12) and the solids are isolated by filtration and washed with one additional portion of hexanes (50 mL). After drying under vacuum, 19.6–19.8 g (98.5–99%) of 2-benzyloxy-1-methylpyridinium trifluoromethanesulfonate is isolated as a white, microcrystalline solid (Note 13).

C. Methyl (R)-(–)-3-benzyloxy-2-methyl propanoate. An oven-dried, 250-mL, three-necked, round-bottomed flask is equipped with a reflux condenser with an inert gas bubbler, a magnetic stir bar, a thermocouple probe and a glass stopper. 2-Benzyloxy-1-methylpyridinium trifluoromethanesulfonate (16.7 g, 48.0 mmol, 2.0 equiv) (Note 14), magnesium oxide (1.94 g, 48.0 mmol, 2.0 equiv) (Note 15), α,α,α-trifluorotoluene (48 mL) (Note 16) and methyl (*R*)-(–)-3-hydroxy-2-methyl propionate (2.65 mL, 23.8 mmol, 1.0 equiv) (Note 17) are charged to the reaction flask by temporarily removing the glass stopper. The heterogeneous reaction mixture is then heated in an oil bath (83 °C bath temperature, 80–82 °C internal temperature) for 24 h (Note 18). The reaction mixture is allowed to cool to room temperature and suction filtered through a pad of Celite on a sintered glass funnel. The Celite is washed with methylene chloride (2 x 20 mL). The filtrate is concentrated by rotary evaporation (15 mmHg, 25 °C bath temperature), and the light brown oil is purified by flash column chromatography (Note 19) to provide 3.77–4.10 g (76–82%) of methyl (*R*)-(–)-3-benzyloxy-2-methyl propanoate as a colorless liquid (Notes 20, 21).

2. Notes

1. The submitters conducted this procedure under an atmosphere of argon; the checkers found substitution of nitrogen to have no adverse impact on yield or purity.
2. Benzyl alcohol (99%), 2-chloropyridine (99%), and 18-crown-6 (99%) were purchased from Sigma-Aldrich and used without further purification.
3. Potassium hydroxide pellets were purchased from J. T. Baker and ground with a mortar and pestle immediately before use.
4. The submitters purchased toluene from VWR and passed it through neutral alumina and 4Å molecular sieves columns under argon atmosphere before use. The checkers obtained ACS grade toluene from Sigma-Aldrich and used it as received.
5. The checkers utilized a heating mantle and a solid state heating controller set to 125 °C at approximately 80% power to maintain reflux. The submitters utilized an oil bath (150 °C bath temperature).
6. At the end of the reaction, 1.4 mL of water was collected in the Dean-Stark trap.
7. The checkers performed a simple distillation under vacuum with the distillation pot heated in an oil bath set at 120 °C. The submitters employed fractional vacuum distillation using a short-path distillation apparatus fitted with a 3-neck pig adapter and thermometer and connected to a vacuum oil pump (0.25 mmHg).
8. The compound had the following physical data: bp: 93–95 °C/ 1.0 mmHg. IR (thin film): 1594, 1569, 1472, 1430, 1270, 988, 777, 733, 695 cm^{-1}. ^{1}H NMR (400 MHz, $CDCl_3$) δ: 5.36 (s, 2 H), 6.80 (dt, *J* = 8.4, 0.9 Hz, 1 H), 6.87 (ddd, *J* = 7.1, 5.1, 0.9 Hz, 1 H), 7.30 (m, 1 H), 7.34–7.39 (m, 2 H), 7.44–7.49 (m, 2 H), 7.57 (ddd, *J* = 8.4, 7.1, 2.0 Hz, 1 H), 8.17 (ddd, *J* = 5.1, 2.0, 0.8 Hz, 1 H); ^{13}C NMR (100 MHz, $CDCl_3$) δ: 67.5, 111.3, 116.9,

127.8, 127.9, 128.4, 137.3, 138.6, 146.8, 163.6; Anal: Calcd for $C_{12}H_{11}NO$: C, 77.81; H, 5.99; N, 7.56. Found: C, 77.56; H, 6.07; N, 7.32. R_f = 0.64 (250 μm silica gel on glass, hexanes:ethyl acetate = 3:1, UV visualization). The spectral data for 2-benzyloxypyridine matched that of the literature data.[3]

9. The submitters used magnetic stirring for this reaction. The checkers observed that cooling of the thick slurry was much less efficient with magnetic stirring than with mechanical stirring at the scale of this preparation.

10. Methyl trifluoromethanesulfonate (MeOTf) was purchased from Sigma-Aldrich, stored and handled under an atmosphere of argon, and used without further purification.

11. TLC analysis showed disappearance of 2-benzyloxypyridine within 1 h by TLC (R_f (reaction product) = 0; R_f (starting material) = 0.6; 3:1 hexanes:EtOAc; UV visualization) and by HPLC (product t_R 2.7 min; starting material t_R 2.4 min; Agilent Zorbax SB-CN column, 4.2 × 75 mm, 3.5 μ silica; mobile phases 0.1% $HClO_4/H_2O$ and acetonitrile; isocratic 95:5 aqueous:acetonitrile for 0.25 min, ramp to 10:90 over 5 min; isocratic at 10:90 for 0.75 min, flow 2 mL/min, UV detection at 210 nm, 30 °C column temperature).

12. The checkers obtained ACS grade hexanes from Sigma-Aldrich and used it as received.

13. Further purification of the material thus obtained was not necessary. The compound had the following physical data: mp: 92–93 °C. IR (thin film): 1639, 1587, 1518, 1254, 1148, 1028, 772, 751, 699, 633 cm^{-1}. ^1H NMR (400 MHz, $CDCl_3$) δ: 4.07 (s, 3 H), 5.54 (s, 2 H), 7.38–7.52 (m, 6 H), 7.60 (d, J = 8.8 Hz, 1 H), 8.31 (ddd, J = 1.8, 7.5, 9.2 Hz, 1 H), 8.47 (dd, J = 6.4, 1.5 Hz, 1 H); ^{13}C NMR (100 MHz, $CDCl_3$) δ: 41.8, 74.3, 112.0, 119.0, 120.6 (q, J_{CF} = 320 Hz), 128.5, 128.9, 129.5, 132.5, 143.7, 148.0, 159.4; ^{19}F NMR (376 MHz, $CDCl_3$, external reference to TFA) δ: –78.8 (s); MS (APCI+) 200.2 $(M–OTf)^+$. Anal: Calcd for $C_{14}H_{14}F_3NO_4S$: C, 48.14; H, 4.04; N, 4.01; Found: C, 48.27; H, 3.73; N, 3.95. The submitters also reported the following data: HRMS (ESI+) found 200.1070 $(M-OTf)^+$ (calcd for $C_{13}H_{14}NO^+$: 200.1075) CHFNOS analysis: Calcd for $C_{14}H_{14}F_3NO_4S$: C, 48.14%; H, 4.04%; F, 16.31%; N, 4.01%; O, 18.32%; S, 9.18%. Found: C, 48.24%; H, 4.09%; F, 16.27%; N, 4.10%; O, 18.53%; S, 9.31%.

14. A minimum of 1.8 equiv of 2-benzyloxy-1-methylpyridinium trifluoromethanesulfonate was found to be necessary for complete consumption of the alcohol substrates. Dibenzyl ether (Bn_2O) was formed

as a by-product during the course of the reaction, perhaps through reaction of the pyridinium salt with MgO and/or adventitious moisture to generate benzyl alcohol *in situ*.

15. The checkers obtained magnesium oxide (light) from Sigma-Aldrich and dried it in a vacuum oven at 60 °C for 12 h. The submitters dried magnesium oxide under vacuum in a flask immersed in an oil bath for 6 h and stored under an atmosphere of argon.

16. α,α,α-Trifluorotoluene in a Sure/Seal bottle was purchased from Sigma-Aldrich and used without further purification.

17. Methyl (*R*)-(–)-3-hydroxy-2-methyl propanoate (99% ee) was purchased from Sigma-Aldrich and used without further purification.

18. The reaction mixture was analyzed by thin layer chromatography (TLC, 250 μm silica gel on glass, hexanes/ethyl acetate = 9:1); visualization was accomplished with 254 nm UV light and anisaldehyde stain with heating: R_f (starting alcohol) = 0.08; R_f (benzyl ether product) = 0.36; R_f (dibenzyl ether by-product) = 0.61.

19. The checkers utilized 120 g of silica gel (60Å, 230–400 mesh) loading the sample neat and used a gradient (19:1 heptanes:EtOAc to elute Bn_2O, to 9:1 heptanes:EtOAc to elute the desired product). The submitters utilized a silica gel column (16 cm length × 4.5 cm width, 90 g of Silica Gel 60 Geduran® 40-63 μM) wet-loaded as a slurry, loading the sample in a minimum volume of eluent, and eluting the product with 750 mL 19:1 hexanes:EtOAc.

20. Careful analysis of the ^{1}H NMR spectrum prior to chromatography and LC/MS analysis of the purified product indicated that impurities were formed from bis- and tris-benzylation (i.e. M+Bn and M+2Bn), presumably by addition to the aromatic ring of the primary benzyl ether product. These impurities seemed to have a greater response factor than the desired product, as their apparent level by LC/MS analysis was significantly higher than that indicated by ^{1}H NMR or elemental analysis. For example, the elemental analysis in Note 21 corresponded to less than 5% of the bis-benzyl product ($C_{19}H_{22}O_3$, vs. $C_{12}H_{16}O_3$ for the desired), and the ^{1}H NMR spectrum showed no discernable quantities of any impurity. This level of impurity should not be an issue for the vast majority of applications of this methodology, but the readers should be aware of the potential presence of these side products.

21. The compound had the following physical data, which matched the data in the literature.[4] ^{1}H NMR (400 MHz, $CDCl_3$) δ:1.16 (d, J = 7.1 Hz, 3 H), 2.79 (X of $ABXY_3$, ddq, J_{BX} = 5.9 Hz, J_{AX} = 7.4 Hz, J_{XY} = 7.1 Hz, 1

H), 3.47 (B of $ABXY_3$, dd, J_{AB} = 9.1 Hz, J_{BX} = 5.9 Hz, 1 H), 3.64 (A of $ABXY_3$, dd, J_{AB} = 9.1 Hz, J_{AX} = 7.4 Hz, 1 H), 3.68 (s, 2H), 4.50 (B of AB, *J* = 12.4 Hz, 1 H), 4.51 (A of AB, *J* = 12.4 Hz, 1 H), 7.26–7.35 (m, 5 H); ^{13}C NMR (100 MHz, $CDCl_3$) δ:13.9, 40.1, 51.7, 71.9, 73.0, 127.51, 127.53, 128.3, 138.1, 175.2; IR (thin film, cm^{-1}): 1736, 1454, 1363, 1198, 1176, 1092, 1028, 736, 698, 608. Anal: Calc'd for $C_{12}H_{16}O_3$: C, 69.21; H, 7.74. Found: C, 69.57; H, 7.86. $[\alpha]_D$ –10.4° (c 2.8, $CHCl_3$); lit $[\alpha]_D$ –10.6° (c 1.7, $CHCl_3$).[5] R_f = 0.36 (250 μm silica gel on glass, hexanes/ethyl acetate = 9:1). Chiral HLPC analysis (Chiracal OD-H 0.46 × 25 cm, 98:2 heptane:2-propanol mobile phase at 1.5 mL/min flow) indicated >99% ee. The (*R*) enantiomer eluted at 3.88 min; the (*S*) enantiomer eluted at 4.17 min.

Waste Disposal Information

All hazardous materials should be handled and disposed of in accordance with "Prudent Practices in the Laboratory"; National Academy Press; Washington, DC, 1995.

3. Discussion

Benzyl ethers are important protecting groups in organic synthesis due to their stability and the minimal electronic impact that they impart on the oxygen atom to which they are attached.[6] For example, benzyl ethers do not interfere with chelation-controlled addition of external nucleophiles to chiral aldehydes, in contrast to silyl ethers.[7] Benzyl (and modified arylmethyl) ethers are perhaps the most versatile of the alkyl ethers with respect to modes of cleavage, which include hydrogenolysis, oxidation, dissolving metal reduction, and acidic decomposition under a range of experimental protocols.

Standard conditions for the formation of benzyl ethers from the corresponding alcohols fall into two main categories: (1) the Williamson ether synthesis, an S_N2-type reaction between benzyl bromide and alkali metal alkoxides, and (2) coupling using benzyl trichloroacetimidate, which is generally promoted by trifluoromethanesulfonic acid (triflic acid, TfOH).[8] Typical benzylation reactions are thus limited to substrates that tolerate either strongly acidic or basic conditions.

2-Benzyloxy-1-methyl-pyridinium trifluoromethanesulfonate (**1**) provides benzyl ethers upon warming in the presence of a free alcohol.[9] The overall balanced equation for the benzylation of alcohols (**2** → **3**) is shown in Eq 1. In contrast to activated acetimidates, oxypyridinium salt **1** is stable to storage and handling. At room temperature, **1** is soluble in chlorinated solvents (dichloromethane, chloroform, dichloroethane), partially soluble in ethereal solvents (THF and ether), and insoluble in aromatic hydrocarbons (benzene, toluene). However, mixtures of **1** and aromatic solvents become homogeneous at elevated temperatures.

(1)

Table 1 illustrates the benzylation reactions of representative alcohols. Primary (entries 1-4) and secondary (entries 5-6) alcohols and certain tertiary alcohols (entry 7) provided the desired benzyl ethers (**3a-g**) in good to excellent yield. Note that phenols are not included in Table 1; aromatic alcohols have yet to be demonstrated as viable substrates in this reaction. However, Mitsunobu conditions[10] can be applied to the benzylation of most phenols.

In conclusion, 2-benzyloxy-1-methylpyridinium triflate (**1**) is a novel benzylation reagent for alcohols. Salt **1** is easy to prepare, bench-stable, and pre-activated. No acidic or basic promoters are needed for benzyl transfer, which occurs upon warming in the presence of the alcohol substrate under mild conditions. New arylmethylation reagents should find use in organic synthesis for the protection of sensitive alcohol substrates.

Table 1. Scope of the Benzylation Reaction

1.0 equiv *R*–OH (**2**) + 2.0 equiv **1** —(2.0 equiv MgO; $PhCF_3$, 83 °C, 1 d)→ *R*-O**Bn** **3**

Entry	ROH (2)	ROBn (3)	Yield[a]
1.	 **2a**	 **3a**	>95%
2.	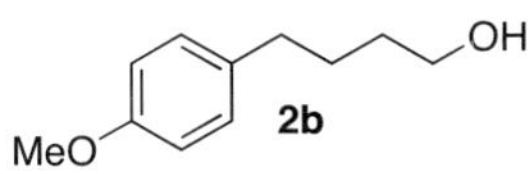 **2b**	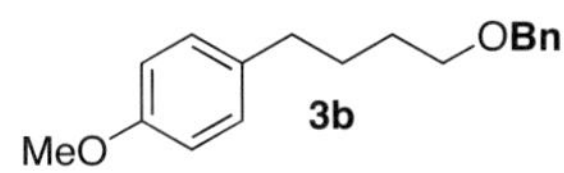 **3b**	>95%
3.	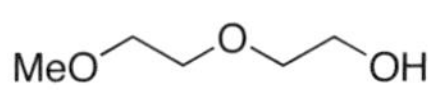 **2c**	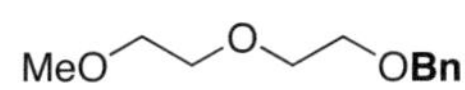 **3c**	>95% 93%[b]
4.	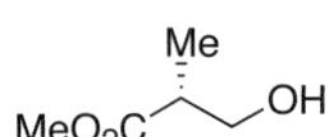 **2d**	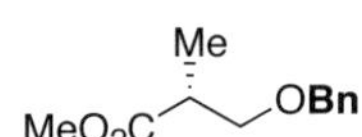 **3d**	85% 76–82%[b]
5.	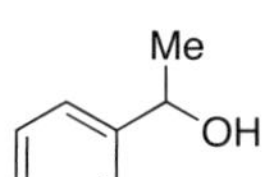 **2e**	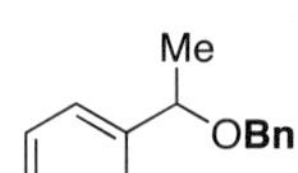 **3e**	83%
6.	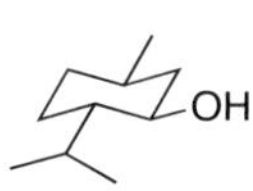 **2f**	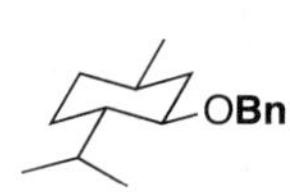 **3f**	88%
7.	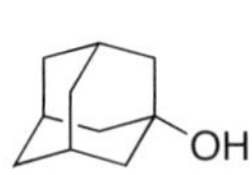 **2g**	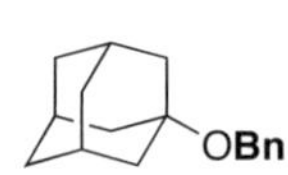 **3g**	80%

[a]Yields are estimated by ^{1}H NMR spectroscopy, unless otherwise indicated.
[b]Isolated yield of pure product.

1. Department of Chemistry and Biochemistry, Florida State University, Tallahassee, FL 32306-4390, USA. E-Mail: gdudley@chem.fsu.edu

2. Chemical Research and Development, Pfizer Global Research and Development, Groton, CT, 06340.

3. Serio Duggan, A. J.; Grabowski, E. J. J.; Russ, W. K. *Synthesis* **1980**, 573–575.

4. White, J. D. and Kawasaki, M. *J. Org. Chem.* **1992,** *57*, 5292–5300.

5. Paquette, L. A.; Guevel, R.; Sakamoto, S.; Kim, I.H.; Crawford, J. *J. Org. Chem.* **2003**, *68*, 6096–6107.

6. (a) Wuts, P. G. M. and Greene, T. W., *Greene's Protective Groups in Organic Synthesis*, 4th ed., John Wiley and Sons: Hoboken, New Jersey, 2007. (b) Kocienski, P. J. *Protecting Groups***,** 3rd ed., Thieme: Stuttgart, 2003**.**

7. Gawley, R. E.; Aubé, J. *Principles of Asymmetric Synthesis*; Baldwin, J. E., Magnus, P. D., Eds.; Tetrahedron Organic Chemistry Series 14; Pergamon: Tarrytown, NY, 1996; pp 121–134.

8. Boa, A. N.; Jenkins, P. R. Benzyl 2,2,2-Trichloroacetimidate. In *Encyclopedia of Reagents for Organic Synthesis*; Paquette, L. A., Ed.; John Wiley and Sons: New York, 1995; Vol. 1, pp 374–375.

9. (a) Poon, K. W. C.; Dudley, G. B. *J. Org. Chem.* **2006**, *71*, 3923–3927. (b) Poon, K. W. C.; House, S. E.; Dudley, G. B. *Synlett* **2005**, 3142–3144.

10. Hughes, D. L. *Org. React.* **1992**, *42*, 335–656.

A CONVENIENT PREPARATION OF AN ORTHOGONALLY PROTECTED C^{α},C^{α}-DISUBSTITUTED AMINO ACID ANALOG OF LYSINE: 1-*tert*-BUTYLOXYCARBONYL-4-((9-FLUORENYLMETHYLOXYCARBONYL)AMINO)-PIPERIDINE-4-CARBOXYLIC ACID

[1,4-Piperidinedicarboxylic acid, 4-[[(9*H*-fluoren-9-ylmethyloxy)carbonyl]amino]-, (1,1-dimethylethyl) ester]

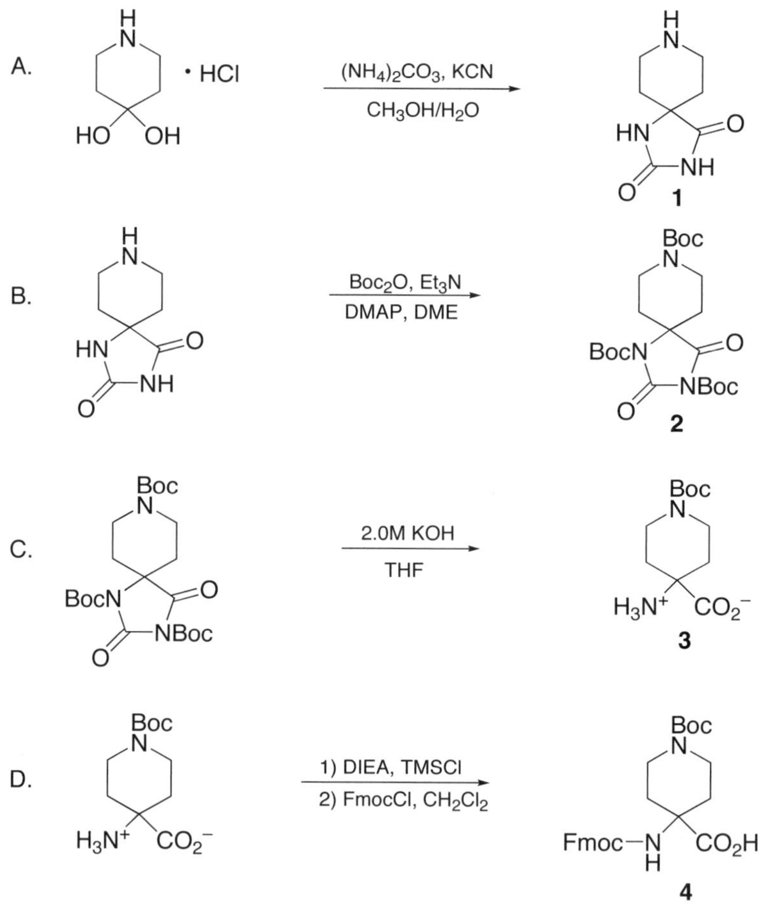

Submitted by Lars G. J. Hammarström, Yanwen Fu, Sidney Vail, Robert P. Hammer, and Mark L. McLaughlin.[1]
Checked by Rick L. Danheiser and Martin E. Hayes.

1. Procedure

Caution: Potassium cyanide is a potent poison, which should always be handled with gloves in a well-ventilated hood. Contact with acid releases toxic hydrogen cyanide gas.

A. Piperidine-4-spiro-5'-hydantoin (**1**). A 1000-mL, single-necked, round-bottomed flask equipped with a magnetic stirbar and an addition funnel fitted with an argon inlet is charged with 4-piperidone monohydrate hydrochloride (30.0 g, 195 mmol), ammonium carbonate (41.3 g, 420 mmol), 250 mL of methanol, and 150 mL of deionized water (Note 1). The mixture is allowed to stir at room temperature until all solids dissolve and then a solution of potassium cyanide (26.7 g, 410 mmol) (Note 2) in 100 mL of deionized water is added dropwise to the reaction mixture over 10 min. The reaction flask is sealed with a glass stopper and the reaction mixture is stirred at room temperature for 48 h. The resulting suspension is concentrated to a volume of 300 mL by rotary evaporation at 40 °C, after which the solution is cooled to 10 °C. The white solid which precipitates is collected in a Büchner funnel using suction filtration. Concentration of the filtrate to a volume of 200 mL results in precipitation of additional product that is separated by filtration and combined with the first crop. The resulting light yellow solid is washed with four 25-mL portions of deionized water to yield a white solid (Note 3). The product is allowed to dry for 2 h in the air and then is dried under reduced pressure (85 °C; 0.5 mm) for 48 h (Note 4) to yield 25.7-26.2 g (77-79%) of the desired hydantoin as a white solid (Note 5).

*B. 1-tert-Butyloxycarbonylpiperidine-4-spiro-5'-(1',3'-bis(tert-butyloxy-carbonyl)) hydantoin (**2**).* A 2000-mL, three-necked, round-bottomed flask equipped with an argon inlet adapter, glass stopper, and an overhead mechanical stirrer is charged with a suspension of the hydantoin **1** (26.0 g, 154 mmol) in 1000 mL of 1,2-dimethoxyethane (Note 6). Triethylamine (15.7 g, 154 mmol) (Note 7) is added in one portion, and the resulting white suspension is stirred for 30 min. Di-*tert*-butyl dicarbonate (168.0 g, 770 mmol) is then added by pipette, followed by 4-dimethylaminopyridine (DMAP) (0.2 g, 1.5 mmol) (Note 8). Six additional 0.2 g-portions of DMAP are added at 12 h intervals during the course of the reaction. The reaction mixture is stirred vigorously for a total of 72 h, and the resulting light yellow solid is then collected in a Büchner funnel using suction filtration. The filtrate is concentrated to a volume of 60 mL by rotary evaporation, and the resulting solution is cooled to 15 °C. The precipitate which appears is collected using suction filtration, added to the first crop, and the combined solids are dissolved in 500 mL of chloroform. This solution is washed with three 200-mL portions of 1.0N HCl, and the combined aqueous phases are extracted with 100 mL of chloroform. The combined organic layers are washed with 100 mL of saturated aq $NaHCO_3$ solution and 100 mL of brine (Note 9), dried over anhydrous $MgSO_4$, filtered, and concentrated by rotary evaporation. The resulting solid is dried at room temperature at 0.01 mm for 24 h. The resulting finely ground light yellow solid is suspended in 400 mL of diethyl ether in a 1000-mL, round-bottomed flask equipped with a magnetic stirbar, stirred for 2 h, and filtered on a Büchner funnel washing with four 50-mL portions of diethyl ether (Note 10). The product is dried under vacuum (85 °C; 0.5 mm) for 24 h to give 60.0–65.3 g (83-90%) of **2** as a ivory-colored solid (Note 11).

C. 4-Amino-1-tert-butyloxycarbonylpiperidine-4-carboxylic acid (**3**). A 2000-mL, round-bottomed flask equipped with a magnetic stirbar is charged with a suspension of the hydantoin **2** (40.0 g, 0.8 mol) in 340 mL of THF (Note 12), and 340 mL of 2.0M potassium hydroxide solution (Note 13) is added in one portion. The flask is stoppered and the reaction mixture is stirred for 4 h (Note 14) and then poured into a 1000-mL separatory funnel. The layers are allowed to separate over 45 min and the aqueous layer is then drained into a 1000-mL round-bottomed flask. This solution is cooled at 0 °C while the pH is adjusted to 8.0 by the slow addition of ca. 100 mL of 6.0N HCl solution. The resulting solution is further acidified to pH 6.5 by slow addition of 2.0 N HCl solution (Note 15). The white precipitate which appears is collected by filtration on a Büchner funnel and the filtrate is concentrated to a volume of 60 mL to furnish additional precipitate which is collected by filtration. The combined portions of white solid are dried at room temperature under reduced pressure (65 °C; 0.5 mm) for 12 h and then suspended in 100 mL of chloroform (Note 16) and stirred for 45 min. The white solid is filtered and then dried under reduced pressure (85 °C; 0.5 mm) for 24 h to yield 13.4–14.1 g (64-68%) (Note 17) of the amino acid **3** as a white solid (Note 18).

D. 1-tert-Butyloxycarbonyl-4-(9-fluorenylmethyloxycarbonylamino) piperidine-4-carboxylic acid (**4**). A 1000-mL, three-necked, round-bottomed flask is equipped with a magnetic stirbar, argon inlet adapter, rubber septum, and an exit tube submerged in a 6.0M KOH solution. The apparatus is flame-dried under a flow of argon and then charged with finely ground amino acid **3** (17.0 g, 69.6 mmol). Anhydrous dichloromethane (500 mL) is added, followed by diisopropylethylamine (22.5 g, 30.3 mL, 174 mmol) which is added via syringe over 10 min (Note 19). The reaction mixture is stirred for 30 min and then chlorotrimethylsilane (15.1 g, 17.6 mL, 140 mmol) (Note 20) is added dropwise over 6 min (Note 21). After 30 min, the septum is replaced with a condenser and the reaction mixture is heated at reflux for 3 h during which time the mixture becomes homogeneous. At 30 min intervals, the reaction vessel is flushed with argon for 30 sec to remove HCl formed in the reaction. The solution is then cooled to –10 °C and 9-fluorenylmethyl chloroformate (Fmoc-Cl) (16.7 g, 64.6 mmol) (Note 22) is added in one portion. The resulting solution is stirred for 3 h under a constant slow stream of argon and then concentrated by rotary evaporation. The resulting paste is taken up in 400 mL of diethyl ether and 1000 mL of aqueous 2.5% Na_2CO_3 solution (Note 23), and the aqueous layer is separated (Note 24) and washed with two 100-mL portions of ether. The aqueous layer is transferred to a 1-L beaker and cooled in an ice bath while 2.0N HCl is added to adjust the pH to 2.0. The resulting suspension is transferred to a 2-L separatory funnel (Note 25), and the precipitated acid is extracted with one 300-mL portion and two 150-mL portions of ethyl acetate. The combined organic phases are dried over $MgSO_4$, filtered, and concentrated by rotary evaporation and then at 60 °C under reduced pressure (0.5 mm) for 10 h. The resulting light yellow solid is suspended in 200 mL of hexane, stirred for 45 min, and then collected by filtration on a Büchner funnel (Note 26). The product is dried under reduced pressure (60 °C; 0.5 mm) for 24 h to afford 22.0–24.4 g (73-76%) (Note 27) of the desired product **4** (Note 28) as a white solid.

2. Notes

1. 4-Piperidone monohydrate hydrochloride and ammonium carbonate were purchased from Alfa-Aesar Chemical Company and used without further purification. Methanol was purchased from Mallinckrodt Chemical Company and used as received.

2. Potassium cyanide was purchased from Alfa Aesar Chemical Company and used without further purification.

3. Precipitated hydantoin must be washed with sufficient water to remove any residual ammonia, which can interfere with subsequent reactions. If an ammonia odor or yellow color persists after the initial water washes, additional washes should be performed.

4. The submitters dried their product in a vacuum oven for 8 h, then ground the resulting solid to a fine powder and returned it to the oven for 40 h further.

5. The hydantoin **1** is only slightly soluble in CD_3SOCD_3 and displays the following spectroscopic properties: ^{1}H NMR (500 MHz, CD_3SOCD_3) δ: 1.37 (d, J = 12.5 Hz, 2 H), 1.68 (dt, J = 4.6, 12.2 Hz, 2 H), 2.64 (app t, J = 12 Hz, 2 H), 2.83 (app d, J = 12.2 Hz, 2 H), 8.47 (s, 1 H), 10.3-10.6 (s, 2 H). ^{13}C NMR (125 MHz, CD_3SOCD_3) δ: 33.8, 41.2, 61.1, 156.4, 178.1.

6. 1,2-Dimethoxyethane (Dri-Solv) was purchased from EM Science Chemical Company and used as received.

7. Triethylamine was purchased from Alfa-Aesar Chemical Company and used without further purification.

8. Di-*tert*-butyl dicarbonate and 4-dimethylaminopyridine were purchased from Alfa-Aesar Chemical Company and used without further purification.

9. The initial HCl wash results in an emulsion and up to 2 h may be required for phase separation. At that point, any remaining emulsion should be separated and added to 100 mL of chloroform and 100 mL of 1.0N HCl solution. The chloroform layer is then combined with the other organic phases.

10. Triturating with diethyl ether was found to be necessary to successfully remove residual di-*tert*-butyl dicarbonate and di-*tert*-butyl iminodicarboxylate which may be produced as a result of residual traces of ammonia from step A.

11. Spectroscopic data for **2**: ^{1}H NMR (500 MHz, $CDCl_3$) δ: 1.46 (s, 9 H), 1.53 (s, 9 H), 1.57 (s, 9 H), 1.73 (m, 2 H), 2.67 (dt, J = 5.2, 13.6 Hz, 2 H), 3.32-3.50 (m, 2 H), 4.00-4.23 (m, 2 H). ^{13}C NMR (125 MHz, $CDCl_3$) δ: 27.9, 28.2, 28.6, 29.7, 29.9, 39.2, 40.2, 62.5, 80.1, 85.3, 87.1, 145.2, 147.3, 148.0, 154.5, 169.8.

12. Tetrahydrofuran was purchased from Mallinckrodt Chemical Company and used as received.

13. Potassium hydroxide was purchased from Mallinckrodt Chemical Company and used without further purification.

14. Two phases appear within the first 15 min due to the heavy ionic content of the aqueous layer. Di-*tert*-butyl iminodicarboxylate, which is produced as a result of the hydrolysis, is selectively soluble in the THF layer, while the amino carboxylate salt of the product remains in solution in the aqueous layer.

15. Reaching the equilibrium where the amino acid zwitterion predominates is a slow process. After acidifying to pH 6.5, the solution is allowed to stir at 0 °C for 25 min during which time the pH of the solution slowly increases. The pH is readjusted to pH 6.5 by slow addition of 2.0N HCl at 0 °C. Repetition of this procedure as many as ten times may be necessary to insure the pH value of the aqueous solution remains at 6.5. Triturating with chloroform is necessary to successfully remove a trace amount of 1-*tert*-butyloxycarbonylpiperidine-4-spiro-5'-(1'-*tert*-butyloxy-carbonyl)-hydantoin.

16. The submitters report obtaining **3** in 87% yield.

17. Compound **3** is only slightly soluble in MeOH-d_4. Spectroscopic data for **3**: ^{1}H NMR (500 MHz, CD_3OD) δ: 1.44 (s, 9 H), 1.58-1.66 (m, 2 H), 2.07-2.14 (m, 2 H), 3.44-3.54 (m, 2 H), 3.64-3.72 (m, 2 H). ^{13}C NMR (125 MHz, $CDCl_3$) δ: 28.8, 33.3, 40.7, 41.7, 59.9, 81.5, 156.5, 175.9.

18. Diisopropylethylamine was purchased from Alfa-Aesar chemical Company and used without further purification. Dichloromethane was purified by pressure filtration through activated alumina.

19. Chlorotrimethylsilane was distilled from calcium hydride immediately prior to use.

20. Upon addition of chlorotrimethylsilane, vigorous HCl production is observed.

21. The checkers purchased Fmoc-Cl from Alfa-Aesar Chemical Company. Impurities are sometimes found in Fmoc-Cl purchased from various sources, and it is necessary to verify its purity by TLC and ^{1}H-NMR analysis. If necessary, Fmoc-Cl can be purified by recrystallization from hexane.

22. Efficient magnetic stirring for up to 2 h is necessary to dissolve the crude product in the mixture of ether and aqueous 2.5% sodium carbonate solution.

23. The solution is allowed to stand in a separatory funnel for at least 30 min before the aqueous layer is drained out. It is sometimes observed that a portion of the sodium carboxylate salt of **4** forms a third layer below the aqueous layer, in which case this layer should be combined with the sodium carbonate layer.

24. The precipitated amino acid forms a gum on the inside walls of the beaker. Magnetic stirring and manual agitation with excess ethyl acetate is necessary to transfer the residual precipitate.

25. Triturating with hexane is necessary to remove trace amounts of solvents (ethyl acetate and ether) in the product.

26. The submitters obtained 26.9 g (83%) of the product.

27. The product displays the following spectroscopic data: ^{1}H NMR (500 MHz, $CDCl_3$) δ: 1.46 (s, 9 H), 1.90-2.12 (br s, 4 H), 3.00-3.13 (br s, 2 H), 3.72-3.92 (br s, 2 H), 4.16-4.22 (m, 1 H), 4.38-4.48 (br s, 2 H), 5.06-5.14 (br s, 1 H), 7.30 (t, J = 7.4 Hz, 2 H), 7.39 (t, J = 7.4 Hz, 2 H), 7.56 (d, J = 7.0 Hz, 2 H), 7.75 (d, J = 7.4 Hz, 2 H). ^{13}C NMR (125 MHz, $(CD_3)_2SO$) δ: 28.1, 31.3, 46.7, 56.7, 65.4, 78.8, 120.2, 125.3, 127.1, 127.7, 140.8, 143.8, 153.9, 155.4, 175.1.

Waste Disposal Information

All toxic materials were disposed of in accordance with "Prudent Practices in the Laboratory"; National Academy Press; Washington, DC, 1995.

3. Discussion

The effect of C^{α},C^{α}-disubstituted amino acids (ααAAs) on peptide secondary structure has been studied in recent years.[2a-d] While longer side-chain C^{α},C^{α}-di-*n*-alkyl amino acids promote extended peptide conformation,[2a] alicyclic ααAAs, in which the C^{α} carbon forms a cyclic bridge with itself, such a 1-aminocyclopentane-1-carboxylic acid (Ac_5c) and 1-aminocyclohexane-1-carboxylic acid (Ac_6c), have helix-forming characteristics similar to those of 1-aminoisobutyric acid (Aib).[2a,c]

Since most ααAAs are hydrophobic in nature, peptides rich in ααAAs are generally restricted to study in organic solvents due to their low solubility in aqueous media. There have been very few examples of side-chain functionalized ααAAs that would allow for the synthesis of highly water-soluble peptides rich in ααAA content.[3] This is primarily due to difficulty of synthesis, since side-chain functionalized derivatives must be orthogonally protected to allow for incorporation into solid-phase peptide synthesis. The harsh conditions, under which standard methods of ααAA synthesis are performed, make this a difficult task.

Despite recent advances in synthetic methods to mildly generate polyfunctional ααAAs,[4] the Bucherer-Berg approach is still the most convenient method to generate simple ααAAs in good yields.[5] However, hydantoins suffer from the limitation of requiring harsh conditions for hydrolysis, thus compromising the ability to generate side-chain protected polyfunctional amino acids. This problem was partially overcome by Rebek and coworkers, who discovered that *N,N'*-bis-(*t*-butyloxycarbonyl)-hydantoins can be hydrolyzed under much milder conditions.[6] The progress made in ααAA synthesis by mild hydantoin hydrolysis allowed the submitters to develop the synthesis of the orthogonally protected ααAA Fmoc-Api(Boc)-OH. This ααAA analog of lysine allows for the preparation of alicyclic ααAAs, while retaining high water-solubility.[6] As with other alicyclic ααAAs, Api has been found to strongly favor helical conformations of resulting peptides.[7b]

It has been found that the *tris*(*tert*-butyloxycarbonyl) protected hydantoin of 4-piperidone **2**, selectively hydrolyses in alkali to yield the *N-tert*-butyloxycarbonylated piperidine amino acid **3**. The hydrolysis, which is performed in a biphasic mixture of THF and 2.0M KOH at room temperature, cleanly partitions the deprotonated 4-amino-*N'*-(*tert*-butyloxycarbonyl)piperidine-4-carboxylic acid into the aqueous phase of the reaction with minimal contamination of the hydrolysis product, di-*tert*-butyl iminodicarboxylate, which partitions into the THF layer. Upon neutralization of the aqueous phase with aqueous hydrochloric acid, the zwitterion of the amino acid is isolated. The Bolin procedure to introduce the 9-fluorenylmethyloxycarbonyl protecting group efficiently produces **4**.[8] This synthesis is a significant improvement over the previously described method[9] where the final protection step was complicated by contamination of the hydrolysis side-product, di-*tert*-butyl iminodicarboxylate, which is very difficult to separate from **4**, even by chromatographic means.

1. Department of Chemistry, Louisiana State University, Baton Rouge, LA 70803.

2. (a) Paul, P. K. C.; Sukumar, M.; Bardi, R.; Piazzesi, A. M.; Valle, G.; Toniolo, C.; Balaram, P. *J. Am. Chem. Soc.* **1986**, *108*, 6363-6370; (b) Toniolo, C.; Benedetti, E. *Macromolecules* **1991**, *24*, 4004-4009; (c) Benedetti, E.; DiBlasio, B.; Iacovino, R.; Menchise, V.; Saviao, M.; Pedone, C.; Bonora, G. M.; Ettore, A.; Graci, L.; Formaggio, F.; Crisma, M.; Valle, G.; Toniolo, C. *J. Chem. Soc., Perkin Trans. 2* **1997**, 2023-2032; (d) Toniolo, C.; Crisma, M.; Formaggio, F.; Benedetti, E.; Santini, A.; Iacovino, R.; Saviano, M.; DiBlasio, B.; Pendone, C.; Kamphius, J. *Biopolymers* **1996**, *40*, 519-522.
3. Yokum, T. S.; Bursavich, M. G.; Piha-Paul, S. A.; Hall, D. A.; McLaughlin, M. L. *Tetrahedron Lett.* **1997**, *38*, 4013-4016.
4. Fu, Y.; Hammarström, L. G. J.; Miller, T. J.; Fronczek, F. R.; McLaughlin, M. L.; Hammer, R. P. *J. Org. Chem.* **2001**, *66*, 7118-7124.

5. (a) Bucherer, H. T.; Steiner, W. *J. Prakt. Chem.* **1934**, *140*, 291-316; (b) Bergs, H.; German patent 566,094 (May 26, 1929); *Chem Abst.* **1993**, *27*, 1001.

6. Kubik, S.; Meissner, R. S. Rebek, Jr.; J. *Tetrahedron Lett.* **1994**, *35*, 6635-6638.

7. (a) Yokum, T. S.; Gauthier, T. J.; Hammer, R. P.; McLaughlin, M. L. *J. Am. Chem. Soc.* **1997**, *119*, 1167-1168; (b) Hammarström, L. G. J.; Gauthier, T. J.; Hammer, R. P.; McLaughlin, M. L. *J. Peptide Res.* **2001**, *58*, 108-116.

8. Bolin, D. R.; Sytwu, I.–I.; Humiec, F.; Meienhofer, J. *Int. J. Pept. Protein Res.* **1989**, 353.

9. Wysong, C. L.; Yokum, T. S.; Morales, G. A.; Gundry, R. L.; McLaughlin, M. L.; Hammer, R. P. *J. Org. Chem.* **1996**, *61*, 7650-7651.

SYNTHESIS OF ELECTRON-DEFICIENT SECONDARY PHOSPHINE OXIDES AND SECONDARY PHOSPHINES: BIS[3,5-BIS(TRIFLUOROMETHYL)PHENYL]PHOSPHINE OXIDE and BIS[3,5-BIS(TRIFLUOROMETHYL)PHENYL]PHOSPHINE

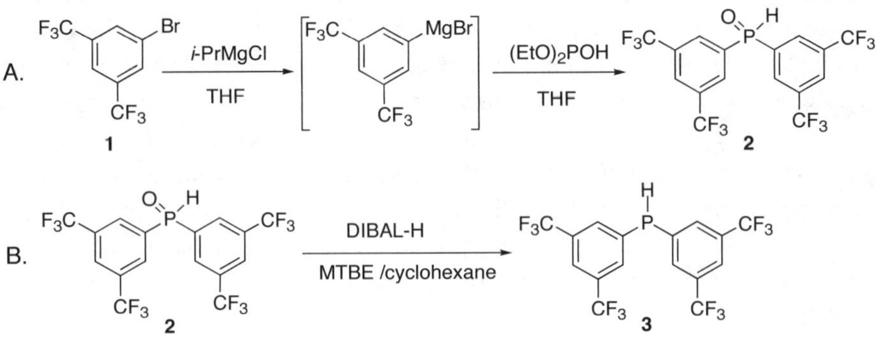

Submitted by Carl A. Busacca, Jon C. Lorenz, Paul Sabila, Nizar Haddad, and Chris H. Senanyake.[1]
Checked by Scott E. Denmark and Christopher S. Regens.

1. Procedure

A. Bis[3,5-bis(trifluoromethyl)phenyl]-phosphine oxide (**2**). To an oven-dried, 1-L, three-necked, round-bottomed flask equipped with a magnetic stir-bar, and fitted with a graduated 250-mL pressure-equalizing addition funnel (fitted with a septum) in the center neck, a thermocouple inserted through a rubber septum in the left neck, and an inert gas inlet valve in the right neck, is evacuated (0.03 mmHg vacuum) and backfilled three times with argon. The flask is charged with 3,5-bis(trifluoromethyl)bromobenzene **1** (59.2 mL, 343 mmol, 3.3 equiv, Note 1) followed by anhydrous THF (60 mL, Note 2) through the addition funnel septum cap with a syringe. The addition funnel is then charged with isopropylmagnesium chloride (171.5 mL, 343 mmol, 3.3 equiv, 2.0 M solution in THF, Note 3) via cannula under argon pressure. The flask is then placed in a dry-ice/acetone bath. When the internal temperature reaches –30 °C, the solution of isopropylmagnesium chloride is added rapidly over 10 min. The internal temperature at the conclusion of the addition is –60 °C. Tetrahydrofuran (10 mL, added through the addition funnel septum cap with a syringe) is used to wash the walls of the addition funnel into the flask. The dry-ice/acetone bath is then replaced by an ice/H_2O bath. The internal temperature rapidly rises to a maximum of 9 °C in approximately 5 min, to give a clear, brown solution, then the internal temperature rapidly falls to 2 °C (Note 4). At this time a 50 μL aliquot is removed by syringe through the septum in the left neck of the flask, and is quenched into a biphasic mixture of Et_2O/2 M HCl (~ 2 mL). The aqueous layer is removed and the organic solution is back extracted with H_2O (1 mL). The organic layer is filtered through a SiO_2 plug (1 mm x 2 cm), which is eluted with Et_2O (~2 mL). The filtrate is analyzed by gas chromatography to monitor reaction progress (Notes 5 and 6). An additional portion of isopropylmagnesium chloride solution (5 mL, 10 mmol, 3% of the original Grignard charge) is added at once by syringe at 1 °C through the septum in the left neck of the flask. After an additional 0.5 h, GC analysis of a quenched aliquot indicates that Grignard exchange is complete (total reaction time 1 h after the initial exotherm).

The addition funnel is then charged with THF (30 mL) followed by diethyl phosphite (13.7 mL, 104 mmol, 1 equiv, Note 7) through the septum cap with a syringe. The reaction solution is again cooled in a dry-ice/acetone bath. When the internal temperature reaches –60 °C, the phosphite solution is added drop-wise over 10 min, causing the internal temperature to rise to –50 °C. The cold bath is removed, and the reaction is allowed to warm to ambient temperature (~18-20 °C). After one hour at ambient temperature, GC analysis of a quenched aliquot indicates the reaction is complete to give phosphine oxide **2** (t_R = 6.90 min, GC Method 2). The reaction mixture is then cooled to 5 °C internal temperature in an ice/H_2O bath, and 3 N HCl (175 mL) is added by addition funnel over 15 min. The addition funnel is removed and saturated aqueous NaCl solution (150 mL) is then added at once. Then EtOAc (250 mL) is added and the biphasic mixture is transferred to a 1-L separatory funnel. The phases are separated and the aqueous phase is back-extracted with EtOAc (250 mL). The combined organic extracts are then dried ($MgSO_4$, 50 g, Note 8), filtered into a 1-L round-bottomed flask and the solvent is removed with rotary evaporation under reduced pressure. The residue is further dried under high vacuum (0.02 mmHg, 15 min) to remove the 1,3-bis(trifluoromethyl)benzene by-product (Note 9). The resulting orange-brown solid is dissolved in boiling MTBE (90 – 110 mL) in a 1-L round-bottomed flask, equipped a magnetic stir-bar, and a 125-mL pressure-equalizing addition funnel with an argon adaptor at the top. The contents are allowed to cool to ambient temperature under argon, with stirring, causing a thick, brown slurry to form (Note 10). Hexanes (140 mL) are added over 10 min to this slurry through the addition funnel (Note 11). The mixture is then stirred for an additional 10 min to break up any large solids, and the mixture is filtered through a medium-fritted filter funnel, washing the flask and filter cake with MTBE (5 mL). The solid is dried under a vigorous stream of argon, from an inverted funnel, flowing over the filter cake. After 2 h, compound **2** (36.0–37.1 g, 73–75%) is obtained as a powdery, white solid (Note 12).

B. Bis[3,5-bis(trifluoromethyl)phenyl]phosphine (**3**). A 500 mL, three-necked, round-bottomed flask containing a magnetic stir-bar is fitted with a 125-mL pressure-equalizing addition funnel (fitted with a rubber septum) in the center neck, a thermocouple through a rubber septum in the left neck, and an inert gas valve in the right neck. Phosphine oxide **2** (10.0 g, 21.0 mmol, 1 equiv) is added under a flow of argon by removal of the inert gas valve. After replacement of the inert gas valve, MTBE (50 mL, Note 13) is added via syringe to the flask through the septum on the addition funnel. The resulting slurry is stirred under argon at ambient temperature for 5 min. The addition funnel is then charged with DIBAL-H (42.2 mL, 42.2 mmol, 1.0 M solution in cyclohexane, 2 equiv, Note 14). This solution is then added drop-wise to the substrate slurry over 10 min. The internal temperature rises to a maximum temperature of 36 °C and a brown solution forms (Fig. 1). The addition funnel is rinsed with MTBE (5 mL). The resulting mixture is then stirred for 1 h, and then is placed in a dry-ice/acetone bath. A needle connected to an argon source is inserted through the flask septum and positioned beneath the solution surface, so that a slow stream of argon is bubbled through the solution (Note 15). The addition funnel is charged with an aqueous 2 N NaOH solution (80 mL). This solution is added drop-wise to the reaction mixture over ~ 15 min. The internal temperature rises to –27 °C during the addition (Note 16). The cold bath is replaced by a water bath (30–35 °C) to raise the internal temperature to approximately 20 °C, and the mixture is stirred for an additional 15 min at this temperature (Note 17). Stirring is stopped, and two clear phases are observed (See Fig. 2). A 32-cm, 18-gauge needle attached to a 60-mL polyethylene syringe is inserted through the flask septum with the needle tip pressed against the bottom of the flask. The lower (aqueous) phase is removed by syringe and discarded as waste. The addition funnel is charged with H_2O (40 mL) by syringe through the addition funnel septum cap and added to the reaction flask by slow addition over approximately 5 min. The mixture is stirred for 5 min, then the lower (aqueous) phase is removed by syringe through flask septum and discarded as waste, leaving the solution of crude secondary phosphine **3** in the reaction flask.

An Airfree filter is charged with Merck silica gel 60 (32 g) followed by anhydrous $MgSO_4$ (28 g), and topped with sand (30 g) (Note 18, see Fig. 1). The filter is then sealed with a rubber septum that is further secured with a piece of copper wire. The bottom of the Airfree filter is connected to a vacuum take-off adapter (Note 19), which in turn is connected to a 500-mL, one–necked, round-bottomed flask used as reccciver. The side arm of the vacuum take-off adapter is connected with a short piece of rubber tubing to a Florence flask cold trap (Note 20) that is cooled to –78 °C in a dry-ice/acetone bath to collect the evaporated organic solvents. The cold trap is vented to an argon bubbler. All joints of the Airfree filter apparatus are greased with Dow silicon grease (Note 21). A mixture of cyclohexane/MTBE (5:1, 100 mL) is then injected by syringe through the septum cap of the Airfree funnel containing the silica gel/$MgSO_4$ column. A slight argon pressure is applied from the upper valve of the Airfree filter, and the solvent is passed through to wet the solids. The receiver flask is removed and the wash solvent discarded as waste. A new, oven-dried, 500-mL one-necked receiver flask is then installed, and argon is purged gently through the system for approximately 5 min. The organic phase from the reaction flask containing the crude secondary phosphine is removed by syringe and injected through the septum cap of the Airfree filter onto the column (see Fig. 3). At the same time, a cyclohexane/MTBE mixture (5:1, 100 mL) is charged to the reaction flask by syringe (Notes 15, 22). A slight positive pressure of argon is applied to the upper valve of the Airfree filter as before, slowly flowing the secondary phosphine solution through the column. Then, 100 mL of the solvent mixture is injected through the septum cap of the Airfree filter and is passed through the solids under a slight argon pressure to wash the remaining column contents into the receiver flask. The entire product is now contained in the 500-mL receiver flask (Fig. 4, Note 23).

The receiver flask is placed in a warm water bath (32 – 40 °C) (Note 24). The rate of argon flow through the Airfree filter side arm is then increased, causing the solvents to be removed from the secondary phosphine product by evaporation, with solvent condensation in the attached cold trap (Note 25). After approximately 8 h, all the solvent is removed and the product secondary phosphine **3** (7.7–7.8 g, 79–83% yield) is obtained as a white crystalline solid in the receiver (Fig. 5, Note 26).

The secondary phosphine can be recrystallized by working in a dry box. The solid is transferred to a one-necked, 100-mL, round-bottomed flask equipped with a magnetic stir-bar, and a septum-sealed, Teflon-taped three-way adaptor. The flask is removed from the dry-box and the adaptor is attached to an argon inert gas line. A 20-gauge needle is inserted through the septum as a vent, then degassed MeOH (18 mL) is added via syringe. The contents are then magnetically stirred and heated with a heat gun until all the solids dissolved. The needle is removed and the flask is isolated from the septum by turning the three-way stopcock. The contents are allowed to cool to ambient temperature. When the contents became turbid (Fig. 6), the flask is placed in an ice bath for 1 h. While still in the ice-bath, the mother-liquor is cannula-transferred to another septum–sealed, one-necked, 100-mL, round-bottomed flask that is purged three times with argon. The crystals remaining are washed with chilled (0 °C) degassed MeOH (5 mL) and the wash is transferred to the flask containing the mother liquor. The crystals are further dried under reduced pressure (0.03 mmHg) for 30 min and then are transferred to the dry-box for isolation (crop 1, 6.0–6.3 g, 61–64%). The flask containing the mother liquor is transferred to the dry-box and placed in a –20 °C freezer for 12 h. A second crop is collected by vacuum filtration in the dry box to give 0.90–0.36 g (combined crops 6.7–6.9 g, 68–70%) of **3** as colorless needles (Note 27).

2. Notes

1. 3,5-Bis(trifluromethyl)bromobenzene (98%) was purchased from Matrix Scientifics and used as received.

2. Anhydrous THF (99.9%, stabilized with ~25 ppm BHT) was purchased from EMD and used as received.

3. Formation of this Grignard reagent from Mg metal is hazardous and must not be done. Old solutions of isopropylmagnesium chloride of uncertain concentration should not be used. Isopropylmagnesium chloride in THF was purchased from Aldrich Chemical Co. in a ®-SurePac cylinder. The solution of isopropylmagnesium chloride *was* titrated as 2.0 M, using both no-D NMR[12] and the procedure of Watson and Eastham.[13]

4. The temperature of the Grignard exchange reaction was critical: If the batch temperature was not allowed to exotherm as described, exchange was incomplete, and products derived from addition of unreacted isopropylmagnesium chloride to the phosphite were ultimately formed. An

undercharge of isopropylmagnesium chloride was also less problematic than an overcharge of isopropylmagnesium chloride for the same reason.

5. Analytical gas chromatography (GC) was performed using a Hewlett-Packard 5890 Gas Chromatograph fitted with a flame ionization detector. Two GC methods were used. GC Method 1: Injections were made onto a Hewlett-Packard HP1 (30 m x 0.32 mm) capillary column. Injector temperature was 250 °C; the detector temperature was 300 °C with a H_2 carrier gas flow of 16 mL/min. The column temperature program was as follows: 50 °C for 1 min, 50 °C to 150 °C at 25 °C/min, hold at 150 °C for 5 minutes. Total run time 12 min. GC Method 2: Injector temperature is 250 °C; the detector temperature is 300 °C with a H_2 carrier gas flow of 16 mL/min. The column temperature program is as follows: 50 °C for 1 min, 50 °C to 250 °C at 25 °C/min, hold at 250 °C for 3 minutes. Total run time 12 min. Retention times (t_R) are obtained using Agilent Chemstation software or a Hewlett-Packard 3396 Series II Integrator.

6. At this point, 4% of aryl bromide **1** (retention time, t_R, = 2.88 min, GC Method 1) still remained, and 1,3-bis(trifluoromethyl)benzene was present at t_R = 1.47 min.

7. Diethyl phosphite (98%) was purchased from Aldrich Chemical Co. and used as received.

8. Some phosphine oxides can crystallize as hydrates, so the organic solution had to be well dried to avoid these forms.

9. Rotary evaporation was preformed at 25 mmHg and the water bath temperature was 40 °C. The boiling point of 1,3-bis(trifluoromethyl)benzene is 116 °C. If the aromatic impurity was not removed, it lowered the yield of the crystallization by partially solubilizing phosphine oxide **2**.

10. The large amounts of solids were difficult to dissolve in MTBE initially. Multiple heating cycles (3-4) using an oil bath (65 °C) were used to completely dissolve all solids. In one case 110 mL of MTBE was needed to complete solubilize the solid.

11. Hexanes was added to maximize the yield of the crystallization. It could be omitted with a ~5% reduction in yield.

12. Data for **2** were as follows: mp 127–128 °C; ^{1}H NMR (500 MHz, $CDCl_3$) δ: 8.12 (s, 2 H), 8.21 (d, J = 13.1 Hz, 4 H), 8.33 (d, $^1J_{HP}$ = 507 Hz, 1 H); ^{13}C NMR (125 MHz, $CDCl_3$) δ: 122.5 (C(5), q, $^1J_{CF}$ = 273.0 Hz), 127.2 (C(4), m), 130.8 (C(2), dq, $^2J_{CP}$ = 13.1 Hz, $^3J_{CF}$ = 4.5 Hz), 133.1 (C(1), d, $^1J_{CP}$ = 100.4 Hz), 133.2 (C(3), dq, $^2J_{CF}$ = 34.0 Hz, $^3J_{CP}$ = 13.0 Hz); ^{31}P NMR (202 MHz, $CDCl_3$) δ: 15.47 ($^1J_{PH}$ = 505 Hz); ^{19}F NMR (469 MHz,

$CDCl_3$) δ: –63.65; IR (Nujol) cm^{-1}: 3080 (w), 3057(w), 3031(w), 2389 (w), 2210 (w), 1619 (m), 1329 (m), 1277 (s), 1196 (s), 1134 (s), 1002 (w), 960 (s), 914 (s), 845 (s), 681 (s); MS (EI, 70 ev) *m/z* (%): 474 (M^+ (100)), 472 (40), 456 (11), 455 (59), 453 (21), 407 (20), 261 (35), 195 (71); Anal. Calcd for $C_{16}H_7F_{12}OP$: C, 40.53; H, 1.49; F, 48.08; Found C, 40.45; H, 1.45; F, 47.82.

13. Anhydrous MTBE (99.8%) was purchased from Aldrich Chemical Co. and was used as received. DIBAL-H/cyclohexane was purchased from Aldrich Chemical Co. and was titrated to 1.0 M using no-D NMR[12] and gas titration according to the method of Brown.[14]

14. DIBAL-H is a pyrophoric material, and standard inert atmosphere syringe and cannula transfer methods were used.

15. All secondary phosphines are air-sensitive compounds. Careful maintenance of inert conditions was critical to successful isolation of these compounds.

16. The initial portion of the quench was very exothermic, and slow addition of the aqueous base was required. Some ice (H_2O) formed, yet was of no consequence. Gas (isobutane and possibly H_2) was formed during the quench, also requiring controlled addition.

17. Full hydrolysis of all carbon-aluminum bonds required ambient temperature conditions. If hydrolysis was incomplete, clogging of the silica gel column by the organic phase occurred.

18. A Chemglass Airfree filter (AF-0542-25) was used. The order of addition of solids to the Airfree filter was critical: The phosphine solution needed to first encounter $MgSO_4$ (top of column) so that the *dried* solution flowed through the silica gel. The solvent system was chosen to allow for elution of non-polar phosphines and retention of more polar impurities.

19. A Chemglass vacuum take-off adapter (Chemglass CG-1049-01) was used.

20. A Florence flask cold trap (Kontes 926400-0500) was used.

21. Dow Silicone grease part no. 2021846-0702.

22. Oxygen, a non-polar gas, has a higher solubility in non-polar solvents such as these than in polar solvents. All organic solvents were degassed for 30 min prior to use by bubbling argon through the solution in a two-necked, 1-L round-bottom flask, equipped with septa in both joints. One septum was pierced and a gas dispersion tube was used to bubble argon through the solution. The other septum contained a 20-gauge needle for a bleed.

23. For many applications of secondary phosphines, this solution was used immediately for the next step (nucleophilic addition, cross-coupling, phosphine-borane formation, etc.), rather than isolating the phosphine in neat form. The solution was collected in a three-neck flask, for example, which served as the reactor for the next step. The solvents used here (MTBE/C_6H_{12}) were chosen to be compatible with these further transformations.

24. The water bath was essential for rapid evaporation of the solvent. The water bath temperature was not allowed to exceed 55 °C, because product sublimation was observed at higher temperatures, which lowered the overall yield of phosphine **3**.

25. The cold trap improved the efficiency of the solvent removal, and avoided venting the solvents to the atmosphere. A combination of this argon sweep with an applied vacuum can be used to remove solvent, though increased product oxidation was observed under these conditions.

26. The yield was determined by weighing the solid in a dry-box. From the evaporation procedure, the secondary phosphine was obtained in 90–95% purity, where phosphine oxide was the major contaminant. The submitters noted that the amount of phosphine oxide did not affect subsequent transformations with the secondary phosphine.

27. Data for **3**: mp 67–70 °C. ^{1}H NMR (500 MHz, C_6D_6) δ: 4.61 (d, $^1J_{HP}$ = 223.3 Hz, 1H), 7.53 (d, J = 6.1 Hz, 4H), 7.65 (s, 2H); ^{13}C NMR (125 MHz, C_6D_6) δ: 123.0 (q, $^1J_{CF}$ = 273.6 Hz), 123.5 (m), 132.6 (dq, J = 5.9, 34.1 Hz), 134.0 (d, J = 18.4 Hz), 137.4 (d, J = 17.5 Hz); ^{31}P NMR (202 MHz, C_6D_6) δ: –41.66 ($^1J_{PH}$ = 223 Hz); ^{19}F NMR (471 MHz, C_6D_6) δ: –63.29; IR (Nujol) cm^{-1}: 2319 (w), 1839 (w), 1820 (w), 1616 (w), 1280 (s), 1173 (s), 1125 (s), 1093 (s), 923 (m), 901 (s), 896 (s), 848 (m), 704 (s), 684 (s); MS (EI, 70 ev) *m/z* (%): 458 (M^+ (51)), 439 (28), 244 (46), 195 (100), 175 (68); Anal. Calcd for $C_{16}H_7F_{12}P$: C, 41.94; H, 1.54; F, 49.76; Found C, 41.82; H, 1.49; F, 50.27.

3. Discussion

This procedure can be used to prepare a variety of secondary phosphines by reduction of secondary phosphine oxides (SPO's), as we have recently reported.[2] The SPO's are in turn prepared from the combination of Grignard reagents and diethyl phosphite. The arylmagnesium reagent prepared in situ here must be prepared by Knochel Grignard exchange,[3] not from Mg metal.[4] This method successfully reduces all classes of SPO's from electron-deficient to electron-rich, even extremely hindered species such as di-*tert*-butylphosphine oxide and dimesitylphosphine oxide (Table 1). The conditions are mild and the reducing agents are among the cheapest available. The reductions are chemoselective, and according to adiabatic calorimetry, are also safe, with typical adiabatic temperature rises of only ~17 °C.

Diphenyldisilane has been used for SPO reductions,[5] but reaction temperatures exceed 200 °C and the silane currently costs ~ $90/mol, more than 50 times the cost of DIBAL-H. Lithium aluminum hydride has been used for this reduction,[6] yet material handling and quenching is hazardous, and primary phosphine impurities are generated with electron-deficient SPO's. Phenylsilane has been employed,[7] but a non-aqueous workup with distillation of the product is required. Many secondary phosphines, such as the one described here, are solids that cannot be readily distilled. Furthermore, when using phenylsilane as the reductant, polysiloxanes are known by-products that must be removed. Phosphorus trichloride in combination with lithium aluminum hydride has been used to generate secondary phosphines from SPO's.[8] This method has the disadvantages of taking two steps rather than one (usually requiring the distillation of the intermediate chlorophosphine, which is often difficult) along with the problems associated with lithium aluminum hydride mentioned above. Trichlorosilane has been used for this reduction,[9] but it has a very low boiling point, 31 °C, and is then routinely used at temperatures 80–100 °C *above* its boiling point (refluxing toluene or xylene), creating a hazard. In addition, trichlorosilane works only for electron-rich SPO's that are capable of nucleophilically attacking silicon, and it is commonly used in very large excess (5–20 equivalents). Trichlorosilane is also about twice as expensive as DIBAL-H per mole on a metric ton basis. Finally, borane has been utilized for this reduction.[10] Disadvantages of this reagent include the generation of phosphinous acid borane impurities (a form of unreacted starting material), the need to use a large excess, and a recent severe industrial accident on storage.[11] In summary, the organoaluminum reduction method described here is superior to existing methods in substrate scope, mildness of conditions, chemoselectivity, number of steps, operability, safety, and cost.

Table 1. Reduction of Secondary Phosphine Oxides by iisobutylaluminum Hydride and Triisobutylaluminum.

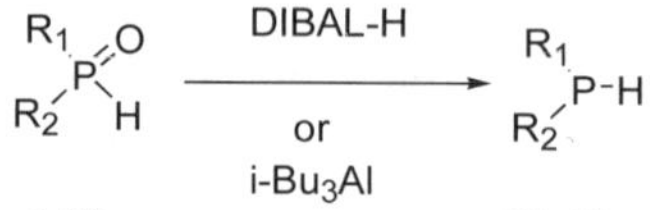

Entry	R_1	SM	R_2	Time/ Temp. (°C)	Al species [a]	Prod	Yield
1	C_6H_5	**1**	C_6H_5	10m/25	**D**	**22**	86
2	4-F-C_6H_4	**2**	4-F-C_6H_4	10m/25	**D**	**23**	90
3	4-Cl-C_6H_4	**3**	4-Cl-C_6H_4	10m/25	**D**	**24**	83
4	4-Me-C_6H_4	**4**	4-Me-C_6H_4	10m/25	**D**	**25**	80[b]
5	3-F-C_6H_4	**5**	3-F-C_6H_4	10m/25	**D**	**26**	89
6	3-Cl-C_6H_4	**6**	3-Cl-C_6H_4	10m/25	**D**	**27**	81
7	3-F, 5-Me-C_6H_3	**7**	3-F, 5-Me-C_6H_3	10m/25	**D**	**28**	82
8	3,5-F_2, 4-OMe-C_6H_2	**8**	3,5-F_2, 4-OMe-C_6H_2	10m/25	**D**	**29**	90
9	3,5-Cl_2-C_6H_3	**9**	3,5-Cl_2-C_6H_3	1h/-20	**D**	**30**	90
10	3,5-F_2-C_6H_3	**10**	3,5-F_2-C_6H_3	1h/-20	**D**	**31**	80
11	2-Me-C_6H_4	**11**	2-Me-C_6H_4	8h/25	**D**	**32**	80
12	4-NMe_2-C_6H_4	**12**	4-NMe_2-C_6H_4	45m/25[c]	**D**	**33**	92
13	4-OMe-C_6H_4	**13**	4-OMe-C_6H_4	40m/25 [c]	**D**	**34**	91
14	2-OMe-C_6H_4	**14**	2-OMe-C_6H_4	8h/35 [c]	**D**	**35**	75
15	C_6H_5	**15**	*i*-Pr	10m/25	**D**	**36**	90
16	C_6H_5	**16**	*t*-Bu	4h/50	**D**	**37**	86
17	n-Bu	**17**	*n*-Bu	1h/25	**D**	**38**	85
18	C_6H_{11}	**18**	C_6H_{11}	4h/50	**D**	**39**	88
19	t-Bu	**19**	*t*-Bu	4h/50	**D**	**40**	87
20	3,5-$(CF_3)_2$-C_6H_3	**20**	3,5-$(CF_3)_2$-C_6H_3	15m/25	**D**	**41**	72
21	3-F, 5-Me-C_6H_3 + PhI [d]	7	3-F, 5-Me-C_6H_3 + PhI [d]	30m/25	**T**	**28**	82
22	3-F, 5-Me-C_6H_3 + $PhCONMe_2$ [d]	7	3-F, 5-Me-C_6H_3 + $PhCONMe_2$ [d]	30m/25	**T**	**28**	81
23	2,4,6-Me_3-C_6H_2	**21**	2,4,6-Me_3-C_6H_2	24h/70 [e]	**T**	**42**	75

[a] **D** = DIBAL-H; **T** = *i*-Bu_3Al. [b] Yield of phosphine borane complex [c] 5 eq. DIBAL-H used [d] competition experiment [e] 8 eq. i-Bu_3Al used

Figure 1: DIBAL-H Reduction Prior to Quench

Figure 2: Reaction Mixture After Aqueous Quench

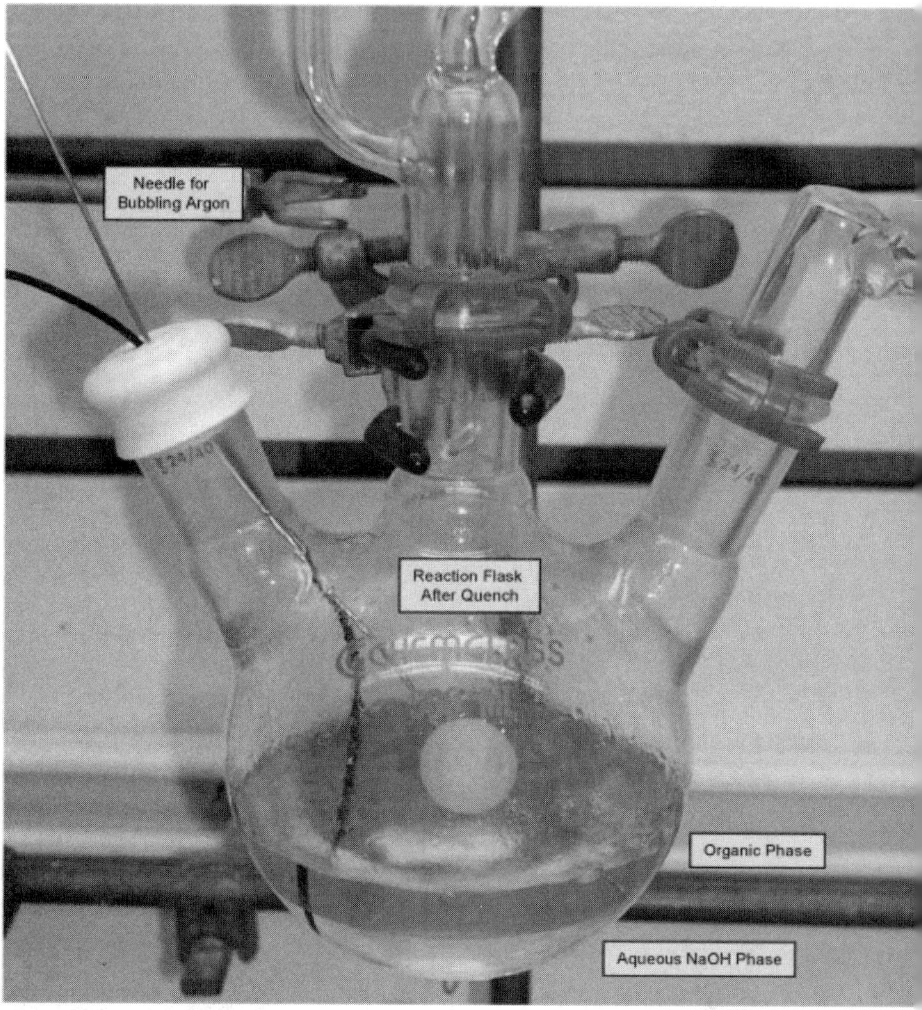

Figure 3: Filtration of Secondary Phosphine Solution

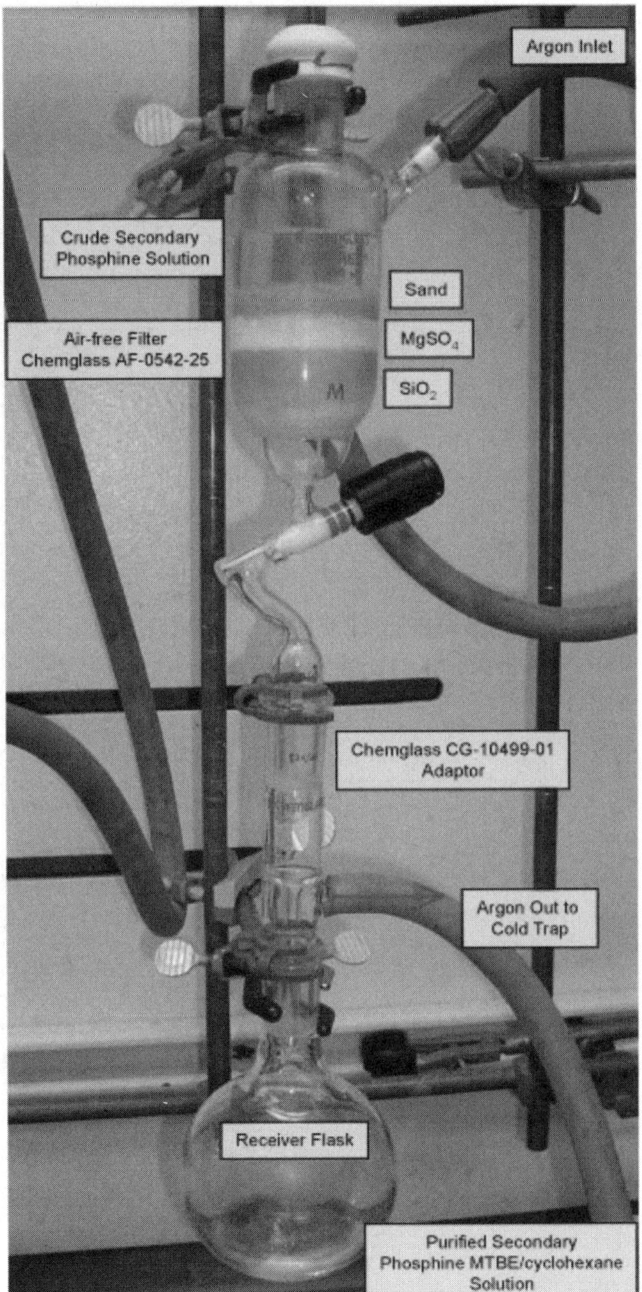

Figure 4: Concentration of Purified Secondary Phosphine

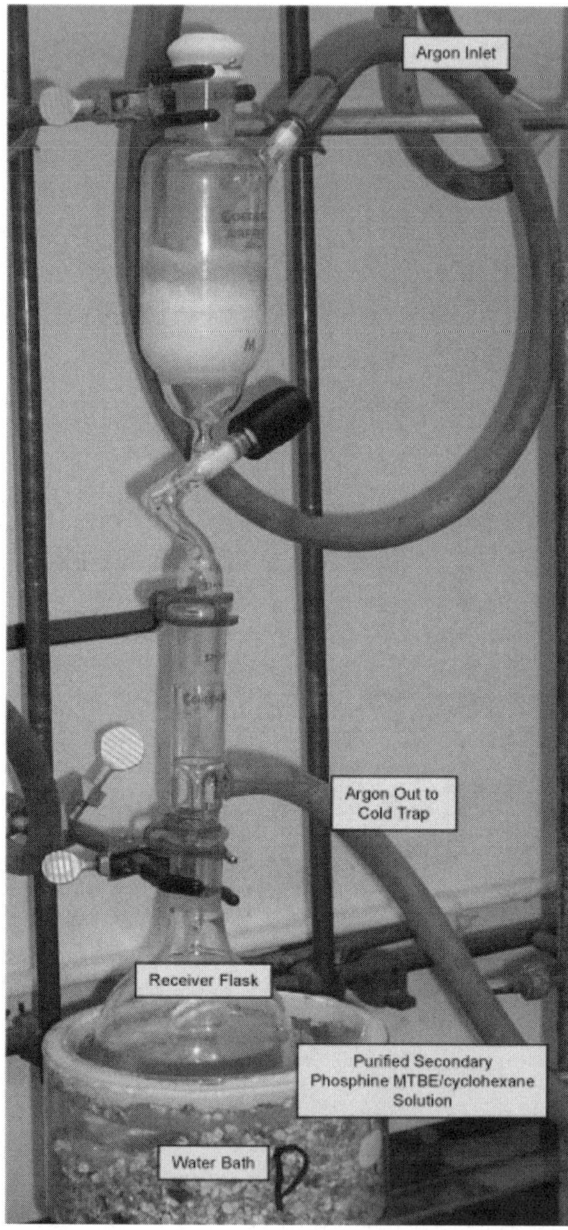

Figure 5: Concentrated Secondary Phosphine

Figure 6: Recrystalization of Secondary Phosphine

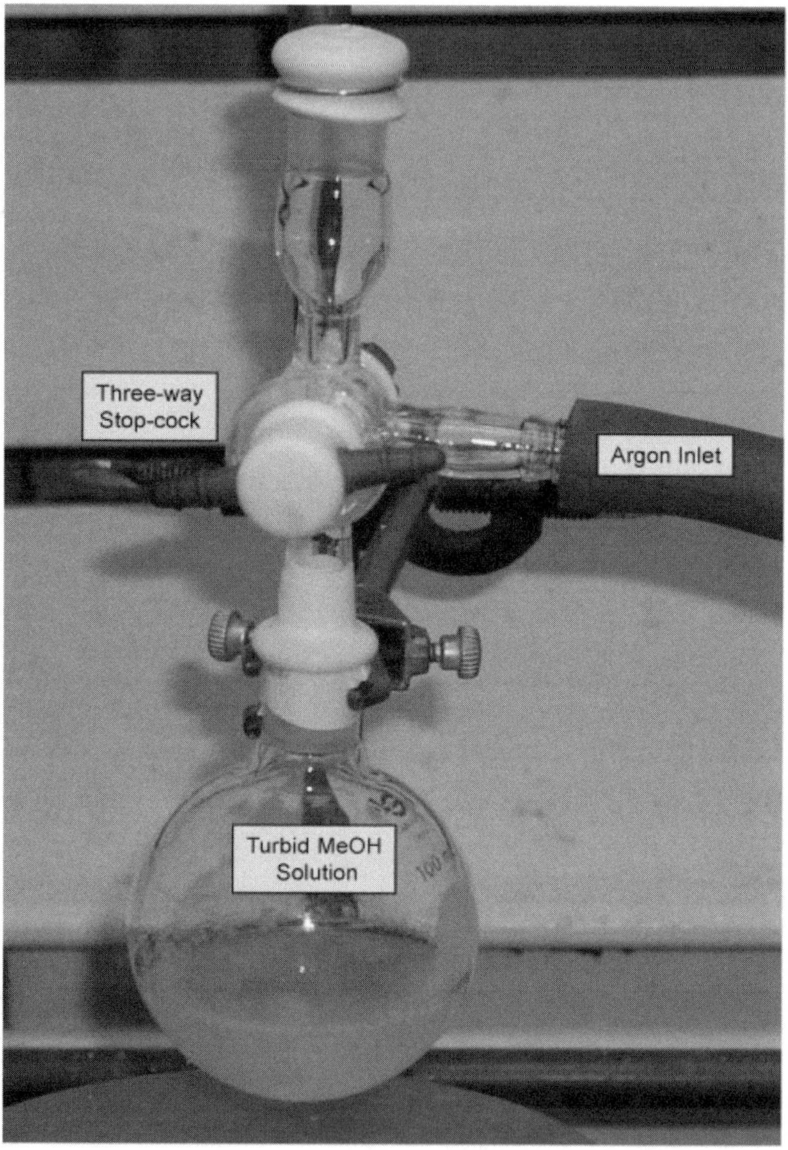

1. Dept. of Chemical Development, Boehringer-Ingelheim Pharmaceuticals, Inc., 900 Ridgebury Rd., Ridgefield, CT 06877.

2. a) Busacca, C. A. U.S. Patent 7 256 314, 2007. b) Busacca, C. A.; Lorenz, J. C.; Grinberg, N.; Haddad, N.; Hrapchak, M.; Latli, B.; Lee, H.; Sabila, P.; Saha, A.; Sarvestani, M.; Shen, S.; Varsolona, R.; Wei, X.; Senanayake, C. H. *Org. Lett.* **2005**, *7*, 4277-4280. c) For the reduction of *tertiary* phosphine oxides with this methodology, see: Busacca, A. A.; Raju, R.; Grinberg, N.; Haddad, N.; Jones, P.-J.; Lee, H.; Lorenz, J. C.; Saha, A.; Senanayake, C. H. *J. Org. Chem,* **2008**, *73* (4) , 1524-1531.

3. Abarbi, M.; Dehmel, F.; Knochel, P. *Tetrahedron Lett.* **1999**, *40*, 7449-7453.

4. a) Leazer, J. L.; Cvetovich, R.; Tsay, F.-R.; Dolling, U.; Vickery, T.; Bachert, D. *J. Org. Chem.* **2003**, *68*, 3695-3698. b) Yakelis, N. A.; Bergman, R. G. *Organometallics* **2005**, *24*, 3579-3581.

5. Livinghouse, T.; McKinstry, L. *Tetrahedron* **1994**, *50*, 6145-6154.

6. Kapoor, P. N.; Venanzi, L. M. *Helv. Chim. Acta* **1977**, *60*, 2824-2829.

7. Pilard, J.-F.; Baba, G.; Gaumont, A.-C.; Denis, J.-M. *Synlett* **1995**, 1168-1170.

8. Rajanbabu, T. V.; Casalnuovo, A. L.; Ayers, T. A.; Warren, T. H. *J. Am. Chem. Soc.* **1994**, *116*, 9869-9882.

9. Dong, X.; Erkey, C. *J. Mol. Catalysis A* **2004**, *211*, 73-81.

10. Pietrusiewicz, K. M.; Stankevic, M. *Synlett* **2003**, *7*, 1012-1016.

11. Reisch, M. *Chem. Eng. News* **2002**, *80*(26), 7.

12. Hoye, T. R.; Aspass, A. W.; Eklov, B. M.; Ryba, T. D. *Org. Lett.* **2005**, *7*, 2205-2208.

13. Watson, S. C.; Eastham, J. F. *J. Organomet. Chem.***1967**, *9,* 165-168.

14. Brown, H. C.; Kramer, G. W.; Levy, A. B.; Midland, M. M. *Organic Synthesis via Boranes*; John Wiley and Sons; New York, **1973**, 21-244.

PREPARATION OF ENANTIOMERICALLY PURE (*R,R*)-BozPHOS [(*2R,5R*)-1-{2-[(*2R,5R*)-2,5-dimethylphospholan-1-yl]phenyl}-2,5-dimethylphospholane 1-oxide]

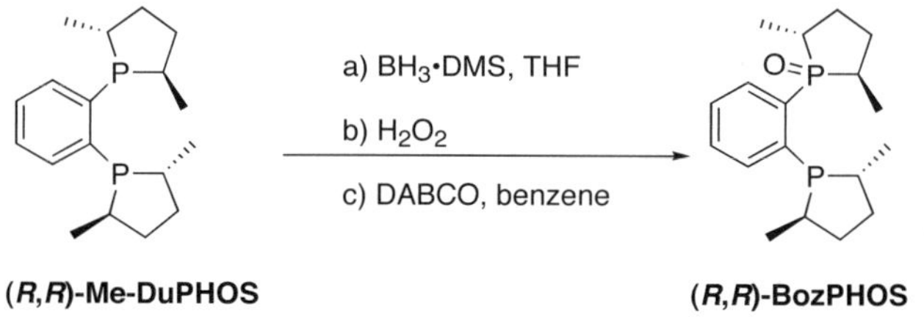

Submitted by Alexandre Côté, Jean-Nicolas Desrosiers, Alessandro A. Boezio and André B. Charette.[1]
Checked by Scott E. Denmark and Justin I. Montgomery.

1. Procedure

(2R,5R)-1-{2-[(2R,5R)-2,5-dimethylphospholan-1-yl]phenyl}-2,5-dimethylphospholane 1-oxide **(*R,R*)-BozPHOS**. A flame-dried, 100-mL, one-necked, round-bottomed flask equipped with an egg-shaped magnetic stirring bar and a rubber septum is charged with **(*R,R*)-Me-DuPHOS** (1 g, 3.26 mmol) (Note 1) in a glove-box under argon. Anhydrous THF (32 mL) (Note 2) is added to the flask via a syringe under argon and the resulting colorless solution is cooled in a 0 °C ice bath for 15 min before BH_3•DMS 10 M (360 μL, 3.6 mmol) (Note 3) is added dropwise (Note 4) via a syringe. The colorless mixture is stirred for 45 min under argon at 0 °C and H_2O_2 35% wt. (3 x 1.15 mL, 39.2 mmol) (Note 5) is added dropwise via syringe (Note 6) at 0 °C under argon in three portions with an interval of 30 min between each addition. The mixture is stirred for an additional 45 min at room temperature, and the reaction is quenched by the dropwise addition via cannula of a saturated aqueous solution of Na_2SO_3 (40 mL) at 0 °C over 30 min (Note 7). The mixture is transferred to a 250-mL separatory funnel and the aqueous layer is extracted with ethyl acetate (3 x 100 mL), and the combined extracts are dried over Na_2SO_4 (50 g), filtered and concentrated by rotary evaporation (30 °C, 50 mmHg) to afford a white foam (1.115 g) (Note 8). The crude product and DABCO (550 mg, 4.9 mmol) (Note 9) are placed in a 100-mL, one-necked, round-bottomed flask equipped with an egg-shaped magnetic stirring bar and a rubber septum. The flask is purged under

argon, then anhydrous benzene (32 mL) (Note 10) is added. The colorless reaction mixture is stirred for 5 h in a 50 °C oil bath and then the solvent is removed by rotary evaporation (25 °C, 40 mmHg) to afford a white residue. The crude product is purified on silica gel (Notes 11, 12) to afford 977 mg (93% yield, >99% ee) of the title compound as a white solid (Notes 13 and 14).

2. Notes

1. (*R,R*)-Me-DuPHOS was purchased from Strem Chemicals, Inc. Although, it could be briefly manipulated under moisture and oxygen, it was stored under argon atmosphere in a glove-box to prevent any undesired oxidation. It was used without any prior purification.

2. Anhydrous THF was obtained by filtration through a drying column on a GlassContour system (Irvine, CA).

3. Borane•dimethyl sulfide (BH_3•DMS) was purchased from Aldrich Chemical Company, Inc. and was used without prior purification.

4. The addition lasted 1 minute.

5. H_2O_2 35% wt. was purchased from Aldrich Chemical Company, Inc. and was used without prior purification.

6. One minute was needed for each addition.

7. The reaction is exothermic and a gas is evolved.

8. The crude product can be stored overnight under argon in a −20 °C freezer with no degradation.

9. DABCO was purchased from Aldrich Chemical Company, Inc. and was recrystallized from a 1:1 mixture of MeOH and hexanes.

10. Anhydrous benzene was obtained by filtration through a drying column on a GlassContour system (Irvine, CA).

11. The product was dissolved in dichloromethane (2 mL) and charged on a column (3 x 15 cm) of 60 g of UltraPure silica gel (40–63 μm) purchased from Silicycle. The product was eluted with 600 mL of 5% MeOH in EtOAc and collected with 8 mL fractions. The desired product was obtained in fractions 16–30, which were combined and concentrated by rotary evaporation (30 °C, 30–50 mmHg). The desired product can be visualized on TLC with a UV lamp or by developing with a $KMnO_4$ solution.

12. No over-oxidation occurred during the chromatography, but the pure compound is typically stored under argon.

13. The physical properties are as follows: R_f 0.33 (5% MeOH in EtOAc) mp 125–127 °C; HRMS (ESI$^+$) *m/z* calc. for $C_{18}H_{29}P_2O$ [M^+ + 1]: 323.1694, found: 323.1688; Elemental Analysis calc. for $C_{18}H_{28}P_2O$: C, 67.07; H, 8.75, found: C, 67.13; H, 9.00; IR (neat) cm^{-1} 730, 738, 758, 1116, 1130, 1159, 1253, 1374, 1455, 2860, 2925; ^1H NMR (400 MHz, C_6D_6) δ: 0.85 (dd, *J* = 17.3, 7.4 Hz, 3 H), 0.91–1.03 (m, 1 H), 1.06 (dd, *J* = 9.1, 7.1 Hz, 3 H), 1.26–1.37 (m, 2 H), 1.28 (dd, *J* = 17.6, 8.3 Hz, 3 H), 1.30 (dd, *J* = 13.7, 6.9 Hz, 3 H), 1.64–1.76 (m, 2 H), 1.84–2.06 (m, 4 H), 2.41–2.52 (m, 1 H), 2.58–2.71 (m, 1 H), 2.71–2.83 (m, 1 H), 7.05 (tddd, *J* = 7.5, 2.6, 1.3, 0.5 Hz, 1 H), 7.13 (tt, *J* = 7.4, 1.5 Hz, 1 H), 7.27–7.34 (m, 1 H), 7.45–7.50 (m, 1 H); ^{13}C NMR (100 MHz, C_6D_6) δ: 13.1 (dd, $J_{C\text{-}P}$ = 2.9, <1 Hz), 17.5 (d, $J_{C\text{-}P}$ = 8.8 Hz), 18.6 (d, $J_{C\text{-}P}$ = 3.9 Hz), 20.6 (d, $J_{C\text{-}P}$ = 36.1 Hz), 31.7 (dd, $J_{C\text{-}P}$ = 67.9, <1 Hz), 31.8 (d, $J_{C\text{-}P}$ = 9.8 Hz), 32.2 (dd, $J_{C\text{-}P}$ = 8.8, 1.9 Hz), 34.4 (d, $J_{C\text{-}P}$ = 11.7 Hz), 35.0 (d, $J_{C\text{-}P}$ = 13.6 Hz), 36.5 (d, $J_{C\text{-}P}$ = 6.8 Hz), 36.8 (dd, $J_{C\text{-}P}$ = 66.9, 4.9 Hz), 36.9 (d, $J_{C\text{-}P}$ = 2.0 Hz), 128.5 (d, $J_{C\text{-}P}$ = 10.8 Hz), 130.9 (dd, $J_{C\text{-}P}$ = 2.9, <1 Hz), 131.5 (dd, $J_{C\text{-}P}$ = 11.2, 9.3 Hz), 134.2 (dd, $J_{C\text{-}P}$ = 10.7, 2.9 Hz), 140.3 (dd, $J_{C\text{-}P}$ = 82.0, 33.2 Hz), 144.0 (dd, $J_{C\text{-}P}$ = 37.1, 9.3 Hz); ^{31}P NMR (202 MHz, C_6D_6) δ: 8.99 (d, *J* = 4.6 Hz), 62.05 (d, *J* = 4.6 Hz); $[\alpha]_D^{20}$ –221.6 (c = 0.836, EtOH). The enantiomeric excess of the product is determined by HPLC analysis at 254 nm [Chiralpak AD, 95:5 hexanes: *i*-PrOH, 1mL/min: (R,R) t_r (major) = 7.3 min, (S,S) t_r (minor) = 10.1 min)] or determined by SFC analysis [Chiralpak AD, 20% i-PrOH, 150 bar CO_2, 1 mL/min, 65° Cl.

14. The checkers found that the appearance of the NMR spectra was highly concentration dependent. The spectra reported were taken as follows: ^1H NMR, 4 mg/1.0 mL; ^{13}C NMR, 309 mg/1.5 mL.

Safety and Waste Disposal Information

All hazardous materials should be handled and disposed of in accordance with "Prudent Practices in the Laboratory"; National Academy Press; Washington, DC, 1995.

3. Discussion

The hemilabile ligand **BozPHOS** is useful for the copper-catalyzed asymmetric addition of diorganozinc reagents to *N*-diphenylphosphinoylimines.[2–5] This bis-phosphine monoxide ligand can be prepared from both enantiomers of the commercially available **Me-DuPHOS** by mono-oxidation. Various methods have been reported to oxidize bidentate phosphines.[6] Among them, Grushin's method involving palladium acetate[7] was unsuitable because it afforded a mixture of the unreacted bis-phosphine, its mono-oxide (40%), its bis-oxide and an unidentified phosphonium salt. The procedure described above employs a selective mono-protection of the bis-phosphine with BH_3•DMS followed by a mono-oxidation with H_2O_2. During this process, no bis-oxidation is obtained even if an excess of H_2O_2 is used. Traces of unreacted **Me-DuPHOS** can be separated easily by silica gel flash chromatography since **BozPHOS** is air stable. The procedure described above can be accomplished on a 0.5 to 10 g-scale to obtain **BozPHOS** in an 87 to 93% yield.

1. Département de Chimie, Université de Montréal, P.O. Box 6128, Station Downtown, Montréal (Québec) Canada, H3C 3J7.

2. Desrosiers, J.-N.; Côté, A.; Boezio, A. A.; Charette, A. B. *Subsequent procedure in this volume,* **2004**.
3. Boezio, A. A.; Pytkowicz, J.; Côté, A.; Charette, A. B. *J. Am. Chem. Soc.* **2003**, *125*, 14260–14261.
4. Côté, A.; Boezio, A. A.; Charette, A. B. *Angew. Chem. Int. Ed.* **2004**, *43*, 6525–6528.
5. Côté, A.; Boezio, A. A.; Charette, A. B. *Proc. Natl. Acad. Sci. USA* **2004**, *101*, 5405–5410.
6. Grushin, V. V. *Chem. Rev.* **2004**, *104*, 1629–1662.
7. Grushin, V. V. *J. Am. Chem. Soc.* **1999**, *121*, 5831–5832.

PREPARATION OF (TRIPHENYLPHOSPHORANYLIDENE)-KETENE FROM (METHOXYCARBONYLMETHYLENE)-TRIPHENYLPHOSPHORANE

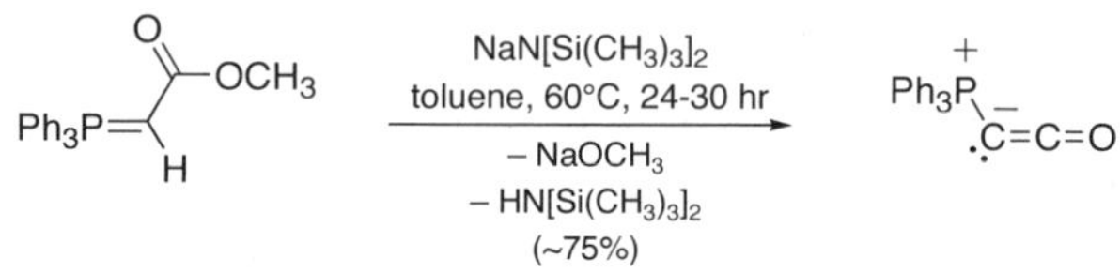

Submitted by Rainer Schobert.[1]
Checked by Robert K. Boeckman Jr. and Joseph E. Pero.

1. Procedure[2]

A 3-L, three-necked, round-bottomed flask, equipped with a mechanical stirrer and a lateral gas inlet is purged with dry argon (Note 1) and charged with sodium amide (19.5 g, 0.5 mol) (Note 2), toluene dried by passage through alumina (1300 mL) (Note 3), and bis(trimethylsilyl)amine (81 g, 105 mL, 0.5 mol) (Note 4). The flask is fitted with a reflux condenser, then flushed once more with argon, and immersed in an oil bath. The mixture is heated at 70-80 °C (bath temperature) until a clear and colorless solution of sodium hexamethyldisilazanide is obtained and the evolution of ammonia has ceased (2-4 h) (Notes 5, 6). The reaction mixture is cooled to room temperature and the condenser is removed briefly, with protection from air by a stream of argon, to allow the addition of (methoxycarbonylmethylene)-triphenylphosphorane (167 g, 0.5 mol) in several large portions (Note 7). The mixture is then heated again to ca. 60-70 °C (bath temperature) and stirring is continued for a further 24-30 h by which time the solution should take on a bright yellow hue (Note 8). While still hot and blanketed with a stream of argon, the reaction mixture is rapidly suction filtered using a 24/40 standard taper vacuum filtration adapter and a 350 mL fritted funnel (course porosity) open to the atmosphere through a 1 inch thick pad of Celite™ into a 2-L 24/40 single-necked round-bottomed flask (Notes 9 and 10).

Once filtration is complete, the receiving flask is flushed with argon and attached to a rotary evaporator (Note 11). The colorless filtrate is evaporated to dryness and the residue recrystallized from dry toluene (~1g / 5mL) by dissolution in hot toluene followed by cooling to -20 °C in a freezer overnight. While the mixture was cold, the resulting solids are collected by suction filtration using a 350 mL coarse fritted funnel and washed on the filter with three 100 mL portions of cold (0 °C) toluene affording the first crop of phosphorane (Note 3). The mother liquor is again concentrated to dryness under vacuum, redissolved in ~100 mL of hot dry toluene, and again cooled to -20 °C in a freezer overnight. The resulting solids are isolated and washed as above. The first and second crops of phosphorane are then combined and dried under high vacuum (ca 0.01 Torr) to a constant weight to afford 106.8-114.9 g (74-76%) (Note 12) of the phosphorane product as a very pale yellow, flaky powder having mp 173 °C (Note 13).

2. Notes

1. An atmosphere of dry argon is maintained throughout the entire process. While neither the starting nor the product ylide are particularly air- or moisture-sensitive in the solid state, hot solutions of the latter are sensitive.

2. Sodium amide should be of 95% purity. Powdery batches were found to react more readily than suspensions in mineral oil or pellets. Sodium amide can be purchased from Aldrich Chemical Company, Inc. The checkers employed sodium amide purchased from Acros Organics CAUTION: Sodium amide is highly flammable and corrosive and can cause severe burns.

3. The checkers employed toluene dried by passage through a column containing anhydrous alumina under nitrogen pressure. The submitters dried toluene by distillation from sodium metal. Attempts by the checkers to employ toluene dried by distillation from CaH_2 led to inferior results during attempts to form sodium hexamethyldisilazide (NaHMDS).

4. Bis(trimethylsilyl)amine (hexamethyldisilazane, HMDS), 99+%, was purchased from Aldrich Chemical Company, Inc. CAUTION: bis(trimethylsilyl)amine is harmful and corrosive.

5. Purging the set-up with argon every now and then cuts down reaction times considerably.

6. Commercially available sodium hexamethyldisilazide (e.g. Aldrich Chemical Company) can be used instead. However, it is costly and the reaction time and yields of product crucially depend on its age and quality.

7. (Methoxycarbonylmethylene)triphenylphosphorane can be bought from Aldrich Chemical Company, Inc. The checkers employed Aldrich material as received. The phosphorane can also be readily prepared at low cost in two steps from triphenylphosphine and methyl bromoacetate according to a literature procedure.[3]

8. Monitoring reaction progress is possible by IR spectroscopy of aliquots of the reaction mixture measured as films on KBr plates. A strong product band at 2090 cm^{-1} of constant intensity and the disappearance of a starting ylide band at 1616 cm^{-1} are indicative of a complete reaction.

9. The submitters employed a funnel with a sintered frit (fine porosity) where the frit is covered with a 2 cm layer of basic alumina (III) and 1 cm of celite atop. Through side-arm stopcocks, the filter funnel is connected to an argon line and the receiving flask to a source of vacuum (water-jet or membrane pump). The whole set-up is positioned upright or slightly slanted and submitted to several cycles of evacuating and flushing with inert gas prior to use. During filtration occasional removal or resuspending of sedimented sodium methoxide from the surface of the celite/alumina plug by scraping with a long spatula may be necessary in order to maintain a continuous flow of filtrate.

10. The submitters filtered the reaction mixture by pouring through a tapered adaptor into a 2-L Schlenk-type filter funnel which is joined to the only neck of a round-bottomed 2-L receiving flask (Note 9). Filtration is then carried out by application of a slight overpressure from an argon line to the stoppered funnel and by gentle suction at the receiving flask (the stopcock to the vacuum source should be opened at regular intervals but only briefly at each time). During filtration which typically takes 30 to 60 min, the temperature of the reaction mixture in the filter funnel should be kept between 30 °C and 50 °C by means of a heating basket, mantle or tape, or at least insulating foil to avoid premature product crystallization.

11. The submitters concentrated the clear filtrate to 500 mL, flushed once more with argon and then placed in a refrigerator for 24 h. The faintly yellow precipitate (ca. 70-80 g) is collected on a second Schlenk-type filter funnel, washed with dry diethyl ether (300 mL) and dried on an oil pump at ca. 0.01 Torr. A second, darker crop of product (ca. 20-35 g) is obtained by concentrating the toluene mother liquor to ca. 250 mL, layering it with diethyl ether (50 mL) and cooling the mixture to 4 °C for 24 h. The resulting precipitate is collected, washed and dried like the first batch to which it is by then identical in color and all analytical data.

12. The submitters obtained 100-115g (66-75%) of pure phosphorane in two crops with identical mp and spectroscopic properties using the above described direct crystallization procedure (Note 11).

13. The product is pure by usual microanalytical standards; IR (KBr, ν): 2090 (s), 1625 (m), 1436 (m), 1110 (m) cm^{-1}; ^{1}H NMR ($CDCl_3$, 400 MHz δ: 7.77-7.64 (m, 6H), 7.63-7.55 (m, 3H), 7.54-7.44 (m, 6H); ^{13}C NMR

($CDCl_3$, 100 MHz δ: 145.6 (d, $^2J_{PC}$ 43.0 Hz, C^{β}), 132.3 (s, C-*meta*), 132.2 (s, C-*para*), 129.6 (d, $^1J_{PC}$ 98.5 Hz, C-*ipso*), 128.8 (d, $^3J_{PC}$ 12.9 Hz, C-*ortho*), -10.5 (d, $^1J_{PC}$ 185.4 Hz, C^{α}); ^{31}P NMR ($CDCl_3$, 162 MHz δ: 6.0. Analysis performed on an unrecrystallized sample of the bulk material: Anal. Calcd for $C_{20}H_{15}OP$: C, 79.46; H, 5.00; P, 10.25. Found. C, 78.66; H, 5.30. Recrystallization of this sample from toluene and reanalysis gave: Anal. Calcd for $C_{20}H_{15}OP$: C, 79.46; H, 5.00; P, 10.25. Found. C, 79.60; H, 5.06. The only detectable impurities by ^1H NMR are the starting ester ylide and toluene (δ 3.60, s and 2.36, s). Based on this intergration, the purity of product is calculated to be ~99.97%. Prior analytical results have been reported.[4-6]

Waste Disposal Information

All toxic materials were disposed of in accordance with "Prudent Practices in the Laboratory"; National Academy Press; Washington, DC 1995.

3. Discussion

Alternative methods for the preparation of Ph_3PCCO: Strong bases other than sodium hexamethyldisilazanide (NaHMDS) have been employed for the elimination of methanol from (methoxycarbonylmethylene)triphenylphosphorane. With crystalline sodium amide, drastic reaction conditions (refluxing in benzene or toluene for several days) are required.[7] However, mixtures of sodium amide and catalytic quantities of HMDS react as readily as stoichiometric amounts of preformed NaHMDS.[8] Alkyllithium bases (e.g. *n*BuLi, PhLi) give low yields of Ph_3PCCO contaminated with various byproducts stemming from attack of the base on the phenyl rings.[9] The title ylide has also been prepared by pyrolysis of the disilylated ester ylide obtained from reaction of [1,1-bis(trimethylsilyl)methylene]triphenylphosphorane with carbon dioxide,[10] and by pyrolysis of the betaine obtained from the reaction of carbodiphosphorane, $Ph_3P{=}C{=}PPh_3$, with carbon dioxide.[6] However, neither protocol is suitable for large scale preparation.

Properties and reactions of Ph_3PCCO:[11] It is remarkably stable and solid samples can be stored for months at room temperature under an atmosphere of argon and handled even under ambient conditions without notable decomposition. This fact is due to its electronic structure featuring an orthogonal pair of filled π-systems ($\pi^4 \perp \pi^4$). Hence Ph_3PCCO shows little propensity to dimerize like the strongly electrophilic ketenes. It does not even enter readily into Wittig alkenation reactions which are so typical of common phosphorus ylides. It represents a nucleophilic-only C_2-building block reacting with a wide range of electrophiles in a variety of ways and undergoing cycloadditions both at the polar P–C^α and at the C^α=C^β bond. For example, acidic compounds, RXH, such as alcohols, thiols, amines, and CH-active 1,3-dicarbonyls add across the C=C bond of Ph_3PCCO to give the corresponding Wittig-active acyl ylides, Ph_3P=CH–C(O)XR. Multicomponent and domino reactions between the cumulated ylide, an

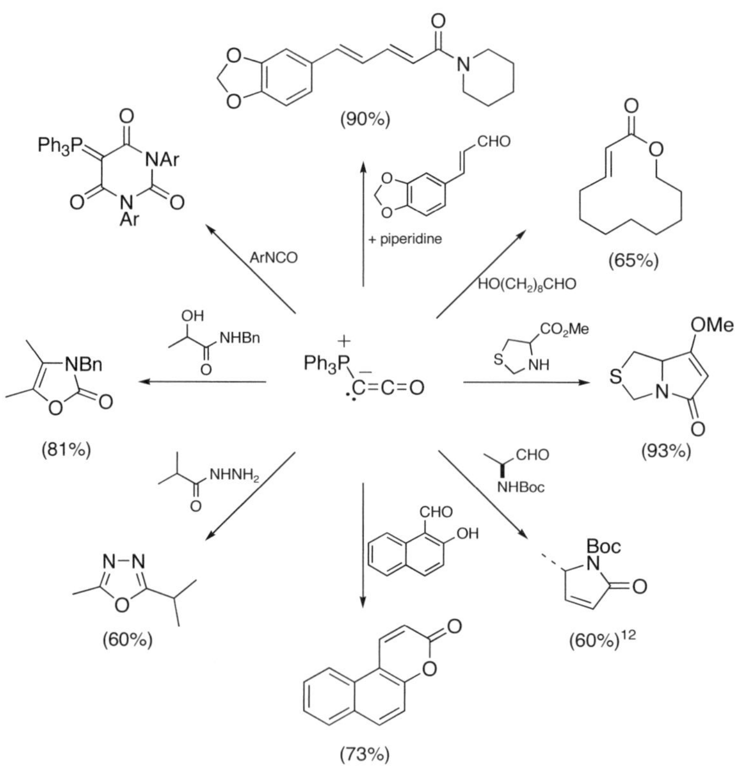

acidic compound, and a ketone or an aldehyde lead to various α,β-unsaturated carbonyl compounds. Intramolecular variants employing difunctional carbonyl derivatives, including esters and amides, bearing such acidic functionalities furnish (hetero)cyclic products with ring sizes ranging from five to twenty. Grignard compounds, RMgX, react with Ph_3PCCO to yield acyl ylides $Ph_3P{=}CHC(O)R$ upon hydrolytic work-up. Both [2+2]- and [2+4]-reactions are known of Ph_3PCCO, e.g. with ketenes to form cyclobutan-1,3-diones, and with iso(thio)cyanates to furnish six-membered pyrimidinetriones, or hexahydro-1,3-dithiazines, respectively.

1. Universität Bayreuth, Organisch-chemisches Laboratorium, Universitätsstrasse 30, D-95440 Bayreuth (Germany). E-mail: Rainer.Schobert@uni-bayreuth.de, Fax: +49 (0)921 552672.

2. Bestmann, H. J.; Sandmeier, D. *Angew. Chem.* **1975**, *87*, 630; *Angew. Chem., Int. Ed. Engl.* **1975**, *14*, 634.

3. Isler, O.; Gutmann, H.; Montavon, M.; Ruegg, R.; Ryser, G.; Zeller, P. *Helv.Chim. Acta* **1957**, *40*, 1242-1249.

4. Daly, J. J.; Wheatley, P. J. *J. Chem. Soc. A* **1966**, 1703-1706.

5. Matthews, C. N.; Birum, G. H. *Chem. Ind. (London)* **1968**, 653.

6. (a) Matthews, C. N.; Birum, G. H. *Tetrahedron Lett.* **1966**, 5707-5710; (b) Matthews, C. N.; Birum, G. H. *Acc. Chem. Res.* **1969**, *2*, 373-379.

7. Bestmann, H. J.; Schmidt, M.; Schobert, R. *Synthesis* **1988**, 49-53.

8. Haas, W. *PhD Thesis*, University of Erlangen-Nürnberg, Germany **1992**, 20.

9. (a) Bestmann, H. J.; Besold, R.; Sandmeier, D. *Tetrahedron Lett.* **1975**, 2293-2294; (b) Buckle, J.; Harrison, P. G. *J. Organomet. Chem.* **1974**, *77*, C22-C24.

10. Bestmann, H. J.; Dostalek, R.; Zimmermann, R. *Chem. Ber.* **1992**, *125*, 2081-2084.

11. (a) Bestmann, H. J. *Angew. Chem.* **1977**, *89*, 361-376; *Angew. Chem., Int. Ed. Engl.* **1977**, *16*, 349-364; (b) Schobert, R.; Gordon, G. J. *Science of Synthesis: Houben-Weyl Methods of Molecular Transformations*; Thieme: Stuttgart, **2004**, Vol. 27, p. 1042.

12. Andrus, M. B.; Li, W.; Keyes, R. F. *J. Org. Chem.* **1997**, *62*, 5542-5549.

PREPARATION OF [1-(METHOXYMETHYLCARBAMOYL)ETHYL] PHOSPHONIC ACID BIS-(2,2,2-TRIFLUOROETHYL) ESTER: A USEFUL INTERMEDIATE IN THE SYNTHESIS OF *Z*-UNSATURATED *N*-METHOXY-*N*-METHYLAMIDES

A.

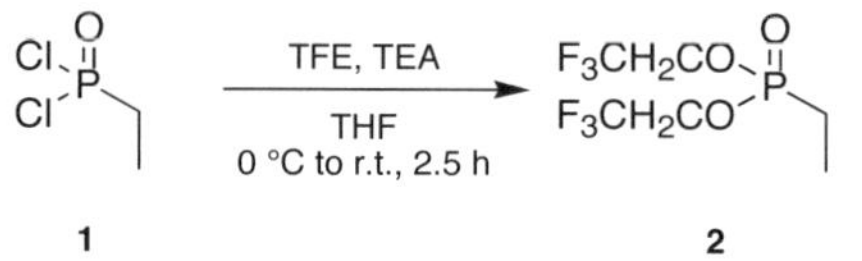

B.

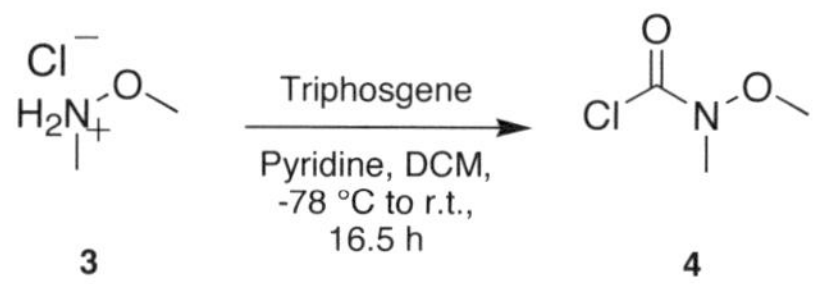

C.

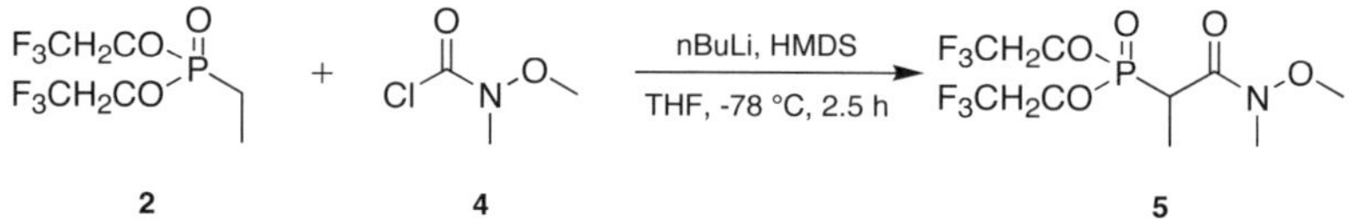

Submitted by Amos B. Smith, III, Jason J. Beiger, Akin H. Davulcu, and Jason M. Cox.[1]
Checked by Mark Lautens and Catherine Taillier.

1. Procedure

A. Ethylphosphonic acid bis-(2,2,2-trifluoroethyl) ester (2).[2] A flame-dried 2-L three-necked round-bottom flask equipped with a Teflon-coated magnetic stir bar, thermometer, rubber septum and argon inlet (Note 1) is charged with 24.1 mL (336 mmol, 2.01 eq.) of 2,2,2-trifluoroethanol (Notes 2,3) and 500 mL of anhydrous THF (Note 4) under an argon atmosphere, and cooled to an internal temperature of 0 °C with an ice-NaCl bath for 10 min. The above flask is then charged dropwise with 51.3 mL (369 mmol, 2.20 eq.) of triethylamine (Note 5) over 10 min., and stirred at 0 °C for 15 min. A separate flame-dried 250-mL one-necked round-bottom flask equipped with a rubber septum and argon inlet is charged with 17.9 mL (167 mmol, 1.00 eq.) of ethylphosphonic dichloride 1 (Note 2) and 65 mL of

THF under an argon atmosphere. The mixture is stirred, and transferred dropwise *via* cannula (Notes 3,6) into the trifluoroethanol/triethylamine solution over 15 min while maintaining the internal temperature below 12 °C. The 250-mL flask is then rinsed with THF (2 x 10 mL), and the THF washes transferred by cannula to the larger flask over a five-minute period. The resulting white slurry is warmed to room temperature over 15 min., stirred at room temperature for 2.25 h, filtered through a sand-covered frit (Note 7), and concentrated by rotary evaporation (35 °C, 20 mmHg) to afford 43 g of a cloudy, nearly colorless liquid. The latter is transferred to a 100-mL one-necked round-bottom flask equipped with a Teflon-coated magnetic stir bar and a short path distillation head fitted with a distribution adapter and three 50-mL pear-shaped receiving flasks. After collecting a forerun (ca. 1 mL), the product is distilled under an aspirator vacuum (bp = 92–95 °C at 25 mmHg) to yield 40.85 g (89%) of ethylphosphonic acid bis-(2,2,2-trifluoroethyl) ester (2) as a clear colorless liquid (Note 8).

B. N-Methoxy-N-methylcarbamoyl chloride (4).[3] A flame-dried 500-mL three-necked round-bottom flask equipped with a Teflon-coated magnetic stir bar, thermometer, rubber septum and argon inlet is charged with 10.1 g (34.1 mmol, 0.400 eq.) of triphosgene (Note 2) and 57 mL of anhydrous dichloromethane (Note 4) under an argon atmosphere, and cooled to an internal temperature of –78 °C with a dry ice-acetone bath for 20 min. The suspension is charged with 8.33 g (85.4 mmol, 1.00 eq.) of *N,O*-dimethylhydroxylamine hydrochloride 3 (Note 2) in three portions by briefly removing the septum and using a powder funnel with 5 min. elapsing between each portion while maintaining the internal temperature below –70 °C. The suspension is recooled to –78 °C and stirred for 30 min. A separate flame-dried, 100-mL round-bottomed flask equipped with a rubber septum and argon inlet is charged with 13.7 mL (171 mmol, 2.00 eq.) of pyridine (Note 5) and 28 mL of dichloromethane under an argon atmosphere. The mixture is stirred and transferred dropwise *via* cannula to the above hydroxylamine/triphosgene suspension over 1.5 h while maintaining the internal temperature below –72 °C (Note 6). The 100-mL flask is rinsed with dichloromethane (2 x 5 mL), and the dichloromethane washes transferred by cannula to the larger flask over 5 min. The resulting yellow-orange slurry is slowly warmed to room temperature over 4 h, stirred at room temperature for 12.5 h, quenched with 100 mL of distilled water, and transferred to a 500-mL separatory funnel. The layers are separated, and the aqueous layer extracted with dichloromethane (3 x 50 mL). The combined organic phases are sequentially washed (Note 5) with aqueous 0.5 *M* HCl (2 x 100 mL), saturated aqueous $NaHCO_3$ (1 x 75 mL), and brine (1 x 75 mL), dried (Na_2SO_4), filtered, and concentrated by rotary evaporation (35 °C, 20 mmHg) to afford 11.3 g of a yellow liquid. The yellow liquid is transferred to a 50-mL one-necked round-bottom flask equipped with a Teflon-coated magnetic stir bar and a short path distillation head fitted with a distribution adapter and three 25-mL pear-shaped receiving flasks. After collecting a forerun (*ca.* 0.5 mL), the product is distilled under vacuum aspirator (bp = 67–69 °C at 25 mmHg) to afford 8.87-9.18 g (84-87%) of *N*-methoxy- *N*-methylcarbamoyl chloride 4 as a clear colorless liquid (Note 9).

C. *[1-(Methoxymethylcarbamoyl)ethyl] phosphonic acid bis-(2,2,2-trifluoroethyl) ester* (**5**).[4] A flame-dried 1-L three-necked round-bottom flask equipped with a Teflon-coated magnetic stir bar, thermometer, rubber septum and argon inlet is charged with 51.0 mL (128 mmol, 2.41 eq.) of a 2.5 *M* solution of *n*-butyllithium in hexanes (Note 2) and 40 mL of THF under an argon atmosphere, and cooled to an internal temperature of –20 ± 3 °C with a dry ice-isopropanol-water bath for 20 min. In a separate flame-dried 100-mL one-necked round-bottom flask equipped with a rubber septum and argon inlet is charged with 29.2 mL (139 mmol, 2.64 eq.) of 1,1,1,3,3,3-hexamethyldisilazane (HMDS) (Note 2) and 40 mL of THF under an argon atmosphere. The mixture is stirred and transferred dropwise *via* cannula to the above *n*-butyllithium solution over 20 min while maintaining the internal temperature below –15 °C (Note 6). The 100-mL flask is then rinsed with THF (2 X 10 mL), and the THF washes cannulated to the larger flask over 5 min. The resulting clear solution is stirred at –20 ± 3 °C for 20 min., and cooled to an internal temperature of –75 °C with a dry ice-acetone bath. A separate flame-dried 100-mL one-necked round-bottomed flask equipped with a rubber septum and argon inlet is then charged with 14.5 g (52.9 mmol, 1.00 eq.) of ethylphosphonic acid bis-(2,2,2-trifluoroethyl) ester **2**,[2] 9.00 g (72.9 mmol, 1.38 eq.) of *N*-methoxy-*N*-methylcarbamoyl chloride **4**,[3] and 50 mL of THF, and transferred dropwise *via* cannula to the above lithium HMDS solution over 30 min. while maintaining the internal temperature below –68 °C (Note 6). The smaller flask is rinsed with THF (2 x 10 mL), and the THF washes transferred by cannula to the larger flask over 5 min. The resulting pale yellow solution is recooled to –75 °C and stirred for 2.5 h (Note 10), then slowly acidified over 5 min. with 130 mL of a 1.0 *M* solution of HCl, and gradually warmed to an internal temperature of 0 °C over 30 min. The solution is diluted with 100 mL of distilled water, and transferred to a 1-L separatory funnel. The flask is rinsed with diethyl ether (3 x 50 mL), transferred to the funnel, shaken, the layers separated, and the aqueous layer extracted with dichloromethane (4 x 100 mL) (Note 11). The combined organic phases are dried ($MgSO_4$), filtered, and concentrated by rotary evaporation (35 °C, 20 mmHg) to afford 19 g of a yellow oil (Note 12). The yellow oil is loaded onto a 80-mm diameter column, wet-packed (4:1 hexanes:ethyl acetate) with 450 grams (25 cm) of silica gel (Note 13), and sequentially eluted with a gradient of hexanes and ethyl acetate (2 L of 2:1, 2 L of 1:1, 1 L of 1:2). The desired product is collected in fractions of 75-mL volume, concentrated by rotary evaporation (35 °C, 20 mmHg), and dried under vacuum (25 °C, 0.01 mmHg) until a constant mass is obtained. The above described procedure affords 15.6 - 16.8 g (82–88%) of [1-(methoxymethylcarbamoyl)ethyl] phosphonic acid bis-(2,2,2-trifluoroethyl) ester **5** as a clear pale yellow oil (Note 14).

2. Notes

1. All flasks were flame-dried and maintained under an argon atmosphere during the course of the reactions. The argon was dried and purified by passing through Drierite® and then an Oxiclear™ disposable gas purifier, which may be purchased from Aldrich Chemical Co., Inc. A gas manifold was then used to distribute argon to each of the attached flasks. An argon inlet was affixed to a flask by inserting its needle through that flask's rubber septum while maintaining slightly positive pressure.

2. 2,2,2-Trifluoroethanol (≥99%), ethylphosphonic dichloride (98%), triphosgene (98%), *N,O*-dimethylhydroxylamine hydrochloride (98%), 1,1,1,3,3,3-hexamethyldisilazane (99.9%) and the 2.5 *M* solution of *n*-butyllithium in hexanes were purchased from Aldrich Chemical Co., Inc. and used as received.

3. Liquids were added to flasks from plastic disposable syringes through stainless steel needles. Liquids were transferred between flasks *via* stainless steel cannulas. All needles and cannulas were oven-dried for at least one hour and cooled to room temperature in a dessicator prior to use.

4. HPLC grade dichloromethane (99.9%) and tetrahydrofuran (THF) (99.9%) were purchased from Fisher Scientific. Dichloromethane and THF were dried by purging with argon over activated molecular sieves and were stored under argon at room temperature. The checkers used THF distilled

from Na/benzophenone ketyl and dichloromethane purified with a MBRAUN® Solvent Purification System.

5. Triethylamine (99%), pyridine (99.9%), dichloromethane, hexanes, ethyl acetate, sodium chloride, sodium bicarbonate, anhydrous magnesium sulfate, anhydrous sodium sulfate, and concentrated hydrochloric acid were purchased from Fisher Scientific and used as received. Triethylamine and pyridine were used by the checkers after distillation over KOH.

6. Alternatively, a flame-dried 100-mL additional funnel may be used to regulate the rate at which the reaction flask is charged with solution.

7. Alternatively, filtration of the resulting white slurry through a short plug of Celite could be performed to remove the ammonium salts.

8. Ethylphosphonic acid bis-(2,2,2-trifluoroethyl) ester (**2**) displays the following spectroscopic properties: IR (film): 2974 (weak), 1463, 1419, 1287, 1255, 1174, 1109, 1079, 1036, 964, 868, 844 cm^{-1}; ^{1}H NMR (400 MHz, $CDCl_3$) δ: 4.30-4.47 (m, 4H), 1.93 (dq, J_{PH} = 18.7 and J_{HH} = 7.7 Hz, 2H),1.22 (dt, J_{PH} = 21.5 and J_{HH} = 7.7 Hz, 3H); ^{13}C NMR (100 MHz, $CDCl_3$) δ: 122.6 (qd, J_{CF} = 277.6 and J_{CP} = 7.7 Hz), 61.8 (qd, J_{CF} = 37.6 and J_{CP} = 6.1 Hz), 18.9 (d, J_{CP} = 143.4 Hz), 5.9 (d, J_{CP} = 7.7 Hz); The checkers have not been able to get an elemental analysis due to volatility of compound **2**. However, the submitters reported the following data for **2**: Anal. Calcd for $C_6H_9F_6O_3P$: C, 26.29; H, 3.31. Found: C, 26.20; H, 3.06.

9. The following characterization data were obtained for *N*-methoxy-*N*-methylcarbamoyl chloride (**4**): IR (film): 2982, 2941, 1732 (broad), 1460, 1443, 1406, 1352, 1182, 1082, 995, 868, 669, 653 cm^{-1}; ^{1}H NMR (500 MHz, $CDCl_3$) δ: 3.78 (s, 3H), 3.33 (s, 3H). The checkers have not been able to get an elemental analysis due to volatility of compound **4**. However, the submitters reported the following data for **4**: Anal. Calcd for $C_3H_6ClNO_2$: C, 29.17; H, 4.90. Found: C, 29.02; H, 4.89.

10. The reaction was monitored by thin layer chromatography (TLC) with 0.25-mm E. Merck pre-coated silica gel plates. The plates were eluted with a 1:1 mixture of hexanes and ethyl acetate. The R_f values of **2**, **4**, and **5**, are 0.55, 0.70, and 0.21, respectively. A potassium permanganate-based stain that was prepared by dissolving 3 g of $KMnO_4$, 20 g of K_2CO_3, and 5 mL of 5% aqueous NaOH in 300 mL of distilled water was suitable for visualizing the product on TLC.

11. If an emulsion occurs during the dichloromethane extractions, it may be broken with the addition of approximately 15 mL of brine. Checkers found that filtration of the biphasic mixture through a short plug of Celite was necessary to break the emulsion.

12. The crude mixture may be distilled under vacuum through a short path apparatus (bp = 105–107 °C at 0.01 mmHg) to yield phosphonate **5** as a slightly pale yellow oil. Purification by this method, however, resulted in a significantly lower yield (40–50%) due to thermal decomposition of **5**.

13. Silica gel may be purchased from Silicycle Chemical Division. The silica gel used by the submitters had the following specifications: pH: 6.5–7.0, particle size: 40–63 μm (230–400 mesh), spec. surface area: 500 m^2/g, pore diameter: 60 Å.

14. Phosphonate **5** exhibits the following spectroscopic properties: IR (film): 2976, 2949, 1651, 1455, 1418, 1393, 1259, 1173; 1072, 987, 962, 845 cm^{-1}; ^{1}H NMR (400 MHz, $CDCl_3$) δ: 4.43-4.57 (m, 2H), 4.35-4.47 (m, 2H), 3.78 (s, 3H), 3.77 (m, 1H), 3.28 (s, 3H), 1.50 (dd, J = 19.8 and J = 7.3 Hz, 3H); ^{13}C NMR (100 MHz, $CDCl_3$) δ: 169.3, 122.6 (qdd, J_{CF} = 277.6, J_{CP} = 22.6 and J_{CP} = 8.8 Hz), 63.1 (qd, J_{CP} = 37.6 and J_{CP} = 5.0 Hz), 61.8 (qd, J_{CP} = 38.3 and J_{CP} = 6.1 Hz), 61.7, 35.7 (d, J_{CP} = 142.6 Hz), 32.0, 12.4 (d, J_{CP} = 7.7 Hz); MS-ESI *m/z* (relative intensity): 284 ($M+Na^+$, 100%), 362 ($M+H^+$, 81%), 319 (27%), 301 (22%), 273 (25%), 245 (8%); High resolution mass spectrum (ES+) *m/z* 384.0393 [$(M+Na)^+$; calcd for $C_9H_{14}F_6NNaO_5P$: 384.0411]. Anal. Calcd for $C_9H_{14}F_6NO_5P$: C, 29.93; H, 3.91; N, 3.88. Found: C, 29.97; H, 3.93; N, 3.92.

Waste Disposal Information

All toxic materials were disposed of in accordance with "Prudent Practices in the Laboratory"; National Academy Press; Washington, DC, 1995.

3. Discussion

The methods described herein illustrate a practical and convenient three-step synthesis of bis(2,2,2-trifluoroethyl)phosphonates in good overall yield (>66%). This class of phosphonates, bearing an *N*-methoxy-*N*-methylamide functionality, is synthetically useful for the construction of *Z*-unsaturated *N*-methoxy-*N*-methylamides.[4] It is well established that the Weinreb amide functionality may then be converted to aldehydes or ketones in good yields.[5]

The formation of bis(2,2,2-trifluoroethyl)phosphonate **5** was previously reported by Deslongchamps et al.[4] following procedures developed by Tius and Busch-Petersen that employed KF/alumina to construct α-heterosubstituted Weinreb amides.[6] They discovered that tris(2,2,2-trifluoroethyl) phosphite reacts with bromides bearing the *N*-methoxy-*N*-methylamide moiety (**6**, **7**) to furnish the respective phosphonates **8** and **5** in modest yield (eq. 1).

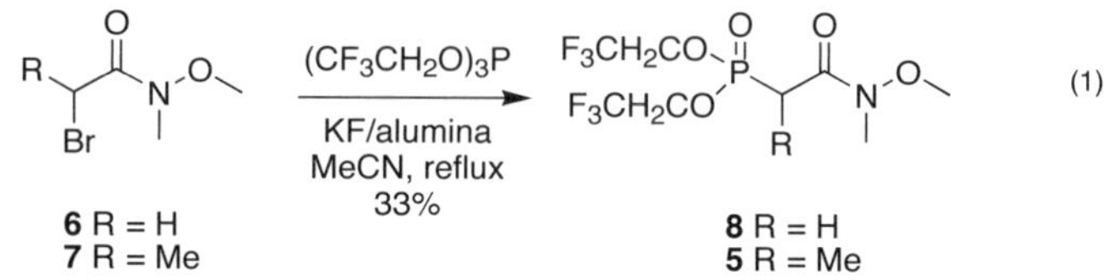

The procedure described herein proceeds in higher yield and circumvents the difficulty of handling and preparing the KF/alumina reagent. For example, although reactivity of the halide increases as the ratio of fluoride salt versus alumina is increased, the reagents became very hygroscopic and difficult to handle.[7] Additionally, Ando et al. discovered that a trace amount of water was essential for promoting the reaction, and thorough drying of the KF/alumina reagent led to a significant reduction in reactivity.[6,8] On the other hand, too much residual water was found to hydrolyze the chlorine analog of Weinreb amide **6**.[6] Tius and Busch-Petersen also reported that continuous sonication of the reaction mixture was necessary to enhance the reactivity of the KF/alumina reagent.[6,9]

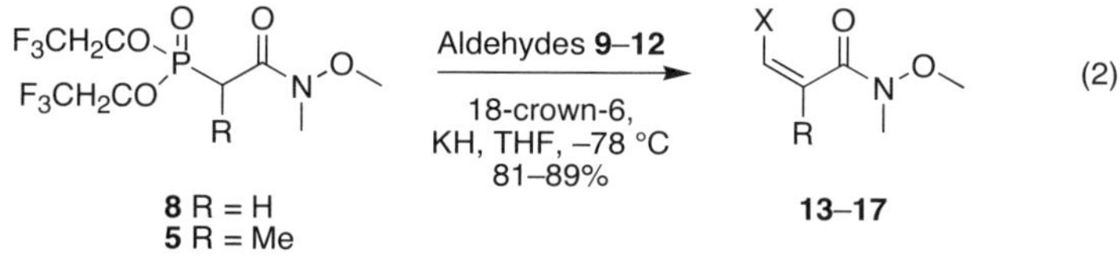

Deslongchamps et al. subsequently demonstrated the utility of phosphonates **5** and **8** in the stereoselective formation of *Z*-unsaturated alkenes using the modified Still conditions.[4,10] Bis(2,2,2-trifluoroethyl)phosphonates **8** and **5** react with aldehydes **9**–**12** to furnish the respective unsaturated *N*-methoxy-*N*-methylamides **13**–**17** in 81–89% yield, with only the *Z* isomer detectable by ^{1}H-NMR (eq. 2).[4]

Table 1 illustrates the several aldehydes (**9**–**12**) that Deslongchamps et al. employed to demonstrate the versatility of bis(2,2,2-trifluoroethyl)phosphonates **5** and **8** in the stereoselective synthesis of *Z*-unsaturated *N*-methoxy-*N*-methylamides **13**–**17**.[4]

Table 1. Synthesis of *Z*-Unsaturated *N*-methoxy-*N*-methylamides.

Entry	Aldehyde	Phosphonate	Product	Yield
1	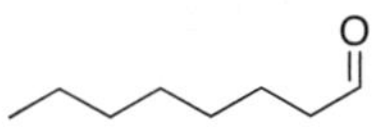 **9**	5	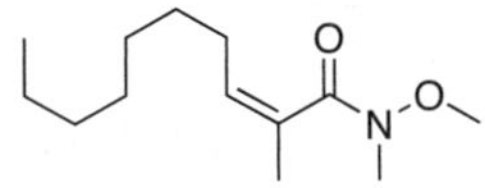 **13**	83
2	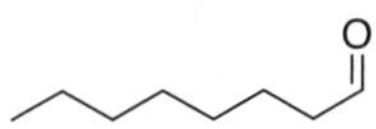 **9**	8	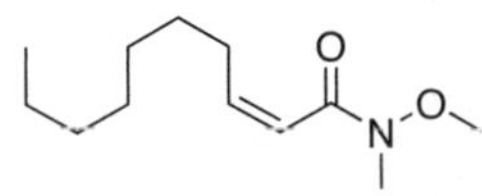 **14**	89
3	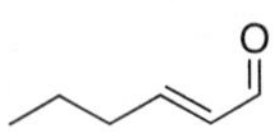 **10**	5	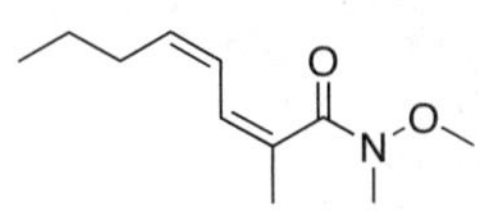 **15**	83
4	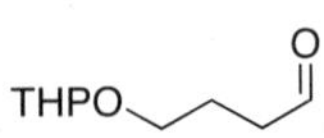 **11**	5	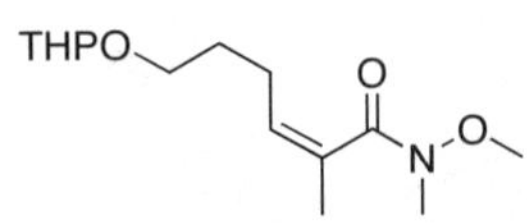 **16**	83
5	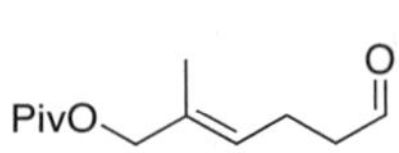 **12**	5	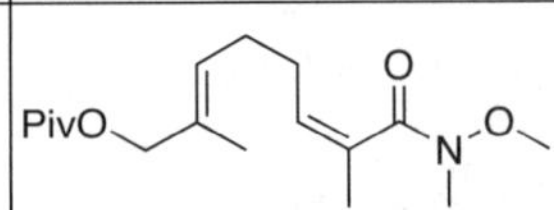 **17**	81

1. Department of Chemistry, University of Pennsylvania, Philadelphia, PA 19104.

2. For leading references on the preparation of ethylphosphonic acid bis-(2,2,2-trifluoroethyl) ester (**2**), see: Patois, C.; Savignac, P.; About-Jaudet, E.; Collignon, N. *Synth. Commun.*.**1991**, *21*(22), 2391–2396. Patois, C.; Savignac, P.; About-Jaudet, E.; Collignon, N. *Org. Synth.* **1998**, *CV 9*, 88–91.

3. For leading references on the preparation of *N*-methoxy-*N*-methylcarbamoyl chloride (**4**), see: Tius, M.A.; Busch-Petersen, J.; Yamashita, M. *Tetrahedron Lett.* **1998**, *39*, 4219–4222. Murakami, M.; Hoshino, Y.; Ito, H.; Ito, Y. *Chem. Lett.* **1998**, *2*, 163–164.

4. For leading references on the preparation of [1-(methoxymethylcarbamoyl)ethyl] phosphonic acid bis-(2,2,2-trifluoroethyl) ester **5**, and its use in the synthesis of *Z*-unsaturated *N*-methoxy-*N*-methylamides, see: Fortin, S.; Dupont, F.; Deslongchamps, P. *J. Org. Chem.* **2002**, *67*, 5437–5439.

5. Nahm, S.; Weinreb, S.M. *Tetrahedron Lett.* **1981**, *22*, 3815–3818.

6. Tius, M.A.; Busch-Petersen, J. *Synlett* **1997**, 531–532.

7. Ando, T.; Yamawaki, J.; Kawate, T.; Sumi, S.; Hanafusa, T. *Bull. Chem. Soc. Jpn.* **1982**, *55*, 2504–2507.

8. Ando, T.; Kawate, T.; Yamawaki, J.; Hanafusa, T. *Chem. Lett.* **1982**, 935–938.

9. Ando, T.; Kawate, T.; Ichihara, J.; Hanafusa, T. *Chem. Lett.* **1984**, 725-728.

10. Still, W.C.; Gennari, C. *Tetrahedron Lett.* **1983**, *24*(41), 4405–4408.

TETRABENZYL PYROPHOSPHATE
[[Diphosphoric acid, tetrakis(phenylmethyl) ester]]

DBP TBPP

Submitted by Todd D. Nelson,[1] Jonathan D. Rosen,[1] M. Bhupathy,[1] James McNamara[1], Michael J. Sowa,[2] Chad Rush,[2] and Louis S. Crocker.[3]
Checked by Stuart J. Conway and Andrew B. Holmes.

1. Procedure

A flame-dried, 250-mL, three-necked, round-bottomed flask is equipped with a magnetic stir bar, low-temperature thermometer, and two rubber septa (Note 1). A balloon filled with argon is connected via an inlet needle through the central rubber septum. The vessel is charged with 15.24 g (54.8 mmol) of dibenzyl phosphate (DBP) (Notes 2, 3) and flushed with argon from the balloon, then 60 mL of isopropyl acetate is introduced through the side septum by syringe. The resulting slurry is stirred and cooled to 3 ± 3 °C, and the balloon is temporarily removed from the central inlet needle while 26 mL of a 1.08M solution (28.2 mmol) of 1,3-dicyclohexylcarbodiimide (DCC) in isopropyl acetate (Notes 4, 5) *(Caution!)* is added via cannula from a flame-dried, 100-mL, round-bottomed flask equipped with another argon-inflated balloon. The reaction temperature is maintained at 3 ± 3 °C by controlling the addition through the cannula by simply removing and re-inserting the needle in the flask containing the DCC (Notes 6, 7). The cold slurry is filtered through a sintered glass filter funnel under aspirator vacuum and the 1,3-dicyclohexylurea (DCU) waste cake is rinsed with 4 × 12-mL portions of isopropyl acetate. The filtrate and rinses are combined and concentrated at 25 °C in vacuo on a rotary evaporator to a final volume of 30 mL (Notes 8, 9).

The concentrated solution is transferred to a 250-mL, three-necked, round-bottomed flask equipped with a magnetic stir bar, low-temperature thermometer and two rubber septa (Note 1). A balloon filled with argon is connected via an inlet needle through the central rubber septum. The solution is diluted by adding 10 mL of heptane through the septum using a syringe (Note 10). The resulting slurry is stirred at room temperature for 15 min, then 80 mL of heptane is added via syringe to the stirred slurry over 10 min (Note 11). The slurry is then cooled to 3 ± 3 °C and stirred for 1 h. The solids are collected using a sintered glass filter funnel with aspirator vacuum and the filter cake is washed with 3 × 20-mL portions of 20% (v/v) isopropyl acetate/heptane. The filter cake is dried under reduced pressure, then transferred to a 250-mL, round-bottomed flask and flushed with a blanket of argon four times at room temperature. The product was allowed to stand under an atmosphere of argon at room temperature overnight (Note 12) to afford 13.0-14.15 g (88-96%) of tetrabenzyl pyrophosphate as a white crystalline solid (Note 13).

2. Notes

1. The submitters used a 12-L, round-bottomed flask equipped with an overhead stirrer, thermocouple, N_2 inlet, and addition funnel to process 762 g of DBP.

2. The submitters carried out this preparation on fifty times the scale reported here.

3. Dibenzyl phosphate was purchased from Aldrich Chemical Company, Inc. The submitters purchased dibenzyl phosphate from Digital Specialty Company, Inc.

4. The checkers dissolved 5.8 g of DCC in 26 mL of isopropyl acetate (both purchased from Aldrich Chemical Company, Inc.). The submitters obtained a prepared solution of DCC (1.08M in isopropyl acetate) from Schweizerhall, Inc.

5. DCC, a strong sensitizer, is readily absorbed through the skin and may cause an allergic skin reaction. This compound is extremely destructive to tissue of the mucous membranes, upper respiratory tract, eyes and skin.

6. Although the dibenzyl phosphate is not completely soluble in isopropyl acetate, it is apparently sufficiently soluble to allow formation of tetrabenzyl pyrophosphate which remains in solution while a slurry of 1,3-dicyclohexylurea (DCU) forms as a white precipitate.

7. Typical addition times are between 10-20 min and the reaction is complete within 90 min (HPLC). The submitters found that addition times of 25-35 min (with 762 g of dibenzyl phosphate) were typical and the reaction was complete within 30 min. The checkers used the following HPLC conditions:

Column:	Dynamax C18, 5 × 250 mm, 8 μm particle size
Guard Tube:	Dynamax C18 Guard, 8 μm particle size
Mobile phase:	Acetonitrile:water 70:30 (v/v)
Run time:	20 min
Flow rate:	1.5 mL/min, at room temperature

Detection at 220 nm and 260 nm (diode array detector)
Retention time for tetrabenzyl pyrophosphate = 12.0 min

The checkers found it advantageous to monitor the reaction by detection at 260 nm as well as 220 nm because at the longer wavelength there was no absorption peak due to isopropyl acetate. The final purity of the sample and the absence of isopropyl acetate were then verified by repeating the HPLC analysis using detection at 220 nm. The submitters used the following HPLC conditions:

Column:	Zorbax RX-C8, 4.6 × 250 mm, 5 μm particle size
Mobile phase:	A) Acetonitrile B) 0.1% H_3PO_4 buffered water Step gradient: 40:60 A:B to 70:30 A:B over 10 min, hold for 5 min, 70:30 A:B to 40:60 A:B over 2 min, hold for 3 min.
Flow rate:	1.5 mL/min, at room temperature

Detection at 220 nm
Retention time for tetrabenzyl pyrophosphate = 12.8 min

8. Concentration of the solution on the rotary evaporator must be ended before the solution becomes saturated. The submitters carried out this process by reduced pressure distillation. The solubility of tetrabenzyl pyrophosphate in isopropyl acetate at 25 °C is ca. 676 mg/g.

9. All operating temperatures should be kept at room temperature or below due to the very large exothermic potential exhibited by dibenzyl phosphate and tetrabenzyl pyrophosphate.

10. The submitters seeded the batch with 1 mol% of tetrabenzyl pyrophosphate. The checkers found it unnecessary to use seed crystals.

11. The solvent composition of the system is approximately 4:1 heptane:isopropyl acetate. This was predetermined by using solubility data of tetrabenzyl pyrophosphate under different conditions (listed below) at 5 °C.

Heptane:isopropyl acetate (v/v)	Solubility (mg TBPP/g solvent)[a]
1:4	82
2:3	34
3:2	10.6
4:1	2.8

[a]Measured at 5 °C

12. The submitters dried the product cake in vacuo and under a blanket of argon overnight at room temperature.

13. The submitters obtained 671 g (91%) from 762 g of dibenzyl phosphate. Characterization data for tetrabenzyl pyrophosphate include the following: TLC on Merck kieselgel 60 F_{254}, $R_f = 0.42$ (1:1 EtOAc:hexane); mp 59-60 °C (lit.[4] 61-62 °C, from cyclohexane); ^{1}H NMR (250 MHz, $CDCl_3$) δ: 5.10-5.13 (m, 8 H), 7.33 (s, 20 H); ^{13}C NMR (100 MHz, $CDCl_3$) δ: 70.3 (t, 4 C, $J = 3$), 127.9 (s, 8 C), 128.4 (s, 8 C), 128.6 (s, 4 C), 134.8 (t, 4 C, $J = 4$); ^{31}P NMR (162 MHz, $CDCl_3$) δ: −12.3 (s, 2 P); IR ($CHCl_3$) cm^{-1}: 3012, 1498, 1456, 1381, 1292, 1021, 956. MS (ES^+) *m/z* (rel intensity) 561 $(100, M + Na)^+$, 539 (50, M + H)+; *m/z* 539.1390 ($[M + H]^+$, calcd for $C_{28}H_{29}O_7P_2$ 539.1389. Anal. Calcd for $C_{28}H_{28}O_7P_2$: C, 62.5; H, 5.2; Found: C, 62.4; H, 5.2.

Waste Disposal Information

All toxic materials were disposed of in accordance with "Prudent Practices in the Laboratory"; National Academy Press; Washington, DC, 1995.

3. Discussion

Recently, tetrabenzyl pyrophosphate (TBPP) has become increasingly utilized for either *O*- or *N*-phosphorylations in complex molecules. A wide variety of substrates have been successfully phosphorylated by the title reagent and this transformation has often been cited as a key step in many syntheses. Debenzylation of the resulting dibenzyl phosphoryl moiety is usually facile and cleanly affords the corresponding phosphate.

In recent years, tetrabenzyl pyrophosphate has been used to synthesize a wide variety of phosphate-containing biologically active molecules,[5] which include prodrugs of tachykinin receptor antagonists,[6] angiotensin II antagonists,[7] 5-phosphatase inhibitors,[8] DHQ synthase inhibitors,[9] EPSP synthase inhibitors,[10] HIV protease inhibitors,[11] SH2 domain inhibitors,[12] antifungal agents[13] and antitumor agents.[14]

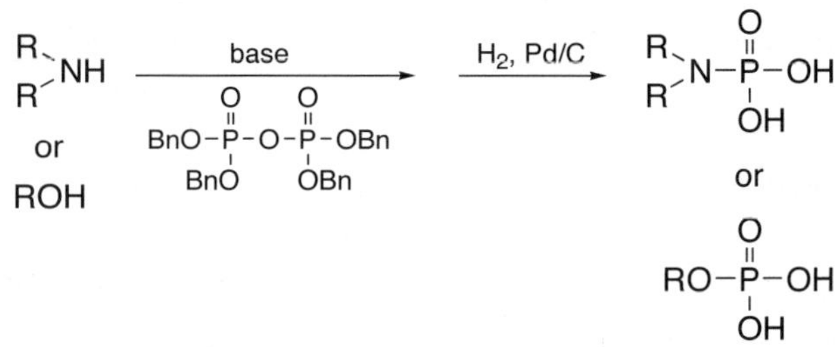

With the increasing use of tetrabenzyl pyrophosphate,[15] especially in the pharmaceutical industry, a cost-effective and convenient multi-kilogram preparation of tetrabenzyl pyrophosphate was desirable. The current procedure is a modification of an earlier synthesis by Todd.[4] Tetrabenzyl pyrophosphate has also been synthesized from: 1) *N*-hydroxysuccinimide and *N*-mercaptosuccinimide esters[16] and 2) the corresponding silver phosphate in carbon disulfide.[17]

Tetrabenzyl pyrophosphate should be stored in a freezer as a dry solid. Stability data on TBPP are tabulated below [reported in Liquid Chromatography Area Percent (LCAP)]. The only impurity detected was dibenzyl phosphate. For this study, solid TBPP was sealed in scintillation vials with Parafilm, which were in turn enclosed in desiccators with indicating Drierite.

Table. Stability Data for TBPP

Temperature	0 days	19 days	31 days	62 days
25 °C	99.9	97.1	96.9	95.5
0 °C	99.9	99.5	99.0	98.4
−25 °C	99.9	99.7	99.7	99.3

1. Merck Research Laboratories, Department of Process Research, P.O. Box 2000, Rahway, NJ 07065.

2. Merck Research Laboratories, Department of Chemical Engineering Research and Development, P.O. Box 2000, Rahway, NJ 07065.

3. Merck Research Laboratories, Department of Analytical Research, P.O. Box 2000, Rahway, NJ 07065.

4. Khorana, H. G.; Todd, A. R. *J. Chem. Soc.* **1953**, 2257.

5. (a) Nicolaou, M. G.; Yuan, C.-S.; Borchardt, R. T. *J. Org. Chem.* **1996**, *61*, 8636; (b) Vacca, J. P.; de Solms, S. J.; Young, S. D.; Huff, J. R.; Billington, D. C.; Baker, R.; Kulagowski, J. J.; Mawer, I. M. *ACS Symposium Series 463*; **1991**; Chapter 5, p. 66.

6. Dorn, C. P.; Hale, J. J.; Maccoss, M.; Mills, S. G. U.S. Patent 5 691 336, 1997.

7. De Laszlo, S. E.; Glinka, T. W.; Greenlee, W. J.; Chakravarty, P. K.; Patchett, A. A. U.S. Patent 5 385 894, 1995.

8. Kozikowski, A. P.; Fauq, A. H.; Wilcox, R. A.; Nahorski, S. R. *J. Org. Chem.* **1994**, *59*, 2279.

9. Montchamp, J.-L.; Peng, J.; Frost, J. W. *J. Org. Chem.* **1994**, *59*, 6999.

10. Marzabadi, M. R.; Font, J. L.; Gruys, K. J.; Pansegrau, P. D.; Sikorski, J. A. *Bioorg. Med. Chem. Lett.* **1992**, *2*, 1435.

11. Dressman, B. A.; Fritz, J. E.; Hammond, M.; Hornback, W. J.; Kaldor, S. W.; Kalish, V. J.; Munroe, J. E.; Reich, S. H.; Tatlock, J. H.; Shepherd, T. A.; Rodriguez, M. J. U.S. Patent 5 484 926, 1996.

12. Pacofsky, G. J.; Lackey, K.; Alligood, K. J.; Berman, J.; Charifson, P. S.; Crosby, R. M.; Dorsey, Jr., G. F.; Feldman, P. F.; Gilmer, T. M.; Hummel, C. W.; Jordan, S. R.; Mohr, C.; Shewchuk, L. M.; Sternbach, D. D.; Rodriguez, M. *J. Med. Chem.* **1998**, *41*, 1894.

13. (a) Balkovec, J. M.; Black, R. M.; Hammond, M. L.; Heck, J. V.; Zambias, R. A.; Abruzzo, G.; Bartizal, K.; Kropp, H.; Trainor, C.; Schwartz, R. E.; McFadden, D. C.; Nollstadt, K. H.; Pittarelli, L. A.; Powles, M. A.; Schmatz, D. M. *J. Med. Chem.* **1992**, *35*, 194; (b) Balkovec, J. M.; Loewe, M. F.; Mathre, D. J. Eur. Patent 525 889, 1991.

14. Ueda, Y.; Mikkilineni, A. B.; Knipe, J. O.; Rose, W. C.; Casazza, A. M.; Vyas, D. M. *Bioorg. Med. Chem. Lett.* **1993**, *3*, 1761.

15. A *Chemical Abstracts* search regarding tetrabenzyl pyrophosphate (on February 1, 2002) revealed a significant increase in the use of this reagent, as indicated by the following data:

Period	Number of Citations
2002-1991	73
1990-1981	10
1980-1971	1

16. Chapman, T. M.; Kleid, D. G. *J. Org. Chem.* **1973**, *38*, 250.

17. Atkinson, R. E.; Cadogan, J. I. G. *J. Chem. Soc. (C)* **1967**, 1356.

IRIDIUM-CATALYZED C-H BORYLATION OF ARENES AND HETEROARENES: 1-CHLORO-3-IODO-5-(4,4,5,5-TETRAMETHYL-1,3,2-DIOXABOROLAN-2-YL)BENZENE AND 2-(4,4,5,5,-TETRAMETHYL-1,3,2-DIOXABOROLAN-2-YL)INDOLE (2-(3-Chloro-5-iodophenyl)-4,4,5,5-tetramethyl-1,3,2-dioxaborolane) (2-(4,4,5,5-Tetramethyl-1,3,2-dioxaborolan-2-yl)-1H-indole)

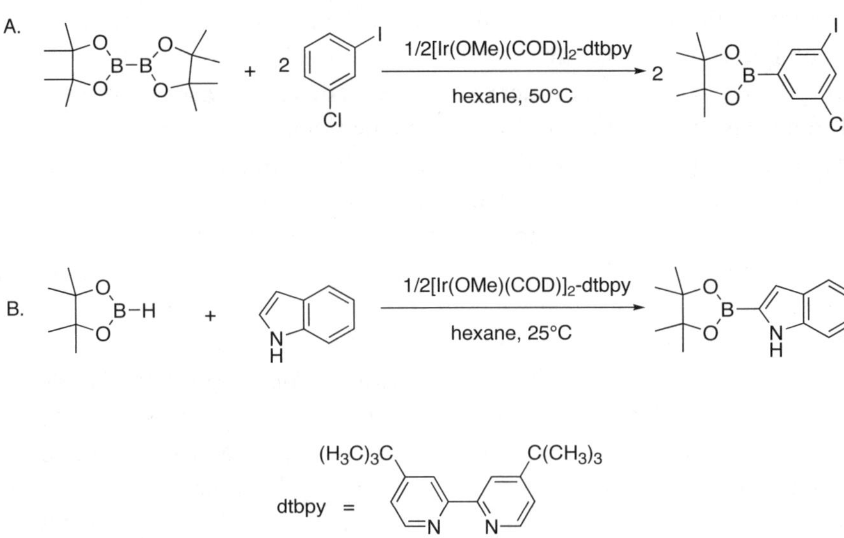

Submitted by Tatsuo Ishiyama, Jun Takagi, Yusuke Nobuta and Norio Miyaura.[1]
Checked by Yili Shi, Daniel J. Weix and Jonathan A. Ellman.

1. Procedure

Caution! The reactions produce hydrogen gas and should be conducted in a well-ventilated hood.

A. 1-Chloro-3-iodo-5-(4,4,5,5-tetramethyl-1,3,2,-dioxaborolan-2-yl)benzene. A 50-mL, two-necked, round-bottomed flask is fitted with a magnetic stirring bar, a rubber septum, and a condenser to which a nitrogen inlet and an oil bubbler are attached, and flushed with nitrogen (Note 1). The septum is removed and the flask is charged with bis(η^4-1,5-cyclooctadiene)-di-μ-methoxy-diiridium(I) ([Ir(OMe)(COD)]$_2$) (33 mg, 0.050 mmol) (Note 2), 4,4'-di-*tert*-butyl-2,2'-bipyridine (dtbpy) (27 mg,

0.10 mmol) (Note 3), and bis(pinacolato)diboron (2.67 g, 10.5 mmol) (Note 4). The septum is again placed on the flask, and the flask is purged with nitrogen for 1 min. Hexane (30 mL) (Note 5) is added by syringe, and the flask is immersed in an oil bath that is maintained at 50 °C. The mixture is stirred for 10 min to give a dark red solution. The flask is charged with 1-chloro-3-iodobenzene (4.75 g, 19.9 mmol) by syringe (Note 6), and then the resulting dark red solution is stirred at 50 °C for 6 h (Note 7). The mixture is removed from the oil bath, allowed to cool to room temperature, and poured into a separatory funnel. The reaction flask is rinsed with hexane (2 x 10 mL). The rinses and water (30 mL) are added to the separatory funnel, the funnel is shaken, the layers are separated, and the organic extracts are dried over anhydrous magnesium sulfate. The drying agent is removed by filtration and is washed with hexane (3 x 10 mL), and the filtrate is concentrated on a rotary evaporator to give a dark brown oil. The oil is distilled under reduced pressure (Note 8) to afford 6.11-6.20 g (84-86%) of 1-chloro-3-iodo-5-(4,4,5,5-tetramethyl-1,3,2-dioxaborolan-2-yl)benzene as a white solid, mp 58.6-60.7 °C (Note 9).

B. 2-(4,4,5,5-Tetramethyl-1,3,2-dioxaborolan-2-yl)indole. A 50-mL, two-necked, round-bottomed flask is fitted with a magnetic stirring bar, a rubber septum, and a condenser to which a nitrogen inlet and an oil bubbler are attached, and flushed with nitrogen (Note 1). The septum is removed and the flask is charged with bis(η^4-1,5-cyclooctadiene)-di-μ-methoxy-diiridium(I) ([Ir(OMe) (COD)]$_2$) (50 mg, 0.075 mmol) (Note 2) and 4,4'-di-*tert*-butyl-2,2'-bipyridine (dtbpy) (40 mg, 0.15 mmol) (Note 3). The septum is again placed on the flask, and the flask is purged with nitrogen for 1 min. Hexane (30 mL) (Note 5) and pinacolborane (4.79 mL, 33.0 mmol) (Note 10) are added by syringe, and the flask is immersed in an oil bath that is maintained at 25 °C. The mixture is stirred for 10 min. to give a dark red solution. The flask is charged with indole (3.51 g, 30.0 mmol) (Note 11), and then the resulting dark red suspension is stirred at 25 °C for 4 h. The mixture is removed from the oil bath, allowed to cool to room temperature, and poured into a separatory funnel. The reaction flask is rinsed with ether (2 x 20 mL). The rinses and water (30 mL) are added to the separatory funnel, the funnel is shaken, the layers are separated, and the organic extracts are dried over anhydrous magnesium sulfate. The drying agent is removed by filtration and is washed with ether (3 x 10 mL), and the filtrate is concentrated on a rotary evaporator to give a dark brown oil. The oil is distilled under reduced pressure (Note 12) to afford 5.33-5.40 g (73-74%) of

2-(4,4,5,5-tetramethyl-1,3,2-dioxaborolan-2-yl)indole as a colorless viscous oil, bp 149-153 °C (0.15 mm) (Note 13).

2. Notes

1. All glassware is pre-dried in an oven at 120 °C for 1 h, assembled while hot, and allowed to cool under a stream of nitrogen.

2. $[Ir(OMe)(COD)]_2$ is prepared from $[IrCl(COD)]_2$ by the reported procedure.[2] The complex can be handled in air, and stored in a tightly closed bottle and in a freezer. Checkers stored the complex in a vial in a nitrogen-filled glove-bag at rt.

3. 4,4'-Di-*tert*-butyl-2,2'-bipyridine (98%) was purchased from Aldrich Chemical Company, Inc., and used without further purification.

4. Bis(pinacolato)diboron is prepared according to the literature procedure,[3] and dried under reduced pressure (0.1 mm) at room temperature for 16 h. prior to use. The reagent is air-stable and commercially available. The checkers purchased bis(pinacolato)diboron (98%) from Aldrich and used it without further purification.

5. Hexane is distilled from lithium aluminum hydride under nitrogen before use. The checkers purified hexanes by passage through a column of activated alumina (type A2, size 12 × 32, Purifry Co.) followed by nitrogen sparge. The hexanes were stored over MS 3Å in a nitrogen-filled glove-box.

6. 1-Chloro-3-iodobenzene was purchased from Tokyo Kasei Kogyo Co., Ltd., and distilled from molecular sieves 4Å under nitrogen before use. The checkers purchased 1-chloro-3-iodobenzene (98%) from Aldrich.

7. The reaction using 1.5 mol% of the catalyst is completed within 4 h at room temperature, but the decreasing the amount of catalyst to 0.5 mol% sufficiently slows the rate of the reaction that heating to 50 °C is required to complete the reaction in a reasonable amount of time.

8. Bp 145-148 °C (0.3 mm) 1-Chloro-3-iodobenzene (28 mg) that is unreacted or is generated by protodeboration of the product is also obtained, bp 60 °C (0.05 mm). The checkers found that either simple distillation with a short-path distillation head or Kugelrohr distillation can be used. To avoid contamination of the product with lower boiling materials (eg starting materials or pinacol), the mixture should be heated until 1-chloro-3-iodobenzene and pinacol have distilled. The apparatus is then allowed to cool, is disassembled, and is cleaned, before the product is distilled over.

9. Gas chromatographic analysis of the product (Hitachi G-3500, OV-101 on UniportB, a glass column, 3 mm x 2 m) shows that the chemical

purity is 97-99%. The spectral data are as follows: ^{1}H NMR (400 MHz, $CDCl_3$) δ 1.30 (s, 12H), 7.72 (d, 1H, J = 2.0), 7.78 (t, 1H, J = 1.8), 7.99 (t, 1H, J = 0.8); ^{13}C NMR (100 MHz, $CDCl_3$) δ 24.8, 84.5, 94.2, 133.7, 134.7, 139.4, 141.4; ^{11}B NMR (128.3 MHz, $CDCl_3$) δ 29.58 (BF_3•OEt_2 as external reference, δ 0.00); HRMS calcd for $C_{12}H_{15}BClIO_2[M^+]$, 363.9899, found 363.9890.

10. Pinacolborane (97%) was purchased from Aldrich Chemical Company, Inc., and used without further purification. Pinacolborane is very sensitive to moisture and was troublesome to store for any period of time. The checkers found it could be stored under nitrogen at -30 °C in a sealable tube.

11. Indole was purchased from Tokyo Kasei Kogyo Co., Ltd., and dried under reduced pressure (0.1 mm) at room temperature for 16 h prior to use. The checkers purchased indole (99%+) from Aldrich.

12. The checkers accomplished the distillation with the apparatus below.

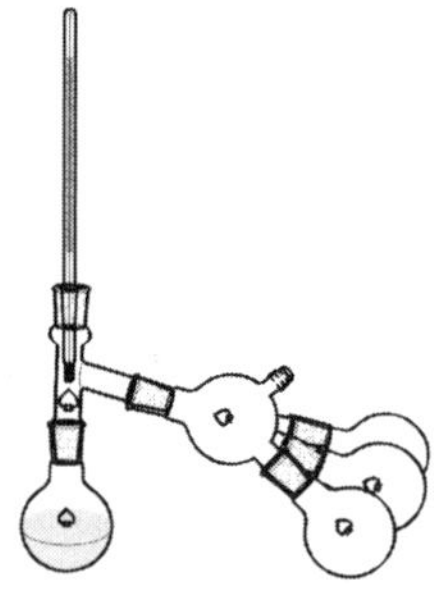

It was found that distillation with a traditional short-path led to clogging of the apparatus by both remaining indole starting material and the highly viscous product. Indole (29 mg) that is unreacted or resulted by protodeboration of the desired product is also obtained, bp 120 °C (0.7 mm). To avoid contamination of the product with lower boiling materials (eg starting materials or pinacol), the mixture should be heated until indole and pinacol have distilled over. The apparatus is then allowed to cool, is disassembled, and is cleaned, before the product is distilled over.

13. Gas chromatographic analysis of the product (Hitachi G-3500, OV-101 on Uniport B, a glass column, 3 mm x 2m) shows that the chemical purity is 96-99%. The spectral data are as follows: ^{1}H NMR (400 MHz, $CDCl_3$) δ 1.37 (s, 12H), 7.10 (dt, 1H, J= 0.8, 8.7), 7.12 (d, H, J= 0.7), 7.23 (dt, 1H, J= 1.1, 7.6), 7.39 (dd, 1H, J= 1.0, 8.3), 7.68 (dd, 1H, J= 1.0, 8.1), 8.56 (br s, 1H); ^{13}C NMR (100 MHz, $CDCl_3$) δ 24.8, 84.1, 111.2, 113.8,

119.8, 121.6, 123.6, 128.3, 138.2; ^{11}B NMR (128.3 MHz, $CDCl_3$) δ 28.45 ($BF_3 \bullet OEt_2$ as external reference, δ 0.00); HRMS calcd for $C_{14}H_{18}BNO_2[M^+]$, 243.1431, found 243.1423.

Safety and Waste Disposal Information

All hazardous materials should be handled and disposed of in accordance with "Prudent Practices in the Laboratory"; National Academy Press; Washington, DC, 1995.

3. Discussion

Aryl- and heteroarylboron derivatives are important class of compounds that have been applied to various fields of chemistry.[4] Traditional methods for their synthesis are based on the reactions of trialkylborates with aryl- and heteroarylmagnesium or lithium reagents.[5] Pd-catalyzed cross-coupling of aryl and heteroaryl halides with tetra(alkoxo)diborons[6] or di(alkoxo)boranes[7] is a milder variant where the preparation of magnesium and lithium reagents is avoided.

Alternatively, the catalytic C-H borylation of arenes and heteroarenes, first reported by Smith,[8] is highly attractive as a halide-free process for the synthesis of aryl- and heteroarylboron compounds.[9] Among the catalysts developed to date, the combination of $1/2[Ir(OMe)(COD)]_2$ and dtbpy described here exhibits high activity, which allows the formation of aryl- and heteroarylboronates in high yields at room temperature from an equimolar equivalent of bis(pinacolato)diboron (pin_2B_2, pin = $Me_4C_2O_2$) or pinacolborane (pinBH) and arenes or heteroarenes (Table 1).[9j-l]

The regiochemistry of arene borylation is primarily controlled by the steric effects of substituents. The reaction occurs at C-H bonds located meta or para to a substituent in preference to those located ortho. Thus, 1,2- and 1,4-disubstituted arenes bearing identical substituents yield arylboronates as single isomers. The borylation of 1,3-disubstituted arenes proceeds at the common meta position; therefore, regioisomerically pure products are obtained even for two distinct substituents on the arenes. In the case of five-membered heteroarenes, the electronegative heteroatom causes the C-H bonds at the α-positions to be active so that the borylation occurs at the α-positions. Thus, the regioselective monoborylation of 2-substituted or benzo-fused substrates can be possible. Although a mixture of 2-borylated and 2,5-diborylated products is formed from unsubstituted substrates, both products are selectively obtained by reactions with the appropriate ratio of substrate and reagent. On the other hand, six-membered heteroarenes such as pyridine shows significantly lower reactivity due to strong coordinating ability of the basic nitrogen for the catalyst. Exceptionally, 2,6-disubstituted pyridines undergo smooth borylation at the 4-position.

TABLE 1.

SYNTHESIS OF ARYL- AND HETEROARYLBORONATES[a]

Product	(%)[b] A[c]	(%)[b] B[d]	Product	(%)[b] A[c]	(%)[b] B[d]
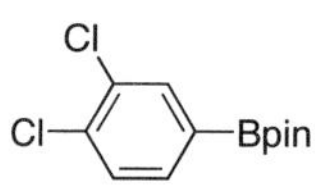	82 (8 h)	73 (8 h)	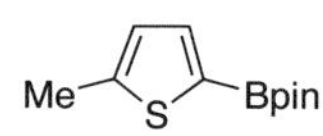	95 (2 h)	91 (2 h)
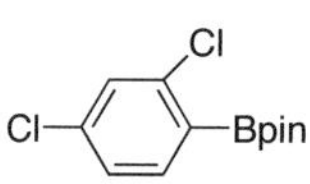	53 (24 h)	22 (24 h)	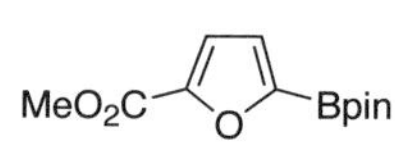	80 (2 h)	95 (0.5 h)
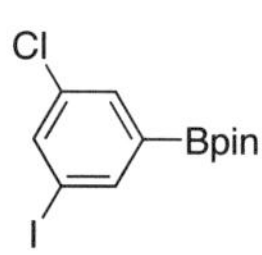	82 (4 h)	67 (8 h)	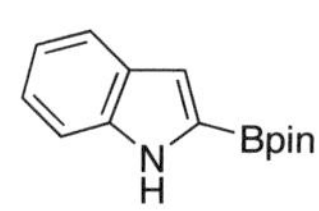	88 (0.5 h)	99 (0.5 h)
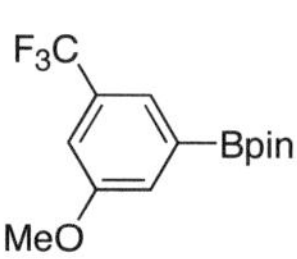	81 (8 h)	73 (24 h)	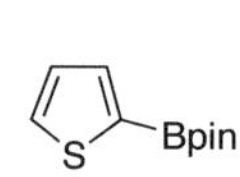	91[e] (1 h)	75[f] (0.5 h)
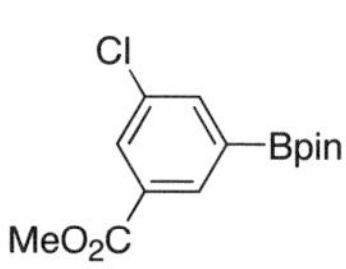	80 (8 h)	70 (24 h)	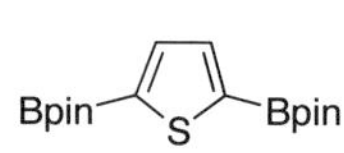	83[g] (0.5 h)	86[h] (2 h)
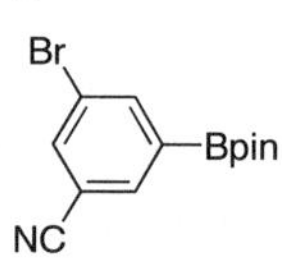	83 (2 h)	74 (2 h)	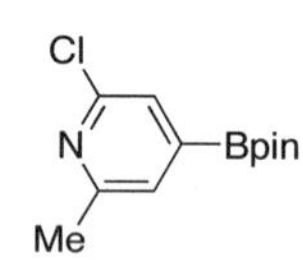	84 (4 h)	75 (2 h)

[a] Reactions were carried out at 25 °C in hexane (6 mL) by using substrate (1.0 mmol) and 1/2[Ir(OMe)(COD)]2-dtbpy catalyst. [b] GC yields based on substrates. [c] Method A: pin_2B_2 (0.50 mmol)/Ir cat. (0.015 mmol). [d] Method B: pinBH (1.1 mmol)/Ir cat (0.03 mmol). [e] 5.0 mmol of thiophene and 1.0 mmol of pinBH were used. The yield is based on pin_2B_2. [f] 5.0 mmol of thiophene was used. The yield is based on pinBH. [g] 1.0 mmol of pin_2B_2 was used. [h] 2.2 mmol of pinBH was used.

Functional group tolerance of the borylation is quite high. The reaction selectively occurs at the C-H bond for substrates possessing Cl, Br, I, CF_3, OMe, CO_2Me, and CN groups. The reaction occurs only at the aromatic C-H bonds even when the substrate has weaker benzylic C-H bonds.

Aryl- and heteroarylboronic acids and esters have been used for the synthesis of biaryls via the palladium-catalyzed cross-coupling reaction with aryl electrophiles.[10] Sequential reactions involving aromatic C-H borylation and cross-coupling with aryl electrophiles in the same flask provide an efficient and convenient route to unsymmetrical biaryls (eq. 1).

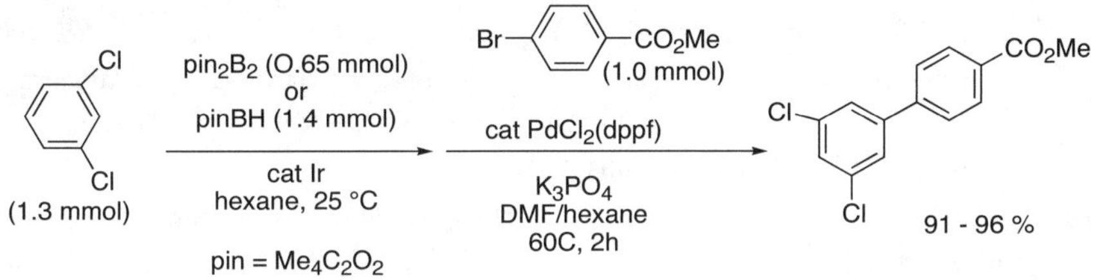

1. Division of Molecular Chemistry, Graduate School of Engineering, Hokkaido University, Sapporo 060-8628, Japan.
2. Uson, R.; Oro, L. A.; Cabeza, J. A., *Inorg. Synth.* **1985**, *23*, 126.
3. (a) Nöth, H. Z. *Naturforsch. B: Anorg. Chem., Org. Chem.* **1984**, *39B*, 1463; (b) Ishiyama, T.; Murata, M.: Ahiko, T.-a; Miyaura, N. *Org. Synth.* **2000**, *77*, 176.
4. Reviews see: (a) Vaultier, M.; Carboni, B. in *Comprehensive Organometallic Chemistry II*; Abel, E. W., Stone, F. G. A., Wilkinson, G., Eds.; Pergamon Press: Oxford, 1995; Vol. 11, p 191; (b) Ishihara, K.; Yamamoto, H. *Eur. J. Org. Chem.* **1999**, 527; (c) Shinkai, S.; Ikeda, M.; Sugasaki, A.; Takeuchi, M. *Acc. Chem. Res.* **2001**, *34*, 494; (d) Entwistle, C.D.; Marder, T.B. *Angew. Chem., Int. Ed.* **2002**, *41*, 2927; (e) Soloway, A. H.; Tjarks, W.; Barnum, B. A.; Rong, F.-G.; Barth, R. F.; Codogni, I. M.; Wilson, J. G. *Chem. Rev.* **1998**, *98*, 1515; (f) Yang, W.; Gao, X.; Wang, B. *Med. Res. Rev.* **2003**, *23*, 346.
5. Nesmeyanov, A. N.; Sokolik, R. A. *Methods of Elemento-Organic Chemistry*; North-Holland: Amsterdam, 1967; Vol. 1.
6. (a) A review see: Ishiyama, T.; Miyaura, N. *J. Synth. Org. Chem., Jpn.* **1999**, *57*, 503; (b) A review see: Ishiyama, T.; Miyaura, N. *J. Organomet. Chem.* **2000**, 611, 392; (c) Isiyama, T.; Murata, M.; Miyaura, N. *J. Org. Chem.* **1995**, *60*, 7508; (d) Isiyama, T.; Itoh, Y.; Kitano, T.; Miyaura, N.; *Tetrahedron Lett.* **1997**, *38*, 3447; (e) Ishiyama, T.; Ishida, K.; Miyaura, N. *Tetrahedron* **2001**, 57, 9813.
7. Murata, M.; Oyama, T.; Watanabe, S.; Masuda, Y. *J. Org. Chem.* **2000**, *65*, 164.

8. Iverson, C. N.; Smith, M. R., III. *J. Am. Chem. Soc.* **1999**, *121*, 7696.

9. (a) A review see: Ishiyama, T.; Miyaura, N. *J. Organomet. Chem.* **2003**, *680*, 3; (b) Chen, H.; Hartwig, J. F. *Angew. Chem. Int. Ed.* **1999**, *38*, 3391; (c) Chen, H.; Schlecht, S.; Semple, T. C.; Hartwig, *J. F. Science* **2000**, *287*, 1995; (d) Cho, J.-Y.; Iverson, C. N.; Smith, M. R., III. *J. Am. Chem. Soc.* **2000**, *122,* 12868; (e) Tse, M. K.; Cho, J. Y.; Smith, M. R., III. *Org. Lett.* **2001**, *3*, 2831; (f) Cho, J.-Y.; Tse, M. K. Holmes, D.; Maleczka, R. E., Jr.; Smith, M. R., III. *Science* **2002**, *295*, 305; (g) Shimada, S.; Batsanov, A. S.; Howard, J. A. K.; Marader, T. B. *Angew. Chem., Int. Ed.* **2001**, *40*, 2168; (h) Isiyama, T.; Takagi, J.; Ishida, K,; Miyaura, N.; Anastasi, N. R.; Hartwig, J. F. *J. Am. Chem. Soc.* **2002**, *124*, 390; (i) Takagi, J.; Sato, K.; Hartwig, J. F.; Ishiyama, T.; Miyaura, N. *Tetrahedron Lett.* **2002**, *43*, 5649; (j) Isiyama, T.; Takagi, J.; Hartwig, J. F.; Miyaura, N. *Angew. Chem., Int. Ed.* **2002**, *41*, 3056; (k) Ishiyama, T.; Takagi, J.; Yonekawa, Y.; Hartwig, J. F.; Miyaura, N. *Adv. Synth. Catal.* **2003**, *345*, 1103; (l) Ishiyama, T.; Nobuta, Y.; Hartwig, J. F.; Miyaura, N. *Chem. Commun.* **2003**, 2924.

10. (a) Miyaura, N.; Suzuki, A. *Chem, Rev.* **1995**, *95*, 2457; (b) Miyaura, N. *Top. Curr. Chem.* **2002**, *219*, 11; (c) Suzuki, A.; Brown, H. C. *Organic Syntheses Via Boranes*; Aldrich Chemical Company, Inc.: Milwaukee, 2003; Vol. 3.

SYNTHESIS OF ORTHO SUBSTITUTED ARYLBORONIC ESTERS BY IN SITU TRAPPING OF UNSTABLE LITHIO INTERMEDIATES: 2-(5,5-DIMETHYL-1,3,2-DIOXABORINAN-2-YL)BENZOIC ACID ETHYL ESTER

[Benzoic acid, 2-(5,5-dimethyl-1,3,2-dioxaborinan-2-yl)-, ethyl ester]

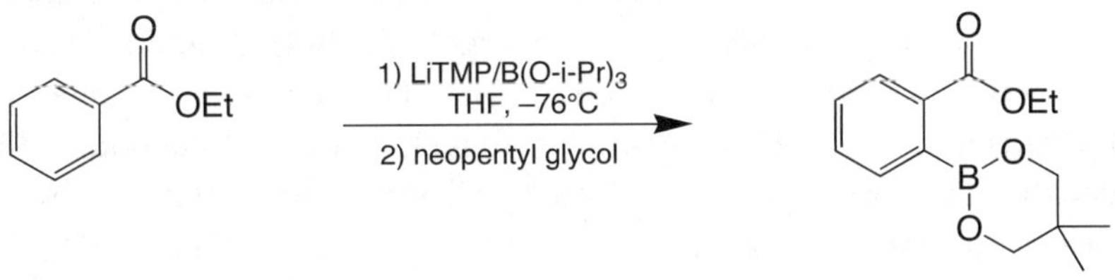

Submitted by Jesper Langgaard Kristensen, Morten Lysén, Per Vedsø and Mikael Begtrup.[1]
Checked by Günter Seidel and Alois Fürstner.

1. Procedure

2-(5,5-Dimethyl-1,2,3-dioxaborinan-2-yl)benzoic acid ethyl ester. An oven-dried, three-necked, 2-L, round-bottomed flask fitted with a thermometer, mechanical stirrer, and a 500-mL pressure-equalizing addition funnel (with volume graduation) capped with a rubber septum and nitrogen inlet is charged with 2,2,6,6-tetramethylpiperidine (51.2 g, 367 mmol) (Note 1) and 400 mL of anhydrous tetrahydrofuran (Note 2). The mixture is cooled to –30 °C (internal temperature) in a dry ice-acetone bath and 243 mL of *n*-butyllithium solution (1.49M in hexanes, 362 mmol) (Note 3) is cannulated directly into the addition funnel from a 1-L sure-seal flask. The *n*-butyllithium solution is added dropwise to the stirred reaction mixture over 30 min while the temperature is maintained between –30 and –35 °C resulting in an orange-yellow solution. The addition funnel is washed with two 10-mL portions of tetrahydrofuran and the reaction mixture is stirred an additional 10 min at –30 °C before being cooled to –76 °C (Note 4). Triisopropyl borate (112 mL, 483 mmol) (Note 5) is cannulated into the addition funnel directly from a sure-seal flask and then added dropwise (Note 6) to the creamy yellow suspension over 20 min while the internal temperature is maintained below –73 °C. Ethyl benzoate (35.7 g, 238 mmol) (Note 7) is added via syringe to the addition funnel and added dropwise to the reaction mixture over 10 min, causing a slight reddening of the suspension. The addition funnel is washed with two 10-mL portions of tetrahydrofuran, and the reaction mixture is stirred for 3.5 h at –73 °C. The resulting deep red suspension is taken out of the cooling bath, and after approximately 20 min the internal temperature rises to –30 °C. Glacial acetic acid (20.7 mL, 362 mmol) is then added dropwise over 5 min via the addition funnel, causing the internal temperature to rise to –10 °C while the color of the reaction mixture changes from red to yellow. The addition funnel is removed, neopentyl glycol (37.1 g, 356 mmol) (Note 8) is added in one portion, and stirring is continued for 2 h. The mixture is decanted from a white precipitate into a 3-L separatory funnel, and 1 L of dichloromethane is added to the separatory funnel (Note 9). The resulting solution is washed with three 300-mL portions of a 1:1 mixture of saturated aqueous NH_4Cl and water, two 300-mL portions of saturated $NaHCO_3$ solution, and two 300-mL portions of water, dried over 100 g of Na_2SO_4, filtered into a 1-L round-bottomed flask, and concentrated by rotary evaporation at 40 °C. When no more solvent can be distilled, a large teflon-coated magnetic stirbar is added to the flask, and the oily residue is stirred overnight at room temperature under reduced pressure (0.3 mm) to give 58.12 g (95%) of 2-(5,5-dimethyl-[1,3,2]dioxaborinan-2-yl)benzoic acid ethyl ester as a golden-brown oil which is pure enough for most purposes (Note 10). Analytically pure material is obtained as a colorless oil by distillation under high vacuum to give 50.3 g (82%), bp 105-110 °C ($1x10^{-5}$ Torr) (Notes 11, 12).

2. Notes

1. 2,2,6,6-Tetramethylpiperidine (99+%) was purchased from Aldrich Chemical Company, Inc. and used as received.
2. Tetrahydrofuran (THF) was distilled from sodium/benzophenone ketyl under a nitrogen atmosphere.
3. Butyllithium (1.6M solution in hexanes) was purchased from Aldrich Chemical Company, Inc. Three sequential titrations using *N*-pivaloyltoluidine (see: Suffert, J. *J. Org. Chem.* **1989**, *54*, 509) gave titers of 1.47M, 1.48M, and 1.51M, and on that basis the titer was assumed to be 1.49M.
4. At approximately –60 °C the mixture becomes cloudy.
5. Triisopropyl borate (98+%) was purchased from Aldrich Chemical Company, Inc. and used as received.

6. Care should be taken that the triisopropyl borate drops directly into the solution and does not run down the side of the flask, as this will cause the triisopropyl borate to precipitate from the reaction mixture.

7. Ethyl benzoate (99+%) was purchased from Aldrich Chemical Company, Inc. and used as received.

8. Neopentyl glycol (2,2-dimethyl-1,3-propanldiol) (99%) was purchased from Aldrich Chemical Company, Inc. and used as received.

9. Ca. 200 mL of the dichloromethane is used to rinse the 2-L flask before being added to the separatory funnel.

10. GC-MS analysis showed the crude material to be 95% pure. The title compound is the only detectable species in the NMR-spectrum. The submitters obtained 61.8 g (99%)of the product which was 99.5% pure by GC-MS analysis.

11. High vacuum is necessary for the success of the distillation. Attempted distillation at 0.3 mm results in decomposition of the material at a bath temperature of approximately 150 °C.

12. The product exhibits the following properties: ^{1}H NMR (300 MHz, $CDCl_3$) δ: 1.11 (s, 6H), 1.39 (t, 3H, J = 7.1 Hz), 3.79 (s, 4H), 4.38 (q, 2H, J = 7.1 Hz), 7.42-7.35 (m, 1H), 7.52-7.48 (m, 2H), 7.93 (td, 1H, J = 7.8, 0.9 Hz); ^{13}C NMR (75 MHz, $CDCl_3$) δ: 14.3, 21.9, 31.6, 61.6, 72.4, 128.3, 128.6, 131.2, 131.9, 133.1, 168.6; ^{11}B NMR (96 MHz, $CDCl_3$) δ 28.1 ppm. Anal. Calcd for $C_{14}H_{19}BO_4$: C 64.15, H 7.31. Found: C 64.02, H 7.46.

2. Waste Disposal Information

All toxic materials were disposed of in accordance with "Prudent Practices in the Laboratory"; National Academy Press; Washington, DC, 1995.

3. Discussion

The procedure described herein is an improved version of our previously reported synthesis,[2] circumventing the need for an intermediate aqueous work-up, thereby providing the title compound in a one-pot procedure.

The transition metal catalyzed cross coupling of an organohalide with a boronic acid derivative, the Suzuki-Miyaura coupling, has become one of the most popular ways of preparing biaryls.[3] The reaction is very robust and can easily be scaled to provide multigrams of material.[4]

Arylboronic acids have traditionally been prepared via the addition of an organomagnesium or organolithium intermediate to a trialkyl borate. Subsequent acidic hydrolysis produces the free arylboronic acid. This limits the type of arylboronic acids one can access via this method, as many functional groups are not compatible with the conditions necessary to generate the required organometallic species, or these species may not be stable intermediates.

Complete characterization of arylboronic acids is often difficult because they are readily transformed into stable cyclic anhydrides called boroxines[5] and other polymeric species. Arylboronic acids are also known to be hygroscopic. Thus, arylboronic acids are often prepared and used directly as a mixture of different entities. Commercial arylboronic acids will very often contain varying amount of anhydrides.

The procedure described herein illustrates two important points:

(1) The in situ trapping of unstable lithio-intermediates as a way of circumventing the above mentioned limitations, and

(2) The direct isolation of the well defined and stable neopentyl glycol arylboronic esters, without the need for an intermediate aqueous work-up.

As first described by Krizan and Martin,[6] the in situ trapping protocol, i.e., having the base and electrophile present in solution simultaneously, makes it possible to lithiate substrates that are not applicable in classical *ortho*-lithiation reactions.[7] Later, Caron and Hawkins utilized the compatibility of lithium diisopropylamide and triisopropyl borate to synthesize arylboronic acid derivatives of bulky, electron deficient neopentyl benzoic acid esters.[8] As this preparation illustrates, the use of lithium tetramethylpiperidide instead of lithium diisopropylamide broadens the scope of the reaction, and makes it possible to functionalize a simple alkyl benzoate.[2]

The conversion of arylboronic acids to the corresponding neopentyl glycol arylboronic esters has several advantages: The esters are readily soluble in organic solvents, shelf stable, non-hygroscopic and easily characterized as a single entity.[9] Furthermore, boronic esters can be utilized in many of the transformations where arylboronic acids usually are employed, making them an attractive alternative from a practical point of view.

The described procedure can also be applied for *ortho*-borylation of chloro, fluoro and cyanonaphthalenes[10] or pyridines.[11]

1. Department of Medicinal Chemistry, The Danish University of Pharmaceutical Sciences, Universitetsparken 2, DK-2100 Copenhagen, Denmark.
2. Kristensen, J.; Lysén, M.; Vedsø, P.; Begtrup, M. *Org. Lett.* **2001**, *3*, 1435.
3. (a) Miyaura, N.; Yanagi, T.; Suzuki, A. *Synth. Commun.* **1981**, *11*, 513; Reviews: (b) Miyaura, N.; Suzuki, A. *Chem. Rev.* **1995**, *95*, 2457; (c) Suzuki, A. *J. Organomet. Chem.* **1999**, *576*, 147.
4. (a) Huff, B.E.; Koenig, T.M.; Mitchell, D.; Staszak, M.A. *Org. Synth., Coll. Vol. X* **2004**, 102. (b) Goodson, F.E.; Wallow, T.I.; Novak, B.M. *Org. Synth., Coll. Vol. X* **2004**, 501. (c) Ruel, F.S.; Braun, M.P.; Johnson, C.R. *Org. Synth., Coll. Vol. X* **2004**, 467.
5. Beckmann, J.; Dakternieks, D.; Duthie, A.; Lim, A.E.K.; Tiekink, E.R.T. *J. Organomet. Chem.* **2001**, *633*, 149 and references cited therein.
6. Krizan, T.D.; Martin, J.C. *J. Am. Chem. Soc.* **1983**, *105*, 6155.
7. (a) Gschwend, H.W.; Rodriguez, H.R. *Org. React.* **1979**, *26*, 1; (b) Snieckus, V. *Chem. Rev.* **1990**, *90*, 879.
8. Caron, S.; Hawkins, J.M. *J. Org. Chem.* **1998**, *63*, 2054.
9. For a related approach to the isolation of arylboronic acids as the corresponding esters, see: Wong, K.-T.; Chien, Y.-Y.; Liao, Y.-L.; Lin, C.-C.; Chou, M.-Y.; Leung, M. *J. Org. Chem.* **2002**, *67*, 1041.
10. Lysén, M.; Madden, M.; Kristensen, J. K. VedsØ, P.; ZØllner, K.; Begtrup, M. *Synthesis* **2006**, *70*, 3478.
11. Hansen, H. M.; Lysén, M.; Begtrup, M.; Kristensen, J. L. *Tetrahedron* **2005**, 9955.

PREPARATION OF (*S,S*)-1,2-BIS(*tert*-BUTYLMETHYLPHOSPHINO)ETHANE ((*S,S*)-*t*-Bu-BISP*) AS A RHODIUM COMPLEX

[Rhodium(1+), [(2,3,5,6-η)-bicyclo[2.2.1]hepta-2,5-diene] (S,S)-1,2-bis(*M*, -butymethylphosphino)ethane, trafluoroborate(1-)]

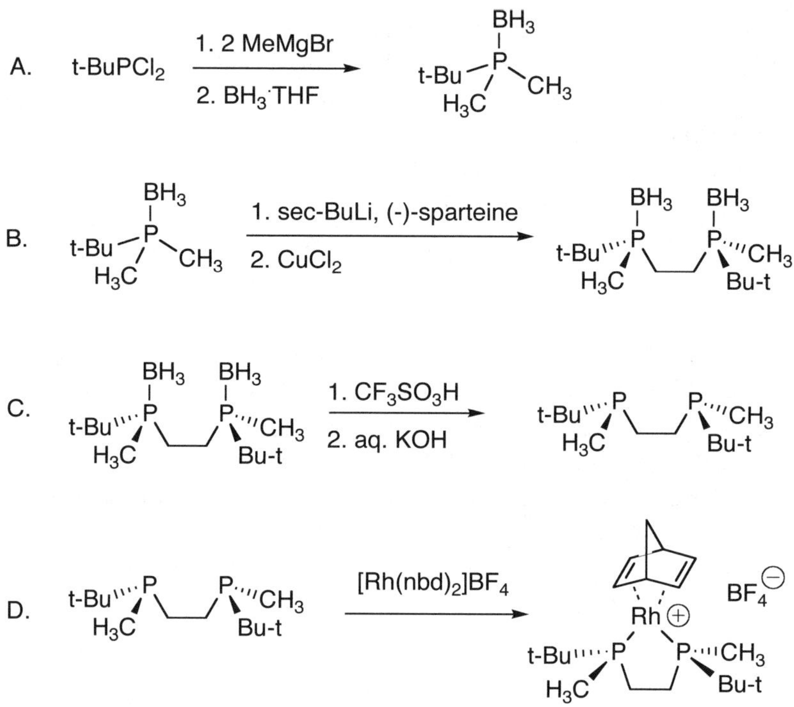

Submitted by Karen V. L. Crépy and Tsuneo Imamoto.[1]
Checked and substantially modified by Günter Seidel and Alois Fürstner.

1. Procedure

Caution! All reactions must be carried out in a well-ventilated hood.

A. tert-Butyl(dimethyl)phosphine–Borane. A 500-mL, three-necked flask equipped with a large football-shaped magnetic stirring bar, a thermometer, a three-way tap connected to an argon line, and a 250-mL pressure-equalizing dropping funnel fitted with a glass stopper is flame dried under vacuum and purged with argon (Note 1). The flask is charged with *tert*-butyldichlorophosphine (12.18 g, 77 mmol) (Note 2) and THF (120 mL) (Note 3) under argon and the resulting solution is cooled to −10 °C using a cryostat. A solution of methylmagnesium bromide (3.0 M in Et_2O, 57 mL, 171 mmol) (Note 4) diluted with THF (100 mL) is added dropwise over 0.5 h at such a rate as to maintain the internal temperature below 0 °C, and the resulting heterogeneous mixture is stirred for 5 h at room temperature. A solution of borane–THF complex (1.0 M in THF, 93 mL, 93 mmol) (Note 5) is then added dropwise over 20 min at −10 °C and stirring is continued at room temperature overnight (18 h). The mixture is cautiously poured into ice/water (120 mL) containing concentrated HCl (30 mL) (Note 6) and the resulting solution is vigorously stirred for 15 min. The layers are separated, the aqueous phase is extracted with ethyl acetate (2 x 100 mL), the combined organic layers are washed with brine (150 mL), dried over Na_2SO_4, filtered and evaporated under reduced pressure. The residue is recrystallized from hot hexane (40 mL) and dried under vacuum to give the *tert*-butyl(dimethyl)phosphine–borane adduct as a white solid (6.1 g, 61 %) (Notes 7,8).

B. (S,S)-1,2-Bis(boranato(tert-butyl)methylphosphino)ethane. A 500-mL, three-necked flask equipped with a football-shaped magnetic stirring bar, a thermometer, a three-way tap connected to an argon line, and a 100-mL pressure-equilizing dropping funnel fitted with a glass stopper is flame-dried under vacuum and filled with argon. After cooling to ambient temperature, the flask is charged with (−)-sparteine (8.18 g, 35 mmol) and diethyl ether (40 mL) (Note 3) and the resulting solution is cooled to −78 °C using a cryostat. *sec*-Butyllithium (1.3 M in cyclohexane, 26.8 mL, 35 mmol) is added dropwise at that temperature over 5 min and stirring is continued for 30 min before a solution of the *tert*-butyl(dimethyl)phosphine–borane adduct (4.21 g, 31.9 mmol) in diethyl ether (40 mL) is added dropwise over 30 min. The mixture is stirred at −78 °C for 1 h and −50 °C for 4 h. Copper(II) chloride (5.6 g, 41 mmol) (Note 9) is added with vigorous stirring and the resulting mixture is gradually warmed to room temperature overnight. Aqueous ammonia (25-28%, 40 mL) is added, the layers are separated, the bright blue aqueous phase is extracted with ethyl acetate (2 x 50 mL), the combined organic layers are successively washed with water (30 mL), aqueous HCl (3 M, 2 x 30 mL), water (30 mL), and brine (30 mL) before being dried over Na_2SO_4 and evaporated under reduced pressure. The residue is recrystallized twice from toluene at 80 °C (20 mL each) (Note 10) to afford diastereomerically pure (*S,S*)-1,2-bis(boranato(*tert*-butyl)methylphosphino)ethane (2.0 g, 48 %) as white needles (Note 11). The enantiomeric purity is checked by HPLC analysis using a chiral column (Note 12).

C. (S,S)-1,2-Bis-((tert-butyl)methylphosphino)ethane ((S,S)-t-Bu-BisP)).* A 100-mL, two-necked flask equipped with a magnetic stirring bar, a three-way tap connected to an argon line, and a glass stopper is flame-dried under vacuum and filled with argon after reaching ambient temperature. The flask is charged with (*S,S*)-1,2-bis(boranato(*tert*-butyl)methylphosphino)ethane (1.14 g, 4.4 mmol) and toluene (35 mL) (Note 3) and the resulting mixture is cooled to 0 °C. Trifluoromethanesulfonic acid (1.92 mL, 22 mmol) is added dropwise and the mixture is stirred for 15 min at 0 °C and for 45 min at ambient temperature. The solvent is evaporated under reduced pressure before a solution of potassium hydroxide (2.44 g, 43 mmol) in freshly degassed ethanol/water (9:1 ratio, 15 mL) (Note 13) is added, and the resulting mixture is stirred at 50 °C for 2 h under argon. After reaching ambient temperature, the mixture is extracted four times with diethyl ether (20 mL each) while keeping the positive pressure of argon. The combined organic phases are dried over Na_2SO_4 and passed through a short column of basic alumina (80 g) under argon (Note 14). The column is carefully rinsed with degassed diethyl ether (total volume 180 mL) and the combined eluents are evaporated to give a colorless pasty oil which solidifies on cooling to 0 °C (980 mg, 95%). The free diphosphine is immediately subjected to complexation with a transition metal precursor.

D. [Rh((S,S)-t-Bu-BisP))(nbd)]BF$_4$.* A 250-mL, two-necked flask equipped with a magnetic stirring bar, a glass stopper, and a three-way tap connected to an argon line is flame-dried under vacuum and filled with argon. After reaching ambient temperature, the flask is charged with $[Rh(nbd)_2]BF_4$ (1.17 g, 3.1 mmol) (Note 15) and THF (100 mL). A solution of (*S,S*)-1,2-bis-((*tert*-butyl)methylphosphino)ethane (0.77 g, 3.3 mmol) in THF (40 mL) is added via syringe to this suspension and stirring is continued for 2 h leading to the formation of an almost clear solution. This mixture is filtered under argon and the filtrate is evaporated. The orange powder is dissolved in hot THF until a clear solution is formed. This solution is slowly cooled to 0 °C, leading to the precipitation of red cubes that are removed through filtration under argon. The filtrate is evaporated and the residue is recrystallized again by the same procedure. The combined crops of five recrystallizations are dried under vacuum to give the rhodium complex in analytically pure form (687 mg, 43 %) (Notes 16, 17).

2. Notes

1. Argon (> 99.999%) was used by the checkers.

2. *tert*-Butyldichlorophosphine was purchased from Aldrich and used as received. *tert*-Butyldichlorophosphine is moderately air-sensitive and must be handled under argon.

3. THF, Et_2O, and toluene were freshly distilled from sodium/ benzophenone at atmospheric pressure under argon immediately prior to use.

4. The submitters used methylmagnesium bromide in THF (1.0 M in THF) purchased from the Tokyo Kasei Company. The checkers purchased methylmagnesium bromide (3 M in Et_2O) from Aldrich. It was transferred from the commercial bottle to the dropping funnel under argon *via* cannula.

5. Borane-THF complex was purchased from Kanto Chemical Company. The checkers purchased this reagent from Aldrich. Borane-THF is air-sensitive and must be handled under argon.

6. *CAUTION*: Excess methylmagnesium bromide and borane–THF complex react violently with water while liberating a large amount of gas. Slow addition under stirring is necessary. The use of a large Erlenmeyer flask is recommended.

7. The submitters reported a yield of 77-81% (19.2-20.4 g scale). The *tert*-butyl(dimethyl)phosphine–borane adduct prepared by this procedure shows the following physical and spectroscopic data: mp 164-165 °C (hexane); R_f 0.44 (5:1 hexane/ethyl acetate); ^{1}H NMR (400 MHz, $CDCl_3$): δ 1.23 [d, 6H, ^2J(P,H) = 9.9 Hz], 1.16 [d, 9H, ^3J (P,H) = 14 Hz], 0.44 [dq, 3H, *J*(B,H) = 95 Hz, ^2J (P,H) = 15 Hz]; ^{13}C NMR (100 MHz, $CDCl_3$): δ 26.6 [d, *J*(P,C) = 35 Hz], 24.7 [d, ^2J (P,C) = 2.2 Hz], 7.3 [d, *J*(P,C) = 35.6 Hz]; ^{11}B NMR (96 MHz, $CDCl_3$): δ –39.9 [dq, *J*(B,H) = 95 Hz, *J*(P,B) = 63 Hz]; ^{31}P NMR (121 MHz, $CDCl_3$): δ 20.9 [q, *J*(P,B) = 62 Hz]. IR (KBr): 2970, 2370, 1070, 945, 920 cm^{-1}.

8. The malodorous smell of the reaction/work-up glassware is removed by immersing all vessels for 1 h in a mixture of domestic bleach and water (ca 1/10 ratio).

9. Finely powdered copper(II) chloride was dried for 2 h at 130–140 °C under vacuum prior to use.

10. The submitters report that 4-5 recrystallizations were necessary to obtain the desired (*S,S*)-t-Bu-BisP*–borane complex in pure form. It is important that the solid is carefully dried before proceeding to the next recrystallization.

11. The submitters reported yields of 60% (2.15 g scale) and 51-55% (5.3-5.6 g scale).

12. (*S,S*)-1,2-bis(boranato(*tert*-butyl)methylphosphino)ethane shows the following analytical and spectroscopic data: mp = 186-187.5 °C; the submitters report a mp = 182-184 °C. R_f 0.22 (5:1 hexane/ethyl acetate); $[\alpha]^{28}_D$ −9.1 (c 1.21, chloroform); ee = 99.1% (250 mm Chiralcel OD-H, 0.5 mL/min; n-heptane/i-propanol = 9:1; RI detection; t = 15.3 min); IR (KBr): 2960, 2380, 2350, 1185, 1065, 765 cm^{-1}; ^1H NMR (400 MHz, $CDCl_3$): δ 1.99 (m, 2H), 1.60 (m, 2H), 1.21 [d, 6H, ^2J(P,H) = 10 Hz], 1.17 [d, 18 H, ^3J(P,H) = 14 Hz], 0.38 [q, 6H, J(B,H) = 95 Hz]; ^{11}B NMR (128 MHz, $CDCl_3$): δ −41.1 [dq, *J*(P,B) = 63 Hz, *J*(B,H) = 94 Hz]; ^{13}C NMR (100 MHz, $CDCl_3$): δ 27.7 [d, *J*(P,C) = 34 Hz], 25.1, 15.9 [d, *J*(P,C) = 30 Hz], 5.6 [d, *J*(P,C) = 34 Hz]; ^{31}P-NMR (162 MHz, $CDCl_3$): δ 30.1. MS (EI) *m/z* (rel. intensity): 262 ([M^+], 4), 261 (25), 260 (15), 259 (27), 258 (15), 257 (20), 256 (9), 247 (100), 203 (24), 191 (21), 189 (55), 133 (17), 108 (5), 93 (6), 75 (5), 57 (42), 41 (41), 29 (17); HR-MS (ESIpos, [M^+ + NH_4]): calcd. 280.26656, found 280.26674.

13. The checkers prepared the degassed solvents by two freeze/thaw cycles, whereas the submitters recommended the following procedure: The solvent is introduced in a two-necked flask fitted with a septum and a three-way tap connected to an argon balloon. It is placed in an ultra-sound bath and vacuum/argon cycles are applied three times.

14. Free phosphines are usually air-sensitive compounds and must be handled under argon. Therefore, all flasks must be evacuated and purged with argon three times before introducing any free phoshines. The alumina column must also be the subject of the same treatment (a thin pressure-equalizing dropping funnel is best used).

15. The checkers prepared the complex $[Rh(nbd)_2]BF_4$ in the following way: $AgBF_4$ (931 mg, 4.8 mmol) is added to a solution of $[Rh(nbd)Cl]_2$ (1.001 g, 2.17 mmol) and norbornadiene (1.8 mL, 16 mmol) in CH_2Cl_2 (30 mL) and the resulting mixture is stirred at ambient temperature for 4 h under argon. The precipitate is filtered through a pad of Celite, the filtrate is evaporated, and the residue is triturated with Et_2O (50 mL). The solid material formed is filtered, rinsed with Et_2O (5 mL) and dried in vacuum to give $[Rh(nbd)_2]BF_4$ as a dark red, air sensitive, crystalline solid. Anal. calcd. for $C_{14}H_{16}RhBF_4$ (373.99): C 44.96, H 4.31; found: C 45.09, H 4.35.

16. The submitters report a yield of 62% (2.24 g scale). The rhodium complex shows the following spectroscopic properties: ^1H NMR (400 MHz,

$CDCl_3$): δ 5.73 (m, 4H), 4.15 (m, 2H), 2.04 (m, 1H), 1.93 (m, 1H), 1.83 (m, 2H), 1.54 (m, 2H), 1.39 [d, 6H, X-part of an $AA'X_3X'_3$ spin system (A = ^{31}P, X = ^{1}H, ^{2}J(P,H) = 8.3 Hz], 1.10 [d, 18H, X-part of an $AA'X_9X'_9$ spin system, ^{3}J(P,H) = 14.5 Hz]; ^{13}C NMR (75 MHz, $CDCl_3$, X-parts of an ABMX spin system (A, B = ^{31}P, M = ^{103}Rh, X = ^{13}C): δ 89.5 {*J*(Rh,C) = 6.7 Hz, [$J(P_A,C)$ + $J(P_B,C)$] = 9.2 Hz}, 85.0 {*J*(Rh,C) = 6.2 Hz, [$J(P_A,C)$ + $J(P_B,C)$] = 8.7 Hz}, 72.3 {*J*(Rh,C) = 3.9 Hz, [$J(P_A,C$ + $J(P_B,C)$] = 5.7 Hz}, 56.7 {*J*(Rh,C) = 1.8 Hz, [$J(P_A,C)$ + $J(P_B,C)$] = 3.4 Hz}, 32.3 [*J*(Rh,C) = 1.6 Hz, $J(P_A,P_B)$ = 21.4 Hz, $J(P_A,C)$ = 25.7 Hz, $J(P_B,C)$ = 0.8 Hz], 26.4 [$J(P_A,C)$ + $J(P_B,C)$] = 4.2 Hz, 21.5 [*J*(Rh,C) = 2.8 Hz, $J(P_A,P_B)$ = 21.4 Hz, $J(P_A,C)$ = 27.8 Hz, $J(P_B,C)$ = 10.9 Hz], 6.0 [*J*(Rh,C) = 1.3 Hz, $J(P_A,P_B)$ = 21.5 Hz, $J(P_A,C)$ = 20.8 Hz, $J(P_B,C)$ = -0.6 Hz]; ^{11}B NMR (128 MHz, $CDCl_3$): δ 0.0; ^{31}P NMR (121 MHz, $CDCl_3$): δ 62.0 [d, *J*(Rh,P) = 152 Hz]. The structure was determined by X-ray crystallographic analysis.[2]

17. The rhodium complex is not readily oxidized on contact with air at room temperature, but it gradually decomposes on prolonged exposure to air. Therefore, storage under argon in a freezer is recommended. The checkers, however, found the complex to be unstable when kept in CD_3OD solution.

Waste Disposal Information

All hazardous materials should be handled and disposed of in accordance with "Prudent Practices in the Laboratory"; National Academy Press; Washington, DC, 1995.

3. Discussion

In contrast to optically active diphosphine ligands bearing a chiral-backbone, C_2-symmetric, P-stereogenic diphosphine ligands, such as (*S,S*)-1,2-bis(alkylmethyl- phosphino)ethane (also known as (*S,S*)-BisP*),[2] have been used only recently as ligands in enantioselective reactions. BisP* was found to be an excellent ligand for enantioselective hydrogenations of olefins or ketones. Its synthesis consists of only three steps starting from *tert*-butyldichlorophosphine. The asymmetric P-center is formed during the deprotonation step using *sec*-BuLi in the presence of (–)-sparteine.[3] Given the fact that the copper-promoted oxidative coupling[4] of the generated anion proceeds through a radical intermediate, the reaction leading to the (*S,S*) or (*R,R*)-isomers has almost the same rate as that leading to the *meso* compound.[5] Thus, (*S,S*)-BisP*–borane (>99% ee) is contaminated by a small amount of (*R,S*)-BisP*–borane which can be removed by recrystallization.

The (*R,R*)-enantionmer ((*R,R*)-t-Bu-BisP*)) may also be prepared via one of three protocols that we have developed.[6] The recommended method[6c] is reported in the scheme below:

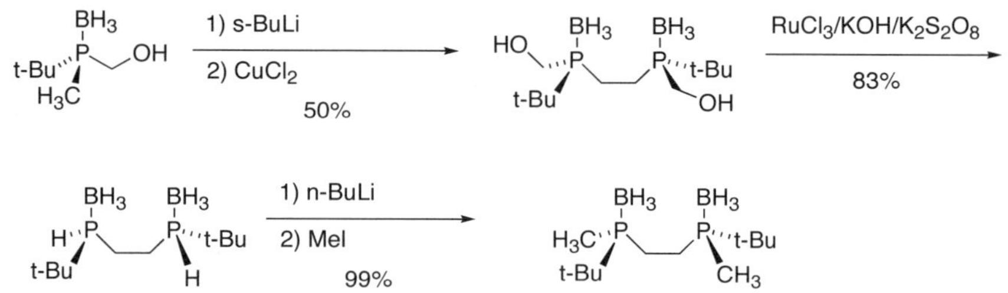

(R,R)-t-Bu-BisP*-borane
98% ee (99% ee after recrystallization)

It should be noted that tricoordinate phosphorus compounds in low oxidation states are usually air-sensitive, making their handling and storage difficult. Moreover, chiral phosphines bearing stereogenic phosphorus atoms are prone to racemization especially at higher temperatures[7] and therefore require additional stabilization. Unlike diaryl- and triarylphosphines, however, optically active trialkylphosphines hardly racemize even at elevated temperatures.[8] Temporary protection of the phosphine with BH_3 usually prevents all problems of this kind.[9] The phosphine–borane adducts are air-stable compounds which can be conveniently isolated, purified and stored. Owing to their remarkable inertness and resistance to a wide range of reactions conditions phosphine-borane complexes have emerged as indispensable intermediates for the preparation of P-stereogenic compounds.[10] Cleavage of the P–B bond releases the chiral phosphines with retention of configuration.[5] Complexation with a transition-metal prevents ready oxidation of the ligand.

1. Department of Chemistry, Faculty of Science, Chiba University, Yayoi-cho, Inage-ku, Chiba 263-8522, Japan.
2. (a) Imamoto, T.; Watanabe, J.; Wada, Y.; Masuda, H.; Yamada, H.; Tsuruta, H.; Matsukawa, S.; Yamaguchi, K. *J. Am. Chem. Soc.* **1998,** *120,* 1635; (b) Gridnev, I. D.; Yamanoi, Y.; Higashi, N.; Tsuruta, H.; Yasutake, M.; Imamoto, T. *Adv. Synth. Catal.* **2001**, *343*, 118.

3. Muci, A. R.; Campos, K. R.; Evans, D. A. *J. Am. Chem. Soc.* **1995**, *117*, 9075.

4. Maryanoff, C. A.; Maryanoff, B. E.; Tang, R.; Mislow, K. *J. Am. Chem. Soc.* **1973**, *95*, 5839.

5. Imamoto, T.; Oshiki, T.; Onozawa, T.; Kusumoto, T.; Sato, K. *J. Am. Chem. Soc.* **1990**, *112*, 5244.

6. (a) Miura, T.; Yamada, H.; Kikuchi, S.; Imamoto, T. *J. Org. Chem.* **2000**, *65*, 1877; (b) Imamoto, T.; Kikuchi, S.; Miura, T.; Wada, Y. *Org. Lett.* **2001**, *3*, 87; (c) Crépy, K. V. L.; Imamoto, T. *Tetrahedron Lett.* **2002**, *43*, 7735.

7. (a) Pietrusiewicz, K. M.; Zablocka, M. *Chem. Rev.* **1994**, *94*, 1375; (b) Valentine, D., Jr. In *Asymmetric Synthesis*; Morrison, J. D., Scott, J. W., Eds; Academic Press, Orlando, 1984; Vol. 4; Chapter 3.

8. Baechler, R. D.; Mislow, K. *J. Am. Chem. Soc.* **1970**, *92*, 3090.

9. (a) Ohff, M.; Holz, J.; Quirmbach, M.; Börner, A. *Synthesis* **1998**, 1391; (b) Carboni, B.; Monnier, L. *Tetrahedron* **1999**, *55*, 1197.

10. Imamoto, T. *Pure Appl. Chem.* **1993**, *65*, 655.

CHIRAL LITHIUM AMIDE BASE DESYMMETRIZATION OF A RING FUSED IMIDE: FORMATION OF (3a*S*,7a*S*)-2-[2-(3,4-DIMETHOXYPHENYL)-ETHYL]-1,3-DIOXO-OCTAHYDRO-ISOINDOLE-3a-CARBOXYLIC ACID METHYL ESTER

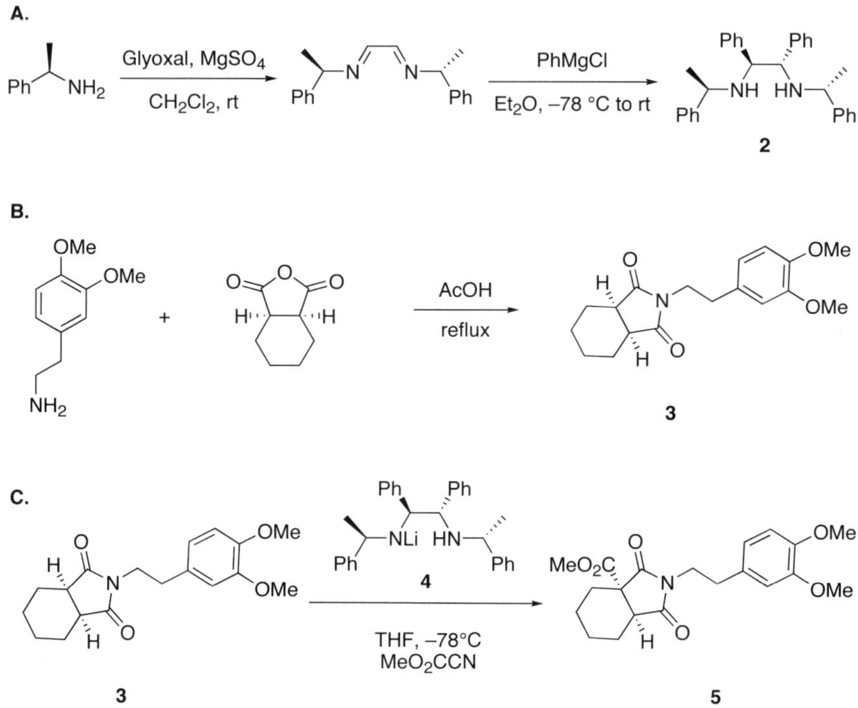

Submitted by Vincent Rodeschini, Nigel S. Simpkins and Fengzhi Zhang.[1]
Checked by Melissa A. Beenen and Jonathan A. Ellman.

1. Procedure

A. 1(S),2(S)-Diphenyl-N,N'-bis-[1(R)-phenyl-ethyl]-ethane-1,2-diamine (2). In a 500-mL, three-necked, round-bottomed flask equipped with a N_2 inlet, two septa and a magnetic stirring bar is introduced successively CH_2Cl_2 (180 mL) (Note 1) and an aqueous glyoxal solution (9.40 mL, 82.5 mmol, 1.0 equiv) (Note 2) by syringe through one of the septa. One septum is temporarily removed and solid magnesium sulfate (40.0 g) (Note 3) is then added portion-wise over 20 min, while the mixture is stirred at room temperature. The resulting suspension is stirred for an additional 10 min, and then (*R*)-(+)-phenylethylamine (21.0 mL, 20.0 g, 165.0 mmol, 2.0 equiv) (Note 4) is introduced dropwise *via* syringe over a 5 min period. The resulting mixture is stirred overnight at room temperature. Magnesium sulfate is removed by filtration and rinsed with CH_2Cl_2 (2 × 25 mL). The filtrate is then concentrated by rotary evaporation and then further dried under high vacuum for 1 h (Note 5) to provide crude bis-imine (**1**) (Note 6) as an orange oil (21.3 g, 98.0%), which is used directly for the next step without further purification. Into a 1-L, three-necked, round-bottomed flask (Note 7) equipped with one low temperature thermometer, one N_2 inlet and one septum is introduced crude bis-imine **1** (21.3 g) as a solution in Et_2O (300 mL) by syringe (Note 8). This solution is cooled to –78 °C (dry ice-acetone bath), and PhMgCl (162 mL, 2.0 M in THF, 325 mmol, 4.0 equiv) (Note 9) is added drop-wise *via* syringe pump (Note 10) while maintaining the internal temperature between –78 °C and –75 °C over 2 h (Note 11). The resulting dark brown mixture is allowed to warm to room temperature gradually over 4 h, and stirred for an additional 2 h at room temperature. The mixture is then cooled to 4 °C (ice/water bath), and carefully quenched by the addition of a saturated aqueous NH_4Cl solution (200 mL) after removal of the septum over 30 min. The solid that is formed is dissolved by the addition of de-ionized water (100 mL), and the two phases are separated in a 1-L separatory funnel. The aqueous phase is then further extracted with ethyl acetate (3 × 150 mL). The combined organic phases are washed with brine (50.0 mL), then dried over $MgSO_4$ (30 g, 5 min) and filtered. The filtrate is concentrated by rotary evaporation and further dried under high vacuum for 1 h. The crude product (33.5 g) is purified by flash column chromatography. Thus, the crude brown solid is dissolved in the minimum amount of CH_2Cl_2 (approximately 10 mL), charged on a column (8 × 16 cm) containing 400 g of silica gel (Note 12) and eluted with Et_2O-petroleum ether (5:95, 2 L, then 10:90, 2 L). After the first apolar impurity (this spot is only evident by UV), the title compound elutes, before a mixture of other isomers (Note 13). The combined fractions of the major isomer are concentrated by rotary evaporation to provide a slightly yellow solid (16.8 g). This solid is dissolved in boiling petroleum ether (250 mL), with addition of a small amount of CH_2Cl_2 (5 mL) to complete the dissolution. The solvent is allowed to evaporate to 1/5 of the initial volume at room temperature over 2 days, allowing the crystallisation to occur (Note 14). These crystals are collected by filtration, and rinsed with a small amount (15.0 mL) of cold (0 °C) petroleum ether. Thus, 13.8–14.7 g (40–43%) of pure diamine **2** are obtained (Note 15).

B. *(3aS,7aR)-2-[2-(3,4-Dimethoxyphenyl)-ethyl]-hexahydro-isoindole-1,3-dione (**3**).* Into a 250-mL, one-necked, round-bottomed flask, equipped with a septum and a magnetic stirring bar, containing cyclohexanedicarboxylic anhydride (8.50 g, 55.1 mmol, 1.0 equiv) (Note 16) is added glacial acetic acid (100 mL), followed by 2-(3,4-dimethoxyphenyl)ethylamine (10.1 g, 55.4 mmol, 1.0 equiv) (Note 17). The septum is removed and the flask is equipped with a Liebig condenser. The reaction is then heated at reflux (oil bath temperature 120 °C) for 13 h. The resulting mixture is allowed to cool to room temperature and poured into a 1-L beaker containing 400 mL of de-ionized water. The aqueous phase is extracted with diethyl ether (5 × 100 mL). The combined organic phases are transferred into a 1-L Erlenmeyer flask, cooled to 4 °C (water/ice bath) and neutralized by the slow addition (30 min) of a saturated aqueous Na_2CO_3 solution (150 mL) under vigorous stirring. After the addition is finished, the two phases are separated and the organic phase is further washed with saturated aqueous Na_2CO_3 (2 × 50 mL). The organic phase is then dried over anhydrous $MgSO_4$ (20.0 g, 5 min). After removal of the solid by filtration, the organic phase is concentrated by rotary evaporation and further dried under high vacuum for 1 h. The resulting yellow solid is recrystallized from boiling light petroleum ether/ethyl acetate (100 mL:25 mL). Crystallization is allowed to occur at room temperature over 2 h, then at 4 °C overnight. The crystals are collected by filtration and washed with cold (0 °C) petroleum ether (25 mL) to give 14.2 g (81%) (Note 18) of the title compound **3** (Note 19).

C. *(3aS,7aS)-2-[2-(3,4-Dimethoxyphenyl)-ethyl]-1,3-dioxo-octahydro-isoindole-3a-carboxylic acid methyl ester (**5**).* Into a 100-mL, two-necked, round-bottomed flask (Note 7), equipped with one septum and one low temperature thermometer, containing a solution of diamine **2** (8.79 g, 20.9 mmol, 1.1 equiv) in THF (50 mL) (Note 20) cooled to –78 °C (dry ice-acetone bath) is introduced *n*-BuLi (8.50 mL, 2.5 M in hexanes, 21.3 mmol, 1.1 equiv) (Note 21) dropwise *via* syringe over 25 min maintaining the internal temperature between –78 °C and –75 °C. The resulting pink solution is allowed to warm to room temperature (23 °C) by removing the cooling bath, and then stirred at this temperature for 30 min (Note 22). This solution is then cooled to –78 °C and transferred over 2 h *via* cannula into a 250-mL two-necked, round-bottomed flask equipped with one septum and one low temperature thermometer containing a solution of imide **3** (6.00 g, 18.9 mmol) in THF (80 mL) while maintaining the internal temperature at –78 °C. The resulting mixture is then stirred for 1 h at –78 °C. A solution of methyl cyanoformate (3.00 mL, 37.8 mmol, 2.0 equiv) (Note 23) in THF (10 mL) is then added dropwise *via* cannula over 15 min. The resulting yellow solution is stirred for 1 h at –78 °C, then the cooling bath is removed, and saturated aqueous $NaHCO_3$ (80.0 mL) is added slowly, followed by de-ionized water (20 mL). The phases are separated, and the aqueous phase is extracted with ethyl acetate (3 × 50 mL). The combined organic phases are washed with brine (50 mL), then dried over $MgSO_4$ (20.0 g, 5 min). After removal of the solid by filtration, the filtrate is concentrated by rotary evaporation. The crude solid residue is dissolved in the minimum amount of CH_2Cl_2 (approximately 15 mL), charged on a column (8 × 16 cm) containing 400 g of silica gel (Note 12) and eluted with ethyl acetate-petroleum ether (30:70, 1 L) to afford recovered diamine **2** (8.40 g, 96%) (Note 24), and then with AcOEt-petroleum ether (50:50, 2 L) to afford title compound **5** (6.20 g, 87 %) (Note 25).

2. Notes

1. The checkers used HPLC grade CH_2Cl_2 purchased from Fisher and passed through two columns of neutral alumina. The submitters used CH_2Cl_2 freshly distilled over CaH_2 under an Ar atmosphere.

2. Glyoxal (40% aqueous solution, ~ 8.8 M) was purchased from Fluka Chemical Company.

3. $MgSO_4$ was purchased from Fisher Company and dried in an oven at 120 °C for 24 h.

4. (*R*)-(+)-Phenylethylamine was purchased from Lancaster Chemical Company {99%, 99% e.e. (HPLC)} and used as received.

5. Throughout this procedure, rotary evaporation refers to a vacuum of 20 mmHg, and high vacuum refers to a vacuum of 1 mmHg.

6. Crude bis-imine (**1**) (R_f = 0.5; petroleum ether:Et_2O 1:1, visualization with $KMnO_4$)

7. The apparatus was dried in an oven (120 °C) overnight and maintained under an atmosphere of dry N_2 during the course of the reaction.

8. The checkers used ACS grade Et_2O stabilized with BHT purchased from Fisher and passed through two columns of neutral alumina. The submitters used anhydrous Et_2O purchased from Fisher Chemicals (water < 0.03%) and purified by pressure filtration under N_2 through activated alumina.

9. Phenylmagnesium chloride was obtained from Aldrich Chemical Company, Inc.

10. The submitters performed this addition via cannula.

11. A white precipitate appeared over the course of the addition.

12. The checkers used silica gel 60A (32- 63D) purchased from Bodman Industries. The submitters used silica gel 60A (35-70μ) purchased from Fluorochem.

13. The ^{1}H NMR of the crude reaction material indicated a ratio of the title compound (1*S*, 2*S*)-**2** to the minor isomer (1*R*, 2*R*)-**2** of 8:2 (R_f (major) = 0.4, (minor) = 0.2, petroleum ether:Et_2O 4:1, visualization with $KMnO_4$). The minor isomer was identified by a characteristic signal at 3.88 ppm in the ^{1}H NMR,[2] and was isolated containing traces of another isomer (3.83 g, brown oil).

14. Alternatively, the solution was simply cooled to room temperature, and then left at 4 °C overnight allowing the crystallisation to occur. After collection of a first batch of crystals, the mother liquor was evaporated to 1/5 of its volume to provide a second batch of crystals.

15. Large colorless crystals. mp = 113–115 °C; $[\alpha]_D^{25}$ +190.0 (*c* = 1.1, $CHCl_3$); FTIR ($CHCl_3$): 3322, 2923, 2859, 1602, 1492, 1454, 1363, 1106, 921, 862 cm^{-1}; ^{1}H NMR (500 MHz, $CDCl_3$) δ: 1.25 (d, *J* = 6.6 Hz, 6 H), 2.25 (br s, 2 H), 3.37 (s, 2 H), 3.43 (q, *J* = 6.4 Hz, 2 H), 6.91–7.24 (m, 20 H); ^{13}C NMR (125 MHz, $CDCl_3$) δ: 25.5, 55.1, 65.9, 126.7 (**2**), 126.8, 128.0, 128.1, 128.5, 141.8, 145.7; HRMS (ES+) *m/z* calcd for $C_{30}H_{33}N_2$ 421.2638, found 421.2644; Found: C, 85.46; H, 7.77; N, 6.64. $C_{30}H_{32}N_2$ requires C, 85.67; H, 7.67; N, 6.66%;

16. Cyclohexanedicarboxylic anhydride (95%) was purchased from Aldrich Chemical Company, Inc.

17. 2-(3,4-Dimethoxyphenyl)ethylamine (98%) was purchased from Alfa Aesar Company.

18. The checkers obtained an 82 % yield when the reaction was run on half-scale.

19. Small white crystals; R_f= 0.4 (petroleum ether: CH_2Cl_2:AcOEt 2:2:1, visualisation with $KMnO_4$); mp = 86.7–88.1 °C; FTIR ($CHCl_3$) 3004, 2922, 2860, 1693, 1514, 1399, 1233, 1150, 1025, 807 cm^{-1}. ^{1}H NMR (400 MHz, $CDCl_3$) δ: 1.28 (m, 2 H), 1.37 (m, 2 H), 1.55 (m, 2 H), 1.74 (m, 2 H), 2.72 (m, 2 H), 2.84 (app. t, *J* = 7.7 Hz, 2 H), 3.70 (app. t, *J* = 7.4 Hz, 2 H), 3.81 (s, 3 H), 3.84 (s, 3 H), 6.71–6.76 (m, 3 H); ^{13}C NMR (100 MHz, $CDCl_3$) δ: 21.6, 23.6, 32.9, 39.2, 39.6, 55.8, 55.9, 111.2, 112.0, 121.0, 130.3, 147.7, 148.8, 179.7; MS (EI) *m/z* 317 (M^+, 28%), 164 ($C_{10}H_{12}O_2$, 28%), 151 ($C_9H_{11}O_2$, 28%). HRMS: found 317.1634, calcd for $C_{18}H_{23}NO_4$ 317.1627.

Found: C, 68.22; H, 7.56; N, 4.40. $C_{18}H_{23}NO_4$ requires C, 68.10; H, 7.31; N, 4.42%.

20. The checkers used HPLC grade THF purchased from Fisher Chemicals and passed though two columns of neutral alumina. The submitters used anhydrous THF purchased from Fisher Chemicals (water < 0.03%) and purified by pressure filtration under N_2 through activated alumina.

21. *n*-BuLi (2.5 M in hexanes) was purchased from Aldrich Chemical Company, Inc.

22. The initial color of this solution varied from orange to pink, and after being stirred at room temperature, from pink to purple.

23. Methyl cyanoformate (99%) purchased from Aldrich Chemical Company, Inc

24. The submitters isolated **2** as small white crystals, $[\alpha]_D^{25}$ +204 (c = 1.05, $CHCl_3$).

25. The checkers obtained an 81% yield when the reaction was run on half-scale. $[\alpha]_D^{25}$ –52.9 (c = 1.1, $CHCl_3$). The submitters obtained an optical rotation of $[\alpha]_D^{25}$ –62.0 (c = 1.1, $CHCl_3$). R_f = 0.4 (petroleum ether:CH_2Cl_2:AcOEt 2:2:1, visualization with $KMnO_4$); FTIR ($CHCl_3$) 2940, 1742, 1701, 1515, 1349, 1234, 1026, 805 cm^{-1}; ^1H NMR (400 MHz, $CDCl_3$) δ: 1.10 (m, 1 H), 1.30–1.37 (m, 3 H), 1.48–1.60 (m, 2 H), 1.98 (ddd, J = 14.2, 8.5, 4.2 Hz, 1 H), 2.26 (m, 1 H), 2.86 (app. td, J = 7.7, 2.2 Hz, 2 H), 3.19 (dd, J = 6.5, 3.7 Hz, 1 H), 3.72 (s, 3 H), 3.73 (m, 2 H), 3.80 (s, 3 H), 3.83 (s, 3 H), 6.69–6.75 (m, 3 H); ^{13}C NMR (125 MHz, $CDCl_3$) δ: 20.5, 20.8, 21.4, 28.5, 32.7, 39.8, 43.8, 53.3, 54.1, 55.9, 56.0, 111.1, 112.0, 121.1, 129.9, 147.8, 148.9, 170.2, 176.1, 177.6; MS (FAB) m/z 375 (M^+, 100%), 164 ($C_{10}H_{12}O_2$, 90%), 151 ($C_9H_{11}O_2$, 35%); HRMS: found 375.1683, calcd for $C_{20}H_{25}NO_6$ 375.1682. Found: C, 63.69; H, 6.75; N, 3.65. $C_{20}H_{25}NO_6$ requires C, 63.99; H, 6.71; N, 3.73. The ee was determined as 86–88% by HPLC (OD column, EtOH:hexanes 10:90, 0.6 mL/min), the retention times were 22.5 min (major) and 27 min (minor). The submitters determined the ee was 93–95% by HPLC (OD column, EtOH:hexanes 10:90, 0.6 mL/min), the retention times were 35 min (major) and 47 min (minor).

Waste Disposal Information

All toxic materials were disposed of in accordance with "Destruction of Hazardous Chemicals in the Laboratory"; Lunn, G.; Sansone, E.B. 2nd Ed.; John Wiley & Sons, Inc.

3. Discussion

Over the past two decades, the use of chiral lithium amide bases in synthesis has become widespread, enabling the synthesis of chiral building blocks with high selectivity.[3] The three main classes of reactions where chiral lithium amide bases have been applied successfully are: (i) deprotonation of prochiral cyclic ketones, (ii) rearrangement of epoxides to allylic alcohols, and (iii) aromatic and benzylic functionalization of tricarbonyl (η^6-arene)chromium complexes. In all these examples, the chiral base selects between enantiotopic protons in kinetically controlled deprotonations of achiral or prochiral substrates. Another fundamentally different process also exists, in which the chiral lithium amide base first acts as a strong base to produce a prochiral carbanion, e.g. an enolate. The stereochemical outcome of the subsequent reaction of the anion with an electrophile is then controlled by the complexed chiral secondary amine. More recently, catalytic variants of some of the reactions described above have emerged. In this case, a substoichiometric amount of the chiral base is used in conjunction with a stoichiometric amount of an achiral base.[4]

The procedure described herein illustrates the efficient use of a chiral lithium amide base to mediate desymmetrization of a prochiral cyclic imide. Imide **5** has been used successfully in a synthesis of the proposed structure of the alkaloid Jamtine.[5] Apart from chiral diamine **2**, commercially available bisphenylethylamine in the form of its lithium amide **6** has been used for the desymmetrization of certain imides.[6] However, in many cases chiral base **2** has been shown to provide better selectivity. For example, in the case of imide **9** (Table 1, entry 2), the use of lithium amide **6** proved less selective, providing **9** in only 70% ee. It is also noteworthy that **2** can be used as its mono-lithiated form **4**, or as its bis-lithiated derivative **7**, both species leading to high enantioselectivities. Table 1 gives examples of other types of imides that have been successfully desymmetrized using chiral base **7**.

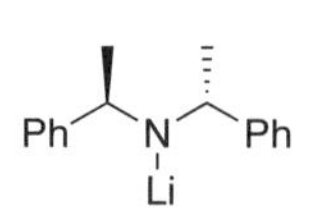

6

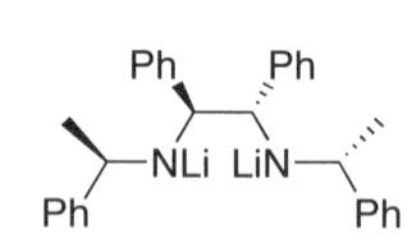

7

Chiral diamine **2** has also been found to be useful for the desymmetrization of prochiral ketones,[7] piperidines,[8] various tricarbonyl(η^6-

arene)chromium complexes,[9] asymmetric rearrangement of episulfoxides,[10] asymmetric thia-Sommelet dearomatization,[11] and asymmetric hydrosilylation[12] with good to excellent levels of enantioselectivities.

The synthesis of diamine **2** suffers from a somewhat low yield; however, it can be easily prepared in large scale.[13] Moreover, it can easily be recovered in essentially pure form by column chromatography after the reaction. Two alternative syntheses of **2** have been reported, either using phenyl lithium instead of a Grignard derivative[2] or *via* pinacol coupling of imines.[14] Neither of these methods, however, provided better yields of the product.

In summary, the chiral base desymmetrization of imides provides an efficient method to access highly functionalised molecules with high enantioselectivities. This strategy has been applied to the synthesis of natural products,[5,15,17] as well as a drug molecule.[17]

Table 1: Examples of Imide Desymmetrization using Base **7**.

Entry	Electrophile	Conditions	Product	Yield	ee
1[15]	TMSCl	THF, –78 °C	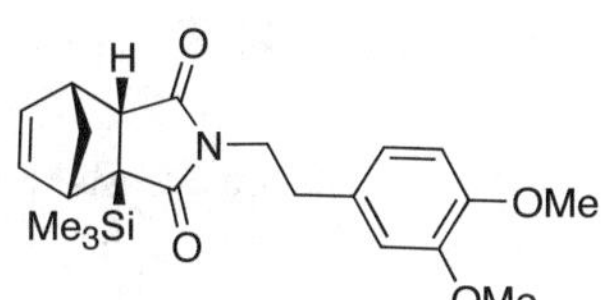 **8**	82%	92–94%
2[16]	TMSCl	THF, –105 °C LiCl	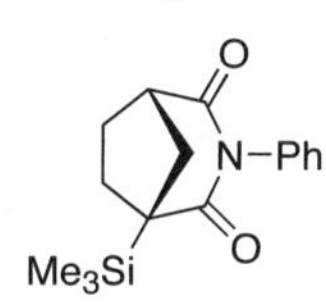 **9**	47%	94%
3[17]	MeI	THF, –78 °C	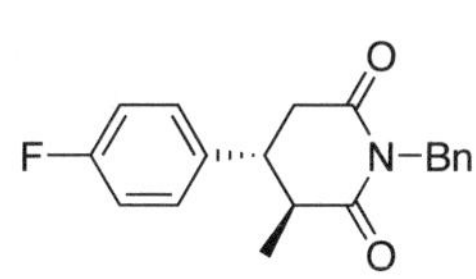 **10**	65%	97%

1. School of Chemistry, University of Birmingham, Edgbaston, Birmingham, UK, B15 2TT, *n.simpkins@bham.ac.uk.*

2. Martelli, G.; Morri, S.; Savoia, D. *Tetrahedron*, **2000**, *56*, 8367-8374.
3. For reviews see (a) Petterson, D.; Amedjkouh, M., Ahlberg, P. In *The Chemistry of Organolithium Compounds*; Rappoport, Z.; Marek, I. Ed.s; John Wiley and Sons: Chichester, 2006; pp 411–476. (b) O'Brien, P. *J. Chem. Soc., Perkin Trans. 1*, **1998**, 1439-1457; (c) Cox, P.J. ; Simpkins, N.S. *Tetrahedron : Asymmetry*, **1991**, *2*, 1-26; (d) Simpkins, N.S., Asymmetric deprotonation reactions using enantiopure lithium amide bases. In *Advanced Asymmetric Synthesis;* Stephenson, G.R. (Ed.), Blackie Academic, 1996; 111- 125. (e) Koga, K. *Pure Appl. Chem.* **1994**, *66*, 1487.
4. Eames, J. *Eur. J. Org. Chem.* **2002**, 393-401 ; see also ref 3.
5. Simpkins, N.S.; Gill, C.D. *Org. Lett.* **2003**, *5*, 535-537.
6. Adams, D.J.; Blake, A.J.; Cooke, P.A.; Gill, C.D.; Simpkins, N.S. *Tetrahedron* **2002**, *58*, 4603-4615.
7. For a recent example see: Armstrong, A.; Ahmed, G.; Dominguez-Fernandez, B.; Hayter, B.R.; Walles, J.S. *J. Org. Chem.* **2002**, *67*, 8610-8617; see also: Newcombe, N.J.; Simpkins, N.S. *J. Chem. Soc., Chem. Commun.* **1995**, 831-833.
8. See for example: Yu, M.; Clive, D.L.J.; Yeh, V.S.C.; Kang, S.; Wang, J. *Tetrahedron Lett.* **2004**, *45*, 2879-2881; Huxford, T.; Simpkins, N.S. *Synlett* **2004**, *13*, 2295-2298; Goldspink, N.J.; Simpkins, N.S.; Beckman, M. *Synlett* **1999**, *8*, 1292-1294.
9. For a recent contribution in this area: Castaldi, M.P.; Gibson, S.E.; Rudd, M.; White, A.J.P. *Angew. Chem. Int. Ed.* **2005**, *44*, 3432-3435.
10. Blake, A.J.; Cooke, P.A.; Kendall, J.D.; Simpkins, N.S.; Westaway, S.J. *J. Chem. Soc., Perkin Trans. 1* **2000**, 153-163.
11. McComas, C.C.; Van Vranken, D.L. *Tetrahedron Lett.* **2003**, *44*, 8203-8205.
12. Bette, V.; Mortreux, A.; Ferioli, F.; Martelli, G.; Savoia, D.; Carpentier, J.F. *Eur. J. Org. Chem.* **2004**, 3040-3045.
13. Bambridge, K.; Begley, M.J.; Simpkins N.S. *Tetrahedron Lett.*, **1994**, *35*, 3391-3394
14. Annunziata, R.; Benaglia, M.; Caporale, M.; Raimondi, L. *Tetrahedron*: *Asymmetry*, **2002**, *13*, 2727-2734.
15. Blake, A.J.; Gill, C.D.; Greenhalgh, D.A.; Simpkins, N.S.; Zhang, F. *Synthesis* **2005**, *19*, 3287-3292
16. Giblin, G.M.P.; Kirk, D.T.; Mitchell, L.; Simpkins, N.S. *Org. Lett.* **2003**, *5*, 1673-1675
17. Gill, C.D.; Greenhalgh, D.A.; Simpkins, N.S. *Tetrahedron* **2003**, *59*, 9213-9230.

CATALYTIC ASYMMETRIC ACYLATION OF ALCOHOLS USING A CHIRAL 1,2-DIAMINE DERIVED FROM (*S*)-PROLINE: (1*S*,2*S*)-*trans*-1-BENZOYLOXY-2-BROMOCYCLOHEXANE

A.

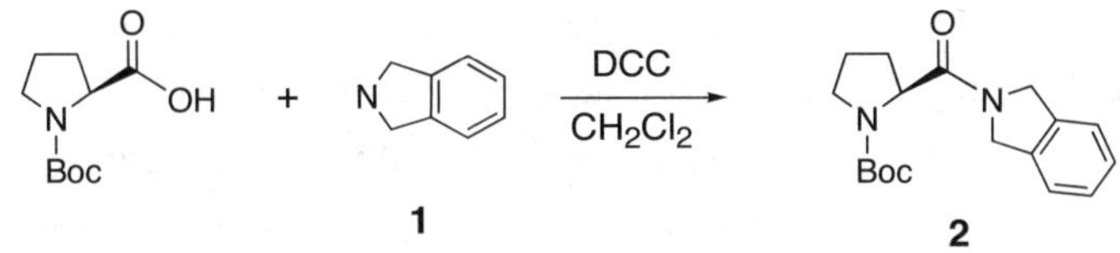

B.

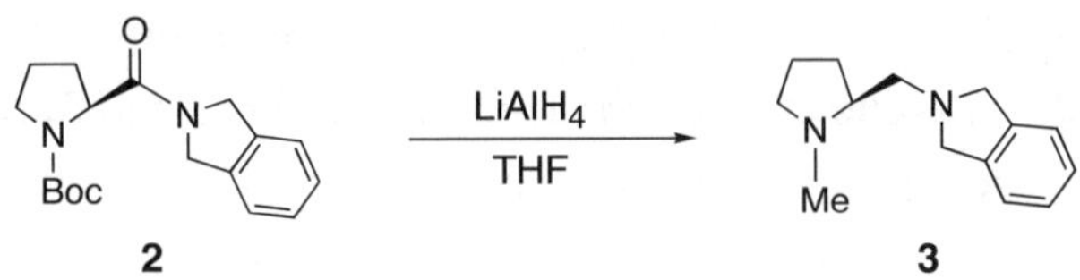

C.

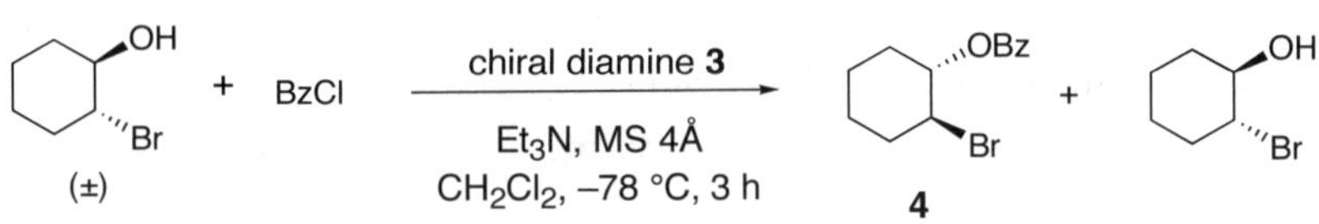

Submitted by Dai Terakado and Takeshi Oriyama.[1]
Checked by Jing Zhang, Fangzheng Li and Marvin J. Miller.

1. Procedure

A. (S)-N-(N-tert-Butoxycarbonylprolyl)dihydroisoindole (**2**). A dry, 100-mL, two-necked flask equipped with a Teflon-coated magnetic stirring bar and a septum cap is charged with (*S*)-*N*-*tert*-butoxycarbonylproline (5.05 g, 23.5 mmol) (Note 1), dihydroisoindole (2.54 g, 21.3 mmol) (Note 2), and dichloromethane (25 mL) (Note 3) under an argon atmosphere. After cooling to 0 °C with the aid of an ice-water bath, a solution of dicyclohexylcarbodiimide (DCC, 5.1 g, 24.7 mmol) (Note 4) in dichloromethane (20 mL) (Note 3) is added and the reaction mixture is allowed to warm to room temperature while stirring overnight. The mixture is filtered through Celite, concentrated, and purified by column chromatography (Note 5) (ethyl acetate/hexanes:1/3) to afford

(*S*)-*N*-(*N*-*tert*-butoxycarbonylprolyl)dihydroisoindole (4.1 g, 61%) as a white solid (Note 6).

B. (S)-1-Methyl-2-[(dihydroisoindol-2-yl)methyl]pyrrolidine (**3**) A dry, 200-mL, three-necked, round-bottomed flask equipped with a Teflon-coated magnetic stirring bar, a reflux condenser, a septum cap, and an argon inlet is charged with lithium aluminum hydride (LAH, 0.96 g, 25.3 mmol) in THF (10 mL). After cooling to 0 °C with the aid of an ice-water bath, a solution of (*S*)-*N*-(*N*-*tert*-butoxycarbonylprolyl)dihydroisoindole (3.96 g, 12.5 mmol) in THF (15 mL) was added under an argon atmosphere. The cooling bath is removed and the reaction mixture is refluxed for 3 h. After cooling the mixture to 0 °C, the reaction is quenched carefully by the slow addition of saturated aqueous sodium sulfate (approx 5 mL). The liquid is decanted away from the precipitate, and the precipitate was washed with THF (2 x 20 mL). The combined liquid is dried over sodium sulfate, filtered and concentrated under reduced pressure. The residue is purified by column chromatography on silica gel (CH_2Cl_2/MeOH/Et_3N:95/5/1), to afford (*S*)-1-methyl-2- [(dihydroisoindol-2-yl)methyl]pyrrolidine (**3**) (1.56 g, 58%) (Note 7).

C. (1S,2S)-trans-1-Benzoyloxy-2-bromocyclohexane (**4**). A 100-mL, two-necked flask equipped with a Teflon-coated magnetic stirring bar and a septum cap is charged with 1 g of molecular sieves 4 Å (Note 8), and flame-dried under reduced pressure. After being allowed to warm to ambient temperature, the apparatus is flushed with argon. The flask is charged with (*S*)-1-methyl-2-[(dihydroisoindol-2-yl)methyl]pyrrolidine (65 mg, 0.28 mmol) in dichloromethane (5 mL) (Note 3), triethylamine (5.56 g, 55 mmol) (Note 9) in dichloromethane (15 mL), and racemic *trans*-2-bromocyclohexanol (17.91 g, 100 mmol) (Note 10) in dichloromethane (40 mL) by means of an oven-dried syringe and needle. After cooling to –78 °C by immersion in a dry-ice bath, benzoyl chloride (9.14 g, 65 mmol) (Note 11) in dichloromethane (20 mL) is added slowly over 30 min by means of an oven-dried syringe and needle. The solution is stirred for 3 h at –78 °C, and then quenched with a phosphate buffer (pH 7) (Note 12). The layers are separated and the aqueous layer extracted with diethyl ether (3 x 50 mL). The combined organic phase is washed with water, dried over anhydrous sodium sulfate, filtered, and concentrated under vacuum. The crude products are purified by column chromatography (Note 13) (ethyl acetate/hexanes:1/50) to give (1*S*, 2*S*)-*trans*-1-benzoyloxy-2-bromocyclohexane (14.25 g, 50%) (Note 14) and unreacted alcohol (1*R*, 2*R*)-*trans*-2-bromocyclohexanol (6.60 g, 37%) (Note 15).

2. Notes

1. The checkers purchased (*S*)-*N*-*tert*-butoxycarbonylproline from Aldrich Chemical Company.

2. The checkers purchased dihydroisoindole (isoindoline) from Aldrich Chemical Company.

3. Dichloromethane was purchased by the submitters as anhydrous solvent from Kanto Chemical Company, Inc., and used without further purification. The dichloromethane used by the checkers was purchased from Fisher Scientific and distilled from CaH_2 prior to use.

4. Dicyclohexylcarbodiimide (DCC) used by the submitters was purchased from Tokyo Kasei Kogyo Co. The checkers used DCC purchased from Aldrich Chemical Co. The submitters reversed the addition by adding the substrates to the DCC mixture. However, the checkers found that the substrates were incompletely soluble in methylene chloride and that transfer was then incomplete.

5. Column chromatography was performed (38 mm x 600 mm column) on Wakogel C-200 that purchased from Wako Chemical Company, Inc. The checkers used Merck silica gel with a 45 mm x 200 mm column.

6. The submitters reported 100% yield. The analytical and spectral data of (*S*)-*N*-(*N*-*tert*-butoxycarbonylprolyl)dihydroisoindole are as follows: mp 150–152 °C (decompose), $[\alpha]_D^{24}$ −18.4 (*c* 1.0, EtOH). The checkers obtained $[\alpha]_D$ −19.8 (*c* 1.0, EtOH); The NMR spectra show the presence of two rotameric forms in an approximate 1:1 ratio. Resonances for both rotamers are included in the following characterization data. ^{1}H NMR (300 MHz, $CDCl_3$) δ: 1.35 (s, 9 H), 1.46 (s, 9 H), 1.82–2.02 (m, 4 H), 2.08–2.29 (m, 4 H), 3.42–3.68 (m, 4 H), 4.47 (dd, 1 H, *J* = 8.1, 4.8 Hz), 4.59 (dd, 1 H, *J* = 7.7, 3.3 Hz), 4.72–4.91 (m, 6 H), 4.99 (d, 1 H, *J* = 13.6 Hz), 5.18 (d, 1 H, *J* = 13.6 Hz), 7.22–7.32 (m, 8 H); ^{13}C NMR (75 MHz, $CDCl_3$) δ: 23.94 (CH_2), 24.55 (CH_2), 28.50 (CH_3), 28.63 (CH_3), 29.64 (CH_2), 30.49 (CH_2), 46.82 (CH_2), 47.03 (CH_2), 52.35 (CH_2), 52.46 (CH_2), 52.56 (CH_2), 57.55 (CH), 57.89 (CH), 79.69 (C), 79.77 (C), 122.70 (CH), 122.75 (CH), 123.02 (CH), 123.20 (CH), 127.53 (CH), 127.73 (CH), 127.83 (CH), 128.07 (CH),

136.13 (C), 136.26 (C), 136.42 (C), 136.50 (C), 153.92 (C), 154.71 (C), 171.73 (C), 171.89 (C); IR (neat) cm^{-1}: 2967, 2867, 1694, 1653, 1405, 1358, 1163, 1127, 888, 755; MS (FAB) exact mass, (*m/z*) Calcd for $C_{18}H_{24}N_2O_3$ 317.1865, Found 317.1862; Anal. Calcd for $C_{18}H_{24}N_2O_3$: C, 68.33; H, 7.65; N, 8.85. Found: C, 68.16; H, 7.60; N, 8.65.

7. The submitters reported 60% yield after distillation under reduced pressure. The checkers had difficulty with reproducibility using distillation. The yield obtained by the checkers after chromatography varied between 58% and 65%. The checkers used Merck silica gel with a 45 mm x 200 mm column. The analytical and spectral data of (*S*)-1-methyl-2-[(dihydroisoindol-2-yl)methyl]pyrrolidine are as follows: bp 112–114 °C / 0.6 mmHg, $[\alpha]_D^{24}$ –70.4 (*c* 1.1, EtOH), ^{1}H NMR (300 MHz, $CDCl_3$) δ: 1.58–1.84 (m, 3 H), 1.94–2.06 (m, 1 H), 2.16 (dt, 1 H, *J* = 9.6, 7.8 Hz), 2.30 (m, 1 H), 2.40 (s, 3 H), 2.62 (dd, 1 H, *J* = 11.7, 7.8 Hz), 2.88 (dd, 1 H, *J* = 12, 4.8 Hz), 3.03 (m, 1 H), 3.89 (s, 4 H), 7.11 (s, 4 H); ^{13}C NMR (75 MHz, $CDCl_3$) δ: 22.60 (CH_2), 30.53 (CH_2), 41.35 (CH_3), 57.67 (CH_2), 59.96 (CH_2), 60.94 (CH_2), 64.96 (CH), 122.04 (CH), 126.50 (CH), 140.28 (C); IR (neat) cm^{-1}: 2938, 2771, 1463, 1149, 742; MS (FAB) exact mass (*m/z*) Calcd for $C_{14}H_{20}N_2$ 217.1705, Found 217.1701. Anal. Calcd for $C_{14}H_{20}N_2$: C, 77.72; H, 9.32; N, 12.95. Found: C, 77.34; H, 9.53; N, 12.83.

8. Molecular sieves 4Å were purchased from Wako Chemical Company, Inc., and dried at 100 °C for 3 h as a powder under reduced pressure before use. The checkers used molecular sieves purchased from Aldrich Chemical Company.

9. Triethylamine was purchased from Tokyo Kasei Kogyo Co., and distilled before use. The triethylamine used by the checkers was purchased from Aldrich Chemical Company and was distilled from CaH_2 prior to use.

10. *trans*-2-Bromocyclohexanol was prepared from cyclohexene oxide and hydrobromic acid according to the following procedure: A 200-mL, round-bottomed flask equipped with a Teflon-coated magnetic stirring bar is charged with hydrobromic acid (47%, 40 mL, 346 mmol) and cooled at 0 °C by immersion in an ice-water bath. Cyclohexene oxide (20 mL, 198 mmol) is added dropwise and the mixture is stirred at room temperature for 8 h. After being cooled to 0 °C, the solution is neutralized by addition of saturated aqueous Na_2CO_3 (approx 30 mL) (*Caution: slow addition of the Na_2CO_3 solution is recommended due to excessive bubbling of the solution*), and extracted with diethyl ether (3 x 30 mL). The combined organic phases are dried over anhydrous sodium sulfate, filtered, concentrated, and distilled under reduced pressure to give

trans-2-bromocyclohexanol (30.0 g, 85%) (bp 92 °C / 11 mmHg).

11. BzCl was purchased from Tokyo Kasei Kogyo Co., and distilled before use.

12. The buffer was prepared by dissolving 33.4 g of disodium hydrogenphosphate dodecahydrate and 6.4 g of potassium dihydrogenphosphate into 300 mL of water. The buffer solution was diluted to a final volume of 700 mL and stored in a glass bottle.

13. The checkers used Merck silica gel with a 70 mm x 200 mm column.

14. The checkers obtained 13.65 g (48%) of the benzoate (**4**) and 7.59 g (42%) of the alcohol. The submitters determined the enantiomeric excess of the benzoate (95% *ee*) by HPLC analysis using a Daicel CHIRALCEL OD column (*i*-PrOH:hexanes = 1:1000, 1.0 mL/min, 254 nm). The retention times for the (1*S*,2*S*)-*trans*-1-benzoyloxy-2-bromocyclohexane are 14.6 min ((+)-1*S*,2*S*) and 16.7 min ((–)-1*R*,2*R*). The checkers determined the enantiomeric excess of the benzoate to be 90% using a 25 x 0.46 cm Daicel CHIRALPAK AD-H column (*i*-PrOH:hexanes = 1:9, 1.0 mL/min, 254 nm). The retention times are 5.05 and 5.31 min. The analytical and spectral data of pure (1*S*, 2*S*)-*trans*-1-benzoyloxy-2-bromocyclohexane are as follows: $[\alpha]_D^{24}$ +104.6 (*c* 1.0, $CHCl_3$, 90.0% ee); ^{1}H NMR (500 MHz, $CDCl_3$) δ: 1.35–1.44 (m, 1 H), 1.48–1.57 (m, 2 H), 1.75–1.85 (m, 2 H), 1.91–1.99 (m, 1 H), 2.25–2.33 (m, 1 H), 2.39–2.45 (m, 1 H), 4.16 (m, 1 H), 5.14 (dt, 1 H, *J* = 9.0, 4.5 Hz), 7.46 (t, 2 H, *J* = 7.5 Hz), 7.57 (m, 1 H), 8.07 (m, 2 H); ^{13}C NMR (125 MHz, $CDCl_3$) δ: 23.5 (CH_2), 25.6 (CH_2), 31.3 (CH_2), 35.7 (CH_2), 52.8 (CH), 76.5 (CH), 128.5 (CH), 129.9 (CH), 130.4 (C), 133.2 (CH), 165.8 (C); IR (neat) cm^{-1}: 2941, 1720, 1450, 1274, 1104, 1027, 945, 711; MS (FAB) exact mass (*m/z*) Calcd for $C_{13}H_{15}BrO_2$ 283.0334, Found 283.0308. The submitters report that the product gave the following elemental anlysis: Anal. Calcd for $C_{13}H_{15}BrO_2$: C, 55.14; H, 5.34. Found: C, 55.05; H, 5.42.

15. The enantiomeric excess was determined by the submitters to be >99% by HPLC analysis using Daicel CHIRALCEL OD (*i*-PrOH:hexanes = 1:1000, 1.0 mL/min, 254 nm) after conversion to the corresponding benzoate. The retention times for the (1*R*, 2*R*)-*trans*-1-benzoyloxy-2-bromocyclohexane are 14.3 min ((+)-1*S*,2*S*) and 16.8 min ((–)-1*R*,2*R*). The analytical and spectral data of (1*R*, 2*R*)-*trans*-2-bromocyclohexanol are as follows: $[\alpha]_D^{24}$ –33.0 (*c* 1.0, $CHCl_3$, 99.3% ee); ^{1}H NMR (500 MHz, $CDCl_3$) δ: 1.22–1.40 (m, 3 H), 1.65–1.70 (m, 1 H), 1.77–1.86 (m, 2 H), 2.10–2.16 (m, 1 H), 2.30–2.36 (m, 1 H), 2.62 (brs, 1 H), 3.60 (dt, 1 H, *J* =

9.9, 4.5 Hz), 3.89 (ddd, 1 H, J = 12.1, 9.5, 4.5 Hz); ^{13}C NMR (125 MHz, $CDCl_3$) δ: 24.26 (CH_2), 26.8 (CH_2), 33.7 (CH_2), 36.4 (CH_2), 61.96 (CH), 75.4 (CH); IR (neat) cm^{-1}: 3350, 2835, 1450, 1070, 955.

Waste Disposal Information

All hazardous materials should be handled and disposed of in accordance with "Prudent Practices in the Laboratory"; National Academy Press; Washington, DC, 1995.

3. Discussion

Asymmetric acylation of alcohols is divided into two types of reactions. These are kinetic resolution of racemic alcohols and desymmetrization of meso-polyols. Although most methods reported so far employ an enzyme such as a lipase or an esterase,[2] some outstanding asymmetric acylations of alcohols by using organocatalysts[3] have recently emerged as reliable alternatives to the well established enzyme-catalyzed reactions.

Kinetic resolution of racemic alcohols via asymmetric acylation has been widely used to construct various useful chiral building blocks in the synthesis of complex natural products.[4] The submitters have demonstrated highly enantioselective desymmetrization of *meso*-diols[5] and highly efficient kinetic resolution of racemic secondary alcohols[6] catalyzed by a chiral 1,2-diamine derived from (*S*)-proline.

Chiral diamine catalysts can be readily prepared in three steps from (*S*)-proline. This type of a 1,2-diamine is frequently used in important and fundamental asymmetric syntheses.[7] This diamine has an advantage in that the non-proline amine portion can be easily modified to include, for example, cyclic pyrrolidine, piperidine, indoline skeletons, or the acyclic benzyl-methylamino skeleton. More efficient reactions and higher enantioselectivities during asymmetric acylation of alcohols are achieved by using a chiral diamine containing a dihydroisoindole or a benzyl-methylamino component. Representative results of asymmetric acylation of alcohols catalyzed by chiral 1,2-diamines are shown in Tables I and II.

TABLE I

CHIRAL 1,2-DIAMINE CATALYZED KINETIC RESOLUTIONS OF RACEMIC SECONDARY ALCOHOLS[a]

Entry	Alcohol	Amine	Ester Yield[b] / %	Ester ee[c] / %	Unreacted alcohol Yield[b] / %	Unreacted alcohol ee[d] / %	s^e
1	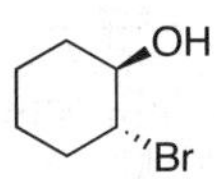	i-Pr_2NEt	47	96	39	95	130
2	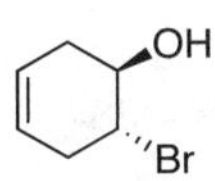	i-Pr_2NEt	46	97	43	91	170
3	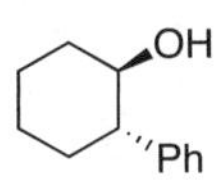	Et_3N	49	96	48	95	160
4[f]	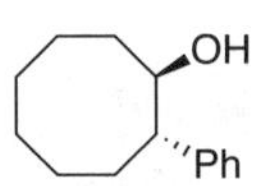	Et_3N	44	95	47	79	88
5	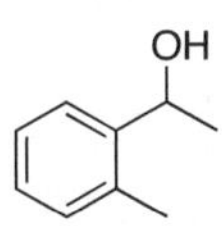	Et_3N	45	82	49	78	20

a) Reaction conditions : MS 4Å (40 mg), (*S*)-1-methyl-2-[(dihydroisoindol-2-yl)methyl]pyrrolidine (0.9 μmol), amine (0.15 mmol), alcohol (0.3 mmol), BzCl (0.23 mmol), –78 °C , 3 h. b) Isolated yield of purified product. c) Determined by chiral HPLC analysis. d) Determined by chiral HPLC analysis after conversion to corresponding benzoate. e) Selectivity factor, calculated from the conversion (isolated yield) and ee of the acylated product. f) Reaction was performed for 24 h.

(1*S*,2*S*)-*trans*-1-Benzoyloxy-2-bromocyclohexane is a valuable synthetic precursor for allylic alcohol derivatives. An enantio-enriched allylic benzoate can be provided via β-elimination. Treatment of **4** with DBN (1,5-diazabicyclo[4.3.0]non-5-ene) in refluxing toluene gives the corresponding allylic benzoate, (*S*)-1-benzoyloxy-2-cyclohexene, without loss of enantio-purity.[6c]

We have presented a promising small organocatalyst for asymmetric acylation of various alcohols including racemic secondary alcohols and *meso*-1,2-, 1,3-, and 1,4-diols. Catalytic asymmetric acylation of alcohols by using a chiral 1,2-diamine has the following distinguishing synthetic features: 1) high enantioselectivity, 2) high efficiency, 3) operational simplicity, 4) widespread applicability, and 5) the absence of a metal.

TABLE II

CHIRAL 1,2-DIAMINE CATALYZED ASYMMETRIC DESYMMETRIZATION OF *meso*-DIOLS[a]

Entry	*meso*-Diol	Amine	Solvent	Monoester Yield[b] / %	Monoester ee[c] / %
1[d]	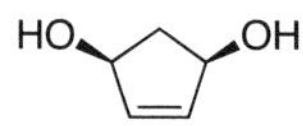	Et_3N	PrCN	38	98
2[d,e]	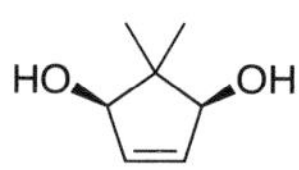	Et_3N	PrCN	87	>99
3[f]		*i*-Pr_2NEt	CH_2Cl_2 - DMF (9 : 1)	57	>99
4	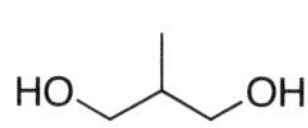	*i*-Pr_2NEt	PrCN	33	96
5[g,h]	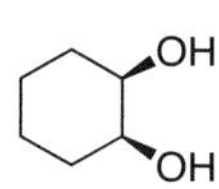	Et_3N	CH_2Cl_2	92	98
6[g,h]	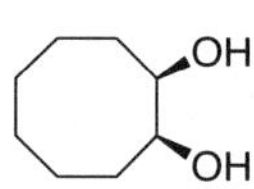	Et_3N	CH_2Cl_2	92	99

a) Reaction conditions : MS 4Å (40 mg), chiral 1,2-diamine (0.0015 mmol), amine (0.45 mmol), diol (0.3 mmol), *p*-*t*-BuC_6H_4COCl (0.45 mmol), –78 °C, 3 h. b) Isolated yield of purified product. c) Determined by chiral HPLC analysis. d) 0.51 mmol of BzCl and Et_3N were used. e) Reaction was performed for 8 h. f) Chiral 1,2-diamine **A** was used. g) 0.003 mmol of chiral 1,2-diamine and 0.3 mmol of Et_3N were used and silylation of monoalcohol was performed in one-pot after acylation for 3 h. h) *p*-$CH_3C_6H_4COCl$ was used instead of *p*-*t*-BuC_6H_4COCl.

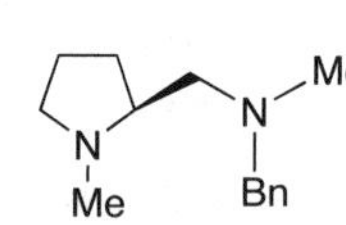

A

1. Faculty of Science, Ibaraki University, Bunkyo, Mito 310-8512, Japan.

2. (a) Klibanov, A. M. *Acc. Chem. Res.* **1990**, *23*, 114; (b) Santaniello, E.; Ferraboschi, P.; Grisenti, P.; Manzocchi, A. *Chem. Rev.* **1992**, *92*, 1071; (c) Otera, J. *Chem. Rev.* **1993**, *93*, 1449; (d) Schoffers, E.; Golebiowski, A.; Johnson, C. R. *Tetrahedron*,

1996, *52*, 3769.

3. (a) Vedejs, E.; MacKay, J. A. *Org. Lett.* **2001**, *3*, 535; (b) Vedejs, E.; Daugulis, O.; Harper, L. A.; MacKay, J. A.; Powell, D. R. *J. Org. Chem.* **2003**, *68*, 5020; (c) Tao, B.; Ruble, J. C.; Hoic, D. A.; Fu, G. C. *J. Am. Chem. Soc.* **1999**, *121*, 5091; (d) Fu, G. C. *Acc. Chem. Res.* **2000**, *33*, 412; (e) Miller, S. J.; Copeland, G. T.; Papaioannou, N.; Horstmann, T. E.; Ruel, E. M. *J. Am. Chem. Soc.* **1998**, *120*, 1629; (f) Spivey, A. C.; Fekner, T.; Spey, S. E. *J. Org. Chem.* **2000**, *65*, 3154; (g) Kawabata, T.; Stragies, R.; Fukaya, T.; Nagaoka, Y.; Schedel, H.; Fuji, K. *Tetrahedron Lett.* **2003**, *44*, 1545; (h) Yamada, S; Katsumata, H. *Chem. Lett.* **1998**, 995; (i) Matsumura, Y; Maki, T; Murakami, S; Onomura, O. *J. Am. Chem. Soc.* **2003**, *125*, 2052; (j) Trost, B. M.; Mino, T. *J. Am. Chem. Soc.* **2003**, *125*, 2410.
4. Theil, F. *Chem. Rev.* **1995**, *95*, 2203. and references cited therein.
5. (a) Oriyama, T.; Imai, K.; Sano, T.; Hosoya, T. *Tetrahedron Lett.* **1998**, *39*, 3529; (b) Oriyama, T.; Hosoya, T.; Sano, T. *Heterocycles* **2000**, *52*, 1065; (c) Oriyama, T.; Taguchi, H.; Terakado, D.; Sano, T. *Chem. Lett.* **2002**, 26.
6. (a) Sano, T.; Oriyama, T. *Yuki Gosei Kagaku Kyokaishi*, **1999**, *57*, 598; (b) Sano, T.; Imai, K.; Ohashi, K.; Oriyama, T. *Chem. Lett.* **1999**, 265; (c) Sano, T.; Miyata, H.; Oriyama, T. *Enantiomer*, **2000**, *5*, 119.
7. (a) Mukaiyama, T.; Tomimori, K.; Oriyama, T. *Chem. Lett.* **1985**, 813; (b) Mukaiyama, T.; Minowa, N.; Oriyama, T.; Narasaka, K. *Chem. Lett.* **1986**, 97; (c) Asami, M. *Yuki Gosei Kagaku Kyokaishi*, **1996**, *54*, 188; (d) Kobayashi, S.; Horibe, M. *Chem. Eur. J.* **1997**, *3*, 1472; (e) Lucet, D.; Gall, T. L.; Mioskowski, C. *Angew. Chem. Int. Ed.* **1998**, *37*, 2580.

(*R*,*R*)-2,2'-BISPYRROLIDINE and (*S*,*S*)-2,2'-BISPYRROLIDINE: USEFUL LIGANDS FOR ASYMMETRIC SYNTHESIS

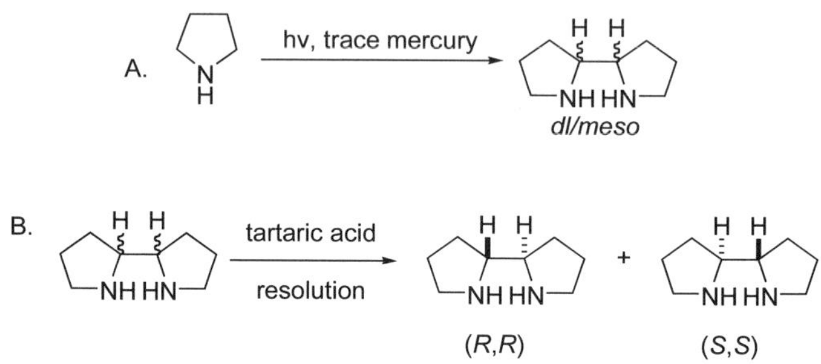

Submitted by Scott E. Denmark, Jiping Fu and Michael J. Lawler.[1]
Checked by Sandra Lee, Elliott Huntsman and Edward J.J. Grabowski.

Caution! This procedure should be carried out in a well-ventilated hood because of the stench of pyrrolidine and 2,2'-bispyrrolidine. The hood doors should be covered with an opaque sheet to shield the UV light.

1. Procedure

A. (R,R*) and (R*,S*)-2,2'-Bispyrrolidine.*[2,3] A three-necked, 500-mL flask equipped with three quartz refluxing columns and three water condensers (Figure 1) is charged with pyrrolidine (160 mL) and a drop of mercury (Notes 1 and 2). The water condensers are equipped with gas inlets connected to a single nitrogen source leading to an oil bubbler. The flask is placed in a Rayonet Photoreactor fitted with 14 x 8 Watt low pressure Hg lamps (254 nm). The reaction mixture is heated to reflux with a heating mantle. The lamps are turned on, after which the mixture is heated at reflux for 7 days. After the lamps are turned off, the system is allowed to cool to room temperature (Note 2). The liquid is then carefully decanted to a distillation flask and the mercury is recovered. Unreacted pyrrolidine and side products are removed by distillation at atmospheric pressure (Note 3). The residue is distilled to provide 67.6 g (50%) of a mixture of (*R**,*R**) and (*R**,*S**)-2,2'-bispyrrolidine as a clear, light yellow liquid (bp 79–81 °C at 3.0 mmHg) (Note 4). The product is of sufficient purity to be used in the resolution step (Note 5).

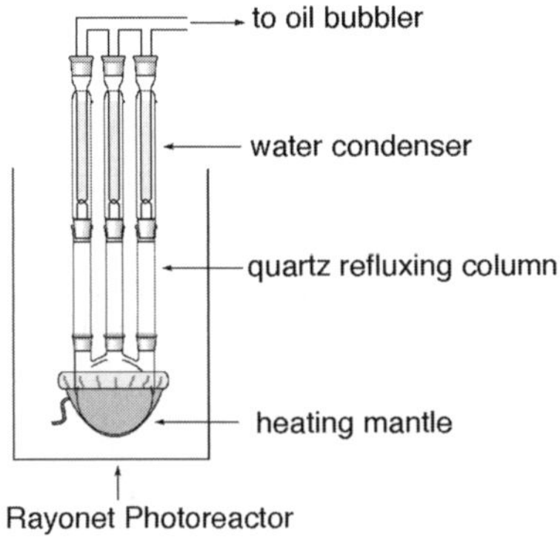

Figure 1. Apparatus for photodimerization of pyrrolidine.

B. Resolution of 2,2'-Bispyrrolidine[3,4]

1. *Preparation of (R,R)-2,2'-bispyrrolidine•(L)-tartrate.* To a solution of a 1/1 mixture of *d,l*- and *meso*-2,2'-bispyrrolidine (67.0 g, 479 mmol) in H_2O (240 mL) is added (L)-(+)-tartaric acid (36.0 g, 240 mmol, 0.5 equiv) and acetic acid (27.4 mL, 479 mmol, 1.0 equiv) (Note 6). The mixture is heated to 90 °C and the homogenous solution is allowed to cool to room temperature slowly before it is placed in an ice bath. After the solid precipitates (Note 7), the mixture is kept in an ice bath for another 2 h. The precipitate is then filtered and is washed with ice-cold water (20 mL). The mother liquor is saved for the recovery of the (*S,S*)-2,2'-bispyrrolidine. The collected solid is dried under high vacuum (0.5 mmHg) at 80 °C for 2 h to give 31.6 g of a light yellow powder. This solid is dissolved in hot water (90 mL) and the solution is allowed to cool slowly to room temperature before it is placed in an ice bath for 1 h. The precipitate is filtered and the solid is washed with ice-cold water (10 mL). The solid is dried under high vacuum (0.5 mmHg) at 80 °C for 2 h to give 23.0 g of white crystals. This recrystallization process is repeated using water (50 mL) to give 21.4 g (62% based on isomer content) of (*R,R*)-2,2'-bispyrrolidine•(L)-tartrate as white prismatic crystals (Note 8).

2. *Preparation of (R,R)-2,2'-Bispyrrolidine.* To a mixture of the tartrate salt (9.1 g) in water (15 mL) at 0 °C are added KOH (20 g) pellets (Note 9). The mixture is then stirred at 0 °C for 10 min before diethyl ether (80 mL) is added, whereupon the solution is stirred at 0 °C for another 30

min. The aqueous layer is then separated and extracted with diethyl ether (6 x 50 mL) (Note 10). The diethyl ether extracts are combined, dried (K_2CO_3) and then are concentrated under vacuum (Note 11). The residue is transferred to a dry 25-mL, round-bottomed flask. To this is added a small piece of sodium (Note 12) and the mixture is stirred at room temperature under nitrogen for 30 min. The residue is distilled under vacuum to give 3.61 g (83%) of (*R*,*R*)-2,2'-bispyrrolidine as a clear colorless oil (bp 97–98 °C at 8.0 mmHg) (Note 13). The enantiomeric purity of product is determined by CSP-SFC and CSP-HPLC analysis of the corresponding dibenzoyl amide derivative (Note 14).

3. *Preparation of (S,S)-2,2'-bispyrrolidine•(D)-tartrate.* The mother liquor from initial resolution is cooled to 0 °C and KOH pellets (80 g) are added slowly. The mixture is stirred vigorously at 0 °C for 10 min (Note 15). To this solution is added diethyl ether (500 mL) and the mixture is stirred at room temperature for 20 min. The aqueous layer is separated and then is extracted with diethyl ether (4 x 500 mL). The diethyl ether extracts are combined, dried (K_2CO_3), and then are concentrated under vacuum to give 48.4 g of a yellow oil. The oil is dissolved in H_2O (150 mL), then (D)-(–)-tartaric acid (34.5 g, 230 mmol) and acetic acid (27.0 mL, 473 mmol) are added (Note 16). The mixture is heated to 90 °C and the homogenous solution is allowed to cool to room temperature slowly before it is cooled in an ice bath. After the solid precipitates, the mixture is kept in an ice bath for another 2 h. The precipitate is filtered and the solid is washed with ice-cold water (10 mL), and then is dried under high vacuum (0.5 mmHg) at 80 °C for 2 h to give 23.5 g of a light-yellow powder. The solid is dissolved in hot water (60 mL) and the solution is allowed to cool slowly to room temperature before it is placed in an ice bath for another 2 h. The precipitate is filtered, washed with 10 mL of ice-cold water, and dried under high vacuum (0.5 mmHg) at 80 °C for 2 h to give 20.16 g of white prismatic crystals. This recrystallization process is repeated using water (55 mL) to give 18.4 g (55% based on isomer content) of (*S*,*S*)-2,2'-bispyrrolidine•(D)-(–)-tartrate as white prismatic crystals (Note 17).

4. *Preparation of (S,S)-2,2'-Bispyrrolidine.* To a solution of the tartrate salt (9.3 g) in water (15 mL) at 0 °C are added KOH pellets (20 g). The mixture is stirred at 0 °C for 10 min before diethyl ether (80 mL) is added, whereupon it is stirred at 0 °C for another 30 min. The aqueous layer is separated and then is extracted with diethyl ether (6 x 50 mL). The diethyl ether extracts are combined, dried (K_2CO_3) and then are concentrated *in vacuo* (Note 11). The residue is then transferred to a dry 25-mL, round-

bottomed flask. To this is added a piece of sodium (Note 12) and the mixture is stirred at room temperature under nitrogen for 30 min. The residue is then distilled under vacuum to give 3.51 g (80%) of (*S,S*)-2,2′-bispyrrolidine as a clear, colorless oil (bp 97−98 °C at 8.0 mmHg) (Note 18). The enantiomeric purity of product is determined by CSP-SFC and CSP-HPLC analysis of corresponding dibenzoyl amide derivative (Note 19).

2. Notes

1. The three quartz-refluxing columns are 34 cm long and 4.0 cm in diameter. All joints are well sealed with high vacuum grease to avoid leakage. Pyrrolidine (99 %) was purchased from Aldrich Chemical Company, Inc., and was used without further purification.

Variables that are difficult to control relative to the photolysis include the age and actual power of the Hg lamps, the actual length of the quartz tubes receiving the UV light and the rate of reflux. All affect the overall rate of photolysis. The checkers found it convenient to follow the progress of the photolysis by periodically sampling the reaction and analyzing it by ^{13}C-NMR spectroscopy in $CDCl_3$. The peak heights for the methylene groups for pyrrolidine (47.1 ppm) and the bis-pyrrolidines (47.0 and 46.6) were used as measures of the relative ratios of these species. When the pyrrolidine is almost consumed in the photolysis, the reflux ceases. The submitters were able to recover ~50% of the bis-pyrrolidines after distillation following a seven-day photolysis (See Note 5). The checkers achieved the following results: 39% in seven days; 60% in nine days; 72% in 11.4 days and 60% in five days at half-scale.

2. The crude reaction mixture can be analyzed by ^{1}H NMR. The ratio of starting material to products can be estimated from the NMR spectrum.

3. The distillate contains 46.6 g of colorless liquid.

4. The distillation should be done carefully to avoid solidification of the diamine in the condenser.

5. The crude material (67.6 g, approximately 50% based on the pyrrolidine charged) contains a ca. 1/1 mixture of *d,l* and *meso* isomers: ^{1}H NMR (500 MHz, $CDCl_3$) d: 1.32−1.46 (m, 2 H), 1.61−1.92 (m, 8 H), 2.82−2.98 (m, 6 H). The checkers obtained 52.6 g from the seven-day photolysis; 81.0 g from the nine day photolysis; 97.0 g from the 11.4 day photolysis and 41.5 g from the five day photolysis at half-scale.

6. L-(+)-Tartaric acid (99% GLC) was purchased from Aldrich

Chemical Company, Inc., and was used without further purification. Acetic acid (glacial) was purchased from Fisher Scientific Company and was used without further purification. In completing the checking of this procedure, the subsequent reactions were scaled to reflect the quantity of bispyrrolidines obtained in the distillations. The reactions checked at the yields indicated.

7. The initial formation of crystals may take up to 16 h. The process can be facilitated by stirring the mixture with glass rod or by addition of small amount of seed crystals.

8. The analytical data for (*R*,*R*)-2,2'-bispyrrolidine•(L)-tartrate are as follows:[4] mp 212–216 °C; ^{1}H NMR (500 MHz, D_2O/DSS) d: 1.88–1.98 (m, 2 H), 2.08–2.28 (m, 4 H), 2.40–2.48 (m, 2 H), 3.51–3.55 (m, 4 H), 3.92–4.00 (m, 2 H), 4.43 (s, 2 H), 4.86 (br, 6 H); ^{13}C NMR (126 MHz, D_2O) d: 25.5, 31.0, 49.1, 63.2, 76.6, 181.4; IR (KBr) cm^{-1}: 3220, 2717, 2516, 1693, 1612, 1583, 1450, 1386, 1321, 1124, 1072, 709; $[\alpha]_D^{24}$ +17.9 (*c* = 1.00, H_2O); Anal. Calcd for $C_{12}H_{22}N_2O_6$: C, 49.65; H, 7.64, N, 9.65. Found: C, 49.80; H, 7.63; N, 9.65. The checkers noted slight chemical shift variations in the NMR spectra of this material, and attribute these to slight differences in concentration and apparent pH in the different samples.

9. Potassium hydroxide (87.7%) was purchased from Fisher Scientific Company and was used without further purification.

10. Diethyl ether was purchased from Mallinckrodt Inc. and was used without purification.

11. The water bath is kept at 0 °C to avoid loss of (*R*,*R*)-2,2′-bispyrrolidine.

12. The piece of sodium is about 0.5 cm^3 and it is washed with hexane before use. After distillation, the sodium was destroyed by the careful addition of isopropyl alcohol.

13. The analytical data for (*R*,*R*)-2,2'-bisyrrolidine are as follows:[3,4] ^{1}H NMR (500 MHz, $CDCl_3$) δ: 1.31–1.38 (m, 2 H), 1.65–1.84 (m, 6 H), 2.06 (br, 2 H), 2.82–2.97 (m, 6 H); ^{13}C NMR (126 MHz, $CDCl_3$) δ: 25.4, 29.0, 46.4, 63.8; IR (KBr) cm^{-1}: 3270, 2956, 2867, 1282, 1118, 1076; MS (FAB) (*m/z*): 141; HRMS (*m/z*) $C_8H_{17}N_2$ (M+H): Calc.: 141.1386; Found: 141.1375; $[\alpha]_D^{24}$ –14.91 (*c* = 1.03, MeOH); Anal. Calcd for $C_8H_{16}N$: C, 68.52; H, 11.50, N, 19.98. Found: C, 68.38; H, 11.64; N, 19.92. The free diamine is extremely hygroscopic, oxygen sensitive and absorbs CO_2 rapidly in air.

14. Procedure for derivatization is as follows (eq 1)[3,4]: To a solution of (*R*,*R*)-**2** (140 mg, 1.0 mmol) in 1.0 mL of methylene chloride (purchased from Fisher Scientific Company and distilled from P_2O_5) at 0 °C is added

triethylamine (278 mL, 2.0 mmol, 2.0 equiv, purchased from Aldrich Chemical Company, Inc., and distilled from CaH_2) and benzoyl chloride (232 mL, 2.0 mmol, 2.0 equiv, purchased from Aldrich Chemical Company, Inc., and distilled before use). The mixture is stirred at room temperature for

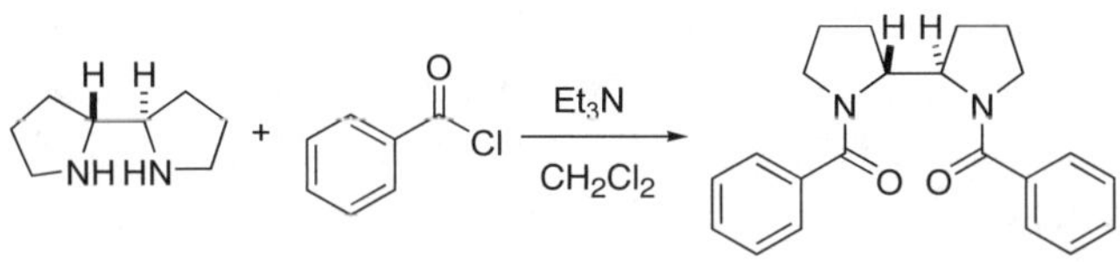

(*R,R*)-**2** (*R,R*)-**3**

4 h and then EtOAc (50 mL) and H_2O (10 mL) are added. The aqueous layer is separated and then is extracted with EtOAc (3 x 15 mL). The organic layers are combined, washed with 15 mL of saturated, aqueous sodium bicarbonate solution, dried over Na_2SO_4 and then concentrated under vacuum. The residue is purified by column chromatography (SiO_2, hexane/*i*-PrOH, 6/1) to give 295 mg (85%) of (*R,R*)-**3** as a white solid. The analytical data for (*R,R*)-**3** are as follows:[4] ^{1}H NMR (500 MHz, $CDCl_3$) δ: 1.76–2.05 (m, 6 H) 2.20–2.28 (m, 2 H), 3.19 (dt, *J* = 10.3, 7.8, 2 H), 3.79 (ddd, *J* = 10.5, 8.8, 5.1, 2 H), 4.59–4.64 (m, 2 H), 7.22–7.36 (m, 6 H), 7.38–7.42 (m, 4 H); ^{13}C NMR (126 MHz, $CDCl_3$) δ: 24.1, 28.2, 49.1, 58.8, 127.1, 128.2, 129.5, 137.2, 164.5; IR ($CHCl_3$) cm^{-1}: 2997, 2881, 1624, 1576, 1427, 700; MS (EI, 70 eV): 348, 175, 174, 105; HRMS (*m/z*): Calc. $C_{22}H_{25}N_2O_2$ (M+H): 349.1911; Found: 349.1900; Supercritical Fluid Chromatography: t_R (*R,R*)-**3** 2.86 min (100 %); t_R (*S,S*)-**3** 3.26 min (0 %) (Chiralpak AS, 40 °C, 150 bar, 15 % MeOH in CO_2, 3.0 mL/min, 220 nm); HPLC: t_R (*S,S*)-**3** 8.2 min (0%); t_R (*R,R*)-**3** 13.2 min (100%) (Chiralpak AD, *i*-PrOH/hexane, 95/5, 0.7 mL/min)

15. The submitters initially used 40 g of KOH pellets for the neutralization. The checkers found that the use of 80 g was necessary to assure extraction of all of the diamine.

16. D-(–)-Tartaric acid (97% GLC) was purchased from Aldrich Chemical Company, Inc., and was used without further purification.

17. The analytical data for (*S,S*)-2,2'-bispyrrolidine•(D)-(–)-tartrate are as follows: mp 214–218 °C; ^{1}H NMR (500 MHz, D_2O/DSS) δ: 1.88–1.99 (m, 2 H), 2.08–2.29 (m, 4 H), 2.41–2.49 (br, 2 H), 3.52–3.57 (m, 4 H), 3.94–4.01 (m, 2 H), 4.44 (s, 2 H), 4.84 (br, 6 H); ^{13}C NMR (126 MHz, D_2O) δ: 25.5, 31.0, 49.1, 63.16, 76.6, 181.4; IR (KBr) cm^{-1}: 3384, 3242, 2997, 2885,

2717, 2517, 1693, 1610, 1583, 1387, 1124, 1072, 710; $[\alpha]_D^{24}$ −17.7 (*c* = 1.02, H_2O); Anal. Calcd for $C_{12}H_{22}N_2O_6$: C, 49.65; H, 7.64, N, 9.65. Found: C, 49.98; H, 7.43; N, 9.37. The checkers noted slight chemical shift variations in the NMR spectra of this material, and attribute these to slight differences in concentration and apparent pH in the different samples.

18. The analytical data for (*S,S*)-2,2′-bisyrrolidine are as follows: ^1H NMR (500 MHz, $CDCl_3$) δ: 1.31-1.38 (m, 2 H), 1.65-1.84 (m, 6 H), 2.06 (br, 2 H), 2.82-2.97 (m, 6 H); ^{13}C NMR (126 MHz, $CDCl_3$) δ: 25.4, 29.0, 46.4, 63.8; IR (KBr) cm^{-1}: 3263, 2954, 2867, 2821, 1457, 1442, 1280, 1118, 1076, 892, 869; MS (FAB) (*m/z*) 141; HRMS (*m/z*) $C_8H_{17}N_2$ (M+H): Calc: 141.1386; Found: 141.1373; $[\alpha]_D^{24}$ 14.82 (*c* = 1.01, MeOH); Anal. Calcd for $C_8H_{16}N$: C, 68.52; H, 11.50, N, 19.98. Found: C, 68.45; H, 11.64; N, 19.79.

19. For the derivatization procedure see Note 14. The analytical data for (*S,S*)-**3** are as follows: ^1H NMR (500 MHz, $CDCl_3$) δ: 1.78–2.05 (m, 6 H), 2.20–2.27 (m, 2 H), 3.22 (dt, *J* = 10.4, 7.8, 2 H), 3.80 (ddd, *J* = 10.6, 8.8, 5.1, 2 H), 4.60–4.64 (m, 2 H), 7.23–7.33 (m, 6 H), 7.38–7.44 (m, 4 H); ^{13}C NMR (126 MHz, $CDCl_3$) δ: 24.3, 28.4, 49.3, 59.0, 127.4, 128.4, 129.8, 137.4, 171.2; MS (EI, 70 eV): 348, 175, 174, 105; HRMS (*m/z*) Calc. $C_{22}H_{25}N_2O_2$ (M+H): 349.1911; Found: 349.1917; Supercritical Fluid Chromatography: t_R (*R,R*)-**3** 2.86 min (0%); t_R (*S,S*)-**3** 3.26 min (100%) (Chiralpak AS, 40°C, 150 bar, 15% MeOH in CO_2, 3.0 mL/min, 220 nm); HPLC: (*S,S*)-**3** t_R 8.2 min (100%); (*R,R*)-**3** t_R 13.2 min (0%) (Chiralpak AD, *i*-PrOH/hexane, 95/5, 0.7 mL/min)

Safety and Waste Disposal Information

All hazardous materials should be handled and disposed of in accordance with "Prudent Practices in the Laboratory"; National Academy Press; Washington, DC, 1995.

3. Discussion

C_2-Symmetric chiral diamines have found extensive application as additives, auxiliaries and catalysts in asymmetric synthesis.[5] (*R,R*)-2,2'-Bispyrrolidine, initially developed by Hirama, has been successfully applied as a ligand for osmium tetraoxide in the asymmetric dihydroxylation of olefins[6] (eq 2) and as a ligand in asymmetric hydrogenation.[7]

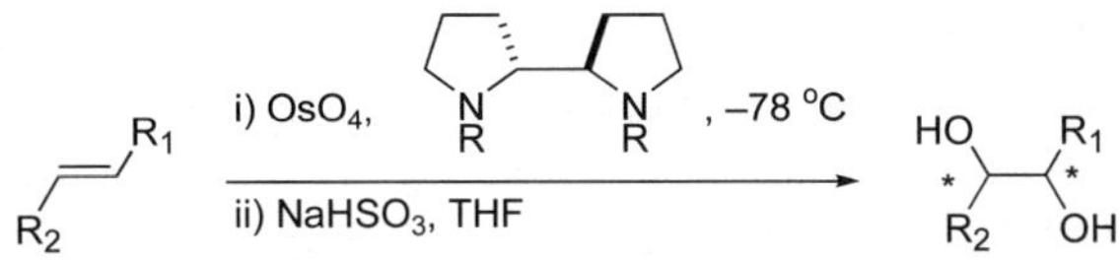

Several syntheses of enantiopure 2,2'-bispyrrolidine have been reported.[4,8,9] The first synthesis described by Masamune and coworkers requires only two steps, but produces a *d,l/meso* mixture of isomers in a sluggish and irreproducible heterogeneous hydrogenation. This short synthesis arrives as the final product by direct resolution of the *d,l/meso* mixture of 2,2'-bispyrrolines.[4] The other routes produce enantiopure 2,2'-bispyrrolines without resolution, but they require multiple-step syntheses from chiral starting materials. For example, the synthesis developed by Kotsuki and coworkers takes 11 steps from (D)-tartaric acid.[8] Most recently, Alexakis reported a five-step synthesis of (*R*,*R*)-2,2'-bispyrrolidine by asymmetric addition to a chiral imine.[9] In the procedure described herein, the *d,l/meso* mixture of 2,2'-bispyrrolidines is easily synthesized on a large scale by the photodimerization of pyrrolidine developed by Crabtree.[2] The previously reported resolution[4] has been modified such that both enantiomers can be obtained.

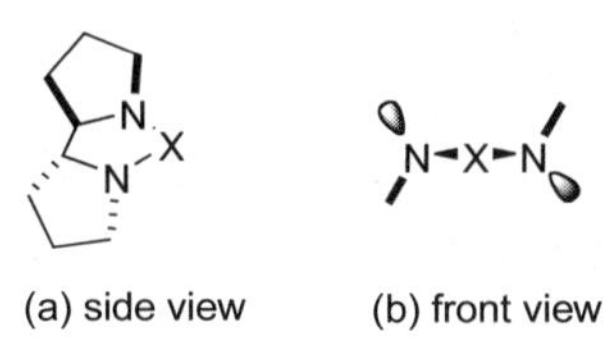

Figure 2: Stair-like structure of two pyrrolidine rings

This diamine possesses very interesting structural features that impart useful characteristics as a bidentate ligand. When the two nitrogen atoms function either in a chelate or are covalently bonded to another atom, the two pyrrolidines adopted a stair-like structure, which creates a highly asymmetric environment (Figure 2). This feature was recently exploited in the development of a highly selective catalyst for asymmetric allylations (Figure 3).[3] The addition of allylic trichlorosilanes to unsaturated aldehydes can be catalyzed by chiral bisphosphoramide **4** derived from 2,2'-bispyrrolidine to give homoallylic alcohols with excellent diastereo- and enantioselectivities. Of particular note is the catalytic enantioselective construction of quaternary centers by the use of γ-disubstituted allylic silanes. The unique structural features of this diamine together with the ease of preparation bode well for further application in asymmetric synthesis.

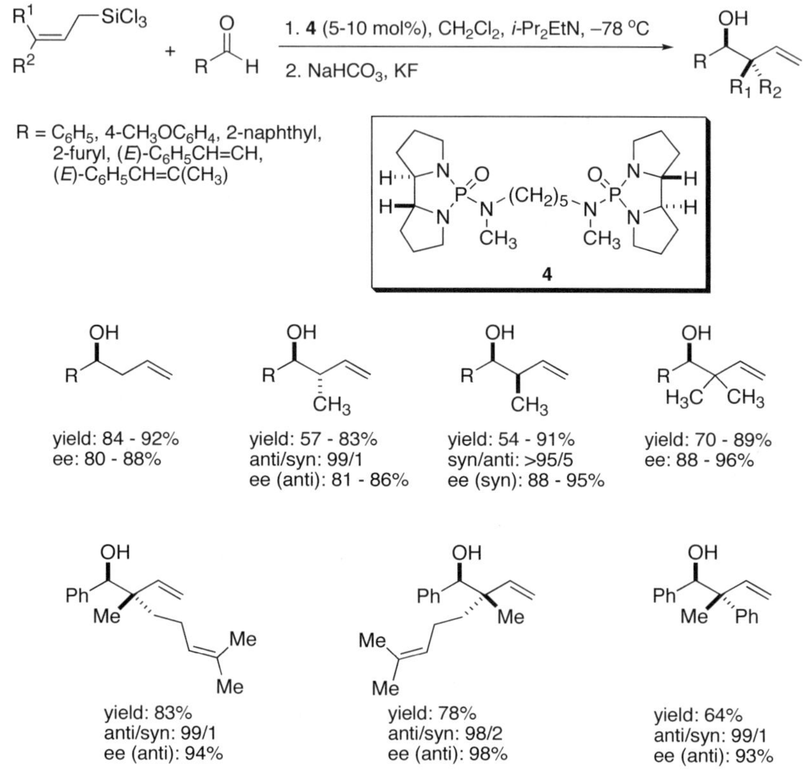

Figure 3. Enantioselective addition of allylic trichlorosilanes catalyzed by 2,2-bispyrrolidine-derived bisphosphoramide **4**.

1. Department of Chemistry, University of Illinois at Urbana-Champaign, Urbana, IL, 61801.
2. Ferguson, R. R.; Boojamra, C. G.; Brown, S. H.; Crabtree, R. H. *Heterocycles* **1989**, *28*, 121.
3. Denmark, S. E.; Fu, J. *J. Am. Chem. Soc.* **2001**, *123*, 9488.
4. Oishi, T.; Hirama, M.; Sita, L. R.; Masamune, S. *Synthesis* **1991**, 789.
5. (a) Bennani, Y. L.; Hannessian, S. *Chem. Rev.* **1997**, *97*, 3161. (b) Lucet, D.; Le Gell, T.; Mioskowski, C. *Angew. Chem. Int. Ed.* **1998**, *37*, 2580.
6. (a) Hirama, M.; Oishi, T.; Ito, S. *J. Chem. Soc., Chem. Commun.* **1989**, 665. (b) Oishi, T.; Hirama, M. *J. Org. Chem.* **1989**, 54, 5834.
7. Hamada, T.; Izawa, K. *Eur. Pat. Appl.* 987271, **2000**.

8. Kotsuki, H.; Kuzume, H.; Gohda, T.; Fukuhara, M.; Ochi, M.; Oishi, T.; Hirama, M.; Shiro, M. *Tetrahedron: Asymmetry,* **1995**, *6*, 2227.

9. Alexakis, A.; Tomassini, A.; Chouillet, C.; Roland, S.; Mangeney, P.; Bernardinelli, G. *Angew. Chem. Int. Ed.* **2000**, *39*, 4093.

LIPASE-CATALYZED RESOLUTION OF 4-TRIMETHYLSILYL-3-BUTYN-2-OL AND CONVERSION OF THE (R)-ENANTIOMER TO (R)-3-BUTYN-2-YL MESYLATE AND (P)-1-TRIBUTYLSTANNYL-1,2-BUTADIENE

[3-Butyn-2-ol, 4-(trimethylsilyl)-, (2*R*)- and Stannane, 1,2-butadienyltributyl-, *(P)*-]

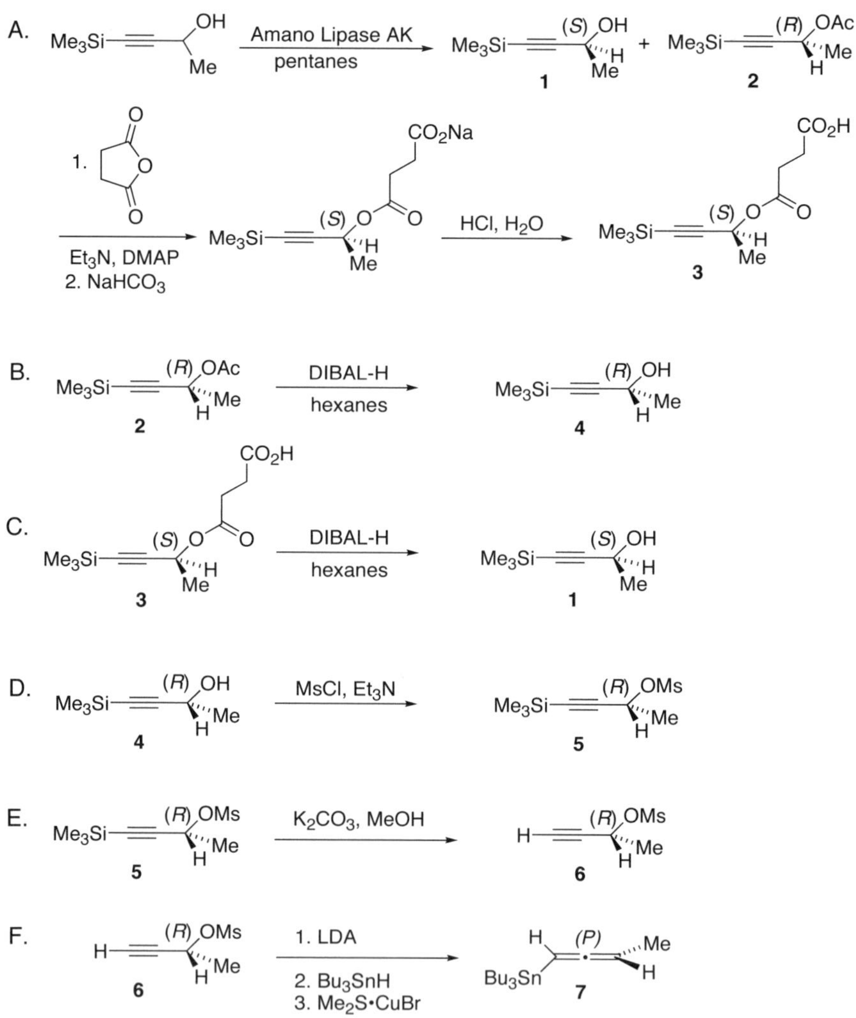

Submitted by James A. Marshall and Harry Chobanian.[1]
Checked by Peter Wipf and Joshua Pierce.

1. Procedure

A. (R)-4-Trimethylsilyl-3-butyn-2-yl Acetate (**2**) *and (S)-4-Trimethylsilyl-3-butyn-2-yl Succinate* (**3**). An oven dried, 1-L, round-bottomed flask, equipped with a large magnetic stir bar and rubber septum with nitrogen inlet, is charged with racemic 4-trimethylsilyl-3-butyn-2-ol (**1**) (10.0 g, 69.9 mmol) (Note 1) and pentane (250 mL) (Note 2). To this stirred solution is added Amano Lipase AK (Note 3) (2.00 g), freshly distilled vinyl acetate (50 mL), and pulverized, activated 4Å molecular sieves (1 g). The mixture is stirred at room temperature for 72 h at which point analysis by GC (Note 4) indicated the reaction had proceeded to 50% completion. The mixture is filtered through a medium porosity sintered glass funnel, washed with additional pentane, and concentrated via rotary evaporation affording 11.5 g of a nearly 1:1 mixture of alcohol and acetate by ^{1}H NMR analysis. To this mixture in THF (50 mL) is added Et_3N (9.1 mL, 65.0 mmol), DMAP (92 mg, 0.75 mmol) and succinic anhydride (4.15 g, 41.1 mmol) successively. The mixture is heated to reflux for 4 h, cooled, and quenched with 40 mL of sat. aq. $NaHCO_3$ to adjust the pH to ≥9. The solution is stirred vigorously for 1 h and diluted with ethyl acetate (EtOAc) (75 mL). The EtOAc solution is separated, washed with 10% HCl (200 mL) and brine (200 mL), dried over $MgSO_4$, filtered, and concentrated by rotary evaporation. The resulting oil is purified by bulb-to-bulb distillation (65 °C at 0.5 mmHg) (Note 5) to yield 6.10-6.11 g (94%) of (*R*)-4-trimethylsilyl-3-butyn-2-yl acetate (**2**) as a clear oil (Note 6). The aqueous phase is carefully acidified with 12 M HCl (~5 mL) to pH ~ 1 and extracted with Et_2O (4 x 150 mL). The combined Et_2O extracts are dried over $MgSO_4$, filtered, and concentrated by rotary evaporation affording 8.47–8.48 g (99%) of (*S*)-4-trimethylsilyl-3-butyn-2-yl succinate (**3**) as a light yellow oil which solidified upon cooling to 0 °C. The acid is carried on without further purification (Note 7).

B. (R)-4-Trimethylsilyl-3-butyn-2-ol (**4**). To an oven dried, 100-mL, round-bottomed flask, equipped with a magnetic stir bar and rubber septum with nitrogen inlet is added acetate **2** (6.11 g, 33.2 mmol) in 40 mL of hexanes at -78 °C followed by 50 mL of DIBAL-H (1.0 M in hexanes). The solution is stirred for 10 min and poured into a rapidly stirred mixture of 300 mL of aqueous Rochelle's salt and 200 mL of Et_2O. Once the Et_2O layer clarifies, it is separated, dried over $MgSO_4$, filtered, and concentrated by distillation at atmospheric pressure to yield an oil. The oil is purified by bulb-to-bulb distillation (80 °C at 0.5 mmHg) to yield 4.27-4.25 g (87%) of (*R*)-4-trimethylsilyl-3-butyn-2-ol (**4**) (Note 8).

C. (S)-4-Trimethylsilyl-3-butyn-2-ol (**1**). To an oven dried, 250-mL, round-bottomed flask, equipped with a magnetic stir bar and rubber septum with nitrogen inlet is added a solution of the foregoing succinate **3** (8.48 g, 34.8 mmol) in 100 mL of CH_2Cl_2 at -78 °C followed by 77 mL of DIBAL-H (1.0 M in hexanes). The solution is stirred for 10 min and the product is isolated as described for (*R*)-4-trimethylsilyl-3-butyn-2-ol (**4**). The resulting oil is purified by bulb-to-bulb distillation (75-80 °C at 0.5 mmHg) to yield 4.26-4.28 g (86%) of (*S*)-4-trimethylsilyl-3-butyn-2-ol (**1**) (Notes 9 and 10).

D. (R)-4-Trimethylsilyl-3-butyn-2-yl Mesylate (**5**). An oven dried, 500-mL, round bottom flask equipped with a magnetic stir bar and rubber septum with nitrogen inlet is charged with (*R*)-4-trimethylsilyl-3-butyn-2-ol (**4**) (4.28 g, 30.1 mmol) and 350 mL of CH_2Cl_2 at -78 °C. To this solution is added Et_3N (8.50 mL, 61.0 mmol) followed by methanesulfonyl chloride (5.23 g, 46.0 mmol). The solution is stirred for 1 h at -78 °C before being quenched with 10 mL of sat. aq. $NaHCO_3$ solution. The solution is allowed to warm to room temperature before being concentrated at reduced pressure by rotary evaporation. The residual oil is diluted with distilled H_2O (200 mL) and Et_2O (100 mL). The Et_2O layer is separated, dried over $MgSO_4$, filtered, and concentrated to afford 6.07-6.09 g (91%) of (*R*)-4-trimethylsilyl-3-butyn-2-yl mesylate (**5**), which can be used without further purification (Note 11).

E. (R)-3-Butyn-2-yl Mesylate (**6**). To a 100-mL round-bottomed flask, equipped with a magnetic stir bar, is added a solution of (*R*)-4-trimethylsilyl-3-butyn-2-yI mesylate (6.09 g, 27.6 mmol) in methanol (20 mL) followed by addition of anhydrous K_2CO_3 (5.76 g, 42.0 mmol). After 15 min, TLC analysis (1:1; EtOAc:hexanes) of the mixture indicates complete cleavage of the TMS group. The solution is diluted with 40 mL of sat NaCl solution and extracted with diethyl ether (Et_2O) (3 x 50 mL). The combined ether layers are washed with distilled H_2O (3 x 50 mL) to remove residual methanol. The ether layer is separated, dried over $MgSO_4$, filtered, and concentrated under reduced pressure (Note 12) to yield 3.48 g (85%) of (*R*)-3-butyn-2-yl mesylate (**6**) as a clear oil that can be used without further purification (Note 13).

F. (P)-(+)-3-(Tributylstannyl)-1,2-butadiene (**7**). An oven dried, 250-mL, round-bottomed flask equipped with a magnetic stir bar and a rubber septum with a nitrogen inlet is charged with diisopropylamine (4.0 mL, 28.0 mmol) and THF (100 mL). To this solution is added 11.2 mL (28.0 mmol) of n-BuLi (2.5 M in hexanes) at 0 °C. After 30 min, tributyltin hydride (8.15 g, 28.0 mmol) (Note 14) is added dropwise and the mixture is stirred an additional 30 min. The yellow solution is then cooled to -78 °C and 5.76 g (28.0 mmol) of $CuBr{\cdot}SMe_2$ is added portion-wise (Note 15). Once addition is complete, the dark solution is stirred an additional 30 min before (*R*)-3-butyn-2-yl mesylate (**6**) (Note 16) is added. After 15 min, the solution is poured into a rapidly stirred solution of 400 mL of 9:1 sat. aq. NH_4Cl/NH_4OH solution and 300 mL of ether. Once the ether layer clarifies, it is separated, dried over $MgSO_4$, filtered, and concentrated under reduced pressure. The resulting oil is purified by bulb-to-bulb distillation (120 °C/ 0.5 mmHg) to yield 5.77-5.79 g (88%) of (*P*)-(+)-3-(tributylstannyl)-1,2-butadiene as a yellow oil (Note 17).

2. Notes

1. 4-Trimethylsilyl-3-butyn-2-ol was obtained from Gelest, Inc. (Submitters) or from Lancaster Research Chemicals (Checkers). All other chemicals were purchased from Aldrich Chemical Company, Inc. and used as received.

2. The selectivity of the resolution diminished at a higher concentration of butynol. A decrease in the volume of pentane from 250 mL to 125 mL led to acetate with er = 88:12 at 50% conversion.

3. Amano Lipase AK from *Pseudomonas fluorescens* was purchased from Aldrich Chemical Company, Inc.

4. The reaction progress was monitored by GC (Carbowax, 110 °C, 1 °C ramp/min; acetate 5.75 min; alcohol 8.22 min. or β-DEX, 89 °C, 0.2 °C ramp/min, alcohol 31.30 min, acetate 32.37 min).

5. Bulb-to-bulb (short-path) distillation was performed with an Aldrich Kugelrohr distillation apparatus.

6. Physical characteristics of (*R*)-2-acetoxy-4-trimethylsilyl-3-butyne (**2**): $[\alpha]_D^{20}$ +119 (c = 2.2, $CHCl_3$); IR (film) cm^{-1}: 2181, 1747; ^1H NMR (300 MHz, $CDCl_3$) δ 0.15 (s, 9H), 1.45 (d, *J* = 6.7, 3H), 2.05 (s, 3H), 5.44 (q, *J* = 6.7, 1H). ^{13}C NMR (75 MHz, $CDCl_3$) δ -0.27, 21.1, 21.4, 60.6, 89.4, 103.5, 169.7. Analysis by GC on a β-Dex column showed a single peak at 32.0 min (80 °C, 0.2 °C ramp/min). The racemic acetate gave rise to peaks at 31.66 and 32.42 min under these conditions.

7. Spectral analysis for (*S*)-4-trimethylsilyl-3-butyn-2-yl succinate (**3**): ^1H NMR (300 MHz, $CDCl_3$) δ 0.16 (s, 9H), 1.47 (d, *J* = 6.6, 3H), 2.50-2.80 (m, 4H), 5.49 (q, *J* = 6.6, 1H).

8. Physical characteristics of (*R*)-4-trimethylsilyl-3-butyn-2-ol (**4**): $[\alpha]_D^{20}$+22.4 (c = 2.01, $CHCl_3$); IR (film) cm^{-1} υ: 3334, 2175; ^1H NMR (300 MHz, $CDCl_3$) δ 0.17 (s, 9H), 1.45 (d, *J* = 6.6, 3H), 1.83 (d, *J* = 5.3, 1H), 4.52 (app p, *J* = 6.3, 1H); ^{13}C NMR (75 MHz, $CDCl_3$) δ -0.20, 24.1, 58.5, 88.2, 107.7.

9. (*S*)-4-Trimethylsilyl-3-butyn-2-ol (**1**): $[\alpha]_D^{20}$–22.3 (c = 2.55, $CHCl_3$); lit. (–25.9, c = 3.12).[2]

10. The acetate derivative of (*S*)-4-trimethylsilyl-3-butyn-2-ol was prepared by treatment with excess acetic anhydride, Et_3N, and DMAP in CH_2Cl_2. Analysis of an aliquot by GC on a β-Dex column (80 °C, 0.2 °C ramp/min) revealed a 98:2 mixture of enantiomers.

11. Physical characteristics of (*R*)-4-trimethylsilyl-3-butyn-2-yl mesylate (**5**): $[\alpha]_D^{20}$ +98.4 (c = 2.30, $CHCl_3$); IR (film) cm^{-1}: 2961, 2175; ^1H NMR (300 MHz, $CDCl_3$) δ 0.15 (s, 9H), 1.58 (d, *J* = 6.7, 3H), 3.07 (s, 3H), 5.20 (q, *J* = 6.7, 1H); ^{13}C NMR (75 MHz, $CDCl_3$) δ –0.63, 22.3, 38.9, 68.3, 93.4, 101.1.

12. Care must be taken not to heat the bath during removal of the solvent from the relatively volatile mesylate solution on the rotary evaporator.

13. Physical characteristics of (*R*)-3-butyn-2-ol mesylate (**6**): $[\alpha]_D^{20}$ +92.9 (c = 2.49, $CHCl_3$); IR (film) cm^{-1}: 3282, 2942, 2124; ^1H NMR (300 MHz, $CDCl_3$) δ 1.63 (d, *J* = 6.6, 3H), 2.72 (d, *J* = 1.8, 1H), 3.10 (s, 3H), 5.27 (dq, *J* = 1.7, 6.4 Hz, 1H); ^{13}C NMR (75 MHz, $CDCl_3$) δ 22.3, 39.0, 67.4, 76.3, 80.0.

14. Tributyltin hydride was purchased from Lancaster Synthesis, Inc. With certain batches of tributyltin hydride the submitters found that extending the stirring time with LDA to 3 hours improved the yield.

15. Addition of the $CuBr{\cdot}SMe_2$ over 5 min in small portions is important to the success of the reaction.

16. After addition of the mesylate to the stirred reaction mixture, it is necessary to quench the reaction within 10 min to minimize copper-catalyzed racemization of the allenylstannane. In our original communication of this methodology[3] we noted that, contrary to the findings of others[4], who reported partial racemization of allenes prepared through additions of alkylcuprates to propargylic esters, the acetoxymethyl stannane (eq 5, R^1 = $AcOCH_2$, R^2 = Me) was not racemized by prolonged exposure to the stannylcupration conditions. It appears that, perhaps not surprisingly, structural factors can influence the configurational stability of allenes

prepared in this manner. It is also possible that contaminants in the copper salts or impurities of an undetermined nature could be responsible for the racemization, which is mechanistically obscure at this time.

17. Physical characteristics of (*P*)-(+)-3-(tributylstannyl)-1,2-butadiene (**7**): $[\alpha]_D^{20}$ +61.2 (c = 3.31, $CHCl_3$); IR (film) cm^{-1}: 2924, 1928; ^1H NMR (300 MHz, $CDCl_3$) δ 0.80-1.70 (m, 30 H), 4.45-4.71 (m, 1H), 4.87-5.12 (m, 1H). ^{13}C NMR (75 MHz, $CDCl_3$) δ 10.3, 13.7, 27.2, 28.9, 74.3, 75.2, 209.0. The ee of the stannane was determined by analysis of the product obtained through $BF_3{\cdot}OEt_2$-promoted addition to isobutyraldehyde. To a stirred solution of (*P*)-(+)-3-(tributylstannyl)-1,2-butadiene (1.97 g, 57.5 mmol) and isobutyraldehyde (300 mg, 42 mmol) in CH_2Cl_2 at -78 °C was added $BF_3{\cdot}OEt_2$ (2.18 mL, 172 mmol). The solution was maintained at -78 °C for 1 h then quenched by the addition of sat. aq. $NaHCO_3$ and warmed to rt. The CH_2Cl_2 layer was removed and stirred vigorously with KF-on-Celite[5] and $MgSO_4$. After 1 h, analysis of an aliquot by GC on a β-Dex column indicated a 98:2 ratio of enantiomers, and a >99:1 ratio of diastereomers (80 °C, 0.2 °C ramp/min; alcohol 21.8 min).

Waste Disposal Information

All hazardous materials should be handled and disposed of in accordance with "Prudent Practices in the Laboratory"; National Academy Press; Washington, DC, 1995.

3. Discussion

The present resolution procedure is based on a report by Burgess and Jennings,[2] but with several key modifications.

1) To minimize losses of the volatile products during solvent removal, pentane was substituted for hexane.

2) For the same reason, and to facilitate larger scale resolutions, the substrate concentration was increased from 0.1 to ~0.2M. However, an increase to 0.4 M led to diminished efficiency.

3) In the original report, the (*S*)-alcohol **1** and (*R*)-acetate **2** were separated by column chromatography which led to isolated yields of 90% for the acetate and only 54% for the alcohol. In the present procedure the alcohol is separated from the acetate by conversion to the half succinate, which is extracted with aqueous sodium bicarbonate. Purification of the acetate is achieved by short-path distillation. After cleavage of the ester

groupings through reduction with DIBAL-H, both the (*S*) and (*R*)-alcohols are isolated following distillation in ~86% overall yield.

The (*R*)-TMS butynyl mesylate **5**, or its enantiomer, are converted *in situ* to the (*P*)-allenyl indium or zinc reagents by reaction with InI or Et_2Zn utilizing $Pd(OAc)_2{\cdot}PPh_3$ as the precatalyst.[6] These metallations are conducted in the presence of various aldehydes to produce *anti* homopropargylic alcohol adducts **8** or **9** of high diastereomeric and enantiomeric purity (eq 1). Additions to chiral α and β-oxygenated aldehydes proceed with a high degree of reagent control to yield *syn*, *anti*

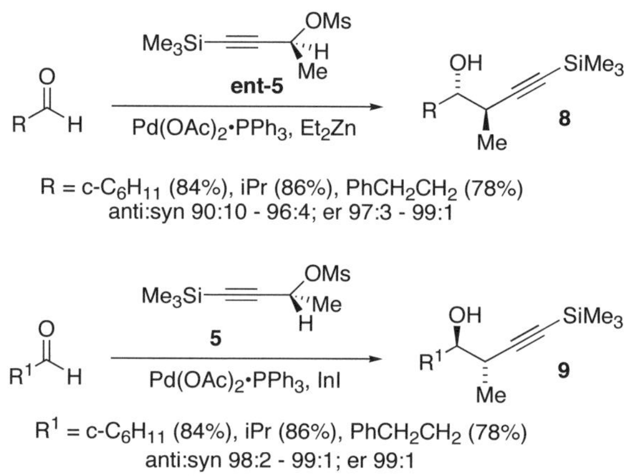

(1)

and *anti*, *anti* stereotriads (eq 2, 3).[7] These additions can also be effected with the corresponding allenyl indium and zinc reagents derived from the terminal alkynyl mesylate **6**, or its enantiomer (eq 4).[8] However, those additions proceed with significantly lower diastereoselectivity, most notably with unbranched aldehydes.

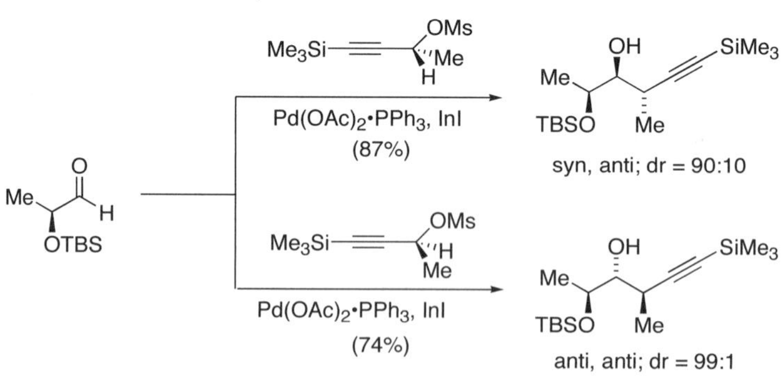

(2)

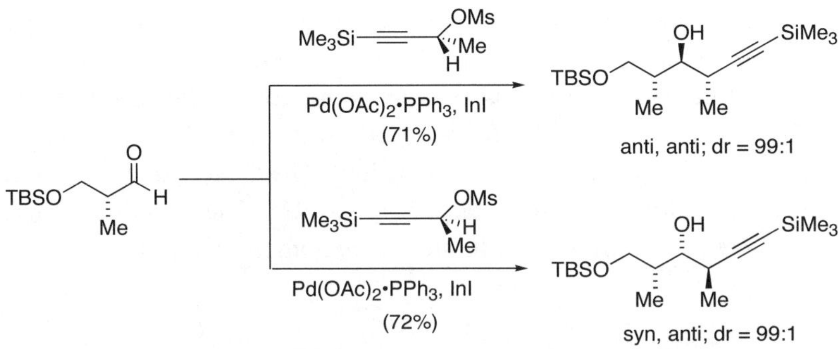

(3)

Propargylic mesylates, such as **6**, can be converted to (*P*)-allenylstannanes through S_N2' displacement of the mesylate with a tributyltin cuprate reagent (eq 5).[9,10] These reagents are formed with inversion of stereochemistry and most are sufficiently stable to be purified by distillation. The reaction proceeds with high levels of enantioselectivity, provided the

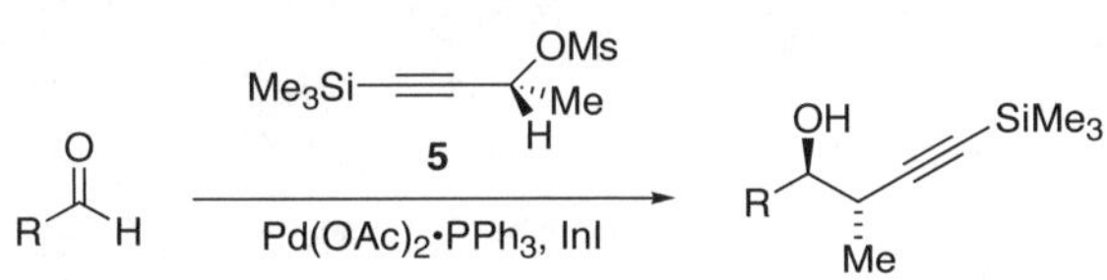

(4)

R = c-C_6H_{11} (86%), iPr (71%), $PhCH_2CH_2$ (71%)
anti:syn 77:23 - 95:5; er 99:1

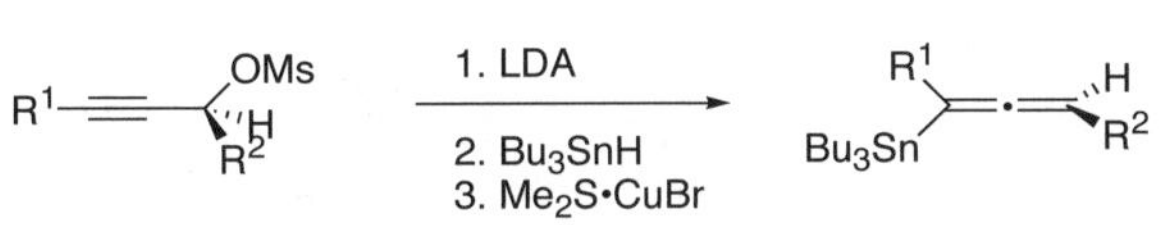

(5)

product is removed from the cuprate salts within 10 min. The allenyl tin reagents afford *syn* homopropargylic alcohol adducts upon addition to

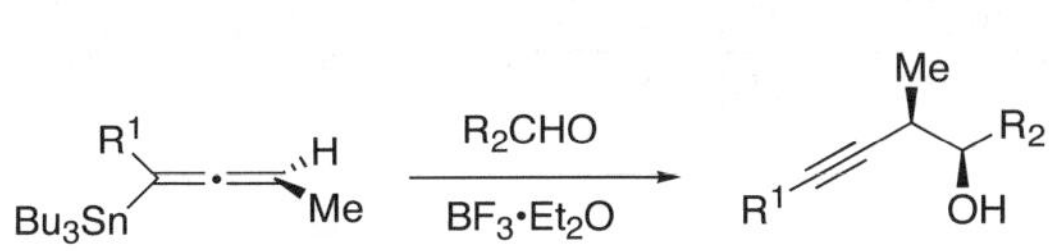

(6)

R^1	R^2	Yield (%)	syn:anti
C_7H_{15}	C_6H_{13}	83	37:63
C_7H_{15}	t-Bu	92	99:1
C_7H_{15}	i-Pr	95	95:5
CH_2OAc	i-Pr	98	95:5

aldehydes in the presence of excess $BF_3{\cdot}OEt_2$ (eq 6). Matching/mismatching occurs in additions to α-chiral aldehydes (eq 7). Allenylstannanes undergo transmetallation with $BuSnCl_3$, $SnCl_4$, or InX_3 salts in the presence of aldehydes to afford allenylcarbinols or *anti* homopropargylic alcohols (eq 8).[11-13] The $SnCl_4$ transmetallations proceed with overall inversion of stereochemistry whereas the InX_3 reactions lead to allenylindium intermediates with retention of stereochemistry (eq 9).[14] The resulting allenyl $SnCl_3$ and InX_2 reagents afford enantiomeric *anti* homopropargylic

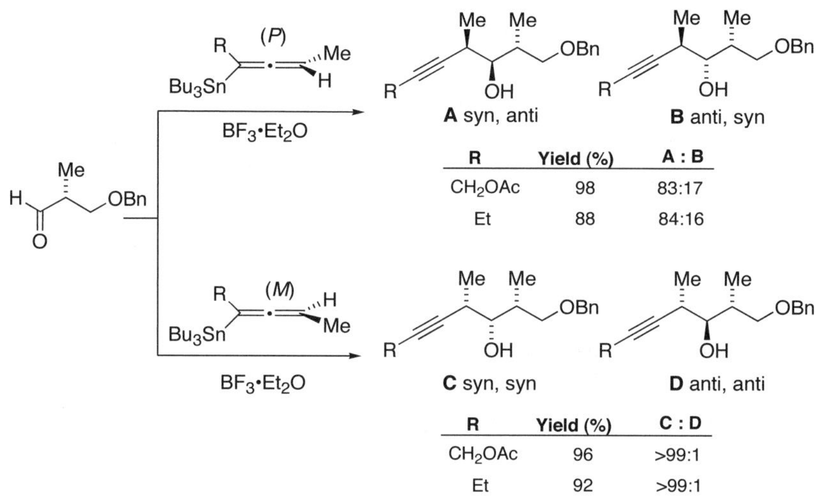

R	Yield (%)	A : B
CH_2OAc	98	83:17
Et	88	84:16

R	Yield (%)	C : D
CH_2OAc	96	>99:1
Et	92	>99:1

(7)

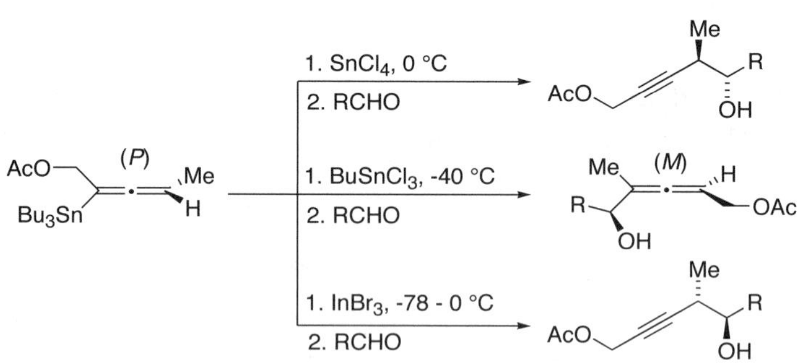

(8)

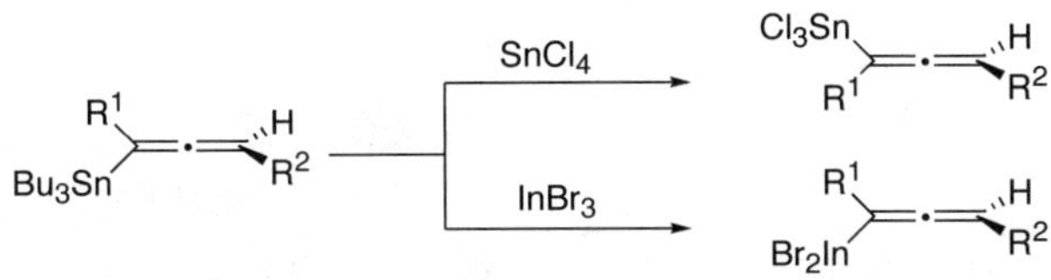

(9)

alcohols upon addition to aldehydes (eq 8). The $BuSnCl_3$ reaction proceeds via the transmetallated propargylic stannanes (eq 10). All three of the foregoing additions proceed through cyclic transition states.

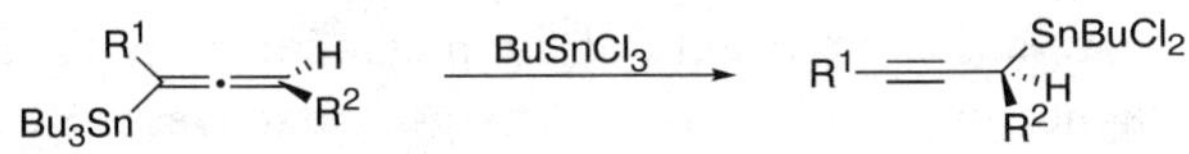

(10)

Additions to α-chiral aldehydes exhibit matching/mismatching characteristics (eq 11-13). Through appropriate selection of chiral partners and allenylmetal reagents, any of the eight possible stereotriads can be prepared by this methodology. The allenylcarbinol products are stereospecifically converted to dihydrofurans upon treatment with 5-10 mol% $AgNO_3$.[13]

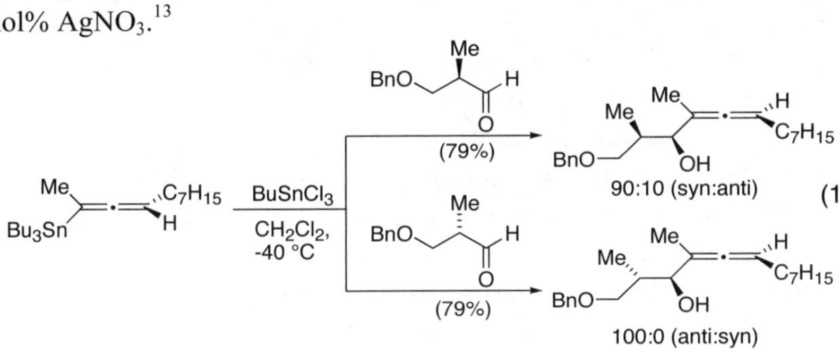

(11)

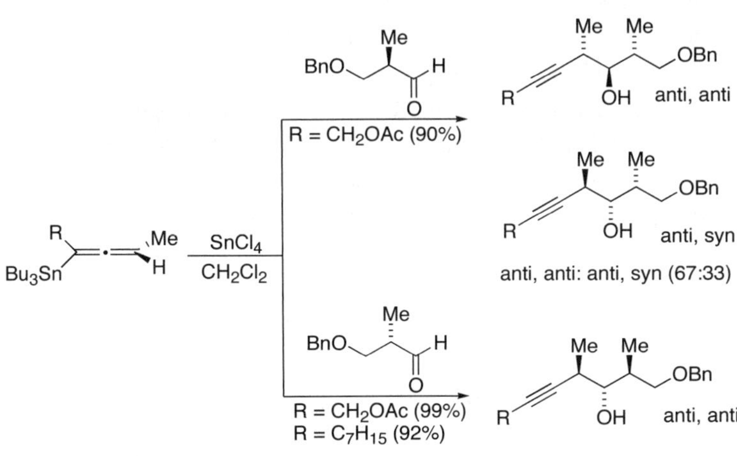

(12)

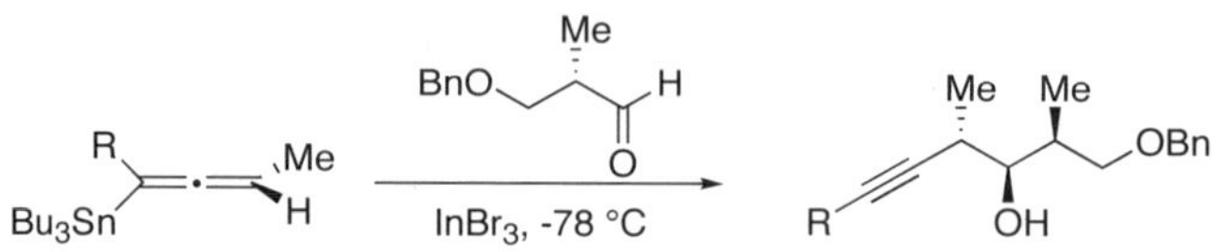

(13)

R = CH_2OAc, anti, syn:anti, anti = 95:5
R = H, anti, syn:anti, anti = 92:8

Chiral allenylmetal reagents have been utilized with excellent effectiveness in the synthesis of a variety of polyketide natural products of marine origin.[15, 16] An attractive feature of the methodology is the ability to incorporate the alkyne function of the adducts as a reactive component for elaboration of the polypropionate and polyacetate backbone of these compounds.

1. Department of Chemistry, McCormick Road, P.O. Box 400319, University of Virginia, Charlottesville, VA 22904.

2. Burgess, K.; Jennings, L. D. *J. Am. Chem. Soc.* **1991**, *113*, 6129.

3. Marshall, J. A.; Wang, X-J. *J. Org. Chem.* **1991**, *56*, 3211.

4. (a) LeQuan, M.; Cadiot, P. *Bull. Soc. Chem. Fr.* **1965**, *45*. (b) LeQuan, M.; Guillem, G. J. *Organomet. Chem.* **1973**, *54*, 153. (c) Claesson, A.; Olsson, L. I. *J. Chem. Soc. Chem. Commun.* **1979**, 524.

5. Savall, B. M.; Powell, N. A.; Roush, W. R. *Org. Lett.* **2001**, *3*, 3057.

6. (a) Marshall, J. A.; Adams, N. D. *J. Org. Chem.* **1999**, *64*, 5201. (b) Marshall, J. A.; Grant, C. M. *J. Org. Chem.* **1999**, *64*, 8214.

7. Marshall, J. A.; Chobanian, H. R. *J. Org. Chem.* **2000**, *65*, 8357

8. Marshall, J. A.; Chobanian, H. R.; Yanik, M. M. *Org. Lett.* **2001**, *3*, 3369.

9. Marshall, J. A.; Wang, X-J. *J. Org. Chem.* **1990**, *55*, 6246.

10. Marshall, J. A.; Wang, X-J. *J. Org. Chem.* **1992**, *57*, 1242.

11. Marshall, J. A.; Perkins, J. *J. Org. Chem.* **1994**, *59*, 3509.

12. Marshall, J. A.; Perkins, J. F.; Wolf, M. A. *J. Org. Chem.* **1995**, *60*, 5556.

13. Marshall, J. A.; Yu, R. H.; Perkins, J. F. *J. Org. Chem.* **1995**, *60*, 5550.

14. Marshall, J. A.; Palovich, M. R. *J. Org. Chem.* **1997**, *62*, 6001.

15. (a) Marshall, J. A.; Johns, B. A. *J. Org. Chem.* **1998**, *63*, 7885. (b) Marshall, J. A.; Fitzgerald, R. N. *J. Org. Chem.* **1999**, *64*, 4477. (c) Marshall, J. A.; Johns, B. A. *J. Org. Chem.* **2000**, *65*, 1501. (d) Marshall, J. A.; Adams, N. D. *Org. Lett.* **2000**, *2*, 2897. (e) Marshall, J. A.; Yanik, M. M. *J. Org. Chem.* **2001**, *66*, 1373. (f) Marshall, J. A.; Schaaf, G. M. *J. Org. Chem.* **2001**, *66*, 7825. (g) Marshall, J. A.; Adams, N. D. *J. Org. Chem.* **2002**, *67*, 733.

16. Several alternative non-resolution, large-scale routes to (*R*)- and (S)-3-butyn-2-ol have also been developed. The first two procedures require stoichiometric amounts of triphenylphosphine, whereas the latter employs lithiodichloromethane as the alkyne precursor; (a) Ku, Y.-Y.; Patel, R. R.; Elisseou, E. M.; Sawich, D. P. *Tetrahedron Lett.* **1995**, *36*, 2733. (b) Marshall, J. A.; Xie, S. *J. Org. Chem.* **1995**, *60*, 7230. (c) Marshall, J. A.; Yanik, M. M.; Adams, N. D.; Ellis, K. C.; Chobanian, H. R. *Org. Syn.* **2004**, *81*, 157.

ASYMMETRIC ALCOHOLYSIS OF MESO-ANHYDRIDES MEDIATED BY ALKALOIDS (Bicyclo[2.2.1]hept-5-ene-2-endo-carboxylic acid, 3-endo-benzyloxycarbonyl, (2*R*,3*S*)-)

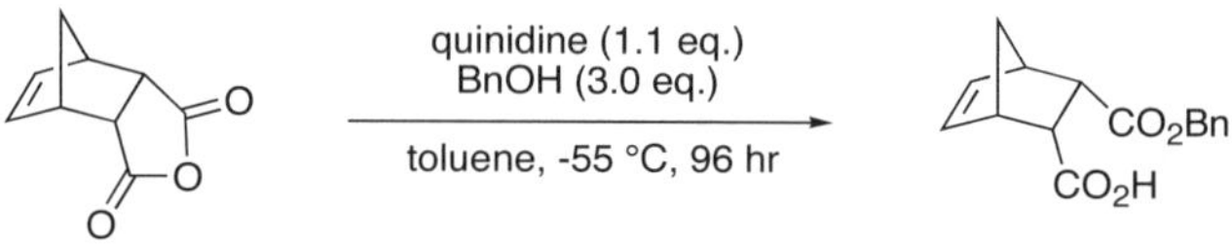

Submitted by Carsten Bolm,[1] Iuliana Atodiresei and Ingo Schiffers.
Checked by Motomu Kanai and Masakatsu Shibasaki.

1. Procedure

A flame-dried 250-mL single-necked, round-bottomed flask equipped with a magnetic stirring bar and charged with quinidine (7.14 g, 22 mmol) (Note 1) and *endo*-bicyclo[2.2.1]hept-5-ene-2,3-dicarboxylic anhydride (3.28 g, 20 mmol) (Note 2) is placed under vacuum for 2 h (Note 3). The evacuated flask is flushed with argon, charged with 100 mL of dry toluene (Note 4), equipped with a rubber septum and the mixture is cooled to -55 °C (Note 5). Benzyl alcohol (6.2 mL, 6.49 g, 60 mmol) (Note 6) is added dropwise via syringe (over a period of 10 min) to the cooled suspension (Note 7), and the reaction mixture is stirred at the indicated temperature for 96 h (Note 8), during which the solid material gradually dissolves. The resulting clear solution is concentrated *in vacuo* to dryness, and the resulting residue is dissolved in diethyl ether (125 mL). The solution is washed with 2 N HCl (3 × 30 mL), and the aqueous layer is back-extracted with ether (5 × 50 mL). The combined organic layers are extracted with a saturated solution of sodium bicarbonate (5 × 75 mL), and the resulting aqueous phase is washed with diethyl ether (1 × 100 mL) in order to remove the traces of benzyl alcohol. The aqueous phase is acidified with 8 N HCl, extracted with CH_2Cl_2 (5 × 100 mL), and the combined organic layers are dried ($MgSO_4$), filtered and concentrated to provide 4.97-5.20 g (91-95%, 97-98% ee) of the benzyl hemiester as a white solid (Note 9). The enantiomeric excess of the half ester was analyzed by chiral HPLC analysis (Note 10). Alternatively, the enantiomeric excess of the benzyl ester[2] could be determined by GC analysis of the corresponding lactone,[3] which was obtained by selective reduction of the ester group with lithium triethylborohydride ($LiBEt_3H$) followed by acid-catalyzed lactonization (Note 11).[4]

Use of quinine in the ring opening of *endo*-bicyclo[2.2.1]hept-5-ene-2,3-dicarboxylic anhydride provides 5.13 g (94%) of the corresponding enantiomeric benzyl hemiester as a white solid (Note 16).

2. Notes

1. Anhydrous quinidine (95%) and quinine (99%) were purchased from Acros Organics and used as supplied.

2. *Endo*-bicyclo[2.2.1]hept-5-ene-2,3-dicarboxylic anhydride (97%) was purchased from Fluka and used as received.

3. High vacuum is used in order to remove traces of moisture.

4. Toluene (Merck, >99%), distilled from sodium benzophenone ketyl radical under argon, was stored over 4 Å molecular sieves.

5. An RL 6 CP type cooling machine was used in order to maintained the low temperature.

6. Anhydrous benzyl alcohol (99.8%) was purchased from Sigma-Aldrich and used as supplied.

7. The mixture is stirred for at least 1 h at the indicated temperature before the benzyl alcohol addition.

8. The reaction has been studied extensively on a 1-mmol scale, and the best asymmetric induction was achieved when the reactions were performed at low temperature. Slightly lower enantioselectivities have been observed when the desymmetrizations were carried out at room temperature.

9. The product has the following characteristics: mp 120 °C (racemate), 88-90 °C (enantiomer); $[\alpha]_D^{rt}$ = +6.6 (*c* = 1.90, $CHCl_3$); ee = 97-98% (GC-analysis of the lactone: Lipodex E, t_1 = 89.3, t_2 = 89.8 major); ^1H NMR (400 MHz, $CDCl_3$) δ 1.33 (d, *J* = 8.6 Hz, 1H), 1.48 (dt, *J* = 8.6, 1.8 Hz, 1H), 3.18 (br s, 1H), 3.30 (br s, 1H), 3.31-3.35 (m, 2H), 4.91 (d, *J* = 12.6 Hz, 1H), 5.09 (d, *J* = 12.6 Hz, 1H), 6.22 (dd, *J* = 3.0, 5.7 Hz, 1H), 6.30 (dd, *J* = 3.0, 5.7 Hz, 1H), 7.27-7.36 (m, 5H); ^{13}C NMR (100 MHz, $CDCl_3$) δ 46.1, 46.5, 48.1, 48.2, 48.7, 66.3, 128.0, 128.2, 128.4, 134.3, 135.5, 135.8, 172.2, 178.7; IR (KBr) 3065, 3032, 2979, 1740, 1707, 1172 cm^{-1}; EI-MS *m/z* = 272 (M^+, 2), 254 (3), 226 (3), 181 (58), 163 (3), 137 (5), 119 (2), 91 (100), 66 (20). Anal. Calcd for $C_{16}H_{16}O_4$ (272.30): C 70.57; H 5.92. Found: C 70.55; H 6.01.

10. Conditions: Daicel CHIRALPAK AS-H, eluent = ^{i}PrOH/hexane = 1/1, flow = 0.5 mL/min, detection = 254 nm, retention time = 10 min

(minor isomer using quinidine) and 12 min (major isomer using quinidine).

11. General procedure for the lactone formation. A 25-mL flame-dried Schlenk-flask equipped with a magnetic stirring bar and rubber septum is purged with argon and charged with (2*R*,3*S*)-3-*endo*-benzyloxycarbonyl bicyclo[2.2.1]hept-5-ene-2-*endo*-carboxylic acid (80 mg, 0.29 mmol) in THF (2 mL) (Note 12). The reaction mixture is cooled to 0 °C with an ice-bath and 6 eq. of $LiBEt_3H$ (2 mL, 1 M solution in THF) (Note 13) are slowly added. After stirring for 1 h at room temperature 2 N HCl (5 mL) is slowly added and the mixture is stirred for additional 2 h. The aqueous phase is extracted with ethyl acetate (3 × 5 mL), and the combined organic phases are dried over $MgSO_4$ and filtered. Evaporation of the solvent yields the corresponding lactone, which is analyzed by GC (Note 14).

In order to recover the alkaloid, the acidic aqueous phase, obtained after the first extraction, is neutralized with Na_2CO_3 and extracted with CH_2Cl_2 (5 × 100 mL). The combined organic phases are dried over $MgSO_4$ and filtered. Evaporation of the solvent yields the alkaloid almost quantitatively (6.93 g, 21.36 mmol, 97%) (Note 15).

12. THF was distilled from sodium benzophenone ketyl radical under argon.

13. $LiBEt_3H$ was purchased from Acros Organics and used as received.

14. Capillary gas chromatograms were obtained using the following column and temperature program: Lipodex E: 2,6-*O*-Dipentyl-3-*O*-butyryl-γ-CD. Column head pressure: 1.0 bar N_2; 100 °C (50 min), heating rate 3.0 °C/min up to 180 °C (60 min). Injector temperature 200 °C, detector temperature 250 °C.

15. The spectral properties of the recovered alkaloid were in accordance with the data published in the literature.[5]

16. The reaction was performed in a 0.1 M solution (with respect to the anhydride) to give (2*S*,3*R*)-3-*endo*-benzyloxycarbonyl-bicyclo [2.2.1] hept-5-ene-2-*endo*-carboxylic acid with 96% ee (GC-analysis of the lactone: Lipodex E, $t_1 = 89.3$, major, $t_2 = 89.8$); $[\alpha]^{rt}_D = -7.4$ ($c = 1.00$, $CHCl_3$).

Safety and Waste Disposal Information

All hazardous materials should be handled and disposed of in accordance with "Prudent Practices in the Laboratory"; National Academy Press; Washington, DC, 1995.

Table 1. Quinidine-mediated ring opening of various meso-anhydrides with benzyl alcohol.[a]

Entry	Substrate[b]	Product	Yield (%)	ee (%)[c]
1	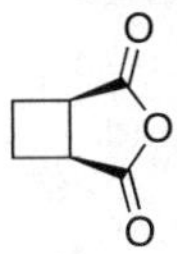	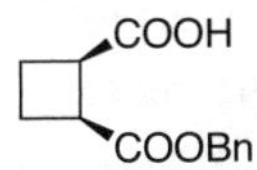	97	96
2	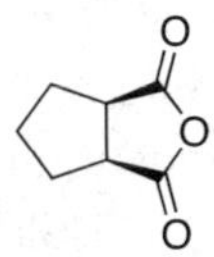	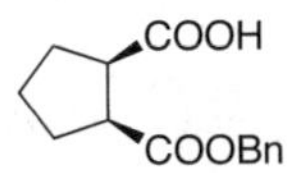	93	97
3	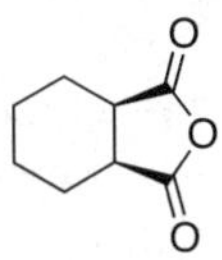	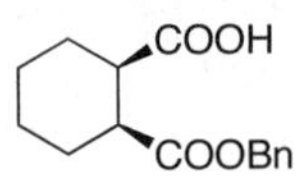	95	97
4	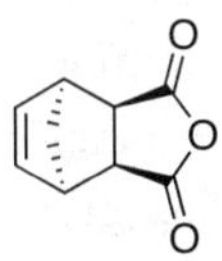	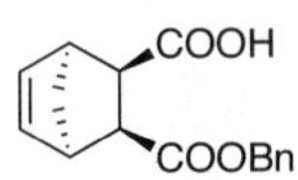	95	97
5[d]	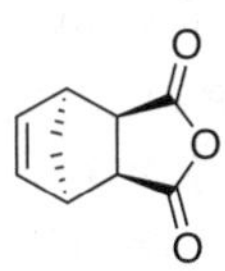	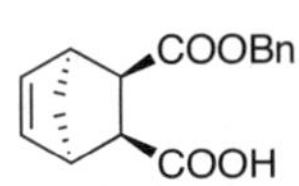	94	96
6	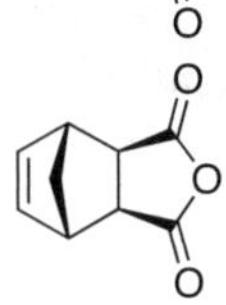	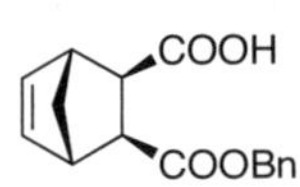	97	96
7	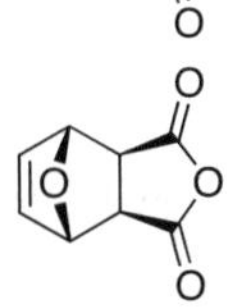	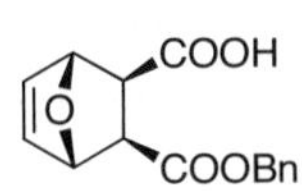	89[e]	99

[a] All reactions were performed in toluene, at –55 °C for 96 hr using 1.1 eq. of quinidine and 3 eq. of benzyl alcohol in a 0.2 M solution related to anhydride. All products have been fully characterized.
[b] All meso-anhydrides were prepared by application of literature procedures or were commercially available. For specific procedures, see ref. 10-14.
[c] Determined by GC-analysis of the corresponding lactones using a chiral stationary phase. For retention times, see ref. 3b.
[d] Quinine was used as chiral mediator (0.1 M solution related to anhydride).
[e] After chromatographic purification.

3. Discussion

The procedure described above is an improved version of previously reported alkaloid-mediated asymmetric anhydride openings.[2,3,6,7] Structurally diverse anhydrides have been converted into their corresponding benzyl monoesters with very high enantiomeric excesses and excellent yields. (Table 1) Both enantiomers are available by using either quinidine or quinine as directing additive. An advantage of the present protocol using benzyl alcohol as the nucleophile is that the reactions can be performed using toluene as solvent, avoiding the use of the previously utilized carbon tetrachloride. A simple aqueous work-up permits the isolation of the products in analytically pure form. The synthetic usefulness of the method was demonstrated by the preparation of optically active β-amino acids[2] and unsymmetrical norbornane scaffolds as inducers for hydrogen bond interactions in peptides.[8] In these applications, the benzyl hemiesters were converted into the corresponding *N*-Cbz-protected β-amino acid benzyl esters by Curtius degradation, which proceeded with neither racemization nor epimerization. Subsequent cleavage of both protecting groups by simple hydrogenation yielded the corresponding free β-amino acids in excellent yields in a single step. Finally, the method is also highly selective for the preparation of the corresponding methyl hemiesters, which have been isolated with up to 99% yield and 99% ee (on a 1 mmol scale). The usefulness of such products for the synthesis of optically active β-amino acids, 1,2-diamines, and polymeric materials has also already been demonstrated.[9]

1. Institut für Organische Chemie der RWTH Aachen, Landoltweg 1, D-52056 Aachen, Germany.

2. Bolm, C.; Schiffers, I.; Atodiresei, I.; Hackenberger, C. P. R. *Tetrahedron: Asymmetry* **2003**, *14*, 3455.
3. (a) Bolm, C.; Gerlach, A.; Dinter, C. L. *Synlett* **1999**, 195. (b) Bolm, C.; Schiffers, I.; Dinter, C. L.; Gerlach, A. *J. Org. Chem.* **2000**, *65*, 6984.
4. Jaeschke, G.; Seebach, D. *J. Org. Chem.* **1998**, *63*, 1190.
5. Raheem, I. T.; Goodman, S. N.; Jacobsen, E. N. *J. Am. Chem. Soc.*, **2004**, *126*, 706.
6. (a) Hiratake, J.; Inagaki, M.; Yamamoto, Y.; Oda, J. *J. Chem. Soc., Perkin Trans. 1* **1987**, 1053. (b) Hiratake, J.; Yamamoto, Y.; Oda, J. *J. Chem. Soc., Chem. Commun.* **1985**, 1717. (c) Aitken, R. A.; Gopal, J. *Tetrahedron: Asymmetry* **1990**,

1, 517. (d) Aitken, R. A.; Gopal, J.; Hirst, J. A. *J. Chem Soc., Chem. Commun.* **1988**, 632.

7. For recent reviews, see: (a) Chen, Y.; McDaid, P.; Deng, L. *Chem. Rev.* **2003**, *103*, 2965. (b) Tian, S.-K.; Chen, Y.; Hang, J.; Tang, L.; McDaid, P.; Deng, L. *Acc. Chem. Res.* **2004**, *37*, 621; (c) Atodiresei, I.; Schiffers, I.; Bolm, C. *Chem Rev.*, accepted for publication.

8. Hackenberger, C. P. R.; Schiffers, I.; Runsink, J.; Bolm, C. *J. Org. Chem.* **2004**, *69*, 739.

9. (a) Bolm, C.; Schiffers, I.; Dinter, C. L.; Defrère, L.; Gerlach, A.; Raabe, G. *Synthesis* **2001**, 1719. (b) Bolm, C.; Dinter, C. L.; Schiffers, I.; Defrère, L. *Synlett* **2001**, 1875. (c) Bolm, C.; Schiffers, I.; Atodiresei, I.; Ozcubukcu, S.; Raabe, G. *New J. Chem.* **2003**, *27*, 14.

10. *cis*-1,2-Cyclobutanedicarboxylic anhydride was obtained by refluxing *cis*-cyclobutane-1,2-dicarboxylic acid (Fluka, >97%) in trifluoroacetic anhydride (Acros Organics, 99+%) for 16 h.

11. *cis*-1,2-Cyclopentanedicarboxylic anhydride was prepared in a 3-step synthesis according to a literature procedure. Padwa, A.; Hornbuckle, S. F.; Fryxell, G. E.; Stull, P. D. *J. Org. Chem.* **1989**, *54*, 817.

12. *cis*-1,2-Cyclohexanedicarboxylic anhydride (99%) was purchased from Acros Organics and used as received.

13. *Exo*-bicyclo[2.2.1]hept-5-ene-2,3-dicarboxylic anhydride was prepared according to a literature procedure. Canonne, P.; Belanger, D.; Lemay, G. *J. Org. Chem.* **1982**, *47*, 3953. It is also available from Sigma-Aldrich (95%).

14. *Exo*-7-oxabicyclo[2.2.1]hept-5-ene-2,3-dicarboxylic anhydride was prepared by Diels-Alder reaction of maleic anhydride and furan. It is also available from different commercial suppliers (Fluka, Sigma-Aldrich, Acros Organics, Lancaster Synthesis).

TYPE OF REACTION INDEX

The index lists the preparations contained in this volume in accordance with general types of reactions. The arrangement of types and of preparations is alphabetical. The reference page is to the title page of the procedure where the preparations may be found.

ACETAL FORMATION

2-Acetamido-2-deoxy-1,3,4,6-tetra-*O*-acetyl-β-D-glucopyranose 787
1,1-Bis(benzyloxy)propan-2-one 473
2,2-Bis-(chloromethyl)-5,5-dimethyl-1,3-dioxane 457
2-Chloro-3,3-diethoxyprop-1-ene 586
2,3-Dibromo-1,1,4,4-tetramethoxybutane 607
1,2:4,5-Di-*O*-isopropylidene-β-D-fructopyranose 177
N-(3,3-Dimethoxy-1-trifluoromethyl-propyl)-acetamide 212
(5*S*)-(d-Menthyloxy)-2(5*H*)-furanone 128
1-Methoxymethyl-1-phenylethane 934
1,2:3,4:5,6-Tri-*O*-isopropylidene-L-mannitol 164

ACYLATION

Acetic acid 5-phenylpent-2-ynyl ester 561
(2-Aminoethyl)carbamic acid *tert*-butyl ester 942
(1S,2S)-trans-1-Benzoyloxy-2-bromocyclohexane 1037
N-Benzyl-4-phenylbutyramide 733
(–)-(S,S)-N,N'-Bis(1-hydroxymethyl-2-methylpropyl)-2,2-dimethyl-malonamide 821
1-[3,5-Bis(trifluoromethyl)phenyl]-ethanone 343
(2*S-trans*)-2-*tert*-Butoxycarbonylacetyl-4-hydroxypyrrolidine-1-carboxylic acid, *tert*-butyl ester 38
5*S*,6*R*)-4-*tert*-Butoxycarbonyl-5,6-diphenylmorpholin-2-one 879
1-*tert*-Butyloxycarbonyl-4-(9-fluorenylmethyloxycarbonylamino)piperidine-4-carboxylic acid 956
1-*tert*-Butyloxycarbonylpiperidine-4-spiro-5'-(1',3'-bis(*tert*-butyloxycarbonyl))- hydantoin 956
(η^5-Carbomethoxycyclopentadienyl)-(η^4-tetraphenyl-cyclobutadiene)cobalt 741
(3,6-Diacetyl-9,10-dimethoxyphenanthrene 519
Dibenzofuran-4,6-dicarboxylic acid 888
(*R,R*)-Dibenzofuran-4,6-dicarboxylic acid bis(2-hydroxy-1-phenylethyl) amide 888
N,N-Dibenzylformamide 425
Dibutyl 3,4,6-tri-*O*-benzyl-2-*O*-pivaloyl-D-glucopyranosyl phosphate 169
N-(2,2-Diethoxy)ethyl formamide 778
(3a*S*,7a*R*)-2-[2-(3,4-Dimethoxyphenyl)-ethyl]-hexahydro-isoindole-1,3-dione 1028
(3a*S*,7a*S*)-2-[2-(3,4-Dimethoxyphenyl)-ethyl]-1,3-dioxo-octahydro-isoindole-3a-carboxylic acid methyl ester 1028
N,N'-Dimethyl-*N*-acetylurea 782
1-(Dimethyl(phenyl)silyl)propan-1-one 197
Dimethylsulfoxonium (*S*)-2-oxo-3-(Boc-amino)-4-phenylbutylide 149
(Ethanol, 1-[[(1*R*,2*S*,4*R*)-1,7,7-trimethylbicyclo[2.2.1]hept-2-yl]oxy]-, acetate) 52
Ethyl (1'*S*,2'*R*)-*N*-*tert*-butyloxycarbonyl-*N*-(1',2'-diphenyl-2'-hydroxyethyl-glycinate 879
Ethyl 6-Oxotridecanoate 281
N-[(9*H*-Fluoren-9-ylmethoxy)carbonyl]-L-aspartic acid, α-*tert*-butyl ester, β-(2-ethyl(1*E*)-[[[(4-nitrophenyl)azo]phenyl]amino]ethyl ester 635
N-Methoxy-*N*-methylcarbamoyl chloride 992
N-(Methoxycarbonyl)-2-propenylamine 244
1-(Methoxymethylcarbamoyl)ethyl] phosphonic acid bis-(2,2,2-trifluoroethyl) ester 992
N-Methyl-*O*-benzoylhydroxylamine hydrochloride 70
(–)-Methyl (*1R,9R*)-6-oxo-7,11-diazatricyclo[7.3.1.0^{2,7}]trideca-2,4-diene-11-carboxylate 25
(4*S*)-4-(1-Methylethyl)-5,5-diphenyl-2-oxazolidinone 896

Methyl 2-oxo-1,2,3,4-tetrahydronaphthalene-1-carboxylate 811
1-Morpholinopropan-1-one 197
Pivaloyl hydrazide 874
Propargyl benzoate 552

ADDITION
Acetic acid 2-methylene-3-phenethylbut-3-enyl ester 561
3-Acetylamino-1-ethoxythiocarbonylsulfanyl-4,4,4-trifluorobutyl acetate 212
(*2S,6S*)-4-Benzyl-1,7-bis(trifluoromethylsulfonyl)-2,6-diisopropyl-1,4,7 –triazaheptane 506
(2*R*,3*S*)-3-*endo*-Benzyloxycarbonyl-Bicyclo[2.2.1]hept-5-ene-2-endo-carboxylic acid 1068
1-[3,5-Bis(trifluoromethyl)phenyl]-ethanone 343
(*S*)-4-(*t*-Butyldimethylsilyloxy)-2-pentynol 615
1-(*tert*-Butylimino-methyl)-1,3-dimethyl-urea hydrochloride 782
2-Cyclopentylacetophenone 433
(4,4-Dibromobut-3-enyl)benzene 561
1,1-Dichloro-(3*S*)-(*t*-butyldimethylsilyloxy)-2-butanol *p*-toluenesulfonate 615
(3*R*)-3,7-Dimethyl-2-(3-oxobutyl)-6-octenal 416
1(*S*),2(*S*)-Diphenyl-*N,N'*-bis-[1(*R*)-phenyl-ethyl]-ethane-1,2-diamine 1028
1,3-Diphenyl-2-(1-phenylethyl)propane-1,3-dione 272
Ethyl 2-{[(*1S,2R,3R,5S*)-2,6,6-trimethylbicyclo[3.1.1]hept-3-yl]methyl}acrylate 308
(*E*)-1-(1-Heptenyl)-4-methoxybenzene 367
(*Z*)-1-(1-Heptenyl)-4-methoxybenzene 359
(*S*)-[1-(Hydroxydiphenylmethyl)-2-methylpropyl]carbamic acid, *tert*-butyl ester 896
5-Methyl-1-phenyl-6-hepten-3-ol 327
3-Methyl-1-phenyl-4-penten-1-ol 327
N-[(4-Methylphenyl)sulfonyl(phenyl)methyl]-*P,P*-diphenyl-phosphinic amide 749
6,9,12,15,18-Pentamethyl-1,6,9,12,15,18-hexahydro(c_{60}-i_h)[5,6]fullerene 319
1-Phenylhex-5-en-3-amine 691
5-Phenylpent-2-yn-1-ol 561
N-[(1S)-1-Phenylpropyl*]-P,P*-diphenylphosphinic amide 749
3-Phenyl-2-(propylthio)-2-propen-1-ol 642
4-Phenyl-3-trifluoromethyl-2-butanone 336
Piperidine-4-spiro-5'-hydantoin 956
(2-Propenoic acid, 3-iodo-, (2*E*)-) 662
Camphorquinone oxime 702
3,4-Dibromo-2,5-dimethoxytetrahydrofuran 607
(6*R*)-(+)-1,6-Dimethylcyclohex-2-en-1-ol 100
5-Oxaspiro[3.4]octan-1-one 256
4-Phenyl-5(*E*)-decen-2-one 261
(*E*)-2,2,2-Trichloroacetimidic acid hex-2-enyl ester 720
(*rac*)-4-Triisopropylsilyl-3-butyn-2-ol 9

ALKENYLATION
Acetic acid 2-methylene-3-phenethylbut-3-enyl ester 561
(*R*)-(*E*)-2-Benzylidiene-1-cyclohexyl-butan-1-ol 205
1-Chloro-4-(2,2-dichloro-1-methylvinyl)benzene 594
(4,4-Dibromobut-3-enyl)benzene 561
Methyl (2*Z*)-2-azido-3-(2-chloropyridin-3-yl)acrylate 802
(*E*)-Methyl 2-diazo-4-phenylbut-3-enoate 296
1-Methyl-3,4-dihydronaphthalene 296
Methyl 2-ethenyl-3-nitrobenzoate 903
(*Z*)-2-Methyl-3-trimethylsilyl-2-butenoic acid 577

ALKYLATION
1-(4-Acetylphenyl)-2-phenylethane 289
O-Allyl-*N*-(9-Anthracenylmethyl)cinchonidinium bromide 404
6-Amino-3,4-dihydro-1(2*H*)-naphthalenone 670
N-(9-Anthracenylmethyl)cinchonidinium chloride 404

(4*S*)-2-(Benzhydrylidenamino)pentanedioic Acid, 1-*tert*-Butyl Ester-5-Methyl Ester 404
3,6-Bis(benzoylbiphenyl)ethynyl-9-tetradecylcarbazole 540
3-Benzyloxy-2-iodophenyl trifluoromethanesulfonate 488
2-Benzyloxy-1-methylpyridinium trifluoromethanesulfonate 947
1-(4-Bromophenyl)-4-methylphenyl)but-3-yn-1-ol 794
(3*R*,5*R*,6*S*)-4-*tert*-Butoxycarbonyl-5,6-diphenyl-3-(1'-prop-2'-enyl)morpholin-2-one 399
3-(3′-Chloropropyl)-3,4-dihydronaphthalen-2(1*H*)-one 811
1,4:5,8-Dimethano-1,2,3,4,5,6,7,8-octahydro-9,10-dimethoxy-anthracene 62
9,10-Dimethoxyphenanthrene 519
2-(2,2-Dimethylpropoxy)-2,3-dihydro-*1H*-indene 46
Ethyl (1'*S*,2'*R*)-*N*-(1',2'-diphenyl-2'-hydroxyethyl)glycinate 879
(6*E*, 10*E*)-Ethyl 7,11,15-Trimethyl-3-oxohexa-deca-6,10,14-trienoate 221
4-(4-Methoxyphenyl)-2-phenyl-1*H*-imidazole 862
Methyl (*R*)-(–)-3-benzyloxy-2-methyl propanoate 947
(*E*)-Methyl 2-diazo-4-phenylbut-3-enoate 296
1-Methyl-3,4-dihydronaphthalene 296
(*S,E*)-Methyl 2-((*R*)-4-methyl-1,2-dihydronaphthalen-2-yl)-4-phenylbut-3-enoate 296
4-Nonylbenzoic acid methyl ester 353
6,9,12,15,18-Pentamethyl-1,6,9,12,15,18-hexahydro(c_{60}-i_h)[5,6]fullerene 319
(*S*)-[(1-Phenylethyl)amino]acetonitrile 139
(*R*)-4-Phenyl-3-(2-propynyl)oxazolidin-2-one 867
N-(2-Propenyl)-*N*-(2-propynyl)-4-methylbenzenesulfonamide 442
1-(2-(1-Propynyl)phenyl)pyrrole 826
8-[2-(Triethoxysilyl)ethyl]-1-tetralone 410
1,6,6-Trimethyl-4,8-dioxaspiro[2.5]oct-1-ene 457
9-Tetradecylcarbazole 540

ALKYNYLATION
3,6-Bis[(trimethylsilyl)ethynyl]-9-tetradecylcarbazole 540
(*R*)-(+)-*N,N*-Dibenzyl-*N*-[1-cyclohexyl-3-(trimethylsilyl)-2-propynyl]-amine 192

AMIDE FORMATION
(–)-(S,S)-N,N'-Bis(1-hydroxymethyl-2-methylpropyl)-2,2-dimethyl-malonamide 821
2-Bromo-2-methylpropanamide 670
(S)-N-(N-tert-Butoxycarbonylprolyl)dihydroisoindole 1037
1-Morpholinopropan-1-one 197

AMINATION
(*R*)-(+)-*N,N*-Dibenzyl-*N*-[1-cyclohexyl-3-(trimethylsilyl)-2-propynyl]-amine 192
N-(3,5-Dimethylphenyl)-*tert*-butylamine 529
Ethyl 4-[(6-chloropyridin-3-yl)amino]benzoate 665
N-(Methoxycarbonyl)-*N*-(2-propenyl)-1,3-decadiynylamide 244
(*R*)-4-Phenyl-3-(2-triisopropylsilyl-ethynyl)oxazolidin-2-one 677

ANNULATION
5-Benzyloxybicyclo[4.2.0]octa-1,3,5-trien-7-one 488
N,N-Dibenzyl-*N*-(2-ethenylcyclopropyl)amine 425

CONDENSATION
(1R,2S)-1-[3-Adamantyl)-2-hydroxy-5-methylbenzyliden-amino]indan-2-ol 498
N-Benzylidene-*p*-anisidine 920
(–)-(*S,S*)-Bis(4-isopropyloxazoline) 821
(2*S-trans*)-2-*tert*-Butoxycarbonylacetyl-4-hydroxypyrrolidine-1-carboxylic acid, *tert*-butyl ester 38
1-(4-Chlorophenyl)ethanone hydrazone 594
(*R,R*)-4,6-Dibenzofurandiyl-2,2'-bis(4-phenyloxazoline) 888
1(*S*),2(*S*)-Diphenyl-*N,N'*-bis-[1(*R*)-phenyl-ethyl]-ethane-1,2-diamine 1028
Dithieno[3,2-*b*:2′,3′-*d*]thiophene-2,6-dicarboxylic acid diethyl ester 843
(1*S*,2*R*)-1-[(2-Hydroxy-3,5-di-*tert*-butylbenzylidene)amino]indan-2-ol 770

1-(2-Iodophenyl)pyrrole 826
[[Methanaminium, *N*-[2-chloro-3-(dimethylamino)-2-propenylidene]-*N*-methyl-, hexafluorophosphate(1-)]] 929
Methyl (2*Z*)-2-azido-3-(2-chloropyridin-3-yl)acrylate 802
Methyl 2-ethenyl-3-nitrobenzoate 903
(4*S*)-4-(2-Phenethyl)oxetan-2-one 506
N-p-Tolylsulfinyl-(*E*)-1-phenylethylideneimine 741
5-(Tosyloxyimino)-2,2-dimethyl-1,3-dioxane-4,6-dione 911

COUPLING
1-(4-Acetylphenyl)-2-phenylethane 289
3,6-Bis(benzoylbiphenyl)ethynyl-9-tetradecylcarbazole 540
(S,S)-1,2-Bis(boranato(tert-butyl)methylphosphino)ethane 1020
3,6-Bis[(trimethylsilyl)ethynyl]-9-tetradecylcarbazole 540
N-(*tert*-Butoxycarbonyl)-β-[4-(methoxycarbonyl)phenyl]alanine methyl ester 384
(S)-N-(N-tert-Butoxycarbonylprolyl)dihydroisoindole 1037
2-Chloro-1-decen-3-yne 234
dl-1,2-Dicyclohexylethanediol 348
o-(1,1-Difluorohex-1-en-2-yl)aniline 834
N-(3,5-Dimethylphenyl)-*tert*-butylamine 529
Ethyl 4-[(6-chloropyridin-3-yl)amino]benzoate 665
S-Ethyl Octanethioate 281
Ethyl 6-Oxotridecanoate 281
(*E*)-1-(1-Heptenyl)-4-methoxybenzene 367
(*Z*)-1-(1-Heptenyl)-4-methoxybenzene 359
N-(Methoxycarbonyl)-*N*-(2-propenyl)-1,3-decadiynylamide 244
1-Methoxy-2-(4-methoxyphenoxy)benzene 626
(*E*)-2-Methyl-3-phenylacrylic acid butyl ester 372
4-Nonylbenzoic acid methyl ester 353
4-Phenyl-5(*E*)-decen-2-one 261
(*R*)-4-Phenyl-3-(2-triisopropylsilyl-ethynyl)oxazolidin-2-one 677
(*E*)-4-(2-Phenylethenyl)benzonitrile 372
4-Phenylmorpholine 684
1-(2-(1-Propynyl)phenyl)pyrrole 826
3-Pyridin-3-yl-quinoline 393
2'-Vinylacetanilide 315

CYCLIZATION
trans-2-Benzyl-1-methylcyclopropan-1-ol 450
N-Benzylpiperidine 697
Bicyclo[3.1.0]hexan-1-ol 450
(–)-(*S,S*)-Bis(4-isopropyloxazoline) 821
N-Boc-3-pyrroline 570
5-(4-Bromophenyl)-3-iodo-2-(4-methylphenyl)furan 794
3-Butyl-2-fluoro-1-tosylindole 834
1,2:5,6-Dianhydro-3,4-*O*-isopropylidene-L-mannitol 164
(*R,R*)-4,6-Dibenzofurandiyl-2,2'-bis(4-phenyloxazoline) 888
3,4-Dimethylcyclopent-2-enone 420
Dithieno[3,2-*b*:2′,3′-*d*]thiophene-2,6-dicarboxylic acid diethyl ester 843
5-Hexenyllithium 244
1-(2-Iodophenyl)pyrrole 826
(4*S*)-4-(1-Methylethyl)-5,5-diphenyl-2-oxazolidinone 896
4-(4-Methoxyphenyl)-2-phenyl-1*H*-imidazole 862
Methyl 4-chloro-1*H*-pyrrolo[3,2-c]pyridine-2-carboxylate 802
Methyl indole-4-carboxylate 903
4-Methylpyrrolo[1,2-a]quinoline 826
5-Oxaspiro[3.4]octan-1-one 256
N-Phenyl-(5*R*)-hydroxymethyl-2-oxazolidinone 867
Piperidine-4-spiro-5'-hydantoin 956

2,3,11,11a-Tetrahydro-1*H*-benzo[d]pyrrolo[1,2-a]azepin-5(6H)-one 811
3-(4,4,5,5-Tetramethyl-[1,3,2]dioxaborolan-2-yl)pyridine 393
(*S*)-N-Trifluoromethylsulfonyl-2-isopropylaziridine 506
1,6,6-Trimethyl-4,8-dioxaspiro[2.5]oct-1-ene 457

CYCLOADDITION
5-Benzyloxybicyclo[4.2.0]octa-1,3,5-trien-7-one 488
2α-Benzyloxy-8-oxabicyclo[3.2.1]oct-6-en-3-one 473
6,13-Bis(triisopropylsiloxy)-9,10-dimethoxy[7]helicenebisquinone 524
1-(*tert*-Butyldimethylsilyloxy)-8-(methoxycarbonyl)-6-methyl- bicyclo[4.2.0]octane 481
(2S,6R)-6-(*tert*-Butyldimethylsilyloxymethyl)-2-methoxy-2,5-dihydro-pyran 498
1,4:5,8-Dimethano-1,4,4a,5,8,8a,9a,10a-octahydroanthracene-9,10-dione 62
cis-5-(5,5-Dimethyl-1,3-dioxan-2-ylidene)hexahydro-1(2*H*)-pentalen-2-one 457
trans-1-(4-Methoxyphenyl)-4-phenyl-3-(phenylthio)azetidin-2-one 920
Methyl-5-methylpyridine-2-carboxylate 911
9-Spiroepoxy-*endo*-tricyclo[5.2.2.$0^{2,6}$]undeca-4,10-dien-8-one 514
2,3,3α,4-Tetrahydro-2-[(4-methylbenzene)sulfonyl]cyclopenta[c]pyrrol-5(1H)-one 442

CYCLOPROPANATION
trans-2-Benzyl-1-methylcyclopropan-1-ol 450
Bicyclo[3.1.0]hexan-1-ol 450
n-Butyl 2,2-difluorocyclopropanecarboxylate 469
N,N-Dibenzyl-*N*-(2-ethenylcyclopropyl)amine 425
1,1-Dichloro-2-ethoxycyclopropane 586

DEHYDRATION
(–)-(*S,S*)-Bis(4-isopropyloxazoline) 821
(*R,R*)-4,6-Dibenzofurandiyl-2,2'-bis(4-phenyloxazoline) 888
2,2-Diethoxy-1-isocyanoethane 778
Diphosphoric acid, tetrakis(phenylmethyl) ester 1001
(η^5-(*S*)-2-(4-Methylethyl)oxazolinylcyclo-pentadienyl)-(η^4-tetraphenylcyclobutadiene)cobalt 708

DEPROTECTION
5-Benzyloxybicyclo[4.2.0]octa-1,3,5-trien-7-one 488
(S,S)-1,2-Bis-((*tert*-butyl)methylphosphino)ethane 1020
(*R*)-(*N-tert*-Butoxycarbonyl)allylglycine 399
(R)-3-Butyn-2-yl mesylate 1056
(*R*)-(+)-*N,N*-Dibenzyl-*N*-(1-cyclohexyl-2-propynyl)-amine 192
3,6-Diethynyl-9-tetradecylcarbazole 540
4,4-Dimethoxybut-2-ynal 607
Methyl-5-methylpyridine-2-carboxylate 911

DESYMMETRIZATION
(3a*S*,7a*S*)-2-[2-(3,4-Dimethoxyphenyl)-ethyl]-1,3-dioxo-octahydro-isoindole-3a-carboxylic acid methyl ester 1028

DIAZOTIZATION
(*E*)-Methyl 2-diazo-4-phenylbut-3-enoate 296
S-Phenyl diazothioacetate 920

ELIMINATION
(*S*)-4-(*t*-Butyldimethylsilyloxy)-2-pentynol 615
2-Chloro-3,3-diethoxyprop-1-ene 586
1,3-Decadiyne 234
1,2:5,6-Dianhydro-3,4-*O*-isopropylidene-L-mannitol 164
o-(1,1-Difluorohex-1-en-2-yl)aniline 834
1,4-Dioxene 850
1-Methyl-3,4-dihydronaphthalene 296
Methyl-5-methylpyridine-2-carboxylate 911

N-[(Phenyl)methylene]-*P,P*-diphenylphosphinic amide 749
5-Phenylpent-2-yn-1-ol 561
1,1,4,4-Tetramethoxybut-2-yne 607
(Triphenylphosphoranylidene)-ketene 986

ENZYMATIC PROCESSES
(R)-4-Trimethylsilyl-3-butyn-2-yl Acetate 1056

EPOXIDATION
Dibutyl 3,4,6-tri-*O*-benzyl-2-*O*-pivaloyl-D-glucopyranosyl phosphate 169
(2*S*,3*S*)-1,2-Epoxy-3-(Boc-amino)-4-phenylbutane 149
(±)-Methyl glycidate 157
(*R,R*)-*trans*-β-Methylstyrene oxide 183
(*R,R*)-1-Phenylcyclohexene oxide 183
9-Spiroepoxy-*endo*-tricyclo[5.2.2.0^{2,6}]undeca-4,10-dien-8-one 514

ESTERIFICATION
Acetic acid 5-phenylpent-2-ynyl ester 561
(1S,2S)-trans-1-Benzoyloxy-2-bromocyclohexane 1037
(2*R*,3*S*)-3-*endo*-Benzyloxycarbonyl-Bicyclo[2.2.1]hept-5-ene-2-endo-carboxylic acid 1068
(5*S*,6*R*)-4-*tert*-Butoxycarbonyl-5,6-diphenylmorpholin-2-one 879
N-(*tert*-Butoxycarbonyl)-L-valine methyl ester 896
2-(5,5-Dimethyl-1,3,2-dioxaborinan-2-yl)benzoic acid ethyl ester 1015
S-Ethyl Octanethioate 281
N-[(9*H*-Fluoren-9-ylmethoxy)carbonyl]-L-aspartic acid, α-*tert*-butyl ester, β-(2-ethyl(1*E*)-[[[(4-nitrophenyl)azo]phenyl]amino]ethyl ester 635
Isopropyl (*E*)-but-2-enoate 420
(*E*)-Methyl 2-diazo-4-phenylbut-3-enoate 296
3-(4,4,5,5-Tetramethyl-[1,3,2]dioxaborolan-2-yl)pyridine 393
(R)-4-Trimethylsilyl-3-butyn-2-yl Acetate 1056
(S)-4-Trimethylsilyl-3-butyn-2-yl Succinate 1056

FORMYLATION
(1R,2S)-1-[3-Adamantyl)-2-hydroxy-5-methylbenzyliden-amino]indan-2-ol 498
3-Bromo-2-hydroxybenzaldehyde 267
3,4-Dibromo-2,5-diformylthiophene 843

GLYCOSYLATION
2-Acetamido-2-deoxy-1,3,4,6-tetra-O-acetyl-β-D-glucopyranose 787
3,4,6-Tri-*O*-benzyl-2-*O*-pivaloyl-β-D-glucopyranosyl-(1→6)-1,2:3,4-di-*O*-isopropylidene-α-D-galactopyranoside 169

HALOGENATION
1-Bromo-1,3-decadiyne 244
5-Bromoisoquinoline 656
5-Bromo-8-nitroisoquinoline 656
5-(4-Bromophenyl)-3-iodo-2-(4-methylphenyl)furan 794
1-Bromo-2-triisopropylsilyl-ethyne 677
n-Butyl 2,2-difluorocyclopropanecarboxylate 469
Dibenzofuran-4,6-dicarbonyl chloride 888
3,4-Dibromo-2,5-dimethoxytetrahydrofuran 607
3,6-Diiodo-9-tetradecylcarbazole 540
Ethyl 3-chloro-3-phenylpropanoate 646
Ethyl 2,2-dibromopropionate 577
Hexakis(4-bromophenyl)benzene 652
1-Iodo-1-heptyne 359
(2*E*)-3-Iodo-2-propenoic acid 662
2-Iodoresorcinol 488
Methyl 2-bromomethyl-3-nitrobenzoate 903

HYDRATION
2,8-Nonanedione 82

HYDROBORATION
Ethyl 2-{[(*1S,2R,3R,5S*)-2,6,6-trimethylbicyclo[3.1.1]hept-3-yl]methyl}acrylate 308

HYDROLYSIS
4-Amino-1-*tert*-butyloxycarbonylpiperidine-4-carboxylic acid 956
6-Amino-3,4-dihydro-1(2*H*)-naphthalenone 670
(2*R*,3*S*)-3-*endo*-Benzyloxycarbonyl-Bicyclo[2.2.1]hept-5-ene-2-endo-carboxylic acid 1068
3-(3′-Chloropropyl)-3,4-dihydronaphthalen-2(1*H*)-one 811
4,4-Dimethoxybut-2-ynal 607
Dithieno[3,2-*b*:2′,3′-*d*]thiophene-2,6-dicarboxylic acid 843
1-Hydroxy-3-phenyl-2-propanone 642
N-Hydroxy-(*S*)-1-phenylethylamine oxalate 139
3,4-*O*-Isopropylidene-L-mannitol 164
4-Nonylbenzoic acid 353
(1S)-1-Phenylpropan-1-amine hydrochloride 749

ION EXCHANGE
1-Butyl-3-methylimidazolium tetrafluoroborate 859

ISOLATION
(–)-Cytisine 25

ISOMERIZATION
6-Amino-3,4-dihydro-1(2*H*)-naphthalenone 670
6,6-Dimethyl-1-methylene-4,8-dioxaspiro[2.5]octane 457
4-Methylpyrrolo[1,2-a]quinoline 826

METALLATION
1-[3,5-Bis(trifluoromethyl)phenyl]-ethanone 343
1-Bromo-2-triisopropylsilyl-ethyne 677
(*S*)-4-(*t*-Butyldimethylsilyloxy)-2-pentynol 615
1-Chloro-3-iodo-5-(4,4,5,5-tetramethyl-1,3,2,-dioxaborolan-2-yl)benzene 1007
3-(3′-Chloropropyl)-3,4-dihydronaphthalen-2(1*H*)-one 811
(Cyclooctadienyl)ruthenium dichloride polymer-(*R*)-BINAP catalyst, $[Et_2NH_2]^+[Ru_2Cl_5(BINAP)_2]^-$ 38
Di-μ-acetatobis[(η^5-(*S*)-(p*R*)-2-(2’-(4’-methylethyl)oxazolinyl)cyclopentadienyl, 1-C, 3’-*N*)(η^4-tetraphenylcyclo-butadiene)cobalt]dipalladium ($[COP\text{-}OAc]_2$) 714
Dibenzofuran-4,6-dicarboxylic acid 888
3,4-Dibromo-2,5-diformylthiophene 843
(3a*S*,7a*S*)-2-[2-(3,4-Dimethoxyphenyl)-ethyl]-1,3-dioxo-octahydro-isoindole-3a-carboxylic acid methyl ester 1028
2-(5,5-Dimethyl-1,3,2-dioxaborinan-2-yl)benzoic acid ethyl ester 1015
Dimethylphenylsilyllithium 197
5-Ethoxy-5-oxopentylzinc iodide 281
Ethyl 2,2-dibromopropionate 577
(6*E*, 10*E*)-Ethyl 7,11,15-Trimethyl-3-oxohexa-deca-6,10,14-trienoate 221
(*Z*)-1-(1-Heptenyl)-1,1-dimethylsilanol 359
(*E*)-1-(1-Heptenyl)-4-methoxybenzene 359
5-Hexenyllithium 433
1-Iodo-1-heptyne 359
Lithium *N*-(3,5-dimethylphenyl)-*tert*-butylamide etherate 529
1-(Methoxymethylcarbamoyl)ethyl] phosphonic acid bis-(2,2,2-trifluoroethyl) ester 992
(*Z*)-2-Methyl-3-trimethylsilyl-2-butenoic acid 577
Methyl tris(isopropoxy)titanium 425

5-Oxaspiro[3.4]octan-1-one 256
Phenylmagnesium bromide 896
5-Phenylpent-2-yn-1-ol 561
3-Pyridylboronic acid [Tris(3-pyridyl)boroxin] 393
2-(4,4,5,5-Tetramethyl-1,3,2-dioxaborolan-2-yl)indole 1007
(Triphenylphosphoranylidene)-ketene 986

METAL-CATALYZED REACTIONS
ALUMINUM
(*4S*)-4-(2-Phenethyl)oxetan-2-one 506

BORON
N-Benzyl-4-phenylbutyramide 733
o-(1,1-Difluorohex-1-en-2-yl)aniline 834

COBALT
2,3,3α,4-Tetrahydro-2-[(4-methylbenzene)sulfonyl]cyclopenta[c]pyrrol-5(1*H*)-one 442

COPPER
2-Chloro-1-decen-3-yne 234
1-Chloro-4-(2,2-dichloro-1-methylvinyl)benzene 594
(*R*)-(+)-*N,N*-Dibenzyl-*N*-[1-cyclohexyl-3-(trimethylsilyl)-2-propynyl]-amine 192
(2*E*)-3-Iodo-2-Propenoic acid 662
N-(Methoxycarbonyl)-*N*-(2-propenyl)-1,3-decadiynylamide 244
1-Methoxy-2-(4-methoxyphenoxy)benzene 626
6,9,12,15,18-Pentamethyl-1,6,9,12,15,18-hexahydro(c_{60}-i_h)[5,6]fullerene 319
4-Phenylmorpholine 684
(*R*)-4-Phenyl-3-(2-triisopropylsilyl-ethynyl)oxazolidin-2-one 677

GOLD
1,3-Diphenyl-2-(1-phenylethyl)propane-1,3-dione 272
2,8-Nonanedione 82

INDIUM
4-Methylpyrrolo[1,2-a]quinoline 826

IRIDIUM
1-Chloro-3-iodo-5-(4,4,5,5-tetramethyl-1,3,2,-dioxaborolan-2-yl)benzene 1007
1-Methoxy-4-vinyloxybenzene 631
2-(4,4,5,5-Tetramethyl-1,3,2-dioxaborolan-2-yl)indole 1007

IRON
4-Nonylbenzoic acid methyl ester 353

MERCURY
(*R*,R**)-2,2'-Bispyrrolidine 1046
(*R*,S**)-2,2'-Bispyrrolidine 1046

NICKEL
(*R*)-(*E*)-2-Benzylidiene-1-cyclohexyl-butan-1-ol 205
5-Methyl-1-phenyl-6-hepten-3-ol 327
3-Methyl-1-pheny-4-penten-1-ol 327

PALLADIUM
1-(4-Acetylphenyl)-2-phenylethane 289
3,6-Bis(benzoylbiphenyl)ethynyl-9-tetradecylcarbazole 540
3,6-Bis[(trimethylsilyl)ethynyl]-9-tetradecylcarbazole 540
N-(*tert*-Butoxycarbonyl)-β-[4-(methoxycarbonyl)phenyl]alanine methyl ester 384
2-Chloro-1-decen-3-yne 234

o-(1,1-Difluorohex-1-en-2-yl)aniline 834
N-(3,5-Dimethylphenyl)-*tert*-butylamine 529
Ethyl 4-[(6-chloropyridin-3-yl)amino]benzoate 665
Ethyl 6-Oxotridecanoate 281
(*E*)-1-(1-Heptenyl)-4-methoxybenzene 359
(*Z*)-1-(1-Heptenyl)-4-methoxybenzene 367
(*E*)-2-Methyl-3-phenylacrylic acid butyl ester 372
(*E*)-4-(2-Phenylethenyl)benzonitrile 372
1-(2-(1-Propynyl)phenyl)pyrrole 826
3-Pyridin-3-yl-quinoline 393
(*S*)-2,2,2-Trichloro-*N*-(1-propylallyl)acetamide 720
2'-Vinylacetanilide 315

RHODIUM
trans-1-(4-Methoxyphenyl)-4-phenyl-3-(phenylthio)azetidin-2-one 920
(*S,E*)-Methyl 2-((*R*)-4-methyl-1,2-dihydronaphthalen-2-yl)-4-phenylbut-3-enoate 296
4-Phenyl-3-trifluoromethyl-2-butanone 336

RUTHENIUM
Acetic acid 2-methylene-3-phenethylbut-3-enyl ester 561
2(*S*)-(β-*tert*-Butoxycarbonyl-α-(*R*)- and (*S*)-hydroxyethyl-4(*R*)-hydroxypyrrolidine-1-carboxylic acid, *tert*-butyl ester 38
N,N-Dimethyl-3-phenylpropylamine 1
(*R,R*)-Hydrobenzoin 17
8-[2-(Triethoxysilyl)ethyl]-1-tetralone 410

SILVER
1,3-Diphenyl-2-(1-phenylethyl)propane-1,3-dione 272

TITANIUM
N,N-Dibenzyl-*N*-(2-ethenylcyclopropyl)amine 348

VANDIUM
(R_S)-(+)-*tert*-Butyl *tert*-butanethiosulfinate 770

METAL COMPLEXATION
(μ_3, η^2: η^3: η^5-Acenaphthylene)$Ru_3(CO)_7$ 1
(η^5-Carbomethoxycyclopentadienyl)-(η^4-tetraphenyl-cyclobutadiene)cobalt 708
1-Chloro-3-iodo-5-(4,4,5,5-tetramethyl-1,3,2,-dioxaborolan-2-yl)benzene 1007
Chromium(III) Cl Complex 498
Di-μ-acetatobis[(η^5-(*S*)-(p*R*)-2-(2'-(4'-methylethyl)oxazolinyl)cyclopentadienyl, 1-*C*, 3'-*N*)(η^4-tetraphenylcyclo-butadiene)cobalt]dipalladium ($[COP\text{-}OAc]_2$) 714
Di-μ-chlorobis[(η^5-(*S*)-(pR)-2-(2'-(4'-methylethyl)oxazolinyl)cyclopentadienyl, 1-*C*, 3'-*N*)(η^4-tetraphenylcyclo-butadiene)cobalt]dipalladium ($[COP\text{-}Cl]_2$) 714
Hexafluoroacetylacetonate[(η^5-(*S*)-(p*R*)-2-(2'-(4'-methylethyl)oxazolinyl)cyclopentadienyl, 1-*C*, 3'-*N*)(η^4-tetraphenylcyclo-butadiene)cobalt]palladium (COP-hfacac) 714
Hexacarbonyl[μ-[(3,4-η:3,4-η)-2-methyl-3-butyn-2-ol]]dicobalt 442
$[MoCl_3(THF)_3]$ 529
[Rh((S,S)-*t*-Bu-BisP*))(nbd)]BF_4 1020
RuCl[(*S,S*)-NTsCH(C_6H_5)CH(C_6H_5)NH_2(η^6-cymene) 9
(Silane, (1,2-dideuterio-1-hexenyl)trimethyl-, (*Z*)-) 600
Tris[*N*-(*tert*-butyl)(3,5-dimethylphenyl)-amido]molybdenum(III) 529
Tris[*N*-(*tert*-butyl)(3,5-dimethylphenyl)-amido]molybdenum(VI) propylidyne 529

METATHESIS
Acetic acid 2-methylene-3-phenethylbut-3-enyl ester 561
2-(Benzoyloxymethyl)-1,3-cyclohexadiene 552
N-Boc-3-pyrroline 570

Carbazole-based tetrakismacrocycle 540

OXIDATION
2-Acetoxy-1,4-dioxane 850
(2-Benzoyloxy)-1,4-cyclohexanedione (mono)ethylene ketal 70
4-Benzoyloxymorpholine 684
N-Benzylidene-benzylamine *N*-oxide 114
1-(4-Bromophenyl)-4-methylphenyl)but-3-yn-1-one 794
5-(4-Bromophenyl)-3-iodo-2-(4-methylphenyl)furan 794
(R_S)-(+)-*tert*-Butyl *tert*-butanethiosulfinate 770
Camphorquinone oxime 702
(*R*)-1-Cyclohexyl-1-hydroxy-butan-2-one 205
1,2:4,5-Di-*O*-isopropylidene-D-erythro-2,3-hexodiulo-2,6-pyranose 177
3,4-Dibromo-2,5-dimethoxytetrahydrofuran 607
1,4:5,8-Dimethano-1,2,3,4,5,6,7,8-octahydroanthracene-9,10-diol 62
1,4:5,8-Dimethano-1,2,3,4,5,6,7,8-octahydro-9,10-dimethoxyanthracenium hexachloroantimonate 62
(*R*)-(+)-3,4-Dimethylcyclohex-2-en-1-one 100
(2*E*)-3,7-Dimethyl-2,6-octadienal 93
(*Z*)-3,7-Dimethyl-2,6-octadienal 77
(2R,5R)-1-{2-[*(2R,5R)*-2,5-Dimethylphospholan-1-yl]phenyl}-2,5-dimethylphospholane 1-oxide 982
Diphenyl disulfide 135
2-(Ethoxycarbonyl)prop-2-en-1-yl phenyl sulfone 308
Ethyl 2,2-dibromopropionate 577
5-Hydroxy-2(5*H*)-furanone 128
4-Methoxyphenylacetic acid 107
(*R*)-(+)-6-Methylcyclohex-2-en-1-one 100
(*R,R*)-*trans*-β-Methylstyrene oxide 183
Methyl-5-methylpyridine-2-carboxylate 911
Methyl phenyl sulfoxide 135
[(1*S*)-1-Phenylethyl]imino]acetonitrile *N*-oxide 139
(*R,R*)-1-Phenylcyclohexene oxide 183
D-Ribonolactone 88
9-Spiroepoxy-*endo*-tricyclo[5.2.2.0^{2,6}]undeca-4,10-dien-8-one 514
2,3,11,11a-Tetrahydro-1H-benzo[d]pyrrolo[1,2-a]azepin-5(6*H*)-one 811
N-*p*-Tolylsulfonyl (*E*)-1-phenylethylideneimine 942
4-Triisopropylsilyl-3-butyn-2-one 9

PHOSPHORYLATION
Dibutyl 3,4,6-tri-*O*-benzyl-2-*O*-pivaloyl-D-glucopyranosyl phosphate 169
Diphosphoric acid, tetrakis(phenylmethyl) ester 1001
(2*Z*,6*E*,10*E*)-Ethyl 7,11,15-Trimethyl-3-diethylphosphoryloxy-6,10,14-trienoate 221

PHOTOCHEMICAL REACTIONS
(*R*,R**)-2,2'-Bispyrrolidine 1046
(*R**,*S**)-2,2'-Bispyrrolidine 1046
5-Hydroxy-2(5*H*)-furanone 128
Methyl 2-bromomethyl-3-nitrobenzoate 903
2-Phenylthio-5-heptanol 121

PROTECTION
(2-Aminoethyl)carbamic acid *tert*-butyl ester 942
tert-Butyl(dimethyl)phosphine–Borane 1020
1-*tert*-Butyldimethylsilyloxy-2-methyl-1-cyclohexene 481
1-*tert*-Butyloxycarbonyl-4-(9-fluorenylmethyloxycarbonylamino)piperidine-4-carboxylic acid 956
1-*tert*-Butyloxycarbonylpiperidine-4-spiro-5'-(1',3'-bis(*tert*-butyloxycarbonyl))-hydantoin 956

Dibutyl 3,4,6-tri-*O*-benzyl-2-*O*-pivaloyl-D-glucopyranosyl phosphate 169
2,3-Dihydro-2-*tert*-butyldimethysiloxy-*1H*-indene 46
(*S*)-Ethyl 2-(*tert*-butyldimethylsilyloxy)propanoate 625
Ethyl (1'*S*,2'*R*)-*N*-*tert*-butyloxycarbonyl-*N*-(1',2'-diphenyl-2'-hydroxyethyl-glycinate 879
3,4-*O*-Isopropylidene-L-mannitol 264
2,3-Isopropylidene(D-ribonolactone) 88
N-(Methoxycarbonyl)-2-propenylamine 244
1-Methoxymethyl-1-phenylethane 934
Methyl (*R*)-(–)-3-benzyloxy-2-methyl propanoate 947

RADICAL CHAIN
3-Acetylamino-1-ethoxythiocarbonylsulfanyl-4,4,4-trifluorobutyl acetate 212

REARRANGEMENTS
6-Amino-3,4-dihydro-1(2*H*)-naphthalenone 670
2-Cyclopentylacetophenone 433
(*R*)-(+)-3,4-Dimethylcyclohex-2-en-1-one 100
5-Hydroxy-2(5*H*)-furanone 128
(*S,E*)-Methyl 2-((*R*)-4-methyl-1,2-dihydronaphthalen-2-yl)-4-phenylbut-3-enoate 296
5-Oxaspiro[3.4]octan-1-one 256
(*R*)-4-Phenyl-3-((1,2 -propadienyl-2-oxazolidinone 726
2-Phenylthio-5-heptanol 121
2,3,11,11a-Tetrahydro-1*H*-benzo[d]pyrrolo[1,2-a]azepin-5(6*H*)-one 811
(*S*)-2,2,2-Trichloro-*N*-(1-propylallyl)acetamide 720

REDUCTION
(2S)-(-)-3-exo-Aminoisoborneol 702
Bis[3,5-bis(trifluoromethyl)phenyl]phosphine 965
(*R*)-(*N-tert*-Butoxycarbonyl)allylglycine 399
2(*S*)-(β-*tert*-Butoxycarbonyl-α-(*R*)- and (*S*)-hydroxyethyl-4(*R*)-hydroxypyrrolidine-1-carboxylic acid, *tert*-butyl ester 38
(*S*)-2-(*t*-Butyldimethylsilyloxy)propanal 615
1,4:5,8-Dimethanododecahydroanthracene-9,10-dione 62
9,10-Dimethoxyphenanthrene 519
N,N-Dimethyl-3-phenylpropylamine 1
2-(2,2-Dimethylpropoxy)-2,3-dihydro-*1H*-indene 46
Dithieno[3,2-*b*:2′,3′-*d*]thiophene 843
(Ethanol, 1-[[(1*R*,2*S*,4*R*)-1,7,7-trimethylbicyclo[2.2.1]hept-2-yl]oxy]-, acetate) 52
(2*S*,3*S*)-1,2-Epoxy-3-(Boc-amino)-4-phenylbutane 149
Ethyl 4-aminobenzoate 57
(2*E*,6*E*,10*E*)-Ethyl 3,7,11,15-Tetramethylhexadeca-2,6,10,14-tetraenoate 221
(*E, E, E*)-Geranylgeraniol 221
(*R,R*)-Hydrobenzoin 17
(*Z*)-1-Iodo-1-heptene 359
1,2:3,4:5,6-Tri-*O*-Isopropylidene-L-mannitol 164
Methyl 4-chloro-1*H*-pyrrolo[3,2-c]pyridine-2-carboxylate 802
(+)-(*1R,2S,9S*)-11-Methyl-7,11-diazatricyclo[7.3.1.$0^{2,7}$]tridecane 25
(S)-1-Methyl-2-[(dihydroisoindol-2-yl)methyl]pyrrolidine 1037
Methyl indole-4-carboxylate 903
(Silane, (1,2-dideuterio-1-hexenyl)trimethyl-, (*Z*)-) 600
(*S*)-4-Triisopropylsilyl-3-butyn-2-ol 9
1,6,6-Trimethyl-4,8-dioxaspiro[2.5]oct-1-ene 457
(R)-4-Trimethylsilyl-3-butyn-2-ol 1056
(S)-4-Trimethylsilyl-3-butyn-2-ol 1056
Tris[*N*-(*tert*-butyl)(3,5-dimethylphenyl)-amido]molybdenum(VI) propylidyne 529

REPLACEMENT REACTIONS
Halogen by Boron

3-Pyridylboronic acid 393

Halogen by Carbon
1-(4-Acetylphenyl)-2-phenylethane 289
1-[3,5-Bis(trifluoromethyl)phenyl]-ethanone 343
N-(*tert*-Butoxycarbonyl)-β-[4-(methoxycarbonyl)phenyl]alanine methyl ester 384
tert-Butyl(dimethyl)phosphine–Borane 1020
(*E*)-1-(1-Heptenyl)-4-methoxybenzene 367
(*Z*)-1-(1-Heptenyl)-4-methoxybenzene 359
(*E*)-2-Methyl-3-phenylacrylic acid butyl ester 372
4-Nonylbenzoic acid methyl ester 353
(*E*)-4-(2-Phenylethenyl)benzonitrile 372
3-Pyridin-3-ylquinoline 393
2'-Vinylacetanilide 315

Halogen by Nitrogen
3-(3′-Azidopropyl)-3,4-dihydronaphthalen-2(1*H*)-one 811
2-Bromo-2-methylpropanamide 670
Methyl azidoacetate 802
(2S)-(-)-exo-(Morpholino)isoborneol 702
1-Morpholinopropan-1-one 197

Halogen by Oxygen
6-Amino-3,4-dihydro-1(2*H*)-naphthalenone 670
3-Benzyloxy-2-iodophenyl trifluoromethanesulfonate 488
2-Benzyloxypyridine 947
Ethylphosphonic acid bis-(2,2,2-trifluoroethyl) ester 992
1-Methoxy-2-(4-methoxyphenoxy)benzene 626

Halogen by Phosphorus
(2-Carbomethoxy-6-nitrobenzyl)triphenylphosphonium bromide 903
2-Carboxyethyltriphenylphosphonium chloride 296

Halogen by Silicon
(*Z*)-1-(1-Heptenyl)-1,1-dimethylsilanol 359

Halogen by Sulfur
Dithieno[3,2-*b*:2′,3′-*d*]thiophene-2,6-dicarboxylic acid diethyl ester 843

Hydrogen by Sulfur
S-(1-Acetylamino-2,2,2-trifluoroethyl) *O*-ethyl dithiocarbonate 212

Hydroxy by Halogen
N-(*tert*-Butoxycarbonyl)-β-iodoalanine methyl ester 384
N-1-(Chloro-2,2,2-trifluoro-ethyl)acetamide 212
Ethyl 3-chloro-3-phenylpropanoate 646
(*E, E*)-Farnesyl bromide 221
6-Iodo-1-hexene 433

Hydroxy by Oxygen
1,2:5,6-Dianhydro-3,4-*O*-isopropylidene-L-mannitol 164a

Oxygen by Halogen
Potassium 2-Phenethyltrifluoroborate 289

Oxygen by Phosphine
Dibutyl 3,4,6-tri-*O*-benzyl-2-*O*-pivaloyl-D-glucopyranosyl phosphate 169

Oxygen by Nitrogen
2-Benzyloxy-1-methylpyridinium trifluoromethanesulfonate 947
N-Benzylpiperidine 697
N-(2,2,2-Trifluoro-1-hydroxyethyl)-acetamide 212
(*S*)-(+)-2,4,6-Trimethylphenylsulfinamide 761

Oxygen by Sulfur
GlcNAc-thiazoline triacetate 787

Phosphine with Oxygen
3,4,6-Tri-*O*-benzyl-2-*O*-pivaloyl-β-D-glucopyranosyl-(1→6)-1,2:3,4-di-*O*-isopropylidene-α-D-galactopyranoside 169

Sulfonate by Tin
(P)-(+)-3-(Tributylstannyl)-1,2-butadiene 1056

Sulfur by Halogen
(*S*)-1-Chloro-3-(Boc-amino)-4-phenyl-2-butanone 149

Sulfur by Nitrogen
(R_S)-(+)-*tert*-Butanesulfinamide 770

RESOLUTION
(1S,2S)-trans-1-Benzoyloxy-2-bromocyclohexane 1037
(R,R)-2,2'-Bispyrrolidine 1046
(*S,S*)-2,2'-Bispyrrolidine 1046
(*S*)-Methyl glycidate 157

SAPONIFICATION
6-Amino-3,4-dihydro-1(2*H*)-naphthalenone 670
Dithieno[3,2-*b*:2′,3′-*d*]thiophene-2,6-dicarboxylic acid 843

SILYLATION
O-Silylation
[1,1-(Bis-benzyloxymethyl)vinyloxy]triethylsilane 473
3,6-Bis[1-(triisopropylsiloxy)ethenyl]-9,10-dimethoxyphenanthrene 524
2,3-Dihydro-2-*tert*-butyldimethysiloxy-*1H*-indene 46
(*S*)-Ethyl 2-(*t*-Butyldimethylsilyloxypropanoate 615
Trimethylsilyl 2-fluorosulfonyl-2,2-difluoroacetate 469

SUBSTITUTION
3-(3′-Azidopropyl)-3,4-dihydronaphthalen-2(1*H*)-one 811
3-Benzyloxy-2-iodophenyl trifluoromethanesulfonate 488
2-Benzyloxypyridine 947
Bis[3,5-bis(trifluoromethyl)phenyl]-phosphine oxide 965
1-(4-Bromophenyl)-4-methylphenyl)but-3-yn-1-ol 794
2-Carboxyethyltriphenylphosphonium chloride 296
3-(3′-Chloropropyl)-3,4-dihydronaphthalen-2(1*H*)-one 811
3,6-Diacetyl-9,10-dimethoxyphenanthrene 519
2-Iodoresorcinol 488
Methyl azidoacetate 802
(*E*)-Methyl 2-diazo-4-phenylbut-3-enoate 296
(*R*)-4-Phenyl-3-(2-triisopropylsilyl-ethynyl)oxazolidin-2-one 677
Potassium 2-Phenethyltrifluoroborate 289
9-Tetradecylcarbazole 540
(R_S,*1S,2R*)-(−)-2,4,6-Trimethylbenzenesulfinic acid 1-(2,4,6-trimethylbenzenesulfonylamino)-2-indan-2-yl ester 761

SULFINATION
3-(2,4,6-Trimethylbenzenesulfonyl)-3,3*a*,8,8*a*-tetrahydro-*2H*-1-oxa-2λ^4-thia-3-aza-cyclopenta[*a*]inden-2-ol 761
(*S*)-(+)-2,4,6-Trimethylphenylsulfinamide 761

SULFONATION
N-Sulfonation
o'-(1,1-Difluorohex-1-en-2-yl)-*p*-toluenesulfonanilide 834
(*1S,2R*)-(+)-*N*-(2-Hydroxyindan-1-yl)-2,4,6-trimethylbenzenesulfonamide 761
N-(2-Propenyl)-4-methylbenzenesulfonamide 442
(*S*)-*N*-Trifluoromethylsulfonyl-2-isopropylaziridine 506

O-Sulfonation
3-Benzyloxy-2-iodophenyl trifluoromethanesulfonate 488
N-(*tert*-Butoxycarbonyl)-*O*-(*p*-toluenesulfonyl)-L-serine methyl ester 384
1-*tert*-Butyldimethylsilyloxy-2-methyl-1-cyclohexene 481
1,1-Dichloro-(3*S*)-(t-butyldimethylsilyloxy)-2-butanol *p*-toluenesulfonate 615
5-Heptenyl methanesulfonate 433
(R)-4-Trimethylsilyl-3-butyn-2-yl Mesylate 1056

TYPE OF COMPOUND INDEX

Preparations are listed by functional group or ring system. Many compounds with multiple functional groups are multi-listed, although not all of the compounds have multiple listings. This choice represents an arbitrary judgment by the editor as to the most likely place a user would look for specific preparations.

ACETALS and HEMIACETALS

2-Acetamido-2-deoxy-1,3,4,6-tetra-*O*-acetyl-β-D-glucopyranose 787
3-Acetylamino-1-ethoxythiocarbonylsulfanyl-4,4,4-trifluorobutyl acetate 212
2-Acetoxy-1,4-dioxane 850
(2-Benzoyloxy)-1,4-cyclohexanedione (mono)ethylene ketal 70
[1,1-(Bis-benzyloxymethyl)vinyloxy]triethylsilane 473
1,1-Bis(benzyloxy)propan-2-one 473
2-Chloro-3,3-diethoxyprop-1-ene 586
1,2:5,6-Dianhydro-3,4-*O*-isopropylidene-L-mannitol 164
3,4-Dibromo-2,5-dimethoxytetrahydrofuran 607
2,3-Dibromo-1,1,4,4-tetramethoxybutane 607
N-(2,2-Diethoxy)ethyl formamide 778
2,2-Diethoxy-1-isocyanoethane 778
1,2:4,5-Di-*O*-isopropylidene-β-D-fructopyranose 177
1,2:4,5-Di-*O*-isopropylidene-D-erythro-2,3-hexodiulo-2,6-pyranose 177
4,4-Dimethoxybut-2-ynal 607
N-(3,3-Dimethoxy-1-trifluoromethyl-propyl)-acetamide 212
5-Hydroxy-2(5*H*)-furanone 128
O-Isopropylidene-α-D-galactopyranoside 169
3,4-*O*-Isopropylidene-L-mannitol 164
2,3-Isopropylidene(D-ribonolactone) 88
1-Methoxymethyl-1-phenylethane 934
(5*S*)-(*d*-Menthyloxy)-2(5*H*)-furanone 128
1,1,4,4-Tetramethoxybut-2-yne 607
3,4,6-Tri-*O*-benzyl-2-*O*-pivaloyl-β-D-glucopyranosyl-(1→6)-1,2:3,4-di-
1,2:3,4:5,6-Tri-*O*-isopropylidene-L-mannitol 164
N-(2,2,2-Trifluoro-1-hydroxyethyl)-acetamide 212
1-[[(1*R*,2*S*,4*R*)-1,7,7-Trimethylbicyclo[2.2.1]hept-2-yl]oxy]-ethanol acetate 52

ACIDS

4-Amino-1-*tert*-butyloxycarbonylpiperidine-4-carboxylic acid 956
(2*R*,3*S*)-3-*endo*-Benzyloxycarbonyl-bicyclo[2.2.1]hept-5-ene-2-endo-carboxylic acid 1068
(*R*)-(*N*-*tert*-Butoxycarbonyl)allylglycine 399
1-*tert*-Butyloxycarbonyl-4-(9-fluorenylmethyloxycarbonylamino)piperidine-4-carboxylic acid 956
2-Carboxyethyltriphenylphosphonium chloride 296
Dibenzofuran-4,6-dicarboxylic acid 888
Dithieno[3,2-*b*:2′,3′-*d*]thiophene-2,6-dicarboxylic acid 843
(2*E*)-3-Iodo-2-propenoic acid 662
4-Methoxyphenylacetic acid 107
(*Z*)-2-Methyl-3-trimethylsilyl-2-butenoic acid 577
4-Nonylbenzoic acid 353

ALCOHOLS

(1R,2S)-1-[3-Adamantyl)-2-hydroxy-5-methylbenzyliden-amino]indan-2-ol 498
(2S)-(-)-3-exo-Aminoisoborneol 702
N-(9-Anthracenylmethyl)cinchonidinium chloride 404
(*R*)-(*E*)-2-Benzylidiene-1-cyclohexyl-butan-1-ol 205
trans-2-Benzyl-1-methylcyclopropan-1-ol 450
Bicyclo[3.1.0]hexan-1-ol 450
3-Bromo-2-hydroxybenzaldehyde 267
1-(4-Bromophenyl)-4-methylphenyl)but-3-yn-1-ol 794

(2*S*-*trans*)-2-*tert*-Butoxycarbonylacetyl-4-hydroxypyrrolidine-1-carboxylic acid, *tert*-butyl ester 38
2(*S*)-(β-*tert*-Butoxycarbonyl-α-(*R*)- and α-(*S*)-hydroxyethyl)-4-(*R*)-hydroxypyrrolidine-1-carboxylic acid, *tert*-butyl ester 38
(*S*)-4-(*t*-Butyldimethylsilyloxy)-2-pentynol 615
(*R*)-1-Cyclohexyl-1-hydroxy-butan-2-one 205
(*R,R*)-Dibenzofuran-4,6-dicarboxylic acid bis(2-hydroxy-1-phenylethyl) amide 888
dl-1,2-Dicyclohexylethanediol 348
(6*R*)-(+)-1,6-Dimethylcyclohex-2-en-1-ol 100
(1*R*,2*R*)-1,2-Diphenyl-1,2-ethanediol 17
Ethyl (1'*S*,2'*R*)-*N*-(1',2'-diphenyl-2'-hydroxyethyl)glycinate 879
Ethyl (1'*S*,2'*R*)-*N*-*tert*-butyloxycarbonyl-*N*-(1',2'-diphenyl-2'-hydroxyethyl-glycinate 879
(*E, E, E*)-Geranylgeraniol 221
(*Z*)-1-(1-Heptenyl)-1,1-dimethylsilanol 359
(1*S*,2*R*)-1-[(2-Hydroxy-3,5-di-*tert*-butylbenzylidene)amino]indan-2-ol 770
(*S*)-[1-(Hydroxydiphenylmethyl)-2-methylpropyl]carbamic acid, *tert*-butyl ester 896
1-Hydroxy-3-phenyl-2-propanone 642
3,4-*O*-Isopropylidene-L-mannitol 164
5-Methyl-1-phenyl-6-hepten-3-ol 327
3-Methyl-1-pheny-4-penten-1-ol 327
(2S)-(-)-exo-(Morpholino)isoborneol 702
N-Phenyl-(5*R*)-hydroxymethyl-2-oxazolidinone 867
5-Phenylpent-2-yn-1-ol 561
3-Phenyl-2-(propylthio)-2-propen-1-ol 642
2-Phenylthio-5-heptanol 121
(*rac*)-4-Triisopropylsilyl-3-butyn-2-ol 9
(*S*)-4-Triisopropylsilyl-3-butyn-2-ol 9
(R)-4-Trimethylsilyl-3-butyn-2-ol 1056
(S)-4-Trimethylsilyl-3-butyn-2-ol 1056

ALDEHYDES
3-Bromo-2-hydroxybenzaldehyde 267
(*S*)-2-(*t*-Butyldimethylsilyloxy)propanal 615
3,4-Dibromo-2,5-diformylthiophene 843
4,4-Dimethoxybut-2-ynal 607
(2*E*)-3,7-Dimethyl-2,6-octadienal 93
(Z)-3,7-Dimethyl-2,6-octadienal 77
(3*R*)-3,7-Dimethyl-2-(3-oxobutyl)-6-octenal 416

ALKENES
2-(Benzoyloxymethyl)-1,3-cyclohexadiene 552
(*R*)-(*E*)-2-Benzylidiene-1-cyclohexyl-butan-1-ol 205
2-Chloro-1-decen-3-yne 234
1-Chloro-4-(2,2-dichloro-1-methylvinyl)benzene 594
2-Chloro-3,3-diethoxyprop-1-ene 586
(*Z*)-(1,2-Dideuterio-1-hexenyl)trimethylsilane 600
o-(1,1-Difluorohex-1-en-2-yl)aniline 834
o'-(1,1-Difluorohex-1-en-2-yl)-*p*-toluenesulfonanilide 834
1,4:5,8-Dimethano-1,4,4a,5,8,8a,9a,10a-octahydroanthracene-9,10-dione 62
(*R*)-(+)-3,4-Dimethylcyclohex-2-en-1-one 100
3,4-Dimethylcyclopent-2-enone 420
1,4-Dioxene 850
(2*E*,6*E*,10*E*)-Ethyl 3,7,11,15-Tetramethylhexadeca-2,6,10,14-tetraenoate 221
Ethyl 2-{[(*1S,2R,3R,5S)*-2,6,6-trimethylbicyclo[3.1.1]hept-3-yl]methyl}acrylate 308
(2*Z*,6*E*,10*E*)-Ethyl 7,11,15-Trimethyl-3-diethylphosphoryloxy-6,10,14-trienoate 221
(6*E*, 10*E*)-Ethyl 7,11,15-Trimethyl-3-oxohexa-deca-6,10,14-trienoate 221
(*E, E*)-Farnesyl bromide 221
(*E, E, E*)-Geranylgeraniol 221
N-(Methoxycarbonyl)-*N*-(2-propenyl)-1,3-decadiynylamide 244
N-(Methoxycarbonyl)-2-propenylamine 244

(*E*)-Methyl 2-diazo-4-phenylbut-3-enoate 296
1-Methyl-3,4-dihydronaphthalene 296
(*S,E*)-Methyl 2-((*R*)-4-methyl-1,2-dihydronaphthalen-2-yl)-4-phenylbut-3-enoate 296
(Z)-2-Methyl-3-trimethylsilyl-2-butenoic acid 577
Methyl 2-ethenyl-3-nitrobenzoate 903
(*R*)-(+)-6-Methylcyclohex-2-en-1-one 100
5-Methyl-1-phenyl-6-hepten-3-ol 327
3-Methyl-1-phenyl-4-penten-1-ol 327
6,9,12,15,18-Pentamethyl-1,6,9,12,15,18-hexahydro(c_{60}-i_h)[5,6]fullerene 319
4-Phenyl-5(*E*)-decen-2-one 261
1-Phenylhex-5-en-3-amine 691
N-(2-Propenyl)-4-methylbenzenesulfonamide 442
N-(2-Propenyl)-*N*-(2-propynyl)-4-methylbenzenesulfonamide 442
(2-Propenoic acid, 3-iodo-, (2*E*)-) 662
2'-Vinylacetanilide 315

ALKYNES

3,6-Bis(benzoylbiphenyl)ethynyl-9-tetradecylcarbazole 540
3,6-Bis[(trimethylsilyl)ethynyl]-9-tetradecylcarbazole 540
1-Bromo-1,3-decadiyne 244
1-(4-Bromophenyl)-4-methylphenyl)but-3-yn-1-ol 794
1-(4-Bromophenyl)-4-methylphenyl)but-3-yn-1-one 794
1-Bromo-2-triisopropylsilyl-ethyne 677
(R)-3-Butyn-2-yl Mesylate 1056
2-Chloro-1-decen-3-yne 234
1,3-Decadiyne 234
(±)-*N,N*-Dibenzyl-*N*-[1-cyclohexyl-3-(trimethylsilyl)-2-propynyl]-amine 192
(*R*)-(+)-*N,N*-Dibenzyl-*N*-[1-cyclohexyl-3-(trimethylsilyl)-2-propynyl]-amine 192
(*R*)-(+)-*N,N*-Dibenzyl-*N*-(1-cyclohexyl-2-propynyl)-amine 192
3,6-Diethynyl-9-tetradecylcarbazole 540
4,4-Dimethoxybut-2-ynal 607
N-(Methoxycarbonyl)-*N*-(2-propenyl)-1,3-decadiynylamide 244
(*R*)-4-Phenyl-3-(2-triisopropylsilyl-ethynyl)oxazolidin-2-one 677
Propargyl benzoate 552
N-(2-Propenyl)-*N*-(2-propynyl)-4-methylbenzenesulfonamide 442
1-(2-(1-Propynyl)phenyl)pyrrole 826
1,1,4,4-Tetramethoxybut-2-yne 607
(*rac*)-4-Triisopropylsilyl-3-butyn-2-ol 9
4-Triisopropylsilyl-3-butyn-2-one 9
(*S*)-4-Triisopropylsilyl-3-butyn-2-ol 9
(R)-4-Trimethylsilyl-3-butyn-2-ol 1056
(S)-4-Trimethylsilyl-3-butyn-2-ol 1056
(R)-4-Trimethylsilyl-3-butyn-2-yl Acetate 1056
(R)-4-Trimethylsilyl-3-butyn-2-yl Mesylate 1056
(S)-4-Trimethylsilyl-3-butyn-2-yl Succinate 1056

AMIDES AND LACTAMS

3-Acetylamino-1-ethoxythiocarbonylsulfanyl-4,4,4-trifluorobutyl acetate 212
S-(1-Acetylamino-2,2,2-trifluoroethyl) *O*-ethyl dithiocarbonate 212
N-Benzyl-4-phenylbutyramide 733
(–)-(S,S)-N,N'-Bis(1-hydroxymethyl-2-methylpropyl)-2,2-dimethyl-malonamide 821
(S)-N-(N-tert-Butoxycarbonylprolyl)dihydroisoindole 1037
2-Bromo-2-methylpropanamide 670
N-1-(Chloro-2,2,2-trifluoro-ethyl)acetamide 212
(*R,R*)-Dibenzofuran-4,6-dicarboxylic acid bis(2-hydroxy-1-phenylethyl) amide 888
N,N-Dibenzylformamide 425
N-(2,2-Diethoxy)ethyl formamide 778
N-(3,3-Dimethoxy-1-trifluoromethyl-propyl)-acetamide 212
N-Methoxy-*N*-methylcarbamoyl chloride 992
[1-(Methoxymethylcarbamoyl)ethyl] phosphonic acid bis-(2,2,2-trifluoroethyl) ester 992

trans-1-(4-Methoxyphenyl)-4-phenyl-3-(phenylthio)azetidin-2-one 920
1-Morpholinopropan-1-one 197
2,3,11,11a-Tetrahydro-1*H*-benzo[d]pyrrolo[1,2-a]azepin-5(6*H*)-one 934
(*S*)-2,2,2-Trichloro-N-(1-propylallyl)acetamide 720
N-(2,2,2-Trifluoro-1-hydroxyethyl)-acetamide 212
2'-Vinylacetanilide 315

AMINES
6-Amino-3,4-dihydro-1(2*H*)-naphthalenone 670
(2-Aminoethyl)carbamic acid *tert*-butyl ester 942
(2S)-(-)-3-exo-Aminoisoborneol 702
(*2S,6S*)-4-Benzyl-1,7-bis(trifluoromethylsulfonyl)-2,6-diisopropyl-1,4,7-triazaheptane 506
N-Benzylpiperidine 697
(*R,R*)-2,2'-Bispyrrolidine 1046
(5*S*,6*R*)-4-*tert*-Butoxycarbonyl-5,6-diphenylmorpholin-2-one 879
(–)-Cytisine 25
(*R*)-(+)-*N,N*-Dibenzyl-*N*-(1-cyclohexyl-2-propynyl)-amine 192
(±)-*N,N*-Dibenzyl-*N*-[1-cyclohexyl-3-(trimethylsilyl)-2-propynyl]-amine 192
(*R*)-(+)-*N,N*-Dibenzyl-*N*-[1-cyclohexyl-3-(trimethylsilyl)-2-propynyl]-amine 192
N-(3,5-Dimethylphenyl)-*tert*-butylamine 529
N,N-Dimethyl-3-phenylpropylamine 1
1(*S*),2(*S*)-Diphenyl-*N,N'*-bis-[1(*R*)-phenyl-ethyl]-ethane-1,2-diamine 1028
Ethyl 4-[(6-chloropyridin-3-yl)amino]benzoate 665
Ethyl (1'*S*,2'*R*)-*N*-(1',2'-diphenyl-2'-hydroxyethyl)glycinate 879
(+)-(*1R,2S,9S*)-11-Methyl-7,11-diazatricyclo[7.3.1.$0^{2,7}$]tridecane 25
(S)-1-Methyl-2-[(dihydroisoindol-2-yl)methyl]pyrrolidine 826
(2S)-(-)-exo-(Morpholino)isoborneol 702
(*S*)-[(1-Phenylethyl)amino]acetonitrile 139
1-Phenylhex-5-en-3-amine 691
4-Phenylmorpholine 684
(1S)-1-Phenylpropan-1-amine hydrochloride 749
N-[(1S)-1-Phenylpropyl]-*P,P*-diphenylphosphinic amide 749

AMINO ACIDS
(*R*)-(*N-tert*-Butoxycarbonyl)allylglycine 399

AZIDES
3-(3′-Azidopropyl)-3,4-dihydronaphthalen-2(1*H*)-one 934
Methyl azidoacetate 802
Methyl (2*Z*)-2-azido-3-(2-chloropyridin-3-yl)acrylate 802

AZO COMPOUNDS
N-[(9*H*-Fluoren-9-ylmethoxy)carbonyl]-L-aspartic acid, α-*tert*-butyl ester, β-(2-ethyl(1*E*)-[[[(4-nitrophenyl)azo]phenyl]amino]ethyl ester 635

CARBAMATES
4-Amino-1-*tert*-butyloxycarbonylpiperidine-4-carboxylic acid 956
(2-Aminoethyl)carbamic acid *tert*-butyl ester 942
N-Boc-3-pyrroline 570
(5*S*,6*R*)-4-*tert*-Butoxycarbonyl-5,6-diphenylmorpholin-2-one 879
(3*R*,5R,6*S*)-4-*tert*-Butoxycarbonyl-5,6-diphenyl-3-(1'-prop-2'-enyl)morpholin-2-one 399
(2*S*-*trans*)-2-*tert*-Butoxycarbonylacetyl-4-hydroxypyrrolidine-1-carboxylic acid, *tert*-butyl ester 38
(*R*)-(*N-tert*-Butoxycarbonyl)allylglycine 399
2(*S*)-(β-*tert*-Butoxycarbonyl-α-(*R*)- and α-(*S*)-hydroxyethyl)- 4-(*R*)-hydroxypyrrolidine-1-carboxylic acid, *tert*-butyl ester 38
1-*tert*-Butyloxycarbonyl-4-(9-fluorenylmethyloxycarbonylamino)piperidine-4-carboxylic acid 956
N-(*tert*-Butoxycarbonyl)-L-iodoalanine methyl ester 384
N-(*tert*-Butoxycarbonyl)-L-[4-(methoxycarbonyl)phenyl]alanine methyl ester 384

1-*tert*-Butyloxycarbonylpiperidine-4-spiro-5'-(1',3'-bis(*tert*-butyloxycarbonyl))hydantoin 956
N-(*tert*-Butoxycarbonyl)-*O*-(p-toluenesulfonyl)-L-serine methyl ester 384
N-(*tert*-Butoxycarbonyl)-L-valine methyl ester 896
(*S*)-1-Chloro-3-(Boc-amino)-4-phenyl-2-butanone 149
Dimethylsulfoxonium (*S*)-2-oxo-3-(Boc-amino)-4-phenylbutylide 149
(2*S*,3*S*)-1,2-Epoxy-3-(Boc-amino)-4-phenylbutane 149
Ethyl (1'*S*,2'*R*)-*N*-*tert*-butyloxycarbonyl-*N*-(1',2'-diphenyl-2'-hydroxyethyl-glycinate 879
N-[(9*H*-Fluoren-9-ylmethoxy)carbonyl]-L-aspartic acid, α-*tert*-butyl ester, β-(2-ethyl(1*E*)-[[[(4-nitrophenyl)azo]phenyl]amino]ethyl ester 635
(*S*)-[1-(Hydroxydiphenylmethyl)-2-methylpropyl]carbamic acid, *tert*-butyl ester 896
N-(Methoxycarbonyl)-2-propenylamine 244
N-(Methoxycarbonyl)-*N*-(2-propenyl)-1,3-decadiynylamide 244
(*R*)-4-Phenyl-3-((1,2 -propadienyl-2-oxazolidinone 726
(*R*)-4-Phenyl-3-(2-propynyl)oxazolidin-2-one 736
(*R*)-4-Phenyl-3-(2-triisopropylsilyl-ethynyl)oxazolidin-2-one 677
(–)-Methyl (*1R,9R*)-6-Oxo-7,11-diazatricyclo[7.3.1.$0^{2,7}$]trideca-2,4-diene-11-carboxylate 25

CARBOHYDRATES
2-Acetamido-2-deoxy-1,3,4,6-tetra-*O*-acetyl-β-D-glucopyranose 787
1,2:5,6-Dianhydro-3,4-*O*-isopropylidene-L-mannitol 164
Dibutyl 3,4,6-tri-*O*-benzyl-2-*O*-pivaloyl-D-glucopyranosyl phosphate 169
1,2:4,5-Di-*O*-isopropylidene-β-D-fructopyranose 177
1,2:4,5-Di-*O*-isopropylidene-D-erythro-2,3-hexodiulo-2,6-pyranose 177
GlcNAc-thiazoline triacetate 787
2,3-Isopropylidene(D-ribonolactone) 88
D-Ribonolactone 88
3,4,6-Tri-*O*-benzyl-2-*O*-pivaloyl-β-D-glucopyranosyl-(1→6)-1,2:3,4-di-*O*-isopropylidene-α-D-galactopyranoside 169
1,2:3,4:5,6-Tri-*O*-isopropylidene-L-mannitol 164

CATALYSTS
Chromium
Chromium (III) complex of *(1R,2S)*-1-[3-Adamantyl)-2-hydroxy-5-methylbenzylidenamino]indan-2-ol 498

Iron
Ferric acetylacetonate 353

Molybdenum
Tris[*N*-(*tert*-butyl)(3,5-dimethylphenyl)-amido]molybdenum(VI) propylidyne 529

Palladium
Bis(tri-*tert*-butylphosphine)palladium 372
Palladium acetate 393
Tris(dibenzylideneacetone)dipalladium 359, 367, 374, 384

Platinum
Tri-*tert*-butylphosphine-Pt(0) complex 367

Rhodium
[Rh*((S,S)*-t-Bu-BisP*))(nbd)]BF_4 1020

Ruthenium
(μ_3, η^2: η^3: η^5-Acenaphthylene)$Ru_3(CO)_7$ 1
Ru(II)-BINAP complex 38

Vanadium
Dichlorodicyclopentadienylvanadium 348

CYCLOBUTANE
5-Benzyloxybicyclo[4.2.0]octa-1,3,5-trien-7-one 488

CYCLOPROPANE
trans-2-Benzyl-1-methylcyclopropan-1-ol 450
Bicyclo[3.1.0]hexan-1-ol 450
n-Butyl 2,2-difluorocyclopropanecarboxylate 469
1,1-Dichloro-2-ethoxycyclopropane 586
6,6-Dimethyl-1-methylene-4,8-dioxaspiro[2.5]octane 457
1,6,6-Trimethyl-4,8-dioxaspiro[2.5]oct-1-ene 457

CUMULLENE

DIAZO
(*E*)-Methyl 2-diazo-4-phenylbut-3-enoate 296
S-Phenyl diazothioacetate 920

DIENES
2-(Benzoyloxymethyl)-1,3-cyclohexadiene 552

ENAMINE
N-[2-Chloro-3-(dimethylamino)-2-propenylidene]-*N*-methyl-methanaminium hexafluorophosphate 929

ENOL
Methyl 2-oxo-1,2,3,4-tetrahydronaphthalene-1-carboxylate 934
3-Phenyl-2-(propylthio)-2-propen-1-ol 642

ESTERS AND LACTONES
2-Acetamido-2-deoxy-1,3,4,6-tetra-*O*-acetyl-β-D-glucopyranose 787
Acetic acid 2-methylene-3-phenethylbut-3-enyl ester 561
Acetic acid 5-phenylpent-2-ynyl ester 561
2-Acetoxy-1,4-dioxane 850
3-Acetylamino-1-ethoxythiocarbonylsulfanyl-4,4,4-trifluorobutyl acetate 212
(2-Aminoethyl)carbamic acid *tert*-butyl ester 942
(4*S*)-2-(Benzhydrylidenamino)pentanedioic Acid, 1-*tert*-Butyl Ester-5-Methyl Ester 404
(1S,2S)-trans-1-Benzoyloxy-2-bromocyclohexane 1037
(2-Benzoyloxy)-1,4-cyclohexanedione (mono)ethylene ketal 70
(2*R*,3*S*)-3-endo-Benzyloxycarbonyl-bicyclo[2.2.1]hept-5-ene-2-endo-carboxylic acid 1068
(3*R*,5*R*,6*S*)-4-*tert*-Butoxycarbonyl-5,6-diphenyl-3-(1'-prop-2'-enyl)morpholin-2-one 399
(2*S*-*trans*)-2-*tert*-Butoxycarbonylacetyl-4-hydroxypyrrolidine-1-carboxylic acid, *tert*-butyl ester 38
2(*S*)-(β-*tert*-Butoxycarbonyl-α-(*R*)- and α-(*S*)-hydroxyethyl)-4-(*R*)-hydroxypyrrolidine-1-carboxylic acid, *tert*-butyl ester 38
N-(*tert*-Butoxycarbonyl)-L-iodoalanine methyl ester 384
N-(*tert*-Butoxycarbonyl)-L-[4-(methoxycarbonyl)phenyl]alanine methyl ester 384
N-(*tert*-Butoxycarbonyl)-*O*-(p-toluenesulfonyl)-L-serine methyl ester 384
N-(*tert*-Butoxycarbonyl)-L-valine methyl ester 896
n-Butyl 2,2-difluorocyclopropanecarboxylate 469
(η^5-Carbomethoxycyclopentadienyl)-(η^4-tetraphenyl-cyclobutadiene)cobalt 708
Dibutyl 3,4,6-tri-*O*-benzyl-2-*O*-pivaloyl-D-glucopyranosyl phosphate 169
(3a*S*,7a*S*)-2-[2-(3,4-Dimethoxyphenyl)-ethyl]-1,3-dioxo-octahydro-isoindole-3a-carboxylic acid methyl ester 1028
2-(5,5-Dimethyl-1,3,2-dioxaborinan-2-yl) benzoic acid ethyl ester 1015
Dimethylsulfoxonium (*S*)-2-oxo-3-(Boc-amino)-4-phenylbutylide 149
Dithieno[3,2-*b*:2',3'-*d*]thiophene-2,6-dicarboxylic acid diethyl ester 843
5-Ethoxy-5-oxopentylzinc iodide 281
Ethyl 4-aminobenzoate 57
(*S*)-Ethyl 2-(*t*-Butyldimethylsilyloxy)propanoate 615
Ethyl (1'*S*,2'*R*)-*N*-*tert*-butyloxycarbonyl-*N*-(1',2'-diphenyl-2'-hydroxyethyl-glycinate 879

Ethyl 4-[(6-chloropyridin-3-yl)amino]benzoate 665
Ethyl 2,2-dibromopropionate 577
Ethyl (1'*S*,2'*R*)-*N*-(1',2'-diphenyl-2'-hydroxyethyl)glycinate 879
S-Ethyl Octanethioate 281
Ethyl 6-Oxotridecanoate 281
(2*Z*,6*E*,10*E*)-Ethyl 7,11,15-Trimethyl-3-diethylphosphoryloxy-6,10,14-trienoate 221
Ethylphosphonic acid bis-(2,2,2-trifluoroethyl) ester 992
(2*E*,6*E*,10*E*)-Ethyl 3,7,11,15-Tetramethylhexadeca-2,6,10,14-tetraenoate 221
(6*E*, 10*E*)-Ethyl 7,11,15-Trimethyl-3-oxohexa-deca-6,10,14-trienoate 221
N-[(9*H*-Fluoren-9-ylmethoxy)carbonyl]-L-aspartic acid, α-*tert*-butyl ester, β-(2-ethyl(1*E*)-[[[(4-nitrophenyl)azo]phenyl]amino]ethyl ester 635
5-Hydroxy-2(5*H*)-furanone 128
Isopropyl *(E)*-but-2-enoate 420
2,3-Isopropylidene(D-ribonolactone) 88
Methyl azidoacetate 802
Methyl (2*Z*)-2-azido-3-(2-chloropyridin-3-yl)acrylate 802
N-Methyl-*O*-benzoylhydroxylamine hydrochloride 70
Methyl (*R*)-(–)-3-benzyloxy-2-methyl propanoate 947
Methyl 2-bromomethyl-3-nitrobenzoate 903
(*E*)-Methyl 2-diazo-4-phenylbut-3-enoate 296
(*S,E*)-Methyl 2-((*R*)-4-methyl-1,2-dihydronaphthalen-2-yl)-4-phenylbut-3-enoate 296
Methyl-5-methylpyridine-2-carboxylate 911
Methyl 2-oxo-1,2,3,4-tetrahydronaphthalene-1-carboxylate 934
(5*S*)-(*d*-Menthyloxy)-2(5*H*)-furanone 128
(*E*)-2-Methyl-3-phenylacrylic acid butyl ester 372
4-Nonylbenzoic acid methyl ester 353
(*4S*)-4-(2-Phenethyl)oxetan-2-one 506
Propargyl benzoate 552
D-Ribonolactone 88
(*E*)-2,2,2-Trichloroacetimidic acid hex-2-enyl ester 1007
1-[[(1*R*,2*S*,4*R*)-1,7,7-Trimethylbicyclo[2.2.1]hept-2-yl]oxy]-ethanol acetate 52
(R)-4-Trimethylsilyl-3-butyn-2-yl Acetate 1056
*(*S*)*-4-Trimethylsilyl-3-butyn-2-yl Succinate 1056

ETHERS
Enol Ethers
3,6-Bis[1-(triisopropylsiloxy)ethenyl]-9,10-dimethoxyphenanthrene 524
[1,1-(Bis-benzyloxymethyl)vinyloxy]triethylsilane 473
1-*tert*-Butyldimethylsilyloxy-2-methyl-1-cyclohexene 481
1-Methoxy-4-vinyloxybenzene 631

Epoxides
1,2:5,6-Dianhydro-3,4-*O*-isopropylidene-L-mannitol 164
(±)-Methyl glycidate 157
(*S*)-Methyl glycidate 157
(*R,R*)-*trans*-β-Methylstyrene oxide 183
(*R,R*)-1-Phenylcyclohexene oxide 183
9-Spiroepoxy-*endo*-tricyclo[5.2.2.0^{2,6}]undeca-4,10-dien-8-one 514

Silyl ethers
6,13-Bis(triisopropylsiloxy)-9,10-dimethoxy[7]helicenebisquinone 524
[1,1-(Bis-benzyloxymethyl)vinyloxy]triethylsilane 473
1-(*tert*-Butyldimethylsilyloxy)-8-(methoxycarbonyl)-6-methyl-bicyclo[4.2.0]octane 481
1-*tert*-Butyldimethylsilyloxy-2-methyl-1-cyclohexene 481
(*S*)-4-(*t*-Butyldimethylsilyloxy)-2-pentynol 615
(*S*)-2-(*t*-Butyldimethylsilyloxy)propanal 615
1,1-Dichloro-(3*S*)-(*tert*-butyldimethylsilyloxy)-2-butanol *p*-toluenesulfonate 615
2,3-Dihydro-2-*tert*-butyldimethysiloxy-*1H*-indene 46
(*S*)-Ethyl 2-(*t*-Butyldimethylsilyloxy)propanoate 615

Miscellaneous
O-Allyl-*N*-(9-Anthracenylmethyl)cinchonidinium bromide 404
2α-Benzyloxy-8-oxabicyclo[3.2.1]oct-6-en-3-one 473
3,6-Diacetyl-9,10-dimethoxyphenanthrene 519
Dibutyl 3,4,6-tri-*O*-benzyl-2-*O*-pivaloyl-D-glucopyranosyl phosphate 169
1,1-Dichloro-2-ethoxycyclopropane 586
1,4:5,8-Dimethano-1,2,3,4,5,6,7,8-octahydro-9,10-dimethoxy-anthracene 62
9,10-Dimethoxyphenanthrene 519
2-(2,2-Dimethylpropoxy)-2,3-dihydro-*1H*-indene 46
(*E*)-1-(1-Heptenyl)-4-methoxybenzene 367
(*Z*)-1-(1-Heptenyl)-4-methoxybenzene 359
1-Methoxy-2-(4-methoxyphenoxy)benzene 626
4-Methoxyphenylacetic acid 107
4-(4-Methoxyphenyl)-2-phenyl-1*H*-imidazole 862
1-Methoxy-4-vinyloxybenzene 631
Methyl (*R*)-(–)-3-benzyloxy-2-methyl propanoate 947

FORMAMIDINE
1-(*tert*-Butylimino-methyl)-1,3-dimethyl-urea hydrochloride 782

FULLERENES
6,9,12,15,18-Pentamethyl-1,6,9,12,15,18-hexahydro(c_{60}-i_h)[5,6]fullerene 319

HALOGENATED COMPOUNDS
Bromine
(1S,2S)-trans-1-Benzoyloxy-2-bromocyclohexane 1037
1-Bromo-1,3-decadiyne 244
3-Bromo-2-hydroxybenzaldehyde 267
5-Bromoisoquinoline 656
2-Bromo-2-methylpropanamide 670
5-Bromo-8-nitroisoquinoline 656
1-(4-Bromophenyl)-4-methylphenyl)but-3-yn-1-ol 794
1-(4-Bromophenyl)-4-methylphenyl)but-3-yn-1-one 794
5-(4-Bromophenyl)-3-iodo-2-(4-methylphenyl)furan 794
1-Bromo-2-triisopropylsilyl-ethyne 677
(4,4-Dibromobut-3-enyl)benzene 561
3,4-Dibromo-2,5-diformylthiophene 843
3,4-Dibromo-2,5-dimethoxytetrahydrofuran 607
2,3-Dibromo-1,1,4,4-tetramethoxybutane 607
Ethyl 2,2-dibromopropionate 577
(*E, E*)-Farnesyl bromide 221
Hexakis(4-Bromophenyl)benzene 652
Methyl 2-bromomethyl-3-nitrobenzoate 903

Chlorine
2,2-Bis-(chloromethyl)-5,5-dimethyl-1,3-dioxane 457
(*S*)-1-Chloro-3-(Boc-amino)-4-phenyl-2-butanone 149
2-Chloro-1-decen-3-yne 234
1-Chloro-4-(2,2-dichloro-1-methylvinyl)benzene 594
2-Chloro-3,3-diethoxyprop-1-ene 586
N-[2-Chloro-3-(dimethylamino)-2-propenylidene]-*N*-methyl-methanaminium hexafluorophosphate 929
1-Chloro-3-iodo-5-(4,4,5,5-tetramethyl-1,3,2,-dioxaborolan-2-yl)benzene 1007
3-(3′-Chloropropyl)-3,4-dihydronaphthalen-2(1*H*)-one 934
N-1-(Chloro-2,2,2-trifluoro-ethyl)acetamide 212
Dibenzofuran-4,6-dicarbonyl chloride 888
1,1-Dichloro-(3*S*)-(*tert*-butyldimethylsilyloxy)-2-butanol *p*-toluenesulfonate 615
1,1-Dichloro-2-ethoxycyclopropane 586
Ethyl 3-chloro-3-phenylpropanoate 646
Ethyl 4-[(6-chloropyridin-3-yl)amino]benzoate 665

N-Methoxy-*N*-methylcarbamoyl chloride 992
Methyl (2*Z*)-2-azido-3-(2-chloropyridin-3-yl)acrylate 802
Methyl 4-chloro-1*H*-pyrrolo[3,2-c]pyridine-2-carboxylate 802
(*E*)-2,2,2-Trichloroacetimidic acid hex-2-enyl ester 720
(*S*)-2,2,2-Trichloro-*N*-(1-propylallyl)acetamide 720

Fluorine
S-(1-Acetylamino-2,2,2-trifluoroethyl) *O*-ethyl dithiocarbonate 212
3-Acetylamino-1-ethoxythiocarbonylsulfanyl-4,4,4-trifluorobutyl acetate 212
Bis[3,5-bis(trifluoromethyl)phenyl]-phosphine oxide 965
Bis[3,5-bis(trifluoromethyl)phenyl]phosphine 965
n-Butyl 2,2-difluorocyclopropanecarboxylate 469
3-Butyl-2-fluoro-1-tosylindole 834
N-1-(Chloro-2,2,2-trifluoro-ethyl)acetamide 212
o-(1,1-Difluorohex-1-en-2-yl)aniline 834
o′-(1,1-Difluorohex-1-en-2-yl)-*p*-toluenesulfonanilide 834
N-(3,3-Dimethoxy-1-trifluoromethyl-propyl)-acetamide 212
Ethylphosphonic acid bis-(2,2,2-trifluoroethyl) ester 992
(2*E*)-3-Iodo-2-Propenoic acid 662
[1-(Methoxymethylcarbamoyl)ethyl] phosphonic acid bis-(2,2,2-trifluoroethyl) ester 992
4-Phenyl-3-trifluoromethyl-2-butanone 336
N-(2,2,2-Trifluoro-1-hydroxyethyl)-acetamide 212
Trimethylsilyl 2-fluorosulfonyl-2,2-difluoroacetate 469

Iodine
3-Benzyloxy-2-iodophenyl trifluoromethanesulfonate 272
5-(4-Bromophenyl)-3-iodo-2-(4-methylphenyl)furan 794
N-(*tert*-Butoxycarbonyl)-L-iodoalanine methyl ester 384
1-Chloro-3-iodo-5-(4,4,5,5-tetramethyl-1,3,2,-dioxaborolan-2-yl)benzene 1007
3,6-Diiodo-9-tetradecylcarbazole 540
(*Z*)-1-Iodo-1-heptene 359
1-Iodo-1-heptyne 359
6-Iodo-1-hexene 433
1-(2-Iodophenyl)pyrrole 826
2-Iodoresorcinol 272

HETEROCYCLES
Three-membered, one N
(*S*)-*N*-Trifluoromethylsulfonyl-2-isopropylaziridine 506

Three-membered, one O
1,2:5,6-Dianhydro-3,4-*O*-isopropylidene-L-mannitol 164
(2*S*,3*S*)-1,2-Epoxy-3-(Boc-amino)-4-phenylbutane 149
9-Spiroepoxy-*endo*-tricyclo[5.2.2.0^{2,6}]undeca-4,10-dien-8-one 514

Four-membered, one N
trans-1-(4-Methoxyphenyl)-4-phenyl-3-(phenylthio)azetidin-2-one 920

Five-membered, one N
(*R,R*)-2,2'-Bispyrrolidine 1046
N-Boc-3-pyrroline 570
5-Bromoisoquinoline 656
5-Bromo-8-nitroisoquinoline 656
(2*S-trans*)-2-*tert*-Butoxycarbonylacetyl-4-hydroxypyrrolidine-1-carboxylic acid, *tert*-butyl ester 38
2(*S*)-(β-*tert*-Butoxycarbonyl-α-(*R*)- and α-(*S*)-hydroxyethyl)-4-(*R*)-hydroxypyrrolidine-1-carboxylic acid, *tert*-butyl ester 38
(S)-N-(N-tert-Butoxycarbonylprolyl)dihydroisoindole 1037
3-Butyl-2-fluoro-1-tosylindole 834
1-(2-Iodophenyl)pyrrole 826

(S)-1-Methyl-2-[(dihydroisoindol-2-yl)methyl]pyrrolidine 1037
4-Methylpyrrolo[1,2-a]quinolines 826
1-(2-(1-Propynyl)phenyl)pyrrole 826
2,3,3α,4-Tetrahydro-2-[(4-methylbenzene)sulfonyl]cyclopenta[c]pyrrol-5(1*H*)-one 442

Five-membered, one O
2*α*-Benzyloxy-8-oxabicyclo[3.2.1]oct-6-en-3-one 473
5-(4-Bromophenyl)-3-iodo-2-(4-methylphenyl)furan 794
3,4-Dibromo-2,5-dimethoxytetrahydrofuran 607
5-Hydroxy-2(5*H*)-furanone 128
(5*S*)-(d-Menthyloxy)-2(5*H*)-furanone 128

Five-membered, one N, one O
(–)-(*S,S*)-*Bis(4-isopropyloxazoline)* 821
Di-μ-acetatobis[(η^5-(*S*)-(p*R*)-2-(2'-(4'-methylethyl)oxazolinyl)cyclopentadienyl, 1-C, 3'-N)(η^4-tetraphenylcyclo-butadiene)cobalt]dipalladium ([COP-OAc]$_2$) 714
(*R,R*)-4,6-Dibenzofurandiyl-2,2'-bis(4-phenyloxazoline) 888
Di-μ-chlorobis[(η^5-(*S*)-(p*R*)-2-(2'-(4'-methylethyl)oxazolinyl)cyclopentadienyl, 1-C, 3'-N)(η^4-tetraphenylcyclo-butadiene)cobalt]dipalladium ([COP-Cl]$_2$) 714
Hexafluoroacetylacetonate[(η^5-(*S*)-(pR)-2-(2'-(4'-methylethyl)oxazolinyl)cyclopentadienyl, 1-C, 3'-*N*)(η^4-tetraphenylcyclo-butadiene)cobalt]palladium (COP-hfacac) 714
(4*S*)-4-(1-Methylethyl)-5,5-diphenyl-2-oxazolidinone 896
(η^5-(*S*)-2-(4-Methylethyl)oxazolinylcyclo-pentadienyl)-(η^4-tetraphenylcyclobutadiene)cobalt 708
N-Phenyl-(5*R*)-hydroxymethyl-2-oxazolidinone 867
(*R*)-4-Phenyl-3-((1,2-propadienyl-2-oxazolidinone 726
(*R*)-4-Phenyl-3-(2-propynyl)oxazolidin-2-one 726

Five-membered, one N, one S
GlcNAc-thiazoline triacetate 787

Five-membered, two N
1-Butyl-3-methylimidazolium tetrafluoroborate 859
1-*tert*-Butyloxycarbonylpiperidine-4-spiro-5'-(1',3'-bis(*tert*-butyloxycarbonyl)) hydantoin 956
4-(4-Methoxyphenyl)-2-phenyl-1*H*-imidazole 862
Piperidine-4-spiro-5'-hydantoin 956

Five-membered, two O
1,2:5,6-Dianhydro-3,4-*O*-isopropylidene-L-mannitol 164
3,4-*O*-Isopropylidene-L-mannitol 164
2,3-Isopropylidene(D-ribonolactone) 88
3,4,6-Tri-*O*-benzyl-2-*O*-pivaloyl-β-D-glucopyranosyl-(1→6)-1,2:3,4-di-*O*-isopropylidene-α-D-galactopyranoside 169
1,2:3,4:5,6-Tri-*O*-isopropylidene-L-mannitol 164

Six-membered, one N
4-Acetylamino-2,2,6,6-tetramethylpiperidine-1-oxoammonium tetrafluoroborate 93
4-Amino-1-*tert*-butyloxycarbonylpiperidine-4-carboxylic acid 956
2-Benzyloxy-1-methylpyridinium trifluoromethanesulfonate 947
2-Benzyloxypyridine 947
N-Benzylpiperidine 697
5-Bromoisoquinoline 656
5-Bromo-8-nitroisoquinoline 656
1-*tert*-Butyloxycarbonyl-4-(9-fluorenylmethyloxycarbonylamino)-piperidine-4-carboxylic acid 956
1-*tert*-Butyloxycarbonylpiperidine-4-spiro-5'-(1',3'-bis(*tert*-butyloxycarbonyl))hydantoin 956

Ethyl 4-[(6-chloropyridin-3-yl)amino]benzoate 665
Methyl (2*Z*)-2-azido-3-(2-chloropyridin-3-yl)acrylate 802
Methyl-5-methylpyridine-2-carboxylate 911
4-Methylpyrrolo[1,2-a]quinolines 826
Piperidine-4-spiro-5'-hydantoin 956
3-Pyridin-3-yl-quinoline 393
3-Pyridylboronic acid 393
3-(4,4,5,5-Tetramethyl-[1,3,2]dioxaborolan-2-yl)-pyridine 393
Tris(3-pyridyl)boroxin 393

Six-membered, one O
2*α*-Benzyloxy-8-oxabicyclo[3.2.1]oct-6-en-3-one 473
(2S,6R)-6-(*tert*-Butyldimethylsilyloxymethyl)-2-methoxy-2,5-dihydro-pyran 498

Six-membered, one N, one O
1-Morpholinopropan-1-one 197
4-Phenylmorpholine 684

Six-membered, two O
2-Acetoxy-1,4-dioxane 850
2,2-Bis-(chloromethyl)-5,5-dimethyl-1,3-dioxane 457
Dibutyl 3,4,6-tri-*O*-benzyl-2-*O*-pivaloyl-D-glucopyranosyl phosphate 169
cis-5-(5,5-Dimethyl-1,3-dioxan-2-ylidene)hexahydro-1(2*H*)-pentalen-2-one 457
6,6-Dimethyl-1-methylene-4,8-dioxaspiro[2.5]octane 457
1,4-Dioxene 850
5-(Tosyloxyimino)-2,2-dimethyl-1,3-dioxane-4,6-dione 911
3,4,6-Tri-*O*-benzyl-2-*O*-pivaloyl-β-D-glucopyranosyl-(1→6)-1,2:3,4-di-*O*-isopropylidene-α-D-galactopyranoside 169
1,6,6-Trimethyl-4,8-dioxaspiro[2.5]oct-1-ene 457

Miscellaneous
O-Allyl-*N*-(9-Anthracenylmethyl)cinchonidinium bromide 404
N-(9-Anthracenylmethyl)cinchonidinium chloride 404
3,6-Bis(benzoylbiphenyl)ethynyl-9-tetradecylcarbazole 540
3,6-Bis[(trimethylsilyl)ethynyl]-9-tetradecylcarbazole 540
1-Chloro-3-iodo-5-(4,4,5,5-tetramethyl-1,3,2,-dioxaborolan-2-yl)benzene 1007
(–)-Cytisine 25
3,4-Dibromo-2,5-diformylthiophene 843
3,6-Diethynyl-9-tetradecylcarbazole 540
3,6-Diiodo-9-tetradecylcarbazole 540
Dithieno[3,2-*b*:2′,3′-*d*]thiophene 843
Dithieno[3,2-*b*:2′,3′-*d*]thiophene-2,6-dicarboxylic acid 843
Dithieno[3,2-*b*:2′,3′-*d*]thiophene-2,6-dicarboxylic acid diethyl ester 843
Dibenzofuran-4,6-dicarboxylic acid 888
Dibenzofuran-4,6-dicarbonyl chloride 888
(*R,R*)-Dibenzofuran-4,6-dicarboxylic acid bis(2-hydroxy-1-phenylethyl) amide 888
(*R,R*)-4,6-Dibenzofurandiyl-2,2'-bis(4-phenyloxazoline) 888
2-(5,5-Dimethyl-1,3,2-dioxaborinan-2-yl) benzoic acid ethyl ester 1015
Methyl 4-chloro-1*H*-pyrrolo[3,2-c]pyridine-2-carboxylate 802
Methyl indole-4-carboxylate 903
(+)-(*1R,2S,9S*)-11-Methyl-7,11-diazatricyclo[7.3.1.$0^{2,7}$]tridecane 25
(–)-Methyl (*1R,9R*)-6-Oxo-7,11-diazatricyclo[7.3.1.$0^{2,7}$]trideca-2,4-diene-11-carboxylate 25
(2S)-(-)-exo-(Morpholino)isoborneol 702
9-Tetradecylcarbazole 540
2,3,11,11a-Tetrahydro-1*H*-benzo[d]pyrrolo[1,2-a]azepin-5(6*H*)-one 934
3-(4,4,5,5-Tetramethyl-[1,3,2]dioxaborolan-2-yl)-pyridine 393
2-(4,4,5,5-Tetramethyl-1,3,2-dioxaborolan-2-yl)indole 1007
Tris(3-pyridyl)boroxin 393

HYDRAZONES
1-(4-Chlorophenyl)ethanone hydrazone 594

HYDROXYLAMINES
4-Benzoyloxymorpholine 684
N-Hydroxy-(*S*)-1-phenylethylamine oxalate 139
N-Methyl-*O*-benzoylhydroxylamine hydrochloride 70

IMIDATES
(*E*)-2,2,2-Trichloroacetimidic acid hex-2-enyl ester 1007

IMIDES
(3a*S*,7a*S*)-2-[2-(3,4-Dimethoxyphenyl)-ethyl]-1,3-dioxo-octahydro-isoindole-3a-carboxylic acid methyl ester 1028
(3a*S*,7a*R*)-2-[2-(3,4-Dimethoxyphenyl)-ethyl]-hexahydro-isoindole-1,3-dione 1028

IMINES
(1R,2S)-1-[3-Adamantyl)-2-hydroxy-5-methylbenzyliden-amino]indan-2-ol 498
(4*S*)-2-(Benzhydrylidenamino)pentanedioic Acid, 1-*tert*-Butyl Ester-5-Methyl Ester 404
N-Benzylidene-*p*-anisidine 920
N-[2-Chloro-3-(dimethylamino)-2-propenylidene]-*N*-methyl-methanaminium hexafluorophosphate 929
(1*S*,2*R*)-1-[(2-Hydroxy-3,5-di-*tert*-butylbenzylidene)amino]indan-2-ol 770
N-[(Phenyl)methylene]-*P,P*-diphenylphosphinic amide 749
N-*p*-Tolylsulfinyl-(*E*)-1-phenylethylideneimine 741
N-*p*-Tolylsulfonyl (*E*)-1-phenylethylideneimine 741

ISOCYANIDE
2,2-Diethoxy-1-isocyanoethane 778

KETONES
1-(4-Acetylphenyl)-2-phenylethane 289
6-Amino-3,4-dihydro-1(2*H*)-naphthalenone 670
3-(3′-Azidopropyl)-3,4-dihydronaphthalen-2(1*H*)-one 934
(2-Benzoyloxy)-1,4-cyclohexanedione (mono)ethylene ketal 70
5-Benzyloxybicyclo[4.2.0]octa-1,3,5-trien-7-one 272
2*α*-Benzyloxy-8-oxabicyclo[3.2.1]oct-6-en-3-one 473
1-[3,5-Bis(trifluoromethyl)phenyl]-ethanone 343
6,13-Bis(triisopropylsiloxy)-9,10-dimethoxy[7]helicenebisquinone 524
1-(4-Bromophenyl)-4-methylphenyl)but-3-yn-1-one 794
(2*S*-*trans*)-2-*tert*-Butoxycarbonylacetyl-4-hydroxypyrrolidine-1-carboxylic acid, *tert*-butyl ester 38
Camphorquinone oxime 702
(*S*)-1-Chloro-3-(Boc-amino)-4-phenyl-2-butanone 149
3-(3′-Chloropropyl)-3,4-dihydronaphthalen-2(1*H*)-one 934
(*R*)-1-Cyclohexyl-1-hydroxy-butan-2-one 205
2-Cyclopentylacetophenone 433
3,6-Diacetyl-9,10-dimethoxyphenanthrene 519
1,2:4,5-Di-*O*-isopropylidene-D-erythro-2,3-hexodiulo-2,6-pyranose 177
1,4:5,8-Dimethanododecahydroanthracene-9,10-dione 62
1,4:5,8-Dimethano-1,4,4a,5,8,8a,9a,10a-octahydroanthracene-9,10-dione 62
(*R*)-(+)-3,4-Dimethylcyclohex-2-en-1-one 100
3,4-Dimethylcyclopent-2-enone 420
cis-5-(5,5-Dimethyl-1,3-dioxan-2-ylidene)hexahydro-1(2*H*)-pentalen-2-one 457
(3*R*)-3,7-Dimethyl-2-(3-oxobutyl)-6-octenal 416
1-(Dimethyl(phenyl)silyl)propan-1-one 197
1,3-Diphenyl-2-(1-phenylethyl)propane-1,3-dione 272
Ethyl 6-Oxotridecanoate 281
(6*E*, 10*E*)-Ethyl 7,11,15-Trimethyl-3-oxohexa-deca-6,10,14-trienoate 221
1-Hydroxy-3-phenyl-2-propanone 642

(*R*)-(+)-6-Methylcyclohex-2-en-1-one 100
Methyl 2-oxo-1,2,3,4-tetrahydronaphthalene-1-carboxylate 934
2,8-Nonanedione 82
5-Oxaspiro[3.4]octan-1-one 256
4-Phenyl-5(*E*)-decen-2-one 261
4-Phenyl-3-trifluoromethyl-2-butanone 336
9-Spiroepoxy-*endo*-tricyclo[5.2.2.$0^{2,6}$]undeca-4,10-dien-8-one 514
2,3,3α,4-Tetrahydro-2-[(4-methylbenzene)sulfonyl]cyclopenta[c]pyrrol-5(1*H*)-one 442
4-Triisopropylsilyl-3-butyn-2-one 9

NITRILES
(*S*)-[(1-Phenylethyl)amino]acetonitrile 139
(1*S*)-1-Phenylethyl]imino]acetonitrile *N*-oxide 139
(*E*)-4-(2-Phenylethenyl)benzonitrile 372

NITRO COMPOUNDS
5-Bromo-8-nitroisoquinoline 656
Methyl 2-bromomethyl-3-nitrobenzoate 903
Methyl 2-ethenyl-3-nitrobenzoate 903

NITRONES
N-Benzylidene-benzylamine *N*-oxide 114
[(1*S*)-1-Phenylethyl]imino]acetonitrile *N*-oxide 139

ORGANOMETALLIC COMPOUNDS
Boron
1-Chloro-3-iodo-5-(4,4,5,5-tetramethyl-1,3,2,-dioxaborolan-2-yl)benzene 1007
2-(5,5-Dimethyl-1,3,2-dioxaborinan-2-yl)benzoic acid ethyl ester 1015
Potassium 2-Phenethyltrifluoroborate 289
3-Pyridylboronic acid 393
2-(4,4,5,5-Tetramethyl-1,3,2-dioxaborolan-2-yl)indole 1007
3-(4,4,5,5-Tetramethyl-[1,3,2]dioxaborolan-2-yl)-pyridine 393
Tris(3-pyridyl)boroxin 393

Cobalt
(η^5-Carbomethoxycyclopentadienyl)-(η^4-tetraphenyl-cyclobutadiene)cobalt 708
Di-μ-acetatobis[(η^5-(*S*)-(p*R*)-2-(2'-(4'-methylethyl)oxazolinyl)cyclopentadienyl, 1-C, 3'-*N*)(η^4-tetraphenylcyclo-butadiene)cobalt]dipalladium ([COP-OAc]$_2$) 714
Di-μ-chlorobis[(η^5-(*S*)-(p*R*)-2-(2'-(4'-methylethyl)oxazolinyl)cyclopentadienyl, 1-*C*, 3'-*N*)(η^4-tetraphenylcyclo-butadiene)cobalt]dipalladium ([COP-Cl]$_2$) 714
Hexacarbonyl[μ-[(3,4-η:3,4-η)-2-methyl-3-butyn-2-ol]]dicobalt 442
Hexafluoroacetylacetonate[(η^5-(*S*)-(p*R*)-2-(2'-(4'-methylethyl)oxazolinyl)cyclopentadienyl, 1-*C*, 3'-*N*)(η^4-tetraphenylcyclo-butadiene)cobalt]palladium (COP-hfacac) 714
(η^5-(*S*)-2-(4-Methylethyl)oxazolinylcyclo-pentadienyl)-(η^4-tetraphenylcyclobutadiene)cobalt 708

Lithium
Dimethylphenysilyllithium 197
5-Hexenyllithium 433
Lithium bis(trimethylsilyl)amide 399

Magnesium
3-Butenylmagnesium bromide 425
Nonylmagnesium bromide 353
Phenylmagnesium bromide 896

Molybdenum
Tris[*N*-(*tert*-butyl)(3,5-dimethylphenyl)-amido]molybdenum(III) 529

Tris[*N*-(*tert*-butyl)(3,5-dimethylphenyl)-amido]molybdenum(VI) propylidyne 529

Palladium
Di-μ-acetatobis[(η^5-(*S*)-(p*R*)-2-(2'-(4'-methylethyl)oxazolinyl)cyclopentadienyl, 1-*C*, 3'-*N*)(η^4-tetraphenylcyclo-butadiene)cobalt]dipalladium ([COP-OAc]$_2$) 714
Di-μ-chlorobis[(η^5-(*S*)-(p*R*)-2-(2'-(4'-methylethyl)oxazolinyl)cyclopentadienyl, 1-*C*, 3'-*N*)(η^4-tetraphenylcyclo-butadiene)cobalt]dipalladium ([COP-Cl]$_2$) 714
Hexafluoroacetylacetonate[(η^5-(*S*)-(p*R*)-2-(2'-(4'-methylethyl)oxazolinyl)cyclopentadienyl, 1-*C*, 3'-*N*)(η^4-tetraphenylcyclo-butadiene)cobalt]palladium (COP-hfacac) 714

Ruthenium
(μ_3, η^2: η^3: η^5-Acenaphthylene)$Ru_3(CO)_7$ 1
Ru(II)-BINAP complex 38
RuCl[(*S,S*)-NTsCH(C_6H_5)CH(C_6H_5)NH_2(η^6-cymene) 9

Silicon
(*Z*)-1-(1-Heptenyl)-1,1-dimethylsilanol 359

Tin
(P)-(+)-3-(Tributylstannyl)-1,2-butadiene 1056

Titanium
Methyl tris(isopropoxy)titanium 425

Zinc
5-Ethoxy-5-oxopentylzinc iodide 281

OXIMES
Camphorquinone oxime 87
5-(Tosyloxyimino)-2,2-dimethyl-1,3-dioxane-4,6-dione 911

PHENOLS
3-Benzyloxy-2-iodophenyl trifluoromethanesulfonate 272
1,4:5,8-Dimethano-1,2,3,4,5,6,7,8-octahydroanthracene-9,10-diol 62
(1*S*,2*R*)-1-[(2-Hydroxy-3,5-di-*tert*-butylbenzylidene)amino]indan-2-ol 770
2-Iodoresorcinol 272

PHOSPHORUS COMPOUNDS
(S,S)-1,2-Bis(boranato(*tert*-butyl)methylphosphino)ethane 1020
(S,S)-1,2-Bis-((*tert*-butyl)methylphosphino)ethane *((S,S)-t*-Bu-BisP*)) 1020
Bis[3,5-bis(trifluoromethyl)phenyl]-phosphine oxide 965
Bis[3,5-bis(trifluoromethyl)phenyl]phosphine 965
tert-Butyl(dimethyl)phosphine–Borane 1020
2-Carboxyethyltriphenylphosphonium chloride 296
(2-Carbomethoxy-6-nitrobenzyl)triphenylphosphonium bromide 903
Dibutyl 3,4,6-tri-*O*-benzyl-2-*O*-pivaloyl-D-glucopyranosyl phosphate 169
(2*R*,5*R*)-1-{2-[(2*R*,5*R*)-2,5-Dimethylphospholan-1-yl]phenyl}-2,5-dimethylphospholane 1-oxide 982
Diphosphoric acid, tetrakis(phenylmethyl) ester 1001
Ethylphosphonic acid bis-(2,2,2-trifluoroethyl) ester 992
(2*Z*,6*E*,10*E*)-Ethyl 7,11,15-Trimethyl-3-diethylphosphoryloxy-6,10,14-trienoate 221
[1-(Methoxymethylcarbamoyl)ethyl] phosphonic acid bis-(2,2,2-trifluoroethyl) ester 992
N-[(4-Methylphenyl)sulfonyl(phenyl)methyl*]-P,P*-diphenyl-phosphinic amide 749
N-[(Phenyl)methylene]-*P,P*-diphenylphosphinic amide 749
N-[(1S)-1-Phenylpropyl]-*P,P*-diphenylphosphinic amide 749
(Triphenylphosphoranylidene)-ketene 986

RADICAL CATIONS
1,4:5,8-Dimethano-1,2,3,4,5,6,7,8-octahydro-9,10-dimethoxyanthracenium hexachloroantimonate 62

SILANES
3,6-Bis[(trimethylsilyl)ethynyl]-9-tetradecylcarbazole 540
1-Bromo-2-triisopropylsilyl-ethyne 677
(±)-*N,N*-Dibenzyl-*N*-[1-cyclohexyl-3-(trimethylsilyl)-2-propynyl]-amine 192
(*R*)-(+)-*N,N*-Dibenzyl-*N*-[1-cyclohexyl-3-(trimethylsilyl)-2-propynyl]-amine 192
(*Z*)-1,2-Dideuterio-1-hexenyl)trimethylsilane 600
1-(Dimethyl(phenyl)silyl)propan-1-one 197
(*Z*)-2-Methyl-3-trimethylsilyl-2-butenoic acid 577
8-[2-(Triethoxysilyl)ethyl]-1-tetralone 410
4-Triisopropylsilyl-3-butyn-2-one 9
(*rac*)-4-Triisopropylsilyl-3-butyn-2-ol 9
(*S*)-4-Triisopropylsilyl-3-butyn-2-ol 9
Trimethylsilyl 2-fluorosulfonyl-2,2-difluoroacetate 469
(*R*)-4-Phenyl-3-(2-triisopropylsilyl-ethynyl)oxazolidin-2-one 677

SULFUR COMPOUNDS
Sulfenates
3-Heptyl benzenesulfenate 121
N-*p*-Tolylsulfinyl-(*E*)-1-phenylethylideneimine 741
(R_S,*1S,2R*)-(–)-2,4,6-Trimethylbenzenesulfinic acid 1-(2,4,6-trimethylbenzenesulfonylamino)-2-indan-2-yl ester 761
S)-(+)-2,4,6-Trimethylphenylsulfinamide 761

Sulfides
Diphenyl disulfide 135
trans-1-(4-Methoxyphenyl)-4-phenyl-3-(phenylthio)azetidin-2-one 920
3-Phenyl-2-(propylthio)-2-propen-1-ol 642
2-Phenylthio-5-heptanol 121

Sulfones
2-(Ethoxycarbonyl)prop-2-en-1-yl phenyl sulfone 308
N-[(4-Methylphenyl)sulfonyl(phenyl)methyl]-*P,P*-diphenyl-phosphinic amide 749

Sulfonic acids and derivatives
(*2S,6S*)-4-Benzyl-1,7-bis(trifluoromethylsulfonyl)-2,6-diisopropyl-1,4,7-triazaheptane 506
3-Benzyloxy-2-iodophenyl trifluoromethanesulfonate 272
2-Benzyloxy-1-methylpyridinium trifluoromethanesulfonate 947
N-(*tert*-Butoxycarbonyl)-*O*-(*p*-toluenesulfonyl)-L-serine methyl ester 384
3-Butyl-2-fluoro-1-tosylindole 834
(R)-3-Butyn-2-yl Mesylate 1056
1,1-Dichloro-(3*S*)-(*tert*-butyldimethylsilyloxy)-2-butanol *p*-toluenesulfonate 615
o'-(1,1-Difluorohex-1-en-2-yl)-*p*-toluenesulfonanilide 834
5-Hexen-1-ol, methanesulfonate 433
(*1S,2R*)-(+)-*N*-(2-Hydroxyindan-1-yl)-2,4,6-trimethylbenzenesulfonamide 761
N-(2-Propenyl)-4-methylbenzenesulfonamide 442
N-(2-Propenyl)-*N*-(2-propynyl)-4-methylbenzenesulfonamide 442
2,3,3α,4-Tetrahydro-2-[(4-methylbenzene)sulfonyl]cyclopenta[c]pyrrol-5(1*H*)-one 442
N-*p*-Tolylsulfonyl (*E*)-1-phenylethylideneimine 741
5-(Tosyloxyimino)-2,2-dimethyl-1,3-dioxane-4,6-dione 911
(R_S,*1S,2R*)-(–)-2,4,6-Trimethylbenzenesulfinic acid 1-(2,4,6-trimethylbenzenesulfonylamino)-2-indan-2-yl ester 761
3-(2,4,6-Trimethylbenzenesulfonyl)-3,3*a*,8,8*a*-tetrahydro-2*H*-1-oxa-2λ^4-thia-3-aza-cyclopenta[*a*]inden-2-ol 761
Trimethylsilyl 2-fluorosulfonyl-2,2-difluoroacetate 469
(R)-4-Trimethylsilyl-3-butyn-2-yl Mesylate 1056

Sulfoxides
(R$_S$)-(+)-*tert*-Butyl *tert*-butanethiosulfinate 770
(*R$_S$*)-(+)-*tert*-Butanesulfinamide 770
Methyl phenyl sulfoxide 135
3-(2,4,6-Trimethylbenzenesulfonyl)-3,3*a*,8,8*a*-tetrahydro-*2H*-1-oxa-2λ^4-thia-3-aza-cyclopenta[*a*]inden-2-ol 761

Thiol esters
S-Ethyl Octanethioate 281
S-Phenyl diazothioacetate 920

Thiophenes
3,4-Dibromo-2,5-diformylthiophene 843
Dithieno[3,2-*b*:2′,3′-*d*]thiophene 843
Dithieno[3,2-*b*:2′,3′-*d*]thiophene-2,6-dicarboxylic acid 843
Dithieno[3,2-*b*:2′,3′-*d*]thiophene-2,6-dicarboxylic acid diethyl ester 843

Xanthates
3-Acetylamino-1-ethoxythiocarbonylsulfanyl-4,4,4-trifluorobutyl acetate 212
S-(1-Acetylamino-2,2,2-trifluoroethyl) *O*-ethyl dithiocarbonate 212

FORMULA INDEX

All compounds for which preparations are provided are included in this Index. The system of indexing is that which is used *by Chemical Abstracts*. The essential principles involved are as follows: (1) The arrangement of symbols in formulas is alphabetical except that C (carbon) is always listed first, followed immediately by H if hydrogen is present; (2) The arrangement of formulas is also alphabetical except that the number of atoms of any specific kind influences the order of compounds; for example, all formulas with one carbon precede those with two carbons; (3) The arrangement of entries under any one heading is strictly alphabetical according to the names of the isomers.

C_2H_5ClO Chloromethyl methyl ether as a solution in toluene 934
$C_3H_3IO_2$ (*E*)-3-iodoprop-2-enoic acid 662
$C_3H_5N_3O_2$ Methyl azidoacetate 802
$C_3H_6ClNO_2$ *N*-Methoxy-*N*-methylcarbamoyl chloride 992
$C_4H_4O_3$ 5-Hydroxy-2(5*H*)-furanone 128
$C_4H_5ClF_3NO$ *N*-1-(Chloro-2,2,2-trifluoro-ethyl)acetamide 212
$C_4H_6F_3NO_2$ *N*-(2,2,2-Trifluoro-1-hydroxyethyl)-acetamide 212
$C_4H_6O_2$ 1,4-Dioxene 850
$C_4H_6O_3$ (±)-Methyl glycidate 157
$C_4H_6O_3$ (*S*)-Methyl glycidate 157
C_4H_7BrMg 3-Butenylmagnesium bromide 425
C_4H_8BrNO 2-Bromo-2-methylpropanamide 670
$C_4H_{11}NOS$ (R_S)-(+)-*tert*-Butanesulfinamide 770
$C_5H_6BNO_2$ 3-Pyridylboronic acid 393
$C_5H_8Br_2O_2$ Ethyl 2,2-dibromopropionate 577
$C_5H_8Cl_2O$ 1,1-Dichloro-2-ethoxycyclopropane 586
$C_5H_8O_3S$ (*R*)-3-Butyn-2-yl Mesylate 1056
$C_5H_8O_5$ *D*-Ribonolactone 88
$C_5H_9F_3O_4SSi$ Trimethylsilyl 2-fluorosulfonyl-2,2-difluoroacetate 469
$C_5H_9NO_2$ *N*-(Methoxycarbonyl)-2-propenylamine 244
$C_5H_{12}N_2O$ Pivaloyl hydrazide 874
$C_6H_2Br_2O_2S$ 3,4-Dibromo-2,5-diformylthiophene 843
C_6H_5BrMg Phenylmagnesium bromide 896
$C_6H_5IO_2$ 2-Iodoresorcinol 488
$C_6H_8O_3$ 4,4-Dimethoxybut-2-ynal 607
$C_6H_9F_6O_3P$ Ethylphosphonic acid bis-(2,2,2-trifluoroethyl) ester 992
$C_6H_{10}Br_2O_3$ 3,4-Dibromo-2,5-dimethoxytetrahydrofuran 607
$C_6H_{10}F_3NO_2S$ (*S*)-*N*-Trifluoromethylsulfonyl-2-isopropylaziridine 506
$C_6H_{10}O$ Bicyclo[3.1.0]hexan-1-ol 450
$C_6H_{10}O$ *trans*-2-Benzyl-1-methylcyclopropan-1-ol 450
$C_6H_{10}O_4$ 2-Acetoxy-1,4-dioxane 850
$C_6H_{11}BrO$ (1*R*, 2*R*)-*trans*-2-bromocyclohexanol 1037
$C_6H_{11}I$ 6-Iodo-1-hexene 433
$C_6H_{11}Li$ 5-Hexenyllithium 433
$C_6H_{18}BP$ *tert*-Butyl(dimethyl)phosphine–Borane 1020
$C_7H_5BrO_2$ 3-Bromo-2-hydroxybenzaldehyde 267
C_7H_8OS Methyl phenyl sulfoxide 135
$C_7H_{10}F_3NO_2S_2$ *S*-(1-Acetylamino-2,2,2-trifluoroethyl) *O*-ethyl dithiocarbonate 212
$C_7H_{10}O$ 3,4-Dimethylcyclopent-2-enone 420
$C_7H_{10}O$ (*R*)-(+)-6-Methylcyclohex-2-en-1-one 100
$C_7H_{10}O_2$ 5-Oxaspiro[3.4]octan-1-one 256
$C_7H_{11}I$ 1-Iodo-1-heptyne 359
$C_7H_{11}N_3O_2$ Piperidine-4-spiro-5'-hydantoin 956
$C_7H_{12}O_2$ Isopropyl (*E*)-but-2-enoate 420

$C_7H_{13}ClO_2$ 2-Chloro-3,3-diethoxyprop-1-ene 586
$C_7H_{13}I$ (*Z*)-1-Iodo-1-heptene 359
$C_7H_{13}IO_2Zn$ 5-Ethoxy-5-oxopentylzinc iodide 281
$C_7H_{13}NO_2$ 2,2-Diethoxy-1-isocyanoethane 778
$C_7H_{13}NO_2$ 1-Morpholinopropan-1-one 197
$C_7H_{14}ClF_6N_2P$ 2-Chloro-1,3-bis(dimethylamino)trimethinium hexafluorophosphate 929
$C_7H_{14}OSi$ (*R*)-4-Trimethylsilyl-3-butyn-2-ol 1056
$C_7H_{14}OSi$ (*S*)-4-Trimethylsilyl-3-butyn-2-ol 1056
$C_7H_{14}O_3S$ 5-Hexen-1-ol, methanesulfonate 433
$C_7H_{15}NO_3$ *N*-(2,2-Diethoxy)ethyl formamide 778
$C_7H_{16}N_2O_2$ (2-Aminoethyl)carbamic acid *tert*-butyl ester 942
$C_8H_4S_3$ Dithieno[3,2-b:2′,3′-d]thiophene 843
$C_8H_6N_2OS$ *S*-Phenyl diazothioacetate 920
$C_8H_9BF_3K$ Potassium 2-Phenethyltrifluoroborate 289
$C_8H_9NO_2$ Methyl-5-methylpyridine-2-carboxylate 911
$C_8H_{10}ClNO_2$ *N*-Methyl-*O*-benzoylhydroxylamine hydrochloride 70
$C_8H_{11}LiSi$ Dimethylphenysilyllithium 197
$C_8H_{12}F_2O_2$ *n*-Butyl 2,2-difluorocyclopropanecarboxylate 469
$C_8H_{14}Cl_2O_2$ 2,2-Bis-(chloromethyl)-5,5-dimethyl-1,3-dioxane 457
$C_8H_{14}F_3NO_3$ *N*-(3,3-Dimethoxy-1-trifluoromethyl-propyl)-acetamide 212
$C_8H_{16}N_2$ (*R,R*)-2,2'-Bispyrrolidine 1046
$C_8H_{16}N_2$ (*S,S*)-2,2'-Bispyrrolidine 1046
$C_8H_{16}O_2Si$ (*Z*)-2-Methyl-3-trimethylsilyl-2-butenoic acid 577
$C_8H_9ClN_2$ 1-(4-Chlorophenyl)ethanone hydrazone 594
$C_8H_{12}Cl_3NO$ (*E*)-2,2,2-Trichloroacetimidic acid hex-2-enyl ester 720
$C_8H_{12}Cl_3NO$ (*S*)-2,2,2-Trichloro-*N*-(1-propylallyl)acetamide 720
$C_8H_{12}O$ (*R*)-(+)-3,4-Dimethylcyclohex-2-en-1-one 100
$C_8H_{12}O_5$ 2,3-Isopropylidene(*D*-ribonolactone) 88
$C_8H_{14}O$ (6*R*)-(+)-1,6-Dimethylcyclohex-2-en-1-ol 100
$C_8H_{14}O_4$ 1,1,4,4-Tetramethoxybut-2-yne 607
$C_8H_{15}BF_4N_2$ 1-Butyl-3-methylimidazolium tetrafluoroborate 859
$C_8H_{16}Br_2O_4$ 2,3-Dibromo-1,1,4,4-tetramethoxybutane 607
$C_8H_{16}O_3SSi$ (*R*)-4-Trimethylsilyl-3-butyn-2-yl Mesylate 1056
$C_8H_{18}ClN_3O$ 1-(*tert*-Butylimino-methyl)-1,3-dimethyl-urea hydrochloride 782
$C_8H_{18}OS_2$ (R_S)-(+)-*tert*-Butyl *tert*-butanethiosulfinate 770
$C_9H_5BrN_2O_2$ 5-Bromo-8-nitroisoquinoline 656
C_9H_6BrN 5-Bromoisoquinoline 656
$C_9H_7ClN_2O_2$ Methyl 4-chloro-1*H*-pyrrolo[3,2-c]pyridine-2-carboxylate 802
$C_9H_7ClN_4O_2$ Methyl (2*Z*)-2-azido-3-(2-chloropyridin-3-yl)acrylate 802
$C_9H_7Cl_3$ 1-Chloro-4-(2,2-dichloro-1-methylvinyl)benzene 594
$C_9H_8BrNO_4$ Methyl 2-bromomethyl-3-nitrobenzoate 903
$C_9H_{10}O$ (*R,R*)-*trans*-β-Methylstyrene oxide 183
$C_9H_{10}O_2$ 1-Methoxy-4-vinyloxybenzene 631
$C_9H_{10}O_2$ 1-Hydroxy-3-phenyl-2-propanone 642
$C_9H_{10}O_3$ 4-Methoxyphenylacetic acid 107
$C_9H_{11}NO_2$ Ethyl 4-aminobenzoate 57
$C_9H_{14}ClN$ (1*S*)-1-Phenylpropan-1-amine hydrochloride 749
$C_9H_{14}F_6NO_5P$ [1-(Methoxymethylcarbamoyl)ethyl] phosphonic acid bis-(2,2,2-trifluoroethyl) ester 992
$C_9H_{14}O_2$ 1,6,6-Trimethyl-4,8-dioxaspiro[2.5]oct-1-ene 457
$C_9H_{14}O_2$ 6,6-Dimethyl-1-methylene-4,8-dioxaspiro[2.5]octane 457
$C_9H_{14}O_4$ 1,2:5,6-Dianhydro-3,4-*O*-isopropylidene-L-mannitol 164
$C_9H_{15}NO_2$ *N*-Boc-3-pyrroline 570
$C_9H_{16}INO_4$ *N*-(*tert*-Butoxycarbonyl)-L-iodoalanine methyl ester 384
$C_9H_{16}O_2$ 2,8-Nonanedione 82
$C_9H_{16}O_2Si$ (*R*)-4-Trimethylsilyl-3-butyn-2-yl Acetate 1056
$C_9H_{18}D_2Si$ (*Z*)-1,2-Dideuterio-1-hexenyl)trimethylsilane 600
$C_9H_{18}O_6$ 3,4-*O*-Isopropylidene-L-mannitol 164
$C_9H_{19}BrMg$ Nonylmagnesium bromide 353
$C_9H_{20}OSi$ (*Z*)-1-(1-Heptenyl)-1,1-dimethylsilanol 359

$C_9H_{20}O_2Si$ (*S*)-2-(*tert*-butyldimethylsilyloxy)propanal 615
$C_{10}H_4O_4S_3$ Dithieno[3,2-b:2′,3′-d]thiophene-2,6-dicarboxylic acid 843
$C_{10}H_6F_6O$ 1-[3,5-Bis(trifluoromethyl)phenyl]-ethanone 343
$C_{10}H_8IN$ 1-(2-Iodophenyl)pyrrole 826
$C_{10}H_8O_2$ Propargyl benzoate 552
$C_{10}H_9NO_2$ Methyl indole-4-carboxylate 903
$C_{10}H_9NO_4$ Methyl 2-ethenyl-3-nitrobenzoate 903
$C_{10}H_{10}Br_2$ (4,4-Dibromobut-3-enyl)benzene 561
$C_{10}H_{10}N_2O$ [(1*S*)-1-Phenylethyl]imino]acetonitrile *N*-oxide 139
$C_{10}H_{11}NO$ 6-Amino-3,4-dihydro-1(2*H*)-naphthalenone 670
$C_{10}H_{11}NO$ 2'-Vinylacetanilide 315
$C_{10}H_{11}NO_3$ *N*-Phenyl-(5*R*)-hydroxymethyl-2-oxazolidinone 867
$C_{10}H_{12}N_2$ (*S*)-[(1-Phenylethyl)amino]acetonitrile 139
$C_{10}H_{13}Br$ 1-Bromo-1,3-decadiyne 244
$C_{10}H_{13}NO$ 4-Phenylmorpholine 684
$C_{10}H_{13}NO_2S$ *N*-(2-Propenyl)-4-methylbenzenesulfonamide 442
$C_{10}H_{13}NO_5$ *N*-Hydroxy-(*S*)-1-phenylethylamine oxalate 139
$C_{10}H_{14}$ 1,3-Decadiyne 234
$C_{10}H_{14}O_2$ 1-Methoxymethyl-1-phenylethane 934
$C_{10}H_{15}Cl$ 2-Chloro-1-decen-3-yne 234
$C_{10}H_{15}NO_2$ Camphorquinone oxime 702
$C_{10}H_{16}BNO_2$ 3-(4,4,5,5-Tetramethyl-[1.3.2]dioxaborolan-2-yl)pyridine 393
$C_{10}H_{16}NO_4$ (*R*)-(*N*-*tert*-Butoxycarbonyl)allylglycine 399
$C_{10}H_{16}O$ (*Z*)-3,7-Dimethyl-2,6-octadienal 77
$C_{10}H_{16}O$ (2*E*)-3,7-Dimethyl-2,6-octadienal 93
$C_{10}H_{18}O_2$ (*R*)-1-Cyclohexyl-1-hydroxy-butan-2-one 205
$C_{10}H_{19}NO$ *(2S)*-(-)-3-*exo*-Aminoisoborneol 702
$C_{10}H_{20}OS$ *S*-Ethyl Octanethioate 281
$C_{10}H_{24}O_3Ti$ Methyl tris(isopropoxy)titanium 425
$C_{11}H_8Co_2O_7$ Hexacarbonyl[μ-[(3,4-η:3,4-η)-2-methyl-3-butyn-2-ol]]dicobalt 442
$C_{11}H_{10}N_2O_2$ (*E*)-Methyl 2-diazo-4-phenylbut-3-enoate 296
$C_{11}H_{11}F_3O$ 4-Phenyl-3-trifluoromethyl-2-butanone 336
$C_{11}H_{12}$ 1-Methyl-3,4-dihydronaphthalene 296
$C_{11}H_{12}O$ 5-Phenylpent-2-yn-1-ol 561
$C_{11}H_{12}O_2$ (*4S*)-4-(2-Phenethyl)oxetan-2-one 506
$C_{11}H_{13}ClO_2$ Ethyl 3-chloro-3-phenylpropanoate 646
$C_{11}H_{13}NO_3$ 4-Benzoyloxymorpholine 684
$C_{11}H_{14}N_2O$ (–)-Cytisine 25
$C_{11}H_{16}F_3NO_4S_2$ 3-Acetylamino-1-ethoxythiocarbonylsulfanyl-4,4,4-trifluorobutyl acetate 212
$C_{11}H_{16}OSi$ 1-(Dimethyl(phenyl)silyl)propan-1-one 197
$C_{11}H_{17}N$ *N,N*-Dimethyl-3-phenylpropylamine 1
$C_{11}H_{18}O_4Si$ *(S)*-4-Trimethylsilyl-3-butyn-2-yl Succinate 1056
$C_{11}H_{20}N_2O_4$ 4-Amino-1-*tert*-butyloxycarbonylpiperidine-4-carboxylic acid 956
$C_{11}H_{21}BF_4N_2O_2$ 4-Acetylamino-2,2,6,6-tetramethylpiperidine-1-oxoammonium tetrafluoroborate 93
$C_{11}H_{21}BrSi$ 1-Bromo-2-triisopropylsilyl-ethyne 677
$C_{11}H_{21}NO_4$ *N*-(*tert*-Butoxycarbonyl)-L-valine methyl ester 896
$C_{11}H_{22}O_2Si$ (*S*)-4-(*tert*-Butyldimethylsilyloxy)-2-pentyn-1-ol 615
$C_{11}H_{24}O_3Si$ (*S*)-Ethyl 2-(*tert*-butyldimethylsilyloxy)propanoate 615
$C_{12}H_{10}S_2$ Diphenyl disulfide 135
$C_{12}H_{11}NO$ 2-Benzyloxypyridine 947
$C_{12}H_{11}NO_2$ (*R*)-4-Phenyl-3-(2-propadienyl)oxazolidin-2-one 726
$C_{12}H_{12}O_2$ 9-Spiroepoxy-*endo*-tricyclo[5.2.2.$0^{2,6}$]undeca-4,10-dien-8-one 514
$C_{12}H_{12}O_3$ Methyl 2-oxo-1,2,3,4-tetrahydronaphthalene-1-carboxylate 811
$C_{12}H_{14}O$ (*R,R*)-1-Phenylcyclohexene oxide 183
$C_{12}H_{14}O_4S$ 2-(Ethoxycarbonyl)prop-2-en-1-yl phenyl sulfone 308
$C_{12}H_{15}BClIO_2$ 1-Chloro-3-iodo-5-(4,4,5,5-tetramethyl-1,3,2,-dioxaborolan-2-yl)benzene 1007
$C_{12}H_{15}F_2N$ *o*-(1,1-Difluorohex-1-en-2-yl)aniline 834
$C_{12}H_{16}O$ 3-Methyl-1-pheny-4-penten-1-ol 327
$C_{12}H_{16}OS$ 3-Phenyl-2-(propylthio)-2-propen-1-ol 642

$C_{12}H_{16}O_3$	Methyl (*R*)-(–)-3-benzyloxy-2-methyl propanoate 947
$C_{12}H_{17}N$	1-Phenylhex-5-en-3-amine 691
$C_{12}H_{17}N$	*N*-Benzylpiperidine 697
$C_{12}H_{18}O_6$	1,2:4,5-Di-*O*-isopropylidene-D-erythro-2,3-hexodiulo-2,6-pyranose 177
$C_{12}H_{19}N$	*N*-(3,5-Dimethylphenyl)-*tert*-butylamine 529
$C_{12}H_{20}O_6$	1,2:4,5-Di-*O*-isopropylidene-β-D-fructopyranose 177
$C_{12}H_{22}N_2$	(+)-(1*R*,2*S*,9*S*)-11-Methyl-7,11-diazatricyclo[7.3.1.0^{2,7}]tridecane 25
$C_{12}H_{24}Cl_3MoO_3$	[$MoCl_3(THF)_3$] 529
$C_{12}H_{27}P$-Pt(0)	Tri(*tert*-butyl)phosphine-platinum(0) complex 367
$C_{12}H_{28}P_2$	*(S,S)*-1,2-Bis-((*tert*-butyl)methylphosphino)ethane ((*S,S)*-*t*-Bu-BisP*)) 1020
$C_{12}H_{34}B_2P_2$	*(S,S)*-1,2-Bis(boranato(*tert*-butyl)methylphosphino)ethane 1020
$C_{13}H_{11}N$	1-(2-(1-Propynyl)phenyl)pyrrole 826
$C_{13}H_{11}N$	4-Methylpyrrolo[1,2-a]quinoline 926
$C_{13}H_{11}NO$	*N,N*-Dibenzylformamide 425
$C_{13}H_{13}NO_7S$	5-(Tosyloxyimino)-2,2-dimethyl-1,3-dioxane-4,6-dione 911
$C_{13}H_{14}O_2$	Acetic acid 5-phenylpent-2-ynyl ester 561
$C_{13}H_{15}BrO_2$	(1*S*,2*S*)-*trans*-1-Benzoyloxy-2-bromocyclohexane 1037
$C_{13}H_{15}ClO$	3-(3′-Chloropropyl)-3,4-dihydronaphthalen-2(1*H*)-one 811
$C_{13}H_{15}NO$	2,3,11,11*a*-Tetrahydro-1H-benzo[d]pyrrolo[1,2-*a*]azepin-5(6*H*)-one 811
$C_{13}H_{15}NO_2S$	*N*-(2-Propenyl)-*N*-(2-propynyl)-4-methylbenzenesulfonamide 442
$C_{13}H_{15}N_3O$	3-(3′-Azidopropyl)-3,4-dihydronaphthalen-2(1*H*)-one 811
$C_{13}H_{16}O$	2-Cyclopentylacetophenone 433
$C_{13}H_{16}N_2O_3$	(–)-Methyl (1*R*,9*R*)-6-oxo-7,11-diazatricyclo[7.3.1.0^{2,7}]trideca-2,4-diene-11-carboxylate 25
$C_{13}H_{20}OS$	3-Heptyl benzenesulfenate 121
	2-Phenylthio-5-heptanol 121
$C_{13}H_{24}OSi$	4-Triisopropylsilyl-3-butyn-2-one 9
$C_{13}H_{26}OSi$	(±)-4-Triisopropylsilyl-3-butyn-2-ol 9
$C_{13}H_{26}OSi$	(*S*)-4-Triisopropylsilyl-3-butyn-2-ol 9
$C_{13}H_{26}OSi$	1-*tert*-Butyldimethylsilyloxy-2-methyl-1-cyclohexene 481
$C_{13}H_{26}O_3Si$	*(2S,6R)*-6-(*tert*-Butyldimethylsilyloxymethyl)-2-methoxy-2,5-dihydropyran 498
$C_{14}H_6Cl_2O_3$	Dibenzofuran-4,6-dicarbonyl chloride 888
$C_{14}H_9O_5$	Dibenzofuran-4,6-dicarboxylic acid 888
$C_{14}H_{10}F_3IO_4S$	3-Benzyloxy-2-iodophenyl trifluoromethanesulfonate 488
$C_{14}H_{10}N_2$	3-Pyridin-3-ylquinoline 393
$C_{14}H_{12}O_4S_3$	Dithieno[3,2-b:2′,3′-d]thiophene-2,6-dicarboxylic acid diethyl ester 843
$C_{14}H_{13}ClN_2O_2$	Ethyl 4-[(6-chloropyridin-3-yl)amino]benzoate 665
$C_{14}H_{13}NO$	*N*-Benzylidene-*p*-anisidine 920
$C_{14}H_{13}NO$	*N*-Benzylidene-benzylamine *N*-oxide 114
$C_{14}H_{14}F_3NO_4S$	2-Benzyloxy-1-methylpyridinium trifluoromethanesulfonate 947
$C_{14}H_{14}O_2$	2-(Benzoyloxymethyl)-1,3-cyclohexadiene 552
$C_{14}H_{14}O_2$	*(1R,2R)*-1,2-Diphenyl-1,2-ethanediol 17
$C_{14}H_{14}O_3$	1-Methoxy-2-(4-methoxyphenoxy)-benzene 626
$C_{14}H_{14}O_3$	2α-Benzyloxy-8-oxabicyclo[3.2.1]oct-6-en-3-one 473
$C_{14}H_{15}NO_3S$	2,3,3α,4-Tetrahydro-2-[(4-methylbenzene)sulfonyl]cyclopenta[c]pyrrol-5(1*H*)-one 442
$C_{14}H_{18}BNO_2$	2-(4,4,5,5-Tetramethyl-1,3,2-dioxaborolan-2-yl)indole 107
$C_{14}H_{18}O_2$	(*E*)-2-Methyl-3-phenylacrylic acid butyl ester 372
$C_{14}H_{19}BO_4$	2-(5,5-Dimethyl-1,3,2-dioxaborinan-2-yl)benzoic acid ethyl ester, 1015
$C_{14}H_{19}NO_7S$	GlcNAc-thiazoline triacetate 787
$C_{14}H_{20}N_2$	(*S*)-1-Methyl-2-[(dihydroisoindol-2-yl)methyl]pyrrolidine 1037
$C_{14}H_{20}O$	(*Z*)-1-Heptenyl-4-methoxybenzene 359
	(*E*)-1-Heptenyl-4-methoxybenzene 367
$C_{14}H_{20}O$	2-(2,2-Dimethylpropoxy)-2,3-dihydro-1*H*-indene 46
$C_{14}H_{20}O$	5-Methyl-1-phenyl-6-hepten-3-ol 327
$C_{14}H_{20}O_3$	*cis*-5-(5,5-Dimethyl-1,3-dioxan-2-ylidene)hexahydro-1(2*H*)-pentalen-2-one 457
$C_{14}H_{22}O_3$	(5*S*)-(*d*-Menthyloxy)-2(5*H*)-furanone 128
$C_{14}H_{24}O_2$	(3*R*)-3,7-Dimethyl-2-(3-oxobutyl)-6-octenal 416
$C_{14}H_{24}O_3$	1-[[(1*R*,2S,4R)-1,7,7-Trimethylbicyclo[2.2.1]hept-2-yl]oxy]-ethanol acetate 52

$C_{14}H_{25}NO_2$ *(2S)-(-)-exo*-(Morpholino)isoborneol 702
$C_{14}H_{26}O_2$ *dl*-1,2-Dicyclohexylethanediol 348
$C_{15}H_{11}N$ (*E*)-4-(2-Phenylethenyl)benzonitrile 372
$C_{15}H_{12}B_3N_3O_3$ Tris(3-pyridyl)boroxin 393
$C_{15}H_{12}O_2$ 5-Benzyloxybicyclo[4.2.0]octa-1,3,5-trien-7-one 488
$C_{15}H_{15}NOS$ *N-p*-Tolylsulfinyl (*E*)-1-phenylethylideneimine 741
$C_{15}H_{15}NO_2S$ *N-p*-Tolylsulfonyl (*E*)-1-phenylethylideneimine 741
$C_{15}H_{16}O_5$ (2-Benzoyloxy)-1,4-cyclohexanedione (mono)ethylene ketal 70
$C_{15}H_{18}O_2$ Acetic acid 2-methylene-3-phenethylbut-3-enyl ester 561
$C_{15}H_{20}ClNO_3$ (*S*)-1-Chloro-3-(Boc-amino)-4-phenyl-2-butanone 149
$C_{15}H_{21}NO_2$ *N*-(Methoxycarbonyl)-*N*-(2-propenyl)-1,3-decadiynylamide 244
$C_{15}H_{21}NO_3$ (2*S*,3*S*)-1,2-Epoxy-3-(Boc-amino)-4-phenylbutane 149
$C_{15}H_{24}OSi$ 2,3-Dihydro-2-*tert*-butyldimethysiloxy-1*H*-indene 46
$C_{15}H_{25}Br$ (*E, E*)-Farnesyl bromide 221
$C_{15}H_{26}N_2O_2$ (–)-(*S,S*)-Bis(4-isopropyloxazoline) 821
$C_{15}H_{26}O_6$ 1,2:3,4:5,6-Tri-*O*-isopropylidene-L-mannitol 164
$C_{15}H_{28}O_3$ Ethyl 6-Oxotridecanoate 281
$C_{15}H_{30}N_2O_4$ (–)-(*S,S*)-*N,N'*-Bis(1-hydroxymethyl-2-methylpropyl)-2,2-dimethyl-malonamide 821
$C_{16}H_7F_{12}OP$ Bis[3,5-bis(trifluoromethyl)phenyl]phosphine oxide 965
$C_{16}H_7F_{12}P$ Bis[3,5-bis(trifluoromethyl)phenyl]phosphine 965
$C_{16}H_{14}N_2O$ 4-(4-Methoxyphenyl)-2-phenyl-1*H*-imidazole 862
$C_{16}H_{14}O_2$ 3,6-Diacetyl-9,10-dimethoxyphenanthrene 519
$C_{16}H_{14}O_2$ 9,10-Dimethoxyphenanthrene 519
$C_{16}H_{16}O$ 1-(4-Acetylphenyl)-2-phenylethane 289
$C_{16}H_{16}O_2$ 1,4:5,8-Dimethano-1,4,4a,5,8,8a,9a,10a-octahydroanthracene-9,10-dione 62
$C_{16}H_{16}O_4$ (2*R*,3*S*)- 3-Benzyloxycarbonyl-Bicyclo[2.2.1]hept-5-ene-2-carboxylic acid 1068
$C_{16}H_{18}O_2$ 1,4:5,8-Dimethano-1,2,3,4,5,6,7,8-octahydroanthracene-9,10-diol 62
$C_{16}H_{20}O_2$ 1,4:5,8-Dimethanododecahydroanthracene-9,10-dione 62
$C_{16}H_{22}O$ 4-Phenyl-5(*E*)-decen-2-one 261
$C_{16}H_{23}NO_{10}$ 2-Acetamido-2-deoxy-1,3,4,6-tetra-*O*-acetyl-β-D-glucopyranose 787
$C_{16}H_{23}NO_7S$ *N*-(*tert*-Butoxycarbonyl)-*O*-(*p*-toluenesulfonyl)-L-serine methyl ester 384
$C_{16}H_{24}O_2$ 4-Nonylbenzoic acid 353
$C_{16}H_{26}O_2$ Ethyl 2-{[(1S,2*R*,3*R*,5*S*)-2,6,6-trimethylbicyclo[3.1.1]hept-3-yl]methyl}–acrylate 308
$C_{16}H_{27}NO_6$ (2*S*,4*R*)-2-*tert*-Butoxycarbonylacetyl-4-hydroxypyrrolidine-1-carboxylic acid, *tert*-butyl ester 38
$C_{16}H_{28}LiNO$ Lithium *N*-(3,5-dimethylphenyl)-*tert*-butylamide etherate 529
$C_{16}H_{29}NO_6$ 2(*S*)-(β-*tert*-ButoxycarbonyL-α-(*R*)- and α-(*S*)-hydroxyethyl-4(*R*)-hydroxypyrrolidine-1-carboxylic acid, *tert*-butyl ester 38
$C_{16}H_{32}Sn$ (*P*)-(+)-3-(Tributylstannyl)-1,2-butadiene 1056
$C_{17}H_{12}BrIO$ 5-(4-Bromophenyl)-3-iodo-2-(4-methylphenyl)furan 794
$C_{17}H_{13}BrO$ 1-(4-Bromophenyl)-4-methylphenyl)but-3-yn-1-one 794
$C_{17}H_{15}BrO$ 1-(4-Bromophenyl)-4-methylphenyl)but-3-yn-1-ol 794
$C_{17}H_{17}N$ *N,N*-Dibenzyl-*N*-(2-ethenylcyclopropyl)amine 425
$C_{17}H_{17}NO$ *N*-Benzyl-4-phenylbutyramide 733
$C_{17}H_{18}O_3$ 1,1-Bis(benzyloxy)propan-2-one 473
$C_{17}H_{23}NO_6$ *N*-(*tert*-Butoxycarbonyl)-β-[4-(methoxycarbonyl)phenyl] alanine methyl ester 384
$C_{17}H_{24}O$ (*R*)-(*E*)-2-Benzylidiene-1-cyclohexyl-butan-1-ol 205
$C_{17}H_{25}NO_4S$ Dimethylsulfoxonium (*S*)-2-oxo-3-(Boc-amino)-4-phenylbutylide 149
$C_{17}H_{26}O_2$ 4-Nonylbenzoic acid methyl ester 353
$C_{17}H_{28}Cl_2O_4Ssi$ 1,1-Dichloro-(3*S*)-(*tert*-butyldimethylsilyloxy)-2-butyl *p*-toluenesulfonate 615
$C_{17}H_{32}O_3Si$ 1-(*tert*-Butyldimethylsilyloxy)-8-(methoxycarbonyl)-6-methyl-bicyclo[4.2.0]octane 481
$C_{18}H_{19}NO_2$ (4*S*)-4-(1-Methylethyl)-5,5-diphenyl-2-oxazolidinone 896
$C_{18}H_{19}NO_4S_2$ 3-(2,4,6-Trimethylbenzenesulfonyl)-3,3a,8,8a-tetrahydro-2*H*-1-oxa-2λ^4-thia-3-aza-cyclopenta[a]inden-2-ol 761
$C_{18}H_{21}NO_3$ Ethyl (1'S,2'*R*)-*N*-(1',2'-diphenyl-2'-hydroxyethyl)glycinate 879

$C_{18}H_{21}NO_3S$ (1*S*,2*R*)-(+)-*N*-(2-Hydroxyindan-1-yl)-2,4,6-trimethylbenzenesulfonamide 761
$C_{18}H_{21}NO_3S$ (*S*)-(+)-2,4,6-Trimethylphenylsulfinamide 761
$C_{18}H_{22}Cl_6O_2Sb$ 1,2,3,4,5,6,7,8-Octahydro-9,10-dimethoxy-1,4:5,8-dimethanoanthracene, radical ion, hexachloroantimonate 62
$C_{18}H_{22}O_2$ 1,4:5,8-Dimethano-1,2,3,4,5,6,7,8-octahydro-9,10-dimethoxy-anthracene 62
$C_{18}H_{23}NO_4$ (3a*S*,7a*R*)-2-[2-(3,4-Dimethoxyphenyl)-ethyl]-hexahydro-isoindole-1,3-dione 1028
$C_{18}H_{24}N_2O_3$ (*S*)-*N*-(*N*-*tert*-Butoxycarbonylprolyl)dihydroisoindole 1037
$C_{18}H_{28}OP_2$ (2*R*,5*R*)-1-{2-[(2*R*,5*R*)-2,5-dimethylphospholan-1-yl]phenyl}-2,5-dimethylphospholane 1-oxide 982
$C_{18}H_{28}O_4Si$ 8-[2-(Triethoxysilyl)ethyl]-1-tetralone 410
$C_{19}H_8O_7Ru_3$ Ruthenium, [μ3-[(1,2,2a,8a,8b-η:3,4-η:5,5a,6-η)-acenaphthylene]]-μ-carbonylhexacarbonyltri-, triangulo 1
$C_{19}H_{16}NOP$ *N*-[(Phenyl)methylene]-*P,P*-diphenylphosphinic amide 749
$C_{19}H_{20}FNO_2S$ 3-Butyl-2-fluoro-1-tosylindole 834
$C_{19}H_{21}F_2NO_2S$ *o*′-(1,1-Difluorohex-1-en-2-yl)-*p*-toluenesulfonanilide 834
$C_{19}H_{29}F_6N_3O_4S_2$ (*2S,6S*)-4-Benzyl-1,7-bis(trifluoromethylsulfonyl)-2,6-diisopropyl-1,4,7-triazaheptane 506
$C_{19}H_{36}BF_4P_2Rh$ [(2,3,5,6-η)-Bicyclo[2.2.1]hepta-2,5-diene]bis(methyldiphenylphosphine)-rhodium tetrafluoroborate 1020
$C_{20}H_{15}OP$ (Triphenylphosphoranylidene)-ketene 986
$C_{20}H_{25}NO_6$ (3a*S*,7a*S*)-2-[2-(3,4-Dimethoxyphenyl)-ethyl]-1,3-dioxo-octahydro-isoindole-3a-carboxylic acid methyl ester 1028
$C_{20}H_{29}NO_2Si$ (*R*)-4-Phenyl-3-(2-triisopropylsilyl-ethynyl)oxazolidin-2-one 677
$C_{20}H_{34}O$ (*E, E, E*)-Geranylgeraniol 221
$C_{21}H_{20}ClO_2P$ 2-Carboxyethyltriphenylphosphonium chloride 296
$C_{21}H_{22}NOP$ *N*-[(1*S*)-1-Phenylpropyl]-*P,P*-diphenylphosphinic amide 749
$C_{21}H_{23}NO_4$ (5*S*,6*R*)-4-*tert*-Butoxycarbonyl-5,6-diphenylmorpholin-2-one 879
$C_{21}H_{34}O_3$ (6*E*, 10*E*)-Ethyl 7,11,15-Trimethyl-3-oxohexa-deca-6,10,14-trienoate 221
$C_{22}H_{19}NO_2S$ *trans*-1-(4-Methoxyphenyl)-4-phenyl-3-(phenylthio)azetidin-2-one 920
$C_{22}H_{22}O_2$ (*S,E*)-Methyl 2-((*R*)-4-methyl-1,2-dihydronaphthalen-2-yl)-4-phenylbut-3-enoate 296
$C_{22}H_{29}NO_3$ (*S*)-[1-(Hydroxydiphenylmethyl)-2-methylpropyl]carbamic acid, *tert*-butyl ester 896
$C_{22}H_{35}N_3O_8$ 1-*tert*-Butyloxycarbonylpiperidine-4-spiro-5'-(1',3'-bis(*tert*-butyloxycarbonyl))hydantoin 956
$C_{22}H_{36}O_2$ (2*E*,6*E*,10*E*)-Ethyl 3,7,11,15-Tetramethylhexadeca-2,6,10,14-tetraenoate 221
$C_{23}H_{20}O_2$ 1,3-Diphenyl-2-(1-phenylethyl)propane-1,3-dione 272
$C_{23}H_{27}N$ (*R*)-(+)-*N,N*-Dibenzyl-*N*-(1-cyclohexyl-2-propynyl)-amine 192
$C_{23}H_{27}NO_4$ (4*S*)-2-(Benzhydrylidenamino)pentanedioic Acid, 1-*tert*-Butyl Ester-5-Methyl Ester 404
$C_{23}H_{29}NO_5$ Ethyl (1'*S*,2'*R*)-*N*-*tert*-butyloxycarbonyl-*N*-(1',2'-diphenyl-2'-hydroxyethyl-glycinate 879
$C_{23}H_{32}O_3Si$ [1,1-(Bis-benzyloxymethyl)vinyloxy]triethylsilane 473
$C_{24}H_{27}NO_4$ (3*R*,5*R*,6*S*)-4-*tert*-Butoxycarbonyl-5,6-diphenyl-3-(1'-prop-2'-enyl)morpholin-2-one 399
$C_{25}H_{28}N_2O_6$ 1-*tert*-Butyloxycarbonyl-4-((9-fluorenylmethyloxy-carbonyl)amino)piperidine-4-carboxylic acid 956
$C_{24}H_{31}NO_2$ (1*S*,2*R*)-1-[(2-Hydroxy-3,5-di-*tert*-butylbenzylidene)amino]indan-2-ol 770
$C_{25}H_{43}O_6P$ (2*Z*,6*E*,10*E*)-Ethyl 7,11,15-Trimethyl-3-diethylphosphoryloxy-6,10,14-trienoate 221
$C_{26}H_{24}NO_3PS$ *N*-[(4-Methylphenyl)sulfonyl(phenyl)methyl]-*P,P*-diphenyl-phosphinic amide 749
$C_{26}H_{35}I_2N$ 3,6-Diiodo-9-tetradecylcarbazole 540
$C_{26}H_{35}NSi$ (±)-*N,N*-Dibenzyl-*N*-[1-cyclohexyl-3-(trimethylsilyl)-2-propynyl]-amine 192
$C_{26}H_{35}NSi$ (*R*)-(+)-*N,N*-Dibenzyl-*N*-[1-cyclohexyl-3-(trimethylsilyl)-2-propynyl]-amine 192
$C_{26}H_{37}N$ 9-Tetradecylcarbazole 540
$C_{27}H_{23}BrNO_4P$ (2-Carbomethoxy-6-nitrobenzyl)triphenylphosphonium bromide 903

$C_{27}H_{29}ClCrNO_2$ Chromium(III) of *(1R,2S)*-1-[3-Adamantyl)-2-hydroxy-5-methylbenzyliden-amino]indan-2-ol 498

$C_{27}H_{31}NO_2$ *(1R,2S)*-1-[3-Adamantyl)-2-hydroxy-5-methylbenzyliden-amino]indan-2-ol 498

$C_{27}H_{31}NO_4S_2$ (R_S,1*S*,2*R*)-(–)-2,4,6-Trimethylbenzenesulfinic acid 1-(2,4,6-trimethylbenzenesulfonylamino)-2-indan-2-yl ester 761

$C_{28}H_{28}O_7P_2$ Diphosphoric acid, tetrakis(phenylmethyl) ester 1001

$C_{30}H_{22}N_2O_3$ (*R,R*)-4,6-Dibenzofurandiyl-2,2'-bis(4-phenyloxazoline) 888

$C_{30}H_{26}N_2O_5$ (*R,R*)-Dibenzofuran-4,6-dicarboxylic acid bis(2-hydroxy-1-phenylethyl) amide 888

$C_{30}H_{32}N_2$ 1(*S*),2(*S*)-Diphenyl-*N,N'*-bis-[1(*R*)-phenyl-ethyl]-ethane-1,2-diamine 1028

$C_{30}H_{37}N$ 3,6-Diethynyl-9-tetradecylcarbazole 540

$C_{31}H_{34}N_2O_2RuS$ [*N*-[(1*S*,2*S*)-2-(Amino-κ*N*)-1,2-diphenylethyl]-4-methylbenzenesulfonamidato(2-)-κ*N*][(1,2,3,4,5,6-η)-1-methyl-4-(1-methylethyl)benzene]-ruthenium 9

$C_{34}H_{28}O_2Pd$ Bis(dibenzylideneacetone)palladium 367

$C_{35}H_{27}CoO_2$ (η5-Carbomethoxycyclopentadienyl)-(η4-tetraphenyl-cyclobutadiene)-cobalt 708

$C_{35}H_{36}ClN_2O$ *N*-(9-Anthracenylmethyl)cinchonidinium chloride 404

$C_{36}H_{53}NSi$ 3,6-Bis[(trimethylsilyl)ethynyl]-9-tetradecylcarbazole 540

$C_{36}H_{54}MoN_3$ Tris[*N*-(*tert*-butyl)(3,5-dimethylphenyl)-amido]molybdenum(III) 529

$C_{38}H_{40}BrN_2O$ *O*-Allyl-*N*-(9-Anthracenylmethyl)cinchonidinium bromide 404

$C_{38}H_{41}N_5O_8$ *N*-[(9-Fluorenylmethoxy)carbonyl]-L-aspartic acid, α-*tert*-butyl ester, β-(2-ethyl[(1*E*)-4-nitrophenyl)azo]phenyl]amino] ethyl ester 635

$C_{38}H_{58}O_4Si_2$ 3,6-Bis[1-(triisopropylsiloxy)ethenyl]-9,10-dimethoxyphenanthrene 524

$C_{39}H_{34}CoNO$ (η5-(*S*)-2-(4-methylethyl)oxazolinylcyclo-pentadienyl)-(η4-tetraphenylcyclobutadiene)cobalt 708

$C_{39}H_{59}MoN_3$ Tris[*N*-(*tert*-butyl)(3,5-dimethylphenyl)-amido]molybdenum(VI) propylidyne 529

$C_{40}H_{55}O_{10}P$ Dibutyl 3,4,6-tri-*O*-benzyl-2-*O*-pivaloyl-D-glucopyranosyl phosphate 169

$C_{42}H_{24}Br_6$ Hexakis(4-bromophenyl)benzene 652

$C_{44}H_{56}O_{12}$ 3,4,6-Tri-*O*-benzyl-2-pivaloyl-β-D-glucopyranosyl-(1→6)-1,2:3,4-di-*O*-isopropylidene-α-D-galactopyranoside 169

$C_{54}H_{74}O_8Si_2$ 6,13-Bis(triisopropylsiloxy)-9,10-dimethoxy[7]helicenebisquinone 524

$C_{65}H_{16}$ 6,9,12,15,18-Pentamethyl-1,6,9,12,15,18-hexahydro(c_{60}-i_h)[5,6]fullerene 319

$C_{68}H_{61}NO_2$ 3,6-Bis(benzoylbiphenyl)ethynyl-9-tetradecylcarbazole 540

$C_{78}H_{66}Cl_2Co_2N_2O_2Pd_2$ Di-μ-chlorobis[(η5-(*S*)-(*pR*)-2-(2'-(4'-methylethyl)oxazolinyl)cyclopentadienyl, 1-C, 3'-*N*)(η4-tetraphenylcyclo-butadiene)cobalt]dipalladium ([COP-Cl]$_2$) 714

$C_{82}H_{72}Co_2N_2O_6Pd_2$ Di-μ-acetatobis[(η5-(*S*)-(*pR*)-2-(2'-(4'-methylethyl)oxazolinyl)cyclopentadienyl, 1-C, 3'-*N*)(η4-tetraphenylcyclo-butadiene)cobalt]dipalladium ([COP-OAc]$_2$) 714

$C_{88}H_{68}Co_2F_{12}N_2O_6Pd_2$ Hexafluoroacetylacetonate[(η5-(*S*)-(*pR*)-2-(2'-(4'-methylethyl)oxazolinyl)cyclopentadienyl, 1-C, 3'-*N*)(η4-tetraphenylcyclo-butadiene)cobalt]palladium (COP-hfacac) 714

$C_{92}H_{76}Cl_5NP_4Ru_2$ Ruthenium-BINAP catalyst 38

$C_{112}H_{140}N_4$ 1,35:3,5:8,10:12,14:17,19:21,23:26,28:30,32-Octaethenodotriaconta[1,2-c:9,10-c':17,18-c'':25,26-c''']tetrapyrrole, 6,7,15,16,24,25,33,34-octadehydro-2,11,20,29-tetrahydro-2,11,20,29-tetratetradecyl-; 540

AUTHOR INDEX

This index comprises the names of contributors to Volumes **80**, **81**, **82**, **83** and **84**. For authors of previous volumes, see either indices in Collective Volumes I through X, or the single volume entitled *Organic Syntheses, Collective Volumes I-VIII, Cumulative Indices,* edited by J. P. Freeman.

Adams, N. D., 615
Akué-Gédu, R., 607
Albiniak, P. A., 947
Alemán, J., 741
Amato, F., 778
Amorelli, B., 787
Amos, D. T., 911
Anderson, C. E., 708, 714, 720
Ando, A., 336
Armitage, M. A., 843
Armstrong, J., 38
Asahara, M., 348
Atodiresei, I., 1068
Aubé, J., 811

Baba, A., 646
Bailey, W. F., 433
Bajwa, J. S., 46
Balenkova, E. S., 594
Banik, B. K., 57
Banik, I., 57
Becker, F. F., 57
Begtrup, M., 1015
Bégué, J.-P., 135
Beiger, J. J., 992
Belecki, K., 934
Bemish, R. J., 874
Bergdahl, M., 261
Berliner, M., 934
Berman, A. M., 684
Bhupathy, M., 1001
Bill, D. R., 874
Bobbitt, J. M., 93
Boezio, A. A., 749, 982
Boisvert, M., 802
Bolm, C., 1068
Bonnet-Delpon, D., 135
Brenek, S., 874
Brenner, M., 896
Brickner, S. J., 867
Brown, W. D., 656
Buck, E., 626
Burns, C. L., 1, 652
Busacca, C. A., 965
Buson, R. A., 874

Cai, D., 393
Cardona, F., 114
Cekovic, Z., 121
Cha, J. K., 450
Chandra, D., 82
Charette, A. B., 749, 982
Chavez, D. E., 498
Chen, D., 879
Chen, Q.-Y., 469
Chen, Y. K., 702
Chhor, R. B., 821
Chiu, C. K.-F., 862, 874
Cho, H. M., 540
Chobanian, H., 615; 1056
Christensen, J. B., 942
Cid, M. B., 741
Coates, R. M., 212
Corey, E. J. 404
Côté, A., 749, 982
Cottineau, B., 315
Cox, J. M., 992
Creary, X., 859
Crépy, K. V. L., 1020
Crocker, L. S., 1001
Crousse, B., 135
Cui, D.-M., 82
Curran, D. P., 888
Cvetovich, R., 343

Danheiser, R. L., 234, 244, 911, 920
Darmency, V., 308
Davies, H. M. L., 296
Davies, I. W., 929
Davis, F. A., 761
Davulcu, A. H., 992
Dembinski, R., 794
de Meijere, A., 425
DeMong, D. E., 399, 879
Denmark, S. E., 359, 367, 1046
Deselnicu, M. I., 62
Desrosiers, J. -N., 749, 982
Díaz, D. D., 782
Diver, S. T., 552
Dixon, A. J., 25
Dixon, D., 662
Dolbier, Jr., W. R., 469
Duan, J.-X., 469
Dudley, G. B., 947
Dunetz, J. R., 244

Eidam, H. S., 9
Eidam, P., 9

El-Batta, A., 261
Ellis, K. C., 615
Ellman, J. A., 770
Enoki, Y., 697
Eppich, J. C., 850
Evans, D. A., 821

Ferguson, M. L., 570
Finn, M. G., 782
Frank, R., 635
Frey, J., 843
Frohn, M., 177, 183
Fu, G. C., 372
Fu, J., 1046
Fu, Y., 956
Fujita, K., 697
Fukuyama, T., 139
Fürstner, A., 353; 826

Gagosz, F., 212
Gommermann, N., 192
Goti, A., 114
Gouliaev, A. H., 656
Grebe, T., 1015
Grecian, S., 811
Grether, U. M., 100
Grubbs, R. H., 570

Haddad, N., 965
Hagiwara, H., 416
Hammer, R. P., 956
Hammarström, L. G. J., 956
Hank, R. F., 862
Hansen, T. V., 267
Haque, M. M., 821
Hashiguchi, S., 17
Hayashi, T., 82
Hilmey, D. G., 256
Hirabayashi, T., 631
Hirano, K., 691
Hirao, T., 348
Hoerrner, R. S., 393
Hoffmann, H. M. R., 473
Holmes, A. B., 843
Hoshi, T., 416
Hsung, R. P., 677, 726
Huhn, R. A., 787

Ichikawa, J., 834
Ihara, M., 481
Ikariya, T., 17
Imamoto, T., 1020
Inanaga, K., 481
Isaka, M., 457
Iserloh, U., 888
Ishida, T., 1
Ishii, T., 481
Ishii, Y., 631
Ishiyama, T., 1007
Itonaga, C., 1

Jackson, R. F. W., 384
Jacobs, K., 164
Jacobsen, E. N., 157, 498
Jamison, T. F., 205
Jensen, M. S., 393
Jeon, S-J., 702
Jezek, E., 821
Jin, Y., 221
Johnson, J. S., 684
Jonckers, T. H. M., 665
Jones, T. C., 70

Kakiuchi, F., 410
Kamath, V. P., 88
Kanemasa, S., 888
Katz, T. J., 519, 524
Kesavan, V., 135
Keller, J., 38
Kessler, A., 315
Kim, S.-H., 450
Kimura, M., 327
King, S. A., 38
Kinoshita, A., 561
Kirsch, S. F., 84, 148
Knapp, S., 787
Knochel, P., 192
Kobayashi, S., 691
Kohnen, A. L., 234, 244
Koide, M., 319
Kopecky, D. J., 52
Korotchenko, V. N., 594
Kreilein, M. M., 850
Kristensen, J. L., 1015
Kuboyama, T., 139
Kulkarni, A. A., 552
Kumadaki, I., 336
Kurtz, K. C. M., 677
Kvernenes, O. H., 586

Larsen, R. D., 393
Laurich, D., 826
LaVecchia, L., 896
Lawler, M. J., 1046
Lawlor, M. D., 920
Leazer, J. L., 343
Leblanc, Y., 802
Lee, C.-S., 100
Lee, T. W., 920
Leitner, A., 353
Leonelli, F., 77
Lettan, II, R. B., 197
Leutert, T., 896
Lewinsky, R., 942
Ley, S. V., 662
Li, B., 862, 874
Li, C.-J., 272
Li, J., 107
Li, W., 393
Littke, A. F., 372

Livinghouse, T., 442
Longbottom, D. A., 662
Loones, K. T. J., 665
Lorenz, J. C., 965
Love, K. R., 169
Lu, Y., 529
Luderer, M. R., 433
Lysén, M., 1015

Maes, B. U. W., 665
Maligres, P., 642
Mamane, V., 826
Manninen, P. R., 867
Manning, J. R., 296
Mano, E., 107
Marcaccini, S., 778
Marcoux, J.-F., 929
Marshall, J. A., 9, 615, 1056
Martínez-Lamenca, C., 473
Matsumoto, K., 577
Matsuo, Y., 319
McGrath, M. J., 25
McLaughlin, M. L., 956
McNamara, J., 1001
Mealy, M.J., 433
Merbouh, N., 93
Milgram, B. C., 197
Miller, K. M., 205
Mills, P. A., 506
Miyaura, N., 1007
Mizuno, M., 670
Mizushima, E., 82
Molander, G. A., 289
Moore, J. S., 529, 540
Moradei, O. M., 128
Mori, M., 561
Mori, T., 834
Mori, Y., 281
Morreale, M. S., 794
Morris, P. E., 88
Motoyama, Y., 1
Muguruma, Y., 348
Mulder, J. A., 726
Murai, S., 410
Muramatsu, A., 319
Murata, K., 17
Murry, J., 862

Nadano, R., 834
Nagashima, H., 1
Nakamura, E., 319, 457
Nakamura, M., 457
Nath, D., 82
Nelson, D. P., 393
Nelson, S. G., 506
Nelson, T. D., 1001
Nenajdenko, V. G., 594
Newell, L., 874
Nielsen, L. P. C., 157
Nobuta, Y., 1007
Noe, M. C., 404
Nosse, B., 821
Noyori, R., 17
Nugent, W. A., 149, 702
Nugiel, D. A., 164

O'Brien, P., 25
Oderaotoshi, Y., 888
Ogawa, A., 348
Okamoto, I., 920
O'Leary, D. J., 570
Omote, M., 336
Onishi, Y., 646
Ono, H., 416
Oriyama, T., 1037
O'Shea, D. F., 315
Overman, L. E., 708, 714, 720

Pagendorf, B. L., 442
Paquette, L. A., 128, 256, 850
Parra, A., 741
Paruch, K., 519, 524
Patel, M. C., 442
Perez-Gonzalez, M., 384
Petrillo, D. E., 289
Petrovic, G., 121
Piancatelli, G., 77
Pittelkow, M., 942
Poon, K. W. C., 947
Porinchu, M., 514
Prasad, K., 46
Proemmel, S., 843
Punzalan, E. R., 433

Ramachandar, T., 761
Rathore, R., 62, 652
Ravikumar, K. S., 135
Reiser, O., 821
Renaud, P., 308
Renslo, A. R., 911
Repič, O., 46
Richards, C. J., 708, 714
Rigo, B., 607
Ripka, A. S., 782
Roberts, F. G., 212
Rodeschini, V., 1028
Rombouts, G., 665
Rose, J. D., 1001
Roth J., 862
Roy, P., 802
Ruano, J. L. G., 741
Rush, C., 1001
Rychnovsky, S, D., 52

Sabila, P., 965
Sagamanova, I. K., 677
Sahu, P. K., 514
Saicic, R. N., 121
Sakaguchi, S., 631
Sakurai, H., 348

Sato, F., 336, 600
Scanlan, E. M., 308
Schaffner, A. P., 308
Schall, A., 821
Scheidt, K. A., 197
Schiffers, I., 1068
Schobert, R., 986
Schwartz, K. D., 420
Seebach, D., 896
Seeberger, P. H., 169
Seidel, G., 353, 826
Seki, M., 281
Senanyake, C. H., 965
Shastin, A. V., 594
Shi, Y., 177, 183
Shinde, Y., 821
Shindo, M., 577
Shishido, K., 577
Shriver, J. A., 903
Shu, L., 183
Simpkins, N. S., 1028
Sinclair, P. J., 399, 879
Singh, V., 514
Skattebøl, L., 267
Smith, A. B., 992
Snelgrove, K., 642
Sniady, A., 794
Söderberg, B. C., 903
Soldaini, G., 114
Song, Z. J., 107, 626
Sowa, M. J., 1001
Stecker, B., 425
Stevenson, C. P., 157
Sugiura, M., 691
Sung, M. J., 450
Suzuki, D., 600
Suzuki, K., 488
Sydnes, L. K., 586

Tabaka, A. C., 164
Tahara, K., 319
Takagi, J., 1007
Takasu, K., 481
Tamaru, Y., 327
Tanaka, M., 82
Takasaki, M., 1
Tang, P., 733
Taylor, J., 929
Teleha, C. A., 164
Terakado, D., 1037
Tian, F., 469
Tobiassen, H., 862
Tokuyama, H., 139
Tomkinson, N. C. O., 70
Tonogaki, K., 561
Townsend, L. B., 88
Tracey, M. R., 726
Tschaen, D. M., 107
Tsujiyama, S., 488
Tu, Y., 177, 183

Urabe, H., 600

Vail, S., 956
Van Dyke, A. R., 205
Vedantham, P., 514
Vedsø, P., 1015
Vidal-Pascual, M., 473
Vyklicky, L., 519, 524

Wada, Y., 834
Wallace, J. A., 903
Walsh, P. J., 702
Wang, X. Q., 457
Wang, Z., 359, 367
Wang, Z.-X., 177, 183
Waters, M. S., 642
Watson, M. P., 720
Weix, D. J., 770
White, J. D., 100, 420

Wang, D., 149
Watson, M. P., 708, 714
White, N., 708
Williams, J. D., 88
Williams, R. M., 399, 879
Willis, E. D., 859
Winsel, H., 425
Woerpel, K. A., 821
Wright, G. T., 911
Wu, Y., 761

Xiong, H., 726

Yamago, S., 457
Yamano, M., 670
Yamaguchi, R., 697
Yamasaki, S., 646
Yanik, M. M., 615
Yao, X., 272
Yasuda, M., 646

Zander, N., 635
Zard, S. Z., 212
Zhai, D., 879
Zhang, F., 1028
Zhang, J., 761
Zhang, W., 529, 540
Zhao, M. M., 107

GENERAL INDEX

This index comprises subject matter for Volumes **80**, **81**, **82**, **83** and **84**. For subjects in previous volumes, see either the indices in Collective Volumes I through X or the single volume entitled *Organic Syntheses, Collective Volumes I-VIII, Cumulative Indices,* edited by J. P. Freeman.

The index lists the names of compounds in two forms. The first is the name used commonly in procedures. The second is the systematic name according to Chemical Abstracts nomenclature, accompanied by its registry number in parentheses. Also included are general terms for classes of compounds, types of reactions, special apparatus, and unfamiliar methods.

Most chemicals used in the procedure will appear in the index as written in the text. There generally will be entries for all starting materials, reagents, intermediates, important by-products, and final products; however common solvents are generally excluded. Entries in capital letters indicate compounds, reactions, or methods appearing in the title of the preparation.

Acenaphthylene; (208-96-8) 1

(μ_3, η^2: η^3: η^5-acenaphthylene)$Ru_3(CO)_7$; Ruthenium, [μ3-[(1,2,2a,8a,8b-η:3,4-η:5,5a,6-η)-acenaphthylene]]-μ-carbonylhexacarbonyltri-, triangulo; (151364-75-9) 1

Acetaldehyde; (75-07-0) 9

p-Acetamidobenzenesulfonyl azide: Benzenesulfonyl azide, 4-(acetylamino)-; (2158-14-7) 296

2-Acetamido-2-deoxy-1,3,4,6-tetra-*O*-acetyl-β-D-glucopyranose: β-D-Glucopyranose, 2-(acetylamino)-2-deoxy-, 1,3,4,6-tetraacetate; (7772-79-4) 787

Acetamido-TEMPO: 1-Piperidinyloxy, 4-(acetylamino)-2,2,6,6-tetramethyl-; (14691-89-5) 93

ACETIC ACID 2-METHYLENE-3-PHENETHYLBUT-3-ENYL ESTER:: Benzenepentanol, β,γ-bis(methylene)-, acetate; (445234-76-4) 561

Acetic acid 5-phenylpent-2-ynyl ester: 2-Pentyn-1-ol, 5-phenyl-, acetate; (445234-71-9) 561

Acetic anhydride: Acetic acid, anhydride; (108-24-7) 52, 343, 561, 787

Acetophenone: Ethanone, 1-phenyl-; (98-86-2) 741

2-Acetoxy-1,4-dioxane: 1,4-Dioxan-2-ol, acetate; (1743-23-3) 850

α-ACETOXY ETHER SYNTHESIS, 52

2-(N-ACETYLAMINO)-4,4-DIMETHOXY-1,1,1-TRIFLUOROBUTANE; N-(3,3-Dimethoxy-1-trifluoromethyl-propyl)-acetamide: Acetamide, *N*-[3,3-dimethoxy-1-(trifluoromethyl)propyl]-; (583029-14-5) 212

3-Acetylamino-1-ethoxythiocarbonylsulfanyl-4,4,4-trifluorobutyl acetate: Carbonodithioic acid, *S*-[3-(acetylamino)-1-(acetyloxy)-4,4,4-trifluorobutyl] *O*-ethyl ester: (583028-99-3) 212

4-Acetylamino-2,2,6,6-tetramethylpiperidine-1-oxoammonium perchlorate: Piperidinium, 4-(acetylamino)-2,2,6,6-tetramethyl-1-oxo-, perchlorate (9); (219543-08-5) 93

4-Acetylamino-2,2,6,6-tetramethylpiperidine-1-oxoammonium tetrafluoroborate (219543-09-6) 93

4-Acetylamino-2,2,6,6-tetramethylpiperidinium acetate: Acetamide, *N*-(2,2,6,6-tetramethyl-4-piperidinyl)-, monoacetate (9); (136708-43-5) 93

S-(1-Acetylamino-2,2,2-trifluoroethyl) *O*-ethyl dithiocarbonate: Carbonodithioic acid, *S*-[1-(acetylamino)-2,2,2-trifluoroethyl] *O*-ethyl ester; (583029-16-7) 212

Acetylation, of alcohols, 561
  reductive, 52

Acetyl bromide; (506-96-7) 506

Acetyl chloride; (75-36-5), 519, 782, 934

ACETYLENE DICARBOXALDEHYDE DIMETHYL ACETAL 607

ACETYLENE DICARBOXALDEHYDE TETRAMETHYL ACETAL 607

N-Acetyl-D-glucosamine: D-Glucose, 2-(acetylamino)-2-deoxy-; (7512-17-6) 787

1-(4-ACETYLPHENYL)-2-PHENYLETHANE: 1-[4-(2-Phenylethyl)phenyl]-ethanone; (785-78-4) 289

Acetyltrimethylsilane; (13411-48-8) 577

Acylation 25, 70, 149, 197, 244, 281, 552, 741, 811, 942, 1028, 1037
Acyl hydrazide synthesis, 874
(1*R*,2*S*)-1-[(3-Adamantyl)-2-hydroxy-5-methylbenzylidenamino]indan-2-ol: *1H*-Inden-2-ol, 2,3-dihydro-1-[[(2-hydroxy-5-methyl-3-tricyclo[3.3.1.13,7]dec-1-decylphenyl)methylene]amino]-, (1*R*,2*S*)-; (231963-92-1) 498
2-(1-Adamantyl)-4-methylphenol: Phenol, 4-methyl-2-tricyclo[3.3.1.13,7]dec-1-yl-; (41031-50-9) 498
Alkyl iodide synthesis, 433
Alkyne-cobalt complex, for Pauson-Khand reaction, 442
Alkyne-hydration 82
Alkyne-titanium alkoxide complex, 600
Alkyne synthesis 561, 615
Alkynyl iodide synthesis, 359
Allene synthesis, 726
Allenylamide synthesis, 726
Allylamine: 2-Propen-1-amine; (107-11-9) 244, 442
O-ALLYL-*N*-(9-ANTHRACENYLMETHYL)-CINCHONIDINIUM BROMIDE: Cinchonanium, 1-(9-anthracenylmethyl)-9-(2-propenyloxy)-, bromide, (8*a*,9*R*)-; (200132-54-3), 404
Allylbenzene: Benzene, 2-propenyl-; (300-57-2), 450
Allyl bromide:1-Propene, 3-bromo-; (106-95-6), 404
Allyl iodide: 1-Propene, 3-iodo-; (556-56-9), 399
N-ALLYL-*N*-(METHOXYCARBONYL)-1,3-DECADIYNYLAMINE: 244
2-Allyl-4,4,5,5-tetramethyl-1,3,2-dioxaborolane: 1,3,2-Dioxaborolane, 4,4,5,5-tetramethyl-2-(2-propenyl)-; (72824-04-5) 691
Aluminum chloride ($AlCl_3$); (7446-70-0), 519
Amide synthesis, 197, 670, 733, 821, 1037
Aminoacetaldehyde diethyl acetal: Ethanamine, 2,2-diethoxy-; (645-36-3) 778
Amino acid synthesis, 956
4-Amino-1-*tert*-butyloxycarbonylpiperidine-4-carboxylic acid: 1,4-Piperidine dicarboxylic acid, 4-Amino-, 1-(1,1-dimethylethyl)ester (9); (183673-71-4), 956
6-AMINO-3,4-DIHYDRO-1(2*H*)-NAPHTHALENONE: 6-amino-1-tetralone: 1(2*H*)-Naphthalenone, 6-amino-3,4-dihydro-; (3470-53-9) 670
(2-AMINOETHYL)CARBAMIC ACID *TERT*-BUTYL ESTER: Carbamic acid, *N*-(2-aminoethyl)-, 1,1-dimethylethyl ester; (57260-73-8) 942
(1*S*,2*R*)-(–)-*cis*-1-Amino-2-indanol: *1H*-Inden-2-ol, 1-amino-2,3-dihydro-, (1*S*,2*R*)-; (126456-43-7) 498, 761, 770
(2*S*)-(–)-3-exo-Aminoisoborneol: 3-Amino-1,7,7-trimethyl-bicyclo[2.2.1]heptan-2-ol; (417199-73-7) 702
4-Amino-2,2,6,6-tetramethylpiperidine: 4-Piperidinamine, 2,2,6,6-tetramethyl- (9); (36768-62-4) 93
Ammonia; (7664-41-7), 399, 770
Ammonium carbonate: Carbonic acid, diammonium salt (8,9); (506-87-6), 956
Ammonium chloride (NH_4Cl) (9); (12125-02-9), 57
p-Anisidine: Benzenamine, 4-methoxy-; (104-94-9), 920
Antimony pentachloride; Antimony chloride; (7647-18-9) 62
Arylboronate synthesis, 393, 1015
Asymmetric hydrogenation, 38
Aziridine Formation; 506

Benzalacetone: 3-buten-2-one, 4-phenyl- (8, 9); (122-57-6) 261
Benzaldehyde; (100-52-7) 296, 749, 920
Benzamidine hydrochloride: Benzenecarboximidamide, monohydrochloride (9); (1670-14-0) 862
Benzenesulfenyl chloride; (931-59-9) 121
Benzenesulfinic acid, sodium salt: Sodium benzenesulfinate; (873-55-2) 308
Benzenethiol; Thiophenol; (108-98-5) 135
(4*S*)-2-(BENZHYDRYLIDENAMINO)-PENTANEDIOIC ACID, 1-*tert*-BUTYL ESTER-5-METHYL ESTER: L-Glutamic acid, *N*-(diphenylmethylene)-, 1-(1,1-dimethylethyl)-5-methyl ester; (212121-62-5) 404
Benzil: Ethanedione, diphenyl-; (134-81-6) 17; 646
Benzoic acid; (65-85-0) 70
rac-Benzoin: Ethanone, 2-hydroxy-1,2-diphenyl-; (19-53-9) 17
Benzonitrile; (100-47-0) 433
1,4-Benzoquinone: 2,5-Cyclohexadiene-1,4-dione; (106-51-4) 524; 62
4-Benzoyl-4'-bromobiphenyl: ethanone, (4'-bromo[1,1'-biphenyl]-4-yl)phenyl-; (63242-14-8) 540
Benzoyl chloride; (98-88-4) 552
5-BENZYLOXYBICYCLO[4.2.0]OCTA-1,3,5-TRIEN-7-ONE: Bicyclo[4.2.0]octa-1,3,5-trien-7-one, 5-(phenylmethoxy)- (9); (169615-68-3) 488
(2-BENZOYLOXY)-1,4-CYCLOHEXANEDIONE (MONO)ETHYLENE KETAL: 1,4-Dioxaspiro[4.5]decan-8-one, 7-(benzoyloxy)-; (872312-37-3) 70
(1*S*, 2*S*)-1-BENZOYLOXY-2-BROMOCYCLOHEXANE; Cyclohexanol, 2-bromo-, benzoate, (1*S*,2*S*)-; (222851-77-6) 1037

4-Benzoyloxymorpholine; (5765-65-1) 684
Benzyl alcohol: Benzenemethanol; (100-51-6) 947, 1068
Benzyl amine: Benzenemethanamine; (100-46-9) 506, 697, 733
(*2S,6S*)-4-Benzyl-1,7-bis(trifluoromethylsulfonyl)-2,6-diisopropyl-1,4,7-triazaheptane: *N,N'*-[[(Phenylmethyl)imino]bis[(1*S*)-1-(1-methylethyl)-2,1-ethanediyl]]bis[1,1,1-trifluoro]-methanesulfonamide; (200351-80-0) 506
Benzyl bromide: CORROSIVE. LACHRYMATOR: Toluene, α-bromo- (8); Benzene, (bromomethyl)- (9); (100-39-0) 488
(*R*)-(*E*)-2-Benzylidiene-1-cyclohexyl-butan-1-ol: Cyclohexanemethanol, α-[(1*E*)-1-(phenylmethylene)propyl]-, (α*R*)-; (512785-86-3) 205
N-Benzylidene-*p*-anisidine: Benzenamine, 4-methoxy-*N*-(phenylmethylene)-; (783-08-4) 920
N-Benzylidene-benzylamine *N*-oxide; Benzenemethanamine, *N*-(phenylmethylene)-, *N*-oxide; (3376-26-9) 114
trans-2-BENZYL-1-METHYLCYCLOPROPAN-1-OL; 1-Cyclopropanol, 1-methyl-2-phenylmethyl-; 450
3-Benzyloxy-2-iodophenyl trifluoromethanesulfonate: Methanesulfonic acid, trifluoro-, 2-iodo-3-(phenylmethoxy)phenyl ester (9); (138720-02-2) 488
2-Benzyloxy-1-methylpyridinium trifluoromethanesulfonate; (882980-43-0) 947
2-Benzyloxypyridine: Pyridine, 2-(phenylmethoxy)-; (40864-08-2) 947
N-BENZYL-4-PHENYLBUTYRAMIDE: Benzenebutanamide, *N*-(phenylmethyl)-; (179923-27-4) 733
N-BENZYLPIPERIDINE: Piperidine, 1-(phenylmethyl)-; (2905-56-8) 697
Benzyltriethylammonium chloride; Benzenemethanaminium, *N,N,N*-triethyl-, chloride; (56-37-1) 586
BICYCLO[2.2.1]HEPT-5-ENE-2-ENDO-CARBOXYLIC ACID, 3-ENDO-BENZYLOXYCARBONYL, (2*R*,3*S*)-: Bicyclo[2.2.1]hept-5-ene-2,3-dicarboxylic acid, mono(phenylmethyl) ester, (1*R*,2*S*,3*R*,4*S*)-; (581100-26-7) 1068
endo-Bicyclo[2.2.1]hept-5-ene-2,3-dicarboxylic anhydride: 4,7-Methanoisobenzofuran-1,3-dione, 3*a*,4,7,7*a*-tetrahydro-, (3a*R*,4*S*,7*R*,7a*S*)-rel-; (129-64-6) 1068
BICYCLO[3.1.0]HEXAN-1-OL; (7422-09-5) 450
rac. BINAP: Phosphine, [1,1'-binaphthalene]-2,2'-diylbis[diphenyl-; (98327-87-8) 665
(*R*)-BINAP: Phosphine, (1*R*)-[1,1: binaphthalene]-2,2'-diayhdris (diphenyl)-; (76189-55-4) 38
3,6-Bis(benzoylbiphenyl)ethynyl-9-tetradecylcarbazole: Methanone, [(9-tetradecyl-9H-carbazole-3,6-diyl)bis(2,1-ethynediyl[1,1'-biphenyl]-4',4-diyl)]bis[phenyl-; (791090-33-0) 540
Bis(2-bromoethyl) ether: Ethane, 1,1'-oxybis[2-bromo-; (5414-19-7) 702
2,2-Bis(chloromethyl)-5,5-dimethyl-1,3-dioxane: 1,3-Dioxane, 2,2-bis(chloromethyl)-5,5-dimethyl-; (133961-12-3) 457
(*S,S*)-1,2-BIS-(*tert*-BUTYLMETHYLPHOSPHINO)ETHANE ((*S,S*)-*t*-Bu-BISP*) 1020
Bis (tri-*tert*-butylphosphine)palladium: Palladium, bis[tris(1,1,-dimethylethyl)phosphine]- (9); (53199-31-8) 372
Bis(η^4-1,5-cyclooctadiene)-di-μ-methoxy-diiridium(I) ($[Ir(OMe)(COD)]_2$): Bis[(1,2,5,6-η)-1,5-cyclooctadiene]di-μ-methoxydiiridium; (12148-71-9) 1007
(*R,R*)-(–)-*N,N'*-Bis(3,5-di-*tert*-butylsalicylidene)-1,2-cyclohexanediaminocobalt(II): Cobalt, [[2,2'-[(1*R*,2*R*)-1,2-cyclohexanediylbis[(nitrilo-κ*N*)methylidyne]]bis[4,6-bis(1,1-dimethylethyl)phenolato-κ*O*]](2-)]-, (SP-4-2)-; (176763-62-5) 157
Bismuth bromide: Bismuthine, tribromo-; (7787-58-8) 46
Bis(pinacolato)diboron: 4,4,4',4',5,5,5',5'-octamethyl-2,2'-Bi-1,3,2-dioxaborolane; (73183-34-3) 1007
2,2-Bispyrrolidine; 2,2-Bipyrrolidine; (74295-58-2) 1046
(*R,R*)-2,2'-BISPYRROLIDINE: 2,2'-Bipyrrolidine, (2*R*,2'*R*)-; (137037-20-8) 1046
(*R,R*)-2,2'-Bispyrrolidine•(L)-tartrate: 2,2'-Bipyrrolidine, (2*R*,2'*R*)-, (2*R*,3*R*)-2,3-dihydroxybutanedioate (1:1)-; (137037-21-9) 1046
(*S,S*)-2,2'-BISPYRROLIDINE: 2,2'-Bipyrrolidine, (2*S*,2'*S*)-; (124779-66-4) 1046
(*S,S*)-2,2'-Bispyrrolidine•(D)-tartrate: 2,2'-Bipyrrolidine, [*S*-(*R**,*R**)]-, [S-(*R**,*R**)]-2,3-dihydroxybutanedioate (1:1); (136937-03-6) 1046
Bis(tricyclohexylphosphine)benzylidine ruthenium(IV) dichloride: Ruthenium, dichloro(phenylmethylene)bis(tricyclohexylphosphine)-; (172222-30-9) 570

3,5-Bis(trifluoromethyl)-acetophenone: 1-[3,5-Bis(trifluoromethyl)phenyl] ethanone; (30071-93-3) 343
3,5-Bis(trifluoromethyl)bromobenzene: 1-Bromo-3,5-bis(trifluoromethyl) benzene; (328-70-1) 343, 965
3,5-BIS(TRIFLUOROMETHYL)-ACETOPHENONE: (1-[3,5-Bis(trifluoromethyl)phenyl]-ethanone); (30071-93-3) 343
BIS[3,5-BIS(TRIFLUOROMETHYL)-PHENYL]PHOSPHINE (166172-69-6) 965
BIS[3,5-BIS(TRIFLUOROMETHYL)-PHENYL]-PHOSPHINE OXIDE (15979-14-3) 965
6,13-BIS(TRIISOPROPYLSILOXY)-9,10-DIMETHOXY[7]HELICENEBISQUINONE: Dinaphtho[2,1-c:1',2'-g]phenanthrene-1,4,15,18-tetrone, 9,10-dimethoxy-6,13-bis[[tris(1-methylethyl)silyl]oxy]-; (310899-14-0) 524
3,6-Bis[1-(triisopropylsiloxy)ethenyl]-9,10-dimethoxyphenanthrene: Silane,[(9,10-dimethoxy-3,6-phenenthrenediyl)bis(ethenylideneoxy)]tris(1-methylethyl)-; 524
3,6-Bis[(trimethylsilyl)ethynyl]-9-tetradecylcarbazole: 9H-Carbazole, 9-tetradecyl-3,6-bis[(trimethylsilyl)ethynyl]-; (197860-65-4) 540
Boc-diallylamine: Carbamic acid, di-2-propenyl-, 1,1-dimethylethyl ester; (151259-38-0) 570
N-Boc hydroxyproline methyl ester; 1,2-Pyrrolidinedicarboxylic acid, 4-hydroxy, 1-(1,1-dimethylethyl) 2-methyl ester, (*2S,4R*); (74844-91-0) 38
Boc-L-phenylalanine 4-nitrophenyl ester: L-Phenylalanine, *N*-[(1,1-dimethylethoxy)carbonyl]-, 4-nitrophenyl ester; (7535-56-0) 149
N-Boc-3-PYRROLINE: *1H*-Pyrrole-1-carboxylic acid, 2,5-dihydro-, 1,1-dimethylethyl ester; (73286-70-1) 570
endo-1-BORNYLOXYETHYL ACETATE: Ethanol, 1-[[(*1R,2S,4R*)-1,7,7-trimethylbicyclo[2.2.1]-hept-2-yl]oxy]-, acetate; (284036-61-9) 52
Borane-dimethylsulfide complex: Boron, trihydro[thiobis[methane]]-(T-4)-; (13292-87-0) 359
Borane-tetrahydrofuran: Boron, trihydro(tetrahydrofuran)-, (14044-65-6) 1020
Boric acid: Boric acid (H_3BO_3); (10043-35-3) 733
BORONIC ACID SYNTHESIS, 393
(*R,R*)-BozPHOS: Phospholane, 1-[2-[(2*R*,5*R*)-2,5-dimethyl-1-oxido-1-phospholanyl]phenyl]-2,5-dimethyl-, (2*R*,5*R*)-; (38132-66-8) 749, 982
Bromination, 244, 577, 656, 677, 903
Bromine; (7726-95-6) 62, 88, 607, 652, 903
Bromoacetonitrile; Acetonitrile, bromo-; (590-17-0), 139
4-Bromoacetophenone: Ethanone, 1-(4-bromophenyl)-: (99-90-1) 289
2-Bromoanisole: Benzene, 1-bromo-2-methoxy-; (578-57-4) 626
Bromobenzene: Benzene, bromo-; (100-86-1) 896
trans-2-Bromocyclohexanol; (2425-33-4) 1037
1-Bromo-1,3-decadiyne: 244
1-Bromo-2-propyne: 1-propyne, 3-bromo-; (106-96-7) 442
4-Bromo-1-butene: 1-Butene, 4-bromo-; (5162-44-7) 425
5-BROMOISOQUINOLINE: Isoquinoline, 5-bromo- (8,11); (34784-04-8) 656
2-Bromo-2-methylpropanamide: Propanamide, 2-bromo-2-methyl-; (7462-74-0) 670
2-Bromo-2-methylpropanoyl bromide: Propanoyl bromide, 2-bromo-2-methyl-; (20769-85-1) 670
5-BROMO-8-NITROISOQUINOLINE: Isoquinoline, 5-bromo-8-nitro-; (63927-23-1) 656
1-Bromononane: Nonane, 1-bromo-; (693-58-3) 353
2-Bromophenol: benzene, 2-bromo-1-hydroxy-(9); (95-56-7) 267
N-(2-Bromophenyl)acetamide; (614-76-6) 315
5-(4-Bromophenyl)-3-iodo-2-(4-methylphenyl)furan; (852211-52-0) 794
1-(4-Bromophenyl)-4-methylphenyl)but-3-yn-1-ol; (852211-40-6) 794
1-(4-Bromophenyl)-4-methylphenyl)but-3-yn-1-one; (852211-45-1) 794
2-(4-Bromophenyl)oxirane: (4-Bromostyrene oxide); (32017-76-8) 794
3-Bromopyridine: Pyridine, 3-bromo- (8, 9); (626-55-1) 393
3-Bromoquinoline: Quinoline, 3-bromo- (8, 9); (5332-24-1) 393
3-BROMOSALICYLALDEHYDE 267
3-Bromosalicylaldehyde: benzaldehyde, 2-hydroxy-3-bromo- (9); (1829-34-1) 267
N-Bromosuccinimide: Succinimide, *N*-bromo-(8); 2,5-Pyrrolidinedione, 1-bromo-; (128-08-5) 244, 656
1-Bromotetradecane; (112-71-0) 540
1-Bromo-2-triisopropylsilyl-ethyne: Silane, (bromoethynyl)tris(1-methylethyl)-; (111409-79-1) 677
5-Bromo-*m*-xylene: Benzene, 1-bromo-3,5-dimethyl-; (556-96-7) 529
(*R*)-BUTAN-2-ONE, 1-CYCLOHEXYL, 1-HYDROXY-; (512785-95-4) 205

3-Butenylmagnesium bromide: Magnesium, bromo-3-butenyl-(8,9); (7013-09-5) 425
(2S-*trans*)-2-*tert*-Butoxycarbonylacetyl-4-hydroxypyrrolidine-1-carboxylic acid, *tert*-butyl ester: 2-Pyrrolidinepropanoic acid, 1-[(1,1-dimethylethoxy)carbonyl]-4-hydroxy-β-oxo-,1,1-dimethylethyl ester, [2*S*-*trans*]- (9); (167963-29-3) 38
(*R*)-(*N*-*tert*-BUTOXYCARBONYL)--ALLYLGLYCINE: 4-Pentenoic acid, 2-[[(1,1-dimethylethoxy)carbonyl]amino]-, (*2R*)-; (170899-08-8) 399
(3*R*,5*R*,6*S*)-4-*tert*-Butoxycarbonyl-5,6-diphenyl-3-(1'-prop-2'-enyl)morpholin-2-one: 4-Morpholinecarboxylic acid, 2-oxo-5,6-diphenyl-3-(2-propenyl)-, 1,1-dimethylethyl ester, [3*R*-(3*a*,5*b*,6*b*)]-; (143140-32-3) 399
2(*S*)-(β-*tert*-BUTOXYCARBONYL-α-(*R*)-HYDROXY)ETHYL-4(*R*)-HYDROXYPYRROLIDINE-1-CARBOXYLIC ACID, *tert*-BUTYL ESTER: (2-Pyrrolidinepropanoic acid, 1-[(1,1-dimethylethoxy)carbonyl]-β,4-dihydroxy-, 1,1-dimethylethyl ester, [2*S*-[2α(*S**),4*b*]]-; (167963-30-6) 38
2(*S*)-(β-*tert*-BUTOXYCARBONYL-α-(*S*)-HYDROXY)ETHYL-4(*R*)-HYDROXYPYRROLIDINE-1-CARBOXYLIC ACID, *tert*-BUTYL ESTER, 38
N-(*tert*-BUTOXYCARBONYL)- L-IODOALANINE METHYL ESTER: L-Alanine, *N*-[(1,1-dimethylethoxy)carbonyl]-3-iodo-, methyl ester; (93267-04-0) 384
N-(*tert*-BUTOXYCARBONYL)- L-4-(METHOXYCARBONYL)PHENYL]ALANINE METHYL ESTER: L-Phenylalanine,*N*-[(1,1-dimethylethoxy)carbonyl]-4-(methoxycarbonyl)-, methyl ester (9); (160168-19-4) 384
tert-Butoxycarbonyl- L-proline: 1,2-Pyrrolidinedicarboxylic acid, 1-(1,1-dimethylethyl)ester, (*S*)-*tert*-Butoxycarbonyl- L-proline (15761-39-4) 1037
(*S*)-*N*-(*N*-*tert*-Butoxycarbonylprolyl)dihydro-isoindole (188122-39-6) 1037
N-(*tert*-Butoxycarbonyl)- L-serine methyl ester: L-Serine, *N*-[(1,1-dimethylethoxy)carbonyl]-, methyl ester (9); (2766-43-0) 384
(*S*)-*N*-(*tert*-Butoxycarbonyl)valine methyl ester: Valine, *N*-[(1,1-dimethylethoxy)-carbonyl]-, methyl ester, (*S*)-; (58561-04-9) 896
(*S*)-N-(*tert*-Butoxycarbonyl)valine: Valine, *N*-[(1,1-dimethylethoxy)carbonyl]-; (13734-41-3) 896
tert-Butyl acetate: Acetic acid, 1,1-dimethylethyl ester; (540-88-5) 38
n-Butyl acrylate: 2-Propenoic acid, butyl ester; (141-32-2) 469
n-Butylamine: 1-Butanamine; (109-73-9) 234
tert-Butylamine: 2-Propanamine, 2-methyl-; (75-64-9) 529
(R_S)-(+)-*tert*-Butyl *tert*-butanethiosulfinate: 2-Propanesulfinothioic acid, 2-methyl-, *S*-(1,1-dimethylethyl) ester, [*S*(*R*)]-; (67734-35-4) 770
tert-Butyldichlorophosphine: Phosphonous dichloride, (1,1-dimethylethyl)-; (25979-07-1) 1020
n-BUTYL 2,2-DIFLUOROCYCLO-PROPANECARBOXYLATE: Cyclopropanecarboxylic acid, 2,2-difluoro-, butyl ester; (260352-79-2) 469
tert-Butyldimethylphosphine)trihydroboron: Boron, [(1,1-dimethylethyl)dimethyl-phosphine]-trihydro-; (203000-43-5) 1020
tert-Butyldimethylsilyl chloride; Silane, chloro(1,1-dimethylethyl)dimethyl- (9); (18162-48-6) 46, 615
(*tert*-Butyldimethylsilyloxy)acetaldehyde: Acetaldehyde, [[(1,1-dimethylethyldimethylsilyl]oxy]-; (102191-92-4) 498
1-(*tert*-BUTYLDIMETHYLSILYLOXY)-8-(METHOXYCARBONYL)-6-METHYLBICYCLO [4.2.0]OCTANE: Bicyclo[4.2.0]octane-7-carboxylic acid, 6-[[(1,1-dimethylethyl)dimethylsilyl]oxy]-1-methyl-, methyl ester, (1*R*,6*R*,7*S*)-rel-: (657428-75-6) 481
1-(*tert*-Butyldimethylsilyloxy)-1-methoxyethene: Silane, (1,1-dimethylethyl)[(1-methoxyethenyl)oxy]dimethyl- (9); (77086-38-5) 488
1-*tert*-Butyldimethylsilyloxy-2-methyl-1-cyclohexene: Silane, (1,1-dimethyl-ethyl)dimethyl[(2-methyl-1-cyclohexen-1-yl)oxy]-; (20152-33-4) 481
(*S*)-4-(*tert*-BUTYLDIMETHYLSILYLOXY)-2-PENTYN-1-OL, 615
(*S*)-2-(*tert*-Butyldimethylsilyloxy)propanal: Propanal, 2-[[(1,1-dimethylethyl)dimethyl silyl]oxy]-, (2*S*)-; (87727-28-4), 615
3-BUTYL-2-FLUORO-1-TOSYLINDOLE: 1*H*-Indole, 3-butyl-2-fluoro-1-[(4-methylphenyl)sulfonyl]-; (195734-36-2) 834
1-(*TERT*-BUTYLIMINO-METHYL)-1,3-DIMETHYL-UREA HYDROCHLORIDE 782
tert-Butyl isocyanide: Propane, 2-isocyano-2-methyl-; (7188-38-7) 782

n-Butyllithium: Lithium, butyl-; (109-72-8) 359, 393, 488, 529, 561, 615, 677, 811, 867, 920, 992, 1015, 1028, 1056
sec-Butyllithium: Lithium, (1-methylpropyl)-; (598-30-1), 888, 1020
tert-Butyllithium; Lithium, (1,1-dimethylethyl)-; (594-19-4) 9, 433
n-Butylmagnesium chloride: Magnesium, butylchloro-; (693-04-9) 450
Butyl methacrylate: 1-Propenoic acid, 2-methyl-, butyl ester; (97-88-1) 372
1-Butyl-3-methylimidazolium chloride: *1H*-Imidazolium, 1-butyl-3-methyl-, chloride; (79917-90-1) 859
1-Butyl-3-methylimidazolium tetrafluoroborate: *1H*-Imidazolium, 1-butyl-3-methyl, tetrafluoroborate (1–); (174501-65-6) 859
(5*S*,6*R*)-4-*tert*-BUTYLOXYCARBONYL-5,6-DIPHENYLMORPHOLIN-2-ONE: 4-Morpholinecarboxylic acid, 6-oxo-2,3-diphenyl-, 1,1-dimethylethyl ester, (2*R*-*cis*)-;(173397-90-5) 31, 879
(5*R*,6*S*)-4-*tert*-BUTYLOXYCARBONYL-5,6-DIPHENYLMORPHOLIN-2-ONE: 4-Morpholinecarboxylic acid, 6-oxo-2,3-diphenyl-, 1,1-dimethylethyl ester, (2*S*-*cis*)-;(112741-50-1) 879
1-*tert*-BUTYLOXYCARBONYL-4-(9-FLUORENYLMETHYLOXYCARBONYLAMINO)PIPERIDINE-4-CARBOXYLIC ACID: 1,4,Piperidinedicarboxylic acid, 4-[[(*9H*-fluoren-9-ylmethyloxy)-carbonyl]amino]-, 1-(1,1-dimethylethyl)ester; (183673-66-7) 956
1-*tert*-Butyloxycarbonylpiperidine-4-spiro-5'-(1',3'-bis(*tert*-butyloxycarbonyl) hydantoin: 1,3,8-Triazaspiro[4.5]decane-1,3,8-tricarboxylic acid, 2,4-diiso-, tris(1-dimethylethyl)ester; (183673-68-9) 956
tert-Butyl phenyl carbonate: Carbonic acid, 1,1-dimethylethyl phenyl ester; (6627-89-0) 942
(*R*)-3-Butyn-2-yl mesylate: 3-Butyn-2-ol, methanesulfonate, (2*R*)-; (121882-95-4) 1056

anti-(1*R*)-(1)-Camphorquinone 3-oxime: 1,7,7-Trimethyl-bicyclo[2.2.1]heptane-2,3-dione 3-oxime; (31571-14-9) 702
(±)-10-Camphorsulfonic acid; (5872-08-2) 212
D-(+)-Camphorsulfonic acid: Bicyclo[2.2.1]heptane-1-methanesulfonic acid, 7,7-dimethyl-2-oxo-, (1*S*,4*R*)-; (3144-16-9) 128
(1*R*)-(+)-Camphor: 1,7,7-Trimethyl-bicyclo[2.2.1]heptan-2-one; (464-49-3) 702
Carbazole: 9H-Carbazole; (86-74-8) 540
CARBAZOLE-BASED TETRAKISMACROCYCLE; (245648-36-6) 540
(η^5-Carbomethoxycyclopentadienyl)-(η^4-tetraphenylcyclobutadiene)cobalt; (182627-81-2) 708
(2-Carbomethoxy-6-nitrobenzyl)triphenylphosphonium bromide: Phosphonium, [[2-(methoxycarbonyl)-6-nitrophenyl]methyl]triphenyl-, bromide; (195992-09-7) 903
Carbon dioxide; (124-38-9) 888
Carbon monoxide; (630-08-0) 442, 903
Carbon tetrabromide: Methane, tetrabromo-; (558-13-4) 561
Carbon tetrachloride: Tetrachloromethane; (56-23-5) 594
Carbonyldihydridotris(triphenylphosphine)ruthenium(II): Ruthenium, carbonyldihydridotris(triphenylphosphine); (25360-32-1) 410
Carbonyldiimidazole: 1*H*-Imidazole, 1,1'-carbonylbis-; (530-62-1) 70
2-Carboxyethyltriphenylphosphonium chloride; (36626-29-6) 296
Catalytic reduction, 62
Catecholborane: 1,3,2-Benzodioxaborole; (274-07-7) 308
Cesium carbonate: Carbonic acid, dicesium salt; (534-17-8) 488, 626, 665
Cesium hydroxide: Cesium hydroxide (CsOH); (21351-79-1) 404
CHIRAL LEWIS ACID TRIDENTATE LIGAND, 888
Chloride formation, 646
Chloroacetic acid; (79-11-8) 770
Chloroacetyl chloride: Acetyl chloride, chloro-; (79-04-9) 929
Chlorobenzene: Benzene, chloro-(8,9); (108-90-7) 372
4-Chlorobenzoic acid methyl ester: Benzoic acid, 4-chloro-, methyl ester; (1126-46-1) 353
4-Chlorobenzonitrile: Benzonitrile, 4-chloro-(9); (623-03-0) 372
2-CHLORO-1,3-BIS(DIMETHYLAMINO)-TRIMETHINIUM HEXAFLUOROPHOSPHATE: Methanaminium, *N*-[2-chloro-3-(dimethylamino)- 2-propenylidene]-*N*-methyl-, hexafluorophosphate(1-); (249561-98-6) 139
(*S*)-1-Chloro-3-(Boc-amino)-4-phenyl-2-butanone: Carbamic acid, [(*1S*)-3-chloro-2-oxo-1-(phenylmethyl)propyl]-, 1,1-dimethylethyl ester; (102123-74-0) 149
2-Chloro-1-decen-3-yne; (534600-74-3) 234
1-CHLORO-4-(2,2-DICHLORO-1-METHYLVINYL)BENZENE: Benzene,

1-chloro-4-(2,2-dichloro-1-methylethenyl)-; (73644-88-9) 594
Chlorodimethylsilane; (1066-35-9) 646
Chlorodimethylphenylsilane; (768-33-2) 197
1-Chloro-3-iodobenzene; (625-99-0) 1007
2-Chloro-5-iodopyridine; (69045-79-0) 665
1-CHLORO-3-IODO-5-(4,4,5,5-TETRAMETHYL-1,3,2-DIOXABOROLAN-2-YL)BENZENE: (2-(3-Chloro-5-iodophenyl)-4,4,5,5-tetramethyl-1,3,2-dioxaborolane; (479411-94-4) 1007
m-Chloroperbenzoic acid; Benzenecarboperoxoic acid, 3-chloro-; (937-14-4), 139
2-Chloropyridine; (109-09-1) 947
2-Chloro-3-pyridinecarboxaldehyde; (36404-88-3) 802
9-Chloromethylanthracene: Anthracene, 9-(chloromethyl)-; (24463-19-2), 404
CHLOROMETHYL METHYL ETHER: Methane, chloromethoxy-; (107-30-2) 934
1-(4-Chlorophenyl)ethanone; (99-91-2) 594
1-(4-Chlorophenyl)ethanone hydrazone; (40137-41-5) 594
3-Chloropropionic acid: Propanoic acid, 3-chloro-; (107-94-8) 296
N-Chlorosuccinimide: 2,5-Pyrrolidinedione, 1-chloro-; (128-09-6), 911
Chlorotitanium triisopropoxide: Titanium, chlorotris(2-propanolato)-, (T-4)-; (20717-86-6) 450
Chlorotriethylsilane: Silane, chlorotriethyl-; (994-30-9) 473
N-1-(Chloro-2,2,2-trifluoroethyl)acetamide; (6776-46-1) 212
Chlorotrimethylsilane: Silane, chlorotrimethyl-(9); (75-77-4) 281, 348, 469, 473, 956
Chlorotris(triphenylphosphine)cobalt(I); (26305-75-9) 708
Chromium(III) Cl Complex: Chromium, chloro[(1*R*,2*S*)-2,3-dihydro-1-[[[2-(hydroxy-κ*O*)-5-methyl-3-tricyclo[3.3.1.13,7dec-1-ylphenyl]methyl-ene]amino-κ*N*]-*1H*-indene-2-olato-(2-)-κ*O*],(SP-4-4); (231963-76-1) 498
Chromium trioxide; (1333-82-0) 794
Cinchonidine: Cinchonan-9-ol, (8a, 9*R*)-; (485-71-2) 404
R-(+)-Citronellal: 6-Octenal, 3,7-dimethyl-, (3*R*)-; (2385-77-5) 416
Claisen condensation, 38
2,4,6-Collidine: Pyridine, 2,4,6-trimethyl-; (108-75-8) 708
(*S*)-COP-Cl: COBALT, BIS[1,1',1'',1'''-(η4-1,3-CYCLOBUTADIENE-1,2,3,4-TETRAYL)TETRAKIS[BENZENE]](di-μ-CHLORODIPALLADIUM)BIS[μ-[(1-η:1,2,3,4,5-η)-2-[(4*S*)-4,5-DIHYDRO-4-(1-METHYLETHYL)-2-OXAZOLYL-κ*N*3]-2,4-CYCLOPENTADIEN-1-YL]]DI-; (581093-92-7) 714, 720
(*S*)-COP-hfacac: HEXAFLUOROACETYLACETONATE[(η^5-(*S*)-(*pR*)-2-(2'-(4'-METHYLETHYL)OXAZOLINYL)CYCLOPENTADIENYL, 1-*C*, 3'-*N*)(η^4-TETRAPHENYLCYCLOBUTADIENE)COBALT] PALLADIUM; (805315-08-6) 714
(*S*)-COP-OAc: Di-μ-ACETATOBIS[(η^5-(*S*)-(*pR*)-2-(2'-(4'-METHYLETHYL)OXAZOLINYL)CYCLOPENTADIENYL, 1-*C*, 3'-N)(η^4-TETRAPHENYLCYCLOBUTADIENE)COBALT] DIPALLADIUM; (849592-74-1) 714
Condensation 741, 802, 821, 826, 843, 1028
Copper(II) acetate monohydrate: Acetic acid, copper(2+) salt, monohydrate; (6046-93-1) 100
Copper(I) bromide; (7787-70-4) 192
Copper (I) chloride: Copper chloride: (7758-89-6) 594, 626
Copper (II) chloride: Copper chloride; (7447-39-4) 1020
Copper(I) Iodide: Copper iodide (CuI); (7681-65-4) 234, 244, 261, 540, 662
Copper-mediated phenol coupling 626
Copper sulfate pentahydrate; Sulfuric acid copper(2+) salt (1:1), pentahydrate; (7758-99-8) 586, 677
Cross coupling, 234, 244, 281, 289, 315, 353, 529, 540, 665, 677, 684, 826
Cross enyne metathesis, 552, 561
Crotonic acid: 2-Butenic Acid; (3724-65-0) 420
Cuprous bromide-dimethylsulfide complex: Copper, [thiobis(methane)]; (54678-23-8) 1056
Cu(OTf)$_2$: Methanesulfonic acid, trifluoro-, copper(2+) salt; (34946-82-2) 749
Cycloaddition, of difluorocarbene, 469
of dichlorocarbene 586
[2+2} Cycloaddition, 481, 488, 920
[3+2} Cycloaddition, 457
[4+3} Cycloaddition, 473
Cyclocondensation, 420, 506, 821, 826, 843
Cyclohexanecarboxaldehyde; (2043-61-0), 192, 348
Cyclohexanedicarboxylic anhydride: 1,3-Isobenzofurandione, hexahydro-; (85-42-7) 1028
1,4-Cyclohexanedione mono-ethylene ketal: 1,4-Dioxaspiro[4.5]decan-8-one; (4746-97-8) 70
Cyclohexene; (110-83-8) 359
Cyclohexylamine: Cyclohexanamine; (108-91-8) 442

2-(Dicyclohexylphosphino)biphenyl: Phosphine, [1,1'-biphenyl]-2-yldicyclohexyl-; (247940-06-3) 529
Cycloisomerization, 826
Cyclooctadienyl)ruthenium dichloride polymer, 38
(Cyclooctadienyl)ruthenium dichloride polymer-(*R*)-BINAP catalyst, $[Et_2NH_2]^+[Ru_2Cl_5(BINAP)_2]^-$, 38
Cyclopentadiene: 1,3-Cyclopentadiene (8,9); (542-92-7) 62, 514
2-Cyclopenten-1-one; (930-30-3), 457
2-CYCLOPENTYLACETOPHENONE: Ethanone, 2-cyclopentyl-1-phenyl-; (23033-65-0), 433
Cyclopentylmagnesium chloride: Magnesium, chlorocyclopentyl-; (32916-51-1), 450
Cyclopropane synthesis, 450; 425; 586
Cyclopropene synthesis, 457
Cyclopropanol synthesis, 450
Cyclopropylamine synthesis, 425
(–)-Cytisine: 1,5-Methano-8*H*-pyrido[1,2-*a*][1,5]diazocin-8-one, 1,2,3,4,5,6-hexahydro-, (1*R*,5*S*)-; (485-35-8) 25

1,3-DECADIYNE; (55682-66-1) 234, 244
1-Decanol: 1-Decanol (9); (112-30-1) 93
Decanal: Decanal (8,9); (112-31-2) 93
Desymmetrization, 1028
3,6-DIACETYL-9,10-DIMETHOXYPHENANTHRENE: Ethanone, 1,1'-(9,10-dimethoxy-3,6-phenanthrenediyl)bis-; (310899-08-2) 519, 524
1,2:5,6-DIANHYDRO-3,4-*O*-ISOPROPYLIDENE-L-MANNITOL: L-MANNITOL,1,2:5,6-DIANHYDRO-3,4-*O*-(1-METHYLETHYLIDENE)-; (153059-37-1) 164
1,4-Diazabicyclo[2.2.2]octane (DABCO); (280-57-9) 982
1,8-Diazabicyclo[5.4.0]undec-7-ene (DBU): Pyrimido[1,2-*a*]azepine, 2,3,4,6,7,8,9,10-octahydro-; (6674-22-2) 296, 720
Deuterium oxide: Water-d2; (7789-20-0) 600
Diazo transfer reaction, 296, 920
Dibenzofuran; (132-64-9) 888
Dibenzofuran-4,6-dicarbonyl chloride: 4,6-Dibenzofurandicarbonyl dichloride; (151412-73-8) 888
Dibenzofuran-4,6-dicarboxylic acid: 4,6-Dibenzofurandicarboxylic acid; (88818-47-7) 888
(*R,R*)-Dibenzofuran-4,6-dicarboxylic acid bis(2-hydroxy-1-phenylethyl)amide: 4,6-Dibenzofurandicarboxamide, *N,N'*-bis[(1*R*)-2-hydroxy-1-phenylethyl]-; (247097-79-6) 888
(*R,R*)-4,6-DIBENZOFURANDIYL-2,2'-BIS(4-PHENYLOXAZOLINE) (DBFOX/PH): Oxazole, 2,2'-(4,6-dibenzofurandiyl)bis(4,5-dihydro-4-phenyl-, (4*R*,4'*R*)-; (195433-00-2) 888
Dibenzoylmethane: 1,3-Propanedione, 1,3-diphenyl-: (120-46-7) 272
Dibenzoyl peroxide: Peroxide, dibenzoyl; (94-36-0) 903
Dibenzylamine: Benzenemethanamine, *N*-(phenylmethyl)-; (103-49-1) 114, 192, 425
N,N-DIBENZYL-*N*-(1-CYCLOHEXYL-2-PROPYNYL)-AMINE: Benzenemethanamine, *N*-[(1*R*)-1-cyclohexyl-2-propynyl]-*N*-(phenylmethyl)-; (872357-86-3) 192
(*R*)-(+)-*N,N*-Dibenzyl-*N*-[1-cyclohexyl-3-(trimethylsilyl)-2-propynyl]-amine: Benzenemethanamine, *N*-[(1*R*)-1-cyclohexyl-3-(trimethylsilyl)-2-propynyl]-*N*-(phenylmethyl)-; (872357-80-7) 192
N,N-DIBENZYL-*N*-(2-ETHENYLCYCLOPROPYL)AMINE: BENZENEMETHANAMINE, N-(2-ETHENYLCYCLOPROPYL)-*N*-(PHENYLMETHYL)-; (220247-75-5) 425
N,N-Dibenzylformamide: Formamide, *N,N*-bis(phenylmethyl)-; (5464-77-7) 425
Dibenzylideneacetone: 1,4-Pentadien-3-one, 1,5-diphenyl-; (538-58-9) 367
Dibenzyl phosphate: Phosphoric acid, bis(phenylmethyl) ester; (1623-08-1) 1001
(4,4-Dibromobut-3-enyl)benene: Benzene, (4,4-dibromo-3-butenyl)-; (119405-97-9) 561
3,4-Dibromo-2,5-diformylthiophene: 2,5-Thiophenedicarboxaldehyde, 3,4-dibromo-; (25373-20-0) 843
Dibromo-2,5-dimethoxytetrahydrofuran: Furan, 3,4-dibromotetrahydro-2,5-dimethoxy-; (91468-55-2) 607
1,2-Dibromoethane: CANCER SUSPECT AGENT; (106-93-4) 281
1,2-Dibromo-1,1,2,2-tetrachloroethane; (630-25-1) 577
Di-*tert*-butyl dicarbonate: Dicarbonic acid, bis(1,1-dimethylethyl) ester; (24424-99-5) 879, 956, 1037
Dibutyl phosphate: Phosphoric acid, dibutyl ester (8,9); (107-66-4) 169
2,3-Dibromo-1,1,4,4-tetramethoxybutane: Butane, 2,3-dibromo-1,1,4,4-tetramethoxy-; (25537-21-7) 607
4,4'-Di-*tert*-butyl-2,2'-bipyridine (dtbpy): 4,4'-Bis(1,1-dimethylethyl)-2,2'-bipyridine: (72914-19-3) 1007
Di-*tert*-butyl disulfide: Disulfide, bis(1,1-dimethylethyl); (110-06-5) 770

3,5-Di-*tert*-butyl salicylaldehyde: Benzaldehyde, 3,5-bis(1,1-dimethylethyl)-2-hydroxy-; (37942-07-7) 770

DIBUTYL 3,4,6-TRI-*O*-BENZYL-2-*O*-PIVALOYL-D-GLUCOPYRANOSYL PHOSPHATE: β-D-GLUCOPYRANOSE, 3,4,5-TRIS-*O*-(PHENYLMETHYL)-, 1-(DIBUTYL PHOSPHATE) 2-(2,2-DIMETHYLPROPANOATE); (223919-63-7) 169

1,3-Dichloroacetone: 2-Propanone, 1,3-dichloro-; (534-07-6) 256, 457

Di-μ-chloro-bis(1,5-cyclooctadiene)diiridium(I): Iridium, di-μ-chlorobis[(1,2,5,6-η)-1,5-cyclooctadiene]di-; (12112-67-3) 631

Dichlorobis(triphenylphosphine)palladium(II); (13965-03-2) 540

1,1-Dichloro-(3*S*)- (*tert*-butyldimethylsilyloxy)-2-butanol *p*-toluenesulfonate: 2-Butanol, 1,1-dichloro-3-[[(1,1-dimethylethyl)dimethylsilyl]oxy]-, 4-methylbenzene sulfonate, (3*S*)-; (329914-17-2) 615

Dichloro(*p*-cymene)ruthenium(II) dimer; (52462-29-0) 9

Di-μ-chloro-dichlorobis(η^5-pentamethylcyclopentadienyl)diiridium $[Cp^*IrCl_2]_2$ Iridium, di-μ-chlorodichlorobis[(1,2,3,4,5-η)-1,2,3,4,5-pentamethyl-2,4-cyclopentadien-1-yl]di-; (12354-84-6) 697

Dichlorodicyclopentadienylvanadium; Vanadium, dichlorobis (η^5-2,4-cyclopentadien-1-yl)-; (12083-48-6) 348

1,1-Dichloro-2-ethoxycyclopropane; cyclopropane, 1,1-dichloro-2-ethoxy-; (7363-99-7) 586

1,1-Dichloropropane; (78-99-9) 529

Dicobalt octacarbonyl: Cobalt, di-μ-carbonylhexacarbonyl di-, (Co-Co); (10210-68-1) 442

1,3-Dicyclohexylcarbodiimide: Cyclohexanamine, *N,N'*-methanetetraylbis-; (538-75-0) 1001, 1037

dl-1,2-DICYCLOHEXYLETHANEDIOL: 1,2-ETHANEDIOL, 1,2-DICYCLOHEXYL-(9); (92319-61-4) 348

Dicyclopentadiene: 4,7-Methano-*1H*-indene, 3*a*,4,7,7*a*-tetrahydro-; (77-73-6), 514

(*Z*)-1,2-DIDEUTERIO-1-(TRIMETHYLSILYL)-1-HEXENE: Silane, (1,2-dideuterio-1-hexenyl)trimethyl-, (*Z*)-; 600

Diels-Alder reaction, 62, 514, 524, 911

1, 3-Diene synthesis, 552, 561

2,2-DIETHOXY-1-ISOCYANOETHANE 778

Diethylamine (8); Ethanamine, *N*-ethyl- (9); (109-89-7) 488

Diethylaminosulfur trifluoride (DAST): Sulfur, (*N*-ethylethanaminato)trifluoro-, (T-4); (38078-09-0) 888

Diethylaminotrimethylsilane: Silanamine, *N,N*-diethyl-1,1,1-trimethyl-; (996-50-9) 416

Diethyl azodicarboxylate: Azodicarboxylic acid diethyl ester; (1972-58-3) 164

Diethyl chlorophosphate: Phosphorochloridic acid, diethyl ester; (814-49-3) 221

Diethyl phosphite: Phosphonic acid, diethyl ester: (762-04-9) 965

Diethylzinc; (557-20-0) 327, 336, 749

3,6-Diethynyl-9-tetradecylcarbazole; (188740-71-8) 540

DIFLUOROCARBENE REAGENT, 469

o-(1,1-Difluorohex-1-en-2-yl)aniline: Benzenamine, 2-[1-(difluoromethylene)pentyl]-; (134810-59-6) 834

o'-(1,1-Difluorohex-1-en-2-yl)-*p*-toluenesulfonanilide: Benzenesulfonamide, *N*-[2-[1-(difluoromethylene)pentyl]phenyl]-4-methyl-; (195734-33-9) 834

(2*R*,5*R*)-(+)-*trans*-Dihydrocarvone: (2*R*,5*R*)-2-Methyl-5-(1-methylethenyl)-cyclohexanone; (5948-04-9) 100

2,3-Dihydrofuran; (1191-99-7) 256

3,4-Dihydro-6-hydroxy-1(2*H*)-naphthalenone, 6-hydroxy-1-tetralone: 1(2*H*)-Naphthalenone, 3,4-dihydro-6-hydroxy-; (3470-50-6) 670

3,6-Diiodo-9-tetradecylcarbazole: 3,6-Diiodo-9-tetradecyl-9H-carbazole: (197860-64-3) 540

DIBAL-H (Diisobutylaluminum hydride): Aluminum, hydrobis (2-methylpropyl)- (9); (1191-15-7) 52, 221, 615, 965, 1056

Diisopropylamine: 2-Propanamine, *N*-(1-methylethyl)-; (109-72-8) 234, 615, 811

Diisopropylethylamine: 2-Propanamine, *N*-ethyl-*N*-(1-methylethyl)-; (7087-68-5) 139, 488, 506, 934, 956

1,2:4,5-DI-*O*-ISOPROPYLIDENE-D-erythro-2,3-HEXODIULO-2,6-PYRANOSE: β-D-erythro-2,3-Hexodiulo-2,6-pyranose, 1,2:4,5-bis-*O*-(1-methylethylidene)-; (18422-53-2) 177, 183

1,2:4,5-Di-*O*-isopropylidene-β-D-fructopyranose: β-D-Fructopyranose, 1,2;4,5-bis-*O*-(1-methylethylidene)-; (25018-67-1) 177

1,2:3,4-Di-*O*-isopropylidene-α-D-galactopyranose: α-D-Galactopyranose, 1,2:3,4-bis-*O*-(1-methylethylidene)-; [4064-06-6] 169

1,4:5,8-Dimethanododecahydroanthracene-9,10-dione: 1,4:5,8-Dimethanoanthracene-9,10-dione, dodecahydro-,

(1a,4a,4aa,5b,8b,8ab,9aa,10ab)-; (2065-48-7) 62
1,4:5,8-Dimethano-1,2,3,4,5,6,7,8-octahydroanthracene-9,10-diol: 1,4:5,8-Dimethanoanthracene-9,10-diol, 1,2,3,4,5,6,7,8-octahydro-, (1a,4a,5b,8b)-; (130778-68-6) 62
1,4:5,8-Dimethano-1,4,4a,5,8,8a,9a,10a-octahydroanthracene-9,10-dione: 1,4:5,8-Dimethanoanthracene-9,10-dione, 1,4,4a,5,8,8a,9a,10a-octahydro-, (1*R*,4*S*,4a*R*,5*S*,8*R*,8a*S*,9a*S*,10a*R*)-; (78548-82-0) 62
1,4:5,8-dimethano-1,2,3,4,5,6,7,8-octahydro-9,10-1,4:5,8-Dimethano-1,2,3,4,5,6,7,8-octahydro-9,10-dimethoxy-anthracene: 1,4:5,8-Dimethanoanthracene, 1,2,3,4,5,6,7,8-octahydro-9,10-dimethoxy-, (1*R*,4*S*,5*S*,8*R*)-; (322733-47-1) 62
Dimethoxyanthracenium Hexachloroantimonate 62
4,4-Dimethoxybut-2-ynal: 2-Butynal, 4,4-dimethoxy-; (124744-10-1) 607
2,5-Dimethoxy-2,5-dihydrofuran: Furan, 2,5-dihydro-2,5-dimethoxy-; (332-77-4) 607
1,2-Dimethoxyethane: Ethane, 1,2-dimethoxy-; (110-71-4) 442
Dimethoxymethane: Methane, dimethoxy-; (109-87-5), 183, 934
9,10-DIMETHOXYPHENANTHRENE: Phenanthrene, 9,10-dimethoxy-; (13935-65-4), 519
2-(3,4-Dimethoxyphenyl)ethylamine: Benzeneethanamine, 3,4-dimethoxy-; (120-20-7) 1028
(3a*S*,7a*S*)-2-[2-(3,4-DIMETHOXYPHENYL)-ETHYL]-1,3-DIOXO-OCTAHYDRO-ISOINDOLE-3A-CARBOXYLIC ACID METHYL ESTER; (501085-21-8) 1028
(3a*S*,7a*R*)-2-[2-(3,4-Dimethoxyphenyl)-ethyl]-hexahydro-isoindole-1,3-dione : 1H-Isoindole-1,3(2H)-dione, 2-[2-(3,4-dimethoxyphenyl)ethyl]-hexahydro-, (3aR,7aS)-rel-; (501085-17-2) 1028
2,2-Dimethoxypropane: Propane, 2,2-dimethoxy-(9); (77-76-9) 88, 164, 177
1,1-Dimethoxy-2-propanone: α,α-Dimethoxyacetone; (6342-56-9) 473
2,5-Dimethoxytetrahydrofuran: Furan, tetrahydro-2,5-dimethoxy-; (696-59-3) 826
4-(Dimethylamino)pyridine (DMAP): 4-Pyridinamine, *N,N*-dimethyl-; (1122-58-3) 52, 384, 956
Dimethyl biphenyl-4,4'-dicarboxylate: [1,1'-Biphenyl]-4,4'- dicarboxylic acid, dimethyl ester; (792-74-5) 384
Dimethyl carbonate: Carbonic acid, dimethyl ester; (616-38-6) 811
(6*R*)-(+)-1,6-Dimethylcyclohex-2-en-1-ol: *trans*-1,6-Dimethyl-2-Cyclohexen-1-ol; (114644-29-0), *cis*-1,6-Dimethyl-2-Cyclohexen-1-ol; (114644-28-9) 100
(*R*)-(+)-3,4-DIMETHYLCYCLOHEX-2-EN-1-ONE: ((*R*)-(+)-3,4-Dimethyl-2-cyclohexen-1-one) (10463-42-0) 100
3,4-DIMETHYLCYCLOPENT-2-ENONE: 3,4-Dimethyl-2-cyclopenten-1-one; (30434-64-1) 420
2-(5,5-DIMETHYL-1,3,2-DIOXABORINAN-2-YL)BENZOIC ACID ETHYL ESTER: BENZOIC ACID, 2-(5,5-DIMETHYL-1,3,2-DIOXABORINAN-2-YL)-, ETHYL ESTER; (346656-34-6) 1015
cis-5-(5,5-DIMETHYL-1,3-DIOXAN-2-YLIDENE)HEXAHYDRO-1(*2H*)-PENTALENONE, 457
N,N-Dimethylformamide: Formamide, *N,N*-dimethyl-; (68-12-2) 200, 708, 888
2,5-Dimethylfuran; (625-86-5) 473
(*Z*)-1,1-Dimethyl-1-heptenylsilanol: Silanol, (*1Z*)-1-heptenyldimethyl-; (261717-40-2) 359
N,O-Dimethylhydroxylamine hydrochloride: Methanamine, *N*-methoxy-, hydrochloride; (6638-79-5) 992
Dimethylimidizolidinone (DMI): 1,3-Dimethyl-2-imidazolidinone; (80-73-9) 205
6,6-DIMETHYL-1-METHYLENE-4,8-DIOXASPIRO[2.5]OCTANE: 4,8-Dioxaspiro[2.5]octane, 6,6-dimethyl-1-methylene; (122968-05-2) 457
(*Z*)-3,7-DIMETHYL-2,6-OCTADIEN-1-AL (NERAL): (2*Z*)-3,7-Dimethyl-2,6-octadienal; (106-26-3) 77
(*Z*)-3,7-Dimethyl-2,6-octadien-1-ol (nerol); (106-25-2) 77
3*R*,7-DIMETHYL-2-(2-OXOBUTYL)-6-OCTENAL: 6-Octenal, 3,7-dimethyl-2-(3-oxo-butyl)-, (3*R*)-; (131308-24-2) 416
N-(3,5-Dimethylphenyl)-*tert*-butylamine: Benzenamine, *N*-(1,1-dimethylethyl)-3,5-dimethyl-; (110993-40-3) 529
N,N-Dimethyl-3-phenylpropionamide: Benzenepropanamide, *N,N*-dimethyl-; (5830-31-9) 1
N,N-Dimethyl-3-phenylpropylamine: Benzenepropanamine, *N,N*-dimethyl-; (1199-99-1) 1
1-(DIMETHYL(PHENYL)SILYL)PROPAN-1-ONE: Dimethyl(1-oxopropyl)phenylsilane; (128084-44-6) 197
2,2-Dimethyl-propanedioyl dichloride : Propanedioyl dichloride, dimethyl-; (5659-93-8) 821
Dimethyl sulfate: Sulfuric acid, dimethyl ester; (77-78-1) 62, 296, 519
Dimethylsulfide: Methane, thiobis-; (75-18-3)

192, 205
Dimethylsulfoxonium (*S*)-2-oxo-3-(Boc-amino)-4-phenylbutylide: Sulfoxonium, dimethyl-, (*3S*)-3-[[(1,1-dimethylethoxy)carbonyl]amino]-2-oxo-4-phenylbutylide: (400611-25-8) 149
1,3-Dimethylurea: Urea, *N,N'*-dimethyl-; (96-31-1) 782
Diol synthesis, 348
Dioxane: 1,4-Dioxane; (123-91-1) 850
1,4-DIOXENE: (2,3-Dihydro-1,4-dioxin) (543-75-9) 850
Diphenylacetylene: Benzene, 1,1'-(1,2-ethynediyl)bis-; (501-65-5) 708
DIPHENYL DISULFIDE: Disulfide, diphenyl; (882-33-7) 135
1(*S*),2(*S*)-Diphenyl-*N,N'*-bis-[1(*R*)-phenyl-ethyl]-ethane-1,2-diamine; (156730-49-3) 1028
(1*S*,2*R*)-1,2-Diphenyl-2-hydroxyethylamine: Benzeneethanol, β-amino-α-phenyl-, (α*R*,β*S*)-rel-; (23412-95-5) 879
(1*R*,2*S*)-1,2-Diphenyl-2-hydroxyethylamine: Benzeneethanol, β-amino-α-phenyl-, (α*S*,β*R*)-; (23364-44-5) 879
N-(Diphenylmethylene)glycine *tert*-butyl ester: Glycine, (diphenylmethylene)-, 1,1-dimethylethyl ester; (81477-94-3) 404
1,3-DIPHENYL-2-(1-PHENYLETHYL)PROPANE-1,3-DIONE: 1,3-Propanedione, 1,3-diphenyl-2-(1-phenylethyl)-; (116140-58-0) 272
P,P-Diphenylphosphinic amide; (5994-87-6) 749
Diphosphine Ligands 1020
Directed metallation, 1015
Disodium ethylenediaminetetraacetate: Glycine, *N,N'*-1,2-ethanediylbis[*N*-(carboxymethyl)-, disodium salt; (139-33-3) 183
Disperse Red 1; Ethanol, 2-[ethyl[4-[4-nitrophenyl)azo]phenyl]amino]-; (2872-52-8), 635
DITHIENO[3,2-*b*:2',3'-*d*]THIOPHENE; (593-75-7) 843
Dithieno[3,2-*b*:2',3'-*d*]thiophene-2,6-dicarboxylic acid; (502764-53-6) 843
Dithieno[3,2-*b*:2',3'-*d*]thiophene-2,6-dicarboxylic acid diethyl ester; (502764-52-5) 843
Dodecylbenzenesulfonic acid; (27176-87-0) 691
4-Dodecylbenzenesulfonyl azide: Benzenesulfonyl azide, 4-dodecyl-; (79791-38-1) 920
(*R,R*)-Me-DuPHOS: Phospholane 1,1'-(1,2-phenylene)bis[2,5-dimethyl-, [2*R*-[1(2'*R**,5'*R**),2*a*,5*b*]]-; (147253-67-6) 982

Electrophilic aromatic substitution, 519, 656, 652
Electrophilic bromination 652
Enol silane synthesis, 193, 473, 524
Epoxide synthesis, 149, 157, 164, 183
(*2S,3S*)-1,2-EPOXY-3-(BOC-AMINO)-4-PHENYLBUTANE: Carbamic acid, [(*1S*)-1-(*2S*)-oxiranyl-2-phenylethyl]-, 1,1-dimethylethyl ester; (98737-29-2) 149
Ester hydrolysis, 353
Esterification, 281, 296, 420, 561, 635, 1037
1,2-Ethanediamine; (107-15-3) 942
Ethanethiol: STENCH; (75-08-1) 281
Ethenone, (triphenylphosphoranylidene)-; (15596-07-3) 986
Ethyl acetoacetate: Butanoic acid, 3-oxo-, ethyl ester; (141-97-9) 221
ETHYL 4-AMINOBENZOATE: BENZOIC ACID, 4-AMINO-2, ETHYL ESTER; (94-09-7) 57, 665
Ethyl benzoate: Benzoic acid ethyl ester; (93-89-0) 1015
Ethyl bromoacetate: Acetic acid, bromo-, ethyl ester; (105-36-2) 879
Ethyl 2-bromopropanoate: Propanoic acid, 2-bromo-, ethyl ester: (535-11-5) 577
(*S*)-Ethyl 2-(*tert*-butyldimethylsilyloxy)lactate: Propanoic acid, 2-[[(1,1-dimethylethyl) dimethyl]oxy], ethyl ester; (106513-42-2) 615
Ethyl (1'*R*,2'*S*)-*N*-*tert*-butyloxycarbonyl-*N*-(1',2'-diphenyl-2'-hydroxyethyl) glycinate: Glycine, *N*-[(1,1-dimethylethoxy)carbonyl]-*N*-(2-hydroxy-1,2-diphenylethyl)-, ethyl ester, [*S*-(*R**,*S**)]-; (112741-73-8) 879
Ethyl (1'*S*,2'*R*)-*N*-*tert*-butyloxycarbonyl-*N*-(1',2'-diphenyl-2'-hydroxyethyl)glycinate: Glycine, *N*-[(1,1-dimethylethoxy)carbonyl]-*N*-(2-hydroxy-1,2-diphenylethyl)-, ethyl ester, [*R*-(*R**,*S**)]-; (112741-70-5) 879
ETHYL 3-CHLORO-3-PHENYLPROPANOATE: Benzenepropanoic acid, β-chloro-, ethyl ester; (77085-24-6) 646
ETHYL 4-[(6-CHLOROPYRIDIN-3-YL)AMINO]BENZOATE; (509953-61-1) 665
Ethyl 2,2-dibromopropionate: Propanoic acid, 2,2-dibromo-, ethyl ester; (34757-17-0) 577
Ethyl (1'*S*,2'*R*)-*N*-(1',2'-diphenyl-2'-hydroxyethyl)glycinate: Glycine, *N*-(2-hydroxy-1,2-diphenylethyl)-, ethyl ester, [*R*-(*R**,*S**)]-; (100678-82-8) 879
Ethyl (1'*R*,2'*S*)-*N*-(1',2'-diphenyl-2'-hydroxyethyl)glycinate: Glycine, *N*-(2-hydroxy-1,2-diphenylethyl)-, ethyl ester, [*S*-(*R**,*S**)]-; (112835-62-8) 879
Ethyl 3-hydroxy-3-phenylpropanoate: Benzenepropanoic acid, β-hydroxy-, ethyl ester; (5764-85-2) 646

Ethylene: Ethene; ((74-85-1) 561
purification of, 561
reaction apparatus for addition of, 561
Ethylenediaminetetraacetic acid (EDTA): Glycine, *N,N'*-1,2-ethanediylbis[*N*-(carboxymethyl)-;(60-00-4) 183
Ethyl 5-iodopentanoate: CANCER SUSPECT AGENT: Pentanoic acid, 5-iodo-, ethyl ester; (41302-32-3) 281
(*S*)-Ethyl lactate: Propanoic acid, 2-hydroxy-, ethyl ester, (2*S*)-; (687-47-8) 615
Ethyl 2-mercaptoacetate: Acetic acid, mercapto-, ethyl ester; (623-51-8) 843
Ethyl methacrylate: 2-Methyl-2-propenoic acid, ethyl ester; (97-63-2) 308
Ethyl 4-nitrobenzoate: Benzoic acid, 4-nitro-, ethyl ester (9); 99-77-4) 57
ETHYL 6-OXOTRIDECANOATE; (333355-32-1) 281
Ethylphosphonic acid bis-(2,2,2-trifluoroethyl) ester: Phosphonic acid, ethyl-, bis(2,2,2-trifluoroethyl) ester; (650-16-8) 992
Ethylphosphonic dichloride; (1066-50-8) 992
Ethyl pentanoate: Pentanoic acid, ethyl ester; (539-82-2) 281
ETHYL 2-[(2,6,6-TRIMETHYLBICYCLO[3.1.1]HEPT-3-YL)METHYL]PROPENOATE; (183623-93-0) 308
Ethyl vinyl ether: Ethene, ethoxy-; (109-92-2) 552, 561, 586
4–Ethynyltoluene; (9766-97-2) 794
2-(Ethoxycarbonyl)prop-2-en-1-yl phenyl sulfone; (89295-32-9) 308

(*E, E*)-Farnesol; (106-28-5) 221
(*E, E*)-Farnesyl bromide: 2,6,10-Dodecatriene, 1-bromo-3,7,11-trimethyl-; (6874-67-5) 221
Ferric acetylacetonate: Tris (2,4-pentanedionato)iron (lll); (14024-18-1) 353
Ferric nitrate nonahydrate: Nitric acid, iron(3+) salt, nonahydrate; (7782-61-8) 457
Ferrous sulfate heptahydrate: Sulfuric acid, iron(2+) salt (1:1), heptahydrate; (7782-63-0) 100
Finkelstein reaction, 433
9-Fluorenylmethyl chloroformate: Carbonochloridic acid, *9H*-fluoren-9-yl-, methyl ester; (28920-43-6) 956
2-Fluorosulfonyl-2,2-difluoroacetic acid: Acetic acid, difluoro(fluorosulfonyl)-; (1717-59-5) 469
Formamidine Urea, 782
Formic acid; (64-18-6) 17, 607
Formylation 843
N-Formylpiperidine: 1-Piperidinecarboxaldehyde; (2591-86-8) 843
D-Fructose; (57-48-7) 177
[60]Fullerene: [5,6]Fullerene-C_{60}-I_h; (99685-96-8) 319
Furfural: 2-Furancarboxaldehyde; (98-01-1) 128

Geranial: 2,6-Octadienal, 3,7-dimethyl-, (2*E*)-; (141-27-5) 93
Geraniol: 2,6-Octadien-1-ol, 3,7-dimethyl-, (2*E*)-; (106-24-1) 93
(*E, E, E*)-GERANYLGERANIOL: 2,6,10,14-Hexadecatetraen-1-ol, 3,7,11,15-tetramethyl-, (2*E*,6*E*,10*E*)- (24034-73-9*)* 221
GLCNAC-THIAZOLINE TRIACETATE: (3a*R*,5*R*,6*S*,7*R*,7a*R*)-5-Acetoxymethyl-6,7-diacetoxy-2-methyl-5,6,7,7a-tetrahydro-3a*H*-pyrano[3,2-*d*]thiazole: 5*H*-Pyrano[3,2-*d*]thiazole-6,7-diol, 5-[(acetyloxy)methyl]-3a,6,7,7a-tetrahydro-2-methyl-, diacetate (ester), (3a*R*,5*R*,6*S*,7*R*,7a*R*)-; (67109-74-4) 787
(*R*)-(–)-Glycidyl butyrate: Butanoic acid, (2*R*)-oxiranylmethyl ester; (60456-26-0) 867
Glycosylation, 169, 787
Glyoxal (aqueous solution): Ethandial; (107-22-2) 1028
Gold(III) chloride; (13453-07-1) 272
Grignard reagents, 353
Grubbs catalyst, 561, 570
Grubbs' second generation catalyst: Ruthenium [1,3-bis-(2,4,6-trimethylphenyl)-2-imidazolidinylidine]-dichloro(phenylmethylene)(tricyclohexylphosphine); (246047-72-3) 552

Heck reaction, 372
HELICENEBISQUINONES, 524
3-Heptanol; (589-82-2) 121
(*Z*)- 1-HEPTENYLDIMETHYLSILANOL: Silanol, (1*Z*)-1-heptenyldimethyl-; (261717-40-2) 359
(Z)-1-HEPTENYL-4-METHOXYBENZENE: BENZENE, 1-(1*Z*)-1-HEPTENYL-4-METHOXY-; (80638-85-3) 359, 367
3-Heptyl benzenesulfenate: Benzenesulfenic acid, 1-ethylpentyl ester; (198778-69-7), 121
1-Heptyne: 1-Heptyne; (628-71-7) 359, 367
N-Heteroannulation, 826, 834, 903
Hetero-Diels Alder 498
Hexabutylditin: Distannane, hexabutyl; (813-19-4), 121
Hexacarbonyl[μ[(3,4-η:3,4-η)-2-methyl-3-butyn-2-ol]]dicobalt: Cobalt, hexacarbonyl[μ-[(3,4-η:3,4-η)-2-methyl-3-butyn-2-ol]]di-, (Co-Co); (40754-33-4), 442
Hexafluorophosphoric acid: Phosphate (1-), hexafluoro-, hydrogen; (16940-81-1), 929
HEXAKIS(4-BROMOPHENYL)BENZENE (HBB) 652

hexakis(4-Bromophenyl)benzene: 1,1':2',1"-Terphenyl, 4,4"-dibromo-3',4',5',6'-tetrakis(4-bromophenyl)-; (19057-50-2) 652
Hexamethylcyclotrisiloxane: Cyclotrisiloxane, hexamethyl-; (54-05-9), 359
1,1,1,3,3,3-Hexamethyldisilazane: Silanamine,1,1,1-trimethyl-*N*-(trimethylsilyl)-; (999-97-3) 920, 986, 992
Hexamethyldisiloxane; (107-46-0) 787
Hexaphenylbenzene: 1,1':2',1"-Terphenyl, 3',4',5',6'-tetraphenyl-; (992-04-1) 652
1-Hexene, 6-iodo-; (18922-04-8) 433
trans-2-Hexen-1-ol: 2-Hexen-1-ol, (2*E*)-; (928-95-0) 720
1-Hexyne (8, 9); (693-02-7) 261
5-Hexen-1-ol; (821-41-0) 433
5-Hexen-1-ol, methanesulfonate; (64818-36-6) 433
Hydantoin synthesis, 956
Hydration, 82
Hydrazine; (10217-52-4) 874
Hydrazine hydrate: Hydrazine, monohydrate; (7803-57-8) 594
Hydriodic acid; (10034-85-2) 662
(*R*,*R*)-Hydrobenzoin: 1,2-Ethanediol, 1,2-diphenyl-, (1*R*,2*R*)-; (52340-78-0) 17
Hydroboration, 308
Hydrocinnamaldehyde: Benzenepropanal; (104-53-0) 506
Hydrofluoric acid: CORROSIVE. (9); (7664-39-3) 488
Hydrogen peroxide (H_2O_2); (7722-84-1) 93, 135, 183, 770
Hydrosilylation, 367
(1*S*,2*R*)-1-[(2-Hydroxy-3,5-di-*tert*-butylbenzylidene)amino]indan-2-ol: *1H*-Inden-2-ol, 1-[[[3,5-bis(1,1-dimethylethyl)-2-hydroxyphenyl]methylene]amino]-2,3-dihydro-, (1*S*,2*R*)-; (212378-89-7) 770
(*S*)-[1-(Hydroxydiphenylmethyl)-2-methylpropyl]carbamic acid, *tert*-butyl ester: Carbamic acid, [1-(hydroxydiphenylmethyl)-2-methylpropyl]-, 1,1-dimethylethyl ester, (*S*)-; (157035-82-0) 896
2-Hydroxy-2(*5H*)-furanone: 2(*5H*)-Furanone, 5-hydroxy-; (14032-66-7) 128
(1*S*,2*R*)-(–)-*N*-(2-Hydroxy-indan-1-yl)-2,4,6-trimethyl-benzenesulfonamide: Benzenesulfonamide, *N*-[(1*S*,2*R*)-2,3-dihydro-2-hydroxy-1*H*-inden-1-yl]-2,4,6-trimethyl-; (473554-01-7) 761
3-Hydroxy-2-iodophenyl trifluoromethanesulfonate: Methanesulfonic acid, trifluoro-, 3-hydroxy-2-iodophenyl ester (9); (514826-79-0) 488
Hydroxylamine hydrochloride: Hydroxylamine, hydrochloride; (5470-11-1) 139
Hydroxylaminolysis, 139
2-Hydroxy-2-methyl-*N*-(5-oxo-5,6,7,8-tetrahydro-2-naphthalenyl)propanamide: Propanamide, 2-hydroxy-2-methyl-*N*-(5,6,7,8-tetrahydro-5-oxo- 2-naphthalenyl)-; (418761-92-9) 670
N-HYDROXY-(*S*)-1-PHENYLETHYLAMINE OXALATE; Benzenemethanamine, *N*-hydroxy-α-methyl-, (α*S*)-, ethanedioate (1:1) salt; (78798-33-1) 139
1-HYDROXY-3-PHENYL-2-PROPANONE: 2-Propanone, 1-hydroxy-3-phenyl-; (4982-08-5) 642

Imidazole: *1H*-Imidazole; (288-32-4) 46, 615
Imidazole synthesis, 862
Imine formation, 749, 920
1*H*-Indene, 2-(2,2-dimethylpropoxy)-2,3-dihydro-2-Indanol: 1*H*-Inden-2-ol, 2,3-dihydro-; (4254-29-9) 46
Indium; (7440-74-6), 57
Indium chloride ($InCl_3$); (10025-82-8) 646, 826
Indole: *1H*-Indole; (120-72-9) 1007
Iodination, 488, 540, 794
of alkynes, 359
Iodine; (7553-56-2) 359, 384, 488
2-Iodoaniline: Benzenamine, 2-iodo-; (615-43-0) 826, 834
4-Iodoanisole: Benzene, 1-iodo-4-methoxy-; (696-62-8), 359, 367
Iodobenzene diacetate (IBD): Bis(acetato-kO)phenyliodine; (3240-34-4) 77
(*Z*)-1-Iodo-1-Heptene: 1-Heptene, 1-iodo-, (1*Z*)-(9); (63318-29-6), 359
1-Iodo-1-heptyne: 1-Heptyne, 1-iodo-; (54573-13-6), 359
Iodomethane; (74-88-4) 296
1-(2-Iodophenyl)pyrrole: 1*H*-Pyrrole, 1-(2-iodophenyl)-; (157017-41-9) 826
(*E*)-3-IODOPROP-2-ENOIC ACID: 2-Propenoic acid, 3-iodo-, (2*E*)-; (6372-02-7), 662
2-Iodoresorcinol: 1,3-Benzenediol, 2-iodo- (9); (41046-67-7) 488
2-Iodoresorcinol bis(trifluoromethanesulfonate): Methanesulfonic acid, trifluoro-, 2-iodo-1,3-phenylene ester (9); (514826-78-9) 488
N-Iodosuccinimide: 1-Iodo-2,5-Pyrrolidinedione; (516-12-1) 540, 794
Ionic liquid 859
Iridium-catalysis 631
Iron(III) nitrate nonahydrate: Nitric acid, iron(3+) salt, nonahydrate; (7782-61-8) 770
Isoamyl nitrite: isopentyl nitrite; (110-46-3) 702
(*R*)-Isobutyl 2-(*tert*-butyldimethylsilyloxy)propanoate, 615
Isocyanate formation, 778

Isoindoline, 1,3-Dihydroisoindole: 1*H*-Isoindole, 2,3-dihydro-; (496-12-8) 1037
Isomerization, 670
of alkynes to allenes, 726
of a methylcyclopropene to a methylenecyclopropane, 457
cycloisomerization, 826
Isoprene: 1,3-Butadiene, 2-methyl-; (78-79-5) 911
Isopropyl acetate: Acetic acid, 1-methylethyl ester; (108-21-4) 1001
Isopropyl (*E*)-but-2-enoate: 2-Butenoic acid, 1-methylethyl ester, (2*E*)-; (6284-46-4) 420
3,4-*O*-Isopropylidene-L-mannitol: L-Mannitol, 3,4-*O*-(1-methylethylidene)-; (153059-36-0) 164
2,3-ISOPROPYLIDENE(D-RIBONOLACTONE) 88
Isopropylmagnesium chloride: Magnesium, chloro(1-methylethyl)-; (1068-55-9) 343, 600
Isoquinoline; (119-65-3) 656

Ketal formation, 128, 177, 212, 457, 473, 586, 787, 934
Ketal hydrolysis, 164
Kulinkovich reaction, 450, 425

Lactone formation, 879, 506
Lauroyl peroxide: Peroxide, bis(1-oxododecyl); (105-74-8) 212
Lead tetraacetate: Acetic acid, lead(4+) salt; (546-67-8) 850
LIPASE-CATALYZED RESOLUTION 1056
Lithium; (7439-93-2) 399, 770
Lithium aluminum hydride; (16863-85-3) 702
Lithium borohydride: Borate(-1), tetrahydro-, lithium; (16949-15-8) 164
Lithium bis(trimethylsilyl)amide; Lithium hexamethyldisilazide: Silanamine, 1,1,1-trimethyl-*N*-(trimethylsilyl)-, lithium salt; (4039-32-1) 38, 399, 920
Lithium diisopropylamide: 2-Propanamine, *N*-(1-methylethyl)-, lithium salt; (4111-54-0) 615, 794
Lithium *N*-(3,5-dimethylphenyl)-*tert*-butylamide etherate: Lithium, [*N*-(3,5-dimethylphenyl)tricyclo[3.3.1.13,7]decan-1-aminato][1,1'-oxybis[ethane]]-; (215180-13-5) 529
Lithium iodide; (10377-51-2) 708
Lithium naphthalenide: Naphthalene, radical ion(1-), lithium; (7308-67-0) 256
Lithium tri-*tert*-butoxyaluminum hydride; (17476-04-9) 149
Lithium triethylborohydride: Borate(1-), triethylhydro-, lithium, (T-4)-; (22560-16-3) 1068

Magnesium; (7439-95-4) 296, 896
Magnesium chloride (8); (7786-30-3) 267
Manganese dioxide; (1313-13-9) 9
Magnesium oxide; (1309-48-4) 947
L-Mannonic acid δ-lactone: L-Mannonic acid, δ-lactone; (22430-23-5) 164
m-CPBA: Benzenecarboperoxoic acid, 3-chloro-; (937-14-4) 741
Meldrum's acid: 1,3-Dioxane-4,6-dione, 2,2-dimethyl-; (2033-24-1) 911
d-Menthol: Cyclohexanol, 5-methyl-2-(1-methylethyl)-, (1*S*,2*R*,5*S*)-; (15356-60-2) 128
(5*S*)-(d-MENTHYLOXY)-2(5H)-FURANONE: 2(*5H*)-Furanone, 5-[[(1*S*,2*R*,5*S*)-5-methyl-2-(1-methylethyl)cyclohexyl]oxy]-, (5*S*)-; (122079-41-8) 128
Mesitylene: 1,2,3-Trimethylbenzene; (108-67-8) 802
2-Mesitylmagnesium bromide: Magnesium, bromo(2,4,6-trimethylphenyl)-; (2633-66-1) 761
Mesyl chloride: Methanesulfonyl chloride; (124-63-0) 1056
Metathesis, 540, 552, 561, 570
Methanesulfonyl chloride; (124-63-0) 433, 708
Methanol; (67-56-1) 607
1-Methoxy-1,3-butadiene: 1,3-Butadiene, 1-methoxy-; (3036-66-6) 498
(Methoxycarbonylmethylene)triphenylphosphorane: Phosphonium, triphenyl-, 2-methoxy-2-oxoethylide; (21204-67-1) 986
N-(Methoxycarbonyl)-2-propenylamine: Carbamic acid, 2-propenyl-, methyl ester; (19364-21-7) 244
1-METHOXY-2-(4-METHOXYPHENOXY)-BENZENE, 626
1-METHOXYMETHOXY-1-PHENYLETHANE: Benzene, [1-(methoxymethoxy)ethyl]-; (94073-86-6) 934
N-Methoxy-*N*-methylcarbamoyl chloride: Carbamic chloride, methoxymethyl-; (30289-28-2) 992
[1-(METHOXYMETHYLCARBAMOYL)ETHYL]PHOSPHONIC ACID BIS-(2,2,2-TRIFLUOROETHYL) ESTER: Phosphonic acid, [2-(methoxymethylamino)-1-methyl-2-oxoethyl]-, bis(2,2,2-trifluoroethyl) ester; (448219-33-8) 992
4-Methoxyphenacyl bromide; Ethanone, 2-bromo-1-(4-methoxyphenyl)-; (2632-13-5) 862
4-Methoxyphenethyl alcohol: Benzeneethanol, 4-methoxy-; (702-23-8) 107
4-Methoxyphenol: Phenol, 4-methoxy-; (150-76-5) 626, 631
4-METHOXYPHENYLACETIC ACID: BENZENEACETIC ACID, 4-METHOXY-; (104-01-8) 107

4-(4-METHOXYPHENYL)-2-PHENYL-1H-IMIDAZOLE: *1H*-IMIDAZOLE, 4-(METHOXYPHENYL)- 2-PHENYL-; (53458-08-5) 862
trans-1-(4-METHOXYPHENYL)-4-PHENYL-3-PHENYLTHIO)AZETIDIN-2-ONE: 2-Azetidinone, 1-(4-methoxyphenyl)-4-phenyl-3-(phenylthio)-*trans*-; (94612-48-3) 920
1-Methoxy-4-vinyloxybenzene, 631
Methyl acrylate: 2-Propenoic acid, methyl ester; (96-33-3) 157, 404, 481
Methyl azidoacetate: Acetic acid, azido-, methyl ester; (1816-92-8) 802
Methyl (2*Z*)-2-azido-3-(2-chloropyridin-3-yl)acrylate: 2-Propenoic acid, 2-azido-3-(2-chloro-3-pyridinyl)-, methyl ester; (688357-18-8) 802
N-Methyl-*O*-benzoylhydroxylamine hydrochloride: Methanamine, *N*-(benzoyloxy)-, hydrochloride; (27130-46-7) 70
METHYL (*R*)-(–)-3-BENZYLOXY-2-METHYL PROPANOATE: Propanoic acid, 2-methyl-3-(phenylmethoxy)-, methyl ester, (2*R*)-; (112068-34-5) 947
Methyl bromoacetate: Acetic acid, bromo-, methyl ester; (96-32-2) 802
Methyl 2-bromomethyl-3-nitrobenzoate: Benzoic acid, 2-bromomethyl-3-nitro-, methyl ester; (98475-07-1) 903
Methyl *tert*-butyl ether; Propane, 2-methoxy-2-methyl-; (1634-04-4) 107, 626
2-Methyl-3-butyn-2-ol: 3-Butyn-2-ol, 2-methyl-; (115-19-5) 442
Methyl chloroformate: Carbonochloridic acid, methyl ester; (79-22-1) 244
METHYL 4-CHLORO-1*H*-PYRROLO[3,2-C]PYRIDINE-2-CARBOXYLATE: 1*H*-Pyrrolo[3,2-c]pyridine-2-carboxylic acid, 4-chloro-, methyl ester; (688357-19-9) 802
Methyl cyanoformate: Carbonocyanidic acid, methyl ester (9CI); (17640-15-2) 1028
(*R*)-(+)-6-Methylcyclohex-2-en-1-one: (6*R*)-6-Methyl-2-cyclohexen-1-one; (62392-84-1) 100
(+)-(1*R*,2*S*,9*S*)-11-METHYL-7,11-DIAZATRICYCLO[7.3.1.0^{2,7}]TRIDECANE: 1,5-Methano-2*H*-pyrido[1,2-*a*][1,5]diazocine, decahydro-3-methyl-, (1*R*,5*S*,11a*S*)-; (475301-86-1) 25
(*E*)-METHYL 2-DIAZO-4-PHENYLBUT-3-ENOATE: (119987-21-2) 296
N-Methyldicyclohexylamine: Cyclohexanamine, *N*-cyclohexyl-*N*-methyl-; (7560-83-0) 372
(*S*)-1-Methyl-2-[(dihydroisoindol-2-yl)methyl]pyrrolidine (159497-37-7) 1037
1-Methyl-3,4-dihydronaphthalene: Naphthalene, 1,2-dihydro-4-methyl-; (4373-13-1) 296
Methyl 2-ethenyl-3-nitrobenzoate: Benzoic acid, 2-ethenyl-3-nitro-, methyl ester; (195992-04-2) 903
(4*S*)-(1-METHYLETHYL)-5,5-DIPHENYLOXAZOLIDIN-2-ONE: 2-Oxazolidinone, 4-(1-methylethyl)-5,5-diphenyl-, (4*S*)-; (184346-45-0) 896
(η^5-(*S*)-2-(4-METHYLETHYL)OXAZOLINYLCYCLO-PENTADIENYL)-(η^4-TETRAPHENYLCYCLOBUTADIENE) COBALT: Cobalt, [1,1',1'',1'''-(η4-1,3-cyclobutadiene-1,2,3,4-tetrayl)tetrakis[benzene]][(1,2,3,4,5-η)-1-[[[(1*S*)-1-(hydroxymethyl)-2-methylpropyl]amino]carbonyl]-2,4-cyclopentadien-1-yl]-; (222400-02-4) 708, 714
(±)-Methyl glycidate: Oxiranecarboxylic acid, methyl ester; (4538-50-5) 157
(*S*)-METHYL GLYCIDATE: Oxiranecarboxylic acid, methyl ester, (2*S*)-; (118712-39-3) 157
Methyl 5-hexenoate: 5-Hexenoic acid, methyl ester; (2396-80-7) 450
N-Methylhydroxylamine hydrochloride: Methanamine, *N*-hydroxy-, hydrochloride; (4229-44-1) 70
Methyl (*R*)-(–)-3-hydroxy-2-methyl propionate; Propanoic acid, 3-hydroxy-2-methyl-, methyl ester, (2*R*)-; (72657-23-9) 947
N-Methylimidazole: *1H*-lmidazole, 1-methyl-; (616-47-7) 635
METHYL INDOLE-4-CARBOXYLATE: *1H*-Indole-4-carboxylic acid, methyl ester; (39830-66-5) 903
Methyl iodide: Methane, iodo-; (74-88-4) 457, 896
Methyl 4-iodobenzoate: Benzoic acid, 4-iodo-, methyl ester; (619-44-3) 384
Methyllithium: Lithium, methyl-; (917-54-4) 425
Methyllithium-lithium bromide; (332360-06-2) 100
Methylmagnesium bromide: Magnesium, bromomethyl-; (75-16-1) 1020
(*S,E*)-METHYL 2-((*R*)-4-METHYL-1,2-DIHYDRONAPHTHALEN-2-YL)-4-PHENYLBUT-3-ENOATE: 2-Naphthaleneacetic acid, 1,2-dihydro-4-methyl-α-[(1E)-2-phenylethenyl]-, methyl ester, (αS,2R)-; (76531-50-5) 296
Methyl 2-methyl-3-nitrobenzoate: Benzoic acid, 2-methyl-3-nitro-, methyl ester; (59382-59-1) 903
METHYL 5-METHYLPYRIDINE-2-CARBOXYLATE: 2-Pyridinecarboxylic acid, 5-methyl-, methyl ester; (260998-85-4) 911

(–)-Methyl (1*R*,9*R*)-6-Oxo-7,11-diazatricyclo[7.3.1.0^{2,7}]trideca-2,4-diene-11-carboxylate: 1,5-Methano-2*H*-pyrido[1,2-*a*][1,5]diazocine-3(4*H*)-carboxylic acid, 1,5,6,8-tetrahydro-8-oxo-, methyl ester, (1*R*,5*R*)-; (125109-97-9) 25
Methyl 2-oxo-1,2,3,4-tetrahydronaphthalene-1-carboxylate: 1-Naphthalenecarboxylic acid, 1,2,3,4-tetrahydro-2-oxo-, methyl ester; (31202-23-0) 811
2-Methyl-2-[(5-oxo-5,6,7,8-tetrahydro-2-naphthalenyl)oxy]propanamide: Propanamide, 2-methyl-2-[(5,6,7,8-tetrahydro-5-oxo-2- naphthalenyl)oxy]-; (418761-91-8) 670
(*E*)-2-METHYL-3-PHENYLACRYLIC ACID BUTYL ESTER: 2-PROPENOIC ACID, 2-METHYL-3-PHENYL-, BUTYL ESTER, (2*E*)-; (21511-00-5) 372
N-[(4-Methylphenyl)sulfonyl(phenyl)methyl] *P,P*-diphenylphosphinic amide; (701291-86-3) 749
METHYL PHENYL SULFOXIDE: Benzene, (methylsulfinyl)-; (1193-82-4) 135
(R_S)-(+)-2-METHYL-2-PROPANESULFINAMIDE [*TERT*-BUTANESULFINAMIDE]: 2-Propanesulfinamide, 2-Methyl-, [*S*(*R*)]-; (196929-78-9) 770
N-Methylpyrrolidinone: 2-Pyrrolidinone, 1-methyl-; (872-50-4) 353, 626
4-METHYLPYRROLO[1,2-*a*]QUINOLINE; (796843-24-8) 826
trans-β-Methylstyrene: Benzene, (1*E*)-1-propenyl-; (873-66-5) 183
(*R,R*)-*trans*-β-METHYLSTYRENE OXIDE: Oxirane, 2-methyl-3-phenyl-, (2*R*,3*R*)-; (14212-54-5) 183
Methyl trifluoromethanesulfonate: Methyl trifluoromethanesulfonate; (333-27-7) 947
Methyltrioxorhenium: Rhenium, methyltrioxo-, (T-4); (70197-13-6) 114
Methyl(triphenylphosphine)gold (CH_3AuPPh_3); (23108-72-7) 82
Methyl tris(isopropoxy)titanium: Titanium, methyltris2-propanolato)-, (T-4)-; (18006-13-8) 425
Methyl vinyl ketone: 3-Buten-2-one; (78-94-4) 416
Mitsunobu reaction, 164
Michael addition, 404
Michael addition, of aldehydes to vinyl ketones, 416
Molybdenum(V) chloride; (10241-05-1) 529
Morpholine; (110-91-8) 197
Morpholinone glycine synthon, 399
    reductive removal of auxiliary, 399
1-Morpholinopropan-1-one: 4-(1-oxopropyl)-morpholine; (30668-14-5) 197
(2*S*)-(–)-3-exo-(MORPHOLINO)ISOBORNEOL [(–)-MIB] 702

Natural product isolation 25
Negishi coupling, 384
(*S*)-(+)-Neomenthyldiphenylphosphine: Phosphine, [(1*S*,2*S*,5*R*)-5-methyl-2-(1-methylethyl)cyclohexyl]diphenyl-; (43077-29-8) 205
Neopentyl glycol: 1,3-Propanediol, 2,2-dimethyl-; (126-30-7) 457, 1015
Nickel(II) acetylacetonate: Bis(2,4-pentanedionate), nickel(II); (3264-82-2) 327
Ni(cod)$_2$: Nickel, bis[(1,2,5,6-η)-1,5-cyclooctadiene]-; (1295-35-8) 205
Nitration, 656
Nitrone synthesis, 114
p-Nitrophenol: 4-Nitrophenol; (100-02-7) 529, 540
1,8-Nonadiyne; (2396-65-8) 82
2,8-NONANEDIONE; (30502-73-9) 82
4-NONYLBENZOIC ACID: Benzoic acid, 4-nonyl-; (38289-46-2) 353
(Nonylmagnesium bromide: Magnesium, bromononyl-; (39691-62-8) 353
(Norbornadiene)rhodium chloride dimer; Rhodium, bis[(2,3,5,6-η)-bicyclo[2.2.1]hepta-2,5-diene]di-μ-chlorodi-; (12257-42-0) 1020

Octanoic acid; (124-07-2) 281
1-Octyne; (629-05-0) 234
Organolithium cyclization, 433
Orthoformylation, 267
Oxalic acid: Ethanedioic acid; (144-62-7) 139
Oxalyl Chloride: Ethanedioyl dichloride; (79-37-8) 708
5-OXASPIRO[3.4]OCTAN-1-ONE; (881389-73-7) 256
Oxazolidinone synthesis, 867
Oxazoline formation, 888
Oxidation, of alcohols to carboxylic acids, 107
    of alcohols to aldehydes, 77
    of alcohols to ketones, 9, 177, 794
    of amines, 114, 139, 684
    of phenols, 514
    of phosphines, 982
    of sulfur compounds, 135, 741
    photolytic, 128
Oximation, 911
Oxone: Peroxymonosulfuric acid, monopotassium salt, mixture with dipotassium sulfate and potassium hydrogen sulfate; (37222-66-5) 183
Oxygen; (7782-44-7) 128

Pd(dba)$_2$: Palladium, bis[(1,2,4,5-η)-1,5-diphenyl-1,4-pentadien-3-one]- (9); (32005-36-0) 359, 367
Palladium (II) acetate: Acetic acid, palladium(2+) salt; (3375-31-3) 289, 393, 665, 714, 903
Palladium chloride: Palladium chloride ($PdCl_2$); (7647-10-1) 367
Palladium, dichlorobis(triphenylphosphine)- ($(PPh_3)_2PdCl_2$); (13965-03-2) 826
Palladium hydroxide; (12135-22-7) 281
Paraformaldehyde; (30525-89-4) 267, 561, 615, 903
PAUSON-KHAND REACTION, 442
6,9,12,15,18-PENTAMETHYL-1,6,9,12,15,18-HEXAHYDRO(C_{60}-I_h)[5,6]FULLERENE, 319
1,5-Pentanediol; (111-29-5) 697
Perchloric acid; (7601-90-3), 177
Phase transfer catalyst for asymmetric alkylation, 404
Phase transfer catalysis, 607
9,10-Phenanthrenequinone; (84-11-7) 519
1,10-Phenanthroline; (66-71-7) 677
α-Phenethyl alcohol: Benzenemethanol, α-methyl-; (98-85-1) 934
Phenethylmagnesium chloride: Magnesium, chloro(2-phenylethyl)-; (90878-19-6) 289
4-Phenyl-3-buten-2-one; (122-57-6) 336
1-Phenyl-1-butyne: 1-Butynylbenzene; (622-76-4) 205, 529
4-Phenylbutyric acid: Benzenebutanoic acid; (1821-12-1) 733
N-Phenylcarbamic acid methyl ester: Carbamic acid, phenyl-methyl ester; (2603-10-3) 867
1-Phenylcyclohexene: Benzene, 1-cyclohexen-1-yl-; (771-98-2) 183
(*R,R*)-1-PHENYLCYCLOHEXENE OXIDE: 7-Oxabicyclo[4.1.0]heptane, 1-phenyl-, (1*R*,6*R*)-; (17540-04-4) 183
4-PHENYL-5(*E*)-DECEN-2-ONE; (42521-58-4) 261
S-Phenyl diazothioacetate: Ethanethioic acid, diazo-, *S*-phenyl ester; (72228-26-3) 920
Phenyldimethylsilane: Silane, dimethylphenyl-; (766-77-8) 1
(*E*)-4-(2-PHENYLETHENYL)BENZONITRILE: BENZONITRILE, 4-[(1*E*)-2-PHENYLETHENYL]- (9); (13041-79-7) 372
(*S*)-1-Phenylethylamine: Benzenemethanamine, α-methyl-, αS)-; (2627-86-3) 139
(*R*)-(+)-Phenylethylamine: Benzenemethanamine, α-methyl-, (αR)-; (3886-69-9) 1028
(*S*)-[(1-Phenylethyl)amino]acetonitrile; Acetonitrile, [[(1*S*)-1-phenylethyl]amino]-; (35341-76-5) 139
[(1*S*)-1-Phenylethyl]imino]acetonitrile *N*-oxide; Acetonitrile, [oxido[(1*S*)-1-phenylethyl]imino]-; (300843-73-6) 139
(4*S*)-4-(2-PHENYLETHYL)-2-OXETANONE: (4*S*)-4-(2-phenylethyl)-2-oxetanone; (214853-90-4) 506
(*R*)-(−)-2-Phenylglycinol: Benzeneethanol, β-amino-, (β*R*)-; (56613-80-0) 888
1-PHENYLHEX-5-EN-3-AMINE: Benzenepropanamine, α-2-propenyl-; (17125-07-4) 691
N-PHENYL-5*R*-HYDROXYMETHYL-2-OXAZOLIDINONE: 2-Oxazolidinone, 5-(hydroxymethyl)-3-phenyl-, (5*R*); (875080-42-7) 867
Phenylmagnesium bromide: Magnesium, bromophenyl-; (100-58-3) 684, 896
Phenylmagnesium chloride: Magnesium, chlorophenyl-; (100-59-4) 1028
N-[(Phenyl)methylene]-*P,P*-diphenylphosphinic amide: Phosphinic amide, *P,P*-diphenyl-*N*-(phenylmethylene)-; (67764-52-7) 749
4-PHENYLMORPHOLINE: (92-53-5) 684
(*R*)-4-Phenyl-2-oxazolidinone: 2-Oxazolidinone, 4-phenyl-, (4*R*)-; (90319-52-1) 677, 726
5-PHENYLPENT-2-YN-1-OL: 2-Pentyn-1-ol, 5-phenyl-; (16900-77-9) 561
R-4-PHENYL-3-(1,2-PROPADIENY)-2-OXAZOLIDINONE: 2-OXAZOLIDINONE, 4-PHENYL-3-(1,2-PROPADIENYL)-, (4*R*)-; (256382-50-0) 726
(1*S*)-1-Phenylpropan-1-amine hydrochloride: Benzenemethanamine, α-ethyl-, hydrochloride, (α*S*)-: (19146-52-2) 749
3-Phenylpropionaldehyde: Benzenepropanal; (104-53-0) 561, 691
N-[(*1S*)-1-Phenylpropyl]-*P,P*-diphenylphosphinic amide: Phosphinic amide, *P,P*-diphenyl-*N*-[(*1S*)-1-phenylpropyl]-; (106651-15-4) 749
3-Phenyl-2-propylthio-2-propen-1-ol, 642
3-Phenyl-2-propyn-1-ol: 2-Propyn-1-ol, 3-phenyl-; (1504-58-1) 642
R-4-Phenyl-3-(2-propynyl)-2-oxazolidinone: 2-Oxazolidinone, 4-phenyl-3-(2-propynyl); (4*R*)-; (256382-74-8) 726
S-Phenyl thioacetate: Ethanethioic acid, *S*-phenyl ester; (934-87-2) 920
2-PHENYLTHIO-5-HEPTANOL: 3-Heptanol, 6-(phenylthio); (198778-75-5) 121
4-PHENYL-3-(TRIFLUOROMETHYL)BUTAN-2-ONE: 2-Butanone, 4,4,4-trifluoro-3-(phenylmethyl)-; (808105-43-3) 336
(*R*)-4-PHENYL-3-(2-TRIISOPROPYLSILYL-ETHYNYL) OXAZOLIDIN-2-ONE: [2-Oxazolidinone, 4-phenyl-3-(2-triisopropylsilyl-ethynyl)-, (4*R*)-];

(503590-28-1) 677
Phosphorus oxychloride: Phophoric trichloride; (10025-87-3) 929
Phosphorus pentasulfide; (1314-80-3) 787
Photolysis, 121, 1046
Pinacol: 2,3-Butanediol, 2,3-dimethyl-; (76-09-5) 393
Pinacolborane: 4,4,5,5-Tetramethyl-1,3,2-dioxaborolane; (25015-63-8) 1007
Pinacolic coupling, 348
(+)-α-Pinene: (1*R*)-2,6,6-Trimethyl-bicyclo[3.1.1]hept-2-ene; (7785-70-8) 308
Piperidine; (110-89-4) 540
Piperidine-4-spiro-5'-hydantoin: 1,3,8-Triazaspiro [4.5]decane-2,4-dione; (13625-39-3) 956
4-Piperidone monohydrate hydrochloride: 4-Piperidinone, hydrochloride; (41979-39-9) 956
Pivaloyl chloride: Propanoyl chloride, 2,2-dimethyl-; (3282-30-2) 169, 874
Platinum(0)-1,3-divinyl-1,1,3,3-tetramethyldisiloxane complex: Platinum, 1,3-diethenyl-1,1,3,3-tetramethyldisiloxane complex; (68478-92-2) 367
cis-Polybutadiene; (40022-03-5) 552
Potassium *tert*-butoxide: 2-Propanol, 2-methyl-, potassium salt; (865-47-4) 164, 296, 457, 702, 896
Potassium cyanide: Potassium cyanide [K(CN)]; (151-50-8) 956
Potassium *O*-ethyl xanthate: Carbonodithioic acid, *O*-ethyl ester, potassium salt; (140-89-6) 212
Potassium hexamethyldisilazide: Silanamine, 1,1,1-trimethyl-*N*-(trimethylsilyl)-, potassium salt; (40949-94-8) 244
Potassium hydrogen fluoride: (7789-29-9) 289
Potassium hydroxide; (1310-58-3) 62, 607, 1020
Potassium nitrate:; (7757-79-1) 656
Potassium 2-phenethyltrifluoroborate: Borate(1-), trifluoro(2-phenylethyl)-, potassium, (T-4)-: (329976-74-1) 289
Propanethiol; (79869-58-2) 642
Propanoic acid, 2-hydroxy-, ethyl ester, (2*S*)-; (687-47-8) 615
2-Propanol; (67-63-0) 17
Propanoyl chloride; (79-03-8) 197
Propargyl alcohol: 2-Propyn-1-ol; (107-19-7) 552
Propargyl benzoate: 2-Propyn-1-ol, benzoate; (6750-04-5) 552
Propargyl bromide: 1-Propyne, 3-bromo-; (106-96-7) 726
Propargylic alcohol synthesis, 561, 615
N-(2-Propenyl-4-methylbenzenesulfonamide: Benzenesulfonamide, 4-methyl-*N*-2-propenyl-; (50487-71-3) 442
N-(2-Propenyl)-*N*-(2-propynyl)-4-methylbenzenesulfonamide: Benzenesulfonamide, 4-methyl-*N*-2-propenyl-*N*-2-propynyl-; (133886-40-5) 442
Propiolic acid: 2-Propynoic acid; (471-25-0) 662
Propyl formate: Formic acid, propyl ester; (110-74-7) 778
1-(2-(1-Propynyl)phenyl)pyrrole: *1H*-Pyrrole, 1-[2-(1-propynyl)phenyl]-; (796843-21-5) 826
Pyridine; (110-86-1) 52, 442, 561, 992
Pyridinium chlorochromate: Chromate(1-), chlorotrioxo-, (T-4)-, hydrogen compound with pyridine (1:1); (26299-14-9) 100, 177
4-Pyridinamine, *N,N*-dimethyl- (DMAP); (1122-58-3) 552
3-PYRIDIN-3-YLQUINOLINE: Quinoline, 3-(3-pyridinyl)-; (96546-80-4) 393
3-PYRIDYLBORONIC ACID: Boronic acid, 3-pyridinyl-; (1692-25-7) 393
Pyrophosphate synthesis, 1001

(*S*)-Quinap: Isoquinoline, 1-[2-(diphenylphosphino)-1-naphthalenyl]-, (1*S*)-; (149341-33-3) 192
Quinidine: Cinchonan-9-ol, 6'-methoxy-, (9*S*)-; (56-54-2) 1068
Quinine: Cinchonan-9-ol, 6'-methoxy-, (8a,9*R*)-; (130-95-0) 1068

Radical cation, 62
Reduction, of alkynes, 359
of aromatic nitro compounds, 57
of carbamates ,25
of amides, 1037
of esters to aldehydes, 615
of esters to alcohols, 221
of ketones, 9, 149
of phosphorus, 965
of quinones, 519
Reductive alkylation, 46
Reductive amination, 498
Reductive decarboxylation, 843
Reductive deuteration, 600
Regioselective alkylation of aromatic ketones, 410
Remote functionalization, 121
Resolution, 157, 1037, 1046
Resorcinol: 1,3-Benzenediol (9); (108-46-3) 488
Rhodium(II) acetate dimer: Acetic acid, rhodium(2+) salt; (5503-41-3) 920
Rhodium(1+), [(2,3,5,6-η)-bicyclo[2.2.1]hepta-2,5-diene]bis(methyldi-phenylphosphine)-, tetrafluoroborate(1-); (34664-31-8) 1020
Rhodium, chlorotris(triphenylphosphine)-, ($RhCl(PPh_3)_3$) (SP-4-2)-; (14694-95-2) 336
Rh_2(*S*-DOSP)$_4$; (179162-34-6) 296

D-RIBONOLACTONE 88
D-Ribonolactone: D-Ribonic acid, γ-lactone; (5336-08-3) 88
D-Ribose; (50-69-1) 88
Rochelle salt: Butanedioic acid, 2,3-dihydroxy-(*2R,3R*)-, monopotassium monosodium salt; (304-59-6) 615
Rose Bengal; (11121-48-5) 128
Ruthenium, dodecacarbonyltri-, triangulo; (15243-33-1) 1
Ruthenium-mediated reduction, 17
Ruthenium [1,3-bis-(2,4,6-trimethylphenyl)-2-imidazolidinylidine]-dichloro(phenylmethylene)(tricyclohexylphosphine); (246047-72-3) 552
RuCl[(1*S*,2*S*)-*p*-TsNCH(C_6H_5)CH(C_6H_5)NH_2](η^6-*p*-cymene): Ruthenium, [*N*-[(*1S,2S*)-2-(amino-κ*N*)-1,2-diphenylethyl]-4-methyl-benzenesulfonamidato-κ*N*]chloro-[(1,2,3,4,5,6-η)-1-methyl-4-(1-methylethyl)benzene]-; (192139-90-5) 9, 17

Salicyl alcohol: Benzenemethanol, 2-hydroxy-; (90-01-7) 514
Saponification 670, 843
Schwartz's reagent: Zirconium, chlorobis(η^5-2,4-cyclopentadien-1-yl)hydro- (9); (37342-97-5) 261
Silane, [(2,3-dihydro-1*H*-inden-2-yl)oxy](1,1-dimethylethyl)dimethyl-; (216884-03-6) 46
Silver carbonate: Carbonic acid, disilver(1+) salt; (534-16-7) 88
Silver trifluoromethanesulfonate: Methanesulfonic acid, trifluoro-, silver(1+) salt; (2923-28-6) 272
Silylation of alcohols, 46, 615
Sodium; (7440-23-5) 457
Sodium amide: Sodium amide ($NaNH_2$); (7782-92-5) 457, 986
Sodium azide; (26628-22-8) 347, 802
Sodium benzophenone ketyl: Benzophenone, radical ion (1-), sodium (8); Methane, diphenyl-, radical ion (1-), sodium (9); (3463-17-0) 261
Sodium bicarbonate: Carbonic acid monosodium salt; (144-55-8) 88
Sodium bisulfite: Sulfurous acid, monosodium salt; (7631-90-5) 75, 652
Sodium *tert*-butoxide: 2-Propanol, 2-methyl-, sodium salt; (865-48-5) 529
Sodium carbonate: Carbonic acid disodium salt; (497-19-8) 631
Sodium chlorite: Chlorous acid, sodium salt; (7758-19-2) 107
Sodium cyclopentadienide: Sodium, 2,4-cyclopentadien-1-yl-; (4984-82-1) 708
Sodium dithionite: Dithionous acid, disodium salt; (7775-14-6) 519
Sodium fluoride: Sodium fluoride (NaF); (7681-49-4) 469
Sodium hexafluoroacetylacetonate: 2,4-Pentanedione, 1,1,1,5,5,5-hexafluoro-, ion(1-), sodium; (22466-49-5) 714
Sodium hydride: Sodium hydride (NaH); (7646-69-7) 540, 726, 811
Sodium hydrogen sulfite: Sulfurous acid, monosodium salt: (7631-90-5) 62
Sodium hypochlorite: Hypochlorous acid, sodium salt (8,9); (7681-52-9), 107; 93
Sodium iodide: Sodium iodide (NaI); (7681-82-5) 384, 433, 811
Sodium metaperiodate: Periodic acid (HIO_4), sodium salt; (7790-28-5) 514
Sodium methoxide: Methanol, sodium salt; (124-41-4) 911
Sodium nitrite: Nitrous acid, sodium salt; (7632-00-0) 911
Sodium tetrafluoroborate; (13755-29-8) 859
Sodium thiosulfate: Thiosulfuric acid ($H_2S_2O_3$), disodium salt; (7772-98-7) 384, 770
Solid supports, 635
(–)-Sparteine: 7,14-Methano-2*H*,6*H*-dipyrido[1,2-*a*:1',2'-*e*][1,5]diazocine, dodecahydro-, (7*S*,7a*R*,14*S*,14a*S*)-; (90-39-1) 1020
9-SPIROEPOXY-*endo*-TRICYCLO[5.2.2.$0^{2,6}$]UNDECA-4,10-DIEN-8-ONE: Spiro[4,7-ethano-*1H*-indene-8,2'-oxiran]-9-one, 3*a*,4,7,7*a*-tetrahydro-; (146924-02-9), 514
Styrene: Benzene, ethenyl-; (100-42-5) 272, 372
Sulfenate synthesis, 121
Sulfide synthesis, 121
Sulfuryl chloride; (7791-25-5) 121
Suzuki synthesis, 315, 393

TETRABENZYL PYROPHOSPHATE: Diphosphoric acid, tetrakis(phenylmethyl) ester; (990-91-0) 1001
Tetrabutylammonium bromide: 1-Butanaminium, *N,N,N*-tributyl-, bromide; (1643-19-2) 519
Tetrabutylammonium fluoride: 1-Butanaminium, *N,N,N*-tributyl-, fluoride; (429-41-4) 192, 834
Tetrabutylammonium fluoride trihydrate: 1-Butanaminium, *N,N,N*-tributyl, fluoride, trihydrate; (87749-50-6) 359, 367
Tetrabutylammonium hydrogen sulfate: 1-Butanaminium, *N,N,N*-tributyl-, sulfate (1:1); (32503-27-8) 183
Tetrabromothiophene; (3958-03-0) 843
Tetrachloromethane: Methane, tetrachloro-; (56-23-5) 778
9-Tetradecylcarbazole: 9-Tetradecyl-9*H*-carbazole; (20863-25-6) 540

Tetrafluoroboric acid: Borate(1-), tetrafluoro-, hydrogen (8,9); (16872-11-0) 93
2,3,11,11a-TETRAHYDRO-1*H*-BENZO[d]PYRROLO[1,2-a]AZEPIN-5(6*H*)-ONE: 5*H*-Pyrrolo[2,1-b][3]benzazepin-5-one, 1,2,3,6,11,11a-hexahydro-; (137125-37-2) 811
2,3,3α,4-TETRAHYDRO-2-[(4-METHYLBENZENE)SULFONYL]CYCLOPENTA-[C]PYRROL-5(*1H*)-ONE: Cyclopenta[*b*]pyrrol-5(1*H*)-one, 2,3,3*a*,4-tetrahydro-1-[(4-methylphenyl)sulfonyl]-; (205885-50-3) 442
Tetrakis(hydroxymethyl)phosphonium sulfate ("Pyroset TKOW"): Phosphonium, tetrakis(hydroxymethyl)-, sulfate (2:1); (55566-30-8) 570
Tetrakis(triphenylphosphine)palladium(0); (14221-01-3) 315
1-Tetralone: 1(2*H*)-Naphthalenone, 3,4-dihydro-; (529-34-0) 296, 410
β-Tetralone: 2(1*H*)-Naphthalenone, 3,4-dihydro-; (530-93-8) 811
1,1,4,4-Tetramethoxybut-2-yne: 2-Butyne, 1,1,4,4-tetramethoxy-; (53281-53-1) 607
2-(4,4,5,5,-TETRAMETHYL-1,3,2-DIOXABOROLAN-2-YL)INDOLE: 2-(4,4,5,5-Tetramethyl-1,3,2-dioxaborolan-2-yl)-*1H*-indole; (476004-81-6) 1007
3-(4,4,5,5-Tetramethyl-[1,3,2]dioxaborolan-2-yl)-pyridine: Pyridine, 3-(4,4,5,5-tetramethyl-1,3,2-dioxaborolan-2-yl)-; (329214-79-1) 393
1,1,3,3,-Tetramethyldisiloxane: Disiloxane, 1,1,3,3,-tetramethyl-; (3277-26-7) 367
N,N,N',N'-Tetramethylethylenediamine: 1,2-Ethanediamine, *N,N,N',N'*-tetramethyl-; (110-18-9) 888
2,2,6,6-Tetramethylheptane-3,5-dione: 3,5-Heptanedione, 2,2,6,6-tetramethyl-: (1118-71-4) 626
2,2,6,6-Tetramethylpiperidine; Piperidine, 2,2,6,6-tetramethyl-; (768-66-1) 1015
2,2,6,6-Tetramethyl-1-piperidinyloxy (TEMPO): 1-Piperidinyoxy, 2,2,6,6-tetramethyl-; (2564-83-2) 77, 107
"Thia-Wolff" rearrangement, 920
Thioanisole: Benzene, (methylthio)-; (100-68-5) 135
Thiophenol: Benzenethiol; (108-98-5) 121
Thionyl chloride; (7719-09-7) 888; 212
Titanium tetrachloride: Titanium chloride ($TiCl_4$)(T-4); (7550-45-0) 425
$Ti(OEt)_4$: Ethanol, titanium(4+) salt; (3087-36-3) 741
Titanium tetraisoproproxide: 2-Propanol, titanium (4+)salt; (546-68-9) 425, 600
Toluene: Benzene, methyl-; (108-88-3) 631
N-p-Toluenesulfinamide: Benzenesulfinamide, 4-methyl-; (6873-55-8) 741
p-Toluenesulfinic acid: Benzenesulfinic acid, 4-methyl-; (536-57-2) 749
N-p-Tolylsulfinyl-(*E*)-1-phenylethylideneimine: Benzenesulfinamide, 4-methyl-*N*-(1-phenylethylidene)-, [*N*(*E*)]-; (177468-94-9) 741
p-Toluenesulfonyl chloride: Benzenesulfonyl chloride, 4-methyl-; (98-59-9) 384, 442, 615, 911
(1*S*,2*S*)-(+)-*N-p*-Toluenesulfonyl-1,2-diphenylethylenediamine: Benzenesulfonamide, *N,N'*-(1,2-diphenyl-1,2-ethanediyl)bis[4-methyl-, [*S*-(*R**,*R**)]-; (170709-41-8) 9
N-p-TOLYLSULFONYL-(*E*)-1-PHENYLETHYLIDENEIMINE: Benzenesulfonamide, 4-methyl-*N*-(1-phenylethylidene)-, (*E*)-; (163586-87-6) 741
5-(Tosyloxyimino)-2,2-dimethyl-1,3-dioxane-4,6-dione: 1,3-Dioxane-4,5,6-trione, 2,2-dimethyl-, 5-O-[(4-methylphenyl)sulfonyl]oxime; (215436-24-1) 911
3,4,6-Tri-*O*-benzyl-D-glucal: D-*arabino*-Hex-1-enitol, 1,5-anhydro-2-deoxy-3,4,6-tris-*O*-(phenylmethyl)-; (55628-54-1) 169
3,4,6-TRI-*O*-BENZYL-2-*O*-PIVALOYL-β-D-GLUCOPYRANOSYL-(1→6)-1,2:3,4-DI-*O*-ISOPROPYLIDENE-α-D-GALACTOPYRANOSIDE: α-D-Galactopyranose, 6-*O*-[2-*O*-(2,2-dimethyl-1-oxopropyl)-3,4,6-tris-*O*-(phenylmethyl)-β-D-glucopyranosyl}-1,2:3,4-bis-*O*-(1-methylethylidene)-; [219122-26-6] 169
Tri-*tert*-butylphosphiine: Phosphine, tris(1,1-dimethylethyl)-; (13716-12-6) 367, 372
(P)-1-Tributylstannyl-1,2-butadiene: Stannane, 1,2-butadienyltributyl-; (202119-26-4) 1056
Tributyltin hydride: Stannane, tributyl-; (688-73-3) 1056
(*E*)-2,2,2-Trichloroacetimidic acid hex-2-enyl ester: Ethanimidic acid, 2,2,2-trichloro-, (2*E*)-2-hexenyl ester; (51479-70-0) 720
Trichloroacetonitrile; (545-06-2) 720
(*S*)-2,2,2-TRICHLORO-*N*-(1-PROPYLALLYL)ACETAMIDE: Acetamide, 2,2,2-trichloro-*N*-[(1*S*)-1-ethenylbutyl]-; (611182-43-5) 720
8-[2-(TRIETHOXYSILYL)ETHYL-1-TETRALONE: 1(*2H*)-Naphthalenone, 3,4-dihydro-8-[2-(triethoxysilyl)ethyl]-; (154735-94-1) 410
Triethoxyvinylsilane: Silane, ethenyltriethoxy-; (78-08-0) 410
Triethylamine; Ethanamine, *N,N*-diethyl-; (121-44-8) 17, 87, 121, 267, 384, 433, 506,

524, 552, 570, 607, 708, 778, 879, 888, 903, 920, 956, 992, 1056
Triethylborane (1.0M solution in hexanes): Borane, triethyl; (97-94-9) 205, 327
Triethylsilane: Silane, triethyl-; (617-86-7) 46, 442
Trifluoroacetic acid; (76-05-1) 811
Trifluoroethanol: Ethanol, 2,2,2-trifluoro-; (75-89-8) 135, 992
2,2,2-Trifluoroethyl *p*-toluenesulfonate: Ethanol, 2,2,2-trifluoro-, 4-methylbenzenesulfonate; (433-06-7) 834
2,2,2-Trifluoroethyl trifluoroacetate: Acetic acid, trifluoro-, 2,2,2-trifluoroethyl ester; (407-38-5) 920
N-(2,2,2-Trifluoro-1-hydroxyethyl)-acetamide; (6776-45-0) 212
Trifluoroiodomethane; (2314-97-8) 336
Trifluoromethylsulfonic acid: Methanesulfonic acid, trifluoro-; (1493-13-6) 1020
Trifluoromethanesulfonic anhydride; (358-23-6) 488, 506
Trifluoromethanesulfonimide: Methanesulfonamide, 1,1,1-trifluoro-*N*-phenyl-*N*-[(trifluoromethyl)sulfonyl]-; (37595-74-7) 481
2,2,2-Trifluoro-1-methoxyethanol; (431-46-9) 212
(*S*)-*N*-Trifluoromethylsulfonyl-2-isopropylaziridine: (*S*)-2-(1-Methylethyl)-1-[(trifluoromethyl)sulfonyl]-aziridine; (196520-85-1) 506
Triisopropyl borate: Boric acid (H_3BO_3), tris(1-methylethyl) ester; (5419-55-6) 393, 1015
1,2:3,4:5,6-Tri-*O*-isopropylidene-L-mannitol: L-Mannitol,1,2:3,4:5,6-tris-*O*-(1-methylethylidene)-; (153059-35-9) 164
Triisopropylsilyl acetylene: Silane, ethynyltris(1-methylethyl)-; (89343-06-6) 9, 677
(*rac*)-4-Triisopropylsilyl-3-butyn-2-ol: 3-Butyn-2-ol, 4-[tris(1-methylethyl)silyl]-: (726202-65-9) 9
(*S*)-4-TRIISOPROPYLSILYL-3-BUTYN-2-OL: (*S*)-3-Butyn-2-ol, 4-[tris(1-methylethyl)silyl]-; (901126-37-2) 9
4-Triisopropylsilyl-3-butyn-2-one: 3-Butyn-2-one, 4-[tris(1-methylethyl)silyl]-: (183852-48-4) 9
Triisopropylsilyl triflate: Methanesulfonic acid, trifluoro-, tris(1-methylethyl)silyl ester; (80522-42-5) 524
Trimethylacetaldehyde: Propanal, 2,2-dimethyl-; (630-19-3) 46
Trimethylamine hydrochloride: Methanamine, *N,N*-dimethyl-, hydrochloride; (593-81-7) 384
Trimethylaluminum; (75-24-1) 506
(*S*)-(+)-2,4,6-TRIMETHYLBENZENESULFINAMIDE: Benzenesulfinamide, 2,4,6-trimethyl-, [$S_{(S)}$]-; (607729-50-0) 761
(R_S,1*S*,2*R*)-(−)-2,4,6-Trimethylbenzenesulfinic acid 1-(2,4,6-trimethylbenzenesulfonylamino)-2-indan-2-yl ester: Benzenesulfinic acid, 2,4,6-trimethyl-, (1*S*,2*R*)-2,3-dihydro-1-[[(2,4,6-trimethylphenyl)sulfonyl]amino]-1*H*-inden-2-yl ester, [$S_{(S)}$]-; (607729-49-7) 761
2,4,6-Trimethylbenzenesulfonyl chloride; (773-64-8) 761
3-(2,4,6-Trimethylbenzenesulfonyl)-3,3*a*,8,8*a*-tetrahydro-2*H*-1-oxa-2λ^4-thia-3-aza-cyclopenta[*a*]inden-2-ol: Indeno[1,2-*d*]-1,2,3-oxathiazole, 3,3*a*,8,8*a*-tetrahydro-3-[(2,4,6-trimethylphenyl)sulfonyl]-, 2-oxide, (2*R*,3*aS*,8*aR*)-; (473554-02-8) 761
Trimethyl borate; (121-43-7) 289
1,6,6-Trimethyl-4,8-dioxaspiro[2.5]oct-1-ene: 4,8-Dioxaspiro[2.5]oct-1-ene, 1,6,6-trimethyl-; (122762-81-6) 457
Trimethylsilylacetylene: Silane, ethynyltrimethyl-; (1066-54-2) 192, 540
(*R*)-4-Trimethylsilyl-3-butyn-2-yl acetate: 3-Butyn-2-ol, 4-(trimethylsilyl)-, acetate; (129571-78-4) 1056
(*R*)-4-Trimethylsilyl-3-butyn-2-yl mesylate: 3-Butyn-2-ol, 4-(trimethylsilyl)-, methanesulfonate, (2*R*)-; (200440-90-0) 1056
(*S*)-4-Trimethylsilyl-3-butyn-2-yl succinate: Butanedioic acid, mono [(1*S*)-1-methyl-3-(trimethylsilyl)-2-propynyl] ester; (375395-73-6) 1056
4-Trimethylsilyl-3-butyn-2-ol: 3-Butyn-2-ol, 4-(trimethylsilyl)-; (6999-19-5); (2*R*)-(121522-26-7); (2S)-(12155-27-8) 1056
TRIMETHYLSILYL 2-FLUOROSULFONYL-2,2-DIFLUOROACETATE: Acetic acid, difluoro(fluorosulfonyl)-, trimethylsilyl ester; (120801-75-4) 469
(*Z*)-1-(Trimethylsilyl)-1-hexene: Silane, 1-hexenyltrimethyl-, (*Z*)-; (52835-06-0) 600
1-(Trimethylsilyl)-1-hexyne: Silane, 1-hexynyltrimethyl-; (3844-94-8) 600
Trimethylsilyl triflate: Methanesulfonic acid, trifluoro-, trimethylsilyl ester; [27607-77-8] 169, 473
Trimethylsulfoxonium iodide; (1774-47-6) 149
Triphenylphosphine: Phosphine, triphenyl-; (603-35-0) 164, 289, 296, 393, 540, 561, 778, 834, 903
Triphenylphosphine oxide: Phosphine oxide, triphenyl-; (791-28-6) 778
(TRIPHENYLPHOSPHORANYLIDENE)KETENE: Phosphonium, triphenyl-, oxoethenylide; (73818-55-0) 986
Triphosgene: Methanol, trichloro-, carbonate (2:1); (32315-10-9) 992

Tris(dibenzylideneacetone)dipalladium: Palladium,tris[μ-[(1,2-η:4,5-η)-(1*E*,4*E*)-1,5,-diphenyl-1,4-pentadien-3-one]]di-; (51364-51-3) 372, 384
Trisamidomolybdenum(VI) propylidyne: MOLYBDENUM, TRIS[*N*-(1,1-DIMETHYLETHYL)-3,5-DIMETHYLBENZENAMINATO] PROPYLIDYNE-, (T-4)-; (616886-28-3) 529, 540
Tris[*N*-(*tert*-butyl)(3,5-dimethylphenyl)-amido]molybdenum(III): (236740-70-8) 529
Tris(dibenzylideneacetone)dipalladium(0)–chloroform (1/1): Palladium, tris[μ-[(1,2-η:4,5-η)-(1*E*,4*E*)-1,5-diphenyl-1,4-pentadien-3-one]]di-, compound with trichloromethane (1:1); (52522-40-4) 834
Tris(dibenzylideneacetone)dipalladium(0) [$Pd_2(dba)_3$]: Palladium, tris[μ-[(1,2-η:4,5-η)-(1*E*,4*E*)-1,5-diphenyl-1,4-pentadien-3-one]]di-; (52409-22-0) 529, 540
Tris(hydroxymethyl)phosphine: Methanol, phosphinidynetris-; (2767-80-8) 570
Tris[2-(2-methoxyethoxy)-ethyl]amine: Ethanamine, 2-(2-methoxyethoxy)-*N,N*-bis[2-(2-methoxyethoxy)ethyl]-; (70384-51-9) 607
Tris(3-pyridyl)boroxin: Pyridine, 3,3',3''-(2,4,6-boroxintriyl)tris-; (160688-99-3) 393
Tri-*o*-tolylphosphine: Phosphine, tris(2-methylphenyl)-; (6163-58-2) 384
2,4,6-Trivinylcyclotriboroxane-pyridine complex: Boron, ethenyl[(ethenylboronic acid-k*O*) bimol. monoanhydridato(2-)] (pyridine)-; (95010-17-6) 315

Urea hydrogen peroxide: Urea, compd. with hydrogen peroxide (H_2O_2) (1:1); (124-43-6) 114

(*S*)-Valinol: (2*S*)-2-Amino-3-methyl-1-butanol; (2026-48-4) 506, 708, 821
Vanadyl bis-acetylacetonate: Vanadium, oxobis(2,4-pentanedionato-κ*O*,κ*O*')-; (3153-26-2) 770
Vinamidinium salts, 929
2′-VINYLACETANILIDE: *N*-(2-Vinylphenyl)acetamide: Acetamide, *N*-(2-ethenylphenyl)-; (29124-68-3) 315
Vinyl acetate: Acetic acid ethenyl ester; (108-05-4) 212, 631, 1056
Vinyl ether formation, 631
Vinylidene chloride: 1,1-Dichloroethene; (75-35-4) 234
Vinyl iodide synthesis, 359
Vinylsilane synthesis, 359
Vinyl sulfide, synthesis, 642
hydrolysis, 642

Wittig reaction, 561, 903

Zinc; (7440-66-6) 348, 384
Zinc bromide; (7699-45-8) 934

CONCORDANCE INDEX

The annual volume (in bold-faced type) and page number where each procedure first appeared is given on the left; the page number for the revised procedure is given on the right.

Annual Volume Number	Page	This Volume Page
80	1	177-182
	9	183-188
	18	879-887
	31	399-403
	38	404-409
	46	888-895
	57	896-902
	66	128-134
	75	903-910
	85	570-576
	93	442-449
	104	410-415
	111	450-456
	120	600-606
	129	662-664
	133	911-919
	144	457-468
	160	920-928
	172	469-472
	177	52-56
	184	135-138
	190	642-645
	195	416-419
	200	929-933
	207	139-148
	219	1001-1006
	227	519-523
	233	524-528
81	1	561-569
	14	425-432
	26	348-352
	33	353-358
	42	359-366
	54	367-371
	63	372-383
	77	384-392
81	89	393-398
	98	656-661
	105	862-866
	112	867-873
	121	433-441
	134	1015-1019
	140	164-168
	147	726-732
	157	615-625
	171	514-518
	178	38-45
	188	57-61
	195	107-113
	204	114-120
	213	956-964
	225	169-176
	235	635-641
	244	121-127
	254	874-878
	262	733-740
82	1	62-69
	10	17-24
	18	778-781
	22	1020-1027
	30	652-655
	34	498-505
	43	1056-1067
	55	631-634
	59	782-786
	64	267-271
	69	626-630
	75	88-92
	80	93-99
	87	702-707
	93	594-599
	99	850-858

Annual Volume Number	Page	This Volume Page
82	108	100-106
	115	343-347
	120	1068-1073
	126	1007-1014
	134	720-725
	140	986-991
	147	992-1000
	157	770-777
	166	859-861
	170	506-513
	179	607-614
	188	1-8
83	1	982-985
	5	749-760
	18	77-81
	24	308-314
	31	684-690
	38	646-651
	45	315-318
	49	420-424
	55	82-87
	61	473-480
	70	1037-1045
	80	319-326
	88	327-335
	97	821-825
	103	826-833
	111	834-842
	121	1046-1055
	131	761-769
	141	25-37
	155	46-51
	162	157-163
	170	691-696
	177	336-342
	184	586-593
	193	481-487
	200	552-560
	209	843-849
	217	697-701
84	1	192-199
	11	577-585
	22	197-204
	32	212-220
	43	221-233
	58	149-156
	68	787-793
	77	234-243
	88	244-255
	102	934-941
	111	205-211
	120	9-16
	129	741-748
	139	708-713
	148	714-719
	156	256-260
	163	529-539
	177	540-551
	192	261-266
	199	794-801
	209	942-946
	215	665-669
	222	272-280
	233	70-76
	242	965-981
	262	802-810
	272	488-497
	285	281-288
	295	947-955
	306	1028-1036
	317	289-295
	325	670-676
	334	296-307
	347	811-820
	359	677-683